P9-COP-831

200

$7470

-ia, -ia, state or condition: insomnia
idi-, idios, one's own: idiopathic
ile-, ileum: ileocolic sphincter
ili-, ilio-, ilium: iliac
in-, in-, in, within, or denoting negative effect: inactivate
infra-, infra, beneath: infraorbital
inter-, inter, between: interventricular
intra-, intra, within: intracapsular
ipsi-, ipse, itself: ipsilateral
iso-, isos, equal: isotonic
-itis, -itis, inflammation: dermatitis
karyo-, karyon, body: megakaryocyte
kerato-, keros, horn: keratin
kino-, -kinin, kinein, to move: bradykinin
lact-, lacto-, -lactin, lac, milk: prolactin
lapar-, lapara, flank or loins: laparoscopy
-lemma, lemma, husk: plasmalemma
leuko-, leukos, white: leukocyte
liga-, ligare, to bind together: ligase
lip-, lipo-, lipos, fat: lipoid
lith-, lithos, stone: cholelithiasis
lyso-, -lysis, -lyze, lysis, dissolution: hydrolysis
macr-, makros, large: macrophage
mal-, mal, abnormal: malabsorption
mamilla-, mamilla, little breast: mamillary
mast-, masto-, mastos, breast: mastoid
mega-, megas, big: megakaryocyte
melan-, melas, black: melanocyte
men-, men, month: menstrual
mero-, meros, part: merocrine
meso-, mesos, middle: mesoderm
meta-, meta, after, beyond: metaphase
micr-, mikros, small: microscope
mono-, monos, single: monocyte
morpho-, morphe, form: morphology
multi-, multus, much, many: multicellular
-mural, murus, wall: intramural
myelo-, myelos, marrow: myeloblast
myo-, mys, muscle: myofilament
narc-, narkoun, to numb or deaden: narcotics
nas-, nasus, nose: nasolacrimal duct
natri-, natrium, sodium: natriuretic
necr-, nekros, corpse: necrosis
nephr-, nephros, kidney: nephron
neur-, neuro-, neuron, nerve: neuromuscular
oculo-, oculus, eye: oculomotor
odont-, odontos, tooth: odontoid process
-oid, eidos, form, resemblance: odontoid process
oligo-, oligos, little, few: oligopeptide
-ology, logos, the study of: physiology
-oma, -oma, swelling: carcinoma
onco-, onkos, mass, tumor: oncology
oo-, oon, egg: oocyte
ophthalm-, ophthalmos, eye: ophthalmic nerve
-opia, ops, eye: myopia
orb-, orbita, a circle: orbicularis oris
orchi-, orchis, testis: orchiectomy
orth-, orthos, correct, straight: orthopedist
-osis, -osis, state, condition: neurosis
osteon, osteo-, os, bone: osteocyte
oto-, otikos, ear: otolith
para-, para, beyond: paraplegia
patho-, -path, -pathy, pathos, disease:
    pathology
pedia-, paidos, child: pediatrician
per-, per, through, throughout: percutaneous
peri-, peri, around: perineurium

phag-, phagein, to eat: phagocyte
-phasia, phasis, speech: aphasia
-phil, -philia, philus, love: hydrophilic
phleb-, phleps, a vein: phlebitis
-phobe, -phobia, phobos, fear: hydrophobic
phot-, phos, light: photoreceptor
-phylaxis, phylax, a guard: prophylaxis
physio-, physis, nature: physiology
-plasia, plasis, formation: dysplasia
platy-, platys, flat: platysma
-plegia, plege, a blow, paralysis: paraplegia
-plexy, plessein, to strike: apoplexy
pneum-, pneuma, air: pneumotaxic center
podo-, podo, podon, foot: podocyte
-poiesis, poiesis, making: hemopoiesis
poly-, polys, many: polysaccharide
post-, post, after: postanal
pre-, prae, before: precapillary sphincter
presby-, presbys, old: presbyopia
pro-, pro, before: prophase
proct-, proktos, anus: proctology
pterygo-, pteryx, wing: pterygoid
pulmo-, pulmo, lung: pulmonary
pulp-, pulpa, flesh: pulpitis
pyel-, pyelos, trough or pelvis: pyelitis
quadr-, quadrans, one quarter: quadriplegia
re-, re-, back, again: reinfection
retro-, retro, backward: retroperitoneal
rhin-, rhis, nose: rhinitis
-rrhage, rhegnymi, to burst forth: hemorrhage
-rrhea, rhein, flow, discharge: amenorrhea
sarco-, sarkos, flesh: sarcomere
scler-, sclero-, skleros, hard: sclera
-scope, skopeo, to view: colonoscope
-sect, sectio, to cut: transect
semi-, semis, half: semitendinosus
-septic, septikos, putrid: antiseptic
-sis, -sis, state or condition: metastasis
som-, -some, soma, body: somatic
spino-, spina, spine, vertebral column:
    spinothalamic pathway
-stalsis, staltikos, contractile: peristalsis
sten-, stenos, a narrowing: stenosis
-stomy, stoma, mouth, opening: colostomy
stylo-, stylus, stake, pole: styloid process
sub-, sub, below; subcutaneous
super-, super, above or beyond: superficial
supra-, supra, on the upper side: supraspinous fossa
syn-, syn, together: synthesis
tachy-, tachys, swift: tachycardia
telo-, telos, end: telophase
tetra-, tettares, four: tetralogy of Fallot
therm-, thermo-, therme, heat: thermoregulation
thorac-, thorax, chest: thoracentesis
thromb-, thrombos, clot: thrombocyte
-tomy, temnein, to cut: appendectomy
tox-, toxikon, poison: toxemia
trans-, trans, through: transudate
tri-, tres, three: trimester
-tropic, trope, turning: adrenocorticotropic
tropho-, trophe, nutrition: trophoblast
-trophy, trophikos, nourishing: atrophy
tropo-, tropikos, turning: troponin
uni-, unus, one: unicellular
uro-, -uria, ouron, urine: glycosuria
vas-, vas, vessel: vascular
zyg-, zygotos, yoked: zygote

FIFTH EDITION

Fundamentals of
Anatomy
&
Physiology

FREDERIC H. MARTINI, PH.D.

with

WILLIAM C. OBER, M.D.
ART COORDINATOR AND ILLUSTRATOR

CLAIRE W. GARRISON, R.N.
ILLUSTRATOR

KATHLEEN WELCH, M.D.
CLINICAL CONSULTANT

RALPH T. HUTCHINGS
BIOMEDICAL PHOTOGRAPHER

PRENTICE HALL, *Upper Saddle River, New Jersey 07458*

Library of Congress Cataloging-in-Publication Data

Martini, Frederic.
    Fundamentals of anatomy & physiology / Frederic H. Martini; with William C. Ober
…[et al.].–5th ed.
      p. ; cm.
    Includes index
    ISBN 0-13-017292-8 (casebound)
    1. Human physiology. 2. Human anatomy. I. Title: Fundamentals of anatomy and
physiology. II. Title.
    [DNLM: 1. Anatomy. 2. Physiology. QS 4 M3855f2001]
    QP34.5 .M27 2001
    612--dc21

                              00-039968

Senior Acquisitions Editor: *Halee Dinsey*
Senior Development Editor: *Karen Karlin*
Production Editor: *Shari Toron*
Project Manager: *Don O'Neal*
Editorial Assistant: *Susan Zeigler*
Assistant Vice President of Production & Manufacturing:
   David W. Riccardi
Executive Managing Editor: *Kathleen Schiaparelli*
Assistant Managing Editor, Science: *Beth Sturla*
Editorial Director: *Paul F. Corey*
Editor in Chief for Biology: *Sheri Snavely*
Editors in Chief, Development: *Ray Mullaney,*
   Carol Trueheart
Senior Marketing Manager: *Martha McDonald*
Executive Marketing Manager for Biology: *Jennifer Welchans*
Director of Marketing: *John Tweedale*
Marketing Assistant: *Andrew Gilfillan*
Manufacturing Manager: *Trudy Pisciotti*
Manufacturing Buyer: *Michael Bell*
Copy Editors: *Margo Quinto, Jennifer Vecchione*
Photo Researcher: *Tobi Zausner*
Director, Creative Services: *Paul Belfanti*
Art Director: *Heather Scott*
Interior Design: *Anne Flanagan*
Cover Designer: *John Christiana*
Art Manager: *Gus Vibal*
Art Editor: *Karen Branson*
Special Projects Manager: *Barbara A. Murray*

Supplements Production Editor: *Meaghan Forbes*
Illustrators: *William C. Ober, M.D.; Claire Garrison, R.N.*
Clinical Consultant: *Kathleen Welch, M.D.*
Biomedical Photographer: *Ralph T. Hutchings*
Composition: *Preparè, Inc.*
Cover Photograph: *Jamey Hampton, Daniel Ezralow,*
   Ashley Roland, and Sheila Lehner
   ISO Dance Company Photo © Lois Greenfield, 1991
Assistant Managing Editor, Science Media: *Alison Lorber*
Media Product Development Editor: *Andrew Stull*
Media Content Team: *Kathleen Flickinger, Michael McKinley,*
   Sharon Simpson, Susan Finazzo, Leslie Carlson, and Sandy
   Amundson
Media Content Consultant: *Frederic H. Martini*
Project Manager, CD-ROM: *Rich Barnes*
Project Manager, Companion Website: *Elizabeth Wright*
Assistant Media Editor: *Anthony M. Maffia*
Web Development: *Allyson Graesser, Jeff Henn, William Johnson,*
   Lawrence La Raia, Anthony M. Maffia, and Elizabeth Wright
Media Consultant: *Dean Fiorino*
Formatting Manager: *Jim Sullivan*
CD-ROM Development/Animations: Visible Productions—
   Thomas McCracken, Dave Carlson, Steve Hade, Sherra Cook,
   Kevin Ryan, Phil Guzy, Frank Forney, and Sean McCracken
Tutorials: Blausen Medical Communications—*Bruce Blausen,*
   David Hansen, David Jonah, Daniela Weil, and Paige Phillips
Manta Consulting: *George Pople*

Notice: Our knowledge in clinical sciences is constantly changing. The author and the publisher of this volume have taken care that the information contained herein is accurate and compatible with the standards generally accepted at the time of publication. Nevertheless, it is difficult to ensure that all information given is entirely accurate for all circumstances. The author and the publisher disclaim any liability, loss, or damage incurred as a consequence, directly or indirectly, of the use and application of any of the contents of this volume.

© 2001, 1998, 1995, 1992, 1989 by Frederic H. Martini, Inc.
Published by Prentice-Hall, Inc.
Upper Saddle River, New Jersey 07458

All rights reserved. No part of this book may be reproduced, in any form or by any means, without permission in writing from the publisher.

Printed in the United States of America

10 9 8 7 6 5 4 3 2 1

ISBN 0-13-017292-8

Prentice-Hall International (UK) Limited, *London*
Prentice-Hall of Australia Pty. Limited, *Sydney*
Prentice-Hall Canada Inc., *Toronto*
Prentice-Hall Hispanoamericana, S.A., *Mexico*
Prentice-Hall of India Private Limited, *New Delhi*
Prentice-Hall of Japan, Inc., *Tokyo*
Pearson Education Asia Pte. Ltd
Editora Prentice-Hall do Brasil, Ltda., *Rio de Janeiro*

Text and Illustration Team

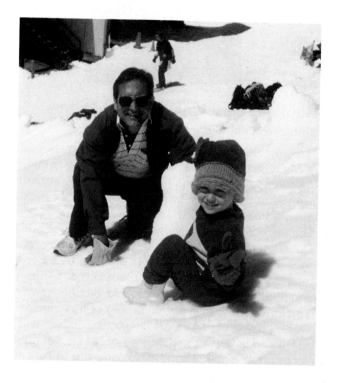

Dr. Frederic (Ric) Martini
Author

Dr. Martini received his Ph.D. from Cornell University in comparative and functional anatomy. His publications include journal articles, technical reports, magazine articles, and a book for naturalists about the biology and geology of tropical islands. He is the coauthor of five other undergraduate texts on anatomy or anatomy and physiology. He is currently on the faculty of the University of Hawaii and remains affiliated with the Shoals Marine Laboratory, a joint venture between Cornell University and the University of New Hampshire. Dr. Martini is a member of the Human Anatomy and Physiology Society, the American Physiological Society, and the American Association of Anatomists. He is also a member of the National Association of Biology Teachers, the Society for College Science Teachers, the Society for Integrative and Comparative Biology, the Western Society of Naturalists, and the International Society of Vertebrate Morphologists.

William C. Ober, M.D.
Art coordinator and illustrator

Dr. Ober received his undergraduate degree from Washington and Lee University and his M.D. from the University of Virginia in Charlottesville. While in medical school, he also studied in the Department of Art as Applied to Medicine at Johns Hopkins University. Dr. Ober is currently on the faculty of the University of Virginia in the Department of Sports Medicine. He is also part of the Core Faculty at the Shoals Marine Laboratory.

Claire W. Garrison, R.N.
Illustrator

Ms. Garrison practiced pediatric and obstetric nursing for nearly 20 years before turning to medical illustration as a full-time career. She has been Dr. Ober's associate since 1986, following a five-year apprenticeship. Ms. Garrison is also a Core Faculty member at the Shoals Marine Laboratory.

Dr. Kathleen Welch
Clinical consultant

Dr. Welch received her M.D. from the University of Washington in Seattle and did her residency at the University of North Carolina in Chapel Hill. For two years, she served as Director of Maternal and Child Health at the LBJ Tropical Medical Center in American Samoa and subsequently was a member of the Department of Family Practice at the Kaiser Permanente Clinic in Lahaina, Hawaii. She has been in private practice since 1987. Dr. Welch is a Fellow of the American Academy of Family Practice and a member of the Hawaii Medical Association and the Human Anatomy and Physiology Society.

Ralph T. Hutchings
Biomedical photographer

Mr. Hutchings was associated with the Royal College of Surgeons for 20 years. An engineer by training, he has focused for years on photographing the structure of the human body. The result has been a series of color atlases, including the *Color Atlas of Human Anatomy, The Color Atlas of Surface Anatomy,* and *The Human Skeleton* (all published by Mosby-Yearbook Publishing, St. Louis, Missouri, USA). For his anatomical portrayal of the human body, the International Photographers Association has chosen Mr. Hutchings as the best photographer of humans in the twentieth century. He lives in North London, where he tries to balance the demands of his photographic assignments with his hobbies of early motor cars and airplanes.

Dr. Kathleen (Ireland) Flickinger
Media advisor

Dr. Flickinger teaches anatomy and physiology on campus and via multimedia distance technologies for the University of Hawaii–Maui Community College. She has worked in medical research for LSU MC, Drake University, the Lion's Eye Clinic in New Orleans, and DuPont Biochemical Corporation. She began teaching anatomy and physiology in Louisiana in the mid-1980s and received her Ph.D. in curriculum and instructional technologies from Iowa State University. Her current research involves distance education and the effective use of technology in higher education.

To my family, whose support makes this project possible, and to my readers, whose thanks and suggestions for improvement were deeply appreciated.

Brief Contents

1 An Introduction to Anatomy and Physiology 1

2 The Chemical Level of Organization 29

3 The Cellular Level of Organization 64

4 The Tissue Level of Organization 106

5 The Integumentary System 143

6 Osseous Tissue and Skeletal Structure 167

7 The Axial Skeleton 193

8 The Appendicular Skeleton 226

9 Articulations 245

10 Muscle Tissue 270

11 The Muscular System 311

12 Neural Tissue 360

13 The Spinal Cord and Spinal Nerves 406

14 The Brain and Cranial Nerves 435

15 Integrative Functions 480

16 The Autonomic Nervous System 503

17 Sensory Function 527

18 The Endocrine System 576

19 Blood 623

20 The Heart 654

21 Blood Vessels and Circulation 691

22 The Lymphatic System and Immunity 751

23 The Respiratory System 797

24 The Digestive System 845

25 Metabolism and Energetics 900

26 The Urinary System 942

27 Fluid, Electrolyte, and Acid–Base Balance 984

28 The Reproductive System 1016

29 Development and Inheritance 1064

Contents

Preface   xix
To the Student   xxix

CHAPTER 1
An Introduction to Anatomy and Physiology   1

INTRODUCTION   2

THE SCIENCES OF ANATOMY AND PHYSIOLOGY   2
Anatomy   3
Physiology   3

LEVELS OF ORGANIZATION   4

HOMEOSTASIS AND SYSTEM INTEGRATION   12
Negative Feedback   12
Positive Feedback   14

A FRAME OF REFERENCE FOR ANATOMICAL STUDIES   15
Superficial Anatomy   15
Sectional Anatomy   19
   ■ FOCUS: Sectional Anatomy and Clinical Technology   23

MEDIALAB   25

CHAPTER REVIEW   26

CHAPTER 2
The Chemical Level of Organization   29

INTRODUCTION   30

ATOMS AND MOLECULES   30
Atomic Structure   30
Chemical Bonds   32

CHEMICAL REACTIONS   36
Basic Energy Concepts   36
FOCUS: Chemical Notation   37
Types of Reactions   38
Reversible Reactions   38
Enzymes and Chemical Reactions   39

INORGANIC COMPOUNDS   39
Water and Its Properties   40
Inorganic Acids and Bases   42
Salts   43
Buffers and pH Control   43

ORGANIC COMPOUNDS   44
Carbohydrates   44
Lipids   46
Proteins   51

Nucleic Acids   56
High-Energy Compounds   57

CHEMICALS AND CELLS   59

MEDIALAB   60

CHAPTER REVIEW   60

CHAPTER 3
The Cellular Level of Organization   64

INTRODUCTION   65

STUDYING CELLS   65
An Overview of Cellular Anatomy   66

THE CELL MEMBRANE   68
Membrane Structure   68
Membrane Permeability   71

THE CYTOPLASM   80
The Cytosol   80
Organelles   80

THE NUCLEUS   89
Chromosome Structure   90
The Genetic Code   91
Gene Activation and Protein Synthesis   91
DNA and the Control of Cell Structure and Function   95

THE CELL LIFE CYCLE   96
Interphase   96
Mitosis   98
Cytokinesis   99
The Mitotic Rate and Energetics   99
Regulation of the Cell Life Cycle   100

CELL DIVERSITY AND DIFFERENTIATION   101

MEDIALAB   102

CHAPTER REVIEW   102

CHAPTER 4
The Tissue Level of Organization   106

INTRODUCTION   107

EPITHELIAL TISSUE   107
Functions of Epithelial Tissue   107
Specializations of Epithelial Cells   108
Maintaining the Integrity of Epithelia   108
Classification of Epithelia   110
Glandular Epithelia   115

CONNECTIVE TISSUES   117
Classification of Connective Tissues   118
Connective Tissue Proper   118
Fluid Connective Tissues   124
Supporting Connective Tissues   125

MEMBRANES   129
Mucous Membranes   129
Serous Membranes   129
The Cutaneous Membrane   131
Synovial Membranes   131

THE CONNECTIVE TISSUE FRAMEWORK OF THE BODY   131

MUSCLE TISSUE   132
Skeletal Muscle Tissue   132
Cardiac Muscle Tissue   132
Smooth Muscle Tissue   132

NEURAL TISSUE   134

TISSUE INJURIES AND AGING   135
Inflammation and Regeneration   135
Aging and Tissue Repair   136
Aging and Cancer Incidence   136

MEDIALAB   138

CHAPTER REVIEW   138

CHAPTER 5
The Integumentary
System   143

INTRODUCTION   144

THE EPIDERMIS   145
Layers of the Epidermis   145
Skin Color   147
The Epidermis and Vitamin D$_3$ 149
Epidermal Growth Factor   149

THE DERMIS   150
Dermal Organization   150
Dermal Circulation and Innervation   151

THE SUBCUTANEOUS LAYER   152

ACCESSORY STRUCTURES   152
Hair Follicles and Hair   152
Glands in the Skin   155
Nails   158

LOCAL CONTROL OF INTEGUMENTARY FUNCTION   158
Injury and Repair   158

AGING AND THE INTEGUMENTARY SYSTEM   160

INTEGRATION WITH OTHER SYSTEMS   160

MEDIALAB   163

CHAPTER REVIEW   163

CHAPTER 6
Osseous Tissue and Skeletal
Structure   167

INTRODUCTION   168

A CLASSIFICATION OF BONES   168
Bone Shapes   168
Bone Structure   168

BONE HISTOLOGY   170
The Matrix of Bone   170
Cells in Bone   171
Compact Bone and Spongy Bone   172
The Periosteum and Endosteum   174

BONE DEVELOPMENT AND GROWTH   175
Intramembranous Ossification   175
Endochondral Ossification   176
The Blood and Nerve Supplies   178

THE DYNAMIC NATURE OF BONE   180
Effects of Exercise on Bone   180
Hormonal and Nutritional Effects on Bone   181
The Skeleton as a Calcium Reserve   182
Fracture Repair   184
Aging and the Skeletal System   185
FOCUS: A Classification of Fractures   186

BONE MARKINGS (SURFACE FEATURES)   187

INTEGRATION WITH OTHER SYSTEMS   187

MEDIALAB   190

CHAPTER REVIEW   190

CHAPTER 7
The Axial Skeleton   193

INTRODUCTION   194

THE SKULL   195
FOCUS: The Individual Bones of the Skull   200
Cranial Bones   200
Facial Bones   205
Summary: The Foramina and Fissures of the Skull   208
The Orbital and Nasal Complexes   208
The Skulls of Infants and Children   211

THE VERTEBRAL COLUMN   212
Spinal Curvature   212
Vertebral Anatomy   213
Vertebral Regions   214

THE THORACIC CAGE   219
The Ribs   219
The Sternum   222

MEDIALAB   222

CHAPTER REVIEW   223

CHAPTER **8**
The Appendicular
Skeleton   226

INTRODUCTION   227

THE PECTORAL GIRDLE AND UPPER LIMBS   228

The Pectoral Girdle   228
The Upper Limbs   230

THE PELVIC GIRDLE AND LOWER LIMBS   233

The Pelvic Girdle   234
The Lower Limbs   237

INDIVIDUAL VARIATION IN THE SKELETAL
SYSTEM   241

MEDIALAB   242

CHAPTER REVIEW   243

CHAPTER **9**
Articulations   245

INTRODUCTION   246

A CLASSIFICATION OF JOINTS   246

Synarthroses (Immovable Joints)   246
Amphiarthroses (Slightly Movable Joints)   247
Diarthroses (Freely Movable Joints)   247

ARTICULAR FORM AND FUNCTION   250

Describing Dynamic Motion   250
Types of Movements   251
A Structural Classification of Synovial Joints   254

REPRESENTATIVE ARTICULATIONS   256

Intervertebral Articulations   256
The Shoulder Joint   258
The Elbow Joint   260
The Hip Joint   260
The Knee Joint   261

AGING AND ARTICULATIONS   265

BONES AND MUSCLES   265

MEDIALAB   266

CHAPTER REVIEW   266

CHAPTER **10**
Muscle Tissue   270

INTRODUCTION   271

SKELETAL MUSCLE TISSUE AND THE MUSCULAR
SYSTEM   271

ANATOMY OF SKELETAL MUSCLE   271

Organization of Connective Tissues   271
Blood Vessels and Nerves   273
Microanatomy of Skeletal Muscle Fibers   273

CONTRACTION OF SKELETAL MUSCLE   279

The Sliding Filament Theory   279
The Control of Skeletal Muscle Activity   280
Relaxation   281

FOCUS: The Contraction Cycle   282

Length–Tension Relationships   285

MUSCLE MECHANICS   286

Tension Production   286
Energetics of Muscular Activity   293
Muscle Performance   296

AGING AND THE MUSCULAR SYSTEM   299

INTEGRATION WITH OTHER SYSTEMS   300

CARDIAC MUSCLE TISSUE   302

Differences between Cardiac and Skeletal Muscle
Tissues   302

SMOOTH MUSCLE TISSUE   303

Differences between Smooth Muscle Tissue and Other
Muscle Tissues   304

MEDIALAB   306

CHAPTER REVIEW   306

CHAPTER **11**
The Muscular
System   311

INTRODUCTION   312

BIOMECHANICS AND MUSCLE ANATOMY   312

Organization of Skeletal Muscle Fibers   312
Levers   313

MUSCLE TERMINOLOGY   315

Origins and Insertions   315
Actions   315
Names of Skeletal Muscles   316
Axial and Appendicular Muscles   317

THE AXIAL MUSCLES   320

Muscles of the Head and Neck   320
Muscles of the Vertebral Column   328
Oblique and Rectus Muscles   330
Muscles of the Pelvic Floor   332

THE APPENDICULAR MUSCLES   335

Muscles of the Shoulders and Upper Limbs   335
Muscles of the Pelvis and Lower Limbs   346

MEDIALAB   356

CHAPTER REVIEW   356

CHAPTER 12
Neural Tissue   360

INTRODUCTION   361

AN OVERVIEW OF THE NERVOUS SYSTEM   361

NEURONS   362
Neuron Structure   362
Neuron Classification   365

NEUROGLIA   367
Neuroglia of the Central Nervous System   367
Neuroglia of the Peripheral Nervous System   370
FOCUS: Neural Responses to Injuries   372

NEUROPHYSIOLOGY   373
The Transmembrane Potential   373
Changes in the Transmembrane Potential   376
Graded Potentials   377
Action Potentials   380

SYNAPTIC ACTIVITY   386
Electrical Synapses   386
Chemical Synapses   387

INFORMATION PROCESSING   392
Postsynaptic Potentials   392
Presynaptic Inhibition and Facilitation   395
The Rate of Action Potential Generation   395

**HIGHER LEVELS OF ORGANIZATION AND
    PROCESSING   396**
Functional Organization   396
Anatomical Organization   399

INTEGRATION WITH OTHER SYSTEMS   399

MEDIALAB   401
CHAPTER REVIEW   401

CHAPTER 13
The Spinal Cord and Spinal Nerves   406

INTRODUCTION   407

GROSS ANATOMY OF THE SPINAL CORD   407
Spinal Meninges   409
Sectional Anatomy of the Spinal Cord   412

SPINAL NERVES   414
Peripheral Distribution of Spinal Nerves   414
Nerve Plexuses   416

AN INTRODUCTION TO REFLEXES   422
The Reflex Arc   422
Classification of Reflexes   423

SPINAL REFLEXES   425
Monosynaptic Reflexes   425
Polysynaptic Reflexes   428

INTEGRATION AND CONTROL OF SPINAL REFLEXES   430
Voluntary Movements and Reflex Motor Patterns   430

MEDIALAB   431
CHAPTER REVIEW   431

CHAPTER 14
The Brain and Cranial Nerves   435

INTRODUCTION   436

**AN INTRODUCTION TO THE ORGANIZATION
    OF THE BRAIN   436**
A Preview of Major Regions and Landmarks   436
Embryology of the Brain   438
Ventricles of the Brain   439

PROTECTION AND SUPPORT OF THE BRAIN   440
The Cranial Meninges   440
Cerebrospinal Fluid   441
The Blood Supply to the Brain   443

THE CEREBRUM   445
The Cerebral Cortex   445
The Cerebral Nuclei   451

THE LIMBIC SYSTEM   452

THE DIENCEPHALON   454
The Thalamus   456
The Hypothalamus   457

THE MESENCEPHALON   459

THE CEREBELLUM   460

THE PONS   462

THE MEDULLA OBLONGATA   462
FOCUS: Cranial Nerves   464
A Summary of Cranial Nerve Branches and Functions   474

MEDIALAB   475
CHAPTER REVIEW   475

CHAPTER 15
Integrative Functions   480

INTRODUCTION   481

SOMATIC SENSORY AND MOTOR PATHWAYS   481
Sensory Pathways   481

Motor Pathways and Motor Control   487

MONITORING BRAIN ACTIVITY: THE ELECTROENCEPHALOGRAM   493

HIGHER-ORDER FUNCTIONS   494

Memory   494
Consciousness   496

BRAIN CHEMISTRY AND BEHAVIOR   498

Personality and Other Mysteries   498

AGING AND THE NERVOUS SYSTEM   499

Age-Related Anatomical and Functional
   Changes   499
Alzheimer's Disease   499

MEDIALAB   500

CHAPTER REVIEW   500

CHAPTER 16
The Autonomic Nervous System   503

INTRODUCTION   504

AN OVERVIEW OF THE ANS   504

Divisions of the ANS   504

THE SYMPATHETIC DIVISION   506

The Sympathetic Chain   507
Collateral Ganglia   510
The Adrenal Medullae   510
Sympathetic Activation   510
Neurotransmitters and Sympathetic Function   511
Summary: The Sympathetic Division   512

THE PARASYMPATHETIC DIVISION   512

Organization and Anatomy of the Parasympathetic
   Division   513
General Functions of the Parasympathetic
   Division   513
Parasympathetic Activation and Neurotransmitter
   Release   514
Summary: The Parasympathetic Division   516

INTERACTIONS OF THE SYMPATHETIC AND PARASYMPATHETIC DIVISIONS   516

Anatomy of Dual Innervation   516
A Comparison of the Sympathetic and
   Parasympathetic Divisions   517
Autonomic Tone   518

INTEGRATION AND CONTROL OF AUTONOMIC FUNCTIONS   519

Visceral Reflexes   520
Higher Levels of Autonomic Control   521

MEDIALAB   524

CHAPTER REVIEW   524

CHAPTER 17
Sensory Function   527

INTRODUCTION   528

RECEPTORS AND RECEPTOR FUNCTION   528

The Detection of Stimuli   528
The Interpretation of Sensory Information   529
Central Processing and Adaptation   529

THE GENERAL SENSES   530

Nociceptors   530
Thermoreceptors   531
Mechanoreceptors   531
Chemoreceptors   533

OLFACTION   534

Olfactory Receptors   535
Olfactory Pathways   535
Olfactory Discrimination   536

GUSTATION   536

Taste Receptors   536
Gustatory Pathways   536
Gustatory Discrimination   537

VISION   538

Accessory Structures of the Eye   538
The Eye   541
Visual Physiology   550
The Visual Pathway   554

EQUILIBRIUM AND HEARING   557

Anatomy of the Ear   557
Equilibrium   561
Hearing   564

MEDIALAB   571

CHAPTER REVIEW   572

CHAPTER 18
The Endocrine System   576

INTRODUCTION   577

AN OVERVIEW OF THE ENDOCRINE SYSTEM   578

Hormone Structure   578
Hormone Distribution and Transport   580
Hormone Function and Mechanisms of
   Action   580
Control of Endocrine Activity   584

THE PITUITARY GLAND   586

The Anterior Pituitary   587
The Posterior Pituitary   591

THE THYROID GLAND   593

Thyroid Follicles and Thyroid Hormones   594
The C Cells of the Thyroid Gland:
    Calcitonin   597

THE PARATHYROID GLANDS   597

Parathyroid Hormone Secretion   598

THE ADRENAL GLANDS   600

The Adrenal Cortex   600
The Adrenal Medullae   603

THE PINEAL GLAND   603

THE ENDOCRINE TISSUES OF OTHER SYSTEMS   604

The Intestines   604
The Kidneys   604
The Heart   605
The Thymus   605
The Pancreas   605
The Gonads   608

PATTERNS OF HORMONAL INTERACTION   610

Hormones and Growth   610
Hormones and Stress   612
Hormones and Behavior   614
Aging and Hormone Production   614

INTEGRATION WITH OTHER SYSTEMS   614

MEDIALAB   618

CHAPTER REVIEW   618

CHAPTER 19
Blood   623

INTRODUCTION   624

FUNCTIONS OF BLOOD   624

COMPOSITION OF BLOOD   624

PLASMA   626

Plasma Proteins   626

FORMED ELEMENTS   627

Hemopoiesis   627
Red Blood Cells   628
White Blood Cells   637
Platelets   644

HEMOSTASIS   645

The Vascular Phase   645
The Platelet Phase   645
The Coagulation Phase   646
Clot Retraction   648
Fibrinolysis   648

MEDIALAB   650

CHAPTER REVIEW   650

CHAPTER 20
The Heart   654

INTRODUCTION   655

AN OVERVIEW OF THE CARDIOVASCULAR SYSTEM   655

ANATOMY OF THE HEART   655

The Pericardium   656
Superficial Anatomy of the Heart   657
The Heart Wall   658
Internal Anatomy and Organization   660
Connective Tissues and the Fibrous Skeleton   664
The Blood Supply to the Heart   664

THE HEARTBEAT   668

Contractile Cells   668
The Conducting System   670
The Electrocardiogram   673
The Cardiac Cycle   674

CARDIODYNAMICS   680

Overview: The Control of Cardiac Output   681
Factors Affecting the Stroke Volume   681
Factors Affecting the Heart Rate   683
Exercise and Cardiac Output   685

**THE HEART AND THE CARDIOVASCULAR
    SYSTEM   687**

MEDIALAB   687

CHAPTER REVIEW   687

CHAPTER 21
Blood Vessels and
    Circulation   691

INTRODUCTION   692

THE ANATOMY OF BLOOD VESSELS   692

The Structure of Vessel Walls   692
Differences between Arteries and Veins   692
Arteries   693
Capillaries   695
Veins   699
The Distribution of Blood   700

CARDIOVASCULAR PHYSIOLOGY   701

Pressure   702
Resistance   702
An Overview of Circulatory Pressures   704
Arterial Blood Pressure   705
Capillary Exchange   706
Venous Pressure and Venous Return   710

CARDIOVASCULAR REGULATION   711

Autoregulation of Blood Flow within
   Tissues   712
Neural Mechanisms   712
Hormones and Cardiovascular Regulation   715

PATTERNS OF CARDIOVASCULAR RESPONSE   717

Exercise and the Cardiovascular System   717
Cardiovascular Response to Hemorrhaging   719

FOCUS: Special Circulation   722

THE BLOOD VESSELS   723

The Pulmonary Circuit   723
The Systemic Circuit   724
Fetal Circulation   742

**AGING AND THE CARDIOVASCULAR
   SYSTEM   744**

INTEGRATION WITH OTHER SYSTEMS   745

MEDIALAB   745

CHAPTER REVIEW   745

CHAPTER 22
The Lymphatic System and
Immunity   751

INTRODUCTION   752

AN OVERVIEW OF THE LYMPHATIC SYSTEM   752

ORGANIZATION OF THE LYMPHATIC SYSTEM   752

Functions of the Lymphatic System   752
Lymphatic Vessels   752
Lymphocytes   756
Lymphoid Tissues   758
Lymphoid Organs   759
The Lymphatic System and Body Defenses   763

NONSPECIFIC DEFENSES   763

Physical Barriers   765
Phagocytes   765
Immunological Surveillance   766
Interferons   767
Complement   767
Inflammation   769
Fever   770

SPECIFIC DEFENSES: THE IMMUNE RESPONSE   771

Forms of Immunity   771
Properties of Immunity   772
Overview: The Immune Response   773
T Cells and Cell-Mediated Immunity   773
B Cells and Antibody-Mediated Immunity   777
Primary and Secondary Responses to Antigen
   Exposure   781
Hormones of the Immune System   782

SUMMARY: THE IMMUNE RESPONSE   784

THE DEVELOPMENT OF RESISTANCE   785

IMMUNE DISORDERS   787

Autoimmune Disorders   787
Immunodeficiency Diseases   788
Allergies   788

STRESS AND THE IMMUNE RESPONSE   789

AGING AND THE IMMUNE RESPONSE   789

INTEGRATION WITH OTHER SYSTEMS   791

MEDIALAB   791

CHAPTER REVIEW   793

CHAPTER 23
The Respiratory System   797

INTRODUCTION   798

FUNCTIONS OF THE RESPIRATORY SYSTEM   798

ORGANIZATION OF THE RESPIRATORY SYSTEM   798

The Respiratory Mucosa   798

THE UPPER RESPIRATORY SYSTEM   801

The Nose and Nasal Cavity   801
The Pharynx   802

THE LARYNX   803

Cartilages and Ligaments of the Larynx   803
The Laryngeal Musculature   805

THE TRACHEA   805

THE PRIMARY BRONCHI   806

THE LUNGS   806

Lobes and Surfaces of the Lungs   806
The Bronchi   808
The Bronchioles   809
Alveolar Ducts and Alveoli   811
The Blood Supply to the Lungs   813

**THE PLEURAL CAVITIES AND PLEURAL
   MEMBRANES   813**

RESPIRATORY PHYSIOLOGY   814

Pulmonary Ventilation   814
Gas Exchange at the Respiratory
   Membrane   822
Gas Pickup and Delivery   826
Summary: Gas Transport   830

CONTROL OF RESPIRATION   830

Local Regulation of Gas Transport and Alveolar
   Function   830
The Respiratory Centers of the Brain   831
Voluntary Control of Respiration   836

CHANGES IN THE RESPIRATORY SYSTEM
   AT BIRTH   836

AGING AND THE RESPIRATORY SYSTEM   837

INTEGRATION WITH OTHER SYSTEMS   837

MEDIALAB   839

CHAPTER REVIEW   840

CHAPTER 24
The Digestive System   845

INTRODUCTION   846

AN OVERVIEW OF THE STRUCTURE AND FUNCTION OF THE
   DIGESTIVE TRACT   846

Histological Organization of the Digestive Tract   846
The Movement of Digestive Materials   849
The Control of Digestive Function   849
The Peritoneum   851

THE ORAL CAVITY   853

The Tongue   854
Salivary Glands   854
The Teeth   856

THE PHARYNX   858

THE ESOPHAGUS   858

Histology of the Esophagus   860
Swallowing   860

THE STOMACH   861

Anatomy of the Stomach   861
Regulation of Gastric Activity   865
Digestion and Absorption in the Stomach   868

THE SMALL INTESTINE AND ASSOCIATED GLANDULAR
   ORGANS   868

The Small Intestine   869
Intestinal Movements   872
The Pancreas   872
The Liver   874
The Gallbladder   879
The Coordination of Secretion and Absorption   880

THE LARGE INTESTINE   882

The Cecum   882
The Colon   882
The Rectum   884
Histology of the Large Intestine   884
Physiology of the Large Intestine   885

DIGESTION AND ABSORPTION   887

The Processing and Absorption of Nutrients   887
Carbohydrate Digestion and Absorption   888

Lipid Digestion and Absorption   890
Protein Digestion and Absorption   890
Water Absorption   891
Ion Absorption   891
Vitamin Absorption   892

AGING AND THE DIGESTIVE SYSTEM   893

INTEGRATION WITH OTHER SYSTEMS   893

MEDIALAB   895

CHAPTER REVIEW   895

CHAPTER 25
Metabolism and
Energetics   900

INTRODUCTION   901

AN OVERVIEW OF METABOLISM   901

CARBOHYDRATE METABOLISM   903

Glycolysis   904
Mitochondrial ATP Production   904
Energy Yield of Glycolysis and Cellular
   Respiration   910
Other Catabolic Pathways   911
Gluconeogenesis   912

LIPID METABOLISM   913

Lipid Catabolism   913
Lipid Synthesis   914
Lipid Transport and Distribution   914

PROTEIN METABOLISM   916

Amino Acid Catabolism   916
Protein Synthesis   919

NUCLEIC ACID METABOLISM   919

RNA Catabolism   919
Nucleic Acid Synthesis   920

METABOLIC INTERACTIONS   920

The Absorptive State   921
The Postabsorptive State   923

DIET AND NUTRITION   926

Food Groups and Food Pyramids   926
Nitrogen Balance   926
Minerals   928
Vitamins   929
Diet and Disease   931

BIOENERGETICS   932

Food and Energy   933
Metabolic Rate   933
Thermoregulation   933

MEDIALAB   937

CHAPTER REVIEW   938

CHAPTER **26**
The Urinary System   942

INTRODUCTION   943

ORGANIZATION OF THE URINARY SYSTEM   943

THE KIDNEYS   943

Superficial Anatomy of the Kidneys   944
Sectional Anatomy of the Kidneys   944
The Blood Supply to the Kidneys   946
Innervation of the Kidneys   946
The Nephron   946

RENAL PHYSIOLOGY   953

Basic Principles of Urine Formation   953
Filtration   954
Reabsorption and Secretion   958
The Function of the Vasa Recta   967
Summary: Renal Function   968
The Composition of Normal Urine   969

URINE TRANSPORT, STORAGE, AND
    ELIMINATION   972

The Ureters   972
The Urinary Bladder   974
The Urethra   976
The Micturition Reflex and Urination   976

AGING AND THE URINARY SYSTEM   977

INTEGRATION WITH OTHER SYSTEMS   978

MEDIALAB   978
CHAPTER REVIEW   980

CHAPTER **27**
Fluid, Electrolyte, and
Acid–Base Balance   984

INTRODUCTION   985

FLUID BALANCE AND ELECTROLYTE BALANCE   985

Basic Concepts in the Regulation of Fluids and
    Electrolytes   987
An Overview of the Primary Regulatory
    Hormones   988
The Interplay between Fluid Balance and Electrolyte
    Balance   989
Fluid Balance   989
Electrolyte Balance   991

ACID–BASE BALANCE   997

The Importance of pH Control   997
Types of Acids in the Body   998

Mechanisms of pH Control   998
Maintenance of Acid–Base Balance   1001

DISTURBANCES OF ACID–BASE BALANCE   1004

Respiratory Acidosis   1005
Respiratory Alkalosis   1006
Metabolic Acidosis   1007
Metabolic Alkalosis   1008
The Detection of Acidosis and
    Alkalosis   1008

AGING AND FLUID, ELECTROLYTE, AND ACID–BASE
    BALANCE   1009

MEDIALAB   1011
CHAPTER REVIEW   1011

CHAPTER **28**
The Reproductive
System   1016

INTRODUCTION   1017

AN OVERVIEW OF THE REPRODUCTIVE
    SYSTEM   1017

THE REPRODUCTIVE SYSTEM OF THE MALE   1017

The Testes   1017
The Anatomy of a Spermatozoon   1025
The Male Reproductive Tract   1025
The Accessory Glands   1028
Semen   1029
The Penis   1030
Hormones and Male Reproductive
    Function   1032

THE REPRODUCTIVE SYSTEM OF THE
    FEMALE   1034

The Ovaries   1035
The Uterine Tubes   1039
The Uterus   1040
The Vagina   1045
The External Genitalia   1046
The Mammary Glands   1047
Hormones and the Female Reproductive
    Cycle   1049

THE PHYSIOLOGY OF SEXUAL INTERCOURSE   1053

Male Sexual Function   1053
Female Sexual Function   1056

AGING AND THE REPRODUCTIVE SYSTEM   1056

Menopause   1056
The Male Climacteric   1057

INTEGRATION WITH OTHER SYSTEMS   1057

MEDIALAB   1059
CHAPTER REVIEW   1059

CHAPTER 29
Development and Inheritance   1064

INTRODUCTION 1065

AN OVERVIEW OF TOPICS IN DEVELOPMENT   1065

FERTILIZATION 1065

The Oocyte at Ovulation   1066
Oocyte Activation   1067

AN OVERVIEW OF PRENATAL DEVELOPMENT   1067

THE FIRST TRIMESTER   1068

Cleavage and Blastocyst Formation   1068
Implantation   1068
Placentation   1072
Embryogenesis   1075

THE SECOND AND THIRD TRIMESTERS   1081

Pregnancy and Maternal Systems   1081
Structural and Functional Changes in the Uterus   1083

LABOR AND DELIVERY   1084

Stages of Labor   1084
Premature Labor   1084
Multiple Births   1086

POSTNATAL DEVELOPMENT   1086

The Neonatal Period, Infancy, and Childhood   1086
Adolescence and Maturity   1088
Senescence   1089

GENETICS, DEVELOPMENT, AND INHERITANCE   1090

Genes and Chromosomes   1090
The Human Genome Project   1095

MEDIALAB   1098

CHAPTER REVIEW   1098

APPENDIXES A-1

Appendix I: Answers to Concept Check Questions   A-1
    Answers to End-of-Chapter Questions   A-11
Appendix II: Weights and Measures   A-32
Appendix III: Periodic Table   A-35
Appendix IV: Condensed Structural Formulas of the
    Amino Acids   A-36
Appendix V: Eponyms in Common Use   A-37
Appendix VI: Normal Physiological Values   A-39

GLOSSARY OF KEY TERMS   G-1

ILLUSTRATION CREDITS   IC-1

INDEX   I-1

LIST OF CLINICAL DISCUSSIONS

Homeostasis and Disease 14
Fatty Acids and Health 48
Tonicity and Normal Saline 74
DNA Fingerprinting 91
Mutations 97
Exfoliative Cytology 115
Marfan's Syndrome 120
Adipose Tissue and Weight Control 122
Fibrocartilages and Joint Injuries 128
Problems with Serous Membranes 130
Tissue Structure and Disease 137
Variations in Skin Pigmentation 149
Drug Administration through the Skin 150
Problems with Dermal Circulation 151
Dermatitis 152
Hair Loss 155
Problems with Sebaceous Glands 156
Burns and Grafts 162
Abnormal Bone Growth and Development 181
TMJ Syndrome 208
Sinus Problems and Septal Defects 210
Craniostenosis and Microcephaly 211
Thoracentesis 222
Bursitis 249
Problems with Intervertebral Discs 257
Knee Injuries 263
The Muscular Dystrophies 277
Interference with Neural Control Mechanisms 281
Rigor Mortis 283
Tetanus 289
Muscle Atrophy 298
Delayed-Onset Muscle Soreness 300
What's New in Anatomy? 325
Hernias 335
Sports Injuries 339
Carpal Tunnel Syndrome 344
Intramuscular Injections 349
Compartment Syndrome 355
CNS Tumors and Neuroglia 370
Neurotoxins 382
Drugs and Synaptic Function 390
Factors That Disrupt Neural Function 398
Spinal Taps and Myelography 411
Damage to Spinal Tracts 413
Nerve Palsies and Spinal Injuries 416
Reinforcement and Inhibition 430
Epidural and Subdural Hemorrhages 443
Hydrocephalus 444
Penetrating the Blood–Brain and Blood–CSF
   Barriers 445
Aphasia and Dyslexia 449
Disconnection Syndrome 451
Cerebral Palsy 489
Amyotrophic Lateral Sclerosis 492
Anencephaly 493
Modifying Autonomic Nervous System Function 522

Biofeedback 523
Tickling and Itching 533
Corneal Transplants 541
Diabetic Retinopathy 545
Cataracts 546
Glaucoma 546
Accommodation Problems 548
Scotomas 550
Retinitis Pigmentosa 550
Night Blindness 554
Color Blindness 555
Motion Sickness 563
Hearing Deficits 571
Growth Hormone Abnormalities 590
Disorders of ADH Production 591
Thyroid Gland Disorders 597
Disorders of Parathyroid Function 598
Disorders of Aldosterone Secretion 601
Abnormal Glucocorticoid Production 602
Abnormal Activity in the Zona Reticularis 602
Diabetes Mellitus 607
Leptin: A Different Hormone 609
Endocrinology and Athletic Performance 611
Endocrine Disorders 616
Blood Collection for Analysis 626
Plasma Expanders 627
Abnormal Hemoglobin 631
Blood Tests and RBCs 634
Transfusions 637
Hemolytic Disease of the Newborn 638
Abnormal Hemostasis 649
Pericarditis 657
Valvular Heart Disease 664
Coronary Artery Disease 667
ECG Analysis 674
Heart Attacks 678
Mitral Valve Prolapse 679
Effects of Drugs on Contractility 683
Abnormal Conditions Affecting Cardiac Output 686
Aneurysms 695
Arteriosclerosis 696
Hypertension and Hypotension 706
Checking the Pulse and Blood Pressure 707
Edema 710
Circulatory Shock 720
Strokes 729
Congenital Circulatory Problems 743
Lymphedema 756
Infected Lymphoid Nodules 759
Cancer and the Lymphatic System 760
Injury to the Spleen 762
NK Cells, Cancer, and Viral Infections 766
Necrosis 770
Graft Rejection and Immunosuppression 777
Drug Reactions 780
Manipulating the Immune Response 784

Treatment of Anaphylaxis   789
AIDS   790
Cystic Fibrosis   800
Nosebleeds   802
Laryngeal Problems   805
Tracheal Blockage   805
Tuberculosis   808
Bronchoscopy   809
Pneumonia   813
Pulmonary Embolisms   813
Pneumothorax   818
Artificial Respiration   819
Pulmonary Function Tests   822
Decompression Sickness   824
Blood Gas Analysis   825
Carbon Monoxide Poisoning   827
Adaptations to High Altitude   829
SIDS   834
Emphysema and Lung Cancer   839
Epithelial Renewal and Repair   848
Peritonitis   851
Mumps   855
Dental Problems and Solutions   858
Esophageal Varices   860
Gastritis and Peptic Ulcers   865
Stomach Cancer   868
Vomiting   872
Pancreatitis   874
Portal Hypertension   877
Liver Disease   878
Problems with Bile Storage and Excretion   880
Colon Cancer   884
Diverticulosis   884
Diarrhea and Constipation   887
Malabsorption Syndromes   893
Carbohydrate Loading   910
Dietary Fats and Cholesterol   915
Obesity   923
Ketoacidosis   925
Hypervitaminosis   930
Alcohol: A Risky Diversion   931
Eating Disorders   932

Heat Exhaustion and Heat Stroke   935
Induced Hypothermia   936
Analysis of Renal Blood Flow   946
Glomerulonephritis   952
Polycystic Kidney Disease   952
Expressing Solute Concentrations   954
Blood Pressure and Glomerular Filtration   955
Renal Failure   958
Diuretics   968
Advances in the Treatment of Renal Failure   970
Urinalysis   972
Problems with the Conducting System   974
Bladder Cancer   976
Urinary Tract Infections   976
Incontinence   977
Athletes and Salt Loss   994
Hypokalemia and Hyperkalemia   995
Hypercalcemia and Hypocalcemia   995
Fluid, Electrolyte, and Acid–Base Balance
   in Infants   1010
Cryptorchidism   1018
Testicular Cancer   1025
Prostatitis   1029
Semen Analysis   1030
The Circumcision Debate   1031
Prostatic Hypertrophy and Prostate Cancer   1032
DHEA   1034
Ovarian Cancer   1039
Pelvic Inflammatory Disease (PID)   1040
Cervical Cancer   1042
Endometriosis   1045
Breast Cancer   1048
Premenstrual Syndrome   1052
Impotence   1053
Birth Control Strategies   1054
Sexually Transmitted Diseases   1056
Gestational Trophoblastic Neoplasias   1070
Technology and the Treatment of Infertility   1079
Abortion   1083
Forceps Deliveries and Breech Births   1086
Chromosomal Abnormalities and Genetic
   Analysis   1097

Preface

To the Instructor

CHANGING TIMES

When the First Edition of *Fundamentals of Anatomy and Physiology* was published in 1989, the options available for presenting, studying, and applying principles of anatomy and physiology (A & P) were much more limited than they are today. Our knowledge base has also expanded tremendously. As we proceed into the new millennium, we know much more than ever before about the biochemical and molecular foundations of many normal physiological processes, we understand the molecular origins of many diseases, and we are close to completing an initial survey of the human genome. This enhanced understanding has led to the development of effective new treatments, protocols, and interventions to combat disease, reduce suffering, and promote good health. Many of these advances could not have been made if they were not preceded by improvements in the speed and reliability of computers. Modern technology, especially computers and the Internet, has changed our daily lives and is reshaping our societies in many ways.

The emerging medical technologies allow us to perform complex surgical procedures on individual organs or to target specific cells within an organ. We can manipulate the DNA within cells, and we can even clone genetically identical animals, although we are still unable to repair even the simplest genetic defect in humans. Many people are horrified by what we can now do with technology; others are intent on pushing ahead without a clear idea of the ultimate destination. Nevertheless, all of us are affected by modern technology. Technology has become a means for acquiring and distributing information, and medical professionals have learned to use technology to increase their effectiveness as well as to improve the quality of medical care.

By enhancing the availability of *specific* information, modern technology has affected how each of us deals with information *in general*. In many cases, knowing where to look for information is preferable to trying to memorize specific data. That is certainly the trend in the training of medical and allied health professionals today. Students begin by mastering the terminology and memorizing a substantial core of basic conceptual information. In the process, they are also given (1) a "mental framework" for organizing new information, (2) the ability to access additional information when needed, by consulting relevant print or electronic data sources, and (3) an understanding of how to apply their knowledge to solve particular problems. The same skills are equally important to people in other career paths. To be effective in almost any job today, you must know how to access and absorb new information, to use (or learn to use) available technology, and to solve problems.

The ultimate goal of any anatomy and physiology course should be to empower students to use their conceptual understanding to solve problems. Preparation for this goal involves a substantial amount of memorization, but the burden of memorization is reduced if the relevance of the information is evident. *Fundamentals of Anatomy and Physiology* has been designed to place the information in a meaningful context and to help students develop their problem-solving skills. The electronic enhancements, including the enclosed CD-ROM and the Companion Website, make it easy for students and instructors to use current technology to access and manage information.

A & P TODAY

Over the last 5 years I have visited instructors and students at campuses throughout North America, Australia, and New Zealand. I have found that most instructors are aware of this relationship among information, technology, and problem solving. I have also found them seeking answers to similar questions, which could be summarized as follows:

- How do I find the time?
- How much material should I teach?
- Which media should I use?

These questions are asked by college faculty members in all courses; in fact, a very similar set of questions is asked by students. (See "To the Student" on page xxix.) The situation is probably more troublesome in A & P than in other courses, because this is the first college-level science course many of these students will take. They must learn not only a new vocabulary but also new ways of studying and organizing information. For most instructors, teaching anatomy and physiology has never been more challenging; in addition to teaching terminology, facts, and concepts, instructors must help their students be good problem solvers. Mastering terminology, facts, and concepts is of little value if that mastery cannot be used to solve problems. It would be analogous to a person's knowing all the parts of an automobile and the concepts behind the internal combustion engine without being able to back the car out of the garage and drive it away.

The focus of this text and its support materials has been to simplify the teaching and learning of anatomy and physiology. In the revision process, I've tried to address each of the above questions.

■ *How do I find the time?* Nothing I can do can give any of us more time. But I have tried to help you make better use of the time you do have. My approach has been (1) to provide a range of options for reducing preparation time for classroom presentations, and (2) to improve the delivery of information in the text and supplements package so that it can assume more of the teaching load.

■ *How much material should I teach?* The answer will vary from class to class and from instructor to instructor. Some courses have prerequisites, many others do not; some courses are taught to prenursing students, others to physical education students, and still others to general biology students. Few instructors cover all the material in the textbook. Therefore, I have tried to assist both instructors and students by making the chapter organization clear at the outset and by providing an internal structure that helps instructors decide—before they read the section—whether the material in it is relevant for their class. Sections can be assigned or skipped as needed without interrupting the flow of information.

■ *Which media should I use?* Instructors vary widely in their need for media products. This variety reflects differences in teaching styles and in classroom facilities. I have visited schools where "multimedia" means colored chalk and other schools where every seat in the lecture hall has a response pad and a data port for a laptop computer. My answer to this question, then, is "How do you want to deliver the key information?" All lecturers rely on visual support of one kind or another. The resources available with this textbook run the gamut from transparencies to a videotape and laser disk, to digital video, and a CD-ROM with an Image Bank and PowerPoint Gallery of slides for computer presentations. For those interested in Web courseware, you can start simply and easily with the Syllabus Manager at the Companion Website to supplement the lecture or lab work, or use WebCT or Blackboard to design distance-learning or Web-based courses. The content is consistent across the various formats. For example, all art in the textbook is prepared by a team of two medical illustrators, so the art is consistent from figure to figure and chapter to chapter. That same art appears on the image bank, in the PowerPoint slides, and on the website, and it forms the basis for the animations and tutorials. In short, the key information is available to you, however you choose to deliver it.

I have often heard the statement "all A & P books are the same." Well, yes and no. All books in this market are about the same size and length, and they all have a similar organization, colorful illustrations, and an assortment of supplements. However, that does not mean that all A & P books are exactly alike, any more than all cars are alike simply because they all have engines and tires and can transport you from one place to another. This textbook has been designed to meet specific needs. How were those needs determined? Personal experience has helped, but no one person has a complete view of any situation. So, wherever and whenever possible, I have met with instructors and students to learn more about shared problems, solutions, and perspectives. Prentice Hall has sponsored student focus groups and instructor focus groups, giving me the opportunity to get direct feedback and to "test-drive" new features. I have received hundreds of letters, phone calls, and e-mail messages from instructors and students with comments and suggestions about this textbook and its supplements. The result is a package that is more than just "reliable transportation"—it will help you negotiate the curves and avoid the potholes. The textbook's important and distinctive features are explained in "To the Student" (page xxix). I urge you to read it before your course gets under way.

WHAT'S DIFFERENT ABOUT THE FIFTH EDITION

Each new edition requires some revising and updating, and that was certainly the case here. In the process, I reexamined the material within each chapter to see what steps could be taken to improve the delivery of information and clarify key concepts. This reexamination was the key to demonstrating the relevancy and applicability of the material, to integrating the components of the learning system more effectively, and to providing additional opportunities for enhancing student understanding during lectures, lab periods, and study times through the use of appropriate technologies. To give you an idea of the scope and nature of the changes, I will begin with changes to the textbook and then consider changes to the teaching and learning system as a whole.

CHANGES TO THE NARRATIVE

The basic chapter sequence and organization of the text remains unchanged. However, in response to instructor and student feedback, I have streamlined the delivery of key physiological concepts and developed a new feature, called a *Navigator*, to help students keep track of their progress through the material. Simply put, a Navigator is a flow chart that includes the key topics covered within a section. This new feature is presented first as a full-size figure, accompanied by a narrative overview. It then reappears in simplifed form each time the reader "steps" from one key topic to another. The Navigator concept, which sounds simple, actually entailed more than just creating new artwork and adding overviews. In some cases, it involved the resequencing of material and led to the creation of summary tables that review key concepts before the next "step" is taken. This approach is suitable neither for all topics nor for all chapters, but it is a big help in getting through tough sections that deal with abstract physiological concepts. For instance, Navigators are used in Chapters 12 (neurophysiology), 21 (cardiovascular physiology), 22 (immunity), 23 (respiratory physiology), and 25 (cellular metabolism). A comparable pattern of presentation—overview followed by blocks of text and art accompanied by summary tables—has been used throughout, even in chapters that do not contain Navigators.

CHANGES TO THE ART PROGRAM

Both anatomical structures and physiological processes must be visualized if students are to understand them. One of the greatest strengths of the art program in this textbook has been that all illustrations in all chapters have been created by the same two medical illustrators: Bill Ober, M.D., and Claire Garrison, R.N. Their efforts in previous editions have resulted in awards from the Association of Medical Illustrators and from the Text and Academic Authors Association. Having one team responsible for the visual presentation of information ensures that structures and processes are depicted in a consistent manner from figure to figure and from chapter to chapter; *Fundamentals of Anatomy and Physiology* is the only A & P textbook that has a unified art program. (To learn more about Bill and Claire, please see their brief biographies on page iii.) We work together as the manuscript evolves, and they read key sections of the manuscript to make sure that the visual and narrative presentations are in alignment. In addition, we have been able to work with Ralph Hutchings, a biomedical photographer recently cited as one of the best photographers of the twentieth century for his anatomical portrayals of the human body. Bill and Claire have incorporated many new photos into the textbook and the Companion Atlas of the *Applications Manual,* and Ralph's photos appear in the textbook, and in the media, Companion Atlas, and lab manuals that accompany the Fifth Edition.

A detailed examination of virtually any chapter in the Fifth Edition will reveal both subtle and dramatic improvements and additions to the illustration program. Dramatic improvements in the art program for this edition include the following:

- Chapters 7–9: New bone photographs have been supplied by Ralph Hutchings, and several photos have been enlarged. New views have been added, providing additional perspectives on the structure of the elbow, hip, and knee joints.

- Chapter 10: Several new illustrations have been added, and the sequence has been reordered to create visual links between succeeding figures and to help students better visualize the key aspects of muscle physiology.
- Chapter 11: New anatomical paintings have been included, depicting the muscles of the forearm and hip, and many figures have been enhanced and improved.
- Chapter 12: The art program for the neurophysiology section has been modified, with new Navigators and new art and illustrated tables linked to the discussions of graded potentials, action potentials, synaptic function, and neural processing.
- Chapters 13, 14, 20, 21, 22, 26, and 28 (among others): New anatomical art and new physiological diagrams have been added to accompany a revised coverage of reflexes, cardiovascular function, immune function, and renal function.

MEDIALABS

Each chapter of the Fifth Edition contains an integrated learning tool called a MediaLab. These innovative sections have been developed with the assistance of Kate Flickinger, Ph.D., of Maui Community College. Kate has extensive research and teaching experience in anatomy and physiology and specializes in media use and instructional design. (Her brief biography appears on page iv.)

Although computers continue to become more available to both faculty and students, relatively few instructors have used Web-based assignments in an A&P course. The reasons are that (1) instructors don't have time to cover the material already in the course syllabus and text; (2) there is no demonstrable link between the URL provided and the course material that will affect student grades; and (3) there are no guidelines for time expenditure—students could wander on the Internet for hours, when they might be better off spending that time studying their textbook and lecture notes. MediaLabs address these concerns by providing specific, time-constrained exercises with clearly stated objectives that relate the Web-based activities to the core chapter content. It is our hope that this feature will enable instructors to capitalize on the wealth of Web-based content as it provides students with alternative learning opportunities.

FORMAT AND DESIGN CHANGES

Introductory A & P students often need a lot of help with organizing and integrating the material, so this textbook offers a variety of pedagogical aids. The red figure locator dots, red checkmarks in Concept Check Questions, and Concept Link icons that distinguish this textbook from all others are important and effective learning aids that were, in many cases, suggested by students themselves. Focus-group participants and reviewers who used the Fourth Edition—as students or as instructors—were virtually unanimous in telling me to retain these features in the Fifth Edition. These pedagogical aids were retained, but the design has been changed to soften their visual impact. As a bonus, this new design has enabled me to coordinate the narrative and illustrations more closely.

CHANGES TO THE LEARNING SYSTEM

One of the most important aspects of this new edition, from my perspective, is that I had the chance to look at the functionality and integration of all the components of the learning system developed over the last 12 years. That means examining not just the textbook and *Applications Manual,* for which I am largely responsible, but the study guides, CD-ROMs, websites, and other components that are created by others. As you know, each medium has its strengths and weaknesses. So, when I started dealing with content issues, reviewer comments, and so forth for the textbook revision, I also began working with the talented editorial and media staff at Prentice Hall to examine each element in the learning system. On the basis of feedback from faculty, students, and technical experts, we made decisions as to how the various components might be rearranged to make better use of the available media and to improve the delivery of the revised text content. I can give you a few specific examples here; others will become apparent as you become more familiar with the new edition.

We began by looking at the material in the textbook, the *Applications Manual,* student CD-ROMs, and Companion Website as though it were a single pool of resources. We then allocated content to take advantage of the strengths of each medium. After distributing assets, we evaluated each to determine how it could be enhanced and integrated with the whole to support an improved learning system. This evaluation led to several major changes from the approach used in the Fourth Edition.

- The Companion Website has been enhanced to include roughly four times as many exercises, many of them graphics-based. It also includes animations, the eBook version of the textbook, and more extensive links to other sites. The Companion Website now incorporates many components that were formerly distributed among different media, making it more convenient and accessible to students.

- Critical-Thinking Questions and Clinical Problems, formerly included in the *Applications Manual,* now appear in an interactive framework on the Companion Website. Adaptation and integration with the Companion Website have made it easier to seek answers, share results, and submit conclusions and opinions for review by study groups or to instructors.

- The content and approach of the interactive exercises used in the Fourth Edition CD-Study Guide has been adapted to the Companion Website for the Fifth Edition. Although the CD-Study Guide has been discontinued, the questions have been merged into a common resource on the Companion Website.

- The eBook version of the textbook, which in the Fourth Edition was part of the CD-ROM, has been moved to the Companion Website, where it can be linked to animations and related websites and used as a resource while you work through the review exercises. Like the Fourth Edition's eBook, this one includes art from the textbook as well as more than 75 animations developed directly from that art. The eBook also has linked Glossary terms and Concept Check Questions (with links to the answers).

- The CD-ROM accompanying the Fifth Edition contains animations, simulations, and tutorials that are either interactive or so graphics-intensive that, given the connect speeds available to most students, it would be impractical to deliver across the Internet. The CD-ROM also contains an audio-glossary to enhance the vocabulary of anatomy and physiology and Case Studies to support problem-based learning. Additionally, a search tool is available to help users easily locate the material most important to them. These resources are fully operational without an Internet portal, but if one is available, the CD-ROM includes a direct link to the Companion Website.

- With the exception of the application questions, which have moved to the Companion Website, and the Case Studies, which have moved to the CD-ROM, the scope and coverage of topics discussed in the *Applications Manual* remains comparable to that of the Fourth Edition. (Additional details are provided in a later section of this Preface.) However, the clinical content has been improved and updated, many new illustrations have been added, and the tables have been extensively revised.

- Images from the Fourth Edition *Applications Manual* have been combined with new images to create a Companion Atlas. This atlas, which appears at the back of the *Applications Manual,* contains many new photographs by Ralph Hutchings. Each image is cross-referenced to figures in *Fundamentals of Anatomy and Physiology.* By placing these images in the Companion Atlas, they can be presented in a larger format than would be possible were they included in the textbook itself, and they can be referenced as needed from many chapters. The Companion Atlas includes (1) color Embryology Summaries dealing with the developmental origins of each body system; (2) color plates of surface anatomy, skeletal anatomy, and cadaver prosections; and (3) a series of high-quality black and white images produced by CT, MRI, and contrast X-ray techniques.

Each of these components will be considered individually in the Supplements section next. But this brief overview should give you a sense of the nature and goals of the reorganization.

THE SUPPLEMENTS PACKAGE

The supplements team consists of talented A & P instructors, most of whom have been using this textbook since its First or Second Edition. Their efforts are coordinated by the supplements editors and multimedia editors at Prentice Hall. During the development of the Fifth Edition, they have worked with me as an extra team of reviewers. In return, I have provided them with additional suggestions and comments while their revisions were under way.

FOR THE INSTRUCTOR

Instructor's Preparation Box

Prentice Hall and I have tried to simplify the task of preparing lectures for this course. The Instructor's Preparation Box contains three sets of materials for each chapter. The first set contains traditional Instructor's Manual materials, including detailed lecture notes and an organizational grid of visual resources by chapter. This set also contains suggested demonstrations, analogies, answers to end-of-chapter questions, and mnemonic activities designed to engage students' interest and to provide some additional flavor for your lectures. The second set contains student handouts—including crossword puzzles and Quiz Art (line art without labels) for each chapter. The third set contains all the transparency acetates by chapter. Also included is a Lecture Presentation Notebook to help you organize your lecture material.
(0-13-019675-4)

Instructor's Manual

This useful resource includes a wealth of materials to help you prepare and organize your lectures, such as lecture notes, vocabulary aids, applications, and classroom demonstrations. To help you organize all the visual resources provided in the supplements package, this manual also includes a detailed grid that correlates each section of the textbook with the visual aids that support that section. At one glance you can determine which transparencies, animations, and CD-ROM images are available for your lecture. New to this edition are MediaLab Instructor Notes, designed to ensure successful and meaningful use of the MediaLab activities in each chapter of the textbook.
(0-13-019676-2)

Transparencies

The transparency set has been significantly improved and expanded. More than 800 acetates are included, with art and labels enlarged for use in large lecture halls. Where possible, complex figures have been divided into separate transparencies for improved clarity and teaching effectiveness. The selection of acetates and the quality of each image have been assessed by faculty reviewers to ensure that we offer the best possible presentation.
(0-13-019677-0)

Test Item File and Computer Test Manager

The Test Item File has been thoroughly revised and reorganized for easier selection of questions for every objective in the textbook. A test bank of more than 3000 questions organized around the three-level learning system (as seen in the textbook and Study Guide) will help you design a variety of tests and quizzes.
(Print format: 0-13-019678-9)
(Software: Windows 0-13-019679-7, Mac 0-13-019681-9)

Image Bank and PowerPoint Gallery

Available in both Windows and Macintosh formats, this CD-ROM includes both an Image Bank and a PowerPoint Gallery. The Image Bank contains more than 1100 illustrations, including line art and photos from the textbook, cadaver and cat dissection photos, MRI and CT scans, X-rays, and laboratory model photos. Of special note is the inclusion of a library of unlabeled art. This material can easily be used to develop art-based quizzes, customized to serve individual presentation needs, or printed within your course support material. In addition, the CD-ROM includes animations of more than 75 of the most difficult concepts in anatomy and physiology. To provide yet another resource, Prentice Hall has assembled a complete series of PowerPoint slides for the entire course. These slides can be customized to fit your particular needs, but the basic—and time-consuming—work has been done for you.
(0-13-019682-7)

Laser Disk

The Prentice Hall laser disk, *Anatomy and Physiology in Motion*, features a collection of professionally narrated animations of the most complex topics and processes in anatomy and physiology.
(0-13-751777-7)

The Martini Companion Website
http://www.prenhall.com/martini/fap5

The new Companion Website has been designed to make it easier to teach anatomy and physiology and to make it easier to incorporate Web-based activities into the core curriculum. One of the key advances is the inclusion of chapter-based MediaLabs. MediaLabs provide specific, time-constrained exercises, called Web Explorations, with clearly stated objectives that relate the Web-based activities to the chapter content. The Web Explorations have been designed (1) to enable instructors to capitalize on the wealth of Web-based content and (2) to provide students with alternative learning opportunities. To help you implement these MediaLabs into your course, the Instructor's Manual offers instructions for their use.

The Companion Website also provides a wealth of other materials to help instructors enhance the course. One of these is the Syllabus Manager tool, which enables instructors to build easily and manage custom online syllabuses. This tool provides a seamless way for you to link to content provided by Prentice Hall and offered by other sources on the Web. If you've ever thought about offering Web-based or distance-learning options to your course, then you should consider this free resource.

In addition to the basic tools available on the Companion Website for this textbook, Prentice Hall provides rich and thorough course cartridges for both the WebCT and the Blackboard course management systems. Many campuses have already selected one of these course management tools to help instructors develop and deliver content for Web-based and distance learning.

I hope that as the year passes, you will use the Companion Website or my email address, listed at the end of the Preface, to contact me with suggestions for future improvements, new illustrations, or news of interesting research developments. Distance is no longer a limiting factor in our "global A & P village," and I hope that even more instructors and students will participate in the development of future editions.

FOR THE STUDENT

Applications Manual

Each new copy of the text comes packaged with an *Applications Manual,* which is written in collaboration with Kathleen Welch, M.D., my wife, who is also the clinical consultant for the text. This unique supplement provides access to interesting and relevant clinical and diagnostic information. It presents that information in a framework that helps students develop problem-solving skills. The book begins with an overview of the way diseases are diagnosed, and this leads to a discussion of the scientific method and to the use of logic and reasoning to evaluate health-related statements and claims. It then provides background information about chemical, cellular, and molecular disorders and techniques. Finally, it presents the major categories of disorders and important diagnostic techniques for each body system. Although it can be read separately, each major topic in the *Applications Manual* is cross-referenced to specific pages in the textbook.

The *Applications Manual* is designed to demonstrate the relevance of key concepts presented in the textbook. By placing the material in a separate volume, topics can be covered in greater depth—or skipped—without interrupting the flow of information in the textbook. Few instructors will require their students to read the entire *Applications Manual,* but each student will refer to segments of particular importance to their career plans or their personal or family health. In subsequent courses, it becomes an invaluable reference that links the topics covered in more-advanced courses with the basic anatomy and physiology covered in their introductory-level course.

The *Applications Manual* also contains a Companion Atlas, which consists of Embryology Summaries for all organ systems; photographs of skeletal anatomy, surface anatomy, cadaver prosections, and plastic models; and representative images produced by CT, MRI, and contrast X-ray procedures. (0-13-031117-0)

Study Guide

This very popular Study Guide, written by Charles Seiger, is an excellent way to review basic facts and concepts as well as to develop problem-solving skills. A variety of questions, including labeling and concept mapping, are keyed to every learning objective in the textbook and are organized around the three-level learning system used in the textbook. (0-13-019691-6)

The Martini Companion Website
http://www.prenhall.com/martini/fap5

Not only will the new design and content of the Companion Website for *Fundamentals of Anatomy and Physiology* make it easier for instructors to use and assign material, but this website has also been redesigned to make it easier for students to learn anatomy and physiology. Notice that I did not say that the website makes it *easy* to learn this subject; rather, it makes the process of learning anatomy and physiology less stressful and more interesting. To do this, the Companion Website must be engaging, appealing to students with diverse learning styles and backgrounds. It must not just provide reinforcement and feedback, although those are important; it must also provide insights that help clarify difficult topics. The benefits of working with this website must be apparent and real. Students should come away with a better understanding of the material, and that understanding should be reflected in their performance on the next exam. To accomplish these goals, the media team has redesigned the website and integrated multi-level questions, image banks, links to other websites, guided activities, and the eBook version of the textbook. For logistical and copyright reasons, the more-comprehensive Companion Website resources are passcode-protected. Each new copy of the Fifth Edition comes with a passcode that, when activated by a student, remains "live" for 18 months. Passcodes can also be purchased online for a small fee. Each instructor receives a passcode good for the life of the edition.

Martini Interactive Student CD-ROM

This CD-ROM was designed to accompany your textbook. It is fully integrated with both the textbook and its Companion Website. The CD-ROM contains interactive tools to help you visualize anatomy and physiology; three-dimensional animations; fly-through panoramas of the heart and gastrointestinal tract; interactive tutorials; clinical case studies; an audio glossary; and a link to the Companion Website.

The animations were developed with data from the Visible Human Project, sponsored by the National Library of Medicine. An easy-to-use navigation bar allows you to rotate each structure 360°, peel away the layers of each image, and view the structures with the labels on or off. The interactive tutorials present the most difficult topics in anatomy and physiology so as to help you discover the underlying principles while you control the flow of information. Each tutorial includes audio narration throughout. The clinical case studies are organized into steps that give you the opportunity to practice and apply your knowledge of anatomy and physiology in a practical and realistic way. These case studies, derived from actual case histories, support problem-based learning.
(0-13-019692-4)

Science on the Internet: A Student's Guide 2000–2001

This guide is the latest in a series written by Andrew Stull and Harry Nickla to help you use the Internet. It is a handy reference to get you up to speed on the World Wide Web's vast resources, including the Companion Website for this book. The guide, a unique resource, gives clear steps to help you access regularly updated biology resources as well as an overview of general navigation strategies and techniques to help you critique material that you find on the Web.
(0-13-028253-7)

Video Tutor for Anatomy and Physiology

This videotape offers high-quality tutorials for anyone with access to a VCR. Each segment includes a professionally narrated animation that walks you through the most difficult physiological concepts and offers self-check questions that allow you to test your understanding of the material. Segments of the Video Tutor include membrane transport, protein synthesis, muscle contraction, action potential propagation, vision, auditory function, heart function, urine formation, and the immune response.
(0-13-751843-9)

The New York Times "Themes of the Times"

Prentice Hall's unique alliance with *The New York Times* enhances your access to current, relevant information and applications. Articles are selected by the textbook author and are compiled into a free supplement that helps you make the connection between your classroom and the outside world.
(0-13-089311-0)

FOR THE LABORATORY

Prentice Hall publishes a variety of laboratory manuals to meet the diverse needs of anatomy and physiology labs. Please see your Prentice Hall sales representative for more details. Here is a list of those manuals:

Laboratory Textbook of Anatomy and Physiology, Second Edition, by Michael G. Wood. This full-color, 700-page lab manual has been designed to complement *Fundamentals of Anatomy and Physiology* by using compatible line art and terminology. It features cat dissections.
(0-13-019694-0)

Laboratory Manual for Fundamentals of Anatomy and Physiology, Fifth Edition, by Roberta M. Meehan. Cross-referenced to the *Fundamentals* textbook and using compatible line art, this manual features both cat and fetal pig dissections.
(0-13-019693-2)

Laboratory Exercises in Anatomy and Physiology with Cat Dissections, Sixth Edition, by Gerard J. Tortora and Robert B. Tallitsch. This full-color manual features cat dissections and graphics.
(0-13-920323-0)

Anatomy and Physiology Laboratory Manual, Sixth Edition, by Gerard J. Tortora and Robert J. Amitrano. This full-color manual offers a variety of laboratory exercises with minimal mammalian dissections.
(0-13-089670-5)

ACKNOWLEDGMENTS

This textbook is not the product of any single individual; rather, it represents a group effort, and the members of the group deserve to be acknowledged. Foremost on my thank-you list are the faculty and reviewers throughout the world who offered suggestions that helped guide me through the revision process. Their interest in the subject, concern for the accuracy and method of presentation, and experience with students of widely varying abilities and backgrounds made the revision process an educational experience for me. To them, I express my sincere thanks and best wishes:

Michèle Bertholf, *Metropolitan State College* and *Front Range Community College*
Michael Bonnert, *University of Toronto*
Charles Dick, *Pasco–Hernando Community College*
Lori K. Garrett, *Danville Area Community College*
Anthony J. Gaudin, *Ivy Tech State College*
Nancy G. Kincaid, *Troy State University Montgomery*
William Mautz, *University of Hawaii at Hilo*

Michael P. McKinley, *Glendale Community College*
Ann Murphy, *Sydney University*
Claire R. Oakley, *Rocky Mountain College*
Betsy Ott, *Tyler Community College*
Michael Palladino, *Monmouth University*
John M. Ripper, *Butler County Community College*
Carolyn J. Rivard, *Fanshawe College*
Frank Schwartz, *Cuyahoga Community College*
Sherrie Shupe, *Delaware Technical and Community College*
Sandra Stewart, *Vincennes University*
Ann Stoeckmann, *Pennsylvania State University–Worthington Scranton*
Diane G. Tice, *SUNY Morrisville*
Bruce Wingerd, *San Diego State University*
Nancy L. Young, *Seattle Pacific University*

After the initial drafts were completed, two dedicated A & P instructors functioned as technical editors, assisting with the review and correction of page proofs:

Michael P. McKinley, *Glendale Community College*
Bruce Wingerd, *San Diego State University*

I extend special thanks to them not only for their work in reviewing pages, but also for providing suggestions during the manuscript stage. They have been immensely helpful to me during the revision of this edition.

Focus groups and casual meetings with students around the world helped me concentrate on the needs of individual students. I first undertook this project in 1983 to address the needs of my own students, and their feedback continues to be very important to me. I also thank the many instructors and students who took the time to contact me by phone or e-mail with specific suggestions, kudos, questions, or problems.

Over time, a textbook evolves. As with organisms, each evolutionary step builds on a preexisting framework. Thus, I also thank the individuals who helped with the development of previous editions. That list includes the following:

Maxine A'Hearn, *Prince George's Community College*
Ahmed Naguy Ali, *Alexandria, Egypt*
Paul Anderson, *Massachusetts Bay Community College*
Timothy Alan Ballard, *University of North Carolina–Wilmington*
Debra Joan Barnes, *Contra Costa College*
Edwin Bartholomew, *Maui Community College*
CeCe Barto, *Tomball College*
Steven Bassett, *Southeast Community College*
Robert Bauman, Jr., *Amarillo College*
Dean Beckwith, *Illinois Central College*
Doris Benfer, *Montgomery Community College*
Latsy Best, *Palm Beach Community College–North*

Charles Biggers, *University of Memphis*
Cynthia Bottrell, *Scott Community College*
Spencer R. Bowers, *Oakton Community College*
Mimi Bres, *Prince George's Community College*
Alan Bretag, *University of South Australia School of Pharmacy*
C. David Bridges, *Purdue University*
Sandra Bruner, *Polk Community College*
Gene Carella, *Niagara County Community College*
Robert M. Carey, *University of Arizona*
Wayne Carley, *Lamar University*
Lucia Cepriano, *SUNY Farmingdale*
William M. Chamberlain, *Indiana State University*
William D. Chapple, *University of Connecticut*
Beng Cheah, *University of Newcastle*
Anthony Chee, *Houston Community College*
Suzette Chopin, *Texas A & M University at Corpus Christi*
Chin Moi Chow, *Cumberland College of Health Sciences, University of Sydney*
O. D. Cockrum, *Texas State Technical College*
Kim Cooper, *Arizona State University*
Richard Coppings, *Chattanooga State College*
William F. Crowley, *Harvard Medical School*
Grant Dahmer, *University of Arizona*
Charles Daniels, *Kapiolani Community College*
Brent DeMars, *Lakeland Community College*
Gerald R. Dotson, *Front Range Community College*
Ellen Dupre, *Indian River Community College*
John Dziak, *Community College of Allegheny County*
Lee Famiano, *Cuyahoga Community College*
Lee Farello, *Niagara County Community College*
Marion Fintel, *Jefferson State Community College*
Kathleen A. Flickinger, *Maui Community College*
Ruby Fogg, *New Hampshire Tech*
Mildred Fowler, *Tidewater Community College*
Sharon Fowler, *Dutchess Community College*
Denise Friedman, *Hudson Valley Community College*
Ann Funkhouser, *University of the Pacific*
Mildred Galliher, *Cochise College*
Jeff Gerst, *North Dakota State University*
Louis Giacinti, *Milwaukee Area Technical College*
Delaine Gilcrease, *Mesa Community College*
Freda Glaser, *University of Maryland–Baltimore County*
Kathleen M. Gorczyca, *North Shore Community College*
Bonnie Gordon, *Memphis State University*
Mac E. Hadley, *University of Arizona*
William Hairston, *Harrisburg Area Community College*
Cecil Hampton, *Jefferson College*
Ernest Harber, *San Antonio College*
John P. Harley, *Eastern Kentucky University*
Ann Harmer, *Orange Coast Community College*
Linden Haynes, *Hinds Community College*
Ruth Lanier Hays, *Clemson University*
Mary Healey, *Springfield College*
Jean Helgeson, *Collin County Community College*

Vickie S. Hennessy, *Sinclair Community College*
Donna Hoel, *Stark Community College*
Elvis J. Holt, *Purdue University*
Jacqueline A. Homan, *South Plains College*
James Horwitz, *Palm Beach Community College*
Beth Howard, *Rutgers University*
Yvette Huet-Hudson, *University of North Carolina–Charlotte*
Angie Huxley, *Pima Community College*
Aaron James, *Gateway Community College*
Desiree Jett, *Essex County College*
Drusilla Jolly, *Forsyth Technical Community College*
David Kalichstein, *Ocean County College*
Eileen Kalmar, *St. John Fisher College*
George Karleskint, *St. Louis Community College*
C. Ward Kischer, *University of Arizona*
Frank Kitakis, *Wayne County Community College*
William Kleinelp, *Middlesex County College*
Michael Kokkinn, *University of South Australia School of Pharmacy*
Bob Kucera, *University of Newcastle*
Jerry K. Lindsey, *Tarrant County Junior College*
Mary Lockwood, *University of New Hampshire*
Susan Lustick, *San Jacinto College North*
Greg Maravellas, *Bristol Community College*
Dan Mark, *Penn Valley Community College*
Jane Marks, *Scottsdale Community College*
Alice Gerke McAfee, *University of Toledo Technical College*
Thomas W. McCort, *Cuyahoga Community College*
Mike McCusker, *Eastern College*
Bob McDonough, *Georgia Perimeter College*
Ruth McFarland, *Mt. Hood Community College*
Paul McGrath, *University of Newcastle*
Mike McKinley, *Glendale Community College*
Roberta Meehan
Richard F. Meginniss, *College of Lake County*
Ann Miller, *Middlesex Community College*
Alice Mills, *Middle Tennessee State University*
Melvin Mills, *Scottsdale Community College*
Ron Mobley, *Wake Technical Community College*
Rose Morgan, *Minot State University*
Aubrey Morris, *Pensacola Junior College*
Robert L. Moskowitz, *Community College of Philadelphia*
Richard Mostardi, *Akron University*
Elizabeth Naugle, *Lane Community College*
Martha Newsome, *Tomball College*
Bill Nicholson, *University of Arizona*
Richard Northrup, *Delta College*
Erik Nyholm, *Umea University*
John M. Olson, *Villanova University*
Betty Orr, *Sinclair Community College*
Mary Theresa Ortiz, *Kinsborough Community College*
David L. Parker, *Northern Virginia Community College*
Mark Paulissen, *Slippery Rock University*
Brian K. Paulsen, *California University of Pennsylvania*

Lois Peck, *Philadelphia College of Pharmacy*
Philip Penner, *Borough of Manhattan Community College*
Ingrid Persson, *Umea University*
Clifford Pohl, *Duquesne University*
Robert Pollack, *Nassau Community College*
Robert L. Preston, *Illinois State University*
Gary Quick, *Paradise Valley Community College*
Anil Rao, *Metro State College of Denver*
Joel Reicherter, *SUNY Farmingdale*
Peta Reid, *University of Sydney, Faculty of Nursing*
Jean Rigden, *Scottsdale Community College*
Carolyn C. Robertson, *Tarrant County Junior College*
Kevin Ryan, *Stark Technical College*
Kristi Sather-Smith, *Hinds Junior College*
Frank Schwartz, *Cuyahoga Community College*
Charles Seiger, *Atlantic Community College*
Mark Shoop, *Macon College*
P. George Simone, *Eastern Michigan University*
Robert A. Sinclair, *San Antonio College*
Tom Smeaton, *University of South Australia School of Pharmacy*
David S. Smith, *San Antonio College*
Jeffery Smith, *Delgado Community College*
Philip Sokolove, *University of Maryland–Baltimore County*
Michael Soules, *University of Washington*
Thomas S. Spurgeon, *Colorado State University*
Jenna Sullivan, *University of Arizona*
Eric Sun, *Macon College*
Richard Symmons, *California State University at Hayward*
Dennis Taylor, *Hiram College*
Kathy Taylor, *University of Arizona*
Jay Templin, *Widener University*
Caryl Tickner, *Stark State Technical College*
Marge Torode, *Cumberland College of Health Sciences, University of Sydney*
Lucia Tranel, *St. Louis Community College* and *St. Louis College*
Steve Trautwein, *Southeast Missouri State University*
Pat Turner, *Howard Community College*
Kent M. Van De Graaff, *Brigham Young University*
Sheila Van Holst, *University of Sydney, Faculty of Nursing*
Michael Vennig, *University of South Australia School of Pharmacy*
Jane Wallace, *Chattanooga State Technical Community College*
Eva Weinreb, *Community College of Philadelphia*
Debra A. Wellner, *Wichita State University*
Rosamund Wendt, *Community College of Philadelphia*
J. Wilkinson, *University of Sydney, School of Biological Sciences*
Stephen Williams, *Glendale Community College*
Eric Wise, *University of California at Santa Barbara*
Michael G. Wood, *Del Mar College*
Jamie Young, *Chattanooga State Technical Community College*

Karen Karlin, Senior Development Editor at Prentice Hall, also played a vital role in the shaping of the Fifth Edition. Throughout the process, she helped me keep the general tone and level of presentation consistent, assisted with the analysis of reviews, and sometimes tolerated my grumpiness when the going got rough. I could not have survived this process without her good-natured support and steadfast determination to make this the best edition yet. The accuracy and currency of the clinical material in this edition and in the *Applications Manual* in large part reflect the detailed clinical reviews performed by my wife, Kathleen Welch, M.D. Virtually without exception, reviewers stressed the importance of accurate, integrated, and visually attractive illustrations in aiding the students to understand essential material. The revision of the art program was directed by Bill Ober, M.D., and Claire Garrison, R.N. Their suggestions about topics of clinical importance, presentation sequence, and revisions to the proposed art were of incalculable value to me and to the project. I also owe thanks to Kate Flickinger for her creative efforts on the MediaLabs and for generating new media components incorporated on the CD-ROM and Companion Website. Special thanks are also due to Ann Murphy, at the University of Sydney, for taking time out from her Christmas holiday to put together a list of alternative international pronunciations for use in the Glossary.

Many of the text's illustrations include color photographs or micrographs collected from a variety of sources. The striking anatomical photos in the text, Companion Atlas of the *Applications Manual,* and lab manuals were the work of Ralph Hutchings, whose efforts on this project are deeply appreciated. Much of the work in tracking down micrographs and other visual materials was performed by Tobi Zausner; I greatly value her help. Dr. Eugene C. Wasson, III, and the staff of Maui Radiology Consultants, Inc., provided valuable assistance in the selection and printing of the CT and MRI scans that are included in the *Applications Manual.*

I also express my appreciation to the editors and support staff at Prentice Hall who made the entire project possible. First on the list is Shari Toron, the Production Editor for my projects. She somehow managed to handle crises and keep things moving in the right direction while maintaining her sense of humor. Her support, hard work, and patience are deeply appreciated.

I owe special thanks to Halee Dinsey, Senior Acquisitions Editor, for her efforts as project coordinator; Don O'Neal, Project Manager, for coordinating and managing the varied print supplements; Andrew Stull, Media Product Development Editor, for doing an outstanding job with a highly complex media supplements program; Martha McDonald, Senior Marketing Manager for Applied Biology, and Jennifer Welchans, Director of Marketing, for their enthusiasm and tireless efforts in promoting this book and its supplements; and Marketing Assistant Andrew Gilfillan and Editorial Assistant Susan Zeigler.

I would also like to thank Tim Bozik, President of the Engineering, Science, and Math Division (ESM), Paul Corey, Vice President and Editorial Director of Math and Science, and Sheri Snavely, Editor in Chief, Biology, for their support of this project; Carol Trueheart, Editor in Chief of ESM Development, for her calm and patience; David Riccardi, Assistant Vice President of ESM Production, Kathleen Schiaparelli, Executive Managing Editor for Science and Mathematics Production, and Beth Sturla, Assistant Managing Editor, Science Production, for providing the extraordinary resources (human as well as technological) required to produce a book of this size and complexity; Paul Belfanti, Creative Director, Heather Scott, Art Director, and Gus Vibal, Art Manager, for their supervision of the design, the art program, and much else; and Barbara Murray, Special Projects Manager, Meaghan Forbes, Supplements Production Editor, and David Munger of The Davidson Group for their work on the transparencies and on the *Applications Manual,* which is a text unto itself.

No one person could expect to produce a flawless textbook of this scope and complexity. Any errors or oversights are strictly my own rather than those of the reviewers, artists, or editors. To help improve future editions, I encourage you to send any pertinent information, suggestions, or comments about the organization or content of this textbook to me directly, using the e-mail address below. I will deeply appreciate any and all comments and suggestions and will carefully consider them in the preparation of the Sixth Edition.

Frederic H. Martini

Frederic H. Martini
Haiku, Hawaii
martini@maui.net

To the Student

Dear student,

Over the last few years, I've met with students in the United States and abroad. They've all expressed similar concerns, which can be boiled down to two basic questions: (1) *How do I find the time?* and (2) *What do I really need to know?* In preparing the Fifth Edition, I've tried to address both questions.

- *How do I find the time?* Although there's nothing I can do to give you more time to study A & P, I've tried to organize the material clearly and concisely and provide tips, checks, and aids—such as short summary tables within the chapters and Navigators (icons described in the walkthrough that follows). These features should help you understand the text discussions and make it easier to review the material as exam time draws near. In other words, I've tried to make sure that you get the most out of the time you *do* have.

- *What do I really need to know?* As you start the course, you may be intimidated by the sheer volume of material you will be expected to learn and understand. If you look at the text as just a block of facts you must memorize, you'd have every right to feel overwhelmed. But that's not the case, because there are general themes and patterns that appear over and over, regardless of the structure or body system you are considering. Once you recognize those themes and patterns, you will have a framework that you can use to organize information. Throughout the text, I have highlighted those patterns for you and present new information in that context. Let's face it—a year after you complete this course, you will probably have forgotten a lot of the details you memorized. But you won't forget the themes and patterns, and new (or rediscovered) details can easily fit into that framework.

A variation on the second question is *What do I need to know to be a* nurse, physical therapist, radiation technologist, and so on? Throughout this book are short boxes that relate the basic information in the text to specific diagnostic procedures or clinical conditions. For students who are especially interested in clinical applications, or who have personal questions regarding specific clinical conditions, my wife and I have prepared the *Applications Manual* that accompanies this book. This manual should be a useful reference that will form a bridge to later coursework, should you pursue a career in the health sciences.

Each of us has a learning style that fits our individual strengths and weaknesses. For some, it is the narrative; for others, the illustrations. Some people need to hear the information, whereas others need to see it, especially when dynamic processes are involved. With this edition, I've given you a lot of options to help you learn key concepts. For example, this textbook devotes more space to the illustration program than other texts do, giving you a lot of visual support. The accompanying CD-ROM offers ani-mated tutorials. If you need more practice and study drills, you can visit the Companion Website, which has interactive review questions tied to an electronic version of the text and hyperlinked to other tutorial sites on the World Wide Web. Study guides, video-tapes, and other supplements are also available. You don't have to use any or all of these supplements; you just need to know that if you cannot quite grasp a concept, plenty of options are at hand.

The next section walks you through the various features of this book. The learning aids introduced there have been built into the textbook to simplify your study of this material and to make that process more rewarding. They have evolved over the last 12 years through brainstorming sessions with students in campuses across the United States. Each year I hear from students, in person or by phone or e-mail, that this system really does work. Yet, it can help you only if you learn how to use it (preferably well before your first exam). This book will help you get the most from your course, so please take a close look at the walkthrough that follows and consult your instructor if you have questions about it.

Good luck and best wishes,

Frederic H. Martini

CHAPTER-OPENING PHOTOGRAPH AND CAPTION

Each chapter begins with a **photograph** that dramatizes some aspect of the material covered in the chapter.

The accompanying **caption** relates the photograph to themes that are developed within the chapter.

CHAPTER OUTLINE AND OBJECTIVES

Before you begin reading, you should know what each chapter covers and what you need to learn from it. For this reason, every chapter opens with a **chapter outline** that gives you an overview of its contents. Each entry in this outline corresponds to a major heading in the text. Page numbers are provided so that you can easily find these headings in the chapter.

The basic themes of the chapter are set out in a series of **learning objectives**. These objectives are designed to help you structure your reading and keep you focused on the key points.

CHAPTER **23**

The Respiratory System

Take a deep breath, and blow very hard to put out the candles. It sounds simple, but the process is very complex. Not only must a wide variety of skeletal muscles work together, but also the usually subconscious regulation of breathing must be consciously and forcefully overridden. In this chapter, we shall examine the homeostatic mechanisms responsible for the conscious and subconscious regulation of respiration. We shall also discuss the functional integration of respiratory and cardiovascular activities.

CHAPTER OUTLINE AND OBJECTIVES

INTRODUCTION   p. 798

FUNCTIONS OF THE RESPIRATORY SYSTEM   p. 798
① Describe the primary functions of the respiratory system.

ORGANIZATION OF THE RESPIRATORY SYSTEM   p. 798
   The Respiratory Mucosa   p. 798
② Explain how the delicate respiratory exchange surfaces are protected from pathogens, debris, and other hazards.

THE UPPER RESPIRATORY SYSTEM   p. 801
   The Nose and Nasal Cavity   p. 801
   The Pharynx   p. 802
③ Identify the organs of the upper respiratory system, and describe their functions.

THE LARYNX   p. 803
   Cartilages and Ligaments of the Larynx   p. 803
   The Laryngeal Musculature   p. 805
④ Describe the structure of the larynx, and discuss its role in normal breathing and sound production.

THE TRACHEA   p. 805

THE PRIMARY BRONCHI   p. 806
⑤ Discuss the structure of the extrapulmonary airways.

THE LUNGS   p. 806
   Lobes and Surfaces of the Lungs   p. 806
   The Bronchi   p. 808
   The Bronchioles   p. 809
   Alveolar Ducts and Alveoli   p. 811
   The Blood Supply to the Lungs   p. 813
⑥ Describe the superficial anatomy of the lungs, the structure of a pulmonary lobule, and the functional anatomy of the alveoli.

THE PLEURAL CAVITIES AND PLEURAL MEMBRANES   p. 813

RESPIRATORY PHYSIOLOGY   p. 814
   Pulmonary Ventilation   p. 814
   Gas Exchange at the Respiratory Membrane   p. 822
   Gas Pickup and Delivery   p. 826
   Summary: Gas Transport   p. 830
⑦ Describe the physical principles governing the movement of air into the lungs and the diffusion of gases into and out of the blood.

⑧ Describe the origins and actions of the respiratory muscles responsible for respiratory movements.
⑨ Differentiate between pulmonary ventilation and alveolar ventilation.
⑩ Describe how oxygen and carbon dioxide are transported in blood.

CONTROL OF RESPIRATION   p. 830
   Local Regulation of Gas Transport and Alveolar Function   p. 830
   The Respiratory Centers of the Brain   p. 831
   Voluntary Control of Respiration   p. 836
⑪ Describe the factors that influence the respiration rate.
⑫ Identify and discuss reflex respiratory activity and the brain centers involved in the control of respiration.

CHANGES IN THE RESPIRATORY SYSTEM AT BIRTH   p. 836

AGING AND THE RESPIRATORY SYSTEM   p. 837

INTEGRATION WITH OTHER SYSTEMS   p. 837
⑬ Give examples of interactions between the respiratory system and each of the other organ systems.

MEDIALAB and CHAPTER REVIEW   pp. 839, 840

A REFINED FRAMEWORK

It can be difficult to organize and synthesize the volume of material you are required to master in an anatomy and physiology course. This textbook provides a highly refined pedagogical framework to help you manage the learning process.

CONCEPT CHECK QUESTIONS

Concept Check Questions are located at the end of major sections in each chapter. You can stop and check your understanding before you move on to the next concept. **Answers** to Concept Check Questions are provided in Appendix I.

KEY TERMS

Throughout the text, the most important new terms are presented in **boldface** type.

WORD ROOTS AND DEFINITIONS

For many terms, a derivation is offered as a memory aid and to help you recognize **word roots** that occur again and again in the language of anatomy and physiology.

The Nucleus 89

generate ATP. Cells throughout the body may be affected, but symptoms involving muscle cells, neurons, and the receptor cells in the eye are most common, because these cells have especially high energy demands.
AM Mitochondrial DNA, Disease, and Evolution

☑ Certain cells in the ovaries and testes contain large amounts of smooth endoplasmic reticulum (SER). Why?
☑ Microscopic examination of a cell's contents reveals many mitochondria. What does this observation imply about the cell's energy requirements?

THE NUCLEUS

The **nucleus** is the control center of cellular operations. A single nucleus stores all the information needed to control the synthesis of the approximately 100,000 different proteins in the human body. The nucleus determines the structural and functional characteristics of the cell by controlling which proteins are

synthesized, and in what amounts. A cell without a nucleus could be compared to a car without a driver. However, a car can sit idle for years, but a cell without a nucleus will disintegrate within 3 or 4 months.

Most cells contain a single nucleus, but exceptions exist. For example, skeletal muscle cells have many nuclei, and mature red blood cells have none. Figure 3-21• details the structure of a typical nucleus. A **nuclear envelope** surrounds the nucleus and separates it from the cytosol. The nuclear envelope is a double membrane, with its two layers separated by a narrow **perinuclear space** (*peri*, around). At several locations, the nuclear envelope is connected to the rough endoplasmic reticulum, as shown in Figure 3-2•, p. 66.

For it to direct processes that take place in the cytosol, the nucleus must receive information about conditions and activities in other parts of the cell. Chemical communication between the nucleus and the cytosol occurs through **nuclear pores**. These pores, which cover about 10 percent of the surface of the nucleus, are large enough to permit the movement of ions and small molecules but too small for the free passage

Nucleoplasm

Nucleolus

Chromatin

Nuclear envelope

Nuclear pores

(a)

Inner membrane of nuclear envelope

Broken edge of outer membrane

Outer membrane of nuclear envelope

(b)

•FIGURE 3-21 The Nucleus. (a) Important nuclear structures. (TEM × 4828) (b) This cell was frozen and then broken apart to make internal structures visible. This technique, called freeze-fracture, provides a unique perspective on the internal organization of cells. The nuclear envelope and nuclear pores are visible. The fracturing process broke away part of the outer membrane of the nuclear envelope, and the cut edge of the nucleus can be seen. (SEM × 9240)

444 Chapter 14 ■ THE BRAIN AND CRANIAL NERVES

Hydrocephalus

The adult brain is surrounded by the relatively inflexible bones of the cranium. The enclosed cranial cavity contains two fluids—blood and cerebrospinal fluid—and the soft tissues of the brain. Because the total volume cannot change, when the volume of blood or CSF increases, the volume of the brain must decrease. For example, if a dural or subarachnoid blood vessel ruptures, the fluid volume increases as blood collects in the cranial cavity. The rising intracranial pressure compresses the brain, leading to neural dysfunction that can end in unconsciousness and ultimately death.

Any change in the rate of CSF production is normally matched by an increase in the rate of CSF removal at the arachnoid granulations. If this equilibrium is disturbed, clinical problems appear as the intracranial pressure changes. The volume of CSF will increase if the rate of formation accelerates or the rate of removal decreases. In either event, the increased fluid volume compresses and distorts the brain. Increased rates of formation can accompany head injuries, but the most common problems arise from masses, such as tumors or abscesses, or from developmental abnormalities. These conditions restrict the normal circulation and reabsorption of CSF. Because CSF production continues, the ventricles gradually expand, distorting the surrounding neural tissues and causing brain function to deteriorate.

Infants are especially sensitive to changes in intracranial pressure, because the arachnoid granulations do not appear until roughly 3 years of age. (Until then, CSF is reabsorbed into small vessels in the subarachnoid space and beneath the ependyma lining the ventricles.) As in adults, if intracranial pressure becomes abnormally high, the ventricles will expand. But in infants, the cranial sutures have yet to fuse, so the skull can enlarge to accommodate the extra fluid volume. The result is an enormously expanded skull, a condition called **hydrocephalus**, or "water on the

•FIGURE 14-6 Hydrocephalus. This infant has hydrocephalus, a condition generally caused by impaired circulation and impaired removal of cerebrospinal fluid. The buildup of CSF leads to the distortion of the brain and enlargement of the cranium.

brain" (Figure 14-6•). Infant hydrocephalus typically results from some interference with normal CSF circulation, such as a blockage of the mesencephalic aqueduct or a constriction of the connection between the subarachnoid spaces of the cranial and spinal meninges. Untreated infants commonly suffer some degree of mental retardation.

Successful treatment generally involves the installation of a *shunt*, a bypass that either avoids the blockage site or drains the excess CSF. The goal is reduction of the intracranial pressure. The shunt may be removed if (1) further growth of the brain eliminates the blockage or (2) the intracranial pressure decreases after the arachnoid granulations develop when the child reaches 3 years of age.

The Blood–Brain Barrier

Neural tissue in the CNS is isolated from the general circulation by the **blood–brain barrier**. This barrier exists because the endothelial cells that line the capillaries of the CNS are extensively interconnected by tight junctions. These junctions prevent the diffusion of materials between adjacent endothelial cells. In general, only lipid-soluble compounds (including carbon dioxide; oxygen; ammonia; lipids, such as steroids or prostaglandins; and small alcohols) can diffuse across

the membranes of endothelial cells into the interstitial fluid of the brain and spinal cord. Water and ions must pass through channels in the apical and basal cell membranes. Larger, water-soluble compounds can cross the capillary walls only by active or passive transport.

The restricted permeability characteristics of the endothelial lining of brain capillaries are in some way dependent on chemicals secreted by astrocytes ∞ p. 368 that are in close contact with CNS capillaries. ∞ p. 368 The outer surfaces of the endothelial cells are covered by the processes of astrocytes. Because the astrocytes

CLINICAL DISCUSSIONS

Important **clinical topics** are presented in context, set off by a caduceus icon. These topics have been selected not only for their medical importance, but also for their value in helping you learn essential principles. Many of these discussions deal with diseases or therapies and show how an understanding of abnormal conditions can shed light on normal function and vice versa. Other clinical discussions deal with personal health and fitness.

FIGURE REFERENCE LOCATORS

Red dots next to figure references serve as place markers, making it easy for you to return to the narrative after you have studied an illustration.

CONCEPT LINK ICONS

Concept link icons provide a quick visual signal notifying you that the material currently being presented is related to or builds on a previous discussion. Each **"link" icon,** ∞, is followed by a page reference to the relevant material from an earlier chapter.

VISUALIZATION IS THE KEY TO SUCCESS

Central to your success in anatomy and physiology is the ability to visualize the structures and functions of the human body. The award-winning medical illustrators William Ober, M.D., and Claire Garrison, R.N., are the single illustration team for *Fundamentals of Anatomy and Physiology*. They create **artwork** known for its clarity and accuracy in showing anatomical structures and physiological processes. The same art is also used in the teaching and learning resources available with the textbook so that students can benefit from a consistent treatment.

NAVIGATOR ICONS

Navigator icons have been inserted alongside the artwork for the most difficult topics in anatomy and physiology. These icons show you where you are in the process, where you have been, and where you are going.

(1) This picture shows what happens when an action potential is generated. The same overview is used as the basis for a series of Navigator icons.

The overview figure is repeated, in reduced and simplified form, in subsequent key figures as topics change. Here, the Navigators indicate that (2) the resting potential is shown and discussed first, then (3) the graded potential, and so on. The highlighted step lets you know that "you are here."

PROCESS ART

Process art breaks down complex physiological processes into **manageable steps** and organizes information to highlight important relationships.

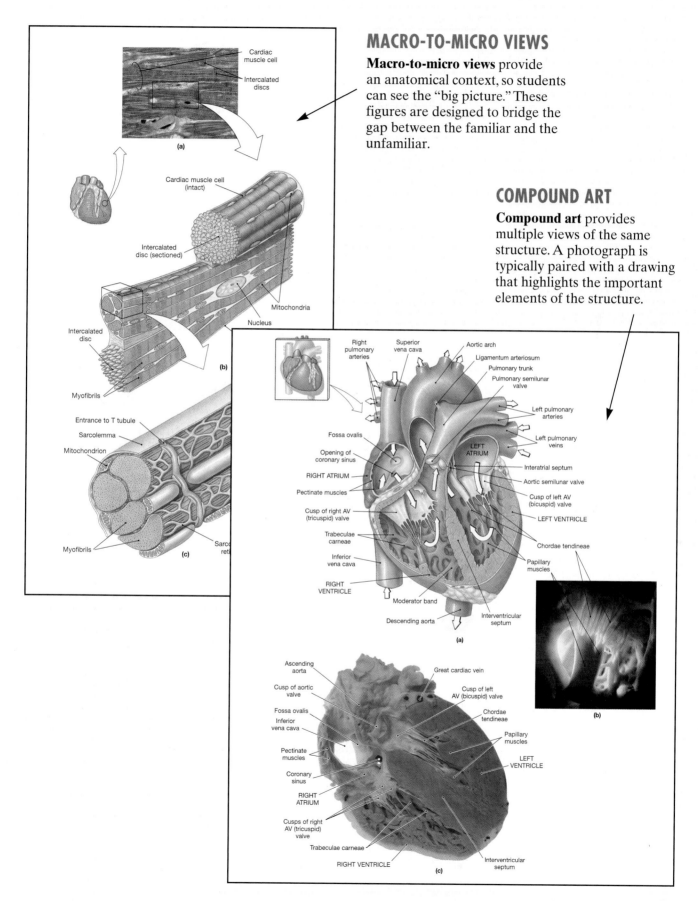

MACRO-TO-MICRO VIEWS

Macro-to-micro views provide an anatomical context, so students can see the "big picture." These figures are designed to bridge the gap between the familiar and the unfamiliar.

COMPOUND ART

Compound art provides multiple views of the same structure. A photograph is typically paired with a drawing that highlights the important elements of the structure.

(labels, top figure)

Cardiac muscle cell

Intercalated discs

(a)

Cardiac muscle cell (intact)

Intercalated disc (sectioned)

Mitochondria

Nucleus

Intercalated disc

(b)

Myofibrils

Entrance to T tubule

Sarcolemma

Mitochondrion

Myofibrils

Sarco... reti...

(c)

(labels, heart figure a)

Right pulmonary arteries

Superior vena cava

Aortic arch

Ligamentum arteriosum

Pulmonary trunk

Pulmonary semilunar valve

Left pulmonary arteries

Left pulmonary veins

Fossa ovalis

Opening of coronary sinus

RIGHT ATRIUM

Pectinate muscles

Cusp of right AV (tricuspid) valve

Trabeculae carneae

Inferior vena cava

RIGHT VENTRICLE

Moderator band

Descending aorta

LEFT ATRIUM

Interatrial septum

Aortic semilunar valve

Cusp of left AV (bicuspid) valve

LEFT VENTRICLE

Chordae tendineae

Papillary muscles

Interventricular septum

(a)

(b)

(labels, heart figure c)

Ascending aorta

Cusp of aortic valve

Fossa ovalis

Inferior vena cava

Pectinate muscles

Coronary sinus

RIGHT ATRIUM

Cusps of right AV (tricuspid) valve

Trabeculae carneae

RIGHT VENTRICLE

Great cardiac vein

Cusp of left AV (bicuspid) valve

Chordae tendineae

Papillary muscles

LEFT VENTRICLE

Interventricular septum

(c)

FOCUS FEATURES

The FOCUS format presents related text, art, and tabular material as an **integrated package.** This feature can help you assimilate complex information as efficiently as possible.

APPLICATIONS MANUAL REFERENCES

Many additional topics of clinical significance are treated in the *Applications Manual*, which accompanies this textbook. Topics that are covered in the *Applications Manual* are identified in the textbook by the **AM icon,** AM , followed by the title of the relevant discussion (in blue). Embryology Summaries, Plates from the Surface Anatomy and Cadaver Atlas, and Scans from the Scanning Atlas are referenced as well.

FACIAL BONES

The Maxillary Bones (Figure 7-10a,b•)

General Functions: The **maxillary bones,** or *maxillae,* support the upper teeth and form the inferior orbital rim, the lateral margins of the external nares, the upper jaw, and most of the hard palate. The *maxillary sinuses* in these bones produce mucous secretions that flush the inferior surfaces of the nasal cavities. These are the largest facial bones, and the maxillary sinuses are the largest sinuses.

Articulations: The maxillary bones articulate with the frontal and ethmoid bones, with one another, and with all the other facial bones except the mandible (Figures 7-3c–e and 7-4a•).

Regions/Landmarks: The **orbital rim** protects the eye and other structures in the orbit.

The **alveolar processes** that border the mouth support the upper teeth.

The **palatine processes** form most of the **hard palate,** or bony roof of the mouth. One type of *cleft palate,* a developmental disorder, results when the maxillary bones fail to meet along the midline of the hard palate. AM *EMBRYOLOGY SUMMARY 6: The Development of the Skull*

The **maxillary sinuses** lighten the portion of the maxillary bone superior to the teeth.

The **nasolacrimal canal,** formed by the maxillary and lacrimal bones, protects the *lacrimal sac* and the *nasolacrimal duct,* which carries tears from the orbit to the nasal cavity.

Foramina: The **infraorbital foramen** marks the path of a major sensory nerve; that nerve reaches the brain by way of the foramen rotundum of the sphenoid bone.

The **inferior orbital fissure** (Fig. 7-3d•), which lies between the maxillary bone and the sphenoid bone, permits passage of cranial nerves and blood vessels.

The Palatine Bones (Figure 7-10b,c•)

General Functions: The **palatine bones** form the posterior portion of the hard palate and contribute to the floor of each orbit.

Articulations: The palatine bones articulate with one another, with the maxillary, sphenoid, and ethmoid bones, with the inferior nasal conchae, and with the vomer (Figures 7-3e and 7-4a•).

Regions/Landmarks: These bones are L-shaped. The **horizontal plate** forms the posterior part of the hard palate; the **perpendicular plate** extends from the horizontal plate to the orbital process. The **orbital process** forms part of the floor of the orbit. This process contains a small sinus that normally opens into the sphenoidal sinus.

Foramina: Small blood vessels and nerves supplying the roof of the mouth penetrate the lateral portion of the horizontal plate.

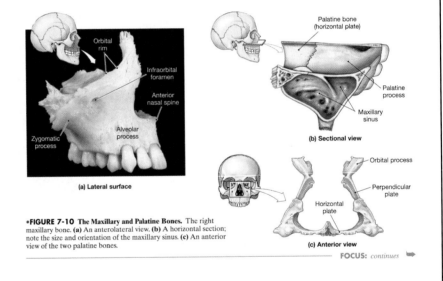

•**FIGURE 7-10** **The Maxillary and Palatine Bones.** The right maxillary bone. **(a)** An anterolateral view. **(b)** A horizontal section; note the size and orientation of the maxillary sinus. **(c)** An anterior view of the two palatine bones.

FOCUS: *continues* ➡

AGING ICONS

Thanks to advances in medicine, life expectancies are increasing. On the rise as well is our interest in the aging process. These **aging icons** highlight text discussions that describe this process for a given body system.

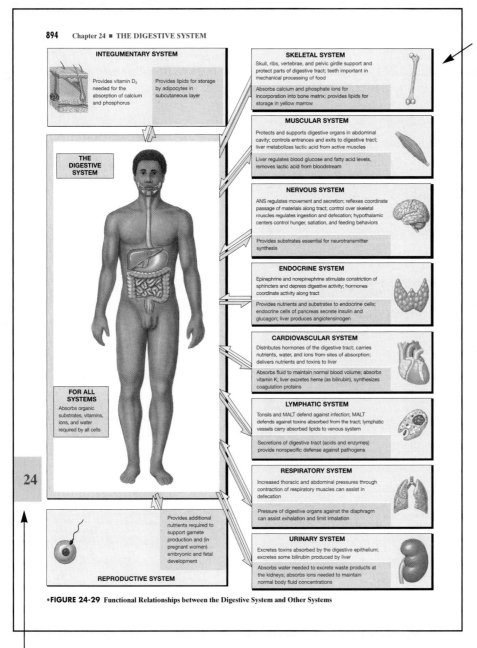

INTEGUMENTARY SYSTEM

Provides vitamin D₃ needed for the absorption of calcium and phosphorus

Provides lipids for storage by adipocytes in subcutaneous layer

THE DIGESTIVE SYSTEM

FOR ALL SYSTEMS
Absorbs organic substrates, vitamins, ions, and water required by all cells

24

SKELETAL SYSTEM
Skull, ribs, vertebrae, and pelvic girdle support and protect parts of digestive tract; teeth important in mechanical processing of food

Absorbs calcium and phosphate ions for incorporation into bone matrix; provides lipids for storage in yellow marrow

MUSCULAR SYSTEM
Protects and supports digestive organs in abdominal cavity; controls entrances and exits to digestive tract; liver metabolizes lactic acid from active muscles

Liver regulates blood glucose and fatty acid levels, removes lactic acid from bloodstream

NERVOUS SYSTEM
ANS regulates movement and secretion; reflexes coordinate passage of materials along tract; control over skeletal muscles regulates ingestion and defecation; hypothalamic centers control hunger, satiation, and feeding behaviors

Provides substrates essential for neurotransmitter synthesis

ENDOCRINE SYSTEM
Epinephrine and norepinephrine stimulate constriction of sphincters and depress digestive activity; hormones coordinate activity along tract

Provides nutrients and substrates to endocrine cells; endocrine cells of pancreas secrete insulin and glucagon; liver produces angiotensinogen

CARDIOVASCULAR SYSTEM
Distributes hormones of the digestive tract; carries nutrients, water, and ions from sites of absorption; delivers nutrients and toxins to liver

Absorbs fluid to maintain normal blood volume; absorbs vitamin K; liver excretes heme (as bilirubin), synthesizes coagulation proteins

LYMPHATIC SYSTEM
Tonsils and MALT defend against infection; MALT defends against toxins absorbed from the tract; lymphatic vessels carry absorbed lipids to venous system

Secretions of digestive tract (acids and enzymes) provide nonspecific defense against pathogens

RESPIRATORY SYSTEM
Increased thoracic and abdominal pressures through contraction of respiratory muscles can assist in defecation

Pressure of digestive organs against the diaphragm can assist exhalation and limit inhalation

Provides additional nutrients required to support gamete production and (in pregnant women) embryonic and fetal development

URINARY SYSTEM
Excretes toxins absorbed by the digestive epithelium; excretes some bilirubin produced by liver

Absorbs water needed to excrete waste products at the kidneys; absorbs ions needed to maintain normal body fluid concentrations

REPRODUCTIVE SYSTEM

•FIGURE 24-29 Functional Relationships between the Digestive System and Other Systems

SYSTEM INTEGRATORS

These figures, located at the end of each body system chapter, show how the system under consideration interacts with each of the body's other systems. They reinforce the concept that the body functions as an **integrative unit.**

Interactions among systems are color-coded:

● Beige is used to highlight the contributions of the system being discussed to each of the body's other systems.

● Blue is used to indicate the contributions other systems make to the system being studied.

THUMB TABS

For easy reference, each chapter is color-coded by body system. The **thumb tabs** in the outer margins provide the chapter number.

PRONUNCIATION GUIDE

Whenever the pronunciation of a new term may not be obvious, a **pronunciation guide** is given in parentheses. Each word is broken into syllables by hyphens, and the primary accented syllable of multisyllabic words appears in capital letters. Overbars are used to indicate long vowel sounds:

"ā" as in "tray" (trā)
"ē" as in "ear" (ēr)
"ī" as in "fly" (flī)
"ō" as in "bone" (bōn)
"ū" as in "you" (ū)

SELECTED CLINICAL TERMINOLOGY

The first part of this section reviews the definitions and (where necessary) the pronunciations of the most important medical terms discussed **in the chapter.** Page references are provided for easy review. The second part of this section gives definitions of medical terms that appear in related discussions **in the Applications Manual.**

STUDY OUTLINE

The study outline provides you with a concise but detailed **summary** of the chapter's most important facts and concepts.
● Short numbered statements are organized under the major chapter headings, which appear with their **page references.** You can easily review the organization of the chapter as well as its content.
● Key terms that are in **boldface** in the text appear in boldface in the summary.
● The **figures and tables** relevant to each section are also listed.

CHAPTER REVIEW

1

SELECTED CLINICAL TERMINOLOGY

Terms Discussed in This Chapter

abdominopelvic quadrant: One of four divisions of the anterior abdominal surface. *(p. 15)*
abdominopelvic region: One of nine divisions of the anterior abdominal surface. *(p. 17)*
CT, CAT (computerized [axial] tomography): An imaging technique that uses X-rays to reconstruct the body's three-dimensional structure. *(p. 24)*
disease: A malfunction of organs or organ systems resulting from a failure of homeostatic regulation. *(p. 14)*
DSA (digital subtraction angiography): A technique used to monitor blood flow through specific organs, such as the brain, heart, lungs, or kidneys. X-rays are taken before and after radiopaque dye is administered, and a computer "subtracts" details common to both images. The result is a high-contrast image showing the distribution of the dye. *(p. 25)*
echogram: An image created by ultrasound. *(p. 24)*

embryology: The study of structural changes during the first 2 months of development. *(p. 3)*
histology (his-TOL-o-jē): The study of tissues. *(p. 3)*
MRI (magnetic resonance imaging): An imaging technique that employs a magnetic field and radio waves to portray subtle structural differences. *(p. 24)*
radiologist: A physician who specializes in performing and analyzing radiological procedures. *(p. 23)*
spiral-CT: A method of processing computerized tomography data to provide rapid, three-dimensional images of internal organs. *(p. 25)*
ultrasound: An imaging technique that uses brief bursts of high-frequency sound reflected by internal structures. *(p. 24)*
X-rays: High-energy radiation that can penetrate living tissues. *(p. 23)*

AM Additional Terms Discussed in the *Applications Manual*

auscultation (aws-kul-TĀ-shun): Listening to a patient's body sounds by means of a stethoscope.
DSR (dynamic spatial reconstruction): A research tool that provides three-dimensional images in real time. The computer-generated image can be rotated or sectioned.
inspection: A careful observation of a patient's appearance and actions.
palpation: Using hands and fingers to feel the patient's body as part of a physical exam.

percussion: Tapping with the fingers or hand to obtain information about the densities of a patient's underlying tissues.
PET (positron emission tomography) scan: A scan taken after radioisotopes are injected or inhaled; the released radiation is monitored, and computers use the data to visualize physiological activity.

STUDY OUTLINE

INTRODUCTION   p. 2

1. **Biology** is the study of life. One of its goals is to discover the unity and patterns that underlie the diversity of organisms.
2. All living things have certain common characteristics, including **organization, responsiveness, growth** and **differentiation, reproduction, movement,** and **metabolism** and **excretion.**

THE SCIENCES OF ANATOMY AND PHYSIOLOGY   p. 2

1. **Anatomy** is the study of internal and external structures and the physical relationships among body parts. **Physiology** is the study of how living organisms perform vital functions. All specific functions are performed by specific structures.

Anatomy   p. 3

2. The boundaries of *microscopic anatomy* are established by the equipment used. In **cytology,** we analyze the internal structure of individual cells. In **histology,** we examine **tissues,** groups of cells that have specific functional roles. Tissues combine to form **organs,** anatomical units with multiple functions. Organs combine to form **organ systems,** groups of organs that function together.
3. In *gross (macroscopic) anatomy,* we consider features that are visible without a microscope. It includes *surface anatomy* (general form and superficial markings); *regional*

anatomy (superficial and internal features in a specific area of the body); and *systemic anatomy* (structure of major organ systems).
4. In *developmental anatomy,* we examine the changes in form that occur between conception and physical maturity. In *embryology,* we study processes that occur during the first 2 months of development.

Physiology   p. 3

5. **Human physiology** is the study of the functions of the human body. It is based on *cell physiology,* the study of the functions of cells. In *special physiology,* we study the physiology of specific organs. In *systemic physiology,* we consider all aspects of the function of specific organ systems. In *pathological physiology,* we study the effects of diseases on organ or system functions.

LEVELS OF ORGANIZATION   p. 4

1. Anatomical structures and physiological mechanisms are arranged in a series of interacting levels of organization. *(Figure 1-1)*
2. The 11 organ systems of the body are the integumentary, skeletal, muscular, nervous, endocrine, cardiovascular, lymphatic, respiratory, digestive, urinary, and reproductive systems. *(Figure 1-2)*

REVIEW QUESTIONS

LEVEL 1 REVIEWING FACTS AND TERMS

Match each numbered item with the most closely related lettered item. Use letters for answers in the spaces provided.

Set 1: The Brain

___ **1.** prosencephalon
___ **2.** rhombencephalon
___ **3.** brain stem
___ **4.** ventricle in diencephalon
___ **5.** dural fold
___ **6.** choroid plexus
___ **7.** limbic system
___ **8.** pineal gland
___ **9.** supraoptic nucleus
___ **10.** paraventricular nucleus
___ **11.** substantia nigra
___ **12.** red nucleus
___ **13.** Purkinje cells
___ **14.** arbor vitae
___ **15.** ataxia
___ **16.** reticular formation
___ **17.** nucleus gracilis
___ **18.** hypothalamus

a. third ventricle
b. oxytocin
c. adjusts muscle tone and limb position
d. melatonin
e. white matter of cerebellum
f. mesencephalon, pons, medulla oblongata
g. link between nervous and endocrine systems
h. releases the neurotransmitter dopamine
i. forebrain
j. cerebellar cortex
k. falx cerebri
l. a disturbance in balance
m. apneustic, pneumotaxic centers
n. hindbrain
o. amygdaloid body
p. medulla oblongata
q. production of CSF
r. antidiuretic hormone

Set 2: The Cranial Nerves

___ **19.** facial nerve
___ **20.** vagus nerve
___ **21.** olfactory nerve
___ **22.** trochlear nerve
___ **23.** hypoglossal nerve
___ **24.** accessory nerve
___ **25.** trigeminal nerve
___ **26.** optic nerve
___ **27.** glossopharyngeal nerve
___ **28.** oculomotor nerve
___ **29.** vestibulocochlear nerve
___ **30.** abducens nerve

a. sensory, smell
b. N V
c. motor, tongue movements
d. sensory, vision
e. N X
f. N III
g. equilibrium, hearing
h. N VII
i. N VI
j. N IX
k. motor, eye movements
l. N XI

31. The primary link between the nervous and endocrine systems is the
(a) hypothalamus   (b) mes
(c) pons   (d) med

32. Immediate reflexive responses to l
are directed by nuclei in the
(a) pons   (b) dier
(c) mesencephalon   (d) mec

33. The regulation of autonomic functi
and blood pressure, originates in the
(a) cerebrum   (b) cer
(c) diencephalon   (d) mec

34. The smooth surface that covers the brain but does not follow the underlying neural convolutions or sulci is the
(a) neural cortex   (b) dura mater
(c) pia mater   (d) arachnoid membrane

35. The meningeal layer that adheres to the surface contour of the brain, extending into every fold and curve, is the
(a) pia mater   (b) dura mater
(c) arachnoid membrane   (d) neural cortex

36. The dural fold that divides the two cerebellar hemispheres is the
(a) transverse sinus   (b) falx cerebri
(c) tentorium cerebelli   (d) falx cerebelli

37. Cerebrospinal fluid is produced and secreted in the
(a) hypothalamus   (b) choroid plexus
(c) medulla oblongata   (d) crista galli

38. The primary purpose of the blood–brain barrier is to
(a) provide the brain with oxygenated blood
(b) drain venous blood via the internal jugular veins
(c) isolate neural tissue in the CNS from the general circulation
(d) a, b, and c are correct

39. The two cerebral hemispheres are functionally different even though anatomically they appear the same.
(a) true   (b) false

40. Reading, writing, and speaking are dependent on processing in the
(a) right cerebral hemisphere
(b) left cerebral hemisphere
(c) prefrontal cortex
(d) postcentral gyrus

41. The establishment of emotional states is a function of the
(a) limbic system   (b) tectum
(c) mamillary bodies   (d) thalamus

42. The final relay point for ascending sensory information that will be projected to the primary sensory cortex is the
(a) hypothalamus   (b) thalamus
(c) spinal cord   (d) medulla oblongata

43. The two hormones secreted by the hypothalamus are
(a) epinephrine and norepinephrine
(b) antidiuretic hormone and oxytocin
(c) melatonin and serotonin
(d) FSH and ATP

44. The centers in the pons that modify the activity of the respiratory rhythmicity centers in the medulla oblongata are the
(a) apneustic and pneumotaxic centers
(b) inferior and superior peduncles
(c) cardiac and vasomotor centers

LEVEL 2 REVIEWING CONCEPTS

51. A person with a damaged visual association area may be
(a) unable to scan the lines of a page or see rows of clear symbols
(b) declared legally blind
(c) unable to see letters but able to identify words and their meanings
(d) able to see letters quite clearly but unable to recognize or interpret them

52. If the corpus callosum is cut,
(a) cross-referencing of sensory information is inhibited
(b) the individual talks constantly but uses the wrong words
(c) objects touched with the left hand cannot be recognized but can be verbally identified
(d) objects touched with the right hand cannot be verbally identified

53. Which structure in the brain would your A & P instructor be referring to when talking about a nucleus that resembles a sea horse and that appears to be important in the storage and retrieval of long-term memories? In which functional system of the brain is it located?

54. Stimulation of which part of the brain would produce sensations of hunger and thirst?

55. Damage to the corpora quadrigemina would interfere with the
(a) control of autonomic function
(b) regulation of body temperature
(c) processing of visual and auditory sensations
(d) conscious control of skeletal muscles

56. If symptoms characteristic of Parkinson's disease appear, which part of the mesencephalon is inhibited from secreting a neurotransmitter? Which neurotransmitter is it?

57. What are the principal functional differences between the right and left hemispheres of the cerebrum?

58. Which major part of the brain is associated with identifying the clinical signs of life?

59. Why can the brain respond to stimuli with greater versatility than the spinal cord can?

60. What kinds of problems are associated with the presence of lesions in Wernicke's area and Broca's area?

LEVEL 3 CRITICAL THINKING AND CLINICAL APPLICATIONS

61. Smelling salts can sometimes help restore consciousness after a person has fainted. The active ingredient of smelling salts is ammonia, and it acts by irritating the lining of the nasal cavity. Propose a mechanism by which smelling salts would raise a person from the unconscious state to the conscious state.

62. A police officer has just stopped Bill on suspicion of driving while intoxicated. The officer asks Bill to walk the yellow line on the road and then to place the tip of his index finger on the tip of his nose. How would these activities indicate Bill's level of sobriety? Which part of the brain is being tested by these activities?

63. While having some dental work performed, Tyler is given an injection of local anesthetic in his lower jaw. His dentist tells him not to eat until the anesthetic wears off, not because of his teeth but because of his tongue. Why is the dentist giving Tyler this advice?

REVIEW QUESTIONS

A **three-level learning system** is consistently used throughout the teaching and learning program. It helps build your confidence and understanding by allowing you to check yourself as you move from one level to another. **Answers** to all Review Questions are given in Appendix I.

Level 1 questions allow students to test their recall of the chapter's basic information and terminology.

Level 2 questions help students check their conceptual understanding.

Level 3 questions enable students to develop their reasoning and analytical skills by applying chapter material to plausible real-world and clinical situations.

INNOVATIVE MEDIA RESOURCES

The **Martini Media Program** uses technology to bring anatomy and physiology to life for students. All the media resources are easy to use and work in concert with the textbook. Many **chapter-specific resources**—animations, self-grading quizzes, tutorials, and so on—are available with the Martini Companion Website and *Martini Interactive* Student CD-ROM.

MEDIA LAUNCHED FROM THE PAGES OF THE TEXTBOOK

MediaLabs offer an opportunity for students to explore relevant material online. The Web Explorations in each MediaLab are exciting, concept-specific activities that are the gateways into the Martini Media Program.

[EXPLORE] [MediaLab]

The bones of the axial skeleton are numerous and include many markings, processes, and functions. To gain a handle on the complexity of the axial skeleton, use the resources available on both the CD-ROM and the Companion Website. The CD-ROM includes animations and tutorials; the website builds on those resources with text content and interactive problems as well as communication avenues for sharing ideas with your class. To complete the following Web Explorations, visit the MediaLab in Chapter 7 of your Companion Website.

Web Exploration 1

Estimated time for completion = 10 minutes

The best way to learn the intricacies of the skull bones is to practice identifying them in various illustrations. After studying the illustrations provided in your text, select the keyword **TUTORIAL** on the Companion Website. Using the text as a reference, try to identify the components of the skull indicated on each of the five questions. If you wish, share your answers with your instructor or classmates by using the communication tools on the Companion Website.

Web Exploration 2

Estimated time for completion = 10 minutes

The 26 bones of the vertebral column, the central portion of the axial skeleton, fit together to support and protect the spinal cord. Sometimes this fit is altered, resulting in scoliosis, kyphosis, or lordosis. To understand how the vertebrae usually fit together and what happens when they don't, select the keyword **VERTEBRAL COLUMN** on the Companion Website. Click on the footprints to move through the pages. When you have finished, draw a typical vertebra and demonstrate how adjacent vertebrae would normally articulate with it. Then try to depict the same articulations in a person with scoliosis, one with kyphosis, and one with lordosis.

WEB EXPLORATIONS

The MediaLab at the end of each chapter asks you to investigate a text topic further on the Web. All activities are focused and have been designed to optimize your time investment: **Estimated completion times** are given for each Web Exploration.

TOOLS FOR SUCCESS

A powerful companion to the text is provided with each new copy of the textbook. The *Martini Interactive* **Student CD-ROM** contains visual tools to help you understand anatomy and physiology: dissectable 3-D animations, tutorials, interactive clinical case studies, an audio glossary, and much more. It is linked to Martini's text-specific **Companion Website.**

COMPANION WEBSITE (CW)
www.prenhall.com/martini/fap5

Martini's **Companion Website** is an innovative online resource center, organized by chapter, that is designed specifically to support and enhance your study of anatomy and physiology. It provides all the resources necessary to complete MediaLab activities. The site offers three levels of interactive, self-grading study questions; multiple-choice, fill-in, matching, labeling, and essay questions are included. Actual passages with animated art from the textbook are provided as support.

The Companion Website also provides access to Prentice Hall's online **Syllabus Manager,** a course management tool for instructors.

ADDITIONAL RESOURCES TO HELP YOU SUCCEED

The new edition of the *Applications Manual* includes **An Introduction to Diagnostics, Applied Research Topics,** and **The Body Systems: Clinical and Applied Topics.** The *Applications Manual* also offers a **Companion Atlas** with diagnostic scans; full-color photographs of surface anatomy, skeletal anatomy, cadaver prosections, and models; and illustrated overviews of the embryological and fetal development of each body system. Illustrations throughout the textbook are cross-referenced to related images in the Companion Atlas.

Scanning Atlas

Embryology Summaries

Surface Anatomy and Cadaver Atlas

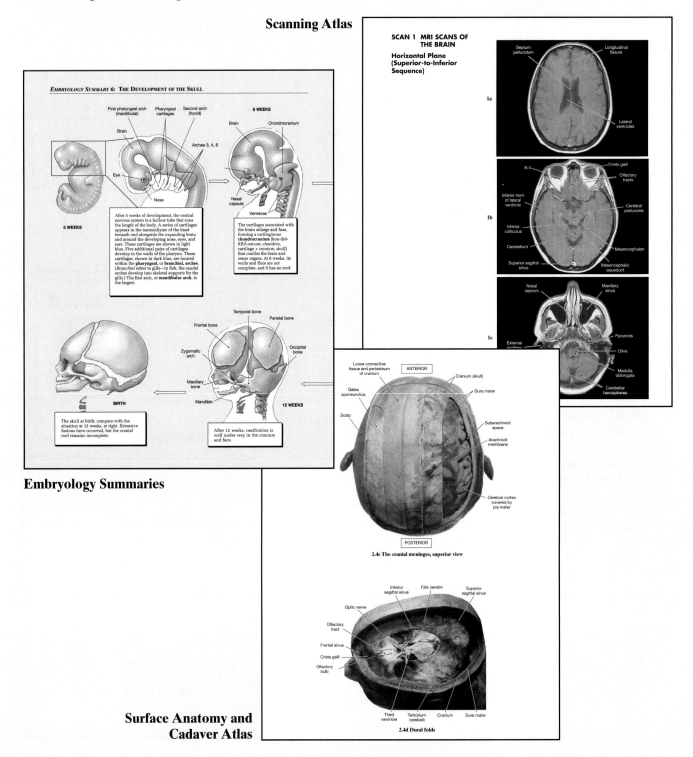

CHAPTER 1

An Introduction to Anatomy and Physiology

Every day we perform a balancing act. It is an amazingly complex performance, and it goes on continuously, without a break or intermission. Every part of our physical being is involved in this balance, from individual atoms to the largest organs and organ systems. Our bodies face constant challenges: stresses (both physical and mental), injury, and disease. Every change, whether it occurs inside us or outside, threatens this delicate balance. Adjusting to these changes is crucial, for if the body loses its balance it may plunge out of control, into dysfunction, illness, and even death.

CHAPTER OUTLINE AND OBJECTIVES

INTRODUCTION   p. 2
① Describe the basic functions of organisms.

THE SCIENCES OF ANATOMY AND PHYSIOLOGY   p. 2
Anatomy   p. 3
Physiology   p. 3
② Define anatomy and physiology, and describe various specialties of each discipline.

LEVELS OF ORGANIZATION   p. 4
③ Identify the major levels of organization in organisms, from the simplest to the most complex.
④ Identify the organ systems of the human body and the major components of each system.

HOMEOSTASIS AND SYSTEM INTEGRATION   p. 12
Negative Feedback   p. 12
Positive Feedback   p. 14
⑤ Explain the concept of homeostasis and its significance for organisms.
⑥ Describe how positive feedback and negative feedback are involved in homeostatic regulation.

A FRAME OF REFERENCE FOR ANATOMICAL STUDIES   p. 15

Superficial Anatomy   p. 15
Sectional Anatomy   p. 19

⑦ Use anatomical terms to describe body sections, body regions, and relative positions.
⑧ Identify the major body cavities and their subdivisions.

■ FOCUS: Sectional Anatomy and Clinical Technology   p. 23

MEDIALAB and CHAPTER REVIEW   pp. 25, 26

Our world contains an enormous diversity of living organisms that vary widely in appearance and lifestyle. One aim of **biology,** the study of life, is to discover the unity and patterns that underlie this diversity. Biologists have found that all living things share basic characteristics, including the following:

■ **Organization:** Every organism has a characteristic pattern of organization that differs from that of inanimate objects. Each organism has a discrete boundary separating it from its environment, and conditions within the organism are often very different from those in the outside world. Life continues only as long as the organism can maintain its physical integrity and can keep its internal organization, and its internal environment, within tolerable limits. We will look more closely at the organization of living things and how they regulate internal conditions in later sections.

■ **Responsiveness:** Organisms respond to changes in their immediate environment; this property is also called *irritability.* You move your hand away from a hot stove; your dog barks at approaching strangers; fish are scared by loud noises; and amoebas glide toward potential prey. Organisms also make longer-term changes as they adjust to their environments. For example, an animal may either grow a heavier coat as winter approaches or migrate to a warmer climate. The potential to make such adjustments is termed *adaptability.*

■ **Growth** and **differentiation:** Over a lifetime, organisms grow larger, increasing in size by increasing the size or number of their cells. In multicellular organisms, the individual cells become specialized to perform particular functions. This specialization is called differentiation. Growth and differentiation often produce changes in form and function. For example, the body proportions and functional capabilities of an adult human are quite different from those of an infant.

■ **Reproduction:** Organisms reproduce, creating subsequent generations of similar organisms.

■ **Movement:** Organisms are capable of producing movement, which may be internal (transporting food, blood, or other materials inside the body) or external (moving through the environment).

■ **Metabolism** and **excretion:** Organisms rely on complex chemical reactions to provide the energy for responsiveness, growth, reproduction, and movement. They must also synthesize complex chemicals, such as proteins. *Metabolism* refers to all the chemical operations under way in the body. Many normal metabolic operations require the absorption of certain materials from the environment. These materials, called *nutrients,* are used for growth and maintenance; many nutrients are also important as energy sources. To generate energy efficiently, most cells require oxygen, an atmospheric gas. *Respiration* is the absorption, transport, and use of oxygen by cells. Metabolic operations often generate unneeded or potentially harmful waste products that must be removed from body fluids through the process of *excretion.*

These are basic characteristics of living things. Several additional functions can be distinguished when you consider animals as complex as fish, cats, or humans. Whereas very small organisms can absorb materials across exposed surfaces, creatures larger than a few millimeters seldom absorb nutrients directly from their environment. For example, humans cannot absorb steaks, apples, or ice cream without processing them first. That processing, called *digestion,* occurs in specialized areas where complex foods are broken down into simpler components that can be absorbed easily. Respiration and excretion are also more complicated for large organisms than for small ones. Humans have specialized structures responsible for gas exchange (lungs) and for waste elimination (kidneys). Finally, because absorption, respiration, and excretion are performed in different regions of the body, humans require an internal transportation system, or *cardiovascular system.*

Biology includes subspecialties, each with a slightly different perspective. This text considers two biological subjects, *anatomy* (ah-NAT-o-mē) and *physiology* (fiz-ē-OL-o-jē). In your journey through these 29 chapters, you will become familiar with the basic anatomy and physiology of the human body.

THE SCIENCES OF ANATOMY AND PHYSIOLOGY

The word *anatomy* has Greek origins, as do many other anatomical terms and phrases. A literal translation would be "a cutting open." **Anatomy** is the study of internal and external structures of the body and the physical relationships among body parts. For example, someone studying anatomy might examine how a particular muscle attaches to the skeleton. **Physiology,** another word adopted from Greek, is the study of how organisms perform their vital functions. Someone studying physiology might consider how a muscle contracts or what forces a contracting muscle exerts on the skeleton.

Anatomy and physiology are closely integrated both theoretically and practically. Anatomical information provides clues about probable functions, and physiological mechanisms can be explained only in terms of the underlying anatomy. This observation leads to a very important concept: *All specific functions are performed by specific structures.*

Anatomists and physiologists approach the relationship between structure and function from different perspectives. To understand the difference, consider a simple nonbiological analogy. Suppose that an anatomist and a physiologist were asked to examine and report on a pickup truck. The anatomist might begin by measuring and photographing the various parts of the truck and, if possible, taking it apart and putting it back together. The anatomist could then explain its key structural relationships—for example, how the movement of the pistons in the engine cylinders causes the drive shaft to rotate and how the transmission then conveys this motion to the wheels. The physiologist also would note the relationships among the components, but the primary focus would be on functional characteristics, such as the amount of power that the engine could generate, the amount of force transmitted to the wheels in different gears, and so on.

The link between structure and function is always present but not always understood. For example, the anatomy of the heart was clearly described in the fifteenth century, but almost 200 years passed before the pumping action of the heart was demonstrated.

This text will introduce anatomical structures and the physiological processes that make human life possible. The information will enable you to understand important mechanisms of disease and will help you make intelligent decisions about personal health.

ANATOMY

Anatomy can be divided into *microscopic anatomy* and *gross (macroscopic) anatomy* on the basis of the degree of structural detail under consideration. Other anatomical specialties focus on specific processes or medical applications.

Microscopic Anatomy

Microscopic anatomy deals with structures that cannot be seen without magnification. The boundaries of microscopic anatomy are established by the limits of the equipment used. With a light microscope, you can see basic details of cell structure; with an electron microscope, you can see individual molecules that are only a few nanometers across. As we proceed through the text, we will be considering details at all levels, from macroscopic to microscopic. (Readers unfamiliar with the terms used to describe measurements and weights should consult the reference tables in Appendix II.)

Microscopic anatomy includes cytology and histology. **Cytology** (sī-TOL-o-jē) is the analysis of the structure of individual **cells,** the simplest units of life. Cells are composed of chemical substances in various combinations, and our lives depend on the chemical processes occurring in the trillions of cells in the body. For this reason we will consider basic chemistry (Chapter 2) before we examine cell structure (Chapter 3).

Histology (his-TOL-o-jē) is the examination of **tissues,** groups of specialized cells and cell products that work together to perform specific functions (Chapter 4). Tissues combine to form **organs,** such as the heart, kidney, liver, or brain. Many organs are easily examined without a microscope, so at the organ level we cross the boundary from microscopic anatomy to gross anatomy.

Gross Anatomy

Gross anatomy, or *macroscopic anatomy,* involves the examination of relatively large structures and features usually visible with the unaided eye. There are many ways to approach gross anatomy:

- *Surface anatomy* is the study of general form and superficial markings.
- *Regional anatomy* focuses on the anatomical organization of specific areas of the body, such as the head, neck, or trunk. Many advanced courses in anatomy stress a regional approach, because it emphasizes the spatial relationships between structures already familiar to students.
- *Systemic anatomy* is the study of the structure of organ systems, such as the *skeletal system* or the *muscular system.* **Organ systems** are groups of organs that function together in a coordinated manner. For example, the heart, blood, and blood vessels form the *cardiovascular system,* which distributes oxygen and nutrients throughout the body. Introductory texts present systemic anatomy because that approach clarifies functional relationships among the component organs. The human body has 11 organ systems, and we will introduce them later in the chapter.
- *Developmental anatomy* deals with the changes in form that occur during the period between conception and physical maturity. Because developmental anatomy considers anatomical structures over such a broad range of sizes (from a single cell to an adult human), techniques used in it are similar to those used in both microscopic anatomy and gross anatomy. The most extensive structural changes occur during the first 2 months of development. The study of these early developmental processes is called *embryology* (em-brē-OL-o-jē).

Several anatomical specialties are important in a clinical setting. Examples include *medical anatomy* (anatomical features that change during illness), *radiographic anatomy* (anatomical structures as seen by using specialized imaging techniques, discussed later in this chapter), and *surgical anatomy* (anatomical landmarks important in surgery). Physicians normally use a combination of anatomical, physiological, and psychological information when they evaluate patients. The evaluation process follows a logical sequence; this sequence is an example of what is known as the *scientific method.* AM *The Scientific Method*

PHYSIOLOGY

As we noted earlier, physiology is the study of the function of anatomical structures; **human physiology** is the study of the functions of the human body. These functions are complex and much more difficult to examine than most anatomical structures. As a result, there are even more specialties in physiology than in anatomy:

1

■ *Cell physiology,* the study of the functions of cells, is the cornerstone of human physiology. Cell physiology deals with events at the chemical and molecular levels: both chemical processes within cells and chemical interactions between cells. Chapters 2–4 focus on the chemical structure, internal organization, and control mechanisms of cells and tissues.

■ *Special physiology* is the study of the physiology of specific organs. For example, *cardiac physiology* is the study of heart function.

■ *Systemic physiology* includes all aspects of the function of specific organ systems. Cardiovascular physiology, respiratory physiology, and reproductive physiology are examples of systemic physiology.

■ *Pathological physiology* is the study of the effects of diseases on organ or system functions. (*Pathos* is the Greek word for "disease.") Modern medicine depends on an understanding of both normal physiology and pathological physiology. You will find extensive information on clinically important topics in subsequent chapters and in the *Applications Manual* that accompanies this text.

LEVELS OF ORGANIZATION

Our study of the human body will begin with an overview of microscopic anatomy and then proceed to the gross anatomy and microscopic anatomy of each organ system. When considering events from the microscopic to the macroscopic scale, we will examine several interdependent levels of organization. Figure 1-1• presents an example of the relationships among these various levels of organization:

■ *The chemical, or molecular, level. Atoms,* the smallest stable units of matter, can combine to form *molecules* with complex shapes. Even at this simplest level, the specialized shape of a molecule determines its function.

■ *The cellular level.* Molecules can interact to form *organelles,* such as the protein filaments found in muscle cells. Each type of organelle has specific functions. For example, interactions among protein filaments produce the contractions of muscle cells in the heart. Cells are the smallest living units in the body; organelles are their structural and functional components.

■ *The tissue level.* A *tissue* is a group of cells working together to perform one or more specific functions. Heart muscle cells, or cardiac muscle cells (*cardium,* heart), interact with other cell types and with extracellular materials to form *cardiac muscle tissue.*

■ *The organ level. Organs* consist of two or more tissues working in combination to perform several functions. Layers of cardiac muscle tissue in combination with *connective tissue,* another tissue type, form the bulk of the wall of the *heart,* a hollow, three-dimensional organ.

■ *The organ system level.* Organs interact in *organ systems.* Each time it contracts, the heart pushes blood into a network of blood vessels. Together the heart, blood, and blood vessels form the cardiovascular system, one of 11 organ systems in the body.

■ *The organism level.* All organ systems of the body work together to maintain life and health. This brings us to the highest level of organization, that of the *organism*—in this case, a human being.

The organization at each level determines the characteristics and functions of higher levels. For example, the arrangement of atoms and molecules at the chemical level creates the protein filaments that, at the cellular level, give cardiac muscle cells the ability to contract powerfully. At the tissue level, these cells are linked, forming cardiac muscle tissue. The structure of the tissue ensures that the contractions are coordinated, producing a heartbeat. When that beat occurs, the internal anatomy of the heart, an organ, enables it to function as a pump. The heart is filled with blood and connected to the blood vessels, and the pumping action circulates blood through the vessels of the cardiovascular system. Through interactions with the respiratory, digestive, urinary, and other systems, the cardiovascular system performs a variety of functions essential to the survival of the organism.

Something that affects a system will ultimately affect each component. For example, the heart cannot pump blood effectively after a massive blood loss. If the heart cannot pump and blood cannot flow, oxygen and nutrients cannot be distributed. Very soon, the cardiac muscle tissue begins to break down as individual muscle cells die from oxygen and nutrient starvation. These changes will not be restricted to the cardiovascular system; all cells, tissues, and organs in the body will be damaged.

Figure 1-2• (pp. 6–11) introduces the 11 organ systems in the human body. These organ systems are interdependent, interconnected, and packaged together in a relatively small space. The cells, tissues, organs, and organ systems of the body live together in a shared environment, like the inhabitants of a large city. Just as city dwellers breathe the city air and drink the water provided by the local water company, cells in the human body absorb oxygen and nutrients from the fluids that surround them. If a city is blanketed in smog or its water supply is contaminated, the inhabitants will become ill. Similarly, if body fluid composition becomes abnormal, cells will be injured or destroyed. Suppose the temperature or salt content of the blood changes. The effect on the heart could range from a minor adjustment (heart muscle tissue contracts more often, so the heart rate goes up) to a total disaster (the heart stops beating, so the individual dies).

Various physiological mechanisms act to prevent potentially damaging changes in the composition of body fluids and the environment inside our cells. **Homeostasis** (*homeo,* unchanging + *stasis,* standing) refers to the existence of a stable internal environment. To survive, every organism must maintain homeostasis.

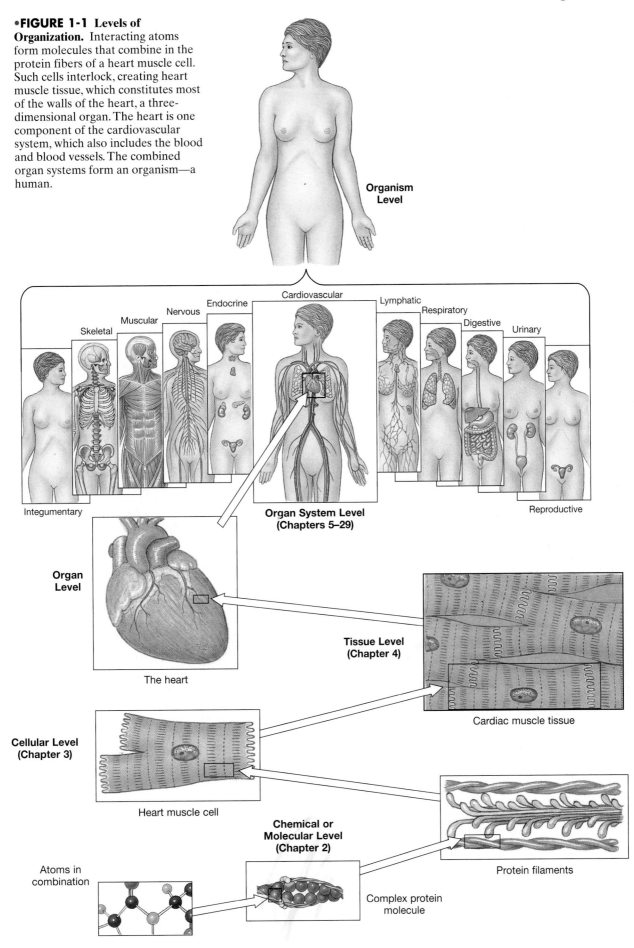

•**FIGURE 1-1** **Levels of Organization.** Interacting atoms form molecules that combine in the protein fibers of a heart muscle cell. Such cells interlock, creating heart muscle tissue, which constitutes most of the walls of the heart, a three-dimensional organ. The heart is one component of the cardiovascular system, which also includes the blood and blood vessels. The combined organ systems form an organism—a human.

Organism Level

Skeletal Muscular Nervous Endocrine Cardiovascular Lymphatic Respiratory Digestive Urinary

Integumentary

Organ System Level (Chapters 5–29)

Reproductive

Organ Level

The heart

Tissue Level (Chapter 4)

Cardiac muscle tissue

Cellular Level (Chapter 3)

Heart muscle cell

Chemical or Molecular Level (Chapter 2)

Atoms in combination

Complex protein molecule

Protein filaments

1

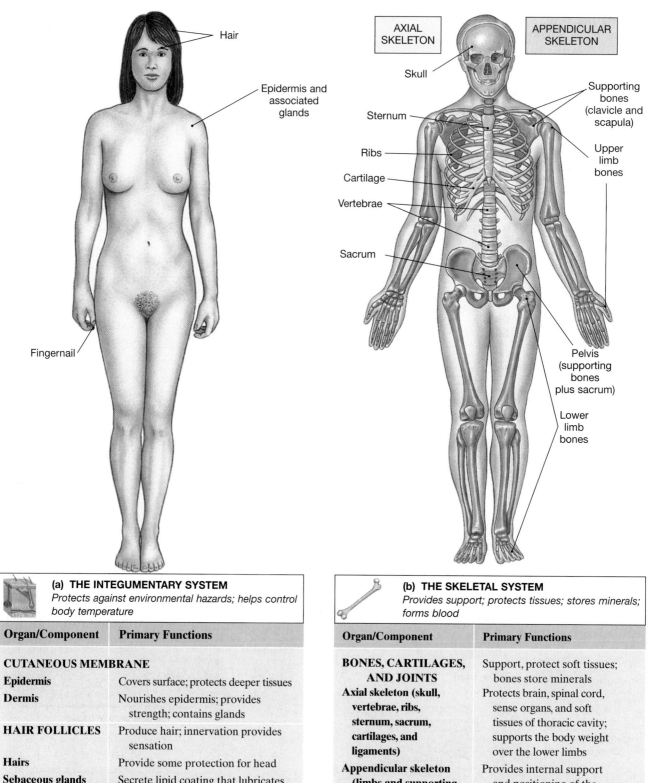

AXIAL SKELETON

APPENDICULAR SKELETON

Hair

Epidermis and associated glands

Skull

Supporting bones (clavicle and scapula)

Sternum

Ribs

Upper limb bones

Cartilage

Vertebrae

Sacrum

Fingernail

Pelvis (supporting bones plus sacrum)

Lower limb bones

(a) THE INTEGUMENTARY SYSTEM
Protects against environmental hazards; helps control body temperature

Organ/Component	Primary Functions
CUTANEOUS MEMBRANE	
Epidermis	Covers surface; protects deeper tissues
Dermis	Nourishes epidermis; provides strength; contains glands
HAIR FOLLICLES	Produce hair; innervation provides sensation
Hairs	Provide some protection for head
Sebaceous glands	Secrete lipid coating that lubricates hair shaft and epidermis
SWEAT GLANDS	Produce perspiration for evaporative cooling
NAILS	Protect and stiffen distal tips of digits
SENSORY RECEPTORS	Provide sensations of touch, pressure, temperature, pain
SUBCUTANEOUS LAYER	Stores lipids; attaches skin to deeper structures

(b) THE SKELETAL SYSTEM
Provides support; protects tissues; stores minerals; forms blood

Organ/Component	Primary Functions
BONES, CARTILAGES, AND JOINTS	Support, protect soft tissues; bones store minerals
Axial skeleton (skull, vertebrae, ribs, sternum, sacrum, cartilages, and ligaments)	Protects brain, spinal cord, sense organs, and soft tissues of thoracic cavity; supports the body weight over the lower limbs
Appendicular skeleton (limbs and supporting bones and ligaments)	Provides internal support and positioning of the limbs; supports and moves axial skeleton
BONE MARROW	Acts as primary site of blood cell production (red blood cells, white blood cells)

•**FIGURE 1-2** An Introduction to Organ Systems

Axial muscles

Appendicular muscles

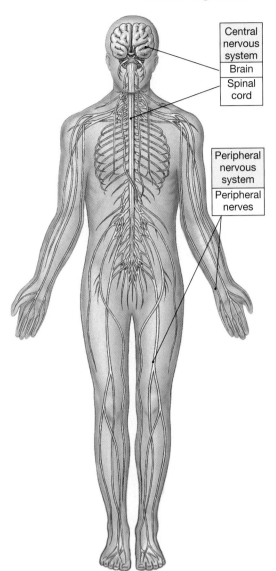

Central nervous system
Brain
Spinal cord

Peripheral nervous system
Peripheral nerves

(c) THE MUSCULAR SYSTEM
Allows for locomotion; provides support; produces heat

Organ/Component	Primary Functions
SKELETAL MUSCLES (700)	Provide skeletal movement; control entrances and exits of digestive tract; produce heat; support skeletal position; protect soft tissues
Axial muscles	Support and position axial skeleton
Appendicular muscles	Support, move, and brace limbs
TENDONS, APONEUROSES	Harness forces of contraction to perform specific tasks

(d) THE NERVOUS SYSTEM
Directs immediate responses to stimuli, usually by coordinating the activities of other organ systems

Organ/Component	Primary Functions
CENTRAL NERVOUS SYSTEM (CNS)	Acts as control center for nervous system: processes information; provides short-term control over activities of other systems
Brain	Performs complex integrative functions; controls both voluntary and autonomic activities
Spinal cord	Relays information to and from brain; performs less-complex integrative functions; directs many simple involuntary activities
PERIPHERAL NERVOUS SYSTEM (PNS)	Links CNS with other systems and with sense organs

1

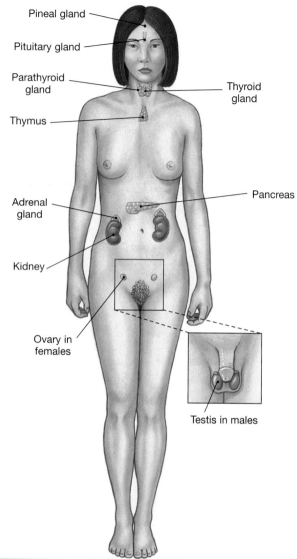

Pineal gland
Pituitary gland
Parathyroid gland
Thyroid gland
Thymus
Adrenal gland
Pancreas
Kidney
Ovary in females
Testis in males

Heart
Capillaries
Artery
Vein

(e) THE ENDOCRINE SYSTEM	
Directs long-term changes in activities of other organ systems	

Organ/Component	Primary Functions
PINEAL GLAND	May control timing of reproduction and set day–night rhythms
PITUITARY GLAND	Controls other endocrine glands; regulates growth and fluid balance
THYROID GLAND	Controls tissue metabolic rate; regulates calcium levels
PARATHYROID GLANDS	Regulate calcium levels (with thyroid)
THYMUS	Controls maturation of lymphocytes
ADRENAL GLANDS	Adjust water balance, tissue metabolism, cardiovascular and respiratory activity
KIDNEYS	Control red blood cell production and elevate blood pressure
PANCREAS	Regulates blood glucose levels
GONADS	
Testes	Support male sexual characteristics and reproductive functions *(see Figure 1-2k)*
Ovaries	Support female sexual characteristics and reproductive functions *(see Figure 1-2l)*

(f) THE CARDIOVASCULAR SYSTEM	
Transports cells and dissolved materials, including nutrients, wastes, and gases	

Organ/Component	Primary Functions
HEART	Propels blood; maintains blood pressure
BLOOD VESSELS	Distribute blood around the body
Arteries	Carry blood from heart to capillaries
Capillaries	Permit diffusion between blood and interstitial fluids
Veins	Return blood from capillaries to the heart
BLOOD	Transports oxygen, carbon dioxide, and blood cells; delivers nutrients and hormones; removes waste products; assists in temperature regulation and defense against disease

•**FIGURE 1-2** An Introduction to Organ Systems *(continued)*

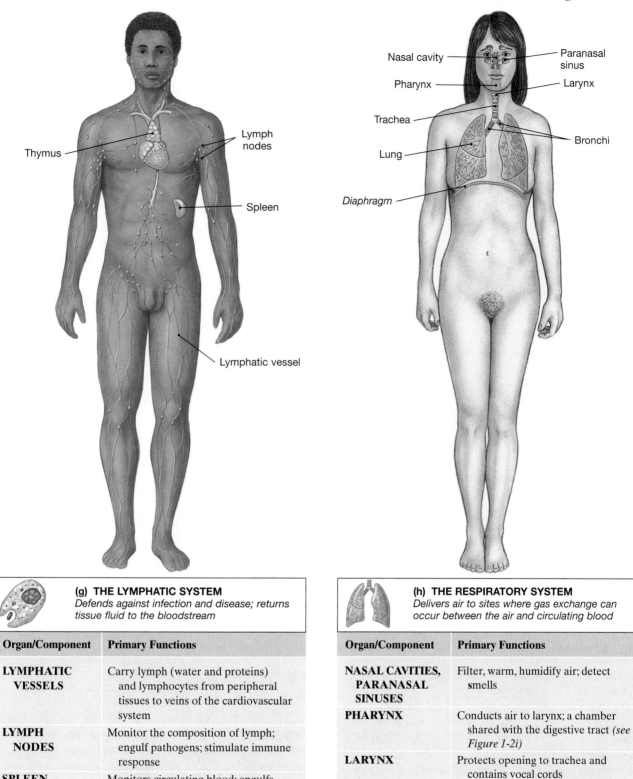

Thymus

Lymph nodes

Spleen

Lymphatic vessel

Nasal cavity

Paranasal sinus

Pharynx

Larynx

Trachea

Lung

Bronchi

Diaphragm

(g) THE LYMPHATIC SYSTEM
Defends against infection and disease; returns tissue fluid to the bloodstream

Organ/Component	Primary Functions
LYMPHATIC VESSELS	Carry lymph (water and proteins) and lymphocytes from peripheral tissues to veins of the cardiovascular system
LYMPH NODES	Monitor the composition of lymph; engulf pathogens; stimulate immune response
SPLEEN	Monitors circulating blood; engulfs pathogens; stimulates immune response
THYMUS	Controls development and maintenance of one class of lymphocytes (T cells)

(h) THE RESPIRATORY SYSTEM
Delivers air to sites where gas exchange can occur between the air and circulating blood

Organ/Component	Primary Functions
NASAL CAVITIES, PARANASAL SINUSES	Filter, warm, humidify air; detect smells
PHARYNX	Conducts air to larynx; a chamber shared with the digestive tract (*see Figure 1-2i*)
LARYNX	Protects opening to trachea and contains vocal cords
TRACHEA	Filters air, traps particles in mucus; cartilages keep airway open
BRONCHI	(Same functions as trachea)
LUNGS	Responsible for air movement through volume changes during movements of ribs and diaphragm; include airways and alveoli
Alveoli	Act as sites of gas exchange between air and blood

1

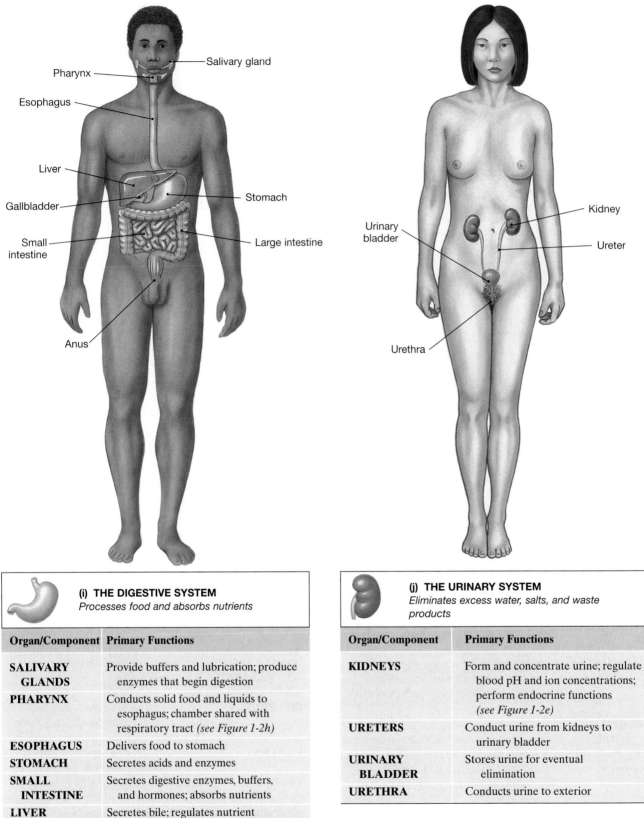

(i) THE DIGESTIVE SYSTEM
Processes food and absorbs nutrients

Organ/Component	Primary Functions
SALIVARY GLANDS	Provide buffers and lubrication; produce enzymes that begin digestion
PHARYNX	Conducts solid food and liquids to esophagus; chamber shared with respiratory tract *(see Figure 1-2h)*
ESOPHAGUS	Delivers food to stomach
STOMACH	Secretes acids and enzymes
SMALL INTESTINE	Secretes digestive enzymes, buffers, and hormones; absorbs nutrients
LIVER	Secretes bile; regulates nutrient composition of blood
GALLBLADDER	Stores bile for release into small intestine
PANCREAS	Secretes digestive enzymes and buffers; contains endocrine cells *(see Figure 1-2e)*
LARGE INTESTINE	Removes water from fecal material; stores wastes

(j) THE URINARY SYSTEM
Eliminates excess water, salts, and waste products

Organ/Component	Primary Functions
KIDNEYS	Form and concentrate urine; regulate blood pH and ion concentrations; perform endocrine functions *(see Figure 1-2e)*
URETERS	Conduct urine from kidneys to urinary bladder
URINARY BLADDER	Stores urine for eventual elimination
URETHRA	Conducts urine to exterior

•FIGURE 1-2 An Introduction to Organ Systems *(continued)*

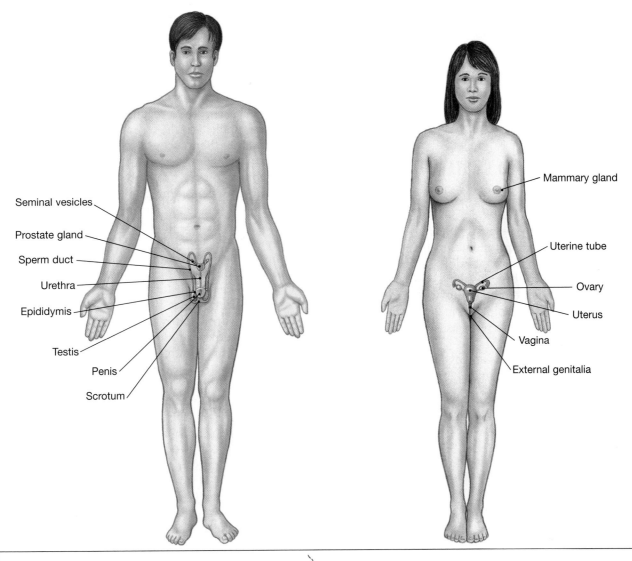

Seminal vesicles
Prostate gland
Sperm duct
Urethra
Epididymis
Testis
Penis
Scrotum

Mammary gland
Uterine tube
Ovary
Uterus
Vagina
External genitalia

(k) THE MALE REPRODUCTIVE SYSTEM
Produces sex cells and hormones

(l) THE FEMALE REPRODUCTIVE SYSTEM
Produces sex cells and hormones

Organ/Component	Primary Functions
TESTES	Produce sperm and hormones (*see Figure 1-2e*)
ACCESSORY ORGANS	
Epididymis	Acts as site of sperm maturation
Ductus deferens (sperm duct)	Conducts sperm between epididymis and prostate gland
Seminal vesicles	Secrete fluid that makes up much of the volume of semen
Prostate gland	Secretes fluid and enzymes
Urethra	Conducts semen to exterior
EXTERNAL GENITALIA	
Penis	Contains erectile tissue; deposits sperm in vagina of female; produces pleasurable sensations during sexual activities
Scrotum	Surrounds the testes and controls their temperature

Organ/Component	Primary Functions
OVARIES	Produce oocytes and hormones (*see Figure 1-2e*)
UTERINE TUBES	Deliver oocyte or embryo to uterus; normal site of fertilization
UTERUS	Site of embryonic development and exchange between maternal and embryonic bloodstreams
VAGINA	Site of sperm deposition; acts as birth canal at delivery; provides passageway for fluids during menstruation
EXTERNAL GENITALIA	
Clitoris	Contains erectile tissue; produces pleasurable sensations during sexual activities
Labia	Contain glands that lubricate entrance to vagina
MAMMARY GLANDS	Produce milk that nourishes newborn infant

☑ Which characteristics of life do white blood cells exhibit when they migrate to an injury site in response to chemicals released from damaged cells?

☑ A histologist investigates structures at which level of organization?

☑ A researcher studies the factors that cause heart failure. What might this person's specialty be called?

HOMEOSTASIS AND SYSTEM INTEGRATION

Homeostasis is absolutely vital; a failure to maintain it soon leads to illness or even death. The principle of homeostasis is the central theme of this text and the foundation of all modern physiology. **Homeostatic regulation** is the adjustment of physiological systems to preserve homeostasis. An understanding of homeostatic regulation will help you make accurate predictions about the body's responses to both normal and abnormal conditions.

Two general mechanisms are involved in homeostatic regulation: autoregulation and extrinsic regulation:

1. **Autoregulation,** or *intrinsic regulation,* occurs when the activities of a cell, tissue, organ, or organ system adjust automatically in response to some environmental change. For example, when oxygen levels decline in a tissue, the cells release chemicals that dilate local blood vessels. This dilation increases the rate of blood flow and provides more oxygen to the region.

2. **Extrinsic regulation** results from the activities of the nervous system or endocrine system, two organ systems that control or adjust the activities of many systems simultaneously. For example, when you exercise, your nervous system issues commands that increase your heart rate so that blood will circulate faster. Your nervous system also reduces blood flow to less-active organs, such as the digestive tract. The oxygen in circulating blood is thus saved for the active muscles.

In general, the nervous system performs crisis management by directing rapid, short-term, and very specific responses. When you accidentally set your hand on a hot stove, the heat produces a painful, localized disturbance of homeostasis. Your nervous system responds by ordering the contraction of specific muscles that will pull your hand away from the stove. The effects last only as long as the neural activity continues, usually a matter of seconds.

In contrast, the endocrine system releases chemical messengers, called *hormones,* that affect tissues and organs throughout the body. The responses may not be immediately apparent, but when the effects appear, they may persist for days or weeks. Examples of endocrine function include the long-term regulation of blood volume and composition and the adjustment of organ system function during development or during starvation.

Regardless of the system involved, homeostatic regulation always attempts to keep the characteristics of the internal environment within desirable limits. The regulatory mechanism consists of three parts: (1) a **receptor,** a sensor that is sensitive to a particular environmental change, or *stimulus;* (2) a **control center,** or *integration center,* which receives and processes the information supplied by the receptor; and (3) an **effector,** a cell or organ that responds to the commands of the control center and whose activity opposes or enhances the stimulus. You are probably already familiar with comparable regulatory mechanisms, such as the thermostat in your home or apartment (Figure 1-3●).

A thermostat is a control center that receives information about room temperature from an internal or remote thermometer (a receptor). The dial on the thermostat establishes the *set point,* or desired value, which in this case is the temperature you select. In our example, the set point is 22°C (about 72°F). The function of the thermostat is to keep room temperature within acceptable limits, usually within a degree or so of the set point. In summer the thermostat accomplishes this function by controlling an air conditioner (an effector). When the temperature at the thermometer rises above the acceptable range, the thermostat turns on the air conditioner. The air conditioner then cools the room (Figure 1-3a●). When the temperature at the thermometer reaches the set point, the thermostat turns off the air conditioner (Figure 1-3b●). The control is not precise; the room is large, and the thermostat is located on one wall. Over time, the temperature in the center of the room oscillates around the set point.

NEGATIVE FEEDBACK

The essential feature of temperature control by thermostat can be summarized very simply: A variation outside the desired range triggers an automatic response that corrects the situation. This method of homeostatic regulation is called **negative feedback,** because an effector activated by the control center either opposes or eliminates the stimulus.

Most homeostatic regulatory mechanisms involve negative feedback. Consider the control of body temperature, a process called *thermoregulation.* In thermoregulation, the relationship between heat loss, which occurs primarily at the body surface, and heat production, which occurs in all active tissues, is altered.

The control center for thermoregulation is in the *hypothalamus,* a region of the brain (Figure 1-4a●). This control center receives information from two sets of temperature receptors, one in the skin and the other inside the hypothalamus. At the normal set point, the body temperature, as measured with an oral thermometer, will be approximately 37°C (98.6°F). If body temperature rises above 37.2°C,

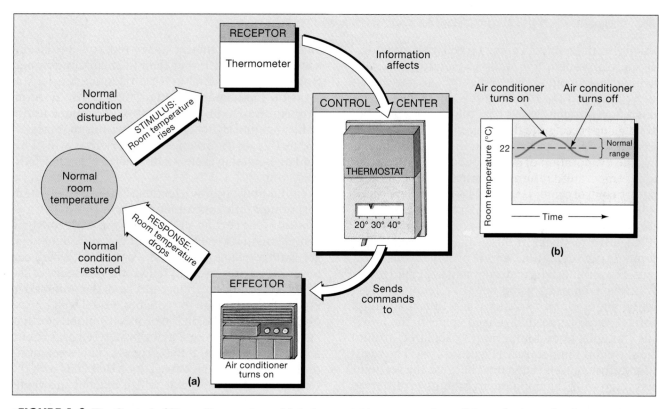

•**FIGURE 1-3** **The Control of Room Temperature.** **(a)** A thermostat turns on an air conditioner (or heater) to keep room temperature near the desired set point. Whether room temperature rises or falls, the thermostat (a control center) triggers an effector response that restores normal temperature. When room temperature rises, the thermostat turns on the air conditioner, and the temperature returns to normal. **(b)** With this regulatory system, room temperature oscillates around the set point.

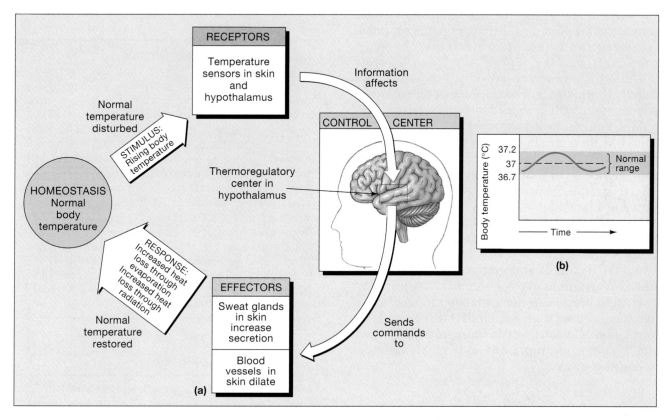

•**FIGURE 1-4** **Negative Feedback: The Control of Body Temperature.** In negative feedback, a stimulus produces a response that opposes or eliminates the stimulus. **(a)** Events comparable to those shown in Figure 1-3 occur in the regulation of body temperature. A control center in the brain functions as a thermostat with a set point of 37°C. If body temperature exceeds 37.2°C, heat loss is increased through enhanced blood flow to the skin and increased sweating. **(b)** The thermoregulatory center keeps body temperature oscillating within an acceptable range, usually between 36.7 and 37.2°C.

1

activity in the control center targets two effectors: (1) muscle tissue in the walls of blood vessels supplying the skin and (2) sweat glands. The muscle tissue relaxes and the blood vessels dilate, increasing blood flow through vessels near the body surface; the sweat glands accelerate their secretion. The skin then acts like a radiator by losing heat to the environment, and the evaporation of sweat speeds the process. As body temperature returns to normal, the thermoregulatory control center becomes less active. Superficial blood flow and sweat gland activity then decrease to previous levels.

Negative feedback is the primary mechanism of homeostatic regulation, and it provides long-term control over internal conditions and systems. Homeostatic mechanisms using negative feedback normally ignore minor variations, and they maintain a normal *range* rather than a fixed value. In the previous example, body temperature oscillated around the set-point temperature (Figure 1-4b•). The regulatory process itself is dynamic, because the set point may vary with changing environments or differing activity levels. For example, when you are asleep, your thermoregulatory set point lowers, whereas when you work outside on a hot day (or when you have a fever), it climbs. Thus body temperature can vary from moment to moment or from day to day for any individual due to (1) small oscillations around the set point or (2) changes in the set point. Comparable variations occur in all other physiological characteristics.

The variability among individuals is even greater than that within an individual. Each of us has homeostatic set points determined by genetic factors, age, gender, general health, and environmental conditions. It is therefore impractical to define "normal" homeostatic conditions very precisely. By convention, physiological values are reported either as averages—average values obtained by sampling a large number of individuals—or as a range that includes 95 percent or more of the sample population. For instance, 5 percent of healthy adults have resting body temperatures that are below 36.7°C or above 37.2°C. These temperatures are perfectly normal for them, and the variations have no clinical significance. Physicians must keep this variability in mind when they review lab reports or clinical discussions, because unusual values—even those outside the normal range—may represent individual variation rather than disease.

POSITIVE FEEDBACK

In **positive feedback,** the initial stimulus produces a response that exaggerates or enhances its effects.

Suppose that the thermostat in Figure 1-3a• is connected to a heater rather than to an air conditioner. Now when room temperature rises, the thermostat turns on the heater, causing a further rise in room temperature. Room temperature will continue to rise until someone switches off the thermostat, turns off the heater, or intervenes in some other way. This kind of escalating cycle is often called a *positive feedback loop.*

In the body, positive feedback loops are often incorporated into control mechanisms in which a potentially dangerous or stressful process must be completed quickly. For example, the immediate danger from a severe cut is the loss of blood, which can lower blood pressure and reduce the efficiency of the heart. The body's response has been diagrammed in Figure 1-5•. Damage to the blood vessel wall releases chemicals that begin the process of blood clotting. As clotting gets under way, each step releases chemicals that accelerate the process. This escalating process is a positive feedback loop that ends with the formation of a blood clot, which patches the vessel wall and stops the bleeding. Blood clotting will be examined more closely in Chapter 19. Labor and delivery, another example of positive feedback in action, will be discussed in Chapter 29.

Homeostasis and Disease

When homeostatic regulation fails, whether as a result of infection, injury, or genetic abnormality, organ systems begin to malfunction. The result is a state known as illness, or **disease.** Chapters 2–29 devote considerable attention to the mechanisms responsible for a variety of human diseases. An understanding of normal homeostatic mechanisms can usually enable you to draw conclusions about what might be responsible for the signs and symptoms characteristic of many diseases. A **sign** is an objective physical indication of a disease. Examples include swellings, fevers, or sounds of abnormal breathing. **Symptoms** are subjective—things that a person experiences and describes but that aren't otherwise detectable. Examples include pain, nausea, and anxiety. Major organizational and functional patterns relevant to clinical practice are discussed in the *Applications Manual.* AM *Homeostasis and Disease*

☑ Why is homeostatic regulation important to humans?

☑ What happens to the body when homeostasis breaks down?

☑ Why is positive feedback helpful in blood clotting but unsuitable for the regulation of body temperature?

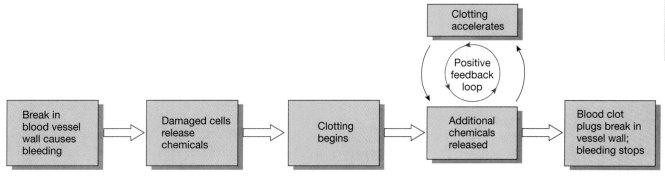

•**FIGURE 1-5** **Positive Feedback: Blood Clotting.** Positive feedback loops are important in accelerating processes that must proceed to completion rapidly. In this example, positive feedback accelerates the clotting process until a blood clot forms and stops the bleeding.

A FRAME OF REFERENCE FOR ANATOMICAL STUDIES

Early anatomists faced serious problems in communication. Stating that a bump is "on the back" does not give very precise information about its location. So anatomists created maps of the human body. The landmarks are prominent anatomical structures; distances are measured in centimeters or inches; and specialized directional terms are used. In effect, anatomy uses a special language that must be learned almost at the start of your study.

A familiarity with Latin and Greek word roots and patterns makes anatomical terms more understandable. As new terms are introduced, notes on pronunciation and relevant word roots will be provided. The front endpapers contain additional information on word roots, prefixes, suffixes, and combining forms.

Latin and Greek terms are not the only ones that have been imported into the anatomical vocabulary over the centuries, and the vocabulary continues to expand. Many anatomical structures and clinical conditions were initially named after either the discoverer or, in the case of diseases, the most famous victim. Over the last 100 years, most of these commemorative names, or *eponyms*, have been replaced by more precise terms. Appendix V lists the most important eponyms and related historical details.

SUPERFICIAL ANATOMY

A familiarity with anatomical landmarks, regions, and directional references will make subsequent chapters more understandable. As you encounter new terms, create your own mental maps with the in-

formation provided in the accompanying anatomical illustrations.

Anatomical Landmarks

Important anatomical landmarks are presented in Figure 1-6•. Anatomical terms are given in boldface type, common names in plain type, and anatomical adjectives in parentheses. Understanding the terms and their origins will help you remember the location of a particular structure as well as its name. For example, *brachium* refers to the arm; later we will consider the *brachialis muscle* and the *brachial artery*, which are (as their names suggest) in the arm.

Standard anatomical illustrations show the human form in the **anatomical position.** When the body is in this position, the hands are at the sides with the palms facing forward. Figure 1-6a• shows an individual in the anatomical position as seen from the front (an *anterior view*), and Figure 1-6b• from the back (a *posterior view*). Unless otherwise noted, all descriptions in this text refer to the body in the anatomical position. A person lying down in the anatomical position is said to be **supine** (soo-PĪN) when face up and **prone** when face down.

Anatomical Regions

Major regions of the body are listed in Table 1-1 (p. 17) and shown in Figure 1-6•. To describe a general area of interest or injury, anatomists and clinicians often need broader terms as well as specific landmarks. Two methods are used, both concerned with mapping the surface of the abdomen and pelvis. Clinicians refer to four **abdominopelvic quadrants.** The area is divided into four segments by using a pair of imaginary lines that intersect at the umbilicus (navel). This simple

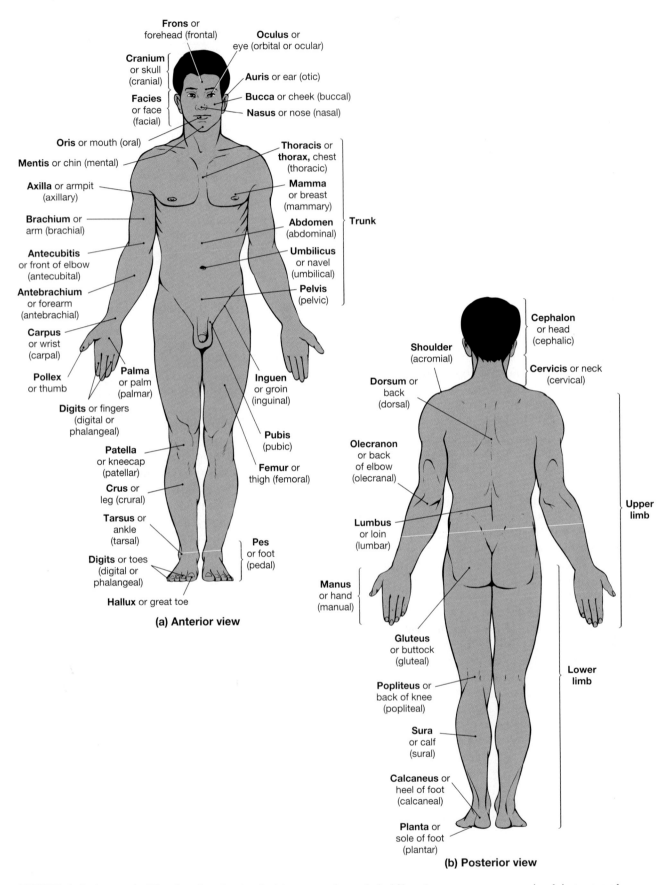

Frons or forehead (frontal)

Oculus or eye (orbital or ocular)

Cranium or skull (cranial)

Auris or ear (otic)

Facies or face (facial)

Bucca or cheek (buccal)

Nasus or nose (nasal)

Oris or mouth (oral)

Thoracis or **thorax**, chest (thoracic)

Mentis or chin (mental)

Axilla or armpit (axillary)

Mamma or breast (mammary)

Brachium or arm (brachial)

Abdomen (abdominal)

Antecubitis or front of elbow (antecubital)

Umbilicus or navel (umbilical)

Antebrachium or forearm (antebrachial)

Pelvis (pelvic)

Trunk

Carpus or wrist (carpal)

Pollex or thumb

Palma or palm (palmar)

Inguen or groin (inguinal)

Digits or fingers (digital or phalangeal)

Pubis (pubic)

Patella or kneecap (patellar)

Femur or thigh (femoral)

Crus or leg (crural)

Tarsus or ankle (tarsal)

Pes or foot (pedal)

Digits or toes (digital or phalangeal)

Hallux or great toe

(a) Anterior view

Cephalon or head (cephalic)

Shoulder (acromial)

Cervicis or neck (cervical)

Dorsum or back (dorsal)

Upper limb

Olecranon or back of elbow (olecranal)

Lumbus or loin (lumbar)

Manus or hand (manual)

Gluteus or buttock (gluteal)

Popliteus or back of knee (popliteal)

Lower limb

Sura or calf (sural)

Calcaneus or heel of foot (calcaneal)

Planta or sole of foot (plantar)

(b) Posterior view

•**FIGURE 1-6 Anatomical Landmarks.** Anatomical terms are shown in boldface type, common names in plain type, and anatomical adjectives in parentheses.

TABLE 1-1 **Regions of the Human Body** *(see Figure 1-6)*

Structure	Region
Cephalon (head)	Cephalic region
Cervicis (neck)	Cervical region
Thoracis (thorax, or chest)	Thoracic region
Brachium (arm)	Brachial region
Antebrachium (forearm)	Antebrachial region
Carpus (wrist)	Carpal region
Manus (hand)	Manual region
Abdomen	Abdominal region
Lumbus (loin)	Lumbar region
Gluteus (buttock)	Gluteal region
Pelvis	Pelvic region
Pubis (anterior pelvis)	Pubic region
Inguen (groin)	Inguinal region
Femur (thigh)	Femoral region
Crus (anterior leg)	Crural region
Sura (calf)	Sural region
Tarsus (ankle)	Tarsal region
Pes (foot)	Pedal region
Planta (sole)	Plantar region

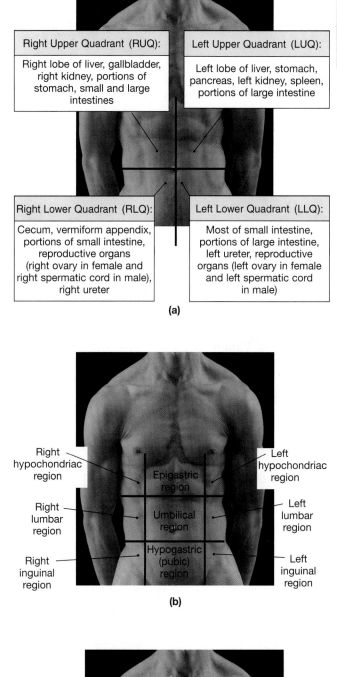

(a)

method, shown in Figure 1-7a•, provides useful references for the description of aches, pains, and injuries. The location can help the physician determine the possible cause; for example, tenderness in the right lower quadrant (RLQ) is a symptom of appendicitis, whereas tenderness in the right upper quadrant (RUQ) may indicate gallbladder or liver problems.

Anatomists like to use more-precise terms to describe the location and orientation of internal organs. They recognize nine **abdominopelvic regions** (Figure 1-7b•). Figure 1-7c• shows the relationships among quadrants, regions, and internal organs.

Anatomical Directions

Figure 1-8• and Table 1-2 show the principal directional terms and examples of their use. There are many

(b)

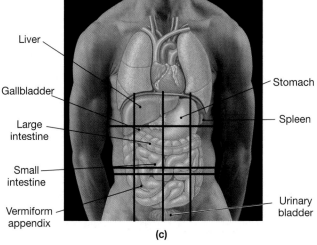

•**FIGURE 1-7 Abdominopelvic Quadrants and Regions.**
(a) Abdominopelvic quadrants divide the area into four sections. These terms, or their abbreviations, are most often used in clinical discussions. **(b)** Abdominopelvic regions divide the same area into nine sections, providing more-precise regional descriptions. **(c)** Overlapping quadrants and regions and the relationship between superficial anatomical landmarks and underlying organs.

(c)

1

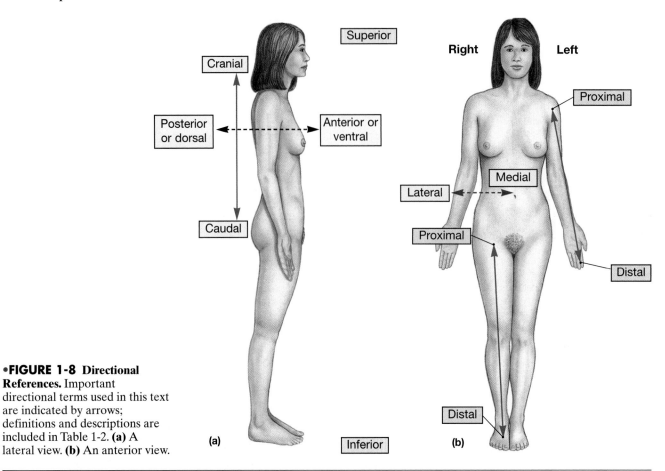

•**FIGURE 1-8 Directional References.** Important directional terms used in this text are indicated by arrows; definitions and descriptions are included in Table 1-2. **(a)** A lateral view. **(b)** An anterior view.

TABLE 1-2 Directional Terms *(see Figure 1-8)*

Term	Region or Reference	Example
Anterior	The front; before	The navel is on the *anterior* surface of the trunk.
Ventral	The belly side (equivalent to anterior when referring to human body)	In humans, the navel is on the *ventral* surface.
Posterior	The back; behind	The shoulder blade is located *posterior* to the rib cage.
Dorsal	The back (equivalent to posterior when referring to human body)	The *dorsal* body cavity encloses the brain and spinal cord.
Cranial or cephalic	The head	The *cranial,* or *cephalic,* border of the pelvis is on the side toward the head rather than toward the thigh.
Superior	Above; at a higher level (in human body, toward the head)	In humans, the cranial border of the pelvis is *superior* to the thigh.
Caudal	The tail (coccyx in humans)	The hips are *caudal* to the waist.
Inferior	Below; at a lower level	The knees are *inferior* to the hips.
Medial	Toward the body's longitudinal axis; toward the midsagittal plane	The *medial* surfaces of the thighs may be in contact; moving medially from the arm across the chest surface brings you to the sternum.
Lateral	Away from the body's longitudinal axis; away from the midsagittal plane	The thigh articulates with the *lateral* surface of the pelvis; moving laterally from the nose brings you to the eyes.
Proximal	Toward an attached base	The thigh is *proximal* to the foot; moving proximally from the wrist brings you to the elbow.
Distal	Away from an attached base	The fingers are *distal* to the wrist; moving distally from the elbow brings you to the wrist.
Superficial	At, near, or relatively close to the body surface	The skin is *superficial* to underlying structures.
Deep	Farther from the body surface	The bone of the thigh is *deep* to the surrounding skeletal muscles.

different terms, and some can be used interchangeably. For example, *anterior* refers to the front of the body when viewed in the anatomical position; in humans, this term is equivalent to *ventral*, which refers to the belly. The terms that appear frequently in later chapters have been emphasized. Before you read further, analyze Table 1-2 in detail, and practice using these terms. If you are familiar with the basic vocabulary, the descriptions in future chapters will be easier to follow. When reading anatomical descriptions, you will find it useful to remember that the terms *left* and *right* always refer to the left and right sides of the subject, not of the observer.

SECTIONAL ANATOMY

A presentation in sectional view is sometimes the only way to illustrate the relationships among the parts of a three-dimensional object. An understanding of sectional views has become increasingly important since the development of electronic imaging techniques that enable us to see inside the living body without resorting to surgery.

Planes and Sections

Any slice through a three-dimensional object can be described with reference to three **sectional planes,** indicated in Figure 1-9• and Table 1-3. The **transverse plane** lies at right angles to the long axis of the body, dividing it into *superior* and *inferior* sections. A cut in this plane is called a **transverse section,** or *cross section.* The **frontal plane,** or *coronal plane,* and the **sagittal plane** are parallel to the long axis of the body.

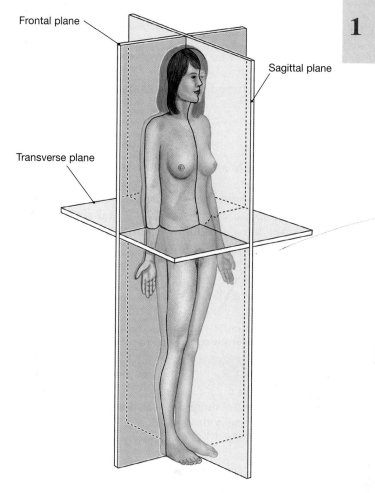

•**FIGURE 1-9 Planes of Section.** The three primary planes of section, which are defined and described in Table 1-3.

TABLE 1-3 Terms That Indicate Planes of Section *(see Figure 1-9)*

Orientation of Plane	Plane	Directional Reference	Description
Perpendicular to long axis	Transverse or horizontal	Transversely or horizontally	A *transverse,* or *horizontal,* section separates superior and inferior portions of the body.
Parallel to long axis	Sagittal	Sagittally	A *sagittal section* separates right and left portions. You examine a sagittal section, but you section sagittally.
	Midsagittal		In a *midsagittal section,* the plane passes through the midline, dividing the body in half and separating right and left sides.
	Parasagittal		A *parasagittal section* misses the midline, separating right and left portions of unequal size.
	Frontal or coronal	Frontally or coronally	A *frontal,* or *coronal, section* separates anterior and posterior portions of the body; *coronal* usually refers to sections passing through the skull.

1

The frontal plane extends from side to side, dividing the body into *anterior* and *posterior* sections. The sagittal plane extends from front to back, dividing the body into left and right sections. A cut that passes along the midline and divides the body into left and right halves is a *midsagittal section,* or *median section;* a cut parallel to the midsagittal line is a *parasagittal section.*

Body Cavities

Many vital organs are suspended in internal chambers called *body cavities.* These cavities have two essential functions: (1) They protect delicate organs, such as the brain and spinal cord, from accidental shocks and cushion them from the thumps and bumps that occur when we walk, jump, or run; and (2) they permit significant changes in the size and shape of internal organs. For example, because they are inside body cavities, the lungs, heart, stomach, intestines, urinary bladder, and many other organs can expand and contract without distorting surrounding tissues or disrupting the activities of nearby organs.

The *dorsal body cavity* contains the brain and spinal cord. The much larger *ventral body cavity* contains organs of the respiratory, cardiovascular, digestive, urinary, and reproductive systems. The relationships among the dorsal and ventral body cavities and their subdivisions are indicated in Figure 1-10• and shown in Figure 1-11•.

Dorsal Body Cavity   The **dorsal body cavity** (Figure 1-11a,c•) is a fluid-filled space whose limits are established by the cranium, the bones of the skull that surround the brain, and the *vertebrae,* the bones of the vertebral column that surround the spinal cord. The dorsal body cavity is subdivided into the **cranial cavity,** which contains the brain, and the **spinal cavity,** which contains the spinal cord.

Ventral Body Cavity   As embryological development proceeds, internal organs grow and their relative positions change. These changes lead to the subdivision of the **ventral body cavity,** or *coelom* (SĒ-lōm; *koila,* cavity). The **diaphragm** (DĪ-ah-fram), a flat muscular sheet, divides the ventral body cavity into a superior **thoracic cavity,** bounded by the chest wall, and an inferior **abdominopelvic cavity,** enclosed by the abdominal wall and by the bones and muscles of the pelvis.

Many of the organs in these cavities change size and shape as they perform their functions. For example, the lungs inflate and deflate as you breathe, and your stomach swells at each meal and shrinks between meals. These organs are surrounded by moist internal spaces that permit expansion and limited movement but prevent friction. The thoracic cavity is subdivided into three separate spaces, whereas the abdominopelvic cavity contains a single, extensive chamber. The internal organs that are partially or completely enclosed by these cavities are called

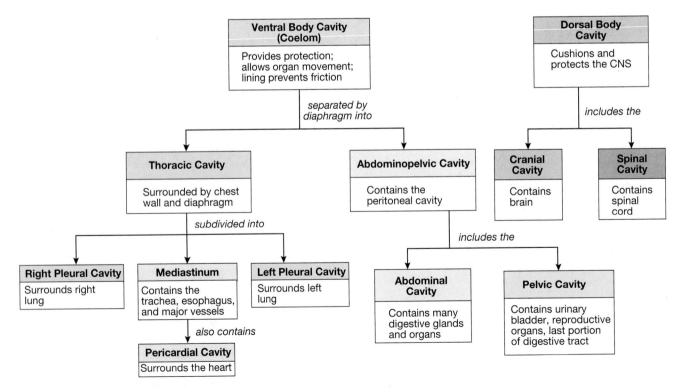

•**FIGURE 1-10** **Relationships of the Various Body Cavities**

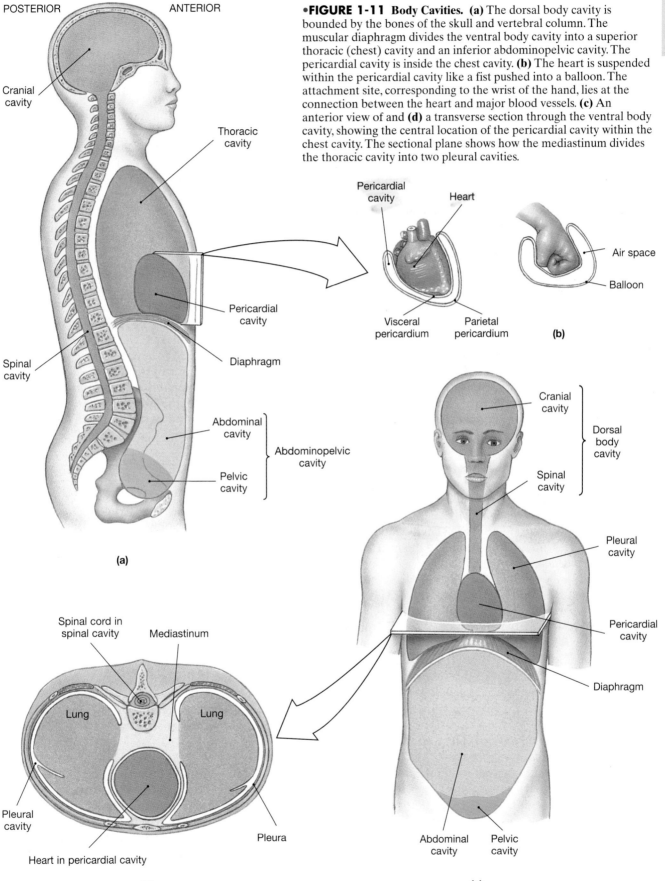

•FIGURE 1-11 Body Cavities. (a) The dorsal body cavity is bounded by the bones of the skull and vertebral column. The muscular diaphragm divides the ventral body cavity into a superior thoracic (chest) cavity and an inferior abdominopelvic cavity. The pericardial cavity is inside the chest cavity. **(b)** The heart is suspended within the pericardial cavity like a fist pushed into a balloon. The attachment site, corresponding to the wrist of the hand, lies at the connection between the heart and major blood vessels. **(c)** An anterior view of and **(d)** a transverse section through the ventral body cavity, showing the central location of the pericardial cavity within the chest cavity. The sectional plane shows how the mediastinum divides the thoracic cavity into two pleural cavities.

POSTERIOR   ANTERIOR

Cranial cavity

Thoracic cavity

Pericardial cavity

Diaphragm

Spinal cavity

Abdominal cavity

Abdominopelvic cavity

Pelvic cavity

(a)

Pericardial cavity   Heart

Air space

Balloon

Visceral pericardium   Parietal pericardium   **(b)**

Cranial cavity

Dorsal body cavity

Spinal cavity

Pleural cavity

Pericardial cavity

Diaphragm

Abdominal cavity   Pelvic cavity

(c)

Spinal cord in spinal cavity   Mediastinum

Lung   Lung

Pleural cavity

Pleura

Heart in pericardial cavity

(d)

1

viscera (VIS-e-ruh). A delicate layer called a *serous membrane* lines the walls of these internal cavities and covers the surfaces of the enclosed viscera. Serous membranes secrete a watery fluid that coats the opposing surfaces and reduces friction.

The Thoracic Cavity The walls of the thoracic cavity surround the lungs and heart; associated organs of the respiratory, cardiovascular, and lymphatic systems; the inferior portions of the esophagus; and the thymus. The thoracic cavity contains three internal chambers: a single *pericardial cavity* and a pair of *pleural cavities* (Figure 1-11a,c•). These cavities are lined by shiny, slippery serous membranes.

The heart is surrounded by the **pericardial cavity.** The relationship between the heart and the cavity resembles that of a fist pushing into a balloon (Figure 1-11b•). The wrist corresponds to the *base* (attached portion) of the heart, and the balloon corresponds to the serous membrane that lines the pericardial cavity. The serous membrane is called the *pericardium (peri-,* around + *kardia,* heart). The layer covering the heart is the *visceral pericardium,* and the opposing surface is the *parietal pericardium.* The space between the visceral pericardium and the parietal pericardium is the pericardial cavity. During each beat, the heart changes in size and shape. The pericardial cavity permits these changes, and the slippery pericardial lining prevents friction between the heart and adjacent structures in the thoracic cavity.

The pericardium lies within the **mediastinum** (mē-dē-as-TĪ-num or mē-dē-AS-ti-num), the portion of the thoracic cavity between the left and right pleural cavities (Figure 1-11d•). The connective tissue of the mediastinum surrounds the pericardial cavity and heart, the large arteries and veins attached to the heart, and the thymus, trachea, and esophagus.

One **pleural cavity** lies on each side of the mediastinum. Each pleural cavity encloses a lung. The relationship between a lung and its pleural cavity is comparable to that between the heart and the pericardial cavity. The serous membrane lining a pleural cavity is called a *pleura* (PLOO-rah). The *visceral pleura* covers the outer surfaces of a lung, and the *parietal pleura* covers the opposing mediastinal surface and the inner body wall.

The Abdominopelvic Cavity The abdominopelvic cavity extends from the diaphragm to the pelvis. It is subdivided into a superior **abdominal cavity** and an inferior **pelvic cavity** (Figure 1-11a,c•). The abdominopelvic cavity contains the *peritoneal* (per-i-tō-NĒ-al) *cavity,* a chamber lined by a serous membrane known as the *peritoneum* (per-i-tō-NĒ-um). The *parietal peritoneum* lines the inner surface of the body wall. A narrow space containing a small amount of fluid separates the parietal peritoneum

from the *visceral peritoneum,* which covers the enclosed organs.

The abdominal cavity extends from the inferior surface of the diaphragm to the level of the superior margins of the pelvis. This cavity contains the liver, stomach, spleen, small intestine, and most of the large intestine. (The positions of most of these organs are shown in Figure 1-7c•, p. 17.) These organs are partially or completely enclosed by the peritoneal cavity, much as the heart or lungs are enclosed by the pericardial or pleural cavities, respectively. A few organs, such as the kidneys and pancreas, lie between the peritoneal lining and the muscular wall of the abdominal cavity. Those organs are said to be *retroperitoneal* (*retro,* behind).

The pelvic cavity is the portion of the ventral body cavity inferior to the abdominal cavity. The bones of the pelvis form the walls of the pelvic cavity, and a layer of muscle forms its floor. The pelvic cavity contains the last portion of the large intestine, the urinary bladder, and various reproductive organs. For example, the pelvic cavity of females contains the ovaries, uterine tubes, and uterus; in males, it contains the prostate gland and seminal vesicles. The pelvic cavity contains the inferior portion of the peritoneal cavity. The superior portion of the urinary bladder in both sexes, as well as the uterine tubes, the ovaries, and the superior portion of the uterus in females, are covered by peritoneum.

☑ Which type of section would separate the two eyes?

☑ If a surgeon makes an incision just inferior to the diaphragm, which body cavity will be opened?

This chapter provided an overview of the locations and functions of the major components of each organ system. It also introduced the anatomical vocabulary needed for you to follow more-detailed anatomical descriptions in later chapters. Many of the figures in later chapters contain images produced by the procedures outlined in Figures 1-12• through 1-14• (see "FOCUS: Sectional Anatomy and Clinical Technology," pp. 23–25), which summarize the most common methods of visualizing anatomical structures in living individuals.

Chapters 2–4 will take you on a tour of the principal levels of organization, from individual atoms to individual humans. As we proceed through the text, we will emphasize major structural and functional patterns. To sharpen your analytical skills, we have included critical-thinking questions at the end of each chapter and on your Companion Website. Chapters 5–29 will focus on the individual organ systems and their interrelationships.

*R*adiological procedures include (1) scanning techniques that involve the use of beams of radiation, such as X-rays, to create a photographic or computer-generated image of internal structures and (2) methods that involve the administration of radioactive materials.

Physicians who specialize in the performance of these procedures and the analysis of the resulting images are called *radiologists.* Radiological procedures can provide detailed information about internal systems. Figures 1-12 through 1-14• compare the views provided by several techniques radiologists and other clinicians use. These figures include *X-rays, CT scans, MRI scans, ultrasound images, spiral-CT scans,* and *DSA images.* These and several other clinical technologies, such as those that produce *PET (positron emission tomography) scans* and *DSR (dynamic spatial reconstruction) images* and more-specialized MRI procedures, are described in the *Applications Manual;* many examples are included as figures later in this text. ⬛AM⬛ *The Steps in Diagnosis*

Whenever you see anatomical diagrams or clinical procedures that present cross-sectional views of the body, remember that the sections are oriented as though the observer is standing at the feet of the subject and looking toward the head.

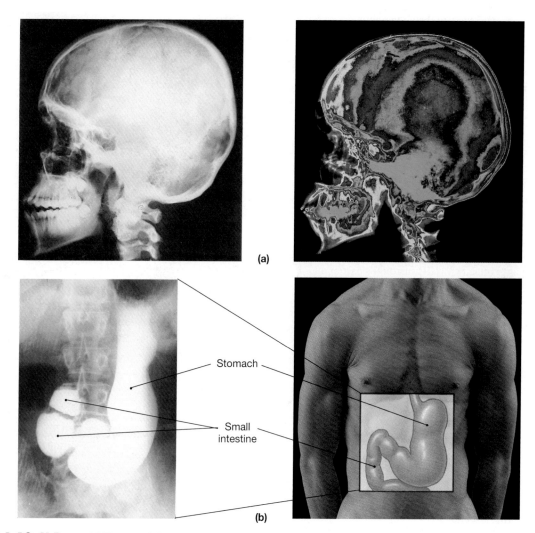

(a)

Stomach

Small
intestine

(b)

•**FIGURE 1-12 X-Rays (a)** X-rays of the skull, taken from the left side. **X-rays** are a form of high-energy radiation that can penetrate tissues. In the most familiar procedure, a beam of X-rays travels through the body and strikes a photographic plate. Not all of the projected X-rays reach the film; some are absorbed or deflected as they pass through the body. The resistance to X-ray penetration is called **radiodensity.** Radiodensity increases in the following sequence: air, fat, liver, blood, muscle, bone. The usual result is an image with radiodense tissues, such as bone, appearing in white, and less-dense tissues in shades of gray to black. (The image on the right was color enhanced.) **(b)** A **barium-contrast X-ray** of the upper digestive tract. Such an X-ray is produced by introducing a radiodense material into the body to provide sharp outlines and contrast and to check the distribution of fluids or the movements of internal organs. In this instance, the patient swallowed a solution of barium. Barium is very dense; the contours of the stomach and intestinal linings are clearly outlined against the white of the barium solution.

●**FIGURE 1-13** **Common Scanning Techniques**

(a) The relative position and orientation of the scans shown in parts (b)–(d).

(b) A color-enhanced CT scan of the abdomen. **C**omputerized **t**omography **(CT)**, formerly called **c**omputerized **a**xial **t**omography **(CAT)**, uses computers to reconstruct sectional views. A single X-ray source rotates around the body, and the X-ray beam strikes a sensor monitored by the computer. The source completes one revolution around the body every few seconds; it then moves a short distance and repeats the process. The result is usually displayed as a sectional view in black and white, but it can be colorized for visual effect. CT scans show three-dimensional relationships and soft-tissue structure more clearly than do standard X-rays.

(c) A color-enhanced MRI scan of the abdomen. **M**agnetic **r**esonance **i**maging **(MRI)** surrounds part or all of the body with a magnetic field about 3000 times as strong as that of Earth. This field affects protons within atomic nuclei throughout the body. The protons line up along the magnetic lines of force like compass needles in Earth's magnetic field. When struck by a radio wave of a certain frequency, a proton will absorb energy. When the wave pulse ends, that energy is released and the source of the radiation is detected. Each element differs in terms of the radio frequency required to affect its protons.

(d) An ultrasound scan of the abdomen. In **ultrasound** procedures, a small transmitter contacting the skin broadcasts a brief, narrow burst of high-frequency sound and then picks up the echoes. The sound waves are reflected by internal structures. An **echogram,** or ultrasound picture, can be assembled from the pattern of echoes. These images lack the clarity of other procedures, but no adverse effects have been reported, and fetal development can be monitored without a significant risk of birth defects. Special methods of transmission and processing permit analysis of the beating heart, without the complications that can accompany dye injections. Note the differences in detail among this image, the CT scan, and the MRI image.

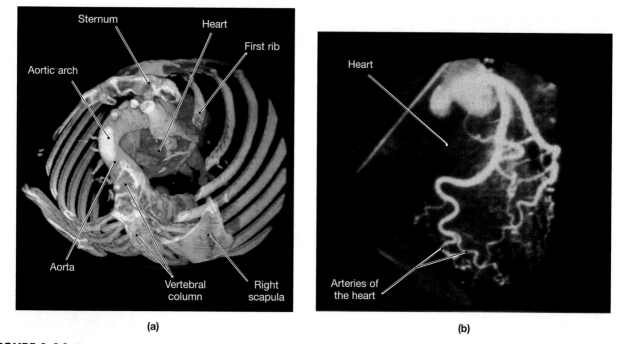

(a)   (b)

•FIGURE 1-14 Special Scanning Methods. (a) A **spiral-CT** scan of the chest. Such an image is created by special processing of CT data to permit rapid three-dimensional visualization of internal organs. Spiral-CT scans are becoming increasingly important in clinical settings. **(b) D**igital **s**ubtraction **a**ngiography **(DSA)** is used to monitor blood flow through specific organs, such as the brain, heart, lungs, or kidneys. X-rays are taken before and after radiopaque dye is administered, and a computer "subtracts" details common to both images. The result is a high-contrast image showing the distribution of the dye.

[EXPLORE]
MediaLab

Anatomy and physiology are two closely related fields of study, each with its own rules and terminology. This chapter introduces you to those fields and provides you with an overview of their organization. The text has both a CD-ROM and a Companion Website to aid you in your studies. The CD-ROM includes animations and tutorials; the website builds on those resources with text content and interactive problems as well as communication avenues for sharing ideas with your class. Additionally, each text chapter ends with a MediaLab. These MediaLabs contain Web Explorations, which are designed to give you practice in using the concepts discussed in the chapter. To see how these study tools work, visit the MediaLab in Chapter 1 of the Companion Website and complete the exercises available there.

Web Exploration 1

Estimated time for completion = 5 minutes

As is true of any scientific field, anatomy and physiology use terminology that may not be familiar to the average person. The best way to approach this unfamiliarity is to examine the formation of the new words and to try to make sense of their roots, prefixes, or suffixes. You can find an excellent

source for understanding medical terminology by selecting the keyword **TERMINOLOGY** on the Companion Website. Read the material on that site, and visit the links for roots, prefixes, and suffixes. Create your own vocabulary list that you can add to as you progress through the text. If you wish, create a shared class list of terms, using the communications tools on the Companion Website.

Web Exploration 2

Estimated time for completion = 10 minutes

As you move through your study of the human body, you will be presented with various sectional views of the body and of specific organs. It is important to be able to relate what you see in a sectional view to the spatial organization of the body as a whole. Select the keyword **PLANES** on the Companion Website to explore the human body "cut" in each of the three sectional planes. Try to locate the brain in each section. Which sectional view gives you the most useful information? Complete the exercise at the bottom of the website page. Can you visualize three-dimensional organs? If you wish, share your hints for this visualization with your peers, using the communications tools on the Companion Website.

1

CHAPTER REVIEW

SELECTED CLINICAL TERMINOLOGY
Terms Discussed in This Chapter

abdominopelvic quadrant: One of four divisions of the anterior abdominal surface. *(p. 15)*

abdominopelvic region: One of nine divisions of the anterior abdominal surface. *(p. 17)*

CT, CAT (computerized [axial] tomography): An imaging technique that uses X-rays to reconstruct the body's three-dimensional structure. *(p. 24)*

disease: A malfunction of organs or organ systems resulting from a failure of homeostatic regulation. *(p. 14)*

DSA (digital subtraction angiography): A technique used to monitor blood flow through specific organs, such as the brain, heart, lungs, or kidneys. X-rays are taken before and after radiopaque dye is administered, and a computer "subtracts" details common to both images. The result is a high-contrast image showing the distribution of the dye. *(p. 25)*

echogram: An image created by ultrasound. *(p. 24)*

embryology: The study of structural changes during the first 2 months of development. *(p. 3)*

histology (his-TOL-o-jē): The study of tissues. *(p. 3)*

MRI (magnetic resonance imaging): An imaging technique that employs a magnetic field and radio waves to portray subtle structural differences. *(p. 24)*

radiologist: A physician who specializes in performing and analyzing radiological procedures. *(p. 23)*

spiral-CT: A method of processing computerized tomography data to provide rapid, three-dimensional images of internal organs. *(p. 25)*

ultrasound: An imaging technique that uses brief bursts of high-frequency sound reflected by internal structures. *(p. 24)*

X-rays: High-energy radiation that can penetrate living tissues. *(p. 23)*

AM Additional Terms Discussed in the *Applications Manual*

auscultation (aws-kul-TĀ-shun): Listening to a patient's body sounds by means of a stethoscope.

DSR (dynamic spatial reconstruction): A research tool that provides three-dimensional images in real time. The computer-generated image can be rotated or sectioned.

inspection: A careful observation of a patient's appearance and actions.

palpation: Using hands and fingers to feel the patient's body as part of a physical exam.

percussion: Tapping with the fingers or hand to obtain information about the densities of a patient's underlying tissues.

PET (positron emission tomography) scan: A scan taken after radioisotopes are injected or inhaled; the released radiation is monitored, and computers use the data to visualize physiological activity.

STUDY OUTLINE

INTRODUCTION p. 2
1. **Biology** is the study of life. One of its goals is to discover the unity and patterns that underlie the diversity of organisms.
2. All living things have certain common characteristics, including **organization, responsiveness, growth** and **differentiation, reproduction, movement,** and **metabolism** and **excretion.**

THE SCIENCES OF ANATOMY AND PHYSIOLOGY p. 2
1. **Anatomy** is the study of internal and external structures and the physical relationships among body parts. **Physiology** is the study of how living organisms perform vital functions. All specific functions are performed by specific structures.

Anatomy p. 3
2. The boundaries of *microscopic anatomy* are established by the equipment used. In **cytology,** we analyze the internal structure of individual cells. In **histology,** we examine **tissues,** groups of cells that have specific functional roles. Tissues combine to form **organs,** anatomical units with multiple functions. Organs combine to form **organ systems,** groups of organs that function together.
3. In *gross (macroscopic) anatomy,* we consider features that are visible without a microscope. It includes *surface anatomy* (general form and superficial markings); *regional anatomy* (superficial and internal features in a specific area of the body); and *systemic anatomy* (structure of major organ systems).
4. In *developmental anatomy,* we examine the changes in form that occur between conception and physical maturity. In *embryology,* we study processes that occur during the first 2 months of development.

Physiology p. 3
5. **Human physiology** is the study of the functions of the human body. It is based on *cell physiology,* the study of the functions of cells. In *special physiology,* we study the physiology of specific organs. In *systemic physiology,* we consider all aspects of the function of specific organ systems. In *pathological physiology,* we study the effects of diseases on organ or system functions.

LEVELS OF ORGANIZATION p. 4
1. Anatomical structures and physiological mechanisms are arranged in a series of interacting levels of organization. *(Figure 1-1)*
2. The 11 organ systems of the body are the integumentary, skeletal, muscular, nervous, endocrine, cardiovascular, lymphatic, respiratory, digestive, urinary, and reproductive systems. *(Figure 1-2)*

3. Homeostasis is the presence of a stable environment within the body.

HOMEOSTASIS AND SYSTEM INTEGRATION   p. 12

1. Physiological systems preserve homeostasis through **homeostatic regulation.**
2. **Autoregulation** occurs when the activities of a cell, tissue, organ, or organ system change automatically in response to an environmental change. **Extrinsic regulation** results from the activities of the nervous or endocrine system.
3. Homeostatic regulation usually involves a **receptor** sensitive to a particular stimulus, a **control center** that receives and processes the information from the receptor, and an **effector** whose activities are regulated by the control center and whose actions have a direct or indirect effect on the same stimulus.

Negative Feedback   p. 12

4. **Negative feedback** is a corrective mechanism involving an action that directly opposes a variation from normal limits. *(Figures 1-3, 1-4)*

Positive Feedback   p. 14

5. In **positive feedback,** a stimulus produces a response that exaggerates the stimulus, creating a *positive feedback loop.* *(Figure 1-5)*

A FRAME OF REFERENCE FOR ANATOMICAL STUDIES   p. 15

Superficial Anatomy   p. 15

1. Standard anatomical illustrations show the body in the **anatomical position.** If the figure is shown lying down, it can be either **supine** (face up) or **prone** (face down). *(Figure 1-6)*

2. **Abdominopelvic quadrants** and **abdominopelvic regions** represent two approaches to describing anatomical regions of the body. *(Figure 1-7; Table 1-1)*
3. The use of special directional terms provides clarity for the description of anatomical structures. *(Figure 1-8; Table 1-2)*

Sectional Anatomy   p. 19

4. The three **sectional planes** (**frontal,** or *coronal,* **plane; sagittal plane;** and **transverse plane**) describe relationships among the parts of the three-dimensional human body. *(Figure 1-9; Table 1-3)*
5. *Body cavities* protect delicate organs and permit changes in the size and shape of internal organs. The **dorsal body cavity** contains the **cranial cavity** (enclosing the brain) and the **spinal cavity** (surrounding the spinal cord). The **ventral body cavity,** or *coelom,* surrounds developing respiratory, cardiovascular, digestive, urinary, and reproductive organs. *(Figures 1-10, 1-11)*
6. During development, the **diaphragm** divides the ventral body cavity into the (superior) **thoracic** and (inferior) **abdominopelvic cavities.** By birth, the thoracic cavity contains two **pleural cavities** (each containing a lung) and a **pericardial cavity** (which surrounds the heart). The abdominopelvic cavity consists of the **abdominal cavity** and the **pelvic cavity.** It contains the *peritoneal cavity,* an internal chamber lined by the *peritoneum,* a *serous membrane. (Figure 1-11)*

FOCUS: Sectional Anatomy and Clinical Technology   p. 23

7. Important *radiological procedures* (which can provide detailed information about internal systems) include **CT, MRI, ultrasound, X-rays, spiral-CT,** and **DSA.** *(Figures 1-12, 1-13)*

REVIEW QUESTIONS

LEVEL 1 REVIEWING FACTS AND TERMS

Match each numbered item with the most closely related lettered item. Use letters for answers in the spaces provided.

___ **1.** cytology
___ **2.** physiology
___ **3.** histology
___ **4.** metabolism
___ **5.** homeostasis
___ **6.** muscle
___ **7.** heart
___ **8.** endocrine
___ **9.** temperature regulation
___ **10.** labor and delivery
___ **11.** supine
___ **12.** prone
___ **13.** ventral body cavity
___ **14.** dorsal body cavity
___ **15.** pericardium

  a. study of tissues
  b. constant internal environment
  c. face up
  d. study of functions
  e. positive feedback
  f. organ system
  g. study of cells
  h. negative feedback
  i. brain and spinal cord
  j. all chemical activity in body
  k. thoracic and abdominopelvic
  l. tissue
  m. serous membrane
  n. organ
  o. face down

16. The process by which an organism increases the size and/or number of cells is
  (a) reproduction
  (b) adaptation
  (c) growth
  (d) metabolism

17. The following is a list of six levels of organization that make up the human body:
  1. tissue
  2. cell
  3. organ
  4. molecule
  5. organism
  6. organ system

The correct order from the smallest to the largest level is
  (a) 2, 4, 1, 3, 6, 5
  (b) 4, 2, 1, 3, 6, 5
  (c) 4, 2, 1, 6, 3, 5
  (d) 4, 2, 3, 1, 6, 5
  (e) 2, 1, 4, 3, 5, 6

18. The study of internal and external structures and the physical relationships among body parts is
  (a) histology
  (b) anatomy
  (c) physiology
  (d) embryology

19. The specialist who attempts to determine the physical and chemical processes responsible for vital functions is a
  (a) physiologist
  (b) physician
  (c) pathologist
  (d) cytologist

20. The relative stability of an organism's internal environment is
  (a) homeopathy
  (b) uniformity
  (c) equilibrium
  (d) homeostasis

21. The automatic change that occurs when the activities of a cell, tissue, organ, or system are faced with some environmental variation is
  (a) autoregulation
  (b) extrinsic regulation
  (c) deregulation
  (d) disease

22. Crisis management by directing rapid, short-term, specific responses is a function of the
   (a) endocrine system
   (b) nervous system
   (c) hormones
   (d) autoregulatory system

23. Growth and sexual maturation are regulated primarily by the
   (a) nervous system   (b) circulatory system
   (c) endocrine system   (d) digestive system

24. When a variation outside normal limits triggers a response that restores the normal condition, the regulatory mechanism involves
   (a) negative feedback   (b) positive feedback
   (c) compensation   (d) adaptation

25. In positive feedback, the initial stimulus produces a response that
   (a) suppresses the stimulus
   (b) has no effect on the stimulus
   (c) interferes with completion of the process
   (d) exaggerates the stimulus

26. Failure of homeostatic regulation in the body results in
   (a) autoregulation
   (b) extrinsic regulation
   (c) disease
   (d) positive feedback

27. The terms that apply to the front of the body when in the anatomical position are
   (a) posterior, dorsal
   (b) back, front
   (c) medial, lateral
   (d) anterior, ventral

28. A plane through the body that passes perpendicular to the long axis of the body and divides the body into a superior and an inferior section is a
   (a) sagittal section   (b) transverse section
   (c) coronal section   (d) frontal section

29. The cranial and spinal cavities are in the
   (a) ventral body cavity
   (b) thoracic cavity
   (c) dorsal body cavity
   (d) abdominopelvic cavity

30. The diaphragm divides the ventral body cavity into a superior _____ cavity and an inferior _____ cavity.
   (a) pleural, pericardial
   (b) abdominal, pelvic
   (c) thoracic, abdominopelvic
   (d) cranial, thoracic

31. The mediastinum is the region between the
   (a) lungs and heart
   (b) two pleural cavities
   (c) chest and abdomen
   (d) heart and pericardium

LEVEL 2 REVIEWING CONCEPTS

32. What basic functions do all living things perform?

33. (a) Define *anatomy*.
   (b) Define *physiology*.

34. Why is it important to understand the different levels of organization in the human body?

35. Beginning with the chemical level, list in sequence the levels of organization from the simplest level to the most complex level.

36. What is homeostatic regulation, and what is its physiological importance?

37. What distinguishes autoregulation from extrinsic regulation?

38. What necessary components are involved in homeostatic regulation? What is the function of each?

39. How does negative feedback differ from positive feedback?

40. Describe the position of the body when it is in the anatomical position.

41. What is the role of serous membranes in the body?

42. As a surgeon, you perform an invasive (surgical) procedure that requires you to cut through the peritoneum. Are you operating on the heart or on the stomach?

43. In which body cavity would each of the following organs or organ systems be?
   (a) brain, spinal cord
   (b) cardiovascular, digestive, and urinary systems
   (c) heart, lungs
   (d) stomach, intestines

44. Name the serous membrane that lines each of the body cavities containing the following organs.
   (a) heart   (b) lungs
   (c) intestines

LEVEL 3 CRITICAL THINKING AND CLINICAL APPLICATIONS

45. The hormone *calcitonin* is released from the thyroid gland in response to increased levels of calcium ions in the blood. If this hormone is controlled by negative feedback, what effect would calcitonin have on blood calcium levels?

46. During exercise, blood flow to skeletal muscles increases. The initial response that increases blood flow is automatic and independent of the nervous and endocrine systems. Which type of homeostatic regulation is this? Why?

47. A chemical imbalance in a heart muscle cell can cause the heart to cease pumping blood; the cessation of blood flow will in turn cause other tissues and organs to cease functioning. This observation supports the view that
   (a) all organisms are composed of cells
   (b) all levels of organization within an organism are interdependent
   (c) chemical molecules make up cells
   (d) all cells are independent of each other
   (e) congenital defects can be life threatening

CHAPTER 2

The Chemical Level of Organization

Water is so common, and seems so ordinary, that it is often taken for granted. Yet all living things, including humans, are totally dependent on water. You can survive for weeks without food but for only a few days without water. The basic reason may surprise you: In large part, you are water. Water makes up almost two-thirds of your body weight.

Chemically, water is a simple substance: two atoms of hydrogen joined to one atom of oxygen. Yet, when these three atoms are linked by chemical bonds, they produce a substance with many unusual properties. In this chapter, you will learn how atoms are bound together to form molecules, the building blocks of cells. You will also meet the larger molecules that form the structural framework of the body and that enable our cells to grow, divide, and BUFFERS and pH CONTROL perform other essential functions.

CHAPTER OUTLINE AND OBJECTIVES

INTRODUCTION p. 30

ATOMS AND MOLECULES p. 30
  Atomic Structure p. 30
  Chemical Bonds p. 32
① Describe an atom and how atomic structure affects interactions between atoms.
② Compare the ways in which atoms combine to form molecules and compounds.

CHEMICAL REACTIONS p. 36

■ **FOCUS: Chemical Notation** p. 37

  Basic Energy Concepts p. 36
  Types of Reactions p. 38
  Reversible Reactions p. 38
  Enzymes and Chemical Reactions p. 39
③ Use chemical notation to symbolize chemical reactions.
④ Distinguish among the major types of chemical reactions that are important for studying physiology.
⑤ Describe the crucial role of enzymes in metabolism.

INORGANIC COMPOUNDS p. 39
  Water and Its Properties p. 40
  Inorganic Acids and Bases p. 42
  Salts p. 43
  Buffers and pH Control p. 43
⑥ Distinguish between organic and inorganic compounds.
⑦ Explain how the chemical properties of water make life possible.
⑧ Discuss the importance of pH and the role of buffers in body fluids.
⑨ Describe the physiological roles of inorganic compounds.

ORGANIC COMPOUNDS p. 44
  Carbohydrates p. 44

  Lipids p. 46
  Proteins p. 51
  Nucleic Acids p. 56
  High-Energy Compounds p. 57
⑩ Discuss the structures and functions of carbohydrates, lipids, proteins, nucleic acids, and high-energy compounds.

CHEMICALS AND CELLS p. 59

MEDIALAB and CHAPTER REVIEW p. 60

Our study of the human body begins at the chemical level of organization. *Chemistry* is the science that deals with the structure of *matter,* defined as anything that takes up space and has mass. *Mass* is a physical property that determines the weight of an object in Earth's gravitational field. For our purposes, the mass of an object is the same as its weight. If we were to ride on the space shuttle, however, we would find that the two are not always equivalent. In orbit we would be weightless, but our mass would remain unchanged.

Atoms are the smallest stable units of matter. Air, elephants, oranges, oceans, rocks, and people are all composed of atoms in varying combinations. The unique characteristics of each object, living or nonliving, result from the types of atoms involved and the ways those atoms combine and interact. You will need to become familiar with the chemical principles governing those interactions before you can understand the anatomy and physiology of the cells, tissues, organs, and organ systems of the human body.

ATOMS AND MOLECULES

Atoms are composed of **subatomic particles.** Although dozens of different subatomic particles exist, only three are important for understanding chemical properties: protons, neutrons, and electrons. Protons and neutrons are similar in size and mass, but **protons** (p^+) bear a positive electrical charge, whereas **neutrons** (n, or n^0) are electrically *neutral,* or uncharged. **Electrons** (e^-) are much lighter than protons—only 1/1836th as massive—and bear a negative electrical charge. The mass of an atom is therefore determined primarily by the number of protons and neutrons in the nucleus. The mass of a large object, such as your body, is the sum of the masses of all the component atoms.

ATOMIC STRUCTURE

Atoms normally contain equal numbers of protons and electrons. The number of protons in an atom is known as its **atomic number.** *Hydrogen* (H) is the simplest atom, with an atomic number of 1. Thus an atom of hydrogen contains one proton, and it contains one electron as well. The proton is located in the center of the atom and forms the nucleus. Hydrogen atoms seldom contain neutrons, but when neutrons are present, they are also located in the nucleus. The electron orbits the nucleus at high speed, forming a spherical **electron cloud.** For convenience, we often illustrate atomic structure in the simplified form shown in Figure 2-1a●. In this two-dimensional representation, the electrons occupy a circular **electron shell.** One reason an electron tends to remain in its electron shell is that the negatively charged electron is attracted to the positively charged proton. The attraction between opposite electrical charges is an example of an *electrical force.* As you shall see in later chapters, electrical forces are involved in many physiological processes.

The dimensions of the electron cloud determine the overall size of the atom. To get an idea of the scale involved, consider that if the nucleus were the size of a tennis ball, the electron cloud would have a radius of 10 km (roughly 6 miles). In reality, atoms are so small that atomic measurements are most conveniently reported in *nanometers* (NA-nō-mē-terz) (nm). One nanometer is 10^{-9} meters (0.000000001 m). The very largest atoms approach 0.5 nm in diameter (0.0000000005 m, or 0.00000002 in.).

All atoms other than the simple hydrogen atom described above have at least one neutron in the nucleus. The number of neutrons present does not affect the properties of an atom, other than its mass.

Elements and Isotopes

Atoms can be assigned to groups called **elements.** Each element includes all the atoms that have the same number of protons and thus the same atomic number. Only 92 elements exist in nature, although 25 additional elements have been created through nuclear reactions in research laboratories. Each element has a chemical symbol, an abbreviation recognized by scientists everywhere. Most of the symbols are easily

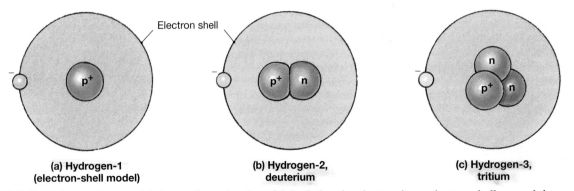

(a) Hydrogen-1
(electron-shell model)

(b) Hydrogen-2,
deuterium

(c) Hydrogen-3,
tritium

Electron shell

●**FIGURE 2-1 Hydrogen Atoms. (a)** A two-dimensional model depicting the electron in an electron shell around the nucleus lets us visualize the components of the atom. A typical hydrogen nucleus contains a proton and no neutrons. **(b)** A deuterium (^{2}H) nucleus contains a proton and a neutron. **(c)** A tritium (^{3}H) nucleus contains a pair of neutrons in addition to the proton.

connected with the English names of the elements, but a few are abbreviations of their Latin names. For example, the symbol for sodium, Na, comes from the Latin word *natrium*. (Appendix III, the *periodic table,* gives the chemical symbols and atomic numbers of each element.) The relative contributions of the 13 most abundant elements in the human body to the total body weight are shown in Table 2-1. The human

body also contains atoms of another 13 elements—called *trace elements*—that are present in such small amounts that percentage values are not shown.

The atoms of a single element can differ in terms of the number of neutrons in the nucleus. For example, although most hydrogen nuclei consist of a single proton, 0.015 percent also contain one neutron, and a very small percentage contain two neutrons. The **isotopes** of an element are atoms whose nuclei contain the same number of protons but different numbers of neutrons. Isotopes are usually indistinguishable except on the basis of mass. The **mass number**—the total number of protons plus neutrons in the nucleus—is used to designate a particular isotope. Thus, the three isotopes of hydrogen are hydrogen-1, or ^{1}H, with one proton and one electron (Figure 2-1a•); hydrogen-2, or ^{2}H, also known as *deuterium,* with one proton, one electron, and one neutron (Figure 2-1b•); and hydrogen-3, or ^{3}H, also known as *tritium,* with one proton, one electron, and two neutrons (Figure 2-1c•).

Radioisotopes are isotopes with nuclei that spontaneously emit subatomic particles or radiation in measurable amounts. After emission, the nuclei are stable; the emission process is called *decay.* Strongly radioactive isotopes are dangerous, because the emissions can break molecules apart, destroy cells, and otherwise damage living tissues. Weakly radioactive isotopes are sometimes used in diagnostic procedures to monitor the structural or functional characteristics of internal organs. The **half-life** of any radioactive isotope is the time required for half of a given amount of the isotope to decay. The half-lives of radioisotopes range from fractions of a second to billions of years.

AM *The Medical Importance of Radioisotopes*

Atomic Weights

A typical atom of *oxygen,* which has an atomic number of 8, contains eight protons, but it also contains eight neutrons. The mass number of this isotope is therefore 16. The mass numbers of other isotopes of oxygen depend on the number of neutrons present. Mass numbers are useful because they tell us the number of subatomic particles in the nuclei of different atoms. However, they do not tell us the *actual* mass of the atoms. For example, they do not take into account the masses of the electrons or the slight difference between the mass of a proton and that of a neutron. The actual mass of an atom is known as its **atomic weight.**

The unit used to express the atomic weight is the **dalton** (also known as the *atomic mass unit,* or *amu*). One dalton is very close to the weight of a single proton. Thus, the atomic weight of the most common isotope of hydrogen is very close to 1, and that of the most common isotope of oxygen is very close to 16.

The atomic weight of an element is an average mass number that reflects the proportions of different isotopes. For example, the atomic number of

TABLE 2-1 Principal Elements in the Human Body

Element (% of total body weight)	Significance
Oxygen (65)	A component of water and other compounds; gaseous form is essential for respiration
Carbon (18.6)	Found in all organic molecules
Hydrogen (9.7)	A component of water and most other compounds in the body
Nitrogen (3.2)	Found in proteins, nucleic acids, and other organic compounds
Calcium (1.8)	Found in bones and teeth; important for membrane function, nerve impulses, muscle contraction, and blood clotting
Phosphorus (1.0)	Found in bones and teeth, nucleic acids, and high-energy compounds
Potassium (0.4)	Important for proper membrane function, nerve impulses, and muscle contraction
Sodium (0.2)	Important for membrane function, nerve impulses, and muscle contraction
Chlorine (0.2)	Important for membrane function and water absorption
Magnesium (0.06)	A cofactor for several enzymes
Sulfur (0.04)	Found in many proteins
Iron (0.007)	Essential for oxygen transport and energy capture
Iodine (0.0002)	A component of hormones of the thyroid gland
Trace elements: silicon (Si), fluorine (F), copper (Cu), manganese (Mn), zinc (Zn), selenium (Se), cobalt (Co), molybdenum (Mo), cadmium (Cd), chromium (Cr), tin (Sn), aluminum (Al), and boron (B)	Some function as cofactors; the functions of many trace elements are poorly understood

2

hydrogen is 1, but the atomic weight of hydrogen is 1.0079, primarily because some hydrogen atoms (0.015 percent) have a mass number of 2, and even fewer have a mass number of 3. The atomic weight of an element is usually very close to the mass number of the most common isotope of that element. The atomic weights of the elements are included in Appendix III.

For any element, a **mole** (abbreviated *mol*) is a quantity with a weight in grams equal to that element's atomic weight. One mole of a given element contains the same number of atoms as a mole of any other element. That number (called *Avogadro's number*) is 6.023×10^{23}, or about 600 billion trillion. Expressing relationships in moles rather than in grams makes it much easier to keep track of the relative numbers of atoms in chemical samples and processes. For example, if a report stated that a sample contains 0.5 mol of hydrogen atoms and 0.5 mol of oxygen atoms, you would know immediately that they were present in equal numbers. That would not be so evident if the report stated that there were 0.505 g of hydrogen atoms and 8 g of oxygen atoms. Most chemical analyses and clinical laboratory tests report data in moles (mol), millimoles (mmol), or micromoles (μmol).

Electrons and Energy Levels

Atoms are electrically neutral; every positively charged proton is balanced by a negatively charged electron. Thus, each increase in the atomic number is accompanied by a comparable increase in the number of electrons orbiting the nucleus. Electrons occupy an orderly series of **energy levels** within the electron cloud. Although the electrons in an energy level may travel in complex orbits around the nucleus, for our purposes they can be diagrammed as a series of concentric electron shells. The first electron shell (the one closest to the nucleus) occupies the lowest energy level.

Each energy level is limited in the number of electrons it can hold. The first energy level can hold at most two electrons, and the second level can hold up to eight electrons. The electrons in an atom are stored in an orderly manner: The first energy level is filled before electrons enter the second, and the second energy level is filled before any electrons enter the third.

The outermost energy level forms the "surface" of the atom. The number of electrons in this level determines the chemical properties of the element. Atoms with unfilled energy levels are unstable—that is, they will react with other atoms, usually in ways that give them full outer electron shells. In contrast, atoms with a filled outermost energy level are stable and therefore do not readily react with other atoms.

As indicated in Figure 2-2•, a hydrogen atom has one electron in the first energy level, and the level is thus unfilled. A hydrogen atom readily reacts with other hydrogen atoms or with the atoms of other elements. A *helium* atom has two electrons in its first en-

•FIGURE 2-2 **Atoms and Energy Levels.** **(a)** A typical hydrogen atom has one proton and one electron. The electron orbiting the nucleus occupies the first energy level, diagrammed as an electron shell. **(b)** An atom of helium has two protons, two neutrons, and two electrons. The two electrons orbit in the same energy level. **(c)** The first energy level can hold only two electrons. In a lithium atom, with three protons, four neutrons, and three electrons, the third electron occupies a second energy level. **(d)** The second level can hold up to eight electrons. A neon atom has 10 protons, 10 neutrons, and 10 electrons; thus both the first and second energy levels are filled.

ergy level. Because its outer energy level is filled, a helium atom is very stable; it will not ordinarily react with other atoms. A *lithium* atom has three electrons. Its first energy level can hold only two of them, so lithium has a single electron in a second, unfilled energy level. As you would expect, lithium is extremely reactive. The second energy level is filled in a *neon* atom, with an atomic number of 10. Neon atoms, like helium atoms, are thus very stable. The atoms most important to biological systems are unstable, because those are the atoms that interact to form larger structures. The elements most important to the human body are listed in Table 2-1. (Further information on these and other elements is given in Appendix III.)

CHEMICAL BONDS

Elements that do not readily participate in chemical processes are said to be *inert*. Helium, neon, and argon have filled outermost energy levels. They are called *inert gases*, because their atoms neither react with one another nor combine with atoms of other elements.

Elements with unfilled outermost energy levels, such as hydrogen and lithium, are called *reactive,* because they readily interact or combine with other atoms. In doing so, these atoms achieve stability by gaining, losing, or sharing electrons to fill their outermost energy level. The interactions often involve the formation of **chemical bonds,** which hold the participating atoms together once the reaction has ended.

We will consider three basic types of chemical bonds: *ionic bonds, covalent bonds,* and *hydrogen bonds.* The term **molecule** refers to any chemical structure consisting of atoms held together by covalent bonds. A **compound** is a chemical structure that contains two or more elements. (Whether the compound is a molecule depends on the type of bonding involved.) A compound has properties that can be quite different from those of its component elements. For example, a mixture of hydrogen and oxygen gases is highly flammable, but the chemical combination of hydrogen and oxygen atoms produces water, a compound used to put out fires.

The human body consists of countless molecules and compounds, so it is a challenge to describe these substances, and their varied interactions. Fortunately, chemists rely on a standardized system of *chemical notation.* The rules of this system are given in "FOCUS: Chemical Notation" on p. 37, and you will find them useful in this chapter and in later chapters.

Ionic Bonds

An atom or molecule that has a charge is called an **ion.** Ions with a positive charge are **cations** (KAT-ī-onz). Ions with a negative charge are **anions** (AN-ī-onz). **Ionic bonds** are chemical bonds created by the electrical attraction between anions and cations.

Ions have an unequal number of protons and electrons. Atoms become ions by losing or gaining electrons. If we assign a value of +1 to the charge on a proton, the charge on an electron is −1. If the number of protons is equal to the number of electrons, an atom is electrically neutral. An atom that loses an electron becomes a cation with a charge of +1, because it then has one proton that lacks a corresponding electron. Losing a second electron would give the cation a charge of +2. Adding an extra electron to a neutral atom produces an anion with a charge of −1; adding a second electron gives the anion a charge of −2.

In the formation of an ionic bond:

- One atom—the *electron donor*—loses one or more electrons and becomes a cation, with a positive (+) charge.
- Another atom—the *electron acceptor*—gains those same electrons and becomes an anion, with a negative (−) charge.
- Attraction between the opposite charges then draws the two ions together.

The formation of an ionic bond is illustrated in Figure 2-3a•. The sodium atom diagrammed in step 1 has an atomic number of 11, so this atom normally

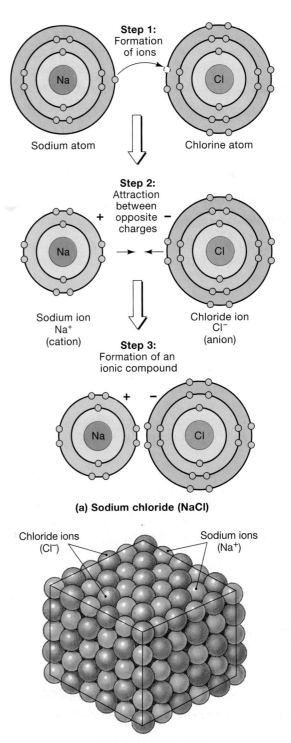

(a) Sodium chloride (NaCl)

(b) Sodium chloride crystal

•**FIGURE 2-3 Ionic Bonding. (a)** Step 1: A sodium (Na) atom loses an electron, which is accepted by a chlorine (Cl) atom. Step 2: Because the sodium (Na^+) and chloride (Cl^-) ions have opposite charges, they are attracted to one another. Step 3: The association of sodium and chloride ions forms the ionic compound sodium chloride. **(b)** Large numbers of sodium and chloride ions form a crystal of sodium chloride (table salt).

2

contains 11 protons and 11 electrons. (Because neutrons are electrically neutral, their presence has no effect on the formation of ions or ionic bonds.) Electrons fill the first and second energy levels, and a single electron occupies the outermost level. Losing that one electron would give the sodium atom a full outermost energy level—the second level—and would produce a **sodium ion,** with a +1 charge. (The chemical shorthand for sodium abbreviates this ion as Na$^+$.) But a sodium atom cannot simply throw away the electron; the electron must be donated to an electron acceptor. A chlorine atom has seven electrons in its outermost energy level, so it needs only one electron to achieve stability. A sodium atom can provide the extra electron. In the process (step 2), the chlorine atom becomes a **chloride ion** (Cl$^-$), with a −1 charge.

Both atoms have now become stable ions with filled outermost energy levels. But the two ions do not move apart after the electron transfer, because the positively charged sodium ion is attracted to the negatively charged chloride ion (step 3). The combination of oppositely charged ions forms the *ionic compound* **sodium chloride,** otherwise known as table salt. Large numbers of sodium and chloride ions interact to form highly structured crystals, held together only by the electrical attraction of oppositely charged ions (Figure 2-3b•). Although sodium chloride and other ionic compounds are common in body fluids, they are not present as intact crystals. When placed in water, ionic compounds dissolve, and the component anions and cations separate.

Covalent Bonds

Atoms complete their outer electron shells not only by gaining or losing electrons but also by sharing electrons with other atoms. The result is a molecule—two or more atoms held together by **covalent** (KŌ-vā-lent) **bonds.**

Individual hydrogen atoms, as diagrammed in Figure 2-2a•, do not exist in nature. Instead, we find hydrogen molecules (Figure 2-4a•). Molecular hydrogen consists of a pair of hydrogen atoms. In chemical shorthand, molecular hydrogen is indicated by H$_2$, where H is the chemical symbol for hydrogen and the subscript 2 indicates the number of atoms. Molecular hydrogen is a gas present in the atmosphere in very small quantities. The two hydrogen atoms share their electrons, and each electron whirls around both nuclei. The sharing of one pair of electrons creates a **single covalent bond** (—).

Oxygen, with an atomic number of 8, has two electrons in its first energy level and six in its second. The oxygen atoms diagrammed in Figure 2-4b• attain a stable electron configuration by sharing two pairs of electrons and thereby forming a **double covalent bond.** In a structural formula, a double covalent bond is represented by two lines (=). Molecular oxygen (O$_2$) is an atmospheric gas that is very important to most organisms. Our cells would die without a relatively constant supply of oxygen.

MOLECULE	ELECTRON-SHELL MODEL AND STRUCTURAL FORMULA
(a) Hydrogen (H$_2$)	H–H
(b) Oxygen (O$_2$)	O=O
(c) Carbon dioxide (CO$_2$)	O=C=O
(d) Nitric oxide (NO)	N=O

•**FIGURE 2-4 Covalent Bonds. (a)** In a hydrogen molecule, two hydrogen atoms share electrons such that each atom has a filled outermost electron shell. This sharing creates a single covalent bond. **(b)** In an oxygen molecule, two oxygen atoms that share two pairs of electrons. The result is a double covalent bond. **(c)** In a carbon dioxide molecule, a central carbon atom forms double covalent bonds with two oxygen atoms. **(d)** A nitric oxide molecule is held together by a double covalent bond, but the outer electron shell of the nitrogen atom requires an additional electron to be complete. Thus, nitric oxide is a free radical, which readily reacts with another atom or molecule.

In our bodies, chemical processes that consume oxygen generally also produce **carbon dioxide** (CO$_2$) as a waste product. The oxygen atoms in a carbon dioxide molecule form double covalent bonds with the carbon atom (Figure 2-4c•).

A triple covalent bond, such as the one joining two nitrogen molecules (N$_2$), is indicated by three lines (≡). Molecular nitrogen accounts for roughly 79 percent of our planet's atmosphere, but our cells ignore it completely. In fact, deep-sea divers live for long periods while breathing artificial air that does not contain nitrogen. (We shall discuss the reasons for eliminating nitrogen under these conditions in Chapter 23.)

Covalent bonds usually form molecules that complete the outer energy levels of the atoms involved. An ion or molecule that contains unpaired electrons in its outermost energy level is called a *free radical.* Free radicals are highly reactive. Almost as fast as it forms, a free radical enters additional reactions that are typically destructive. For example, free radicals can damage or destroy vital compounds, such as proteins. Free radicals sometimes form in the course of normal metabolism, but cells have several methods of removing or inactivating them. However, *nitric oxide* (NO) (Figure 2-4d•)

is a free radical that has important functions in the body. It is involved in chemical communication in the nervous system, in the control of blood vessel diameter, in blood clotting, and in the defense against bacteria and other pathogens. It has been suggested that the cumulative damage produced by free radicals inside and outside our cells is a major factor in the aging process.

Nonpolar Covalent Bonds Covalent bonds are very strong, because the shared electrons hold the atoms together. In typical covalent bonds the atoms remain electrically neutral, because each shared electron spends just as much time "at home" as away. (If you and a friend were tossing a pair of baseballs back and forth as fast as you could, on average, each of you would have just one baseball.) Many covalent bonds involve an equal sharing of electrons. Such bonds, which occur, for instance, between two atoms of the same type, are called **nonpolar covalent bonds.** Nonpolar covalent bonds, especially those between carbon atoms, create the stable framework of the large molecules that make up most of the structural components of the human body.

Polar Covalent Bonds Covalent bonds involving different types of atoms may instead involve a very unequal sharing of electrons, because the elements differ in how strongly they attract electrons. An unequal sharing of electrons creates a **polar covalent bond.** For example, in a molecule of water (Figure 2-5•), an oxygen atom forms covalent bonds with two hydrogen atoms. The oxygen nucleus has a much stronger attraction for the shared electrons than the hydrogen

atoms do. As a result, the electrons spend more time orbiting the oxygen nucleus than orbiting the hydrogen nuclei. Because it has two extra electrons most of the time, the oxygen atom develops a slight (partial) negative charge, indicated by δ^-. At the same time, each hydrogen atom develops a slight (partial) positive charge, δ^+, because its electron is away much of the time. (Suppose you and a friend were tossing a pair of baseballs back and forth, but one of you returned them back as fast as possible, and the other held onto them for a while before throwing them back. One of you would now, on average, have more than one baseball, and the other would have fewer than one.) The unequal sharing of electrons makes polar covalent bonds somewhat weaker than nonpolar covalent bonds.

Hydrogen Bonds

Covalent and ionic bonds tie atoms together to form molecules and compounds. There are also comparatively weak forces that act between adjacent molecules and even between atoms within a large molecule. **Hydrogen bonds** are the most important of these weak attractive forces. A hydrogen bond is the attraction between a δ^+ on the hydrogen atom of a polar covalent bond and a δ^- on an oxygen or nitrogen atom of another polar covalent bond. The polar covalent bond containing the oxygen or nitrogen atom can be in a different molecule or in the same molecule as the hydrogen atom.

Hydrogen bonds are too weak to create molecules, but they can change molecular shapes or pull molecules together. For example, hydrogen bonding occurs between water molecules (Figure 2-6•). At the

•**FIGURE 2-5 Polar Covalent Bonds and the Structure of Water. (a)** In forming a water molecule, an oxygen atom completes its outermost energy level by sharing electrons with a pair of hydrogen atoms. The sharing is unequal, because the oxygen atom holds the electrons more tightly than do the hydrogen atoms. **(b)** Because the oxygen atom has two extra electrons much of the time, it develops a slight negative charge and the hydrogen atoms become weakly positive. The bonds in a water molecule are polar covalent bonds.

•**FIGURE 2-6 Hydrogen Bonds.** The hydrogen atoms of a water molecule have a slight positive charge, and the oxygen atom has a slight negative charge (see Figure 2-5b). The attraction between a hydrogen atom of one water molecule and the oxygen atom of another is a hydrogen bond.

water surface, the attraction between molecules slows the rate of evaporation and creates the phenomenon known as surface tension. **Surface tension** acts as a barrier that keeps small objects from entering the water; it is the reason that insects can walk across the surface of a pond or puddle. Surface tension in a layer of tears keeps small objects such as dust particles from touching the surface of the eye. At the cellular level, hydrogen bonds affect the shapes and properties of complex molecules, such as proteins and nucleic acids (including DNA), and they may also determine the three-dimensional relationships between molecules.

States of Matter

Matter in our environment exists in one of three states: solid, liquid, or gaseous. *Solids* maintain their volume and shape at ordinary temperatures and pressures. A lump of granite, a brick, and a textbook are solid objects. *Liquids* have a constant volume but no fixed shape. The shape of a liquid is determined by the shape of its container. Water, coffee, and soda are liquids. A *gas* has neither a constant volume nor a fixed shape. Gases can be compressed or expanded, and they will fill a container of any shape. The air in our atmosphere is the gas with which we are most familiar.

Whether a particular substance is a solid, a liquid, or a gas depends on the degree of interaction among its atoms or molecules. For example, hydrogen molecules have little attraction for one another; in our environment, molecular hydrogen exists as a gas. Water is the only substance that occurs as a solid (ice), a liquid (water), and a gas (water vapor) at temperatures compatible with life. Water exists in the liquid state over a broad range of temperatures primarily due to hydrogen bonding among the water molecules. We will talk more about the unusual properties of water in a later section.

Molecular Weights

The **molecular weight** of a molecule is the sum of the atomic weights of its component atoms. In other words, the molecular weight in grams is equal to the weight of one mole of molecules. Molecular weights are important because you cannot handle individual molecules; nor could you easily count the billions of molecules involved in chemical reactions in the body. Using molecular weights, you can calculate the quantities of reactants needed to perform a specific reaction and determine the amount of product generated. For example, suppose you want to create water from hydrogen and oxygen, according to the equation

$$2H_2 + O_2 \longrightarrow 2H_2O$$

The first step in performing such an experiment would be to calculate the molecular weights involved. The atomic weight of hydrogen is close to 1.0, so one hydrogen molecule (H_2) has a molecular weight near 2.0. Oxygen has an atomic weight of about 16, so the molecular weight of an oxygen molecule (O_2) is roughly 32. To perform the experiment, you would combine 4 g of hydrogen with 32 g of oxygen to produce 36 g of water. You could also work with ounces, pounds, or tons, as long as the proportions remained the same.

☑ Both oxygen and neon are gases at room temperature. Oxygen combines readily with other elements, but neon does not. Why?

☑ How is it possible for two samples of hydrogen to contain the same number of atoms yet have different weights?

☑ Which kind of bond holds atoms in a water molecule together? What attracts water molecules to one another?

CHEMICAL REACTIONS

Cells remain alive and functional by controlling chemical reactions. In a **chemical reaction,** new chemical bonds form between atoms or existing bonds between atoms are broken. These changes occur as atoms in the reacting substances, or **reactants,** are rearranged to form different substances, or **products.** (See "FOCUS: Chemical Notation," p. 37.)

In effect, each cell is a chemical factory. For example, growth, maintenance and repair, secretion, and contraction all involve complex chemical reactions. **Metabolism** (me-TAB-ō-lizm) refers to all the chemical reactions that occur in an organism. Cells use chemical reactions to provide the energy needed to maintain homeostasis and to perform essential functions.

BASIC ENERGY CONCEPTS

An understanding of some basic relationships between matter and energy is essential for any discussion of chemical reactions. **Work** is movement or a change in the physical structure of matter. **Energy** is the capacity to perform work; movement or physical change cannot occur unless energy is provided. The two major types of energy are kinetic energy and potential energy:

1. **Kinetic energy** is the energy of motion—energy that is doing work. When you fall off a ladder, it is kinetic energy that does the damage.

2. **Potential energy** is stored energy—energy that has the potential to do work. It may result from the position of an object (a book sitting on a high shelf, or you standing on a ladder) or from an object's physical or chemical structure (a stretched spring or a charged battery).

efore we can consider the specific compounds that occur in the human body, we must be able to describe chemical compounds and reactions effectively. The use of sentences to describe chemical structures and events often leads to confusion. A simple form of "chemical shorthand" makes communication much more efficient. The chemical shorthand we will use is known as **chemical notation.** Chemical notation enables us to describe complex events briefly and precisely. The rules of chemical notation are summarized in Table 2-2.

TABLE 2-2 Rules of Chemical Notation

1. The symbol of an element indicates one atom of that element:

    H = one atom of hydrogen
    O = one atom of oxygen

2. A number preceding the symbol of an element indicates more than one atom:

    2H = two atoms of hydrogen
    2O = two atoms of oxygen

3. A subscript following the symbol of an element indicates a molecule with that number of atoms of that element:

    H_2 = one hydrogen molecule,
         composed of two hydrogen atoms
    O_2 = one oxygen molecule, composed
         of two oxygen atoms
    H_2O = one water molecule, composed of two
         hydrogen atoms and one oxygen atom

4. In a description of a chemical reaction, the participants at the start of the reaction are called reactants, and the reaction generates one or more products. An arrow indicates the direction of the reaction, from reactants (usually on the left) to products (usually on the right). In the following reaction, two atoms of hydrogen combine with one atom of oxygen to produce a single molecule of water:

    $$2H + O \longrightarrow H_2O$$

5. A superscript plus or minus sign following the symbol of an element indicates an ion. A single plus sign indicates a cation with a charge of +1 (the original atom has lost one electron). A single minus sign indicates an anion with a charge of −1 (a gain of one electron). If more than one electron has been lost or gained, the charge on the ion is indicated by a number preceding the plus or minus:

    Na^+ = one sodium ion (the sodium
         atom has lost 1 electron)
    Cl^- = one chloride ion (the chlorine atom
         has gained 1 electron)
    Ca^{2+} = one calcium ion (the calcium atom
         has lost 2 electrons)

6. Chemical reactions neither create nor destroy atoms; they merely rearrange atoms into new combinations. Therefore, the numbers of atoms of each element must always be the same on both sides of the equation for a chemical reaction. When this is the case, the equation is balanced:

    Unbalanced: $H_2 + O_2 \longrightarrow H_2O$
    Balanced: $H_2 + O_2 \longrightarrow 2H_2O$

Kinetic energy must be used in lifting the book to the shelf, in climbing the ladder, in stretching the spring, or in charging the battery. The potential energy is converted back into kinetic energy when the book falls (or you fall), the spring recoils, or the battery discharges. The kinetic energy can then be used to perform work. For example, the potential energy stored in small batteries is converted to kinetic energy in powering flashlights, portable radios, and other small appliances.

Energy cannot be lost; it can only be converted from one form to another. A conversion between potential energy and kinetic energy is never 100 percent efficient. Each time an energy exchange occurs, some of the energy is released in the form of heat. *Heat* is an increase in random molecular motion. The temperature of an object is proportional to its heat content.

Cells may perform work as movement or physical change or both. For example, a skeletal muscle at rest contains potential energy in the form of the positions of protein filaments and the covalent bonds between molecules inside the cell. When a muscle contracts, it performs work; potential energy is converted into kinetic energy, and heat is released. The amount of heat is proportional to the amount of work done. As a result, when you exercise, your body temperature rises.

2

TYPES OF REACTIONS

Three types of chemical reactions are important to the study of physiology: (1) decomposition reactions, (2) synthesis reactions, and (3) exchange reactions.

Decomposition

Decomposition is a reaction that breaks a molecule into smaller fragments. You could diagram a typical *decomposition reaction* as

$$AB \longrightarrow A + B$$

Decomposition reactions occur outside cells as well as inside them. For example, a typical meal contains molecules of fats, sugars, and proteins that are too large and too complex to be absorbed and used by your body. Decomposition reactions in the digestive tract break these down into smaller fragments before absorption begins.

Decomposition reactions involving water are important in the breakdown of complex molecules in the body. In **hydrolysis** (hī-DROL-i-sis; *hydro-,* water + *lysis,* dissolution), one of the bonds in a complex molecule is broken and the components of a water molecule (H and OH) are added to the resulting fragments:

$$A-B-C-D-E + H_2O \longrightarrow$$
$$A-B-C-H + HO-D-E$$

Catabolism (ka-TAB-ō-lizm; *katabole,* a throwing down) is the decomposition of molecules within cells. When a covalent bond—a form of potential energy—is broken, it releases kinetic energy that can perform work. By harnessing the energy released in this way, cells perform vital functions such as growth, movement, and reproduction.

Synthesis

Synthesis (SIN-the-sis) is the opposite of decomposition. A *synthesis reaction* assembles larger molecules from smaller components. Simple synthetic reactions could be diagrammed as

$$A + B \longrightarrow AB$$

Synthesis reactions may involve individual atoms or the combination of molecules to form even larger products. The formation of water from hydrogen and oxygen molecules is a synthesis reaction. Synthesis always involves the formation of new chemical bonds, whether the reactants are atoms or molecules.

Dehydration synthesis, or *condensation,* is the formation of a complex molecule by the removal of water:

$$A-B-C-H + HO-D-E \longrightarrow$$
$$A-B-C-D-E + H_2O$$

It is therefore the opposite of hydrolysis. We will encounter examples of both dehydration synthesis and hydrolysis in later sections.

Anabolism (a-NAB-ō-lizm; *anabole,* a throwing upward) is the synthesis of new compounds within the body. Because it takes energy to create a chemical bond, anabolism is usually an "uphill" process. Cells must balance their energy budgets, with catabolism providing the energy to support anabolism as well as other vital functions.

Exchange

In an **exchange reaction,** parts of the reacting molecules are shuffled around. The following equation represents a simple exchange reaction:

$$AB + CD \longrightarrow AD + CB$$

This equation has two reactants and two products. Although the reactants and products contain the same components (A, B, C, and D), those components are present in different combinations. In an exchange reaction, the reactant molecules AB and CD must break apart (a decomposition) before they can interact with each other to form AD and CB (a synthesis).

Specific chemical reactions may absorb or release energy, generally in the form of heat. If more energy is released by the breaking of old bonds than by the formation of new ones, the exchange reaction will release energy. Reactions that release energy are said to be **exergonic** (*exo-,* outside). If more energy is required for the synthesis part of the reaction than is released by the associated decomposition, additional energy must be provided. Reactions that absorb energy are **endergonic** (*endo-,* inside). Exergonic reactions are relatively common in the body; they are responsible for generating the heat that maintains your body temperature.

REVERSIBLE REACTIONS

Chemical reactions are (at least theoretically) reversible, so if A + B $\longrightarrow$ AB, then AB $\longrightarrow$ A + B. Many important biological reactions are freely reversible. Such reactions can be represented as an equation:

$$A + B \rightleftharpoons AB$$

This equation reminds you that, in a sense, two reactions are occurring simultaneously: one, a synthesis (A + B $\longrightarrow$ AB), and the other, a decomposition (AB $\longrightarrow$ A + B). At **equilibrium** (ē-kwi-LIB-rē-um), the rates at which the two equations proceed are in balance. As fast as one molecule of AB forms, another degrades into A + B.

The result of a disturbance in the equilibrium condition can be predicted. In our example, the rate at which the synthesis reaction proceeds is directly proportional to the frequency of encounters between A and B. The frequency of encounters depends on the degree of crowding—you are much more likely to run into another person in a crowded room than in a room that is almost empty. The addition of more AB molecules will increase the rate of conversion of AB to A and B. The amounts of A and B will then increase, leading to an increase in the rate of the reverse reaction—the formation of AB from A and B. Eventually, an equilibrium is again established.

Not all chemical reactions, however, are easily reversed. The requirements for two paired reactions may differ, so at any time and place the overall reaction will proceed chiefly in one direction only. For example, the synthesis reaction may occur only when A and B molecules are heated; the decomposition reaction may occur only when AB molecules are placed in water. In this case, the reaction would be written as

$$A + B \underset{H_2O}{\overset{heat}{\rightleftharpoons}} AB$$

ENZYMES AND CHEMICAL REACTIONS

Most chemical reactions do not occur spontaneously, or they occur so slowly that they would be of little value to cells. Before a reaction can proceed, enough energy must be provided to activate the reactants. **Activation energy** is the amount of energy required to start a reaction. Although many reactions can be activated by changes in temperature or acidity, such changes are deadly to cells. For example, every day your cells break down complex sugars as part of your normal metabolism. Yet, to break down a complex sugar in a laboratory, you must boil it in an acidic solution. Your cells don't have that option; temperatures that high and solutions that corrosive would immediately destroy living tissues. Instead, your cells use special proteins called *enzymes* to perform most of the complex synthesis and decomposition reactions in your body.

Enzymes promote chemical reactions by lowering the activation energy requirements (Figure 2-7●). In doing so, they make it possible for chemical reactions, such as the breakdown of sugars, to proceed under conditions compatible with life. Enzymes belong to a class of substances called **catalysts** (KAT-uh-lists; *katalysis,* dissolution), compounds that accelerate chemical reactions without themselves being permanently changed or consumed. A cell makes an enzyme molecule to promote a specific reaction. Enzymatic reactions, which are reversible, can be written as

$$A + B \overset{enzyme}{\rightleftharpoons} AB$$

Although the presence of an appropriate enzyme can accelerate a reaction, an enzyme affects only the rate of the reaction, not the direction of the reaction or the products that will be formed. An enzyme cannot bring about a reaction that would otherwise be impossible. Enzymatic reactions are generally reversible, and they proceed until an equilibrium becomes established.

The complex reactions that support life proceed in a series of interlocking steps, and each step is controlled by a specific enzyme. Such a reaction sequence is called a *pathway*. A synthetic pathway can be diagrammed as

$$A \underset{Step\ 1}{\overset{enzyme\ 1}{\rightleftharpoons}} B \underset{Step\ 2}{\overset{enzyme\ 2}{\rightleftharpoons}} C \underset{Step\ 3}{\overset{enzyme\ 3}{\rightleftharpoons}} \text{and so on}$$

In some cases, the steps in the synthetic pathway differ from those of the decomposition pathway, and separate enzymes may be involved.

☑ In cells, glucose, a six-carbon molecule, is converted into two three-carbon molecules by a reaction that releases energy. How would you classify this reaction?

☑ What is the significance of enzymes in our cells?

INORGANIC COMPOUNDS

Although the human body is very complex, it contains relatively few elements, as indicated in Table 2-1, p. 31. But knowing the identity and quantity of each element will not help you understand a human being any more than studying the alphabet will help you understand this textbook. The rest of this chapter focuses on nutrients and metabolites. **Nutrients** are the essential elements and molecules normally obtained from the diet. **Metabolites** (me-TAB-ō-līts; *metabole,* change) include all the molecules synthesized or broken down by chemical reactions inside our bodies. Nutrients and metabolites can be broadly categorized as inorganic or organic. **Inorganic compounds** generally do not contain carbon and hydrogen atoms as the primary structural ingredients, whereas carbon and hydrogen *always* form the basis for **organic compounds.**

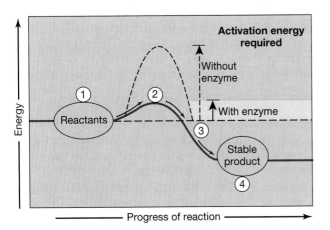

●**FIGURE 2-7** **Enzymes and Activation Energy.** Enzymes lower the activation energy requirements so a reaction can occur readily, in order from 1–4, under conditions in the body.

The most important inorganic compounds in the body are (1) *carbon dioxide,* a byproduct of cell metabolism; (2) *oxygen,* an atmospheric gas required in important metabolic reactions; (3) *water,* which accounts for most of our body weight; and (4) *inorganic acids, bases,* and *salts,* compounds held together partially or completely by ionic bonds. In this section, we will be concerned primarily with water, its properties, and how those properties establish the conditions necessary for life. Most of the other inorganic molecules and compounds in the body exist in association with water, the primary component of our body fluids. For example, carbon dioxide and oxygen are gas molecules that occur both in the atmosphere and in body fluids; all the inorganic acids, bases, and salts we will discuss are dissolved in body fluids.

WATER AND ITS PROPERTIES

Water, H_2O, is the single most important constituent of the body, accounting for up to two-thirds of total body weight. A change in body water content can have fatal consequences because virtually all physiological systems will be affected.

Although water is familiar to everyone, it has some very unusual properties. These properties are a direct result of the hydrogen bonding that occurs between adjacent water molecules:

1. *Solubility.* A remarkable number of inorganic and organic molecules will *dissolve* (break up) in water, creating a solution. Every **solution** is a uniform mixture of two or more substances—whether in solid, liquid, or gaseous form. The **solvent** is the medium in which other atoms, ions, or molecules are dispersed. The dispersed substances are called **solutes.** In *aqueous solutions,* water is the solvent. The solvent properties of water are so important that we will consider them in detail in the next section.

2. *Reactivity.* In our bodies, chemical reactions occur in water, and water molecules are also participants in some reactions. Hydrolysis and dehydration synthesis are two examples noted earlier in the chapter.

3. *High heat capacity.* Heat capacity is the ability to absorb and retain heat. Water has an unusually high heat capacity, because water molecules in the liquid state are attracted to one another through hydrogen bonding. Important consequences of this attraction include the following:

   ■ The temperature must be quite high before individual molecules have enough energy to break free and become water vapor, a gas. Consequently, water stays in the liquid state over a broad range of environmental temperatures, and the freezing and boiling points of water are far apart.

   ■ Water carries a great deal of heat away with it when it finally does change from a liquid to a gas. This feature accounts for the cooling effect of perspiration on the skin.

   ■ An unusually large amount of heat energy is required to change the temperature of 1 g of water by 1°C. Thus, once a quantity of water has reached a particular temperature, it will change temperature only slowly. This property is called *thermal inertia.* Because water accounts for up to two-thirds of the weight of the human body, thermal inertia helps stabilize body temperature. Similarly, water's high heat capacity allows the blood plasma to transport and redistribute large amounts of heat as it circulates within the body. For example, heat absorbed as the blood flows through active muscles will be released when the blood reaches vessels in the relatively cool body surface.

4. *Lubrication.* Water is an effective lubricant. There is little friction between water molecules. Thus, if two opposing surfaces are separated by even a thin layer of water, friction between those surfaces will be greatly reduced. (That is why driving on wet roads can be tricky; your tires may start sliding on a layer of water rather than maintaining contact with the road.) Within joints such as the knee, an aqueous solution prevents friction between the opposing surfaces. Similarly, a small amount of fluid in the ventral body cavities prevents friction between internal organs, such as the heart or lungs, and the body wall. ∞ *p. 20*

Aqueous Solutions

Its chemical structure makes water an unusually effective solvent. The bonds in a water molecule are oriented such that the hydrogen atoms are relatively close together (Figure 2-8a•). As a result, the water molecule has positive and negative poles. A water molecule is therefore called a *polar molecule,* or a *dipole.*

Many inorganic compounds are held together partially or completely by ionic bonds. In water, these compounds undergo *ionization* (ī-on-i-ZĀ-shun), or *dissociation* (di-sō-sē-Ā-shun). In this process (Figure 2-8b•), ionic bonds are broken as the individual ions interact with the positive or negative ends of polar water molecules. The result is a mixture of cations and anions surrounded by water molecules. The water molecules around each ion form a *hydration sphere.*

An aqueous solution containing anions and cations will conduct an electrical current. Cations (+) move toward the negative side, or negative *terminal,* and anions (−) move toward the positive terminal. Electrical forces across cell membranes affect the functioning of all cells, and small electrical currents carried by ions are essential to muscle contraction and nerve function. Chapters 10 and 12 will discuss these processes in more detail.

Soluble inorganic molecules whose ions will conduct an electrical current in solution are called **electrolytes** (e-LEK-trō-līts). Sodium chloride is an

(a) Water molecule

(b) Sodium chloride in solution

(c) Glucose in solution

•**FIGURE 2-8** **Water Molecules and Solutions.** **(a)** In a water molecule, oxygen forms polar covalent bonds with two hydrogen atoms. Because both hydrogen atoms are at one end, the molecule has an uneven distribution of charges, creating positive and negative poles. **(b)** Ionic compounds, such as sodium chloride, dissociate in water as the polar water molecules break the ionic bonds. Each ion in solution is surrounded by water molecules, creating hydration spheres. **(c)** Hydration spheres also form around an organic molecule containing polar covalent bonds. If the molecule is small, as is glucose, it will be carried into solution.

electrolyte. The dissociation of electrolytes in blood and other body fluids releases sodium ions (Na^+), potassium ions (K^+), calcium ions (Ca^{2+}), and chloride ions (Cl^-). Table 2-3 lists important electrolytes and the ions released by their dissociation.

Changes in the body fluid concentrations of electrolytes will disturb almost every vital function. For example, declining potassium levels will lead to a general muscular paralysis, and rising concentrations will cause weak and irregular heartbeats. The concentrations of ions in body fluids are carefully regulated, primarily by the coordination of activities at the kidneys (ion excretion), the digestive tract (ion absorption), and the skeletal system (ion storage or release).

Organic molecules typically contain polar covalent bonds, which also attract water molecules. The hydration spheres that form then carry these molecules into solution (Figure 2-8c•). Molecules such as *glucose,* an important soluble sugar, that readily interact with water molecules in this way are called **hydrophilic** (hī-drō-FI-lik; *hydro-,* water + *philos,* loving).

Molecules that do not readily interact with water are called **hydrophobic** (hī-drō-FŌ-bik; *hydro-,* water + *phobos,* fear). Hydrophobic molecules have few if any polar covalent bonds; they are said to be *nonpolar molecules.* When nonpolar molecules are exposed to water, hydration spheres do not form and the molecules do not dissolve. Body fat deposits consist of large, insoluble droplets of hydrophobic molecules trapped in the watery interior of cells. Gasoline, heating oil, and diesel fuel are hydrophobic molecules not found in the body.

Solute Concentrations The **concentration** of a substance is the amount of that substance present in a specified volume of solvent. Physiologists and clinicians often monitor inorganic and organic solute concentrations in body fluids such as blood or urine. Each solute has a normal range of values (see Appendix VI), and variations outside this range may indicate disease. Many solutes are participants in biochemical reactions, and the concentrations of reactants and products in a chemical reaction directly affect reaction rates.

Solute concentrations can be reported in several ways, and we will introduce two methods here. In the first method, we report the number of solute atoms, molecules, or ions in a specific volume of solution. Values are reported in moles per liter (mol/l, or M) or millimoles per liter (mmol/l, or mM). The term *molarity* refers to the concentration of solutes, in moles or millimoles. As we discussed earlier in the chapter, a mole is the number of atoms or molecules in a sample that has a weight in grams equal to the atomic weight or molecular weight, respectively. Physiological concentrations today are most often reported in millimoles per liter.

You can report concentrations in moles or millimoles only when you know the molecular weight of

TABLE 2-3 **Important Electrolytes That Dissociate in Body Fluids**

Electrolyte		Ions Released
NaCl (sodium chloride)	→	Na^+ + Cl^-
KCl (potassium chloride)	→	K^+ + Cl^-
CaCl$_2$ (calcium chloride)	→	Ca^{2+} + $2\,Cl^-$
NaHCO$_3$ (sodium bicarbonate)	→	Na^+ + HCO_3^-
MgCl$_2$ (magnesium chloride)	→	Mg^{2+} + $2\,Cl^-$
Na$_2$HPO$_4$ (disodium phosphate)	→	$2\,Na^+$ + HPO_4^{2-}
Na$_2$SO$_4$ (sodium sulfate)	→	$2\,Na^+$ + SO_4^{2-}

the ion or molecule in question. When the chemical structure is unknown or you are dealing with a complex mixture of materials, concentration is reported in terms of the weight of material dissolved in a unit volume of solution. Values are most often expressed in milligrams (mg) or grams (g) per deciliter (dl, or 100 ml). This is the method used, for example, for reporting the concentration of plasma proteins in a blood sample. [AM] *Solutions and Concentrations*

Colloids and Suspensions

Body fluids may contain large and complex organic molecules, such as proteins and protein complexes, that are held in solution by their association with water molecules (Figure 2-8c•). A solution containing dispersed proteins or other large molecules is called a **colloid.** Liquid Jell-O™ is a familiar viscous (thick) colloid.

The particles or molecules in a colloid will remain in solution indefinitely. A **suspension** contains even larger particles that will, if undisturbed, settle out of solution due to the force of gravity. Stirring beach sand into a bucket of water creates a temporary suspension that will last only until the sand settles to the bottom. Whole blood is another temporary suspension, because the blood cells are suspended in the blood plasma. If clotting is prevented, the cells in a blood sample will gradually settle to the bottom of the container. That settling rate, or "sedimentation rate," is a common laboratory test.

Hydrogen Ions in Body Fluids

A hydrogen atom can achieve stability either by sharing an electron with another atom in a covalent bond or by giving away its electrons and becoming a hydrogen ion, H^+. Hydrogen ions are extremely reactive in solution. In excessive numbers, they will break chemical bonds, change the shapes of complex molecules, and generally disrupt cell and tissue functions. As a result, the concentration of hydrogen ions in body fluids must be precisely regulated.

A few hydrogen ions are normally present, even in a sample of pure water, because some of the water molecules dissociate spontaneously, releasing cations and anions. The dissociation of water is a reversible reaction that can be represented as

$$H_2O \rightleftharpoons H^+ + OH^-$$

The dissociation of one water molecule yields a hydrogen ion and a *hydroxide* (hī-DROK-sīd) *ion,* OH^-.

Very few water molecules ionize in pure water, and the number of hydrogen and hydroxide ions is very small. The quantities are usually reported in moles, making it easy to keep track of the relative numbers of hydrogen and hydroxide ions. One liter of pure water contains about 0.0000001 mol of hydrogen ions and an equal number of hydroxide ions. In other words, the concentration of hydrogen ions in a solution of pure water is 0.0000001 mol per liter. This can be written as

$$[H^+] = 1 \times 10^{-7} \text{ mol/l}$$

The brackets around H^+ indicate "the concentration of," another example of chemical notation.

pH The hydrogen ion concentration in body fluids is so important to physiological processes that a special shorthand is used to express it. The **pH** of a solution is defined as the negative logarithm of the hydrogen ion concentration in moles per liter. Thus, instead of using the equation $[H^+] = 1 \times 10^{-7}$ mol/l, we say that the pH of pure water is $-(-7)$, or 7. Using pH values saves space, but always remember that the pH number is an *exponent* and that the pH scale is logarithmic. For instance, a pH of 6 ($[H^+] = 1 \times 10^{-6}$, or 0.000001) means that the concentration of hydrogen ions is *10 times as great* as it is at a pH of 7 ($[H^+] = 1 \times 10^{-7}$, or 0.0000001). For common liquids, the pH scale, included in Figure 2-9•, ranges from 0 to 14.

Although pure water has a pH of 7, solutions display a wide range of pH values, depending on the nature of the solutes involved. A solution with a pH of 7 is said to be **neutral,** because it contains equal numbers of hydrogen and hydroxide ions. A solution with a pH below 7 is **acidic** (a-SI-dik), meaning that it contains more hydrogen ions than hydroxide ions. A pH above 7 is **basic,** or *alkaline* (AL-kuh-lin), and has more hydroxide ions than hydrogen ions.

The pH of blood normally ranges from 7.35 to 7.45. Abnormal fluctuations in pH can damage cells and tissues by breaking chemical bonds, changing the shapes of proteins, and altering cellular functions. *Acidosis* is an abnormal physiological state caused by low blood pH (below 7.35); a pH below 7 can produce coma. *Alkalosis* results from an abnormally high pH (above 7.45); a blood pH above 7.8 generally causes uncontrollable, sustained skeletal muscle contractions.

INORGANIC ACIDS AND BASES

The body contains both inorganic and organic *acids* and *bases* that may cause acidosis or alkalosis, respectively. An **acid** is any solute that dissociates in solution and releases hydrogen ions, thereby lowering the pH. Because a hydrogen atom that loses its electron consists solely of a proton, hydrogen ions are often referred to simply as protons, and acids as *proton donors.*

A *strong acid* dissociates completely in solution, and the reaction occurs essentially in one direction only. *Hydrochloric acid* (HCl) is a representative strong acid; in water it ionizes as

$$HCl \longrightarrow H^+ + Cl^-$$

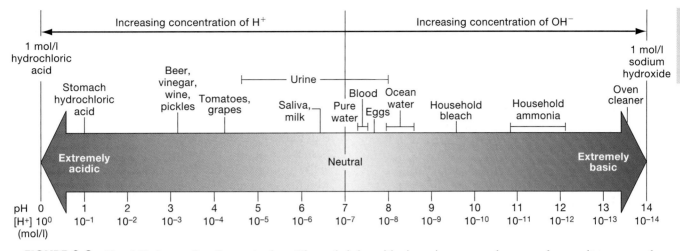

•**FIGURE 2-9** **pH and Hydrogen Ion Concentration.** The scale is logarithmic; an increase or decrease of one unit corresponds to a 10-fold change in H^+ concentration.

The stomach produces this powerful acid to assist in the breakdown of food. Hardware stores sell HCl under the name muriatic acid, for cleaning sidewalks and swimming pools.

A **base** is a solute that removes hydrogen ions from a solution and thereby acts as a *proton acceptor*. In solution, many bases release a hydroxide ion (OH^-). Hydroxide ions have a strong affinity for hydrogen ions and quickly react with them to form water molecules. A *strong base* dissociates completely in solution. *Sodium hydroxide,* NaOH, is a strong base; in solution, it releases sodium ions and hydroxide ions:

$$NaOH \longrightarrow Na^+ + OH^-$$

Strong bases have a variety of industrial and household uses. Drain openers (Drano™) and lye are two familiar examples.

Weak acids and *weak bases* fail to dissociate completely. At equilibrium, a significant number of molecules remain intact in the solution. For the same number of molecules in solution, weak acids and weak bases therefore have less of an impact on pH than do strong acids and strong bases. *Carbonic acid* (H_2CO_3) is a weak acid found in body fluids. In solution, carbonic acid reversibly dissociates into a hydrogen ion and a *bicarbonate ion,* HCO_3^-:

$$H_2CO_3 \rightleftharpoons H^+ + HCO_3^-$$

SALTS

A **salt** is an ionic compound consisting of any cation except a hydrogen ion and any anion except a hydroxide ion. Because they are held together by ionic bonds, many salts dissociate completely in water, releasing cations and anions. For example, sodium chloride (table salt) immediately dissociates in water, releasing Na^+ and Cl^- ions. Sodium and chloride are the most abundant ions in body fluids. However, many other ions are present in lesser amounts as a result of the dissociation of other inorganic compounds. Ionic concentrations in the body are regulated by mechanisms we shall describe in Chapters 26 and 27.

The ionization of sodium chloride does not affect the local concentrations of hydrogen ions or hydroxide ions, so NaCl, like many salts, is a "neutral" solute. Through their interactions with water molecules, however, other salts may indirectly affect the concentrations of H^+ and OH^- ions. Thus, the dissociation of some salts make a solution slightly acidic or slightly basic.

BUFFERS AND pH CONTROL

Buffers are compounds that stabilize the pH of a solution by removing or replacing hydrogen ions. *Buffer systems* typically involve a weak acid and its related salt, which functions as a weak base. For example, the carbonic acid–bicarbonate buffer system (which shall be detailed in Chapter 27) consists of carbonic acid and sodium bicarbonate, $NaHCO_3$, otherwise known as baking soda. Buffers and buffer systems in body fluids maintain pH within normal limits. Figure 2-9• includes the pH of several body fluids. Antacids such as Alka-Seltzer® and Rolaids® use sodium bicarbonate to neutralize excess hydrochloric acid in the stomach. The effects of neutralization are most evident when you add a strong acid to a strong base. For example, by adding hydrochloric acid to sodium hydroxide, you neutralize both the strong acid and the strong base:

$$HCl + NaOH \longrightarrow H_2O + NaCl$$

This reaction produces water and a salt—in this case, the neutral salt sodium chloride.

☑ Why does a solution of table salt conduct electricity but a sugar solution does not?

☑ How does an antacid help decrease stomach discomfort?

ORGANIC COMPOUNDS

Organic compounds always contain the elements carbon and hydrogen and generally oxygen as well. Many organic molecules contain long chains of carbon atoms linked by covalent bonds. These carbon atoms typically form additional covalent bonds with hydrogen or oxygen atoms and, less commonly, with nitrogen, phosphorus, sulfur, iron, or other elements.

Many organic molecules are soluble in water. Although the previous discussion focused on inorganic acids and bases, there are also important organic acids and bases. For example, *lactic acid* is an organic acid, generated by active muscle tissues, that must be neutralized by the buffers in body fluids.

In this discussion, we consider four major classes of organic compounds: *carbohydrates, lipids, proteins,* and *nucleic acids.* We also introduce *high-energy compounds,* which are insignificant in terms of their abundance in the body but vital to the survival of our cells. In addition, the human body contains small quantities of many other organic compounds whose structures and functions we will consider in later chapters.

Although organic compounds are diverse, structural patterns are common among all classes. *Functional groups* are groups of atoms that help determine the properties of the entire molecule. Table 2-4 details the functional groups you will encounter in this chapter.

CARBOHYDRATES

A **carbohydrate** is an organic molecule that contains carbon, hydrogen, and oxygen in a ratio near 1:2:1. Familiar carbohydrates include the *sugars* and *starches* that make up roughly half of the typical U.S. diet. Our tissues can break down most carbohydrates, which typically account for less than 1 percent of the total body weight. Although they may have other functions, carbohydrates are most important as sources of energy that are catabolized rather than stored. We will focus on *monosaccharides, disaccharides,* and *polysaccharides.*

Monosaccharides

A *simple sugar,* or **monosaccharide** (mon-ō-SAK-uh-rīd; *mono-,* single + *sakcharon,* sugar), is a carbohydrate containing from three to seven carbon atoms. A monosaccharide can be called a *triose* (three-carbon), *tetrose* (four-carbon), *pentose* (five-carbon), *hexose* (six-carbon), or *heptose* (seven-carbon). The hexose **glucose** (GLOO-kōs), $C_6H_{12}O_6$, is the most important metabolic "fuel" in the body. The atoms in a glucose molecule may form a straight chain (Figure 2-10a●) or a ring (Figure 2-10b●). In the body, the ring form is more common. A three-dimensional model shows the arrangement of atoms in the ring most clearly (Figure 2-10c●).

The three-dimensional structure of an organic molecule is an important characteristic, because it usually

TABLE 2-4 Important Functional Groups

Functional Group	Structural Formula*	Importance	Examples
Carboxyl group, —COOH	R...—C(OH)=O	Acts as an acid, releasing H^+ to become R-COO—	Fatty acids, amino acids
Amino group, —NH_2	R—N(H)(H)	Can accept or release H^+, depending on pH; can form bonds with other molecules	Amino acids
Hydroxyl group, —OH	R—O—H	Strong bases dissociate to release hydroxide ions (OH^-); may link molecules through dehydration synthesis	Carbohydrates, fatty acids, amino acids
Phosphate group, —PO_4	R—O—P(=O)(O⁻)—O⁻	May link other molecules to form larger structures; may store energy in high-energy bonds	Phospholipids, nucleic acids, high-energy compounds

*The term *R-group* is used to denote the rest of the molecule, whatever that might be. The R-group is also known as a *side chain.*

(a)   (b)

(c)

•**FIGURE 2-10** **Glucose.** **(a)** The straight-chain structural
formula. **(b)** The ring form that is most common in nature.
(c) A three-dimensional model that shows the organization
of the atoms in the ring.

= Carbon

= Oxygen

= Hydrogen

determines the molecule's fate or function. **Isomers** are
molecules that have the same molecular formula—in
other words, the same types and numbers of atoms—but
different structures. The body usually treats different
isomers as distinct molecules. For example, the monosaccharides glucose and fructose are isomers. *Fructose*
is a hexose found in many fruits and in secretions of the
male reproductive tract. Although its chemical formula—$C_6H_{12}O_6$—is the same as that of glucose, separate
enzymes and reaction sequences control its breakdown
and synthesis. Monosaccharides such as glucose and
fructose dissolve readily in water and are rapidly distributed throughout the body by blood and other body
fluids. [AM] *The Pharmaceutical Use of Isomers*

Disaccharides and Polysaccharides

Carbohydrates other than simple sugars are complex
molecules composed of monosaccharide building
blocks. Two monosaccharides joined together form a
disaccharide (dī-SAK-uh-rīd; *di-*, two). Disaccharides
such as *sucrose* (table sugar) have a sweet taste
and, like monosaccharides, are quite water-soluble.
The formation of sucrose (Figure 2-11a•) involves dehydration synthesis, a process introduced on page 38.
Dehydration synthesis, or condensation, links molecules together by the removal of a water molecule. The
breakdown of sucrose into simple sugars is an example
of hydrolysis, the functional opposite of dehydration
synthesis (Figure 2-11b•).

(a) During dehydration synthesis, two molecules are joined by the removal of a water molecule.

(b) Hydrolysis reverses the steps of dehydration synthesis; a complex
molecule is broken down by the addition of a water molecule.

•**FIGURE 2-11** **The Formation and Breakdown of Complex Sugars.** These reactions are performed by enzymes in the cell.

Many foods contain disaccharides, but all carbohydrates except monosaccharides must be disassembled through hydrolysis before they can provide useful energy. Most popular junk foods, such as candies and sodas, abound in monosaccharides (commonly fructose) and disaccharides (generally sucrose). Some people cannot tolerate sugar for medical reasons; others avoid it in an effort to control their weight. (Excess sugars are converted to fat for long-term storage.) Many such people use *artificial sweeteners* in their foods and beverages. These compounds have a very sweet taste, but they either cannot be broken down in the body or are used in insignificant amounts. [AM] *Artificial Sweeteners*

Dehydration synthesis can continue adding monosaccharides or combining disaccharides to create very complex carbohydrates. These large molecules are called **polysaccharides** (pol-ē-SAK-uh-rīdz; *poly-*, many). Polysaccharide chains can be straight or highly branched. *Starches* are glucose-based polysaccharides. Most starches are manufactured by plants. Our digestive tract can break these molecules into monosaccharides. Starches such as those in potatoes and grains are a major dietary energy source. *Cellulose,* a structural component of many plants, is a polysaccharide that our bodies cannot digest. Foods such as celery—which contains cellulose, water, and little else—contribute to the bulk of digestive wastes but are useless as an energy source. (In fact, you expend more energy when you chew a stalk of celery than you obtain by digesting it.)

A branched polysaccharide, **glycogen** (GLĪ-kō-jen), or *animal starch,* is composed of interconnected glucose molecules (Figure 2-12•). Like most other polysaccharides, glycogen does not dissolve in water or other body fluids. Liver tissues and muscle tissues make and store glycogen. When these tissues have a high demand for glucose, glycogen molecules are broken down; when the demand is low, these tissues absorb glucose from the bloodstream and rebuild glycogen reserves.

Despite their metabolic importance, carbohydrates account for less than 3 percent of our total body weight. Table 2-5 summarizes information about the carbohydrates.

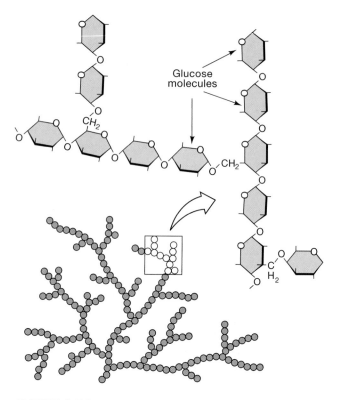

•**FIGURE 2-12** **The Structure of a Polysaccharide.** Liver and muscle cells store glucose as the polysaccharide glycogen, a long, branching chain of glucose molecules. This figure uses a different method of representing a carbon ring structure: At each corner of the solid hexagon is a carbon atom. The position of an oxygen atom in each glucose ring is shown.

LIPIDS

Like carbohydrates, **lipids** (*lipos,* fat) contain carbon, hydrogen, and oxygen, and the carbon to hydrogen ratio is commonly 1:2. However, lipids contain much less oxygen than do carbohydrates with the same number of carbon atoms. The hydrogen-to-oxygen ratio is therefore very large; a representative lipid, such as lauric acid, has a formula of $C_{12}H_{24}O_2$. Lipids may also contain small quantities of other elements, such as phosphorus, nitrogen, or sulfur. Familiar lipids include *fats, oils,* and *waxes.* Most lipids are insoluble in water, but special transport mechanisms carry them in the circulating blood.

TABLE 2-5 Carbohydrates in the Body

Structure	Example(s)	Primary Function(s)	Remarks
Monosaccharides (simple sugars)	Glucose, fructose	Energy source	Manufactured in the body and obtained from food; distributed in body fluids
Disaccharides	Sucrose, lactose, maltose	Energy source	Sucrose is table sugar, lactose is in milk, maltose is malt sugar; all must be broken down to monosaccharides before absorption
Polysaccharides	Glycogen	Storage of glucose molecules	Glycogen is in animal cells; other starches and cellulose are in plant cells

TABLE 2-6 Representative Lipids and Their Functions in the Body

Lipid Type	Example(s)	Primary Function(s)	Remarks
Fatty acids	Lauric acid	Energy source	Absorbed from food or synthesized in cells; transported in the blood
Eicosanoids	Prostaglandins, leukotrienes	Chemical messengers coordinating local cellular activities	Prostaglandins: produced in most body tissues
Glycerides	Monoglycerides, diglycerides, triglycerides	Energy source, energy storage, insulation, and physical protection	Stored in fat deposits; must be broken down to fatty acids and glycerol before they can be used as an energy source
Steroids	Cholesterol	Structural component of cell membranes, hormones, digestive secretions in bile	All have the same carbon ring framework
Phospholipids, glycolipids	Lecithin (a phospholipid)	Structural components of cell membranes	Derived from fatty acids and nonlipid components

Lipids form essential structural components of all cells. In addition, lipid deposits are important as energy reserves. On average, lipids provide roughly twice as much energy as carbohydrates do, gram for gram, when broken down in the body. When the supply of lipids exceeds the demand for energy, the excess is stored in fat deposits. For this reason, there has been great interest in developing *fat substitutes* that provide less energy but have the same taste and texture as lipids. AM *Fat Substitutes*

Lipids normally account for 12–18 percent of the total body weight of adult males and 18–24 percent of that of adult women. Many kinds of lipids exist in the body, and major groups are presented in Table 2-6. We will consider five classes of lipid molecules: *fatty acids, eicosanoids, glycerides, steroids,* and *phospholipids and glycolipids.*

Fatty Acids

Fatty acids are long carbon chains with hydrogen atoms attached. One end of the carbon chain always bears a *carboxylic* (kar-bok-SIL-ik) *acid group,* COOH (Table 2-4). The name *carboxyl* should help you remember that a carbon and a hydroxyl (—OH) group are the important structural features of fatty acids. The end opposite the carboxylic acid group is known as a hydrocarbon *tail* of the fatty acid. Figure 2-13a• shows a representative fatty acid, *lauric acid.*

When a fatty acid is in solution, only the carboxyl end associates with water molecules, because that is the only hydrophilic portion of the molecule. The hydrocarbon tail is hydrophobic, so fatty acids have a very limited solubility in water. In general, the longer the hydrocarbon tail, the lower the solubility.

In a *saturated* fatty acid, each carbon atom in the hydrocarbon tail has four single covalent bonds. If some of the carbon-to-carbon bonds are double covalent bonds, the fatty acid is said to be *unsaturated.*

Lauric acid ($C_{12}H_{24}O_2$)

(a)

Saturated

Unsaturated

(b)

•**FIGURE 2-13 Fatty Acids. (a)** Lauric acid demonstrates two structural characteristics common to all fatty acids: a long chain of carbon atoms and a carboxylic acid group (—COOH) at one end. **(b)** A fatty acid is saturated or unsaturated. Unsaturated fatty acids have double covalent bonds; their presence causes a sharp bend in the molecule.

2

These terms refer to the ratio of hydrogen atoms to carbon atoms in the fatty acid. Replacing a double bond between carbon atoms with a single bond allows the molecule to accept two more hydrogen atoms; hence, a double-bonded chain is unsaturated (Figure 2-13b•). A *monounsaturated* fatty acid has a single unsaturated bond in the hydrocarbon tail. A *polyunsaturated* fatty acid contains multiple unsaturated bonds.

•**FIGURE 2-14 Prostaglandins.** Prostaglandins are unusual short-chain fatty acids.

Fatty Acids and Health

Both saturated and unsaturated fatty acids can be broken down for energy, but a diet containing large amounts of saturated fatty acids increases the risk of heart disease and other circulatory problems. Butter, fatty meat, and ice cream are popular dietary sources of saturated fatty acids. Vegetable oils, such as olive oil or corn oil, contain a mixture of monounsaturated and polyunsaturated fatty acids. Current research indicates that the monounsaturated fats may be more effective than polyunsaturated fats in lowering the risk of heart disease. In fact, the manufacture of margarine and vegetable shortening from polyunsaturated fats produces compounds called *trans* fatty acids, which may actually increase the risk of heart disease. Increasing the proportion of oleic acid, an 18-carbon monounsaturated fatty acid, in cooking oils and other products could therefore yield health benefits.

The carbons in a fatty acid molecule are numbered, beginning at the carboxylic acid end. The last carbon in the chain is called the *omega* carbon. Eskimos have lower rates of heart disease than other populations do, even though the Eskimo diet is high in fats and cholesterol. The fatty acids in the Eskimo diet have an unsaturated bond three carbons before the omega carbon, a position known as "omega minus 3," or *omega-3*. Fish flesh and fish oils contain omega-3 fatty acids. For unknown reasons, the presence of omega-3 fatty acids (or some unidentified component of fish flesh) in the diet reduces the risks of heart disease, rheumatoid arthritis, and other inflammatory diseases.

Eicosanoids

Eicosanoids (ī-KŌ-sa-noydz) are lipids derived from *arachidonic* (ah-rak-i-DON-ik) *acid,* a fatty acid that must be absorbed in the diet because it cannot be synthesized by the body. Two major classes of eicosanoids exist: (1) *prostaglandins* and (2) *leukotrienes.* We will consider only prostaglandins here, because virtually all tissues synthesize and respond to prostaglandins. Leukotrienes are primarily produced by cells involved with coordinating the responses to injury or disease. We shall consider leukotrienes in Chapters 18 and 22.

Prostaglandins (pros-tuh-GLAN-dinz) are short-chain fatty acids in which five of the carbon atoms are joined in a ring (Figure 2-14•). Cells release prostaglandins to coordinate or direct local cellular activities. Prostaglandins are extremely powerful and effective in minute quantities. Almost every tissue in the body releases and responds to prostaglandins. Their effects vary with the nature of the prostaglandin and the site of release. Consider two examples:

1. Prostaglandins released by damaged tissues stimulate nerve endings and produce the sensation of pain (Chapter 17).
2. Prostaglandins released in the uterus help trigger the start of labor contractions (Chapter 29).

Hormones are chemical messengers produced at one site that journey through the bloodstream to regulate activities under way in a different portion of the body. Prostaglandins are chemical messengers released to coordinate local activities; they usually do not travel through the circulatory system to reach their target cells. As a result, prostaglandins are often called *local hormones.*

Glycerides

Individual fatty acids cannot be strung together in a chain by dehydration synthesis, as monosaccharides can. But they can be attached to another compound, **glycerol** (GLI-se-rol), through a similar reaction. The result is a lipid known as a **glyceride** (GLI-se-rīd). Dehydration synthesis can produce a **monoglyceride** (mo-nō-GLI-se-rīd), consisting of glycerol plus one fatty acid. Subsequent reactions can yield a **diglyceride** (glycerol + two fatty acids) and then a **triglyceride** (glycerol + three fatty acids), as in Figure 2-15•. Hydrolysis breaks the glycerides into fatty acids and glycerol. Compare Figure 2-15 with Figure 2-11• to convince yourself that dehydration synthesis and hydrolysis operate the same way, whether the molecules involved are carbohydrates or lipids.

Triglycerides, also known as *neutral fats,* have three important functions:

1. *Energy source.* Fatty deposits in specialized sites of the body represent a significant energy reserve. In times of need, the triglycerides are disassembled, yielding fatty acids that can be broken down to provide energy.
2. *Insulation.* Fat deposits under the skin serve as insulation, preventing heat loss to the environment. Heat loss across a layer of lipids is only about one-third that through other tissues.

•**FIGURE 2-15** **Triglyceride Formation.** The formation of a triglyceride involves the attachment of fatty acids to the carbons of a glycerol molecule. In this example, a triglyceride is formed by the attachment of one unsaturated and two saturated fatty acids to a glycerol molecule.

3. *Protection.* A fat deposit around a delicate organ such as a kidney provides a cushion that protects against shocks or blows.

Triglycerides are stored in the body as lipid droplets within cells. These lipids absorb and accumulate lipid-soluble vitamins, drugs, or toxins that appear in body fluids. This accumulation has both positive and negative effects. For example, the body's lipid reserves retain both valuable lipid-soluble vitamins (A, D, E, K) and potentially dangerous lipid-soluble pesticides, such as DDT.

Steroids

Steroids are large lipid molecules that share a distinctive carbon framework. They differ in the carbon chains attached to this basic structure. Figure 2-16a• shows the structural formula of the steroid **cholesterol** (ko-LES-ter-ol; *chole-*, bile + *stereos*, solid). Cholesterol and related steroids are important for the following reasons:

▪ All animal cell membranes contain cholesterol. Cells need cholesterol to maintain their cell membranes as well as for cell growth and division.

▪ Steroid hormones are involved in the regulation of sexual function. Examples include the sex hormones, such as *testosterone* and the *estrogens* (Figure 2-16b,c•).

▪ Steroid hormones are important in the regulation of tissue metabolism and mineral balance. Examples include the hormones of the adrenal cortex, called *corticosteroids*, and *calcitriol*, a hormone important in calcium ion regulation.

▪ Steroid derivatives called *bile salts* are required for the normal processing of dietary fats. Bile salts, which are produced in the liver and secreted in bile, interact with lipids in the intestinal tract and facilitate the digestion and absorption of lipids.

Cholesterol is obtained from two sources: (1) by absorption from animal products in the diet and (2) by synthesis within the body. Meat, cream, and egg yolks are especially rich dietary sources of cholesterol. A diet high in cholesterol can be harmful, because a strong link exists between high blood cholesterol levels and heart disease. Current nutritional advice suggests that you limit cholesterol intake to less than 300 mg per day. This amount represents a 40 percent reduction for the average adult in the United States. Unfortunately, because the body can synthesize cholesterol as well, blood cholesterol levels can be difficult to control by dietary restriction alone. We shall examine the connection between blood cholesterol levels and heart disease more closely in later chapters.

•**FIGURE 2-16** **Steroids.** All steroids share a complex four-ring structure. Individual steroids differ in the side chains attached to the carbon rings.

Phospholipids and Glycolipids

Phospholipids (FOS-fō-lip-idz) and **glycolipids** (GLĪ-kō-lip-idz) are structurally related, and our cells can synthesize both types of lipids, primarily from fatty acids. In a *phospho*lipid, a *phosphate group* (PO_4^{3-}) links a diglyceride to a nonlipid group (Figure 2-17a•).

In a *glyco*lipid, a carbohydrate is attached to a diglyceride (Figure 2-17b•). Note that placing *-lipid* last in these names indicates that the molecule consists primarily of lipid.

The long hydrocarbon tails of phospholipids and glycolipids are hydrophobic, but the opposite ends, the nonlipid *heads,* are hydrophilic. In water, large num-

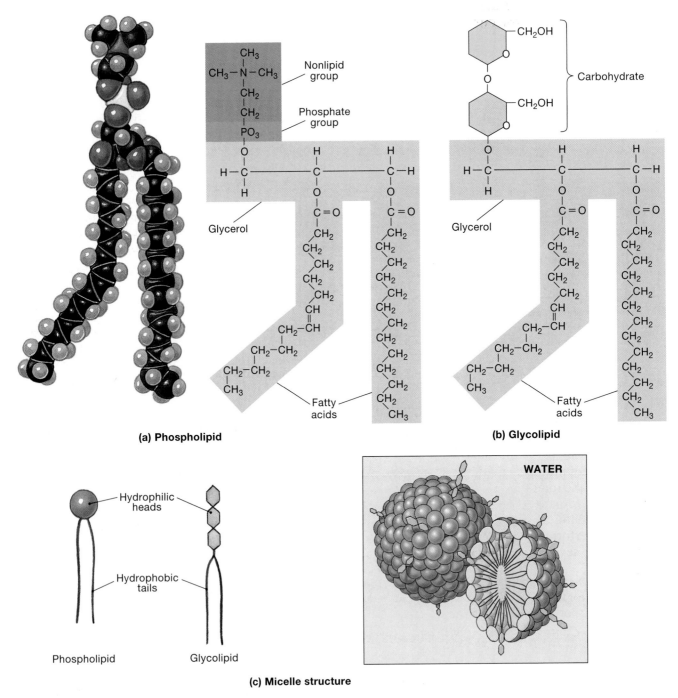

(a) Phospholipid

(b) Glycolipid

(c) Micelle structure

•**FIGURE 2-17 Phospholipids and Glycolipids. (a)** The phospholipid *lecithin.* In a phospholipid, a phosphate group links a nonlipid molecule to a diglyceride. **(b)** In a glycolipid, a carbohydrate is attached to a diglyceride. **(c)** When present in large numbers, these molecules will form micelles, with the hydrophilic heads facing the water molecules and the hydrophobic tails on the inside of each droplet.

bers of these molecules tend to form droplets, or *micelles* (mī-SELZ), with the hydrophilic portions on the outside (Figure 2-17c•). Most meals contain a mixture of lipids and other organic molecules, and micelles form as the food breaks down in your digestive tract. In addition to phospholipids and glycolipids, micelles may contain other insoluble lipids, such as steroids, glycerides, and long-chain fatty acids.

Cholesterol, phospholipids, and glycolipids are called *structural lipids,* because they help form and maintain intracellular structures called membranes. At the cellular level, *membranes* are sheets or layers composed primarily of hydrophobic lipids. A membrane surrounds each cell and separates the aqueous solution inside the cell from the aqueous solution in the extracellular environment. Because the two solutions are separated by a membrane, their compositions can be very different. A variety of internal membranes subdivide the interior of the cell into specialized compartments, each with a distinctive chemical nature and, as a result, a different function.

☑ A food contains organic molecules with the elements C, H, and O in a ratio of 1:2:1. Which type of compound is this, and what is its major function in the body?

☑ When two monosaccharides undergo a dehydration synthesis reaction, which type of molecule is formed?

☑ Which kind of lipid would be found in a sample of fatty tissue taken from beneath the skin?

☑ Which lipids would you find in human cell membranes?

PROTEINS

Proteins are the most abundant organic components of the human body and in many ways the most important. The human body may have as many as 140,000 types of proteins, and they account for about 20 percent of the total body weight. All proteins contain carbon, hydrogen, oxygen, and nitrogen; smaller quantities of sulfur may also be present.

Proteins perform a variety of essential functions. These include seven major functional categories:

1. *Support.* *Structural proteins* create a three-dimensional framework for the body, providing strength, organization, and support for cells, tissues, and organs.

2. *Movement.* *Contractile proteins* are responsible for muscular contraction; related proteins are responsible for the movement of individual cells.

3. *Transport.* Insoluble lipids, respiratory gases, special minerals, such as iron, and several hormones must be bound to special *transport proteins* before they can be transported in the blood. Other specialized proteins transport materials from one part of a cell to another.

4. *Buffering.* Proteins provide a considerable buffering action and thereby help prevent potentially dangerous changes in pH in our cells and tissues.

5. *Metabolic regulation.* *Enzymes* accelerate chemical reactions in cells. The sensitivity of enzymes to environmental factors is extremely important in controlling the pace and direction of metabolic operations.

6. *Coordination and control.* Protein hormones can influence the metabolic activities of every cell in the body or affect the function of specific organs or organ systems.

7. *Defense.* The tough, waterproof proteins of the skin, hair, and nails protect the body from environmental hazards. Proteins called *antibodies,* components of the *immune response,* help protect us from disease. Special *clotting proteins* restrict bleeding after an injury to the cardiovascular system.

Structure of Proteins

Proteins consist of long chains of organic molecules called **amino acids.** Twenty types of amino acids occur in significant quantities in the body. A typical protein contains 1000 amino acids; the largest protein complexes have 100,000 or more. Each amino acid consists of a central carbon atom to which four groups are attached: (1) a hydrogen atom, (2) an *amino group* ($-NH_2$), (3) a carboxylic acid group ($-COOH$), and (4) a variable group, known as an *R group,* or *side chain* (Figure 2-18•). Different R groups distinguish one amino acid from another, giving each its own chemical properties. The name *amino acid* refers to the presence of the *amino* group and the carboxylic *acid* group. Amino acids are relatively small, water-soluble molecules. The 20 amino acids common in proteins are shown in Appendix IV; compare R groups. In the normal pH range of body fluids, the carboxylic acid groups on many amino acids give up their hydrogens. When the carboxylic acid group changes from $-COOH$ to $-COO^-$, these amino acids become negatively charged.

Structure of an amino acid

•**FIGURE 2-18 Amino Acids.** Each amino acid consists of a central carbon atom to which four different groups are attached: a hydrogen atom, an amino group ($-NH_2$), a carboxylic acid group ($-COOH$), and a variable group designated R.

Glycine (gly)   Alanine (ala)

Dehydration   Hydrolysis
synthesis

Peptide bond

Peptide bond formation

•**FIGURE 2-19 Peptide Bonds.** The amino acids glycine and alanine, linked to form a dipeptide. Peptides form as dehydration synthesis creates a peptide bond between the carboxylic acid group of one amino acid and the amino group of another.

Figure 2-19• shows two representative amino acids, *glycine* and *alanine*. As the figure indicates, dehydration synthesis can link two amino acids. This process creates a covalent bond between the carboxylic acid group of one amino acid and the amino group of another. Such a bond is known as a **peptide bond.** Molecules consisting of amino acids held together by peptide bonds are called **peptides.** The molecule created in this example is called a *dipeptide,* because it contains two amino acids.

The chain can be lengthened by the addition of more amino acids. Attaching a third amino acid produces a *tripeptide;* next are *tetrapeptides, pentapeptides,* and so forth. Tripeptides and larger peptide chains are called **polypeptides.** Polypeptides containing more than 100 amino acids are usually called proteins. You are probably familiar with the names of several important proteins, including *hemoglobin* in red blood cells and *keratin* in fingernails and hair. Because each protein contains amino acids that are negatively charged, the entire protein has a net negative charge. For this reason, proteins are often indicated by the abbreviation Pr⁻.

Protein Shape

The characteristics of a particular protein are determined in part by the R groups on its component amino acids. But the properties of a protein are more than just the sum of the properties of its parts, for polypeptides can have very complex shapes. Proteins have four levels of structural complexity:

1. **Primary structure** is the sequence of amino acids along the length of a single polypeptide. The primary structure of a short peptide chain is diagrammed in Figure 2-20a•.

2. **Secondary structure** results from bonds that develop between atoms at different parts of the polypeptide chain. Hydrogen bonding, for example, may create a simple spiral, known as an *alpha-helix,* or a flat *pleated sheet* (Figure 2-20b•). Which one forms depends on the sequence of amino acids in the peptide chain and where hydrogen bonding occurs along the length of the peptide. The alpha-helix is the most common form, but a single polypeptide chain may have both helical and pleated sections.

3. **Tertiary structure** is the complex coiling and folding that gives the protein its final three-dimensional shape. Tertiary structure results primarily from interactions between the polypeptide chain and the surrounding water molecules and to a lesser extent from interactions between the R groups of amino acids in different parts of the molecule. *Disulfide bonds* are covalent bonds that may form between two molecules of the amino acid *cysteine,* each molecule located at a different site along the chain. Disulfide bonds create permanent loops or coils in a polypeptide chain. **Globular proteins** (Figure 2-20c•) are compact, generally rounded, and readily enter solution. *Myoglobin,* a protein in muscle cells, is a globular protein.

4. **Quaternary structure** is the interaction between individual polypeptide chains to form a protein complex (Figure 2-20d•). Each of the polypeptide subunits has its own secondary and tertiary structures. The protein *hemoglobin* contains four globular subunits. Hemoglobin is found within red blood cells, where it binds and transports oxygen. In *keratin* and *collagen,* three alpha-helical polypeptides are wound together like the strands of a rope. Keratin is the tough, water-resistant protein at the surface of the skin and in nails and hair. Collagen is the most abundant structural protein, as collagen fibers form the framework that supports cells in most tissues. Keratin and collagen are **fibrous proteins.** Fibrous proteins are tough, durable, and generally insoluble.

Shape and Function Proteins are very versatile and have a variety of functions. The shape of a protein determines its functional properties, and the ultimate determinant of shape is the sequence of amino acids. The 20 common amino acids can be linked in an astonishing number of combinations, creating proteins of enormously varied shape and function. Changing the identity of a single amino acid out of the 10,000 or more amino acids in a protein can significantly alter the protein's functional properties. For example, several cancers and *sickle cell anemia,* a blood disorder, result from single changes in the amino acid sequences of complex proteins.

(a) Primary structure

Linear chain of amino acids

Bond formation

A2 • • • • A6

A1 • A3 A5 A7 A9

Alpha-helix

(b) Secondary structure

Bond formation

A1 A2 A3 A4

A5

A9 A8 A7 A6

A10

A11 A12 A13 A14

OR

Pleated sheet

Alpha-helix

(c) Tertiary structure

Hemoglobin (globular protein)

OR

Keratin or collagen (fibrous protein)

(d) Quaternary structure

•**FIGURE 2-20 Protein Structure. (a)** The primary structure of a polypeptide is the sequence of amino acids (A1, A2, A3, and so on) along its length. **(b)** Secondary structure is primarily the result of hydrogen bonding along the length of the polypeptide chain. Such bonding often produces a simple spiral (an alpha-helix) or a flattened arrangement known as a pleated sheet. **(c)** Tertiary structure is the coiling and folding of a polypeptide. Within the cylindrical segments of this globular protein, the polypeptide chain is arranged in an alpha-helix. **(d)** Quaternary structure develops when separate polypeptide subunits interact to form a larger molecule. A single hemoglobin molecule contains four globular subunits. Hemoglobin transports oxygen in the blood; the oxygen binds reversibly to the heme units. In keratin and collagen, three fibrous subunits intertwine. Keratin is a tough, water-resistant protein in skin, hair, and nails. Collagen is the principal extracellular protein in most organs.

The tertiary and quaternary shapes of complex proteins depend not only on their amino acid sequence, but also on the local environmental characteristics. Small changes in the ionic composition, temperature, or pH of their surroundings can thus affect the function of proteins. Protein shape can also be affected by hydrogen bonding to other molecules in solution. The significance of these factors is most striking when we consider the function of enzymes, for these proteins are essential to the metabolic operations under way in every cell in the body. Table 2-7 identifies the major types of proteins in the body.

We will now take a closer look at enzyme function, a topic introduced on p. 39.

TABLE 2-7 **Representative Proteins and Their Functions**

Protein Type	Example(s)	Representative Location(s)	Function(s)
Structural proteins	Keratin	Skin surface, hair, nails	Provides strength and waterproofing
	Collagen	Dermis of skin, tendons	Provides strength
Contractile proteins	Actin, myosin	Muscle cells	Perform contraction and movement
Transport proteins	Albumin	Circulating blood	Transports fatty acids and steroid and thyroid hormones
	Transferrin	Circulating blood	Transports iron
	Apolipoproteins	Circulating blood	Transport glycerides
	Hemoglobin	Circulating blood	Transports oxygen in blood (in red blood cells)
Buffers	Intracellular and extracellular proteins	In cells and body fluids	Stabilize pH
Enzymes	Hydrolases	All cells	Catalyze hydrolysis of organic molecules
	Kinases	All cells	Attach phosphate groups to organic substrates
	Proteases	All cells; digestive secretions of stomach, pancreas	Break down proteins
	Carbohydrases	All cells; digestive secretions of salivary glands, pancreas	Break down carbohydrates
	Lipases	All cells; digestive secretions of pancreas	Break down lipids
Hormones	Insulin, glucagon	Circulating blood	Coordinate and/or control metabolic activities
Antibodies (immunoglobulins)	Gamma globulins	Circulating blood	Attack foreign proteins and pathogens

Enzyme Function

The reactants in enzymatic reactions are called **substrates.** As in other types of chemical reactions, the interactions among substrates yield specific products. Before an enzyme can function as a catalyst—to accelerate a chemical reaction without itself being permanently changed or consumed—the substrates must bind to a special region of the enzyme. Called the **active site,** such a region is typically a groove or pocket into which one or more substrates nestle, like a key fitting into a lock. This physical fit is reinforced by weak electrical attractive forces, such as hydrogen bonding. The tertiary or quaternary structure of the enzyme molecule determines the shape of the active site. Although enzymes are proteins, any organic or inorganic compound that will bind to the active site can be a substrate.

Figure 2-21• diagrams enzyme function in a synthesis reaction. Substrates bind to the enzyme at its active site (step 1). It appears likely that substrate binding results in a temporary, reversible change in the shape of the protein and that this shape change promotes the reaction by placing physical stresses on the substrate molecules. The enzyme then promotes

product formation (step 2). The completed product then detaches from the active site (step 3), and the enzyme is free to repeat the process. Enzymes work quickly, cycling rapidly between substrates and products. For example, an enzyme providing energy during a muscular contraction performs its reaction sequence 100 times per second.

Our example in Figure 2-21• is an enzyme that catalyzes a synthesis reaction. Other enzymes may catalyze decomposition reactions or exchange reactions. All enzymes share three basic characteristics:

1. *Specificity.* Each enzyme catalyzes only one type of reaction and can accommodate only one type of substrate molecule. This property is called **specificity.** The specificity of an enzyme is determined by the ability of a particular substrate to bind to the active site, a relatively small portion of the entire protein. Differences in the quaternary structure of the enzyme that do not affect the active site or change the response of the enzyme to substrate binding do not affect enzyme function. *Isozymes* are enzymes that differ in structure but catalyze the same reaction. Different tissues typically contain different isozymes.

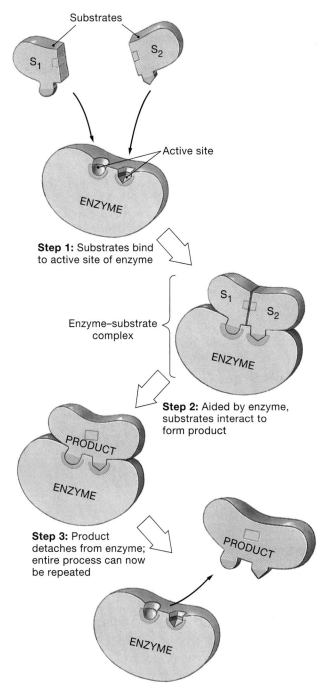

Substrates

S_1 S_2

Active site

ENZYME

Step 1: Substrates bind to active site of enzyme

Enzyme–substrate complex

S_1 S_2

ENZYME

Step 2: Aided by enzyme, substrates interact to form product

PRODUCT

ENZYME

Step 3: Product detaches from enzyme; entire process can now be repeated

PRODUCT

ENZYME

•**FIGURE 2-21** **A Simplified View of Enzyme Structure and Function.** Each enzyme contains a specific active site somewhere on its exposed surface.

not affect the rate of reaction unless additional enzyme molecules are provided. The substrate concentration required to have the maximum rate of reaction is called the *saturation limit.* An enzyme that has reached its saturation limit is said to be **saturated.** To increase the reaction rate further, the cell must increase the number of enzyme molecules available. This is one important way that cells promote specific reactions.

3. *Regulation.* A variety of factors can turn enzymes "on" or "off" and thereby control reaction rates inside the cell. We will consider only one example here, the presence or absence of *cofactors* (considered next). In fact, virtually anything that changes the tertiary or quaternary shape of an enzyme may turn enzyme function on or off. Each cell contains an assortment of enzymes, and any particular enzyme may be active under one set of conditions and inactive under another. Because the change is immediate, enzyme activation or inactivation is an important method of short-term control over reaction rates and pathways.

   [AM] *A Closer Look: The Control of Enzyme Activity*

Cofactors and Enzyme Function A **cofactor** is an ion or a molecule that must bind to the enzyme before substrates can also bind. Without a cofactor, the enzyme is intact but nonfunctional; with the cofactor, the enzyme can catalyze a specific reaction. Examples of cofactors include ions such as calcium (Ca^{2+}) and magnesium (Mg^{2+}), which bind at the enzyme's active site. Cofactors may also bind at other sites, as long as they produce a change in the shape of the active site that makes substrate binding possible.

Coenzymes are nonprotein organic molecules that function as cofactors. Our bodies convert many vitamins into essential coenzymes. *Vitamins,* detailed in Chapter 25, are structurally related to lipids or carbohydrates but have unique functional roles. The human body cannot synthesize most of the vitamins it needs, so you must obtain them from your diet.

Temperature and pH Each enzyme works best at specific temperature and pH values. As temperatures rise, protein shape changes and enzyme function deteriorates. Death occurs at very high body temperatures (above 43°C, or 110°F), because proteins undergo **denaturation,** a change in their tertiary or quaternary structure. Denatured proteins are nonfunctional, and the loss of structural proteins and enzymes soon causes irreparable damage to organs and organ systems.

Denaturation can be temporary or permanent. You see denaturation when you fry an egg. As the temperature rises, the proteins in the clear white denature. Eventually, the proteins become completely and irreversibly denatured, forming an insoluble white mass.

Enzymes are equally sensitive to pH changes. *Pepsin,* an enzyme that breaks down proteins in the contents of your stomach, works best at a pH of 2.0 (strongly acidic). Your small intestine contains *trypsin,*

2. *Saturation limits.* The rate of an enzymatic reaction is directly proportional to the concentration of substrate molecules and enzymes. An enzyme molecule must encounter appropriate substrates before it can catalyze a reaction; the higher the substrate concentration, the more frequent encounters will be. When substrate concentrations are high enough that every enzyme molecule is cycling through its reaction sequence at top speed, further increases in substrate concentration will

2

another enzyme that attacks proteins. Trypsin works only in an alkaline environment, with an optimum pH of 7.7 (weakly basic).

Glycoproteins and Proteoglycans

Glycoproteins (GLĪ-kō-prō-tēnz) and **proteoglycans** (prō-tē-ō-GLĪ-kanz) are combinations of protein and carbohydrate molecules. Glyco*proteins* are large proteins with small carbohydrate groups attached. These molecules may function as enzymes, antibodies, hormones, or protein components of cell membranes. Glycoproteins in cell membranes play a major role in the identification of normal versus abnormal cells as well as in the initiation and coordination of the immune response (Chapter 22). Glycoprotein secretions called *mucins* absorb water to form **mucus.** Mucus coats the surfaces of the respiratory and digestive tracts, providing lubrication. Proteo*glycans* are large polysaccharide molecules linked by polypeptide chains. The proteoglycans in tissue fluids give them a syrupy consistency.

☑ Proteins are chains of which small organic molecules?

☑ Which level of protein structure would be affected by an agent that breaks hydrogen bonds?

☑ Why does boiling a protein affect its structural and functional properties?

☑ How might a change in an enzyme's active site affect its function?

NUCLEIC ACIDS

Nucleic (noo-KLĀ-ik) **acids** are large organic molecules composed of carbon, hydrogen, oxygen, nitrogen, and phosphorus. Nucleic acids store and process information at the molecular level, inside cells. Two classes of nucleic acid molecules exist: (1) **deoxyribonucleic** (dē-ok-sē-rī-bō-noo-KLĀ-ik) **acid,** or **DNA,** and (2) **ribonucleic** (rī-bō-noo-KLĀ-ik) **acid,** or **RNA.** As we shall see, these two classes of nucleic acids differ in composition, structure, and function.

The DNA in our cells determines our inherited characteristics, such as eye color, hair color, and blood type. DNA affects all aspects of body structure and function, because DNA molecules encode the information needed to build proteins. By directing the synthesis of structural proteins, DNA controls the shape and physical characteristics of our bodies. By controlling the manufacture of enzymes, DNA regulates not only protein synthesis but all aspects of cellular metabolism, including the creation and destruction of lipids, carbohydrates, and other vital molecules.

Several forms of RNA cooperate to manufacture specific proteins by using the information provided by DNA. We shall detail the functional relationships between DNA and RNA in Chapter 3.

Structure of Nucleic Acids

A nucleic acid consists of nucleotides linked by dehydration synthesis. Each **nucleotide** has three components: (1) a sugar, (2) a phosphate group, and (3) a **nitrogenous** (nitrogen-containing) **base.** The sugar is a *pentose* (five-carbon sugar), either *ribose* (in RNA) or *deoxyribose* (in DNA). Each pentose is attached to a phosphate group and to a nitrogenous base. Five nitrogenous bases occur in nucleic acids: **adenine (A), guanine (G), cytosine (C), thymine (T),** and **uracil (U)** (Figure 2-22●). Adenine and guanine are double-ringed molecules called *purines;* the other three bases are single-ringed molecules called *pyrimidines.* Both RNA and DNA contain adenine, guanine, and cytosine. Uracil occurs only in RNA, and thymine only in DNA.

In the formation of a nucleic acid, a nitrogenous base is attached to a pentose molecule. A nucleotide

•**FIGURE 2-22 Purines and Pyrimidines. (a)** Purines and **(b)** pyrimidines are the nitrogenous bases in nucleic acids.

forms when a phosphate group binds to the sugar. Dehydration synthesis then attaches the phosphate group of one nucleotide to the carbohydrate of another. The "backbone" of a nucleic acid molecule is thus a linear sugar-to-phosphate-to-sugar sequence, with the nitrogenous bases projecting to one side.

RNA and DNA

Important structural differences distinguish RNA from DNA. A molecule of RNA consists of a single chain of nucleotides (Figure 2-23a•). Its shape depends on the order of the nucleotides and the interactions

(a) RNA molecule

(b) DNA molecule

•**FIGURE 2-23** **Nucleic Acids: RNA and DNA.** Nucleic acids are long chains of nucleotides. Each molecule starts at the sugar of the first nucleotide and ends at the phosphate group of the last member of the chain. **(a)** An RNA molecule has a single nucleotide chain. Its shape is determined by the sequence of nucleotides and by the interactions among them. **(b)** A DNA molecule has a pair of nucleotide chains linked by hydrogen bonding between complementary base pairs.

among them. Our cells have three types of RNA: (1) *messenger RNA (mRNA)*, (2) *transfer RNA (tRNA)*, and (3) *ribosomal RNA (rRNA)*. These types have different shapes and functions, but all three are required for the synthesis of proteins (Chapter 3).

A DNA molecule consists of a *pair* of nucleotide chains (Figure 2-23b•). Hydrogen bonding between opposing nitrogenous bases holds the two strands together. The shapes of the nitrogenous bases allow adenine to bond only to thymine and cytosine to bond only to guanine. As a result, the combinations adenine–thymine and cytosine–guanine are known as **complementary base pairs.**

The two strands of DNA twist around one another in a double helix that resembles a spiral staircase. Each step of the staircase corresponds to one complementary base pair. Table 2-8 compares RNA and DNA.

HIGH-ENERGY COMPOUNDS

To perform their vital functions, cells must use energy, obtained by the catabolism of organic substrates. To be useful, that energy must be captured by enzymes and transferred from molecule to molecule or from one part of the cell to another. The usual method of energy transfer involves the creation of *high-energy bonds.* A high-energy bond is a covalent bond whose breakdown releases energy the cell can harness. In your cells, a high-energy bond generally connects a phosphate group (PO_4^{3-}) to an organic molecule. The resulting complex is called a **high-energy compound.**

Phosphorylation (fos-for-i-LĀ-shun) is the attachment of a phosphate group to another molecule. This process does not necessarily produce high-energy bonds. The creation of a high-energy compound requires (1) a phosphate group, (2) appropriate enzymes, and (3) suitable organic substrates. The most important substrate is *adenosine,* a combination of adenine and ribose, with two phosphate groups attached. This compound is **adenosine diphosphate (ADP).** It is created by the phosphorylation of the nucleotide *adenosine monophosphate (AMP),* a building block of nucleic acids. A significant energy input is required to convert AMP to ADP, and the second phosphate is attached by a high-energy bond. Even more energy is required to add a third phosphate and thereby create the high-energy compound **adenosine triphosphate (ATP).** Figure 2-24• shows the structure of ATP.

The conversion of ADP to ATP and the reversion of ATP to ADP are the most common methods of energy transfer in our cells:

ADP + phosphate group + energy $\rightleftharpoons$ ATP + H_2O

The conversion of ATP to ADP requires an enzyme known as **adenosine triphosphatase,** or **ATPase.** Throughout life, our cells continuously generate ATP from ADP and use the energy provided by that ATP

TABLE 2-8 A Comparison of RNA and DNA

Characteristic	RNA	DNA
Sugar	Ribose	Deoxyribose
Nitrogenous bases	Adenine Guanine Cytosine Uracil	Adenine Guanine Cytosine Thymine
Number of nucleotides in typical molecule	Varies from fewer than 100 to about 50,000	Always more than 45 million
Shape of molecule	Varies with hydrogen bonding along the length of the strand; 3 main types (mRNA, rRNA, tRNA)	Paired strands coiled in a double helix
Function	Performs protein synthesis as directed by DNA	Stores genetic information that controls protein synthesis

•**FIGURE 2-24 The Structure of ATP.** A molecule of ATP consists of adenosine (adenine plus ribose) to which three phosphate groups have been joined. Cells most often transfer energy by attaching a third phosphate group to ADP with a high-energy bond and then removing that phosphate group at another site, where the associated release of energy performs cellular work.

TABLE 2-9 Structure and Function of Biologically Important Compounds

Class	Building Blocks	Sources	Functions
Inorganic			
Water (pp. 40–42)	Hydrogen and oxygen atoms	Absorbed as liquid water or generated by metabolism	Solvent; transport medium for dissolved materials and heat; cooling through evaporation; medium for chemical reactions; reactant in hydrolysis
Acids, bases, salts (pp. 42–43)	H^+, OH^-, various anions and cations	Obtained from the diet or generated by metabolism	Structural components; buffers; sources of ions
Dissolved gases (p. 40)	O, C, N, and other atoms	Atmosphere	O_2: required for cellular metabolism CO_2: generated by cells as a waste product NO: chemical messenger involved in cardiovascular, nervous, and lymphatic systems
Organic			
Carbohydrates (pp. 44–46)	C, H, O, in some cases N; CHO in a 1:2:1 ratio	Obtained from the diet or manufactured in the body	Energy source; some structural role when attached to lipids or proteins; energy storage
Lipids (pp. 46–51)	C, H, O, in some cases N or P; CHO not in 1:2:1 ratio	Obtained from the diet or manufactured in the body	Energy source; energy storage; insulation; structural components; chemical messengers; protection
Proteins (pp. 51–56)	C, H, O, N, commonly S	20 common amino acids; roughly half can be manufactured in the body, others must be obtained from the diet	Catalysts for metabolic reactions; structural components; movement; transport; buffers; defense; control and coordination of activities
Nucleic acids (pp. 56–57)	C, H, O, N, and P; nucleotides composed of phosphates, sugars, and nitrogenous bases	Obtained from the diet or manufactured in the body	Storage and processing of genetic information
High-energy compounds (pp. 57–59)	Nucleotides joined to phosphates by high-energy bonds	Synthesized by all cells	Storage or transfer of energy

to perform vital functions, such as the synthesis of proteins or the contraction of muscles.

Although ATP is the most abundant high-energy compound, there are others—typically other nucleotides that have undergone phosphorylation. For example, *guanosine triphosphate* (*GTP*) and *uridine triphosphate* (*UTP*) are nucleotide-based high-energy compounds that transfer energy in specific enzymatic reactions.

Table 2-9 summarizes information about the inorganic and organic compounds that we covered in this chapter.

CHEMICALS AND CELLS

The human body is more than a collection of chemicals. Biochemical building blocks form functional units called **cells.** ∞ p. 3 Each cell behaves like a minia-

ture organism, responding to internal and external stimuli. A phospholipid membrane separates the cell from its environment, and internal membranes create compartments with specific functions. Proteins form an internal supporting framework and, as enzymes, accelerate and control the chemical reactions that maintain homeostasis. Nucleic acids direct the synthesis of all cellular proteins, including the enzymes that enable the cell to synthesize a wide variety of other substances. Carbohydrates provide energy (transferred by high-energy compounds) to support vital activities, and they form part of specialized compounds such as proteoglycans and glycolipids.

Cells are dynamic structures that adapt to changes in their environment. That adaptation may involve changes in the chemical organization of the cell. Such changes are easily made because organic molecules other than DNA are temporary rather than permanent components of the cell. Their continuous removal and replacement are part of the process of **metabolic turnover.**

Most of the organic molecules in the cell are replaced at intervals ranging from hours to months. The average time between synthesis and recycling is known as the *turnover rate*. Table 2-10 indicates the turnover rates of the organic components of representative cells.

☑ A large organic molecule is composed of the sugar ribose, nitrogenous bases, and phosphate groups. Which nucleic acid is this?

☑ What molecule is produced by the phosphorylation of ADP?

TABLE 2-10 Turnover Rates

Cell Type	Component	Average Recycling Time*
Liver	Total protein	5–6 days
	Enzymes	1 hour to several days, depending on the enzyme
	Glycogen	1–2 days
	Cholesterol	5–7 days
Muscle cell	Total protein	30 days
	Glycogen	12–24 hours
Neuron	Phospholipids	200 days
	Cholesterol	100+ days
Fat cell	Triglycerides	15–20 days

*Most values were obtained from studies on mammals other than humans.

[EXPLORE] MediaLab

A grasp of anatomy and physiology requires the ability to deal with basic chemistry. To gain a better understanding of chemistry, use the resources available on both the CD-ROM and the Companion Website. The CD-ROM includes animations and tutorials; the website builds on those resources with text content and interactive problems as well as communication avenues for sharing ideas with your class. To complete the following Web Explorations, visit the MediaLab in Chapter 2 of your Companion Website.

Web Exploration 1

Estimated time for completion = 5 minutes

Many physiological processes function properly only under a narrow range of pH values. Serious consequences result from an acid–base imbalance in the body. After reading the subsection "pH" in the text, select the keyword **pH** on the Companion Website. Follow the directions there to adjust the concentration of hydrogen ions in solution; note what happens to the pH. This website is maintained by the Anesthesiology Department of Tulane Medical School. Explain why pH chemistry is important to anesthesiologists.

Web Exploration 2

Estimated time for completion = 10 minutes

ATP stores energy in two high-energy phosphate bonds. In the body, ATP is broken down to ADP, which is then broken down to AMP; energy is released at both steps. Select the keyword **ATP** on the Companion Website to obtain an image of the three compounds and information about their function. John Walker, Paul Boyer, and Jens Skou were awarded the Nobel Prize for Chemistry in 1997 for their discovery of the way in which ATP works. After reading the website, describe this discovery. Why was it considered so important?

CHAPTER REVIEW

SELECTED CLINICAL TERMINOLOGY
Terms Discussed in This Chapter

cholesterol: A steroid, important in the structure of cellular membranes, that in high concentrations increases the risk of heart disease. *(p. 49)*

mole (mol), **millimole** (mmol): The number of atoms or molecules that has a weight in grams equal to the atomic weight or the molecular weight, respectively; 1 mmol = 0.001 mol. *(p. 32)*

omega-3 fatty acids: Fatty acids, abundant in fish flesh and fish oils, that have a double bond three carbons away from the end of the hydrocarbon chain. Their presence in the diet has been linked to reduced risks of heart disease and other conditions. *(p. 48)*

radioisotopes: Isotopes with unstable nuclei, which spontaneously emit subatomic particles or radiation in measurable amounts. *(p. 31)*

AM **Additional Terms Discussed in the** *Applications Manual*

albinism: A genetic disorder resulting in a lack of skin pigment.

allosteric effect: A change in the shape of an enzyme's active site caused by interaction between the enzyme and another molecule.

apoenzyme: An enzyme that is inactive when a cofactor is absent.

equivalent (Eq): A mole of positive or negative charges.

familial hypercholesterolemia: A genetic disorder resulting in high cholesterol levels in blood and cholesterol buildup in body tissue, especially the walls of blood vessels.

galactosemia: A disorder resulting from an inability to metabolize galactose, a monosaccharide.

holoenzyme: An enzyme activated by an appropriate cofactor.

nuclear imaging: A procedure in which an image is created on a photographic plate or video screen by the radiation emitted by injected radioisotopes.

phenylketonuria: A metabolic disorder resulting from an inability to convert phenylalanine to tyrosine.

radiopharmaceuticals: Drugs that incorporate radioactive atoms; administered to expose specific target tissues to radiation.

tracer: A compound labeled with a radioisotope that can be tracked in the body by the radiation it releases.

STUDY OUTLINE

INTRODUCTION p. 30

ATOMS AND MOLECULES p. 30

1. **Atoms** are the smallest units of matter. They consist of **protons, neutrons,** and **electrons.** *(Figure 2-1)*

Atomic Structure p. 30

2. Within an atom, an **electron cloud** surrounds the nucleus. *(Figure 2-1; Table 2-1)*

3. The number of protons in an atom is its **atomic number.** Each **element** includes all the atoms that have the same number of protons and thus the same atomic number.

4. The **mass number** of an atom is the total number of protons and neutrons in its nucleus. **Isotopes** are atoms of the same element whose nuclei contain different numbers of neutrons.

5. Electrons occupy a series of **energy levels,** which are commonly illustrated as **electron shells.** The electrons in the outermost energy level determine an atom's chemical properties. *(Figure 2-2)*

Chemical Bonds p. 32

6. Atoms can combine through chemical reactions that create **chemical bonds. A molecule** is any chemical structure consisting of atoms held together by covalent bonds. A **compound** is a chemical structure containing atoms of two or more elements.

7. An **ionic bond** results from the attraction between **ions,** atoms that have gained or lost electrons. **Cations** are positively charged; **anions** are negatively charged. *(Figure 2-3)*

8. Atoms that share electrons and form a molecule are held together by **covalent bonds.** A sharing of one pair of electrons is a **single covalent bond;** a sharing of two pairs is a **double covalent bond.** A bond with equal sharing of electrons is a **nonpolar covalent bond;** a bond with unequal sharing of electrons is a **polar covalent bond.** *(Figures 2-4, 2-5)*

9. A **hydrogen bond** is a weak but important force that can affect the shapes and properties of molecules. *(Figure 2-6)*

10. Matter can exist as a *solid,* a *liquid,* or a *gas,* depending on the nature of the interactions among the component atoms or molecules.

CHEMICAL REACTIONS p. 36

1. **Metabolism** is all the **chemical reactions** in the body. Through metabolism, cells capture, store, and use energy to maintain homeostasis and to support essential functions.

FOCUS: Chemical Notation p. 37

2. **Chemical notation** allows us to describe reactions between **reactants** that generate one or more **products.** *(Table 2-2)*

Basic Energy Concepts p. 36

3. **Work** is movement of an object or a change in its physical structure. **Energy** is the capacity to perform work.

4. **Kinetic energy** is the energy of motion. **Potential energy** is stored energy that results from the position or structure of an object. Conversions from potential to kinetic energy are not 100 percent efficient; every such energy conversion releases *heat.*

Types of Reactions p. 38

5. Cells gain energy to power their functions by **catabolism,** the breakdown of complex molecules. Much of this energy supports **anabolism,** the synthesis of new molecules.

6. A chemical reaction is classified as a **decomposition, synthesis,** or **exchange reaction. Exergonic reactions** release energy; **endergonic reactions** require energy.

Reversible Reactions p. 38

7. At **equilibrium,** the rates of two opposing reactions are in balance.

Enzymes and Chemical Reactions p. 39

8. **Activation energy** is the amount of energy required to start a reaction. **Enzymes** control many chemical reactions in our bodies. Enzymes are **catalysts**—substances that accelerate chemical reactions without themselves being permanently changed or used up. *(Figure 2-7)*

9. The complex reactions that support life proceed in a series of interlocking steps called *pathways.* Each step in a pathway is controlled by a different enzyme.

INORGANIC COMPOUNDS p. 39

1. **Nutrients** and **metabolites** are broadly classified as **inorganic** or **organic.**

Water and Its Properties p. 40

2. Water is the most important component of the body.

3. A **solution** is a uniform mixture of two or more substances. It consists of a medium, or **solvent,** in which atoms, ions, or molecules of another substance, or **solute,** are dispersed. In *aqueous solutions,* water is the solvent.

4. Many inorganic compounds, called **electrolytes,** undergo *ionization,* or *dissociation,* in water to form ions. *(Figure 2-8; Table 2-3)*

5. The **pH** of a solution indicates the concentration of hydrogen ions it contains. Solutions are classified as **neutral, acidic,** or **basic** *(alkaline)* on the basis of pH. *(Figure 2-9)*

Inorganic Acids and Bases p. 42

6. An **acid** releases hydrogen ions; a **base** removes hydrogen ions from a solution. *Strong acids* and *strong bases* ionize completely, whereas *weak acids* and *weak bases* do not.

Salts p. 43

7. A **salt** is an electrolyte whose cation is not hydrogen (H^+) and whose anion is not hydroxide (OH^-).

Buffers and pH Control p. 43

8. **Buffers** remove or replace hydrogen ions in solution. Buffers and *buffer systems* maintain the pH of body fluids within normal limits.

ORGANIC COMPOUNDS p. 44

1. Organic compounds contain carbon and hydrogen and generally oxygen as well. Major classes of organic compounds include *carbohydrates, lipids, proteins,* and *nucleic acids. High-energy compounds* are small in terms of abundance but absolutely vital to the survival of our cells. *(Table 2-4)*

Carbohydrates p. 44

2. **Carbohydrates** are most important as an energy source for metabolic processes. The three major types of carbohydrates are **monosaccharides** (*simple sugars*), **disaccharides,** and **polysaccharides.** Disaccharides and polysaccharides form from monosaccharides by **dehydration synthesis.** *(Figures 2-10–2-12; Table 2-5)*

Lipids p. 46

3. **Lipids** are water-insoluble molecules that include *fats, oils,* and *waxes.* The five important classes of lipids are **fatty acids, eicosanoids, glycerides, steroids,** and **phospholipids** and **glycolipids.** *(Figures 2-13–2-17; Table 2-6)*
4. **Triglycerides** (*neutral fats*) consist of three fatty acid molecules attached by dehydration synthesis to a molecule of **glycerol. Diglycerides** consist of two fatty acids and glycerol. **Monoglycerides** consist of one fatty acid plus glycerol.
5. **Steroids** (1) are involved in the structure of cell membranes, (2) include sex hormones and hormones regulating metabolic activities, and (3) are important in lipid digestion.

Proteins p. 51

6. **Proteins** perform a great variety of functions in the body. Important types of proteins include *structural proteins, contractile proteins, transport proteins, enzymes,* and *antibodies. (Table 2-7)*
7. Proteins are chains of **amino acids.** Each amino acid consists of an *amino group,* a *carboxylic acid group,* and an *R group* (*side chain*). A protein, or **polypeptide,** is a linear sequence of amino acids held together by **peptide bonds.** *(Figures 2-18, 2-19)*
8. The four levels of protein structure are **primary structure** (amino acid sequence), **secondary structure** (amino acid interactions, such as hydrogen bonds), **tertiary structure** (complex folding, *disulfide bonds,* and interaction with water molecules), and **quaternary structure** (formation of protein complexes from individual subunits). **Globular proteins,** such as *myoglobin,* are generally rounded and water-soluble. **Fibrous proteins,** such as *keratin* and *collagen,* are tough, durable, and generally insoluble. *(Figure 2-20)*

9. The reactants in an enzymatic reaction, called **substrates,** interact to yield a product by binding to the enzyme's **active site. Cofactors** are ions or molecules that must bind to the enzyme before substrate binding can occur. **Coenzymes** are organic cofactors commonly derived from *vitamins. (Figure 2-21)*
10. The shape of a protein determines its functional characteristics. Each protein works best at an optimal combination of temperature and pH and will undergo temporary or permanent **denaturation** at temperatures or pH values outside the normal range.

Nucleic Acids p. 56

11. **Nucleic acids** store and process information at the molecular level. The two kinds of nucleic acids are **deoxyribonucleic acid (DNA)** and **ribonucleic acid (RNA).** *(Figures 2-22, 2-23; Table 2-8)*
12. Nucleic acids are chains of **nucleotides.** Each nucleotide contains a sugar, a phosphate group, and a **nitrogenous base.** The sugar is *ribose* in RNA and *deoxyribose* in DNA. DNA, which is a two-stranded double helix, contains the nitrogenous bases **adenine, guanine, cytosine,** and **thymine.** RNA, which consists of a single strand, contains **uracil** instead of thymine.

High-Energy Compounds p. 57

13. Cells store energy in the *high-energy bonds* of **high-energy compounds** for later use. The most important high-energy compound is **ATP (adenosine triphosphate).** Cells make ATP by adding a phosphate group to **ADP (adenosine diphosphate)** through **phosphorylation.** When ATP is broken down to ADP and phosphate, energy is released. The cell can use this energy to power essential activities. *(Figure 2-24; Table 2-9)*

CHEMICALS AND CELLS p. 59

1. Biochemical building blocks form functional units called **cells.**
2. The continuous removal and replacement of cellular organic molecules (other than DNA), a process called **metabolic turnover,** allows cells to change and adapt to changes in their environment. *(Table 2-10)*

REVIEW QUESTIONS

LEVEL 1 REVIEWING FACTS AND TERMS

1. In atoms, protons and neutrons are
   (a) only in the nucleus
   (b) outside the nucleus
   (c) inside and outside the nucleus
   (d) in the electron shells
2. Isotopes of an element differ from each other in the number of
   (a) protons in the nucleus
   (b) neutrons in the nucleus
   (c) electrons in the outer shells
   (d) protons and electrons in the nucleus

3. The number and arrangement of electrons in an atom's outer energy level determines the atom's
   (a) atomic weight   (b) atomic number
   (c) electrical properties (d) chemical properties
4. The bond between sodium and chlorine in the compound sodium chloride (NaCl) is
   (a) an ionic bond
   (b) a single covalent bond
   (c) a nonpolar covalent bond
   (d) a double covalent bond

Match each numbered item with the most closely related lettered item. Use letters for answers in the spaces provided.

___ **5.** atomic number   **a.** sum of atomic weights
___ **6.** mass number   **b.** water-insoluble
___ **7.** covalent bond   **c.** synthesis
___ **8.** ionic bond   **d.** catalyst
___ **9.** molecular weight   **e.** sharing of electrons
___ **10.** catabolism   **f.** $A + B \rightleftharpoons AB$
___ **11.** anabolism   **g.** stabilize pH
___ **12.** exchange reaction   **h.** number of protons
___ **13.** reversible reaction   **i.** decomposition
___ **14.** hydrophobic   **j.** carbohydrates, lipids, proteins
___ **15.** acid   **k.** loss or gain of electrons
___ **16.** enzyme   **l.** water, salts
___ **17.** buffer   **m.** H^+ donor
___ **18.** organic compounds   **n.** number of protons plus number of neutrons
___ **19.** inorganic compounds   **o.** $AB + CD \longrightarrow AD + CB$

20. The oxygen atoms in a molecule of oxygen are held together by
    (a) a single covalent bond   **(b)** a double covalent bond
    (c) a polar covalent bond   **(d)** an ionic bond

21. All the chemical reactions that occur in the human body are collectively referred to as
    (a) anabolism   **(b)** catabolism
    (c) metabolism   **(d)** homeostasis

22. Which of the following equations illustrates a typical decomposition reaction?
    (a) $A + B \longrightarrow AB$ **(b)** $AB + CD \longrightarrow AD + CB$
    (c) $2 A_2 + B_2 \longrightarrow 2 A_2 B$ **(d)** $AB \longrightarrow A + B$

23. Of the following choices, the pH of the least acidic solution is
    (a) 6.0   **(b)** 4.5
    (c) 2.3   **(d)** 1.0

24. A pH of 7.8 in the human body typifies a condition referred to as
    (a) acidosis   **(b)** alkalosis
    (c) metabolic acidosis   **(d)** homeostasis

25. A(n) _____ is a solute that dissociates to release hydrogen ions, and a(n) _____ is a solute that removes hydrogen ions from solution.
    (a) base, acid   **(b)** salt, base
    (c) acid, salt   **(d)** acid, base

26. Carbohydrates, lipids, and proteins are formed from their basic building blocks by the
    (a) removal of a water molecule between the building blocks
    (b) addition of a water molecule between the building blocks
    (c) addition of carbon to each molecule
    (d) addition of oxygen to each molecule

27. Complementary base pairing in DNA includes the pairs
    (a) adenine–uracil; cytosine–guanine
    (b) adenine–thymine; cytosine–guanine
    (c) adenine–guanine; cytosine–thymine
    (d) guanine–uracil; cytosine–thymine

28. What are the three stable fundamental particles in atoms?
29. What is the difference between potential energy and kinetic energy?
30. What is the role of enzymes in chemical reactions?

31. List six inorganic molecules that are abundant in the human body.
32. What four major classes of organic compounds are in the body?
33. List three important functions of triglycerides (neutral fats) in the body.
34. List seven major functions performed by proteins.
35. **(a)** What three basic components make up a nucleotide of DNA?
    (b) What three basic components make up a nucleotide of RNA?
36. What three components are required to create the high-energy compound ATP?

LEVEL 2 REVIEWING CONCEPTS

37. If an isotope of oxygen has 8 protons, 10 neutrons, and 8 electrons, its mass number is
    (a) 26   **(b)** 16   **(c)** 18   **(d)** 8

38. Hydrophilic molecules readily associate with
    (a) lipid molecules
    (b) insoluble droplets of hydrophobic molecules
    (c) water molecules
    (d) a, b, and c are all correct

39. The speed, or rate, of a chemical reaction is influenced by
    (a) the presence of catalysts
    (b) the temperature
    (c) the concentration of the reactants
    (d) a, b, and c are all correct

40. Using the periodic table of the elements (Appendix III), determine the following information about calcium:
    (a) number of protons
    (b) number of electrons
    (c) number of neutrons
    (d) atomic number
    (e) atomic weight
    (f) number of electrons in each shell

41. Explain the differences among nonpolar covalent bonds, polar covalent bonds, and ionic bonds.
42. How does thermal inertia help stabilize body temperature?
43. Why does pure water have a neutral pH?
44. What role do buffer systems play in the human body?
45. A sample that contains an organic molecule has the following constituents: carbon, hydrogen, oxygen, nitrogen, and phosphorus. Is the molecule a carbohydrate, a lipid, a protein, or a nucleic acid?

LEVEL 3 CRITICAL THINKING AND CLINICAL APPLICATIONS

46. The element sulfur has an atomic number of 16 and an atomic weight of 32. How many neutrons are in the nucleus of a sulfur atom? If sulfur forms covalent bonds with hydrogen, how many hydrogen atoms can bond to one sulfur atom?
47. An important buffer system in the human body involves carbon dioxide (CO_2) and bicarbonate ion (HCO_3^-) as shown:

$$CO_2 + H_2O \rightleftharpoons H_2CO_3 \rightleftharpoons H^+ + HCO_3^-$$

If a person becomes excited and exhales large amounts of CO_2, how will the pH of the person's body be affected?
48. A student needs to prepare 1 liter of a solution of NaCl that contains 1.2 moles of ions per liter of solution. How many moles of NaCl should the student use to prepare the solution? What will that quantity weigh, in grams?

CHAPTER 3

The Cellular Level of Organization

Many familiar structures, from pyramids to patchwork quilts, are made up of numerous small, similar components. The human body is built on the same principle; it is made up of several trillion tiny units called cells. Cells, however, differ from many other building blocks. For instance, they vary widely in size and appearance. Here, small blood cells lie on a layer of large, flattened cells that line a blood vessel. An individual brick can perform few (if any) of the functions of a building. But a cell can perform many of the functions of the body of which it is a part. Indeed, cells are the smallest entities that can perform all the basic life functions discussed in Chapter 1. In addition, each cell in the body performs specialized functions and helps maintain homeostasis. What internal structures do cells contain? How are those structures arranged, and how does that arrangement enable them to perform their specific functions? We shall see in this chapter.

CHAPTER OUTLINE AND OBJECTIVES

INTRODUCTION   p. 65

STUDYING CELLS   p. 65
An Overview of Cellular Anatomy   p. 66

THE CELL MEMBRANE   p. 68
Membrane Structure   p. 68
Membrane Permeability   p. 71
① List the functions of the cell membrane and the structural features that enable it to perform those functions.
② Specify the routes by which different ions and molecules can enter or leave a cell and the factors that may restrict such movement.
③ Describe the various transport mechanisms that cells use to facilitate the absorption or removal of specific substances.
④ Explain the origin and significance of the transmembrane potential.

THE CYTOPLASM   p. 80
The Cytosol   p. 80
Organelles   p. 80
⑤ Describe the organelles of a typical cell, and indicate the specific functions of each.

THE NUCLEUS   p. 89
Chromosome Structure   p. 90
The Genetic Code   p. 91
Gene Activation and Protein Synthesis   p. 91
DNA and the Control of Cell Structure and Function   p. 95
⑥ Explain the functions of the cell nucleus.
⑦ Discuss the nature and importance of the genetic code.
⑧ Summarize the process of protein synthesis.

THE CELL LIFE CYCLE   p. 96
Interphase   p. 96
Mitosis   p. 98
Cytokinesis   p. 99
The Mitotic Rate and Energetics   p. 99
Regulation of the Cell Life Cycle   p. 100
⑨ Describe the stages of the cell life cycle.
⑩ Describe the process of mitosis, and explain its significance.

CELL DIVERSITY AND DIFFERENTIATION   p. 101
⑪ Define differentiation, and explain its importance.

MEDIALAB and CHAPTER REVIEW   p. 102

A toms are the building blocks of molecules; cells are the building blocks of the human body. Cells were first described by the English scientist Robert Hooke around 1665. Hooke used an early light microscope to examine a dried piece of cork, a material from the interior of woody plants. He observed in the cork thousands of tiny, empty chambers, which he named *cells.* Later that decade, other scientists, observing the structure of living plants, realized that in life these chambers were filled with a gelatinous material. Research over the next 175 years led to the *cell theory,* the concept that **cells** are the fundamental units of all plant and animal tissues. Since that time, the cell theory has been expanded to incorporate five basic concepts relevant to our discussion of the human body:

1. Cells are the building blocks of all plants and animals.
2. Cells are produced by the division of preexisting cells.
3. Cells are the smallest units that perform all vital physiological functions.
4. Each cell maintains homeostasis at the cellular level.
5. Homeostasis at the tissue, organ, organ system, and organism levels reflects the combined and coordinated actions of many cells.

Cells have a variety of forms and functions. Figure 3-1● gives examples of the range of sizes and shapes of cells in the human body. The relative proportions of the cells in this figure are correct, but all have been magnified roughly 500 times. Together, these and other types of cells create and maintain all anatomical structures and perform all vital physiological functions.

The human body contains trillions of cells, and all our activities—from running to thinking—result from the combined and coordinated responses of millions or even billions of cells. Yet each cell also functions as an individual entity, responding to a variety of environmental cues. As a result, anyone interested in understanding how the human body functions must first become familiar with basic concepts of cell biology.

AM *A Closer Look: The Nature of Pathogens*

STUDYING CELLS

Cytology (sī-TOL-o-jē; *cyto-,* cell + *-ology,* the study of) is the study of the structure and function of cells. What we have learned since the 1950s has given us new insights into cellular physiology and the mechanisms of homeostatic control. Today, cytology is part of the broader discipline **cell biology,** which incorporates aspects of biology, chemistry, and physics. The two most common methods used to study cell and tissue structure are *light microscopy* and *electron microscopy.*

AM *Methods of Microanatomy*

In this chapter, we describe the structure of a typical animal cell, some of the ways in which cells interact with their environment, and how cells reproduce. Later chapters deal with the coordination of cellular activities in physiological systems.

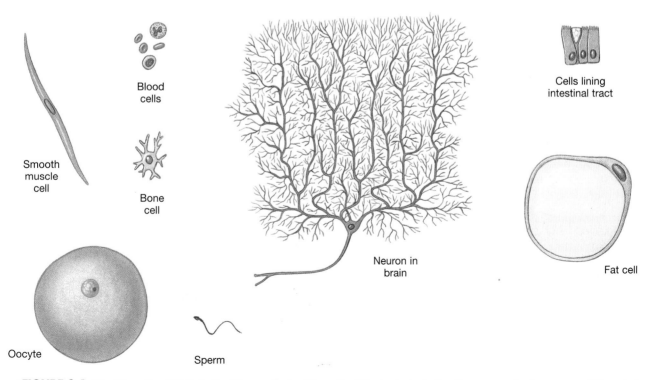

Smooth muscle cell

Blood cells

Bone cell

Neuron in brain

Cells lining intestinal tract

Fat cell

Oocyte

Sperm

●**FIGURE 3-1 The Diversity of Cells in the Human Body.** The cells of the body, including these examples, have many different shapes and a variety of special functions. All the cells are shown with the dimensions they would have if magnified approximately 500 times.

3

AN OVERVIEW OF CELLULAR ANATOMY

The human body contains two classes of cells: sex cells and somatic cells. **Sex cells** (also called *germ cells* or *reproductive cells*) are either (1) the *sperm* of males or (2) the *oocytes* of females. **Somatic cells** (*soma*, body) include all the other cells in the human body. In this chapter, we will examine somatic cells; we shall discuss sex cells in chapters dealing with the reproductive system.

The "typical" somatic cell is like the "average" person. Any description masks enormous individual variations. Our model cell will share features with most cells of the body without being identical to any one. Figure 3-2● shows such a cell, and Table 3-1 summarizes the structures and functions of its parts.

Our cell is surrounded by a watery medium known as the **extracellular fluid.** The extracellular fluid in most tissues is called **interstitial** (in-ter-STISH-ul) **fluid** (*interstitium,* something standing between). A *cell membrane* separates the cell contents, or *cytoplasm,* from the extracellular fluid. The cytoplasm surrounds the membranous *nucleus.* The cytoplasm can itself be subdivided into (1) the *cytosol,* a liquid, and (2) intracellular structures collectively known as *organelles* (or-gan-ELZ; "little organs").

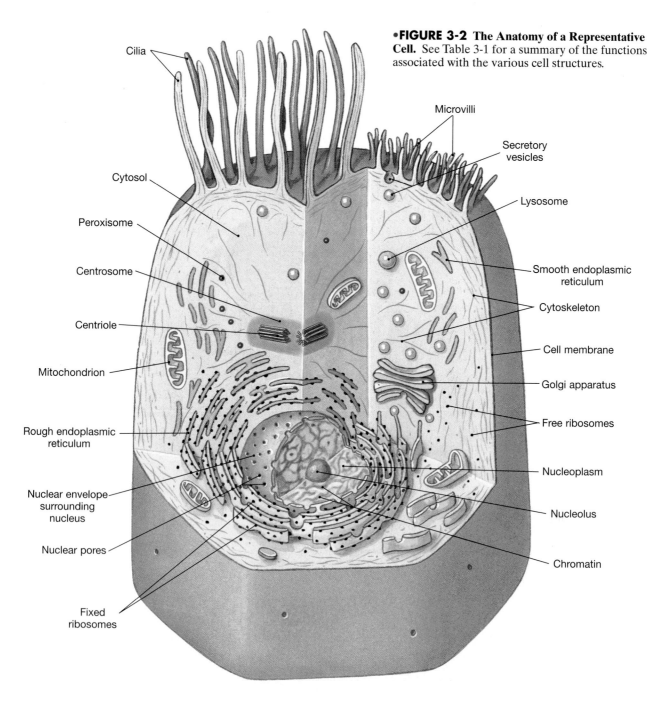

●FIGURE 3-2 The Anatomy of a Representative Cell. See Table 3-1 for a summary of the functions associated with the various cell structures.

Cilia

Microvilli

Secretory vesicles

Cytosol

Lysosome

Peroxisome

Smooth endoplasmic reticulum

Centrosome

Cytoskeleton

Centriole

Cell membrane

Mitochondrion

Golgi apparatus

Rough endoplasmic reticulum

Free ribosomes

Nucleoplasm

Nuclear envelope surrounding nucleus

Nucleolus

Nuclear pores

Chromatin

Fixed ribosomes

TABLE 3-1 The Organelles of a Representative Cell

Appearance	Structure	Composition	Function(s)
	CELL MEMBRANE	Lipid bilayer, containing phospholipids, steroids, proteins, and carbohydrates	Isolation; protection; sensitivity; support; control of entrance/exit of materials
	CYTOSOL	Fluid component of cytoplasm	Distributes materials by diffusion
	NONMEMBRANOUS ORGANELLES		
	Cytoskeleton Microtubule Microfilament	Proteins organized in fine filaments or slender tubes	Strength and support; movement of cellular structures and materials
	Microvilli	Membrane extensions containing microfilaments	Increase surface area to facilitate absorption of extracellular materials
	Centrosome Centriole	Cytoplasm containing two centrioles, at right angles; each centriole is composed of 9 microtubule triplets	Essential for movement of chromosomes during cell division; organization of microtubules in cytoskeleton
	Cilia	Membrane extensions containing microtubule doublets in a 9 + 2 array	Movement of materials over cell surface
	Ribosomes	RNA + proteins; fixed ribosomes bound to rough endoplasmic reticulum, free ribosomes scattered in cytoplasm	Protein synthesis
	MEMBRANOUS ORGANELLES		
	Endoplasmic reticulum (ER)	Network of membranous channels extending throughout the cytoplasm	Synthesis of secretory products; intracellular storage and transport
	Rough ER (RER)	Has ribosomes bound to membranes	Modification and packaging of newly synthesized proteins
	Smooth ER (SER)	Lacks attached ribosomes	Lipid and carbohydrate synthesis
	Golgi apparatus	Stacks of flattened membranes (cisternae) containing chambers	Storage, alteration, and packaging of secretory products and lysosomal enzymes
	Lysosomes	Vesicles containing digestive enzymes	Intracellular removal of damaged organelles or of pathogens
	Peroxisomes	Vesicles containing degradative enzymes	Catabolism of fats and other organic compounds; neutralization of toxic compounds generated in the process
	Mitochondria	Double membrane, with inner membrane folds (cristae) enclosing important metabolic enzymes	Produce 95% of the ATP required by the cell
Nuclear pore / Nuclear envelope	**NUCLEUS**	Nucleoplasm containing nucleotides, enzymes, nucleoproteins, and chromatin; surrounded by double membrane (nuclear envelope)	Control of metabolism; storage and processing of genetic information; control of protein synthesis
	Nucleolus	Dense region in nucleoplasm containing DNA and RNA	Site of rRNA synthesis and assembly of ribosomal subunits

THE CELL MEMBRANE

The **cell membrane,** also called the **plasma membrane** or *plasmalemma* (*lemma,* husk), forms the outer boundary of the cell. Its general functions include:

■ *Physical isolation.* The cell membrane is a physical barrier that separates the inside of the cell from the surrounding extracellular fluid. Conditions inside and outside the cell are very different, and those differences must be maintained to preserve homeostasis. For example, the cell membrane keeps enzymes and structural proteins inside the cell.

■ *Regulation of exchange with the environment.* The cell membrane controls the entry of ions and nutrients, such as glucose; the elimination of wastes; and the release of secretions.

■ *Sensitivity.* The cell membrane is the first part of the cell affected by changes in the composition, concentration, or pH of the extracellular fluid. It also contains a variety of receptors that allow the cell to recognize and respond to specific molecules in its environment. For instance, the cell membrane may bind chemical signals from other cells. The binding of just one molecule may trigger the activation or deactivation of enzymes that affect many cellular activities.

■ *Structural support.* Specialized connections between cell membranes or between membranes and extracellular materials give tissues a stable structure. For example, the cells at the surface of the skin are bound together, and those in the deepest layers are attached to extracellular protein fibers in underlying tissues.

MEMBRANE STRUCTURE

The cell membrane is extremely thin and delicate, ranging from 6 to 10 nm in thickness. This membrane contains lipids, proteins, and carbohydrates.

Membrane Lipids

Figure 3-3• shows cell membrane structure. Lipids form most of the surface but account for only about 42 percent of its weight. Proteins, which are relatively dense, account for more than half the weight of the membrane. The cell membrane is called a **phospholipid bilayer,** because the phospholipids in it form two layers. In each layer, the phospholipids lie with their hydrophilic heads at the membrane surface and their hydrophobic tails on the inside, in association with cholesterol and small quantities of other lipids.

Note the similarities in lipid organization between the cell membrane and a micelle (Figure 2-17c•, p. 50). Ions and water-soluble compounds cannot enter the interior of a micelle, because the lipid tails of the phospholipid molecules are hydrophobic and will not associate with water molecules. For the same reason, water and solutes

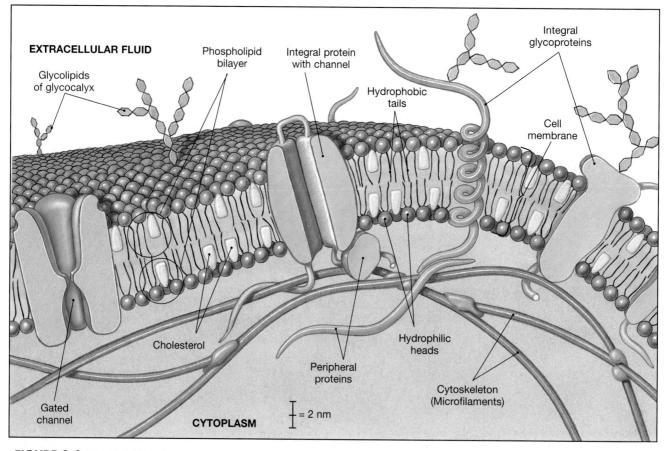

•**FIGURE 3-3** **The Cell Membrane**

cannot cross the lipid portion of a cell membrane. Thus, the membrane isolates the cytoplasm from the surrounding fluid environment. Such isolation is important because the composition of cytoplasm is very different from that of extracellular fluid, and the cell cannot survive if those differences are eliminated.

Membrane Proteins

Proteins account for roughly 55 percent of the weight of a cell membrane. Several types of proteins are in-

volved. **Integral proteins** are part of the membrane structure. Most integral proteins span the width of the membrane one or more times and are therefore known as *transmembrane proteins.* **Peripheral proteins** are bound to the inner or outer surface of the membrane and are easily separated from it. Integral proteins greatly outnumber peripheral proteins. Membrane proteins (Figure 3-4•) may function specifically as anchoring proteins, recognition proteins, enzymes, receptors, carriers, or channels:

•**FIGURE 3-4** Membrane Proteins

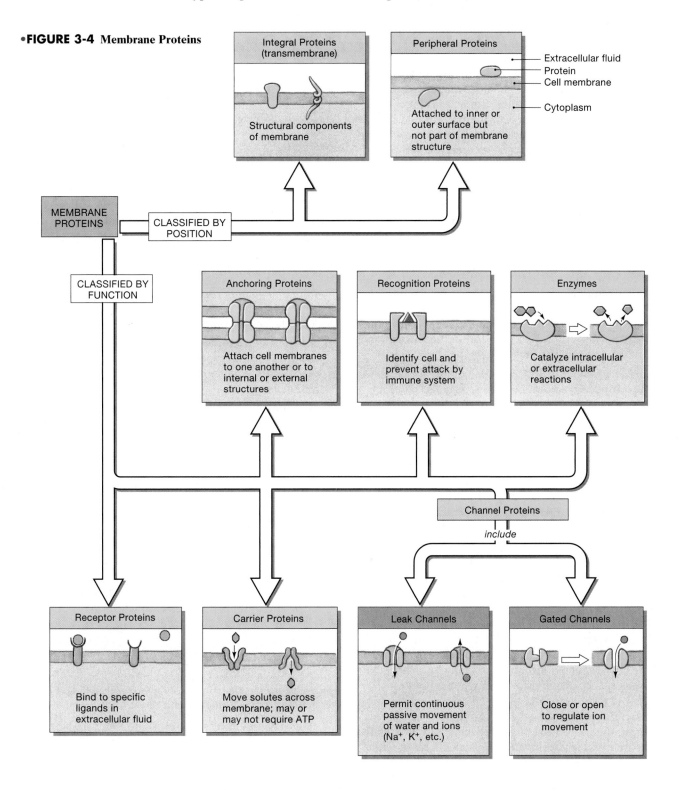

1. *Anchoring proteins.* Membrane proteins called **anchoring proteins** attach the cell membrane to other structures and stabilize its position. Inside the cell, membrane proteins are bound to the *cytoskeleton,* a network of supporting filaments in the cytoplasm. Outside the cell, other membrane proteins may attach the cell to extracellular protein fibers or to another cell.

2. *Recognition proteins (identifiers).* The cells of the immune system recognize other cells as normal or abnormal on the basis of the presence or absence of characteristic **recognition proteins.** Many important recognition proteins are glycoproteins. (We shall discuss one group, the *MHC proteins* involved in the immune response, in Chapter 22.)

3. *Enzymes.* Enzymes in cell membranes may be integral or peripheral proteins. These enzymes catalyze reactions in the extracellular fluid or in the cytosol, depending on the location of the protein and its active site. For example, dipeptides are broken down into amino acids by enzymes on the exposed membranes of cells that line the intestinal tract.

4. *Receptor proteins.* **Receptor proteins** in the cell membrane are sensitive to the presence of specific extracellular molecules called **ligands** (LĪ-gandz). A receptor protein exposed to an appropriate ligand will bind to it, and that binding may trigger changes in the activity of the cell. For example, the binding of the hormone *insulin* to a specific membrane receptor is the key step that leads to an increase in the rate of glucose absorption by the cell. Cell membranes differ in the type and number of receptor proteins they contain; these differences account for their differing sensitivities to hormones and other solutes.

5. *Carrier proteins.* **Carrier proteins** bind solutes and transport them across the cell membrane. The transport process involves a change in the shape of the carrier protein; the shape changes when solute binding occurs, and the protein returns to its original shape when the solute is released. Carrier proteins may require ATP as an energy source. For example, virtually all cells have carrier proteins that can bring glucose into the cytoplasm without expending ATP, but these cells must expend ATP to transport ions such as sodium and calcium across the cell membrane and out of the cytoplasm.

6. *Channels.* Some integral proteins contain a central pore, or **channel,** that forms a passageway. The channel permits the movement of water and small solutes across the cell membrane. Ions do not dissolve in lipids, so they cannot cross the phospholipid bilayer. Thus, ions and other small water-soluble materials can cross the membrane only by passing through channels. The movement of ions through channels is involved in a variety of physiological mechanisms. Physiologists speak of sodium channels, calcium channels, potassium channels, and so forth when referring to channels that permit the movement of only specific ions. There are two major kinds of channels: (1) **leak channels,** which permit water and ion movement at all times (although the rate can vary), and (2) **gated channels,** which open or close to regulate ion passage. Channels account for about 0.2 percent of the total membrane surface area.

Membrane structure is not rigid. The embedded proteins drift across the surface of the membrane like ice cubes in a bowl of punch. In addition, the composition of the cell membrane can change over time as components of the membrane are removed and recycled in the process of metabolic turnover. ∞ *p. 60*

The inner and outer surfaces of the cell membrane differ in their protein and lipid compositions. For example, some cytoplasmic enzymes are found only on the inner surface of the membrane, and some receptors are found exclusively on its outer surface.

Membrane Carbohydrates

Carbohydrates account for roughly 3 percent of the weight of a cell membrane. The carbohydrates in the cell membrane are components of complex molecules such as *proteoglycans, glycoproteins,* and *glycolipids.* ∞ *pp. 50, 56* The carbohydrate portions of these large molecules extend beyond the outer surface of the membrane, forming a layer known as the **glycocalyx** (glī-kō-KĀ-liks; *calyx,* cup). The glycocalyx has a variety of important functions, including the following:

■ *Lubrication and protection.* The glycoproteins and glycolipids form a viscous layer that lubricates and protects the cell membrane.

■ *Anchoring and locomotion.* Because the components are sticky, the glycocalyx can help anchor the cell in place. It also participates in the locomotion of specialized cells.

■ *Specificity in binding.* Glycoproteins and glycolipids can function as receptors, binding specific extracellular compounds. Such binding can alter the properties of the cell surface and indirectly affect the cell's behavior.

■ *Recognition.* Glycoproteins and glycolipids are recognized as normal or abnormal by cells involved with the immune response. The characteristics of the glycocalyx are genetically determined. For example, your blood type (A, B, AB, or O) is determined by the presence or absence of membrane glycolipids on circulating red blood cells. The body's immune system can recognize normal membrane glycoproteins and glycolipids as "self" rather than as "foreign." This recognition system keeps your immune system from attacking your blood cells but enables it to recognize and destroy foreign blood cells, should they appear in the bloodstream.

☑ Which component of the cell membrane is primarily responsible for the membrane's ability to form a physical barrier between the cell's internal and external environments?

☑ Which type of integral protein allows water and small ions to pass through the cell membrane?

MEMBRANE PERMEABILITY

The **permeability** of the cell membrane is the property that determines precisely which substances can enter or leave the cytoplasm. A membrane through which nothing can pass is described as **impermeable.** A membrane through which any substance can pass without difficulty is **freely permeable.** The permeability of cell membranes lies somewhere between those extremes, so cell membranes are said to be **selectively permeable.** A selectively permeable membrane permits the free passage of some materials and restricts the passage of others. The distinction may be based on size, electrical charge, molecular shape, lipid solubility, or a combination of factors. Cells differ in their permeabilities according to variations in the organization and identity of membrane lipids and proteins.

Passage across the membrane is either passive or active. *Passive processes* move ions or molecules across the cell membrane with no expenditure of energy by the cell. *Active processes* require that the cell expend energy, generally in the form of ATP.

Transport processes are also categorized by the mechanism involved. The four major categories are:

1. *Diffusion,* which results from the random motion and collisions of ions and molecules. Diffusion is a passive process.

2. *Filtration,* which occurs when hydrostatic pressure forces fluids and solutes across a membrane barrier. Filtration is also a passive process.

3. *Carrier-mediated transport,* which requires the presence of specialized integral membrane proteins. Car-rier-mediated transport can be passive or active, depending on the substance transported and the nature of the transport mechanism.

4. *Vesicular transport,* which involves the movement of materials within small membranous sacs, or *vesicles.* Vesicular transport is an active process.

Diffusion

Ions and molecules are constantly in motion, colliding and bouncing off one another and off obstacles in their paths. The movement is random: A molecule can bounce in any direction. One result of this continuous random motion is that, over time, the molecules in any given space will tend to become evenly distributed. This distribution process is **diffusion.** As the molecules move around, there will be a net movement of material from areas of higher concentration to areas of lower concentration. The difference between the high and low concentrations is a **concentration gradient.** Diffusion tends to eliminate that gradient.

After the gradient has been eliminated, the molecular motion continues, but net movement no longer occurs in any particular direction. For convenience, we restrict use of the term *diffusion* to the directional movement that eliminates concentration gradients—a process sometimes called *net diffusion.* Because diffusion tends to spread materials from a region of higher concentration to one of lower concentration, it is often described as proceeding "down a concentration gradient" or "downhill."

All of us have experienced the effects of diffusion, which occurs in air as well as in water. The scent of fresh flowers in a vase sweetens the air in a large room; a drop of ink spreads to color an entire glass of water. Each case begins with a very high concentration of molecules in a localized area. Consider a colored sugar cube dropped in water (Figure 3-5•). Placing the cube in a large volume of clear water establishes a steep

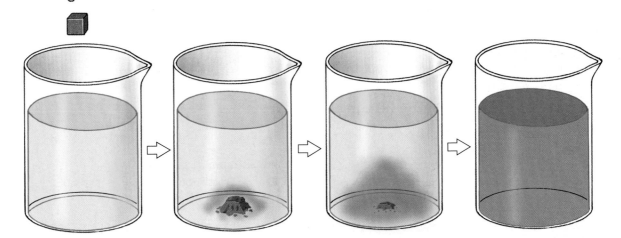

•FIGURE 3-5 Diffusion. Placing a colored sugar cube in a glass of water establishes a steep concentration gradient. As the cube dissolves, many sugar molecules are in one location and none are elsewhere. As diffusion occurs, the molecules spread through the solution. Eventually, diffusion eliminates the concentration gradient. The sugar cube has dissolved completely, and the molecules are distributed evenly. Molecular motion continues, but there is no directional movement.

concentration gradient for the ingredients as they dissolve: The sugar and dye concentration is high near the cube and negligible elsewhere. As diffusion proceeds, the sugar and dye molecules spread through the solution until they are distributed evenly.

Diffusion is important in body fluids because it tends to eliminate local concentration gradients. For example, every cell in the body generates carbon dioxide, and the intracellular concentration is relatively high. Carbon dioxide concentrations are lower in the surrounding interstitial fluid and lower still in the circulating blood. Because cell membranes are freely permeable to carbon dioxide, it can diffuse down its concentration gradient—traveling from the cell's interior into the interstitial fluid, and from the interstitial fluid into the bloodstream, for eventual delivery to the lungs.

To be effective, the diffusion of nutrients, waste products, and dissolved gases must be able to keep pace with the demands of active cells. Important factors that influence diffusion rates include:

■ *Distance.* Concentration gradients are eliminated quickly over short distances. The greater the distance, the longer the time required. In the human body, diffusion distances are generally small. For example, few cells are farther than 125 μm from a blood vessel.

■ *Gradient size.* The larger the concentration gradient, the faster diffusion proceeds. When cells become more active, the intracellular concentration of oxygen declines. This change increases the concentration gradient for oxygen between the inside of the cell (relatively low) and the interstitial fluid outside (relatively high). The rate of oxygen diffusion into the cell then increases.

■ *Molecule size.* Ions and small organic molecules such as glucose diffuse more rapidly than do large proteins.

■ *Temperature.* The higher the temperature, the faster the diffusion rate. The human body maintains a temperature of about 37°C (98.6°F), and diffusion proceeds more rapidly at this temperature than at normal environmental temperatures.

■ *Electrical forces.* The interior of the cell membrane has a net negative charge relative to the exterior surface, in part due to the high concentration of proteins in the cell. Opposite charges (+ and −) are attracted to each other; similar charges (+ and + or − and −) repel each other. The negative charge on the inside of the cell membrane tends to pull positive ions from the extracellular fluid into the cell and to oppose the entry of negative ions. For example, interstitial fluid contains higher concentrations of sodium ions (Na^+) and chloride ions (Cl^-) than does cytosol. Both the concentration gradient, or *chemical gradient,* and the electrical gradient favor the diffusion of the positively charged sodium ions into the cell. In contrast, diffusion of the negatively charged chloride ions is favored by the chemical gradient but opposed by the electrical gradient. For any ion, the net result of the chemical and electrical forces acting on it is called the *electrochemical gradient.*

Diffusion Across Cell Membranes In extracellular fluids, water and dissolved solutes diffuse freely. A cell membrane, however, acts as a barrier that selectively restricts diffusion: Some substances pass through easily; others cannot penetrate the membrane. An ion or molecule can diffuse across a cell membrane only by (1) passing through a membrane channel or (2) crossing the lipid portion of the membrane (Figure 3-6●).

EXTRACELLULAR FLUID

Lipid-soluble molecules diffuse through membrane lipids

Cell membrane

Channel protein

Large molecules that cannot diffuse through lipids cannot cross the membrane unless they are transported by a carrier mechanism

Small soluble molecules and ions diffuse through membrane channels

CYTOPLASM

●**FIGURE 3-6** Diffusion across the Cell Membrane

If there is an electrochemical gradient for a particular ion or molecule, whether that substance moves across the membrane will depend on its lipid solubility and its size relative to the sizes of membrane channels.

Alcohol, fatty acids, and steroids can enter cells easily, because they can diffuse through the lipid portions of the membrane. Dissolved gases, such as oxygen and carbon dioxide, and lipid-soluble drugs also enter and leave our cells by diffusing through the phospholipid bilayer. **AM** *Drugs and the Cell Membrane*

Ions and most water-soluble compounds are not lipid-soluble, so they must pass through a membrane channel to enter the cytoplasm. These channels are very small—on average, about 0.8 nm in diameter. Water molecules can enter or exit freely, as can ions such as sodium, potassium, calcium, hydrogen, and chloride, but even a small organic molecule, such as glucose, is too big to fit through the channels.

Osmosis: A Special Case of Diffusion   The net diffusion of water across a membrane is so important that it is given a special name, **osmosis** (oz-MŌ-sis; *osmos*, thrust). For convenience, we will always use the term *osmosis* when we consider water movement and will restrict use of the term *diffusion* to the movement of solutes.

Intracellular and extracellular fluids are solutions that contain a variety of dissolved materials. Each solute tends to diffuse as though it were the only material in solution. For example, the diffusion of sodium ions occurs only in response to the existence of a concentration gradient for sodium. Thus, a concentration gradient for another ion will have no effect on the rate or direction of sodium ion diffusion. Some solutes diffuse into the cytoplasm, others diffuse out, and a few,

such as proteins, are unable to diffuse across the cell membrane at all. But if we ignore the individual identities and simply count ions and molecules, we find that the *total* concentration of dissolved ions and molecules on either side of the cell membrane stays the same.

This state of equilibrium persists because a typical cell membrane is freely permeable to water. Whenever a solute concentration gradient exists, a concentration gradient for water exists also. Because dissolved solute molecules occupy space that would otherwise be taken up by water molecules, the higher the solute concentration, the lower the water concentration. As a result, water molecules tend to diffuse across a membrane toward the solution containing the higher solute concentration, because this movement is down the concentration gradient for water.

Remember these three characteristics of osmosis:

1. Osmosis is the diffusion of water molecules across a membrane.
2. Osmosis occurs across a selectively permeable membrane that is freely permeable to water but not freely permeable to solutes.
3. In osmosis, water will flow across a membrane toward the solution that has the higher concentration of solutes, because that is where the concentration of water is lowest.

Osmosis and Osmotic Pressure   Figure 3-7• diagrams the process of osmosis. Step 1 shows two solutions (A and B), with different solute concentrations, separated by a selectively permeable membrane. As osmosis occurs, water molecules cross the membrane until the solute concentrations in the two solutions are identical (step 2a). Thus, the volume of solution B

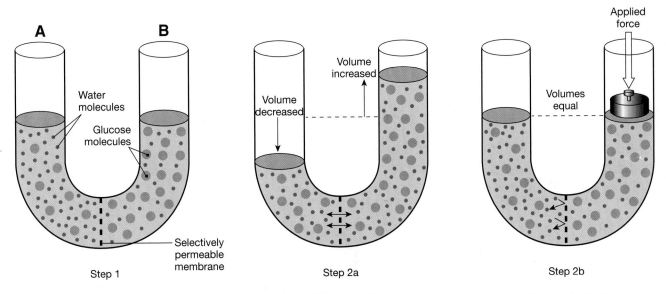

•**FIGURE 3-7 Osmosis.** *Step 1:* Two solutions containing different solute concentrations are separated by a selectively permeable membrane. Water molecules (small blue dots) begin to cross the membrane toward solution B, the solution with the higher concentration of solutes (larger pink circles). *Step 2a:* At equilibrium, the solute concentrations on the two sides of the membrane are equal. The volume of solution B has increased at the expense of that of solution A. *Step 2b:* Osmosis can be prevented by resisting the volume change. The osmotic pressure of solution B is equal to the amount of hydrostatic pressure required to stop the osmotic flow.

increases while that of solution A decreases. The greater the initial difference in solute concentrations, the stronger the osmotic flow. The **osmotic pressure** of a solution is an indication of the force of water movement *into that solution* as a result of its solute concentration. We can measure a solution's osmotic pressure in several ways. For example, an opposing pressure can prevent the osmotic flow of water into the solution. Pushing against a fluid generates **hydrostatic pressure.** In step 2b, hydrostatic pressure opposes the osmotic pressure of solution B, so no net osmotic flow occurs.

Osmosis eliminates solute concentration differences much more quickly than we might predict on the basis of diffusion rates for other molecules. When water molecules cross a membrane, they move in groups held together by hydrogen bonding. So, whereas solute molecules usually diffuse through membrane channels one at a time, water molecules move together in large numbers. This phenomenon is called *bulk flow.*

Osmolarity and Tonicity   The total solute concentration in an aqueous solution is the solution's **osmolarity, or osmotic concentration.** The nature of the solutes, however, is often as important as the total osmolarity (see the clinical discussion "Tonicity and Normal Saline"). Therefore, when we describe the effects of various osmotic solutions on cells, we use the term **tonicity** instead of osmolarity. If a solution does not cause an osmotic flow of water into or out of a cell, the solution is called **isotonic** (*iso,* same + *tonos,* tension).

Figure 3-8a● shows the appearance of a red blood cell in an isotonic solution. If a red blood cell is in a hypotonic solution, water will flow into the cell, causing it to swell up like a balloon (Figure 3-8b●). The cell may rupture, releasing its contents. This event is **hemolysis** (*hemo-,* blood + *lysis,* dissolution). A cell in a **hypertonic** solution will lose water by osmosis. As it does, the cell shrivels and dehydrates. The shrinking of red blood cells is called **crenation** (Figure 3-8c●).

Tonicity and Normal Saline

Although often used interchangeably, the terms *osmolarity* and *tonicity* do not always mean the same thing. For example, consider a solution that has the same osmolarity as the intracellular fluid but a higher concentration of one or more individual ions. If any of those ions can diffuse into the cell, the osmolarity of the intracellular fluid will increase and that of the extracellular solution will decrease. Osmosis will then occur, moving water into the cell. If the process continues, the cell will eventually burst. In this case, the extracellular solution and the intracellular fluid were equal in osmolarity, but the solution was not isotonic.

It is often necessary to give patients large volumes of fluid after severe blood loss or dehydration. One fluid often administered is a 0.9 percent (0.9 g/dl) solution of sodium chloride (NaCl). This solution, which approximates the normal osmotic concentration of the extracellular fluids, is called *normal saline.* It is used because sodium and chloride are the most abundant ions in the extracellular fluid. Little net movement of either ion across cell membranes occurs; thus, normal saline is essentially isotonic with respect to body cells. An alternative treatment involves the use of an isotonic saline solution containing *dextran,* a carbohydrate that cannot cross cell membranes.

(a) Isotonic   **(b) Hypotonic**   **(c) Hypertonic**

●**FIGURE 3-8 Osmotic Flow across a Cell Membrane.** Black arrows indicate an equilibrium with no net water movement. Blue arrows indicate the direction of osmotic water movement. **(a)** Because these red blood cells are immersed in an isotonic saline solution, no osmotic flow occurs and the cells appear normal. **(b)** Immersion in a hypotonic saline solution results in the osmotic flow of water into the cells. The swelling may continue until the cell membrane ruptures, or lyses. **(c)** Exposure to a hypertonic solution results in the movement of water out of the cells. The red blood cells shrivel and become crenated. (SEM × 833)

Filtration

In **filtration,** hydrostatic pressure forces water across a membrane, and solute molecules may be transported with the water. If they are smaller than the membrane pores, molecules of solute will be carried along with the water. We can see filtration in action in a coffeemaker. Gravity forces hot water through the filter, and the water carries with it a variety of dissolved compounds. The large coffee grounds never reach the pot, because they cannot fit through the fine pores in the filter.

In the body, the heart pushes blood through the cardiovascular system while generating hydrostatic pressure. Filtration occurs across the walls of *capillaries,* the most delicate blood vessels, pushing water and dissolved solutes into the tissues of the body. Filtration across specialized blood vessels in the kidneys is an essential step in the production of urine.

☑ How would a decrease in the concentration of oxygen in the lungs affect the diffusion of oxygen into the blood?

☑ Some pediatricians recommend the use of a 10 percent salt solution to relieve congestion for infants with stuffy noses. What effect would such a solution have on the cells lining the nasal cavity, and why?

Carrier-Mediated Transport

In **carrier-mediated transport,** integral proteins bind specific ions or organic substrates and carry these substances across the cell membrane. All forms of carrier-mediated transport share the following characteristics with enzymes: ∞ *pp. 54–55*

1. *Specificity.* Each carrier protein in the cell membrane will bind and transport only certain substances. For example, the carrier protein that transports glucose will not transport other simple sugars.

2. *Saturation limits.* The availability of substrate molecules and carrier proteins limits the rate of transport into or out of the cell, just as enzymatic reaction rates are limited by the availability of substrates and enzymes. When all the available carrier proteins are operating at maximum speed, the proteins are said to be *saturated.* The rate of transport cannot increase further, regardless of the size of the concentration gradient.

3. *Regulation.* Just as enzyme activity often depends on the presence of cofactors, the binding of other molecules, such as hormones, can affect the activity of carrier proteins. Hormones thus provide an important means of coordinating carrier protein activity throughout the body. The interplay between hormones and cell membranes will be examined when we deal with the endocrine system (Chapter 18) and metabolism (Chapter 25).

Many examples of carrier-mediated transport involve the movement of a single substrate molecule across the cell membrane. A few carrier mechanisms transport more than one substrate at a time. In **cotransport,** or *symport,* the carrier transports two substances in the same direction simultaneously, either into or out of the cell. In **countertransport,** or *antiport,* one substance moves into the cell and the other moves out.

We will consider two major examples of carrier-mediated transport here: (1) *facilitated diffusion* and (2) *active transport.*

Facilitated Diffusion Many essential nutrients, such as glucose and amino acids, are insoluble in lipids and too large to fit through membrane channels. These substances can be passively transported across the membrane by carrier proteins in a process called **facilitated diffusion** (Figure 3-9•). The molecule to be transported must first bind to a **receptor site** on the protein. The shape of the protein then changes, moving

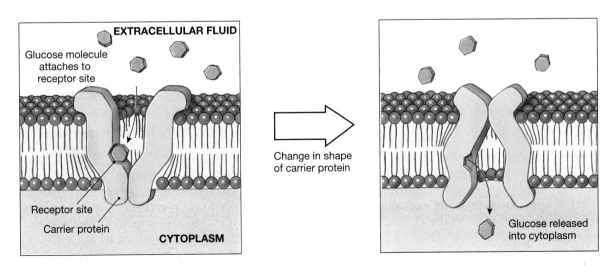

•**FIGURE 3-9 Facilitated Diffusion.** In facilitated diffusion, an extracellular molecule, such as glucose, binds to a receptor site on a carrier protein. The binding alters the shape of the protein, which then releases the molecule to diffuse into the cytoplasm.

3

the molecule to the inside of the cell membrane, where it is released into the cytoplasm.

As in the case of simple diffusion, no ATP is expended in facilitated diffusion, and the molecules move from an area of higher concentration to one of lower concentration. However, facilitated diffusion differs from simple diffusion because once the carrier proteins are saturated, the rate of transport cannot increase, regardless of further changes in the concentration gradient.

All cells move glucose across their membranes through facilitated diffusion. However, several different carrier proteins are involved. In muscle cells, fat cells, and many other cell types, the glucose transporter functions only when stimulated by the hormone *insulin*. Inadequate production of this hormone is one cause of *diabetes mellitus*, a metabolic disorder that we shall discuss in Chapter 18.

Active Transport In **active transport,** the high-energy bond in ATP (or another high-energy compound) provides the energy needed to move ions or molecules across the membrane. Despite the energy cost, active transport offers one great advantage: It is not dependent on a concentration gradient. As a result, the cell can import or export specific substrates *regardless of their intracellular or extracellular concentrations.*

All cells contain carrier proteins called **ion pumps,** which actively transport the cations sodium (Na^+), potassium (K^+), calcium (Ca^{2+}), and magnesium (Mg^{2+}) across their cell membranes. Specialized cells can transport additional ions such as iodide (I^-), chloride (Cl^-), and iron (Fe^{2+}). Many of these carrier proteins move a specific cation or anion in one direction only, either into or out of the cell. In a few instances, one carrier protein will move more than one kind of ion at the same time. If one kind of ion moves in one direction and the other moves in the opposite direction, the carrier protein is called an **exchange pump.**

The Sodium–Potassium Exchange Pump Sodium and potassium ions are the principal cations in body fluids. Sodium ion concentrations are high in the extracellular fluids but low in the cytoplasm. The distribution of potassium in the body is just the opposite—low in the extracellular fluids and high in the cytoplasm. As a result, sodium ions slowly diffuse into the cell, and potassium ions diffuse out through leak channels. Homeostasis within the cell depends on the ejection of sodium ions and recapture of lost potassium ions. This exchange is accomplished through the activity of a **sodium–potassium exchange pump.** The carrier protein involved in this process is called *sodium–potassium ATPase.*

The sodium–potassium exchange pump exchanges intracellular sodium for extracellular potassium (Figure 3-10•). On average, for each ATP molecule

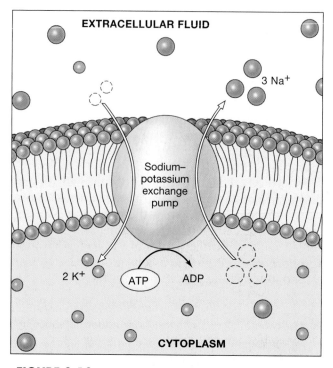

EXTRACELLULAR FLUID

3 Na$^+$

Sodium–potassium exchange pump

2 K$^+$ ATP   ADP

CYTOPLASM

•**FIGURE 3-10** The Sodium–Potassium Exchange Pump. The operation of the sodium–potassium exchange pump is an example of active transport. For each ATP converted to ADP, the protein called sodium–potassium ATPase carries three Na^+ out of the cell and two K^+ into the cell.

consumed, three sodium ions are ejected and two potassium ions are reclaimed by the cell. If ATP is readily available, the rate of transport depends on the concentration of sodium ions in the cytoplasm. When that concentration rises, the pump becomes more active. The energy demands are impressive; sodium–potassium ATPase may use up to 40 percent of the ATP produced by a resting cell.

Secondary Active Transport In **secondary active transport,** the transport mechanism itself does not require energy, but the cell often needs to expend ATP at a later time to preserve homeostasis. As does facilitated transport, a secondary active transport mechanism moves a specific substrate down a concentration gradient. Unlike the proteins in facilitated transport, however, these carrier proteins can also move another substrate at the same time, without regard to its concentration gradient. In effect, the concentration gradient for one substance provides the driving force needed by the carrier protein, and the second substance gets a free ride.

The concentration gradient for sodium ions most often provides the driving force for cotransport mechanisms that move materials into the cell. For example, sodium-linked cotransport is important in the absorption of glucose and amino acids along the intestinal tract. Although the transport activity proceeds

•**FIGURE 3-11** Secondary Active Transport. In secondary active transport, glucose transport by this carrier protein will occur only after the carrier has bound a sodium ion. In three cycles, three glucose molecules and three sodium ions are transported into the cytoplasm. The cell now pumps the sodium ions across the cell membrane via the sodium–potassium exchange pump, at a cost of one ATP.

without directly expending energy, the cell must expend ATP to pump the arriving sodium ions out of the cell by using the sodium–potassium exchange pump (Figure 3-11•). Sodium ions are also involved with many countertransport mechanisms. Sodium–calcium countertransport is responsible for keeping intracellular calcium ion concentrations very low.

Vesicular Transport

In **vesicular transport,** materials move into or out of the cell by means of **vesicles,** small membranous sacs that form at or fuse with the cell membrane. Because large volumes of fluid and solutes are transported in this way, this process is also known as *bulk transport.* Two major categories of vesicular transport exist: (1) *endocytosis* and (2) *exocytosis.*

Endocytosis **Endocytosis** is the packaging of extracellular materials in a vesicle at the cell surface for import into the cell. This process involves relatively large volumes of extracellular material. Three major types of endocytosis exist: (1) *receptor-mediated endocytosis,* (2) *pinocytosis,* and (3) *phagocytosis.* All three are active processes that require energy in the form of ATP.

Vesicles produced by receptor-mediated endocytosis or by pinocytosis are called *endosomes;* those produced by phagocytosis are called *phagosomes.* The contents of endosomes and phagosomes remain isolated within the vesicle. The movement of materials into the surrounding cytoplasm may involve active transport, simple or facilitated diffusion, or the destruction of the vesicle membrane.

Receptor-Mediated Endocytosis A selective process, **receptor-mediated endocytosis** involves the formation of small vesicles at the membrane surface. This process produces vesicles that contain a specific target molecule in high concentrations. Receptor-mediated endocytosis begins when materials in the extracellular fluid bind to receptors on the membrane surface (Figure 3-12•). Most receptor molecules are glycoproteins, and

each binds to a specific ligand, or target, such as a transport protein or a hormone. Some receptors are distributed widely over the surface of the cell membrane. Others are restricted to pits on the cell surface.

Receptors bound to ligands cluster together. Once an area of the cell membrane has become covered with ligands, it forms grooves or pockets that pinch off to form an endosome. Such endosomes are called **coated vesicles,** because they are surrounded by a protein–fiber network. This coating is essential to their formation and movement. Inside the cell, the coated vesicles fuse with *primary lysosomes,* vesicles containing digestive enzymes. The fusion of one or more primary lysosomes with another vesicle creates a *secondary lysosome.* Lysosomal enzymes then free the ligands from their receptors, and the ligands enter the cytosol by diffusion or active transport. After the vesicle membrane pinches off from the secondary lysosome and returns to the cell surface, its receptors are ready to bind more ligands.

Many important substances, including cholesterol and iron ions (Fe^{2+}), are distributed through the body attached to special transport proteins. These proteins are too large to pass through membrane pores but can enter the cell by receptor-mediated endocytosis.

Pinocytosis and Phagocytosis Cell drinking, or **pinocytosis** (pi-nō-sī-TŌ-sis), is the formation of endosomes filled with extracellular fluid. This process is not as selective as receptor-mediated endocytosis, because no receptor proteins are involved. The target appears to be the fluid contents in general rather than specific bound ligands. In pinocytosis, a deep groove or pocket forms in the cell membrane and then pinches off (Figure 3-13a•). The steps involved in the formation and fate of an endosome created by pinocytosis are similar to the steps in receptor-mediated endocytosis, except that ligand binding is not the trigger.

Cell eating, or **phagocytosis** (fa-gō-sī-TŌ-sis), produces phagosomes containing solid objects that may be as large as the cell itself. In this process, cytoplasmic

3

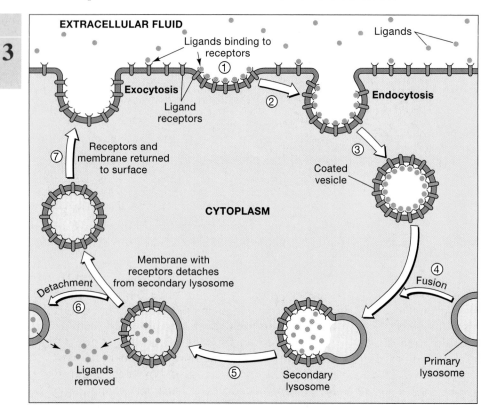

•**FIGURE 3-12 Receptor-Mediated Endocytosis.** In this process, ① specific target molecules called ligands bind to receptors, generally glycoproteins, in the membrane. ② Membrane areas coated with ligands pinch off to form ③ vesicles that ④ fuse with primary lysosomes. ⑤ The ligands are freed from the receptors and, if necessary, broken down by enzymes before diffusing or being transported into the surrounding cytoplasm. ⑥ The membrane containing the receptor molecules separates from the membrane of the lysosome and ⑦ returns to the cell surface to bind additional ligands.

(a)

•**FIGURE 3-13 Pinocytosis and Phagocytosis.** **(a)** An electron micrograph showing pinocytosis at the bases of microvilli in a cell lining the intestinal tract. **(b)** Material brought into the cell by phagocytosis is enclosed in a phagosome and subsequently exposed to lysosomal enzymes. After nutrients are absorbed from the vesicle, the residue is discharged by exocytosis.

(b)

TABLE 3-2 Summary: Mechanisms Involved in Movement across Cell Membranes

Mechanism	Process	Factors Affecting Rate	Substances Involved (location)
Diffusion	Molecular movement of solutes; direction determined by relative concentrations	Size of gradient; size of molecules; charge; lipid solubility, temperature	Small inorganic ions; lipid-soluble materials (all cells)
Osmosis	Movement of water molecules toward solution containing relatively higher solute concentration; requires selectively permeable membrane	Concentration gradient; opposing osmotic or hydrostatic pressure	Water only (all cells)
Filtration	Movement of water, usually with solute, by hydrostatic pressure; requires filtration membrane	Amount of pressure; size of pores in filter	Water and small ions (blood vessels)
Carrier-Mediated Transport			
Facilitated diffusion	Carrier proteins passively transport solutes across a membrane down a concentration gradient	Size of gradient, temperature and availability of carrier protein	Glucose and amino acids (all cells, but several different regulatory mechanisms exist)
Active transport	Carrier proteins actively transport solutes across a membrane regardless of any concentration gradients	Availability of carrier, substrate, and ATP	Na^+, K^+, Ca^{2+}, Mg^{2+} (all cells); other solutes by specialized cells
Secondary active transport	Carrier proteins passively transport two solutes, with one (normally Na^+) moving down its concentration gradient; the cell must later expend ATP to eject the Na^+	Availability of carrier, substrates, and ATP	Glucose and amino acids (specialized cells)
Vesicular Transport			
Endocytosis	Creation of membranous vesicles containing fluid or solid material	Stimulus and mechanics incompletely understood; requires ATP	Fluids, nutrients (all cells); debris, pathogens (specialized cells)
Exocytosis	Fusion of vesicles containing fluids and/or solids within the cell membrane	Stimulus and mechanics incompletely understood; requires ATP	Fluids, debris (all cells)

extensions called **pseudopodia** (soo-dō-PŌ-dē-ah; *pseudo-,* false + *podon,* foot; singular *pseudopodium*) surround the object, and their membranes fuse to form a vesicle (Figure 3-13b•). The phagosome then fuses with many lysosomes, whereupon its contents are digested by lysosomal enzymes. Most cells display pinocytosis, but phagocytosis is performed only by specialized cells, such as the macrophages that protect tissues by engulfing bacteria, cell debris, and other abnormal materials.

Exocytosis Exocytosis (ek-sō-sī-TŌ-sis) is the functional reverse of endocytosis. In exocytosis (Figure 3-13b•), a vesicle created inside the cell fuses with the cell membrane and discharges its contents into the extracellular environment. The ejected material may be secretory products, such as mucins or hormones, or waste products, such as those accumulating in endocytic vesicles.

In a few specialized cells, endocytosis produces vesicles on one side of the cell that are discharged through exocytosis on the opposite side. This method of bulk transport is common in cells lining capillaries. These cells use a combination of pinocytosis and exocytosis to transfer fluid and solutes from the bloodstream into the surrounding tissues.

Many different mechanisms are moving materials into and out of the cell at any moment. Before proceeding further, review and compare the mechanisms summarized in Table 3-2.

The Transmembrane Potential

As we noted previously (p. 72), the inside of the cell membrane has a slight negative charge with respect to the outside. The cause is a slight excess of positively charged ions outside the cell membrane and a slight excess of negatively charged ions inside the cell membrane. This unequal charge distribution is created by differences in the permeability of the membrane to various ions as well as by active transport mechanisms.

Although the positive and negative charges are attracted to each other, they are separated by the phospholipid membrane. When positive and negative charges are held apart, a **potential difference** exists—like the energy stored in a stretched spring—and the opposite charges would rush together if they were not separated by the cell membrane. We will use the term **transmembrane potential** when we refer to the potential difference across a cell membrane.

The *volt* (V) is the unit of measurement for potential difference. Most cars have 12-V batteries. The transmembrane potentials of cells are much smaller, averaging about 0.07 V for a neuron. This value is usually reported as −70 mV, or −70 *millivolts* (thousandths of a volt), with the minus sign signifying that the inside of the cell membrane contains an excess of negative charges as compared with the outside.

The **resting potential** is the transmembrane potential in an undisturbed cell. Each cell type has a characteristic resting potential between −10 mV and −100 mV. Examples include fat cells (−40 mV), thyroid cells (−50 mV), neurons (−70 mV), skeletal muscle cells (−85 mV), and cardiac muscle cells (−90 mV).

If the lipid barrier were removed, the positive and negative charges would move together. The cell membrane thus acts like a dam across a stream. A dam resists the water pressure that builds up on the upstream side; a cell membrane resists electrochemical forces that would otherwise drive ions into or out of the cell. The water retained behind a dam and the ions held on either side of the cell membrane have *potential energy.* People have designed many ways to make use of the potential energy stored behind a dam—for example, turning a mill wheel or a turbine. Similarly, cells have ways of utilizing the potential energy stored in the transmembrane potential. For example, it is the transmembrane potential that makes possible the transmission of information in the nervous system, and thus our perceptions and thoughts. As we shall see in later chapters, changes in the transmembrane potential also trigger the contraction of muscles and the secretion of glands.

☑ During digestion in the stomach, the concentration of hydrogen ions (H^+) rises to many times that in cells of the stomach. Which transport process could be responsible?

☑ If the cell membrane were freely permeable to sodium ions (Na^+), how would the transmembrane potential be affected?

☑ When they encounter bacteria, certain types of white blood cells engulf the bacteria and bring them into the cell. What is this process called?

THE CYTOPLASM

Cytoplasm is a general term for the material located inside the cell membrane and outside the membrane surrounding the nucleus. Cytoplasm contains many more proteins than does extracellular fluid, and proteins account for 15–30 percent of the weight of the cell. The cytoplasm includes two major subdivisions:

1. **Cytosol,** or **intracellular fluid,** which contains dissolved nutrients, ions, soluble and insoluble proteins, and waste products. The cell membrane separates the cytosol from the surrounding extracellular fluid.
2. **Organelles,** which are structures that perform specific functions within the cell.

THE CYTOSOL

The three most important differences between cytosol and extracellular fluid are as follows:

1. The cytosol contains a higher concentration of potassium ions and a lower sodium ion concentration; extracellular fluid contains a higher concentration of sodium ions and a lower potassium ion concentration.
2. The cytosol contains a high concentration of suspended proteins. Many of the proteins are enzymes that regulate metabolic operations; others are associated with the various organelles. The cytosol is a colloid with a consistency that varies between that of thin maple syrup and almost-set gelatin. ∞ *p. 42*
3. The cytosol usually contains small quantities of carbohydrates and large reserves of amino acids and lipids. The carbohydrates are broken down to provide energy, and the amino acids are used to manufacture proteins. Lipids are used primarily as an energy source when carbohydrates are unavailable.

The cytosol may also contain masses of insoluble materials known as **inclusions.** Among the most common inclusions are stored nutrients. For example, glycogen granules in liver or in skeletal muscle cells, and lipid droplets in fat cells are inclusions.

ORGANELLES

Each organelle performs specific functions that are essential to normal cell structure, maintenance, and metabolism. Cellular organelles can be divided into two broad categories. (1) **Nonmembranous organelles** are not completely enclosed by membranes, and all of their components are in direct contact with the cytosol. (2) **Membranous organelles** are isolated from the cytosol by their membranes, just as the cell membrane isolates the cytosol from the extracellular fluid.

The cell's nonmembranous organelles include the *cytoskeleton, microvilli, centrioles, cilia, flagella,* and *ribosomes.* Membranous organelles include the *endoplasmic reticulum,* the *Golgi apparatus, lysosomes, peroxisomes,* and *mitochondria.* (The *nucleus,* which is also surrounded by a membranous envelope and therefore sometimes included with the membranous organelles, will be the focus of a separate section.)

The Cytoskeleton

The **cytoskeleton** is an internal protein framework that gives the cytoplasm strength and flexibility. It includes *microfilaments, intermediate filaments, microtubules,* and in some cases *thick filaments.* Thick filaments appear only in muscle cells. All cells contain the other types of filaments in varying numbers. In a typical cell, the microfilaments, intermediate filaments, and microtubules form a dynamic network. The organizational details are as yet poorly understood, because the network is extremely delicate and thus hard to study in an intact state. Figure 3-14a• is based on our current knowledge of cytoskeletal structure.

Microfilaments   **Microfilaments,** or *thin filaments,* are slender protein strands, generally less than 6 nm in diameter. Typical microfilaments are composed of the protein **actin.** In most cells, the actin filaments are common in the periphery of the cell but relatively rare in the region immediately surrounding the nucleus. In cells that form a layer or lining, such as the lining of the intestinal tract, actin filaments also form a layer, the *terminal web,* just inside the membrane at the exposed surface of the cell.

Microfilaments have three major functions:

1. Microfilaments anchor the cytoskeleton to integral proteins of the cell membrane. They provide additional mechanical strength to the cell and attach the cell membrane to the enclosed cytoplasm.

2. Microfilaments, interacting with other proteins, determine the consistency of the cytoplasm. The consistency can vary by region, from a viscous gel to a fluid, depending on the nature of these interactions. For example, actin filaments in a dense, flexible network give the cytoplasm a gelatinous consistency, whereas fluid cytoplasm contains dispersed microfilaments.

3. Actin can interact with the protein **myosin** to produce active movement of a portion of a cell or to change the shape of the entire cell.

Intermediate Filaments   The protein composition of **intermediate filaments** varies from one cell type to another. These filaments, which range from 7 to 11 nm in diameter, are intermediate in size between microfilaments and thick filaments. Intermediate

Microvillus

Cell membrane

Microfilaments

Terminal web

Mitochondrion

Ribosome cluster (polyribosome)

Intermediate filament

Endoplasmic reticulum

Microtubule

Secretory vesicle

(a)

(b)

•**FIGURE 3-14   The Cytoskeleton. (a)** The cytoskeleton provides strength and structural support for the cell and its organelles. Interactions between cytoskeletal components are also important in moving organelles and in changing the shape of the cell. **(b)** An SEM image of the microfilaments and microvilli of an intestinal cell.

filaments (1) provide strength and stability to cell shape, (2) stabilize the positions of organelles, and (3) stabilize the position of the cell with respect to surrounding cells through specialized attachment to the cell membrane. Intermediate filaments are insoluble fibers, and these are the most durable of the cytoskeletal elements. Many cells contain specialized intermediate filaments with unique functions. For example, the keratin fibers in superficial layers of the skin are intermediate filaments that make these layers tough and water-resistant. ∞ *p. 52*

Microtubules All our cells contain **microtubules,** hollow tubes built from the globular protein **tubulin.** Microtubules are the largest components of the cytoskeleton, with diameters of about 25 nm. The number and distribution of microtubules in a cell change over time. A microtubule forms by the aggregation of tubulin molecules. It continuously undergoes remodeling, with tubulin molecules dispersing at one end of the microtubule and attaching at the other. The entire structure persists for a time and then disassembles into individual tubulin molecules. The microtubular array in a cell is centered near the nucleus, in a region known as the *centrosome* (see Figure 3-2•, p. 66). From the centrosome, microtubules extend outward into the periphery of the cell.

Microtubules have the following functions:

1. Microtubules form the primary components of the cytoskeleton, giving the cell strength and rigidity and anchoring the position of major organelles.

2. Disassembly of microtubules provides a mechanism for changing the shape of the cell, perhaps assisting in cell movement.

3. Microtubules can attach to organelles and other intracellular materials that then move along the axis of the microtubule. The movement is performed by proteins called *molecular motors.* These proteins, which attach to the microtubule and to the vesicle or other organelle, "walk" their way along the microtubule. Whether the transport occurs toward one end of the microtubule or the other depends on the identity of the molecular motor. For example, if *kinesin* moves in one direction, *dynein* will move in the opposite direction. Regardless of the direction of transport or the nature of the motor, the process requires ATP and is essential to normal cellular function.

4. During cell division, microtubules form the *spindle apparatus,* which distributes the duplicated chromosomes to opposite ends of the dividing cell. We shall consider this process in more detail in a later section.

5. Microtubules form structural components of organelles such as *centrioles, cilia,* and *flagella.*

In addition to those functions, the cytoskeleton has other functions that are only now being investigated. For example, many intracellular enzymes, especially those involved with metabolism and energy production, and the ribosomes and RNA molecules responsible for the synthesis of proteins are attached to the microfilaments and microtubules of the cytoskeleton. Thus, the cytoskeleton plays a role in the metabolic organization of the cell by determining where in the cytoplasm key enzymatic reactions occur and where specific proteins are synthesized.

Thick Filaments **Thick filaments** are relatively massive bundles of myosin protein subunits. Thick filaments may reach 15 nm in diameter. They appear only in muscle cells, where they interact with actin filaments to produce powerful contractions.

Microvilli

Microvilli are small, finger-shaped projections of the cell membrane (Figure 3-14b•) that greatly increase the surface area of the cell exposed to the extracellular environment. They cover the surfaces of cells that are actively absorbing materials from the extracellular fluid, such as the cells lining the digestive tract. Microvilli differ from other membranous projections, such as pseudopodia, in their degree of integration with the cytoskeleton. A core of microfilaments stiffens each microvillus and anchors it to the cytoskeleton at the terminal web.

Centrioles

A **centriole** is a cylindrical structure composed of short microtubules (Figure 3-15a•). The microtubules form nine groups, with three in each group. Each of the nine triplets is connected to its nearest neighbors on either side. With no central microtubules, this organization is called a *9 + 0 array.* Some preparations show an axial structure, with radial spokes leading toward the microtubular groups. The function of this axial complex is not known.

All animal cells capable of reproducing themselves contain a pair of centrioles. During cell division, the centrioles form the spindle apparatus associated with the movement of DNA strands. (Cell division is detailed later in this chapter.) Mature red blood cells, skeletal muscle cells, cardiac muscle cells, and typical neurons have no centrioles; as a result, these cells are not capable of dividing.

The **centrosome,** the cytoplasm surrounding the centrioles, is the heart of the cytoskeletal system. Microtubules of the cytoskeleton generally begin at the centrosome and radiate through the cytoplasm.

Cilia

Cilia (singular, *cilium*) contain nine pairs of microtubules surrounding a central pair (Figure 3-15b•). This organization is known as a *9 + 2 array.* Cilia are anchored to a compact **basal body** situated just

•**FIGURE 3-15** **Centrioles and Cilia.** **(a)** A centriole consists of nine microtubule triplets (known as a 9 + 0 array). The centrosome contains a pair of centrioles oriented at right angles to one another. **(b)** A cilium contains nine pairs of microtubules surrounding a central pair (9 + 2 array). The basal body to which the cilium is anchored has a structure similar to that of a centriole. **(c)** A single cilium swings forward and then returns to its original position. During the power stroke, the cilium is relatively stiff; during the return stroke, it bends and moves back to its original position.

beneath the cell surface. In section, the organization of microtubules in the basal body resembles that of a centriole.

The exposed portion of the cilium is completely covered by the cell membrane. Cilia "beat" rhythmically (Figure 3-15c•). The cilium is relatively stiff during the effective *power stroke* and flexible during the *return stroke*. The coordinated action of cilia moves fluids or secretions across the cell surface. For example, cilia lining the respiratory tract beat in a synchronized manner to move sticky mucus and trapped dust particles toward the throat and away from delicate respiratory surfaces. If the cilia are damaged or immobilized by heavy smoking or a metabolic problem, the cleansing action is lost and the irritants will no longer be removed. As a result, chronic respiratory infections develop.

Flagella

Flagella (fla-JEL-uh; singular, *flagellum,* whip) resemble cilia, but they are much longer. Flagella move a cell through the surrounding fluid rather than moving the fluid past a stationary cell. Sperm are the only human cells that have flagella, which are used to move the cells along the female reproductive tract. If the flagella of sperm are paralyzed or otherwise abnormal, the individual will be sterile because immobile sperm cannot perform fertilization.

Ribosomes

Ribosomes are organelles that manufacture proteins, using information provided by the DNA of the nucleus. The number of ribosomes in a particular cell varies with the type of cell. For example, liver cells, which

3

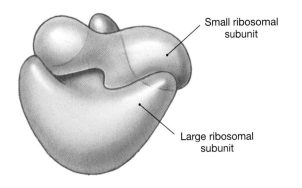

Small ribosomal
subunit

Large ribosomal
subunit

•**FIGURE 3-16 Ribosomes.** A diagrammatic view of the structure of an intact ribosome. The subunits are separate unless the ribosome is engaged in protein synthesis.

manufacture blood proteins, contain far more ribosomes than do fat cells, which synthesize triglycerides.

Ribosomes are not visible with the light microscope. In an electron micrograph, ribosomes appear as dense granules roughly 25 nm in diameter. Each ribosome consists of roughly 60 percent RNA and 40 percent protein.

A functional ribosome consists of two subunits that are normally separate and distinct. The subunits differ in size; one is a **small ribosomal subunit,** and the other is a **large ribosomal subunit** (Figure 3-16•). These subunits contain special proteins and **ribosomal RNA (rRNA),** one of the RNA types introduced in Chapter 2. ∞ *p. 57* Before protein synthesis can begin, a small and a large ribosomal subunit must join together with a strand of *messenger RNA* (mRNA, another type of RNA) to create a functional ribosome.

Two major types of functional ribosomes exist: (1) free ribosomes and (2) fixed ribosomes (Figure 3-2•, p. 66). **Free ribosomes** are scattered throughout the

cytoplasm. The proteins they manufacture enter the cytosol. **Fixed ribosomes** are attached to the *endoplasmic reticulum (ER),* a membranous organelle. Proteins manufactured by fixed ribosomes enter the ER, where they are modified and packaged for secretion. We shall examine ribosomal structure and functions in later sections when we deal with the endoplasmic reticulum and protein synthesis.

☑ Cells lining the small intestine have numerous fingerlike projections on their free surface. What are these structures, and what is their function?

☑ How would the absence of a flagellum affect a sperm?

The Endoplasmic Reticulum

The **endoplasmic reticulum** (en-dō-PLAZ-mik re-TIK-ū-lum), or **ER,** is a network of intracellular membranes that is connected to the *nuclear envelope,* which surrounds the nucleus. This network has four major functions:

1. *Synthesis.* Specialized regions of the ER synthesize proteins, carbohydrates, and lipids.
2. *Storage.* The ER can store synthesized molecules or materials absorbed from the cytosol without affecting other cellular operations.
3. *Transport.* Materials can travel from place to place in the ER.
4. *Detoxification.* Drugs or toxins can be absorbed by the ER and neutralized by enzymes within it.

The ER (Figure 3-17•) forms hollow tubes, flattened sheets, and chambers called **cisternae** (sis-TUR-nē; singular, *cisterna,* a reservoir for water). Two types of ER exist: (1) the *smooth endoplasmic reticulum* and (2) the *rough endoplasmic reticulum.*

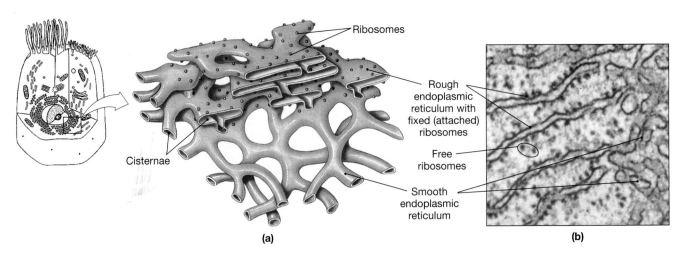

Ribosomes

Rough
endoplasmic
reticulum with
fixed (attached)
ribosomes

Free
ribosomes

Smooth
endoplasmic
reticulum

Cisternae

(a)

(b)

•**FIGURE 3-17 The Endoplasmic Reticulum. (a)** The three-dimensional relationships between the rough and smooth endoplasmic reticula. **(b)** Rough endoplasmic reticulum and free ribosomes in the cytoplasm of a cell. (TEM × 111,000)

The Smooth Endoplasmic Reticulum No ribosomes are associated with the **smooth endoplasmic reticulum (SER).** The SER has a variety of functions that center around the synthesis of lipids and carbohydrates. For example, the SER is responsible for:

- Synthesizing phospholipids and cholesterol needed for maintenance and growth of the cell membrane, ER, nuclear membrane, and Golgi apparatus in all cells.
- Synthesizing steroid hormones, such as *androgens* and *estrogens* (the dominant sex hormones in males and in females, respectively) in the reproductive organs and the steroid hormones of the adrenal glands.
- Synthesizing and storing glycerides, especially triglycerides, in liver cells and fat cells.
- Synthesizing and storing glycogen in skeletal muscle and liver cells.

In skeletal muscle cells, neurons, and many other cell types, the SER also adjusts the composition of the cytosol by absorbing and storing ions, such as Ca^{2+}, or larger molecules. In addition, the SER is responsible for the detoxification or inactivation of drugs by liver and kidney cells.

The Rough Endoplasmic Reticulum The **rough endoplasmic reticulum (RER)** functions as a combination workshop and shipping depot. It is where many newly synthesized proteins undergo chemical modification and where they are packaged for export to their next destination, the *Golgi apparatus.*

The outer surface of the rough endoplasmic reticulum contains fixed ribosomes (Figure 3-17•). Ribosomes synthesize proteins by using instructions provided by messenger RNA. As the polypeptide chains grow, they enter the cisternae of the endoplasmic reticulum. Inside the ER, each protein assumes its secondary and tertiary structures. Some of the proteins are enzymes that will function inside the ER. Other proteins are chemically modified in the ER by the attachment of carbohydrates, creating glycoproteins. Most of the proteins and glycoproteins produced by the RER are packaged into small membranous sacs that pinch off the tips of the cisternae. These **transport vesicles** deliver the products to the Golgi apparatus.

The RER and SER in Specialized Cells The amount of endoplasmic reticulum and the proportion of RER to SER vary with the type of cell and its ongoing activities. For example, pancreatic cells that manufacture digestive enzymes contain an extensive RER, but the SER is relatively small. The situation is just the reverse in the cells that synthesize steroid hormones in the reproductive system.

The Golgi Apparatus

The **Golgi** (GŌL-jē) **apparatus** (Figure 3-18•) consists of flattened membrane discs called *cisternae.* A typical Golgi apparatus consists of five or six cisternae. A single cell may contain several sets of cisternae, each resembling a stack of dinner plates. Most often, these stacks lie near the nucleus of the cell. The Golgi cisternae communicate with the ER and with the cell surface by the formation, movement, and fusion of vesicles.

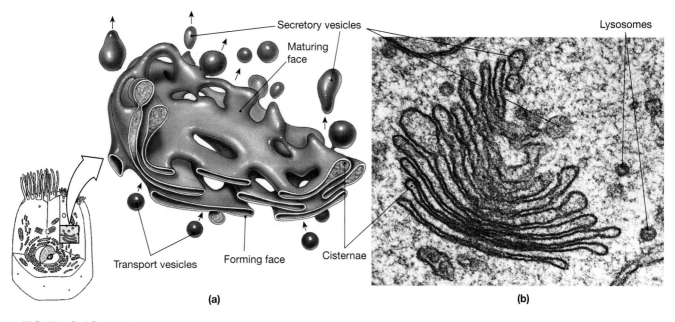

•**FIGURE 3-18 The Golgi Apparatus. (a)** A three-dimensional view of the Golgi apparatus with a cut edge corresponding to **(b),** a sectional view of the Golgi apparatus of an active secretory cell. (TEM × 57,660)

3

(a)

•**FIGURE 3-19** **Functions of the Golgi Apparatus.** **(a)** Transport vesicles carry the secretory product from the endoplasmic reticulum to the Golgi apparatus (simplified to clarify the relationships between the membranes). Small vesicles move membrane and materials between the Golgi cisternae. At the maturing face, three functional categories of vesicles develop. Secretory vesicles carry the secretion from the Golgi to the cell surface, where exocytosis releases the contents into the extracellular fluid. Other vesicles add surface area and integral proteins to the cell membrane. Lysosomes, which remain in the cytoplasm, are vesicles filled with enzymes. **(b)** Exocytosis at the surface of a cell.

The Golgi apparatus has three major functions:

1. *It modifies and packages secretions, such as hormones or enzymes, for release through exocytosis.*
2. *It renews or modifies the cell membrane.*
3. *It packages special enzymes within vesicles for use in the cytosol.*

Secretory Vesicles and Exocytosis Figure 3-19a• diagrams the role of the Golgi apparatus in packaging secretions. Proteins and glycoproteins that will reach the Golgi apparatus are synthesized in the RER. Transport vesicles then deliver these products to the Golgi apparatus. The vesicles generally arrive at a cisterna known as the *forming* (or *cis*) *face.* The transport vesicles then fuse with the Golgi membrane, emptying their contents into the cisterna. Inside the Golgi apparatus, enzymes modify the arriving proteins and glycoproteins. For example, the enzymes may change the carbohydrate structure of a glycoprotein, or they may attach a phosphate group, sugar, or fatty acid to a protein.

Small vesicles move material from one cisterna to the next. Ultimately, the product arrives at the *maturing* (or *trans*) *face,* which usually faces the cell surface. At the maturing face, three types of vesicles form that carry materials away from the Golgi apparatus:

1. *Secretory vesicles.* **Secretory vesicles** are vesicles containing secretions that will be discharged from the cell by exocytosis (Figure 3-19b•).
2. *Membrane renewal vesicles.* When vesicles produced at the Golgi apparatus fuse with the surface of the cell, they are adding new lipids and proteins to the cell membrane. At the same time, other areas of the cell membrane are being removed by endocytosis. The Golgi apparatus can thus change the properties of the cell membrane over time. For example, new glycoprotein receptors can be added, making the cell more sensitive to a particular stimulus; alternatively, receptors can be removed, making the cell less sensitive. Such changes can profoundly alter the sensitivity and functions of the cell.
3. *Lysosomes.* Vesicles that remain in the cytoplasm, called *lysosomes,* contain digestive enzymes. Their varied functions will be detailed next.

Lysosomes

Lysosomes (LĪ-sō-sōmz; *lyso-,* dissolution + *soma,* body) are vesicles filled with digestive enzymes. They are small, often spherical, and their contents look dense and dark in electron micrographs (Figure 3-18b•). Lysosomes are produced at the Golgi apparatus (Figure 3-19a•). *Primary lysosomes* contain inactive enzymes. Enzyme activation occurs when the lysosome fuses with the membranes of damaged organelles, such as mitochondria or fragments of the ER. This fusion is one way to create a *secondary lysosome,* which contains active enzymes. These enzymes then break down the lysosomal contents. Nutrients reenter the cytosol, and the remaining material is eliminated by exocytosis.

Lysosomes also function in the defense against disease. Cells may remove bacteria as well as liquids and organic debris from the extracellular fluid through endocytosis. (The role of lysosomes in receptor-mediated endocytosis was detailed in Figures 3-12 and 3-13b•, p. 78.) Lysosomes fuse with the vesicles created through endocytosis, and the digestive enzymes then break down the contents and release usable substances, such as sugars or amino acids. In this way, the cell both protects itself against pathogenic organisms and obtains valuable nutrients.

Lysosomes perform essential cleanup and recycling functions inside the cell. For example, when muscle cells are inactive, lysosomes gradually break down their contractile proteins; if the cells become active once again, this destruction ceases. This regulatory mechanism fails in a damaged or dead cell. Lysosomes then disintegrate, releasing active enzymes into the cytosol. These enzymes rapidly destroy the proteins and organelles of the cell, a process called **autolysis** (aw-TAH-li-sis; *auto-,* self). We do not know how to control lysosomal activities or why the enclosed enzymes do not digest the lysosomal walls unless the cell is damaged.

Problems with lysosomal enzyme production cause more than 30 serious diseases affecting children. In these conditions, called *lysosomal storage diseases,* the lack of a specific lysosomal enzyme results in the buildup of waste products and debris normally removed and recycled by lysosomes. Affected individuals may die when vital cells, such as those of the heart, can no longer function. AM *Lysosomal Storage Diseases*

Peroxisomes

Peroxisomes are smaller than lysosomes and carry a different group of enzymes. In contrast to lysosomes, which are produced at the Golgi apparatus, new peroxisomes are produced by the growth and subdivison of existing peroxisomes. Their enzymes are produced at free ribosomes and later are transported from the cytosol into the peroxisomes by carrier proteins. Peroxisomes absorb and break down fatty acids and other organic compounds. As they do so, peroxisomes generate hydrogen peroxide (H_2O_2), a potentially dangerous free radical. ∞ *p. 34* Other enzymes then break down the hydrogen peroxide to oxygen and water. Peroxisomes thus protect the cell from the potentially damaging effects of free radicals produced during catabolism. These organelles are present in all cells; their numbers are highest in metabolically active cells, such as liver cells.

Membrane Flow

When the temperature changes markedly, you change your clothes. When a cell's environment changes, it alters the structure and properties of its cell membrane. With the exception of mitochondria, all membranous organelles in the cell are either interconnected or in communication through the movement of vesicles. The RER and SER are continuous and are connected to the nuclear envelope. Transport vesicles connect the ER with the Golgi apparatus, and secretory vesicles link the Golgi apparatus with the cell membrane. Finally, vesicles forming at the exposed surface of the cell remove and recycle segments of the cell membrane. This continuous movement and exchange is called **membrane flow.** In an actively secreting cell, an area equal to the entire membrane surface may be replaced each hour.

Membrane flow is another example of the dynamic nature of cells. It provides a mechanism by which cells change the characteristics of their cell membranes—lipids, receptors, channels, anchors, and enzymes—as they grow, mature, or respond to a specific environmental stimulus.

Mitochondria

Mitochondria (mī-tō-KON-drē-uh; singular, *mitochondrion; mitos,* thread + *chondrion,* granule) are small organelles that can have a variety of shapes, from long and slender to short and fat. The number of mitochondria in a particular cell varies with the cell's energy demands. Red blood cells lack mitochondria, but these organelles may account for 20 percent of the volume of an active liver cell.

Mitochondria have an unusual double membrane (Figure 3-20a•). The outer membrane surrounds the organelle. The inner membrane contains numerous folds called **cristae.** Cristae increase the surface area exposed to the fluid contents, or **matrix,** of the mitochondrion. Metabolic enzymes in the matrix catalyze the reactions that provide energy for cellular functions.

Most of the chemical reactions that release energy occur in the mitochondria, but most of the cellular activities that require energy occur in the surrounding cytoplasm. Cells must therefore store energy in a form that can be moved from place to place. Energy is stored and transferred in the form of *high-energy bonds.* Such a bond generally attaches a phosphate group (PO_4^{3-}) to a suitable molecule, creating a *high-energy compound.* ∞ *p. 57* The most important high-energy compound is *adenosine triphosphate (ATP).* Cells break the high-energy bond under controlled conditions, reconverting ATP to ADP and phosphate and thereby releasing energy for the cell's use.

Mitochondrial Energy Production Most cells generate ATP and other high-energy compounds through the breakdown of carbohydrates, especially glucose. We shall examine the entire process in Chapter 25, but a few basic concepts now will help you follow discussions of muscle contraction, neuron function, and endocrine function in Chapters 10–18.

3

(a)

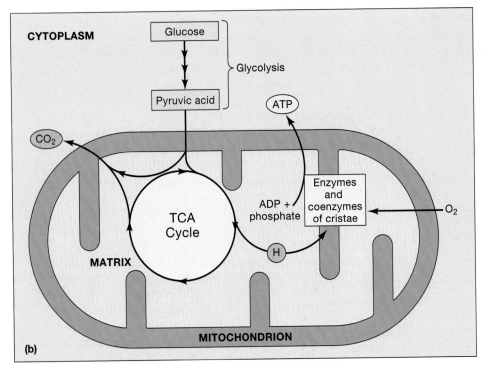

(b)

•**FIGURE 3-20** **Mitochondria.** (a) The three-dimensional organization, and a color-enhanced TEM of a typical mitochondrion in section. (TEM × 46,332). (b) An overview of the role of mitochondria in energy production. Mitochondria absorb short carbon chains (such as pyruvic acid) and oxygen and generate carbon dioxide and ATP.

Although most of the ATP production occurs inside mitochondria, the first steps take place in the cytosol. In this reaction sequence, called **glycolysis,** each glucose molecule is broken down into two molecules of *pyruvic acid.* The pyruvic acid molecules are then absorbed by mitochondria (Figure 3-20b•).

In the mitochondrial matrix, a CO_2 molecule is removed from each absorbed pyruvic acid molecule; the remainder enters the **tricarboxylic acid cycle,** or **TCA cycle.** The TCA cycle is an enzymatic pathway that breaks down organic molecules. The remnants of pyruvic acid molecules contain carbon, oxygen, and hydrogen atoms. The carbon and oxygen atoms are released as carbon dioxide, which diffuses out of the cell. The hydrogen atoms are delivered to carrier proteins in the cristae. The electrons are then removed and passed along a chain of coenzymes. The energy released during these steps performs the enzymatic conversion of ADP to ATP. ∞ *p. 57* Because mitochondrial activity requires oxygen, this method of ATP production is known as **aerobic metabolism** (*aero-,* air + *bios,* life), or *cellular respiration.* Aerobic metabolism in mitochondria produces about 95 percent of the ATP needed to keep a cell alive. (Enzymatic reactions in the cytosol produce the rest.)

Several inheritable disorders result from abnormal mitochondrial activity. The mitochondria involved have defective enzymes that reduce their ability to generate ATP. Cells throughout the body may be affected, but symptoms involving muscle cells, neurons, and the receptor cells in the eye are most common, because these cells have especially high energy demands.

[AM] *Mitochondrial DNA, Disease, and Evolution*

☑ Certain cells in the ovaries and testes contain large amounts of smooth endoplasmic reticulum (SER). Why?

☑ Microscopic examination of a cell's contents reveals many mitochondria. What does this observation imply about the cell's energy requirements?

THE NUCLEUS

The **nucleus** is the control center of cellular operations. A single nucleus stores all the information needed to control the synthesis of the approximately 100,000 different proteins in the human body. The nucleus determines the structural and functional characteristics of the cell by controlling which proteins are

synthesized, and in what amounts. A cell without a nucleus could be compared to a car without a driver. However, a car can sit idle for years, but a cell without a nucleus will disintegrate within 3 or 4 months.

Most cells contain a single nucleus, but exceptions exist. For example, skeletal muscle cells have many nuclei, and mature red blood cells have none. Figure 3-21• details the structure of a typical nucleus. A **nuclear envelope** surrounds the nucleus and separates it from the cytosol. The nuclear envelope is a double membrane, with its two layers separated by a narrow **perinuclear space** (*peri-*, around). At several locations, the nuclear envelope is connected to the rough endoplasmic reticulum, as shown in Figure 3-2•, p. 66.

For it to direct processes that take place in the cytosol, the nucleus must receive information about conditions and activities in other parts of the cell. Chemical communication between the nucleus and the cytosol occurs through **nuclear pores.** These pores, which cover about 10 percent of the surface of the nucleus, are large enough to permit the movement of ions and small molecules but too small for the free passage

Nucleoplasm

Nucleolus

Chromatin

Nuclear envelope

Nuclear pores

(a)

Inner membrane of nuclear envelope

Broken edge of outer membrane

Outer membrane of nuclear envelope

(b)

•**FIGURE 3-21 The Nucleus. (a)** Important nuclear structures. (TEM × 4828) **(b)** This cell was frozen and then broken apart to make internal structures visible. This technique, called freeze-fracture, provides a unique perspective on the internal organization of cells. The nuclear envelope and nuclear pores are visible. The fracturing process broke away part of the outer membrane of the nuclear envelope, and the cut edge of the nucleus can be seen. (SEM × 9240)

3

of proteins or DNA. Each nuclear pore contains regulatory proteins that transport specific proteins and RNA into or out of the nucleus.

The term *nucleoplasm* refers to the fluid contents of the nucleus. The nucleoplasm contains the **nuclear matrix,** a network of fine filaments that provides structural support and may be involved in the regulation of genetic activity. The nucleoplasm also contains ions, enzymes, RNA and DNA nucleotides, small amounts of RNA, and DNA.

Most nuclei contain several dark-staining areas called **nucleoli** (noo-KLĒ-ō-lī); singular, *nucleolus*). Nucleoli are nuclear organelles that synthesize ribosomal RNA. They also assemble the ribosomal subunits, which reach the cytoplasm by carrier-mediated transport at the nuclear pores. Nucleoli are composed of RNA, enzymes, and proteins called **histones.** The nucleoli form around areas of DNA that contain the instructions for producing ribosomal proteins and RNA. Nucleoli are most prominent in cells that manufacture large amounts of proteins, because such cells need large numbers of ribosomes. Examples include liver cells and muscle cells.

CHROMOSOME STRUCTURE

It is the DNA in the nucleus that stores the instructions for protein synthesis, and this DNA is contained in **chromosomes** (*chroma*, color). The nuclei of somatic cells (all cells other than sex cells) in humans contain 23 pairs of chromosomes. One member of each pair is derived from the mother, and one from the father. Each chromosome contains DNA strands bound to histones. The DNA strands coil around the histones, allowing a great deal of DNA to be packaged in a small space. Interactions between the DNA and the histones help determine what information is available to the cell at any moment.

Figure 3-22• illustrates chromosome structure. At intervals, the DNA strands wind around the histones, forming a complex known as a **nucleosome.** The entire chain of nucleosomes may coil around other proteins. The degree of coiling determines whether the chromosome is long and thin or short and fat. In cells that are not dividing, the chromosomal material is loosely coiled, forming a tangle of fine filaments known as **chromatin.** Chromatin gives the nucleus a clumped, grainy appearance.

•**FIGURE 3-22 Chromosome Structure.**
DNA strands are coiled around histones to form nucleosomes. Nucleosomes form coils that may be very tight or rather loose. In cells that are not dividing, the DNA is loosely coiled, forming a tangled network known as chromatin. When the coiling becomes tighter, as it does in preparation for cell division, the DNA becomes visible as distinct structures called chromosomes.

Chromosomes don't become visible until just before cell division begins. At that time, two copies of each chromosome exist. Each copy is called a **chromatid** (KRŌ-ma-tid). The two chromatids are joined together at a single point, the **centromere** (SEN-trō-mēr), which is surrounded by a protein complex called a **kinetochore** (ki-NE-tō-kor). You will learn more about these structures later in the chapter, when we discuss cell division.

THE GENETIC CODE

The **genetic code** is the method of information storage in the DNA strands of the nucleus. An understanding of the genetic code has enabled us to determine how cells build proteins and how various structural and functional characteristics are inherited from generation to generation. Researchers are experimenting with the manipulation of the genetic information in human cells; the techniques may revolutionize the treatment of some serious diseases.

We described the basic structure of nucleic acids in Chapter 2. ⚬ *pp. 56–57* A single DNA molecule consists of a pair of DNA strands held together by hydrogen bonding between complementary nitrogenous bases. Information is stored in the sequence of nitrogenous bases along the length of the DNA strands. Those nitrogenous bases are adenine, A; thymine, T; cytosine, C; and guanine, G.

The genetic code is called a *triplet code,* because a sequence of three nitrogenous bases can specify the identity of a single amino acid. For example, the DNA triplet thymine–guanine–thymine (TGT) codes for the amino acid cysteine. More than one triplet may represent the same amino acid, however. The DNA triplet thymine–guanine–cytosine (TGC) also codes for cysteine.

A **gene** is the functional unit of heredity; it contains all the DNA triplets needed to produce one specific protein. The number of triplets varies from gene to gene, depending on the size of the polypeptide represented. A relatively short polypeptide chain might require fewer than 100 triplets, whereas the instructions for building a large protein might involve 1000 or more triplets. Not all of the DNA molecule is devoted to carrying instructions for proteins; some segments contain instructions for the synthesis of transfer RNA or ribosomal RNA, and others have no apparent function.

Every gene contains segments responsible for regulating its own activity. In effect, these are triplets that say "do (or do not) read this message," "message starts here," or "message ends here." The "read me," "don't read me," and "start" signals form a special region of DNA called the **promoter,** or *control segment,* at the start of each gene. Each gene ends with a "stop" signal.

 DNA Fingerprinting
Every nucleated somatic cell in the body carries a set of 46 chromosomes identical to the set formed at fertilization. Not all the DNA of these chromosomes codes for proteins, however, and a significant percentage of DNA segments have no known function. Some of the "useless" segments contain the same nucleotide sequence repeated over and over. The number of segments and the number of repetitions vary from individual to individual. The chance that any two individuals, other than identical twins, will have the same pattern of repeating DNA segments is less than one in 9 billion. The identification of individuals can therefore be made on the basis of a DNA pattern analysis, just as it can on the basis of a fingerprint. Skin scrapings, blood, semen, hair, or other tissues can be used as a sample source. Information from **DNA fingerprinting** is used to convict (and to acquit) persons accused of committing violent crimes, such as rape or murder.

GENE ACTIVATION AND PROTEIN SYNTHESIS

Each DNA molecule contains thousands of genes and therefore holds the information needed to synthesize thousands of proteins. These genes are normally tightly coiled, and histones bound to the promoters keep the genes inactive. Before a gene can affect the cell, the portion of the DNA molecule containing that gene must be uncoiled and the histones temporarily removed. The factors controlling this process, called **gene activation,** are only partially understood. Gene activation begins with the temporary disruption of the weak bonds between the nitrogenous bases of the two DNA strands and the removal of the histone that guards the promoter.

After the complementary strands have separated and the histone has been removed, the enzyme **RNA polymerase** binds to the promoter of the gene. This binding is the first step in the process of **transcription,** the production of RNA from a DNA template. The term *transcription* is appropriate, because the RNA polymerase is transcribing, or "taking notes," from the gene. Transcription can produce any of the three forms of RNA, but we will focus on the transcription of mRNA, which carries the information needed to synthesize proteins. **Messenger RNA (mRNA)** is absolutely vital, because activated genes neither leave the nucleus nor lose their connections with other genes along the length of the DNA strand. Instead, they pass their information to messenger RNA, which carries that information to the cytoplasm, where protein synthesis occurs.

The Transcription of mRNA

The two DNA strands in a gene are complementary. One of the two strands contains the triplets that specify the sequence of amino acids in the polypeptide.

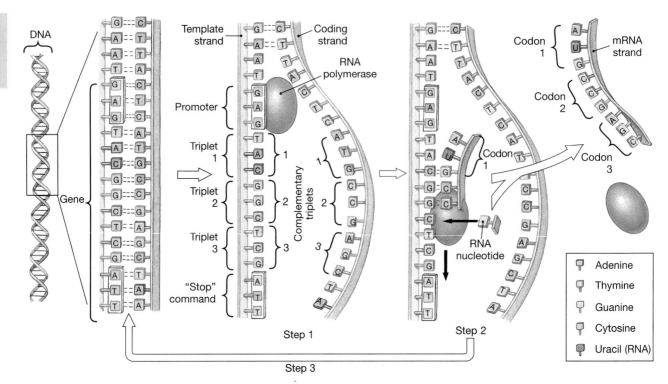

•**FIGURE 3-23 mRNA Transcription.** A small portion of a single DNA molecule, containing a single gene available for transcription. *Step 1:* The two DNA strands separate, and RNA polymerase binds to the promoter of the gene. *Step 2:* The RNA polymerase moves from one nucleotide to another along the length of the template strand. At each site, complementary RNA nucleotides form hydrogen bonds with the DNA nucleotides of the template strand. The RNA polymerase then strings the arriving nucleotides together into a strand of mRNA. *Step 3:* On reaching the stop signal at the end of the gene, the RNA polymerase and the mRNA strand detach, and the two DNA strands reassociate.

This is the **coding strand.** The other strand, called the **template strand,** contains complementary triplets that will be used as a template for mRNA production. The resulting mRNA will have a nucleotide sequence identical to that of the coding strand but with uracil substituted for thymine. Figure 3-23• details transcription:

STEP 1. Once the DNA strands have separated and the promoter has been exposed, transcription can begin. The key event is the attachment of RNA polymerase to the template strand.

STEP 2. RNA polymerase promotes hydrogen bonding between the nitrogenous bases of the template strand and complementary nucleotides in the nucleoplasm. This enzyme begins at a "start" signal in the promoter region. It then strings nucleotides together by covalent bonding. The RNA polymerase interacts with only a small portion of the template strand at any one time as it travels along the DNA strand. The complementary strands separate in front of the enzyme as it moves one nucleotide at a time, and they reassociate behind it. The enzyme collects additional nucleotides and attaches them to the growing chain. The nucleotides involved are those characteristic of RNA, not of DNA; RNA polymerase can attach adenine, guanine, cytosine, or uracil but never thymine. Thus, wherever an A occurs in the DNA strand,

the polymerase will attach a U rather than a T to the growing mRNA strand. In this way, RNA polymerase assembles a complete strand of mRNA. The nucleotide sequence of the template strand determines the nucleotide sequence of the mRNA strand. Thus, each DNA triplet will correspond to a sequence of three nucleotide bases in the mRNA strand. Such a three-base mRNA sequence is called a **codon** (KŌ-don). Codons contain nitrogenous bases that are complementary to those of the triplets in the template strand. For example, if the DNA triplet is TCG, the corresponding mRNA codon will be AGC.

STEP 3. At the "stop" signal, the enzyme and the mRNA strand detach from the DNA strand, and transcription ends. The complementary DNA strands now complete their reassociation as hydrogen bonding occurs between complementary base pairs.

Each gene includes a number of triplets that do not contain information needed to build a functional protein. As a result, the mRNA strand assembled during transcription, sometimes called immature mRNA or *pre-mRNA,* must be "edited" before it leaves the nucleus to direct protein synthesis. In this **RNA processing,** nonsense regions, called **introns,** are snipped out, and the remaining segments, or **exons,** are spliced together. This process creates a much shorter, functional strand of mRNA that then enters the cytoplasm through one of the nuclear pores.

Translation

Protein synthesis is the assembling of functional polypeptides in the cytoplasm. Figure 3-24• provides an overview of this process. Protein synthesis occurs through translation. **Translation** is the formation of a linear chain of amino acids by using the information an mRNA strand provides. This process occurs at a ribosome. The codons of an mRNA strand carry the message spelled out in the triplets of the coding strand of the gene. Each mRNA codon designates a particular amino acid to be incorporated into the polypeptide chain. During translation, the sequence of codons will determine the sequence of amino acids in the polypeptide. The amino acids are provided by **transfer RNA (tRNA),** a relatively small and mobile type of RNA. Each tRNA molecule binds and delivers an amino acid of a specific type. More than 20 kinds of transfer RNA exist—at least one for each of the amino acids used in protein synthesis.

Each tRNA molecule has a tail that binds an amino acid. Roughly midway along its length, the nucleotide chain of the tRNA forms a tight loop. It is this loop that can interact with an mRNA strand. The loop contains three nitrogenous bases that form an **anticodon.** The anticodon can bond complementarily with an appropriate mRNA codon during translation. The base sequence of the anticodon indicates the type of amino acid carried by the tRNA. For example, a tRNA with the anticodon GGG always carries the amino acid *proline,* whereas a tRNA with the anticodon CGG carries *alanine.* Table 3-3 provides examples of several codons and anticodons that specify individual amino acids and summarizes the relationships among DNA, codons, and anticodons.

The tRNA molecules thus provide the physical link between codons and amino acids. During translation, each codon along the mRNA strand will bind a

TABLE 3-3 Examples of the Triplet Code

DNA Triplets		mRNA Codon	tRNA Anticodon	Amino Acid
Coding Strand	Template Strand			
TTT	AAA	UUU	AAA	Phenylalanine
TTA	AAT	UUA	AAU	Leucine
TGT	ACA	UGU	ACA	Cysteine
GTT	CAA	GUU	CAA	Valine
CCC	GGG	CCC	GGG	Proline
GCC	CGG	GCC	CGG	Alanine

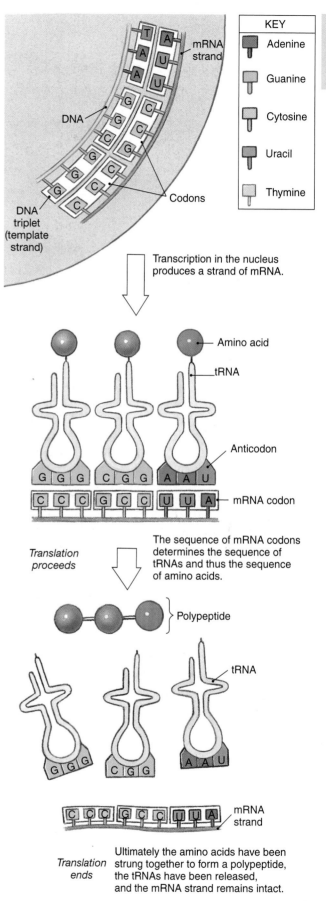

Transcription in the nucleus produces a strand of mRNA.

Translation proceeds

The sequence of mRNA codons determines the sequence of tRNAs and thus the sequence of amino acids.

Translation ends Ultimately the amino acids have been strung together to form a polypeptide, the tRNAs have been released, and the mRNA strand remains intact.

•FIGURE 3-24 An Overview of Protein Synthesis

3

Step 1: The mRNA strand binds to the small ribosomal subunit and is joined at the start codon by the first tRNA, which carries the amino acid methionine. Binding occurs between complementary base pairs of the codon and anticodon.

Step 2: The small and large ribosomal subunits interlock around the mRNA strand.

•FIGURE 3-25 The Process of Translation. For clarity, the components are not drawn to scale and their three-dimensional relationships have been simplified.

complementary anticodon on a tRNA molecule. Thus, if the mRNA has the codons CCC–GCC–UUA, it will bind to tRNAs with anticodons GGG–CGG–AAU. The amino acid sequence of the polypeptide chain created is determined by the sequence of delivery by tRNAs, and that sequence depends on the arrangement of codons along the mRNA strand. In this case, the amino acid sequence in the polypeptide would be proline–alanine–leucine.

The translation process is illustrated in Figure 3-25•:

STEP 1. Translation begins as the mRNA strand binds to a small ribosomal subunit. The first codon, or *start codon,* of the mRNA strand always has the base sequence AUG. It binds a tRNA with the complementary anticodon sequence UAC. This tRNA, which carries the amino acid *methionine,* attaches to the first of two tRNA binding sites on the small ribosomal subunit. (The initial methionine will be removed from the finished protein.)

STEP 2. When this tRNA binding occurs, a large ribosomal subunit joins the complex to create a complete ribosome. The mRNA strand nestles in the gap between the small and the large ribosomal subunits.

STEP 3. A second tRNA now arrives at the second tRNA binding site of the ribosome, and its anticodon binds to the next codon of the mRNA strand.

STEP 4. Enzymes of the large ribosomal subunit then break the linkage between the tRNA and its amino acid. At the same time, the enzymes attach the amino acid to

its neighbor by means of a peptide bond. The ribosome then moves one codon down the mRNA strand.

STEP 5. The cycle is then repeated with the arrival of another molecule of tRNA. The tRNA already stripped of its amino acid drifts away. It will soon bind to another amino acid and be ready to participate in protein synthesis again.

STEP 6. The polypeptide chain continues to grow by the addition of amino acids until the ribosome reaches a "stop" signal, or *stop codon,* at the end of the mRNA strand. The ribosomal subunits now detach, leaving an intact strand of mRNA and a completed polypeptide.

Translation proceeds swiftly, producing a typical protein in about 20 seconds. The mRNA strand remains intact, and it can interact with other ribosomes to create additional copies of the same polypeptide chain. The process does not continue indefinitely, however, because after a few minutes to a few hours, mRNA strands are broken down and the nucleotides are recycled. However, large numbers of protein chains can be produced during that time. Although only two mRNA codons are "read" by a ribosome at any one time, the entire strand may contain thousands of codons. As a result, many ribosomes can bind to a single mRNA strand. At any moment, each ribosome will be reading a different part of the same message, but each will end up constructing a copy of the same protein as the others. The arrangement is similar to a line of people at a

Step 3: A second tRNA arrives at the adjacent binding site of the ribosome. The anticodon of the second tRNA binds to the next mRNA codon.

Peptide bond

Step 4: The first amino acid is detached from its tRNA and is joined to the second amino acid by a peptide bond. The ribosome moves one codon farther along the mRNA strand; the first tRNA detaches as another tRNA arrives.

buffet lunch; all the people will assemble the same meal, but each person is always a step behind the person ahead. A series of ribosomes attached to the same mRNA strand is called a *polyribosome,* or *polysome* (Figure 3-14a•, p. 81).

DNA AND THE CONTROL OF CELL STRUCTURE AND FUNCTION

The DNA of the nucleus controls the cell by directing the synthesis of specific proteins. Through control of protein synthesis, virtually every aspect of cell structure and function can be regulated. Two levels of control are involved:

1. The DNA of the nucleus has *direct* control over the synthesis of structural proteins, such as cytoskeletal components, membrane proteins (including receptors), and secretory products. By issuing appropriate orders, the nucleus can alter the internal structure of the cell, its sensitivity to substances in its environment, or its secretory functions to meet changing needs.

2. The DNA of the nucleus has *indirect* control over all other aspects of cellular metabolism, because it regulates the synthesis of enzymes. By ordering the production or stopping the production of appropriate enzymes, the nucleus can regulate all metabolic activities and functions of the cell. For example, the nucleus can accelerate the rate of glycolysis by increasing the number of needed enzymes in the cytoplasm.

☑ How does the nucleus control the activities of a cell?

☑ What process would be affected by the lack of the enzyme RNA polymerase?

Step 5: This cycle is repeated as the ribosome moves along the length of the mRNA strand, binds new tRNAs, and incorporates their amino acids into the polypeptide chain.

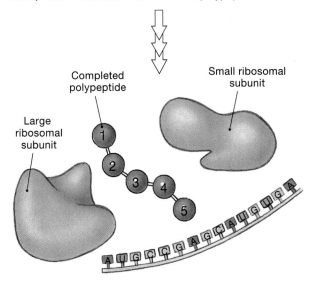

Completed polypeptide

Small ribosomal subunit

Large ribosomal subunit

Step 6: The chain elongates until the stop codon is reached; the components then separate.

3

THE CELL LIFE CYCLE

The period between fertilization and physical maturity involves tremendous changes in organization and complexity. At fertilization a single cell is all there is; at maturity an individual has roughly 75 trillion cells. This amazing transformation involves a form of cellular reproduction called **cell division.** The division of a single cell produces a pair of **daughter cells,** each half the size of the original. Before dividing, each of the daughter cells will grow to the size of the original cell.

Even when development is complete, cell division continues to be essential to survival. Cells are highly adaptable, but they can be damaged by physical wear and tear, toxic chemicals, temperature changes, and other environmental stresses. And cells, like individuals, are subject to aging. The life span of a cell varies from hours to decades, depending on the type of cell and the stresses involved. Many cells appear to be programmed to self-destruct after a certain period of time as a result of the activation of specific "suicide genes" in the nucleus. The genetically controlled death of cells is called **apoptosis** (ap-op-TŌ-sis or ap-ō-TŌ-sis; *ptosis,* a falling away). A gene involved in the regulation of this process has been identified. This gene, called *bcl-2,* appears to prevent apoptosis and to keep a cell alive and functional. If something interferes with the function of this gene, the cell self-destructs.

Because a typical cell does not live nearly as long as a typical person, cell populations must be maintained over time by cell division. For cell division to be successful, the genetic material in the nucleus must be duplicated accurately and one copy must be distributed to each daughter cell. The duplication of the cell's genetic material is called **DNA replication,** and nuclear division is called **mitosis** (mī-TŌ-sis). Mitosis occurs during the division of somatic cells. The production of sex cells involves a different process, **meiosis** (mī-Ō-sis), to be described in Chapter 28.

Figure 3-26● presents the life cycle of a typical cell. That life cycle includes a fairly brief period of mitosis alternating with an *interphase* of variable duration.

INTERPHASE

Most cells spend only a small part of their time actively engaged in cell division. Somatic cells spend the majority of their functional lives in a state known as **interphase.** During interphase, a cell performs all its normal functions and, if necessary, prepares for cell division. In a cell preparing to divide, interphase can be divided into the G_1, S, and G_2 *phases.* An interphase cell in the G_0 *phase* is not preparing for division but is performing all other normal cell functions. Some mature cells, such as skeletal muscle cells and most neurons, remain in G_0 indefinitely and never divide. In contrast, *stem cells,* which divide repeatedly with very brief interphase periods, never enter G_0.

●**FIGURE 3-26** The Cell Life Cycle

The G_1 Phase

A cell ready to divide first enters the **G_1 phase.** In this phase, the cell makes enough mitochondria, cytoskeletal elements, endoplasmic reticulum, ribosomes, Golgi membranes, and cytosol for two functional cells. Centriole replication begins in G_1 and commonly continues until G_2. In cells dividing at top speed, G_1 may last just 8–12 hours. Such cells pour all their energy into mitosis, and all other activities cease. If G_1 lasts for days, weeks, or months, preparation for mitosis occurs as the cells perform their normal functions.

The S Phase

When the activities of G_1 have been completed, the cell enters the **S phase.** Over the next 6–8 hours, the cell duplicates its chromosomes. This involves DNA replication and the synthesis of histones and other proteins in the nucleus. The goal of DNA replication, which occurs in cells preparing to undergo either mitosis or meiosis, is to copy the genetic information in the nucleus. In mitosis, one chromosome copy is given to each of the two daughter cells.

DNA Replication Each DNA molecule consists of a pair of DNA strands joined by hydrogen bonding between complementary nitrogenous bases. ∞ *p. 57* Figure 3-27● diagrams DNA replication. It begins when enzymes unwind the strands and disrupt the weak bonds between the bases. As the strands unwind, molecules of **DNA polymerase** bind to the exposed nitrogenous bases. This enzyme (1) promotes bonding between the

•**FIGURE 3-27 DNA Replication.** In DNA replication, the DNA strands unwind and DNA polymerase begins attaching complementary DNA nucleotides along each strand. On one original strand, the complementary copy is produced as a continuous strand. Along the other original strand, the copy begins as a series of short segments that are spliced together by ligases. This process ultimately produces two identical copies of the original DNA molecule.

nitrogenous bases of the DNA strand and complementary DNA nucleotides dissolved in the nucleoplasm and (2) links the nucleotides by covalent bonds.

Many molecules of DNA polymerase work simultaneously along the DNA strands (Figure 3-27•). DNA polymerase can work in only one direction along a strand of DNA, but the two strands in a DNA molecule are oriented differently. As a result, the DNA polymerase on one strand works toward the site where the strands are unzipping, but those on the other strand work away from it. As the two original strands gradually separate, the DNA polymerase bound to one strand adds nucleotides to make a single, continuous complementary copy of that strand. Refer to the upper strand in Figure 3-27•. This copy grows toward the "zipper" from right to left, adding nucleotides in the sequence

$$9 \leftarrow 8 \leftarrow 7 \leftarrow 6 \leftarrow 5 \leftarrow 4 \leftarrow 3 \leftarrow 2 \leftarrow 1$$

The 1 is added first, then 2 to the left of 1, and so on.

But DNA polymerase on the other original strand can work only toward the free end of the DNA strand, away from the unzipping site. Consider the lower strand in Figure 3-27•. The first DNA polymerase to bind to it must work from left to right, adding nucleotides in the sequence $1 \rightarrow 2 \rightarrow 3 \rightarrow 4 \rightarrow 5$. But as the original strands continue to unzip, additional nucleotides are continuously exposed. This molecule of DNA polymerase cannot go into reverse; it can only continue working from left to right. Thus, a second molecule of DNA polymerase must bind closer to the point of unzipping and assemble a complementary copy that grows in the sequence $6 \rightarrow 7 \rightarrow 8 \rightarrow 9$ until it bumps into the segment created by the first DNA polymerase. The two segments are then spliced together by enzymes called **ligases** (LĪ-gās-ez; *liga,* to tie).

Eventually, the unzipping completely separates the original strands. The copying ends, the last splicing is done, and two identical DNA molecules have formed.

The G₂ Phase

Once DNA replication has ended, there is a brief (2–5 hour) **G₂ phase** devoted to last-minute protein synthesis and to the completion of centriole replication. The cell then enters the **M phase,** and mitosis begins.

Mutations

Mutations are permanent changes in a cell's DNA that affect the nucleotide sequence of one or more genes. The simplest is a **point mutation,** a change in a single nucleotide that affects one codon. The triplet code has some flexibility, because several different codons can specify the same amino acid. But a point mutation that produces a codon that specifies a different amino acid will usually change the structure of the completed protein. A single change in the amino acid sequence of a structural protein or enzyme can prove fatal. Certain cancers and two potentially lethal blood disorders, *thalassemia* and *sickle cell anemia,* result from variations in a single nucleotide.

More than 100 inherited disorders have been traced to abnormalities in enzyme or protein structure that reflect single changes in nucleotide sequence. More-elaborate mutations, such as additions or deletions of nucleotides, can affect multiple codons in one gene or in several adjacent genes, or they can affect the structure of one or more chromosomes.

Because most mutations occur during DNA replication, they are most likely to involve cells that are undergoing cell division. A single cell, a group of cells, or an entire individual may be affected. This last prospect occurs when the changes are made early in development. For example, a mutation affecting the DNA of an individual's sex cells will be inherited by that individual's children. Our understanding of genetic structure is opening the possibility of diagnosing and correcting some of these problems.

AM *Genetic Engineering and Gene Therapy*

3

•**FIGURE 3-28** **Interphase, Mitosis, and Cytokinesis.** Diagrammatic and microscopic views of representative cells undergoing cell division. (LM × 450)

MITOSIS

Mitosis is a process that separates the duplicated chromosomes of a cell into two identical nuclei. Mitosis specifically refers to the division and duplication of the cell's nucleus. Division of the cytoplasm to form two distinct new cells involves a separate but related process known as **cytokinesis** (sī-tō-ki-NĒ-sis; *cyto-*, cell + *kinesis*, motion). Figure 3-28a• depicts interphase, and Figure 3-28b–f• summarizes the four stages of mitosis: *prophase* (early and late), *metaphase, anaphase,* and *telophase.*

Stage 1: Prophase

Prophase (PRŌ-fāz; *pro,* before) begins when the chromosomes coil so tightly that they become visible as individual structures. As a result of DNA replication during the S phase, two copies of each chromosome now exist. As noted earlier (p. 91), each copy is called a chromatid. A chromatid is connected to its duplicate copy at the centromere, the region surrounded by the protein complex of the kinetochore.

As the chromosomes appear, the nucleoli disappear. This disappearance occurs in late prophase, often called *prometaphase.* Around this time, the two pairs of centrioles replicated during the G₁–G₂ period move toward opposite poles of the nucleus. An array of microtubules called **spindle fibers** extends between the centriole pairs. Smaller microtubules called *astral*

rays radiate into the surrounding cytoplasm. Late in prophase, the nuclear envelope disappears. The spindle fibers now form among the chromosomes, and the kinetochore of each chromatid becomes attached to a spindle fiber called a *chromosomal microtubule.*

Stage 2: Metaphase

Metaphase (MET-a-fāz; *meta,* after) begins as the chromatids move to a narrow central zone called the **metaphase plate.** Metaphase ends when all the chromatids are aligned in the plane of the metaphase plate.

Stage 3: Anaphase

Anaphase (AN-a-fāz; *ana,* apart) begins when the centromere of each chromatid pair splits and the chromatids separate. The two **daughter chromosomes** are now pulled toward opposite ends of the cell along the chromosomal microtubules. This movement involves an interaction between the kinetochore and the microtubule. Anaphase ends when the daughter chromosomes arrive near the centrioles at opposite ends of the cell.

Stage 4: Telophase

During **telophase** (TĒL-ō-fāz; *telo,* end), each cell prepares to return to the interphase state. The nuclear membranes form, the nuclei enlarge, and the chromosomes gradually uncoil. Once the chromosomes have relaxed and the fine filaments of chromatin become

(e) Anaphase

(f) Telophase

Daughter chromosomes

Cytokinesis

Daughter cells

visible, nucleoli reappear and the nuclei resemble those of interphase cells. This stage marks the end of mitosis.

CYTOKINESIS

Cytokinesis is the cytoplasmic division of the daughter cells. This process usually begins in late anaphase. As the daughter chromosomes approach the ends of the spindle apparatus, the cytoplasm constricts along the plane of the metaphase plate. This process continues throughout telophase and is usually completed sometime after a nuclear membrane has reformed around each daughter nucleus. The completion of cytokinesis marks the end of cell division.

THE MITOTIC RATE AND ENERGETICS

The preparations for cell division that occur between G_1 and the M phase are difficult to recognize in a light micrograph. However, the start of mitosis is easy to recognize, because the chromosomes become condensed and highly visible. The frequency of cell division can thus be estimated by the number of cells in mitosis at any time. As a result, we often use the term **mitotic rate** when we discuss rates of cell division. In general, the longer the life expectancy of a cell type, the slower the

mitotic rate. Long-lived cells, such as muscle cells and neurons, either never divide or do so only under special circumstances. Other cells, such as those covering the surface of the skin or the lining of the digestive tract, are subject to attack by chemicals, pathogens, and abrasion. They survive for only days or even hours. Special cells called **stem cells** maintain these cell populations through repeated cycles of cell division.

Stem cells are relatively unspecialized, and their only function is the production of daughter cells. Each time a stem cell divides, one of its daughter cells develops functional specializations while the other prepares for further stem cell divisions. The rate of stem cell division can vary with the tissue type and the demand for new cells. In heavily abraded skin, stem cells may divide more than once a day, but stem cells in adult connective tissues may remain inactive for years.

Dividing cells use an unusually large amount of energy. For example, they must synthesize new organic materials and move organelles and chromosomes within the cell. All these processes require ATP in substantial amounts. Cells that do not have adequate energy sources cannot divide. In a person who is starving, normal cell growth and maintenance grind to a halt. For this reason, prolonged starvation slows wound healing, lowers resistance to disease, thins the skin, and changes the lining of the digestive tract.

3

REGULATION OF THE CELL LIFE CYCLE

In normal tissues, the rate of cell division balances the rate of cell loss or destruction. Mitotic rates are genetically controlled, and many different stimuli may be responsible for activating genes that promote cell division. Some of the stimuli are internal, and many cells set their own pace of mitosis and cell division. An important internal trigger is the level of **m-phase promoting factor (MPF),** also known as *maturation-promoting factor.* MPF is assembled from two parts: a *cell division cycle protein* called *Cdc2* and a second protein, called *cyclin.* Cyclin levels climb as the cell life cycle proceeds. When levels are high enough, MPF appears in the cytoplasm and mitosis gets under way.

Various extracellular compounds, generally peptides, can stimulate the division of specific cell types. These compounds include several hormones and a variety of **growth factors.** Table 3-4 lists some of these stimulatory compounds and their target tissues; we shall discuss these hormones and factors in later chapters. Each of these compounds appears to exert its effects by binding to receptors on the cell membrane. This binding initiates a series of biochemical events that ultimately trigger cell division. The effects of these stimulatory factors may be opposed by a poorly understood class of peptides called *chalones* (KĀ-lōnz).

Many of the peptide growth factors bind to membrane receptors at the cell surface. Binding activates enzymes inside the cell, starting a chain reaction leading to the activation of genes that prepare the cell to undergo cell division. The presence or absence of the appropriate binding protein determines whether a particular cell will respond to a particular growth factor. The mechanism inside the cell appears to be very similar, however, regardless of the type of membrane receptor. Binding at the membrane surface triggers the activation and release of intermediaries known as *Ras proteins.* These proteins in turn activate intracellular enzymes and promote gene activation. Growth factors that do not use Ras proteins may use other intermediaries or may enter the cell and exert their effects directly on the nucleus.

Genetic mechanisms for inhibiting cell division have recently been identified. The genes involved are known as *repressor genes.* One gene, called *p53,* controls a protein that resides in the nucleus and activates genes that direct the production of growth-inhibiting factors inside the cell. Roughly half of all cancers are associated with abnormal forms of the *p53* gene.

Cell Division and Cancer

When the rate of cell division and growth exceeds the rate of cell death, a tissue begins to enlarge. A **tumor,** or **neoplasm,** is a mass or swelling produced by abnormal cell growth and division. In a **benign tumor,**

TABLE 3-4 Representative Chemical Factors Affecting Cell Division

Factor	Source(s)	Effect(s)	Target(s)
M-phase promoting factor (maturation-promoting factor)	Forms within cytoplasm from Cdc2 and cyclin	Triggers start of mitosis	Regulatory mechanism active in all dividing cells
Growth hormone	Anterior pituitary gland	Stimulation of growth, cell division, differentiation	All cells, especially in epithelia and connective tissues
Prolactin	Anterior pituitary gland	Stimulation of cell growth, division, development	Gland and duct cells of mammary glands
Nerve growth factor (NGF)	Salivary glands; other sources suspected	Stimulation of nerve cell repair and development	Neurons and neuroglia
Epidermal growth factor (EGF)	Duodenal glands; other sources suspected	Stimulation of stem cell divisions and epithelial repairs	Epidermis of skin
Fibroblast growth factor (FGF)	Unknown	Division and differentiation of fibroblasts and related cells	Connective tissues
Erythropoietin	Kidneys (primary source)	Stimulation of stem cell divisions and maturation of red blood cells	Bone marrow
Thymosins and related compounds	Thymus	Stimulation of division and differentiation of lymphocytes (especially T cells)	Thymus and other lymphoid tissues and organs
Chalones	Many tissues	Inhibition of cell division	Cells in the immediate area

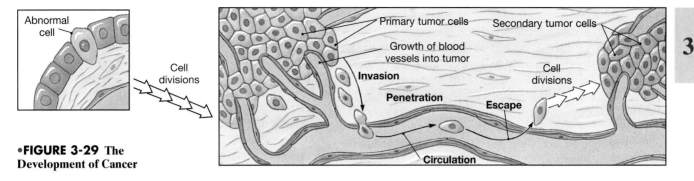

•**FIGURE 3-29** The Development of Cancer

the cells usually remain within a connective tissue capsule. Such a tumor seldom threatens an individual's life. The tumor can commonly be surgically removed if its size or position disturbs tissue function.

Cells in a **malignant tumor** no longer respond to normal control mechanisms. These cells spread into surrounding tissues from the *primary tumor* (or *primary neoplasm*). This spreading is called **invasion.** Malignant cells may also travel to distant tissues and organs and establish *secondary tumors.* This dispersion is called **metastasis** (me-TAS-ta-sis; *meta,* after + *stasis,* standing still). Metastasis is dangerous and difficult to control.

Cancer is an illness characterized by malignant cells. Cancer develops in the series of steps diagrammed in Figure 3-29•. Initially, the cancer cells are restricted to the primary tumor. In most cases, all the cells in the tumor are the daughter cells of a single malignant cell. Normal cells often become malignant when a mutation occurs in a gene involved with cell growth, differentiation, or division. The modified genes are called **oncogenes** (ON-kō-gēnz). [AM] *A Closer Look: Cancer*

Cancer cells gradually lose their resemblance to normal cells. They change size and shape, typically becoming abnormally large or small. At first, the growth of the primary tumor distorts the tissue, but the basic tissue organization remains intact. Metastasis begins with invasion as tumor cells "break out" of the primary tumor and invade the surrounding tissue. They may then enter the lymphatic system and accumulate in nearby lymph nodes. When metastasis involves the penetration of blood vessels, the cancer cells circulate throughout the body.

Responding to cues that are as yet unknown, cancer cells in the circulatory system ultimately escape out of the blood vessels to establish secondary tumors at other sites. These tumors are extremely active metabolically, and their presence stimulates the growth of blood vessels into the area. The increased circulatory supply provides additional nutrients to the cancer cells and further accelerates tumor growth and metastasis.

Organ function begins to deteriorate as the number of cancer cells increases. These cells may no longer perform their original functions, or they may perform normal functions in an unusual way. For example, endocrine cancer cells may produce normal hormones but in excessively large amounts. Cancer cells do not use energy very efficiently. They grow and multiply at the expense of healthy tissues, competing for space and nutrients with normal cells. This competition accounts for the starved appearance of many patients in the late stages of cancer. Death may occur as a result of the compression of vital organs when nonfunctional cancer cells have killed or replaced the healthy cells in those organs or when the cancer cells have starved normal tissues of essential nutrients.

The growth of blood vessels into the tumor is a vital step in the development and spread of the cancer. Without those vessels, the growth and metastasis of the cancer cells would be limited by the availability of oxygen and nutrients. A polypeptide called *antiangiogenesis factor* can prevent the growth of blood vessels and can slow the growth of cancers. This chemical, produced in normal human cartilage, can be extracted in large quantities from sharks, whose skeletons are entirely cartilaginous. Sharks are now being collected to obtain antiangiogenesis factor for use in experimental cancer therapies. We will return to the subject of cancer in later chapters that deal with specific systems.

CELL DIVERSITY AND DIFFERENTIATION

The liver cells, fat cells, and neurons of an individual contain the same chromosomes and genes, but in each case a different set of genes has been turned *off.* In other words, liver cells differ from other cells because different genes are accessible for transcription.

When a gene is functionally eliminated, the cell loses the ability to create a particular protein—and thus to perform any functions involving that protein. Each time another gene switches off, the cell's functional abilities become more restricted. This specialization process is called **differentiation.**

Fertilization produces a single cell with all its genetic potential intact. Repeated cell divisions follow, and differentiation begins as the number of cells increases. Differentiation produces specialized cells with limited capabilities. These cells form organized collections known as *tissues,* each with discrete functional roles. In Chapter 4, we shall examine the structure and function of tissues and shall consider the role of tissue interactions in the maintenance of homeostasis.

☑ A cell is actively manufacturing enough organelles to serve two functional cells. This cell is probably in which phase of its life cycle?

☑ During DNA replication, a nucleotide is deleted from a sequence that normally codes for a polypeptide. What effect will this deletion have on the amino acid sequence of the polypeptide?

☑ What would happen if spindle fibers failed to form in a cell during mitosis?

According to the cell theory, cells are the fundamental units of all plant and animal tissues. An understanding of cells is critical to a grasp of the anatomy and physiology of higher levels of organization. To gain a better understanding of cells, use the resources available on both the CD-ROM and the Companion Website. The CD-ROM includes animations and tutorials; the website builds on those resources with text content and interactive problems as well as communication avenues for sharing ideas with your class. To complete the following Web Explorations, visit the MediaLab in Chapter 3 of your Companion Website.

Web Exploration 1

Estimated time for completion = 15 minutes

The nucleus houses deoxyribonucleic acid (DNA), which stores all of a cell's genetic information. That information is needed to synthesize proteins through transcription and translation. To visualize the steps involved, select the keyword **REPLICATION** on the Companion Website. Click on

the DNA Workshop activity to transcribe and translate a series of nucleic acids. Reflect on how the cell carries out these processes. What is a mutation? Relate what you have experienced to genetic change. If you wish, share your ideas with your classmates, using the communications tools of the Companion Website.

Web Exploration 2

Estimated time for completion = 15 minutes

When a cell undergoes mitosis, it forms two identical daughter cells. The subsection "Mitosis" in the text explains this process. Draw a diagram of the cell cycle. What triggers a cell to begin and continue mitosis? Select the keyword **MITOSIS** on the Companion Website to read the article presented there. Focus on the roles of promoting factors and cyclins. Then add the promoting factors and cyclins at the appropriate spots on your diagram. You might share your diagram, in the form of a poster, with your classmates or use the communication tools on the Companion Website to describe your findings.

CHAPTER REVIEW

SELECTED CLINICAL TERMINOLOGY
Terms Discussed in This Chapter

benign tumor: A mass or swelling in which the cells usually remain within a connective tissue capsule; rarely life-threatening. *(p. 100)*

cancer: An illness characterized by the metastasis of malignant tumor cells. *(p. 101)*

dextran: A carbohydrate that cannot cross cell membranes; commonly administered in solution to patients after blood loss or dehydration. *(p. 74)*

DNA fingerprinting: Identifying an individual on the basis of repeating nucleotide sequences in his or her DNA. *(p. 91)*

invasion: The spread of cancer cells from a primary tumor into surrounding tissues. *(p. 101)*

malignant tumor: A mass or swelling in which the cells no longer respond to normal control mechanisms but divide rapidly. *(p. 101)*

metastasis (me-TAS-ta-sis): The spread of malignant cells into distant tissues and organs, where secondary tumors subsequently develop. *(p. 101)*

normal saline: A solution that approximates the normal osmotic concentration of extracellular fluids. *(p. 74)*

oncogene (ON-kō-jēn): A cancer-causing gene created by a somatic mutation in a normal gene involved with growth, differentiation, or cell division. *(p. 101)*

primary tumor (*primary neoplasm*): The site at which a cancer cell initially develops. *(p. 101)*

secondary tumor: A colony of cancerous cells formed by metastasis. *(p. 101)*

tumor (neoplasm): A mass or swelling produced by abnormal cell growth and division. *(p. 100)*

AM Additional Terms Discussed in the *Applications Manual*

carcinogen (kar-SIN-ō-jen): An environmental factor that stimulates the conversion of a normal cell to a cancer cell.
genetic engineering: The research and experiments related to changing the genetic makeup of an organism.
karyotyping (KAR-ē-ō-tī-ping): The determination of the shape, size, and centromere position of a person's chromosomes.

mutagen (MŪ-ta-jen): A factor that can damage DNA strands and may cause chromosomal breakage, stimulating the development of cancer cells.
recombinant DNA: DNA created by splicing a specific gene from one organism into the DNA strand of another organism.

STUDY OUTLINE

INTRODUCTION p. 65

1. Contemporary *cell theory* incorporates several basic concepts: (1) **Cells** are the building blocks of all plants and animals; (2) cells are produced by the division of preexisting cells; (3) cells are the smallest units that perform all vital physiological functions; (4) each cell maintains homeostasis at the cellular level; and (5) homeostasis at the tissue, organ, organ system, and organism levels reflects the combined and coordinated actions of many cells. *(Figure 3-1)*

STUDYING CELLS p. 65

1. **Cytology** is the study of the structure and function of cells; it is part of **cell biology.**

An Overview of Cellular Anatomy p. 66

2. The human body contains two types of cells: **sex cells** (*sperm* and *oocytes*) and **somatic cells** (all other cells).

THE CELL MEMBRANE p. 68

1. A typical cell is surrounded by **extracellular fluid,** specifically the **interstitial fluid** of a tissue. The cell's outer boundary is the **cell membrane,** or **plasma membrane.** The membrane's functions include physical isolation, regulation of exchange with the environment, sensitivity, and structural support. *(Figure 3-2; Table 3-1)*

Membrane Structure p. 68

2. The cell membrane, which is a **phospholipid bilayer,** contains lipids, proteins, and carbohydrates. Phospholipids are the largest contributors to membrane structure. *(Figure 3-3)*
3. **Integral proteins** are part of the membrane itself; **peripheral proteins** are attached but can separate from it.
4. Membrane proteins can act as anchors **(anchoring proteins),** identifiers **(recognition proteins),** enzymes, receptors **(receptor proteins),** carriers **(carrier proteins),** or **channels.** **Leak channels** allow water and ions to move across the membrane at all times; **gated channels** open or close to regulate ion passage. *(Figure 3-4)*

Membrane Permeability p. 71

5. **Permeability** is the ease with which substances can cross the cell membrane. Nothing can pass through an **impermeable** barrier; anything can pass through a **freely permeable** barrier. Cell membranes are **selectively permeable.**
6. **Diffusion** is the net movement of material from an area where its concentration is higher to an area where its concentration is lower. Diffusion occurs until the **concentration gradient** is eliminated. *(Figures 3-5, 3-6)*
7. **Osmosis** is the diffusion of water across a membrane in response to differences in solute concentration. **Osmotic pressure** is the force of water movement into a solution as the result of solute concentration. **Hydrostatic pressure** can oppose osmotic pressure. Water molecules undergo *bulk flow,* movement in groups across a membrane. *(Figure 3-7)*

8. **Tonicity** describes the effects of osmotic solutions on cells. A solution that does not cause an osmotic flow is **isotonic.** One that causes water to flow into a cell is **hypotonic** and can lead to **hemolysis.** One that causes water to flow out of a cell is **hypertonic** and can lead to **crenation.** *(Figure 3-8)*
9. In **filtration,** hydrostatic pressure forces water across a membrane. If membrane pores are large enough, molecules of solute will be carried along.
10. In **facilitated diffusion,** a type of **carrier-mediated transport,** compounds are transported across a membrane after binding to a **receptor site** of a carrier protein. *(Figure 3-9)*
11. **Active transport** mechanisms consume ATP but are independent of concentration gradients. Some **ion pumps** are **exchange pumps. Secondary active transport** may involve **cotransport** or **countertransport.** *(Figures 3-10, 3-11)*
12. In **vesicular transport,** material moves into or out of the cell in membranous sacs called **vesicles.** Movement into the cell is accomplished through **endocytosis,** an active process that can take three forms: **receptor-mediated endocytosis** (by means of **coated vesicles**), **pinocytosis,** or **phagocytosis** (using **pseudopodia**). The ejection of materials from the cytoplasm is accomplished by **exocytosis.** *(Figures 3-12, 3-13, 3-19; Table 3-2)*
13. The **potential difference** between the two sides of a cell membrane is a **transmembrane potential.** The transmembrane potential in an undisturbed cell is its **resting potential.**

THE CYTOPLASM p. 80

The Cytosol p. 80

1. The **cytoplasm** contains the fluid **cytosol** and the **organelles** the cytosol surrounds.

Organelles p. 80

2. **Nonmembranous organelles** are not enclosed by membranes, and they are always in contact with the cytosol. They include the *cytoskeleton, microvilli, centrioles, cilia, flagella,* and *ribosomes. (Table 3-1)*
3. **Membranous organelles** are surrounded by lipid membranes that isolate them from the cytosol. They include the *endoplasmic reticulum,* the *Golgi apparatus, lysosomes, peroxisomes,* and *mitochondria. (Table 3-1)*
4. The **cytoskeleton** gives the cytoplasm strength and flexibility. It has four components: **microfilaments** (typically made of **actin**), **intermediate filaments, microtubules** (made of **tubulin**), and **thick filaments** (made of **myosin**). *(Figure 3-14)*
5. **Microvilli** are small projections of the cell membrane that increase the surface area exposed to the extracellular environment. *(Figure 3-14)*
6. **Centrioles** direct the movement of chromosomes during cell division and organize the cytoskeleton. The **centrosome** is the cytoplasm surrounding the centrioles. *(Figure 3-15)*
7. **Cilia,** anchored by a **basal body,** beat rhythmically to move fluids or secretions across the cell surface. *(Figure 3-15)*

3

8. **Flagella** move a cell through surrounding fluid rather than moving fluid past a stationary cell.

9. A **ribosome** is an intracellular factory that manufactures proteins. Ribosomes are composed of a **small** and a **large ribosomal subunit,** both of which contain **ribosomal RNA (rRNA). Free ribosomes** are in the cytoplasm, and **fixed ribosomes** are attached to the rough endoplasmic reticulum. *(Figure 3-16)*

10. The **endoplasmic reticulum (ER)** is a network of intracellular membranes that function in synthesis, storage, transport, and detoxification. The ER forms hollow tubes, flattened sheets, and round chambers called **cisternae. Rough endoplasmic reticulum (RER)** contains ribosomes on its outer surface and forms **transport vesicles; smooth endoplasmic reticulum (SER)** does neither. *(Figure 3-17)*

11. The **Golgi apparatus** forms **secretory vesicles** and new membrane components, and packages *lysosomes.* Secretions are discharged from the cell by exocytosis. *(Figures 3-18, 3-19)*

12. **Lysosomes** are vesicles filled with digestive enzymes.

13. **Peroxisomes** carry enzymes that neutralize toxins.

14. **Mitochondria** are responsible for 95 percent of the ATP production in a typical cell. The **matrix,** or fluid contents of a mitochondrion, lie inside the **cristae,** or folds of an inner membrane. *(Figure 3-20)*

THE NUCLEUS   p. 89

1. The **nucleus** is the control center of cellular operations. It is surrounded by a **nuclear envelope** (a double membrane with a **perinuclear space**), through which it communicates with the cytosol by way of **nuclear pores.** The nucleus contains a supportive **nuclear matrix.** *(Figure 3-21)*

Chromosome Structure   p. 90

2. The nucleus controls the cell by directing the synthesis of specific proteins, using information stored in **chromosomes.** Chromosomes consist of DNA bound to **histones.** In nondividing cells, chromosomes form a tangle of filaments called **chromatin.** The nucleus may also contain **nucleoli.** Dividing cells contain two copies **(chromatids)** of each chromosome. The chromatids are joined at the **centromere,** a region surrounded by the **kinetochore.** *(Figure 3-22)*

The Genetic Code   p. 91

3. The cell's information storage system, the **genetic code,** is called a *triplet code* because a sequence of three nitrogenous bases codes for a single amino acid. Each **gene** contains all the triplets needed to produce a specific polypeptide chain. *(Table 3-3)*

Gene Activation and Protein Synthesis   p. 91

4. As **gene activation** begins, **RNA polymerase** must bind to the gene.

5. **Transcription** is the formation of RNA from a DNA template. After mRNA transcription, a strand of **messenger RNA (mRNA)** carries instructions from the nucleus to the cytoplasm. *(Figure 3-23)*

6. During **translation,** a functional polypeptide is constructed by using the information contained in the sequence of **codons** along an mRNA strand. The sequence of codons determines the sequence of amino acids in the polypeptide. *(Figures 3-24, 3-25)*

7. By complementary base pairing of **anticodons** to mRNA codons, **transfer RNA (tRNA)** molecules bring amino acids to the ribosomal complex. *(Figures 3-24, 3-25; Table 3-3)*

DNA and the Control of Cell Structure and Function   p. 95

8. The DNA of the nucleus has both direct and indirect control over protein synthesis.

THE CELL LIFE CYCLE   p. 96

1. **Cell division** is the reproduction of cells. **Apoptosis** is the genetically controlled death of cells. **Mitosis** is the nuclear division of somatic cells. Sex cells are produced by **meiosis.**

Interphase   p. 96

2. Most somatic cells spend most of their time in **interphase,** which includes **DNA replication.** *(Figures 3-26, 3-27)*

Mitosis   p. 98

3. Mitosis proceeds in four stages: **prophase, metaphase, anaphase,** and **telophase.** *(Figure 3-28)*

Cytokinesis   p. 99

4. During **cytokinesis,** the cytoplasm is divided, generally producing two identical **daughter cells.** *(Figure 3-28)*

The Mitotic Rate and Energetics   p. 99

5. In general, the longer the life expectancy of a cell type, the slower the **mitotic rate. Stem cells** undergo frequent mitoses to replace other, more-specialized cells.

6. Cell division is genetically controlled.

Regulation of the Cell Life Cycle   p. 100

7. A variety of **growth factors** can stimulate cell division and growth. *(Table 3-4)*

8. Produced by abnormal cell growth and division, a **tumor,** or **neoplasm,** can be **benign** or **malignant.** Malignant cells may spread locally **(invasion)** or to distant tissues and organs **(metastasis). Cancer** is the illness caused by malignant cells. *(Figure 3-29)*

CELL DIVERSITY AND DIFFERENTIATION   p. 101

1. **Differentiation** is the process of specialization that is due to gene regulation and produces cells with limited capabilities. These specialized cells form organized collections called *tissues,* each of which has certain functional roles.

REVIEW QUESTIONS

LEVEL 1 REVIEWING FACTS AND TERMS

1. All the following membrane transport mechanisms are passive processes except
   (a) diffusion
   (b) facilitated diffusion
   (c) vesicular transport
   (d) filtration

2. The following pairs are the principal cations in human body fluids:
   (a) calcium and magnesium
   (b) chloride and bicarbonate
   (c) sodium and potassium
   (d) sodium and chloride

3. _____ ion concentrations are high in the extracellular fluids, and _____ ion concentrations are high in the cytoplasm.
- **(a)** Calcium, magnesium
- **(b)** Chloride, sodium
- **(c)** Potassium, sodium
- **(d)** Sodium, potassium

4. In a resting transmembrane potential, the inside of the cell is _____, and the cell exterior is _____:
- **(a)** slightly negative, slightly positive
- **(b)** slightly positive, slightly negative
- **(c)** slightly positive, neutral
- **(d)** slightly negative, neutral

5. The process that generates carbon dioxide while removing hydrogen atoms from organic molecules is
- **(a)** the TCA cycle
- **(b)** active transport
- **(c)** glycolysis
- **(d)** phosphorylation

6. The reaction sequence in which glucose is broken down into pyruvic acid is
- **(a)** aerobic metabolism
- **(b)** the TCA cycle
- **(c)** mitochondrial energy production
- **(d)** glycolysis

7. The construction of a functional polypeptide by using the information in an mRNA strand is
- **(a)** translation
- **(b)** transcription
- **(c)** replication
- **(d)** gene activation

8. Our somatic cell nuclei contain _____ pairs of chromosomes.
- **(a)** 8
- **(b)** 16
- **(c)** 23
- **(d)** 46

9. Translation ends in the production of a(n)
- **(a)** DNA molecule
- **(b)** mRNA molecule
- **(c)** tRNA molecule
- **(d)** protein

10. Differentiation is
- **(a)** the loss of genes from cells
- **(b)** the acquisition of new functional capabilities by cells
- **(c)** the production of functionally specialized cells
- **(d)** the division of genes among types of cells

11. The interphase of the cell life cycle is divided into
- **(a)** prophase, metaphase, anaphase, and telophase
- **(b)** G_0, G_1, S, and G_2
- **(c)** mitosis and cytokinesis
- **(d)** a, b, and c are correct

12. List the five basic concepts that make up the modern-day cell theory.

13. What are four general functions of the cell membrane?

14. What are the primary functions of membrane proteins?

15. By what four major transport mechanisms do substances get into and out of cells?

16. List four important factors that influence diffusion rates.

17. State three general characteristics of osmosis that define its role in the body.

18. List **(a)** the nonmembranous organelles and **(b)** the membranous organelles of a typical cell.

19. What are the four major functions of the endoplasmic reticulum?

20. List the four stages of mitosis in their order of occurrence.

LEVEL 2 REVIEWING CONCEPTS

21. Diffusion is important in body fluids because it tends to
- **(a)** increase local concentration gradients
- **(b)** eliminate local concentration gradients
- **(c)** move substances against concentration gradients
- **(d)** create concentration gradients

22. Osmotic pressure differs from hydrostatic pressure because the osmotic pressure of a solution is an indication of the force of water movement resulting from
- **(a)** its solute concentration
- **(b)** the volume of water
- **(c)** the permeability of the membrane
- **(d)** a, b, and c are correct

23. When a cell is placed in a _____ solution, the cell will lose water through osmosis. This process results in the _____ of red blood cells.
- **(a)** hypotonic, crenation
- **(b)** hypertonic, crenation
- **(c)** isotonic, hemolysis
- **(d)** hypotonic, hemolysis

24. Suppose that a DNA segment has the following nucleotide sequence: CTC–ATA–CGA–TTC–AAG–TTA. Which nucleotide sequences would a complementary mRNA strand have?
- **(a)** GAG–UAU–GAU–AAC–UUG–AAU
- **(b)** GAG–TAT–GCT–AAG–TTC–AAT
- **(c)** GAG–UAU–GCU–AAG–UUC–AAU
- **(d)** GUG–UAU–GGA–UUG–AAC–GGU

25. How many amino acids are coded in the DNA segment in Review Question 24?
- **(a)** 18
- **(b)** 9
- **(c)** 6
- **(d)** 3

26. What general characteristics are important in carrier-mediated transport mechanisms?

27. (a) What are the similarities between facilitated diffusion and active transport?
- **(b)** What are the differences?

28. What role does the sodium–potassium exchange pump play in stabilizing the resting membrane potential?

29. How does the cytosol differ in composition from interstitial fluid?

30. Differentiate between transcription and translation.

31. List in sequence the phases of the interphase stage of the cell life cycle, and briefly describe what happens in each.

32. List the stages of mitosis, and briefly describe the events that occur in each.

33. (a) What is cytokinesis?
- **(b)** What is its role in the cell cycle?

LEVEL 3 CRITICAL THINKING AND CLINICAL APPLICATIONS

34. The transport of a certain molecule exhibits the following characteristics: (1) The molecule moves down its concentration gradient; (2) at concentrations above a given level, the rate of transport does not increase; and (3) cellular energy is not required for transport to occur. Which transport process is at work?

35. Solutions A and B are separated by a selectively permeable barrier. Over time, the level of fluid on side A increases. Which solution initially had the higher concentration of solute?

36. In kidney dialysis, a person's blood is passed through a bath that contains several ions and molecules. The blood is separated from the dialysis fluid by a membrane that allows water, small ions, and small molecules to pass but does not allow large proteins or blood cells to pass. What should the composition of dialysis fluid be for it to remove urea (a small molecule) without changing blood volume (removing water from the blood)?

CHAPTER 4

The Tissue Level of Organization

Exotic creatures on the deep-sea floor? No. These "creatures" are much smaller than the strange life forms photographed in the ocean depths, and they are much closer to home. These are cells that line the trachea, the airway leading to your lungs. The "tentacles" are cilia that help remove dirt and harmful microbes from air you inhale. Notice that the tracheal lining contains more than one type of cell; some cells have long cilia, and others have stumpy microvilli. Groups of cells specialized to perform a particular set of functions are called tissues. Which type of tissue is this, and what are its functions? You will find out in this chapter. Each of your body's tissues has its own distinctive structure and plays its own role in maintaining homeostasis.

CHAPTER OUTLINE AND OBJECTIVES

INTRODUCTION   p. 107
① Identify the four major tissue types of the body and their roles.

EPITHELIAL TISSUE   p. 107
Functions of Epithelial Tissue   p. 107
Specializations of Epithelial Cells   p. 108
Maintaining the Integrity of Epithelia   p. 108
Classification of Epithelia   p. 110
Glandular Epithelia   p. 115
② Discuss the types and functions of epithelial cells.
③ Describe the relationship between form and function for each epithelial type.

CONNECTIVE TISSUES   p. 117
Classification of Connective Tissues   p. 118
Connective Tissue Proper   p. 118
Fluid Connective Tissues   p. 124
Supporting Connective Tissues   p. 125
④ Compare the structures and functions of the various types of connective tissues.

MEMBRANES   p. 129
Mucous Membranes   p. 129
Serous Membranes   p. 129
The Cutaneous Membrane   p. 131
Synovial Membranes   p. 131
⑤ Explain how epithelial and connective tissues combine to form four types of membranes, and specify the functions of each.

THE CONNECTIVE TISSUE FRAMEWORK OF THE BODY   p. 131
⑥ Describe how connective tissue establishes the framework of the body.

MUSCLE TISSUE   p. 132
Skeletal Muscle Tissue   p. 132
Cardiac Muscle Tissue   p. 132
Smooth Muscle Tissue   p. 132
⑦ Describe the three types of muscle tissue and the special structural features of each type.

NEURAL TISSUE   p. 134
⑧ Discuss the basic structure and role of neural tissue.

TISSUE INJURIES AND AGING   p. 135
Inflammation and Regeneration   p. 135
Aging and Tissue Repair   p. 136
Aging and Cancer Incidence   p. 136
⑨ Describe how injuries and aging affect the tissues of the body.

MEDIALAB and CHAPTER REVIEW   p. 138

No single cell contains the metabolic machinery and organelles needed to perform all the many functions of the human body. Instead, through differentiation, each cell develops a characteristic set of structural features and a limited number of functions. ∞ *p. 101* These structures and functions can be quite distinct from those of nearby cells. Nevertheless, all the cells in a given location work together.

A detailed examination of the body reveals a number of patterns at the cellular level. Although the human body contains trillions of cells, there are only about 200 types of cells. These cell types combine to form **tissues,** collections of specialized cells and cell products that perform a relatively limited number of functions. **Histology** is the study of tissues.

Four basic *tissue types* exist: epithelial tissue, connective tissue, muscle tissue, and neural tissue:

1. *Epithelial tissue* covers exposed surfaces, lines internal passageways and chambers, and forms glands.

2. *Connective tissue* fills internal spaces, provides structural support for other tissues, transports materials within the body, and stores energy reserves.

3. *Muscle tissue* contracts to perform specific movements and in the process generates heat that warms the body.

4. *Neural tissue* carries information from one part of the body to another in the form of electrical impulses.

In this chapter, we will discuss the characteristics of each tissue type, focusing on the relationship between cellular organization and tissue function. In later chapters, our primary interest will be the patterns of tissue interaction in various organs and organ systems. AM *EMBRYOLOGY SUMMARY 1: The Formation of Tissues*

EPITHELIAL TISSUE

Epithelial tissue includes epithelia and glands. **Epithelia** (e-pi-THĒ-lē-a; singular, *epithelium*) are layers of cells that cover internal or external surfaces. **Glands** are secretory structures derived from epithelia. Epithelia cover every exposed body surface. The skin surface is a good example, and epithelia line the digestive, respiratory, reproductive, and urinary tracts—passageways that communicate with the outside world. Epithelia also line internal cavities and passageways, such as the chest cavity, fluid-filled chambers in the brain, and the inner surfaces of blood vessels and heart.

Epithelia have several important characteristics:

- *Cellularity.* Epithelia are composed almost entirely of cells bound closely together by one or more types of cell junctions. In other tissue types, the cells are often widely separated by extracellular materials.

- *Polarity.* An epithelium has an exposed surface that faces the exterior of the body or some internal space. It

also has a base, where the epithelium is attached to underlying tissues. The surfaces differ in membrane structure (and function), and organelles and other cytoplasmic components are not evenly distributed between these surfaces. **Polarity** is the term for this uneven distribution of membrane functions and organelles between the exposed and attached surfaces.

- *Attachment.* The base of an epithelium is bound to a thin **basement membrane.** The basement membrane is a complex structure produced by the basal surface of the epithelium and the underlying connective tissue.

- *Avascularity.* Epithelia are **avascular** (ā-VAS-kū-lar; *a-,* without + *vas,* vessel); that is, they do not contain blood vessels. Epithelial cells must therefore obtain nutrients by diffusion or absorption across the exposed or attached surface.

- *Regeneration.* Epithelial cells damaged or lost at the exposed surface are continuously replaced through the divisions of stem cells in the epithelium. Although regeneration is a characteristic of other tissues as well, the rates of cell division and replacement are typically much higher in epithelia than in other tissues.

FUNCTIONS OF EPITHELIAL TISSUE

Epithelia perform four essential functions:

1. *Provide physical protection.* Epithelia protect exposed and internal surfaces from abrasion, dehydration, and destruction by chemical or biological agents.

2. *Control permeability.* Any substance that enters or leaves your body has to cross an epithelium. Some epithelia are relatively impermeable; others are easily crossed by compounds as large as proteins. Many epithelia contain the molecular "machinery" needed for selective absorption or secretion. The epithelial barrier can be regulated and modified in response to stimuli. For example, hormones can affect the transport of ions and nutrients through epithelial cells. Even physical stress can alter the structure and properties of epithelia; think of calluses that form on your hands when you do rough physical work for a while.

3. *Provide sensation.* Most epithelia are extensively innervated by sensory nerves; that is, they have a large sensory nerve supply. Specialized epithelial cells can detect changes in their environment and convey information about such changes to the nervous system. For example, touch receptors in the deepest epithelial layers of the skin respond to pressure by stimulating adjacent sensory nerves. A *neuroepithelium* is an epithelium containing sensory cells that provide sensations of smell, taste, sight, equilibrium, or hearing.

4. *Produce specialized secretions.* Epithelial cells that produce secretions are called *gland cells.* Individual gland cells are typically scattered among other cell types in an epithelium. In a **glandular epithelium,** most or all of the epithelial cells produce secretions. Those secretions are either discharged onto the surface of

4

the skin or other epithelium to provide physical protection or are released into the surrounding interstitial fluid and blood to act as chemical messengers.

SPECIALIZATIONS OF EPITHELIAL CELLS

Epithelial cells have several structural specializations that distinguish them from other body cells. For the epithelium as a whole to perform the functions listed in the previous section, individual epithelial cells may be specialized for (1) the movement of fluids over the epithelial surface, providing protection and lubrication; (2) the movement of fluids through the epithelium, to control permeability; or (3) the production of secretions that provide physical protection or act as chemical messengers. These epithelial cells generally show a strong polarity. The arrangement of their organelles varies with the functions of the epithelium. Figure 4-1• shows a common type of polarity.

The cell is often divided into two functional regions: (1) the *apical surface,* where the cell is exposed to an internal or external environment; and (2) the *basolateral surface,* which includes the base, where the cell attaches to underlying epithelial cells or deeper tissues, and the sides, where the cell contacts its neighbors.

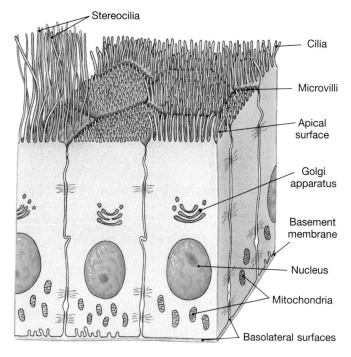

•**FIGURE 4-1 The Polarity of Epithelial Cells.** Many epithelial cells have an uneven distribution of organelles between the free surface (here, the top) and the basement membrane. In many cases, the free surface bears microvilli; in some cases, this surface has cilia or (very rarely) stereocilia. *(All three would not normally be on the same group of cells but are depicted here for purposes of illustration.)* In some epithelia, such as the lining of the kidney tubules, mitochondria are concentrated near the base of the cell, probably to provide energy for the cell's transport activities.

Many epithelial cells that line internal passageways have microvilli on their exposed surfaces. ∞ *p. 82* Just a few may be present, or microvilli may carpet the entire surface. Microvilli are especially abundant on epithelial surfaces where absorption and secretion take place, such as along portions of the digestive and urinary tracts. The epithelial cells in these locations are transport specialists. A cell with microvilli has at least 20 times more surface area available for material transport than a cell without them.

Stereocilia are structurally similar to microvilli but are very long (up to 250 μm) and incapable of active movement. These structures occur in two places only: (1) along portions of the male reproductive tract and (2) on receptor cells of the inner ear.

The photo on page 106 shows the surface of a **ciliated epithelium.** A typical ciliated cell contains about 250 cilia that beat in a coordinated fashion. As though on an escalator, substances are moved over the epithelial surface by the synchronized beating of cilia. For example, the ciliated epithelium that lines the respiratory tract moves mucus up from the lungs and toward the throat. The mucus traps foreign particles such as dust, pollen, and pathogens and carries them away from more-delicate surfaces deeper in the lungs. Injury to the cilia or to the epithelial cells in general can stop ciliary movement and block the protective flow of mucus. This is one consequence of smoking.

MAINTAINING THE INTEGRITY OF EPITHELIA

Three factors help maintain the physical integrity of an epithelium: (1) intercellular connections, (2) attachment to the basement membrane, and (3) epithelial maintenance and repair.

Intercellular Connections

Cells in an epithelium are firmly attached to one another, and the epithelium as a unit is attached to extracellular fibers of the basement membrane. Many other cells in your body form permanent or temporary bonds to other cells or extracellular material. Epithelial cells, however, are specialists in intercellular connection (Figure 4-2a•).

Intercellular connections may involve extensive areas of opposing cell membranes, or they may be concentrated at specialized attachment sites. Large areas of opposing cell membranes are interconnected by transmembrane proteins called **cell adhesion molecules (CAMs),** which bind to each other and to other extracellular materials. For example, CAMs on the attached base of an epithelium help bind the cell to the underlying basement membrane. The membranes of adjacent cells may also be bonded by **intercellular cement,** a thin layer of proteoglycans. These proteoglycans contain

Interlocking
junctional proteins

Tight junction

Belt desmosome

Terminal web

Button
desmosome

(b) Tight junction

Hemidesmosome

(a)

Gap junctions

Embedded
proteins
(connexons)

(c) Gap junction

Intermediate
filaments

Dense area

Basal
lamina
Reticular
lamina
} Basement
membrane

Intercellular
cement

(e) Hemidesmosome

(d) Desmosome

•**FIGURE 4-2 Intercellular Connections. (a)** A diagrammatic view of an epithelial cell, showing the major types of intercellular connections. **(b)** A tight junction is formed by the fusion of the outer layers of two cell membranes. Bands of tight junctions encircle the apical portion of many epithelial cells, preventing the diffusion of fluids and solutes between the cells. **(c)** Gap junctions permit the free diffusion of ions and small molecules between two cells. **(d)** A desmosome has a more organized network of intermediate filaments. Desmosomes attach one cell to another. A continuous belt of desmosomes lies deep to the tight junctions. This belt is tied to the microfilaments of the terminal web. **(e)** Hemidesmosomes attach a cell to extracellular structures, such as the protein fibers in the basement membrane.

polysaccharide derivatives known as *glycosaminogly-cans,* most notably **hyaluronan** *(hyaluronic acid).*

Cell junctions are specialized areas of the cell membrane that attach a cell to another cell or to extracellular materials. Three types of cell junctions are (1) tight junctions, (2) gap junctions, and (3) desmosomes.

At a **tight junction,** the lipid portions of the two cell membranes are tightly bound together by interlocking membrane proteins (Figure 4-2b•). That attachment is so tight that these junctions prevent the passage of water and solutes between the cells. When

the epithelium forms a tube, such as the lining of the intestinal tract, the space in contact with the apical surface is called the **lumen** (LOO-men). Tight junctions effectively isolate the contents of the lumen from the basolateral surfaces of the cell. For example, tight junctions near the apical surfaces of cells that line the digestive tract keep enzymes, acids, and wastes in the lumen from reaching the basolateral surfaces and then damaging delicate underlying tissues.

At a **gap junction** (Figure 4-2c•), two cells are held together by an interlocking of membrane proteins

4

called *connexons*. Because these are channel proteins, the result is a narrow passageway that lets small molecules and ions pass from cell to cell. Gap junctions are common among epithelial cells, where they help coordinate functions such as the beating of cilia. Gap junctions are most abundant in cardiac muscle tissue and in smooth muscle tissue, in which they are essential to the coordination of muscle cell contractions.

At a **desmosome** (DEZ-mō-sōm; *desmos*, ligament + *soma*, body), CAMs and proteoglycans link the opposing cell membranes. A desmosome is formed by two cells. Within each cell is a complex known as a *dense area*, which is connected to the cytoskeleton (Figure 4-2d•). A *belt desmosome* forms a continuous band that encircles the cell and binds it to its neighbors. This band is attached to the microfilaments of the terminal web. ∞ *p. 81 Button desmosomes* are small discs connected to bands of intermediate fibers. The intermediate fibers function as cross-braces to stabilize the shape of the cell. Desmosomes are very strong, and they can resist stretching and twisting. These connections are most abundant between cells in the superficial layers of the skin, where desmosomes create links so strong that dead skin cells are shed in thick sheets rather than individually. Desmosomes are also common in cardiac muscle tissue, interconnecting the muscle cells. *Hemidesmosomes* resemble half of a button desmosome. Rather than attaching one cell to another, a hemidesmosome attaches a cell to extracellular filaments in the basement membrane (Figure 4-2e•). This attachment helps stabilize the position of the epithelial cell and anchors the cell to underlying tissues.

Attachment to the Basement Membrane

Not only do epithelial cells hold on to one another, but they also remain firmly connected to the rest of the body. The basal surface of each epithelium is attached to a special two-part basement membrane. The layer closer to the epithelium, the **basal lamina** (LA-mi-nah; *lamina*, thin layer), contains glycoproteins and a network of fine protein filaments (Figure 4-2e•). The basal lamina, secreted by the adjacent layer of epithelial cells, is a barrier that restricts the movement of proteins and other large molecules from the underlying connective tissue into the epithelium. The deeper layer of the basement membrane, the **reticular lamina,** contains bundles of coarse protein fibers produced by connective tissue cells. The reticular lamina gives the basement membrane its strength. Attachments between the fibers of the basal lamina and those of the reticular lamina hold the two layers together, and hemidesmosomes attach the epithelial cells to the composite basement membrane.

Epithelial Maintenance and Repair

An epithelium must continuously repair and renew itself. Epithelial cells lead hard lives, for they are exposed to disruptive enzymes, toxic chemicals, path-

ogenic bacteria, or mechanical abrasion. Consider the lining of the small intestine, where epithelial cells are exposed to a variety of enzymes and abrasion from partially digested food. In this extreme environment, an epithelial cell may last just a day or two before it is shed or destroyed. The only way the epithelium can maintain its structure over time is by the continual division of *stem cells.* ∞ *p. 99* Most stem cells, also called **germinative cells,** are located in the deepest layers of the epithelium, near the basement membrane. AM *EMBRYOLOGICAL SUMMARY 2: The Development of Epithelia*

☑ An epithelial surface bears many microvilli. What is the probable function of this epithelium?

☑ What is the functional significance of gap junctions?

☑ Why do the pharynx, esophagus, anus, and vagina have the same epithelial organization?

CLASSIFICATION OF EPITHELIA

One way to classify epithelia is by cell shape. Three cell shapes are identified: *squamous, cuboidal,* and *columnar.* This classification is based on the appearance of the cell when seen in a section that is perpendicular to the apical surface and to the basement membrane. A second way to classify epithelia is by the number of cell layers. This classification scheme recognizes two types of layering: *simple* and *stratified.*

If only one layer of cells covers the basement membrane, the epithelium is a **simple epithelium.** Simple epithelia are thin. All the cells have the same polarity, so the distance from the nucleus to the basement membrane does not change from one cell to the next. Because they are so thin, simple epithelia are also fragile. A single layer of cells cannot provide much mechanical protection, so simple epithelia are located only in protected areas inside the body. They line internal compartments and passageways, including the ventral body cavities, the heart chambers, and blood vessels.

Simple epithelia are also characteristic of regions where secretion or absorption occurs, such as the lining of the intestines and the gas-exchange surfaces of the lungs. In these places, the thinness of simple epithelia is an advantage, for it reduces the time required for materials to cross the epithelial barrier.

In a **stratified epithelium,** several layers of cells cover the basement membrane. Stratified epithelia are generally located in areas subject to mechanical or chemical stresses, such as the surface of the skin and the lining of the mouth.

We can describe almost every epithelium in the body by combining the two ways to classify epithelia— by cell shape (squamous, cuboidal, or columnar) and by number of cell layers (simple or stratified). In most stratified epithelia, the shapes of the cells change as they approach the exposed surface. As a result, the classification scheme for stratified epithelia is based only on the appearance of the cells in the superficial layer.

Squamous Epithelia

The cells in a **squamous epithelium** (SKWĀ-mus; *squama,* plate or scale) are thin, flat, and somewhat irregular in shape, like puzzle pieces (Figure 4-3●). In sectional view, the disc-shaped nucleus occupies the thickest portion of each cell. From the surface, the cells look like fried eggs laid side by side.

A **simple squamous epithelium** is the body's most delicate type of epithelium (Figure 4-3a●). This type is located in protected regions where absorption takes place or where a slick, slippery surface reduces friction. Examples are the respiratory exchange surfaces *(alveoli)*

of the lungs, the lining of the ventral body cavities, and the lining of the heart and blood vessels.

Special names have been given to simple squamous epithelia that line chambers and passageways that do not communicate with the outside world. The simple squamous epithelium that lines the ventral body cavities is a **mesothelium** (mez-ō-THĒ-lē-um; *mesos,* middle). The pleura, peritoneum, and pericardium each contain a superficial layer of mesothelium. The simple squamous epithelium lining the inner surface of the heart and all blood vessels is an **endothelium** (en-dō-THĒ-lē-um).

A **stratified squamous epithelium** (Figure 4-3b●) is generally located where mechanical stresses are severe. The cells form a series of layers, like a stack of

●**FIGURE 4-3 Squamous Epithelia. (a)** A superficial view of the simple squamous epithelium (mesothelium) that lines the peritoneal cavity. The three-dimensional drawing shows the epithelium in superficial and sectional views. **(b)** Sectional and diagrammatic views of the stratified squamous epithelium that covers the tongue.

4

plywood sheets. The surface of the skin and the lining of the mouth, esophagus, and anus are areas where this epithelial type protects against physical and chemical attacks. On exposed body surfaces, where mechanical stress and dehydration are potential problems, apical layers of epithelial cells are packed with filaments of the protein *keratin.* As a result, superficial layers are both tough and water-resistant; the epithelium is said to be *keratinized.* A *nonkeratinized* stratified squamous epithelium resists abrasion but will dry out and deteriorate unless kept moist. Nonkeratinized stratified squamous epithelia are situated in the oral cavity, pharynx, esophagus, anus, and vagina.

Cuboidal Epithelia

The cells of a **cuboidal epithelium** resemble hexagonal boxes; they appear square in typical sectional views. The spherical nuclei are near the center of each cell. The distance between adjacent nuclei is roughly equal to the height of the epithelium. A **simple cuboidal epithelium** provides limited protection and occurs where secretion or absorption takes place. Such an epithelium lines portions of the kidney tubules (Figure 4-4a•). In the pancreas and salivary glands, simple cuboidal epithelia secrete enzymes and buffers and line portions of the ducts that discharge those secretions. The thyroid gland contains chambers, called *thyroid follicles,* that are lined by a cuboidal secretory epithelium. Thyroid hormones are stored in the follicles and released as needed into the bloodstream.

    Stratified cuboidal epithelia are relatively rare; they are located along the ducts of sweat glands (Figure 4-4b•) and in the larger ducts of the mammary glands. A **transitional epithelium** (Figure 4-4c•) is unusual because, unlike most epithelia, it tolerates considerable stretching. It is called transitional because the appearance of the epithelium changes as the stretching occurs. A transitional epithelium is situated in regions of the urinary system, such as the urinary bladder, where large changes in volume occur. In an empty urinary bladder, the epithelium seems to have many layers, and the superficial cells are typically plump cuboidal cells. The layered appearance results from overcrowding; the actual structure of the epithelium can be seen in the full urinary bladder, when the volume of urine has stretched the lining to its natural thickness.

Columnar Epithelia

In a typical sectional view, **columnar epithelial cells** appear rectangular. In reality, the densely packed columnar cells are hexagonal, but they are taller and more slender than cells in a cuboidal epithelium (Figure 4-5•, p. 114). The elongated nuclei are crowded into a narrow band close to the basement membrane. The height of the epithelium is several times the distance between adjacent nuclei. A **simple columnar epithelium** typically

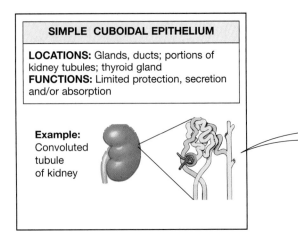

SIMPLE CUBOIDAL EPITHELIUM

LOCATIONS: Glands, ducts; portions of kidney tubules; thyroid gland
FUNCTIONS: Limited protection, secretion and/or absorption

Example:
Convoluted tubule of kidney

STRATIFIED CUBOIDAL EPITHELIUM

LOCATIONS: Lining of some ducts (rare)
FUNCTIONS: Protection, secretion, absorption

Example:
Duct of sweat gland

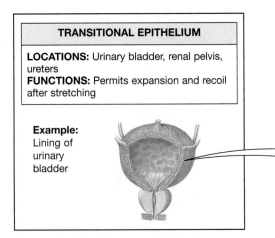

TRANSITIONAL EPITHELIUM

LOCATIONS: Urinary bladder, renal pelvis, ureters
FUNCTIONS: Permits expansion and recoil after stretching

Example:
Lining of urinary bladder

•**FIGURE 4-4 Cuboidal Epithelia. (a)** A section through the simple cuboidal epithelial cells of a kidney tubule. **(b)** A sectional view of the stratified cuboidal epithelium that lines a sweat gland duct in the skin. **(c)** Left: The lining of the empty urinary bladder, showing a transitional epithelium in the relaxed state. Right: The lining of the full bladder, showing the effects of stretching on the appearance of cells in the epithelium.

4

Lumen of tubule

Cuboidal cells

Connective tissue

Kidney tubule × 2000

(a)

Basement membrane   Connective tissue

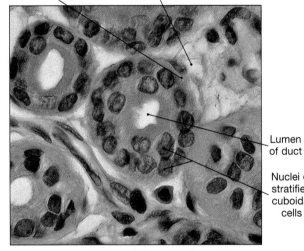

Lumen of duct

Nuclei of stratified cuboidal cells

Stratified cuboidal cells

Lumen of duct

Nuclei

Sweat gland duct × 1413

(b)

Relaxed epithelium

Basement membrane

Connective tissue and smooth muscle layers

Epithelium (stretched)   Basement membrane   Connective tissue and smooth muscle layers

Empty bladder × 454

(c)

Full bladder × 454

4

Example:
Lining
of small
intestine

SIMPLE COLUMNAR EPITHELIUM

LOCATIONS: Lining of
stomach, intestine, gallbladder,
uterine tubes, collecting ducts
of kidneys
FUNCTIONS: Protection,
secretion, absorption

Intestinal lining × 350   **(a)**

Microvilli
Cytoplasm
Nucleus
Basement
membrane
Loose
connective
tissue

Example:
Lining
of trachea
and bronchi

PSEUDOSTRATIFIED CILIATED COLUMNAR EPITHELIUM

LOCATIONS: Lining of nasal
cavity, trachea, bronchi;
portions of male reproductive
tract
FUNCTIONS: Protection,
secretion

Trachea × 251   **(b)**

Cilia
Cytoplasm
Nuclei
Basement
membrane
Loose
connective
tissue

Example:
Duct of
a salivary
gland

STRATIFIED COLUMNAR EPITHELIUM

LOCATIONS: Small areas of
the pharynx, epiglottis, anus,
mammary gland and salivary
gland ducts, urethra
FUNCTION: Protection

Superficial
columnar cells
Lumen
Salivary gland duct   **(c)**

Cytoplasm
Deeper
basal
cells
Nuclei
Basement
membrane
Loose
connective
tissue

•**FIGURE 4-5 Columnar Epithelia.** Note the thickness of the epithelium and the location and orientation of the nuclei. **(a)** The simple columnar epithelium lining the small intestine. **(b)** The pseudostratified ciliated columnar epithelium of the respiratory tract. Note the uneven layering of the nuclei. **(c)** A stratified columnar epithelium occurs along some large ducts, such as this salivary gland duct.

occurs where absorption or secretion is under way, such as in the small intestine (Figure 4-5a•). In the stomach and large intestine, the secretions of simple columnar epithelia protect against chemical stresses.

Portions of the respiratory tract contain a **pseudostratified columnar epithelium,** a columnar epithelium that includes several cell types with varying shapes and functions. The distances between the cell nuclei and the exposed surface vary, so the epithelium appears to be layered, or stratified (Figure 4-5b•). But it is not truly stratified, because all the epithelial cells contact the basement membrane. Pseudostratified columnar epithelial cells typically possess cilia. This type of epithelium lines most of the nasal cavity, the trachea (windpipe), bronchi, and portions of the male reproductive tract.

Stratified columnar epithelia are relatively rare, providing protection along portions of the pharynx, epiglottis, anus, and urethra as well as along a few large excretory ducts. The epithelium has either two layers (Figure 4-5c•) or multiple layers. When multiple layers exist, only the superficial cells are columnar.

Exfoliative Cytology

Exfoliative cytology (eks-FŌ-lē-a-tiv; *ex,* from + *folium,* leaf) is the study of cells shed or collected from epithelial surfaces. The cells are examined for a variety of reasons—for example, to check for cellular changes that indicate cancer or to identify the pathogens involved in an infection. The cells are collected by sampling the fluids that cover the epithelia lining the respiratory, digestive, urinary, or reproductive tract or by removing fluid from one of the ventral body cavities. The sampling procedure is often called a *Pap test,* named after Dr. George Papanicolaou, who pioneered its use. The most familiar Pap test is that for cervical cancer, which involves the scraping of cells from the tip of the *cervix,* the portion of the uterus that projects into the vagina.

Amniocentesis is another important test based on exfoliative cytology. In this procedure, exfoliated epithelial cells are collected from a sample of *amniotic fluid,* the fluid that surrounds and protects a developing fetus. Examination of these cells can determine if the fetus has a genetic abnormality, such as *Down syndrome,* that affects the number or structure of chromosomes.

GLANDULAR EPITHELIA

Many epithelia contain gland cells that produce secretions. We classify these secretions by the location to which they are discharged:

1. *Endocrine (endo-,* inside + *krinein,* to secrete) *secretions* are released into the surrounding interstitial fluid and blood. These secretions, also called *hormones,* regulate or coordinate the activities of other tissues, organs, and organ systems. Endocrine secretions are produced in organs such as the pancreas, thyroid, and anterior pituitary gland.

2. *Exocrine (exo-,* outside) *secretions* are discharged onto the skin or other epithelial surface. Enzymes entering the digestive tract, perspiration on the skin, and milk produced by mammary glands are exocrine secretions.

Endocrine and exocrine secretions are released by endocrine glands and exocrine glands, respectively.

Only a few complex glands produce both exocrine and endocrine secretions. For example, the pancreas contains endocrine cells that secrete hormones and exocrine cells that produce digestive enzymes and buffers.

Endocrine Glands

Endocrine glands release their secretions into the surrounding interstitial fluid. Because the secretions are not released into *ducts,* or tubes that open onto a free surface, endocrine glands are often called *ductless glands.* From the interstitial fluids, these secretions, hormones, enter the bloodstream for distribution throughout the body.

Endocrine cells may be part of an epithelial surface, such as the lining of the digestive tract, or they may be separate, as in the pancreas, thyroid gland, thymus, and pituitary gland. We shall consider endocrine cells, organs, and hormones further in Chapter 18.

Exocrine Glands

Exocrine glands produce secretions that are discharged onto epithelial surfaces. Most exocrine glands release their secretions into tubular passageways called **ducts.** Ducts then empty onto the skin surface or onto an epithelium lining an internal passageway that communicates with the exterior. Exocrine secretions include sweat, saliva, and tears.

Exocrine glands are classified by their mode of secretion, the type of secretion, and the structure of the gland cells and the associated ducts.

Modes of Secretion   A glandular epithelial cell releases its secretions by (1) merocrine secretion, (2) apocrine secretion, or (3) holocrine secretion. In **merocrine secretion** (MER-ō-krin; *meros,* part), the product is released from secretory vesicles by exocytosis (Figure 4-6a•). This is the most common mode of secretion. One type of merocrine secretion, mucin, mixes with water to form **mucus,** an effective lubricant, a protective barrier, and a sticky trap for foreign particles and microorganisms. The mucous secretions of merocrine glands coat digestive and respiratory tract passageways. In the skin, merocrine sweat glands produce the watery perspiration that helps cool you on a hot day.

Apocrine secretion (AP-ō-krin; *apo-,* off) involves the loss of cytoplasm as well as the secretory product (Figure 4-6b•). The apical portion of the cytoplasm becomes packed with secretory vesicles and is then shed. The thick, sticky underarm perspiration targeted by the deodorant industry results from apocrine secretion.

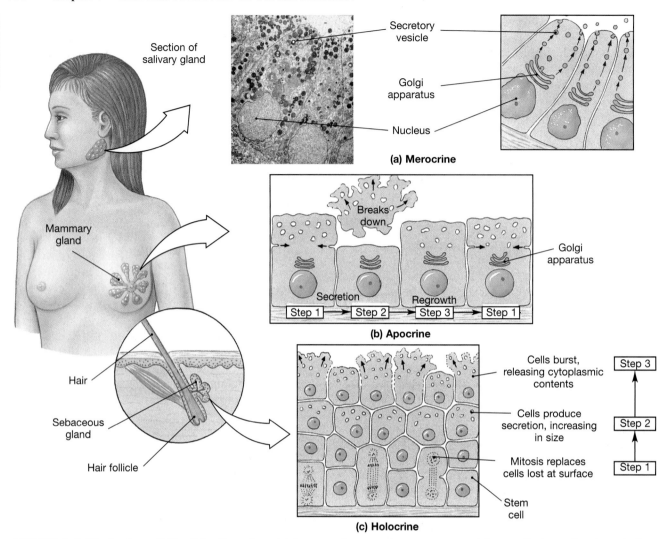

•**FIGURE 4-6 Mechanisms of Glandular Secretion. (a)** In merocrine secretion, secretory vesicles are discharged at the apical surface of the gland cell by exocytosis. **(b)** Apocrine secretion involves the loss of apical cytoplasm. Inclusions, secretory vesicles, and other cytoplasmic components are shed in the process. The gland cell then undergoes growth and repair before it releases additional secretions. **(c)** Holocrine secretion occurs as superficial gland cells burst. Continued secretion involves the replacement of these cells through the mitotic division of underlying stem cells.

Milk production in the mammary glands involves a combination of merocrine and apocrine secretions.

Merocrine and apocrine secretions leave a cell relatively intact and able to continue secreting. **Holocrine secretion** (HOL-ō-krin; *holos,* entire) destroys the gland cell. During holocrine secretion, the entire cell becomes packed with secretory products and then bursts (Figure 4-6c•), releasing the secretion but killing the cell. Further secretion depends on the replacement of destroyed gland cells by the division of stem cells. Sebaceous glands, associated with hair follicles, produce an oily hair coating by means of holocrine secretion.

Types of Secretions Exocrine glands are also categorized into one of three types of secretion produced:

1. **Serous glands** secrete a watery solution that contains enzymes. The parotid salivary glands are serous glands.

2. **Mucous glands** secrete *mucins,* glycoproteins that, upon hydration, form mucus, a slippery lubricant. The sublingual salivary glands and the submucosal glands of the small intestine are mucous glands.

3. **Mixed exocrine glands** contain more than one type of gland cell and may produce two different exocrine secretions—one serous and the other mucous. The submandibular salivary glands are mixed exocrine glands.

Gland Structure The final classification method is by structure (Figure 4-7•). In epithelia that have independent, scattered gland cells, the individual secretory cells are called **unicellular glands. Multicellular glands** include glandular epithelia and aggregations of gland cells that produce exocrine or endocrine secretions.

The only **unicellular exocrine glands** in the body are goblet cells. **Goblet cells,** which secrete mucins, are scat-

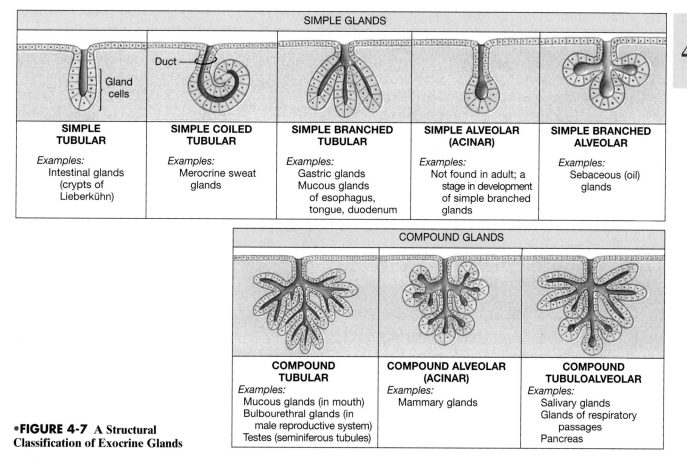

•FIGURE 4-7 A Structural Classification of Exocrine Glands

tered among other epithelial cells. For example, both the pseudostratified ciliated columnar epithelium that lines the trachea and the columnar epithelium of the small and large intestines have an abundance of goblet cells.

The simplest **multicellular exocrine gland** is a **secretory sheet,** in which gland cells dominate the epithelium and release their secretions into an inner compartment. The continuous secretion of mucin-secreting cells that line the stomach, for instance, protects the stomach from its own acids and enzymes. Most other multicellular exocrine glands are in pockets set back from the epithelial surface; their secretions travel through one or more ducts to the epithelial surface. Examples include the salivary glands, which produce mucins and digestive enzymes.

Three characteristics are used to describe the structure of multicellular exocrine glands:

1. *The shape of the secretory portion of the gland.* Glands whose glandular cells form tubes are *tubular.* Those that form blind pockets are *alveolar* (al-VĒ-ō-lar; *alveolus,* sac) or *acinar* (A-si-nar; *acinus,* chamber). Glands whose secretory cells form both tubes and pockets are called *tubuloalveolar* or *tubuloacinar.*

2. *The branching pattern of the duct.* A gland is *simple* if it has a single duct that does not divide on its way to the gland cells. The gland is *compound* if the duct divides repeatedly.

3. *The relationship between the ducts and the glandular areas.* A gland is *branched* if several secretory areas (tubular or acinar) share a duct. "Branched" refers to the glandular areas and not to the duct.

☑ With a light microscope, you examine a tissue and see a simple squamous epithelium on the outer surface. Can this be a sample of the skin surface?

☑ The secretory cells of sebaceous glands fill with secretions and then rupture, releasing their contents. Which type of secretion is this?

☑ A gland has no ducts to carry the glandular secretions, and the gland's secretions are released directly into the extracellular fluid. Which type of gland is this?

CONNECTIVE TISSUES

Connective tissues include tissues such as bone, fat, and blood. They are quite different in appearance and function but share three basic components: (1) specialized cells, (2) extracellular protein fibers, and (3) a fluid known as **ground substance.** The extracellular fibers and ground substance constitute the **matrix,** which surrounds the cells. Whereas cells make up the bulk of epithelial tissue, the matrix typically accounts for most of the volume of connective tissues. [AM] *EMBRYOLOGY SUMMARY 3: The Origins of Connective Tissues*

Connective tissues are situated throughout the body but are never exposed to the outside environment. Many connective tissues are highly vascular (that is, they have many blood vessels) and contain sensory receptors that provide pain, pressure, temperature, and other sensations. The functions of connective tissues include:

- Establishing a structural framework for the body.
- Transporting fluids and dissolved materials.
- Protecting delicate organs.
- Supporting, surrounding, and interconnecting other tissue types.
- Storing energy reserves, especially in the form of lipids.
- Defending the body from invading microorganisms.

CLASSIFICATION OF CONNECTIVE TISSUES

The three general categories of connective tissue are connective tissue proper, fluid connective tissues, and supporting connective tissues (Figure 4-8•):

1. **Connective tissue proper** includes those connective tissues with many types of cells and extracellular fibers in a syrupy ground substance. Such connective tissues differ in the number of cell types they contain and in the relative properties and proportions of fibers and ground substance. For example, both *adipose tissue* (fat) and *tendons* are connective tissue proper, but they have very different structural and functional characteristics. Connective tissue proper is divided into *loose connective tissues* and *dense connective tissues* on the basis of the relative proportions of cells, fibers, and ground substance.

2. **Fluid connective tissues** have distinctive populations of cells suspended in a watery matrix that contains dissolved proteins. Two types exist: *blood* and *lymph.*

3. **Supporting connective tissues** differ from connective tissue proper in having a less diverse cell population and a matrix containing much more densely packed

fibers. The two types of supporting connective tissues are *cartilage* and *bone.* The matrix of cartilage is a gel whose characteristics vary with the predominant fiber type. The matrix of bone is said to be **calcified,** because it contains mineral deposits (primarily calcium salts). These minerals give bone its rigidity.

CONNECTIVE TISSUE PROPER

Connective tissue proper contains extracellular fibers, a viscous (syrupy) ground substance, and a varied cell population. Some of these cells, including *fibroblasts, adipocytes,* and *mesenchymal cells,* are involved with local maintenance, repair, and energy storage. These cells are permanent residents of the connective tissue. Other cells, including *macrophages, mast cells, lymphocytes, plasma cells,* and *microphages,* defend and repair damaged tissues. Such cells are not permanent residents; they migrate through healthy connective tissues and aggregate at sites of tissue injury.

The number of cells and cell types in a tissue at any moment varies with local conditions. Figure 4-9• summarizes the cells and fibers of connective tissue proper.

The Cell Population

Fibroblasts (FĪ-brō-blasts) are the most abundant permanent residents of connective tissue proper and are the only cells *always* present in it. Fibroblasts secrete hyaluronan (a polysaccharide derivative) and proteins. These two components interact in the extracellular fluid to form the proteoglycans that make ground substance viscous. Each fibroblast also secretes protein subunits that interact to form large extracellular fibers.

In addition to fibroblasts, connective tissues proper may contain several other cell types:

- **Macrophages** (MAK-rō-fā-jez; *phagein,* to eat) are large, amoeboid cells scattered throughout the matrix. They engulf pathogens or damaged cells that enter the tissue. Although not abundant, they are important in

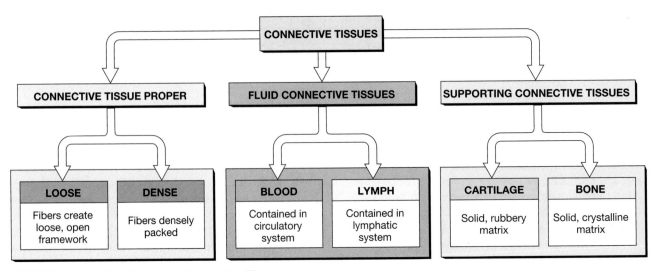

•**FIGURE 4-8** A Classification of Connective Tissues

Connective tissue proper (areolar tissue) × 384

•**FIGURE 4-9** **The Cells and Fibers of Connective Tissue Proper.** Diagrammatic and histological views of the cell types and fibers of connective tissue proper. (Microphages, not shown, are common only in damaged or abnormal tissues.)

mobilizing the body's defenses. When stimulated, they release chemicals that activate the immune system and attract large numbers of additional macrophages and other cells involved in tissue defense. Macrophages are often categorized as *fixed macrophages,* which spend long periods in a tissue, or *free macrophages,* which migrate rapidly through the tissue. In effect, fixed macrophages provide a "front-line" defense that can be reinforced by the arrival of free macrophages and other specialized cells.

- **Adipocytes** (AD-i-pō-sīts) are also known as *adipose cells,* or fat cells. A typical adipocyte contains a single, enormous lipid droplet. The nucleus and other organelles are squeezed to one side, making the cell in section resemble a class ring. The number of fat cells varies from one type of connective tissue to another, from one region of the body to another, and among individuals.

- **Mesenchymal cells** are stem cells present in many connective tissues. These cells respond to local injury or infection by dividing to produce daughter cells that differentiate into fibroblasts, macrophages, or other connective tissue cells.

- **Melanocytes** (me-LAN-ō-sīts) synthesize and store the brown pigment **melanin** (ME-la-nin). Melanin gives the tissue a dark color. Melanocytes are common in the epithelium of the skin, where they play a major role in determining skin color. However, melanocytes are also abundant in connective tissues of the eye, and they are in the dermis of the skin, although the number present differs by body region and among individuals.

- **Mast cells** are small, mobile connective tissue cells common near blood vessels. The cytoplasm of a mast cell is filled with **histamine** (HIS-tuh-mēn) and **heparin** (HEP-

uh-rin) granules. These chemicals, released after injury or infection, stimulate local inflammation. *Basophils,* blood cells that enter damaged tissues and enhance the inflammation process, also contain those granules.

- **Lymphocytes** (LIM-fō-sīts) migrate throughout the body. Their numbers increase markedly wherever tissue damage occurs, and some lymphocytes may then develop into plasma cells. **Plasma cells** produce *antibodies,* proteins involved in defending the body against disease.

- **Microphages** (*neutrophils* and *eosinophils*) are phagocytic blood cells that move through connective tissues in small numbers. When an infection or injury occurs, chemicals released by macrophages and mast cells attract microphages in large numbers.

Connective Tissue Fibers

Three types of fibers occur in connective tissue: collagen, reticular, and elastic. Fibroblasts form all three by secreting protein subunits that interact in the matrix.

1. **Collagen fibers** are long, straight, and unbranched. They are the most common fibers in connective tissue proper. Each collagen fiber consists of a bundle of fibrous protein subunits wound together like the strands of a rope. Like a rope, a collagen fiber is flexible, but it is stronger than steel when pulled from either end. *Tendons,* which connect skeletal muscles to bones, consist almost entirely of collagen fibers. Typical *ligaments* are similar to tendons, but they connect one bone to another. Tendons and ligaments can withstand tremendous forces. Uncontrolled muscle contractions or skeletal movements are more likely to break a bone than to snap a tendon or a ligament.

2. **Reticular fibers** (*reticulum,* network) contain the same protein subunits as do collagen fibers, but arranged differently. Thinner than collagen fibers, they form a tough but flexible branching, interwoven framework. Because they form a network rather than share a common alignment, reticular fibers resist forces applied from many directions. This interwoven network is called a *stroma.* It stabilizes the relative positions of the functional cells, or **parenchyma** (pa-RENG-ki-ma) of the organ. Reticular fibers also stabilize the positions of an organ's blood vessels, nerves, and other structures despite changing positions and the pull of gravity.

3. **Elastic fibers** contain the protein *elastin.* Elastic fibers are branched and wavy. After stretching, they will return to their original length. **Elastic ligaments** are dominated by elastic fibers. They are relatively rare but have important functions, such as interconnecting vertebrae.

Ground Substance

Ground substance fills the spaces between cells and surrounds connective tissue fibers (Figure 4-9•). In connective tissue proper, it is clear, viscous, and colorless. Its typical consistency is like that of maple syrup, due to the presence of proteoglycans and glycoproteins. ∞ *p. 56* Ground substance is dense enough that bacteria have trouble moving through it—imagine swimming in molasses. This density slows the spread of pathogens and makes them easier for phagocytes to catch.

Marfan's Syndrome

Marfan's syndrome is an inherited condition caused by the production of an abnormal form of *fibrillin,* a glycoprotein important to the strength and elasticity of connective tissues. Because most organs contain connective tissues, the effects of this defect are widespread. The most visible sign of Marfan's syndrome involves the skeleton; most individuals with this condition are tall and have abnormally long limbs and fingers. The most serious consequences involve the cardiovascular system. Roughly 90 percent of people with Marfan's syndrome have structural abnormalities in their cardiovascular system. The most dangerous potential result is that the weakened elastic connective tissues in the walls of major arteries, such as the aorta, may burst, causing a sudden, fatal loss of blood.

Embryonic Connective Tissues

Mesenchyme, or *embryonic connective tissue,* is the first connective tissue to appear in a developing embryo. Mesenchyme contains star-shaped stem cells (mesenchymal cells) separated by a matrix with very fine protein filaments (Figure 4-10a•). This tissue gives rise to all other connective tissues. **Mucous connective tissue** (Figure 4-10b•) is a loose connective tissue in many parts of the embryo, including the umbilical cord.

Adults have neither of these embryonic forms of connective tissue. However, many adult connective tis-

Mesenchymal cells

(a) Mesenchyme ×136

Mesenchymal cells   Blood vessel

(b) Mucous connective tissue × 136

•**FIGURE 4-10** Connective Tissues in Embryos. **(a)** Mesenchyme, the first connective tissue to appear in an embryo. **(b)** Mucous connective tissue from the umbilical cord of a fetus. Mucous connective tissue in this location is also known as *Wharton's jelly.*

sues contain scattered mesenchymal stem cells that can assist in tissue repair after an injury.

Loose Connective Tissues

Loose connective tissues are the "packing materials" of the body. They fill spaces between organs, cushion and stabilize specialized cells in many organs, and support epithelia. These tissues surround and support blood vessels and nerves, store lipids, and provide a route for the diffusion of materials. Loose connective tissues include mucous connective tissue in embryos and *areolar tissue, adipose tissue,* and *reticular tissue* in adults.

Areolar Tissue   **Areolar tissue** (*areola,* little space) is the least specialized connective tissue in adults. It may contain all the cells and fibers of any connective tissue proper (Figure 4-9•). Areolar tissue has an open framework; ground substance accounts for most of its volume. This syrupy fluid absorbs shocks. The fibers are loosely organized, so areolar tissue can distort without damage. Elastic fibers make it resilient, so this tissue returns to its original shape after external pressure is relieved.

Areolar tissue forms a layer that separates the skin from deeper structures. In addition to providing padding, the elastic properties of this layer allow a considerable amount of independent movement. Thus, if you pinch the skin of your arm, you will not affect the underlying muscle. Conversely, contractions of the underlying muscles do not pull against your skin; as the muscle bulges, the areolar tissue stretches. Because this tissue has an extensive circulatory supply, the areolar tissue layer under the skin is a common injection site for drugs.

The capillaries (thin-walled blood vessels) in areolar tissue deliver oxygen and nutrients and remove carbon dioxide and waste products and also carry wandering cells to and from the tissue. Epithelia commonly cover a layer of areolar tissue; fibroblasts maintain the reticular lamina of the basement membrane that separates these tissues. The epithelial cells rely on the diffusion of oxygen and nutrients across that membrane from capillaries in the underlying connective tissue.

Adipose Tissue   The distinction between areolar tissue and fat, or **adipose tissue,** is somewhat arbitrary. Adipocytes account for most of the volume of adipose tissue (Figure 4-11a•) but only a fraction of the volume of areolar tissue. Adipose tissue provides padding, absorbs shocks, acts as an insulator to slow heat loss through the skin, and serves as packing or filler around structures. Adipose tissue is common under the skin of the groin, sides, buttocks, and breasts. It fills the bony sockets behind the eyes, surrounds the kidneys, and is common beneath the mesothelial lining of the pericardial and abdominal cavities. Most of the adipose tissue in the body is called **white fat,** because it has a pale, yellow-white color.

In infants and young children, the adipose tissue between the shoulder blades, around the neck, and

Adipose tissue × 118   **(a)**   SEM (color-enhanced) × 236

Adipocytes (fat cells)

Reticular tissue × 330   **(b)**

Reticular fibers

•**FIGURE 4-11   Adipose and Reticular Tissues.** **(a)** Adipose tissue is a loose connective tissue dominated by adipocytes. In standard histological preparations, the tissue looks empty because the lipids in the fat cells dissolve in the alcohol used in tissue processing. **(b)** Reticular tissue has an open framework of reticular fibers. These fibers are usually very difficult to see because of the large numbers of cells around them.

4

possibly elsewhere in the upper body is highly vascularized, and the individual adipocytes contain numerous mitochondria. Together, these characteristics give the tissue a deep, rich color from which the name **brown fat** is derived. When these cells are stimulated by the nervous system, lipid breakdown accelerates. The cells do not capture the energy released, and it is absorbed by the surrounding tissues as heat. This heat warms the circulating blood, which distributes the heat throughout the body. In this way, an infant can increase metabolic heat generation by 100 percent very quickly. Adults have little if any brown fat. Shivering is significantly more effective in elevating the body temperature of adults, who have a larger body size and more skeletal muscle mass and insulation than infants do.

Adipose Tissue and Weight Control

Adipocytes are metabolically active cells; their lipids are constantly being broken down and replaced. When nutrients are scarce, adipocytes deflate like collapsing balloons. This deflation occurs during a weight-loss program. Because the cells are not killed but merely reduced in size, the lost weight can easily be regained in the same areas of the body. In adults, adipocytes are incapable of dividing. The number of fat cells in peripheral tissues is established in the first few weeks of a newborn's life, perhaps in response to the amount of fats in the diet. However, this is not the end of the story, because loose connective tissues also contain mesenchymal cells. If circulating lipid levels are chronically elevated, the mesenchymal cells will divide, giving rise to cells that differentiate into fat cells. As a result, areas of areolar tissue can become adipose tissue in times of nutritional plenty, even in adults. In the procedure known as *liposuction,* unwanted adipose tissue is surgically removed. Because adipose tissue can regenerate through the differentiation of mesenchymal cells, liposuction provides only a temporary solution to the problem of excess weight.

Reticular Tissue Organs such as the spleen and liver contain **reticular tissue,** in which reticular fibers create a complex three-dimensional network, or stroma (Figure 4-11b•). The stroma supports the functional cells of these organs. This fibrous framework is also located in the lymph nodes and bone marrow. Fixed macrophages and fibroblasts are associated with the reticular fibers, but these cells are seldom visible because the organs are dominated by specialized cells with other functions.

Dense Connective Tissues

Most of the volume of **dense connective tissues** is occupied by fibers. Dense connective tissues are often called **collagenous** (ko-LA-jin-us) **tissues,** because collagen fibers are the dominant fiber type. The body has two types of dense connective tissues: (1) dense regular connective tissue and (2) dense irregular connective tissue.

In **dense regular connective tissue,** the collagen fibers are parallel to each other, packed tightly, and aligned with the forces applied to the tissue. **Tendons**

•**FIGURE 4-12 Dense Connective Tissues. (a)** The dense regular connective tissue in a tendon. Notice the densely packed, parallel bundles of collagen fibers. The fibroblast nuclei are flattened between the bundles. **(b)** The deep dermis of the skin contains a thick layer of dense irregular connective tissue. **(c)** An elastic ligament, an example of elastic tissue. Elastic ligaments extend between the vertebrae of the vertebral column. The bundles are fatter than those of a tendon or a ligament composed of collagen.

DENSE REGULAR CONNECTIVE TISSUE

LOCATIONS: Between skeletal muscles and skeleton (tendons and aponeuroses); between bones (ligaments); covering skeletal muscles; deep fasciae
FUNCTIONS: Provides firm attachment; conducts pull of muscles; reduces friction between muscles; stabilizes relative positions of bones

DENSE IRREGULAR CONNECTIVE TISSUE

LOCATIONS: Capsules of visceral organs; periostea and perichondria; nerve and muscle sheaths; dermis
FUNCTIONS: Provides strength to resist forces applied from many directions; helps prevent overexpansion of organs such as the urinary bladder

ELASTIC TISSUE

LOCATIONS: Between vertebrae of the spinal column (ligamentum flavum and ligamentum nuchae); ligaments supporting penis; ligaments supporting transitional epithelia; in walls of blood vessels
FUNCTIONS: Stabilizes positions of vertebrae and penis; cushions shocks; permits expansion and contraction of organs

are cords of dense regular connective tissue that attach skeletal muscles to bones (Figure 4-12a•). The collagen fibers run along the longitudinal axis of the tendon and transfer the pull of the contracting muscle to the bone. An **aponeurosis** (AP-ō-noo-RŌ-sis) is a tendinous sheet. **Ligaments** resemble tendons but connect one bone to another. Large numbers of fibroblasts are scattered among the collagen fibers of tendons, aponeuroses, and ligaments.

The fibers in **dense irregular connective tissue** form an interwoven meshwork in no consistent pattern (Figure 4-12b•). These tissues strengthen and support areas subjected to stresses from many directions.

Example:
Tendon

Collagen
fibers

Fibroblast
nuclei

Tendon × 440

(a)

Example:
Deep dermis
of skin

Collagen
fiber
bundles

Deep dermis × 111

(b)

Example:
Elastic
ligament

Elastic
fibers

Fibroblast
nuclei

Elastic ligament × 887

(c)

4

A layer of dense irregular connective tissue gives skin its strength. Cured leather (animal skin) is an excellent illustration of the interwoven nature of this tissue. Except at joints, dense irregular connective tissue forms a sheath around cartilages (the *perichondrium*) and bones (the *periosteum*). Dense irregular connective tissue also forms a thick fibrous layer called a **capsule,** which surrounds internal organs, such as liver, kidneys, and spleen, and encloses the cavities of joints.

Dense regular and dense irregular connective tissues contain variable amounts of elastic fibers. When elastic fibers outnumber collagen fibers, the tissue has a springy, resilient nature that allows it to tolerate cycles of extension and recoil. For instance, many elastic fibers are present in the connective tissue that supports transitional epithelia; in the walls of large blood vessels, such as the aorta; and around the respiratory passageways.

Elastic tissue is a dense regular connective tissue dominated by elastic fibers. **Elastic ligaments,** which are almost completely dominated by elastic fibers, help stabilize the positions of the vertebrae of the spinal column (Figure 4-12c●).

☑ Lack of vitamin C in the diet interferes with the ability of fibroblasts to produce collagen. What effect might this interference have on connective tissue?

☑ Many allergy sufferers take antihistamines to relieve their allergy symptoms. Which type of cell produces the molecule that this medication blocks?

☑ Which type of connective tissue contains primarily triglycerides?

FLUID CONNECTIVE TISSUES

Blood and *lymph* are connective tissues that contain distinctive collections of cells in a fluid matrix. The watery matrix of blood and lymph contains cells and many types of suspended proteins that do not form insoluble fibers under normal conditions.

Blood contains blood cells and fragments of cells, collectively known as *formed elements* (Figure 4-13●). Three types of formed elements exist: (1) red blood cells, (2) white blood cells, and (3) platelets. A single cell type, the **red blood cell,** or **erythrocyte** (e-RITH-rō-sīt; *erythros,* red), accounts for almost half the volume of blood. Red blood cells are responsible for the transport of oxygen and, to a lesser degree, of carbon dioxide in the blood. The watery matrix of blood, called **plasma,** also contains small numbers of **white blood cells,** or **leukocytes** (LOO-kō-sīts; *leuko,* white). White blood cells include the phagocytic microphages (*neutrophils* and *eosinophils*), *basophils, lymphocytes,* and *monocytes.* The white blood cells are important components of the immune system, which protects the body from infection and disease. Tiny membrane-enclosed packets of cytoplasm called **platelets** contain enzymes and special proteins. Platelets function in the clotting response that seals breaks in the endothelial lining.

Extracellular fluid includes three major subdivisions: *plasma, interstitial fluid,* and *lymph.* Plasma is normally confined to the vessels of the circulatory system, and contractions of the heart keep it in motion. **Arteries** are vessels that carry blood away from the heart toward fine, thin-walled vessels called **capillaries. Veins** are vessels that drain the capillaries and return blood to the heart, completing the circuit of blood. In tissues, filtration moves water and small solutes out of the capillaries and into the interstitial fluid, which bathes the body's cells. The major difference between plasma and interstitial fluid is that plasma contains large numbers of suspended proteins.

Lymph forms as interstitial fluid enters **lymphatic vessels,** small passageways that return it to the cardiovascular system. Along the way, cells of the immune system monitor the composition of the lymph and respond to signs of injury or infection. The number of cells in lymph may vary, but ordinarily 99 percent of them are lymphocytes. The rest are primarily macrophages or microphages.

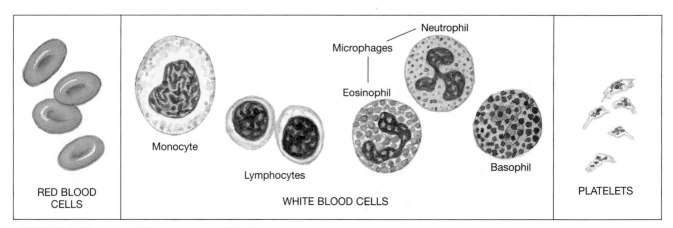

●**FIGURE 4-13** Formed Elements of the Blood

SUPPORTING CONNECTIVE TISSUES

Cartilage and bone are called supporting connective tissues because they provide a strong framework that supports the rest of the body. In these connective tissues, the matrix contains numerous fibers and, in some cases, deposits of insoluble calcium salts.

Cartilage

The matrix of **cartilage** is a firm gel that contains polysaccharide derivatives called **chondroitin sulfates** (kon-DROY-tin; *chondros,* cartilage). Chondroitin sulfates form complexes with proteins in the ground substance, forming proteoglycans. Cartilage cells, or **chondrocytes** (KON-drō-sīts), are the only cells in the matrix. They occupy small chambers known as **lacunae** (la-KOO-nē; *lacus,* pool). The physical properties of cartilage depend on the type and abundance of extracellular fibers as well as the proteoglycans.

Cartilage is avascular, and all nutrient and waste product exchange must occur by diffusion through the matrix. Blood vessels do not grow into cartilage because chondrocytes produce a chemical that discourages their formation. This chemical, named **antiangiogenesis factor** (*anti-,* against + *angeion,* vessel + *genno,* to produce), is now being tested as a potential anticancer agent. Tumors enlarge rapidly because blood vessels usually grow into areas where cells are crowded and active, improving nutrient and oxygen delivery. In theory, antiangiogenesis factor could prevent this growth, but the quantities produced in normal human cartilage are extremely small.

A cartilage is generally set apart from surrounding tissues by a fibrous **perichondrium** (pe-rē-KON-drē-um; *peri-,* around). The perichondrium contains two distinct layers: an outer, fibrous region of dense irregular connective tissue and an inner, cellular layer. The fibrous layer provides mechanical support and protection and attaches the cartilage to other structures. The cellular layer is important to the growth and maintenance of the cartilage.

Cartilage Growth Cartilage grows by two mechanisms: (1) interstitial growth and (2) appositional growth. In **interstitial growth,** chondrocytes in the cartilage matrix undergo cell division and the daughter cells produce additional matrix (Figure 4-14a•). This process enlarges the cartilage from within, like inflating a balloon by

Matrix — Chondrocyte — Lacuna

Chondrocyte undergoes division within a lacuna surrounded by cartilage matrix.

As daughter cells secrete additional matrix, they are pushed apart, expanding the cartilage from within.

New matrix

(a) Interstitial growth

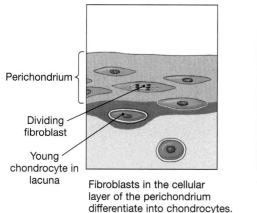

Perichondrium

Dividing fibroblast

Young chondrocyte in lacuna

Fibroblasts in the cellular layer of the perichondrium differentiate into chondrocytes.

These chondrocytes secrete new matrix.

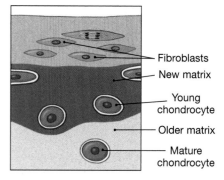

Fibroblasts — New matrix — Young chondrocyte — Older matrix — Mature chondrocyte

As matrix enlarges, more fibroblasts are incorporated; they are replaced by divisions of cells in the perichondrium.

(b) Appositional growth

•**FIGURE 4-14** **The Formation and Growth of Cartilage.** **(a)** Interstitial growth. The cartilage expands from within as chondrocytes in the matrix divide, grow, and produce new matrix. **(b)** Appositional growth. The cartilage grows at its external surface through the differentiation of fibroblasts into chondrocytes in the cellular layer of the perichondrium.

forcing more air into it. In **appositional growth,** new layers of cartilage are added to the surface (Figure 4-14b•). In this process, cells of the inner layer of the perichondrium undergo repeated cycles of division. The innermost cells then begin producing cartilage matrix and differentiate into chondrocytes. This process gradually increases the size of the cartilage by adding to its outer surface. Interstitial growth is most important during embryonic development. Appositional growth begins early in development and continues through adolescence. Neither interstitial nor appositional growth occurs in the cartilages of normal adults. However, appositional growth may occur in unusual circumstances, such as after cartilage has been damaged or under excessive stimulation with *growth hormone* from the pituitary gland. Minor damage to cartilage can be repaired by appositional growth at the damaged surface. After more severe damage, the injured portion of the cartilage will be replaced by a dense fibrous patch.

Types of Cartilage   The body contains three major types of cartilage: hyaline cartilage, elastic cartilage, and fibrocartilage:

1. **Hyaline cartilage** (HĪ-uh-lin; *hyalos,* glass) is the most common type of cartilage. Except inside joint cavities, hyaline cartilage is covered by a dense perichondrium (Figure 4-15a•). The matrix of hyaline cartilage contains closely packed collagen fibers making it tough but somewhat flexible. Because the fibers are not in large bundles and do not stain darkly, they are not always apparent in light microscopy (Figure 4-15b•). Examples in adults include the connections between the ribs and the sternum; the nasal cartilages and the supporting cartilages along the conducting passageways of the respiratory tract; and the *articular cartilages,* which cover opposing bone surfaces within many joints, such as the elbow and knee.

2. **Elastic cartilage** (Figure 4-15c•) contains numerous elastic fibers that make it extremely resilient and flexible. These cartilages usually have a yellowish color on gross dissection. Elastic cartilage forms the external flap (the *auricle* or *pinna*) of the outer ear, the epiglottis, an airway to the middle ear cavity (the *auditory tube*), and small cartilages in the larynx (the *cuneiform cartilages*).

3. **Fibrocartilage** has little ground substance, and its matrix is dominated by collagen fibers (Figure 4-15d•). The collagen fibers are densely interwoven, making this tissue extremely durable and tough. Fibrocartilaginous pads lie between the spinal vertebrae, between the pubic bones of the pelvis, and around or in a few joints and tendons. In these positions, fibrocartilage resists compression, absorbs shocks, and prevents damaging bone-to-bone contact. Cartilages heal

Perichondrium

Hyaline cartilage

(a)

HYALINE CARTILAGE

LOCATIONS: Between tips of ribs and bones of sternum; covering bone surfaces at synovial joints; supporting larynx (voicebox), trachea, and bronchi; forming part of nasal septum
FUNCTIONS: Provides stiff but somewhat flexible support; reduces friction between bony surfaces

---

ELASTIC CARTILAGE

LOCATIONS: Auricle of external ear; epiglottis; auditory canal; cuneiform cartilages of larynx
FUNCTIONS: Provides support but tolerates distortion without damage and returns to original shape

---

FIBROCARTILAGE

LOCATIONS: Pads within knee joint; between pubic bones of pelvis; intervertebral discs
FUNCTIONS: Resists compression; prevents bone-to-bone contact; limits relative movement

---

•**FIGURE 4-15 The Perichondrium and Types of Cartilage. (a)** A perichondrium separates cartilage from other tissues. **(b)** Hyaline cartilage. Note the translucent matrix and the absence of prominent fibers. **(c)** Elastic cartilage. The closely packed elastic fibers are visible between the chondrocytes. **(d)** Fibrocartilage. The collagen fibers are extremely dense, and the chondrocytes are relatively far apart.

Example:
Shoulder joint

Chondrocytes
in lacunae

Matrix

Hyaline cartilage × 500

(b)

Example:
Auricle of
external ear

Nucleus of
chondrocyte

Lacuna

Elastic fibers
in matrix

Matrix

Elastic cartilage × 320

(c)

Example:
Discs
separating
vertebrae

Collagen
fibers in
matrix

Lacuna

Chondrocyte

Fibrocartilage × 750

(d)

4

poorly, and damaged fibrocartilages in joints such as the knee can interfere with normal movements.

Fibrocartilages and Joint Injuries

Several complex joints, including the knee, contain both hyaline cartilage and fibrocartilage. The hyaline cartilage covers bony surfaces, and fibrocartilage pads in the joint prevent bone contact during movement. Injuries to these joints can produce tears in the fibrocartilage pads, and these tears do not heal. Eventually, joint mobility is severely reduced. Surgery generally produces only a temporary or incomplete repair.

Recent advances in tissue culture have enabled researchers to grow fibrocartilages in the laboratory. Chondrocytes removed from the knees of injured dogs are cultured in an artificial framework of collagen fibers. They eventually produce masses of fibrocartilage that can be inserted into the damaged joints. Over time, the pads change shape and grow, restoring normal joint function. In the future, this technique may be used to treat severe joint injuries in humans. ▪

Bone

Because we shall examine the detailed histology of **bone,** or **osseous** (OS-ē-us) **tissue** (*os,* bone), in Chapter 6, here we will focus on significant differences between cartilage and bone. The volume of ground substance in bone is very small. Roughly two-thirds of the matrix of bone consists of a mixture of calcium salts, primarily calcium phosphate with lesser amounts of calcium carbonate. The rest of the matrix is dominated by collagen fibers. This combination gives bone truly remarkable properties. By themselves, calcium salts are hard but rather brittle. Collagen fibers are stronger but relatively flexible. In bone, the minerals are organized around the collagen fibers. The result is a strong, somewhat flexible combination that is very resistant to shattering. In its overall properties, bone can compete with the best steel-reinforced concrete.

Figure 4-16● shows the general organization of osseous tissue. Lacunae in the matrix contain **osteocytes** (OS-tē-ō-sīts), or bone cells. The lacunae are typically organized around blood vessels that branch through the bony matrix. Although diffusion cannot occur through the hard matrix, osteocytes communicate with the blood vessels and with one another by means of slender cytoplasmic extensions. These extensions run through long, slender passageways in the matrix. Called **canaliculi** (kan-a-LIK-ū-lē; little canals), these passageways form a branching network for the exchange of materials between blood vessels and osteocytes.

Except in joint cavities, where they are covered by a layer of hyaline cartilage, bone surfaces are sheathed by a **periosteum** (pe-rē-OS-tē-um), a layer composed of fibrous (outer) and cellular (inner) layers. The periosteum assists in the attachment of a bone to surrounding tissues and to associated tendons and ligaments. The cellular layer functions in appositional bone growth and participates in repairs after an injury. Unlike cartilage, bone undergoes extensive remodeling throughout life, and complete repairs can be made even after severe damage has occurred. Bones also respond to the stresses placed on them, growing thicker and stronger with exercise and becoming thin and brittle with inactivity.

Table 4-1 summarizes the similarities and differences between cartilage and bone.

☑ Why does cartilage heal so slowly?

☑ If a person has a herniated intervertebral disc, which type of cartilage has been damaged?

☑ Which two types of connective tissue have a matrix that is fluid?

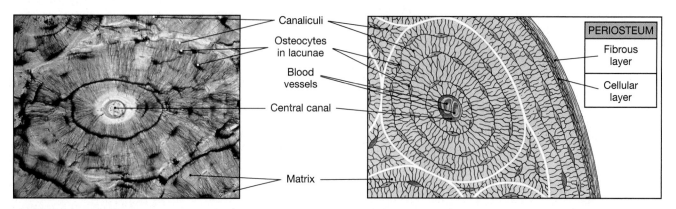

Canaliculi

Osteocytes in lacunae

Blood vessels

Central canal

Matrix

PERIOSTEUM

Fibrous layer

Cellular layer

●**FIGURE 4-16 Bone.** The osteocytes in bone are generally organized in groups around a central space that contains blood vessels. Bone dust produced during grinding fills the lacunae and the central canal, making them appear dark. (LM × 362)

TABLE 4-1   A Comparison of Cartilage and Bone

Characteristic	Cartilage	Bone
Structural Features		
Cells	Chondrocytes in lacunae	Osteocytes in lacunae
Ground substance	Chondroitin sulfate (in proteoglycan) and water	A small volume of liquid surrounding insoluble crystals of calcium salts (calcium phosphate and calcium carbonate)
Fibers	Collagen, elastic, and reticular fibers (proportions vary)	Collagen fibers predominate
Vascularity	None	Extensive
Covering	Perichondrium, two-part	Periosteum, two-part
Strength	Limited: bends easily but hard to break	Strong: resists distortion until breaking point
Metabolic Features		
Oxygen demands	Low	High
Nutrient delivery	By diffusion through matrix	By diffusion through cytoplasm and fluid in canaliculi
Growth	Interstitial and appositional	Appositional only
Repair capabilities	Limited	Extensive

MEMBRANES

Epithelia and connective tissues combine to form membranes that cover and protect other structures and tissues. Four such membranes occur in the body: (1) *mucous membranes,* (2) *serous membranes,* (3) the *cutaneous membrane,* and (4) *synovial membranes.*

MUCOUS MEMBRANES

Mucous membranes, or **mucosae** (mū-KŌ-sē), line cavities that communicate with the exterior, including the digestive, respiratory, reproductive, and urinary tracts (Figure 4-17a•). The epithelial surfaces are kept moist at all times. They are lubricated either by mucus produced by goblet cells or multicellular glands or by exposure to fluids such as urine or semen. The areolar tissue component of a mucous membrane is called the **lamina propria** (PRŌ-prē-uh). We will consider the organization of specific mucous membranes in greater detail in later chapters.

Many mucous membranes are lined by simple epithelia that perform absorptive or secretory functions, such as the simple columnar epithelium of the digestive tract. Other types of epithelia may be involved, however. For example, a stratified squamous epithelium is part of the mucous membrane of the mouth, and the mucous membrane along most of the urinary tract has a transitional epithelium.

SEROUS MEMBRANES

Serous membranes line the sealed, internal divisions of the ventral body cavity. These membranes consist of a mesothelium supported by areolar tissue (Figure 4-17b•). Recall from Chapter 1 that the three types of serous membranes are (1) the *pleura,* which lines the pleural cavities and covers the lungs; (2) the *peritoneum,* which lines the peritoneal cavity and covers the surfaces of the enclosed organs; and (3) the *pericardium,* which lines the pericardial cavity and covers the heart. ∞ *p. 22* Serous membranes are very thin, but they are firmly attached to the body wall and to the organs they cover. When looking at an organ such as the heart or stomach, you are really seeing the tissues of the organ through a transparent serous membrane.

Each serous membrane can be divided into a *parietal portion,* which lines the inner surface of the cavity, and an opposing *visceral portion,* or **serosa,** which covers the outer surfaces of visceral organs. The parietal and visceral surfaces of a serous membrane are in close contact at all times. Minimizing friction between these opposing surfaces is the primary function of serous membranes. Because the mesothelia are very thin, serous membranes are relatively permeable. Tissue fluids diffuse onto the exposed surface, keeping it moist and slippery.

The fluid formed on the surfaces of a serous membrane is called a *transudate* (TRAN-sū-dāt; *trans,* across). In healthy individuals, the total volume of transudate is extremely small, just enough to prevent friction between the walls of the cavities and the surfaces of internal organs. But after an injury or in certain disease states, the volume of transudate may increase dramatically, complicating existing medical problems or producing new ones.

●**FIGURE 4-17** Membranes.
(a) Mucous membranes are coated with the secretions of mucous glands. These membranes line the digestive, respiratory, urinary, and reproductive tracts. **(b)** Serous membranes line the ventral body cavities (the peritoneal, pleural, and pericardial cavities). **(c)** The cutaneous membrane, or skin, covers the outer surface of the body. **(d)** Synovial membranes line joint cavities and produce the fluid within the joint.

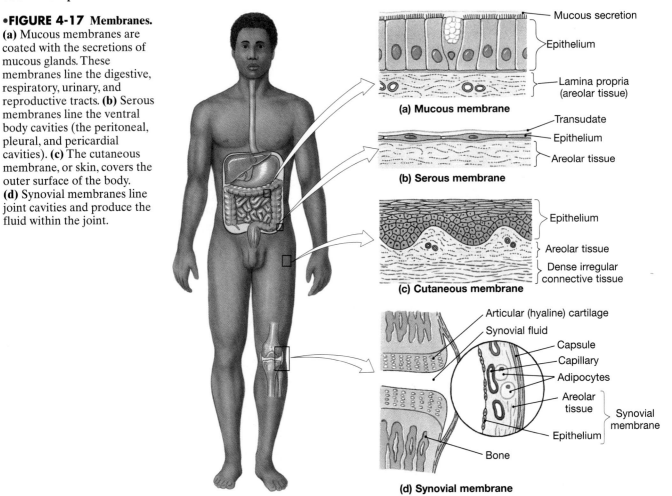

Problems with Serous Membranes

Several clinical conditions, including infection and chronic irritation, can cause the abnormal buildup of fluid in a ventral body cavity. *Pleuritis,* or *pleurisy,* is an inflammation of the pleural cavities. At first the membranes become dry, and the opposing membranes may scratch against one another, producing a sound known as a *pleural rub.* Friction between opposing layers of serous membranes may promote the formation of **adhesions,** fibrous connections that lock the membranes together and eliminate the friction. They also severely restrict the movement of the affected organ or organs and may compress blood vessels or nerves. Adhesions seldom form between the serous membranes of the pleural cavities. More commonly, continued inflammation and rubbing lead to a gradual increase in fluid production to levels well above normal. Fluid then accumulates in the pleural cavities, producing a condition known as *pleural effusion.* Pleural effusion is also caused by heart conditions that elevate the pressure in blood vessels of the lungs. As fluids build up in the pleural cavities, the lungs are compressed, making breathing difficult. The combination of severe pleural effusion and heart disease can be lethal.

Pericarditis is an inflammation of the pericardium. This condition typically leads to a *pericardial effusion,* an abnormal accumulation of the fluid in the pericardial cavity. When sudden or severe, the fluid buildup can seriously reduce the efficiency of the heart and restrict blood flow through major vessels.

Peritonitis, an inflammation of the peritoneum, can follow infection of or injury to the peritoneal lining. Peritonitis is a potential complication of any surgical procedure in which the peritoneal cavity is opened. Liver disease, kidney disease, or heart failure can cause an increase in the rate of fluid movement through the peritoneal lining. *Ascites* (a-SĪ-tēz), the accumulation of peritoneal fluid, creates a characteristic abdominal swelling. The distortion of internal organs by the contained fluid can lead to symptoms such as heartburn, indigestion, and low-back pain.

THE CUTANEOUS MEMBRANE

The **cutaneous membrane,** or skin, covers the surface of the body. It consists of a stratified squamous epithelium and a layer of areolar tissue reinforced by underlying dense connective tissue (Figure 4-17c•). In contrast to serous and mucous membranes, the cutaneous membrane is thick, relatively waterproof, and usually dry.

SYNOVIAL MEMBRANES

Bones contact one another at joints, or *articulations.* Joints that permit significant movement are surrounded by a fibrous capsule and contain a joint cavity lined by a synovial membrane (Figure 4-17d•). **Synovial** (sin-Ō-vē-ul) **membranes** consist of extensive areas of specialized areolar tissue containing a matrix of interwoven collagen fibers, proteoglycans, and glycoproteins. They surround the joint cavity. This space is filled with **synovial fluid,** a solution similar to the ground substance in loose connective tissues. The lining of the joint cavity consists of an incomplete layer of macrophages and specialized fibroblasts, which regulate the composition of the synovial fluid. Although this lining is often called an epithelium, it differs from true epithelia in four respects: (1) The lining cells develop within a connective tissue; (2) no basement membrane is present; (3) gaps of up to 1 mm may separate adjacent cells; and (4) fluid and solutes are continuously exchanged between the synovial fluid and capillaries in the underlying connective tissue.

THE CONNECTIVE TISSUE FRAMEWORK OF THE BODY

4

Connective tissues create the internal framework of the body. Layers of connective tissue connect the organs within the dorsal and ventral body cavities with the rest of the body. These layers (1) provide strength and stability; (2) maintain the relative positions of internal organs; and (3) provide a route for the distribution of blood vessels, lymphatic vessels, and nerves. **Fasciae** (FASH-ē-ē); singular, *fascia*) are connective tissue layers and wrappings that support and surround organs. We can divide the fasciae into three types of layers: the superficial fascia, the deep fascia, and the subserous fascia. The functional anatomy of these layers is illustrated in Figure 4-18•.

1. The **superficial fascia,** or **subcutaneous layer** (*sub,* below + *cutis,* skin) is also termed the **hypodermis** (*hypo,* below + *dermis,* skin). This layer of areolar tissue and fat separates the skin from underlying tissues and organs. It provides insulation and padding and lets the skin and underlying structures move independently.

2. The **deep fascia** consists of dense connective tissue. The organization of the fibers resembles that of plywood: All the fibers in one layer run in the same direction, but the orientation of the fibers changes from layer to layer. This arrangement helps the tissue resist forces applied from many directions. The tough capsules that surround most organs, including the kidneys and the organs in the thoracic and peritoneal cavities, are bound to the deep fascia. The perichondrium

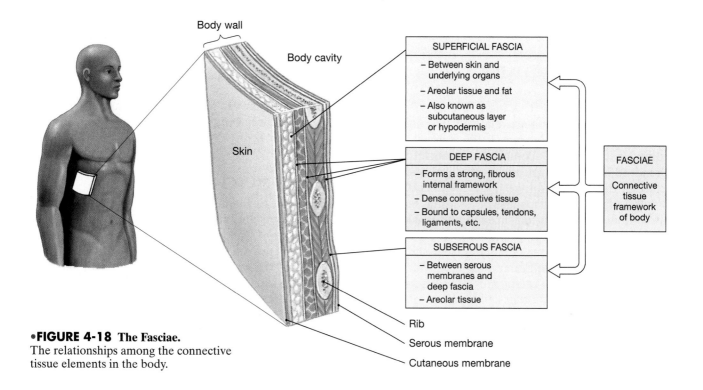

•**FIGURE 4-18** The Fasciae.
The relationships among the connective tissue elements in the body.

4

around cartilages, the periosteum around bones and the ligaments that interconnect them, and the connective tissues of muscle, including tendons, are also connected to the deep fascia. The dense connective tissue components are interwoven. For example, the deep fascia around a muscle blends into the tendon, whose fibers intermingle with those of the periosteum. This arrangement creates a strong, fibrous network and ties structural elements together.

3. The **subserous fascia** is a layer of areolar tissue that lies between the deep fascia and the serous membranes that line body cavities. Because this layer separates the serous membranes from the deep fascia, movements of muscles or muscular organs do not severely distort the delicate lining.

☑ Which cavities in the body are lined by serous membranes?

☑ The lining of the nasal cavity is normally moist, contains numerous goblet cells, and rests on a layer of connective tissue called the lamina propria. Which type of membrane is this?

☑ A sheet of tissue has many layers of collagen fibers that run in different directions in successive layers. Which type of tissue is this?

MUSCLE TISSUE

Muscle tissue is specialized for contraction. Muscle cells possess organelles and properties distinct from those of other cells. The body has three types of muscle tissue: (1) *skeletal muscle,* (2) *cardiac muscle,* and (3) *smooth muscle.* The contraction mechanism is similar in all three, but they differ in internal organization. We will describe each muscle type more fully in Chapter 10. Here we focus on general characteristics.

SKELETAL MUSCLE TISSUE

Skeletal muscle tissue contains very large muscle cells—up to 0.3 meter (1 ft) or more in length. Because the individual muscle cells are relatively long and slender, they are usually called **muscle fibers.** Each fiber is **multinucleated**; that is, it contains many nuclei. Several hundred nuclei are typically located just under the surface of the cell membrane (Figure 4-19a●). Skeletal muscle fibers are incapable of dividing, but new muscle fibers are produced through the divisions of **satellite cells,** stem cells that persist in adult skeletal muscle tissue. As a result, skeletal muscle tissue can at least partially repair itself after an injury.

The cytoskeleton of skeletal muscle fibers contains actin and myosin, as do other cell types. ∞ *p. 81* In skeletal muscle fibers, however, these filaments are organized in repeating groups that give the cells a *striated,* or banded, appearance. These *striations,* or bands, are readily apparent in light micrographs. Skele-

tal muscle fibers do not usually contract unless stimulated by nerves, and the nervous system provides voluntary control over their activities. Thus, skeletal muscle is called **striated voluntary muscle.**

Skeletal muscle fibers are tied together by collagen and elastic fibers. These fibers surround each cell and each group of cells and blend into the attached tendon. The tendon fibers, which conduct the force of contraction, are typically connected to a bone. Thus, when the muscle tissue contracts, it pulls on the attached bone.

CARDIAC MUSCLE TISSUE

Cardiac muscle tissue is located only in the heart. A typical cardiac muscle cell, also known as a **cardiocyte,** or *cardiac myocyte,* is smaller than a skeletal muscle cell. A typical cardiac muscle cell has one centrally positioned nucleus, but some cardiocytes have as many as five. Prominent striations (Figure 4-19b●) resemble those of skeletal muscle; the actin and myosin filaments are arranged the same way in both cell types.

Cardiac muscle cells form extensive connections with one another. The connections occur at specialized regions known as **intercalated discs.** At an intercalated disc, the membranes are locked together by desmosomes, intercellular cement, and gap junctions. As a result, cardiac muscle tissue consists of a branching network of interconnected muscle cells. The desmosomes and intercellular cement lock the cells together during a contraction, and ion movement through gap junctions helps coordinate the contractions of the cardiac muscle cells. Like skeletal muscle fibers, cardiac muscle cells cannot divide. Because this tissue lacks satellite cells, cardiac muscle tissue damaged by injury or disease cannot regenerate.

Cardiac muscle cells do not rely on nerve activity to start a contraction. Instead, specialized cardiac muscle cells called *pacemaker cells* establish a regular rate of contraction. Although the nervous system can alter the rate of pacemaker cell activity, it does not provide voluntary control over individual cardiac muscle cells. Therefore, cardiac muscle is called **striated involuntary muscle.**

SMOOTH MUSCLE TISSUE

Smooth muscle tissue is located (1) in the walls of blood vessels; (2) around hollow organs such as the urinary bladder; and (3) in layers around the respiratory, circulatory, digestive, and reproductive tracts. A smooth muscle cell is a small, spindle-shaped cell with tapering ends and a single, oval nucleus (Figure 4-19c●). Smooth muscle cells can divide, so smooth muscle tissue can regenerate after an injury. The actin and myosin filaments in smooth muscle cells are organized differently from those of skeletal and cardiac muscles, and smooth muscle tissue has no striations. Smooth muscle

4

SKELETAL MUSCLE TISSUE

Cells are long, cylindrical, striated, and multinucleate.

LOCATIONS: Combined with connective tissues and nervous tissue in skeletal muscles

FUNCTIONS: Moves or stabilizes the position of the skeleton; guards entrances and exits to the digestive, respiratory, and urinary tracts; generates heat; protects internal organs

Nuclei

Striations

Muscle fiber

Skeletal muscle × 155   **(a)**

CARDIAC MUSCLE TISSUE

Cells are short, branched, and striated, usually with a single nucleus; cells are interconnected by intercalated discs.

LOCATION: Heart

FUNCTIONS: Circulates blood; maintains blood (hydrostatic) pressure

Intercalated disc

Nucleus

Cardiocytes

Striations

Cardiac muscle × 425   **(b)**

SMOOTH MUSCLE TISSUE

Cells are short, spindle-shaped, and nonstriated, with a single, central nucleus.

LOCATIONS: Encircles blood vessels; in the walls of digestive, respiratory, urinary, and reproductive organs

FUNCTIONS: Moves food, urine, and reproductive tract secretions; controls diameters of respiratory passageways and blood vessels

Smooth muscle cell

Nucleus

Smooth muscle × 330   **(c)**

•**FIGURE 4-19 Muscle Tissue. (a)** Skeletal muscle fibers are large, have multiple, peripherally located nuclei, and exhibit prominent striations (banding) and an unbranched arrangement. **(b)** Cardiac muscle cells differ from skeletal muscle fibers in three major ways: size (cardiac muscle cells are smaller), organization (cardiac muscle cells branch), and number and location of nuclei (a typical cardiac muscle cell has one centrally placed nucleus). Both skeletal and cardiac muscle cells contain actin and myosin filaments in an organized array that produces striations. **(c)** Smooth muscle cells are small and spindle-shaped, with a central nucleus. They do not branch or have striations.

4

cells may contract on their own, with gap junctions between adjacent cells coordinating the contractions of individual cells. The contraction of some smooth muscle tissue can be controlled by the nervous system, but the contractile activity is not under your voluntary control. Because your nervous system usually does not provide voluntary control over smooth muscle contractions, smooth muscle is known as **nonstriated involuntary muscle.**

NEURAL TISSUE

Neural tissue, which is also known as *nervous tissue* or *nerve tissue,* is specialized for the conduction of electrical impulses from one region of the body to another. Ninety-eight percent of the neural tissue in the body is concentrated in the brain and spinal cord, which are the control centers of the nervous system.

Neural tissue contains two basic types of cells: (1) **neurons** (NOO-ronz; *neuro,* nerve) and (2) several kinds of supporting cells, collectively called **neuroglia** (noo-ROG-lē-uh or noo-rō-GLĔ-uh), or

glial cells (*glia,* glue). Our conscious and unconscious thought processes reflect the communication among neurons in the brain. Such communication involves the propagation of electrical impulses in the form of changes in the transmembrane potential. Information is conveyed by the frequency and pattern of impulse generation. Neuroglia have different functions, such as supporting neural tissue and supplying nutrients to neurons.

The longest cells in your body are neurons; many neurons are a meter (39 in.) long. Most neurons cannot divide under normal circumstances, so they have a very limited ability to repair themselves after injury. A typical neuron has a large **cell body** with a large nucleus and a prominent nucleolus (Figure 4-20•). Extending from the cell body are various branching processes (projections or outgrowths) termed **dendrites** (DEN-drīts; *dendron,* a tree) and one **axon.** Dendrites receive information, typically from other neurons, and axons carry that information to other cells. Because axons tend to be very long and slender, they are also called **nerve fibers.** In Chapter 12, we shall further examine the properties of neural tissue.

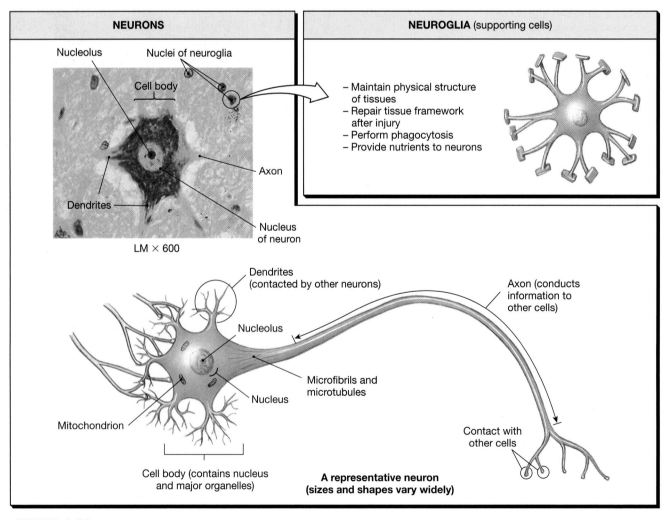

•FIGURE 4-20 **Neural Tissue**

☑ Which type of muscle tissue has small, tapering cells with single nuclei and no obvious striations?

☑ A tissue contains irregularly shaped cells with many fibrous projections, some several centimeters long. These are probably which type of cell?

☑ If skeletal muscle cells in adults are incapable of dividing, how is new skeletal muscle formed?

TISSUE INJURIES AND AGING

Tissues are not isolated; they combine to form organs with diverse functions. Any injury affects several tissue types simultaneously. These tissues must respond in a coordinated way to preserve homeostasis.

INFLAMMATION AND REGENERATION

The restoration of homeostasis after a tissue has been injured involves two related processes: inflammation and repair. First, immediately after the injury, the area is isolated while damaged cells, tissue components, and any dangerous microorganisms are cleaned up. This phase, which coordinates the activities of several types of tissues, is called **inflammation,** or the **inflammatory response.** It produces several familiar sensations, including swelling, redness, warmth, and pain. An **infection** is an inflammation resulting from the presence of pathogens, such as harmful bacteria.

Second, the damaged tissues are replaced or repaired to restore normal function. This repair process is called **regeneration.** Inflammation and regeneration are controlled at the tissue level. The two phases overlap; isolation establishes a framework that guides the cells responsible for reconstruction, and repairs are under way well before cleanup operations have ended. We will now consider the basics of the repair process after an injury. At this time, we will focus on the interaction among different tissues. Our example includes

two connective tissues (areolar tissue and blood), an epithelium (the endothelia of blood vessels), a muscle tissue (smooth muscle in the vessel walls), and neural tissue (sensory nerve endings). In later chapters, especially Chapters 5 and 22, we shall examine inflammation and regeneration in more detail.

First Phase: Inflammation

Many stimuli—including impact, abrasion, distortion, chemical irritation, infection by pathogenic organisms (such as bacteria or viruses), and extreme temperatures (hot or cold)—can produce inflammation. Each of these stimuli kills cells, damages fibers, or injures the tissue in some other way. These changes alter the chemical composition of the interstitial fluid: Damaged cells release prostaglandins, proteins, and potassium ions, and the injury itself may have introduced foreign proteins or pathogens.

Immediately after the injury, tissue conditions become even more abnormal. **Necrosis** (ne-KRŌ-sis) is the tissue degeneration that occurs after cells have been injured or destroyed. The process begins several hours after the original injury, and the damage is caused by lysosomal enzymes. Through widespread autolysis, lysosomes release enzymes that first destroy the injured cells and then attack surrounding tissues. ∞ *p. 87* The result may be an accumulation of debris, fluid, dead and dying cells, and necrotic tissue components collectively known as **pus.** An accumulation of pus in an enclosed tissue space is an **abscess.**

These tissue changes trigger the inflammatory response by stimulating mast cells, connective tissue cells introduced on page 119. Figure 4-21• follows the events set in motion by the activation of mast cells. Although in this example the injury has occurred in an areolar tissue, the process would be basically the same after an injury to any connective tissue proper.

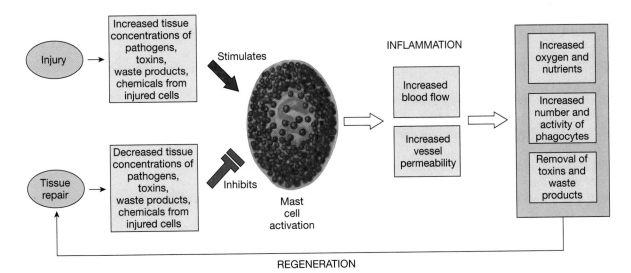

•**FIGURE 4-21** An Introduction to Inflammation

4

Because all organs have connective tissues, inflammation can occur anywhere in the body.

When an injury occurs that damages fibers and cells, mast cells release a variety of chemicals. These chemicals, including histamine and prostaglandins, trigger changes in local circulation. In response, the smooth muscle tissue that surrounds local blood vessels relaxes and the vessels *dilate,* or enlarge in diameter. This dilation increases blood flow through the tissue, turning the region red and making it warm to the touch. The combination of abnormal tissue conditions and chemicals released by mast cells stimulates sensory nerve endings that produce sensations of pain. At the same time, the chemicals released by mast cells make the endothelial cells of local capillaries more permeable. Plasma, including blood proteins, now diffuses into the injured tissue, so the area becomes swollen.

The increased blood flow accelerates the delivery of nutrients and oxygen and the removal of dissolved waste products and toxic chemicals. It also brings white blood cells to the region. These phagocytic cells migrate to the injury site and assist in the defense and cleanup operations. Macrophages and microphages protect the tissue from infection and perform cleanup by engulfing both debris and bacteria. Over a period of hours to days, this cleanup process generally succeeds in eliminating the inflammatory stimulus.

Second Phase: Regeneration

Although the situation is under control and no further damage will occur, many cells in the area have either already died or will soon die as a consequence of the original injury. As tissue conditions return to normal, fibroblasts move into the necrotic area, laying down a network of collagen fibers that stabilizes the injury site. This process produces a dense, collagenous framework known as *scar tissue* or *fibrous tissue.* Over time, scar tissue is usually remodeled and gradually assumes a more normal appearance. The cell population in the area gradually increases; some cells migrate to the site, and others are produced by the division of mesenchymal stem cells.

Each organ has a different ability to regenerate after injury. That ability can be directly linked to the pattern of tissue organization in the injured organ. Epithelia and connective tissues regenerate well; smooth and skeletal muscle tissues do so relatively poorly; and cardiac muscle tissue and neural tissue cannot regenerate at all. Your skin, which is dominated by epithelia and connective tissues, regenerates rapidly and completely after injury. (We shall describe this process in Chapter 5.) In contrast, damage to the heart is much more serious, because, although the connective tissues of the heart can be repaired, lost cardiac muscle cells are replaced only by fibrous tissue. The permanent replacement of normal tissue by fibrous tissue is called *fibrosis* (fī-BRŌ-sis). Fibrosis in muscle and other tissues may occur in response to injury, disease, or aging.

AGING AND TISSUE REPAIR

Tissues change with age, and the speed and effectiveness of tissue repairs decrease. Repair and maintenance activities throughout the body slow down, and a combination of hormonal changes and changes in lifestyle affect the structure and chemical composition of many tissues. Epithelia get thinner and connective tissues more fragile. Individuals bruise easily and bones become brittle; joint pain and broken bones are common in the elderly. Because cardiac muscle cells and neurons cannot be replaced, cumulative damage can eventually cause major health problems, such as cardiovascular disease or deterioration in mental function.

In later chapters, we will consider the effects of aging on specific organs and systems. Some of these effects are genetically programmed. For example, the chondrocytes of older individuals produce a slightly different form of proteoglycan than do those of younger people. The difference probably accounts for the thinner and less resilient cartilage of older people. In some cases, the tissue degeneration can be temporarily slowed or even reversed. The age-related reduction in bone strength, a condition called *osteoporosis,* typically results from a combination of inactivity, low dietary calcium levels, and a reduction in circulating sex hormones. A program of exercise, calcium supplements, and hormone replacement therapies can generally maintain normal bone structure for many years.

AGING AND CANCER INCIDENCE

Cancer rates increase with age, and roughly 25 percent of all people in the United States develop cancer at some point in their lives. It has been estimated that 70–80 percent of cancer cases result from chemical exposure, environmental factors, or some combination of the two, and 40 percent of those cancers are caused by cigarette smoke. Each year in the United States, more than 500,000 individuals die of cancer, making it second only to heart disease. We discussed cancer development and growth in Chapter 3. *pp. 100–101* AM *A Closer Look: Cancer*

This chapter introduced the four basic tissue types that occur in the body. In terms of their contribution to total body weight, muscle tissue is most important (about 50 percent), followed by connective tissues (45 percent), epithelium and glands (3 percent), and neural tissue (2 percent). In combination, these tissues form all the organs and systems we shall discuss in subsequent chapters. In Chapter 5 we shall consider the *integument,* an easily accessible structure with structural characteristics intermediate between those of organs and organ systems. AM *EMBRYOLOGY SUMMARY 4: The Development of Organ Systems*

Tissue Structure and Disease

Pathologists (pa-THOL-o-jists) are physicians who specialize in the study of disease processes. Diagnosis, rather than treatment, is usually the main focus of their activities. In their analyses, pathologists integrate anatomical and histological observations to determine the nature and severity of the disease. Because disease processes affect the histological organization of tissues and organs, **biopsies,** or tissue samples, often play a key role in their diagnoses.

Figure 4-22• diagrams the histological changes induced by one relatively common irritating stimulus, cigarette smoke. The first abnormality to be observed is **dysplasia** (dis-PLĀ-zē-uh), a change in the normal shape, size, and organization of tissue cells. Dysplasia is generally a response to chronic irritation or inflammation, and the changes are reversible. The normal trachea (windpipe) and its branches are lined by a pseudostratified ciliated columnar epithelium. The cilia move a mucous layer that traps foreign particles and moistens incoming air. The drying and chemical effects of smoking first paralyze the cilia, halting the movement of mucus (Figure 4-22a•). As mucus builds up, the individual coughs to dislodge it (the well-known "smoker's cough").

Epithelia and connective tissues may undergo more-radical changes in structure, caused by the division and differentiation of stem cells. **Metaplasia** (me-tuh-PLĀ-zē-uh) is a structural change that dramatically alters the character of the tissue. In our heavy smoking example, over time the epithelial cells lose their cilia altogether.

As metaplasia occurs, the epithelial cells produced by stem cell divisions no longer differentiate into ciliated columnar cells. Instead, they form a stratified squamous epithelium that provides greater resistance to drying and chemical irritation (Figure 4-22b•). This epithelium protects the underlying tissues more effectively, but it eliminates the moisturization and cleaning properties of the epithelium. Cigarette smoke will now have an even greater effect on more-delicate portions of the respiratory tract. Fortunately, metaplasia is reversible, and the epithelium gradually returns to normal if the individual quits smoking.

In **anaplasia** (a-nuh-PLĀ-zē-uh), tissue organization breaks down. Tissue cells change size and shape, typically becoming unusually large or small (Figure 4-22c•). Anaplasia occurs in smokers who develop one form of lung cancer; the cells divide more frequently, but not all divisions proceed in the normal way. Many of the tumor cells have abnormal chromosomes. Unlike dysplasia and metaplasia, anaplasia is irreversible.

Irritant chemicals and particles in smoke

NORMAL RESPIRATORY EPITHELIUM

Reversible

(a) The cilia of respiratory epithelial cells are damaged and paralyzed by exposure to cigarette smoke. These changes cause the local buildup of mucus and reduce the effectiveness of the epithelium in protecting deeper, more delicate portions of the respiratory tract.

DYSPLASIA

Reversible

(b) In metaplasia, a tissue changes its structure. In this case the stressed respiratory surface converts to a stratified epithelium that protects underlying connective tissues but does nothing for other areas of the respiratory tract.

METAPLASIA

Irreversible

(c) In anaplasia, the tissue cells become tumor cells; anaplasia produces a cancerous tumor.

ANAPLASIA

•**FIGURE 4-22** Changes in a Tissue under Stress

[EXPLORE] [MediaLab]

Each of the four basic tissue types in the body has individual characteristics that relate to its function. A firm grounding in histology is essential to a grasp of anatomy and physiology. To gain a better understanding of tissues, use the resources available on both the CD-ROM and the Companion Website. The CD-ROM includes animations and tutorials; the website builds on those resources with text content and interactive problems as well as communication avenues for sharing ideas with your class. To complete the following Web Explorations, visit the MediaLab in Chapter 4 of your Companion Website.

Web Exploration 1

Estimated time for completion = 15 minutes

One of the four basic tissue types is connective tissue. Connective tissue is classified in part by the type of matrix—the noncellular material—it contains. The identification of connective tissue is an important skill to master; once you have identified such a tissue, you can surmise its general function.

This skill can be difficult, however, without practice. Select the keyword **CONNECTIVE TISSUE** on the Companion Website to gain some practice. Read the available material; then click on "Quiz" to test your understanding.

Web Exploration 2

Estimated time for completion = 15 minutes

A good way to try to assess the function of an organ is to identify the major tissue type it contains and to relate the functions of that particular tissue to the organ itself. To hone your tissue identification skills, select the keyword **HISTOLOGY** on the Companion Website. This website consists of five pages: an introduction to tissues, and a separate linked page for each basic tissue type. As you read these pages, create a study sheet that includes the function of each tissue type and a description of the tissue's appearance. You may wish to share your study sheet with your classmates by using the communications tools on the Companion Website.

CHAPTER REVIEW

SELECTED CLINICAL TERMINOLOGY
Terms Discussed in This Chapter

abscess: The accumulation of pus within an enclosed tissue space. *(p. 135)*
adhesions: Restrictive fibrous connections that can result from surgery, infection, or other injuries to serous membranes. *(p. 130)*
anaplasia (a-nuh-PLĀ-zē-uh): An irreversible change in the size and shape of tissue cells. *(p. 137)*
antiangiogenesis factor: A secretion, produced by chondrocytes, that inhibits the growth of blood vessels. *(p. 125)*
ascites (a-SĪ-tēz): The accumulation of fluid in the peritoneal cavity, usually caused by liver or kidney disease or heart failure. *(p. 130)*
dysplasia (dis-PLĀ-zē-uh): A change in the normal shape, size, and organization of tissue cells. *(p. 137)*
exfoliative cytology: The study of cells shed or collected from epithelial surfaces. *(p. 115)*
liposuction: A surgical procedure to remove unwanted adipose tissue by sucking it out through a tube. *(p. 122)*
metaplasia (me-tuh-PLĀ-zē-uh): A structural change that alters the character of a tissue. *(p. 137)*

necrosis: Tissue degeneration that occurs after cells have been injured or destroyed; a result of the release of lysosomal enzymes through autolysis. *(p. 135)*
pathologists (pa-THOL-o-jists): Physicians who specialize in the study of disease processes. *(p. 137)*
pericarditis: An inflammation of the pericardial lining that may lead to the accumulation of pericardial fluid (a *pericardial effusion*). *(p. 130)*
peritonitis: An inflammation of the peritoneum after infection or injury. *(p. 130)*
pleural effusion: The accumulation of fluid within the pleural cavities as a result of chronic infection or inflammation of the pleura. *(p. 130)*
pleuritis (*pleurisy*): An inflammation of the pleural cavities. This condition may cause the production of a sound known as a *pleural rub*. *(p. 130)*
regeneration: The repairing of injured tissues that follows inflammation. *(p. 135)*

AM Additional Terms Discussed in the *Applications Manual*

chemotherapy: The administration of drugs that either kill cancerous tissues or prevent mitotic divisions.
immunotherapy: The administration of drugs that help the immune system recognize and attack cancer cells.

oncologists (on-KOL-o-jists): Physicians who specialize in identifying and treating cancers.
remission: A stage in which a tumor stops growing or becomes smaller; the goal of cancer treatment.

STUDY OUTLINE

INTRODUCTION p. 107

1. **Tissues** are collections of specialized cells and cell products that are organized to perform a relatively limited number of functions. The four *tissue types* are *epithelial tissue, connective tissue, muscle tissue,* and *neural tissue.*

2. **Histology** is the study of tissues.

EPITHELIAL TISSUE p. 107

1. **Epithelial tissue** includes epithelia and glands. An **epithelium** is an **avascular** layer of cells that forms a barrier that provides protection and regulates permeability. **Glands** are secretory structures derived from epithelia. Epithelial cells may show **polarity,** an uneven distribution of cytoplasmic components.

Functions of Epithelial Tissue p. 107

2. Epithelia provide physical protection, control permeability, provide sensation, and produce specialized secretions. Gland cells are epithelial cells that produce secretions. In **glandular epithelia,** most cells produce secretions.

3. A **basement membrane** attaches epithelia to underlying connective tissues.

Specializations of Epithelial Cells p. 108

4. Epithelial cells are specialized to allow them to perform secretory or transport functions and to maintain the physical integrity of the epithelium. *(Figure 4-1)*

5. Many epithelial cells have microvilli, and some have **stereocilia.**

6. The coordinated beating of the cilia on a **ciliated epithelium** moves materials across the epithelial surface.

Maintaining the Integrity of Epithelia p. 108

7. Cells can attach to other cells or to extracellular protein fibers by means of **cell adhesion molecules (CAMs)** or at specialized attachment sites called **cell junctions.** The three major types of cell junctions are **tight junctions, desmosomes,** and **gap junctions.** *(Figure 4-2)*

8. The inner surface of each epithelium is connected to a two-part basement membrane consisting of a **basal lamina** and a **reticular lamina.** Divisions by **germinative cells** continually replace the short-lived epithelial cells.

Classification of Epithelia p. 110

9. Epithelia are classified on the basis of the number of cell layers and the shape of the cells at the apical surface.

10. A **simple epithelium** has a single layer of cells covering the basement membrane; a **stratified epithelium** has several layers. The cells in a **squamous epithelium** are thin and flat. Cells in a **cuboidal epithelium** resemble little hexagonal boxes; those in a **columnar epithelium** are taller and more slender. *(Figures 4-3, 4-4, 4-5)*

Glandular Epithelia p. 115

11. **Exocrine glands** discharge secretions onto the body surface or into **ducts,** which communicate with the exterior. *Hormones,* the secretions of **endocrine glands,** are released by gland cells into the surrounding interstitial fluid.

12. A glandular epithelial cell may release its secretions by merocrine, apocrine, or holocrine modes. In **merocrine secretion,** the most common secretion mode, the product is released through exocytosis. **Apocrine secretion** involves the loss of both secretory product and cytoplasm. Unlike the other two methods, **holocrine secretion** destroys the

cell, which becomes packed with secretions and bursts. *(Figure 4-6)*

13. In epithelia that contain scattered gland cells, individual secretory cells are called **unicellular glands. Multicellular glands** are organs that contain glandular epithelia that produce exocrine or endocrine secretions.

14. Exocrine glands can be classified on the basis of structure as **unicellular exocrine glands (goblet cells)** or as **multicellular exocrine glands.** Multicellular exocrine glands can be further classified according to structure. *(Figure 4-7)*

CONNECTIVE TISSUES p. 117

1. **Connective tissues** are internal tissues with many important functions: establishing a structural framework; transporting fluids and dissolved materials; protecting delicate organs; supporting, surrounding, and interconnecting tissues; storing energy reserves; and defending the body from microorganisms. *(Figure 4-8)*

2. All connective tissues contain both specialized cells and a **matrix,** composed of extracellular protein fibers and a **ground substance.**

Classification of Connective Tissues p. 118

3. **Connective tissue proper** is connective tissue that contains varied cell populations and fiber types surrounded by a syrupy ground substance.

4. **Fluid connective tissues** have a distinctive population of cells suspended in a watery ground substance that contains dissolved proteins. The two types are *blood* and *lymph.*

5. **Supporting connective tissues** have a less diverse cell population than connective tissue proper does and a dense ground substance with closely packed fibers. The two types of supporting connective tissues are *cartilage* and *bone.*

Connective Tissue Proper p. 118

6. Connective tissue proper contains fibers, a viscous ground substance, and a varied population of cells, including **fibroblasts, macrophages, adipocytes, mesenchymal cells, melanocytes, mast cells, lymphocytes,** and **microphages.**

7. The three types of fibers in connective tissue are **collagen fibers, reticular fibers,** and **elastic fibers.**

8. The first connective tissue to appear in an embryo is **mesenchyme,** or *embryonic connective tissue.*

9. Connective tissue proper is classified as **loose connective tissue** or **dense connective tissue.** The loose connective tissues are mucous connective tissue in the embryo, **areolar tissue, adipose tissue,** including **white fat** and **brown fat,** and **reticular tissue.** Most of the volume in dense connective tissue consists of fibers. The two types of dense connective tissue are **dense regular connective tissue** and **dense irregular connective tissue.** *(Figures 4-9–4-12)*

Fluid Connective Tissues p. 124

10. **Blood** and **lymph** are connective tissues that contain distinctive collections of cells in a fluid matrix.

11. Blood contains *formed elements:* **red blood cells (erythrocytes), white blood cells (leukocytes),** and **platelets.** Its watery matrix is called **plasma.** *(Figure 4-13)*

12. **Arteries** carry blood from the heart and toward **capillaries,** where water and small solutes move into the interstitial fluid of surrounding tissues. **Veins** return blood to the heart.

4

13. Lymph forms as interstitial fluid enters the **lymphatic vessels,** which return lymph to the cardiovascular system.

Supporting Connective Tissues   p. 125

14. Cartilage and bone are called supporting connective tissues because they support the rest of the body.

15. **Cartilage** grows by two mechanisms: **interstitial growth** and **appositional growth.** *(Figure 4-14)*

16. The matrix of cartilage is a firm gel that contains **chondroitin sulfates** (used to form proteoglycans) and cells called **chondrocytes.** Chondrocytes occupy chambers called **lacunae.** A fibrous **perichondrium** separates cartilage from surrounding tissues. The three types of cartilage are **hyaline cartilage, elastic cartilage,** and **fibrocartilage.** *(Figure 4-15)*

17. Chondrocytes rely on diffusion through the avascular matrix to obtain nutrients.

18. **Bone,** or **osseous tissue,** has **osteocytes,** little ground substance, and a dense, mineralized matrix. Osteocytes are situated in lacunae. The matrix consists of calcium salts and collagen fibers, giving it unique properties. *(Figure 4-16; Table 4-1)*

19. Osteocytes depend on diffusion through **canaliculi** for nutrient intake.

20. Each bone is surrounded by a **periosteum** with fibrous and cellular layers.

MEMBRANES   p. 129

1. Membranes form a barrier or interface. Epithelia and connective tissues combine to form membranes that cover and protect other structures and tissues. The four types of membranes are *mucous, serous, cutaneous,* and *synovial.* *(Figure 4-17)*

Mucous Membranes   p. 129

2. **Mucous membranes** line cavities that communicate with the exterior. They contain areolar tissue called the **lamina propria.** *(Figure 4-17)*

Serous Membranes   p. 129

3. **Serous membranes** line the body's sealed internal cavities. They form a fluid called a *transudate. (Figure 4-17)*

The Cutaneous Membrane   p. 131

4. The **cutaneous membrane,** or skin, covers the body surface. *(Figure 4-17)*

Synovial Membranes   p. 131

5. **Synovial membranes** form an incomplete lining within the cavities of synovial joints. *(Figure 4-17)*

THE CONNECTIVE TISSUE FRAMEWORK OF THE BODY   p. 131

1. Internal organs and systems are tied together by a network of connective tissue proper. It consists of the **superficial fascia** (the **subcutaneous layer,** or **hypodermis,** separating the skin from underlying tissues and organs), the **deep fascia** (dense connective tissue), and the **subserous fascia** (the layer between the deep fascia and the serous membranes that line body cavities). *(Figure 4-18)*

MUSCLE TISSUE   p. 132

1. **Muscle tissue** is specialized for contraction. The three types of muscle tissue are *skeletal muscle, cardiac muscle,* and *smooth muscle. (Figure 4-19)*

Skeletal Muscle Tissue   p. 132

2. The cells of **skeletal muscle tissue** are **multinucleated.** Skeletal muscle, or **striated voluntary muscle,** produces new fibers by the division of **satellite cells.**

Cardiac Muscle Tissue   p. 132

3. **Cardiocytes,** the cells of **cardiac muscle tissue,** occur only in the heart. Cardiac muscle, or **striated involuntary muscle,** relies on *pacemaker cells* for regular contraction.

Smooth Muscle Tissue   p. 132

4. **Smooth muscle tissue,** or **nonstriated involuntary muscle,** is not striated. Smooth muscle cells can divide and therefore regenerate after injury has occurred.

NEURAL TISSUE   p. 134

1. **Neural tissue** conducts electrical impulses, which convey information from one area of the body to another.

2. Cells in neural tissue are either neurons or neuroglia. **Neurons** transmit information as electrical impulses. Several kinds of **neuroglia** exist, but their basic functions include supporting neural tissue and helping supply nutrients to neurons. *(Figure 4-20)*

3. A typical neuron has a **cell body, dendrites,** and an **axon,** or **nerve fiber.** The axon carries information to other cells.

TISSUE INJURIES AND AGING   p. 135

Inflammation and Regeneration   p. 135

1. Any injury affects several tissue types simultaneously, and they respond in a coordinated manner. Homeostasis is restored by two processes: inflammation and regeneration.

2. **Inflammation,** or the **inflammatory response,** isolates the injured area while damaged cells, tissue components, and any dangerous microorganisms (which could cause **infection**) are cleaned up. **Regeneration** is the repair process that restores normal function. *(Figure 4-21)*

Aging and Tissue Repair   p. 136

3. Tissues change with age. Repair and maintenance become less efficient, and the structure and chemical composition of many tissues are altered.

Aging and Cancer Incidence   p. 136

4. Cancer incidence increases with age, with roughly three-quarters of all cases caused by exposure to chemicals or by other environmental factors, such as cigarette smoke. *(Figure 4-22)*

REVIEW QUESTIONS

LEVEL 1 REVIEWING FACTS AND TERMS

Match each numbered item with the most closely related lettered item. Use letters for answers in the spaces provided.

___ **1.** histology
___ **2.** microvilli
___ **3.** gap junction
___ **4.** tight junction
___ **5.** ground substance
___ **6.** basal and reticular laminae
___ **7.** germinative cells
___ **8.** mesothelium
___ **9.** endothelium
___ **10.** mucus
___ **11.** destroys gland cells
___ **12.** hormones
___ **13.** goblet cells
___ **14.** adipocytes
___ **15.** macrophages
___ **16.** mast cells
___ **17.** bone-to-bone attachment
___ **18.** muscle-to-bone attachment
___ **19.** skeletal muscle
___ **20.** cardiac muscle

a. hyaluronan
b. lines heart and blood vessels
c. repair and renewal
d. ligament
e. endocrine secretion
f. merocrine secretion
g. absorption and secretion
h. fat cells
i. holocrine secretion
j. study of tissues
k. unicellular exocrine glands
l. tendon
m. intercellular connection
n. histamine and heparin
o. interlocking of membrane proteins
p. lines ventral body cavities
q. intercalated discs
r. striated, voluntary
s. defense and repair
t. basement membrane

21. The four basic tissue types in the body are
  (a) epithelial, connective, muscle, and neural
  (b) simple, cuboidal, squamous, and stratified
  (c) fibroblasts, adipocytes, melanocytes, and mesenchyme
  (d) lymphocytes, macrophages, microphages, and adipocytes

22. Long microvilli incapable of movement are
  (a) cilia   **(b)** flagella
  (c) stereocilia   **(d)** a, b, and c are correct

23. A type of junction common in cardiac and smooth muscle tissues is the
  (a) desmosome   **(b)** basement membrane
  (c) tight junction   **(d)** gap junction

24. The most abundant connections between cells in the superficial layers of the skin are
  (a) connexons   **(b)** gap junctions
  (c) desmosomes   **(d)** tight junctions

25. _____ membranes have an epithelium that is incomplete and atypical.
  (a) Cutaneous   **(b)** Serous
  (c) Synovial   **(d)** Mucous

26. Mucous secretions that coat the passageways of the digestive and respiratory tracts result from _____ secretion.
  (a) apocrine   **(b)** merocrine
  (c) holocrine   **(d)** endocrine

27. The type of tissue that contains a fluid known as the ground substance is
  (a) epithelial   **(b)** neural
  (c) muscle   **(d)** connective

28. The three basic types of fibers in connective tissue are
  (a) tendons, ligaments, and elastic ligaments
  (b) loose, dense, and irregular
  (c) cartilage, bone, and collagen
  (d) collagen, reticular, and elastic

29. The three types of loose connective tissue are
  (a) collagen, reticular, and elastic
  (b) areolar, adipose, and reticular
  (c) collagen, bone, and cartilage
  (d) fluid, supporting, and connective tissue proper

30. Two major examples of dense regular connective tissue are
  (a) cartilage and bone
  (b) elastic tissue and bone
  (c) tendons and elastic tissue
  (d) collagen and tendons

31. The three major types of cartilage in the body are
  (a) collagen, reticular, and elastic
  (b) areolar, adipose, and reticular
  (c) hyaline, elastic, and fibrocartilage
  (d) tendons, reticular, and elastic

32. The primary function of serous membranes in the body is
  (a) to minimize friction between opposing surfaces
  (b) to line cavities that communicate with the exterior
  (c) to perform absorptive and secretory functions
  (d) to cover the surface of the body

33. The layer of areolar tissue that separates the skin from underlying tissues and organs is the
  (a) superficial fascia   **(b)** subcutaneous layer
  (c) hypodermis   **(d)** a, b, and c are correct

34. Intercalated discs and pacemaker cells are characteristic of
  (a) smooth muscle tissue
  (b) cardiac muscle tissue
  (c) skeletal muscle tissue
  (d) a, b, and c are correct

35. Axons, dendrites, and a cell body are characteristics of cells located in
  (a) neural tissue   **(b)** muscle tissue
  (c) connective tissue   **(d)** epithelial tissue

36. The repair process necessary to restore normal function in damaged tissues is
  (a) isolation   **(b)** regeneration
  (c) reconstruction   **(d)** a, b, and c are correct

37. What are the four major characteristics of epithelial tissue?

38. What are the four essential functions of epithelial tissue?

39. Which three types of cell junctions serve as cellular attachments in the body?

40. What three types of layering make epithelial tissue recognizable?

41. What three cell shapes describe almost every epithelium in the body?

42. What three methods do various glandular epithelial cells use to release their secretions?

43. List three basic components of connective tissues.

44. What six basic functions do connective tissues perform in the body?

45. Describe the fluid connective tissues and supporting connective tissues in the human body.

46. What are the three types of fibers in connective tissues?

47. Give four major examples of dense regular connective tissue.

48. What are the three major types of cartilage in the body?

49. What are the four kinds of membranes composed of epithelial and connective tissue that cover and protect other structures and tissues in the body?

50. What is a transudate? Where might one form, and what is its function?

51. What are the three types of muscle tissue in the body?

52. What two cell populations make up neural tissue? What is the function of each?

LEVEL 2 REVIEWING CONCEPTS

53. In surfaces of the body where mechanical stresses are severe, the dominant epithelium is
  (a) stratified squamous epithelium
  (b) simple cuboidal epithelium
  (c) simple columnar epithelium
  (d) stratified cuboidal epithelium

54. Why is the presence of cilia on the respiratory surfaces important?

55. Why does holocrine secretion require continuous cell division?

56. What is the difference between an exocrine and an endocrine secretion?

57. A significant structural feature in the digestive system is the presence of tight junctions near the exposed surfaces of cells lining the digestive tract. Why are these junctions so important?

58. Why are dead skin cells generally shed in thick sheets rather than individually?

59. Why are infections always a serious threat after a severe burn or abrasion?

60. Why can't adults generate metabolic heat as readily as infants and young children can?

61. After a weight-loss program, why is the lost weight often regained quickly in the same areas of the body?

62. Cartilage heals poorly and in many instances does not heal or recover at all after a severe injury. Why not?

63. What characteristics make the cutaneous membrane different from the serous and mucous membranes?

64. Why is cardiac muscle tissue that has been damaged by injury or disease incapable of regeneration?

65. Where are you most likely to find elastic ligaments in the body, and why?

LEVEL 3 CRITICAL THINKING AND CLINICAL APPLICATIONS

66. Assuming that you had the necessary materials to perform a detailed chemical analysis of body secretions, how could you determine whether a secretion was merocrine or apocrine?

67. After many years of chronic smoking, Mr. Butts is plagued by a hacking cough. Explain the causes of this cough.

68. A biology student accidentally loses the labels of two prepared slides she is studying. One is a slide of animal intestine; the other, of animal esophagus. You volunteer to help her sort them out. How would you decide which slide is which?

69. You are asked to develop a two-step scheme that can be used to identify the three types of muscle tissue. What would the two steps be?

CHAPTER 5

The Integumentary System

How many sun-lovers ever stop to consider the contributions and sacrifices made by their skin? This remarkable structure absorbs ultraviolet radiation, prevents dehydration, preserves normal body temperature, and tolerates the chafing and abrasion of the sand. Although few of us think of it in these terms, the skin is an organ—the largest organ of the body. In this chapter, we will examine its varied functions.

CHAPTER OUTLINE AND OBJECTIVES

INTRODUCTION   p. 144

THE EPIDERMIS   p. 145
  Layers of the Epidermis   p. 145
  Skin Color   p. 147
  The Epidermis and Vitamin D₃   p. 149
  Epidermal Growth Factor   p. 149
① Describe the main structural features of the epidermis, and explain their functional significance.
② Explain what accounts for individual and racial differences in skin, such as skin color.
③ Discuss the effects of ultraviolet radiation on the skin and the role played by melanocytes.

THE DERMIS   p. 150
  Dermal Organization   p. 150
  Dermal Circulation and Innervation   p. 151
④ Describe the structure and functions of the dermis.

THE SUBCUTANEOUS LAYER   p. 152
⑤ Describe the structure and functions of the subcutaneous layer.

ACCESSORY STRUCTURES   p. 152
  Hair Follicles and Hair   p. 152
  Glands in the Skin   p. 155
  Nails   p. 158
⑥ Explain the mechanisms that produce hair, and determine hair texture and color.
⑦ Discuss the various kinds of glands in the skin and the secretions of those glands.
⑧ Explain how the sweat glands of the integumentary system play a major role in regulating body temperature.
⑨ Describe the anatomical structure of nails and how they are formed.

LOCAL CONTROL OF INTEGUMENTARY FUNCTION   p. 158
  Injury and Repair   p. 158
⑩ Explain how the skin responds to injury and repairs itself.

AGING AND THE INTEGUMENTARY SYSTEM   p. 160
⑪ Summarize the effects of the aging process on the skin.

INTEGRATION WITH OTHER SYSTEMS   p. 160
⑫ Give examples of interactions between the integumentary system and each of the other organ systems.

MEDIALAB and CHAPTER REVIEW   p. 163

The **integument** (in-TEG-ū-ment) can be considered to be either a large, highly complex organ or a structurally integrated organ system. Most people recognize both possibilities and consider the terms *integument* and **integumentary system** to be interchangeable. The components of the integumentary system include the **cutaneous membrane,** or *skin,* and the associated hairs, nails, and exocrine glands.

Few organs are as accessible, large, varied in function, and underappreciated. Your integument accounts for about 16 percent of your total body weight. Its surface, 1.5–2 square meters in area, is continuously abused, abraded, attacked by microorganisms, irradiated by sunlight, and exposed to environmental chemicals. Of all the body systems, the integumentary system is the only one you see every day.

Because others see this system as well, you devote a lot of time to improving the appearance of the skin and associated structures. Washing your face, brushing and trimming your hair, showering, and applying deodorant are activities that modify the appearance or properties of the integument. And when something goes wrong with your integument, the effects are immediately apparent. You will notice even a minor skin condition or blemish at once, whereas you may ignore more serious problems in other systems. In addition to watching for abnormal skin signs, physicians note the general appearance of the skin, because changes in the color, flexibility, elasticity, or sensitivity of the integument may indicate a dysfunction in another system. [AM] *Examination of the Skin*

The integumentary system has two major components, the cutaneous membrane and the accessory structures:

1. The cutaneous membrane itself has two components: the **epidermis** (*epi-,* above), or superficial epithelium, and the **dermis,** an underlying area of connective tissues.
2. The **accessory structures** include hair, nails, and multicellular exocrine glands. These structures are located primarily in the dermis and protrude through the epidermis to the skin surface.

The integument does not function in isolation. An extensive network of blood vessels branches through the dermis, and sensory receptors that monitor touch, pressure, temperature, and pain provide valuable information to the central nervous system about the state of the body. The general structure of the integument is shown in Figure 5-1•. Deep to the dermis, the loose connective tissue of the **subcutaneous layer,** also known as the superficial fascia or *hypodermis,* separates the integument from the deep fascia around other organs, such as muscles and bones. ∞ *p. 131* Although it is often not considered a part of the integument, we will consider the subcutaneous layer in this chapter because its connective tissue fibers are interwoven with those of the dermis.

•**FIGURE 5-1 The Components of the Integumentary System.** The relationships among the major components of the integumentary system (with the exception of nails, shown in Figure 5-10).

The general functions of the skin and subcutaneous layer include:

- *Protection* of underlying tissues and organs against shocks, abrasion, and chemical attack.
- *Excretion* of salts, water, and organic wastes by integumentary glands.
- *Maintenance* of normal body temperature through either insulation or evaporative cooling, as needed.
- *Synthesis* of *vitamin D_3,* a steroid that is subsequently converted to the hormone calcitriol, important to normal calcium metabolism.
- *Storage* of nutrients. Lipids are stored in adipocytes in the dermis and in adipose tissue in the subcutaneous layer.
- *Detection* of touch, pressure, pain, and temperature stimuli and the relay of that information to the nervous system.

We shall explore each of those functions more fully as we discuss each component of the integument.

THE EPIDERMIS

The epidermis consists of a stratified squamous epithelium (Figures 5-1 and 5-2•). It provides mechanical protection and helps keep microorganisms outside the body.

The body's most abundant epithelial cells, called **keratinocytes** (ker-A-tin-ō-sīts), form several layers. The precise boundaries between these layers can be difficult to see in a light micrograph. In **thick skin,** on the palms of the hands and soles of the feet, five layers can be distinguished. Only four layers are distinguished in the **thin skin** that covers the rest of the body. The terms *thick* and *thin* refer to the relative thickness of the epidermis, not to the integument as a whole.

Most of the body is covered by thin skin. In a sample of thin skin (Figure 5-2a,b•), the epidermis is a mere 0.08 mm thick. The epidermis in a sample of thick skin (Figure 5-2c•) can be as thick as 0.5 mm.

LAYERS OF THE EPIDERMIS

Figure 5-3• shows the layers in a section of the epidermis of thick skin. In order, from the basement membrane toward the free surface, are the *stratum germinativum,* the *stratum spinosum,* the *stratum granulosum,* the *stratum lucidum,* and the *stratum corneum.*

Stratum Germinativum

The innermost epidermal layer is the **stratum germinativum** (STRA-tum jer-mi-na-TĒ-vum), or *stratum basale* (Figure 5-3•). Hemidesmosomes attach the cells of this layer to the basement membrane that separates the epidermis from the loose connective tissue of the adjacent dermis. ∞ *p. 110* The stratum germinativum forms **epidermal ridges,** which extend into the dermis, increasing the area of contact between the two regions. Dermal projections called **dermal papillae** (singular, *papilla;* a nipple-shaped mound) extend between adjacent epidermal ridges (Figure 5-2a•).

The contours of the skin surface follow the ridge patterns, which vary from small conical pegs (in thin skin) to the complex whorls seen on the thick skin of the palms and soles. Ridges on the palms and soles increase the surface area of the skin and increase friction, ensuring a secure grip. Ridge shapes are genetically determined: Your epidermal ridges are unique to you and do not change during your lifetime. Fingerprints are ridge patterns on the tips of the fingers that can be used to identify individuals. These patterns have been used in criminal investigations for more than a century.

•**FIGURE 5-2 Thin Skin and Thick Skin. (a)** The basic organization of the epidermis. The proportions of the various layers vary with the location sampled. **(b)** Thin skin covers most of the exposed body surface. (LM × 144) **(c)** Thick skin covers the surfaces of the palms and soles. (LM × 154)

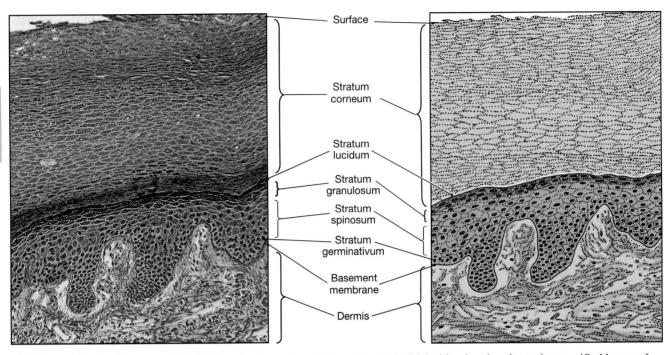

•**FIGURE 5-3** **The Structure of the Epidermis.** A portion of the epidermis in thick skin, showing the major stratified layers of epidermal cells. Langerhans cells cannot be distinguished in standard histological preparations. (LM × 210)

Large **basal cells,** or *germinative cells,* dominate the stratum germinativum. They are stem cells whose divisions replace the more superficial keratinocytes that are lost or shed at the epithelial surface. Skin surfaces that lack hair contain specialized epithelial cells known as **Merkel cells.** These cells are located among the cells of the stratum germinativum. They are sensitive to touch; when compressed, Merkel cells release chemicals that stimulate sensory nerve endings. (The skin contains many other kinds of sensory receptors, as we will see in later sections.) The brown tones of skin result from the synthetic activities of *melanocytes,* which are pigment cells. ∞ *p. 119* Melanocytes are scattered throughout the stratum germinativum, with cell processes extending into more-superficial layers.

Stratum Spinosum

Each time a stem cell divides, one of the daughter cells is pushed superficial to the stratum germinativum into the **stratum spinosum,** which means "spiny layer" (Figure 5-3•). The stratum spinosum consists of 8–10 layers of cells, and the keratinocytes are bound together by desmosomes. ∞ *p. 110* Standard histological procedures shrink the cytoplasm, but the cytoskeletal elements and desmosomes remain intact, so the cells look like miniature pincushions. Some of the cells entering this layer from the stratum germinativum continue to divide, further increasing the thickness of the epithelium. The stratum spinosum also contains **Langerhans cells,** participants in the immune response. These cells are responsible for stimulating a defense against (1) microorganisms that manage to penetrate

the superficial layers of the epidermis and (2) superficial skin cancers.

Stratum Granulosum

The region superficial to the stratum spinosum is the **stratum granulosum,** or "grainy layer" (Figure 5-3•). The stratum granulosum consists of 3–5 layers of keratinocytes displaced from the stratum spinosum. By the time cells reach this layer, most have stopped dividing. They begin making large amounts of the proteins **keratin** (KER-a-tin; *keros,* horn) and **keratohyalin** (ker-a-tō-HĪ-a-lin). In humans, keratin, a fibrous protein, is the basic structural component of hair and nails. ∞ *p. 52* As keratin fibers develop, the cells become thinner and flatter. The cell membranes thicken and become less permeable. Keratohyalin forms dense granules in the cytoplasm that promote the dehydration of the cell and the aggregation and cross-linking of the keratin fibers. The nuclei and other organelles then disintegrate, and the cells die. Further dehydration creates a tightly interlocked layer of cells that consist of keratin fibers surrounded by keratohyalin.

Stratum Lucidum

In the thick skin of the palms and soles, a glassy **stratum lucidum** (clear layer) covers the stratum granulosum (Figure 5-3•). The cells in the stratum lucidum are flattened, densely packed, and filled with keratin.

Stratum Corneum

At the exposed surface of both thick skin and thin skin is the **stratum corneum** (KOR-nē-um; *cornu,* horn)

(Figure 5-3•). The stratum corneum normally contains 15–30 layers of keratinized cells. **Keratinization,** or *cornification,* is the formation of protective, superficial layers of cells filled with keratin. This process occurs on all exposed skin surfaces except the anterior surfaces of the eyes. The dead cells in each layer of the stratum corneum remain tightly interconnected by desmosomes. Those connections are so secure that keratinized cells are generally shed in large groups or sheets rather than individually.

It takes 15–30 days for a cell to move from the stratum germinativum to the stratum corneum. The dead cells generally remain in the exposed stratum corneum layer for an additional 2 weeks before they are shed or washed away. This arrangement places the deeper portions of the epithelium and underlying tissues beneath a protective barrier of dead, durable, and expendable cells. Normally, the surface of the stratum corneum is relatively dry, so it is unsuitable for the growth of many microorganisms. Maintenance of this barrier involves coating the surface with lipid secretions from sebaceous glands.

The stratum corneum is water-resistant but not waterproof. Water from interstitial fluids slowly penetrates the surface, to be evaporated into the surrounding air. You lose roughly 500 ml (about 1 pt) of water in this way each day. This process is called **insensible perspiration,** to distinguish it from the **sensible perspiration** produced by active sweat glands. Damage to the epidermis can increase the rate of insensible perspiration. If the damage breaks connections between superficial and deeper layers of the epidermis, this fluid will accumulate in pockets, or *blisters,* within the epidermis. (Blisters also form between the epidermis and dermis if that attachment is disrupted.) If the damage to the epidermis causes the stratum corneum to lose its effectiveness as a water barrier, the rate of insensible perspiration skyrockets and a potentially dangerous fluid loss occurs. This is a serious consequence of severe burns and a complication in the condition known as *xerosis* (excessively dry skin). [AM] *Environmental Stress, Inflammation, and Infections of the Skin*

When the skin is immersed in water, osmotic forces may move water into or out of the epithelium. ∞ *p. 73* Sitting in a freshwater bath causes water to move into the epidermis, because fresh water is hypotonic (has fewer dissolved materials) compared with body fluids. The epithelial cells may swell to four times their normal volumes, a phenomenon particularly noticeable in the thickly keratinized areas of the palms and soles. Swimming in the ocean reverses the direction of osmotic flow, because the ocean is a hypertonic solution. As a result, water leaves the body, crossing the epidermis from the underlying tissues. The process is slow, but long-term exposure to seawater endangers survivors of a shipwreck by accelerating dehydration.

☑ Excessive shedding of cells from the outer layer of skin in the scalp causes dandruff. What is the name of this skin layer?

☑ A splinter pierces the palm of your hand and lodges in the third layer of the epidermis. Identify this layer.

☑ Why does swimming in fresh water for an extended period cause epidermal swelling?

☑ Some criminals sand the tips of their fingers so as not to leave recognizable fingerprints. Would this practice permanently remove fingerprints? Why or why not?

SKIN COLOR

The color of your skin is a result of an interaction between (1) the epidermal pigmentation and (2) the dermal circulation.

Epidermal Pigmentation

The epidermis contains variable quantities of two pigments, carotene and melanin. **Carotene** (KAR-ō-tēn) is an orange-yellow pigment that normally accumulates in epidermal cells. It is most apparent in cells of the stratum corneum of light-skinned individuals. This pigment is also distributed in fatty tissues in the dermis. Carotene is found in a variety of orange vegetables, such as carrots and squashes. The skin of a Caucasian vegetarian with a special fondness for carrots can actually turn orange from an overabundance of carotenes. The skin of people of other races will pick up the orange tone, but the color change is less striking. Carotene can be converted to vitamin A, which is required for (1) the normal maintenance of epithelia and (2) the synthesis of photoreceptor pigments in the eye. **Melanin** is a brown, yellow-brown, or black pigment produced by melanocytes.

Melanocytes **Melanocytes** are located in the stratum germinativum, squeezed between or deep to the epithelial cells (Figure 5-4•). Melanocytes manufacture the pigment melanin from molecules of the amino acid *tyrosine*. The melanin is packaged in intracellular vesicles called *melanosomes*. These vesicles travel within the processes of melanocytes and are transferred intact to keratinocytes. The transfer of pigmentation colors the keratinocyte temporarily, until the melanosomes are destroyed by fusion with lysosomes. In Caucasians, this transfer occurs in the stratum germinativum and stratum spinosum, and the cells of more-superficial layers lose their pigmentation. In dark-skinned people, the melanosomes are larger and the transfer may occur in the stratum granulosum as well; the pigmentation is thus darker and more persistent.

The melanin in keratinocytes protects your epidermis and dermis from the harmful effects of sunlight, which contains significant amounts of **ultraviolet (UV) radiation.** A small amount of UV radiation is beneficial, because it stimulates synthetic activity in the epidermis (a process discussed in a later section). However, UV

(a)

Keratinocyte

Melanosome

Melanin pigment

Melanocyte

Basement membrane

(b)

•**FIGURE 5-4 Melanocytes. (a)** The location and orientation of melanocytes in the stratum germinativum of a dark-skinned person. **(b)** Comparable micrograph of the skin of a Caucasian person.

radiation can damage DNA, causing mutations and promoting cancer development. In the keratinocytes, melanosomes become concentrated in the region around the nucleus, where the melanin pigments provide some UV protection for the DNA in these cells. In addition, UV radiation can produce the immediate effects of mild or even serious burns, damaging both the epidermis and the dermis. Thus, the presence of a pigment layer in the deep layers of the epidermis helps protect both epidermal and dermal tissues. However, although melanocytes respond to UV exposure by increasing their activity, the response is not rapid enough to prevent sunburn the first day you spend at the beach. Melanin synthesis accelerates slowly, peaking about 10 days after the initial exposure. Individuals of any skin color can suffer sun damage to the integument, but dark-skinned individuals have greater initial protection against the effects of UV radiation.

Over time, cumulative damage to the integument by UV exposure can harm fibroblasts, leading to impaired maintenance of the dermis. The resulting structural alterations lead to premature wrinkling. In addition, skin cancers can result from chromosomal damage in germinative cells or melanocytes. One of the major consequences of the global depletion of the ozone layer in Earth's upper atmosphere is likely the sharp increase in the rate of skin cancers, such as *malignant melanoma.* Such an increase has been reported in Australia, which has already experienced a significant loss of ozone, as

well as in the United States, Canada, and parts of Europe, which have experienced a more moderate ozone loss. For this reason, limiting UV exposure through a combination of protective clothing and sunscreens (or, better yet, sunblocks) is recommended during outdoor activities. AM *Skin Tumors and Skin Cancers*

The ratio of melanocytes to germinative cells ranges between 1:4 and 1:20, depending on the region of the body. For example, the skin covering most areas of the body has about 1000 melanocytes per square millimeter. The cheeks and forehead, the nipples, and the genital region (the scrotum of males and the labia majora of females) have higher concentrations (about $2000/mm^2$). Observed differences in skin color among individuals and even among races do not reflect different numbers of melanocytes but merely different levels of synthetic activity. Even the melanocytes of *albino* individuals are distributed normally, although these cells are incapable of producing melanin.

Dermal Circulation

Blood contains red blood cells filled with the pigment *hemoglobin.* Hemoglobin binds and transports oxygen in the bloodstream. When bound to oxygen, hemoglobin is bright red, giving blood vessels in the dermis a reddish tint that is most apparent in lightly pigmented individuals. When those vessels are dilated, as during inflammation, the red tones become much more pronounced.

When the circulatory supply is temporarily reduced, the skin becomes relatively pale; when frightened, a Caucasian may "turn white" as a result of a sudden drop in blood supply to the skin. During a sustained reduction in circulatory supply, the tissue oxygen levels decline; the hemoglobin in the tissues releases oxygen and turns a much darker red. Seen from the surface, the skin takes on a bluish coloration called **cyanosis** (sī-uh-NŌ-sis; *kyanos,* blue). In individuals of any skin color, cyanosis is most apparent in areas of thin skin, such as the lips or beneath the nails. It can be a response to extreme cold or a result of circulatory or respiratory disorders, such as heart failure or severe asthma.

Variations in Skin Pigmentation

Freckles are small pigmented spots that appear on the skin of pale-skinned individuals. These spots, which typically have an irregular border, represent the area serviced by melanocytes that are producing larger-than-average amounts of melanin. Freckles tend to be most abundant on surfaces such as the face, probably owing to sun exposure. *Lentigos* are similar to freckles but have regular borders and contain abnormal melanocytes. *Senile lentigos,* or *liver spots,* are variably pigmented areas that develop on exposed skin in older Caucasians.

Several diseases that have primary impacts on other systems may have secondary effects on skin color and pigmentation. Because the skin is easily observed, these color changes can be useful in diagnosis. For example:

- In *jaundice* (JAWN-dis), the liver is unable to excrete bile, so a yellowish pigment accumulates in body fluids. In advanced stages, the skin and whites of the eyes turn yellow.
- Some tumors affecting the pituitary gland result in the secretion of large amounts of *melanocyte-stimulating hormone (MSH).* This hormone causes a darkening of the skin, as if the individual has an extremely deep bronze tan.
- In *Addison's disease,* the pituitary gland secretes large quantities of *adrenocorticotropic hormone (ACTH),* a hormone that is structurally similar to MSH. The effect of ACTH on skin coloration is similar to that of MSH.
- In *vitiligo* (vi-ti-LĪ-gō), individuals lose their melanocytes. The condition develops in about 1 percent of the population, and the incidence increases among individuals with thyroid gland disorders, Addison's disease, and several other disorders. It is suspected that this disorder develops when the immune defenses malfunction and antibodies attack normal melanocytes. The primary problem with vitiligo is cosmetic, especially for individuals with darkly pigmented skin. Pop-music star Michael Jackson is said to have vitiligo.

THE EPIDERMIS AND VITAMIN D₃

Although strong sunlight can damage epithelial cells and deeper tissues, limited exposure to sunlight is very beneficial. When exposed to ultraviolet radiation, epidermal cells in the stratum spinosum and stratum germinativum convert a cholesterol-related steroid into **vitamin D₃,** or **cholecalciferol** (kō-le-kal-SIF-e-rol). The liver converts cholecalciferol into an intermediary product used by the kidneys to synthesize the hormone **calcitriol** (kal-si-TRĪ-ol). Calcitriol is essential for the normal absorption of calcium and phosphorus by the small intestine. An inadequate supply of calcitriol leads to impaired bone maintenance and growth.

If present in the diet, cholecalciferol can also be absorbed by the digestive tract. This fact accounts for the use of the term *vitamin,* even though the body can synthesize its own cholecalciferol. (*Vitamin* usually refers to an essential organic nutrient that the body either cannot make or makes in insufficient amounts.) Children who live in areas with overcast skies and whose diet lacks cholecalciferol can have abnormal bone development. This condition has largely been eliminated in the United States, because dairy companies add cholecalciferol, usually identified as "vitamin D," to the milk sold in grocery stores. In Chapter 6, we shall consider the hormonal control of bone growth in greater detail.

EPIDERMAL GROWTH FACTOR

Epidermal growth factor (EGF) is one of the peptide growth factors introduced in Chapter 3. ∞ *p. 100* Produced by the salivary glands and glands of the duodenum, EGF has widespread effects on epithelia, especially the epidermis. Its effects include:

- Promoting the divisions of germinative cells in the stratum germinativum and stratum spinosum.
- Accelerating the production of keratin in differentiating epidermal cells.
- Stimulating epidermal development and epidermal repair after injury.
- Stimulating synthetic activity and secretion by epithelial glands.

Epidermal growth factor has such a pronounced effect that it can be used to stimulate the growth and division of epidermal cells outside the body. In this technique, known as *tissue culture,* sheets of epidermal cells are produced for use in the treatment of extensive burns. Burned areas can be covered by epidermal sheets "grown" from a small sample of intact skin from another part of the burn victim's body. (We will consider this treatment later in this chapter in the discussion of burns.)

☑ Why does exposure to sunlight or sunlamps darken skin?

☑ Why does the skin appear red when it is warm?

☑ In some cultures, women must be covered completely, except for their eyes, when they go outside. These women exhibit a high incidence of problems with their bones. Why?

Drug Administration through the Skin

If you dissolve a drug in an oil or other solvent that is lipid-soluble, the drug can be carried across the cell membranes of the epidermis. The movement is slow, primarily owing to the density of the stratum corneum. But once a drug reaches the underlying connective tissues, it will be absorbed into the circulation. Drugs can also be administered by packaging them in *liposomes,* artificially produced lipid droplets. Liposomes containing DNA fragments have been used experimentally to introduce normal genes into abnormal human cells. For example, genes carried by liposomes have been used to alter the cell membranes of skin cancer cells so that the tumor cells will be attacked by the immune system. Another experimental procedure involves the creation of a transient change in skin permeability by administering a brief pulse of electricity. This electrical pulse temporarily changes the positions of the cells in the stratum corneum, creating channels that allow the drug to penetrate.

A useful technique for long-term drug administration is the placement of a sticky, drug-containing patch over an area of thin skin. To overcome the slow rate of diffusion, the patch must contain an extremely high concentration of the drug. This procedure is called *transdermal administration.* A single patch may work for several days, making daily pills unnecessary. Several drugs are now routinely administered transdermally:

- Transdermal scopolamine, a drug that affects the nervous system, can control the nausea associated with motion sickness.
- Transdermal nitroglycerin is used to improve blood flow within heart muscle to prevent a heart attack.
- Transdermal estrogens are administered to women to reduce symptoms of menopause.
- Transdermal pain medications and drugs are used to control high blood pressure.
- Transdermal nicotine is used to suppress the urge to smoke cigarettes. Nicotine is the addictive compound in tobacco. The nicotine from the patch depresses the craving for a cigarette; the transdermal dosage can be gradually reduced in small, controlled steps.

Dimethyl sulfoxide (DMSO) is a transdermal drug intended for the treatment of injuries to the muscles and joints of domesticated animals, such as horses or cows. It is a solvent that rapidly crosses the skin; drugs dissolved in DMSO are carried into the body along with it. Although DMSO has not been tested or approved for the treatment of humans in the United States, either for joint or muscle injuries or as a transdermal solvent, it can be prescribed in Canada and Europe. The long-term risks associated with it are unknown. Reported short-term side effects include nausea, vomiting, cramps, and chills.

THE DERMIS

The dermis lies between the epidermis and the subcutaneous layer (Figure 5-1•, p. 144). It has two major components: (1) a superficial *papillary layer* and (2) a deeper *reticular layer.*

DERMAL ORGANIZATION

The **papillary layer** consists of areolar tissue. This region contains the capillaries and sensory neurons that supply the surface of the skin. The papillary layer derives its name from the dermal papillae that project between the epidermal ridges (Figure 5-2a•, p. 145).

The **reticular layer,** deep to the papillary layer, consists of an interwoven meshwork of dense irregular connective tissue. ∞ *p. 122* Bundles of collagen fibers extend beyond the reticular layer to blend into those of the superficial papillary layer, so the boundary between the two layers is indistinct. Collagen fibers of the reticular layer also extend into the deeper subcutaneous layer.

In addition to extracellular protein fibers, the dermis contains all the cells of connective tissue proper.

∞ *p. 118* Accessory organs of epidermal origin, such as hair follicles and sweat glands, extend into the dermis. In addition, the reticular and papillary layers of the dermis contain networks of blood vessels, lymph vessels, and nerve fibers (Figure 5-1•, p. 144).

Wrinkles and Stretch Marks

Individual collagen fibers are very strong and resist stretching, but they are easily bent or twisted. Elastic fibers permit stretching and will then recoil to their original length. These fiber types are interwoven within the dermis. As a result, the dermis tolerates limited stretching. Your skin remains flexible and somewhat elastic, but the collagen fibers stop the distortion before tissue damage occurs. The water content of the skin also helps maintain its flexibility and resilience, properties known as *skin turgor.* As a result, dehydration temporarily reduces skin flexibility. Aging, hormones, and the destructive effects of ultraviolet radiation permanently reduce the amount of elastin in the dermis, producing wrinkles and sagging skin. The extensive distortion of the dermis that occurs over the abdomen during pregnancy or after a substantial weight gain can exceed the elastic capabilities of the skin. The resulting damage to the dermis prevents it from recoiling to its original size after delivery or weight loss. The skin then wrinkles and creases, creating a network of **stretch marks.**

Tretinoin (Retin-A™) is a derivative of vitamin A that can be applied to the skin as a cream or gel. This drug was originally developed to treat acne, but it also increases blood flow to the dermis and stimulates dermal repair. As a result, the rate of wrinkle formation decreases, and existing wrinkles become smaller. The degree of improvement varies among individuals.

Lines of Cleavage

Most of the collagen and elastic fibers at any location are arranged in parallel bundles. Their orientation depends on the stress placed on the skin during normal movement: The bundles are aligned to resist applied forces. The resulting pattern of fiber bundles establishes the **lines of cleavage** of the skin (Figure 5-5•). Lines of cleavage are clinically significant, because a cut parallel to a cleavage line will usually remain closed, whereas a cut at right angles to a cleavage line will be pulled open as cut elastic fibers recoil. Surgeons choose their incision patterns accordingly, for a neatly closed incision made parallel to the lines of cleavage will heal faster and with less scarring than will an incision that crosses lines of cleavage.

DERMAL CIRCULATION AND INNERVATION

The Dermal Blood Supply

Arteries supplying the skin form a network in the subcutaneous layer along its border with the reticular layer of the dermis. This network is called the *cutaneous plexus.* Tributaries of these arteries supply the adipose tissues of the subcutaneous layer and the tissues of the integument. As small arteries travel toward the epidermis, branches supply the hair follicles, sweat glands, and other structures in the dermis. On reaching the papillary layer, these small arteries form another branching network, the *papillary plexus,* which provides arterial blood to capillary loops that follow the contours of the epidermis–dermis boundary. These capillaries empty into a network of small veins connected to larger veins in the subcutaneous layer.

Front   Back

•**FIGURE 5-5 Lines of Cleavage of the Skin.** Lines of cleavage follow the pattern of fiber bundles in the skin. They reflect the orientation of collagen fiber bundles in the dermis.

 Problems with Dermal Circulation
Problems with dermal circulation affect the epidermis and the dermis. An **ulcer** is a localized shedding of an epithelium. **Decubitis ulcers,** or *bedsores,* affect bedridden or mobile patients with circulatory restrictions, especially when splints, casts, or bedding continuously presses against superficial blood vessels. Such sores most commonly affect the skin near joints or projecting bones, where dermal blood vessels are pressed against deeper structures. The chronic lack of circulation kills epidermal cells, removing a barrier to bacterial infection. Eventually, dermal tissues deteriorate as well. (This type of tissue degeneration, or *necrosis,* occurs in any tissue deprived of adequate blood flow. ∞ *p. 135*) Bedsores can be prevented or treated by frequently changing body position to vary the pressures applied to specific blood vessels.

Tumors sometimes form in the dermal blood vessels during development. The result is a temporary or permanent *birthmark. A capillary hemangioma* is a birthmark that involves capillaries of the papillary layer of the dermis. It generally enlarges after birth but subsequently fades and disappears. *Cavernous hemangiomas,* or "port-wine stains," affect larger vessels in the dermis. Most such birthmarks last a lifetime, but some can be removed by laser surgery.

Trauma to the skin often results in a *contusion,* the rupture of dermal blood vessels. Blood leaks into the dermis, and the area develops a dark blue color responsible for the term "black and blue."

The Innervation of the Skin

Nerve fibers in the skin control blood flow, adjust gland secretion rates, and monitor sensory receptors in the dermis and the deeper layers of the epidermis. We have already noted the presence of Merkel cells in the deeper layers of the epidermis. These cells are monitored by sensory terminals known as *tactile discs,* or *Merkel's discs.* The epidermis also contains the extensions of sensory neurons that provide sensations of pain and temperature. The dermis contains similar receptors as well as other, more specialized receptors. Examples discussed in Chapter 17 and shown in Figure 5-1•, p. 144, include receptors sensitive to light touch—*tactile (Meissner's) corpuscles,* located in dermal papillae—and to deep pressure and vibration—*lamellated (pacinian) corpuscles,* in the reticular layer.

Dermatitis

Due to the abundance of sensory receptors in the skin, regional infection or inflammation can be very painful. **Dermatitis** (der-muh-TĪ-tis) is an inflammation of the skin that primarily involves the papillary layer. The inflammation typically begins in a part of the skin exposed to infection or irritated by chemicals, radiation, or mechanical stimuli. Dermatitis may cause no discomfort, or it may produce an annoying itch, as in poison ivy. Other forms of this condition can be quite painful, and the inflammation can spread rapidly across the entire integument.

Dermatitis has many forms, some of them common:

- **Contact dermatitis** generally occurs in response to strong chemical irritants. It produces an itchy rash that may spread to other areas; poison ivy is an example.

- **Eczema** (EK-se-muh) is a dermatitis that can be triggered by temperature changes, fungi, chemical irritants, greases, detergents, or stress. Hereditary factors, environmental factors, or both can promote its development.

- **Diaper rash** is a localized dermatitis caused by a combination of moisture, irritating chemicals from fecal or urinary wastes, and flourishing microorganisms.

- **Urticaria** (ur-ti-KAR-ē-uh), or *hives,* is an extensive allergic response to a food, drugs, an insect bite, infection, stress, or other stimulus.

☑ Where are the capillaries and sensory neurons that supply the epidermis located?

☑ What accounts for the ability of the dermis to undergo repeated stretching?

THE SUBCUTANEOUS LAYER

The connective tissue fibers of the reticular layer are extensively interwoven with those of the subcutaneous layer, or **hypodermis.** The boundary between the two is generally indistinct (Figure 5-1•, p. 144).

Although the subcutaneous layer is not a part of the integument, it is important in stabilizing the position of the skin in relation to underlying tissues, such as skeletal muscles or other organs, while permitting independent movement.

The subcutaneous layer consists of areolar and adipose tissues. Most infants and small children have extensive "baby fat," which helps them reduce heat loss. Subcutaneous fat also serves as a substantial energy reserve and a shock absorber for the rough-and-tumble activities of our early years.

As we grow, the distribution of our subcutaneous fat changes. The greatest changes occur in response to circulating sex hormones. Beginning at puberty, men accumulate subcutaneous fat at the neck, on the arms, along the lower back, and over the buttocks. In contrast, women accumulate subcutaneous fat at the breasts, buttocks, hips, and thighs. In adults of either gender, the subcutaneous layer of the backs of the hands and the upper surfaces of the feet contain few fat cells, whereas distressing amounts of adipose tissue can accumulate in the abdominal region, producing a prominent "potbelly."

The subcutaneous layer is quite elastic. Only the superficial region contains large arteries and veins. The venous circulation of this layer contains a substantial amount of blood, and much of this volume can be shifted to the general circulation by constricting the veins. For this reason, the skin is often described as a *blood reservoir.* The rest of the subcutaneous layer contains a limited number of capillaries and no vital organs. This last characteristic makes **subcutaneous injection**—by means of a **hypodermic needle**—a useful method of administering drugs.

ACCESSORY STRUCTURES

The accessory structures of the integument include hair follicles, sebaceous glands, sweat glands, and nails. During embryological development, these structures originate from the epidermis, so they are also known as *epidermal derivatives.* Although located in the dermis, they project through the epidermis to the integumentary surface. AM *EMBRYOLOGY SUMMARY 5: The Development of the Integumentary System*

HAIR FOLLICLES AND HAIR

Hairs project above the surface of the skin almost everywhere except over the sides and soles of the feet, the palms of the hands, the sides of the fingers and toes, the lips, and portions of the external genitalia. The human body has about 5 million hairs, and 98 percent of them are on the general body surface, not on the head. Hairs are nonliving structures produced in organs called **hair follicles.** Hair production is a complex process involving cooperation between the dermis and epidermis.

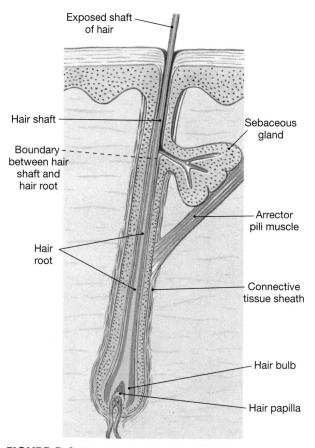

Exposed shaft of hair

Hair shaft

Boundary between hair shaft and hair root

Hair root

Sebaceous gland

Arrector pili muscle

Connective tissue sheath

Hair bulb

Hair papilla

•**FIGURE 5-6** The Anatomy of a Single Hair

5

Hair Structure

Each hair is a long, cylindrical structure that extends outward, past the epidermal surface (Figure 5-6•). The **hair root**—the portion that anchors the hair into the skin—begins at the base of the hair, at the *hair bulb,* and extends distally to the point at which the internal organization of the hair is complete, about halfway to the skin surface. The **hair shaft**—the part we see on the surface—extends from this point to the exposed tip of the hair (Figures 5-6, 5-7•). The **cuticle** is the layer that forms the outer surface of the shaft. It consists of an overlapping layer of dead, keratinized cells. The cuticle covers the **cortex,** an intermediate layer. The cells in the cortex and cuticle contain thick layers of **hard keratin,** which give the hair its stiffness. The **medulla,** or core, of the hair contains flexible **soft keratin.**

Hair follicles, which produce hairs, extend deep into the dermis, typically projecting into the underlying subcutaneous layer. At the base of a hair follicle is a small **hair papilla,** a peg of connective tissue containing capillaries and nerves. The **hair bulb** consists of epithelial cells that surround the hair papilla (Figure 5-6•).

Types of Hairs

Hairs first appear after roughly 3 months of embryonic development. These hairs, collectively known as *lanugo* (la-NOO-gō), are extremely fine and unpig-

mented. Most lanugo hairs are shed before birth. They are replaced by one of two types of hairs in the adult integument: (1) vellus hairs or (2) terminal hairs. **Vellus hairs** are the fine "peach fuzz" hairs located over much of the body surface. **Terminal hairs** are heavy, more deeply pigmented, and sometimes curly. The hair on your head, including your eyebrows and eyelashes, are terminal hairs present throughout life. Hair follicles may alter the structure of the hairs in response to circulating hormones. For example, vellus hairs are present at the armpits, groin, and limbs until puberty. Thereafter, the follicles produce terminal hairs, in response to circulating sex hormones.

Hair Color

Variations in hair color reflect differences in structure and variations in the pigment produced by melanocytes at the hair papilla. For example, different forms of melanin give dark brown, yellow-brown, or red coloration to the hair. These structural and biochemical characteristics are genetically determined, but hormonal and environmental factors influence the condition of your hair. As pigment production decreases with age, the hair color lightens. White hair results from the combination of a lack of pigment and the presence of air bubbles in the medulla of the hair shaft. As the proportion of white hairs increases, the individual's overall hair color is described as gray. Because the hair itself is dead and inert, any changes in coloration are gradual.

Functions of Hair

The 5 million hairs on your body have important functions. The roughly 100,000 hairs on your head protect your scalp from ultraviolet light, help cushion a (light) blow to the head, and insulate the skull. The hairs guarding the entrances to your nostrils and external ear canals help prevent the entry of foreign particles and insects, and your eyelashes perform a similar function for the surface of the eye. A **root hair plexus** of sensory nerves surrounds the base of each hair follicle (Figure 5-7a•). As a result, you can feel the movement of the shaft of even a single hair. This sensitivity provides an early-warning system that may help prevent injury. For example, you may be able to swat a mosquito before it reaches your skin surface.

Ribbons of smooth muscle called **arrector pili** (a-REK-tor PI-lē; plural, *arrectores pilorum*) muscles extend from the papillary layer of the dermis to the connective tissue sheath surrounding the hair follicle. When stimulated, the arrector pili pulls on the follicles and forces the hairs to stand erect. Contraction may be the result of emotional states, such as fear or rage, or a response to cold, producing "goose bumps." In a furry mammal, this action increases the thickness of the insulating coat, rather like putting on an extra sweater. Although humans do not receive any comparable insulating benefits, the reflex persists.

5

•FIGURE 5-7 **Hair Follicles.** **(a)** A three-dimensional view of a single hair follicle. **(b)** A cross section through a hair follicle and a hair. **(c)** A longitudinal section along two hair follicles, showing the base of the follicle and the matrix and papilla at the hair root.

Labels (a): Exposed shaft of hair; Sebaceous gland; Arrector pili muscle; Connective tissue sheath; Root hair plexus

Labels (b): Cortex of hair; Connective tissue sheath; Glassy membrane; External root sheath; Internal root sheath; Cuticle of hair; Medulla of hair

Labels (c): External root sheath; Internal root sheath; Glassy membrane; Cuticle of hair; Cortex of hair; Medulla of hair; Matrix; Hair bulb; Hair papilla; Subcutaneous adipose tissue

Hair Production and Follicle Structure

Hair production involves a specialization of keratinization. It begins at the deepest part of the hair follicle (Figure 5-7a•). The epithelial layer involved is the **hair matrix** (Figure 5-7c•). Basal cells near the center of the hair matrix divide, producing daughter cells that are gradually pushed toward the surface. Those cells closest to the center of the matrix form the medulla; cells closer to the edge of the developing hair form the cortex. As cell divisions continue at the hair matrix, the daughter cells are pushed toward the skin surface, and the hair gets longer. Keratinization is completed by the time these cells approach the surface. At this point, corresponding to the start of the hair shaft, the cells of the medulla and cortex are dead. Dead cells at the surface form the cuticle.

The cells of the follicle walls are organized into several concentric layers (Figure 5-7b,c•). Outward from the hair cuticle, these layers include:

■ The **internal root sheath,** which surrounds the hair root and the deeper portion of the shaft. It is produced by the cells at the periphery of the hair matrix. The cells of the internal root sheath disintegrate quickly. This layer does not extend the entire length of the follicle.

■ The **external root sheath,** which in longitudinal section extends from the skin surface to the hair matrix. Over most of that distance, it has all the cell layers found in the superficial epidermis. However, where the external root sheath joins the hair matrix, all the cells resemble those of the stratum germinativum.

■ The **glassy membrane,** a thickened basement membrane wrapped in a dense connective tissue sheath.

Growth and Replacement of Hair

Our hairs grow and are shed according to a **hair growth cycle.** A hair in the scalp grows for 2–5 years, at a rate of about 0.33 mm per day. Variations in the growth rate and in the duration of the hair growth cycle account for individual differences in the length of uncut hair.

While hair is growing, the cells of the hair root absorb nutrients and incorporate them into the hair structure. Clipping or collecting hair for analysis can be helpful in diagnosing several disorders. For example, hairs of individuals with lead poisoning or other heavy-metal poisoning contain high quantities of those metal ions. Hair samples can also be used for identification purposes through DNA fingerprinting. ∞ *p. 91*

As it grows, the root is firmly attached to the matrix of the follicle. At the end of the growth cycle, the follicle becomes inactive. The hair is now a **club hair.** The follicle gets smaller, and over time the connections between the hair matrix and the club hair root break down. When another cycle begins, the follicle produces a new hair; the old club hair is pushed to the surface and is shed.

Hair Loss

If you are a healthy adult, you lose about 50 hairs from your head each day. Several factors affect this rate. Sustained losses of more than 100 hairs per day generally indicate that something is wrong. Temporary increases in hair loss can result from drugs, dietary factors, radiation, vitamin A excess, high fever, stress, or hormonal factors related to pregnancy. In males, changes in the level of the sex hormones circulating in the blood can affect the scalp, causing a shift from terminal hair to vellus hair production, beginning at the temples and the crown of the head. This alteration is called **male pattern baldness.** Some cases of male pattern baldness respond to drug therapies, such as topical application of *minoxidyl (Rogaine™).*
AM *Baldness and Hirsutism*

☑ What happens when the arrector pili muscles contract?

☑ A person gets a burn on the forearm that destroys the epidermis and extensive areas of the deep dermis. When the injury heals, will hair grow again in the area of the injury?

GLANDS IN THE SKIN

The skin contains two types of exocrine glands: (1) *sebaceous glands* and (2) *sweat glands.* Sebaceous glands produce an oily lipid that coats hair shafts and the epidermis. Sweat glands produce a watery solution and perform other special functions.

Sebaceous (Oil) Glands

Sebaceous (se-BĀ-shus) **glands,** or *oil glands,* are holocrine glands that discharge a waxy, oily secretion into hair follicles (Figure 5-8●). Sebaceous glands that communicate with a single follicle share a duct

Sebaceous gland (LM × 150)

●**FIGURE 5-8** **Sebaceous Glands and Follicles.** The structure of sebaceous glands and sebaceous follicles in the skin.

5

and thus are classified as simple branched alveolar glands. ∞ *p. 117* The gland cells produce large quantities of lipids as they mature. The lipid product is released through holocrine secretion, a process that involves the rupture of the secretory cells. ∞ *p. 116*

The lipids released from these cells enter the **lumen,** or open passageway of the gland. Contraction of the arrector pili muscles that erect the hair squeezes the sebaceous gland, forcing an oily secretion into the hair follicle and onto the surface of the skin. This secretion, called **sebum** (SĒ-bum), is a mixture of triglycerides, cholesterol, proteins, and electrolytes. Sebum inhibits the growth of bacteria. Sebum also lubricates and protects the keratin of the hair shaft and conditions the surrounding skin. Keratin is a tough protein, but dead, keratinized cells become dry and brittle once exposed to the environment. Shampooing removes the natural oily coating; excessive washing can make hairs stiff and brittle.

Sebaceous follicles are large sebaceous glands that are not associated with hair follicles. Their ducts discharge sebum directly onto the epidermis. Sebaceous follicles are located on the face, back, chest, nipples, and male sex organs.

Surprisingly, sebaceous glands are very active during the last few months of fetal development. Their secretions, mixed with shed epidermal cells, form a protective superficial layer that coats the skin surface. This layer is called the *vernix caseosa*. Sebaceous gland activity all but ceases after birth but increases again at puberty in response to rising levels of sex hormones.

 Problems with Sebaceous Glands

Although sebum has bactericidal (bacteria-killing) properties, under some conditions bacteria can invade sebaceous glands or follicles. The presence of bacteria in sebaceous glands or follicles can produce *folliculitis* (fo-lik-ū-LĪ-tis), a local inflammation. If the duct of a sebaceous gland becomes blocked, a distinctive abscess called a *furuncle* (FUR-ung-kl), or *boil,* develops. The usual treatment for a furuncle is to cut it open, or "lance" it, so that drainage and healing can occur.

Sebaceous glands and sebaceous follicles are very sensitive to changes in the concentrations of sex hormones, and their secretory activities accelerate at puberty. For this reason, an individual with large sebaceous glands may be especially prone to develop acne during adolescence. In **acne,** sebaceous ducts become blocked and secretions accumulate, causing inflammation and a raised "pimple." The trapped secretions provide a fertile environment for bacterial infection. [AM] *Acne*

Seborrheic dermatitis is an inflammation around abnormally active sebaceous glands. The affected area becomes red, and some epidermal scaling usually occurs. Sebaceous glands of the scalp are most often involved. In infants, mild cases are called *cradle cap*. This condition is one cause of dandruff in adults. Anxiety, stress, and food allergies can aggravate the problem.

Sweat Glands

The skin contains two types of sweat glands, or **sudoriferous glands:** (1) *apocrine sweat glands* and (2) *merocrine sweat glands* (Figure 5-9●). These names

(a) Apocrine sweat gland

(b) Merocrine sweat gland

●**FIGURE 5-9 Sweat Glands. (a)** Apocrine sweat glands, located in the axillae, groin, and nipples, produce a thick, odorous fluid by apocrine secretion. (LM × 369) **(b)** Merocrine sweat glands produce a watery fluid by merocrine secretion. (LM × 194) refer to the mechanism of secretion, introduced in Chapter 4. ∞ *p. 115*

Apocrine Sweat Glands

In the armpits (axillae), around the nipples, and in the groin, **apocrine sweat glands** secrete their products into hair follicles (Figure 5-9a•). These coiled, tubular glands produce a sticky, cloudy, and potentially odorous secretion. The name *apocrine* was originally chosen because it was thought that the gland cells use an apocrine method of secretion. Although we now know that they rely on merocrine secretion, the name has not changed.

Apocrine sweat glands begin secreting at puberty. The sweat produced is a nutrient source for bacteria, which intensify its odor. Special **myoepithelial cells** (*myo-,* muscle) surround the secretory cells in these glands. Myoepithelial cells contract, squeezing the gland and thereby discharging the accumulated secretion into the hair follicles. The secretory activities of the gland cells and the contractions of myoepithelial cells are controlled by the nervous system and by circulating hormones.

Merocrine (Eccrine) Sweat Glands

Merocrine sweat glands, also known as **eccrine** (EK-rin) **sweat glands,** are far more numerous and widely distributed than apocrine glands (Figure 5-9b•). The adult integument contains 2–5 million merocrine sweat glands. They are smaller than apocrine sweat glands and do not extend as far into the dermis. Palms and soles have the highest numbers; the palm of the hand has an estimated 500 merocrine sweat glands per square centimeter (3000 per in.^{2}). These coiled, tubular glands discharge their secretions directly onto the surface of the skin.

As we learned earlier, the sweat produced by merocrine sweat glands is called sensible perspiration. Sweat is 99 percent water, but it also contains some electrolytes (chiefly sodium chloride), organic nutrients, and waste products. It has a pH of 4–6.8, and the presence of sodium chloride gives sweat a salty taste. See Appendix VI for a complete analysis of the composition of normal sweat.

The functions of merocrine sweat gland activity include:

- *Cooling the surface of the skin to reduce body temperature.* This is the primary function of sensible perspiration, and the degree of secretory activity is regulated by neural and hormonal mechanisms. When all the merocrine sweat glands are working at their maximum, the rate of perspiration can exceed a gallon per hour; dangerous fluid and electrolyte losses can occur. For this reason, athletes in endurance sports must pause at regular intervals to drink fluids.
- *Excreting water and electrolytes.* A number of ingested drugs are excreted as well.
- *Protecting from environmental hazards.* Sweat dilutes harmful chemicals and discourages the growth of microorganisms.

Other Integumentary Glands

Merocrine sweat glands are widely distributed across the body surface, and sebaceous glands are located wherever there are hair follicles. Apocrine sweat glands are located in relatively restricted areas. The skin also contains a variety of specialized glands that are restricted to specific locations. We shall encounter many of them in later chapters but will cite two important examples here:

1. The **mammary glands** of the breasts are anatomically related to apocrine sweat glands. A complex interaction between sex hormones and pituitary hormones controls their development and secretion. We shall discuss mammary gland structure and function in Chapter 28.

2. **Ceruminous** (se-ROO-mi-nus) **glands** are modified sweat glands in the passageway of the external ear. Their secretions combine with those of nearby sebaceous glands, forming a mixture called **cerumen,** or ear wax. Ear wax, together with tiny hairs along the ear canal, helps trap foreign particles or small insects and keeps them from reaching the eardrum.

Control of Glandular Secretions

Sebaceous glands and apocrine sweat glands can be collectively turned on or off by the autonomic nervous system, but no regional control is possible. When one sebaceous or apocrine gland is activated, so are all the other glands of that type in the body. Merocrine sweat glands are much more precisely controlled, and the amount of secretion and the area of the body involved can vary independently. For example, when you are nervously awaiting an anatomy and physiology exam, only your palms may begin to sweat.

As we have seen, the primary function of sensible perspiration is to cool the surface of the skin and to reduce body temperature. When you sweat in the hot sun, all your merocrine glands are working together. The blood vessels beneath your epidermis are flushed with blood, and your skin reddens. Your skin surface is warm and wet. As the moisture evaporates, your skin cools. If your body temperature then falls below normal, sensible perspiration ceases, blood flow to the skin is reduced, and the cool, dry surface releases little heat into the environment. We introduced the negative feedback mechanisms of thermoregulation (temperature control) in Chapter 1 (∞ *p. 12*) and shall provide additional details in Chapter 25.

☑ What are the functions of sebaceous secretions?

☑ Deodorants are used to mask the effects of secretions from which type of skin gland?

☑ Which type of skin gland is most affected by the hormonal changes that occur during puberty?

NAILS

Nails form on the dorsal surfaces of the tips of the fingers and toes (Figure 5-10a•). The nails protect the exposed tips of the fingers and toes and help limit their distortion when they are subjected to mechanical stress—for example, when you run or grasp objects. The **nail body,** the visible portion of the nail, covers an area of epidermis called the **nail bed** (Figure 5-10b•). The nail body is recessed deep to the level of the surrounding epithelium, and it is bounded on either side by **nail grooves** (depressions) and **nail folds.** The **free edge** of the nail—the distal portion that continues past the nail bed—extends over the **hyponychium** (hī-pō-NIK-ē-um), an area of thickened stratum corneum.

Nail production occurs at the **nail root,** an epidermal fold not visible from the surface (Figure 5-10c•). The deepest portion of the nail root lies very close to the bone of the fingertip. A portion of the stratum corneum of the nail root extends over the exposed nail, forming the **eponychium** (ep-ō-NIK-ē-um; *epi-,* over + *onyx,*

nail), or **cuticle.** Underlying blood vessels give the nail its characteristic pink color. Near the root, these vessels may be obscured, leaving a pale crescent known as the **lunula** (LOO-nū-la; *luna,* moon) (Figure 5-10a,c•).

The body of the nail consists of dead, tightly compressed cells packed with keratin. The cells producing the nails can be affected by conditions that alter body metabolism, so changes in the shape, structure, or appearance of the nails can assist in diagnosis. For example, the nails may turn yellow in individuals who have chronic respiratory disorders, thyroid gland disorders, or AIDS. Nails may become pitted and distorted as a result of *psoriasis* (a condition marked by rapid stem cell division in the stratum germinativum) and concave as a result of some blood disorders.

LOCAL CONTROL OF INTEGUMENTARY FUNCTION

The integumentary system displays a significant degree of functional independence. It often responds directly and automatically to local influences without the involvement of the nervous or endocrine system. For example, when the skin is subjected to mechanical stresses, stem cells in the stratum germinativum divide more rapidly and the depth of the epithelium increases. That is why calluses form on your palms when you perform manual labor. A more dramatic display of local regulation can be seen after an injury to the skin.

INJURY AND REPAIR

The skin can regenerate effectively even after considerable damage has occurred, because stem cells persist in both the epithelial and connective tissue components. Germinative cell divisions replace lost epidermal cells, and mesenchymal cell divisions replace lost dermal cells. This process can be slow. When large surface areas are involved, problems of infection and fluid loss complicate the situation. The relative speed and effectiveness of skin repair vary with the type of wound involved. A slender, straight cut, or *incision,* may heal relatively quickly compared with a deep scrape, or *abrasion,* which involves a much greater surface area of skin to be repaired.

AM *Trauma to the Skin*

Figure 5-11• illustrates the four stages in the regeneration of the skin after an injury. When damage extends through the epidermis and into the dermis, bleeding generally occurs (step 1). The blood clot, or **scab,** that forms at the surface temporarily restores the integrity of the epidermis and restricts the entry of additional microorganisms into the area (step 2). The bulk of the clot consists of an insoluble network of *fibrin,* a fibrous protein that forms from blood proteins during the clotting response. The clot's color reflects the presence of trapped red blood cells. Cells of

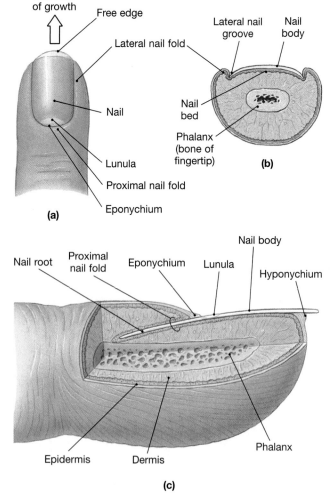

(a)

(b)

(c)

•**FIGURE 5-10 The Structure of a Nail. (a)** A superficial view. **(b)** A cross-sectional view. **(c)** A longitudinal section.

Epidermis

Dermis

Step 1: Bleeding occurs at the injury site immediately after the injury, and mast cells in the region trigger an inflammatory response.

Scab

Macrophages and fibroblasts

Sweat gland

Granulation tissue

Migratory epithelial cells

Step 2: After several hours, a scab has formed and cells of the stratum germinativum are migrating along the edges of the wound. Phagocytic cells are removing debris, and more of these cells are arriving via the enhanced circulation. Clotting around the edges of the affected area partially isolates the region.

Fibroblasts

Step 3: One week after the injury, the scab has been undermined by epidermal cells migrating over the meshwork produced by fibroblast activity. Phagocytic activity around the site has almost ended, and the fibrin clot is disintegrating.

Scar tissue

Step 4: After several weeks, the scab has been shed, and the epidermis is complete. A shallow depression marks the injury site, but fibroblasts in the dermis continue to create scar tissue that will gradually elevate the overlying epidermis.

•**FIGURE 5-11** **Integumentary Repair**

the stratum germinativum undergo rapid divisions and begin to migrate along the sides of the wound in an attempt to replace the missing epidermal cells. Meanwhile, macrophages patrol the damaged area of the dermis, phagocytizing any debris and pathogens.

If the wound covers an extensive area or involves a region covered by thin skin, dermal repairs must be under way before epithelial cells can cover the surface. Fibroblast and mesenchymal cell divisions produce mobile cells that invade the deeper areas of injury. Endothelial cells of damaged blood vessels also begin to divide, and capillaries follow the fibroblasts, providing a circulatory supply. The combination of blood clot, fibroblasts, and an extensive capillary network is called **granulation tissue.**

Over time, the clot dissolves and the number of capillaries declines. Fibroblast activity leads to the appearance of collagen fibers and typical ground substance (step 3). These repairs do not restore the integument to its original condition, however, because the dermis will contain an abnormally large number of collagen fibers and relatively few blood vessels. Severely damaged hair follicles, sebaceous or sweat glands, muscle cells, and nerves are seldom repaired, and they too are replaced by fibrous tissue. The formation of this rather inflexible, fibrous, noncellular **scar tissue** can be considered a practical limit to the healing process (step 4).

We do not know what regulates the extent of scar tissue formation, and the process is highly variable. For example, surgical procedures performed on a fetus do not leave scars, perhaps because damaged fetal tissues do not produce the same types of growth factors that adult tissues do. In some adults, most often those with dark skin, scar tissue formation may continue beyond the requirements of tissue repair. The result is a flattened mass of scar tissue that begins at the injury site and grows into the surrounding dermis. This thickened area of scar tissue, called a **keloid** (KĒ-loyd), is covered by a shiny, smooth epidermal surface. Keloids most commonly develop on the upper back, shoulders, anterior chest, or earlobes. They are harmless, and some aboriginal cultures intentionally produce keloids as a form of body decoration.

AGING AND THE INTEGUMENTARY SYSTEM

Aging affects all the components of the integumentary system. These changes include the following:

- The epidermis thins as germinative cell activity declines, making older people more prone to injury and skin infections.
- The number of Langerhans cells decreases to about 50 percent of levels seen at maturity (roughly, age 21). This decrease may reduce the sensitivity of the immune system and further encourage skin damage and infection.

- Vitamin D_3 production declines by about 75 percent. The result can be reduced calcium and phosphate absorption, eventually leading to muscle weakness and a reduction in bone strength.
- Melanocyte activity declines, and in Caucasians the skin becomes very pale. With less melanin in the skin, people become more sensitive to sun exposure and more likely to experience sunburn.
- Glandular activity declines. The skin becomes dry and often scaly, because sebum production is reduced. Merocrine sweat glands are also less active; with impaired perspiration, older people cannot lose heat as fast as younger people can. Thus, the elderly are at greater risk of overheating in warm environments.
- The blood supply to the dermis is reduced while sweat glands become less active. Because blood flow decreases, the skin becomes cool; this cooling can stimulate thermoreceptors, making a person feel cold, even in a warm room. However, with reduced circulation and sweat gland function, the elderly are less able than younger people to lose body heat. As a result, overexertion or exposure to high temperatures (such as a sauna or hot tub) can cause dangerously high body temperatures.
- Hair follicles stop functioning or produce thinner, finer hairs. With decreased melanocyte activity, these hairs are gray or white.
- The dermis thins, and the elastic fiber network decreases in size. The integument therefore becomes weaker and less resilient; sagging and wrinkling occur. These effects are most pronounced in areas of the body that have been exposed to the sun.
- With changes in levels of sex hormones, secondary sexual characteristics in hair and body-fat distribution begin to fade. In consequence, people age 90–100 of both sexes and all races tend to look alike.
- Skin repairs proceed relatively slowly, and recurring infections may result. Skin repairs proceed most rapidly in young, healthy individuals. For example, barring infection, it might take 3–4 weeks to complete the repairs to a blister site in a young adult; the same repairs at age 65–75 could take 6–8 weeks.

☑ What is the combination of fibrin clot, fibroblasts, and the extensive network of capillaries in healing tissue?

☑ Why can skin regenerate effectively even after considerable damage has occurred?

☑ Older individuals do not tolerate the summer heat as well as they did when they were young, and they are more prone to heat-related illness. What accounts for this change?

INTEGRATION WITH OTHER SYSTEMS

Although the integumentary system can function independently, many of its activities are integrated with those of other systems. Figure 5-12● diagrams the major functional relationships.

5

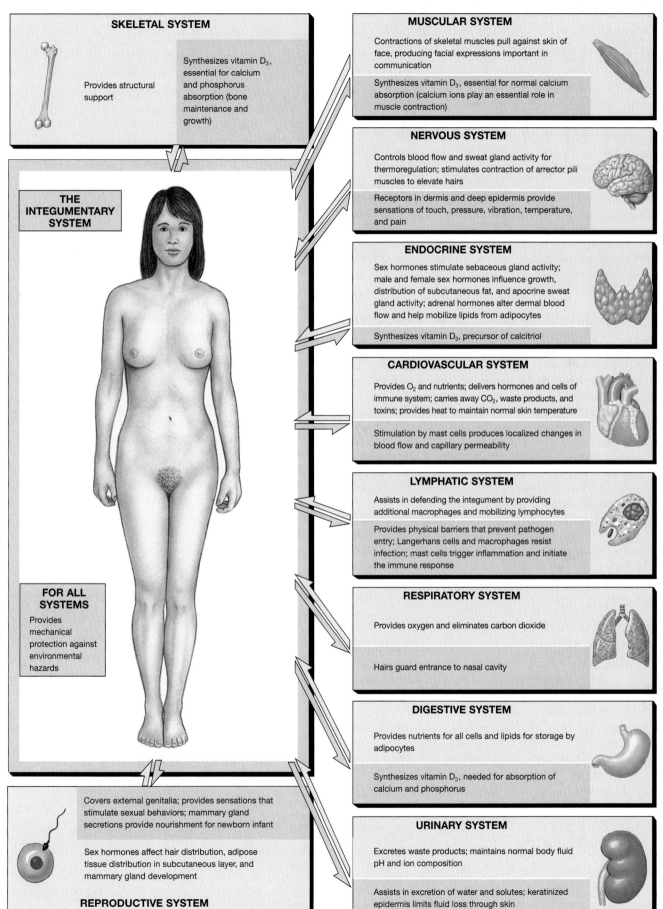

SKELETAL SYSTEM

Provides structural support

Synthesizes vitamin D_3, essential for calcium and phosphorus absorption (bone maintenance and growth)

THE INTEGUMENTARY SYSTEM

FOR ALL SYSTEMS

Provides mechanical protection against environmental hazards

Covers external genitalia; provides sensations that stimulate sexual behaviors; mammary gland secretions provide nourishment for newborn infant

Sex hormones affect hair distribution, adipose tissue distribution in subcutaneous layer, and mammary gland development

REPRODUCTIVE SYSTEM

MUSCULAR SYSTEM

Contractions of skeletal muscles pull against skin of face, producing facial expressions important in communication

Synthesizes vitamin D_3, essential for normal calcium absorption (calcium ions play an essential role in muscle contraction)

NERVOUS SYSTEM

Controls blood flow and sweat gland activity for thermoregulation; stimulates contraction of arrector pili muscles to elevate hairs

Receptors in dermis and deep epidermis provide sensations of touch, pressure, vibration, temperature, and pain

ENDOCRINE SYSTEM

Sex hormones stimulate sebaceous gland activity; male and female sex hormones influence growth, distribution of subcutaneous fat, and apocrine sweat gland activity; adrenal hormones alter dermal blood flow and help mobilize lipids from adipocytes

Synthesizes vitamin D_3, precursor of calcitriol

CARDIOVASCULAR SYSTEM

Provides O_2 and nutrients; delivers hormones and cells of immune system; carries away CO_2, waste products, and toxins; provides heat to maintain normal skin temperature

Stimulation by mast cells produces localized changes in blood flow and capillary permeability

LYMPHATIC SYSTEM

Assists in defending the integument by providing additional macrophages and mobilizing lymphocytes

Provides physical barriers that prevent pathogen entry; Langerhans cells and macrophages resist infection; mast cells trigger inflammation and initiate the immune response

RESPIRATORY SYSTEM

Provides oxygen and eliminates carbon dioxide

Hairs guard entrance to nasal cavity

DIGESTIVE SYSTEM

Provides nutrients for all cells and lipids for storage by adipocytes

Synthesizes vitamin D_3, needed for absorption of calcium and phosphorus

URINARY SYSTEM

Excretes waste products; maintains normal body fluid pH and ion composition

Assists in excretion of water and solutes; keratinized epidermis limits fluid loss through skin

•**FIGURE 5-12** **Functional Relationships between the Integumentary System and Other Systems**

Burns and Grafts

Burns result from the exposure of skin to heat, radiation, electrical shock, or strong chemical agents. The severity of the burn reflects the depth of penetration and the total area affected.

First- and second-degree burns are also called *partial-thickness burns,* because damage is restricted to the superficial layers of the skin. Only the surface of the epidermis is affected by a *first-degree burn.* In this type of burn, including most sunburns, the skin reddens and can be painful. The redness, a sign called **erythema** (er-i-THE-ma), results from inflammation of the sun-damaged tissues. In a *second-degree burn,* the entire epidermis and perhaps some of the dermis are damaged. Accessory structures such as hair follicles and glands are generally not affected, but blistering, pain, and swelling occur. If the blisters rupture at the surface, infection can easily develop. Healing typically takes 1–2 weeks, and some scar tissue may form. *Full-thickness burns,* or *third-degree burns,* destroy the epidermis and dermis, extending into subcutaneous tissues. Despite swelling, these burns are less painful than second-degree burns, because sensory nerves are destroyed along with accessory structures, blood vessels, and other dermal components. Extensive third-degree burns cannot repair themselves, because granulation tissue cannot form and epithelial cells are unable to cover the injury site. As a result, the affected area remains open to potential infection.

Each year in the United States, roughly 10,000 people die from the effects of burns. The larger the area burned, the more significant the effects on integumentary function. Figure 5-13● presents a standard reference for calculating the percentage of total surface area damaged. Burns that cover more than 20 percent of the skin surface represent serious threats to life, because they affect the following functions:

■ *Fluid and electrolyte balance.* Even areas with partial-thickness burns lose their effectiveness as barriers to fluid and electrolyte losses. In full-thickness burns, the rate of fluid loss through the skin may reach five times the normal level.

■ *Thermoregulation.* Increased fluid loss means increased evaporative cooling. More energy must be expended to keep body temperature within acceptable limits.

■ *Protection from attack.* The epidermal surface, damp from uncontrolled fluid losses, encourages bacterial growth. If the skin is broken at a blister or the site of a third-degree burn, infection is likely. Widespread bacterial infection, or **sepsis** (*septikos,* rotting), is the leading cause of death in burn victims.

Effective treatment of full-thickness burns focuses on the following four procedures:

1. Replacing lost fluids and electrolytes.
2. Providing sufficient nutrients to meet increased metabolic demands for thermoregulation and healing.
3. Preventing infection by cleaning and covering the burn while administering antibiotic drugs.
4. Assisting tissue repairs.

Because large full-thickness burns cannot heal unaided, surgical procedures are necessary to encourage healing. In a **skin graft,** areas of intact skin are transplanted to cover the burn site. A *split-thickness graft* takes a shaving of the epidermis and superficial portions of the dermis. A *full-thickness graft* involves the epidermis and both layers of the dermis.

With fluid-replacement therapies, infection control methods, and grafting techniques, young patients with burns over 80 percent of the body have about a 50 percent chance of recovery. Recent advances in cell culturing may improve survival rates further. A small section of undamaged epidermis can be removed and grown in the laboratory. Over time, germinative cell divisions produce large sheets of epidermal cells that can cover the burn area. From initial samples the size of a postage stamp, square meters of epidermis have been grown and transplanted onto body surfaces. Although questions remain about the strength and flexibility of the repairs, skin cultivation is a substantial advance in the treatment of serious burns. AM *Synthetic Skin*

Head 9%

Upper limb 9% (each)

Trunk – 18% (front or back)

Genitalia 1%

Lower limb 18% (each)

Head 15%

Trunk – 16% (front or back)

Upper limb 9% (each)

Genitalia 1%

Lower limb 17% (each)

ADULT   CHILD (5-year-old)

●**FIGURE 5-13 A Quick Method of Estimating the Percentage of Surface Area Affected by Burns.** The method is called the *rule of nines,* because the surface area in adults is in multiples of nine. This rule must be modified for children, because their proportions are quite different.

[EXPLORE] MediaLab

The integumentary system is the most familiar of the 11 body systems. We see and deal with this system daily, but it has many hidden structures and physiological concepts that are not so familiar. To gain a better understanding of the integumentary system, use the resources available on both the CD-ROM and the Companion Website. The CD-ROM includes animations and tutorials; the website builds on those resources with text content and interactive problems as well as communication avenues for sharing ideas with your class. To complete the following Web Explorations, visit the MediaLab in Chapter 5 of your Companion Website.

Web Exploration 1

Estimated time for completion = 15 minutes

The danger of excessive sun exposure is well known. Tanning products advertise their Sun Protection Factor (SPF) on the front label. What do we know about melanoma? How dangerous can a suntan be? Select the keyword **MELANOMA** on the Companion Website. Read each of the five sections prepared by the Skin Cancer Foundation. Describe why melanoma is such a serious health problem. What is being done to educate the public and to treat this disease? What can you do for protection?

Web Exploration 2

Estimated time for completion = 10 minutes

Acne is a skin problem that gets a lot of media exposure. Many potions and lotions are marketed as acne treatments. How can you make an educated decision as to which will work and which are not worth purchasing? Select the keyword **ACNE** on the Companion Website. Read the article by Solomon and Falcon entitled "Taming Your Acne." Then turn to Figure 5-1 in your text. Trace or reproduce the diagram. Add your interpretation of where and how acne cysts develop.

Which of the treatments included in the article would you choose? Why? If you wish, share your thoughts with your classmates by using the communications tools on the Companion Website.

CHAPTER REVIEW

SELECTED CLINICAL TERMINOLOGY
Terms Discussed in This Chapter

acne: A sebaceous gland inflammation caused by an accumulation of secretions. *(p. 156 and AM)*

capillary hemangioma: A birthmark caused by a tumor in the capillaries of the papillary layer of the dermis. It generally enlarges after birth but subsequently fades and disappears. *(p. 151)*

cavernous hemangioma ("port-wine stain"): A birthmark caused by a tumor affecting large vessels in the dermis; generally lasts a lifetime. *(p. 151)*

contact dermatitis: A dermatitis generally caused by strong chemical irritants. It produces an itchy rash that may spread to other areas as scratching distributes the chemical agent; includes poison ivy. *(p. 152)*

cyanosis (sī-uh-NŌ-sis): Bluish skin color as a result of reduced oxygenation of the blood in superficial vessels. *(p. 149)*

decubitis ulcers *(bedsores):* Ulcers that occur in areas subject to restricted circulation, especially common in bedridden persons. *(p. 151)*

dermatitis: An inflammation of the skin that primarily involves the papillary region of the dermis. *(p. 152)*

diaper rash: A localized dermatitis caused by a combination of moisture, irritating chemicals from fecal or urinary wastes, and flourishing microorganisms. *(p. 152)*

eczema (EK-se-muh): A dermatitis that can be triggered by temperature changes, fungi, chemical irritants, greases, detergents, or stress and that can be related to hereditary and/or environmental factors. *(p. 152)*

folliculitis (fo-lik-ū-LĪ-tis): A local inflammation caused by bacterial infection of a sebaceous gland or sebaceous follicle. *(p. 156)*

furuncle (FUR-ung-kl): A boil, an abscess that develops when the duct of a sebaceous gland is blocked. *(p. 156)*

granulation tissue: A combination of fibrin, fibroblasts, and capillaries that forms during tissue repair after inflammation. *(p. 160)*

hypodermic needle: A needle used to administer drugs via subcutaneous injection. *(p. 152)*

keloid (KĒ-loyd): A thickened area of scar tissue covered by a shiny, smooth epidermal surface. *(p. 160)*

male pattern baldness: Hair loss in an adult male due to changes in levels of circulating sex hormones. *(p. 155)*

malignant melanoma (mel-uh-NŌ-muh): A skin cancer originating in malignant melanocytes *(p. 148 and AM)*

psoriasis (sō-RĪ-uh-sis): A painless condition characterized by rapid stem cell divisions in the stratum germinativum of the scalp, elbows, palms, soles, groin, and nails. Affected areas appear dry and scaly. *(p. 158 and AM)*

scab: A blood clot that forms at the surface of a wound to the skin. *(p. 158)*

seborrheic dermatitis: An inflammation around abnormally active sebaceous glands. *(p. 156)*

sepsis: A dangerous, widespread bacterial infection; the leading cause of death in burn patients. *(p. 162)*

skin graft: The transplantation of a section of skin (partial thickness or full thickness) to cover an extensive injury site, such as a third-degree burn. *(p. 162)*

ulcer: A localized shedding of an epithelium. *(p. 151)*

urticaria (ur-ti-KAR-ē-uh) *(hives):* An extensive dermatitis resulting from an allergic reaction to food, drugs, an insect bite, infection, stress, or other stimulus. *(p. 152)*

basal cell carcinoma: A malignant cancer that originates in the stratum germinativum; the most common skin cancer. Roughly two-thirds of cases of this cancer appear in areas subjected to chronic UV exposure. Metastasis seldom occurs.

erysipelas (er-i-SIP-e-lus): A widespread inflammation of the dermis caused by bacterial infection.

pruritis (proo-RĪ-tis): An irritating itching sensation, common in skin conditions.

skin signs: Characteristic abnormalities in the skin surface that can assist in diagnosing skin conditions. They include flat macules, wheals, papules, nodules, vesicles (blisters), pustules, erosions (ulcers), crusts, scales, and fissures.

squamous cell carcinoma: A form of skin cancer less common than basal cell carcinoma, almost totally restricted to areas of sun-exposed skin. Metastasis seldom occurs.

xerosis (ze-RŌ-sis): "Dry skin," a common complaint of older persons and almost anyone living in an arid climate.

STUDY OUTLINE

INTRODUCTION p. 144

1. The **integument,** or **integumentary system,** consists of the **cutaneous membrane,** or *skin,* which includes the **epidermis** and **dermis** and the **accessory structures.** Beneath the dermis lies the **subcutaneous layer.** *(Figure 5-1)*

THE EPIDERMIS p. 145

1. **Thin skin,** formed by four layers of **keratinocytes,** covers most of the body. Heavily abraded body surfaces may be covered by **thick skin,** formed by five layers. *(Figure 5-2)*
2. The epidermis provides mechanical protection, prevents fluid loss, and helps keep microorganisms out of the body.

Layers of the Epidermis p. 145

3. Cell divisions in the **stratum germinativum,** the innermost epidermal layer, replace more-superficial cells. *(Figure 5-3)*
4. As epidermal cells age, they pass through the **stratum spinosum,** the **stratum granulosum,** the **stratum lucidum** (if thick skin), and the **stratum corneum.** In the process, they accumulate large amounts of **keratin.** Ultimately, the cells are shed or lost. *(Figure 5-3)*
5. **Epidermal ridges,** such as those on the palms and soles, improve our gripping ability and increase the skin's sensitivity. They interlock with **dermal papillae** of the underlying dermis.
6. **Langherhans cells** are part of the immune system. **Merkel cells** provide information about objects that touch the skin.

Skin Color p. 147

7. The color of the epidermis depends on two factors: blood supply and pigment composition and concentration.
8. The epidermis contains the pigments **carotene** and **melanin. Melanocytes,** which produce melanin, protect us from **ultraviolet (UV) radiation.** *(Figure 5-4)*
9. Interruptions of the dermal blood supply can lead to such conditions as **cyanosis** or **decubitis ulcers.**

The Epidermis and Vitamin D₃ p. 149

10. Epidermal cells synthesize **vitamin D₃,** or **cholecalciferol,** when exposed to the UV radiation in sunlight.

Epidermal Growth Factor p. 149

11. **Epidermal growth factor (EGF)** stimulates the maintenance and repair of the epidermis and the secretion of epithelial glands.

THE DERMIS p. 150

Dermal Organization p. 150

1. The dermis consists of the superficial **papillary layer** and the deeper **reticular layer.**
2. The papillary layer of the dermis contains blood vessels, lymphatics, and sensory nerves. This layer supports and nourishes the overlying epidermis. The reticular layer consists of a meshwork of collagen and elastic fibers oriented to resist tension in the skin.
3. Extensive distortion of the dermis can cause **stretch marks.**
4. The pattern of collagen and elastic fiber bundles forms **lines of cleavage.** *(Figure 5-5)*

Dermal Circulation and Innervation p. 151

5. **Dermatitis** is an inflammation that generally affects the papillary layer.

THE SUBCUTANEOUS LAYER p. 152

1. The subcutaneous layer, or **hypodermis,** stabilizes the skin's position against underlying organs and tissues. *(Figure 5-1)*

ACCESSORY STRUCTURES p. 152

Hair Follicles and Hair p. 152

1. **Hairs** originate in complex organs called **hair follicles.** Each hair has a **root** and a **shaft.** At the base of the root are a **hair papilla,** surrounded by a **hair bulb,** and a **root hair plexus** of sensory nerves. Hair production involves cell specialization of the **hair matrix** to form a soft core, or **medulla,** surrounded by a **cortex.** The **cuticle** is a hard layer that coats the hair. *(Figures 5-6, 5-7)*
2. The **arrector pili** muscles can erect the hairs. *(Figures 5-6, 5-7)*
3. Our bodies have both **vellus hairs** ("peach fuzz") and heavy **terminal hairs.** A hair that has stopped growing is called a **club hair.**
4. Our hairs grow and are shed according to the **hair growth cycle.** A single hair grows for 2–5 years and is subsequently shed.

Glands in the Skin p. 155

5. A typical **sebaceous gland** discharges the waxy **sebum** into a **lumen** and ultimately into a hair follicle. **Sebaceous follicles** are large sebaceous glands that empty directly onto the surface of the epidermis. Inflammation of sebaceous glands can lead to **acne** or **seborrheic dermatitis.** *(Figure 5-8)*

6. The two classes of sweat glands, or **sudoriferous glands,** are apocrine and merocrine sweat glands. **Apocrine sweat glands** produce an odorous secretion. The more numerous **merocrine,** or **eccrine, sweat glands** produce a watery secretion known as **sensible perspiration.** *(Figure 5-9)*

7. **Mammary glands** of the breasts are related to apocrine sweat glands. **Ceruminous glands** in the ear produce a waxy **cerumen.**

Nails p. 158

8. The **nail body** of a **nail** covers the **nail bed.** Nail production occurs at the **nail root,** which is overlain by the **cuticle,** or **eponychium.** The **free edge** of the nail extends over the **hyponychium.** *(Figure 5-10)*

LOCAL CONTROL OF INTEGUMENTARY FUNCTION p. 158

Injury and Repair p. 158

1. The skin can regenerate effectively even after considerable damage. The process includes the formation of a **scab, granulation tissue,** and **scar tissue.** *(Figure 5-11)*

AGING AND THE INTEGUMENTARY SYSTEM p. 160

1. Aging affects all components of the integumentary system.

INTEGRATION WITH OTHER SYSTEMS p. 160

1. Many activities of the integumentary system are integrated with those of other systems. *(Figure 5-12)*

REVIEW QUESTIONS

LEVEL 1 REVIEWING FACTS AND TERMS

1. The two major components of the integumentary system are
   (a) the cutaneous membrane and the accessory structures
   (b) the epidermis and the hypodermis
   (c) the hair and the nails
   (d) the dermis and the subcutaneous layer

2. Beginning at the basement membrane and traveling toward the free surface, the epidermis includes the following layers:
   (a) corneum, lucidum, granulosum, spinosum, germinativum
   (b) granulosum, lucidum, spinosum, germinativum, corneum
   (c) germinativum, spinosum, granulosum, lucidum, corneum
   (d) lucidum, granulosum, spinosum, germinativum, corneum

3. The fibrous protein that forms the basic structural component of hair and nails is
   (a) collagen   (b) melanin
   (c) elastin   (d) keratin

4. The primary pigments contained in the epidermis are
   (a) carotene and xanthophyll
   (b) carotene and melanin
   (c) melanin and chlorophyll
   (d) xanthophyll and melanin

5. The two major components of the dermis are the
   (a) superficial fascia and cutaneous membrane
   (b) epidermis and hypodermis
   (c) papillary layer and reticular layer
   (d) stratum germinativum and stratum corneum

6. The cutaneous plexus and papillary plexus consist of
   (a) a network of arteries providing the dermal blood supply
   (b) a network of nerves providing dermal sensations
   (c) specialized cells for cutaneous sensations
   (d) gland cells that release cutaneous secretions

7. The accessory structures of the integument include the
   (a) blood vessels, glands, muscles, and nerves
   (b) Merkel cells, lamellated corpuscles, and tactile corpuscles
   (c) hair, skin, and nails
   (d) hair follicles, nails, sebaceous glands, and sweat glands

8. The two major types of hairs in the integument are
   (a) thick, thin   (b) vellus, terminal
   (c) primary, secondary (d) short, long

9. The two types of exocrine glands in the skin are
   (a) merocrine and sweat glands
   (b) sebaceous and sweat glands
   (c) apocrine and sweat glands
   (d) eccrine and sweat glands

10. Sweat glands that communicate with hair follicles in the armpits and produce an odorous secretion are
    (a) apocrine glands   (b) merocrine glands
    (c) sebaceous glands   (d) a, b, and c are correct

11. The primary function of sensible perspiration is to
    (a) get rid of wastes
    (b) protect the skin from dryness
    (c) maintain electrolyte balance
    (d) reduce body temperature

12. The thickened stratum corneum over which the free edge of the nail extends is called the
    (a) eponychium   (b) hyponychium
    (c) cuticle   (d) cerumen

13. Muscle weakness and a reduction in bone strength in the elderly results from decreased
    (a) vitamin D_3 production
    (b) melanin production
    (c) sebum production
    (d) dermal blood supply

14. Older persons are more sensitive to sun exposure and more likely to get sunburned than are younger persons because
    (a) melanocyte activity declines with age
    (b) vitamin D_3 production declines with age
    (c) glandular activity declines with age
    (d) skin thickness decreases with age

15. In which layer(s) of the epidermis does cell division occur?

16. What are the protein precursors of keratin, and in which layer(s) of the epidermis are these proteins produced?

17. What two skin pigments are in the epidermis?

18. What widespread effects does epidermal growth factor (EGF) have on the integument?

19. What two major layers constitute the dermis, and what components are in each layer?

20. Beginning at the hair cuticle, what three cell layers make up the walls of the hair follicle?

21. Which two groups of sweat glands are contained in the skin?

LEVEL 2 REVIEWING CONCEPTS

22. How do insensible perspiration and sensible perspiration differ?

23. During the transdermal administration of drugs, why are fat-soluble drugs more desirable than those drugs that are water-soluble?

24. In our society, a tan body is associated with good health. However, medical research constantly warns about the dangers of excessive exposure to the sun. What are the benefits of a tan?

25. Why is it important for a surgeon to choose an incision pattern according to the lines of cleavage of the skin?

26. Why is regional infection or inflammation of the skin usually very painful?

27. Why is a subcutaneous injection with a hypodermic needle a useful method of administering drugs?

28. How are changes in the shape, structure, or appearance of the nails clinically significant?

29. Why is scab formation important in skin repair?

30. Why does skin sag and wrinkle as a person ages?

LEVEL 3 CRITICAL THINKING AND CLINICAL APPLICATIONS

31. A new mother notices that her 6-month-old child has a yellow-orange complexion. Fearful that the child may have jaundice (a condition caused by a toxic yellow-orange pigment in the blood), she takes him to her pediatrician. After examining the child, the pediatrician declares him perfectly healthy and advises the mother to watch the child's diet. Why?

32. Vanessa's 80-year-old grandmother sets her thermostat at 80°F and wears a sweater on balmy spring days. When asked why, the grandmother says she is cold. Can you explain why?

33. Exposure to optimum amounts of sunlight is necessary for proper bone maintenance and growth in children.

   (a) What does sunlight do to promote bone maintenance and growth?

   (b) If a child lives in an area where exposure to sunlight is rare because of pollution or overcast skies, what can be done to minimize impaired maintenance and growth of bone?

34. One of the factors to which lie detectors respond is an increase in skin conductivity due to the presence of moisture. Explain the physiological basis for the use of this indicator.

CHAPTER 6

Osseous Tissue and Skeletal Structure

The Tin Man wore his skeleton on the outside. The armor plating was useful, but it had drawbacks, such as the need to oil the joints. We avoid such problems by having internal skeletons with self-lubricating joints. Our skeletons are made of bone, a remarkable tissue that is as strong as steel-reinforced concrete, immune to rust (unlike the Tin Man's), and capable of repairing itself even after serious injuries. This chapter focuses on the nature of bone.

CHAPTER OUTLINE AND OBJECTIVES

INTRODUCTION   p. 168
① Describe the functions of the skeletal system.

A CLASSIFICATION OF BONES   p. 168
  Bone Shapes   p. 168
  Bone Structure   p. 168
② Classify bones according to their shapes and internal organization, and give examples of each type.

BONE HISTOLOGY   p. 170
  The Matrix of Bone   p. 170
  Cells in Bone   p. 171
  Compact Bone and Spongy Bone   p. 172
  The Periosteum and Endosteum   p. 174
③ Identify the cell types in bone, and list their major functions.
④ Compare the structures and functions of compact bone and spongy bone.

BONE DEVELOPMENT AND GROWTH   p. 175
  Intramembranous Ossification   p. 175
  Endochondral Ossification   p. 176
  The Blood and Nerve Supplies   p. 178
⑤ Compare the mechanisms of intramembranous ossification and endochondral ossification.
⑥ Discuss the timing of bone development and growth, and account for the differences in the internal structure of the bones of adults.

THE DYNAMIC NATURE OF BONE   p. 180
  Effects of Exercise on Bone   p. 180
  Hormonal and Nutritional Effects on Bone   p. 181
  The Skeleton as a Calcium Reserve   p. 182
  Fracture Repair   p. 184

▬ FOCUS: A Classification of Fractures   p. 186

  Aging and the Skeletal System   p. 185
⑦ Describe the remodeling and homeostatic mechanisms of the skeletal system.
⑧ Discuss the effects of nutrition, hormones, exercise, and aging on bone development and on the skeletal system.
⑨ Describe the types of fractures, and explain how they heal.

BONE MARKINGS (SURFACE FEATURES)   p. 187
⑩ Identify the major types of bone markings, and explain their functional significance.

INTEGRATION WITH OTHER SYSTEMS   p. 187
⑪ Give examples of interactions between the skeletal system and each of the other organ systems.

MEDIALAB and CHAPTER REVIEW   p. 190

6

This chapter begins our examination of the skeletal system. This system includes the bones of the skeleton and the cartilages, ligaments, and other connective tissues that stabilize or connect them. Skeletal elements are more than just racks from which muscles hang; they have a great variety of vital functions. In addition to supporting the weight of the body, bones work together with muscles to maintain body position and to produce controlled, precise movements. Without the skeleton to pull against, contracting muscle fibers could not make us sit, stand, walk, or run. Without something to hold onto, contracting muscles merely get shorter and fatter.

The skeletal system has five primary functions:

1. *Support.* The skeletal system provides structural support for the entire body. Individual bones or groups of bones provide a framework for the attachment of soft tissues and organs.

2. *Storage of minerals and lipids.* As we shall learn in Chapter 25, minerals are inorganic ions that contribute to the osmotic concentration of body fluids. They also participate in various physiological processes, and several minerals are important as enzyme cofactors. The calcium salts of bone are a valuable mineral reserve that maintains normal concentrations of calcium and phosphate ions in body fluids. Calcium is the most abundant mineral in the human body. In addition to acting as a mineral reserve, the bones of the skeleton store energy reserves as lipids in areas filled with *yellow marrow.*

3. *Blood cell production.* Red blood cells, white blood cells, and other blood elements are produced in *red marrow,* which fills the internal cavities of many bones. We shall describe the role of bone marrow in blood cell formation when we deal with the cardiovascular and lymphatic systems (Chapters 19 and 22).

4. *Protection.* Many delicate tissues and organs are surrounded by skeletal elements. The ribs protect the heart and lungs; the skull encloses the brain; the vertebrae shield the spinal cord; and the pelvis cradles delicate digestive and reproductive organs.

5. *Leverage.* Many bones function as levers that can change the magnitude and direction of the forces generated by skeletal muscles. The movements produced range from the dainty motion of a fingertip to powerful changes in the position of the entire body.

A CLASSIFICATION OF BONES

A description of bone may indicate its general shape or its histological organization. Before considering specific bones of the skeleton, you must be familiar with both classification methods.

BONE SHAPES

Every adult skeleton contains 206 major bones. We can divide these bones into six broad categories according to their individual shapes (Figure 6-1•):

1. **Long bones** are relatively long and slender (Figure 6-1a•). Long bones are located in the arm and forearm, thigh and leg, palms, soles, fingers, and toes. The femur, the long bone of the thigh, is the largest and heaviest bone in the body.

2. **Short bones** are boxy (Figure 6-1b•). Examples of short bones include the carpal bones (wrists) and tarsal bones (ankles).

3. **Flat bones** have thin, roughly parallel surfaces. Flat bones form the roof of the skull (Figure 6-1c•), the sternum, the ribs, and the scapula. They provide protection for underlying soft tissues and offer an extensive surface area for the attachment of skeletal muscles.

4. **Irregular bones** have complex shapes with short, flat, notched, or ridged surfaces (Figure 6-1d•). The spinal vertebrae, the bones of the pelvis, and several skull bones are irregular bones.

5. **Sesamoid bones** are generally small, flat, and shaped somewhat like a sesame seed (Figure 6-1e•). They develop inside tendons and are most commonly located near joints at the knees, the hands, and the feet. Everyone has sesamoid *patellae* (pa-TEL-ē), or kneecaps, but individuals vary in terms of the location and abundance of other sesamoid bones. This variation accounts for disparities in the total number of bones in the skeleton. (Sesamoid bones may form in at least 26 locations.)

6. **Sutural bones,** or *Wormian bones,* are small, flat, irregularly shaped bones between the flat bones of the skull (Figure 6-1f•). There are individual variations in the number, shape, and position of the sutural bones. Their borders are like puzzle pieces, and they range in size from a grain of sand to a quarter.

BONE STRUCTURE

Each bone in the skeleton contains two forms of **osseous tissue,** or bone tissue: (1) compact bone and (2) spongy bone. **Compact bone,** or *dense bone,* is relatively solid, whereas **spongy bone,** or *cancellous* (KAN-sel-us) *bone,* forms an open network of struts and plates. Compact bone is always located on the surface of a bone, where it forms a sturdy protective layer. Spongy bone is in the interior of a bone. The relationship between compact bone and spongy bone and their relative proportions vary with the shape of the bone. We will consider two extremes: long bones and flat bones.

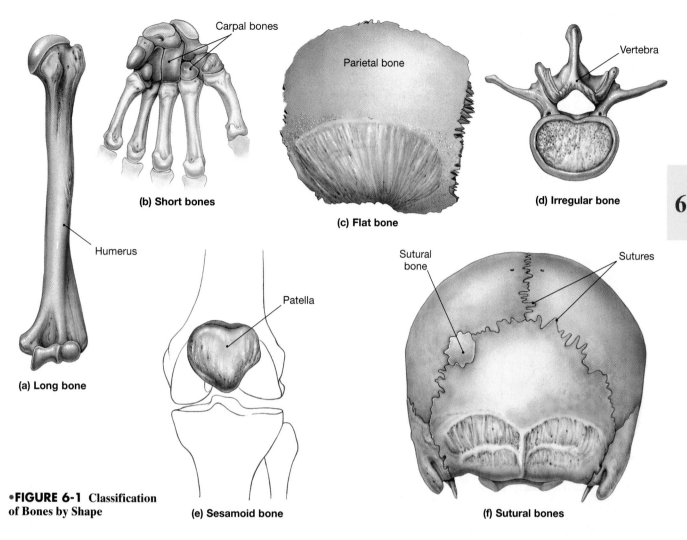

Carpal bones

(b) Short bones

Parietal bone

(c) Flat bone

Vertebra

(d) Irregular bone

6

Humerus

(a) Long bone

Patella

Sutural bone

Sutures

•**FIGURE 6-1** Classification of Bones by Shape

(e) Sesamoid bone

(f) Sutural bones

Figure 6-2• introduces the anatomy of the femur, a representative long bone. It has a long tubular shaft, or **diaphysis** (dī-AF-i-sis). At each end is an expanded area known as the **epiphysis** (ē-PIF-i-sis). The diaphysis is connected to each epiphysis at a narrow zone known as the **metaphysis** (me-TAF-i-sis). The wall of the diaphysis consists of a layer of compact bone that surrounds a central space called the **marrow cavity,** or *medullary cavity* (*medulla,* innermost part) (Figure 6-2a•). The epiphyses consist largely of spongy bone with a thin covering, or **cortex,** of compact bone. The epiphyses of the femur form complex joints at the hip and knee. These joints have large joint cavities lined by synovial membranes and filled with synovial fluid. ∞ *p. 131* The portion of an epiphysis within such a joint cavity is covered by a thin layer of hyaline cartilage called the *articular cartilage.*

The marrow cavity of the diaphysis and the spaces between the struts and plates of the epiphyses contain **bone marrow,** a loose connective tissue. **Yellow bone marrow** is dominated by fat cells. **Red bone marrow** consists largely of a mixture of mature and immature red blood cells, white blood cells, and the stem cells that produce them. Yellow bone marrow is an important energy reserve; areas of red bone marrow are important sites of blood cell formation.

Figure 6-2b• details the structure of a flat bone from the skull, such as one of the *parietal bones.* A flat bone resembles a spongy bone sandwich, with layers of compact bone covering a core of spongy bone. Although bone marrow is present within the spongy bone, there is no marrow cavity. The layer of spongy bone between the layers of compact bone is called the *diploë* (DIP-lō-ē).

Many people think of the skeleton as a rather dull collection of bony props. This is far from the truth. Our bones are complex, dynamic organs that constantly change to adapt to the demands we place on them. We will now consider the internal organization of a typical bone and will see how that seemingly inert structure is remodeled—that is, replaced with new bone—and repaired.

6

(a)

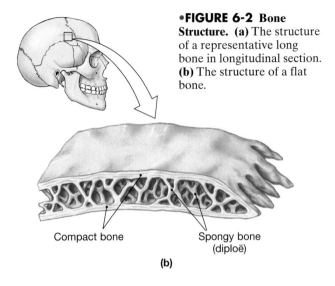

•**FIGURE 6-2** Bone Structure. **(a)** The structure of a representative long bone in longitudinal section. **(b)** The structure of a flat bone.

(b)

3. *Canaliculi,* narrow passageways through the matrix, extend between the lacunae and nearby blood vessels, forming a branching network for the exchange of nutrients, waste products, and gases.

4. Except at joints, the outer surfaces of bones are covered by a *periosteum,* which consists of outer fibrous and inner cellular layers.

We will now take a closer look at the organization of the matrix and cells of bone.

THE MATRIX OF BONE

Calcium phosphate, $Ca_3(PO_4)_2$, accounts for almost two-thirds of the weight of bone. Calcium phosphate interacts with calcium hydro-xide, $Ca(OH)_2$, to form crystals of **hydroxyapatite,** $Ca_{10}(PO_4)_6(OH)_2$. As they form, these crystals also incorporate other calcium salts, such as calcium carbonate ($CaCO_3$), and ions such as sodium, magnesium, and fluoride. Roughly one-third of the weight of bone is contributed by collagen fibers. Cells account for only 2 percent of the mass of a typical bone.

Calcium phosphate crystals are very hard but relatively inflexible and quite brittle. They can withstand compression, but the crystals are likely to shatter when exposed to bending, twisting, or sudden impacts. Collagen fibers are remarkably strong; when subjected to tension (pull), they are stronger than steel. Flexible as well as tough, they can easily tolerate twisting and bending but offer little resistance to compression. When compressed, they simply bend out of the way.

In bone, collagen fibers provide an organic framework on which hydroxyapatite crystals can form. These crystals form small plates and rods that are locked into the collagen fibers can form. The result is a protein–crystal combination with properties intermediate between those of collagen and those of pure mineral crystals. Bone is strong, somewhat flexible, and very resistant to shattering. In its overall properties, bone is on

BONE HISTOLOGY

Osseous tissue is a supporting connective tissue. (You may wish to review the sections on dense connective tissues, cartilage, and bone in Chapter 4.) ∞ *pp. 125–128* Like other connective tissues, osseous tissue contains specialized cells and a matrix consisting of extracellular protein fibers and a ground substance. The matrix of bone tissue is solid and sturdy owing to the deposition of calcium salts around the protein fibers.

In Chapter 4, which introduced the organization of bone tissue, we discussed four characteristics of bone:

1. The matrix of bone is very dense and contains deposits of calcium salts.

2. The matrix contains bone cells, or *osteocytes,* within pockets called *lacunae.* (The spaces that chondrocytes occupy in cartilage are also called lacunae. ∞ *p. 125*) The lacunae of bone are typically organized around blood vessels that branch through the bony matrix.

a par with the best steel-reinforced concrete. In reality, bone is far superior to concrete, because it can be remodeled regularly and can repair itself after injury.

CELLS IN BONE

Although osteocytes are most abundant, bone contains four cell types: osteocytes, osteoblasts, osteoprogenitor cells, and osteoclasts (Figure 6-3a•):

1. **Osteocytes** (*osteon,* bone) are mature bone cells that account for most of the cell population. Each osteocyte occupies a **lacuna,** a pocket sandwiched between layers of matrix (Figure 6-3b,c•). These layers are called **lamellae** (lah-MEL-lē; singular, *lamella,* a thin plate). Osteocytes cannot divide, and a lacuna never contains more than one osteocyte. Narrow passageways called **canaliculi** penetrate the lamellae, radiating through the matrix and connecting lacunae with one another and with nutrient sources. Canaliculi contain cytoplasmic extensions of the osteocytes. Neighboring osteocytes are linked by gap junctions, which permit the exchange of ions and small molecules, including

nutrients and hormones, between the cells. The interstitial fluid that surrounds the osteocytes and their extensions provides an additional route for the diffusion of nutrients and waste products.

Osteocytes have two major functions: (1) They maintain and monitor the protein and mineral content of the surrounding matrix. As osteocytes release chemicals that dissolve the minerals in the adjacent matrix, those minerals enter the circulation. Ultimately the recycled minerals are replaced through the deposition of new hydroxyapatite crystals. (2) They can participate in the repair of damaged bone. If released from their lacunae, these cells can convert to a less specialized type of cell, such as an osteoblast or an osteoprogenitor cell.

2. **Osteoblasts** (OS-tē-ō-blasts; *blast,* precursor) produce new bone matrix, a process called **osteogenesis** (os-tē-ō-JEN-e-sis; *gennan,* to produce). These cells make and release the proteins and other organic components of the matrix. Before calcium salts deposit, the matrix is called **osteoid** (OS-tē-oyd). Osteoblasts also assist in elevating local concentrations of calcium phosphate

6

Osteocyte: Mature bone cell that maintains the bone matrix

Osteoblast: Immature bone cell that secretes organic components of matrix

Osteoprogenitor cell: Stem cell whose divisions produce osteoblasts

Osteoclast: Multinucleate cell that secretes acids and enzymes to dissolve bone matrix

(a) Cells of bone

(b)

(c)

•**FIGURE 6-3** **The Histology of Compact Bone. (a)** Bone cells. **(b)** A thin section through compact bone. By this procedure, the intact matrix and central canals appear white, and the lacunae and canaliculi are shown in black. (LM × 343) **(c)** Several osteons in compact bone. (SEM × 182) Reproduced from R. G. Kessel and R. H. Kardon, *Tissues and Organs: A Text-Atlas of Scanning Electron Microscopy,* W. H. Freeman & Co., 1979.

and promoting the deposition of calcium salts in the organic matrix. This process converts osteoid to bone. Osteocytes develop from osteoblasts that have become completely surrounded by bone matrix.

3. Bone contains small numbers of mesenchymal cells called **osteoprogenitor** (os-tē-ō-prō-JEN-i-tor) **cells** (*progenitor,* ancestor). These stem cells divide to produce daughter cells that differentiate into osteoblasts. Osteoprogenitor cells maintain populations of osteoblasts and are important in the repair of a *fracture* (a break or a crack in a bone). They are located in the inner, cellular layer of the periosteum; in an inner layer, or *endosteum,* that lines marrow cavities; and in the lining of vascular passageways in the matrix.

4. **Osteoclasts** (OS-tē-ō-klasts; *clast,* to break), giant cells with 50 or more nuclei, remove bone matrix. Osteoclasts are not related to osteoprogenitor cells or their descendants. Instead, they are derived from monocytes, a circulating form of macrophage. ∞ *pp. 119, 124* Acids and proteolytic (protein-digesting) enzymes se-

creted by osteoclasts dissolve the matrix and release the stored minerals. This erosion process, called **osteolysis** (os-tē-OL-i-sis) or *resorption,* is important in the regulation of calcium and phosphate concentrations in body fluids. Osteoclasts are always removing matrix, and osteoblasts are always adding to it. The balance between the activities of osteoblasts and osteoclasts is very important. When osteoclasts remove calcium salts faster than osteoblasts deposit them, bones weaken. When osteoblast activity predominates, bones become stronger and more massive.

COMPACT BONE AND SPONGY BONE

The composition of the matrix in compact bone is the same as that in spongy bone, but they differ in the arrangement of osteocytes, canaliculi, and lamellae.

Compact Bone

The basic functional unit of mature compact bone is the **osteon** (OS-tē-on), or *Haversian system* (Figures 6-3c and 6-4a•). In an osteon, the osteocytes are arranged in concentric layers around a **central canal,** or *Haversian canal.* This canal contains one or more

•**FIGURE 6-4** The Structure of Osseous Tissue.
(a) The relationships between spongy bone and compact bone. **(b)** The orientation of collagen fibers in adjacent lamellae. **(c)** The organization of spongy bone.

Central canal

Concentric lamellae

Endosteum

Collagen fiber orientation

(b)

Venule   Capillary

Osteons

Circumferential lamellae

Perforating fibers

Interstitial lamellae

Vein
Artery
Arteriole
Central canal
Perforating canal

Trabeculae of spongy bone

(a)

Trabeculae of spongy bone

Trabeculae of spongy bone

Canaliculi opening on surface

Endosteum   Lamellae

(c)

blood vessels (normally a capillary and a *venule*, a very small vein) that carry blood to and from the osteon. Central canals generally run parallel to the bone surface. Other passageways, known as **perforating canals,** or the *canals of Volkmann*, extend roughly perpendicular to the surface. Blood vessels in these canals supply blood to osteons deeper in the bone and to tissues of the marrow cavity.

The lamellae of each osteon are cylindrical and aligned parallel to the long axis of the central canal. These are *concentric lamellae*. Collectively, the concentric lamellae resemble a bull's-eye around the central canal. Canaliculi radiating through the lamellae interconnect the lacunae of the osteon with one another and with the central canal. *Interstitial lamellae* fill in the spaces between the osteons in compact bone. These lamellae are remnants of osteons whose matrix components have been almost completely recycled by osteoclasts. *Circumferential lamellae* (*circum-*, around + *ferre*, to bear) are covered by the periosteum or endosteum (Figure 6-4a,b•). These lamellae are produced during the growth of the bone.

Function of Compact Bone Compact bone is thickest where stresses arrive from a limited range of directions. All osteons in compact bone are aligned the same way, making such bones very strong when stressed along the axis of alignment. You might envision a single osteon as a drinking straw with very thick walls. When you attempt to push the ends of a straw together or to pull them apart, the straw is quite strong. But if you hold the ends and push from the side, it will break relatively easily.

The osteons in the diaphysis are parallel to the long axis of the shaft. Thus, the shaft does not bend even when extreme forces are applied to either end. (The femur can withstand 10–15 times the body's weight without breaking.) Yet a much smaller force applied to the side of the shaft can break the femur.

Figure 6-5• shows the distribution of forces applied to the compact bone and spongy bone of the femur. The hip joint consists of the head of the femur and a corresponding socket on the lateral surface of the hip bone. The femoral head projects medially, and body weight compresses the medial side of the diaphysis. Because the force is applied off-center, the bone must resist the tendency to bend into a lateral bow. The other side of the shaft, which resists this bending, is placed under a stretching load, or *tension*. Notice that the central part of the bone is subjected to neither compression nor tension. The powerful forces are restricted to the compact bone of the shaft, so the presence of the marrow cavity does not reduce the bone's strength.

Spongy Bone

In spongy bone, lamellae are not arranged in osteons. The matrix in spongy bone forms struts and plates called **trabeculae** (tra-BEK-ū-lē) (Figure 6-4a,c•). The

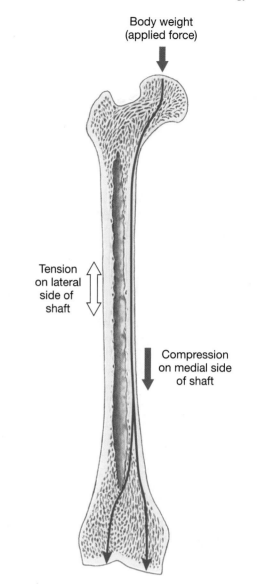

•**FIGURE 6-5 The Distribution of Forces on a Long Bone.** The femur, or thigh bone, has a diaphysis (shaft) with walls of compact bone and epiphyses filled with spongy bone. The body weight is transferred to the femur at the hip joint. Because the hip joint is off-center relative to the axis of the shaft, the body weight is distributed along the bone such that the medial (inner) portion of the shaft is compressed and the lateral (outer) portion is stretched.

thin trabeculae branch, creating an open network. Nutrients reach the osteocytes by diffusion along canaliculi that open onto the surfaces of trabeculae.

Function of Spongy Bone Spongy bone is located where bones are not heavily stressed or where stresses arrive from many directions. The trabeculae are oriented along stress lines but with extensive cross-bracing. At the proximal epiphysis, trabeculae transfer forces from the hip to the compact bone of the femoral shaft; at the distal epiphysis, trabeculae transfer weight from the shaft to the leg, across the knee joint.

6

In addition to being able to withstand stresses applied from many directions, spongy bone is much lighter than compact bone. Spongy bone reduces the weight of the skeleton and makes it easier for muscles to move the bones. Finally, the framework of trabeculae supports and protects the cells of the bone marrow. The red bone marrow within the spongy bone of the femoral epiphyses is a key site of blood cell formation.

THE PERIOSTEUM AND ENDOSTEUM

Except within joint cavities, the superficial layer of compact bone that covers all bones is wrapped by a **periosteum.** The periosteum has a fibrous outer layer and a cellular inner layer (Figure 6-6a•). The periosteum (1) isolates the bone from surrounding tissues, (2) provides a route for circulatory and nervous supply, and (3) actively participates in bone growth and repair.

Near joints, the periosteum becomes continuous with the connective tissues that lock the bones together. At a synovial joint, the periosteum is continuous with the joint capsule. The fibers of the periosteum are also interwoven with those of the tendons attached to the bone. As the bone grows, these tendon fibers are cemented into the superficial lamellae by osteoblasts from the cellular layer of the periosteum. Collagen fibers incorporated into bone tissue from tendons and ligaments, as well as from the superficial

periosteum, are called *perforating fibers (Sharpey's fibers)*. This method of attachment bonds the tendons and ligaments into the general structure of the bone, providing a much stronger attachment than would otherwise be possible. An extremely powerful pull on a tendon or ligament will usually break a bone rather than snap the collagen fibers at the bone surface.

The **endosteum,** an incomplete cellular layer, lines the marrow cavity (Figure 6-6b•). This layer covers the trabeculae of spongy bone and lines the inner surfaces of the central canals. It is active during bone growth, repair, and remodeling. It consists of a simple flattened layer of osteoprogenitor cells that covers the bone matrix, generally without intervening connective tissue fibers. Where the cellular layer is not complete, the matrix is exposed. At these exposed sites, osteoclasts and osteoblasts can remove or deposit matrix components. The osteoclasts generally occur in shallow depressions *(Howship's lacunae)* they have eroded into the matrix.

☑ How would the strength of a bone be affected if the ratio of collagen to hydroxyapatite increased?

☑ A sample of bone has concentric lamellae surrounding a central canal. Is the sample from the cortex or the marrow cavity of a long bone?

☑ If the activity of osteoclasts exceeds the activity of osteoblasts in a bone, how will the mass of the bone be affected?

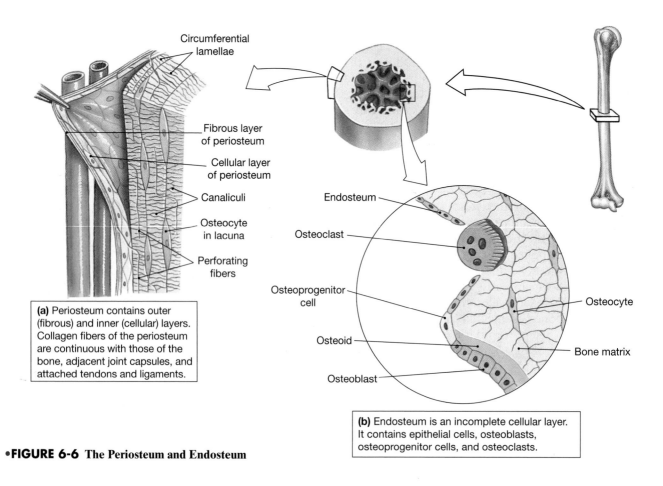

Circumferential lamellae

Fibrous layer of periosteum

Cellular layer of periosteum

Canaliculi

Osteocyte in lacuna

Perforating fibers

(a) Periosteum contains outer (fibrous) and inner (cellular) layers. Collagen fibers of the periosteum are continuous with those of the bone, adjacent joint capsules, and attached tendons and ligaments.

Endosteum

Osteoclast

Osteoprogenitor cell

Osteoid

Osteoblast

Osteocyte

Bone matrix

(b) Endosteum is an incomplete cellular layer. It contains epithelial cells, osteoblasts, osteoprogenitor cells, and osteoclasts.

•**FIGURE 6-6** The Periosteum and Endosteum

BONE DEVELOPMENT AND GROWTH

The growth of the skeleton determines the size and proportions of your body. The bony skeleton begins to form about 6 weeks after fertilization, when the embryo is approximately 12 mm (0.5 in.) long. (At this stage the existing skeletal elements are cartilaginous.) During subsequent development, the bones undergo a tremendous increase in size. Bone growth continues through adolescence, and portions of the skeleton generally do not stop growing until roughly age 25. The entire process is carefully regulated, and a breakdown in regulation will ultimately affect all body systems. In this section, we will consider the physical process of osteogenesis (bone formation) and bone growth. In the next section, we will examine the maintenance and replacement of mineral reserves in the adult skeleton.

AM *Examination of the Skeletal System*

The process of replacing other tissues with bone is called **ossification.** Ossification refers specifically to the formation of bone. The process of **calcification,** the deposition of calcium salts, occurs during ossification, but it can also occur in other tissues. (The result is a calcified tissue, such as calcified cartilage, that does not resemble bone.) Two major forms of ossification exist: (1) intramembranous and (2) endochondral ossifications. In *intramembranous ossification,* bone develops directly from mesenchyme or fibrous connective tissue. In *endochondral ossification,* bone replaces existing cartilage.

INTRAMEMBRANOUS OSSIFICATION

Intramembranous (in-tra-MEM-bra-nus) **ossification,** or *dermal ossification,* begins when osteoblasts differentiate within a mesenchymal or fibrous connective tissue. This type of ossification normally occurs in the deeper layers of the dermis. The bones that result are often called **dermal bones.** Examples of dermal bones include the flat bones of the skull, the mandible (lower jaw), and the clavicle (collarbone). In response to abnormal stresses, bone may form in other dermal areas or in connective tissues within tendons, around joints, in the kidneys, or in skeletal muscles. Dermal bones forming in abnormal locations are called *heterotopic bones* (*hetero-,* different + *topos,* place), or *ectopic bones* (*ektos,* outside). AM *Heterotopic Bone Formation*

The steps in the process of intramembranous ossification (Figure 6-7●) can be summarized as follows:

STEP 1: Mesenchymal cells first cluster together and start to secrete the organic components of the matrix. The resulting osteoid then becomes mineralized through the crystallization of calcium salts. (The enzyme *alkaline phosphatase* plays a role in this process.) As calcification

●**FIGURE 6-7** Intramembranous Ossification

Step 1: Mesenchymal cells aggregate, differentiate, and begin the ossification process. The bone expands as a series of spicules that spread into surrounding tissues. (LM × 22)

Step 2: As the spicules interconnect, they trap blood vessels within the bone. (LM × 23)

Step 3: Over time, the bone assumes the structure of spongy bone. Areas of spongy bone may later be removed, creating marrow cavities. Through remodeling, spongy bone formed in this way can be converted to compact bone.

occurs, the mesenchymal cells differentiate into osteoblasts. The location in a tissue where ossification begins is called an **ossification center.** The developing bone grows outward from the ossification center in small struts called **spicules.** As ossification proceeds, it traps some osteoblasts inside bony pockets; these cells differentiate into osteocytes. Meanwhile, mesenchymal cell divisions continue to produce additional osteoblasts.

STEP 2: Bone growth is an active process, and osteoblasts require oxygen and a reliable supply of nutrients. Blood vessels begin to grow into the area. As spicules meet and fuse together, some of these blood vessels become trapped within the developing bone.

STEP 3: Initially, the intramembranous bone consists only of spongy bone. Subsequent remodeling around trapped blood vessels can produce osteons typical of compact bone. As the rate of growth slows, the connective tissue around the bone becomes organized into the fibrous layer of the periosteum. The osteoblasts closest to the bone surface become less active but remain as the inner, cellular layer of the periosteum.

ENDOCHONDRAL OSSIFICATION

Endochondral (en-dō-KON-drul) **ossification** (*endo-*, inside + *chondros*, cartilage) begins with the formation of a hyaline cartilage model. Most bones in the body form in this way. As an example, we will follow the steps in limb bone development. By the time an embryo is 6 weeks old, the proximal bone of the limb—either the humerus (arm) or femur (thigh)—is present, but it is composed entirely of cartilage. This cartilage model continues to grow by expansion of the cartilage matrix (*interstitial growth*) and the production of new cartilage at the outer surface (*appositional growth*). ∞ *pp. 125–126* Steps in the growth and ossification of a limb bone are diagrammed in Figure 6-8a●:

STEP 1: As the cartilage enlarges, chondrocytes near the center of the shaft begin to increase greatly in size. As these cells enlarge, their lacunae expand and the matrix is reduced to a series of thin struts. These struts soon begin to calcify. The enlarged chondrocytes are now deprived of nutrients, because diffusion cannot occur through calcified cartilage. They soon die and disintegrate.

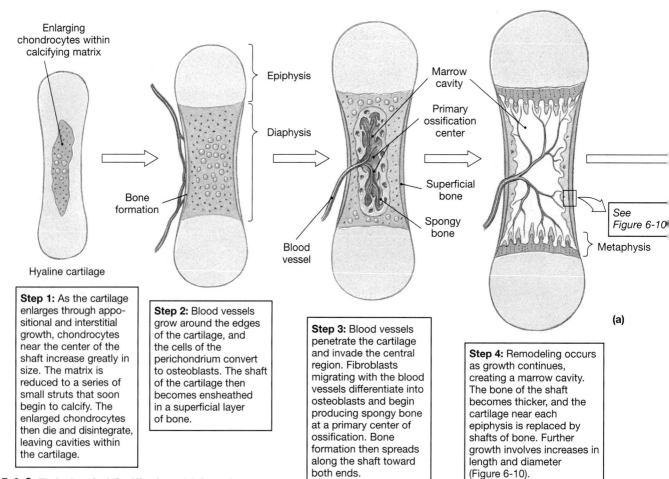

Step 1: As the cartilage enlarges through appositional and interstitial growth, chondrocytes near the center of the shaft increase greatly in size. The matrix is reduced to a series of small struts that soon begin to calcify. The enlarged chondrocytes then die and disintegrate, leaving cavities within the cartilage.

Step 2: Blood vessels grow around the edges of the cartilage, and the cells of the perichondrium convert to osteoblasts. The shaft of the cartilage then becomes ensheathed in a superficial layer of bone.

Step 3: Blood vessels penetrate the cartilage and invade the central region. Fibroblasts migrating with the blood vessels differentiate into osteoblasts and begin producing spongy bone at a primary center of ossification. Bone formation then spreads along the shaft toward both ends.

Step 4: Remodeling occurs as growth continues, creating a marrow cavity. The bone of the shaft becomes thicker, and the cartilage near each epiphysis is replaced by shafts of bone. Further growth involves increases in length and diameter (Figure 6-10).

●**FIGURE 6-8 Endochondral Ossification.** **(a)** Steps in endochondral ossification.

STEP 2: Blood vessels grow into the perichondrium surrounding the shaft of the cartilage. (We introduced the structure of the perichondrium and its role in cartilage formation in Chapter 4. ∞ *p. 125*) The cells of the inner layer of the perichondrium in this region then differentiate into osteoblasts. The perichondrium is now a periosteum, and the inner cellular layer soon produces a thin layer of bone around the shaft of the cartilage.

STEP 3: While these changes are under way, the blood supply to the periosteum increases, and capillaries and fibroblasts migrate into the heart of the cartilage, invading the spaces left by the disintegrating chondrocytes. The calcified cartilaginous matrix breaks down; the fibroblasts differentiate into osteoblasts that replace it with spongy bone. Bone development begins at this site, called the **primary center of ossification,** and spreads toward both ends of the cartilaginous model. While the diameter is small, the entire diaphysis is filled with spongy bone.

STEP 4: As the bone enlarges, osteoclasts appear and begin eroding the trabeculae in the center of the diaphysis, creating a marrow cavity. Further growth involves two distinct processes: (1) an increase in length, and (2) an enlargement in diameter by appositional growth. (We will consider appositional growth in a later section.)

During the initial stages of osteogenesis, osteoblasts move away from the center of ossification toward the epiphyses. But they do not complete the ossification of the model immediately, because the epiphyseal cartilages continue to enlarge. The region where the cartilage is being replaced by bone lies at the metaphysis. On the shaft side of the metaphysis, osteoblasts continuously invade the cartilage and replace it with bone. But on the epiphyseal side, new cartilage is continuously added. The osteoblasts are therefore moving toward the epiphysis, which is being pushed ahead by the growth of the cartilage. The situation is like a pair of joggers, one in front of the other. As long as they are running at the same speed, they can run for miles without colliding. The osteoblasts don't catch up to the epiphysis, but both the osteoblasts and the epiphysis "run away" from the primary ossification center. As they do so, the bone grows longer and longer.

6

Step 5: Capillaries and osteoblasts migrate into the epiphyses, creating secondary ossification centers.

Step 6: Soon the epiphyses are filled with spongy bone. An articular cartilage remains exposed to the joint cavity; over time it will be reduced to a thin superficial layer. At each metaphysis, an epiphyseal cartilage separates the epiphysis from the diaphysis.

(b)

•**FIGURE 6-8** Endochondral Ossification *(continued).*
(b) A light micrograph showing the interface between the degenerating cartilage and the advancing osteoblasts.

6

STEP 5: The next major change occurs when the centers of the epiphyses begin to calcify. Capillaries and osteoblasts then migrate into these areas, creating **secondary ossification centers.** The time of appearance of secondary ossification centers varies from one bone to another and from individual to individual. Secondary ossification centers may occur at birth in both ends of the humerus (arm), femur (thigh), and tibia (leg), but the ends of some other bones remain cartilaginous until early adulthood.

STEP 6: The epiphyses eventually become filled with spongy bone. A thin cap of the original cartilage model remains exposed to the joint cavity as the **articular cartilage.** This cartilage prevents damaging bone-to-bone contact within the joint. At the metaphysis, a relatively narrow cartilaginous region called the **epiphyseal cartilage,** or *epiphyseal plate,* now separates the epiphysis from the diaphysis. Figure 6-8b• shows the interface between the degenerating cartilage and the advancing osteoblasts. As long as the rate of cartilage growth keeps pace with the rate of osteoblast invasion, the shaft grows longer but the epiphyseal cartilage survives.

At puberty, the combination of rising levels of sex hormones, growth hormone, and thyroid hormones stimulates bone growth dramatically. Osteoblasts now begin producing bone at a rate faster than that of epiphyseal cartilage expansion. As a result, the epiphyseal cartilage gets narrower and narrower until it ultimately disappears. In adults, the former location of this cartilage is detected in X-rays as a distinct **epiphyseal line,** which remains after epiphyseal growth has ended (Figure 6-9•). The completion of epiphyseal growth is called *closure.*

Appositional Growth

A superficial layer of bone forms early in endochondral ossification (Figure 6-8•, step 2). Thereafter, the developing bone increases in diameter through appositional growth at the outer surface (Figure 6-10a•). In this process, cells of the inner layer of the periosteum differentiate into osteoblasts and contribute to the growth of the bone matrix. Eventually they become surrounded by matrix and differentiate into osteocytes. Over much of the surface, appositional growth adds a series of layers that form circumferential lamellae. In time, the deeper lamellae are recycled and replaced with osteons typical of compact bone. However, blood vessels and collagen fibers of the periosteum can become enclosed within the matrix. Osteons then form around the smaller vessels, as Figure 6-10a• shows.

While bone matrix is being added to the outer surface of the growing bone, osteoclasts are removing bone matrix at the inner surface. As a result, the marrow cavity gradually enlarges as the bone gets larger in diameter (Figure 6-10b•).

☑ During intramembranous ossification, which type(s) of tissue is (are) replaced by bone?

☑ In endochondral ossification, what is the original source of osteoblasts?

☑ How could X-rays of the femur be used to determine whether a person has reached full height?

THE BLOOD AND NERVE SUPPLIES

Osseous tissue is very vascular, and the bones of the skeleton have an extensive blood supply. In a typical

(a) Epiphyseal cartilages

(b) Epiphyseal lines

•**FIGURE 6-9** **Bone Growth at an Epiphyseal Cartilage.** **(a)** An X-ray of growing epiphyseal cartilages (arrows). **(b)** Epiphyseal lines in an adult (arrows).

Artery   Ridge   Periosteum

Step 1: Bone formation at the surface of the bone produces ridges that parallel a blood vessel.

Perforating canal

Step 2: The ridges enlarge and create a deep pocket.

Step 3: The ridges meet and fuse, trapping the vessel inside the bone.

Central canal of new osteon

Steps 4–6: Bone deposition then proceeds inward toward the vessel, creating a typical osteon. Meanwhile, additional circumferential lamellae are deposited and the bone continues to increase in diameter. As it does so, additional blood vessels will be enclosed.

(a) Steps in appositional bone growth

Bone deposited by osteoblasts   Bone resorbed by osteoclasts

Infant

Child

Adult

(b) Appositional growth and remodeling

•**FIGURE 6-10 Appositional Bone Growth. (a)** Three-dimensional diagrams illustrating the mechanism responsible for increasing the diameter of a growing bone. **(b)** A bone grows in diameter as new bone is added to the outer surface. At the same time, osteoclasts resorb bone on the inside, enlarging the marrow cavity.

6

bone such as the humerus, three major sets of blood vessels develop (Figure 6-11●):

1. *The nutrient artery and vein.* The blood vessels that supply the diaphysis form by invading the cartilage model as endochondral ossification begins. Most bones have only one *nutrient artery* and one *nutrient vein,* but a few bones, including the femur, have more than one of each. The vessels enter the bone through one or more round passageways called *nutrient foramina* in the diaphysis. Branches of these large vessels form smaller perforating canals and extend along the length of the shaft into the osteons of the surrounding cortex.

2. *Metaphyseal vessels.* The *metaphyseal vessels* supply blood to the inner (diaphyseal) surface of each epiphyseal cartilage, where that cartilage is being replaced by bone.

3. *Periosteal vessels.* Blood vessels from the periosteum are incorporated into the developing bone surface as described previously (Figure 6-10a●). These *periosteal vessels* provide blood to the superficial osteons of the shaft. During endochondral bone formation, branches of periosteal vessels also enter the epiphyses, providing blood to the secondary ossification centers.

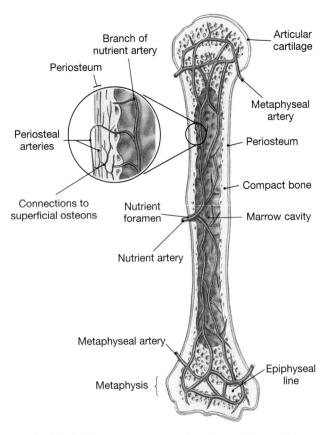

●**FIGURE 6-11 The Circulatory Supply to a Mature Bone**

Following the closure of the epiphyses, all three sets of vessels become extensively interconnected.

The periosteum contains an extensive network of lymphatic vessels and sensory nerves. The lymphatics collect lymph from branches that enter the bone and reach individual osteons via the perforating canals. The sensory nerves penetrate the cortex with the nutrient artery to innervate the endosteum, marrow cavity, and epiphyses. Because of the rich sensory innervation, injuries to bones are usually very painful.

THE DYNAMIC NATURE OF BONE

The organic and mineral components of the bone matrix are continuously being recycled and renewed through the process of **remodeling.** Bone remodeling is under way throughout life, as part of normal bone maintenance. Remodeling can replace the matrix but leave the bone as a whole unchanged, or it may change the shape, internal architecture, or mineral content of the bone.

Bone remodeling involves an interplay among the activities of osteocytes, osteoblasts, and osteoclasts. In adults, osteocytes are continuously removing and replacing the surrounding calcium salts. But osteoclasts and osteoblasts also remain active, even after the epiphyseal cartilages have closed. Normally, their activities are balanced: As quickly as osteoblasts form one osteon, osteoclasts remove another osteon by osteolysis. The turnover rate of bone is quite high. In young adults, each year almost one-fifth of the adult skeleton is recycled and replaced. Not every part of every bone is affected; the rate of turnover differs regionally and even locally. For example, the spongy bone in the head of the femur may be replaced two or three times each year, whereas the compact bone along the shaft remains largely unchanged.

EFFECTS OF EXERCISE ON BONE

The turnover and recycling of minerals give each bone the ability to adapt to new stresses. The sensitivity of osteoblasts to electrical events has been theorized as the mechanism that controls the internal organization and structure of bone. Whenever a bone is stressed, the mineral crystals generate minute electrical fields. Osteoblasts are apparently attracted to these electrical fields and, once in the area, begin to produce bone. (Electrical fields are also used to stimulate the repair of severe fractures.)

Because bones are adaptable, their shapes reflect the forces applied to them. For example, bumps and ridges on the surface of a bone mark the sites where tendons attach to the bone. If muscles become more powerful, the corresponding bumps and ridges enlarge to withstand the increased forces. Heavily stressed bones become thicker and stronger, whereas bones not subjected to ordinary stresses become thin and brittle. Regular exercise is therefore important as a stimulus that maintains normal bone structure. Champion weight lifters have massive bones with thick, prominent ridges where muscles attach. In nonathletes (especially couch potatoes), moderate amounts of physical activity and weight-bearing activities are essential to stimulate normal bone maintenance and to maintain adequate bone strength.

Degenerative changes in the skeleton occur after relatively brief periods of inactivity. For example, you may use a crutch to take weight off an injured leg while you wear a cast. After a few weeks, your unstressed bones will lose up to a third of their mass. The bones rebuild just as quickly when you resume normal weight loading. However, the removal of calcium salts can be a serious health hazard both for astronauts remaining in a weightless environment and for bedridden or paralyzed patients who spend months or years without stressing their skeleton.

HORMONAL AND NUTRITIONAL EFFECTS ON BONE

Normal bone growth and maintenance depend on a combination of nutritional and hormonal factors:

- Normal bone growth and maintenance cannot occur without a constant dietary source of calcium and phosphate salts. Lesser amounts of other minerals, such as magnesium, fluoride, iron, and manganese, are also required.

- The hormone *calcitriol*, synthesized in the kidneys, is essential for normal calcium and phosphate ion absorption in the digestive tract. Calcitriol synthesis is dependent on the availability of a related steroid, *cholecalciferol* (vitamin D_3), which may be synthesized in the skin or absorbed from the diet. ∞ *p. 149*

- Adequate levels of vitamin C must be present in the diet. This vitamin is required by certain key enzymatic reactions in collagen synthesis and stimulates osteoblast differentiation. One of the signs of vitamin C deficiency, a condition called *scurvy,* is a loss of bone mass and strength.

- Three other vitamins have significant effects on bone structure. Vitamin A, which stimulates osteoblast activity, is particularly important for normal bone growth in children. Vitamins K and B_{12} are required for the synthesis of proteins in normal bone.

- *Growth hormone,* produced by the pituitary gland, and *thyroxine,* from the thyroid gland, stimulate bone growth. Growth hormone stimulates protein synthesis and cell growth throughout the body. Thyroxine stimulates cell metabolism and increases the rate of osteoblast activity. In proper balance, these hormones maintain normal activity at the epiphyseal cartilages until roughly the time of puberty.

- At puberty, rising levels of sex hormones (*estrogens* in females and *androgens* in males) stimulate osteoblasts to produce bone faster than the rate at which epiphyseal cartilage expands. Over time, the epiphyseal cartilages narrow and eventually close. The timing of epiphyseal closure differs from bone to bone and from individual to individual. The toes may complete ossification by age 11, but parts of the pelvis or the wrist may continue to enlarge until roughly age 25. Differences in the male and female sex hormones account for the gender variation and for related variations in body size and proportions. (For example, because estrogens cause faster epiphyseal closure than do androgens, women are generally shorter than men at maturity.)

Two other hormones—*calcitonin* (kal-si-TŌ-nin) from the thyroid gland and *parathyroid hormone* from the parathyroid gland—are important in the homeostatic control of calcium and phosphate levels in body fluids. We will consider the interactions of these hormones in the next section. The major hormones affecting the growth and maintenance of the skeletal system are summarized in Table 6-1.

Abnormal Bone Growth and Development

A variety of endocrine or metabolic problems can result in characteristic skeletal changes. In **pituitary growth failure** (*pituitary dwarfism*), inadequate growth hormone production leads to reduced epiphyseal cartilage activity and abnormally short bones. This condition is becoming increasingly rare in the United States, because children can be treated with human growth hormone.

Gigantism results from an overproduction of growth hormone before puberty. (The world record for height is 272 cm, or 8 ft 11 in., with a weight of 216 kg, or 475 lb.) If growth hormone levels rise abnormally after epiphyseal cartilages close, the skeleton does not grow larger but cartilage growth and alterations in soft tissue structure lead to changes in bone density and in physical features, such as the contours of the face. These physical changes occur in the disorder called **acromegaly.**

Several inherited metabolic problems that affect many systems influence the growth and development of the skeletal system. These conditions produce characteristic variations in body proportions. For example, many individuals with **Marfan's syndrome** are very tall and have long, slender limbs, due to excessive cartilage formation at the epiphyseal cartilages. ∞ *p. 120* Although this is an obvious physical distinction, the characteristic body proportions are not in themselves dangerous. However, the underlying mutation, which affects connective tissue structure, commonly causes life-threatening problems with the cardiovascular system. [AM] *Congenital Disorders of the Skeleton*

TABLE 6-1 **Hormones Involved in the Regulation of Bone Growth and Maintenance**

Hormone	Primary Source	Effects on Skeletal System
Calcitriol	Kidneys	Promotes calcium and phosphate ion absorption along the digestive tract
Growth hormone	Pituitary gland	Stimulates osteoblast activity and the synthesis of bone matrix
Thyroxine	Thyroid gland (follicle cells)	With growth hormone, stimulates osteoblast activity and synthesis of bone matrix
Sex hormones	Ovaries (estrogens) Testes (androgens)	Stimulate osteoblast activity and synthesis of bone matrix
Parathyroid hormone	Parathyroid glands	Stimulates osteoclast activity; elevates calcium ion concentrations in body fluids
Calcitonin	Thyroid gland (C cells)	Inhibits osteoclast activity; reduces calcium ion concentrations in body fluids

☑ Why would you expect the arm bones of a weight lifter to be thicker and heavier than those of a jogger?

☑ A child who enters puberty several years later than the average age is generally taller than average as an adult. Why?

☑ A 7-year-old child has a pituitary tumor involving the cells that secrete growth hormone (GH), resulting in increased levels of GH. How will this condition affect the child's growth?

THE SKELETON AS A CALCIUM RESERVE

The bones of the skeleton are important mineral reservoirs (Figure 6-12•). For the moment, we will focus on the homeostatic regulation of the calcium ion concentration in body fluids; we shall consider other minerals in later chapters. Calcium is the most abundant mineral in the human body. A typical human body contains 1–2 kg (2.2–4.4 lb.) of calcium, with roughly 99 percent of it deposited in the skeleton.

Calcium ions play a role in a variety of physiological processes, so the body must closely control calcium ion concentrations in order to prevent damage to essential physiological systems. Even small variations from the normal concentration will affect cellular operations. Larger changes can cause a clinical crisis. Calcium ions are particularly important to the membranes and to the intracellular activities of neurons and muscle cells, especially cardiac muscle cells. If the calcium concentration of body fluids increases by 30 percent, neurons and muscle cells become relatively unresponsive. If calcium levels decrease by 35 percent, neurons become so excitable that convulsions can occur. A 50 percent reduction in calcium concentration

generally causes death. Calcium ion concentration is so closely regulated, however, that daily fluctuations of more than 10 percent are very unusual.

Hormones and Calcium Balance

Calcium ion homeostasis is maintained by a negative feedback system involving a pair of hormones with opposing effects. These hormones, parathyroid hormone and calcitonin, coordinate the storage, absorption, and excretion of calcium ions. Three target sites are involved: (1) the bones (storage), (2) the digestive tract (absorption), and (3) the kidneys (excretion). Figure 6-13a• indicates factors that elevate calcium levels; Figure 6-13b• indicates factors that depress calcium ion concentrations.

When calcium ion concentrations in the blood fall below normal, cells of the **parathyroid glands,** embedded in the thyroid gland in the neck, release **parathyroid hormone (PTH)** into the bloodstream. Parathyroid hormone has three major effects:

1. *Stimulating osteoclast activity* and enhancing the recycling of minerals by osteocytes.

2. *Increasing the rate of intestinal absorption of calcium ions* by enhancing the action of calcitriol. Under normal circumstances, calcitriol is always present, and parathyroid hormone controls its impact on the intestinal epithelium.

3. *Decreasing the rate of excretion of calcium ions at the kidneys.*

Under these conditions, more calcium ions enter body fluids, and losses are restricted. The calcium ion concentration increases to normal levels, and homeostasis is restored.

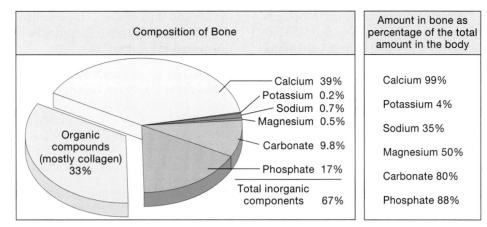

•**FIGURE 6-12** A Chemical Analysis of Bone

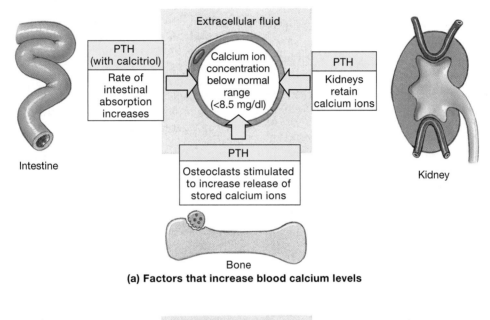

(a) Factors that increase blood calcium levels

(b) Factors that decrease blood calcium levels

•**FIGURE 6-13** Factors That Alter the Concentration of Calcium Ions in Body Fluids

If the calcium ion concentration of the blood instead rises above normal, special cells (*parafollicular cells,* or *C cells*) in the thyroid gland secrete **calcitonin.** Calcitonin has two major functions:

1. *Inhibiting osteoclast activity.*
2. *Increasing the rate of excretion of calcium ions at the kidneys.*

Under these conditions, less calcium *enters* body fluids because osteoclasts leave the mineral matrix alone. More calcium *leaves* body fluids because osteoblasts continue to produce new bone matrix while calcium ion excretion at the kidneys accelerates. The net result is a decline in the calcium ion concentration of body fluids, restoring homeostasis.

By providing a calcium reserve, the skeleton plays the primary role in the homeostatic maintenance of normal calcium ion concentrations of body fluids. This function can have a direct effect on the shape and strength of the bones in the skeleton. When large numbers of calcium ions are mobilized in body fluids, the bones become weaker; when calcium salts are deposited, the bones become denser and stronger.

Because the bone matrix contains protein fibers as well as mineral deposits, changes in mineral content do not necessarily affect the shape of the bone. In *osteomalacia* (os-tē-ō-ma-LĀ-shē-uh; *malakia,* softness), the bones appear normal although they are weak and flexible owing to poor mineralization. *Rickets,* a form of osteomalacia affecting children, generally results from a vitamin D_3 deficiency caused by inadequate exposure to sunlight and inadequate dietary supply. The bones of children with rickets are so poorly mineralized that they become very flexible. Affected individuals develop a bowlegged appearance as the thigh and leg bones bend under the weight of the body.

☑ Why does a child who has rickets have difficulty walking?

☑ What effect would increased PTH secretion have on blood calcium levels?

☑ How does calcitonin help lower the calcium ion concentration of blood?

FRACTURE REPAIR

Despite its mineral strength, bone can crack or even break if subjected to extreme loads, sudden impacts, or stresses from unusual directions. The damage produced constitutes a **fracture.** (See "FOCUS: A Classification of Fractures" on page 186.) Most fractures heal even after severe damage, provided that the blood supply and the cellular components of the endosteum and periosteum survive. Steps in the repair process are illustrated in Figure 6-14●:

STEP 1: In even a small fracture, many blood vessels are broken and extensive bleeding occurs. A large blood clot, or **fracture hematoma,** soon closes off the injured vessels and leaves a fibrous meshwork in the damaged area. The disruption of circulation kills osteocytes around the fracture, broadening the area affected. Dead bone soon extends along the shaft in either direction from the break.

STEP 2: In adults, the cells of the periosteum and endosteum are relatively inactive. When a fracture occurs, the cells of the intact endosteum and periosteum undergo rapid mitoses, and the daughter cells migrate into the fracture zone. An **external callus** (*callum,* hard skin), or enlarged collar of cartilage and bone, forms and encircles the bone at the level of the fracture. An extensive **internal callus** organizes within the marrow cavity and between the broken ends of the shaft. At the center of the external callus, cells differentiate into chondrocytes and produce blocks of cartilage. At the edges of each callus, the cells differentiate into osteoblasts and begin creating a bridge between the bone fragments on either side of the fracture.

STEP 3: As the repair continues, osteoblasts replace the central cartilage of the external callus with spongy bone. When this conversion is complete, the external and internal calluses form an extensive and continuous brace at the fracture site. Struts of spongy bone now unite the broken ends. The surrounding area is gradually reshaped as fragments of dead bone are removed and replaced.

STEP 4: Osteoclasts and osteoblasts continue to remodel the region of the fracture for a period ranging from 4 months to well over a year. When the remodeling is complete, the bone of the calluses is gone and only living compact bone remains. The repair may be "good as new," and it may leave no indications that a fracture ever occurred, or the bone may be slightly thicker and stronger than normal at the fracture site. Under comparable stresses, a second fracture will generally occur at a different site.

Comparable events occur after more complex breaks or even after the transplantation of a bone fragment. Each year in the United States, roughly 200,000 people receive *bone grafts,* transplants of bone, to stimulate bone repair. The bone fragments are commonly taken from another part of the body, such as the hip. However, because the inserted bone is ultimately destroyed and replaced, dead and sterilized bone fragments from other donors or even from other species of animals can be used to establish a framework for the repair process. In situations in which the fracture is

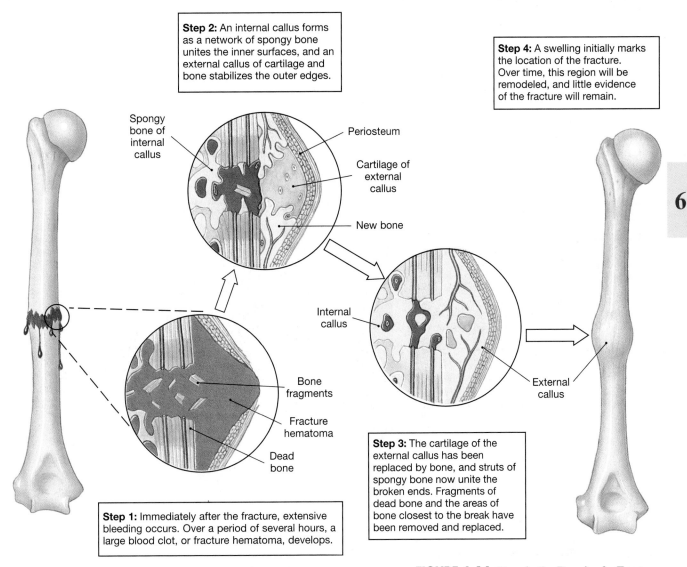

Step 2: An internal callus forms as a network of spongy bone unites the inner surfaces, and an external callus of cartilage and bone stabilizes the outer edges.

Step 4: A swelling initially marks the location of the fracture. Over time, this region will be remodeled, and little evidence of the fracture will remain.

Spongy bone of internal callus

Periosteum

Cartilage of external callus

New bone

Internal callus

External callus

Bone fragments

Fracture hematoma

Dead bone

Step 3: The cartilage of the external callus has been replaced by bone, and struts of spongy bone now unite the broken ends. Fragments of dead bone and the areas of bone closest to the break have been removed and replaced.

Step 1: Immediately after the fracture, extensive bleeding occurs. Over a period of several hours, a large blood clot, or fracture hematoma, develops.

•**FIGURE 6-14** Steps in the Repair of a Fracture

so severe that normal repair processes cannot occur, even with bone grafts, surgeons may rely on synthetic bone or the use of strong electrical fields to stimulate osteoblast activity. AM *Stimulation of Bone Growth and Repair*

AGING AND THE SKELETAL SYSTEM

The bones of the skeleton become thinner and weaker as a normal part of the aging process. Inadequate ossification is called **osteopenia** (os-tē-ō-PĒ-nē-uh; *penia,* lacking), and all of us become slightly osteopenic as we age. This reduction in bone mass begins between the ages of 30 and 40. Over this period, osteoblast activity begins to decline while osteoclast activity continues at previous levels. Once the reduc-

tion begins, women lose roughly 8 percent of their skeletal mass every decade, whereas the skeletons of men deteriorate at about 3 percent per decade. All parts of the skeleton are not equally affected. Epiphyses, vertebrae, and the jaws lose more than their share, resulting in fragile limbs, a reduction in height, and the loss of teeth.

When the reduction in bone mass is sufficient to compromise normal function, the condition is known as **osteoporosis** (os-tē-ō-por-Ō-sis; *porosus,* porous). The fragile bones that result are likely to break when exposed to stresses that younger individuals could easily tolerate. For example, a hip fracture can occur when a woman in her nineties simply tries to stand. Any fractures that do occur lead to a loss of independence and an immobility that further weakens the skeleton.

A Classification of Fractures

Fractures are classified according to their external appearance, location, and the nature of the crack or break in the bone. Important fracture types are indicated here with representative X-rays. **Closed,** or *simple,* fractures are completely internal. They do not involve a break in the skin. **Open,** or *compound,* fractures project through the skin. They are more dangerous than closed fractures due to the possibility of infection or uncontrolled bleeding. Many fractures fall into more than one category. For example, a *Colles' fracture* is a transverse fracture that, depending on the injury, may also be an open or closed *comminuted fracture.* Representative examples of the common types of fractures are shown in Figure 6-15•.

A **Pott's fracture** occurs at the ankle and affects both bones of the leg.

Comminuted fractures, such as this fracture of the femur, shatter the affected area into a multitude of bony fragments.

Transverse fractures, such as this fracture of the radius, break a shaft bone across its long axis.

Spiral fractures, such as this fracture of the tibia, are produced by twisting stresses that spread along the length of the bone.

Displaced fractures produce new and abnormal bone arrangements; **nondisplaced fractures** retain the normal alignment of the bones or fragments.

A **Colles' fracture,** a break in the distal portion of the radius, is typically the result of reaching out to cushion a fall.

In a **greenstick fracture,** such as this fracture of the radius, only one side of the shaft is broken, and the other is bent. This type generally occurs in children, whose long bones have yet to ossify fully.

Epiphyseal fractures, such as this fracture of the femur, tend to occur where the bone matrix is undergoing calcification and chondrocytes are dying. A clean transverse fracture along this line generally heals well. Unless carefully treated, fractures between the epiphysis and the epiphyseal cartilage can permanently stop growth at this site.

Compression fractures occur in vertebrae subjected to extreme stresses, as when you land on your seat in a fall.

•**FIGURE 6-15** Major Types of Fractures

(a)   (b)

•**FIGURE 6-16** **The Effects of Osteoporosis.** **(a)** Normal spongy bone from the epiphysis of a young adult. (SEM × 25) **(b)** Spongy bone from a person with osteoporosis. (SEM × 21)

The extent of the loss of spongy bone mass due to osteoporosis is shown in Figure 6-16•; the reduction in compact bone mass is equally severe.

Sex hormones are important in maintaining normal rates of bone deposition. Over age 45, an estimated 29 percent of women and 18 percent of men have osteoporosis. The condition accelerates after menopause owing to a decline in circulating estrogens. Because men continue to produce androgens until relatively late in life, severe osteoporosis is less common in males below age 60 than in females in that same age group.

Osteoporosis can also develop as a secondary effect of many cancers. Cancers of the bone marrow, breast, or other tissues release a chemical known as **osteoclast-activating factor.** This compound increases both the number and activity of osteoclasts and produces a severe osteoporosis. [AM] *Osteoporosis and Age-Related Skeletal Abnormalities*

BONE MARKINGS (SURFACE FEATURES)

Each bone in the body has characteristic external and internal features. Elevations or projections form where tendons and ligaments attach and where adjacent bones articulate (that is, form joints). Depressions, grooves, and tunnels in bone indicate sites where blood vessels or nerves lie alongside or penetrate the bone. Detailed examination of these **bone markings,** or *surface features,* can yield an abundance of anatomical information. For example, anthropologists, criminologists, and pathologists can often determine the size, weight, sex, and general appearance of an individual on the basis of incomplete skeletal remains. (We shall discuss this topic further in Chapter 8.)

We will ignore minor variations of individual bones to focus on prominent features that identify the bone (Table 6-2). These markings provide fixed landmarks that can help us determine the position of the soft tissue components of other systems. Specific anatomical terms are used to describe the various elevations and depressions.

INTEGRATION WITH OTHER SYSTEMS

Although bones may seem inert, you should now realize that they are dynamic structures. The entire skeletal system is intimately associated with other systems. For instance, the bones of the skeleton are attached to the muscular system, extensively connected to the cardiovascular and lymphatic systems, and largely under the physiological control of the endocrine system. The digestive and urinary systems also play important roles in providing the calcium and phosphate minerals needed for bone growth. In return, the skeleton represents a reserve of calcium, phosphate, and other minerals that can compensate for changes in the dietary supply of those ions. The functional relationships between the skeletal system and other systems are diagrammed in Figure 6-17• (p. 189).

☑ At which point in fracture repair would you find an external callus?

☑ Why is osteoporosis more common in older women than in older men?

☑ X-rays of a child's skeleton indicate bones forming in the tendons of several joints. Which kind of bones are these?

TABLE 6-2 An Introduction to Skeletal Terminology

General Description	Anatomical Term	Definition
Elevations and projections (general)	**Process**	Any projection or bump
	Ramus	An extension of a bone making an angle to the rest of the structure
Processes formed where tendons or ligaments attach	**Trochanter**	A large, rough projection
	Tuberosity	A smaller, rough projection
	Tubercle	A small, rounded projection
	Crest	A prominent ridge
	Line	A low ridge
	Spine	A pointed process
Processes formed for articulation with adjacent bones	**Head**	The expanded articular end of an epiphysis, separated from the shaft by the neck
	Neck	A narrow connection between the epiphysis and the diaphysis
	Condyle	A smooth, rounded articular process
	Trochlea	A smooth, grooved articular process shaped like a pulley
	Facet	A small, flat articular surface
Depressions	**Fossa**	A shallow depression
	Sulcus	A narrow groove
Openings	**Foramen**	A rounded passageway for blood vessels and/or nerves
	Canal	A large-diameter passageway through the substance of a bone
	Fissure	An elongate cleft
	Sinus or **antrum**	A chamber within a bone, normally filled with air

INTEGUMENTARY SYSTEM

Synthesizes vitamin D₃, essential for calcium and phosphorus absorption (bone maintenance and growth)

Provides structural support

THE SKELETAL SYSTEM

6

FOR ALL SYSTEMS

Provides mechanical support
Stores energy reserves
Stores calcium and phosphate reserves

Sex hormones stimulate growth and maintenance of bones; surge of sex hormones at puberty causes acceleration of growth and closure of epiphyseal cartilages

Pelvis protects reproductive organs of female; protects portion of ductus deferens and accessory glands in male

REPRODUCTIVE SYSTEM

MUSCULAR SYSTEM

Stabilizes bone positions; tension in tendons stimulates bone growth and maintenance

Provides calcium needed for normal muscle contraction; bones act as levers to produce body movements

NERVOUS SYSTEM

Regulates bone position by controlling muscle contractions

Provides calcium for neural function; protects brain, spinal cord; receptors at joints provide information about body position

ENDOCRINE SYSTEM

Skeletal growth regulated by growth hormone, thyroid hormones, and sex hormones; calcium mobilization regulated by parathyroid hormone and calcitonin

Protects endocrine organs, especially in brain, chest, and pelvic cavity

CARDIOVASCULAR SYSTEM

Provides oxygen, nutrients, hormones, blood cells; removes waste products and carbon dioxide

Provides calcium needed for cardiac muscle contraction; blood cells produced in bone marrow

LYMPHATIC SYSTEM

Lymphocytes assist in the defense and repair of bone following injuries

Lymphocytes and other cells of the immune response are produced and stored in bone marrow

RESPIRATORY SYSTEM

Provides oxygen and eliminates carbon dioxide

Movements of ribs important in breathing; axial skeleton surrounds and protects lungs

DIGESTIVE SYSTEM

Provides nutrients, calcium, and phosphate

Ribs protect portions of liver, stomach, and intestines

URINARY SYSTEM

Conserves calcium and phosphate needed for bone growth; disposes of waste products

Axial skeleton provides some protection for kidneys and ureters; pelvis protects urinary bladder and proximal urethra

•**FIGURE 6-17** Functional Relationships between the Skeletal System and Other Systems

6

[EXPLORE MediaLab]

Far from being dry, static structures, bones are living, changing organs that perform a vital role in your body's homeostasis. To gain a better understanding of bone markings, bone structure, and bone formation, use the resources available on both the CD-ROM and the Companion Website. The CD-ROM includes animations and tutorials; the website builds on those resources with text content and interactive problems as well as communication avenues for sharing ideas with your class. To complete the following Web Explorations, visit the MediaLab in Chapter 6 of your Companion Website.

Web Exploration 1

Estimated time for completion = 10 minutes

Bone remodeling occurs continuously in the healthy body. When bones are immobilized for a long time, such as after hip replacement surgery, bone remodeling can lead to a loss of bone density, similar to that from osteoporosis. What happens to the femur and hip area during this procedure? Select the keyword **REMODELING** on the Companion Website to read a report about bone density changes with hip replacement surgery. Pay special attention to the conclusions. What does this study mean to you? Is there a message here for general skeletal health?

Web Exploration 2

Estimated time for completion = 10 minutes

The American Society for Bone and Mineral Research published a series of articles about studies of dietary intake and bone density. Select the keyword **ASBMR** on the Companion Website to help you determine whether the findings make sense. Read and take notes on the article entitled "Low Levels of Magnesium in College Students Could Affect Future Rates of Osteoporosis." Back at the Companion Website, select the keyword **BROCCOLI** to read the article "Broccoli May Be Good for Bones." Reflect on your lifestyle and your knowledge of bone physiology. Do these studies make sense? Will you change your diet or lifestyle as a consequence?

CHAPTER REVIEW

SELECTED CLINICAL TERMINOLOGY
Terms Discussed in This Chapter

acromegaly: A condition caused by excess secretion of growth hormone after puberty. Skeletal abnormalities develop, affecting the cartilages and various small bones. *(p. 181 and AM)*

external callus: A toughened layer of connective tissue that encircles and stabilizes a bone at a fracture site. *(p. 184)*

fracture: A crack or break in a bone. *(p. 184)*

fracture hematoma: A large blood clot that closes off the injured vessels around a fracture and leaves a fibrous meshwork in the damaged area of bone; the first step in fracture repair. *(p. 184)*

gigantism: A condition resulting from an overproduction of growth hormone before puberty. *(p. 181)*

internal callus: A bridgework of bone trabeculae that unites the broken ends of a bone on the marrow side of a fracture. *(p. 184)*

Marfan's syndrome: An inherited condition linked to defective production of a connective tissue glycoprotein. Extreme height and long, slender limbs are the most obvious physical indications. *(p. 181 and AM)*

osteoclast-activating factor: A compound released by cancers of the bone marrow, breast, or other tissues. It produces a severe osteoporosis. *(p. 187)*

osteomalacia (os-tē-ō-ma-LĀ-shē-uh): A softening of bone due to a decrease in the mineral content. *(p. 184)*

osteopenia (os-tē-ō-PĒ-nē-uh): Inadequate ossification, leading to thinner, weaker bones. *(p. 185)*

osteoporosis (os-tē-ō-por-Ō-sis): A reduction in bone mass to a degree that compromises normal function. *(p. 185 and AM)*

pituitary growth failure (*pituitary dwarfism*): A type of dwarfism caused by inadequate growth hormone production. *(p. 181)*

rickets: A childhood disorder that reduces the amount of calcium salts in the skeleton; typically characterized by a bowlegged appearance, because the leg bones bend under the body's weight. *(p. 184)*

scurvy: A condition involving weak, brittle bones as a result of a vitamin C deficiency. *(p. 181)*

AM Additional Terms Discussed in the *Applications Manual*

achondroplasia (ā-kon-drō-PLĀ-sē-uh): A condition resulting from abnormal epiphyseal activity; the epiphyseal cartilages grow unusually slowly, and the individual develops short, stocky limbs. The trunk is normal in size, and sexual and mental development remain unaffected.

hyperostosis (hī-per-os-TŌ-sis): The excessive formation of bone tissue.

osteogenesis imperfecta (im-per-FEK-tuh): An inherited condition affecting the organization of collagen fibers. Osteoblast function is impaired, growth is abnormal, and the bones are very fragile, leading to progressive skeletal deformation and repeated fractures.

osteomyelitis (os-tē-ō-mī-e-LĪ-tis): A painful infection in a bone, generally caused by bacteria.

osteopetrosis (os-tē-ō-pe-TRŌ-sis): A condition caused by a decrease in osteoclast activity, causing increased bone mass and various skeletal deformities.

Paget's disease, or **osteitis deformans** (os-tē-Ī-tis de-FOR-manz): A condition characterized by a gradual deformation of the skeleton.

STUDY OUTLINE

INTRODUCTION   p. 168

1. The skeletal system includes the bones of the skeleton and the cartilages, ligaments, and other connective tissues that stabilize or interconnect bones. Its functions include structural support, storage of minerals and lipids, blood cell production, protection, and leverage.

A CLASSIFICATION OF BONES   p. 168

Bone Shapes   p. 168

1. Categories of bones include **long bones, short bones, flat bones, irregular bones, sesamoid bones,** and **sutural bones** (*Wormian bones*). *(Figure 6-1)*

Bone Structure   p. 168

2. A representative long bone has a **diaphysis, epiphyses, metaphyses, articular cartilages,** and a **marrow cavity.**
3. The two types of bone are **compact** *(dense)* **bone** and **spongy** *(cancellous)* **bone.** The marrow cavity and spaces within spongy bone contain **bone marrow,** either **yellow bone marrow** (for lipid storage) or **red bone marrow** (for blood cell formation). *(Figure 6-2)*

BONE HISTOLOGY   p. 170

1. **Osseous tissue** is a supporting connective tissue with a solid matrix.

The Matrix of Bone   p. 170

2. Bone matrix consists largely of crystals of **hydroxyapatite.**

Cells in Bone   p. 171

3. **Osteocytes,** located in **lacunae,** are the most abundant bone cells. **Osteoblasts** synthesize the bony matrix by **osteogenesis. Osteoclasts** dissolve the bony matrix through **osteolysis. Osteoprogenitor cells** differentiate into osteoblasts. *(Figure 6-3)*

Compact Bone and Spongy Bone   p. 172

4. The basic functional unit of compact bone is the **osteon,** containing osteocytes arranged around a **central canal. Lamellae** are layers of calcified matrix. **Canaliculi** within and between lamellae interconnect the lacunae. *(Figures 6-3, 6-4)*
5. Spongy bone contains **trabeculae,** typically in an open network. *(Figure 6-4)*
6. Compact bone is located where stresses come from a limited range of directions, such as along the diaphysis of some bones. Spongy bone is located where stresses are few or come from many directions, such as at the epiphyses of some bones. *(Figure 6-5)*

The Periosteum and Endosteum   p. 174

7. A bone is covered by a **periosteum** and lined with an **endosteum.** *(Figure 6-6)*

BONE DEVELOPMENT AND GROWTH   p. 175

1. **Ossification** is the process of converting other tissues to bone. **Calcification** is the process of depositing calcium salts within a tissue.

Intramembranous Ossification   p. 175

2. **Intramembranous ossification** begins when osteoblasts differentiate within connective tissue and can produce spongy or compact bone. Such ossification begins at an **ossification center.** *(Figure 6-7)*

Endochondral Ossification   p. 176

3. **Endochondral ossification** begins by forming a cartilage model that is gradually replaced by bone at the metaphysis. In this way, bone length increases. *(Figure 6-8)*
4. The timing of *closure* of the **epiphyseal cartilage** differs among bones and among individuals. *(Figure 6-9)*
5. Bone diameter increases through *appositional growth.* *(Figure 6-10)*

The Blood and Nerve Supplies   p. 178

6. Three major sets of blood vessels provide an extensive supply of blood to bone. *(Figure 6-11)*

THE DYNAMIC NATURE OF BONE   p. 180

1. The organic and mineral components of bone are continuously recycled and renewed through **remodeling.**

Effects of Exercise on Bone   p. 180

2. The shapes and thicknesses of bones reflect the stresses applied to them.

Hormonal and Nutritional Effects on Bone   p. 181

3. Normal osteogenesis requires a reliable source of minerals, vitamins, and hormones.
4. *Growth hormone* and *thyroxine* stimulate bone growth. *(Table 6-1)*

The Skeleton as a Calcium Reserve   p. 182

5. Calcium is the most common mineral in the human body, with roughly 99 percent of it located in the skeleton. *(Figure 6-12)*
6. Interactions among the bones, intestinal tract, and kidneys affect the calcium ion concentration. *(Figure 6-13)*
7. Two hormones, **calcitonin** and **parathyroid hormone (PTH),** regulate calcium ion homeostasis. Calcitonin leads to a decline in the calcium concentration, whereas parathyroid hormone increases the calcium concentration. *(Figure 6-14)*

Fracture Repair   p. 184

8. A **fracture** is a crack or a break in a bone. The repair of a fracture involves the formation of a **fracture hematoma,** an **external callus,** and an **internal callus.** *(Figure 6-14)*

FOCUS: A Classification of Fractures   p. 186

9. Various criteria are used to classify fractures. *(Figure 6-15)*

Aging and the Skeletal System   p. 185

10. The effects of aging on the skeleton can include **osteopenia** and **osteoporosis.** *(Figure 6-16)*

BONE MARKINGS (SURFACE FEATURES)   p. 187

1. **Bone markings** can be used to identify specific bones. *(Table 6-2)*

INTEGRATION WITH OTHER SYSTEMS   p. 187

1. The skeletal system is dynamically associated with all the other systems of the body. *(Figure 6-17)*

REVIEW QUESTIONS

LEVEL 1 REVIEWING FACTS AND TERMS

1. The bones of the skeleton store energy reserves as lipids in areas of
 (a) red marrow
 (b) yellow marrow
 (c) the matrix of bone tissue
 (d) the ground substance

2. Two-thirds of the weight of bone is accounted for by
 (a) crystals of calcium phosphate
 (b) collagen fibers
 (c) osteocytes
 (d) calcium carbonate

3. The two types of osseous tissue are
 (a) compact bone and spongy bone
 (b) dense bone and compact bone
 (c) spongy bone and cancellous bone
 (d) a, b, and c are correct

4. The basic functional units of mature compact bone are called
 (a) lacunae   (b) osteocytes
 (c) osteons   (d) canaliculi

5. Unlike compact bone, spongy bone contains concentric lamellae that form struts or plates called
 (a) canaliculi   (b) canals of Volkmann
 (c) osteons   (d) trabeculae

6. The vitamins essential for normal adult bone maintenance and repair are
 (a) A and E   (b) C and D
 (c) B and E   (d) B complex and K

7. The hormones that coordinate the storage, absorption, and excretion of calcium ions are
 (a) growth hormone and thyroxine
 (b) calcitonin and parathyroid hormone
 (c) calcitriol and cholecalciferol
 (d) estrogens and androgens

8. Inadequate ossification due to the aging process is
 (a) osteomalacia   (b) osteolysis
 (c) osteopenia   (d) osteogenesis

9. The primary reason that osteoporosis accelerates after menopause in women is
 (a) reduced levels of circulating estrogens
 (b) reduced levels of vitamin C
 (c) diminished osteoclast activity
 (d) increased osteoblast activity

10. A child with rickets would have
 (a) oversized facial bones
 (b) long arms and legs
 (c) weak, brittle bones
 (d) bowlegs

11. What are the five primary functions of the skeletal system?

12. List the four distinctive cell populations of osseous tissue.

13. What is the functional difference between an osteoblast and an osteoclast?

14. What are the three primary functions of the periosteum?

15. What is the primary difference between intramembranous ossification and endochondral ossification?

16. What three major sets of blood vessels develop in typical bones in the skeleton?

17. (a) What nutritional factors are essential for normal bone growth and maintenance?
 (b) What hormonal factors are necessary for normal bone growth and maintenance?

18. Which three organs and/or tissues interact to assist in the regulation of calcium ion concentration in body fluids?

19. What major effects of parathyroid hormone oppose those of calcitonin?

20. What are the major functions of the hormone calcitonin?

LEVEL 2 REVIEWING CONCEPTS

21. If spongy bone has no osteons, how do nutrients reach the osteocytes?

22. Why are stresses or impacts to the side of the shaft in a long bone more dangerous than stress applied to the long axis of the shaft?

23. Why do extended periods of inactivity cause degenerative changes in the skeleton?

24. What are the functional relationships between the skeleton and both the digestive and urinary systems?

25. During the growth of a long bone, how is the epiphysis forced farther from the shaft?

26. What role do sex hormones in males and females play in bone growth and development?

27. How might damage to the thyroid gland influence calcium regulation in the body?

28. Why does a second fracture in the same bone tend to occur at a site different from that of the first fracture?

29. Why are individuals in their eighties shorter than they were in their forties?

30. How might bone markings be useful in identifying the remains of a criminal who has been shot and killed?

31. What purpose do elevations or projections serve on bones?

LEVEL 3 CRITICAL THINKING AND CLINICAL APPLICATIONS

32. While playing on her swing set, 10-year-old Sally falls and breaks her right leg. At the emergency room, the doctor tells her parents that the proximal end of the tibia where the epiphysis meets the diaphysis is fractured. The fracture is properly set and eventually heals. During a routine physical when she is 18, Sally learns that her right leg is 2 cm shorter than her left, probably because of her accident. What might account for this difference?

33. Sherry is a pregnant teenager. Her diet before she was pregnant consisted mostly of junk food, and that hasn't changed since she became pregnant. Approximately 8–10 weeks into her pregnancy, she falls and breaks her arm. She doesn't understand why the bone broke, because it wasn't a hard fall. Test results determine that a significant amount of bone demineralization is occurring. Explain what is happening to Sherry.

34. Would you expect to see changes in blood levels of the hormones calcitonin and PTH as a result of vitamin D_3 deficiency? Explain.

35. Why might a person in kidney failure exhibit symptoms similar to those of osteoporosis?

CHAPTER **7**

The Axial Skeleton

CHAPTER OUTLINE AND OBJECTIVES

INTRODUCTION   p. 194
① Identify the bones of the axial skeleton, and specify their functions.

THE SKULL   p. 195

▦ FOCUS: The Individual Bones of the Skull   p. 200

Cranial Bones   p. 200

Facial Bones   p. 205

Summary: The Foramina and Fissures of the Skull   p. 208
The Orbital and Nasal Complexes   p. 208
The Skulls of Infants and Children   p. 211
② Identify the bones of the cranium and face, and explain the significance of the markings on the individual bones.
③ Describe the structure of the nasal complex and the functions of the individual bones.
④ Explain the functions of the paranasal sinuses.
⑤ Describe key structural differences among the skulls of infants, children, and adults.

THE VERTEBRAL COLUMN   p. 212
Spinal Curvature   p. 212
Vertebral Anatomy   p. 213
Vertebral Regions   p. 214
⑥ Identify and describe the curvatures of the spinal column and their functions.
⑦ Identify the vertebral regions, and describe the distinctive structural and functional characteristics of each vertebral group.

THE THORACIC CAGE   p. 219
The Ribs   p. 219
The Sternum   p. 222
⑧ Explain the significance of the articulations between the thoracic vertebrae and ribs and between the ribs and sternum.

MEDIALAB and CHAPTER REVIEW   p. 222, 223

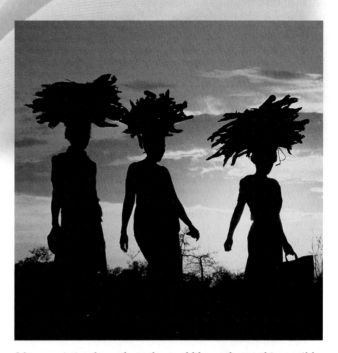

Many societies throughout the world have chosen this sensible method of carrying heavy loads. Rather than bearing the load in their arms and possibly straining their back, people place bundles on their head. The weight is thus distributed along the longitudinal axis of the body via the axial skeleton. One or both hands are free to perform other tasks. Weight bearing is a major function of the axial skeleton, but, as you will see, these bones have many other important roles.

7

Bone first appeared more than 500 million years ago in primitive fishes. It became very common in many groups of fishes within 100 million years. This was dermal bone (∞ *p. 175*) that provided a mineral reserve and armor plating to protect against predators. The internal skeletons of these fishes were composed of cartilage. In the evolutionary branch that led toward humans, bones replaced most of the cartilages.

As we saw in Chapter 6, bones are remarkable structures. They are as strong as reinforced concrete but are considerably lighter. ∞ *p. 171* Better yet, bones are remodeled and reshaped to meet metabolic demands or to adapt to changing activity patterns. The basic features of the human skeleton have been shaped by evolution, but the detailed characteristics of each bone reflect the stresses placed on it. As a result, the skeleton changes in the course of a lifetime. Examples discussed in Chapter 6 included the proportional changes at puberty and the gradual osteopenia of aging. In this chapter, we will consider additional examples, such as the changes that occurred in the shape of your vertebral column as you made the transition from crawling to walking.

The skeletal system typically includes 206 separate bones and a number of associated cartilages and ligaments. This system is divided into axial and appendicular divisions (Figure 7-1●). The **axial skeleton** (shown in Figure 7-1a● and highlighted in Figure 7-1b●) forms the longitudinal axis of the body. The axial skeleton has 80 bones, roughly 40 percent of the bones in the human body. The axial components are:

- The *skull* (8 *cranial bones* and 14 *facial bones*).
- Bones associated with the skull (6 *auditory ossicles* and the *hyoid bone*).
- The *thoracic cage* (the *sternum* and 24 *ribs*).
- The *vertebral column* (24 *vertebrae,* the *sacrum,* and the *coccyx*).

The axial skeleton provides a framework that supports and protects organs in the dorsal and ventral body cavities. It also provides an extensive surface area for the attachment of muscles that (1) adjust the positions of the head, neck, and trunk; (2) perform respiratory movements; and (3) stabilize or position parts of the appendicular skeleton. The joints of the axial skeleton permit limited movement, but they are very strong and heavily reinforced with ligaments.

The **appendicular skeleton** consists of 126 bones. This division includes the bones of the limbs and the pectoral and pelvic girdles that attach the limbs to the trunk. We shall examine the appendicular skeleton in Chapter 8. This chapter describes the functional anatomy of the axial skeleton; we will begin with the skull. You may wish

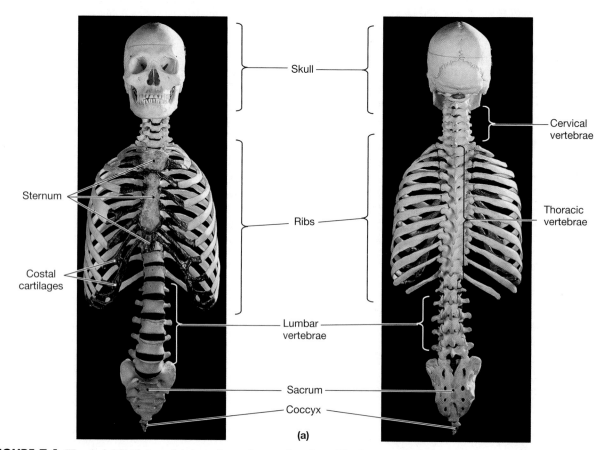

(a)

●**FIGURE 7-1 The Axial Skeleton. (a)** Anterior and posterior views. The bones associated with the skull are not visible.

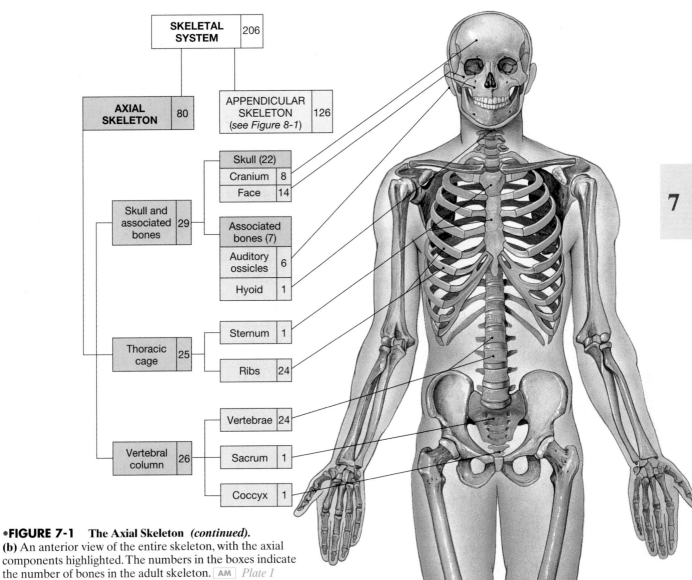

•FIGURE 7-1   The Axial Skeleton (continued).
(b) An anterior view of the entire skeleton, with the axial
components highlighted. The numbers in the boxes indicate
the number of bones in the adult skeleton. [AM]   *Plate 1*

to review the directional references in Tables 1-2 and 1-3 and the terms
in Table 6-2. ∞ *pp. 18, 19, 188*

THE SKULL

The bones of the **skull** protect the brain and guard the entrances to the
digestive and respiratory systems. The skull contains 22 bones: 8 form
the **cranium,** or *braincase,* and 14 are associated with the face (Figure
7-2•). Seven additional bones are associated with the skull: The 6 auditory ossicles are situated within the *temporal bones* of the cranium,
and the hyoid bone is connected to the inferior surfaces of the temporal bones by a pair of ligaments.

The cranium consists of the 8 **cranial bones:** the *occipital, parietal,
frontal, temporal, sphenoid,* and *ethmoid bones.* Together the cranial
bones enclose the **cranial cavity,** a fluid-filled chamber that cushions
and supports the brain. Blood vessels, nerves, and membranes that stabilize the position of the brain are attached to the inner surface of the
cranium. Its outer surface provides an extensive area for the attachment of muscles that move the eyes, jaws, and head. A joint between

(b)

Cranial bones

ASSOCIATED BONES	7		CRANIUM	8		FACE	14
			Occipital bone	1		Maxillary bones	2
			Parietal bones	2		Palatine bones	2
Auditory ossicles (detailed in Chapter 17)	6	Within temporal bones	Frontal bone	1		Nasal bones	2
			Temporal bones	2		Vomer	1
			Sphenoid bone	1		Inferior nasal conchae	2
Hyoid bone	1	Suspended from temporal bones	Ethmoid bone	1		Zygomatic bones	2
						Lacrimal bones	2
						Mandible	1

Facial bones

•**FIGURE 7-2 Cranial and Facial Subdivisions of the Skull.** The seven associated bones are not illustrated.

the occipital bone and the first vertebra of the neck stabilizes the positions of the brain and spinal cord while permitting a range of head movements.

If the cranium is the house where the brain resides, the *facial complex* is the front porch. **Facial bones** protect and support the entrances to the digestive and respiratory tracts. The superficial facial bones (the *maxillary, lacrimal, nasal,* and *zygomatic bones* and the *mandible*) (Figure 7-2•) provide areas for the attachment of muscles that control facial expressions and assist in food manipulation. The deeper facial bones (the *palatine bones, inferior nasal conchae,* and *vomer*) help separate the oral and nasal cavities, increase the surface area of the nasal cavities, or help form the **nasal septum** (*septum,* wall), which subdivides the nasal cavity.

Several bones of the skull contain air-filled chambers called **sinuses.** Sinuses have two major functions: (1) The presence of a sinus makes a bone much lighter than it would be otherwise, and (2) the mucous membrane lining the sinuses produces mucus that moistens and cleans the air in and adjacent to the sinus. We will consider the sinuses as we discuss specific bones.

Joints, or *articulations,* form where two bones interconnect. Except where the mandible contacts the cranium, the connections between the skull bones of adults are immovable joints called **sutures.** At a suture, bones are tied firmly together with dense fibrous connective tissue. Each suture of the skull has a name, but now you need to know only four major sutures:

1. *Lambdoidal suture.* The **lambdoidal** (lam-DOYD-ul) **suture** (Greek *lambda,* L + *eidos,* shape) arches across the posterior surface of the skull (Figure 7-3a•). This suture separates the occipital bone from the two parietal bones. One or more **sutural bones** (*Wormian bones*) may be present along this suture. ∞ *p. 168*

2. *Coronal suture.* The **coronal suture** attaches the frontal bone to the parietal bones of either side (Figure 7-3b•). The occipital, parietal, and frontal bones form the **calvaria** (kal-VAR-ē-uh), or skullcap.

3. *Sagittal suture.* The **sagittal suture** extends from the lambdoidal suture to the coronal suture, between the parietal bones (Figure 7-3b•).

4. *Squamous sutures.* A **squamous suture** on each side of the skull is the boundary between the temporal bone and the parietal bone of that side. Figure 7-3a• shows the intersection between the squamous sutures and the lambdoidal suture; Figure 7-3c• shows the path of the squamous suture on the right side of the skull.

"FOCUS: The Individual Bones of the Skull" on pages 200–207 considers each skull bone. Refer to the figures there and to Figures 7-3 and 7-4• (the adult skull in superficial and sectional views, respectively) to develop a three-dimensional perspective of the relationships among the bones. The ridges and *foramina* (fō-RAM-in-uh; passageways; singular, *foramen*) detailed here mark the attachment sites of muscles discussed in Chapter 11 or the passage of nerves and blood vessels discussed in Chapters 14 and 21, respectively.

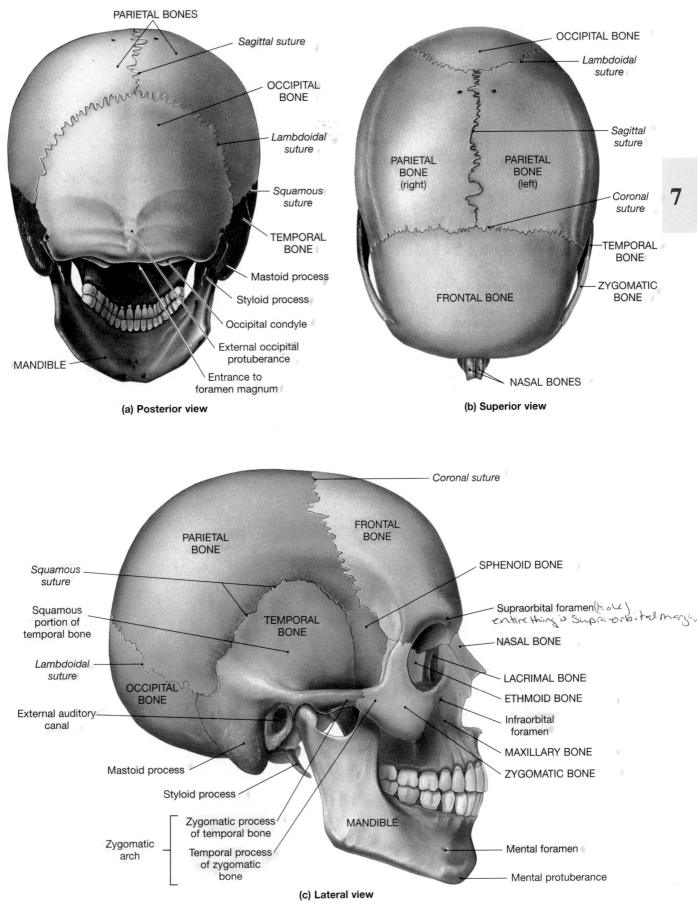

PARIETAL BONES

Sagittal suture

OCCIPITAL
BONE

Lambdoidal
suture

Squamous
suture

TEMPORAL
BONE

Mastoid process

Styloid process

Occipital condyle

External occipital
protuberance

MANDIBLE

Entrance to
foramen magnum

(a) Posterior view

OCCIPITAL BONE

Lambdoidal
suture

7

Sagittal
suture

PARIETAL
BONE
(right)

PARIETAL
BONE
(left)

Coronal
suture

TEMPORAL
BONE

ZYGOMATIC
BONE

FRONTAL BONE

NASAL BONES

(b) Superior view

Coronal suture

PARIETAL
BONE

FRONTAL
BONE

Squamous
suture

SPHENOID BONE

Supraorbital foramen (hole)
entire thing is supra-orbital margin

Squamous
portion of
temporal bone

TEMPORAL
BONE

NASAL BONE

LACRIMAL BONE

Lambdoidal
suture

ETHMOID BONE

OCCIPITAL
BONE

Infraorbital
foramen

External auditory
canal

MAXILLARY BONE

ZYGOMATIC BONE

Mastoid process

Styloid process

MANDIBLE

Zygomatic process
of temporal bone

Zygomatic
arch

Temporal process
of zygomatic
bone

Mental foramen

Mental protuberance

(c) Lateral view

•**FIGURE 7-3** **The Adult Skull.** AM *Plates 2.1, 2.2*

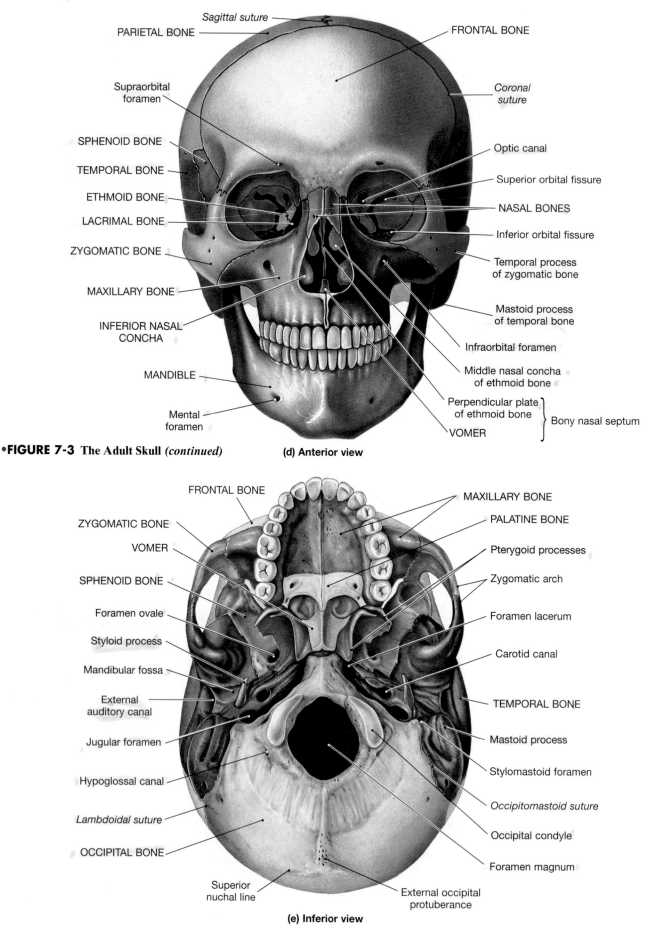

Sagittal suture

PARIETAL BONE

FRONTAL BONE

Supraorbital foramen

Coronal suture

SPHENOID BONE

Optic canal

TEMPORAL BONE

Superior orbital fissure

ETHMOID BONE

NASAL BONES

LACRIMAL BONE

Inferior orbital fissure

ZYGOMATIC BONE

Temporal process of zygomatic bone

MAXILLARY BONE

Mastoid process of temporal bone

INFERIOR NASAL CONCHA

Infraorbital foramen

Middle nasal concha of ethmoid bone

MANDIBLE

Perpendicular plate of ethmoid bone
} Bony nasal septum

Mental foramen

VOMER

•**FIGURE 7-3** The Adult Skull *(continued)*   **(d) Anterior view**

FRONTAL BONE

MAXILLARY BONE

ZYGOMATIC BONE

PALATINE BONE

VOMER

Pterygoid processes

SPHENOID BONE

Zygomatic arch

Foramen ovale

Foramen lacerum

Styloid process

Carotid canal

Mandibular fossa

External auditory canal

TEMPORAL BONE

Jugular foramen

Mastoid process

Hypoglossal canal

Stylomastoid foramen

Lambdoidal suture

Occipitomastoid suture

OCCIPITAL BONE

Occipital condyle

Foramen magnum

Superior nuchal line

External occipital protuberance

(e) Inferior view

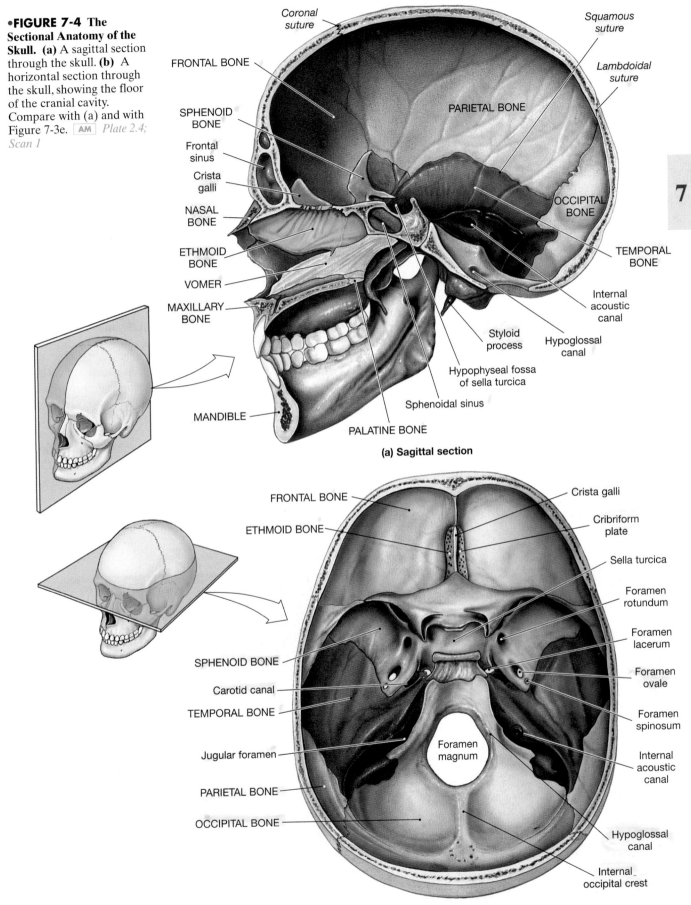

●**FIGURE 7-4** The **Sectional Anatomy of the Skull. (a)** A sagittal section through the skull. **(b)** A horizontal section through the skull, showing the floor of the cranial cavity. Compare with (a) and with Figure 7-3e. AM *Plate 2.4; Scan 1*

7

Coronal suture

Squamous suture

Lambdoidal suture

FRONTAL BONE

PARIETAL BONE

SPHENOID BONE

Frontal sinus

Crista galli

NASAL BONE

ETHMOID BONE

VOMER

MAXILLARY BONE

OCCIPITAL BONE

TEMPORAL BONE

Internal acoustic canal

Hypoglossal canal

Styloid process

Hypophyseal fossa of sella turcica

Sphenoidal sinus

MANDIBLE

PALATINE BONE

(a) Sagittal section

FRONTAL BONE

ETHMOID BONE

Crista galli

Cribriform plate

Sella turcica

Foramen rotundum

Foramen lacerum

Foramen ovale

Foramen spinosum

Internal acoustic canal

Hypoglossal canal

Internal occipital crest

SPHENOID BONE

Carotid canal

TEMPORAL BONE

Jugular foramen

PARIETAL BONE

OCCIPITAL BONE

Foramen magnum

(b) Horizontal section

CRANIAL BONES

The Occipital Bone (Figure 7-5a•)

General Functions: The **occipital bone** forms much of the posterior and inferior surfaces of the cranium.

Articulations: The occipital bone articulates with the parietal bones, the temporal bones, the sphenoid bone, and the first cervical vertebra (the atlas) (Figures 7-3a–c,e and 7-4•).

Regions/Landmarks: The **external occipital protuberance** is a small bump at the midline on the inferior surface.

The **occipital crest,** which begins at the external occipital protuberance, marks the attachment of a ligament that helps stabilize the vertebrae of the neck.

The **occipital condyles** are the site of articulation between the skull and the first vertebra of the neck.

The *inferior* and *superior nuchal* (NOO-kul) *lines* are ridges that intersect the occipital crest. They mark the attachment sites of muscles and ligaments that stabilize the articulation at the occipital condyles and balance the weight of the head over the vertebrae of the neck.

The concave internal surface of the occipital bone (Figure 7-4a•) closely follows the contours of the brain. The grooves follow the paths of major blood vessels, and the ridges mark the attachment sites of membranes that stabilize the position of the brain.

Foramina: The **foramen magnum** connects the cranial cavity with the spinal cavity, which is enclosed by the vertebral column (Figure 7-4b•). This foramen surrounds the connection between the brain and spinal cord.

The **jugular foramen** lies between the occipital bone and the temporal bone (Figure 7-3e•). The *internal jugular vein* passes through this foramen, carrying venous blood from the brain.

The **hypoglossal canals** (Figure 7-3e•) begin at the lateral base of each occipital condyle and end on the inner surface of the occipital bone near the foramen magnum. The *hypoglossal nerves,* cranial nerves that control the tongue muscles, pass through these canals.

The Parietal Bones (Figure 7-5b•)

General Functions: The **parietal bones** form part of the superior and lateral surfaces of the cranium.

Articulations: The parietal bones articulate with one another and with the occipital, temporal, frontal, and sphenoid bones (Figures 7-3a–d and 7-4•).

Regions/Landmarks: The *superior* and *inferior temporal lines* are low ridges that mark the attachment sites of the *temporalis muscle,* a large muscle that closes the mouth.

Grooves on the inner surface of the parietal bones mark the paths of cranial blood vessels (Figure 7-4a•).

(a) Occipital bone, inferior view

(b) Right parietal bone, lateral view

•**FIGURE 7-5** The Occipital and Parietal Bones

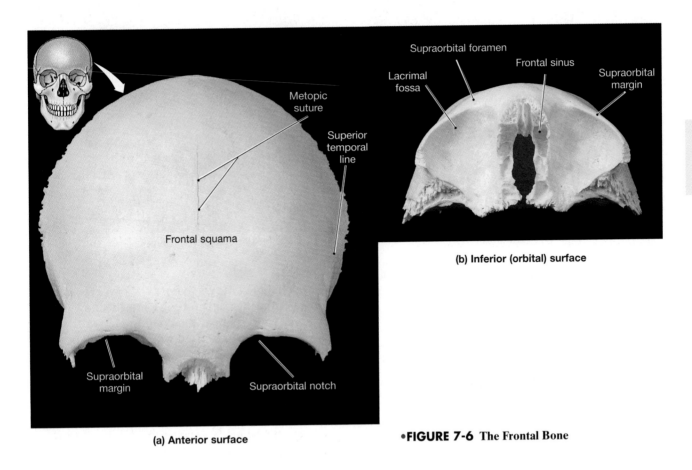

(a) Anterior surface

(b) Inferior (orbital) surface

•**FIGURE 7-6** The Frontal Bone

The Frontal Bone (Figure 7-6a,b•)

General Functions: The **frontal bone** forms the anterior portion of the cranium and the roof of the *orbits* (eye sockets). Mucous secretions of the *frontal sinuses* within this bone help flush the surfaces of the nasal cavities.

Articulations: The frontal bone articulates with the parietal, sphenoid, ethmoid, nasal, lacrimal, maxillary, and zygomatic bones (Figures 7-3b–e and 7-4•).

Regions/Landmarks: The **frontal squama,** or forehead, forms the anterior, superior portion of the cranium and provides surface area for the attachment of facial muscles.

　The *superior temporal line* is continuous with the superior temporal line of the parietal bone.

　The **supraorbital margin** is a thickening of the frontal bone that helps protect the eye.

　The **lacrimal fossa** on the anterior and medial surface of the orbit is a shallow depression that marks

the location of the *lacrimal* (tear) *gland,* which lubricates the surface of the eye.

　The **frontal sinuses** are extremely variable in size and time of appearance. They generally appear after age 6, but some people never develop them. We will describe the frontal sinuses and other sinuses of the cranium and face in a later section.

Foramina: The **supraorbital foramen** provides passage for blood vessels that supply the eyebrow, eyelids, and frontal sinuses. In some cases this foramen is incomplete; the vessels then cross the orbital rim within a **supraorbital notch.**

Remarks: During development, the bones of the cranium form by the fusion of separate centers of ossification. At birth, the fusions have not been completed—two frontal bones articulate along the *metopic suture.* Although the suture generally disappears by age 8 as the bones fuse, the adult skull commonly retains traces of the suture line. This suture, or what remains of it, runs down the center of the frontal squama.

FOCUS: *continues* ➡

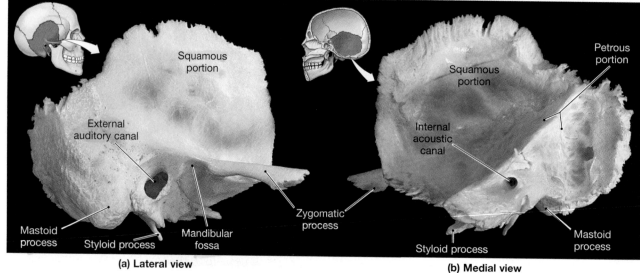

(a) Lateral view

(b) Medial view

(c) Mastoid air cells

•**FIGURE 7-7** **The Temporal Bones. (a,b)** The right temporal bone. **(c)** A cutaway view of the mastoid air cells.

The Temporal Bones (Figure 7-7a,b•)

General Functions: The **temporal bones** form part of both the lateral walls of the cranium and the *zygomatic arches;* form the only articulations with the mandible; surround and protect the sense organs of the inner ear; and are attachment sites for muscles that close the jaws and move the head.

Articulations: The temporal bones articulate with the zygomatic, sphenoid, parietal, and occipital bones and with the mandible (Figures 7-3 and 7-4•).

Regions/Landmarks: The **squamous portion,** or *squama,* of the temporal bone is the convex, irregular surface that borders the squamous suture.

The **zygomatic process,** inferior to the squamous portion, articulates with the *temporal process* of the zygomatic bone. Together these processes form the **zygomatic arch,** or cheekbone.

The **mandibular fossa** on the inferior surface marks the site of articulation with the mandible.

The **mastoid process** is an attachment site for muscles that rotate or extend the head. *Mastoid air cells* within the mastoid process are connected to the middle ear cavity (Figure 7-7c•). If pathogens invade the mastoid air cells, *mastoiditis* develops. Symptoms include severe earaches, fever, and swelling behind the ear.

The **styloid** (STĪ-loyd; *stylos,* pillar) **process,** near the base of the mastoid process, is attached to ligaments that support the hyoid bone and to the tendons of several muscles associated with the hyoid bone, the tongue, and the pharynx.

The **petrous portion** of the temporal bone, located on its internal surface, encloses the structures of the *inner ear*—sense organs that provide information about hearing and balance. The **auditory ossicles** are located in the *tympanic cavity,* or *middle ear,* a cavity within the petrous portion. These tiny bones—three on each side—transfer sound vibrations from the delicate *tympanic membrane,* or *eardrum,* to the inner ear. (We shall discuss each bone in Chapter 17.)

Foramina (Figure 7-3e•): The jugular foramen, between the temporal and occipital bones, provides passage for the internal jugular vein.

The **carotid canal** provides passage for the internal carotid artery, a major artery to the brain.

The **foramen lacerum** (LA-se-rum; *lacerare,* to tear) is a jagged slit extending between the sphenoid bone and the petrous portion of the temporal bone. It contains hyaline cartilage and small arteries that supply the inner surface of the cranium.

The **external auditory canal** on the lateral surface ends at the tympanic membrane (which disintegrates during the preparation of a dried skull).[1]

The **stylomastoid foramen** lies posterior to the base of the styloid process. The *facial nerve* passes through this foramen to control the facial muscles.

The **internal acoustic canal**[1] begins on the medial surface of the petrous portion of the temporal bone. It carries blood vessels and nerves to the inner ear, and the facial nerve to the stylomastoid foramen.

---

[1] The names for these passageways vary widely; the terms *auditory* and *acoustic* are used interchangeably, as are *canal* and *meatus.*

The Sphenoid Bone (Figure 7-8a,b•)

General Functions: The **sphenoid bone** forms part of the floor of the cranium; unites the cranial and facial bones; and acts as a cross-brace that strengthens the sides of the skull. Mucous secretions of the *sphenoidal sinuses* within this bone help clean the surfaces of the nasal cavities.

Articulations: The sphenoid bone articulates with the frontal, occipital, parietal, ethmoid, and temporal bones of the cranium and the palatine bones, zygomatic bones, maxillary bones, and vomer of the face (Figures 7-3c–e and 7-4•).

Regions/Landmarks: The shape of the sphenoid bone has been compared to a bat with its wings extended. Although this bone is relatively large, much of it is hidden by more-superficial bones.

The **body** forms the central axis of the sphenoid bone.

The **sella turcica** (TUR-si-kuh), or Turkish saddle, is a bony, saddle-shaped enclosure on the superior surface of the body.

The **hypophyseal** (hī-po-FIZ-ē-ul) **fossa** is the depression within the sella turcica. The *pituitary gland* occupies this fossa.

The **sphenoidal sinuses** are on either side of the body, inferior to the sella turcica.

The **lesser wings** extend horizontally anterior to the sella turcica.

The **greater wings** extend laterally from the body and form part of the cranial floor. A sharp *sphenoidal spine* lies at the posterior, lateral corner of each greater wing. Anteriorly, each greater wing contributes to the posterior wall of the orbit.

The **pterygoid** (TER-i-goyd; *pterygion,* wing) **processes** are vertical projections that originate on either side of the body. Each pterygoid process forms a pair of *pterygoid plates,* which are attachment sites for muscles that move the lower jaw and soft palate.

Foramina: The **optic canals** permit passage of the optic nerves from the eyes to the brain.

A **superior orbital fissure, foramen rotundum, foramen ovale** (ō-VAH-lē), and **foramen spinosum** penetrate each greater wing. These passages carry blood vessels and nerves to the orbit, face, and jaws, respectively.

7

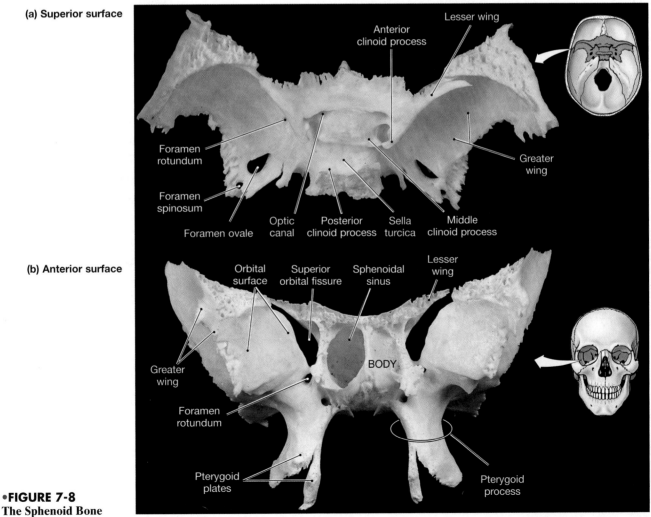

(a) Superior surface

Lesser wing
Anterior clinoid process
Foramen rotundum
Foramen spinosum
Greater wing
Foramen ovale
Optic canal
Posterior clinoid process
Sella turcica
Middle clinoid process

(b) Anterior surface

Orbital surface
Superior orbital fissure
Sphenoidal sinus
Lesser wing
Greater wing
BODY
Foramen rotundum
Pterygoid plates
Pterygoid process

•**FIGURE 7-8**
The Sphenoid Bone

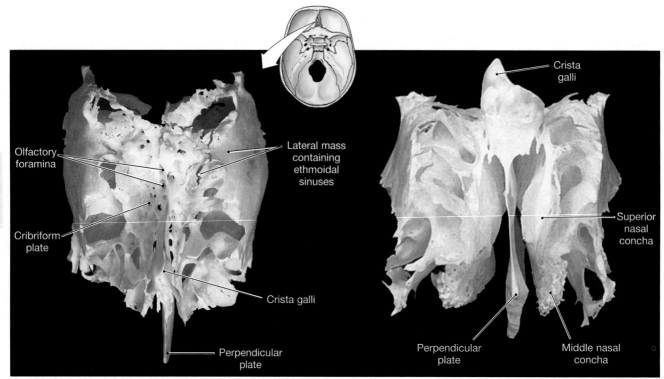

(a) Superior surface

(b) Posterior surface

•**FIGURE 7-9** The Ethmoid Bone

The Ethmoid Bone (Figure 7-9a,b•)

General Functions: The **ethmoid bone** forms the anteromedial floor of the cranium, the roof of the nasal cavity, and part of the nasal septum and medial orbital wall. Mucous secretions from the *ethmoidal sinuses* within this bone flush the surfaces of the nasal cavities.

Articulations: The ethmoid bone articulates with the frontal and sphenoid bones of the cranium and with the nasal, lacrimal, palatine, and maxillary bones and the inferior nasal conchae and vomer of the face (Figures 7-3c,d and 7-4•).

Regions/Landmarks: The ethmoid has three parts: (1) the cribriform plate, (2) the paired lateral masses, and (3) the perpendicular plate.

The **cribriform plate** (*cribrum*, sieve) forms the anteromedial floor of the cranium and the roof of the nasal cavity. The **crista galli** (*crista*, crest + *gallus*, chicken; cock's comb) is a bony ridge that projects superior to the cribriform plate. The *falx cerebri*, a membrane that stabilizes the position of the brain, attaches to this ridge.

The **lateral masses** contain the **ethmoidal sinuses,** or *ethmoidal air cells,* which open into the nasal cavity on each side. The **superior nasal conchae** (KONG-kē; sin-

gular, *concha,* a snail shell) and the **middle nasal conchae** are delicate projections of the lateral masses.

The **perpendicular plate** forms part of the nasal septum, along with the vomer and a piece of hyaline cartilage.

Foramina: The **olfactory foramina** in the cribriform plate permit passage of the olfactory nerves, which provide your sense of smell.

Remarks: *Olfactory* (smell) *receptors* are located in the epithelium that covers the inferior surfaces of the cribriform plate, the medial surfaces of the superior nasal conchae, and the superior portion of the perpendicular plate.

The nasal conchae break up the airflow in the nasal cavity, creating swirls, turbulence, and eddies that have three major functions: (1) The swirling throws any particles in the air against the sticky mucus that covers the walls of the nasal cavity; (2) the turbulence slows air movement, providing time for warming, humidification, and dust removal before the air reaches more-delicate portions of the respiratory tract; and (3) the eddies direct air toward the superior portion of the nasal cavity, adjacent to the cribriform plate, where the olfactory receptors are located.

FACIAL BONES

The Maxillary Bones (Figure 7-10a,b•)

General Functions: The **maxillary bones,** or *maxillae,* support the upper teeth and form the inferior orbital rim, the lateral margins of the external nares, the upper jaw, and most of the hard palate. The *maxillary sinuses* in these bones produce mucous secretions that flush the inferior surfaces of the nasal cavities. These are the largest facial bones, and the maxillary sinuses are the largest sinuses.

Articulations: The maxillary bones articulate with the frontal and ethmoid bones, with one another, and with all the other facial bones except the mandible (Figures 7-3c–e and 7-4a•).

Regions/Landmarks: The **orbital rim** protects the eye and other structures in the orbit.

The **alveolar processes** that border the mouth support the upper teeth.

The **palatine processes** form most of the **hard palate,** or bony roof of the mouth. One type of *cleft palate,* a developmental disorder, results when the maxillary bones fail to meet along the midline of the hard palate. **AM** *EMBRYOLOGY SUMMARY 6: The Development of the Skull*

The **maxillary sinuses** lighten the portion of the maxillary bone superior to the teeth.

The **nasolacrimal canal,** formed by the maxillary and lacrimal bones, protects the *lacrimal sac* and the *nasolacrimal duct,* which carries tears from the orbit to the nasal cavity.

Foramina: The **infraorbital foramen** marks the path of a major sensory nerve; that nerve reaches the brain by way of the foramen rotundum of the sphenoid bone.

The **inferior orbital fissure** (Fig. 7-3d•), which lies between the maxillary bone and the sphenoid bone, permits passage of cranial nerves and blood vessels.

The Palatine Bones (Figure 7-10b,c•)

General Functions: The **palatine bones** form the posterior portion of the hard palate and contribute to the floor of each orbit.

Articulations: The palatine bones articulate with one another, with the maxillary, sphenoid, and ethmoid bones, with the inferior nasal conchae, and with the vomer (Figures 7-3e and 7-4a•).

Regions/Landmarks: These bones are L-shaped. The **horizontal plate** forms the posterior part of the hard palate; the **perpendicular plate** extends from the horizontal plate to the orbital process. The **orbital process** forms part of the floor of the orbit. This process contains a small sinus that normally opens into the sphenoidal sinus.

Foramina: Small blood vessels and nerves supplying the roof of the mouth penetrate the lateral portion of the horizontal plate.

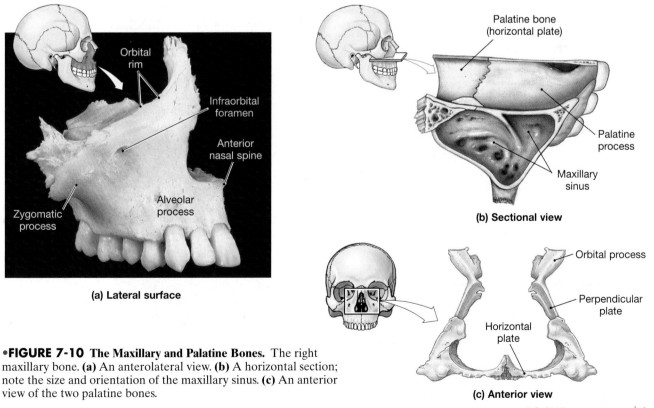

Orbital rim

Infraorbital foramen

Anterior nasal spine

Zygomatic process

Alveolar process

(a) Lateral surface

Palatine bone (horizontal plate)

Palatine process

Maxillary sinus

(b) Sectional view

Orbital process

Perpendicular plate

Horizontal plate

(c) Anterior view

•**FIGURE 7-10** **The Maxillary and Palatine Bones.** The right maxillary bone. **(a)** An anterolateral view. **(b)** A horizontal section; note the size and orientation of the maxillary sinus. **(c)** An anterior view of the two palatine bones.

The Nasal Bones (Figure 7-11•)

General Functions: The **nasal bones** support the superior portion of the bridge of the nose and form the superior border of the **external nares** (NA-rēz), the entrance to the nasal cavity.

Articulations: The paired nasal bones articulate with one another and with the frontal, ethmoid, and maxillary bones (Figures 7-3b–d and 7-4a•).

The Vomer (Figure 7-11•)

General Functions: The **vomer** forms the inferior portion of the bony nasal septum.

Articulations: The vomer articulates with the sphenoid, ethmoid, palatine, and maxillary bones and with the cartilaginous part of the nasal septum, which extends into the fleshy part of the nose (Figures 7-3d,e and 7-4a•).

The Inferior Nasal Conchae (Figure 7-11•)

General Functions: The **inferior nasal conchae** create turbulence in air passing through the nasal cavity and increase the epithelial surface area to promote warming and humidification of inhaled air.

Articulations: The inferior nasal conchae articulate with the ethmoid, maxillary, palatine, and lacrimal bones (Figure 7-3d•).

The Zygomatic Bones (Figure 7-11•)

General Functions: The **zygomatic bones** contribute to the rim and lateral wall of the orbit and form part of the zygomatic arch.

Articulations: The zygomatic bones articulate with the frontal, temporal, sphenoid, and maxillary bones (Figure 7-3b–e•).

Regions/Landmarks: The **temporal process** curves posteriorly to meet the zygomatic process of the temporal bone.

Foramina: The **zygomaticofacial foramen** on the anterior surface of each zygomatic bone carries a sensory nerve that innervates the cheek.

The Lacrimal Bones (Figure 7-11•)

General Functions: The **lacrimal bones** form part of the medial wall of the orbit.

Articulations: These are the smallest facial bones. The lacrimal bones articulate with the frontal, maxillary, and ethmoid bones (Figure 7-3c,d•).

Regions/Landmarks: The **lacrimal sulcus** is a groove along the anterior, lateral surface of the lacrimal bone. This sulcus leads to the nasolacrimal canal, which begins at the orbit and opens into the nasal cavity. As noted earlier, this canal is formed by the lacrimal bone and the maxillary bone.

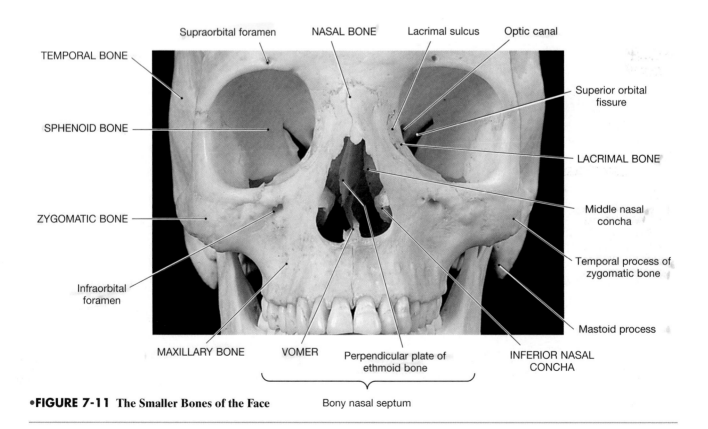

•**FIGURE 7-11** The Smaller Bones of the Face

The Mandible (Figure 7-12a,b•)

General Functions: The **mandible** forms the lower jaw.

Articulations: The mandible articulates with the mandibular fossae of the temporal bones (Figures 7-3c,e and 7-7a•).

Regions/Landmarks: The **body** of the mandible is the horizontal portion of that bone.

The **alveolar processes** support the lower teeth.

The *mental protuberance,* or chin, is the attachment site for several facial muscles.

A prominent depression on the medial surface marks the position of the *submandibular salivary gland.*

The *mylohyoid line* marks the insertion of the *mylohyoid muscle,* which supports the floor of the mouth.

The **ramus** of the mandible is the ascending part that begins at the *mandibular angle* on either side. On each ramus:

The **condylar process** articulates with the temporal bone at the *temporomandibular joint.*

The **coronoid** (ko-RŌ-noyd) **process** is the insertion point for the *temporalis muscle,* a powerful muscle that closes the jaws.

The **mandibular notch** is the depression that separates the condylar and coronoid processes.

Foramina: The **mental foramina** (*mentalis,* chin) are openings for nerves that carry sensory information from the lips and chin to the brain.

The **mandibular foramen** is the entrance to the *mandibular canal,* a passageway for blood vessels and nerves that service the lower teeth. Before they work on the lower teeth, dentists typically anesthetize the sensory nerve that enters this canal.

The Hyoid Bone (Figure 7-12c•)

General Functions: The **hyoid bone** supports the larynx and is the attachment site for muscles of the larynx, pharynx, and tongue.

Articulations: *Stylohyoid ligaments* connect the *lesser horn* to the styloid processes of the temporal bones.

Regions/Processes: The **body** of the hyoid is an attachment site for muscles of larynx, tongue, and pharynx.

The **greater horns,** or *greater cornua* (*cornua,* horns; singular, *cornu*), help support the larynx and are attached to muscles that move the tongue.

The **lesser horns,** or *lesser cornua,* are attached to the stylohyoid ligaments; from these ligaments, the hyoid and larynx hang beneath the skull like a child's swing from the limb of a tree.

7

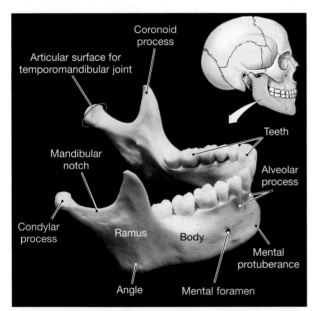

Coronoid process

Articular surface for temporomandibular joint

Teeth

Mandibular notch

Alveolar process

Condylar process

Ramus

Body

Mental protuberance

Angle

Mental foramen

(a) Lateral view

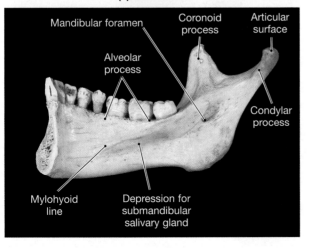

Mandibular foramen

Coronoid process

Articular surface

Alveolar process

Condylar process

Mylohyoid line

Depression for submandibular salivary gland

(b) Medial view

•FIGURE 7-12 The Mandible and Hyoid Bone. (a) A lateral view of the mandible. **(b)** A medial view of the right mandible. **(c)** An anterior view of the hyoid bone.

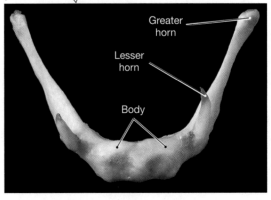

Greater horn

Lesser horn

Body

(c) Anterior view

SUMMARY: THE FORAMINA AND FISSURES OF THE SKULL

Table 7-1 summarizes information about the foramina and fissures introduced thus far. This reference will be especially important to you in later chapters when you deal with the nervous and cardiovascular systems.

TMJ Syndrome

The *temporomandibular joint (TMJ)*, between each temporal bone and the mandible, is quite mobile; it allows your jaw to move while you chew or talk. The disadvantage of such mobility is that your jaw can easily be dislocated by forceful forward or lateral displacement. The connective tissue sheath, or *capsule*, that surrounds the joint is relatively loose, and the opposing bone surfaces are separated by a fibrocartilage pad. In **TMJ syndrome**, or *myofacial pain syndrome*, the mandible is pulled slightly out of alignment, generally by spasms in one of the jaw muscles. The individual experiences (1) facial pain that radiates around the ear on the affected side and (2) an inability to open the mouth fully.

TMJ syndrome is a repeating cycle of muscle spasm → misalignment → pain → muscle spasm. It has been linked to involuntary behaviors, such as grinding of the teeth during sleep, and to emotional stress. Treatment focuses on breaking the cycle of muscle spasm and pain and, when necessary, providing emotional support. The application of heat to the affected joint, coupled with the use of anti-inflammatory drugs, local anesthetics, or both, may help. If teeth grinding is suspected, special mouth guards may be worn at night.

☑ In which bone is the foramen magnum located?

☑ Tomás suffers a blow to the skull that fractures the right superior lateral surface of his cranium. Which bone is fractured?

☑ The internal jugular veins are important blood vessels of the head. Between which bones do these blood vessels pass?

☑ Which bone contains the depression called the sella turcica? What is located in this depression?

☑ Which of the five senses would be affected if the cribriform plate of the ethmoid bone failed to form?

THE ORBITAL AND NASAL COMPLEXES

Together, certain cranial bones and facial bones form the *orbital complex*, which surrounds each eye, and the *nasal complex*, which surrounds the nasal cavities.

The **orbits** are the bony recesses that contain the eyes. Each orbit is formed by the seven bones of the **orbital complex** (Figure 7-13•). The frontal bone forms the roof, and the maxillary bone provides most of the orbital floor. The orbital rim and the first portion of the medial wall are formed by the maxillary bone, the lacrimal bone, and the lateral mass of the ethmoid bone. The lateral mass articulates with the sphenoid bone and a small process of the palatine bone. Several prominent foramina and fissures penetrate the sphenoid or lie between it and the maxillary bone. Laterally, the sphenoid bone and maxillary bone articulate with the zygomatic bone, which forms the lateral wall and rim of the orbit.

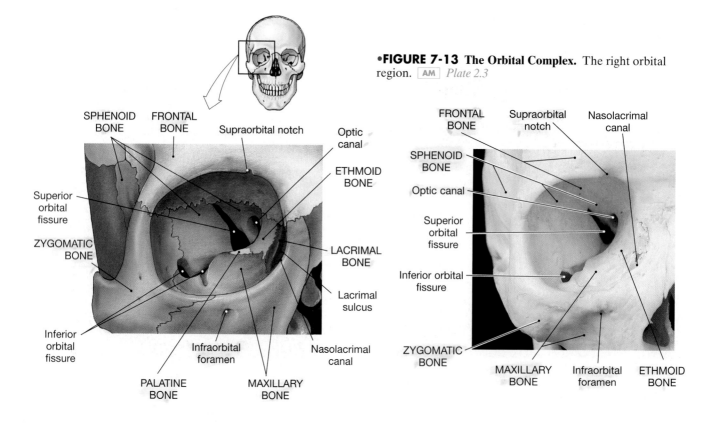

•**FIGURE 7-13** **The Orbital Complex.** The right orbital region. [AM] *Plate 2.3*

TABLE 7-1   A Key to the Foramina and Fissures of the Skull

Bone	Foramen/Fissure	Major Structures Using Passageway	
		Neural Tissue*	**Vessels and Other Structures**
Occipital bone	Foramen magnum	Medulla oblongata (most caudal portion of brain) and accessory nerve (XI), which provides motor control over several neck and back muscles	Vertebral arteries to brain; supporting membranes around central nervous system
	Hypoglossal canal	Hypoglossal nerve (XII) provides motor control to muscles of the tongue	
With temporal bone	Jugular foramen	Glossopharyngeal nerve (IX), vagus nerve (X), accessory nerve (XI). Nerve IX provides taste sensation; X is important for visceral functions; XI innervates important muscles of the back and neck	Internal jugular vein; important vein returning blood from brain to heart
Frontal bone	Supraorbital foramen (or notch)	Supraorbital nerve, sensory branch of ophthalmic nerve, innervating the eyebrow, eyelid, and frontal sinus	Supraorbital artery delivers blood to same region
Lacrimal bone	Lacrimal sulcus, nasolacrimal canal (with maxillary bone)		Tear duct; drains into nasal cavity
Temporal bone	Stylomastoid foramen	Facial nerve (VII) provides motor control of facial muscles	
	Carotid canal		Internal carotid artery; supplying blood to brain
	External auditory canal		Air in canal conducts sound to eardrum
	Internal acoustic canal	Vestibulocochlear nerve (VIII) from sense organs for hearing and balance. Facial nerve (VII) enters here, exits at stylomastoid foramen	Internal acoustic artery to inner ear
Sphenoid bone	Optic canal	Optic nerve (II) brings information from the eye to the brain	Ophthalmic artery brings blood into orbit
	Superior orbital fissure	Oculomotor nerve (III), trochlear nerve (IV), ophthalmic branch of trigeminal nerve (V), abducens nerve (VI). Ophthalmic nerve provides sensory information about eye and orbit; other nerves control muscles that move the eye	Ophthalmic vein returns blood from orbit
	Foramen rotundum	Maxillary branch of trigeminal nerve (V) provides sensation from the face	
	Foramen ovale	Mandibular branch of trigeminal nerve (V) controls the muscles that move the lower jaw and provides sensory information from that area	
	Foramen spinosum		Vessels to membranes around central nervous system
With temporal and occipital bones	Foramen lacerum		Internal carotid artery leaves carotid canal, enters cranium through foramen lacerum
With maxillary bone	Inferior orbital fissure	Maxillary branch of trigeminal nerve (V). *See Foramen rotundum*	
Ethmoid bone	Cribriform plate	Olfactory nerve (I) provides sense of smell	
Maxillary bone	Infraorbital foramen	Infraorbital nerve, maxillary branch of trigeminal nerve (V) from the inferior orbital fissure to face	Infraorbital artery with same distribution
Mandible	Mental foramen	Mental nerve, sensory branch of the mandibular nerve, provides sensation from the chin and lips	Mental vessels to chin and lips
	Mandibular foramen	Inferior alveolar nerve, sensory branch of mandibular nerve, provides sensation from the gums, teeth	Inferior alveolar vessels supply same region
Zygomatic bone	Zygomaticofacial foramen	Zygomaticofacial nerve, sensory branch of mandibular nerve to cheek	

*Twelve pairs of cranial nerves exist, numbered I–XII. Their functions and distribution are detailed in Chapter 14, pp. 464–474.

7

The **nasal complex** includes the bones that enclose the nasal cavities and the **paranasal sinuses,** air-filled chambers connected to the nasal cavities. The frontal, sphenoid, and ethmoid bones form the superior wall of the nasal cavities. The lateral walls are formed by the maxillary and lacrimal bones, the ethmoid bone (the superior and middle nasal conchae), and the inferior nasal conchae (Figure 7-14•). Much of the anterior margin of the nasal cavity is formed by the soft tissues of the nose, but the bridge of the nose is supported by the maxillary and nasal bones.

Paranasal Sinuses

The frontal, sphenoid, ethmoid, palatine, and maxillary bones contain the paranasal sinuses. Figure 7-14a• shows the location of the frontal and sphenoidal sinuses. The ethmoidal and maxillary sinuses are shown in Figure 7-14b•. (The tiny palatine sinuses, not shown, generally open into the sphenoidal sinus.) The paranasal sinuses lighten the skull bones and provide an extensive area of mucous epithelium. The mucous secretions are released into the nasal cavities. The ciliated epithelium passes the mucus back toward the throat, where it is eventually swallowed. Incoming air is humidified and warmed as it flows across this thick carpet of mucus. Foreign particulate matter, such as dust or microorganisms, becomes trapped in this sticky mucus and is then swallowed.

This mechanism helps protect more-delicate portions of the respiratory tract.

Sinus Problems and Septal Defects

Flushing the nasal epithelium with mucus produced in the paranasal sinuses often succeeds in removing a mild irritant. But a viral or bacterial infection produces an inflammation of the mucous membrane of the nasal cavity. As swelling occurs, the communicating passageways narrow. Drainage of mucus slows, congestion increases, and the individual experiences headaches and a feeling of pressure within the facial bones. This condition of sinus inflammation and congestion is called **sinusitis.** The maxillary sinuses are commonly involved. Because gravity does little to assist mucus drainage from these sinuses, the effectiveness of the flushing action is reduced and pressure on the sinus walls typically increases.

Temporary sinus problems may accompany allergies or the exposure of the mucous epithelium to chemical irritants or invading microorganisms. Chronic sinusitis may occur as the result of a **deviated** (nasal) **septum.** In this condition, the nasal septum has a bend in it, generally at the junction between the bony and cartilaginous regions. Septal deviation often blocks the drainage of one or more sinuses, producing chronic cycles of infection and inflammation. A deviated septum can result from developmental abnormalities or from injuries to the nose. The condition can usually be corrected or improved by surgery.

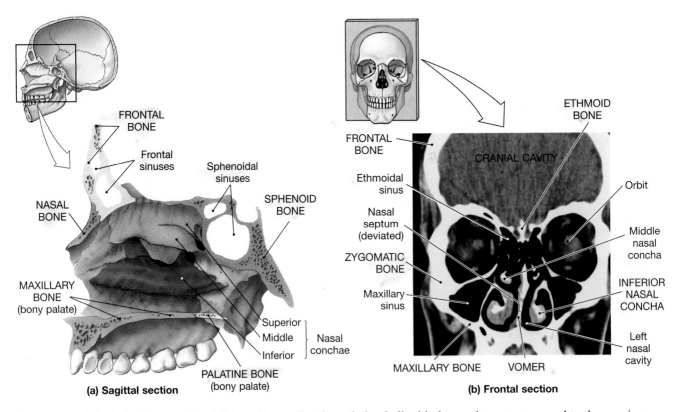

•**FIGURE 7-14 The Nasal Complex. (a)** A sagittal section through the skull, with the nasal septum removed to show major features of the wall of the right nasal cavity. **(b)** An MRI scan showing a frontal section through the ethmoidal and maxillary sinuses, part of the paranasal sinuses. The sphenoidal sinuses are visible in (a). AM *Scans 1, 2*

THE SKULLS OF INFANTS AND CHILDREN

Many different centers of ossification are involved in the formation of the skull. As development proceeds, the centers fuse, producing a smaller number of composite bones. For example, the sphenoid bone begins as 14 separate ossification centers. At birth, fusion has not been completed: There are two frontal bones, four occipital bones, and several sphenoid and temporal elements.

The skull organizes around the developing brain. As the time of birth approaches, the brain enlarges rapidly. Although the bones of the skull are also growing, they fail to keep pace. At birth, the cranial bones are connected by areas of fibrous connective tissue. These connections are quite flexible, and the skull can be distorted without damage. Such distortion normally occurs during delivery and eases the passage of the infant through the birth canal. The largest fibrous areas between the cranial bones are known as **fontanels** (fon-tuh-NELZ; sometimes spelled *fontanelles*) (Figure 7-15●):

- The *anterior fontanel* is the largest fontanel. It lies at the intersection of the frontal, sagittal, and coronal sutures.

- The *occipital fontanel* is at the junction between the lambdoidal and sagittal sutures.

- The *sphenoidal fontanels* are at the junctions between the squamous sutures and the coronal suture.

- The *mastoid fontanels* are at the junctions between the squamous sutures and the lambdoidal suture.

The occipital, sphenoidal, and mastoid fontanels disappear within a month or two after birth. The anterior fontanel generally persists until a child is nearly 2 years old. Even after the fontanels disappear, the bones of the skull remain separated by fibrous connections.

The skulls of infants and adults differ in terms of the shape and structure of cranial elements. This difference accounts for variations in proportions as well as in size. The most significant growth in the skull occurs before age 5, because at that time the brain stops growing and the cranial sutures develop. As a result, the cranium of a young child, when compared with the skull as a whole, is relatively larger than that of an adult. The growth of the cranium is generally coordinated with the expansion of the brain. If one or more sutures form before the brain stops growing, the skull will be abnormal in shape, size, or both.

Craniostenosis and Microcephaly

Unusual distortions of the skull result from *craniostenosis* (krā-nē-ō-sten-Ō-sis; *stenosis*, narrowing), the premature closure of one or more fontanels. As the brain continues to enlarge, the rest of the skull distorts to accommodate it. A long and narrow head is produced by early closure of the sagittal suture, whereas a very broad skull results if the coronal suture forms prematurely. Early closure of all cranial sutures restricts the development of the brain, and surgery must be performed to prevent brain damage. If brain enlargement stops due to genetic or developmental abnormalities, however, skull growth ceases as well. This condition, which results in a very undersized head, is called *microcephaly* (mī-krō-SEF-uh-lē).

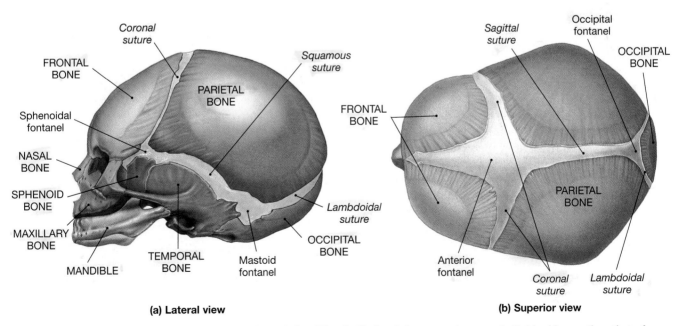

(a) Lateral view

(b) Superior view

●**FIGURE 7-15 The Skull of an Infant. (a)** A lateral view. The skull of an infant contains more individual bones than that of an adult. Many of the bones eventually fuse; thus the adult skull has fewer bones. The flat bones of the skull are separated by areas of fibrous connective tissue, allowing for cranial expansion and the distortion of the skull during birth. The large fibrous areas are called fontanels. By about age 4, these areas will disappear. **(b)** A superior view.

THE VERTEBRAL COLUMN

The rest of the axial skeleton is subdivided on the basis of vertebral structure. The adult **vertebral column,** or **spine,** consists of 26 bones: the **vertebrae** (24), the **sacrum,** and the **coccyx** (KOK-siks), or tailbone. The vertebrae provide a column of support, bearing the weight of the head, neck, and trunk and ultimately transferring the weight to the appendicular skeleton of the lower limbs. The vertebrae also protect the spinal cord and help maintain an upright body position, as in sitting or standing.

The vertebral column is divided into cervical, thoracic, lumbar, sacral, and coccygeal regions (Figure 7-16●). Seven **cervical vertebrae** constitute the neck and extend inferiorly to the trunk. Twelve **thoracic vertebrae** form the superior portion of the back; each articulates with one or more pairs of ribs. Five **lumbar vertebrae** form the inferior portion of the back; the fifth articulates with the sacrum, which in turn articulates with the coccyx. The cervical, thoracic, and lumbar regions consist of individual vertebrae. During development, the sacrum originates as a group of five vertebrae, and the coccyx begins as three to five very small vertebrae. In general, the vertebrae of the sacrum are completely fused by age 25–30. Ossification of the distal coccygeal vertebrae is not complete before puberty, and thereafter fusion occurs at a variable pace. The total length of the vertebral column of an adult averages 71 cm (28 in.). [AM] *EMBRYOLOGY SUMMARY 7: The Development of the Vertebral Column*

SPINAL CURVATURE

The vertebral column is not straight and rigid. A lateral view of this column shows four **spinal curves** (Figure 7-16●): the (1) **cervical curve,** (2) **thoracic curve,** (3) **lumbar curve,** and (4) **sacral curve.**

The thoracic and sacral curves are called **primary curves,** because they appear late in fetal development, or **accommodation curves,** because they accommodate the thoracic and abdominopelvic viscera. Only the primary curves are present in the vertebral column of a newborn. The lumbar and cervical curves, known as **secondary curves,** do not appear until several months after birth. These are also called **compensation curves,** because they help shift the trunk weight over the lower limbs. The cervical curve develops as the infant learns to balance the head upright. The lumbar curve develops with the ability to stand. Both compensations become accentuated as the toddler learns to walk and run. All four curves are fully developed by age 10.

Several abnormal distortions of spinal curvature may appear during childhood and adolescence. Examples include *kyphosis* (kī-FŌ-sis), exaggerated thoracic curvature; *lordosis* (lor-DŌ-sis), exaggerated lumbar curvature; and *scoliosis* (skō-lē-Ō-sis), an ab-

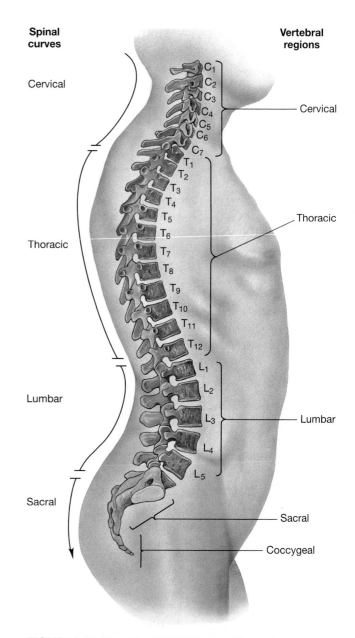

Spinal curves / Vertebral regions

Cervical — Cervical

Thoracic — Thoracic

Lumbar — Lumbar

Sacral — Sacral

— Coccygeal

●**FIGURE 7-16** **The Vertebral Column.** The major divisions of the adult vertebral column; notice the four spinal curves. [AM] *Scan 3*

normal lateral curvature. [AM] *Kyphosis, Lordosis, and Scoliosis*

When you stand, the weight of your body must be transmitted through the vertebral column to the hips and ultimately to the lower limbs. Yet most of your body weight lies anterior to the vertebral column. The various curves bring that weight in line with the body axis. Consider what you do automatically when standing with a heavy object hugged to your chest. You avoid toppling forward by exaggerating the lumbar curve and moving the weight back toward the body axis. This posture can lead to discomfort at the base of the spinal column. For example, many women in the last 3 months of pregnancy develop chronic back pain from the changes in lumbar curvature that must adjust for the increasing weight of the fetus. African and South American natives often balance heavy objects on their head, as seen in the chapter-opening photograph on page 193. This practice increases the load on the vertebral column, but the spinal curves are not affected because the weight is aligned with the axis of the spine.

VERTEBRAL ANATOMY

Each vertebra consists of three basic parts: (1) a *vertebral body,* (2) a *vertebral arch,* and (3) *articular processes* (Figure 7-17•).

The Vertebral Body

The **vertebral body,** or *centrum* (plural, *centra*), is the part of a vertebra that transfers weight along the axis of the vertebral column. The bodies of adjacent vertebrae are interconnected by ligaments but are separated by pads of fibrocartilage, the **intervertebral discs.**

The Vertebral Arch

The **vertebral arch** forms the posterior margin of each **vertebral foramen.** Together, the vertebral foramina of successive vertebrae form the **vertebral canal,** which encloses the spinal cord. The vertebral arch has walls, called **pedicles** (PE-di-kulz), and a roof, formed by flat layers called **laminae** (LA-mi-nē; singular, *lamina,* a

(a) Lateral view

(b) Inferior view

(c) Posterior view

(d) Lateral view

•**FIGURE 7-17** **Vertebral Anatomy.** The anatomy of a typical vertebra and the arrangement of articulations between vertebrae. **(a)** A lateral and slightly inferior view of a vertebra. **(b)** An inferior view of a vertebra. **(c)** A posterior view of three articulated vertebrae. **(d)** A lateral and sectional view of three articulated vertebrae.

thin plate). The pedicles arise along the posterior and lateral margins of the body. The laminae on either side extend dorsally and medially to complete the roof.

In the condition called *spina bifida* (SPĪ-nuh BI-fi-duh), the vertebral laminae fail to unite during development. The vertebral arch is incomplete, and the membranes, or *meninges,* that line the dorsal body cavity bulge outward. Mild cases involving the sacral and lumbar regions may pass unnoticed; in severe cases, the entire spinal column and skull are affected.

A **spinous process** projects posteriorly from the point where the vertebral laminae fuse to complete the vertebral arch. You can see—and feel—the spinous processes through the skin of the back when the spine is flexed. **Transverse processes** project laterally or dorsolaterally on both sides from the point where the laminae join the pedicles. These processes are sites of muscle attachment, and they may also articulate with the ribs.

The Articular Processes

Like the transverse processes, the **articular processes** arise at the junction between the pedicles and laminae. A **superior** and an **inferior articular process** lie on each side of the vertebra.

Vertebral Articulation

The inferior articular processes of one vertebra articulate with the superior articular processes of the next vertebra. Each articular process has a smooth concave surface called an **articular facet.** The superior processes have articular facets on their dorsal surfaces, whereas the inferior processes articulate along their ventral surfaces.

Adjacent vertebral bodies are separated by intervertebral discs, and gaps separate the pedicles of successive vertebrae. These gaps, called **intervertebral foramina,** permit the passage of nerves running to or from the enclosed spinal cord.

☑ Why does the vertebral column of an adult have fewer vertebrae than that of a newborn?

☑ What is the importance of the secondary curves of the spine?

☑ When you run your finger along a person's spine, what part of the vertebrae are you feeling just beneath the skin?

VERTEBRAL REGIONS

When referring to a specific vertebra, we use a capital letter to indicate the vertebral region: C, T, L, S, and Co indicate the cervical, thoracic, lumbar, sacral, and coccygeal regions, respectively. In addition, we use a subscript number to indicate the relative position of the vertebra within that region, with 1 indicating the vertebra closest to the skull. For example, C_3 is the third cervical vertebra, with C_1 in contact with the skull; L_4 is the fourth lumbar vertebra, with L_1 in contact with the last thoracic vertebra (Figure 7-16, p. 212●). We shall use this shorthand throughout the text.

Although each vertebra bears characteristic markings and articulations, we will focus on the general characteristics of each region and on how regional variations determine the vertebral group's function.

Cervical Vertebrae

The seven cervical vertebrae (Figure 7-18a●) are the smallest of the vertebrae. They extend from the occipital bone of the skull to the thorax. The body of a cervical vertebra is small compared with the size of the vertebral foramen (Figure 7-18b●). At this level, the spinal cord still contains most of the axons that connect the brain to the rest of the body. As you proceed caudally along the vertebral canal, the diameter of the spinal cord decreases, and so does the diameter of the vertebral arch. However, cervical vertebrae support only the weight of the head, so the vertebral body can be relatively small and light. As you continue toward the sacrum, the loading increases and the vertebral bodies gradually enlarge.

Compared with the cervical vertebrae, your head is relatively massive. It sits atop the cervical vertebrae like a soup bowl on the tip of a finger. With this arrangement, small muscles can produce significant effects by tipping the balance one way or another. But if you suddenly change position, as in a fall or during rapid acceleration (a jet takeoff) or deceleration (a car crash), the balancing muscles are not strong enough to stabilize the head. A dangerous partial or complete dislocation of the cervical vertebrae can result, with injury to muscles and ligaments and potential injury to the spinal cord. The term **whiplash** is used to describe such an injury, because the movement of the head resembles the cracking of a whip.

In a cervical vertebra, the superior surface of the body is concave from side to side, and it slopes, with the anterior edge inferior to the posterior edge (Figure 7-18c●). The spinous process is relatively stumpy, generally shorter than the diameter of the vertebral foramen. The tip of each process other than that of C_7 bears a prominent notch (Figure 7-18b●). A notched spinous process is said to be **bifid** (BĪ-fid; *bifidus,* cut into two parts).

Laterally, the transverse processes are fused to the **costal processes,** which originate near the ventrolateral portion of the body. The costal and transverse processes encircle prominent, round, **transverse foramina.** These passageways protect the *vertebral arteries* and *vertebral veins,* important blood vessels that service the brain.

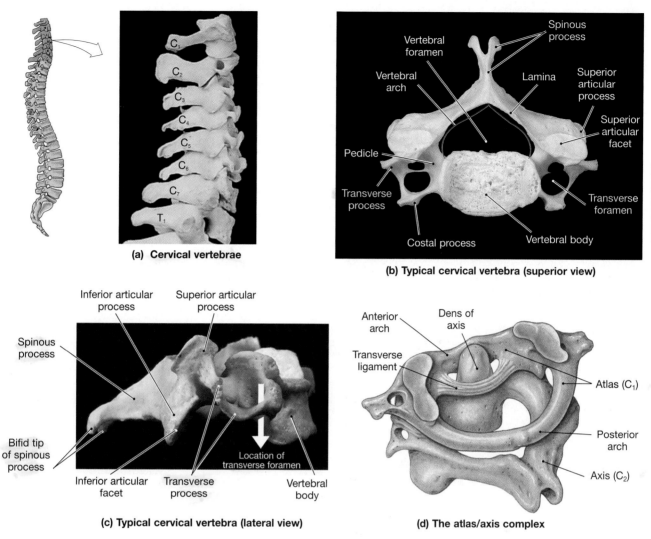

(a) Cervical vertebrae

(b) Typical cervical vertebra (superior view)

(c) Typical cervical vertebra (lateral view)

(d) The atlas/axis complex

•FIGURE 7-18 The Cervical Vertebrae. (a) A lateral view of the cervical vertebrae, C_1–C_7. **(b)** A superior view of a typical cervical vertebra (C_3–C_6). Notice the characteristic features listed in Table 7-2. **(c)** A lateral view of the same vertebra. **(d)** The atlas (C_1) and axis (C_2) AM *Scan 3*

This description would be adequate to identify all but the first two cervical vertebrae. When cervical vertebrae C_3–C_7 articulate, their interlocking bodies permit more flexibility than do those of other regions. The first two cervical vertebrae are unique, and the seventh is modified. Table 7-2 summarizes the features of cervical vertebrae.

The Atlas (C_1)   The **atlas,** cervical vertebra C_1, holds up the head, articulating with the occipital condyles of the skull (Figure 7-18d•). It is named after Atlas, who, according to Greek myth, holds the world on his shoulders. The articulation between the occipital condyles and the atlas is a joint that permits you to nod (such as when you indicate "yes"). The atlas can easily be distinguished from other vertebrae by (1) the lack of a body and spinous process and (2) the presence of a large, round vertebral foramen bounded by **anterior** and **posterior arches.**

The atlas articulates with the second cervical vertebra, the *axis.* This articulation permits rotation (such as when you shake your head to indicate "no").

The Axis (C_2)   During development, the body of the atlas fuses to the body of the second cervical vertebra, called the **axis** (C_2) (Figure 7-18d•). This fusion creates the prominent **dens** (DENZ; *dens,* tooth), or *odontoid* (ō-DON-toyd; *odontos,* tooth) *process,* of the axis. A transverse ligament binds the dens to the inner surface of the atlas, forming a pivot for rotation of the atlas and skull. Important muscles controlling the position of the head and neck attach to the especially robust spinous process of the axis.

In children, the fusion between the dens and axis is incomplete. Impacts or even severe shaking can cause dislocation of the dens and severe damage to the spinal cord. In adults, a blow to the base of the skull can be equally dangerous, because a dislocation of the

TABLE 7-2 Regional Differences in Vertebral Structure and Function

Feature	Type (Number)		
	Cervical Vertebrae (7)	Thoracic Vertebrae (12)	Lumbar Vertebrae (5)
Location	Neck	Chest	Inferior portion of back
Body	Small, oval, curved faces	Medium, heart-shaped, flat faces; facets for rib articulations	Massive, oval, flat faces
Vertebral foramen	Large	Smaller	Smallest
Spinous process	Long; split tip; points inferiorly	Long, slender; not split; points inferiorly	Blunt, broad, points posteriorly
Transverse process	Has transverse foramen	All but two (T_{11}, T_{12}) have facets for rib articulations	Short; no articular facets or transverse foramen
Functions	Support skull, stabilize relative positions of brain and spinal cord, and allow controlled head movement	Support weight of head, neck, upper limbs, chest; articulate with ribs to allow changes in volume of thoracic cage	Support weight of head, neck, upper limbs, and trunk

axis–atlas joint can force the dens into the base of the brain, with fatal results.

The Vertebra Prominens (C₇)   The transition from one vertebral region to another is not abrupt, and the last vertebra of one region generally resembles the first vertebra of the next. The **vertebra prominens,** or seventh cervical vertebra (C_7), has a long, slender spinous process that ends in a broad tubercle that you can feel through the skin at the base of the neck. This vertebra is the interface between the cervical curve, which arches anteriorly, and the thoracic curve, which arches posteriorly (see Figures 7-16 and 7-19a•). The transverse processes are large, providing additional surface area for muscle attachment. The **ligamentum nuchae** (li-guh-MEN-tum NOO-kē; *nucha,* nape), a large elastic ligament, begins at the vertebra prominens and extends to an insertion along the occipital crest of the skull. Along the way, it attaches to the spinous processes of the other cervical vertebrae. When your head is upright, this ligament acts like the string on a bow, maintaining the cervical curvature without muscular effort. If you have bent your neck forward, the elas-

ticity in this ligament helps return your head to an upright position.

Thoracic Vertebrae

There are 12 thoracic vertebrae. A typical thoracic vertebra (Figure 7-19b•) has a distinctive heart-shaped body that is more massive than that of a cervical vertebra. The vertebral foramen is relatively smaller, and the long, slender spinous process projects posteriorly and inferiorly. The spinous processes of T_{10}, T_{11}, and T_{12} increasingly resemble those of the lumbar series, as the transition between the thoracic and lumbar curves approaches. Because the inferior thoracic and lumbar vertebrae carry so much weight, the transition between the thoracic and lumbar curves is difficult to stabilize. As a result, compression fractures or compression–dislocation fractures after a hard fall tend to involve the last thoracic and first two lumbar vertebrae.

Each thoracic vertebra articulates with ribs along the dorsolateral surfaces of the body. The location and structure of the articulations vary somewhat from vertebra to vertebra (Figure 7-19a•). Rib pairs 2

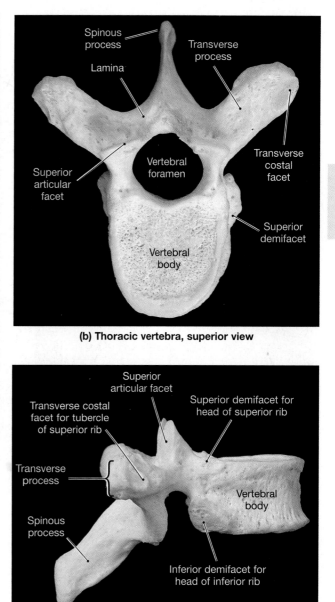

7

(b) Thoracic vertebra, superior view

(a) Thoracic vertebrae, lateral view

•**FIGURE 7-19** **The Thoracic Vertebrae.** **(a)** A lateral view of the thoracic region of the vertebral column. The vertebra prominens (C_7) resembles T_1 but lacks facets for rib articulation. Vertebra T_{12} resembles the first lumbar vertebra (L_1) but has a facet for rib articulation. **(b)** Thoracic vertebra, superior view. **(c)** Thoracic vertebra, lateral view. Notice the characteristic features listed in Table 7-2.

(c) Thoracic vertebra, lateral view

through 8 originate between adjacent vertebrae, so vertebrae T_2–T_8 have superior and inferior **demifacets** (*partial facets*) on each side. The first rib originates at the body of T_1. That vertebra contacts a **whole facet**— a complete articular surface on the body of a single vertebra. Vertebra T_1 also has an inferior demifacet on each side. Vertebra T_9 has only a superior demifacet on each side, whereas T_{10}, T_{11}, and T_{12} have a single whole facet on each side.

The transverse processes of vertebrae T_1–T_{10} are relatively thick, and they contain **transverse costal facets** for rib articulation (Figure 7-19b,c•). Thus, rib pairs 1 through 10 contact their vertebrae at two points—at a whole facet or demifacet and at a transverse costal facet. Table 7-2 summarizes the features of the thoracic vertebrae.

Lumbar Vertebrae

The five lumbar vertebrae are the largest vertebrae. The body of a typical lumbar vertebra (Figure 7-20•) is thicker than that of a thoracic vertebra, and the superior and inferior surfaces are oval rather than heart-shaped. Other noteworthy features include the following: (1) Lumbar vertebrae have neither whole facets nor demifacets; (2) the slender transverse processes, which lack costal facets, project dorsolaterally; (3) the vertebral foramen is triangular; (4) the stumpy spinous processes project dorsally; (5) the superior articular processes face medially ("up and in"); and (6) the inferior articular processes face laterally ("down and out").

The lumbar vertebrae bear the most weight. Their massive spinous processes provide surface area for the

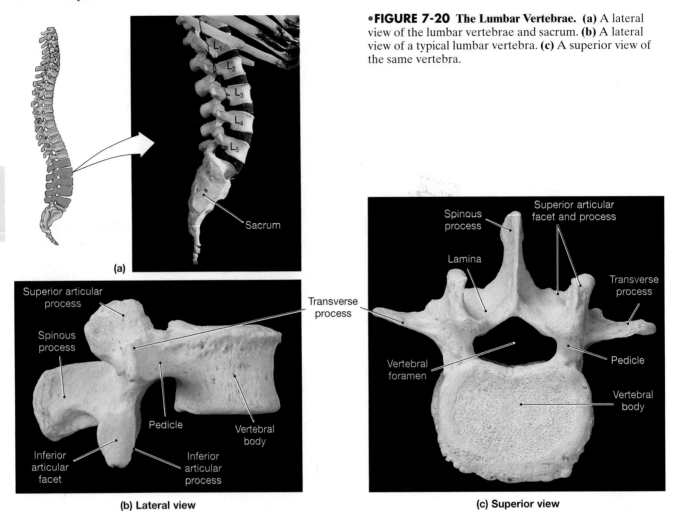

•FIGURE 7-20 The Lumbar Vertebrae. **(a)** A lateral view of the lumbar vertebrae and sacrum. **(b)** A lateral view of a typical lumbar vertebra. **(c)** A superior view of the same vertebra.

(a)

(b) Lateral view

(c) Superior view

attachment of lower back muscles that reinforce or adjust the lumbar curve. Table 7-2 summarizes the characteristics of lumbar vertebrae.

The Sacrum

The sacrum consists of the fused components of five sacral vertebrae. These vertebrae begin fusing shortly after puberty and in general are completely fused at age 25–30. The sacrum protects the reproductive, digestive, and urinary organs and, via paired articulations, attaches the axial skeleton to the pelvic girdle of the appendicular skeleton (Figure 7-1b•, p. 195).

The broad posterior surface of the sacrum (Figure 7-21a•) provides an extensive area for the attachment of muscles, especially those that move the thigh. The superior articular processes of the first of the five fused sacral vertebrae articulate with the last lumbar vertebra. The **sacral canal** is a passageway that begins between these articular processes and extends the length of the sacrum. Nerves and membranes that line the vertebral canal in the spinal cord continue into the sacral canal.

The **median sacral crest** is a ridge formed by the fused spinous processes of the sacral vertebrae. The laminae of the fifth sacral vertebra fail to contact one another at the midline; they form the **sacral cornua** (KORN-ū-uh; singular, *cornu*). These ridges form the margins of the **sacral hiatus** (hī-Ā-tus), the opening at the inferior end of the sacral canal. This opening is covered by connective tissues. Four pairs of **sacral foramina** open on either side of the median sacral crest. The intervertebral foramina of the fused sacral vertebrae open into these passageways. The **lateral sacral crest** on each side is a ridge that represents the fused transverse processes of the sacral vertebrae. The sacral crests provide surface area for the attachment of muscles of the lower back and hip.

The sacrum is curved, with a convex posterior surface (Figure 7-21b•). The degree of curvature is more pronounced in males than in females. The **auricular surface** is a thickened, flattened area lateral to the superior portion of the lateral sacral crest. The auricular surface is the site of articulation with the pelvic girdle (the *sacroiliac joint*). The **sacral tuberosity** is a rough-

(a) Posterior surface (b) Lateral surface (c) Anterior surface

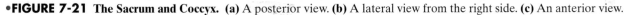

•**FIGURE 7-21** **The Sacrum and Coccyx.** **(a)** A posterior view. **(b)** A lateral view from the right side. **(c)** An anterior view.

ened area between the lateral sacral crest and the auricular surface. It marks the attachment site of ligaments that stabilize the sacroiliac joint.

The subdivisions of the sacrum are most clearly seen in anterior view (Figure 7-21c•). The narrow, inferior portion is the sacral **apex,** whereas the broad superior surface forms the **base.** The **sacral promontory,** a prominent bulge at the anterior tip of the base, is an important landmark in females during pelvic examinations and during labor and delivery. The region corresponding to the fused vertebral bodies is called the **sacral body.** Once this fusion is complete, prominent *transverse lines* mark the former boundaries of individual vertebrae. At the base, a broad sacral **ala,** or *wing,* extends on either side lateral to the sacral body. The anterior and superior surfaces of each ala provide an extensive area for muscle attachment. At the apex, a flattened area marks the site of articulation with the coccyx.

The Coccyx

The small coccyx consists of 3–5 (typically 4) coccygeal vertebrae that have generally begun fusing by age 26 (Figure 7-21•). The coccyx provides an attachment site for a number of ligaments and for a muscle that constricts the anal opening. The first two coccygeal vertebrae have transverse processes and unfused vertebral arches. The prominent laminae of the first coccygeal vertebrae are known as the **coccygeal cornua.** They curve to meet the cornua of the sacrum. The coccygeal vertebrae do not fuse completely until late in adulthood. In very old people, the coccyx may fuse with the sacrum.

☑ Joe suffered a hairline fracture at the base of the dens. Which bone is fractured, and where is it located?

☑ Examining a human vertebra, you notice that in addition to the large foramen for the spinal cord, two smaller foramina are on either side of the bone in the region of the transverse processes. From which region of the vertebral column is this vertebra?

☑ Why are the bodies of the lumbar vertebrae so large?

THE THORACIC CAGE

The **thoracic cage,** or skeleton of the chest, consists of the thoracic vertebrae, the ribs, and the sternum (breastbone) (Figure 7-22•). The ribs and the sternum form the *rib cage* and support the walls of the thoracic cavity. The thoracic cage serves two functions:

1. It protects the heart, lungs, thymus, and other structures in the thoracic cavity.

2. It serves as an attachment point for muscles involved in (1) respiration, (2) the position of the vertebral column, and (3) movements of the pectoral girdle and upper limbs.

THE RIBS

Ribs, or *costae,* are elongate, curved, flattened bones that (1) originate on or between the thoracic vertebrae and (2) end in the wall of the thoracic cavity. Each of us has 12 pairs of ribs (Figure 7-22•). The first seven pairs are called **true ribs,** or *vertebrosternal ribs.* They

(a) Anterior view

(b) Posterior view

•**FIGURE 7-22** **The Thoracic Cage.**
(a) An anterior view of the thoracic cage and sternum. **(b)** A posterior view of the thoracic cage.

(a) Superior view

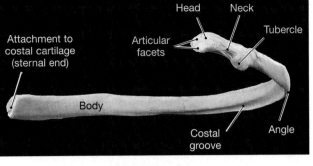

(b) Posterior view

7

•**FIGURE 7-23** **The Ribs.** **(a)** Details of rib structure and the articulations between the ribs and thoracic vertebrae. **(b)** A posterior view of the head of a representative rib from the right side (ribs 2–9). **(c)** The effect of rib movement on the thoracic cavity, similar to the movement of a bucket handle.

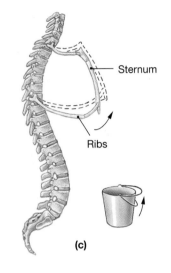

(c)

reach the anterior body wall and are connected to the sternum by separate cartilaginous extensions, the **costal cartilages.** Beginning with the first rib, the vertebrosternal ribs gradually increase in length and in the radius of curvature.

Ribs 8–12 are called **false ribs,** because they do not attach directly to the sternum. The costal cartilages of ribs 8–10, the *vertebrochondral ribs,* fuse together and merge with the cartilages of rib pair 7 before they reach the sternum (Figure 7-22a•). The last two pairs of ribs (11 and 12) are called *floating ribs,* because they have no connection with the sternum, or *vertebral ribs,* because they are attached only to the vertebrae (Figure 7-22a,b•).

Figure 7-23a• shows the superior surface of a typical rib. The *vertebral end* of the rib articulates with the vertebral column at the **head,** or *capitulum* (ka-PIT-ū-lum). A ridge divides the articular surface of the head into superior and inferior articular facets (Figure 7-23b•). From the head, a short **neck** leads to the **tubercle,** or *tuberculum* (too-BER-kū-lum), a small elevation that projects dorsally. The inferior portion of the tubercle contains an articular facet that contacts the transverse process of the thoracic vertebra.

Ribs 1 and 10 originate at whole facets on vertebrae T_1 and T_{10}, respectively, and their tubercular facets articulate with those vertebrae. Ribs 2–9 originate at demifacets, and their tubercular facets articulate with the inferior member of the vertebral pair. Ribs 11–12 originate at whole facets on T_{11} and T_{12}. These ribs do not have tubercular facets and do not contact the transverse processes of T_{11} or T_{12}. The difference in rib orientation can be seen by comparing Figure 7-19a, p. 217, with Figure 7-22b•.

The bend, or *angle,* of the rib is the site where the tubular **body,** or *shaft,* begins curving toward the sternum. The internal rib surface is concave, and a promi-

nent *costal groove* along its inferior border marks the path of nerves and blood vessels. The superficial surface is convex and provides an attachment site for muscles of the pectoral girdle and trunk. The *intercostal muscles,* which move the ribs, are attached to the superior and inferior surfaces.

With their complex musculature, dual articulations at the vertebrae, and flexible connection to the sternum, the ribs are quite mobile. Note how the ribs curve away from the vertebral column to angle inferiorly (Figure 7-22b•). A typical rib acts as if it were the handle on a bucket, lying just below the horizontal plane. Pushing the handle down forces it inward; pulling it up swings it outward (Figure 7-23c•). In addition, because of the curvature of the ribs, the same movements change the position of the sternum. Depressing the ribs pulls the sternum inward, whereas elevation moves it outward. As a result, movements of the ribs affect both the width and the depth of the thoracic cage, increasing or decreasing its volume accordingly.

The ribs can bend and move to cushion shocks and absorb blows, but severe or sudden impacts can cause painful rib fractures. Because the ribs are tightly bound in connective tissues, a cracked rib can heal without a cast or splint. But compound fractures of the ribs can send bone splinters or fragments into the thoracic cavity, with potential damage to internal organs.

Surgery on the heart, lungs, or other organs in the thorax typically involves entering the thoracic cavity. The mobility of the ribs and the cartilaginous connections with the sternum allow the ribs to be temporarily moved out of the way. "Rib-spreaders" are used to push the ribs apart in much the same way that a jack lifts a car off the ground for a tire change. If more extensive access is required, the cartilages of the sternum can be cut and the entire sternum folded out of the way. Once replaced, scar tissue reunites the cartilages, and the ribs heal fairly rapidly.

Thoracentesis

After thoracic surgery, *chest tubes* may be inserted through the wall of the thoracic cavity to permit drainage of fluids. To install a chest tube or obtain a sample of pleural fluid, a physician must penetrate the wall of the thorax. This process, called *thoracentesis* (tho-ra-sen-TĒ-sis) or *thoracocentesis* (tho-ra-kō-sen-TĒ-sis; *kentesis,* perforating), involves the penetration of the thoracic wall along the superior border of one of the ribs. Penetration at this location avoids damage to vessels and nerves within the costal groove.

THE STERNUM

The adult **sternum,** or breastbone, is a flat bone that forms in the anterior midline of the thoracic wall (Figure 7-22a•). The sternum has three components:

1. The broad, triangular **manubrium** (ma-NOO-brē-um) articulates with the *clavicles* (collarbones) and the cartilages of the first pair of ribs. This is the widest and most superior portion of the sternum. Only the first pair of ribs is attached by cartilage to this portion of the sternum. The **jugular notch,** located between the clavicular articulations, is a shallow indentation on the superior surface of the manubrium.

2. The tongue-shaped **body** attaches to the inferior surface of the manubrium and extends inferiorly along the midline. Individual costal cartilages from rib pairs 2–7 are attached to this portion of the sternum.

3. The **xiphoid** (ZĪ-foyd) **process,** the smallest part of the sternum, is attached to the inferior surface of the body. The muscular *diaphragm* and *rectus abdominis muscles* attach to the xiphoid process.

Ossification of the sternum begins at 6 to 10 centers, and fusion is not completed until at least age 25. Before age 25, the sternal body consists of four separate bones. In adults, their boundaries appear as a series of transverse lines crossing the sternum. The xiphoid process is generally the last sternal component to ossify and fuse. Its connection to the sternal body can be broken by impact or by strong pressure, creating a spear of bone that can severely damage the liver. Cardiopulmonary resuscitation (CPR) training strongly emphasizes proper hand positioning to reduce the chances of breaking ribs or the xiphoid process.

☑ How could you distinguish between true ribs and false ribs?

☑ Improper administration of CPR (cardiopulmonary resuscitation) can result in a fracture of which bone(s)?

[EXPLORE] MediaLab

The bones of the axial skeleton are numerous and include many markings, processes, and functions. To gain a handle on the complexity of the axial skeleton, use the resources available on both the CD-ROM and the Companion Website. The CD-ROM includes animations and tutorials; the website builds on those resources with text content and interactive problems as well as communication avenues for sharing ideas with your class. To complete the following Web Explorations, visit the MediaLab in Chapter 7 of your Companion Website.

Web Exploration 1

Estimated time for completion = 10 minutes

The best way to learn the intricacies of the skull bones is to practice identifying them in various illustrations. After studying the illustrations provided in your text, select the keyword **TUTORIAL** on the Companion Website. Using the text as a reference, try to identify the components of the skull indicated on each of the five questions. If you wish, share your answers with your instructor or classmates by using the communication tools on the Companion Website.

Web Exploration 2

Estimated time for completion = 10 minutes

The 26 bones of the vertebral column, the central portion of the axial skeleton, fit together to support and protect the spinal cord. Sometimes this fit is altered, resulting in scoliosis, kyphosis, or lordosis. To understand how the vertebrae usually fit together and what happens when they don't, select the keyword **VERTEBRAL COLUMN** on the Companion Website. Click on the footprints to move through the pages. When you have finished, draw a typical vertebra and demonstrate how adjacent vertebrae would normally articulate with it. Then try to depict the same articulations in a person with scoliosis, one with kyphosis, and one with lordosis.

CHAPTER REVIEW

SELECTED CLINICAL TERMINOLOGY
Terms Discussed in This Chapter

chest tube: A drain installed after thoracic surgery to permit removal of blood and pleural fluid. *(p. 222)*

craniostenosis (krā-nē-ō-sten-Ō-sis): The premature closure of one or more fontanels, which can lead to unusual distortions of the skull. *(p. 211)*

deviated (nasal) septum: A bent nasal septum that slows or prevents sinus drainage. *(p. 210)*

kyphosis (kī-FŌ-sis): An abnormal exaggeration of the thoracic curve that produces a humpback appearance. *(p. 212 and AM)*

lordosis (lor-DŌ-sis): An abnormal lumbar curve that gives a swayback appearance. *(p. 212 and AM)*

microcephaly (mī-krō-SEF-uh-lē): An undersized head resulting from genetic or developmental abnormalities. *(p. 211)*

scoliosis (skō-lē-Ō-sis): An abnormal lateral curvature of the spine. *(p. 212 and AM)*

sinusitis: Inflammation and congestion of the sinuses. *(p. 210)*

spina bifida (SPĪ-nuh BI-fi-duh): A condition resulting from the failure of the vertebral laminae to unite during development; commonly associated with developmental abnormalities of the brain and spinal cord. *(p. 214 and AM)*

thoracentesis (tho-ra-sen-TĒ-sis), or **thoracocentesis** (tho-ra-kō-sen-TĒ-sis): The penetration of the thoracic wall along the superior border of one of the ribs. *(p. 222)*

TMJ syndrome: A painful condition resulting from a misalignment of the mandible at the temporomandibular joint. *(p. 208)*

whiplash: An injury resulting from a sudden change in the body position that can injure the cervical vertebrae. *(p. 214)*

7

STUDY OUTLINE

INTRODUCTION p. 194

1. The skeletal system consists of the axial skeleton and the appendicular skeleton. The **axial skeleton** can be divided into the *skull*, the *auditory ossicles* and *hyoid bone*, the *vertebral column*, and the *thoracic cage*. *(Figure 7-1)*

2. The **appendicular skeleton** includes the pectoral and pelvic girdles, which support the upper and lower limbs.

THE SKULL p. 195

1. The **skull** consists of the **cranium** and the bones of the face. The cranium, composed of **cranial bones**, encloses the **cranial cavity**, a division of the dorsal body cavity. The **facial bones** protect and support the entrances to the digestive and respiratory tracts. *(Figure 7-2)*

2. Prominent superficial landmarks on the skull include the **lambdoidal, coronal, sagittal,** and **squamous sutures.** *(Figure 7-3)*

FOCUS: The Individual Bones of the Skull p. 200

Cranial Bones p. 200

3. The cranial bones are the **occipital bone;** the two **parietal bones;** the **frontal bone;** the two **temporal bones;** the **sphenoid bone;** and the **ethmoid bone.** *(Figures 7-3–7-9)*

4. The occipital bone surrounds the **foramen magnum.** *(Figures 7-3–7-5)*

5. The frontal bone contains the **frontal sinuses.** *(Figures 7-4, 7-6)*

6. The **auditory ossicles** are located in a cavity within the temporal bone.

Facial Bones p. 205

7. The bones of the face are the **maxillary bones,** the **palatine bones,** the **nasal bones,** the **vomer,** the **inferior nasal conchae,** the **zygomatic bones,** the **lacrimal bones,** the **mandible,** and the **hyoid bone.** *(Figures 7-3, 7-4, 7-10–7-12)*

8. The left and right maxillary bones, or *maxillae,* are the largest facial bones; they form the upper jaw. *(Figures 7-3, 7-4, 7-10)*

9. The palatine bones are small, L-shaped bones that form the posterior portions of the **hard palate** and contribute to the floor of the orbital cavities. *(Figures 7-3, 7-4, 7-10)*

10. The paired nasal bones extend to the superior border of the **external nares.** *(Figures 7-3, 7-4, 7-11)*

11. The vomer forms the inferior portion of the **nasal septum.** *(Figures 7-3, 7-4, 7-11)*

12. The **temporal process** of the zygomatic bone articulates with the **zygomatic process** of the temporal bone to form the **zygomatic arch.** *(Figures 7-3, 7-7, 7-11)*

13. The paired lacrimal bones, the smallest bones of the face, are situated medially in each **orbit.** *(Figures 7-3, 7-11)*

14. The mandible is the bone of the lower jaw. *(Figures 7-3, 7-4, 7-12)*

15. The **hyoid bone,** suspended by *stylohyoid ligaments,* does not articulate with other bones. *(Figure 7-12)*

Summary: The Foramina and Fissures of the Skull p. 208

16. Features of the adult skull are summarized in *Table 7-1.*

The Orbital and Nasal Complexes p. 208

17. Seven bones form each **orbital complex.** *(Figure 7-13)*

18. The **nasal complex** includes the bones that enclose the nasal cavities and the **paranasal sinuses,** hollow airways that connect with the nasal passages. *(Figure 7-14)*

The Skulls of Infants and Children p. 211

19. Fibrous connections at **fontanels** permit the skulls of infants and children to continue growing. *(Figure 7-15)*

THE VERTEBRAL COLUMN p. 212

1. The **vertebral column** consists of the vertebrae, sacrum, and coccyx. We have 7 **cervical vertebrae** (the first articulates with the skull), 12 **thoracic vertebrae** (which articulate with ribs), and 5 **lumbar vertebrae** (the last articulates with the sacrum). The **sacrum** and **coccyx** consist of fused vertebrae. *(Figure 7-16)*

Spinal Curvature p. 212
2. The spinal column has four **spinal curves**. The **thoracic** and **sacral curves** are called **primary**, or **accommodation, curves**; the **lumbar** and **cervical curves** are known as **secondary**, or **compensation, curves**. *(Figure 7-16)*

Vertebral Anatomy p. 213
3. A typical vertebra has a **body** (*centrum*) and a **vertebral arch** and articulates with adjacent vertebrae at the **superior** and **inferior articular processes**. *(Figure 7-17)*
4. Adjacent vertebrae are separated by **intervertebral discs**. Spaces between successive **pedicles** form the **intervertebral foramina**. *(Figure 7-17)*

Vertebral Regions p. 214
5. Cervical vertebrae are distinguished by the shape of the body, the relative size of the vertebral foramen, the presence of **costal processes** with **transverse foramina**, and notched **spinous processes**. These vertebrae include the **atlas, axis,** and **vertebra prominens**. *(Figure 7-18; Table 7-2)*
6. Thoracic vertebrae have a distinctive heart-shaped body; long, slender spinous processes; and articulations for the ribs. *(Figures 7-19, 7-22; Table 7-2)*

7. The lumbar vertebrae are the most massive and least mobile; they are subjected to the greatest strains. *(Figure 7-20; Table 7-2)*
8. The sacrum protects reproductive, digestive, and urinary organs and articulates with the pelvic girdle and with the fused elements of the coccyx. *(Figure 7-21)*

THE THORACIC CAGE p. 219
1. The skeleton of the **thoracic cage** consists of the thoracic vertebrae, the ribs, and the sternum. The **ribs** and **sternum** form the *rib cage*. *(Figure 7-22)*

The Ribs p. 219
2. Ribs 1–7 are **true**, or *vertebrosternal*, **ribs**. Ribs 8–12 are called **false ribs**; they include the *vertebrochondral ribs* and two pairs of *floating (vertebral) ribs*. A typical rib has a **head**, or *capitulum*; a **neck**; a **tubercle**, or *tuberculum*; an *angle*; and a **body**, or *shaft*. A *costal groove* marks the path of nerves and blood vessels. *(Figures 7-22, 7-23)*

The Sternum p. 222
3. The sternum consists of the **manubrium, body,** and **xiphoid process**. *(Figure 7-22)*

REVIEW QUESTIONS

LEVEL 1 REVIEWING FACTS AND TERMS

Match each numbered item with the most closely related lettered item. Use letters for answers in the spaces provided.

___ 1. foramina   **a.** skullcap
___ 2. sinuses   **b.** tear ducts
___ 3. sutures   **c.** atlas
___ 4. calvaria   **d.** lumbar and cervical
___ 5. auditory ossicles   **e.** air-filled chambers
___ 6. hypophyseal fossa   **f.** axis
___ 7. lacrimal bones   **g.** ear bones
___ 8. accommodation curve   **h.** ribs
___ 9. compensation curve   **i.** passageways
___ 10. costae   **j.** sella turcica
___ 11. C_1 **k.** thoracic and sacral
___ 12. C_2 **l.** immovable joints

13. The axial skeleton consists of the bones of the
 (a) pectoral and pelvic girdles
 (b) skull, thorax, and vertebral column
 (c) arms, legs, hands, and feet
 (d) limbs, pectoral girdle, and pelvic girdle

14. The appendicular skeleton consists of the bones of the
 (a) pectoral and pelvic girdles
 (b) skull, thorax, and vertebral column
 (c) arms, legs, hands, and feet
 (d) limbs, pectoral girdle, and pelvic girdle

15. Which list contains *only* bones of the cranium?
 (a) frontal, parietal, occipital, sphenoid
 (b) frontal, occipital, zygomatic, parietal
 (c) occipital, sphenoid, temporal, lacrimal
 (d) mandible, maxillary, nasal, zygomatic

16. The facial bones include the
 (a) lacrimal, nasal, maxillary, mandible
 (b) frontal, lacrimal, zygomatic, sphenoid
 (c) maxillary, mandible, ethmoid, sphenoid
 (d) ethmoid, sphenoid, temporal, parietal

17. The boundaries between skull bones are immovable joints called
 (a) foramina   (b) fontanels
 (c) lacunae   (d) sutures

18. The major sutures of the skull are the
 (a) frontal, parietal, occipital, sphenoid
 (b) frontal, lambdoidal, occipital, coronal
 (c) lambdoidal, coronal, sagittal, squamous
 (d) coronal, sagittal, frontal, parietal

19. Of the following bones, which is unpaired?
 (a) vomer   (b) maxillary
 (c) palatine   (d) nasal

20. The cribriform plate, crista galli, and superior conchae are parts of the
 (a) parietal bone   (b) occipital bone
 (c) sphenoid bone   (d) ethmoid bone

21. The bone that houses the inner ear structures associated with hearing and balance is the
 (a) temporal   (b) occipital
 (c) parietal   (d) sphenoid

22. The bony enclosure that forms a prominent depression on the superior surface of the sphenoid bone and houses the pituitary gland is the
 (a) clinoid process
 (b) sella turcica
 (c) external auditory canal
 (d) auditory tube

23. The hyoid bone
 (a) forms the inferior portion of the skull
 (b) is attached to the zygomatic bone
 (c) does not articulate with any other bone
 (d) is a cartilaginous structure

24. The membranous areas between the cranial bones of the fetal skull are
 (a) fontanels   (b) sutures
 (c) Wormian bones   (d) foramina

25. The part(s) of the vertebra that transfer(s) weight along the axis of the vertebral column is (are) the

    (a) vertebral arch   **(b)** lamina

    (c) pedicles   **(d)** body

26. Progressing inferiorly, the components of the sternum include the

    (a) xiphoid process, body, manubrium

    (b) body, manubrium, xiphoid process

    (c) manubrium, body, xiphoid process

    (d) manubrium, xiphoid process, body

27. Which five bones make up the facial complex?

28. Which four major sutures make up the boundaries between the skull bones?

29. What is the primary function of the vomer?

30. Which bones contain the paranasal sinuses?

31. What characteristic of the hyoid bone makes that bone unique?

32. What purpose do the fontanels serve during delivery?

33. Beginning at the skull, cite the regions of the vertebral column and list the number of vertebrae in each region of the adult vertebral column.

34. List the four spinal curves that are apparent in the adult spinal column when viewed from the side.

35. What two primary functions does the thoracic cage perform?

36. Why are the upper seven pairs of ribs called *true* ribs and the lower five pairs called *false* ribs?

LEVEL 2 REVIEWING CONCEPTS

37. The atlas (C_1) can be distinguished from the other vertebrae by

    (a) the presence of anterior and posterior vertebral arches

    (b) the lack of a body

    (c) the presence of superior facets and inferior articular facets

    (d) a, b, and c are correct

38. What is the relationship between the temporal bone and the ear?

39. What is the relationship between the sphenoid bone and the pituitary gland?

40. Unlike that of other animals, the skull of humans is balanced above the vertebral column, allowing an upright posture. What anatomical adaptations are necessary in humans to maintain the balance of the skull?

41. The structural features and skeletal components of the sternum make it a part of the axial skeleton that is important in a variety of clinical situations. If you were teaching this information to prospective nursing students, what clinical applications would you cite?

42. Why are ruptured intervertebral discs more common in lumbar vertebrae and dislocations and fractures more common in cervical vertebrae?

43. We generally associate a "runny" nose with a cold. Does this nasal condition serve a purpose? What is the cause of this type of reaction?

44. Why is it important to keep your back straight when you lift a heavy object?

7

LEVEL 3 CRITICAL THINKING AND CLINICAL APPLICATIONS

45. Tess is diagnosed with a disease that affects the membranes surrounding the brain. The physician tells Tess's family that the disease is caused by an airborne virus. Explain how this virus could have entered the cranium.

46. Joe is 40 years old and 30 pounds overweight. Like many middle-aged men, Joe carries most of this extra weight in his abdomen and jokes with his friends about his "beer gut." During an annual physical, Joe's physician advises Joe that his spine is developing an abnormal curvature. Why is the curvature of Joe's spine changing, and what is this condition called?

47. A babysitter is on trial for the death of a 10-month-old infant. The prosecutor contends that the child died as the result of being violently shaken. The defense claims that the child's head became stuck in the slats of his crib and, in trying to twist free, the child broke his neck. The medical examiner testifies that the child died as the result of a compression of the spinal cord between the sixth and seventh cervical vertebrae. The superior articular processes of the seventh cervical vertebra and the inferior articular processes of the sixth cervical vertebra were fractured, and the processes on the right side were laterally displaced, causing the sixth vertebra to slide laterally across the seventh, damaging the spinal cord. On the basis of this evidence, whom do you believe? Why?

48. While working at an excavation, an archaeologist finds several small skull bones. She examines the frontal, parietal, and occipital bones and concludes that the skulls are those of children not yet 1 year old. How can she tell their ages from examining the bones?

CHAPTER **8**

The Appendicular Skeleton

CHAPTER OUTLINE AND OBJECTIVES

INTRODUCTION   p. 227
① Identify each bone of the appendicular skeleton.

THE PECTORAL GIRDLE AND UPPER LIMBS   p. 228
   The Pectoral Girdle   p. 228
   The Upper Limbs   p. 230
② Identify the bones that form the pectoral girdle,
   their functions, and their superficial features.
③ Identify the bones of the upper limbs, their functions,
   and their superficial features.

THE PELVIC GIRDLE AND LOWER LIMBS   p. 233
   The Pelvic Girdle   p. 234
   The Lower Limbs   p. 237
④ Identify the bones that form the pelvic girdle,
   their functions, and their superficial features.
⑤ Identify the bones of the lower limbs, their functions,
   and their superficial features.
⑥ Discuss structural and functional differences between
   the pelvis of females and that of males.

INDIVIDUAL VARIATION IN THE SKELETAL SYSTEM   p. 241
⑦ Explain how study of the skeleton can reveal
   significant information about an individual.
⑧ Summarize the skeletal differences between males
   and females.
⑨ Briefly describe how the aging process affects
   the skeletal system.

MEDIALAB and CHAPTER REVIEW   p. 242, 243

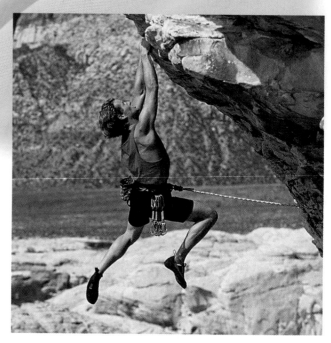

Clinging to a cliff face by your fingers and toes places unusually severe stresses on the appendicular skeleton and on the associated muscles and supporting connective tissues. Even if you do not climb mountains, your appendicular skeleton plays a major role in your daily life. Make a mental list of all the things you have done with your arms or legs in the past day. Standing, walking, writing, turning pages, eating, dressing, shaking hands, waving—the list quickly becomes unwieldy. Your axial skeleton protects and supports internal organs, and it participates in vital functions, such as respiration. But it is your appendicular skeleton that gives you control over your environment, changes your position in space, and provides mobility. In this chapter, we examine the components of the appendicular skeleton; in Chapter 9, we will consider the ways those elements interact during normal movements.

The **appendicular skeleton** includes the bones of the limbs and the supporting elements, or *girdles,* that connect them to the trunk (Figure 8-1•). The descriptions in this chapter emphasize surface features that have functional importance, such as the attach- ment sites for skeletal muscles or the paths of major nerves and blood vessels.

The appendicular skeleton is dominated by the long bones that support the limbs. ∞ *p. 168* Each long bone shares common features with other long bones.

8

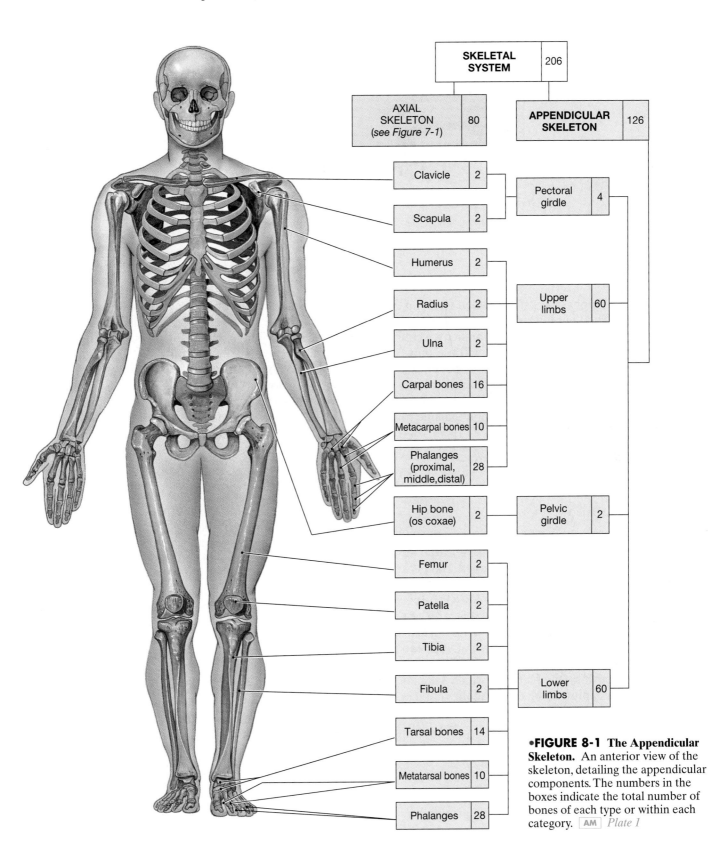

•**FIGURE 8-1 The Appendicular Skeleton.** An anterior view of the skeleton, detailing the appendicular components. The numbers in the boxes indicate the total number of bones of each type or within each category. AM *Plate 1*

(a) Anterior view

•FIGURE 8-2 **The Clavicle. (a)** The position of the clavicle, anterior view.

For example, one epiphysis is usually called the *head*, the diaphysis is called the *shaft*, and the head and shaft are normally separated by a *neck*. For simplicity, when an illustration shows a single bone, we will use those terms without including the name of the bone. Thus a photo of the humerus will be labeled *head* rather than *head of the humerus* or *humeral head*. When more than one bone is shown, we will give the complete name to avoid confusion.

THE PECTORAL GIRDLE AND UPPER LIMBS

Each arm articulates (that is, forms a joint) with the trunk at the **pectoral girdle,** or *shoulder girdle*. The pectoral girdle consists of two S-shaped *clavicles* (KLAV-i-kulz; collarbones) and two broad, flat *scapulae* (SKAP-ū-lē; singular, *scapula,* SKAP-ū-luh; shoulder blade). Each clavicle articulates with the manubrium of the sternum. ∞ *p. 222* These articulations are the *only* direct connections between the pectoral girdle and the axial skeleton. Skeletal muscles support and position the scapulae, which have no direct bony or ligamentous connections to the thoracic cage. As a result, the shoulders are extremely mobile but not very strong.

THE PECTORAL GIRDLE

Movements of the clavicles and scapulae position the shoulder joints and provide a base for arm move-

ment. Once the shoulder joints are in position, muscles that originate on the pectoral girdle help move the upper limbs. The surfaces of the scapulae and clavicles are extremely important as sites for muscle attachment. Where they attach, major muscles leave marks in the form of bony ridges and flanges. Other bone markings, such as sulci or foramina, indicate the position of nerves or the passage of blood vessels that control the muscles and nourish the muscles and bones, respectively.

The Clavicles

The **clavicles** are S-shaped bones that originate at the superior, lateral border of the manubrium of the sternum, lateral to the jugular notch (Figure 8-2a•). From the roughly pyramidal **sternal end,** each clavicle curves laterally and dorsally until it articulates with a process of the scapula, the *acromion* (a-KRŌ-mē-on). The broad, flat, **acromial end** of the clavicle is broader than the sternal end (Figure 8-2b,c•).

The smooth, superior surface of the clavicle lies just beneath the skin. The rough, inferior surface of the acromial end is marked by prominent lines and tubercles. These surface features are attachment sites for muscles and ligaments of the shoulder.

You can explore the interaction between scapulae and clavicles. With your fingers in your jugular notch, locate the clavicle on either side. When you move your shoulders, you can feel the clavicles change their positions. Because the clavicles are so close to the skin, you can trace one laterally until it articulates with the scapula.

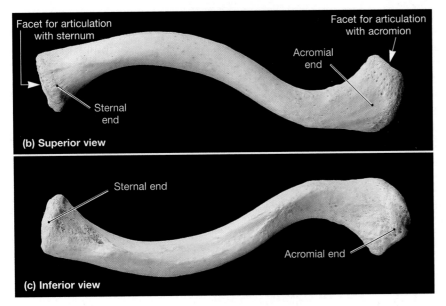

Facet for articulation with sternum

Facet for articulation with acromion

Acromial end

Sternal end

(b) Superior view

Sternal end

Acromial end

(c) Inferior view

•**FIGURE 8-2** The Clavicle *(continued).* **(b)** Superior and **(c)** inferior views of the right clavicle.

8

The clavicles are relatively small and fragile, and therefore fractures of the clavicle are fairly common. For example, you can fracture a clavicle as the result of a simple fall if you land on your hand with your arm outstretched. Fortunately, in view of the clavicle's vulnerability, most clavicular fractures heal rapidly without a cast.

The Scapulae

The anterior surface of the **body** of each **scapula** forms a broad triangle (Figure 8-3a•). The three sides of that triangle are the **superior border;** the **medial border,** or *vertebral border;* and the **lateral border,** or *axillary border* (*axilla*, armpit). Muscles that position the scapula attach along these edges. The corners of the triangle

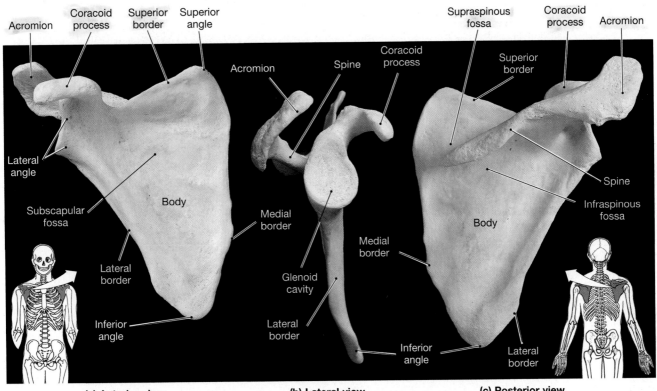

Acromion

Coracoid process

Superior border

Superior angle

Acromion

Spine

Coracoid process

Supraspinous fossa

Coracoid process

Acromion

Superior border

Lateral angle

Subscapular fossa

Body

Medial border

Medial border

Glenoid cavity

Lateral border

Inferior angle

Spine

Infraspinous fossa

Body

Lateral border

Lateral border

Inferior angle

Lateral border

Inferior angle

(a) Anterior view

(b) Lateral view

(c) Posterior view

•**FIGURE 8-3** The Scapula. **(a)** Anterior, **(b)** lateral, and **(c)** posterior views of the right scapula.

are called the *superior angle,* the *inferior angle,* and the *lateral angle.* The lateral angle, or *head* of the scapula, forms a broad process that supports the cup-shaped **glenoid cavity.** At the glenoid cavity, the scapula articulates with the *humerus,* the proximal bone of the upper limb. This articulation is the shoulder joint, also known as the *glenohumeral joint.* The anterior surface of the body of the scapula is relatively smooth and concave. The depression in the anterior surface is called the **subscapular fossa.**

Two large scapular processes extend beyond the margin of the glenoid cavity (Figure 8-3b•) superior to the head of the humerus. The smaller, anterior projection is the **coracoid** (KOR-uh-koyd) **process.** The **acromion** is the larger, posterior process. If you run your fingers along the superior surface of the shoulder joint, you will feel this process. The acromion articulates with the clavicle at the *acromioclavicular joint.* Both the acromion and the coracoid process are attached to ligaments and tendons associated with the shoulder joint.

The acromion is continuous with the **scapular spine** (Figure 8-3c•), a ridge crosses that scapular body before ending at the medial border. The scapular spine divides the convex dorsal surface of the body into two regions. The area superior to this spine constitutes the **supraspinous fossa** (*supra,* above); the region inferior to the spine is the **infraspinous fossa** (*infra,* beneath). The entire posterior surface is mar-ked by small ridges and lines where smaller muscles attach to the scapula.

☑ Why would a broken clavicle affect the mobility of the scapula?

☑ Which bone articulates with the scapula at the glenoid cavity?

THE UPPER LIMBS

The skeleton of the upper limbs consists of the bones of the arms, forearms, wrists, and hands. Notice that in anatomical descriptions, the term *arm* refers only to the proximal portion of the upper limb, not to the entire limb. The anatomical usage and common usage are thus quite different. We will examine the bones of the right upper limb. The arm, or *brachium,* contains one bone, the **humerus.** The humerus extends from the scapula to the elbow. At the proximal end of the humerus, the round head articulates with the scapula (Figure 8-4•). At its distal end, the humerus articulates with the *radius* and the *ulna,* the bones of the forearm (or *antebrachium*).

The Humerus

The prominent **greater tubercle** of the humerus is a rounded projection near the lateral surface of the humeral head (Figure 8-4•). The greater tubercle establishes the lateral contour of the shoulder. You can verify its position by feeling for a bump situated a few centimeters from the tip of the acromion. The **lesser tubercle** is a smaller projection that lies on the anterior, medial surface of the epiphysis, separated from the greater tubercle by the **intertubercular groove,** or *intertubercular sulcus.* Both tubercles are important sites for muscle attachment; a large tendon runs along the groove. The **anatomical neck** marks the extent of the joint capsule. The anatomical neck lies between the tubercles and the articular surface of the head. The narrower **surgical neck** corresponds to the metaphysis of the growing bone. This name reflects the fact that fractures typically occur at this site.

The proximal shaft of the humerus is round in section. The **deltoid tuberosity** is a large, rough elevation on the lateral surface of the shaft, roughly halfway along its length. It is named after the *deltoid muscle,* which attaches to it.

On the posterior surface, the deltoid tuberosity ends at the **radial groove** (Figure 8-4b•). This depression marks the path of the *radial nerve,* a large nerve that provides sensory information from the posterior surface of the limb and motor control over the large muscles that straighten the elbow. Distal to the radial groove, the posterior surface of the humerus is relatively flat. Near the distal articulation with the bones of the forearm, the shaft expands to either side at the **medial** and **lateral epicondyles.** *Epicondyles* are processes that develop proximal to an articulation and provide additional surface area for muscle attachment. The *ulnar nerve* crosses the posterior surface of the medial epicondyle. A blow at the humeral side of the elbow joint can strike this nerve and produce a temporary numbness and paralysis of muscles on the anterior surface of the forearm. It causes an odd sensation, so this area is sometimes called the *funny bone.*

At the articular **condyle,** the humerus articulates with the bones of the forearm—the radius and the ulna. The articular condyle is divided into two articular regions, the trochlea and the capitulum (Figure 8-4•). The **trochlea** (*trochlea,* a pulley) is the spool-shaped medial portion of the condyle. The trochlea extends from the base of the **coronoid** (*corona,* crown) **fossa** on the anterior surface to the **olecranon** (ō-LEK-ruh-non) **fossa** on the posterior surface (Figure 8-4b•). These depressions accept projections from the ulnar surface as the elbow approaches the limits of its range of motion. The rounded **capitulum** forms the lateral surface of the condyle. A shallow **radial fossa** superior to the capitulum accommodates a portion of the radial head as the forearm approaches the humerus.

The Ulna

The **ulna** and **radius** are parallel bones that support the forearm. In the anatomical position, the ulna lies medial to the radius.

The **olecranon** is the point of the elbow (Figure 8-5a•). On the anterior surface of the proximal epiphysis (Figure 8-5b•), the **trochlear notch** articulates with the trochlea of the humerus at the elbow joint.

The olecranon forms the superior lip of the trochlear notch, and the **coronoid process** forms its inferior lip. At the limit of *extension,* with the forearm and arm forming a straight line, the olecranon swings into the olecranon fossa on the posterior surface of the humerus. At the limit of *flexion,* with the arm and forearm forming a V, the coronoid process projects into the coronoid fossa on the anterior humeral surface. Lateral to the coronoid process, a smooth **radial notch** accommodates the head of the radius at the *proximal radioulnar joint.*

When viewed in cross section, the shaft of the ulna is roughly triangular. The **interosseous membrane,** a fibrous sheet, connects the lateral margin of the ulna to the radius. Near the wrist, the shaft of the ulna narrows before ending at a disc-shaped **ulnar head,** or *head of the ulna.* The posterior, lateral surface of the ulnar head bears a short **styloid process** (*styloid,* long and pointed). A triangular *articular disc* attaches to the styloid process; this cartilage separates the ulnar head from the bones of the wrist. The lateral surface of the ulnar head articulates with the distal end of the radius to form the *distal radioulnar joint.*

The Radius

The radius is the lateral bone of the forearm. The disc-shaped **radial head,** or *head of the radius,* articulates with the capitulum of the humerus. During flexion, the radial head swings into the radial fossa of the humerus. A narrow neck extends from the radial head to the **radial tuberosity,** which marks the attachment site of the *biceps brachii* muscle, a large muscle on the anterior surface of the arm. The shaft of the radius curves along its length. It also enlarges, and the distal portion of the radius is considerably larger than the distal portion of the ulna. The **ulnar notch** on the medial surface of the distal end of the radius marks the site of articulation with the head of the ulna. The distal end of the radius articulates with the bones of the

•**FIGURE 8-4** **The Humerus.** **(a)** The anterior and **(b)** posterior surfaces of the right humerus.

•**FIGURE 8-5 The Radius and Ulna.** The right radius and ulna **(a)** in posterior and **(b)** anterior views.

wrist. The **styloid process** on the lateral surface of the radius helps stabilize this joint.

The Carpal Bones

The **carpus,** or wrist, contains eight **carpal bones.** These bones form two rows: four **proximal carpal bones** and four **distal carpal bones.**

The proximal carpal bones are the scaphoid bone, lunate bone, triquetrum, and pisiform bone (Figure 8-6a•). The **scaphoid bone** is the proximal carpal bone on the lateral border of the wrist; it is the carpal bone closest to the styloid process of the radius. The comma-shaped **lunate** (*luna,* moon) **bone** lies medial to the scaphoid bone. Like the scaphoid bone, the lunate bone articulates with the radius. The **triquetrum** is medial to the lunate bone. It has the shape of a small pyramid. The triquetrum articulates with the articular disc that separates the ulnar head from the wrist. The small, pea-shaped **pisiform** (PIS-i-form) **bone** sits anterior to the triquetrum.

The distal carpal bones are the trapezium, trapezoid bone, capitate bone, and hamate bone (Figure 8-6b•). The **trapezium** is the lateral bone of the distal row; its proximal surface articulates with the scaphoid bone. The wedge-shaped **trapezoid bone** lies medial to the trapezium. Like the trapezium, it has a proximal articulation with the scaphoid bone. The **capitate bone,** the largest carpal bone, sits between the trapezoid bone and the hamate bone. The **hamate** (*hamatum,* hooked) **bone** is the medial distal carpal bone.

It may help you to remember "Sam Likes To Push The Toy Car Hard." In lateral-to-medial order, the first four words stand for the proximal carpal bones (scaphoid, lunate, triquetrum, pisiform) and the last four, for the distal carpal bones (trapezium, trapezoid, capitate, hamate).

The carpal bones articulate with one another at joints that permit limited sliding and twisting. Ligaments interconnect the carpal bones and help stabilize the wrist joint. The tendons of muscles that flex

RADIUS
Lunate bone
Scaphoid bone
Trapezium
Trapezoid bone
Capitate bone
Metacarpal bones
Proximal phalanx
Distal phalanx

ULNA
Pisiform bone
Triquetrum
Pisiform bone
Hamate bone

I
II
III
IV
V

(a) Anterior view

RADIUS
Lunate bone
Scaphoid bone
Trapezium
Trapezoid bone
Capitate bone
Metacarpal bones
Proximal phalanx
Middle phalanx
Distal phalanx

V
IV
III
II
I

(b) Posterior view

8

•**FIGURE 8-6** **Bones of the Wrist and Hand.** **(a)** Anterior and **(b)** posterior views of the right hand.

the fingers pass across the anterior surface of the wrist, sandwiched between the intercarpal ligaments and a broad, superficial transverse ligament called the *flexor retinaculum.* Inflammation of the connective tissues between the flexor retinaculum and the carpal bones can compress the tendons and adjacent sensory and motor nerves, producing pain and a loss of wrist mobility. This condition, called *carpal tunnel syndrome,* is discussed further in Chapter 11.

The Hand

Five **metacarpal** (met-uh-KAR-pul; *metacarpus,* hand) **bones** articulate with the distal carpal bones and support the hand (Figure 8-6•). Roman numerals I–V are used to identify the metacarpal bones, beginning with the lateral metacarpal bone that articulates with the trapezium.

Distally, the metacarpal bones articulate with the proximal finger bones. Each hand has 14 finger bones, or **phalanges** (fa-LAN-jēz; singular, *phalanx*). The first finger, known as the **pollex** (POL-eks), or thumb, has two phalanges (*proximal* and *distal*). Each of the other fingers has three phalanges (*proximal, middle, and distal*).

☑ The rounded projections on either side of the elbow are parts of which bone?

☑ Which bone of the forearm is lateral in the anatomical position?

☑ Bill accidentally fractures his first distal phalanx with a hammer. Which finger is broken?

THE PELVIC GIRDLE AND LOWER LIMBS

Because they must withstand the stresses involved in weight bearing and locomotion, the bones of the **pelvic girdle** are more massive than those of the pectoral girdle. For similar reasons, the bones of the lower limbs are more massive than those of the upper limbs. The pelvic girdle consists of the two fused *ossa coxae,* or *innominate bones.* The *pelvis* is a composite structure that includes the ossa coxae of the appendicular skeleton and the sacrum and coccyx of the axial skeleton. *pp. 218–219* Each lower limb consists of the *femur* (thigh), the *patella* (kneecap), the *tibia* and *fibula* (leg), and the bones of the ankle (*tarsal bones*) and foot.

THE PELVIC GIRDLE

The pelvic girdle consists of the paired hip bones. Each hip bone, or **os coxae** (plural, *ossa coxae*), forms by the fusion of three bones: an **ilium** (IL-ē-um; plural, *ilia*), an **ischium** (IS-kē-um; plural, *ischia*), and a **pubis** (PŪ-bis) (Figure 8-7•). The ilia have a sturdy articulation with the auricular surfaces of the sacrum, attaching the pelvic girdle to the axial skeleton. Ventrally, the hip bones are connected by a pad of fibrocartilage at a joint called the **pubic symphysis.** On the lateral surface of each hip bone, the **acetabulum** (a-se-TAB-ū-lum; *acetabulum*, vinegar cup), a socket with a curved surface, articulates with the head of the femur (Figure 8-7a•). The acetabulum has a diameter of about 5 cm (2 in.). A ridge of bone forms the lateral and superior margins of the acetabulum. The anterior and inferior portion of the ridge is incomplete; the gap is called the *acetabular notch.* The smooth, C-shaped articular surface of the acetabulum is the **lunate surface.**

The ilium, ischium, and pubis meet inside the acetabulum, as though it were a pie sliced into three pieces. Superior to the acetabulum, the ilium forms a broad, curved surface that provides an extensive area for the attachment of muscles, tendons, and ligaments (Figure 8-7a•). Landmarks along the margin of the ilium include the *iliac spines,* which mark the attachment sites of important muscles and ligaments; the *gluteal lines,* which mark the attachment of large hip muscles; and the **greater sciatic** (sī-A-tik) **notch,** through which a major nerve (the *sciatic nerve*) reaches the lower limb.

Near the edge of the acetabulum, the ilium fuses with the ischium. Posterior to the acetabulum, the prominent **ischial spine** projects superior to the *lesser sciatic notch,* through which blood vessels, nerves, and a small muscle pass. The **ischial tuberosity,** a roughened projection, is located at the posterior and lateral edge of the ischium. When you are seated, the ischial tuberosities bear your body's weight.

The narrow **ischial ramus** (branch) of the ischium continues until it meets the **inferior ramus** of the pubis. The inferior pubic ramus extends between the ischial ramus and the *pubic tubercle,* a small, elevated area anterior and lateral to the

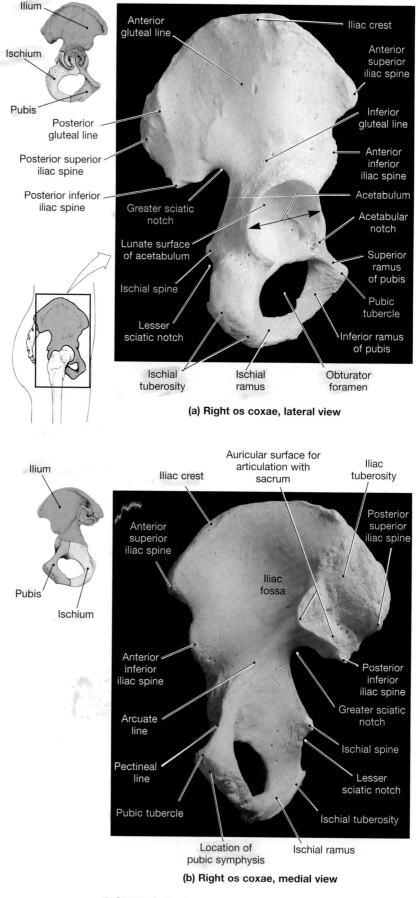

(a) Right os coxae, lateral view

(b) Right os coxae, medial view

•**FIGURE 8-7** The Os Coxae

pubic symphysis. There the inferior pubic ramus meets the **superior ramus** of the pubis, which originates near the acetabulum. The anterior, superior surface of the superior ramus bears the *pectineal line,* a ridge that ends at the pubic tubercle. The pubic and ischial rami encircle the **obturator** (OB-tū-rā-tor) **foramen.** This space is closed by a sheet of collagen fibers whose inner and outer surfaces provide a firm base for the attachment of muscles of the hip.

The broadest part of the ilium extends between the **arcuate line,** which is continuous with the pectineal line, and the **iliac crest.** These prominent ridges mark the attachments of ligaments and muscles (Figure 8-7b●). The area between the arcuate line and the iliac crest forms a shallow depression known as the **iliac fossa.** The concave surface of the iliac fossa helps support the abdominal organs and provides additional area for muscle attachment.

In medial view, the anterior and medial surface of the pubis contains a roughened area that marks the site of articulation with the pubis of the opposite side (Figure 8-7b●). At this articulation, the pubic symphysis, the two pubic bones are attached to a median fibrocartilage pad. Posteriorly, the **auricular surface** of the ilium articulates with the auricular surface of the sacrum at the *sacroiliac joint.* ∞ *p. 218* Ligaments arising at the **iliac tuberosity,** a roughened area superior to the auricular surface, stabilize this joint.

The Pelvis

Figure 8-8● shows anterior and posterior views of the **pelvis,** which consists of the two hip bones, the sacrum, and the coccyx. An extensive network of ligaments connects the lateral borders of the sacrum with the iliac crest, the ischial tuberosity, the ischial spine, and the arcuate line. Other ligaments tie the ilia to the posterior lumbar vertebrae. These interconnections increase the stability of the pelvis.

The pelvis may be divided into the **false** (*greater*) **pelvis** and the **true** (*lesser*) **pelvis** (Figure 8-9a,b●). The true pelvis encloses the *pelvic cavity,* a subdivision of the abdominopelvic cavity. ∞ *p. 22* The superior limit of the true pelvis is a line that extends from either side of the base of the sacrum, along the arcuate line and pectineal line to the pubic symphysis. The bony edge of the true pelvis is called the **pelvic brim,** or *linea terminalis,* and the enclosed space is the **pelvic inlet.** The false pelvis consists of the

expanded, bladelike portions of each ilium superior to the pelvic brim.

The **pelvic outlet** is the opening bounded by the coccyx, the ischial tuberosities, and the inferior border of the pubic symphysis (Figure 8-9b,c●). The region bounded by the inferior edges of the pelvis is called the **perineum** (per-i-NĒ-um). Perineal muscles form the floor of the pelvic cavity and support the organs in the true pelvis.

(a) Anterior view

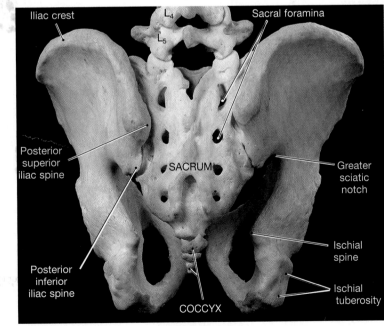

(b) Posterior view

●**FIGURE 8-8 The Pelvis.** The pelvis of an adult male. (*See Figure 7-21, p. 219, for a detailed view of the sacrum and coccyx.*)

8

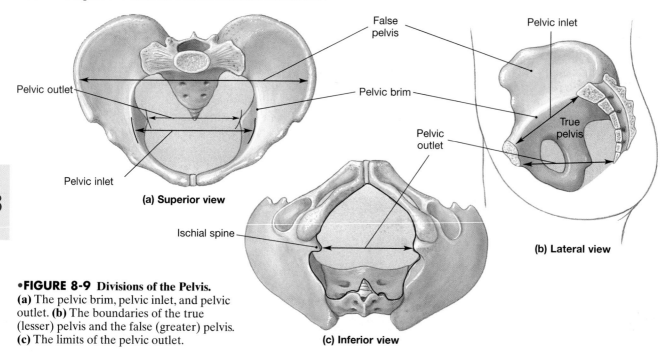

•FIGURE 8-9 Divisions of the Pelvis.
(a) The pelvic brim, pelvic inlet, and pelvic outlet. **(b)** The boundaries of the true (lesser) pelvis and the false (greater) pelvis. **(c)** The limits of the pelvic outlet.

The shape of the pelvis of a female is somewhat different from that of a male (Figure 8-10•). Some of these differences are the result of variations in body size and in muscle mass. For example, in females, the pelvis is generally smoother, lighter, and has less prominent markings. Females have other adaptations for childbearing, including:

■ An enlarged pelvic outlet.
■ Less curvature on the sacrum and coccyx, which in males arc into the pelvic outlet.
■ A wider, more circular pelvic inlet.
■ A relatively broad, low pelvis.

■ Ilia that project farther laterally but do not extend as far superior to the sacrum.
■ A broader *pubic angle* (the inferior angle between the pubic bones); in females, it is greater than 100°.

These adaptations are related to (1) the support of the weight of the developing fetus and uterus and (2) the passage of the newborn through the pelvic outlet during delivery. In addition, the hormone *relaxin,* produced during pregnancy, loosens the pubic symphysis, allowing relative movement between the hip bones that can further increase the size of the pelvic inlet and outlet.

•FIGURE 8-10 Anatomical Differences in the Pelvis of a Male and a Female. Notice the much sharper pubic angle in **(a)** the pelvis of a male than in **(b)** that of a female.

☑ Which three bones make up the os coxae?

☑ How is the pelvis of females adapted for childbearing?

☑ When you are seated, which part of the pelvis bears your body's weight?

THE LOWER LIMBS

The skeleton of each lower limb consists of a femur, a patella, a tibia and fibula, and the bones of the ankle and foot. Once again, anatomical terminology differs from common usage. In anatomical terms, *leg* refers only to the distal portion of the limb, not to the entire lower limb. Thus we will use *thigh* and *leg* when people unfamiliar with proper terminology might use *upper leg* and *lower leg,* respectively.

The functional anatomy of the lower limbs is different from that of the upper limbs, primarily because the lower limbs transfer the body weight to the ground. We now examine the bones of the right lower limb.

The Femur

The **femur** is the longest and heaviest bone in the body (Figure 8-11●). It articulates with the os coxae at the hip joint and with the tibia of the leg at the knee joint. The rounded epiphysis, or **femoral head,** articulates with the pelvis at the acetabulum. A ligament attaches the acetabulum to the femur at the **fovea capitis,** a small pit in the center of the femoral head. The **neck** of the femur joins the **shaft** at an angle of about 125°. The **greater** and **lesser trochanters** are large, rough

8

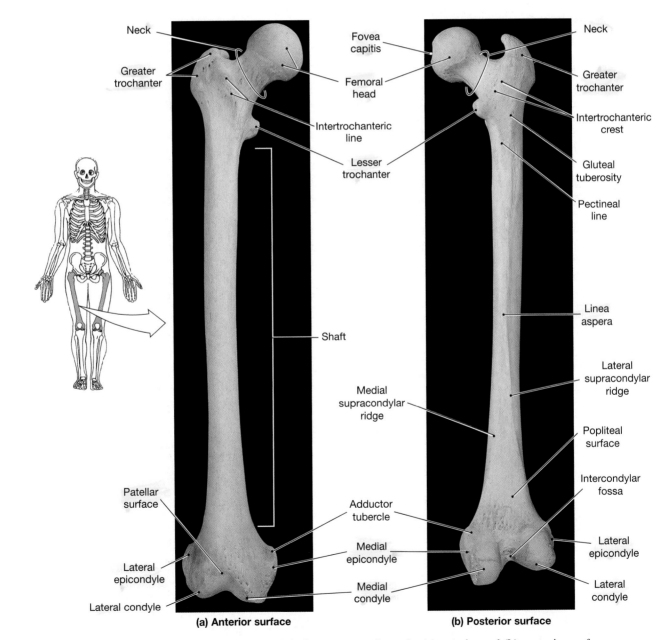

(a) Anterior surface   **(b) Posterior surface**

●**FIGURE 8-11 The Femur.** Bone markings on the right femur as seen from the **(a)** anterior and **(b)** posterior surfaces.

projections that extend laterally from the junction of the neck and shaft. Both trochanters develop where large tendons attach to the femur. On the anterior surface of the femur, the raised **intertrochanteric** (in-ter-trō-kan-TER-ik) **line** marks the edge of the articular capsule. This line continues around to the posterior surface as the **intertrochanteric crest.**

The **linea aspera** (*aspera,* rough), a prominent elevation, runs along the center of the posterior surface, marking the attachment site of powerful hip muscles (Figure 8-11b•). As it approaches the knee joint, the linea aspera divides into a pair of ridges that continue to the **medial** and **lateral epicondyles.** These smoothly rounded projections form superior to the **medial** and **lateral condyles,** which participate in the knee joint. The two condyles are separated by a deep **intercondylar fossa.**

The medial and lateral condyles extend across the inferior surface of the femur, but the intercondylar fossa does not reach the anterior surface (Figure 8-11a•). The anterior and inferior surfaces of the two condyles are separated by the **patellar surface,** a smooth articular surface over which the patella glides.

The Patella

The **patella** is a large sesamoid bone that forms within the tendon of the *quadriceps femoris,* a group of muscles that extend (straighten) the knee. The patella has a rough, convex anterior surface and a broad **base** (Figure 8-12a•). The roughened surface reflects the attachment of the quadriceps tendon (anterior and superior surfaces) and the *patellar ligament* (anterior and inferior surfaces). The patellar ligament connects the **apex** of the patella to the tibia. The posterior surface (Figure 8-12b•) presents two concave facets for articulation with the medial and lateral condyles of the femur.

The Tibia

The **tibia** (TI-bē-uh), or shinbone, is the large medial bone of the leg (Figure 8-13a•). The medial and lateral condyles of the femur articulate with the **medial** and **lateral tibial condyles** at the proximal end of the tibia. The **intercondylar eminence,** is a ridge that, separates the condyles (Figure 8-13b•). The anterior surface of the tibia near the condyles bears a prominent, rough **tibial tuberosity,** which you can feel through the skin. This tuberosity marks the attachment of the patellar ligament.

The **anterior margin** is a ridge that begins at the tibial tuberosity and extends distally along the anterior tibial surface. You can also easily feel the anterior margin of the tibia through the skin. As it approaches the ankle joint, the tibia broadens, and the medial border ends in the **medial malleolus** (ma-LĒ-ō-lus; *malleolus,* hammer), a large process. The inferior surface of the tibia articulates with the proximal bone of the ankle; the medial malleolus provides medial support for this joint.

The Fibula

The slender **fibula** (FIB-ū-luh) parallels the lateral border of the tibia (Figure 8-13a,b•). The head of the fibula articulates with the tibia. The articular facet is located on the anterior, inferior surface of the lateral tibial condyle. The medial border of the thin shaft is bound to the tibia by the **interosseous membrane,** which extends to the lateral margin of the tibia. This membrane helps stabilize the positions of the two bones and provides additional surface area for muscle attachment.

The fibula does not articulate with the femur or help transfer weight to the ankle and foot. However, it is important as a site for muscle attachment. In addition, the distal tip of the fibula extends lateral to the ankle joint. This fibular process, the **lateral malleolus,** provides lateral stability to the ankle. Yet, forceful

Attachment area for quadriceps tendon

Base of patella

Medial facet, for medial condyle of femur

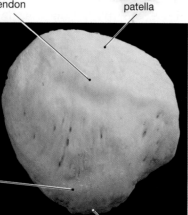

Lateral facet, for lateral condyle of femur

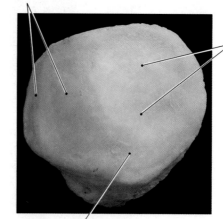

Attachment site for patellar ligament

Apex of patella

Articular surface of patella

•**FIGURE 8-12**
The Right Patella

(a) **Anterior view**

(b) **Posterior view**

•**FIGURE 8-13** **The Tibia and Fibula.** **(a)** Anterior and **(b)** posterior views of the right tibia and fibula.

movement of the foot outward and backward can dislocate the ankle, breaking both the lateral malleolus of the fibula and the medial malleolus of the tibia. This injury is called a *Pott's fracture.* ∞ *p. 186*

The Tarsal Bones

The ankle, or **tarsus,** consists of seven **tarsal bones** (Figure 8-14•). The large **talus** transmits the weight of the body from the tibia toward the toes. The articulation between the talus and the tibia occurs across the superior and medial surfaces of the **trochlea,** a pulley-shaped articular process. The lateral surface of the trochlea articulates with the lateral malleolus of the fibula.

The **calcaneus** (kal-KĀ-nē-us), or heel bone, is the largest of the tarsal bones. When you stand normally, most of your weight is transmitted from the tibia to the talus to the calcaneus, and then to the ground. The posterior portion of the calcaneus is a rough, knob-shaped projection. This is the attachment site for the *calcanean tendon (Achilles tendon),* which arises at strong calf muscles. These muscles raise the heel and depress the sole, such as when you stand on tiptoes. The superior

and anterior surfaces of the calcaneus bear smooth facets for articulation with other tarsal bones.

The **cuboid bone** articulates with the anterior surface of the calcaneus. The **navicular bone** is anterior to the talus, on the medial side of the ankle. It articulates with the talus and with the cuneiform bones. The three **cuneiform bones** are wedge-shaped bones arranged in a row, with articulations between them. They are named according to their position: **medial cuneiform, intermediate cuneiform,** and **lateral cuneiform.** Proximally, the cuneiform bones articulate with the anterior surface of the navicular bone. The lateral cuneiform bone also articulates with the medial surface of the cuboid bone. The distal surfaces of the cuboid bone and the cuneiform bones articulate with the metatarsal bones of the foot. To remember the names of the tarsal bones in the order presented, try the memory aid "Tom Can Control Not Much In Life."

The Foot

The **metatarsal bones** are five long bones that form the distal portion of the foot, or *metatarsus* (Figure 8-14•).

8

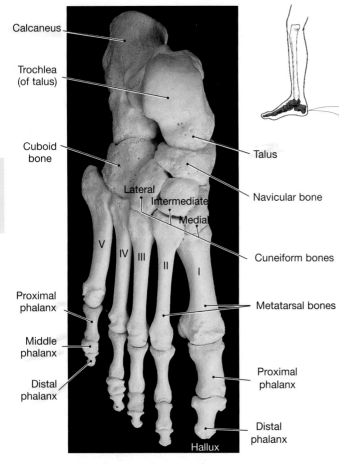

(a) Superior view, right foot

Calcaneus

Trochlea (of talus)

Cuboid bone

Lateral
Intermediate
Medial

V IV III II I

Proximal phalanx

Middle phalanx

Distal phalanx

Hallux

Talus

Navicular bone

Cuneiform bones

Metatarsal bones

Proximal phalanx

Distal phalanx

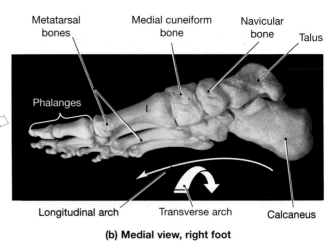

Metatarsal bones

Medial cuneiform bone

Navicular bone

Talus

Phalanges

I

Longitudinal arch

Transverse arch

Calcaneus

(b) Medial view, right foot

•**FIGURE 8-14** Bones of the Ankle and Foot. *Plate 8.14d, Scan 8*

The metatarsal bones are identified by Roman numerals I–V, proceeding from medial to lateral across the sole. Proximally, metatarsal bones I–III articulate with the three cuneiform bones, and metatarsal bones IV and V articulate with the cuboid bone. Distally, each metatarsal bone articulates with a different proximal phalanx. The **phalanges,** or toe bones (Figure 8-14•), have the same anatomical organization as the fingers. The toes contain 14 phalanges. The **hallux,** or great toe, has two phalanges (*proximal* and *distal),* and the other four toes have three phalanges apiece (*proximal, middle,* and *distal).*

Arches of the Foot   Weight transfer occurs along the **longitudinal arch** of the foot (Figure 8-14b•). Ligaments and tendons maintain this arch by tying the calcaneus to the distal portions of the metatarsal bones. However, the lateral, or *calcaneal,* portion of the longitudinal arch has much less curvature than the medial, *talar* portion, in part because the talar portion has considerably more elasticity. As a result, the medial plantar surface of the foot remains elevated and the muscles, nerves, and blood vessels that supply the inferior surface are not squeezed between the metatarsal bones and the ground. In the condition known as *flatfeet,* normal arches are lost ("fall") or never form. In-

dividuals with this condition cannot walk long distances without discomfort; for this reason they cannot enlist in the U.S. Army.

The elasticity of the talar portion of the longitudinal arch absorbs the shocks from sudden changes in weight loading. For example, the stresses that running or ballet dancing place on the toes are cushioned by the elasticity of this portion of the arch. The degree of curvature changes from the medial to the lateral borders of the foot, so a **transverse arch** also exists.

While you stand normally, your body weight is distributed evenly between the calcaneus and the distal ends of the metatarsal bones. The amount of weight transferred forward depends on the position of the foot and the placement of body weight. During *dorsiflexion* of the foot, such as when you "dig in your heels," all your body weight rests on the calcaneus. During *plantar flexion,* when you "stand on tiptoe," the talus and calcaneus transfer your weight to the metatarsal bones and phalanges through the more anterior tarsal bones.

Congenital Talipes Equinovarus

Congenital talipes equinovarus *(clubfoot)* results from an inherited developmental abnormality that affects 2 in 1000 births. Boys are affected roughly twice as often as girls. One or both feet may be involved, and the condition can be mild, moderate, or severe. The underlying problem is abnormal muscle development that distorts growing bones and joints. In most cases, the tibia, ankle, and foot are affected, and the feet are turned medially and inverted. The longitudinal arch is exaggerated. If both feet are involved, the soles face one another. Prompt treatment with casts or other supports in infancy helps alleviate the problem, and fewer than half the cases require surgery. Kristi Yamaguchi, an Olympic gold medalist in figure skating, was born with this condition. AM *Problems with the Ankle and Foot*

☑ The fibula does not participate in the knee joint or bear weight. When it is fractured, however, walking becomes difficult. Why?

☑ While jumping off the back steps at his house, 10-year-old Joey lands on his right heel and breaks his foot. Which foot bone is most likely broken?

☑ Which foot bone transmits the weight of the body from the tibia toward the toes?

☑ Ballet dancers strengthen their thigh muscles by placing one leg laterally on a waist-high bar and contracting against the resistance. How would the increased strength of these muscles affect the shape of the femur?

INDIVIDUAL VARIATION IN THE SKELETAL SYSTEM

A comprehensive study of a human skeleton can reveal important information about the individual. We can estimate a person's muscular development and body weight from the development of various ridges and general bone mass. Details such as the condition of the teeth or the presence of healed fractures give an indication of the individual's medical history. Two important details, gender and age, can be determined or closely estimated on the basis of measurements indicated in Tables 8-1 and 8-2. AM *EMBRYOLOGY SUMMARY 8: The Development of the Appendicular Skeleton*

Table 8-1 identifies characteristic differences between the skeletons of males and females, but not every skeleton shows every feature in classic detail. Many differences, including markings on the skull, cranial capacity, and general skeletal features, reflect differences in average body size, muscle mass, and muscular strength. The general changes in the skeletal system that take place with age are summarized in Table 8-2. Note that these changes begin at age 1 and continue throughout life. For example, the epiphyseal cartilages begin to fuse at about age 3, and degenerative changes in the normal skeletal system, such as a reduction in mineral content in the bony matrix, typically do not begin until age 30–45.

An understanding of individual variation and of the normal timing of skeletal development is important in clinical diagnosis and treatment. Several professions focus on specific aspects of skeletal form and function. Each specialty has a different perspective with its own techniques, traditions, and biases. For example, a person with back pain may consult an *orthopedist,* an *osteopath,* or a *chiropractor.* AM *A Matter of Perspective*

TABLE 8-1 Gender Differences in the Human Skeleton

Region/Feature	Male (as compared with female)	Female (as compared with male)
Skull		
General appearance	Heavier, rougher	Lighter, smoother
Forehead	More sloping	More vertical
Sinuses	Larger	Smaller
Cranium	About 10% larger (average)	About 10% smaller
Mandible	Larger, more robust	Smaller, lighter
Teeth	Larger	Smaller
Pelvis		
General appearance	Narrower, more robust, rougher	Broader, lighter, smoother
Pelvic inlet	Heart-shaped	Oval to round
Iliac fossa	Deeper	Shallower
Ilium	Extends farther superior to sacral articulation	More vertical; less extension superior to sacroiliac joint
Angle inferior to pubic symphysis	Under 90°	100° or more *(see Figure 8-10, p. 236)*
Acetabulum	Directed laterally	Faces slightly anteriorly as well as laterally
Obturator foramen	Oval	Triangular
Ischial spine	Points medially	Points posteriorly
Sacrum	Long, narrow triangle with pronounced sacral curvature	Broad, short triangle with less curvature
Coccyx	Points anteriorly	Points inferiorly
Other skeletal elements		
Bone weight	Heavier	Lighter
Bone markings	More prominent	Less prominent

8

TABLE 8-2 Age-Related Changes in the Skeleton

Region/Feature	Event(s)	Age (Years)
General skeleton		
Bony matrix	Reduction in mineral content; increased risk of osteoporosis	Begins at ages 30–45; values differ for males versus females between ages 45 and 65; similar reductions occur in both genders after age 65
Markings	Reduction in size, roughness	Gradual reduction with increasing age and decreasing muscular strength and mass
Skull		
Fontanels	Closure	Completed by age 2
Metopic suture	Fusion	2–8
Occipital bone	Fusion of ossification centers	1–4
Styloid process	Fusion with temporal bone	12–16
Hyoid bone	Complete ossification and fusion	25–30
Teeth	Loss of "baby teeth"; appearance of secondary dentition; eruption of posterior molars	Detailed in Chapter 24 (digestive system)
Mandible	Loss of teeth; reduction in bone mass; change in angle at mandibular notch	Accelerates in later years (60+)
Vertebrae		
Curvature	Development of major curves	3 months–10 years
Intervertebral discs	Reduction in size, percentage contribution to height	Accelerates in later years (60+)
Long bones		
Epiphyseal cartilages	Fusion	Begins about age 3; ranges vary, but general analysis permits determination of approximate age
Pectoral and pelvic girdles		
Epiphyses	Fusion	Relatively narrow ranges (14–16, 16–18, 22–25) that increase accuracy of age estimates

[EXPLORE] MediaLab

Although you are now familiar with the basic parts of the appendicular skeleton, these bones of the pectoral and pelvic girdles and the limbs have many markings, processes, and functions. To gain a better understanding of these bones and their features, use the resources available on both the CD-ROM and the Companion Website. The CD-ROM includes animations and tutorials; the website builds on those resources with text content and interactive problems as well as communication avenues for sharing ideas with your class. To complete the following Web Exercises, visit the Media-Lab in Chapter 8 of your Companion Website.

Web Exploration 1

Estimated time for completion = 10 minutes

How did the bones of the appendicular skeleton get their names? What interesting facts are known about these bones? After reading about each bone in the text, select the keyword **FUN FACTS** on the Companion Website. From the menu, choose five bones and the pelvis. Click on each bone name, and read the information provided there. Report any new information you have learned; be sure to reference the source correctly. If you wish, use the communications tools on the Companion Website to post your findings to a chat room so that others can learn from your explorations.

Web Exploration 2

Estimated time for completion = 10 minutes

Having completed your study of the skeletal system, you might find it fun to test your understanding of that system. Select the keyword **BONE QUIZ** on the Companion Website to get a series of bone-related questions. Write your answers down; then click on the Answer button to see how you did. Now go back to the Companion Website, and select the keyword **BROKEN BONE QUIZ.** Try the questions there. Were you able to apply your knowledge of bones to this situation? If you wish, use the communications tools on the Companion Website to send your quiz scores to your instructor.

CHAPTER REVIEW

SELECTED CLINICAL TERMINOLOGY
Terms Discussed in This Chapter

carpal tunnel syndrome: An inflammation of the tissues at the anterior wrist, causing compression of adjacent tendons and nerves. Symptoms are pain and a loss of wrist mobility. *(p. 233)*

congenital talipes equinovarus, or *clubfoot:* A congenital deformity affecting one or both feet. It develops secondary to abnormalities in muscular development. *(p. 240)*
flatfeet: The loss or absence of a longitudinal arch. *(p. 240)*

8

AM **Additional Terms Discussed in the *Applications Manual***

dancer's fracture: A fracture of the fifth metatarsal bone, generally near its proximal articulation.

STUDY OUTLINE

INTRODUCTION   p. 227
1. The **appendicular skeleton** includes the bones of the upper and lower limbs and the pectoral and pelvic girdles, which connect the limbs to the trunk. *(Figure 8-1)*

THE PECTORAL GIRDLE AND UPPER LIMBS   p. 228
1. Each upper limb articulates with the trunk via the **pectoral girdle,** or *shoulder girdle,* which consists of the **scapulae** and **clavicles.**

The Pectoral Girdle   p. 228
2. On each side, a clavicle and scapula position the shoulder joint, help move the upper limb, and provide a base for muscle attachment. *(Figures 8-2, 8-3)*
3. Both the **coracoid process** and the **acromion** are attached to ligaments and tendons. The **scapular spine** crosses the scapular **body.** *(Figure 8-3)*

The Upper Limbs   p. 230
4. The scapula articulates with the **humerus** at the shoulder *(glenohumeral)* joint. The **greater** and **lesser tubercles** are important sites of muscle attachment. *(Figure 8-4)*
5. The humerus articulates with the **radius** and **ulna,** the bones of the forearm, at the elbow joint. *(Figure 8-5)*
6. The **carpal bones** of the wrist, or **carpus,** form two rows. The distal row articulates with the five **metacarpal bones.** Four of the fingers contain three **phalanges;** the **pollex** (thumb) has only two phalanges. *(Figure 8-6)*

THE PELVIC GIRDLE AND LOWER LIMBS   p. 233
1. The bones of the **pelvic girdle** are more massive than those of the pectoral girdle.

The Pelvic Girdle   p. 234
2. The pelvic girdle consists of two **ossa coxae.** Each os coxae forms through the fusion of an **ilium,** an **ischium,** and a **pubis.** *(Figure 8-7)*
3. The ilium is the largest hip bone. Inside the **acetabulum,** the ilium is fused to the ischium (posteriorly) and the pubis (anteriorly). The **pubic symphysis** limits movement between the pubic bones of the left and right hip bones. *(Figures 8-7, 8-8)*
4. The **pelvis** consists of the hip bones, the sacrum, and the coccyx. It is subdivided into the **false** *(greater)* **pelvis** and the **true** *(lesser)* **pelvis.** *(Figures 8-8–8-10)*

The Lower Limbs   p. 237
5. The **femur** is the longest bone in the body. It articulates with the **tibia** at the knee joint. *(Figures 8-11, 8-13)*
6. The **patella** is a large sesamoid bone. *(Figure 8-12)*
7. The **fibula** parallels the tibia laterally. *(Figure 8-13)*
8. The **tarsus,** or ankle, has seven **tarsal bones.** *(Figure 8-14)*
9. The basic organizational pattern of the **metatarsal bones** and **phalanges** of the foot resembles that of the hand. All the toes have three phalanges except for the **hallux,** which has two phalanges. *(Figure 8-14)*
10. When a person stands normally, most of the body weight is transferred to the **calcaneus** and the rest is passed on to the five metatarsal bones. Weight transfer occurs along the **longitudinal arch;** there is also a **transverse arch.** *(Figure 8-14)*

INDIVIDUAL VARIATION IN THE SKELETAL SYSTEM   p. 241
1. Studying a human skeleton can reveal important information such as race, medical history, weight, gender, body size, muscle mass, and age. *(Tables 8-1, 8-2)*
2. Age-related changes and events take place in the skeletal system. These changes begin at about age 1 and continue throughout life. *(Table 8-2)*

REVIEW QUESTIONS

LEVEL 1 REVIEWING FACTS AND TERMS
1. The only direct connection between the pectoral girdle and the axial skeleton is where the
   (a) clavicle articulates with the humerus
   (b) clavicle articulates with the manubrium of the sternum
   (c) os coxae articulates with the femur
   (d) vertebral column articulates with the skull

2. The presence of foramina in bones indicates the position of
   (a) tendons and ligaments   (b) nerves and blood vessels
   (c) ridges and flanges   (d) a, b, and c are correct
3. At the glenoid cavity, the scapula articulates with the proximal end of the
   (a) humerus   (b) radius
   (c) ulna   (d) femur

4. In anatomical position, the ulna lies
   (a) medial to the radius   (b) lateral to the radius
   (c) inferior to the radius   (d) superior to the radius
5. The proximal carpal bones include the
   (a) trapezium, trapezoid, capitate, hamate
   (b) scaphoid, capitate, lunate, hamate
   (c) trapezium, triquetrum, trapezoid, pisiform
   (d) scaphoid, lunate, triquetrum, pisiform
6. The distal carpal bones are the
   (a) scaphoid, pisiform, capitate, hamate
   (b) trapezium, trapezoid, capitate, hamate
   (c) scaphoid, lunate, triquetrum, pisiform
   (d) trapezium, lunate, triquetrum, hamate
7. The bones of the hand articulate distally with the
   (a) carpal bones   (b) ulna and radius
   (c) metacarpal bones   (d) phalanges
8. Each os coxae of the pelvic girdle consists of three fused bones:
   (a) ulna, radius, humerus
   (b) ilium, ischium, pubis
   (c) femur, tibia, fibula
   (d) hamate, capitate, trapezium
9. The large foramen between the pubic and ischial rami is the
   (a) foramen magnum   (b) suborbital foramen
   (c) acetabulum   (d) obturator foramen
10. Which of the following is an adaptation for childbearing?
   (a) inferior angle between the pubic bones 100° or more
   (b) a relatively broad, low pelvis
   (c) less curvature of the sacrum and coccyx
   (d) a, b, and c are correct
11. The epiphysis of the femur articulates with the pelvis at the
   (a) pubic symphysis   (b) acetabulum
   (c) sciatic notch   (d) obturator foramen
12. The large medial bone of the leg is the
   (a) tibia   (b) femur
   (c) fibula   (d) humerus
13. The selection that includes *only* tarsal bones is
   (a) scaphoid, pisiform, capitate, hamate, talus
   (b) calcaneus, navicular, lunate, capitate, talus
   (c) talus, calcaneus, cuboid, navicular, cuneiforms
   (d) navicular, scaphoid, capitate, talus, calcaneus
14. The calcaneus is the attachment site of the
   (a) calcanean tendon   (b) muscles of the calf
   (c) talus   (d) a, b, and c are correct

15. Name the skeletal components of the pectoral girdle.
16. Which two large scapular processes are associated with the shoulder joint?
17. Which bones make up the arm and forearm?
18. Which two movements are associated with the proximal radioulnar articulation?
19. List the eight carpal bones.
20. Which bones constitute the lower limb?
21. Name the components of each os coxae.
22. How do anatomists distinguish between the false (greater) pelvis and the true (lesser) pelvis?
23. Which seven bones make up the ankle (tarsus)?
24. Distinguish between the pollex and the hallux.

LEVEL 2 REVIEWING CONCEPTS

25. Why are injuries of the clavicle common?
26. What is the difference in skeletal structure between the pelvic girdle and the pelvis?
27. Why is the tibia but not the fibula involved in the transfer of weight to the ankle and foot?
28. What are the direct anatomical connections between the skeletal and muscular systems?
29. Why would an instructor teaching self-defense advise a student to strike an assailant's clavicle in an attack?
30. Why is it necessary for the bones of the pelvic girdle to be more massive than the bones of the pectoral girdle?

LEVEL 3 CRITICAL THINKING AND CLINICAL APPLICATIONS

31. To settle a bet, you need to measure the length of your lower limb (femur and tibia). What landmarks would you use to make the measurement?
32. While Fred, a fireman, is fighting a fire in a building, part of the ceiling collapses and a beam strikes him on his left shoulder. He is rescued but has a great deal of pain in his shoulder. He cannot move his arm properly, especially in the anterior direction. His clavicle is not broken, and his humerus is intact. What is the probable nature of Fred's injury?
33. Cindy is anxiously awaiting the birth of her first child. As she gets closer to term, she has an ultrasound scan. After seeing the results, her physician makes some calculations and informs Cindy that she will probably have to have a cesarean section. What clues might tell the physician that Cindy can't deliver the baby by natural childbirth?

CHAPTER 9

Articulations

CHAPTER OUTLINE AND OBJECTIVES

INTRODUCTION   p. 246

A CLASSIFICATION OF JOINTS   p. 246
  Synarthroses (Immovable Joints)   p. 246
  Amphiarthroses (Slightly Movable Joints)   p. 247
  Diarthroses (Freely Movable Joints)   p. 247
① Contrast the major categories of joints, and explain the relationship between structure and function for each category.
② Describe the basic structure of a synovial joint, identifying possible accessory structures and their functions.

ARTICULAR FORM AND FUNCTION   p. 250
  Describing Dynamic Motion   p. 250
  Types of Movements   p. 251
  A Structural Classification of Synovial Joints   p. 254
③ Describe the dynamic movements of the skeleton.
④ List the types of synovial joints, and discuss how the characteristic motions of each type are related to its anatomical structure.

REPRESENTATIVE ARTICULATIONS   p. 256
  Intervertebral Articulations   p. 256
  The Shoulder Joint   p. 258
  The Elbow Joint   p. 260
  The Hip Joint   p. 260
  The Knee Joint   p. 261
⑤ Describe the articulations between the vertebrae of the vertebral column.
⑥ Describe the structure and function of the shoulder, elbow, hip, and knee joints.
⑦ Explain the relationship between joint strength and mobility, using specific examples.

Our movements must conform to the limitations of the human body. For example, it is not possible to bend the shaft of the humerus or femur; movements are restricted to joints, where bones articulate. The same is true of artificial limbs, which use mechanical equivalents of human joints. Very early in life, we learn to work within those limitations. Each joint—natural or artificial—will tolerate a specific range of motion. A variety of bony surfaces, cartilages, ligaments, tendons, and muscles— or metal and plastic balls, shafts, and screws—work together to keep movement within the normal range. The stronger the joint, the more restrictive the limitations are likely to be. In this chapter, we will examine representative joints and explore the relationship between structural stability and the range of motion.

AGING AND ARTICULATIONS   p. 265

BONES AND MUSCLES   p. 265

MEDIALAB and CHAPTER REVIEW   p. 266

In the last two chapters, you have become familiar with the individual bones of the skeleton. Your understanding of the skeletal system is still incomplete, however, because in life the skeleton is always in motion. Most of our activities—from breathing or speaking to writing or running—involve movements of the skeleton. The bones of the skeleton are rigid, and movements can occur only at **articulations,** or joints, where two bones interconnect. The characteristic structure of a joint determines the type of movement that may occur. Each joint reflects a compromise between the need for strength and the need for mobility.

This chapter compares the relationships between articular form and function. We will use several examples that range from relatively immobile but very strong (the intervertebral articulations) to highly mobile but relatively weak (the shoulder).

A CLASSIFICATION OF JOINTS

Two classification methods are used to categorize joints. The first, and the one we will use in this chapter, considers the amount of movement permitted, a property known as the *range of motion* (Table 9-1). These functional categories are further subdivided on the basis of the anatomical structure of the joint or the range of motion permitted:

1. An *immovable joint* is a **synarthrosis** (sin-ar-THRŌ-sis; *syn,* together + *arthros,* joint). A synarthrosis can be *fibrous* or *cartilaginous,* depending on the nature of the connection. Over time, the two bones may fuse.

2. A *slightly movable joint* is an **amphiarthrosis** (am-fē-ar-THRŌ-sis; *amphi,* on both sides). An amphiarthrosis is *fibrous* or *cartilaginous,* depending on the nature of the connection between the opposing bones.

3. A *freely movable joint* is a **diarthrosis** (dī-ar-THRŌ-sis; *dia,* through), or *synovial joint.* Diarthroses are subdivided according to the amount or range of movement permitted.

The second classification scheme relies solely on the anatomical organization of the joint, without regard to the degree of permitted movement. In this framework, joints are classified as *bony, fibrous, cartilaginous,* or *synovial* (Table 9-2). We will use the functional classification rather than the anatomical one, because we are primarily interested in how joints work.

SYNARTHROSES (IMMOVABLE JOINTS)

At a synarthrosis, the bony edges are quite close together and may even interlock. These extremely strong joints are located where movement between the bones must be prevented. There are four major types of synarthrotic joints:

TABLE 9-1 A Functional Classification of Articulations

Functional Category	Structural Category	Description	Example
Synarthrosis (no movement)	**Fibrous**		
	Suture	Fibrous connections plus interlocking projections	Between the bones of the skull
	Gomphosis	Fibrous connections plus insertion in alveolar process	Between the teeth and jaws
	Cartilaginous		
	Synchondrosis	Interposition of cartilage plate	Epiphyseal cartilages
	Bony fusion		
	Synostosis	Conversion of other articular form to solid mass of bone	Portions of the skull, epiphyseal lines
Amphiarthrosis (little movement)	**Fibrous**		
	Syndesmosis	Ligamentous connection	Between the tibia and fibula
	Cartilaginous		
	Symphysis	Connection by a fibrocartilage pad	Between right and left pubic bones of pelvis; between adjacent vertebral bodies along vertebral column
Diarthrosis (free movement)	**Synovial**	Complex joint bounded by joint capsule and containing synovial fluid	Numerous; subdivided by range of movement *(see Figure 9-6)*
	Monaxial	Permits movement in one plane	Elbow, ankle
	Biaxial	Permits movement in two planes	Ribs, wrist
	Triaxial	Permits movement in all three planes	Shoulder, hip

TABLE 9-2 A Structural Classification of Articulations

Structure	Type	Functional Category	Example
Bony fusion	Synostosis (illustrated)	Synarthrosis	Frontal bone — Metopic suture (fusion)
Fibrous joint	Suture (illustrated)	Synarthrosis	Skull — Lambdoidal suture
	Gomphosis	Synarthrosis	
	Syndesmosis	Amphiarthrosis	
Cartilaginous joint	Synchondrosis	Synarthrosis	Pelvis — Pubic symphysis
	Symphysis (illustrated)	Amphiarthrosis	
Synovial joint	Monaxial Biaxial Triaxial (illustrated)	Diarthroses	Synovial joint

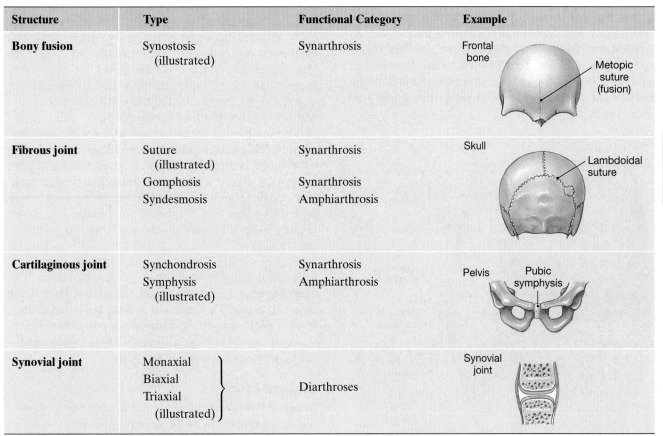

9

1. *Sutures.* A **suture** (*sutura,* a sewing together) is a synarthrotic joint located only between the bones of the skull. The edges of the bones are interlocked and bound together at the suture by dense connective tissue.

2. *Gomphoses.* A **gomphosis** (gom-FŌ-sis; *gomphosis,* a bolting together) is a synarthrosis that binds the teeth to bony sockets in the maxillary bone and mandible. The fibrous connection between a tooth and its socket is a *periodontal* (pe-rē-ō-DON-tal) *ligament* (*peri,* around + *odontos,* tooth).

3. *Synchondroses.* A **synchondrosis** (sin-kon-DRŌ-sis; *syn,* together + *chondros,* cartilage) is a rigid, cartilaginous bridge between two articulating bones. The epiphyseal cartilage is a synchondrosis that connects the diaphysis of a long bone with an epiphysis, even though the two bones are part of the same skeletal element. ∞ *p. 178* Another example is the cartilaginous connection between the ends of the first pair of vertebrosternal ribs and the sternum.

4. *Synostoses.* A **synostosis** (sin-os-TŌ-sis) is a totally rigid, immovable joint created when two separate bones fuse and the boundary between them disappears. The *metopic suture* of the frontal bone and the epiphyseal lines of mature bones are synostoses. ∞ *pp. 201, 178*

AMPHIARTHROSES (SLIGHTLY MOVABLE JOINTS)

An amphiarthrosis is another compromise between mobility and strength. It permits more movement than a synarthrosis but is much stronger than a freely movable joint. The articulating bones are connected by collagen fibers or cartilage. There are two major types of amphiarthrotic joints:

1. At a **syndesmosis** (sin-dez-MŌ-sis; *desmos,* a band or ligament), bones are connected by a ligament. One example is the distal articulation between the tibia and fibula.

2. At a **symphysis,** or *symphyseal joint,* the articulating bones are separated by a wedge or pad of fibrocartilage. The articulation between the bodies of vertebrae (at the *intervertebral disc*) and the connection between the two pubic bones (the *pubic symphysis*) are examples of this type of joint.

DIARTHROSES (FREELY MOVABLE JOINTS)

Diarthroses, or **synovial** (si-NŌ-vē-ul) **joints,** permit a wider range of motion than do other types of joints. A synovial joint is surrounded by a fibrous

9

articular capsule, and a *synovial membrane* lines the articular cavity. Recall that a synovial membrane consists of areolar tissue covered by an incomplete epithelial layer. The synovial fluid that fills the joint cavity originates in the areolar tissue of the synovial membrane. ∞ *p. 131* Figure 9-1• surveys the structure of synovial joints. These joints are typically located at the ends of long bones, such as those of the upper and lower limbs.

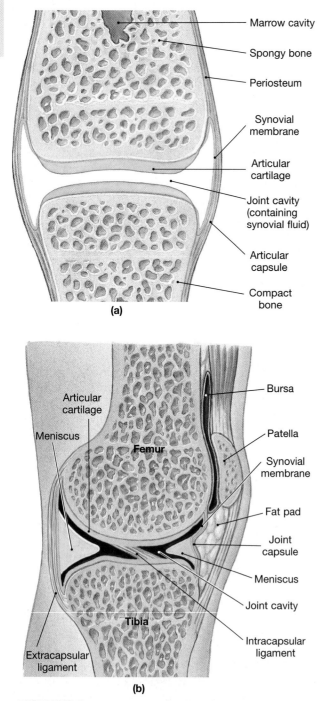

(a)

(b)

•**FIGURE 9-1** **The Structure of a Synovial Joint. (a)** A diagrammatic view of a simple articulation. **(b)** A simplified sectional view of the knee joint.

Articular Cartilages

Under normal conditions, the bony surfaces at a synovial joint cannot contact one another, because the articulating surfaces are covered by special **articular cartilages.** Articular cartilages resemble hyaline cartilages elsewhere in the body. However, articular cartilages have no perichondrium (the fibrous sheath described in Chapter 4), and the matrix contains more water than that of other cartilages. ∞ *p. 125*

The surfaces of the articular cartilages are slick and smooth. This feature alone can reduce friction during movement at the joint. However, even when pressure is applied across a joint, the smooth articular cartilages do not touch one another, because they are separated by a thin film of synovial fluid in the joint cavity (Figure 9-1a•). This fluid acts as a lubricant, minimizing friction.

Proper synovial function can continue only if the articular cartilages retain their normal structure. If an articular cartilage is damaged, the matrix may begin to break down. The exposed surface will then change from a slick, smooth, gliding surface to a rough feltwork of bristly collagen fibers. This feltwork drastically increases friction at the joint.

Synovial Fluid

Synovial fluid resembles interstitial fluid but contains a high concentration of proteoglycans secreted by fibroblasts of the synovial membrane. It is a thick, viscous solution with the consistency of heavy molasses. The synovial fluid within a joint has three primary functions:

1. *Lubrication.* The articular cartilages are like sponges filled with synovial fluid. When part of an articular cartilage is compressed, some of the synovial fluid is squeezed out of the cartilage and into the space between the opposing surfaces. This thin layer of fluid markedly reduces friction between moving surfaces, just as a thin film of water reduces friction between a car's tires and a highway. When the compression stops, synovial fluid is sucked back into the articular cartilages.

2. *Nutrient distribution.* The total quantity of synovial fluid in a joint is normally less than 3 ml, even in a large joint such as the knee. This small volume of fluid must circulate continuously to provide nutrients and a waste-disposal route for the chondrocytes of the articular cartilages. The synovial fluid circulates whenever the joint moves, and the compression and reexpansion of the articular cartilages pump synovial fluid into and out of the cartilage matrix. As the synovial fluid flows through the areolar tissue of the synovial membrane, waste products are absorbed and additional nutrients are obtained by diffusion across capillary walls.

3. *Shock absorption.* Synovial fluid cushions shocks in joints that are subjected to compression. For example, your hip, knee, and ankle joints are compressed as you

walk and are more severely compressed when you jog or run. When the pressure across a joint suddenly increases, the synovial fluid lessens the shock by distributing it evenly across the articular surfaces.

Accessory Structures

Synovial joints may have a variety of accessory structures, including pads of cartilage or fat, ligaments, tendons, and bursae (Figure 9-1b•).

Cartilages and Fat Pads   In several joints, including the knee (Figure 9-1b•), accessory structures may lie between the opposing articular surfaces. These include menisci and fat pads. A **meniscus** (men-IS-kus; a crescent; plural, *menisci*) is a pad of fibrocartilage situated between opposing bones within a synovial joint. Menisci, or *articular discs,* may subdivide a synovial cavity, channel the flow of synovial fluid, or allow for variations in the shapes of the articular surfaces.

Fat pads are localized masses of adipose tissue covered by a layer of synovial membrane. They are commonly superficial to the joint capsule (Figure 9-1b•). Fat pads protect the articular cartilages and act as packing material for the joint. When the bones move, the fat pads fill in the spaces created as the joint cavity changes shape.

Ligaments   The joint capsule that surrounds the entire joint is continuous with the periostea of the articulating bones. **Accessory ligaments** are localized thickenings of the capsule. These ligaments reinforce and strengthen the capsule, and they may also limit rotation at the joint. **Extracapsular ligaments** interconnect the articulating bones and pass across the outside of the capsule. These ligaments provide additional support to the wall of the joint. **Intracapsular ligaments,** inside the capsule, help prevent extreme movements that might otherwise damage the joint. In a **sprain,** a ligament is stretched to the point at which some of the collagen fibers are torn, but the ligament as a whole survives and the joint is not damaged. Ligaments are very strong, and one of the attached bones typically breaks before the ligament tears. In general, a broken bone heals much more quickly and effectively than does a torn ligament.

Tendons   Although not part of the articulation itself, tendons passing across or around a joint may limit the range of motion and provide mechanical support. For example, tendons associated with the muscles of the arm provide much of the bracing for the shoulder joint.

Bursae   **Bursae** are small, fluid-filled pockets in connective tissue. They contain synovial fluid and are lined by a synovial membrane. Bursae may be connected to the joint cavity or may be separate from it. They form where a tendon or ligament rubs against other tissues. They reduce friction and act as shock absorbers. Bursae are located around most synovial joints, such as the shoulder joint. **Synovial tendon sheaths** are tubular bursae that surround tendons where they cross bony surfaces. Bursae may also appear deep to the skin covering a bone or within other connective tissues exposed to friction or pressure. Bursae that develop in abnormal locations, or because of abnormal stresses, are called *adventitious bursae.*

Bursitis

When bursae become inflamed, causing pain in the affected area whenever the tendon or ligament moves, the condition of **bursitis** exists. Inflammation can result from the friction of repetitive motion, pressure over the joint, irritation by chemical stimuli, infection, or trauma. Bursitis associated with repetitive motion typically occurs at the shoulder; musicians, golfers, pitchers, and tennis players may develop bursitis there. The most common pressure-related bursitis is a **bunion.** Bunions form over the base of the great toe as a result of friction and distortion of the first metatarsophalangeal joint by tight shoes, especially narrow shoes with pointed toes.

We have special names for bursitis at other locations, indicating the occupations most often associated with them. In "housemaid's knee," which accompanies prolonged kneeling, the affected bursa lies between the patella and the skin. The condition of "student's elbow" is a form of bursitis that can result from propping your head up with your arm on a desk while you read your A & P textbook.

Factors That Stabilize Joints

A joint cannot be both highly mobile and very strong. The greater the range of motion at a joint, the weaker it becomes. A synarthrosis, the strongest type of joint, permits no movement, whereas a diarthrosis, such as the shoulder, permits a broad range of movements. Any mobile diarthrosis will be damaged by movement beyond its normal range of motion. Several factors are responsible for limiting the range of motion, stabilizing the joint, and reducing the chance of injury:

- The collagen fibers of the joint capsule and any accessory, extracapsular, or intracapsular ligaments.
- The shapes of the articulating surfaces, which may prevent movement in specific directions.
- The presence of other bones, skeletal muscles, or fat pads around the joint.
- Tension in tendons attached to the articulating bones. When a skeletal muscle contracts and pulls on a tendon, movement in a specific direction may be either encouraged or opposed.

The pattern of stabilizing structures varies among joints. For example, the hip joint is stabilized by the shapes of the bones (the head of the femur projects into the acetabulum), a heavy capsule, intracapsular and extracapsular ligaments, tendons, and massive muscles. It is therefore very strong and very stable. In contrast, the elbow, another very stable joint, gains its stability primarily from the interlocking of the articulating bones, with additional support of the capsule and associated ligaments. In general, the more stable the joint, the more restricted its range of motion. The shoulder joint, the most mobile synovial joint, relies only on the surrounding ligaments, muscles, and tendons for stability. It is thus fairly weak.

When reinforcing structures cannot protect a joint from extreme stresses, a **dislocation,** or **luxation** (luks-Ā-shun), results. In a dislocation, the articulating surfaces are forced out of position. This displacement can damage the articular cartilages, tear ligaments, or distort the joint capsule. Although the *inside* of a joint has no pain receptors, nerves that monitor the capsule, ligaments, and tendons are quite sensitive, so dislocations are very painful. The damage accompanying a partial dislocation, or **subluxation** (sub-luks-Ā-shun), is less severe. People who are "double-jointed" have joints that are weakly stabilized. Although their joints permit a greater range of motion than do those of other individuals, they are more likely to suffer partial or complete dislocations.

☑ What common characteristics do typical synarthrotic and amphiarthrotic joints share?

☑ In a newborn infant, the large bones of the skull are joined by fibrous connective tissue. Which type of joints are these? These bones later grow, interlock, and form immovable joints. Which type of joints are these?

☑ Why would improper circulation of synovial fluid lead to the degeneration of articular cartilages in the affected joint?

ARTICULAR FORM AND FUNCTION

To *understand* human movement, you must be aware of the relationship between structure and function at each articulation. To *describe* human movement, you need a frame of reference that permits accurate and precise communication. We can classify the synovial joints according to their anatomical and functional properties. To demonstrate the basis for that classification, we will use a simple model to describe the movements that occur at a typical synovial joint.

DESCRIBING DYNAMIC MOTION

Take a pencil (or a pen) as your model and stand it upright on the surface of a desk or table (Figure 9-2a●). The pencil represents a bone, and the desk is an articular surface. A little imagination and a lot of twisting, pushing, and pulling will demonstrate that there are only three ways to move the model. Considering them one at a time will provide a frame of reference for us to analyze complex movements:

(a) **Initial position**
Pencil at right angles to surface.

(b) **Gliding movement**
Pencil remains vertical, but tip moves away from point of origin.

(c) **Angular motion**
Tip remains stationary, but shaft changes angle relative to the surface.

(d) **Circumduction (a type of angular motion)**
Tip remains stationary while the shaft, held at an angle less than 90°, describes a complete circle.

(e) **Rotation**
With tip at same point, the angle of the shaft remains unchanged as the shaft spins around its longitudinal axis.

●**FIGURE 9-2** A Simple Model of Articular Motion

Possible movement 1: *The point can move.* If you hold the pencil upright but do not secure the point, you can push the pencil across the surface. This kind of motion, *gliding* (Figure 9-2b●), is an example of **linear motion.** You could slide the point forward or backward, from side to side, or diagonally. However you move the pencil, the motion can be described by using two lines of reference. One line represents forward/backward motion, and the other left/right movement. For example, a simple movement along one axis could be described as "forward 1 cm" or "left 2 cm." A diagonal movement could be described with both axes, as in "backward 1 cm and to the right 2.5 cm."

Possible movement 2: *The shaft can change its angle with the surface.* With the tip held in position, you can move the free (eraser) end forward and backward, from side to side, or at some intermediate angle. These movements, which change the angle between the shaft and the articular surface, are examples of **angular motion** (Figure 9-2c•). We can describe such motion by the angle the shaft makes with the surface.

Any angular movement can be described with reference to the same two axes (forward/backward, left/right) and the angular change (in degrees). In one instance, however, a special term is used to describe a complex angular movement. Grasp the free end of the pencil and move it in any direction until the shaft is no longer vertical. Now swing that end through a complete circle (Figure 9-2d•). This movement, which corresponds to the path of your arm when you draw a large circle on a chalkboard, is very difficult to describe. Anatomists avoid the problem by using a special term, **circumduction** (sir-kum-DUK-shun; *circum,* around), for this type of angular motion.

> *Possible movement 3:* The shaft can rotate. If you prevent movement of the base and keep the shaft vertical, you can spin the shaft around its longitudinal axis. This movement is called **rotation** (Figure 9-2e•). Several articulations permit partial rotation, but none can rotate freely. Such a movement would hopelessly tangle the blood vessels, nerves, and muscles that cross the joint.

An articulation that permits movement along only one axis is called **monaxial** (mon-AKS-ē-ul). In the above model, if an articulation permits only angular movement in the forward/backward plane or prevents any movement other than rotation around its longitudinal axis, it is monaxial. If movement can occur along two axes, the articulation is **biaxial** (bī-AKS-ē-ul). If the pencil could undergo angular motion in the forward/backward *and* left/right planes, but not rotation, it would be biaxial. The most mobile joints permit a combination of angular movement and rotation. These are said to be **triaxial** (trī-AKS-ē-ul).

Joints that permit gliding allow only small amounts of movement. These joints may be called *nonaxial,* because they permit only small sliding movements, or *multiaxial,* because sliding may occur in any direction.

TYPES OF MOVEMENTS

In descriptions of motion at synovial joints, phrases such as "bend the leg" or "raise the arm" are not sufficiently precise. Anatomists use descriptive terms that have specific meanings. We will consider these motions with reference to the basic categories of movement we discussed previously: gliding, angular motion, and rotation.

Linear Motion (Gliding)

In **gliding,** two opposing surfaces slide past one another, as in possible movement 1. Gliding occurs between the surfaces of articulating carpal bones, between tarsal bones, and between the clavicles and the sternum. The movement can occur in almost any direction, but the amount of movement is slight, and

rotation is generally prevented by the capsule and associated ligaments.

Angular Motion

Examples of angular motion include *flexion, extension, abduction, adduction,* and *circumduction.* The descriptions of the movements are based on reference to an individual in the anatomical position.

Flexion and Extension **Flexion** (FLEK-shun) is movement in the anterior/posterior plane that reduces the angle between the articulating elements. **Extension** occurs in the same plane, but it increases the angle between articulating elements (Figure 9-3a•). These terms are usually applied to the movements of the long bones of the limbs, but they are also used to describe movements of the axial skeleton. For example, when you bring your head toward your chest, you flex the intervertebral joints of the neck. When you bend down to touch your toes, you flex the intervertebral joints of the spine. Extension reverses these movements, returning you to the anatomical position. When a person is in the anatomical position, all of the major joints of the axial and appendicular skeletons other than the ankle are at full extension. (Special terms used to describe movements of the ankle joint are introduced below.)

Flexion at the shoulder joint or hip joint moves the limbs anteriorly, whereas extension moves them posteriorly. Flexion of the wrist joint moves the hand anteriorly, and extension moves it posteriorly. In each of these examples, extension can be continued past the anatomical position. Extension past the anatomical position is called **hyperextension** (Figure 9-3a•). When you hyperextend your neck, you can gaze at the ceiling. Hyperextension of many joints is prevented by ligaments, bony processes, or soft tissues.

Abduction and Adduction **Abduction** (*ab,* from) is movement *away from the longitudinal axis of the body* in the frontal plane (Figure 9-3b•). For example, swinging the upper limb to the side is abduction of the limb. Moving it back to the anatomical position constitutes **adduction** (*ad,* to). Adduction of the wrist moves the heel of the hand *toward* the body, whereas abduction moves it farther away. Spreading the fingers or toes apart abducts them, because they move *away* from a central digit (Figure 9-3c•). Bringing them together constitutes adduction. (Fingers move toward or away from the middle finger; toes move toward or away from the second toe.) Abduction and adduction always refer to movements of the appendicular skeleton, not to those of the axial skeleton.

Circumduction We introduced a special type of angular motion, circumduction, in our model. Moving your arm in a loop is circumduction (Figure 9-3d•), as when you draw a large circle on a chalkboard. Your hand moves in a circle, but your arm does not rotate.

9

•**FIGURE 9-3** **Angular Movements.** The red dots indicate the locations of the joints involved in the illustrated movement.

Rotation

Rotational movements are also described with reference to a figure in the anatomical position. Rotation of the head may involve **left rotation** or **right rotation** (Figure 9-4a,b•). Limb rotation may be described by reference to the longitudinal axis of the trunk. During **medial rotation,** also known as *internal rotation* or *inward rotation,* the anterior surface of a limb turns toward the long axis of the trunk. The reverse movement is called **lateral rotation,** *external rotation,* or *outward rotation* (Figure 9-4b•).

The proximal articulation between the radius and ulna permits rotation of the radial head. As the shaft of the radius rotates, the distal epiphysis of the radius rolls across the anterior surface of the ulna. This movement, called **pronation** (prō-NĀ-shun), turns the wrist and hand from palm-facing-front to palm-facing-back (Figure 9-4c•). The opposing movement, in which the palm is turned anteriorly, is **supination** (soo-pi-NĀ-shun).

Special Movements

Several special terms apply to specific articulations or unusual types of movement (Figure 9-5•):

- **Inversion** (*in,* into + *vertere,* to turn) is a twisting motion of the foot that turns the sole inward, elevating the medial edge of the sole. The opposite movement is called **eversion** (ē-VER-zhun; *e,* out).

- **Dorsiflexion** is flexion at the ankle joint and elevation of the sole, as when you dig in your heel. **Plantar flexion** (*planta,* sole), the opposite movement, extends the ankle joint and elevates the heel, as when you stand on tiptoe.

- **Opposition** is movement of the thumb toward the surface of the palm. Opposition permits you to grasp and hold objects between your thumb and palm. It involves movement at the first carpometacarpal and metacarpophalangeal joints.

- **Protraction** entails moving a part of the body anteriorly in the horizontal plane. **Retraction** is the reverse movement. You protract your jaw when you grasp your upper lip with your lower teeth, and you protract your clavicles when you cross your arms.

- **Elevation** and **depression** occur when a structure moves in a superior or inferior direction, respectively. You depress your mandible when you open your mouth; you elevate it as you close it. Another familiar elevation occurs when you shrug your shoulders.

- **Lateral flexion** occurs when your vertebral column bends to the side. This movement is most pronounced in the cervical and thoracic regions.

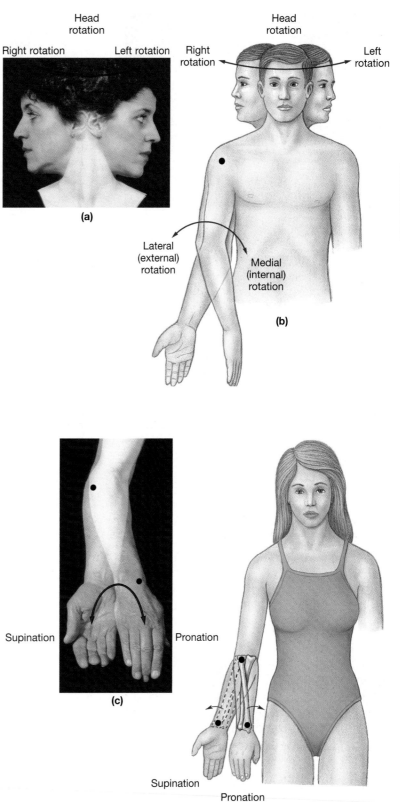

•**FIGURE 9-4** Rotational Movements

9

Eversion   Inversion   Opposition

Retraction   Protraction   Depression   Elevation   Lateral flexion

•**FIGURE 9-5** **Special Movements**

A STRUCTURAL CLASSIFICATION OF SYNOVIAL JOINTS

Synovial joints are described as *gliding, hinge, pivot, ellipsoidal, saddle,* or *ball-and-socket joints* on the basis of the shapes of the articulating surfaces. Each type of joint permits a different type and range of motion.

Gliding Joints

Gliding joints (Figure 9-6a•), also called *planar joints,* have flattened or slightly curved faces. The relatively flat articular surfaces slide across one another, but the amount of movement is very slight. Although rotation is theoretically possible at such a joint, ligaments usually prevent or restrict such movement. Gliding joints are nonaxial or multiaxial. These joints are located at the ends of the clavicles, between the carpal bones, between the tarsal bones, between the cartilages of ribs 2–7 and the sternum, and between the articular facets of adjacent vertebrae.

Hinge Joints

Hinge joints (Figure 9-6b•) permit angular motion in a single plane, like the opening and closing of a door. A hinge joint is a monaxial joint. Examples include

the elbow, knee, ankle, and interphalangeal joints of the appendicular skeleton.

Pivot Joints

Pivot joints (Figure 9-6c•) are also monaxial, but they permit only rotation. A pivot joint between the atlas and axis allows you to rotate your head to either side, and another between the head of the radius and the proximal shaft of the ulna permits pronation and supination of the palm.

Ellipsoidal Joints

In an **ellipsoidal joint** (Figure 9-6d•), or *condyloid joint,* an oval articular face nestles within a depression in the opposing surface. With such an arrangement, angular motion occurs in two planes, along or across the length of the oval. It is thus a biaxial joint. Any form of angular motion, including circumduction, is permitted, but rotation cannot occur. Ellipsoidal joints connect the radius with the proximal carpal bones, the phalanges of the fingers with the metacarpal bones, and the phalanges of the toes with the metatarsal bones.

Saddle Joints

Saddle joints (Figure 9-6e•), or *sellaris joints,* have saddle-shaped articular faces. Each face is concave on one

•**FIGURE 9-6** A Functional Classification of Synovial Joints

(a) Gliding joint

(b) Hinge joint

(c) Pivot joint

(d) Ellipsoidal joint

(e) Saddle joint

(f) Ball-and-socket joint

axis and convex on the other, and the opposing faces nest together. This arrangement permits angular motion, including circumduction, but prevents rotation. Saddle joints are usually considered to be biaxial. The carpometacarpal joint at the base of the thumb is the best example of a saddle joint, and twiddling your thumbs will demonstrate the possible movements.

Ball-and-Socket Joints

In a **ball-and-socket joint** (Figure 9-6f•), the round head of one bone rests within a cup-shaped depression in another. All combinations of angular and rotational movements, including circumduction and rotation, can be performed at ball-and-socket joints. These are triaxial joints. Examples include the shoulder and hip joints.

☑ Give the proper term for each of the following types of motion: (a) moving the humerus away from the midline of the body; (b) turning the palms so that they face forward; (c) bending the elbow.

☑ When you do jumping jacks, which lower limb movements are necessary?

☑ Which movements are associated with hinge joints?

REPRESENTATIVE ARTICULATIONS

Here we consider articulations that demonstrate important functional principles. We will first consider the *intervertebral articulations* of the axial skeleton. The articulations between adjacent vertebrae include gliding articulations and cartilaginous symphyses. We will then discuss the synovial articulations of the appendicular skeleton. The shoulder has great mobility; the elbow has great strength; and the wrist makes fine adjustments in the orientation of the palm and fingers. The functional requirements of the joints in the lower limb are very different from those of the upper limb. Articulations at the hip, knee, and ankle must transfer the body weight to the ground; during movements such as running, jumping, or twisting, the applied forces are much greater than the weight of the body. We consider only representative articulations here, but Tables 9-3 and 9-4 (pp. 258 and 264) summarize data on most articulations in the body.

INTERVERTEBRAL ARTICULATIONS

The articulations between the superior and inferior articular processes of adjacent vertebrae are gliding joints that permit small movements associated with flexion and rotation of the vertebral column. Little gliding occurs between adjacent vertebral bodies. From axis to sacrum, the vertebrae are separated and cushioned by pads of fibrocartilage called **intervertebral discs** (Figure 9-7•). Thus the bodies of vertebrae form symphyseal joints. Intervertebral discs and symphyseal joints are not found in the sacrum or coccyx, where vertebrae have fused, or between the first and second cervical

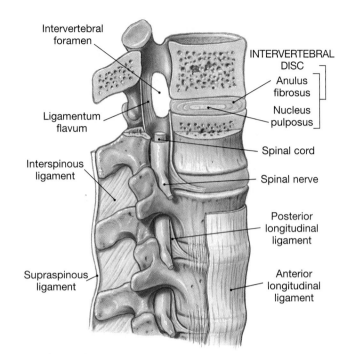

•**FIGURE 9-7** **Intervertebral Articulations.** *Scan 3*

vertebrae. The first cervical vertebra has no vertebral body and no intervertebral disc; the only articulation between the first two cervical vertebrae is a pivot joint that permits much more rotation than do the symphyseal joints between other cervical vertebrae.

The Intervertebral Discs

Each intervertebral disc has a tough outer layer of fibrocartilage, the **anulus fibrosus** (AN-ū-lus fī-BRŌ-sus). The collagen fibers of this layer attach the disc to the bodies of adjacent vertebra. The anulus fibrosus surrounds the **nucleus pulposus** (pul-PŌ-sus), a soft, elastic, and gelatinous core. The nucleus pulposus gives the disc resiliency and lets it absorb shocks.

Movement of the vertebral column compresses the nucleus pulposus and displaces it in the opposite direction. This displacement permits smooth gliding movements by each vertebra while maintaining their alignment. The discs make a significant contribution to an individual's height: They account for roughly one-quarter the length of the vertebral column superior to the sacrum. As we grow older, the water content of the nucleus pulposus in each disc decreases. The discs gradually become less effective as a cushion, and the chances for vertebral injury increase. Loss of water by the discs also causes shortening of the vertebral column; this loss accounts for the characteristic decrease in height with advancing age.

Intervertebral Ligaments

Numerous ligaments are attached to the bodies and processes of all vertebrae to bind them together and stabilize the vertebral column. Ligaments interconnecting adjacent vertebrae include:

- The *anterior longitudinal ligament,* which connects the anterior surfaces of adjacent vertebral bodies.
- The *posterior longitudinal ligament,* which parallels the anterior longitudinal ligament and connects the posterior surfaces of adjacent vertebral bodies.
- The *ligamentum flavum* (plural, *ligamenta flava*), which connects the laminae of adjacent vertebrae.
- The *interspinous ligament,* which connects the spinous processes of adjacent vertebrae.
- The *supraspinous ligament,* which interconnects the tips of the spinous processes from C_7 to the sacrum. The *ligamentum nuchae* extends from vertebra C_7 to the base of the skull. It is continuous with the supraspinous ligament. ∞ *p. 216*

Vertebral Movements

The following movements can occur across the intervertebral joints of the vertebral column: (1) flexion, bending anteriorly; (2) extension, bending posteriorly; (3) lateral flexion, bending laterally; and (4) rotation. Table 9-3 summarizes information about other articulations of the axial skeleton.

☑ Which regions of the vertebral column do not contain intervertebral discs? Why is the lack of discs significant?

☑ Which vertebral movements are involved in (a) bending forward, (b) bending to the side, and (c) moving the head to signify "no"?

9

Problems with Intervertebral Discs

An intervertebral disc compressed beyond its normal limits can be temporarily or permanently damaged. If the posterior longitudinal ligaments are weakened, as often occurs with advancing age, the compressed nucleus pulposus may distort the anulus fibrosus, forcing it partway into the vertebral canal. This condition is called a **slipped disc** (Figure 9-8a•), although the disc does not actually slip. A disc problem can occur at any age as a result of an accidental injury, such as a hard fall or a "whiplash" injury to the neck. But, with advanced age, the supporting ligaments may become so weak that the problem could occur without warning or apparent cause.

If the nucleus pulposus breaks through the anulus fibrosis, it too may protrude into the vertebral canal. This condition is called a **herniated disc** (Figure 9-8b•). When a disc herniates, sensory nerves are distorted and the protruding mass can also compress the nerves passing through the adjacent intervertebral foramen. The result is severe backache, abnormal posture, a burning or tingling sensation from the lower back and lower limbs, and in some cases a partial loss of control over skeletal muscles innervated by the compressed nerves. [AM] *Diagnosing and Treating Disc Problems*

| (a) | (b) |

•**FIGURE 9-8 Damage to the Intervertebral Discs. (a)** A lateral view of the lumbar region of the spinal column, showing a distorted intervertebral disc (a "slipped" disc). **(b)** A sectional view through a herniated disc, showing the release of the nucleus pulposus and its effect on the spinal cord and adjacent nerves.

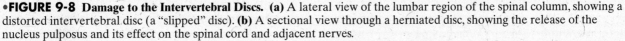

TABLE 9-3   Articulations of the Axial Skeleton

Element	Joint	Type of Articulation	Movements
Skull			
Cranial and facial bones of skull	Various	Synarthroses (suture or synostosis)	None
Maxillary bone/teeth and mandible/teeth		Synarthrosis (gomphosis)	None
Temporal bone/mandible	Temporomandibular	Combined gliding joint and hinge diarthrosis	Elevation, depression, and lateral gliding
Vertebral column			
Occipital bone/atlas	Atlanto-occipital	Ellipsoidal diarthrosis	Flexion/extension
Atlas/axis	Atlanto-axial	Pivot diarthrosis	Rotation
Other vertebral elements	Intervertebral (between vertebral bodies)	Amphiarthrosis (symphysis)	Slight movement
	Intervertebral (between articular processes)	Gliding diarthrosis	Slight rotation and flexion/extension
L_5/sacrum	Between L_5 body and sacral body	Amphiarthrosis (symphysis)	Slight movement
	Between inferior articular processes of L_5 and articular processes of sacrum	Gliding diarthrosis	Slight flexion/extension
Sacrum/os coxae	Sacroiliac	Gliding diarthrosis	Slight movement
Sacrum/coccyx	Sacrococcygeal	Gliding diarthrosis (may become fused)	Slight movement
Coccygeal bones		Synarthrosis (synostosis)	No movement
Thoracic cage			
Bodies of T_1–T_{12} and heads of ribs	Costovertebral	Gliding diarthrosis	Slight movement
Transverse processes of T_1–T_{10}	Costovertebral	Gliding diarthrosis	Slight movement
Ribs and costal cartilages		Synarthrosis (synchondrosis)	No movement
Sternum and first costal cartilage	Sternocostal (1st)	Synarthrosis (synchondrosis)	No movement
Sternum and costal cartilages 2–7	Sternocostal (2nd–7th)	Gliding diarthrosis*	Slight movement

*Commonly converts to synchondrosis in elderly individuals.

THE SHOULDER JOINT

The shoulder joint, or *glenohumeral joint,* permits the greatest range of motion of any joint. Because it is also the most frequently dislocated joint, it provides an excellent demonstration of the principle that stability must be sacrificed to obtain mobility.

This joint is a ball-and-socket diarthrosis formed by the articulation of the head of the humerus with the glenoid cavity of the scapula (Figure 9-9a•). The surface of the glenoid cavity is covered by a fibrocartilaginous **glenoid labrum** (*labrum,* lip or edge), which extends past the bony rim and deepens the socket (Figure 9-9b•). The relatively loose articular capsule extends from the scapula, proximal to the glenoid labrum, to the anatomical neck of the humerus. It is a somewhat oversized capsule that permits an extensive range of motion. The bones of the pectoral girdle provide some stability to the superior surface, because the acromion and coracoid process project laterally superior to the head of the humerus. However, most of the stability at this joint is provided by the surrounding skeletal muscles, with help from their associated tendons and various ligaments. Bursae reduce friction between the tendons and other tissues at the joint.

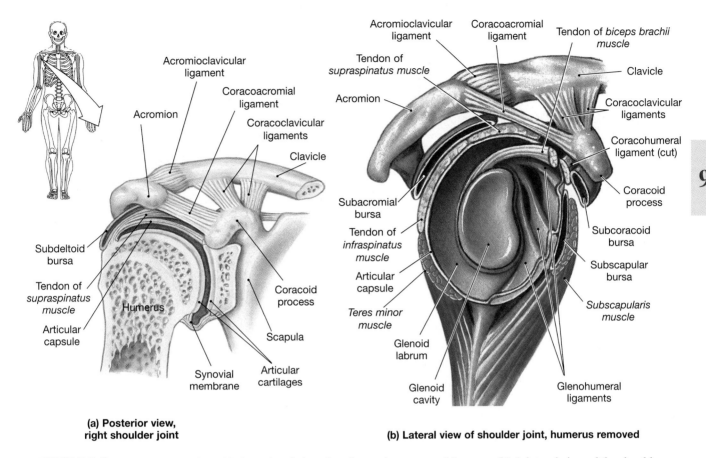

(a) Posterior view,
right shoulder joint

(b) Lateral view of shoulder joint, humerus removed

•**FIGURE 9-9** **The Shoulder Joint.** (a) A sectional view showing major structural features. (b) A lateral view of the shoulder joint with the humerus removed.

The major ligaments that help stabilize the shoulder joint are the *glenohumeral, coracohumeral, coracoacromial, coracoclavicular,* and *acromioclavicular ligaments.* The acromioclavicular ligament reinforces the capsule of the acromioclavicular joint and supports the superior surface of the shoulder. A **shoulder separation** is a relatively common injury involving partial or complete dislocation of the acromioclavicular joint. This injury can result from a blow to the superior surface of the shoulder. The acromion is forcibly depressed, but the clavicle is held back by strong muscles.

The muscles that move the humerus do more to stabilize the shoulder joint than do all the ligaments and capsular fibers combined. Muscles originating on the trunk, pectoral girdle, and humerus cover the anterior, superior, and posterior surfaces of the capsule. The tendons of the *supraspinatus, infraspinatus, subscapularis,* and *teres minor* muscles reinforce the joint capsule and limit the range of movement. These muscles, known as the muscles of the *rotator cuff,* are the primary mechanism for supporting the shoulder joint and limiting the range of movement. Damage to

the rotator cuff typically occurs when individuals are engaged in sports that place severe strains on the shoulder. White-water kayakers, baseball pitchers, and quarterbacks are all at high risk for rotator cuff injuries.

The anterior, superior, and posterior surfaces of the shoulder joint are reinforced by ligaments, muscles, and tendons, but the inferior capsule is poorly reinforced. As a result, a dislocation caused by an impact or violent muscle contraction is most likely to occur at this site. Such a dislocation can tear the inferior capsular wall and the glenoid labrum. The healing process typically leaves a weakness that increases the chances for future dislocations.

As at other joints, bursae at the shoulder reduce friction where large muscles and tendons pass across the joint capsule. The shoulder has a relatively large number of important bursae, such as the *subacromial bursa,* the *subcoracoid bursa,* the *subdeltoid bursa,* and the *subscapular bursa* (Figure 9-9•). Inflammation of one or more of these bursae can restrict motion and produce the painful symptoms of bursitis (p. 249).

9

Humerus — Medial epicondyle

Radius — Radial tuberosity — Annular ligament — Ulnar collateral ligament

Ulna

(a) Right elbow, medial view

Humerus — Radial collateral ligament — Annular ligament (covering head and neck of radius) — Radial tuberosity — Interosseous membrane

Capitulum — Olecranon of ulna — Ulna

(b) Right elbow, lateral view

•**FIGURE 9-10** **The Elbow Joint.** The right elbow joint. **(a)** A medial view, showing ligaments that stabilize the joint. **(b)** A lateral view.

THE ELBOW JOINT

The elbow joint is a complex hinge joint that involves the humerus, radius, and ulna (Figure 9-10•). The largest and strongest articulation at the elbow is the *humeroulnar joint,* where the trochlea of the humerus articulates with the trochlear notch of the ulna. ∞ *pp. 230, 231* At the smaller *humeroradial joint,* the capitulum of the humerus articulates with the head of the radius.

Muscles that extend the elbow attach to the rough surface of the olecranon. These muscles are primarily under the control of the *radial nerve,* which passes along the radial groove of the humerus. ∞ *p. 230* The large *biceps brachii muscle* covers the anterior surface of the arm. Its tendon is attached to the radius at the radial tuberosity. Contraction of this muscle produces supination of the forearm and flexion at the elbow.

The elbow joint is extremely stable because (1) the bony surfaces of the humerus and ulna interlock; (2) the articular capsule is very thick; and (3) the capsule is reinforced by strong ligaments. The *radial collateral ligament* stabilizes the lateral surface of the joint. It extends between the lateral epicondyle of the humerus and the *annular ligament,* which binds the head of the radius to the ulna. The medial surface of the joint is stabilized by the *ulnar collateral ligaments.* This ligament extends from the medial epicondyle of the humerus anteriorly to the coronoid processes of the ulna and posteriorly to the olecranon (Figure 9-10•).

Despite the strength of the capsule and ligaments, the elbow joint can be damaged by severe impacts or unusual stresses. For example, if you fall on your hand with a partially flexed elbow, contractions of muscles that extend the elbow may break the ulna at the center of the trochlear notch. Less-violent stresses can produce dislocations or other injuries to the elbow, es-

pecially if epiphyseal growth has not been completed. For example, parents in a hurry may drag a toddler along behind them, exerting an upward, twisting pull on the elbow joint that can result in a partial dislocation known as *nursemaid's elbow.*

Table 9-4, p. 264, summarizes the characteristics of the joints of the upper limb.

☑ Which tissues or structures provide most of the stability for the shoulder joint?

☑ Would a tennis player or a jogger be more likely to develop inflammation of the subscapular bursa? Why?

☑ A football player received a blow to the upper surface of his shoulder, causing a shoulder separation. What does this mean?

☑ Terry suffers an injury to his forearm and elbow. After the injury, he notices an unusually large degree of motion between the radius and the ulna at the elbow. Which ligament did Terry most likely damage?

THE HIP JOINT

The hip joint, or *coxal joint,* is a sturdy ball-and-socket diarthrosis that permits flexion and extension, adduction and abduction, circumduction, and rotation. Figure 9-11• introduces the structure of the hip joint. The acetabulum, a deep fossa, accommodates the head of the femur. ∞ *p. 234* Within the acetabulum, a fibrocartilage pad extends like a horseshoe to either side of the acetabular notch (Figure 9-11a•). The *acetabular labrum,* a projecting rim of fibrocartilage, increases the depth of the joint cavity.

The articular capsule of the hip joint is extremely dense and strong. It extends from the lateral and inferior surfaces of the pelvic girdle to the

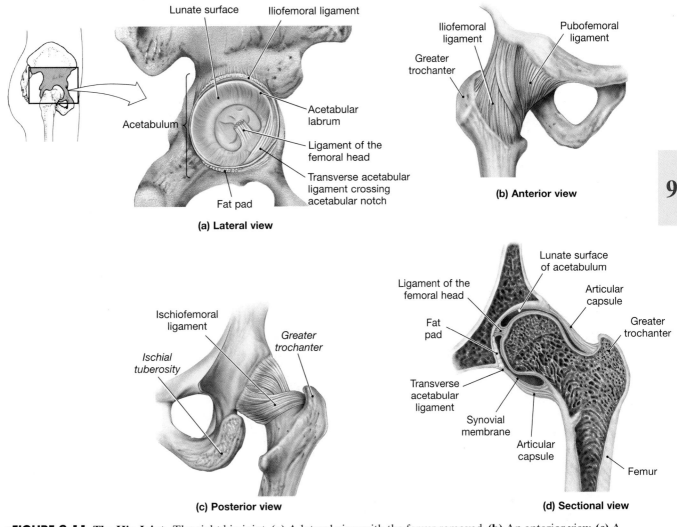

Lunate surface   Iliofemoral ligament

Acetabulum

Acetabular labrum

Ligament of the femoral head

Transverse acetabular ligament crossing acetabular notch

Fat pad

(a) Lateral view

Iliofemoral ligament   Pubofemoral ligament

Greater trochanter

(b) Anterior view

9

Ischiofemoral ligament

Ischial tuberosity

Greater trochanter

(c) Posterior view

Ligament of the femoral head

Fat pad

Transverse acetabular ligament

Synovial membrane

Lunate surface of acetabulum

Articular capsule

Greater trochanter

Articular capsule

Femur

(d) Sectional view

•**FIGURE 9-11** **The Hip Joint.** The right hip joint. **(a)** A lateral view with the femur removed. **(b)** An anterior view. **(c)** A posterior view, showing additional ligaments that add strength to the capsule. **(d)** A sectional view.   AM   *Plate 8.4; Scan 4*

---

intertrochanteric line and intertrochanteric crest of the femur, enclosing both the head and neck of the femur. ∞ *p. 238* This arrangement helps keep the femoral head from moving too far from the acetabulum.

Four broad ligaments reinforce the articular capsule (Figure 9-11•). Three of them—the *iliofemoral, pubofemoral,* and *ischiofemoral ligaments*—are regional thickenings of the capsule. The *transverse acetabular ligament* crosses the acetabular notch, filling in the gap in the inferior border of the acetabulum. A fifth ligament, the *ligament of the femoral head,* or *ligamentum teres* (*teres,* long and round), originates along the transverse acetabular ligament (Figure 9-11a•) and attaches to the fovea capitis, a small pit at the center of the femoral head. ∞ *p. 237* This ligament tenses only when the hip is flexed and the thigh is undergoing lateral rotation. Much more important stabilization is provided by the bulk of the

surrounding muscles, aided by ligaments and capsular fibers.

The combination of an almost complete bony socket, a strong articular capsule, supporting ligaments, and muscular padding makes the hip joint an extremely stable joint. Fractures of the femoral neck or between the greater and lesser trochanters of the femur are more common than hip dislocations. As we noted in Chapter 6, however, hip fractures are relatively common in elderly individuals with severe osteoporosis. ∞ *p. 185*

THE KNEE JOINT

The hip joint passes weight to the femur, and the knee joint transfers the weight from the femur to the tibia. The shoulder is mobile; the hip, stable; and the knee, …? If you had to choose one word, it would probably be "complicated." Although the knee functions as a hinge joint, the articulation is far more complex than that of the elbow or even the ankle. The rounded condyles of the femur roll across the superior surface of the tibia, so the points of contact are constantly changing. The joint permits flexion and extension and very limited rotation.

Structurally, the knee resembles three separate joints: two between the femur and tibia (medial condyle to medial condyle and lateral condyle to lateral condyle) and one between the patella and the patellar surface of the femur (Figure 9-12•).

The Articular Capsule and Joint Cavity

The articular capsule at the knee joint is thin and in some areas incomplete, but it is strengthened by various ligaments and the tendons of associated mus-

•**FIGURE 9-12** **The Knee Joint.** The right knee. **(a)** Superficial posterior and anterior views of the extended knee joint. **(b)** A deep posterior view, at full extension. **(c)** An anterior view, at full flexion.

AM *Plates 8.9, 8.11; Scans 5, 6, 7*

(a) Superficial view, extended

(b) Deep posterior view, extended

(c) Deep anterior view, flexed

cles. A pair of fibrocartilage pads, the **medial** and **lateral menisci,** lie between the femoral and tibial surfaces (Figures 9-1b, p.248, and 9-12b,c•). The menisci (1) act as cushions, (2) conform to the shape of the articulating surfaces as the femur changes position, and (3) provide lateral stability to the joint. Prominent fat pads cushion the margins of the joint and assist the many bursae in reducing friction between the patella and other tissues.

Supporting Ligaments

A complete dislocation of the knee is very rare, largely because seven major ligaments stabilize the knee joint:

1. The tendon from the muscles responsible for extending the knee passes over the anterior surface of the joint. The patella is embedded in this tendon, and the *patellar ligament* continues to its attachment on the anterior surface of the tibia. The patellar ligament and two ligamentous bands known as the *patellar retinaculae* support the anterior surface of the knee joint (Figure 9-12a•).

2, 3. Two *popliteal ligaments* extend between the femur and the heads of the tibia and fibula (Figure 9-12a•). These ligaments reinforce the knee joint's posterior surface.

4, 5. Inside the joint capsule, the *anterior cruciate ligament* (ACL) and *posterior cruciate ligament* (PCL) attach the intercondylar area of the tibia to the condyles of the femur. *Anterior* and *posterior* refer to their sites of origin on the tibia. They cross one another as they proceed to their destinations on the femur (Figure 9-12b,c•). (The term *cruciate* is derived from the Latin word *crucialis*, meaning a cross.) These ligaments limit the anterior and posterior movement of the femur and maintain the alignment of the femoral and tibial condyles.

6, 7. The *tibial collateral ligament* reinforces the medial surface of the knee joint, and the *fibular collateral ligament* reinforces the lateral surface (Figure 9-12•). These ligaments tighten only at full extension. In this position, they stabilize the joint.

At full extension, a slight lateral rotation of the tibia tightens the anterior cruciate ligament and jams the lateral meniscus between the tibia and femur. The knee joint is essentially locked in the extended position. With the joint locked, you can stand for prolonged periods without using (and tiring) the muscles that extend the knee. Unlocking the knee joint requires muscular contractions that medially rotate the tibia or laterally rotate the femur. If the

locked knee is struck from the side, the lateral meniscus can tear and serious damage can occur to the supporting ligaments.

The knee is structurally complex and subjected to severe stresses in the course of normal activities. Painful knee injuries are all too familiar to both amateur and professional athletes. Treatment is often costly and prolonged, and repairs seldom make the joint "good as new."

Table 9-4 summarizes information about the articulations of the pelvic girdle and lower limb.

9

Knee Injuries

Athletes place tremendous stresses on their knees. Ordinarily, the medial and lateral menisci move as the position of the femur changes. Placing a lot of weight on the knee while it is partially flexed can trap a meniscus between the tibia and femur, resulting in a break or tear in the cartilage. In the most common injury, the lateral surface of the leg is driven medially, tearing the medial meniscus. In addition to being quite painful, the torn cartilage may restrict movement at the joint. It can also lead to chronic problems and the development of a "trick knee," a knee that feels unstable. Sometimes the meniscus can be heard and felt popping in and out of position when the knee is extended.

Less common knee injuries involve tearing one or more stabilizing ligaments or damaging the patella. Torn ligaments can be difficult to correct surgically, and healing is slow. The patella can be injured in a number of ways. If the leg is immobilized (as it might be in a football pileup) while you try to extend the knee, the muscles are powerful enough to pull the patella apart. Impacts to the anterior surface of the knee can also shatter the patella. Treatment of a fractured patella is difficult and time-consuming. The fragments must be surgically removed; and the tendons and ligaments, repaired. The joint must then be immobilized. Total knee replacements are rarely performed on young people, but they are becoming increasingly common among elderly patients with severe arthritis. [AM] *Arthroscopic Surgery and the Diagnosis of Joint Injuries*

☑ Where would you find the following ligaments: iliofemoral ligament, pubofemoral ligament, and ischiofemoral ligament?

☑ What symptoms would you expect to see in an individual who has damaged the menisci of the knee joint?

☑ Why is "clergyman's knee" (a type of bursitis) common among carpet layers and roofers?

9

TABLE 9-4 Articulations of the Appendicular Skeleton

Element	Joint	Type of Articulation	Movements
ARTICULATIONS OF THE PECTORAL GIRDLE AND UPPER LIMB			
Sternum/clavicle	Sternoclavicular	Gliding diarthrosis*	Protraction/retraction, elevation/depression, slight rotation
Scapula/clavicle	Acromioclavicular	Gliding diarthrosis	Slight movement
Scapula/humerus	Shoulder, or glenohumeral	Ball-and-socket diarthrosis	Flexion/extension, adduction/abduction, circumduction, rotation
Humerus/ulna and humerus/radius	Elbow (humeroulnar and humeroradial)	Hinge diarthrosis	Flexion/extension
Radius/ulna	Proximal radioulnar	Pivot diarthrosis	Rotation
	Distal radioulnar	Pivot diarthrosis	Pronation/supination
Radius/carpal bones	Radiocarpal	Ellipsoidal diarthrosis	Flexion/extension, adduction/abduction, circumduction
Carpal bone to carpal bone	Intercarpal	Gliding diarthrosis	Slight movement
Carpal bone to metacarpal bone (I)	Carpometacarpal of thumb	Saddle diarthrosis	Flexion/extension, adduction/abduction, circumduction, opposition
Carpal bone to metacarpal bone (II–V)	Carpometacarpal	Gliding diarthrosis	Slight flexion/extension, adduction/abduction
Metacarpal bone to phalanx	Metacarpophalangeal	Ellipsoidal diarthrosis	Flexion/extension, adduction/abduction, circumduction
Phalanx/phalanx	Interphalangeal	Hinge diarthrosis	Flexion/extension
ARTICULATIONS OF THE PELVIC GIRDLE AND LOWER LIMB			
Sacrum/ilium of os coxae	Sacroiliac	Gliding diarthrosis	Slight movement
Os coxae/os coxae	Pubic symphysis	Amphiarthrotic symphysis	None†
Os coxae/femur	Hip	Ball-and-socket diarthrosis	Flexion/extension, adduction/abduction, circumduction, rotation
Femur/tibia	Knee	Complex, functions as hinge	Flexion/extension, limited rotation
Tibia/fibula	Tibiofibular (proximal)	Gliding diarthrosis	Slight movement
	Tibiofibular (distal)	Gliding diarthrosis and amphiarthrotic syndesmosis	Slight movement
Tibia and fibula with talus	Ankle, or talocrural	Hinge diarthrosis	Dorsiflexion/plantar flexion
Tarsal bone to tarsal bone	Intertarsal	Gliding diarthrosis	Slight movement
Tarsal bone to metatarsal bone	Tarsometatarsal	Gliding diarthrosis	Slight movement
Metatarsal bone to phalanx	Metatarsophalangeal	Ellipsoidal diarthrosis	Flexion/extension, adduction/abduction
Phalanx/phalanx	Interphalangeal	Hinge diarthrosis	Flexion/extension

* A "double gliding joint," with two joint cavities separated by an articular cartilage.
† During pregnancy, hormones weaken the symphysis and permit movement important to childbirth; see Chapter 29.

⧗ AGING AND ARTICULATIONS

Joints are subjected to heavy wear and tear throughout our lifetimes, and problems with joint function are relatively common, especially in older individuals. **Rheumatism** (ROO-muh-tizm) is a general term that indicates pain and stiffness affecting the skeletal system, the muscular system, or both. Several major forms of rheumatism exist. **Arthritis** (ar-THRĪ-tis) encompasses all the rheumatic diseases that affect synovial joints. Arthritis always involves damage to the articular cartilages, but the specific cause can vary. For example, arthritis can result from bacterial or viral infection, injury to the joint, metabolic problems, or severe physical stresses.

Osteoarthritis (os-tē-ō-ar-THRĪ-tis), also known as *degenerative arthritis* or *degenerative joint disease (DJD)*, generally affects individuals age 60 or older. Osteoarthritis can result from cumulative wear and tear at the joint surfaces or from genetic factors affecting collagen formation. In the U.S. population, 25 percent of women and 15 percent of men over age 60 show signs of this disease.

Rheumatoid arthritis is an inflammatory condition that affects roughly 2.5 percent of the adult population. At least some cases occur when the immune response mistakenly attacks the joint tissues. Such a condition, in which the body attacks its own tissues, is called an *autoimmune disease*. Allergies, bacteria, viruses, and genetic factors have all been proposed as contributing to or triggering the destructive inflammation.

Regular exercise, physical therapy, and drugs that reduce inflammation, such as aspirin, can slow the progress of arthritis. Surgical procedures can realign or redesign the affected joint. In extreme cases involving the hip, knee, elbow, or shoulder, the defective joint can be replaced by an artificial one. AM *Rheumatism, Arthritis, and Synovial Function*

Degenerative changes comparable to those seen in arthritis may result from joint immobilization. When motion ceases, so does the circulation of synovial fluid, and the cartilages begin to suffer. **Continuous passive motion (CPM)** of any injured joint appears to encourage the repair process by improving the circulation of synovial fluid. The movement is often performed by a physical therapist or a machine during the recovery process.

With age, bone mass decreases and bones become weaker, so the risk of fractures increases. ∞ *p. 185* If osteoporosis develops, the bones may weaken to the point at which fractures occur in response to stresses that could easily be tolerated by normal bones. Hip fractures are among the most dangerous fractures seen in elderly persons, with or without osteoporosis. These fractures, most often involving individuals over age 60, may be accompanied by hip dislocation or by pelvic fractures.

Healing proceeds very slowly, and the powerful muscles that surround the hip joint often prevent proper alignment of the bone fragments. Fractures at the greater or lesser trochanter generally heal well if the joint can be stabilized; steel frames, pins, screws, or some combination of these devices may be needed to preserve alignment and to permit healing to proceed normally.

Although severe hip fractures are most common among those over age 60, in recent years the frequency of hip fractures has increased dramatically among young, healthy professional athletes. AM *Hip Fractures, Aging, and Professional Athletes*

BONES AND MUSCLES

The skeletal and muscular systems are structurally and functionally interdependent. Their interactions are so extensive that they are often considered to be parts of a single *musculoskeletal system.* The two systems have direct physical connections. Many of the anatomical landmarks identified in this chapter are attachment sites of skeletal muscles, and the connective tissues that surround individual muscle fibers are continuous with those that form the organic framework of an attached bone. Muscles and bones are also physiologically linked through mutual dependence on stable calcium ion concentrations in body fluids. With most of the body's calcium tied up in the skeleton, bone abnormalities can have a direct effect on the muscle.

The bones, in turn, are directly affected by muscular activity. When a muscle enlarges due to regular exercise, the increased forces exerted on the skeleton will make bones become stronger and more massive. The reverse holds true as well, for both muscle and bones decrease in size and strength after periods of inactivity.

In the next two chapters, we will examine the structure and function of the muscular system and shall discuss how muscular contractions perform specific movements.

9

[EXPLORE]
[MediaLab]

The skeletal system provides support and protection, but it must also allow for movement. This need is met by means of three types of articulation. To help you study the various structures and functions of these articulations, use the resources available on both the CD-ROM and the Companion Website. The CD-ROM includes animations and tutorials; the website builds on those resources with text content and interactive problems as well as communication avenues for sharing ideas with your class. To complete the following Web Explorations, visit the MediaLab in Chapter 9 of your Companion Website.

Web Exploration 1

Estimated time for completion = 15 minutes

The knee has a unique structure, combining three articulations in one joint. It carries the weight of the body yet seems hardly stable enough to function. To examine this joint more fully, select the keyword **KNEE** on the Companion Website. Read the information presented, and study the labeled photographs of a dissected knee. Prepare your own description of the knee joint, including the placement of structures that are typically associated with synovial joints. How can the knee both support your weight and allow for a range of motions?

Web Exploration 2

Estimated time for completion = 10 minutes

Arthritis is a degenerative disease of synovial joints. Read the discussion of arthritis in the "Aging and Articulations" section of the text and the related material in the *Applications Manual*. Research is currently under way to identify any genetic factors that correlate with the development of this disease. Select the keyword **RHEUMATOID ARTHRITIS** on the Companion Website to learn about the work of the Arthritis Foundation. Scroll past the list of participating research centers to find a description of the disease process. Do you think this project will provide the answers that scientists are hoping for? Would you be willing to assist in this project as a donor, a laboratory technician, or an investor? Explain.

CHAPTER REVIEW

SELECTED CLINICAL TERMINOLOGY
Terms Discussed in This Chapter

arthritis (ar-THRĪ-tis): Rheumatic diseases that affect synovial joints. Arthritis always involves damage to the articular cartilages, but the specific cause can vary. The diseases of arthritis are usually classified as either *degenerative* or *inflammatory*. *(p. 265 and AM)*

bunion: The most common pressure-related bursitis, involving a tender nodule formed around bursae over the base of the great toe. *(p. 249)*

bursitis: An inflammation of a bursa, causing pain whenever the associated tendon or ligament moves. *(p. 249)*

continuous passive motion (CPM): A therapeutic procedure involving the passive movement of an injured joint to stimulate the circulation of synovial fluid. The goal is to prevent degeneration of the articular cartilages. *(p. 265 and AM)*

herniated disc: A condition caused by intervertebral compression severe enough to rupture the anulus fibrosus and release the nucleus pulposus, which may protrude beyond the intervertebral space. *(p. 257)*

luxation (luks-Ā-shun): A dislocation; a condition in which the articulating surfaces are forced out of position. *(p. 250)*

osteoarthritis (os-tē-ō-ar-THRĪ-tis) (*degenerative arthritis* or *degenerative joint disease, DJD*): An arthritic condition resulting from (1) cumulative wear and tear on joint surfaces or (2) genetic predisposition. In the U.S. population, 25 percent of women and 15 percent of men over age 60 show signs of this disease. *(p. 265 and AM)*

rheumatism (ROO-muh-tizm): A general term that indicates pain and stiffness affecting the skeletal system, the muscular system, or both. *(p. 265 and AM)*

rheumatoid arthritis: An inflammatory arthritis that affects roughly 2.5 percent of the adult U.S. population. The cause is uncertain, although allergies, bacteria, viruses, and genetic factors have all been proposed. The primary symptom is *synovitis*, swelling and inflammation of the synovial membrane. *(p. 265 and AM)*

shoulder separation: The partial or complete dislocation of the acromioclavicular joint. *(p. 259)*

slipped disc: A common name for a condition caused by the distortion of an intervertebral disc. The distortion applies pressure to spinal nerves, causing pain and limiting range of motion. *(p. 257)*

sprain: A condition in which a ligament is stretched to the point at which some of the collagen fibers are torn. The ligament remains functional, and the structure of the joint is not affected. *(p. 249)*

subluxation (sub-luks-Ā-shun): A partial dislocation; the displacement of articulating surfaces sufficient to cause discomfort but resulting in less physical damage to the joint than occurs during a dislocation. *(p. 250)*

AM Additional Terms Discussed in the *Applications Manual*

ankylosis (ang-ki-LŌ-sis): An abnormal fusion between articulating bones in response to trauma and friction within a joint.

arthroscope: A fiber-optic instrument used to view the inside of joint cavities.

arthroscopic surgery: The surgical modification of a joint by using an arthroscope.

bony crepitus: A crackling or popping sound produced during movement of an abnormal joint.

laminectomy (la-mi-NEK-to-mē): The removal of vertebral laminae; may be performed to access the vertebral canal and relieve symptoms of a herniated disc.

meniscectomy: The surgical removal of a damaged meniscus.

monoarthritic: An arthritic condition that affects a single articulation.

polyarthritic: An arthritic condition that affects multiple articulations.

STUDY OUTLINE

9

INTRODUCTION   p. 246

1. **Articulations** (joints) exist wherever two bones interact.

A CLASSIFICATION OF JOINTS   p. 246

1. *Immovable joints* are **synarthroses;** *slightly movable joints* are **amphiarthroses;** and joints that are *freely movable* are called **diarthroses,** or **synovial joints.** (*Table 9-1*)
2. Alternatively, joints are classified structurally as *bony, fibrous, cartilaginous,* or *synovial.* (*Table 9-2*)

Synarthroses (Immovable Joints)   p. 246

3. The four major types of synarthroses are a **suture** (skull bones bound together by dense connective tissue), a **gomphosis** (teeth bound to bony sockets by *periodontal ligaments*), a **synchondrosis** (two bones joined by a rigid cartilaginous bridge), and a **synostosis** (two bones completely fused).

Amphiarthroses (Slightly Movable Joints)   p. 247

4. The two major types of amphiarthroses are a **syndesmosis** (bones connected by a ligament) and a **symphysis** (bones separated by fibrocartilage).

Diarthroses (Freely Movable Joints)   p. 247

5. The bony surfaces at diarthroses are enclosed within an **articular capsule,** covered by **articular cartilages,** and lubricated by **synovial fluid.**
6. Other synovial structures can include **menisci,** or *articular discs;* **fat pads; accessory ligaments;** and **bursae.** (*Figure 9-1*)
7. A **dislocation** occurs when articulating surfaces are forced out of position.

ARTICULAR FORM AND FUNCTION   p. 250

Describing Dynamic Motion   p. 250

1. Possible movements are classified as **linear motion, angular motion,** or **rotation.** (*Figure 9-2*)
2. Joints are called **monoaxial, biaxial,** or **triaxial,** depending on the planes of movement they allow.

Types of Movements   p. 251

3. In **gliding,** two surfaces slide past one another.
4. Important terms that describe angular motion are **flexion, extension, hyperextension, circumduction, abduction,** and **adduction.** (*Figure 9-3*)
5. Rotational movement can be **left** or **right rotation, medial** (*internal*) or **lateral** (*external*) **rotation,** or, in the bones of the forearm, **pronation** or **supination.** (*Figure 9-4*)

6. Movements of the foot include **inversion** and **eversion.** The ankle undergoes **dorsiflexion** and **plantar flexion. Opposition** is the thumb movement that enables us to grasp objects. (*Figure 9-5*)
7. **Protraction** involves moving something anteriorly; **retraction** involves moving it posteriorly. **Depression** and **elevation** occur when we move a structure inferiorly and superiorly, respectively. **Lateral flexion** occurs when the vertebral column bends to one side. (*Figure 9-5*)

A Structural Classification of Synovial Joints   p. 254

8. **Gliding joints** permit limited movement, generally in a single plane. (*Figure 9-6*)
9. **Hinge joints** are monaxial joints that permit only angular movement in one plane. (*Figure 9-6*)
10. **Pivot joints** are monaxial joints that permit only rotation. (*Figure 9-6*)
11. **Ellipsoidal joints** are biaxial joints with an oval articular face that nestles within a depression in the opposing surface. (*Figure 9-6*)
12. **Saddle joints** are biaxial joints with articular faces that are concave on one axis and convex on the other. (*Figure 9-6*)
13. **Ball-and-socket joints** are triaxial joints that permit rotation as well as other movements. (*Figure 9-6*)

REPRESENTATIVE ARTICULATIONS   p. 256

Intervertebral Articulations   p. 256

1. The articular processes of vertebrae form gliding joints with those of adjacent vertebrae. The bodies form symphyseal joints. They are separated and cushioned by **intervertebral discs,** which contain an inner **nucleus pulposus** and an outer **anulus fibrosus.** Several ligaments stabilize the vertebral column. (*Figures 9-7, 9-8; Table 9-3*)

The Shoulder Joint   p. 258

2. The shoulder joint, or *glenohumeral joint,* is formed by the glenoid cavity and the head of the humerus. This articulation permits the greatest range of motion of any joint. It is a ball-and-socket diarthrosis with various stabilizing ligaments. Strength and stability are sacrificed to obtain mobility. (*Figure 9-9; Table 9-4*)

The Elbow Joint   p. 260

3. The elbow joint permits only flexion/extension. It is a hinge diarthrosis whose capsule is reinforced by strong ligaments. (*Figure 9-10; Table 9-4*)

The Hip Joint p. 260

4. The hip joint is a ball-and-socket diarthrosis formed by the union of the acetabulum with the head of the femur. The joint permits flexion/extension, adduction/abduction, circumduction, and rotation; it is stabilized by numerous ligaments. *(Figure 9-11; Table 9-4)*

The Knee Joint p. 261

5. The knee joint is a hinge joint formed by the union of the condyles of the femur with the superior condylar surfaces of the tibia. The joint permits flexion/extension and limited rotation, and it has various supporting ligaments. *(Figure 9-12; Table 9-4)*

AGING AND ARTICULATIONS p. 265

1. Problems with joint function are relatively common, especially in older individuals. **Rheumatism** is a general term for pain and stiffness affecting the skeletal system, the muscular system, or both; several major forms exist. **Arthritis** encompasses all the rheumatic diseases that affect synovial joints. Both conditions become increasingly common with age.

BONES AND MUSCLES p. 265

1. The interactions of the skeletal and muscular systems are so extensive that we often refer to a single *musculoskeletal system*.

REVIEW QUESTIONS

LEVEL 1 REVIEWING FACTS AND TERMS

1. A synarthrosis located only between the bones of the skull is a
   - (a) symphysis
   - (b) syndesmosis
   - (c) synchondrosis
   - (d) suture

2. The distal articulation between the tibia and fibula is a
   - (a) syndesmosis
   - (b) symphysis
   - (c) synchondrosis
   - (d) synostosis

3. The anterior articulation between the two pubic bones is a
   - (a) synchondrosis
   - (b) synostosis
   - (c) symphysis
   - (d) synarthrosis

4. Joints typically located at the end of long bones are
   - (a) synarthroses
   - (b) amphiarthroses
   - (c) diarthroses
   - (d) symphyses

5. The function of the synovial fluid is
   - (a) to nourish chondrocytes
   - (b) to provide lubrication
   - (c) to absorb shock
   - (d) a, b, and c are correct

6. The structures responsible for channeling the flow of synovial fluid are
   - (a) menisci
   - (b) bursae
   - (c) carpal tunnels
   - (d) articular cartilages

7. The structures that limit the range of motion and provide mechanical support across or around a joint are
   - (a) bursae
   - (b) tendons
   - (c) menisci
   - (d) a, b, and c are correct

8. A partial dislocation of an articulating surface is a
   - (a) circumduction
   - (b) hyperextension
   - (c) subluxation
   - (d) supination

9. Abduction and adduction always refer to movements of the
   - (a) axial skeleton
   - (b) appendicular skeleton
   - (c) skull
   - (d) vertebral column

10. Rotation of the forearm that makes the palm face posteriorly is
    - (a) supination
    - (b) pronation
    - (c) proliferation
    - (d) projection

11. Spreading the fingers or toes apart is
    - (a) dorsiflexion
    - (b) circumduction
    - (c) adduction
    - (d) abduction

12. Standing on tiptoe is an example of
    - (a) elevation
    - (b) dorsiflexion
    - (c) plantar flexion
    - (d) retraction

13. Examples of monaxial joints, which permit angular movement in a single plane, are the
    - (a) carpals and tarsals
    - (b) shoulder and hip
    - (c) elbow and knee
    - (d) a, b, and c are correct

14. Joints that connect the fingers and toes with the metacarpals and metatarsals are
    - (a) ellipsoidal joints
    - (b) saddle joints
    - (c) pivot joints
    - (d) hinge joints

15. Movements that occur at the shoulder and the hip represent the action that occurs at a _____ joint.
    - (a) hinge
    - (b) ball-and-socket
    - (c) pivot
    - (d) gliding

16. The anulus fibrosus and nucleus pulposus are structures associated with the
    - (a) intervertebral discs
    - (b) knee and elbow
    - (c) shoulder and hip
    - (d) carpals and tarsals

17. Subacromial, subcoracoid, and subscapular bursae reduce friction in the _____ joint.
    - (a) hip
    - (b) knee
    - (c) elbow
    - (d) shoulder

18. The functional anatomy of the lower limbs is very different from that of the upper limbs primarily because
    - (a) the upper limbs are used for grasping
    - (b) the upper limbs contain more bones than the lower limbs do
    - (c) the lower limbs must transfer the body weight to the ground
    - (d) the lower limbs are used for bipedal mobility

LEVEL 2 REVIEWING CONCEPTS

19. The hip is an extremely stable joint because it has
    - (a) a complete bony socket
    - (b) a strong articular capsule
    - (c) supporting ligaments
    - (d) a, b, and c are correct

20. Complete dislocation of the knee is an extremely rare event because
    - (a) the knee is protected by the patella
    - (b) the femur articulates with the tibia at the knee
    - (c) the knee contains seven major ligaments
    - (d) the knee contains fat pads to absorb shocks

21. How does a meniscus (articular disc) function in a joint?

22. The greater the range of motion at a joint, the weaker the joint becomes. Why?

23. How do articular cartilages differ from other cartilages in the body?

24. What is the significance of the fact that the pubic symphysis is a slightly movable joint?

25. How would you explain to your grandmother the characteristic decrease in height with advancing age?

26. When the biceps brachii muscle contracts, what movements does it produce?

27. When you stand for a long period of time, why should you "lock" your knee in the extended position? How does the knee lock?

LEVEL 3 CRITICAL THINKING AND CLINICAL APPLICATIONS

28. While playing tennis, Dave "overturns" his ankle. He experiences swelling and pain. After being examined, he is told that he has no torn ligaments and that the structure of the ankle is not affected. On the basis of the symptoms and the examination results, what happened to Dave's ankle?

29. During a basketball game, Bob injured his right knee when he jumped to retrieve the ball and then landed off-balance on his right leg. Since then he has pain and limited mobility of his right knee joint. What type of injury did Bob sustain?

30. A high-school student comes to the emergency room complaining of persistent pain beneath her right shoulder blade. In talking with her, you discover that she has been spending many hours trying to improve her pitching skills for her school's softball team. What is likely causing the pain?

9

CHAPTER 10

Muscle Tissue

CHAPTER OUTLINE AND OBJECTIVES

INTRODUCTION   p. 271

**SKELETAL MUSCLE TISSUE
AND THE MUSCULAR SYSTEM**   p. 271
① Describe the characteristics and functions of skeletal
   muscle tissue.

ANATOMY OF SKELETAL MUSCLE   p. 271
   Organization of Connective Tissues   p. 271
   Blood Vessels and Nerves   p. 273
   Microanatomy of Skeletal Muscle Fibers   p. 273
② Describe the organization of muscle at the tissue level.
③ Explain the unique characteristics of skeletal muscle
   fibers.
④ Identify the structural components of a sarcomere.

CONTRACTION OF SKELETAL MUSCLE   p. 279
   The Sliding Filament Theory   p. 279
   The Control of Skeletal Muscle Activity   p. 280

■ **FOCUS: The Contraction Cycle**   p. 282

   Relaxation   p. 281
   Length–Tension Relationships   p. 285
⑤ Identify the components of the neuromuscular junction
   and summarize the events involved in the neural
   control of skeletal muscles.
⑥ Explain the key steps involved in the contraction of a
   skeletal muscle fiber.

MUSCLE MECHANICS   p. 286
   Tension Production   p. 286
   Energetics of Muscular Activity   p. 293
   Muscle Performance   p. 296
⑦ Compare the types of muscle contractions.
⑧ Describe the mechanisms by which muscle fibers
   obtain the energy to power contractions.
⑨ Relate the types of muscle fibers to muscle
   performance.
⑩ Distinguish between aerobic and anaerobic endurance,
   and explain their implications for muscular
   performance.

AGING AND THE MUSCULAR SYSTEM   p. 299
⑪ Describe the effects of aging on muscles.

INTEGRATION WITH OTHER SYSTEMS   p. 300
⑫ Give examples of interactions between the muscular
   system and each of the other organ systems.

CARDIAC MUSCLE TISSUE   p. 302
   **Differences between Cardiac and Skeletal Muscle
   Tissues**   p. 302

SMOOTH MUSCLE TISSUE   p. 303
   **Differences between Smooth Muscle Tissue and Other
   Muscle Tissues**   p. 304
⑬ Identify the structural and functional differences
   among skeletal, cardiac, and smooth muscle cells.
⑭ Discuss the role that smooth muscle tissue plays in
   systems throughout the body.

MEDIALAB and CHAPTER REVIEW   p. 306

*Astronaut Susan Still seems to enjoy the experience of zero
gravity during weightlessness training with astronaut
candidates, but the situation is less than ideal for her bones
and muscles. Our bones and muscles work together. Under
weightless conditions, bones are not stressed by the weight of
the body, and muscles no longer work to oppose the force of
gravity. "What you don't use, you lose," and prolonged
weightlessness results in a loss of bone and muscle mass.
Because this loss can be potentially dangerous on return to
normal gravity, astronauts in orbit have exercise periods with
workouts to stress their muscles and prevent or reduce the loss
of muscle mass. This chapter addresses two related questions:
How do muscles work, and how does exercise—in space or on
Earth—affect strength and endurance?*

Think for a moment what life would be like without muscle tissue. Imagine being unable to sit, stand, walk, speak, or grasp objects. Imagine, too, how your internal functions would be affected. Blood would not circulate, because you would have no heartbeat to propel it through the vessels. You would be unable to breathe, speak, or eat, and food could not move along your digestive tract. In fact, there would be practically no movement along any of your internal passageways.

This is not to say that all life depends on muscle tissue. Some large organisms get by very nicely without it; we call them plants. But life as *we* live it would be impossible, because virtually all of our dynamic interactions with the environment involve muscle tissue.

Muscle cells are specialized to contract. The individual cells in muscle tissue are tied together, primarily by collagen fibers. When the muscle cells contract, they pull on those fibers the way a group of people might pull on a rope. The pull, called **tension**, is an active force—energy must be expended to produce it. Tension is applied *to* some object, whether a rope, a rubber band, or a book on a tabletop.

Tension applied to an object tends to pull the object toward the source of the tension. However, before movement can occur, the applied tension must overcome the object's **resistance**, a passive force that opposes movement. The amount of resistance can depend on the weight of the object, its shape, friction, and other factors. When the applied tension exceeds the resistance, the object moves. In contrast, **compression**, or a push applied to an object, tends to force the object away from the source of the compression. Again, no movement can occur until the applied compression exceeds the resistance of the object. Muscle cells can use energy to shorten and generate tension but not to lengthen and generate compression.

As we noted in Chapter 6, collagen fibers cannot withstand compression; when compressed, they simply bend out of the way. ∞ *p. 170* But collagen fibers can resist considerable tension, so when a muscle pulls on collagen fibers, those fibers transmit the force and pull on something else. What happens next depends largely on what the collagen fibers are attached to and on how the muscle cells are arranged. We will encounter several examples in this chapter as well as in Chapters 11, 20, and 24.

SKELETAL MUSCLE TISSUE AND THE MUSCULAR SYSTEM

Muscle tissue, one of the four primary tissue types, consists chiefly of muscle cells that are highly specialized for contraction. Three types of muscle tissue exist: (1) *skeletal muscle,* (2) *cardiac muscle,* and (3) *smooth muscle.* ∞ *p. 132* Without these muscle tissues, nothing in the body would move, and no body movement could occur. Skeletal muscle tissue moves the body by pulling on bones of the skeleton, making it possible for us to walk, dance, bite an apple, or play the ukulele. Cardiac muscle tissue pushes blood through the circulatory system. Smooth muscle tissue pushes fluids and solids along the digestive tract and regulates the diameters of small arteries, among other functions.

This chapter deals primarily with the structure and function of skeletal muscle tissue, in preparation for our discussion of the muscular system (Chapter 11). This chapter will also provide an overview of the differences among skeletal, cardiac, and smooth muscle tissues.

Skeletal muscle tissue forms **skeletal muscles**, organs that also contain connective tissues, nerves, and blood vessels. Each cell in skeletal muscle tissue is a single muscle *fiber.* Skeletal muscles are directly or indirectly attached to the bones of the skeleton. Our skeletal muscles perform the following five functions:

1. *Produce skeletal movement.* Skeletal muscle contractions pull on tendons and move the bones of the skeleton. The effects range from simple motions such as extending the arm or breathing to the highly coordinated movements of swimming, skiing, or typing.

2. *Maintain posture and body position.* Tension in our skeletal muscles maintains body posture—for example, holding your head in position when you read a book or balancing your body weight above your feet when you walk. Without constant muscular activity, we could not sit upright or stand.

3. *Support soft tissues.* The abdominal wall and the floor of the pelvic cavity consist of layers of skeletal muscle. These muscles support the weight of visceral organs and shield internal tissues from injury.

4. *Guard entrances and exits.* The openings of the digestive and urinary tracts are encircled by skeletal muscles. These muscles provide voluntary control over swallowing, defecation, and urination.

5. *Maintain body temperature.* Muscle contractions require energy; whenever energy is used in the body, some of it is converted to heat. The heat released by working muscles keeps our body temperature in the range required for normal functioning.

We will begin our discussion with the gross anatomy of a typical skeletal muscle. We will then consider at the microscopic level the structural features that make contractions possible.

ANATOMY OF SKELETAL MUSCLE

Figure 10-1• illustrates the appearance and organization of a representative skeletal muscle. As we noted, a skeletal muscle contains connective tissues, blood vessels, and nerves as well as skeletal muscle tissue.

ORGANIZATION OF CONNECTIVE TISSUES

Three layers of connective tissue are part of each muscle: (1) an outer epimysium, (2) a central perimysium,

10

•FIGURE 10-1 **The Organization of Skeletal Muscles.** A skeletal muscle consists of fascicles (bundles of muscle fibers) enclosed by the epimysium. The bundles are separated by connective tissue fibers of the perimysium, and within each bundle the muscle fibers are surrounded by the endomysium. Each muscle fiber has many superficial nuclei as well as mitochondria and other organelles (detailed in Figure 10-3).

and (3) an inner endomysium. These layers and the relationships among them are diagrammed in Figure 10-1•.

The entire muscle is surrounded by the **epimysium** (ep-i-MĪZ-ē-um; *epi-,* on + *mys,* muscle), a dense layer of collagen fibers. The epimysium separates the muscle from surrounding tissues and organs. It is connected to the deep fascia, a dense connective tissue layer (see Figure 4-18•). ∞ *p. 131*

The connective tissue fibers of the **perimysium** (per-i-MĪZ-ē-um; *peri-,* around) divide the skeletal muscle into a series of compartments, each containing a bundle of muscle fibers called a **fascicle** (FAS-si-kl; *fasciculus,*

a bundle). In addition to collagen and elastic fibers, the perimysium contains blood vessels and nerves that maintain blood flow and innervate the fascicles. Each fascicle receives branches of these blood vessels and nerves.

Within a fascicle, the delicate connective tissue of the **endomysium** (en-dō-MĪZ-ē-um; *endo-,* inside) surrounds the individual skeletal muscle fibers and interconnects adjacent muscle fibers. Scattered between the endomysium and the muscle fibers are **satellite cells,** embryonic stem cells that function in the repair of damaged muscle tissue. ∞ *p. 132*

AM *Disruption of Normal Muscle Organization*

The collagen fibers of the endomysium and perimysium are interwoven and blend into one another. At each end of the muscle, the collagen fibers of the epimysium, perimysium, and endomysium come together to form a bundle known as a **tendon** or a broad sheet called an **aponeurosis**. Tendons and aponeuroses usually attach skeletal muscles to bones.[1] Where a tendon attaches to a bone, the tendon fibers extend into the bone matrix, providing a firm attachment. As a result, any contraction of the muscle will exert a pull on its tendon and thereby on the attached bone (or bones).

BLOOD VESSELS AND NERVES

The connective tissues of the epimysium and perimysium contain the blood vessels and nerves that supply the muscle fibers. Muscle contraction requires tremendous quantities of energy. An extensive vascular network delivers the necessary oxygen and nutrients and carries away the metabolic wastes generated by active skeletal muscles. The blood vessels and the nerve supply generally enter the muscle together and follow the same branching pattern through the perimysium. Within the endomysium, arterioles supply blood to a capillary network that surrounds each individual muscle fiber.

Skeletal muscles contract only under stimulation from the central nervous system. Axons, or *nerve fibers,* penetrate the epimysium, branch through the perimysium, and enter the endomysium to innervate individual muscle fibers. Skeletal muscles are often called voluntary muscles, because we have voluntary control over their contractions. Many skeletal muscles may also be controlled at a subconscious level. For example, skeletal muscles involved with breathing, such as the *diaphragm,* usually work outside our conscious awareness.

Next we will examine the microscopic structure of a typical skeletal muscle fiber and will relate that microstructure to the physiology of the contraction process.

MICROANATOMY OF SKELETAL MUSCLE FIBERS

Skeletal muscle fibers are quite different from the "typical" cells we described in Chapter 3. One obvious difference is size: Skeletal muscle fibers are enormous. A muscle fiber from a thigh muscle could have a diameter of 100 μm and a length equal to that of the entire muscle (up to 30 cm, or 12 in.). A second obvious difference is that skeletal muscle fibers are *multinucleate:* Each skeletal muscle fiber contains hundreds of nuclei just internal to the cell membrane (Figure 10-2•). The genes in these nuclei direct the production of enzymes and structural proteins required for normal muscle contraction, and the presence of multiple copies of these genes speeds up the process. This feature is important because metabolic turnover tends to be very rapid in skeletal muscle fibers. ∞ *p. 60*

The distinctive features of size and multiple nuclei are related. During development, groups of embryonic cells called **myoblasts** fuse, forming individual skeletal muscle fibers (Figure 10-2a•). Each nucleus in a skeletal

[1] In a few cases, they connect skeletal muscles; specific examples will be discussed in Chapter 11.

Myoblasts

Muscle fibers develop through the fusion of mesodermal cells called *myoblasts.*

(a)

Satellite cell

Nuclei

Immature muscle fiber

(b)

•**FIGURE 10-2 The Formation and Nuclei of a Skeletal Muscle. (a)** The formation of muscle fibers by the fusion of myoblasts. Notice the multiple nuclei. **(b)** A micrograph and diagrammatic view of one muscle fiber. (LM × 612) muscle fiber reflects the contribution of a single myoblast. Some myoblasts, however, do not fuse with developing muscle fibers. These unfused cells remain in adult skeletal muscle tissue as the satellite cells seen in Figures 10-1 and 10-2a•. After an injury, satellite cells may enlarge, divide, and fuse with damaged muscle fibers, thereby assisting in the regeneration of the tissue.

The Sarcolemma and Transverse Tubules

The **sarcolemma** (sar-kō-LEM-uh; *sarkos,* flesh + *lemma,* husk), or cell membrane of a muscle fiber, surrounds the **sarcoplasm** (SAR-kō-plazm), or cytoplasm of the muscle fiber (Figure 10-3a•). Like other cell membranes, the sarcolemma has a characteristic transmembrane potential due to the unequal distribution of positive and negative charges across the membrane. ∞ *p. 79* In a skeletal muscle fiber, a sudden change in the transmembrane potential is the first step that leads to a contraction.

A skeletal muscle fiber is very large, but all regions of the cell must contract simultaneously. Thus, the signal to contract must be distributed quickly throughout the interior of the cell. This signal is conducted through the transverse tubules. **Transverse tubules,** or **T tubules,** are narrow tubes that are continuous with the sarcolemma and extend into the sarcoplasm at right angles to the cell surface (Figure 10-3b•). Filled with extracellular fluid, T tubules form passageways through the muscle fiber, like a series of tunnels through a mountain. The T tubules have the same general properties as the sarcolemma. Electrical impulses conducted by the sarcolemma travel along the T tubules. These

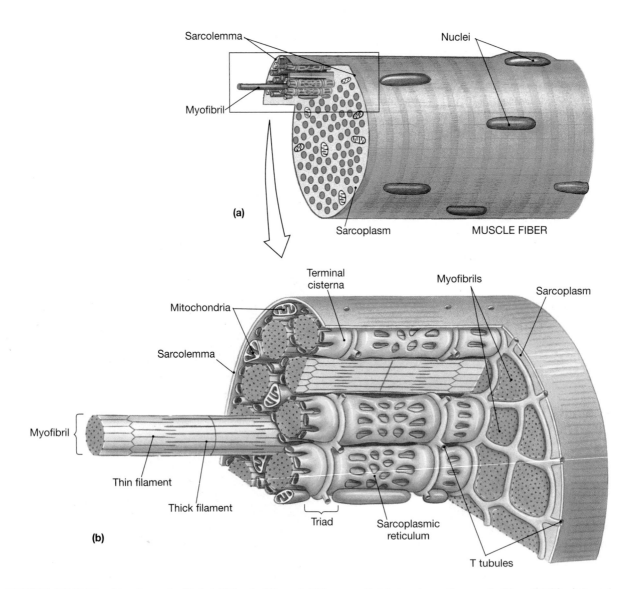

•**FIGURE 10-3 The Structure of a Skeletal Muscle Fiber. (a)** The superficial structure of a muscle fiber. **(b)** The internal organization of a muscle fiber.

impulses, or *action potentials,* are the trigger for muscle fiber contraction.

Myofibrils

Inside the muscle fiber, branches of the transverse tubules encircle cylindrical structures called **myofibrils** (Figure 10-3•). A myofibril is 1–2 μm in diameter and as long as the entire cell. Each skeletal muscle fiber contains hundreds to thousands of myofibrils.

Myofibrils consist of bundles of **myofilaments,** protein filaments composed primarily of actin and myosin. The actin forms the bulk of **thin filaments,** and the myosin forms **thick filaments.** We introduced both types of filaments in Chapter 3. ∞ *pp. 81, 82* Myofibrils, which can actively shorten, are responsible for skeletal muscle fiber contraction. At each end of the skeletal muscle fiber, the myofibrils are anchored to the inner surface of the sarcolemma. In turn, the outer surface of the sarcolemma is attached to collagen fibers of the tendon or aponeurosis of the skeletal muscle. As a result, when the myofibrils contract, the entire cell shortens. In doing so, it pulls on the tendon. Scattered among and around the myofibrils are mitochondria and granules of glycogen, the storage form of glucose. Glucose breakdown through glycolysis and mitochondrial activity provides the ATP needed to power muscular contractions. ∞ *p. 88*

The Sarcoplasmic Reticulum

Wherever a transverse tubule encircles a myofibril, the tubule is tightly bound to the membranes of the sarcoplasmic reticulum. The **sarcoplasmic reticulum (SR)** is a membrane complex similar to the smooth endoplasmic reticulum of other cells. In skeletal muscle fibers, the SR forms a tubular network around each individual myofibril (Figure 10-3b•). On either side of a T tubule, the tubules of the SR enlarge, fuse, and form expanded chambers called **terminal cisternae.** The combination of a pair of terminal cisternae plus a transverse tubule is known as a **triad.** Although the membranes of the triad are tightly bound together, their fluid contents are separate and distinct.

In Chapter 3, we noted the existence of special ion pumps that keep the intracellular concentration of calcium ions (Ca^{2+}) very low. ∞ *p. 76* Most cells pump the calcium ions across their cell membranes and into the extracellular fluid. Although skeletal muscle fibers do pump Ca^{2+} out of the cell in this way, they also remove calcium ions from the sarcoplasm by actively transporting them into the terminal cisternae of the sarcoplasmic reticulum. The sarcoplasm of a resting skeletal muscle fiber contains very low concentrations of Ca^{2+}, around 10^{-7} mmol/l. The free Ca^{2+} concentration inside the terminal cisternae may be as much as 1000 times higher. In addition, cister-

nae contain the protein *calsequestrin,* which reversibly binds Ca^{2+}. Including both the free calcium and the bound calcium, the total concentration of Ca^{2+} inside cisternae can be 40,000 times that of the surrounding sarcoplasm.

A muscle contraction begins when stored calcium ions are released into the sarcoplasm. These ions then diffuse into individual contractile units called *sarcomeres.*

Sarcomeres

As we have seen, myofibrils are bundles of thin and thick filaments. These myofilaments are organized into repeating functional units called **sarcomeres** (SAR-kō-mērz; *sarkos,* flesh + *meros,* part).

Sarcomere Organization A myofibril consists of approximately 10,000 sarcomeres end to end. Each sarcomere has a resting length of 1.6–2.6 μm. Sarcomeres are the smallest functional units of the muscle fiber. Interactions between the thick and thin filaments of sarcomeres are responsible for muscle contraction. A sarcomere contains (1) thick filaments, (2) thin filaments, (3) proteins that stabilize the positions of the thick and thin filaments, and (4) proteins that regulate the interactions between thick and thin filaments.

Differences in the size, density, and distribution of thick filaments and thin filaments account for the banded appearance of each myofibril (Figure 10-4•). There are dark bands (**A bands**) and light bands (**I bands**). The names of these bands are derived from *anisotropic* and *isotropic,* which refer to the appearance of these bands when they are viewed under polarized light. You may find it helpful to remember that A bands are d*A*rk and that I bands are l*I*ght in a typical light micrograph.

The A Band The thick filaments are located at the center of a sarcomere, in the A band. The length of the A band is equal to the length of a typical thick filament. The A band, which also includes portions of thin filaments, contains the following three subdivisions (Figure 10-4•):

1. *The M line.* The central portion of each thick filament is connected to its neighbors by proteins of the **M line.** These dark-staining proteins help stabilize the positions of the thick filaments.

2. *The H zone.* In a resting sarcomere, the **H zone,** or *H band,* is a lighter region on either side of the M line. The H zone contains thick filaments but no thin filaments.

3. *The zone of overlap.* In the **zone of overlap,** thin filaments are situated between the thick filaments. In this region, each thin filament is surrounded by three thick filaments, and each thick filament is surrounded by six thin filaments.

•**FIGURE 10-4 Sarcomere Structure, Part I. (a)** A longitudinal section of a sarcomere. **(b)** A corresponding view of a sarcomere in a myofibril from a muscle fiber in the *gastrocnemius* muscle of the calf. (TEM ×64,000)

The cross-sectional views in Figure 10-5• should help you visualize these features of the three-dimensional structure of the sarcomere.

The I Band Each I band, which contains thin filaments but not thick filaments, extends from the A band of one sarcomere to the A band of the next sarcomere. **Z lines** mark the boundary between adjacent sarcomeres. The Z lines consist of proteins called *connectins,* which interconnect thin filaments of adjacent sarcomeres. From the Z lines at either end of the sarcomere, thin filaments extend toward the M line and into the zone of overlap. Strands of the protein **titin** extend from the tips of the thick filaments to attachment sites at the Z line (Figure 10-5b•). Titin helps keep the thick and thin filaments in proper alignment; it also helps the muscle fiber resist extreme stretching that would otherwise disrupt the contraction mechanism.

Two transverse tubules encircle each sarcomere, and triads are located on each side of the M line at the zone of overlap. As a result, calcium ions released by the SR enter the regions where thick and thin filaments can interact.

•FIGURE 10-5 Sarcomere Structure, Part II. **(a)** A superficial view of a sarcomere. **(b)** Cross-sectional views of different portions of a sarcomere. Dashed lines show the relationships between thick and thin filaments in the zone of overlap.

Sarcomere

Myofibril

(a)

Thick filament

Thin filament

Connectin filaments

Attachment of titin

Titin filament

(b)   Z line   I band   M line   H zone   Zone of overlap

Each Z line is surrounded by a meshwork of intermediate filaments that interconnect adjacent myofibrils. The myofibrils closest to the sarcolemma are bound to attachment sites on the inside of the membrane. Because the Z lines of all the myofibrils are aligned in this way, the muscle fiber as a whole has a banded appearance (Figure 10-2b•). These bands, or *striations*, are visible with the light microscope, so skeletal muscle tissue is also known as striated muscle. ∞ *p. 132*

Figure 10-6• reviews the levels of organization we have discussed so far. We shall now consider the molecular structure of the myofilaments responsible for muscle contraction.

The Muscular Dystrophies

Abnormalities in the genes that code for structural and functional proteins in muscle fibers are responsible for a number of inherited diseases collectively known as the **muscular dystrophies** (DIS-trō-fēz). These conditions, which cause a progressive muscular weakness and deterioration, are the result of abnormalities in the sarcolemma or in the structure of internal proteins. The best-known example is *Duchenne's muscular dystrophy,* which typically develops in males from 3 to 7 years old.

AM *The Muscular Dystrophies*

Thin Filaments A typical thin filament is 5–6 nm in diameter and 1 μm in length (Figure 10-7a•, p. 279). A single thin filament contains four proteins: F actin, nebulin, tropomyosin, and troponin:

1. **F actin** is a twisted strand composed of two rows of 300–400 individual globular molecules of **G actin** (Figure 10-7b•). A long strand of **nebulin** spirals along the F actin strand in the cleft between the rows of G actin molecules. Nebulin holds the F actin strand together; as thin filaments develop, the length of the nebulin molecule probably determines the length of the F actin strand. Each molecule of G actin contains an **active site** that can bind to a thick filament much as a substrate molecule binds to the active site of an enzyme. ∞ *p. 54* Under resting conditions, myosin binding is prevented by the *troponin–tropomyosin complex.*

2. Strands of **tropomyosin** (trō-pō-MĪ-ō-sin) cover the active sites and prevent actin–myosin interaction (Figure 10-7b•). A tropomyosin molecule is a double-stranded protein that covers seven active sites. It is bound to one molecule of troponin midway along its length.

3. A **troponin** (TRŌ-pō-nin; *trope,* turning) molecule consists of three globular subunits. One subunit binds to tropomyosin, locking them together as a troponin–tryopomyosin complex; a second subunit binds to one

SKELETAL MUSCLE

Surrounded by:
Epimysium

Contains:
Muscle fascicles

(a)

MUSCLE FASCICLE

Surrounded by:
Perimysium

Contains:
Muscle fibers

(b)

MUSCLE FIBER

Surrounded by:
Endomysium

Contains:
Myofibrils

(c)

MYOFIBRIL

Surrounded by:
Sarcoplasmic
reticulum

Consists of:
Sarcomeres
(Z line to Z line)

(d)

I band   A band

SARCOMERE

Contains:
Thick filaments

Thin filaments

Z line   M line   Titin   Z line

(e)

•FIGURE 10-6 Levels of Functional Organization in a Skeletal Muscle Fiber

G actin, holding the troponin–tropomyosin complex in position; the third subunit has a receptor that binds a calcium ion. In a resting muscle, intracellular Ca^{2+} concentrations are very low and that binding site is empty.

A contraction cannot occur unless the position of the troponin–tropomyosin complex changes, exposing the active sites on F actin. The necessary change in position occurs when calcium ions bind to receptors on the troponin molecules.

At either end of the sarcomere, the thin filaments are attached to the Z line (Figure 10-7a•). Although it is called a *line* because it looks like a dark line on the surface of the myofibril, the Z line in sectional view is more like an open meshwork (Figure 10-5b•). For this reason, the Z line is often called the *Z disc*.

Thick Filaments   Thick filaments are 10–12 nm in diameter and 1.6 μm long (Figure 10-7c•). Each thick filament consists of roughly 500 myosin molecules. Each myosin molecule consists of a pair of myosin subunits twisted around one another (Figure 10-7d•). The long, attached **tail** is bound to other myosin molecules in the thick filament. The free **head**, which projects outward toward the nearest thin filament, consists of two globular protein subunits. Because myosin heads interact with thin filaments during a contraction, they are also known as **cross-bridges**. The connection between the head and the tail functions as a hinge that lets the head pivot at its base. When pivoting occurs, the head swings toward or away from the M line. As we will see in a later section, this pivoting is the key step in muscle contraction.

All the myosin molecules are arranged with their tails pointing toward the M line (Figure 10-7c•). The H zone includes a central region where there are no myosin heads. Elsewhere, the myosin heads are arranged in a spiral, each facing one of the surrounding thin filaments.

Each thick filament has a core of titin. On either side of the M line, a strand of titin extends the length of the thick filament and then continues across the I band to the Z line on that side. The portion of the titin strand exposed within the I band is highly elastic and will recoil after stretching. In the normal resting sarcomere, the titin strands are completely relaxed; they become tense only when some external force stretches the sarcomere.

☑ How would severing the tendon attached to a muscle affect the muscle's ability to move a body part?

☑ Why does skeletal muscle appear striated when viewed through a microscope?

☑ Where would you expect the greatest concentration of Ca^{2+} in resting skeletal muscle to be?

•**FIGURE 10-7 Thick and Thin Filaments. (a)** The gross structure of a thin filament, showing the attachment at the Z line. **(b)** The organization of G actin subunits in an F actin strand, and the position of the troponin–tropomyosin complex. **(c)** The structure of thick filaments, showing the orientation of the myosin molecules. **(d)** The structure of a myosin molecule.

CONTRACTION OF SKELETAL MUSCLE

THE SLIDING FILAMENT THEORY

When a skeletal muscle fiber contracts, (1) the H bands and I bands get smaller, (2) the zones of overlap get larger, (3) the Z lines move closer together, and (4) the width of the A bands remains constant. The contraction ends once the fiber has shortened by about 30 percent, which coincides with the elimination of the I bands. These observations make sense only if the thin filaments are sliding toward the center of the sarcomere, alongside the thick filaments (Figure 10-8•). This explanation for the physical changes that occur during contraction is the **sliding filament theory.**

The sliding filament theory explains what happens to the sarcomere during a contraction, but it does not explain the mechanism involved. To do so, we must examine the contraction sequence in greater detail.

•**FIGURE 10-8 Changes in the Appearance of a Sarcomere during the Contraction of a Skeletal Muscle Fiber.** During a contraction, the A band stays the same width, but the Z lines move closer together and the I band gets smaller.

THE CONTROL OF SKELETAL MUSCLE ACTIVITY

Skeletal muscle fibers contract only under the control of the nervous system. Communication between the nervous system and a skeletal muscle fiber occurs at a specialized intercellular connection known as a **neuromuscular junction (NMJ),** or *myoneural junction.* One such junction is shown in Figure 10-9a•.

The Neuromuscular Junction

Each skeletal muscle fiber is controlled by a neuron at a single neuromuscular junction midway along the fiber's length. Figure 10-9b• summarizes key features of this structure. A single axon branches with-

in the perimysium to form a number of fine branches. Each branch ends at an expanded **synaptic terminal.** The cytoplasm of the synaptic terminal contains mitochondria and vesicles filled with molecules of **acetylcholine** (as-ē-til-KŌ-lēn), or **ACh.** Acetylcholine is a *neurotransmitter,* a chemical released by a neuron to change the membrane properties of another cell. In this case, the release of ACh from the synaptic terminal can alter the permeability of the sarcolemma and trigger the contraction of the muscle fiber.

The **synaptic cleft,** a narrow space, separates the synaptic terminal of the neuron from the opposing sarcolemmal surface. This surface, which contains membrane receptors that bind ACh, is known as the

•**FIGURE 10-9** **Skeletal Muscle Innervation.**
(a) A diagrammatic view of a neuromuscular junction. **(b)** Details of the neuromuscular junction. **(c,d)** Changes at the motor end plate that trigger an action potential in the sarcolemma.

motor end plate. The motor end plate has deep creases called *junctional folds,* which increase its surface area and thus the number of available ACh receptors. The synaptic cleft and sarcolemma also contain molecules of the enzyme **acetylcholinesterase** (**AChE,** or *cholinesterase*), which breaks down ACh.

When a neuron stimulates a muscle fiber, the stimulus for ACh release is the arrival of an electrical impulse, or **action potential,** at the synaptic terminal. An action potential is a sudden change in the transmembrane potential propagated along the length of the axon. When that impulse reaches the synaptic terminal, permeability changes in the membrane trigger the exocytosis of ACh into the synaptic cleft. This exocytosis is accomplished when vesicles in the synaptic terminal fuse with the membrane of the neuron.

Molecules of ACh diffuse across the synaptic cleft and bind to ACh receptors on the motor end plate. The binding of ACh changes the permeability of the motor end plate to sodium ions (Figure 10-9c●). As we learned in Chapter 3, the extracellular fluid contains a high concentration of sodium ions, whereas sodium ion concentrations inside the cell are very low. ∞ *p. 76* When the membrane permeability to sodium increases, sodium ions rush into the sarcoplasm. This influx continues until AChE removes the ACh from the receptors.

The sudden influx of sodium ions results in the generation of an action potential in the sarcolemma at the edges of the motor end plate. This electrical impulse sweeps across the entire membrane surface and travels along each T tubule. The arrival of an action potential at the synaptic terminal thus leads to the appearance of an action potential in the sarcolemma. Even before the action potential has spread across the entire membrane, the ACh has been broken down by AChE. This sequence of events can now be repeated if another action potential arrives at the synaptic terminal.

Excitation–Contraction Coupling

The link between the generation of an action potential in the sarcolemma and the start of a muscle contraction is called **excitation–contraction coupling.** This coupling occurs at the triads. On reaching a triad, an action potential triggers the release of Ca^{2+} from the cisternae of the sarcoplasmic reticulum. The change in the permeability of the SR to Ca^{2+} is temporary, lasting only about 0.03 second. Yet within a millisecond the Ca^{2+} concentration in and around the sarcomere reaches 100 times resting levels. Because the terminal cisternae are situated at the zones of overlap, where the thick and thin filaments interact, the effect of calcium release on the sarcomere is almost instantaneous. The binding of Ca^{2+} to troponin exposes the active sites along the thin filaments, initiating the contraction. The **contraction cycle** then begins.

During the contraction cycle, cross-bridges bind to exposed active sites and the myosin heads pivot, shortening the sarcomeres. Each myosin head continues to attach, pivot, and detach as long as Ca^{2+} and ATP are available. This process is detailed in Figure 10-10●, in "FOCUS: The Contraction Cycle" (pp. 282–283).

Interference with Neural Control Mechanisms

Anything that interferes with neural function or with excitation–contraction coupling will cause muscular paralysis. Two examples are worth noting:

1. *Botulism* results from the consumption of contaminated canned or smoked foods that contain a toxin. The toxin, produced by bacteria, prevents the release of ACh at the synaptic terminals, leading to a potentially fatal muscular paralysis.

2. The progressive muscular paralysis of *myasthenia gravis* results from the loss of ACh receptors at the junctional folds. The primary cause is a misguided attack on the ACh receptors by the immune system. Genetic factors play a role in predisposing individuals to this condition. **AM** *Problems with the Control of Muscle Activity*

RELAXATION

The duration of a contraction depends on the duration of stimulation at the neuromuscular junction. The ACh released after a single action potential arrives at the synaptic terminal does not remain intact for long. Whether it is bound to the sarcolemma or is free in the synaptic cleft, the released ACh is rapidly broken down and inactivated by AChE. Inside the muscle fiber, the permeability changes in the SR are also very brief. A contraction will therefore continue only if additional action potentials arrive at the synaptic terminal in rapid succession. When they do, the continual release of ACh into the synaptic cleft produces a series of action potentials in the sarcolemma that keeps Ca^{2+} levels elevated in the sarcoplasm. Under these conditions, the contraction cycle will be repeated over and over.

If just one action potential arrives at the neuromuscular junction, Ca^{2+} concentrations in the sarcoplasm will quickly return to normal resting levels. Two mechanisms are involved in this process: (1) active Ca^{2+} transport across the cell membrane into the extracellular fluid and (2) active Ca^{2+} transport into the SR. Of the two, transport into the SR is the more important. Virtually as soon as the calcium ions have been released, the SR returns to its normal permeability and begins actively absorbing Ca^{2+} from the surrounding sarcoplasm. As Ca^{2+} concentrations in the sarcoplasm fall, (1) calcium ions detach from troponin, (2) troponin returns to its original position, and (3) the active sites are re-covered by tropomyosin. The contraction has ended.

Once the contraction has ended, the sarcomere does not automatically return to its original length.

igure 10-10• details the molecular events that occur during the contraction cycle. In the resting sarcomere, each cross-bridge is already "energized"—charged with the energy that will be used to power a contraction. The cross-bridge functions as ATPase, an enzyme that can break down ATP. ∞ *p. 57* At the start of the contraction cycle, each cross-bridge has already split a molecule of ATP and stored the energy released in the process. The breakdown products, ADP and phosphate (PO_4^{3-}, often represented as (P)), remain bound to the cross-bridge.

The contraction cycle involves five interlocking steps (Figure 10-10•):

STEP 1: *Exposure of active sites.* The calcium ions entering the sarcoplasm bind to troponin. This binding weakens the bond between the troponin–tropomyosin complex and actin. The troponin molecule then changes position, pulling the tropomyosin molecule away from the active sites and allowing cross-bridges to form.

STEP 2: *Attachment of cross-bridges.* When the active sites are exposed, the myosin cross-bridges bind to them.

STEP 3: *Pivoting.* In the resting sarcomere, each cross-bridge points away from the M line. In this position, the myosin head is "cocked" like the spring in a mousetrap. Cocking the myosin head requires energy, and the energy is obtained by breaking down ATP into ADP and a phosphate group. In the cocked position, both the ADP and the phosphate are still bound to the myosin head. After cross-bridge attachment has occurred, the stored energy is released as the myosin head pivots toward the M line. This action is called the *power stroke*. When this occurs, the ADP and phosphate group are released.

STEP 4: *Detachment of cross-bridges.* When an ATP binds to the myosin head, the link between the active

Resting sarcomere

Step 1: Active-site exposure

Step 2: Cross-bridge attachment

•**FIGURE 10-10** The Molecular Events of the Contraction Process site on the actin molecule and the myosin head is broken. The active site is now exposed and able to interact with another cross-bridge.

STEP 5: *Reactivation of myosin.* Myosin reactivation occurs when the free myosin head splits the ATP into ADP and a phosphate group. The energy released in this process is used to recock the myosin head. The entire cycle can now be repeated. If calcium ion concentrations remain elevated and ATP reserves are sufficient, each myosin head will repeat this cycle about five times per second. Each power stroke shortens the sarcomere by about 1 percent, so each second the sarcomere can shorten by roughly 5 percent. Because all the sarcomeres contract together, the entire muscle shortens at the same rate.

To appreciate the overall effect, imagine that you are pulling on a large rope. You are the myosin head, and the rope is a thin filament. You reach forward, grab the rope with both hands, and pull it toward you. This action corresponds to cross-bridge attachment and pivoting. You now release the rope, reach forward, and grab it once again. By repeating the cycle over and over, you can gradually pull in the rope.

Now consider several people lined up, all pulling on the same rope, as in a tug-of-war team. Each person reaches forward, grabs the rope, pulls it, releases it, and then grabs it again to repeat the cycle. The individual actions are not coordinated: At any one moment, some people are grabbing, some are pulling, and others are letting go. The amount of tension produced is a function of how many people are pulling at any given instant. A comparable situation applies to tension in a muscle fiber, where the myosin heads along a thick filament work together to pull a thin filament toward the center of the sarcomere.

Rigor Mortis

When death occurs, circulation ceases and the skeletal muscles are deprived of nutrients and oxygen. Within a few hours, the skeletal muscle fibers have run out of ATP and the sarcoplasmic reticulum becomes unable to pump Ca^{2+} out of the sarcoplasm. Calcium ions diffusing into the sarcoplasm from the extracellular fluid or leaking out of the sarcoplasmic reticulum then trigger a sustained contraction. Without ATP, the cross-bridges cannot detach from the active sites. Skeletal muscles throughout the body become locked in the contracted position. Because all the skeletal muscles are involved, the individual becomes "stiff as a board." This physical state, **rigor mortis,** lasts until the lysosomal enzymes released by autolysis break down the myofilaments 15–25 hours later. The timing is dependent on environmental factors, such as temperature. Forensic pathologists can estimate the time of death on the basis of the degree of rigor mortis and the characteristics of the local environment.

10

Step 5: Myosin reactivation

Step 3: Pivoting of myosin head

Step 4: Cross-bridge detachment

Sarcomeres shorten actively, but there is no active mechanism for reversing the process. External forces must act on the contracted muscle fiber to stretch the myofibrils and sarcomeres to their original dimensions. We will describe those forces in a later section.

Before you proceed, review the entire sequence of events from neural activation through excitation–contraction coupling to the completion of a contraction. Table 10-1 provides a summary of the contraction process, from ACh release to the end of the contraction.

TABLE 10-1 A Summary of the Steps Involved in Skeletal Muscle Contraction

Key steps in the initiation of a contraction include the following:

1. At the neuromuscular junction (NMJ), ACh released by the synaptic terminal binds to receptors on the sarcolemma.

2. The resulting change in the transmembrane potential of the muscle fiber leads to the production of an action potential that spreads across its entire surface and along the T tubules.

3. The sarcoplasmic reticulum (SR) releases stored calcium ions, increasing the calcium concentration of the sarcoplasm in and around the sarcomeres.

4. Calcium ions bind to troponin, producing a change in the orientation of the troponin–tropomyosin complex that exposes active sites on the thin (actin) filaments. Myosin cross-bridges form when myosin heads bind to active sites.

5. The contraction begins as repeated cycles of cross-bridge binding, pivoting, and detachment occur, powered by the hydrolysis of ATP. These events produce filament sliding, and the muscle fiber shortens.

This process continues for a brief period, until:

6. Action potential generation ceases as ACh is broken down by acetylcholinesterase (AChE).

7. The SR reabsorbs calcium ions, and the concentration of calcium ions in the sarcoplasm declines.

8. When calcium ion concentrations approach normal resting levels, the troponin–tropomyosin complex returns to its normal position. This change re-covers the active sites and prevents further cross-bridge interaction.

9. Without cross-bridge interactions, further sliding cannot take place and the contraction ends.

10. Muscle relaxation occurs, and the muscle returns passively to resting length.

Steps in the initiation of a contraction   Steps that end the contraction

LENGTH–TENSION RELATIONSHIPS

When many people pull on a rope, the amount of tension produced is proportional to the number of people pulling. In a skeletal muscle fiber, the amount of tension generated during a contraction depends on the number of pivoting cross-bridges in all the sarcomeres along all the myofibrils. The number of cross-bridges that can form depends on the degree of overlap between thick filaments and thin filaments. When the muscle fiber is stimulated to contract, only myosin heads in the zone of overlap can bind to active sites and produce tension. The tension produced by the entire muscle fiber can thus be related to the structure of an individual sarcomere.

When the sarcomeres are as short as they can be, the thick filaments are jammed against the Z lines. Although cross-bridge binding can occur, the myosin heads cannot pivot, and no tension is produced (Figure 10-11a•). Even if the sarcomeres are somewhat longer, the thin filaments that extend across the center of the sarcomere collide with or overlap the thin filaments of the opposite side (Figure 10-11b•). This disruption of the normal arrangement of thick and thin filaments interferes with the binding of cross-bridges to active sites. As a result, little tension is produced when the muscle fiber is stimulated.

Within the optimal range of sarcomere lengths (Figure 10-11c•), the maximum number of cross-bridges can form and the tension produced is highest. Any further increase in sarcomere length reduces the tension produced by reducing the size of the zone of overlap

and the number of potential cross-bridge interactions (Figure 10-11d•). When the zone of overlap is reduced to zero, thin and thick filaments cannot interact (Figure 10-11e•). The muscle fiber cannot produce any tension, and a contraction cannot occur. Such extreme stretching of a muscle fiber is normally prevented by the titin filaments in the muscle fiber, which tie the thick filaments to the Z lines, and by the surrounding connective tissues, which limit the degree of muscle stretch.

In summary, muscle fibers contract most forcefully when stimulated over a narrow range of resting lengths. The normal range of sarcomere lengths in the body (Figure 10-11•) is 75 to 130 percent of the optimal length. The arrangement of skeletal muscles, connective tissues, and bones normally prevents the extreme compression or excessive stretching that corresponds to Figure 10-11a or 10-11e•. For example, straightening your elbow stretches your *biceps brachii muscle,* but the bones and ligaments of the elbow end this movement before the muscle fibers stretch too far.

During normal movements, your muscle fibers perform over a broad range of intermediate lengths; the tension produced varies with the initial length of the muscle fibers. During an activity such as walking, in which muscles contract and relax cyclically, muscle fibers are stretched to a length very close to "ideal" before they are stimulated to contract. Some muscles must contract over a large range of resting lengths. When they contract at inefficient resting lengths, these muscles are generally assisted by the contractions of other muscles whose lengths are closer to ideal. (In Chapter 11, we will discuss the mechanical principles involved.)

10

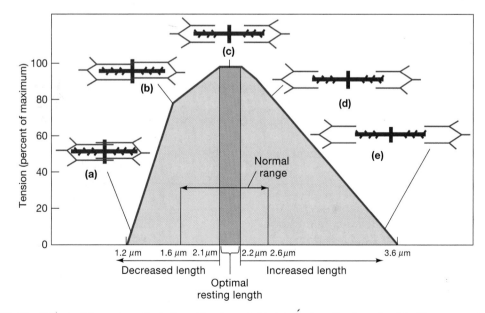

•**FIGURE 10-11 The Effect of Sarcomere Length on Tension. (a,b)** At short resting lengths, thin filaments extending across the center of the sarcomere interfere with the normal orientation of thick and thin filaments, reducing tension production. In addition, very little contraction can occur before the thick filaments crash into the Z lines. **(c)** The maximum tension is produced when the zone of overlap is large but the thin filaments do not extend across the sarcomere's center. If the sarcomeres are stretched too far, the zone of overlap **(d)** is reduced or **(e)** disappears, and cross-bridge interactions are reduced or cannot occur. The light purple area represents the normal range of sarcomere lengths.

☑ How would a drug that interferes with cross-bridge formation affect muscle contraction?

☑ What would you expect to happen to a resting skeletal muscle if the sarcolemma suddenly became very permeable to Ca^{2+}?

☑ Predict what would happen to a muscle if the motor end plate failed to produce acetylcholinesterase.

MUSCLE MECHANICS

Next, we will consider the coordination, metabolism, and performance of skeletal muscles as functional organs.

TENSION PRODUCTION

Now that you are familiar with the basic mechanisms of muscle contraction at the level of the individual muscle fiber, we can begin to examine the performance of skeletal muscles—organs of the muscular system. In this section, we will consider the coordinated contractions of an entire population of skeletal muscle fibers.

The amount of tension produced by an individual muscle fiber depends solely on the number of pivoting cross-bridges. If a muscle fiber at a given resting length is stimulated to contract, it will always produce the same amount of tension. There is no mechanism to regulate the amount of tension produced in that contraction by changing the number of contracting sarcomeres. When calcium ions are released, they are released from all triads in the muscle fiber. As a result, all the myosin heads within zones of overlap will interact with thin filaments. Thus a muscle fiber is either "ON" (producing as much tension as possible at that resting length) or "OFF" (relaxed). This feature of muscle mechanics is known as the **all-or-none principle.** The amount of tension produced in the skeletal muscle *as a whole* is therefore determined by (1) the frequency of stimulation and (2) the number of muscle fibers stimulated.

The Frequency of Muscle Stimulation

A **twitch** is a single stimulus–contraction–relaxation sequence in a muscle fiber. Twitches vary in duration. Twitches in one eye muscle fiber can be as brief as 7.5 msec, but a twitch in a muscle fiber from the *soleus,* a small calf muscle, lasts about 100 msec. Figure 10-12a• is a **myogram,** or graph of tension development in various muscles during a twitch contraction.

Figure 10-12b• details the phases of a 40-msec twitch in the *gastrocnemius muscle,* a prominent calf muscle. A single twitch can be divided into a *latent period,* a *contraction phase,* and a *relaxation phase:*

1. The **latent period** begins at stimulation and typically lasts about 2 msec. Over this period, the action potential sweeps across the sarcolemma and the sarcoplasmic reticulum releases calcium ions. The muscle fiber does not produce tension during the latent period, because the contraction cycle has yet to begin.

2. In the **contraction phase,** tension rises to a peak. As tension rises, calcium ions are binding to troponin, active sites on thin filaments are being exposed, and cross-bridge interactions are occurring. The contraction phase ends roughly 15 msec after stimulation.

3. The **relaxation phase** then continues for about another 25 msec. During this period, calcium levels are falling, active sites are being covered by tropomyosin, and the number of active cross-bridges is declining.

•**FIGURE 10-12 The Twitch and Development of Tension.** **(a)** A myogram showing differences in tension over time for a twitch contraction in different skeletal muscles. **(b)** The details of tension over time for a single twitch contraction in the gastrocnemius muscle. Notice the presence of a latent period, which corresponds to the time needed for the conduction of an action potential and the subsequent release of calcium ions by the sarcoplasmic reticulum.

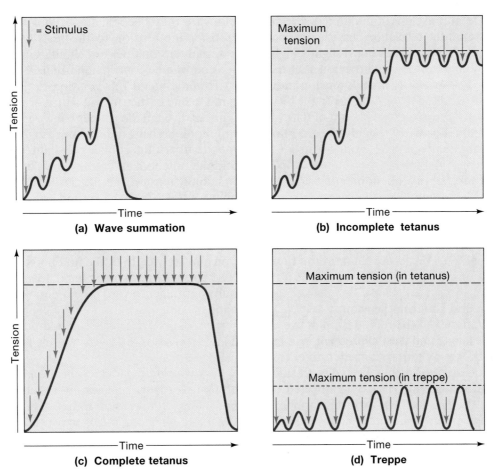

= Stimulus

Tension

Time

(a) Wave summation

Maximum
tension

Time

(b) Incomplete tetanus

Tension

Time

(c) Complete tetanus

Maximum tension (in tetanus)

Maximum tension (in treppe)

Time

(d) Treppe

•FIGURE 10-13 Effects of Repeated Stimulations. (a) Wave summation occurs when successive stimuli arrive before the relaxation phase (the downturn of the curve) has been completed. **(b)** Incomplete tetanus occurs if the rate of stimulation increases further. Tension production will rise to a peak, and the periods of relaxation will be very brief. **(c)** During complete tetanus, the frequency of stimulation is so high that the relaxation phase is eliminated; tension plateaus at maximal levels. **(d)** Treppe is an increase in peak tension with each successive stimulus delivered shortly after the completion of the relaxation phase of the preceding twitch.

A single stimulation produces a single twitch, but twitches in a skeletal muscle do not accomplish anything useful. *All normal activities involve sustained muscle contractions.* The mechanism involved is most easily understood by examining the responses of an isolated skeletal muscle when it is stimulated under laboratory conditions. Under these conditions, the myogram of tension production changes dramatically as the rate of stimulation increases (Figure 10-13•).

Wave Summation and Incomplete Tetanus If a second stimulus arrives before the relaxation phase has ended, a second, more powerful contraction occurs. The addition of one twitch to another in this way constitutes the **summation of twitches**, or **wave summation** (Figure 10-13a•). The duration of a single twitch determines the maximum time available to produce wave summation. For example, if a twitch lasts 20 msec (1/50 sec), subsequent stimuli must be separated by less than 20 msec—a stimulation rate of more than 50 stimuli per second.

Rather than refer to stimulation rate, we usually use *frequency,* which is a number per unit time. In this instance, a stimulus frequency of greater than 50 per second produces wave summation, whereas a stimulus frequency below 50 per second will produce individual twitches.

If the stimulation continues and the muscle is never allowed to relax completely, tension will rise to a peak (Figure 10-13b•). A muscle producing peak tension during rapid cycles of contraction and relaxation is in **incomplete tetanus** (*tetanos,* convulsive tension).

Complete Tetanus **Complete tetanus** is obtained by increasing the stimulation rate until the relaxation phase is eliminated (Figure 10-13c•). During complete tetanus, action potentials arrive so rapidly that the sarcoplasmic reticulum does not have time to reclaim the calcium ions. The high Ca^{2+} concentration in the cytoplasm prolongs the contraction state, making it continuous. Virtually all normal muscular contractions involve complete tetanus of the participating muscle fibers.

Treppe   If a skeletal muscle is stimulated a second time immediately after the relaxation phase has ended, the contraction that occurs will develop a slightly higher maximum tension than did the contraction after the first stimulation. The increase in peak tension indicated in Figure 10-13d• will continue over the first 30–50 stimulations. Thereafter, the amount of tension produced will remain constant at roughly 25 percent of the maximal tension that would be produced in complete tetanus. Because the tension rises in stages, like the steps in a staircase, this phenomenon is called **treppe** (TREP-ē), a German word meaning "stairs." The rise is thought to result from a gradual increase in the concentration of calcium ions in the sarcoplasm, in part because the ion pumps in the sarcoplasmic reticulum are unable to recapture them in the time between stimulations.

Internal Tension and External Tension

We have already discussed the relationship between resting sarcomere length and tension production in individual muscle fibers. When we consider an entire skeletal muscle, the situation is complicated by the fact that the muscle fibers are not directly connected to the structures they pull against. The extracellular fibers of the endomysium, perimysium, epimysium, and tendons are flexible and somewhat elastic. When a skeletal muscle contracts, the myofibrils in the muscle fibers generate *internal tension*. This internal tension is applied to the extracellular fibers. The tension in the extracellular fibers is called *external tension.*

As the external tension rises, the extracellular fibers stretch; for this reason, they are called *series elastic elements.* The series elastic elements behave like fat rubber bands. They stretch easily at first, but as they elongate they become stiffer and more effective at transferring tension to the resistance. As a result, external tension does not climb as quickly as internal tension when a contraction occurs.

To understand this relationship, attach a rubber band to one of the rings in a three-ring notebook. Put a finger through the rubber band and use it to pull the notebook across a table. Your finger represents the muscle fibers; the rubber band, the attached tendon; and the notebook, a bone of the skeleton. When you first apply tension, the rubber band stretches and becomes stiffer. Over this period, external tension rises. The notebook starts to move when the rubber band becomes sufficiently taut—that is, when the tension in the rubber band (the external tension) overcomes friction and the weight of the notebook (the resistance).

If you now relax your hand, the rubber band will pull your finger toward the notebook. The same thing happens in a muscle: When the contraction ends, the series elastic elements recoil and pull on the muscle. This recoil helps return the muscle to its original resting length.

A myogram performed in the laboratory generally measures the tension in a tendon and so is reporting external tension rather than internal tension. Figure 10-14a• compares the internal and external tensions during a twitch contraction. A single twitch is so short in duration that there isn't enough time for the external tension to rise as high as the internal tension. Twitches are therefore ineffective in terms of performing useful work. Notice that the external tension remains elevated until the relaxation phase has ended. If a second twitch occurs before the external tension returns to zero, the external tension will peak at a higher level, because more of the

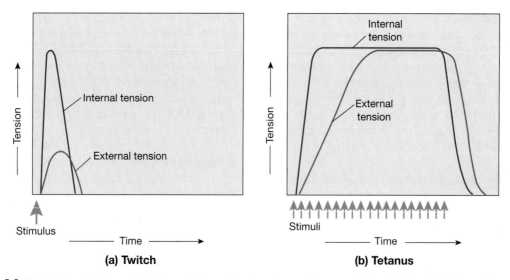

•**FIGURE 10-14 Internal and External Tensions.** Internal tension rises as the muscle fiber contracts. External tension rises more slowly as the series elastic elements are stretched. **(a)** During a single-twitch contraction, external tension cannot rise as high as internal tension before the relaxation phase begins. **(b)** During a tetanic contraction, external tension soon plateaus at a level roughly equivalent to internal tension. External tension remains elevated for the duration of the contraction.

Tetanus

Children are often told to be careful of rusty nails. Parents should worry most not about the rust or the nail but about infection with the very common bacterium *Clostridium tetani*, which can cause **tetanus.** Although they have the same name, the disease tetanus has no relation to the normal muscle response to neural stimulation. The *Clostridium* bacterium occurs virtually everywhere in the environment, but it can thrive only in tissues that contain low amounts of oxygen. For this reason, a deep puncture wound, such as that from a nail, carries a much greater risk of producing tetanus than does a shallow, open cut that bleeds freely.

When active in body tissues, these bacteria release a powerful toxin that affects the central nervous system. Motor neurons, which control skeletal muscles throughout the body, are particularly sensitive to it. The toxin suppresses the mechanism that regulates motor neuron activity. The result is a sustained, powerful contraction of skeletal muscles throughout the body.

The incubation period (the time from exposure to the development of symptoms) is generally less than 2 weeks. The most common complaints are headache, muscle stiffness, and difficulty in swallowing. Because it soon becomes difficult to open the mouth, this disease is also called *lockjaw.* Widespread muscle spasms typically develop within 2 or 3 days of the initial symptoms and continue for a week before subsiding. After 2–4 weeks, patients who survive recover with no after-effects.

Severe tetanus has a 40–60 percent mortality rate; that is, for every 100 people who develop severe tetanus, 40 to 60 die. Fortunately, immunization is effective in preventing the disease. Approximately 500,000 cases of tetanus occur worldwide each year, but only about 100 of them occur in the United States, thanks to an effective immunization program. ("Tetanus shots" are recommended, with booster shots every 10 years.) Severe symptoms in unimmunized patients can be prevented by early administration of an antitoxin, in most cases *human tetanus immune globulin.* Such treatment does not reduce symptoms that have already appeared, however.

internal tension will be conveyed to the series elastic elements. Think of pushing a child on a swing: You push once to start the swing moving; if you push a second time, the child swings higher because the energy of the second push is added to the energy remaining from the first. This mechanism of elevated external tension is now thought to be the primary basis of wave summation. During a tetanic contraction, there is sufficient time for internal and external tensions to equalize (Figure 10-14b•).

Motor Units and Tension Production

You have a remarkable ability to control the amount of tension exerted by your skeletal muscles. During a normal contraction, tension rises smoothly, not jerkily, because activated muscle fibers are stimulated to complete tetanus. The *total force* exerted by the skeletal muscle depends on how many muscle fibers are activated.

A typical skeletal muscle contains thousands of muscle fibers. Although some motor neurons control a few muscle fibers, most control hundreds of them. All the muscle fibers controlled by a single motor neuron constitute a **motor unit.** The size of a motor unit is an indication of how fine the control of movement can be. In the muscles of the eye, where precise control is extremely important, a motor neuron may control 4–6 muscle fibers. We have much less precise control over our leg muscles, where a single motor neuron may control 1000–2000 muscle fibers (depending on the specific muscle and the reference consulted). The muscle fibers of each motor unit are intermingled with those of other motor units (Figure 10-15a•). Because of this intermingling, the direction of pull exerted on the tendon does not change despite variations in the numbers of activated motor units.

When you decide to perform a specific arm movement, specific groups of motor neurons in the spinal cord are stimulated. The contraction begins with the activation of the smallest motor units in the stimulated muscle. These motor units generally contain muscle fibers that contract relatively slowly. Over time, larger motor units containing faster and more powerful muscle fibers are activated, and tension production rises steeply. The smooth but steady increase in muscular tension produced by increasing the number of active motor units is called **recruitment,** or **multiple motor unit summation.**

Peak tension production occurs when all motor units in the muscle contract in a state of complete tetanus. Such powerful contractions do not last long, however, because the individual muscle fibers soon use up their available energy reserves. During a sustained tetanic contraction, motor units are activated on a rotating basis, so some of them are resting and recovering while others are actively contracting. This "relay team" approach, called *asynchronous motor unit summation,* lets each motor unit recover somewhat

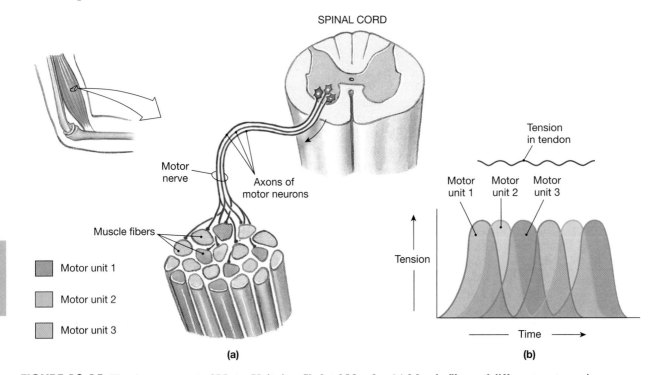

•**FIGURE 10-15** **The Arrangement of Motor Units in a Skeletal Muscle.** (a) Muscle fibers of different motor units are intermingled, so the forces applied to the tendon remain roughly balanced regardless of which muscle groups are stimulated. (b) The tension applied to the tendon remains relatively constant even though individual motor units cycle between contraction and relaxation.

before it is stimulated again (Figure 10-15b•). As a result, when your muscles contract for sustained periods, they produce slightly less than maximal tension.

Muscle Tone   In any skeletal muscle, some motor units are always active, even when the entire muscle is not contracting. Their contractions do not produce enough tension to cause movement, but they do tense and firm the muscle. This resting tension in a skeletal muscle is called **muscle tone.** A muscle with little muscle tone appears limp and flaccid, whereas one with moderate muscle tone is firm and solid. The identity of the stimulated motor units changes constantly, so a constant tension in the attached tendon is maintained, but individual muscle fibers can relax.

Resting muscle tone stabilizes the position of bones and joints. For example, in muscles involved with balance and posture, enough motor units are stimulated to produce the tension needed to maintain body position. Muscle tone also helps prevent sudden, uncontrolled changes in the position of bones and joints. In addition to bracing the skeleton, the elastic nature of muscles and tendons lets skeletal muscles act as shock absorbers that cushion the impact of a sudden bump or shock. Heightened muscle tone accelerates the recruitment process during a voluntary contraction, because some of the motor units are already stimulated. Strong muscle tone also makes skeletal muscles appear firm and well defined, even at rest.

Isotonic and Isometric Contractions

We can classify muscle contractions as *isotonic* or *isometric* on the basis of the pattern of tension production.

Isotonic Contractions   In an **isotonic contraction** (*iso-,* equal + *tonos,* tension), tension rises and the skeletal muscle's length changes. Lifting an object off a desk, walking, and running involve isotonic contractions.

Two types of isotonic contractions exist: (1) concentric and (2) eccentric. In a **concentric contraction,** the muscle tension exceeds the resistance and the muscle shortens. Consider the experiment summarized in Figure 10-16•. A skeletal muscle 1 cm^2 in cross-sectional area can produce roughly 4 kg of tension in complete tetanus. If we hang a 2-kg weight from that muscle and stimulate it, the muscle will shorten (Figure 10-16a•). Before the muscle can shorten, the cross-bridges must produce enough tension to overcome the resistance—in this case, the 2-kg weight. Over this period, internal tension in the muscle fibers rises until the external tension in the tendon exceeds the amount of resistance. As the muscle shortens, the internal and external tensions in the skeletal muscle remain constant at a value that just exceeds the resistance (Figure 10-16b•). The term *isotonic* originated from this type of experiment.

In the body, however, the situation is more complicated. For example, muscles are not always positioned

•FIGURE 10-16 Isotonic and Isometric Contractions. (a,b) This muscle is attached to a weight less than its peak tension capabilities. On stimulation, it develops enough tension to lift the weight. Tension remains constant for the duration of the contraction, although the length of the muscle changes. This is an example of isotonic contraction. **(c,d)** The same muscle is attached to a weight that exceeds its peak tension capabilities. On stimulation, tension will rise to a peak, but the muscle as a whole cannot shorten. This is an isometric contraction.

directly above the resistance, and they are attached to bones rather than to static weights. Changes in the relative positions of the muscle and the articulating bones, the effects of gravity, and other mechanical and physical factors interact to increase or decrease the amount of resistance the muscle must overcome as a movement proceeds. Nevertheless, at any time during a concentric contraction, the tension produced exceeds that resistance.

The speed of shortening varies with the difference between the amount of tension produced and the amount of resistance. If all the muscle units are stimulated and the resistance is relatively small, the muscle will shorten very quickly. In contrast, if the muscle barely produces enough tension to overcome the resistance, it will shorten very slowly.

In an **eccentric contraction,** the peak tension developed is less than the resistance, and the muscle

elongates owing to the contraction of another muscle or the pull of gravity. Think of a tug-of-war team trying to stop a moving car. Although everyone pulls as hard as they can, the rope slips through their fingers. The speed of elongation depends on the difference between the amount of tension developed by the active muscle fibers and the amount of resistance. In our analogy, the team might slow down a small car but would have little effect on a large truck.

Eccentric contractions are very common, and they are an important part of a variety of movements. In these movements, you exert precise control over the amount of tension produced. By varying the tension in an eccentric contraction, you can control the rate of elongation, just as you can vary the tension in a concentric contraction. For example, precisely controlled eccentric contractions occur each time you walk down stairs or settle into a chair. During physical training, people commonly perform cycles of concentric and eccentric contractions, as when you hold a weight in your hand and flex and extend your elbow.

Isometric Contractions   In an **isometric contraction** (*metric*, measure), the muscle as a whole does not change length, and the tension produced never exceeds the resistance. Figure 10-16c• shows what happens if we attach a weight heavier than 4 kg to the experimental muscle and then stimulate the muscle. Although cross-bridges form and tension rises to peak values, the muscle cannot overcome the resistance of the weight and so cannot shorten (Figure 10-16d•). Examples of isometric contractions include holding a heavy weight above the ground, pushing against a locked door, or trying to pick up a car. These are rather unusual movements. However, many of the reflexive muscle contractions that keep your body upright when you stand or sit involve the isometric contractions of muscles that oppose the force of gravity.

You may have noticed that when you perform an isometric contraction, the contracting muscle bulges (but not as much as it does during an isotonic contraction). In an isometric contraction, although the muscle *as a whole* does not shorten, the individual muscle fibers shorten until the tendons are taut and the external tension equals the internal tension generated by the muscle fibers. The muscle fibers cannot shorten further, because the external tension does not exceed the resistance.

Normal daily activities therefore involve a combination of isotonic and isometric muscular contractions. As you sit and read this text, isometric contractions of postural muscles stabilize your vertebrae and maintain your upright position. When you turn a page, the movements of your arm, forearm, hand, and fingers are produced by a combination of concentric and eccentric isotonic contractions.

Resistance and Speed of Contraction

You can lift a light object more rapidly than you can lift a heavy one because the resistance and the speed of contraction are inversely related. If the resistance is less than the tension produced, an isotonic, concentric contraction will occur; the muscle will shorten. The heavier the resistance, the longer it takes for the movement to begin, because muscle tension, which increases gradually, must exceed the resistance before shortening can occur (Figure 10-17•). The contraction itself proceeds more slowly. At the cellular level, the speed of cross-bridge pivoting is reduced as the load increases.

For each muscle, an optimal combination of tension and speed exists for any given resistance. If you have ever ridden a 10-speed bicycle, you are probably already aware of this fact. When you are cruising along comfortably, your thigh and leg muscles are working at

•**FIGURE 10-17** **The Resistance and Speed of Contraction.** The heavier the resistance on a muscle, the longer it will take for the muscle to begin to shorten and the less the muscle will shorten.

an optimal combination of speed and tension. When you come to a hill, the resistance increases. Your muscles must now develop more tension, and they move more slowly; they are no longer working at optimal efficiency. You then shift to a lower gear. The load on your muscles decreases, the speed increases, and the muscles are once again working efficiently.

Muscle Relaxation and the Return to Resting Length

As we noted earlier, there is no active mechanism for muscle fiber elongation. The sarcomeres in a muscle fiber can shorten and develop tension, but the power stroke cannot be reversed to push the Z lines farther apart. After a contraction, a muscle fiber returns to its original length through a combination of elastic forces, opposing muscle contractions, and gravity.

Elastic Forces   When the contraction ends, some of the energy initially "spent" in stretching the series elastic elements is recovered as they recoil. The recoil of the series elastic elements gradually helps return the muscle fiber to its original resting length.

Opposing Muscle Contractions   The contraction of opposing muscles can return a muscle to its resting length more quickly than elastic forces can. Consider the muscles of the arm that flex or extend the elbow. Contraction of the *biceps brachii muscle* on the anterior part of the arm flexes the elbow; contraction of the *triceps brachii muscle* on the posterior part of the arm extends the elbow. When the biceps brachii contracts, the triceps brachii is stretched. When the biceps brachii relaxes, contraction of the triceps brachii extends the elbow and stretches the muscle fibers of the biceps brachii to their original length.

Gravity Gravity may assist opposing muscle groups in quickly returning a muscle to its resting length after a contraction. For example, imagine the biceps brachii muscle fully contracted with the elbow pointed at the ground. When the muscle relaxes, gravity will pull the forearm down and stretch the muscle. Although gravity can provide assistance in stretching muscles, some active muscle tension is needed to control the rate of movement and to prevent damage to the joint. In the previous example, eccentric contraction of the biceps brachii muscle can control the movement.

☑ Why is it difficult to contract a muscle that has been over-stretched?

☑ During treppe, why does tension in a muscle gradually increase even though the stimulus strength and frequency are constant?

☑ Can a skeletal muscle contract without shortening? Explain.

ENERGETICS OF MUSCULAR ACTIVITY

A single muscle fiber may contain 15 billion thick filaments. When that muscle fiber is actively contracting, each thick filament breaks down roughly 2500 ATP molecules per second. Because even a small skeletal muscle contains thousands of muscle fibers, the ATP demands of a contracting skeletal muscle are enormous. In practical terms, the demand for ATP in a contracting muscle fiber is so high that it would be impossible to have all the necessary energy available as ATP before the contraction begins. Instead, a resting muscle fiber contains only enough ATP and other high-energy compounds to sustain a contraction until additional ATP can be generated. Throughout the rest of the contraction, the muscle fiber will generate ATP at roughly the same rate as it is used.

ATP and CP Reserves

The primary function of ATP is the transfer of energy from one location to another rather than the long-term storage of energy. At rest, a skeletal muscle fiber produces more ATP than it needs. Under these conditions, ATP transfers energy to creatine. *Creatine* is a small molecule that muscle cells assemble from fragments of amino acids. The energy transfer creates another high-energy compound, **creatine phosphate (CP)**, or *phosphorylcreatine:*

$$ATP + creatine \rightarrow ADP + creatine\ phosphate$$

During a contraction, each myosin cross-bridge breaks down ATP, producing ADP and a phosphate group. The energy stored in creatine phosphate is then used to "recharge" ADP, converting it back to ATP through the reverse reaction:

$$ADP + creatine\ phosphate \rightarrow ATP + creatine$$

The enzyme that facilitates this reaction is **creatine phosphokinase (CPK or CK).** When muscle cells are damaged, CPK leaks across the cell membranes and into the bloodstream. Thus, a high blood concentration of CPK usually indicates serious muscle damage.

The energy reserves of a representative muscle fiber are indicated in Table 10-2. A resting skeletal muscle fiber contains about six times as much creatine phosphate as ATP. But when a muscle fiber is undergoing a sustained contraction, these energy reserves are exhausted in only about 15 seconds. The muscle fiber must then rely on other mechanisms to convert ADP to ATP.

ATP Generation

We learned in Chapter 3 that most cells in the body generate ATP through aerobic metabolism in mitochondria and through glycolysis in the cytoplasm. ∞ *p. 88*

Aerobic Metabolism **Aerobic metabolism** normally provides 95 percent of the ATP demands of a resting cell. In this process, mitochondria absorb oxygen, ADP, phosphate ions, and organic substrates from the surrounding cytoplasm. The substrates then enter the *TCA (tricarboxylic acid) cycle* (also known as the *citric acid cycle* or the *Krebs cycle*), an enzymatic pathway that breaks down organic molecules. The carbon atoms are released as carbon dioxide. The hydrogen atoms are shuttled to respiratory enzymes in the inner mitochondrial membrane, where their electrons are

10

TABLE 10-2 **Sources of Energy Stored in a Typical Muscle Fiber**

Energy Stored As	Utilized through	Initial Quantity	Number of Twitches Supported by Each Energy Source Alone	Duration of Isometric Tetanic Contraction Supported by Each Energy Source Alone
ATP	ATP → ADP + P	3 mmol	10	2 sec
CP	ADP + CP → ATP + C	20 mmol	70	15 sec
Glycogen	Glycolysis (anaerobic)	100 mmol	670	130 sec
	Aerobic metabolism		12,000	2400 sec (40 min)

removed. After a series of intermediate steps, the protons and electrons are combined with oxygen to form water. Along the way, large amounts of energy are released and used to make ATP. The entire process is very efficient; for each organic molecule "fed" to the TCA cycle, the cell will gain 17 ATP molecules.

Resting skeletal muscle fibers rely almost exclusively on the aerobic metabolism of fatty acids to generate ATP. When the muscle starts contracting, the mitochondria begin breaking down molecules of pyruvic acid instead of fatty acids. The pyruvic acid is provided by the enzymatic pathway of glycolysis.

Glycolysis **Glycolysis** is the breakdown of glucose to pyruvic acid in the cytoplasm of a cell. It is called an **anaerobic** process, because it does not require oxygen. Glycolysis provides a net gain of 2 ATP and generates 2 pyruvic acid molecules from each glucose molecule. The ATP produced by glycolysis is thus only a small fraction of that produced by aerobic metabolism, in which the breakdown of the 2 pyruvic acid molecules in mitochondria would generate 34 ATP. Yet, because it can proceed in the absence of oxygen, glycolysis can be important when the availability of oxygen limits the rate of mitochondrial ATP production.

In most skeletal muscles, glycolysis is the primary source of ATP during peak periods of activity. The glucose broken down under these conditions is obtained primarily from the reserves of glycogen in the sarcoplasm. Glycogen, diagrammed in Figure 2-12•, page 16, is a polysaccharide chain of glucose molecules. Typical skeletal muscle fibers contain large glycogen reserves, which may account for 1.5 percent of the total muscle weight. When the muscle fiber begins to run short of ATP and CP, enzymes split the glycogen molecules apart, releasing glucose, which can be used to generate more ATP. As the level of muscular activity increases and these reserves are mobilized, the pattern of energy production and use changes.

Energy Use and the Level of Muscle Activity

In a resting skeletal muscle (Figure 10-18a•), the demand for ATP is low. More than enough oxygen is available for the mitochondria to meet that demand, and they produce a surplus of ATP. The extra ATP is used to build up reserves of CP and glycogen. Resting muscle fibers absorb fatty acids and glucose that are delivered by the bloodstream. The fatty acids are broken down in the mitochondria, and the ATP generated is used to convert creatine to creatine phosphate and glucose to glycogen.

At moderate levels of activity (Figure 10-18b•), the demand for ATP increases. This demand is met by the mitochondria. As the rate of mitochondrial ATP production rises, so does the rate of oxygen consumption. Oxygen availability is not a limiting factor, because

(a) Resting muscle: Fatty acids are catabolized; the ATP produced is used to build energy reserves of ATP, CP, and glycogen.

(b) Moderate activity: Glucose and fatty acids are catabolized; the ATP produced is used to power contraction.

(c) Peak activity: Most ATP is produced through glycolysis, with lactic acid as a byproduct. Mitochondrial activity (not shown) now provides only about one-third of the ATP consumed.

•**FIGURE 10-18** Muscle Metabolism. **(a)** A resting muscle breaks down fatty acids by aerobic metabolism to make ATP. Surplus ATP is used to build reserves of creatine phosphate (CP) and glycogen. **(b)** At modest activity levels, mitochondria can meet the ATP demands through the aerobic metabolism of fatty acids and glucose. **(c)** At peak activity levels, mitochondria cannot get enough oxygen to meet ATP demands. Most of the ATP is provided by glycolysis, leading to the production of lactic acid.

oxygen can diffuse into the muscle fiber fast enough to meet mitochondrial needs. But all the ATP produced is needed by the muscle fiber, and no surplus is available. The skeletal muscle now relies primarily on the aerobic metabolism of pyruvic acid to generate ATP. The pyruvic acid is provided by glycolysis, which breaks down glucose molecules obtained from glycogen in the muscle fiber. If glycogen reserves are low, the muscle fiber can also break down other substrates, such as lipids or amino acids. As long as the demand for ATP can be met by mitochondrial activity, the ATP provided by glycolysis makes a relatively minor contribution to the total energy budget of the muscle fiber.

At peak levels of activity, the ATP demands are enormous and mitochondrial ATP production rises to a maximum. This maximum rate is determined by the availability of oxygen, and oxygen cannot diffuse into the muscle fiber fast enough to enable the mitochondria to produce the required ATP. At peak levels of exertion, mitochondrial activity can provide only about one-third of the ATP needed. The remainder is produced through glycolysis (Figure 10-18c•).

When glycolysis produces pyruvic acid faster than it can be utilized by the mitochondria, pyruvic acid levels rise in the sarcoplasm. Under these conditions, pyruvic acid is converted to **lactic acid,** a related three-carbon molecule.

The anaerobic process of glycolysis enables the cell to generate additional ATP when the mitochondria are unable to meet the current energy demands. However, anaerobic energy production has its drawbacks:

- Lactic acid is an organic acid that in body fluids dissociates into a hydrogen ion and a negatively charged *lactate ion.* The production of lactic acid can therefore lower the intracellular pH. Buffers in the sarcoplasm can resist pH shifts, but these defenses are limited. Eventually, changes in pH will alter the functional characteristics of key enzymes. The muscle fiber then cannot continue to contract.

- Glycolysis is a relatively inefficient way to generate ATP. Under anaerobic conditions, each glucose molecule generates 2 pyruvic acid molecules, which are converted to lactic acid. In return, the cell gains 2 ATP through glycolysis. Had those 2 pyruvic acid molecules been catabolized aerobically in a mitochondrion, the cell would have gained 34 additional ATP.

Muscle Fatigue

A skeletal muscle fiber is said to be **fatigued** when it can no longer contract despite continued neural stimulation. The cause of muscle fatigue varies with the level of muscle activity. After short peak levels of activity, such as running a 100-meter dash, fatigue may result from the exhaustion of ATP and CP reserves or from the drop in pH that accompanies the buildup of lactic acid. After prolonged exertion, such as running a

marathon, fatigue may involve physical damage to the sarcoplasmic reticulum that interferes with the regulation of intracellular Ca^{2+} concentrations. Muscle fatigue is cumulative—the effects become more pronounced as more muscle fibers are affected. The result is a gradual reduction in the capabilities of the entire skeletal muscle.

If the muscle fiber is contracting at moderate levels and ATP demands can be met through aerobic metabolism, fatigue will not occur until glycogen, lipid, and amino acid reserves are depleted. This type of fatigue affects the muscles of long-distance athletes, such as marathon runners, after hours of exertion.

When a muscle produces a sudden, intense burst of activity at peak levels, most of the ATP is provided by glycolysis. After just seconds to minutes, the rising lactic acid levels lower the tissue pH, and the muscle can no longer function normally. Athletes who run sprints, such as the 100-meter dash, get this type of muscle fatigue. We will return to the topics of fatigue, athletic training, and metabolic activity later in the chapter.

Normal muscle function requires (1) substantial intracellular energy reserves, (2) a normal circulatory supply, and (3) a normal blood oxygen concentration. Anything that interferes with one or more of those factors will promote premature muscle fatigue. For example, reduced blood flow from tight clothing, a circulatory disorder, or loss of blood slows the delivery of oxygen and nutrients, accelerates the buildup of lactic acid, and promotes muscle fatigue.

The Recovery Period

When a muscle fiber contracts, the conditions in the sarcoplasm are changed. Energy reserves are consumed, heat is released, and, if the contraction was at peak levels, lactic acid is generated. In the **recovery period,** the conditions in muscle fibers are returned to normal, preexertion levels. It may take several hours for muscle fibers to recover from a period of moderate activity. After sustained activity at higher levels, complete recovery can take a week.

Lactic Acid Removal and Recycling   Glycolysis enables a skeletal muscle to continue contracting even when mitochondrial activity is limited by the availability of oxygen. As we have seen, however, lactic acid production is not an ideal way to generate ATP. It squanders the glucose reserves of the muscle fibers, and it is potentially dangerous because lactic acid can alter the pH of the blood and tissues.

During the recovery period, when oxygen is available in abundance, that lactic acid can be recycled by conversion back to pyruvic acid. The pyruvic acid can then be used either by mitochondria to generate ATP or as a substrate for enzyme pathways that synthesize glucose and rebuild glycogen reserves.

During a period of exertion, lactic acid diffuses out of the muscle fibers and into the bloodstream. This process continues after the exertion has ended, because intracellular lactic acid concentrations are still relatively high. The liver absorbs the lactic acid and converts it to pyruvic acid. Roughly 30 percent of these pyruvic acid molecules are broken down in the TCA cycle, providing the ATP needed to convert the other pyruvic acid molecules to glucose. (We shall cover these processes more fully in Chapter 25.) The glucose molecules are then released into the circulation, where they are absorbed by skeletal muscle fibers and used to rebuild their glycogen reserves. This shuffling of lactic acid to the liver and glucose back to muscle cells is called the **Cori cycle.**

The Oxygen Debt   During the recovery period, oxygen is readily available and the body's oxygen demand remains elevated above normal resting levels. The recovery period is powered by the ATP that aerobic metabolism generates. The more ATP required, the more oxygen will be needed. The **oxygen debt,** or *excess postexercise oxygen consumption (EPOC),* created during exercise is the amount of oxygen needed to restore normal, preexertion conditions. Skeletal muscle fibers, which must restore ATP, creatine phosphate, and glycogen concentrations to their former levels, and liver cells, which generate the ATP needed to convert excess lactic acid to glucose, are responsible for most of the additional oxygen consumption. However, several other tissues also increase their rate of oxygen consumption and ATP generation during the recovery period. For example, sweat glands increase their secretory activity until normal body temperature is restored. While the oxygen debt is being repaid, the breathing rate and depth are increased. As a result, you continue to breathe heavily long after you stop exercising.

Heat Loss   Muscular activity generates substantial amounts of heat. When a catabolic reaction occurs, such as the breakdown of glycogen or the reactions of glycolysis, the muscle fiber captures only a portion of the released energy. ∞ *p. 37* The rest is released as heat. A resting muscle fiber relying on aerobic metabolism captures about 42 percent of the energy released in catabolism. The other 58 percent warms the sarcoplasm, interstitial fluid, and circulating blood. Active skeletal muscles release roughly 85 percent of the heat needed to maintain normal body temperature.

When muscles become active, their consumption of energy skyrockets. As anaerobic energy production becomes the primary method of ATP generation, muscle fibers become less efficient at capturing energy. At peak levels of exertion, only about 30 percent of the released energy is captured as ATP; the remaining 70 percent warms the muscle and surrounding tissues.

Body temperature soon climbs, and heat loss at the skin accelerates through mechanisms introduced in Chapters 1 and 5. ∞ *pp. 12, 157*

Hormones and Muscle Metabolism

Metabolic activities in skeletal muscle fibers are adjusted by hormones of the endocrine system. *Growth hormone* from the pituitary gland and *testosterone* (the primary sex hormone in males) stimulate the synthesis of contractile proteins and the enlargement of skeletal muscles. *Thyroid hormones* elevate the rate of energy consumption by resting and active skeletal muscles. During a sudden crisis, hormones of the adrenal gland, notably *epinephrine* (adrenaline), stimulate muscle metabolism and increase both the duration of stimulation and the force of contraction. We shall further examine the effects of hormones on muscle and other tissues in Chapter 18.

MUSCLE PERFORMANCE

Muscle performance can be considered in terms of **power,** the maximum amount of tension produced by a particular muscle or muscle group, and **endurance,** the amount of time for which the individual can perform a particular activity. Two major factors determine the performance capabilities of any skeletal muscle: (1) the types of muscle fibers in the muscle and (2) physical conditioning or training.

Types of Skeletal Muscle Fibers

The human body has three major types of skeletal muscle fibers: *fast fibers, slow fibers,* and *intermediate fibers.*

Fast Fibers   Most of the skeletal muscle fibers in the body are called **fast fibers,** because they can contract in 0.01 sec or less after stimulation. Fast fibers are large in diameter; they contain densely packed mofibrils, large glycogen reserves, and relatively few mito-chondria. The tension produced by a muscle fiber is directly proportional to the number of sarcomeres, so muscles dominated by fast fibers produce powerful contractions. However, fast fibers fatigue rapidly because their contractions use ATP in massive amounts, so prolonged activity is supported primarily by anaerobic metabolism. Several other names are used to refer to these muscle fibers, including *white muscle fibers, fast-twitch glycolytic fibers,* and *Type II-A fibers.*

Slow Fibers   **Slow fibers** are only about half the diameter of fast fibers and take three times as long to contract after stimulation. Slow fibers are specialized to enable them to continue contracting for extended periods, long after a fast muscle would have become fatigued. The most important specializations improve mitochondrial performance. Slow muscle tissue contains a more extensive network of capillaries than is typical of fast muscle tissue and so has a dramatically higher oxygen supply. In addition, slow fibers contain the red pigment **myoglobin** (MĪ-ō-glō-bin). This globular protein is structurally related to hemoglobin, the oxygen-carrying pigment in blood. Both myoglobin and hemoglobin are red pigments that reversibly bind oxygen molecules. Although other muscle fiber types contain small amounts of myoglobin, it is most abundant in slow fibers. As a result, resting slow fibers contain substantial oxygen reserves that can be mobilized during a contraction. Because slow fibers have both an extensive capillary supply and a high concentration of myoglobin, skeletal muscles dominated by slow fibers are dark red. They are also known as *red muscle fibers, slow-twitch oxidative fibers,* and *Type I fibers.*

With oxygen reserves and a more efficient blood supply, the mitochondria of slow fibers can contribute more ATP during contraction. Thus, slow fibers are less dependent on anaerobic metabolism than are fast fibers. Some of the mitochondrial energy production involves the breakdown of stored lipids rather than glycogen, so glycogen reserves of slow fibers are smaller than those of fast fibers. Slow fibers also contain more mitochondria than do fast fibers. Figure 10-19• compares the appearance of fast and slow fibers.

Intermediate Fibers The properties of **intermediate fibers** are intermediate between those of fast fibers and slow fibers. In appearance, intermediate fibers most closely resemble fast fibers, for they contain little myoglobin and are relatively pale. They have a more extensive capillary network around them, however, and are more resistant to fatigue than are fast fibers. Intermediate fibers are also known as *fast-twitch oxidative fibers* and *Type II-B fibers.*

The three types of muscle fibers are compared in Table 10-3. In muscles that contain a mixture of fast and intermediate fibers, the proportion can change with physical conditioning. For example, if a muscle is used repeatedly for endurance events, some of the fast fibers will develop the appearance and functional capabilities of intermediate fibers. The muscle as a whole will thus become more resistant to fatigue.

Muscle Performance and the Distribution of Muscle Fibers

The percentages of fast, intermediate, and slow fibers in a skeletal muscle can be quite variable. Muscles dominated by fast fibers appear pale and are often called **white muscles.** Chicken breasts contain "white meat" because chickens use their wings only for brief intervals, as when fleeing from a predator, and the power for flight comes from fast fibers in their breast muscles. As we learned earlier, the extensive blood

Slow fibers
Smaller diameter, darker color due to myoglobin; fatigue-resistant

Slow

Fast fibers
Larger diameter, paler color; easily fatigued

Fast

(a)

(b)

•**FIGURE 10-19 Fast versus Slow Fibers.** **(a)** The slender, slow fiber (R for red) has more mitochondria (M) and a more extensive capillary supply (cap) than does the fast fiber (W for white). (LM × 783) **(b)** Notice the difference in the sizes of slow fibers, above, and of fast fibers, below. (LM × 171)

TABLE 10-3 Properties of Skeletal Muscle Fiber Types

Property	Slow	Intermediate	Fast
Cross-sectional diameter	Small	Intermediate	Large
Tension	Low	Intermediate	High
Contraction speed	Slow	Fast	Fast
Fatigue resistance	High	Intermediate	Low
Color	Red	White	White
Myoglobin content	High	Low	Low
Capillary supply	Dense	Intermediate	Scarce
Mitochondria	Many	Intermediate	Few
Glycolytic enzyme concentration in sarcoplasm	Low	High	High
Substrates used for ATP generation during contraction	Lipids, carbohydrates, amino acids (aerobic)	Primarily carbohydrates (anaerobic)	Carbohydrates (anaerobic)
Alternative names	Type I, S (slow), red, SO (slow oxidizing), slow-twitch oxidative	Type II-B, FR (fast resistant), fast-twitch oxidative	Type II-A, FF (fast fatigue), white, fast-twitch glycolytic

vessels and myoglobin in slow fibers give these fibers a reddish color; muscles dominated by slow fibers are therefore known as **red muscles.** Chickens walk around all day, and the movements are performed by the slow fibers in the "dark meat" of their legs.

Most human muscles contain a mixture of fiber types and so appear pink. However, there are no slow fibers in muscles of the eye or hand, where swift but brief contractions are required. Many back and calf muscles are dominated by slow fibers; these muscles contract almost continuously to maintain an upright posture. The percentage of fast versus slow fibers in each muscle is genetically determined. As we noted earlier, the proportion of intermediate fibers to fast fibers can increase as a result of athletic training.

Muscle Hypertrophy

As a result of repeated, exhaustive stimulation, muscle fibers develop more mitochondria, a higher concentration of glycolytic enzymes, and larger glycogen reserves. Such muscle fibers have more myofibrils than do muscles that are less used, and each myofibril contains more thick and thin filaments. The net effect is **hypertrophy,** or an enlargement of the stimulated muscle. The number of muscle fibers does not change significantly, but the muscle as a whole enlarges because each muscle fiber increases in diameter. Hypertrophy occurs in muscles that have been repeatedly stimulated to produce near-maximal tension. The intracellular changes that occur increase the amount of tension produced when these muscles contract. A champion weight lifter or bodybuilder is an excellent example of hypertrophied muscular development.

Muscle Atrophy

A skeletal muscle that is not regularly stimulated by a motor neuron loses muscle tone and mass. The muscle becomes flaccid, and the muscle fibers become smaller and weaker. This reduction in muscle size, tone, and power is called **atrophy.** Individuals paralyzed by spinal injuries or other damage to the nervous system will gradually lose muscle tone and size in the areas affected. Even a temporary reduction in muscle use can lead to muscular atrophy; you can easily observe this effect by comparing "before and after" limb muscles in someone who has worn a cast. Muscle atrophy is initially reversible, but dying muscle fibers are not replaced. In extreme atrophy, the functional losses are permanent. That is why physical therapy is crucial for people who are temporarily unable to move normally.

Because skeletal muscles depend on their motor neurons for stimulation, disorders that affect the nervous system can indirectly affect the muscular system. In *polio,* a virus attacks motor neurons in the spinal cord and brain, producing muscular paralysis and atrophy. AM *Problems with the Control of Muscle Activity*

Physical Conditioning

Physical conditioning and training schedules enable athletes to improve both power and endurance. In practice, the training schedule varies depending on whether the activity is supported primarily by aerobic or anaerobic energy production.

Anaerobic endurance is the length of time muscular contraction can continue to be supported by glycolysis and by the existing energy reserves of ATP and CP. Anaerobic endurance is limited by (1) the amount of ATP and CP on hand, (2) the amount of glycogen available for breakdown, and (3) the ability of the muscle to tolerate the lactic acid generated during the anaerobic period. Typically, the onset of muscle fatigue occurs within 2 minutes of the start of maximal activity.

Activities that require above-average levels of anaerobic endurance include a 50-meter dash or swim, a pole vault, and a weight-lifting competition. These activities involve the contractions of fast fibers. The energy for the first 10–20 seconds of activity comes from the ATP and CP reserves of the cytoplasm. As these reserves dwindle, glycogen breakdown and glycolysis provide additional energy. Athletes training to improve anaerobic endurance perform frequent, brief, intensive workouts that stimulate muscle hypertrophy.

Aerobic endurance is the length of time a muscle can continue to contract while supported by mitochondrial activities. Aerobic endurance is determined primarily by the availability of substrates for aerobic respiration, which the muscle fibers can obtain by breaking down carbohydrates, lipids, or amino acids. Initially, many of the nutrients catabolized by the muscle fiber are obtained from reserves in the sarcoplasm. Prolonged aerobic activity, however, must be supported by nutrients provided by the circulating blood. During exercise, blood vessels in the skeletal muscles dilate, increasing blood flow and thus bringing oxygen and nutrients to the active muscle tissue. Warm-up periods are therefore important not only in that they take advantage of treppe, the increase in tension production noted on page 288, but also because they stimulate circulation in the muscles before the serious workout begins. Because glucose is a preferred energy source, aerobic athletes such as marathon runners typically "load" or "bulk up" on carbohydrates for the last three days before an event. They may also consume glucose-rich "sports drinks" during a competition. (We shall consider the risks and benefits of these practices in Chapter 25.)

Training to improve aerobic endurance generally involves sustained low levels of muscular activity. Examples include jogging, distance swimming, and other exercises that do not require peak tension production. Improvements in aerobic endurance result from altering the characteristics of muscle fibers and improving the performance of the cardiovascular system:

1. *Altering the characteristics of muscle fibers.* The composition of fast and slow fibers in each muscle is genetically determined, and individual differences are significant. These variations affect aerobic endurance, because a person with more slow fibers in a particular muscle will be better able to perform under aerobic conditions than will a person with fewer. However, skeletal muscle cells respond to changes in the pattern of neural stimulation. Fast fibers trained for aerobic competition develop the characteristics of intermediate fibers, and this change improves aerobic endurance.

2. *Improving cardiovascular performance.* Cardiovascular activity affects muscular performance by delivering oxygen and nutrients to active muscles. Physical training alters cardiovascular function by accelerating blood flow, thus improving oxygen and nutrient availability. We shall examine factors involved in improving cardiovascular performance in Chapter 21.

Aerobic activities do not promote muscle hypertrophy. Many athletes train using a combination of aerobic and anaerobic exercises so that their muscles will enlarge and both anaerobic and aerobic endurance will improve. These athletes alternate an aerobic activity, such as swimming, with sprinting or weight lifting. The combination is known as *interval training* or *cross-training.* Interval training is particularly useful for persons engaged in racquet sports, such as tennis or squash, which are dominated by aerobic activities but are punctuated by brief periods of anaerobic effort.

AGING AND THE MUSCULAR SYSTEM

As the body ages, the size and power of all muscle tissues decrease. The effects on the muscular system can be summarized as follows:

- *Skeletal muscle fibers become smaller in diameter.* This reduction in size reflects primarily a decrease in the number of myofibrils. In addition, the muscle fibers contain smaller ATP, CP, and glycogen reserves and less myoglobin. The overall effect is a reduction in skeletal muscle size, strength, and endurance, combined with a tendency to fatigue rapidly. Because cardiovascular performance also decreases with age, blood flow to active muscles does not increase with exercise as rapidly as it does in younger people. These factors interact to produce decreases in anaerobic and aerobic performance of 30–50 percent by age 65.

- *Skeletal muscles become less elastic.* Aging skeletal muscles develop increasing amounts of fibrous connective tissue, a process called **fibrosis.** Fibrosis makes the muscle less flexible, and the collagen fibers can restrict movement and circulation.

- *Tolerance for exercise decreases.* A lower tolerance for exercise results in part from the tendency for rapid fatigue and in part from the reduction in thermoregulatory ability described in Chapter 5. ∞ *p. 160* Individuals over age 65 cannot eliminate the heat their muscles generate during contraction as effectively as younger people can and thus are subject to overheating.

10

Delayed-Onset Muscle Soreness

You have probably experienced muscle soreness the day after a period of physical exertion. Considerable controversy exists over the source and significance of this pain, which is known as *delayed-onset muscle soreness* (DOMS) and has several interesting characteristics:

■ The soreness of DOMS is distinct from the soreness you experience immediately after you stop exercising. The short-term soreness is probably related to the biochemical events associated with muscle fatigue.

■ The soreness generally begins several hours after the exercise period ends and may last 3 or 4 days.

■ The amount of soreness is highest when the activity involves eccentric contractions. Activities dominated by concentric or isometric contractions produce less soreness.

■ Levels of CPK and myoglobin are elevated in the blood, indicating damage to muscle cell membranes. The nature of the activity (eccentric versus concentric versus isometric) has no effect on these levels, nor can the levels be used to predict the degree of soreness experienced.

Three mechanisms have been proposed to explain DOMS:

1. Small tears may exist in the muscle tissue, leaving muscle fibers with damaged membranes. The sarcolemma of each damaged muscle fiber permits the loss of enzymes, myoglobin, and other chemicals that may stimulate pain receptors in the region.

2. The pain may result from muscle spasms in the affected skeletal muscles. In some studies, stretching the muscle involved can reduce the degree of soreness.

3. The pain may result from tears in the connective tissue framework and tendons of the skeletal muscle.

Some evidence supports each of those mechanisms, but it is unlikely that any one tells the entire story. For example, muscle fiber damage is certainly supported by the biochemical findings, but if that were the only factor, the type of activity would be correlated with the level of pain experienced.

■ *The ability to recover from muscular injuries decreases.* The number of satellite cells steadily decreases with age, and the amount of fibrous tissue increases. As a result, when an injury occurs, repair capabilities are limited. Scar tissue formation is the usual result.

The *rate* of decline in muscular performance is the same in all individuals, regardless of their exercise patterns or lifestyle. Therefore, to be in good shape late in life, you must be in *very* good shape early in life. Regular exercise helps control body weight, strengthens bones, and generally improves the quality of life at all ages. Extremely demanding exercise is not as important as regular exercise. In fact, extreme exercise in the elderly can damage tendons, bones, and joints. Although it has obvious effects on the quality of life, there is no clear evidence that exercise prolongs life expectancy.

INTEGRATION WITH OTHER SYSTEMS

To operate at maximum efficiency, the muscular system must be supported by many other systems. The changes that occur during exercise provide a good example of such interaction. As we noted earlier, active muscles consume oxygen and generate carbon dioxide and heat. Responses of other systems include:

■ *Cardiovascular system.* Blood vessels in the active muscles and in the skin dilate, and the heart rate increases. These adjustments accelerate oxygen delivery and carbon dioxide removal at the muscle and bring heat to the skin for radiation into the environment.

■ *Respiratory system.* The respiratory rate and depth of respiration increase. Air moves into and out of the lungs more quickly, keeping pace with the increased rate of blood flow through the lungs.

■ *Integumentary system.* Blood vessels dilate, and sweat gland secretion increases. This combination helps promote evaporation at the skin surface and removes the excess heat generated by muscular activity.

■ *Nervous and endocrine systems.* The responses of other systems are directed by controlling the heart rate, respiratory rate, and sweat gland activity.

Even when the body is at rest, the muscular system has extensive interactions with other systems. Figure 10-20● summarizes the range of interactions between the muscular system and other vital systems.

☑ Why would a sprinter experience muscle fatigue before a marathon runner would?

☑ Which activity would be more likely to create an oxygen debt: swimming laps or lifting weights?

☑ Which type of muscle fibers would you expect to predominate in the large leg muscles of someone who excels at endurance activities, such as cycling or long-distance running?

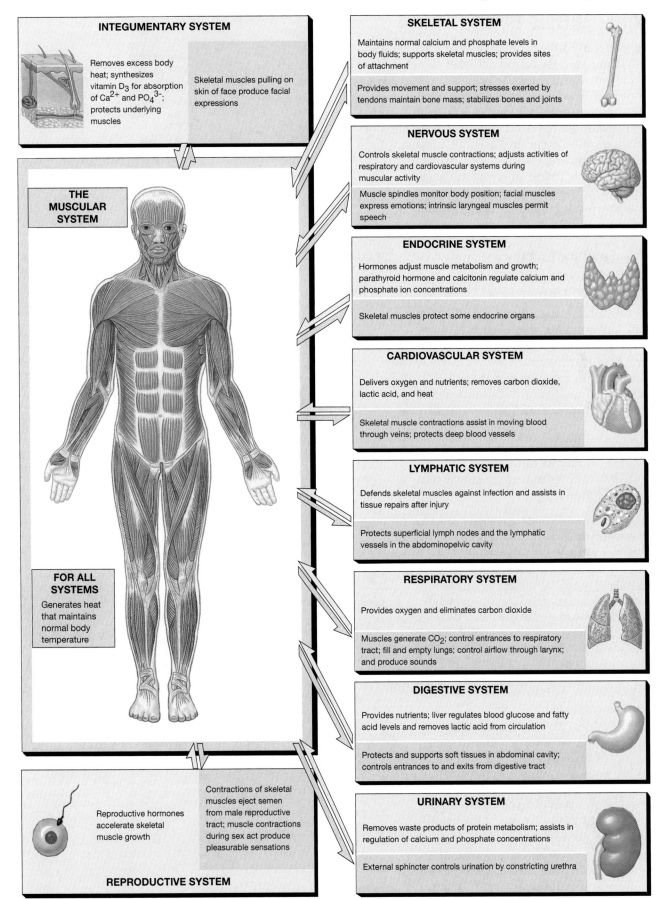

INTEGUMENTARY SYSTEM

Removes excess body heat; synthesizes vitamin D_3 for absorption of Ca^{2+} and PO_4^{3-}; protects underlying muscles

Skeletal muscles pulling on skin of face produce facial expressions

THE MUSCULAR SYSTEM

FOR ALL SYSTEMS
Generates heat that maintains normal body temperature

REPRODUCTIVE SYSTEM

Reproductive hormones accelerate skeletal muscle growth

Contractions of skeletal muscles eject semen from male reproductive tract; muscle contractions during sex act produce pleasurable sensations

SKELETAL SYSTEM

Maintains normal calcium and phosphate levels in body fluids; supports skeletal muscles; provides sites of attachment

Provides movement and support; stresses exerted by tendons maintain bone mass; stabilizes bones and joints

NERVOUS SYSTEM

Controls skeletal muscle contractions; adjusts activities of respiratory and cardiovascular systems during muscular activity

Muscle spindles monitor body position; facial muscles express emotions; intrinsic laryngeal muscles permit speech

ENDOCRINE SYSTEM

Hormones adjust muscle metabolism and growth; parathyroid hormone and calcitonin regulate calcium and phosphate ion concentrations

Skeletal muscles protect some endocrine organs

CARDIOVASCULAR SYSTEM

Delivers oxygen and nutrients; removes carbon dioxide, lactic acid, and heat

Skeletal muscle contractions assist in moving blood through veins; protects deep blood vessels

LYMPHATIC SYSTEM

Defends skeletal muscles against infection and assists in tissue repairs after injury

Protects superficial lymph nodes and the lymphatic vessels in the abdominopelvic cavity

RESPIRATORY SYSTEM

Provides oxygen and eliminates carbon dioxide

Muscles generate CO_2; control entrances to respiratory tract; fill and empty lungs; control airflow through larynx; and produce sounds

DIGESTIVE SYSTEM

Provides nutrients; liver regulates blood glucose and fatty acid levels and removes lactic acid from circulation

Protects and supports soft tissues in abdominal cavity; controls entrances to and exits from digestive tract

URINARY SYSTEM

Removes waste products of protein metabolism; assists in regulation of calcium and phosphate concentrations

External sphincter controls urination by constricting urethra

10

•**FIGURE 10-20** Functional Relationships between the Muscular System and Other Systems

CARDIAC MUSCLE TISSUE

We introduced **cardiac muscle tissue** in Chapter 4 and briefly compared its properties with those of other muscle types. **Cardiac muscle cells,** also called *cardiocytes* or *cardiac myocytes,* are relatively small, averaging 10–20 μm in diameter and 50–100 μm in length. A typical cardiac muscle cell (Figure 10-21a,b•) has a single, centrally placed nucleus, although a few may have two or more. As the name implies, cardiac muscle tissue is found only in the heart. Table 10-4 (p. 304) compares skeletal, cardiac, and smooth muscle tissues in greater detail.

DIFFERENCES BETWEEN CARDIAC AND SKELETAL MUSCLE TISSUES

As do skeletal muscle fibers, each cardiac muscle cell contains organized myofibrils, and the presence of many aligned sarcomeres gives it striations. However, cardiac muscle cells are much smaller than skeletal muscle fibers, and significant structural and functional differences exist between the two.

Structural Differences

Important structural differences between skeletal muscle fibers and cardiac muscle cells include the following:

- The T tubules in a cardiac muscle cell are short and broad, and there are no triads. The T tubules encircle the sarcomeres at the Z lines rather than at the zone of overlap.

- The SR of a cardiac muscle cell lacks terminal cisternae, and its tubules contact the cell membrane as well as the T tubules (Figure 10-21c•). As in skeletal muscle fibers, the appearance of an action potential triggers calcium release from the SR and the contraction of sarcomeres; it also increases the permeability of the sarcolemma to extracellular calcium ions.

- Cardiac muscle cells are almost totally dependent on aerobic metabolism to obtain the energy needed to continue contracting. The sarcoplasm of a cardiac muscle cell thus contains large numbers of mitochondria and abundant reserves of myoglobin (to store oxygen). Energy reserves are maintained in the form of glycogen and lipid inclusions.

- Each cardiac muscle cell contacts several others at specialized sites known as **intercalated** (in-TER-ka-lā-ted) **discs.** ∞ *p. 132* Intercalated discs play a vital role in the function of cardiac muscle, as we shall learn next.

•**FIGURE 10-21 Cardiac Muscle Tissue. (a)** A light micrograph of a cardiac muscle tissue. Notice the striations and the intercalated discs. **(b,c)** The structure of a cardiac muscle cell; compare with Figure 10-3.

Intercalated Discs   At an intercalated disc (Figure 10-21a,b•), the cell membranes of two adjacent cardiac muscle cells are extensively intertwined and bound together by gap junctions and desmosomes. ∞ *pp. 109, 110* These connections help stabilize the relative positions of adjacent cells and maintain the three-dimensional structure of the tissue. The gap junctions allow ions and small molecules to move from one cell to another. This arrangement creates a direct electrical connection between the two muscle cells. An action potential can travel across an intercalated disc, moving quickly from one cardiac muscle cell to another.

Myofibrils in the two interlocking muscle cells are firmly anchored to the membrane at the intercalated disc. Because their myofibrils are essentially locked together, the two muscle cells can "pull together" with maximum efficiency. Because the cardiac muscle cells are mechanically, chemically, and electrically connected to one another, the entire tissue resembles a single, enormous muscle cell. For this reason, cardiac muscle has been called a *functional syncytium* (sin-SISH-ē-um; a fused mass of cells).

Functional Differences

In Chapter 20, we will examine cardiac muscle physiology in detail; here we will briefly summarize four major functional specialties of cardiac muscle:

1. Cardiac muscle tissue contracts without neural stimulation. This property is called **automaticity.** The timing of contractions is normally determined by specialized cardiac muscle cells called **pacemaker cells.**

2. Innervation by the nervous system can alter the pace established by the pacemaker cells and adjust the amount of tension produced during a contraction.

3. Cardiac muscle cell contractions last roughly 10 times longer than do those of skeletal muscle fibers.

4. The properties of cardiac muscle cell membranes differ from those of skeletal muscle fiber membranes. As a result, individual twitches cannot undergo wave summation, and cardiac muscle tissue cannot produce tetanic contractions. This difference is important because a heart in a sustained tetanic contraction could not pump blood.

SMOOTH MUSCLE TISSUE

Smooth muscle cells range from 5 to 10 μm in diameter and from 30 to 200 μm in length. Each cell is spindle-shaped and has a single, centrally located nucleus. Figure 10-22a• shows typical smooth muscle fibers as seen by light microscopy, and Figure 10-22b• shows diagrammatic views of relaxed and contracted smooth muscle cells.

Smooth muscle tissue occurs within almost every organ, forming sheets, bundles, or sheaths around other tissues. Smooth muscles around blood vessels regulate blood flow through vital organs. In the digestive and urinary systems, rings of smooth muscle, called *sphincters,* regulate the movement of materials along internal passageways. Smooth muscles in bundles, layers, or sheets play a variety of other roles:

(a)

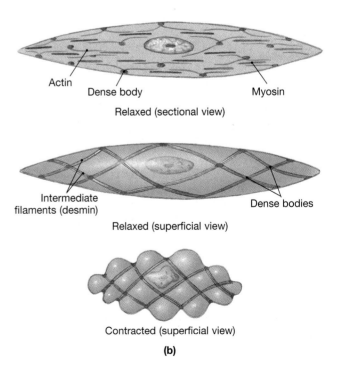

Relaxed (sectional view)

Relaxed (superficial view)

Contracted (superficial view)

(b)

•**FIGURE 10-22 Smooth Muscle Tissue.** **(a)** Many visceral organs contain several layers of smooth muscle tissue oriented in different directions. As a result, a single sectional view shows a smooth muscle cell in both longitudinal (L) and transverse (T) sections. **(b)** A single relaxed smooth muscle cell is spindle-shaped and has no striations. Note the changes in cell shape as contraction occurs.

10

- *Integumentary system.* Smooth muscles around blood vessels regulate the flow of blood to the superficial dermis; smooth muscles of the arrector pili elevate hairs. ∞ *p. 153*
- *Cardiovascular system.* Smooth muscles encircling blood vessels control the distribution of blood and help regulate blood pressure.
- *Respiratory system.* Smooth muscle contraction or relaxation alters the diameters of the respiratory passageways and changes the resistance to airflow.
- *Digestive system.* Extensive layers of smooth muscle in the walls of the digestive tract play an essential role in moving materials along the tract. Smooth muscle in the walls of the gallbladder contract to eject bile into the digestive tract.
- *Urinary system.* Smooth muscle tissue in the walls of small blood vessels alters the rate of filtration at the kidneys. Layers of smooth muscle in the walls of the ureters transport urine to the urinary bladder; the contraction of the smooth muscle in the wall of the urinary bladder forces urine out of the body.
- *Reproductive system.* Layers of smooth muscle help move sperm along the reproductive tract in males and cause the ejection of glandular secretions from the accessory glands into the reproductive tract. In females, layers of smooth muscle help move oocytes (and perhaps sperm) along the reproductive tract, and contraction of the smooth muscle in the walls of the uterus expels the fetus at delivery.

DIFFERENCES BETWEEN SMOOTH MUSCLE TISSUE AND OTHER MUSCLE TISSUES

Smooth muscle tissue differs from both skeletal and cardiac muscle tissues in structure and function (Table 10-4).

Structural Differences

Actin and myosin are present in all three muscle types. In skeletal and cardiac muscle cells, these proteins are organized in sarcomeres, with thin and thick filaments. The internal organization of a smooth muscle cell is very different:

- A smooth muscle fiber has no T tubules, and the sarcoplasmic reticulum forms a loose network throughout the sarcoplasm. Smooth muscle tissue has no myofibrils or sarcomeres. As a result, this tissue also has no striations and is called **nonstriated** muscle.
- Thick filaments are scattered throughout the sarcoplasm of a smooth muscle cell. The myosin proteins are organized differently than in skeletal or cardiac muscle cells, and smooth muscle cells have more crossbridges per thick filament.
- The thin filaments in a smooth muscle cell are attached to **dense bodies,** structures distributed throughout the sarcoplasm in a network of intermediate filaments composed of the protein *desmin* (Figure 10-22b•). Some of the dense bodies are firmly attached to the sarcolemma. The dense bodies and intermediate filaments anchor the thin filaments such that, when sliding occurs between thin and thick filaments, the cell shortens. Dense bodies are not arranged in straight lines, so when a contraction occurs, the muscle cell twists like a corkscrew.
- Adjacent smooth muscle cells are bound together at dense bodies, transmitting the contractile forces from cell to cell throughout the tissue.
- Although smooth muscle cells are surrounded by connective tissue, the collagen fibers never unite to form tendons or aponeuroses as they do in skeletal muscles.

TABLE 10-4 A Comparison of Skeletal, Cardiac, and Smooth Muscle Tissues

Property	Skeletal Muscle	Cardiac Muscle	Smooth Muscle
Fiber dimensions (diameter × length)	100 μm × up to 30 cm	10–20 μm × 50–100 μm	5–10 μm × 30–200 μm
Nuclei	Multiple, near sarcolemma	Generally single, centrally located	Single, centrally located
Filament organization	In sarcomeres along myofibrils	In sarcomeres along myofibrils	Scattered throughout sarcoplasm
SR	Terminal cisternae in triads at zones of overlap	SR tubules contact T tubules at Z lines	Dispersed throughout sarcoplasm, no T tubules
Control mechanism	Neural, at single neuromuscular junction	Automaticity (pacemaker cells)	Automaticity (pacesetter cells), neural or hormonal control
Ca²⁺ source	Release from SR	Extracellular fluid and release from SR	Extracellular fluid
Contraction	Rapid onset; may be tetanized; rapid fatigue	Slower onset; cannot be tetanized; resistant to fatigue	Slow onset; may be tetanized; resistant to fatigue
Energy source	Aerobic metabolism at moderate levels of activity; glycolysis (anaerobic) during peak activity	Aerobic metabolism, usually lipid or carbohydrate substrates	Primarily aerobic metabolism

Functional Differences

Smooth muscle tissue differs from other muscle types in terms of (1) excitation–contraction coupling, (2) length–tension relationships, (3) control of contractions, and (4) smooth muscle tone.

Excitation–Contraction Coupling   The trigger for smooth muscle contraction is the appearance of free calcium ions in the cytoplasm. Most of these calcium ions enter the cell from the extracellular fluid. Once in the sarcoplasm, the calcium ions interact with **calmodulin,** a calcium-binding protein. Calmodulin then activates the enzyme **myosin light chain kinase,** which breaks down ATP and initiates the contraction. This situation is quite different from the one in skeletal and cardiac muscles, in which the trigger for contraction is the binding of calcium ions to troponin.

Length–Tension Relationships   Because the thick and thin filaments are scattered and are not organized into sarcomeres, tension development and resting length in smooth muscle are not directly related. A stretched smooth muscle soon adapts to its new length and retains the ability to contract on demand. This ability to function over a wide range of lengths is called **plasticity.** Smooth muscle can contract over a range of lengths four times greater than that of skeletal muscle. This ability is especially important for digestive organs that undergo great changes in volume, such as the stomach. Despite the lack of sarcomere organization, smooth muscle contractions can be just as powerful as those of skeletal muscles. Like skeletal muscle fibers, smooth muscle cells often undergo sustained tetanic contractions.

Control of Contractions   Many smooth muscle cells are not innervated by motor neurons, and the neurons that do innervate smooth muscles are not under voluntary control. The nature of the connection with the nervous system provides a means of categorizing smooth muscle cells as either multiunit or visceral. **Multiunit smooth muscle cells** are innervated in motor units comparable to those of skeletal muscles, but each smooth muscle cell may be connected to several motor neurons rather than to just one. In contrast, many **visceral smooth muscle cells** lack a direct contact with any motor neuron.

Multiunit smooth muscle cells resemble skeletal muscle fibers and cardiac muscle cells, in that neural activity produces an action potential that is propagated over the sarcolemma. However, the contractions of these smooth muscle cells are more leisurely than are those of skeletal or cardiac muscle cells. Multiunit smooth muscle cells are located in the iris of the eye, where they regulate the diameter of the pupil; along portions of the male reproductive tract; within the walls of large arteries; and in the arrector pili muscles of the skin. Multiunit smooth muscle cells do not typically occur in the digestive tract.

Visceral smooth muscle cells are arranged in sheets or layers. Within each layer, adjacent muscle cells are connected by gap junctions. Because they are connected in this way, whenever one muscle cell contracts, the electrical impulse that triggered the contraction can travel to adjacent smooth muscle cells. The contraction therefore spreads in a wave that soon involves every smooth muscle cell in the layer. The initial stimulus may be the activation of a motor neuron that contacts one of the muscle cells in the region. But smooth muscle cells will also contract or relax in response to chemicals, hormones, local concentrations of oxygen or carbon dioxide, or physical factors such as extreme stretching or irritation.

Many visceral smooth muscle networks show rhythmic cycles of activity in the absence of neural stimulation. These cycles are characteristic of the smooth muscle cells in the wall of the digestive tract, where **pacesetter cells** undergo spontaneous depolarization and trigger the contraction of entire muscular sheets. Visceral smooth muscle cells are located in the walls of the digestive tract, the gallbladder, the urinary bladder, and many other internal organs.

Smooth Muscle Tone   Both multiunit and visceral smooth muscle tissues show a normal background level of activity, or smooth muscle tone. The regulatory mechanisms detailed above stimulate contraction and increase muscle tone. Neural, hormonal, or chemical factors can also stimulate smooth muscle relaxation, producing a decrease in muscle tone. For example, smooth muscle cells at the entrances to capillaries regulate the amount of blood flow into each vessel. If the tissue becomes oxygen-starved, the smooth muscle cells relax. Blood flow increases, delivering additional oxygen. As conditions return to normal, the smooth muscle regains its normal muscle tone.

☑ What feature of cardiac muscle tissue allows the heart to act as a functional syncytium?

☑ Why are cardiac and smooth muscle contractions more affected by changes in extracellular Ca^{2+} than are skeletal muscle contractions?

☑ Smooth muscle can contract over a wider range of resting lengths than skeletal muscle can. Why?

[EXPLORE]
MediaLab

To gain a better understanding of muscle physiology, use the resources available on both the CD-ROM and the Companion Website. The CD-ROM includes animations and tutorials; the website builds on those resources with text content and interactive problems as well as communication avenues for sharing ideas with your class. To complete the following Web Explorations, visit the Media-Lab in Chapter 10 of your Companion Website.

Web Exploration 1

Estimated time for completion = 15 minutes

Skeletal muscle fibers can contract quickly (fast fibers) or slowly (slow fibers). How do these two cell types differ? First reread the subsection "Types of Skeletal Muscle Fibers" in the text. Then draw a preliminary sketch of these two fiber types. A good way to refine that sketch is to read others' descriptions. In an abstract, Hoffer, Cresswell, and Thorstensson give a brief explanation of their research on the anatomy of fast and slow fibers. To learn of their discoveries, select the keyword **MUSCLE FIBERS** on the Companion Website. Finally, refine your sketch and compare your ideas with those of your classmates by using the communications tools on the Companion Website.

Web Exploration 2

Estimated time for completion = 10 minutes

After having read this chapter, you should have a good understanding of the anatomy of a single muscle fiber. Look at Figures 10-2 and 10-3, and reread the subsection "Muscle Hypertrophy" in the text. Although hypertrophy is considered to be the primary mechanism of muscle enlargement in humans, controversy exists as to whether the *number* of muscle fibers changes during hypertrophy and, if so, by how much. This process has been called muscle *hyperplasia*. By what mechanisms might the number of muscle fibers in an adult skeletal muscle increase? On the Companion Website, select the keyword **HYPERTROPHY** to read an article about the debate over the existence and importance of hyperplasia. Then summarize your thoughts. Discuss factors that might make it difficult to detect and prove hyperplasia in a human skeletal muscle. If you wish, share your ideas with your classmates by using the communications tools on the Companion Website. Compare your own conclusions with the contributions of your peers. Is this debate resolvable?

CHAPTER REVIEW

SELECTED CLINICAL TERMINOLOGY
Terms Discussed in This Chapter

botulism: A severe, potentially fatal paralysis of skeletal muscles, resulting from the consumption of a bacterial toxin. *(p. 281 and AM)*

Duchenne's muscular dystrophy (DMD): One of the most common and best understood of the muscular dystrophies. *(p. 277 and AM)*

fibrosis: A process in which muscle tissue is replaced by fibrous connective tissue, making muscles weaker and less flexible. *(p. 299)*

muscular dystrophies (DIS-trō-fēz): A varied collection of congenital diseases that produce progressive muscle weakness and deterioration. *(p. 277 and AM)*

myasthenia gravis (mī-as-THĒ-nē-uh GRA-vis): A general muscular weakness resulting from a reduction in the number of ACh receptors on the motor end plate. *(p. 281 and AM)*

polio: A disease resulting from the destruction of motor neurons by a certain virus and characterized by the paralysis and atrophy of motor units. *(p. 298 and AM)*

rigor mortis: A state following death during which muscles are locked in the contracted position, making the body extremely stiff. *(p. 283)*

tetanus: A disease caused by a bacterial toxin that leads to the production of sustained, powerful contractions of skeletal muscles. *(p. 289)*

Additional Terms Discussed in the *Applications Manual*

cholinesterase inhibitor: A compound that "ties up" the active sites at which acetylcholinesterase normally binds ACh.
chronic fatigue syndrome: A condition, resembling fibromyalgia, that may suddenly develop soon after a viral infection.

fibromyalgia syndrome: An inflammatory disorder characterized by a distinctive pattern of symptoms, including tender points on the body surface.
trichinosis (trik-i-NŌ-sis): A condition resulting from a parasitic infection by a nematode worm.

STUDY OUTLINE

INTRODUCTION   p. 271

1. Muscle cells are specialized to contract by pulling on collagen fibers. In muscle tissue, muscle cells exert **tension,** or a pull, on those fibers.

SKELETAL MUSCLE TISSUE AND THE MUSCULAR SYSTEM   p. 271

1. The three types of muscle tissue are skeletal muscle, cardiac muscle, and smooth muscle.
2. **Skeletal muscles** attach to bones directly or indirectly. Their functions are to (1) produce skeletal movement, (2) maintain posture and body position, (3) support soft tissues, (4) guard entrances and exits, and (5) maintain body temperature.

ANATOMY OF SKELETAL MUSCLE   p. 271

Organization of Connective Tissues   p. 271

1. Each muscle fiber is surrounded by an **endomysium.** Bundles of muscle fibers are sheathed by a **perimysium,** and the entire muscle is covered by an **epimysium.** At the ends of the muscle are **tendons** or **aponeuroses** that attach the muscle to bones. *(Figure 10-1)*

Blood Vessels and Nerves   p. 273

2. The epimysium and perimysium contain the blood vessels and nerves that supply the muscle fibers.

Microanatomy of Skeletal Muscle Fibers   p. 273

3. A skeletal muscle fiber has a **sarcolemma,** or cell membrane; **sarcoplasm** (cytoplasm); and **sarcoplasmic reticulum (SR),** similar to the smooth endoplasmic reticulum of other cells. **Transverse (T) tubules** and **myofibrils** aid in contraction. Filaments in a myofibril are organized into repeating functional units called **sarcomeres.** *(Figures 10-2–10-6)*
4. **Myofilaments** form myofibrils and consist of **thin filaments** and **thick filaments.** *(Figures 10-2–10-6)*
5. Thin filaments consist of **F actin, tropomyosin,** and **troponin.** Tropomyosin molecules cover **active sites** on the **G actin** subunits that form the F actin strand. Troponin binds to G actin and tropomyosin and holds the tropomyosin in position. *(Figure 10-7)*
6. Thick filaments consist of a bundle of myosin molecules, each consisting of an elongate **tail** and a globular **head,** or **cross-bridge.** In a resting muscle cell, the interactions between the active sites on G actin and myosin cross-bridges are prevented by tropomyosin. *(Figures 10-5, 10-7)*

CONTRACTION OF SKELETAL MUSCLE   p. 279

The Sliding Filament Theory   p. 279

1. The **sliding filament theory** explains how the relationship between thick and thin filaments changes as the muscle contracts and shortens. *(Figure 10-8)*

The Control of Skeletal Muscle Activity   p. 280

2. Neural control of muscle function involves a link between electrical activity in the sarcolemma and the initiation of a contraction.
3. The activity of a muscle fiber is controlled by a neuron at a **neuromuscular** *(myoneural)* **junction (NMJ).** *(Figure 10-9)*
4. When an **action potential** arrives at the **synaptic terminal, acetylcholine (ACh)** is released into the **synaptic cleft.** The binding of ACh to receptors on the opposing *junctional folds* leads to the generation of an action potential in the sarcolemma. *(Figure 10-9)*
5. **Excitation–contraction coupling** occurs as the passage of an action potential along a T tubule triggers the release of Ca^{2+} from the cisternae of the SR.

FOCUS: The Contraction Cycle   p. 282

6. Each contraction involves a cycle of attachment, pivoting, detachment, and return. It begins when the SR releases calcium ions. The calcium ions bind to troponin, which changes position and moves tropomyosin away from the active sites of actin. Cross-bridges of myosin heads then bind to actin. Next, each myosin head pivots at its base, pulling the actin filament toward the center of the sarcomere. *(Figure 10-10)*

Relaxation   p. 281

7. **Acetylcholinesterase (AChE)** breaks down ACh and limits the duration of stimulation. *(Table 10-1)*
8. The contraction continues until either (1) Ca^{2+} concentrations return to resting levels or (2) the muscle fiber runs out of ATP.

Length–Tension Relationships   p. 285

9. The number of cross-bridge interactions in a muscle fiber determines the amount of tension produced during a contraction. The fibers can contract most forcefully when stimulated over a narrow range of resting lengths. *(Figure 10-11)*

MUSCLE MECHANICS   p. 286

Tension Production   p. 286

1. A **twitch** is a cycle of contraction and relaxation produced by a single stimulus. *(Figure 10-12)*
2. Repeated stimulation before the relaxation phase ends may produce **summation of twitches (wave summation),** in

which one twitch is added to another; **incomplete tetanus,** in which tension peaks because the muscle is never allowed to relax completely; or **complete tetanus,** in which the relaxation phase is eliminated. *(Figure 10-13)*

3. Repeated stimulation after the relaxation phase has been completed produces **treppe.** *(Figure 10-13)*

4. *Internal tension* is generated inside contracting skeletal muscle fibers; *external tension* is generated in the extracellular fibers. *(Figure 10-14)*

5. The number and size of a muscle's **motor units** indicate how precisely controlled its movements are. *(Figure 10-15)*

6. Resting **muscle tone** stabilizes bones and joints.

7. Normal activities generally include both **isotonic contractions** (in which the tension in a muscle rises and the length of the muscle changes) and **isometric contractions** (in which tension rises but the length of the muscle remains constant). *(Figure 10-16)*

8. The return to resting length after a contraction may involve *series elastic elements,* the contraction of opposing muscle groups, and gravity. *(Figure 10-17)*

Energetics of Muscular Activity p. 293

9. Muscle contractions require large amounts of energy. A resting muscle generates ATP through **aerobic metabolism** in mitochondria. *(Table 10-2)*

10. **Creatine phosphate (CP)** can release stored energy to convert ADP to ATP.

11. When a muscle fiber runs short of ATP and CP, enzymes can break down glycogen molecules to release glucose, which can in turn be broken down via glycolysis.

12. At rest or at moderate levels of activity, aerobic metabolism can provide most of the ATP needed to support muscle contractions. *(Figure 10-18)*

13. At peak levels of activity, the cell relies heavily on the **anaerobic** process of **glycolysis** to generate ATP, because the mitochondria cannot obtain enough oxygen to meet the existing ATP demands. *(Figure 10-18)*

14. A **fatigued** muscle can no longer contract, because of changes in pH due to the buildup of **lactic acid,** a lack of energy, or other problems.

15. The **recovery period** begins immediately after a period of muscle activity and continues until conditions inside the muscle have returned to preexertion levels. The **oxygen debt,** or *excess postexercise oxygen consumption (EPOC),* created during exercise is the amount of oxygen used in the recovery period to restore normal conditions.

16. Circulating hormones may alter metabolic activities in skeletal muscle fibers.

Muscle Performance p. 296

17. The three types of skeletal muscle fibers are **fast fibers, slow fibers,** and **intermediate fibers.** *(Table 10-3)*

18. Fast fibers, which are large in diameter, contain densely packed myofibrils, large glycogen reserves, and relatively few mitochondria. They produce rapid and powerful contractions of relatively brief duration. *(Figure 10-19)*

19. Slow fibers are about half the diameter of fast fibers and take three times as long to contract after stimulation. Specializations such as abundant mitochondria, an extensive capillary supply, and high concentrations of **myoglobin** enable slow fibers to continue contracting for extended periods. *(Figure 10-19)*

20. Intermediate fibers are very similar to fast fibers but have a greater resistance to fatigue.

21. **Anaerobic endurance** is the time over which a muscle can support sustained, powerful muscle contractions through anaerobic mechanisms. Training to develop anaerobic endurance can lead to **hypertrophy** (enlargement) of the stimulated muscles.

22. **Aerobic endurance** is the time over which a muscle can continue to contract while supported by mitochondrial activities.

AGING AND THE MUSCULAR SYSTEM   p. 299

1. The aging process reduces the size, elasticity, and power of all muscle tissues. Both exercise tolerance and the ability to recover from muscular injuries decrease.

INTEGRATION WITH OTHER SYSTEMS   p. 300

1. To operate at maximum efficiency, the muscular system must be supported by many other systems. *(Figure 10-20)*

CARDIAC MUSCLE TISSUE   p. 302

Differences between Cardiac and Skeletal Muscle Tissues   p. 302

1. **Cardiac muscle tissue** is located only in the heart. **Cardiac muscle cells** and skeletal muscle fibers differ structurally in size, the number and location of nuclei, the size and location of the T tubules at the sarcomeres, their dependence on aerobic metabolism when contracting at peak levels, the nature of their metabolic reserves, and the presence or absence of **intercalated discs.** *(Figure 10-21; Table 10-4)*

2. Cardiac muscle cells and skeletal muscle fibers differ functionally in (1) the source of the stimulus for contraction, (2) the relationship with the autonomic nervous system, (3) the duration of contractions, and (4) the ability to undergo tetanic contractions.

SMOOTH MUSCLE TISSUE   p. 303

Differences between Smooth Muscle Tissue and Other Muscle Tissues   p. 304

1. **Smooth muscle tissue** is nonstriated, involuntary muscle tissue that exhibits **plasticity.**

2. Smooth muscle cells lack sarcomeres and the resulting striations. The thin filaments are anchored to **dense bodies.** *(Figure 10-22; Table 10-4)*

3. **Multiunit smooth muscle cells** are innervated by more than one motor neuron. **Visceral smooth muscle cells** are not always innervated by motor neurons. Neurons that innervate smooth muscle cells are not under voluntary control.

REVIEW QUESTIONS

1 REVIEWING FACTS AND TERMS

1. A skeletal muscle contains
   (a) connective tissues
   (b) blood vessels and nerves
   (c) skeletal muscle tissue
   (d) a, b, and c are correct

2. The contraction of a muscle exerts a pull on a bone because
   (a) muscles are attached to bones by ligaments
   (b) muscles are directly attached to bones
   (c) muscles are attached to bones by tendons
   (d) a, b, and c are correct

3. The detachment of the myosin cross-bridges is directly triggered by
   (a) the repolarization of T tubules
   (b) the attachment of ATP to myosin heads
   (c) the hydrolysis of ATP
   (d) calcium ions

4. A muscle producing peak tension during rapid cycles of contraction and relaxation is said to be in
   (a) incomplete tetanus   (b) treppe
   (c) complete tetanus   (d) a twitch

5. The smooth, steady increase in muscular tension produced by increasing the number of active motor units is
   (a) complete tetanus   (b) treppe
   (c) wave summation   (d) recruitment

6. The type of contraction in which the tension rises but the resistance does not change is
   (a) a wave summation
   (b) a twitch
   (c) an isotonic contraction
   (d) an isometric contraction

7. Large-diameter, densely packed myofibrils, large glycogen reserves, and few mitochondria are characteristics of
   (a) slow fibers   (b) intermediate fibers
   (c) fast fibers   (d) red muscles

8. An action potential can travel quickly from one cardiac muscle cell to another because of the presence of
   (a) gap junctions   (b) tight junctions
   (c) intercalated discs   (d) a and c are correct

9. List the three types of muscle tissue in the body.

10. What are the five functions of skeletal muscle?

11. What three layers of connective tissue are part of each muscle? What functional role does each layer play?

12. What is a motor unit?

13. What structural feature of a skeletal muscle fiber is responsible for conducting action potentials into the interior of the cell?

14. Under resting conditions, what two proteins prevent interactions between cross-bridges and active sites?

15. What five interlocking steps are involved in the contraction process?

16. What two factors affect the amount of tension produced when a skeletal muscle contracts?

17. Which specialized type of muscle fiber plays an important role in the reflex control of position and posture?

18. What forms of energy reserves do resting skeletal muscle fibers contain?

19. What two mechanisms are used to generate ATP from glucose in muscle cells?

20. What three factors are necessary to prevent premature muscle fatigue?

21. What is the calcium-binding protein in smooth muscle tissue?

LEVEL 2 REVIEWING CONCEPTS

22. An activity that would require anaerobic endurance is
   (a) a 50-meter dash
   (b) a pole vault
   (c) a weight-lifting competition
   (d) a, b, and c are correct

23. Areas of the body where you would not expect to find slow fibers include the
   (a) back and calf muscles   (b) eye and hand
   (c) chest and abdomen   (d) a, b, and c are correct

24. When contraction occurs,
   (a) the H and I bands get smaller
   (b) the Z lines move closer together
   (c) the width of the A band remains constant
   (d) a, b, and c are correct

25. The amount of tension produced during a contraction is *not* affected by
   (a) the amount of ATP available to the muscle
   (b) the degree of overlap between thick and thin filaments
   (c) the interaction of troponin and tropomyosin
   (d) the number of cross-bridge interactions in a muscle fiber

26. Describe the basic sequence of events that occurs at a neuromuscular junction.

27. Why is the multinucleate condition important in skeletal muscle fibers?

28. Describe the graphic events seen on a myogram as tension is developed in a stimulated calf muscle fiber during a twitch.

29. What three processes are involved in repaying the oxygen debt during the recovery period?

30. What are the effects of aging on the muscular system?

31. Why is cardiac muscle an example of a functional syncytium?

32. How does cardiac muscle tissue contract without neural stimulation?

33. Atracurium is a drug that blocks the binding of ACh. Give an example of a site where such binding normally occurs, and predict the physical effect of this drug.

10

34. Describe what happens to muscles that are not "used" on a regular basis. What can be done to offset this effect?

35. The time of a murder victim's death is commonly estimated by the flexibility of the body. Explain why.

36. Many visceral smooth muscle cells lack motor neuron innervation. How are their contractions coordinated and controlled?

37. A motor unit in a skeletal muscle contains 1500 muscle fibers. Would this muscle be involved in fine, delicate movements or powerful, gross movements? Explain.

LEVEL 3 CRITICAL THINKING AND CLINICAL APPLICATIONS

38. Many potent insecticides contain toxins, called organophosphates, that interfere with the action of the enzyme acetylcholinesterase. Ivan is using an insecticide containing organophosphates and is very careless. He does not use gloves or a dust mask and absorbs some of the chemical through his skin. He inhales a large amount as well. What symptoms would you expect to observe in Ivan as a result of the organophosphate poisoning?

39. A rare (hypothetical) genetic disease causes the body to produce antibodies that compete with acetylcholine for receptors on the motor end plate. Patients with this disease exhibit varying degrees of muscle weakness and flaccid paralysis in the affected muscles. If you could administer a drug that inhibits acetylcholinesterase or a drug that inhibits acetylcholine, which would you use to alleviate these symptoms?

40. Thirty minutes after Mary has completed a 10-km race, she begins to notice soreness and stiffness in her leg muscles. She wonders whether she may have damaged the muscles during the race. She visits her physician, who orders a blood test. How could the physician tell from a blood test whether muscle damage had occurred?

CHAPTER 11

The Muscular System

CHAPTER OUTLINE AND OBJECTIVES

INTRODUCTION   p. 312

BIOMECHANICS AND MUSCLE ANATOMY   p. 312
   Organization of Skeletal Muscle Fibers   p. 312
   Levers   p. 313
① Describe the arrangement of fascicles in the various
   types of muscles, and explain the resulting functional
   differences.
② Describe the classes of levers and how they make
   muscles more efficient.

MUSCLE TERMINOLOGY   p. 315
   Origins and Insertions   p. 315
   Actions   p. 315
   Names of Skeletal Muscles   p. 316
   Axial and Appendicular Muscles   p. 317
③ Predict the actions of a muscle on the basis
   of the relative positions of its origin and insertion.
④ Explain how muscles interact to produce or oppose
   movements.
⑤ Explain how the name of a muscle can help identify
   its location, appearance, and/or function.

THE AXIAL MUSCLES   p. 320
   Muscles of the Head and Neck   p. 320
   Muscles of the Vertebral Column   p. 328
   Oblique and Rectus Muscles   p. 330
   Muscles of the Pelvic Floor   p. 332
⑥ Identify the principal axial muscles of the body
   together with their origins, insertions, actions,
   and innervation.

*Skeletal muscles have a remarkable ability to adapt to
the demands placed on them. Intensive weight training
and physical conditioning will lead to an increase in the size
and strength of the muscles being exercised. However, even
the most dedicated bodybuilders cannot work on developing
all of the 700 or so skeletal muscles in the human body.
Instead, they concentrate on the largest, most prominent
muscles responsible for powerful movements of the axial
and appendicular skeletons. This chapter will introduce those
skeletal muscles and the movements they perform.*

THE APPENDICULAR MUSCLES   p. 335
   Muscles of the Shoulders and Upper Limbs   p. 335
   Muscles of the Pelvis and Lower Limbs   p. 346
⑦ Identify the principal appendicular muscles of the body
   together with their origins, insertions, actions, and
   innervation.
⑧ Compare the major muscle groups of the upper and
   lower limbs, and relate their differences to their
   functional roles.

MEDIALAB and CHAPTER REVIEW   p. 356

The **muscular system** includes all the skeletal muscles that can be controlled voluntarily. Most of the muscle tissue in the body is part of this system, and approximately 700 skeletal muscles have been identified. Some are attached to bony processes and others to broad sheets of connective tissue, but all are directly or indirectly associated with the skeletal system. Rather than attempt to survey all 700 skeletal muscles, we will focus on a relatively small but representative number of muscles, about 20 percent of the total. To simplify the memorization, we have organized these muscles into anatomical and functional groups.

The shape or appearance of each muscle provides clues to its primary function. Muscles involved with locomotion and posture work across joints, producing skeletal movement. Those that support soft tissue form slings or sheets between relatively stable bony elements, whereas those that guard an entrance or exit completely encircle the opening.

At the level of the individual skeletal muscle, two factors interact to determine the effects of its contraction: (1) the anatomical arrangement of the muscle fibers and (2) the way the muscle attaches to the bones of the skeletal system. We can understand the performance of muscles in the body in terms of basic mechanical laws. The analysis of biological systems in mechanical terms is the study of *biomechanics*. In this chapter, we examine the biomechanics and gross anatomy of the muscular system.

BIOMECHANICS AND MUSCLE ANATOMY

Although most skeletal muscle fibers contract at comparable rates and shorten to the same degree, variations in microscopic and macroscopic organization can dramatically affect the power, range, and speed of movement produced when a muscle contracts.

ORGANIZATION OF SKELETAL MUSCLE FIBERS

Muscle fibers in a skeletal muscle form bundles called *fascicles*. ∞ *p. 272* The muscle fibers in a single fascicle are parallel, but the organization of fascicles in the skeletal muscle can vary, as can the relationship between the fascicles and the associated tendon. Four patterns of fascicle organization form *parallel muscles, convergent muscles, pennate muscles,* and *circular muscles.*

Parallel Muscles

In a **parallel muscle,** the fascicles are parallel to the long axis of the muscle. Most of the skeletal muscles in the body are parallel muscles. Some are flat bands with broad attachments *(aponeuroses)* at each end; others are plump and cylindrical with tendons at one or both ends. In the latter case, the muscle is spindle-shaped (Figure 11-1a▪), with a central **body,** also known as the *belly,* or *gaster* (GAS-ter; stomach). The *biceps brachii muscle* of the arm is a parallel muscle with a central body. When a parallel muscle contracts, it gets shorter and larger in diameter. You can see the bulge of the contracting biceps brachii on the anterior surface of your arm when you flex your elbow.

A skeletal muscle cell can contract until it has shortened by roughly 30 percent. Because the fibers in a parallel muscle are parallel to the long axis of the muscle, when the fibers contract together, the entire muscle shortens by the same amount. If the muscle is 10 cm long, the end of the tendon will move 3 cm when the muscle contracts. The tension developed during this contraction depends on the total number of myofibrils the muscle contains. Because the myofibrils are distributed evenly through the sarcoplasm of each cell, we can use the cross-sectional area of the resting muscle to estimate the tension. A parallel muscle 6.25 cm^2 (1 in.^{2}) in cross-sectional area can develop approximately 23 kg (50 lb) of tension.

Convergent Muscles

In a **convergent muscle,** the muscle fibers are spread over a broad area, but all the fibers converge at one common attachment site. They may pull on a tendon, an aponeurosis (tendinous sheet), or a slender band of collagen fibers known as a **raphe** (RĀ-fē; seam). The muscle fibers typically spread out, like a fan or a broad triangle, with a tendon at the apex. The prominent chest muscles of the *pectoralis group* have this shape (Figure 11-1b▪). A convergent muscle has versatility, because the stimulation of only one portion of the muscle can change the direction of pull. However, when the entire muscle contracts, the muscle fibers do not pull as hard on the attachment site as would a parallel muscle of the same size. The reason is that the convergent muscle fibers pull in different directions rather than all pulling in the same direction.

Pennate Muscles

In a **pennate muscle** (*penna*, feather), the fascicles form a common angle with the tendon. Because the muscle cells pull at an angle, contracting pennate muscles do not move their tendons as far as parallel muscles do. But a pennate muscle contains more muscle fibers—and, as a result, produces more tension—than does a parallel muscle of the same size. (Tension production is proportional to the number of contracting sarcomeres; the more muscle fibers, the more myofibrils and sarcomeres.)

If all the muscle fibers are on the same side of the tendon, the pennate muscle is *unipennate.* The *extensor digitorum muscle,* a forearm muscle that extends the finger joints, is unipennate (Figure 11-1c▪). More

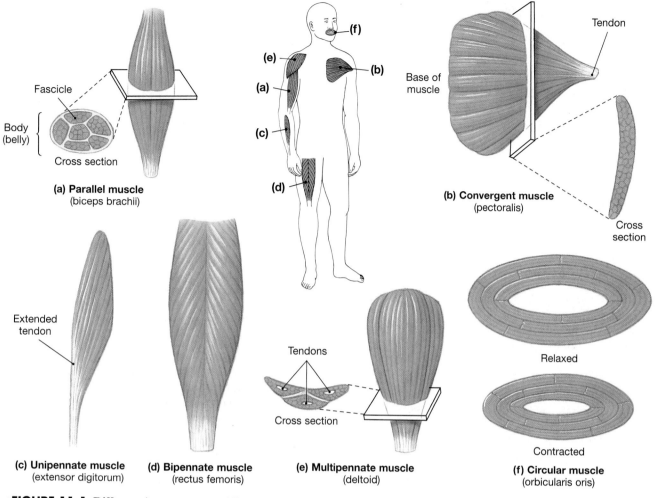

(a) Parallel muscle
(biceps brachii)

(b) Convergent muscle
(pectoralis)

(c) Unipennate muscle
(extensor digitorum)

(d) Bipennate muscle
(rectus femoris)

(e) Multipennate muscle
(deltoid)

(f) Circular muscle
(orbicularis oris)

•**FIGURE 11-1** Different Arrangements of Skeletal Muscle Fibers

commonly, a pennate muscle has fibers on both sides of the tendon. Such a muscle is called *bipennate.* The *rectus femoris muscle,* a prominent muscle that extends the knee, is bipennate (Figure 11-1d•). If the tendon branches within a pennate muscle, the muscle is said to be *multipennate.* The triangular *deltoid muscle* of the shoulder is multipennate (Figure 11-1e•).

Circular Muscles

In a **circular muscle,** or **sphincter** (SFINK-ter), the fibers are concentrically arranged around an opening or a recess. When the muscle contracts, the diameter of the opening decreases. Circular muscles guard entrances and exits of internal passageways such as the digestive and urinary tracts. An example is the *orbicularis oris muscle* of the mouth (Figure 11-1f•).

LEVERS

Skeletal muscles do not work in isolation. When a muscle is attached to the skeleton, the nature and site of the connection will determine the force, speed, and range of the movement produced. These characteris-

tics are interdependent, and the relationships can explain a great deal about the general organization of the muscular and skeletal systems.

The force, speed, or direction of movement produced by contraction of a muscle can be modified by attaching the muscle to a lever. A **lever** is a rigid structure—such as a board, a crowbar, or a bone—that moves on a fixed point called the **fulcrum.** A child's teeter-totter, or seesaw, provides a familiar example of lever action. In the body, each bone is a lever, and each joint is a fulcrum. Levers can change (1) the direction of an applied force, (2) the distance and speed of movement produced by an applied force, and (3) the effective strength of an applied force.

Classes of Levers

Three classes of levers are found in the human body. The seesaw is a **first-class lever.** In such a lever, the fulcrum lies between the applied force (AF) and the resistance (R). The body has few first-class levers. One, involved with extension of the neck, is shown in Figure 11-2a•.

(a) First-class lever

(b) Second-class lever

Movement completed

(c) Third-class lever

Movement completed

•**FIGURE 11-2** **The Three Classes of Levers.** **(a)** In a first-class lever, the applied force and the resistance are on opposite sides of the fulcrum. **(b)** In a second-class lever, the resistance lies between the applied force and the fulcrum. **(c)** In a third-class lever, the force is applied between the resistance and the fulcrum.

In a **second-class lever** (Figure 11-2b•), the resistance is located between the applied force and the fulcrum. A familiar example is a loaded wheelbarrow. The weight of the load is the resistance, and the upward lift on the handle is the applied force. Because in this arrangement the force is always farther from the fulcrum than the resistance is, a small force can balance a larger weight. That is, the effective force is increased. Notice, however, that when a force moves the handle, the load moves more slowly and covers a

shorter distance. The body has few second-class levers. In performing plantar flexion, the calf muscles use a second-class lever (Figure 11-2b•).

Third-class levers are the most common levers in the body. In this lever system, a force is applied between the resistance and the fulcrum (Figure 11-2c•). An example is a ladder that you raise to lean against a building. The fulcrum is the base of the ladder, in contact with the ground. Force is applied where you grasp the ladder, and the resistance is the weight of the ladder between your hands and the free end. The effect is the reverse of that for a second-class lever: Speed and distance traveled are increased at the expense of effective force. In the example shown (the biceps brachii muscle, which flexes the elbow), the resistance is six times farther from the fulcrum than is the applied force. The effective force is reduced to the same degree. The muscle must generate 180 kg of tension at its attachment to the forearm to support 30 kg held in the hand. However, the distance traveled and the speed of movement are increased by that same 6:1 ratio: The load will travel 45 cm when the point of attachment moves 7.5 cm.

Although not every muscle operates as part of a lever system, the presence of levers provides speed and versatility far in excess of what we would predict on the basis of muscle physiology alone. Skeletal muscle fibers resemble one another closely, and their abilities to contract and generate tension are quite similar. Consider a skeletal muscle that can contract in 500 msec and shorten 1 cm while it exerts a 10-kg pull. Without using a lever, this muscle would be performing efficiently only when moving a 10-kg weight a distance of 1 cm. But, by using a lever, the same muscle operating at the same efficiency could move 20 kg a distance of 0.5 cm, 5 kg a distance of 2 cm, or 1 kg a distance of 10 cm.

☑ Why does a pennate muscle generate more tension than does a parallel muscle of the same size?

☑ Which type of muscle would you expect to be guarding the opening between the stomach and the small intestine?

☑ The joint between the occipital bone of the skull and the first cervical vertebra (atlas) is part of which type of lever system?

MUSCLE TERMINOLOGY

This chapter focuses on the functional anatomy of skeletal muscles and of muscle groups. You must learn a number of new terms, and this section attempts to give you assistance in understanding them.

ORIGINS AND INSERTIONS

The site at which a muscle begins is its **origin,** and the site at which it ends is its **insertion.** Each muscle contracts to produce a specific **action,** or movement. In general, the origin remains stationary— attached to a bone or other structure that doesn't move—whereas the insertion moves. In most cases, the origin is proximal to the insertion. For example, consider the *gastrocnemius muscle* (in the calf), which extends from the distal portion of the femur to the calcaneus. As Figure 11-2b• shows, when the gastrocnemius muscle contracts, it pulls the calcaneus toward the knee. As a result, we say that the gastrocnemius muscle has its origin at the femur and its insertion at the calcaneus.

The decision as to which end is the origin and which is the insertion is usually based on movement from the anatomical position. Part of the fun of studying the muscular system is that you can actually do the movements and think about the muscles involved. As a result, laboratory discussions of the muscular system tend to resemble disorganized aerobics classes.

When the origins and insertions cannot be determined easily on the basis of common movement or position, other rules are used. If a muscle extends between a broad aponeurosis and a narrow tendon, the aponeurosis is the origin and the tendon is the insertion. If several tendons are at one end and just one is at the other, the muscle has multiple origins and a single insertion. These simple rules cannot cover every situation. Knowing which end is the origin and which is the insertion is ultimately less important than knowing where the two ends attach and what the muscle accomplishes when it contracts.

Most muscles originate at a bone, but some originate at a connective tissue sheath or band. Examples are *intermuscular septa* (components of the deep fascia that may separate adjacent skeletal muscles), the interosseous membranes of the forearm or leg, and the fibrous sheet that spans the obturator foramen of the pelvis.

ACTIONS

Almost all skeletal muscles either originate or insert on the skeleton. When a muscle moves a portion of the skeleton, that movement may involve flexion, extension, adduction, abduction, protraction, retraction, elevation, depression, rotation, circumduction, prona-

tion, supination, inversion, eversion, lateral flexion, or opposition. Before proceeding, you may want to review the discussions of planes of motion and Figures 9-2 to 9-5. ∞ *pp. 250–254*

Actions can be described by one of two methods. The first, used by most undergraduate textbooks and references such as *Gray's Anatomy,* describes actions in terms of the bone affected. Thus a muscle such as the biceps brachii muscle is said to perform "flexion of the forearm." The second method, of increasing use among specialists such as kinesiologists, identifies the joint involved. With this method, the action of the biceps brachii muscle would be "flexion at (or of) the elbow." Both methods are valid, and each has its advantages. We shall primarily use the latter method.

When complex movements occur, muscles commonly work in groups rather than individually. Their cooperation improves the efficiency of a particular movement. For example, large muscles of the limbs produce flexion or extension over an extended range of motion. Although these muscles cannot develop much tension at full extension, they are generally paired with one or more smaller muscles that provide assistance until the larger muscle can perform at maximum efficiency. At the start of the movement, the smaller muscle is producing maximum tension while tension production by the larger muscle is at a minimum. The importance of this smaller "assistant" decreases as the movement proceeds and the efficiency of the primary muscle increases.

On the basis of size and range of motion, muscles are described as follows:

- An **agonist,** or **prime mover,** is a muscle whose contraction is chiefly responsible for producing a particular movement. The biceps brachii muscle is an agonist that produces flexion at the elbow.

- **Antagonists** are muscles whose actions oppose that of the agonist under consideration. The *triceps brachii muscle* is an agonist that extends the elbow. It is therefore an antagonist of the biceps brachii muscle, and the biceps brachii is an antagonist of the triceps brachii. Agonists and antagonists are functional opposites; if one produces flexion, the other will produce extension. When an agonist contracts to produce a particular movement, the corresponding antagonist will be stretched, but it will usually not relax completely. Instead, it will contract eccentrically, with the tension adjusted to control the speed of the movement and ensure its smoothness. ∞ *p. 291* You may find it easiest to learn about muscles in agonist/antagonist pairs (flexors/extensors, abductors/adductors) that act at a specific joint. This method highlights the functions of the muscles involved, and it can help organize the information in a logical framework. The tables in this chapter are arranged to facilitate this approach.

- When a **synergist** (*syn-,* together + *ergon,* work) contracts, it helps a larger agonist work efficiently.

Synergists may provide additional pull near the insertion or may stabilize the point of origin. Their importance in assisting a particular movement may change as the movement progresses. In many cases, they are most useful at the start, when the agonist is stretched and unable to develop maximum tension. For example, the *latissimus dorsi muscle* is a large trunk muscle that extends, adducts, and medially rotates the arm at the shoulder joint. A much smaller muscle, the *teres* (TER-ēz) *major muscle*, assists in starting such movements when the shoulder joint is at full flexion. Synergists may also assist an agonist by preventing movement at another joint and thereby stabilizing the origin of the agonist. Such synergists are called **fixators.**

NAMES OF SKELETAL MUSCLES

Except for the *platysma* and the *diaphragm,* the complete names of all skeletal muscles include the term *muscle.* Although the full name, such as the biceps brachii muscle, will usually appear in the text, for simplicity only the descriptive name (biceps brachii) will be used in figures and tables.

You need not learn every one of the approximately 700 muscles in the human body, but you will have to become familiar with the most important ones. Fortunately, anatomists assigned names to the muscles that provide clues to their identification. If you can learn to recognize the clues, you will find it easier to remember the names and identify the muscles. The name of a muscle may include information about its fascicle organization, location, relative position, structure, size, shape, origin and insertion, or action.

Fascicle Organization

A muscle name may refer to the orientation of the muscle fibers within a particular skeletal muscle. **Rectus** means "straight," and rectus muscles are parallel muscles whose fibers generally run along the long axis of the body. Because we have several rectus muscles, the name typically includes a second term that refers to a precise region of the body. For example, the *rectus abdominis muscle* is located on the abdomen, and the rectus femoris muscle on the thigh. Other directional indicators include **transversus** and **oblique** for muscles whose fibers run across or at an oblique angle to the longitudinal axis of the body, respectively.

Location

Table 11-1 includes a useful summary of terms that designate specific regions of the body. They are common as modifiers that help identify individual muscles, as in the case of the rectus muscles. In a few cases, the muscle is such a prominent feature of the region that the regional name alone will identify it. Examples include the *temporalis muscle* of the head and the *brachialis* (brā-kē-A-lis) *muscle* of the arm.

Relative Position

Muscles visible at the body surface are often called **externus** or **superficialis,** whereas deeper muscles are termed **internus** or **profundus.** Superficial muscles that position or stabilize an organ are called **extrinsic;** muscles located entirely within an organ are **intrinsic.**

Structure, Size, and Shape

Some muscles are named after distinctive structural features. The biceps brachii muscle, for example, has two tendons of origin (*bi-,* two + *caput,* head); the triceps brachii muscle has three; and the *quadriceps group,* four. Shape is sometimes an important clue to the name of a muscle. For example, the *trapezius* (tra-PĒ-zē-us), *deltoid,* *rhomboideus* (rom-BOY-dē-us), and *orbicularis* (or-bik-ū-LAR-is) muscles look like a trapezoid, a triangle, a rhomboid, and a circle, respectively. Long muscles are called **longus** (long) or **longissimus** (longest), and **teres** muscles are both long and round. Short muscles are called **brevis.** Large ones are called **magnus** (big), **major** (bigger), or **maximus** (biggest); and small ones are called **minor** (smaller) or **minimus** (smallest).

Origin and Insertion

Many names tell you the specific origin and insertion of each muscle. In such cases, the first part of the name indicates the origin, and the second part the insertion. The *genioglossus muscle,* for example, originates at the chin *(geneion)* and inserts in the tongue *(glossus).* The names may be long and difficult to pronounce, but Table 11-1 and the anatomical terms introduced in Chapter 1 can help you identify and remember them. ∞ *pp. 15–19*

Action

Many muscles are named *flexor, extensor, retractor, abductor,* and so on. These are such common actions that the names almost always include other clues as to the appearance or location of the muscle. For example, the *extensor carpi radialis longus muscle* is a long muscle along the radial (lateral) border of the forearm. When it contracts, its primary function is extension at the carpus (wrist).

A few muscles are named after the specific movements associated with special occupations or habits. The *sartorius* (sar-TO-rē-us) *muscle* is active when you cross your legs. Before sewing machines were invented, a tailor would sit on the floor cross-legged, and the name of this muscle was derived from *sartor,* the Latin word for "tailor." On the face, the *buccinator* (BUK-si-nā-tor) *muscle* compresses the cheeks, as when you purse your lips and blow forcefully. *Buccinator* translates as "trumpet player." Another facial muscle, the *risorius* (ri-SO-rē-us) *muscle,* was supposedly named after the mood expressed. The Latin term *risor,* however, means "laugher"; a more appropriate description for the effect would be "grimace."

TABLE 11-1 Muscle Terminology

Terms Indicating Direction Relative to Axes of the Body	Terms Indicating Specific Regions of the Body*	Terms Indicating Structural Characteristics of the Muscle	Terms Indicating Actions
Anterior (front)	Abdominis (abdomen)	**Origin**	**General**
Externus (superficial)	Anconeus (elbow)	Biceps (two heads)	Abductor
Extrinsic (outside)	Auricularis (auricle of ear)	Triceps (three heads)	Adductor
Inferioris (inferior)	Brachialis (brachium)	Quadriceps (four heads)	Depressor
Internus (deep, internal)	Capitis (head)		Extensor
Intrinsic (inside)	Carpi (wrist)	**Shape**	Flexor
Lateralis (lateral)	Cervicis (neck)	Deltoid (triangle)	Levator
Medialis/medius (medial, middle)	Cleido-/-clavius (clavicle)	Orbicularis (circle)	Pronator
	Coccygeus (coccyx)	Pectinate (comblike)	Rotator
Oblique	Costalis (ribs)	Piriformis (pear-shaped)	Supinator
Posterior (back)	Cutaneous (skin)	Platys- (flat)	Tensor
Profundus (deep)	Femoris (femur)	Pyramidal (pyramid)	
Rectus (straight, parallel)	Genio- (chin)	Rhomboideus (rhomboid)	**Specific**
Superficialis (superficial)	Glosso-/-glossal (tongue)	Serratus (serrated)	Buccinator (trumpeter)
Superioris (superior)	Hallucis (great toe)	Splenius (bandage)	Risorius (laugher)
Transversus (transverse)	Ilio- (ilium)	Teres (long and round)	Sartorius (like a tailor)
	Inguinal (groin)	Trapezius (trapezoid)	
	Lumborum (lumbar region)		
	Nasalis (nose)	**Other Striking Features**	
	Nuchal (back of neck)	Alba (white)	
	Oculo- (eye)	Brevis (short)	
	Oris (mouth)	Gracilis (slender)	
	Palpebrae (eyelid)	Lata (wide)	
	Pollicis (thumb)	Latissimus (widest)	
	Popliteus (posterior to knee)	Longissimus (longest)	
	Psoas (loin)	Longus (long)	
	Radialis (radius)	Magnus (large)	
	Scapularis (scapula)	Major (larger)	
	Temporalis (temples)	Maximus (largest)	
	Thoracis (thoracic region)	Minimus (smallest)	
	Tibialis (tibia)	Minor (smaller)	
	Ulnaris (ulna)	-tendinosus (tendinous)	
	Uro- (urinary)	Vastus (great)	

*For other regional terms, refer to Figure 1-6, p. 16, which deals with anatomical landmarks.

11

AXIAL AND APPENDICULAR MUSCLES

The separation of the skeletal system into axial and appendicular divisions provides a useful guideline for subdividing the muscular system as well:

1. The **axial musculature** arises on the axial skeleton. It positions the head and spinal column and also moves the rib cage, assisting in the movements that make breathing possible. It does not play a role in movement or support of either the pectoral or pelvic girdle or the limbs. This category encompasses roughly 60 percent of the skeletal muscles in the body.

2. The **appendicular musculature** stabilizes or moves components of the appendicular skeleton and includes the remaining 40 percent of all skeletal muscles.

Figure 11-3• provides an overview of the major axial and appendicular muscles of the human body. These are the superficial muscles, which tend to be relatively large. The superficial muscles cover deeper, smaller muscles that cannot be seen unless the overlying muscles are either removed or *reflected*—that is, cut and pulled out of the way. Later figures that show deep muscles in specific regions will indicate whether

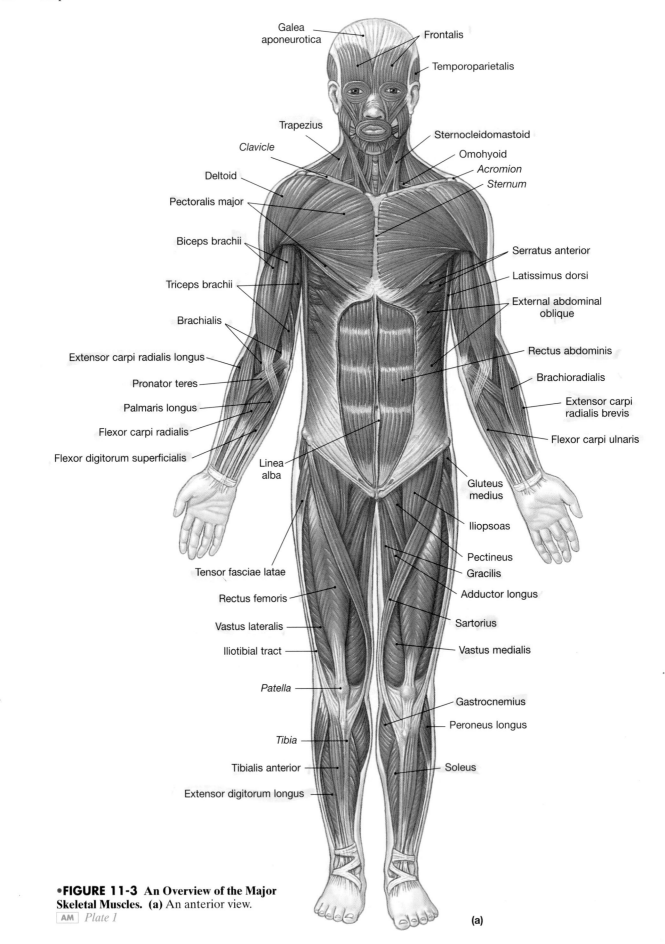

•**FIGURE 11-3** **An Overview of the Major Skeletal Muscles. (a)** An anterior view.

AM *Plate 1*

(a)

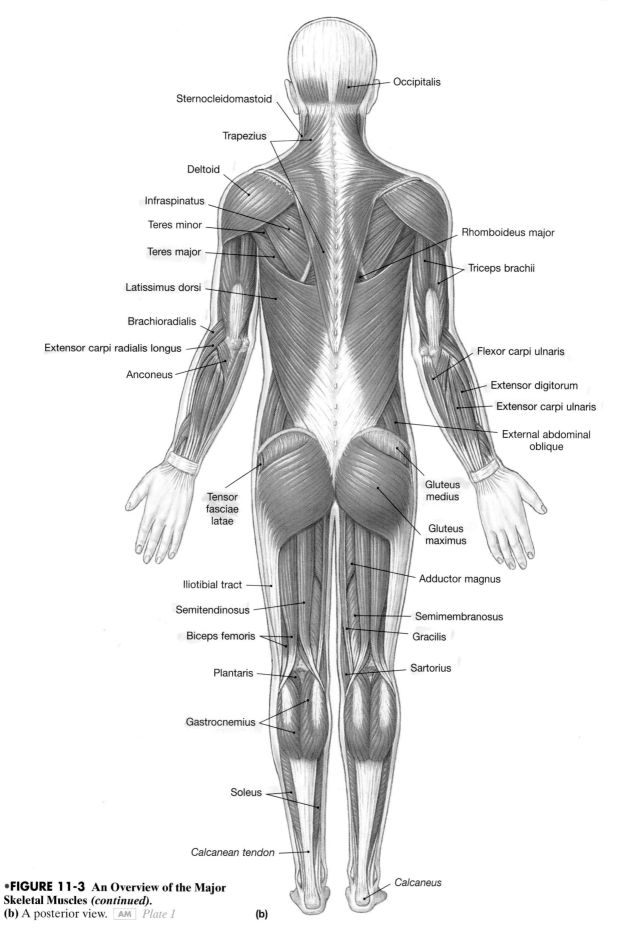

Occipitalis

Sternocleidomastoid

Trapezius

Deltoid

Infraspinatus

Teres minor

Teres major

Latissimus dorsi

Brachioradialis

Extensor carpi radialis longus

Anconeus

Rhomboideus major

Triceps brachii

Flexor carpi ulnaris

Extensor digitorum

Extensor carpi ulnaris

External abdominal
oblique

Gluteus
medius

Gluteus
maximus

Tensor
fasciae
latae

Adductor magnus

Iliotibial tract

Semitendinosus

Biceps femoris

Plantaris

Gastrocnemius

Semimembranosus

Gracilis

Sartorius

Soleus

Calcanean tendon

Calcaneus

11

•**FIGURE 11-3** **An Overview of the Major
Skeletal Muscles** *(continued)*.
(b) A posterior view.   AM   *Plate 1*   **(b)**

superficial muscles have been removed or reflected for the sake of clarity.

Paying attention to patterns of origin, insertion, and action, we will now examine representatives of both muscular divisions. This discussion assumes that you already understand skeletal anatomy. As you examine the figures in this chapter, you will find that some bony and cartilaginous landmarks are labeled for orientation purposes. These labels are shown in italics, to differentiate them from the muscles that are the primary focus of each figure. Should you need further review of skeletal anatomy, figure captions in this chapter indicate the relevant figure numbers in Chapters 7, 8, or 9.

The muscle tables that follow also contain information about the innervation of the individual muscles. **Innervation** is the distribution of nerves to a region or organ; the tables indicate the nerves that control each muscle. Many of the muscles of the head and neck are innervated by cranial nerves, such as the *facial nerve,* or seventh cranial nerve (N VII), which innervates the facial musculature. Cranial nerves originate at the brain and pass through the foramina of the skull. Spinal nerves are connected to the spinal cord and pass through the intervertebral foramina. For example, spinal nerve L_1 passes between vertebrae L_1 and L_2. Spinal nerves may form a complex network, or *plexus;* one branch, such as the *sciatic nerve* of the thigh, may contain axons from several spinal nerves. Thus, many tables include the spinal nerves as well as the names of the peripheral nerves.

☑ The *gracilis* muscle is attached to the anterior surface of the tibia at one end and to the pubis and ischium of the pelvis at the other. When the muscle contracts, hip flexion occurs. Which attachment point is the muscle's origin?

☑ Muscle A abducts the humerus, and muscle B adducts the humerus. What is the relationship between these two muscles?

☑ What does the name *flexor carpi radialis longus* tell you about this muscle?

THE AXIAL MUSCLES

The axial muscles fall into logical groups on the basis of location, function, or both. The groups do not always have distinct anatomical boundaries. For example, a function such as extension of the vertebral column involves muscles along its entire length and flexion at each of the intervertebral joints. We will discuss the axial muscles in four groups:

1. *The muscles of the head and neck.* This group includes muscles that move the face, tongue, and larynx. They are therefore responsible for verbal and nonverbal communication—laughing, talking, frowning, smiling, whistling, and so on. You also use this group of muscles when you eat—especially in sucking and chewing—and even when you look for something to eat, by controlling your eye movements. This group does not include muscles of the neck that are involved with movements of the vertebral column.

2. *The muscles of the spine.* This group includes numerous flexors, extensors, and rotators of the vertebral column.

3. *The oblique and rectus muscles.* This group forms the muscular walls of the thoracic and abdominopelvic cavities between the first thoracic vertebra and the pelvis. In the thoracic area these muscles are partitioned by the ribs, but over the abdominal surface they form broad muscular sheets. The neck also has oblique and rectus muscles. Although they do not form a complete muscular wall, they share a common developmental origin with the oblique and rectus muscles of the trunk. AM *EMBRYOLOGY SUMMARY 9: The Development of the Muscular System*

4. *The muscles of the pelvic floor.* These muscles extend between the sacrum and pelvic girdle. This group forms the *perineum,* a muscular sheet that closes the pelvic outlet.

MUSCLES OF THE HEAD AND NECK

We can divide the muscles of the head and neck into several functional groups. The *muscles of facial expression,* the *muscles of mastication* (chewing), the *muscles of the tongue,* and the *muscles of the pharynx* originate on the skull or hyoid bone. Muscles involved with sight and hearing are also based on the skull. We will consider here the *extrinsic eye muscles*—those associated with movements of the eye. We shall discuss the intrinsic eye muscles, which control the diameter of the pupil and the shape of the lens and the tiny skeletal muscles associated with the auditory ossicles, in Chapter 17. In the neck, the *extrinsic muscles of the larynx* adjust the position of the hyoid bone and larynx. We shall examine the intrinsic laryngeal muscles, including those of the vocal cords, in Chapter 23 (the respiratory system).

Muscles of Facial Expression

The muscles of facial expression originate on the surface of the skull (Figure 11-4• and Table 11-2, p. 322). At their insertions, the fibers of the epimysium are woven into those of the superficial fascia and the dermis of the skin: When they contract, the skin moves. These muscles are innervated by the facial nerve.

The largest group of facial muscles is associated with the mouth. The **orbicularis oris** muscle constricts

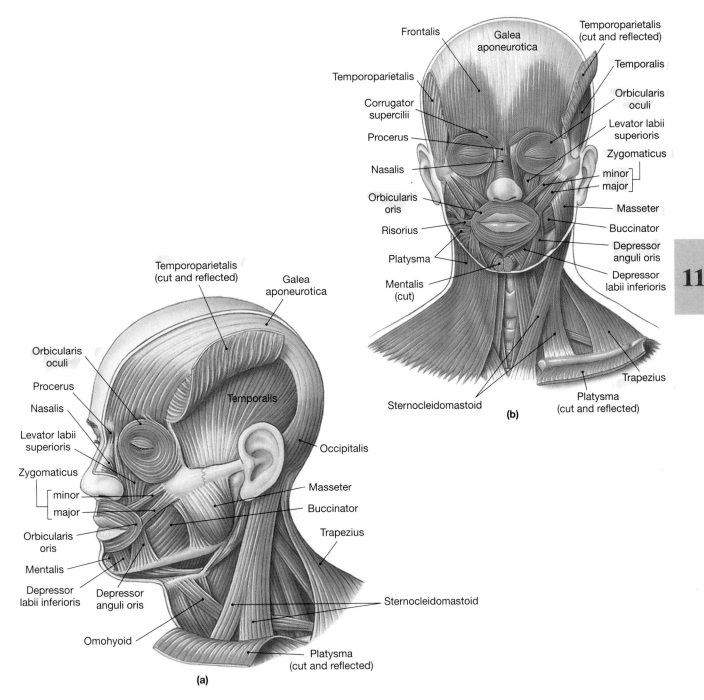

•**FIGURE 11-4 Muscles of Facial Expression.** *See also Figure 7-3, pp. 197–198.* **(a)** An anterolateral view. **(b)** An anterior view.
AM   *Plates 5.1, 5.2*

the opening, and other muscles move the lips or the corners of the mouth. The **buccinator** muscle has two functions related to eating (in addition to its importance to musicians). During chewing, it cooperates with the masticatory muscles by moving food back across the teeth from the space inside the cheeks. In infants, the buccinator provides suction for suckling at the breast.

Smaller groups of muscles control movements of the eyebrows and eyelids, the scalp, the nose, and the external ear. The **epicranius** (ep-i-KRĀ-nē-us; *epi-*, on + *kranion*, skull), or scalp, contains two muscles, the **frontalis** and the **occipitalis** muscles. These muscles are separated by the **galea aponeurotica** (GĀ-lē-uh ap-ō-nū-RO-ti-kuh; *galea*, a helmet + aponeurosis), a collagenous sheet. The **platysma** (pla-TIZ-muh; *platys*, flat) covers the ventral surface of the neck, extending from the base of the neck to the periosteum of the mandible and the fascia at the corner of the mouth.

TABLE 11-2 Muscles of Facial Expression *(Figure 11-4)*

Region/Muscle	Origin	Insertion	Action	Innervation
Mouth				
Buccinator	Alveolar processes of maxillary bone and mandible	Blends into fibers of orbicularis oris	Compresses cheeks	Facial nerve (N VII)
Depressor labii inferioris	Mandible between the anterior midline and the mental foramen	Skin of lower lip	Depresses lip	As above
Levator labii superioris	Lower margin of orbit, superior to the infraorbital foramen	Orbicularis oris	Elevates upper lip	As above
Mentalis	Incisive fossa of mandible	Skin of chin	Elevates and protrudes lower lip	As above
Orbicularis oris	Maxillary bone and mandible	Lips	Compresses, purses lips	As above
Risorius	Fascia surrounding parotid salivary gland	Angle of mouth	Draws corner of mouth to the side	As above
Depressor anguli oris	Anterolateral surface of mandibular body	Skin at angle of mouth	Depresses corner of mouth	As above
Zygomaticus major	Zygomatic bone near zygomaticomaxillary suture	Angle of mouth	Retracts and elevates corner of mouth	As above
Zygomaticus minor	Zygomatic bone posterior to zygomaticotemporal suture	Upper lip	As above	As above
Eye				
Corrugator supercilii	Orbital rim of frontal bone near nasal suture	Eyebrow	Pulls skin inferiorly and anteriorly; wrinkles brow	As above
Levator palpebrae superioris (not shown)	Tendinous band around optic foramen	Upper eyelid	Elevates upper eyelid	Oculomotor nerve (N III)*
Orbicularis oculi	Medial margin of orbit	Skin around eyelids	Closes eye	Facial nerve (N VII)
Nose				
Procerus	Nasal bones and lateral nasal cartilages	Aponeurosis at bridge of nose and skin of forehead	Moves nose, changes position and shape of nostrils	As above
Nasalis	Maxillary bone and alar cartilage of nose	Bridge of nose	Compresses bridge, depresses tip of nose; elevates corners of nostrils	As above
Ear (extrinsic)				
Temporoparietalis	Fascia around external ear	Galea aponeurotica	Tenses scalp, moves auricle of ear	As above
Scalp (Epicranius)				
Frontalis	Galea aponeurotica	Skin of eyebrow and bridge of nose	Raises eyebrows, wrinkles forehead	As above
Occipitalis	Superior nuchal line	Galea aponeurotica	Tenses and retracts scalp	As above
Neck				
Platysma	Superior thorax between cartilage of 2nd rib and acromion of scapula	Mandible and skin of cheek	Tenses skin of neck; depresses mandible	As above

*This muscle originates in association with the extrinsic eye muscles, so its innervation is unusual. *(See Figure 14-22, p. 467.)*

(a) Medial surface, right eye   (b) Lateral surface, right eye

•**FIGURE 11-5** Extrinsic Eye Muscles. *See also Figure 7-13, p. 208.* [AM] *Plates 3.2, 3.3*

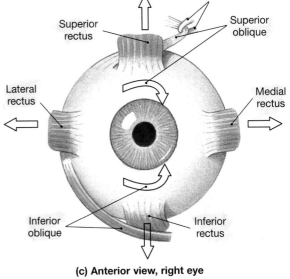

(c) Anterior view, right eye

Extrinsic Eye Muscles

Six **extrinsic eye muscles,** or **oculomotor** (ok-ū-lō-MŌ-ter) **muscles,** originating on the surface of the orbit control the position of each eye. These muscles, shown in Figure 11-5• and detailed in Table 11-3, are the **inferior rectus, medial rectus, superior rectus, lateral rectus, inferior oblique,** and **superior oblique** muscles. The extrinsic eye muscles are innervated by the third *(oculomotor),* fourth *(trochlear),* and sixth *(abducens)* cranial nerves.

TABLE 11-3 Extrinsic Eye Muscles (Figure 11-5)

Muscle	Origin	Insertion	Action	Innervation
Inferior rectus	Sphenoid bone around optic canal	Inferior, medial surface of eyeball	Eye looks down	Oculomotor nerve (N III)
Medial rectus	As above	Medial surface of eyeball	Eye rotates medially	As above
Superior rectus	As above	Superior surface of eyeball	Eye looks up	As above
Lateral rectus	As above	Lateral surface of eyeball	Eye rotates laterally	Abducens nerve (N VI)
Inferior oblique	Maxillary bone at anterior portion of orbit	Inferior, lateral surface of eyeball	Eye rolls, looks up and to the side	Oculomotor nerve (N III)
Superior oblique	Sphenoid bone around optic canal	Superior, lateral surface of eyeball	Eye rolls, looks down and to the side	Trochlear nerve (N IV)

11

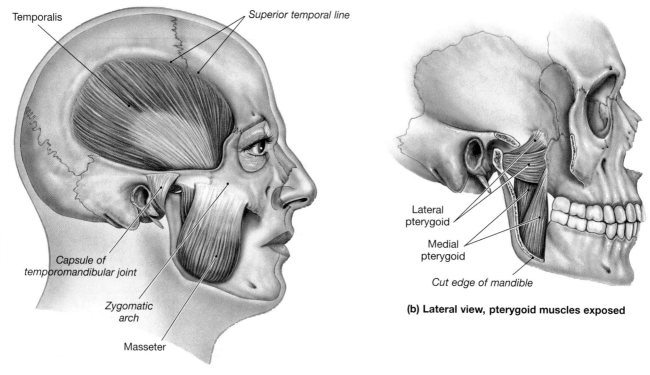

(a) Lateral view

(b) Lateral view, pterygoid muscles exposed

•**FIGURE 11-6** **Muscles of Mastication.** **(a)** The temporalis passes medial to the zygomatic arch to insert on the coronoid process of the mandible. The masseter inserts on the angle and lateral surface of the mandible. **(b)** The location and orientation of the pterygoid muscles can be seen after the overlying muscles, along with a portion of the mandible, are removed. *See also Figures 7-3 and 7-12, pp. 197–198, 207.* [AM] *Plate 5.2*

Muscles of Mastication

The muscles of mastication (Figure 11-6• and Table 11-4) move the mandible at the temporomandibular joint. The large **masseter** muscle is the strongest jaw muscle. The **temporalis** muscle assists in elevation of the mandible. The **pterygoid** muscles, used in various combinations, can elevate, depress, or protract the mandible or slide it from side to side, a movement called *lateral excursion*. These movements are important in making efficient use of your teeth while you chew foods of various consistencies. The muscles of mastication are innervated by the fifth cranial nerve, the *trigeminal nerve*.

TABLE 11-4 Muscles of Mastication (Figure 11-6)

Muscle	Origin	Insertion	Action	Innervation
Masseter	Zygomatic arch	Lateral surface of mandibular ramus	Elevates mandible	Trigeminal nerve (N V), mandibular branch
Temporalis	Along temporal lines of skull	Coronoid process of mandible	Elevates mandible	As above
Pterygoids (medial and lateral)	Lateral pterygoid plate	Medial surface of mandibular ramus	*Medial:* Elevates the mandible and closes the jaws, or performs lateral excursion	As above
			Lateral: Opens jaws, protrudes mandible, or performs lateral excursion	As above

What's New in Anatomy?

Like most people, you might assume that every anatomical structure in the human body was described centuries ago and that nothing is "new" in the field of anatomy. Many people were surprised in 1996, however, when anatomical researchers at the University of Maryland documented the existence of a previously unknown skeletal muscle. This "new" muscle, named the *sphenomandibularis* muscle, extends from the lateral surface of the sphenoid bone to the mandible (Figure 11-7•). The sphenomandibularis muscle assists the muscles of mastication. The research began with the computer analysis of the Virtual Human database, a digitized photographic atlas of cross-sectional anatomy. That initial work was then supported by careful cadaver dissections. Although the discovery of this muscle remains controversial (for example, it may have been described previously as a portion of the temporalis muscle), it is a good example of how modern technologies are providing new perspectives on the human body.

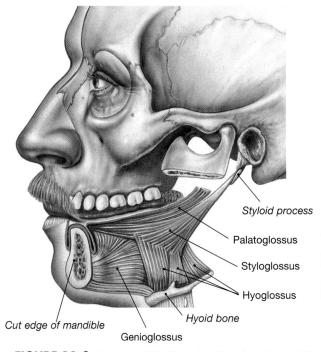

Styloid process
Palatoglossus
Styloglossus
Hyoglossus
Cut edge of mandible
Hyoid bone
Genioglossus

11

•**FIGURE 11-8** Muscles of the Tongue. *See also Figure 7-3, pp. 197–198.*

Muscles of the Tongue

The muscles of the tongue have names ending in *glossus*, the Greek word for "tongue." Once you can recall the structures referred to by *palato-, stylo-, genio-,* and *hyo-,* the names will be self-explanatory. The **palatoglossus** muscle originates at the palate, the **styloglossus** muscle at the styloid process of the temporal bone, the **genioglossus** muscle at the chin, and the **hyoglossus** muscle at the hyoid bone (Figure 11-8•). These muscles, used in various combinations, move the tongue in the delicate and complex patterns necessary for speech, and they manipulate food within the mouth in preparation for swallowing. Most are innervated by the *hypoglossal nerve* (N XII), a cranial nerve whose name indicates its function and its location (Table 11-5).

Origin of superior portion of lateral pterygoid
Sphenomandibularis

•**FIGURE 11-7** The Sphenomandibularis Muscle

TABLE 11-5 Muscles of the Tongue *(Figure 11-8)*

Muscle	Origin	Insertion	Action	Innervation
Genioglossus	Medial surface of mandible around chin	Body of tongue, hyoid bone	Depresses and protracts tongue	Hypoglossal nerve (N XII)
Hyoglossus	Body and greater cornu of hyoid bone	Side of tongue	Depresses and retracts tongue	As above
Palatoglossus	Anterior surface of soft palate	As above	Elevates tongue, depresses soft palate	Cranial branch of accessory nerve (N XI)
Styloglossus	Styloid process of temporal bone	Along the side to tip and base of tongue	Retracts tongue, elevates side	Hypoglossal nerve (N XII)

Muscles of the Pharynx

The muscles of the pharynx (Figure 11-9• and Table 11-6) are responsible for initiating the swallowing process. The **pharyngeal constrictors** (*superior, middle,* and *inferior*) move materials into the esophagus. The **laryngeal elevators** elevate the larynx. The **palatal muscles,** the *tensor veli palatini* and the *levator veli palatini,* elevate the soft palate and adjacent portions of the pharyngeal wall and also pull open the entrance to the auditory tube. As a result, swallowing repeatedly can open the entrance to the auditory tube to help you adjust to pressure changes when you fly or dive.

Anterior Muscles of the Neck

The anterior muscles of the neck include (1) muscles that control the position of the larynx, (2) muscles that depress the mandible and tense the floor of the mouth, and (3) muscles that provide a stable foundation for muscles of the tongue and pharynx (Figure 11-10• and Table 11-7). The **digastric** (dī-GAS-trik) muscle has two bellies, as the name implies (*di-,* two + *gaster,* stomach). One belly extends from the chin to the hyoid bone; the other continues from the hyoid bone to the mastoid portion of the temporal bone. Depending on which belly contracts and whether fixator muscles are stabilizing the position of the hyoid bone, the digastric muscle can open the mouth by depressing the mandible, or it can elevate the larynx by raising the

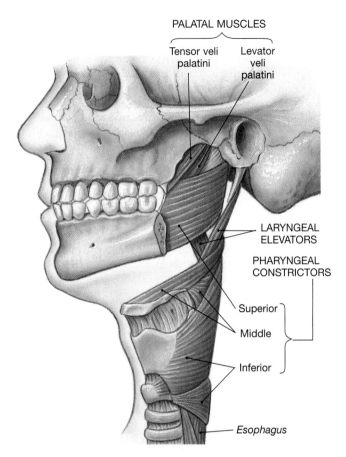

•**FIGURE 11-9** Muscles of the Pharynx. A lateral view. *See also Figure 7-4, p. 199.*

TABLE 11-6 Muscles of the Pharynx *(Figure 11-9)*

Muscle	Origin	Insertion	Action	Innervation
Pharyngeal constrictors				
Superior constrictor	Pterygoid process of sphenoid bone, medial surfaces of mandible	Median raphe attached to occipital bone	Constrict pharynx to propel bolus into esophagus	Branches of pharyngeal plexus (N X)
Middle constrictor	Cornu of hyoid bone	Median raphe	As above	As above
Inferior constrictor	Cricoid and thyroid cartilages of larynx	As above	As above	As above
Laryngeal elevators*	Ranges from soft palate, to cartilage around inferior portion of auditory tube, to styloid process of temporal bone	Thyroid cartilage	Elevate larynx	Branches of pharyngeal plexus (N IX and N X)
Palatal muscles				
Levator veli palatini	Petrous portion of temporal bone; tissues around the auditory tube	Soft palate	Elevates soft palate	N XI
Tensor veli palatini	Spine of sphenoid bone; tissues around the auditory tube	As above	As above	N V

*The palatopharyngeus, salpingopharyngeus, and stylopharyngeus, assisted by the thyrohyoid, geniohyoid, stylohyoid, and hyoglossus muscles, discussed in Tables 11-5 and 11-7.

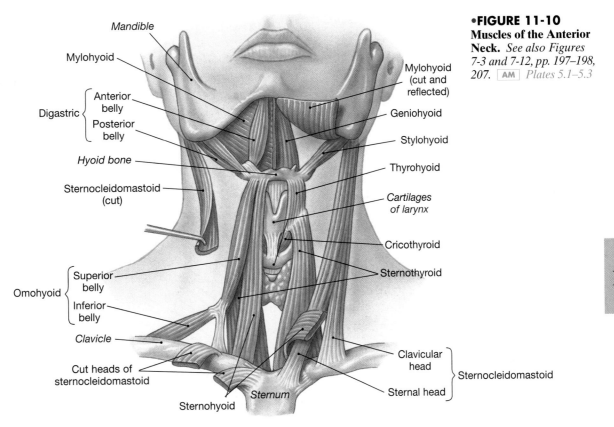

•**FIGURE 11-10**
Muscles of the Anterior Neck. *See also Figures 7-3 and 7-12, pp. 197–198, 207.* [AM] *Plates 5.1–5.3*

11

TABLE 11-7 Anterior Muscles of the Neck (Figure 11-10)

Muscle	Origin	Insertion	Action	Innervation
Digastric	Two bellies: *posterior* from mastoid region of temporal; *anterior* from inferior surface of mandible at chin	Hyoid bone	Depresses mandible and/or elevates larynx	Trigeminal nerve (N V), mandibular branch, to anterior belly Facial nerve (N VII) to posterior belly
Geniohyoid	Medial surface of mandible at chin	Hyoid bone	As above and pulls hyoid bone anteriorly	Cervical nerve C_1 via hypoglossal nerve (N XII)
Mylohyoid	Mylohyoid line of mandible	Median connective tissue band (raphe) that runs to hyoid bone	Elevates floor of mouth and hyoid bone and/or depresses mandible	Trigeminal nerve (N V), mandibular branch
Omohyoid	Central tendon attaches to clavicle and 1st rib	Two bellies: *superior* attaches to hyoid bone; *inferior* to superior margin of scapula	Depresses hyoid bone and larynx	Cervical spinal nerves C_2–C_3
Sternohyoid	Clavicle and manubrium	Hyoid bone	As above	Cervical spinal nerves C_1–C_3
Sternothyroid	Dorsal surface of manubrium and 1st costal cartilage	Thyroid cartilage of larynx	As above	As above
Stylohyoid	Styloid process of temporal bone	Hyoid bone	Elevates larynx	Facial nerve (N VII)
Thyrohyoid	Thyroid cartilage of larynx	Hyoid bone	Elevates thyroid, depresses hyoid bone	Cervical spinal nerves C_1–C_2 via hypoglossal nerve (N XII)
Sternocleido-mastoid	Two bellies: *clavicular head* attaches to sternal end of clavicle; *sternal head* attaches to manubrium	Mastoid region of skull	Together, they flex the neck; alone, one side bends head toward shoulder and turns face to opposite side	Accessory nerve (N XI) and cervical spinal nerves (C_2–C_3) of cervical plexus

hyoid bone. The digastric muscle overlies the broad, flat **mylohyoid** muscle, which provides a muscular floor to the mouth. The **stylohyoid** muscle forms a muscular connection between the hyoid bone and the styloid process of the skull. The **sternocleidomastoid** (ster-nō-klī-dō-MAS-toyd) muscle extends from the clavicle and the sternum to the mastoid region of the skull (Figures 11-4, p. 321, and 11-10•). The **omohyoid** (ō-mō-HĪ-oyd) muscle attaches to the scapula, the clavicle and first rib, and the hyoid bone. The other members of this group are straplike muscles that extend between the sternum and larynx *(sternothyroid)* or hyoid bone *(sternohyoid)*, between the larynx and hyoid bone *(thyrohyoid),* and between the hyoid bone and chin *(geniohyoid).*

☑ If you were contracting and relaxing your masseter muscle, what would you probably be doing?

☑ Which facial muscle would you expect to be well developed in a trumpet player?

☑ Why can swallowing help alleviate the pressure sensations at the eardrum when you are in an airplane that is changing altitude?

MUSCLES OF THE VERTEBRAL COLUMN

The muscles of the vertebral column are covered by more-superficial back muscles, such as the trapezius and latissimus dorsi muscles (Figure 11-3b•, p. 319). The **erector spinae,** or *spinal extensors,* include superficial and deep layers. The superficial layer can be divided into **spinalis, longissimus,** and **iliocostalis** divisions (Figure 11-11• and Table 11-8). In the inferior lumbar and sacral regions, the boundary between the longissimus and iliocostalis muscles is indistinct. When contracting together, the erector spinae extend the vertebral column. When the muscles on only one side contract, the result is lateral flexion of the vertebral column.

Deep to the spinalis muscles, the muscles of the vertebral column interconnect and stabilize the vertebrae. These muscles include the **semispinalis** muscles and the **multifidus, interspinales, intertransversarii,** and **rotatores** muscles (Figure 11-11a,b•). In various combinations, they produce slight extension or rotation of the vertebral column. They are also important in making delicate adjustments in the positions of individual vertebrae, and they stabilize adjacent vertebrae.

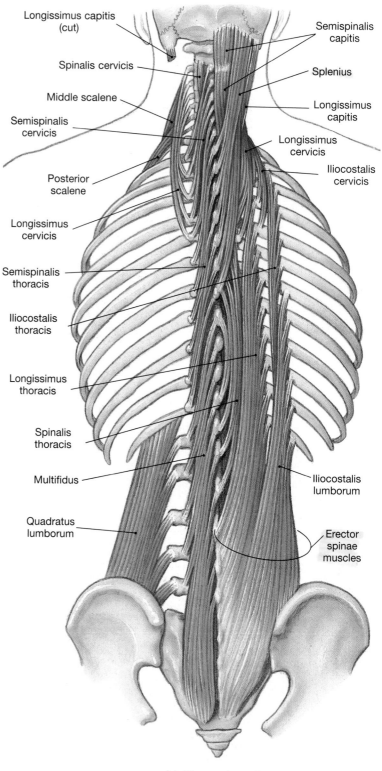

(a) The erector spinae

•**FIGURE 11-11** Muscles of the Vertebral Column. *See also Figures 7-1a and 7-22, pp. 194, 220.*

•**FIGURE 11-11** Muscles of the Vertebral Column *(continued)*

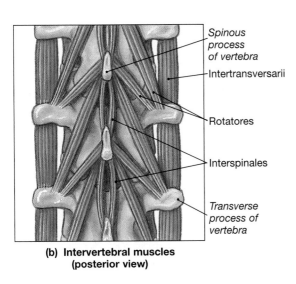

(b) Intervertebral muscles (posterior view)

(c) Muscles arising from the anterior surfaces of the superior vertebrae

11

TABLE 11-8 Muscles of the Vertebral Column *(Figure 11-11)*

Group/Muscle	Origin	Insertion	Action	Innervation
SUPERFICIAL LAYER				
Splenius (Splenius capitis, splenius cervicis)	Spinous processes and ligaments connecting inferior cervical and superior thoracic vertebrae	Mastoid process, occipital bone of skull, and superior cervical vertebrae	Together, the two sides extend neck; alone, each rotates and laterally flexes neck to that side	Cervical spinal nerves
Erector spinae				
Spinalis group				
Spinalis cervicis	Inferior portion of ligamentum nuchae and spinous process of C_7	Spinous process of axis	Extends neck	As above
Spinalis thoracis	Spinous processes of inferior thoracic and superior lumbar vertebrae	Spinous processes of superior thoracic vertebrae	Extends vertebral column	Thoracic and lumbar spinal nerves
Longissimus group				
Longissimus capitis	Processes of inferior cervical and superior thoracic vertebrae	Mastoid process of temporal bone	Together, the two sides extend head; alone, each rotates and laterally flexes neck to that side	Cervical and thoracic spinal nerves
Longissimus cervicis	Transverse processes of superior thoracic vertebrae	Transverse processes of middle and superior cervical vertebrae	As above	As above
Longissimus thoracis	Broad aponeurosis and transverse processes of inferior thoracic and superior lumbar vertebrae; joins iliocostalis	Transverse processes of superior vertebrae and inferior surfaces of ribs	Extends vertebral column; alone, each produces lateral flexion to that side	Thoracic and lumbar spinal nerves
Iliocostalis group				
Iliocostalis cervicis	Superior borders of vertebrosternal ribs near the angles	Transverse processes of middle and inferior cervical vertebrae	Extends neck, elevates ribs	Cervical and superior thoracic spinal nerves
Iliocostalis thoracis	Superior borders of inferior 7 ribs medial to the angles	Upper ribs and transverse process of last cervical vertebra	Stabilizes thoracic vertebrae in extension	Thoracic spinal nerves
Iliocostalis lumborum	Iliac crest, sacral crests, and spinous processes	Inferior surfaces of inferior 7 ribs near their angles	Extends vertebral column, depresses ribs	Inferior thoracic and lumbar spinal nerves

TABLE 11-8 Muscles of the Vertebral Column *(continued)*

Group/Muscle	Origin	Insertion	Action	Innervation
DEEP LAYER (TRANSVERSOSPINALIS)				
Semispinalis group				
Semispinalis capitis	Processes of inferior cervical and superior thoracic vertebrae	Occipital bone, between nuchal lines	Together, the two sides extend head; alone, each extends and laterally flexes neck	Cervical spinal nerves
Semispinalis cervicis	Transverse processes of $T_1–T_5/T_6$	Spinous processes of $C_2–C_5$	Extends vertebral column and rotates toward opposite side	As above
Semispinalis thoracis	Transverse processes of $T_6–T_{10}$	Spinous processes of $C_5–T_4$	As above	Thoracic spinal nerves
Multifidus	Sacrum and transverse processes of each vertebra	Spinous processes of the third or fourth more-superior vertebrae	As above	Cervical, thoracic, and lumbar spinal nerves
Rotatores	Transverse processes of each vertebra	Spinous processes of adjacent, more-superior vertebra	As above	As above
Interspinales	Spinous processes of each vertebra	Spinous processes of preceding vertebra	Extends vertebral column	As above
Intertransversarii	Transverse processes of each vertebra	Transverse process of preceding vertebra	Laterally flexes the vertebral column	As above
SPINAL FLEXORS				
Longus capitis	Transverse processes of cervical vertebrae	Base of the occipital bone	Together, the two sides bend head forward; alone, each rotates head to that side	Cervical spinal nerves
Longus colli	Anterior surfaces of cervical and superior thoracic vertebrae	Transverse processes of superior cervical vertebrae	Flexes and/or rotates neck; limits hyperextension	As above
Quadratus lumborum	Iliac crest and iliolumbar ligament	Last rib and transverse processes of lumbar vertebrae	Together, they depress ribs; alone, each side laterally flexes vertebral column	Thoracic and lumbar spinal nerves

If injured, these muscles can start a cycle of pain → muscle stimulation → contraction → pain. Pressure on adjacent spinal nerves results, leading to sensory losses as well as limiting mobility. Many of the warm-up and stretching exercises recommended before athletic events are intended to prepare these small but very important muscles for their supporting role.

The muscles of the vertebral column include many dorsal extensors but few ventral flexors. The vertebral column does not need a massive series of flexor muscles, because (1) many of the large trunk muscles flex the vertebral column when they contract, and (2) most of the body weight lies anterior to the vertebral column, so gravity tends to flex the spine. However, a few spinal flexors are associated with the anterior surface of the vertebral column. In the neck, the **longus capitis** and the **longus colli** muscles rotate or flex the neck, depending on whether the muscles of one or both sides are contracting (Figure 11-11c•). In the lumbar region, the large **quadratus lumborum** muscles flex the vertebral column and depress the ribs (Figure 11-11a•).

OBLIQUE AND RECTUS MUSCLES

The oblique and rectus muscles lie within the body wall, between the spinous processes of vertebrae and the ventral midline (Figures 11-3, pp. 318–319, and 11-12• and Table 11-9, p. 332). The oblique muscles compress underlying structures or rotate the vertebral column, depending on whether one or both sides contract. The rectus muscles are important flexors of the vertebral column, acting in opposition to the erector spinae. The oblique and rectus muscles share embryological origins. We can divide them into cervical, thoracic, and abdominal groups.

The oblique group includes the **scalene** muscles of the neck (Figure 11-11c•) and the **intercostal** and **transversus** muscles of the thorax (Figure 11-12a,b•). The scalene muscles (*anterior, middle,* and *posterior*) elevate the first two ribs and assist in flexion of the neck. In the thorax, the oblique muscles extend between the ribs, and the **external intercostal** muscles cover the **internal intercostal** muscles. Both groups of intercostal muscles aid in respiratory movements of the ribs. A small **transversus thoracis** muscle crosses the inner surface of the rib cage and is covered by the *pleura,* the serous membrane that lines the pleural cavities. ∞ *p. 22* The sternum occupies the place where we might expect thoracic rectus muscles to be.

The same basic pattern of musculature extends unbroken across the abdominopelvic surface (Figure 11-12a,c•). Here the muscles are called the **external abdominal oblique, internal abdominal oblique, transversus abdominis,** and **rectus abdominis**

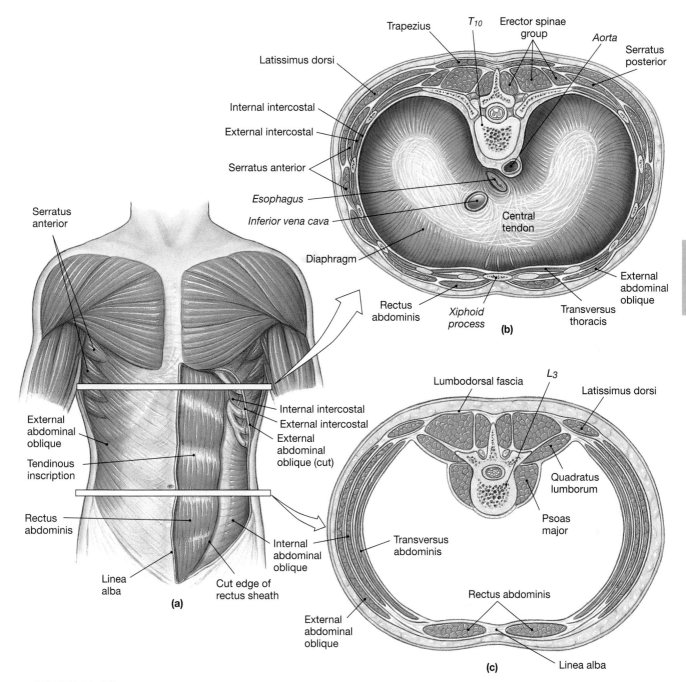

•**FIGURE 11-12 Oblique and Rectus Muscles and the Diaphragm.** (a) An anterior view. (b) A sectional view at the level of the diaphragm. (c) A sectional view at the level of the umbilicus. *See also Figures 7-1a and 7-22, pp. 194, 220.* AM *Plate 7.9*

11

muscles. The rectus abdominis muscle inserts at the xiphoid process and originates near the pubic symphysis. This muscle is longitudinally divided by the **linea alba** (white line), a median collagenous partition (Figure 11-3a•, p. 318). The rectus abdominis muscle is separated into segments by transverse bands of collagen fibers called **tendinous inscriptions.** Each segment contains muscle fibers that extend longitudinally, originating and inserting on the tendinous inscriptions.

The Diaphragm

The term *diaphragm* refers to any muscular sheet that forms a wall. When used without a modifier, however, **diaphragm,** or *diaphragmatic muscle,* specifies the muscular partition that separates the abdominopelvic and thoracic cavities (Figure 11-12b•). We include this muscle here because it develops in association with the other muscles of the chest wall. The diaphragm is a major muscle of respiration.

TABLE 11-9 **Oblique and Rectus Muscles*** *(Figure 11-12)*

Group/Muscle	Origin	Insertion	Action	Innervation
OBLIQUE GROUP				
Cervical region				
Scalenes	Transverse and costal processes of cervical vertebrae	Superior surfaces of first two ribs	Elevates ribs and/or flexes neck	Cervical spinal nerves
Thoracic region				
External intercostals	Inferior border of each rib	Superior border of more-inferior rib	Elevate ribs	Intercostal nerves (branches of thoracic spinal nerves)
Internal intercostals	Superior border of each rib	Inferior border of the preceding rib	Depress ribs	As above
Transversus thoracis	Medial surface of sternum	Cartilages of ribs	As above	As above
Abdominal region				
External abdominal oblique	Lower eight ribs	Linea alba and iliac crest	Compresses abdomen, depresses ribs, flexes or bends spine	Intercostal, iliohypogastric, and ilioinguinal nerves
Internal abdominal oblique	Lumbodorsal fascia and iliac crest	Inferior ribs, xiphoid process, and linea alba	As above	As above
Transversus abdominis	Cartilages of inferior ribs, iliac crest, and lumbodorsal fascia	Linea alba and pubis	Compresses abdomen	As above
RECTUS GROUP				
Cervical region	*See muscles in Table 11-6*			
Thoracic region				
Diaphragm	Xiphoid process, cartilages of ribs 4–10, and anterior surfaces of lumbar vertebrae	Central tendinous sheet	Contraction expands thoracic cavity, compresses abdominopelvic cavity	Phrenic nerves (C_3–C_5)
Abdominal region				
Rectus abdominis	Superior surface of pubis around symphysis	Inferior surfaces of costal cartilages (ribs 5–7) and xiphoid process	Depresses ribs, flexes vertebral column	Intercostal nerves (T_7–T_{12})

* Where appropriate, spinal nerves involved are given in parentheses.

MUSCLES OF THE PELVIC FLOOR

The muscles of the pelvic floor extend from the sacrum and coccyx to the ischium and pubis. These muscles (1) support the organs of the pelvic cavity, (2) flex the sacrum and coccyx, and (3) control the movement of materials through the urethra and anus (Figure 11-13• and Table 11-10, p. 334).

The boundaries of the **perineum,** the muscular sheet that forms the pelvic floor, are established by the inferior margins of the pelvis. If you draw a line between the ischial tuberosities, you will divide the perineum into two triangles—an anterior **urogenital triangle** and a posterior **anal triangle.** The superficial muscles of the urogenital triangle are the muscles of the external genitalia. They cover deeper muscles that strengthen the pelvic floor and encircle the urethra. These muscles constitute the **urogenital diaphragm,** a deep muscular layer that extends between the pubic bones.

An even more extensive muscular sheet, the **pelvic diaphragm,** forms the muscular foundation of the anal triangle. This layer, covered by the urogenital diaphragm, extends as far as the pubic symphysis.

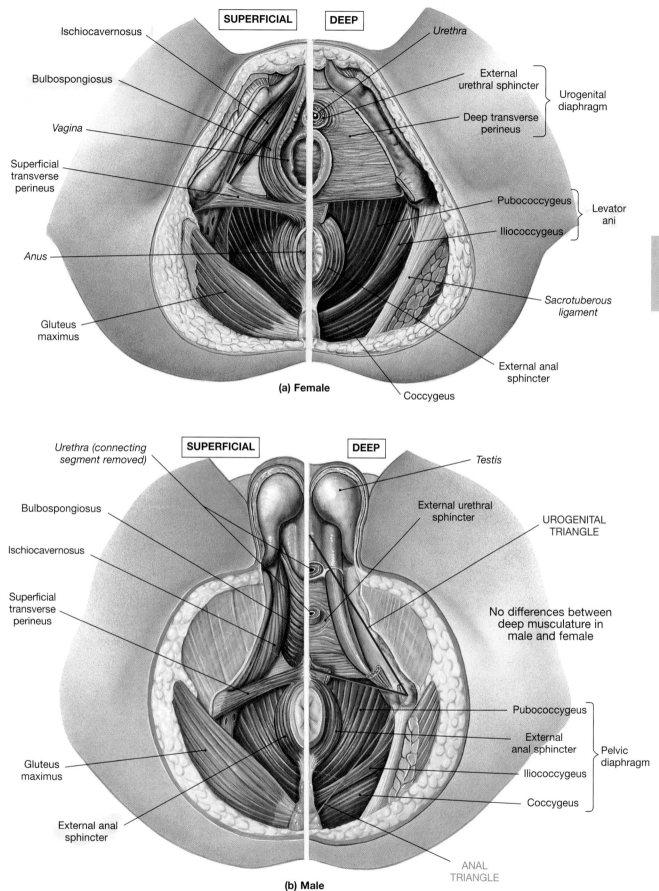

SUPERFICIAL | DEEP

Ischiocavernosus
Bulbospongiosus
Vagina
Superficial transverse perineus
Anus
Gluteus maximus

Urethra
External urethral sphincter ⎤ Urogenital
Deep transverse perineus ⎦ diaphragm
Pubococcygeus ⎤ Levator
Iliococcygeus ⎦ ani
Sacrotuberous ligament
External anal sphincter
Coccygeus

(a) Female

SUPERFICIAL | DEEP

Urethra (connecting segment removed)
Bulbospongiosus
Ischiocavernosus
Superficial transverse perineus
Gluteus maximus
External anal sphincter

Testis
External urethral sphincter
UROGENITAL TRIANGLE
No differences between deep musculature in male and female
Pubococcygeus
External anal sphincter ⎤
Iliococcygeus ⎥ Pelvic
Coccygeus ⎦ diaphragm
ANAL TRIANGLE

(b) Male

•**FIGURE 11-13** **Muscles of the Pelvic Floor.** *See also Figures 7-1a, 8-8, and 8-9, pp. 194, 235, 236.*

11

TABLE 11-10 Muscles of the Pelvic Floor *(Figure 11-13)*

Group/Muscle	Origin	Insertion	Action	Innervation
UROGENITAL TRIANGLE				
Superficial muscles				
Bulbospongiosus:				
Males	Collagen sheath at base of penis; fibers cross over urethra	Median raphe and central tendon of perineum	Compresses base and stiffens penis; ejects urine or semen	Pudendal nerve, perineal branch (S_2–S_4)
Females	Collagen sheath at base of clitoris; fibers run on either side of urethral and vaginal opening	Central tendon of perineum	Compresses and stiffens clitoris; narrows vaginal opening	As above
Ischiocavernosus	Ischial ramus and tuberosity	Pubic symphysis anterior to base of penis or clitoris	Compresses and stiffens penis or clitoris	As above
Superficial transverse perineus	Ischial ramus	Central tendon of perineum	Stabilizes central tendon of perineum	As above
Deep muscles:				
Urogenital diaphragm				
Deep transverse perineus	Ischial ramus	Median raphe of urogenital diaphragm	As above	As above
External urethral sphincter:				
Males	Ischial and pubic rami	To median raphe at base of penis; inner fibers encircle urethra	Closes urethra; compresses prostate and bulbourethral glands	As above
Females	Ischial and pubic rami	To median raphe; inner fibers encircle urethra	Closes urethra; compresses vagina and greater vestibular glands	As above
ANAL TRIANGLE				
Pelvic diaphragm				
Coccygeus	Ischial spine	Lateral, inferior borders of sacrum	Flexes coccygeal joints	Inferior sacral nerves (S_4–S_5)
External anal sphincter	Via tendon from coccyx	Encircles anal opening	Closes anal opening	Pudendal nerve, hemorrhoidal branch (S_2–S_4)
Levator ani:				
Iliococcygeus	Ischial spine, pubis	Coccyx and median raphe	Tenses floor of pelvis; flexes coccygeal joints; elevates and retracts anus	Pudendal nerve (S_2–S_4)
Pubococcygeus	Inner margins of pubis	As above	As above	As above

The urogenital and pelvic diaphragms do not completely close the pelvic outlet, for the urethra, vagina, and anus pass through them to open on the external surface. Muscular sphincters surround their openings and permit voluntary control of urination and defecation. Muscles, nerves, and blood vessels also pass through the pelvic outlet as they travel to or from the lower limbs.

☑ Damage to the external intercostal muscles would interfere with what important process?

☑ If someone hit you in your rectus abdominis muscle, how would your body position change?

☑ After spending an afternoon carrying heavy boxes from his basement to his attic, Joe complains that the muscles in his back hurt. Which muscle(s) is (are) most likely sore?

Hernias

When your abdominal muscles contract forcefully, pressures in your abdominopelvic cavity can skyrocket, and those pressures are applied to internal organs. If you exhale at the same time, the pressure is relieved, because your diaphragm can move upward as your lungs collapse. But during vigorous isometric exercises or when you lift a weight while holding your breath, pressure in the abdominopelvic cavity can rise high enough to cause a variety of problems, among them a hernia. A **hernia** develops when an organ protrudes through an abnormal opening. The most common hernias are inguinal and diaphragmatic hernias. *Inguinal*

hernias typically occur in males, at the *inguinal canal,* the site where blood vessels, nerves, and reproductive ducts pass through the abdominal wall to reach the testes. Elevated abdominal pressure can force open the inguinal canal and push a portion of the intestine into the pocket created. *Diaphragmatic hernias* develop when visceral organs, such as a portion of the stomach, are forced into the left pleural cavity. If herniated structures become trapped or twisted, surgery may be required to prevent serious complications, such as intestinal blockage or tissue degeneration due to the interruption of blood flow. AM *Hernias*

11

THE APPENDICULAR MUSCLES

The appendicular musculature positions and stabilizes the pectoral and pelvic girdles and moves the upper and lower limbs. There are two major groups of appendicular muscles: (1) *the muscles of the shoulders and upper limbs* and (2) *the muscles of the pelvic girdle and lower limbs.* The functions and required ranges of motion are very different from one group to another. In addition to increasing the mobility of the arms, the muscular connections between the pectoral girdle and the axial skeleton must act as shock absorbers. For example, while you jog, you can still perform delicate hand movements, because the muscular connections between the axial and appendicular skeletons smooth out the bounces in your stride. In contrast, the pelvic girdle has evolved to transfer weight from the axial to the appendicular skeleton. A muscular connection would reduce the efficiency of the transfer, and the emphasis is on strength rather than versatility. Figure 11-14• provides an introduction to the organization of the appendicular muscles.

MUSCLES OF THE SHOULDERS AND UPPER LIMBS

Muscles associated with the shoulders and upper limbs can be divided into four groups: (1) *muscles that position the pectoral girdle,* (2) *muscles that move the arm,* (3) *muscles that move the forearm and hand,* and (4) *muscles that move the hand and fingers.*

Muscles That Position the Pectoral Girdle

The large, superficial **trapezius** muscles cover the back and portions of the neck, reaching to the base of the skull. These muscles originate along the midline of the neck and back and insert on the clavicles and the scapular spines (Figures 11-14 and 11-15a•, p. 338). The trapezius muscles are innervated by more than one nerve (Table 11-11), and specific regions can be made to contract independently. As a result, their actions are quite varied.

Removing the trapezius muscle reveals the **rhomboideus** and **levator scapulae** muscles (Figure 11-15a•). These muscles are attached to the dorsal surfaces of the cervical and thoracic vertebrae. They insert along the vertebral border of each scapula, between the superior and inferior angles. Contraction of a rhomboideus muscle adducts (retracts) the scapula on that side. The levator scapulae muscle, as its name implies, elevates the scapula.

On the chest, the **serratus anterior** muscle originates along the anterior surfaces of several ribs (Figures 11-3, pp. 318–319, and 11-15a,b•). This fan-shaped muscle inserts along the anterior margin of the vertebral border of the scapula. When the serratus anterior muscle contracts, it abducts (protracts) the scapula and swings the shoulder anteriorly.

Two other deep chest muscles arise along the ventral surfaces of the ribs on either side. The **subclavius** (sub-KLĀ-vē-us; *sub-,* below + *clavius,* clavicle) muscle inserts on the inferior border of the clavicle (Figure 11-15b•). When it contracts, it depresses and protracts the scapular end of the clavicle. Because ligaments connect this end to the shoulder joint and scapula, those structures move as well. The **pectoralis** (pek-to-RA-lis) **minor** muscle attaches to the coracoid process of the scapula. The contraction of this muscle generally complements that of the subclavius muscle.

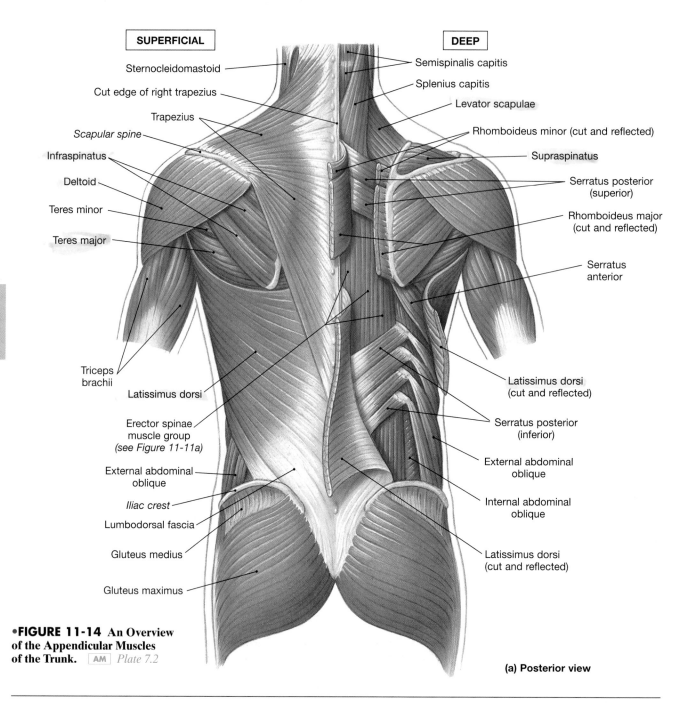

SUPERFICIAL

DEEP

Sternocleidomastoid

Cut edge of right trapezius

Trapezius

Scapular spine

Infraspinatus

Deltoid

Teres minor

Teres major

Triceps brachii

Latissimus dorsi

Erector spinae muscle group *(see Figure 11-11a)*

External abdominal oblique

Iliac crest

Lumbodorsal fascia

Gluteus medius

Gluteus maximus

Semispinalis capitis

Splenius capitis

Levator scapulae

Rhomboideus minor (cut and reflected)

Supraspinatus

Serratus posterior (superior)

Rhomboideus major (cut and reflected)

Serratus anterior

Latissimus dorsi (cut and reflected)

Serratus posterior (inferior)

External abdominal oblique

Internal abdominal oblique

Latissimus dorsi (cut and reflected)

•**FIGURE 11-14** An Overview of the Appendicular Muscles of the Trunk. **AM** *Plate 7.2*

(a) Posterior view

TABLE 11-11 Muscles That Position the Pectoral Girdle *(Figures 11-14, 11-15)*

Muscle	Origin	Insertion	Action	Innervation
Levator scapulae	Transverse processes of first 4 cervical vertebrae	Vertebral border of scapula near superior angle	Elevates scapula	Cervical nerves C_3–C_4 and dorsal scapular nerve (C_5)
Pectoralis minor	Ventral surfaces of ribs 3–5	Coracoid process of scapula	Depresses and protracts shoulder; rotates scapula so glenoid cavity moves inferiorly (downward rotation); elevates ribs if scapula is stationary	Medial pectoral nerve (C_8, T_1)
Rhomboideus major	Spinous processes of superior thoracic vertebrae	Vertebral border of scapula from spine to inferior angle	Adducts scapula and performs downward rotation	Dorsal scapular nerve (C_5)
Rhomboideus minor	Spinous processes of vertebrae C_7–T_1	Vertebral border of scapula near spine	As above	As above

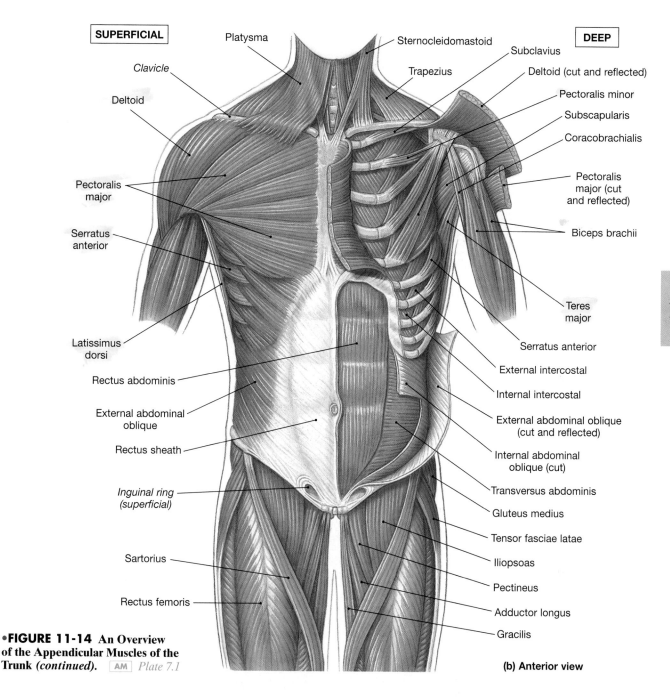

SUPERFICIAL

DEEP

Platysma
Clavicle
Deltoid
Pectoralis major
Serratus anterior
Latissimus dorsi
Rectus abdominis
External abdominal oblique
Rectus sheath
Inguinal ring (superficial)
Sartorius
Rectus femoris

Sternocleidomastoid
Trapezius
Subclavius
Deltoid (cut and reflected)
Pectoralis minor
Subscapularis
Coracobrachialis
Pectoralis major (cut and reflected)
Biceps brachii
Teres major
Serratus anterior
External intercostal
Internal intercostal
External abdominal oblique (cut and reflected)
Internal abdominal oblique (cut)
Transversus abdominis
Gluteus medius
Tensor fasciae latae
Iliopsoas
Pectineus
Adductor longus
Gracilis

11

•FIGURE 11-14 An Overview of the Appendicular Muscles of the Trunk *(continued)*. AM *Plate 7.1*

(b) Anterior view

TABLE 11-11 Muscles That Position the Pectoral Girdle *(continued)*

Muscle	Origin	Insertion	Action	Innervation
Serratus anterior	Anterior and superior margins of ribs 1–9	Anterior surface of vertebral border of scapula	Protracts shoulder; rotates scapula so glenoid cavity moves superiorly (upward rotation)	Long thoracic nerve (C_5–C_7)
Subclavius	First rib	Clavicle (inferior border)	Depresses and protracts shoulder	Subclavian nerve
Trapezius	Occipital bone, ligamentum nuchae, and spinous processes of thoracic vertebrae	Clavicle and scapula (acromion and scapular spine)	Depends on active region and state of other muscles; may elevate, retract, depress, or rotate scapula upward and/or elevate clavicle or extend neck	Accessory nerve (N XI) and cervical spinal nerves (C_3–C_4)

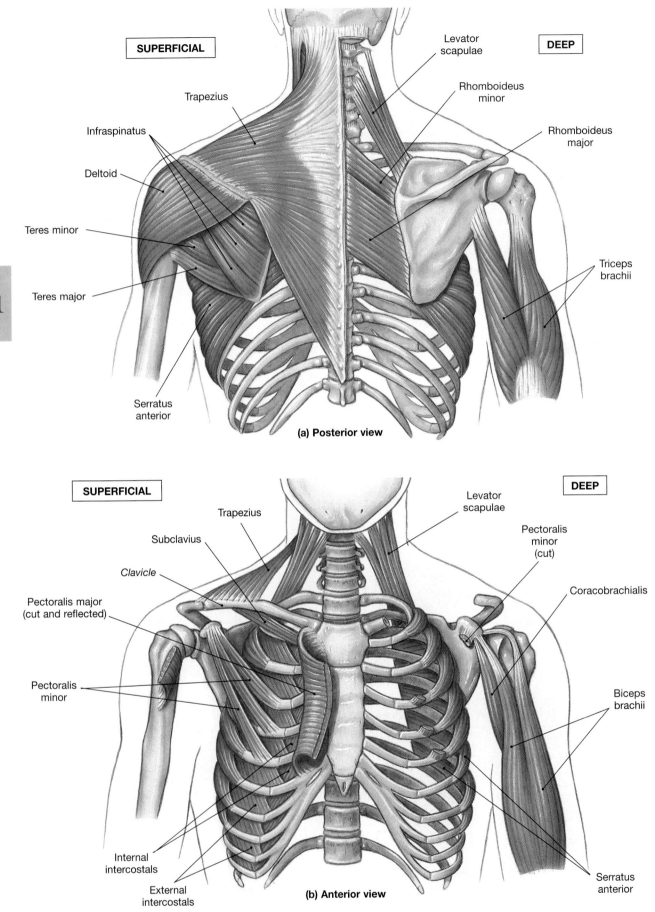

SUPERFICIAL

Trapezius

Infraspinatus

Deltoid

Teres minor

Teres major

Serratus
anterior

Levator
scapulae

DEEP

Rhomboideus
minor

Rhomboideus
major

Triceps
brachii

(a) Posterior view

SUPERFICIAL

Trapezius

Subclavius

Clavicle

Pectoralis major
(cut and reflected)

Pectoralis
minor

Internal
intercostals

External
intercostals

Levator
scapulae

DEEP

Pectoralis
minor
(cut)

Coracobrachialis

Biceps
brachii

Serratus
anterior

(b) Anterior view

•**FIGURE 11-15** **Muscles That Position the Pectoral Girdle.** *See also Figures 8-2, 8-3, 8-4, and 9-9, pp. 229, 231, 259.*
AM *Plates 6.1, 7.1–7.3*

Muscles That Move the Arm

The muscles that move the arm (Figures 11-14–11-16•) are easiest to remember when grouped by their actions at the shoulder joint (Table 11-12). The **deltoid** muscle is the major abductor, but the **supraspinatus** (soo-pra-spī-NA-tus) muscle assists at the start of this movement. The **subscapularis** and **teres major** muscles produce medial rotation at the shoulder, whereas the **infraspinatus** and the **teres minor** muscles produce lateral rotation. All these muscles originate on the scapula. The small **coracobrachialis** (KOR-uh-kō-brā-kē-A-lis) muscle is the only muscle attached to the scapula that produces flexion and adduction at the shoulder.

The **pectoralis major** muscle extends between the anterior portion of the chest and the crest of the greater tubercle of the humerus. The **latissimus dorsi** (la-TIS-i-mus DOR-sē) muscle extends between the thoracic vertebrae at the posterior midline and the intertubercular groove of the humerus (Figure 11-16b•). The pectoralis major muscle produces flexion at the shoulder joint, and the latissimus dorsi muscle produces extension. These two muscles can also work together to produce adduction and medial rotation of the humerus at the shoulder.

Collectively, the supraspinatus, infraspinatus, subscapularis, and teres minor muscles and their associated tendons form the **rotator cuff,** a common site of sports injuries.

Sports Injuries

Exercise carries risks due to the stresses placed on muscles, joints, and connective tissues. Many of us participate in exercise programs and sports on a regular basis. More than 30 million people jog in the United States, and millions more participate in various amateur and professional sports. As a result, *sports injuries* are very common, and sports medicine has become an active area of professional and academic research interest. AM *Sports Injuries*

Muscles That Move the Forearm and Hand

Although most of the muscles that insert on the forearm and hand originate on the humerus, the biceps brachii and triceps brachii muscles are noteworthy exceptions. The **biceps brachii** muscle and the *long head* of the

11

TABLE 11-12 **Muscles That Move the Arm** *(Figures 11-14 to 11-16)*

Muscle	Origin	Insertion	Action	Innervation
Coracobrachialis	Coracoid process	Medial margin of shaft of humerus	Adduction and flexion at shoulder	Musculocutaneous nerve (C_5–C_7)
Deltoid	Clavicle and scapula (acromion and adjacent scapular spine)	Deltoid tuberosity of humerus	Abduction at shoulder	Axillary nerve (C_5–C_6)
Supraspinatus	Supraspinous fossa of scapula	Greater tubercle of humerus	As above	Suprascapular nerve (C_5)
Infraspinatus	Infraspinous fossa of scapula	Greater tubercle of humerus	Lateral rotation at shoulder	Suprascapular nerve (C_5–C_6)
Subscapularis	Subscapular fossa of scapula	Lesser tubercle of humerus	Medial rotation at shoulder	Subscapular nerves (C_5–C_6)
Teres major	Inferior angle of scapula	Medial lip of intertubercular groove of humerus	Extension, adduction, and medial rotation at shoulder	Lower subscapular nerve (C_5–C_6)
Teres minor	Lateral border of scapula	Greater tubercle of humerus	Lateral rotation at shoulder	Axillary nerve (C_5)
Triceps brachii (long head)	*See Table 11-13*			
Latissimus dorsi	Spinous processes of inferior thoracic and all lumbar vertebrae, ribs 8–12, and lumbodorsal fascia	Floor of intertubercular groove of the humerus	Extension, adduction, and medial rotation at shoulder	Thoracodorsal nerve (C_6–C_8)
Pectoralis major	Cartilages of ribs 2–6, body of sternum, and inferior, medial portion of clavicle	Crest of greater tubercle and lateral lip of intertubercular groove of humerus	Flexion, adduction, and medial rotation at shoulder	Pectoral nerves (C_5–T_1)

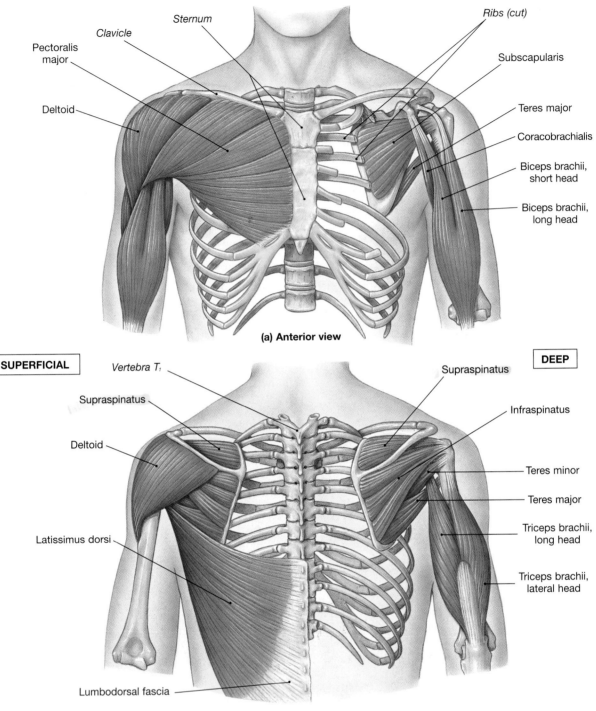

(a) Anterior view

SUPERFICIAL

DEEP

(b) Posterior view

•**FIGURE 11-16** **Muscles That Move the Arm.** *See also Figures 7-22, 8-3, and 9-9, pp. 220, 229, 259.* ⟨AM⟩ *Plates 6.1, 7.1–7.3*

triceps brachii muscle originate on the scapula and insert on the bones of the forearm (Figure 11-17•). The triceps brachii muscle inserts on the olecranon. Contraction of the triceps brachii muscle extends the elbow, as when you do push-ups. The biceps brachii muscle inserts on the radial tuberosity, a roughened area on the anterior surface of the radius. ∞ *p. 231* Contraction of the biceps brachii muscle flexes the elbow and supinates the forearm. With the forearm pronated (palm

facing back), the biceps brachii muscle cannot function effectively. As a result, you are strongest when you flex your elbow with a supinated forearm; the biceps brachii muscle then makes a prominent bulge.

More muscles are shown in Figure 11-17• and listed in Table 11-13 (p. 342). The **brachialis** and **brachioradialis** (brā-kē-ō-rā-dē-A-lis) muscles flex the elbow and are opposed by the **anconeus** muscle and the triceps brachii muscle, respectively.

•FIGURE 11-17 Muscles That Move the Forearm and Hand. Superficial muscles are shown in (a) posterior and (b) anterior views. Deeper muscles are shown in the sectional views and in Figure 11-18. *See also Figures 8-4, 8-5, and 9-10, pp. 231, 232, 260.*
AM *Plates 6.1–6.5*

Posterior

Triceps brachii

lateral head   medial head   long head

Humerus

Brachialis

Artery

Vein

Nerve

Biceps brachii

Anterior

Humerus

Biceps brachii, long head

Coracobrachialis

Biceps brachii, short head

Triceps brachii, long head

Triceps brachii, medial head

Brachialis

Pronator teres

Brachioradialis

Flexor carpi radialis

Flexor digitorum superficialis

Palmaris longus

Flexor carpi ulnaris

Flexor retinaculum

Pronator quadratus

(b) Anterior view, superficial layer

Triceps brachii, long head

Triceps brachii, lateral head

Brachioradialis

Anconeus

Extensor carpi radialis longus

Olecranon of ulna

Extensor carpi radialis brevis

Radius

Flexor carpi ulnaris

Extensor carpi ulnaris

Extensor digitorum

Ulna

Abductor pollicis longus

Extensor retinaculum

(a) Posterior view, superficial layer

Anterior

Palmaris longus   Flexor pollicis longus   Flexor carpi radialis

Flexor digitorum superficialis

Brachioradialis

Radius

Flexor carpi ulnaris

Extensor carpi radialis longus

Extensor carpi radialis brevis

Flexor digitorum profundus

Abductor pollicis longus

Ulna

Extensor digitorum

Extensor carpi ulnaris

Extensor pollicis

Extensor digiti minimi

Posterior

11

341

TABLE 11-13 Muscles That Move the Forearm and Hand *(Figure 11-17)*

Muscle	Origin	Insertion	Action	Innervation
ACTION AT THE ELBOW				
Flexors				
Biceps brachii	*Short head* from the coracoid process; *long head* from the supraglenoid tubercle (both on the scapula)	Tuberosity of radius	Flexion at elbow and shoulder; supination	Musculo-cutaneous nerve (C_5–C_6)
Brachialis	Anterior, distal surface of humerus	Tuberosity of ulna	Flexion at elbow	As above and radial nerve (C_7–C_8)
Brachioradialis	Lateral epicondyle of humerus	Lateral aspect of styloid process of radius	As above	Radial nerve (C_5–C_6)
Extensors				
Anconeus	Posterior surface of lateral epicondyle of humerus	Lateral margin of olecranon on ulna	Extension at elbow	Radial nerve (C_7–C_8)
Triceps brachii lateral head	Superior, lateral margin of humerus	Olecranon of ulna	As above	As above
long head	Infraglenoid tubercle of scapula	As above	As above	As above
medial head	Posterior surface of humerus inferior to radial groove	As above	As above	As above
PRONATORS/SUPINATORS				
Pronator quadratus	Medial surface of distal portion of ulna	Anterolateral surface of distal portion of radius	Pronation	Median nerve (C_8–T_1)
Pronator teres	Medial epicondyle of humerus and coronoid process of ulna	Distal lateral surface of radius	As above	Median nerve (C_6–C_7)
Supinator	Lateral epicondyle of humerus and ridge of proximal ulna	Anterolateral surface of radius distal to the radial tuberosity	Supination	Deep radial nerve (C_6–C_8)
ACTION AT THE HAND				
Flexors				
Flexor carpi radialis	Medial epicondyle of humerus	Bases of 2nd and 3rd metacarpal bones	Flexion and abduction at wrist	Median nerve (C_6–C_7)
Flexor carpi ulnaris	Medial epicondyle of humerus; adjacent medial surface of olecranon and anteromedial portion of ulna	Pisiform bone, hamate bone, and base of 5th metacarpal bone	Flexion and adduction at wrist	Ulnar nerve (C_8–T_1)
Palmaris longus	Medial epicondyle of humerus	Palmar aponeurosis and flexor retinaculum	Flexion at wrist	Median nerve (C_6–C_7)
Extensors				
Extensor carpi radialis longus	Lateral supracondylar ridge of humerus	Base of 2nd metacarpal bone	Extension and abduction at wrist	Radial nerve (C_6–C_7)
Extensor carpi radialis brevis	Lateral epicondyle of humerus	Base of 3rd metacarpal bone	As above	As above
Extensor carpi ulnaris	Lateral epicondyle of humerus; adjacent dorsal surface of ulna	Base of 5th metacarpal bone	Extension and adduction at wrist	Deep radial nerve (C_6–C_8)

The **flexor carpi ulnaris, flexor carpi radialis,** and **palmaris longus** muscles are superficial muscles that work together to produce flexion of the wrist. The flexor carpi radialis muscle flexes and *ab*ducts, and the flexor carpi ulnaris muscle flexes and *ad*ducts. *Pitcher's arm* is an inflammation at the origins of the flexor carpi muscles at the medial epicondyle of the humerus. This condition results from forcibly flexing the wrist just before releasing a baseball.

The **extensor carpi radialis** muscles and the **extensor carpi ulnaris** muscle have a similar relationship to that between the flexor carpi muscles. The extensor carpi radialis muscles produce extension and *ab*duction, whereas the extensor carpi ulnaris muscle produces extension and *ad*duction.

The **pronator teres** and **supinator** muscles originates on both the humerus and ulna. These muscles rotate the radius without either flexing or extending the elbow. The **pronator quadratus** muscle originates on the ulna and assists the pronator teres muscle in opposing the actions of the supinator or biceps brachii muscles. The muscles involved in pronation and supination are shown in Figure 11-18•. During pronation, the tendon of the biceps brachii muscle rotates with the radius. As a result, this muscle cannot assist in flexion of the elbow when the forearm is pronated.

As you study the muscles included in Table 11-13, notice that, in general, the extensor muscles lie along the posterior and lateral surfaces of the arm, whereas the flexors are on the anterior and medial surfaces.

Muscles That Move the Hand and Fingers

Several superficial and deep muscles of the forearm flex and extend the finger joints (Figure 11-18• and Table 11-14). These muscles stop before reaching the wrist, and only their tendons cross the articulation. The muscles are relatively large, and keeping them clear of the joints ensures maximum mobility at both the wrist and hand. The tendons that cross the dorsal and ventral surfaces of the wrist pass through **synovial tendon sheaths,** elongate bursae that reduce friction. ∞ *p. 249*

The muscles of the forearm provide strength and crude control of the hand and fingers. These muscles are known as

(a) Anterior view, middle layer

(b) Anterior view, deepest layer

(c) Posterior view, middle layer

(d) Posterior view, deepest layer

•**FIGURE 11-18 Muscles That Move the Hand and Fingers.** Middle and deep muscle layers of the right forearm; for superficial muscles, see Figure 11-17. *See also Figure 8-5, p. 232.* [AM] *Plates 6.3–6.5*

TABLE 11-14 **Muscles That Move the Hand and Fingers** *(Figure 11-18)*

Muscle	Origin	Insertion	Action	Innervation
Abductor pollicis longus	Proximal dorsal surfaces of ulna and radius	Lateral margin of 1st metacarpal bone	Abduction at joints of thumb and wrist	Deep radial nerve (C_6–C_7)
Extensor digitorum	Lateral epicondyle of humerus	Posterior surfaces of the phalanges, fingers 2–5	Extension at finger joints and wrist	Deep radial nerve (C_6–C_8)
Extensor pollicis brevis	Shaft of radius distal to origin of adductor pollicis longus	Base of proximal phalanx of thumb	Extension at joints of thumb; abduction at wrist	Deep radial nerve (C_6–C_7)
Extensor pollicis longus	Posterior and lateral surfaces of ulna and interosseous membrane	Base of distal phalanx of thumb	As above	Deep radial nerve (C_6–C_8)
Extensor indicis	Posterior surface of ulna and interosseous membrane	Posterior surface of phalanges of index finger (2), with tendon of extensor digitorum	Extension and adduction at joints of index finger	As above
Extensor digiti minimi	Via extensor tendon to lateral epicondyle of humerus, and from intermuscular septa	Posterior surface of proximal phalanx of little finger (5)	Extension at joints of little finger	As above
Flexor digitorum superficialis	Medial epicondyle of humerus; adjacent anterior surfaces of ulna and radius	Midlateral surfaces of middle phalanges of fingers 2–5	Flexion at proximal interphalangeal, metacarpophalangeal, and wrist joints	Median nerve (C_7–T_1)
Flexor digitorum profundus	Medial and posterior surfaces of ulna, medial surface of coronoid process, and interosseus membrane	Bases of distal phalanges of fingers 2–5	Flexion at distal interphalangeal joints and, to lesser degree, proximal interphalangeal joints and wrist	Palmar interosseous nerve, from median nerve and ulnar nerve (C_8–T_1)
Flexor pollicis longus	Anterior shaft of radius, interosseous membrane	Base of distal phalanx of thumb	Flexion at joints of thumb	Median nerve (C_8–T_1)

the *extrinsic muscles of the hand*. Fine control of the hand involves small *intrinsic muscles,* which originate on the carpal and metacarpal bones. No muscles originate on the phalanges, and only tendons extend across the distal joints of the fingers. The intrinsic muscles of the hand are detailed in Figure 11-19• and Table 11-15 (p. 346).

The fascia of the forearm thickens on the posterior surface of the wrist, forming the **extensor retinaculum** (ret-i-NAK-ū-lum), a wide band of connective tissue. The extensor retinaculum holds the tendons of the extensor muscles in place. On the anterior surface, the fascia also thickens to form another wide band of connective tissue, the **flexor retinaculum,** which stabilizes the tendons of the flexor muscles. Inflammation of the retinacula and synovial tendon sheaths can restrict movement and irritate the *median nerve,* a mixed (sensory and motor) nerve that innervates the hand. This condition, known as *carpal tunnel syndrome,* causes chronic pain.

Carpal Tunnel Syndrome

Tenosynovitis is the inflammation of a tendon sheath. **Carpal tunnel syndrome** results from tenosynovitis of the synovial tendon sheath surrounding the flexor tendons of the palm. The inflammation leads to compression of the median nerve. Symptoms include pain, especially on palmar flexion, a tingling sensation or numbness on the palm, and weakness in the *abductor pollicis brevis.* This common condition often strikes persons engaged in repetitive hand movements, such as typing, working at a computer keyboard, or playing the piano. Treatment involves the administration of anti-inflammatory drugs such as aspirin, the injection of anti-inflammatory agents such as *glucocorticoids* (steroid hormones produced by the adrenal cortex), and the use of a splint to prevent wrist flexion and to stabilize the region.

Carpal tunnel syndrome is an example of a *cumulative trauma disorder,* or *overuse syndrome.* These disorders are caused by repetitive movements of the arms, hands, and fingers. Such musculoskeletal problems now account for over 50 percent of all work-related injuries in the United States.

•**FIGURE 11-19** Intrinsic Muscles of the Hand.
See also Figure 8-6, p. 233. AM *Plate 6.6*

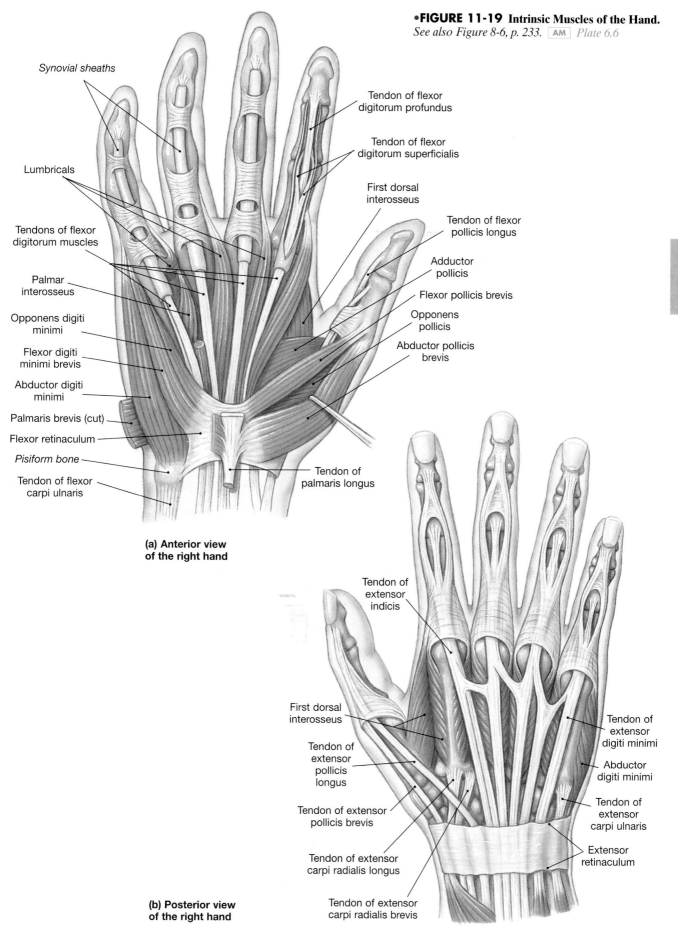

Synovial sheaths

Lumbricals

Tendons of flexor
digitorum muscles

Palmar
interosseus

Opponens digiti
minimi

Flexor digiti
minimi brevis

Abductor digiti
minimi

Palmaris brevis (cut)

Flexor retinaculum

Pisiform bone

Tendon of flexor
carpi ulnaris

Tendon of flexor
digitorum profundus

Tendon of flexor
digitorum superficialis

First dorsal
interosseus

Tendon of flexor
pollicis longus

Adductor
pollicis

Flexor pollicis brevis

Opponens
pollicis

Abductor pollicis
brevis

Tendon of
palmaris longus

**(a) Anterior view
of the right hand**

Tendon of
extensor
indicis

First dorsal
interosseus

Tendon of
extensor
pollicis
longus

Tendon of extensor
pollicis brevis

Tendon of extensor
carpi radialis longus

Tendon of extensor
carpi radialis brevis

Tendon of
extensor
digiti minimi

Abductor
digiti minimi

Tendon of
extensor
carpi ulnaris

Extensor
retinaculum

**(b) Posterior view
of the right hand**

TABLE 11-15 Intrinsic Muscles of the Hand *(Figure 11-19)*

Muscle	Origin	Insertion	Action	Innervation
Adductor pollicis	Metacarpal and carpal bones	Proximal phalanx of thumb	Adduction of thumb	Ulnar nerve, deep branch (C_8–T_1)
Opponens pollicis	Trapezium	First metacarpal bone	Opposition of thumb	Median nerve (C_6–C_7)
Palmaris brevis	Palmar aponeurosis	Skin of medial border of hand	Moves skin on medial border toward midline of palm	Ulnar nerve, superficial branch (C_8)
Abductor digiti minimi	Pisiform bone	Proximal phalanx of little finger	Abduction of little finger and flexion at its metacarpophalangeal joint	Ulnar nerve, deep branch (C_8–T_1)
Abductor pollicis brevis	Transverse carpal ligament, scaphoid bone, and trapezium	Radial side of base of proximal phalanx of thumb	Abduction of thumb	Median nerve (C_6–C_7)
Flexor pollicis brevis*	Flexor retinaculum, trapezium, capitate bone, and ulnar side of 1st metacarpal bone	Ulnar side of proximal phalanx of thumb	Flexion and adduction of thumb	Branches of median and ulnar nerves
Flexor digiti minimi brevis	Hamate bone	Proximal phalanx of little finger	Flexion at joints of little finger	Ulnar nerve, deep branch (C_8–T_1)
Opponens digiti minimi	As above	Fifth metacarpal bone	Opposition of 5th metacarpal bone	As above
Lumbricals (4)	Tendons of flexor digitorum profundus	Tendons of extensor digitorum	Flexion at metacarpophalangeal joints; extension at proximal and distal interphalangeal joints	No. 1 and no. 2 by median nerve; no. 3 and no. 4 by ulnar nerve, deep branch
Dorsal interossei (4)	Each originates from opposing faces of two metacarpal bones (I and II, II and III, III and IV, IV and V)	Bases of proximal phalanges of fingers 2–4	Abduction at metacarpophalangeal joints of fingers 2 and 4; flexion at metacarpophalangeal joints; extension at interphalangeal joints	Ulnar nerve, deep branch (C_8–T_1)
Palmar interossei (3–4)*	Sides of metacarpal bones II, IV, and V	Bases of proximal phalanges of fingers 2, 4, and 5	Adduction at metacarpophalangeal joints of fingers 2, 4, and 5; flexion at metacarpophalangeal joints; extension at interphalangeal joints	As above

*The portion of the flexor pollicis brevis originating on the first metacarpal bone is sometimes called the *first palmar interosseus muscle.*

☑ Which muscle are you using when you shrug your shoulders?

☑ Baseball pitchers sometimes suffer from rotator cuff injuries. Which muscles are involved in this type of injury?

☑ Injury to the flexor carpi ulnaris muscle would impair which two movements?

MUSCLES OF THE PELVIS AND LOWER LIMBS

The pelvic girdle is tightly bound to the axial skeleton, and little relative movement is permitted. In our discussion of the axial musculature, we therefore encountered few muscles that can influence the position of the pelvis. The muscles that position the lower limbs can be divided into three functional groups: (1) *muscles that move the thigh,* (2) *muscles that move the leg,* and (3) *muscles that move the foot and toes.*

Muscles That Move the Thigh

Table 11-16 lists the muscles that move the thigh. **Gluteal muscles** cover the lateral surfaces of the ilia

TABLE 11-16 **Muscles That Move the Thigh** *(Figure 11-20)*

Group/Muscle	Origin	Insertion	Action	Innervation
Gluteal group				
Gluteus maximus	Iliac crest, posterior gluteal line, and lateral surface of ilium; sacrum, coccyx, and lumbodorsal fascia	Iliotibial tract and gluteal tuberosity of femur	Extension and lateral rotation at hip	Inferior gluteal nerve (L_5–S_2)
Gluteus medius	Anterior iliac crest of ilium, lateral surface between posterior and anterior gluteal lines	Greater trochanter of femur	Abduction and medial rotation at hip	Superior gluteal nerve (L_4–S_1)
Gluteus minimus	Lateral surface of ilium between inferior and anterior gluteal lines	As above	As above	As above
Tensor fasciae latae	Iliac crest and lateral surface of anterior superior iliac spine	Iliotibial tract	Flexion and medial rotation at hip; tenses fascia lata, which laterally supports the knee	As above
Lateral rotator group				
Obturators (externus and internus)	Lateral and medial margins of obturator foramen	Trochanteric fossa of femur (externus); medial surface of greater trochanter (internus)	Lateral rotation at hip	Obturator nerve (externus: L_3–L_4) and special nerve from sacral plexus (internus: L_5–S_2)
Piriformis	Anterolateral surface of sacrum	Greater trochanter of femur	Lateral rotation and abduction at hip	Branches of sacral nerves (S_1–S_2)
Gemelli (superior and inferior)	Ischial spine and tuberosity	Medial surface of greater trochanter	Lateral rotation at hip	Nerves to obturator internus and quadratus femoris
Quadratus femoris	Lateral border of ischial tuberosity	Intertrochanteric crest of femur	As above	Special nerve from sacral plexus (L_4–S_1)
Adductor group				
Adductor brevis	Inferior ramus of pubis	Linea aspera of femur	Adduction, flexion, and medial rotation at hip	Obturator nerve (L_3–L_4)
Adductor longus	Inferior ramus of pubis anterior to adductor brevis	As above	As above	As above
Adductor magnus	Inferior ramus of pubis posterior to adductor brevis and ischial tuberosity	Linea aspera and adductor tubercle of femur	Abduction at hip; superior part produces flexion and medial rotation; inferior part produces extension and lateral rotation	Obturator and sciatic nerves
Pectineus	Superior ramus of pubis	Pectineal line inferior to lesser trochanter of femur	Flexion, medial rotation, and adduction at hip	Femoral nerve (L_2–L_4)
Gracilis	Inferior ramus of pubis	Medial surface of tibia inferior to medial condyle	Flexion at knee; adduction and medial rotation at hip	Obturator nerve (L_3–L_4)
Iliopsoas group				
Iliacus	Iliac fossa of ilium	Femur distal to lesser trochanter; tendon fused with that of psoas	Flexion at hip and/or lumbar intervertebral joints	Femoral nerve (L_2–L_3)
Psoas major	Anterior surfaces and transverse processes of vertebrae T_{12}–L_5	Lesser trochanter in company with iliacus	As above	Branches of the lumbar plexus (L_2–L_3)

11

(a) Lateral view, gluteal muscle group of the right side

Tensor fasciae latae

Gluteus maximus

Iliotibial tract

(b) Posterior view, gluteal muscle group

Gluteus medius (cut)

Gluteus minimus

Obturator internus

Gluteus medius

Gluteus maximus

(c) Posterior view, deep muscles

Gluteus maximus (cut)

Sacrum

Iliac crest

Gluteus medius (cut)

Gluteus minimus

Piriformis

Superior gemellus

Obturator internus

Tensor fasciae latae

Gluteus medius (cut)

Greater trochanter of femur

Quadratus femoris

Gluteus maximus (cut)

Inferior gemellus

Ischial tuberosity

Gracilis

Adductor magnus

Semitendinosus

Iliotibial tract

Adductor magnus

Biceps femoris (long head)

(d) The iliopsoas muscle and the adductor group

Iliopsoas

Iliacus Psoas major

Piriformis

Pectineus

Adductor brevis

Obturator externus

Adductor longus

Adductor magnus

Gracilis

Sartorius *(Table 11-17)*

•**FIGURE 11-20 Muscles That Move the Thigh.** *See also Figures 8-7, 8-8, 8-11, and 9-11, pp. 234, 235, 237, 261.* [AM] *Plates 7.10d,e, 8.1–8.6*

(Figures 11-14, pp. 336–337, and 11-20a,b,c•). The **gluteus maximus** muscle is the largest and most posterior of the gluteal muscles. Its origin includes parts of the ilium; the sacrum, coccyx, and associated ligaments; and the lumbodorsal fascia. Acting alone, this massive muscle produces extension and lateral rotation at the

hip joint. The gluteus maximus shares an insertion with the **tensor fasciae latae** (FASH-ē-ē LĀ-tā) muscle, which originates on the iliac crest and the anterior superior iliac spine. Together these muscles pull on the **iliotibial** (il-ē-ō-TIB-ē-ul) **tract,** a band of collagen fibers that extends along the lateral surface of the thigh and inserts on the tibia. This tract provides a lateral brace for the knee that becomes particularly important when you balance on one foot.

The **gluteus medius** and **gluteus minimus** muscles (Figure 11-20b,c•) originate anterior to the origin of the gluteus maximus muscle and insert on the greater trochanter of the femur. The anterior gluteal line on the lateral surface of the ilium marks the boundary between these muscles.

The **lateral rotators** originate at or inferior to the horizontal axis of the acetabulum. There are six lateral rotator muscles in all, of which the **piriformis** (pir-i-FOR-mis) muscle and the **obturator** muscles are dominant (Figure 11-20c,d•).

The **adductors** (Figure 11-20c,d•) originate inferior to the horizontal axis of the acetabulum. This muscle group includes the **adductor magnus, adductor brevis, adductor longus, pectineus** (pek-TIN-ē-us), and **gracilis** (GRAS-i-lis) muscles. All but the adductor magnus originate both anterior and inferior to the joint, so they perform hip flexion as well as adduction. The adductor magnus muscle can produce either adduction and flexion or adduction and extension, depending on the region stimulated. The adductor magnus muscle may also produce medial or lateral rotation at the hip. The other muscles produce medial rotation. These muscles insert on low ridges along the posterior surface of the femur. When an athlete suffers a *pulled groin,* the problem is a *strain*—a muscle tear or break—in one of these adductor muscles.

The medial surface of the pelvis is dominated by a pair of muscles. The large **psoas** (SŌ-us) **major** muscle originates alongside the inferior thoracic and lumbar vertebrae, and its insertion lies on the lesser trochanter of the femur. Before reaching this insertion, its tendon merges with that of the **iliacus** (il-Ē-uh-kus) muscle, which nestles within the iliac fossa. These two muscles are powerful hip flexors and are often referred to collectively as the **iliopsoas** (il-ē-ō-SŌ-us) muscle.

Muscles That Move the Leg

As in the upper limb, muscle distribution in the lower limb, exhibits a pattern (Figure 11-21• and Table 11-17, p. 351). Extensor muscles are located along the anterior and lateral surfaces of the leg, and flexors lie along the posterior and medial surfaces. Although the flexors and adductors originate on the pelvic girdle, most extensors originate on the femoral surface.

The *flexors of the knee* include the **biceps femoris, semimembranosus** (sem-ē-mem-bra-NŌ-sus), **semitendinosus** (sem-ē-ten-di-NŌ-sus), and **sartorius** muscles (Figure 11-21•). These muscles originate along the edges of the pelvis and insert on the tibia and fibula, and their contractions produce flexion at the knee. The sartorius muscle is the only knee flexor that originates superior to the acetabulum, and its insertion lies along the medial surface of the tibia. When the sartorius con-

tracts, it produces flexion at the knee and lateral rotation at the hip—as when you cross your legs.

Because the biceps femoris, semimembranosus, and semitendinosus muscles originate on the pelvic surface inferior and posterior to the acetabulum, their contractions also produce extension of the hip. These three muscles are often called the **hamstrings.** A *pulled hamstring* is a relatively common sports injury caused by a strain affecting one of the hamstring muscles.

The knee joint can be locked at full extension by a slight lateral rotation of the tibia. ∞ *p. 263* The small **popliteus** (pop-LI-tē-us) muscle originates on the femur near the lateral condyle and inserts on the posterior tibial shaft. When flexion is initiated, this muscle contracts to produce a slight medial rotation of the tibia that unlocks the knee joint.

Collectively, the four *knee extensors* make up the **quadriceps femoris:** the three **vastus** muscles, which originate along the shaft of the femur, and the **rectus femoris.** Together, the vastus muscles cradle the rectus femoris muscle the way a bun surrounds a hot dog (Figure 11-21c•). All four muscles insert on the patella; the force of their contraction is relayed to the tibial tuberosity by way of the patellar ligament. The rectus femoris muscle originates on the anterior inferior iliac spine and the superior acetabular rim—so in addition to extending the knee, it assists in flexion of the hip.

11

Intramuscular Injections

Drugs are commonly injected into tissues rather than directly into the circulation. This method enables the physician to introduce a large amount of a drug at one time yet have it enter the circulation gradually. In an **intramuscular (IM) injection,** the drug is introduced into the mass of a large skeletal muscle. Uptake is generally faster and accompanied by less tissue irritation than when drugs are administered *intradermally* or *subcutaneously* (injected into the dermis or subcutaneous layer, respectively). ∞ *p. 152* Up to 5 ml of fluid may be injected at one time, and multiple injections are possible.

The most common complications involve accidental injection into a blood vessel or piercing of a nerve. The sudden entry of massive quantities of a drug into the bloodstream can have fatal consequences, and damage to a nerve can cause motor paralysis or sensory loss. As a result, the site of injection must be selected with care. Bulky muscles that contain few large vessels or nerves make ideal targets. The gluteus medius muscle or the posterior, lateral, superior part of the gluteus maximus muscle is commonly selected. The deltoid muscle of the arm, about 2.5 cm (1 in.) distal to the acromion, is another popular site. Probably the most satisfactory from a technical point of view is the vastus lateralis muscle of the thigh; an injection into this thick muscle will not encounter vessels or nerves. This is the preferred injection site in infants and young children, whose gluteal and deltoid muscles are relatively small. This site is also used in elderly patients or others with atrophied gluteal and deltoid muscles.

11

(a) Posterior view of right thigh

(b) Anterior view of right thigh

(c) Sectional view

•**FIGURE 11-21 Muscles That Move the Leg.** *See also Figures 8-11–8-13, and 9-12, pp. 237–239, 262.* AM *Plates 8.3, 8.7, 8.8, 8.10; Scans 4, 5, 6, 7*

TABLE 11-17 Muscles That Move the Leg *(Figure 11-21)*

Muscle	Origin	Insertion	Action	Innervation
Flexors of the knee				
Biceps femoris	Ischial tuberosity and linea aspera of femur	Head of fibula, lateral condyle of tibia	Flexion at knee; extension and lateral rotation at hip	Sciatic nerve; tibial portion (S_1–S_3; to long head) and common peroneal branch (L_5–S_2; to short head)
Semimembranosus	Ischial tuberosity	Posterior surface of medial condyle of tibia	Flexion at knee; flexion and medial rotation at hip	Sciatic nerve (tibial portion; L_5–S_2)
Semitendinosus	As above	Proximal, medial surface of tibia near insertion of gracilis	As above	As above
Sartorius	Anterior superior iliac spine	Medial surface of tibia near tibial tuberosity	Flexion at knee; flexion and lateral rotation at hip	Femoral nerve (L_2–L_3)
Popliteus	Lateral condyle of femur	Posterior surface of proximal tibial shaft	Medial rotation of tibia (or lateral rotation of femur) at knee	Tibial nerve (L_4–S_1)
Extensors of the knee				
Rectus femoris	Anterior inferior iliac spine and superior acetabular rim of ilium	Tibial tuberosity via patellar ligament	Extension at knee; flexion at hip	Femoral nerve (L_2–L_4)
Vastus intermedius	Anterolateral surface of femur and linea aspera (distal half)	As above	Extension at knee	As above
Vastus lateralis	Anterior and inferior to greater trochanter of femur and along linea aspera (proximal half)	As above	As above	As above
Vastus medialis	Entire length of linea aspera of femur	As above	As above	As above

11

Muscles That Move the Foot and Toes

The extrinsic muscles that move the foot and toes are shown in Figure 11-22• and listed in Table 11-18 (p. 353). Most of the muscles that move the ankle produce the plantar flexion involved with walking and running movements. The **gastrocnemius** (gas-trok-NĒ-mē-us; *gaster*, stomach + *kneme*, knee) muscle of the calf is an important plantar flexor, but the slow muscle fibers of the underlying **soleus** (SŌ-lē-us) muscle are more powerful. These muscles are best seen in posterior and lateral views (Figure 11-22a,b•). The gastrocnemius muscle arises from two heads located on the medial and lateral epicondyles of the femur just proximal to the knee. The **fabella,** a sesamoid bone, is generally present within the lateral head of the gastrocnemius muscle. The gastrocnemius and soleus muscles share a common tendon, the **calcanean tendon,** commonly known as the *Achilles tendon.*

Deep to the gastrocnemius and soleus muscles lie a pair of **peroneus** muscles (Figure 11-22b,c•). The peroneus muscles produce eversion and plantar flexion at the ankle. Inversion is caused by the contraction of the **tibialis** (tib-ē-A-lis) muscles. The large **tibialis anterior** muscle (Figure 11-22b,d•) dorsiflexes the ankle and opposes the gastrocnemius muscle.

SUPERFICIAL MUSCLES

Gastrocnemius

Soleus

Calcanean tendon

(a) Posterior view

Popliteus

Gastrocnemius, cut and removed

SECOND LAYER

Head of fibula

Lateral head of gastrocnemius

Tibialis anterior

Peroneus longus

Soleus

Peroneus brevis

Extensor digitorum longus

Superior extensor retinaculum

Inferior extensor retinaculum

Calcanean tendon

Lateral malleolus

(b) Lateral view

Head of fibula

Flexor hallucis longus

Peroneus brevis

Tendon of peroneus brevis

Tendon of peroneus longus

Tibialis posterior

Peroneus longus

Flexor digitorum longus

THIRD LAYER

DEEPEST LAYER

Tibialis posterior

(c) Posterior view

Patella

Patellar ligament

Tibia

Gastrocnemius

Tibialis anterior

Superior extensor retinaculum

Inferior extensor retinaculum

Tibialis anterior tendon

Soleus

Tibialis posterior

Medial malleolus

Calcanean tendon

(d) Medial view

•FIGURE 11-22 **Extrinsic Muscles That Move the Foot and Toes.** *See also Figures 8-13 and 8-14, pp. 239, 240.*

AM *Plates 8.12–8.14; Scans 5, 6, 7*

11

TABLE 11-18 Extrinsic Muscles That Move the Foot and Toes *(Figure 11-22)*

Muscle	Origin	Insertion	Action	Innervation
ACTION AT THE ANKLE				
Dorsiflexors				
Tibialis anterior	Lateral condyle and proximal shaft of tibia	Base of 1st metatarsal bone and medial cuneiform bone	Dorsiflexion at ankle; inversion of foot	Deep peroneal nerve (L_4–S_1)
Plantar flexors				
Gastrocnemius	Femoral condyles	Calcaneus via calcanean tendon	Plantar flexion at ankle; inversion and adduction of foot; flexion at knee	Tibial nerve (S_1–S_2)
Peroneus brevis	Midlateral margin of fibula	Base of 5th metatarsal bone	Eversion of foot and plantar flexion at ankle	Superficial peroneal nerve (L_4–S_1)
Peroneus longus	Lateral condyle of tibia, head and proximal shaft of fibula	Base of 1st metatarsal bone and medial cuneiform bone	Eversion of foot and plantar flexion at ankle; supports longitudinal arch	As above
Plantaris	Lateral supracondylar ridge	Posterior portion of calcaneus	Plantar flexion at ankle; flexion at knee	Tibial nerve (L_4–S_1)
Soleus	Head and proximal shaft of fibula, and adjacent posteromedial shaft of tibia	Calcaneus via calcanean tendon (with gastrocnemius)	Plantar flexion at ankle; adduction of foot	Sciatic nerve, tibial branch (S_1–S_2)
Tibialis posterior	Interosseous membrane and adjacent shafts of tibia and fibula	Tarsal and metatarsal bones	Adduction and inversion of foot; plantar flexion at ankle	As above
ACTION AT THE TOES				
Digital flexors				
Flexor digitorum longus	Posteromedial surface of tibia	Inferior surfaces of distal phalanges, toes 2–5	Flexion at joints of toes 2–5	Sciatic nerve, tibial branch (L_5–S_1)
Flexor hallucis longus	Posterior surface of fibula	Inferior surface, distal phalanx of great toe	Flexion at joints of great toe	As above
Digital extensors				
Extensor digitorum longus	Lateral condyle of tibia, anterior surface of fibula	Superior surfaces of phalanges, toes 2–5	Extension at joints of toes 2–5	Deep peroneal nerve (L_4–S_1)
Extensor hallucis longus	Anterior surface of fibula	Superior surface, distal phalanx of great toe	Extension at joints of great toe	As above

Important digital muscles originate on the surface of the tibia, the fibula, or both (Figure 11-22b,c,d•). Large synovial tendon sheaths surround the tendons of the tibialis anterior, **extensor digitorum longus,** and **extensor hallucis longus** muscles, where they cross the ankle joint. The positions of these sheaths are stabilized by superior and inferior **extensor retinacula** (Figure 11-22b,d•).

Intrinsic muscles of the foot originate on the tarsal and metatarsal bones (Figure 11-23• and Table 11-19). Their contractions move the toes and contribute to the maintenance of the longitudinal arch of the foot. ∞ *p. 240*

☑ Which leg movement would be impaired by injury to the obturator muscle?

☑ You often hear of athletes who suffer a pulled hamstring. To what does this phrase refer?

☑ How would you expect a torn calcanean tendon to affect movement of the foot?

11

Tendon of peroneus brevis

Superior extensor retinaculum

Lateral malleolus of fibula

Inferior extensor retinaculum

Tendons of extensor digitorum longus

Dorsal interosseus muscles

Tendons of extensor digitorum brevis

Medial malleolus of tibia

Tendon of tibialis anterior

Extensor digitorum brevis

Tendon of extensor hallucis longus

Abductor hallucis

(a) Dorsal view

TABLE 11-19 Intrinsic Muscles of the Foot *(Figure 11-23)*

Muscle	Origin
Extensor digitorum brevis	Calcaneus (superior and lateral surfaces)
Abductor hallucis	Calcaneus (tuberosity on inferior surface)
Flexor digitorum brevis	As above
Abductor digiti minimi	As above
Quadratus plantae	Calcaneus (medial, inferior surfaces)
Lumbricals (4)	Tendons of flexor digitorum longus
Flexor hallucis brevis	Cuboid and lateral cuneiform bones
Adductor hallucis	Bases of metatarsal bones II–IV and plantar ligaments
Flexor digiti minimi brevis	Base of metatarsal bone V
Interossei dorsal (4)	Sides of metatarsal bones
Interossei plantar (3)	Bases and medial sides of metatarsal bones

•**FIGURE 11-23** **Intrinsic Muscles of the Foot.** *See also Figure 8-14, p. 240.* AM *Plate 8.14, Scan 8*

Lumbricals

Tendons of flexor digitorum brevis overlying tendons of flexor digitorum longus

Flexor digiti minimi brevis

Abductor digiti minimi

Plantar aponeurosis (cut)

Fibrous tendon sheaths

Flexor hallucis brevis

Abductor hallucis

Flexor digitorum brevis

Calcaneus

(b) Plantar view, superficial layer

Tendons of flexor digitorum longus

Tendons of flexor digitorum brevis (cut)

Lumbricals

Abductor digiti minimi (cut)

Flexor digiti minimi brevis

Tendon of peroneus brevis

Tendon of peroneus longus

Abductor digiti minimi (cut)

Tendon of flexor hallucis longus

Flexor hallucis brevis

Abductor hallucis (cut and retracted)

Tendon of flexor digitorum longus

Tendon of tibialis posterior

Quadratus plantae

Flexor digitorum brevis (cut)

Calcaneus

(c) Plantar view, deep layer

Insertion	Action	Innervation
Dorsal surfaces of toes 1–5	Extension at metatarsophalangeal joints of toes 1–4	Deep peroneal nerve (L_5–S_1)
Medial side of proximal phalanx of great toe	Abduction at metatarsophalangeal joint of great toe	Medial plantar nerve (L_4–L_5)
Sides of middle phalanges, toes 2–5	Flexion at proximal interphalangeal joints of toes 2–5	As above
Lateral side of proximal phalanx, toe 5	Abduction at metatarsophalangeal joints of toe 5	Lateral plantar nerve (L_4–L_5)
Tendon of flexor digitorum longus	Flexion at joints of toes 2–5	As above
Insertions of extensor digitorum longus	Flexion at metatarsophalangeal joints; extension at proximal interphalangeal joints of toes 2–5	Medial plantar nerve (1), lateral plantar nerve (2–4)
Proximal phalanx of great toe	Flexion at metatarsophalangeal joint of great toe	Medial plantar nerve (L_4–L_5)
As above	Adduction at metatarsophalangeal joint of great toe	Lateral plantar nerve (S_1–S_2)
Lateral side of proximal phalanx of toe 5	Flexion at metatarsophalangeal joint of toe 5	As above
Sides of toes 2–4	Abduction at metatarsophalangeal joints of toes 3 and 4	As above
Sides of toes 3–5	Adduction at metatarsophalangeal joints of toes 3–5	As above

11

Compartment Syndrome

In the limbs, the interconnections between the superficial fascia, the deep fascia of the muscles, and the periostea of the appendicular skeleton are quite substantial. The muscles within a limb are in effect isolated in **compartments** formed by dense collagenous sheets (Figure 11-24•). Blood vessels and nerves traveling to specific muscles within the limb enter and branch within the appropriate compartments.

When a crushing injury, severe contusion, or muscle strain occurs, the blood vessels in one or more compartments may be damaged. These compartments become swollen with blood and fluid leaked from damaged vessels. The connective tissue partitions are very strong; the accumulated fluid cannot escape, so pressure rises within the affected compartments. Eventually, the pressures can become so high that they compress regional blood vessels and eliminate the circulatory supply to the muscles and nerves of the compartment. This compression produces **ischemia** (is-KĒ-mē-uh), or "blood starvation," known as **compartment syndrome.**

Slicing into the compartment along its longitudinal axis or implanting a drain are emergency measures used to relieve the pressure. If such steps are not taken, the contents of the compartment will suffer severe damage. Nerves in the affected compartment will be destroyed after 2–4 hours of ischemia, although they can regenerate to some degree if the circulation is restored. After 6 hours or more, the muscle tissue will also be destroyed, and no regeneration can occur. The

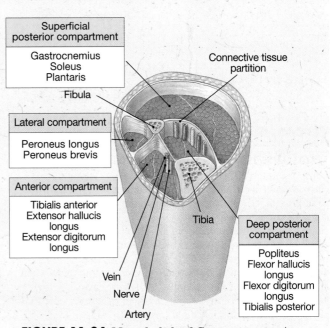

•**FIGURE 11-24 Musculoskeletal Compartments.** A section through the leg, with the muscles partially removed. A section through the thigh or arm would show a comparable arrangement of dense connective tissue partitions. The anterior and lateral compartments of the leg contain extensor/dorsiflexor muscles, and the posterior compartments contain the flexor/plantar flexor muscles.

muscles will be replaced by scar tissue, and shortening of the connective tissue fibers may result in *contracture,* a permanent contraction of an entire muscle following the atrophy of individual muscle fibers.

[EXPLORE]
[MediaLab]

The muscular system must allow for both strong and delicate movements. To gain an understanding of the anatomy of this dynamic system, use the resources available on both the CD-ROM and the Companion Website. The CD-ROM includes animations and tutorials; the website builds on those resources with text content and interactive problems as well as communication avenues for sharing ideas with your class. To complete the following Web Explorations, visit the MediaLab in Chapter 11 of your Companion Website.

Web Exploration 1
Estimated time for completion = 15 minutes
To learn even a fraction of the more than 700 muscles in the body requires study and practice. For practice in locating and discussing the major skeletal muscles, select the keyword **MUSCLES** on the Companion Website. Clicking on any area of the body shown there will lead you to a Web page about the skeletal muscles that lie in the chosen area. Using the text and this page, prepare a chart of the major skeletal muscles. Include muscle name, location, origin, insertion, action (movement), and one suggested exercise for strengthening the muscle or muscle group.

Web Exploration 2
Estimated time for completion = 10 minutes
The news reports about anabolic steroids as drugs that are abused by athletes at all levels of competition. What are anabolic steroids, and why are they federally regulated as controlled substances? Select the keyword **STEROID LAWS** on the Companion Website to read what Rick Collins, a criminal trial lawyer, has compiled about performance-enhancing steroids. After reading the history section, click on the link to current laws; read that portion as well. What are your thoughts? Do you agree with the current legislation? Is it based on scientifically correct information? If you wish, use the communications tools on the Companion Website to share your thoughts.

CHAPTER REVIEW

SELECTED CLINICAL TERMINOLOGY
Terms Discussed in This Chapter

carpal tunnel syndrome: An inflammation of the sheath surrounding the flexor tendons of the palm that leads to nerve compression and pain. *(p. 344)*
compartment syndrome: Ischemia resulting from accumulated blood and fluid trapped within a musculoskeletal compartment. *(p. 355)*
diaphragmatic hernia *(hiatal hernia):* A hernia that occurs when abdominal organs are forced into the thoracic cavity. *(p. 335)*
hernia: A condition involving an organ or a body part that protrudes through an abnormal opening. *(p. 335)*

inguinal hernia: A condition in which the inguinal canal enlarges and abdominal contents are forced into it. *(p. 335)*
intramuscular (IM) injection: The administration of a drug by injecting it into the mass of a large skeletal muscle. *(p. 349)*
ischemia (is-KĒ-mē-uh): A condition of "blood starvation" resulting from compression of regional blood vessels. *(p. 355)*
rotator cuff: The muscles that surround the shoulder joint; a common site of sports injuries. *(p. 339)*

AM Additional Terms Discussed in the *Applications Manual*

bone bruise: Bleeding within the periosteum of a bone.
bursitis: The inflammation of the bursae around one or more joints.
muscle cramps: Prolonged, involuntary, painful muscular contractions.
sprains: Tears or breaks in ligaments or tendons.

strains: Tears or breaks in muscles.
stress fractures: Cracks or breaks in bones subjected to repeated stresses or trauma.
tendinitis: The inflammation of the connective tissue surrounding a tendon.

STUDY OUTLINE

INTRODUCTION   p. 312

1. The **muscular system** includes all the skeletal muscle tissue that can be controlled voluntarily.

BIOMECHANICS AND MUSCLE ANATOMY   p. 312

Organization of Skeletal Muscle Fibers   p. 312

1. A muscle can be classified according to the arrangement of fibers and fascicles as a **parallel muscle, convergent muscle, pennate muscle,** or **circular muscle (sphincter).** A pennate muscle may be *unipennate, bipennate,* or *multipennate. (Figure 11-1)*

Levers   p. 313

2. A **lever** is a rigid structure that moves on a fixed point called the **fulcrum.** Levers can change the direction, speed, or distance of muscle movements and can modify the force applied to muscles.

3. Levers are classified as **first-class, second-class,** or **third-class levers.** Third-class levers are the most common type in the body. *(Figure 11-2)*

MUSCLE TERMINOLOGY   p. 315

Origins and Insertions   p. 315

1. Each muscle can be identified by its **origin, insertion,** and **action.**

Actions   p. 315

2. According to function, a muscle can be classified as an **agonist,** or **prime mover;** an **antagonist;** a **synergist;** or a **fixator.**

Names of Skeletal Muscles   p. 316

3. The names of muscles commonly provide clues to their fascicle organization, location, relative position, structure, size, shape, origin and insertion, or action. *(Table 11-1)*

Axial and Appendicular Muscles   p. 317

4. The **axial musculature** arises on the axial skeleton; it positions the head and spinal column and moves the rib cage. The **appendicular musculature** stabilizes or moves components of the appendicular skeleton. *(Figure 11-3)*

THE AXIAL MUSCLES   p. 320

1. The axial muscles fall into logical groups on the basis of location, function, or both.

2. **Innervation** refers to the identity of the nerve that controls a given muscle.

Muscles of the Head and Neck   p. 320

3. The muscles of facial expression are the **orbicularis oris, buccinator,** and **epicranius (frontalis** and **occipitalis)** muscles and the **platysma.** *(Figure 11-4; Table 11-2)*

4. Six **extrinsic eye muscles (oculomotor muscles)** control external eye movements: the **inferior** and **superior rectus** muscles, the **lateral** and **medial rectus** muscles, and the **inferior** and **superior oblique** muscles. *(Figure 11-5; Table 11-3)*

5. The muscles of mastication (chewing) are the **masseter, temporalis,** and **pterygoid** muscles. *(Figure 11-6; Table 11-4)*

6. The muscles of the tongue are necessary for speech and swallowing and assist in mastication. They are the **genioglossus, hyoglossus, palatoglossus,** and **styloglossus** muscles. *(Figure 11-8; Table 11-5)*

7. The muscles of the pharynx constrict the pharyngeal walls **(pharyngeal constrictors),** elevate the larynx **(laryngeal elevators),** or raise the soft palate **(palatal muscles).** *(Figure 11-9; Table 11-6)*

8. The muscles of the neck control the position of the larynx, depress the mandible, and provide a foundation for the muscles of the tongue and pharynx. *(Figure 11-10; Table 11-7)*

Muscles of the Vertebral Column   p. 328

9. The superficial muscles of the spine can be classified into the **spinalis, longissimus,** and **iliocostalis** divisions. *(Figure 11-11; Table 11-8)*

10. Other muscles of the spine include the **longus capitis** and **longus colli** muscles of the neck and the **quadratus lumborum** muscle of the lumbar region. *(Figure 11-11; Table 11-8)*

Oblique and Rectus Muscles   p. 330

11. The oblique muscles include the **scalene** muscles and the **intercostal** and **transversus** muscles. The **external** and **internal intercostal** muscles are important in respiratory movements of the ribs. Also important to respiration is the **diaphragm.** *(Figure 11-12; Table 11-9)*

Muscles of the Pelvic Floor   p. 332

12. The **perineum** can be divided into an anterior **urogenital triangle** and a posterior **anal triangle.** The pelvic floor consists of the **urogenital diaphragm** and the **pelvic diaphragm.** *(Figure 11-13; Table 11-10)*

THE APPENDICULAR MUSCLES   p. 335

Muscles of the Shoulders and Upper Limbs   p. 335

1. The **trapezius** muscle affects the positions of the shoulder girdle, head, and neck. Other muscles inserting on the scapula include the **rhomboideus, levator scapulae, serratus anterior, subclavius,** and **pectoralis minor** muscles. *(Figures 11-14, 11-15; Table 11-11)*

2. The **deltoid** and the **supraspinatus** muscles are important abductors. The **subscapularis** and **teres major** muscles produce medial rotation at the shoulder; the **infraspinatus** and **teres minor** muscles perform lateral rotation; and the **coracobrachialis** muscle produces flexion and adduction at the shoulder joint. *(Figures 11-14–11-16; Table 11-12)*

3. The **pectoralis major** muscle flexes the shoulder joint, and the **latissimus dorsi** muscle extends it. *(Figures 11-14–11-16; Table 11-12)*

4. The actions of the **biceps brachii** muscle and the **triceps brachii** muscle (long head) affect the elbow joint. The **brachialis** and **brachioradialis** muscles flex the elbow, opposed by the **anconeus** muscle. The **flexor carpi ulnaris, flexor carpi radialis,** and **palmaris longus** muscles cooperate to flex the wrist. They are opposed by the **extensor carpi radialis** muscles and the **extensor carpi ulnaris** muscle. The **pronator teres** and **pronator quadratus** muscles pronate the forearm and are opposed by the **supinator** muscle. *(Figures 11-16–11-19; Tables 11-13–11-15)*

Muscles of the Pelvis and Lower Limbs   p. 346

5. **Gluteal muscles** cover the lateral surfaces of the ilia. The largest is the **gluteus maximus** muscle, which shares an insertion with the **tensor fasciae latae.** Together these muscles pull on the **iliotibial tract.** *(Figures 11-14, 11-20; Table 11-16)*

11

6. The **piriformis** muscle and the **obturator** muscles are the most important **lateral rotators.** The **adductors** can produce a variety of movements. *(Figure 11-20; Table 11-16)*

7. The **psoas major** and **iliacus** muscles merge to form the **iliopsoas** muscle, a powerful flexor of the hip. *(Figures 11-20, 11-21; Table 11-16)*

8. The *flexors of the knee* include the **biceps femoris, semimembranosus,** and **semitendinosus** muscles (the three **hamstrings**) and the **sartorius** muscle. The **popliteus** muscle unlocks the knee joint. *(Figures 11-21, 11-22; Table 11-17)*

9. Collectively, the *knee extensors* are known as the **quadriceps femoris.** This group consists of the three **vastus** muscles and the **rectus femoris** muscle. *(Figure 11-21; Table 11-17)*

10. The **gastrocnemius** and **soleus** muscles produce plantar flexion. A pair of **peroneus** muscles produces eversion as well as plantar flexion at the ankle. *(Figure 11-22; Table 11-18)*

11. Smaller muscles of the calf and shin position the foot and move the toes. Precise control of the phalanges is provided by muscles originating at the tarsal and metatarsal bones. *(Figure 11-23; Table 11-19)*

REVIEW QUESTIONS

LEVEL 1 REVIEWING FACTS AND TERMS

11

1. Muscle fibers in a skeletal muscle form bundles called
   (a) aponeuroses   (b) fascicles
   (c) sarcomeres   (d) myofibrils

2. Most muscles in the body are _____ muscles.
   (a) convergent   (b) pennate
   (c) circular   (d) parallel

3. Skeletal muscle fibers produce maximum tension over a relatively narrow range of
   (a) complex movements   (b) extensions
   (c) flexions   (d) sarcomere lengths

4. The bones, which serve as levers in the body, change
   (a) the direction of an applied force
   (b) the distance and speed of movement produced by a force
   (c) the effective strength of a force
   (d) a, b, and c are correct

5. A general rule about the difference between an origin and insertion is
   (a) the origin moves while the insertion remains stationary
   (b) each muscle begins and ends at the origin
   (c) the origin remains stationary while the insertion moves
   (d) each muscle begins and ends at the insertion

6. If an agonist produces flexion, the action of the antagonist will be
   (a) pronation   (b) adduction
   (c) abduction   (d) extension

7. The muscles of facial expression are innervated by
   (a) cranial nerve VII   (b) cranial nerve V
   (c) cranial nerve IV   (d) cranial nerve VI

8. The strongest masticatory muscle is the _____ muscle.
   (a) pterygoid   (b) masseter
   (c) temporalis   (d) mandible

9. The muscles of the tongue are innervated by the
   (a) hypoglossal nerve (N XII)
   (b) trochlear nerve (N IV)
   (c) abducens nerve (N VI)
   (d) a, b, and c are correct

10. The muscle that opens the mouth by depressing the mandible is the _____ muscle.
    (a) stylopharyngeus   (b) digastric
    (c) sternocleidomastoid   (d) hypoglossus

11. Important flexors of the spinal column that act in opposition to the erector spinae are the _____ muscles.
    (a) rectus   (b) longus capitis
    (c) longus colli   (d) scalene

12. The linea alba is a median collagenous partition that longitudinally divides the _____ muscle.
    (a) external abdominal oblique   (b) rectus abdominis
    (c) external intercostal   (d) rectus femoris

13. The muscular partition that separates the abdominopelvic and thoracic cavities is the
    (a) masseter   (b) perineum
    (c) diaphragm   (d) transversus abdominis muscle

14. The major abductor of the arm is the _____ muscle.
    (a) deltoid   (b) biceps brachii
    (c) triceps brachii   (d) subscapularis

15. The muscles that rotate the radius without producing either flexion or extension of the elbow are the _____ muscles.
    (a) brachialis and brachioradialis
    (b) pronator teres and supinator
    (c) biceps brachii and triceps brachii
    (d) a, b, and c are correct

16. The powerful flexors of the hip are the _____ muscles.
    (a) piriformis   (b) obturators
    (c) pectineus   (d) iliopsoas

17. Knee extensors known as the quadriceps consist of the _____.
    (a) three vastus muscles and the rectus femoris muscle
    (b) biceps femoris, gracilis, and sartorius muscles
    (c) popliteus, iliopsoas, and gracilis muscles
    (d) gastrocnemius, tibialis, and peroneus muscles

18. What two factors interact to determine the effects of a muscle contraction?

19. List the four fascicle organizations that produce the different patterns of skeletal muscles.

20. What are the distinguishing characteristics of unipennate, bipennate, and multipennate muscles?

21. How do first-, second-, and third-class levers differ?

22. What three actions are used to identify muscle groups? Give a brief description of each action.

23. What is the functional difference between the axial musculature and the appendicular musculature?

24. Which four muscle groups make up the axial musculature?

25. Which axial muscles are used in mastication?

26. What three functions are accomplished by the muscles of the pelvic floor?

27. Which four groups of muscles are associated with the shoulders and upper limbs?

28. What three functional groups make up the muscles of the lower limbs?

29. Which three muscular sites are most desirable for intramuscular injections? Why?

LEVEL 2 REVIEWING CONCEPTS

30. Of the following examples, the one that illustrates the action of a second-class lever is
  (a) knee extension
  (b) plantar flexion
  (c) flexion at the elbow
  (d) a, b, and c are correct

31. An example of an agonist that produces flexion at the elbow is the _____ muscle.
  (a) brachioradialis
  (b) biceps brachii
  (c) brachialis
  (d) biceps femoris

32. The removal of the trapezius muscle exposes the
  (a) serratus anterior and subclavian muscles
  (b) infraspinatus and teres minor muscles
  (c) deltoid and supraspinatus muscles
  (d) rhomboideus and levator scapulae muscles

33. The muscles of the spine include many dorsal extensors but few ventral flexors. Why?

34. What specific structural characteristic makes the voluntary control of urination and defecation possible?

35. Why does a convergent muscle exhibit more versatility when contracting than does a parallel muscle?

36. Why can a pennate muscle generate more tension than can a parallel muscle of the same size?

37. Why is it difficult to lift a heavy object when the elbow is at full extension?

38. If cranial nerve V (trigeminal) is damaged or severed, what function in the body will be affected?

39. Which types of movements are affected when the hamstrings are injured?

LEVEL 3 CRITICAL THINKING AND CLINICAL APPLICATIONS

40. Mary sees Jill coming toward her and immediately contracts her frontalis and procerus muscles. She also contracts her levator labii on the right side. Is Mary glad to see Jill? How can you tell?

41. Jeff is interested in building up his thigh muscles, specifically the quadriceps group. What exercises would you recommend to help Jeff accomplish his goal?

42. Shelly gives her son an ice-cream cone. The boy grasps the cone with his right hand, opens his mouth, and begins to lick at the ice cream. Which muscles does he use to perform these actions?

11

CHAPTER 12

Neural Tissue

CHAPTER OUTLINE AND OBJECTIVES

INTRODUCTION   p. 361

AN OVERVIEW OF THE NERVOUS SYSTEM   p. 361
① List the two major anatomical divisions of the nervous system, and describe the characteristics of each division.

NEURONS   p. 362
  Neuron Structure   p. 362
  Neuron Classification   p. 365
② Sketch and label the structure of a typical neuron, and describe the functions of each component.
③ Classify neurons on the basis of their structure and function.

NEUROGLIA   p. 367
  Neuroglia of the Central Nervous System   p. 367
  Neuroglia of the Peripheral Nervous System   p. 370

  ■ FOCUS: Neural Responses to Injuries   p. 372

④ Describe the locations and functions of neuroglia.

NEUROPHYSIOLOGY   p. 373
  The Transmembrane Potential   p. 373
  Changes in the Transmembrane Potential   p. 376
  Graded Potentials   p. 377
  Action Potentials   p. 380
⑤ Explain how the resting potential is created and maintained.
⑥ Describe the events involved in the generation and propagation of an action potential.
⑦ Discuss the factors that affect the speed with which action potentials are propagated.

SYNAPTIC ACTIVITY   p. 386
  Electrical Synapses   p. 386
  Chemical Synapses   p. 387
⑧ Describe the structure of a synapse, and explain the mechanism involved in synaptic activity.
⑨ Describe the major types of neurotransmitters and neuromodulators, and discuss their effects on postsynaptic membranes.

INFORMATION PROCESSING   p. 392
  Postsynaptic Potentials   p. 392
  Presynaptic Inhibition and Facilitation   p. 395
  The Rate of Action Potential Generation   p. 395
⑩ Discuss the interactions that make possible the processing of information in neural tissue.

HIGHER LEVELS OF ORGANIZATION AND PROCESSING   p. 396
  Functional Organization   p. 396
  Anatomical Organization   p. 399
⑪ Describe the patterns of interaction between neurons that are involved in the processing of information at higher levels.

INTEGRATION WITH OTHER SYSTEMS   p. 399
⑫ Give examples of interactions between the nervous system and each of the other organ systems.

MEDIALAB and CHAPTER REVIEW   p. 401

If you think of the nervous system as an organic computer, individual neurons are the computer chips that make it work. This scanning electron micrograph shows a human neuron growing on the surface of a silicon chip. The circuits visible on the chip are quite small, but the branches of the neuron—the organic circuits—are considerably smaller. Any comparison between the nervous system and a computer is misleading, because even the most sophisticated computer lacks the versatility and adaptability of a single neuron. A neuron may process information from 100,000 different sources, and the nervous system has more than 20 billion neurons. How is information carried and transferred from cell to cell? How can simple cells outperform sophisticated computers? You will find answers to these questions as you read this chapter.

In the next seven chapters, our attention will shift to mechanisms that coordinate the activities of the body's organ systems. These activities are adjusted to meet changing situations and environmental conditions. You sit, stand, or walk by controlling muscular activities; your body temperature remains stable on a cold winter day or in a warm kitchen because your rates of heat generation and heat loss are closely regulated.

Two organ systems, the *endocrine system* and the *nervous system,* provide the necessary regulation. These systems share many characteristics, and they usually act in a complementary fashion. The endocrine system adjusts the metabolic operations of other systems in response to changes in the availability of nutrients and the demand for energy. It also directs activities that continue for extended periods, such as growth and maturation, sexual development, pregnancy, and responses to chronic environmental stresses. Most endocrine responses develop more slowly but last much longer than responses of the nervous system. The nervous system provides swifter but generally briefer responses to stimuli by temporarily modifying the activities of other organ systems. The modifications may appear in a matter of milliseconds, but the effects disappear soon after neural activity ceases.

The nervous system, which accounts for a mere 3 percent of the total body weight, is the most complex organ system. It is vital not only to life but also to our appreciation of life. This chapter details the structure and function of neural tissue and introduces principles of neurophysiology that are vital to an understanding of the nervous system's capabilities and limitations. In subsequent chapters, we will examine increasing levels of structural and functional complexity. [AM] *The Neurological Examination*

AN OVERVIEW OF THE NERVOUS SYSTEM

The nervous system includes all the **neural tissue** in the body. ∞ *p. 134* The basic functional units of the nervous system are individual cells called **neurons**. Supporting cells, or **neuroglia** (noo-RŌG-lē-uh or noo-rō-GLĒ-uh; *glia,* glue), separate and protect the neurons, provide a supportive framework for neural tissue, act as phagocytes, and help regulate the composition of the interstitial fluid. Neuroglia, also called *glial cells,* far outnumber neurons and account for roughly half the volume of the nervous system.

Neural tissue, with supporting blood vessels and connective tissues, forms the organs of the nervous system: the brain; the spinal cord; the receptors in complex sense organs, such as the eye and ear; and the *nerves* that interconnect these organs and link the ner-

vous system with other systems. In Chapter 1, we introduced the two major anatomical divisions of the nervous system: (1) the central nervous system and (2) the peripheral nervous system. ∞ *p. 7*

The **central nervous system (CNS)** consists of the spinal cord and brain. These are complex organs that include not only neural tissue but also blood vessels and the various connective tissues that provide physical protection and support. The CNS is responsible for integrating, processing, and coordinating sensory data and motor commands. Sensory data convey information about conditions inside or outside the body. Motor commands control or adjust the activities of peripheral organs, such as skeletal muscles. When you stumble, the CNS integrates information regarding balance and limb position and then coordinates your recovery by sending motor commands to appropriate skeletal muscles—all in a split second and without conscious effort. The CNS—specifically, the brain—is also the seat of higher functions, such as intelligence, memory, learning, and emotion.

The **peripheral nervous system (PNS)** includes all the neural tissue outside the CNS. The PNS delivers sensory information to the CNS and carries motor commands to peripheral tissues and systems. Bundles of axons, or *nerve fibers,* carry sensory information and motor commands in the PNS. Such bundles, with associated blood vessels and connective tissues, are called *peripheral nerves,* or simply **nerves.** Nerves connected to the brain are called **cranial nerves;** those attached to the spinal cord are called **spinal nerves.**

Figure 12-1• diagrams the functional divisions of the nervous system. The PNS is subdivided into afferent and efferent divisions. The **afferent division** (*ad,* to + *ferre,* to carry) of the PNS brings sensory information to the CNS from receptors in peripheral tissues and organs. **Receptors** are sensory structures that either detect changes in the internal environment or respond to the presence of specific stimuli. These structures range from the dendrites (slender cytoplasmic extensions) of single cells to complex organs. Receptors may be neurons or specialized cells of other tissues, such as the Merkel cells of the epidermis. ∞ *p. 146*

The **efferent division** (*ex,* from) of the PNS carries motor commands from the CNS to muscles and glands. These target organs, which respond by *doing* something, are called **effectors.** The efferent division has both somatic and autonomic components. The **somatic nervous system (SNS)** controls skeletal muscle contractions. *Voluntary* contractions are under conscious control; you exert conscious control over your arm as you raise a full glass of water to your lips. *Involuntary* contractions may be simple, automatic responses or complex movements, but they are directed at the subconscious level, outside your conscious awareness. If you accidentally place your hand on a hot stove, you

12

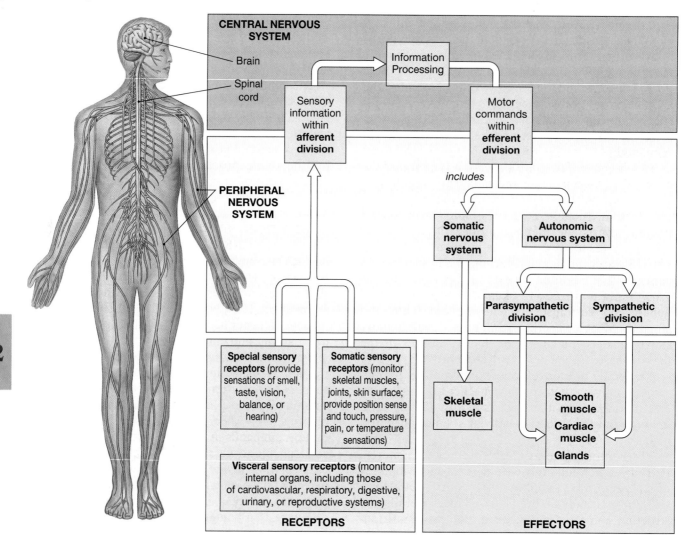

•**FIGURE 12-1** A Functional Overview of the Nervous System

will withdraw it immediately, usually before you notice the pain. This type of automatic response is called a **reflex**. The **autonomic nervous system (ANS),** or *visceral motor system,* provides automatic regulation of smooth muscle, cardiac muscle, and glandular secretions at the subconscious level. The ANS includes a *sympathetic division* and a *parasympathetic division*, which commonly have antagonistic effects. For example, activity of the sympathetic division accelerates the heart rate, whereas parasympathetic activity slows the heart rate.

NEURONS

Before we consider how neurons relay information, we must take a closer look at the structure of an individual neuron. In this section, we will consider (1) the structure of a "model" neuron and (2) the structural and functional classifications of neurons.

NEURON STRUCTURE

Figure 12-2• shows the structure of a representative neuron. Neurons have a variety of shapes. The one shown here is a *multipolar neuron,* the most common type of neuron in the central nervous system. The basic characteristics of a multipolar neuron were introduced in Chapter 4. ∞ *p. 134* Each multipolar neuron has a large *cell body,* or *soma,* that is connected to a single, elongate *axon* and several short, branched *dendrites*.

The Cell Body

The **cell body,** or **soma** (plural, *somata*), contains a relatively large, round nucleus with a prominent nucleolus (Figure 12-2•). The cytoplasm surrounding the nucleus constitutes the **perikaryon** (per-i-KAR-ē-on; *karyon,* nucleus). The cytoskeleton of the perikaryon contains **neurofilaments** and **neurotubules,** which are similar to the microfilaments and microtubules of other types of cells. Bundles of neurofilaments,

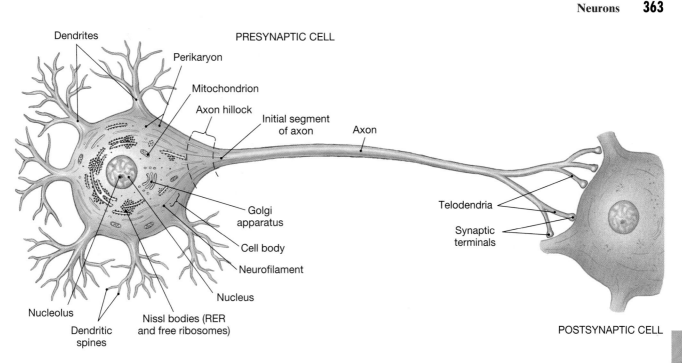

Dendrites

Perikaryon

Mitochondrion

Axon hillock

Initial segment of axon

Axon

Golgi apparatus

Cell body

Neurofilament

Nucleus

Nucleolus

Dendritic spines

Nissl bodies (RER and free ribosomes)

Telodendria

Synaptic terminals

POSTSYNAPTIC CELL

•**FIGURE 12-2** **The Anatomy of a Multipolar Neuron.** A diagrammatic view of a neuron, showing major organelles.

12

called **neurofibrils,** extend into the dendrites and axon, providing internal support for these slender processes.

The perikaryon contains organelles that provide energy and synthesize organic materials, especially the chemical neurotransmitters that are important in cell-to-cell communication. ∞ *p. 280* The numerous mitochondria, free and fixed ribosomes, and membranes of the rough endoplasmic reticulum (RER) give the perikaryon a coarse, grainy appearance. Mitochondria generate ATP to meet the high energy demands of an active neuron. The ribosomes and RER synthesize proteins. Some areas of the perikaryon contain clusters of RER and free ribosomes. These regions, which stain darkly, are called *Nissl bodies,* because they were first described by the German microscopist Franz Nissl. Nissl bodies account for the gray color of areas containing neuron cell bodies—the *gray matter* seen in gross dissection.

Most neurons lack centrioles, important organelles involved in the organization of the cytoskeleton and the movement of chromosomes during mitosis. ∞ *p. 98* As a result, typical CNS neurons cannot divide. Thus they cannot be replaced if lost to injury or disease. Although neural stem cells persist in the adult nervous system, they are inactive except in the nose, where the regeneration of olfactory (smell) receptors maintains our sense of smell, and in the *hippocampus,* a portion of the brain involved with memory storage. The control mechanisms that trigger neural stem cell activity are now being investigated with the goal of preventing or reversing neuron loss due to trauma, disease, or aging.

The Dendrites

A variable number of slender, sensitive processes known as **dendrites** extend out from the cell body (Figure 12-2•). Typical dendrites are highly branched, and each branch bears fine processes called *dendritic spines.* In the CNS, a neuron receives information from other neurons primarily at the dendritic spines, which represent 80–90 percent of the neuron's total surface area.

The Axon

An **axon** is a long cytoplasmic process capable of propagating an electrical impulse known as an *action potential.* ∞ *p. 281* The **axoplasm** (AK-sō-plazm), or cytoplasm of the axon, contains neurofibrils, neurotubules, small vesicles, lysosomes, mitochondria, and various enzymes. The axoplasm is surrounded by the **axolemma** (*lemma,* husk), a specialized portion of the cell membrane. In the CNS, the axolemma may be exposed to the interstitial fluid or covered by the processes of neuroglia. The base, or **initial segment,** of the axon in a multipolar neuron is attached to the cell body at a thickened region known as the **axon hillock** (Figure 12-2•).

An axon may branch along its length, producing side branches collectively known as **collaterals.** Collaterals enable a single neuron to communicate with several other cells. The main axon trunk and any collaterals end in a series of fine extensions, or **telodendria** (te-lō-DEN-drē-uh; *telo-,* end + *dendron,* tree) (Figure 12-2•). The telodendria of an axon end at **synaptic terminals.**

The Synapse

Each synaptic terminal is part of a **synapse,** a specialized site where the neuron communicates with another cell. Two cells meet at every synapse: (1) the *presynaptic cell,* which has the synaptic terminal and sends a message, and (2) the *postsynaptic cell,* which receives the message (Figure 12-3•). The communication between cells at a synapse most commonly involves the release of chemicals called **neurotransmitters** by the synaptic terminal. These chemicals, released by the presynaptic cell, affect the activity of the postsynaptic cell. As we saw in Chapter 10, this release is triggered by electrical events, such as the arrival of an action potential. ∞ *p. 281* The neurotransmitters are typically packaged in *synaptic vesicles.*

The presynaptic cell is usually a neuron. (Specialized receptor cells may form synaptic connections with the dendrites of neurons; this process will be described in Chapter 17.) The postsynaptic cell can be either a neuron or another cell type. When one neuron communicates with another, the synapse may occur on a dendrite, on the cell body, or along the length of the axon of the receiving cell. A synapse between a neuron and a muscle cell is called a **neuromuscular junction.** ∞ *p. 280* At a **neuroglandular junction,** a neuron controls or regulates the activity of a secretory (gland) cell. Neurons also innervate a variety of other cell types, such as adipocytes (fat cells). We shall consider the nature of that innervation in later chapters.

The structure of the synaptic terminal varies with the type of postsynaptic cell. A relatively simple, round **synaptic knob** occurs where the postsynaptic cell is another neuron.[1] The synaptic terminal at a neuromuscular junction is much more complex. ∞ *p. 280* We will deal primarily with the structure of synaptic knobs here; we shall detail other types of synaptic terminals in later chapters.

Axoplasmic Transport Each synaptic knob contains mitochondria, portions of the endoplasmic reticulum, and thousands of vesicles filled with neurotransmitter molecules. Breakdown products of neurotransmitter released at the synapse are reabsorbed and reassembled at the synaptic knob. The synaptic knob also receives a continuous supply of neurotransmitter synthesized in the cell body, along with enzymes

---

[1] The term *synaptic knob* is widely recognized and will be used throughout this text. However, the same structures are also called terminal buttons, terminal boutons, end bulbs, or neuropodia.

•**FIGURE 12-3 The Structure of a Typical Synapse.** A diagrammatic view of the synapse and a micrograph of a typical synapse between two neurons. (TEM, color-enhanced, ×222,000)

and lysosomes. These materials travel the length of the axon along neurotubules and involve *kinesins,* molecular motors that run on ATP. The movement of materials between the cell body and synaptic knobs is called **axoplasmic transport**. Some materials travel slowly, at rates of a few millimeters per day. This transport mechanism is known as the "slow stream." Vesicles containing neurotransmitter move much more rapidly, traveling in the "fast stream" at 5–10 mm per hour.

Axoplasmic transport occurs in both directions. The flow of materials from the cell body to the synaptic knob is *anterograde* (AN-ter-ō-grād) *flow.* At the same time, other substances are being transported toward the cell body in *retrograde* (RET-rō-grād) *flow* (*retro,* backward). If debris or unusual chemicals appear in the synaptic knob, retrograde flow soon delivers them to the cell body. The arriving materials may then alter the activity of the cell by turning appropriate genes on or off.

Rabies is perhaps the most dramatic example of a clinical condition directly related to retrograde flow. A bite from a rabid animal injects the rabies virus into peripheral tissues, where virus particles quickly enter synaptic knobs and peripheral axons. Retrograde flow then carries the virus into the CNS, with fatal results. Many toxins, including heavy metals, some pathogenic bacteria, and other viruses also rely on some form of axoplasmic transport to bypass CNS defenses. AM

Axoplasmic Transport and Disease

NEURON CLASSIFICATION

The billions of neurons in the nervous system are quite variable in form. They can be grouped by structure or by function.

Structural Classification

Neurons are classified as anaxonic, bipolar, unipolar, or multipolar on the basis of the relationship of the dendrites to the cell body and the axon:

1. **Anaxonic** (an-ak-SON-ik) **neurons** are small and have no anatomical clues to distinguish dendrites from axons; all the cell processes look alike (Figure 12-4a•). Anaxonic neurons are located in the brain and in special sense organs. Their functions are poorly understood.

2. **Bipolar neurons** have two distinct processes—one dendritic process that branches extensively at its distal tip, and one axon, with the cell body between them (Figure 12-4b•). Bipolar neurons are rare but occur in special sense organs, where they relay information about sight, smell, or hearing from the receptor cells to other neurons. They are smaller than unipolar or multipolar neurons; the largest measure less than 30 mm from end to end.

3. In a **unipolar neuron,** or *pseudounipolar neuron,* the dendrites and axon are continuous—basically, fused—and the cell body lies off to one side (Figure 12-4c•). In such a neuron, the initial segment lies where the dendrites converge. The rest of the process, which carries action potentials, is usually considered to be an axon.

12

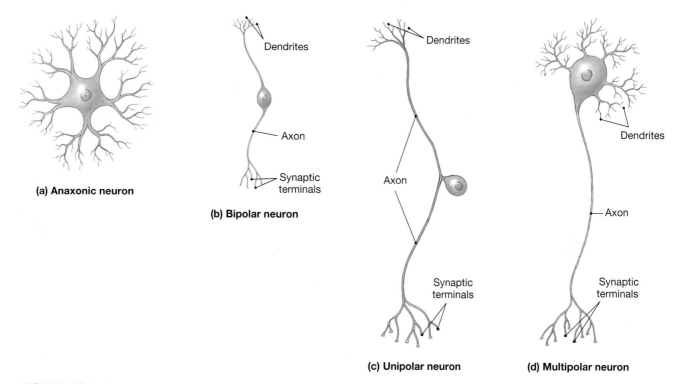

(a) **Anaxonic neuron**

(b) **Bipolar neuron**

Dendrites

Axon

Synaptic terminals

(c) **Unipolar neuron**

Dendrites

Axon

Synaptic terminals

(d) **Multipolar neuron**

Dendrites

Axon

Synaptic terminals

•**FIGURE 12-4 A Structural Classification of Neurons.** The neurons are not drawn to scale; typical anaxonic neurons and bipolar neurons are many times smaller than typical unipolar or multipolar neurons.

Most sensory neurons of the peripheral nervous system are unipolar. Their axons may extend a meter or more, ending at synapses in the central nervous system.

4. **Multipolar neurons** have two or more dendrites and a single axon. Multipolar neurons (Figure 12-4d•) are the most common type of neuron in the CNS. For example, all the motor neurons that control skeletal muscles are multipolar neurons. Their axons can be as long as those of unipolar neurons.

Functional Classification

Alternatively, neurons are categorized by function as (1) sensory neurons, (2) motor neurons, and (3) interneurons. Their relationships are diagrammed in Figure 12-5•.

Sensory Neurons   **Sensory neurons,** or *afferent neurons*, form the afferent division of the PNS. They deliver information from sensory receptors to the CNS. The cell bodies of sensory neurons are located in peripheral *sensory ganglia.* (A *ganglion* is a collection of neuron cell bodies in the PNS.) Sensory neurons are unipolar neurons with processes, known as **afferent fibers,** that extend between a sensory receptor and the spinal cord or brain. Sensory neurons collect information concerning the external or internal environment. The human body has about 10 million sensory neurons. **Somatic sensory neurons** monitor the outside world and our position within it. **Visceral sensory neurons** monitor internal conditions and the status of other organ systems.

Sensory receptors may be the processes of specialized sensory neurons or cells monitored by sensory neurons. Receptors are broadly categorized as follows:

1. **Exteroceptors** (*extero-,* outside) provide information about the external environment in the form of touch, temperature, or pressure sensations and the more complex senses of sight, smell, and hearing.

2. **Proprioceptors** (prō-prē-ō-SEP-torz) monitor the position and movement of skeletal muscles and joints.

3. **Interoceptors** (*intero-,* inside) monitor the digestive, respiratory, cardiovascular, urinary, and reproductive systems and provide sensations of taste, deep pressure, and pain.

Motor Neurons   **Motor neurons,** or *efferent neurons*, form the efferent division of the PNS. These neurons carry instructions from the CNS to peripheral effectors in a peripheral tissue, organ, or organ system. Your body has about half a million motor neurons. Axons traveling away from the CNS are called **efferent fibers.** As we learned earlier, the two major efferent systems are the somatic nervous system (SNS) and the autonomic (visceral) nervous system (ANS). The somatic nervous system includes all the **somatic motor neurons** that innervate skeletal muscles. You have conscious control over the activity of the SNS. The cell body of a somatic motor neuron lies in the CNS, and its axon extends into the periphery to innervate skeletal muscle fibers at neuromuscular junctions.

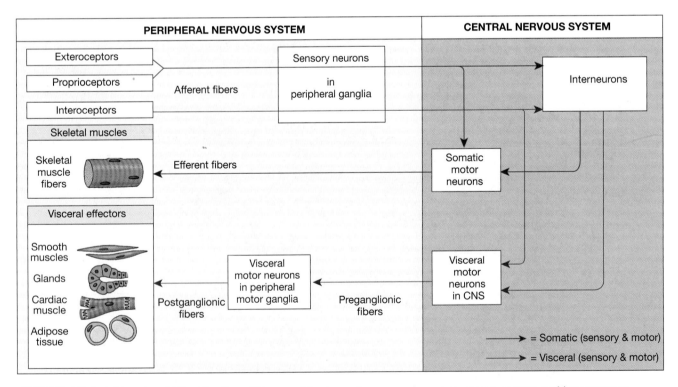

•**FIGURE 12-5 A Functional Classification of Neurons.** The boxes for sensory neurons, motor neurons, and interneurons indicate the locations of the cell bodies.

You do not have conscious control over the activities of the ANS. **Visceral motor neurons** innervate all peripheral effectors other than skeletal muscles. Thus, the ANS innervates smooth muscle, cardiac muscle, glands, and adipose tissue throughout the body. The axons of visceral motor neurons in the CNS innervate a second set of visceral motor neurons in peripheral *autonomic ganglia.* The neurons whose cell bodies are located in those ganglia innervate and control peripheral effectors.

To get from the CNS to a visceral effector, such as a smooth muscle cell, the signal must travel along one axon, be relayed across a synapse, and then travel along a second axon to its final destination. The axons extending from the CNS to an autonomic ganglion are called *preganglionic fibers.* Axons connecting the ganglion cells with the peripheral effectors are known as *postganglionic fibers.*

Interneurons   The 20 billion or so **interneurons,** or *association neurons,* outnumber all other types of neurons combined. Most are located within the brain and spinal cord; some are in autonomic ganglia. Interneurons are responsible for the distribution of sensory information and the coordination of motor activity. For example, one or more interneurons are situated between sensory neurons and motor neurons. The more complex the response to a given stimulus, the greater the number of interneurons involved. Interneurons are also involved with all higher functions, such as memory, planning, and learning.

NEUROGLIA

The organization of neural tissue in the CNS differs significantly from that in the PNS, owing primarily to their distinctive neuroglia populations. The CNS has a greater variety of neuroglia than the PNS has. Although histological descriptions have been available for the past century, the technical problems involved in isolating and manipulating individual glial cells have limited our understanding of their functions. Figure 12-6• summarizes information known about the major neuroglia populations in the CNS and PNS.

NEUROGLIA OF THE CENTRAL NERVOUS SYSTEM

The central nervous system has four types of neuroglia: (1) *ependymal cells,* (2) *astrocytes,* (3) *oligodendrocytes,* and (4) *microglia.*

12

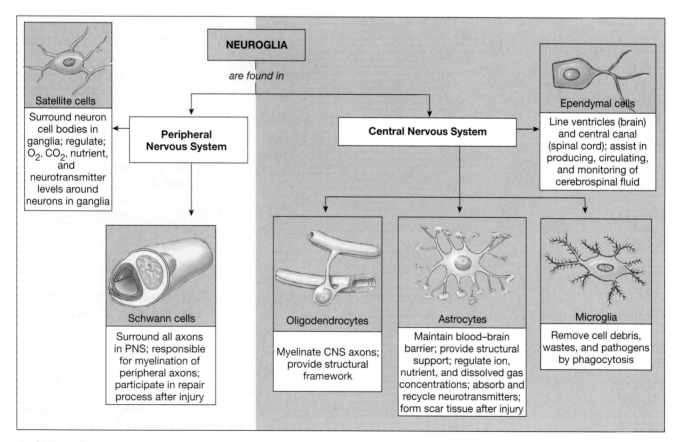

•**FIGURE 12-6** An Introduction to Neuroglia

Ependymal Cells

A fluid-filled central passageway extends along the longitudinal axis of the spinal cord and brain. The wall thickness and the diameter of this passageway vary from one region to another. The narrow passageway in the spinal cord is called the *central canal;* in the brain, the passageway forms enlarged chambers called *ventricles.* The central canal and ventricles are lined by a cellular layer of epithelial cells called the **ependyma** (e-PEN-di-muh) and are filled with **cerebrospinal fluid (CSF).** This fluid, which also surrounds the brain and spinal cord, provides a protective cushion and transports dissolved gases, nutrients, wastes, and other materials.

The ependyma consists of **ependymal cells,** which are cuboidal to columnar in sectional view (Figure 12-7a,b•). During embryonic development and early childhood, the free surfaces of ependymal cells are covered with cilia. The cilia persist in adults only within the ventricles of the brain, where they assist in the circulation of the CSF. In other areas, the ependymal cells typically have scattered microvilli. In a few parts of the brain, specialized ependymal cells participate in the secretion of the CSF. Other regions of the ependyma may have sensory functions, such as monitoring the composition of the CSF.

Unlike typical epithelial cells, ependymal cells have slender processes that branch extensively and make direct contact with neuroglia in the surrounding neural tissue. The functions of these connections are not known. During early embryonic development, stem cells line the central canal and ventricles; the divisions of these stem cells give rise to neurons and all CNS neuroglia other than microglia. [AM] *EMBRYOLOGY SUMMARY 10: An Introduction to the Development of the Nervous System*

Evidence indicates that the ependyma in adults contains stem cells that can divide to produce additional neurons. The specific regulatory mechanisms are now being investigated. If we could control this process, the treatment of strokes and several other debilitating nervous system disorders would be revolutionized.

Astrocytes

Astrocytes (AS-trō-sīts; *astro-,* star + *cyte,* cell) are the largest and most numerous neuroglia in the CNS (Figure 12-7b•). Astrocytes have a variety of functions, many of them poorly understood. These functions include the following:

■ *Maintaining the blood–brain barrier.* Compounds dissolved in the circulating blood do not have free access to the interstitial fluid of the CNS. Neural tissue must be isolated from the general circulation, because hormones or other chemicals in the blood can alter neuron function. The endothelial cells lining CNS capillaries control the chemical exchange between the blood and interstitial fluid. These cells create a **blood–brain barrier,** which isolates the CNS from the general circulation. The slender cytoplasmic extensions of astrocytes end in expanded "feet," processes that wrap around capillaries. Astrocytic processes form a complete blanket around the capillaries, interrupted only where other neuroglia come in contact with the capillary walls. Chemicals secreted by astrocytes are somehow responsible for maintaining the special permeability characteristics of the endothelial cells. (We shall discuss the blood–brain barrier further in Chapter 14.)

■ *Creating a three-dimensional framework for the CNS.* Astrocytes are packed with microfilaments that extend across the breadth of the cell and its processes. This extensive cytoskeletal reinforcement assists them in providing a structural framework for the neurons of the brain and spinal cord.

■ *Repairing damaged neural tissue.* In the CNS, damaged neural tissue seldom regains normal function. However, astrocytes moving into the injury site can make structural repairs that stabilize the tissue and prevent further injury. We will detail neural damage and subsequent repair in a later section.

■ *Guiding neuron development.* Astrocytes in the embryonic brain appear to be involved in directing both the growth and interconnection of developing neurons.

■ *Controlling the interstitial environment.* Although much remains to be learned about astrocyte physiology, evidence indicates that astrocytes adjust the composition of interstitial fluid by several means: (1) regulating the concentration of sodium ions, potassium ions, and carbon dioxide; (2) providing a rapid-transit system for the transport of nutrients, ions, and dissolved gases between capillaries and neurons; (3) controlling the volume of blood flow through the capillaries; and (4) absorbing and recycling some neurotransmitters.

Oligodendrocytes

Like astrocytes, **oligodendrocytes** (o-li-gō-DEN-drō-sīts; *oligo-,* few) possess slender cytoplasmic extensions, but the cell bodies of oligodendrocytes are smaller, and they have fewer processes than do astrocytes (Figure 12-7b•). The processes of oligodendrocytes generally are in contact with the exposed surfaces of neurons. The functions of processes ending at the neuron cell body have yet to be determined; much more is known about the

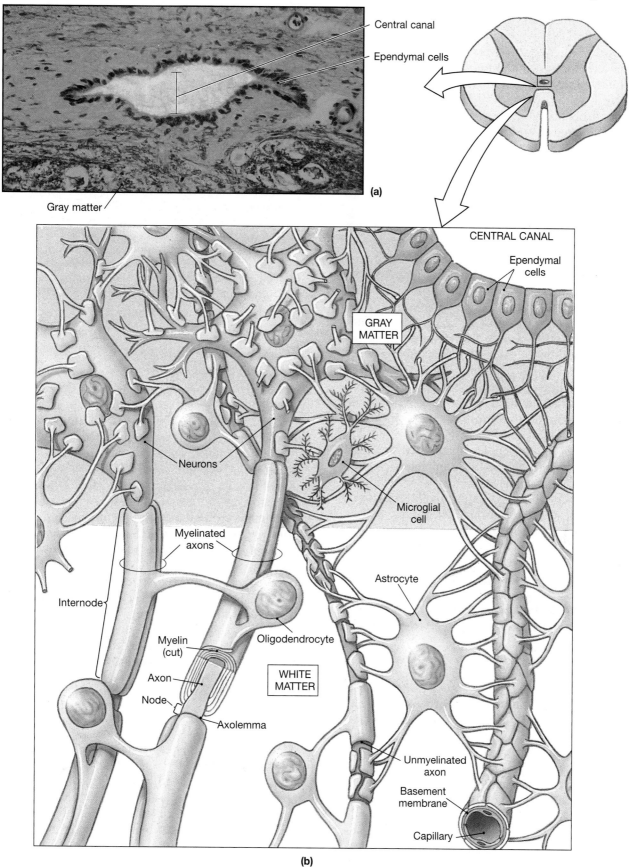

●*FIGURE 12-7* **Neuroglia in the CNS.** **(a)** Light micrograph, showing the ependymal lining of the central canal of the spinal cord. (LM × 236) **(b)** A diagrammatic view of neural tissue in the CNS, showing relationships between neuroglia and neurons.

processes that end on the surfaces of axons. Many axons in the CNS are completely sheathed in the processes of oligodendrocytes.

The axolemma, or cell membrane of the axon, may be insulated from contact with the extracellular fluid by the processes of neuroglia, especially oligodendrocytes. Near the tip of each process, the axolemma expands to form an enormous membranous pad and the cytoplasm there becomes very thin. This flattened pancake somehow gets wound around the axon, forming concentric layers of cell membrane (Figure 12-7b•). This membranous wrapping of insulation, called **myelin** (MĪ-e-lin), increases the speed at which an action potential travels along the axon.

Many oligodendrocytes cooperate in the formation of a **myelin sheath** along the length of an axon. The axon is then said to be **myelinated.** Each oligodendrocyte myelinates short segments of several axons. The relatively large areas of the axon that are thus wrapped in myelin are called **internodes** (*inter,* between). Between myelinated segments are small gaps called **nodes,** or *nodes of Ranvier* (RAHN-vē-ā). In dissection, myelinated axons appear glossy white, primarily because of the lipids present. Regions dominated by myelinated axons constitute the **white matter** of the CNS.

Not all axons in the CNS are myelinated, however. **Unmyelinated** axons may not be completely covered by the processes of neuroglia. Such axons are common where relatively short axons and collaterals form synapses with densely packed neuron cell bodies. These areas have a dusky gray color, primarily due to the concentration of Nissl bodies; they constitute the **gray matter** of the CNS.

In summary, oligodendrocytes play a role in structural organization by tying clusters of axons together, and these neuroglia also improve the functional performance of neurons by wrapping axons within a myelin sheath.

Microglia

The least numerous and smallest neuroglia in the CNS are **microglia** (mī-KRŌG-lē-uh). Their slender cytoplasmic processes have many fine branches (Figure 12-7b•). These cells are capable of migrating through neural tissue. Microglia appear early in embryonic development. They originate from mesodermal stem cells that are related to those stem cells that produce monocytes and macrophages. ∞ *pp. 124, 118* Microglia migrate into the CNS as the nervous system forms. Thereafter, they remain isolated in the neural tissue, where they act as a wandering police force and janitorial service—engulfing cellular debris, waste products, and pathogens.

CNS Tumors and Neuroglia

Tumors of the brain, spinal cord, and associated membranes result in approximately 90,000 deaths in the United States each year. Tumors that originate in the central nervous system are called *primary CNS tumors,* to distinguish them from *secondary CNS tumors,* which arise from the metastasis of cancer cells that originate elsewhere. Roughly 75 percent of CNS tumors are primary tumors. In adults, primary CNS tumors result from the divisions of abnormal neuroglia rather than from the divisions of abnormal neurons, because typical neurons in adults cannot divide. However, through the divisions of stem cells, neurons increase in number until children reach age 4. As a result, primary CNS tumors involving abnormal neurons can occur in young children. Symptoms of CNS tumors vary with the location affected. Treatment may involve surgery, radiation, chemotherapy, or a combination of these procedures.

NEUROGLIA OF THE PERIPHERAL NERVOUS SYSTEM

Neuron cell bodies in the PNS are clustered in masses called **ganglia** (singular, *ganglion*). Neuron cell bodies and axons in the PNS are completely insulated from their surroundings by the processes of neuroglia. The two types of neuroglia involved are called *satellite cells* and *Schwann cells.*

Satellite cells, or *amphicytes* (AM-fi-sīts), surround the neuron cell bodies in ganglia. They regulate the environment around the neurons, much as astrocytes do in the CNS. **Schwann cells,** or *neurilemmal cells (neurilemmocytes),* form a sheath around every peripheral axon, whether unmyelinated or myelinated. Wherever a Schwann cell covers an axon, the outer surface of the Schwann cell is called the **neurilemma** (noo-ri-LEM-uh).

Whereas an oligodendrocyte in the CNS may myelinate portions of several adjacent axons (Figure 12-7•), a Schwann cell can myelinate only one segment of a single axon (Figure 12-8a•). However, a Schwann cell can *enclose* segments of several unmyelinated axons (Figure 12-8b•).

Demyelination is the progressive destruction of myelin sheaths both in the CNS and PNS. The result is a loss of sensation and motor control that leaves affected regions numb and paralyzed. Many unrelated conditions that result in the destruction of myelin can cause symptoms of demyelination. Several important demyelinating desorders are *heavy metal poisoning, diphtheria, multiple sclerosis (MS),* and *Guillain–Barré syndrome.* [AM] *Demyelination Disorders*

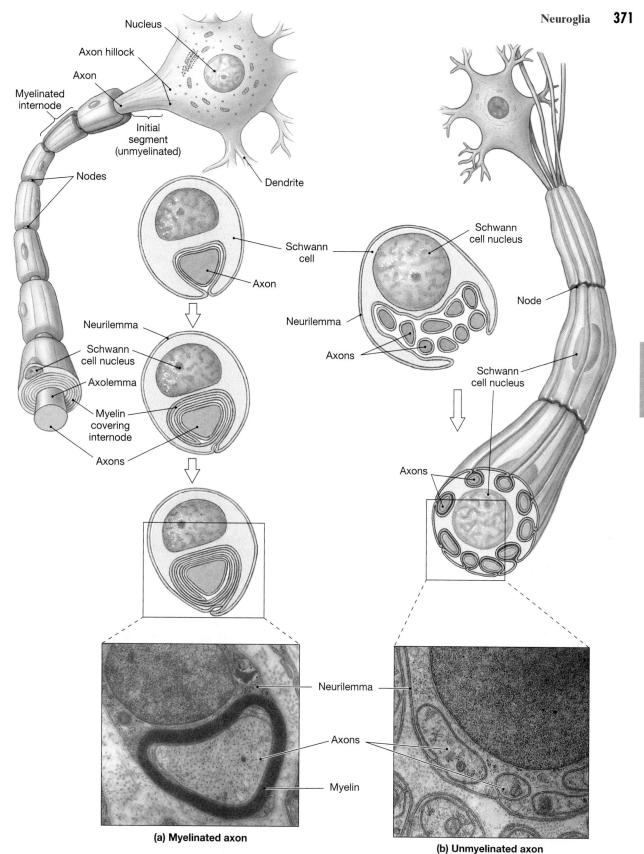

(a) Myelinated axon

(b) Unmyelinated axon

•FIGURE 12-8 Schwann Cells and Peripheral Axons. (a) A myelinated axon, showing the organization of Schwann cells along the length of the axon. Also shown are stages in the formation of a myelin sheath by a single Schwann cell, which myelinates a portion of a single axon. This situation differs from the way myelin forms in the CNS; compare with Figure 12-7b. **(b)** A single Schwann cell can enfold a group of unmyelinated axons. A series of Schwann cells is required to cover the axons along their entire length. Every PNS axon, whether myelinated or unmyelinated, is shielded from contact with the interstitial fluid by Schwann cells.

12

•**FIGURE 12-9** Peripheral Nerve Regeneration
after Injury

A neuron responds to injury in a very limited, stereotyped fashion. In the cell body, the Nissl bodies disperse and the nucleus moves away from its centralized location as the cell increases its rate of protein synthesis. If the neuron recovers its functional abilities, it will return to a normal appearance. The key to recovery seems to be the events under way in the axon. Consider the response of a neuron to mechanical stresses, such as the pressure applied during a crushing injury. The pressure produces a local decrease in blood flow and oxygen availability, and the affected membrane becomes unexcitable. If the pressure is released after an hour or two, the neuron will recover within a few weeks. More severe or prolonged pressure will produce effects similar to those caused by cutting off the distal portion of the axon.

In the PNS, the Schwann cells participate in the repair of damaged nerves. In the process known as **Wallerian degeneration** (Figure 12-9•), the axon distal to the injury site degenerates and macrophages migrate into the area to phagocytize the debris. The Schwann cells do not degenerate. Instead, they proliferate and form a solid cellular cord that follows the path of the original axon. As the neuron recovers, its axon grows into the injury site and the Schwann cells wrap around the axon.

If it continues to grow into the periphery alongside the appropriate cord of Schwann cells, the axon may eventually reestablish its normal synaptic contacts. If it stops growing or wanders off in some new direction, normal function will not return. The growing axon is most likely to arrive at its appropriate destination if the cut edges of the original nerve bundle remain in contact.

Limited regeneration can occur in the CNS, but the situation is more complicated because (1) many more axons are likely to be involved, (2) astrocytes produce scar tissue that can prevent axon growth across the damaged area, and (3) astrocytes release chemicals that block the regrowth of axons.

STEP 1:
Fragmentation of axon and myelin occurs in distal stump.

STEP 2:
Schwann cells form cord, grow into cut, and unite stumps.
Macrophages engulf degenerating axon and myelin.

STEP 3:
Axon sends buds into network of Schwann cells and
then starts growing along cord of Schwann cells.

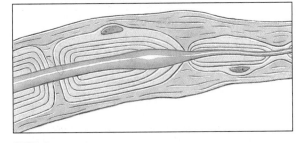

STEP 4:
Axon continues to grow into distal stump
and is enfolded by Schwann cells.

☑ What would damage to the afferent division of the PNS affect?

☑ A tissue sample shows unipolar neurons. Are these more likely to be sensory neurons or motor neurons?

☑ Which type of neuroglia would occur in larger than normal numbers in the brain tissue of a person with a CNS infection?

NEUROPHYSIOLOGY

In Chapter 3, we introduced the concepts of *transmembrane potential* and *resting potential,* two characteristic features of all cells. ∞ *pp. 79–80* In this section, we will focus on the membrane properties of neurons; many of the principles discussed also apply to other cell types. Figure 12-10• introduces the important membrane processes we will be discussing. All neural activities begin with a change in the resting potential of a neuron. A stimulus that disturbs the resting potential produces a temporary, localized change in the transmembrane potential; this change is called a *graded potential.* If the graded potential is sufficiently large, it produces an *action potential* in the membrane of the axon. This action potential is then propagated along the axon to one or more synapses. Activity at each synapse produces graded potentials at the postsynaptic cell. The effects then depend on the *information processing* that occurs at the post-synaptic cell. When you understand each step, you will know how neurons process information and communicate with one another and with peripheral effectors.

THE TRANSMEMBRANE POTENTIAL

Chapter 3 introduced three important concepts regarding the transmembrane potential:

1. The intracellular fluid (cytosol) and extracellular fluid differ markedly in ionic composition. The extracellular fluid (ECF) contains high concentrations of sodium ions (Na^+) and chloride ions (Cl^-), whereas the cytosol contains high concentrations of potassium ions (K^+) and negatively charged proteins.

2. If the cell membrane were freely permeable, diffusion would continue until those ions were evenly distributed across the membrane, at equilibrium. But an even distribution does not occur, because cells have selectively permeable membranes. ∞ *p. 71* Ions cannot freely cross the lipid portions of the cell membrane; they can enter or leave the cell only through membrane channels. Many kinds of membrane channels exist, each with its own properties. At the resting potential, or transmembrane potential of an undisturbed cell, ion movement occurs through *leak channels,* membrane channels that are always open. ∞ *p. 70* There are also active transport mechanisms that move specific ions into or out of the cell.

3. These passive and active mechanisms do not ensure the equal distribution of charges across the cell membrane. Membrane permeability varies by ion. For example, negatively charged proteins inside the cell cannot cross the membrane, and it is easier for K^+ to diffuse out of the cell through a potassium channel than it is for Na^+ to enter the cell through a sodium channel. As a result, the inner surface has an excess of negative charges with respect to the outer surface.

12

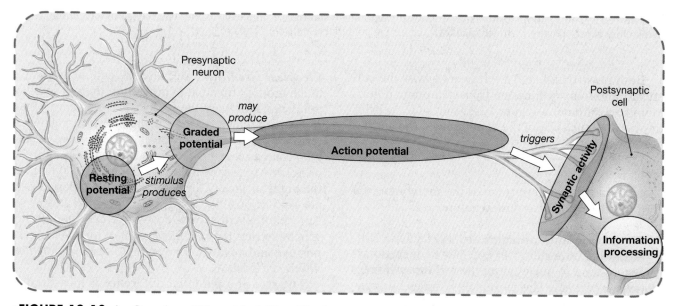

•**FIGURE 12-10 An Overview of Neural Activities.** The important membrane processes are shown in order of their presentation in the text. This figure will be repeated, in reduced and simplified form as Navigator icons, in key figures throughout this chapter as we change topics.

•**FIGURE 12-11 An Introduction to the Resting Potential.** The resting potential is the transmembrane potential of an undisturbed cell. The phospholipid bilayer of the cell membrane is shown as a simple blue band. The Navigator icon in the dashed box repeats Figure 12-10 but highlights the resting potential to indicate "You are here!"

Both passive and active forces act across the cell membrane to determine the transmembrane potential at any moment. Figure 12-11• provides a brief overview of the state of the membrane at the normal resting potential.

Passive Forces

The passive forces acting across the membrane are both chemical and electrical in nature.

Chemical Gradients Because the intracellular concentration of potassium ions is relatively high, potassium ions tend to move out of the cell through open potassium channels. This movement is driven by a concentration gradient, or *chemical gradient*. Similarly, a chemical gradient for sodium ions tends to drive those ions into the cell.

Electrical Gradients Because the cell membrane is much more permeable to potassium than to sodium, potassium ions leave the cytoplasm more rapidly than sodium ions enter. As a result, the cytosol along the interior of the membrane experiences a net loss of positive charges, leaving an excess of negatively charged proteins. At the same time, the extracellular fluid near the outer surface of the cell membrane experiences a net gain of positive charges. The positive and negative charges are separated by the cell membrane, which restricts the free movement of ions. Whenever positive and negative ions are held apart, a *potential difference* exists.

The size of a potential difference is measured in volts (V) or millivolts (mV; thousandths of a volt). The resting potential varies widely, depending on the cell type, but averages about 0.07 V for many cells, including most neurons. We will use this value in our discussion. We usually express this value as −70 mV (Figure 12-11•). The minus sign signifies that the inner surface of the cell membrane is negatively charged with respect to the exterior.

Positive and negative charges attract one another. If nothing separates them, oppositely charged ions will move together and eliminate the potential difference. A movement of charges to eliminate a potential difference is called a **current.** If a barrier, such as a cell membrane, separates the oppositely charged ions, the amount of current depends on how easily the ions can cross the membrane. The **resistance** of the membrane is a measure of how much the membrane restricts ion movement. If the resistance is high, the current is very small, because few ions can cross the membrane. If the resistance is low, the current is very large, because ions flood across the membrane. The resistance of a cell membrane can be changed by the opening or closing of ion channels. These changes result in currents that bring ions into or out of the cytoplasm.

The Electrochemical Gradient
Electrical gradients can reinforce or oppose the chemical gradient for each ion. The **electrochemical gradient** for a specific ion is the sum of the chemical and electrical forces acting across the cell membrane. The electrochemical gradients for K^+ and Na^+ are the primary factors affecting the resting potential of most cells, including neurons.[2]

Sodium ions in the extracellular fluid are attracted by the excess of negative charges on the inner surface of the cell membrane, so both chemical and electrical forces drive Na^+ into the cell. If the cell membrane were freely permeable to Na^+, these ions would continue to enter the cell until the interior of the cell membrane contained enough excess positive charges to reverse the electrical gradient. The interior would then be positively charged, and the repulsion between positive charges would prevent any further net movement of Na^+ into the cell. The equilibrium potential of an ion is the transmembrane potential at which no net movement occurs across the cell membrane. The equilibrium potential of Na^+ is about +66 mV. The resting potential is nowhere near that value, because the resting membrane permeability to Na^+ is very low.

The chemical gradient for potassium ions tends to drive them out of the cell, but the movement is opposed by (1) the attraction between K^+ and the negative charges on the inside of the cell membrane and (2) the repulsion between K^+ and the positive charges on

the outside of the membrane. Thus the chemical gradient is opposed by the electrical gradient, reducing the net force that drives K^+ out of the cytoplasm. If the cell membrane were freely permeable to K^+ but impermeable to other positively charged ions, potassium ions would continue to leave the cell until the electrical gradient (preventing K^+ from leaving the cell) was as large as the chemical gradient (pushing K^+ out of the cell). This equilibrium would occur at a transmembrane potential of about −90 mV, the *equilibrium potential* for potassium. (In reality, the continuous leakage of Na^+ into the cell keeps the resting potential near −70 mV rather than −90 mV.)

An electrochemical gradient is a form of *potential energy.* ∞ *p. 36* Potential energy is stored energy—the energy of position, as in a stretched spring, a charged battery, or water behind a dam. Without a cell membrane, diffusion would eliminate these electrochemical gradients. In effect, the cell membrane acts like a dam across a river. Without the dam, water would simply respond to gravity and flow downstream, gradually losing energy. With the dam in place, even a small opening will release water under tremendous pressure. Similarly, any stimulus that increases the membrane permeability to sodium or potassium ions will produce sudden and dramatic ion movement. For example, a stimulus that opens sodium ion channels will trigger an immediate rush of Na^+ into the cell. The nature of the stimulus does not determine the amount of ion movement. If the stimulus opens the door, the electrochemical gradient will do the rest.

Active Forces: The Sodium–Potassium Exchange Pump

We can compare a cell to a leaky fishing boat with a deckload of fish. The hull represents the cell membrane; the fish, K^+; and the ocean water, Na^+. As the boat rumbles and rolls, water comes in through the cracks and fish fall overboard. To stay afloat and keep the catch, the fisher has to work hard, bailing out the water and recapturing the lost fish.

Similarly, at the normal resting potential, the cell must expend energy, in the form of ATP, to bail out sodium ions that leak in and to recapture potassium ions that leak out. The ATP is used by an ion pump, introduced in Chapter 3, that depends on the carrier protein *sodium–potassium ATPase.* ∞ *p. 76* This pump exchanges 3 intracellular sodium ions for 2 extracellular potassium ions. At the normal resting potential, the pump's primary significance is that it ejects sodium ions as quickly as they enter the cell. Thus, the activity of this exchange pump exactly balances the passive forces of diffusion, and the resting potential remains stable.

Table 12-1 provides a summary of what we have learned about the resting potential.

[2] We can ignore the contributions of the major negatively charged ions (proteins and Cl^-), because (1) proteins are too large to pass through the cell membrane, (2) the normal resting potential is near the equilibrium potential for Cl^-, and (3) the negative charges carried by the interior proteins repel the extracellular Cl^-.

TABLE 12-1 Summary: The Resting Potential

- The resting potential reflects primarily the high membrane permeability to K^+. It is therefore fairly close to -90 mV, the equilibrium potential for K^+.

- Although the electrochemical gradient for Na^+ is very large, sodium permeability is very low. Consequently, sodium ions have only a small effect on the normal resting potential, making it just slightly less negative than it would otherwise be.

- The sodium–potassium exchange pump stabilizes the resting potential when the ratio of passive Na^+ entry to passive K^+ loss is 3:2.

- At the normal resting potential, these passive and active mechanisms are in balance. The resting potential varies widely with the cell type. A typical neuron has a resting potential of approximately -70 mV.

CHANGES IN THE TRANSMEMBRANE POTENTIAL

The resting potential is the transmembrane potential of an "undisturbed" cell. Yet cells are dynamic structures that continually modify their activities either in response to external stimuli or to perform specific functions. The transmembrane potential is equally dynamic, rising or falling in response to temporary changes in membrane permeability. Those changes result from the opening or closing of specific membrane channels.

Membrane Channels

Membrane channels control the movement of ions across the cell membrane. Our discussion will focus on the membrane permeability to sodium and potassium ions, which are the primary determinants of the transmembrane potential of many cell types, including neurons. Sodium and potassium ion channels are either passive or active. **Passive channels,** or **leak channels,** are always open. However, their permeability can vary from moment to moment as the proteins that make up the channel change shape in response to local conditions. Sodium and potassium leak channels are important in establishing the normal resting potential of the cell, as we noted earlier (Figure 12-11●).

Cell membranes also contain **active channels,** which are often called **gated channels.** Gated channels open or close in response to specific stimuli. Each gated channel can be in one of three states: (1) closed but capable of opening, (2) open **(activated),** or (3) closed and incapable of opening **(inactivated).**

Three classes of gated channels exist: chemically regulated channels, voltage-regulated channels, and mechanically regulated channels:

1. **Chemically regulated channels** open or close when they bind specific chemicals (Figure 12-12a●). The receptors that bind acetylcholine (ACh) at the neuromuscular junction are chemically regulated channels. ∞ *p. 281* Chemically regulated channels are most abundant on the dendrites and cell body of a neuron, the areas where most synaptic communication occurs.

Chemically regulated channels are also important in specialized receptor cells of the nose and eye. Olfactory receptors are stimulated by the opening of ion channels in response to the presence of specific extracellular chemicals. When light strikes a visual receptor, chemicals are released that trigger the closing of chemically regulated ion channels. In Chapter 17, we will see how the resulting changes in permeability are the basis for our senses of smell and vision.

2. **Voltage-regulated channels** are characteristic of areas of **excitable membrane,** a membrane capable of generating and conducting an action potential. Examples of excitable membranes include the axons of unipolar and multipolar neurons and the sarcolemma (including T tubules) of skeletal muscle fibers and cardiac muscle cells. ∞ *pp. 274, 302* Voltage-regulated channels open or close in response to changes in the transmembrane potential (Figure 12-12b●). The most important, for our purposes, are voltage-regulated sodium channels, potassium channels, and calcium channels. The opening of voltage-regulated sodium channels is a key step in the generation of an action potential. These channels have two gates that function independently: an *activation gate* and an *inactivation gate.* In the resting membrane, the activation gate is closed and the inactivation gate is open. The activation gate opens when the proper chemical stimulation has occurred; sodium ions can then enter. Sodium ion entry stops when the inactivation gate closes. During the recovery period that follows, the activation gate closes and the inactivation gate reopens. Until the recovery period has passed, the ion channel cannot respond to voltage changes in the surrounding membrane.

3. **Mechanically regulated channels** open or close in response to physical distortion of the membrane surface (Figure 12-12c●). Such channels are important in sensory receptors that respond to touch, pressure, or vibration. We shall discuss these receptors in more detail in Chapter 17.

At the resting potential, most gated channels are closed. The opening of gated channels alters the rate of ion movement across the cell membrane and thus changes the transmembrane potential. The distribu-

•**FIGURE 12-12** **Gated Channels.** We use Na$^+$ channels here, but comparable gated channels regulate the movements of other cations and anions. **(a)** A chemically regulated Na$^+$ channel that opens in response to the presence of ACh at a binding site. **(b)** A voltage-regulated Na$^+$ channel that responds to changes in the transmembrane potential. At the normal resting potential, the channel is closed. At a membrane potential of −60 mV, the channel opens. At +30 mV, it is inactivated. Later we will consider factors that alter the membrane potential. **(c)** Mechanically regulated channels open in response to membrane distortion.

tion of membrane channels can vary from one region of the cell membrane to another; this variation can affect how and where a cell responds to specific stimuli. For example, chemically regulated sodium channels are widespread on the surfaces of a neuron, but voltage-regulated sodium channels are restricted to the axon, its branches, and the synaptic terminals; mechanically regulated channels are typically located only on the dendrites of sensory neurons. The functional implications of these distribution differences will become apparent in later sections.

GRADED POTENTIALS

Graded potentials, or *local potentials,* are changes in the transmembrane potential that cannot spread far from the area surrounding the site of stimulation. Any stimulus that opens a gated channel will produce a graded potential. Figure 12-13a,b• shows what happens when a membrane is exposed to a chemical that opens chemically regulated sodium channels. (For clarity, only gated channels are shown; leak channels are present, but they are not involved in the production

(a) Resting membrane with closed chemically regulated sodium ion channels

(b) Membrane exposed to chemical that opens these sodium ion channels

(c) Spread of sodium ions inside cell membrane produces a local current that depolarizes adjacent portions of the cell membrane

•**FIGURE 12-13 Graded Potentials.** The depolarization radiates in all directions away from the source of stimulation. For clarity, only gated channels are shown; leak channels are present but are not responsible for the production of graded potentials. Color changes in the phospholipid bilayer indicate that the resting potential has been disturbed and that the transmembrane potential is no longer −70 mV. Notice that the dashed box now highlights the graded potential.

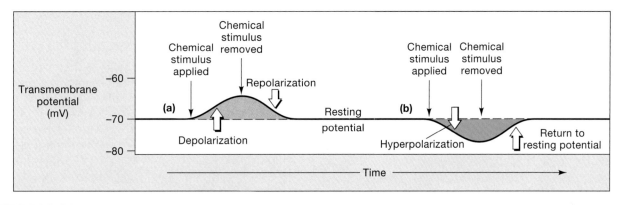

•**FIGURE 12-14 Depolarization and Hyperpolarization. (a)** Depolarization and repolarization in response to the application and removal of a stimulus that opens chemically regulated sodium channels. **(b)** Hyperpolarization in response to the application of a stimulus that opens chemically regulated potassium channels. When the stimulus is removed, the membrane potential returns to the resting level.

of graded potentials.) Sodium ions enter the cell and are attracted to the negative charges along the inner surface of the membrane. The arrival and spreading out of additional positive charges shift the transmembrane potential toward 0 mV. Any shift from the resting potential toward 0 mV is called a **depolarization** (Figure 12-14a•). This term applies to changes in potential from −70 mV to smaller negative values (−65 mV, −45 mV, −10 mV) as well as to membrane potentials above 0 mV (+10 mV, +30 mV).

At the resting potential, sodium ions are drawn to the outer membrane surface, attracted by the excess of negative ions on the inside of the membrane. As the cell membrane depolarizes, sodium ions are released from the outer membrane surface. These ions, accompanied by other extracellular sodium ions, then move toward the open channels, replacing ions that have already entered the cell. This movement of positive charges parallel to the inner and outer surfaces of a membrane is called a **local current** (Figure 12-13c•).

The degree of depolarization decreases with distance, because the cytosol offers considerable resistance to ion movement and because some of the sodium ions entering the cell then move back across the membrane through sodium leak channels. At some distance from the entry point, the effects on the transmembrane potential are undetectable (Figure 12-13c•). The maximum change in the transmembrane potential is proportional to the size of the stimulus, because that determines the number of open sodium channels. The more open channels, the more sodium ions enter the cell, the greater the membrane area affected, and the greater the degree of depolarization.

Opening a gated potassium channel would have the opposite effect on the transmembrane potential as opening a gated sodium channel: The rate of potassium outflow would increase, and the interior of the cell would lose positive ions. The loss of positive ions produces **hyperpolarization,** an increase in the negativity of the resting potential, from −70 mV perhaps to −80 mV or more (Figure 12-14b•). Again, a local current would distribute the effect to adjacent portions of the cell membrane, and the effect would decrease with distance from the open channel (or channels).

Both depolarization and hyperpolarization move the transmembrane potential away from resting levels. When a chemical stimulus is removed and normal membrane permeability is restored, the transmembrane potential soon returns to resting levels. The process of restoring the normal resting potential after depolarization is called **repolarization** (Figure 12-14a•). Repolarization typically involves a combination of ion movement through membrane channels and the activities of ion pumps, especially the sodium–potassium exchange pump. Table 12-2 summarizes information about graded potentials.

TABLE 12-2 Summary: Graded Potentials

Graded potentials, whether depolarizing or hyperpolarizing, share four basic characteristics:

1. The transmembrane potential is most affected at the site of stimulation, and the effect decreases with distance.

2. The effect spreads passively owing to local currents.

3. The graded potential change may involve either depolarization or hyperpolarization. The nature of the change is determined by the properties of the membrane channels involved. For example, in a resting membrane, the opening of sodium channels will cause depolarization, whereas the opening of potassium channels will cause hyperpolarization.

4. The stronger the stimulus, the greater the change in the transmembrane potential and the larger the area affected.

The Distribution and Importance of Graded Potentials

Graded potentials occur in the membranes of many cell types, such as epithelial cells, gland cells, adipocytes, and a variety of sensory receptors. They are often the trigger for specific cell functions; for example, a graded potential at the surface of a gland cell may trigger the exocytosis of secretory vesicles. Similarly, a graded potential at the synaptic terminal is the trigger for the exocytosis of a neurotransmitter, such as ACh. Finally, we saw in Chapter 10 that the graded depolarization of the motor end plate by ACh can result in the appearance of an action potential in adjacent portions of the sarcolemma. ∞ *p. 281* The motor end plate supports graded potentials, whereas the rest of the sarcolemma consists of excitable membrane.

This brings us to an interesting observation: Each neuron receives information in the form of graded potentials on the dendrites and cell body, and graded potentials at its synaptic terminals trigger the release of neurotransmitter. Yet the largest graded potentials affect an area perhaps only 1 mm in diameter. In a small neuron with short processes, such as an anaxonic neuron or a bipolar neuron of the eye, a graded potential in any part of the cell membrane can spread to the synaptic terminals and alter the rate of neurotransmitter release. This is not the case for most neurons, however. In a typical sensory or motor neuron, the axon is so long that graded potentials on the dendrites and cell body have no direct impact on the synaptic terminals. Action potentials propagated along the axon thus link graded potentials at the dendrites or cell body to the cell membrane at the synaptic terminals.

ACTION POTENTIALS

Action potentials are propagated changes in the transmembrane potential that, once initiated, affect an entire excitable membrane. In a representative neuron, an action potential generally begins at the initial segment of the axon. The action potential is then propagated along the length of the axon, ultimately reaching the synaptic terminals. The action potential begins with the opening of voltage-regulated sodium ion channels at one site. The movement of sodium ions into the cell depolarizes adjacent sites, triggering the opening of additional voltage-regulated channels. The result is a chain reaction that spreads across the membrane surface like a line of falling dominoes.

The All-or-None Principle

The stimulus that initiates an action potential is a depolarization large enough to open voltage-regulated sodium channels. That opening occurs at a transmembrane potential known as the **threshold.** Threshold for an axon is typically between −60 mV and −55 mV, a depolarization of 10–15 mV. A stimulus that shifts the resting membrane potential from −70 mV to −62 mV will not produce an action potential, only a graded depolarization. When such a stimulus is removed, the transmembrane potential returns to the resting level. The depolarization of the initial segment of the axon is caused by local currents resulting from the graded depolarization of the axon hillock.

The initial depolarization acts like pressure on the trigger of a gun. If a slight pressure is applied, the gun will not fire. It will fire only when a certain minimum pressure is applied to the trigger. Once the trigger pressure reaches this threshold, the firing pin drops and the gun discharges. At that point, it no longer matters whether the pressure was applied gradually or suddenly, or whether it was caused by the precise movement of just one finger or by the clenching of the entire hand. The speed and range of the bullet that leaves the gun do not change, regardless of the forces that were applied to the trigger.

In the case of an axon or another area of excitable membrane, a graded depolarization is the pressure on the trigger, and the action potential is the firing of the gun. All stimuli that bring the membrane to threshold generate identical action potentials. In other words, the properties of the action potential are independent of the relative strength of the depolarizing stimulus as long as that stimulus exceeds threshold. This concept is called the **all-or-none principle,** because a given stimulus either triggers a typical action potential or does not produce one at all. If an action potential *is* produced, it will be propagated over the entire surface of the excitable membrane. The all-or-none principle applies to all excitable membranes. An action potential is *generated* at one site as the result of a localized depolarization. It is then *propagated* away from that site. We will consider each process individually.

Action Potential Generation

Figure 12-15● diagrams the steps involved in the generation of an action potential from the resting state. At the normal resting potential, the activation gates of the voltage-regulated sodium channels are closed:

STEP 1: *Depolarization to threshold.* Before an action potential can begin, an area of excitable membrane must be depolarized to threshold by local currents.

STEP 2: *The activation of sodium channels and rapid depolarization.* At threshold, the sodium activation gates open and the cell membrane becomes much more permeable to Na^+. Driven by the large electrochemical gradient, sodium ions rush into the cytoplasm, and rapid depolarization occurs at this site. In less than a millisecond, the inner membrane surface has changed; it now contains more positive ions than negative ones, and the transmembrane potential has changed from −60 mV to positive values closer to the equilibrium potential for sodium (+66 mV).

STEP 1:
Depolarization
to threshold

STEP 2:
Activation of sodium
channels and rapid
depolarization

STEP 3:
Inactivation of sodium
channels and activation
of potassium channels

STEP 4:
The return to
normal permeability

STEP 5:
Resting state

12

Notice that the first two steps in action potential generation are an example of positive feedback: A small depolarization triggers a larger depolarization.

STEP 3: *The inactivation of sodium channels and the activation of potassium channels.* As the transmembrane potential approaches +30 mV, the inactivation gates of the voltage-regulated sodium channels begin closing. This step is known as **sodium channel inactivation.** While sodium channel inactivation is under way, voltage-regulated potassium channels are opening. At a transmembrane potential of +30 mV, the cytosol along the interior of the membrane contains an excess of positive charges. Thus, in contrast to the situation in the resting membrane (p. 375), both the electrical *and* chemical gradients favor the movement of K⁺ out of the cell. The sudden loss of positive charges then shifts the transmembrane potential back toward resting levels, and repolarization begins.

STEP 4: *The return to normal permeability.* The voltage-regulated sodium channels remain inactivated until the membrane has repolarized partway toward threshold. At this time, they regain their normal status—closed but capable of opening. The voltage-regulated potassium channels begin closing as the membrane reaches the normal resting potential (about −70 mV), but the process takes at least a millisecond. Over that period, potassium ions continue to move out of the cell at a faster rate than when at rest, producing a brief hyperpolarization that brings the transmembrane potential very close to the equilibrium potential for potassium (−90 mV). As the voltage-regulated potassium channels close, the transmembrane potential returns to normal resting levels. The membrane is now in a prestimulation condition, and the action potential is over.

The Refractory Period From the time an action potential begins until the normal resting potential has stabilized, the membrane will not respond normally to additional depolarizing stimuli. This period is known as the **refractory period** of the membrane. From the moment the voltage-regulated sodium channels open at threshold until sodium channel inactivation ends, the membrane cannot respond to further stimulation, because all the voltage-regulated sodium channels are either already open or are inactivated.

•**FIGURE 12-15** **The Generation of an Action Potential.**
For clarity, only gated channels are shown.

This portion of the refractory period, the **absolute refractory period,** lasts 0.4–1.0 msec; the smaller the axon diameter, the longer the duration. The **relative refractory period** begins when the sodium channels regain their normal resting condition and continues until the transmembrane potential stabilizes at resting levels. Another action potential can occur over this period if the membrane is sufficiently depolarized.

That depolarization requires a larger-than-normal stimulus because (1) the local current must deliver enough Na^+ to counteract the loss of positively charged K^+ through voltage-regulated K^+ channels, and (2) through most of the relative refractory period, the membrane is hyperpolarized to some degree.

The Role of the Sodium–Potassium Exchange Pump   In an action potential, depolarization results from the influx of Na^+ and repolarization involves the loss of K^+. Over time, the sodium–potassium exchange pump returns intracellular and extracellular ion concentrations to prestimulation levels. Compared with the total number of ions inside and outside the cell, however, the number involved in a single action potential is insignificant. Tens of thousands of action potentials can occur before intracellular ion concentrations change enough to disrupt the entire mechanism. Thus the exchange pump is not essential to any single action potential.

However, a maximally stimulated neuron can generate action potentials at a rate of 1000 per second. Under these circumstances, the exchange pump is needed if ion concentrations are to remain within acceptable limits over a prolonged period. The sodium–potassium exchange pump requires energy in the form of ATP. Each time the pump exchanges two extracellular potassium ions for three intracellular sodium ions, one molecule of ATP is broken down to ADP. The transmembrane protein of the exchange pump is called sodium–potassium ATPase because it splits ATP to ADP. If sodium–potassium ATPase is inactivated by a metabolic poison or if the cell runs out of ATP, a neuron will soon lose its ability to function.

Table 12-3 summarizes the process of action potential generation.

Neurotoxins

Potentially deadly forms of poisoning result from eating seafood containing *neurotoxins,* poisons that affect primarily neurons. Several neurotoxins, such as *tetrodotoxin* (TTX), prevent sodium from entering voltage-regulated sodium channels. The result is an inability to generate action potentials. Motor neurons cannot function under these conditions, and death by suffocation can result from paralysis of the respiratory muscles. AM *Neurotoxins in Seafood*

TABLE 12-3 Summary: Action Potentials

STEP 1: Depolarization to threshold
- A graded depolarization brings an area of excitable membrane to threshold (−60 mV).

STEP 2: Activation of sodium channels and rapid depolarization
- The voltage-regulated sodium channels open (sodium channel activation).
- Sodium ions, driven by charge attraction and the concentration gradient, flood into the cell.
- The transmembrane potential goes from −60 mV, the threshold level, toward +30 mV.

STEP 3: Inactivation of sodium channels and activation of potassium channels
- The voltage-regulated sodium channels close (sodium channel inactivation) occurs at +30 mV.
- Voltage-regulated potassium channels are now open, and potassium ions move out of the cell.
- Repolarization begins.

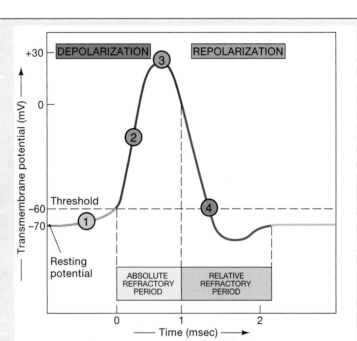

STEP 4: Return to normal permeability
- Voltage-regulated sodium channels regain their normal properties in 0.4–1.0 msec. The membrane is now capable of generating another action potential if a larger-than-normal stimulus is provided.
- The voltage-regulated potassium channels begin closing at −70 mV. Because they do not all close at the same time, potassium loss continues and a temporary hyperpolarization to approximately −90 mV occurs.
- At the end of the relative refractory period, all voltage-regulated channels have closed and the membrane is back to its resting state.

Action Potential Propagation

The sequence of events described earlier occurs in a relatively small portion of the total membrane surface. But we have already noted that, unlike graded potentials, which diminish rapidly with distance, action potentials spread to affect the entire excitable membrane. To understand the basic principle involved, imagine that you are standing by the doors of a movie theater at the start of a long line. Everyone is waiting for the doors to open. The manager steps outside and says to you, "Let everyone know that we're opening in 15 minutes." How would you spread the news? If you treated the line as an inexcitable membrane, you would shout "The doors open in 15 minutes!" as loudly as you could. The closest people in the line would hear the news very clearly, but those farther away might not hear the entire message, and those at the end of the line might not hear you at all. If you treated the crowd as an excitable membrane, you would give the message to another person in line, with instructions to pass it on. In this way, the message would travel along the line until everyone had heard the news. Such a message "moves" as each person repeats it to someone else. Distance is not a factor; the line can contain 50 people or 5000.

This situation is comparable to the way an action potential spreads across an excitable membrane. An action potential (message) is relayed from one location to another in a series of steps. At each step, the message is repeated. Because the same events take place over and over, the term **propagation** is preferable to the term *conduction*.[3]

Continuous Propagation The basic mechanism of action potential propagation along an unmyelinated axon is shown in Figure 12-16•. For convenience, we will consider the membrane as a series of adjacent segments. The action potential begins at the initial segment. For a brief moment at the peak of the action potential, the transmembrane potential becomes positive rather than negative (Figure 12-16a•). A local current then develops, and sodium ions begin moving in the cytosol and in the extracellular fluid (Figure 12-16b•). The local current spreads in all directions, depolarizing adjacent portions of the membrane. The axon hillock cannot respond with an action potential (like the rest of the cell body, it lacks voltage-gated channels). But when the adjacent segment of the axon is depolarized to threshold, an action potential develops there. The process then continues in a chain reaction (Figure 12-16c,d•). Eventually, the most distant portions of the cell membrane will be affected. As in our "movie

line" model, the message is being relayed from one location to another. At each step along the way, the message is retold, so distance has no effect on the process. The action potential reaching the synaptic knob is identical to the one generated at the initial segment, and the net effect is the same as if a single action potential had traveled across the membrane surface. This form of action potential propagation is known as **continuous propagation**.

(a) As an action potential develops in the initial segment, the transmembrane potential depolarizes to +30 mV.

(b) A local current depolarizes the adjacent portion of the membrane to threshold.

(c) An action potential develops at this location, and the initial segment enters the refractory period.

(d) A local current depolarizes the adjacent portion of the membrane to threshold, and the cycle is repeated.

•**FIGURE 12-16 Action Potential Propagation along an Unmyelinated Axon.** The axon can be viewed as a series of adjacent segments.

---

[3] In electrical conduction, electrons flow along a *conductor*, such as a copper wire. Axons are relatively poor conductors of electricity.

Each time a local current develops, the action potential moves forward, not backward, because the previous segment of the axon is still in the absolute refractory period. As a result, an action potential always proceeds away from the site of generation and cannot reverse direction. For a second action potential to occur at the same site, a second stimulus must be applied.

In continuous propagation, an action potential appears to move across the membrane surface in a series of tiny steps. Even though the events at any one location take only about a millisecond, the sequence must be repeated at each step along the way. Continuous propagation along unmyelinated axons occurs at speeds of about 1 meter per second (approximately 2 mph).

Saltatory Propagation   In a myelinated axon, the axolemma is wrapped in a myelin sheath that is complete except at the nodes. Continuous propagation cannot occur along a myelinated axon, because myelin increases resistance to the flow of ions across the membrane. Ions can readily cross the cell membrane only at the nodes. As a result, only the nodes can respond to a depolarizing stimulus.

When an action potential appears at the initial segment of a myelinated axon, the local current skips the internodes and depolarizes the closest node to threshold (Figure 12-17•). Because the nodes may be 1–2 mm apart in a large myelinated axon, the action potential "jumps" from node to node rather than moving along the axon in a series of tiny steps. This process is **saltatory propagation** (*saltare,* leaping). Imagine relaying a message along a line of people spaced 5 meters apart. Each person shouts the message to the next person in line; by the time the message has been repeated four times, it has moved 20 meters. In our model of continuous propagation, in which people were closely packed, the message would have moved only a few meters by the time

it had been repeated four times. Saltatory propagation in the CNS and PNS carries nerve impulses along an axon much more rapidly than does continuous propagation. It also uses proportionately less energy, because less surface area is involved and fewer sodium ions need to be pumped out of the cytoplasm.

Table 12-4 reviews key differences between graded potentials and action potentials.

Axon Diameter and Propagation Speed

Axon diameter, as well as myelination, affects propagation velocity. To depolarize adjacent portions of the cell membrane, ions must move through the cytoplasm. Cytoplasm offers resistance to ion movement, although the resistance is much less than that of the cell membrane. In this instance, an axon behaves like an electrical cable: The larger the diameter, the lower the resistance. (That is why motors with large current demands, such as the starter on a car, an electric stove, or a big air conditioner, use such thick wires.)

Axons are classified into three groups according to the relationships among diameter, myelination, and propagation speed:

1. **Type A fibers** are the largest axons, with diameters ranging from 4 to 20 μm. These are myelinated axons that carry action potentials at speeds of up to 140 meters per second, or over 300 mph.

2. **Type B fibers** are smaller myelinated axons, with diameters of 2–4 μm. Their propagation speeds average around 18 meters per second, or roughly 40 mph.

3. **Type C fibers** are unmyelinated and less than 2 μm in diameter. These axons propagate action potentials at the leisurely pace of 1 meter per second, or a mere 2 mph.

TABLE 12-4 A Comparison of Graded Potentials and Action Potentials

Graded Potentials	Action Potentials
Depolarizing or hyperpolarizing	Always depolarizing
No threshold value	Must depolarize to threshold before action potential begins
Amount of depolarization or hyperpolarization depends on intensity of stimulus	All-or-none phenomenon; all stimuli that exceed threshold will produce identical action potentials
Passive spread from site of stimulation	Action potential at one site depolarizes adjacent sites to threshold
Effect on membrane potential decreases with distance from stimulation site	Propagated across entire membrane surface without decrease in strength
No refractory period	Refractory period
Occur in most cell membranes	Occur only in excitable membranes of specialized cells such as neurons and muscle cells

(a) Action potential at initial segment

(b) Depolarization to threshold at node 1

(c) Action potential at node 1

(d) Depolarization to threshold at node 2

•**FIGURE 12-17** **Saltatory Propagation along a Myelinated Axon.** This process will continue along the entire length of the axon.

We can understand the relative importance of myelin by noting that in going from Type C to Type A fibers, the diameter increases tenfold but the propagation speed increases by 140 times!

Type A fibers carry to the CNS sensory information about position, balance, and delicate touch and pressure sensations from the skin surface. The motor neurons that control skeletal muscles also send their commands over large, myelinated Type A axons. Type B fibers and Type C fibers carry to the CNS information about temperature, pain, and general touch and pressure sensations and carry instructions to smooth muscle, cardiac muscle, glands, and other peripheral effectors.

When we need to tell a friend urgent news or receive an immediate response, we usually pick up the telephone. For general correspondence, we usually send a first-class letter or an e-mail message, both of which produce fast but not immediate responses. If we have to distribute an enormous volume of information to a huge number of people and are in no particular rush, bulk mail offers efficiency at a considerable savings. Instead of representing a compromise between time and money, information transfer in the nervous system reflects a compromise between time and space. Axons carry the information, and the larger the axon, the faster the rate of transmission. But if all sensory information were carried by Type A fibers, your peripheral nerves would be the size of garden hoses and your spinal cord would have the diameter of a garbage can. Instead, only around one-third of all axons carrying sensory information are myelinated, and most sensory information arrives over slender Type C fibers.

Myelination improves coordination and control by decreasing the time between the reception of a sensation and the initiation of an appropriate response. Myelination begins relatively late in development, and the myelination of sensory and motor fibers is not completed until early adolescence. In growing children, the pace of myelination and the pathways involved can be quite variable. This variability contributes to the observed range of physical capabilities in a given age group.

Action Potentials in Muscle Tissues

Although neurons, skeletal muscle fibers, multiunit smooth muscle cells, and cardiac muscle cells have excitable membranes, their action potentials differ in several important ways:

- *The potential difference is larger in muscle cells than in axons.* The resting potential of a skeletal, cardiac, or multiunit smooth muscle cell is −85 mV to −90 mV, close to the potassium equilibrium potential. Threshold values in these cells are comparable to those of axons.

- *Action potentials last longer in muscle cells than in axons.* An action potential in an axon may last 0.5 msec, versus 7.5 msec in a skeletal muscle fiber, 50 msec in a multiunit smooth muscle cell, and 250–300 msec in a cardiac muscle cell from one of the ventricles.

- *Despite the large diameters of muscle cells, action potentials are propagated relatively slowly.* Action potentials of muscle cells travel by continuous propagation at speeds of 3–5 meters per second (6–10 mph).

Visceral smooth muscle cells do not have typical excitable membranes, and most have neither stable resting potentials nor action potentials comparable to those of other muscle types. When visceral smooth muscle cells are stimulated to contract, changes in the transmembrane potential reflect primarily changes in the membrane permeability to calcium ions.

☑ How would a chemical that blocks the sodium channels in neuron cell membranes affect a neuron's ability to depolarize?

☑ What effect would decreasing the concentration of extracellular potassium ions have on the transmembrane potential of a neuron?

☑ One axon propagates action potentials at 50 meters per second; another carries them at 1 meter per second. Which axon is myelinated?

SYNAPTIC ACTIVITY

In the nervous system, messages move from one location to another in the form of action potentials along axons. These electrical events are also known as **nerve impulses.** To be effective, a message must not only be propagated along an axon but must also be transferred in some way to another cell. At a synapse between two neurons, the impulse passes from the **presynaptic neuron** to the **postsynaptic neuron.** A synapse may also involve other types of postsynaptic cells. For example, the neuromuscular junction is a synapse where the postsynaptic cell is a skeletal muscle fiber.

At a synapse, a change in the transmembrane potential of the synaptic terminal affects the activity of another cell. A synapse may be *electrical,* with direct physical contact between the cells, or *chemical,* involving a neurotransmitter.

ELECTRICAL SYNAPSES

Electrical synapses are located in the CNS and PNS, but they are extremely rare. They are present in some areas of the brain, including the *vestibular nuclei,* in the eye, and in at least one pair of PNS ganglia (the *ciliary ganglia*). At an electrical synapse, the presynaptic and postsynaptic membranes are locked together at gap junctions (Figure 4-2•, p. 109). The lipid portions of the opposing membranes, separated by only 2 nm, are held in position by binding between integral membrane proteins called *connexons*. These proteins contain pores that permit the passage of ions between the cells. Because the two cells are linked in this way, changes in the transmembrane potential of one cell will produce local currents that affect the other cell as if the two shared a common cell membrane. As a result, an electrical synapse quickly and efficiently propagates action potentials from one cell to the next.

CHEMICAL SYNAPSES

The situation at a **chemical synapse** is far more dynamic than that at an electrical synapse, because the cells are not directly coupled. For example, an action potential that reaches an electrical synapse will *always* be propagated to the next cell. But at a chemical synapse, an arriving action potential *may or may not* release enough neurotransmitter to bring the postsynaptic neuron to threshold. In addition, other factors may intervene and make the postsynaptic cell more or less sensitive to arriving stimuli. In essence, the postsynaptic cell at a chemical synapse is not a slave to the presynaptic neuron; its activity can be adjusted, or "tuned," by a variety of factors.

Chemical synapses are by far the most abundant type of synapse. Most synapses between neurons and all communications between neurons and other cell types involve chemical synapses. Communication across a chemical synapse can normally occur in only one direction: from the presynaptic membrane to the postsynaptic membrane.

Although acetylcholine is the neurotransmitter that has received the most attention, there are other important chemical transmitters. Neurotransmitters are often classified as excitatory or inhibitory on the basis of their effects on postsynaptic membranes. **Excitatory neurotransmitters** cause depolarization and promote action potential generation, whereas **inhibitory neurotransmitters** cause hyperpolarization and suppress action potential generation.

This classification is useful but not always precise. For example, acetylcholine typically produces a depolarization in the postsynaptic membrane. But the acetylcholine released at neuromuscular junctions in the heart has an inhibitory effect, producing a transient hyperpolarization of the postsynaptic membrane. This situation highlights an important aspect of neurotransmitter function: The effect of a neurotransmitter on the postsynaptic membrane depends on the properties of the receptor, not on the nature of the neurotransmitter.

Cholinergic Synapses

We shall focus on chemical synapses that release the neurotransmitter **acetylcholine (ACh).** These are known as **cholinergic synapses.** The neuromuscular junction is a cholinergic synapse. ∞ *p. 280* ACh, the most widespread (and best-studied) neurotransmitter, is released (1) at all neuromuscular junctions involving skeletal muscle fibers, (2) at many synapses in the CNS, (3) at all neuron-to-neuron synapses in the PNS, and (4) at all neuromuscular and neuroglandular junctions of the parasympathetic division of the ANS.

At a cholinergic synapse between two neurons, the presynaptic and postsynaptic membranes are separated by a synaptic cleft that averages 20 nm (0.02 μm) in width. Most of the ACh in the synaptic knob is packaged in synaptic vesicles, each containing several thousand molecules of neurotransmitter. A single synaptic knob may contain a million such vesicles.

Figure 12-18• diagrams the events that occur at a cholinergic synapse after an action potential arrives at a synaptic knob. For convenience, we assume that this synapse is adjacent to the initial segment of the axon.

STEP 1: *The arrival of an action potential and depolarization of the synaptic knob* (Figure 12-18a•). The normal stimulus for neurotransmitter release is the depolarization of the synaptic knob by the arrival of an action potential.

STEP 2: *The entry of extracellular calcium ions into the synaptic knob, triggering the exocytosis of ACh* (Figure 12-18b•). The depolarization of the synaptic knob opens voltage-regulated calcium channels. In the brief period during which these channels remain open, calcium ions rush into the synaptic knob. Their arrival triggers exocytosis and the release of ACh into the synaptic cleft. The ACh is released in packets of roughly 3000 molecules, the average number of ACh molecules in a single vesicle. The release of ACh stops very soon, because the calcium ions that triggered exocytosis are rapidly removed from the cytoplasm by active transport mechanisms. These mechanisms either pump the calcium ions out of the cell or transfer them into mitochondria, vesicles, or the endoplasmic reticulum.

STEP 3: *The binding of ACh and the depolarization of the postsynaptic membrane* (Figure 12-18c•). The ACh released through exocytosis diffuses across the synaptic cleft toward receptors on the postsynaptic membrane. These receptors are chemically regulated ion channels. The primary response is an increased permeability to Na^+, producing a depolarization that lasts about 20 msec.[4] This depolarization is a graded potential: The more ACh released at the presynaptic membrane, the larger the depolarization. If the depolarization brings an adjacent area of excitable membrane to threshold, an action potential will appear in the postsynaptic neuron.

---

[4] These channels also let potassium ions out of the cell. But, because sodium ions are driven by a much stronger electrochemical gradient, the net effect is a slight depolarization of the postsynaptic membrane.

12

(a) **STEP 1:** Arrival of action potential at synaptic knob

(b) **STEP 2:** Entry of extracellular Ca^{2+} and exocytosis of ACh

(c) **STEP 3:** Binding of ACh to receptors and depolarization of postsynaptic membrane may bring initial segment to threshold

(d) **STEP 4:** Removal of ACh by acetylcholinesterase (AChE)

●FIGURE 12-18 The Function of a Cholinergic Synapse

STEP 4: *The removal of ACh by AChE* (Figure 12-18d●). The effects on the postsynaptic membrane are temporary, because the synaptic cleft and the postsynaptic membrane contain *acetylcholinesterase (AChE,* or *cholinesterase).* Roughly half of the ACh released at the presynaptic membrane is broken down before it reaches receptors on the postsynaptic membrane. ACh molecules that succeed in binding to receptor sites are generally broken down within 20 msec of their arrival.

The enzyme AChE breaks down molecules of ACh into **acetate** and **choline.** The choline is actively absorbed by the synaptic knob and is used to synthesize more ACh, using acetate provided by *coenzyme A (CoA).* (As you may recall from Chapter 2, coenzymes derived from vitamins are required in many enzymatic reactions. ∞ *p. 55*) Acetate diffusing away from the synapse can be absorbed and metabolized by the postsynaptic cell or by other cells and tissues.

Table 12-5 summarizes the events that occur at a cholinergic synapse.

Synaptic Delay   A 0.2–0.5 msec **synaptic delay** occurs between the arrival of the action potential at the synaptic knob and the effect on the postsynaptic membrane. Most of that delay reflects the time involved in calcium influx and neurotransmitter release, not in neurotransmitter diffusion—the synaptic cleft is narrow, and neurotransmitters can diffuse across it in very little time. Although a delay of 0.5 msec is not very long, in that time an action potential may travel more than 7 cm (about 3 in.) along a myelinated axon. When information is being passed along a chain of interneurons in the CNS, the cumulative synaptic delay may exceed the propagation time along the axons. Reflexes, which provide rapid and automatic responses to stimuli, involve a small number of synapses. The fewer synapses involved, the shorter the total synaptic delay and the faster the response. The fastest reflexes have just one synapse, with a sensory neuron directly controlling a motor neuron. The muscle spindle reflexes, discussed in Chapter 13, are an important example.

TABLE 12-5 Summary: Synaptic Activity

The sequence of events at a typical cholinergic synapse:

STEP 1:
- An arriving action potential depolarizes the synaptic knob.

STEP 2:
- Calcium ions enter the cytoplasm of the synaptic knob.
- ACh is released through exocytosis of neurotransmitter vesicles.

STEP 3:
- ACh diffuses across the synaptic cleft and binds to receptors on the postsynaptic membrane.
- Chemically regulated sodium channels on the postsynaptic surface are activated, producing a graded depolarization.
- ACh release ceases because calcium ions are removed from the cytoplasm of the synaptic knob.

STEP 4:
- The depolarization ends as ACh is broken down into acetate and choline by AChE.
- The synaptic knob reabsorbs choline from the synaptic cleft and uses it to resynthesize ACh.

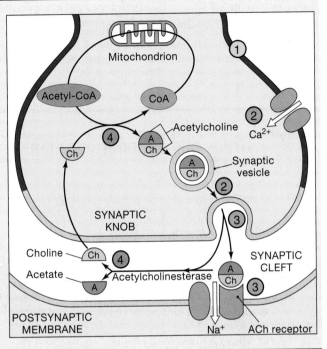

12

Synaptic Fatigue Because ACh molecules are recycled, the synaptic knob is not totally dependent on the ACh synthesized in the cell body and delivered by axoplasmic transport. But under intensive stimulation, resynthesis and transport mechanisms may be unable to keep pace with the demand for neurotransmitter. **Synaptic fatigue** then occurs; the synapse remains inactive until ACh has been replenished.

Other Neurotransmitters

Neurotransmitters include *biogenic amines, amino acids, neuropeptides,* and *dissolved gases* as well as a variety of other compounds. We will consider only a few of the most important neurotransmitters here; we shall encounter additional examples in later chapters:

- **Norepinephrine** (nor-ep-i-NEF-rin), or **NE,** is a neurotransmitter that is widely distributed in the brain and in portions of the ANS. Norepinephrine is also called *noradrenaline,* and synapses that release NE are known as **adrenergic synapses.** Norepinephrine typically has an excitatory, depolarizing effect on the postsynaptic membrane, but the mechanism is quite distinct from that of ACh. We will consider specifics in Chapter 16.

- **Dopamine** (DŌ-puh-mēn) is a CNS neurotransmitter released in many areas of the brain. It may have inhibitory or excitatory effects. For example, inadequate dopamine production and release in one portion of the brain can lead to the overstimulation of neurons that control skeletal muscle tone, owing to a lack of inhibitory control. The result can be the characteristic rigidity and stiffness of *Parkinson's disease,* a condition we shall describe in Chapter 14. At other sites, dopamine can have excitatory effects. One of the effects of cocaine use is the inhibition of dopamine removal from synapses in specific areas of the brain. The resulting rise in dopamine concentrations at these synapses is responsible for the "high" experienced.

- **Serotonin** (ser-o-TŌ-nin) is another important CNS neurotransmitter. Inadequate serotonin production can have widespread effects on attention and emotional states and may be responsible for many cases of severe chronic depression. *Fluoxetine (Prozac™), Paxil™, Zoloft™,* and related antidepressant drugs inhibit the reabsorption of serotonin by synaptic knobs. This inhibition leads to increased serotonin concentrations at synapses; over time, the increase may relieve the symptoms of depression. Interactions among serotonin, norepinephrine, and other neurotransmitters are thought to be involved in the regulation of asleep–awake cycles.

- **Gamma aminobutyric** (a-MĒ-nō-bū-TĒR-ik) **acid,** or **GABA,** generally has an inhibitory effect. Its release appears to reduce anxiety, and some anti-anxiety drugs work by enhancing these effects.

Many neurotransmitters exist whose functions are less well understood. In a clear demonstration of the principle "the more you look, the more you see," at least 50 neurotransmitters have been identified, including certain amino acids, peptides, polypeptides, prostaglandins, and ATP. In addition, two gases have recently been identified as important neurotransmitters: nitric oxide and carbon monoxide. Nitric oxide (NO) is generated by synaptic terminals that innervate smooth muscle in the walls of blood vessels in the PNS and at synapses in several regions of the brain. Carbon monoxide (CO), best known as a component of automobile exhaust, is also generated by specialized synaptic knobs in the brain, where it functions as a neurotransmitter. Our knowledge of the significance of these compounds and the mechanisms involved in their synthesis and release remains incomplete. Table 12-6 lists major neurotransmitters of the brain and spinal cord and their primary effects (if known).

Drugs and Synaptic Function

Many drugs interfere with key steps in synaptic activity. These drugs (1) interfere with neurotransmitter synthesis, (2) alter the rate of neurotransmitter release, (3) prevent neurotransmitter inactivation, or (4) prevent neurotransmitter binding to receptors. Our discussion here is limited to clinically important compounds that act at cholinergic synapses (Figure 12-19●).

Some drugs affect the synaptic terminals. *Botulinus toxin* is responsible for the primary symptom of *botulism*, a widespread paralysis of skeletal muscles. ∞ *p. 281 Botulinus* toxin blocks the release of ACh at the presynaptic membrane of cholinergic neurons. The venom of the black widow spider has the opposite effect. It causes a massive release of ACh that produces intense muscular cramps and spasms.

Other drugs primarily affect the postsynaptic membrane. **Anticholinesterase drugs,** or *cholinesterase inhibitors,* block the breakdown of ACh by acetylcholinesterase. The result is an exaggerated and prolonged stimulation of the postsynaptic membrane. At the neuromuscular junctions, this abnormal stimulation produces an extended and extreme state of contraction. Military nerve gases block cholinesterase activity for weeks, although few persons exposed are likely to live long enough to regain normal synaptic function. Most animals utilize ACh as a neurotransmitter, and anticholinesterase (AChE) drugs, such as *malathion,* are in widespread use in insect-control projects.

Drugs such as **atropine** and **d-tubocurarine** prevent ACh from binding to the postsynaptic receptors. They work on different types of ACh receptors. Atropine can be administered to counteract the effects of AChE poisoning on ACh receptors in smooth and cardiac muscles; d-tubocurarine is a derivative of *curare,* a plant extract used by certain South American tribes to paralyze their prey. Curare and related compounds induce paralysis by preventing the stimulation of the neuromuscular junction by ACh. Other compounds, including **nicotine,** an active ingredient in cigarette smoke, bind to the receptor sites and stimulate the postsynaptic membrane. No enzymes exist that remove these compounds, and the effects are relatively prolonged. [AM] *Neuroactive Drugs*

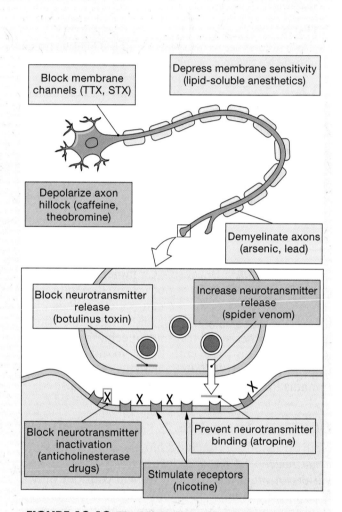

•**FIGURE 12-19** **The Mechanism of Drug Action at a Cholinergic Synapse.** Factors that facilitate neural function and make neurons more excitable are shown in violet. Factors that inhibit or depress neural function are in blue.

Figure 12-19● also indicates the sites of action for other chemicals mentioned in this chapter. Several toxins found in seafood (p. 000 and the *Applications Manual*), including *tetrodotoxin* (TTX), *saxitoxin* (STX), and *ciguatoxin* (CTX), block sodium ion channels. At low doses, these toxins can produce abnormal sensations and interfere with muscle control. Higher doses cause death, generally through the paralysis of respiratory muscles.

TABLE 12-6 Representative Neurotransmitters and Their Effects

Neurotransmitter	Mechanism of Action	Distribution (Examples)
Acetylcholine	Primarily direct, through binding to chemically regulated channels	*Widespread in CNS and PNS* *CNS:* Synapses throughout brain and spinal cord *PNS:* Neuromuscular junctions; preganglionic synapses of ANS; neuroglandular junctions of parasympathetic division and (rarely) sympathetic division of ANS
Biogenic Amines		
Norepinephrine (NE)	Indirect, using second messengers	*CNS and PNS* *CNS:* Cerebral cortex, hypothalamus, brain stem, cerebellum, spinal cord *PNS:* Most neuromuscular and neuroglandular junctions of sympathetic division of ANS
Epinephrine	Indirect, using second messengers	*CNS:* Thalamus, hypothalamus, midbrain, spinal cord
Dopamine	Direct or indirect, depending on receptor type	*CNS:* Hypothalamus, limbic system
Serotonin	Direct or indirect, depending on receptor type	*CNS:* Limbic system, hypothalamus, cerebellum, spinal cord
Histamine	Indirect, using second messengers	*CNS:* Hypothalamus
Excitatory Amino Acids		
Glutamate	Direct or indirect, depending on receptor type	*CNS:* Cerebral cortex, brain stem
Aspartate	Direct or indirect, depending on receptor type	*CNS:* Spinal cord
Inhibitory Amino Acids		
Gamma aminobutyric acid (GABA)	Direct or indirect, depending on receptor type	*CNS:* Cerebellum, cerebral cortex, interneurons throughout brain and spinal cord
Glycine	Direct	*CNS:* Interneurons throughout brain and spinal cord
Neuropeptides		
Substance P	Indirect	*CNS:* Synapses of sensory axons within spinal cord, hypothalamus, other areas of brain
Opioids (endorphins, enkephalins, dynorphins)	Indirect, using second messengers	*CNS:* Hypothalamus, thalamus, brain stem
Other Synaptic Chemicals		
High-energy compounds (ATP, GTP)	Indirect, using second messengers	*CNS:* Spinal cord
Hormones (ADH, oxytocin, insulin, glucagon, secretin, CCK, and many others)	Generally indirect	*CNS:* Brain; widespread
Prostaglandins	Indirect, using second messengers	*CNS:* Brain; widespread
Dissolved Gases		
Carbon monoxide (CO)	Indirect, by diffusion, using second messengers	*CNS:* Brain; some neuromuscular and neuroglandular junctions
Nitric oxide (NO)	Indirect, by diffusion, using second messengers	*CNS:* Brain; some neuromuscular and neuroglandular junctions *PNS:* Some sympathetic neuromuscular and neuroglandular junctions (rare)

12

Neuromodulators

Chemical synapses are always active to some degree, and small numbers of neurotransmitter molecules continuously leak through the presynaptic membranes. Other chemicals may also be released by the synaptic knob, through gradual diffusion or exocytosis, in the company of neurotransmitter molecules. **Neuromodulators** (noo-rō-MOD-ū-lā-torz) are compounds that influence either the release of neurotransmitter by the presynaptic neuron or the postsynaptic cell's response to the neurotransmitter. Neuromodulators are typically **neuropeptides,** small peptide chains synthesized and released by the synaptic knob. Many neuromodulators act by binding to receptors in the presynaptic or postsynaptic membranes and activating cytoplasmic enzymes.

Neuromodulators called **opioids** (Ō-pē-oydz) have effects similar to those of the drugs *opium* and *morphine,* because they bind to the same postsynaptic receptors. Three classes of opioids are identified in the CNS: (1) the **endorphins** (en-DOR-finz), (2) the **enkephalins** (en-KEF-a-linz), (3) and the **dynorphins** (dī-NOR-finz). The primary function of opioids is probably the relief of pain. Pain relief occurs through the inhibition of the release of the neurotransmitter *substance P* at synapses that relay pain sensations. Dynorphins have far more powerful analgesic (pain-relieving) effects than morphine or the other opioids.

How Neurotransmitters and Neuromodulators Work

Functionally, neurotransmitters and neuromodulators fall into one of three groups:

1. *Compounds that have a direct effect on membrane potential by opening or closing chemically regulated channels.* Examples include ACh and the amino acids *glutamate* and *aspartate.* In addition, some chemically regulated channels are sensitive to GABA and norepinephrine, although other membrane receptors for these neurotransmitters produce indirect effects.

2. *Compounds that have an indirect effect on membrane potential through the activities of second messengers.* **Second messengers** are intracellular intermediaries produced or released when neurotransmitters or other extracellular chemicals bond to specific membrane receptors. When epinephrine, norepinephrine, dopamine, serotonin, histamine, GABA, or any of the endorphins bind to membrane receptors, the receptors activate the intracellular enzyme **adenylate cyclase,** also known as *adenylyl cyclase.* This enzyme catalyzes the formation of cyclic-AMP from ATP at the inner surface of the cell membrane. **Cyclic-AMP (cAMP)** is a second messenger that may open membrane channels, may activate intracellular enzymes, or may do both, depending on the nature of the postsynaptic cell. We shall encounter other second messengers, including *cyclic-GMP (cGMP)*, calcium ions, and *inositol triphosphate*, in later chapters.

3. *Lipid-soluble gases.* Gases such as NO and CO are lipid-soluble, so they can diffuse through the membrane lipids and bind to enzymes on the inner surface of the membrane. These enzymes then promote the appearance of second messengers that can affect cellular activity.

It can be very difficult to distinguish neurotransmitters from neuromodulators on biochemical or functional grounds. In general, neuromodulators (1) have long-term effects that are relatively slow to appear; (2) trigger responses that involve second messengers; (3) may affect the presynaptic membrane, the postsynaptic membrane, or both; and (4) can be released alone or in the company of a neurotransmitter. However, the same compound may function in one site as a neurotransmitter and in another as a neuromodulator. For this reason, Table 12-6 does not distinguish between neurotransmitters and neuromodulators.

☑ What effect would blocking voltage-regulated calcium channels at a cholinergic synapse have on synaptic communication?

☑ One pathway in the central nervous system consists of three neurons, and another consists of five neurons. If the neurons in the two pathways are identical, which pathway will transmit impulses more rapidly?

☑ Norepinephrine causes the hyperpolarization of smooth muscle cells in the blood vessels that serve skeletal muscles, but it causes the depolarization of smooth muscle cells in blood vessels that supply the intestines. How is this possible?

INFORMATION PROCESSING

At each synapse, an action potential arriving at a synaptic knob triggers chemical or electrical events that affect another cell. Some of the neurotransmitters arriving at the postsynaptic cell at any moment may be excitatory, whereas others, from different presynaptic neurons, may be inhibitory. The net result may be a depolarization, a hyperpolarization, or no appreciable change in the transmembrane potential at the initial segment. The transmembrane potential at the initial segment therefore represents an integration of all the excitatory and inhibitory stimuli affecting the neuron at that moment. Excitatory and inhibitory stimuli are integrated through interactions between *postsynaptic potentials*. This interaction is the simplest level of **information processing** in the nervous system.

POSTSYNAPTIC POTENTIALS

Postsynaptic potentials are graded potentials that develop in the postsynaptic membrane in response to a neurotransmitter. Two major types of postsynaptic potentials develop at neuron-to-neuron synapses: (1) excitatory postsynaptic potentials and (2) inhibitory postsynaptic potentials.

An **excitatory postsynaptic potential,** or **EPSP,** is a graded depolarization caused by the arrival of a neurotransmitter at the postsynaptic membrane. An EPSP results from the opening of chemically regulated membrane channels. The depolarization produced by the binding of ACh is an EPSP. Because it is a graded potential, an EPSP affects only the area immediately surrounding the synapse, as in Figure 12-13•, p. 378.

We have already learned that not all neurotransmitters have an excitatory (depolarizing) effect. An **inhibitory postsynaptic potential,** or **IPSP,** is a transient hyperpolarization of the postsynaptic membrane. For example, an IPSP may result from the opening of chemically regulated potassium channels. While the hyperpolarization continues, the neuron is **inhibited,** because a larger-than-usual depolarizing stimulus must be provided to bring the membrane potential to threshold. A stimulus sufficient to shift the transmembrane potential by 10 mV (from −70 mV to −60 mV) would normally produce an action potential. But if the transmembrane potential were reset at −85 mV by an IPSP, the same stimulus would depolarize it to only −75 mV— well below threshold.

Summation

Summation is the mechanism responsible for the integration of EPSPs, IPSPs, or both in the postsynaptic neuron. We will use the summation of EPSPs to demonstrate this process. An individual EPSP has a small effect on the transmembrane potential; a typical EPSP produces a depolarization of about 0.5 mV at the postsynaptic membrane. Before an action potential will appear in the initial segment, local currents must depolarize that region by at least 10 mV. A single EPSP will therefore not result in an action potential, even if the synapse is on the axon hillock. But individual EPSPs combine through the process of summation. Two forms of summation exist: (1) temporal summation and (2) spatial summation.

Temporal summation (*tempus,* time) is the addition of stimuli occurring in rapid succession. It occurs at a *single synapse* that is active *repeatedly.* Consider that you cannot fill a bathtub with a single bucket of water, but you can fill it if you keep using the bucket over and over. The water in each bucketful corresponds to the sodium ions that enter

the cytoplasm during an EPSP. Suppose that a second EPSP arrives before the effects of the first have disappeared (Figure 12-20a•). A typical EPSP lasts about 20 msec, but under maximum stimulation one action potential can reach the synaptic knob each millisecond. Every time an action potential arrives, a group of vesicles discharges ACh into the synaptic cleft. Every time more ACh molecules arrive at the postsynaptic

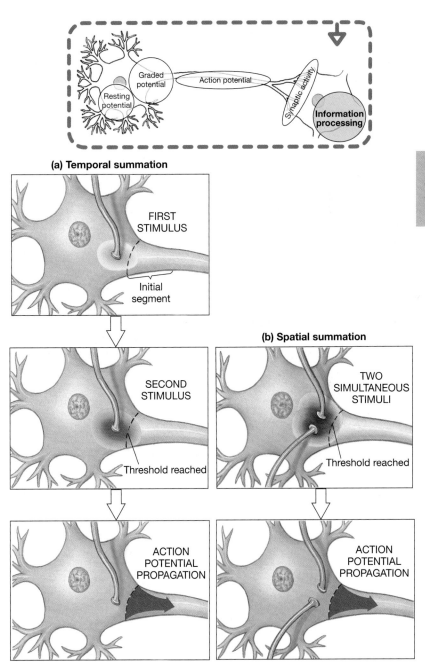

(a) Temporal summation

FIRST STIMULUS

Initial segment

SECOND STIMULUS

Threshold reached

ACTION POTENTIAL PROPAGATION

(b) Spatial summation

TWO SIMULTANEOUS STIMULI

Threshold reached

ACTION POTENTIAL PROPAGATION

•**FIGURE 12-20** **Temporal and Spatial Summation.** **(a)** Temporal summation occurs on a membrane that receives two depolarizing stimuli separated in time. The effects of the second stimulus are added to those of the first. **(b)** Spatial summation occurs when sources of stimulation arrive simultaneously but at different locations. Local currents spread the depolarizing effects, and areas of overlap experience the combined effects.

membrane, more chemically regulated channels open and the degree of depolarization increases. In this way, a series of small steps can eventually bring the initial segment to threshold.

Spatial summation occurs when simultaneous stimuli at different locations have a cumulative effect on the transmembrane potential. Spatial summation involves *multiple synapses* that are active *simultaneously.* In terms of our bucket analogy, you could fill a bathtub immediately if 50 friends emptied their buckets into the bathtub at the same time.

The activity of one synapse produces a graded potential with localized effects. If more than one synapse is active at the same time, all will "pour" sodium ions across the postsynaptic membrane. At each active synapse, the sodium ions that produce the EPSP spread out along the inner surface of the membrane and mingle with those entering at other synapses. As a result, the effects on the initial segment are cumulative (Figure 12-20b•). The degree of depolarization dep-ends on how many synapses are active at any moment and on their distance from the initial segment. As in temporal summation, an action potential appears when the transmembrane potential at the initial segment reaches threshold.

Facilitation   Spatial or temporal summation of EPSPs may not depolarize the initial segment to threshold, but every step closer to threshold makes it easier for the *next* stimulus to trigger an action potential. A neuron that is brought closer to threshold is said to be **facilitated.** The larger the degree of facilitation, the smaller the additional stimulus needed to trigger an action potential. In a highly facilitated neuron, even a small depolarizing stimulus produces an action potential.

Facilitation also results from neuron exposure to certain drugs in the extracellular fluid. For example, nicotine stimulates postsynaptic ACh receptors, producing prolonged EPSPs that facilitate CNS neurons. The active ingredients of coffee and colas (caffeine), cocoa (theobromine), and tea (theophylline) also cause facilitation, but they act in a different way. These compounds lower the threshold at the initial segment, so a smaller-than-usual depolarization will cause an action potential. Once an action potential reaches the synaptic knob, these compounds also increase the amount of ACh released. The nervous system becomes more excitable, and you feel more alert. "Coffee makes you jumpy" is not an exaggeration; after several cups, you may jump at a sudden noise because your reflexes are primed to respond to the slightest stimulus. Because nicotine and caffeine stimulate in different ways, the combination of cigarettes and coffee has a dramatic effect on the CNS.

Summation of EPSPs and IPSPs   Like EPSPs, IPSPs summate spatially and temporally. EPSPs and IPSPs reflect the activation of different types of chemically regulated channels; they have opposing effects on the transmembrane potential. The antagonism between IPSPs and EPSPs is an important mechanism for cellular information processing. In terms of our bucket analogy, EPSPs put water into the bathtub, and IPSPs take water out. If more buckets add water than remove water, the water level in the tub will rise. If more buckets remove water, the level will fall. If a bucket of water is removed every time another bucket is dumped in, the level will remain stable. Comparable interactions between EPSPs and IPSPs (Figure 12-21•) determine the transmembrane potential at the boundary between the axon hillock and the initial segment.

•**FIGURE 12-21 EPSP–IPSP Interactions.** At time 1, a small depolarizing stimulus produces an EPSP. At time 2, a small hyperpolarizing stimulus produces an IPSP of comparable magnitude. If the two stimuli are applied simultaneously, as at time 3, summation occurs. Because the two are equal in size, the membrane potential remains at the resting level. If the EPSP were larger, a net depolarization would result; if the IPSP were larger, a net hyperpolarization would result instead.

Neuromodulators, hormones, or both can change the postsynaptic membrane's sensitivity to excitatory or inhibitory neurotransmitters. By shifting the balance between EPSPs and IPSPs, these compounds promote facilitation or inhibition of CNS and PNS neurons.

Chloride Channels and EPSP–IPSP Summation Many chemicals that inhibit EPSP formation open chemically regulated chloride channels rather than potassium channels. The opening of chemically regulated chloride channels in a resting membrane is difficult to detect, because the equilibrium potential for chloride ions is -70 mV, the same as the typical resting potential. However, with the chloride channels open, Cl^- is free to cross the membrane if the transmembrane potential changes. Consider what happens if the cell receives a depolarizing stimulus that would, under other circumstances, shift the transmembrane potential from -70 mV to -65 mV. As soon as the potential climbs above -70 mV, chloride ions start rushing into the cell. The situation now is like a bathtub with a large hole just above the normal water level; to raise the water level, you must add much more water—enough to overcome the water loss through the hole. In a similar way, once the chloride ion channels are open, a much larger-than-normal stimulus is needed to depolarize the membrane to threshold.

PRESYNAPTIC INHIBITION AND FACILITATION

Inhibitory or excitatory responses may occur at an *axoaxonal synapse,* a synapse between the axon of one neuron and the axon of another. An axoaxonal synapse that occurs on the synaptic knob can modify the rate of neurotransmitter release at the presynaptic membrane. In one form of **presynaptic inhibition,** the release of GABA inhibits the opening of voltage-regulated calcium channels in the synaptic knob (Figure 12-22a•). This inhibition reduces the amount of neurotransmitter released when an action potential arrives at the synaptic knob and thus limits the effects on the postsynaptic membrane.

In **presynaptic facilitation,** activity at an axoaxonal synapse increases the amount of neurotransmitter released when an action potential arrives at the synaptic knob (Figure 12-22b•). This increase enhances and prolongs the neurotransmitter's effects on the postsynaptic membrane. The neurotransmitter serotonin is involved in presynaptic facilitation. In the presence of serotonin released at an axoaxonal synapse, voltage-regulated calcium channels remain open longer.

12

THE RATE OF ACTION POTENTIAL GENERATION

In a computer, complex information is translated into a simple binary code of ones and zeros. In the nervous system, all sensory information and motor commands must be translated into action potentials that can be propagated along axons. On arrival, the message is

(a) Presynaptic inhibition

(b) Presynaptic facilitation

•**FIGURE 12-22 Presynaptic Inhibition and Facilitation. (a)** Steps in presynaptic inhibition. **(b)** Steps in presynaptic facilitation.

often interpreted solely on the basis of the frequency of action potentials. For example, action potentials arriving at a neuromuscular junction at the rate of 1 per second may produce a series of isolated twitches in the associated skeletal muscle fiber, whereas at the rate of 100 per second they will cause a sustained tetanic contraction. A few action potentials per second along a sensory fiber may be perceived as a feather-light touch, whereas hundreds of action potentials per second along that same axon may be perceived as unbearable pressure. In this section, we will examine factors that vary the rate of action potential generation; we shall consider the functional significance of these changes in later chapters.

If a graded potential briefly depolarizes the axon hillock such that the initial segment reaches threshold, an action potential will be propagated along the axon. Now consider what happens if the axon hillock *remains* depolarized past threshold levels for an extended period. The longer the initial segment remains above threshold, the more action potentials it will produce. The *frequency* of action potentials depends on the degree of depolarization above threshold: The greater the degree of depolarization, the higher the frequency of action potentials. The membrane can respond to a second stimulus as soon as the absolute refractory period ends. Holding the membrane above threshold has the same effect as applyng a second, larger-than-normal stimulus.

The rate of action potential generation reaches a maximum when the relative refractory period has been completely eliminated. The maximum theoretical frequency is therefore established by the duration of the absolute refractory period. The absolute refractory period is shortest in large-diameter axons, in which the *theoretical* maximum frequency of action potentials is 2500 per second. However, the highest frequencies recorded from axons in the body range between 500 and 1000 per second.

Table 12-7 summarizes the basic principles of information processing.

HIGHER LEVELS OF ORGANIZATION AND PROCESSING

FUNCTIONAL ORGANIZATION

The human body has about 10 million sensory neurons, 20 billion interneurons, and one-half million motor neurons. The interneurons are organized into a smaller number of neuronal pools. A **neuronal pool** is a group of interconnected neurons with specific functions. Neuronal pools are defined on the basis of function. A pool may be diffuse, involving neurons in several regions of the brain, or localized, with neurons restricted to one specific location in the brain or spinal cord. Estimates of the actual number of pools range between a few hundred and a few thousand. Each pool has a limited number of input sources and output destinations, and each may contain both excitatory and inhibitory neurons. The output of the entire pool may stimulate or depress the activity of other pools, or it may exert direct control over motor neurons or peripheral effectors.

The pattern of interaction among neurons provides clues to the functional characteristics of a neuronal pool. It is customary to refer to such a "wiring diagram" as a neural circuit, just as we refer to electrical circuits in the wiring of a house. We can distinguish five circuit patterns:

1. **Divergence** is the spread of information from one neuron to several neurons (Figure 12-23a•) or from one pool to multiple pools. Divergence permits the broad distribution of a specific input. Considerable divergence occurs when sensory neurons bring information into the CNS, for the information is distributed to neuronal pools throughout the spinal cord and brain. For example, visual information arriving from the eyes reaches your conscious awareness at the same time it is distributed to areas of the brain that control posture and balance at the subconscious level.

2. In **convergence,** several neurons synapse on the same postsynaptic neuron (Figure 12-23b•). Several patterns of activity in the presynaptic neurons can there-

---

TABLE 12-7 Summary: Information Processing

- The neurotransmitters released at a synapse may have excitatory or inhibitory effects. The effect on the axon's initial segment reflects a summation of the stimuli that arrive at any moment. The frequency of action potential generation is an indication of the degree of sustained depolarization at the axon hillock.

- Neuromodulators can alter the rate of neurotransmitter release or the response of a postsynaptic neuron to specific neurotransmitters.

- Neurons may be facilitated or inhibited by extracellular chemicals other than neurotransmitters or neuromodulators.

- The effect of a presynaptic neuron's activation on a postsynaptic neuron at a synapse may be altered by other neurons, through presynaptic inhibition or presynaptic facilitation.

(a) Divergence

(c) Serial processing

(b) Convergence

(d) Parallel processing

(e) Reverberation

•**FIGURE 12-23** The Organization of Neuronal Pools. **(a)** Divergence, a mechanism for spreading stimulation to multiple neurons or neuronal pools in the CNS. **(b)** Convergence, a mechanism providing input to a single neuron from multiple sources. **(c)** Serial processing, in which neurons or pools work sequentially. **(d)** Parallel processing, in which neurons or pools process information simultaneously. **(e)** Reverberation, a positive feedback mechanism.

12

fore have the same effect on the postsynaptic neuron. Through convergence, the same motor neurons can be subject to both conscious and subconscious control. For example, the movements of your diaphragm and ribs are now being controlled by your brain at the subconscious level. But the same motor neurons can also be controlled consciously, as when you take a deep breath and hold it. Two neuronal pools are involved, both synapsing on the same motor neurons.

3. Information may be relayed in a stepwise fashion, from one neuron to another or from one neuronal pool to the next. This pattern is called **serial processing** (Figure 12-23c•). Serial processing occurs as sensory information is relayed from one part of the brain to another. For example, pain sensations en route to your conscious awareness make two stops along the way, at neuronal pools along the pain pathway.

4. **Parallel processing** occurs when several neurons or neuronal pools process the same information at one time (Figure 12-23d•). Divergence must take place before parallel processing can occur. Thanks to parallel processing, many responses can occur simulta-

neously. For example, when you step on a sharp object, sensory neurons that distribute the information to a number of neuronal pools are stimulated. As a result of parallel processing, you might withdraw your foot, shift your weight, move your arms, feel the pain, and shout "Ouch!" all at the same time.

5. Some neural circuits utilize positive feedback to produce **reverberation.** In this arrangement, collateral branches of axons somewhere along the sequence extend back toward the source of an impulse and further stimulate the presynaptic neurons. Once a reverberating circuit has been activated, it will continue to function until synaptic fatigue or inhibitory stimuli break the cycle. As can convergence and divergence, reverberation can occur within a single neuronal pool, or it may involve a series of interconnected pools. A simple example of reverberation is shown in Figure 12-23e•. Reverberation is like a positive feedback loop involving neurons; once the circuit is activated, it continues to stimulate itself. Very complicated examples of reverberation among neuronal pools in the brain may help maintain consciousness, muscular coordination, and normal breathing.

Factors That Disrupt Neural Function

Two key factors can disrupt normal CNS and PNS functions: (1) changes in the extracellular environment and (2) an inability to meet the metabolic demands of active neurons.

Environmental Factors

Environmental factors, such as changes in pH, ionic composition, or temperature, can alter the resting membrane potential or disrupt the metabolic operations that support action potential generation:

■ *Changes in hydrogen ion concentration (pH) can have dramatic effects on neural activity.* The normal extracellular pH averages 7.35–7.45. If the pH rises, neurons are facilitated: At a pH near 7.8, they begin to generate action potentials spontaneously, producing severe convulsions. If the pH drops, neurons are inhibited: At a pH around 7.0, the rate of action potential generation is so low that the nervous system shuts down, and the individual becomes completely unresponsive.

■ *Changes in the ionic composition of the extracellular fluids have a direct effect on neural function.* Fluctuations in Na^+ or K^+ concentrations, such as those caused by dehydration or kidney disease, may facilitate or inhibit neural activity by depolarizing or hyperpolarizing the cell membrane. For example, an elevated extracellular K^+ concentration, a condition called **hyperkalemia** (hī-per-ka-LĒ-mē-uh; *kalium*, potassium + *haima*, blood) reduces the chemical gradient for K^+ across the cell membrane. The rate of potassium loss decreases as a result, and the retention of positive charges gradually depolarizes the membrane. Hyperkalemia has damaging effects on all excitable membranes. A slight hyperkalemia causes facilitation, but extreme hyperkalemia interferes with repolarization and suppresses action potential generation. The result is a general paralysis of skeletal muscles and death by cardiac arrest. Abnormally high or low extracellular Ca^{2+} concentrations affect synaptic function by reducing or enhancing calcium entry into the synaptic knob, thereby modifying the amount of neurotransmitter released. There are also direct effects on the excitability of membranes.

■ *Changes in body temperature have a direct effect on the activity of neurons.* This is one reason the regulation of body temperature is so important. If your body temperature rises, your neurons become more excitable; an individual with a high fever may experience hallucinations or convulsions. If your body temperature falls, your neurons become inhibited; an individual whose body temperature has fallen outside normal limits will be lethargic and confused and may become unconscious.

Metabolic Processes

The brain accounts for just 2 percent of your body weight, but it accounts for 18 percent of your resting energy consumption. Active neurons need ATP to support (1) the synthesis, release, and recycling of neurotransmitter molecules; (2) the movement of materials to and from the cell body by axoplasmic transport; (3) the maintenance of the normal resting potential; and (4) the recovery from action potentials through the activity of the sodium–potassium exchange pump. Each time an action potential occurs, sodium ions enter and potassium ions leave the cell; over time, ATP must be expended to maintain normal cytoplasmic ion concentrations. When impulses are generated at high frequencies, the energy demands are enormous.

Neurons normally derive ATP solely through aerobic mechanisms. Because their cytoplasm does not contain glycogen reserves, these cells are totally dependent on a continuous and reliable supply of both oxygen and glucose from the blood. In cases of severe malnutrition, neural function deteriorates as the body's energy reserves are exhausted.

Neural function is also impaired if the local circulation is restricted or, worse yet, shut off. If the circulation to a region is interrupted for just a few seconds, neurons in that region will be injured. The longer the interruption, the more severe the injury will be. In a **stroke,** the blood supply to the brain is interrupted by a circulatory blockage or by vascular rupture. The degree of functional impairment after a stroke is determined by (1) the location and size of the region deprived of circulation and (2) the duration of the circulatory interruption. In subsequent chapters, we will consider the origins and treatment of strokes in greater detail.

Several inherited abnormalities in neural function are caused by metabolic problems in neurons. For example, *Tay-Sachs disease* is a genetic abnormality involving the metabolism of *gangliosides*, glycolipids that are important components of neuron cell membranes.

AM   *Tay-Sachs Disease*

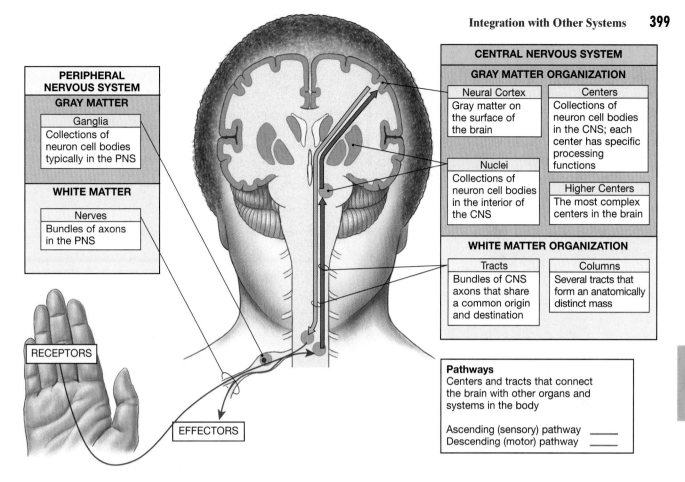

•FIGURE 12-24 An Introduction to the Anatomical Organization of the Nervous System

ANATOMICAL ORGANIZATION

The functions of the nervous system depend on the interactions among neurons in neuronal pools; the most complex neural processing steps occur in the spinal cord and brain. Neurons and their axons are not randomly scattered in the CNS and PNS. Instead, they form masses or bundles with distinct anatomical boundaries, and they are identified by specific terms. We shall use these terms in all later chapters, so we offer a brief overview here (Figure 12-24•):

In the PNS:

- Neuron cell bodies are located in *ganglia.*

- Axons are bundled together in *nerves,* with *spinal nerves* connected to the spinal cord and *cranial nerves* connected to the brain.

In the CNS:

- A collection of neuron cell bodies with a common function is called a **center.** A center with a discrete anatomical boundary is a **nucleus.** Portions of the brain surface are covered by a thick layer of gray matter called the **neural cortex.** The term *higher centers* refers to the most complex integration centers, nuclei, and cortical areas of the brain.

- The white matter of the CNS contains bundles of axons that share common origins, destinations, and

functions. These bundles are called **tracts.** Tracts in the spinal cord form larger groups called **columns.**

- The centers and tracts that link the brain with the rest of the body are called **pathways.** For example, **sensory pathways** distribute information from peripheral receptors to processing centers in the brain, and **motor pathways** begin at CNS centers concerned with motor control and end at the effectors they control.

☑ A neuron from a cutaneous receptor synapses with a motor neuron located in the spinal cord. A motor neuron from the brain also synapses with the motor neuron in the spinal cord. Which type of neuronal pool does this arrangement represent?

☑ In an injury involving a peripheral nerve, why is it important for the two ends of the damaged nerve to be closely aligned?

☑ How would damage to interneurons in the spinal cord affect nervous system function?

INTEGRATION WITH OTHER SYSTEMS

Figure 12-25• diagrams the relationships between the nervous system and other physiological systems. We shall explore many of these relationships in greater detail in subsequent chapters.

INTEGUMENTARY SYSTEM

Provides sensations of touch, pressure, pain, vibration, and temperature; hair provides some protection and insulation for skull and brain; protects peripheral nerves

Controls contraction of arrector pili muscles and secretion of sweat glands

THE NERVOUS SYSTEM

FOR ALL SYSTEMS

Monitors pressure, pain, and temperature; adjusts tissue blood flow patterns

Sex hormones affect CNS development and sexual behaviors

Controls sexual behaviors and sexual function

REPRODUCTIVE SYSTEM

SKELETAL SYSTEM

Provides calcium for neural function; protects brain and spinal cord

Controls skeletal muscle contractions that promote bone thickening and maintenance and determine bone position

MUSCULAR SYSTEM

Facial muscles express emotional state; intrinsic laryngeal muscles permit communication; muscle spindles provide proprioceptive sensations

Controls skeletal muscle contractions; coordinates respiratory and cardiovascular activities

ENDOCRINE SYSTEM

Many hormones affect CNS neural metabolism; reproductive hormones and thyroid hormone influence CNS development and behavior

Controls pituitary gland and many other endocrine organs; secretes ADH and oxytocin

CARDIOVASCULAR SYSTEM

Endothelial cells maintain blood–brain barrier; blood vessels (with ependymal cells) produce CSF

Modifies heart rate and blood pressure

LYMPHATIC SYSTEM

Defends against infection and assists in tissue repairs

Release of neurotransmitters and hormones affects sensitivity of immune response

RESPIRATORY SYSTEM

Provides oxygen and eliminates carbon dioxide

Controls pace and depth of respiration

DIGESTIVE SYSTEM

Provides nutrients for energy production and neurotransmitter synthesis

Regulates digestive tract movement and secretion

URINARY SYSTEM

Eliminates metabolic wastes; regulates body fluid pH and electrolyte concentrations

Adjusts renal blood pressure; controls urination

•**FIGURE 12-25** Functional Relationships between the Nervous System and Other Systems

[EXPLORE] [MediaLab]

As you have seen, nervous tissue is amazing. The maintenance of membrane potentials and the generation of action potentials through ion movement happen continuously in the body. To help you understand these processes, the CD-ROM and the Companion Website offer some interactive resources. The CD-ROM includes animations and tutorials; the website builds on those resources with text content and interactive problems as well as communication avenues for sharing ideas with your class. To complete the following Web Explorations, visit the MediaLab in Chapter 12 of your Companion Website.

Web Exploration 1

Estimated time for completion = 30 minutes

One of the central concepts in this chapter is the mechanism of action potential generation and propagation in a neuron. The two ions involved (sodium and potassium) must remain in the proper concentrations on either side of the cell membrane for this process to function properly. For a wonderful simulation of a laboratory experiment on action potentials, select the keyword **NERVE IMPULSE** on the Companion Website. There you can alter the concentrations of sodium

and potassium on either side of the membrane of an active neuron. Design your own experiment to determine what happens to the action potential as the sodium ion concentration declines outside the neuron. Report your findings, using the communications tools on the Companion Website. Design a second experiment and post it as well.

Web Exploration 2

Estimated time for completion = 10 minutes

Neuroglia are an interesting group of cells in neural tissue. Figure 12-7 and the accompanying text in the subsection "Neuroglia of the Central Nervous System" describe these cells and their functions. What does actor/director Christopher Reeve know about neuroglia? Since damaging his spinal cord in a horseback-riding accident, he has been working to find a cure for spinal cord injury. Select the keyword **CHRISTOPHER REEVE** on the Companion Website to learn the current state of research on neuroglia. Read the section of the news release titled "Brain and Spinal Cord Injury." What does this research seem to indicate for future spinal-cord injury victims? Prepare an outline of your thoughts to present to your classmates as a discussion topic.

12

CHAPTER REVIEW

SELECTED CLINICAL TERMINOLOGY
Terms Discussed in This Chapter

anticholinesterase drug: A drug that blocks the breakdown of ACh by AChE. *(p. 390)*

atropine: A drug that prevents ACh from binding to the postsynaptic membrane of cardiac muscle and smooth muscle cells. *(p. 390)*

demyelination: The destruction of the myelin sheaths around axons in the CNS and PNS. *(p. 370 and AM)*

d-tubocurarine: A drug, derived from curare, that prevents ACh from binding to the postsynaptic membrane of skeletal muscle fibers. *(p. 390)*

endorphins (en-DOR-finz):   Neuropeptides produced in the brain and spinal cord that appear to relieve pain and to affect mood. *(p. 392)*

hyperkalemia (hī-per-ka-LĒ-mē-uh):   An abnormal physiological state resulting from a high extracellular concentration of potassium. *(p. 398)*

neurotoxin: A compound that disrupts normal nervous system function by interfering with the generation or propagation of action potentials. Examples include *tetrodotoxin (TTX), saxitoxin (STX), paralytic shellfish poisoning (PSP),* and *ciguatoxin (CTX)*. *(p. 382 and AM)*

nicotine: A compound, found in tobacco, that binds to specific ACh receptor sites and stimulates the postsynaptic membrane. *(p. 390)*

rabies: A fatal disease caused by a virus that reaches the CNS by retrograde flow along peripheral axons. *(p. 365 and AM)*

Tay-Sachs disease: A genetic abnormality involving the metabolism of gangliosides, important components of neuron cell membranes. The result is a gradual deterioration of neurons due to the buildup of metabolic byproducts and the release of lysosomal enzymes. *(p. 398 and AM)*

[AM] Additional Terms Discussed in the *Applications Manual*

diphtheria (dif-THĒ-rē-uh):   A disease that results from a bacterial infection of the respiratory tract. Among other effects, the bacterial toxins damage Schwann cells and cause PNS demyelination.

Guillain–Barré syndrome: A progressive demyelination that appears to be linked with the immune response to a viral infection.

multiple sclerosis (skler-Ō-sis) **(MS):**   A disease marked by recurrent incidents of demyelination affecting axons in the optic nerve, brain, and/or spinal cord.

STUDY OUTLINE

INTRODUCTION   p. 361

1. The nervous and endocrine systems coordinate the activities of other organ systems. The *nervous system* provides swift but brief responses to stimuli; the *endocrine system* adjusts metabolic operations and directs long-term changes.

AN OVERVIEW OF THE NERVOUS SYSTEM   p. 361

1. The nervous system includes all the **neural tissue** in the body. The basic functional unit is the **neuron.** The anatomical divisions of the nervous system are the **central nervous system (CNS)** (the brain and spinal cord) and the **peripheral nervous system (PNS)** (all the neural tissue outside the CNS). Bundles of *axons* (*nerve fibers*) in the PNS are called **nerves.**

2. Functionally, the PNS can be divided into an **afferent division,** which brings sensory information to the CNS, and an **efferent division,** which carries motor commands to muscles and glands, called **effectors.** *(Figure 12-1)*

3. The efferent division of the PNS includes the **somatic nervous system (SNS)** (which controls skeletal muscle contractions) and the **autonomic nervous system (ANS)** (which regulates smooth muscle, cardiac muscle, and glandular activity). *(Figure 12-1)*

NEURONS   p. 362

Neuron Structure   p. 362

1. The **perikaryon** of a multipolar neuron contains organelles, including **neurofilaments, neurotubules,** and **neurofibrils.** The **axon hillock** connects the **initial segment** of the **axon** to the **cell body,** or **soma.** The **axoplasm** contains numerous organelles. *(Figure 12-2)*

2. **Collaterals** may branch from an axon, with **telodendria** branching from the axon's tip.

3. A **synapse** is a site of intercellular communication. *(Figures 12-2, 12-3)*

Neuron Classification   p. 365

4. Neurons are described histologically as **anaxonic, unipolar, bipolar,** or **multipolar.** *(Figure 12-4)*

5. The three functional categories of neurons are sensory neurons, motor neurons, and interneurons. *(Figure 12-5)*

6. **Sensory neurons** form the afferent division of the PNS. They deliver information to the CNS. *(Figure 12-5)*

7. **Motor neurons** form the efferent division of the PNS. These neurons stimulate or modify the activity of a peripheral tissue, organ, or organ system. *(Figure 12-5)*

8. **Interneurons** (*association neurons*) are always located in the CNS and may be situated between sensory and motor neurons. They analyze sensory inputs and coordinate motor outputs. *(Figure 12-5)*

NEUROGLIA   p. 367

Neuroglia of the Central Nervous System   p. 367

1. The four types of **neuroglia,** or *glial cells,* in the CNS are (1) **ependymal cells,** with functions related to the **cerebrospinal fluid (CSF);** (2) **astrocytes,** the largest and most numerous neuroglia; (3) **oligodendrocytes,** which are responsible for the **myelination** of CNS axons; and (4) **microglia,** or phagocytic cells. *(Figures 12-6, 12-7)*

Neuroglia of the Peripheral Nervous System   p. 370

2. Neuron cell bodies in the PNS are clustered into **ganglia,** *(Figure 12-8)*

3. **Satellite cells,** or *amphicytes,* surround neuron cell bodies within ganglia. **Schwann cells** ensheath axons in the PNS. A single Schwann cell may myelinate one segment of an axon or enfold segments of several unmyelinated axons. *(Figure 12-8)*

FOCUS: Neural Responses to Injuries   p. 372

4. In the PNS, functional repair may follow **Wallerian degeneration.** In the CNS, many factors complicate the repair process and reduce the chances of functional recovery. *(Figure 12-9)*

NEUROPHYSIOLOGY   p. 373

The Transmembrane Potential   p. 373

1. The **electrochemical gradient** is the sum of all chemical and electrical forces acting across the membrane. *(Figures 12-10, 12-11)*

2. The *resting potential* of a neuron, about -70 mV, is determined chiefly by the membrane permeability to potassium ions. *(Figure 12-11; Table 12-1)*

Changes in the Transmembrane Potential   p. 376

3. The cell membrane contains **passive (leak) channels,** which are always open, and **active (gated) channels,** which open or close in response to specific stimuli.

4. The three types of gated channels are **chemically regulated channels, voltage-regulated channels,** and **mechanically regulated channels.** *(Figure 12-12)*

Graded Potentials   p. 377

5. A localized **depolarization** or **hyperpolarization** is a **graded potential** (a change in potential that decreases with distance). *(Figures 12-13, 12-14; Table 12-2)*

Action Potentials   p. 380

6. An **action potential** appears when a region of excitable membrane depolarizes to **threshold.** The steps involved include membrane depolarization and the activation of sodium channels, sodium channel inactivation, potassium channel activation, and the return to normal permeability. *(Figure 12-15; Tables 12-3, 12-4)*

7. The activity of the sodium–potassium exchange pump is necessary to maintain ion concentrations within acceptable limits over time.

8. In **continuous propagation** (*continuous conduction*), an action potential spreads across the entire excitable membrane surface in a series of small steps. *(Figure 12-16)*

9. In **saltatory propagation** (*saltatory conduction*), the action potential appears to leap from node to node, skipping the intervening membrane surface. Saltatory propagation carries nerve impulses many times more rapidly than does continuous propagation. *(Figure 12-17)*

10. Axons are classified as **Type A fibers, Type B fibers,** or **Type C fibers** on the basis of diameter, myelination, and propagation speed.

11. Compared with action potentials in neural tissue, those in muscle tissue have (1) higher resting potentials, (2) longer-lasting action potentials, and (3) slower action potential propagation.

SYNAPTIC ACTIVITY   p. 386

1. An action potential traveling along an axon is a **nerve impulse.** At a synapse between two neurons, information passes from the **presynaptic neuron** to the **postsynaptic neuron.** A synapse may also involve other types of postsynaptic effector cells.

2. A synapse is either electrical (with direct physical contact between cells) or chemical (involving a neurotransmitter).

Electrical Synapses   p. 386

3. **Electrical synapses** occur in the CNS and PNS, but they are rare. At an electrical synapse, the presynaptic and postsynaptic cell membranes are bound by interlocking membrane proteins at a gap junction. Pores within these proteins permit the passage of local currents, and the two neurons act as if they share a common cell membrane.

Chemical Synapses   p. 387

4. **Chemical synapses** are more common than electrical synapses. **Excitatory neurotransmitters** cause depolarization and promote action potential generation, whereas **inhibitory neurotransmitters** cause hyperpolarization and depress action potential generation.

5. The effect of a neurotransmitter on the postsynaptic membrane depends on the properties of the receptor, not on the nature of the neurotransmitter.

6. **Cholinergic synapses** release the neurotransmitter **acetylcholine (ACh).** Communication moves from the presynaptic neuron to the postsynaptic neuron. The separation of the presynaptic membrane and postsynaptic membrane by a synaptic cleft causes a **synaptic delay.** *(Figures 12-18, 12-19; Table 12-5)*

7. The **choline** released during the breakdown of ACh in the synaptic cleft is reabsorbed and recycled by the synaptic knob. If stores of ACh are exhausted, **synaptic fatigue** can occur. *(Table 12-5)*

8. **Adrenergic synapses** release **norepinephrine (NE),** also called *noradrenaline.* Other important neurotransmitters include **dopamine, serotonin,** and **gamma aminobutyric acid (GABA).** *(Table 12-6)*

9. **Neuromodulators** influence the postsynaptic cell's response to neurotransmitters.

INFORMATION PROCESSING   p. 392

1. Excitatory and inhibitory stimuli are integrated through interactions between **postsynaptic potentials.** This interaction is the simplest level of **information processing** in the nervous system.

Postsynaptic Potentials   p. 392

2. A depolarization caused by a neurotransmitter is an **excitatory postsynaptic potential (EPSP).** Individual EPSPs can combine through **summation.** Summation can be **temporal** (occurring at a single synapse when a second EPSP arrives before the effects of the first have disappeared) or **spatial** (resulting from the cumulative effects of multiple synapses at various locations). *(Figure 12-20)*

3. Hyperpolarization of the postsynaptic membrane is an **inhibitory postsynaptic potential (IPSP).**

4. The most important determinants of neural activity are EPSP–IPSP interactions. *(Figure 12-21)*

Presynaptic Inhibition and Facilitation   p. 395

5. In **presynaptic inhibition,** GABA release at an *axoaxonal synapse* inhibits the opening of voltage-regulated calcium channels in the synaptic knob. This inhibition reduces the amount of neurotransmitter released when an action potential arrives at the synaptic knob. *(Figure 12-22)*

6. In **presynaptic facilitation,** activity at an axoaxonal synapse increases the amount of neurotransmitter released when an action potential arrives at the synaptic

knob. This increase enhances and prolongs the effects of the neurotransmitter on the postsynaptic membrane. *(Figure 12-22)*

7. The neurotransmitters released at a synapse have excitatory or inhibitory effects. The effect on the initial segment reflects an integration of the stimuli arriving at any moment. The frequency of action potential generation is an indication of the degree of depolarization at the axon hillock. *(Table 12-7)*

8. Neuromodulators can alter the rate of neurotransmitter release or the response of a postsynaptic neuron to specific neurotransmitters. Neurons may be facilitated or inhibited by extracellular chemicals other than neurotransmitters or neuromodulators. *(Table 12-7)*

9. The effect of a presynaptic neuron's activation on a postsynaptic neuron at a synapse may be altered by other neurons. *(Table 12-7)*

The Rate of Action Potential Generation   p. 395

10. The greater the degree of sustained depolarization at the axon hillock, the higher the frequency of action potential generation. At a frequency of about 1000 per second, the relative refractory period has been eliminated and further depolarization will have no effect.

HIGHER LEVELS OF ORGANIZATION AND PROCESSING   p. 396

Functional Organization   p. 396

1. The roughly 20 billion interneurons are classified into **neuronal pools,** groups of interconnected neurons with specific functions.

2. **Divergence** is the spread of information from one neuron or neuronal pool to several neurons or pools. In **convergence,** several neurons synapse on the same postsynaptic neuron. Neurons or pools may also function in sequence **(serial processing),** or they may process the same information at one time **(parallel processing).** In **reverberation,** collateral axons establish a circuit that further stimulates presynaptic neurons. *(Figure 12-23)*

Anatomical Organization   p. 399

3. The functions of the nervous system as a whole depend on interactions among neurons in neuronal pools. In the PNS, *spinal nerves* communicate with the spinal cord, and *cranial nerves* are connected to the brain.

4. In the CNS, a collection of neuron cell bodies that share a particular function is a **center.** A center with a discrete anatomical boundary is a **nucleus.** Portions of the brain surface are covered by a thick layer of gray matter called the **neural cortex.** The term *higher centers* refers to the most complex integration centers, nuclei, and cortical areas of the brain. *(Figure 12-24)*

5. The white matter of the CNS contains bundles of axons called **tracts,** which share common origins, destinations, and functions. Tracts in the spinal cord form larger groups called **columns. Sensory pathways** carry information from peripheral receptors to the brain; **motor pathways** extend from CNS centers concerned with motor control to the associated effectors. *(Figure 12-24)*

INTEGRATION WITH OTHER SYSTEMS   p. 399

1. The nervous system interacts with every other body system. *(Figure 12-25)*

12

REVIEW QUESTIONS

LEVEL 1 REVIEWING FACTS AND TERMS

1. Regulation by the nervous system provides
   (a) relatively slow but long-lasting responses to stimuli
   (b) swift, long-lasting responses to stimuli
   (c) swift but brief responses to stimuli
   (d) relatively slow, short-lived responses to stimuli

2. The efferent division of the PNS
   (a) brings sensory information to the CNS
   (b) carries motor commands to muscles and glands
   (c) processes and integrates sensory data
   (d) is the seat of higher functions in the body

3. The part of the nervous system that controls voluntary contractions of skeletal muscles is the
   (a) somatic nervous system
   (b) autonomic nervous system
   (c) visceral motor system
   (d) sympathetic division of the ANS

4. Smooth muscle, cardiac muscle, and glands are among the targets of the
   (a) somatic nervous system
   (b) sensory neurons
   (c) afferent division of the PNS
   (d) autonomic nervous system

5. The cellular layer that lines the ventricles of the brain and the central canal of the spinal cord is the
   (a) microglia   **(b)** ganglia
   (c) ependyma   **(d)** oligodendrocyte

6. Neuroglia responsible for maintaining the blood–brain barrier are the
   (a) microglia   **(b)** ependymal cells
   (c) astrocytes   **(d)** oligodendrocytes

7. Phagocytic cells in neural tissue of the CNS are
   (a) astrocytes   **(b)** ependymal cells
   (c) oligodendrocytes **(d)** microglia

8. Substances being transported from an axon terminal to the cell body of the same neuron are delivered by
   (a) axoplasmic transport
   (b) synaptic conduction
   (c) retrograde flow
   (d) active transport

9. All the motor neurons that control skeletal muscles are
   (a) multipolar neurons
   (b) myelinated bipolar neurons
   (c) unipolar, unmyelinated sensory neurons
   (d) anaxonic neurons

10. The type of neural cells responsible for the analysis of sensory inputs and coordination of motor outputs are
   (a) neuroglia   **(b)** interneurons
   (c) sensory neurons **(d)** motor neurons

11. Depolarization of a neuron cell membrane will shift the membrane potential toward
   (a) 0 mV   **(b)** −70 mV
   (c) −90 mV   **(d)** a, b, and c are correct

12. The primary determinant of the resting membrane potential is
   (a) the membrane permeability to sodium
   (b) the membrane permeability to potassium
   (c) intracellular negatively charged proteins
   (d) negatively charged chloride ions in the ECF

13. Receptors that bind acetylcholine at the postsynaptic membrane are
   (a) chemically regulated channels
   (b) voltage-regulated channels
   (c) passive channels
   (d) mechanically regulated channels

14. Gated channels that open or close in response to a change in the transmembrane potential are
   (a) mechanically regulated channels
   (b) voltage-regulated channels
   (c) chemically regulated channels
   (d) a, b, and c are correct

15. Changes in the transmembrane potential that are not restricted to the area surrounding the site of stimulation are
   (a) action potentials
   (b) graded potentials
   (c) inhibitory potentials
   (d) hyperpolarizing potentials

16. Neuromodulators are compounds that influence the
   (a) postsynaptic cell's response to a neurotransmitter
   (b) synaptic vesicles in the synaptic knob
   (c) release of calcium ions into the axoplasm
   (d) a, b, and c are correct

17. A transient hyperpolarization of the postsynaptic membrane is
   (a) a refractory period   **(b)** an EPSP
   (c) an IPSP   **(d)** threshold

18. When simultaneous stimuli have a cumulative effect on the transmembrane potential, the process is
   (a) reverberation   **(b)** spatial summation
   (c) divergence   **(d)** temporal summation

19. The synapsing of several neurons on the same postsynaptic neuron is
   (a) serial processing   **(b)** reverberation
   (c) divergence   **(d)** convergence

20. Positive feedback in neural circuits produces
   (a) reverberation   **(b)** divergence
   (c) convergence   **(d)** temporal summation

21. What are the major components of **(a)** the central nervous system? **(b)** the peripheral nervous system?

22. What two major cell populations are found in the nervous system? What is the primary function of each cell type?

23. Which neuroglia **(a)** maintain the blood–brain barrier? **(b)** serve a phagocytic function in the CNS?

24. Which two types of neuroglia insulate neuron cell bodies and axons in the PNS from their surroundings?

25. Classifying neurons on the basis of *structure,* what four types are found in the nervous system?

26. What three *functional* groups of neurons are found in the nervous system? What is the function of each type of neuron?

27. What two integrated steps are necessary for the transfer of nerve impulses from neuron to neuron?

28. What is the *functional* difference among voltage-regulated, chemically regulated, and mechanically regulated channels?

29. What four basic characteristics are associated with graded potentials?

30. State the all-or-none principle of action potentials.

31. Describe the steps involved in the generation of an action potential.

32. Describe the steps that take place at a typical cholinergic synapse.

33. What environmental factors play a role in altering the function of neurons in the CNS and PNS?

LEVEL 2 REVIEWING CONCEPTS

34. If the resting membrane potential is -70 mV and the threshold is -55 mV, a membrane potential of -60 mV will
  (a) produce an action potential
  (b) make it easier to produce an action potential
  (c) make it harder to produce an action potential
  (d) hyperpolarize the membrane

35. A graded potential
  (a) decreases with distance from the point of stimulation
  (b) spreads passively because of local currents
  (c) may involve either depolarization or hyperpolarization
  (d) a, b, and c are correct

36. For an action potential to begin, an area of excitable membrane must
  (a) have its voltage-regulated gates inactivated
  (b) be hyperpolarized
  (c) be depolarized to threshold level
  (d) not be in a relative refractory period

37. During an absolute refractory period, the membrane
  (a) continues to hyperpolarize
  (b) cannot respond to further stimulation
  (c) can respond to a larger-than-normal depolarizing stimulus
  (d) will respond to summated stimulation

38. Action potentials are restricted to areas of excitable membranes that contain voltage-regulated channels:
  (a) true   (b) false

39. The all-or-none principle states that
  (a) the properties of an action potential are independent of the strength of the depolarizing stimulus
  (b) all stimuli will produce action potentials
  (c) all graded potentials will generate action potentials
  (d) if stimulated to threshold, any cell membrane can generate and propagate an action potential

40. The loss of positive ions from the interior of a neuron produces
  (a) depolarization   (b) threshold
  (c) hyperpolarization   (d) an action potential

41. The continuous propagation of an action potential cannot occur in
  (a) myelinated axons   (b) unmyelinated axons
  (c) Type A fibers   (d) Type B fibers

42. Why can't most neurons in the CNS be replaced when they are lost to injury or disease?

43. What purpose do collaterals serve in the nervous system?
44. What is the difference between axoplasmic transport and retrograde flow?
45. What is the difference between continuous propagation and saltatory propagation?
46. How does an action potential in a skeletal muscle fiber differ from that in a neuron?
47. How does the action of a neurotransmitter differ from that of a neuromodulator?
48. What is the difference between temporal summation and spatial summation?
49. How does the neuronal activity of divergence differ from that of convergence?
50. Which functions of neurons necessitate the support of energy from ATP?
51. Why is an electrical synapse a more efficient carrier of nerve impulses from cell to cell than is a chemical synapse?
52. When a runner experiences the "runner's high," why is the suppression of pain common?
53. Multiple sclerosis (MS) is a demyelination disorder. How does this condition produce muscular paralysis and sensory losses?

LEVEL 3 CRITICAL THINKING AND CLINICAL APPLICATIONS

54. If neurons in the CNS lack centrioles and are unable to divide, how can a person develop brain cancer?
55. Harry has a kidney condition that causes changes in his body's electrolyte levels (concentration of ions in the extracellular fluid). As a result, he is exhibiting tachycardia, an abnormally fast heart rate. Which ion is involved, and how does a change in its concentration cause Harry's symptoms?
56. Twenty neurons synapse with a single receptor neuron. Fifteen of these neurons release neurotransmitters that produce EPSPs at the postsynaptic membrane, and the other five release neurotransmitters that produce IPSPs. Each time one of the neurons is stimulated, it releases enough neurotransmitter to produce a 2-mV change in potential at the postsynaptic membrane. If the threshold of the postsynaptic neuron is 10 mV, how many of the excitatory neurons must be stimulated to produce an action potential in the receptor neuron if all five inhibitory neurons are stimulated? (Assume that spatial summation occurs.)
57. The myelination of peripheral neurons occurs rapidly throughout the first year of life. How can this process explain the increased abilities of infants during their first year?
58. A drug that blocks ATPase is introduced into an experimental neuron preparation. The neuron is then repeatedly stimulated, and recordings are made of the response. What effect would you expect to observe?

12

CHAPTER 13

The Spinal Cord and Spinal Nerves

CHAPTER OUTLINE AND OBJECTIVES

INTRODUCTION   p. 407

GROSS ANATOMY OF THE SPINAL CORD   p. 407
Spinal Meninges   p. 409
Sectional Anatomy of the Spinal Cord   p. 412
① Discuss the structure and functions of the spinal cord.
② Describe the three meningeal layers that surround the central nervous system.
③ Explain the roles of white matter and gray matter in processing and relaying sensory information and motor commands.

SPINAL NERVES   p. 414
Peripheral Distribution of Spinal Nerves   p. 414
Nerve Plexuses   p. 416
④ Describe the major components of a spinal nerve.
⑤ Relate the distribution pattern of spinal nerves to the regions they innervate.

AN INTRODUCTION TO REFLEXES   p. 422
The Reflex Arc   p. 422
Classification of Reflexes   p. 423
⑥ Describe the process of a neural reflex.
⑦ Classify the types of reflexes, and explain the functions of each.

SPINAL REFLEXES   p. 425
Monosynaptic Reflexes   p. 425
Polysynaptic Reflexes   p. 428
⑧ Distinguish between the types of motor responses produced by various reflexes.
⑨ Explain how reflexes interact to produce complex behaviors.

Your conscious thoughts and intentions originate in your brain, but your awareness of your limbs and trunk and your control over posture and movements depend on continuous communication between your brain and spinal cord. If something disrupts that communication, sensory and motor losses result. The degree of loss depends on the area of the spinal cord affected. In the case of actor Christopher Reeve, a fall from a horse resulted in a fracture of cervical vertebrae C_1 and C_2. This injury disrupted communication between the brain and the spinal cord inferior to the fracture. Yet most of his somatic and visceral reflexes remain intact and functional, because these reflexes are based in the spinal cord rather than in the brain. In this chapter, we shall examine the spinal cord and its control over a variety of reflexes and motor patterns.

INTEGRATION AND CONTROL OF SPINAL REFLEXES   p. 430
Voluntary Movements and Reflex Motor Patterns   p. 430
⑩ Explain how higher centers control and modify reflex responses.

MEDIALAB and CHAPTER REVIEW   p. 431

Over this and the next three chapters, we will examine the functional organization of the nervous system at increasing levels of complexity. We will begin with the spinal cord and spinal nerves and will consider the interactions between neuronal pools that direct relatively simple spinal reflexes. *Spinal reflexes* are stereotyped motor responses that are triggered by specific stimuli and controlled in the spinal cord. For example, a reflex controlled in the spinal cord, not in the brain, makes you drop a scalding-hot frying pan after you have accidentally grabbed it.

Your spinal cord is structurally and functionally integrated with your brain. Chapter 14 provides an overview of the major components and functions of the brain and cranial nerves. It also discusses the *cranial reflexes,* comparable to those of the spinal cord. We then consider the nervous system as an integrated unit. Chapter 15 covers complex *higher-order functions,* such as memory and learning, that involve many regions of the brain and affect all the activities of the nervous system. That chapter also deals with the interplay between centers in the brain and spinal cord that occurs in the processing of sensory information and in both the reflex control and the voluntary control of skeletal muscle activity. Chapter 16 completes the sequence with an examination of the autonomic nervous system. This system, which has processing centers in the brain, spinal cord, and peripheral nervous system, is responsible for the control of visceral effectors, such as peripheral smooth muscles, cardiac muscle, and glands.

GROSS ANATOMY OF THE SPINAL CORD

The adult spinal cord (Figure 13-1a•) measures approximately 45 cm (18 in.) in length and has a maximum width of roughly 14 mm (0.55 in.). The posterior (dorsal) surface of the spinal cord bears a shallow longitudinal groove, the **posterior median sulcus.** The **anterior median fissure** is a deeper groove along the anterior (ventral) surface.

The amount of gray matter is greatest in segments of the spinal cord that deal with the sensory and motor control of the limbs. These areas are expanded, forming the **enlargements** of the spinal cord. The **cervical enlargement** supplies nerves to the shoulder girdles and upper limbs; the **lumbar enlargement** provides innervation to structures of the pelvis and lower limbs. Inferior to the lumbar enlargement, the spinal cord becomes tapered and conical; this region is the **conus medullaris.** The **filum terminale** ("terminal thread"), a slender strand of fibrous tissue, extends from the inferior tip of the conus medullaris. It continues along the length of the vertebral canal as far as the second sacral vertebra. There it provides longitudinal support

to the spinal cord as a component of the *coccygeal ligament.*

Figure 13-1b• provides a series of sectional views that demonstrate the variations in the relative mass of gray matter and white matter in the cervical, thoracic, lumbar, and sacral regions of the spinal cord. The entire spinal cord can be divided into 31 segments on the basis of the origins of the spinal nerves. Each segment is identified by a letter and number designation, the same method used to identify vertebrae. For example, C_3, the segment in the uppermost section of Figure 13-1•, is the third cervical segment.

Every spinal segment is associated with a pair of **dorsal root ganglia** (Figure 13-1b•). These ganglia, situated near the spinal cord, contain the cell bodies of sensory neurons. The axons of these neurons form the **dorsal roots,** which bring sensory information into the spinal cord. A pair of **ventral roots** contains the axons of motor neurons that extend into the periphery to control somatic and visceral effectors. On both sides, the dorsal and ventral roots of each segment pass between the vertebral canal and the periphery at the *intervertebral foramen* between successive vertebrae. The dorsal root ganglion lies between the pedicles of the adjacent vertebrae. (You may wish to review the description of vertebral anatomy in Chapter 7, Figure 7-17•, p. 213.)

Distal to each dorsal root ganglion, the sensory and motor roots are bound together into a single **spinal nerve.** Spinal nerves are classified as **mixed nerves**—that is, they contain both afferent (sensory) and efferent (motor) fibers. There are 31 pairs of spinal nerves, each identified by its association with adjacent vertebrae. For example, we may speak of "cervical spinal nerves" or even "cervical nerves" when we make a general reference to spinal nerves of the neck. When we indicate specific spinal nerves, it is customary to give them a regional number, as indicated in Figure 13-1•. Each spinal nerve inferior to the first thoracic vertebra takes its name from the vertebra immediately preceding it. Thus, spinal nerve T_1 emerges immediately inferior to vertebra T_1, spinal nerve T_2 follows vertebra T_2, and so forth.

The arrangement differs in the cervical region, because the first pair of spinal nerves, C_1, passes between the skull and the first cervical vertebra. For this reason, each cervical nerve takes its name from the vertebra immediately *following* it. In other words, cervical nerve C_2 *precedes* vertebra C_2, and the same system is used for the rest of the cervical series. The transition from one numbering system to another occurs between the last cervical vertebra and first thoracic vertebra. Because the spinal nerve lying between them has been designated C_8, there are only seven cervical vertebrae but *eight* cervical nerves.

The spinal cord continues to enlarge and elongate until an individual is approximately 4 years old. Up to

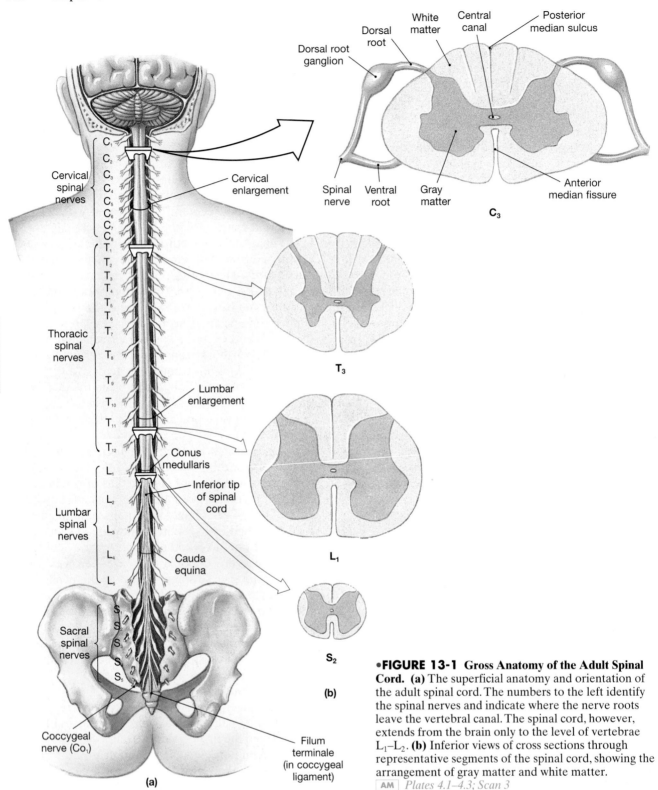

•FIGURE 13-1 Gross Anatomy of the Adult Spinal Cord. (a) The superficial anatomy and orientation of the adult spinal cord. The numbers to the left identify the spinal nerves and indicate where the nerve roots leave the vertebral canal. The spinal cord, however, extends from the brain only to the level of vertebrae L_1–L_2. (b) Inferior views of cross sections through representative segments of the spinal cord, showing the arrangement of gray matter and white matter.
AM *Plates 4.1–4.3; Scan 3*

that time, enlargement of the spinal cord keeps pace with the growth of the vertebral column. Throughout this period, the ventral and dorsal roots are very short, and they enter the intervertebral foramina immediately adjacent to their spinal segment. After age 4, the vertebral column continues to elongate but the spinal cord does not. This vertebral growth carries the dorsal roots and spinal nerves farther and farther from their original positions relative to the spinal cord. As a result, the dorsal and ventral roots gradually elongate, and the correspondence between spinal segment and the vertebral segment is lost. For example, in adults the sacral segments of the spinal cord are at the level of vertebrae L_1–L_2.

Because the adult spinal cord extends only to the level of the first or second lumbar vertebra, the dorsal and ventral roots of spinal segments L_2 to S_5 extend inferiorly, past the inferior tip of the conus medullaris. When seen in gross dissection, the filum terminale and the long ventral and dorsal roots resemble a horse's tail. As a result, early anatomists called this complex the **cauda equina** (KAW-duh ek-WĪ-nuh; *cauda,* tail + *equus,* horse).

SPINAL MENINGES

The vertebral column and its surrounding ligaments, tendons, and muscles isolate the spinal cord from the external environment. The delicate neural tissues must also be protected from damaging contacts with the surrounding bony walls of the vertebral canal. The **spinal meninges** (men-IN-jēz; singular, *meninx,* membrane), a series of specialized membranes surrounding the spinal cord, provide the necessary physical stability and shock absorption. Blood vessels branching within these layers deliver oxygen and nutrients to the spinal cord.

The relationships among the spinal meninges are shown in Figure 13-2a•. The spinal meninges consist of three layers: (1) the *dura mater,* (2) the *arachnoid,* and (3) the *pia mater.* At the foramen magnum of the skull, the spinal meninges are continuous with the **cranial meninges,** which surround the brain. (We shall discuss the cranial meninges, which have the same three layers, in Chapter 14.)

Bacterial or viral infection can cause **meningitis,** or inflammation of the meningeal membranes. Meningitis is dangerous because it can disrupt the normal circulatory and cerebrospinal fluid supplies, damaging or killing neurons and neuroglia in the affected areas. Although an initial diagnosis may specify the meninges of the spinal cord (*spinal meningitis*) or brain (*cerebral meningitis*), in later stages the entire meningeal system is usually affected.

AM *Meningitis*

The Dura Mater

The tough, fibrous **dura mater** (DOO-ruh MĀ-ter; *dura,* hard + *mater,* mother) is the layer that forms the outermost covering of the spinal cord (Figure 13-2a•).

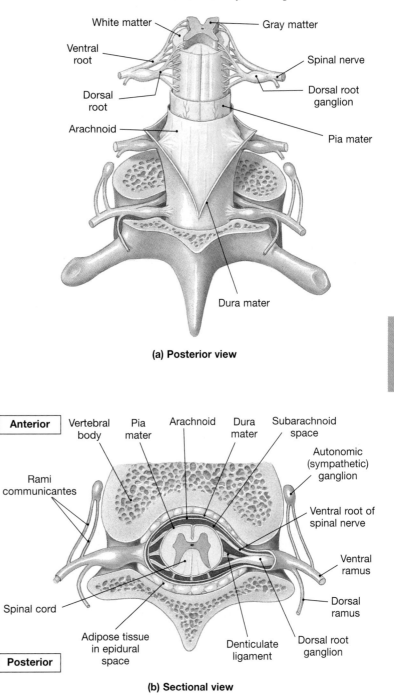

(a) Posterior view

13

(b) Sectional view

•**FIGURE 13-2 The Spinal Cord and Spinal Meninges. (a)** A posterior view of the spinal cord, showing the meningeal layers, superficial landmarks, and the distribution of gray matter and white matter. **(b)** A sectional view through the spinal cord and meninges, showing the peripheral distribution of spinal nerves.

This layer contains dense collagen fibers that are oriented along the longitudinal axis of the spinal cord. Between the dura mater and the walls of the vertebral canal lies the **epidural space,** a region that contains loose connective tissue, blood vessels, and a protective padding of adipose tissue (Figure 13-2b•).

The spinal dura mater does not have extensive, firm connections to the surrounding vertebrae. Longitudinal stability is provided by localized attachment sites at either end of the vertebral canal. Cranially, the outer layer of the spinal dura mater fuses with the periosteum of the occipital bone around the margins of the foramen magnum. There the spinal dura mater becomes continuous with the cranial dura mater. Within the sacral canal, the spinal dura mater tapers from a sheath to a dense cord of collagen fibers that blend with components of the filum terminale to form the **coccygeal ligament.**

The coccygeal ligament continues along the sacral canal, ultimately blending into the periosteum of the coccyx. Lateral support for the spinal dura mater is provided by loose connective tissue and adipose tissue within the epidural space. In addition, this dura mater extends between adjacent vertebrae at each intervertebral foramen, fusing with the connective tissues that surround the spinal nerves.

Injecting an anesthetic into the epidural space of the spinal cord will affect only the spinal nerves in the immediate area of the injection. The result is an **epidural block**—a temporary sensory loss or a sensory and motor paralysis, depending on the anesthetic selected. Epidural blocks in the inferior lumbar or sacral regions may be used to control pain during childbirth. AM *Spinal Anesthesia*

The Arachnoid

In most anatomical and histological preparations, a narrow **subdural space** separates the dura mater from deeper meningeal layers. It is likely, however, that in life no such space exists and that the inner surface of the dura mater is in contact with the outer surface of the **arachnoid** (a-RAK-noyd; *arachne*, spider), the middle meningeal layer (Figure 13-2b•). The inner surface of the dura mater and the outer surface of the arachnoid are covered by simple squamous epithelia. The arachnoid includes this epithelium, called the *arachnoid membrane*, and the *arachnoid trabeculae*, a delicate network of collagen and elastic fibers that extends between the arachnoid membrane and the outer surface of the pia mater. The intervening region is called the **subarachnoid space.** It is filled with **cerebrospinal fluid (CSF),** which acts as a shock absorber and a diffusion medium for dissolved gases, nutrients, chemical messengers, and waste products.

The spinal arachnoid extends inferiorly as far as the filum terminale, and the dorsal and ventral roots of the cauda equina lie within the fluid-filled subarachnoid space. In adults, the withdrawal of cerebrospinal fluid, a procedure known as a **spinal tap,** involves the insertion of a needle into the subarachnoid space in the inferior lumbar region. (See the clinical discussion "Spinal Taps and Myelography.") This

placement avoids damage to the spinal cord. Spinal taps are performed when CNS infection is suspected or to diagnose severe back pain, headaches, disc problems, and some types of strokes.

The Pia Mater

The subarachnoid space bridges the gap between the arachnoid epithelium and the innermost meningeal layer, the **pia mater** (*pia*, delicate + *mater*, mother). The pia mater consists of a meshwork of elastic and collagen fibers that is firmly bound to the underlying neural tissue (Figure 13-2b•). These connective tissue fibers are extensively interwoven with those that span the subarachnoid space, firmly binding the arachnoid to the pia mater. The blood vessels servicing the spinal cord run along the surface of the spinal pia mater, within the subarachnoid space.

Along the length of the spinal cord, paired **denticulate ligaments** extend from the pia mater through the arachnoid to the dura mater (Figures 13-2b and 13-3•). Denticulate ligaments, which originate along either side of the spinal cord, prevent lateral (side-to-side) movement. The dural connections at the foramen magnum and the coccygeal ligament prevent longitudinal (superior/inferior) movement.

Spinal cord

Anterior median fissure

Pia mater

Denticulate ligaments

Arachnoid and dura mater (reflected)

Spinal blood vessel

Dorsal root of cervical nerve C_6

Ventral root of cervical nerve C_6

Anterior view

•**FIGURE 13-3 The Cervical Spinal Cord.** An anterior view of the spinal cord and spinal nerve roots in the vertebral canal. The dura mater and arachnoid membrane have been cut and reflected; notice the blood vessels that run in the subarachnoid space, bound to the outer surface of the delicate pia mater.

Spinal Taps and Myelography

Tissue samples, or **biopsies,** are taken from many organs to assist in diagnosis. For example, when a liver or skin disorder is suspected, small plugs of tissue are removed and examined for signs of infection or cell damage or are used to identify the bacteria that cause an infection. Unlike many other tissues, however, neural tissue consists largely of cells rather than extracellular fluids or fibers. Neural tissue samples are seldom removed for analysis, because the body usually does not replace extracted or damaged neurons. Instead, small volumes of cerebrospinal fluid are collected and analyzed. Cerebrospinal fluid is intimately associated with the neural tissue of the CNS. When present, pathogens, cell debris, or metabolic wastes in the CNS are therefore detectable in the cerebrospinal fluid.

The withdrawal of CSF, a procedure called a *spinal tap,* must be done with care to avoid injuring the spinal cord. The adult spinal cord extends inferiorly only as far as vertebra L_1 or L_2. Between vertebra L_2 and the sacrum, the spinal meninges remain intact, but they enclose only the relatively sturdy components of the cauda equina and a significant quantity of CSF. When the vertebral column is flexed, a needle can be inserted between the inferior lumbar vertebrae and into the subarachnoid space with minimal risk to the cauda equina. In this procedure, known as a **lumbar puncture,** 3–9 ml of fluid is taken from the subarachnoid space between vertebrae L_3 and L_4 (Figure 13-4a•). Spinal taps are performed when CNS infection is suspected or when severe back pain, headaches, disc problems, and some types of strokes are diagnosed.

Myelography is the introduction of radiopaque dyes into the CSF of the subarachnoid space. Because the dyes are opaque to X-rays, the CSF appears white on an X-ray image (Figure 13-4b•). Any tumors, inflammation, or adhesions that distort or divert CSF circulation appear in silhouette. In the event of severe infection, inflammation, or leukemia (cancer of the white blood cells), antibiotics, steroids, or anti-cancer drugs can be injected into the subarachnoid space.

(a)

(b)

•**FIGURE 13-4 Spinal Taps and Myelography. (a)** The lumbar puncture needle is in the subarachnoid space, between the nerves of the cauda equina. The needle has been inserted between the third and fourth lumbar spinous processes, pointing at a superior angle. **(b)** A myelogram—an X-ray image of the spinal cord after a radiopaque dye has been introduced into the CSF—of the inferior lumbar region.

13

The spinal meninges accompany the dorsal and ventral roots as these roots pass through the intervertebral foramina. As the sectional view of Figures 13-2b• indicates, the meningeal membranes are continuous with the connective tissues that surround the spinal nerves and their peripheral branches.

☑ Damage to which root of a spinal nerve would interfere with motor function?

☑ Where is the cerebrospinal fluid that surrounds the spinal cord located?

SECTIONAL ANATOMY OF THE SPINAL CORD

To understand the functional organization of the spinal cord, you must become familiar with its sectional organization. The anterior median fissure and the posterior median sulcus mark the division between left and right sides of the spinal cord. The superficial white matter contains large numbers of myelinated and unmyelinated axons. The gray matter, dominated by the cell bodies of neurons, neuroglia, and unmyelinated axons, surrounds the narrow **central canal** and forms an H or butterfly shape (Figure 13-5•). The projections of gray matter toward the outer surface of the spinal cord are called **horns.**

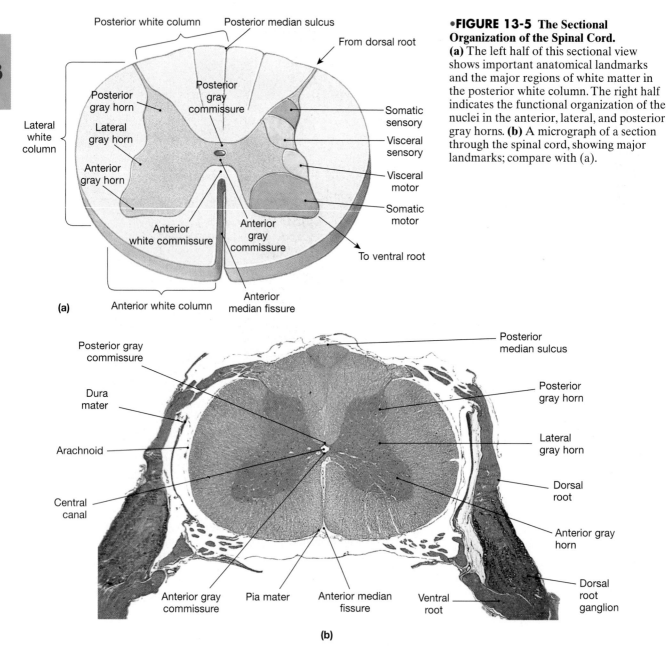

•**FIGURE 13-5 The Sectional Organization of the Spinal Cord.** **(a)** The left half of this sectional view shows important anatomical landmarks and the major regions of white matter in the posterior white column. The right half indicates the functional organization of the nuclei in the anterior, lateral, and posterior gray horns. **(b)** A micrograph of a section through the spinal cord, showing major landmarks; compare with (a).

Organization of Gray Matter

The cell bodies of neurons in the gray matter of the spinal cord are organized into functional groups called *nuclei*. **Sensory nuclei** receive and relay sensory information from peripheral receptors. **Motor nuclei** issue motor commands to peripheral effectors. Although sensory and motor nuclei appear rather small in transverse section, they may extend for a considerable distance along the length of the spinal cord.

A frontal section along the length of the central canal of the spinal cord separates the sensory (posterior, or dorsal) nuclei from the motor (anterior, or ventral) nuclei. The **posterior gray horns** contain somatic and visceral sensory nuclei, whereas the **anterior gray horns** contain somatic motor nuclei. The **lateral gray horns**, located only in the thoracic and lumbar segments, contain visceral motor nuclei. The **gray commissures** (*commissura*, a joining together) posterior to and anterior to the central canal contain axons that cross from one side of the cord to the other before they reach a destination in the gray matter.

Figure 13-5a• shows the relationship between the function of a particular nucleus (sensory or motor) and its relative position in the gray matter of the spinal cord. The nuclei within each gray horn are also organized. For example, the anterior gray horns of the cervical enlargement contain nuclei whose motor neurons control the muscles of the upper limbs. On each side of the spinal cord, in medial to lateral sequence, are somatic motor nuclei that control (1) muscles that position the pectoral girdle, (2) muscles that move the arm, (3) muscles that move the forearm and hand, and (4) muscles that move the hand and fingers. Within each of these regions, the motor neurons that control flexor muscles are grouped separately from those that control extensor muscles. Because the spinal cord is so highly organized, we can predict the muscles that will be affected by damage to a specific area of gray matter.

Organization of White Matter

The white matter on each side of the spinal cord can be divided into three regions called **columns,** or *funiculi* (Figure 13-5a•). The **posterior white columns** lie between the posterior gray horns and the posterior median sulcus. The **anterior white columns** lie between the anterior gray horns and the anterior median fissure. The anterior white columns are interconnected by the **anterior white commissure,** a region where axons cross from one side of the spinal cord to the other. The white matter between the anterior and posterior columns on each side makes up the **lateral white column.**

Each column contains tracts whose axons share functional and structural characteristics. A **tract,** or *fasciculus* (fa-SIK-ū-lus; bundle), is a bundle of axons in the CNS that are relatively uniform with respect to diameter, myelination, and conduction speed. All the axons within a tract relay the same type of information (sensory or motor) in the same direction. Short tracts carry sensory or motor signals between segments of the spinal cord, and longer tracts connect the spinal cord with the brain. **Ascending tracts** carry sensory information toward the brain, and **descending tracts** convey motor commands to the spinal cord. We shall examine specific tracts and their functions in Chapter 15.

Damage to Spinal Tracts

All the ascending sensory information and descending somatic motor commands associated with the trunk and limbs involve nuclei and tracts in the spinal cord. *Multiple sclerosis (MS)*, a disorder that we mentioned in Chapter 12, produces muscular paralysis and sensory losses through demyelination. ∞ *p. 370* The initial symptoms appear as the result of myelin degeneration within the white matter of the lateral and posterior columns of the spinal cord or along tracts within the brain. During subsequent attacks, the effects may become more widespread. The cumulative sensory and motor losses can eventually lead to a generalized muscular paralysis. [AM] *Multiple Sclerosis*

The viral disease *polio* causes paralysis due to the destruction of somatic motor neurons. This disorder, introduced in Chapter 10, has been almost eliminated in the Western Hemisphere by an aggressive immunization program. Immunization continues because polio remains relatively common in other areas of the world. The disease could be brought into the United States at any time, leading to an epidemic among unimmunized children.

Paralysis can also result from physical damage to the spinal cord due to a severe auto crash or other accident. The damaged tracts seldom undergo even partial repairs. Extensive damage to the spinal cord at or superior to the fifth cervical vertebra eliminates sensation and motor control of the upper and lower limbs. The extensive paralysis produced is called **quadriplegia. Paraplegia,** the loss of motor control of the lower limbs, may follow damage to the thoracic spinal cord. [AM] *Spinal Cord Injuries and Experimental Treatments*

Less severe injuries affecting the spinal cord or cauda equina produce symptoms of sensory loss or motor paralysis that reflect the specific nuclei, tracts, or spinal nerves involved. We will explain one example, the loss of peripheral sensation along the distribution of a spinal nerve, in a later section.

☑ A person with polio has lost the use of his leg muscles. In which area of his spinal cord would you expect the virus-infected motor neurons to be?

☑ Which portion of the spinal cord would be affected by a disease that damages myelin sheaths?

13

SPINAL NERVES

A series of connective tissue layers surrounds each spinal nerve and continues along all its peripheral branches. These layers, best seen in sectional view (Figure 13-6•), are comparable to those associated with skeletal muscles (Figure 10-1•, p. 272). The **epineurium,** or outermost layer, consists of a dense network of collagen fibers. The fibers of the **perineurium,** the middle layer, extend inward from the epineurium. These connective tissue partitions divide the nerve into a series of compartments that contain bundles of axons, or *fascicles.* Delicate connective tissue fibers of the **endoneurium,** the innermost layer, extend from the perineurium and surround individual axons.

Arteries and veins penetrate the epineurium and branch within the perineurium. Capillaries leaving the perineurium branch in the endoneurium and supply the axons and Schwann cells of the nerve and the fibroblasts of the connective tissues.

PERIPHERAL DISTRIBUTION OF SPINAL NERVES

Figure 13-7• shows the distribution, or pathway, of a typical spinal nerve that originates from the thoracic or superior lumbar segments of the spinal cord. The spinal nerve forms just lateral to the intervertebral foramen, where the dorsal and ventral roots unite. Let us follow the nerve's distribution in the periphery.

In the thoracic and superior lumbar regions (segments T_1–L_2), the first branch from the spinal nerve carries visceral motor fibers to a nearby **autonomic ganglion.** This ganglion is part of the sympathetic division of the autonomic nervous system (ANS). (The *sympathetic division* is, among its other functions, responsible for elevating the metabolic rate and for increasing alertness.) Because preganglionic axons are myelinated, this branch has a light color and is known as the **white ramus** ("branch"). Postganglionic fibers innervating glands and smooth muscles in the body wall or limbs return from the ganglia to rejoin the spinal nerve. These fibers are unmyelinated and have a darker color; they form the **gray ramus.** The gray ramus is typically proximal to the white ramus. The gray and white rami are collectively termed the *rami communicantes,* or "communicating branches." Postganglionic fibers inner- vating visceral organs in the thoracic cavity form a series of separate **sympathetic nerves.**

The **dorsal ramus** of each spinal nerve is a bundle of axons that provide sensory and motor innervation to the skin and muscles of the back. The axons in the relatively large **ventral ramus** supply the ventrolateral body surface, structures in the body wall, and the limbs.

(a)

Blood vessels

Perineurium (around one fascicle)

Endoneurium (around one axon)

Schwann cell

Myelinated axon

(b)

Fascicle

Epineurium (around peripheral nerve)

•**FIGURE 13-6 A Peripheral Nerve. (a)** Electron micrograph and **(b)** diagrammatic views of a typical peripheral nerve and its connective tissue wrappings, the perineurium, endoneurium, and epineurium. (SEM × 340) Reproduced from R. G. Kessel and R. H. Kardon, *Tissues and Organs: A Text-Atlas of Scanning Electron Microscopy,* W. H. Freeman & Co., 1979.

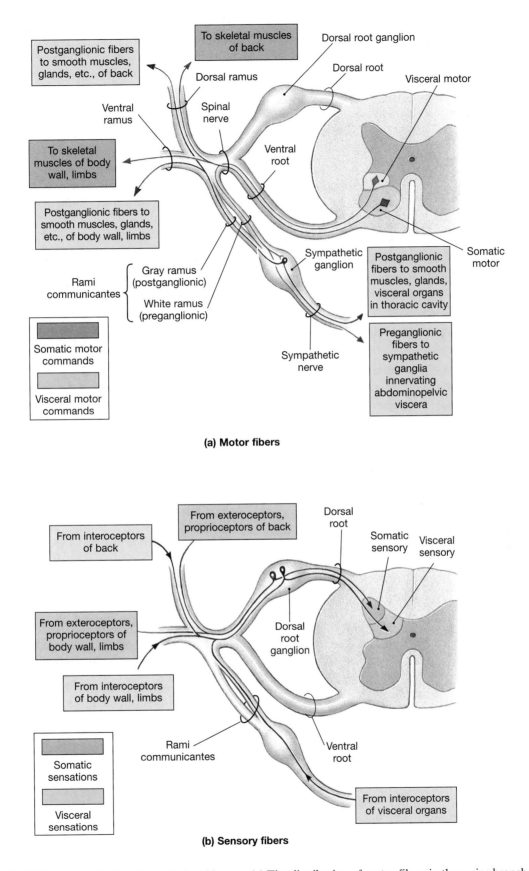

(a) Motor fibers

(b) Sensory fibers

•**FIGURE 13-7 Peripheral Distribution of Spinal Nerves. (a)** The distribution of motor fibers in the major branches of a representative thoracic or superior lumbar spinal nerve. (Although the gray ramus is normally proximal to the white ramus, this diagrammatic view makes it easier to follow the relationships between preganglionic and postganglionic fibers.) **(b)** A comparable view of the distribution of sensory fibers.

•FIGURE 13-8 Dermatomes. (a) Anterior and **(b)** posterior distributions of dermatomes on the surface of the skin.

The specific region of the skin surface monitored by a single pair of spinal nerves is known as a **dermatome.** Each pair of spinal nerves services its own dermatome (Figure 13-8•), although the precise boundaries of adjacent dermatomes overlap to some degree. Dermatomes are clinically important, because damage or infection of a spinal nerve or dorsal root ganglion will produce a characteristic loss of sensation in the skin. For example, in the condition *shingles,* a virus infects dorsal root ganglia, causing a painful rash whose distribution corresponds to that of the affected sensory nerves. AM *Shingles and Hansen's Disease; EMBRYOLOGY SUMMARY 11: The Development of the Spinal Cord and Spinal Nerves*

NERVE PLEXUSES

The simple distribution pattern of dorsal and ventral rami in Figure 13-7• applies to spinal nerves T_2–T_{12}. But in segments controlling the skeletal musculature of the neck, upper limbs, or lower limbs, the situation is more complicated. During development, small skeletal muscles innervated by different ventral rami typically fuse to form larger muscles with compound origins. The anatomical distinctions between the component muscles may disappear, but separate ventral rami continue to provide sensory innervation and motor control to each part of the compound muscle. As they converge, the ventral rami of adjacent spinal nerves blend their fibers, producing a series of compound nerve trunks. Such a complex interwoven network of nerves is a **nerve plexus** (PLEK-sus; *plexus,* braid). There are four major plexuses: (1) the *cervical plexus,* (2) the *brachial plexus,* (3) the *lumbar plexus,* and (4) the *sacral plexus* (Figure 13-9•).

In Chapter 11, we introduced the peripheral nerves that control the major axial and appendicular muscles. You may find it helpful to refer to the tables in that chapter as we proceed to review the innervation of the skeletal muscle groups. ∞ *pp. 320–355*

Nerve Palsies and Spinal Injuries
Peripheral *nerve palsies,* or **peripheral neuropathies,** are regional losses of sensory and motor function as a result of nerve trauma or compression. You have experienced a mild, temporary palsy if your arm or leg has ever "fallen asleep" after you leaned or sat in an uncomfortable position. AM *Peripheral Neuropathies* Although the location of the affected dermatomes provides clues to the location of injuries along the spinal cord, the information is not precise, because adjacent dermatomes overlap to a variable extent. More-exact conclusions can be drawn from the loss of motor control, on the basis of the origin and distribution of the peripheral nerves originating at nerve plexuses.

If a peripheral axon is damaged but not displaced, normal function may eventually return as the cut stump grows across the injury site away from the cell body and along its former path. ∞ *p. 372* Repairs made after an entire peripheral *nerve* has been damaged are generally incomplete, primarily because of problems with axon alignment and regrowth. Various technologically sophisticated procedures designed to improve nerve regeneration and repair are currently under evaluation. An entire family of *nerve growth factors* has been discovered in recent years. Their use alone or in combination with other therapies may ultimately revolutionize the treatment of damaged neural tissue inside and outside the CNS. AM *Damage and Repair of Peripheral Nerves*

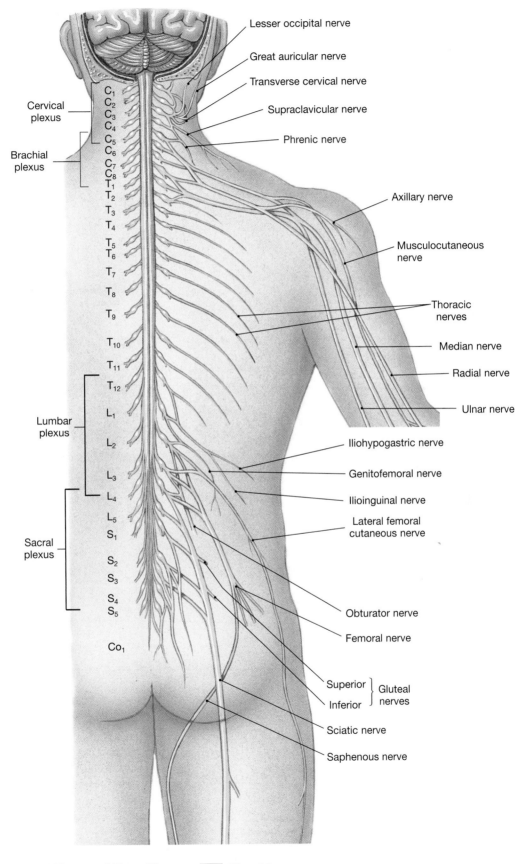

Cervical plexus

Brachial plexus

C₁
C₂
C₃
C₄
C₅
C₆
C₇
C₈
T₁
T₂
T₃
T₄
T₅
T₆
T₇
T₈
T₉
T₁₀
T₁₁
T₁₂

Lumbar plexus

L₁
L₂
L₃
L₄
L₅
S₁

Sacral plexus

S₂
S₃
S₄
S₅

Co₁

Lesser occipital nerve
Great auricular nerve
Transverse cervical nerve
Supraclavicular nerve
Phrenic nerve
Axillary nerve
Musculocutaneous nerve
Thoracic nerves
Median nerve
Radial nerve
Ulnar nerve
Iliohypogastric nerve
Genitofemoral nerve
Ilioinguinal nerve
Lateral femoral cutaneous nerve
Obturator nerve
Femoral nerve
Superior
Inferior } Gluteal nerves
Sciatic nerve
Saphenous nerve

13

•**FIGURE 13-9** **Peripheral Nerves and Nerve Plexuses.** AM *Plate 4.1*

The Cervical Plexus

The **cervical plexus** consists of the ventral rami of spinal nerves C_1–C_5 (Figures 13-9, 13-10●; Table 13-1). Its branches innervate the muscles of the neck and extend into the thoracic cavity, where they control the diaphragmatic muscles. The **phrenic nerve,** the major nerve of the cervical plexus, provides the entire nerve supply to the diaphragm, a key respiratory muscle. Other branches of this nerve plexus are distributed to the skin of the neck and the upper part of the chest.

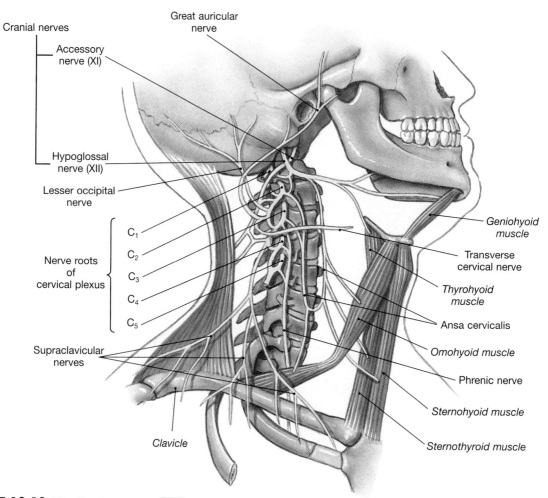

●**FIGURE 13-10** The Cervical Plexus. [AM] *Plates 5.2, 5.3*

TABLE 13-1 The Cervical Plexus

Nerve(s)	Spinal Segments	Distribution
Ansa cervicalis (superior and inferior branches)	C_1–C_4	Five of the extrinsic laryngeal muscles: sternothyroid, sternohyoid, omohyoid, geniohyoid, and thyrohyoid (via N XII)
Lesser occipital, transverse cervical, supraclavicular, and great auricular nerves	C_2–C_3	Skin of upper chest, shoulder, neck, and ear
Phrenic nerve	C_3–C_5	Diaphragm
Cervical nerves	C_1–C_5	Levator scapulae, scalenes, sternocleidomastoid, and trapezius (with N XI)

The Brachial Plexus

The **brachial plexus** innervates the pectoral girdle and upper limb, with contributions from the ventral rami of spinal nerves C_5–T_1 (Figures 13-9 and 13-11•; Table 13-2). The nerves that form this plexus originate from trunks and cords. **Trunks** are large bundles of axons contributed by several spinal nerves. **Cords** are smaller branches that originate at trunks. Both trunks and cords are named according to their location relative to the *axillary artery*, a large artery supplying the upper limb. Hence we have *superior, middle,* and *inferior trunks* and *lateral, medial,* and *posterior cords*. The lateral cord forms the **musculocutaneous nerve** exclusively and, together with the medial cord, contributes to the **median nerve.** The **ulnar nerve** is the other major nerve of the medial cord. The posterior cord gives rise to the **axillary nerve** and the **radial nerve.** Table 13-2 provides

13

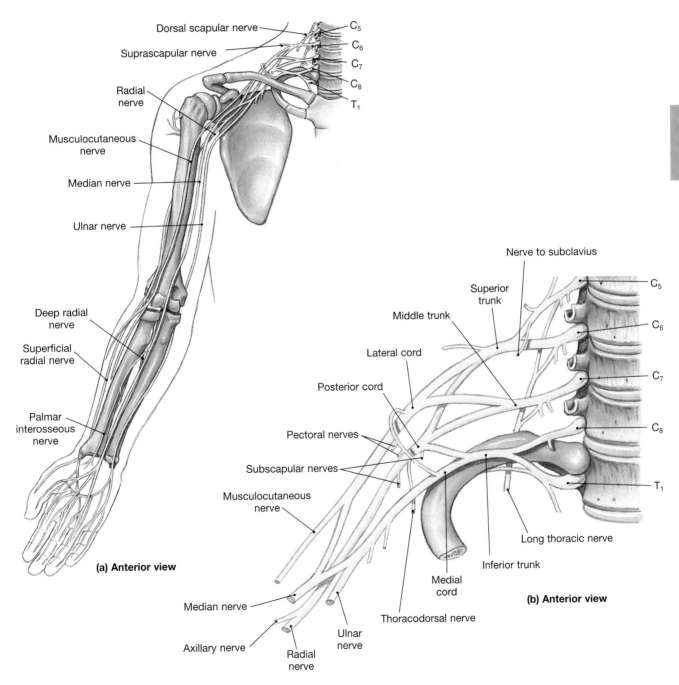

Dorsal scapular nerve
Suprascapular nerve
C_5
C_6
C_7
C_8
T_1
Radial nerve
Musculocutaneous nerve
Median nerve
Ulnar nerve
Deep radial nerve
Superficial radial nerve
Palmar interosseous nerve

(a) Anterior view

Nerve to subclavius
Superior trunk
Middle trunk
Lateral cord
Posterior cord
Pectoral nerves
Subscapular nerves
Musculocutaneous nerve
Median nerve
Axillary nerve
Radial nerve
Ulnar nerve
Thoracodorsal nerve
Medial cord
Inferior trunk
Long thoracic nerve
C_5
C_6
C_7
C_8
T_1

(b) Anterior view

•**FIGURE 13-11 The Brachial Plexus. (a)** Major nerves originating at the right brachial plexus. **(b)** The right brachial plexus.

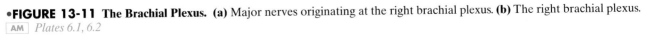
AM *Plates 6.1, 6.2*

TABLE 13-2 The Brachial Plexus

Nerve(s)	Spinal Segments	Distribution
Nerve to subclavius	C_4–C_6	Subclavius muscle
Dorsal scapular nerve	C_5	Rhomboideus muscles and levator scapulae
Long thoracic nerve	C_5–C_7	Serratus anterior
Suprascapular nerve	C_5, C_6	Supraspinatus and infraspinatus muscles
Pectoral nerves (medial and lateral)	C_5–T_1	Pectoralis muscles
Subscapular nerves	C_5, C_6	Subscapularis and teres major muscles
Thoracodorsal nerve	C_6–C_8	Latissimus dorsi muscle
Axillary nerve	C_5, C_6	Deltoid and teres minor muscles; skin of shoulder
Radial nerve	C_5–T_1	Extensor muscles on the arm and forearm (triceps brachii, brachioradialis, extensor carpi radialis, and extensor carpi ulnaris muscles)
		Digital extensors and abductor pollicis muscle; skin over the posterolateral surface of the arm
Musculocutaneous nerve	C_5–C_7	Flexor muscles on the arm (biceps brachii, brachialis, and coracobrachialis muscles); skin over lateral surface of the forearm
Median nerve	C_6–T_1	Flexor muscles on the forearm (flexor carpi radialis and palmaris longus muscles); pronator quadratus and pronator teres muscles; digital flexors (through the *palmar interosseous nerve*); skin over anterolateral surface of the hand
Ulnar nerve	C_8, T_1	Flexor carpi ulnaris muscle; adductor pollicis muscle and small digital muscles; skin over medial surface of the hand

further information about these and other major nerves of the brachial plexus.

The Lumbar and Sacral Plexuses

The **lumbar plexus** and the **sacral plexus** arise from the lumbar and sacral segments of the spinal cord. The nerves arising at these plexuses innervate the pelvic girdle and lower limbs (Figures 13-9 and 13-12●). The individual nerves that form the lumbar and sacral plexuses are listed in Table 13-3 (p. 422).

The lumbar plexus contains axons from the ventral rami of spinal nerves T_{12}–L_4. The major nerves of this plexus are the **genitofemoral nerve,** the **lateral femoral cutaneous nerve,** and the **femoral nerve.** The sacral plexus contains axons from the ventral rami of spinal nerves L_4–S_4. Two major nerves arise at this plexus: the **sciatic nerve** and the **pudendal nerve.** The sciatic

nerve passes posterior to the femur, deep to the long head of the biceps femoris muscle. As it approaches the knee, the sciatic nerve divides into two branches—the **peroneal nerve** and the **tibial nerve.** The *sural nerve,* formed by branches of the peroneal nerve, is a sensory nerve innervating the lateral portion of the foot. A section of this nerve is often removed for use in nerve grafts.

☑ An anesthetic blocks the function of the dorsal rami of the cervical spinal nerves. Which areas of the body will be affected?

☑ Injury to which of the nerve plexuses would interfere with the ability to breathe?

☑ Compression of which nerve produces the sensation that your leg has "fallen asleep"?

(a) The lumbar plexus, anterior view

Intercostal nerve
Iliohypogastric nerve
Ilioinguinal nerve
Genitofemoral nerve
Lateral femoral cutaneous nerve
Femoral nerve
Obturator nerve
T_{12}
L_1
L_2
L_3
L_4
L_5
LUMBAR PLEXUS
Lumbosacral trunk

(b) The sacral plexus, anterior view

Lumbosacral trunk
Superior gluteal nerve
Inferior gluteal nerve
Sciatic nerve
Posterior femoral cutaneous nerve
L_4
L_5
S_1
S_2
S_3
S_4
S_5
Co_1
SACRAL PLEXUS
Pudendal nerve

(c) Nerves of the lumbar plexus, anterior view

Iliohypogastric nerve
Ilioinguinal nerve
Genitofemoral nerve
Lateral femoral cutaneous nerve
Femoral nerve
Superior gluteal nerve
Pudendal nerve
Sciatic nerve (cut)
Posterior femoral cutaneous nerve (cut)
Intercostal nerve
Inferior gluteal nerve
Obturator nerve
Saphenous nerve

(d) Nerves of the sacral plexus, posterior view

Pudendal nerve
Tibial nerve
Superior gluteal nerve
Inferior gluteal nerve
Sciatic nerve
Posterior femoral cutaneous nerve
Peroneal nerve
Sural nerve

13

•FIGURE 13-12 The Lumbar and Sacral Plexuses. (a) The right lumbar plexus. (b) The right sacral plexus. (c) The major branches of the right lumbar plexus. (d) The major branches of the right sacral plexus. [AM] Plates 8.7c, 8.8a, 8.10, 8.12c, 8.13a

TABLE 13-3 The Lumbar and Sacral Plexuses

Nerve(s)	Spinal Segment(s)	Distribution
LUMBAR PLEXUS		
Iliohypogastric nerve	T_{12}, L_1	Abdominal muscles (external and internal abdominal oblique muscles, transversus abdominis muscle); skin over inferior abdomen and buttocks
Ilioinguinal nerve	L_1	Abdominal muscles (with iliohypogastric nerve); skin over superior, medial thigh and portions of external genitalia
Genitofemoral nerve	L_1, L_2	Skin over anteromedial surface of thigh and portions of external genitalia
Lateral femoral cutaneous nerve	L_2, L_3	Skin over anterior, lateral, and posterior surfaces of thigh
Femoral nerve	L_2–L_4	Anterior muscles of thigh (sartorius muscle and quadriceps group); adductors of thigh (pectineus and iliopsoas muscles); skin over anteromedial surface of thigh, medial surface of leg and foot
Obturator nerve	L_2–L_4	Adductors of thigh (adductors magnus, brevis, and longus); gracilis muscle; skin over medial surface of thigh
Saphenous nerve	L_2–L_4	Skin over medial surface of leg
SACRAL PLEXUS		
Gluteal nerves: Superior	L_4–S_2	Abductors of thigh (gluteus minimus, gluteus medius, and tensor fasciae latae)
Inferior		Extensor of thigh (gluteus maximus)
Posterior femoral cutaneous nerve	S_1–S_3	Skin of perineum and posterior surface of thigh and leg
Sciatic nerve:	L_4–S_3	Two of the hamstrings (semimembranosus and semitendinosus); adductor magnus (with obturator nerve)
Tibial nerve		Flexors of leg and plantar flexors of foot (popliteus, gastrocnemius, soleus, and tibialis posterior muscles and long head of biceps femoris muscle); flexors of toes; skin over posterior surface of leg, plantar surface of foot
Peroneal nerve		Biceps femoris muscle (short head); peroneus (brevis and longus) and tibialis anterior muscles; extensors of toes; skin over anterior surface of leg and dorsal surface of foot; skin over lateral portion of foot (through the *sural nerve*)
Pudendal nerve	S_2–S_4	Muscles of perineum, including urogenital diaphragm and external anal and urethral sphincter muscles; skin of external genitalia and related skeletal muscles (bulbospongiosus and ischiocavernosus muscles)

AN INTRODUCTION TO REFLEXES

Conditions inside or outside the body can change rapidly and unexpectedly. **Reflexes** are rapid, automatic responses to specific stimuli. Reflexes preserve homeostasis by making rapid adjustments in the function of organs or organ systems. The response shows little variability; each time a particular reflex is activated, it usually produces the same motor response. In Chapter 1, we introduced the basic functional components involved in all types of homeostatic regulation: a *receptor*, an *integration center,* and an *effector.* ∞ *p. 12* Here we consider *neural reflexes,* in which sensory fibers deliver information to the CNS and motor fibers carry motor commands to peripheral effectors. In Chapter 18, we will examine *endocrine reflexes,* in which the commands to peripheral tissues and organs are delivered by hormones in the bloodstream.

THE REFLEX ARC

The "wiring" of a single reflex is called a **reflex arc.** A reflex arc begins at a receptor and ends at a peripher-

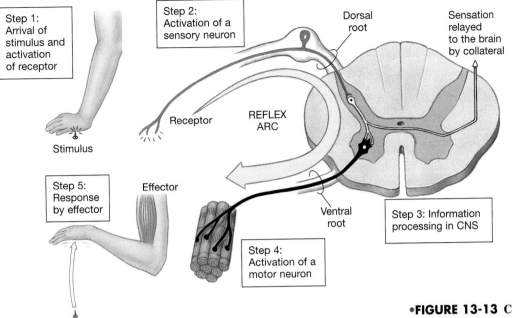

Step 1:
Arrival of
stimulus and
activation
of receptor

Step 2:
Activation of a
sensory neuron

Dorsal
root

Sensation
relayed
to the brain
by collateral

Receptor

REFLEX
ARC

Stimulus

Step 5:
Response
by effector

Effector

Step 3: Information
processing in CNS

Ventral
root

Step 4:
Activation of a
motor neuron

13

•**FIGURE 13-13** Components of a Reflex Arc

al effector, such as a muscle fiber or a gland cell. Figure 13-13• diagrams the five steps in a neural reflex:

STEP 1: *The Arrival of a Stimulus and Activation of a Receptor.* A *receptor* is a specialized cell or the dendrites of a sensory neuron. Receptors are sensitive to physical or chemical changes in the body or to the external environment. We introduced general categories of sensory receptors in Chapter 12. ∞ *p. 366* When you lean on a tack, pain receptors in the palm of your hand are activated (Figure 13-13•). Pain receptors are the dendrites of sensory neurons; they respond to stimuli that cause or accompany tissue damage. (We shall discuss the link between receptor stimulation and sensory neuron activation further in Chapter 17.)

STEP 2: *The Activation of a Sensory Neuron.* The stimulation of pain receptors leads to the formation and propagation of action potentials along the axons of sensory neurons. This information reaches your spinal cord by way of a dorsal root. In our example, steps 1 and 2 involve the same cell. However, the two steps may involve different cells. For example, reflexes triggered by loud sounds begin when receptor cells in the inner ear release neurotransmitters that stimulate sensory neurons.

STEP 3: *Information Processing.* Information processing begins when a neurotransmitter released by synaptic knobs of each sensory neuron affects the postsynaptic membrane of an interneuron, producing an excitatory postsynaptic potential (EPSP). ∞ *p. 393* This stimulus is integrated with other stimuli arriving at the postsynaptic neuron at that moment. In a neuronal pool, the next step may coordinate an appropriate motor response by stimulating some motor neurons and inhibiting others. Interneurons are not always involved in a reflex arc;

the sensory neuron may innervate a motor neuron directly. In that case, the motor neuron performs the information processing as summation occurs at the axon hillock.

STEP 4: *The Activation of a Motor Neuron.* The axons of the stimulated motor neurons carry action potentials into the periphery—in this example, through the ventral root of a spinal nerve.

STEP 5: *The Response of a Peripheral Effector.* The release of neurotransmitter by the motor neurons at synaptic knobs then leads to a response by a peripheral effector—in this case, a skeletal muscle whose contraction pulls your hand away from the tack.

A reflex response generally removes or opposes the original stimulus; here, the contracting muscle pulls your hand away from the painful stimulus. This reflex arc is therefore an example of *negative feedback.* ∞ *p. 12* By opposing potentially harmful changes in the internal or external environment, reflexes play an important role in homeostatic maintenance. The immediate reflex response is typically not the only response to a stimulus. In this example, you might wince, say "Ouch," and shake your hand. These other responses, which are directed by your brain, involve multiple synapses and take longer to organize and coordinate.

CLASSIFICATION OF REFLEXES

Reflexes are classified by (1) their development, (2) the site of information processing, (3) the nature of the resulting motor response, or (4) the complexity of the

13

•FIGURE 13-14 Methods of Classifying Reflexes

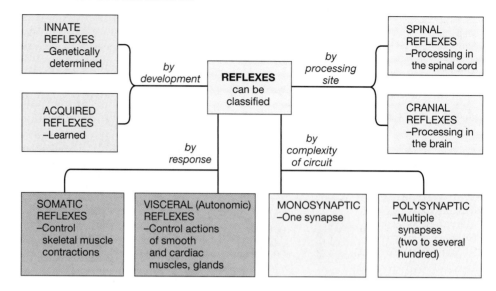

neural circuit involved. These categories are not mutually exclusive. They represent different ways of describing a single reflex (Figure 13-14•).

Development of Reflexes

Innate reflexes result from the connections that form between neurons during development. Such reflexes generally appear in a predictable sequence, from the simplest reflex responses (withdrawal from pain) to more-complex motor patterns (chewing, suckling, or tracking objects with the eyes). The neural connections responsible for basic motor patterns of an innate reflex are genetically or developmentally programmed. Examples include the reflexive removal of your hand from a hot stovetop and blinking when your eyelashes are touched.

More-complex, learned motor patterns are called **acquired reflexes.** An experienced driver steps on the brake when trouble appears ahead; a professional skier makes rapid, automatic adjustments in body position while racing. These motor responses are rapid and automatic, but they were learned rather than preestablished. Such reflexes are enhanced by repetition. The distinction between innate and acquired reflexes is not absolute. Some people can learn motor patterns more quickly than others can, and the differences probably have a genetic basis.

Most reflexes, whether innate or acquired, can be modified over time or suppressed through conscious effort. For example, while walking a tightrope over the Grand Canyon, you might ignore a bee sting on your hand, although under other circumstances you would probably withdraw your hand immediately, with shouting and thrashing.

Processing Sites

In a **spinal reflex,** the important interconnections and processing events occur in the spinal cord. We will dis-

cuss these reflexes in detail in the next section. Reflexes processed in the brain are called **cranial reflexes.** For example, the reflex movements in response to a sudden loud noise or bright light are cranial reflexes directed by nuclei in the brain. We shall encounter many other cranial reflexes in Chapters 15 and 16.

Nature of the Response

Somatic reflexes provide a mechanism for the involuntary control of the muscular system. **Visceral reflexes,** or *autonomic reflexes,* control the activities of other systems. We will consider somatic reflexes in detail in this chapter; we shall examine visceral reflexes in Chapter 16.

The movements directed by somatic reflexes are not delicate or precise. You might therefore wonder why they exist at all, because we have voluntary control over the same muscles. Somatic reflexes are absolutely vital, primarily because they are *immediate.* Making decisions and coordinating voluntary responses take time, and in an emergency situation—you slip while descending a flight of stairs or lean your hand against a knife edge—any delay increases the likelihood of a severe injury. Thus, somatic reflexes provide a rapid response that, if necessary, can later be modified by voluntary motor commands.

Complexity of the Circuit

In the simplest reflex arc, a sensory neuron synapses directly on a motor neuron, which serves as the processing center. Such a reflex is a **monosynaptic reflex.** Transmission across a chemical synapse always involves a synaptic delay; but, with only one synapse, the delay between the stimulus and the response is minimized. Circuit 1 in Figure 13-15a• is a typical monosynaptic reflex.

Most other types of reflexes have at least one interneuron between the sensory neuron and the motor neuron, as does circuit 2 of Figure 13-15b•.

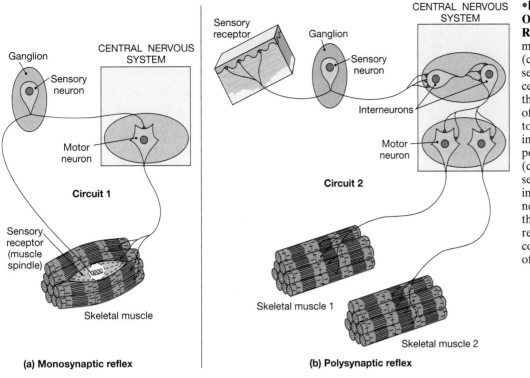

13

These **polysynaptic reflexes** have a longer delay between stimulus and response. The length of the delay is proportional to the number of synapses involved. Polysynaptic reflexes can produce far more complicated responses than monosynaptic reflexes can, because the interneurons can control several muscle groups.

SPINAL REFLEXES

Spinal reflexes range in complexity from simple monosynaptic reflexes involving a single segment of the spinal cord to polysynaptic reflexes that involve many segments. In the most complicated spinal reflexes, called **intersegmental reflex arcs,** many segments interact to produce a coordinated, highly variable motor response.

MONOSYNAPTIC REFLEXES

In a monosynaptic reflex, there is little delay between sensory input and motor output. These reflexes control the most-rapid, stereotyped motor responses of the nervous system to specific stimuli. The best-known monosynaptic reflex is the *stretch reflex.*

The Stretch Reflex

The **stretch reflex** (Figure 13-16•) provides automatic regulation of skeletal muscle length. The stimulus (increasing muscle length) activates a sensory neuron, which triggers an immediate motor response (contraction of the stretched muscle) that counteracts the stimulus. Action potentials traveling toward and away

•**FIGURE 13-16** Components of the Stretch Reflex. A diagram of the steps in a stretch reflex.

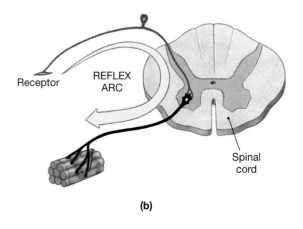

•FIGURE 13-17 The Patellar Reflex. (a) The patellar reflex is controlled by muscle spindles in the muscles that straighten the knee. A reflex hammer striking the muscle tendon provides the stimulus that stretches the spindle fibers. The response is an immediate increase in muscle tone and a reflexive kick. **(b)** The basic wiring of a monosynaptic reflex arc.

from the spinal cord are conducted along large, myelinated, Type A fibers. ∞ *p. 384* The entire reflex is completed within 20–40 msec.

The **patellar reflex,** or **knee jerk reflex** (Figure 13-17•), is a familiar example of a stretch reflex. This reflex is triggered by stretching the quadriceps muscles, which extend the knee. You can demonstrate this reflex by tapping on the patellar tendon of a flexed knee when the thigh and leg muscles are relaxed. A tap on the patellar ligament stretches muscle spindles in the quadriceps group. This stretching produces a sudden burst of activity in the sensory neurons that monitor these muscle spindles, so muscle tone in the quadriceps group increases rapidly. The rise is so sudden that it usually produces a noticeable kick. As the muscles contract, the muscle spindles return to their resting length; the rate of action potential generation in the sensory neurons declines. This decline causes a drop in muscle tone, and the leg then swings back.

Muscle Spindles The sensory receptors involved in the stretch reflex are **muscle spindles.** Each muscle spindle consists of a bundle of small, specialized skeletal muscle fibers called **intrafusal muscle fibers** (Figure 13-18a•). The muscle spindle is surrounded by larger **extrafusal muscle fibers,** responsible for the resting muscle tone and, at greater levels of stimulation, for the contraction of the entire muscle.

Each intrafusal fiber is innervated by both sensory and motor neurons. The dendrites of the sensory neuron spiral around the central portion of the intrafusal fiber. Axons from a separate population of spinal motor neurons form neuromuscular junctions on either end of this fiber. Motor neurons innervating intrafusal fibers are called **gamma (γ) motor neurons;** their axons are called **gamma efferents.** An intrafusal fiber has one set of myofibrils at each end. Instead of running the length of the fiber, each set of myofibrils is attached to the sarcolemma in the central region that is monitored by the sensory neuron.

The sensory neuron is always active, conducting impulses to the CNS. The axon enters the CNS in a dorsal root and synapses on motor neurons in the anterior gray horn of the spinal cord. Collaterals distribute the information to the brain, providing information about the state of the muscle spindle. Stretching the central portion of the intrafusal fiber distorts the dendrites and stimulates the sensory neuron, increasing the frequency of action potential generation. Compressing the central portion inhibits the sensory neuron, decreasing the frequency of action potential generation.

The axon of the sensory neuron synapses on CNS motor neurons that control the extrafusal muscle fibers of the same muscle. An increase in stimulation of the sensory neuron, caused by stretching of the intrafusal fiber, will increase stimulation to the surrounding extrafusal fibers, so muscle tone increases. A decrease in the stimulation of the sensory neuron, due to compression of the intrafusal fiber, will lead to a decrease in the stimulation of extrafusal fibers and in turn to a drop in muscle tone.

•**FIGURE 13-18 Intrafusal Fibers. (a)** The structure of a muscle spindle. **(b)** The effect of changes in muscle spindle length on muscle tone in the surrounding extrafusal fibers.

When a skeletal muscle is stretched, muscle spindles elongate and muscle tone increases (Figure 13-18a•). This increase provides automatic resistance that reduces the chance of muscle damage due to overstretching. When a skeletal muscle is compressed, muscle spindles are also compressed, and muscle tone decreases (Figure 13-18b•). This decrease helps reduce resistance to the movement under way.

Many stretch reflexes are **postural reflexes,** reflexes that maintain normal upright posture. For example, standing involves a cooperative effort of many muscle groups. Some of these muscles work in opposition to one another, exerting forces that keep the body's weight balanced over the feet. If the body leans forward, stretch receptors in the calf muscles are stimulated; those muscles respond by contracting. This contraction returns the body to an upright position. If the muscles overcompensate and the body begins to lean back, the calf muscles relax. But then stretch receptors in muscles of the shins and thighs are stimulated, and the problem is corrected immediately.

Postural muscles generally have a firm muscle tone and extremely sensitive stretch receptors. As a result,

very fine adjustments are continually being made, and you are not aware of the cycles of contraction and relaxation that occur. Stretch reflexes are only one type of postural reflex; there are many complex polysynaptic postural reflexes.

The CNS can adjust the sensitivity of muscle spindles by stimulating gamma motor neurons. This ability is important whenever voluntary contractions change the length of a muscle. For example, suppose you partially flex your elbow by contracting the biceps brachii muscle. You would expect that, in the new position, the muscle spindles would be compressed and muscle tone decreased. That is not the case, because impulses arriving over gamma efferents cause the contraction of myofibrils in the intrafusal fibers in the biceps brachii muscle. The myofibrils pull on the membrane in the sensory portion of the intrafusal fiber until that membrane is stretched to its normal resting length. As a result, the muscle spindles remain sensitive to any externally imposed changes in muscle length. For instance, if someone drops a ball into your palm with your arm in this position, the muscle spindles will automatically adjust muscle tone to compensate for the increased load.

☑ What is the minimum number of neurons in a reflex arc?

☑ One of the first somatic reflexes to develop is the suckling reflex. Which type of reflex is this?

☑ How would the stimulation of the muscle spindles involved in the patellar (knee jerk) reflex by gamma motor neurons affect the speed of the reflex?

POLYSYNAPTIC REFLEXES

Polysynaptic reflexes can produce far more complicated responses than can monosynaptic reflexes. One reason is that the interneurons involved can control several muscle groups. Moreover, these interneurons may produce either excitatory or inhibitory postsynaptic potentials (EPSPs or IPSPs) at CNS motor nuclei, so the response can involve the stimulation of some muscles and the inhibition of others.

The Tendon Reflex

The stretch reflex regulates the length of a skeletal muscle. The **tendon reflex** monitors the external tension produced during a muscular contraction and prevents tearing or breaking of the tendons. The sensory receptors for this reflex have not been identified, but they are distinct from both muscle spindles and *Golgi tendon organs,* proprioceptors in tendons. The receptor is stimulated when the collagen fibers are stretched to a dangerous degree. In the spinal cord, these neurons stimulate inhibitory interneurons that innervate the motor neurons controlling the skeletal muscle. The greater the tension in the tendon, the greater the inhibitory effect on the motor neurons. As a result, skeletal muscles generally cannot develop enough tension to break their tendons.

Withdrawal Reflexes

Withdrawal reflexes move affected parts of the body away from a source of stimulation. The strongest withdrawal reflexes are triggered by painful stimuli, but these reflexes are sometimes initiated by the stimulation of touch receptors or pressure receptors.

The **flexor reflex,** a representative withdrawal reflex, affects the muscles of a limb (Figure 13-19a•). Recall from Chapters 9 and 11 that flexion is a reduction in the angle between two articulating bones and that the contractions of flexor muscles perform this movement. ∞ *pp. 251, 316* When you step on a tack, a dramatic flexor reflex is produced in the affected leg. When the pain receptors in your foot are stimulated, the sensory neurons activate interneurons in the spinal cord that stimulate motor neurons in the anterior gray horns. The result is a contraction of flexor muscles that yanks your foot off the ground.

When a specific muscle contracts, opposing muscles must relax to permit the movement. For example, the flexor muscles that bend the leg are opposed by extensor muscles that straighten it out. A potential conflict exists: In theory, the contraction of a flexor muscle should trigger a stretch reflex in the extensors that would cause them to contract, opposing the movement. Interneurons in the spinal cord prevent such competition through **reciprocal inhibition.** When one set of motor neurons is stimulated, those neurons that control antagonistic muscles are inhibited. The term *reciprocal* refers to the fact that the system works both ways. When the flexors contract, the extensors relax; when the extensors contract, the flexors relax.

Withdrawal reflexes are much more complex than any monosynaptic reflex. They also show tremendous versatility, because the sensory neurons activate many pools of interneurons. If the stimuli are strong, interneurons will carry excitatory and inhibitory impulses up and down the spinal cord, affecting motor neurons in many segments. The end result is always the same: a coordinated movement away from the stimulus. But the distribution of the effects and the strength and character of the motor responses depend on the intensity and location of the stimulus. When you step on something sharp, mild discomfort might provoke a brief contraction in muscles of your ankle and foot. More-powerful stimuli would produce coordinated muscular contractions affecting the positions of your ankle, foot, and leg. Severe pain would also stimulate contractions of your shoulder, trunk, and arm muscles. These contractions could persist for several seconds owing to the activation of reverberating circuits. ∞ *p. 397* In contrast, monosynaptic reflexes are relatively invariable and brief; the patellar reflex is completed in roughly 20 msec.

Crossed Extensor Reflexes

The stretch, tendon, and withdrawal reflexes are **ipsilateral reflex arcs** (*ipsi,* same + *lateral,* side), in which the sensory stimulus and the motor response occur on the same side of the body. The **crossed extensor reflex** (Figure 13-19b•) is called a **contralateral reflex arc** (*contra,* opposite), because the motor response occurs on the side opposite the stimulus.

The crossed extensor reflex complements the flexor reflex, and the two occur simultaneously. When you step on a tack, while the flexor reflex pulls that foot away from the ground, the crossed extensor reflex stiffens your other leg to support your body weight. In the crossed extensor reflex, the axons of interneurons re-

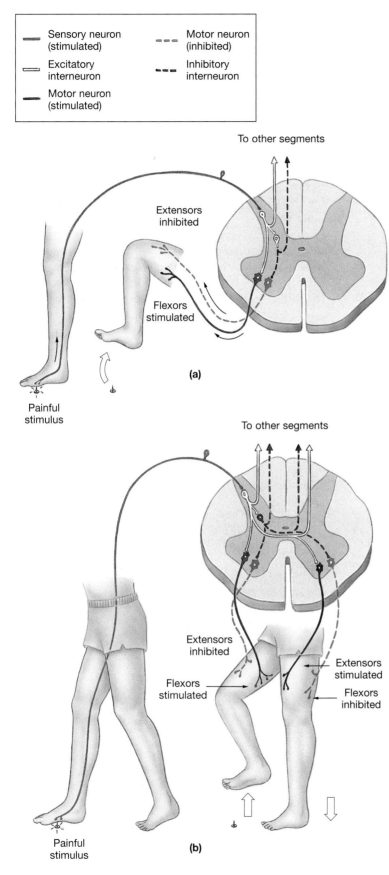

Legend:
- ── Sensory neuron (stimulated)
- ── Excitatory interneuron
- ── Motor neuron (stimulated)
- --- Motor neuron (inhibited)
- --- Inhibitory interneuron

To other segments

Extensors inhibited

Flexors stimulated

(a)

Painful stimulus

To other segments

Extensors inhibited

Flexors stimulated

Extensors stimulated

Flexors inhibited

(b)

Painful stimulus

•**FIGURE 13-19 The Flexor and Crossed Extensor Reflexes.**
(a) The flexor reflex, an example of a withdrawal reflex. **(b)** The crossed extensor reflex.

sponding to the pain cross to the other side of the spinal cord and stimulate motor neurons that control the extensor muscles of the uninjured leg. Your opposite leg straightens to support the shifting weight. Reverberating circuits use positive feedback to ensure that the movement lasts long enough to be effective, despite the absence of motor commands from higher centers of the brain.

Polysynaptic reflexes range in complexity from a simple tendon reflex to the complex and variable reflexes associated with standing, walking, and running. Yet all polysynaptic reflexes share the same five basic characteristics:

1. *They involve pools of interneurons.* Processing occurs in pools of interneurons before motor neurons are activated. The result may be excitation or inhibition; the tendon reflex produces inhibition of motor neurons, whereas the flexor and crossed extensor reflexes direct specific muscle contractions.

2. *They are intersegmental in distribution.* The interneuron pools extend across spinal segments and may activate muscle groups in many parts of the body.

3. *They involve reciprocal inhibition.* Reciprocal inhibition coordinates muscular contractions and reduces resistance to movement. In the flexor and crossed extensor reflexes, the contraction of one muscle group is associated with the inhibition of opposing muscles.

4. *They have reverberating circuits, which prolong the reflexive motor response.* Positive feedback between interneurons that innervate motor neurons and the processing pool maintains the stimulation even after the initial stimulus has faded.

5. *Several reflexes may cooperate to produce a coordinated, controlled response.* As a reflex movement is under way, any antagonistic reflexes are inhibited. For example, during the stretch reflex, antagonistic muscles are inhibited; in the tendon reflex, antagonistic muscles are stimulated. In complex polysynaptic reflexes, commands may be distributed along the length of the spinal cord, producing a well-coordinated response.

13

INTEGRATION AND CONTROL OF SPINAL REFLEXES

Reflex motor behaviors occur automatically, without instructions from higher centers. However, higher centers can have a profound effect on reflex performance. Processing centers in the brain can facilitate or inhibit reflex motor patterns based in the spinal cord. Descending tracts originating in the brain synapse on interneurons and motor neurons throughout the spinal cord. These synapses are continuously active, producing EPSPs or IPSPs at the postsynaptic membrane.

VOLUNTARY MOVEMENTS AND REFLEX MOTOR PATTERNS

Centers in the brain can interact with the stereotyped motor patterns programmed into the spinal cord. By making use of preexisting patterns, relatively few descending fibers can control complex motor functions. For example, the motor patterns for walking, running, and jumping are primarily directed by neuronal pools in the spinal cord. The descending pathways provide appropriate facilitation, inhibition, or "fine-tuning" of the established patterns.

When complicated voluntary movements are under way, the neurons involved with spinal reflexes assist with muscular coordination and control. For instance, the descending tracts that stimulate motor neurons controlling the biceps brachii muscle, brachialis muscle, and other elbow flexors also stimulate (1) the gamma efferents to the intrafusal fibers in these muscles and (2) inhibitory interneurons to relax antagonistic muscle groups, such as the triceps brachii muscle. The gamma efferents regulate the sensitivity of the stretch reflex, and the interneurons are the same as those activated in the withdrawal and crossed extensor reflexes. As muscle contraction pro-

---

 ## Reinforcement and Inhibition

Synapses producing EPSPs make postsynaptic neurons more sensitive to other excitatory stimuli. The resulting level of generalized facilitation rises and falls, depending on the activity in higher centers. For example, a voluntary effort to pull apart clasped hands elevates the general state of facilitation along the spinal cord. This elevated facilitation leads to **reinforcement,** an enhancement of spinal reflexes. Clinicians use reinforcement to decide whether a particular reflex is weak (or voluntarily suppressed) or absent.

Other descending fibers have an inhibitory effect on spinal reflexes. In adults, stroking the foot on the side of the sole produces a curling of the toes, called a **plantar reflex,** or *negative Babinski reflex,* after about a 1-second delay (Figure 13-20a•). Stroking an infant's foot on the side of the sole produces a fanning of the toes known as the **Babinski sign,** or *positive Babinski reflex.* This response disappears as descending motor pathways develop. If either the higher centers or the descending tracts are damaged, the Babinski sign will reappear in an adult (Figure 13-20b•). As a result, this reflex is often tested if CNS injury is suspected.

Many reflexes can be assessed through careful observation and the use of simple tools. The procedures are easy to perform, and the results can provide valuable information about damage to the spinal cord or spinal nerves. By testing a series of spinal and cranial reflexes, a physician can assess the function of sensory pathways and motor centers throughout the spinal cord and brain. AM *Reflexes and Diagnostic Testing*

•**FIGURE 13-20 The Babinski Reflexes. (a)** The plantar reflex (negative Babinski reflex), a curling of the toes, is seen in healthy adults. **(b)** The Babinski sign (positive Babinski reflex) occurs in the absence of descending inhibition. It is normal in infants but pathological in adults.

ceeds, tendon reflexes keep tension production within tolerable limits.

Motor control therefore involves a series of interacting levels. At the lowest level are monosynaptic reflexes that are rapid but stereotyped and relatively inflexible. At the highest level are centers in the brain that can modulate or build on reflexive motor patterns.

☑ A weight lifter is straining to lift a 200-kg barbell. Shortly after he lifts it to chest height, his muscles appear to relax and he drops the barbell. Which reflex has occurred?

☑ During a withdrawal reflex, what happens to the limb on the side opposite the stimulus? What is this response called?

☑ After injuring her back, Tina exhibits a positive Babinski reflex. What does this imply about Tina's injury?

[EXPLORE] MediaLab

The anatomical simplicity of the spinal cord and spinal nerves is in direct contrast to the complex physiological processes they perform. To further your understanding of this portion of the central nervous system, use the resources available on both the CD-ROM and the Companion Website. The CD-ROM includes animations and tutorials; the website builds on those resources with text content and interactive problems as well as communication avenues for sharing ideas with your class. To complete the following Web Explorations, visit the MediaLab in Chapter 13 of your Companion Website.

Web Exploration 1

Estimated time for completion = 10 minutes

Reflect on the lives of individuals with a spinal cord injury. Using what you know about the spinal cord, spinal nerves, and their functions, decide how a spinal cord injury would affect you. Select the keyword **SPINAL INJURY** on the Companion Website to see various statistics about spinal cord injuries. How does spinal cord injury change the life ex-

pectancy of a person your age? What are the physiological and anatomical bases for the changes? Working alone or in a group, relate the information on this Web page to the information in the text to describe spinal cord function.

Web Exploration 2

Estimated time for completion = 10 minutes

A good way to check what you have learned is to practice answering questions. To access 13 questions about the general anatomy of the spinal cord, select the keyword **SPINAL CORD QUIZ** on the Companion Website. Once you answer a question, you will get immediate, useful feedback. Take note of all questions you got wrong. You can return to the Companion Website and select the same keyword to retake the quiz. Do not get discouraged; this is a very thorough quiz prepared for pharmacology students at the University of Kansas Medical Center. If you wish, share your score with your instructor by using the communication tools on the Companion Website.

13

CHAPTER REVIEW

SELECTED CLINICAL TERMINOLOGY
Terms Discussed in This Chapter

Babinski sign *(positive Babinski reflex):* A spinal reflex in infants, consisting of a fanning of the toes and produced by stroking the foot on the side of the sole; in adults, a sign of CNS injury. *(p. 430)*

epidural block: The injection of anesthetic into the epidural space to eliminate sensory and motor innervation via spinal nerves in the area of injection. *(p. 410 and AM)*

lumbar puncture: A spinal tap performed between adjacent lumbar vertebrae inferior to the conus medullaris. *(p. 411)*

meningitis: An inflammation of the meninges involving the spinal cord *(spinal meningitis)* and/or brain *(cerebral meningitis)*; generally caused by or viral pathogens. *(p. 409 and AM)*

myelography: A diagnostic procedure in which a radiopaque dye is introduced into the cerebrospinal fluid to obtain an X-ray of the spinal cord and cauda equina. *(p. 411)*

nerve growth factor: A peptide factor that promotes the growth and maintenance of neurons. Other factors important to neuron growth and repair include BDNF, NT-3, NT-4, GAP-43. *(p. 416 and AM)*

nerve palsies (peripheral neuropathies): Regional losses of sensory and motor function as a result of nerve trauma or com-

pression. Common palsies include *radial nerve palsy, ulnar palsy, sciatica,* and *peroneal palsy.* *(p. 416 and AM)*

paraplegia: Paralysis involving a loss of motor control of the lower but not the upper limbs. *(p. 413)*

patellar reflex (knee jerk reflex): A stretch reflex resulting from the stimulation of stretch receptors in the quadriceps muscles. *(p. 426)*

plantar reflex *(negative Babinski reflex):* A spinal reflex in adults, consisting of a curling of the toes and produced by stroking the foot on the side of the sole. *(p. 430)*

quadriplegia: Paralysis involving the loss of sensation and motor control of the upper and lower limbs. *(p. 413 and AM)*

shingles: A condition caused by the infection of neurons in dorsal root ganglia by the virus *Herpes varicella-zoster.* The primary symptom is a painful rash along the sensory distribution of the affected spinal nerves. *(p. 416 and AM)*

spinal tap: A procedure in which cerebrospinal fluid is removed from the subarachnoid space through a needle, generally inserted between the lumbar vertebrae. *(p. 410)*

AM **Additional Terms Discussed in the *Applications Manual***

abdominal reflex: A reflexive twitch in abdominal muscles that moves the navel toward a stimulus; often used to provide information about descending tracts in the spinal cord.

areflexia (ā-re-FLEK-sē-uh): A condition in which normal reflexes fail to appear, even with reinforcement.

caudal anesthesia: The introduction of anesthetics into the epidural space of the sacrum to paralyze lower abdominal and perineal structures.

clonus (KLŌ-nus): Oscillations that develop between opposing stretch reflexes in individuals with hyperreflexia.

Hansen's disease *(leprosy):* A bacterial infection that begins in sensory nerves of the skin and gradually progresses to a motor paralysis of the same regions.

hyperreflexia: A condition resulting from a high degree of facilitation along the spinal cord; reflexes are easily triggered, and the responses may be grossly exaggerated.

hyporeflexia: A condition in which normal reflexes are weak but demonstrable with reinforcement.

mass reflex: A sudden, abnormal, and prolonged activation of somatic and visceral reflexes throughout the spinal cord, resulting from an extreme state of motor neuron facilitation.

paresthesia: An abnormal tingling sensation, usually described as "pins and needles," that accompanies sensory return after a temporary palsy.

sciatica (sī-AT-i-kuh): The painful result of compression of the roots of the sciatic nerve.

spinal shock: A period of depressed sensory and motor function following any severe injury to the spinal cord. Spinal shock may accompany *spinal concussion, spinal contusion, spinal laceration, spinal compression,* or *spinal transection.*

STUDY OUTLINE

13

INTRODUCTION p. 407

1. In addition to relaying information to and from the brain, the spinal cord integrates and processes information on its own.

GROSS ANATOMY OF THE SPINAL CORD p. 407

1. The adult spinal cord includes localized **enlargements,** which provide innervation to the limbs. The spinal cord has 31 segments, each associated with a pair of **dorsal roots** and a pair of **ventral roots.** *(Figure 13-1)*

2. The **filum terminale** (a strand of fibrous tissue), which originates at the **conus medullaris,** ultimately becomes part of the *coccygeal ligament. (Figure 13-1)*

3. **Spinal nerves** are **mixed nerves:** They contain both afferent (sensory) and efferent (motor) fibers.

Spinal Meninges p. 409

4. The **spinal meninges** provide physical stability and shock absorption for neural tissues of the spinal cord; the **cranial meninges** surround the brain. *(Figure 13-2)*

5. The **dura mater** covers the spinal cord; inferiorly it tapers into the **coccygeal ligament.** The **epidural space** separates the dura mater from the walls of the vertebral canal. *(Figures 13-2, 13-3)*

6. Interior to the inner surface of the dura mater are the **subdural space,** the **arachnoid** (the second meningeal layer), and the **subarachnoid space.** The subarachnoid space contains **cerebrospinal fluid (CSF),** which acts as a shock absorber and a diffusion medium for dissolved gases, nutrients, chemical messengers, and waste products. *(Figures 13-2, 13-3)*

7. The **pia mater,** a meshwork of elastic and collagen fibers, is the innermost meningeal layer. Unlike more-superficial meninges, it is bound to the underlying neural tissue. *(Figures 13-2, 13-3)*

Sectional Anatomy of the Spinal Cord p. 412

8. The white matter contains myelinated and unmyelinated axons, whereas the gray matter contains cell bodies of neurons and neuroglia and unmyelinated axons. The projections of gray matter toward the outer surface of the spinal cord are called **horns.** *(Figure 13-5)*

9. The **posterior gray horns** contain somatic and visceral sensory nuclei; nuclei in the **anterior gray horns** deal with somatic motor control. The **lateral gray horns** contain visceral motor neurons. The **gray commissures** contain axons that cross from one side of the spinal cord to the other. *(Figure 13-5)*

10. The white matter can be divided into six **columns** *(funiculi),* each of which contains **tracts** *(fasciculi).* **Ascending tracts** relay information from the spinal cord to the brain, and **descending tracts** carry information from the brain to the spinal cord. *(Figure 13-5)*

SPINAL NERVES p. 414

1. There are 31 pairs of spinal nerves. Each has an **epineurium** (outermost layer), **perineurium,** and **endoneurium** (innermost layer). *(Figure 13-6)*

Peripheral Distribution of Spinal Nerves p. 414

2. A typical spinal nerve has a **white ramus** (containing myelinated axons), a **gray ramus** (containing unmyelinated fibers that innervate glands and smooth muscles in the body wall or limbs), a **dorsal ramus** (providing sensory and motor innervation to the skin and muscles of the back), and a **ventral ramus** (supplying the ventrolateral body surface, structures in the body wall, and the limbs). Each pair of nerves monitors a region of the body surface called a **dermatome.** *(Figures 13-7, 13-8)*

Nerve Plexuses p. 416

3. A complex, interwoven network of nerves is a **nerve plexus.** The four large plexuses are the **cervical plexus,** the **brachial plexus,** the **lumbar plexus,** and the **sacral plexus.** *(Figures 13-9–13-12; Tables 13-1–13-3)*

AN INTRODUCTION TO REFLEXES p. 422

1. **Reflexes** are rapid, automatic responses to stimuli. A *neural reflex* is an automatic, involuntary motor response by the nervous system that helps preserve homeostasis by rapidly adjusting the functions of organs or organ systems.

The Reflex Arc p. 422

2. A **reflex arc** is the neural "wiring" of a single reflex. *(Figure 13-13)*
3. The five steps involved in a neural reflex are (1) the arrival of a stimulus and activation of a receptor, (2) the activation of a sensory neuron, (3) information processing, (4) the activation of a motor neuron, and (5) a response by an effector. *(Figure 13-13)*
4. A *receptor* is a specialized cell that monitors conditions in the body or external environment. Each receptor has a characteristic range of sensitivity.

Classification of Reflexes p. 423

5. Reflexes are classified according to (1) their development, (2) the site of information processing, (3) the nature of the resulting motor response, and (4) the complexity of the neural circuit. *(Figure 13-14)*
6. **Innate reflexes** result from the connections that form between neurons during development. **Acquired reflexes** are learned and typically are more complex.
7. Reflexes processed in the brain are **cranial reflexes.** In a **spinal reflex,** the important interconnections and processing events occur in the spinal cord.
8. **Somatic reflexes** control skeletal muscles; **visceral reflexes** *(autonomic reflexes)* control the activities of other systems.
9. In a **monosynaptic reflex,** the simplest reflex arc, a sensory neuron synapses directly on a motor neuron, which acts as the processing center. **Polysynaptic reflexes,** which have at least one interneuron between the sensory afferent and the motor efferent, have a longer delay between stimulus and response. *(Figure 13-15)*

SPINAL REFLEXES p. 425

1. Spinal reflexes range from simple monosynaptic reflexes to more-complex polysynaptic reflexes, in which many segments interact to produce a coordinated motor response.

Monosynaptic Reflexes p. 425

2. The **stretch reflex** (such as the **patellar,** or **knee jerk, reflex**) is a monosynaptic reflex that automatically regulates skeletal muscle length and muscle tone. The sensory receptors involved are **muscle spindles.** *(Figures 13-16–13-18)*
3. A **postural reflex** maintains normal upright posture.

Polysynaptic Reflexes p. 428

4. Polysynaptic reflexes can produce more-complicated responses than can monosynaptic reflexes. Examples include the **tendon reflex** (which monitors the tension produced during muscular contractions and prevents damage to tendons) and **withdrawal reflexes** (which move affected portions of the body away from a source of stimulation). The **flexor reflex** is a withdrawal reflex affecting the muscles of a limb. The **crossed extensor reflex** complements withdrawal reflexes. *(Figure 13-19)*
5. In an **ipsilateral reflex arc,** the sensory stimulus and motor response occur on the same side of the body. In a **contralateral reflex arc,** the motor response occurs on the side opposite the stimulus. *(Figure 13–19)*
6. All polysynaptic reflexes (1) involve pools of interneurons, (2) are intersegmental in distribution, (3) involve reciprocal inhibition, and (4) have reverberating circuits, which prolong the reflexive motor response. (5) Several reflexes may cooperate to produce a coordinated response.

INTEGRATION AND CONTROL OF SPINAL REFLEXES p. 430

1. The brain can facilitate or inhibit reflex motor patterns based in the spinal cord.
2. The enhancement of spinal reflexes is **reinforcement.** Spinal reflexes may instead be inhibited, as in the **plantar reflex** in adults versus the **Babinski sign** in infants. *(Figure 13-20)*

Voluntary Movements and Reflex Motor Patterns p. 430

3. Motor control involves a series of interacting levels. Monosynaptic reflexes form the lowest level; at the highest level are the centers in the brain that can modulate or build on reflexive motor patterns.

13

REVIEW QUESTIONS

LEVEL 1 REVIEWING FACTS AND TERMS

1. The expanded area of the spinal cord that supplies nerves to the pectoral girdle and arms is the
   (a) conus medullaris   (b) filum terminale
   (c) lumbar enlargement   (d) cervical enlargement
2. The dorsal roots of each spinal segment
   (a) bring sensory information into the spinal cord
   (b) control peripheral effectors
   (c) contain somatic motor and visceral motor neurons
   (d) a, b, and c are correct
3. Spinal nerves are called mixed nerves because they
   (a) contain sensory and motor fibers
   (b) exit at intervertebral foramina
   (c) are associated with a pair of dorsal root ganglia
   (d) are associated with dorsal and ventral roots
4. The spinal cord continues to enlarge and elongate until an individual is
   (a) born   (b) 1 year old
   (c) 4 years old   (d) 10 years old
5. The tough, fibrous outermost covering of the spinal cord is the
   (a) arachnoid   (b) pia mater
   (c) dura mater   (d) epidural block
6. The gray matter of the spinal cord is dominated by
   (a) myelinated axons only
   (b) unmyelinated axons only
   (c) Schwann cells and satellite cells
   (d) cell bodies of neurons, neuroglia, and unmyelinated axons
7. The outermost layer of connective tissue that surrounds each spinal nerve is the
   (a) perineurium   (b) epineurium
   (c) endoneurium   (d) epimysium
8. A sensory region monitored by the dorsal rami of a single spinal segment is
   (a) a ganglion   (b) a fascicle
   (c) a dermatome   (d) a ramus
9. The major nerve of the cervical plexus that innervates the diaphragm is the
   (a) median nerve   (b) axillary nerve
   (c) phrenic nerve (d) peroneal nerve

10. The genitofemoral, femoral, and lateral femoral cutaneous nerves are major nerves of the
  (a) lumbar plexus
  (b) sacral plexus
  (c) brachial plexus
  (d) cervical plexus

11. The final step in a neural reflex is
  (a) activation of a receptor
  (b) response by an effector
  (c) activation of a motor neuron
  (d) information processing

12. The reflexes that control the most-rapid, stereotyped motor responses of the nervous system to stimuli are
  (a) monosynaptic reflexes
  (b) polysynaptic reflexes
  (c) tendon reflexes

13. An example of a stretch reflex triggered by passive muscle movement is the
  (a) tendon reflex   **(b)** patellar reflex
  (c) flexor reflex   **(d)** ipsilateral reflex

14. The contraction of flexor muscles and the relaxation of extensor muscles illustrates the principle of
  (a) reverberating circuitry   **(b)** generalized facilitation
  (c) reciprocal inhibition   **(d)** reinforcement

15. Reflex arcs in which the sensory stimulus and the motor response occur on the same side of the body are
  (a) contralateral   **(b)** paraesthetic
  (c) ipsilateral   **(d)** monosynaptic

16. All polysynaptic reflexes
  (a) have reverberating circuits that prolong the reflexive motor response
  (b) involve pools of interneurons
  (c) involve reciprocal inhibition
  (d) a, b, and c are correct

17. Which three meningeal layers provide physical stability and shock absorption for the spinal cord?

18. Beginning with the outermost layer, list the connective tissue layers that surround each spinal nerve.

LEVEL 2 REVIEWING CONCEPTS

19. Polysynaptic reflexes can produce far more complicated responses than can monosynaptic reflexes because
  (a) the response time is quicker
  (b) the response is initiated by highly sensitive receptors
  (c) motor neurons carry impulses at a faster rate than do sensory neurons
  (d) the interneurons involved can control several muscle groups

20. Why do cervical nerves outnumber cervical vertebrae?

21. If the anterior gray horns of the spinal cord were damaged, what type of control would be affected?

22. Proceeding inward from the outermost layer, number the following in the correct sequence:
  ___ walls of vertebral canal   ___ subdural space
  ___ pia mater   ___ subarachnoid space
  ___ dura mater   ___ epidural space
  ___ arachnoid membrane   ___ spinal cord

23. In which of the structures in Review Question 22 is cerebrospinal fluid (CSF) located? What are the functions of CSF?

24. What five characteristics are common to all polysynaptic reflexes?

25. Which ramus would be involved if you felt something brushing against your _____?
  (a) abdomen   **(b)** left forearm
  (c) upper back   **(d)** right ankle

26. Which plexus has primary responsibility for innervating the following muscles?
  (a) diaphragm   **(b)** pelvic diaphragm
  (c) deltoid   **(d)** gastrocnemius

27. Why is response time in a monosynaptic reflex much faster than response time in a polysynaptic reflex?

28. List, in sequence, the five steps involved in a neural reflex.

29. What would happen if the ventral root of a spinal nerve were damaged or transected?

30. How does the stimulation of a sensory neuron that innervates an extrafusal muscle fiber affect muscle tone?

31. Why is it important that a spinal tap be done between the third and fourth lumbar vertebrae?

LEVEL 3 CRITICAL THINKING AND CLINICAL APPLICATIONS

32. Mary complains that when she wakes up in the morning, her thumb and forefinger are always "asleep." She mentions this condition to her physician, who asks if Mary sleeps with her arm above her head. She replies that she does. The physician tells Mary that sleeping in that position compresses a portion of one of the nerves of the brachial plexus, producing her symptoms. Which nerve is involved?

33. The improper use of crutches can produce a condition known as "crutch paralysis," which is characterized by a lack of response by the extensor muscles of the arm, and a condition known as "wrist drop," an inability to extend the fingers and wrist. Which nerve is involved?

34. During childbirth, some women are given a local anesthetic that temporarily deadens sensory neurons in the region of the genitals. Which nerve does this anesthetic affect?

35. While playing football, Ramon is tackled hard and injures his left leg. As he tries to get up, he cannot flex his left thigh or extend the leg. Which nerve is damaged, and how would this damage affect sensory perception in his left leg?

CHAPTER 14

The Brain and Cranial Nerves

CHAPTER OUTLINE AND OBJECTIVES

INTRODUCTION   p. 436

AN INTRODUCTION TO THE ORGANIZATION
OF THE BRAIN   p. 436
  A Preview of Major Regions and Landmarks   p. 436
  Embryology of the Brain   p. 438
  Ventricles of the Brain   p. 439
① Name the major regions of the brain, and describe
   their functions.
② Name the ventricles of the brain, and describe their
   locations and the connections between them.

PROTECTION AND SUPPORT OF THE BRAIN   p. 440
  The Cranial Meninges   p. 440
  Cerebrospinal Fluid   p. 441
  The Blood Supply to the Brain   p. 443
③ Explain how the brain is protected and supported.
④ Discuss the formation, circulation, and functions
   of the cerebrospinal fluid.

THE CEREBRUM   p. 445
  The Cerebral Cortex   p. 445
  The Cerebral Nuclei   p. 451
⑤ Locate the motor, sensory, and association areas
   of the cerebral cortex, and discuss their functions.

THE LIMBIC SYSTEM   p. 452

THE DIENCEPHALON   p. 454
  The Thalamus   p. 456
  The Hypothalamus   p. 457

THE MESENCEPHALON   p. 459

THE CEREBELLUM   p. 460

THE PONS   p. 462

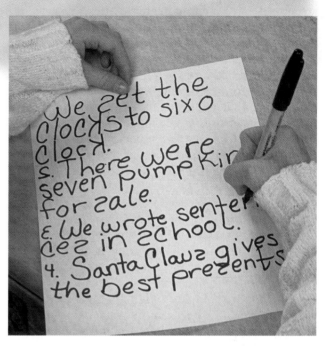

Our perceptions of the world around us depend on thousands of interactions among neurons in the central nervous system. We seldom realize how complex these processes are unless they go wrong in some way. This child has dyslexia, a condition characterized by difficulties with the recognition and use of words. Although the cause of dyslexia remains a mystery, there is general agreement that it results from problems with the integration and processing of visual or auditory information. There is much that we still do not understand about such activities, which ultimately create our consciousness and our unique personalities. This chapter introduces the brain regions involved in our conscious and subconscious thought processes and considers complex neural functions, such as memory and learning.

THE MEDULLA OBLONGATA   p. 462
⑥ Identify important structures within each region
   of the brain, and explain their functions.

■ FOCUS: Cranial Nerves   p. 464

  A Summary of Cranial Nerve Branches
    and Functions   p. 474
⑦ Identify the cranial nerves, and relate each nerve
   to its principal destinations and functions.

MEDIALAB and CHAPTER REVIEW   p. 475

The brain is probably the most fascinating organ in the body, yet we know relatively little about its structural and functional complexities. What we do know is that all our dreams, passions, plans, and memories result from brain activity. The brain contains tens of billions of neurons organized into hundreds of neuronal pools; it has a complex three-dimensional structure and performs a bewildering array of functions. If an individual's heart, liver, lung, or kidney stops working, heroic measures, including organ transplantation, can be taken to preserve the person's life. But if the brain permanently stops working, the person is classified as "brain dead" and for all intents and purposes ceases to exist.

The brain is far more complex than the spinal cord, and it can respond to stimuli with greater versatility. That versatility results from (1) the tremendous number of neurons and neuronal pools in the brain and (2) the complexity of the interconnections between those neurons and neuronal pools. Because the interconnections are complex and the pools are large, the response can be varied to meet changing circumstances. But this versatility has a price: A response cannot be immediate, precise, and adaptable all at the same time. Adaptability requires multiple processing steps, and every synapse adds to the delay between stimulus and response. One of the major functions of spinal reflexes is to provide an *immediate* response that can be fine-tuned or elaborated on by more-versatile—but slower—processing centers in the brain. For example, the flexor reflex will pull your hand from a hot stove before you become consciously aware that something is wrong. ∞ *p. 428* In effect, this spinal reflex prevents further injury and gives the brain time to decide what to do next.

In this chapter, we consider the brain and the cranial nerves and their primary functional roles. The chapter organization, which parallels that of Chapter 13, will make it easy for you to compare the functional organization of the brain with that of the spinal cord.

AN INTRODUCTION TO THE ORGANIZATION OF THE BRAIN

The adult human brain contains almost 98 percent of the body's neural tissue. A "typical" brain weighs 1.4 kg (3 lb) and has a volume of 1200 cc (71 in.^{3}). Brain size varies considerably among individuals. The brains of males average about 10 percent larger than those of females, owing to differences in average body size. No correlation exists between brain size and intelligence. Individuals with the smallest brains (750 cc) and the largest brains (2100 cc) are functionally normal.

A PREVIEW OF MAJOR REGIONS AND LANDMARKS

The adult brain is dominated in size by the cerebrum:

■ *Cerebrum.* Viewed from the anterior and superior surfaces (Figure 14-1a,b●), the **cerebrum** (SER-e-brum, or se-RĒ-brum) of the adult brain can be divided into large, paired **cerebral hemispheres.** Conscious thoughts, sensations, intellect, memory, and complex movements originate in the cerebrum. The surfaces of both the cerebral hemispheres and the cerebellum are highly folded and covered by **neural cortex** (*cortex*, rind or bark), a superficial layer of gray matter. The term *cerebral cortex* refers to the neural cortex of the cerebral hemispheres, as opposed to the *cerebellar cortex* of the cerebellar hemispheres.

Second in size to the cerebrum is the cerebellum:

■ *Cerebellum.* The hemispheres of the **cerebellum** (ser-e-BEL-um) are partially hidden by the cerebral hemispheres (Figure 14-1b–d●). The cerebellum adjusts ongoing movements on the basis of comparisons between arriving sensations and sensations experienced previously, allowing you to perform the same movements over and over.

The other major regions of the brain can best be examined after the cerebral hemispheres have been removed (Figure 14-2●, p. 438):

■ *Diencephalon.* The walls of the **diencephalon** (dī-en-SEF-a-lon; *dia,* through) are composed of the **left thalamus** and **right thalamus** (THAL-a-mus; plural, *thalami*). Each thalamus contains relay and processing centers for sensory information. The **hypothalamus** (*hypo-*, below), or floor of the diencephalon, contains centers involved with emotions, autonomic function, and hormone production. The *infundibulum,* a narrow stalk, connects the hypothalamus to the **pituitary gland,** a component of the endocrine system. As we shall see in Chapter 18, it is the primary link between the nervous and endocrine systems.

The diencephalon is a structural and functional link between the cerebral hemispheres and the components of the brain stem. The **brain stem** includes the mesencephalon, pons, and medulla oblongata.[1] It contains a variety of important processing centers and nuclei that relay information headed to or from the cerebrum or cerebellum:

■ *Mesencephalon.* Sensory nuclei in the **mesencephalon** (*mesos*, middle), or midbrain, process visual and auditory information and control reflexes triggered by

---

[1] Some sources consider the brain stem to include the diencephalon. We will use the more restrictive definition here.

(b) Superior view

(a) Anterior view

(c) Posterior view

(d) Lateral view

•**FIGURE 14-1** **The Adult Brain.** **(a)** An anterior view of the brain. **(b)** A superior view, which is dominated by the paired cerebral hemispheres. **(c)** A posterior view. **(d)** A lateral view.

14

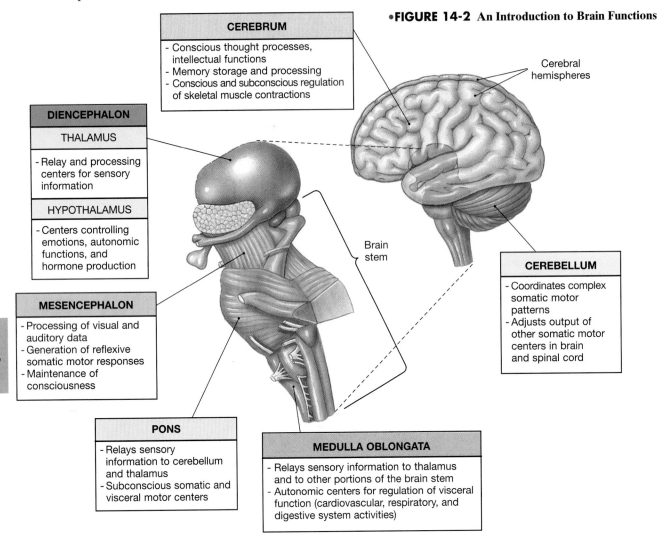

•**FIGURE 14-2** An Introduction to Brain Functions

CEREBRUM

- Conscious thought processes, intellectual functions
- Memory storage and processing
- Conscious and subconscious regulation of skeletal muscle contractions

DIENCEPHALON

THALAMUS

- Relay and processing centers for sensory information

HYPOTHALAMUS

- Centers controlling emotions, autonomic functions, and hormone production

MESENCEPHALON

- Processing of visual and auditory data
- Generation of reflexive somatic motor responses
- Maintenance of consciousness

PONS

- Relays sensory information to cerebellum and thalamus
- Subconscious somatic and visceral motor centers

MEDULLA OBLONGATA

- Relays sensory information to thalamus and to other portions of the brain stem
- Autonomic centers for regulation of visceral function (cardiovascular, respiratory, and digestive system activities)

CEREBELLUM

- Coordinates complex somatic motor patterns
- Adjusts output of other somatic motor centers in brain and spinal cord

Cerebral hemispheres

Brain stem

these stimuli. For example, your immediate, reflexive responses to a loud, unexpected noise (eye movements and head turning) are directed by nuclei in the midbrain. This region also contains centers that help maintain consciousness.

- *Pons.* The term *pons* is Latin for "bridge"; the **pons** of the brain connects the cerebellum to the brain stem. In addition to tracts and relay centers, the pons also contains nuclei involved with somatic and visceral motor control.

- *Medulla oblongata.* The spinal cord connects to the brain at the **medulla oblongata.** Near the pons, the posterior wall of the medulla oblongata is thin and membranous. The inferior portion of the medulla oblongata resembles the spinal cord in that it has a narrow central canal. The medulla oblongata relays sensory information to the thalamus and to centers in other portions of the brain stem. The medulla oblongata also contains major centers that regulate autonomic function, such as heart rate, blood pressure, and digestion.

The boundaries and general functions of the diencephalon and brain stem are listed in Figure 14-2•.

☑ What are the three primary brain vesicles, and what does each contribute to the structure of the adult brain?

☑ In response to a loud noise, your head automatically turns toward the source of the sound. Which part of the brain directs this response?

EMBRYOLOGY OF THE BRAIN

To understand the internal organization of the adult brain, we must consider its embryological origins. The central nervous system (CNS) begins as a hollow tube known as the *neural tube.* This tube has a fluid-filled internal cavity, the *neurocoel.* In the cephalic portion of the neural tube, three areas enlarge rapidly through expansion of the neurocoel. This enlargement creates three prominent divisions called **primary brain vesicles.** The primary brain vesicles are named for their relative positions: the **prosencephalon** (prō-zen-SEF-a-lon; *proso,* forward + *enkephalos,* brain), or "forebrain"; the mesencephalon, or "midbrain"; and the **rhomben-cephalon** (rom-ben-SEF-a-lon), or "hindbrain."

TABLE 14-1 **Development of the Brain**

Primary Brain Vesicles (3 weeks)	Secondary Brain Vesicles (6 weeks)	Brain Regions at Birth

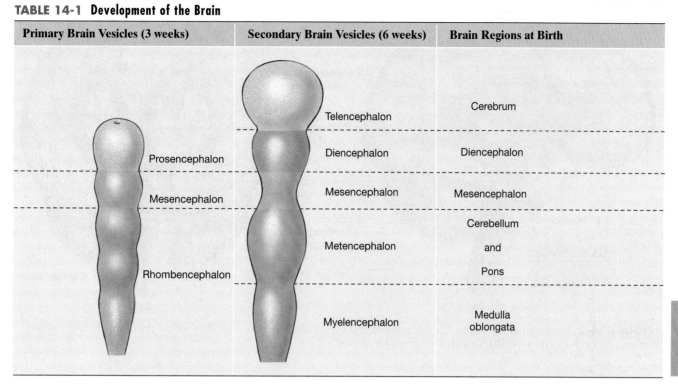

Primary Brain Vesicles (3 weeks)	Secondary Brain Vesicles (6 weeks)	Brain Regions at Birth
Prosencephalon	Telencephalon	Cerebrum
	Diencephalon	Diencephalon
Mesencephalon	Mesencephalon	Mesencephalon
Rhombencephalon	Metencephalon	Cerebellum and Pons
	Myelencephalon	Medulla oblongata

14

The fate of the three primary divisions of the brain is summarized in Table 14-1. The prosencephalon and rhombencephalon are subdivided further, forming **secondary brain vesicles.** The prosencephalon forms the **telencephalon** (tel-en-SEF-a-lon; *telos,* end) and the diencephalon. The telencephalon will ultimately form the cerebrum of the adult brain. The walls of the mesencephalon thicken, and the neurocoel becomes a relatively narrow passageway comparable to the central canal of the spinal cord. The portion of the rhombencephalon adjacent to the mesencephalon forms the **metencephalon** (met-en-SEF-a-lon; *meta,* after). The dorsal portion of the metencephalon will become the cerebellum, and the ventral portion will develop into the pons. The portion of the rhombencephalon closer to the spinal cord forms the **myelencephalon** (mī-el- en-SEF-a-n; *myelon,* spinal cord), which will become the medulla oblongata. ▢AM *EMBRYOLOGY SUMMARY 12: The Development of the Brain and Cranial Nerves*

VENTRICLES OF THE BRAIN

During development, the neurocoel within the cerebral hemispheres, diencephalon, metencephalon, and medulla oblongata expands to form chambers called ventricles (VEN-tri-kls). The ventricles are lined by cells of the *ependyma.* ∞ *p. 368*

Each cerebral hemisphere contains a large **lateral ventricle. The septum pellucidum,** a thin medial partition, separates the two lateral ventricles. Because there are two lateral ventricles (first and second), the ventricle in the diencephalon is called the **third ventricle.** Although the two lateral ventricles are not directly connected, each communicates with the third ventricle of the diencephalon through an **interventricular foramen** (*foramen of Monro*) (Figure 14-3●).

The mesencephalon has a slender canal known as the **mesencephalic aqueduct** (*aqueduct of Sylvius* or *cerebral aqueduct*). This passageway connects the third ventricle with the **fourth ventricle.** The superior portion of the fourth ventricle lies between the posterior surface of the pons and the anterior surface of the cerebellum. The fourth ventricle extends into the superior portion of the medulla oblongata. This ventricle then narrows and becomes continuous with the central canal of the spinal cord.

The ventricles are filled with cerebrospinal fluid (CSF). The CSF continuously circulates from the ventricles and central canal into the *subarachnoid space* of the surrounding cranial meninges. The CSF passes between the interior and exterior of the CNS through foramina in the roof of the fourth ventricle.

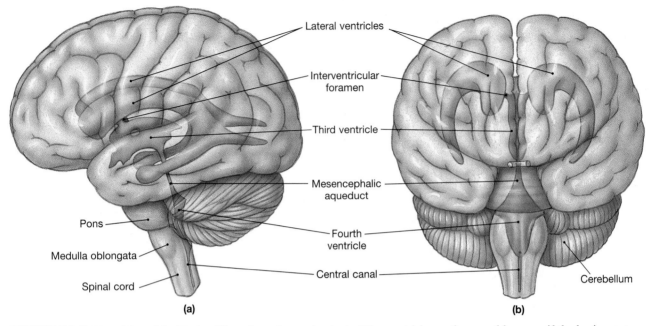

•**FIGURE 14-3** **Ventricles of the Brain.** The orientation and extent of the ventricles as they would appear if the brain were transparent. **(a)** A lateral view. **(b)** An anterior view.   AM   *Scans 1, 2*

14

PROTECTION AND SUPPORT OF THE BRAIN

The delicate tissues of the brain are protected from mechanical forces by (1) the bones of the cranium, (2) the *cranial meninges,* and (3) cerebrospinal fluid. In addition, the neural tissue of the brain is biochemically isolated from the general circulation by the *blood–brain barrier.* Refer to Figures 7-3 and 7-4• (pp. 197–199) for a review of the bones of the cranium. We will discuss the other protective factors here.

THE CRANIAL MENINGES

The layers that make up the cranial meninges—the cranial dura mater, arachnoid, and pia mater—are continuous with those of the spinal meninges. ∞ *p. 409* However, the cranial meninges have distinctive anatomical and functional characteristics:

■ The cranial *dura mater* consists of outer and inner fibrous layers. The outer layer is fused to the periosteum of the cranial bones. As a result, there is no epidural space superficial to the dura mater, as there is along the spinal cord. The outer, or *endosteal,* and inner, or *meningeal,* layers of the cranial dura mater are typically separated by a slender gap that contains tissue fluids and blood vessels, including several large venous sinuses. The veins of the brain open into these sinuses, which deliver the venous blood to the *internal jugular veins* of the neck.

■ The cranial *arachnoid* consists of the arachnoid membrane, an epithelial layer, and the cells and fibers of the arachnoid trabeculae that cross the subarachnoid space to the pia mater. The arachnoid membrane covers the brain, providing a smooth surface that does not follow the brain's underlying folds. This membrane is in contact with the inner epithelial layer of the dura mater. The subarachnoid space extends between the arachnoid membrane and the pia mater.

■ The *pia mater* sticks to the surface of the brain, anchored by the processes of astrocytes. It extends into every fold and accompanies the branches of cerebral blood vessels as they penetrate the surface of the brain to reach internal structures.

Functions of the Cranial Meninges

The brain is cradled within the cranium. The overall shape of the brain roughly corresponds to that of the cranial cavity (Figure 14-4a•). The massive cranial bones provide mechanical protection, but they also pose a threat. The brain is like a person driving a car. If the car hits a tree, the car protects the driver from contact with the tree, but serious injury will occur unless a seat belt or airbag protects the driver from contact with the car. The cranial meninges protect the brain effectively. Tough, fibrous extensions of the dura mater, called *dural folds,* act like safety belts that hold the brain in position. The cerebrospinal fluid in the subarachnoid space acts like an airbag that cushions against sudden jolts and shocks.

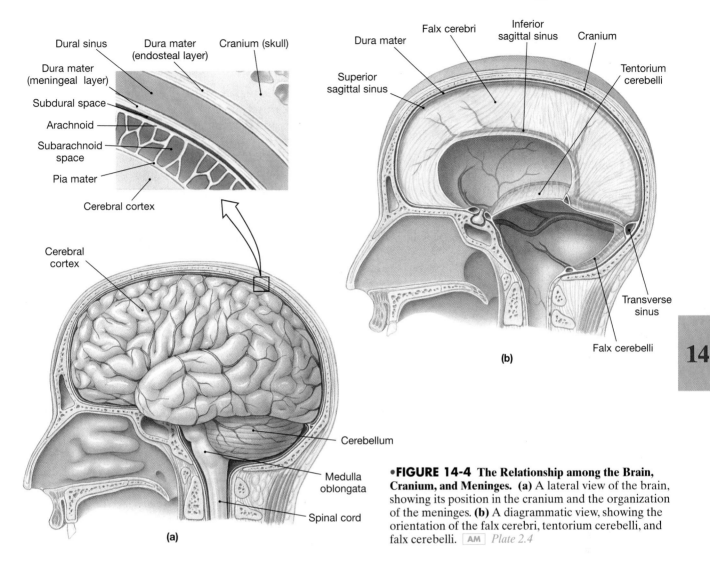

•FIGURE 14-4 The Relationship among the Brain, Cranium, and Meninges. (a) A lateral view of the brain, showing its position in the cranium and the organization of the meninges. **(b)** A diagrammatic view, showing the orientation of the falx cerebri, tentorium cerebelli, and falx cerebelli. AM *Plate 2.4*

14

Cranial trauma is a head injury resulting from impact with another object. Each year in the United States, roughly 8 million cases of cranial trauma occur, but only 1 case in 8 results in serious brain damage. The percentage is relatively low because the cranial meninges and CSF are so effective in protecting the brain. AM *Cranial Trauma*

Dural Folds In several locations, the inner layer of the dura mater extends into the cranial cavity, forming a sheet that dips inward and then returns. These **dural folds** provide additional stabilization and support to the brain. **Dural sinuses** are large collecting veins located between the two layers of a dural fold. The three largest dural folds are called the falx cerebri, the tentorium cerebelli, and the falx cerebelli (Figure 14-4b•):

1. The **falx cerebri** (FALKS ser-Ē-brē; *falx,* curving or sickle-shaped) is a fold of dura mater that projects between the cerebral hemispheres in the longitudinal fissure. Its inferior portions attach anteriorly to the crista galli and posteriorly to the *internal occipital crest,* a ridge along the inner surface of the occipital bone. The **superior sagittal sinus** and the **inferior sagittal sinus,** two large venous sinuses, lie within this dural fold. The posterior margin of the falx cerebri intersects the tentorium cerebelli.

2. The **tentorium cerebelli** (ten-TOR-ē-um ser-e-BEL-ē; *tentorium,* a covering) separates and protects the cerebellar hemispheres from those of the cerebrum. It extends across the cranium at right angles to the falx cerebri. The **transverse sinus** lies within the tentorium cerebelli.

3. The **falx cerebelli** divides the two cerebellar hemispheres along the midsagittal line inferior to the tentorium cerebelli.

CEREBROSPINAL FLUID

Cerebrospinal fluid (CSF) completely surrounds and bathes the exposed surfaces of the CNS. The CSF has several important functions, including the following:

■ *Cushioning delicate neural structures.*

■ *Supporting the brain.* In essence, the brain is suspended inside the cranium and floats in the CSF. A human brain weighs about 1400 g in air but only about 50 g when supported by CSF.

■ *Transporting nutrients, chemical messengers, and waste products.* Except at the choroid plexus, where CSF is produced, the ependymal lining is freely permeable and the CSF is in constant chemical communication with the interstitial fluid of the CNS.

Because free exchange occurs between the interstitial fluid and the CSF, changes in CNS function can produce changes in the composition of the CSF. As we noted in Chapter 13, a *spinal tap* can provide useful clinical information about CNS injury, infection, or disease. ∞ *pp. 410, 411*

The Formation of CSF

The **choroid plexus** (*choroid,* vascular coat; *plexus,* network) consists of a combination of specialized ependymal cells and permeable capillaries for the production of cerebrospinal fluid. Two extensive folds of the choroid plexus originate in the roof of the third ventricle and extend through the interventricular foramina. These folds cover the floors of the lateral ventricles (Figure 14-5a●). In the inferior brain stem, a region of the choroid plexus in the roof of the fourth ventricle projects between the cerebellum and the pons.

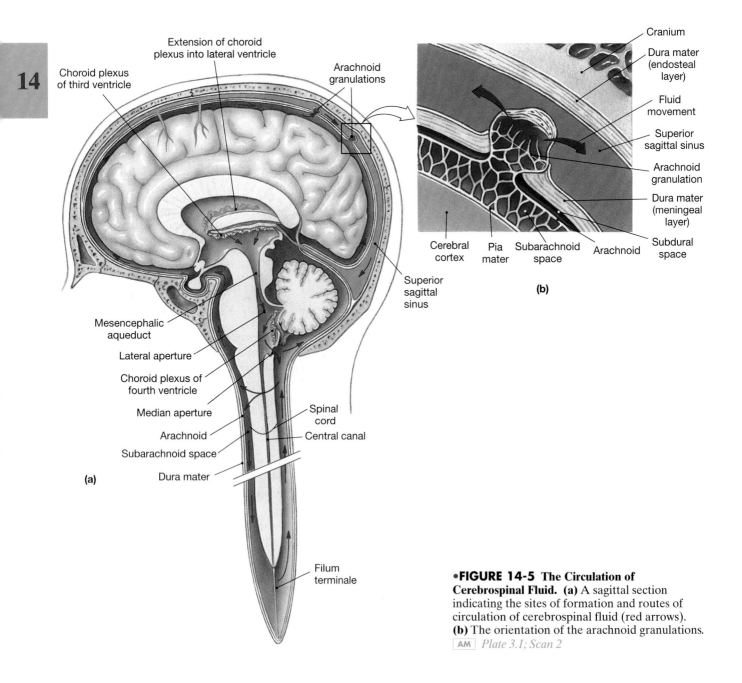

•**FIGURE 14-5 The Circulation of Cerebrospinal Fluid. (a)** A sagittal section indicating the sites of formation and routes of circulation of cerebrospinal fluid (red arrows). **(b)** The orientation of the arachnoid granulations.
AM *Plate 3.1; Scan 2*

Specialized ependymal cells, interconnected by tight junctions, surround the capillaries of the choroid plexus. The ependymal cells secrete CSF into the ventricles; they also remove waste products from the CSF and adjust its composition over time. The differences in composition between CSF and blood plasma (blood with the cellular elements removed) are quite pronounced. For example, the blood contains high concentrations of soluble proteins, but the CSF does not. The concentrations of individual ions and the levels of amino acids, lipids, and waste products are also different.

Circulation of CSF

The choroid plexus produces CSF at a rate of about 500 ml/day. The total volume of CSF at any moment is approximately 150 ml; thus, the entire volume of CSF is replaced roughly every 8 hours. Despite this rapid turnover, the composition of CSF is closely regulated and the rate of removal normally keeps pace with the rate of production.

The CSF circulates from the choroid plexus through the ventricles and the central canal of the spinal cord (Figure 14-5a•). As the CSF circulates, diffusion between it and the interstitial fluid of the CNS is unrestricted between and across the ependymal cells. The CSF reaches the subarachnoid space through two **lateral apertures** and a single **median aperture,** openings in the roof of the fourth ventricle. Cerebrospinal fluid then flows through the subarachnoid space surrounding the brain, spinal cord, and cauda equina.

Fingerlike extensions of the arachnoid membrane, called the *arachnoid villi,* penetrate the meningeal layer of the dura mater and extend into the superior sagittal sinus. In adults, clusters of villi form large **arachnoid granulations** (Figure 14-5b•). Cerebrospinal fluid is absorbed into the venous circulation at the arachnoid granulations.

If the normal circulation or reabsorption of CSF is interrupted, a variety of clinical problems may appear. For example, a problem with the reabsorption of CSF in infancy is responsible for symptoms of *hydrocephalus,* or "water on the brain." Infants with this condition have enormously expanded skulls due to the presence of an abnormally large volume of CSF. (See the clinical discussion "Hydrocephalus" on p. 444.) In adults, a failure of reabsorption or a blockage of CSF circulation can distort and damage the brain.

THE BLOOD SUPPLY TO THE BRAIN

As we noted in Chapter 12, neurons have a high demand for energy, but they have neither energy reserves, in the form of carbohydrates or lipids, nor oxygen reserves, in the form of myoglobin. ∞ *p. 398*

Your brain, with billions of neurons, is an extremely active organ with a continuous demand for nutrients and oxygen. These demands are met by an extensive circulatory supply. Arterial blood reaches the brain through the *internal carotid arteries* and the *vertebral arteries.* Most of the venous blood from the brain leaves the cranium in the *internal jugular veins,* which drain the dural sinuses. A head injury that damages cerebral blood vessels may cause bleeding into the dura mater, either near the dural epithelium or between the outer layer of the dura mater and the bones of the skull. These are serious conditions, because the blood entering these spaces compresses and distorts the relatively soft tissues of the brain.

Cerebrovascular diseases are circulatory disorders that interfere with the normal circulatory supply to the brain. The particular distribution of the vessel involved determines the symptoms, and the degree of oxygen or nutrient starvation determines their severity. A **cerebrovascular accident (CVA),** or *stroke,* occurs when the blood supply to a portion of the brain is shut off. Affected neurons begin to die in a matter of minutes.

14

 Epidural and Subdural Hemorrhages
A severe head injury may damage meningeal vessels and cause bleeding into the cranial cavity. The most serious cases involve an arterial break, because arterial blood pressure is relatively high. If blood is forced between the dura mater and the cranium, the condition is known as an **epidural hemorrhage.** The elevated fluid pressure then distorts the underlying tissues of the brain. The individual loses consciousness from minutes to hours after the injury, and death follows in untreated cases. An epidural hemorrhage involving a damaged vein does not produce massive symptoms immediately, and the individual may become unconscious several hours to several days or even weeks after the original injury. As a result, the problem may not be noticed until the nervous tissue has been severely damaged by distortion, compression, and secondary hemorrhaging. Epidural hemorrhages are rare, occurring in fewer than 1 percent of head injuries. However, the mortality rate is 100 percent in untreated cases and over 50 percent even after the blood pool has been removed and the damaged vessels have been closed.

The term **subdural hemorrhage** may be misleading, because in many cases blood enters the meningeal layer of the dura mater, flowing deep to the epithelium that contacts the arachnoid membrane, rather than between the dura mater and the arachnoid. Subdural hemorrhages are roughly twice as common as epidural hemorrhages. The most common source of blood is a small vein or one of the dural sinuses. Because the venous blood pressure in a subdural hemorrhage is lower than that in an arterial epidural hemorrhage, the extent of distortion produced and the effects on brain function can be quite variable.

Hydrocephalus

The adult brain is surrounded by the relatively inflexible bones of the cranium. The enclosed cranial cavity contains two fluids—blood and cerebrospinal fluid—and the soft tissues of the brain. Because the total volume cannot change, when the volume of blood or CSF increases, the volume of the brain must decrease. For example, if a dural or subarachnoid blood vessel ruptures, the fluid volume increases as blood collects in the cranial cavity. The rising intracranial pressure compresses the brain, leading to neural dysfunction that can end in unconsciousness and ultimately death.

Any change in the rate of CSF production is normally matched by an increase in the rate of CSF removal at the arachnoid granulations. If this equilibrium is disturbed, clinical problems appear as the intracranial pressure changes. The volume of CSF will increase if the rate of formation accelerates or the rate of removal decreases. In either event, the increased fluid volume compresses and distorts the brain. Increased rates of formation can accompany head injuries, but the most common problems arise from masses, such as tumors or abscesses, or from developmental abnormalities. These conditions restrict the normal circulation and reabsorption of CSF. Because CSF production continues, the ventricles gradually expand, distorting the surrounding neural tissues and causing brain function to deteriorate.

Infants are especially sensitive to changes in intracranial pressure, because the arachnoid granulations do not appear until roughly 3 years of age. (Until then, CSF is reabsorbed into small vessels in the subarachnoid space and beneath the ependyma lining the ventricles.) As in adults, if intracranial pressure becomes abnormally high, the ventricles will expand. But in infants, the cranial sutures have yet to fuse, so the skull can enlarge to accommodate the extra fluid volume. The result is an enormously expanded skull, a condition called **hydrocephalus,** or "water on the

•FIGURE 14-6 Hydrocephalus. This infant has hydrocephalus, a condition generally caused by impaired circulation and impaired removal of cerebrospinal fluid. The buildup of CSF leads to the distortion of the brain and enlargement of the cranium.

brain" (Figure 14-6•). Infant hydrocephalus typically results from some interference with normal CSF circulation, such as a blockage of the mesencephalic aqueduct or a constriction of the connection between the subarachnoid spaces of the cranial and spinal meninges. Untreated infants commonly suffer some degree of mental retardation.

Successful treatment generally involves the installation of a *shunt,* a bypass that either avoids the blockage site or drains the excess CSF. The goal is reduction of the intracranial pressure. The shunt may be removed if (1) further growth of the brain eliminates the blockage or (2) the intracranial pressure decreases after the arachnoid granulations develop when the child reaches 3 years of age.

The Blood–Brain Barrier

Neural tissue in the CNS is isolated from the general circulation by the **blood–brain barrier.** This barrier exists because the endothelial cells that line the capillaries of the CNS are extensively interconnected by tight junctions. These junctions prevent the diffusion of materials between adjacent endothelial cells. In general, only lipid-soluble compounds (including carbon dioxide; oxygen; ammonia; lipids, such as steroids or prostaglandins; and small alcohols) can diffuse across the membranes of endothelial cells into the interstitial fluid of the brain and spinal cord. Water and ions must pass through channels in the apical and basal cell membranes. Larger, water-soluble compounds can cross the capillary walls only by active or passive transport.

The restricted permeability characteristics of the endothelial lining of brain capillaries are in some way dependent on chemicals secreted by astrocytes—cells that are in close contact with CNS capillaries. ∞ *p. 368* The outer surfaces of the endothelial cells are covered by the processes of astrocytes. Because the astrocytes release chemicals that control the permeabilities of the endothelium, these cells play a key supporting role in the blood–brain barrier. If the astrocytes are damaged or stop stimulating the endothelial cells, the blood–brain barrier disappears.

The choroid plexus is not part of the neural tissue of the brain, so no astrocytes are in contact with the endothelial cells there. As a result, capillaries in the choroid plexus are highly permeable. Substances do not have free access to the CNS, however, because a **blood–CSF barrier** is created by specialized ependymal cells. These cells, interconnected by tight junctions, surround the capillaries of the choroid plexus.

Transport across the blood–brain and blood–CSF barriers is selective and directional. Even the passage of small ions, such as sodium, hydrogen, potassium, or chloride, is controlled. As a result, the pH and concentrations of sodium, potassium, calcium, and magnesium ions in the blood and CSF are different. Some organic compounds are readily transported, and others cross only in minute amounts. Neurons have a constant need for glucose. This need must be met regardless of the relative concentrations in the blood and interstitial fluid. Even when circulating glucose levels are low, endothelial cells continue to transport glucose from the blood to the interstitial fluid of the brain. In contrast, only trace amounts of circulating norepinephrine, epinephrine, dopamine, or serotonin pass into the interstitial fluid or CSF of the brain. This limitation is important, because these compounds are neurotransmitters—their entry from the bloodstream (where concentrations can be relatively high) could result in the uncontrolled stimulation of neurons throughout the brain.

The blood–brain barrier remains intact throughout the CNS, with four noteworthy exceptions:

1. In portions of the hypothalamus, the capillary endothelium is extremely permeable. This permeability exposes hypothalamic nuclei to circulating hormones and permits the diffusion of hypothalamic hormones into the circulation.

2. Capillaries in the posterior pituitary gland, which is continuous with the floor of the hypothalamus, are highly permeable. At this site, the hormones *ADH* and *oxytocin,* produced by hypothalamic neurons, are released into the circulation.

3. Capillaries in the *pineal gland* are also very permeable. The pineal gland, an endocrine structure, is located in the posterior, superior surface of the diencephalon. The capillary permeability allows pineal secretions into the general circulation.

4. Capillaries at the choroid plexus are extremely permeable. Although the capillary characteristics of the blood–brain barrier are lost there, the transport activities of specialized ependymal cells in the choroid plexus maintain the blood–CSF barrier.

Penetrating the Blood–Brain and Blood–CSF Barriers

Physicians must sometimes get specific compounds into the interstitial fluid of the brain to fight CNS infections or to treat other neural disorders. To do this, they must understand the limitations of the blood–brain and blood–CSF barriers. For example, *Parkinson's disease,* discussed on page 460, results from inadequate release of the neurotransmitter dopamine within the cerebral nuclei. Although dopamine will not cross the blood–brain or blood–CSF barrier, a related compound, *L-dopa,* passes readily and is converted to dopamine inside the brain. In treatments for meningitis or other CNS infections, the antibiotics *penicillin* and *tetracycline* are excluded from the brain, but *erythromycin, sulfisoxazole,* and *sulfadiazine* enter the CNS very rapidly.

When the CNS is damaged, the blood–brain barrier breaks down. As a result, the locations of injury sites, infections, and tumors can be determined by injecting into the bloodstream tracers that normally cannot enter the CNS tissues. One common method is to use the plasma protein *albumin* labeled with radioactive iodine. Any radioactivity in the brain is easily detected in a brain scan.

14

☑ What would happen if an interventricular foramen became blocked?

☑ How would decreased diffusion across the arachnoid granulations affect the volume of cerebrospinal fluid in the ventricles?

☑ Many water-soluble molecules that are relatively abundant in the blood occur in small amounts or not at all in the extracellular fluid of the brain. Why?

THE CEREBRUM

The cerebrum is the largest region of the brain. Conscious thoughts and all intellectual functions originate in the cerebral hemispheres. Much of the cerebrum is involved in the processing of somatic sensory and motor information.

Figure 14-1b,d•, p. 437, provides a perspective on the cerebrum and its relationships with other regions of the brain. Gray matter in the cerebrum is located in the *cerebral cortex* and in deeper *cerebral nuclei.* The white matter of the cerebrum lies deep to the neural cortex and around the cerebral nuclei.

THE CEREBRAL CORTEX

A blanket of neural cortex covers the paired cerebral hemispheres, which dominate the superior and lateral surfaces of the cerebrum (Figure 14-1•). The cortical surface forms a series of elevated ridges, or **gyri** (JĪ-rī; singular, *gyrus*). The gyri are separated by shallow depressions called **sulci** (SUL-sī) or by deeper grooves called **fissures.** Gyri increase the surface area of the cerebral hemispheres. The greater the area of neural cortex, the greater the number of cortical neurons. The total surface area of the cerebral hemispheres is roughly equivalent to 2200 cm^2 (2.5 ft^2) of flat surface. An area that large can be packed into the skull only when folded like a crumpled piece of paper. Complex analytical and integrative functions require large numbers of neurons. The entire brain has enlarged over the course of human evolution, but the cerebral hemispheres have enlarged at a much faster rate than has the rest of the brain.

Cerebral Landmarks and Lobes

The two cerebral hemispheres are almost completely separated by a deep **longitudinal fissure** (Figure 14-1a•, p. 437). They remain connected by a thick band of white matter called the *corpus callosum*. Each cerebral hemisphere can be divided into *lobes*, or regions, named after the overlying bones of the skull. Your brain has a unique pattern of sulci and gyri, as individual as a fingerprint, but the boundaries between lobes are reliable landmarks. On each hemisphere, the **central sulcus,** a deep groove, divides the anterior **frontal lobe** from the more posterior **parietal lobe.** The roughly horizontal **lateral sulcus** separates the frontal lobe from the **temporal lobe.** Pushing the temporal lobe to the side exposes the **insula** (IN-sū-luh; *insula,* island), an "island" of cortex that is otherwise invisible. The more-posterior **parieto-occipital sulcus** separates the parietal lobe from the **occipital lobe.**

Each lobe contains functional regions whose boundaries are less clearly defined. Some of these regions deal with sensory information and others with motor commands. Keep in mind three points about the cerebral lobes:

1. *Each cerebral hemisphere receives sensory information from, and sends motor commands to, the opposite side of the body.* For example, the motor areas of the left cerebral hemisphere control muscles on the right side, and the right cerebral hemisphere controls muscles on the left side. This crossing over has no known functional significance.

2. *The two hemispheres have different functions, even though they look almost identical.* We will discuss these differences in a later section.

3. *The assignment of a specific function to a specific region of the cerebral cortex is imprecise.* Because the boundaries are indistinct and have considerable overlap, one region may have several functions. Some aspects of cortical function, such as consciousness, cannot easily be assigned to any single region. However, we know that normal individuals use all portions of the brain.

The White Matter of the Cerebrum

The interior of the cerebrum consists primarily of white matter. The axons can be roughly classified as association fibers, commissural fibers, and projection fibers (Figure 14-7•). **Association fibers** interconnect areas of neural cortex within a single cerebral hemisphere. Shorter association fibers are called **arcuate** (AR-kū-āt) **fibers,** because they curve in an arc to pass from one gyrus to another. Longer association fibers are organized into discrete bundles, or *fasciculi*. The **longitudinal fasciculi** connect the frontal lobe to the other lobes of the same hemisphere.

Commissural (kom-MIS-ū-rul) **fibers** (*commissura,* crossing over) interconnect and permit communication between the cerebral hemispheres. Bands of commissural fibers linking the hemispheres include the **corpus callosum** and the **anterior commissure.** The corpus callosum alone contains more than 200 million axons, carrying some 4 billion impulses per second!

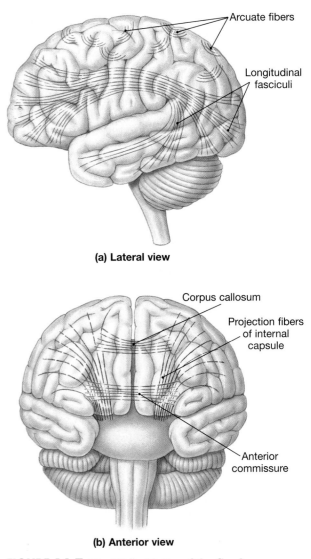

(a) Lateral view

(b) Anterior view

•**FIGURE 14-7** **The White Matter of the Cerebrum**

Projection fibers link the cerebral cortex to the diencephalon, brain stem, cerebellum, and spinal cord. All projection fibers must pass through the diencephalon, where axons heading to sensory areas of the cerebral cortex pass among the axons descending from motor areas of the cortex. In gross dissection, the ascending fibers and descending fibers look alike. The entire collection of fibers is known as the **internal capsule.**

Motor and Sensory Areas of the Cortex

The major motor and sensory regions of the cerebral cortex are listed in Table 14-2 and shown in Figure 14-8•. The central sulcus separates the motor and sensory areas of the cortex. The **precentral gyrus** of the frontal lobe forms the anterior border of the central sulcus. The surface of this gyrus is the **primary motor cortex.** Neurons of the primary motor cortex direct voluntary movements by controlling somatic motor neurons in the brain stem and spinal cord. These cortical neurons are called **pyramidal cells,** because their cell bodies resemble little pyramids.

The primary motor cortex is like the keyboard of a piano. If you strike a specific piano key, you produce a specific sound; if you stimulate a specific motor neuron in the primary motor cortex, you generate a contraction in a specific skeletal muscle.

TABLE 14-2 The Cerebral Cortex

Lobe/Region	Function
Frontal lobe Primary motor cortex	Voluntary control of skeletal muscles
Parietal lobe Primary sensory cortex	Conscious perception of touch, pressure, vibration, pain, temperature, and taste
Occipital lobe Visual cortex	Conscious perception of visual stimuli
Temporal lobe Auditory cortex and olfactory cortex	Conscious perception of auditory and olfactory stimuli
All lobes Association areas	Integration and processing of sensory data; processing and initiation of motor activities

Like the gauges in the dashboard of a car, the sensory areas of the cerebral cortex report key information. At each location, sensory information is reported in the pattern of neuron activity in the cortex. The **postcentral gyrus** of the parietal lobe forms the posterior border of the central sulcus, and its surface contains the **primary sensory cortex.** Neurons in this region receive somatic sensory information from receptors for touch, pressure, pain, vibration, taste, or temperature. We are consciously aware of these sensations only when nuclei in the thalamus relay the information to the primary sensory cortex.

Sensations of sight, sound, smell, and taste arrive at other portions of the cerebral cortex (Figure 14-8a•). The **visual cortex** of the occipital lobe receives visual information, and the **auditory cortex** and **olfactory cortex** of the temporal lobe receive information about hearing and smell, respectively. The **gustatory cortex,** which receives information from taste receptors of the tongue and pharynx, lies in the anterior portion of the insula and adjacent portions of the frontal lobe.

Association Areas

The sensory and motor regions of the cortex are connected to nearby **association areas,** regions of the cortex that interpret incoming data or coordinate a motor response (Figure 14-8a•). Like the information provided by the gauges in a car, the arriving information must be noticed and interpreted before it can be acted on. *Sensory association areas* are cortical regions that monitor and interpret the information that arrives at the sensory areas of the cortex. Examples include the somatic sensory association area, visual association area, and auditory association area.

The **somatic sensory association area** monitors activity in the primary sensory cortex. It is the somatic sensory association area that allows you to recognize a touch as light as the arrival of a mosquito on your arm (and gives you a chance to swat the mosquito before it bites).

The special senses of smell, sight, and hearing involve separate areas of the sensory cortex, and each has its own association area. These areas monitor and interpret arriving sensations. For example, the **visual association area** monitors the patterns of activity in the visual cortex and interprets the results. You see the symbols *c, a,* and *r* when the stimulation of receptors in your eyes leads to the stimulation of neurons in your visual cortex. Your visual association area recognizes that these are letters and that *c + a + r = car.* An individual with a damaged visual association area could scan the lines of a printed page and see rows of symbols that are clear but would perceive no meaning from the symbols. Similarly, the **auditory association area** monitors sensory activity in the auditory cortex; word recognition occurs in this association area.

14

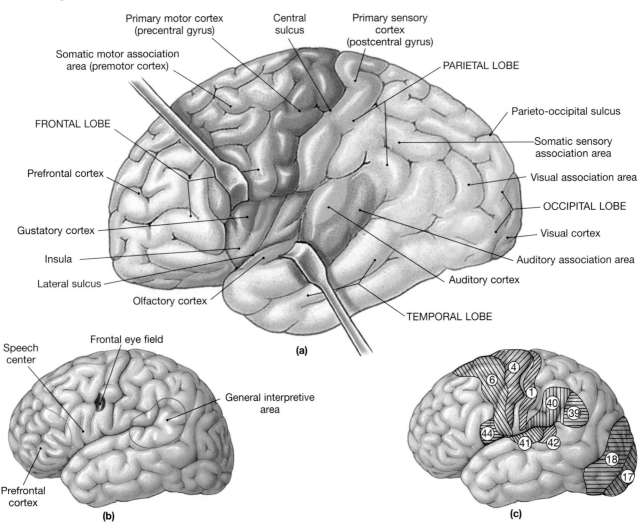

•**FIGURE 14-8 The Cerebral Hemispheres. (a)** Major anatomical landmarks on the surface of the left cerebral hemisphere. The lateral sulcus has been pulled apart to expose the insula. **(b)** The left hemisphere generally contains the general interpretive area and the speech center. The prefrontal cortex of each hemisphere is involved with conscious intellectual functions. **(c)** Regions of the cerebral cortex as determined by histological analysis. Several of the 47 regions described by Brodmann are shown for comparison with the results of functional mapping.

The **somatic motor association area,** or **premotor cortex,** is responsible for the coordination of learned movements. The primary motor cortex does nothing on its own, any more than a piano keyboard can play itself. The neurons in the primary motor cortex must be stimulated by neurons in other parts of the cerebrum. When you perform a voluntary movement, the instructions are relayed to the primary motor cortex by the premotor cortex. With repetition, the proper pattern of stimulation becomes stored in your premotor cortex. You can then perform the movement smoothly and easily by triggering the *pattern* rather than by controlling the individual neurons. This principle applies to any learned movement, from something as simple as picking up a glass to something as complex as playing the piano. One area of the premotor cortex, the *frontal eye field,* controls learned eye movements, such as when you scan these lines of type. Someone with damage to the frontal eye field can understand written letters and words but cannot read, because his or her eyes cannot follow the lines on a printed page.

Integrative Centers

Integrative centers are areas that receive information from many association areas and direct extremely complex motor activities. These centers also perform complicated analytical functions. For example, the *prefrontal cortex* of the frontal lobe (Figure 14-8b•) integrates information from sensory association areas and performs abstract intellectual functions, such as predicting the consequences of possible responses.

Integrative centers are located in the lobes and cortical areas of both cerebral hemispheres. Integrative centers concerned with the performance of complex processes, such as speech, writing, mathematical computation, or understanding spatial relationships, are restricted to either the left or the right hemisphere. These centers include the *general interpretive area* and the *speech center*. The corresponding regions on the opposite hemisphere are also active, but their functions are less well defined.

The General Interpretive Area   The **general interpretive area,** or *Wernicke's area* (Figure 14-8b•), also called the *gnostic area,* receives information from all the sensory association areas. This analytical center is present in only one hemisphere (typically the left). This region plays an essential role in your personality by integrating sensory information and coordinating access to complex visual and auditory memories. Damage to the general interpretive area affects the ability to interpret what is seen or heard, even though the words are understood as individual entities. For example, if your general intrerpretive area were damaged, you might still understand the meaning of the spoken words *sit* and *here,* because word recognition occurs in the auditory association areas. But you would be totally bewildered by the request *sit here.* Damage to another portion of the general interpretive area might leave you able to see a chair clearly and to know that you recognize it, but you would be unable to name it because the connection to your visual association area has been disrupted.

Aphasia and Dyslexia

Aphasia (*a-*, without + *phasia*, speech) is a disorder affecting the ability to speak or read. **Global aphasia** results from extensive damage to the general interpretive area or to the associated sensory tracts. Affected individuals are unable to speak, to read, or to understand the speech of others. Global aphasia often accompanies a severe stroke or tumor that affects a large area of cortex, including the speech and language areas. Recovery is possible when the condition results from edema or hemorrhage, but the process often takes months or even years. AM *Aphasia*

Dyslexia (*lexis,* diction) is a disorder affecting the comprehension and use of words. **Developmental dyslexia** affects children; estimates indicate that up to 15 percent of children in the United States have some degree of dyslexia. For unknown reasons, the problem is much more common among left-handed children than among right-handed children. Children with dyslexia have difficulty reading and writing, although their other intellectual functions may be normal or above normal. Their writing looks uneven and unorganized; letters are typically written in the wrong order (*dig* becomes *gid*) or reversed (*E* becomes Ǝ). Recent evidence suggests that at least some forms of dyslexia result from problems in processing, sorting, and integrating visual or auditory information.

The Speech Center   Some of the neurons in the general interpretive area innervate the **speech center,** also called *Broca's area* or the *motor speech area* (Figure 14-8b•). This center lies along the edge of the premotor cortex in the same hemisphere as the general interpretive area (usually the left). The speech center regulates the patterns of breathing and vocalization needed for normal speech. This regulation involves coordinating the activities of the respiratory muscles, the laryngeal and pharyngeal muscles, and the muscles of the tongue, cheeks, lips, and jaws. A person with a damaged speech center can make sounds but not words.

The motor commands issued by the speech center are adjusted by feedback from the auditory association area, also called the *receptive speech area.* Damage to the related sensory areas can cause a variety of speech-related problems. Some affected individuals have difficulty speaking although they know exactly which words to use; others talk constantly but use all the wrong words.

The Prefrontal Cortex   The **prefrontal cortex** of the frontal lobe coordinates information relayed from the association areas of the entire cortex. In doing so, it performs such abstract intellectual functions as predicting the consequences of events or actions. Damage to the prefrontal cortex leads to difficulties in estimating temporal relationships between events. Questions such as "How long ago did this happen?" or "What happened first?" become difficult to answer.

The prefrontal cortex has extensive connections with other cortical areas and with other portions of the brain. Feelings of frustration, tension, and anxiety are generated at the prefrontal cortex as it interprets ongoing events and makes predictions about future situations or consequences. If the connections between the prefrontal cortex and other brain regions are severed, the frustrations, tensions, and anxieties are removed. During the middle of the twentieth century, this rather drastic procedure, called **prefrontal lobotomy,** was used to "cure" a variety of mental illnesses, especially those associated with violent or antisocial behavior. After a lobotomy, the patient would no longer be concerned about what had previously been a major problem, whether psychological (hallucinations) or physical (severe pain). However, the individual was often equally unconcerned about tact, decorum, and toilet training. Drugs have been developed to target specific pathways and regions of the CNS, so lobotomies are no longer used to change behavior.

Brodmann's Areas   Early in the twentieth century, several attempts were made to describe and classify regional differences in the histological organization of the cerebral cortex. It was hoped that the patterns of

14

cellular organization could be correlated to specific functions. By 1919, at least 200 patterns had been described, but most of the classification schemes have since been abandoned. However, the cortical map prepared by Korbinian Brodmann in 1909 has proved useful to neuroanatomists. Brodmann, a German neurologist, described 47 patterns of cellular organization in the cerebral cortex. Several of these *Brodmann's areas* are shown in Figure 14-8c•. Some correspond to known functional areas. For example, Brodmann's area 44 corresponds to the speech center and area 41 to the auditory cortex; area 4 follows the contours of the primary motor cortex. In other cases, the correspondence is less precise. For instance, Brodmann's area 42 forms only a small portion of the auditory association area.

Hemispheric Lateralization

Each of the two cerebral hemispheres is responsible for specific functions that are not ordinarily performed by the opposite hemisphere. This regional specialization has been called *hemispheric lateralization.* Figure 14-9• indicates the major functional differences between the hemispheres. In most people, the left hemisphere contains the general interpretive and speech centers and is responsible for language-based skills. For example, reading, writing, and speaking are dependent on processing done in the left cerebral hemisphere. In addition, the premotor cortex involved with the control of hand movements is larger on the left side for right-handed individuals than for left-handed ones. The left hemisphere is also important in performing analytical tasks, such as

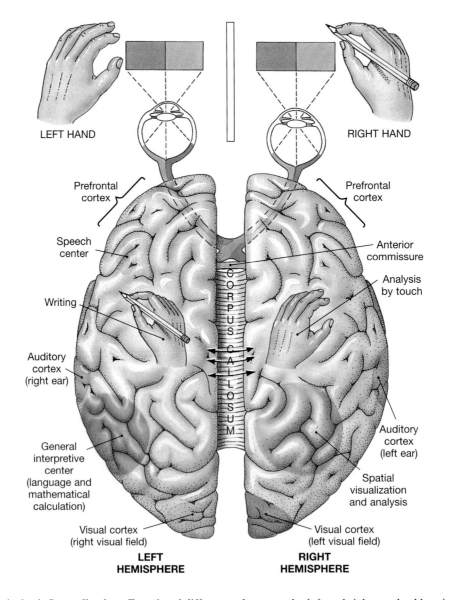

•**FIGURE 14-9** **Hemispheric Lateralization.** Functional differences between the left and right cerebral hemispheres.

mathematical calculations and logical decision making. For these reasons, the left cerebral hemisphere has been called the *dominant hemisphere,* or the *categorical hemisphere.*

The right cerebral hemisphere analyzes sensory information and relates the body to the sensory environment. Interpretive centers in this hemisphere permit you to identify familiar objects by touch, smell, sight, taste, or feel. For example, the right hemisphere plays a dominant role in recognizing faces and in understanding three-dimensional relationships. It is also important in analyzing the emotional context of a conversation—for instance, determining whether the phrase "get lost" was intended as a threat or a question. Individuals with a damaged right hemisphere may be unable to add emotional inflections to their own words.

Left-handed persons represent 9 percent of the human population; in most cases, although the primary motor cortex of the right hemisphere controls motor function for the dominant hand, the centers involved with speech and analytical function are in the left hemisphere. Interestingly, an unusually high percentage of musicians and artists are left-handed. The complex motor activities performed by these individuals are directed by the primary motor cortex and association areas on the right cerebral hemisphere, near the association areas involved with spatial visualization and emotions.

Disconnection Syndrome

The functional differences between the hemispheres become apparent if the corpus callosum is cut, a procedure sometimes performed to treat epileptic seizures that cannot be controlled by other methods. This surgery produces symptoms of **disconnection syndrome.** In this condition, the two hemispheres function independently, each "unaware" of stimuli or motor commands that involve its counterpart. The result is some rather interesting changes in the individual's abilities. For example, objects touched by the left hand can be recognized but not verbally identified, because the sensory information arrives at the right hemisphere but the speech center is on the left. The object can be verbally identified if felt with the right hand, but the person cannot say whether it is the same object previously touched with the left hand. Sensory information from the left side of the body arrives at the right hemisphere and cannot reach the general interpretive area. Thus, conscious decisions are made without regard to sensations from the left side. Two years after a surgical sectioning of the corpus callosum, the most striking behavioral abnormalities have disappeared and the person may test normally. In addition, individuals born without a functional corpus callosum do not have sensory, motor, or intellectual problems. In some way, the CNS adapts to these situations, probably by increasing the amount of information transferred across the anterior commissure.

☑ Shelly suffers a head injury that damages her primary motor cortex. Where is this area located?

☑ Which senses would be affected by damage to the temporal lobes of the cerebrum?

☑ After suffering a stroke, Jake is unable to speak. He can understand what is said to him, and he can understand written messages, but he cannot express himself verbally. Which part of his brain has been affected by the stroke?

THE CEREBRAL NUCLEI

While your cerebral cortex is consciously directing a complex movement or solving some intellectual puzzle, other centers of your cerebrum, diencephalon, and brain stem are processing sensory information and issuing motor commands outside your conscious awareness. Many of these activities, which occur at the *subconscious* level, are directed by the cerebral nuclei. The **cerebral nuclei** (Figure 14-10•) are masses of gray matter that lie within each hemisphere deep to the floor of the lateral ventricle. They are embedded in the white matter of the cerebrum, and the radiating projection fibers and commissural fibers travel around or between these nuclei. The cerebral nuclei are components of a functional group known as the **basal nuclei,** or *basal ganglia.*[2] The latter term has persisted despite the fact that ganglia are restricted to the PNS.

The **caudate nucleus** has a massive head and a slender, curving tail that follows the curve of the lateral ventricle. The head of the caudate nucleus lies anterior to the **lentiform** (lens-shaped) **nucleus.** The lentiform nucleus consists of a medial **globus pallidus** (GLŌ-bus PAL-i-dus; pale globe) and a lateral **putamen** (pū-TĀ-men). The term *corpus striatum* (striated body) has been used to refer to the caudate and lentiform nuclei, or to the caudate and putamen. The name refers to the striated (striped) appearance of the internal capsule as it passes among these nuclei. The **amygdaloid** (ah-MIG-da-loyd) **body** (*amygdale,* almond) is a separate nucleus located anterior to the tail of the caudate nucleus and inferior to the lentiform nucleus. The amygdaloid body is a component of the *limbic system,* which we will consider in a later section.

Functions of the Cerebral Nuclei

The cerebral nuclei are involved with the subconscious control of skeletal muscle tone and the coordination of learned movement patterns. Under normal conditions, these nuclei do not initiate particular movements. But once a movement is under way, the cerebral nuclei provide the general pattern and rhythm, especially for movements of the trunk and proximal limb muscles.

---

[2] *Basal ganglia,* like the term *brain stem,* has historically been used in several ways. In some sources, it is synonymous with *cerebral nuclei;* others use it to refer to a functional group that consists of the cerebral nuclei and nuclei in the diencephalon and mesencephalon.

•**FIGURE 14-10** The Cerebral Nuclei.
(a) The relative positions of the cerebral nuclei in the intact brain. (b) Frontal sections.

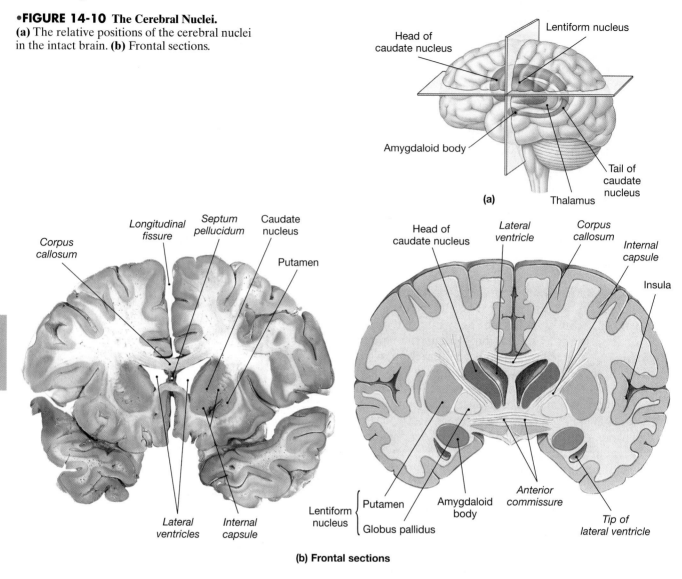

(a)

(b) Frontal sections

Information arrives at the caudate nucleus and putamen from sensory, motor, and association areas of the cerebral cortex. Processing and integration occur in these nuclei and in the adjacent globus pallidus. Most of the output of the cerebral nuclei leaves the globus pallidus and synapses in the thalamus. Nuclei in the thalamus then project the information to appropriate areas of the cerebral cortex. The cerebral nuclei alter the motor commands issued by the cerebral cortex through this feedback loop. For example:

- When you walk, your cerebral nuclei control the cycles of arm and thigh movements that occur between the time you decide to "start" walking and the time you give the "stop" order.

- As you begin a voluntary movement, your cerebral nuclei control and adjust muscle tone, particularly in the appendicular muscles, to set your body position. When you decide to pick up a pencil, you consciously reach and grasp with your forearm, wrist, and hand while the cerebral nuclei position your shoulder and stabilize your arm.

THE LIMBIC SYSTEM

The **limbic system** (*limbus*, border) includes nuclei and tracts along the border between the cerebrum and diencephalon. This system is a functional grouping rather than an anatomical one and includes portions of the cerebrum, diencephalon, and mesencephalon. The functions of the limbic system include (1) establishing emotional states; (2) linking the conscious, intellectual functions of the cerebral cortex with the unconscious and autonomic functions of the brain stem; and (3) facilitating memory storage and retrieval. Whereas the sensory cortex, motor cortex, and association areas of the cerebral cortex enable you to perform complex tasks, it is largely the limbic system that makes you *want* to do them. For this reason, the limbic system is also known as the *motivational system*.

Figure 14-11• focuses on major components of the limbic system. The amygdaloid body at the tail end of the caudate nucleus is a cerebral contribution to the

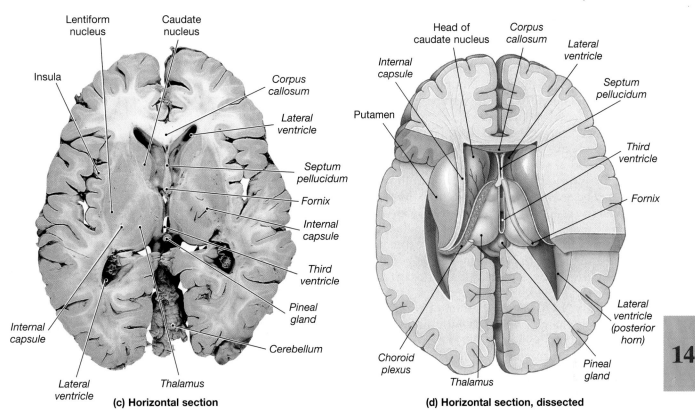

(c) Horizontal section

(d) Horizontal section, dissected

•**FIGURE 14-10** **The Cerebral Nuclei** *(continued)*. **(c)** A horizontal section; compare this with **(d)**, the view in dissection.

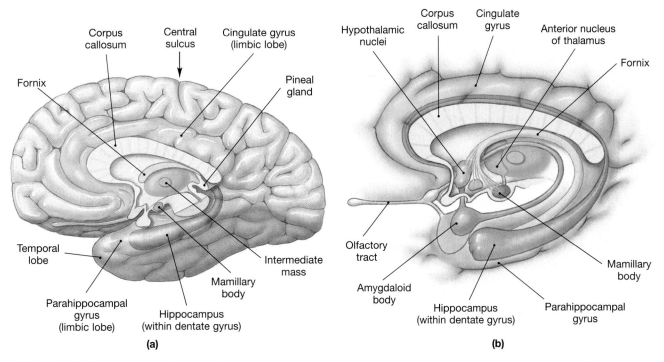

(a)

(b)

•**FIGURE 14-11** **The Limbic System.** **(a)** A diagrammatic sagittal section through the cerebrum, showing the cortical areas associated with the limbic system. The parahippocampal gyrus is shown as though transparent to make deeper limbic components visible. **(b)** A three-dimensional reconstruction of the limbic system, showing the relationships among the major components.

14

limbic system. The amygdaloid body appears to act as an interface between the limbic system, the cerebrum, and various sensory systems. It plays a role in the regulation of heart rate, in the control of the "flight or fight" response, and in linking emotions with specific memories. The **limbic lobe** of the cerebral hemisphere consists of the gyri and underlying structures adjacent to the diencephalon. The gyri curve along the corpus callosum and onto the medial surface of the temporal lobe (Figure 14-11a•). The **cingulate** (SIN-gū-lāt) **gyrus** (*cingulum,* girdle or belt) sits superior to the corpus callosum. The **dentate gyrus** is adjacent to the **parahippocampal** (pa-ra-hip-ō-KAM-pal) **gyrus.** Both conceal the **hippocampus,** a nucleus inferior to the floor of the lateral ventricle. To early anatomists, this structure resembled a sea horse (*hippocampus*); it appears to be important in learning, especially in the storage and retrieval of new long-term memories.

The **fornix** (FOR-niks) is a tract of white matter that connects the hippocampus with the hypothalamus (Figures 14-10c,d and 14-11•). From the hippocampus, the fornix curves medially, meeting its counterpart from the opposing hemisphere. The fornix proceeds anteriorly, inferior to the corpus callosum and septum pellucidum, before curving toward the hypothalamus. Many fibers of the fornix end in the **mamillary** (MAM-i-lar-ē) **bodies** (*mamilla,* or *mammilla,* breast or nipple), prominent nuclei in the floor of the hypothalamus (Figures 14-11 and 14-12a•). The mamillary bodies process sensory information, including olfactory sensations. They also contain motor nuclei that control reflex movements associated with eating, such as chewing, licking, and swallowing.

Several other nuclei in the wall (thalamus) and floor (hypothalamus) of the diencephalon are components of the limbic system. The *anterior nucleus* of the thalamus relays information from the mamillary body (of the hypothalamus) to the cingulate gyrus on that side. The boundaries between some of the hypothalamic nuclei tend to be poorly defined, but experimental stimulation has outlined a number of important hypothalamic centers responsible for the emotions of rage, fear, pain, sexual arousal, and pleasure. The stimulation of specific regions of the hypothalamus can also produce heightened alertness and a generalized excitement or generalized lethargy and sleep. These responses are caused by the stimulation or inhibition of the **reticular formation,** an interconnected network of neurons and nuclei. The reticular formation extends the length of the brain stem, but its headquarters lies in the mesencephalon.

Table 14-3 summarizes the organization and functions of the limbic system. Figure 14-12•, which shows the brain in midsagittal and frontal sections, includes many of the limbic structures of the cerebrum, diencephalon, and brain stem.

TABLE 14-3 The Limbic System

FUNCTIONS
Processing of memories; creation of emotional states, drives, and associated behaviors

CEREBRAL COMPONENTS
Cortical areas: limbic lobe (cingulate gyrus, dentate gyrus, and parahippocampal gyrus)
Nuclei: hippocampus, amygdaloid body
Tracts: fornix

DIENCEPHALIC COMPONENTS
Thalamus: anterior nuclear group
Hypothalamus: centers concerned with emotions, appetites (thirst, hunger), and related behaviors (*see Table 14-5*)

OTHER COMPONENTS
Reticular formation: network of interconnected nuclei throughout brain stem

---

☑ What symptoms would you expect to observe in an individual who has damage to the cerebral nuclei?

☑ Paul is having a difficult time remembering facts and recalling long-term memories. Which part of his cerebrum is probably involved?

☑ After suffering a head injury in an automobile accident, Terri is having trouble chewing and swallowing. This difficulty suggests that Terri may have sustained damage to which part of her brain?

THE DIENCEPHALON

The diencephalon plays a vital role in integrating conscious and unconscious sensory information and motor commands. It consists of the epithalamus, thalamus, and hypothalamus. Figures 14-10c,d through 14-13• show the position of the diencephalon and its relationship to landmarks on the brain stem. Figure 14-13a• (p. 456) includes the origins of 11 of the 12 pairs of cranial nerves. The individual cranial nerves are identified by Roman numerals. We will introduce the full names of these nerves in a later section.

The *epithalamus* is the roof of the diencephalon superior to the third ventricle. The anterior portion of the epithalamus contains an extensive area of choroid plexus that extends through the interventricular foramina into the lateral ventricles. The posterior portion of the epithalamus contains the **pineal gland** (Figure 14-13b•), an endocrine structure that secretes the hormone **melatonin.** Melatonin is important in the regulation of day–night cycles and also

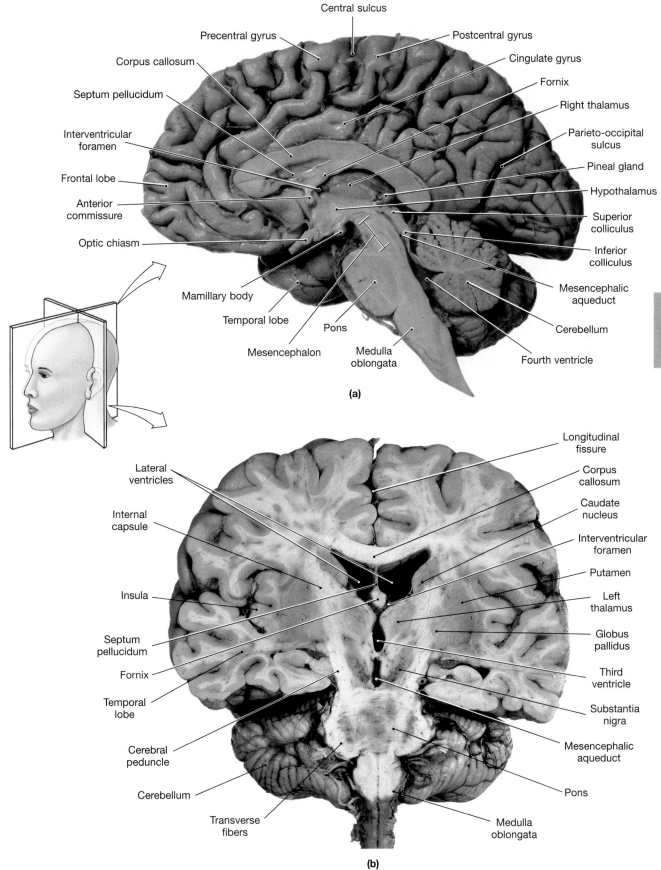

Central sulcus

Precentral gyrus

Corpus callosum

Septum pellucidum

Interventricular foramen

Frontal lobe

Anterior commissure

Optic chiasm

Postcentral gyrus

Cingulate gyrus

Fornix

Right thalamus

Parieto-occipital sulcus

Pineal gland

Hypothalamus

Superior colliculus

Inferior colliculus

Mesencephalic aqueduct

Cerebellum

Fourth ventricle

Mamillary body

Temporal lobe

Pons

Mesencephalon

Medulla oblongata

(a)

Longitudinal fissure

Corpus callosum

Caudate nucleus

Interventricular foramen

Putamen

Left thalamus

Globus pallidus

Third ventricle

Substantia nigra

Mesencephalic aqueduct

Pons

Medulla oblongata

Lateral ventricles

Internal capsule

Insula

Septum pellucidum

Fornix

Temporal lobe

Cerebral peduncle

Cerebellum

Transverse fibers

(b)

•FIGURE 14-12 The Brain in Section. (a) A midsagittal section. **(b)** A frontal section. AM *Plate 3.1; Scan 1e*

14

(a) Lateral view

(b) Posterior view

•**FIGURE 14-13** **The Diencephalon and Brain Stem.** **(a)** A lateral view, as seen from the left side. **(b)** A posterior view.

in the regulation of reproductive functions. (We shall describe the role of melatonin in Chapter 18.)

Most of the neural tissue in the diencephalon is concentrated in two structures: (1) the thalami (the wall), and (2) the hypothalamus (the floor). Ascending sensory information from the spinal cord and cranial nerves (other than the olfactory tract) synapses in a nucleus in the left or right thalamus before reaching the cerebral cortex and our conscious awareness. The hypothalamus contains centers involved with emotions and visceral processes that affect the cerebrum as well as other components of the brain stem. It also controls a variety of autonomic functions and forms the link between the nervous and endocrine systems.

THE THALAMUS

The thalamus is the final relay point for ascending sensory information that will be projected to the primary sensory cortex. It acts as a filter, passing on only a small portion of the arriving sensory information. The thalamus also coordinates the activities of the cerebral nuclei and the cerebral cortex by relaying information between them.

The left thalamus and right thalamus are separated by the third ventricle. Each thalamus consists of a rounded mass of *thalamic nuclei* (Figure 14-14•). The lateral border of the thalamus is established by the fibers of the internal capsule (p. 447). Viewed in mid-sagittal section (Figure 14-12a•), each thalamus extends from the anterior commissure to the inferior

•**FIGURE 14-14** **The Thalamus.** **(a)** A lateral view of the brain, color coded to indicate the regions that receive input from the thalamic nuclei. **(b)** An enlarged view of the thalamic nuclei of the left side.

base of the pineal gland. A projection of gray matter called an **intermediate mass** extends into the ventricle from the thalamus on either side (Figure 14-11a•, p. 453). In roughly 70 percent of the human population, the two thalami fuse in the midline.

Functions of Thalamic Nuclei

The thalamic nuclei deal primarily with the relay of sensory information to the cerebral nuclei and cerebral cortex. The five major groups of thalamic nuclei, listed in Table 14-4 and shown in Figure 14-14•, are the anterior, medial, ventral, posterior, and lateral groups:

1. The *anterior group* includes the **anterior nuclei,** part of the limbic system.

2. The nuclei of the *medial group* provide a conscious awareness of emotional states by connecting emotional centers in the hypothalamus with the frontal lobes of the cerebral hemispheres. The medial group also receives and relays sensory information from other portions of the thalamus.

3. The nuclei of the *ventral group* relay information from the cerebral nuclei and cerebellum to the primary motor cortex and motor association areas. Ventral group nuclei also relay sensory information about touch, pressure, pain, temperature, and proprioception (position) to the primary sensory cortex.

4. The *posterior group* includes the pulvinar and the geniculate nuclei. The **pulvinar** integrates sensory information for projection to the association areas of the cerebral cortex. The **lateral geniculate** (je-NIK-ū-lāt) **nucleus** (*genicula*, little knee) of each thalamus receives visual information over the *optic tract*, which originates at the eyes. The output of the lateral geniculate nucleus goes to the visual cortex and to the mesencephalon. The **medial geniculate nucleus** relays auditory information to the auditory cortex from specialized receptors of the inner ear.

5. The nuclei of the *lateral group* form feedback loops with the cingulate gyrus of the limbic system and the parietal lobe of the cerebral hemisphere on that side. They thus affect emotional states and the integration of sensory information.

THE HYPOTHALAMUS

The hypothalamus extends from the area superior to the *optic chiasm,* a cross-over where the optic tracts from the eyes arrive at the brain, to the posterior margins of the mamillary bodies (Figure 14-12a•). Posterior to the optic chiasm, a narrow stalk called the **infundibulum** (in-fun-DIB-ū-lum; *infundibulum*, funnel) extends inferiorly, connecting the floor of the hypothalamus to the pituitary gland (Figure 14-15•).

Viewed in sagittal section (Figure 14-15•), the floor of the hypothalamus between the infundibulum and the mamillary bodies is the **tuberal area** (*tuber,* swelling). The tuberal area contains nuclei that are involved with the control of pituitary gland function. The region adjacent to the infundibulum is called the **median eminence.**

Functions of the Hypothalamus

The hypothalamus contains important control and integrative centers, in addition to those associated with the limbic system. These centers and their functions are shown in Figure 14-15a• and summarized in Table 14-5. Hypothalamic centers may be stimulated by (1) sensory information from the cerebrum, brain stem, and spinal cord; (2) changes in the compositions of the CSF and interstitial fluid; or (3) chemical stimuli in the circulating blood that enter the hypothalamus across highly permeable capillaries, because the hypothalamus has no blood–brain barrier.

The hypothalamus performs the following functions:

1. *The subconscious control of skeletal muscle contractions.* The hypothalamus directs somatic motor patterns associated with rage, pleasure, pain, and sexual arousal by stimulating centers in other portions of the brain. For example, the changes in facial expression that accompany rage and the basic movements associated with sexual activity are controlled by hypothalamic centers.

2. *The control of autonomic function.* The hypothalamus adjusts and coordinates the activities of autonomic centers in the pons and medulla oblongata that regulate heart rate, blood pressure, respiration, and digestive functions.

Group/Nuclei	Function(s)
Anterior group	Part of the limbic system
Medial group	Integrates sensory information for projection to the frontal lobes
Ventral group	Projects sensory information to the primary sensory cortex; relays information from cerebellum and cerebral nuclei to motor area of cerebral cortex
Posterior group	
Pulvinar	Integrates sensory information for projection to association areas of cerebral cortex
Lateral geniculate nuclei	Project visual information to the visual cortex
Medial geniculate nuclei	Project auditory information to the auditory cortex
Lateral group	Integrates sensory information and influences emotional states

14

•FIGURE 14-15 The Hypothalamus in Sagittal Section.
(a) A diagrammatic view of the hypothalamus, showing the locations of major nuclei and centers. (b) The hypothalamus and adjacent portions of the brain.

TABLE 14-5 Components and Functions of the Hypothalamus

Region/Nucleus	Function
Supraoptic nucleus	Secretes ADH, restricting water loss at the kidneys
Paraventricular nucleus	Secretes oxytocin
Preoptic areas	Regulate body temperature
Suprachiasmatic nucleus	Coordinates day–night cycles of activity
Tuberal area and median eminence	Release hormones that control endocrine cells of the anterior pituitary gland
Autonomic centers	Control medullary nuclei that regulate heart rate and blood pressure
Mamillary bodies	Control feeding reflexes (licking, swallowing, etc.)

Autonomic centers (sympathetic)
Paraventricular nucleus
Preoptic area
Autonomic centers (parasympathetic)
Supraoptic nucleus
Suprachiasmatic nucleus
Optic chiasm
Tuberal area
Infundibulum
Pituitary gland
Thalamus
Hypothalamus
Mamillary body
Midbrain
Median eminence
Pons

(a)

Corpus callosum
Septum pellucidum
Anterior commissure
Anterior cerebral artery
Frontal lobe
Optic chiasm
Optic nerve
Fornix
Choroid plexus
Right thalamus
Pineal gland
Hypothalamus
Cerebellum
Mesencephalic aqueduct
Fourth ventricle
Infundibulum (cut)
Tuberal area and median eminence
Mamillary body

(b)

3. *The coordination of activities of the nervous and endocrine systems.* The hypothalamus coordinates neural and endocrine activities by inhibiting or stimulating endocrine cells in the pituitary gland through the production of *regulatory hormones.* These hormones are produced at the tuberal area and are released into local capillaries for transport to the anterior pituitary gland.

4. *The secretion of hormones.* The hypothalamus secretes *antidiuretic hormone (ADH),* which is produced by the **supraoptic nucleus** and restricts water loss at the kidneys, and *oxytocin,* which is produced by the **paraventricular nucleus** and stimulates smooth muscle contractions in the uterus and mammary glands of females and the prostate gland of males. These hormones are transported along axons that pass through the infundibulum to the posterior pituitary gland. There the hormones are released into the blood for distribution throughout the body.

5. *The production of emotions and behavioral drives.* Specific hypothalamic centers produce sensations that lead to conscious or subconscious changes in behavior. For example, stimulation of the **feeding center** produces the sensation of hunger, and stimulation of the **thirst center** produces the sensation of thirst. These unfocused "impressions" originating in the hypothalamus are called **drives.** The conscious sensations are only part of the hypothalamic response. For instance, the thirst center also orders the release of ADH by neurons in the supraoptic nucleus.

6. *Coordination between voluntary and autonomic functions.* When you think about a dangerous or stressful situation, your heart rate and respiratory rate go up and your body prepares for an emergency. These autonomic adjustments are made by the hypothalamus.

7. *The regulation of body temperature.* The **preoptic area** of the hypothalamus coordinates the activities of other CNS centers and regulates other physiological systems to maintain normal body temperature. If body temperature falls, the preoptic area sends instructions to the *vasomotor center,* an autonomic center in the medulla oblongata that controls blood flow by regulating the diameter of peripheral blood vessels. In response, the vasomotor center decreases the blood supply to the skin, reducing the rate of heat loss.

8. *The control of circadian rhythms.* The **suprachiasmatic nucleus** coordinates daily cycles of activity that are linked to the day–night cycle. This nucleus receives input from the retina of the eye, and its output adjusts the activities of other hypothalamic nuclei, the pineal gland, and the reticular formation.

☑ Damage to the lateral geniculate nuclei of the thalamus would interfere with the functions of which senses?

☑ Which area of the diencephalon would be stimulated by changes in body temperature?

THE MESENCEPHALON

The external anatomy of the mesencephalon, or midbrain, is shown in Figure 14-13•, and the major nuclei are listed in Table 14-6 and shown in Figure 14-16•. The **tectum,** or roof of the mesencephalon, contains two pairs of sensory nuclei known collectively as the **corpora quadrigemina** (KOR-po-ra quad-ri-JEM-i-nuh). These nuclei, the superior and inferior colliculi, process visual and auditory sensations. Each **superior colliculus** (kol-IK-ū-lus; *colliculus,* a small hill) receives

TABLE 14-6 Components and Functions of the Mesencephalon

Subdivision	Region/Nuclei	Function(s)
GRAY MATTER		
Tectum (roof)	Superior colliculi	Integrate visual information with other sensory inputs; initiate reflex responses to visual stimuli
	Inferior colliculi	Relay auditory information to medial geniculate nuclei; initiate reflex responses to auditory stimuli
Walls and floor	Red nuclei	Involuntary control of background muscle tone and limb position
	Substantia nigra	Regulates activity in the cerebral nuclei
	Reticular formation (headquarters)	Automatic processing of incoming sensations and outgoing motor commands; can initiate involuntary motor responses to stimuli; helps maintain consciousness (RAS)
	Other nuclei/centers	Nuclei associated with two cranial nerves (N III, N IV)
WHITE MATTER		
	Cerebral peduncles	Connect primary motor cortex with motor neurons in brain and spinal cord; carry ascending sensory information to thalamus

•FIGURE 14-16 The Mesencephalon. (a) A posterior view. Superficial structures are labeled on the left; underlying nuclei are colored and labeled on the right. **(b)** A superior view of a section at the level of the mesencephalon.   AM   *Plate 3.3; Scan 1b*

visual inputs from the lateral geniculate nucleus of the thalamus on that side. The **inferior colliculus** receives auditory data from nuclei in the medulla oblongata and pons. Some of this information may be forwarded to the medial geniculate on the same side. The superior colliculi control the reflex movements of the eyes, head, and neck in response to visual stimuli, such as a bright light. The inferior colliculi control reflex movements of the head, neck, and trunk in response to auditory stimuli, such as a loud noise.

On each side, the mesencephalon contains a red nucleus and the substantia nigra. The **red nucleus** contains numerous blood vessels, which give it a rich red color. This nucleus receives information from the cerebrum and cerebellum. The red nucleus issues subconscious motor commands that affect background muscle tone and limb position. The **substantia nigra** (NĪ-gruh; black) is a nucleus that lies lateral to the red nucleus (Figure 14-16•). The gray matter in this region contains darkly pigmented cells, giving it a black color.

Neurons in the substantia nigra inhibit the activity of the cerebral nuclei by releasing the neurotransmitter *dopamine.* ∞ *p. 389* If the substantia nigra is damaged or the neurons secrete less dopamine, the cerebral nuclei become more active. The result is a gradual, generalized increase in muscle tone and the appearance of symptoms characteristic of **Parkinson's disease.** Persons with Parkinson's disease have difficulty starting voluntary movements, because opposing muscle groups do not relax; they must be overpowered. Once a movement is under way, every

aspect must be voluntarily controlled through intense effort and concentration.   AM   *The Cerebral Nuclei and Parkinson's Disease*

The nerve fiber bundles on the ventrolateral surfaces of the mesencephalon (Figures 14-13, p. 456, and 14-16•) are the **cerebral peduncles** (*peduncles,* little feet). They contain (1) descending fibers that go to the cerebellum by way of the pons and (2) descending fibers that carry voluntary motor commands from the primary motor cortex of each cerebral hemisphere.

The mesencephalon also contains the headquarters of the **reticular activating system (RAS),** a specialized component of the reticular formation. Stimulation of the mesencephalic portion of the RAS makes you more alert and attentive. We shall consider the role of the RAS in the maintenance of consciousness in Chapter 15.

THE CEREBELLUM

The cerebellum (Figure 14-17• and Table 14-7, p. 462) is an automatic processing center. It has two primary functions:

1. *Adjusting the postural muscles of the body.* The cerebellum coordinates rapid, automatic adjustments that maintain balance and equilibrium. These alterations in muscle tone and position are made by modifying the activities of the red nuclei.

2. *Programming and fine-tuning movements controlled at the conscious and subconscious levels.* The cerebellum refines learned movement patterns. This func-

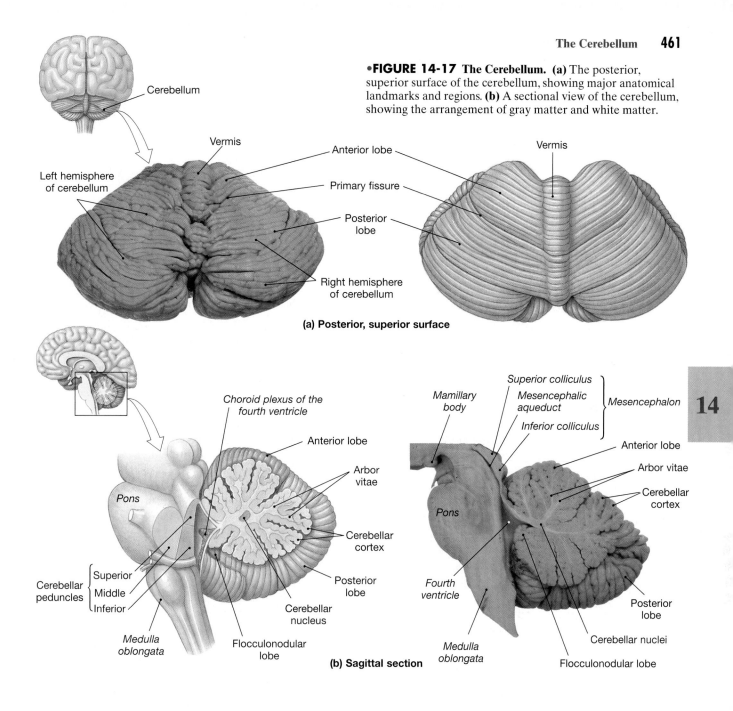

Cerebellum

Vermis

Left hemisphere of cerebellum

Anterior lobe

Primary fissure

Posterior lobe

Right hemisphere of cerebellum

Vermis

•FIGURE 14-17 The Cerebellum. (a) The posterior, superior surface of the cerebellum, showing major anatomical landmarks and regions. **(b)** A sectional view of the cerebellum, showing the arrangement of gray matter and white matter.

(a) Posterior, superior surface

Choroid plexus of the fourth ventricle

Anterior lobe

Arbor vitae

Cerebellar cortex

Posterior lobe

Cerebellar nucleus

Flocculonodular lobe

Pons

Cerebellar peduncles { Superior Middle Inferior }

Medulla oblongata

Superior colliculus

Mamillary body

Mesencephalic aqueduct

Inferior colliculus

Mesencephalon

14

Anterior lobe

Arbor vitae

Cerebellar cortex

Pons

Fourth ventricle

Medulla oblongata

Flocculonodular lobe

Posterior lobe

Cerebellar nuclei

(b) Sagittal section

tion is performed indirectly by regulating activity along motor pathways at the cerebral cortex, cerebral nuclei, and motor centers in the brain stem. The cerebellum compares the motor commands with proprioceptive information (position sense) and performs any adjustments needed to make the movement smooth.

The cerebellum has a complex, highly convoluted surface composed of neural cortex. The **folia** (FŌ-lē-uh; leaves), or folds of the cerebellum surface, are less prominent than the gyri of the cerebral hemispheres. The **anterior** and **posterior lobes** are separated by the **primary fissure** (Figure 14-17a•). Along the midline, a narrow band of cortex known as the **vermis** (VER-mis; worm) separates the **cerebellar hemispheres.** The slender **flocculonodular** (flok-ū-lō-NOD-ū-ler) **lobe** lies be-

tween the roof of the fourth ventricle and the cerebellar hemispheres and vermis (Figure 14-17b•).

Like the cerebrum, the cerebellum has a superficial layer of neural cortex. The cerebellar cortex contains huge, highly branched **Purkinje** (pur-KIN-jē) **cells.** The extensive dendrites of each Purkinje cell receive input from up to 200,000 synapses. Internally, the white matter of the cerebellum forms a branching array that in sectional view resembles a tree. Anatomists call it the **arbor vitae,** or "tree of life." The cerebellum receives proprioceptive information from the spinal cord and monitors all proprioceptive, visual, tactile, balance, and auditory sensations received by the brain. Information about the motor commands issued at the conscious and subconscious levels reaches the cerebellum indirectly, relayed by nuclei in the

TABLE 14-7 Components of the Cerebellum

Subdivision	Region/Nucleus	Function(s)
Gray matter	Cerebellar cortex	Involuntary coordination and control of ongoing body movements
	Cerebellar nuclei	As above
White matter	Arbor vitae	Connects cerebellar cortex and nuclei with cerebellar peduncles
	Cerebellar peduncles	
	Superior	Link the cerebellum with mesencephalon, diencephalon, and cerebrum
	Middle	Contain transverse fibers and carry communications between the cerebellum and pons
	Inferior	Link the cerebellum with the medulla oblongata and spinal cord
	Transverse fibers	Interconnect pontine nuclei with the cerebellar hemispheres on the opposite side

14

pons. A relatively small portion of the afferent fibers synapse in **cerebellar nuclei** before projecting to the cerebellar cortex. Most axons that carry sensory information do not synapse in the cerebellar nuclei but pass through the deeper layers of the cerebellum on their way to the cerebellar cortex.

Tracts that link the cerebellum with the brain stem, cerebrum, and spinal cord leave the cerebellar hemispheres as the superior, middle, and inferior cerebellar peduncles. The **superior cerebellar peduncles** link the cerebellum with nuclei in the midbrain, diencephalon, and cerebrum. The **middle cerebellar peduncles** are connected to a broad band of fibers that cross the ventral surface of the pons at right angles to the axis of the brain stem. The middle cerebellar peduncles also connect the cerebellar hemispheres with sensory and motor nuclei in the pons. The **inferior cerebellar peduncles** permit communication between the cerebellum and nuclei in the medulla oblongata and carry ascending and descending cerebellar tracts from the spinal cord.

The cerebellum can be permanently damaged by trauma or stroke or temporarily affected by drugs such as alcohol. The result is **ataxia** (a-TAK-sē-uh; *ataxia*, lack of order), a disturbance in balance. In severe ataxia, the individual cannot sit or stand without assistance.
 AM *Cerebellar Dysfunction*

THE PONS

The pons links the cerebellar hemispheres with the mesencephalon, diencephalon, cerebrum, and spinal cord. Important features and regions of the pons are indicated in Figures 14-13, p. 456, and 14-18a•. The pons contains the following:

- *Sensory and motor nuclei for four cranial nerves* (N V, N VI, N VII, and N VIII). These cranial nerves innervate the jaw muscles, the anterior surface of the face, one of the extrinsic eye muscles (the lateral rectus), and the sense organs of the inner ear (the *vestibular* and *cochlear nuclei*).

- *Nuclei involved with the control of respiration.* On each side of the brain, the reticular formation in this region contains two respiratory centers: the *apneustic center* and the *pneumotaxic center*. These centers modify the activity of the *respiratory rhythmicity center* in the medulla oblongata.

- *Nuclei and tracts that process and relay information heading to or from the cerebellum.* The pons links the cerebellum with the brain stem, cerebrum, and spinal cord.

- *Ascending, descending, and transverse tracts.* Longitudinal tracts interconnect other portions of the CNS. The middle cerebellar peduncles are connected to the **transverse fibers,** which cross the anterior surface of the pons. These fibers are axons that link nuclei of the pons (*pontine nuclei*) with the cerebellar hemisphere of the opposite side.

THE MEDULLA OBLONGATA

The medulla oblongata is continuous with the spinal cord. Figure 14-13•, p. 456, shows the external surface of the medulla oblongata; Figure 14-18b• and Table 14-8 summarize its major components.

The medulla oblongata connects the brain with the spinal cord, and many of its functions are directly related to that fact. All communication between the brain and spinal cord involves tracts that ascend or descend through the medulla oblongata.

The medulla oblongata includes three groups of nuclei that we will encounter in later chapters:

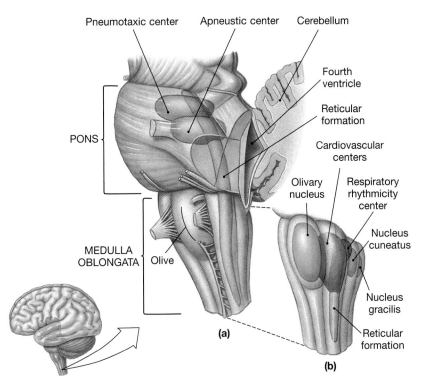

•**FIGURE 14-18** **The Pons and Medulla Oblongata.** **(a)** Nuclei in the pons. **(b)** Nuclei in the medulla oblongata.

TABLE 14-8 **Components and Functions of the Medulla Oblongata**

Subdivision	Region/Nucleus	Function(s)
Gray matter	Nucleus gracilis Nucleus cuneatus }	Relay somatic sensory information to the thalamus
	Olivary nuclei	Relay information from the red nucleus, other nuclei of the mesencephalon, and the cerebral cortex to the cerebellum
	Reflex centers	
	Cardiac centers	Regulate heart rate and force of contraction
	Vasomotor centers	Regulate distribution of blood flow
	Respiratory rhythmicity centers	Set the pace of respiratory movements
	Other nuclei/centers	Contain sensory and motor nuclei of five cranial nerves; relay ascending sensory information from the spinal cord to higher centers
White matter	Ascending and descending tracts	Link the brain with the spinal cord

1. *Autonomic nuclei controlling visceral activities.* The portion of the reticular formation in the medulla oblongata contains nuclei and centers responsible for the regulation of vital autonomic functions. These **reflex centers** receive inputs from cranial nerves, the cerebral cortex, and the brain stem. The output of these centers controls or adjusts the activities of one or more peripheral systems. Major reflex centers regulating autonomic functions include the following:

- The **cardiovascular centers,** which adjust the heart rate, the strength of cardiac contractions, and the flow of blood through peripheral tissues. In terms of function, the cardiovascular centers are subdivided into **cardiac** (*kardia,* heart) and **vasomotor** (*vas,* canal) **centers,** but their anatomical boundaries are difficult to determine.

■ The **respiratory rhythmicity centers,** which set the basic pace for respiratory movements. Their activity is regulated by inputs from the apneustic and pneumotaxic centers of the pons.

2. *Sensory or motor nuclei of cranial nerves.* The medulla oblongata contains sensory and motor nuclei associated with five of the cranial nerves (N VIII, N IX, N X, N XI, and N XII). These cranial nerves provide motor commands to muscles of the pharynx, neck, and back as well as to the visceral organs of the thoracic and peritoneal cavities. Cranial nerve VIII carries sensory information from receptors in the inner ear to the vestibular and cochlear nuclei, which extend from the pons into the medulla oblongata.

3. *Relay stations along sensory or motor pathways.* The **nucleus gracilis** and the **nucleus cuneatus** pass somatic sensory information to the thalamus, and the **olivary nuclei** relay information from the spinal cord, cerebral cortex, and brain stem to the cerebellar cortex. The bulk of these nuclei create the **olives,** prominent olive-shaped bulges along the ventrolateral surface of the medulla oblongata. The tracts leaving these nuclei cross to the opposite side of the brain before reaching their destinations. This crossing over is called *decussation* (dē-kuh-SĀ-shun; *decussatio,* crossing over).

☑ Which part of the brain has a worm (vermis) and a tree (arbor vitae)?

☑ The medulla oblongata is one of the smallest sections of the brain, yet damage there can cause death, whereas similar damage in the cerebrum might go unnoticed. Why?

☑ How would a decrease in the release of neurotransmitter from cells of the substantia nigra affect the activity of cerebral neurons?

Cranial Nerves

Cranial nerves are PNS components that connect directly to the brain rather than to the spinal cord. The twelve pairs of cranial nerves are located on the ventrolateral surface of the brain (Figure 14-19•), each with a name related to its appearance or function.

The number assigned to a cranial nerve roughly corresponds to the nerve's position along the longitudinal axis of the brain, beginning at the cerebrum. Roman numerals are usually used, either alone or with the prefix N or CN. We will use the abbreviation N, which is generally preferred by neuroanatomists and clinical neurologists. The abbreviation CN is preferred by comparative anatomists.

Each cranial nerve attaches to the brain near the associated sensory or motor nuclei. The sensory nuclei act as switching centers, with the postsynaptic neurons relaying the information to other nuclei or to processing centers in the cerebral or cerebellar cortex. In a similar fashion, the motor nuclei receive convergent inputs from higher centers or from other nuclei along the brain stem.

In this section, we classify cranial nerves as primarily sensory, special sensory, motor, or mixed (sensory and motor). In this classification, sensory nerves carry somatic sensory information, including touch, pressure, vibration, temperature, or pain. Special sensory nerves carry the sensations of smell, sight, hearing, or balance.

Motor nerves are dominated by the axons of somatic motor neurons; mixed nerves have a mixture of sensory and motor fibers. This is a useful method of classification, but it is based on the primary function, and a cranial nerve can have important secondary functions. Three examples are worth noting:

1. The olfactory receptors, the visual receptors, and the receptors of the inner ear are innervated by cranial nerves that are dedicated almost entirely to carrying special sensory information. The sensation of taste, considered to be one of the special senses, is carried by axons that form only a small part of large cranial nerves that have other primary functions.

2. As elsewhere in the PNS, a nerve containing tens of thousands of motor fibers that lead to a skeletal muscle will also contain sensory fibers from muscle spindles and tendon organs in that muscle. We assume that these sensory fibers are present but ignore them in the classification of the nerve.

3. Regardless of their other functions, several cranial nerves (N III, VII, IX, and X) distribute autonomic fibers to peripheral ganglia, just as spinal nerves deliver them to ganglia along the spinal cord. We will note the presence of small numbers of autonomic fibers (and shall discuss them further in Chapter 16) but ignore them in the classification of the nerve.

•FIGURE 14-19 Origins of the Cranial Nerves. (a) An inferior view of the brain. **(b)** A diagrammatic view, showing the attachment of the 12 pairs of cranial nerves.

The Olfactory Nerves (N I)

Primary function: Special sensory (smell)
Origin: Receptors of olfactory epithelium
Pass through: Cribriform plate of ethmoid bone
  ∞ *pp. 199, 204*
Destination: Olfactory bulbs

The first pair of cranial nerves (Figure 14-19•) carries special sensory information responsible for the sense of smell. The olfactory receptors are specialized neurons in the epithelium covering the roof of the nasal cavity, the superior nasal conchae, and the superior parts of the nasal septum. Axons from these sensory neurons collect to form 20 or more bundles that penetrate the cribriform plate of the ethmoid bone. These bundles are components of the **olfactory nerves** (N I). Almost at once these bundles enter the **olfactory bulbs,** neural masses on either side of the crista galli. The olfactory afferents synapse within the olfactory bulbs. The axons of the postsynaptic neurons proceed to the cerebrum along the slender **olfactory tracts** (Figures 14-19 and 14-20•).

Because the olfactory tracts look like typical peripheral nerves, anatomists about one hundred years ago misidentified these tracts as the first cranial nerve. Later studies demonstrated that the olfactory tracts and bulbs are part of the cerebrum, but by then the numbering system was already firmly established. Anatomists were left with a forest of tiny olfactory nerve bundles lumped together as N I.

The olfactory nerves are the only cranial nerves attached directly to the cerebrum. The rest originate or terminate within nuclei of the diencephalon or brain stem, and the ascending sensory information synapses in the thalamus before reaching the cerebrum.

The Optic Nerves (N II)

Primary function: Special sensory (vision)
Origin: Retina of eye
Pass through: Optic canals of sphenoid bone
  ∞ *p. 203*
Destination: Diencephalon via the optic chiasm

The **optic nerves** (N II) carry visual information from special sensory ganglia in the eyes. These nerves (Figure 14-21•) contain about 1 million sensory nerve fibers. The optic nerves pass through the optic canals of the sphenoid bone. Then they converge at the ventral, anterior margin of the diencephalon, at the **optic chiasm** (*chiasma,* a crossing). At the optic chiasm, fibers from the nasal half of each retina cross over to the opposite side of the brain.

The reorganized axons continue toward the lateral geniculate nuclei of the thalamus as the **optic tracts** (Figures 14-19 and 14-21•). After synapsing in the lateral geniculates, projection fibers deliver the information to the visual cortex of the occipital lobe. With this arrangement, each cerebral hemisphere receives visual information from the lateral half of the retina of the eye on that side and from the medial half of the retina of the eye of the opposite side. Relatively few axons in the optic tracts bypass the lateral geniculates and synapse in the superior colliculus of the midbrain. We shall consider that pathway in Chapter 17.

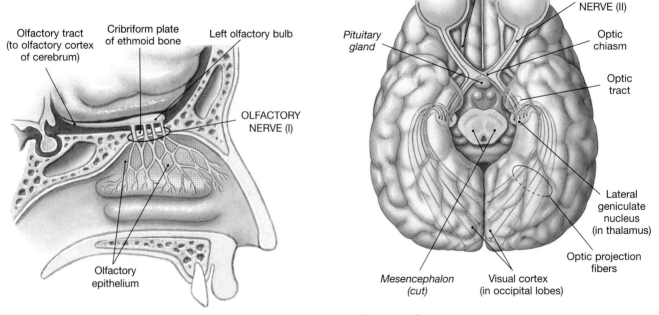

•**FIGURE 14-20** The Olfactory Nerve

•**FIGURE 14-21** The Optic Nerve

The Oculomotor Nerves (N III)

Primary function: Motor (eye movements)
Origin: Mesencephalon
Pass through: Superior orbital fissures of sphenoid bone ∞ *pp. 198, 203, 208*
Destination: *Somatic motor:* superior, inferior, and medial rectus muscles; inferior oblique muscle; levator palpebrae superioris muscle. *Visceral motor:* intrinsic eye muscles

The mesencephalon contains the motor nuclei controlling the third and fourth cranial nerves. Each **oculomotor nerve** (N III) innervates four of the six extrinsic eye muscles and the levator palpebrae superioris, the muscle that raises the upper eyelid (Figure 14-22●). On each side of the brain, N III emerges from the ventral surface of the mesencephalon and penetrates the posterior wall of the orbit at the superior orbital fissure. Individuals with damage to this nerve often complain of pain over the eye and double vision, because the movements of the left and right eyes cannot be coordinated properly.

The oculomotor nerve also delivers preganglionic autonomic fibers to neurons of the **ciliary ganglion.** The neurons of the ciliary ganglion control intrinsic eye muscles. These muscles change the diameter of the pupil, adjusting the amount of light entering the eye, and change the shape of the lens to focus images on the retina.

The Trochlear Nerves (N IV)

Primary function: Motor (eye movements)
Origin: Mesencephalon
Pass through: Superior orbital fissures of sphenoid bone ∞ *pp. 198, 203, 208*
Destination: Superior oblique muscle

A **trochlear** (TRŌK-lē-ar; *trochlea,* a pulley) **nerve** (N IV), the smallest cranial nerve, innervates the superior oblique muscle of each eye (Figure 14-22●). The trochlea is a pulley-shaped, ligamentous sling. Each superior oblique muscle passes through a trochlea on its way to its insertion on the surface of the eye. An individual with damage to N IV or to its nucleus will have difficulty looking down and to the side.

The Abducens Nerves (N VI)

Primary function: Motor (eye movements)
Origin: Pons
Pass through: Superior orbital fissures of sphenoid bone ∞ *pp. 198, 203, 208*
Destination: Lateral rectus muscle

The **abducens** (ab-DŪ-senz) **nerves** (N VI) innervate the lateral rectus muscles, the sixth pair of extrinsic eye muscles. This muscle makes the eye look to the side; in essence, the *abdu*cens causes *abdu*ction of the eye. Each abducens nerve emerges from the inferior surface of the brain stem at the border between the pons and the medulla oblongata (Figure 14-22●). Along with the oculomotor and trochlear nerves from that side, it reaches the orbit through the superior orbital fissure.

14

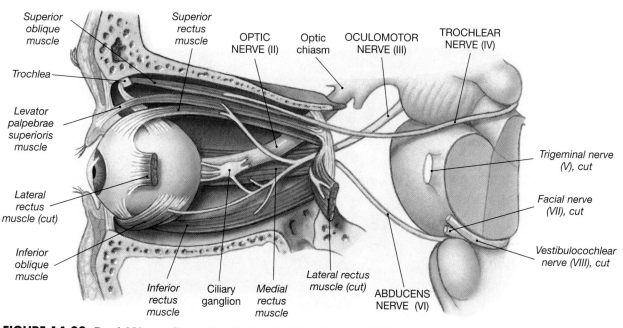

Superior oblique muscle
Superior rectus muscle
OPTIC NERVE (II)
Optic chiasm
OCULOMOTOR NERVE (III)
TROCHLEAR NERVE (IV)
Trochlea
Levator palpebrae superioris muscle
Trigeminal nerve (V), cut
Lateral rectus muscle (cut)
Facial nerve (VII), cut
Inferior oblique muscle
Vestibulocochlear nerve (VIII), cut
Inferior rectus muscle
Ciliary ganglion
Medial rectus muscle
Lateral rectus muscle (cut)
ABDUCENS NERVE (VI)

●FIGURE 14-22 Cranial Nerves Controlling the Extrinsic Eye Muscles. AM *Plate 3.2*

The Trigeminal Nerves (N V)

Primary function: Mixed (sensory and motor) to face

Origin: *Ophthalmic branch* (sensory): orbital structures, nasal cavity, skin of forehead, upper eyelid, eyebrow, nose (part). *Maxillary branch* (sensory): lower eyelid, upper lip, gums, and teeth; cheek; nose, palate, and pharynx (part). *Mandibular branch* (mixed): sensory from lower gums, teeth, and lips; palate and tongue (part); motor from motor nuclei of pons

Pass through (on each side): Ophthalmic branch through superior orbital fissure, maxillary branch through foramen rotundum, mandibular branch through foramen ovale ∞ *pp. 198, 199, 203, 208*

Destination: Ophthalmic and maxillary branches to sensory nuclei in pons; mandibular branch innervates muscles of mastication ∞ *p. 324*

The pons contains the nuclei associated with three cranial nerves (N V, VI, and VII) and contributes to a fourth (N VIII). The **trigeminal** (trī-JEM-i-nal) **nerves** (N V), the largest cranial nerves, are mixed nerves. Each provides both somatic sensory information from the head and face and motor control over the muscles of mastication. Sensory (dorsal) and motor (ventral) roots originate on the lateral surface of the pons (Figure 14-23●). The sensory branch is larger, and the enormous **semilunar ganglion** contains the cell bodies of the sensory neurons. As the name implies, the trigeminal has three major branches; the relatively small motor root contributes to only one of the three. **Tic douloureux** (doo-loo-ROO; *douloureux,* painful) is a painful condition affecting the area innervated by the maxillary and mandibular branches of the trigeminal nerve. Sufferers complain of debilitating pain triggered by contact with the lip, tongue, or gums. The cause of the condition is unknown. [AM] *Tic Douloureux*

The trigeminal nerve branches are associated with the *ciliary, sphenopalatine, submandibular,* and *otic ganglia.* These are autonomic (parasympathetic) ganglia whose neurons innervate structures of the face. However, although its nerve fibers may pass around or through these ganglia, the trigeminal nerve does not contain visceral motor fibers. We discussed the ciliary ganglion on page 467 and will describe the other ganglia next, with the branches of the *facial nerves* (N VII) and the *glossopharyngeal nerves* (N IX).

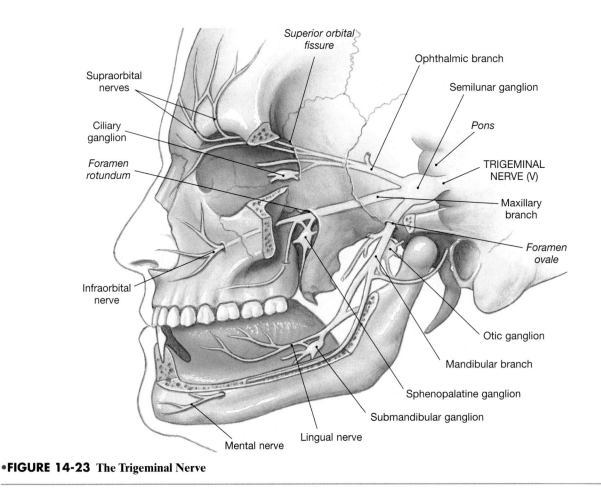

●**FIGURE 14-23** The Trigeminal Nerve

The Facial Nerves (N VII)

Primary function: Mixed (sensory and motor) to face

Origin: *Sensory:* taste receptors on anterior two thirds of tongue. *Motor:* motor nuclei of pons

Pass through: Internal acoustic canals to the *facial canals,* which end at the stylomastoid foramina ∞ *pp. 198, 199, 202*

Destination: *Sensory:* sensory nuclei of pons. *Somatic motor:* muscles of facial expression. ∞ *p. 320 Visceral motor:* lacrimal (tear) gland and nasal mucous glands by way of the sphenopalatine ganglion; submandibular and sublingual salivary glands by way of the submandibular ganglion

The **facial nerves** (N VII) are mixed nerves. The cell bodies of the sensory neurons are located in the **geniculate ganglia,** and the motor nuclei are in the pons. On each side, the sensory and motor roots emerge from the pons and enter the internal acoustic canal of the temporal bone. Each facial nerve then passes through the facial canal to reach the face by way of the stylomastoid foramen. The nerve then splits to form the temporal, zygomatic, buccal, mandibular, and cervical branches (Figure 14-24●).

The sensory neurons monitor proprioceptors in the facial muscles, provide deep pressure sensations over the face, and receive taste information from receptors along the anterior two-thirds of the tongue. Somatic motor fibers control the superficial muscles of the scalp and face and deep muscles near the ear.

The facial nerves carry preganglionic autonomic fibers to the sphenopalatine and submandibular ganglia. Postganglionic fibers from the **sphenopalatine ganglia** innervate the lacrimal glands and small glands of the nasal cavity and pharynx. The **submandibular ganglia** innervate the *submandibular* and *sublingual* (*sub-*, under, + *lingua,* tongue) salivary glands.

Bell's palsy is a cranial nerve disorder that results from an inflammation of a facial nerve. The condition is probably due to a viral infection. Symptoms include paralysis of facial muscles on the affected side and loss of taste sensations from the anterior two-thirds of the tongue. The condition is usually painless and in most cases goes away after a few weeks or months.

14

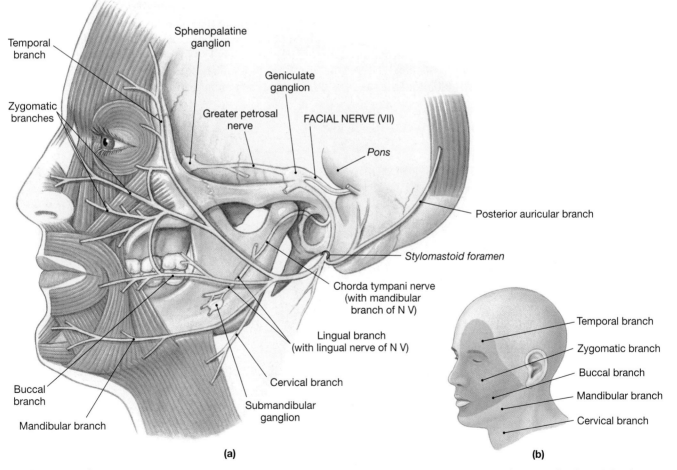

•**FIGURE 14-24 The Facial Nerve (a)** The origin and branches of the facial nerve. **(b)** The superficial distribution of the five major branches of the facial nerve.

The Vestibulocochlear Nerves (N VIII)

Primary function: Special sensory: balance and equilibrium (vestibular branch) and hearing (cochlear branch)

Origin: Monitor receptors of the inner ear (vestibule and cochlea)

Pass through: Internal acoustic canals of temporal bones ∞ *pp. 199, 202*

Destination: Vestibular and cochlear nuclei of pons and medulla oblongata

The **vestibulocochlear nerves** (N VIII) are also known as the *acoustic nerves,* the *auditory nerves,* and the *stato-acoustic nerves.* We will use *vestibulocochlear,* because this term indicates the names of the two major branches: the vestibular branch and the cochlear branch. Each vestibulocochlear nerve lies posterior to the origin of the facial nerve, straddling the boundary between the pons and the medulla oblongata (Figure 14-25●). This nerve reaches the sensory receptors of the inner ear by entering the internal acoustic canal in company with the facial nerve. Each vestibulocochlear nerve has two distinct bundles of sensory fibers. The **vestibular branch** (*vestibulum,* cavity) originates at the receptors of the *vestibule,* the portion of the inner ear concerned with balance sensations. The sensory neurons are located in an adjacent sensory ganglion, and their axons target the **vestibular nuclei** of the pons and medulla oblongata. These afferents convey information about the orientation and movement of the head. The **cochlear branch** (*cochlea,* snail shell) monitors the receptors in the *cochlea,* the portion of the inner ear that provides the sense of hearing. The cell bodies of the sensory neurons are located within a peripheral ganglion (the *spiral ganglion*), and their axons synapse within the **cochlear nuclei** of the pons and medulla oblongata. Axons leaving the vestibular and cochlear nuclei relay the sensory information to other centers or initiate reflexive motor responses. We shall discuss balance and the sense of hearing in Chapter 17.

The Glossopharyngeal Nerves (N IX)

Primary function: Mixed (sensory and motor) to head and neck

Origin: *Sensory:* posterior one-third of the tongue, part of the pharynx and palate, carotid arteries of the neck. *Motor:* motor nuclei of medulla oblongata

Pass through: Jugular foramina between the occipital bone and the temporal bones ∞ *pp. 198, 199*

Destination: *Sensory:* sensory nuclei of medulla oblongata. *Somatic motor:* pharyngeal muscles involved in swallowing. *Visceral motor:* parotid salivary gland by way of the otic ganglion

The medulla oblongata contains the sensory and motor nuclei of the cranial nerves N IX, N X, N XI, and N XII, in addition to the vestibular nucleus of N VIII. The **glossopharyngeal** (glos-ō-fah-RIN-jē-al; *glossum,* tongue) **nerves** (N IX) innervate the tongue and pharynx. Each glossopharyngeal nerve penetrates the cranium within the jugular foramen, with N X and N XI.

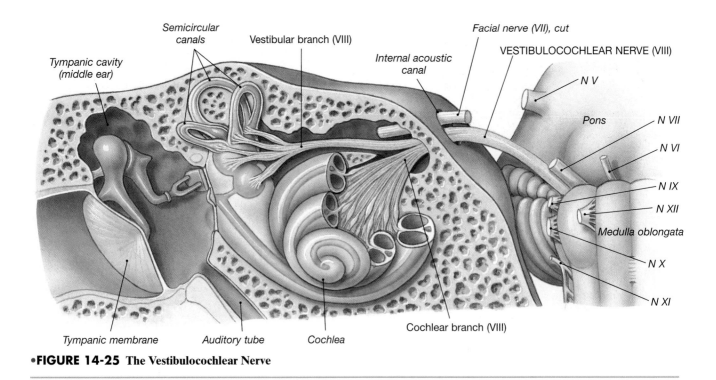

●**FIGURE 14-25** The Vestibulocochlear Nerve

The glossopharyngeal nerves are mixed nerves, but sensory fibers are most abundant. The sensory neurons on each side are in the **superior** (*jugular*) **ganglion** and **inferior** (*petrosal*) **ganglion** (Figure 14-26•). The sensory fibers carry general sensory information from the lining of the pharynx and the soft palate to a nucleus in the medulla oblongata. These nerves also provide taste sensations from the posterior third of the tongue and have special receptors that monitor the blood pressure and dissolved gas concentrations in major blood vessels.

The somatic motor fibers control the pharyngeal muscles involved in swallowing. Visceral motor fibers synapse in the **otic ganglion,** and postganglionic fibers innervate the parotid salivary gland of the cheek.

The Vagus Nerves (N X)

Primary function: Mixed (sensory and motor), widely distributed in the thorax and abdomen

Origin: *Sensory:* pharynx (part), auricle and external auditory canal, diaphragm, and visceral organs in thoracic and abdominopelvic cavities. *Motor:* motor nuclei in medulla oblongata

Pass through: Jugular foramina between the occipital bone and the temporal bones ∞ *pp. 198, 199*

Destination: *Sensory:* sensory nuclei and autonomic centers of medulla oblongata. *Visceral motor:* muscles of the palate, pharynx, digestive, respiratory, and cardiovascular systems in the thoracic and abdominal cavities

The **vagus** (VĀ-gus) **nerves** (N X) arise immediately posterior to the attachment of the glossopharyngeal nerves. Many small rootlets contribute to their formation, and developmental studies indicate that these nerves probably represent the fusion of several smaller cranial nerves during our evolutionary history. As the name suggests (*vagus,* wanderer), the vagus nerves branch and radiate extensively. Figure 14-27• shows only the general pattern of distribution.

Sensory neurons are located in the **jugular** and **nodose** (NŌ-dōs) **ganglia** (*node,* knot). Each vagus nerve provides somatic sensory information about the external auditory canal, a portion of the ear, and the diaphragm and special sensory information from pharyngeal taste receptors. But most of the vagal afferents carry visceral sensory information from receptors along the esophagus, respiratory tract, and abdominal viscera as distant as the last portions of the large intestine. This visceral sensory information is vital to the autonomic control of visceral function.

The motor components of the vagus are equally diverse. Each vagus nerve carries preganglionic autonomic (parasympathetic) fibers that affect the heart and control smooth muscles and glands within the areas monitored by its sensory fibers, including the stomach, intestines, and gallbladder. Difficulty in swallowing is one of the most common signs of damage to either N IX or N X, because damage to either one prevents the coordination of the swallowing reflex.

14

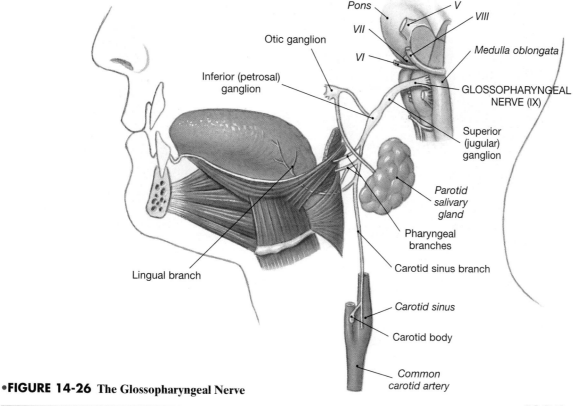

•**FIGURE 14-26** The Glossopharyngeal Nerve

FOCUS: *continues* ➡

14

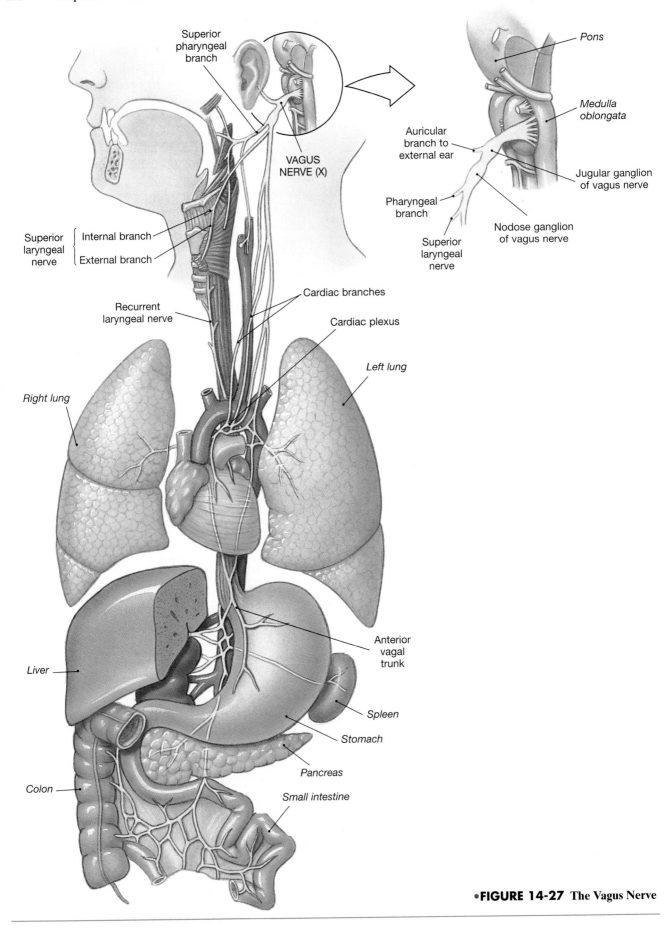

Superior pharyngeal branch

Pons

Medulla oblongata

Auricular branch to external ear

Jugular ganglion of vagus nerve

VAGUS NERVE (X)

Pharyngeal branch

Nodose ganglion of vagus nerve

Superior laryngeal nerve

Superior laryngeal nerve
{ Internal branch
  External branch }

Recurrent laryngeal nerve

Cardiac branches

Cardiac plexus

Left lung

Right lung

Anterior vagal trunk

Liver

Spleen

Stomach

Pancreas

Colon

Small intestine

•**FIGURE 14-27** The Vagus Nerve

The Accessory Nerves (N XI)

Primary function: Motor to muscles of the neck and upper back

Origin: Motor nuclei of spinal cord and medulla oblongata

Pass through: Jugular foramina between the occipital bone and the temporal bones ∞ *pp. 198, 199*

Destination: Medullary branches innervate voluntary muscles of palate, pharynx, and larynx; spinal branches control sternocleidomastoid and trapezius muscles

The **accessory nerves** (N XI) are also known as the *spinal accessory nerves* or the *spinoaccessory nerves.* Unlike other cranial nerves, each accessory nerve has some motor fibers that originate in the lateral gray horns of the first five cervical segments of the spinal cord (Figure 14-28•). These somatic motor fibers enter the cranium through the foramen magnum, join the motor fibers from a nucleus in the medulla oblongata, and leave the cranium through the jugular foramen.

Each accessory nerve has two branches. The **medullary branch** joins the vagus nerve and innervates the voluntary swallowing muscles of the soft palate and pharynx and the intrinsic muscles that control the vocal cords. The **spinal branch** controls the sternocleidomastoid and trapezius muscles of the neck and back. ∞ *pp. 326, 335* The motor fibers of this branch originate in the lateral gray horns of cervical spinal nerves C_1 to C_5.

The Hypoglossal Nerves (N XII)

Primary function: Motor (tongue movements)

Origin: Motor nuclei of medulla oblongata

Pass through: Hypoglossal canals of occipital bone ∞ *pp. 198, 199, 200*

Destination: Muscles of the tongue ∞ *p. 325*

Each **hypoglossal** (hī-pō-GLOS-al) **nerve** (N XII) leaves the cranium through the hypoglossal canal. The nerve then curves to reach the skeletal muscles of the tongue (Figure 14-28•). This cranial nerve provides voluntary motor control over movements of the tongue. Its condition is checked by having you stick out your tongue. Damage to one hypoglossal nerve or to its associated nuclei causes the tongue to veer toward the affected side.

14

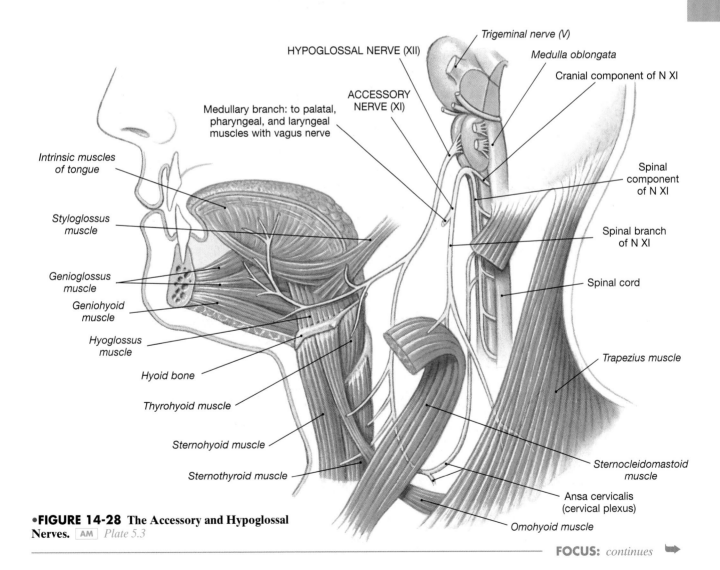

HYPOGLOSSAL NERVE (XII)

ACCESSORY NERVE (XI)

Medullary branch: to palatal, pharyngeal, and laryngeal muscles with vagus nerve

Trigeminal nerve (V)

Medulla oblongata

Cranial component of N XI

Spinal component of N XI

Spinal branch of N XI

Spinal cord

Trapezius muscle

Sternocleidomastoid muscle

Ansa cervicalis (cervical plexus)

Omohyoid muscle

Intrinsic muscles of tongue

Styloglossus muscle

Genioglossus muscle

Geniohyoid muscle

Hyoglossus muscle

Hyoid bone

Thyrohyoid muscle

Sternohyoid muscle

Sternothyroid muscle

•**FIGURE 14-28** The Accessory and Hypoglossal Nerves. [AM] *Plate 5.3*

A SUMMARY OF CRANIAL NERVE BRANCHES AND FUNCTIONS

Few people can easily recall the names, numbers, and functions of the cranial nerves. Mnemonic devices may prove useful. For example, you may find *"Oh, Once One Takes The Anatomy Final, Very Good Vacations Are Heavenly"* easy to remember; the first letter of each word corresponds to the first letter of a cranial nerve, proceeding from N I to N XII. Table 14-9 summarizes the basic distribution and function of each cranial nerve.

Cranial nerves are clinically important, in part because they can provide clues to underlying CNS problems. As a result, a number of standardized tests for cranial nerve function are used. AM *Cranial Reflexes and Cranial Nerve Tests*

TABLE 14-9 **The Cranial Nerves**

Cranial Nerve (number)	Sensory Ganglion	Branch	Primary Function	Foramen	Innervation
Olfactory (I)			Special sensory	Cribriform plate of ethmoid bone	Olfactory epithelium
Optic (II)			Special sensory	Optic canal	Retina of eye
Oculomotor (III)			Motor	Superior orbital fissure	Inferior, medial, superior rectus, inferior oblique and levator palpebrae superioris muscles; intrinsic eye muscles
Trochlear (IV)			Motor	Superior orbital fissure	Superior oblique muscle
Trigeminal (V)	Semilunar		Mixed		Areas associated with the jaws
		Ophthalmic	Sensory	Superior orbital fissure	Orbital structures, nasal cavity, skin of forehead, upper eyelid, eyebrows, nose (part)
		Maxillary	Sensory	Foramen rotundum	Lower eyelid; superior lip, gums, and teeth; cheek, nose (part), palate, and pharynx (part)
		Mandibular	Mixed	Foramen ovale	*Sensory:* inferior gums, teeth, lips, palate (part), and tongue (part) *Motor:* muscles of mastication
Abducens (VI)			Motor	Superior orbital fissure	Lateral rectus muscle
Facial (VII)	Geniculate		Mixed	Internal acoustic canal to facial canal; exits at stylomastoid foramen	*Sensory:* taste receptors on anterior 2/3 of tongue *Motor:* muscles of facial expression, lacrimal gland, submandibular gland, sublingual salivary glands
Vestibulocochlear (Acoustic) (VIII)			Special sensory	Internal acoustic canal	
		Cochlear			Cochlea (receptors for hearing)
		Vestibular			Vestibule (receptors for motion and balance)
Glossopharyngeal (IX)	Superior (jugular) and inferior (petrosal)		Mixed	Jugular foramen	*Sensory:* posterior 1/3 of tongue; pharynx and palate (part); receptors for blood pressure, pH, oxygen, and carbon dioxide concentrations *Motor:* pharyngeal muscles and parotid salivary gland
Vagus (X)	Jugular and nodose		Mixed	Jugular foramen	*Sensory:* pharynx; auricle and external auditory canal; diaphragm; visceral organs in thoracic and abdominopelvic cavities *Motor:* palatal and pharyngeal muscles and visceral organs in thoracic and abdominopelvic cavities
Accessory (XI)		Medullary	Motor	Jugular foramen	Skeletal muscles of palate, pharynx, and larynx (with vagus nerve)
		Spinal	Motor	Jugular foramen	Sternocleidomastoid and trapezius muscles
Hypoglossal (XII)			Motor	Hypoglossal canal	Tongue musculature

[EXPLORE]
[MediaLab]

"The human head weighs 5 pounds" reports the young actor in the movie *Jerry McGuire*. Most of that weight is attributable to the brain. This organ, along with its associated cranial nerves, performs many vital functions. To gain a better understanding of the brain and cranial nerves, use the resources available on both the CD-ROM and the Companion Website. The CD-ROM includes animations and tutorials; the website builds on those resources with text content and interactive problems as well as communication avenues for sharing ideas with your class. To complete the following Web Explorations, visit the MediaLab in Chapter 14 of your Companion Website.

Web Exploration 1

Estimated time for completion = 10 minutes

The frontal lobe of the brain includes the primary motor cortex, which controls skeletal muscle movements. How was the primary motor cortex mapped? Which parts of the frontal lobe control muscles in any one area of the body? To investigate this part of the brain, select the keyword **PROBE** on the Companion Website. Then click on "Mapping the Motor Cortex: A History." Next, return to the previous page and click on "Probe the Brain activity." Follow the directions and map the primary motor cortex, creating a reference for yourself.

Web Exploration 2

Estimated time for completion = 10 minutes

As we learned in the text, a stroke occurs when the blood supply to part of the brain is shut off; neurons die as a result. The symptoms of stroke vary with the area of the brain affected. Select the keyword **STROKE** on the Companion Website to read a short description of the brain, and study the diagram and table that are shown there. Click on each label of the diagram for a list of that area's functions. Create your own table, similar to the one on the Web page. Include all symptoms that would result from a blockage of the left middle cerebral artery, circled in the diagram.

14

CHAPTER REVIEW

SELECTED CLINICAL TERMINOLOGY
Terms Discussed in This Chapter

ataxia: A disturbance of balance that in severe cases leaves the individual unable to stand without assistance. It is caused by problems affecting the cerebellum. *(p.462)*

Bell's palsy: A condition resulting from an inflammation of the facial nerve; symptoms include a paralysis of facial muscles on the affected side and the loss of taste sensations from the anterior two-thirds of the tongue. *(p. 469)*

cerebrovascular accident (CVA), or *stroke:* A condition in which the blood supply to a portion of the brain is blocked off. *(p. 443)*

cranial trauma: A head injury resulting from violent contact with another object. Cranial trauma may cause a *concussion,* a condition characterized by a temporary loss of consciousness and a variable period of amnesia. *(p. 441 and AM)*

dyslexia: A disorder affecting the comprehension and use of words. *(p. 449)*

epidural hemorrhage: A condition involving bleeding between the dura mater and the cranium, generally resulting from cranial trauma. *(p. 443)*

hydrocephalus, or "water on the brain": A condition resulting from interference with the normal circulation and/or reabsorption of cerebrospinal fluid. *(p. 444 and AM)*

Parkinson's disease *(paralysis agitans):* A condition characterized by a pronounced increase in muscle tone, resulting from the excitation of neurons in the cerebral nuclei. *(p. 460 and AM)*

subdural hemorrhage: A condition in which blood accumulates under the dural epithelium in contact with the arachnoid membrane. *(p. 443)*

tic douloureux (doo-loo-ROO): A disorder of the maxillary and mandibular branches of N V characterized by almost totally debilitating pain triggered by contact with the lip, tongue, or gums. *(p. 468 and AM)*

AM Additional Terms Discussed in the *Applications Manual*

decerebrate rigidity: A generalized state of muscular contraction resulting from a loss of CNS inhibitory control.

dysmetria (dis-MET-rē-a): An inability to stop a movement at a precise, predetermined position; it commonly leads to an **intention tremor** in the affected individual. Dysmetria generally reflects cerebellar dysfunction.

spasticity: A condition characterized by hesitant, jerky voluntary movements, increased muscle tone, and hyperactive stretch reflexes.

tremor: Cyclic oscillations in limb position resulting from a "tug of war" between antagonistic muscle groups.

STUDY OUTLINE

INTRODUCTION p. 436

1. The brain is far more complex and adaptable than the spinal cord.

AN INTRODUCTION TO THE ORGANIZATION OF THE BRAIN p. 436

A Preview of Major Regions and Landmarks p. 436

1. The six regions in the adult brain are the cerebrum, diencephalon, mesencephalon ("midbrain"), pons, cerebellum, and medulla oblongata. (*Figures 14-1, 14-2; Table 14-1*)

2. Conscious thought, intellectual functions, memory, and complex motor patterns originate in the **cerebrum.** The **cerebellum** adjusts ongoing motor activities on the basis of sensory data and stored memories. (*Figure 14-2*)

3. Each wall of the **diencephalon** forms a **thalamus,** which contains relay and processing centers for sensory data. The **hypothalamus** contains centers involved with emotions, autonomic function, and hormone production. (*Figure 14-2*)

4. The **mesencephalon** processes visual and auditory information and generates reflexive somatic motor responses. (*Figure 14-2*)

5. The **pons** connects the cerebellum to the brain stem; it is also involved with somatic and visceral motor control. The spinal cord connects to the brain at the **medulla oblongata,** which relays sensory information and regulates autonomic functions. (*Figure 14-2*)

6. The brain contains extensive areas of **neural cortex,** a layer of gray matter on the surfaces of the cerebrum and cerebellum.

Embryology of the Brain p. 438

7. The brain forms from three swellings at the superior tip of the developing *neural tube:* the **prosencephalon,** the mesencephalon, and the **rhombencephalon.** The prosencephalon ("forebrain") forms the **telencephalon** (which becomes the cerebrum) and diencephalon; the rhombencephalon ("hindbrain") forms the **metencephalon** (cerebellum and pons) and **myelencephalon** (medulla oblongata). (*Table 14-1*)

Ventricles of the Brain p. 439

8. The central passageway of the brain expands to form chambers called **ventricles.** Cerebrospinal fluid continuously circulates from the ventricles and central canal of the spinal cord into the subarachnoid space of the meninges that surround the CNS. (*Figure 14-3*)

PROTECTION AND SUPPORT OF THE BRAIN p. 440

The Cranial Meninges p. 440

1. The cranial meninges (the *dura mater, arachnoid,* and *pia mater*) are continuous with those of the spinal cord but have anatomical and functional differences.

2. Folds of dura mater, including the **falx cerebri, tentorium cerebelli,** and **falx cerebelli,** stabilize the position of the brain. (*Figure 14-4*)

Cerebrospinal Fluid p. 441

3. Cerebrospinal fluid (CSF) is produced at the **choroid plexus.**

4. Cerebrospinal fluid (1) cushions delicate neural structures, (2) supports the brain, and (3) transports nutrients, chemical messengers, and waste products. Cerebrospinal fluid reaches the subarachnoid space through the **lateral apertures** and a **median aperture.** Diffusion across the **arachnoid granulations** into the **superior sagittal sinus** returns CSF to the venous circulation. (*Figure 14-5*)

The Blood Supply to the Brain p. 443

5. The **blood–brain barrier** isolates neural tissue from the general circulation.

6. The blood–brain barrier is intact throughout the CNS except in parts of the hypothalamus, the pineal gland, and at the choroid plexus in the diencephalon and medulla oblongata.

THE CEREBRUM p. 445

The Cerebral Cortex p. 445

1. The cortical surface contains **gyri** (elevated ridges) separated by **sulci** (shallow depressions) or **fissures** (deeper grooves). The **longitudinal fissure** separates the two **cerebral hemispheres.** The **central sulcus** separates the **frontal** and **parietal lobes.** Other sulci form the boundaries of the **temporal** and **occipital lobes.** (*Figures 14-1, 14-8*)

2. Each cerebral hemisphere receives sensory information and generates motor commands that concern the opposite side of the body. The two hemispheres have significant functional differences.

3. The white matter of the cerebrum contains **association fibers, commissural fibers,** and **projection fibers.** (*Figure 14-7*)

4. The **primary motor cortex** of the **precentral gyrus** directs voluntary movements. The **primary sensory cortex** of the **postcentral gyrus** receives somatic sensory information from touch, pressure, pain, taste, and temperature receptors. (*Figure 14-8; Table 14-2*)

5. **Association areas,** such as the **somatic sensory association area, visual association area,** and **somatic motor association area (premotor cortex),** control our ability to understand sensory information and coordinate a motor response. (*Figure 14-8; Table 14-2*)

6. The **general interpretive area** receives information from all the sensory association areas. It is present in only one hemisphere—generally the left. (*Figure 14-8*)

7. The **speech center** regulates the patterns of breathing and vocalization needed for normal speech. (*Figure 14-8*)

8. The **prefrontal cortex** coordinates information from the secondary and special association areas of the entire cortex and performs abstract intellectual functions. (*Figure 14-8*)

9. The left hemisphere typically contains the general interpretive and speech centers and is responsible for language-based skills. The right hemisphere is typically responsible for spatial relationships and analyses. (*Figure 14-9*)

The Cerebral Nuclei p. 451

10. The **cerebral nuclei** in the white matter of the cerebrum include the **caudate nucleus, amygdaloid body, globus pallidus,** and **putamen;** they control muscle tone and coordinate learned movement patterns and other somatic motor activities. *(Figure 14-10)*

THE LIMBIC SYSTEM p. 452

1. The **limbic system,** or *motivational system,* includes the amygdaloid body, **cingulate gyrus, parahippocampal gyrus, hippocampus, mamillary bodies,** and **fornix.** The functions of the limbic system involve emotional states and related behavioral drives. *(Figures 14-10–14-12; Table 14-3)*

THE DIENCEPHALON p. 454

1. The diencephalon provides the switching and relay centers necessary to integrate the conscious and subconscious sensory and motor pathways. *(Figures 14-10–14-13)*

The Thalamus p. 456

2. The thalamus is the final relay point for ascending sensory information and coordinates the activities of the cerebral cortex and cerebral nuclei. *(Figures 14-12–14-14; Table 14-4)*

The Hypothalamus p. 457

3. The hypothalamus contains important control and integrative centers. It can (1) control somatic motor activities at the subconscious level, (2) control autonomic function, (3) coordinate activities of the nervous and endocrine systems, (4) secrete hormones, (5) produce emotions and behavioral **drives,** (6) coordinate voluntary and autonomic functions, (7) regulate body temperature, and (8) coordinate circadian cycles of activity. *(Figure 14-15; Table 14-5)*

THE MESENCEPHALON p. 459

1. The **tectum** (roof of the mesencephalon) contains two pairs of sensory nuclei, the **corpora quadrigemina.** On each side, the **superior colliculus** receives visual inputs from the thalamus, and the **inferior colliculus** receives auditory data from the medulla oblongata. The **red nucleus** integrates information from the cerebrum and cerebellum and is involved with the control of muscle tone and posture. The **substantia nigra** regulates the motor output of the cerebral nuclei. The **cerebral peduncles** contain ascending fibers headed for thalamic nuclei and descending pyramidal fibers that carry voluntary motor commands from the primary motor cortex. *(Figures 14-13, 14-16; Table 14-6)*

THE CEREBELLUM p. 460

1. The cerebellum adjusts postural muscles and programs and tunes ongoing movements. The **cerebellar hemispheres** consist of neural cortex formed into folds, or **folia.** The surface consists of the **anterior** and **posterior lobes,** the **vermis,** and the **flocculonodular lobe.** *(Figure 14-17; Table 14-7)*

THE PONS p. 462

1. The pons contains (1) sensory and motor nuclei for four cranial nerves; (2) nuclei that help control respiration; (3) nuclei and tracts linking the cerebellum with the brain stem, cerebrum, and spinal cord; and (4) ascending and descending tracts. *(Figures 14-13, 14-18)*

THE MEDULLA OBLONGATA p. 462

1. The medulla oblongata connects the brain and spinal cord. It contains **olivary nuclei,** which relay information from the spinal cord, cerebral cortex, and brain stem to the cerebellar cortex. Its **reflex centers,** including the **cardiovascular** and **respiratory rhythmicity centers,** control or adjust activities of peripheral systems. *(Figures 14-13, 14-18; Table 14-8)*

FOCUS: Cranial Nerves p. 464

1. We have 12 pairs of cranial nerves. Each nerve attaches to the brain's ventrolateral surface near the associated sensory or motor nuclei. *(Figure 14-19)*

2. The **olfactory nerves** (N I) carry sensory information responsible for the sense of smell. The olfactory afferents synapse within the **olfactory bulbs.** *(Figures 14-19, 14-20)*

3. The **optic nerves** (N II) carry visual information from special sensory receptors in the eyes. *(Figures 14-19, 14-21)*

4. The **oculomotor nerves** (N III) are the primary source of innervation for four of the extrinsic oculomotor muscles. *(Figure 14-22)*

5. The **trochlear nerves** (N IV), the smallest cranial nerves, innervate the eyes' superior oblique muscles. *(Figure 14-22)*

6. The **trigeminal nerves** (N V), the largest cranial nerves, are mixed nerves with *ophthalmic, maxillary,* and *mandibular branches. (Figure 14-23)*

7. The **abducens nerves** (N VI) innervate the lateral rectus muscles. *(Figure 14-22)*

8. The **facial nerves** (N VII) are mixed nerves that control muscles of the scalp and face. They provide pressure sensations over the face and receive taste information from the tongue. *(Figure 14-24)*

9. The **vestibulocochlear nerves** (N VIII) contain the **vestibular branch,** which monitors sensations of balance, position, and movement, and the **cochlear branch,** which monitors hearing receptors. *(Figure 14-25)*

10. The **glossopharyngeal nerves** (N IX) are mixed nerves that innervate the tongue and pharynx and control the action of swallowing. *(Figure 14-26)*

11. The **vagus nerves** (N X) are mixed nerves that are vital to the autonomic control of visceral function. *(Figure 14-27)*

12. The **accessory nerves** (N XI) have **medullary branches,** which innervate voluntary swallowing muscles of the soft palate and pharynx, and **spinal branches,** which control muscles associated with the pectoral girdle. *(Figure 14-28)*

13. The **hypoglossal nerves** (N XII) provide voluntary motor control over tongue movements. *(Figure 14-28)*

A Summary of Cranial Nerve Branches and Functions p. 474

14. The branches and functions of the cranial nerves are summarized in *Table 14-9.*

14

REVIEW QUESTIONS

LEVEL 1 REVIEWING FACTS AND TERMS

Match each numbered item with the most closely related lettered item. Use letters for answers in the spaces provided.

Set 1: The Brain

___ 1. prosencephalon
___ 2. rhombencephalon
___ 3. brain stem
___ 4. ventricle in diencephalon
___ 5. dural fold
___ 6. choroid plexus
___ 7. limbic system
___ 8. pineal gland
___ 9. supraoptic nucleus
___ 10. paraventricular nucleus
___ 11. substantia nigra
___ 12. red nucleus
___ 13. Purkinje cells
___ 14. arbor vitae
___ 15. ataxia
___ 16. reticular formation
___ 17. nucleus gracilis
___ 18. hypothalamus

a. third ventricle
b. oxytocin
c. adjusts muscle tone and limb position
d. melatonin
e. white matter of cerebellum
f. mesencephalon, pons, medulla oblongata
g. link between nervous and endocrine systems
h. releases the neurotransmitter dopamine
i. forebrain
j. cerebellar cortex
k. falx cerebri
l. a disturbance in balance
m. apneustic, pneumotaxic centers
n. hindbrain
o. amygdaloid body
p. medulla oblongata
q. production of CSF
r. antidiuretic hormone

Set 2: The Cranial Nerves

___ 19. facial nerve
___ 20. vagus nerve
___ 21. olfactory nerve
___ 22. trochlear nerve
___ 23. hypoglossal nerve
___ 24. accessory nerve
___ 25. trigeminal nerve
___ 26. optic nerve
___ 27. glossopharyngeal nerve
___ 28. oculomotor nerve
___ 29. vestibulocochlear nerve
___ 30. abducens nerve

a. sensory, smell
b. N V
c. motor, tongue movements
d. sensory, vision
e. N X
f. N III
g. equilibrium, hearing
h. N VII
i. N VI
j. N IX
k. motor, eye movements
l. N XI

31. The primary link between the nervous and endocrine systems is the
(a) hypothalamus   (b) mesencephalon
(c) pons   (d) medulla oblongata

32. Immediate reflexive responses to a loud, unexpected noise are directed by nuclei in the
(a) pons   (b) diencephalon
(c) mesencephalon   (d) medulla oblongata

33. The regulation of autonomic function, such as heart rate and blood pressure, originates in the
(a) cerebrum   (b) cerebellum
(c) diencephalon   (d) medulla oblongata

34. The smooth surface that covers the brain but does not follow the underlying neural convolutions or sulci is the
(a) neural cortex   (b) dura mater
(c) pia mater   (d) arachnoid membrane

35. The meningeal layer that adheres to the surface contour of the brain, extending into every fold and curve, is the
(a) pia mater   (b) dura mater
(c) arachnoid membrane   (d) neural cortex

36. The dural fold that divides the two cerebellar hemispheres is the
(a) transverse sinus   (b) falx cerebri
(c) tentorium cerebelli   (d) falx cerebelli

37. Cerebrospinal fluid is produced and secreted in the
(a) hypothalamus   (b) choroid plexus
(c) medulla oblongata   (d) crista galli

38. The primary purpose of the blood–brain barrier is to
(a) provide the brain with oxygenated blood
(b) drain venous blood via the internal jugular veins
(c) isolate neural tissue in the CNS from the general circulation
(d) a, b, and c are correct

39. The two cerebral hemispheres are functionally different even though anatomically they appear the same.
(a) true   (b) false

40. Reading, writing, and speaking are dependent on processing in the
(a) right cerebral hemisphere
(b) left cerebral hemisphere
(c) prefrontal cortex
(d) postcentral gyrus

41. The establishment of emotional states is a function of the
(a) limbic system   (b) tectum
(c) mamillary bodies   (d) thalamus

42. The final relay point for ascending sensory information that will be projected to the primary sensory cortex is the
(a) hypothalamus   (b) thalamus
(c) spinal cord   (d) medulla oblongata

43. The two hormones secreted by the hypothalamus are
(a) epinephrine and norepinephrine
(b) antidiuretic hormone and oxytocin
(c) melatonin and serotonin
(d) FSH and ATP

44. The centers in the pons that modify the activity of the respiratory rhythmicity centers in the medulla oblongata are the
(a) apneustic and pneumotaxic centers
(b) inferior and superior peduncles
(c) cardiac and vasomotor centers
(d) nucleus gracilis and nucleus cuneatus

45. Briefly summarize the overall function of the cerebellum.

46. What role does the hypothalamus play in the body?

47. What are the three important functions of the CSF?

48. Which three areas in the brain are not isolated from the general circulation by the blood–brain barrier?

49. Where is the limbic system located, and what are its three functions?

50. Using the mnemonic device "Oh, Once One Takes The Anatomy Final, Very Good Vacations Are Heavenly," list the 12 pairs of cranial nerves and their functions.

LEVEL 2 REVIEWING CONCEPTS

51. A person with a damaged visual association area may be
  (a) unable to scan the lines of a page or see rows of clear symbols
  (b) declared legally blind
  (c) unable to see letters but able to identify words and their meanings
  (d) able to see letters quite clearly but unable to recognize or interpret them

52. If the corpus callosum is cut,
  (a) cross-referencing of sensory information is inhibited
  (b) the individual talks constantly but uses the wrong words
  (c) objects touched with the left hand cannot be recognized but can be verbally identified
  (d) objects touched with the right hand cannot be verbally identified

53. Which structure in the brain would your A & P instructor be referring to when talking about a nucleus that resembles a sea horse and that appears to be important in the storage and retrieval of long-term memories? In which functional system of the brain is it located?

54. Stimulation of which part of the brain would produce sensations of hunger and thirst?

55. Damage to the corpora quadrigemina would interfere with the
  (a) control of autonomic function
  (b) regulation of body temperature
  (c) processing of visual and auditory sensations
  (d) conscious control of skeletal muscles

56. If symptoms characteristic of Parkinson's disease appear, which part of the mesencephalon is inhibited from secreting a neurotransmitter? Which neurotransmitter is it?

57. What are the principal functional differences between the right and left hemispheres of the cerebrum?

58. Which major part of the brain is associated with identifying the clinical signs of life?

59. Why can the brain respond to stimuli with greater versatility than the spinal cord can?

60. What kinds of problems are associated with the presence of lesions in Wernicke's area and Broca's area?

LEVEL 3 CRITICAL THINKING AND CLINICAL APPLICATIONS

61. Smelling salts can sometimes help restore consciousness after a person has fainted. The active ingredient of smelling salts is ammonia, and it acts by irritating the lining of the nasal cavity. Propose a mechanism by which smelling salts would raise a person from the unconscious state to the conscious state.

62. A police officer has just stopped Bill on suspicion of driving while intoxicated. The officer asks Bill to walk the yellow line on the road and then to place the tip of his index finger on the tip of his nose. How would these activities indicate Bill's level of sobriety? Which part of the brain is being tested by these activities?

63. While having some dental work performed, Tyler is given an injection of local anesthetic in his lower jaw. His dentist tells him not to eat until the anesthetic wears off, not because of his teeth but because of his tongue. Why is the dentist giving Tyler this advice?

14

CHAPTER 15

Integrative Functions

CHAPTER OUTLINE AND OBJECTIVES

INTRODUCTION   p. 481

SOMATIC SENSORY AND MOTOR PATHWAYS   p. 481
 Sensory Pathways   p. 481
 Motor Pathways and Motor Control   p. 487
① Identify the principal sensory and motor pathways.
② Compare the components, processes, and functions of the various motor pathways.
③ Explain how we can distinguish among sensations that originate in different areas of the body.
④ Describe the levels of information processing involved in motor control.
⑤ Discuss how the brain integrates sensory information and coordinates responses.

MONITORING BRAIN ACTIVITY: THE ELECTROENCEPHALOGRAM   p. 493

HIGHER-ORDER FUNCTIONS   p. 494
 Memory   p. 494
 Consciousness   p. 496
⑥ Explain how memories are created, stored, and recalled.
⑦ Distinguish between the levels of consciousness and unconsciousness, and identify the characteristics of brain activity associated with the different levels of sleep.

BRAIN CHEMISTRY AND BEHAVIOR   p. 498
 Personality and Other Mysteries   p. 498
⑧ Describe drug-related alterations in brain function.

AGING AND THE NERVOUS SYSTEM   p. 499
 Age-Related Anatomical and Functional Changes   p. 499
 Alzheimer's Disease   p. 499
⑨ Summarize the effects of aging on the nervous system.

MEDIALAB and CHAPTER REVIEW   p. 500

It is said that "big cities never sleep." In New York City or Los Angeles at 3:00 A.M., shops are open, deliveries are made, people are on the street, and traffic moves briskly. The central nervous system (CNS) is much more complex than any city and far busier. Information continuously flows among the brain, spinal cord, and peripheral nerves. At any moment, millions of sensory neurons are delivering information to processing centers in the CNS, and millions of motor neurons are controlling or adjusting the activities of peripheral effectors. This process continues 24 hours a day, whether we are awake or asleep. We may sleep soundly, but our nervous system does not; many brain stem centers are active throughout our lives, performing vital autonomic functions. This chapter examines the flow of sensory information and motor commands between the PNS and the CNS. It also examines complex aspects of human consciousness, such as memory and learning, that involve multiple regions of the brain.

Chapters 13 and 14 introduced the major components and functions of the central nervous system (CNS) and peripheral nervous system (PNS). Those chapters, however, could not tell the entire story, for the CNS is more than a collection of independent parts. Think of the nervous system as a symphony orchestra. Thus far, you have studied the appearance and sound of the individual cellos, clarinets, and other instruments. Although essential to an understanding of an orchestra, this information does not tell you how the orchestra performs or what a symphony sounds like. The neural "orchestra" includes processing centers in the brain and spinal cord.

Like any orchestra, your nervous system has many individual components, all working at the same time. For instance, while you are reading this text and trying to memorize important information, you are also breathing, balancing upright, controlling your body temperature, and possibly listening to the radio or television. Each of these seemingly independent actions must be coordinated. Rather than having a single conductor, the nervous system relies on several integrated control centers, such as the cerebral cortex, the cerebellum, and the hypothalamus.

This chapter and the next consider the ways this orchestra monitors and controls somatic and visceral functions. In this chapter we will examine the pathways and the integrative centers that control or adjust skeletal muscle activity. Chapter 16 studies the pathways and integrative centers that control or adjust visceral effectors, such as smooth muscle, cardiac muscle, glands, and fat cells.

Figure 15-1• provides an overview of the pathways we will cover in this chapter. *Pathways* consist of the nuclei and tracts that link the brain with the rest of the body. Somatic sensory pathways deliver sensory information from the head, neck, body wall, and limbs to the CNS. These pathways end in either the primary sensory cortex of the cerebral hemispheres or appropriate areas of the cerebellar hemispheres. The motor commands issued by these higher centers may be triggered by arriving sensory information and may be modified on the basis of planning, memories, and learning—the so-called *higher-order functions* of the brain. Somatic motor commands—commands that are either voluntary or involuntary—travel along somatic motor pathways. They leave the CNS as the somatic nervous system (SNS). ∞ *p. 361*

All sensory and motor pathways involve a series of synapses, one after the other. Along the way, the information is distributed to processing centers operating at the subconscious level. For example, while your cerebral cortex struggles to decide on a voluntary response to a stumble, monosynaptic and polysynaptic reflexes, aided by commands issued by your brain stem and cerebellum, may prevent a fall.

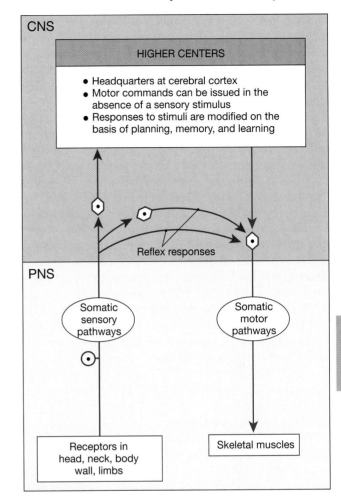

•**FIGURE 15-1** An Overview of the Pathways Involved in Somatic Motor Control

SOMATIC SENSORY AND MOTOR PATHWAYS

We will begin by considering the pathways that utilize the major ascending (sensory) and descending (motor) tracts of the spinal cord. It may help you to remember that tract names indicate the origins and destinations of the axons:

- If the name of a tract begins with *spino-*, the tract must *start* in the spinal cord and *end* in the brain. The rest of the name indicates the tract's destination. For example, a *spinothalamic tract* begins in the spinal cord and carries sensory information to the thalamus of the brain.

- If the name of a tract ends in -*spinal,* the tract's axons *start* in a higher center of the brain and *end* in the spinal cord. The rest of the name indicates the nucleus or cortical area of the brain where the tract originates. For example, a *corticospinal tract* carries motor commands from the cerebral cortex to the spinal cord.

SENSORY PATHWAYS

Sensory receptors represent the interface between the nervous system and the internal and external environments. A **sensory receptor** is a specialized cell or dendrite that monitors specific conditions in the body or the external environment. When stimulated, a receptor passes information to the CNS in the form of action potentials along the axon of a sensory neuron. The arriving information is called a **sensation.** We introduced three general categories of sensory receptors in Chapter 12: *exteroceptors, proprioceptors,* and *interoceptors.* ∞ *p. 366* These categories reflect the source of the stimulation: Exteroceptors provide information about the external environment; proprioceptors track the positions of bones and joints; and interoceptors monitor internal tissues and organs.

The term **general senses** refers to sensations of temperature, pain, touch, pressure, vibration, and proprioception. General sensory receptors are distributed throughout the body, and they are relatively simple in structure. Some of the information they send to the CNS reaches the primary sensory cortex and our conscious awareness. As we noted in Chapter 12, sensory information is interpreted on the basis of the frequency of arriving action potentials. ∞ *pp. 395–396* For example, when pressure sensations are arriving, the harder the pressure, the higher the frequency of action potentials. The conscious awareness of a sensation is called a **perception.**

In Chapter 17, we will consider the origins of sensations and the pathways involved in relaying the information to conscious and subconscious processing centers in the CNS. Most of the processing occurs in centers along the sensory pathways in the spinal cord or brain stem. Only about 1 percent of the information provided by afferent fibers reaches the cerebral cortex and our conscious awareness. For example, we usually do not feel the clothes we wear or hear the hum of the engine in our car.

Receptors for the General Senses

Receptors for the general senses are classified according to the nature of the stimulus that excites them:

1. **Nociceptors** (nō-sē-SEP-torz; *noceo,* hurt) respond to a variety of stimuli generally associated with tissue damage. Receptor activation causes the sensation of pain.

2. **Thermoreceptors** respond to temperature changes.

3. **Mechanoreceptors** are stimulated or inhibited by physical distortion, contact, or pressure on their cell membranes. There are many types of mechanoreceptors. Examples introduced in earlier chapters include *proprioceptors,* such as muscle spindles and Golgi tendon organs, and *tactile receptors,* such as touch receptors in the skin.

4. **Chemoreceptors** monitor the chemical composition of body fluids and respond to the presence of specific molecules. Chemoreceptors are important components of several visceral reflexes.

In the next section, we discuss ascending pathways that carry somatic sensory information to the primary sensory cortex of the cerebral hemispheres. Visceral sensory information, which arises from receptors in visceral organs, is generally routed to processing centers in the brain stem and diencephalon rather than to the cerebrum. As a result, those sensations seldom reach our conscious awareness. We shall consider visceral sensory information further in Chapter 16.

The Organization of Sensory Pathways

The sensory neuron that delivers the sensations to the CNS is often called a **first-order neuron.** The cell body of a first-order sensory neuron is located in a dorsal root ganglion. In the CNS, the axon of that sensory neuron synapses on an interneuron known as a **second-order neuron,** which may be located in the spinal cord or brain stem. If the sensation is to reach our conscious awareness, the second-order neuron will synapse on a **third-order neuron** in the thalamus. Somewhere along its length, the axon of the second-order neuron crosses over to the opposite side of the CNS. As a result, the right side of the thalamus receives sensory information from the left side of the body, and vice versa.

The axons of the third-order neurons ascend without crossing over and synapse on neurons of the primary sensory cortex of the cerebral hemisphere. As a result, the right cerebral hemisphere receives sensory information from the left side of the body, and the left cerebral hemisphere receives sensations from the right side.

Three major somatic sensory pathways exist: (1) the *posterior column pathway,* (2) the *spinothalamic pathway,* and (3) the *spinocerebellar pathway.* Two pairs of spinal tracts are associated with each pathway, one pair on each side of the spinal cord. All the axons within a tract share a common origin and destination. Figure 15-2• indicates the relative positions of the spinal tracts involved.

The Posterior Column Pathway

The **posterior column pathway** carries sensations of highly localized ("fine") touch, pressure, vibration, and proprioception (position sense). This pathway begins at a peripheral receptor and ends at the primary sensory cortex of the cerebral hemispheres. The spinal tracts involved are the left and right **fasciculus gracilis** (*gracilis,* delicate) and the left and right **fasciculus cuneatus** (*cuneus,* wedge-shaped). On each side of the posterior median sulcus, the fasciculus gracilis is medial to the fasciculus cuneatus.

The axons of the first-order neurons reach the CNS within the dorsal roots of spinal nerves and the sensory

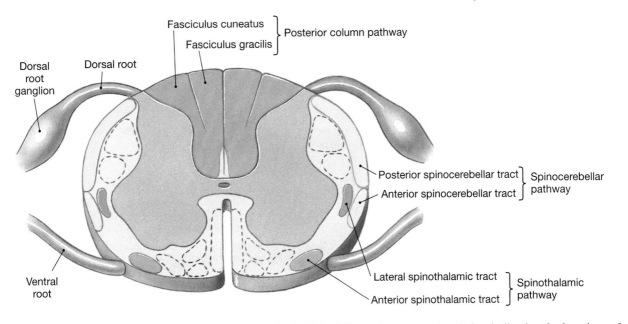

•**FIGURE 15-2** **Ascending Sensory Pathways and Tracts in the Spinal Cord.** A cross-sectional view indicating the locations of the major ascending sensory tracts in the spinal cord. For information about these tracts, see Table 15-1. Descending motor tracts (identified in *Figure 15-6*) are shown in dashed outline.

15

roots of cranial nerves. The axons ascending within the posterior column are organized according to the region innervated. Axons carrying sensations from the inferior half of the body ascend within the fasciculus gracilis and synapse in the *nucleus gracilis* of the medulla oblongata. Axons carrying sensations from the superior half of the trunk, upper limbs, and neck ascend in the fasciculus cuneatus and synapse in the *nucleus cuneatus.* ∞ *p. 464*

Axons of the second-order neurons of the nucleus gracilis and nucleus cuneatus ascend to the thalamus. As they ascend, these axons cross over to the opposite side of the brain stem. The crossing of an axon from the left side to the right side, or vice versa, is called **decussation.** Once on the opposite side of the brain, the axons enter a tract called the **medial lemniscus** (*lemniskos,* ribbon). Before reaching the thalamus, the medial lemniscus receives sensory information collected by cranial nerves V, VII, IX, and X.

The axons in the medial lemniscus synapse on third-order neurons in the ventral nuclei of the thalamus. ∞ *p. 457* These nuclei sort the arriving information according to (1) the nature of the stimulus and (2) the region of the body involved. Processing in the thalamus determines whether we perceive a given sensation as fine touch rather than as pressure or vibration.

Our ability to localize the sensation—to determine precisely where on the body a specific stimulus originated—depends on the projection of information from the thalamus to the primary sensory cortex. Sensory information from the toes arrives at one end of the primary sensory cortex, and information from the head arrives at the other. When neurons in one portion of your primary sensory cortex are stimulated, you become consciously aware of sensations originating at a specific location. If your primary sensory cortex were damaged or the projection fibers were cut, you could detect a light touch but would be unable to determine its source.

The same sensations are reported whether the cortical neurons are activated by axons ascending from the thalamus or by direct electrical stimulation. Researchers have electrically stimulated the primary sensory cortex in awake individuals during brain surgery and asked the subjects where they thought the stimulus originated. The results were used to create a functional map of the primary sensory cortex. This map, shown in Figure 15-3•, is called a **sensory homunculus** ("little man").

The proportions of the sensory homunculus are very different from those of any individual. For example, the face is huge and distorted, with enormous lips and tongue, whereas the back is relatively tiny. These distortions occur because the area of sensory cortex devoted to a particular body region is proportional not to the region's absolute size but to the *number of sensory receptors* it contains. In other words, many more cortical neurons are required to process sensory information arriving from the tongue, which has tens of thousands of taste and touch receptors, than to analyze sensations originating on the back, where touch receptors are few and far between.

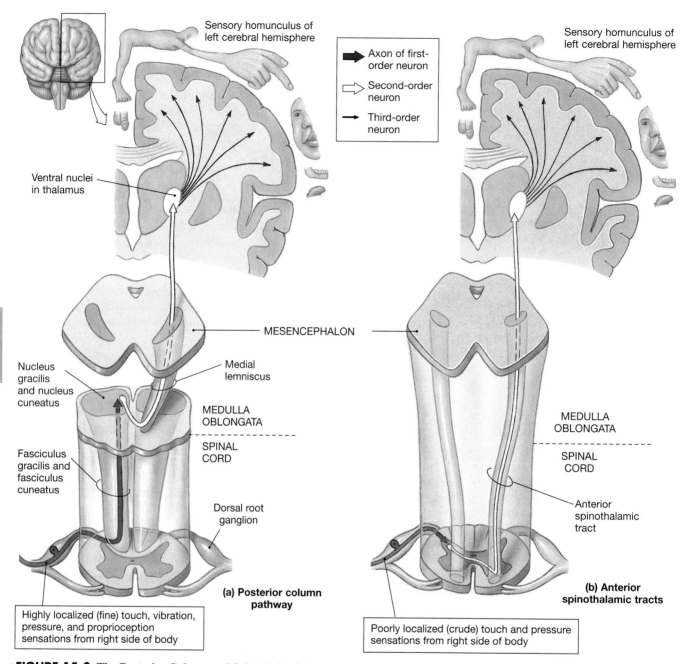

Sensory homunculus of left cerebral hemisphere

| Axon of first-order neuron |
| Second-order neuron |
| Third-order neuron |

Ventral nuclei in thalamus

Sensory homunculus of left cerebral hemisphere

MESENCEPHALON

Nucleus gracilis and nucleus cuneatus

Medial lemniscus

MEDULLA OBLONGATA

SPINAL CORD

Fasciculus gracilis and fasciculus cuneatus

Dorsal root ganglion

MEDULLA OBLONGATA

SPINAL CORD

Anterior spinothalamic tract

(a) Posterior column pathway

Highly localized (fine) touch, vibration, pressure, and proprioception sensations from right side of body

(b) Anterior spinothalamic tracts

Poorly localized (crude) touch and pressure sensations from right side of body

•**FIGURE 15-3** **The Posterior Column and Spinothalamic Pathways.** For clarity, only the pathway for sensations originating on the right side of the body are shown. **(a)** The posterior column pathway delivers fine touch, vibration, and proprioception information to the primary sensory cortex on the opposite side of the body. **(b)** The anterior spinothalamic tracts carry sensations of crude touch and pressure to the primary sensory cortex on the opposite side of the body.

The Spinothalamic Pathway

The **spinothalamic pathway** provides conscious sensations of poorly localized ("crude") touch, pressure, pain, and temperature. This pathway begins at peripheral receptors and ends at the primary sensory cortex of the cerebral hemispheres.

The axons of first-order sensory neurons enter the spinal cord and synapse on second-order neurons within the posterior gray horns. The axons of these in-terneurons cross to the opposite side of the spinal cord and ascend within the anterior or lateral spinothalamic tracts. The left and right **anterior spinothalamic tracts** carry crude touch and pressure sensations (Figure 15-3b•), whereas the left and right **lateral spinothalamic tracts** carry pain and temperature sensations (Figure 15-3c•). These tracts end at third-order neurons in the ventral nucleus group of the thalamus. After the sensations have been sorted and processed, they are relayed to the primary sensory cortex.

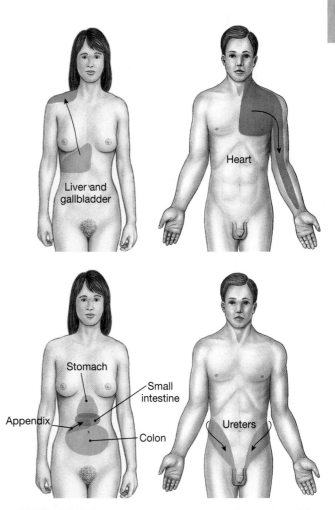

An individual can experience painful sensations that are not real. For example, a person may continue to experience pain in an amputated limb. This *phantom limb pain* is caused by activity in the sensory neurons or interneurons along the spinothalamic pathway. The neurons involved once monitored conditions in the intact limb.

An individual can feel pain in an uninjured part of the body when the pain actually originates at another location. For example, strong visceral pain sensations arriving at a segment of the spinal cord can stimulate interneurons that are normally part of the spinothalamic pathway. Activity in these interneurons leads to the stimulation of the primary sensory cortex, so the individual feels pain in a specific part of the body surface. This phenomenon is called **referred pain.** Two familiar examples are (1) the pain of a heart attack, which is typically felt in the left arm, and (2) the pain of appendicitis, which is generally felt in the right lower quadrant. These and additional examples are shown in Figure 15-4•. We will consider the origins and distribution of pain sensations further in Chapter 17.

15

•**FIGURE 15-3** **The Posterior Column and Spinothalamic Pathways** *(continued).* **(c)** The lateral spinothalamic tracts carry sensations of pain and temperature to the primary sensory cortex on the opposite side of the body.

•**FIGURE 15-4** **Referred Pain.** Pain sensations originating in visceral organs are often perceived as involving specific regions of the body surface innervated by the same spinal segments. Each surface region of perceived pain is labeled according to the organ at which the pain originates.

As in the posterior column pathway, the determination that an arriving stimulus is painful rather than cold, hot, or vibrating is determined by the projection of the sensation to different populations of second-order and third-order neurons. The ability to localize that stimulus depends on the stimulation of an appropriate area of the primary sensory cortex. Any abnormality along the pathway can result in inappropriate sensations. Consider these examples:

The Spinocerebellar Pathway

The cerebellum receives proprioceptive information about the position of skeletal muscles, tendons, and joints from the **spinocerebellar pathway** (Figure 15-5•). This information does not reach our conscious awareness. The axons of first-order sensory neurons synapse on interneurons in the dorsal gray horns of the spinal cord. The axons of these second-order neurons ascend in one of the spinocerebellar tracts. Axons that enter the left or right **posterior spinocerebellar tract** do not cross over to the opposite side of the spinal cord. These axons reach the cerebellar cortex via the inferior cerebellar peduncle of that side. Axons that do cross over to the opposite side of the spinal cord enter the left or right **anterior spinocerebellar tract.**[1] These axons reach the cerebellar cortex via the superior cerebellar peduncle.

Information carried by the axons of the spinocerebellar tracts reaches the *Purkinje cells* of the cerebellar cortex. ∞ *p. 461* These cells are complex and highly branched. Proprioceptive information from each part of the body is relayed to a specific portion of the cerebellar cortex. We will consider the integration of proprioceptive information and the role of the cerebellum in somatic motor control in a later section.

Table 15-1 reviews the three major sensory pathways we discussed in this section.

[1] Each anterior spinocerebellar tract also contains uncrossed axons from interneurons on the same side of the spinal cord as the stimulus.

Axon of first-order neuron

Second-order neuron

PONS

MEDULLA OBLONGATA

Arbor vitae

CEREBELLUM

Spinocerebellar pathway

Posterior spinocerebellar tract

Anterior spinocerebellar tract

SPINAL CORD

Proprioceptive input from Golgi tendon organs, muscle spindles, and joint capsules

•**FIGURE 15-5** The Spinocerebellar Pathway

TABLE 15-1 Principal Ascending (Sensory) Tracts in the

Pathway/Tract	Sensation(s)
POSTERIOR COLUMN PATHWAY	
Fasciculus gracilis	Proprioception and fine touch, pressure, and vibration from inferior half of body
Fasciculus cuneatus	Proprioception and fine touch, and ventral pressure, and vibration from superior half of body
SPINOTHALAMIC PATHWAY	
Lateral spinothalamic tracts	Pain and temperature
Anterior spinothalamic tracts	Crude touch and pressure
SPINOCEREBELLAR PATHWAY	
Posterior spinocerebellar tracts	Proprioception
Anterior spinocerebellar tracts	Proprioception

☑ As a result of pressure on her spinal cord, Jill cannot feel touch or pressure on her lower limbs. Which spinal tract is being compressed?

☑ Which spinal tract carries action potentials generated by nociceptors?

☑ Which cerebral hemisphere receives impulses conducted by the right fasciculus gracilis?

MOTOR PATHWAYS AND MOTOR CONTROL

Motor commands issued by the CNS are distributed by the somatic nervous system (SNS) and the autonomic nervous system (ANS). The somatic nervous system, also called the *somatic motor system,* controls the contractions of skeletal muscles. The output of the SNS is under voluntary control. The autonomic nervous system, or *visceral motor system,* controls visceral effectors, such as smooth muscle, cardiac muscle, and glands. In Chapter 16, we will examine the organization of the ANS; our interest here is the structure of the SNS. Throughout this discussion we will use the terms *motor neuron* and *motor control* to refer specifically to somatic motor neurons and pathways that control skeletal muscles.

Somatic motor pathways always involve at least two motor neurons: an **upper motor neuron,** whose cell body lies in a CNS processing center, and a **lower motor neuron,** whose cell body lies in a nucleus of the brain stem or spinal cord. The upper motor neuron

15

Spinal Cord and the Sensory Information They Provide

	Location of Neurons			
First-Order	**Second-Order**	**Third-Order**	**Final Destination**	**Site of Cross-Over**
Dorsal root ganglia of inferior half of the body; axons enter CNS in dorsal roots and join fasciculus gracilis	Nucleus gracilis of medulla oblongata; axons cross over before entering medial lemniscus	Ventral nuclei of thalamus	Primary sensory cortex on side opposite stimulus	Axons of second-order neurons before entering medial lemniscus
Dorsal root ganglia of superior half of the body; axons enter CNS in dorsal roots and join fasciculus cuneatus	Nucleus cuneatus of medulla oblongata; axons cross over before entering medial lemniscus	As above	As above	As above
Dorsal root ganglia; axons enter CNS in dorsal roots	Interneurons in posterior gray horn; axons enter lateral spinothalamic tract on opposite side	Ventral nuclei of thalamus	Primary sensory cortex on side opposite stimulus	Axons of second-order neurons at level of entry
As above	Interneurons in posterior gray horn; axons enter anterior spinothalamic tract on opposite side	As above	As above	As above
Dorsal root ganglia; axons enter CNS in dorsal roots	Interneurons in posterior gray horn; axons enter posterior spinocerebellar tract on same side	Not present	Cerebellar cortex on side of stimulus	None
As above	Interneurons in same spinal segment; axons enter anterior spinocerebellar tract on same or opposite side	Not present	Cerebellar cortex on side opposite (and side of) stimulus	Axons of most second-order neurons cross before entering tract

•FIGURE 15-6 **Descending Motor Tracts in the Spinal Cord.** A cross-sectional view indicating the locations of the major descending motor tracts that contain the axons of upper motor neurons. The origins and destinations of these tracts are listed in Table 15-2. Sensory tracts (shown in *Figure 15-2*) appear in dashed outline.

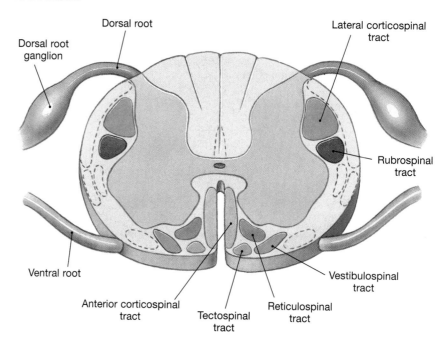

synapses on the lower motor neuron, which in turn innervates a single motor unit in a skeletal muscle. Activity in the upper motor neuron may facilitate or inhibit the lower motor neuron. Activation of the lower motor neuron triggers a contraction in the innervated muscle. Only the axon of the lower motor neuron extends outside the CNS. Destruction of or damage to a lower motor neuron eliminates voluntary and reflex control over the innervated motor unit.

Conscious and subconscious motor commands control skeletal muscles by traveling over three integrated motor pathways: the *corticospinal pathway,* the *medial pathway,* and the *lateral pathway.* Figure 15-6• indicates the positions of the associated motor tracts in the spinal cord. Activity within these motor pathways is monitored and adjusted by the cerebral nuclei and cerebellum. Their output stimulates or inhibits the activity of either (1) motor nuclei or (2) the primary motor cortex.

The Corticospinal Pathway

The **corticospinal pathway,** sometimes called the *pyramidal system* (Figure 15-7•) provides voluntary control over skeletal muscles. This system begins at the *pyramidal cells* of the primary motor cortex. ∞ *p. 447* The axons of these upper motor neurons descend into the brain stem and spinal cord to synapse on lower motor neurons that control skeletal muscles. In general, the corticospinal pathway is direct: The upper motor neurons synapse directly on the lower motor neurons. However, the corticospinal pathway also works indirectly, as it innervates centers of the medial and lateral pathways.

The corticospinal pathway contains three pairs of descending tracts: (1) the *corticobulbar tracts,* (2) the *lateral corticospinal tracts,* and (3) the *anterior corticospinal tracts.* These tracts enter the white matter of the internal capsule, descend into the brain stem, and emerge on either side of the mesencephalon as the *cerebral peduncles* (Figures 15-7, 14-13, and 14-16•, pp. 456, 460).

The Corticobulbar Tracts Axons in the **corticobulbar** (kor-ti-kō-BUL-bar) **tracts** (*bulbar,* brain stem) synapse on lower motor neurons in the motor nuclei of cranial nerves III, IV, V, VI, VII, IX, XI, and XII. The corticobulbar tracts provide conscious control over skeletal muscles that move the eye, jaw, and face and some muscles of the neck and pharynx. The corticobulbar tracts also innervate the motor centers of the medial and lateral pathways.

The Corticospinal Tracts Axons in the **corticospinal tracts** synapse on lower motor neurons in the anterior gray horns of the spinal cord. As they descend, the corticospinal tracts are visible along the ventral surface of the medulla oblongata as a pair of thick bands, the **pyramids.** Along the length of the pyramids, roughly 85 percent of the axons cross the midline (decussate) to enter the descending **lateral corticospinal tracts** on the opposite side of the spinal cord. The other 15 percent continue uncrossed along the spinal cord as the **anterior corticospinal tracts.** At the spinal segment it targets, an axon in the anterior corticospinal tract crosses over to the opposite side of the spinal cord in the anterior white commissure before synapsing on lower motor neurons in the anterior gray horns.

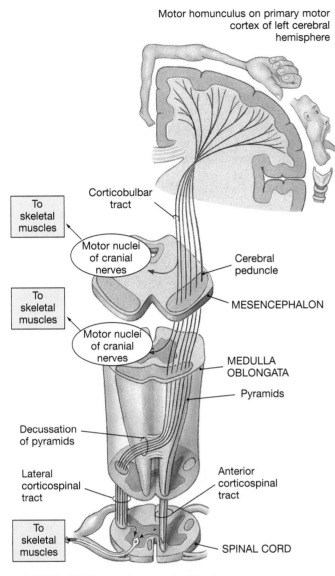

Motor homunculus on primary motor cortex of left cerebral hemisphere

Corticobulbar tract

To skeletal muscles

Motor nuclei of cranial nerves

Cerebral peduncle

MESENCEPHALON

To skeletal muscles

Motor nuclei of cranial nerves

MEDULLA OBLONGATA

Pyramids

Decussation of pyramids

Lateral corticospinal tract

Anterior corticospinal tract

To skeletal muscles

SPINAL CORD

•**FIGURE 15-7** **The Corticospinal Pathway.**
The corticospinal pathway originates at the primary motor cortex. The corticobulbar tracts end at the motor nuclei of cranial nerves on the opposite side of the brain. Most fibers in this pathway cross over in the medulla and enter the lateral corticospinal tracts; the rest descend in the anterior corticospinal tracts and cross over after reaching target segments in the spinal cord.

The Motor Homunculus The activity of pyramidal cells in a specific portion of the primary motor cortex will result in the contraction of specific peripheral muscles. The identities of the stimulated muscles depend on the region of motor cortex that is active. As in the primary sensory cortex, the primary motor cortex corresponds point by point with specific regions of the body. The cortical areas have been mapped out in diagrammatic form, creating a **motor homunculus.** Figure 15-7• shows the motor homunculus of the left cerebral hemisphere and the corticospinal pathway controlling skeletal muscles on the right side of the body.

The proportions of the motor homunculus are quite different from those of the actual body (Figure 15-7•), because the motor area devoted to a specific region of the cortex is proportional to the number of motor units involved in the region's control rather than to its actual size. As a result, the homunculus provides an indication of the degree of fine motor control available. For example, the hands, face, and tongue, all of which are capable of varied and complex movements, appear very large, whereas the trunk is relatively small. These proportions are similar to those of the sensory homunculus (Figure 15-3•, pp. 484–485). The sensory and motor homunculi differ in other respects because some highly sensitive regions, such as the sole of the foot, contain few motor units, and some areas with an abundance of motor units, such as the eye muscles, are not particularly sensitive.

Cerebral Palsy

The term **cerebral palsy** refers to a number of disorders that affect voluntary motor performance; they appear during infancy or childhood and persist throughout the life of the affected individual. The cause may be trauma associated with premature or unusually stressful birth, maternal exposure to drugs, including alcohol, or a genetic defect that causes the improper development of motor pathways. Problems with labor and delivery result from the compression or interruption of placental circulation or oxygen supplies. If the oxygen concentration in fetal blood declines significantly for as little as 5–10 minutes, CNS function can be permanently impaired. The cerebral cortex, cerebellum, cerebral nuclei, hippocampus, and thalamus are likely targets, producing abnormalities in motor skills, posture and balance, memory, speech, and learning abilities.

15

The Medial and Lateral Pathways

Several centers in the cerebrum, diencephalon, and brain stem may issue somatic motor commands as a result of processing performed at a subconscious level. These centers and their associated tracts were long known as the *extrapyramidal system (EPS),* because it was thought that they operated independent of, and in parallel to, the *pyramidal system* (corticospinal pathway). This classification scheme is both inaccurate and misleading, because motor control is integrated at all levels through extensive feedback loops and interconnections. It is more appropriate to group these nuclei and tracts in terms of their primary functions: The components of the **medial pathway** help control gross movements of the trunk and proximal limb muscles, whereas those of the **lateral pathway** help control the distal limb muscles that perform more-precise movements.

The medial and lateral pathways can modify or direct skeletal muscle contractions by stimulating, facilitating, or inhibiting lower motor neurons. It is important to note that the axons of upper motor neurons in the medial and lateral pathways synapse on the same lower motor neurons innervated by the corticospinal pathway. This means that the various motor pathways interact not only within the brain, through interconnections between the primary motor cortex and motor centers in the brain stem, but also through excitatory or inhibitory interactions at the level of the lower motor neuron.

The Medial Pathway The medial pathway is primarily concerned with the control of muscle tone and gross movements of the neck, trunk, and proximal limb muscles. The upper motor neurons of the medial pathway are located in the *vestibular nuclei,* the *superior* and *inferior colliculi,* and the *reticular formation.*

The vestibular nuclei receive information, over the vestibulocochlear nerve (N VIII), from receptors in the inner ear that monitor the position and movement of the head. These nuclei respond to changes in the orientation of the head, sending motor commands that alter the muscle tone, extension, and position of the neck, eyes, head, and limbs. The primary goal is to maintain posture and balance. The descending fibers in the spinal cord constitute the **vestibulospinal tracts.**

The superior and inferior colliculi are located in the *tectum,* or roof of the mesencephalon (Figure 14-16•, p. 460). The colliculi receive visual (superior) and auditory (inferior) sensations. Axons of upper motor neurons in the colliculi descend in the **tectospinal tracts.** These axons cross to the opposite side immediately, before descending to synapse on lower motor neurons in the brain stem or spinal cord. Axons in the tectospinal tracts direct reflexive changes in the position of the head, neck, and upper limbs in response to bright lights, sudden movements, or loud noises.

The reticular formation is a loosely organized network of neurons that extends throughout the brain stem. The reticular formation receives input from almost every ascending and descending pathway. It also has extensive interconnections with the cerebrum, the cerebellum, and brain stem nuclei. Axons of upper motor neurons in the reticular formation descend into the **reticulospinal tracts** without crossing to the opposite side. The effects of reticular formation stimulation are determined by the region stimulated. For example, the stimulation of upper motor neurons in one portion of the reticular formation produces eye movements, whereas the stimulation of another portion activates respiratory muscles.

The Lateral Pathway The lateral pathway is primarily concerned with the control of muscle tone and the more-precise movements of the distal parts of the limbs. The upper motor neurons of the lateral pathway lie within the *red nuclei* of the mesencephalon. Axons of upper motor neurons in the red nuclei cross to the opposite side of the brain and descend into the spinal cord in the **rubrospinal tracts** (*ruber,* red). In humans, the rubrospinal tracts are small and extend only to the cervical spinal cord. There they provide motor control over distal muscles of the upper limbs; normally, their role is insignificant as compared with that of the lateral corticospinal tracts. However, the rubrospinal tracts can be important in maintaining motor control and muscle tone in the upper limbs if the lateral corticospinal tracts are damaged.

Table 15-2 reviews the major motor tracts we discussed in this section.

The Cerebral Nuclei and Cerebellum

The cerebral nuclei and cerebellum are responsible for coordination and feedback control over muscle contractions, whether those contractions are consciously or subconsciously directed.

The Cerebral Nuclei The cerebral nuclei provide the background patterns of movement involved in voluntary motor activities. For example, they may control muscles that determine the background position of the trunk or limbs, or they may direct rhythmic cycles of movement, as in walking or running. The cerebral nuclei do not exert direct control over lower motor neurons; instead, they adjust the activities of upper motor neurons in the various motor pathways. The cerebral nuclei receive input from all portions of the cerebral cortex as well as from the substantia nigra.

The cerebral nuclei use two major pathways to adjust or establish patterns of movement:

1. One group of axons synapses on thalamic neurons, which then send their axons to the premotor cortex, the motor association area that directs activities of the primary motor cortex. This arrangement creates a feedback loop that changes the sensitivity of the pyramidal cells and alters the pattern of instructions carried by the corticospinal tracts.

2. A second group of axons synapses in the reticular formation, altering the excitatory or inhibitory output of the reticulospinal tracts.

Two distinct populations of neurons exist: one that stimulates neurons by releasing acetylcholine (ACh) and another that inhibits neurons by the release of gamma aminobutyric acid (GABA). Under normal conditions, the excitatory interneurons are kept inactive and the tracts leaving the cerebral nuclei have an inhibitory effect on upper motor neurons. In *Parkinson's disease* (Chapter 14), the excitatory neurons become more active, leading to problems with the voluntary control of movement. ∞ *p. 460*

If the primary motor cortex is damaged, the individual loses the ability to exert fine control over skeletal muscles. However, some voluntary movements can

TABLE 15-2   Principal Descending (Motor) Tracts in the Spinal Cord and the Functions of the Associated Nuclei in the Brain

Tract	Location of Upper Motor Neurons	Destination	Site of Cross-Over	Action
CORTICOSPINAL PATHWAY				
Corticobulbar tracts	Primary motor cortex (cerebral hemisphere)	Lower motor neurons of cranial nerve nuclei in brain stem	Brain stem	Conscious motor control of skeletal muscles
Lateral corticospinal tracts	As above	Lower motor neurons of anterior gray horns of spinal cord	Pyramids of medulla oblongata	As above
Anterior corticospinal tracts	As above	As above	Level of lower motor neuron	As above
MEDIAL PATHWAY				
Vestibulospinal tracts	Vestibular nuclei (at border of pons and medulla oblongata)	As above	Uncrossed	Subconscious regulation of balance and muscle tone
Tectospinal tracts	Tectum (mesencephalon: superior and inferior colliculi)	Lower motor neurons of anterior gray horns (cervical spinal cord only)	Brain stem (mesencephalon)	Subconscious regulation of eye, head, neck, and upper limb position in response to visual and auditory stimuli
Reticulospinal tracts	Reticular formation (network of nuclei in brain stem)	Lower motor neurons of anterior gray horns of spinal cord	Uncrossed	Subconscious regulation of reflex activity
LATERAL PATHWAY				
Rubrospinal tracts	Red nuclei of mesencephalon	As above	Brain stem (mesencephalon)	Subconscious regulation of upper limb muscle tone and movement

15

still be controlled by the cerebral nuclei. In effect, the medial and lateral pathways function as they usually do, but the corticospinal pathway cannot fine-tune the movements. For example, after damage to the primary motor cortex, the cerebral nuclei can still receive information about planned movements from the prefrontal cortex and can perform preparatory movements of the trunk and limbs. But, because the corticospinal pathway is inoperative, precise movements of the forearms, wrists, and hands cannot occur. An individual in this condition can stand, maintain balance, and even walk, but all movements are hesitant, awkward, and poorly controlled.

The Cerebellum The cerebellum monitors proprioceptive (position) sensations, visual information from the eyes, and vestibular (balance) sensations from the inner ear as movements are under way. Axons relaying proprioceptive information reach the cerebellar cortex in the spinocerebellar tracts. Visual information is relayed from the superior colliculi, and balance information is relayed from the vestibular nuclei. The output of

the cerebellum affects upper motor neuron activity in the corticospinal, medial, and lateral pathways.

All motor pathways send information to the cerebellum when motor commands are issued. As the movement proceeds, the cerebellum monitors proprioceptive and vestibular information and compares the arriving sensations with those experienced during previous movements. It then adjusts the activities of the upper motor neurons involved. In general, any voluntary movement begins with the activation of far more motor units than are required—or even desirable. The cerebellum provides the necessary inhibition, reducing the number of motor commands to an efficient minimum. As the movement proceeds, the pattern and degree of inhibition change, producing the desired result.

The patterns of cerebellar activity are learned by trial and error, over many repetitions. Many of the basic patterns are established early in life; examples include the fine balancing adjustments you make while standing and walking. The ability to fine-tune a complex pattern of movement improves with practice, until the movements become fluid and automatic. Consider the relaxed, smooth movements of acrobats, golfers, and sushi chefs. These people move without thinking about the details of their movements. This ability is important, because when you concentrate on voluntary control, the rhythm and pattern of the movement usually fall apart as your primary motor cortex starts overriding the commands of the cerebral nuclei and cerebellum.

Amyotrophic Lateral Sclerosis

Amyotrophic lateral sclerosis (ALS), formerly known as *Lou Gehrig's disease,* is a progressive, degenerative disorder that affects motor neurons in the spinal cord, brain stem, and cerebral hemispheres. The degeneration affects both upper and lower motor neurons. Because a motor neuron and its dependent muscle fibers are so intimately related, the destruction of CNS neurons causes atrophy of the associated skeletal muscles. The specificity of this disease is remarkable: Motor neurons are destroyed, but sensory neurons, sensory pathways, and intellectual functions remain unaffected.

Roughly 90 percent of ALS cases have no known cause. About 10 percent of cases can be linked to genetic factors, but the nature of the genetic abnormality is uncertain. Most people with ALS do not develop symptoms until after age 50, although inherited forms may affect individuals in their early teens. Affected individuals typically die within 5 years after symptoms appear. There is no effective treatment. AM *Amyotrophic Lateral Sclerosis*

☑ For what anatomical reason does the left side of the brain control motor function on the right side of the body?

☑ An injury involving the superior portion of the motor cortex would affect which region of the body?

☑ What effect would increased stimulation of the motor neurons of the red nucleus have on muscle tone?

Levels of Processing and Motor Control

When you touch a hot stove, you withdraw your hand before you are consciously aware that you have burned yourself. Voluntary motor responses, such as shaking your hand, stepping back, and crying out, occur somewhat later. This is a general pattern: Spinal and cranial reflexes provide rapid, involuntary, pre-programmed responses that preserve homeostasis over the short term. Voluntary responses are more complex, but they require more time to prepare and execute.

Figure 15-8• reviews the primary sites of somatic motor control. Cranial and spinal reflexes control the most basic motor activities. Integrative centers in the brain perform more-elaborate processing, and as we move from the medulla oblongata to the cerebral cortex, the motor patterns become increasingly complex and variable. The most complex and variable motor activities are directed by the primary motor cortex of the cerebral hemispheres.

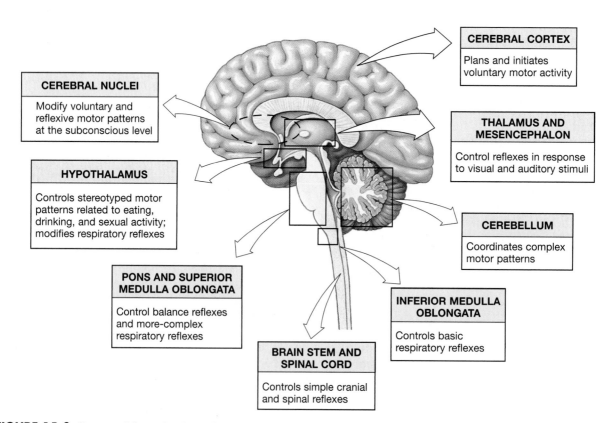

CEREBRAL NUCLEI
Modify voluntary and reflexive motor patterns at the subconscious level

HYPOTHALAMUS
Controls stereotyped motor patterns related to eating, drinking, and sexual activity; modifies respiratory reflexes

PONS AND SUPERIOR MEDULLA OBLONGATA
Control balance reflexes and more-complex respiratory reflexes

BRAIN STEM AND SPINAL CORD
Controls simple cranial and spinal reflexes

CEREBRAL CORTEX
Plans and initiates voluntary motor activity

THALAMUS AND MESENCEPHALON
Control reflexes in response to visual and auditory stimuli

CEREBELLUM
Coordinates complex motor patterns

INFERIOR MEDULLA OBLONGATA
Controls basic respiratory reflexes

•**FIGURE 15-8** **Centers of Somatic Motor Control.** Somatic motor control involves many regions of the brain. As we ascend toward the cerebral cortex, the motor patterns generated become increasingly complex and variable.

During development, the spinal and cranial reflexes are the first to appear. More-complex reflexes and motor patterns develop as CNS neurons multiply, enlarge, and interconnect. The process proceeds relatively slowly, as billions of neurons establish trillions of synaptic connections. At birth, neither the cerebral nor the cerebellar cortex is fully functional. The behavior of newborn infants is directed primarily by centers in the diencephalon and brain stem.

Among the anatomical factors that contribute to the maturation of the CNS and the refinement of motor skills are the following:

- The number of neurons in the cerebral cortex continues to increase until at least age 1.

- The brain as a whole grows in size and complexity until at least age 4.

- The myelination of CNS axons continues at least until age 1–2, and peripheral myelination may continue through puberty. Myelination reduces the delay between stimulus and response and improves motor control.

As these events occur, cortical neurons continue to establish new synaptic interconnections that will have a long-term effect on mental capabilities.

Anencephaly

Although it may sound strange, physicians generally take a newborn infant into a dark room and shine a light against the skull. They are checking for **anencephaly** (an-en-SEF-uh-lē), a rare condition in which the brain fails to develop at levels above the mesencephalon or lower diencephalon. In most such cases, the cranium also fails to develop, and diagnosis is easy, but in some cases, a normal skull forms. In such instances, the cranium is empty and translucent enough to transmit light. Unless the condition is discovered right away, the parents may take the infant home, unaware of the problem. All the normal behavior patterns expected of a newborn are present, including suckling, stretching, yawning, crying, kicking, sticking fingers in the mouth, and tracking movements with the eyes. However, death will occur naturally within days or months. This tragic condition provides a striking demonstration of the role of the brain stem in controlling complex motor patterns. During normal development, these patterns become incorporated into variable and versatile behaviors as control centers and analytical centers appear in the cerebral cortex.

MONITORING BRAIN ACTIVITY: THE ELECTROENCEPHALOGRAM

The primary sensory cortex and the primary motor cortex have been mapped by direct stimulation in patients undergoing brain surgery. The functions of other regions of the cerebrum can be revealed by the behavioral changes that follow localized injuries or strokes, and the activities of specific regions can be examined by a PET scan or sequential MRI scans.

The electrical activity of the brain is commonly monitored to assess brain activity. Neural function depends on electrical events within the cell membrane of neurons. The brain contains billions of neurons, and their activity generates an electrical field that can be measured by placing electrodes on the brain or on the outer surface of the skull. The electrical activity changes constantly, as nuclei and cortical areas are stimulated or quiet down. A printed report of the electrical activity of the brain is called an **electroencephalogram (EEG).** The electrical patterns observed are called *brain waves*.

Typical brain waves are shown in Figure 15-9a–d•. **Alpha waves** occur in the brains of healthy, awake adults who are resting with their eyes closed (Figure 15-9e•). Alpha waves disappear during sleep, but they also vanish when the individual begins to concentrate on some specific task. During attention to stimuli or tasks, alpha waves are replaced by higher-frequency **beta waves.** Beta waves are typical of individuals who are either concentrating on a task, under stress, or in a state of psychological tension. **Theta waves** may appear transiently during sleep in normal adults but are most often observed in children and in intensely frustrated adults. The presence of theta waves under other circumstances may indicate the presence of a brain disorder, such as a tumor. **Delta waves** are very-large-amplitude, low-frequency waves. They are normally seen during deep sleep in individuals of all ages. Delta waves are also seen in the brains of infants (in whom cortical development is still incomplete) and in the awake adult when a tumor, vascular blockage, or inflammation has damaged portions of the brain.

Electrical activity in the two hemispheres is generally synchronized by a "pacemaker" mechanism that appears to involve the thalamus. Asynchrony between the hemispheres can therefore be used to detect localized damage or other cerebral abnormalities. For example, a tumor or injury affecting one hemisphere typically changes the pattern in that hemisphere, and the patterns of the two hemispheres are no longer aligned. A **seizure** is a temporary cerebral disorder accompanied by abnormal movements, unusual sensations, inappropriate behavior, or some combination of these symptoms. Clinical conditions characterized by seizures are known as seizure disorders, or *epilepsies*. Seizures of all kinds are accompanied by a marked change in the pattern of electrical activity monitored in an electroencephalogram. The change begins in one portion of the cerebral cortex but may subsequently spread across the entire cortical surface, like a wave on the surface of a pond.

The nature of the symptoms produced depends on the region of the cortex involved. If a seizure affects the primary motor cortex, movements will occur; if it affects the auditory cortex, the individual will hear strange sounds. AM *Seizures and Epilepsies*

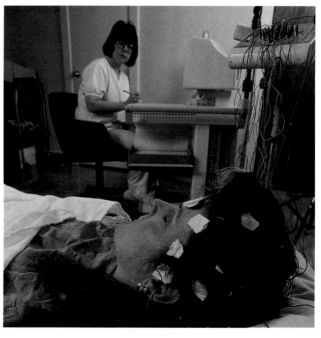

(e)

•**FIGURE 15-9 Brain Waves. (a)** Alpha waves are characteristic of healthy resting adults. **(b)** Beta waves typically accompany intense concentration. **(c)** Theta waves are seen in children and in frustrated adults. **(d)** Delta waves occur during deep sleep and in certain pathological states. (These four waves are not drawn to the same scale.) **(e)** A patient wired for EEG monitoring.

HIGHER-ORDER FUNCTIONS

Higher-order functions share the following three characteristics:

1. The cerebral cortex is required for their performance, and they involve complex interactions among areas of the cortex and between the cerebral cortex and other areas of the brain.
2. They involve both conscious and unconscious information processing.
3. They are not part of the programmed "wiring" of the brain; therefore, the functions are subject to modification and adjustment over time.

In Chapter 14, we considered functional areas of the cerebral cortex and the regional specializations of the left and right cerebral hemispheres. ∞ *pp. 447–451* In this section, we consider the mechanisms of memory and learning and describe the neural interactions responsible for consciousness, sleep, and arousal. In the next section, we will provide an overview of brain chemistry and its effects on behavior and personality.

MEMORY

What was the topic of the last sentence you read? What do your parents look like? What is your Social Security number? When did Columbus reach the New World? What does a red traffic light mean? What does a hot dog taste like? Answering these questions involves accessing *memories,* stored bits of information gathered through experience. **Fact memories** are specific bits of information, such as the color of a stop sign or the smell of a perfume. **Skill memories** are learned motor behaviors. For example, you can probably remember how to light a match or open a screw-top jar. With repetition, skill memories become incorporated at the unconscious level. Examples include the complex motor patterns involved in skiing, playing the violin, and similar activities. Skill memories related to programmed behaviors, such as eating, are stored in appropriate portions of the brain stem. Complex skill memories involve the integration of motor patterns in the cerebral nuclei, cerebral cortex, and cerebellum.

Two classes of memories are identified. **Short-term memories,** or *primary memories,* do not last long, but while they persist, the information can be recalled immediately. Primary memories contain small bits of information, such as a person's name or a telephone number. Repeating a phone number or other bit of information reinforces the original short-term memory and helps ensure its conversion to a long-term memory. **Long-term memories** last much longer, in some cases for an entire lifetime. The conversion from short-term to long-term memory is called **memory consolidation.** There are two types of long-term memory: (1) *Secondary memories* are long-term memories that fade with time and may require considerable effort to recall. (2) *Tertiary memories* are long-term memories that seem to be part of consciousness, such as your name or the contours of your own body. Proposed re-

•FIGURE 15-10 Memory
Storage. Steps in the storage of
memories and the conversion of a
short-term memory to a long-
term memory. Long-term
memories can be temporarily lost,
as often occurs after a mild
stroke. Damage resulting from
more-severe strokes or from
trauma or tumors can lead to
permanent memory loss.

lationships among these memory classes are diagrammed in Figure 15-10•.

Brain Regions Involved in Memory Consolidation and Access

The amygdaloid body and the hippocampus, two components of the limbic system (Figure 14-11•, p. 453),
are essential to memory consolidation. Damage to the
hippocampus leads to an inability to convert short-
term memories to new long-term memories, although
existing long-term memories remain intact and accessible. Tracts leading from the amygdaloid body to the
hypothalamus may link memories to specific emotions.

The **nucleus basalis,** a cerebral nucleus near the
diencephalon, plays an uncertain role in memory
storage and retrieval. Tracts connect this nucleus with
the hippocampus, amygdaloid body, and all areas of
the cerebral cortex. Damage to this nucleus is associated with changes in emotional states, memory, and
intellectual function (as we will see in the discussion
of Alzheimer's disease later in this chapter).

Most long-term memories are stored in the cerebral cortex. Conscious motor and sensory memories
are referred to the appropriate association areas. For
example, visual memories are stored in the visual association area, and memories of voluntary motor activity are stored in the premotor cortex. Special
portions of the occipital and temporal lobes are crucial
to the memories of faces, voices, and words. In at least
some cases, a specific memory probably depends on
the activity of a single neuron. For example, in one
portion of the temporal lobe an individual neuron responds to the sound of one word and ignores others.
A specific neuron may also be activated by the proper combination of sensory stimuli associated with a
particular individual, such as your grandmother. As a
result, these neurons are called "grandmother cells."

Information on one subject is parceled out to many
different regions of the brain. Your memories of cows
are stored in the visual association area (what a cow
looks like, how the letters *c-o-w* mean "cow"), the auditory association area (the "moo" sound and how the
word *cow* sounds), the speech center (how to say the
word *cow*), and the frontal lobes (how big cows are,
what they eat). Related information, such as how you
feel about cows and what milk tastes like, is stored in
other locations. If one of those storage areas is damaged, your memory will be incomplete in some way.
How these memories are accessed and assembled on
demand remains a mystery.

Cellular Mechanisms of Memory Formation and Storage

Memory consolidation at the cellular level involves
anatomical and physiological changes in neurons and
synapses. For legal, ethical, and logistical reasons, it is
not possible to perform much research on these mechanisms with human subjects. Research on other animals, commonly those with relatively simple nervous
systems, has indicated that the following mechanisms
may be involved:

- *Increased neurotransmitter release.* A synapse that is
  frequently active increases the amount of neurotransmitter that it stores, and it releases more on each stimulation. The more neurotransmitter released, the
  greater the effect on the postsynaptic neuron.

- *Facilitation at synapses.* When a neural circuit is repeatedly activated, the synaptic terminals begin releasing
  neurotransmitter in small quantities continuously.
  The neurotransmitter binds to receptors on the postsynaptic membrane, producing a graded depolarization that brings the membrane closer to threshold. The
  facilitation that results affects all neurons in the circuit.

15

■ *The formation of additional synaptic connections.* Evidence indicates that when one neuron repeatedly communicates with another, the axon tip branches and forms additional synapses on the postsynaptic neuron. As a result, the presynaptic neuron will have a greater effect on the transmembrane potential of the postsynaptic neuron.

These processes create anatomical changes that facilitate communication along a specific neural circuit. This facilitated communication is thought to be the basic method of memory storage. A single circuit that corresponds to a single memory has been called a **memory engram.** This definition is based on function rather than structure; we know too little about the organization and storage of memories to be able to describe the neural circuits involved.

Efficient conversion of a short-term memory into a memory engram takes time, usually at least an hour. Whether that conversion will occur depends on several factors, including the nature, intensity, and frequency of the original stimulus. Very strong, repeated, or exceedingly pleasant or unpleasant events are most likely to be converted to long-term memories. Drugs that stimulate the CNS, such as caffeine and nicotine, may enhance memory consolidation through facilitation; we discussed the membrane effects of those drugs in Chapter 12. ∞ *pp. 390, 394*

The hippocampus plays a key role in the consolidation of memories. The mechanism, which remains unknown, is linked to the presence of *NMDA (N-methyl D-aspartate) receptors,* which are chemically gated calcium channels. When activated by the neurotransmitter *glycine,* the gates open and calcium enters the cell. Blocking NMDA receptors in the hippocampus prevents long-term memory formation.

Amnesia is the loss of memory as a result of disease or trauma. The type of memory loss depends on the specific regions of the brain affected. For example, damage to the auditory association areas can make it difficult to remember sounds. AM *Amnesia*

☑ Toward the end of her A & P test, Tina begins to feel a great deal of stress. Which type of brain wave pattern would you expect her to exhibit?

☑ After suffering a head injury in an automobile accident, David has difficulty comprehending what he hears or reads. This symptom might indicate damage to which portion of his brain?

☑ As you recall facts while you take your A & P test, which type of memory are you using?

CONSCIOUSNESS

A conscious individual is alert and attentive; an unconscious individual is not. The difference is obvious, but there are many gradations of both the conscious and unconscious states. For example, a healthy conscious person can be almost asleep, wide awake, or high-strung and jumpy; a healthy sleeping person can be dozing lightly or so deeply asleep that he or she cannot easily be awakened. Healthy individuals cycle between the alert, conscious state and the asleep state each day. The degree of wakefulness at any moment is an indication of the level of ongoing CNS activity. When CNS function becomes abnormal or depressed, the state of wakefulness can be affected. For example, in a state of *coma,* the individual is unconscious and cannot be awakened, even by strong stimuli. As a result, clinicians are quick to note any change in the responsiveness of comatose patients. AM *Altered States*

Sleep

Conscious implies a state of awareness of and attention to external events and stimuli. *Unconscious* can imply a number of conditions, ranging from the deep, unresponsive state induced by anesthesia before major surgery to the light, drifting "nod" that occasionally plagues students who are reading anatomy and physiology textbooks. You are considered to be asleep when you are unconscious but can still be awakened by normal sensory stimuli.

Two levels of sleep are recognized, each typified by characteristic patterns of brain wave activity (Figure 15-11●):

1. In **deep sleep,** also called *slow wave* or *non-REM (NREM) sleep,* your entire body relaxes and activity at the cerebral cortex is at a minimum. Heart rate, blood pressure, respiratory rate, and energy utilization decline by up to 30 percent.

2. During **rapid eye movement (REM) sleep,** active dreaming occurs, accompanied by changes in your blood pressure and respiratory rate. Although the EEG shows traces resembling those of the awake state, you become even less receptive to outside stimuli than in deep sleep, and muscle tone decreases markedly. Intense inhibition of somatic motor neurons probably prevents you from physically producing the responses you envision during the dream sequence. The neurons controlling the eye muscles escape this inhibitory influence, and your eyes move rapidly as the dream events unfold.

Periods of REM and deep sleep alternate throughout the night, beginning with a period of deep sleep that lasts about an hour and a half. Rapid eye movement periods initially average about 5 minutes in length, but they gradually increase to about 20 minutes over an 8-hour night. Each night we probably spend less than 2 hours dreaming, but variation among individuals is significant. For example, children devote more time to REM sleep than do adults, and extremely tired individuals have very short and infrequent REM periods.

Awake

REM sleep

Deep (slow
wave) sleep

(a)

Awake

REM sleep

Transition
period

Deep sleep

10:00 P.M.   Midnight   2:00 A.M.   4:00 A.M.   6:00 A.M.

Time

(b)

•FIGURE 15-11 Levels of Sleep. (a) An EEG from the
awake, REM, and deep (slow wave) sleep states. The EEG
pattern during REM sleep resembles the alpha waves typical
of awake adults. **(b)** The typical pattern of oscillation
between sleep stages of a healthy young adult during a single
night's sleep.

connections with the sensory, motor, and integrative
nuclei and pathways all along the brain stem.

The Reticular Activating System   Your state of consciousness is determined by complex interactions between your brain stem and cerebral cortex. One of the
most important brain stem components is a diffuse
network in the reticular formation known as the **reticular activating system (RAS).** ∞ *p. 460* This network
extends from the medulla oblongata to the mesencephalon (Figure 15-12•). The output of the RAS projects to thalamic nuclei that influence large areas of
the cerebral cortex. When the RAS is inactive, so is
the cerebral cortex; stimulation of the RAS produces
a widespread activation of the cerebral cortex.

The mesencephalic portion of the RAS appears to
be the center of the system. Stimulation of this area produces the most pronounced and long-lasting effects on
the cerebral cortex. Stimulating other portions of the
RAS seems to have an effect only to the degree that it
changes the activity of the mesencephalic region. The
greater the stimulation to the mesencephalic region of
the RAS, the more alert and attentive the individual will
be to incoming sensory information. The thalamic nuclei
associated with the RAS may also play an important
role in focusing attention on specific mental processes.

Sleep may be ended by any stimulus sufficient to
activate the reticular formation and RAS. Arousal

15

Sleep produces only minor changes in the physiological activities of other organs and systems, and none
of these changes appear to be essential to normal function. The significance of sleep must lie in its impact on
the CNS, but the physiological or biochemical basis
remains to be determined. We do know that protein
synthesis in neurons increases during sleep. Extended
periods without sleep will lead to a variety of disturbances in mental function. Roughly 25 percent of the
U.S. population experiences some form of a *sleep disorder.* Examples of such disorders include abnormal
patterns or duration of REM sleep or unusual behaviors performed while sleeping, such as sleepwalk-
ing. In some cases, these problems begin to affect the
individual's conscious activities. Slowed reaction
times, irritability, and behavioral changes may result.
[AM] *Sleep Disorders*

Arousal

Arousal, or awakening from sleep, appears to be one
of the functions of the reticular formation. The reticular formation is especially well suited for providing
"watchdog" services, because it has extensive inter-

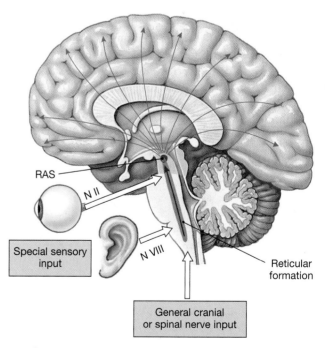

RAS

N II

Special sensory
input

N VIII

General cranial
or spinal nerve input

Reticular
formation

•FIGURE 15-12 The Reticular Activating System. The
mesencephalic headquarters of the reticular formation
receives collateral inputs from a variety of sensory pathways.
Stimulation of this region produces arousal and heightened
states of attentiveness.

15

occurs rapidly, but the effects of a single stimulation of the RAS last less than a minute. Thereafter, consciousness can be maintained by positive feedback, because activity in the cerebral cortex, cerebral nuclei, and sensory and motor pathways will continue to stimulate the RAS.

After many hours of activity, the reticular formation becomes less responsive to stimulation. The individual becomes less alert and more lethargic. The precise mechanism remains unknown, but neural fatigue probably plays a relatively minor role in the reduction of RAS activity. Evidence suggests that the regulation of awake–asleep cycles involves an interplay between brain stem nuclei that use different neurotransmitters. One group of nuclei stimulates the RAS with norepinephrine and maintains the awake, alert state. The other group, which depresses RAS activity with serotonin, promotes deep sleep. These "dueling" nuclei are located in the brain stem.

BRAIN CHEMISTRY AND BEHAVIOR

We discussed the general distribution of neurotransmitters in the central nervous system in Chapter 12. ∞ *pp. 387–391* Of the roughly 50 known neurotransmitters, the most important and widespread are acetylcholine (ACh), norepinephrine (NE), glutamate, aspartate, glycine, gamma aminobutyric acid (GABA), dopamine, serotonin, histamine, Substance P, and the enkephalins and endorphins (see Table 12-6, p. 391). The distribution of neurons that release each of these compounds varies from region to region in the brain. However, tracts originating at each nucleus distribute these neurotransmitters throughout the CNS.

Changes in the normal balance between two or more neurotransmitters can also profoundly affect brain function. For example, an interplay between serotonin and norepinephrine appears to be involved in the regulation of awake–asleep cycles. Another example deals with *Huntington's disease.* The primary problem in this inherited disease is the destruction of ACh-secreting and GABA-secreting neurons in the cerebral nuclei. The reason for this destruction is unknown. Symptoms appear as the cerebral nuclei and frontal lobes slowly degenerate. An individual with Huntington's disease has difficulty controlling movements, and intellectual abilities gradually decline. [AM] *Huntington's Disease*

In many cases, the importance of a specific neurotransmitter has been revealed during the search for a mechanism for the effects of administered drugs. Three examples demonstrate patterns that are emerging:

1. An extensive network of tracts delivers serotonin to nuclei and higher centers throughout the brain. Compounds that enhance the effects of serotonin produce hallucinations; for example, *LSD (lysergic acid diethylamide)* is a powerful hallucinogenic drug that activates serotonin receptors in the brain stem, hypothalamus, limbic system, and spinal cord. Compounds that inhibit serotonin production or block its action cause severe depression and anxiety. The most effective antidepressive drug now in use, *fluoxetine (Prozac®),* slows the removal of serotonin at synapses, causing an increase in serotonin concentrations at the postsynaptic membrane.

2. Norepinephrine is another important neurotransmitter with pathways throughout the brain. Drugs that stimulate NE release cause exhilaration, and those that depress NE release cause depression. One inherited form of depression has been linked to a defective enzyme involved in NE synthesis.

3. Disturbances in dopamine transmission have been linked to several neurological disorders. Inadequate dopamine production causes the motor problems of Parkinson's disease. Excessive production of dopamine may be associated with **schizophrenia,** a psychological disorder marked by pronounced disturbances of mood, thought patterns, and behavior. Amphetamines, or "speed," stimulate dopamine secretion and, in large doses, can produce symptoms resembling those of schizophrenia. (They also affect other neurotransmitter systems, producing a variety of changes in CNS function.) [AM] *Pharmacology and Drug Abuse*

PERSONALITY AND OTHER MYSTERIES

The basic pathways and components of the CNS have been traced, and memory acquisition, learning, and associated mental processes can now be manipulated and explored in animal experiments. But the truly remarkable human characteristics of self-awareness and personality remain phenomena without discrete anatomical foundations. That is not to say that an anatomical basis does not exist, just that self-awareness and personality are more characteristics of the brain as an integrated system than functions of any specific nucleus or region. As a result, the origins of human consciousness and personality may remain a mystery for the immediate future. But our knowledge of the anatomy and biochemistry of specific tracts, nuclei, and regions will continue to provide useful clinical information.

In recent years, significant advances have been made by the correlation of anatomical or physiological deficits with observed behavioral disorders. In some cases, such as the use of L-dopa to treat Parkinson's disease, this procedure has led to new and effective forms of treatment.

AGING AND THE NERVOUS SYSTEM

The aging process affects all body systems, and the nervous system is no exception. Anatomical and physiological changes begin shortly after maturity (probably by age 30) and accumulate over time. Although an estimated 85 percent of people above age 65 lead relatively normal lives, they exhibit noticeable changes in mental performance and in CNS function.

AGE-RELATED ANATOMICAL AND FUNCTIONAL CHANGES

Common age-related anatomical changes in the nervous system include the following:

- *A reduction in brain size and weight.* This reduction results primarily from a decrease in the volume of the cerebral cortex. The brains of elderly individuals have narrower gyri and wider sulci than do those of young persons, and the subarachnoid space is larger.

- *A reduction in the number of neurons.* Brain shrinkage has been linked to a loss of cortical neurons, although evidence exists that neuronal loss does not occur (at least to the same degree) in brain stem nuclei.

- *A decrease in blood flow to the brain.* With age, fatty deposits gradually accumulate in the walls of blood vessels. Just as a kink in a garden hose or a clog in a drain reduces water flow, these deposits reduce the rate of blood flow through arteries. (This process, called *arteriosclerosis,* affects arteries throughout the body; we shall discuss it further in Chapter 21.) The reduction in blood flow may not cause an acute cerebral crisis, but it does increase the chances that a stroke will occur.

- *Changes in the synaptic organization of the brain.* In many areas, the number of dendritic branches, spines, and interconnections appears to decrease. Synaptic connections are lost, and the rate of neurotransmitter production declines.

- *Intracellular and extracellular changes in CNS neurons.* Many neurons in the brain accumulate abnormal intracellular deposits, including lipofuscin and neurofibrillary tangles. **Lipofuscin** is a granular pigment that has no known function. **Neurofibrillary tangles** are masses of neurofibrils that form dense mats inside the cell body and axon. **Plaques** are extracellular accumulations of **amyloid,** an unusual fibrillar protein, surrounded by abnormal dendrites and axons. The significance of these histological abnormalities is unknown. Evidence indicates that they appear in all aging brains, but, when present in excess, they seem to be associated with clinical abnormalities.

These anatomical changes are linked to functional changes. In general, neural processing becomes less efficient with age. Memory consolidation typically becomes more difficult, and secondary memories, especially those of the recent past, become harder to access. The sensory systems of the elderly—notably, hearing, balance, vision, smell, and taste—become less acute. Lights must be brighter, sounds louder, and smells stronger before they are perceived. Reaction times are slowed, and reflexes—even some withdrawal reflexes—weaken or disappear. The precision of motor control decreases, and it takes longer to perform a given motor pattern than it did 20 years earlier.

For roughly 85 percent of the elderly population, these changes do not interfere with their abilities to function in society. But for as yet unknown reasons, some elderly individuals become incapacitated by progressive CNS changes. By far the most common such incapacitating condition is *Alzheimer's disease.*

ALZHEIMER'S DISEASE

Alzheimer's disease is a progressive disorder characterized by the loss of higher-order cerebral functions. It is the most common cause of **senile dementia,** or *senility.* Symptoms generally appear at 50–60 years of age, although the disease occasionally affects younger individuals. Alzheimer's disease has widespread impact. An estimated 2 million people in the United States, including roughly 15 percent of those over age 65, have some form of the condition, and it causes approximately 100,000 deaths each year.

The link is uncertain, but the areas containing plaques and neurofibrillary tangles are the same regions involved with memory, emotions, and intellectual function. It remains to be determined whether these deposits cause Alzheimer's disease or are secondary signs of ongoing metabolic alterations with an environmental, hereditary, or infectious basis.

Genetic factors certainly play a major role. The late-onset form of Alzheimer's disease has been traced to a gene on chromosome 19 that codes for proteins involved in cholesterol transport. Less than 5 percent of people with Alzheimer's disease have the early-onset form; these individuals develop the condition before age 50. The early-onset form of Alzheimer's disease has been linked to genes on chromosomes 21 and 14. Interestingly, the majority of individuals with *Down syndrome* develop Alzheimer's disease relatively early in life. (Down syndrome results from an extra copy of chromosome 21; we shall discuss this condition further in Chapter 29.) [AM] *Alzheimer's Disease*

☑ You are asleep. What would happen to you if your reticular activating system were suddenly stimulated?

☑ What would be the effect of a drug that substantially increases the amount of serotonin released in the brain?

☑ One of the problems associated with aging is difficulty in recalling things or even a total loss of memory. What are some possible reasons for these changes?

15

[EXPLORE]
[MediaLab]

Beginning with sensory and motor pathways and moving through complex integrative capacities such as memory and behavior, you have learned how the brain and spinal cord interact with the peripheral nervous system to maintain homeostasis. To improve your understanding of integrative functions, use the resources available on both the CD-ROM and the Companion Website. The CD-ROM includes animations and tutorials; the website builds on those resources with text content and interactive problems as well as communication avenues for sharing ideas with your class. To complete the following Web Explorations, visit the MediaLab in Chapter 15 of your Companion Website.

Web Exploration 1
Estimated time for completion = 5 minutes

Four of the principal ascending (sensory) tracts in the spinal cord are described as carrying the sensation of proprioception to the primary sensory cortex or to the cerebellar cortex. Review the subsections "The Posterior Column Pathway" and "The Spinocerebellar Pathway" in the text. Then select the keyword **PROPRIOCEPTION** on the Companion Website to discover how discriminatory the information carried in these tracts really is. Perform the simple tasks described on the site, and record your results. Analyze your results in light of what you know about the integrative functions of the nervous system. How much do you rely on your visual system?

Web Exploration 2
Estimated time for completion = 10 minutes

Neuroscience is not a static field of knowledge; advances and new discoveries are being described at an amazing rate. To gain a historical perspective on neuroscience, select the keyword **BREAKTHROUGHS** on the Companion Website. Read the article titled "Brain Breakthroughs: Looking Back on the Decade of the Brain." Think about the discoveries of the past decade. What can be and is being done with the knowledge gained in the past 10 years? Where will research take us in the next decade? If you wish, share your thoughts and concerns with your classmates by using the communication tools on the Companion Website.

CHAPTER REVIEW

SELECTED CLINICAL TERMINOLOGY
Terms Discussed in This Chapter

Alzheimer's disease: A progressive disorder marked by the loss of higher-order cerebral functions. *(p. 499 and AM)*

amnesia: A temporary or permanent loss of memory. *(p. 496 and AM)*

amyotrophic lateral sclerosis (ALS): A progressive, degenerative disorder affecting motor neurons of the spinal cord, brain stem, and cerebral hemispheres. *(p. 492 and AM)*

anencephaly (an-en-SEF-a-lē): A rare condition in which the brain fails to develop at levels above the mesencephalon or inferior part of the diencephalon. *(p. 493)*

cerebral palsy: A disorder that affects voluntary motor performance and arises in infancy or childhood as a result of prenatal trauma, drug exposure, or a congenital defect. *(p. 489)*

electroencephalogram (EEG): A printed record of the brain's electrical activity over time. *(p. 493)*

epilepsy: One of more than 40 conditions characterized by a recurring pattern of seizures over extended periods. *(p. 493 and AM)*

Huntington's disease: An inherited disease marked by a progressive deterioration of mental abilities and by motor disturbances. *(p. 498 and AM)*

schizophrenia: A psychological disorder marked by pronounced disturbances of mood, thought patterns, and behavior. *(p. 498)*

seizure: A temporary disorder of cerebral function, accompanied by abnormal movements, unusual sensations, and/or inappropriate behavior; includes *focal seizures, temporal lobe seizures, convulsive seizures, generalized seizures, grand mal seizures,* and *petit mal seizures.* *(p. 493 and AM)*

senile dementia, or *senility:* A progressive loss of memory, spatial orientation, language, and personality as a consequence of aging. *(p. 499)*

AM | Additional Terms Discussed in the *Applications Manual*

anterograde amnesia: The loss of the ability to store new memories after an injury, although earlier memories remain intact and accessible.

aphasia: A disorder affecting the ability to speak or read.

delirium, or *acute confusional state (ACS):* A state characterized by wild oscillations in the level of wakefulness, little or no grasp of reality, and possible hallucinations.

dementia: A stable, chronic state of consciousness characterized by deficits in memory, spatial orientation, language, or personality.

hypersomnia: Unusually long periods of sleep that is otherwise normal.

insomnia: Shortened sleeping periods and difficulty in getting to sleep.

parasomnias: Abnormal behaviors performed during sleep.

post-traumatic amnesia (PTA): Amnesia caused by head injury, in which the individual can neither remember the past nor consolidate memories of the present.

retrograde amnesia (*retro-*, behind): The loss of memories of events that occurred prior to an injury.

sleep apnea: A condition in which a sleeping individual stops breathing for short intervals throughout the night.

STUDY OUTLINE

INTRODUCTION   p. 481

1. The brain, spinal cord, and peripheral nerves continuously communicate with each other. *(Figure 15-1)*

SOMATIC SENSORY AND MOTOR PATHWAYS   p. 481

Sensory Pathways   p. 481

1. A **sensation** arrives in the form of action potentials in an afferent fiber. The **posterior column pathway** carries fine touch, pressure, and proprioceptive sensations. The axons ascend within the **fasciculus gracilis** and **fasciculus cuneatus** and relay information to the thalamus via the **medial lemniscus**. Before the axons enter the medial lemniscus, they cross over to the opposite side of the brain stem. This crossing over is called **decussation.** *(Figures 15-2, 15-3; Table 15-1)*
2. The **spinothalamic pathway** carries poorly localized sensations of touch, pressure, pain, and temperature. The axons involved decussate in the spinal cord and ascend within the **anterior** and **lateral spinothalamic tracts** to the ventral nuclei of the thalamus. *(Figures 15-3, 15-4; Table 15-1)*
3. The **spinocerebellar pathway,** including the **posterior** and **anterior spinocerebellar tracts,** carries sensations to the cerebellum concerning the position of muscles, tendons, and joints. *(Figure 15-5; Table 15-1)*

Motor Pathways and Motor Control   p. 487

4. Somatic motor pathways always involve an **upper motor neuron** (whose cell body lies in a CNS processing center) and a **lower motor neuron** (whose cell body is located in a motor nucleus of the brain stem or spinal cord). *(Figure 15-6)*
5. The neurons of the primary motor cortex are *pyramidal cells.* The **corticospinal pathway** provides voluntary skeletal muscle control. The **corticobulbar tracts** terminate at the cranial nerve nuclei; the **corticospinal tracts** synapse on motor neurons in the anterior gray horns of the spinal cord. The corticospinal tracts are visible along the medulla as a pair of thick bands, the **pyramids,** where most of the axons decussate to enter the descending **lateral corticospinal tracts.** Those that do not cross over enter the **anterior corticospinal tracts.** The corticospinal pathway provides a rapid, direct mechanism for controlling skeletal muscles. *(Figures 15-6, 15-7; Table 15-2)*
6. The **medial** and **lateral pathways** include several other centers that issue motor commands as a result of processing performed at a subconscious level. *(Figure 15-6; Table 15-2)*
7. The medial pathway primarily controls gross movements of the trunk and proximal limbs. It includes the vestibulospinal, tectospinal, and reticulospinal tracts. The **vestibulospinal tracts** carry information related to maintaining balance and posture. Commands carried by the **tectospinal tracts** change the position of the eyes, head, neck, and arms in response to bright lights, sudden movements, or loud noises. Motor commands carried by the **reticulospinal tracts** vary according to the region stimulated. *(Table 15-2)*
8. The lateral pathway consists of the **rubrospinal tracts,** which primarily control muscle tone and more-precise movements of the distal muscles of the upper limbs. *(Table 15-2)*
9. The cerebral nuclei adjust the motor commands issued in other processing centers.
10. The cerebellum regulates the activity of motor centers and along the motor pathways. The integrative activities performed by neurons in the cortex and nuclei of the cerebellum are essential for the precise control of movements. *(Figure 15-8)*

MONITORING BRAIN ACTIVITY: THE ELECTROENCEPHALOGRAM   p. 493

1. An **electroencephalogram (EEG)** is a printed report of the electrical activity of the brain over time. An EEG can be used to check for a variety of clinical conditions. *(Figure 15-9)*

HIGHER-ORDER FUNCTIONS   p. 494

1. Higher-order functions (1) are performed by the cerebral cortex and involve complex interactions among areas of the cerebral cortex and between that cortex and other areas of the brain, (2) involve conscious and unconscious information processing, and (3) are subject to modification and adjustment over time.

Memory   p. 494

2. Memories can be classified as short-term or long-term. **Short-term,** or *primary,* **memories** are short-lived, but while they persist, the information can be recalled immediately. **Long-term memories** last much longer.
3. Long-term memories can be subdivided into *secondary memories* (which fade with time and may be difficult to recall) and *tertiary memories* (which seem to be part of consciousness). *(Figure 15-10)*
4. The conversion from short-term to long-term memory is **memory consolidation.** The amygdaloid body and the hippocampus are essential to memory consolidation.
5. Cellular mechanisms that seem to be involved in memory formation and storage include increased neurotransmitter release, facilitation at synapses, and the formation of additional synaptic connections.

Consciousness   p. 496

6. In **deep sleep** (*slow wave* or *non-REM sleep*), the body relaxes and cerebral cortex activity is low. In **rapid eye movement (REM) sleep,** active dreaming occurs. Periods of REM and deep sleep alternate throughout the night. *(Figure 15-11)*
7. Consciousness is determined by interactions between the brain stem and the cerebral cortex. The **reticular activating system (RAS),** a network in the reticular formation, is one of the most important brain stem components *(Figure 15-12)*

BRAIN CHEMISTRY AND BEHAVIOR   p. 498

1. Changes in the normal balance between two or more neurotransmitters can profoundly alter brain function. The most important and widespread neurotransmitters are acetylcholine (ACh), norepinephrine (NE), gamma aminobutyric acid (GABA), dopamine, serotonin, histamine, Substance P, and the enkephalins and endorphins.

Personality and Other Mysteries   p. 498

2. Self-awareness and personality appear to be characteristic of the brain as an integrated system rather than functions of any specific component.

AGING AND THE NERVOUS SYSTEM   p. 499

Age-Related Anatomical and Functional Changes   p. 499

1. Age-related changes in the nervous system include a reduction in brain size and weight, a reduction in the number of neurons, a decrease in blood flow to the brain, changes in the synaptic organization of the brain, and intracellular and extracellular changes in CNS neurons.

Alzheimer's Disease   p. 499

2. **Alzheimer's disease** is a progressive disorder characterized by the loss of higher-order cerebral functions. It is the most common cause of **senile dementia.**

15

REVIEW QUESTIONS

LEVEL 1 REVIEWING FACTS AND TERMS

1. The corticospinal tract
   (a) carries motor commands from the cerebral cortex to the spinal cord
   (b) carries sensory information from the spinal cord to the brain
   (c) starts in the spinal cord and ends in the brain
   (d) a, b, and c are correct

2. The ability to localize a specific stimulus depends on the organized distribution of sensory information to the
   (a) spinothalamic pathway
   (b) posterior column tract
   (c) spinocerebellar pathway
   (d) primary sensory cortex

3. Destruction of or damage to a lower motor neuron in the somatic nervous system produces
   (a) the inability to localize a stimulus
   (b) a subconscious response to stimulation
   (c) paralysis of the innervated motor unit
   (d) a stimulation of the innervated muscle

4. A progressive disorder characterized by the loss of higher-order cerebral functions is
   (a) Parkinson's disease
   (b) parasomnia
   (c) Huntington's disease
   (d) Alzheimer's disease

5. What are the four types of sensory receptors for the general senses? What is the nature of the stimuli that excite each type?

6. What are the three major somatic sensory pathways, and what is the function of each pathway?

7. Which three pairs of descending tracts make up the corticospinal pathway?

8. Which three motor pathways make up the medial pathway?

9. What are the two primary functional roles of the cerebellum?

10. What three anatomical factors contribute to the maturation of the CNS and the refinement of motor skills?

11. What three characteristics are shared by higher-order functions?

12. As a result of animal studies, what cellular mechanisms are thought to be involved in memory formation and storage?

13. What physiological activities distinguish non-REM sleep from REM sleep?

14. What anatomical and functional changes in the brain are linked to alterations that occur with aging?

LEVEL 2 REVIEWING CONCEPTS

15. Damage to the hippocampus, a component of the limbic system, leads to
   (a) a loss of emotion due to forgetfulness
   (b) a loss of consciousness
   (c) a loss of long-term memory
   (d) an immediate loss of short-term memory

16. Describe the relationship among first-, second-, and third-order neurons.

17. How is the autonomic nervous system *functionally* different from the somatic nervous system?

18. What is a motor homunculus? How does it differ from a sensory homunculus?

19. What effect does injury to the primary motor cortex have on peripheral muscles?

20. By which structures and in which part of the brain is the level of muscle tone in the body's skeletal muscles controlled? How is this control exerted?

21. One patient suffers a cerebrovascular accident (CVA) in the left hemisphere; another suffers a CVA in the right hemisphere. What functions could be affected in each patient?

LEVEL 3 CRITICAL THINKING AND CLINICAL APPLICATIONS

22. Kelly is having difficulty controlling her eye movements and has lost some control of her facial muscles. After an examination and testing, Kelly's physician tells her that her cranial nerves are perfectly normal but that a small tumor is putting pressure on certain fiber tracts in her brain. This pressure is the cause of Kelly's symptoms. Where is the tumor most likely located?

23. Doris develops a clot that blocks the right branch of the middle cerebral artery, a blood vessel that serves the anterior portion of the right cerebral hemisphere. What symptoms would you expect to observe as a result of this blockage?

24. Researchers studying an illegal drug find that it causes individuals to see strange shapes and to hear sounds that aren't actually present. Perceptions of color are exaggerated, and individuals taking the drug report increased sexual appetites. Which neurotransmitter is this drug mimicking, and which part or parts of the brain are involved?

CHAPTER 16

The Autonomic Nervous System

CHAPTER OUTLINE AND OBJECTIVES

INTRODUCTION   p. 504

AN OVERVIEW OF THE ANS   p. 504
   Divisions of the ANS   p. 504
① Compare the organization of the autonomic nervous system with that of the somatic nervous system.

THE SYMPATHETIC DIVISION   p. 506
   The Sympathetic Chain   p. 507
   Collateral Ganglia   p. 510
   The Adrenal Medullae   p. 510
   Sympathetic Activation   p. 510
   Neurotransmitters and Sympathetic Function   p. 511
   Summary: The Sympathetic Division   p. 512

THE PARASYMPATHETIC DIVISION   p. 512
   Organization and Anatomy of the Parasympathetic Division   p. 513
   General Functions of the Parasympathetic Division   p. 513
   Parasympathetic Activation and Neurotransmitter Release   p. 514
   Summary: The Parasympathetic Division   p. 516
② Contrast the structures and functions of the sympathetic and parasympathetic divisions of the autonomic nervous system.
③ Describe the mechanisms of neurotransmitter release in the autonomic nervous system.
④ Compare the effects of the various autonomic neurotransmitters on target organs and tissues.

INTERACTIONS OF THE SYMPATHETIC AND PARASYMPATHETIC DIVISIONS   p. 516
   Anatomy of Dual Innervation   p. 516
   A Comparison of the Sympathetic and Parasympathetic Divisions   p. 517
   Autonomic Tone   p. 518
⑤ Discuss the relationship between the two divisions of the autonomic nervous system and the significance of dual innervation.
⑥ Explain the importance of autonomic tone.

INTEGRATION AND CONTROL OF AUTONOMIC FUNCTIONS   p. 519
   Visceral Reflexes   p. 520
   Higher Levels of Autonomic Control   p. 521
⑦ Describe the hierarchy of interacting levels of control in the autonomic nervous system.

MEDIALAB and CHAPTER REVIEW   p. 524

This is the control center of a nuclear power station. The generation and harnessing of nuclear energy are complex, potentially dangerous processes. To operate this power station, multiple systems—reactor, cooling, and electrical—must work smoothly and in balance. The systems involved are so complex, and the responses must be so swift and precise, that no individual can manage a single system unaided, let alone the entire facility. Instead, computers are programmed to make adjustments in system operation, and the people look at the "big picture." The human body is an extremely complex biological mechanism. As in the power station, the individual systems are regulated automatically and our conscious minds are aware of only the general patterns and results of those regulatory activities. In this chapter, we will examine the regulatory mechanism by exploring the structure and function of the autonomic nervous system.

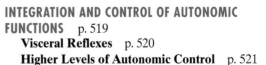

Your conscious thoughts, plans, and actions represent a tiny fraction of the activities of the nervous system. In practical terms, your conscious thoughts and the somatic nervous system (SNS), which operates under conscious control, seldom have a direct effect on your long-term survival. Of course, the somatic nervous system can be important in moving you out of the way of a speeding bus or pulling your hand from a hot stove—but it was your conscious movements that put you in jeopardy in the first place. If all consciousness were eliminated, vital physiological processes would continue virtually unchanged; a night's sleep is not a life-threatening event. Longer, deeper states of unconsciousness are not necessarily more dangerous, as long as nourishment is provided. People who have suffered severe brain injuries can survive in a coma for decades.

Survival under these conditions is possible because routine homeostatic adjustments in physiological systems are made by the **autonomic nervous system (ANS).** It is the ANS that coordinates cardiovascular, respiratory, digestive, urinary, and reproductive functions. In doing so, the ANS adjusts internal water, electrolyte, nutrient, and dissolved gas concentrations in body fluids—and it does so without instructions or interference from the conscious mind.

In this chapter, we examine the anatomical and physiological divisions of the ANS. Our understanding of this system has had a profound effect on the practice of medicine. For example, in 1960, the 5-year survival rate for patients surviving their first heart attack was very low, primarily because it was difficult and sometimes impossible to control high blood pressure. Forty years later, many survivors of heart attacks lead normal lives. This dramatic change occurred as we learned to manipulate the ANS with specific drugs and procedures.

AN OVERVIEW OF THE ANS

It will be useful to compare the organization of the ANS with that of the SNS, which controls our skeletal muscles. We will focus on (1) the neural interactions that direct motor output and (2) the subdivisions of the ANS, based on structural and functional patterns of peripheral innervation.

Figure 16-1• compares the organization of the somatic and autonomic nervous systems. Both are efferent divisions that carry motor commands to peripheral effectors. In the SNS, lower motor neurons exert direct control over skeletal muscles (Figure 16-1a•). In the ANS, a second visceral motor neuron always separates the CNS and the visceral effector (Figure 16-1b•). The visceral motor neurons in the

CNS are known as **preganglionic neurons.** The axons of these neurons are called **preganglionic fibers.**

Preganglionic fibers leave the CNS and synapse on **ganglionic neurons,** visceral motor neurons in peripheral ganglia. These ganglia, which contain hundreds to thousands of visceral motor neurons, are called **autonomic ganglia.** The ganglionic neurons innervate visceral effectors such as cardiac muscle, smooth muscles, glands, and adipose tissues. The axons of ganglionic neurons are called **postganglionic fibers,** because they begin at the autonomic ganglia and extend to the peripheral target organs.[1]

We will begin our discussion with the anatomy and physiology of the ANS. We will then consider the nature of *visceral reflexes,* polysynaptic reflexes that regulate visceral function. Finally, we will discuss the higher centers involved with the coordination of visceral functions and the integration of ANS activity with the other activities of the nervous system.

The ANS contains two subdivisions, the *sympathetic division* and the *parasympathetic division.* Most often, the two divisions have opposing effects; if the sympathetic division causes excitation, the parasympathetic causes inhibition. However, this is not always the case, because (1) the two divisions may work independently, with some structures innervated by only one division or the other; and (2) the two divisions may work together, each controlling one stage of a complex process. In general, the parasympathetic division predominates under resting conditions, and the sympathetic division "kicks in" only during periods of exertion, stress, or emergency.

DIVISIONS OF THE ANS

Preganglionic fibers from the thoracic and superior lumbar segments of the spinal cord synapse in ganglia near the spinal cord. These axons and ganglia are part of the **sympathetic division,** or *thoracolumbar* (tho-ra-kō-LUM-bar) *division,* of the ANS. In the sympathetic division, the preganglionic fibers are short and the postganglionic fibers are long.

The sympathetic division prepares the body for heightened levels of somatic activity. When fully activated, this division produces what is known as the "fight or flight" response, which readies the body for a crisis that may require sudden, intense physical activity. An increase in sympathetic activity generally

---

[1]Ganglionic neurons are sometimes called *postganglionic neurons.* This term is potentially confusing, because (1) only the axons extend beyond the ganglia and (2) the adrenal medulla contains ganglionic neurons that either don't have axons or have short ones that do not leave the ganglion.

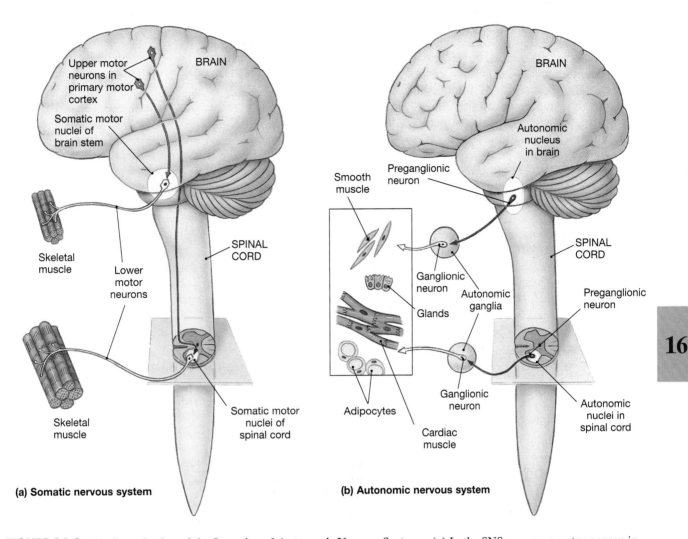

•**FIGURE 16-1** **The Organization of the Somatic and Autonomic Nervous Systems.** **(a)** In the SNS, an upper motor neuron in the CNS controls a lower motor neuron in the brain stem or spinal cord. The axon of the lower motor neuron has direct control over skeletal muscle fibers. The stimulation of the lower motor neuron always has an excitatory effect on the skeletal muscle fibers. **(b)** In the ANS, the axon of a preganglionic neuron in the CNS controls ganglionic neurons in the periphery. The stimulation of ganglionic neurons leads to either excitation or inhibition of the visceral effector innervated.

stimulates tissue metabolism and increases alertness. Imagine walking down a long, dark alley and hearing strange noises in the darkness ahead. Your body responds immediately, and you become more alert and aware of your surroundings. Your metabolic rate rises quickly, to as much as twice the resting level. Your digestive and urinary activities are suspended temporarily, and blood flow to your skeletal muscles increases. You begin breathing more quickly and more deeply. Both your heart rate and blood pressure increase, circulating your blood more rapidly. You feel warm and begin to perspire. The general pattern can be summarized as follows: (1) heightened mental alertness, (2) increased metabolic rate, (3) reduced diges-

tive and urinary functions, (4) activation of energy reserves, (5) increased respiratory rate and dilation of respiratory passageways, (6) increased heart rate and blood pressure, and (7) activation of sweat glands.

Preganglionic fibers originating in the brain stem and the sacral segments of the spinal cord are part of the **parasympathetic division,** or *craniosacral* (krā-nē-ō-SĀ-krul) *division,* of the ANS. In the parasympathetic division, the preganglionic fibers are long and the postganglionic fibers are short. The preganglionic fibers synapse on neurons of **terminal ganglia,** which are located near the target organs, or **intramural ganglia** (*murus,* wall), which are embedded in the tissues of the target organs.

The parasympathetic division stimulates visceral activity. For example, it is responsible for the state of "rest and repose" that follows a big dinner. General parasympathetic activation conserves energy and promotes sedentary activities, such as digestion. Your body relaxes, energy demands are minimal, and both your heart rate and blood pressure are relatively low. Meanwhile, your digestive organs are highly stimulated. Your salivary glands and other secretory glands are active; your stomach is contracting; and smooth muscle contractions move materials along your digestive tract. This movement promotes defecation; at the same time, smooth muscle contractions along your urinary tract promote urination. The overall pattern is as follows: (1) decreased metabolic rate, (2) decreased heart rate and blood pressure, (3) increased secretion by salivary and digestive glands, (4) increased motility and blood flow in the digestive tract, and (5) stimulation of urination and defecation.

Innervation Patterns in the ANS

Figure 16-2• provides an overview of the functional relationships of the ANS. Sensory information triggers visceral reflexes that control both sympathetic and parasympathetic activities. The sympathetic and parasympathetic divisions affect target organs through the controlled release of neurotransmitters by postganglionic fibers. Whether the result is a stimulation or an inhibition of activity depends on the response of the membrane receptors to the presence of the neurotransmitter. We can make three general statements about the neurotransmitters and their effects:

1. All preganglionic autonomic fibers (sympathetic and parasympathetic) release the neurotransmitter *acetylcholine (ACh)* at their synaptic terminals. The effects are always excitatory.

2. Postganglionic parasympathetic fibers also release ACh, but the effects may be excitatory or inhibitory, depending on the nature of the receptor.

3. Most postganglionic sympathetic terminals release the neurotransmitter *norepinephrine (NE)*. The effects are generally excitatory, but the response depends on the nature of the postsynaptic receptor.

☑ How many motor neurons are required to conduct an action potential from the spinal cord to smooth muscles in the wall of the intestine?

☑ While out for a brisk walk, Julie is suddenly confronted by an angry dog. Which division of the autonomic nervous system is responsible for the physiological changes that occur in Julie as she turns and runs?

☑ On the basis of anatomy, how could you distinguish the sympathetic division from the parasympathetic division of the autonomic nervous system?

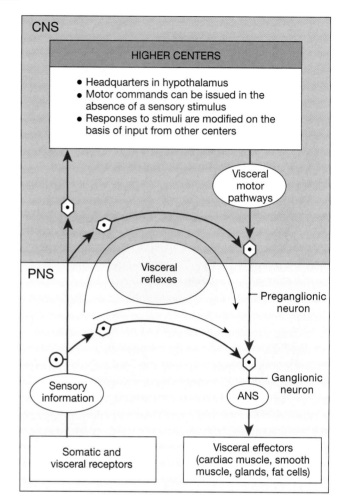

•**FIGURE 16-2** **Functional Relationships of the Autonomic Nervous System (ANS).** The ANS includes higher centers in the brain and motor neurons whose cell bodies are located in the CNS and in PNS ganglia. The output of the ganglionic neurons may be controlled by visceral reflexes or by the commands of higher centers.

THE SYMPATHETIC DIVISION

The sympathetic division of the ANS consists of preganglionic neurons that are located between segments T_1 and L_2 of the spinal cord and ganglionic neurons that are located in ganglia near the vertebral column. The cell bodies of the preganglionic neurons are situated in the lateral gray horns, and their axons enter the ventral roots of these segments. The ganglionic neurons occur in three locations (Figure 16-3•):

1. *Sympathetic chain ganglia.* **Sympathetic chain ganglia,** also called *paravertebral ganglia* or *lateral ganglia,* lie on both sides of the vertebral column. Neurons in these ganglia control effectors in the body wall,

Preganglionic fibers

Postganglionic fibers

Hormones released
into circulation

•**FIGURE 16-3** The Organization of the Sympathetic
Division of the ANS

16

inside the thoracic cavity, and in the head and limbs
(Figure 16-4a•).

2. *Collateral ganglia.* **Collateral ganglia,** also known as
prevertebral ganglia, are anterior to the vertebral bodies (Figure 16-4b•). Collateral ganglia contain ganglionic neurons that innervate tissues and organs in
the abdominopelvic cavity.

3. *The adrenal medullae.* The center of each adrenal
gland, an area known as the *adrenal medulla,* is a modified sympathetic ganglion (Figure 16-4c•). The ganglionic neurons of the adrenal medullae have very
short axons; when stimulated, they release their neurotransmitters into the bloodstream. This change in
the release site—from a synapse to a capillary—allows
the neurotransmitters to function as hormones that
affect target cells throughout the body.

Preganglionic fibers are relatively short, because
the ganglia are located relatively near the spinal cord.
In contrast, postganglionic fibers are relatively long,
except at the adrenal medullae.

THE SYMPATHETIC CHAIN

The ventral roots of spinal segments T_1 to L_2 contain
sympathetic preganglionic fibers. The basic pattern

of sympathetic innervation in these regions was described in Figure 13-7a•, p. 415. After passing through
the intervertebral foramen, each ventral root gives
rise to a myelinated *white ramus,* which carries myelinated preganglionic fibers into a nearby sympathetic chain ganglion. These fibers may synapse within
the sympathetic chain ganglia (Figure 16-4a•), at one
of the collateral ganglia (Figure 16-4b•), or in the
adrenal medullae (Figure 16-4c•). Extensive divergence occurs, with one preganglionic fiber synapsing
on two dozen or more ganglionic neurons. Preganglionic fibers running between the sympathetic chain
ganglia interconnect them, making the chain resemble a string of pearls. Each ganglion in the sympathetic chain innervates a particular body segment or
group of segments.

If a preganglionic fiber carries motor commands
that target structures in the body wall or thoracic
cavity, or in the head, neck, or limbs, it will synapse
in one or more sympathetic chain ganglia. The paths
of the unmyelinated postganglionic fibers differ depending on whether their targets lie in the body wall
or within the thoracic cavity. Postganglionic fibers
that control visceral effectors in the body wall, such
as the sweat glands of the skin or the smooth muscles

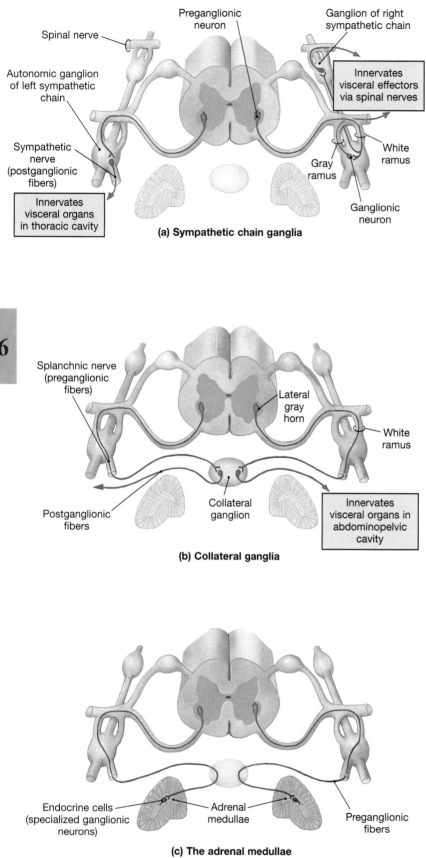

(a) Sympathetic chain ganglia

(b) Collateral ganglia

(c) The adrenal medullae

•**FIGURE 16-4** **Sympathetic Pathways.** Superior views of sections through the thoracic spinal cord, showing the distribution of preganglionic and postganglionic fibers.

in superficial blood vessels, enter the *gray ramus* and return to the spinal nerve for subsequent distribution. However, spinal nerves do not innervate structures in the ventral body cavities. Postganglionic fibers targeting structures in the thoracic cavity, such as the heart and lungs, form **sympathetic nerves,** which proceed directly to their peripheral targets. Although Figure 16-4a• shows sympathetic nerves on the left side and spinal nerve distribution on the right, in reality *both* innervation patterns occur on *each* side of the body.

Anatomy of the Sympathetic Chain

Figure 16-5• provides a more detailed view of the structure of the sympathetic division. The left side represents the distribution to the skin and to skeletal muscles and other tissues of the body wall; the right side depicts the innervation of visceral structures.

Each sympathetic chain contains 3 cervical, 10–12 thoracic, 4–5 lumbar, and 4–5 sacral ganglia, plus 1 coccygeal ganglion. Preganglionic neurons are limited to spinal cord segments T_1–L_2, and these spinal nerves have both white rami (myelinated preganglionic fibers) and gray rami (unmyelinated postganglionic fibers). The neurons in the cervical, inferior lumbar, and sacral sympathetic chain ganglia are innervated by preganglionic fibers that run along the axis of the chain. In turn, these chain ganglia provide postganglionic fibers, through gray rami, to the cervical, lumbar, and sacral spinal nerves. As a result, although only spinal nerves T_1–L_2 have white rami, every spinal nerve has a gray ramus that carries sympathetic postganglionic fibers for distribution in the body wall.

About 8 percent of the axons in each spinal nerve are sympathetic postganglionic fibers. As a result, the spinal nerves, which provide somatic motor innervation to skeletal muscles of the body wall and limbs, also distribute sympathetic postganglionic fibers (Figure 16-5•). In the head, postganglionic sympathetic fibers leaving the cervical sympathetic ganglia supply the regions and structures innervated by cranial nerves III, VII, IX, and X.

∞ *pp. 467, 469, 470, 471*

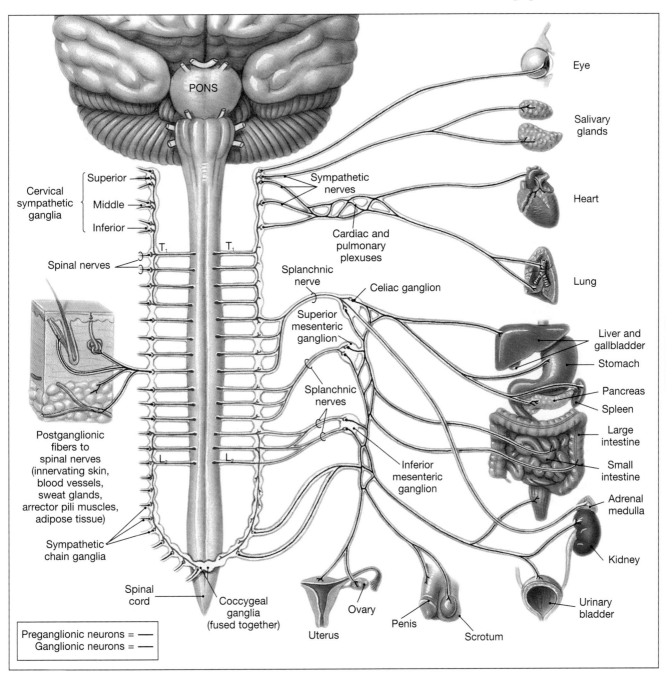

•**FIGURE 16-5 The Distribution of Sympathetic Innervation.** The distribution of sympathetic fibers is the same on both sides of the body. For clarity, the innervation of somatic structures is shown on the left, and the innervation of visceral structures on the right.

In summary:

- Only the thoracic and superior lumbar ganglia receive preganglionic fibers from white rami;

- the cervical, inferior lumbar, and sacral chain ganglia receive preganglionic innervation by preganglionic fibers from spinal segments T_1–L_2; and

- every spinal nerve receives a gray ramus from a ganglion of the sympathetic chain.

This anatomical arrangement means that if the ventral roots of thoracic spinal nerves are damaged, there will be no sympathetic motor function on the affected side of the head, neck, or trunk. Yet damage to the ventral roots of cervical spinal nerves will produce voluntary muscle paralysis on the affected side but will leave sympathetic function intact, because the preganglionic fibers innervating the cervical ganglia originate in the white rami of thoracic segments, which are undamaged.

16

In contrast, damage to the cervical ganglia or thoracic segments can eliminate sympathetic innervation to the face, although sensation and muscle control remain unaffected. The affected side of the face becomes flushed, although sweating does not occur, and the pupil of the eye constricts. This combination of symptoms is known as *Horner's syndrome*. AM *Hypersensitivity and Sympathetic Function*

COLLATERAL GANGLIA

The abdominopelvic viscera receive sympathetic innervation by way of sympathetic preganglionic fibers that pass through the sympathetic chain without synapsing. They synapse in separate collateral ganglia (Figure 16-4b•). Preganglionic fibers that innervate the collateral ganglia form the **splanchnic** (SPLANK-nik) **nerves,** which lie in the dorsal wall of the abdominal cavity. Splanchnic nerves from both sides of the body converge on these ganglia. Although there are two sympathetic chains, one on each side of the vertebral column, most collateral ganglia are single rather than paired.

Postganglionic fibers leaving the collateral ganglia extend throughout the abdominopelvic cavity, innervating a variety of visceral tissues and organs. The general pattern is (1) a reduction of blood flow and energy use by visceral organs that are not important to short-term survival, such as the digestive tract, and (2) the release of stored energy reserves.

The splanchnic nerves innervate three collateral ganglia (Figure 16-5•). Preganglionic fibers from the seven inferior thoracic segments end at the **celiac** (SĒ-lē-ak) **ganglion** or the **superior mesenteric ganglion.** These ganglia are embedded in an extensive network of autonomic nerves. Preganglionic fibers from the lumbar segments form splanchnic nerves that end at the **inferior mesenteric ganglion.** These ganglia are named for their association with adjacent arteries. For example, the celiac ganglion is named after the *celiac trunk.* The celiac ganglion most commonly consists of a pair of interconnected masses of gray matter situated at the base of that artery. The celiac ganglion may also form a single mass or many small, interwoven masses. Postganglionic fibers from this ganglion innervate the stomach, liver, gallbladder, pancreas, and spleen. The superior mesenteric ganglion is located near the base of the *superior mesenteric artery.* Postganglionic fibers leaving the superior mesenteric ganglion innervate the small intestine and the initial segments of the large intestine. The inferior mesenteric ganglion is located near the base of the *inferior mesenteric artery.* Postganglionic fibers from this ganglion provide sympathetic innervation to the terminal portions of the large intestine, the kidney and urinary bladder, and the sex organs.

THE ADRENAL MEDULLAE

Preganglionic fibers entering an adrenal gland proceed to its center, a region called the **adrenal medulla** (Figures 16-4c and 16-5•). The adrenal medulla is a modified sympathetic ganglion where preganglionic fibers synapse on *neuroendocrine cells,* specialized neurons that secrete hormones into the bloodstream. The neuroendocrine cells of the adrenal medullae secrete the neurotransmitters *epinephrine (E)* and *norepinephrine (NE).* Epinephrine, or *adrenaline,* accounts for 75–80 percent of the secretory output; the rest is NE.

The bloodstream then carries the neurotransmitters throughout the body, causing changes in the metabolic activities of many different cells. These effects resemble those produced by the stimulation of sympathetic postganglionic fibers. They differ, however, in two respects: (1) Cells not innervated by sympathetic postganglionic fibers are affected as well; and (2) the effects last much longer than those produced by direct sympathetic innervation, because the hormones continue to diffuse out of the bloodstream for an extended period.

SYMPATHETIC ACTIVATION

The sympathetic division can change the activities of tissues and organs by releasing NE at peripheral synapses and by distributing E and NE throughout the body in the bloodstream. The visceral motor fibers that target specific effectors, such as smooth muscle fibers in blood vessels of the skin, can be activated in reflexes that do not involve other visceral effectors. In a crisis, however, the entire division responds. This event is called **sympathetic activation.** It is controlled by sympathetic centers in the hypothalamus. The effects are not limited to peripheral tissues; sympathetic activation also alters CNS activity. When sympathetic activation occurs, an individual experiences:

- Increased alertness via stimulation of the reticular activating system, causing the individual to feel "on edge."

- A feeling of energy and euphoria, often associated with a disregard for danger and a temporary insensitivity to painful stimuli.

- Increased activity in the cardiovascular and respiratory centers of the pons and medulla oblongata, leading to elevations in blood pressure, heart rate, breathing rate, and depth of respiration.

- A general elevation in muscle tone through stimulation of the medial and lateral pathways, so the person *looks* tense and may begin to shiver.

- The mobilization of energy reserves, through the accelerated breakdown of glycogen in muscle and liver cells and the release of lipids by adipose tissues.

These changes, plus the peripheral changes already noted, complete the preparations necessary for the individual to cope with a stressful situation.

NEUROTRANSMITTERS AND SYMPATHETIC FUNCTION

We have examined the distribution of sympathetic impulses and the general effects of sympathetic activation. We now consider the cellular basis of these effects on peripheral organs. The stimulation of sympathetic preganglionic neurons leads to the release of ACh at synapses with ganglionic neurons. Synapses that use ACh as a transmitter are called *cholinergic.* ∞ *p. 387* The effect on the ganglionic neurons is always excitatory.

The stimulation of these ganglionic neurons leads to the release of neurotransmitter at specific target organs. The synaptic terminals are typically different from neuromuscular junctions of the somatic nervous system. Instead of forming synaptic knobs, telodendria form a branching network. Each branch resembles a string of pearls. Each "pearl," a swollen segment called a **varicosity,** is packed with neurotransmitter vesicles. Chains of varicosities pass along or near the surfaces of the effector cells (Figure 16-6•). There are no specialized postsynaptic membranes, but membrane receptors are scattered across the surfaces of the target cells.

Most sympathetic ganglionic neurons release NE at their varicosities. Neurons that use NE as a neurotransmitter are called *adrenergic.* ∞ *p. 389* The sympathetic division also contains a small but significant number of ganglionic neurons that release ACh rather than NE. Varicosities releasing ACh are located in the body wall, the skin, the brain, and skeletal muscles.

The NE released by varicosities affects its targets until it is reabsorbed or inactivated by enzymes. From 50 to 80 percent of the NE is reabsorbed by varicosities and either reused or broken down by the enzyme *monoamine oxidase (MAO)*. The rest of the NE diffuses out of the area or is broken down by the enzyme *catechol-O-methyltransferase (COMT)* in surrounding tissues. In general, the effects of NE on the postsynaptic membrane persist for a few seconds, significantly longer than the 20-msec duration of ACh effects. (As usual, the responses of the target cells vary with the nature of the receptor on the postsynaptic membrane.) The effects of NE or E released by the adrenal medullae last even longer, because (1) the bloodstream does not contain MAO or COMT, and (2) most tissues contain relatively low concentrations of those enzymes. After the adrenal medullae are stimulated, tissue concentrations of NE and E throughout the body may remain elevated for as long as 30 seconds, and the effects may persist for several minutes.

Sympathetic Stimulation and the Release of NE and E

The effects of sympathetic stimulation result primarily from interactions with membrane receptors sensitive to NE and E. There are two classes of sympathetic receptors: (1) *alpha receptors* and (2) *beta receptors.* In general, norepinephrine stimulates alpha receptors more than it does beta receptors, whereas epinephrine stimulates both classes of receptors.

The stimulation of **alpha (α) receptors** activates enzymes on the inside of the cell membrane. There are two types of alpha receptors: alpha-1 (α_1) and alpha-2 (α_2). Their actions rely on the production of *second messengers,* intracellular intermediaries that have specific functions. ∞ *p. 392* The function of the more common type of alpha receptor, α_1, is the release of intracellular calcium ions from reserves in the endoplasmic reticulum. This action generally has an excitatory effect on the target cell. For example, the stimulation of α_1 receptors on the surfaces of smooth muscle cells is responsible for the constriction of peripheral blood vessels and the closure of sphincters along the digestive tract. Alpha-2 receptor stimulation results in a lowering of cyclic-AMP (cAMP) levels in the cytoplasm. This reduction generally has an inhibitory effect on the cell. The presence of α_2 receptors in the parasympathetic division helps coordinate sympathetic and parasympathetic activities. When the sympathetic division is active, the NE released binds to parasympathetic neuromuscular and neuroglandular junctions and inhibits their activity.

Preganglionic fiber
(myelinated)

Ganglionic
neuron

Postganglionic fiber
(unmyelinated)

Ganglion

Varicosities

Vesicles containing
norepinephrine

Mitochondrion

Schwann cell
cytoplasm

5 μm

Varicosities

Smooth muscle cells

•**FIGURE 16-6** Sympathetic Varicosities

16

16

Beta (β) receptors are located on the membranes of cells in many organs, including skeletal muscles, the lungs, the heart, and the liver. The stimulation of beta receptors triggers changes in the metabolic activity of the target cell. These changes occur indirectly, as the beta receptor causes the formation of cAMP, a second messenger that activates or inactivates key enzymes.

There are two major types of beta receptors: beta-1 (β_1) and beta-2 (β_2). The stimulation of β_1 receptors leads to an increase in metabolic activity. For example, the stimulation of β_1 receptors on skeletal muscles accelerates the metabolic activities of the muscles. The stimulation of β_1 receptors in the heart causes an increase in heart rate and in the force of contraction. The stimulation of β_2 receptors causes inhibition, triggering a relaxation of smooth muscles along the respiratory tract. As a result, the diameter of the respiratory passageways increases, making breathing easier. This response accounts for the effectiveness of the inhalers used to treat asthma.

Sympathetic Stimulation and the Release of ACh and NO

Although the vast majority of sympathetic postganglionic fibers are adrenergic (releasing NE), a few are cholinergic (releasing ACh). These postganglionic fibers innervate sweat glands of the skin and the blood vessels to skeletal muscles and the brain. The activation of these sympathetic fibers stimulates sweat gland secretion and dilates the blood vessels.

It may seem strange that sympathetic terminals release ACh, which is the neurotransmitter used by the parasympathetic nervous system. However, neither the body wall nor skeletal muscles are innervated by the parasympathetic division, and NE and ACh differ significantly in function. For example, the effects of ACh release last for a much shorter period than do those of NE release; ACh stimulates sweat gland secretion much more than does NE; and ACh causes the dilation of most small peripheral arteries *(vasodilation),* whereas NE causes their constriction *(vasoconstriction).* The distribution of cholinergic fibers within the sympathetic division provides a method of stimulating sweat gland secretion and selectively enhancing blood flow to skeletal muscles while the adrenergic terminals reduce the blood flow to other tissues in the body wall.

The sympathetic division also includes *nitroxidergic* synapses, which release *nitric oxide (NO)* as a neurotransmitter. Such synapses occur where neurons innervate smooth muscles in the walls of blood vessels in many regions, notably in skeletal muscles and the brain. The activity of these synapses promotes immediate vasodilation and thereby increased blood flow through the region.

SUMMARY: THE SYMPATHETIC DIVISION

To summarize our discussion of the sympathetic division:

1. The sympathetic division of the ANS includes two sets of sympathetic chain ganglia, one on each side of the vertebral column; three collateral ganglia anterior to the vertebral column; and two adrenal medullae.

2. The preganglionic fibers are short, because the ganglia are close to the spinal cord. The postganglionic fibers are longer and extend a considerable distance before reaching their target organs. (In the case of the adrenal medullae, very short axons end at capillaries that carry their secretions to the bloodstream.)

3. The sympathetic division shows extensive divergence, and a single preganglionic fiber may innervate two dozen or more ganglionic neurons in different ganglia. As a result, a single sympathetic motor neuron in the CNS can control a variety of visceral effectors and can produce a complex and coordinated response.

4. All preganglionic neurons release ACh at their synapses with ganglionic neurons. Most postganglionic fibers release NE, but a few release ACh or NO.

5. The effector response depends on the second messengers activated when NE or E binds to alpha receptors or beta receptors.

☑ Where do the nerves that synapse in collateral ganglia originate?

☑ How would a drug that stimulates acetylcholine receptors affect the sympathetic nervous system?

☑ An individual with high blood pressure is given a medication that blocks beta receptors. How could this medication help correct that person's condition?

THE PARASYMPATHETIC DIVISION

The parasympathetic division of the ANS consists of the following (Figure 16-7•):

1. *Preganglionic neurons in the brain stem and in sacral segments of the spinal cord.* The mesencephalon, pons, and medulla oblongata contain autonomic nuclei associated with cranial nerves III, VII, IX, and X. In sacral segments of the spinal cord, the autonomic nuclei lie in the lateral gray horns of spinal segments S_2–S_4.

2. *Ganglionic neurons in peripheral ganglia within or adjacent to the target organs.* Preganglionic fibers of the parasympathetic division do not diverge as extensively as do those of the sympathetic division. A typical preganglionic fiber synapses on six to eight ganglionic neurons. In contrast to the pattern in the sympathetic division, all these ganglionic neurons are located in the same ganglion and their postganglionic fibers influence the same target organ. Thus, the effects of parasympathetic stimulation are more specific and localized than are those of the sympathetic division.

PREGANGLIONIC NEURONS		GANGLIONIC NEURONS	TARGET ORGANS

PARASYMPATHETIC DIVISION OF ANS

Nuclei in brain stem

N III → Ciliary ganglion → Intrinsic eye muscles (pupil and lens shape)

N VII → Sphenopalatine and submandibular ganglia → Nasal glands, tear glands, and salivary glands

N IX → Otic ganglion → Parotid salivary gland

N X → Intramural ganglia → Visceral organs of neck, thoracic cavity, and most of abdominal cavity

Nuclei in spinal cord segments $S_2 - S_4$ → Pelvic nerves → Intramural ganglia → Visceral organs in inferior portion of abdominopelvic cavity

➡ Preganglionic fibers
➡ Postganglionic fibers

16

ORGANIZATION AND ANATOMY OF THE PARASYMPATHETIC DIVISION

Parasympathetic preganglionic fibers leave the brain as components of cranial nerves III (oculomotor), VII (facial), IX (glossopharyngeal), and X (vagus) (Figure 16-8•). These fibers carry the cranial parasympathetic output. Parasympathetic fibers in the oculomotor, facial, and glossopharyngeal nerves control visceral structures in the head. These fibers synapse in the *ciliary, sphenopalatine, submandibular,* and *otic ganglia.* ∞ *pp. 467, 469, 471* Short postganglionic fibers continue to their peripheral targets. The vagus nerve provides preganglionic parasympathetic innervation to structures in the neck and in the thoracic and abdominopelvic cavity as distant as the distal portion of the large intestine. The vagus nerve alone provides roughly 75 percent of all parasympathetic outflow.

The preganglionic fibers in the sacral segments of the spinal cord carry the sacral parasympathetic output. These fibers do not join the ventral roots of the spinal nerves. Instead, the preganglionic fibers form distinct **pelvic nerves,** which innervate intramural ganglia in the walls of the kidney, urinary bladder, terminal portions of the large intestine, and sex organs.

GENERAL FUNCTIONS OF THE PARASYMPATHETIC DIVISION

The major effects produced by the parasympathetic division include the following:

- Constriction of the pupils, to restrict the amount of light that enters the eyes, and focusing the eyes on nearby objects.
- Secretion by digestive glands, including salivary glands, gastric glands, duodenal glands, intestinal glands, the pancreas, and the liver.
- The secretion of hormones that promote the absorption and utilization of nutrients by peripheral cells.
- An increase in smooth muscle activity along the digestive tract.
- The stimulation and coordination of defecation.
- Contraction of the urinary bladder during urination.
- Constriction of the respiratory passageways.
- A reduction in heart rate and in the force of contraction.
- Sexual arousal and the stimulation of sexual glands in both genders.

These functions center on relaxation, food processing, and energy absorption. The parasympathetic division has been called the *anabolic system,* because

16

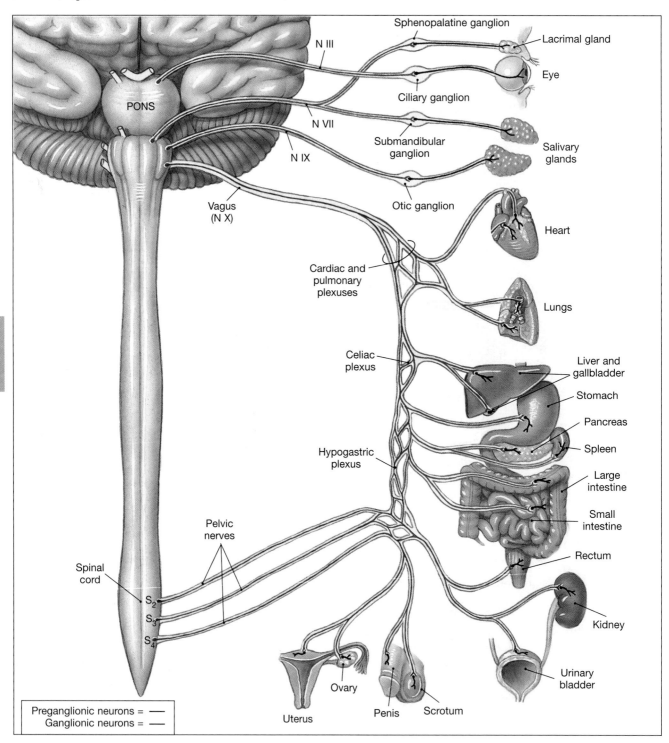

•FIGURE 16-8 The Distribution of Parasympathetic Innervation

its stimulation leads to a general increase in the nutrient content of the blood. (*Anabolic* comes from the Greek word *anabole,* which means "a rising up.") Cells throughout the body respond to this increase by absorbing nutrients and using them to support growth, cell division, and the creation of energy reserves in the form of lipids or glycogen.

PARASYMPATHETIC ACTIVATION AND NEUROTRANSMITTER RELEASE

All the preganglionic and postganglionic fibers in the parasympathetic division release ACh. The neuromuscular and neuroglandular junctions are small and have narrow synaptic clefts. The effects of stimulation are short-lived, because most of the ACh released is inactivated by *acetylcholinesterase (AChE)* at the synapse. Any ACh diffusing into the surrounding tissues will be inactivated by the enzyme *tissue cholinesterase.* As a result, the effects of parasympathetic stimulation are quite localized, and they last a few seconds at most.

Membrane Receptors and Responses

Although all the synapses (neuron to neuron) and neuromuscular or neuroglandular junctions (neuron to effector) of the parasympathetic division use the same transmitter, ACh, two types of ACh receptors occur on the postsynaptic membranes:

1. **Nicotinic** (nik-ō-TIN-ik) **receptors** are located on the surfaces of ganglion cells of both the parasympathetic and sympathetic divisions as well as at neuromuscular junctions of the somatic nervous system. Exposure to ACh always causes excitation of the ganglionic neuron or muscle fiber by the opening of chemically gated channels in the postsynaptic membrane.

2. **Muscarinic** (mus-kar-IN-ik) **receptors** are located at cholinergic neuromuscular or neuroglandular junctions in the parasympathetic division as well as at the few cholinergic junctions in the sympathetic division. The stimulation of muscarinic receptors produces longer-lasting effects than does the stimulation of nicotinic receptors. The response, which reflects the activation or inactivation of specific enzymes, can be excitatory or inhibitory.

The names *nicotinic* and *muscarinic* originated with researchers who found that dangerous environmental toxins bind to these receptor sites. Nicotinic receptors bind *nicotine,* a powerful toxin that can be obtained from a variety of sources, including tobacco leaves. Muscarinic receptors are stimulated by *muscarine,* a toxin produced by some poisonous mushrooms.

These compounds have discrete actions, targeting either the autonomic ganglia and skeletal neuromuscular junctions (nicotine) or the parasympathetic neuromuscular or neuroglandular junctions (muscarine). They produce dangerously exaggerated, uncontrolled responses that parallel those produced by normal receptor stimulation. For example, nicotine poisoning occurs if as little as 50 mg of the compound is ingested or absorbed through the skin. The symptoms reflect widespread autonomic activation: vomiting, diarrhea, high blood pressure, rapid heart rate, sweating, and profuse salivation. Because the neuromuscular junctions of the somatic nervous system are stimulated, convulsions occur. In severe cases, the stimulation of nicotinic receptors inside the CNS can lead to coma and death within minutes. The symptoms of muscarine poisoning are almost entirely restricted to the parasympathetic division: salivation, nausea, vomiting, diarrhea, constriction of respiratory passages, low blood pressure, and an abnormally slow heart rate.

Table 16-1 summarizes details about the adrenergic and cholinergic receptors of the ANS.

16

TABLE 16-1 Adrenergic and Cholinergic Receptors of the ANS

Receptor	Location(s)	Response(s)	Mechanism
Adrenergic			
α_1	Widespread, found in most tissues	Excitation, stimulation of metabolism	Activation of enzymes; release of intracellular Ca^{2+}
α_2	Sympathetic neuromuscular or neuroglandular junctions	Inhibition of effector cell	Reduction of cAMP concentrations
	Parasympathetic neuromuscular or neuroglandular junctions	Inhibition of neurotransmitter release	As above
β_1	Heart, kidneys, liver, adipose tissue*	Stimulation, increased energy consumption	Enzyme activation
β_2	Smooth muscle in vessels of heart and skeletal muscle; smooth muscle layers in intestines, lungs, bronchi	Inhibition, relaxation	As above
Cholinergic			
Nicotinic	All autonomic synapses between preganglionic and ganglionic neurons; neuromuscular junctions of SNS	Stimulation, excitation; muscular contraction	Opening of chemically regulated Na^+ channels
Muscarinic	All parasympathetic and cholinergic sympathetic neuromuscular or neuroglandular junctions	Variable	Enzyme activation causing changes in membrane permeability to K^+

* Adipocytes also contain an additional receptor type, β_3, not found in other tissues.

SUMMARY: THE PARASYMPATHETIC DIVISION

In summary,

- The parasympathetic division includes visceral motor nuclei associated with cranial nerves III, VII, IX, and X and with sacral segments S_2–S_4.
- Ganglionic neurons are located within or next to their target organs.
- The parasympathetic division innervates areas serviced by the cranial nerves and organs in the thoracic and abdominopelvic cavities.
- All parasympathetic neurons are cholinergic. Ganglionic neurons have nicotinic receptors, which are excited by ACh. Muscarinic receptors at neuromuscular or neuroglandular junctions produce either excitation or inhibition, depending on the nature of the enzymes activated when ACh binds to the receptor.
- The effects of parasympathetic stimulation are generally brief and restricted to specific organs and sites.

☑ Which nerve is responsible for the parasympathetic innervation of the lungs, heart, stomach, liver, pancreas, and parts of the small and large intestines?

☑ How would the stimulation of muscarinic receptors in cardiac muscle affect the heart?

☑ Why is the parasympathetic division sometimes referred to as the anabolic system?

INTERACTIONS OF THE SYMPATHETIC AND PARASYMPATHETIC DIVISIONS

The sympathetic division has widespread impact, reaching visceral organs and tissues throughout the body. The parasympathetic division innervates only visceral structures that are serviced by the cranial nerves or that lie within the abdominopelvic cavity. Although some organs are innervated by just one division, most vital organs receive **dual innervation:** they receive instructions from both the sympathetic and the parasympathetic divisions. Where dual innervation exists, the two divisions commonly have opposing effects. Dual innervation with opposing effects is most evident in the digestive tract, heart, and lungs. At other sites, the responses are divergent or complementary. Secretory control of the salivary glands and the

sexual functions of the male reproductive tract are examples.

ANATOMY OF DUAL INNERVATION

Parasympathetic postganglionic fibers from the ciliary, sphenopalatine, submandibular, and otic ganglia of the head accompany the cranial nerves to their peripheral destinations. The sympathetic innervation reaches the same structures by traveling directly from the superior cervical ganglia of the sympathetic chain.

In the thoracic and abdominopelvic cavities, the sympathetic postganglionic fibers mingle with parasympathetic preganglionic fibers at a series of nerve networks collectively called *autonomic plexuses:* the cardiac plexus, the pulmonary plexus, the esophageal plexus, the celiac plexus, the inferior mesenteric plexus, and the hypogastric plexus (Figure 16-9•). Nerves leaving these networks travel with the blood vessels and lymphatic vessels that supply visceral organs.

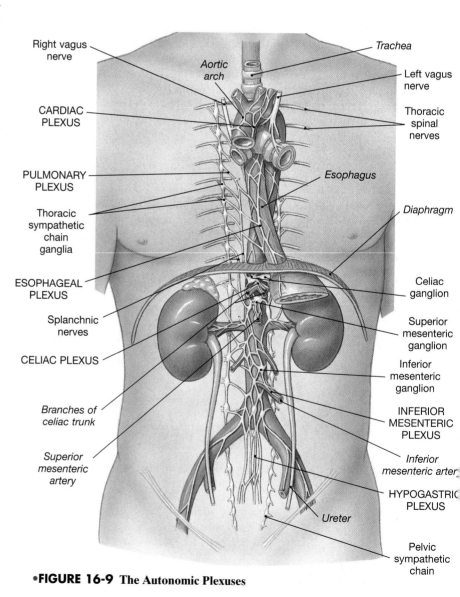

•**FIGURE 16-9** The Autonomic Plexuses

Autonomic fibers entering the thoracic cavity intersect at the **cardiac plexus** and the **pulmonary plexus.** These plexuses contain sympathetic and parasympathetic fibers bound for the heart and lungs, respectively, as well as the parasympathetic ganglia whose output affects those organs. The **esophageal plexus** contains descending branches of the vagus nerve and splanchnic nerves leaving the sympathetic chain on either side.

Parasympathetic preganglionic fibers of the vagus nerve enter the abdominopelvic cavity with the esophagus. There the fibers enter the **celiac plexus,** also known as the *solar plexus.* The celiac plexus and associated smaller plexuses, such as the **inferior mesenteric plexus,** innervate viscera within the abdominal cavity. The **hypogastric plexus** contains the parasympathetic outflow of the pelvic nerves, sympathetic postganglionic fibers from the inferior mesenteric ganglion, and splanchnic nerves from the sacral sympathetic chain. This plexus innervates the digestive, urinary, and reproductive organs of the pelvic cavity.

A COMPARISON OF THE SYMPATHETIC AND PARASYMPATHETIC DIVISIONS

Figure 16-10• and Table 16-2 compare key structural features of the sympathetic and parasympathetic divisions of the ANS. The distinctions have physiological and functional correlates. Table 16-3 provides a functional comparison of the two divisions, taking into account the effect of sympathetic or parasympathetic activity on specific organs and systems.

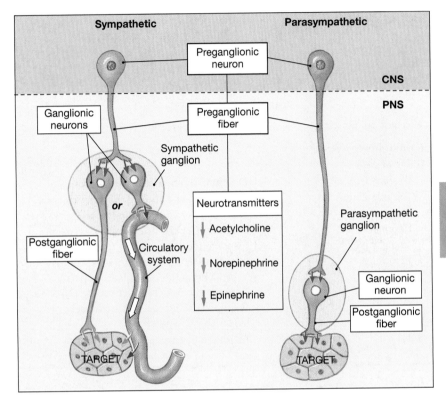

•**FIGURE 16-10** Summary: The Anatomical Differences between the Sympathetic and Parasympathetic Divisions

16

TABLE 16-2 A Structural Comparison of the Sympathetic and Parasympathetic Divisions of the ANS

Characteristic	Sympathetic Division	Parasympathetic Division
Location of CNS visceral motor neurons	Lateral gray horns of spinal segments T_1–L_2	Brain stem and spinal segments S_2–S_4
Location of PNS ganglia	Near vertebral column	Typically intramural
Preganglionic fibers		
Length	Relatively short	Relatively long
Neurotransmitter released	Acetylcholine	Acetylcholine
Postganglionic fibers		
Length	Relatively long	Relatively short
Neurotransmitter released	Normally NE; sometimes ACh	Acetylcholine
Neuromuscular or neuroglandular junction	Varicosities and enlarged terminal knobs that release transmitter near target cells	Junctions that release transmitter to special receptor surface
Degree of divergence from CNS to ganglion cells	Approximately 1:32	Approximately 1:6
General function(s)	Stimulates metabolism; increases alertness; prepares for emergency ("fight or flight")	Promotes relaxation, nutrient uptake, energy storage ("rest and repose")

TABLE 16-3 A Functional Comparison of the Sympathetic and Parasympathetic Divisions of the ANS

Structure	Sympathetic Effects (receptor or synapse type)	Parasympathetic Effects (all muscarinic receptors)
Eye	Dilation of pupil (α_1); accommodation for distance vision (β_2)	Constriction of pupil; accommodation for close vision
Lacrimal glands	None (not innervated)	Secretion
Salivary glands	Production of viscous secretion (α_1, β_1) containing mucins and enzymes	Production of copious, watery secretion
Skin		
Sweat glands	Increased secretion, palms and soles (α_1); generalized increase in secretion (cholinergic)	None (not innervated)
Arrector pili muscles	Contraction; erection of hairs (α_1)	As above
Cardiovascular system		
Blood vessels		As above
To skin	Dilation (cholinergic); constriction (α_1)	
To skeletal muscles	Dilation (β_2 and cholinergic; nitroxidergic)	
To heart	Dilation (β_2); constriction (α_1, α_2)	
To lungs	Dilation (β_2); constriction (α_1)	
To digestive viscera	Constriction (α_1); dilation (β_2)	
To kidneys	Constriction, decreased urine production (α_1, α_2); dilation, increased urine production (β_1, β_2)	
To brain	Dilation (cholinergic and nitroxidergic)	
Veins	Constriction (α_1, β_2)	
Heart	Increased heart rate, force of contraction, and blood pressure (α_1, β_1)	Decreased heart rate, force of contraction, and blood pressure

AUTONOMIC TONE

Even in the absence of stimuli, autonomic motor neurons show a resting level of spontaneous activity. The background level of activation determines an individual's **autonomic tone.** Autonomic tone is an important aspect of ANS function, just as muscle tone is a key aspect of muscular function. If a nerve is absolutely inactive under normal conditions, then all it can do is increase its activity on demand. But if the nerve maintains a background level of activity, it can increase or decrease its activity, providing a range of control options.

Autonomic tone is significant where dual innervation occurs and the two ANS divisions have opposing effects (Table 16-3). It is even more important in situations in which dual innervation does not occur. To demonstrate how autonomic tone affects ANS function, we will consider one example of each situation.

The heart is an organ that receives dual innervation. Recall that the heart consists of cardiac muscle tissue and that its contractions are triggered by specialized pacemaker cells. ∞ *pp. 302, 303* The two autonomic divisions have opposing effects on heart

function. Acetylcholine released by postganglionic fibers of the parasympathetic division causes a reduction in heart rate, whereas NE released by varicosities of the sympathetic division accelerates the heart rate. Because autonomic tone exists, small amounts of both of these neurotransmitters are released continuously. However, parasympathetic innervation dominates under resting conditions. If you remove all autonomic innervation, the heart rate goes up; if you remove the parasympathetic innervation but leave the sympathetic innervation intact, the heart rate goes up a lot. The heart rate can be controlled very precisely to meet the demands of active tissues by making small adjustments in the balance between parasympathetic stimulation and sympathetic stimulation. In a crisis, the stimulation of the sympathetic innervation and the inhibition of the parasympathetic innervation accelerate the heart rate to the maximum extent possible.

Several organs are innervated by one division only. For example, most structures in the body wall, such as blood vessels, sweat glands, adipose tissue, and skeletal muscles, receive only sympathetic innervation.

TABLE 16-3 A Functional Comparison of the Sympathetic and Parasympathetic Divisions of the ANS *(continued)*

Structure	Sympathetic Effects (receptor or synapse type)	Parasympathetic Effects (all muscarinic receptors)
Endocrine system		
Adrenal gland	Secretion of epinephrine, norepinephrine by medulla	None (not innervated)
Posterior pituitary	Secretion of ADH (β_1)	As above
Pancreas	Decreased insulin secretion (α_2)	Increased insulin secretion
Pineal gland	Increased melatonin secretion (β)*	
Respiratory system		
Airways	Increased airway diameter (β_2)	Decreased airway diameter
Digestive System		
Sphincters	Constriction (α_1)	Dilation
General level of activity	Decreased (α_2, β_2)	Increased
Secretory glands	Inhibition (α_2)	Stimulation
Liver	Glycogen breakdown, glucose synthesis and release (α_1, β_2)	Glycogen synthesis
Pancreas	Decreased exocrine secretion (α_1)	Increased exocrine secretion
Skeletal muscles	Increased force of contraction, glycogen breakdown (β_2)	None (not innervated)
	Facilitation of ACh release at neuromuscular junction (α_2)	As above
Adipose tissue	Lipolysis, fatty acid release ($\alpha_1, \beta_1, \beta_3$)	
Urinary system		
Kidneys	Secretion of renin (β_1)	Uncertain effects on urine production
Urinary bladder	Constriction of internal sphincter; relaxation of urinary bladder (α_1, β_2)	Tensing of urinary bladder, relaxation of internal sphincter to eliminate urine
Male reproductive system	Increased glandular secretion and ejaculation (α_1)	Erection
Female reproductive system	Increased glandular secretion; contraction of pregnant uterus (α_1)	Variable (depending on hormones present)
	Relaxation of nonpregnant uterus (β_2)	As above

*The type of beta receptor has not yet been determined.

The sympathetic control of blood vessel diameter demonstrates how autonomic tone allows fine adjustment of peripheral activities when the target organ is not innervated by both ANS divisions. It is very important that blood flow to specific organs is controlled to meet the tissue demands for oxygen and nutrients. When a blood vessel dilates, blood flow through it increases; when it constricts, blood flow is reduced. Sympathetic postganglionic fibers that release NE innervate the smooth muscle cells in the walls of peripheral vessels. The background sympathetic tone keeps these muscles partially contracted, so the vessels are ordinarily at roughly half their maximum diameter. When increased blood flow is needed, the rate of NE release decreases and sympathetic cholinergic fibers are stimulated. As a result, the smooth muscle cells relax, the vessels dilate, and blood flow quickly increases. By adjusting sympathetic tone and the activity of cholinergic fibers, the sympathetic division can exert precise control of vessel diameter over its entire range.

INTEGRATION AND CONTROL OF AUTONOMIC FUNCTIONS

Figure 15-8•, p. 492, diagrammed the relationships among centers involved in somatic motor control. The lowest level of regulatory control consists of the lower motor neurons involved in cranial and spinal reflex arcs. The highest level consists of the pyramidal motor neurons of the primary motor cortex, operating with the feedback from the cerebellum and cerebral nuclei.

The ANS is also organized into a series of interacting levels. At the bottom are visceral motor neurons in the lower brain stem and spinal cord that are involved in cranial and spinal visceral reflexes. **Visceral reflexes** provide automatic motor responses that can be modified, facilitated, or inhibited by higher centers, especially those of the hypothalamus.

For example, when a light is shined in one of your eyes, a visceral reflex constricts the pupils of *both* eyes (the *consensual light reflex*). The visceral motor commands are distributed by parasympathetic fibers. In darkness, your pupils dilate; this *pupillary reflex* is directed by sympathetic postganglionic fibers. However, the motor nuclei directing pupillary constriction or dilation are also controlled by hypothalamic centers concerned with emotional states. When you are queasy or nauseated, your pupils constrict; when you are sexually aroused, your pupils dilate.

VISCERAL REFLEXES

Each **visceral reflex arc** consists of a receptor, a sensory neuron, a processing center (interneuron), and two visceral motor neurons (Figure 16-11●). All visceral reflexes are polysynaptic; they are either long reflexes or short reflexes. **Long reflexes** are the autonomic equivalents of the polysynaptic reflexes introduced in Chapter 13. ∞ *p. 428* Visceral sensory neurons deliver information to the CNS along the

dorsal roots of spinal nerves, within the sensory branches of cranial nerves, and within the autonomic nerves that innervate visceral effectors. The processing steps involve interneurons within the CNS, and the ANS carries the motor commands to the appropriate visceral effectors.

Short reflexes bypass the CNS entirely; they involve sensory neurons and interneurons whose cell bodies are located within autonomic ganglia. The interneurons synapse on ganglionic neurons, and the motor commands are then distributed by postganglionic fibers. Short reflexes control very simple motor responses with localized effects. In general, short reflexes may control patterns of activity in one small part of a target organ, whereas long reflexes coordinate the activities of the entire organ.

In most organs, long reflexes are most important in regulating visceral activities, but this is not the case with the digestive tract and its associated glands. The ganglia in the walls of the digestive tract contain the cell bodies of visceral sensory neurons, interneurons, and visceral motor neurons. Although parasympathetic innervation of the visceral motor neurons can stimulate and coordinate various digestive activities, the digestive system is quite capable of controlling its own functions independent of the central nervous system. For this reason it has been called the *enteric nervous system (ENS)*; we shall consider its functions further in Chapter 24.

As we examine other body systems in later chapters, we will encounter many examples of autonomic

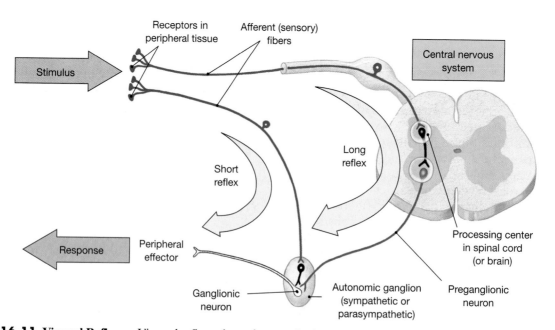

●FIGURE 16-11 Visceral Reflexes. Visceral reflexes have the same basic components as somatic reflexes, but all visceral reflexes are polysynaptic.

TABLE 16-4 Representative Visceral Reflexes

Reflex	Stimulus	Response	Comments
Parasympathetic reflexes			
Gastric and intestinal reflexes (Chapter 24)	Pressure and physical contact	Smooth muscle contractions that propel food materials and mix with secretions	Via vagus nerve
Defecation (Chapter 24)	Distention of rectum	Relaxation of internal anal sphincter	Requires voluntary relaxation of external anal sphincter
Urination (Chapter 26)	Distention of urinary bladder	Contraction of walls of urinary bladder; relaxation of internal urethral sphincter	Requires voluntary relaxation of external urethral sphincter
Direct light and consensual light reflexes (Chapter 14)	Bright light shining in eye(s)	Constriction of pupils of both eyes	
Swallowing reflex (Chapter 24)	Movement of food and liquids into pharynx	Smooth muscle and skeletal muscle contractions	Coordinated by medullary swallowing center
Coughing reflex (Chapter 23)	Irritation of respiratory tract	Sudden explosive ejection of air	Coordinated by medullary coughing center
Baroreceptor reflex (Chapters 17, 20, 21)	Sudden rise in carotid blood pressure	Reduction in heart rate and force of contraction	Coordinated in cardiac center of medulla oblongata
Sexual arousal (Chapter 28)	Erotic stimuli (visual or tactile)	Increased glandular secretions, sensitivity	
Sympathetic reflexes			
Cardioacceleratory reflex	Sudden decline in blood pressure in carotid artery	Increase in heart rate and force of contraction	Coordinated in cardiac center of medulla oblongata
Vasomotor reflexes (Chapter 21)	Changes in blood pressure in major arteries	Changes in diameter of peripheral vessels	Coordinated in vasomotor center in medulla oblongata
Pupillary reflex (Chapter 17)	Low light level reaching visual receptors	Dilation of pupil	
Ejaculation (in males) (Chapter 28)	Erotic stimuli (tactile)	Skeletal muscle contractions ejecting semen	

16

reflexes involved in respiration, cardiovascular function, and other visceral activities. Some of the most important are previewed in Table 16-4. Notice that the parasympathetic division participates in reflexes that affect individual organs and systems. This specialization reflects the relatively specific and restricted pattern of innervation. In contrast, fewer sympathetic reflexes exist. The sympathetic division is typically activated as a whole, in part because it has such a high degree of divergence and in part because the release of hormones by the adrenal medullae produces widespread peripheral effects.

HIGHER LEVELS OF AUTONOMIC CONTROL

The levels of activity in the sympathetic and parasympathetic divisions of the ANS are controlled by centers in the brain stem that deal with specific visceral functions. Figure 16-12• diagrams the levels of autonomic control. As in the SNS, in the ANS simple reflexes based in the spinal cord provide relatively rapid and automatic responses to stimuli. More-complex sympathetic and parasympathetic reflexes are coordinated by processing centers in the medulla oblongata. In

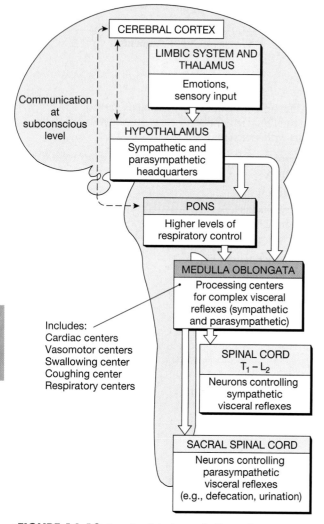

•**FIGURE 16-12** Levels of Autonomic Control

 Modifying Autonomic Nervous System Function

Drugs may be administered to counteract or reduce symptoms caused or aggravated by autonomic activities. These drugs are called **mimetic** if they mimic the activity of one of the normal autonomic neurotransmitters. Drugs that reduce the effects of autonomic stimulation by keeping the neurotransmitter from affecting the postsynaptic membranes are known as **blocking agents.** Mimetic drugs commonly have advantages over natural neurotransmitters. For example, norepinephrine and epinephrine must be administered into the bloodstream by injection or infusion, and the side effects are short-lived. In contrast, **sympathomimetic drugs** may survive oral administration, produce longer-lasting effects, or have more-specific actions than does E or NE. Examples include drugs applied topically to reduce hemorrhaging, by spray to reduce nasal congestion, by inhalation to dilate the respiratory passageways, or in drops to dilate the pupils.

Sympathetic blocking agents bind to the receptor sites and prevent a normal response to the presence of neurotransmitters or sympathomimetic drugs. **Alpha-blockers** eliminate the peripheral vasoconstriction (a reduction of the diameter of blood vessels) that accompanies sympathetic stimulation, and **beta-blockers** are effective and clinically useful for treating chronic high blood pressure and other forms of cardiovascular disease.

Parasympathomimetic drugs may be used to increase the activity along the digestive tract and to encourage defecation and urination. **Parasympathetic blocking agents** target the muscarinic receptors at neuromuscular or neuroglandular junctions. These drugs have diverse effects, but they are often used to control the diarrhea and cramps associated with various forms of food poisoning. `AM`

Pharmacology and the Autonomic Nervous System

---

addition to the cardiovascular and respiratory centers, the medulla oblongata, contains centers and nuclei involved with salivation, swallowing, digestive secretions, peristalsis, and urinary function. These centers are in turn subject to regulation by the hypothalamus. ∞ *pp. 463–464*

The term *autonomic* was originally applied because the regulatory centers involved with the control of visceral function were thought to operate independent of other CNS activities. This view has been drastically revised in light of subsequent research. Because the hypothalamus interacts with all other portions of the brain, activity in the limbic system, thalamus, or cerebral cortex can have dramatic effects on autonomic function. For example, when you become angry, your heart rate accelerates, your blood pressure rises, and your respiratory rate increases; when you remember your last big dinner, your stomach "growls" and your mouth waters.

☑ What effect would the loss of sympathetic tone have on blood flow to a tissue?

☑ What physiological changes would you expect in a patient who is about to undergo a root canal and is quite anxious about the procedure?

☑ Harry has a brain tumor that is interfering with the function of his hypothalamus. Would you expect this tumor to interfere with autonomic function? Why or why not?

The Integration of SNS and ANS Activities

Figure 16-13• and Table 16-5 indicate how the activities of the somatic nervous system, discussed in Chapter 15, and those of the autonomic nervous system are integrated. Although we have considered somatic and visceral portions of the nervous system separately, the two have many parallels, both in terms of organization and function. Integration occurs at the level of the brain stem, and both systems are under the control of higher centers.

•**FIGURE 16-13** **A Comparison of Somatic and Autonomic Function.** The somatic and autonomic nervous systems are organized in parallel and are integrated at the level of the brain stem. Blue arrows indicate ascending sensory information; red arrows, descending motor commands; and dashed lines, communication and feedback.

Biofeedback

Although conscious thought processes affect the ANS, you normally do not perceive the effect because visceral sensory information does not reach your cerebral cortex. Even when a conscious mental process triggers a physiological shift, such as a change in blood pressure, sweat gland activity, skin temperature, or muscle tone, the information never arrives at your primary sensory cortex. **Biofeedback** is an attempt to bridge this gap. In this technique, a person's physiological processes are monitored, and a visual or auditory signal is used to alert the person when a particular change takes place. These signals let the individual know when ongoing conscious thought processes have triggered a desirable change in autonomic function. For example, when biofeedback is used to regulate blood pressure, a light or tone informs the person when blood pressure drops.

With practice, some people can learn to re-create the thoughts, sensations, or mood that will lower their blood pressure. This action is possible because, as Figure 16-13• indicates, the cerebral cortex (conscious thought), thalamus (sensory feedback), and limbic system (emotions) all affect the hypothalamic centers that control autonomic activity.

Biofeedback techniques have been used to promote conscious control of blood pressure, heart rate, circulatory pattern, skin temperature, brain waves, and so forth. By reducing stress, lowering blood pressure, and improving circulation, these techniques can reduce the severity of clinical symptoms. In the process, they also lower the risks for serious complications, such as heart attacks or strokes, in individuals with high blood pressure. Unfortunately, not everyone can learn to influence their own autonomic functions, and the combination of variable results and expensive equipment makes biofeedback unsuitable for widespread application.

16

TABLE 16-5 A Comparison of the ANS and SNS

Characteristic	ANS	SNS
Innervation	Visceral effectors, including cardiac muscle, smooth muscle, glands, fat cells	Skeletal muscles
Activation	In response to sensory stimuli or from commands of higher centers	Same
Relay and processing centers	Brain stem	Brain stem and thalamus
Headquarters	Hypothalamus	Cerebral cortex
Feedback received from	Limbic system and thalamus	Cerebellum and cerebral nuclei
Control method	Adjustment of activity in brain stem processing centers that innervate preganglionic neurons	Direct (corticospinal) and indirect (medial and lateral) pathways that innervate lower motor neurons
Reflexes	Polysynaptic (short and long)	Monosynaptic and polysynaptic

[EXPLORE]
MediaLab

With few exceptions, our organs are controlled primarily by the autonomic division of the peripheral nervous system. This control is established through two very different pathways operating in concert. To reinforce your understanding of these opposing pathways, use the resources available on both the CD-ROM and the Companion Website. The CD-ROM includes animations and tutorials; the website builds on those resources with text content and interactive problems as well as communication avenues for sharing ideas with your class. To complete the following Web Explorations, visit the MediaLab in Chapter 16 of your Companion Website.

Web Exploration 1

Estimated time for completion = 10 minutes

Review the information about cholinergic receptors in Table 16-1; note the functions of nicotinic receptors in particular. To determine whether these receptors relate to the drug nicotine, select the keyword **NICOTINE** on the Companion Web-site. Choose the link "What is nicotine?" and read the first three articles ("What is Nicotine?"; "Nicotine's Effects"; and "Nicotine's Actions"). Summarize what you have learned about the actions and effects of nicotine. If you wish, prepare a short debate article, taking a stance either for or against the continued availability of nicotine for recreational use.

Web Exploration 2

Estimated time for completion = 7 minutes

Read the clinical discussion "Biofeedback" in the text (p. 523). What is this technique, and how is it used today? To learn about one application, the treatment of attention deficit disorder (ADD), select the keyword **ADD** on the Companion Website. How does this application of biofeedback relate to the information in the text? Describe any effects this technique might have on persons with ADD. If you wish, share your response with others by using the communications tools on the Companion Website.

CHAPTER REVIEW

SELECTED CLINICAL TERMINOLOGY
Terms Discussed in This Chapter

alpha-blockers: Drugs that eliminate the peripheral vasoconstriction that accompanies sympathetic stimulation. *(p. 522 and AM)*

beta-blockers: Drugs that decrease the heart rate and force of contraction, lowering peripheral blood pressure. *(p. 522 and AM)*

parasympathetic blocking agents: Drugs that target the muscarinic receptors at neuromuscular or neuroglandular junctions. *(p. 522 and AM)*

parasympathomimetic drugs: Drugs that mimic parasympathetic stimulation and increase the activity along the digestive tract. *(p. 522 and AM)*

sympathetic blocking agents: Drugs that bind to receptor sites, preventing a normal response to neurotransmitters or sympathomimetic drugs. *(p. 522 and AM)*

sympathomimetic drugs: Drugs that mimic the effects of sympathetic stimulation. *(p. 522 and AM)*

STUDY OUTLINE

INTRODUCTION   p. 504

1. The **autonomic nervous system (ANS)** coordinates cardiovascular, respiratory, digestive, urinary, and reproductive functions.

AN OVERVIEW OF THE ANS   p. 504

1. **Preganglionic neurons** in the CNS send axons to synapse on **ganglionic neurons** in **autonomic ganglia** outside the CNS. *(Figure 16-1)*

Divisions of the ANS   p. 504

2. Preganglionic fibers from the thoracic and lumbar segments form the **sympathetic division,** or *thoracolumbar division* ("fight or flight" system), of the ANS. Preganglionic fibers leaving the brain and sacral segments form the **parasympathetic division,** or *craniosacral division* ("rest and repose" system). The ANS includes processing centers in the CNS and two sets of motor neurons, one whose cell bodies are located in the CNS and one whose cell bodies are located in the PNS. *(Figure 16-2)*

THE SYMPATHETIC DIVISION   p. 506

1. The sympathetic division consists of preganglionic neurons between segments T_1 and L_2, ganglionic neurons in ganglia near the vertebral column, and specialized neurons in the adrenal glands. *(Figures 16-3–16-5)*

2. The two types of sympathetic ganglia are **sympathetic chain ganglia** (*paravertebral ganglia*) and **collateral ganglia** (*prevertebral ganglia*). *(Figures 16-3, 16-4)*

The Sympathetic Chain   p. 507

3. **Postganglionic fibers** targeting structures in the body wall and limbs rejoin the spinal nerves and reach their destinations by way of the dorsal and ventral rami. *(Figure 16-4)*

4. Postganglionic fibers targeting structures in the thoracic cavity form **sympathetic nerves,** which go directly to their visceral

destination. Preganglionic fibers run between the sympathetic chain ganglia and interconnect them. *(Figure 16-4)*

Collateral Ganglia p. 510

5. The abdominopelvic viscera receive sympathetic innervation via preganglionic fibers that synapse within collateral ganglia. The preganglionic fibers that innervate the collateral ganglia form the **splanchnic nerves.** *(Figure 16-4)*

6. The **celiac ganglion** innervates the stomach, liver, gallbladder, pancreas, and spleen; the **superior mesenteric ganglion** innervates the small intestine and initial segments of the large intestine; and the **inferior mesenteric ganglion** innervates the kidney, urinary bladder, sex organs, and terminal portions of the large intestine. *(Figures 16-5, 16-9)*

The Adrenal Medullae p. 510

7. Preganglionic fibers entering an adrenal gland synapse within the **adrenal medulla.** *(Figures 16-4, 16-5)*

Sympathetic Activation p. 510

8. In a crisis, the entire sympathetic division responds—an event called **sympathetic activation.** Its effects include increased alertness, a feeling of energy and euphoria, increased cardiovascular and respiratory activities, and general elevation in muscle tone.

Neurotransmitters and Sympathetic Function p. 511

9. The stimulation of the sympathetic division has two distinctive results: the release of *norepinephrine (NE)* at specific locations and the secretion of *epinephrine (E)* and NE into the general circulation. *(Figure 16-6)*

10. The two types of sympathetic receptors are **alpha receptors** (which respond to NE or E by depolarizing the membrane) and **beta receptors** (which are particularly sensitive to E).

11. Most postganglionic fibers are *adrenergic;* a few are *cholinergic* or *nitroxidergic.* Postganglionic fibers innervating sweat glands and blood vessels to skeletal muscles release ACh. Postganglionic fibers innervating smooth muscle in the walls of small arteries in skeletal muscles and the brain release *nitric oxide (NO),* promoting vasodilation.

Summary: The Sympathetic Division p. 512

12. Preganglionic sympathetic fibers are relatively short. Except for those of the adrenal medullae, postganglionic fibers are quite long. Each fiber typically synapses with many ganglionic neurons in different ganglia. A single effector cell may have more than one type of receptor, so the result of sympathetic stimulation depends on interaction among the receptors.

THE PARASYMPATHETIC DIVISION p. 512

Organization and Anatomy of the Parasympathetic Division p. 513

1. The parasympathetic division includes preganglionic neurons in the brain stem and sacral segments of the spinal cord and ganglionic neurons in peripheral ganglia located within or next to target organs. *(Figure 16-7)*

2. Preganglionic fibers leaving the sacral segments form **pelvic nerves.** *(Figure 16-8)*

General Functions of the Parasympathetic Division p. 513

3. The effects produced by the parasympathetic division center on relaxation, food processing, and energy absorption.

Parasympathetic Activation and Neurotransmitter Release p. 514

4. All parasympathetic preganglionic and postganglionic fibers release ACh at synapses and neuromuscular or neuroglandular junctions. The effects are short-lived, because ACh is inactivated by *acetylcholinesterase (AChE)* and by *tissue cholinesterase.*

5. Postsynaptic membranes have two types of ACh receptors. The stimulation of **muscarinic receptors** produces a longer-lasting effect than does the stimulation of **nicotinic receptors.** *(Table 16-1)*

Summary: The Parasympathetic Division p. 516

6. The parasympathetic division innervates areas serviced by cranial nerves and organs in the thoracic and abdominopelvic cavities. All preganglionic and postganglionic parasympathetic neurons are cholinergic, and the effects of stimulation are usually brief and restricted to specific sites.

INTERACTIONS OF THE SYMPATHETIC AND PARASYMPATHETIC DIVISIONS p. 516

1. The sympathetic division has widespread influence on visceral and somatic structures.

2. The parasympathetic division innervates only visceral structures serviced by cranial nerves or lying within the abdominopelvic cavity. Organs with **dual innervation** receive input from both divisions.

Anatomy of Dual Innervation p. 516

3. In body cavities, the parasympathetic and sympathetic nerves intermingle to form a series of characteristic *autonomic plexuses* (nerve networks): the **cardiac, pulmonary, esophageal, celiac, inferior mesenteric,** and **hypogastric plexuses.** *(Figure 16-9)*

A Comparison of the Sympathetic and Parasympathetic Divisions p. 517

4. Important physiological and functional differences exist between the sympathetic and parasympathetic divisions. *(Figure 16-10; Tables 16-2, 16-3)*

5. Sympathetic effects include mental alertness; increased metabolic rate; suspension of digestive and urinary tract functions; activation of energy reserves; increased respiratory rate and efficiency; increased heart rate and blood pressure; and activation of sweat glands.

6. Parasympathetic effects include decreased metabolic rate, heart rate, and blood pressure; salivary and digestive gland secretion; increased digestive tract activity; and urination and defecation.

Autonomic Tone p. 518

7. Even when stimuli are absent, autonomic motor neurons show a resting level of activation, the **autonomic tone.**

INTEGRATION AND CONTROL OF AUTONOMIC FUNCTIONS p. 519

Visceral Reflexes p. 520

1. **Visceral reflexes** are the simplest function of the ANS. *(Figure 16-11)*

2. Parasympathetic reflexes include (1) gastric and intestinal reflexes, (2) defecation, (3) urination, (4) the direct light and consensual light reflexes, (5) the swallowing reflex, (6) the coughing reflex, (7) the baroreceptor reflex, and (8) sexual arousal. Sympathetic reflexes include (1) the cardioacceleratory reflex, (2) vasomotor reflexes, (3) the pupillary reflex, and (4) ejaculation (in males). *(Table 16-4)*

Higher Levels of Autonomic Control p. 521

3. The hypothalamus receives input from many other portions of the brain, including the cerebral cortex. Centers in the hypothalamus deal with the coordination of sympathetic and parasympathetic functions. The regulatory activities occur outside conscious awareness and control. *(Figure 16-12)*

4. The SNS and ANS are organized in parallel. Integration occurs at the level of the brain stem and higher centers. *(Figure 16-13; Table 16-5)*

16

REVIEW QUESTIONS

LEVEL 1 REVIEWING FACTS AND TERMS

1. The CNS is always separated by a synapse from visceral effectors in the
   (a) autonomic nervous system
   (b) somatic nervous system
   (c) reflex arc
   (d) a, b, and c are correct

2. The preganglionic fibers of the sympathetic division of the ANS originate in the
   (a) cerebral cortex of the brain
   (b) medulla oblongata
   (c) craniosacral region of the brain stem and spinal cord
   (d) thoracolumbar region of the spinal cord

3. All preganglionic autonomic fibers release _____ at their synaptic terminals, and the effects are always _____.
   (a) norepinephrine; inhibitory
   (b) norepinephrine; excitatory
   (c) acetylcholine; excitatory
   (d) acetylcholine; inhibitory

4. Ganglionic neurons that innervate tissues and organs in the abdominopelvic cavity are located in
   (a) sympathetic chain ganglia
   (b) collateral ganglia
   (c) paravertebral ganglia
   (d) lateral ganglia

5. In the dorsal wall of the abdominal cavity, preganglionic fibers that innervate the collateral ganglia form the
   (a) splanchnic nerve   (b) phrenic nerve
   (c) sciatic nerve   (d) adrenal medulla

6. Postganglionic fibers from the celiac ganglion innervate the
   (a) heart, lungs, thymus, and diaphragm
   (b) cerebrum, pons, medulla, and cerebellum
   (c) stomach, liver, pancreas, and spleen
   (d) large intestine, kidney, urinary bladder, and sex organs

7. The effect of the neurotransmitter on the target cell depends on the nature of the
   (a) neurotransmitter
   (b) receptor on the presynaptic membrane
   (c) receptor on the postsynaptic membrane
   (d) a, b, and c are correct

8. Approximately 75 percent of parasympathetic outflow is provided by the
   (a) vagus nerve
   (b) sciatic nerve
   (c) glossopharyngeal nerves
   (d) pelvic nerves

9. The neurotransmitter at all synapses and neuromuscular or neuroglandular junctions in the parasympathetic division of the ANS is
   (a) epinephrine   (b) norepinephrine
   (c) cyclic-AMP   (d) acetylcholine

10. How does the emergence of sympathetic fibers from the spinal cord differ from the emergence of parasympathetic fibers?

11. Starting in the spinal cord, trace an impulse through the sympathetic division of the ANS until it reaches a target organ in the abdominopelvic region.

12. Which three collateral ganglia serve as origins for ganglionic neurons that innervate organs or tissues in the abdominopelvic region?

13. What two distinctive results are produced by the stimulation of sympathetic ganglionic neurons?

14. What is the difference between a cholinergic synapse and an adrenergic synapse?

15. Which four pairs of cranial nerves are associated with the cranial segment of the parasympathetic division of the ANS?

16. Which four ganglia serve as origins for postganglionic fibers that deal with the control of visceral structures in the head?

17. Which six plexuses in the thoracic and abdominopelvic cavities innervate visceral organs, and what are the effects of sympathetic versus parasympathetic stimulation?

18. What are the components of a visceral reflex arc?

19. Which control centers in the brain stem influence the levels of activity in the sympathetic and parasympathetic divisions of the ANS?

LEVEL 2 REVIEWING CONCEPTS

20. Autonomic tone in autonomic motor neurons exists because ANS neurons
   (a) from both divisions commonly innervate the same organ
   (b) rarely innervate the same organs as somatic motor neurons do
   (c) are inactive unless stimulated by higher centers
   (d) are always active to some degree

21. Dual innervation refers to situations in which
   (a) vital organs receive instructions from both sympathetic and parasympathetic fibers
   (b) the atria and ventricles of the heart receive autonomic stimulation from the same nerves
   (c) sympathetic and parasympathetic fibers have similar effects
   (d) a, b, and c are correct

22. Compare the general effects of the sympathetic and parasympathetic divisions of the ANS.

23. Why does sympathetic function remain intact even when the ventral roots of the cervical spinal nerves are damaged?

24. Why is the adrenal medulla considered to be a modified sympathetic ganglion?

25. How does the result of alpha-receptor stimulation differ from that of beta-receptor stimulation?

26. Why are the effects of parasympathetic stimulation quite localized and of short duration?

27. Why is autonomic tone a significant part of ANS function?

28. You are home alone at night when you hear what sounds like breaking glass. What physiological effects would this experience probably produce, and what would be their cause?

LEVEL 3 CRITICAL THINKING AND CLINICAL APPLICATIONS

29. In some very severe cases of stomach ulcers, the branches of the vagus nerve (N X) that lead to the stomach are surgically severed. How would this procedure help control the ulcers?

30. Mr. Martin is suffering from a condition known as ventricular tachycardia, in which his heart beats too quickly. Would an alpha-blocker or a beta-blocker help alleviate his problem? Why?

31. Little Billy is stung on his cheek by a wasp. Because Billy is allergic to wasp venom, his throat begins to swell and his respiratory passages constrict. Would acetylcholine or epinephrine be more helpful in relieving his symptoms? Why?

32. While studying the activity of smooth muscle in blood vessels, Shelly discovers that, when applied to muscle membrane, a molecule chemically similar to a neurotransmitter triggers an increase in intracellular calcium ions. Which neurotransmitter is the molecule mimicking, and to which receptors is it binding?

CHAPTER 17

Sensory Function

CHAPTER OUTLINE AND OBJECTIVES

INTRODUCTION p. 528

RECEPTORS AND RECEPTOR FUNCTION p. 528
   The Detection of Stimuli p. 528
   The Interpretation of Sensory Information p. 529
   Central Processing and Adaptation p. 529
① Distinguish between the general and special senses.
② Explain why receptors respond to specific stimuli
   and how the organization of a receptor affects
   its sensitivity.

THE GENERAL SENSES p. 530
   Nociceptors p. 530
   Thermoreceptors p. 531
   Mechanoreceptors p. 531
   Chemoreceptors p. 533
③ Identify the receptors for the general senses,
   and describe how they function.

OLFACTION p. 534
   Olfactory Receptors p. 535
   Olfactory Pathways p. 535
   Olfactory Discrimination p. 536
④ Describe the sensory organs of smell, and trace
   the olfactory pathways to their destinations in the brain.

GUSTATION p. 536
   Taste Receptors p. 536
   Gustatory Pathways p. 536
   Gustatory Discrimination p. 537
⑤ Describe the sensory organs of taste, and trace
   the gustatory pathways to their destinations in the brain.

VISION p. 538
   Accessory Structures of the Eye p. 538
   The Eye p. 541
   Visual Physiology p. 550
   The Visual Pathway p. 554
⑥ Identify the accessory structures of the eye, and explain
   their functions.
⑦ Describe the internal structures of the eye, and explain
   their functions.

Our comprehension of the world around us is based on information provided by our senses. For example, if we can see something, it must be there, because seeing is believing ... or is it? The more you study this image, the more confusing it becomes. We are so used to trusting what we see that when we see something that is physically impossible, it shocks us; some people become dizzy at the sight of an Escher drawing. This chapter examines the way receptors provide sensory information as well as the sensory pathways that distribute this information to provide us with our senses of smell, taste, vision, equilibrium, and hearing.

⑧ Explain how we are able to distinguish colors
   and perceive depth.
⑨ Explain how light stimulates production of nerve
   impulses, and trace the visual pathways to their
   destinations in the brain.

EQUILIBRIUM AND HEARING p. 557
   Anatomy of the Ear p. 557
   Equilibrium p. 561
   Hearing p. 564
⑩ Describe the structures of the external and middle ears,
   and explain how they function.
⑪ Describe the parts of the inner ear and their roles
   in equilibrium and hearing.
⑫ Trace the pathways for the sensations of equilibrium
   and hearing to their respective destinations in the brain.

MEDIALAB and CHAPTER REVIEW pp. 571, 572

Our knowledge of the world around us is limited to those characteristics that stimulate our sensory receptors. Although we may not realize it, our picture of the environment is incomplete. Colors we cannot distinguish guide insects to flowers; sounds we cannot hear and smells we cannot detect provide dolphins, dogs, and cats with important information about their surroundings.

What we *do* perceive varies considerably with the state of our nervous systems. For example, when sympathetic activation occurs, we experience heightened awareness of sensory information and hear sounds that would normally escape our notice. Yet, when concentrating on a difficult problem, we may remain unaware of relatively loud noises. Finally, our perception of any stimulus reflects activity in the cerebral cortex, and that activity can be inappropriate. In cases of phantom limb pain, a person feels pain in a missing limb; during an epileptic seizure, an individual may experience sights, sounds, or smells that have no physical basis.

We begin this chapter by examining receptor function and basic concepts in sensory processing. We will then apply that information to each of the general and special senses.

RECEPTORS AND RECEPTOR FUNCTION

Sensory receptors are specialized cells or cell processes that provide your central nervous system with information about conditions inside or outside your body. A sensory receptor detects an arriving stimulus and translates it into an action potential that can be conducted to the CNS. This translation process is called **transduction.** AM *A Closer Look: Transduction*

Sensory receptors are the interface between your nervous system and your internal and external environments. The **general senses** of pain, temperature, touch, pressure, vibration, and proprioception (position sense) are distributed throughout the body. We introduced the ascending pathways that provide conscious perception of those sensations in Chapter 15. ∞ *pp. 482–485* The **special senses** are **olfaction** (smell), **vision** (sight), **gustation** (taste), **equilibrium** (balance), and **hearing**. These sensations are provided by receptors that are structurally more complex than those of the general senses. Special sensory receptors are located in **sense organs** such as the eye or ear, where the receptors are protected by surrounding tissues. The information these receptors provide is distributed to specific areas of the cerebral cortex (the auditory cortex, the visual cortex, and so forth) and to centers throughout the brain stem. AM *Analyzing Sensory Disorders*

THE DETECTION OF STIMULI

Each receptor has a characteristic sensitivity. For example, a touch receptor is very sensitive to pressure but relatively insensitive to chemical stimuli, whereas a taste receptor is sensitive to dissolved chemicals but insensitive to pressure. This feature is called *receptor specificity.*

Specificity may result from the structure of the receptor cell or from the presence of accessory cells or structures that shield the receptor cell from other stimuli. The simplest receptors are the dendrites of sensory neurons. The branching tips of these dendrites, called **free nerve endings,** are not protected by accessory structures. Free nerve endings extend through a tissue the way grass roots extend into the soil. They can be stimulated by many different stimuli and therefore exhibit little receptor specificity. For example, free nerve endings that respond to tissue damage by providing pain sensations may be stimulated by chemical stimulation, pressure, temperature changes, or trauma. Complex receptors, such as the eye's visual receptors, are protected by accessory cells and connective tissue layers. These cells are seldom exposed to any stimulus other than light and so provide very specific information.

The area monitored by a single receptor cell is its *receptive field* (Figure 17-1●). Whenever a sufficiently strong stimulus arrives in the receptive field, the CNS receives the information "stimulus arriving at receptor X." The larger the receptive field, the poorer your ability to localize a stimulus. For example, a touch receptor on the general body surface may have a receptive field 7 cm (2.5 in.) in diameter. As a result, you can describe a light touch there as affecting only a general area, not an exact spot. On the tongue or fingertips, where the receptive fields are less than a millimeter in diameter, you can be very precise about the location of a stimulus.

An arriving stimulus can take many forms. It may be a physical force (such as pressure), a dissolved chemical, a sound, or a light. Regardless of the nature of the stimulus, however, sensory information must be

●**FIGURE 17-1** **Receptors and Receptive Fields.** Each receptor monitors a specific area known as the receptive field.

sent to the CNS in the form of action potentials, which are electrical events.

As we learned earlier, transduction is the translation of an arriving stimulus into an action potential by a sensory receptor. Transduction begins when a stimulus changes the transmembrane potential of the receptor cell. This change, called a *receptor potential,* is either a graded depolarization or a hyperpolarization. The stronger the stimulus, the larger the receptor potential.

The typical receptors for the general senses are the dendrites of sensory neurons, and the sensory neuron is the receptor cell. Any receptor potential that depolarizes the cell membrane will bring the membrane closer to threshold. A receptor potential large enough to produce an action potential is called a *generator potential.*

Sensations of taste, hearing, equilibrium, and vision are provided by specialized receptor cells that communicate with sensory neurons across chemical synapses. The receptor cells develop graded receptor potentials in response to stimulation, and the change in membrane potential alters the rate of neurotransmitter release at the synapse. The result is a depolarization or hyperpolarization of the sensory neuron. If sufficient depolarization occurs, an action potential appears in the sensory neuron. In this case, the receptor potential and the generator potential occur in different cells: The receptor potential develops in the receptor cell, and the generator potential appears later, in the sensory neuron.

Whenever a generator potential appears, action potentials develop in the axon of a sensory neuron. For reasons discussed in Chapter 12, the greater the degree of sustained depolarization produced by the generator potential, the higher the frequency of action potentials in the afferent fiber. ∞ *p. 396* The arriving information is then processed and interpreted by the CNS at the conscious and subconscious levels.

THE INTERPRETATION OF SENSORY INFORMATION

Sensory information that arrives at the CNS is routed according to the location and nature of the stimulus. In previous chapters, we emphasized the fact that axons in the CNS are organized in bundles with specific origins and destinations. Along sensory pathways, a series of neurons relays information from one point (the receptor) to another (a neuron at a specific site in the cerebral cortex). For example, sensations of touch, pressure, pain, and temperature arrive at the primary sensory cortex; visual, auditory, gustatory, and olfactory sensations reach the visual, auditory, gustatory, and olfactory regions of the cortex, respectively.

The neural link between receptor and cortical neuron is called a *labeled line.* Each labeled line consists of axons carrying information about one modality, or type of stimulus (touch, pressure, light, sound, and so forth). A labeled line begins at receptors in a specific part of the body. The CNS interprets the modality entirely on the basis of the labeled line over which it arrives. As a result, you cannot tell the difference between a true sensation and a false one generated somewhere along the line. For example, when you rub your eyes, you commonly see flashes of light. Although the stimulus is mechanical rather than visual, any activity along the optic nerve is projected to the visual cortex and experienced as a visual perception.

The identity of the active labeled line indicates the type of stimulus. Where it arrives within the sensory cortex determines its perceived location. For example, if activity in a labeled line that carries touch sensations stimulates the facial region of your primary sensory cortex, you perceive a touch on the face. *All other characteristics* of the stimulus—its strength, duration, and variation—are conveyed by the frequency and pattern of action potentials. The translation of complex sensory information into meaningful patterns of action potentials is called *sensory coding.*

Some sensory neurons, called **tonic receptors,** are always active. The frequency with which they generate action potentials indicates the background level of stimulation. When the stimulus increases or decreases, their rate of action potential generation changes accordingly. Other receptors are normally inactive but become active for a short time whenever a change occurs in the conditions they are monitoring. These receptors, called **phasic receptors,** provide information about the intensity and rate of change of a stimulus. Receptors that combine phasic and tonic coding can convey extremely complicated sensory information.

CENTRAL PROCESSING AND ADAPTATION

Adaptation is a reduction in sensitivity in the presence of a constant stimulus. You seldom notice the rumble of the tires when you ride in a car, or the background noise of the air conditioner, because your nervous system quickly adapts to stimuli that are painless and constant. *Peripheral adaptation* occurs when the level of receptor activity changes. The receptor responds strongly at first, but thereafter its activity gradually declines, in part because the size of the generator potential gradually decreases. This response is characteristic of phasic receptors, which are hence also called **fast-adapting receptors.** Temperature receptors *(thermoreceptors)* are phasic receptors; you seldom notice room temperature unless it changes suddenly. Tonic receptors show little peripheral adaptation and so are called **slow-adapting receptors.** Pain receptors *(nociceptors)* are slow-adapting receptors, which is one reason why pain sensations remind you of an injury long after the initial damage has occurred.

Adaptation also occurs inside the CNS along the sensory pathways. For example, a few seconds after you have been exposed to a new smell, conscious awareness of the stimulus virtually disappears, although the sensory neurons are still quite active. This process is known as *central adaptation*. Central adaptation generally involves the inhibition of nuclei along a sensory pathway.

Peripheral adaptation reduces the amount of information that reaches the CNS. Central adaptation at the subconscious level further restricts the amount of detail that arrives at the cerebral cortex. Most of the incoming sensory information is processed in centers along the spinal cord or brain stem, potentially triggering reflex responses, such as the withdrawal reflex. Only about 1 percent of the information provided by afferent fibers reaches the cerebral cortex and our conscious awareness.

The output from higher centers can increase receptor sensitivity or facilitate transmission along a sensory pathway. The reticular activating system in the mesencephalon helps focus our attention and thus heightens or reduces our awareness of arriving sensations. ∞ *p. 497* This adjustment of sensitivity can occur under conscious or subconscious direction. When you "listen carefully," your sensitivity and awareness of auditory stimuli increase. Output from higher centers can also inhibit transmission along a sensory pathway. This inhibition occurs when you enter a noisy factory or walk along a crowded city street, as you automatically tune out the high level of background noise.

In this discussion, we have introduced basic concepts of receptor function and sensory processing. We will next consider how those concepts apply to the general senses.

THE GENERAL SENSES

Receptors for the general senses are scattered throughout the body and are relatively simple in structure. The simple classification scheme introduced in Chapter 12 divides them into exteroceptors, proprioceptors, and interoceptors. ∞ *p. 366* *Exteroceptors* provide information about the external environment; *proprioceptors* report the positions of skeletal muscles and joints; and *interoceptors* monitor visceral organs and functions.

A more detailed classification system divides the general sensory receptors into four types by the nature of the stimulus that excites them: *nociceptors* (pain), *thermoreceptors* (temperature), *mechanoreceptors* (touch, pressure, proprioception), and *chemoreceptors* (chemical concentration). Each class of receptors has distinct structural and functional characteristics.

NOCICEPTORS

Pain receptors, or **nociceptors,** are especially common in the superficial portions of the skin, in joint capsules, within the periostea of bones, and around the walls of blood vessels. Other deep tissues and most visceral organs have few nociceptors. Pain receptors are free nerve endings with large receptive fields (Figure 17-2a•). As a result, it is often difficult to determine the exact source of a painful sensation.

Three populations of nociceptors are identified: those sensitive (1) to extremes of temperature, (2) to mechanical damage, and (3) to dissolved chemicals, such as chemicals released by injured cells. Very strong stimuli, however, will excite all three receptor types. For that reason, people describing very painful sensations—whether caused by acids, heat, or a deep cut—use similar descriptive terms, such as "burning."

Stimulation of the dendrites of a nociceptor causes depolarization. When the initial segment of the axon reaches threshold, an action potential heads toward the CNS. The sensory neurons use the amino acid *glutamate* and the neuropeptide *Substance P* as neurotransmitters. Painful sensations are distributed as follows: The axon synapses on an interneuron, whose axon crosses the spinal cord to ascend within the lateral spinothalamic tract. After a synapse in the thalamus, the information is relayed to the primary sensory cortex.

Two types of axons—Type A and Type C fibers (∞ *p. 384*)—carry painful sensations. Myelinated Type A fibers carry sensations of **fast pain,** or *prickling pain.* An injection or a deep cut produces this type of pain. These sensations very quickly reach the CNS, where they often trigger somatic reflexes. They are also relayed to the primary sensory cortex and so receive conscious attention. In most cases, the arriving information permits the stimulus to be localized to an area several inches in diameter.

Slower, Type C fibers carry sensations of **slow pain,** or *burning and aching pain.* These sensations cause a generalized activation of the reticular formation and thalamus. The individual becomes aware of the pain but has only a general idea of the area affected.

Pain receptors are tonic receptors. Significant peripheral adaptation does not occur, and the receptors continue to respond as long as the painful stimulus remains. Painful sensations cease only after tissue damage has ended. However, central adaptation may reduce the *perception* of the pain while the pain receptors are still stimulated. This effect involves the inhibition of centers in the thalamus, reticular formation, lower brain stem, and spinal cord.

An understanding of the origins of pain sensations and an ability to control or reduce pain levels have always been among the most important aspects of med-

(a) Free nerve endings

(b) Root hair plexus

(c) Merkel cells and tactile discs

(d) Tactile corpuscle LM × 330

(e) Lamellated corpuscle LM × 75

(f) Ruffini corpuscle

•FIGURE 17-2 Tactile Receptors in the Skin

ical treatment. After all, it is usually pain that induces someone to seek treatment; conditions that are not painful are typically ignored or tolerated. Although we often use the term *pain pathways,* it is becoming clear that pain distribution and perception are extremely complex—more so than had previously been imagined. AM *A Closer Look: Pain Mechanisms, Pathways, and Control*

THERMORECEPTORS

Temperature receptors, or **thermoreceptors,** are free nerve endings located in the dermis of the skin, in skeletal muscles, in the liver, and in the hypothalamus. Cold receptors are three or four times more numerous than warm receptors. No structural differences between warm and cold thermoreceptors have been identified.

Temperature sensations are conducted along the same pathways that carry pain sensations. They are sent to the reticular formation, the thalamus, and (to a lesser extent) the primary sensory cortex. Thermoreceptors are phasic receptors: They are very active when the temperature is changing, but they quickly adapt to a stable temperature. When you enter an air-conditioned classroom on a hot summer day or a warm lecture hall on a brisk fall evening, the temperature change seems extreme at first, but you quickly become comfortable as adaptation occurs.

MECHANORECEPTORS

Mechanoreceptors are sensitive to stimuli that distort their cell membranes. These membranes contain *mechanically regulated ion channels* whose gates open or close in response to stretching, compression, twisting, or other distortions of the membrane. There are three classes of mechanoreceptors:

1. **Tactile receptors,** which provide sensations of touch, pressure, and vibration. These sensations are closely related. Touch sensations provide information about shape or texture, whereas pressure sensations indicate the degree of mechanical distortion. Vibration sensations indicate a pulsing or oscillating pressure. The receptors involved may be specialized in some way. For example, rapidly adapting tactile receptors are best suited for detecting vibration. But your interpretation of a sensation as touch rather than pressure is typically a matter of the degree of stimulation and not of differences in the type of receptor stimulated.

2. **Baroreceptors** (bar-ō-rē-SEP-torz; *baro-,* pressure), which detect pressure changes in the walls of blood vessels and in portions of the digestive, reproductive, and urinary tracts.

3. **Proprioceptors,** which monitor the positions of joints and muscles. These are the most structurally and functionally complex of the general sensory receptors.

Tactile Receptors

Fine touch and pressure receptors provide detailed information about a source of stimulation, including its exact location, shape, size, texture, and movement. These receptors are extremely sensitive and have relatively narrow receptive fields. The sensory information reaches our conscious awareness through the posterior column pathway. ∞ *p. 482* **Crude touch and pressure receptors** provide poor localization and, because they have relatively large receptive fields, give little additional information about the stimulus. These sensations ascend within the anterior spinothalamic tract, and the thalamus relays the information to appropriate areas of the primary sensory cortex. ∞ *p. 484*

Tactile receptors range in complexity from free nerve endings to specialized sensory complexes with accessory cells and supporting structures. Figure 17-2• shows six types of tactile receptors in the skin:

1. Free nerve endings sensitive to touch and pressure are situated between epidermal cells (Figure 17-2a•). There appear to be no structural differences between these receptors and the free nerve endings that provide temperature or pain sensations. These are the only sensory receptors on the corneal surface of the eye, but in other portions of the body surface, more-specialized tactile receptors are probably more important. Free nerve endings that provide touch sensations are tonic receptors with small receptive fields.

2. Wherever hairs are located, the nerve endings of the **root hair plexus** monitor distortions and movements across the body surface (Figure 17-2b•). When a hair is displaced, the movement of the follicle distorts the sensory dendrites and produces action potentials. These receptors adapt rapidly, so they are best at detecting initial contact and subsequent movements. For

example, you generally feel your clothing only when you move or when you consciously focus on tactile sensations from the skin.

3. **Tactile discs,** or *Merkel's* (MER-kelz) *discs,* are fine touch and pressure receptors (Figure 17-2c•). They are extremely sensitive tonic receptors, with very small receptive fields. The dendritic processes of a single myelinated afferent fiber make close contact with unusually large epithelial cells in the stratum germinativum of the skin; we described these *Merkel cells* in Chapter 5. ∞ *p. 146*

4. **Tactile corpuscles,** or *Meissner's* (MĪS-nerz) *corpuscles,* perceive sensations of fine touch and pressure and low-frequency vibration. They adapt to stimulation within a second after contact. Tactile corpuscles are fairly large structures, measuring roughly 100 μm in length and 50 μm in width. These receptors are most abundant in the eyelids, lips, fingertips, nipples, and external genitalia. The dendrites are highly coiled and interwoven, and they are surrounded by modified Schwann cells. A fibrous capsule surrounds the entire complex and anchors it within the dermis (Figure 17-2d•).

5. **Lamellated** (LAM-e-lā-ted; *lamella,* a little thin plate) **corpuscles,** or *pacinian* (pa-SIN-ē-an) *corpuscles,* are sensitive to deep pressure. Because they are fast-adapting receptors, they are most sensitive to pulsing or high-frequency vibrating stimuli. A single dendrite lies within a series of concentric layers of collagen fibers and supporting cells (specialized fibroblasts) (Figure 17-2e•). The entire corpuscle may reach 4 mm in length and 1 mm in diameter. The concentric layers, separated by interstitial fluid, shield the dendrite from virtually every source of stimulation other than direct pressure. Lamellated corpuscles adapt quickly because distortion of the capsule soon relieves pressure on the sensory process. These receptors are scattered throughout the dermis, notably in the fingers, mammary glands, and external genitalia. They occur also in the superficial and deep fasciae, in joint capsules, in mesenteries, in the pancreas, and in the walls of the urethra and urinary bladder.

6. **Ruffini** (roo-FĒ-nē) **corpuscles** are also sensitive to pressure and distortion of the skin, but they are located in the reticular (deep) dermis. These receptors are tonic and show little if any adaptation. The capsule surrounds a core of collagen fibers that are continuous with those of the surrounding dermis (Figure 17-2f•). In the capsule, a network of dendrites is intertwined with the collagen fibers. Any tension or distortion of the dermis tugs or twists the capsular fibers, stretching or compressing the attached dendrites and altering the activity in the myelinated afferent fiber.

In Chapter 15, we considered the central distribution of tactile sensations through the posterior column and spinothalamic pathways. ∞ *pp. 482–485* Our sensitivity to tactile sensations may be altered by infection, disease, or damage to sensory neurons or pathways. As a result, mapping tactile responses can

sometimes aid clinical assessment. Sensory losses with clear regional boundaries indicate trauma to spinal nerves. For example, sensory loss along the boundary of a dermatome can help identify the affected spinal nerve or nerves. ∞ *p. 416* Regional sensitivity to light touch can be checked by gentle contact with a fingertip or a slender wisp of cotton. Vibration receptors are tested by applying the base of a tuning fork to the skin. We discuss more-detailed procedures, such as the *two-point discrimination test,* in the *Applications Manual.*
AM *Assessment of Tactile Sensitivities*

Tickling and Itching

Tickle and itch sensations are closely related to the sensations of touch and pain. The receptors involved are free nerve endings, and the information is carried by unmyelinated Type C fibers. Tickle sensations, which are usually (but not always) described as pleasurable, are produced by a light touch that moves across the skin. Psychological factors are involved in the interpretation of tickle sensations, and tickle sensitivity differs greatly among individuals. Little is known about the labeled line involved, but the sensations travel over the spinothalamic pathway. Itching is probably produced by the stimulation of the same receptors. Specific "itch spots" can be mapped in the skin, the inner surfaces of the eyelids, and the mucous membrane of the nose. Itch sensations are absent from other mucous membranes and from deep tissues and viscera. Itching is extremely unpleasant, even more unpleasant than pain. Individuals with extreme itching will scratch even when pain is the result. Itch receptors can be stimulated by the injection of histamine or proteolytic enzymes into the epidermis and superficial dermis. The precise receptor mechanism is unknown.

Baroreceptors

Baroreceptors monitor changes in pressure. A baroreceptor consists of free nerve endings that branch within the elastic tissues in the wall of a distensible organ, such as a blood vessel or a portion of the respiratory, digestive, or urinary tract. When the pressure changes, the elastic walls of the tract recoil or expand. This movement distorts the dendritic branches and alters the rate of action potential generation. Baroreceptors respond immediately to a change in pressure, but they adapt rapidly, and the output along the afferent fibers gradually returns to normal. Table 17-1 surveys baroreceptor distribution and function.

Baroreceptors monitor blood pressure in the walls of major vessels, including the carotid artery (at the *carotid sinus*) and the aorta (at the *aortic sinus*). The information plays a major role in regulating cardiac function and adjusting blood flow to vital tissues. Baroreceptors in the lungs monitor the degree of lung expansion. This information is relayed to the respiratory rhythmicity centers, which set the pace of respi-

ration. Comparable stretch receptors in the digestive and urinary tracts trigger a variety of visceral reflexes, including those of urination and defecation. We shall describe those baroreceptor reflexes in chapters that deal with specific physiological systems.

Proprioceptors

Proprioceptors monitor the position of joints, the tension in tendons and ligaments, and the state of muscular contraction. There are three major groups of proprioceptors (Table 17-1):

1. *Muscle spindles.* Muscle spindles monitor skeletal muscle length and trigger stretch reflexes. ∞ *p. 426*
2. *Golgi tendon organs.* **Golgi tendon organs** are similar in function to Ruffini corpuscles but are located at the junction between a skeletal muscle and its tendon. In a Golgi tendon organ, dendrites branch repeatedly and wind around the densely packed collagen fibers of the tendon. These receptors are stimulated by tension in the tendon; they thus monitor the external tension developed during muscle contraction.
3. *Receptors in joint capsules.* Joint capsules are richly innervated by free nerve endings that detect pressure, tension, and movement at the joint. Your sense of body position results from the integration of information from these receptors with the information provided by muscle spindles, Golgi tendon organs, and the receptors of the inner ear.

Proprioceptors do not adapt to constant stimulation, and each receptor continuously sends information to the CNS. A relatively small proportion of the arriving proprioceptive information reaches your conscious awareness over the posterior column pathway. Most of the proprioceptive information is processed at subconscious levels. The *spinocerebellar pathway* delivers proprioceptive information to the cerebellum outside your conscious awareness. ∞ *p. 486*

CHEMORECEPTORS

Specialized chemoreceptive neurons can detect small changes in the concentration of specific chemicals or compounds. In general, **chemoreceptors** respond only to water-soluble and lipid-soluble substances that are dissolved in the surrounding fluid. These receptors exhibit peripheral adaptation over a period of seconds, and central adaptation may also occur. Table 17-1 surveys the locations and functions of chemoreceptors.

The chemoreceptors included in the general senses do not send information to the primary sensory cortex, so we are not consciously aware of the sensations they provide. The arriving sensory information is routed to brain stem centers that deal with the autonomic control of respiratory and cardiovascular functions. Neurons in the respiratory centers of the brain respond to the concentration of hydrogen ions (pH) and carbon dioxide molecules in the cerebrospinal fluid.

17

TABLE 17-1 **Baroreceptors, Proprioceptors, and Chemoreceptors**

Receptor Location(s)	Function(s)
Baroreceptors	
Carotid sinus and aortic sinus	Monitor arterial blood pressure; information goes to cardiovascular and respiratory control centers
Lung	Monitor lung stretching; information goes to respiratory rhythmicity centers for control of respiratory depth and rate
Intestinal tract (general)	Monitor volume of tract segments; trigger reflex movement of materials along the tract
Intestinal tract (colon)	Monitor volume of fecal material in colon; trigger defecation reflex
Urinary bladder	Monitor volume of urinary bladder; trigger micturition (urination) reflex
Proprioceptors	
Skeletal muscles (muscle spindles)	Monitor degree of stretch; trigger stretch reflexes; information relayed to higher centers is involved in somatic motor control
Muscle/tendon junctions (Golgi tendon organs)	Monitor amount of tension in the tendon; information relayed to higher centers is involved in somatic motor control
Receptors in joint capsules	Monitor position and rate of movement at the joint; information relayed to higher centers is involved in somatic motor control
Chemoreceptors	
Carotid bodies	Monitor pH, carbon dioxide, and oxygen levels in blood; information goes to respiratory and cardiovascular centers
Aortic bodies	Monitor pH, carbon dioxide, and oxygen levels in blood; information goes to respiratory and cardiovascular centers
Surface of medulla oblongata (near respiratory centers)	Monitor pH and carbon dioxide levels in cerebrospinal fluid; information goes to respiratory and cardiovascular centers

Chemoreceptive neurons are also located in the **carotid bodies,** near the origin of the internal carotid arteries on each side of the neck, and in the **aortic bodies,** between the major branches of the aortic arch. These receptors monitor the pH and the carbon dioxide and oxygen concentrations of arterial blood. The afferent fibers leaving the carotid or aortic bodies reach the respiratory centers by traveling within cranial nerves IX (glossopharyngeal) and X (vagus).

☑ Receptor A has a circular receptive field with a diameter of 2.5 cm. Receptor B has a circular receptive field 7.0 cm in diameter. Which receptor will provide more-precise sensory information?

☑ When the nociceptors in your hand are stimulated, what sensation do you perceive?

☑ What would happen to you if the information from proprioceptors in your legs were blocked from reaching the CNS?

We will next turn our attention to the five *special senses:* olfaction, gustation, vision, equilibrium, and hearing. Although the sense organs involved are structurally more complex than those of the general senses, the same basic principles of receptor function apply. ☒ *EMBRYOLOGY SUMMARY 13: The Development of Special Sense Organs*

OLFACTION

The sense of smell, more precisely called *olfaction,* is provided by paired **olfactory organs** (Figure 17-3●). These organs are located in the nasal cavity on either side of the nasal septum. The olfactory organs contain the following two layers:

1. An **olfactory epithelium,** which contains the **olfactory receptors,** supporting cells, and **basal cells** *(stem cells).* The olfactory epithelium covers the inferior surface of the cribriform plate, the superior portion of the perpendicular plate, and the superior nasal conchae of the ethmoid bone. ∞ *p. 204*

2. The *lamina propria,* an underlying layer of areolar tissue. ∞ *p. 129* This layer contains **olfactory glands,** also called *Bowman's glands,* whose secretions absorb water and form a thick, pigmented mucus. The lamina propria also contains numerous blood vessels and nerves.

When you draw air in through your nose, the air swirls and eddies within the nasal cavity. This turbulence brings airborne compounds to your olfactory organs. A normal, relaxed inhalation carries a small sample (about 2 percent) of the inhaled air to the olfactory organs. Sniffing repeatedly increases the flow of air across the olfactory epithelium, intensifying the

(a)

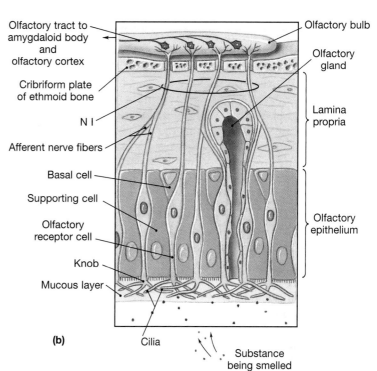

(b)

•**FIGURE 17-3** **The Olfactory Organs.** **(a)** The olfactory organ on the left side of the nasal septum. **(b)** An olfactory receptor is a modified neuron with multiple cilia extending from its free surface.

17

stimulation of the receptors. Once compounds have reached the olfactory organs, water-soluble and lipid-soluble materials must diffuse into the mucus before they can stimulate the olfactory receptors.

OLFACTORY RECEPTORS

Olfactory receptors are highly modified neurons. The exposed tip of each receptor cell forms a prominent knob that projects beyond the epithelial surface (Figure 17-3b•). That knob provides a base for up to 20 cilia that extend into the surrounding mucus. These cilia lie parallel to the epithelial surface, exposing their considerable surface area to dissolved compounds.

Olfactory reception occurs on the surfaces of the olfactory cilia as dissolved chemicals interact with receptors, called *odorant binding proteins,* on the membrane surface. *Odorants* are chemicals that stimulate olfactory receptors. In general, odorants are small organic molecules; the strongest smells are associated with molecules of high solubility both in water and in lipids. The binding of an odorant to its binding protein leads to the activation of adenylate cyclase, the enzyme that converts ATP to cyclic-AMP (cAMP). The cAMP then opens sodium channels in the membrane, which, as a result, begins to depolarize. If sufficient depolarization occurs, an action potential is triggered in the axon and the information is relayed to the CNS.

Between 10 and 20 million olfactory receptors are packed into an area of roughly 5 cm^2 (0.8 in.^{2}). If we take into account the exposed ciliary surfaces, the actual sensory area probably approaches that of the entire body surface. Nevertheless, our olfactory sensitivities cannot compare with those of other vertebrates such as dogs, cats, or fishes. A German shepherd sniffing for smuggled drugs or explosives has an olfactory receptor surface 72 times greater than that of the nearby customs inspector.

OLFACTORY PATHWAYS

The olfactory system is very sensitive. As few as four molecules of an odorous substance can activate an olfactory receptor. However, the activation of an afferent fiber does not guarantee a conscious awareness of the stimulus. Considerable convergence (∞ *pp. 396–397*) occurs along the olfactory pathway, and inhibition at the intervening synapses can prevent the sensations from reaching the *olfactory cortex* of the cerebral hemispheres. (The olfactory cortex is located on the inferior surface of each temporal lobe.) The olfactory receptors adapt very little to a persistent stimulus. It is central adaptation that ensures that you quickly lose awareness of a new smell but retain sensitivity to others.

Axons leaving the olfactory epithelium collect into 20 or more bundles that penetrate the cribriform plate of the ethmoid bone to reach the *olfactory bulbs* of the cerebrum (Figure 17-3•), where the first synapse occurs. Efferent fibers from nuclei elsewhere in the brain also innervate neurons of the olfactory bulbs. This arrangement provides a mechanism for central adaptation or facilitation of olfactory sensitivity. Axons leaving the olfactory bulb travel along the olfactory tract to reach the olfactory cortex, the hypothalamus, and portions of the limbic system.

Olfactory stimulation is the only type of sensory information that reaches the cerebral cortex without first synapsing in the thalamus. The extensive limbic and hypothalamic connections help explain the profound emotional and behavioral responses that can be produced by certain smells. The perfume industry, which understands the practical implications of these connections, expends considerable effort to develop odors that trigger sexual responses.

OLFACTORY DISCRIMINATION

The olfactory system can make subtle distinctions among 2000–4000 chemical stimuli. No apparent structural differences exist among the olfactory cells, but the epithelium as a whole contains receptor populations with distinct sensitivities. At least 50 "primary smells" are known, and our language does not allow us to describe these sensory impressions effectively. It appears likely that the CNS interprets each smell on the basis of the overall pattern of receptor activity.

 Aging and Olfactory Sensitivity

The olfactory receptor population shows considerable turnover, with new receptor cells being produced by the division and differentiation of basal cells in the epithelium. This turnover is one of the few examples of neuronal replacement in adult humans. Despite this process, the total number of receptors declines with age, and the remaining receptors become less sensitive. As a result, elderly individuals have difficulty detecting odors in low concentrations. This decline in the number of receptors accounts for Grandmother's tendency to apply perfume in excessive quantities and explains why Grandfather's aftershave seems so overdone: They must apply more to be able to smell it.

GUSTATION

Gustation, or taste, provides information about the foods and liquids we consume. **Taste receptors,** or *gustatory* (GUS-ta-tor-ē) *receptors,* are distributed over the superior surface of the tongue and adjacent portions of the pharynx and larynx. By the time we reach adulthood, the taste receptors on the pharynx and larynx have decreased in importance and abundance. The most important taste receptors are on the tongue, although a few remain on the epiglottis and adjacent areas of the pharynx. Taste receptors and specialized epithelial cells form sensory structures called **taste buds.** An adult has about 3000 taste buds.

The superior surface of the tongue bears epithelial projections called **lingual papillae** (pa-PIL-lē; *papilla,* a nipple-shaped mound). The human tongue bears three types of lingual papillae: (1) **filiform** (*filum,* thread) **papillae,** (2) **fungiform** (*fungus,* mushroom) **papillae,** and (3) **circumvallate** (sir-kum-VAL-āt) **papillae** (*circum-,* around + *vallum,* wall). The distribution of these lingual papillae varies by region (Figure 17-4a•). Filiform papillae provide friction against objects in the mouth. They do not contain taste buds. Each small fungiform papilla contains about five taste buds; each large circumvallate papilla contains as many as 100 taste buds. The circumvallate papillae form a V near the posterior margin of the tongue.

TASTE RECEPTORS

Each taste bud (Figure 17-4b,c•) contains about 40 slender receptors called **gustatory cells** and many supporting cells. Taste buds are recessed into the surrounding epithelium and are isolated from the relatively unprocessed contents of the mouth. Each gustatory cell extends slender microvilli, sometimes called *taste hairs,* into the surrounding fluids through the **taste pore,** a narrow opening. Despite this relatively protected position, it's still a hard life: A typical gustatory cell survives for only about 10 days before it is replaced by the division of nearby epithelial cells.

GUSTATORY PATHWAYS

Taste buds are monitored by cranial nerves VII (facial), IX (glossopharyngeal), and X (vagus). The facial nerve monitors all the taste buds located on the anterior two-thirds of the tongue, from the tip to the line of circumvallate papillae. The circumvallate papillae and the posterior one-third of the tongue are innervated by the glossopharyngeal nerve. The vagus nerve innervates taste buds scattered on the surface of the epiglottis. The sensory afferents carried by these cranial nerves synapse in the **nucleus solitarius,** or *solitary nucleus,* of the medulla oblongata, and the axons of the postsynaptic neurons enter the medial lemniscus. There the neurons join axons that carry somatic sensory information on touch, pressure, and proprioception. After another synapse in the thalamus, the information is projected to the appropriate portions of the primary sensory cortex.

A conscious perception of taste is produced as the information received from the taste buds is correlated with other sensory data. Information about the texture of food, along with taste-related sensations such as "peppery" or "burning hot," is provided by sensory afferents in the trigeminal nerve (N V). In addition, the level of stimulation from the olfactory receptors plays an overwhelming role in taste perception. You are several thousand times more sensitive to "tastes" when your olfactory organs are fully functional. When you have a cold, airborne molecules cannot reach your olfactory receptors, so meals taste dull and unappealing. This reduction in taste perception occurs even though the taste buds may be responding normally.

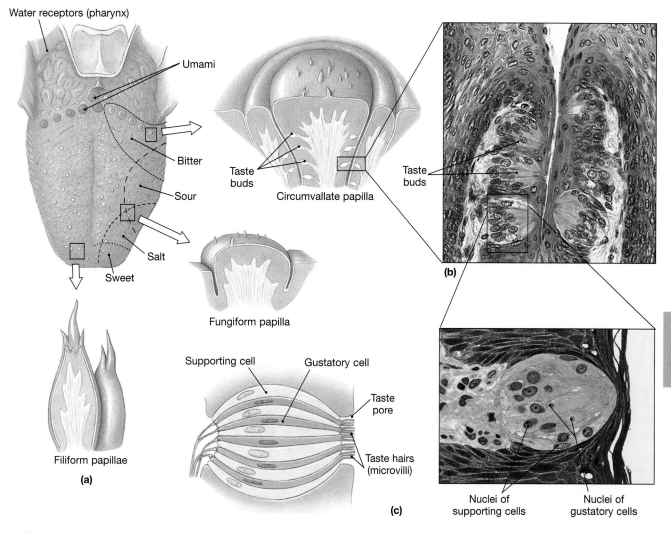

•**FIGURE 17-4 Gustatory Reception. (a)** Taste receptors are located in taste buds, which form pockets in the epithelium of fungiform or circumvallate papillae. **(b)** Taste buds in a circumvallate papilla. (LM × 280) **(c)** A taste bud, showing receptor (gustatory) cells and supporting cells. The diagrammatic view shows details of the taste pore that are not visible in the micrograph. (LM × 650)

GUSTATORY DISCRIMINATION

The mechanism behind gustatory reception resembles that of olfaction. Dissolved chemicals contacting the taste hairs bind to receptor proteins of the gustatory cell. The different tastes involve different receptor mechanisms. For example, salt receptors depolarize after Na^+ channels are opened, whereas sweet receptors depolarize after K^+ channels are closed. The end result of taste receptor stimulation is the release of neurotransmitter by the receptor cell. The dendrites of the sensory afferents are tightly wrapped by folds of the receptor cell membrane, and neurotransmitter release leads to the generation of action potentials in the afferent fiber. Taste receptors adapt slowly, but central adaptation quickly reduces your sensitivity to a new taste.

You are probably already familiar with the four **primary taste sensations:** sweet, salt, sour, and bit-ter. At least two additional tastes have been discovered in humans:

- *Umami.* **Umami** (oo-MAH-mē) is a pleasant taste that is characteristic of beef broth and chicken broth. This taste is produced by receptors sensitive to the presence of amino acids, especially glutamate, small peptides, and nucleotides. The distribution of these receptors is not known in detail, but they are present in taste buds of the circumvallate papillae.

- *Water.* Most people say that water has no flavor. However, research on humans and other vertebrates has demonstrated the presence of **water receptors,** especially in the pharynx. Their sensory output is processed in the hypothalamus and affects several systems that deal with water balance and the regulation of blood volume. For example, a mouthful of distilled water held for 20 minutes will inhibit ADH secretion and promote water loss at the kidneys.

A sensory map of the tongue indicates the regional differences in primary sensitivity (Figure 17-4a•). The threshold for receptor stimulation varies for each of the primary taste sensations, and the taste receptors respond more readily to unpleasant than to pleasant stimuli. For example, we are almost a thousand times more sensitive to acids, which taste sour, than to either sweet or salty chemicals, and we are a hundred times more sensitive to bitter compounds than to acids. This sensitivity has survival value, because acids can damage the mucous membranes of the mouth and pharynx and many potent biological toxins produce an extremely bitter taste.

Taste sensitivity differs significantly among individuals. Many conditions related to taste sensitivity are inherited. The best-known example involves sensitivity to the compound *phenylthiourea,* also known as *phenylthiocarbamide,* or **PTC.** Roughly 70 percent of Caucasians can taste this substance; the other 30 percent are unable to detect it.

⧗ Aging and Gustatory Sensitivity

Our tasting abilities change with age. We begin life with more than 10,000 taste buds, but the number begins declining dramatically by age 50. The sensory loss becomes especially significant because, as we noted, aging individuals also experience a decline in the number of olfactory receptors. As a result, many elderly people find that their food tastes bland and unappetizing, whereas children tend to find the same foods too spicy.

☑ When you first enter the A & P lab for dissection, you are very aware of the odor of preservatives. By the end of the lab period, the smell doesn't seem to be nearly as strong. Why?

☑ If you completely dry the surface of your tongue and then place salt or sugar crystals on it, you can't taste them. Why not?

☑ Your grandfather can't understand why foods he used to enjoy just don't taste the same anymore. Explain it to him.

VISION

We rely more on vision than on any other special sense. Our visual receptors are contained in the eyes, elaborate structures that enable us not only to detect light but also to create detailed visual images. We will begin our discussion of these fascinating organs by considering the *accessory structures* of the eye, which provide protection, lubrication, and support.

ACCESSORY STRUCTURES OF THE EYE

The **accessory structures** of the eye include the eyelids, the superficial epithelium of the eye, and the structures associated with the production, secretion, and removal of tears. Figure 17-5a• shows the superficial anatomy of the eye and its accessory structures.

Eyelids

The **eyelids,** or **palpebrae** (pal-PĒ-brē), are a continuation of the skin. The eyelids act like windshield wipers: Their continual blinking movements keep the surface lubricated and free from dust and debris. They can also close firmly to protect the delicate surface of the eye. The **palpebral fissure** is a gap that separates the free margins of the upper and lower eyelids. The two eyelids are connected, however, at the **medial canthus** (KAN-thus) and the **lateral canthus** (Figure 17-5•). The **eyelashes,** along the margins of the eyelids, are very robust hairs. These hairs help prevent foreign matter and insects from reaching the surface of the eye.

The eyelashes are associated with unusually large sebaceous glands. Along the inner margin of the lid, modified sebaceous glands called **tarsal glands,** or *Meibomian* (mī-BŌ-mē-an) *glands,* secrete a lipid-rich product that helps keep the eyelids from sticking together. At the medial canthus, the **lacrimal caruncle** (KAR-ung-kul), a soft tissue mass, contains glands producing the thick secretions that contribute to the gritty deposits that sometimes appear after a good night's sleep. These various glands are subject to occasional invasion and infection by bacteria. A *chalazion* (kah-LĀ-zē-on; small lump), or cyst, generally results from the infection of a tarsal gland. An infection in a sebaceous gland of one of the eyelashes, a tarsal gland, or one of the many sweat glands that open to the surface between the follicles produces a painful localized swelling known as a *sty.*

The skin covering the visible surface of the eyelid is very thin. Deep to the skin lie the muscle fibers of the *orbicularis oculi* and *levator palpebrae superioris muscles.* ∞ *p. 322* These skeletal muscles are responsible for closing the eyelids and raising the upper eyelid, respectively.

The epithelium covering the inner surfaces of the eyelids and the outer surface of the eye is called the **conjunctiva** (kon-junk-TĪ-vuh). It is a mucous membrane covered by a specialized stratified squamous epithelium. The **palpebral conjunctiva** covers the inner surface of the eyelids, and the **ocular conjunctiva,** or *bulbar conjunctiva,* covers the anterior surface of the eye (Figure 17-5b•). The ocular conjunctiva extends to the edges of the **cornea** (KOR-nē-uh), a transparent part of the outer fibrous layer of the eye. The cornea is covered by a very delicate squamous *corneal epithelium,* 5–7 cells thick, that is continuous with the ocular conjunctiva. A constant supply of fluid washes over the surface of the eyeball, keeping the ocular conjunctiva and cornea moist and clean. Goblet cells in the epithelium assist the accessory glands in providing a superficial lubricant that prevents friction and drying of the opposing conjunctival surfaces.

Conjunctivitis, or pinkeye, results from damage to and irritation of the conjunctival surface. The most obvious symptom, reddening, is due to the dilation of blood vessels deep to the conjunctival epithelium. This condition may be caused by pathogenic infection or by physical or chemical irritation of the conjunctival surface.

The Lacrimal Apparatus

A constant flow of tears keeps conjunctival surfaces moist and clean. Tears reduce friction, remove debris, prevent bacterial infection, and provide nutrients and oxygen to portions of the conjunctival epithelium. The **lacrimal apparatus** produces, distributes, and removes tears. The lacrimal apparatus of each eye consists of (1) a *lacrimal gland* with associated ducts, (2) *lacrimal canaliculi*, (3) a *lacrimal sac*, and (4) a *nasolacrimal duct.*

The pocket created where the palpebral conjunctiva becomes continuous with the ocular conjunctiva is known as the **fornix** (Figure 17-6a•).[1] The lateral portion of the superior fornix receives 10–12 ducts from the **lacrimal gland,** or tear gland (Figure 17-5b•). The lacrimal gland is about the size and shape of an almond, measuring roughly 12–20 mm (0.5–0.75 in.). It nestles within a depression in the frontal bone, just inside the orbit and superior and lateral to the eyeball. ∞ *p. 201* The lacrimal gland normally provides the key ingredients and most of the volume of the tears that bathe the conjunctival surfaces. Its secretions are watery and slightly alkaline, and they contain the enzyme **lysozyme,** which attacks bacteria.

The lacrimal gland produces about 1 ml of tears each day. Once the lacrimal secretions have reached the ocular surface, they mix with the products of

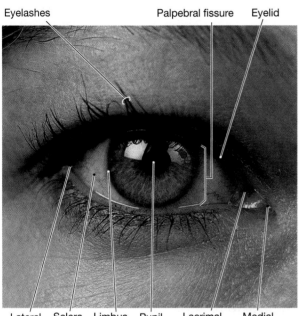

Eyelashes   Palpebral fissure   Eyelid

Lateral   Sclera   Limbus   Pupil   Lacrimal   Medial
canthus   caruncle   canthus

(a)

[1] *Fornix,* like *diaphragm* and *membrane,* is a term that refers to structures in several systems. Regardless of which system is involved, each fornix has an arched or curved shape. We have already discussed the fornix of the brain, a curving component of the limbic system.

17

Superior rectus muscle

Lacrimal gland

Lacrimal gland ducts

Ocular conjunctiva

Lateral canthus

Palpebral conjunctiva

Lower eyelid

Inferior rectus muscle

Inferior oblique muscle

Tendon of superior oblique muscle

Orbital fat

Lacrimal punctum

Superior lacrimal canaliculus

Medial canthus

Inferior lacrimal canaliculus

Lacrimal sac

Nasolacrimal duct

Opening of nasolacrimal duct

Inferior nasal concha

Inferior meatus

(b)

•**FIGURE 17-5** **External Features and Accessory Structures of the Eye.** **(a)** Gross and superficial anatomies of the accessory structures. **(b)** The organization of the lacrimal apparatus.

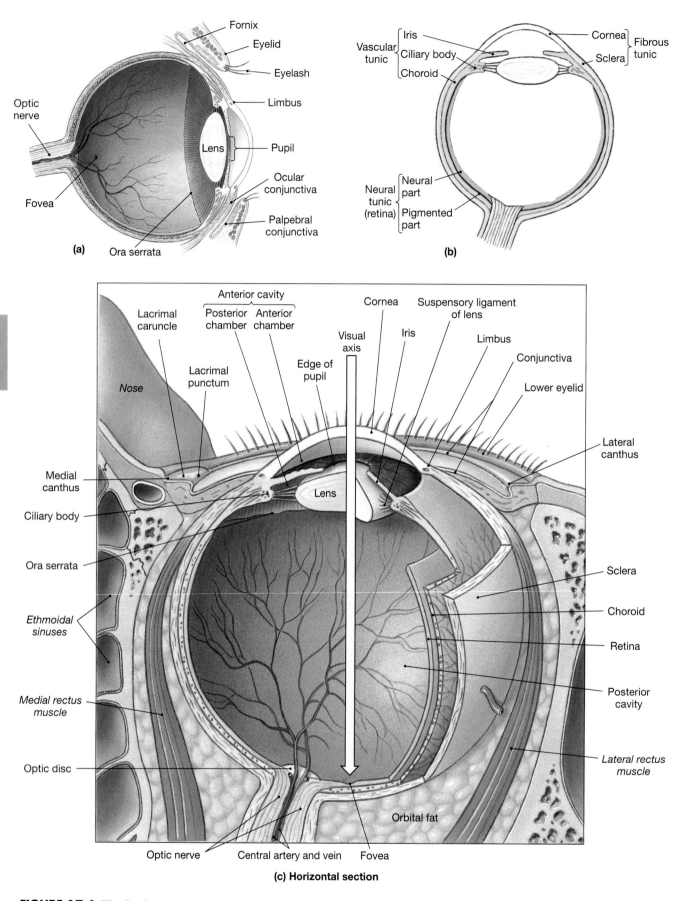

(a)

(b)

(c) Horizontal section

•**FIGURE 17-6 The Sectional Anatomy of the Eye. (a)** A sagittal section through the left eye, showing the position of the fornix. **(b)** A sagittal section, showing the three layers, or tunics, of the eye. **(c)** A horizontal section of the right eye.

accessory glands and the oily secretions of the tarsal glands. The result is a superficial "oil slick" that assists in lubrication and slows evaporation.

Blinking sweeps the tears across the ocular surface, and they accumulate at the medial canthus in an area known as the **lacrimal lake** *(lacus lacrimalis),* or "lake of tears." The lacrimal lake covers the lacrimal caruncle, which bulges anteriorly. The **lacrimal puncta** (singular, *punctum),* two small pores, drain the lacrimal lake. They empty into the **lacrimal canaliculi,** small canals that in turn lead to the **lacrimal sac** (Figure 17-5b●). This sac nestles within the lacrimal sulcus of the orbit. ∞ *p. 206* From the inferior portion of the lacrimal sac, the **nasolacrimal duct** passes through the *nasolacrimal canal,* formed by the lacrimal bone and the maxillary bone. The nasolacrimal duct delivers tears to the nasal cavity on that side. It empties into the *inferior meatus,* a narrow passageway inferior and lateral to the inferior nasal concha.

THE EYE

The eyes are extremely sophisticated visual instruments—more versatile and adaptable than the most expensive cameras, yet light, compact, and durable. Each eye is a slightly irregular spheroid with an average diameter of 24 mm (almost 1 in.), a little smaller than a Ping-Pong ball, and a weight of about 8 g (0.28 oz). Within the orbit, the eyeball shares space with the extrinsic eye muscles, the lacrimal gland, and the cranial nerves and blood vessels that supply the eye and adjacent portions of the orbit and face. **Orbital fat** pads and insulates the eye (Figures 17-5b and 17-6c●).

The wall of the eye contains three distinct layers, or *tunics* (Figure 17-6b●): (1) an outer *fibrous tunic,* (2) an intermediate *vascular tunic,* and (3) an inner *neural tunic.* The visual receptors, or *photoreceptors,* are located in the neural tunic. The eyeball is hollow; its interior can be divided into two cavities (Figure 17-6c●). The large **posterior cavity** is also called the *vitreous chamber,* because it contains the gelatinous *vitreous body.* The smaller **anterior cavity** is subdivided into two *chambers,* anterior and posterior. The shape of the eye is stabilized in part by the vitreous body and the clear *aqueous humor,* which fills the anterior cavity.

The Fibrous Tunic

The **fibrous tunic,** the outermost layer of the eye, consists of the *sclera* and the *cornea.* The fibrous tunic (1) provides mechanical support and some degree of physical protection, (2) serves as an attachment site for the extrinsic eye muscles, and (3) contains structures that assist in the focusing process.

Most of the ocular surface is covered by the **sclera** (SKLER-uh) (Figure 17-6b,c●). The sclera, or "white of the eye," consists of a dense fibrous connective tissue containing both collagen and elastic fibers. This layer is thickest over the posterior surface of the eye, near the exit of the optic nerve, and thinnest over the anterior surface. The six extrinsic eye muscles insert on the sclera, blending their collagen fibers with those of the fibrous tunic. ∞ *p. 323*

The surface of the sclera contains small blood vessels and nerves that penetrate the sclera to reach internal structures. The network of small vessels interior to the ocular conjunctiva generally does not carry enough blood to lend an obvious color to the sclera, but on close inspection the vessels are visible as red lines against the white background of collagen fibers.

The transparent cornea is structurally continuous with the sclera; the **limbus** (Figures 17-5a, 17-6a,c●) is the border between the two. Deep to the delicate corneal epithelium, the cornea consists primarily of a dense matrix containing multiple layers of collagen fibers. These fibers are organized into a series of layers that do not interfere with the passage of light. The cornea has no blood vessels. The superficial epithelial cells must obtain oxygen and nutrients from the tears that flow across their free surfaces. The cornea also has numerous free nerve endings, and it is the most sensitive portion of the eye.

Corneal Transplants

Corneal damage will cause blindness even though the rest of the eye—including the photoreceptors—may be perfectly normal. The cornea has a very restricted ability to repair itself, so corneal injuries must be treated immediately to prevent serious vision losses. Restoring vision after corneal scarring generally requires the replacement of the cornea through a corneal transplant. Corneal replacement is probably the most common form of transplant surgery. Such transplants can be performed between unrelated individuals, because there are no blood vessels to carry white blood cells, which attack foreign tissues, into the area. Corneal grafts are obtained from the eyes of donors who have died from illness or accident. For best results, the tissues must be removed within 5 hours after the donor's death.

The Vascular Tunic (Uvea)

The **vascular tunic,** or **uvea,** contains numerous blood vessels, lymphatic vessels, and the intrinsic eye muscles (Figure 17-6b,c●). The functions of this middle layer include (1) providing a route for blood vessels and lymphatics that supply tissues of the eye; (2) regulating the amount of light that enters the eye; (3) secreting and reabsorbing the *aqueous humor* that circulates within the eye; and (4) controlling the shape of the *lens,* an essential part of the focusing process. The vascular tunic includes the *iris,* the *ciliary body,* and the *choroid.*

17

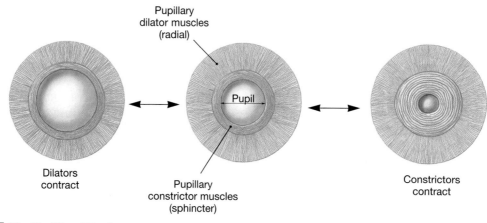

Pupillary
dilator muscles
(radial)

Pupil

Dilators
contract

Pupillary
constrictor muscles
(sphincter)

Constrictors
contract

•**FIGURE 17-7** The Pupillary Muscles

The Iris   The **iris,** which we can see through the transparent corneal surface, contains blood vessels, pigment cells, and two layers of smooth muscle fibers (Figure 17-7•). When these muscles contract, they change the diameter of the **pupil,** or central opening of the iris. One group of smooth muscle fibers, the **pupillary constrictor muscles,** forms a series of concentric circles around the pupil. When these sphincter muscles contract, the diameter of the pupil decreases. A second group of smooth muscles, the **pupillary dilator muscles,** extends radially away from the edge of the pupil. Contraction of these muscles enlarges the pupil. Both muscle groups are controlled by the autonomic nervous system. For example, parasympathetic activation in response to bright light causes the pupils to constrict (the *consensual light reflex*), and sympathetic activation in response to dim light causes the pupils to dilate (the *pupillary reflex*). ∞ *p. 520*

The body of the iris consists of a highly vascular, pigmented, loose connective tissue. The anterior surface has no epithelial covering; instead, it has an incomplete layer of fibroblasts and melanocytes. Melanocytes are also scattered within the body of the iris. The posterior surface is covered by a pigmented epithelium that contains melanin granules. This pigmented epithelium is part of the neural tunic. Your eye color is determined by the density and distribution of melanocytes and by the density of the pigmented epithelium. When the connective tissue of the iris contains few melanocytes, light passes through it and bounces off the pigmented epithelium. The eye then appears blue. Individuals with gray, brown, or black eyes have increasing numbers of melanocytes in the body and surface of the iris. The eyes of human albinos appear a very pale gray or blue-gray.

The Ciliary Body   At its periphery, the iris attaches to the anterior portion of the ciliary body. The **ciliary body** is a thickened region that begins deep to the junction between the cornea and the sclera. It extends posteriorly to the level of the **ora serrata** (Ō-ra ser-RA-tuh; serrated mouth), the serrated anterior edge of the thick, inner portion of the neural tunic (Figure 17-6•). The bulk of the ciliary body consists of the **ciliary muscle,** a smooth muscular ring that projects into the interior of the eye. Its epithelium is thrown into numerous folds called **ciliary processes.** The **suspensory ligaments** of the lens attach to these processes. These connective tissue fibers hold the lens posterior to the iris and centered on the pupil. As a result, any light passing through the pupil and headed for the photoreceptors will pass through the lens.

The Choroid   The **choroid** is a vascular layer that separates the fibrous and neural tunics posterior to the ora serrata (Figure 17-6c•). It is covered by the sclera and attached to the outermost layer of the retina. The choroid contains an extensive capillary network that delivers oxygen and nutrients to the retina. It also contains melanocytes, which are especially numerous near the sclera.

The Neural Tunic

The **neural tunic,** or **retina,** is the innermost layer of the eye. It consists of a thin, outer layer, called the *pigmented part,* and a thick inner layer, called the *neural part.* The pigmented part of the retina absorbs light that passes through the neural part. The pigment cells also have important biochemical interactions with the retina's light receptors, which are located in the neural part of the retina. In addition to light receptors, the neural part of the retina contains supporting cells and neurons that perform preliminary processing and integration of visual information, as well as blood vessels supplying the tissues that line the posterior cavity.

The two layers of the retina are normally very close together but not tightly interconnected. The pigmented part of the retina continues over the ciliary body and iris, although the neural part extends anteriorly only as far as the ora serrata. The neural part of the retina thus forms a cup that establishes the posterior and lateral boundaries of the posterior cavity (Figure 17-6b,c•).

Organization of the Retina In sectional view, the retina contains several layers of cells (Figure 17-8a•). The outermost layer, closest to the pigmented part of the retina, contains the **photoreceptors,** or cells that detect light.

Photoreceptors are entirely dependent on the diffusion of oxygen and nutrients from blood vessels in the choroid. In a **detached retina,** the neural part of

•**FIGURE 17-8 The Organization of the Retina. (a)** The cellular organization of the retina. The photoreceptors are closest to the choroid rather than near the posterior cavity (vitreous chamber). (LM × 290) **(b)** The optic disc in diagrammatic horizontal section. **(c)** A photograph of the retina as seen through the pupil.

the retina becomes separated from the pigmented part. This condition can result from a variety of factors, including a sudden impact to the eye. Unless the two parts of the neural tunic are reattached, the photoreceptors will degenerate and vision will be lost. The reattachment is generally performed by "welding" the two layers together by means of laser beams focused through the cornea. These beams heat the layers and thereby fuse them together at several points around the retina. However, this procedure destroys the photoreceptors and other cells at these locations.

The eye has two types of photoreceptors: rods and cones. **Rods** do not discriminate among colors of light. They are very light-sensitive and enable us to see in dimly lit rooms, at twilight, and in pale moonlight. **Cones** provide us with color vision. Three types of cones are present, and their stimulation in various combinations provides the perception of different colors. Cones give us sharper, clearer images than rods do, but cones require more-intense light. If you sit outside at sunset, you will probably be able to tell when your visual system shifts from cone-based vision (clear images in full color) to rod-based vision (relatively grainy images in black and white).

Rods and cones are not evenly distributed across the outer surface of the retina. Approximately 125 million rods form a broad band around the periphery of the retina. Roughly 6 million cones span the retina's posterior surface. Most of these are concentrated in the area where a visual image arrives after it passes through the cornea and lens. This region, which is known as the **macula lutea** (LOO-tē-uh; yellow spot), contains no rods. The highest concentration of cones occurs at the center of the macula lutea, an area called the **fovea** (FŌ-vē-uh; shallow depression), or *fovea centralis* (Figure 17-8c●). The fovea is the site of sharpest vision: When you look directly at an object, its image falls on this portion of the retina. A line drawn from the center of that object through the center of the lens to the fovea establishes the **visual axis** of the eye (Figure 17-6c●).

You are probably already aware of the visual consequences of this distribution. When you look directly at an object, you are focusing its image on your fovea, the center of color vision. You see a very good image as long as there is enough light to stimulate the cones. In very dim light, cones cannot function. For example, when you try to stare at a dim star, you are unable to see it. But if you look a little to one side rather than directly at the star, you will see it quite clearly. Shifting your gaze moves the image of the star from the fovea, where it does not provide enough light to stimulate the cones, to the periphery, where it can affect the more-sensitive rods.

Rods and cones synapse with roughly 6 million neurons called **bipolar cells** (Figure 17-8a●). Bipolar cells in turn synapse within the layer of neurons called **ganglion cells** adjacent to the posterior cavity. A network of **horizontal cells** extends across the outer portion of the retina at the level of the synapses between photoreceptors and bipolar cells. A comparable layer of **amacrine** (AM-a-krin) **cells** occurs where bipolar cells synapse with ganglion cells. Horizontal and amacrine cells can facilitate or inhibit communication between photoreceptors and ganglion cells, thereby adjusting the sensitivity of the retina. The effect could be compared to adjusting the "contrast" setting on a television. These cells play an important role in the eye's adjustment to dim or brightly lit environments.

The Optic Disc Axons from an estimated 1 million ganglion cells converge on the **optic disc,** a circular region just medial to the fovea. The optic disc is the origin of the optic nerve (N II). From this point, the axons turn, penetrate the wall of the eye, and proceed toward the diencephalon (Figure 17-8b●). The *central retinal artery* and *central retinal vein,* which supply the retina, pass through the center of the optic nerve and emerge on the surface of the optic disc (Figure 17-8b,c●). The optic disc has no photoreceptors or other structures typical of the rest of the retina. Because light striking this area goes unnoticed, the optic disc is commonly called the **blind spot.** You do not notice a blank spot in your field of vision primarily because involuntary eye movements keep the visual image moving and allow your brain to fill in the missing information. However, a simple experiment, shown in Figure 17-9●, will demonstrate the presence and location of the blind spot in your field of vision.

●**FIGURE 17-9** **The Optic Disc.** Close your left eye and stare at the cross with your right eye, keeping the cross in the center of your field of vision. Begin with the page a few inches away from your eye, and gradually increase the distance. The dot will disappear when its image falls on the blind spot, at your optic disc. To check the blind spot in your left eye, close your right eye and repeat this sequence while you stare at the dot.

Diabetic Retinopathy

A *retinopathy* is a disease of the retina. **Diabetic retinopathy** develops in most individuals with *diabetes mellitus*, an endocrine disorder that affects primarily glucose metabolism. Many systems are affected, but serious circulatory problems are very common. Diabetic retinopathy develops over a period of years owing to the degeneration and rupture of blood vessels of the retina. Visual acuity is lost, and over time the photoreceptors are destroyed as a result of oxygen starvation.

The Chambers of the Eye

As we learned earlier, the ciliary body and lens divide the interior of the eye into a large posterior cavity, or vitreous chamber, and a smaller anterior cavity. The anterior cavity is subdivided into the **anterior chamber,** which extends from the cornea to the iris, and a **posterior chamber,** between the iris and the ciliary body and lens. The anterior and posterior chambers are filled with the fluid *aqueous humor.* The posterior cavity is filled with a gelatinous substance known as the *vitreous body,* or *vitreous humor.*

Aqueous Humor   **Aqueous humor** is a fluid that circulates within the anterior cavity, passing from the posterior to the anterior chamber through the pupil (Figure 17-10•). This fluid forms through active secretion by epithelial cells of the ciliary body's ciliary

processes. It is secreted into the posterior chamber at a rate of 1–2 µl per minute. The epithelial cells regulate its composition, which resembles that of cerebrospinal fluid. Because aqueous humor circulates, it provides an important route for nutrient and waste transport in addition to forming a fluid cushion.

The eye is filled with fluid, and fluid pressure in the aqueous humor helps retain the eye's shape. Fluid pressure also stabilizes the position of the retina, pressing the neural part of the retina against the pigmented part. In effect, the aqueous humor acts like the air inside a balloon. The **intraocular pressure** can be measured in the anterior chamber, where the fluid pushes against the inner surface of the cornea. Intraocular pressure is most often checked by bouncing a tiny blast of air off the surface of the eye and then measuring the deflection produced. Normal intraocular pressures range from 12 to 21 mm Hg.

Aqueous humor returns to the circulation from the anterior chamber. After filtering through a network of connective tissues located near the base of the iris, aqueous humor enters the **canal of Schlemm,** or *scleral venous sinus,* a passageway that extends completely around the eye at the level of the limbus. Collecting channels deliver the aqueous humor from this canal to veins in the sclera. The rate of removal normally keeps pace with the rate of generation at the ciliary processes, and aqueous humor is removed and recycled within a few hours of its formation.

17

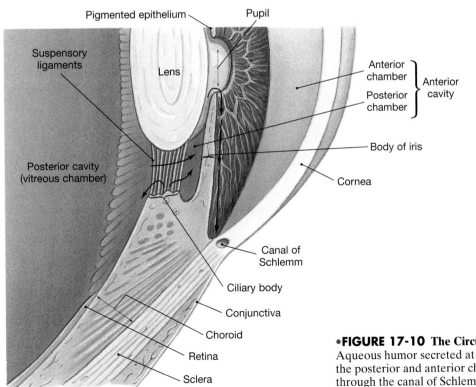

•**FIGURE 17-10 The Circulation of Aqueous Humor.**
Aqueous humor secreted at the ciliary body circulates through the posterior and anterior chambers before it is reabsorbed through the canal of Schlemm.

17

The Vitreous Body The posterior cavity of the eye contains the **vitreous body** (or *vitreous humor*), a gelatinous mass. The vitreous body helps stabilize the shape of the eye and gives additional physical support to the retina. Specialized cells embedded in the vitreous body produce the collagen fibers and proteoglycans that account for the gelatinous consistency of this mass. Unlike the aqueous humor, the vitreous body is not replaced. Aqueous humor produced in the posterior chamber freely diffuses through the vitreous body and across the surface of the retina.

The Lens

The **lens** lies posterior to the cornea, held in place by the suspensory ligaments that originate on the ciliary body of the choroid (Figures 17-6b, p. 540, and 17-10•). The primary function of the lens is to focus the visual image on the photoreceptors. The lens does so by changing its shape.

The lens consists of concentric layers of cells that are precisely organized. A dense fibrous capsule covers the entire lens. Many of the capsular fibers are elastic. Unless an outside force is applied, they will contract and make the lens spherical. Around the edges of the lens, the capsular fibers intermingle with those of the suspensory ligaments. The cells in the interior of the lens are called **lens fibers.** These highly specialized cells have lost their nuclei and other organelles. They are slender and elongate and are filled with transparent proteins called **crystallins.** Crystallins are responsible for both the clarity and the focusing power of the lens. They are extremely stable proteins that remain intact and functional for a lifetime without the need for replacement.

Cataracts

The transparency of the lens depends on a precise combination of structural and biochemical characteristics. When that balance becomes disturbed, the lens loses its transparency; the abnormal lens is known as a **cataract.** Cataracts can result from drug reactions, injuries, or radiation, but **senile cataracts,** a natural consequence of aging, are the most common form.

Over time, the lens turns yellowish and eventually begins to lose its transparency. As the lens becomes "cloudy," the individual needs brighter and brighter light for reading, and visual clarity begins to fade. If the lens becomes completely opaque, the person will be functionally blind, even though the photoreceptors are normal. Surgical procedures involve removal of the lens, either intact or in pieces, after it has been shattered with high-frequency sound. The missing lens is replaced by an artificial substitute, and vision is then fine-tuned with glasses or contact lenses.

Refraction The retina has about 130 million photoreceptors, each monitoring a specific location. A visual image results from the processing of information from all the receptors. The eye is often compared to a camera. To provide useful information, the lens of the eye, like a camera lens, must focus the arriving image. "In focus" means that the rays of light arriving from an object strike the sensitive surface of the retina (or film) in precise order so as to form a miniature image of the original. If the rays are not perfectly focused, the image is blurry. Focusing normally occurs in two steps, as light passes through (1) the cornea and (2) the lens.

Light is **refracted,** or bent, when it passes from one medium to a medium with a different density. You can demonstrate this by sticking a pencil into a glass of

Glaucoma

If aqueous humor cannot enter the canal of Schlemm, the condition of **glaucoma** results. Although drainage is impaired, the production of aqueous humor continues, so the intraocular pressure begins to rise. Because the sclera is a fibrous coat, it cannot expand significantly. The increasing pressure therefore begins to distort soft tissues within the eye. The tough sclera cannot enlarge like an inflating balloon, but it does have one weak point, where the optic nerve penetrates the wall of the eye. Because the optic nerve must penetrate all three tunics, it is not wrapped in connective tissue. When intraocular pressures have risen to roughly twice normal levels, the distortion of the nerve fibers begins to block the propagation of action potentials and vi-

sion begins to deteriorate. If this condition is not corrected, blindness eventually results.

Glaucoma affects roughly 2 percent of the population over age 35. In most cases, the primary factors responsible cannot be determined. Because glaucoma is a relatively common condition—over 2 million cases in the United States alone—most eye exams include a test of intraocular pressure. Glaucoma may be treated by the application of drugs that constrict the pupil and tense the edge of the iris, making the surface more permeable to aqueous humor. Surgical correction involves perforating the wall of the anterior chamber to encourage drainage. This procedure is now performed by laser surgery on an outpatient basis.

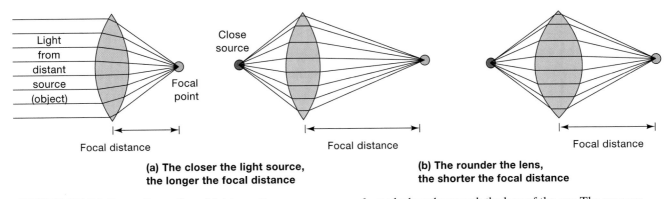

(a) The closer the light source, the longer the focal distance

(b) The rounder the lens, the shorter the focal distance

•**FIGURE 17-11** **Image Formation.** Light rays from a source are refracted when they reach the lens of the eye. The rays are then focused onto a single focal point. **(a)** The focal distance increases as the object nears the lens. **(b)** A rounder lens has a shorter focal distance than a flatter lens does.

water. Because refraction occurs as the light passes into the air from the much denser water, the pencil shaft appears to bend sharply at the air–water interface.

In the human eye, the greatest amount of refraction occurs when light passes from the air into the corneal tissues, which have a density close to that of water. When you open your eyes underwater, you cannot see clearly because the air–water interface has been eliminated; light is passing from one watery medium to another.

Additional refraction takes place when the light passes from the aqueous humor into the relatively dense lens. The lens provides the extra refraction you need to focus the light rays from an object toward a **focal point,** a specific point of intersection on the retina. The distance between the center of the lens and its focal point is the **focal distance.** The focal distance of a lens, whether the lens of your eye or the lens in your camera, is determined by the following two factors:

1. *The distance from the object to the lens.* The focal distance increases as an object moves closer to the lens (Figure 17-11a•).

2. *The shape of the lens.* The rounder the lens, the more refraction occurs, so a very round lens has a shorter focal distance than a flatter one (Figure 17-11b•).

Accommodation Using a camera, we can focus an image by moving the lens toward or away from the film. This method of focusing cannot work in our eyes, because the distance from the lens to the macula lutea cannot change. We focus images on the retina by changing the shape of the lens to keep the focal length constant. This process is **accommodation.** During accommodation, the lens becomes rounder to focus the image of a nearby object on the retina; the lens flattens when we focus on a distant object.

The lens is held in place by the suspensory ligaments that originate at the ciliary body. Smooth muscle fibers in the ciliary body act like sphincter muscles. When the ciliary muscle contracts, the ciliary body moves toward the lens. This movement reduces the tension in the suspensory ligaments, and the elastic capsule pulls the lens into a more spherical shape. The rounder shape increases the refractive power of the lens, enabling it to bring light from nearby objects into focus on the retina (Figure 17-12a•). When the ciliary muscle relaxes, the suspensory ligaments pull at the circumference of the lens, making the lens flatter (Figure 17-12b•).

The greatest amount of refraction is required to view objects that are very close to the lens. The inner

(a) Ciliary muscle contracted, lens rounded for close vision

(b) Ciliary muscle relaxed, lens flattened for distant vision

•**FIGURE 17-12** **Accommodation.** For the eye to form a sharp image, the focal distance must equal the distance between the center of the lens and the retina. **(a)** When the ciliary muscle contracts, the ligaments allow the lens to round up. **(b)** When the ciliary muscle relaxes, the suspensory ligaments pull against the margins of the lens and flatten it.

Accommodation Problems

In the healthy eye, when the ciliary muscle is relaxed and the lens is flattened, a distant image will be focused on the retina's surface (Figure 17-13a•). This condition is called **emmetropia** (*emmetro-,* proper), or normal vision.

Figure 17-13b,c• diagrams two common problems with the accommodation mechanism. If the eyeball is too deep or the resting curvature of the lens is too great, the image of a distant object will form in front of the retina (Figure 17-13b•). The individual will see distant objects as blurry and out of focus. Vision at close range will be normal, because the lens will be able to round up as needed to focus the image on the retina. As a result, such individuals are said to be *nearsighted.* Their condition is more formally termed **myopia** (*myein,* to shut + *ops,* eye). Myopia can be treated by placing a *diverging* lens in front of the eye (Figure 17-13d•). This lens shape, typical of the lenses used in prescription corrective glasses, spreads the light rays apart as if the object were closer to the viewer.

If the eyeball is too shallow or the lens is too flat, **hyperopia** results (Figure 17-13c•). The ciliary muscle must contract to focus even a distant object on the retina, and at close range the lens cannot provide enough refraction to focus an image on the retina. Individuals with this problem are said to be *farsighted,* because they can see distant objects most clearly. Older individuals become farsighted as their lenses lose elasticity; this form of hyperopia is called **presbyopia** (*presbys,* old man). Hyperopia can be corrected by placing a *converging* lens in front of the eye. This lens provides the additional refraction needed to bring nearby objects into focus (Figure 17-13e•).

Variable success at correcting myopia and hyperopia has been achieved by surgically reshaping the cornea to alter its refractive powers. This procedure, called **radial keratotomy,** remains controversial. Although roughly two-thirds of patients are satisfied with the results, corneal healing takes several years. Many ophthalmological surgeons have expressed concerns that the scarring that develops after radial keratotomy creates weak points in the cornea that may increase the chances of a dangerous corneal rupture.

Another controversial procedure is **photorefractive keratectomy (PRK),** in which a computer-guided laser shapes the cornea to exact specifications. Tissue is removed only to a depth of 10–20 μm—no more than about 10 percent of the cornea's thickness. The entire procedure can be done in less than a minute. Each year, an estimated 100,000 people undergo PRK therapy in the United States. Advances in laser design and improvements in accuracy may make the procedure more popular during the next decade.

(a) Emmetropia

(b) Myopia

(c) Hyperopia

Diverging lens

(d) Myopia (corrected)

Converging lens

(e) Hyperopia (corrected)

•**FIGURE 17-13 Visual Abnormalities. (a)** In normal vision, the lens focuses the visual image on the retina. Common problems with the accommodation mechanism involve **(b)** myopia, an inability to lengthen the focal distance enough to focus the image of a distant object on the retina, and **(c)** hyperopia, an inability to shorten the focal distance adequately for nearby objects. These conditions can be corrected by placing appropriately shaped lenses in front of the eyes. **(d)** A diverging lens is used to correct myopia, and **(e)** a converging lens is used to correct hyperopia.

limit of clear vision, known as the *near point of vision,* is determined by the degree of elasticity in the lens. Children can usually focus on something 7–9 cm from the eye, but over time the lens tends to become stiffer and less responsive. A young adult can usually focus on objects 15–20 cm away. As aging proceeds, this distance gradually increases; the near point at age 60 is typically about 83 cm.

If light passing through the cornea and lens fails to refract properly, the visual image will be distorted. In the condition called **astigmatism,** the degree of curvature in the cornea or lens varies from one axis to another. For example, the cornea may be more strongly curved vertically than horizontally. Astigmatism can be corrected by glasses or special contact lenses. Minor astigmatism is very common; the image distortion may be so minimal that people are unaware of the condition.

Image reversal   Thus far, we have considered light that originates at a single point, either near or far from the viewer. An object in view is a complex light source, however, that can be treated as a large number of individual points. Light from each point is focused on the retina as indicated in Figure 17-14a,b•. The result is the creation of a miniature image of the original, but the image arrives upside down and backward.

To understand why, consider Figure 17-14c•, a sagittat section through an eye that is looking at a telephone pole. The image of the top of the pole lands at the bottom of the retina, and the image of the bottom hits the top of the retina. Now consider Figure 17-14d•, a section through an eye that is looking at a picket fence. The image of the left edge of the fence falls on the right side of the retina, and the image of the right edge falls on the left side of the retina. The brain compensates for this image reversal, and we are not consciously aware of any difference between the orientation of the image on the retina and that of the object.

Visual Acuity   The clarity of vision, or **visual acuity,** is rated against a "normal" standard. A person whose vision is rated 20/20 can see details at a distance of 20 feet as clearly as would an individual with normal vision. Vision noted as 20/15 is better than average, because at 20 feet the person is able to see details that would be clear to a normal eye only at a distance of 15 feet. Conversely, a person with 20/30 vision must be 20 feet from an object to discern details that a person with normal vision could make out at a distance of 30 feet.

When visual acuity falls below 20/200, even with the help of glasses or contact lenses, the individual is considered to be legally blind. There are probably fewer than 400,000 legally blind people in the United States; more than half are over 65 years of age. The term *blindness* implies a total absence of vision due to damage to the eyes or to the optic pathways. Common causes of blindness include diabetes mellitus, cataracts, glaucoma, corneal scarring, detachment of the retina, accidental injuries, and hereditary factors that are as yet poorly understood.

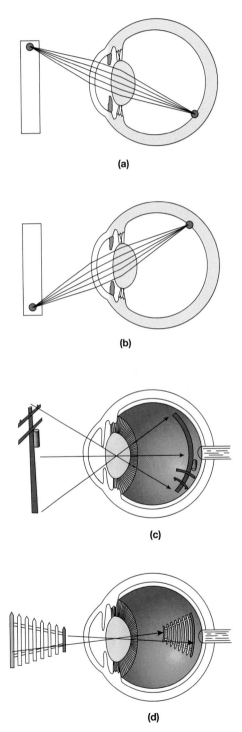

(a)

17

(b)

(c)

(d)

•**FIGURE 17-14** **Image Formation. (a,b)** Light from each portion of an object is focused on a different part of the retina. The resulting image arrives **(c,d)** upside down and backward.

Scotomas

Abnormal blind spots, or **scotomas** (skō-TŌ-muhz), may appear in the field of vision at positions other than at the optic disc. Scotomas are permanent abnormalities that are fixed in position. They may result from a compression of the optic nerve, damage to photoreceptors, or central damage along the visual pathway. *Floaters,* small spots that drift across the field of vision, are generally temporary phenomena that result from blood cells or cellular debris in the vitreous body. They can be detected by staring at a blank wall or a white sheet of paper.

☑ Which layer of the eye would be affected first by the inadequate production of tears?

☑ When the lens of your eye is very round, are you looking at an object that is close to you or far from you?

☑ As Renee enters a dark room, most of the available light becomes focused on the fovea of her eye. Will she be able to see very clearly?

☑ How would a blockage of the canal of Schlemm affect your vision?

VISUAL PHYSIOLOGY

The rods and cones of the retina are called *photoreceptors* because they detect *photons,* basic units of visible light. Light is a form of *radiant energy.* This energy is radiated in waves that have a characteristic *wavelength* (distance between wave peaks). Visible light is one small part of *electromagnetic radiation,* the entire spectrum of radiant energy that ranges from long-wavelength radio waves to short-wavelength gamma rays.

Our eyes are sensitive to wavelengths of 700–400 nm, the spectrum of visible light. This spectrum, seen in a rainbow, can be remembered by the acronym ROY G. BIV (*R*ed, *O*range, *Y*ellow, *G*reen, *B*lue, *I*ndigo, *Vio*let). Photons of red light carry the least energy, and those from the violet portion of the spectrum contain the most. Rods provide the central nervous system with information about the presence or absence of photons, without regard to wavelength. Cones provide information about the wavelength of arriving photons, giving us a perception of color.

Anatomy of Rods and Cones

Figure 17-15a• compares the structures of rods and cones. The elongate **outer segment** of a photoreceptor contains hundreds to thousands of flattened membranous plates, or **discs.** The names *rod* and *cone* refer to the outer segment's shape. In a rod, each disc is an independent entity, and the outer segment forms an elongate cylinder. In a cone, the discs are infoldings of the cell membrane and the outer segment tapers to a blunt tip.

A narrow connecting stalk attaches the outer segment to the **inner segment,** a region that contains all the usual cellular organelles. The inner segment makes synaptic contact with other cells, and it is here that neurotransmitters are released.

Visual Pigments   The discs of the outer segment in both rods and cones contain special organic compounds called **visual pigments.** The absorption of photons by visual pigments is the first key step in the process of *photoreception,* the detection of light. Visual pigments are derivatives of the compound **rhodopsin** (rō-DOP-sin), or *visual purple,* the visual pigment found in rods. Rhodopsin consists of a protein, **opsin,** bound to the pigment **retinal** (RET-i-nal). Retinal, also called *retinene,* is synthesized from **vitamin A.** One form of opsin is characteristic of all rods. Figure 17-15b• diagrams the structure of a rhodopsin molecule.

Cones contain the same retinal pigment that rods do, but in cones the retinal is attached to other forms of opsin. The type of opsin present determines the wavelength of light that can be absorbed by the retinal. Differential stimulation of these cone populations is the basis of color vision.

New discs containing visual pigment are continuously assembled at the base of the outer segment. A completed disc then moves toward the tip of the outer segment. After about 10 days, it will be shed in a small droplet of cytoplasm. These droplets are absorbed by the pigment cells, which break down the membrane components and reconvert the retinal to vitamin A. The vitamin A is then stored within the pigment cells for subsequent transfer to the photoreceptors.

Retinitis Pigmentosa

The term **retinitis pigmentosa** (RP) refers to a collection of inherited retinopathies. Together they are the most common inherited visual pathology, affecting approximately 1 individual in 3000. The visual receptors gradually deteriorate, and blindness eventually results. The mutations responsible change the structure of the photoreceptors—specifically, the visual pigments of the membrane discs. It is not known how the altered pigments lead to photoreceptor destruction.

Photoreception

The cell membrane in the outer segment of the photoreceptor contains chemically regulated sodium ion channels. In darkness, these gated channels are kept

•FIGURE 17-15 Rods and Cones. (a) The structures of rods and cones. Notice the shapes of their outer segments. **(b)** The structure of a rhodopsin molecule.

open in the presence of *cyclic-GMP (cyclic guanosine monophosphate,* or *cGMP),* a derivative of the high-energy compound *guanosine triphosphate* (GTP). Because these channels are open, the transmembrane potential is approximately −40 mV rather than the −70 mV that is typical of resting neurons. At this transmembrane potential, the photoreceptor is continuously releasing neurotransmitters (in this case, glutamate) across synapses at the inner segment. The inner segment also continuously pumps sodium ions out of the cytoplasm. The movement of

sodium ions into the outer segment, to the inner segment, and out of the cell is known as the *dark current* (Figure 17-16a•).

Rhodopsin-based photoreception begins when a photon strikes the retinal portion of a rhodopsin molecule. The bound retinal molecule has two possible configurations: the **11-*cis*** form and the **11-*trans*** form (Figure 17-16b•). It is normally in the 11-*cis* form; on absorbing light, it changes to the more linear 11-*trans* form. This change in shape triggers a chain of enzymatic steps (Figure 17-16c•):

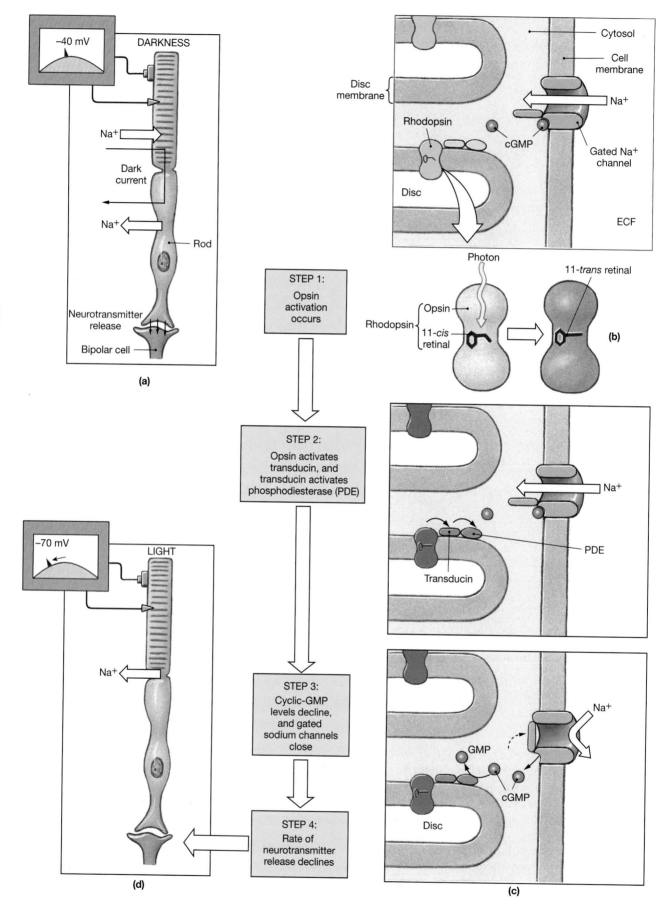

•**FIGURE 17-16** Photoreception

STEP 1: *Opsin activation occurs.* Rhodopsin is a receptor protein embedded in the membrane of the disc. Opsin is an enzyme that is inactive in the dark. When a photon of light strikes the rhodopsin, the pigment retinal in that molecule switches from the 11-*cis* to the 11-*trans* form. This switch activates the opsin portion of rhodopsin.

STEP 2: *Opsin activates a second enzyme, transducin, which in turn activates a third enzyme, phosphodiesterase.* **Transducin** is a **G-protein,** a membrane-bound enzyme activated by interaction with receptor proteins in the cell membranes. In this case, transducin is activated by opsin, and transducin in turn activates **phosphodiesterase (PDE).** Phosphodiesterase breaks down cGMP.

STEP 3: *Cyclic-GMP levels decline, and gated sodium ion channels close.* The removal of cGMP from the gated sodium channels results in their inactivation. The rate of Na^+ entry into the cytoplasm then decreases. The arrival of a photon thus reduces the dark current.

STEP 4: *The rate of neurotransmitter release declines.* Because active transport continues to remove Na^+ from the cytoplasm, when the sodium channels close, the transmembrane potential drops toward 70 mV (Figure 17-16d•). As the membrane hyperpolarizes, the rate of neurotransmitter release decreases. This decrease indicates to the adjacent bipolar cell that the photoreceptor has absorbed a photon.

Recovery After Stimulation After absorbing a photon, retinal does not spontaneously convert back to the 11-*cis* form. Instead, the entire rhodopsin molecule must be broken down and reassembled. Shortly after the shape change occurs, the rhodopsin molecule begins to break down into retinal and opsin, a process known as **bleaching** (Figure 17-17•). Before it can recombine with opsin, the retinal must be enzymatically converted to the 11-*cis* form. This conversion requires energy in the form of ATP (adenosine triphosphate), and it takes time.

Bleaching contributes to the lingering visual impression you have after you see a flashbulb go off. After an intense exposure to light, a photoreceptor cannot respond to further stimulation until its rhodopsin molecules have been regenerated. As a result, a "ghost" image remains on the retina. Bleaching is seldom noticeable under ordinary circumstances, because the eyes are constantly making small, involuntary changes in position that move the image across the retina's surface.

While the rhodopsin molecule is being reassembled, membrane permeability is returning to normal. Opsin is inactivated when bleaching occurs, and the breakdown of cGMP halts as a result. As other enzymes generate cGMP in the cytoplasm, the chemically gated sodium channels reopen.

17

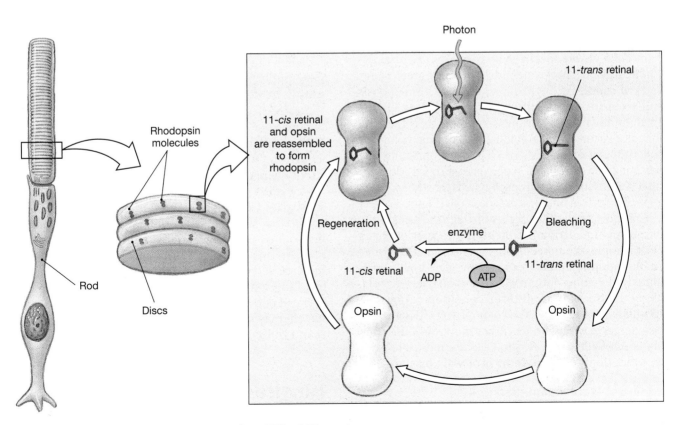

•**FIGURE 17-17** **Bleaching and Regeneration of Visual Pigments**

Night Blindness

The visual pigments of the photoreceptors are synthesized from vitamin A. The body contains vitamin A reserves sufficient for several months, and a significant amount is stored in the cells of the pigmented part of the retina. If dietary sources are inadequate, these reserves are gradually exhausted and the amount of visual pigment in the photoreceptors begins to decline. Daylight vision is affected, but in daytime the light is usually bright enough to stimulate any visual pigments that remain within the densely packed cone population of the fovea. As a result, the problem first becomes apparent at night, when the dim light proves insufficient to activate the rods. This condition, known as **night blindness,** can be treated by the administration of vitamin A. The body can convert the carotene pigments in many vegetables to vitamin A. Carrots are a particularly good source of carotene, which explains the old adage that carrots are good for your eyes.

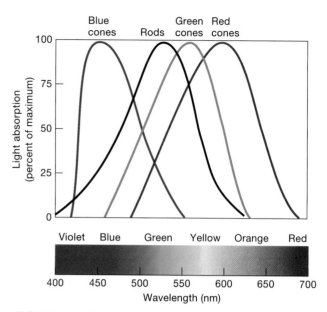

•**FIGURE 17-18** **Cones and Color Vision.** A graph comparing the absorptive characteristics of blue, green, and red cones with those of typical rods. Notice that the rod sensitivities overlap those of the cones and that the three cone types have overlapping sensitivity curves.

Color Vision

An ordinary lightbulb or the sun emits photons of all wavelengths. These photons stimulate both rods and cones. When all three types of cones are stimulated, or when rods alone are stimulated, you see a "white" light. Your eyes also detect photons that reach your retina after the photons bounce off objects around you. If photons of all colors bounce off the object, it will appear white to you; if all the photons are absorbed by the object (and so none reach the retina), it will appear black. An object will appear to have a particular color if it reflects (or transmits) photons from one portion of the visible spectrum and absorbs the rest.

The three types of cones are **blue cones, green cones,** and **red cones.** Each type has a different form of opsin and a sensitivity to a different range of wavelengths. Their stimulation in various combinations is the basis for color vision. In an individual with normal vision, the cone population consists of 16 percent blue cones, 10 percent green cones, and 74 percent red cones. Although their sensitivities overlap, each type is most sensitive to a specific portion of the visual spectrum (Figure 17-18•).

Color discrimination occurs through the integration of information arriving from all three types of cones. For example, the perception of yellow results from a combination of inputs from green cones (highly stimulated), red cones (stimulated), and blue cones (relatively unaffected). If all three cone populations are stimulated, we perceive the color as white. Because we also perceive white if rods, rather than cones, are stimulated, everything appears to be black and white when we enter dimly lit surroundings or walk by starlight.

Light and Dark Adaptation

The sensitivity of your visual system varies with the intensity of illumination. After 30 minutes or more in the dark, almost all visual pigments will be fully receptive to stimulation. This is the **dark-adapted state.** When dark-adapted, the visual system is extremely sensitive. For example, a single rod will hyperpolarize in response to a single photon of light. Even more remarkable, if as few as seven rods absorb photons at one time, you will see a flash of light.

When the lights come on, at first they seem almost unbearably bright, but over the next few minutes your sensitivity decreases as bleaching occurs. Eventually, the rate of visual pigment breakdown is balanced by the rate of reformation. This condition is the **light-adapted state.** If you moved from the depths of a cave to the full light of midday, your receptor sensitivity would decrease by about 25,000 times.

A variety of central responses further adjust light sensitivity. Constriction of the pupil, via the *pupillary constrictor reflex,* reduces the amount of light entering your eye to 1/30 the maximum dark-adapted levels. Dilating the pupil fully can produce a 30-fold increase in the amount of light entering the eye, and facilitating some of the synapses along the visual pathway can perhaps triple its sensitivity. Hence, the entire system may increase its efficiency by a factor of more than 1 million.

THE VISUAL PATHWAY

The visual pathway begins at the photoreceptors and ends at the visual cortex of the cerebral hemispheres.

Color Blindness

Persons unable to distinguish certain colors have a form of **color blindness.** The standard tests for color vision involve picking numbers or letters out of a complex and colorful picture (Figure 17-19•). Color blindness occurs when one or more classes of cones are nonfunctional. The cones may be absent, or they may be present but unable to manufacture the necessary visual pigments. In the most common type of color blindness (red–green color blindness), the red cones are missing, so the individual cannot distinguish red light from green light.

Inherited color blindness involving one or two cone pigments is not unusual. The genes for the red and green cone pigments are located on the X chromosome. As we shall see in Chapter 29, women have two copies of this chromosome *(XX),* whereas men have one X chromosome paired with a Y chromosome *(XY).* As a result, men are much more likely to be color blind than are women, who must have the same abnormal cone pigment genes on both X chromosomes to exhibit color blindness. Ten percent of all men show some color blindness, whereas the incidence among women is only about 0.67 percent. Total color blindness is extremely rare; only 1 person in 300,000 fails to manufacture any cone pigments.

•**FIGURE 17-19 A Standard Test for Color Vision.** If you lack one or more populations of cones, you will be unable to distinguish the patterned image (the number 12).

17

In other sensory pathways we have examined, at most one synapse lies between a receptor and a sensory neuron that delivers information to the CNS. In the visual pathway, the message must cross two synapses (photoreceptor to bipolar cell and bipolar cell to ganglion cell) before it heads toward the brain. The extra synapse adds to the synaptic delay, but it provides an opportunity for the processing and integration of visual information before it leaves the retina.

Processing by the Retina

Each photoreceptor in the retina monitors a specific receptive field. The retina contains about 130 million photoreceptors, 6 million bipolar cells, and 1 million ganglion cells. Thus, a considerable amount of convergence occurs at the start of the visual pathway. The degree of convergence differs between rods and cones. Regardless of the amount of convergence, each ganglion cell monitors a specific portion of the field of vision.

As many as a thousand rods may pass information via their bipolar cells to a single ganglion cell. The ganglion cells that monitor rods are relatively large, and they are called **M cells** *(magnocells).* M cells provide information about general form, motion, and shadows under dim light conditions. Because so much convergence occurs, when an M cell becomes active, it indicates that light has arrived in a general area rather than at a specific location.

The loss of specificity due to convergence is partially overcome by the fact that the activity of ganglion cells varies according to the pattern of activity in their sensory field. The sensory field is generally circular, and a ganglion cell typically responds differently to stimuli that arrive in the center than to stimuli that arrive at the edges of its receptive field (Figure 17-20•). Some ganglion cells **(on-center neurons)** are excited by light arriving in the center of their sensory field and are inhibited when light strikes the edges of their receptive field. Others **(off-center neurons)** are inhibited by light in the central zone but stimulated by illumination at the edges. On-center and off-center neurons provide information about which portion of their sensory field is illuminated.

Cones typically show very little convergence; in the fovea, the ratio of cones to ganglion cells is 1:1. The ganglion cells that monitor cones, called **P cells** *(parvo cells),* are smaller and more numerous than M cells. P cells are active in bright light, and they provide

Surface of retina

Receptive field of ganglion cell

Receptive field

Photoreceptors

Horizontal cell

Bipolar cell

Amacrine cell

Ganglion cell

•FIGURE 17-20 Convergence and Ganglion Cell Function. Photoreceptors are organized in groups within the field of vision. Each ganglion cell monitors a well-defined portion of that field. Some ganglion cells (on-center neurons) respond strongly to light arriving at the center of their receptive field (receptor A). Others (off-center neurons) respond most strongly to illumination of the edges of their receptive field (receptors B).

information about edges, fine detail, and color. Because little convergence occurs, the activation of a P cell means that light has arrived at one specific location. As a result, cones provide more-precise information about a visual image than do rods. In photographic terms, pictures formed by rods have a coarse, grainy appearance that blurs the details; those produced by cones are fine-grained, sharp, and clear.

Central Processing of Visual Information

Axons from the entire population of ganglion cells converge on the optic disc, penetrate the wall of the eye, and proceed toward the diencephalon as the optic nerve (N II). The two optic nerves, one from each eye, reach the diencephalon at the optic chiasm (Figure 17-21•). From this point, approximately half the fibers proceed toward the lateral geniculate nucleus of the same side of the brain, whereas the other half cross over to reach the lateral geniculate nucleus of the opposite side. ∞ *p. 457* The lateral geniculate nuclei act as switching and processing centers that relay visual information to reflex centers in the brain stem as well as to the cerebral cortex. For example, the pupillary reflexes and reflexes that control eye movement are triggered by information relayed to the superior colliculi by the lateral geniculate nuclei.

From each lateral geniculate nucleus, visual information travels to the occipital cortex of the cerebral hemisphere on that side. The bundle of projection fibers linking the lateral geniculates with the visual cortex is known as the **optic radiation.**

The Field of Vision   The perception of a visual image reflects the integration of information that arrives at the visual cortex of the occipital lobes. Each eye receives a slightly different visual image, because (1) the foveas are 5–7.5 cm apart, and (2) the nose and eye socket block the view of the opposite side. **Depth perception,** an interpretation of the three-dimensional relationships among objects in view, is obtained by comparing the relative positions of objects within the images received by the left and right eyes.

When you look straight ahead, the visual images from your left and right eyes overlap (Figure 17-21•). The image received by the fovea of each eye lies in the center of the region of overlap. A vertical line drawn through the center of the region of overlap marks the division of visual information at the optic chiasm. Visual information from the left half of the combined field of vision reaches the visual cortex of your right occipital lobe; visual information from the right half of the combined field of vision arrives at your left visual cortex.

The cerebral hemispheres thus contain a map of the entire field of vision. As in the case of the primary sensory cortex, the map does not faithfully duplicate the relative areas within the sensory field. For example, the area assigned to the macula lutea and fovea covers about 35 times the surface it would cover if the map were proportionally accurate. The map is also upside down and backward, duplicating the orientation of the visual image at the retina.

The Brain Stem and Visual Processing   Many centers in the brain stem receive visual information, either from the lateral geniculate nuclei or through collaterals from the optic tracts. Collaterals that bypass the lateral geniculates synapse in the superior colliculi or in the hypothalamus. The superior colliculus of the mesencephalon issues motor commands that control

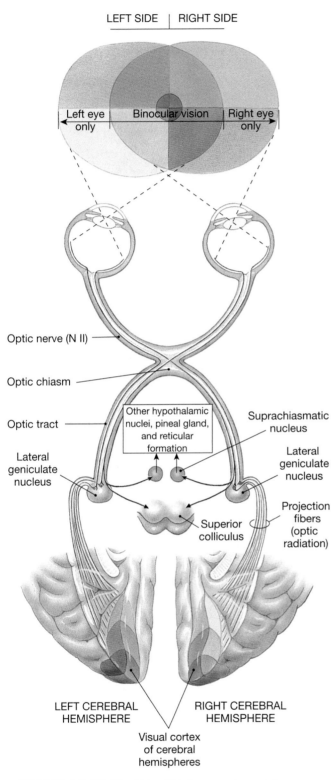

LEFT SIDE | RIGHT SIDE

Left eye only ← Binocular vision → Right eye only

Optic nerve (N II)

Optic chiasm

Optic tract

Other hypothalamic nuclei, pineal gland, and reticular formation

Suprachiasmatic nucleus

Lateral geniculate nucleus

Lateral geniculate nucleus

Superior colliculus

Projection fibers (optic radiation)

LEFT CEREBRAL HEMISPHERE

RIGHT CEREBRAL HEMISPHERE

Visual cortex of cerebral hemispheres

•FIGURE 17-21 The Visual Pathways. A partial crossover of nerve fibers occurs at the optic chiasm. As a result, each hemisphere receives visual information from the medial half of the field of vision of the eye on that side and from the lateral half of the field of vision of the eye on the opposite side. Visual association areas integrate this information to develop a composite picture of the entire field of vision.

unconscious eye, head, or neck movements in response to visual stimuli. Visual inputs to the suprachiasmatic nucleus of the hypothalamus affect the function of other brain stem nuclei. ∞ *p. 459* This nucleus and the *pineal gland* of the epithalamus establish a daily pattern of visceral activity that is tied to the day–night cycle. This **circadian rhythm** (*circa,* about + *dies,* day) affects your metabolic rate, endocrine function, blood pressure, digestive activities, awake–asleep cycle, and other physiological and behavioral processes.

☑ If you had been born without cones in your eyes, would you still be able to see? Explain.

☑ How could a diet deficient in vitamin A affect vision?

☑ What effect would a decrease in phosphodiesterase activity in photoreceptor cells have on vision?

EQUILIBRIUM AND HEARING

The *inner ear,* which consists of receptor complexes in the petrous portion of the temporal bone of the skull, provides two senses: equilibrium and hearing. *Equilibrium* sensations inform us of the position of the head in space by monitoring gravity, linear acceleration, and rotation. *Hearing* enables us to detect and interpret sound waves. The basic receptor mechanism for both senses is the same. The receptors, called *hair cells,* are simple mechanoreceptors. The complex structure of the inner ear and the different arrangement of accessory structures account for the abilities of the hair cells to respond to different stimuli and thus to provide the input for both senses.

ANATOMY OF THE EAR

The ear is divided into three anatomical regions: the external ear, the middle ear, and the inner ear (Figure 17-22•). The *external ear,* the visible portion of the ear, collects and directs sound waves toward the *middle ear,* a chamber located within the petrous portion of the temporal bone. Structures of the middle ear collect sound waves and transmit them to an appropriate portion of the inner ear. The *inner ear* contains the sensory organs for hearing and equilibrium.

The External Ear

The **external ear** includes the fleshy flap and cartilaginous **auricle,** or *pinna,* which surrounds the **external auditory canal,** or *ear canal.* The auricle protects the opening of the canal and provides directional sensitivity to the ear. Sounds coming from behind the head are blocked by the auricle; sounds coming from the side or front are collected and channeled into the external auditory canal. (When you

17

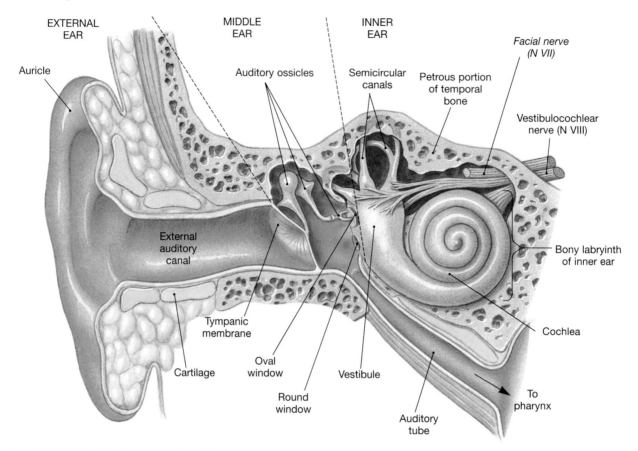

•FIGURE 17-22 The Anatomy of the Ear. The boundaries separating the three regions of the ear (external, middle, and inner) are roughly marked by the dashed lines.

"cup" your ear with your hand to hear a faint sound more clearly, you are exaggerating this effect.) The external auditory canal is a passageway that ends at the **tympanic membrane,** also called the *tympanum* or *eardrum.* The tympanic membrane is a thin, semitransparent sheet that separates the external ear from the middle ear (Figure 17-22•).

The tympanic membrane is very delicate. The auricle and the narrow external auditory canal provide some protection from accidental injury. In addition, **ceruminous glands,** integumentary glands along the external auditory canal, secrete a waxy material that helps deny access to foreign objects or insects, as do many small, outwardly projecting hairs. The waxy secretion of the ceruminous glands, called **cerumen,** also slows the growth of microorganisms in the external auditory canal and reduces the chances of infection.

The Middle Ear

The **middle ear,** or **tympanic cavity,** is filled with air. It is separated from the external auditory canal by the tympanic membrane, but it communicates with the *nasopharynx* (the superior portion of the pharynx) through the **auditory tube** and with the

mastoid air cells through a number of small connections (Figures 17-22 and 17-23a•). ∞ *p.202* The auditory tube is also called the *pharyngotympanic tube* or the *Eustachian tube.* This tube, about 4 cm long, consists of two portions. The portion near the connection to the middle ear is relatively narrow and is supported by elastic cartilage. The portion near the opening into the nasopharynx is relatively broad and funnel-shaped. The auditory tube permits the equalization of pressures inside and outside the tympanic membrane. Unfortunately, the auditory tube can also allow microorganisms to travel from the nasopharynx into the middle ear. Invasion by microorganisms can lead to an unpleasant middle ear infection known as *otitis media.* AM *Otitis Media and Mastoiditis*

The Auditory Ossicles The middle ear contains three tiny ear bones, collectively called **auditory ossicles.** The ear bones connect the tympanic membrane with one of the receptor complexes of the inner ear (Figures 17-22 and 17-23•). The three auditory ossicles are the malleus, the incus, and the stapes. The **malleus** (*malleus,* hammer) attaches at three points to the interior surface of the tympanic membrane. The

Temporal bone

Branch of N VII (cut)

External auditory canal

Tympanic membrane

Malleus

Incus

Footplate of stapes in oval window

Tensor tympani muscle

Stapes

Stapedius muscle

Round window

Auditory tube

(a)

Malleus

Tendon of tensor tympani muscle

Malleus attached to tympanic membrane

Inner surface of tympanic membrane

Incus

Footplate of stapes at oval window

Stapes

Stapedius muscle

(b)

•**FIGURE 17-23** **The Middle Ear.** **(a)** The structures of the middle ear. **(b)** The tympanic membrane and auditory ossicles.

incus (*incus*, anvil), the middle ossicle, attaches the malleus to the **stapes** (*stapes*, stirrup), the inner ossicle. The edges of the base of the stapes are bound to the edges of the *oval window*, an opening in the bone that surrounds the inner ear.

Vibration of the tympanic membrane converts arriving sound waves into mechanical movements. The auditory ossicles act as levers that conduct those vibrations to the inner ear. The ossicles are connected in such a way that an in–out movement of the tympanic membrane produces a rocking motion of the stapes. The tympanic membrane is 22 times larger and heavier than the oval window, so a 1-μm movement of the tympanic membrane produces a 22-μm deflection of the base of the stapes. Thus, the amount of move-

ment increases markedly from tympanic membrane to oval window.

Because this amplification occurs, we can hear very faint sounds. But that degree of amplification can be a problem when we are exposed to very loud noises. In the middle ear, two small muscles protect the tympanic membrane and ossicles from violent movements under very noisy conditions:

1. The **tensor tympani** (TEN-sor tim-PAN-ē) **muscle** is a short ribbon of muscle whose origin is the petrous portion of the temporal bone and the auditory tube and whose insertion is on the "handle" of the malleus. When the tensor tympani contracts, the malleus is pulled medially, stiffening the tympanic membrane.

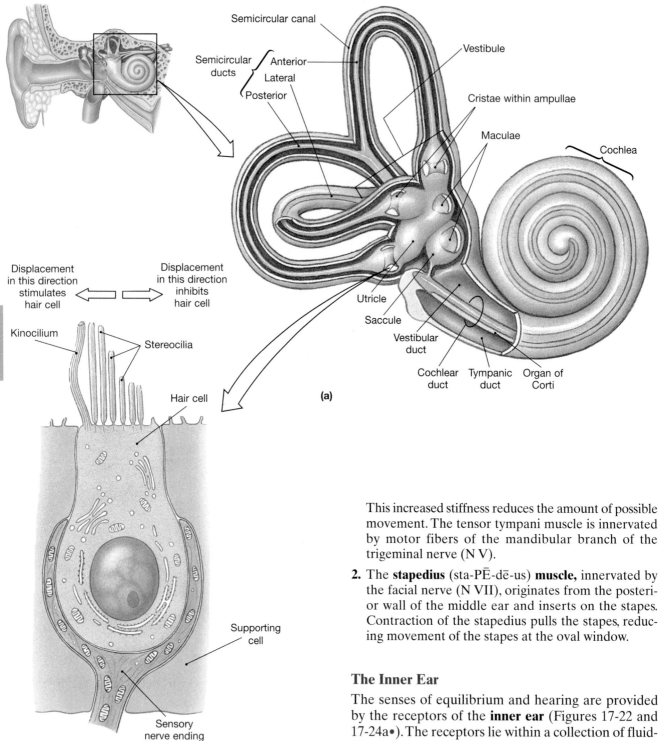

17

Semicircular canal

Semicircular ducts {
Anterior
Lateral
Posterior
}

Vestibule

Cristae within ampullae

Maculae

Cochlea

Displacement in this direction stimulates hair cell ⇦ ⇨ Displacement in this direction inhibits hair cell

Kinocilium

Stereocilia

Hair cell

Utricle

Saccule

Vestibular duct

Cochlear duct   Tympanic duct   Organ of Corti

(a)

Supporting cell

Sensory nerve ending

(b) Hair cell

•**FIGURE 17-24 The Inner Ear. (a)** The bony and membranous labyrinths. Areas of the membranous labyrinth containing sensory receptors (cristae, maculae, and the organ of Corti) are shown in purple. **(b)** A representative hair cell (receptor) from the vestibular complex. Bending the stereocilia toward the kinocilium depolarizes the cell and stimulates the sensory neuron. Displacement in the opposite direction inhibits the sensory neuron.

This increased stiffness reduces the amount of possible movement. The tensor tympani muscle is innervated by motor fibers of the mandibular branch of the trigeminal nerve (N V).

2. The **stapedius** (sta-PĒ-dē-us) **muscle,** innervated by the facial nerve (N VII), originates from the posterior wall of the middle ear and inserts on the stapes. Contraction of the stapedius pulls the stapes, reducing movement of the stapes at the oval window.

The Inner Ear

The senses of equilibrium and hearing are provided by the receptors of the **inner ear** (Figures 17-22 and 17-24a•). The receptors lie within a collection of fluid-filled tubes and chambers known as the **membranous labyrinth** (*labyrinthos,* network of canals). The membranous labyrinth contains **endolymph** (EN-dō-limf), a fluid with electrolyte concentrations different from those of typical body fluids.

The **bony labyrinth** is a shell of dense bone that surrounds and protects the membranous labyrinth. Its inner contours closely follow the contours of the membranous labyrinth, and its outer walls are fused with the surrounding temporal bone. Between the bony and membranous labyrinths flows the perilymph (PER-i-limf), a liquid whose properties closely resemble those of cerebrospinal fluid. (See Appendix VI for a chemical analysis of perilymph, endolymph, and other body fluids.)

The bony labyrinth can be subdivided into the *vestibule,* three *semicircular canals,* and the *cochlea* (Figure 17-24a•). The **vestibule** (VES-ti-būl) consists of a pair of membranous sacs: the **saccule** (SAK-ūl) and the **utricle** (Ū-tri-kul), or the *sacculus* and *utriculus.* Receptors in the saccule and utricle provide sensations of gravity and linear acceleration.

The **semicircular canals** enclose slender **semicircular ducts.** Receptors in the semicircular ducts are stimulated by rotation of the head. The combination of vestibule and semicircular canals is called the **vestibular complex.** The fluid-filled chambers within the vestibule are broadly continuous with those of the semicircular canals.

The **cochlea** (KOK-lē-uh; *cochlea,* a snail shell) is a spiral-shaped, bony chamber that contains the **cochlear duct** of the membranous labyrinth. Receptors within the cochlear duct provide the sense of hearing. The cochlear duct is sandwiched between a pair of perilymph-filled chambers. The entire complex makes $2\frac{1}{2}$ turns around a central bony hub. In sectional view, the spiral arrangement resembles a snail shell.

The walls of the bony labyrinth consist of dense bone everywhere except at two small areas near the base of the cochlear spiral (Figure 17-22•). The **round window** is a thin, membranous partition that separates the perilymph of the cochlear chambers from the air spaces of the middle ear. Collagen fibers connect the bony margins of the opening known as the **oval window** to the base of the stapes. When a sound vibrates the tympanic membrane, the movements are conducted to the stapes by the other auditory ossicles. Movement of the stapes ultimately leads to the stimulation of receptors in the cochlear duct, and we hear the sound.

Receptor Function in the Inner Ear

The sensory receptors of the inner ear are called **hair cells** (Figure 17-24b•). These receptor cells are surrounded by supporting cells and are monitored by sensory afferent fibers. The free surface of each hair cell supports 80–100 long stereocilia, which resemble very long microvilli. Each hair cell in the vestibule also contains a **kinocilium,** a single large cilium. Hair cells do not actively move their kinocilia or stereocilia. However, when an external force pushes against these processes, the distortion of the cell membrane alters the rate at which the hair cell releases chemical transmitter.

Hair cells provide information about the direction and strength of mechanical stimuli. The stimuli involved, however, are quite varied: gravity or accelera-

tion in the vestibule, rotation in the semicircular canals, and sound in the cochlea. The sensitivities of the hair cells differ, because each of these regions has different accessory structures that determine which stimulus will provide the force to deflect the kinocilia and stereocilia. The importance of these accessory structures will become apparent when we consider the way sensations of equilibrium and hearing are produced.

EQUILIBRIUM

As we just learned, equilibrium sensations are provided by receptors of the vestibular complex. The semicircular ducts provide information about rotational movements of the head. For example, when you turn your head to the left, stimulated receptors in the semicircular ducts tell you how rapid the movement is and in which direction. The saccule and the utricle provide information about your position with respect to gravity. If you stand with your head tilted to one side, these receptors will report the angle involved and whether your head tilts forward or backward. These receptors are also stimulated by sudden acceleration. When your car accelerates from a stop, the saccular and utricular receptors give you the impression of increasing speed.

The Semicircular Ducts

Sensory receptors in the semicircular ducts respond to rotational movements of the head. These hair cells are active during a movement but are quiet when the body is motionless. The **anterior, posterior,** and **lateral semicircular ducts** are continuous with the utricle (Figures 17-24a and 17-25a•). Each semicircular duct contains an **ampulla,** an expanded region that contains the hair cells. Hair cells attached to the wall of the ampulla form a raised structure known as a **crista** (Figure 17-25b•). The kinocilia and stereocilia of the hair cells are embedded in the **cupula** (KŪ-pū-luh), a gelatinous structure. Because the cupula has a density very close to that of the surrounding endolymph, it essentially floats above the receptor surface, nearly filling the ampulla. When your head rotates in the plane of the duct, the movement of endolymph along the length of the semicircular duct pushes the cupula and distorts the receptor processes (Figure 17-25c•). Fluid movement in one direction stimulates the hair cells, and movement in the opposite direction inhibits them. When the endolymph stops moving, the elastic nature of the cupula makes it bounce back to its normal position.

Even the most complex movement can be analyzed in terms of motion in three rotational planes. Each semicircular duct responds to one of these rotational movements. A horizontal rotation, as in shaking your

17

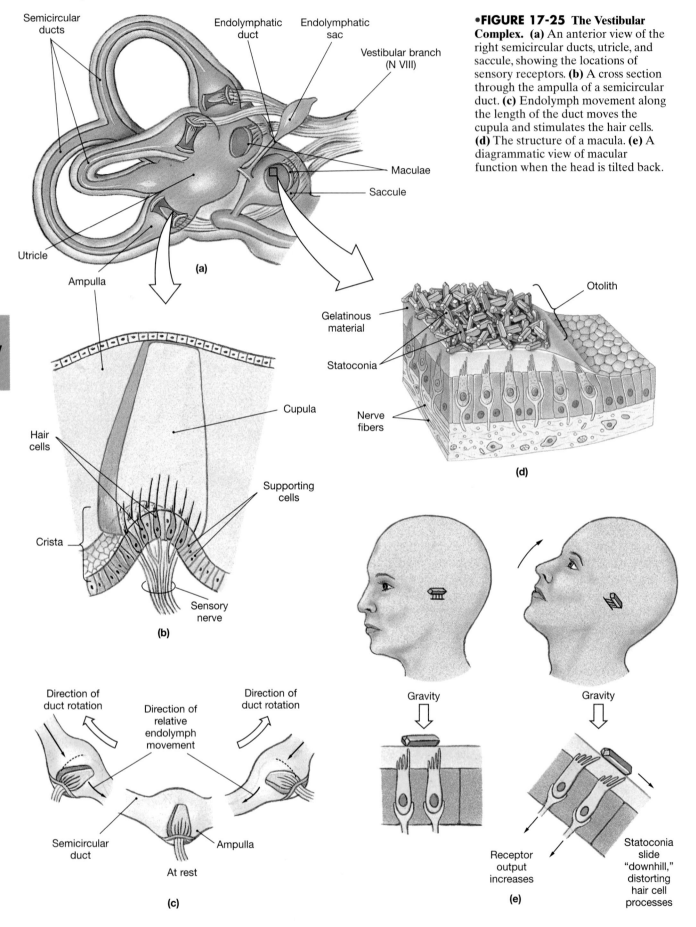

Semicircular ducts

Endolymphatic duct

Endolymphatic sac

Vestibular branch (N VIII)

Maculae

Saccule

Utricle

Ampulla

(a)

•**FIGURE 17-25** **The Vestibular Complex.** **(a)** An anterior view of the right semicircular ducts, utricle, and saccule, showing the locations of sensory receptors. **(b)** A cross section through the ampulla of a semicircular duct. **(c)** Endolymph movement along the length of the duct moves the cupula and stimulates the hair cells. **(d)** The structure of a macula. **(e)** A diagrammatic view of macular function when the head is tilted back.

Gelatinous material

Otolith

Statoconia

Nerve fibers

(d)

Cupula

Hair cells

Supporting cells

Crista

Sensory nerve

(b)

Direction of duct rotation

Direction of relative endolymph movement

Direction of duct rotation

Semicircular duct

Ampulla

At rest

(c)

Gravity

Gravity

Receptor output increases

Statoconia slide "downhill," distorting hair cell processes

(e)

head "no," stimulates the hair cells of the lateral semi-circular duct. Nodding "yes" excites the anterior duct, and tilting your head from side to side activates the receptors in the posterior duct.

The Utricle and Saccule

The utricle and saccule provide equilibrium sensations whether the body is moving or is stationary. These chambers are connected by a slender passageway continuous with the narrow endolymphatic duct. The **endolymphatic duct** ends in a blind pouch called the **endolymphatic sac** (Figure 17-25a•). This sac projects through the dura mater that lines the temporal bone and into the subdural space. Portions of the cochlear duct continuously secrete endolymph, and at the endolymphatic sac excess fluids return to the general circulation.

The hair cells of the utricle and saccule are clustered in oval structures called **maculae** (MAK-ū-lē; *macula,* spot) (Figure 17-25a•). As they are in ampullae, the hair cell processes are embedded in a gelatinous mass. However, the surface of this gelatinous material contains densely packed calcium carbonate crystals known as **statoconia** (*statos,* standing + *conia,* dust). The complex as a whole (gelatinous matrix and statoconia) is called an **otolith** ("ear stone").

The macula of the saccule is diagrammed in Figure 17-25d•. When your head is in the normal, upright position, the statoconia sit atop the macula. Their weight presses down on the macular surface, pushing the hair cell processes down rather than to one side or another. When your head is tilted, the pull of gravity on the statoconia shifts them to the side. This shift distorts the hair cell processes, and the change in receptor activity tells the CNS that your head is no longer level (Figure 17-25e•).

A similar mechanism accounts for your perception of linear acceleration when you are in a car that speeds up suddenly. The statoconia lag behind, and the effect on the hair cells is comparable to tilting your head back. Under normal circumstances, your nervous system distinguishes between the sensations of tilting and linear acceleration by integrating vestibular sensations with visual information.

Pathways for Equilibrium Sensations

Hair cells of the vestibule and semicircular ducts are monitored by sensory neurons located in adjacent **vestibular ganglia.** Sensory fibers from these ganglia form the **vestibular branch** of the vestibulocochlear nerve (N VIII). ∞ *p. 470* These fibers innervate neurons within the pair of **vestibular nuclei** at the boundary between the pons and the medulla oblongata. The two vestibular nuclei have the following four functions:

1. Integrating the sensory information about balance and equilibrium that arrives from both sides of the head.
2. Relaying information from the vestibular complex to the cerebellum.
3. Relaying information from the vestibular complex to the cerebral cortex, providing a conscious sense of head position and movement.
4. Sending commands to motor nuclei in the brain stem and spinal cord.

Motion Sickness

The exceedingly unpleasant symptoms of **motion sickness** include headache, sweating, flushing of the face, nausea, vomiting, and various changes in mental perspective. (Sufferers may go from a state of giddy excitement to almost suicidal despair in a matter of moments.) It has been suggested that the condition results when central processing stations, such as the tectum of the mesencephalon, receive conflicting sensory information. Why and how these conflicting reports result in nausea, vomiting, and other symptoms are not known. Sitting below the deck on a moving boat or reading in a car or airplane tends to provide the necessary conditions. Your eyes (which are tracking lines on a page) report that your position in space is not changing, but your semicircular ducts report that your body is lurching and turning. As a result, seasick sailors watch the horizon rather than their immediate surroundings so that their eyes will provide visual confirmation of the movements detected by their inner ears. It is not known why some individuals are almost immune to motion sickness whereas others find travel by boat or plane almost impossible.

Drugs commonly administered to prevent motion sickness include dimenhydrinate *(Dramamine),* scopolamine, and promethazine. These compounds appear to depress activity at the vestibular nuclei. Sedatives, such as prochlorperazine *(Compazine),* may also be effective. Scopolamine can be administered across the skin surface by using an adhesive patch *(Transderm Scōp™).*

The reflexive motor commands that are issued by the vestibular nuclei are distributed to the motor nuclei for cranial nerves involved with eye, head, and neck movements (N III, IV, VI, and XI). Instructions descending in the *vestibulospinal tracts* of the spinal cord adjust peripheral muscle tone and complement the reflexive movements of the head or neck. ∞ *p. 490* These pathways are indicated in Figure 17-26•.

The automatic movements of the eye that occur in response to sensations of motion are directed by the *superior colliculus* of the mesencephalon. ∞ *pp. 459–460* These movements attempt to keep your gaze focused on a specific point in space despite changes in body position and orientation. If your body is turning or spinning rapidly, your eyes will fix on one point for a moment, then they will jump ahead to another in a series of short, jerky movements.

This type of eye movement can occur even when the body is stationary if either the brain stem or the inner ear is damaged. An individual with this condition, which is called **nystagmus,** has trouble controlling eye movements. Physicians commonly check for nystagmus by asking the patient to watch a small penlight as it is moved across the field of vision. AM *Vertigo and Ménière's Disease*

HEARING

The receptors of the cochlear duct provide us with a sense of hearing that enables us to detect the quietest whisper yet remain functional in a crowded, noisy room. The receptors responsible for auditory sensations are hair cells similar to those of the vestibular complex. However, their placement within the cochlear duct and the organization of the surrounding accessory structures shield them from stimuli other than sound.

In conveying vibrations from the tympanic membrane to the oval window, the auditory ossicles convert pressure fluctuations in air to pressure fluctuations in the perilymph of the cochlea. These fluctuations stimulate hair cells along the cochlear spiral. The *frequency* of the perceived sound is determined by *which part* of the cochlear duct is stimulated. The *intensity* (volume) of the perceived sound is determined by *how many* of the hair cells at that location are stimulated. We will now consider the mechanics of this remarkably elegant process.

The Cochlear Duct

In sectional view (Figures 17-27a,b and 17-28a,b•, p. 566), the cochlear duct, or *scala media,* lies between a pair of perilymphatic chambers: the **vestibular duct**

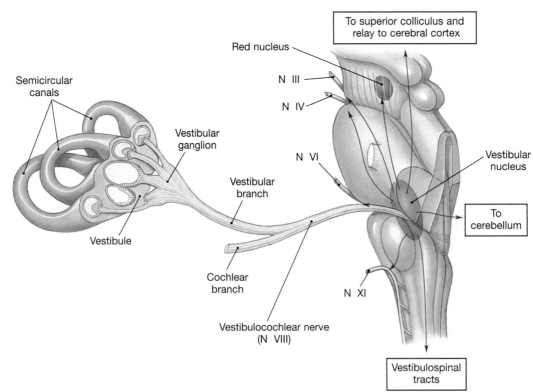

•**FIGURE 17-26** Pathways for Equilibrium Sensations

From oval window
to tip of spiral

From tip of spiral
to round window

Cochlear duct

Round window

Stapes at oval window

Tympanic duct

Vestibular duct

Vestibular branch

Cochlear branch

Vestibulocochlear nerve (N VIII)

Semicircular canals

(a)

17

Vestibular membrane

Tectorial membrane

Basilar membrane

Vestibular duct (contains perilymph)

Cochlear duct (contains endolymph)

Organ of Corti

Tympanic duct (contains perilymph)

Cochlear branch

Spiral ganglion

Vestibulocochlear nerve (N VIII)

From oval window

Spiral ganglion

Vestibular membrane

Basilar membrane

To round window

(b)

•**FIGURE 17-27** **The Cochlea.** **(a)** The structure of the cochlea. **(b)** Diagrammatic and sectional views of the cochlear spiral.

(*scala vestibuli*) and the **tympanic duct** (*scala tympani*). These ducts are interconnected at the tip of the cochlear spiral. The outer surfaces of these ducts are encased by the bony labyrinth everywhere except at the oval window (the base of the vestibular duct) and the round window (the base of the tympanic duct).

The hair cells of the cochlear duct are located in a structure called the **organ of Corti** (Figures 17-27b and 17-28a,b•). This sensory structure sits on the **basilar membrane,** a membrane that separates the cochlear duct from the tympanic duct. The hair cells are arranged in a series of longitudinal rows. They lack kinocilia, and their stereocilia are in contact with the overlying **tectorial** (tek-TOR-ē-al) **membrane**

(*tectum,* roof). This membrane is firmly attached to the inner wall of the cochlear duct. When a portion of the basilar membrane bounces up and down, the stereocilia of the hair cells are pressed against the tectorial membrane and become distorted. The basilar membrane moves in response to pressure fluctuations within the perilymph. These pressure changes are triggered by sound waves arriving at the tympanic membrane. To understand this process, we must consider the basic properties of sound.

An Introduction to Sound

Hearing is the detection of sound, which consists of waves of pressure conducted through a medium such

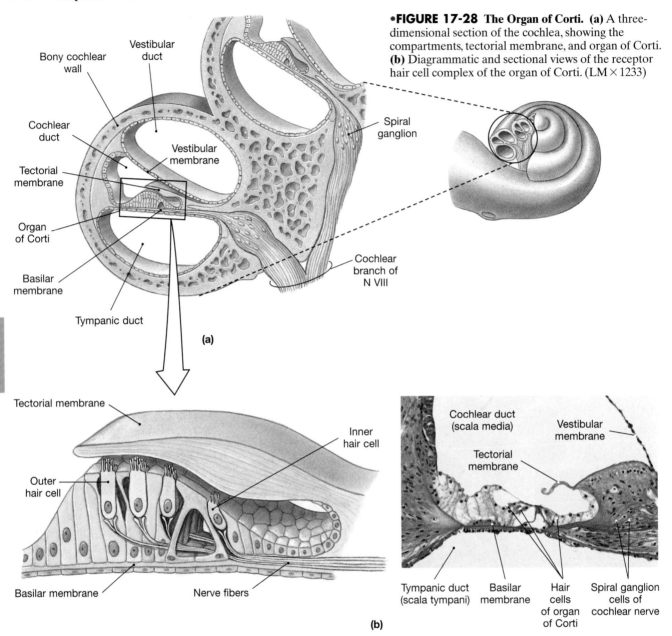

•**FIGURE 17-28 The Organ of Corti. (a)** A three-dimensional section of the cochlea, showing the compartments, tectorial membrane, and organ of Corti. **(b)** Diagrammatic and sectional views of the receptor hair cell complex of the organ of Corti. (LM × 1233)

17

as air or water. In air, each *pressure wave* consists of a region where the air molecules are crowded together and an adjacent zone where they are farther apart (Figure 17-29a•). These waves are sine waves—that is, S-shaped curves that repeat in a regular pattern—and travel through the air at about 1235 km/h (768 mph).

The *wavelength* of sound is the distance between two adjacent wave crests (peaks) or, equivalently, the distance between two adjacent wave troughs (Figure 17-29b•). Wavelength is inversely related to **frequency,** the number of waves that pass a fixed reference point in a given time. Physicists use the term **cycles** rather than *waves.* Hence the frequency of a

sound is measured in terms of the number of cycles per second (cps), a unit called **hertz (Hz).** What we perceive as the **pitch** of a sound is our sensory response to its frequency. A *high-frequency* sound (high pitch, short wavelength) might have a frequency of 15,000 Hz or more; a very *low-frequency* sound (low pitch, long wavelength) could have a frequency of 100 Hz or less.

It takes energy to produce sound waves. When you strike a tuning fork, it vibrates and pushes against the surrounding air, producing sound waves whose frequency depends on the frequency of vibration. The harder you strike the tuning fork, the more energy you provide and the louder the sound.

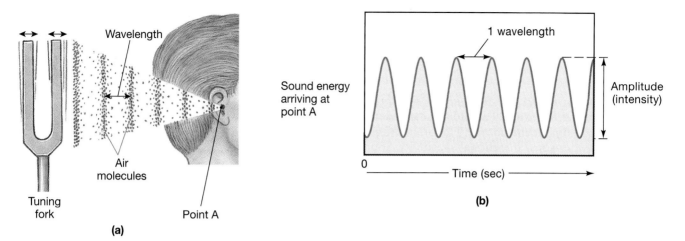

•**FIGURE 17-29** **Sound Production.** **(a)** Sound waves generated by a tuning fork travel through the air as pressure waves. The frequency of a sound wave is the number of waves that pass a fixed reference point in a given time. **(b)** Frequencies are reported in cycles per second (cps), or hertz (Hz).

The loudness increases because the sound waves carry more energy away with them. The energy content, or *power,* of a sound determines its **intensity,** or volume. Intensity is reported in **decibels** (DES-i-belz). Table 17-2 indicates the decibel levels of familiar sounds.

When sound waves strike an object, their energy is a physical pressure. You may have seen windows move in a room in which a stereo is blasting. The more flexible the object, the more easily it will respond to sound pressure. Even soft stereo music will vibrate a sheet of paper held in front of the speaker. Given the right combination of frequencies and intensities of sound, an object will begin to vibrate at the same frequency as the sound. This phenomenon is called *resonance.* The greater the sound intensity, the greater the amount of vibration produced. For you to be able to hear any sound, your thin, flexible tympanic membrane must vibrate in resonance with the sound waves.

TABLE 17-2 **Sound Intensities**

Typical Decibel Level	Example	Dangerous Time Exposure
0	Lowest audible sound	
30	Quiet library; soft whisper	
40	Quiet office; living room; bedroom away from traffic	
50	Light traffic at a distance; refrigerator; gentle breeze	
60	Air conditioner at 20 feet; conversation; sewing machine in operation	
70	Busy traffic; noisy restaurant	Some damage if continuous
80	Subway; heavy city traffic; alarm clock at 2 feet; factory noise	More than 8 hours
90	Truck traffic; noisy home appliances; shop tools; gas lawnmower	Less than 8 hours
100	Chain saw; boiler shop; pneumatic drill	2 hours
120	"Heavy metal" rock concert; sandblasting; thunderclap nearby	Immediate danger
140	Gunshot; jet plane	Immediate danger
160	Rocket launching pad	Hearing loss inevitable

The Hearing Process

The process of hearing can be divided into six basic steps, summarized in Table 17-3 and diagrammed in Figure 17-30a•:

STEP 1: *Sound waves arrive at the tympanic membrane.* Sound waves enter the external auditory canal and travel toward the tympanic membrane. The orientation of the canal provides some directional sensitivity. Sound waves approaching the side of the head have direct access to the tympanic membrane on that side, whereas sounds arriving from another direction must bend around corners or pass through the auricle or other body tissues.

STEP 2: *Movement of the tympanic membrane causes displacement of the auditory ossicles.* The tympanic membrane provides the surface for sound collection, and it vibrates in resonance to sound waves with frequencies between approximately 20 and 20,000 Hz (in a young child). When the tympanic membrane vibrates, so do the malleus and, through their articulations, the incus and stapes. In this way, the sound is amplified.

STEP 3: *Movement of the stapes at the oval window establishes pressure waves in the perilymph of the vestibular duct.* Liquids are incompressible; if you push down on one part of a water bed, the water bed bulges somewhere else. Because the rest of the cochlea is sheathed in bone, pressure applied at the oval window can be relieved only at the round window. When the stapes moves inward, the round window bulges outward. As the stapes moves in and out, vibrating at the frequency of the sound arriving at the tympanic membrane, it creates pressure waves within the perilymph.

STEP 4: *The pressure waves distort the basilar membrane on their way to the round window of the tympanic duct.*

TABLE 17-3 Steps in the Production of an Auditory Sensation

1. Sound waves arrive at the tympanic membrane.

2. Movement of the tympanic membrane causes displacement of the auditory ossicles.

3. Movement of the stapes at the oval window establishes pressure waves in the perilymph of the vestibular duct.

4. The pressure waves distort the basilar membrane on their way to the round window of the tympanic duct.

5. Vibration of the basilar membrane causes vibration of hair cells against the tectorial membrane.

6. Information about the region and intensity of stimulation is relayed to the CNS over the cochlear branch of N VIII.

The pressure waves established by movement of the stapes travel through the perilymph of the vestibular and tympanic ducts to reach the round window. In doing so, these pressure waves distort the basilar membrane. The location of maximum distortion varies with the frequency of the sound, owing to regional differences in the width and flexibility of the basilar membrane along its length. High-frequency sounds, which have a very short wavelength, vibrate the basilar membrane near the oval window. The lower the frequency of the sound, the longer the wavelength, and the farther from the oval window the area of maximum distortion will be (Figure 17-30b–d•). Thus, frequency information is translated into *position* information.

The *amount* of movement at a given location depends on the amount of force applied by the stapes. The amount of force is a function of the intensity of the sound. The louder the sound, the more the basilar membrane moves.

STEP 5: *Vibration of the basilar membrane causes vibration of hair cells against the tectorial membrane.* Vibration of the affected region of the basilar membrane moves hair cells against the tectorial membrane. This movement leads to the displacement of the stereocilia, which in turn opens ion channels in their cell membranes. A subsequent inrush of ions depolarizes the hair cells. The depolarization leads to neurotransmitter release and thus to the stimulation of sensory neurons.

The hair cells of the organ of Corti are arranged in several rows. A very soft sound may stimulate only a few hair cells in a portion of one row. As the intensity of a sound increases, not only do these hair cells become more active, but additional hair cells—at first in the same row, and then in adjacent rows—are stimulated as well. The number of hair cells responding in a given region of the organ of Corti thus provides information on the intensity of the sound.

STEP 6: *Information about the region and intensity of stimulation is relayed to the CNS over the cochlear branch of the vestibulocochlear nerve (N VIII).* The cell bodies of the bipolar sensory neurons that monitor the cochlear hair cells are located at the center of the bony cochlea in the **spiral ganglion** (Figure 17-28a•). From there, the information is carried by N VIII to the cochlear nuclei of the medulla oblongata for subsequent distribution to other centers in the brain.

Auditory Pathways

Hair cell stimulation activates sensory neurons whose cell bodies are in the adjacent spiral ganglion. Their afferent fibers form the **cochlear branch** of the vestibulocochlear nerve (N VIII) (Figure 17-31•, p. 570). These axons enter the medulla oblongata, where they synapse at the **cochlear nucleus.** From there, the information crosses to the opposite side of the brain and

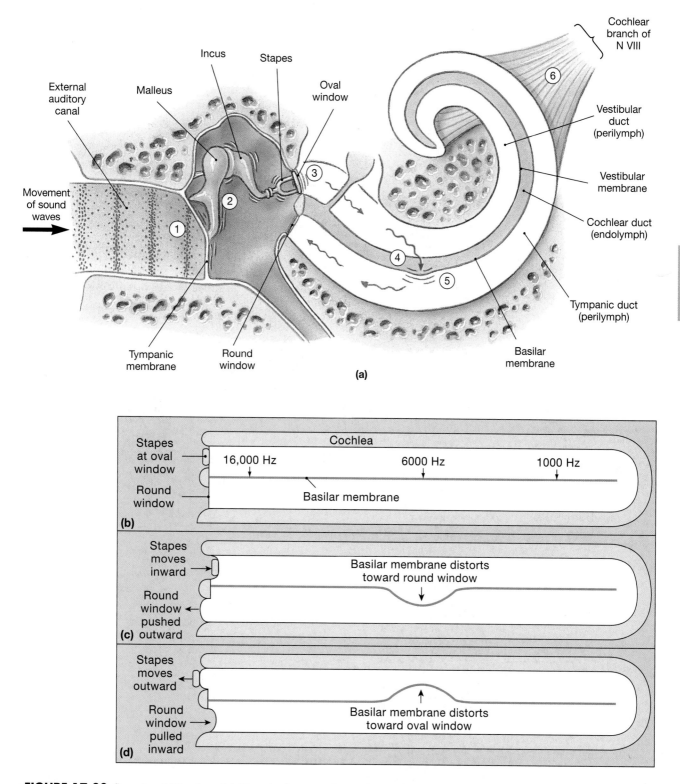

(a)

(b)

(c)

(d)

•**FIGURE 17-30 Sound and Hearing. (a)** Steps in the reception and transduction of sound energy. (See text and Table 17-3.) **(b)** The flexibility of the basilar membrane changes along its length, so pressure waves of different frequencies affect different parts of the membrane. **(c,d)** The effects of a vibration of the stapes at a frequency of 6000 Hz. When the stapes moves inward, as in part (c), the basilar membrane distorts toward the round window, which bulges into the middle ear cavity. When the stapes moves outward, as in part (d), the basilar membrane rebounds and distorts toward the oval window.

ascends to the inferior colliculus of the mesencephalon. This processing center coordinates a number of responses to acoustic stimuli, including auditory reflexes that involve skeletal muscles of the head, face, and trunk. These reflexes automatically change the position of your head in response to a sudden loud noise; you usually turn your head and your eyes toward the source of the sound.

Before reaching the cerebral cortex and your conscious awareness, ascending auditory sensations synapse in the medial geniculate nucleus of the thalamus. Projection fibers then deliver the information to the auditory cortex of the temporal lobe. Information travels to the cortex over labeled lines: High-frequency sounds activate one portion of the cortex, and low-frequency sounds activate another. In effect, the auditory cortex contains a map of the organ of Corti. Thus, *frequency* information, translated into *position* information on the basilar membrane, is projected in this form onto the auditory cortex. There, it is interpreted to produce your subjective sensation of pitch.

An individual whose auditory cortex is damaged will respond to sounds and have normal acoustic reflexes, but sound interpretation and pattern recognition will be difficult or impossible. Damage to the adjacent association area leaves the ability to detect the tones and patterns but produces an inability to comprehend their meaning.

Auditory Sensitivity

Our hearing abilities are remarkable, but it is difficult to assess the absolute sensitivity of the system. The range from the softest audible sound to the loudest tolerable blast represents a trillionfold increase

•**FIGURE 17-31 Pathways for Auditory Sensations.** Auditory sensations are carried by the cochlear branch of N VIII to the cochlear nucleus of the medulla oblongata. From there, the information is relayed to the inferior colliculus, a center that directs a variety of unconscious motor responses to sounds. Ascending acoustic information goes to the medial geniculate nucleus before being forwarded to the auditory cortex of the temporal lobe.

in power. The receptor mechanism is so sensitive that, if we were to remove the stapes, we could in theory hear air molecules bouncing off the oval window. We never use the full potential of this system, because body movements and our internal organs produce squeaks, groans, thumps, and other sounds that are tuned out by central and peripheral adaptation. When other environmental noises fade away, the level of adaptation drops and the system becomes increasingly sensitive. For example, when you relax in a quiet room, your heartbeat seems to get louder and louder as the auditory system adjusts to the level of background noise.

Young children have the greatest hearing range: They can detect sounds ranging from a 20-Hz buzz to a 20,000-Hz whine. With age, damage due to loud noises or other injuries accumulates. The tympanic membrane gets less flexible; the articulations between the ossicles stiffen; and the round window may begin to ossify. As a result, older individuals show some degree of hearing loss.

☑ If the round window were not able to bulge out with increased pressure in the perilymph, how would sound perception be affected?

☑ How would the loss of stereocilia from hair cells of the organ of Corti affect hearing?

☑ Why does a blockage of the auditory tube produce an earache?

Hearing Deficits

Probably more than 6 million people in the United States alone have at least a partial hearing deficit. **Conductive deafness** results from conditions in the outer or middle ear that block the normal transfer of vibrations from the tympanic membrane to the oval window. An external auditory canal plugged with accumulated wax or trapped water can cause a temporary hearing loss. Scarring or perforation of the tympanic membrane and immobilization of one or more of the auditory ossicles are more serious causes of conductive deafness.

In **nerve deafness,** the problem lies within the cochlea or somewhere along the auditory pathway. The vibrations reach the oval window and enter the perilymph, but either the receptors cannot respond or their response cannot reach its central destinations. For example:

- Very loud (high-intensity) sounds can produce nerve deafness by breaking stereocilia off the surfaces of the hair cells. (The reflex contraction of the tensor tympani and stapedius muscles in response to a dangerously loud noise occurs in less than 0.1 second, but this may not be fast enough to prevent damage.)

- Drugs such as the aminoglycoside antibiotics (*neomycin* and *gentamicin*) can diffuse into the endolymph and kill hair cells. Because hair cells and sensory nerves can also be damaged by bacterial infection, the potential side effects must be balanced against the severity of infection.

Many treatment options are available for conductive deafness; treatment options for nerve deafness are relatively limited. Because many of these problems become progressively worse, early diagnosis improves the chances of successful treatment. **AM** *Testing and Treating Hearing Deficits*

Our sensory capabilities are remarkable. Through our sensory organs, we can richly experience our environment. To enhance your understanding of the functioning of these organs, use the resources available on both the CD-ROM and the Companion Website. The CD-ROM includes animations and tutorials; the website builds on those resources with text content and interactive problems as well as communication avenues for sharing ideas with your class. To complete the following Web Explorations, visit the MediaLab in Chapter 17 of your Companion Website.

Web Exploration 1

Estimated time for completion = 10 minutes

The sense of vision is our most acute sense. Unfortunately, many people do not have perfect vision. Refractive surgery is one option for some who have poor vision. To learn about this surgical procedure, select the keyword **VISION** on the Companion Website. Read the page on radial and astigmat-

ic keratotomy, then choose the links at the bottom of the page to explore ALK and LASIK surgeries. What do you think of these new procedures? Would you be willing to undergo either one? Why or why not? Write a short essay describing your thoughts.

Web Exploration 2

Estimated time for completion = 5 minutes

The clinical discussion "Hearing Deficits" in the text (above) describes two types of hearing deficits: conductive deafness and nerve deafness. Until recently, not much could be done about nerve deafness. To learn of a special implant that is now being used to restore partial hearing in some individuals with nerve deafness, select the keyword **IMPLANT** on the Companion Website. Which structure or structures in the inner ear does this implant replace? Why are the external ear and middle ear bypassed? Diagram the pathway of sound through the ear; note in red the areas that are replaced by this implant.

CHAPTER REVIEW

SELECTED CLINICAL TERMINOLOGY
Terms Discussed in This Chapter

cataract: An abnormal lens that has lost its transparency. *(p. 546)*

color blindness: A condition in which a person is unable to distinguish certain colors. *(p. 555)*

conductive deafness: Deafness resulting from conditions in the outer or middle ear that block the transfer of vibrations from the tympanic membrane to the oval window. *(p. 571)*

glaucoma: A condition characterized by increased intraocular pressure due to the impaired reabsorption of aqueous humor; can result in blindness. *(p. 546)*

hyperopia, or *farsightedness:* A condition in which nearby objects are blurry but distant objects are clear. *(p. 548)*

myopia, or *nearsightedness:* A condition in which vision at close range is normal but distant objects appear blurry. *(p. 548)*

nerve deafness: Deafness resulting from problems within the cochlea or along the auditory pathway. *(p. 571)*

nystagmus: Abnormal eye movements that may appear after the brain stem or inner ear is damaged. *(p. 564)*

presbyopia: A type of hyperopia that develops with age as lenses become less elastic. *(p. 548)*

retinitis pigmentosa: A group of inherited retinopathies characterized by the progressive deterioration of photoreceptors, eventually resulting in blindness. *(p. 550)*

AM Additional Terms Discussed in the *Applications Manual*

anesthesia: A total or partial loss of sensation.

hypesthesia: A reduction in sensitivity.

Ménière's disease: A condition in which high fluid pressures rupture the walls of the membranous labyrinth, resulting in

acute vertigo (an inappropriate sense of motion) and inappropriate auditory sensations.

paresthesia: Abnormal sensations, such as a "pins and needles" sensation.

STUDY OUTLINE

INTRODUCTION   p. 528
RECEPTORS AND RECEPTOR FUNCTION   p. 528

1. The **general senses** are pain, temperature, touch, pressure, vibration, and proprioception. Receptors for those sensations are distributed throughout the body. The **special senses** are **olfaction** (smell), **gustation** (taste), **vision** (sight), **equilibrium** (balance), and **hearing.** Receptors for these senses are located in specialized areas called **sense organs.**

2. A *sensory receptor* is a specialized cell that translates a stimulus into an action potential that can be conducted to the CNS. **Transduction** is the translation of a stimulus into an action potential.

The Detection of Stimuli   p. 528

3. *Receptor specificity* allows each receptor to respond to particular stimuli. The simplest receptors are **free nerve endings.** The area monitored by a single receptor cell is the *receptive field. (Figure 17-1)*

4. Transduction involves the development of *receptor potentials,* which can summate to produce a *generator potential* in an afferent fiber. The resulting action potential travels to the CNS.

The Interpretation of Sensory Information   p. 529

5. The identity of a sensation is indicated by the *labeled line* that carries the action potentials into the CNS. The intensity of the stimulus is indicated by the frequency and pattern of action potentials. This phenomenon is called *sensory coding.*

6. **Tonic receptors** always send signals to the CNS; **phasic receptors** become active only when the conditions they monitor change.

Central Processing and Adaptation   p. 529

7. **Adaptation** (a reduction in sensitivity in the presence of a constant stimulus) may involve changes in receptor sensitivity (*peripheral adaptation*) or inhibition along the sensory pathways (*central adaptation*). **Fast-adapting receptors** are phasic; **slow-adapting receptors** are tonic.

THE GENERAL SENSES   p. 530
Nociceptors   p. 530

1. **Nociceptors** respond to a variety of stimuli, generally ones associated with tissue damage. The two types of painful sensations are **fast** *(prickling)* **pain** and **slow** *(burning and aching)* **pain.** *(Figure 17-2)*

Thermoreceptors   p. 531

2. **Thermoreceptors** respond to changes in temperature.

Mechanoreceptors   p. 531

3. **Mechanoreceptors** respond to physical distortion, contact, or pressure on their cell membranes. The three types of mechanoreceptors are **tactile receptors, baroreceptors,** and **proprioceptors.**

4. Tactile receptors respond to touch, pressure, and vibration and can be classified by the fineness of touch they detect. **Fine touch and pressure receptors** provide detailed information about a source of stimulation; **crude touch and pressure receptors** are poorly localized. The six types of tactile receptors are free nerve endings, the **root hair plexus, tactile discs, tactile corpuscles, lamellated corpuscles,** and **Ruffini corpuscles.** *(Figure 17-2)*

5. Baroreceptors monitor changes in blood pressure in the walls of major arteries and veins. Baroreceptors along the digestive tract help coordinate reflex activities of digestion, and others monitor pressure changes in the reproductive and urinary tracts. These mechanoreceptors respond immediately but adapt rapidly. *(Table 17-1)*

6. Proprioceptors monitor the position of joints, tension in tendons and ligaments, and the state of muscular contraction. Proprioceptors include Golgi tendon organs and muscle spindles. *(Table 17-1)*

Chemoreceptors p. 533

7. **Chemoreceptors** respond to water-soluble and lipid-soluble substances dissolved in surrounding fluid. These receptors monitor the chemical makeup of body fluids. *(Table 17-1)*

OLFACTION p. 534

Olfactory Receptors p. 535

1. The **olfactory organs** contain the **olfactory epithelium** with **olfactory receptors** (neurons sensitive to chemicals dissolved in the overlying mucus), supporting cells, and **basal (stem) cells.** Their surfaces are coated with the secretions of the **olfactory glands.** *(Figure 17-3)*
2. The olfactory receptors are modified neurons.

Olfactory Pathways p. 535

3. The highly sensitive olfactory system has extensive limbic and hypothalamic connections.

Olfactory Discrimination p. 536

4. The olfactory system can distinguish thousands of chemical stimuli. The CNS interprets smells by the pattern of receptor activity.
5. The olfactory receptor population shows considerable turnover. The number of olfactory receptors declines with age.

GUSTATION p. 536

1. **Taste** *(gustatory)* **receptors** are clustered in **taste buds.**
2. Taste buds are associated with epithelial projections on the dorsal surface of the tongue (**lingual papillae**). *(Figure 17-4)*

Taste Receptors p. 536

3. Each taste bud contains **gustatory cells,** which extend *taste hairs* through a narrow **taste pore.** *(Figure 17-4)*

Gustatory Pathways p. 536

4. The taste buds are monitored by cranial nerves that synapse within the **nucleus solitarius.**

Gustatory Discrimination p. 537

5. The **primary taste sensations** are sweet, salt, sour, and bitter; **umami** and **water receptors** are also present. *(Figure 17-4)*
6. Taste sensitivity exhibits significant individual differences, some of which are inherited.
7. The number of taste buds declines with age.

VISION p. 538

Accessory Structures of the Eye p. 538

1. The **accessory structures** of the eye include the **eyelids (palpebrae),** which are separated by the **palpebral fissure.** The **eyelashes** line the margins of the eyelids. Along the inner margin of the lid are **tarsal glands,** which secrete a lipid-rich product. Glands at the **lacrimal caruncle** produce other secretions. *(Figure 17-5)*
2. An epithelium called the **conjunctiva** covers most of the exposed surface of the eye. The **ocular** *(bulbar)* **conjunctiva** covers the anterior surface of the eye, and the **palpebral conjunctiva** lines the inner surface of the eyelids. The **cornea** is transparent. *(Figures 17-5, 17-6)*
3. The secretions of the **lacrimal gland** bathe the conjunctiva, are slightly alkaline, and contain **lysozyme** (an enzyme that attacks bacteria). Tears collect in the **lacrimal lake.** The tears reach the inferior meatus of the nose after they pass through the **lacrimal puncta,** the **lacrimal canaliculi,** the **lacrimal sac,** and the **nasolacrimal duct.** *(Figures 17-5, 17-6)*

The Eye p. 541

4. The eye has three layers: an outer **fibrous tunic,** a middle **vascular tunic,** and an inner **neural tunic.**

5. The fibrous tunic consists of the **sclera** (a dense fibrous connective tissue that covers most of the ocular surface), the cornea, and the **limbus** (the border between the sclera and the cornea). *(Figure 17-6)*
6. The vascular tunic, or **uvea,** includes the **iris,** the **ciliary body,** and the **choroid.** The iris contains blood vessels, pigment cells, and muscle fibers that change the diameter of the **pupil.** The ciliary body contains the **ciliary muscle** and the **ciliary processes,** which attach to the **suspensory ligaments** of the **lens.** *(Figures 17-6, 17-7)*
7. The neural tunic, or **retina,** consists of an outer *pigmented part* and an inner *neural part;* the latter contains visual receptors and associated neurons. *(Figures 17-6, 17-8)*
8. The direct line to the CNS proceeds from the **photoreceptors** to **bipolar cells,** then to **ganglion cells,** and to the brain via the optic nerve. The axons of ganglion cells converge at the **optic disc,** or **blind spot. Horizontal cells** and **amacrine cells** modify the signals passed among other components of the retina. *(Figures 17-8, 17-9)*
9. The ciliary body and lens divide the interior of the eye into a large **posterior cavity,** or *vitreous chamber,* and a smaller **anterior cavity.** The anterior cavity is subdivided into the **anterior chamber,** which extends from the cornea to the iris, and a **posterior chamber,** between the iris and the ciliary body and lens. *(Figure 17-10)*
10. The fluid **aqueous humor** circulates within the eye and reenters the circulation after diffusing through the walls of the anterior chamber and into the **canal of Schlemm.** *(Figure 17-10)*
11. The **lens** lies posterior to the cornea and forms the anterior boundary of the posterior cavity. This cavity contains the **vitreous body,** a gelatinous mass that helps stabilize the shape of the eye and support the retina. *(Figures 17-6, 17-10)*
12. The lens focuses a visual image on the photoreceptors. A lens that has lost its transparency is a **cataract.**
13. Light is **refracted** (bent) when it passes through the cornea and lens. During **accommodation,** the shape of the lens changes to focus an image on the retina. "Normal" **visual acuity** is rated 20/20. *(Figures 17-11–17-14)*

Visual Physiology p. 550

14. The two types of photoreceptors (visual receptors of the retina) are **rods,** which respond to almost any photon, regardless of its energy content, and **cones,** which have characteristic ranges of sensitivity. Many cones are densely packed within the **fovea** (the central portion of the **macula lutea**), the site of sharpest vision.
15. Each photoreceptor contains an **outer segment** with membranous **discs.** A narrow stalk connects the outer segment to the **inner segment.** Light absorption occurs in the **visual pigments,** which are derivatives of **rhodopsin** (**opsin** plus the pigment **retinal,** which is synthesized from **vitamin A**). The retinal molecule can be in either the normal **11-*cis*** or (after absorbing light) the **11-*trans*** form. *(Figures 17-15–17-17)*
16. Color sensitivity depends on the integration of information from **red, green,** and **blue** cones. **Color blindness** is the inability to detect certain colors. *(Figures 17-18, 17-19)*
17. In the **dark-adapted state,** most visual pigments are fully receptive to stimulation. In the **light-adapted state,** the pupil constricts and **bleaching** of the visual pigments occurs.

The Visual Pathway p. 554

18. Each photoreceptor monitors a specific receptive field. The ganglion cells that monitor rods, called **M cells** *(magnocells),* are relatively large. The ganglion cells that monitor cones, called **P cells** *(parvo cells),* are smaller and more numerous. *(Figure 17-20)*

17

19. Visual data from the left half of the combined field of vision arrive at the visual cortex of the right occipital lobe; data from the right half of the combined field of vision arrive at the visual cortex of the left occipital lobe. *(Figure 17-21)*

20. Visual inputs to the suprachiasmatic nucleus of the hypothalamus affect the function of other brain stem nuclei. This nucleus establishes a visceral **circadian rhythm,** which is tied to the day–night cycle and affects other metabolic processes.

EQUILIBRIUM AND HEARING p. 557

1. The senses of equilibrium and hearing are provided by the receptors of the inner ear.

Anatomy of the Ear p. 557

2. The ear is divided into the **external ear, middle ear,** and **inner ear.** *(Figure 17-22)*

3. The external ear includes the **auricle,** or *pinna,* which surrounds the entrance to the **external auditory canal.** That canal ends at the **tympanic membrane** *(eardrum). (Figure 17-22)*

4. The middle ear communicates with the nasopharynx via the **auditory** *(pharyngotympanic)* **tube.** The middle ear encloses and protects the **auditory ossicles,** bones that connect the tympanic membrane with the receptors of the inner ear. *(Figures 17-22, 17-23)*

5. The **membranous labyrinth** (the chambers and tubes) of the inner ear contains the fluid **endolymph.** The **bony labyrinth** surrounds and protects the membranous labyrinth. The bony labyrinth can be subdivided into the **vestibule,** the **semicircular canals,** and the **cochlea.** The structures and air spaces of the external and middle ears help capture and transmit sound to the cochlea. *(Figure 17-24)*

6. The vestibule of the inner ear encloses the **saccule** and **utricle,** a pair of membranous sacs whose receptors provide sensations of gravity and linear acceleration. The semicircular canals contain the **semicircular ducts,** whose receptors provide sensations of rotation. The cochlea contains the **cochlear duct,** an elongated portion of the membranous labyrinth. *(Figure 17-24)*

7. The basic receptors of the inner ear are **hair cells.** The surfaces of each hair cell support stereocilia and (except in the cochlear duct) a single **kinocilium.** Hair cells provide information about the direction and strength of varied mechanical stimuli. *(Figure 17-24)*

Equilibrium p. 561

8. The **anterior, posterior,** and **lateral semicircular ducts** are continuous with the utricle. Each duct contains an **ampulla** with sensory receptors. Here the cilia contact a gelatinous **cupula.** *(Figures 17-24, 17-25)*

9. The saccule and utricle are connected by a passageway continuous with the **endolymphatic duct,** which terminates in the **endolymphatic sac.** In the saccule and utricle, hair cells cluster within **maculae,** where their cilia contact the **otolith** (densely packed mineral crystals, called **statoconia,** in a matrix). When the head tilts, the statoconia shift. The resulting distortion in hair cell processes sends a signal to the CNS. *(Figure 17-25)*

10. The vestibular receptors activate sensory neurons of the **vestibular ganglia.** The axons form the **vestibular branch** of the vestibulocochlear nerve (N VIII), synapsing within the **vestibular nuclei.** *(Figure 17-26)*

Hearing p. 564

11. The energy content of a sound determines its **intensity,** measured in **decibels.** Sound waves travel toward the tympanic membrane, which vibrates; the auditory ossicles conduct these vibrations to the inner ear. Movement at the **oval window** applies pressure to the **perilymph** of the **vestibular duct.** *(Tables 17-2, 17-3; Figures 17-27, 17-28)*

12. Pressure waves distort the **basilar membrane** and push the hair cells of the **organ of Corti** against the **tectorial membrane.** The **tensor tympani** and **stapedius muscles** contract to reduce the amount of motion when very loud sounds arrive. *(Figures 17-29, 17-30)*

13. The sensory neurons are located in the **spiral ganglion** of the cochlea. Their afferent fibers form the **cochlear branch** of the vestibulocochlear nerve (N VIII), synapsing at the **cochlear nucleus.** *(Figure 17-31)*

REVIEW QUESTIONS

LEVEL 1 REVIEWING FACTS AND TERMS

1. Regardless of the nature of a stimulus, sensory information must be sent to the CNS in the form of
   - (a) dendritic processes
   - (b) action potentials
   - (c) neurotransmitter molecules
   - (d) generator potentials

2. A membrane depolarization that leads to an action potential in a sensory neuron is
   - (a) a labeled line
   - (b) a sensation
   - (c) a neurotransmitter
   - (d) a generator potential

3. A reduction of sensitivity in the presence of a constant stimulus is
   - (a) transduction
   - (b) sensory coding
   - (c) line labeling
   - (d) adaptation

4. Sensations produced by an injection or a deep cut would reach the CNS very quickly via
   - (a) Type A fibers
   - (b) Type C fibers
   - (c) tactile receptors
   - (d) proprioceptors

5. Mechanoreceptors that detect pressure changes in the walls of blood vessels and in portions of the digestive, reproductive, and urinary tracts are
   - (a) tactile receptors
   - (b) baroreceptors
   - (c) proprioceptors
   - (d) free nerve endings

6. Lamellated corpuscles in the skin detect
   - (a) deep pressure
   - (b) fine touch
   - (c) muscle stretch
   - (d) temperature changes

7. Examples of proprioceptors that monitor the position of joints and the state of muscular contraction are
   - (a) lamellated and tactile corpuscles
   - (b) carotid and aortic sinuses
   - (c) tactile discs and Ruffini corpuscles
   - (d) Golgi tendon organs and muscle spindles

8. When chemicals dissolve in the nasal cavity, they stimulate
   - (a) gustatory cells
   - (b) olfactory cilia
   - (c) rod cells
   - (d) tactile receptors

9. The taste sensation of sweetness is experienced on the
   - (a) posterior part of the tongue
   - (b) anterior part of the tongue
   - (c) right and left lateral sides of the tongue
   - (d) the middle part of the tongue

10. The anterior, transparent part of the fibrous tunic is the
(a) cornea (b) iris
(c) sclera (d) retina

11. Tarsal glands secrete a lipid-rich product that helps
(a) produce and secrete tears in the lacrimal apparatus
(b) keep the cornea moist and clean
(c) dilate the blood vessels beneath the conjunctival epithelium
(d) keep the eyelids from sticking together

12. Tears produced by the lacrimal apparatus
(a) keep conjunctival surfaces moist and clean
(b) reduce friction and remove debris from the eye
(c) provide nutrients and oxygen to the conjunctiva
(d) a, b, and c are correct

13. The thick, gel-like fluid that helps support the structure of the eyeball is the
(a) vitreous humor (b) aqueous humor
(c) ora serrata (d) perilymph

14. Pupillary muscle groups are controlled by the ANS. Parasympathetic activation causes pupillary _____, and sympathetic activation causes _____.
(a) dilation, constriction (b) dilation, dilation
(c) constriction, dilation (d) constriction, constriction

15. The retina is the
(a) vascular tunic (b) fibrous tunic
(c) neural tunic (d) a, b, and c are correct

16. At sunset your visual system adapts to
(a) fovea vision (b) rod-based vision
(c) macular vision (d) cone-based vision

17. The primary function of the lens of the eye is to
(a) stabilize the shape of the eye
(b) facilitate communication between photoreceptors and ganglion cells
(c) focus the visual image on the photoreceptors
(d) a, b, and c are correct

18. A better-than-average visual acuity rating is
(a) 20/20 (b) 20/30
(c) 15/20 (d) 20/15

19. Ceruminous glands along the external auditory canal are responsible for
(a) slowing the growth of microorganisms in the exterior auditory canal
(b) reducing chances of infection in the ear
(c) helping deny access to foreign objects
(d) a, b, and c are correct

20. The malleus, incus, and stapes are the tiny bones located in the
(a) outer ear (b) middle ear
(c) inner ear (d) membranous labyrinth

21. Receptors in the saccule and utricle provide sensations of
(a) angular acceleration
(b) hearing
(c) vibration
(d) gravity and linear acceleration

22. The organ of Corti is located in the _____ of the inner ear.
(a) utricle (b) bony labyrinth
(c) vestibule (d) cochlea

23. Auditory information about the frequency and intensity of stimulation is relayed to the CNS over the cochlear branch of cranial nerve
(a) IV (b) VI
(c) VIII (d) X

24. What three steps are necessary for transduction to occur?

25. What three types of mechanoreceptors respond to stretching, compression, twisting, or other distortions of the cell membrane?

26. Identify six types of tactile receptors located in the skin, and describe their sensitivities.

27. Trace the olfactory pathway from the time an odor reaches the olfactory epithelium until it reaches its final destination in the brain.

28. What are the three types of papillae on the human tongue?

29. (a) What structures make up the fibrous tunic of the eye?
(b) What are the functions of the fibrous tunic?

30. What structures make up the vascular tunic of the eye?

31. Trace the pathway of a nerve impulse from the optic nerve to the visual cortex.

32. What are the three auditory ossicles in the middle ear, and what are their functions?

33. What six basic steps are involved in the process of hearing?

LEVEL 2 REVIEWING CONCEPTS

34. The CNS interprets sensory information on the basis of the
(a) strength of the action potential
(b) number of generator potentials
(c) line over which it arrives
(d) a, b, and c are correct

35. If the auditory cortex is damaged, the individual will respond to sounds and will have normal acoustic reflexes, but
(a) the sounds may produce nerve deafness
(b) the auditory ossicle may be immobilized
(c) sound interpretation and pattern recognition may be impossible
(d) normal transfer of vibration to the oval window is inhibited

36. Distinguish between the general senses and the special senses in the human body.

37. In what form does the CNS receive a stimulus detected by a sensory receptor?

38. What type of information about a stimulus does sensory coding provide?

39. Why are olfactory sensations long-lasting and an important part of our memories and emotions?

40. What is the usual result if a sebaceous gland of an eyelash or a tarsal gland becomes infected?

41. Jane makes an appointment with the optometrist for a vision test. Her test results are reported as 20/15. What does this test result mean? Is a rating of 20/20 better or worse?

LEVEL 3 CRITICAL THINKING AND CLINICAL APPLICATIONS

42. You are at a park watching some deer 35 feet away from you. A friend taps you on the shoulder to ask a question. As you turn to look at your friend, who is standing just 2 feet away, what changes would your eyes undergo?

43. Sally's driver's license indicates that she must wear glasses when she drives. She does not need glasses to read or to see objects that are close. What type of lenses are in Sally's glasses?

44. After attending a Fourth of July fireworks extravaganza, Millie finds it difficult to hear normal conversation and her ears keep "ringing." What is causing her hearing problems?

45. For a few seconds after you ride the express elevator from the twentieth floor to the ground floor, you still feel as if you are descending, even though you have come to a stop. Why?

46. Juan has a disorder involving the saccule and the utricle. He is asked to stand with his feet together and arms extended forward. As long as he keeps his eyes open, he exhibits very little movement. But when he closes his eyes, his body begins to sway a great deal, and his arms tend to drift in the direction of the impaired vestibular receptors. Why does this occur?

17

CHAPTER **18**

The Endocrine System

CHAPTER OUTLINE AND OBJECTIVES

INTRODUCTION p. 577

AN OVERVIEW OF THE ENDOCRINE SYSTEM p. 578
Hormone Structure p. 578
Hormone Distribution and Transport p. 580
Hormone Function and Mechanisms of Action p. 580
Control of Endocrine Activity p. 584
① Compare the endocrine and nervous systems.
② Compare the cellular components of the endocrine system with those of other tissues and systems.
③ Compare the major chemical classes of hormones.
④ Explain the general mechanisms of hormonal action.
⑤ Describe how endocrine organs are controlled.

THE PITUITARY GLAND p. 586
The Anterior Pituitary p. 587
The Posterior Pituitary p. 591

THE THYROID GLAND p. 593
Thyroid Follicles and Thyroid Hormones p. 594
The C Cells of the Thyroid Gland: Calcitonin p. 597

THE PARATHYROID GLANDS p. 597
Parathyroid Hormone Secretion p. 598

THE ADRENAL GLANDS p. 600
The Adrenal Cortex p. 600
The Adrenal Medullae p. 603

THE PINEAL GLAND p. 603

THE ENDOCRINE TISSUES OF OTHER SYSTEMS p. 604
The Intestines p. 604
The Kidneys p. 604
The Heart p. 605
The Thymus p. 605
The Pancreas p. 605
The Gonads p. 608
⑥ Describe the location, hormones, and functions of the pituitary gland, thyroid gland, parathyroid glands, adrenal glands, pineal gland, intestines, kidneys, heart, thymus, pancreas, testes, and ovaries.
⑦ Discuss the results of abnormal levels of hormone production.

Gheorghe Muresan (left) and Tony Massenburg differ in height by about 0.3 m—about 1 foot. Genetically programmed levels of hormones, especially growth hormone, were probably responsible for this difference. Hormone levels during childhood and adolescence gave Muresan a height advantage, but both players have the right blend of coordination, balance, strength, speed, and intelligence. This pattern is common: Hormones direct long-term changes (such as those that determine body form); moment-to-moment adjustments (which make basketball games exciting) are controlled by the nervous system. This chapter introduces the major hormones and the endocrine organs that produce them.

PATTERNS OF HORMONAL INTERACTION p. 610
Hormones and Growth p. 610
Hormones and Stress p. 612
Hormones and Behavior p. 614
Aging and Hormone Production p. 614
⑧ Explain how hormones interact to produce coordinated physiological responses.
⑨ Identify the hormones that are especially important to normal growth, and discuss their roles.
⑩ Define the general adaptation syndrome, and compare homeostatic responses with stress responses.
⑪ Describe the effects that hormones have on behavior.

INTEGRATION WITH OTHER SYSTEMS p. 614
⑫ Give examples of interactions between the endocrine system and each of the other organ systems.

MEDIALAB and CHAPTER REVIEW p. 618

To function effectively, every cell in the body must communicate with its neighbors and with cells and tissues in distant portions of the body. The methods of communication are listed in Table 18-1. In a few specialized cases, cellular activities are coordinated by the exchange of ions and molecules from one cell to the next across gap junctions. This **direct communication** occurs between cells of the same type, and the two cells must be in extensive physical contact. The two cells communicate so closely that they function as a single entity. For example, gap junctions (1) coordinate ciliary movement among epithelial cells, (2) coordinate the contractions of cardiac muscle cells, and (3) facilitate the propagation of action potentials from one neuron to the next at electrical synapses.

Direct communication is highly specialized and relatively rare. Most of the communication between cells involves the release and receipt of chemical messages. Each cell continuously "talks" to its neighbors by releasing chemicals into the extracellular fluid. These chemicals tell cells what their neighbors are doing at any moment. The result is the coordination of tissue function at the local level. The use of chemical messengers to transfer information from cell to cell within a single tissue is called **paracrine communication.** The chemicals involved are called **paracrine factors,** also known as *cytokines,* or *local hormones.* Examples of paracrine factors include the prostaglandins, introduced in Chapter 2, and the various growth factors discussed in Chapter 3. ∞ *pp. 48, 100*

Paracrine factors enter the bloodstream, but the concentrations are usually so low that distant cells and tissues are not affected. However, some paracrine factors, including several of the prostaglandins and related chemicals, have primary effects in their tissues of origin and secondary effects in other tissues and organs. When secondary effects occur, the paracrine factors are also acting as hormones. **Hormones** are chemical messengers that are released in one tissue and transported in the bloodstream to reach specific cells in other tissues. Whereas most cells release paracrine factors, typical hormones are produced only by specialized cells.

In intercellular communication, hormones are letters and the circulatory system is the postal service. A hormone released into the bloodstream will be dis-

18

TABLE 18-1 Mechanisms of Intercellular Communication

Mechanism	Transmission	Chemical Mediators	Distribution of Effects
Direct communication	Through gap junctions from cytoplasm to cytoplasm	Ions, small solutes, lipid-soluble materials	Usually limited to adjacent cells of the same type that are directly interconnected by connexons
Paracrine communication	Through extracellular fluid	Paracrine factors	Primarily limited to local area, where concentrations are relatively high; target cells must have appropriate receptors
Endocrine communication	Through the circulatory system	Hormones	Target cells are primarily in other tissues and organs and must have appropriate receptors
Synaptic communication	Across synaptic clefts	Neurotransmitters	Limited to very specific area; target cells must have appropriate receptors

tributed throughout the body. Each hormone has **target cells,** specific cells that respond to its presence. These cells possess the receptors needed to bind and "read" the hormonal message. Although every cell in the body is exposed to the mixture of hormones in circulation at any moment, each individual cell will respond to only a few of the hormones present. The other hormones are treated like junk mail and ignored, because the cell lacks the receptors to read the messages they contain. The use of hormones to coordinate cellular activities in tissues in distant portions of the body is called **endocrine communication.**

Because the target cells can be anywhere in the body, a single hormone can alter the metabolic activities of multiple tissues and organs simultaneously. These effects may be slow to appear, but they typically persist for days. Consequently, hormones are effective in coordinating cell, tissue, and organ activities on a sustained, long-term basis. For example, circulating hormones keep body water content and levels of electrolytes and organic nutrients within normal limits 24 hours a day throughout our entire lives.

While the effects of a single hormone persist, a cell may receive additional instructions from other hormones. The result will be a further modification of cellular operations. Gradual changes in the quantities and identities of circulating hormones can produce complex changes in physical structure and physiological capabilities. Examples include the processes of embryological and fetal development, growth, and puberty.

The nervous system also relies primarily on chemical communication, but it does not use the bloodstream for message delivery. Instead, neurons release a neurotransmitter at a synapse very close to the target cells that bear the appropriate receptors. The command to release the neurotransmitter rapidly travels from one location to another in the form of action potentials propagated along axons. The nervous system thus acts like a telephone company, carrying high-speed "messages" from one location in the body to another and delivering them to a specific destination. The effects of neural stimulation are generally short-lived, and they tend to be restricted to specific target cells—primarily because the neurotransmitter is rapidly broken down or recycled. This form of **synaptic communication** is ideal for crisis management: If you are in danger of being hit by a speeding bus, the nervous system can coordinate and direct your leap to safety. Once the crisis is over and the neural circuit quiets down, things soon return to normal.

Table 18-1 provides a summary of the ways our cells and tissues communicate with one another. When viewed from a general perspective, the differences between the nervous and endocrine systems seem relatively clear. In fact, these broad organizational and functional distinctions are the basis for treating them

as two separate systems. Yet, when we consider them in detail, the two systems are organized along parallel lines. For example:

- Both systems rely on the release of chemicals that bind to specific receptors on their target cells.
- The two systems share many chemical messengers; for example, norepinephrine and epinephrine are called *hormones* when released into the bloodstream but *neurotransmitters* when released across synapses.
- Both systems are regulated primarily by negative feedback control mechanisms.
- The two systems share a common goal: to preserve homeostasis by coordinating and regulating the activities of other cells, tissues, organs, and systems.

In this chapter, we introduce the components and functions of the endocrine system and explore the interactions between the nervous and endocrine systems. In later chapters, we shall consider specific endocrine organs, hormones, and functions in detail.

AN OVERVIEW OF THE ENDOCRINE SYSTEM

The **endocrine system** includes all the endocrine cells and tissues of the body. As we noted in Chapter 4, *endocrine cells* are glandular secretory cells that release their secretions into the extracellular fluid. This characteristic distinguishes them from *exocrine cells,* which secrete their products onto epithelial surfaces, generally by way of ducts. ∞ *p. 115* The chemicals released by endocrine cells may affect only adjacent cells, as in the case of most paracrine factors, or they may affect cells throughout the body.

The components of the endocrine system are introduced in Figure 18-1●. This figure also lists the major hormones produced in each endocrine tissue and organ. Some of these organs, such as the pituitary gland, have endocrine secretion as a primary function; others, such as the pancreas, have many other functions in addition to endocrine secretion. We consider the structure and functions of these endocrine organs in detail in chapters dealing with other systems. Examples include the hypothalamus (Chapter 14), the adrenal medullae (Chapter 16), the heart (Chapter 20), the thymus (Chapter 22), the pancreas and digestive tract (Chapter 24), the kidneys (Chapter 26), the reproductive organs (Chapter 28), and the placenta (Chapter 29).

HORMONE STRUCTURE

Hormones can be divided into three groups on the basis of chemical structure: (1) *amino acid derivatives,* (2) *peptide hormones,* and (3) *lipid derivatives.*

18

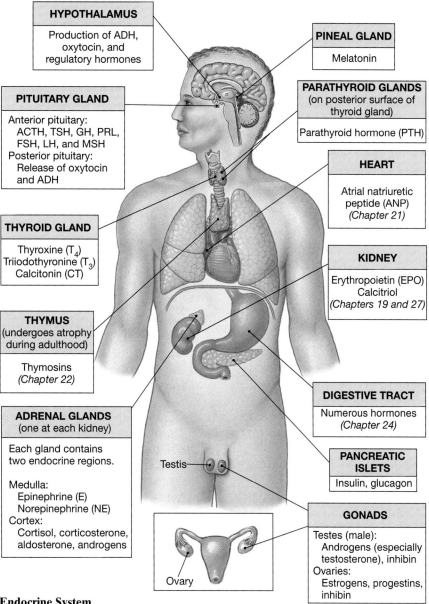

HYPOTHALAMUS
Production of ADH, oxytocin, and regulatory hormones

PINEAL GLAND
Melatonin

PITUITARY GLAND
Anterior pituitary:
ACTH, TSH, GH, PRL, FSH, LH, and MSH
Posterior pituitary:
Release of oxytocin and ADH

PARATHYROID GLANDS
(on posterior surface of thyroid gland)
Parathyroid hormone (PTH)

HEART
Atrial natriuretic peptide (ANP)
(Chapter 21)

THYROID GLAND
Thyroxine (T$_4$)
Triiodothyronine (T$_3$)
Calcitonin (CT)

KIDNEY
Erythropoietin (EPO)
Calcitriol
(Chapters 19 and 27)

THYMUS
(undergoes atrophy during adulthood)
Thymosins
(Chapter 22)

ADRENAL GLANDS
(one at each kidney)
Each gland contains two endocrine regions.
Medulla:
Epinephrine (E)
Norepinephrine (NE)
Cortex:
Cortisol, corticosterone, aldosterone, androgens

DIGESTIVE TRACT
Numerous hormones
(Chapter 24)

PANCREATIC ISLETS
Insulin, glucagon

Testis

GONADS
Testes (male):
Androgens (especially testosterone), inhibin
Ovaries:
Estrogens, progestins, inhibin

Ovary

•**FIGURE 18-1** The Endocrine System

18

Amino Acid Derivatives

Amino acid derivatives are relatively small molecules that are structurally similar to amino acids, the building blocks of proteins. ∞ *p. 51* (The amino acids are shown in Appendix IV.) This group of hormones, sometimes known as the *biogenic amines,* includes epinephrine, norepinephrine, dopamine, the thyroid hormones, and melatonin. Epinephrine (E), norepinephrine (NE), and dopamine are structurally similar; these compounds are sometimes called. *catecholamines* (kat-e-KŌ-la-mēnz). As we saw in Chapter 16, both E and NE are secreted by the adrenal medullae during sympathetic activation. ∞ *p. 510* **Thyroid hormones** are produced by the thyroid gland, and **melatonin** (mel-a-TŌ-nin) is produced by the

pineal gland. Catecholamines and thyroid hormones are synthesized from molecules of the amino acid *tyrosine* (TĪ-rō-sēn). Melatonin is manufactured from molecules of the amino acid *tryptophan.*

Peptide Hormones

Peptide hormones are chains of amino acids. In general, peptide hormones are produced as *prohormones,* inactive molecules that are converted to active hormones either before or after they are secreted.

Peptide hormones can be divided into two groups. One large and diverse group includes hormones that range from short polypeptide chains, such as *antidiuretic hormone (ADH)* and *oxytocin* (9 amino acids apiece), to small proteins, such as *growth hormone* (GH; 191 amino acids) and *prolactin* (PRL; 198 amino acids). This group includes all the hormones secreted by the hypothalamus, posterior pituitary gland, heart, thymus, digestive tract, and pancreas and most of the hormones of the anterior pituitary gland.

The second group of peptide hormones consists of glycoproteins. ∞ *p. 56* These proteins are more than 200 amino acids long and have carbohydrate side chains. The glycoproteins include *thyroid-stimulating hormone (TSH), luteinizing hormone (LH),* and *follicle-stimulating hormone (FSH)* from the anterior pituitary gland as well as several hormones produced in other organs.

Lipid Derivatives

There are two classes of *lipid derivatives:* (1) *steroid hormones,* derived from cholesterol, and (2) *eicosanoids* (ī-KŌ-sa-noydz), derived from *arachidonic acid,* a 20-carbon fatty acid.

Steroid Hormones **Steroid hormones** are lipids structurally similar to cholesterol (Figure 2-16a•, p. 49). Steroid hormones are released by male and female reproductive organs (*androgens* by the testes, *estrogens* and *progestins* by the ovaries), the adrenal glands *(corticosteroids),* and the kidneys *(calcitriol).* The individual hormones differ in the side chains attached to the basic ring structure.

In the blood, steroid hormones are bound to specific transport proteins in the plasma. For this reason, they remain in circulation longer than do secreted peptide hormones. The liver gradually absorbs these steroids and converts them to a soluble form that can be excreted in the bile or urine.

Eicosanoids **Eicosanoids** are small molecules with a five-carbon ring at one end. These compounds are important paracrine factors that coordinate cellular activities and affect enzymatic processes (such as blood clotting) that occur in extracellular fluids. Some of the eicosanoids also have secondary roles as hormones. Examples of important eicosanoids include the following:

■ **Leukotrienes** are released by activated white blood cells, or *leukocytes.* Leukotrienes are important in coordinating tissue responses to injury or disease.

■ **Prostaglandins** are produced in most tissues of the body. Within each tissue, the prostaglandins released are involved primarily in coordinating local cellular activities. In some tissues, prostaglandins are converted to **thromboxanes** (throm-BOX-ānz) and **prostacyclins** (pros-ta-SĪ-klinz), which also have strong paracrine effects.

Our focus in this chapter is on circulating hormones whose primary functions are the coordination of activities in many tissues and organs. We shall consider eicosanoids in chapters that deal with individual tissues and organs, including Chapters 19 (the blood), 22 (the lymphatic system), and 28 (the reproductive system). Examples of each group of hormones discussed here are included in Figure 18-2•.

HORMONE DISTRIBUTION AND TRANSPORT

Hormone release typically occurs where capillaries are abundant, and the hormones quickly enter the bloodstream for distribution throughout the body. Within the blood, hormones may circulate freely, or they may be bound to special carrier proteins. A freely circulating hormone remains functional for less than 1 hour, and sometimes for as little as 2 minutes. It is inactivated when (1) it diffuses out of the bloodstream and binds to receptors in target tissues; (2) it is absorbed and broken down by cells of the liver or kidneys; or (3) it is broken down by enzymes in the plasma or interstitial fluids.

Thyroid hormones and steroid hormones remain in circulation much longer, because when these hormones enter the bloodstream, almost all of them become attached to special transport proteins. Less than 1 percent of these hormones circulate freely. An equilibrium state exists between the free hormones and the bound hormones: As the free hormones are removed and inactivated, they are replaced by the release of bound hormones. Thus the bloodstream contains a substantial reserve (several weeks' supply) of these hormones at any time.

HORMONE FUNCTION AND MECHANISMS OF ACTION

All cellular structures and functions depend on proteins. Structural proteins determine the general shape and internal structure of a cell, and enzymes affect its metabolic activities. Hormones alter cellular operations by changing the types, activities, or quantities of important enzymes and structural proteins. In other words, a hormone may:

■ Stimulate the synthesis of an enzyme or a structural protein not already present in the cytoplasm by activating appropriate genes in the cell nucleus.

■ Turn an existing enzyme "on" or "off" by changing its shape or structure.

■ Increase or decrease the rate of synthesis of a particular enzyme or other protein by changing the rate of transcription or translation.

Through one or more of these mechanisms, a hormone can modify the physical structure or biochemical properties of its target cells.

•**FIGURE 18-2** A Structural Classification of Hormones

To affect a target cell, a hormone must first interact with an appropriate receptor. Each cell has the receptors needed to respond to several different hormones, but cells in different tissues have different combinations of receptors. This arrangement accounts for the differential effects of hormones on specific tissues. For every cell, the presence or absence of a specific receptor determines the cell's hormonal sensitivities. If the cell has a receptor that will bind a particular hormone, that cell will respond to the hormone's presence. If the cell lacks the proper receptor for that hormone, the hormone will have no effect on that cell.

Hormone receptors are located either (1) on the cell membrane or (2) inside the cell. Using a few specific examples, we will now introduce the basic mechanisms involved.

Hormones and the Cell Membrane

The receptors for catecholamines (E, NE, and dopamine), peptide hormones, and eicosanoids are in the cell membranes of their respective target cells. Because catecholamines and peptide hormones are not lipid-soluble, they are unable to penetrate a cell membrane. Instead, these hormones bind to receptor proteins at the outer surface of the cell membrane. Conversely, eicosanoids, which are lipid-soluble, diffuse across the membrane to reach receptor proteins on the inner membrane surface.

Any hormone that binds to receptors in the cell membrane does not have direct effects on the target cell. For example, such a hormone does not begin building a protein or catalyzing a specific reaction. The hormone acts as a **first messenger,** a substance that causes the appearance of a **second messenger** in the cytoplasm. The second messenger functions inside the cell as an enzyme activator, inhibitor, or cofactor, but the net result is a change in the rates of various metabolic reactions. The most important second messengers are (1) *cyclic-AMP (cAMP),* a derivative of ATP; (2) *cyclic-GMP (cGMP),* a derivative of GTP, another high-energy compound; and (3) calcium ions.

The binding of a small number of hormone molecules to membrane receptors may lead to the appearance of thousands of second messengers in a cell. This process, which magnifies the effect of a hormone on the target cell, is *amplification.* In addition, the arrival of a single hormone may promote the release of more than one type of second messenger. Thus, the hormone can alter many aspects of cell function at the same time.

The presence or absence of a hormone can also affect the nature and number of hormone receptor proteins in the cell membrane. **Down-regulation** is a process in which the presence of a hormone triggers a decrease in the number of hormone receptors. In down-regulation, when levels of a particular hormone are high, cells become less sensitive to it. Conversely, **up-regulation** is a process in which the presence of a hormone triggers an increase in the number of hormone receptors. In up-regulation, when levels of a particular hormone are low, cells become more sensitive to it.

G-Proteins   The link between the first messenger and the second messenger generally involves a **G-protein,** an enzyme complex coupled to a membrane receptor. The name *G-protein* refers to the fact that these proteins bind GTP. ∞ *p. 59* There are several types of G-proteins, but in each type the G-protein is activated when a hormone binds to its receptor at the membrane surface. Figure 18-3• diagrams important mechanisms that may be set in motion by the activa-

tion of a G-protein. Each sequence affects levels of a second messenger in the cytoplasm:

1. *The activation of adenylate cyclase.* When activated by a G-protein, the enzyme **adenylate cyclase,** also called *adenylyl cyclase,* converts ATP to a ring-shaped molecule of **cyclic-AMP (cAMP).** Cyclic-AMP then functions as the second messenger. The first step typically involves the activation of a kinase. A *kinase* is an enzyme that performs *phosphorylation,* the attachment of a phosphate group (PO_4^{3-}) to another molecule. ∞ *p. 57* Cyclic-AMP activates kinases that phosphorylate proteins. The effect on the target cell depends on the nature of the proteins affected. The phosphorylation of membrane proteins can open ion channels, and in the cytoplasm many important enzymes can be activated only by phosphorylation. The hormones calcitonin, parathyroid hormone, ADH, ACTH, epinephrine, FSH, LH, TSH, and glucagon produce their effects by increasing intracellular concentrations of cyclic-AMP. The increase is usually short-lived, because the cytoplasm contains another enzyme, **phosphodiesterase (PDE),** which inactivates cyclic-AMP by converting it to AMP (adenosine monophosphate).

2. *The entry and/or release of calcium ions.* An activated G-protein can trigger the opening of calcium ion channels in the membrane or the release of calcium ions from intracellular stores. The G-protein first activates the enzyme **phospholipase C** (PLC). This enzyme generates the second messengers **diacylglycerol** (DAG) and **inositol triphosphate** (IP$_3$) from membrane phospholipids. The mechanism proceeds as follows:

   STEP 1: Inositol triphosphate diffuses into the cytoplasm and triggers the release of Ca^{2+} from intracellular reserves, such as those in the smooth endoplasmic reticulum of many cells.

   STEP 2: The combination of diacylglycerol and intracellular calcium ions activates another membrane protein, **protein kinase C** (PKC). The activation of PKC leads to the phosphorylation of calcium channel proteins. This process opens the channels and permits the entry of extracellular Ca^{2+}.

   STEP 3: The calcium ions then serve as second messengers, generally in combination with an intracellular protein called **calmodulin.** Once it has bound calcium ions, calmodulin can activate specific cytoplasmic enzymes. This chain of events is responsible for the stimulatory effects that follow the activation of α_1 receptors by epinephrine or norepinephrine. ∞ *p. 511* Calmodulin activation is also involved in the responses to oxytocin and to several regulatory hormones secreted by the hypothalamus.

3. *The inhibition of cellular activities.* The activation of some inhibitory G-proteins leads to a reduction in cAMP levels in the cell. In the example shown in Figure 18-3•, the activated G-protein inhibits adenylate cyclase activity and stimulates PDE activity. Levels

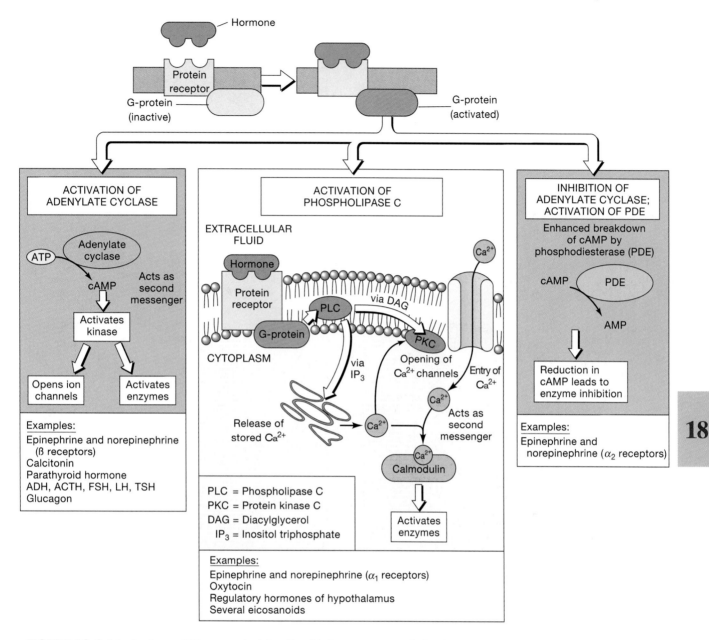

•**FIGURE 18-3** **Mechanisms of Hormone Activity.** Peptide hormones, catecholamines, and eicosanoids bind to membrane receptors and activate G-proteins. Three possible results are shown, each affecting the concentration of second messengers in the cytoplasm. G-protein activation may lead to (1) activation of adenylate cyclase and subsequent formation of cyclic-AMP, (2) opening of calcium ion channels and calcium release into the cytoplasm, or (3) inhibition of cellular activities by reduction of second messenger concentrations.

of cAMP then decline. This decline has an inhibitory effect on the cell, because without phosphorylation, key enzymes remain inactive. This is the mechanism responsible for the inhibitory effects that follow the stimulation of α_2 receptors, as we discussed in Chapter 16. ∞ *p. 511*

Hormones and Intracellular Receptors

Steroid hormones diffuse across the lipid part of the cell membrane and bind to receptors in the cytoplasm or nucleus. The hormone–receptor complexes then activate or deactivate specific genes (Figure 18-4a•). By this mechanism, steroid hormones can alter the rate of DNA transcription in the nucleus, thereby changing the pattern of protein synthesis. In turn, the cell structure or, if enzymes are involved, the metabolic activity of the target cell may be altered. For example, the sex hormone *testosterone* stimulates the production of enzymes and structural proteins in skeletal muscle fibers, causing an increase in muscle size and strength.

Thyroid hormones cross the cell membrane by diffusion or by a carrier mechanism. Once in the

18

•**FIGURE 18-4 Hormone Effects on Gene Activity.** **(a)** Steroid hormones diffuse through the membrane lipids and bind to receptors in the cytoplasm or nucleus. The complex then binds to DNA in the nucleus and activates specific genes. **(b)** Thyroid hormones enter the cytoplasm and bind to receptors in the nucleus to activate specific genes. They also bind to receptors on mitochondria and accelerate ATP production.

cytosol, these hormones bind to receptors within the nucleus and on mitochondria (Figure 18-4b•). The hormone–receptor complexes in the nucleus activate specific genes or change the rate of transcription. This rate change affects the metabolic activities of the cell by increasing or decreasing the concentration of specific enzymes. Thyroid hormones bound to mitochondria increase the mitochondria's rates of ATP production.

CONTROL OF ENDOCRINE ACTIVITY

As we noted earlier, the functional organization of the nervous and endocrine systems is parallel in many ways. In Chapter 13, we considered the basic operation of neural reflex arcs, the simplest organizational units in the nervous system. ∞ *pp. 422–423* The most direct arrangement was a monosynaptic reflex, such as the stretch reflex. Polysynaptic reflexes provide more-complex and variable responses to stimuli, and higher centers, which integrate multiple inputs, can facilitate or inhibit these reflexes as needed.

Endocrine Reflexes

Endocrine reflexes are the functional counterparts of neural reflexes. They can be triggered by (1) *humoral stimuli* (changes in the composition of the extracellular fluid), (2) *hormonal stimuli* (the arrival or removal of a specific hormone), or (3) *neural stimuli* (the arrival of neurotransmitter at neuroglandular junctions). In most cases, endocrine reflexes are controlled by negative feedback mechanisms: A stimulus triggers the production of a hormone whose direct or indirect effects reduce the intensity of the stimulus.

Endocrine cells in a simple endocrine reflex involve only one hormone. The endocrine cells involved respond directly to changes in the composition of the extracellular fluid. The secreted hormone adjusts the activities of target cells and restores homeostasis. Simple endocrine reflexes control hormone secretion by the heart, pancreas, parathyroid gland, and digestive tract.

More-complex endocrine reflexes involve one or more intermediary steps and two or more hormones. The hypothalamus, the highest level of endocrine control, integrates the activities of the nervous and endocrine systems. This integration involves three mechanisms, summarized in Figure 18-5•:

1. The hypothalamus secretes **regulatory hormones,** special hormones that control endocrine cells in the pituitary gland. (By convention, the use of *hormone* in the name indicates that the substance's identity is known; if the identity is not known, the term *factor* is used instead.) The hypothalamic regulatory hormones control the secretory activities of endocrine cells in the anterior portion of the pituitary gland. The hormones released by the anterior pituitary, in turn, control the activities of endocrine cells in the thyroid, adrenal cortex, and reproductive organs.

2. The hypothalamus contains autonomic centers that exert direct neural control over the endocrine cells of the adrenal medullae. When the sympathetic division is activated, the adrenal medullae release hormones into the bloodstream.

3. The hypothalamus acts as an endocrine organ by releasing hormones into the circulation at the posterior pituitary. We introduced two of these hormones, ADH and oxytocin, in Chapter 14. ∞ *p. 459*

The hypothalamus secretes ADH and regulatory hormones in response to changes in the composition of the circulating blood. The pathways leading to the secretion of E, NE, and oxytocin involve both neural and hormonal components. For example, the commands issued to the adrenal medullae that trigger secretion of E and NE arrive in the form of action potentials along efferent fibers rather than as circulating hormones. Such pathways are called *neuroendocrine reflexes,* because they include both neural and

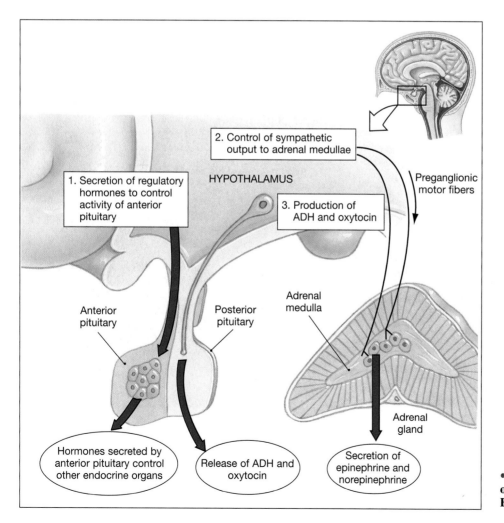

1. Secretion of regulatory hormones to control activity of anterior pituitary

2. Control of sympathetic output to adrenal medullae

HYPOTHALAMUS

3. Production of ADH and oxytocin

Preganglionic motor fibers

Anterior pituitary

Posterior pituitary

Adrenal medulla

Adrenal gland

Hormones secreted by anterior pituitary control other endocrine organs

Release of ADH and oxytocin

Secretion of epinephrine and norepinephrine

•**FIGURE 18-5** Three Methods of Hypothalamic Control over Endocrine Function endocrine components. We will consider these reflex patterns in more detail as we deal with specific endocrine tissues and organs.

In Chapter 17, we noted that receptors provide complex information by varying the frequency and pattern of action potentials in a sensory neuron. In the endocrine system, complex commands are issued by changing the amount of hormone secreted and the pattern of hormone release. In a simple endocrine reflex, hormones are released continuously but the rate of secretion rises and falls in response to humoral stimuli. For example, when blood glucose levels climb, the pancreas increases its secretion of *insulin,* a hormone that stimulates glucose absorption and utilization. As insulin levels rise, glucose levels decline; in turn, the stimulation of the insulin-secreting cells is reduced. As glucose levels return to normal, the rate of insulin secretion reaches resting levels. (We discussed the same pattern in Chapter 1 when we considered the negative feedback control of body temperature. ∞ *pp. 12–14*)

In this example, the responses of the target cells change over time, because the impact of insulin is proportional to its concentration. However, the relationship between hormone concentration and target cell response is not always predictable. For instance, a hormone can have one effect at low concentrations and additional effects—or even different effects—at high concentrations. (We will consider specific examples later in the chapter.)

Several hormones of the hypothalamus and pituitary gland are released in sudden bursts called *pulses* rather than continuously. When hormones arrive in pulses, target cells may vary their response with the frequency of pulses. For example, the target cell response to one pulse every 3 hours can differ from the response when pulses arrive every 30 minutes. The most-complicated hormonal instructions issued by the hypothalamus involve changes in the frequency of pulses *and* in the amount secreted in each pulse.

☑ How could you distinguish between a neural response and an endocrine response on the basis of response time and response duration?

☑ How would the presence of a molecule that blocks adenylate cyclase affect the activity of a hormone that produces its cellular effects by way of the second messenger cAMP?

☑ What primary factor determines each cell's hormonal sensitivities?

THE PITUITARY GLAND

Figure 18-6• shows the anatomical organization of the **pituitary gland,** or **hypophysis** (hī-POF-i-sis). This small, oval gland lies nestled within the *sella turcica,* a depression in the sphenoid bone (Figure 7-8•, p. 203). The pituitary gland hangs inferior to the hypothalamus, connected by the slender, funnel-shaped structure

(a) (b)

•**FIGURE 18-6** The Anatomy and Orientation of the Pituitary Gland. (LM × 62) called the **infundibulum** (in-fun-DIB-ū-lum; funnel). The base of the infundibulum lies between the optic chiasm and the mamillary bodies. The pituitary gland is cradled by the sella turcica and held in position by the *diaphragma sellae,* a dural sheet that encircles the infundibulum. The diaphragma sellae locks the pituitary gland in position and isolates it from the cranial cavity.

The pituitary gland can be divided into posterior and anterior portions on the basis of function and developmental anatomy (AM *EMBRYOLOGY SUMMARY 14: The Development of the Endocrine System*). Nine important peptide hormones are released by the pituitary gland—seven by the anterior pituitary and two by the posterior pituitary. All nine hormones bind to membrane receptors, and all nine use cAMP as a second messenger.

THE ANTERIOR PITUITARY

The **anterior pituitary gland,** or **adenohypophysis** (ad-ē-nō-hī-POF-i-sis), contains a variety of endocrine cell types. The anterior pituitary can be subdivided into three regions: (1) the **pars distalis** (dis-TAL-is; distal part), which is the largest and most anterior portion of the pituitary gland; (2) an extension called the **pars tuberalis,** which wraps around the adjacent portion of the infundibulum; and (3) the slender **pars intermedia,** which forms a narrow band bordering the posterior pituitary (Figure 18-6●). An extensive capillary network radiates through these regions, so every endocrine cell has immediate access to the circulatory system.

The Hypophyseal Portal System

By secreting specific regulatory hormones, the hypothalamus controls the production of hormones in the anterior pituitary. At the median eminence (∞ *p. 457*), a swelling near the attachment of the infundibulum, hypothalamic neurons release regulatory factors into the surrounding interstitial fluids. Their secretions enter the bloodstream quite easily, because the endothelial cells lining the capillaries in this region are unusually permeable.

These **fenestrated** (FEN-es-trā-ted) **capillaries** (*fenestra,* window) allow relatively large molecules to enter or leave the circulatory system. The capillary networks in the median eminence are supplied by the *superior hypophyseal artery.*

Before leaving the hypothalamus, the capillary networks unite to form a series of larg-er vessels that spiral around the infundibulum to reach the anterior pituitary gland. Once within the anterior pituitary, these vessels form a second capillary network that branches among the endocrine cells (Figure 18-7●).

This vascular arrangement is unusual: A typical artery conducts blood from the heart to a capillary network, and a typical vein carries blood from a capillary network back to the heart. The vessels between the median eminence and the anterior pituitary, however, carry blood from one capillary network to another. Blood vessels that link two capillary networks are called **portal vessels;** in this case, they have histological structure of veins. The entire complex is a **portal system.** Portal systems are named after their destinations, so this particular network is known as the **hypophyseal** (hī-pō-FI-sē-al) **portal system.**

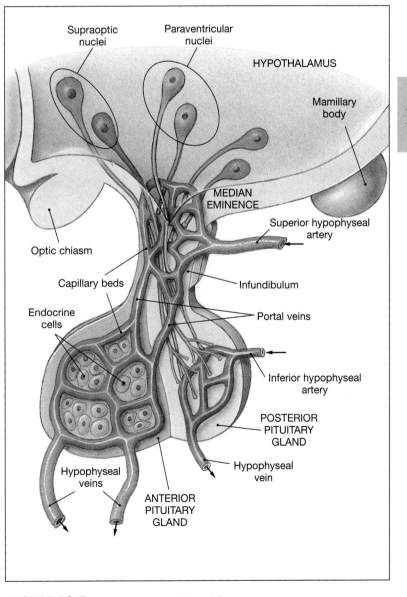

●**FIGURE 18-7** The Hypophyseal Portal System

Portal systems provide an efficient means of chemical communication by ensuring that all the blood entering the portal vessels will reach the intended target cells before it returns to the general circulation. The communication is strictly one-way, however, because any chemicals released by the cells "downstream" must do a complete tour of the circulatory system before they reach the capillaries at the start of the portal system.

Hypothalamic Control of the Anterior Pituitary

Two classes of hypothalamic regulatory hormones exist: (1) releasing hormones and (2) inhibiting hormones. A **releasing hormone (RH)** stimulates the synthesis and secretion of one or more hormones at the anterior pituitary, whereas an **inhibiting hormone (IH)** prevents the synthesis and secretion of hormones from the anterior pituitary. An endocrine cell in the anterior pituitary may be controlled by releasing hormones, inhibiting hormones, or some combination of the two. The regulatory hormones released at the hypothalamus are transported directly to the anterior pituitary by the hypophyseal portal system.

The rate at which the hypothalamus secretes regulatory hormones is controlled by negative feedback. The primary regulatory patterns are diagrammed in Figure 18-8•; we will refer to them as we examine specific pituitary hormones.

Hormones of the Anterior Pituitary

We will discuss seven hormones whose functions and control mechanisms are reasonably well understood:

Regulatory Hormone	Hormone 1 (from Pituitary)	Endocrine Target	Hormone 2 (from Target organ)
TRH	TSH	Thyroid gland	Thyroid hormones
CRH	ACTH	Adrenal cortex	Gluco-corticoids
GnRH	FSH	Testes	Inhibin
		Ovaries	{ Inhibin / Estrogens
	LH	Ovaries	{ Progestins / Estrogens
		Testes	Androgens

•**FIGURE 18-8 Feedback Control of Endocrine Secretion. (a)** A typical pattern of regulation when multiple endocrine organs are involved. The hypothalamus produces a releasing hormone (RH) to stimulate hormone production by other glands, and control is via negative feedback. **(b)** The regulation of prolactin (PRL) production by the anterior pituitary. In this case, the hypothalamus produces both a releasing factor (PRF) and an inhibiting hormone (PIH); when one is stimulated, the other is inhibited. **(c)** The regulation of growth hormone (GH) production by the anterior pituitary; when GH-RH release is inhibited, GH-IH release is stimulated.

thyroid-stimulating hormone, adrenocorticotropic hormone, follicle-stimulating hormone, luteinizing hormone, prolactin, growth hormone, and *melanocyte-stimulating hormone.* Of the six hormones produced by the pars distalis, four regulate the production of hormones by other endocrine glands. The names of these hormones indicate their activities, but many of the phrases are so long that abbreviations are often used instead.

The hormones of the anterior pituitary are also called *tropic hormones* (*tropé,* a turning), because they "turn on" endocrine glands or support the functions of other organs. (Some sources call them *trophic hormones* [*trophé,* nourishment] instead.)

Thyroid-Stimulating Hormone (TSH) **Thyroid-stimulating hormone (TSH),** or *thyrotropin,* targets the thyroid gland and triggers the release of thyroid hormones. TSH is released in response to *thyrotropin-releasing hormone (TRH)* from the hypothalamus. As circulating concentrations of thyroid hormones rise, the rates of TRH and TSH production decline (Figure 18-8a•).

Adrenocorticotropic Hormone (ACTH) **Adrenocorticotropic hormone (ACTH),** also known as *corticotropin,* stimulates the release of steroid hormones by the *adrenal cortex,* the outer portion of the adrenal gland. ACTH specifically targets cells that produce *glucocorticoids* (gloo-kō-KOR-ti-koydz), hormones that affect glucose metabolism. ACTH release occurs under the stimulation of **corticotropin-releasing hormone (CRH)** from the hypothalamus. As glucocorticoid levels increase, the rates of CRH release and ACTH release decline (Figure 18-8a•).

The Gonadotropins Follicle-stimulating hormone and luteinizing hormone are called **gonadotropins** (gō-nad-ō-TRŌ-pinz), because they regulate the activities of the *gonads.* (These organs—the testes and ovaries in males and females, respectively—produce the reproductive cells as well as hormones.) The production of gonadotropins occurs under stimulation by **gonadotropin-releasing hormone (GnRH)** from the hypothalamus. Abnormally low production of gonadotropins produces **hypogonadism.** Children with this condition will not undergo sexual maturation, and adults with hypogonadism cannot produce functional sperm or oocytes.

Follicle-stimulating hormone (FSH) The gonadotropin **follicle-stimulating hormone (FSH),** or *follitropin,* promotes follicle development in females and, in combination with luteinizing hormone, stimulates the secretion of *estrogens* (ES-trō-jenz) by ovarian cells. *Estradiol* is the most important estrogen. In males, FSH stimulates *sustentacular cells,* specialized cells in the tubules where sperm differentiate. In response, the sustentacular cells promote the physical maturation of developing sperm. FSH production is inhibited by *inhibin,* a peptide hormone released by cells in the testes and ovaries (Figure 18-8a•). (The role of inhibin in suppressing the release of GnRH as well as FSH is under debate.)

Luteinizing hormone (LH) **Luteinizing** (LOO-tē-in-ī-zing) **hormone (LH),** or *lutropin,* induces *ovulation,* the production of reproductive cells in females. It also promotes the secretion, by the ovaries, of estrogens and the *progestins* (such as *progesterone*), which prepare the body for possible pregnancy. In males, this gonadotropin is sometimes called *interstitial cell–stimulating hormone* (ICSH), because it stimulates the production of sex hormones by the *interstitial cells* of the testes. These sex hormones are called **androgens** (AN-drō-jenz; *andros,* man), the most important of which is *testosterone.* LH production, like FSH production, is stimulated by GnRH from the hypothalamus. GnRH production is inhibited by estrogens, progestins, and androgens (Figure 18-8a•).

Prolactin (PRL) **Prolactin** (prō-LAK-tin; *pro-,* before + *lac,* milk) **(PRL),** or *mammotropin,* works with other hormones to stimulate mammary gland development. In pregnancy and during the nursing period that follows delivery, PRL also stimulates milk production by the mammary glands. The functions of PRL in males are poorly understood, but evidence indicates that PRL helps regulate androgen production by making interstitial cells more sensitive to LH.

Prolactin production is inhibited by **prolactin-inhibiting hormone (PIH).** This hormone is the neurotransmitter dopamine. The hypothalamus also secretes a prolactin-releasing hormone, but the identity of this *prolactin-releasing factor (PRF)* is a mystery.

Circulating PRL stimulates PIH release and inhibits the secretion of prolactin-releasing factor (Figure 18-8b•).

Although PRL exerts the dominant effect on the glandular cells, normal development of the mammary glands is regulated by the interaction of several hormones. Prolactin, estrogens, progesterone, glucocorticoids, pancreatic hormones, and hormones produced by the placenta cooperate in preparing the mammary glands for secretion, and milk ejection occurs only in response to oxytocin release at the posterior pituitary. We shall describe the functional development of the mammary glands in Chapter 28.

Growth Hormone **Growth hormone (GH),** or **somatotropin** (*soma,* body), stimulates cell growth and replication by accelerating the rate of protein synthesis. Although virtually every tissue responds to some

18

degree, skeletal muscle cells and chondrocytes (cartilage cells) are particularly sensitive to GH levels.

The stimulation of growth by GH involves two mechanisms. The primary mechanism, which is indirect, is best understood. Liver cells respond to the presence of GH by synthesizing and releasing **somatomedins,** or **insulin-like growth factors (IGFs),** which are peptide hormones that bind to receptor sites on a variety of cell membranes (Figure 18-8c•). In skeletal muscle fibers, cartilage cells, and other target cells, somatomedins increase the rate of uptake of amino acids and their incorporation into new proteins. These effects develop almost immediately after GH is released; they are particularly important after a meal, when the blood contains high concentrations of glucose and amino acids. In functional terms, cells can now obtain ATP easily through the aerobic metabolism of glucose, and amino acids are readily available for protein synthesis. Under these conditions, GH, acting through the somatomedins, stimulates protein synthesis and cell growth.

The direct actions of GH are more selective and tend not to appear until after blood glucose and amino acid concentrations have returned to normal levels:

■ In epithelia and connective tissues, GH stimulates stem cell divisions and the differentiation of daughter cells.

The subsequent growth of these daughter cells will be stimulated by somatomedins.

■ In adipose tissue, GH stimulates the breakdown of stored triglycerides by adipocytes (fat cells), which then release fatty acids into the blood. As circulating fatty acid levels rise, many tissues stop breaking down glucose and start breaking down fatty acids to generate ATP. This process is termed a **glucose-sparing effect.**

■ In the liver, GH stimulates the breakdown of glycogen reserves by liver cells. These cells then release glucose into the bloodstream. Because most other tissues are now metabolizing fatty acids rather than glucose, blood glucose concentrations begin to climb, perhaps to levels significantly higher than normal. The elevation of blood glucose levels by GH has been called a **diabetogenic effect,** because *diabetes mellitus,* an endocrine disorder we will consider later in the chapter, is characterized by abnormally high blood glucose concentrations.

The production of GH is regulated by **growth hormone–releasing hormone** (**GH-IH,** or *somatocrinin*) and **growth hormone–inhibiting hormone** (**GH-IH,** or *somatostatin*) from the hypothalamus. Somatomedins stimulate GH-IH and inhibit GH-RH (Figure 18-8c•).

18

Growth Hormone Abnormalities

Growth hormone stimulates muscular and skeletal development. If it is administered before the epiphyseal cartilages have closed, it will cause an increase in height, weight, and muscle mass. In extreme cases, *gigantism* can result. ∞ *p. 181* In **acromegaly** (*akron,* extremity + *megale,* great), an excessive amount of GH is released after puberty, when most of the epiphyseal cartilages have already closed. Cartilages and small bones respond to the hormone, however, resulting in abnormal growth at the hands, feet, lower jaw, skull, and clavicle. Figure 18-19a• (p. 616) shows a typical acromegalic individual.

Children unable to produce adequate concentrations of GH have *pituitary growth failure,* a condition we introduced in Chapter 6. ∞ *p. 181* The steady growth and maturation that typically precede and accompany puberty do not occur in these individuals, who have short stature, slow epiphyseal growth, and larger-than-normal adipose tissue reserves.

Normal growth patterns can be restored by the administration of GH. Before the advent of gene splicing and recombinant DNA techniques, GH had to be carefully extracted and purified from the pituitary

glands of cadavers at considerable expense. Now genetically manipulated bacteria are used to produce human GH in commercial quantities.

The current availability of purified human growth hormone has led to its use under medically questionable circumstances. For example, it is now being praised as an "anti-aging" miracle cure. Although GH supplements do slow or even reverse the losses in bone and muscle mass that accompany aging, little is known about adverse side effects that may accompany long-term use in mature adults. GH is also being sought by some parents of short but otherwise healthy children. These parents view short stature as a handicap that merits treatment by a physician. Whether we are considering GH treatment of adults or children, it is important to remember that GH and the somatomedins affect many different tissues and have widespread metabolic impacts. For example, children exposed to GH may grow faster, but their body fat content declines drastically and sexual maturation is delayed. This decline is associated with metabolic changes in many organs. The range and significance of these metabolic side effects are now the subject of long-term studies.

Melanocyte-Stimulating Hormone   The pars intermedia may secrete two forms of **melanocyte-stimulating hormone (MSH),** or *melanotropin.* As the name indicates, MSH stimulates the melanocytes of the skin, increasing their production of melanin, a brown, black, or yellow-brown pigment. ∞ *p. 147* The release of MSH is inhibited by dopamine, acting as an inhibiting hormone.

Melanocyte-stimulating hormone is important in the control of skin pigmentation in fishes, amphibians, reptiles, and many mammals other than primates. The pars intermedia in adult humans is virtually nonfunctional, and the circulating blood usually does not contain MSH. However, MSH is secreted by the human pars intermedia (1) during fetal development, (2) in very young children, (3) in pregnant women, and (4) in some disease states. The functional significance of MSH secretion under these circumstances is not known. The administration of a synthetic form of MSH causes the skin to darken, so MSH has been suggested as a means of obtaining a "sunless tan."

THE POSTERIOR PITUITARY

The **posterior pituitary** is also called the **neurohypophysis** (noo-rō-hī-POF-isis), or *pars nervosa* (nervous part), because it contains the axons of hypothalamic neurons. Neurons of the **supraoptic** and **paraventricular nuclei** manufacture antidiuretic hormone (ADH) and oxytocin, respectively. These products move along axons in the infundibulum to the basement membranes of capillaries in the posterior pituitary gland by means of axoplasmic transport. ∞ *pp. 364–365*

Antidiuretic Hormone

Antidiuretic hormone (ADH), also known as *arginine vasopressin (AVP),* is released in response to a variety of stimuli, most notably a rise in the electrolyte concentration in the blood or a fall in blood volume or blood pressure. A rise in the electrolyte concentration stimulates the secretory neurons directly. Because they respond to a change in the osmotic concentration of body fluids, these neurons are called *osmoreceptors.*

The primary function of ADH is to decrease the amount of water lost at the kidneys. With losses minimized, any water absorbed from the digestive tract will be retained, reducing the concentration of electrolytes in the extracellular fluid. In high concentrations, ADH also causes *vasoconstriction,* a constriction of peripheral blood vessels that helps elevate blood pressure. ADH release is inhibited by alcohol, which explains the increased fluid excretion that follows the consumption of alcoholic beverages.

Disorders of ADH Production

Diabetes occurs in several forms, all characterized by excessive urine production **(polyuria).** Although diabetes can be caused by physical damage to the kidneys, most forms are the result of endocrine abnormalities. The two most prevalent forms are diabetes mellitus and diabetes insipidus. Diabetes mellitus is described on page 607. **Diabetes insipidus** generally develops because the posterior pituitary no longer releases adequate amounts of ADH. Water conservation at the kidneys is impaired, and excessive amounts of water are lost in the urine. As a result, the individual is constantly thirsty, but the fluids consumed are not retained by the body.

Mild cases of diabetes insipidus may not require treatment if fluid and electrolyte intake keep pace with urinary losses. In severe cases, the fluid losses can reach 10 liters per day, and dehydration and electrolyte imbalances are fatal without treatment. In one innovative treatment, *desmopressin acetate* (DDAVP), a synthetic form of ADH, is administered in a nasal spray. The drug enters the bloodstream after diffusing through the nasal epithelium and may alleviate the symptoms of diabetes insipidus. This method is not effective in treating *nephrogenic diabetes insipidus,* which results from an inability to concentrate urine in the kidneys. In this condition, the circulating levels of ADH are already elevated but the kidney cells are unable to respond appropriately to the hormonal stimulation.

18

Oxytocin

In women, **oxytocin** *(oxy-,* quick + *tokos,* childbirth), or **OT,** stimulates smooth muscle tissue in the wall of the uterus, promoting labor and delivery. After delivery, oxytocin stimulates the contraction of myoepithelial cells (∞ *p. 157*) around the secretory alveoli and the ducts of the mammary glands, promoting the ejection of milk.

Until the last stages of pregnancy, the uterine smooth muscles are relatively insensitive to oxytocin, but sensitivity becomes more pronounced as the time of delivery approaches. The trigger for normal labor and delivery is probably a sudden rise in oxytocin levels at the uterus. There is good evidence, however, that the oxytocin released by the posterior pituitary plays only a supporting role and that most of the oxytocin involved is secreted by the uterus and fetus.

Oxytocin secretion and milk ejection are part of a neuroendocrine reflex. The stimulus is an infant suckling at the breast, and sensory nerves innervating the nipples relay the information to the hypothalamus. Oxytocin is then released into the circulation at the posterior pituitary, and the myoepithelial cells respond by squeezing milk from the secretory alveoli into large collecting ducts. This *milk let-down reflex* can be modified by any factor that affects the hypothalamus. For example, anxiety, stress, and other factors can prevent the flow of milk, even when the mammary glands are fully functional. In contrast, nursing mothers can become conditioned to associate a baby's crying with suckling. These women may begin milk let-down as soon as they hear a baby cry.

Although the functions of oxytocin in sexual activity remain uncertain, it is known that circulating concentrations of oxytocin in both genders rise during sexual arousal and peak at orgasm. Evidence indicates that in men oxytocin stimulates smooth muscle contractions in the walls of the sperm duct *(ductus deferens)* and prostate gland. These actions may be important in *emission,* the ejection of secretions of the prostate gland, sperm, and the secretions of other glands into the male reproductive tract before ejaculation. Studies suggest that the oxytocin released dur-

ing intercourse in females may stimulate smooth muscle contractions in the uterus and vagina that promote sperm transport toward the uterine tubes.

Figure 18-9• and Table 18-2 summarize important information about the hormonal products of the pituitary gland. Review these carefully before considering the structure and function of other endocrine organs.

☑ If a person were dehydrated, how would the level of ADH released by the posterior pituitary gland change?

☑ A blood sample shows elevated levels of somatomedins. Which pituitary hormone would you expect to be elevated as well?

☑ What effect would elevated circulating levels of cortisol, a steroid hormone from the adrenal cortex, have on the pituitary secretion of ACTH?

•**FIGURE 18-9** **Pituitary Hormones and Their Targets**

TABLE 18-2 The Pituitary Hormones

Region/Area	Hormone(s)	Target(s)	Hormonal Effect(s)	Hypothalamic Regulatory Hormone
Anterior Pituitary (Adenohypophysis)				
Pars distalis	Thyroid-stimulating hormone (TSH)	Thyroid gland	Secretion of thyroid hormones	Thyrotropin-releasing hormone (TRH)
	Adrenocorticotropic hormone (ACTH)	Adrenal cortex (zona fasciculata)	Glucocorticoid secretion	Corticotropin-releasing hormone (CRH)
	Gonadotropins:			
	Follicle-stimulating hormone (FSH)	Follicle cells of ovaries	Estrogen secretion, follicle development	Gonadotropin-releasing hormone (GnRH)
		Sustentacular cells of testes	Stimulation of sperm maturation	As above
	Luteinizing hormone (LH)	Follicle cells of ovaries	Ovulation, formation of corpus luteum, progesterone secretion	As above
		Interstitial cells of testes	Testosterone secretion	As above
	Prolactin (PRL)	Mammary glands	Production of milk	Prolactin-releasing factor (PRF) Prolactin-inhibiting hormone (PIH)
	Growth hormone (GH)	All cells	Growth, protein synthesis, lipid mobilization and catabolism	Growth hormone–releasing hormone (GH-RH) Growth hormone–inhibiting hormone (GH-IH)
Pars intermedia (not active in normal adults)	Melanocyte-stimulating hormone (MSH)	Melanocytes	Increased melanin synthesis in epidermis	Melanocyte-stimulating hormone–inhibiting hormone (MSH-IH)
Posterior Pituitary (Neurohypophysis or Pars Nervosa)				
	Antidiuretic hormone (ADH)	Kidneys	Reabsorption of water, elevation of blood volume and pressure	None: Transported along axons from supraoptic nucleus to posterior pituitary gland
	Oxytocin (OT)	Uterus, mammary glands (females)	Labor contractions, milk ejection	None: Transported along axons from paraventricular nucleus to posterior pituitary gland
		Ductus deferens and prostate gland (males)	Contractions of ductus deferens and prostate	

18

THE THYROID GLAND

The **thyroid gland** curves across the anterior surface of the trachea just inferior to the *thyroid* ("shield-shaped") *cartilage,* which forms most of the anterior surface of the larynx (Figure 18-10a•). The two **lobes** of the thyroid gland are united by a slender connection, the **isthmus** (IS-mus). You can easily feel the gland with your fingers. When something goes wrong with it, the thyroid gland typically becomes visible as it swells and distorts the surface of the neck. The size of the gland is quite variable, depending on heredity and environmental and nutritional factors, but its average weight is about 34 g (1.2 oz). An extensive blood supply gives the thyroid gland a deep red color.

18

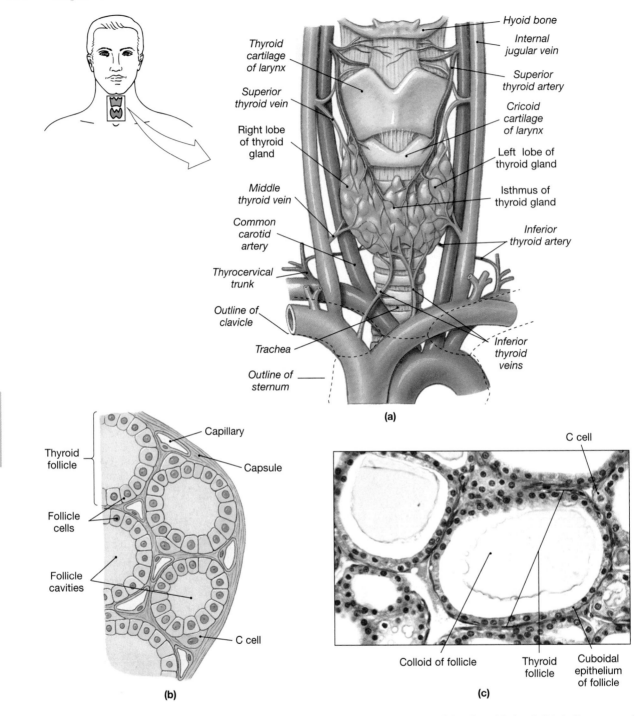

Hyoid bone

Internal jugular vein

Superior thyroid artery

Cricoid cartilage of larynx

Left lobe of thyroid gland

Isthmus of thyroid gland

Inferior thyroid artery

Inferior thyroid veins

Thyroid cartilage of larynx

Superior thyroid vein

Right lobe of thyroid gland

Middle thyroid vein

Common carotid artery

Thyrocervical trunk

Outline of clavicle

Trachea

Outline of sternum

(a)

Thyroid follicle

Capillary

Capsule

Follicle cells

Follicle cavities

C cell

(b)

C cell

Colloid of follicle

Thyroid follicle

Cuboidal epithelium of follicle

(c)

•**FIGURE 18-10** **The Thyroid Gland.** **(a)** The location, anatomy, and blood supply of the thyroid gland. **(b)** A diagrammatic view of a section through the wall of the thyroid gland. **(c)** Histological details, showing thyroid follicles. (LM × 211)

THYROID FOLLICLES AND THYROID HORMONES

The thyroid gland contains large numbers of **thyroid follicles,** spheres lined by a simple cuboidal epithelium (Figure 18-10b,c•). The follicle cells surround a **follicle cavity.** This cavity holds a viscous *colloid,* a fluid containing large quantities of suspended proteins. A net-work of capillaries surrounds each follicle, delivering nutrients and regulatory hormones to the glandular cells and accepting their secretory products and metabolic wastes.

Follicle cells synthesize a globular protein called **thyroglobulin** (thī-rō-GLOB-ū-lin) and secrete it into the colloid of the thyroid follicles (Figures 18-10c and 18-11a•). Each thyroglobulin molecule contains the

•**FIGURE 18-11** **The Functions of the Thyroid Follicles.** **(a)** The synthesis, storage, and secretion of thyroid hormones. For a detailed explanation of the numbered events, see the text. **(b)** The regulation of thyroid secretion.

(a)

(b)

amino acid *tyrosine,* the building block of thyroid hormones. The formation of thyroid hormones involves three basic steps:

1. Iodide ions are absorbed from the diet at the digestive tract and are delivered to the thyroid gland by the bloodstream. Carrier proteins in the basal membrane of the follicle cells transport iodide ions (I^-) into the cytoplasm. The follicle cells normally maintain intracellular concentrations of iodide that are many times higher than those in the extracellular fluid.

2. The iodide ions diffuse to the apical surface of each follicle cell, where they are converted to an activated form of iodide (I^+) by the enzyme *thyroid peroxidase.* This reaction sequence also attaches one or two of these iodide ions to the tyrosine molecules of thyroglobulin.

3. Tyrosine molecules to which iodide ions have been attached are paired, forming molecules of thyroid hormones that remain incorporated into thyroglobulin. The pairing process is probably performed by thyroid peroxidase. The hormone **thyroxine** (thī-ROKS-ēn), also known as *tetraiodothyronine,* or T_4, contains four iodide ions. **Triiodothyronine,** or T_3, is a related molecule containing three iodide ions. Eventually, each molecule of thyroglobulin contains 4–8 molecules of T_3 or T_4 hormones or both.

The major factor controlling the rate of thyroid hormone release is the concentration of TSH in the circulating blood (Figure 18-11b•). TSH stimulates iodide transport into the follicle cells and stimulates the production of thyroglobulin and thyroid peroxidase. TSH also stimulates the release of thyroid hormones. Under the influence of TSH, the following steps occur:

4. Follicle cells remove thyroglobulin from the follicles by endocytosis.

5. Lysosomal enzymes then break the protein down, and the amino acids and thyroid hormones enter the cytoplasm. The amino acids are recycled and used to synthesize thyroglobulin.

6. The released molecules of T_3 and T_4 diffuse across the basement membrane and enter the bloodstream. About 90 percent of all thyroid secretions is T_4; T_3 is secreted in comparatively small amounts.

7. Roughly 75 percent of the T_4 and 70 percent of the T_3 molecules entering the bloodstream become attached to transport proteins called **thyroid-binding globulins (TBGs).** Most of the rest of the T_4 and T_3 in the circulation is attached to **transthyretin,** also known as

18

thyroid-binding prealbumin (TBPA), or to *albumin,* one of the plasma proteins. Only the relatively small quantities of thyroid hormones that remain unbound—roughly 0.3 percent of the circulating T_3 and 0.03 percent of the circulating T_4—are free to diffuse into peripheral tissues.

An equilibrium exists between the bound and unbound thyroid hormones. At any moment, free thyroid hormones are being bound to carriers at the same rate at which bound hormones are being released. When unbound thyroid hormones diffuse out of the bloodstream and into other tissues, the equilibrium is disturbed. The carrier proteins then release additional thyroid hormones until a new equilibrium is reached. The bound thyroid hormones represent a substantial reserve. The bloodstream normally contains more than a week's supply of thyroid hormones.

TSH plays a key role in both the synthesis and the release of thyroid hormones. In the absence of TSH, the thyroid follicles become inactive and neither synthesis nor secretion occurs. TSH binds to membrane receptors and, by stimulating adenylate cyclase, activates key enzymes involved in thyroid hormone production (Figure 18-3•, p. 583).

Functions of Thyroid Hormones

Thyroid hormones readily cross cell membranes, and they affect almost every cell in the body. Inside a cell, they bind to (1) receptors in the nucleus, (2) receptors on the surfaces of mitochondria, and (3) receptors in the cytoplasm. Thyroid hormones bound to cytoplasmic receptors are essentially held in storage. If intracellular levels of thyroid hormones decline, the bound thyroid hormones are released into the cytoplasm.

The thyroid hormones binding to mitochondria increase the rates of mitochondrial ATP production. The binding to receptors in the nucleus activates genes that control the synthesis of enzymes involved with energy transformation and utilization. One specific effect is the accelerated production of sodium–potassium ATPase, the membrane protein responsible for the ejection of intracellular sodium and the recovery of extracellular potassium. As we noted in Chapter 3, this exchange pump consumes large amounts of ATP. ∞ *p. 76*

Thyroid hormones also activate genes that code for the synthesis of enzymes involved in glycolysis and ATP production. This effect, coupled with the direct effect of thyroid hormones on mitochondria, increases the metabolic rate of the cell. Because the cell consumes more energy and because energy use is measured in *calories,* the effect is called the **calorigenic effect** of thyroid hormones. When the metabolic rate increases, more heat is generated. In young children, TSH production increases in cold weather; the calori-

genic effect may help them adapt to cold climates. (This response does not occur in adults.) In growing children, thyroid hormones are also essential to normal development of the skeletal, muscular, and nervous systems.

T_3 Versus T_4 The thyroid gland produces large amounts of T_4, but T_3 is primarily responsible for the observed effects of thyroid hormones: a strong, immediate, and short-lived increase in the rate of cellular metabolism. Peripheral tissues have two sources of T_3:

1. *Release by the thyroid gland.* At any moment, T_3 from the thyroid gland accounts for only 10–15 percent of the T_3 in peripheral tissues.

2. *The conversion of T_4 to T_3.* Enzymes in the liver, kidneys, and other tissues can convert T_4 to T_3. Roughly 85–90 percent of the T_3 that reaches the target cells is produced by the conversion of T_4 within peripheral tissues.

Table 18-3 summarizes the effects of thyroid hormones on major organs and systems.

Iodine and Thyroid Hormones Iodine in the diet is absorbed at the digestive tract as I^-. The follicle cells in the thyroid gland absorb 120–150 μg of I^- each day, the minimum dietary amount needed to maintain normal thyroid function. The iodide ions are actively transported into the thyroid follicle cells, so the concentration of I^- inside thyroid follicle cells is generally about 30 times higher than that in the blood plasma. If plasma I^- levels rise, so do levels inside the follicle cells.

The thyroid follicles contain most of the iodide reserves in the body. The active transport mechanism for iodide is stimulated by TSH, and the increased movement of iodide into the cytoplasm accelerates the formation of thyroid hormones.

TABLE 18-3 Effects of Thyroid Hormones on Peripheral Tissues

1. Elevate rates of oxygen consumption and energy consumption; in children, may cause a rise in body temperature
2. Increase heart rate and force of contraction; generally cause a rise in blood pressure
3. Increase sensitivity to sympathetic stimulation
4. Maintain normal sensitivity of respiratory centers to changes in oxygen and carbon dioxide concentrations
5. Stimulate the formation of red blood cells and thereby enhance oxygen delivery
6. Stimulate the activity of other endocrine tissues
7. Accelerate the turnover of minerals in bone

The typical diet in the United States provides approximately 500 μg of iodide per day, roughly three times the minimum daily requirement. Much of the excess is due to the addition of I⁻ to the table salt sold in grocery stores as "iodized salt." Thus, iodide deficiency is seldom responsible for limiting the rate of thyroid hormone production. (This is not necessarily the case in other countries.) Excess I⁻ is removed from the blood at the kidneys, and each day a small amount of I⁻ (about 20 μg) is excreted by the liver into the bile. *Bile* is an exocrine product stored in the gallbladder. Iodide excreted at the kidneys is eliminated in urine; the I⁻ excreted in bile is eliminated with intestinal wastes. The losses in the bile, which continue even if the diet contains less than the minimum iodide requirement, can gradually deplete the iodide reserves in the thyroid. Thyroid hormone production then declines, regardless of the circulating levels of TSH.

Thyroid Gland Disorders

Normal production of thyroid hormones establishes the background rates of cellular metabolism. These hormones exert their primary effects on metabolically active tissues and organs, including skeletal muscles, the liver, and the kidneys. The inadequate production of thyroid hormones is called **hypothyroidism.** Hypothyroidism in an infant produces *cretinism,* a condition marked by inadequate skeletal and nervous development and a metabolic rate as much as 40 percent below normal levels. This condition affects approximately 1 birth out of every 5000. Cretinism developing later in childhood will retard growth and mental development and delay puberty.

Adults with hypothyroidism are lethargic and unable to tolerate cold temperatures. The symptoms of adult hypothyroidism, collectively known as *myxedema* (miks-e-DĒ-muh), include subcutaneous swelling, dry skin, hair loss, low body temperature, muscular weakness, and slowed reflexes. Hypothyroidism and myxedema as the result of inadequate dietary iodide are very rare in the United States but can be relatively common in many poorer countries.

Hyperthyroidism, or *thyrotoxicosis,* occurs when thyroid hormones are produced in excessive quantities. The metabolic rate climbs, and the skin becomes flushed and moist with perspiration. Blood pressure and heart rate increase, and the heartbeat may become irregular as circulatory demands escalate. The effects on the central nervous system make the individual restless, excitable, and subject to shifts in mood and emotional states. Despite the drive for increased activity, the person has limited energy reserves and fatigues easily. *Graves' disease* is a form of hyperthyroidism that afflicted President George Bush and Barbara Bush during their stay in the White House.

AM *Thyroid Gland Disorders*

THE C CELLS OF THE THYROID GLAND: CALCITONIN

A second population of endocrine cells lies sandwiched between the cuboidal follicle cells and their basement membrane. These cells are larger than those of the follicular epithelium, and they do not stain as clearly. They are the **C (clear) cells,** or *parafollicular cells* (Figure 18-10b,c•). C cells produce the hormone **calcitonin (CT).** Calcitonin aids in the regulation of Ca^{2+} concentrations in body fluids. We introduced the functions of this hormone in Chapter 6. ∞ *p. 184* The net effect of calcitonin release is a drop in the Ca^{2+} concentration in body fluids. This reduction is accomplished by (1) the inhibition of osteoclasts, which slows the rate of Ca^{2+} release from bone, and (2) the stimulation then drop of Ca^{2+} excretion at the kidneys.

The control of calcitonin secretion is a direct endocrine regulation—without the involvement of the hypothalamus or pituitary gland: The C cells respond directly to elevations in the Ca^{2+} concentration of blood. When those concentrations rise, calcitonin secretion increases. The Ca^{2+} concentrations then drop, eliminating the stimulus and "turning off" the C cells.

Calcitonin is probably most important during childhood, when it stimulates bone growth and mineral deposition in the skeleton. It also appears to be important in reducing the loss of bone mass (1) during prolonged starvation and (2) in the late stages of pregnancy, when the maternal skeleton competes with the developing fetus for calcium ions absorbed by the digestive tract. The role of calcitonin in the healthy nonpregnant adult is uncertain.

In several chapters, we have considered the importance of Ca^{2+} in controlling muscle cell and neuron activities. Calcium ion concentrations also affect the sodium permeabilities of excitable membranes. At high Ca^{2+} concentrations, sodium permeability decreases and membranes become less responsive. Such problems are relatively rare. Problems caused by lower-than-normal Ca^{2+} concentrations are equally dangerous and much more common. When calcium ion concentrations decline, sodium permeabilities increase and cells become extremely excitable. If calcium levels fall too far, convulsions or muscular spasms can result. Maintenance of adequate calcium levels involves the *parathyroid glands* and *parathyroid hormone.*

THE PARATHYROID GLANDS

There are normally two pairs of **parathyroid glands** embedded in the posterior surfaces of the thyroid gland (Figure 18-12a•). The gland cells are separated by the dense capsular fibers of the thyroid. All together the four parathyroid glands weigh a mere 1.6 g (0.06 oz). The histological appearance of a

18

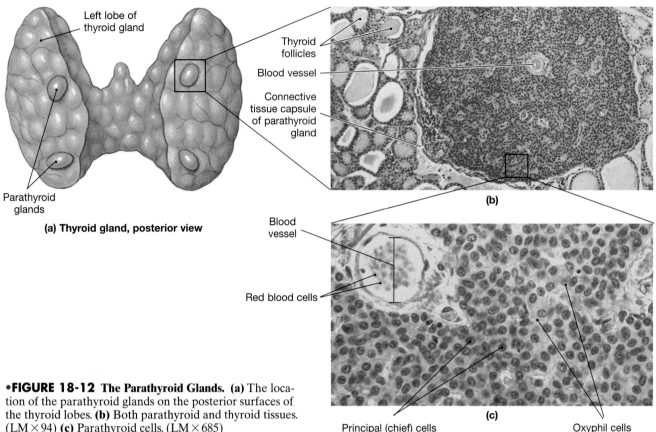

Left lobe of thyroid gland

Thyroid follicles

Blood vessel

Connective tissue capsule of parathyroid gland

Parathyroid glands

(a) Thyroid gland, posterior view

(b)

Blood vessel

Red blood cells

Principal (chief) cells

(c)

Oxyphil cells

•**FIGURE 18-12** **The Parathyroid Glands.** **(a)** The location of the parathyroid glands on the posterior surfaces of the thyroid lobes. **(b)** Both parathyroid and thyroid tissues. (LM × 94) **(c)** Parathyroid cells. (LM × 685)

18

parathyroid gland is shown in Figure 18-12b,c•. The parathyroid glands have at least two cell populations. The **chief cells** produce parathyroid hormone; the functions of the other cells, called *oxyphils,* are unknown.

PARATHYROID HORMONE SECRETION

Like the C cells of the thyroid gland, the chief cells monitor the circulating concentration of calcium ions. When the Ca^{2+} concentration of the blood falls below normal, the chief cells secrete **parathyroid hormone (PTH),** or *parathormone.* The net result of PTH secretion is an increase in Ca^{2+} concentration in body fluids. Parathyroid hormone has four major effects:

1. It stimulates osteoclasts, accelerating mineral turnover and the release of Ca^{2+} from bone.
2. It inhibits osteoblasts, reducing the rate of calcium deposition in bone.
3. It enhances the reabsorption of Ca^{2+} at the kidneys, reducing urinary losses.
4. It stimulates the formation and secretion of *calcitriol* at the kidneys. In general, the effects of calcitriol complement or enhance those of PTH, but one major effect of calcitriol is the enhancement of Ca^{2+} and PO_4^{3-} absorption by the digestive tract. ∞ *pp. 181, 182*

Figure 18-13• compares the functions of calcitonin and PTH. It is likely that PTH, aided by calcitriol, is the primary regulator of circulating calcium ion concentrations in healthy adults. Information about the hormones of the thyroid gland and parathyroid glands is summarized in Table 18-4.

Disorders of Parathyroid Function

When the parathyroid gland secretes inadequate or excessive amounts of parathyroid hormone, Ca^{2+} concentrations move outside normal homeostatic limits. Inadequate parathyroid hormone production, a condition called **hypoparathyroidism,** leads to low Ca^{2+} concentrations in body fluids. The most obvious symptoms involve neural and muscle tissues: The nervous system becomes more excitable, and the affected individual may experience *hypocalcemic tetany,* a dangerous condition characterized by prolonged muscle spasms that initially involve the limbs and face.

In **hyperparathyroidism,** Ca^{2+} concentrations become abnormally high. Bones grow thin and brittle, skeletal muscles become weak, CNS function is depressed, and nausea and vomiting occur. In severe cases, the patient may become comatose. AM *Disorders of Parathyroid Function*

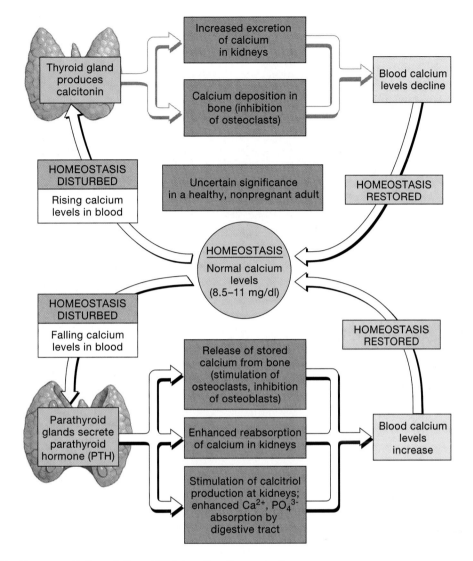

•**FIGURE 18-13** The Homeostatic Regulation of Calcium Ion Concentrations

18

TABLE 18-4 Hormones of the Thyroid Gland and Parathyroid Glands

Gland/Cells	Hormone(s)	Targets	Hormonal Effects	Regulatory Control
Thyroid Gland				
Follicular epithelium	Thyroxine (T_4), triiodothyronine (T_3)	Most cells	Increase energy utilization, oxygen consumption, growth, and development *(see Table 18-3)*	Stimulated by TSH from anterior pituitary *(see Figure 18-8a)*
C cells	Calcitonin (CT)	Bone, kidneys	Decreases Ca^{2+} concentrations in body fluids *(see Figure 18-13)*	Stimulated by elevated blood Ca^{2+} levels; actions opposed by PTH
Parathyroid Glands				
Chief cells	Parathyroid hormone (PTH)	Bone, kidneys	Increases Ca^{2+} concentrations in body fluids *(see Figure 18-13)*	Stimulated by low blood Ca^{2+} levels; PTH effects enhanced by calcitriol and opposed by calcitonin

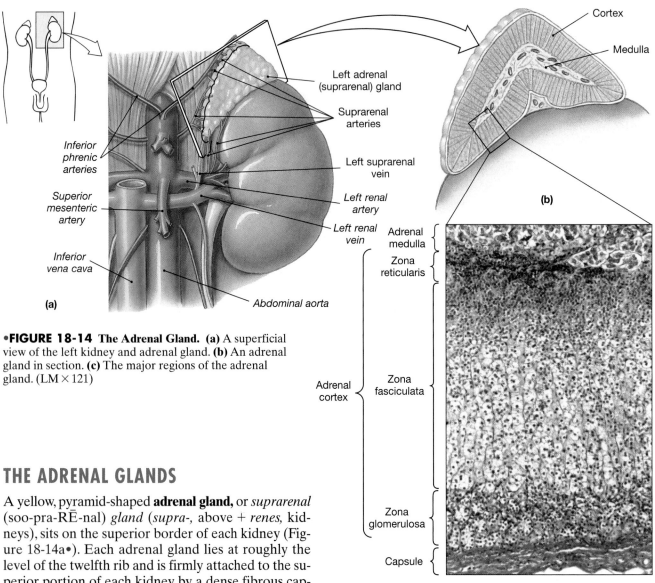

•**FIGURE 18-14** **The Adrenal Gland.** **(a)** A superficial view of the left kidney and adrenal gland. **(b)** An adrenal gland in section. **(c)** The major regions of the adrenal gland. (LM × 121)

18

THE ADRENAL GLANDS

A yellow, pyramid-shaped **adrenal gland,** or *suprarenal* (soo-pra-RĒ-nal) *gland* (*supra-,* above + *renes,* kidneys), sits on the superior border of each kidney (Figure 18-14a•). Each adrenal gland lies at roughly the level of the twelfth rib and is firmly attached to the superior portion of each kidney by a dense fibrous capsule. The adrenal gland on each side nestles between the kidney, the diaphragm, and the major arteries and veins that run along the dorsal wall of the abdominopelvic cavity. The adrenal glands project into the peritoneal cavity, and their anterior surfaces are covered by a layer of parietal peritoneum. Like other endocrine glands, the adrenal glands are highly vascularized.

A typical adrenal gland weighs about 7.5 g (0.19 oz), but adrenal size can vary greatly as secretory demands change. The adrenal gland is divided into two parts: a superficial **adrenal cortex** and an inner **adrenal medulla** (Figure 18-14b•).

THE ADRENAL CORTEX

The yellowish color of the adrenal cortex is due to the presence of stored lipids, especially cholesterol and various fatty acids. The adrenal cortex produces more than two dozen steroid hormones, collectively called **adrenocortical steroids,** or simply **corticosteroids.** In

the bloodstream, these hormones are bound to transport proteins called *transcortins.*

Corticosteroids are vital: If the adrenal glands are destroyed or removed, the individual will die unless corticosteroids are administered. Corticosteroids, like other steroid hormones, exert their effects by determining which genes are transcribed in the nuclei of their target cells and at what rates. The resulting changes in the nature and concentration of enzymes in the cytoplasm affect cellular metabolism.

Deep to the adrenal capsule are three distinct regions, or zones, in the adrenal cortex: (1) an outer, *zona glomerulosa;* (2) a middle, *zona fasciculata;* and (3) an inner, *zona reticularis* (Figure 18-14c•). Each zone synthesizes specific steroid hormones, as listed in Table 18-5.

TABLE 18-5 The Adrenal Hormones

Region/Zone	Hormone(s)	Targets	Hormonal Effects	Regulatory Control
Cortex				
Zona glomerulosa	Mineralocorticoids (MCs), primarily aldosterone	Kidneys	Increase renal reabsorption of Na^+ and water (especially in the presence of ADH) and accelerate urinary loss of K^+	Stimulated by angiotensin II; inhibited by ANP
Zona fasciculata	Glucocorticoids (GCs): cortisol (hydrocortisone), corticosterone, cortisone	Most cells	Release of amino acids from skeletal muscles, and lipids from adipose tissues; promote liver formation of glucose and glycogen; promote peripheral utilization of lipids; anti-inflammatory effects	Stimulated by ACTH from anterior pituitary gland *(see Figure 18-8a)*
Zona reticularis	Androgens		Uncertain significance under normal conditions	Stimulated by ACTH; significance uncertain
Medulla	Epinephrine, norepinephrine	Most cells	Increase cardiac activity, blood pressure, glycogen breakdown, blood glucose levels; release of lipids by adipose tissue	Stimulated during sympathetic activation by sympathetic preganglionic fibers *(see Chapter 16)*

18

The Zona Glomerulosa

The **zona glomerulosa** (glō-mer-ū-LŌ-suh), the outer region of the adrenal cortex, produces **mineralocorticoids (MCs),** steroid hormones that affect the electrolyte composition of body fluids. **Aldosterone** is the principal mineralocorticoid produced by the adrenal cortex.

The zona glomerulosa accounts for about 15 percent of the volume of the adrenal cortex (Figure 18-14c•). A *glomerulus* is a little ball; as the term *zona glomerulosa* implies, the endocrine cells in this region form small, dense knots or clusters. This zone extends from the capsule to the radiating cords of the deeper zona fasciculata.

Aldosterone Aldosterone secretion stimulates the conservation of sodium ions and the elimination of potassium ions. This hormone targets cells that regulate the ionic composition of excreted fluids. It causes the retention of sodium ions at the kidneys, sweat glands, salivary glands, and pancreas, preventing Na^+ loss in urine, sweat, saliva, and digestive secretions. The retention of Na^+ is accompanied by a loss of K^+. As a secondary effect, the reabsorption of Na^+ enhances the osmotic reabsorption of water at the kidneys, sweat glands, salivary glands, and pancreas. The effect at the kidneys is most dramatic when normal levels of ADH are present. In addition, aldosterone increases the sensitivity of salt receptors in the taste buds of the tongue. As a result, interest in (and consumption of) salty food increases.

Aldosterone secretion occurs in response to a drop in blood Na^+ content, blood volume, or blood pressure or to a rise in the blood K^+ concentration. Changes in the Na^+ or K^+ concentration have a direct effect on the zona glomerulosa, stimulating the release of aldosterone. Aldosterone release also occurs in response to *angiotensin II*. We will discuss this hormone, part of the *renin–angiotensin system,* later in this chapter.

Disorders of Aldosterone Secretion

In **hypoaldosteronism,** the zona glomerulosa fails to produce enough aldosterone, generally because the kidneys are not releasing adequate amounts of the enzyme *renin*. Affected individuals lose excessive amounts of water and Na^+ at the kidneys; the water loss leads to low blood volume and low blood pressure. Changes in electrolyte concentrations affect transmembrane potentials, eventually disrupting neural and muscular tissues throughout the body.

Hypersecretion of aldosterone results in the condition of **aldosteronism.** Under continued aldosterone stimulation, the kidneys retain sodium ions very effectively, but potassium ions are lost in large quantities. A crisis eventually ensues when low extracellular K^+ concentrations disrupt normal cardiac, neural, and kidney cell functions.

AM *Disorders of the Adrenal Cortex*

The Zona Fasciculata

The **zona fasciculata** (fa-sik-ū-LA-tuh; *fasciculus,* little bundle) produces steroid hormones collectively known as **glucocorticoids (GCs)** due to their effects on glucose metabolism. This zone begins at the inner border of the zona glomerulosa and extends toward the adrenal medulla. It contributes about 78 percent of the cortical volume. The endocrine cells are larger and contain more lipid than do those of the zona glomerulosa, and the lipid droplets give the cytoplasm a pale, foamy appearance. The cells of the zona fasciculata form individual cords composed of stacks of cells. Adjacent cords are separated by flattened blood vessels with fenestrated walls.

The Glucocorticoids   When stimulated by ACTH from the anterior pituitary, the zona fasciculata secretes primarily **cortisol** (KOR-ti-sol), also called *hydrocortisone,* along with smaller amounts of the related steroid **corticosterone** (kor-ti-KOS-te-rōn). The liver converts some of the circulating cortisol to **cortisone,** another active glucocorticoid. Glucocorticoid secretion is regulated by negative feedback: The glucocorticoids released have an inhibitory effect on the production of corticotropin-releasing hormone (CRH) in the hypothalamus and of ACTH in the anterior pituitary. This relationship was diagrammed in Figure 18-8a•, p. 588.

Effects of Glucocorticoids   Glucocorticoids accelerate the rates of glucose synthesis and glycogen formation, especially in the liver. Adipose tissue responds by releasing fatty acids into the blood, and other tissues begin to break down fatty acids and proteins instead of glucose. This process is another example of a glucose-sparing effect (p. 590).

Glucocorticoids also show **anti-inflammatory** effects; that is, they inhibit the activities of white blood cells and other components of the immune system. "Steroid creams" are commonly used to control irritating allergic rashes, such as those produced by poison ivy, and injections of glucocorticoids may be used to control more-severe allergic reactions. Glucocorticoids slow the migration of phagocytic cells into an injury site and cause phagocytic cells already in the area to become less active. In addition, mast cells exposed to these steroids are less likely to release histamine and other chemicals that promote inflammation. ∞ *pp. 135–136* As a result, swelling and further irritation are dramatically reduced. On the negative side, the rate of wound healing decreases, and the weakening of the region's defenses makes it an easy target for infectious organisms. For that reason, topical steroids are used to treat superficial rashes but are never applied to open wounds.

Abnormal Glucocorticoid Production

Addison's disease is caused by inadequate glucocorticoid production (see Figure 18-19d•, p. 616). The usual cause is a destruction of the zona fasciculata as the result of an *autoimmune response,* in which the body's defenses mistakenly attack healthy tissues, or after infection by the bacteria responsible for *tuberculosis.* Individuals become weak and lose weight; they cannot effectively use their lipid reserves to generate ATP, and blood glucose concentrations fall sharply within hours after a meal. In most cases the zona glomerulosa is affected as well, so symptoms similar to those of hypoaldosteronism may also appear.

Cushing's disease results from an overproduction of glucocorticoids, generally owing to the hypersecretion of ACTH. The symptoms resemble those of a protracted and exaggerated response to stress (see Figure 18-19e•, p. 616). Glucose metabolism is suppressed, lipid reserves are mobilized, and peripheral proteins are broken down. Lipid reserves are redistributed, and the distribution of body fat changes. Individuals develop a "moon-faced" appearance due to the deposition of lipids in the subcutaneous tissues of the face. Skeletal muscles become weak as contractile proteins break down. AM *Disorders of the Adrenal Cortex*

The Zona Reticularis

The **zona reticularis** (re-tik-ū-LAR-is; *reticulum,* network) forms a narrow band bordering each adrenal medulla (Figure 18-14c•). This zone accounts for only about 7 percent of the total volume of the adrenal cortex. The endocrine cells of the zona reticularis form a folded, branching network, and fenestrated blood vessels wind between the cells.

The zona reticularis normally produces small quantities of androgens, the sex hormones produced in large quantities by the testes in males. Once in the bloodstream, some of the androgens released by the zona reticularis are converted to estrogens, the dominant sex hormone in females. When secreted in normal amounts, neither the androgens nor the estrogens affect sexual characteristics, and the significance of adrenal sex hormone production remains uncertain. ACTH stimulates the zona reticularis to a slight degree, but the effects are generally insignificant.

Abnormal Activity in the Zona Reticularis

The zona reticularis ordinarily produces a negligible amount of androgens. If a tumor forms there, androgen secretion may increase dramatically, producing symptoms of **adrenogenital syndrome.** In women, this condition leads to the gradual development of male secondary sex characteristics, including body and facial hair patterns, adipose tissue distribution, and muscle development. Tumors affecting the zona reticularis of males can result in the production of large quantities of estrogens. Affected males develop enlarged breasts and, in some cases, other female secondary sex characteristics. This array of symptoms is called **gynecomastia** (*gynaikos,* woman + *mastos,* breast).

THE ADRENAL MEDULLAE

The boundary between the adrenal cortex and the adrenal medulla is irregular, and the supporting connective tissues and blood vessels are extensively interconnected. The adrenal medulla is reddish brown owing in part to the many blood vessels in this area. It contains large, rounded cells—similar to those in sympathetic ganglia that are innervated by preganglionic sympathetic fibers. The secretory activities of the adrenal medullae are controlled by the sympathetic division of the autonomic nervous system. ⚬⚬ *p. 510*

The adrenal medullae contain two populations of secretory cells: one producing epinephrine (adrenaline) and the other, norepinephrine (noradrenaline). Evidence suggests that the two cell types are distributed in different areas of the medulla and that their secretory activities can be independently controlled. The secretions are packaged in vesicles that form dense clusters just inside cell membranes. The hormones in these vesicles are continuously released at low levels, by exocytosis. Sympathetic stimulation dramatically accelerates the rate of exocytosis and hormone release.

Epinephrine and Norepinephrine

Epinephrine makes up 75–80 percent of the secretions from the adrenal medullae, the rest being norepinephrine. We described the peripheral effects of these hormones, which result from interaction with alpha and beta receptors on cell membranes, in Chapter 16. ⚬⚬ *p. 511* The stimulation of α_1 and β_1 receptors, the most common types, accelerates cellular energy utilization and the mobilization of energy reserves.

When the adrenal medulla is activated:

- Epinephrine and norepinephrine trigger a mobilization of glycogen reserves in skeletal muscles and accelerate the breakdown of glucose to provide ATP. This combination increases both muscular strength and endurance.

- In adipose tissue, stored fats are broken down to fatty acids, which are released into the bloodstream for use by other tissues for ATP production.

- In the liver, glycogen molecules are broken down. The resulting glucose molecules are released into the bloodstream, primarily for use by neural tissues, which cannot shift to fatty acid metabolism.

- In the heart, the stimulation of β_1 receptors triggers an increase in the rate and force of cardiac muscle contraction.

The metabolic changes that follow the release of catecholamines such as E and NE are at their peak 30 seconds after adrenal stimulation, and they linger for several minutes thereafter. As a result, the effects produced by the stimulation of the adrenal medullae outlast the other signs of sympathetic activation. **AM** *Disorders of the Adrenal Medullae*

☑ What symptoms would you expect to see in an individual whose diet lacks iodine?

☑ When a person's thyroid gland is removed, signs of decreased thyroid hormone concentration do not appear until about one week later. Why?

☑ The removal of the parathyroid glands would result in a decrease in the blood concentration of which important mineral?

☑ What effect would elevated cortisol levels have on the level of glucose in the blood?

THE PINEAL GLAND

The **pineal gland,** part of the epithalamus, lies in the posterior portion of the roof of the third ventricle. ⚬⚬ *p. 454* The pineal gland contains neurons, neuroglia, and special secretory cells called **pinealocytes** (pi-NĒ-al-ō-sīts). These cells synthesize the hormone melatonin from molecules of the neurotransmitter *serotonin.* Collaterals from the visual pathways enter the pineal gland and affect the rate of melatonin production. Melatonin production is lowest during daylight hours and highest at night.

Several functions, including the following, have been suggested for melatonin in humans:

- *Inhibiting reproductive functions.* In some other mammals, melatonin slows the maturation of sperm, oocytes, and reproductive organs by reducing the rate of GnRH secretion. The significance of this effect remains uncertain, but circumstantial evidence suggests that melatonin may play a role in the timing of human sexual maturation. For example, melatonin levels in the blood decline at puberty, and pineal tumors that eliminate melatonin production cause premature puberty in young children.

- *Protecting against damage by free radicals.* Melatonin is a very effective *antioxidant;* it may protect CNS neurons from free radicals, such as nitric oxide (NO) or hydrogen peroxide (H_2O_2), that may be generated in active neural tissue.

- *Setting circadian rhythms.* Because pineal activity is cyclical, the pineal gland may also be involved with the maintenance of basic *circadian rhythms,* daily changes in physiological processes that follow a regular pattern. ⚬⚬ *p. 459* Increased melatonin secretion in darkness has been suggested as a primary cause of *seasonal affective disorder (SAD).* This condition, characterized by changes in mood, eating habits, and sleeping patterns, can develop during the winter in people who live at high latitudes, where sunshine is scarce or lacking. **AM** *Light and Behavior*

18

THE ENDOCRINE TISSUES OF OTHER SYSTEMS

As we noted earlier, many organs that are part of other body systems have secondary endocrine functions. Examples include the intestines (digestive system), the kidneys (urinary system), the heart (cardiovascular system), the thymus (lymphatic system), the pancreas (digestive system), and the *gonads*—the testes in males and the ovaries in females (reproductive system). Table 18-6 provides an overview of some of the hormones these organs produce.

THE INTESTINES

The intestines, which process and absorb nutrients, release a variety of hormones that coordinate the activities of the digestive system. Although the pace of digestive activities can be affected by the autonomic nervous system, most digestive processes are controlled locally. These hormones will be described in Chapter 24.

THE KIDNEYS

The kidneys release the steroid hormone *calcitriol*, the peptide hormone *erythropoietin*, and the enzyme *renin*. Calcitriol is important for calcium ion homeostasis. Erythropoietin and renin are involved in the regulation of blood volume and blood pressure.

Calcitriol

Calcitriol is a steroid hormone secreted by the kidneys in response to the presence of parathyroid hormone (PTH). *Cholecalciferol* (vitamin D_3) is a related steroid that is synthesized in the skin or absorbed from the diet. ∞ *p. 149* Cholecalciferol is converted to calcitriol, although not directly. The term *vitamin D* applies to the entire group of related steroids, including calcitriol, cholecalciferol, and various intermediate products.

The best-known function of calcitriol is the stimulation of calcium and phosphate ion absorption along the digestive tract. The effects of PTH on Ca^{2+} absorption result primarily from stimulation of calcitriol release. Calcitriol's other effects on calcium metabolism include (1) stimulating the formation and differentiation of osteoprogenitor cells and osteoclasts, (2) stimulating bone resorption by osteoclasts, (3) stimulating Ca^{2+} reabsorption at the kidneys, and (4) suppressing PTH production. Evidence indicates that calcitriol affects lymphocytes and keratinocytes in the skin; these effects have nothing to do with regulating calcium levels.

Erythropoietin

Erythropoietin (e-rith-rō-poy-Ē-tin; *erythros,* red + *poiesis,* making), or **EPO,** is a peptide hormone released by the kidneys in response to low oxygen levels in kidney tissues. It stimulates the production of red blood cells by bone marrow. The increase in the number of red blood cells elevates blood volume. Because these cells transport oxygen, the increase in their number improves oxygen delivery to peripheral tissues. We shall consider EPO again in Chapter 19.

Renin

Renin is released by specialized kidney cells in response to (1) sympathetic stimulation or (2) a decline in renal blood flow. Once in the bloodstream, renin functions as an enzyme that starts an enzymatic chain reaction known as the *renin–angiotensin system.* First, renin converts **angiotensinogen,** a plasma protein produced by the liver, to **angiotensin I.** In the capillaries of the lungs, angiotensin I is then modified to the hormone **angiotensin II.** Angiotensin II stimulates the secretion of aldosterone by the adrenal cortex and ADH at the

TABLE 18-6 Representative Hormones Produced by Organs of Other Systems

Organ	Hormone(s)	Primary Target(s)	Hormonal Effects
Intestines	Many (secretin, gastrin, cholecystokinin, etc.)	Other regions and organs of the digestive system	Coordinate digestive activities
Kidneys	Erythropoietin (EPO)	Red bone marrow	Stimulates red blood cell production
	Calcitriol	Intestinal lining	Stimulates calcium and phosphate absorption; stimulates Ca^{2+} release from bone; inhibits PTH secretion
Heart	Atrial natriuretic peptide (ANP)	Kidney, hypothalamus, adrenal gland	Increases water and salt loss at kidneys; increases thirst; suppresses secretion of ADH and aldosterone
Thymus	Thymosins (many)	Lymphocytes and other cells of the immune response	Coordinate and regulate immune response
Pancreas	*See Table 18-7*		
Gonads	*See Table 18-8*		

posterior pituitary gland. This combination restricts salt and water losses at the kidneys. Angiotensin II also stimulates thirst and elevates blood pressure.

Because renin plays such a key role in the renin–angiotensin system, many physiological and endocrinological references consider renin to be a hormone. We shall take a closer look at the renin–angiotensin system in our discussion of the control of blood pressure and blood volume in Chapter 21.

THE HEART

The endocrine cells in the heart are cardiac muscle cells in the walls of the *atria*, chambers that receive blood from the veins. If blood volume becomes too great, these cells are excessively stretched and release the hormone **atrial natriuretic** (nā-trē-ū-RET-ik) **peptide (ANP)** *(natrium,* sodium + *ouresis,* making water). In general, the effects of ANP oppose those of angiotensin II: It promotes the loss of Na^+ and water at the kidneys and inhibits renin release and the secretion of ADH and aldosterone. It also suppresses thirst and prevents angiotensin II and norepinephrine from elevating blood pressure. The net result is a reduction in both blood volume and blood pressure, thereby reducing the stretching of the cardiac muscle cells in the atrial walls. We shall consider the actions of this hormone further when we discuss the control of blood pressure and volume in Chapters 21 and 26.

THE THYMUS

The **thymus** is located in the mediastinum, generally just posterior to the sternum. This organ produces several hormones that are important to the development and maintenance of immune defenses. **Thymosin** (THĪ-mō-sin) is the name originally given to an ex-

tract from the thymus that promotes the development and maturation of *lymphocytes,* the white blood cells responsible for immunity. The thymic extract actually contains a blend of several complementary hormones; the term *thymosins* is now sometimes used to refer to all thymic hormones. We shall further consider the histological organization of the thymus and the functions of the thymosins in Chapter 22.

THE PANCREAS

The **pancreas** lies within the abdominopelvic cavity in the J-shaped loop between the stomach and the small intestine. It is a slender, pale organ with a nodular (lumpy) consistency (Figure 18-15a•). The adult pancreas is 20–25 cm (8–10 in.) long and weighs about 80 g (2.8 oz). We shall consider its anatomy further in Chapter 24, because it is primarily a digestive organ that makes digestive enzymes. The **exocrine pancreas,** roughly 99 percent of the pancreatic volume, consists of clusters of gland cells, called *pancreatic acini,* and their attached ducts. Together, the gland and duct cells secrete large quantities of an alkaline, enzyme-rich fluid. This secretion reaches the lumen of the digestive tract by traveling along a network of secretory ducts.

The **endocrine pancreas** consists of small groups of cells scattered among the exocrine cells. The endocrine clusters are known as **pancreatic islets,** or the *islets of Langerhans* (LAN-ger-hanz) (Figure 18-15b•). Pancreatic islets account for only about 1 percent of all cells in the pancreas. Nevertheless, a typical pancreas contains roughly 2 million pancreatic islets.

Like other endocrine tissue, the pancreatic islets are surrounded by an extensive, fenestrated capillary network that carries its hormones into the bloodstream. Each islet contains four cell types:

18

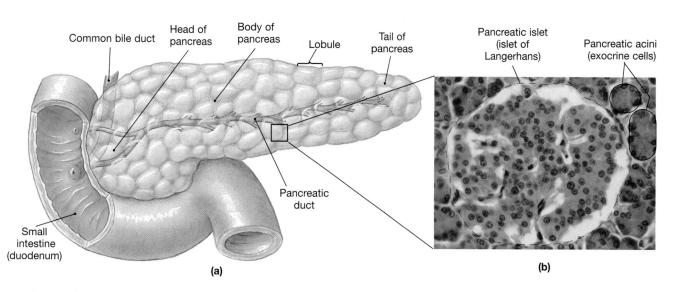

•**FIGURE 18-15 The Endocrine Pancreas. (a)** The gross anatomy of the pancreas. **(b)** A pancreatic islet surrounded by exocrine cells. (LM × 276)

1. **Alpha cells** produce the hormone **glucagon** (GLOO-ka-gon). Glucagon raises blood glucose levels by increasing the rates of glycogen breakdown and glucose release by the liver.

2. **Beta cells** produce the hormone **insulin** (IN-suh-lin). Insulin lowers blood glucose levels by increasing the rate of glucose uptake and utilization by most body cells and increasing glycogen synthesis in skeletal muscles and the liver. Beta cells also secrete *amylin*, a recently discovered peptide hormone whose role is uncertain.

3. **Delta cells** produce a peptide hormone identical to growth hormone–inhibiting hormone (GH-IH), a hypothalamic regulatory hormone. This pancreatic hormone suppresses the release of glucagon and insulin by other islet cells and slows the rates of food absorption and enzyme secretion along the digestive tract.

4. **F cells** produce the hormone **pancreatic polypeptide (PP)**. It inhibits gallbladder contractions and regulates the production of some pancreatic enzymes; it may help control the rate of nutrient absorption by the digestive tract.

We will focus on insulin and glucagon, the hormones responsible for the regulation of blood glucose concentrations. These hormones interact to control blood glucose levels (Figure 18-16•). When blood glucose levels rise, beta cells secrete insulin, which then stimulates the transport of glucose across cell membranes. When blood glucose levels decline, alpha cells secrete glucagon, which stimulates glucose release by the liver.

Insulin

Insulin is a peptide hormone released by beta cells when glucose levels exceed normal levels (70–110 mg/dl).

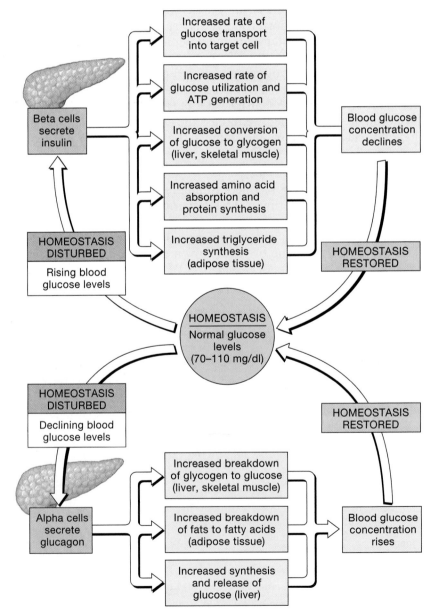

•FIGURE 18-16 The Regulation of Blood Glucose Concentrations

Insulin secretion is also stimulated by elevated levels of some amino acids, including arginine and leucine. Insulin exerts its effects on cellular metabolism in a series of steps that begins when insulin binds to receptor proteins on the cell membrane. Binding leads to the activation of the receptor, which functions as a kinase and attaches phosphate groups to intracellular enzymes. The phosphorylation of enzymes then produces primary and secondary effects in the cell; the biochemical details remain unresolved.

One of the most important effects is the enhancement of glucose absorption and utilization. Insulin receptors are present in most cell membranes; such cells are called *insulin-dependent*. However, cells in the brain and kidneys, cells in the lining of the digestive tract, and red blood cells lack insulin receptors. These cells are called *insulin-independent*, because they can absorb and utilize glucose without insulin stimulation.

The effects of insulin on its target cells include:

- *The acceleration of glucose uptake (all target cells).* This effect results from an increase in the number of glucose transport proteins in the cell membrane. These proteins transport glucose into the cell by facilitated diffusion. This movement follows the concentration gradient for glucose, and ATP is not required.

- *The acceleration of glucose utilization (all target cells) and enhanced ATP production.* This effect occurs for two reasons: (1) The rate of glucose use is proportional to its availability; when more glucose enters the cell, more is used. (2) Second messengers activate a key enzyme involved in the initial steps of glycolysis.

- *The stimulation of glycogen formation (skeletal muscles and liver cells).* When excess glucose enters these cells, it is stored in the form of glycogen.

- *The stimulation of amino acid absorption and protein synthesis.*

- *The stimulation of triglyceride formation in adipose tissues.* Insulin stimulates the absorption of fatty acids and glycerol by adipocytes. The adipose cells then store these components as triglycerides. Adipocytes also increase their absorption of glucose; excess glucose is used in the synthesis of additional triglycerides.

In summary, insulin is secreted when glucose is abundant, and this hormone stimulates glucose utilization to support growth and the establishment of carbohydrate (glycogen) and lipid (triglyceride)

Diabetes Mellitus

Whether glucose is absorbed across the digestive tract or manufactured and released by the liver, very little glucose leaves the body intact once it has entered the bloodstream. Glucose does get filtered out of the blood at the kidneys, but virtually all of it is reabsorbed, so urinary glucose losses are negligible.

Diabetes mellitus (mel-Ī-tus; *mellitum,* honey) is characterized by glucose concentrations that are high enough to overwhelm the reabsorption capabilities of the kidneys. (The term **hyperglycemia** [hī-per-glī-SĒ-mē-ah] refers to the presence of abnormally high glucose levels in the blood.) Glucose appears in the urine (**glycosuria;** glī-kō-SOO-rē-a), and urine production generally becomes excessive *(polyuria).*

Diabetes mellitus can be caused by genetic abnormalities, and some of the genes responsible have been identified. Mutations that result in inadequate insulin production, the synthesis of abnormal insulin molecules, or the production of defective receptor proteins will produce comparable symptoms. Diabetes mellitus can also result from other pathological conditions, injuries, immune disorders, or hormonal imbalances. There are two major types of diabetes mellitus: *insulin-dependent (Type 1) diabetes* and *non-insulin-dependent (Type 2) diabetes.* AM *Diabetes Mellitus*

Probably because glucose levels cannot be stabilized adequately, even with treatment, persons with diabetes mellitus commonly develop chronic medical problems. The most common examples include the following:

- The proliferation of capillaries and hemorrhaging at the retina typically cause partial or complete blindness. This condition is called **diabetic retinopathy.**

- Changes occur in the clarity of the lens of the eye, producing cataracts.

- Small hemorrhages and inflammation at the kidneys cause degenerative changes that can lead to kidney failure. This condition is called **diabetic nephropathy.** Treatment with drugs that improve blood flow to the kidneys can slow the progression to kidney failure.

- A variety of neural problems appear, including *peripheral neuropathies* and abnormal autonomic function. p. 416 These disorders, collectively termed **diabetic neuropathy,** are probably related to disturbances in the blood supply to neural tissues.

- Degenerative changes in cardiac circulation can lead to early heart attacks. For a given age group, heart attacks are 3–5 times more likely in diabetic individuals than in nondiabetic persons.

- Other changes in the vascular system can disrupt normal blood flow to the distal portions of the limbs. For example, a reduction in blood flow to the feet can lead to tissue death, ulceration, infection, and loss of toes or a major portion of one or both feet.

TABLE 18-7 Hormones of the Pancreas

Structure/Cells	Hormone	Primary Targets	Hormonal Effects	Regulatory Control
Pancreatic Islets				
Alpha cells	Glucagon	Liver, adipose tissues	Mobilizes lipid reserves; promotes glucose synthesis and glycogen breakdown in liver; elevates blood glucose concentrations	Stimulated by low blood glucose concentrations; inhibited by GH-IH from delta cells
Beta cells	Insulin	Most cells	Facilitates uptake of glucose by target cells; stimulates formation and storage of lipids and glycogen	Stimulated by high blood glucose concentrations, parasympathetic stimulation, and high levels of some amino acids; inhibited by GH-IH from delta cells and by sympathetic activation
Delta cells	GH-IH	Other islet cells, digestive epithelium	Inhibits insulin and glucagon secretion; slows rates of nutrient absorption and enzyme secretion along digestive tract	Stimulated by protein-rich meal; mechanism uncertain
F cells	Pancreatic polypeptide (PP)	Digestive organs	Inhibits gallbladder contraction; regulates production of pancreatic enzymes; influences rate of nutrient absorption by digestive tract	Stimulated by protein-rich meal and by parasympathetic stimulation

18

reserves. The accelerated use of glucose soon brings circulating glucose levels within normal limits.

Glucagon

When glucose concentrations fall below normal, alpha cells release glucagon and energy reserves are mobilized. When glucagon binds to a receptor in the cell membrane, this hormone activates adenylate cyclase, and cAMP acts as a second messenger that activates cytoplasmic enzymes. The primary effects of glucagon are as follows:

- *The stimulation of glycogen breakdown in skeletal muscle and liver cells.* The glucose molecules released will either be metabolized for energy (skeletal muscle fibers) or released into the bloodstream (liver cells).
- *The stimulation of triglyceride breakdown in adipose tissues.* The adipocytes then release the fatty acids into the bloodstream for use by other tissues.
- *The stimulation of glucose production at the liver.* Liver cells absorb amino acids from the bloodstream, convert them to glucose, and release the glucose into the circulation. This process of glucose synthesis in the liver is called *gluconeogenesis* (gloo-kō-nē-ō-JEN-e-sis).

The results are a reduction in glucose use and the release of more glucose into the bloodstream. Blood glucose concentrations soon rise toward normal levels.

Pancreatic alpha cells and beta cells monitor blood glucose concentrations, and the secretion of glucagon and insulin occur without endocrine or nervous instructions. Yet, because the alpha cells and beta cells are very sensitive to changes in blood glucose levels, any

hormone that affects blood glucose concentrations will indirectly affect the production of both insulin and glucagon. Insulin production is also influenced by autonomic activity. Parasympathetic stimulation enhances insulin release, and sympathetic stimulation inhibits it.

Information about insulin, glucagon, and other pancreatic hormones is summarized in Table 18-7.

THE GONADS

In males, the **interstitial cells** of the testes produce the male hormones known as androgens. **Testosterone** (tes-TOS-ter-ōn) is the most important androgen. This hormone promotes the production of functional sperm, maintains the secretory glands of the male reproductive tract, stimulates growth, and determines male secondary sex characteristics, such as the distribution of facial hair and body fat. Testosterone also affects metabolic operations throughout the body. It stimulates protein synthesis and muscle growth, and it produces aggressive behavioral responses. During embryonic development, the production of testosterone affects the development of CNS structures, including hypothalamic nuclei that will later influence sexual behaviors.

Sustentacular cells in the testes support the differentiation and physical maturation of sperm. Under FSH stimulation, these cells secrete the hormone **inhibin,** which inhibits the secretion of FSH at the anterior pituitary and perhaps suppresses GnRH release at the hypothalamus.

Immature female reproductive cells, called *oocytes,* are produced in specialized structures called **follicles.**

Follicles develop under FSH stimulation. The cells of these follicles, which form a layer around each developing oocyte, produce estrogens under FSH and LH stimulation. **Estrogens** are steroid hormones that support the maturation of the oocytes and stimulate the growth of the lining of the uterus. **Estradiol** is the principal estrogen. Under FSH stimulation, follicle cells secrete inhibin, which suppresses FSH release through a feedback mechanism comparable to that in males.

A surge in LH secretion during the ovarian cycle triggers ovulation, and LH then causes the remaining follicle cells to reorganize into a **corpus luteum** (LOO-tē-um; "yellow body"). These cells release a mixture of estrogens and hormones called **progestins. Progesterone** (prō-JES-ter-ōn), the principal progestin, has several important functions, including the following:

- It prepares the uterus for the arrival of a developing embryo, in case *fertilization* (the fusion of oocyte and sperm) occurs.
- It accelerates the movement of the oocyte or embryo to the uterus.
- It causes an enlargement of the mammary glands, working in combination with other hormones, such as estradiol, growth hormone, and prolactin.

During pregnancy, additional hormones produced by the placenta and uterus interact with those produced by the ovaries and the pituitary gland to promote normal fetal development and delivery. We shall consider the endocrinological aspects of pregnancy in Chapter 29.

Information about the reproductive hormones of the testes and ovaries is summarized in Table 18-8.

Leptin: A Different Hormone

Since 1990, several new hormones have been identified. In most cases, their structures and modes of action remain to be determined. One of the most interesting of these hormones is leptin. Unlike typical hormones, which are produced by endocrine cells in one specific location, **leptin** is secreted by adipose tissues throughout the body. This peptide hormone has several functions, the best known being the feedback control of appetite. When you eat, adipose tissues absorb glucose and lipids and synthesize triglycerides for storage. At the same time, they release leptin into the bloodstream. Leptin binds to neurons in the hypothalamus that deal with emotion and appetite control. The result is a sense of satiation and the suppression of appetite.

Leptin was first discovered in a strain of obese mice. These mice had a defective leptin gene, and the administration of leptin to one of these overweight mice quickly turned it into a slim, athletic animal. Initial hope that leptin could be used to treat human obesity was soon dashed, however. Most obese people appear to have defective leptin receptors (or leptin pathways) in the appetite centers of the CNS. Their circulating leptin levels are already several times higher than in individuals of normal body weight, so the administration of additional leptin would have no effect. Researchers are now working on the structure of the receptor protein and the biochemistry of the pathway triggered by leptin binding.

Leptin also has a permissive effect on GnRH and gonadotropin synthesis. This effect explains (1) why thin girls commonly enter puberty relatively late, (2) why an increase in body fat content can improve fertility, and (3) why women stop menstruating when their body fat content becomes very low.

18

TABLE 18-8 Hormones of the Reproductive System

Structure/Cells	Hormone(s)	Primary Targets	Hormonal Effects	Regulatory Control
Testes				
Interstitial cells	Androgens	Most cells	Support functional maturation of sperm, protein synthesis in skeletal muscles, male secondary sex characteristics, and associated behaviors	Stimulated by LH from anterior pituitary gland (*see Figure 18-8a*)
Sustentacular cells	Inhibin	Anterior pituitary gland	Inhibits secretion of FSH	Stimulated by FSH from anterior pituitary gland (*see Figure 18-8a*)
Ovaries				
Follicular cells	Estrogens	Most cells	Support follicle maturation, female secondary sex characteristics, and associated behaviors	Stimulated by FSH and LH from anterior pituitary gland (*see Figure 18-8a*)
	Inhibin	Anterior pituitary gland	Inhibits secretion of FSH	Stimulated by FSH from anterior pituitary gland (*see Figure 18-8a*)
Corpus luteum	Progestins	Uterus, mammary glands	Prepare uterus for implantation; prepare mammary glands for secretory activity	Stimulated by LH from anterior pituitary gland (*see Figure 18-8a*)

☑ Why does a person with Type 1 or Type 2 diabetes urinate frequently and have a pronounced thirst?

☑ What effect would increased levels of glucagon have on the amount of glycogen stored in the liver?

☑ Increased amounts of light would inhibit the production of which hormone?

PATTERNS OF HORMONAL INTERACTION

Although hormones are usually studied individually, the extracellular fluids contain a mixture of hormones whose concentrations change daily or even hourly. As a result, cells never respond to only one hormone; instead, they respond to multiple hormones simultaneously. When a cell receives instructions from two hormones at the same time, four results are possible:

1. The two hormones may have **antagonistic,** or opposing, effects, as in the case of PTH and calcitonin or insulin and glucagon. The net result will depend on the balance between the two hormones. In general, when antagonistic hormones are present, the observed effects will be weaker than those produced by either hormone acting unopposed.

2. The two hormones may have additive effects. The net result is greater than the effect that each would produce acting alone. In some cases the net result is greater than the *sum* of their individual effects. An example of such a **synergistic** (sin-er-JIS-tik; *synairesis,* a drawing together) effect is the glucose-sparing action of GH and glucocorticoids.

3. One hormone can have a **permissive** effect on another. In such cases, the first hormone is needed for the second to produce its effect. For example, epinephrine does not change energy consumption unless thyroid hormones are also present in normal concentrations.

4. Finally, hormones may produce different but complementary results in specific tissues and organs. These **integrative** effects are important in coordinating the activities of diverse physiological systems. The differing effects of calcitriol and parathyroid hormone on tissues involved in calcium metabolism are an example.

In this section, we shall present three examples of the ways hormones interact to produce complex, well-coordinated results. The patterns introduced will provide the background needed to understand the more detailed discussions found in chapters that deal with cardiovascular function, metabolism, excretion, and reproduction.

HORMONES AND GROWTH

Normal growth requires the cooperation of several endocrine organs. Six hormones—GH, thyroid hormones, insulin, PTH, calcitriol, and reproductive hormones—are especially important, although many others have secondary effects on growth. The circulating concentrations of these hormones are regulated independently. Every time the hormonal mixture changes, metabolic operations are modified to some degree. The modifications vary in duration and intensity, producing unique individual growth patterns.

1. *Growth hormone (GH).* Growth hormone assists in the maintenance of normal blood glucose concentrations and in the mobilization of lipid reserves stored in adipose tissues. It is not the primary hormone involved, however, and an adult with a GH deficiency but normal levels of thyroxine (T_4), insulin, and glucocorticoids will have no physiological problems. The effects of GH on protein synthesis and cellular growth are most apparent in children, in whom GH supports muscular and skeletal development. Undersecretion or oversecretion of GH can lead to pituitary growth failure or gigantism. ∞ *p. 181*

2. *Thyroid hormones.* Normal growth also requires appropriate levels of thyroid hormones. If these hormones are absent for the first year after birth, the nervous system will fail to develop normally and mental retardation and other signs of cretinism will result. If T_4 concentrations decline later in life but before puberty, normal skeletal development will not continue.

3. *Insulin.* Growing cells need adequate supplies of energy and nutrients. Without insulin, the passage of glucose and amino acids across cell membranes will be drastically reduced or eliminated.

4, 5. *Parathyroid hormone (PTH) and calcitriol.* Parathyroid hormone and calcitriol promote the absorption of calcium salts for subsequent deposition in bone. Without adequate levels of both hormones, bones can still enlarge but will be poorly mineralized, weak, and flexible. For example, in *rickets,* a condition typically caused by inadequate calcitriol production in growing children, the limb bones are so weak that they bend under the body's weight. ∞ *p. 184*

6. *Reproductive hormones.* The activity of osteoblasts in key locations and the growth of specific cell populations are affected by the presence or absence of reproductive hormones (androgens in males, estrogens in females). These sex hormones stimulate cell growth and differentiation in their target tissues. The targets differ for androgens and estrogens, and the differential growth induced by these hormones accounts for gender-related differences in skeletal proportions and secondary sex characteristics.

Endocrinology and Athletic Performance

Medical research involving humans is generally subject to tight ethical constraints and meticulous scientific scrutiny. Yet a clandestine, unscientific, and potentially quite dangerous program of "experimentation" with hormones is being pursued by athletes in many countries. The use of hormones to improve performance is banned by the International Olympic Committee, the U.S. Olympic Committee, the National Collegiate Athletic Association, and the National Football League, and it is condemned by the American Medical Association and the American College of Sports Medicine. A significant number of amateur and professional athletes, however, persist in this dangerous practice. Synthetic forms of testosterone are used most often, but athletes may use any combination of testosterone, GH, EPO, and a variety of synthetic hormones.

Androgen Abuse   The use of *anabolic steroids,* or androgens, has become popular with many amateur and professional athletes. The goal of steroid use is to increase muscle mass, endurance, and "competitive spirit." It has been suggested that as many as 30 percent of college and professional athletes and 10–20 percent of male high school athletes are using anabolic steroids (with or without GH) to improve performance. Among U.S. bodybuilders, the proportion who use steroids may be as high as 80 percent. The use of steroids, such as *androstenedione,* which the body can convert to testosterone, was highlighted during 1998–1999 when baseball slugger Mark McGwire admitted using it to improve his performance. (Performance-enhancing drugs are still legal in professional baseball, although they are banned in some other professional sports.)

One supposed justification for this practice has been the unfounded opinion that compounds manufactured in the body are not only safe but are good for you. In reality, the administration of natural or synthetic androgens in abnormal amounts carries unacceptable health risks. Androgens affect many tissues in a variety of ways. Known complications include (1) premature closure of epiphyseal cartilages, (2) various liver dysfunctions (including jaundice and liver tumors), (3) prostate gland enlargement and urinary tract obstruction, and (4) testicular atrophy and infertility. A link to heart attacks, impaired cardiac function, and strokes has also been suggested.

Moreover, the normal regulation of androgen production involves a feedback mechanism comparable to that described for adrenal steroids earlier in this chapter. GnRH stimulates the production of LH, and LH stimulates the secretion of testosterone and other androgens by the interstitial cells of the testes. Circulating androgens, in turn, inhibit the production of both GnRH and LH. Thus, when synthetic androgens are administered in high doses, they can (1) suppress the normal production of testosterone and (2) depress the manufacture of GnRH by the hypothalamus. *This suppression of GnRH release can be permanent.*

The use of androgenic "bulking agents" by female bodybuilders may not only add muscle mass but can alter muscular proportions and secondary sex characteristics as well. For example, women taking steroids can develop irregular menstrual periods and changes in body hair distribution (including baldness). Finally, androgen abuse can cause a generalized depression of the immune system.

EPO Abuse   Because it is now synthesized by recombinant DNA techniques, EPO is readily available. Endurance athletes, such as cyclists or marathon runners, may use it to boost the number of oxygen-carrying red blood cells in the bloodstream. Although this effect improves the oxygen content of blood, it also makes blood denser; the heart must work harder to push it around the circulatory system. Between 1991 and 1994, the deaths of 18 young, otherwise healthy European cyclists were attributed to heart failure related to EPO abuse.

GHB and Clenbuterol   Androgens and EPO are hormones with reasonably well-understood effects. Because drug testing is now widespread in amateur and professional sports, athletes interested in "getting an edge" are experimenting with drugs not easily detected by standard tests. The long-term and short-term effects of these drugs are difficult to predict. Two examples are the recent use of *GHB* and *clenbuterol* by amateur athletes.

Gamma-hydroxybutyrate (GHB) was tested for use as an anesthetic in the 1960s. It was rejected, in part because it was linked to petit mal and grand mal seizures. In 1990, this drug appeared in health-food stores, where it was sold as an anabolic agent and diet aid. During a 5-month period in 1990, 16 cases of severe reaction to GHB were treated in San Francisco alone. Symptoms included confusion, hallucinations, seizures, and coma at doses from 0.25 teaspoons to 4 tablespoons.

Clenbuterol, sometimes used to treat asthma, mimics epinephrine and stimulates β_2 receptors. This effect increases the diameter of the respiratory passageways and accelerates blood flow through active skeletal muscles. Although it is also rumored to have anabolic properties comparable to those of androgens, no evidence supports that rumor. Clenbuterol abuse is reportedly widespread, although exact numbers are difficult to obtain. Heavy usage can cause severe headaches, tremors, insomnia, and potentially dangerous abnormal heartbeats. During the 1992 Olympics in Barcelona, Spain, two American athletes were disqualified because they tested positive for this drug.

18

HORMONES AND STRESS

Any condition—physical or emotional—that threatens homeostasis is a form of **stress.** Many stresses are opposed by specific homeostatic adjustments. For example, a decline in body temperature will lead to shivering or changes in the pattern of blood flow in an attempt to restore normal body temperature.

In addition, the body has a *general* response to stress that can occur while other, more specific, responses are under way. Exposure to a wide variety of stress-causing factors will produce the same general pattern of hormonal and physiological adjustments. These responses are part of the **general adaptation syndrome (GAS),** also known as the **stress response.** The GAS, first described by Hans Selye in 1936, can be divided into three phases: (1) the *alarm phase,* (2) the *resistance phase,* and (3) the *exhaustion phase* (Figure 18-17•).

The Alarm Phase

During the **alarm phase,** an immediate response to the stress occurs. This response is directed by the sympathetic division of the autonomic nervous system. In the alarm phase, (1) energy reserves are mobilized, mainly in the form of glucose, and (2) the body prepares to deal with the stress-causing factor by "fight or flight" responses. ∞ *pp. 504–505*

Epinephrine is the dominant hormone of the alarm phase, and its secretion accompanies a generalized sympathetic activation. The characteristics of the alarm phase include:

■ Increased mental alertness.

■ Increased energy consumption by skeletal muscles and many other tissues.

■ The mobilization of energy reserves (glycogen and lipids).

■ Increased blood flow to skeletal muscles.

■ Decreased blood flow to the skin, kidneys, and digestive organs.

■ A drastic reduction in digestion and urine production.

■ Increased sweat gland secretion.

■ Increased blood pressure, heart rate, and respiratory rate.

Although the effects of epinephrine are most apparent during the alarm phase, other hormones play supporting roles. For example, the reduction of water losses by circulatory changes, ADH production, and aldosterone secretion can be very important if the stress involves a loss of blood.

The Resistance Phase

The temporary adjustments of the alarm phase are often sufficient to remove or overcome the stress. But some stresses, such as starvation, acute illness, or severe anxiety, can persist for hours, days, or weeks. If a stress lasts longer than a few hours, the individual will enter the **resistance phase** of the GAS.

Glucocorticoids are the dominant hormones of the resistance phase. Epinephrine, GH, and thyroid hormones are also involved. Energy demands in the resistance phase remain higher than normal owing to the combined effects of these hormones.

Neural tissue has a high demand for energy, and neurons must have a reliable supply of glucose. If blood glucose concentrations fall too far, neural function will deteriorate. Glycogen reserves are adequate to maintain normal glucose concentrations during the alarm phase but are nearly exhausted after several hours. The endocrine secretions of the resistance phase are coordinated to achieve four integrated results:

1. *The mobilization of lipid and protein reserves.* The hypothalamus produces GH-RH and CRH, stimulating the release of GH and, by means of ACTH, the secretion of glucocorticoids. Adipose tissues respond to GH and glucocorticoids by releasing stored fatty acids. Skeletal muscles respond to glucocorticoids by breaking down proteins and releasing amino acids into the bloodstream.

2. *The conservation of glucose for neural tissues.* Glucocorticoids and GH from the anterior pituitary stimulate lipid metabolism in most tissues. These glucose-sparing effects maintain normal blood glucose levels even after long periods of starvation. Neural tissues do not alter their metabolic activities, however, and they continue to use glucose as an energy source.

3. *The elevation and stabilization of blood glucose concentrations.* As blood glucose levels decline, glucagon and glucocorticoids stimulate the liver to manufacture glucose from other carbohydrates, from glycerol by way of triglycerides, and from amino acids provided by skeletal muscles. The glucose molecules are then released into the bloodstream, and blood glucose concentrations return to normal levels.

4. *The conservation of salts and water and the loss of K^+ and H^+.* Blood volume is conserved through the actions of ADH and aldosterone. As Na^+ is conserved, K^+ and H^+ are lost.

The body's lipid reserves are sufficient to maintain the resistance phase for a period of weeks or even months. (These reserves account for the ability to endure lengthy periods of starvation.) But the resistance phase cannot be sustained indefinitely. If starvation is the primary stress, the resistance phase ends as lipid reserves are exhausted and structural proteins become the primary energy source. If

ALARM PHASE

**"Fight or Flight"
Immediate short-term responses to crises**

Brain

General sympathetic activation

Epinephrine Norepinephrine

Adrenal medulla

→ Mobilization of glucose reserves

→ Changes in circulation

→ Increases in heart rate and respiratory rate

→ Increased energy use by all cells

RESISTANCE PHASE

Long-term metabolic adjustments occur

Brain

Sympathetic stimulation

GH → *Mobilization of remaining energy reserves:* Lipids are released by adipose tissue, amino acids are released by skeletal muscle

GC → *Conservation of glucose:* Peripheral tissue (except neural) breaks down lipids to obtain energy

Glucagon

Pancreas

→ *Elevation of blood glucose concentrations:* Liver synthesizes glucose primarily from other carbohydrates, glycerol, and amino acids

Kidney

ACTH

MC (with ADH) → *Conservation of salts and water, loss of K^+ and H^+*

Adrenal cortex

Renin

Angiotensin

GH	Growth hormone
GC	Glucocorticoids
MC	Mineralocorticoids (aldosterone)
ACTH	Adrenocorticotropic hormone
ADH	Antidiuretic hormone

18

EXHAUSTION PHASE

Collapse of vital systems

Causes may include:
— Exhaustion of lipid reserves
— Inability to produce glucocorticoids
— Failure to maintain electrolyte balance
— Cumulative structural or functional damage to vital organs

•**FIGURE 18-17**
**The General
Adaptation Syndrome**

another factor is the cause, the resistance phase will end due to complications brought about by hormonal side effects. For example:

■ Although the metabolic effects of glucocorticoids are essential to normal resistance, their anti-inflammatory action slows wound healing and increases susceptibility to infection.

■ The continued conservation of fluids under the influence of ADH and aldosterone stresses the cardiovascular system by producing elevated blood volumes and higher-than-normal blood pressures.

■ The adrenal cortex may become unable to continue producing glucocorticoids, quickly eliminating the ability to maintain blood glucose concentrations at acceptable levels.

Poor nutrition, emotional or physical trauma, chronic illness, or damage to key organs such as the heart, liver, or kidneys will hasten the end of the resistance phase.

The Exhaustion Phase

When the resistance phase ends, homeostatic regulation breaks down and the **exhaustion phase** begins. Unless corrective actions are taken almost immediately, the failure of one or more organ systems will prove fatal.

Mineral imbalances contribute to the existing problems with major systems. The production of aldosterone throughout the resistance phase results in a conservation of Na$^+$ at the expense of K$^+$. As the body's K$^+$ content declines, a variety of cells—notably neurons and muscle fibers—begin to malfunction. Although a single cause, such as heart failure, may be listed as the cause of death, the underlying problem is the inability to sustain the endocrine and metabolic adjustments of the resistance phase. Figure 18-17• summarizes the three phases of the GAS.

HORMONES AND BEHAVIOR

As we have seen, many endocrine functions are regulated by the hypothalamus, and hypothalamic neurons monitor the levels of many circulating hormones. Other portions of the CNS are also quite sensitive to hormonal stimulation.

The clearest demonstrations of the effects of specific hormones involve individuals whose endocrine glands are oversecreting or undersecreting. But even normal changes in circulating hormone levels can cause behavioral changes. In *precocious* (premature) *puberty,* sex hormones are produced at an inappropriate time, perhaps as early as age 5 or 6. An affected child not only begins to develop adult secondary

sex characteristics but also undergoes significant behavioral changes. The "nice little kid" disappears, and the child becomes aggressive and assertive. These behavioral changes represent the effects of sex hormones on CNS function. Thus, behaviors that in normal teenagers are usually attributed to environmental stimuli, such as peer pressure, have a physiological basis as well. In adults, changes in the mixture of hormones reaching the CNS can have significant effects on intellectual capabilities, memory, learning, and emotional states.

AGING AND HORMONE PRODUCTION

The endocrine system shows relatively few functional changes with age. The most dramatic exception is the decline in the concentration of reproductive hormones. We noted the effects of these hormonal changes on the skeletal system in Chapter 6 (∞ *pp. 185, 187*); we shall continue the discussion in Chapter 29.

Blood and tissue concentrations of many other hormones, including TSH, thyroid hormones, ADH, PTH, prolactin, and glucocorticoids, remain unchanged with advancing age. Although circulating hormone levels may remain within normal limits, however, some endocrine tissues become less responsive to stimulation. For example, in elderly individuals smaller amounts of GH and insulin are secreted after a carbohydrate-rich meal. The reduction in levels of GH and other tropic hormones affects tissues throughout the body; these hormonal effects are associated with the reductions in bone density and muscle mass noted in earlier chapters.

Finally, age-related changes in other tissues affect their abilities to respond to hormonal stimulation. As a result, peripheral tissues may become less responsive to some hormones. This loss of sensitivity has been documented for glucocorticoids and ADH.

☑ Insulin lowers the level of glucose in the blood, and glucagon causes glucose levels to rise. What is this type of hormonal interaction called?

☑ The lack of which hormones would inhibit skeletal formation?

☑ Why do levels of GH-RH and CRH rise during the resistance phase of the general adaptation syndrome?

INTEGRATION WITH OTHER SYSTEMS

The relationships between the endocrine system and the other systems of the body are summarized in Figure 18-18•.

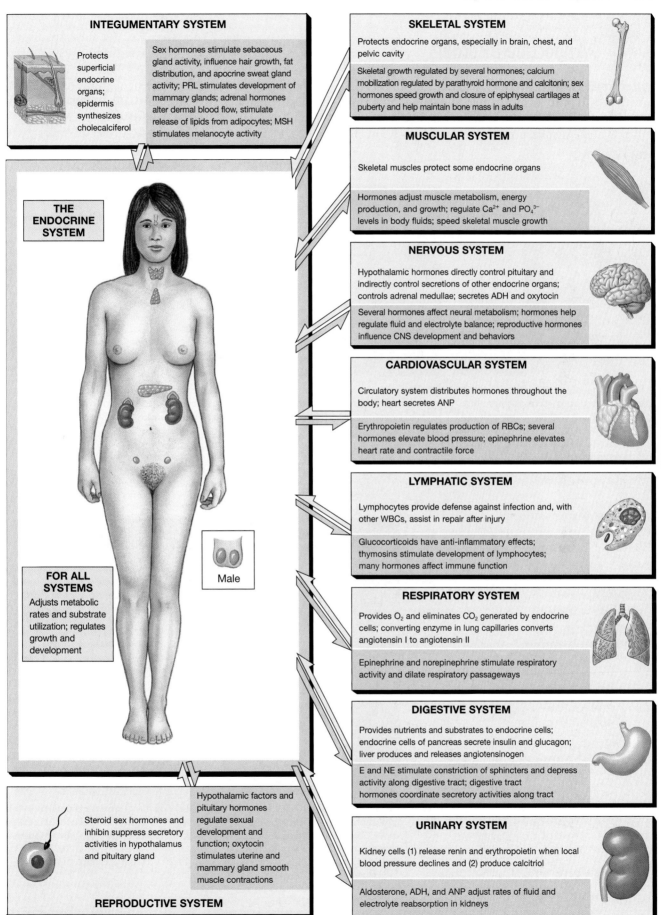

INTEGUMENTARY SYSTEM

Protects superficial endocrine organs; epidermis synthesizes cholecalciferol

Sex hormones stimulate sebaceous gland activity, influence hair growth, fat distribution, and apocrine sweat gland activity; PRL stimulates development of mammary glands; adrenal hormones alter dermal blood flow, stimulate release of lipids from adipocytes; MSH stimulates melanocyte activity

THE ENDOCRINE SYSTEM

FOR ALL SYSTEMS

Adjusts metabolic rates and substrate utilization; regulates growth and development

Male

REPRODUCTIVE SYSTEM

Steroid sex hormones and inhibin suppress secretory activities in hypothalamus and pituitary gland

Hypothalamic factors and pituitary hormones regulate sexual development and function; oxytocin stimulates uterine and mammary gland smooth muscle contractions

SKELETAL SYSTEM

Protects endocrine organs, especially in brain, chest, and pelvic cavity

Skeletal growth regulated by several hormones; calcium mobilization regulated by parathyroid hormone and calcitonin; sex hormones speed growth and closure of epiphyseal cartilages at puberty and help maintain bone mass in adults

MUSCULAR SYSTEM

Skeletal muscles protect some endocrine organs

Hormones adjust muscle metabolism, energy production, and growth; regulate Ca^{2+} and PO_4^{3-} levels in body fluids; speed skeletal muscle growth

NERVOUS SYSTEM

Hypothalamic hormones directly control pituitary and indirectly control secretions of other endocrine organs; controls adrenal medullae; secretes ADH and oxytocin

Several hormones affect neural metabolism; hormones help regulate fluid and electrolyte balance; reproductive hormones influence CNS development and behaviors

CARDIOVASCULAR SYSTEM

Circulatory system distributes hormones throughout the body; heart secretes ANP

Erythropoietin regulates production of RBCs; several hormones elevate blood pressure; epinephrine elevates heart rate and contractile force

LYMPHATIC SYSTEM

Lymphocytes provide defense against infection and, with other WBCs, assist in repair after injury

Glucocorticoids have anti-inflammatory effects; thymosins stimulate development of lymphocytes; many hormones affect immune function

RESPIRATORY SYSTEM

Provides O_2 and eliminates CO_2 generated by endocrine cells; converting enzyme in lung capillaries converts angiotensin I to angiotensin II

Epinephrine and norepinephrine stimulate respiratory activity and dilate respiratory passageways

DIGESTIVE SYSTEM

Provides nutrients and substrates to endocrine cells; endocrine cells of pancreas secrete insulin and glucagon; liver produces and releases angiotensinogen

E and NE stimulate constriction of sphincters and depress activity along digestive tract; digestive tract hormones coordinate secretory activities along tract

URINARY SYSTEM

Kidney cells (1) release renin and erythropoietin when local blood pressure declines and (2) produce calcitriol

Aldosterone, ADH, and ANP adjust rates of fluid and electrolyte reabsorption in kidneys

18

•**FIGURE 18-18** Functional Relationships between the Endocrine System and Other Systems

Endocrine Disorders

Endocrine disorders fall into two basic categories: They reflect either abnormal hormone production or abnormal cellular sensitivity to hormones. The symptoms highlight the significance of normally "silent" hormonal contributions. The characteristic features of some of these conditions are illustrated in Figure 18-19•, and a summary is presented in Table 18-9.

(a) *Acromegaly* results from the over-production of growth hormone after the epiphyseal cartilages have closed. Bone shapes change, and cartilaginous areas of the skeleton enlarge. Notice the broad facial features and the enlarged lower jaw.

(b) *Cretinism,* or *congenital hypothyroidism,* results from thyroid hormone insufficiency in infancy.

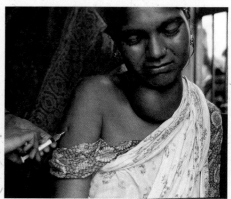

(c) An enlarged thyroid gland, or *goiter,* can be associated with thyroid hyposecretion due to iodine insufficiency in adults.

(d) *Addison's disease* is caused by hyposecretion of corticosteroids, especially glucocorticoids. Pigment changes result from stimulation of melanocytes by ACTH, which is structurally similar to MSH.

(e) *Cushing's disease* is caused by hypersecretion of glucocorticoids. Lipid reserves are mobilized, and adipose tissue accumulates in the cheeks and at the base of the neck.

•**FIGURE 18-19** Endocrine Abnormalities

TABLE 18-9 Clinical Implications of Endocrine Malfunctions

Hormone	Underproduction Syndrome	Principal Symptoms	Overproduction Syndrome	Principal Symptoms
Growth hormone (GH)	Pituitary growth failure (pp. 181, 590)	Retarded growth, abnormal fat distribution, low blood glucose hours after a meal	Gigantism, acromegaly (pp. 181, 590 and AM)	Excessive growth
Antidiuretic hormone (ADH) or arginine vasopressin (AVP)	Diabetes insipidus (p. 591)	Polyuria, dehydration, thirst	SIADH (syndrome of inappropriate ADH secretion) (AM)	Increased body weight and water content
Thyroxine (T_4), triiodothyronine (T_3)	Myxedema, cretinism (p. 597 and AM)	Low metabolic rate; low body temperature; impaired physical and mental development	Hyperthyroidism, Graves' disease, (p. 597 and AM)	High metabolic rate and body temperature
Parathyroid hormone (PTH)	Hypoparathyroidism (p. 598 and AM)	Muscular weakness, neurological problems, formation of dense bones, tetany due to low blood Ca^{2+} concentrations	Hyperparathyroidism (p. 598 and AM)	Neurological, mental, muscular problems due to high blood Ca^{2+} concentrations; weak and brittle bones
Insulin	Diabetes mellitus (Type 1) (p. 607 and AM)	High blood glucose, impaired glucose utilization, dependence on lipids for energy; glycosuria	Excess insulin production or administration	Low blood glucose levels, possibly causing coma
Mineralocorticoids (MCs)	Hypoaldosteronism (p. 601 and AM)	Polyuria, low blood volume, high blood K^+, low blood Na^+ concentrations	Aldosteronism (p. 601 and AM)	Increased body weight due to Na^+ and water retention; low blood K^+ concentration
Glucocorticoids (GCs)	Addison's disease (p. 602 and AM)	Inability to tolerate stress, mobilize energy reserves, or maintain normal blood glucose concentrations	Cushing's disease (p. 602 and AM)	Excessive breakdown of tissue proteins and lipid reserves; impaired glucose metabolism
Epinephrine (E), norepinephrine (NE)	None identified		Pheochromocytoma (AM)	High metabolic rate, body temperature, and heart rate; elevated blood glucose levels
Estrogens (females)	Hypogonadism (p. 589)	Sterility; lack of secondary sex characteristics	Adrenogenital syndrome (p. 602)	Overproduction of androgens by zona reticularis of adrenal cortex leads to masculinization
			Precocious puberty (p. 614)	Premature sexual maturation and related behavioral changes
Androgens (males)	Hypogonadism (p. 589)	Sterility, lack of secondary sex characteristics	Adrenogenital syndrome (gynecomastia) (p. 602)	Abnormal production of estrogen, sometimes due to adrenal or interstitial cell tumors; leads to breast enlargement
			Precocious puberty (p. 614)	Premature sexual maturation and related behavioral changes

18

[EXPLORE] [MediaLab]

The endocrine system is one of the major means of communication within the body. Hormones, which have far-reaching effects, link all the organs of this system. To gain a better understanding of the functions of these organs and their hormones, use the resources available on both the CD-ROM and the Companion Website. The CD-ROM includes animations and tutorials; the website builds on those resources with text content and interactive problems as well as communication avenues for sharing ideas with your class. To complete the following Web Explorations, visit the MediaLab in Chapter 18 of your Companion Website.

Web Exploration 1

Estimated time for completion = 10 minutes

To study the endocrine system, you need the ability to visualize the regulation of hormones. Most hormones function in a negative feedback loop with many regulating factors. To investigate the actions and feedbacks for growth hormone (GH) and prolactin, select the keyword **ANTERIOR PITUITARY** on the Companion Website. Page forward

through the series of six slides about GH and prolactin (pages 6–11). Create your own diagrams of the feedback mechanisms for these hormones. Using the text, do the same for the other anterior pituitary hormones. If you wish, establish a set of visual study aids with your peers.

Web Exploration 2

Estimated time for completion = 5 minutes

Hypersecretion and hyposecretion of hormones can cause serious endocrine imbalances. In hyperparathyroidism, the parathyroid glands are overactive. Review the section "The Parathyroid Glands" in the text. What do you think causes hyperparathyroidism? What would it do to the body? Write down your predictions. To see if your predictions are correct, select the keyword **PARATHYROID** on the Companion Website. Revise your original answers as needed. What can be done to treat this condition? Use your answer sheet as a model for condensing the information in the text about hypersecretion and hyposecretion by other glands.

---

18 CHAPTER REVIEW

SELECTED CLINICAL TERMINOLOGY
Terms Discussed in This Chapter

Addison's disease: A condition caused by the hyposecretion of glucocorticoids and mineralocorticoids; characterized by an inability to mobilize energy reserves and maintain normal blood glucose levels. *(p. 602 and AM)*

cretinism *(congenital hypothyroidism):* A condition caused by hypothyroidism in infancy; marked by inadequate skeletal and nervous development and a metabolic rate as much as 40 percent below normal levels. *(p. 597)*

Cushing's disease: A condition caused by the hypersecretion of glucocorticoids; characterized by the excessive breakdown and relocation of lipid reserves and proteins. *(p. 602 and AM)*

diabetes insipidus: A disorder that develops when the posterior pituitary no longer releases adequate amounts of ADH or when the kidneys cannot respond to ADH. *(p. 591)*

diabetes mellitus (mel-Ī-tus): A disorder characterized by glucose concentrations high enough to overwhelm the kidneys' reabsorption capabilities. *(p. 607 and AM)*

diabetic retinopathy, nephropathy, neuropathy: Disorders of the retina, kidneys, and peripheral nerves, respectively, related to diabetes mellitus; the conditions most often afflict middle-aged or older diabetics. *(p. 607)*

general adaptation syndrome (GAS): The pattern of hormonal and physiological adjustments with which the body responds to all forms of stress. *(p. 612)*

glycosuria: The presence of glucose in the urine. *(p. 607)*

goiter: An abnormal enlargement of the thyroid gland. *(p. 616 and AM)*

hyperglycemia: Abnormally high blood glucose levels. *(p. 607)*

hypocalcemic tetany: Muscle spasms affecting the face and upper extremities; caused by low Ca^{2+} concentrations in body fluids. *(p. 598 and AM)*

insulin-dependent diabetes mellitus (IDDM), or *Type 1 diabetes* or *juvenile-onset diabetes:* A type of diabetes mellitus; the primary cause is inadequate insulin production by the beta cells of the pancreatic islets. *(p. 607 and AM)*

myxedema: Symptoms of hyposecretion of thyroid hormones, including subcutaneous swelling, hair loss, dry skin, low body temperature, muscle weakness, and slowed reflexes. *(p. 597 and AM)*

non-insulin-dependent diabetes mellitus (NIDDM), or *Type 2 diabetes,* or *maturity-onset diabetes:* A type of diabetes mellitus in which insulin levels are normal or elevated but peripheral tissues no longer respond normally. *(p. 607 and AM)*

polyuria: The production of excessive amounts of urine; a symptom of diabetes. *(p. 591)*

seasonal affective disorder (SAD): A condition characterized by depression, lethargy, an inability to concentrate, and altered sleep and eating habits; linked to enhanced melatonin levels in individuals exposed to only short periods of daylight. *(p. 603 and AM)*

thyrotoxicosis: A condition caused by the oversecretion of thyroid hormones (hyperthyroidism). Symptoms include increases in metabolic rate, blood pressure, and heart rate; excitability and emotional instability; and lowered energy reserves. *(p. 597 and AM)*

AM **Additional Terms Discussed in the** *Applications Manual*

exophthalmos: A protrusion of the eyes from their sockets; commonly a symptom of the hypersecretion of thyroid hormones.
hypoglycemia: Abnormally low blood glucose levels.
ketoacidosis: A condition in which large numbers of ketone bodies in the blood lead to a dangerously low blood pH.

ketone bodies: Organic acids produced during the breakdown of fatty acids.
thyrotoxic crisis: An acute state of hyperthyroidism, characterized by high fever, rapid heart rate, and the malfunctioning of multiple organs.

STUDY OUTLINE

INTRODUCTION   p. 577

1. In general, the nervous system performs short-term "crisis management," whereas the endocrine system regulates longer-term, ongoing metabolic processes. *Endocrine cells* release chemicals called **hormones,** which alter the metabolic activities of many tissues and organs simultaneously. *(Figure 18-1; Table 18-1)*
2. Hormones exert their effects by modifying the activities of **target cells.**

AN OVERVIEW OF THE ENDOCRINE SYSTEM   p. 578

Hormone Structure   p. 578
1. Hormones can be divided into three groups on the basis of chemical structure: *amino acid derivatives,* structurally similar to amino acids; *peptide hormones,* chains of amino acids; and *lipid derivatives,* including **steroid hormones** and **eicosanoids.** *(Figure 18-2)*

Hormone Distribution and Transport   p. 580
2. Hormones may circulate freely or bound to transport proteins. Free hormones are rapidly removed from the bloodstream.

Hormone Function and Mechanisms of Action   p. 580
3. Receptors for *catecholamines* and peptide hormones are located on the cell membranes of target cells; the hormone acts as a **first messenger,** which causes a **second messenger** to appear in the cytoplasm. Thyroid and steroid hormones cross the cell membrane and bind to receptors in the cytoplasm or nucleus. The hormone–receptor complex activates or inactivates specific genes. *(Figures 18-3, 18-4)*

Control of Endocrine Activity   p. 584
4. **Endocrine reflexes** are the functional counterparts of neural reflexes. *(Figure 18-5)*
5. The hypothalamus regulates the activities of the nervous and endocrine systems by three mechanisms: (1) It secretes **regulatory hormones,** which control the activities of endocrine cells in the pituitary gland; (2) its autonomic centers exert direct neural control over the endocrine cells of the adrenal medullae; and (3) it acts as an endocrine organ by releasing hormones into the bloodstream at the posterior pituitary. *(Figure 18-5)*

THE PITUITARY GLAND   p. 586

1. The **pituitary gland** releases nine important peptide hormones; all bind to membrane receptors and use **cyclic-AMP** as a second messenger. *(Figure 18-6; Table 18-2)*

The Anterior Pituitary   p. 587
2. The **anterior pituitary gland (adenohypophysis)** can be subdivided into the **pars distalis,** the **pars intermedia,** and the **pars tuberalis.** *(Figure 18-6)*

3. At the median eminence of the hypothalamus, low neurons release regulatory factors into the surrounding interstitial fluids. Their secretions enter the bloodstream easily, because the capillaries in this area are **fenestrated.** *(Figure 18-7)*
4. The **hypophyseal portal system** ensures that all the blood entering the **portal vessels** will reach the intended target cells before it returns to the general circulation. *(Figure 18-7)*
5. **Thyroid-stimulating hormone (TSH)** triggers the release of thyroid hormones. *Thyrotropin-releasing hormone (TRH)* promotes the secretion of TSH.
6. **Adrenocorticotropic hormone (ACTH)** stimulates the release of *glucocorticoids* by the adrenal gland. **Corticotropin-releasing hormone (CRH)** causes the secretion of ACTH.
7. **Follicle-stimulating hormone (FSH)** stimulates follicle development and estrogen secretion in females and sperm production in males. **Luteinizing hormone (LH)** causes *ovulation* and *progestin* production in females and **androgen** production in males. **Gonadotropin-releasing hormone (GnRH)** promotes the secretion of both FSH and LH.
8. **Prolactin (PRL),** with other hormones, stimulates both the development of the mammary glands and milk production. Prolactin secretion is inhibited by **prolactin-inhibiting hormone (PIH)** and is stimulated by a *prolactin-releasing factor (PRF).* *(Figure 18-8)*
9. **Growth hormone (GH,** or **somatotropin)** stimulates cell growth and replication through the release of **somatomedins** from liver cells and has direct metabolic effects. The production of GH is regulated by **growth hormone–releasing hormone (GH-RH)** and **growth hormone–inhibiting hormone (GH-IH).** *(Figure 18-8)*
10. In many nonhuman vertebrates, **melanocyte-stimulating hormone (MSH),** released by the pars intermedia, stimulates melanocytes to produce melanin, but MSH is not normally secreted by nonpregnant adult humans.

The Posterior Pituitary   p. 591
11. The **posterior pituitary (neurohypophysis)** contains the axons of hypothalamic neurons. Neurons of the **supraoptic** and **paraventricular nuclei** manufacture **antidiuretic hormone (ADH)** and **oxytocin,** respectively. ADH decreases the amount of water lost at the kidneys and in higher concentrations elevates blood pressure through *vasoconstriction.* In women, oxytocin stimulates contractile cells in the mammary glands and has a stimulatory effect on smooth muscles in the uterus. *(Figure 18-9)*

THE THYROID GLAND   p. 593

1. The **thyroid gland** lies near the *thyroid cartilage* of the larynx and consists of two **lobes** connected by a narrow **isthmus.** *(Figure 18-10)*

18

Thyroid Follicles and Thyroid Hormones p. 594

2. The thyroid gland contains numerous **thyroid follicles.** Thyroid follicles release several hormones, including **thyroxine (T_4)** and **triiodothyronine (T_3).** *(Figures 18-10, 18-11; Table 18-4)*

3. The follicle cells manufacture **thyroglobulin** and store it as a colloid. *(Figure 18-11)*

4. Most of the thyroid hormones entering the bloodstream are attached to special **thyroid-binding globulins (TBGs),** and the rest are attached to **transthyretin** or to albumin. The unbound hormones affect peripheral tissues immediately; the binding proteins gradually release their hormones over a week or more. *(Figure 18-11; Table 18-3)*

5. Thyroid hormones exert a **calorigenic effect,** which enables us to adapt to cold temperatures.

The C Cells of the Thyroid Gland: Calcitonin p. 597

6. The **C cells** of the thyroid follicles produce **calcitonin (CT),** which helps regulate Ca^{2+} concentrations in body fluids, especially during childhood and pregnancy. *(Figures 18-10, 18-13; Table 18-4)*

THE PARATHYROID GLANDS p. 597

Parathyroid Hormone Secretion p. 598

1. Four **parathyroid glands** are embedded in the posterior surface of the thyroid gland. The **chief cells** of the parathyroid produce **parathyroid hormone (PTH)** in response to lower-than-normal concentrations of Ca^{2+}. The parathyroid glands, aided by *calcitriol* from the kidneys, are the primary regulators of Ca^{2+} levels in healthy adults. *(Figures 18-12, 18-13; Table 18-4)*

THE ADRENAL GLANDS p. 600

1. One **adrenal** *(suprarenal)* **gland** lies along the superior border of each kidney. Each adrenal gland is surrounded by a fibrous capsule. The gland is subdivided into the superficial **adrenal cortex** and the inner **adrenal medulla.** *(Figure 18-14)*

The Adrenal Cortex p. 600

2. The adrenal cortex manufactures steroid hormones called **adrenocortical steroids (corticosteroids).** The cortex can be subdivided into three areas: (1) the **zona glomerulosa,** which releases **mineralocorticoids (MCs;** all of which restrict Na^+ losses at the kidneys, sweat glands, digestive tract, and salivary glands), principally **aldosterone;** (2) the **zona fasciculata,** which produces **glucocorticoids (GCs),** notably **cortisol** and **corticosterone** (which, along with **cortisone,** exert a **glucose-sparing effect** on peripheral tissues); and (3) the **zona reticularis,** which produces androgens of uncertain significance. *(Figure 18-14, Table 18-5)*

The Adrenal Medullae p. 603

3. The adrenal medullae produce epinephrine (75–80 percent of medullary secretions) and norepinephrine (20–25 percent.) *(Figure 18-14; Table 18-5)*

THE PINEAL GLAND p. 603

1. The **pineal gland** contains **pinealocytes,** which synthesize **melatonin.** Although research and debate continue, melatonin appears to (1) slow the maturation of sperm, oocytes, and reproductive organs by reducing the rate of GnRH secretion and (2) establish *circadian rhythms.*

THE ENDOCRINE TISSUES OF OTHER SYSTEMS p. 604

The Intestines p. 604

1. The intestines produce hormones important to the coordination of digestive activities. *(Table 18-6)*

The Kidneys p. 604

2. Endocrine cells in the kidneys produce the hormones *calcitriol* and *erythropoietin* and the enzyme *renin.* These secretions are important for the regulation of calcium metabolism, blood volume, and blood pressure, respectively. *(Table 18-6)*

3. **Calcitriol** stimulates calcium (Ca^{2+}) and phosphate (PO_4^{3-}) ion absorption along the digestive tract.

4. **Erythropoietin (EPO)** stimulates red blood cell production by the bone marrow.

5. **Renin** converts **angiotensinogen** to **angiotensin I.** In the capillaries of the lungs, this compound is converted to **angiotensin II,** the hormone that (1) stimulates the adrenal production of aldosterone, (2) stimulates the pituitary release of ADH, (3) promotes thirst, and (4) constricts blood vessels, elevating blood pressure.

The Heart p. 605

6. Specialized muscle cells in the heart produce **atrial natriuretic peptide (ANP)** when blood volume becomes excessive. In general, its actions oppose those of angiotensin II. *(Table 18-6)*

The Thymus p. 605

7. The **thymus** produces several hormones, collectively known as **thymosins,** which play a role in developing and maintaining normal immune defenses. *(Table 18-6)*

The Pancreas p. 605

8. The **pancreas** contains both exocrine and endocrine cells. The **exocrine pancreas** secretes an enzyme-rich fluid that travels to the lumen of the digestive tract. Cells of the **endocrine pancreas** form clusters called **pancreatic islets** *(islets of Langerhans).* These islets contain **alpha cells** (which secrete the hormone **glucagon**), **beta cells** (which secrete **insulin**), **delta cells** (which secrete GH-IH), and **F cells** (which secrete **pancreatic polypeptide**). *(Figure 18-15; Table 18-7)*

9. Insulin lowers blood glucose by increasing the rate of glucose uptake and utilization; glucagon raises blood glucose by increasing the rates of glycogen breakdown and glucose manufacture in the liver. GH-IH suppresses glucagon and insulin release and slows the rates of food absorption and enzyme secretion in the digestive tract. Pancreatic polypeptide inhibits gallbladder contractions and regulates the production of some pancreatic enzymes. *(Figure 18-16)*

The Gonads p. 608

10. The **interstitial cells** of the testes produce androgens. **Testosterone** is the most important sex hormone in males. *(Table 18-8)*

11. In females, *oocytes* develop in **follicles;** follicle cells produce **estrogens,** especially **estradiol.** After ovulation, the remaining follicle cells reorganize into a **corpus luteum.** Those cells release a mixture of estrogens and **progestins,** especially **progesterone.** *(Table 18-8)*

PATTERNS OF HORMONAL INTERACTION p. 610

1. The endocrine system functions as an integrated unit, and hormones often interact. These interactions may exert several effects: (1) **antagonistic** (opposing) effects; (2) **synergistic** (additive) effects; (3) **permissive** effects, in which one

hormone is necessary for another to produce its effect; or (4) **integrative** effects, in which hormones produce different but complementary results.

Hormones and Growth p. 610

2. Normal growth requires the cooperation of several endocrine organs. Six hormones are especially important: GH, thyroid hormones, insulin, PTH, calcitriol, and reproductive hormones.

Hormones and Stress p. 612

3. Any condition that threatens homeostasis is a **stress.** Our bodies respond to a variety of stress-causing factors by the **general adaptation syndrome (GAS),** or **stress response.** The GAS can be divided into three phases: (1) the **alarm phase** (an immediate, "fight or flight" response, under the direction of the sympathetic division of the ANS); (2) the **resistance phase,** dominated by glucocorticoids (mobilization of lipid and protein reserves, elevation and stabilization

of blood glucose concentrations, and conservation of glucose for neural tissues); and (3) the **exhaustion phase,** the eventual breakdown of homeostatic regulation and failure of one or more organ systems. *(Figure 18-17)*

Hormones and Behavior p. 614

4. Many hormones affect the nervous system, producing changes in mood, emotional states, and behavior.

Aging and Hormone Production p. 614

5. The endocrine system shows few functional changes with advanced age. The chief change is a decline in the concentration of reproductive hormones.

INTEGRATION WITH OTHER SYSTEMS p. 614

1. The endocrine system interacts with each of the other body systems. *(Figure 18-18)*

REVIEW QUESTIONS

LEVEL 1 REVIEWING FACTS AND TERMS

1. Peptide hormones, which are chains of amino acids, generally are produced as
   (a) prehormones (b) prohormones
   (c) hormones (d) liposomes
2. Cyclic-AMP functions as a second messenger to
   (a) build proteins and catalyze specific reactions
   (b) activate adenylate cyclase
   (c) open ion channels and activate key enzymes in the cytoplasm
   (d) bind the hormone–receptor complex to DNA segments
3. Adrenocorticotropic hormone (ACTH) stimulates the release of
   (a) thyroid hormones by the hypothalamus
   (b) gonadotropins by the adrenal glands
   (c) growth hormones by the hypothalamus
   (d) steroid hormones by the adrenal glands
4. FSH production in males supports
   (a) the maturation of sperm by stimulating sustentacular cells
   (b) the development of muscles and strength
   (c) the production of male sex hormones
   (d) an increased desire for sexual activity
5. The hormone that induces ovulation in women and promotes the secretion of progesterone by the ovaries is
   (a) interstitial cell–stimulating hormone
   (b) estradiol
   (c) luteinizing hormone
   (d) prolactin
6. The two hormones released by the posterior pituitary are
   (a) GH and gonadotropin
   (b) estrogen and progesterone
   (c) GH and prolactin
   (d) ADH and oxytocin
7. The primary function of ADH is to
   (a) increase the amount of water lost at the kidneys
   (b) decrease the amount of water lost at the kidneys
   (c) dilate peripheral blood vessels to decrease blood pressure
   (d) increase absorption along the digestive tract

8. The element required for normal thyroid function is
   (a) magnesium (b) calcium
   (c) potassium (d) iodine
9. Reduced fluid losses in urine due to the retention of Na^+ and water is a result of the action of
   (a) antidiuretic hormone (b) calcitonin
   (c) aldosterone (d) cortisone
10. The adrenal medulla produces the hormones
    (a) cortisol and cortisone
    (b) epinephrine and norepinephrine
    (c) corticosterone and testosterone
    (d) androgens and progesterone
11. Hormones released by the kidneys include
    (a) calcitriol and erythropoietin
    (b) ADH and aldosterone
    (c) epinephrine and norepinephrine
    (d) cortisol and cortisone
12. What four events are set in motion by the activation of a G-protein when a hormone binds to its receptor site?
13. What three higher-level mechanisms are involved in integrating the activities of the nervous and endocrine systems?
14. Which seven hormones are released by the anterior pituitary gland?
15. Which two hormones are released by the posterior pituitary gland?
16. What five primary effects result from the action of thyroid hormones?
17. What effects do calcitonin and parathyroid hormone have on blood calcium levels?
18. What three zones make up the adrenal cortex, and what kinds of hormones are produced by each zone?
19. What peripheral effects does the activation of the adrenal medulla have on alpha and beta receptors?
20. Which two hormones are released by the kidneys, and what is the importance of each hormone?
21. What are the four opposing effects of atrial natriuretic peptide and angiotensin II?
22. What four cell populations make up the endocrine pancreas? Which hormone does each cell type produce?

18

23. (a) Describe the three phases of the general adapation syndrome (GAS) that constitute the body's response to stress.

(b) Which endocrine secretions play dominant roles in the alarm and resistance phases?

LEVEL 2 REVIEWING CONCEPTS

24. What is the primary difference in the way the nervous and endocrine systems communicate with their target cells?

25. How can a hormone modify the activities of its target cells?

26. What possible results occur when a cell receives instructions from two hormones at the same time?

27. What is an endocrine reflex? Compare endocrine and neural reflexes.

28. How would blocking the activity of phosphodiesterase affect a cell that responds to hormonal stimulation by the cAMP second messenger system?

LEVEL 3 CRITICAL THINKING AND CLINICAL APPLICATIONS

29. Roger has been extremely thirsty; he drinks numerous glasses of water every day and urinates a great deal. Name two disorders that could produce these symptoms. What test could a clinician perform to determine which disorder Roger has?

30. Julie is pregnant but is not receiving prenatal care. She has a poor diet consisting mostly of fast food. She drinks no milk, preferring colas instead. How would this situation affect Julie's level of parathyroid hormone?

31. Sherry tells her physician that she has been restless and irritable lately. Sherry has a hard time sleeping and complains of diarrhea and shortness of breath. During the examination, her physician notices a higher-than-normal heart rate and decreased circulating levels of cholesterol. What tests could the physician suggest to make a positive diagnosis of Sherry's condition?

32. Patients receiving steroid hormone frequently retain large amounts of water. Why?

CHAPTER **19**

Blood

CHAPTER OUTLINE AND OBJECTIVES

INTRODUCTION   p. 624

FUNCTIONS OF BLOOD   p. 624

COMPOSITION OF BLOOD   p. 624
① Describe the important components and major functions of blood.
② Identify locations on the body used for blood collection, and list the basic physical characteristics of the blood samples drawn from these locations.

PLASMA   p. 626
  Plasma Proteins   p. 626
③ Discuss the composition and functions of plasma.

FORMED ELEMENTS   p. 627
  Hemopoiesis   p. 627
  Red Blood Cells   p. 628
  White Blood Cells   p. 637
  Platelets   p. 644
④ Describe the origin and production of the formed elements in blood.
⑤ List the characteristics and functions of red blood cells.
⑥ Describe the structure of hemoglobin, and indicate its functions.
⑦ Describe the recycling system for aged or damaged red blood cells.
⑧ Define erythropoiesis, identify the stages involved in red blood cell maturation, and describe the homeostatic regulation of red blood cell production.
⑨ List examples of important blood tests, and cite the normal values for each test.
⑩ Explain the importance of blood typing on the basis of ABO and Rh incompatibilities.
⑪ Categorize the various white blood cells on the basis of their structures and functions, and discuss the factors that regulate the production of each class.
⑫ Describe the structure, function, and production of platelets.

Out of sight, out of mind. That saying certainly applies to blood. As long as we don't see it, we don't think much about it. But looking at vials or bags of it in a hospital—or, worse, watching even a drop of our own oozing from a cut or scrape—makes most of us feel uneasy; some people faint at the sight. This anxiety probably comes from being reminded of how dependent we are on our relatively small supply of the precious fluid. If you have three half-gallon milk containers in your refrigerator at home, you probably have more milk than you do blood. You can easily spare a drop of blood or donate a pint, but if you suddenly lose more than 15 to 20 percent of your blood supply, you will need more in a hurry. Fortunately, blood transfusions are now fairly routine; but, as we will see in this chapter, not just anyone's blood will do. We'll also explore the many functions of this vital substance.

HEMOSTASIS   p. 645
  The Vascular Phase   p. 645
  The Platelet Phase   p. 645
  The Coagulation Phase   p. 646
  Clot Retraction   p. 648
  Fibrinolysis   p. 648
⑬ Discuss mechanisms that control blood loss after an injury, and describe the reaction sequences responsible for blood clotting.

MEDIALAB and CHAPTER REVIEW   p. 650

We can compare the cardiovascular system to the cooling system of a car. The basic components of the **cardiovascular system** include a circulating fluid (blood), a pump (the heart), and an assortment of conducting pipes (the blood vessels). Although this system is far more complicated and versatile than that of a car, both mechanical and biological systems can suffer from fluid losses, pump failures, or damaged pipes.

Small embryos don't need cardiovascular systems, because diffusion across their exposed surfaces provides adequate oxygen and removes waste products as rapidly as they are generated. By the time a human embryo reaches a few millimeters in length, however, developing tissues are consuming oxygen and nutrients faster than they can be provided by simple diffusion. At that stage, the cardiovascular system must begin functioning to provide a rapid-transport system for oxygen, nutrients, and waste products. It is the first system to become fully operational: The heart begins beating at the end of the third week of embryonic life, when many other systems have barely begun their development.

When the heart starts beating, blood begins circulating. The embryo can now more efficiently use the nutrients obtained from the maternal bloodstream, and its size doubles in the next week. In adults, circulating blood provides nutrients, oxygen, chemical instructions, and a mechanism for waste removal to each of the roughly 75 trillion cells in the human body. The blood also transports specialized cells that defend peripheral tissues from infection and disease. These services are essential, and a region deprived of circulation dies in a matter of minutes.

In the next three chapters, we shall examine the individual components of the cardiovascular system. This chapter deals with the nature of the circulating blood, Chapter 20 covers the structure and function of the heart, and Chapter 21 examines the organization of blood vessels and the integrated functioning of the cardiovascular system.

The cardiovascular system is extensively integrated with the *lymphatic system*. The two systems are interconnected and interdependent. Fluid leaves the bloodstream, enters the tissues, and returns to the blood within the vessels of the lymphatic system. For this reason, the cardiovascular and lymphatic systems are often said to be components of a single *circulatory system*. The bloodstream carries cells, antibodies, and cytokines of the lymphatic system throughout the body, and the cardiovascular system assists the lymphatic system in defending the body against invasion by pathogens. In Chapter 22, we shall describe the components of the lymphatic system and shall consider the array of defenses that protect the body from internal and external hazards.

FUNCTIONS OF BLOOD

In this chapter, we examine the structure and functions of **blood,** a specialized fluid connective tissue that contains cells suspended in a fluid matrix. ∞ *p. 124* The functions of blood include the following:

- *The transportation of dissolved gases, nutrients, hormones, and metabolic wastes.* Blood carries oxygen from the lungs to peripheral tissues and carbon dioxide from those tissues to the lungs. Blood distributes nutrients absorbed at the digestive tract or released from storage in adipose tissue or in the liver. It carries hormones from endocrine glands toward their target cells, and it absorbs and carries the wastes produced by tissue cells to the kidneys for excretion.

- *The regulation of the pH and ion composition of interstitial fluids.* Diffusion between interstitial fluids and blood eliminates local deficiencies or excesses of ions such as calcium or potassium. Blood also absorbs and neutralizes the acids generated by active tissues, such as the lactic acid produced by skeletal muscles.

- *The restriction of fluid losses at injury sites.* Blood contains enzymes and other substances that respond to breaks in the vessel walls by initiating the process of *blood clotting.* A blood clot acts as a temporary patch and prevents further blood loss.

- *Defense against toxins and pathogens.* Blood transports these *white blood cells,* specialized cells that migrate into peripheral tissues to fight infections or remove debris. It also delivers *antibodies,* special proteins that attack invading organisms or foreign compounds.

- *The stabilization of body temperature.* Blood absorbs the heat generated by active skeletal muscles and redistributes it to other tissues. If the body temperature is already high, that heat will be lost across the surface of the skin. If the body temperature is too low, the warm blood is directed to the brain and to other temperature-sensitive organs.

COMPOSITION OF BLOOD

Blood has a characteristic and unique composition (Figure 19-1•). It is a fluid connective tissue with a matrix called **plasma** (PLAZ-muh). Plasma proteins are in solution rather than forming insoluble fibers like those in other connective tissues, such as loose connective tissue or cartilage. Because these proteins are in solution, plasma is slightly denser than water.

Formed elements are blood cells and cell fragments that are suspended in plasma. Three types of formed elements exist: (1) red blood cells, (2) white blood cells, and (3) platelets. **Red blood cells (RBCs),** or **erythrocytes** (e-RITH-rō-sīts; *erythros,* red + *cyte,* cell), are the most abundant blood cells. These specialized cells are essential for the transport of oxygen in the blood. The less numerous **white blood cells (WBCs),** or **leukocytes** (LOO-kō-sīts; *leukos,* white), are cells

•**FIGURE 19-1 The Composition of Whole Blood.**
(a) Drawing blood. (b) The composition of a typical
sample of plasma (see Appendix VI). (c) The formed
elements (see Table 19-3, p. 642).

involved with the body's defense mechanisms. **Platelets** are small, membrane-bound packets of cytoplasm that contain enzymes and other substances important to the process of blood clotting.

Together, the plasma and formed elements constitute **whole blood.** These components can be **fractionated,** or separated, for analytical or clinical purposes. We shall encounter examples of uses for fractionated blood later in the chapter.

Whole blood from any source—venous blood, blood from peripheral capillaries, or arterial blood—has the same basic physical characteristics:

- *Temperature.* Blood temperature is roughly 38°C (100.4°F), slightly above normal body temperature.
- *Viscosity.* Blood is five times as viscous as water—that is, five times stickier, more cohesive, and resistant to flow than water. The high viscosity results from interactions among dissolved proteins, formed elements, and surrounding water molecules in plasma.
- *pH.* Blood is slightly alkaline, with a pH between 7.35 and 7.45, averaging 7.4.

The cardiovascular system of an adult man contains 5–6 liters (5.3–6.4 quarts) of whole blood; that of an adult woman contains 4–5 liters (4.2–5.3 quarts). The gender differences in blood volume primarily reflect differences in average body size. Blood volume in liters can be estimated for an individual of either gender by calculating 7 percent of the body weight in kilograms. For example, a 75-kg individual would have a blood volume of approximately 5.25 liters (5.4 quarts). More-precise determinations of blood volume can be obtained by using special dyes or radioisotopes, such as phosphorus-32.

Blood Collection for Analysis

Fresh whole blood is generally collected from a superficial vein, such as the *median cubital vein* on the anterior surface of the elbow (Figure 19-1a●). This procedure is **venipuncture** (VĒN-i-punk-chur; *vena,* vein + *punctura,* a piercing). It is a common sampling technique because (1) superficial veins are easy to locate; (2) the walls of veins are thinner than those of comparably sized arteries; and (3) blood pressure in the venous system is relatively low, so the puncture wound seals quickly. The most common clinical procedures examine venous blood.

Blood from peripheral capillaries can be obtained by puncturing the tip of a finger, the lobe of an ear, or (in infants) the great toe or heel. A small drop of capillary blood can be used to prepare a *blood smear,* a thin film of blood on a microscope slide. The blood smear is then stained with special dyes to show each type of formed element.

An **arterial puncture,** or "arterial stick," can be used for checking the efficiency of gas exchange at the lungs. Samples are generally drawn from the *radial artery* at the wrist or the *brachial artery* at the elbow.

PLASMA

The composition of whole blood is summarized in Figure 19-1b●. Plasma makes up 46–63 percent of the volume of whole blood, with water accounting for 92 percent of the plasma volume. Together, plasma and interstitial fluid account for most of the volume of extracellular fluid (ECF) in the body.

In many respects, the composition of the plasma resembles that of interstitial fluid. The concentrations of the major plasma ions, for example, are similar to those of interstitial fluid and differ markedly from those inside cells. This similarity is understandable, as water, ions, and small solutes are continuously exchanged between plasma and interstitial fluids across the walls of capillaries. The capillaries normally deliver more liquid and solutes to a tissue than they remove. The excess flows through the tissue, into vessels of the lymphatic system, and eventually back to the bloodstream. The primary differences between plasma and interstitial fluid involve (1) the concentrations of dissolved proteins, because plasma proteins cannot cross capillary walls; and (2) the levels of respiratory gases (oxygen and carbon dioxide), due to the respiratory activities of tissue cells.

PLASMA PROTEINS

Plasma contains significant quantities of dissolved proteins. On average, each 100 ml of plasma contains 7.6 g (0.3 oz) of protein, almost five times the concentration in interstitial fluid. The large size and globular shapes of most blood proteins prevent them from crossing capillary walls, and they remain trapped within the circulatory system. Three primary classes of plasma proteins exist: (1) *albumins* (al-BŪ-minz), (2) *globulins* (GLOB-ū-linz), and (3) *fibrinogen* (fī-BRIN-ō-jen).

Albumins, Globulins, and Fibrinogen

Albumins constitute roughly 60 percent of the plasma proteins. As the most abundant plasma proteins, they are major contributors to the osmotic pressure of plasma. Albumins are also important in the transport of fatty acids, thyroid hormones, some steroid hormones, and other substances.

Globulins account for approximately 35 percent of the protein population in plasma. Examples of important plasma globulins include antibodies and transport globulins. **Antibodies,** also called **immunoglobulins** (i-mū-nō-GLOB-ū-linz), attack foreign proteins and pathogens. We shall describe several classes of immunoglobulins in Chapter 22. **Transport globulins** bind small ions, hormones, or compounds that might otherwise be lost at the kidneys or that have very low solubility in water. Important examples of transport globulins include the following:

- *Thyroid-binding globulin* and *transthyretin,* which transport thyroid hormones; *transcortin,* which transports ACTH; and *transcalciferin,* which transports calcitriol.
- *Metalloproteins,* which transport metal ions. *Transferrin,* for example, is a metalloprotein that transports iron (Fe^{2+}).
- *Apolipoproteins* (ā-pō-lī-pō-PRŌ-tēnz), which carry triglycerides and other lipids in blood. When bound to lipids, an apolipoprotein becomes a **lipoprotein** (lī-pō-PRŌ-tēn).
- *Steroid-binding proteins,* which transport steroid hormones in blood. For example, *testosterone-binding globulin* (TeBG) binds and transports testosterone.

The third type of plasma protein, **fibrinogen,** functions in blood clotting. Fibrinogen normally accounts for roughly 4 percent of plasma proteins. Under certain conditions, fibrinogen molecules interact, forming large, insoluble strands of **fibrin** (FĪ-brin). These fibers provide the basic framework for a blood clot. If steps are not taken to prevent clotting, the conversion of fibrinogen to fibrin will occur in a sample of plasma. This conversion removes the clotting proteins, leaving a fluid known as **serum.** The clotting process also removes calcium ions and other materials from solution. As a result, plasma and serum differ in several significant ways (see Appendix VI). Because the normal ranges of physiological values differ, the results of a blood test generally indicate whether the sample source is plasma (P) or serum (S).

Other Plasma Proteins

Peptide hormones, including insulin, prolactin (PRL), and the glycoproteins thyroid-stimulating hormone (TSH), follicle-stimulating hormone (FSH), and luteinizing hormone (LH), are normally present in circulating blood. ∞ *pp. 606, 589* Their plasma concentrations rise and fall from day to day or even from hour to hour.

Origins of the Plasma Proteins

The liver synthesizes and releases more than 90 percent of the plasma proteins, including all albumins and fibrinogen, most globulins, and various prohormones. Antibodies are produced by *plasma cells.* Plasma cells are derived from *lymphocytes,* the primary cells of the lymphatic system. Peptide hormones are produced in a variety of endocrine organs. ∞ *pp. 579–580*

Because the liver is the primary source of plasma proteins, liver disorders can alter the composition and functional properties of blood. For example, some forms of liver disease can lead to uncontrolled bleeding due to the inadequate synthesis of fibrinogen and other proteins involved in blood clotting.

Plasma Expanders

Plasma expanders can be used to increase blood volume temporarily, over a period of hours, while preliminary lab work is under way to determine blood type. Isotonic electrolyte solutions such as normal saline can be used as a plasma expander, but its effects are short-lived due to diffusion into interstitial fluid and cytosol. This fluid loss is slowed by the addition of solutes that cannot freely diffuse across cell membranes. One example is *Ringer's solution,* isotonic saline containing lactate ions. The effects of Ringer's solution fade gradually as the liver, skeletal muscles, and other tissues absorb and metabolize the lactate ions. Another option is the administration of isotonic saline solution containing purified human albumin. However, the plasma expanders in clinical use often contain large carbohydrate molecules, rather than proteins, to maintain proper osmotic concentration. (We noted the emergency use of the carbohydrate dextran in sodium chloride solutions in Chapter 3. ∞ *p. 74*) Although these carbohydrates are not metabolized, they are gradually removed from the bloodstream by phagocytes, and the blood volume slowly declines. Plasma expanders are easily stored, and their sterile preparation avoids viral or bacterial contamination, which can be a problem with donated plasma. Although they provide a temporary solution to low blood volume, plasma expanders do not increase the amount of oxygen carried by the blood, however; that function is performed by red blood cells.

☑ Why is venipuncture a common technique for obtaining a blood sample?

☑ What would be the effects of a decrease in the amount of plasma proteins?

☑ Which plasma protein would you expect to be elevated during a viral infection?

19

FORMED ELEMENTS

As we noted earlier in this chapter, the major cellular components of blood are red blood cells and white blood cells. Blood also contains platelets, which are noncellular formed elements. The formed elements in blood are illustrated in Figure 19-1c•.

HEMOPOIESIS

The process of **hemopoiesis** (hēm-ō-poy-Ē-sis), or *hematopoiesis,* produces formed elements. Embryonic blood cells appear in the bloodstream during the third week of development. These cells divide repeatedly, increasing in number. The vessels of the embryonic yolk sac are the primary site of blood formation for the first 8 weeks of development. As other organ systems appear, some of the embryonic blood cells move out of the bloodstream and into the liver, spleen, thymus, and bone marrow. These embryonic cells differentiate into stem cells that produce blood cells by their divisions. The liver and spleen are the primary sites of hemopoiesis from the second to fifth months of development. As the skeleton enlarges, the bone marrow becomes increasingly important, and it predominates after the fifth developmental month. In adults, red bone marrow is the only site of red blood cell production as well as the primary site of white blood cell formation.

Hemocytoblasts, or *pleuripotent stem cells,* divide to produce *myeloid stem cells* and *lymphoid stem cells.* These stem cells remain capable of division, but the daughter cells they generate are involved with the production of specific types of blood cells. Lymphoid stem cells are responsible for lymphocyte production, whereas myeloid stem cells are responsible for the production of all other kinds of formed elements. We will consider the fates of the myeloid and lymphoid stem cells as we discuss the formation of each type of formed element.

RED BLOOD CELLS

Red blood cells (RBCs) contain the red pigment *hemoglobin,* which binds and transports oxygen and carbon dioxide. The most abundant blood cells, RBCs account for 99.9 percent of the formed elements. They give whole blood its deep red color.

Abundance of RBCs

A standard blood test reports the number of RBCs per microliter (μl) of whole blood as the *red blood cell count.* In adult males, one microliter, or one *cubic millimeter* (mm^3), of whole blood contains 4.5–6.3 million RBCs; in adult females, one microliter contains 4.2–5.5 million. A single drop of whole blood contains approximately 260 million RBCs, and the blood of an average adult has 25 trillion. The number of RBCs thus accounts for roughly one-third of all cells in the human body.

The **hematocrit** (he-MA-tō-krit) is the percentage of whole blood occupied by cellular elements. The normal hematocrit in adult males averages 46 (range: 40–54); the average for adult females is 42 (range: 37–47). The gender difference in hematocrit primarily reflects the fact that androgens (male hormones) stimulate red blood cell production, whereas estrogens (female hormones) do not.

The hematocrit is determined by centrifuging a blood sample so that all the formed elements come out of suspension. Whole blood contains roughly 1000 red blood cells for each white blood cell. After centrifugation, the white blood cells and platelets form a very thin *buffy coat* above a thick layer of RBCs.

Because the hematocrit value is due almost entirely to the volume of RBCs, hematocrit is commonly reported as the **volume of packed red cells (VPRC)** or simply the **packed cell volume (PCV).**

Many conditions can affect the hematocrit. For example, the hematocrit increases in cases of dehydration, owing to a reduction in plasma volume, or after *erythropoietin (EPO)* stimulation. ∞ *p. 604* The hematocrit can decrease as a result of internal bleeding or problems with RBC formation. As a result, the hematocrit alone does not provide specific diagnostic information. Yet a change in hematocrit is an indication that other, more specific tests are needed. (We will consider some of those tests later in the chapter.) AM *Polycythemia*

Structure of RBCs

Red blood cells are among the most specialized cells of the body. A red blood cell is very different from the "typical cell" we discussed in Chapter 3. Each RBC is a biconcave disc with a thin central region and a thicker outer margin (Figure 19-2•). An average RBC has a diameter of 7.8 μm and a maximum thickness of 2.6 μm, although the center narrows to about 0.8 μm. (Normal ranges of measurement are given in Figure 19-2d•.)

This unusual shape has three important effects on RBC function:

1. *It gives each RBC a large ratio of surface area to volume.* The RBC carries oxygen bound to intracellular proteins. That oxygen must be absorbed or released quickly as the RBC passes through the capillaries of the lungs or peripheral tissues. The greater the surface area per unit volume, the faster the exchange between the cell's interior and the surrounding plasma. The total surface area of the RBCs in the blood of a typical adult is about 3800 square meters, roughly 2000 times the total surface area of the body.

2. *It enables RBCs to form stacks, like dinner plates, that smooth the flow through narrow blood vessels.* These stacks form and dissociate repeatedly without affecting the cells involved. An entire stack can pass along a blood vessel only slightly larger than the diameter of a single RBC, whereas individual cells would bump the walls, bang together, and form logjams that could restrict or prevent blood flow.

3. *It enables them to bend and flex when entering small capillaries and branches.* Red blood cells are very flexible. By changing shape, individual RBCs can squeeze through capillaries as narrow as 4 μm.

During their differentiation, the RBCs of humans and other mammals lose most of their organelles, including nuclei; these cells retain only the cytoskeleton. (The RBCs of nonmammalian vertebrates have nuclei.) Because they lack nuclei and ribosomes, your

•**FIGURE 19-2** **The Anatomy of Red Blood Cells. (a)** When viewed in a standard blood smear, red blood cells appear as two-dimensional objects, because they are flattened against the surface of the slide. (LM × 320) **(b)** When traveling through relatively narrow capillaries, RBCs may stack like dinner plates. (LM × 1072) **(c)** The three-dimensional structure of red blood cells. (SEM × 1195) **(d)** A sectional view of a mature red blood cell, showing the normal ranges for its dimensions.

19

circulating RBCs cannot divide or synthesize structural proteins or enzymes. As a result, RBCs cannot perform repairs, and their life span is relatively short—normally less than 120 days. Their energy demands are low. In the absence of mitochondria, they obtain the energy they need through the anaerobic metabolism of glucose absorbed from the surrounding plasma. The lack of mitochondria ensures that absorbed oxygen will be carried to peripheral tissues, not "stolen" by mitochondria in the cell.

Hemoglobin

In effect, a developing red blood cell loses any organelle not directly associated with the cell's primary function: the transport of respiratory gases. Molecules of **hemoglobin** (HĒ-mō-glō-bin) **(Hb)** account for over 95 percent of the intracellular proteins. The hemoglobin content of whole blood is reported in grams of Hb per 100 ml of whole blood (g/dl). Normal ranges

are 14–18 g/dl in males and 12–16 g/dl in females. Hemoglobin is responsible for the cell's ability to transport oxygen and carbon dioxide.

Hemoglobin Structure   Each Hb molecule has a complex quaternary shape. The Hb molecule has two *alpha (α) chains* and two *beta (β) chains* of polypeptides. Each individual chain is a globular protein subunit that resembles the myoglobin in skeletal and cardiac muscle cells. Like myoglobin, each Hb chain contains a single molecule of **heme,** a pigment complex (Figure 19-3•). Each heme unit holds an iron ion in such a way that the iron can interact with an oxygen molecule, forming **oxyhemoglobin,** HbO_2. Blood containing RBCs filled with oxyhemoglobin is bright red. The iron–oxygen interaction is very weak; the two can easily be separated without damaging the heme unit or the oxygen molecule. The binding of an oxygen molecule to the iron in a heme unit is therefore completely reversible. A hemoglobin molecule in which the iron

Heme

•**FIGURE 19-3** **The Structure of Hemoglobin.**
Hemoglobin consists of four globular protein subunits.
Each subunit contains a single molecule of heme, a ring
surrounding a single ion of iron.

Hemoglobin molecule

19

has separated from the oxygen molecule is called *de-oxyhemoglobin.* Blood containing RBCs filled with deoxyhemoglobin is dark red, almost burgundy.

The RBCs of an embryo or fetus contain a different form of hemoglobin, known as *fetal hemoglobin,* which binds oxygen more readily than does the hemoglobin of adults. For this reason, the developing fetus can "steal" oxygen from the maternal bloodstream at the placenta. The conversion from fetal hemoglobin to the adult form begins shortly before birth and continues over the next year. The production of fetal hemoglobin can be stimulated in adults by the administration of drugs such as *hydroxyurea* or *butyrate.* This is one method of treatment for conditions, such as *sickle cell anemia* or *thalassemia,* that result from the production of abnormal forms of adult hemoglobin.

Hemoglobin Function There are about 280 million Hb molecules in each red blood cell. Because a Hb molecule contains four heme units, each RBC can potentially carry more than a billion molecules of oxygen. Roughly 98.5 percent of the oxygen carried by the blood travels through the bloodstream bound to Hb molecules inside RBCs.

The amount of oxygen bound to hemoglobin depends primarily on the oxygen content of the plasma. When plasma oxygen levels are low, hemoglobin re-

leases oxygen. Under these conditions, typical of peripheral capillaries, plasma carbon dioxide levels are elevated. The alpha and beta chains of hemoglobin then bind carbon dioxide, forming **carbaminohemoglobin.** In the capillaries of the lungs, plasma oxygen levels are high and carbon dioxide levels are low. Upon reaching these capillaries, RBCs absorb oxygen (which is then bound to hemoglobin) and release carbon dioxide. We shall revisit these processes in Chapter 23.

Normal activity levels can be sustained only when tissue oxygen levels are kept within normal limits. If the hematocrit is low or the Hb content of the RBCs is reduced, the condition of **anemia** exists. Anemia interferes with oxygen delivery to peripheral tissues. Every system is affected as organ function deteriorates owing to oxygen starvation. Anemic individuals become weak, lethargic, and often confused as the brain is affected. Anemia occurs in many forms; we will consider specific examples both in this chapter and in the *Applications Manual.*

RBC Life Span and Circulation

A red blood cell is exposed to severe mechanical stresses. A single round-trip from the heart, through the peripheral tissues, and back to the heart usually takes less than a minute. In that time, an RBC gets

Abnormal Hemoglobin

Several inherited disorders are characterized by the production of abnormal hemoglobin. Two of the best known are thalassemia and sickle cell anemia (SCA).

The various forms of **thalassemia** (thal-ah-SĒ-mē-uh) result from an inability to produce adequate amounts of alpha or beta chains of hemoglobin. As a result, the rate of RBC production is slowed, and the mature RBCs are fragile and short-lived. The scarcity of healthy RBCs reduces the oxygen-carrying capacity of the blood and leads to problems with the development and growth of systems throughout the body. Individuals with severe thalassemia must periodically undergo *transfusions*—the administration of blood components—to keep adequate numbers of RBCs in the bloodstream. [AM] *Thalassemia; Transfusions and Synthetic Blood*

Sickle cell anemia results from a mutation affecting the amino acid sequence of the beta chains of the Hb molecule. When blood contains abundant oxygen, the Hb molecules and the RBCs that carry them appear normal (Figure 19-4a•). But when the defective hemoglobin gives up enough of its bound oxygen, the adjacent Hb molecules interact and the cells become stiff and curved (Figure 19-4b•). This "sickling" makes the RBCs fragile and easily damaged. Moreover, an RBC that has folded to squeeze into a narrow capillary delivers its oxygen to the surrounding tissue, but the cell can become stuck as sickling occurs. A circulatory blockage results, and nearby tissues become oxygen-starved. [AM] *Sickle Cell Anemia*

(a) Normal RBC   (b) Sickled RBC

•**FIGURE 19-4** Sickling in Red Blood Cells. **(a)** When fully oxygenated, the cells of an individual with the sickling trait appear relatively normal. **(b)** At lower oxygen concentrations, the RBCs change shape, becoming more rigid and sharply curved. (SEM × 5200)

19

pumped out of the heart and forced along vessels, where it bounces off the walls and collides with other RBCs. It forms stacks, contorts and squeezes through tiny capillaries, and then joins its comrades in a headlong rush back to the heart for another round.

With all this wear and tear and no repair mechanisms, a typical RBC has a relatively short life span. After traveling about 700 miles in 120 days, either the cell membrane ruptures or the damage is detected by phagocytic cells, which engulf the cell. The continuous elimination of RBCs is usually unnoticed because new ones enter the bloodstream at a comparable rate. About 1 percent of the circulating RBCs are replaced each day, and in the process approximately 3 million new RBCs enter the bloodstream *each second!*

Hemoglobin Conservation and Recycling Phagocytic cells of the liver, spleen, and bone marrow monitor the condition of circulating RBCs; the cells generally recognize and engulf RBCs before they **hemolyze,** or rup-

ture. These phagocytes also detect and remove Hb molecules and cell fragments from the relatively small number of RBCs that hemolyze in the bloodstream (about 10 percent of the total recycled each day).

If the Hb released by hemolysis is not phagocytized, its components will not be recycled. Hemoglobin remains intact only under the conditions inside RBCs. When hemolysis occurs, the Hb breaks down and the alpha and beta chains are filtered by the kidneys and eliminated in urine. When abnormally large numbers of RBCs break down in the bloodstream, urine may turn red or brown. This condition is called **hemoglobinuria. Hematuria** (hē-ma-TOO-rē-uh), the presence of intact RBCs in urine, occurs only after kidney damage or damage to vessels along the urinary tract.

Once an RBC has been engulfed and broken down by a phagocytic cell, each component of the Hb molecule has a different fate (Figure 19-5•). The globular proteins are disassembled into their component amino acids. These amino acids are either metabolized by the cell or

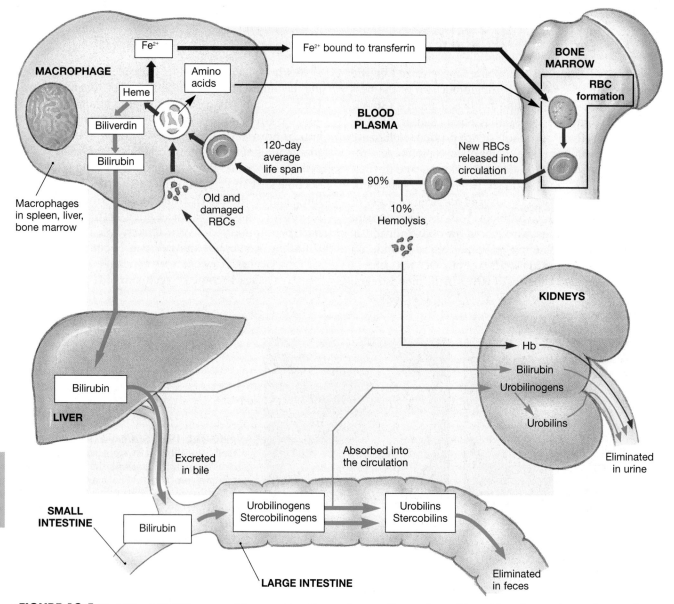

•**FIGURE 19-5 Red Blood Cell Turnover.** The normal pathways for recycling amino acids and iron from aging or damaged RBCs, broken down by macrophages. The amino acids are absorbed, especially by developing cells in bone marrow. The iron is stored in many sites. The rings of the heme units are converted to bilirubin, absorbed by the liver, and excreted in bile or urine; some of the breakdown products produced in the large intestine are recirculated.

released into the bloodstream for use by other cells. Each heme unit is stripped of its iron and converted to **biliverdin** (bil-ē-VER-din), an organic compound with a green color. (Bad bruises commonly develop a green tint due to biliverdin formation in the blood-filled tissues.) Biliverdin is then converted to **bilirubin** (bil-ē-ROO-bin), an orange-yellow pigment, and released into the bloodstream. There, the bilirubin binds to albumin and is transported to the liver for excretion in bile.

If the bile ducts are blocked or the liver cannot absorb or excrete bilirubin, circulating levels of bilirubin climb rapidly. Bilirubin then diffuses into peripheral tissues, giving them a yellow color that is most apparent in the skin and over the sclera of the eyes. This combination of signs (yellow skin and eyes) is called **jaundice** (JAWN-dis). [AM] *Bilirubin Tests and Jaundice*

In the large intestine, bacteria convert bilirubin to related pigments called *urobilinogens* and *stercobilinogens*. Some of the urobilinogens are absorbed into the bloodstream and are subsequently excreted into urine. On exposure to oxygen, some of the urobilinogens and stercobilinogens are converted to **urobilins** and **stercobilins.** Urine is usually yellow because it contains urobilins; feces are yellow-brown or brown owing to the presence of urobilins and stercobilins in varying proportions.

Iron Large quantities of free iron are toxic to cells, so iron in the body is generally bound to transport or storage proteins. Iron extracted from the heme molecules may be bound and stored in a phagocytic cell or released into the bloodstream, where it binds to **transferrin** (tranz-FER-in), a plasma protein. Red blood cells developing in the bone marrow absorb the amino acids and transferrins from the bloodstream and use them to synthesize new Hb molecules. Excess transferrins are removed in the liver and spleen, and the iron is stored in two special protein–iron complexes: **ferritin** (FER-i-tin) and **hemosiderin** (hē-mō-SID-e-rin).

This recycling system is remarkably efficient. Although roughly 26 mg of iron is incorporated into Hb molecules each day, a dietary supply of 1–2 mg can keep pace with the incidental losses that occur at the kidneys and digestive tract.

Any impairment in iron uptake or metabolism can cause serious clinical problems, because RBC formation will be affected. *Iron-deficiency anemia,* which results from a lack of iron in the diet or from problems with iron absorption, is one example. Too much iron can also cause problems owing to excessive buildup in secondary storage sites, such as the liver and cardiac muscle tissue. Excessive iron deposition in cardiac muscle cells has also been linked to heart disease.

AM *Iron Deficiences and Excesses*

RBC Production

Red blood cell formation, or **erythropoiesis** (e-rith-rō-poy-Ē-sis), in adults occurs in the red bone marrow. *Red bone marrow,* or **myeloid** (MĪ-e-loyd) **tissue** (*myelos,* marrow), is located in portions of the vertebrae, sternum, ribs, skull, scapulae, pelvis, and proximal limb bones. Other marrow areas contain a fatty tissue known as *yellow bone marrow.* ∞ *p. 169* Under extreme stimulation, such as severe and sustained blood loss, areas of yellow marrow can convert to red marrow, increasing the rate of RBC formation.

Stages in RBC Maturation During its maturation, a red blood cell passes through a series of stages. **Hematologists** (hē-ma-TOL-o-jists), specialists in blood formation and function, have given specific names to key stages. Divisions of hemocytoblasts in bone marrow produce (1) **myeloid stem cells,** which in turn divide to produce red blood cells and several classes of white blood cells, and (2) **lymphoid stem cells,** which divide to produce the various classes of lymphocytes. Cells destined to become RBCs first differentiate into **proerythroblasts** and then proceed through various **erythroblast** stages (Figure 19-6•). Erythroblasts actively synthesize hemoglobin. They are categorized on the basis of total size, the amount of hemoglobin present, and the size and appearance of the nucleus.

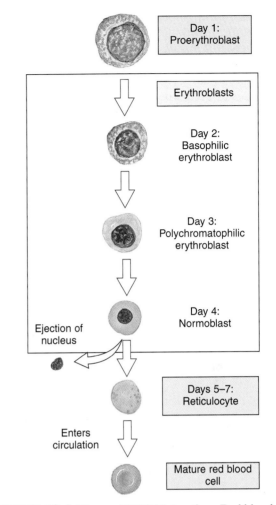

•**FIGURE 19-6 Stages of RBC Maturation.** Red blood cells are produced in the red bone marrow. The color density indicates the abundance of hemoglobin in the cytoplasm. Note the reduction in the size of the cell and the size of the nucleus before a reticulocyte is formed.

19

After roughly 4 days of differentiation, the erythroblast, now called a *normoblast,* sheds its nucleus and becomes a **reticulocyte** (re-TIK-ū-lō-sīt). A reticulocyte contains 80 percent of the Hb of a mature RBC, and Hb synthesis continues for 2–3 more days. During this period, while the cells are synthesizing hemoglobin and other proteins, their cytoplasm still contains RNA, which can be seen under the microscope with certain stains. After 2 days in the bone marrow, reticulocytes enter the bloodstream. At this time, reticulocytes normally account for about 0.8 percent of the RBC population in the blood and can still be detected by staining. After 24 hours in circulation, the reticulocytes complete their maturation and become indistinguishable from other mature RBCs.

Regulation of Erythropoiesis For erythropoiesis to proceed normally, the red bone marrow must receive adequate supplies of amino acids, iron, and vitamins (including B_{12}, B_6, and folic acid) required for protein synthesis. For example, we obtain **vitamin B₁₂** from dairy products and meat, and its absorption requires the presence of *intrinsic factor* produced in the stomach. If vitamin B₁₂ is not obtained from the diet, normal stem cell divisions cannot occur and *pernicious anemia* results. Thus, pernicious anemia is caused by a vitamin B₁₂ deficiency, by a problem with the production of intrinsic factor, or by a problem with the absorption of vitamin B₁₂ bound to intrinsic factor.

Erythropoiesis is stimulated directly by the peptide hormone erythropoietin (∞ *p. 604*) and indirectly by several hormones, including thyroxine, androgens, and growth hormone. As we noted above, estrogens do not stimulate erythropoiesis.

Erythropoietin, also called **EPO** or **erythropoiesis-stimulating hormone,** is a glycoprotein that appears in the plasma when peripheral tissues, especially the kidneys, are exposed to low oxygen concentrations. The state of low tissue oxygen levels is called **hypoxia** (hī-POKS-ē-uh; *hypo-,* below + *ox-,* presence of oxygen). Erythropoietin is released (1) during anemia, (2) when blood flow to the kidneys declines, (3) when the oxygen content of air in the lungs declines, owing to disease or to high altitudes, and (4) when the respiratory surfaces of the lungs are damaged. Once in the bloodstream, EPO travels to areas of red bone marrow, where it stimulates stem cells and developing RBCs.

Erythropoietin has two major effects: (1) It stimulates increased cell division rates in erythroblasts and in the stem cells that produce erythroblasts; and (2) it speeds up the maturation of RBCs, mainly by accelerating the rate of Hb synthesis. Under maximum EPO stimulation, bone marrow can increase the rate of RBC formation tenfold, to about 30 million per second.

This reserve is important to a person recovering from a severe blood loss. But if EPO is administered to a healthy individual, as in the case of the cyclists mentioned in Chapter 18, the hematocrit may rise to 65 or more. ∞ *p. 611* This increase can place an intolerable strain on the heart. Comparable problems can occur after **blood doping.** In this practice, athletes reinfuse packed RBCs removed at an earlier date and stored in hopes of increasing hematocrit to improve performance. [AM] *Erythrocytosis and Blood Doping*

Blood Tests and RBCs

Blood tests provide information about the general health of an individual, usually with a minimum of trouble and expense. Several common blood tests focus on red blood cells, the most abundant formed elements. These *RBC tests* assess the number, size, shape, and maturity of circulating RBCs. Such assessment provides an indication of the erythropoietic activities under way, and it can also be useful in detecting problems, such as internal bleeding, that may not produce other obvious symptoms. Table 19-1 lists examples of important blood tests and related terms. (See the *Applications Manual* for sample calculations.) [AM] *Blood Tests and RBCs*

☑ How would the hematocrit change after an individual suffered a hemorrhage?

☑ Dave develops a blockage in his renal arteries that restricts blood flow to the kidneys. Will his hematocrit change?

☑ How would the level of bilirubin in the blood be affected by a disease that causes damage to the liver?

Blood Types

Antigens are substances that can trigger an *immune response,* a defense mechanism that protects you from infection. Most antigens are proteins, although some

---

TABLE 19-1 **RBC Tests and Related Terminology**

Test	Determines	Terms Associated with Abnormal Values	
		Elevated	**Depressed**
Hematocrit (Hct)	Percentage of formed elements in whole blood Normal = 37–54	Polycythemia (may reflect erythrocytosis or leukocytosis)	Anemia
Reticulocyte count (Retic.)	Circulating percentage of reticulocytes Normal = 0.8%	Reticulocytosis	
Hemoglobin concentration (Hb)	Concentration of hemoglobin in blood Normal = 12–18 g/dl		Anemia
RBC count	Number of RBCs per μl of whole blood Normal = 4.2–6.3 million/μl	Erythrocytosis/polycythemia	Anemia
Mean corpuscular volume (MCV)	Average volume of single RBC Normal = 82–101 μm^3 (normocytic)	Macrocytic	Microcytic
Mean corpuscular hemoglobin concentration (MCHC)	Average amount of Hb in one RBC Normal = 27–34 pg/μl (normochromic)	Hyperchromic	Hypochromic

other types of organic molecules are antigens as well. Your cell membranes contain **surface antigens,** substances that your immune defenses recognize as "normal." In other words, your immune system ignores these substances rather than attacking them as "foreign." Your **blood type** is a classification determined by the presence or absence of specific surface antigens in the RBC cell membranes. The surface antigens involved are integral membrane glycoproteins or glycolipids whose characteristics are genetically determined. The surface antigens of RBCs are often called **agglutinogens** (a-gloo-TIN-ō-jenz). Red blood cells have at least 50 kinds of surface antigens. Three of particular importance have been designated as surface antigens **A, B,** and **Rh** (or **D**).

The RBCs of an individual have (1) either A or B surface antigens, (2) both A and B surface antigens, or (3) neither A nor B surface antigens (Figure 19-7a•).

Type A blood has surface antigen A only; **Type B** has surface antigen B only; **Type AB** has both A and B; and **Type O** has neither A nor B. These blood types are not evenly distributed throughout the population. The average values for the U.S. population are as follows: Type O, 46 percent; Type A, 40 percent; Type B, 10 percent; and Type AB, 4 percent (Table 19-2).

The term **Rh-positive** (Rh$^+$) indicates the presence of the Rh surface antigen, sometimes called the *Rh factor.* The absence of this antigen is indicated as **Rh-negative** (Rh$^-$). When the complete blood type is recorded, the term *Rh* is usually omitted and the data are reported as O-negative (O$^-$), A-positive (A$^+$), and so on. As in the distribution of A and B surface antigens, Rh type differs by race and by region (Table 19-2).

You are probably aware that your blood type must be checked before you can give or receive blood. Your immune system ignores the surface antigens on your

19

(a)

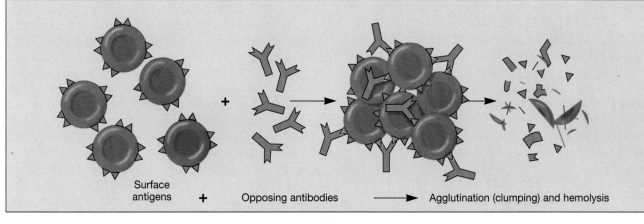

(b)

•**FIGURE 19-7 Blood-Typing and Cross-Reactions.** The blood type depends on the presence of surface antigens (agglutinogens) on RBC surfaces. **(a)** The plasma contains antibodies (agglutinins) that will react with foreign surface antigens. The relative frequencies of each blood type in the U.S. population are indicated in Table 19-2. **(b)** In a cross-reaction, antibodies that encounter their target antigens lead to agglutination and hemolysis of the affected RBCs.

TABLE 19-2 Differences in Blood Group Distribution

Population	Percentage with Each Blood Type				
	O	**A**	**B**	**AB**	**Rh⁺**
U.S. (average)	46	40	10	4	85
African American	49	27	20	4	95
Caucasian	45	40	11	4	85
Chinese American	42	27	25	6	100
Filipino American	44	22	29	6	100
Hawaiian	46	46	5	3	100
Japanese American	31	39	21	10	100
Korean American	32	28	30	10	100
Native North American	79	16	4	<1	100
Native South American	100	0	0	0	100
Australian Aborigine	44	56	0	0	100

own RBCs. However, your plasma contains antibodies, sometimes called **agglutinins** (a-GLOO-ti-ninz), that will attack the antigens on "foreign" RBCs. When the agglutinins attack, the foreign cells **agglutinate,** or clump together; the process is called **agglutination.** The plasma of a Type A, Type B, or Type O individual always contains either anti-A or anti-B antibodies, or both, even if the person has never been exposed to RBCs that carry foreign surface antigens. If you have Type A blood, your plasma contains anti-B antibodies, which will attack Type B surface antigens. If you have Type B blood, your plasma contains anti-A antibodies. The RBCs of an individual with Type O blood have neither A nor B surface antigens, and that person's plasma contains both anti-A and anti-B antibodies. A Type AB individual has RBCs with both A and B surface antigens, and the plasma does not contain anti-A or anti-B antibodies.

In contrast to the situation with surface antigens A and B, the plasma of an Rh-negative individual does not necessarily contain anti-Rh antibodies. These antibodies are present only if the individual has been **sensitized** by previous exposure to Rh-positive RBCs. Such exposure can occur accidentally during a transfusion, but it can also accompany a seemingly normal pregnancy involving an Rh-negative mother and an Rh-positive fetus. (See the box "Hemolytic Disease of the Newborn" on pages 638–639.)

Cross-Reactions   When an antibody meets its specific surface antigen, the RBCs agglutinate and may also hemolyze. This reaction is called a **cross-reaction** (Figure 19-7b●). For instance, an anti-A antibody that encounters A surface antigens will cause the RBCs bearing the surface antigens to clump or even break

up. Clumps and fragments of RBCs under attack form drifting masses that can plug small blood vessels in the kidneys, lungs, heart, or brain, damaging or destroying dependent tissues. Such cross-reactions, or *transfusion reactions,* can be prevented by ensuring that the blood types of the donor and the recipient are **compatible;** that is, the donor's blood cells will not cross-react with the plasma of the recipient. Unless large volumes of whole blood or plasma are transferred, cross-reactions between the donor's plasma and the recipient's blood cells will fail to produce significant agglutination. Packed RBCs, with a minimal amount of plasma, are commonly transfused. Even when whole blood is transferred, the plasma will be diluted quickly through mixing with the relatively large plasma volume of the recipient. (One unit of whole blood, 500 ml, contains roughly 275 ml of plasma, or approximately 10 percent of the normal plasma volume.)

Testing for compatibility   Testing for compatibility normally involves two steps: (1) a determination of blood type and (2) a *cross-match test.* At least 50 surface antigens have been identified on RBCs, but the standard test for blood type considers only the three most likely to produce dangerous cross-reactions: A, B, and Rh. The test involves taking drops of blood and mixing them separately with solutions containing anti-A, anti-B, and anti-Rh (anti-D) antibodies. Any cross-reactions are then recorded. For example, if an individual's RBCs clump together when exposed to anti-A and to anti-B antibodies, the individual has Type AB blood. If no reactions occur after exposure, that person must have Type O blood. The presence or absence of the Rh surface antigen is also noted, and the individual is classified as Rh-positive or Rh-negative on that basis. Type O⁺ is the most common blood type. Although the RBCs of Type O⁺ individuals do not have surface antigens A or B, these cells do have the Rh antigen.

Standard blood-typing of both donor and recipient can be completed in a matter of minutes. However, in an emergency, there may not be time for preliminary testing. For example, a person with a severe gunshot wound may require 5 *liters* or more of blood before the damage can be repaired. Under these circumstances, Type O blood (preferably O⁻) will be administered. Because the donated RBCs lack both A and B surface antigens, the recipient's blood can have anti-A antibodies, anti-B antibodies, or both and still not cross-react with the donor's blood. (More than 100,000 units of Type O⁻ blood were transfused under emergency conditions during the war in Vietnam, with no deaths due to cross-reactions.) Because cross-reactions with Type O blood are so unlikely, Type O individuals are sometimes called *universal donors.* Type AB individuals were once called *universal recipients,* because they do not have anti-A or anti-B antibodies that would attack donated RBCs. This term has largely been dropped, primarily because if the recipient's blood type is known to be AB, Type AB blood can be administered.

It is now possible to use enzymes to strip off the A or B surface antigens from RBCs and create Type O blood in the laboratory. The procedure is expensive and time-consuming and has limited use in emergency treatment. In addition, because at least 48 other surface antigens are present, cross-reactions can still occur, even to Type O⁻ blood. As a result, whenever time and facilities permit, further testing is performed to ensure complete compatibility between donor blood and recipient blood. **Cross-match testing** involves exposing the donor's RBCs to a sample of the recipient's plasma under controlled conditions. This procedure reveals the presence of significant cross-reactions involving surface antigens other than A, B, or Rh. It is also possible to replace lost blood with synthetic blood substitutes.

Because blood groups are inherited, blood tests are also used as paternity tests and in crime detection. The blood collected cannot prove that a particular individual *is* a certain child's father or *is* guilty of a specific crime, but it can prove that the individual is *not* involved. For example, it is impossible for an adult with Type AB blood to be the parent of an infant with Type O blood.

Transfusions

Changes in blood volume have direct effects on other body systems. Clinicians use the terms **hypovolemic** (hī-pō-vō-LĒ-mik), **normovolemic** (nor-mō-vō-LĒ-mik), and **hypervolemic** (hī-per-vō-LĒ-mik) to refer to low, normal, and excessive blood volumes, respectively. The hypovolemic and hypervolemic conditions are potentially dangerous, because variations in blood volume affect other components of the cardiovascular system. For example, an abnormally large blood volume can place a severe stress on the heart, which must keep the extra fluid circulating through the lungs and throughout the body.

Hypovolemia is clinically much more common than hypervolemia. Short-term therapies to treat hypovolemia may include the use of a plasma expander (p. 627) or the infusion of blood, a procedure known as a **transfusion.** In a transfusion, blood components are provided to restore blood volume or to remedy a deficiency in blood composition. Whole blood, packed red blood cells, plasma, platelets, extracted proteins, clotting factors, or enzymes may be administered.

Two major complications from transfusions are (1) cross-reactions between plasma antibodies and donated red blood cells (a transfusion reaction) and (2) infection from bacterial or viral pathogens in the donated blood. Improved cross-match testing and screening for infected blood have helped reduce the frequency of transfusion-related problems.

AM *Transfusions and Synthetic Blood*

☑ What are surface antigens on RBCs?

☑ Which blood type(s) can be transfused into a person with Type O blood?

☑ Why can't a person with Type A blood safely receive blood from a person with Type B blood?

WHITE BLOOD CELLS

Unlike red blood cells, white blood cells have nuclei and other organelles, and these cells lack hemoglobin. White blood cells (WBCs), or leukocytes, help defend the body against invasion by pathogens, and they remove toxins, wastes, and abnormal or damaged cells. Traditionally, WBCs have been divided into two groups on the basis of their appearance after staining:[1] (1) *granular leukocytes,* or *granulocytes* (with abundant stained granules)—the *neutrophils, eosinophils,* and *basophils;* and (2) *agranular leukocytes,* or *agranulocytes* (with few if any stained granules)—the *monocytes* and *lymphocytes.* This categorization is convenient but somewhat misleading, because the granules in "granular leukocytes" are secretory vesicles and lysosomes, and the "agranular leukocytes" contain vesicles and lysosomes as well; they are just smaller and difficult to see with the light microscope.

A typical microliter of blood contains 6000–9000 WBCs. Most of the WBCs in the body at any moment are in connective tissues proper or in organs of the lymphatic system. Circulating WBCs thus represent only a small fraction of the total WBC population.

WBC Circulation and Movement

Unlike RBCs, WBCs circulate for only a short portion of their life span. White blood cells migrate through the loose and dense connective tissues of the body. They use the bloodstream primarily to travel from one organ to another and for rapid transportation to areas of invasion or injury. As they travel along the miles of capillaries, WBCs can detect the chemical signs of damage to surrounding tissues. When problems are detected, these cells leave the bloodstream and enter the damaged area.

Circulating WBCs have four characteristics:

1. *They are capable of amoeboid movement. Amoeboid movement* is a gliding motion accomplished by the flow of cytoplasm into a slender cellular process extended in front of the cell. The mechanism is not fully understood, but it involves the continuous rearrangement of bonds between actin filaments in the cytoskeleton, and it requires calcium ions and ATP. This mobility allows WBCs to move along the walls of

---

[1] The stains used in standard blood work are *Wright's stain* and the *Giemsa stain.*

19

Hemolytic Disease of the Newborn

Genes controlling the presence or absence of any surface antigen in the membrane of a red blood cell are provided by both parents, so a child can have a blood type different from that of either parent. During pregnancy, when fetal and maternal circulatory systems are closely intertwined, the mother's antibodies may cross the placenta, attacking and destroying fetal RBCs. The resulting condition is called **hemolytic disease of the newborn (HDN).**

This disease has many forms, some so mild as to remain undetected. Those involving the Rh surface antigen are potentially quite dangerous, because the anti-Rh antibodies are able to cross the placenta and enter the fetal bloodstream, whereas the anti-A and anti-B antibodies are not. An Rh-positive mother (who lacks anti-Rh antibodies) can carry an Rh-negative fetus without difficulty. Because maternal antibodies can cross the placenta, potential problems appear when an Rh-negative woman carries an Rh-positive fetus. Sensitization generally occurs at delivery, when bleeding takes place at the placenta and uterus. This event can trigger the mother's immune system to produce anti-Rh antibodies. Within 6 months of delivery, roughly 20 percent of Rh-negative mothers who carried Rh-positive children have become sensitized.

Because the anti-Rh antibodies are not produced in significant amounts until after delivery, the first infant is not affected. (Some fetal RBCs cross into the maternal bloodstream during pregnancy but generally not in numbers sufficient to stimulate antibody production.) But if a second pregnancy occurs involving an Rh-positive fetus, maternal anti-Rh antibodies produced after the first delivery cross the placenta and enter the fetal bloodstream (Figure 19-8•). These an-

tibodies destroy fetal RBCs and produce a dangerous anemia. The fetal demand for blood cells increases, and they begin leaving the bone marrow and entering the bloodstream before completing their development. Because these immature RBCs are erythroblasts, HDN is also known as **erythroblastosis fetalis** (e-rith-rō-blas-TŌ-sis fē-TAL-is).

Without treatment, the fetus will probably die before delivery or shortly thereafter. A newborn with severe HDN is anemic, and the high concentration of circulating bilirubin produces jaundice. Because the maternal antibodies will remain active for 1–2 months after delivery, the entire blood volume of the infant may have to be replaced. This blood replacement removes most of the maternal anti-Rh antibodies as well as the affected RBCs, reducing the complications and the chance of death.

If the fetus is in danger of not surviving to full term, premature delivery may be induced after 7–8 months of development. In a severe case affecting a fetus at an earlier stage, one or more transfusions can be given while the fetus continues to develop in the uterus.

The maternal production of anti-Rh antibodies can be prevented by administering anti-Rh antibodies (available under the name *RhoGam*) to the mother in the last 3 months of pregnancy and during and after delivery. (It is also given after a miscarriage or an abortion.) These antibodies will destroy any fetal RBCs that cross the placenta before they can stimulate an immune response in the mother. Sensitization therefore does not occur, so no anti-Rh antibodies are produced. This relatively simple procedure could almost entirely prevent HDN mortality caused by Rh incompatibilities.

blood vessels and, when outside the bloodstream, through surrounding tissues.

2. *They can migrate out of the bloodstream.* When white blood cells in the bloodstream become activated, they contact and adhere to the vessel walls. This process is called *margination.* After further interaction with the endothelial cells, the activated WBCs squeeze between adjacent endothelial cells and enter the surrounding tissue. This process is called *emigration,* or *diapedesis.*

3. *They are attracted to specific chemical stimuli.* This characteristic, called **positive chemotaxis** (kē-mō-TAK-sis), guides WBCs to invading pathogens, damaged tissues, and active WBCs.

4. *Neutrophils, eosinophils, and monocytes are capable of phagocytosis.* These cells may engulf pathogens, cell debris, or other materials. Neutrophils and eosinophils are sometimes called *microphages,* to distinguish them from the larger macrophages in connective tissues. Macrophages are monocytes that have moved out of the bloodstream and have become actively phagocytic.

General Functions

Neutrophils, eosinophils, basophils, and monocytes contribute to the body's *nonspecific defenses.* Such immune defenses are activated by a variety of stimuli, but they do not discriminate between one type of threat

•**FIGURE 19-8 Rh Factors and Pregnancy.** When an Rh-negative woman has her first Rh-positive child, fetal and maternal blood mix at delivery when the placenta breaks down. The appearance of Rh-positive blood cells in the maternal bloodstream sensitizes the mother, stimulating the production of anti-Rh antibodies. If another pregnancy occurs with an Rh-positive fetus, maternal anti-Rh antibodies can cross the placenta and attack fetal blood cells, producing hemolytic disease of the newborn (HDN).

and another. Lymphocytes, in contrast, are responsible for *specific defenses:* the mounting of a counterattack against particular invading pathogens or foreign proteins on an individual basis. We will discuss the interactions among WBCs and the relationships between specific and nonspecific defenses in Chapter 22.

Neutrophils

Fifty to seventy percent of the circulating WBCs are **neutrophils** (NOO-trō-filz). This name reflects the fact that their granules are chemically neutral and thus are difficult to stain with either acidic or basic dyes. A ma-

ture neutrophil has a very dense, segmented nucleus that forms two to five lobes resembling beads on a string (Figure 19-9a•). This trait has given neutrophils another name, **polymorphonuclear** (pol-ē-mor-fō-NOO-klē-ar) **leukocytes** (*poly,* many + *morphe,* form), or **PMNs.** "Polymorphs," or "polys," as they are often called, are roughly 12 μm in diameter. Their cytoplasm is packed with pale granules containing lysosomal enzymes and bactericidal (bacteria-killing) compounds.

Neutrophils are highly mobile; they are generally the first of the WBCs to arrive at an injury site. They are very active cells that specialize in attacking and digesting bacteria that have been "marked" with

(a) Neutrophil

(b) Eosinophil

(c) Basophil

(d) Monocyte

(e) Lymphocyte

•**FIGURE 19-9** White Blood Cells. (LMs × 1500)

antibodies or with *complement proteins,* plasma proteins involved in tissue defenses. (We shall discuss the complement system in Chapter 22.)

Upon encountering a bacterium, the neutrophil quickly engulfs it, and the metabolic rate of the neutrophil increases dramatically. This *respiratory burst* accompanies the production of destructive chemical agents, including *hydrogen peroxide* (H_2O_2) and *superoxide anions* (O_2^-). These highly reactive chemicals can kill bacteria.

Meanwhile, the vesicle containing the engulfed pathogen fuses with lysosomes that contain digestive enzymes and small peptides called **defensins.** This process, which reduces the number of granules in the cytoplasm, is called **degranulation.** Defensins kill a variety of pathogens, including bacteria, fungi, and some viruses, by combining to form large channels in their cell membranes. The digestive enzymes then break down the bacterial remains. While actively engaged in attacking bacteria, a neutrophil releases prostaglandins and leukotrienes. ∞ *p. 580* The prostaglandins increase capillary permeability in the affected region, which contributes to local inflammation and thereby restricts the spread of injury and infection. Leukotrienes are hormones of the immune system that attract other phagocytes and help coordinate the immune response.

Most neutrophils have a short life span, surviving in the bloodstream for only about 10 hours. When actively engulfing debris or pathogens, they may last 30 minutes or less. A neutrophil dies after engulfing 1–2 dozen bacteria, but its breakdown releases additional chemicals that attract other neutrophils to the site.

Eosinophils

Eosinophils (ē-ō-SIN-ō-filz) were so named because their granules stain darkly with *eosin,* a red dye. The granules will also stain with other acid dyes, so the name **acidophils** (a-SID-ō-filz) applies as well. Eosinophils generally represent 2–4 percent of the circulating WBCs. These cells are similar in size to neutrophils, but the combination of deep red granules and a bilobed (two-lobed) nucleus makes an eosinophil easy to identify (Figure 19-9b•).

Eosinophils attack objects that are coated with antibodies. They are phagocytic cells and will engulf antibody-marked bacteria, protozoa, or cellular debris. However, their primary mode of attack is the exocytosis of toxic compounds, including nitric oxide and cytotoxic enzymes, onto the surface of their targets. Eosinophils defend against large multicellular parasites, such as flukes or parasitic worms; these WBCs increase in number dramatically during a parasitic infection.[2] Because they are sensitive to circulating

[2] To review the major classes of pathogens, see "A Closer Look: The Nature of Pathogens" in Unit 2 of the *Applications Manual.*

allergens (materials that trigger allergies), eosinophils increase in number during allergic reactions as well. Eosinophils are also attracted to sites of injury, where they release enzymes that reduce the degree of inflammation and control its spread to adjacent tissues.

Basophils

Basophils (BĀ-sō-filz) have numerous granules that stain darkly with basic dyes. In a standard blood smear, the inclusions are deep purple or blue (Figure 19-9c•). Measuring 8–10 μm in diameter, basophils are smaller than neutrophils or eosinophils. They are also relatively rare, accounting for less than 1 percent of the circulating WBC population.

Basophils migrate to injury sites and cross the capillary endothelium to accumulate in the damaged tissues. There, they discharge their granules into the interstitial fluids. The granules contain *histamine,* which dilates blood vessels, and *heparin,* a compound that prevents blood clotting. Stimulated basophils release these chemicals into the interstitial fluids, and their arrival enhances the local inflammation initiated by mast cells. ∞ *pp. 119, 135* Although the same compounds are released by mast cells in damaged connective tissues, mast cells and basophils are distinct populations with separate origins. Other chemicals released by stimulated basophils attract eosinophils and other basophils to the area.

Monocytes

Monocytes (MON-ō-sīts) in blood are spherical cells that may exceed 15 μm in diameter, nearly twice the diameter of a typical red blood cell. When flattened in a blood smear, they look even larger, so monocytes are relatively easy to identify (Figure 19-9d•). The nucleus is large and tends to be oval or kidney bean–shaped. Monocytes normally account for 2–8 percent of the circulating WBCs.

An individual monocyte uses the bloodstream as a highway, remaining in circulation for only about 24 hours before entering peripheral tissues to become a tissue macrophage. Macrophages are aggressive phagocytes, often attempting to engulf items as large as or larger than themselves. While active, they release chemicals that attract and stimulate neutrophils, monocytes, and other phagocytic cells. Active macrophages also secrete substances that lure fibroblasts into the region. The fibroblasts then begin producing scar tissue, which will wall off the injured area.

Lymphocytes

Typical **lymphocytes** (LIM-fō-sīts) are slightly larger than RBCs and lack abundant, deeply stained granules. In fact, when you see a lymphocyte in a blood smear, you generally see just a thin halo of cytoplasm around a relatively large nucleus (Figure 19-9e•).

Lymphocytes account for 20–30 percent of the WBC population of blood. Lymphocytes continuously migrate from the bloodstream, through peripheral tissues, and back to the bloodstream. Circulating lymphocytes represent only a minute fraction of all lymphocytes, for at any moment most of your body's lymphocytes are in other connective tissues and in organs of the lymphatic system.

The circulating blood contains three classes of lymphocytes, which cannot be distinguished with a light microscope:

1. *T cells.* **T cells** are responsible for *cell-mediated immunity,* a defense mechanism against invading foreign cells and tissues, and for the coordination of the immune response. T cells either enter peripheral tissues and attack foreign cells directly or stimulate or inhibit the activities of other lymphocytes.

2. *B cells.* **B cells** are responsible for *humoral immunity,* a defense mechanism that involves the production and distribution of antibodies, which in turn attack foreign antigens throughout the body. Activated B cells differentiate into **plasma cells,** which are specialized for the synthesis and secretion of antibodies. Whereas the T cells responsible for cellular immunity must migrate to their targets, the antibodies produced by plasma cells in one location can destroy antigens almost anywhere in the body.

3. *NK cells.* **Natural killer (NK) cells** are responsible for *immune surveillance,* the detection and subsequent destruction of abnormal tissue cells. These cells, sometimes known as *large granular lymphocytes,* are important in preventing cancer.

The Differential Count and Changes in WBC Profiles

A variety of disorders, including pathogenic infection, inflammation, and allergic reactions, cause characteristic changes in circulating populations of WBCs. By examining a stained blood smear, we can obtain a **differential count** of the WBC population. The values reported indicate the number of each type of cell in a sample of 100 WBCs.

The normal range for each cell type is indicated in Table 19-3. The term **leukopenia** (loo-kō-PĒ-nē-uh; *penia,* poverty) indicates inadequate numbers of WBCs. **Leukocytosis** (loo-kō-sī-TO-sis) refers to excessive numbers of WBCs. A modest leukocytosis is normal during an infection. Extreme leukocytosis (100,000/μl or more) generally indicates the presence of some form of **leukemia** (loo-KĒ-mē-uh). There are many types of leukemias. Treatment helps in some cases; unless treated, all are fatal. The endings *-penia* and *-osis* can also indicate low or high numbers of specific types of WBCs. For example, *lymphopenia* means too few lymphocytes, and *lymphocytosis* means too many. [AM] *The Leukemias*

TABLE 19-3 A Review of the Formed Elements of the Blood

Cell	Abundance (average number per μl)	Appearance in a Stained Blood Smear	Functions	Remarks
Red Blood Cells	5.2 million (range: 4.4–6.0 million)	Flattened, circular cell; no nucleus, mitochondria, or ribosomes; red	Transport oxygen from lungs to tissues and carbon dioxide from tissues to lungs	Remain in bloodstream; 120-day life expectancy; amino acids and iron recycled; produced in bone marrow
White Blood Cells	7000 (range: 6000–9000)			
Neutrophils	4150 (range: 1800–7300) Differential count: 50–70%	Round cell; nucleus lobed and may resemble a string of beads; cytoplasm contains large, pale inclusions	Phagocytic: Engulf pathogens or debris in tissues, release cytotoxic enzymes and chemicals	Move into tissues after several hours; may survive minutes to days, depending on tissue activity; produced in bone marrow
Eosinophils	165 (range: 0–700) Differential count: 2–4%	Round cell; nucleus generally in two lobes; cytoplasm contains large granules that generally stain bright red	Phagocytic: Engulf antibody-labeled materials, release cytotoxic enzymes, reduce inflammation	Move into tissues after several hours; survive minutes to days, depending on tissue activity; produced in bone marrow
Basophils	44 (range: 0–150) Differential count: <1%	Round cell; nucleus generally cannot be seen through dense, blue-stained granules in cytoplasm	Enter damaged tissues and release histamine and other chemicals that promote inflammation	Survival time unknown; assist mast cells of tissues in producing inflammation; produced in bone marrow
Monocytes	456 (range: 200–950) Differential count: 2–8%	Very large cell; kidney bean–shaped nucleus; abundant pale cytoplasm	Enter tissues to become macrophages; engulf pathogens or debris	Move into tissues after 1–2 days; survive for months or longer; primarily produced in bone marrow
Lymphocytes	2185 (range: 1500–4000) Differential count: 20–30%	Generally round cell, slightly larger than RBC; round nucleus; very little cytoplasm	Cells of lymphatic system, providing defense against specific pathogens or toxins	Survive for months to decades; circulate from blood to tissues and back; produced in bone marrow and lymphoid tissues
Platelets	350,000 (range: 150,000–500,000)	Round to spindle-shaped cytoplasmic fragment; contains enzymes and proenzymes; no nucleus	Hemostasis: Clump together and stick to vessel wall (platelet phase); activate intrinsic pathway of coagulation phase	Remain in bloodstream or in vascular organs; remain intact for 7–12 days; produced by megakaryocytes in bone marrow

WBC Production

Stem cells responsible for the production of WBCs originate in the bone marrow, with the divisions of hemocytoblasts. As we noted earlier, hemocytoblast divisions produce myeloid stem cells and lymphoid stem cells. Myeloid stem cell division creates **progenitor cells,** which give rise to all the formed elements except lymphocytes. One type of progenitor cell produces daughter cells that mature into RBCs; a second type produces cells that manufacture platelets.

Neutrophils, eosinophils, basophils, and monocytes develop from daughter cells produced by a third type of progenitor cell. All but the monocytes complete their development in the bone marrow. Monocytes begin their differentiation in the bone marrow, enter the bloodstream, and complete development when they become free macrophages in peripheral tissues. Each of the other cell types goes through a characteristic series of maturational stages, proceeding from *blast cells* to *myelocytes* to *band cells* before becoming mature WBCs. For example, a cell differentiating into a neutrophil goes from a myeloblast to a *neutrophilic myelocyte* and then becomes a *neutrophilic band cell.*

Some band cells enter the bloodstream before completing their maturation; normally 3–5 percent of all circulating WBCs are band cells. Figure 19-10● summarizes the relationships among the various WBC populations and compares the formation of WBCs and RBCs.

Many of the lymphoid stem cells responsible for the production of lymphocytes migrate from the bone

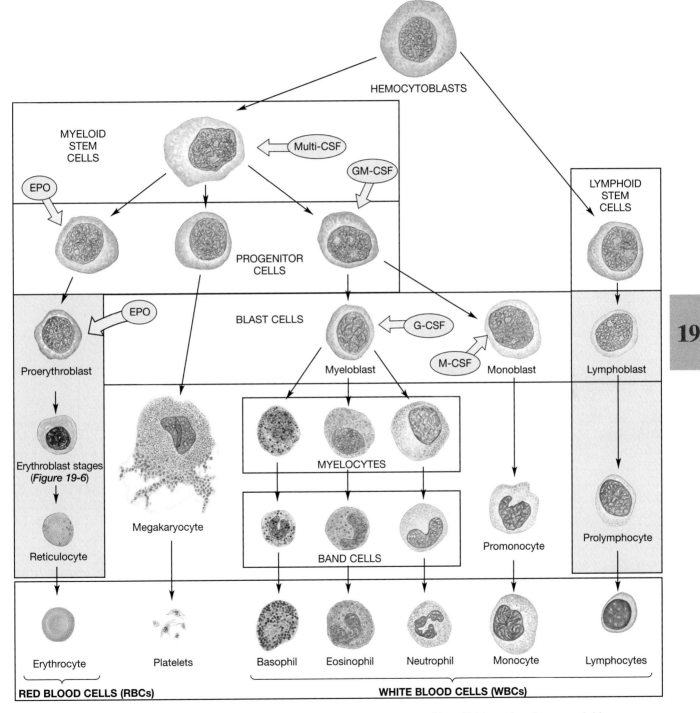

●**FIGURE 19-10 The Origins and Differentiation of Formed Elements.** Hemocytoblast divisions give rise to myeloid or lymphoid stem cells. Lymphoid stem cells produce the various lymphocytes. Myeloid stem cells produce progenitor cells that divide to produce the other classes of formed elements. The targets of EPO and the four colony-stimulating factors (CSFs) are indicated.

marrow to peripheral **lymphoid tissues,** including the thymus, spleen, and lymph nodes. As a result, lymphocytes are produced in these organs as well as in the bone marrow. The process of lymphocyte production is called **lymphopoiesis.**

Regulation of WBC Production   Factors that regulate lymphocyte maturation are as yet incompletely understood. Prior to maturity, hormones produced by the thymus promote the differentiation and maintenance of T cell populations. The importance of the thymus in adulthood, especially in aging, remains controversial. In adults, the production of B and T lymphocytes is regulated primarily by exposure to antigens (foreign proteins, cells, or toxins). When antigens appear, lymphocyte production escalates. We shall describe the control mechanisms in Chapter 22.

Several hormones are involved in the regulation of other WBC populations. Figure 19-10• diagrams the targets of these hormones, called **colony-stimulating factors (CSFs),** and the target of erythropoietin (EPO), discussed on page 634. Four CSFs have been identified, each stimulating the formation of WBCs or both WBCs and RBCs. The abbreviation for each factor indicates its target:

1. **M-CSF** stimulates the production of monocytes.
2. **G-CSF** stimulates the production of granulocytes (neutrophils, eosinophils, and basophils).
3. **GM-CSF** stimulates the production of both granulocytes and monocytes.
4. **Multi-CSF** accelerates the production of granulocytes, monocytes, platelets, and RBCs.

Chemical communication between lymphocytes and other WBCs assists in the coordination of the immune response. For example, active macrophages release chemicals that make lymphocytes more sensitive to antigens and accelerate the development of specific immunity. In turn, active lymphocytes release multi-CSF and GM-CSF, reinforcing nonspecific defenses. The molecular structure of many of the stimulating factors has been determined, and several can be produced by genetic engineering. The U.S. Food and Drug Administration approved the administration of synthesized forms of EPO, G-CSF, and GM-CSF to stimulate the production of specific blood cell lines. For instance, a genetically engineered form of G-CSF, sold under the name *filgrastim (Neupogen),* is used to stimulate the production of neutrophils in patients undergoing cancer chemotherapy.

☑ Which type of white blood cell would you find in the greatest numbers in an infected cut?

☑ Which cell type would you find in elevated numbers in a person who is producing large amounts of circulating antibodies to combat a virus?

☑ How do basophils respond during inflammation?

PLATELETS

Platelets (PLĀT-lets) are flattened discs, round when viewed from above and spindleshaped when seen in section or in a blood smear (Figure 19-9e•). They average about 4 μm in diameter and are roughly 1 μm thick. Platelets in nonmammalian vertebrates are nucleated cells called **thrombocytes** (THROM-bō-sīts; *thrombos,* clot). Because in humans they are cell fragments rather than individual cells, the term *platelet* is preferred when referring to our blood.

Platelets are continuously replaced. Each platelet circulates for 9–12 days before being removed by phagocytes, mainly in the spleen. Each microliter of circulating blood contains 150,000–500,000 platelets; 350,000/μl is the average concentration. Roughly one-third of the platelets in the body at any moment are held in the spleen and other vascular organs rather than in the bloodstream. These reserves are mobilized during a circulatory crisis, such as severe bleeding.

An abnormally low platelet count (80,000/μl or less) is known as **thrombocytopenia** (throm-bō-sī-tō-PĒ-nē-uh). Thrombocytopenia generally indicates excessive platelet destruction or inadequate platelet production. Symptoms include bleeding along the digestive tract, within the skin, and occasionally inside the CNS. In **thrombocytosis** (throm-bō-sī-TŌ-sis), platelet counts can exceed 1,000,000/μl. Thrombocytosis generally results from accelerated platelet formation in response to infection, inflammation, or cancer.

Platelet Functions

Platelets are one participant in a vascular *clotting system* that also includes plasma proteins and the cells and tissues of the circulatory network. The functions of platelets include:

■ *The transport of chemicals important to the clotting process.* By releasing enzymes and other factors at the appropriate times, platelets help initiate and control the clotting process.

■ *The formation of a temporary patch in the walls of damaged blood vessels.* Platelets clump together at an injury site, forming a *platelet plug,* which can slow the rate of blood loss while clotting occurs.

■ *Active contraction after clot formation has occurred.* Platelets contain filaments of actin and myosin. After a blood clot has formed, the contraction of platelet filaments shrinks the clot and reduces the size of the break in the vessel wall.

Platelet Production

Platelet production, or **thrombocytopoiesis,** occurs in the bone marrow. Normal bone marrow contains a number of **megakaryocytes** (meg-a-KAR-ē-ō-sīts; *mega-,* big + *karyon,* nucleus + *-cyte,* cell), enormous cells (up to 160 μm in diameter) with large nuclei. During their development and growth, megakaryocytes manufacture structural proteins, enzymes, and membranes. They then begin shedding cytoplasm in small membrane-enclosed packets. These packets are the platelets that enter the bloodstream. A mature megakaryocyte gradually loses all of its cytoplasm, producing about 4000 platelets before the nucleus is engulfed by phagocytes and broken down for recycling.

The rate of megakaryocyte activity and platelet formation is stimulated by (1) *thrombopoietin* (TPO), or *thrombocyte-stimulating factor,* a peptide hormone produced in the kidneys and perhaps other sites as well, which accelerates platelet formation and stimulates the production of megakaryocytes; (2) *interleukin-6* (IL-6), a hormone of the immune system, which stimulates platelet formation; and (3) multi-CSF, which stimulates platelet production by promoting the formation and growth of megakaryocytes.

HEMOSTASIS

The process of **hemostasis** (*haima,* blood + *stasis,* halt), the cessation of bleeding, prevents the loss of blood through the walls of damaged vessels. At the same time, it establishes a framework for tissue repairs. It consists of three phases: (1) the *vascular phase,* (2) the *platelet phase,* and (3) the *coagulation phase.* Although hemostasis can be analyzed in terms of these three phases, it is more like a chain reaction. Each step modifies the events already under way. As a result, it is easier to say when a particular step begins than when it ends.

THE VASCULAR PHASE

Cutting the wall of a blood vessel triggers a contraction in the smooth muscle fibers of the vessel wall (Figure 19-11•). This local contraction of the vessel is a **vascular spasm,** which decreases the diameter of the vessel at the injury site. Such a constriction can slow or even stop the loss of blood through the wall of a small vessel. The vascular spasm lasts about 30 minutes; this period is called the **vascular phase** of hemostasis.

During the vascular phase, changes occur in the vessel endothelium at the injury site:

- *The endothelial cells contract and expose the underlying basement membrane to the bloodstream.*

- *The endothelial cells begin releasing chemical factors and local hormones.* We shall discuss several of these factors, including *ADP, tissue factor,* and *prostacyclin,* in later sections. Endothelial cells also release **endothelins,** peptide hormones that (1) stimulate smooth muscle contraction and promote vascular spasms and (2) stimulate the division of endothelial cells, smooth muscle fibers, and fibroblasts to accelerate the repair process.

- *The endothelial cell membranes become "sticky."* In small capillaries, endothelial cells on opposite sides of the vessel may stick together and close off the passageway.

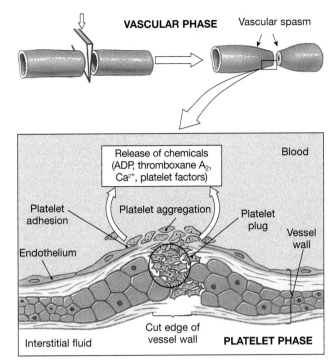

•**FIGURE 19-11** The Vascular and Platelet Phases of Hemostasis

THE PLATELET PHASE

Platelets now begin to attach to sticky endothelial surfaces, to the basement membrane, and to exposed collagen fibers. This attachment marks the start of the **platelet phase** of hemostasis (Figure 19-11•). The attachment of platelets to exposed surfaces is called **platelet adhesion.** As more and more platelets arrive, they begin sticking to one another as well. This process, called **platelet aggregation,** forms a **platelet plug** that may close the break in the vessel wall, if the damage is not severe or the vessel is relatively small. Platelet aggregation begins within 15 seconds after an injury occurs.

As they arrive at the injury site, platelets become activated. The first sign of activation is that they change shape, becoming more spherical and developing cytoplasmic processes that extend toward adjacent platelets. At this time, the platelets begin releasing a wide variety of compounds, including (1) *adenosine diphosphate (ADP),* which stimulates platelet aggregation and secretion; (2) *thromboxane A_2* and *serotonin,* which stimulate vascular spasms; (3) *clotting factors,* proteins that play a role in blood clotting; (4) *platelet-derived growth factor* (PDGF), a peptide that promotes vessel repair; and (5) calcium ions, which are required for platelet aggregation and by several steps in the clotting process.

The platelet phase proceeds rapidly, because the ADP, thromboxane, and calcium ions that each arriving platelet releases stimulate further aggregation. This positive feedback loop ultimately produces a platelet plug that will be reinforced as clotting occurs. However,

19

platelet aggregation must be controlled and restricted to the injury site. Several key factors limit the growth of the platelet plug: (1) **prostacyclin,** a prostaglandin that inhibits platelet aggregation and is released by endothelial cells; (2) inhibitory compounds released by white blood cells entering the area; (3) circulating plasma enzymes that break down ADP near the plug; (4) compounds that, when abundant, inhibit plug formation (for example, serotonin at high concentrations will block the action of ADP); and (5) the development of a blood clot, which reinforces the platelet plug but separates it from the general circulation.

THE COAGULATION PHASE

The vascular and platelet phases begin within a few seconds after the injury. The **coagulation** (cō-ag-ū-LĀ-shun) **phase** does not start until 30 seconds or more after the vessel has been damaged. **Coagulation,** or *blood clotting,* involves a complex sequence of steps leading to the conversion of circulating fibrinogen into the insoluble protein fibrin. As the fibrin network grows, it covers the surface of the platelet plug. Passing blood cells and additional platelets are trapped in the fibrous tangle, forming a **blood clot,** which effectively seals off the damaged portion of the vessel. Figure 19-12• shows the formation and structure of a blood clot.

Clotting Factors

Normal blood clotting cannot occur unless the plasma contains the necessary **clotting factors,** or **procoagulants.** Important clotting factors include Ca^{2+} and 11 different proteins. Many of the proteins are **proenzymes,** which, when converted to active enzymes, direct essential reactions in the clotting response.

For reference, specific clotting factors are identified in Table 19-4 (p. 648). Many are identified by Roman numerals; Ca^{2+}, for example, is also known as clotting Factor IV. All but three of the clotting factors (Factors III, IV, and VIII) are synthesized and released by the liver and are always present in the bloodstream. Activated platelets release five clotting factors (Factors III, IV, V, VIII, and XIII) during the platelet phase. During the coagulation phase, enzymes and proenzymes interact. The activation of one proenzyme commonly creates an enzyme that activates a second proenzyme, and so on in a chain reaction, or *cascade.*

Figure 19-12a• surveys the cascades involved in the *extrinsic, intrinsic,* and *common pathways.* The extrinsic pathway begins outside the bloodstream; the intrinsic pathway begins inside the bloodstream. These two pathways converge at the common pathway. AM

A Closer Look: The Clotting System

The Extrinsic Pathway

The **extrinsic pathway** begins with the release of **Factor III,** also known as **tissue factor** (TF), by damaged endothelial cells or peripheral tissues. The greater the damage, the more tissue factor is released and the faster clotting occurs. Tissue factor then combines with Ca^{2+} and another clotting factor (Factor VII) to form an enzyme complex capable of activating Factor X, the first step in the common pathway.

The Intrinsic Pathway

The **intrinsic pathway** begins with the activation of proenzymes (usually Factor XII) exposed to collagen fibers at the injury site (or a glass surface on a slide or collection tube). This pathway proceeds with the assistance of **PF-3,** a platelet factor released by aggregating platelets. Platelets also release a variety of other factors that accelerate the reactions of the intrinsic pathway. After a series of linked reactions, activated Factors VIII and IX combine to form an enzyme complex capable of activating Factor X.

The Common Pathway

The **common pathway** begins when enzymes from either the extrinsic or the intrinsic pathway activate Factor X, forming the enzyme **prothrombinase.** Prothrombinase converts the proenzyme prothrombin into the enzyme **thrombin** (THROM-bin). Thrombin then completes the clotting process by converting fibrinogen, a plasma protein, to insoluble strands of fibrin.

Interactions among the Pathways

When a blood vessel is damaged, both the extrinsic and the intrinsic pathways respond. The extrinsic pathway is shorter and faster than the intrinsic pathway, and it is usually the first to initiate the clotting process. In essence, the extrinsic pathway produces a small amount of thrombin very quickly. This quick patch is reinforced by the intrinsic pathway, which produces more thrombin, but somewhat later.

The time required to complete clot formation varies with the site and the nature of the injury. In tests of the clotting system, blood held in fine glass tubes normally clots in 8–18 minutes (the *coagulation time*), and a small puncture wound typically stops bleeding in 1–4 minutes (the *bleeding time*).

Feedback Control of Blood Clotting

Thrombin generated in the common pathway stimulates blood clotting by (1) stimulating the formation of tissue factor and (2) stimulating the release of PF-3 by platelets. Thus, the activity of the common pathway stimulates both the intrinsic and extrinsic pathways. This positive feedback loop accelerates the clotting process, and speed can be very important in reducing blood loss after a severe injury.

Blood clotting is restricted by factors that either inactivate or remove clotting factors and other stimulatory agents from the blood. Examples include the following:

(a) The coagulation phase

Platelets

Fibrin network

Trapped RBC

(b) A blood clot

•FIGURE 19-12 The Coagulation Phase of Hemostasis.
(a) Events of the coagulation phase. **(b)** The network of fibrin that forms the framework of a clot. (Color-enhanced SEM × 3561) Red blood cells trapped in those fibers add to the mass of the blood clot and give it a red color.

19

- Normal plasma contains several **anticoagulants,** enzymes that inhibit clotting. One, **antithrombin-III,** inhibits several clotting factors, including thrombin.

- **Heparin,** a compound released by basophils and mast cells, is a cofactor that accelerates the activation of antithrombin-III. Heparin is used clinically to impede or prevent clotting.

- **Thrombomodulin** is released by endothelial cells. This protein binds to thrombin and converts it to an enzyme that activates protein C. **Protein C** is a plasma protein that inactivates several clotting factors and stimulates the formation of *plasmin,* an enzyme that gradually breaks down fibrin strands.

- Prostacyclin released during the platelet phase inhibits platelet aggregation and opposes the stimulatory action of thrombin, ADP, and other factors.

- Other plasma proteins with anticoagulant properties include *alpha-2-macroglobulin,* which inhibits thrombin, and *C₁ inactivator,* which inhibits several clotting factors involved in the intrinsic pathway.

Calcium Ions, Vitamin K, and Blood Clotting

Calcium ions and **vitamin K** affect almost every aspect of the clotting process. All three pathways (intrinsic, extrinsic, and common) require Ca^{2+}. Any disorder that lowers plasma Ca^{2+} concentrations will impair blood clotting.

Adequate amounts of vitamin K must be present for the liver to be able to synthesize four of the clotting factors, including prothrombin. Vitamin K is a fat-soluble vitamin, present in green vegetables, grain, and organ meats, that is absorbed with dietary lipids. It is

TABLE 19-4 Clotting Factors

Factor	Structure	Name	Source	Concentration in Plasma (µg/ml)	Pathway
I	Protein	Fibrinogen	Liver	2500–3500	Common
II	Protein	Prothrombin	Liver, requires vitamin K	100	Common
III	Lipoprotein	Tissue factor (TF)	Damaged tissue, activated platelets	0	Extrinsic
IV	Ion	Calcium ions	Bone, diet, platelets	100	Entire process
V	Protein	Proaccelerin	Liver, platelets	10	Extrinsic and intrinsic
VI	(No longer used)				
VII	Protein	Proconvertin	Liver, requires vitamin K	0.5	Extrinsic
VIII	Protein	Antihemophilic factor (AHF)	Platelets, endothelial cells	15	Intrinsic
IX	Protein	Plasma thromboplastin factor	Liver, requires vitamin K	3	Intrinsic
X	Protein	Stuart–Prower factor	Liver, requires vitamin K	10	Extrinsic and intrinsic
XI	Protein	Plasma thromboplastin antecedent (PTA)	Liver	<5	Intrinsic
XII	Protein	Hageman factor	Liver	<5	Intrinsic; also activates plasmin
XIII	Protein	Fibrin-stabilizing factor (FSF)	Liver, platelets	20	Stabilizes fibrin, slows fibrinolysis

obtained from the diet, and roughly half the daily requirement is manufactured by bacteria in the large intestine. A diet inadequate in fats or in vitamin K, or a disorder that affects fat digestion and absorption (such as problems with bile production) will lead to a vitamin K deficiency. This condition will cause the eventual breakdown of the common pathway due to a lack of clotting factors, and the entire clotting system will be inactivated.

CLOT RETRACTION

Once the fibrin meshwork has appeared, platelets and red blood cells stick to the fibrin strands. The platelets then contract, and the entire clot begins to undergo **clot retraction,** or *syneresis* (sin-ER-ē-sis; "a drawing together"). Clot retraction occurs over a period of 30–60 minutes. It (1) pulls the torn edges of the vessel closer together, reducing residual bleeding and stabilizing the injury site; and (2) reduces the size of the damaged area, making it easier for fibroblasts, smooth muscle cells, and endothelial cells to complete repairs.

FIBRINOLYSIS

As the repairs proceed, the clot gradually dissolves. This process, called **fibrinolysis** (fī-bri-NOL-i-sis), begins with the activation of the proenzyme **plasminogen.**

Plasminogen is activated by two enzymes: thrombin, produced by the common pathway, and **tissue plasminogen activator (t-PA),** released by damaged tissues at the injury site. The activation of plasminogen produces the enzyme **plasmin** (PLAZ-min), which begins digesting the fibrin strands and eroding the foundation of the clot.

☑ A sample of bone marrow has unusually few megakaryocytes. What body process would you expect to be impaired as a result?

☑ About half of our vitamin K is produced by bacteria in the large intestine. How would the extended use of antibiotics affect blood clotting?

☑ Unless chemically treated, blood will coagulate in a test tube. The process begins when Factor XII becomes activated. Which clotting pathway is involved in this process?

To perform its vital functions, blood must be kept in motion. If the circulatory supply is cut off, dependent tissues may be destroyed in a matter of minutes. On average, an RBC completes two trips around the circulatory system each minute. The circulation of blood begins in the third week of embryonic development and continues throughout life. In Chapter 20, we shall examine the structure and function of the heart, the pump that maintains this vital blood flow.

Abnormal Hemostasis

The clotting process involves a complex chain of events, and a disorder that affects any individual clotting factor can disrupt the entire process. As a result, many clinical conditions involve the clotting system, and it is often necessary to control or manipulate the clotting response. [AM] *Testing the Clotting System*

Excessive or Abnormal Blood Clotting

If the clotting response is inadequately controlled, blood clots will begin to form in the bloodstream rather than at an injury site. These blood clots do not stick to the wall of the vessel but continue to drift around until either plasmin digests them or they become stuck in a small blood vessel. A drifting blood clot is a type of **embolus** (EM-bō-lus; *embolos,* plug), an abnormal mass within the bloodstream. An embolus that becomes stuck in a blood vessel blocks circulation to the area downstream, killing the affected tissues. The blockage is called an **embolism,** and the tissue damage caused by the circulatory interruption is an **infarct.** Infarcts at the brain are known as *strokes;* infarcts at the heart are called *myocardial infarctions,* or *heart attacks.*

An embolus in the arterial system can get stuck in capillaries in the brain, causing a stroke. An embolus in the venous system will probably become lodged in one of the capillaries of the lungs, causing a *pulmonary embolism.*

A **thrombus** (*thrombos,* clot), or blood clot attached to a vessel wall, begins to form when platelets stick to the wall of an intact blood vessel. Often, the platelets are attracted to areas called *plaques,* where endothelial and smooth muscle cells contain large quantities of lipids. (The mechanism of plaque formation shall be discussed in Chapter 21.) The thrombus gradually enlarges, projecting into the lumen of the vessel and reducing its diameter. Eventually, the vessel may be completely blocked, or a large chunk of the clot may break off, creating an equally dangerous embolus. [AM] *Disseminated Intravascular Coagulation*

To treat these circulatory blockages, clinicians may attempt surgery to remove the obstruction or may use enzymes to attack blood clots and prevent further clot formation. Important anticoagulant drugs include:

- Heparin, which activates antithrombin-III.
- **Coumadin** (COO-ma-din), or *warfarin* (WAR-fa-rin), and **dicumarol** (dī-KOO-ma-rol), which depress the synthesis of several clotting factors by blocking the action of vitamin K.
- Recombinant DNA–synthesized **tissue plasminogen activator (t-PA),** which stimulates plasmin formation.
- **Streptokinase** (strep-tō-KĪ-nās) and **urokinase** (ū-rō-KĪ-nās), enzymes that convert plasminogen to plasmin.
- Aspirin, which inactivates platelet enzymes involved with the production of thromboxanes and prostaglandins and inhibits the production of

prostacyclin by endothelial cells. Daily ingestion of small quantities of aspirin reduces the sensitivity of the clotting process. This method has been proven to be effective in preventing heart attacks in people with significant heart disease.

It is sometimes necessary to control blood clotting within a blood sample, to avoid changes in plasma composition. Blood samples can be stabilized temporarily by adding heparin or *EDTA* (*e*thylene*d*iamine*t*etroacetic *a*cid) to the sample. The acid EDTA removes Ca^{2+} from plasma, effectively preventing clotting. In units of whole blood held for extended periods in a blood bank, *citratephosphate dextrose* (CPD) is typically added. Like EDTA, CPD ties up plasma Ca^{2+}.

Inadequate Blood Clotting

Hemophilia (hē-mō-FĒL-ē-uh) is one of many inherited disorders characterized by the inadequate production of clotting factors. This condition affects about 1 in 10,000 people, 80–90 percent of whom are males. In hemophilia, the production of a single clotting factor (most commonly Factor VIII) is inadequate; the severity of the condition depends on the degree of underproduction. In severe cases, extensive bleeding accompanies the slightest mechanical stress; hemorrhages occur spontaneously at joints and around muscles.

In many cases, transfusions of clotting factors can reduce or control the symptoms of hemophilia, but plasma samples from many individuals must be pooled (combined) to obtain adequate amounts of clotting factors. This procedure makes the treatment very expensive and increases the risk of blood-borne infections such as hepatitis or AIDS. Gene-splicing techniques have been used to manufacture clotting Factor VIII, an essential component of the intrinsic clotting pathway. As methods are developed to synthesize other clotting factors, treatment of the various forms of hemophilia will become safer and cheaper.

The condition known as **von Willebrand disease** is the most common inherited coagulation disorder. The *von Willebrand factor* (vWF) is a plasma protein that binds and stabilizes Factor VIII. The symptoms and severity of the bleeding vary widely. Many individuals with mild forms of von Willebrand disease remain unaware of any bleeding problems until they have an accident or undergo surgery. Treatment consists of the administration of pooled Factor VIII, rather than synthetic forms, because pooled Factor VIII contains normal vWF. In some forms of this disease in which normal vWF is produced but plasma levels are abnormally low, bleeding can be controlled by the use of nasal sprays containing a synthetic form of antidiuretic hormone (ADH). The absorbed ADH appears to stimulate the release of vWF from endothelial cells.

19

[EXPLORE]
[MediaLab]

Blood maintains the internal environment of the body. In a way, it IS the internal environment. To gain a better understanding of this important connective tissue, use the resources available on both the CD-ROM and the Companion Website. The CD-ROM includes animations and tutorials; the website builds on those resources with text content and interactive problems as well as communication avenues for sharing ideas with your class. To complete the following Web Explorations, visit the MediaLab in Chapter 19 of your Companion Website.

Web Exploration 1

Estimated time for completion = 5 minutes

Hemoglobin is an amazing protein. It responds to changing environmental stimuli by either loading or dumping oxygen, according to the body's needs. What happens to the body's tissues if the structure of hemoglobin is altered, as occurs in thalassemia? Review Figure 19-3 and the clinical discussion "Abnormal Hemoglobin"; then select the keyword **THALASSEMIA** on the Companion Website to read about

this condition. Summarize the information in your own words, paying special attention to changes in the hemoglobin molecule and their effects on the body.

Web Exploration 2

Estimated time for completion = 5 minutes

A healthy body requires healthy blood and a sufficient blood volume. When blood composition is deficient in some way or blood levels are too low, transfusions are prescribed. This practice is common and can save lives. Jehovah's Witnesses, however, have a deep religious conviction against receiving any type of blood product. What is the correct medical procedure in such cases? To learn more, select the keyword **TRANSFUSION** on the Companion Website. If you were an attending physician, how would you handle this situation? Express your thoughts in a short essay. If you wish, share it by using the communications tools on the Companion Website. Can you reach a consensus with your peers as to how best to proceed?

CHAPTER REVIEW

SELECTED CLINICAL TERMINOLOGY
Terms Discussed in This Chapter

19

anemia (a-NĒ-mē-uh): A condition in which the oxygen-carrying capacity of blood is reduced owing to low hematocrit or low blood hemoglobin concentrations. *(p. 630 and AM)*

embolism: A condition in which a drifting blood clot becomes stuck in a blood vessel, blocking circulation to the area downstream. *(p. 649)*

hematocrit (hē-MA-tō-krit): The value that indicates the percentage of whole blood occupied by cellular elements. *(p. 628)*

hematuria (hē-ma-TOO-rē-uh): The presence of red blood cells in urine. *(p. 631)*

hemoglobinuria: The presence of hemoglobin in urine. *(p. 631)*

hemolytic disease of the newborn (HDN): A condition in which fetal red blood cells have been destroyed by maternal antibodies. *(p. 638)*

hemophilia (hē-mō-FĒL-ē-uh): Inherited disorders characterized by the inadequate production of clotting factors. *(p. 649)*

hypervolemic (hī-per-vō-LĒ-mik): Having an excessive blood volume. *(p. 637)*

hypovolemic (hī-pō-vō-LĒ-mik): Having a low blood volume. *(p. 637)*

hypoxia (hī-POKS-ē-uh): Low tissue oxygen levels. *(p. 624 and AM)*

jaundice (JAWN-dis): A condition characterized by yellow skin and eyes, caused by abnormally high levels of plasma bilirubin; examples include *hemolytic jaundice* and *obstructive jaundice. (p. 632 and AM)*

leukemia (loo-KĒ-mē-uh): A condition characterized by extremely elevated levels of circulating white blood cells; includes both *myeloid* and *lymphoid* forms. *(p. 641 and AM)*

leukocytosis (loo-kō-sī-TŌ-sis): Excessive numbers of white blood cells in the bloodstream. *(p. 641)*

leukopenia (loo-kō-PĒ-nē-uh): Inadequate numbers of white blood cells in the bloodstream. *(p. 641)*

normochromic: Having red blood cells that contain normal amounts of hemoglobin. *(p. 634 and AM)*

normocytic: Having cells of normal size. *(p. 634 and AM)*

normovolemic (nor-mō-vō-LĒ-mik): Having a normal blood volume. *(p. 637)*

plaque: An abnormal area within a blood vessel where large quantities of lipids accumulate. *(p. 649)*

RBC tests: These tests include a *reticulocyte count, hematocrit, hemoglobin concentration, RBC count, mean corpuscular volume,* and *mean corpuscular hemoglobin concentration. (p. 634 and AM)*

sickle cell anemia: An anemia resulting from the production of an abnormal form of hemoglobin; causes red blood cells to become sickle-shaped at low oxygen levels. *(p. 631 and AM)*

thalassemia: A disorder resulting from the production of an abnormal form of hemoglobin. *(p. 631 and AM)*

thrombus: A blood clot attached to the lumenal surface of a blood vessel. *(p. 649)*

transfusion: A procedure in which blood components are given to someone whose blood volume has been reduced or whose blood is defective. *(p. 637 and AM)*

venipuncture (VĒN-i-punk-chur): The puncturing of a vein for any purpose, including the withdrawal of blood or the administration of medication. *(p. 626)*

Additional Terms Discussed in the *Applications Manual* `AM`

autologous marrow transplant: The reinfusion of an individual's own bone marrow collected before chemotherapy or radiation treatment.

clotting system tests: These tests include *bleeding time, coagulation time, partial thromboplastin time,* and *plasma prothrombin time.*

erythrocytosis (e-rith-rō-sī-TŌ-sis): A polycythemia affecting only red blood cells.

heterologous marrow transplant: The transplantation of bone marrow from one individual to another to replace bone marrow destroyed during cancer therapy.

packed red cells: Red blood cells from which most of the plasma has been removed.

polycythemia (po-lē-sī-THĒ-mē-uh): A descriptive term indicating an elevated hematocrit with a normal blood volume.

septicemia: A dangerous infection of the bloodstream that distributes bacteria and bacterial toxins throughout the body; can cause *disseminated intravascular coagulation* (DIC).

thrombocytopenic purpura: An immune system disorder resulting in the production of antibodies that attack platelets. The platelet count is low, and the individual bruises easily because the remaining platelets are very fragile.

STUDY OUTLINE

INTRODUCTION   p. 624

1. The **cardiovascular system** provides a mechanism for the rapid transport of nutrients, waste products, respiratory gases, and cells within the body.

FUNCTIONS OF BLOOD   p. 624

1. **Blood** is a specialized fluid connective tissue. Its functions include: (1) transporting dissolved gases, nutrients, hormones, and metabolic wastes; (2) regulating the pH and electrolyte composition of interstitial fluids; (3) restricting fluid losses through damaged vessels or at other injury sites; (4) defending the body against toxins and pathogens; and (5) regulating body temperature by absorbing and redistributing heat.

COMPOSITION OF BLOOD   p. 624

1. Blood contains **plasma** and **formed elements—red blood cells (RBCs), white blood cells (WBCs),** and **platelets.** The plasma and formed elements constitute **whole blood,** which can be **fractionated** for analytical or clinical purposes. *(Figure 19-1; Table 19-3)*

2. Whole blood from any region of the body has roughly the same temperature, viscosity, and pH.

PLASMA   p. 626

1. Plasma accounts for about 55 percent of the volume of blood; roughly 92 percent of plasma is water. *(Figure 19-1)*

2. Plasma has a higher dissolved oxygen concentration and more dissolved proteins than interstitial fluid has.

Plasma Proteins   p. 626

3. The three primary classes of plasma proteins are *albumins, globulins,* and *fibrinogen.*

4. **Albumins** constitute about 60 percent of plasma proteins. **Globulins** constitute roughly 35 percent of plasma proteins; they include **antibodies (immunoglobulins),** which attack foreign proteins and pathogens, and **transport globulins,** which bind ions, hormones, and other compounds. **Fibrinogen** molecules are converted to **fibrin** in the clotting process. The removal of fibrinogen from plasma leaves a fluid called **serum.**

FORMED ELEMENTS   p. 627

Hemopoiesis   p. 627

1. **Hemopoiesis** is the process of blood cell formation. Circulating **stem cells** called **hemocytoblasts** divide to form all types of blood cells.

Red Blood Cells   p. 628

2. Red blood cells **(erythrocytes)** account for slightly less than half the blood volume and 99.9 percent of the formed elements. The **hematocrit** value indicates the percentage of whole blood occupied by formed elements. *(Table 19-1)*

3. Red blood cells transport oxygen and carbon dioxide within the bloodstream. They are highly specialized cells with a relatively large surface area. They typically degenerate after about 120 days in the bloodstream. *(Figure 19-2)*

4. Molecules of **hemoglobin (Hb)** account for over 95 percent of the proteins in RBCs. Hemoglobin is a globular protein formed from two pairs of polypeptide subunits. Each subunit contains a single molecule of **heme** and can reversibly bind an oxygen molecule. Damaged or dead RBCs are recycled by phagocytes. *(Figures 19-3, 19-5)*

5. **Erythropoiesis,** the formation of red blood cells, occurs mainly within the **red bone marrow (myeloid tissue)** in adults. This process increases under stimulation by **erythropoiesis-stimulating hormone (erythropoietin, EPO).** Stages in RBC development include **erythroblasts** and **reticulocytes.** *(Figures 19-6, 19-10)*

6. **Blood type** is determined by the presence or absence of specific **surface antigens (agglutinogens)** in the RBC cell membranes: antigens **A, B,** and **Rh (D).** Antibodies **(agglutinins)** in the plasma will react with RBCs bearing different surface antigens. *(Figure 19-7; Table 19-2)*

White Blood Cells   p. 637

7. White blood cells **(leukocytes)** defend the body against pathogens and remove toxins, wastes, and abnormal or damaged cells. *(Figure 19-9)*

8. White blood cells exhibit (1) *emigration (diapedesis),* or movement into peripheral tissues through vessel walls, and (2) **chemotaxis,** an attraction to specific chemicals.

9. *Granular leukocytes (granulocytes)* are subdivided into **neutrophils, eosinophils,** and **basophils.** Fifty to seventy percent of circulating WBCs are neutrophils, which are highly mobile phagocytes. The much less common eosinophils are phagocytes attracted to foreign compounds that have reacted with circulating antibodies. The relatively rare basophils migrate to damaged tissues and release *histamine* and *heparin,* aiding the inflammation response. *(Figure 19-9)*

10. *Agranular leukocytes (agranulocytes)* are subdivided into **monocytes** and **lymphocytes.** Monocytes that migrate into peripheral tissues become tissue macrophages. Lymphocytes, the primary cells of the lymphatic system, include **T cells** (which enter peripheral tissues and attack foreign cells directly or affect the activities of other lymphocytes), **B cells** (which produce antibodies), and

19

natural killer (NK) cells (which destroy abnormal tissue cells). *(Figure 19-9; Table 19-3)*

11. Granulocytes and monocytes are produced by stem cells in the bone marrow. Stem cells responsible for **lymphopoiesis** (the production of lymphocytes) also originate in the bone marrow, but many migrate to peripheral **lymphoid tissues.** *(Figure 19-10)*

12. Factors that regulate lymphocyte maturation are not completely understood. Several **colony-stimulating factors (CSFs)** are involved in regulating other WBC populations and in coordinating RBC and WBC production. *(Figure 19-10)*

Platelets p. 644

13. **Megakaryocytes** in the bone marrow release packets of cytoplasm (platelets) into the circulating blood. The functions of platelets include (1) transporting chemicals important to the clotting process, (2) forming a temporary patch in the walls of damaged blood vessels, and (3) contracting after a clot has formed in order to reduce the size of the break in the vessel wall.

HEMOSTASIS p. 645

1. **Hemostasis** prevents the loss of blood through the walls of damaged vessels. It consists of three phases: the *vascular phase,* the *platelet phase,* and the *coagulation phase.*

The Vascular Phase p. 645

2. The **vascular phase** is a period of local blood vessel constriction, or **vascular spasm,** at the injury site. *(Figure 19-11)*

The Platelet Phase p. 645

3. The **platelet phase** follows as platelets are activated, aggregate at the site, and adhere to the damaged surfaces. *(Figure 19-11)*

The Coagulation Phase p. 646

4. The **coagulation phase** occurs as factors released by platelets and endothelial cells interact with **clotting factors** to form a **blood clot.** In this reaction sequence, suspended fibrinogen is converted to large, insoluble fibers of fibrin. *(Figure 19-12; Table 19-4)*

Clot Retraction p. 648

5. During **clot retraction,** platelets contract and pull the torn edges of the damaged vessel closer together.

Fibrinolysis p. 648

6. During **fibrinolysis,** the clot gradually dissolves through the action of **plasmin,** the activated form of circulating **plasminogen.**

7. Clotting can be prevented by drugs that depress the clotting response or dissolve existing clots. Important **anticoagulant** drugs include heparin, **coumadin, dicumarol, t-PA, streptokinase, urokinase,** and aspirin.

REVIEW QUESTIONS

LEVEL 1 REVIEWING FACTS AND TERMS

1. The formed elements of the blood include
   (a) plasma, fibrin, serum
   (b) albumins, globulins, fibrinogen
   (c) WBCs, RBCs, platelets
   (d) a, b, and c are correct

2. Blood temperature is approximately _____, and blood pH averages _____.
   (a) 98.6°F, 7.0
   (b) 104°F, 7.8
   (c) 100.4°F, 7.4
   (d) 96.8°F, 7.0

3. Plasma contributes approximately _____ percent of the volume of whole blood, and water accounts for _____ percent of the plasma volume.
   (a) 55, 92
   (b) 25, 55
   (c) 92, 55
   (d) 35, 72

4. When the clotting proteins are removed from plasma, _____ remains.
   (a) fibrinogen
   (b) fibrin
   (c) serum
   (d) heme

5. In adults, the only site of red blood cell production, and the primary site of white blood cell formation, is the
   (a) liver
   (b) spleen
   (c) thymus
   (d) red bone marrow

6. The most numerous WBCs in a differential count of a healthy individual are
   (a) neutrophils
   (b) basophils
   (c) lymphocytes
   (d) monocytes

7. The differential count of a person who has an allergy would indicate a high number of
   (a) neutrophils
   (b) eosinophils
   (c) basophils
   (d) monocytes

8. Stem cells responsible for lymphopoiesis are located in the
   (a) thymus and spleen
   (b) lymph nodes
   (c) red bone marrow
   (d) a, b, and c are correct

9. The first step in hemostasis is
   (a) clotting
   (b) the platelet phase
   (c) fibrinolysis
   (d) vascular spasm

10. The complex sequence of steps leading to the conversion of fibrinogen to fibrin is
    (a) fibrinolysis
    (b) clotting
    (c) retraction
    (d) the platelet phase

11. What five major functions are performed by blood?

12. What three primary classes of plasma proteins are in the blood? What is the major function of each?

13. Which type of antibodies does plasma contain for each of the following blood types?
    (a) Type A
    (b) Type B
    (c) Type AB
    (d) Type O

14. What four characteristics of WBCs are important to their response to tissue invasion or injury?

15. Which kinds of WBCs contribute to the body's nonspecific defenses?

16. Which three classes of lymphocytes are the primary cells of the lymphatic system? What are the functions of each class?

17. Which kinds of WBCs are produced by each of the four colony-stimulating factors (CSFs)?

18. What are the three functions of platelets during the clotting process?

19. What three factors regulate the rate of megakaryocyte activity and platelet formation?

20. What five steps are necessary for hemostasis to be completed after damage to the wall of a blood vessel?

21. What contribution from the intrinsic and the extrinsic pathways is necessary for the common pathway to begin?

22. Distinguish between an embolus and a thrombus.

LEVEL 2 REVIEWING CONCEPTS

23. Dehydration would cause
  (a) an increase in the hematocrit
  (b) a decrease in the hematocrit
  (c) no effect on the hematocrit
  (d) an increase in plasma volume

24. Erythropoietin directly stimulates RBC formation by
  (a) increasing rates of mitotic divisions in erythroblasts
  (b) speeding up the maturation of red blood cells
  (c) accelerating the rate of hemoglobin synthesis
  (d) a, b, and c are correct

25. A person with Type A blood has
  (a) antigen A in the plasma
  (b) anti-B antibodies in the plasma
  (c) anti-A antibodies on the red blood cells
  (d) antigen B on the red blood cell membranes

26. Hemolytic disease of the newborn may result if
  (a) the mother is Rh-positive and the father is Rh-negative
  (b) both the father and the mother are Rh-negative
  (c) both the father and the mother are Rh-positive
  (d) an Rh-negative mother carries an Rh-positive fetus

27. How do red blood cells differ from typical cells in the body?

28. How does blood defend against toxins and pathogens in the body?

29. What is the role of blood in the stabilization and maintenance of body temperature?

30. You are a respiratory therapist, and you need blood to check the efficiency of gas exchange at a patient's lungs. Which type of "stick" should you use, and from which blood vessels might you draw the sample?

31. Linda was given RhoGam during and after the delivery of her baby. What does the administration of RhoGam imply, and what effect does it have on Linda?

32. Why is aspirin sometimes prescribed for the prevention of vascular problems?

LEVEL 3 CRITICAL THINKING AND CLINICAL APPLICATIONS

33. A test for prothrombin time is used to determine deficiencies in the extrinsic clotting pathway and is prolonged if any of the factors are deficient. A test for activated partial thromboplastin time is used in a similar fashion to detect deficiencies in the intrinsic clotting pathway. Which factor would be deficient if a person had a prolonged prothrombin time but a normal partial thromboplastin time?

34. Which of the formed elements would increase after the donation of a pint of blood?

35. Mary took the antibiotic cephalosporin, and her platelet count dropped to $50,000/\mu l$. What signs and symptoms will she exhibit?

36. Why do people with advanced kidney disease commonly become anemic?

37. After Randy was diagnosed with stomach cancer, nearly all of his stomach had to be removed. Postoperative treatment included weekly injections of vitamin B_{12}. Why was this vitamin prescribed, and why was this mode of administration specified?

38. How would the extended use of antibiotics affect blood clotting?

19

CHAPTER 20

The Heart

CHAPTER OUTLINE AND OBJECTIVES

INTRODUCTION   p. 655

AN OVERVIEW OF THE CARDIOVASCULAR SYSTEM   p. 655

ANATOMY OF THE HEART   p. 655
   The Pericardium   p. 656
   Superficial Anatomy of the Heart   p. 657
   The Heart Wall   p. 658
   Internal Anatomy and Organization   p. 660
   Connective Tissues and the Fibrous Skeleton   p. 664
   The Blood Supply to the Heart   p. 664
  ① Describe the location and general features of the heart.
  ② Describe the structure of the pericardium, and explain its functions.
  ③ Trace the flow of blood through the heart, identifying the major blood vessels, chambers, and heart valves.
  ④ Identify the layers of the heart wall.
  ⑤ Describe the vascular supply to the heart.

THE HEARTBEAT   p. 668
   Contractile Cells   p. 668
   The Conducting System   p. 670
   The Electrocardiogram   p. 673
   The Cardiac Cycle   p. 674
  ⑥ Describe the events of an action potential in cardiac muscle, and explain the importance of calcium ions to the contractile process.
  ⑦ Discuss the differences between nodal cells and conducting cells, and describe the components and functions of the conducting system of the heart.
  ⑧ Identify the electrical events associated with a normal electrocardiogram.
  ⑨ Explain the events of the cardiac cycle, including atrial and ventricular systole and diastole, and relate the heart sounds to specific events in this cycle.

CARDIODYNAMICS   p. 680
   Overview: The Control of Cardiac Output   p. 681
   Factors Affecting the Stroke Volume   p. 681

Hard-working muscles require a steady supply of blood to provide them with nutrients and oxygen. This is especially true of the hardest-working muscle of all—your heart. Unlike most other muscles, the heart never rests. Not surprisingly, then, any substantial interruption or reduction in the flow of blood to this organ has grave consequences: what we commonly call a heart attack. Such attacks typically occur when one of the arteries that supply the heart muscle becomes partially or completely blocked. Physicians can check the status of these vessels by using scanning techniques that produce computer-enhanced images such as these. But how does the heart supply blood, not only to itself but to all other tissues and organs? How are its activities adjusted so that the blood supply to active tissues—including the heart—can meet the body's constantly changing demands? We shall see in this chapter.

   Factors Affecting the Heart Rate   p. 683
   Exercise and Cardiac Output   p. 685
  ⑩ Define stroke volume and cardiac output, and describe the factors that influence these variables.
  ⑪ Explain how adjustments in stroke volume and cardiac output are coordinated at different levels of activity.
  ⑫ Describe the effects of autonomic innervation on heart function.
  ⑬ Describe the effects of hormones on the heart.

THE HEART AND THE CARDIOVASCULAR SYSTEM   p. 687

MEDIALAB and CHAPTER REVIEW   p. 687

E very cell relies on the surrounding interstitial fluid for oxygen, nutrients, and waste disposal. The composition of the interstitial fluid in tissues throughout the body is kept stable through continuous exchange between the peripheral tissues and the bloodstream. Yet blood can help maintain homeostasis only as long as it stays in motion. If blood remains stationary, its oxygen and nutrient supplies are quickly exhausted, its capacity to absorb wastes is soon saturated, and neither hormones nor white blood cells can reach their intended targets. Thus all the functions of the cardiovascular system ultimately depend on the **heart.** This muscular organ beats approximately 100,000 times each day, pumping roughly 8000 liters of blood—enough to fill forty 55-gallon drums, or 8800 quart-sized milk cartons. Try transferring a gallon of water by using a squeeze-pump, and you'll appreciate just how hard the heart has to work to keep you alive.

We begin this chapter by examining the structural features that enable the heart to perform so reliably. We will then consider the physiological mechanisms that regulate cardiac activity to meet changing circumstances.

AN OVERVIEW OF THE CARDIOVASCULAR SYSTEM

Blood flows through a network of blood vessels that extend between the heart and peripheral tissues. Those blood vessels can be subdivided into a **pulmonary circuit,** which carries blood to and from the gas exchange surfaces of the lungs, and a **systemic circuit,** which transports blood to and from the rest of the body (Figure 20-1•). Each circuit begins and ends at the heart, and blood travels through these circuits in sequence. For example, blood returning to the heart from the systemic circuit must complete the pulmonary circuit before reentering the systemic circuit.

Arteries, or *efferent vessels,* carry blood away from the heart; **veins,** or *afferent vessels,* return blood to the heart. **Capillaries** are small, thin-walled vessels between the smallest arteries and veins. Capillaries are called **exchange vessels,** because their thin walls permit the exchange of nutrients, dissolved gases, and waste products between the blood and surrounding tissues.

Despite its impressive workload, the heart is a small organ, roughly the size of a clenched fist. The heart contains four muscular chambers, two associated with each circuit. The **right atrium** (Ā-trē-um; chamber; plural, *atria*) receives blood from the systemic circuit and passes it to the **right ventricle** (VEN-tri-kl; little

•**FIGURE 20-1** **An Overview of the Cardiovascular System.** Driven by the pumping of the heart, blood flows through separate pulmonary and systemic circuits. Each circuit begins and ends at the heart and contains arteries, capillaries, and veins.

belly). The right ventricle pumps blood into the pulmonary circuit. The **left atrium** collects blood from the pulmonary circuit and empties it into the **left ventricle.** Contraction of the left ventricle ejects blood into the systemic circuit. When the heart beats, the atria contract first, then the ventricles contract. The two ventricles contract at the same time and eject equal volumes of blood into the pulmonary and systemic circuits.

ANATOMY OF THE HEART

The heart is located near the anterior chest wall, directly posterior to the sternum (Figure 20-2a•). A midsagittal section through the trunk does not divide the heart into two equal halves, because the heart is (1) slightly left of the midline, (2) at an angle to the longitudinal axis of the body, and (3) rotated toward the left side. The heart is surrounded by the **pericardial** (per-i-KAR-dē-al) **cavity,** in the anterior portion of

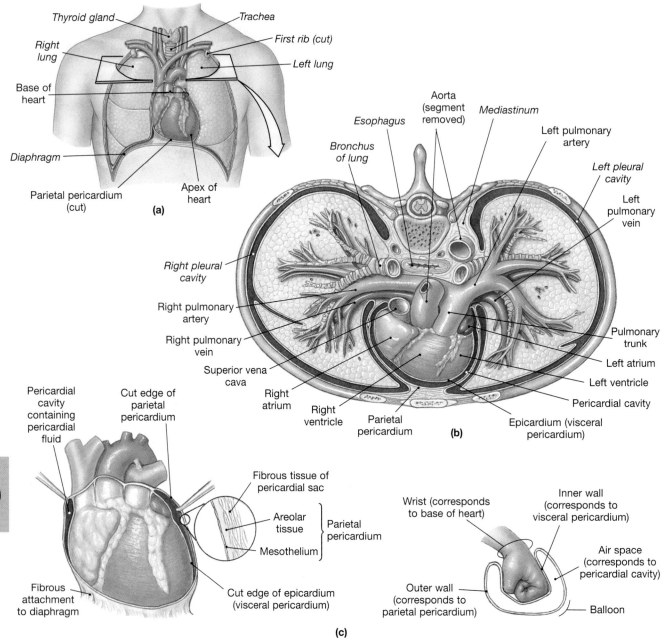

•**FIGURE 20-2** **The Location of the Heart in the Thoracic Cavity.** The heart is situated in the anterior part of the mediastinum, immediately posterior to the sternum. **(a)** An anterior view of the open chest cavity, showing the position of the heart and major vessels relative to the lungs. **(b)** A sectional view of the heart and other organs in the mediastinum. **(c)** The relationship between the heart and the pericardial cavity; compare with the fist and balloon example. [AM] *Plates 7.4a, 7.7a*

the mediastinum. The **mediastinum,** the region between the two pleural cavities, also contains the thymus, esophagus, and trachea. ∞ *p. 22* Figure 20-2b• is a sectional view that illustrates the position of the heart relative to other structures in the mediastinum.

THE PERICARDIUM

The lining of the pericardial cavity is called the **pericardium** ∞ *p. 129* To visualize the relationship between the heart and the pericardial cavity, imag-

ine pushing your fist toward the center of a large balloon (Figure 20-2c•). The balloon represents the pericardium, and your fist is the heart. Your wrist, where the balloon folds back on itself, corresponds to the *base* of the heart, to which the *great vessels,* the largest veins and arteries in the body, are attached. The air space inside the balloon corresponds to the pericardial cavity.

The pericardium is lined by a delicate serous membrane that can be subdivided into the visceral pericardium and the parietal pericardium. The visceral pericardium, or *epicardium*, covers the outer surface of the heart; the **parietal pericardium** lines the inner surface of the **pericardial sac,** which surrounds the heart (Figure 20-2c•). The pericardial sac, or *fibrous pericardium,* which consists of a dense network of collagen fibers, stabilizes the position of the heart and associated vessels within the mediastinum.

The space between the parietal and visceral surfaces is the pericardial cavity. It normally contains 10–20 ml of **pericardial fluid,** secreted by the pericardial membranes. This fluid is a lubricant, reducing friction between the opposing surfaces as the heart beats.

Pericarditis

Pathogens can infect the pericardium, producing the condition **pericarditis.** The inflamed pericardial surfaces rub against one another, producing a distinctive scratching sound. The pericardial irritation and inflammation also commonly result in an increased production of pericardial fluid. Fluid then collects in the pericardial cavity, restricting the movement of the heart. This condition, called **cardiac tamponade** (tam-po-NĀD; *tampon,* plug), can also be caused by traumatic injuries, such as gunshot wounds, that produce bleeding into the pericardial cavity. [AM] *Infection and Inflammation of the Heart*

SUPERFICIAL ANATOMY OF THE HEART

The four cardiac chambers can easily be identified in a superficial view of the heart (Figure 20-3•). The two atria have relatively thin muscular walls and are highly expandable. When not filled with blood, the outer portion of each atrium deflates and becomes a lumpy, wrinkled flap. This expandable extension of an atrium is called an *atrial appendage,* or an **auricle** (AW-ri-kl; *auris,* ear), because it reminded early anatomists of the external ear (Figure 20-3a•). The **coronary sulcus,** a deep groove, marks the border between the atria and the ventricles. The **anterior interventricular sulcus** and the **posterior interventricular sulcus,** shallower depressions, mark the boundary between the left and right ventricles (Figure 20-3a,b•).

The connective tissue of the epicardium at the coronary and interventricular sulci generally contains substantial amounts of fat. In fresh or preserved hearts, this fat must be stripped away to expose the underlying grooves. These sulci also contain the arteries and veins that supply blood to the cardiac muscle.

The great veins and arteries of the circulatory system are connected to the superior end of the heart at the attached **base.** The base sits posterior to the

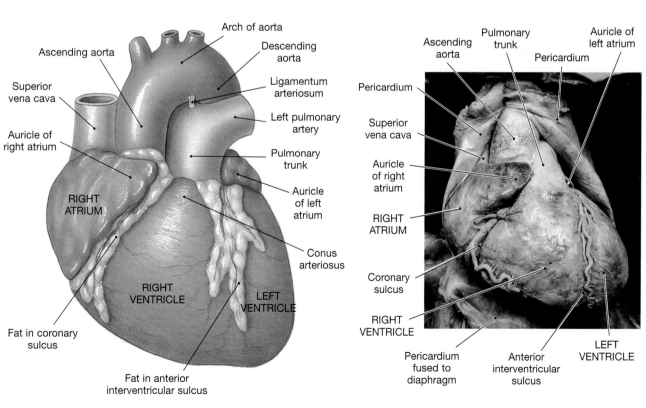

(a) Anterior surface

•**FIGURE 20-3** **The Superficial Anatomy of the Heart.** (a) Major anatomical features on the anterior surface.

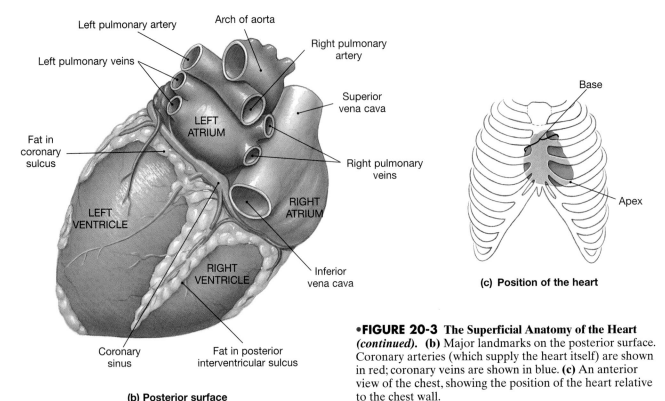

Left pulmonary artery

Arch of aorta

Right pulmonary artery

Left pulmonary veins

LEFT ATRIUM

Superior vena cava

Fat in coronary sulcus

Right pulmonary veins

LEFT VENTRICLE

RIGHT ATRIUM

RIGHT VENTRICLE

Inferior vena cava

Coronary sinus

Fat in posterior interventricular sulcus

(b) Posterior surface

Base

Apex

(c) Position of the heart

•**FIGURE 20-3** **The Superficial Anatomy of the Heart** *(continued).* **(b)** Major landmarks on the posterior surface. Coronary arteries (which supply the heart itself) are shown in red; coronary veins are shown in blue. **(c)** An anterior view of the chest, showing the position of the heart relative to the chest wall.

sternum at the level of the third costal cartilage, centered about 1.2 cm (0.5 in.) to the left side (Figure 20-3c•). The inferior, pointed tip of the heart is the free **apex** (Ā-peks). A typical adult heart measures approximately 12.5 cm (5 in.) from the base to the apex. The apex reaches the fifth intercostal space approximately 7.5 cm (3 in.) to the left of the midline.

THE HEART WALL

A section through the wall of the heart reveals three distinct layers: an outer epicardium, a middle myocardium, and an inner endocardium:

1. The **epicardium** is the visceral pericardium that covers the outer surface of the heart. This serous membrane consists of an exposed mesothelium and an underlying layer of loose connective tissue that is attached to the myocardium.

2. The **myocardium,** or muscular wall of the heart, forms both atria and ventricles. This layer contains cardiac muscle tissue, blood vessels, and nerves. The myocardium consists of concentric layers of cardiac muscle tissue. The atrial myocardium contains muscle bundles that wrap around the atria and form figure eights that pass through the interatrial septum. Superficial ventricular muscles wrap around both ven-

tricles; deeper muscle layers spiral around and between the ventricles toward the apex (Figure 20-4•).

3. The inner surfaces of the heart, including those of the heart valves, are covered by the **endocardium.** It is simple squamous epithelium that is continuous with the endothelium of the attached blood vessels.

Figure 20-5a• illustrates these three layers.

•**FIGURE 20-4** **Layers of the Myocardium.** Cardiac muscle tissue forms concentric layers that wrap around the atria and spiral within the walls of the ventricles.

Parietal pericardium { Areolar tissue / Epithelium

Pericardial cavity

Dense fibrous layer

Connective tissues

MYOCARDIUM (cardiac muscle tissue)

ENDOCARDIUM { Areolar tissue / Endothelium

Epithelium
Areolar tissue
EPICARDIUM (visceral pericardium)

(a)

Cardiac muscle cell (intact)

Intercalated disc (sectioned)

Intercalated disc

Cardiac muscle cell (sectioned)

Bundles of myofibrils

(c)

(b)

Gap junction

Z lines bound to opposing cell membranes

Intercalated disc

Desmosomes

(d)

•**FIGURE 20-5** The Heart Wall and Cardiac Muscle Cells. **(a)** A diagrammatic section through the heart wall, showing the relative positions of the epicardium, myocardium, and endocardium. **(b)** Sectional and **(c)** diagrammatic views of cardiac muscle tissue. (LM × 575) **(d)** The structure of an intercalated disc.

20

Cardiac Muscle Tissue

Recall from Chapter 10 that **cardiac muscle cells** are interconnected by **intercalated discs** (Figure 20-5b,c●). ∞ *pp. 302–303* At an intercalated disc, the interlocking membranes of adjacent cells are held together by desmosomes and linked by gap junctions (Figure 20-5d●). Intercalated discs convey the force of contraction from cell to cell and propagate action potentials. Table 20-1 provides a quick review of the structural and functional differences between cardiac muscle cells and skeletal muscle fibers. Histological characteristics of cardiac muscle cells that differ from those of skeletal muscle fibers include (1) small size; (2) a single, centrally located nucleus; (3) branching interconnections between cells; and (4) the presence of intercalated discs.

INTERNAL ANATOMY AND ORGANIZATION

The right atrium communicates with the right ventricle, and the left atrium with the left ventricle. The atria are separated by the **interatrial septum** (*septum,* wall); the ventricles are separated by the much thicker **interventricular septum** (Figure 20-6a,c●). Each septum is a muscular partition. **Atrioventricular (AV) valves,** folds of fibrous tissue, extend into the openings between the atria and ventricles. These valves permit blood flow in one direction only: from the atria to the ventricles.

The Right Atrium

The right atrium receives blood from the systemic circuit through the two great veins, the **superior vena cava** (VĒ-na KĀ-vuh; plural, *venae cavae*) and the **inferior vena cava.** The superior vena cava delivers blood to the right atrium from the head, neck, upper limbs, and chest. The superior vena cava opens into the posterior and superior portion of the right atrium. The inferior vena cava carries blood to the right atrium from the rest of the trunk, the viscera, and the lower limbs. The inferior vena cava opens into the posterior and inferior portion of the right atrium. The *coronary veins* of the heart return blood to the **coronary sinus,** a large, thin-walled vein that opens into the right atrium inferior to the connection with the inferior vena cava.

The **pectinate muscles** (*pectin,* comb), or *musculi pectinati,* are prominent muscular ridges that run along the inner surface of the auricle and across the adjacent anterior atrial wall (Figure 20-6a,c●). From the fifth week of embryonic development until birth, the

---

TABLE 20-1 Structural and Functional Differences between Cardiac Muscle Cells and Skeletal Muscle Fibers

Feature	Cardiac Muscle Cells	Skeletal Muscle Fibers
Size	10–20 μm $\times$ 50–100 μm	100 μm $\times$ up to 40 cm
Nuclei	Typically 1 (rarely 2–5)	Multiple (hundreds)
Contractile proteins	Sarcomeres along myofibrils	Sarcomeres along myofibrils
Internal membranes	Short T tubules; no triads formed with sarcoplasmic reticulum	Long T tubules form triads with cisternae of the sarcoplasmic reticulum
Mitochondria	Abundant (25% of cell volume)	Relatively scarce
Inclusions	Myoglobin, lipids, glycogen	Little myoglobin, few lipids, but extensive glycogen reserves
Circulatory supply	Very extensive	More extensive than in most connective tissues but sparse compared with supply to cardiac muscle cells
Metabolism (resting)	Not applicable	Aerobic, primarily lipid based
Metabolism (active)	Aerobic, primarily using lipids and carbohydrates	Anaerobic, through breakdown of glycogen reserves
Contractions	Twitches with brief relaxation periods; long refractory period prevents tetanic contractions	Usually sustained tetanic contractions
Stimulus for contraction	Autorhythmicity of pacemaker cells generates action potentials	Activity of somatic motor neuron generates action potentials in sarcolemma
Trigger for contraction	Calcium entry from the ECF and calcium release from the sarcoplasmic reticulum	Calcium release from the sarcoplasmic reticulum
Intercellular connections	Branching network with cell membranes locked together at intercalated discs; connective tissue fibers tie adjacent layers together	Adjacent fibers tied together by connective tissue fibers

20

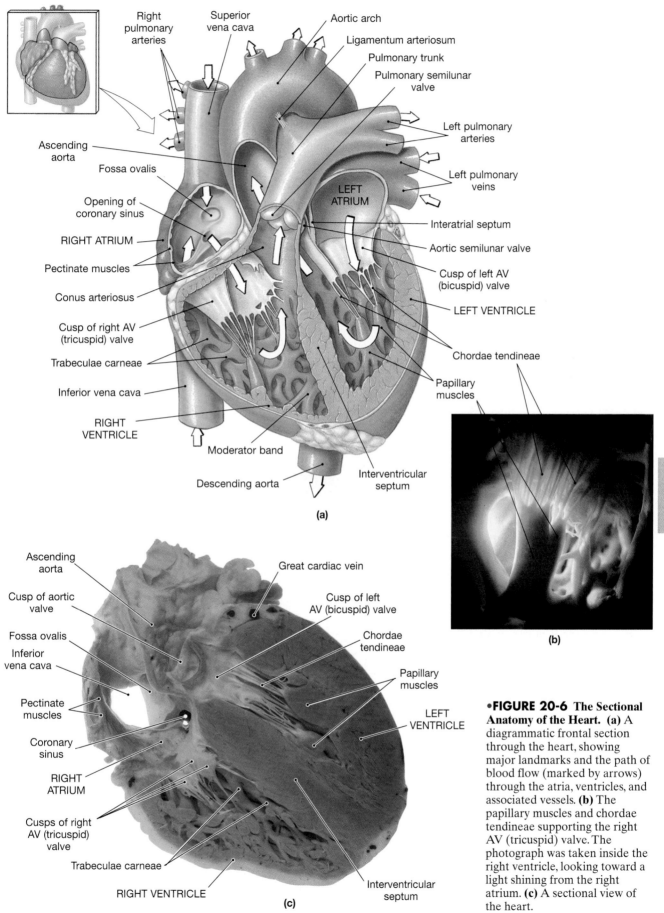

Right pulmonary arteries

Superior vena cava

Aortic arch

Ligamentum arteriosum

Pulmonary trunk

Pulmonary semilunar valve

Left pulmonary arteries

Left pulmonary veins

Ascending aorta

Fossa ovalis

Opening of coronary sinus

RIGHT ATRIUM

Pectinate muscles

Conus arteriosus

Cusp of right AV (tricuspid) valve

Trabeculae carneae

Inferior vena cava

RIGHT VENTRICLE

Descending aorta

Moderator band

LEFT ATRIUM

Interatrial septum

Aortic semilunar valve

Cusp of left AV (bicuspid) valve

LEFT VENTRICLE

Chordae tendineae

Papillary muscles

Interventricular septum

(a)

(b)

20

Ascending aorta

Cusp of aortic valve

Fossa ovalis

Inferior vena cava

Pectinate muscles

Coronary sinus

RIGHT ATRIUM

Cusps of right AV (tricuspid) valve

Trabeculae carneae

RIGHT VENTRICLE

Great cardiac vein

Cusp of left AV (bicuspid) valve

Chordae tendineae

Papillary muscles

LEFT VENTRICLE

Interventricular septum

(c)

•**FIGURE 20-6** **The Sectional Anatomy of the Heart.** **(a)** A diagrammatic frontal section through the heart, showing major landmarks and the path of blood flow (marked by arrows) through the atria, ventricles, and associated vessels. **(b)** The papillary muscles and chordae tendineae supporting the right AV (tricuspid) valve. The photograph was taken inside the right ventricle, looking toward a light shining from the right atrium. **(c)** A sectional view of the heart.

foramen ovale, an oval opening, penetrates the interatrial septum and connects the two atria. The foramen ovale permits blood flow from the right atrium to the left atrium while the lungs are developing before birth. At birth, the foramen ovale closes; after 48 hours, the opening is permanently sealed. The **fossa ovalis,** a small, shallow depression, persists at this site in the adult heart (Figure 20-6a,c•). If the foramen ovale does not close, blood will flow from the left atrium into the right atrium rather than the opposite way, because after birth, blood pressure in the pulmonary circuit is lower than that in the systemic circuit. We will consider the physiological effects of this condition in Chapter 21. *EMBRYOLOGY SUMMARY 15: The Development of the Heart*

The Right Ventricle

Blood travels from the right atrium into the right ventricle through a broad opening bounded by three fibrous flaps. These flaps, or **cusps,** are part of the **right atrioventricular (AV) valve,** also known as the **tricuspid** (trī-KUS-pid; *tri,* three) **valve.** The free edge of each cusp is attached to tendinous connective tissue fibers called the **chordae tendineae** (KOR-dē TEN-di-nē-ē; tendinous cords). These fibers originate at the **papillary** (PAP-i-ler-ē) **muscles,** conical muscular projections that arise from the inner surface of the right ventricle (Figure 20-6a,b•). The right AV valve closes when the right ventricle contracts, preventing the backflow of blood into the right atrium. Without the chordae tendineae, the cusps would be like swinging doors that permitted blood flow in both directions.

The internal surface of the ventricle also contains a series of muscular ridges, the **trabeculae carneae** (tra-BEK-ū-lē KAR-nē-ē; *carneus,* fleshy). The *moderator band* is a muscular ridge that extends horizontally from the inferior portion of the interventricular septum and connects to the anterior papillary muscle. This ridge contains a portion of the *conducting system,* an internal network that coordinates the contractions of cardiac muscle cells. The moderator band delivers the contraction stimulus to the papillary muscles so that they begin tensing the chordae tendineae before the rest of the ventricle contracts.

The superior end of the right ventricle tapers to the **conus arteriosus,** a conical pouch that ends at the **pulmonary semilunar valve.** The pulmonary semilunar valve consists of three semilunar (half-moon–shaped) cusps of thick connective tissue. Blood flowing from the right ventricle passes through this valve to enter the **pulmonary trunk,** the start of the pulmonary circuit. The arrangement of cusps prevents backflow as the right ventricle relaxes. Once in the pulmonary trunk, blood flows into the **left pulmonary arteries** and the **right pulmonary arteries.** These vessels branch repeatedly within the lungs before supplying the capillaries, where gas exchange occurs.

The Left Atrium

From the respiratory capillaries, blood collects into small veins that ultimately unite to form the four pulmonary veins. The posterior wall of the left atrium receives blood from two **left** and two **right pulmonary veins.** Like the right atrium, the left atrium has an auricle. A valve, the **left atrioventricular (AV) valve,** or **bicuspid** (bī-KUS-pid) **valve,** guards the entrance to the attached ventricle (Figure 20-6a,c•). As the name *bicuspid* implies, the left AV valve contains a pair, not a trio, of cusps. Clinicians often call this valve the **mitral** (MĪ-tral; *mitre,* a bishop's hat) **valve.** The left AV valve permits the flow of blood from the left atrium into the left ventricle.

The Left Ventricle

The right and left ventricles contain equal amounts of blood. The left ventricle is much larger than the right, however, because it has thicker walls. The thick, muscular wall enables the left ventricle to develop pressure sufficient to push blood through the large systemic circuit, whereas the right ventricle needs to pump blood, at lower pressure, only about 15 cm (6 in.) to and from the lungs. The internal organization of the left ventricle resembles that of the right ventricle (Figure 20-6a,c•). The trabeculae carneae are prominent, and a pair of large papillary muscles tense the chordae tendineae that brace the cusps of the AV valve and prevent the backflow of blood into the left atrium.

Blood leaves the left ventricle by passing through the **aortic semilunar valve** into the **ascending aorta.** The arrangement of cusps in the aortic semilunar valve is the same as that in the pulmonary semilunar valve. Saclike dilations of the base of the ascending aorta are adjacent to each cusp. These sacs, called **aortic sinuses,** prevent the individual cusps from sticking to the wall of the aorta when the valve opens. The *right* and *left coronary arteries,* which deliver blood to the myocardium, originate at the aortic sinuses. Once the blood has been pumped out of the heart and into the systemic circuit, the aortic semilunar valve prevents backflow into the left ventricle. From the ascending aorta, blood flows on through the **aortic arch** and into the **descending aorta** (Figure 20-6a•). The pulmonary trunk is attached to the aortic arch by the *ligamentum arteriosum,* a ligament that is a remnant of an important fetal blood vessel that once linked the pulmonary and systemic circuits.

Structural Differences between the Left and Right Ventricles

The function of an atrium is to collect blood that is returning to the heart and deliver it to the attached ventricle. The functional demands on the right and left atria are very similar, and the two chambers look almost identical. The demands on the right and left ventricles, however, are very different, and the two have significant structural differences.

Anatomical differences between the left and right ventricles are best seen in a three-dimensional view (Figure 20-7a•). The lungs are close to the heart, and the pulmonary blood vessels are relatively short and wide. Thus the right ventricle normally does not need to push very hard to propel blood through the pulmonary circuit. The wall of the right ventricle is relatively thin. In sectional view, it resembles a pouch attached to the massive wall of the left ventricle. When it contracts, the right ventricle acts like a bellows pump, squeezing the blood against the thick wall of the left ventricle. This mechanism moves blood very efficiently with minimal effort, but it develops relatively low pressures.

A comparable pumping arrangement would not be suitable for the left ventricle, because six to seven times as much force must be exerted to push blood around the systemic circuit as around the pulmonary circuit. The left ventricle has an extremely thick mus-

cular wall and is round in cross section (Figure 20-7a,b•). When this ventricle contracts, (1) the distance between the base and apex decreases, and (2) the diameter of the ventricular chamber decreases. If you imagine the effects of simultaneously squeezing and rolling up the end of a toothpaste tube, you will get the idea. The forces generated are quite powerful, more than enough to open the semilunar valve and eject blood into the ascending aorta. As the powerful left ventricle contracts, it also bulges into the right ventricular cavity (Figure 20-7c•). This dual action improves the efficiency of the right ventricle's efforts. Individuals whose right ventricular musculature has been severely damaged may survive, because the contraction of the left ventricle helps push blood into the pulmonary circuit. We will return to this topic in Chapter 21, as we consider the integrated functioning of the cardiovascular system. [AM] *The Cardiomyopathies*

The Heart Valves

We will now describe the structure and function of the various heart valves.

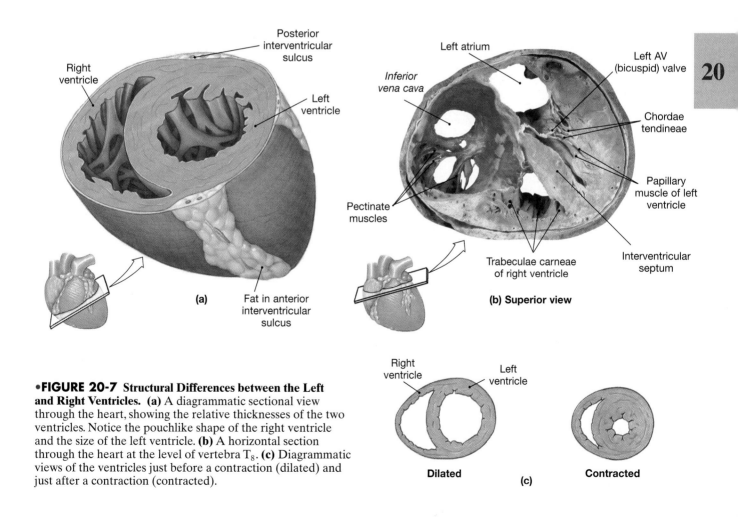

•**FIGURE 20-7 Structural Differences between the Left and Right Ventricles.** (a) A diagrammatic sectional view through the heart, showing the relative thicknesses of the two ventricles. Notice the pouchlike shape of the right ventricle and the size of the left ventricle. (b) A horizontal section through the heart at the level of vertebra T_8. (c) Diagrammatic views of the ventricles just before a contraction (dilated) and just after a contraction (contracted).

The Atrioventricular Valves The atrioventricular (AV) valves prevent the backflow of blood from the ventricles to the atria when the ventricles are contracting. The chordae tendineae and papillary muscles play an important role in the normal function of the AV valves. When the ventricles are relaxed, the chordae tendineae are loose and the AV valves offer no resistance to the flow of blood from the atria to the ventricles (Figure 20-8a•). When the ventricles contract, blood moving back toward the atria swings the cusps together, closing the valves (Figure 20-8b•). At the same time, the contraction of the papillary muscles tenses the chordae tendineae and stops the cusps before they swing into the atria. If the chordae tendineae are cut or the papillary muscles are damaged, there is backflow, or **regurgitation,** of blood into the atria each time the ventricles contract.

The Semilunar Valves The pulmonary and aortic semilunar valves prevent the backflow of blood from the pulmonary trunk and aorta into the right and left ventricles, respectively. The semilunar valves do not require muscular braces as the AV valves do, because the arterial walls do not contract and the relative positions of the cusps are stable. When the semilunar valves close, the three symmetrical cusps support one another like the legs of a tripod (Figure 20-8a,c•).

Valvular Heart Disease
20 Serious valve problems can interfere with cardiac function. If valve function deteriorates to the point at which the heart cannot maintain adequate circulatory flow, symptoms of **valvular heart disease (VHD)** appear. Congenital malformations may be responsible, but in many cases the condition develops after *carditis,* an inflammation of the heart, occurs.

One relatively common cause of carditis is **rheumatic** (roo-MA-tik) **fever,** an acute childhood reaction to infection by streptococcal bacteria. Valve problems serious enough to reduce cardiac function may not appear until 10–20 years after the initial infection. The resulting clinical disorder is known as *rheumatic heart disease (RHD).* Classic cases of rheumatic fever have become less common in the United States owing to the ready availability of antibiotics. However, failure to complete the full course of antibiotic therapy can result in the increased risk of developing carditis and subsequent rheumatic heart disease. AM *RHD and Valvular Stenosis*

CONNECTIVE TISSUES AND THE FIBROUS SKELETON

The connective tissues of the heart include large numbers of collagen and elastic fibers. Each cardiac muscle cell is wrapped in a strong but elastic sheath, and adjacent cells are tied together by fibrous cross-links, or "struts." These fibers are in turn interwoven into sheets that separate the superficial and deep muscle layers. These connective tissue fibers (1) provide physical support for the cardiac muscle fibers, blood vessels, and nerves of the myocardium; (2) help distribute the forces of contraction; (3) add strength and prevent overexpansion of the heart; and (4) provide elasticity that helps return the heart to its original size and shape after a contraction.

The **fibrous skeleton** of the heart consists of four dense bands of tough, elastic tissue that encircle the bases of the pulmonary trunk and aorta and the heart valves (Figure 20-8•). These bands stabilize the positions of the heart valves and ventricular muscle cells and physically isolate the ventricular cells from the atrial cells.

THE BLOOD SUPPLY TO THE HEART

The heart works continuously, and cardiac muscle cells require reliable supplies of oxygen and nutrients. The **coronary circulation** supplies blood to the muscle tissue of the heart. During maximum exertion, the demand for oxygen rises considerably. The blood flow to the heart may then increase to nine times that of resting levels. The coronary circulation includes an extensive network of coronary blood vessels (Figure 20-9•, p. 666).

The Coronary Arteries

The left and right **coronary arteries** originate at the base of the ascending aorta, at the aortic sinuses (Figure 20-9a•, p. 666). Blood pressure here is the highest in the systemic circuit. This high pressure ensures a continuous flow of blood to meet the demands of active cardiac muscle tissue.

The **right coronary artery,** which follows the coronary sulcus around the heart, supplies blood to (1) the right atrium, (2) portions of both ventricles, and (3) portions of the conducting system of the heart, including the *SA (sinoatrial)* and *AV (atrioventricular) nodes.* The cells of these nodes are essential to establishing the normal heart rate. We will focus on their functions and their part in regulation of the heart rate in a later section. Inferior to the right atrium, the right coronary artery generally gives rise to one or more **marginal branches,** which extend across the surface of the right ventricle (Figure 20-9a,b•). It then continues across the posterior surface of the heart, supplying the **posterior interventricular branch,** or *posterior descending artery,* which runs toward the apex within the posterior interventricular sulcus (Figure 20-9b,c•). The posterior interventricular branch supplies blood to the interventricular septum and adjacent portions of the ventricles.

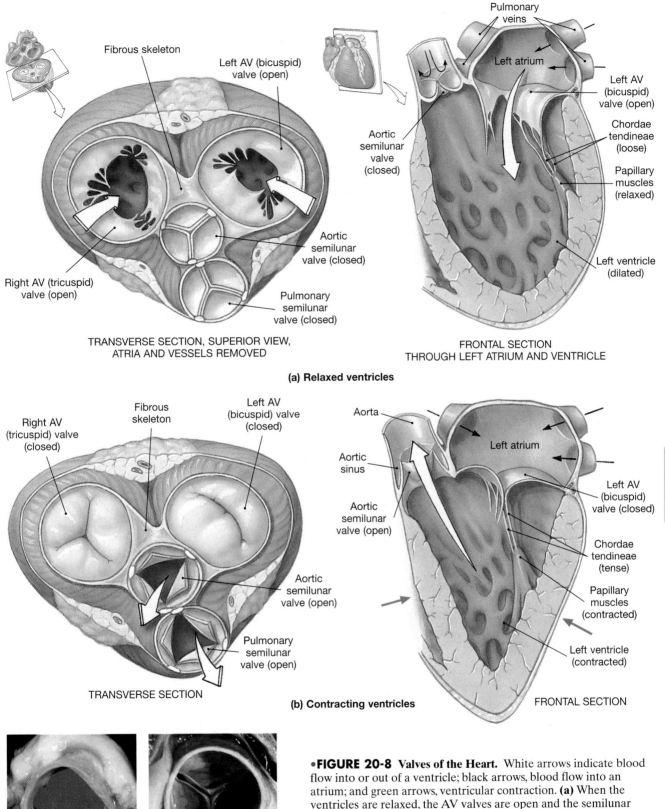

TRANSVERSE SECTION, SUPERIOR VIEW,
ATRIA AND VESSELS REMOVED

FRONTAL SECTION
THROUGH LEFT ATRIUM AND VENTRICLE

(a) Relaxed ventricles

TRANSVERSE SECTION

(b) Contracting ventricles

FRONTAL SECTION

20

Open   Closed

(c) Semilunar valve function

•**FIGURE 20-8 Valves of the Heart.** White arrows indicate blood flow into or out of a ventricle; black arrows, blood flow into an atrium; and green arrows, ventricular contraction. **(a)** When the ventricles are relaxed, the AV valves are open and the semilunar valves are closed. The chordae tendineae are loose, and the papillary muscles are relaxed. **(b)** When the ventricles are contracting, the AV valves are closed and the semilunar valves are open. In the frontal section, notice the attachment of the left AV valve to the chordae tendineae and papillary muscles. **(c)** The aortic semilunar valve in the open (left) and closed (right) positions. The individual cusps brace one another in the closed position.

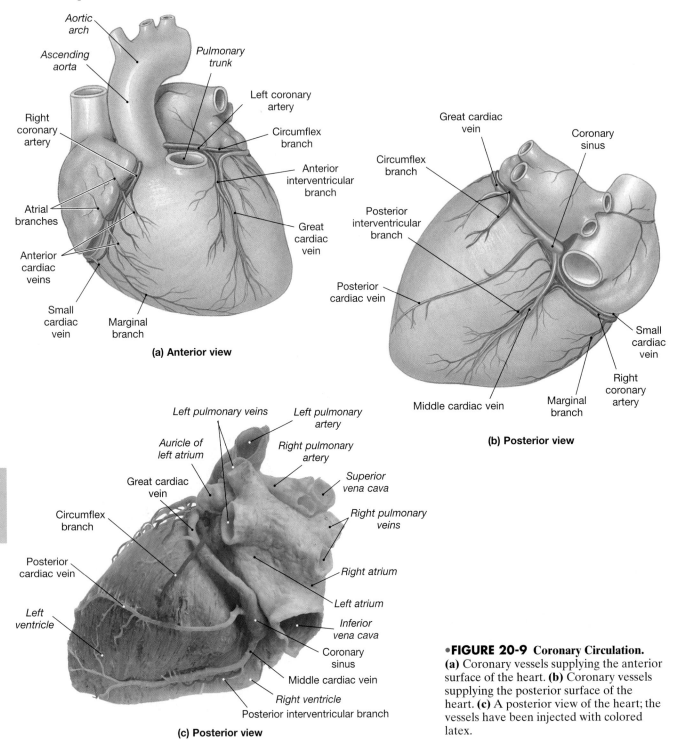

(a) Anterior view

(b) Posterior view

(c) Posterior view

•**FIGURE 20-9** Coronary Circulation. **(a)** Coronary vessels supplying the anterior surface of the heart. **(b)** Coronary vessels supplying the posterior surface of the heart. **(c)** A posterior view of the heart; the vessels have been injected with colored latex.

The **left coronary artery** supplies blood to the left ventricle, left atrium, and interventricular septum. As it reaches the anterior surface of the heart, it gives rise to a circumflex branch and an anterior interventricular branch. The **circumflex branch** curves to the left around the coronary sulcus, eventually meeting and fusing with small branches of the right coronary artery (Figure 20-9a,b,c•). The much larger **anterior interventricular branch,** or *left anterior descending artery,*

swings around the pulmonary trunk and runs along the anterior surface within the anterior interventricular sulcus (Figure 20-9a•).

The anterior interventricular branch supplies small tributaries continuous with those of the posterior interventricular branch of the right coronary artery. Such interconnections between arteries are called **anastomoses** (a-nas-tō-MŌ-sēz; *anastomosis,* outlet). Because the arteries are interconnected in this way,

Coronary Artery Disease

The term **coronary artery disease (CAD)** refers to areas of partial or complete blockage of coronary circulation. Cardiac muscle cells need a constant supply of oxygen and nutrients, so any reduction in coronary circulation produces a corresponding reduction in cardiac performance. Such reduced circulatory supply, known as **coronary ischemia** (is-KĒ-mē-uh), generally results from partial or complete blockage of the coronary arteries. The usual cause is the formation of a fatty deposit, or *plaque,* in the wall of a coronary vessel. The plaque, or an associated thrombus (clot) then narrows the passageway and reduces blood flow. Spasms in the smooth muscles of the vessel wall can further decrease or even stop blood flow. Plaques may be visible by *angiography* or high-resolution ultrasound, and the metabolic effects can be detected in DSA (digital subtraction angiography) scans of the heart (Figure 20-10a,b●). We shall consider plaque development and growth in Chapter 21.

One of the first symptoms of CAD is commonly **angina pectoris** (an-JĪ-nuh PEK-tor-is; *angina,* pain spasm + *pectoris,* of the chest). In the most common form of angina, a temporary ischemia develops when the workload of the heart increases. Although the individual may feel comfortable at rest, exertion or emotional stress can produce a sensation of pressure, chest constriction, and pain that may radiate from the sternal area to the arms, back, and neck.

Angina can typically be controlled by a combination of drug treatment and changes in lifestyle. Lifestyle changes to combat angina include (1) limiting activities known to trigger angina attacks, such as strenuous exercise, and avoiding stressful situations; (2) stopping smoking; and (3) lowering fat consumption. Medications useful for controlling angina include drugs that block sympathetic stimulation *(propranolol* or *metoprolol),* vasodilators such as *nitroglycerin* (nī-trō-GLIS-er-in), and drugs that block calcium movement into the cardiac muscle cells *(calcium channel blockers).*

Angina can also be treated surgically. A single, soft plaque may be reduced with the aid of a long, slender **catheter** (KATH-e-ter), a small-diameter tube. The catheter is inserted into a large artery and guided into a coronary artery to the plaque. A variety of surgical

(a) Normal circulation

(b) Restricted circulation

(c)

●**FIGURE 20-10 Coronary Circulation and Clinical Testing. (a)** A color-enhanced DSA image of a healthy heart. The ventricular walls have an extensive circulatory supply. (The atria are not shown.) **(b)** A color-enhanced DSA image of a damaged heart. Most of the ventricular myocardium is deprived of circulation. **(c)** Balloon angioplasty may be used to remove a circulatory blockage. The catheter is guided through the coronary arteries to the site of blockage and inflated to press the soft plaque against the vessel wall.

tools can be slid into the catheter, and the plaque can then be removed with laser beams or chewed to pieces by a device that resembles a miniature version of a Roto-Rooter® drain-cleaner. Debris created during plaque destruction is sucked up by the catheter, preventing the blockage of smaller vessels.

In **balloon angioplasty** (AN-jē-ō-plas-tē; *angeion,* vessel), the catheter tip contains an inflatable balloon (Figure 20-10c●). Once in position, the balloon is inflated, pressing the plaque against the vessel walls. This procedure works best in small (under 10 mm), soft plaques. Several factors make this a highly attractive treatment: (1) The mortality rate during surgery is only about 1 percent; (2) the success rate is over 90 percent; and (3) it can be performed on an outpatient basis. But plaques commonly rebound or redevelop after angioplasty. A *stent,* a fine wire tube, may then be inserted into the vessel. The stent pushes against the vessel wall, holding it open.

When angioplasty and stents are not feasible, coronary bypass surgery may be done. In a **coronary artery bypass graft (CABG),** a small section is removed from either a small artery (commonly the *internal thoracic artery*) or a peripheral vein (such as the *great saphenous vein* of the leg) and is used to create a detour around the obstructed portion of a coronary artery. As many as four coronary arteries can be rerouted this way during a single operation. The procedures are named according to the number of vessels repaired, so we speak of single, double, triple, or quadruple coronary bypasses. The mortality rate during surgery for such operations performed before significant heart damage has occurred is 1–2 percent. Under these conditions, the procedure eliminates the angina symptoms in 70 percent of the cases and provides partial relief in another 20 percent.

Although coronary bypass surgery does offer certain advantages, recent studies have shown that for mild angina, this surgery does not yield significantly better results than drug therapy. Current recommendations are that coronary bypass surgery be reserved for cases of severe angina that do not respond to other treatment.

the blood supply to the cardiac muscle remains relatively constant despite pressure fluctuations in the left and right coronary arteries as the heart beats.

The Cardiac Veins

The various cardiac veins are shown in Figure 20-9●. The **great cardiac vein** begins on the anterior surface of the ventricles, along the interventricular sulcus. This vein drains blood from the region supplied by the anterior interventricular branch of the left coronary artery. The great cardiac vein reaches the level of the atria and then curves around the left side of the heart within the coronary sulcus. The vein empties into the coronary sinus, which lies in the posterior portion of the coronary sulcus. The coronary sinus communicates with the right atrium near the base of the inferior vena cava. Other cardiac veins that empty into the great cardiac vein or the coronary sinus include (1) the **posterior cardiac vein,** draining the area served by the circumflex branch of the left coronary artery; (2) the **middle cardiac vein,** draining the area supplied by the posterior interventricular branch of the right coronary artery; and (3) the **small cardiac vein,** which receives blood from the posterior surfaces of the right atrium and ventricle. The **anterior cardiac veins,** which drain the anterior surface of the right ventricle, empty directly into the right atrium.

☑ Damage to the semilunar valves on the right side of the heart would interfere with blood flow to which vessel?

☑ What prevents the AV valves from opening back into the atria?

☑ Why is the left ventricle more muscular than the right ventricle?

THE HEARTBEAT

In a single **heartbeat,** the entire heart—atria and ventricles—contracts in a coordinated manner so that blood flows in the right direction at the proper time. Each time your heart beats, the contractions of individual cardiac muscle cells in the atria and ventricles must occur in a specific sequence. Two types of cardiac muscle cells are involved in a normal heartbeat: (1) *contractile cells,* which produce the powerful contractions that propel blood, and (2) specialized muscle cells of the *conducting system,* which control and coordinate the activities of the contractile cells.

CONTRACTILE CELLS

Contractile cells form the bulk of the atrial and ventricular walls. In the discussions of cardiac muscle tissue in earlier chapters, we considered only the structure of contractile cells, which account for roughly 99 percent of the muscle cells in the heart. In both cardiac muscle cells and skeletal muscle fibers, (1) an action potential leads to the appearance of Ca^{2+} among the myofibrils, and (2) the binding of Ca^{2+} to troponin on the thin filaments initiates the contraction. But skeletal and cardiac muscle cells differ in terms of the nature of the action potential, the source of the Ca^{2+}, and the duration of the resulting contraction.

The Action Potential in Cardiac Muscle Cells

In Chapter 12, we compared action potentials in skeletal muscle fibers and in cardiac muscle cells. ∞ *p. 386* We will now take a closer look at the origin and conduction of an action potential in a contractile cell.

The resting potential of a ventricular contractile cell is approximately −90 mV, comparable to that of a resting skeletal muscle fiber (−85 mV). (The resting potential of atrial contractile cells is about −80 mV, but the basic principles described here apply to atrial cells as well.) Although their resting potentials are similar, an action potential in a ventricular contractile cell lasts 250–300 msec, 30 times as long as a typical action potential in a skeletal muscle fiber.

An action potential begins when the membrane of the ventricular muscle cell is brought to threshold, usually at about −75 mV. Threshold is normally reached in a portion of the membrane next to an intercalated disc. The typical stimulus is the excitation of an adjacent muscle cell. Once threshold has been reached, the action potential proceeds in three basic steps (Figure 20-11a):

STEP 1: *Rapid depolarization.* The stage of rapid depolarization in a cardiac muscle cell resembles that in a skeletal muscle fiber. At threshold, voltage-regulated sodium channels open and the membrane suddenly becomes permeable to Na^+. The result is rapid depolarization of the sarcolemma. The channels involved are called **fast channels,** because they open quickly and remain open for only a few milliseconds.

STEP 2: *The plateau.* As the transmembrane potential approaches +30 mV, the voltage-regulated sodium channels close. They will remain closed and inactivated until the membrane potential reaches −60 mV. As the sodium channels are closing, voltage-regulated calcium channels are opening. These channels are called **slow calcium channels** because they open slowly and remain open for a relatively long period—roughly 175 msec. As a result, the transmembrane potential remains near 0 mV for an extended period. This portion of the action potential curve is called the *plateau*. The presence of a plateau is the major difference between action potentials in cardiac muscle cells and in skeletal muscle fibers. In a skeletal muscle fiber, rapid depolarization is immediately followed by a rapid repolarization.

STEP 3: *Repolarization.* As the plateau continues, slow calcium channels begin closing and **slow potassium channels** begin opening. The result is a period of rapid repolarization that restores the resting potential.

The Refractory Period As with skeletal muscle contractions, for some time after an action potential begins, the membrane will not respond normally to a second stimulus. This time is called the refractory period. In the absolute refractory period, the membrane

•**FIGURE 20-11** The Action Potential in Skeletal and Cardiac Muscle. **(a)** An action potential in a ventricular muscle cell. **(b)** The relationship between an action potential and a twitch contraction in skeletal muscle and cardiac muscle.

cannot respond at all, because the sodium channels are either already open or are closed and inactivated. In a ventricular muscle cell, the absolute refractory period lasts approximately 200 msec, spanning the duration of the plateau and the initial period of rapid repolarization.

The absolute refractory period is followed by a shorter (50 msec) relative refractory period. During this period, the voltage-regulated sodium channels are closed but can open. The membrane will respond to a stronger-than-normal stimulus by initiating another action potential.

Calcium Ions and Cardiac Contractions

The appearance of an action potential in the cardiac muscle cell membrane produces a contraction by causing an increase in the concentration of Ca^{2+} around the myofibrils. This process occurs in two steps:

1. Calcium ions entering the cell membrane during the plateau phase of the action potential provide roughly 20 percent of the Ca^{2+} required for a contraction.

2. The arrival of extracellular Ca^{2+} is the trigger for the release of additional Ca^{2+} from reserves in the sarcoplasmic reticulum (SR).

Extracellular calcium ions thus have both direct and indirect effects on cardiac muscle cell contraction. For this reason, cardiac muscle tissue is very sensitive to changes in the Ca^{2+} concentration of the extracellular fluid.

In a skeletal muscle fiber, the action potential is relatively brief and ends as the related twitch contraction begins (Figure 20-11b•). The twitch contraction is short and ends as the SR reclaims the Ca^{2+} it released. In a cardiac muscle cell, as we have seen, the action potential is prolonged and calcium ions continue to enter the cell throughout the plateau. As a result, the period of active muscle cell contraction continues until the plateau

ends. As the slow calcium channels close, the intracellular calcium ions are absorbed by the SR or are pumped out of the cell and the muscle cell relaxes.

In skeletal muscle fibers, the refractory period ends before peak tension develops. As a result, twitches can summate and tetanus can occur. In cardiac muscle cells, the absolute refractory period continues until relaxation is under way. Because summation is not possible, tetanic contractions cannot occur in a normal cardiac muscle cell, regardless of the frequency or intensity of stimulation. This feature is vital: A heart in tetany could not pump blood. With a single twitch lasting 250 msec or longer, a normal cardiac muscle cell can reach 300–400 contractions per minute under maximum stimulation. This rate is not reached in a normal heart due to limitations imposed by the conducting system, which we consider next.

THE CONDUCTING SYSTEM

In contrast to skeletal muscle, cardiac muscle tissue contracts on its own, in the absence of neural or hormonal stimulation. This property is called **automaticity,** or *autorhythmicity.* The cells responsible for initiating and distributing the stimulus to contract are part of the conducting system of the heart. The **conducting system,** also known as the *cardiac conduction system* or the *nodal system,* is a network of specialized cardiac muscle cells that initiates and distributes electrical impulses. The actual contraction lags behind the passage of an electrical impulse, as excitation–contraction coupling occurs and cross-bridge interactions take place. The basic mechanical process of contraction in a cardiac muscle cell resembles that which we described for skeletal muscle fibers in Chapter 10. ∞ *pp. 282–283*

The conducting system (Figure 20-12a•) includes the following:

(a) The conducting system

(b) Depolarization at the SA node

•**FIGURE 20-12 The Conducting System of the Heart.**
(a) Components of the conducting system. (b) The spontaneous changes in the membrane potential of a pacemaker cell in the SA node that is establishing a heart rate of 72 beats per minute.

■ The *sinoatrial (SA) node,* located in the wall of the right atrium.

■ The *atrioventricular (AV) node,* located at the junction between the atria and ventricles.

■ *Conducting cells,* which interconnect the two nodes and distribute the contractile stimulus throughout the myocardium. Conducting cells in the atria are found in *internodal pathways,* which distribute the contractile stimulus to atrial muscle cells as the impulse travels from the SA node to the AV node. (The importance of these pathways in relaying the signal to the AV node remains in dispute, because an impulse spread from contractile cell to contractile cell reaches the AV node at about the same time as an impulse that traverses an internodal pathway.) The ventricular conducting cells include those in the *AV bundle* and the *bundle branches* as well as the *Purkinje* (pur-KIN-jē) *fibers,* which distribute the stimulus to the ventricular myocardium.

Most of the cells of the conducting system are smaller than the contractile cells of the myocardium and contain very few myofibrils. Conducting cells share another important characteristic: They cannot maintain a stable resting potential. Each time repolarization occurs, the membrane again gradually drifts toward threshold. This gradual depolarization is called a **prepotential** (Figure 20-12b•).

The rate of spontaneous depolarization varies in different portions of the conducting system. It is fastest at the SA node, which in the absence of neural or hormonal stimulation will generate action potentials at a rate of 80–100 per minute. Isolated cells of the AV node depolarize more slowly, generating 40–60 action potentials per minute. Under normal conditions, most cells of the AV bundle, the bundle branches, and Purkinje fibers do not depolarize spontaneously.

In a normal heart, the cells of the conducting system are interconnected. An action potential appearing at any one site will be conducted across the entire network. Because the SA node reaches threshold first, it establishes the heart rate. The impulse generated by the SA node brings the AV nodal cells to threshold faster than does the prepotential of the AV nodal cells.

We will now trace the path of an impulse from its initiation at the SA node and will consider its effects on the surrounding myocardium.

The Sinoatrial (SA) Node

The **sinoatrial** (sī-nō-Ā-trē-al) **node (SA node)** is embedded in the posterior wall of the right atrium, near the entrance of the superior vena cava (Figures 20-12a and 20-13 [step 1]•). The SA node contains **pacemaker cells,** which establish the heart rate. As a result, the SA node is also known as the *cardiac pacemaker* or the *natural pacemaker.* It is connected to the larger AV node by the internodal pathways in the atrial walls.

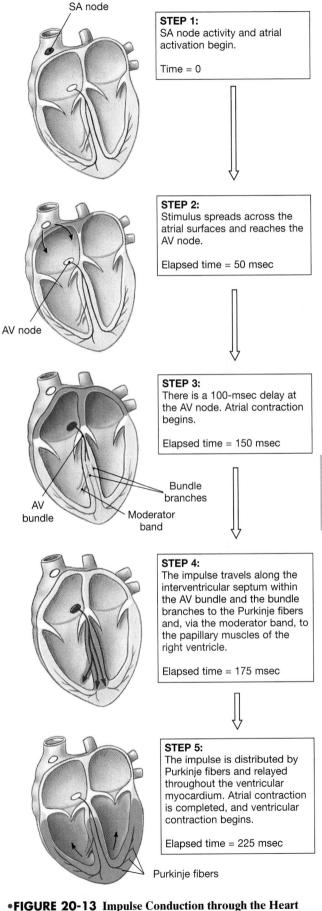

STEP 1:
SA node activity and atrial activation begin.

Time = 0

STEP 2:
Stimulus spreads across the atrial surfaces and reaches the AV node.

Elapsed time = 50 msec

STEP 3:
There is a 100-msec delay at the AV node. Atrial contraction begins.

Elapsed time = 150 msec

STEP 4:
The impulse travels along the interventricular septum within the AV bundle and the bundle branches to the Purkinje fibers and, via the moderator band, to the papillary muscles of the right ventricle.

Elapsed time = 175 msec

STEP 5:
The impulse is distributed by Purkinje fibers and relayed throughout the ventricular myocardium. Atrial contraction is completed, and ventricular contraction begins.

Elapsed time = 225 msec

•**FIGURE 20-13 Impulse Conduction through the Heart**

20

It takes roughly 50 msec for an action potential to travel from the SA node to the AV node along these pathways. Along the way, the conducting cells pass the stimulus to contractile cells of both atria. The action potential then spreads across the atrial surfaces by cell-to-cell contact (Figure 20-13 [step 2]•). The stimulus affects only the atria, because the fibrous skeleton isolates the atrial myocardium from the ventricular myocardium.

The Atrioventricular (AV) Node

The relatively large **atrioventricular (AV) node** (Figures 20-12a and 20-13 [step 2]•) sits within the floor of the right atrium near the opening of the coronary sinus. The rate of propagation slows as the impulse leaves the internodal pathways and enters the AV node, because the nodal cells are smaller in diameter than the conducting cells. (Chapter 12 included a discussion of the relationship between diameter and propagation speed. ∞ *pp. 384, 386*) In addition, the connections between nodal cells are less efficient than those between conducting cells at relaying the impulse from one cell to another. As a result, it takes about 100 msec for the impulse to pass through the AV node and enter the AV bundle (Figure 20-13 [step 3]•).

The delay at the AV node is important, because the atria must contract before the ventricles do. Once the ventricles start contracting, the AV valves close. If all the chambers contracted at the same time, the AV valves would prevent blood flow from the atria and into the ventricles. Because there is a delay at the AV node, the atrial myocardium completes its contraction before ventricular contraction begins.

The cells of the AV node can conduct impulses at a maximum rate of 230 per minute. Because each impulse will result in a ventricular contraction, this value is the maximum normal heart rate. Even if the SA node generates impulses at a faster rate, the ventricles will still contract at 230 beats per minute (bpm). This limitation is important, because mechanical factors, discussed later, begin to decrease the pumping efficiency of the heart at rates above approximately 180 bpm. Rates above 230 bpm occur only when the heart or the conducting system has been damaged or stimulated by drugs. As ventricular rates increase toward their theoretical maximum limit of 300–400 bpm, pumping efficiency becomes dangerously if not fatally reduced.

A number of clinical problems are the result of abnormal pacemaker function. **Bradycardia** (brā-dē-KAR-dē-uh; *bradys,* slow) is a condition in which the heart rate is slower than normal, whereas **tachycardia** (tak-ē-KAR-dē-uh; *tachys,* swift) indicates a faster-than-normal heart rate. These are relative terms, and in clinical practice the definitions vary with the normal resting heart rate of the individual. AM *Problems with Pacemaker Function*

The AV Bundle, Bundle Branches, and Purkinje Fibers

The connection between the AV node and the **AV bundle,** or *bundle of His,* is the only electrical connection between the atria and the ventricles. Once an impulse enters the AV bundle, it travels to the interventricular septum and enters the **right** and **left bundle branches.** The left bundle branch, which supplies the massive left ventricle, is much larger than the right bundle branch. Both bundle branches extend toward the apex of the heart, turn, and fan out deep to the endocardial surface. As the branches diverge, they conduct the impulse to **Purkinje fibers** and, through the moderator band, to the papillary muscles of the right ventricle.

Purkinje fibers conduct action potentials very rapidly—as fast as small myelinated axons. Within about 75 msec, the signal to begin a contraction has reached all the ventricular cardiac muscle cells. The entire process, from the generation of an impulse at the SA node to the complete depolarization of the ventricular myocardium, normally takes around 225 msec. The atria have now completed their contractions, and ventricular contraction can safely occur.

Because the bundle branches deliver the impulse across the moderator band to the papillary muscles directly rather than by way of Purkinje fibers, the papillary muscles begin contracting before the rest of the ventricular musculature does. This contraction applies tension to the chordae tendineae, bracing the AV valves. By limiting the movement of the cusps, tension in the papillary muscles prevents the backflow of blood into the atria when the ventricles contract. The Purkinje fibers radiate from the apex toward the base of the heart. As a result, ventricular contraction proceeds in a wave that begins at the apex and spreads toward the base. Blood is therefore pushed toward the base of the heart, into the aortic and pulmonary trunks.

If the conducting pathways are damaged, the normal rhythm of the heart will be disturbed. The resulting problems are called *conduction deficits.* If the SA node or internodal pathways are damaged, the AV node will assume command. The heart will continue beating normally, although at a slower rate. If an abnormal conducting cell or ventricular muscle cell begins generating action potentials at a more rapid rate, the impulses can override those of the SA or AV node. The origin of these abnormal signals is called an *ectopic* (ek-TOP-ik; out of place) **pacemaker.** The activity of an ectopic pacemaker partially or completely bypasses the conducting system, disrupting the timing of ventricular contraction. The result is a dangerous reduction in the efficiency of the heart. Such conditions are commonly diagnosed with the aid of an *electrocardiogram.* AM *Diagnosing Abnormal Heartbeats*

☑ If the cells of the SA node failed to function, how would the heart rate be affected?

☑ In a person with bradycardia, is cardiac output likely to be greater than or less than normal? Explain.

☑ Why is it important for impulses from the atria to be delayed at the AV node before they pass into the ventricles?

THE ELECTROCARDIOGRAM

The electrical events occurring in the heart are powerful enough to be detected by electrodes on the body surface. A recording of these events is an **electrocardiogram** (ē-lek-trō-KAR-dē-ō-gram), also called an **ECG** or **EKG.** Each time the heart beats, a wave of depolarization radiates through the atria, reaches the AV node, travels down the interventricular septum to the apex, turns, and spreads through the ventricular myocardium toward the base (Figure 20-13●).

By comparing the information obtained from electrodes placed at different locations, a clinician can monitor the electrical activity of the heart, which is directly related to the performance of specific nodal, conducting, and contractile components. For example,

when a portion of the heart has been damaged, the affected muscle cells will no longer conduct action potentials. An ECG will reveal an abnormal pattern of impulse conduction.

The appearance of the ECG varies with the placement of the monitoring electrodes, or *leads.* Figure 20-14a● shows the leads in one of the standard configurations. Figure 20-14b● depicts the important features of an ECG obtained with that configuration. Note the following ECG features:

- The small **P wave** accompanies the depolarization of the atria. The atria begin contracting about 100 msec after the start of the P wave.
- The **QRS complex** appears as the ventricles depolarize. This is a relatively strong electrical signal, because the ventricular muscle is much more massive than that of the atria. It is also a complex signal, in part because it incorporates atrial repolarization as well as ventricular depolarization. The ventricles begin contracting shortly after the peak of the **R wave.**
- The smaller **T wave** indicates ventricular repolarization. You do not see a deflection corresponding to atrial repolarization, because it occurs while the ventricles are depolarizing and the electrical events are masked by the QRS complex.

●**FIGURE 20-14 An Electrocardiogram. (a)** Electrode placement for recording a standard ECG. **(b)** An ECG printout is a strip of graph paper containing a record of the electrical events monitored by the electrodes. The placement of electrodes on the body surface affects the size and shape of the waves recorded. This example is a normal ECG; the enlarged section indicates the major components of the ECG and the measurements most often taken during clinical analysis.

ECG Analysis

To analyze an ECG, you must measure the size of the voltage changes and determine the durations and temporal relationships of the various components. Attention usually is focused on the amount of depolarization occurring during the P wave and the QRS complex. For example, a smaller-than-normal electrical signal may mean that the mass of the heart muscle has decreased; an excessively large QRS complex may indicate that the heart has become enlarged.

The size and shape of the T wave may also be affected by any condition that slows ventricular repolarization. For example, starvation and low cardiac energy reserves, coronary ischemia, or abnormal ion concentrations will reduce the size of the T wave.

You must also measure the time between waves. The values are reported as *segments* or *intervals*. Segments generally extend from the end of one wave to the start of another; intervals are more variable but always include at least one entire wave. Commonly used segments and intervals are indicated in Figure 20-14b•. The names, however, do not always seem to fit. For example:

- The **P–R interval** extends from the start of atrial depolarization to the start of the QRS complex (ventricular depolarization) rather than to R, because in abnormal ECGs, the peak can be difficult to determine. Extension of the P–R interval to more than 0.2 seconds can indicate damage to the conducting pathways or AV node.

- The **Q–T interval** indicates the time required for the ventricles to undergo a single cycle of depolarization and repolarization. It is usually measured from the end of the P–R interval rather than from the bottom of the Q wave. The Q–T interval can be lengthened by conduction problems, coronary ischemia, or myocardial damage. A congenital heart defect that can cause sudden death without warning may be detectable as a prolonged Q–T interval.

Despite the variety of sophisticated equipment available to assess or visualize cardiac function, in the majority of cases the ECG provides the most important diagnostic information. ECG analysis is especially useful in detecting and diagnosing **cardiac arrhythmias** (ā-RITH-mē-az), abnormal patterns of cardiac electrical activity. Momentary arrhythmias are not inherently dangerous; about 5 percent of the healthy population experiences a few abnormal heartbeats each day. Clinical problems appear when the arrhythmias reduce the pumping efficiency of the heart. Serious arrhythmias may indicate damage to the myocardium, injuries to the pacemakers or conduction pathways, exposure to drugs, or variations in the electrolyte composition of extracellular fluids. AM *Diagnosing Abnormal Heartbeats; Monitoring the Heart*

THE CARDIAC CYCLE

The period between the start of one heartbeat and the beginning of the next is a single **cardiac cycle.** The cardiac cycle therefore includes alternating periods of contraction and relaxation. For any one chamber in the heart, the cardiac cycle can be divided into two phases: (1) systole and (2) diastole. During **systole** (SIS-tō-lē), or contraction, the chamber contracts and pushes blood into an adjacent chamber or into an arterial trunk. Systole is followed by **diastole** (dī-AS-tō-lē), or relaxation. During diastole, the chamber fills with blood and prepares for the next cardiac cycle.

Fluids move from an area of high pressure to one of lower pressure. In the course of the cardiac cycle, the pressure within each chamber rises in systole and falls in diastole. Valves between adjacent chambers help ensure that blood flows in the desired direction, but the mere presence of valves is not enough. Blood will flow from one chamber to another only if the pressure in the first chamber exceeds that in the second. This basic principle governs the movement of blood between atria and ventricles, between ventricles and arterial trunks, and between major veins and atria.

The correct pressure relationships are dependent on the careful timing of contractions. For example, blood movement could not occur in the desired direction if an atrium and its attached ventricle contracted at precisely the same moment. The elaborate pacemaking and conducting systems normally provide the required spacing between atrial and ventricular systoles. At a representative heart rate of 75 bpm, a sequence of systole and diastole in either the atria or the ventricles lasts 800 msec. For convenience, we will consider that the cardiac cycle is determined by the atria and that it includes one cycle of atrial systole and atrial diastole. This convention follows our description of the conducting system and the propagation of the stimulus for contraction.

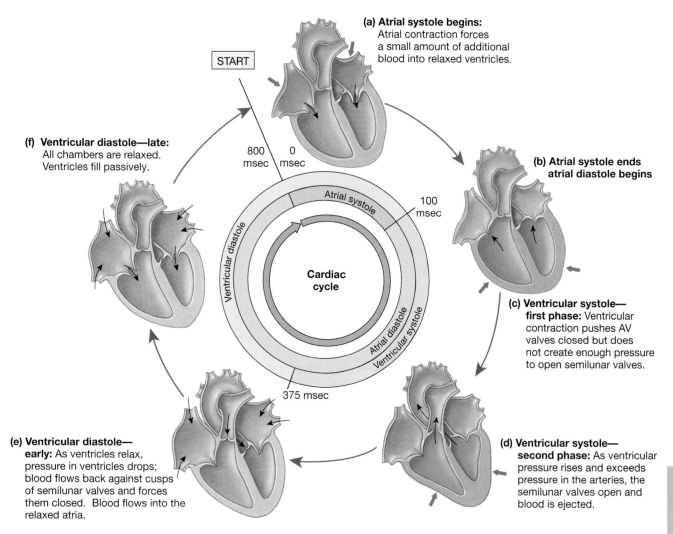

(a) Atrial systole begins: Atrial contraction forces a small amount of additional blood into relaxed ventricles.

START

800 msec   0 msec

(f) Ventricular diastole—late: All chambers are relaxed. Ventricles fill passively.

Atrial systole

100 msec

Ventricular diastole

Cardiac cycle

Atrial diastole

Ventricular systole

(b) Atrial systole ends atrial diastole begins

(c) Ventricular systole— first phase: Ventricular contraction pushes AV valves closed but does not create enough pressure to open semilunar valves.

375 msec

(e) Ventricular diastole— early: As ventricles relax, pressure in ventricles drops; blood flows back against cusps of semilunar valves and forces them closed. Blood flows into the relaxed atria.

(d) Ventricular systole— second phase: As ventricular pressure rises and exceeds pressure in the arteries, the semilunar valves open and blood is ejected.

20

•**FIGURE 20-15** Phases of the Cardiac Cycle. Thin black arrows indicate blood flow, and green arrows indicate contractions.

Phases of the Cardiac Cycle

The phases of atrial systole, atrial diastole, ventricular systole, and ventricular diastole are diagrammed in Figure 20-15• for a heart rate of 75 bpm. As this cardiac cycle begins, the chambers are relaxed and the ventricles are partially filled with blood. During atrial systole, the atria contract and the ventricles become completely filled with blood (Figure 20-15a,b•). Atrial systole lasts 100 msec. Over this period, blood cannot flow into the atria because atrial pressure exceeds venous pressure. Although the connections with the venous system have no valves, very little backflow occurs, because blood takes the path of least resistance. Resistance to blood flow through the broad AV connections and into the ventricles is less than that through the smaller, angled openings of the large veins, which are connected to miles of smaller vessels.

The atria next enter atrial diastole, which continues until the start of the next cardiac cycle. Atrial diastole and ventricular systole begin at the same time. Ventricular systole lasts 270 msec. During this period, blood is pushed through the systemic and pulmonary circuits and toward the atria (Figure 20-15c,d•). The heart then enters ventricular diastole (Figure 20-15e,f•), which lasts 530 msec (the 430 msec remaining in this cardiac cycle, plus the first 100 msec of the next). For the rest of this cycle, filling occurs passively and both the atria and the ventricles are relaxed. We are at the start of the next cardiac cycle, which begins with atrial systole and the completion of ventricular filling.

When the heart rate increases, all the phases of the cardiac cycle are shortened. The greatest reduction occurs in the length of time spent in diastole. When the heart rate climbs from 75 bpm to 200 bpm, the time spent in systole drops by less than 40 percent but the duration of diastole is reduced by almost 75 percent.

Pressure and Volume Changes in the Cardiac Cycle

In considering the pressure and volume changes that occur during the cardiac cycle, we will use reference numbers that match those in Figure 20-16•, which follow pressures and volumes within the left atrium and left ventricle. The narrative description applies to both sides of the heart. Although pressures are lower in the right atrium and right ventricle, both sides of the heart

contract at the same time, and they eject equal volumes of blood.

Atrial Systole The cardiac cycle begins with atrial systole, which lasts about 100 msec in a resting adult:

1. As the atria contract, rising atrial pressures push blood into the ventricles through the open right and left AV valves.

2. At the start of atrial systole, the ventricles are already filled to about 70 percent of capacity, and atrial sys-

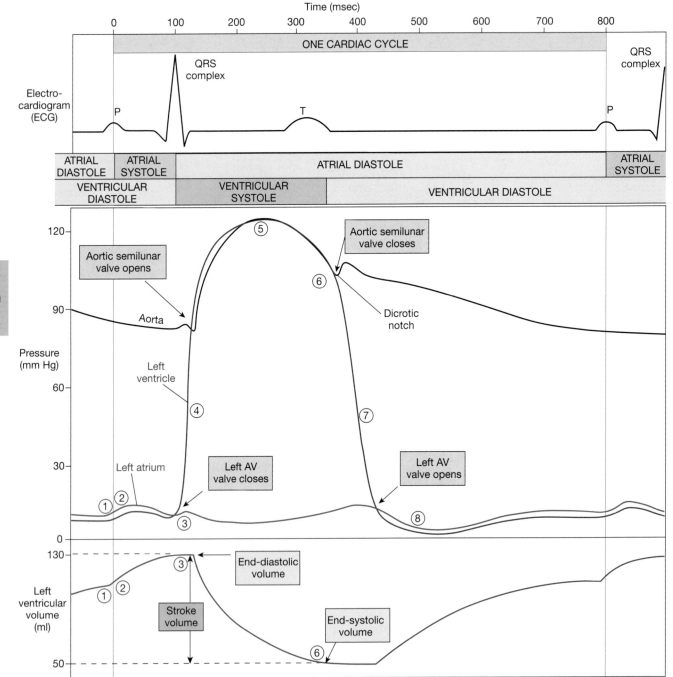

•**FIGURE 20-16 Pressure and Volume Relationships in the Cardiac Cycle.** Major features of the cardiac cycle are shown for a heart rate of 75 bpm. The circled numbers correspond to the numbered list on this page and the next.

tole essentially "tops them off" by providing the additional 30 percent. (The ventricles filled passively during the previous cardiac cycle.)

3. At the end of atrial systole, each ventricle contains the maximum amount of blood that it will hold in this cardiac cycle. That quantity is called the **end-diastolic volume (EDV).** The end-diastolic volume in an adult who is standing at rest is typically about 130 ml.

Ventricular Systole As atrial systole ends, ventricular systole begins. This period lasts approximately 270 msec in a resting adult. As the pressures in the ventricles rise above those in the atria, the AV valves swing shut.

4. During this stage of ventricular systole, the ventricles are contracting. Blood flow has yet to occur, however, because ventricular pressures are not high enough to force open the semilunar valves and push blood into the pulmonary or aortic trunk. Over this period, the ventricles contract isometrically: They generate tension and ventricular pressures rise, but blood flow does not occur. The ventricles are now in the period of **isovolumetric contraction.** During isovolumetric contraction, all the heart valves are closed, the volumes of the ventricles remain constant, and ventricular pressures rise.

5. Once pressure in the ventricles exceeds that in the arterial trunks, the semilunar valves open and blood flows into the pulmonary and aortic trunks. This point marks the beginning of the period of **ventricular ejection.** The ventricles now contract isotonically: The muscle cells shorten, and tension production remains relatively constant. (To review isotonic versus isometric contractions, see Figure 10-16•, p. 291.)

   After reaching a peak, ventricular pressures gradually decline near the end of ventricular systole. Figure 20-16• shows values for the left ventricle and aorta. Although pressures in the right ventricle and pulmonary trunk are much lower, the right ventricle also goes through periods of isovolumetric contraction and ventricular ejection. During this period, each ventricle will eject 70–80 ml of blood. This is the **stroke volume (SV)** of the heart. The stroke volume in this case is roughly 60 percent of the end-diastolic volume. This percentage, known as the *ejection fraction,* can vary in response to changing demands on the heart. (We will discuss the regulatory mechanisms in the next section.)

6. As the end of ventricular systole approaches, ventricular pressures fall rapidly. Blood in the aorta and pulmonary trunk now starts to flow back toward the ventricles, and this movement closes the semilunar valves. As the backflow begins, pressure decreases in the aorta. When the semilunar valves close, pressure rises again as the elastic arterial walls recoil. This small, temporary rise produces a valley in the pressure tracing that is called a *dicrotic* (dī-KRO-tik) *notch* (*dikrotos,* double-beating). The amount of blood remaining in the ventricle

when the semilunar valve closes is the **end-systolic volume (ESV).** The end-systolic volume in this case is 50 ml, about 40 percent of the end-diastolic volume.

Ventricular Diastole The period of ventricular diastole lasts for the 430 msec remaining in this cycle and continues through atrial systole in the next cardiac cycle.

7. All the heart valves are now closed, and the ventricular myocardium is relaxing. Because ventricular pressures are still higher than atrial pressures, blood cannot flow into the ventricles. This is the period of **isovolumetric relaxation.** Ventricular pressures drop rapidly over this period, because the elasticity of the connective tissues of the heart and fibrous skeleton helps reexpand the ventricles toward their resting dimensions.

8. When ventricular pressures fall below those of the atria, the atrial pressures force the AV valves open. Blood now flows from the atria into the ventricles. Both the atria and the ventricles are in diastole, but the ventricular pressures continue to fall as the ventricular chambers expand. Throughout this period, pressures in the ventricles are so far below those in the major veins that blood pours through the relaxed atria and into the ventricles. This passive mechanism is the primary method of ventricular filling. The ventricles will be nearly three-quarters full before the cardiac cycle ends.

The relatively minor contribution that atrial systole makes to ventricular volume explains why individuals can survive quite normally when their atria have been so severely damaged that they can no longer function. In contrast, damage to one or both ventricles can leave the heart unable to maintain adequate blood flow through peripheral tissues and organs. A condition of **heart failure** then exists. [AM] *Heart Failure*

20

Heart Sounds

Listening to the heart, a technique called *auscultation,* is a simple and effective method of cardiac diagnosis. Physicians use an instrument called a **stethoscope** to listen to normal and abnormal heart sounds. Where to place the stethoscope depends on which valve is under examination (Figure 20-17a•). Valve sounds must pass through the pericardium, surrounding tissues, and the chest wall, and some tissues muffle sounds more than others. As a result, the placement of the stethoscope does not always correspond to the position of the valve under review.

There are four **heart sounds,** designated as S_1 through S_4 (Figure 20-17b•). When you listen to your own heart, you usually hear the *first* and *second heart sounds.* These sounds accompany the closing of your heart valves. The first heart sound, known as "lubb" (S_1), lasts a little longer than the second. S_1, which marks the start of ventricular contraction, is produced as the AV valves close. The second heart sound,

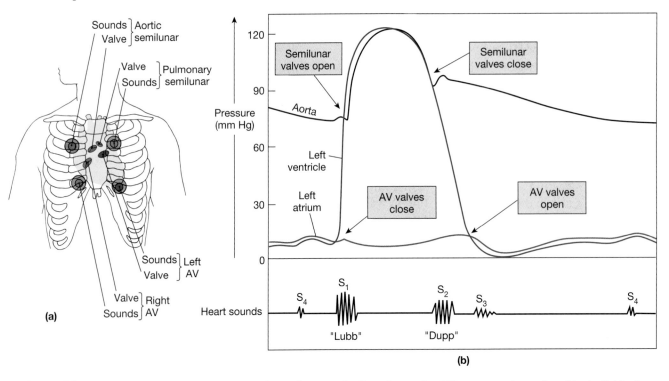

Sounds] Aortic
Valve] semilunar

Valve] Pulmonary
Sounds] semilunar

Sounds] Left
Valve] AV

Valve] Right
Sounds] AV

(a)

Pressure
(mm Hg)

Semilunar
valves open

Semilunar
valves close

Aorta

Left
ventricle

Left
atrium

AV valves
close

AV valves
open

Heart sounds

S_4 S_1 S_2 S_3 S_4

"Lubb" "Dupp"

(b)

•**FIGURE 20-17** Heart Sounds. **(a)** Placements of a stethoscope for listening to the different sounds produced by individual valves. **(b)** The timing of heart sounds in relation to key events in the cardiac cycle.

"dupp" (S_2), occurs at the beginning of ventricular filling, when the semilunar valves close.

Third and *fourth heart sounds* may be audible, but they are usually very faint and seldom are detectable in healthy adults. These sounds are associated with blood flowing into the ventricles (S_3) and atrial contraction (S_4) rather than with valve action.

Mitral Valve Prolapse

Minor abnormalities in valve shape are relatively common. For example, 5 to 10 percent of healthy individuals have some degree of **mitral valve prolapse.** In this condition, the mitral valve cusps do not close properly. The problem may involve abnormally long (or short) chordae tendineae or malfunctioning papillary muscles. Because the valve does not work perfectly, some regurgitation occurs during left ventricular systole. The surges, swirls, and eddies that occur during regurgitation create a rushing, gurgling sound known as a **heart murmur.** Minor heart murmurs are very common. Most individuals with this condition are completely asymptomatic and live normal lives, unaware of any circulatory malfunction. Extreme prolapse and valve failure, as may be caused by breakage of chordae tendineae, can be life-threatening. This condition is known as a *mitral flail.*

The Energy for Cardiac Contractions

When a normal heart is beating, the energy required is obtained by the mitochondrial breakdown of fatty acids (stored as lipid droplets) and glucose (stored as glycogen). These aerobic reactions can occur only when oxygen is readily available.

In addition to obtaining oxygen from their own the coronary circulation, cardiac muscle cells maintain their own sizable reserves of oxygen. In these cells, oxygen molecules are bound to the heme units of myoglobin molecules. (We discussed this globular protein, which reversibly binds oxygen molecules, and its function in muscle fibers in Chapter 10.) ∞ *p. 297* Normally, the combination of circulatory supplies plus myoglobin reserves is enough to meet the oxygen demands of your heart, even when it is working at maximum capacity.

☑ Is the heart always pumping blood when pressure in the left ventricle is rising? Explain.

☑ What factor or factors could cause an increase in the size of the QRS complex of an electrocardiogram recording?

Heart Attacks

In a **myocardial** (mī-ō-KAR-dē-al) **infarction (MI),** or *heart attack,* the coronary circulation becomes blocked and the cardiac muscle cells die from lack of oxygen. The affected tissue then degenerates, creating a nonfunctional area known as an *infarct.* Heart attacks most commonly result from severe coronary artery disease (CAD, p. 667). The consequences depend on the site and nature of the circulatory blockage. If it occurs near the start of one of the coronary arteries, the damage will be widespread and the heart may stop beating. If the blockage involves one of the smaller arterial branches, the individual may survive the immediate crisis but may have many complications, all unpleasant. As scar tissue forms in the damaged area, the heartbeat may become irregular and other vessels can become constricted, creating additional circulatory problems.

Myocardial infarctions are generally associated with fixed blockages, such as those seen in CAD. When the crisis develops as a result of thrombus (clot) formation at a plaque, the condition is called **coronary thrombosis.** A vessel already narrowed by plaque formation may also become blocked by a sudden spasm in the smooth muscles of the vascular wall. The individual then may experience intense pain, similar to that of an angina attack but persisting even at rest. However, pain does not always accompany a heart attack. These *silent heart attacks* may be even more dangerous, because the condition may not be diagnosed and treated before a fatal MI occurs. Roughly 25 percent of heart attacks are not recognized when they occur.

The cytoplasm of a damaged cardiac muscle cell differs from that of a normal muscle cell. As the supply of oxygen decreases, the cells become more dependent on anaerobic metabolism to meet their energy needs. ∞ *p. 294* Over time, the cytoplasm accumulates large numbers of enzymes involved with anaerobic energy production.

As the cell membranes of the cardiac muscle cells deteriorate, these enzymes enter the surrounding intercellular fluids. The appearance of such enzymes in the circulation thus indicates that an infarct has occurred. The enzymes that are tested for in a diagnostic blood test include **lactate dehydrogenase (LDH), serum glutamic oxaloacetic transaminase (SGOT,** also called *aspartate aminotransferase*), **creatine phosphokinase (CPK, or CK),** and a special form of creatine phosphokinase that occurs only in cardiac muscle **(CK-MB).**

About 25 percent of MI patients die before obtaining medical assistance, and 65 percent of MI deaths among those under age 50 occur within an hour after the initial infarct. The goals of treatment are to limit the size of the infarct and to prevent additional complications by preventing irregular contractions, improving circulation with vasodilators, providing additional oxygen, reducing the cardiac workload, and, if possible, eliminating the cause of the circulatory blockage. Anticoagulants (even aspirin chewed and swallowed at the start of an MI) may help prevent the formation of additional thrombi, and clot-dissolving enzymes may reduce the extent of the damage if they are administered within 6 hours after the MI occurred. Current evidence suggests that t-PA (tissue plasminogen activator), which is relatively expensive, is more beneficial than other fibrinolytic agents, such as urokinase or streptokinase. A procedure that combines small amounts of t-PA with streptokinase has been proposed as a workable compromise that produces effects equivalent to high doses of t-PA alone. Follow-up treatment with heparin, aspirin, or both is recommended; without further treatment, the circulatory blockages will reappear in about 20 percent of patients.

Roughly 1.3 million MIs occur in the United States each year, and half the victims die within a year of the incident. The following 10 factors appear to increase the risk of a heart attack: (1) smoking, (2) high blood pressure, (3) high blood cholesterol levels, (4) high circulating levels of low-density lipoproteins (LDLs), (5) diabetes, (6) male gender (below age 70), (7) severe emotional stress, (8) obesity, (9) genetic predisposition, and (10) a sedentary lifestyle. We shall consider the role of lipoproteins and cholesterol in plaque formation and heart disease in Chapter 21. Although the heart attack rate of women under age 70 is lower than that of men, the mortality rate for women is higher—perhaps because heart disease in women is neither diagnosed as early nor treated as aggressively as that in men.

The presence of two risk factors more than doubles the risk of heart attack, so eliminating as many risk factors as possible will improve the chances of preventing or surviving a heart attack. Changing the diet to limit cholesterol, exercising to lower weight, and seeking treatment for high blood pressure are steps in the right direction. It has been estimated that reduction of coronary risk factors could prevent 150,000 deaths each year in the United States alone.

20

CARDIODYNAMICS

The term **cardiodynamics** refers to the movements and forces generated during cardiac contractions. Each time the heart beats, the two ventricles eject equal amounts of blood. Earlier we introduced these terms:

- *End-diastolic volume (EDV)*. The amount of blood in each ventricle at the end of ventricular diastole (the start of ventricular systole).
- *End-systolic volume (ESV)*. The amount of blood remaining in each ventricle at the end of ventricular systole (the start of ventricular diastole).
- *Stroke volume (SV)*. The amount of blood pumped out of each ventricle during a single beat, which can be expressed as EDV − ESV = SV.
- *Ejection fraction*. The percentage of the EDV represented by the SV.

Stroke volume is the most important factor in an examination of a single cardiac cycle. If the heart were an old-fashioned bicycle pump, the stroke volume would be the amount of air pumped in one up–down cycle of the pump handle (Figure 20-18●). Where you stop when you lift the handle determines the end-diastolic volume—how much air the pump contains. How far down you push the handle determines the end-systolic volume—how much air remains in the pump at the end of the cycle. You can increase the amount of air pumped in each cycle by increasing the range of movement of the handle. You pump the maximum amount when the handle moves all the way from the top to the bottom (Figure 20-18b,d●). In other words, you get the largest stroke volume when the EDV is as high as it can be and the ESV is as small as it can be.

●**FIGURE 20-18 A Simple Model of Stroke Volume.** Stroke volume of the heart can be compared to the amount of air pumped from an old-fashioned bicycle pump. That amount varies with the amount of handle movement. The extent of upward movement corresponds to the EDV; the extent of downward movement corresponds to the ESV.

When considering cardiac function over time, physicians generally are most interested in the **cardiac output (CO),** the amount of blood pumped by each ventricle in 1 minute. The cardiac output provides a useful indication of ventricular efficiency over time. We can calculate it by multiplying the average stroke volume (SV) by the heart rate (HR):

$$
\begin{array}{ccccc}
\text{CO} & = & \text{SV} & \times & \text{HR} \\
\text{cardiac} & & \text{stroke} & & \text{heart} \\
\text{output} & & \text{volume} & & \text{rate} \\
\text{(ml/min)} & & \text{(ml/beat)} & & \text{(beats/min)}
\end{array}
$$

For example, if the stroke volume is 80 ml per beat and the heart rate is 75 bpm, the cardiac output will be

$$
\begin{aligned}
\text{CO} &= 80 \text{ ml/beat} \times 75 \text{ bpm} \\
&= 6000 \text{ ml/min (6 l/min)}
\end{aligned}
$$

Cardiac output is precisely adjusted such that peripheral tissues receive an adequate circulatory supply under a variety of conditions. When necessary, stroke volume in a normal heart can almost double and the heart rate can increase by 250 percent.

OVERVIEW: THE CONTROL OF CARDIAC OUTPUT

Figure 20-19• summarizes the factors involved in the normal regulation of cardiac output. Cardiac output can be adjusted by changes in either stroke volume or heart rate. For convenience, we can consider these independently as we discuss the individual factors involved. However, changes in cardiac output generally reflect changes in both stroke volume and heart rate.

The stroke volume can be adjusted by changing the end-diastolic volume (how full the ventricles are when they start to contract), the end-systolic volume (how much blood remains in the ventricle after it contracts), or both. As Figure 20-18• showed, stroke volume peaks when EDV is high and ESV is low. The heart rate can be adjusted by the activities of the autonom-

ic nervous system or by circulating hormones. A variety of other factors can influence cardiac output under abnormal circumstances; we will consider several examples in a separate section.

FACTORS AFFECTING THE STROKE VOLUME

The stroke volume is the difference between the end-diastolic volume and the end-systolic volume. Thus, changes in either the EDV or the ESV can change both the stroke volume and cardiac output.

The EDV

The EDV indicates the amount of blood a ventricle contains at the end of diastole, just before a contraction begins. This volume is affected by two factors: the filling time and the venous return. **Filling time** is the duration of ventricular diastole. As such, it depends entirely on the heart rate. The faster the heart rate, the shorter the available filling time. **Venous return** is the rate of blood flow over this period. Venous return changes in response to alterations in cardiac output, blood volume, patterns of peripheral circulation, skeletal muscle activity, and other factors that affect the rate of blood flow through the venae cavae. (We shall explore these factors in Chapter 21.)

The ESV

After the ventricle contracts and the stroke volume has been ejected, the ESV is the amount of blood that remains in the ventricle at the end of ventricular systole. Three factors that influence the ESV are (1) the *preload,* (2) the *contractility* of the ventricle, and (3) the *afterload.*

Preload The degree of stretching experienced during ventricular diastole is called the **preload.** The preload is directly proportional to the EDV: The greater the EDV, the larger the preload. Preload is significant because it affects the ability of muscle cells to produce tension. In Chapter 10, we considered the length–tension relationship for skeletal muscle fibers; the same principles apply to cardiac muscle cells. ∞ *p. 285* If a muscle cell is stimulated when the resting length is very short, a powerful contraction cannot occur, because the sarcomeres are already at minimum size. The degree of shortening and the amount of tension produced during a contraction increase as the resting length of the muscle increases. The range of optimal resting length is narrow. As a muscle cell stretches *beyond* its optimal length, its abilities to shorten and produce tension decline again, because its sarcomeres no longer function efficiently. If stretched to the point at which the thick and thin filaments in its sarcomeres no longer overlap, a muscle cell cannot contract.

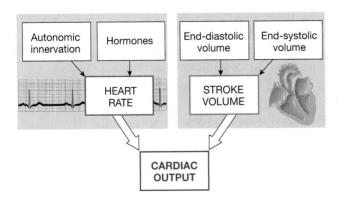

•**FIGURE 20-19** An Introduction to Factors Affecting Cardiac Output

The amount of preload varies with the demands on the heart. When you are standing at rest, your EDV is low; the ventricular muscle is stretched very little, and the sarcomeres are relatively short. During ventricular systole, the cardiac muscle cells develop little power, and the ESV is relatively high because the muscle cells contract only a short distance. If you begin exercising, venous return increases and more blood flows into your heart. Your EDV increases, and the myocardium stretches further. As the sarcomeres approach optimal lengths, the ventricular muscle cells can contract more efficiently and produce more forceful contractions. They also shorten more, and more blood is pumped out of your heart.

In general, the greater the EDV, the larger the stroke volume. Stretching *past* optimal length, which would reduce the force of contraction, does not normally occur because ventricular expansion is limited by myocardial connective tissues, the fibrous skeleton, and the pericardial sac.

The Frank–Starling Principle The relationship between the amount of ventricular stretching and the contractile force means that, within normal physiological limits, increasing the EDV results in a corresponding increase in the stroke volume. This general rule of "more in = more out" was first proposed by Ernest H. Starling on the basis of an analysis of research performed by Otto Frank. The relationship is therefore known as the **Frank–Starling principle,** or *Starling's law of the heart.*

Autonomic adjustments to cardiac output make the effects of the Frank–Starling principle difficult to see. However, it can be demonstrated effectively in individuals who have received a heart transplant, because the implanted heart is not innervated by the ANS. The most obvious effect of the Frank–Starling principle in these hearts is that the outputs of the left and right ventricles remain balanced under a variety of conditions. For example, consider an individual at rest, with the two ventricles ejecting equal volumes of blood. Although the two ventricles contract together, they function in series: When the heart contracts, blood leaving the right ventricle heads to the lungs; during the next ventricular diastole, that volume of blood will pass through the left atrium, to be ejected by the left ventricle at the next contraction. If the venous return decreases, the EDV of the right ventricle will decline. During ventricular systole it will then pump less blood into the pulmonary circuit. In the next cardiac cycle, the EDV of the left ventricle will be reduced, and that ventricle will eject a smaller volume of blood. The output of the two ventricles will again be in balance, but both will have smaller stroke volumes than they did initially.

Contractility **Contractility** is the amount of force produced during a contraction, at a given preload. Under normal circumstances, contractility can be altered by autonomic innervation or circulating hormones. Under special circumstances, contractility can be altered by drugs or as a result of abnormal changes in ion concentrations in the extracellular fluid.

Factors that increase contractility are said to have a *positive inotropic action;* those that decrease contractility have a *negative inotropic action.* Positive inotropic agents typically stimulate Ca^{2+} entry into cardiac muscle cells, thus increasing the force and duration of ventricular contractions. Negative inotropic agents may block Ca^{2+} movement or depress cardiac muscle metabolism. Positive and negative inotropic factors include ANS activity, hormones, and changes in extracelluar ion concentrations.

Autonomic Activity Autonomic activity alters the degree of contraction and changes the ESV in the following ways:

- Sympathetic stimulation has a positive inotropic effect. It causes the release of norepinephrine (NE) by postganglionic fibers of the cardiac nerves and the secretion of epinephrine (E) and NE by the adrenal medullae. In addition to their effects on heart rate, discussed below, these hormones stimulate cardiac muscle cell metabolism and increase the force and degree of contraction. These effects are produced by stimulating alpha and beta receptors in cardiac muscle cell membranes. ∞ *pp. 511–512* The net effect is that the ventricles contract more forcefully, increasing the ejection fraction and decreasing the ESV.

- Parasympathetic stimulation from the vagus nerve has a negative inotropic effect. The primary effect of acetylcholine (ACh) is at the membrane surface, where it produces hyperpolarization and inhibition. The primary result is a decrease in heart rate through effects on the SA and AV nodes. The force of cardiac contractions is also reduced; because the ventricles are not extensively innervated by the parasympathetic division, the atria show the greatest changes in contractile force. However, under strong parasympathetic stimulation or after the administration of drugs that mimic the actions of ACh, the ventricles contract less forcefully, the ejection fraction decreases, and the ESV enlarges.

Hormones Several hormones, including the following, affect the contractility of the heart:

- Epinephrine and norepinephrine have positive inotropic effects.

- Glucagon has a positive inotropic effect. Before synthetic inotropic agents were available, glucagon was widely used to stimulate cardiac function. It is still used in cardiac emergencies and to treat some forms of heart disease.

- Thyroid hormones have positive inotropic effects.

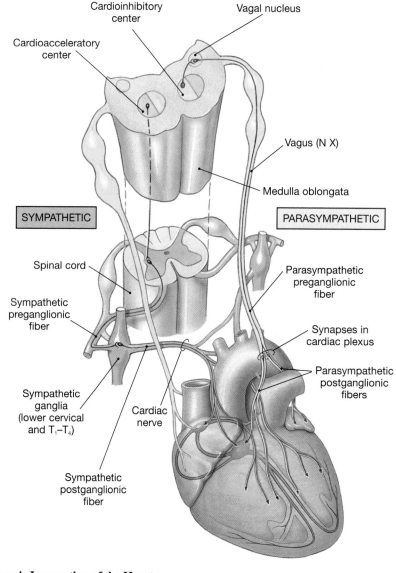

•**FIGURE 20-20** Autonomic Innervation of the Heart

innervation beats at 70–80 bpm due to activity in the parasympathetic nerves innervating the SA node. If parasympathetic activity increases, the heart rate declines further; the heart rate will increase if either parasympathetic activity decreases or sympathetic activation occurs. Through dual innervation and adjustments in autonomic tone, the ANS can make very delicate adjustments in cardiovascular function to meet the demands of other systems.

Effects on the SA Node The sympathetic and parasympathetic divisions alter the heart rate by changing the permeabilities of cells in the conducting system. The most dramatic effects are seen at the SA node, where changes in the rate of impulse generation affect the heart rate. As a secondary effect, autonomic activity also increases or decreases the rate of impulse propa-

gation along the conducting system. We will ignore these changes in propagation rate and will focus on the SA node and the modification of the heart rate.

Consider the SA node of a resting individual whose heart is beating at 75 bpm (Figure 20-21a•). Any factor that changes the rate of spontaneous depolarization or the duration of repolarization will alter the heart rate by changing the time required to reach threshold. Acetylcholine released by parasympathetic neurons opens chemically regulated K^+ channels in the cell membrane. This opening dramatically slows the rate of spontaneous depolarization and also slightly extends the duration of repolarization (Figure 20-21b•). The result is a decline in heart rate.

The NE released by sympathetic neurons binds to beta-1 receptors; this binding leads to the opening of calcium ion channels. The subsequent influx of

(a)

(b)

(c)

— Time (sec) —

•FIGURE 20-21 Pacemaker Function.
(a) Pacemaker cells have membrane potentials lower than those of other cardiac muscle cells (−60 mV versus −90 mV). Their cell membranes undergo spontaneous depolarization to threshold, producing action potentials at a frequency determined by (1) the resting membrane potential and (2) the rate of depolarization (slope of the prepotential). **(b)** Parasympathetic stimulation releases ACh, which extends repolarization and decreases the rate of spontaneous depolarization. The heart rate slows. **(c)** Sympathetic stimulation releases NE, which shortens repolarization and accelerates the rate of spontaneous depolarization. As a result, the heart rate increases.

20

Ca^{2+} increases the rate of depolarization and shortens the period of repolarization. The nodal cells reach threshold more quickly, and the heart rate increases (Figure 20-21c•).

The Atrial Reflex The **atrial reflex,** or *Bainbridge reflex,* involves adjustments in heart rate in response to an increase in the venous return. When the walls of the right atrium are stretched, the stimulation of stretch receptors in the atrial walls triggers a reflexive increase in heart rate caused by increased sympathetic activity (Figure 20-21c•).

Hormones

Just as epinephrine, norepinephrine, and thyroid hormones increase contractility, all these hormones also cause an increase in the heart rate. The effects of epinephrine on the SA node are similar to those of norepinephrine. Epinephrine also affects the contractile cells; after massive sympathetic stimulation of the adrenal medullae, the myocardium may become so excitable that abnormal contractions occur.

EXERCISE AND CARDIAC OUTPUT

In most healthy people, increasing both the stroke volume and the heart rate, as during heavy exercise, can raise the cardiac output by 300–500 percent, to 18–30 l/min. The difference between resting and maximal cardiac outputs is the **cardiac reserve.** Trained athletes exercising at maximal levels may increase cardiac output by nearly 700 percent, to 40 l/min. Cardiac output cannot increase indefinitely, primarily because the available filling time shortens as the heart rate increases. At heart rates up to 160–180 bpm, the combination of an increased rate of venous return and increased contractility compensates for the reduction in filling time. Over this range, cardiac output and heart rate increase together. But if the heart rate continues to climb, the stroke volume begins to drop. Cardiac output first plateaus and then declines.

Figure 20-22• summarizes factors involved in the regulation of heart rate and stroke volume—the factors that interact to determine cardiac output under normal conditions.

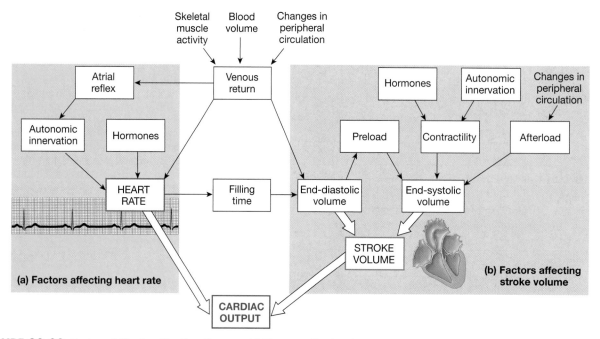

•**FIGURE 20-22** **Factors Affecting Cardiac Output.** **(a)** Factors affecting heart rate. **(b)** Factors affecting stroke volume.

Abnormal Conditions Affecting Cardiac Output

Under unusual conditions, various drugs, abnormal variations in ion concentrations, and changes in body temperature can alter the basic rhythm of contraction established by the SA node. In Chapter 12, we noted that several drugs, including caffeine and nicotine, have a stimulatory effect on excitable membranes in the nervous system. ∞ *p. 394* These drugs also cause an increase in heart rate. Caffeine acts directly on the conducting system and increases the rate of depolarization at the SA node. Nicotine acts indirectly by stimulating the activity of sympathetic neurons that innervate the heart.

Disorders affecting ion concentrations or body temperature can have direct effects on cardiac output by changing stroke volume, heart rate, or both. Abnormal ion concentrations can change both the contractility of the heart, by affecting the cardiac muscle cells, and the heart rate, by affecting the SA nodal cells. The most obvious and clinically important examples of problems with ion concentrations involve K^+ and Ca^{2+}.

Hyperkalemia and Hypokalemia

Changes in the extracellular K^+ concentration affect the transmembrane potential and the rates of depolarization and repolarization. In **hyperkalemia** (hī-per-ka-LĒ-mē-uh), in which K^+ concentrations are high, the muscle cells depolarize and repolarization is inhibited. Cardiac contractions become weak and irregular; in severe cases, the heart eventually stops in diastole. In **hypokalemia** (hī-pō-ka-LĒ-mē-uh), in which K^+ concentrations are

abnormally low, the membranes of cardiac muscle cells hyperpolarize and become less responsive to stimulation. Meanwhile, the hyperpolarization of nodal cells leads to a reduction in the heart rate. Blood pressure falls, and, in severe cases, the heart eventually stops in systole. Severe hyperkalemia and hypokalemia are life threatening; they require immediate corrective action.

Hypercalcemia and Hypocalcemia

In **hypercalcemia** (hī-per-kal-SĒ-mē-uh), a condition in which the extracellular concentration of Ca^{2+} is elevated, cardiac muscle cells become extremely excitable. Their contractions become powerful and prolonged. In extreme cases, the heart goes into an extended state of contraction that is generally fatal. In **hypocalcemia** (hī-pō-kal-SĒ-mē-uh), in which the Ca^{2+} concentration is abnormally low, the contractions become very weak and may cease altogether.

Abnormal Body Temperatures

Temperature changes affect metabolic operations throughout the body. For example, a reduction in temperature slows the rate of depolarization at the SA node, lowers the heart rate, and reduces the strength of cardiac contractions. (In open-heart surgery, the exposed heart may be chilled until it stops beating.) An elevated body temperature accelerates the heart rate and the contractile force. That is one reason your heart seems to race and pound when you have a fever.

THE HEART AND THE CARDIOVASCULAR SYSTEM

The goal of cardiovascular regulation is to maintain adequate blood flow to all body tissues. The heart cannot accomplish this goal by itself, and it does not work in isolation. For example, when blood pressure changes, the cardiovascular centers adjust not only the heart rate but also the diameters of peripheral blood vessels. These adjustments work together to keep blood pressure within normal limits and to maintain circulation to vital tissues and organs. Chapter 21 will complete this story by detailing the cardiovascular responses to changing activity patterns and circulatory emergencies. We will then conclude our discussion of the cardiovascular system by examining the anatomy of the pulmonary and systemic circuits.

☑ What effect would drinking large amounts of caffeinated coffee have on the heart?

☑ If the cardioinhibitory center of the medulla oblongata were damaged, which part of the autonomic nervous system would be affected, and how would the heart be affected?

☑ How would an increase in the extracellular concentration of calcium ions affect the strength of a cardiac contraction?

☑ How would a drug that increases the length of time required for the repolarization of pacemaker cells affect the heart rate?

The heart keeps our blood in motion. Without this push to the circulatory system, we would receive none of the benefits of blood, severely compromising homeostasis. To gain a better understanding of the anatomy and physiology of the heart, use the resources available on both the CD-ROM and the Companion Website. The CD-ROM includes animations and tutorials; the website builds on those resources with text content and interactive problems as well as communication avenues for sharing ideas with your class. To complete the following Web Explorations, visit the MediaLab in Chapter 20 of your Companion Website.

Web Exploration 1

Estimated time for completion = 10 minutes

An electrocardiogram (ECG) allows us to visualize the beating of the heart. Many heart anomalies can be detected by analysis of ECG recordings, but reading these tracings is an art. To learn how to read them, select the keyword **ECG** on the Companion Website. Look at the normal tracing, and link to the abnormal tracings and their explanations. Reconstruct a normal ECG recording; indicate the portions identified on the web page. Choose four representative abnormal tracings, too, and mark the portion of the tracing that is abnormal. Test yourself and your peers by reproducing tracings for later identification.

Web Exploration 2

Estimated time for completion = 5 minutes

What is a heart murmur? What causes this condition? What does it sound like? To explore these questions, select the keyword **MURMUR** on the Companion Website. After reading the page, click on the musical notes next to each heart murmur type to hear typical murmurs. Can you distinguish the sounds? Using the web page and the text as references, list the cause of each murmur, and try to describe the change in sound. If you have access to a stethoscope, listen to actual heartbeats. Can you distinguish what you are hearing?

20

CHAPTER REVIEW

SELECTED CLINICAL TERMINOLOGY
Terms Discussed in This Chapter

angina pectoris (an-JĪ-nuh PEK-tor-is): A condition in which exertion or stress produces severe chest pain, resulting from temporary ischemia when the heart's workload increases. *(p. 667)*

balloon angioplasty: A technique for reducing the size of a coronary plaque by compressing it against the arterial walls, using a catheter with an inflatable collar. *(p. 668)*

bradycardia (brā-dē-KAR-dē-uh): A heart rate that is slower than normal. *(p. 672 and AM)*

calcium channel blockers: Drugs that reduce contractility of the heart by slowing the influx of calcium ions during the plateau phase of the cardiac muscle action potential. *(p. 667)*

cardiac arrhythmias (ā-RITH-mē-az): Abnormal patterns of cardiac electrical activity, indicating abnormal contractions. *(p. 674 and AM)*

cardiac tamponade: A condition, resulting from pericardial irritation and inflammation, in which fluid collects in the pericardial sac and restricts cardiac output. *(p. 657 and AM)*

carditis (kar-DĪ-tis): A general term indicating inflammation of the heart. *(p. 664 and AM)*

conduction deficit: An abnormality in the conducting system of the heart that affects the timing and pacing of cardiac contractions. *(p. 672 and AM)*

coronary artery bypass graft (CABG): The routing of blood around an obstructed coronary artery (or one of its branches) by a vessel transplanted from another part of the body. *(p. 668)*

coronary artery disease (CAD): The obstruction of coronary circulation. *(p. 667)*

coronary ischemia: The restriction of the circulatory supply to the heart, potentially causing tissue damage and a reduction in cardiac efficiency. *(p. 667)*

coronary thrombosis: A blockage due to the formation of a clot (thrombus) at a plaque in a coronary artery. *(p. 679)*

electrocardiogram (ē-lek-trō-KAR-dē-ō-gram) (**ECG** or **EKG**): A recording of the electrical activities of the heart over time. *(p. 673)*

heart failure: A condition in which the heart weakens and peripheral tissues suffer from oxygen and nutrient deprivation. *(p. 677 and AM)*

myocardial (mī-ō-KAR-dē-al) **infarction (MI):** A condition in which the coronary circulation becomes blocked and cardiac muscle cells die from oxygen starvation; also called a *heart attack. (p. 679)*

pericarditis: Inflammation of the pericardium. *(p. 657)*

rheumatic heart disease (RHD): A disorder in which the heart valves become thickened and stiffen into a partially closed position, affecting the efficiency of the heart. *(p. 664 and AM)*

tachycardia (tak-ē-KAR-dē-uh): A heart rate that is faster than normal. *(p. 672 and AM)*

valvular heart disease (VHD): A condition caused by abnormal functioning of one of the cardiac valves. The severity of the condition depends on the degree of damage and the valve involved. *(p. 664 and AM)*

⏍ Additional Terms Discussed in the *Applications Manual*

atrial fibrillation: Rapid, uncontrolled, and uncoordinated spasms in the atrial walls.

atrial flutter: Rapid contractions of the atria that are ineffective in propelling blood.

cardiomyopathies (kar-dē-ō-mī-OP-a-thēz): A group of diseases characterized by the progressive, irreversible degeneration of the myocardium.

coronary arteriography: The production of an X-ray image of coronary circulation after the introduction of a radiopaque dye into one of the coronary arteries via a catheter; the resulting image is a *coronary angiogram.*

defibrillator: A device used to eliminate atrial or ventricular fibrillation and restore normal cardiac rhythm.

echocardiography: Ultrasound analysis of the heart and the blood flow through major vessels.

endocarditis: Inflammation of the endocardial lining.

heart block: A condition affecting the normal rhythm of the heart; characterized by impaired communication between the SA node, the AV node, and the ventricular myocardium as a result of damage to conduction pathways due to mechanical distortion, ischemia, infection, or inflammation.

myocarditis: Inflammation of the myocardium.

paroxysmal atrial tachycardia (PAT): Bursts of rapid atrial depolarization, leading to multiple atrial contractions with little pause between them.

premature atrial contractions (PACs): Atrial contractions that occur in advance of the rhythm established by the SA node.

premature ventricular contractions (PVCs): Premature ventricular beats that are directed by an ectopic pacemaker.

valvular stenosis (ste-NŌ-sis): A condition in which the opening between thickened heart valves becomes narrower.

ventricular escape: The initiation of ventricular contractions that are controlled by the AV node, Purkinje fibers, or other parts of the ventricular conducting system rather than by the SA node.

ventricular fibrillation (VF): The most serious type of cardiac arrhythmia, in which the normal rhythm becomes disrupted because the cardiac muscle fibers are stimulating one another at such a rapid rate. Ventricular contractions are rapid, uncontrolled, and uncoordinated; the ventricles are incapable of pumping blood.

ventricular tachycardia (VT): Rapid ventricular beats that are not coordinated with atrial activity.

STUDY OUTLINE

INTRODUCTION   p. 655

AN OVERVIEW OF THE CARDIOVASCULAR SYSTEM   p. 655

1. The circulatory pathways can be subdivided into the **pulmonary circuit** (which carries blood to and from the lungs) and the **systemic circuit** (which transports blood to and from the rest of the body). **Arteries** carry blood away from the **heart; veins** return blood to the heart. **Capillaries** are thin-walled, narrow-diameter vessels that connect the smallest arteries and veins. *(Figure 20-1)*

2. The heart has four chambers: the **right atrium** and **right ventricle** and the **left atrium** and **left ventricle.**

ANATOMY OF THE HEART   p. 655

1. The heart is surrounded by the **pericardial cavity** and lies within the anterior portion of the **mediastinum,** which separates the two pleural cavities. *(Figure 20-2)*

The Pericardium   p. 656

2. The pericardial cavity is lined by the **pericardium.** The **visceral pericardium (epicardium)** covers the heart's outer surface, and the **parietal pericardium** lines the inner surface of the **pericardial sac,** which surrounds the heart. *(Figure 20-2)*

Superficial Anatomy of the Heart   p. 657

3. The **coronary sulcus,** a deep groove, marks the boundary between the atria and the ventricles. Other surface markings also provide useful reference points in describing the heart and associated structures. *(Figure 20-3)*

The Heart Wall   p. 658

4. The bulk of the heart consists of the muscular **myocardium.** The **endocardium** lines the inner surfaces of the heart, and the **epicardium** covers the outer surface. *(Figures 20-4, 20-5)*

5. **Cardiac muscle cells** are interconnected by **intercalated discs,** which convey the force of contraction from cell to cell and conduct action potentials. *(Figure 20-5; Table 20-1)*

Internal Anatomy and Organization   p. 660

6. The atria are separated by the **interatrial septum,** and the ventricles are divided by the **interventricular septum.** The right atrium receives blood from the systemic circuit via two large veins, the **superior vena cava** and the **inferior vena cava.** (The atrial walls contain the **pectinate muscles,** prominent muscular ridges.) *(Figure 20-6)*

7. Blood flows from the right atrium into the right ventricle via the **right atrioventricular (AV) valve (tricuspid valve).** This opening is bounded by three **cusps** of fibrous tissue braced by the tendinous **chordae tendineae,** which are connected to **papillary muscles.** *(Figure 20-6)*

8. Blood leaving the right ventricle enters the **pulmonary trunk** after passing through the **pulmonary semilunar valve.** The pulmonary trunk divides to form the **left** and **right pulmonary arteries.** The **left** and **right pulmonary veins** return blood to the left atrium. Blood leaving the left atrium flows into the left ventricle via the **left atrioventricular (AV) valve (bicuspid,** or **mitral, valve).** Blood leaving the left ventricle passes through the **aortic semilunar valve** and into the systemic circuit via the **ascending aorta.** *(Figure 20-6)*

9. Anatomical differences between the ventricles reflect the functional demands placed on them. The wall of the right ventricle is relatively thin, whereas the left ventricle has a massive muscular wall. *(Figure 20-7)*

10. Valves normally permit blood flow in only one direction, preventing the **regurgitation** (backflow) of blood. *(Figure 20-8)*

Connective Tissues and the Fibrous Skeleton   p. 664

11. The connective tissues of the heart (mainly collagen and elastic fibers) and the **fibrous skeleton** support the heart's contractile cells and valves. *(Figure 20-8)*

The Blood Supply to the Heart   p. 664

12. The **coronary circulation** meets the high oxygen and nutrient demands of cardiac muscle cells. The **coronary arteries** originate at the base of the ascending aorta. Interconnections between arteries called **anastomoses** ensure a constant blood supply. The **great, posterior, small, anterior,** and **middle cardiac veins** carry blood from the coronary capillaries to the **coronary sinus.** *(Figure 20-9)*

13. In **coronary artery disease (CAD),** areas of the coronary circulation undergo partial or complete blockage. *(Figure 20-10)*

THE HEARTBEAT   p. 668

1. Two general classes of cardiac muscle cells are involved in the normal **heartbeat: contractile cells** and cells of the *conducting system.*

Contractile Cells   p. 668

2. Cardiac muscle cells have a long refractory period, so rapid stimulation produces isolated rather than tetanic contractions. *(Figure 20-11)*

The Conducting System   p. 670

3. The **conducting system** is composed of the *sinoatrial node,* the *atrioventricular node,* and *conducting cells.* This system initiates and distributes electrical impulses within the heart. Nodal cells establish the rate of cardiac contraction, and conducting cells distribute the contractile stimulus from node to node (along *internodal pathways*) and from the nodes to the ventricular myocardium. *(Figure 20-12)*

4. Unlike skeletal muscle, cardiac muscle contracts without neural or hormonal stimulation. **Pacemaker cells** in the **sinoatrial (SA) node** *(cardiac pacemaker)* normally establish the rate of contraction. From the SA node, the stimulus travels to the **atrioventricular (AV) node,** then to the **AV bundle,** which divides into **bundle branches.** From there, **Purkinje fibers** convey the impulses to the ventricular myocardium. *(Figure 20-13)*

The Electrocardiogram   p. 673

5. A recording of electrical activities in the heart is an **electrocardiogram (ECG** or **EKG).** Important landmarks of an ECG include the **P wave** (atrial depolarization), the **QRS complex** (ventricular depolarization), and the **T wave** (ventricular repolarization). *(Figure 20-14)*

The Cardiac Cycle   p. 674

6. The **cardiac cycle** contains periods of **atrial** and **ventricular systole** (contraction) and **diastole** (relaxation). *(Figure 20-15)*

7. When the heart beats, the two ventricles eject equal volumes of blood. *(Figure 20-16)*

8. The closing of valves and rushing of blood through the heart cause characteristic **heart sounds,** which can be heard during *auscultation. (Figure 20-17)*

CARDIODYNAMICS   p. 680

1. The amount of blood ejected by a ventricle during a single beat is the **stroke volume (SV).** The amount of blood pumped by a ventricle each minute is the **cardiac output (CO).** *(Figure 20-18)*

Overview: The Control of Cardiac Output   p. 681

2. Cardiac output can be adjusted by changes in either stroke volume or heart rate. *(Figure 20-19)*

Factors Affecting the Stroke Volume   p. 681

3. The stroke volume is the difference between the **end-diastolic volume (EDV)** and the **end-systolic volume (ESV).** The **filling time** and **venous return** interact to determine the EDV. Normally, the greater the EDV, the more powerful the succeeding contraction (the **Frank–Starling principle**).

Factors Affecting the Heart Rate   p. 683

4. The **cardioacceleratory center** in the medulla oblongata activates sympathetic neurons; the **cardioinhibitory center** governs the activities of the parasympathetic neurons. These cardiac centers receive inputs from higher centers and from receptors monitoring blood pressure and the concentrations of dissolved gases. *(Figure 20-20)*

5. The basic heart rate is established by the pacemaker cells of the SA node, but it can be modified by the autonomic nervous system. The **atrial reflex** accelerates the heart rate when the walls of the right atrium are stretched. *(Figure 20-21)*

6. Sympathetic activity produces more-powerful contractions that reduce the ESV. Parasympathetic stimulation slows the heart rate, reduces the contractile strength, and raises the ESV.

7. Cardiac output is affected by various factors, including autonomic innervation and hormones. *(Figure 20-22)*

Exercise and Cardiac Output   p. 685

8. The difference between resting and maximal cardiac outputs is the **cardiac reserve.**

THE HEART AND THE CARDIOVASCULAR SYSTEM   p. 687

1. The heart does not work in isolation in maintaining adequate blood flow to all tissues.

20

REVIEW QUESTIONS

LEVEL 1 REVIEWING FACTS AND TERMS

1. The blood supply to the muscles of the heart is provided by the
(a) systemic circulation   (b) pulmonary circulation
(c) coronary circulation   (d) coronary portal system

2. The autonomic centers for cardiac function are located in the
(a) myocardial tissue of the heart
(b) cardiac centers of the medulla oblongata
(c) cerebral cortex
(d) a, b, and c are correct

3. The serous membrane covering the outer surface of the heart is the
(a) parietal pericardium   (b) visceral pericardium
(c) myocardium   (d) endocardium

4. The simple squamous epithelium covering the valves of the heart constitutes the
(a) epicardium   (b) endocardium
(c) myocardium   (d) fibrous skeleton

5. The structure that permits blood flow from the right atrium to the left atrium while the lungs are developing before birth is the
(a) foramen ovale   (b) interatrial septum
(c) coronary sinus   (d) fossa ovalis

6. Blood leaves the left ventricle by passing through the
(a) aortic semilunar valve
(b) pulmonary semilunar valve
(c) mitral valve
(d) tricuspid valve

7. The QRS complex of the ECG appears as the
(a) atria depolarize   (b) ventricles depolarize
(c) ventricles repolarize   (d) atria repolarize

8. During diastole, a chamber of the heart
(a) relaxes and fills with blood
(b) contracts and pushes blood into an adjacent chamber
(c) experiences a sharp increase in pressure
(d) reaches a pressure of approximately 120 mm Hg

9. During the cardiac cycle, the amount of blood remaining in the ventricle when the semilunar valve closes is the
(a) stroke volume (SV)
(b) end-diastolic volume (EDV)
(c) end-systolic volume (ESV)
(d) cardiac output (CO)

10. What role do the chordae tendineae and papillary muscles play in the normal function of the AV valves?

11. What is the purpose of each of the three distinct layers that make up the heart wall?

12. What are the principal valves in the heart, and what is the function of each?

13. Trace the normal pathway of an electrical impulse through the conducting system of the heart.

14. What is the cardiac cycle? What phases and events are necessary to complete the cardiac cycle?

15. What three factors regulate stroke volume to ensure that the left and right ventricles pump equal volumes of blood?

LEVEL 2 REVIEWING CONCEPTS

16. During the absolute refractory period, a cell membrane cannot respond because
(a) sodium channels are open or are closed and inactivated
(b) potassium channels are temporarily closed
(c) the membrane is hyperpolarized
(d) a, b, and c are correct

17. Tetanic muscle contractions cannot occur in a normal cardiac muscle cell because
(a) cardiac muscle tissue contracts on its own
(b) there is no neural or hormonal stimulation
(c) the refractory period lasts until the muscle cell relaxes
(d) the refractory period ends before the muscle cell reaches peak tension

18. The amount of blood that is forced out of the heart depends on
(a) the degree of stretching at the end of ventricular diastole
(b) the contractility of the ventricle
(c) the amount of pressure required to eject blood
(d) a, b, and c are correct

19. The cardiac output cannot increase indefinitely because
(a) the available filling time becomes shorter as the heart rate increases
(b) the cardiovascular centers adjust the heart rate
(c) the rate of spontaneous depolarization decreases
(d) the ion concentrations of pacemaker cell membranes decrease

20. Describe the association of the four muscular chambers of the heart with the pulmonary and systemic circuits.

21. What are the source and significance of the heart sounds?

22. Differentiate between stroke volume and cardiac output. How is cardiac output calculated?

23. What factors influence cardiac output?

24. What is the significance of preload to cardiac muscle cells?

25. What effect does sympathetic stimulation have on the heart? What effect does parasympathetic stimulation have on the heart?

26. Describe the effects that epinephrine, norepinephrine, glucagon, and thyroid hormones have on the contractility of the heart.

LEVEL 3 CRITICAL THINKING AND CLINICAL APPLICATIONS

27. Most of the ATP produced in cardiac muscle is derived from the metabolism of fatty acids. During times of exertion, cardiac muscle cells can use lactic acid as an energy source. Why would this adaptation be advantageous to cardiac function?

28. A patient's ECG recording shows a consistent pattern of two P waves followed by a normal QRS complex and T wave. What is the cause of this abnormal wave pattern?

29. The following measurements were made on two individuals. (The values recorded remained stable for 1 hour.)
       Person 1: heart rate, 75 bpm; stroke volume, 60 ml
       Person 2: heart rate, 90 bpm; stroke volume, 95 ml
Which person has the greater venous return? Which person has the longer ventricular filling time?

30. Karen is taking the medication *verapamil,* a drug that blocks the calcium channels in cardiac muscle cells. What effect should this medication have on Karen's stroke volume?

31. After a myocardial infarction, the cells surrounding the damaged tissue frequently become hyperexcitable and act as ectopic pacemakers. This condition can lead to abnormal ventricular rhythms, with fatal consequences. What would cause the excitability of the uninjured cells?

32. Preventricular contractions (PVCs) occur when a Purkinje cell or contractile cell in the ventricle depolarizes to threshold, triggering a premature contraction. A person experiencing a PVC may feel that her heart has "skipped a beat." If the PVC causes an extra contraction, why does the person feel that a beat has been skipped?

20

CHAPTER 21

Blood Vessels and Circulation

When we think of the cardiovascular system, we think first of the heart or of the great blood vessels that service it. But the real work of the cardiovascular system is done in microscopic vessels that permeate most tissues. This is a network of capillaries, delicate vessels that permit diffusion between blood and interstitial fluid. Capillaries weave throughout active tissues, forming intricate networks that surround muscle fibers, radiate through connective tissues, and branch beneath the basement membranes of epithelia. Capillaries provide oxygen and nutrients and remove carbon dioxide and wastes. Homeostatic mechanisms operating at the local, regional, and systemic levels adjust blood flow through the capillaries to meet the demands of peripheral tissues. In this chapter, we consider these mechanisms and the structure and distribution of all the blood vessels.

CHAPTER OUTLINE AND OBJECTIVES

INTRODUCTION   p. 692

THE ANATOMY OF BLOOD VESSELS   p. 692
   The Structure of Vessel Walls   p. 692
   Differences between Arteries and Veins   p. 692
   Arteries   p. 693
   Capillaries   p. 695
   Veins   p. 699
   The Distribution of Blood   p. 700
① Distinguish among the types of blood vessels on the basis of their structure and function.
② Describe how and where fluid and dissolved materials enter and leave the cardiovascular system.

CARDIOVASCULAR PHYSIOLOGY   p. 701
   Pressure   p. 702
   Resistance   p. 702
   An Overview of Circulatory Pressures   p. 704
   Arterial Blood Pressure   p. 705
   Capillary Exchange   p. 706
   Venous Pressure and Venous Return   p. 710
③ Explain the mechanisms that regulate blood flow through arteries, capillaries, and veins.
④ Describe the factors that influence blood pressure and how blood pressure is regulated.
⑤ Discuss the mechanisms and various pressures involved in the movement of fluids between capillaries and interstitial spaces.

CARDIOVASCULAR REGULATION   p. 711
   Autoregulation of Blood Flow within Tissues   p. 712
   Neural Mechanisms   p. 712
   Hormones and Cardiovascular Regulation   p. 715
⑥ Describe how central and local control mechanisms interact to regulate blood flow and pressure in tissues.
⑦ Explain how the activities of the cardiac, vasomotor, and respiratory centers are coordinated to control blood flow through the tissues.

PATTERNS OF CARDIOVASCULAR RESPONSE   p. 717
   Exercise and the Cardiovascular System   p. 717
   Cardiovascular Response to Hemorrhaging   p. 719

■ **FOCUS: Special Circulation**   p. 722

⑧ Explain how the cardiovascular system responds to the demands of exercise, hemorrhaging, and shock.
⑨ Identify the principal blood vessels and the functional characteristics of the special circulation to the brain, heart, and lungs.

THE BLOOD VESSELS   p. 723
   The Pulmonary Circuit   p. 723
   The Systemic Circuit   p. 724
   Fetal Circulation   p. 742
⑩ Identify the major arteries and veins and the areas they serve.

AGING AND THE CARDIOVASCULAR SYSTEM   p. 744
⑪ Discuss the effects of aging on the cardiovascular system.

INTEGRATION WITH OTHER SYSTEMS   p. 745
⑫ Give examples of interactions between the cardiovascular system and each of the other organ systems.

MEDIALAB and CHAPTER REVIEW   p. 745

In the last two chapters, we examined the composition of blood and the structure and function of the heart, whose pumping action keeps the blood in motion. We shall now consider the vessels that carry blood to peripheral tissues and the nature of the exchange that occurs between the blood and interstitial fluids of the body.

Blood leaves the heart in the pulmonary and aortic trunks, each with an internal diameter of about 2.5 cm (1 in.). The pulmonary arteries that branch from the pulmonary trunk carry blood to the lungs. The systemic arteries that branch from the aorta distribute blood to all other organs. Within these organs, further branching occurs, creating several hundred million tiny arteries that provide blood to more than 10 billion capillaries. These capillaries, barely the diameter of a single red blood cell, form extensive, branching networks. If all the capillaries in the body were placed end to end, they would circle the globe with a combined length of over 25,000 miles.

The vital functions of the cardiovascular system depend entirely on events at the capillary level: All chemical and gaseous exchange between blood and interstitial fluid takes place across capillary walls. Tissue cells rely on capillary diffusion to obtain nutrients and oxygen and to remove metabolic wastes, such as carbon dioxide and urea. Diffusion occurs very rapidly, because the distances involved are very small; few cells lie farther than 125 μm (0.005 in.) from a capillary.

Arteries carry blood away from the heart. As they enter peripheral tissues, arteries branch repeatedly, and the branches decrease in diameter. The smallest arterial branches are called **arterioles** (ar-TĒ-rē-ōlz). From the arterioles, blood moves into the **capillaries,** where diffusion occurs between blood and interstitial fluid. From the capillaries, blood enters small **venules** (VĒ-nūlz), which unite to form larger branches, **veins,** that return blood to the heart. This chapter begins with a description of the histological organization of arteries, capillaries, and veins. We then consider the functions of these vessels and basic principles of cardiovascular regulation. We also examine the distribution of major blood vessels in the body. We will then be ready to consider the organization and function of the lymphatic system, the focus of Chapter 22.

THE ANATOMY OF BLOOD VESSELS

THE STRUCTURE OF VESSEL WALLS

The walls of arteries and veins contain three distinct layers—the tunica interna, tunica media, and tunica externa (Figure 21-1•):

1. The **tunica interna** (in-TER-nuh), or *tunica intima,* is the innermost layer of a blood vessel. This layer includes the endothelial lining and an underlying layer of connective tissue with a variable number of elastic fibers. In arteries, the outer margin of the tunica interna contains a thick layer of elastic fibers called the **internal elastic membrane.**

2. The **tunica media,** the middle layer, contains concentric sheets of smooth muscle tissue in a framework of loose connective tissue. The collagen fibers bind the tunica media to the tunica interna and tunica externa. The tunica media is commonly the thickest layer in the wall of a small artery. It is separated from the surrounding tunica externa by the **external elastic membrane,** a thin band of elastic fibers. The smooth muscle cells of the tunica media encircle the endothelium lining the lumen of the blood vessel. When these smooth muscles contract, the vessel decreases in diameter; when they relax, the diameter increases. Large arteries also contain layers of longitudinally arranged smooth muscle cells.

3. The **tunica externa** (eks-TER-nuh), or *tunica adventitia* (ad-ven-TISH-ē-uh), is the outermost layer and forms a connective tissue sheath around the vessel. In arteries, this layer contains collagen fibers with scattered bands of elastic fibers. In veins, this layer, which is generally thicker than the tunica media, contains networks of elastic fibers and bundles of smooth muscle cells. The connective tissue fibers of the tunica externa typically blend into those of adjacent tissues, stabilizing and anchoring the blood vessel.

Their layered walls give arteries and veins considerable strength. The muscular and elastic components also permit controlled alterations in diameter as blood pressure or blood volume changes. However, the walls of arteries and veins are too thick to allow diffusion between the bloodstream and surrounding tissues, or even between the blood and the tissues of the vessel itself. For this reason, the walls of large vessels contain small arteries and veins that supply the smooth muscle cells and fibroblasts of the tunica media and tunica externa. These blood vessels are called the *vasa vasorum* ("vessels of vessels").

DIFFERENCES BETWEEN ARTERIES AND VEINS

Arteries and veins servicing the same region typically lie side by side (Figure 21-1•). In sectional view, arteries and veins may be distinguished by the following characteristics:

- In general, the walls of arteries are thicker than those of veins. The tunica media of an artery contains more smooth muscle and elastic fibers than does that of a vein. These contractile and elastic components resist the pressure generated by the heart as it forces blood into the circuit.

- When not opposed by blood pressure, the elastic fibers in the arterial walls recoil, constricting the lumen. Thus,

21

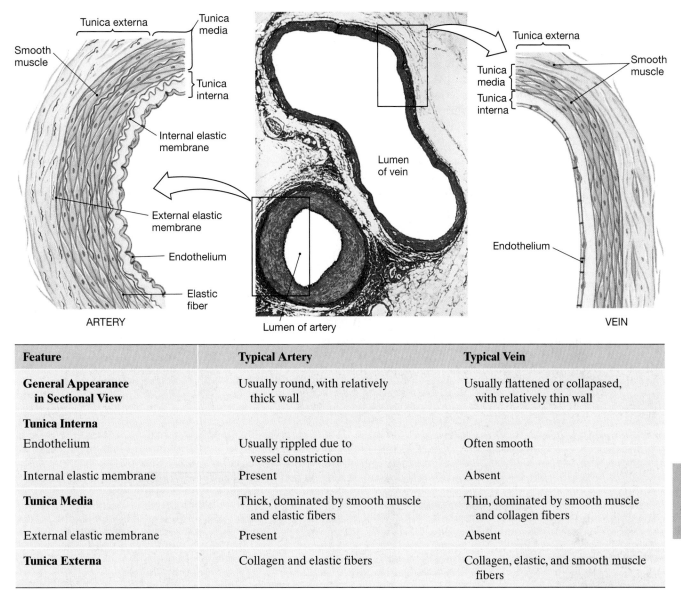

Feature	Typical Artery	Typical Vein
General Appearance in Sectional View	Usually round, with relatively thick wall	Usually flattened or collapased, with relatively thin wall
Tunica Interna		
Endothelium	Usually rippled due to vessel constriction	Often smooth
Internal elastic membrane	Present	Absent
Tunica Media	Thick, dominated by smooth muscle and elastic fibers	Thin, dominated by smooth muscle and collagen fibers
External elastic membrane	Present	Absent
Tunica Externa	Collagen and elastic fibers	Collagen, elastic, and smooth muscle fibers

•FIGURE 21-1 A Comparison of a Typical Artery and a Typical Vein

when seen on dissection or in sectional view, the lumen of an artery often looks smaller than that of the corresponding vein. Because the walls of arteries are relatively thick and strong, they retain their circular shape in section. Cut veins tend to collapse, and in section they often look flattened or grossly distorted.

- The endothelial lining of an artery cannot contract, so when an artery constricts, its endothelium is thrown into folds that give arterial sections a pleated appearance. The lining of a vein lacks these folds.

In gross dissection, arteries and veins can generally be distinguished because:

- The thicker walls of the arteries can be felt if the vessels are compressed.
- Arteries usually retain their cylindrical shape, whereas veins often collapse.

- Arteries are more resilient; when stretched, they keep their shape and elongate, and when released they snap back. A small vein cannot tolerate as much distortion without collapsing or tearing.
- Veins typically contain *valves,* internal structures that prevent the backflow of blood toward the capillaries. In an intact vein, the location of each valve is marked by a slight distension of the vessel wall. (We will consider valve structure in a later section.)

ARTERIES

Their relatively thick, muscular walls make arteries *elastic* and *contractile.* Elasticity permits passive changes in vessel diameter in response to changes in blood pressure. It allows arteries to absorb the pressure pulses that accompany the contractions of the ventricles.

The contractility of the arterial walls gives them the ability to change in diameter actively, primarily under the control of the sympathetic division of the autonomic nervous system. When stimulated, arterial smooth muscles contract and thereby constrict the artery—a process called **vasoconstriction.** Relaxation of these smooth muscles increases the diameter of the lumen—a process called **vasodilation.** Vasoconstriction and vasodilation affect (1) the afterload on the heart, (2) peripheral blood pressure, and (3) capillary blood flow. We will explore these effects in a later section. Contractility is also important during the *vascular phase* of hemostasis, when the contraction of a damaged vessel wall helps reduce bleeding. ∞ *p. 645*

In traveling from the heart to peripheral capillaries, blood passes through *elastic arteries, muscular arteries,* and *arterioles* (Figure 21-2•).

Elastic Arteries

Elastic arteries, or *conducting arteries,* are large vessels with a diameter up to 2.5 cm (1 in.). These vessels transport large volumes of blood away from the heart. The pulmonary and aortic trunks and their major arterial branches (the *pulmonary, common*

•**FIGURE 21-2 Histological Structure of Blood Vessels.** Representative diagrammatic cross-sectional views of the walls of arteries, veins, and capillaries. Notice the relative sizes of the layers in these vessels.

carotid, subclavian, and *common iliac arteries*) are elastic arteries.

The walls of elastic arteries (Figure 21-2●) are extremely resilient. The tunica media of these vessels contains a high density of elastic fibers and relatively few smooth muscle cells. As a result, elastic arteries can tolerate the pressure changes that occur during the cardiac cycle. During ventricular systole, pressures rise rapidly and the elastic arteries expand. During ventricular diastole, blood pressure within the arterial system falls and the elastic fibers recoil to their original dimensions. Their expansion cushions the sudden rise in pressure during ventricular systole, and their recoil slows the drop in pressure during ventricular diastole. This feature is important because blood pressure is the driving force behind blood flow: The greater the pressure oscillations, the greater the changes in blood flow. The elasticity of the arterial system dampens the pressure peaks and valleys that accompany the heartbeat. By the time blood reaches the arterioles, the pressure oscillations have disappeared and blood flow is continuous.

Muscular Arteries

Muscular arteries, also known as *medium-sized arteries* or *distribution arteries,* distribute blood to the body's skeletal muscle and internal organs. A typical muscular artery has a lumen diameter of approximately 0.4 cm (0.15 in.). Muscular arteries are characterized by a thick tunica media that contains more smooth muscle cells than does the tunica media of elastic arteries (Figures 21-1 and 21-2●). The *external carotid arteries* of the neck, the *brachial arteries* of the arms, and the *femoral arteries* of the thighs are muscular arteries.

Arterioles

Arterioles, with an internal diameter of 30 μm or less, are considerably smaller than muscular arteries. Arterioles have a poorly defined tunica externa, and the tunica media in the larger arterioles consists of one or two layers of smooth muscle cells (Figure 21-2●). The tunica media of the smallest arterioles contains scattered smooth muscle cells that do not form a complete layer.

The diameters of smaller muscular arteries and arterioles change in response to local conditions or to sympathetic or endocrine stimulation. For example, arterioles in most tissues vasodilate when oxygen levels are low and, as we saw in Chapter 16, vasoconstrict under sympathetic stimulation. ∞ *p. 512* Changes in their diameter affect the amount of force required to push blood around the cardiovascular system: More pressure is required to push blood through a constricted vessel than through a dilated one. The force opposing blood flow is called **resistance (R),** and arterioles are therefore called **resistance vessels.**

Vessel characteristics change gradually as we travel away from the heart. For example, elastic and muscular arteries are interconnected; the largest muscular arteries contain a considerable amount of elastic tissue, whereas the smallest resemble heavily muscled arterioles.

Aneurysms

An **aneurysm** (AN-ū-rizm) is a bulge in the weakened wall of an artery. This bulge resembles a bubble in the wall of a tire; like a bad tire, the affected artery can suffer a catastrophic blowout. The most dangerous aneurysms are those involving arteries of the brain, where they cause strokes, or of the aorta, where a ruptured aneurysm will cause fatal bleeding in a matter of minutes.

Aneurysms most commonly occur in individuals with *arteriosclerosis.* Over time, arteriosclerosis causes vessel walls to become less elastic, and a weak spot can develop. Aneurysms can also be associated with other conditions that weaken arterial walls, such as arterial inflammation or infection, and with *Marfan's syndrome,* a connective tissue disorder. ∞ *pp. 120, 181* As you might expect, individuals with high blood pressure are most likely to develop dangerous aneurysms, because the elevated arterial pressures place great stresses on the vessel walls. Unfortunately, because they are often painless, aneurysms are likely to go undetected. AM *Aneurysms*

CAPILLARIES

Capillaries are the only blood vessels whose walls permit exchange between the blood and the surrounding interstitial fluids. Because capillary walls are thin, the diffusion distances are small, so exchange can occur quickly. In addition, blood flows through capillaries relatively slowly, allowing sufficient time for the diffusion or active transport of materials across the capillary walls. Thus, the histological structure of capillaries permits a two-way exchange of substances between blood and interstitial fluid.

A typical capillary consists of an endothelial tube inside a delicate basement membrane. Neither a tunica media nor a tunica externa is present. The average diameter of a capillary is a mere 8 μm, very close to that of a single red blood cell. There are two major types of capillaries: (1) *continuous capillaries* and (2) *fenestrated capillaries.*

Continuous Capillaries

Most regions of the body are supplied by continuous capillaries. In a **continuous capillary,** the capillary endothelium is a complete lining. A cross section through a large continuous capillary cuts across several endothelial cells (Figure 21-3a●). In a small continuous capillary, a single endothelial cell may wrap all the way around the lumen, as your hand wraps around a small glass.

21

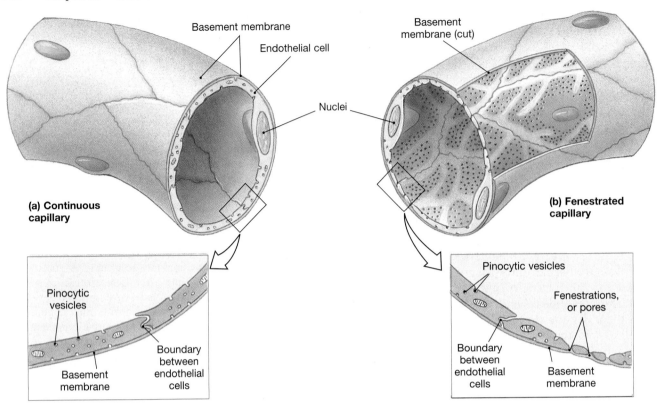

•**FIGURE 21-3 Capillary Structure. (a)** A continuous capillary, showing routes for the diffusion of water and solutes. **(b)** A fenestrated capillary, showing the pores that allow diffusion across the endothelial lining.

Arteriosclerosis

Arteriosclerosis (ar-tē-rē-ō-skle-RŌ-sis) is a thickening and toughening of arterial walls. This condition may not sound life-threatening, but complications related to arteriosclerosis account for roughly half of all deaths in the United States. The effects of arteriosclerosis are varied; for example, arteriosclerosis of coronary vessels is responsible for *coronary artery disease (CAD)*, and arteriosclerosis of arteries supplying the brain can lead to strokes. ∞ *p. 667* Arteriosclerosis takes two major forms, focal calcification and atherosclerosis:

1. **Focal calcification** is the gradual degeneration of smooth muscle in the tunica media and the subsequent deposition of calcium salts. This process typically involves arteries of the limbs and genital organs. Some focal calcification occurs as part of the aging process, and it may develop in association with atherosclerosis. Rapid and severe calcification may occur as a complication of diabetes mellitus, an endocrine disorder. ∞ *p. 607*

2. **Atherosclerosis** (ath-er-ō-skle-RŌ-sis) is associated with damage to the endothelial lining and the formation of lipid deposits in the tunica media. This is the most common form of arteriosclerosis.

Many factors may be involved in the development of atherosclerosis. One major factor is lipid levels in the blood. Atherosclerosis tends to develop in persons whose blood contains elevated levels of plasma lipids, specifically cholesterol. Circulating cholesterol is transported to peripheral tissues in *lipoproteins,* protein–lipid complexes. (We shall discuss the various types of lipoproteins in Chapter 25.) Recent evidence indicates that many forms of atherosclerosis are associated with either (1) low levels of *apolipoprotein-E (ApoE),* a transport protein whose lipids are quickly absorbed by body tissues, or (2) high levels of *lipoprotein(a), a low-density lipoprotein (LDL)* that is absorbed at a much slower rate.

When ApoE levels are low or lipoprotein(a) levels are high, cholesterol-rich lipoproteins remain in circulation for an extended period. Circulating monocytes then begin removing them from the bloodstream. Eventually, the monocytes become filled with lipid droplets. Now called *foam cells,* they attach themselves to the endothelial walls of blood vessels, where they release growth factors. These cytokines stimulate the divisions of smooth muscle cells near the tunica interna, thickening the vessel wall.

Other monocytes then invade the area, migrating between the endothelial cells. As these changes occur, the monocytes, smooth muscle cells, and endothelial cells begin phagocytizing lipids as well. The result is a **plaque,** a fatty mass of tissue that projects into the lumen of the vessel. At this point, the plaque has a relatively simple structure, and evidence suggests that the process can be reversed if appropriate dietary adjustments are made.

If the conditions persist, the endothelial cells become swollen with lipids, and gaps appear in the endothelial lining. Platelets now begin sticking to the exposed collagen fibers. The combination of platelet adhesion and aggregation leads to the formation of a localized blood clot, which will further restrict blood flow through the artery. The structure of the plaque is now relatively complex. Plaque growth can be halted, but the structural changes are generally permanent.

Typical plaques are shown in Figure 21-4•. Elderly individuals, especially elderly men, are most likely to develop atherosclerotic plaques. Evidence suggests that estrogens may slow plaque formation; this may account for the lower incidence of CAD, myocardial infarctions (MIs), and strokes in women. After menopause, when estrogen production declines, the risk of CAD, MIs, and strokes in women increases markedly.

In addition to advanced age and male gender, other important risk factors include high blood cholesterol levels, high blood pressure, and cigarette smoking. Roughly 20 percent of middle-aged men have all three of these risk factors; these individuals are four times more likely to experience an MI or a cardiac arrest than are other men in their age group. Although fewer women develop atherosclerotic plaques, elderly women smokers with high blood cholesterol and high blood pressure are at much greater risk than are other women. Factors that can promote the development of atherosclerosis in both men and women include diabetes mellitus, obesity, and stress. Evidence also indicates that at least some forms of atherosclerosis may be linked to chronic infection with *Chlamydia pneumoniae,* a bacterium responsible for several types of respiratory infections, including some forms of pneumonia.

We discussed potential treatments for atherosclerotic plaques, such as catheterization, balloon angioplasty, and bypass surgery, in Chapter 20. ∞ *p. 668.* In cases in which dietary modifications do not lower circulating LDL levels sufficiently, drug therapies can bring them under control. Genetic engineering techniques have recently been used to treat an inherited form of *hypercholesterolemia* (high blood cholesterol) linked to extensive plaque formation. (Individuals with this condition are unable to absorb and recycle cholesterol in the liver.) In this experimental procedure, circulating cholesterol levels declined after copies of appropriate genes were inserted into some of the individual's liver cells.

Without question, the best approach to atherosclerosis is to try to avoid it by eliminating or reducing associated risk factors. Suggestions include (1) reducing the amount of dietary cholesterol and saturated fats by restricting consumption of fatty meats (such as beef, lamb, and pork), egg yolks, and cream; (2) not smoking; (3) checking your blood pressure and taking steps to lower it if necessary; (4) having your blood cholesterol levels checked annually; (5) controlling your weight; and (6) exercising regularly.

21

(a)

Plaque deposit in vessel wall

(b)

•**FIGURE 21-4 A Plaque Blocking an Artery. (a)** A section of a coronary artery narrowed by plaque formation. **(b)** A sectional view of a large plaque. (LM × 18)

Continuous capillaries are located in all tissues except epithelia and cartilage. Continuous capillaries permit the diffusion of water, small solutes, and lipid-soluble materials into the surrounding interstitial fluid but prevent the loss of blood cells and plasma proteins. In addition, some transport may occur between blood and interstitial fluid by vesicular transport, the movement of vesicles that form at the inner endothelial surface. ∞ *p. 77*

The endothelial cells in specialized continuous capillaries throughout most of the central nervous system and in the thymus are bound together by tight junctions. These capillaries have very restricted permeability characteristics. We discussed one example, the capillaries responsible for the *blood–brain barrier,* in Chapters 12 and 14. ∞ *pp. 368, 444–445*

Fenestrated Capillaries

Fenestrated (FEN-es-trā-ted) **capillaries** (*fenestra,* window) are capillaries that contain "windows," or pores, that span the endothelial lining (Figure 21-3b•). The pores permit the rapid exchange of water and solutes as large as small peptides between plasma and interstitial fluid. Examples of fenestrated capillaries noted in earlier chapters include the *choroid plexus* of the brain and the blood vessels in a variety of endocrine organs, including the hypothalamus and the pituitary, pineal, and thyroid glands. Fenestrated capillaries are also located along absorptive areas of the intestinal tract and at filtration sites in the kidneys. Both the number of pores and their permeability characteristics may vary from one region of the capillary to another.

Sinusoids (SĪ-nus-oydz) resemble fenestrated capillaries that are flattened and irregular. In contrast to fenestrated capillaries, sinusoids commonly have gaps between adjacent endothelial cells, and the basement membrane is either thinner or absent. As a result, sinusoids permit the free exchange of water and solutes as large as plasma proteins between blood and interstitial fluid.

Blood moves through sinusoids relatively slowly, maximizing the time available for exchange across the sinusoidal walls. Sinusoids occur in the liver, bone marrow, spleen, and many endocrine organs, such as the pituitary and adrenal glands. At liver sinusoids, plasma proteins secreted by the liver cells enter the bloodstream. Along sinusoids of the liver, spleen, and bone marrow, phagocytic cells monitor the passing blood, engulfing damaged red blood cells, pathogens, and cellular debris.

Capillary Beds

Capillaries do not function as individual units but as part of an interconnected network called a **capillary bed,** or **capillary plexus** (Figure 21-5•). A single arteriole generally gives rise to dozens of capillaries that empty into several *venules,* the smallest vessels of the venous system. The entrance to each capillary is guarded by a band of smooth muscle called a **precapillary sphincter.** Contraction of the smooth muscle cells constricts and narrows the diameter of the capillary entrance, thereby reducing the flow of blood. Relaxation of the sphincter dilates the opening, allowing blood to enter the capillary faster.

The capillary bed contains several relatively direct connections between arterioles and venules. The wall in the initial part of such a passageway contains smooth muscles capable of changing its diameter. This segment is called a **metarteriole** (met-ar-TĒ-rē-ōl). The rest of the passageway, which resembles a typical capillary in structure, is called a **thoroughfare channel.**

A capillary bed may receive blood from more than one artery. The multiple arteries, called **collaterals,** enter the region and fuse before giving rise to arterioles. The fusion of two collateral arteries that supply a capillary bed is an example of an **arterial anastomosis.** (An *anastomosis* is the joining of two tubes.) The interconnections between the *anterior* and *posterior interventricular branch* arteries of the heart are arterial anastomoses. ∞ *pp. 664, 666* An arterial anastomosis acts like an insurance policy: If one artery is compressed or blocked, capillary circulation will continue.

Arteriovenous (ar-tē-rē-ō-VĒ-nus) **anastomoses** are direct connections between arterioles and venules. When an arteriovenous anastomosis is dilated, blood will bypass the capillary bed and flow directly into the venous circulation. The pattern of blood flow through an arteriovenous anastomosis is regulated primarily by sympathetic innervation under the control of the cardiovascular centers of the medulla oblongata.

Vasomotion

Although blood normally flows from arterioles to venules at a constant rate, the flow within each capillary is quite variable. Each precapillary sphincter alternately contracts and relaxes, perhaps a dozen times per minute. As a result, the blood flow within any capillary occurs in pulses rather than as a steady and constant stream. The net effect is that blood may reach the venules by one route now and by a different route later. The cycling of contraction and relaxation of smooth muscles that changes blood flow through capillary beds is called **vasomotion.**

Vasomotion is controlled locally by changes in the concentrations of chemicals and dissolved gases in the interstitial fluids. For example, when dissolved oxygen concentrations decline within a tissue, the capillary sphincters relax, so blood flow to the area increases. This process, an example of capillary *autoregulation,* will be the focus of a later section.

When you are at rest, blood flows through roughly 25 percent of the vessels within a capillary bed. Your

(b)

(a)

•**FIGURE 21-5** The Organization of a Capillary Bed. **(a)** A typical capillary bed. **(b)** A micrograph of a capillary bed. Solid arrows indicate consistent blood flow; dashed arrows indicate variable or pulsating blood flow.

cardiovascular system does not contain enough blood to maintain adequate blood flow to all the capillaries in all the capillary beds in your body at the same time. As a result, when many tissues become active, the blood flow through capillary beds must be coordinated. We shall describe the mechanisms by which the cardiovascular centers perform this coordination later in the chapter.

VEINS

Veins collect blood from all tissues and organs and return it to the heart. The walls of veins are thinner than those of corresponding arteries. Venous walls need not be as thick as arterial walls because the blood pressure in veins is lower than that in the arteries. Veins are classified on the basis of their size. Even though their walls are thinner, in general, veins are larger in diameter than their corresponding arteries. Review Figures 21-1, p. 693, and 21-2•, p. 694, to compare typical arteries and veins.

Venules

Venules, which collect blood from capillary beds, are the smallest venous vessels. They vary widely in size and character. An average venule has an internal di-

ameter of roughly 20 μm. Venules smaller than 50 μm lack a tunica media, and the smallest venules resemble expanded capillaries.

Medium-Sized Veins

Medium-sized veins range from 2 to 9 mm in internal diameter, comparable in size to muscular arteries. In these veins, the tunica media is thin and contains relatively few smooth muscle cells. The thickest layer of a medium-sized vein is the tunica externa, which contains longitudinal bundles of elastic and collagen fibers.

Large Veins

Large veins include the superior and inferior venae cavae and their tributaries within the abdominopelvic and thoracic cavities. All the tunica layers are present in large veins. The slender tunica media is surrounded by a thick tunica externa composed of a mixture of elastic and collagen fibers.

Venous Valves

The arterial system is a high-pressure system: It takes almost all the force developed by the heart to push blood through the network of arteries and across miles of capillaries. Blood pressure in a peripheral venule is

21

only about 10 percent of that in the ascending aorta, and pressures continue to fall along the venous system.

The blood pressure in venules and medium-sized veins is so low that it cannot oppose the force of gravity. In the limbs, veins of this size contain valves (Figure 21-6•). **Valves** are folds of the tunica interna that project from the vessel wall and point in the direction of blood flow. These valves act like the valves in the heart in that they permit blood flow in one direction only. Venous valves prevent the backflow of blood toward the capillaries.

As long as the valves function normally, any movement that distorts or compresses a vein will push blood toward the heart. This effect improves *venous return,* the rate of blood flow to the heart. ∞ *p. 681* This mechanism is particularly important when you are standing, because blood returning from your feet must overcome the pull of gravity to ascend to the heart. Valves compartmentalize the blood within the veins, thereby dividing the weight of the blood between the compartments. Any movement in the surrounding skeletal muscles squeezes the blood toward the heart. Although you are probably not aware of it, when you stand, rapid cycles of contraction and relaxation are occurring within your leg muscles. These contractions help push blood toward the trunk. When you lie down, venous valves have much less impact on venous return, because your heart and major vessels are at the same level.

If the walls of the veins near the valves weaken or become stretched and distorted, the valves may not work properly. Blood then pools in the veins, and the vessels become grossly distended. The effects range from mild discomfort and a cosmetic problem, as in superficial *varicose veins* in the thighs and legs, to painful distortion of adjacent tissues, as in *hemorrhoids.* AM *Problems with Venous Valve Function*

THE DISTRIBUTION OF BLOOD

The total blood volume is unevenly distributed among arteries, veins, and capillaries (Figure 21-7•). The heart, arteries, and capillaries normally contain 30–35 percent of the blood volume (roughly 1.5 liters of whole blood), and the venous system contains the rest (65–70 percent, or about 3.5 liters). Of the blood in the venous system, roughly one-third (about a liter) is circulating within the liver, bone marrow, and skin. These organs contain extensive venous networks that at any moment contain large volumes of blood.

Because their walls are thinner and contain a lower proportion of smooth muscle, veins are much more distensible (expandable) than arteries. For a given rise in blood pressure, a typical vein will stretch about eight times as much as a corresponding artery. The *capacitance* of a blood vessel is the relationship between the volume of blood it contains and the blood pressure. If the vessel behaves like a child's balloon, expanding easily with little pressure, it has high capacitance. If it behaves more like a truck tire, expanding only when large pressures are applied, it has low capacitance. Veins expand easily, so they are called **capacitance vessels.** Because veins have high capacitance, large changes in blood volume have little effect on blood pressure. If the blood volume rises or falls, the elastic walls stretch or recoil, changing the volume of blood in the venous system.

If serious hemorrhaging occurs, the *vasomotor centers* of the medulla oblongata stimulate sympathetic nerves that innervate smooth muscle cells in the walls of medium-sized veins. This activity has two major effects:

1. *Systemic veins constrict.* This process is called **venoconstriction** (vē-nō-kon-STRIK-shun). Venoconstriction reduces the amount of blood within the venous system and increases the volume within the arterial system and capillaries. This effect can keep the blood volume within the arteries and capillaries at near-normal levels despite a significant blood loss.

Valve closed

Valve open

Valve closed

•**FIGURE 21-6 The Function of Valves in the Venous System.** Valves in the walls of medium-sized veins prevent the backflow of blood. Venous compression caused by the contraction of adjacent skeletal muscles assists in maintaining venous blood flow.

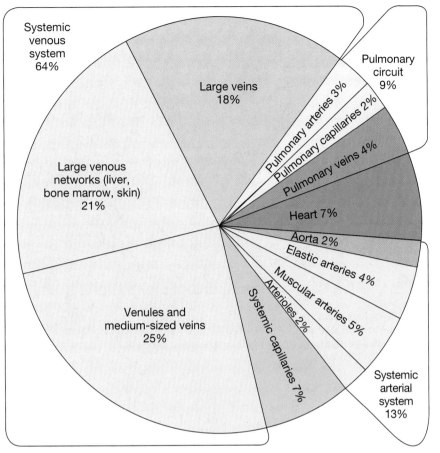

•**FIGURE 21-7** The Distribution of
Blood in the Cardiovascular System

2. *The constriction of veins in the liver, skin, and lungs redistributes a significant proportion of the total blood volume.* As a result, blood flow to delicate organs, such as the brain, and to active skeletal muscles can be increased or maintained after a blood loss. The amount of blood that can be shifted from veins in the liver, skin, and lungs to the general circulation is called the **venous reserve.** The venous reserve normally is about 20 percent of the total blood volume.

☑ A cross section of tissue shows several small, thin-walled vessels with very little smooth muscle tissue in the tunica media. Which type of vessels are these?

☑ Why are valves located in veins but not in arteries?

☑ Where in the body would you find fenestrated capillaries?

CARDIOVASCULAR PHYSIOLOGY

The goal of cardiovascular regulation is the maintenance of adequate blood flow through peripheral tissues and organs. Under normal circumstances, blood flow is equal to cardiac output. When cardiac output goes up, so does the blood flow through capillary beds; when cardiac output declines, capillary blood flow is reduced. We considered the primary factors involved in the regulation of cardiac output in Chapter 20. ∞ *p. 681* The afterload of the heart is determined by the interplay between *pressure* and *resistance* in the cardiovascular network. If there were no resistance to blood flow in the cardiovascular system, the heart would not have to generate pressure to force blood around the pulmonary and systemic circuits.

Figure 21-8• provides an overview of our discussion of cardiovascular physiology. We will begin by considering the nature of blood pressure and resistance. We will pay particular attention to arterial blood pressure, which is responsible for maintaining blood flow within capillaries. Diffusion and osmosis occur within the capillaries, between the blood and the surrounding interstitial fluid. This *capillary exchange* performs vital functions: providing tissues with oxygen and nutrients and removing the carbon dioxide and waste products generated by active cells. Blood then leaves the capillaries and enters the venous system. There, the low venous blood pressure, aided by valves, skeletal muscle contraction, and other factors, returns the blood to the heart. The factors regulating this venous return were considered in Chapter 20. ∞ *pp. 681, 686*

The cardiovascular system must be continuously adjusted to maintain homeostasis. Active tissues

21

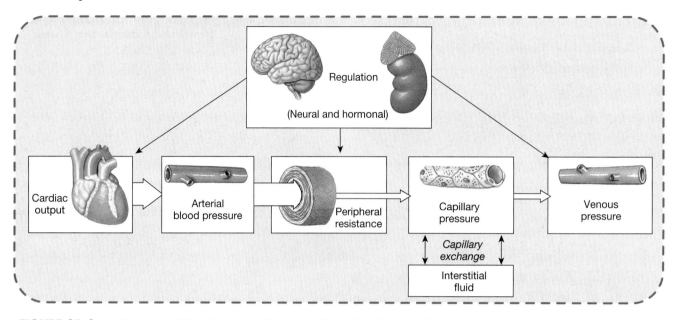

•**FIGURE 21-8 An Overview of Cardiovascular Physiology.** Neural and hormonal activities influence cardiac output, blood pressure, peripheral resistance, and venoconstriction. Peripheral resistance controls blood flow and capillary pressure, which drives the exchange between blood and interstitial fluid. This figure will be repeated, in reduced and simplified form as Navigator icons, in key figures throughout this chapter as we change topics.

require more blood flow than inactive ones; even something as simple as a change in position—going from sitting to standing, for instance—triggers a number of cardiovascular changes. We will end this section with a discussion of what those changes are and how they are coordinated. Some occur automatically, at the local level, but many others are coordinated by the nervous and endocrine systems.

PRESSURE

Liquids, including blood, cannot be compressed. A force exerted against a liquid generates **hydrostatic pressure (HP),** a fluid pressure that is conducted in all directions. ∞ *p. 74* If a pressure gradient exists, hydrostatic pressure will push a liquid from an area of higher pressure toward an area of lower pressure. The water in the pipes of your home is under hydrostatic pressure that is generated by some combination of pressure pumps and gravity. That hydrostatic pressure is higher than atmospheric pressure; hence when you open a faucet, water flows out. The greater the water pressure, the faster the water flow. In other words, the flow rate is directly proportional to the pressure gradient.

In the systemic circuit of the cardiovascular system, the pressure gradient is the **circulatory pressure,** the pressure difference between the base of the ascending aorta and the entrance to the right atrium. Circulatory pressures average about 100 mm Hg. This relatively high pressure is needed primarily to force blood

through the arterioles—resistance vessels—and into peripheral capillaries.

For convenience, the circulatory pressure is often divided into three components:

1. *Blood pressure.* When referring to arterial pressure, we will use the term **blood pressure (BP)** to distinguish it from the total circulatory pressure. Capillary blood flow is directly proportional to blood pressure, which is closely regulated by a combination of neural and hormonal mechanisms. Blood pressure in the systemic arterial system ranges from an average of 100 mm Hg to roughly 35 mm Hg at the start of a capillary network.

2. *Capillary hydrostatic pressure.* **Capillary hydrostatic pressure (CHP),** or *capillary pressure* is the pressure within capillary beds. Along the length of a typical capillary, pressures decline from roughly 35 mm Hg to about 18 mm Hg.

3. *Venous pressure.* **Venous pressure** is the pressure within the venous system. Venous pressure is quite low; the pressure gradient from the venules to the right atrium is only about 18 mm Hg.

RESISTANCE

A resistance is any force that opposes movement. If you put a kink in a garden hose or cover the mouth of the hose with your finger, you will both increase the resistance and decrease the water flow. With the tap closed at the kitchen sink, the resistance is sufficient to stop the water flow. The more open the tap, the lower the resistance and the faster the water flow.

The resistance of the cardiovascular system opposes the movement of blood. The greater the resistance, the slower the blood flow. For circulation to occur, the pressure gradient must be great enough to overcome the **total peripheral resistance,** the resistance of the entire cardiovascular system. Because the resistance of the venous system is very low, for reasons we will describe below, attention focuses on the **peripheral resistance (PR),** the resistance of the arterial system. For blood to flow into peripheral capillaries, the pressure gradient must be great enough to overcome the peripheral resistance. The higher the peripheral resistance, the lower the rate of blood flow: $F \propto \Delta P/R$. In words, the flow, F, is directly proportional to the pressure gradient, ΔP, and inversely proportional to the resistance, R. Sources of peripheral resistance include *vascular resistance, viscosity,* and *turbulence.*

Vascular Resistance

Vascular resistance is the resistance of the blood vessels; it is the largest component of peripheral resistance. The most important factor in vascular resistance is friction between blood and the vessel walls. The amount of friction depends on the length and diameter of the vessel.

Vessel Length Increasing the length of a blood vessel increases friction: The longer the vessel, the larger the surface area in contact with blood. For example, you can easily blow the water out of a snorkel that is 2.5 cm (1 in.) in diameter and 25 cm (10 in.) long, but you cannot blow the water out of a 15-m-long garden hose, because the total friction is too great. The most dramatic changes in blood vessel length occur between birth and maturity, as growth occurs. In adults, vessel length can increase or decrease gradually when the individual gains or loses weight, but on a day-to-day basis this component of vascular resistance can be considered constant.

Vessel Diameter The effects of friction on blood act primarily in a narrow zone closest to the vessel wall. In a small-diameter vessel, nearly all the blood is slowed down by friction with the walls. Resistance is therefore relatively high. Blood near the center of a large-diameter vessel does not encounter resistance from friction with the walls, so the resistance is therefore relatively low.

Differences in diameter have much more significant effects on resistance than do differences in length. If two vessels have equal diameter but one is twice as long as the other, the longer vessel will offer twice as much resistance to blood flow. But with two vessels of equal length, one twice the diameter of the other, the narrower one will offer 16 times as much resistance to blood flow. This relationship, expressed in terms of the vessel radius, r, and resistance, R, can be summarized as $R \propto 1/r^4$.

More significantly, vessel length is constant, but vessel diameter can change. Most of the peripheral resistance occurs in arterioles, the smallest vessels of the arterial system. As we noted earlier in the chapter, arterioles are extremely muscular; the wall of an arteriole with a lumen diameter of 30 μm can have a 20-μm-thick layer of smooth muscle. When these smooth muscles contract or relax, peripheral resistance increases or decreases. Because a small change in diameter produces a large change in resistance, mechanisms that alter the diameters of arterioles provide control over peripheral resistance and blood flow.

Viscosity

Viscosity, introduced in Chapter 19, is the resistance to flow caused by interactions among molecules and suspended materials in a liquid. ∞ *p. 626* Liquids of low viscosity, such as water (viscosity 1.0), flow at low pressures; thick, syrupy fluids, such as molasses (viscosity 300), flow only under higher pressures. Whole blood has a viscosity about five times that of water, owing to the presence of plasma proteins and blood cells. Under normal conditions, the viscosity of blood remains stable, but anemia, polycythemia, and other disorders that affect the hematocrit also change blood viscosity and peripheral resistance.

Turbulence

High flow rates, irregular surfaces, and sudden changes in vessel diameter upset the smooth flow of blood, creating eddies and swirls. This phenomenon, called **turbulence,** slows the flow and increases resistance.

Turbulence normally occurs when blood flows between the atria and the ventricles and between the ventricles and the aortic and pulmonary trunks. It also develops in large arteries, such as the aorta, when cardiac output and arterial flow rates are very high. However, turbulence seldom occurs in smaller vessels unless their walls are damaged. For example, scar tissue formation at an injury site or the development of an atherosclerotic plaque creates abnormal turbulence and restricts blood flow. Because of the distinctive sound, or *bruit* (broo-Ē), produced by this turbulence, the presence of plaques in large blood vessels can often be detected with a stethoscope.

Table 21-1 provides a quick review of the terms and relationships we discussed in this section.

21

TABLE 21-1 Some of the Key Terms and Relationships Pertaining to Blood Circulation

Blood Flow (F): The volume of blood flowing per unit of time through a vessel or group of vessels; may refer to circulation through a capillary, a tissue, an organ, or the entire vascular network. Total blood flow is equal to cardiac output.

Blood Pressure (BP): The hydrostatic pressure in the arterial system that pushes blood through capillary beds.

Circulatory Pressure: The pressure difference between the base of the ascending aorta and the entrance to the right atrium.

Hydrostatic Pressure: A pressure exerted by a liquid in response to an applied force.

Peripheral Resistance (PR): The resistance of the arterial system; affected by such factors as vascular resistance, viscosity, and turbulence.

Resistance (R): A force that opposes movement (in this case, blood flow).

Total Peripheral Resistance: The resistance of the entire cardiovascular system.

Turbulence: A resistance due to the irregular, swirling movement of blood at high flow rates or to exposure to irregular surfaces.

Vascular Resistance: A resistance due to friction within a blood vessel, primarily between the blood and the vessel walls. Increases with increasing length or decreasing diameter; vessel length is constant, but vessel diameter can change.

Venous Pressure: The hydrostatic pressure in the venous system.

Viscosity: A resistance to flow due to interactions among molecules within a liquid.

Relationships Among these Terms:

$F \propto \Delta P$ (flow is proportional to the pressure gradient)

$F \propto 1/R$ (flow is inversely proportional to resistance)

$F \propto \Delta P/R$ (flow is directly proportional to the pressure gradient and inversely proportional to resistance)

$F \propto BP/PR$ (flow is directly proportional to blood pressure and inversely proportional to peripheral resistance)

$R \propto 1/r^4$ (resistance is inversely proportional to the fourth power of the vessel radius)

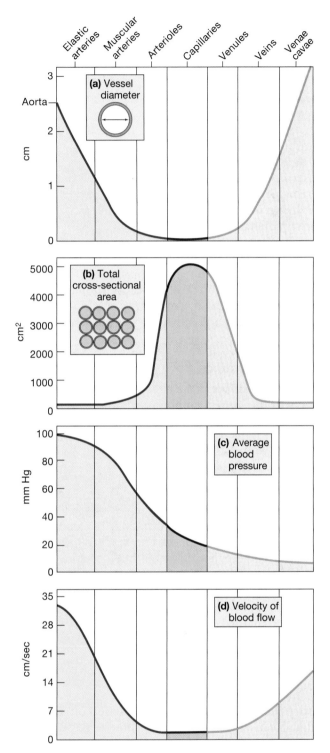

•**FIGURE 21-9** Relationships among Vessel Diameter, Cross-Sectional Area, Blood Pressure, and Blood Velocity

AN OVERVIEW OF CIRCULATORY PRESSURES

The graphs in Figure 21-9• provide an overview of the vessel diameters, areas, pressures, and velocity of blood flow in the systemic circuit. Notice the following:

■ As you proceed from the aorta toward the capillaries, divergence occurs and the arteries branch repeatedly; each branch is smaller in diameter than the preceding one (Figure 21-9a•). As you proceed from the capillaries toward the venae cavae, convergence occurs and the diameters increase as venules combine to form small and medium-sized veins.

■ Although the arterioles, capillaries, and venules are small in diameter, the body has a great many of them (Figure 21-9b•). All the blood flowing through the aorta also flows through peripheral capillaries. In effect, your blood moves from one big pipe into countless tiny ones. The blood pressure and the speed of blood flow are proportional to the cross-sectional area of the vessels involved. What is important is not the cross-sectional area of each individual vessel but their *combined* cross-sectional area. The aorta has a cross-sectional area of 4.5 cm², and all the blood entering the systemic circuit flows through it. That systemic blood also flows through the peripheral capillaries. Collectively, these capillaries have a cross-sectional area of 5000 cm².

- As arterial branching occurs, the cross-sectional area increases and blood pressure falls rapidly. Most of the decline occurs in the small arteries and arterioles of the arterial system; venous pressures are relatively low (Figure 21-9c•).

- Like a fast-flowing river delivering water to a floodplain, blood flow decreases in velocity as the total cross-sectional area of the vessels increases from the aorta toward the capillaries; it then rises as the cross-sectional area drops from the capillaries toward the venae cavae (Figure 21-9d•).

Figure 21-10• graphs the blood pressure throughout the cardiovascular system. Systemic pressures are highest in the aorta, peaking at about 120 mm Hg, and reach a minimum at the entrance to the right atrium. Pressures in the pulmonary circuit are much lower than those in the systemic circuit; the right ventricle is not a high-pressure pump, and the pulmonary vessels are much shorter and more distensible than are systemic vessels and thus provide less resistance to blood flow.

ARTERIAL BLOOD PRESSURE

Arterial pressure is important, because it maintains blood flow through capillary beds. To do this, it must always be high enough to overcome the peripheral resistance. Arterial pressure is not stable; it rises during ventricular systole and falls during ventricular diastole. The peak blood pressure measured during ventricular systole is called **systolic pressure,** and the minimum blood pressure at the end of ventricular diastole is called **diastolic pressure.** In recording blood pressure, we separate systolic and diastolic pressures by a slash mark, as in "120/80" ("one-twenty over eighty") or "110/75."

A *pulse* is a rhythmic pressure oscillation that accompanies each heartbeat. The difference between the systolic and diastolic pressures is the **pulse pressure** (Figure 21-10•). To report a single blood pressure value, we use the **mean arterial pressure (MAP).** The mean arterial pressure is calculated by adding one-third of the pulse pressure to the diastolic pressure:

$$\text{MAP} = \text{diastolic pressure} + \frac{\text{pulse pressure}}{3}$$

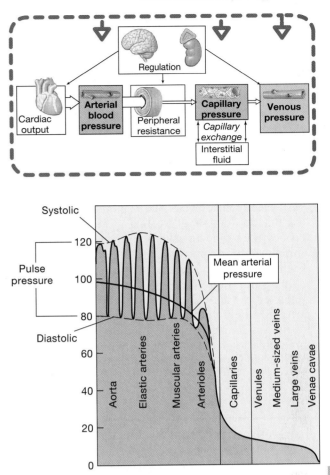

•**FIGURE 21-10** **Pressures within the Cardiovascular System.** Notice the general reduction in circulatory pressure within the systemic circuit and the elimination of the pulse pressure within the arterioles. The Navigator icon in the dashed box repeats Figure 21-8 but highlights the factors that are relevant to *this* figure, so you can see how circulatory pressures relate to the rest of cardiovascular physiology.

21

For a systolic pressure of 120 mm Hg and a diastolic pressure of 90 mm Hg, the MAP is 100 mm Hg:

$$\text{MAP} = 90 + \frac{120 - 90}{3} = 90 + 10 = 100 \text{ mm Hg}$$

Elastic Rebound

As systolic pressure climbs, the arterial walls stretch, just as an extra puff of air expands a partially inflated balloon. This expansion allows the arterial system to accommodate some of the blood provided by ventricular systole. When diastole begins and blood pressures fall, the arteries recoil to their original dimensions. The aortic semilunar valve prevents return of blood to the heart, so arterial recoil pushes blood toward the capillaries. This phenomenon is **elastic rebound.**

Pressures in Small Arteries and Arterioles

The mean arterial pressure and the pulse pressure become smaller as the distance from the heart increases (Figure 21-10•):

- The mean arterial pressure declines as the arterial branches become smaller and more numerous. In essence, the blood pressure decreases as it produces blood flow and overcomes friction.

- The pulse pressure fades as a result of the cumulative effects of elastic rebound. Each arterial segment reduces the magnitude of the pressure change that is experienced by its downstream neighbors. The effect is like that of a loud shout creating a series of ever-softer echoes. Each time an echo is produced, the reflecting surface absorbs some of the sound energy. Eventually, the echo disappears. The pressure surge accompanying ventricular ejection is the shout, and it is reflected by the wall of the aorta, echoing down the arterial system until it finally disappears at the level of the small arterioles. By the time blood reaches a precapillary sphincter, no pressure oscillations remain and the blood pressure remains steady at approximately 35 mm Hg.

Hypertension and Hypotension

Hypertension is the presence of abnormally high blood pressure. The usual criterion for hypertension in adults is a blood pressure greater than 150/90. One study estimated that 20 percent of the Caucasian U.S. population has blood pressures greater than 160/95 and that another 25 percent is on the borderline, with pressures above 140/90. The figures for other racial groups vary; the incidence of hypertension among African Americans is roughly twice that of Caucasian Americans. The extent to which these data reflect genetic rather than environmental factors is not known.

Hypertension significantly increases the workload on the heart, and the left ventricle gradually enlarges. More muscle mass means a greater oxygen demand. When the coronary circulation cannot keep pace, symptoms of coronary ischemia appear. ∞ *p. 667* Increased arterial pressures also place a physical stress on the walls of blood vessels throughout the body. This stress promotes or accelerates the development of arteriosclerosis and increases the risk of aneurysms, heart attacks, and strokes.

Hypotension is low blood pressure. This condition is most often seen in patients who have received overly aggressive treatment for hypertension. [AM] *Hypertension and Hypotension*

CAPILLARY EXCHANGE

As blood flows through peripheral tissues, blood pressure forces water and solutes out of the plasma, across capillary walls. Most of this material is reabsorbed by the capillaries, but about 3.6 liters of water and solutes flows through peripheral tissues each day and enters the *lymphatic system,* which empties into the bloodstream. The continuous movement of water out of the capillaries, through peripheral tissues, then back to the bloodstream by way of the lymphatic system has four important functions:

1. It ensures that plasma and interstitial fluid, two major components of extracellular fluid, are in constant communication.

2. It accelerates the distribution of nutrients, hormones, and dissolved gases throughout tissues.

3. It assists in the transport of insoluble lipids and tissue proteins that cannot enter the bloodstream by crossing the capillary walls.

4. It has a flushing action that carries bacterial toxins and other chemical stimuli to lymphoid tissues and organs responsible for providing immunity from disease.

Capillary exchange thus plays an important role in homeostasis. Next we will consider the factors and mechanisms involved in capillary exchange. The most important processes that move materials across typical capillary walls are *diffusion, filtration,* and *reabsorption.*

Diffusion

As we saw in Chapter 3, *diffusion* is the net movement of ions or molecules from an area where their concentration is higher to an area where their concentration is lower. ∞ *p. 71* The difference between the high and low concentrations represents a *concentration gradient,* and diffusion tends to eliminate that gradient. Diffusion occurs most rapidly when (1) the distances involved are small, (2) the concentration gradient is large, and (3) the ions or molecules involved are small.

Diffusion across capillary walls can occur by five routes:

1. Water, ions, and small organic molecules, such as glucose, amino acids, and urea, can usually enter or leave the bloodstream by diffusion between adjacent endothelial cells or through the pores of fenestrated capillaries.

2. Many ions, including sodium, potassium, calcium, and chloride, can diffuse across the endothelial cells by passing through channels in the cell membranes.

3. Large water-soluble compounds are unable to enter or leave the bloodstream except at fenestrated capillaries, such as those of the hypothalamus, the kidneys, many endocrine organs, and the intestinal tract.

4. Lipids, such as fatty acids and steroids, and lipid-soluble materials, including soluble gases such as oxygen and carbon dioxide, can cross capillary walls by diffusion through the endothelial cell membranes.

5. Plasma proteins are normally unable to cross the endothelial lining anywhere except in sinusoids, such as those of the liver, where plasma proteins enter the bloodstream.

Checking the Pulse and Blood Pressure

You can feel your pulse within any of the large or medium-sized arteries. The usual procedure involves using your fingertips to squeeze an artery against a relatively solid mass, preferably a bone. When the vessel is compressed, you feel your pulse as a pressure against your fingertips. The inside of the wrist is commonly used, because the *radial artery* can easily be pressed against the distal portion of the radius (Figure 21-23•, p. 726). Other accessible arteries include the *external carotid, brachial, temporal, facial, femoral,* and *popliteal arteries* (Figures 21-23, 21-24, and 21-28•, pp. 726, 728, and 733). Firm pressure exerted at these locations, called **pressure points,** can reduce or eliminate arterial bleeding distal to the site.

Blood pressure not only forces blood through the circulatory system, but it also pushes outward against the walls of the containing vessels, just as air pushes against the walls of an inflated balloon. As a result, we can measure blood pressure indirectly by determining how forcefully the blood presses against the vessel walls.

The instrument used to measure blood pressure is called a **sphygmomanometer** (sfig-mō-ma-NOM-e-ter; *sphygmos,* pulse + *manometer,* device for measuring pressure). An inflatable cuff is placed around the arm in such a position that its inflation compresses the brachial artery (Figure 21-11•). A stethoscope is placed over the artery distal to the cuff, and the cuff is then inflated. A tube connects the cuff to a pressure gauge that reports the cuff pressure in millimeters of mercury (mm Hg). Inflation continues until the cuff pressure is roughly 30 mm Hg above the pressure sufficient to collapse the brachial artery completely, stop the flow of blood, and eliminate the sound of the pulse.

The investigator then slowly lets the air out of the cuff. When the pressure in the cuff falls below systolic pressure, blood can again enter the artery. At first, blood enters only at peak systolic pressures, and the stethoscope picks up the sound of blood pulsing through the artery. As the pressure falls further, the sound changes, because the artery is remaining open for longer and longer periods. When the cuff pressure falls below diastolic pressure, blood flow becomes continuous and the sound of the pulse becomes muffled or disappears. Thus the pressure at which the pulse appears corresponds to the peak systolic pressure; when the pulse fades, the pressure has reached diastolic levels. The distinctive sounds heard during this test are called *sounds of Korotkoff* (sometimes spelled *Korotkov* or *Korotkow*). These sounds are produced by turbulence as blood flows past the constricted portion of the artery.

•**FIGURE 21-11** Using a Sphygmomanometer to Measure Blood Pressure

Filtration

The driving force for filtration is hydrostatic pressure. As we saw earlier, hydrostatic pressure pushes water from an area of higher pressure to an area of lower pressure. Blood pressure in a capillary is the capillary hydrostatic pressure (CHP).

In *capillary filtration,* water is forced across a capillary wall and small solute molecules travel with the water (Figure 21-12•). The solute molecules must be small enough to pass between adjacent endothelial cells or through the pores in a fenestrated capillary; larger solutes and suspended proteins are filtered out and remain in the bloodstream.

Along the length of a typical capillary, blood pressure gradually falls from about 35 mm Hg to roughly 18 mm Hg, the pressure at the start of the venous system. Filtration occurs primarily at the arterial end of a capillary, where CHP is highest.

Reabsorption

Reabsorption occurs as the result of osmosis. *Osmosis* is a special term used to refer to the diffusion of water across a selectively permeable membrane separating two solutions of differing solute concentrations. Water molecules tend to diffuse across a membrane toward the solution containing the higher solute concentration (Figure 3-7•, p. 73).

The **osmotic pressure (OP)** of a solution is an indication of the force of osmotic water movement—in other words, the pressure that must be applied to

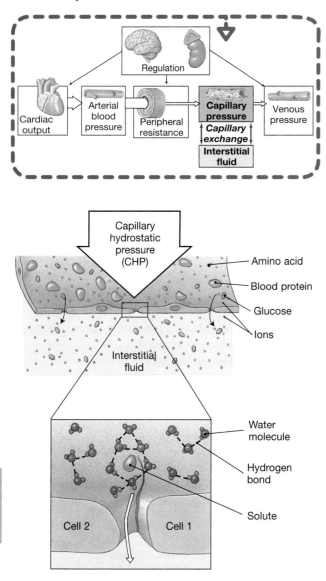

•**FIGURE 21-12 Capillary Filtration.** Capillary hydrostatic pressure forces water and solutes through the gaps between adjacent endothelial cells in continuous capillaries. The size of solutes that move across the capillary wall is determined primarily by the dimensions of the gaps. Notice that the relevant portions of the dashed box are highlighted.

prevent osmotic movement across a membrane. The higher the solute concentration of a solution, the greater the solution's osmotic pressure. The osmotic pressure of the blood is also called *blood colloid osmotic pressure (BCOP),* because only the suspended proteins are unable to cross the capillary walls.[1] Osmotic water movement will continue until either the solute concentrations are equalized or the movement is prevented by an opposing hydrostatic pressure.

---

[1] Clinicians often use the term *oncotic pressure (onkos,* a swelling) when referring to the colloid osmotic pressure of body fluids. The two terms are equivalent.

We will now consider the interplay between filtration and reabsorption along the length of a typical capillary. As the discussion proceeds, remember that hydrostatic pressure forces water *out of* a solution, whereas osmotic pressure draws water *into* a solution.

The Interplay between Filtration and Reabsorption

The rates of filtration and reabsorption gradually change as blood passes along the length of a capillary. The factors involved are diagrammed in Figure 21-13•.

The *net hydrostatic pressure* tends to push water and solutes out of capillaries and into the interstitial fluid. The net hydrostatic pressure is the difference between

1. the capillary hydrostatic pressure (CHP), which ranges from 35 mm Hg at the arterial end of a capillary to 18 mm Hg at the venous end, and

2. the *hydrostatic pressure of the interstitial fluid (IHP).* Measurements of IHP have yielded very small values that differ from tissue to tissue—from +6 mm Hg in the brain to −6 mm Hg in subcutaneous tissues. A positive IHP opposes CHP, and the tissue hydrostatic pressure must be overcome before fluid can move out of a capillary. A negative IHP assists CHP, and additional fluid will be pulled out of the capillary.

The *net colloid osmotic pressure* tends to pull water and solutes into a capillary from the interstitial fluid. The net colloid osmotic pressure is the difference between

1. the *blood colloid osmotic pressure (BCOP),* which is roughly 25 mm Hg, and

2. the *interstitial fluid colloid osmotic pressure (ICOP).* The ICOP is as variable and low as the IHP, because the interstitial fluid in most tissues contains negligible quantities of suspended proteins. Reported values of ICOP are from 0 to 5 mm Hg, within the range of pressures recorded for the IHP.

Because (1) the ICOP and the IHP oppose one another, (2) these pressures are of comparable size, and (3) they are difficult to measure accurately, both values have been set at 0 mm Hg in our model in Figure 21-13•. This value is acceptable because CHP and BCOP are the primary forces acting across a capillary wall under normal circumstances. However, the method of calculation described here would still apply, regardless of the values selected for IHP and ICOP.

The **net filtration pressure (NFP)** is the difference between the net hydrostatic pressure and the net osmotic pressure. In terms of the factors listed above, this means that

CHP = Capillary hydrostatic pressure
BCOP = Blood colloid osmotic pressure
NFP = Net filtration pressure

Lymphatic capillary

Returned to circulation

3.6 l/day reabsorbed into lymphatic capillaries

Arteriole

Venule

24 l/day moves out of capillaries

No net fluid movement

20.4 l/day reabsorbed

35 mm Hg

25 mm Hg

18 mm Hg

NFP = +10 mm Hg

NFP = 0

NFP = −7 mm Hg

25 mm Hg

25 mm Hg

25 mm Hg

CHP > BCOP
Fluid forced out of capillary

CHP = BCOP
No net movement of fluid

BCOP > CHP
Fluid moves into capillary

•**FIGURE 21-13** **Forces Acting across Capillary Walls.** At the arterial end of the capillary, capillary hydrostatic pressure (CHP) is stronger than blood colloid osmotic pressure (BCOP), so fluid moves out of the capillary. Near the venule, CHP is lower than BCOP, so fluid moves into the capillary. In this model, interstitial fluid colloid osmotic pressure (ICOP) and interstitial fluid hydrostatic pressure (IHP) are assumed to be 0 mm Hg and so are not shown.

21

$$\frac{\text{net filtration}}{\text{pressure}} = \frac{\text{net hydrostatic}}{\text{pressure}} - \frac{\text{net colloid}}{\text{osmotic pressure}}$$

$$\text{NFP} = (\text{CHP} - \text{IHP}) - (\text{BCOP} - \text{ICOP})$$

At the arterial end of a capillary, the net filtration pressure is +10 mm Hg:

$$\text{NFP} = (35 - 0) - (25 - 0) = 35 - 25 = 10 \text{ mm Hg}$$

Because this value is positive, it indicates that fluid will tend to move *out of* the capillary and into the interstitial fluid. At the venous end of the capillary, the net filtration pressure is −7 mm Hg:

$$\text{NFP} = (18 - 0) - (25 - 0) = 18 - 25 = -7 \text{ mm Hg}$$

The minus sign indicates that fluid tends to move *into* the capillary; that is, reabsorption is occurring.

The transition between filtration and reabsorption occurs where the CHP is 25 mm Hg, because at that point the hydrostatic and osmotic forces are equal. That is, the NFP is 0 mm Hg. If the maximum filtration pressure at the arterial end of the capillary were

equal to the maximum reabsorption pressure at the venous end, this transition point would lie midway along the length of the capillary. Under these circumstances, filtration would occur along the first half of the capillary, and an identical amount of reabsorption would occur along the second half. However, the maximum filtration pressure is higher than the maximum reabsorption pressure, so the transition point between filtration and reabsorption normally lies closer to the venous end of the capillary than to the arterial end. As a result, more filtration than reabsorption occurs along the capillary. Of the roughly 24 liters of fluid that moves out of the plasma and into the interstitial fluid each day, 85 percent is reabsorbed. The remainder (3.6 liters) flows through the tissues and into lymphatic vessels, for eventual return to the venous system.

Any condition that affects hydrostatic or osmotic pressures in the blood or tissues will shift the balance between hydrostatic and osmotic forces. We can then predict the effects on the basis of an understanding of capillary dynamics. For example:

- If hemorrhaging occurs, both blood volume and blood pressure decline. This reduction in CHP lowers the NFP and increases the amount of reabsorption. The result is a reduction in the volume of interstitial fluid and an increase in the circulating plasma volume. This process is known as a *recall of fluids.*

- If dehydration occurs, the plasma volume decreases owing to water loss, and the concentration of plasma proteins increases. The increase in BCOP accelerates reabsorption and a recall of fluids that delays the onset and severity of clinical symptoms.

- If the CHP rises or the BCOP declines, fluid moves out of the blood and builds up in peripheral tissues, a condition called *edema.*

Edema

Edema (e-DĒ-muh) is an abnormal accumulation of interstitial fluid. Edema has many causes, and we will encounter specific examples in later chapters. The underlying problem in all types of edema is a disturbance in the normal balance between hydrostatic and osmotic forces at the capillary level. For example:

- When a capillary is damaged, plasma proteins can cross the capillary wall and enter the interstitial fluid. The resulting elevation of the ICOP will reduce the rate of capillary reabsorption and produce a localized edema. This is why you usually have swelling at a bruise.

- In acute starvation, the liver cannot synthesize enough plasma proteins to maintain normal concentrations in the blood. BCOP declines, and fluids begin moving from the blood into peripheral tissues. In children, fluid accumulates in the abdominopelvic cavity, producing the swollen bellies typical of starvation victims. A reduction in BCOP is also seen after severe burns and in several types of liver or kidney diseases.

- In the U.S. population, most serious cases of edema result from an increase in arterial blood pressure, in venous pressure, or in total circulatory pressure. The increase may result from heart problems, such as heart failure, venous blood clots that elevate venous pressures, or other circulatory abnormalities. The net result is an increase in CHP that accelerates fluid movement into the tissues.

Edema can also result from problems with other systems, such as the blockage of lymphatic vessels or impaired urine formation:

- If the lymphatic vessels in a region become blocked, the volume of interstitial fluid will continue to rise and the IHP will gradually increase until capillary filtration ceases. In *filariasis,* a condition we shall consider in Chapter 22, parasites can block lymphatic vessels and cause a massive regional edema known as *elephantiasis.*

- If the kidneys are unable to produce urine but the individual continues to drink liquids, the blood volume will rise. This situation ultimately leads to elevated CHP and enhances fluid movement into the peripheral tissues.

VENOUS PRESSURE AND VENOUS RETURN

Venous pressure, although low, determines the venous return, the amount of blood arriving at the right atrium each minute. Venous return has a direct impact on cardiac output. ∞ *p. 681* Although pressure at the start of the venous system is only about one-tenth that at the start of the arterial system, before returning to the heart the blood must still travel through a vascular network as complex as the arterial system.

Pressures at the entrance to the right atrium fluctuate, but they average about 2 mm Hg. Thus the effective pressure in the venous system is roughly 16 mm Hg (from 18 mm Hg in the venules to 2 mm Hg in the venae cavae), as compared with 65 mm Hg in the arterial system (from 100 mm Hg at the aorta to 35 mm Hg at the capillaries). Yet, although venous pressures are low, veins offer comparatively little resistance, so pressure declines very slowly as blood moves through the venous system. As blood continues toward the heart, the veins become larger, resistance drops, and the velocity of blood flow increases (Figure 21-9●, p. 704).

When you stand, the venous blood returning from your body inferior to the heart must overcome gravity as it ascends within the inferior vena cava. Two factors assist the low venous pressures in propelling blood toward your heart: (1) *muscular compression* of peripheral veins and (2) the *respiratory pump.*

Muscular Compression

The contractions of skeletal muscles near a vein compress it, helping push blood toward the heart. The valves in small and medium-sized veins ensure that blood flow occurs in one direction only (Figure 21-6●, p. 700). During normal standing and walking, the cycles of contraction and relaxation that accompany normal movements assist venous return. If you stand at attention, with knees locked and leg muscles immobilized, that assistance is lost. The reduction in venous return then leads to a fall in cardiac output, which reduces the blood supply to the brain. This decline is sometimes enough to cause **fainting,** a temporary loss of consciousness. You would then collapse, but while you were in the horizontal position, both venous return and cardiac output would return to normal.

The Respiratory Pump

As you inhale, your thoracic cavity expands and pressures within the pleural cavities decline. This drop in pressure pulls air into your lungs. At the same time, blood is pulled into the inferior vena cava and right atrium from the smaller veins of your abdominal cavity and lower body. The effect on venous return from the superior vena cava is less pronounced, as blood in that vessel normally has the assistance of gravity.

As you exhale, your thoracic cavity decreases in size. Internal pressures then rise, forcing air out of your lungs and pushing venous blood into the right atrium. This mechanism is called the **respiratory pump,** or *thoracoabdominal pump.* The importance of this pumping action increases during heavy exercise, when respirations are deep and frequent.

☑ In a healthy individual, where would the blood pressure be greater—at the aorta or at the inferior vena cava? Explain.

☑ While standing in the hot sun, Sally begins to feel light-headed and faints. Explain.

☑ Terry's blood pressure is 125/70. At what pressure did the nurse taking Terry's blood pressure first hear the sounds of Korotkoff?

CARDIOVASCULAR REGULATION

Homeostatic mechanisms regulate cardiovascular activity to ensure that **tissue perfusion,** or blood flow through tissues, meets the demand for oxygen and nutrients. The three variable factors are (1) cardiac output, (2) peripheral resistance, and (3) blood pressure. We discussed cardiac output in Chapter 20 (∞ *p. 681*) and considered peripheral resistance and blood pressure earlier in this chapter.

Most cells are relatively close to capillaries. When a group of cells becomes active, the circulation to that region must increase to deliver the necessary oxygen and nutrients and to carry away the waste products and carbon dioxide they generate. The goal of cardiovascular regulation is to ensure that these blood flow changes occur (1) at an appropriate time, (2) in the right area, and (3) without drastically changing blood pressure and blood flow to vital organs.

Factors involved in the regulation of cardiovascular function include local factors, neural mechanisms, and endocrine mechanisms (Figure 21-14•):

- *Local factors.* Local factors change the pattern of blood flow within capillary beds in response to chemical changes in interstitial fluids. This is an example of *autoregulation* at the tissue level. Autoregulation causes

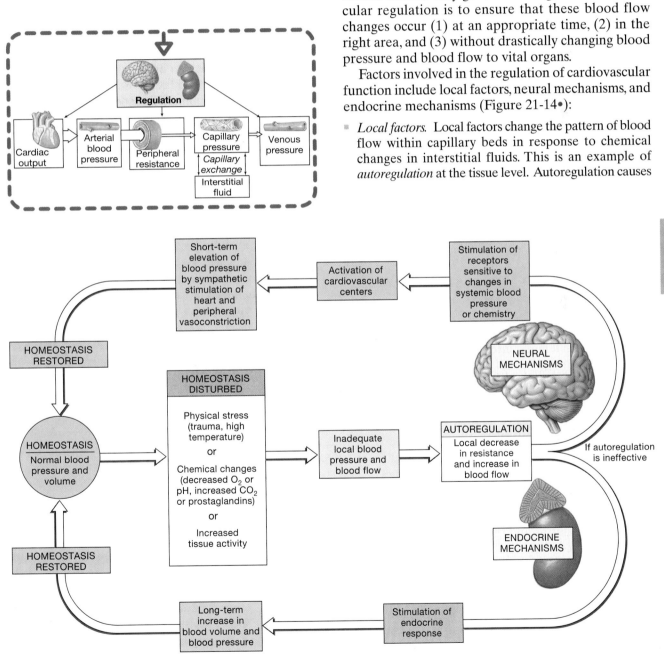

•**FIGURE 21-14** **Homeostatic Adjustments That Maintain Blood Pressure and Blood Flow**

immediate, localized homeostatic adjustments. If autoregulation fails to normalize conditions at the tissue level, neural mechanisms and endocrine factors are activated.

- *Neural mechanisms.* Neural mechanisms respond to changes in arterial pressure or blood gas levels at specific sites. When those changes occur, the cardiovascular centers of the autonomic nervous system adjust cardiac output and peripheral resistance to maintain blood pressure and ensure adequate blood flow.
- *Endocrine mechanisms.* The endocrine system releases hormones that enhance short-term adjustments and direct long-term changes in cardiovascular performance.

We shall next consider each of these regulatory mechanisms individually.

AUTOREGULATION OF BLOOD FLOW WITHIN TISSUES

Under normal resting conditions, cardiac output remains stable and peripheral resistance within individual tissues is adjusted to control local blood flow.

Factors that promote the dilation of precapillary sphincters are called **vasodilators. Local vasodilators** act at the tissue level and accelerate blood flow through the tissue of origin. Examples include:

- Decreased tissue oxygen levels or increased CO_2 levels.
- The generation of lactic acid or other acids by tissue cells.
- The release of nitric oxide (NO), formerly known as *endothelium-derived relaxation factor (EDRF)*, from endothelial cells.
- Rising concentrations of potassium ions or hydrogen ions in interstitial fluid.
- Chemicals released during local inflammation, including histamine and NO.
- Elevated local temperatures.

These factors work by stimulating the relaxation of the smooth muscle cells of the precapillary sphincters. All these factors indicate that conditions in the tissue are abnormal in one way or another. An improvement in blood flow, which will bring oxygen, nutrients, and buffers, may be sufficient to restore homeostasis.

As we noted in Chapter 19, aggregating platelets and damaged tissues produce compounds that stimulate the constriction of precapillary sphincters. These compounds are **local vasoconstrictors.** Examples include prostaglandins and thromboxanes released by activated platelets and white blood cells and the endothelins released by damaged endothelial cells. ∞ *p. 645*
Local vasodilators and vasoconstrictors control blood flow within a single capillary bed (Figure 21-5•, p. 699). When present in high concentrations, these factors also affect arterioles, increasing or decreasing blood flow to all the capillary beds in a given area.

NEURAL MECHANISMS

The nervous system is responsible for adjusting cardiac output and peripheral resistance to maintain adequate blood flow to vital tissues and organs. Centers responsible for these regulatory activities include the *cardiac centers* and the *vasomotor centers* of the medulla oblongata. ∞ *p. 464* It is difficult to distinguish the cardiac and vasomotor centers anatomically, and they are often considered to form complex **cardiovascular (CV) centers.** In functional terms, however, the cardiac and vasomotor centers often act independently. As we noted in Chapter 20, each cardiac center consists of a *cardioacceleratory center,* which increases cardiac output through sympathetic innervation, and a *cardioinhibitory center,* which reduces cardiac output through parasympathetic innervation. ∞ *p. 683*

The vasomotor centers contain two populations of neurons: (1) a very large group responsible for widespread vasoconstriction and (2) a smaller group responsible for the vasodilation of arterioles in skeletal muscles and the brain. The vasomotor centers exert their effects by controlling the activity of sympathetic motor neurons:

1. *Control of vasoconstriction.* The neurons innervating peripheral blood vessels in most tissues are *adrenergic;* that is, they release the neurotransmitter norepinephrine (NE). The response to NE release is the stimulation of smooth muscle in the walls of arterioles, producing vasoconstriction.

2. *Control of vasodilation.* Vasodilator neurons innervate blood vessels in skeletal muscles and in the brain. The stimulation of these neurons will relax smooth muscle cells in the walls of arterioles, producing vasodilation. The relaxation of smooth muscle cells is triggered by the appearance of NO in their surroundings. The vasomotor centers may control NO release indirectly or directly. The most common vasodilator synapses are *cholinergic,* and their synaptic knobs release ACh. ACh stimulates the release of NO by endothelial cells in the area; the NO then causes local vasodilation. Another population of vasodilator synapses is *nitroxidergic;* the synaptic knobs release NO as a neurotransmitter. Nitric oxide has an immediate and direct relaxing effect on the vascular smooth muscle cells in the area.

Vasomotor Tone

In Chapter 16, we discussed the significance of autonomic tone in setting a background level of neural activity that can increase or decrease on demand. ∞ *pp. 518–519* The sympathetic vasoconstrictor nerves are chronically active, producing a significant **vasomotor tone.** Vasoconstrictor activity is normally sufficient to keep the arterioles partially constricted. Under maximal stimulation, arterioles constrict to about half their resting diameter of the vassel, whereas a fully dilated arteriole increases its resting diameter by roughly 1.5 times. Constriction has a significant effect on resistance, because, as we saw earlier (p. 703), the resistance increases sharply as the diameter of the vessel decreases. The resistance of a maximally constricted arteriole is roughly *80 times* that of a fully dilated arteriole. Because blood pressure varies directly with peripheral resistance, the vasomotor centers can very effectively control arterial blood pressure by making modest adjustments in vessel diameter. Extreme stimulation of the vasomotor centers will also produce venoconstriction and a mobilization of the venous reserve.

Reflex Control of Cardiovascular Function

The cardiovascular centers detect changes in tissue demand by monitoring arterial blood, with particular attention to blood pressure and to the pH and dissolved gas concentrations. The *baroreceptor reflexes* respond to changes in blood pressure, and the *chemoreceptor reflexes* monitor changes in the chemical composition of arterial blood. These reflexes are regulated through a negative feedback loop. The stimulation of a receptor by an abnormal condition leads to a response that counteracts the stimulus and restores normal conditions.

Baroreceptor Reflexes *Baroreceptors* are specialized receptors that monitor the degree of stretch in the walls of expandable organs. ∞ *p. 533* The baroreceptors involved with cardiovascular regulation are located in the walls of (1) the **carotid sinuses,** expanded chambers near the bases of the *internal carotid arteries* of the neck (Figure 21-24•, p. 728), (2) the **aortic sinuses,** pockets in the walls of the ascending aorta adjacent to the heart (Figure 20-8b•, p. 665), and (3) the wall of the right atrium. These receptors are components of the **baroreceptor reflexes,** which adjust cardiac output and peripheral resistance to maintain normal arterial pressures.

Aortic baroreceptors monitor blood pressure within the ascending aorta. There, the **aortic reflex** adjusts blood pressure in response to changes in pressure. The goal is to maintain adequate blood pressure and blood flow through the systemic circuit. Carotid sinus baroreceptors respond to changes in blood pressure at the carotid sinus. These receptors trigger reflexes that maintain adequate blood flow to the brain. The blood flow to the brain must remain constant, and the carotid sinus receptors are extremely sensitive. Figure 21-15• presents the basic organization of the baroreceptor reflexes triggered by changes in blood pressure at these locations.

When blood pressure climbs, the increased output from the baroreceptors alters activity in the CV centers and produces two major effects (Figure 21-15a•):

1. *A decrease in cardiac output,* due to parasympathetic stimulation and the inhibition of sympathetic activity.

2. *Widespread peripheral vasodilation,* due to the inhibition of excitatory neurons in the vasomotor centers.

The decrease in cardiac output reflects primarily a reduction in heart rate due to acetylcholine release at the sinoatrial (SA) node. ∞ *p. 684* The widespread vasodilation lowers peripheral resistance, and this effect, combined with a reduction in cardiac output, leads to a decline in blood pressure to normal levels.

When blood pressure falls below normal, baroreceptor output is reduced accordingly (Figure 21-15b•). This change has two major effects:

1. *An increase in cardiac output* through the stimulation of sympathetic innervation to the heart. This results from the stimulation of the cardioacceleratory centers and is accompanied by an inhibition of the cardioinhibitory centers.

2. *Widespread peripheral vasoconstriction,* caused by the stimulation of sympathetic vasoconstrictor neurons by the vasomotor centers.

The effects on the heart result from the release of NE by sympathetic neurons innervating the SA node, the atrioventricular (AV) node, and the general myocardium. In a crisis, sympathetic activation occurs, and its effects will be enhanced by the release of both NE and epinephrine (E) from the adrenal medullae. The net effect is an immediate increase in heart rate and stroke volume and a corresponding rise in cardiac output. The vasoconstriction, which also results from NE release by sympathetic neurons, increases peripheral resistance. These adjustments, increased cardiac output and increased peripheral resistance, work together to elevate blood pressure.

Atrial baroreceptors are receptors that monitor blood pressure at the end of the systemic circuit—at the venae cavae and the right atrium. The **atrial reflex** responds to a stretching of the wall of the right atrium. ∞ *p. 685*

Under normal circumstances, the heart pumps blood into the aorta at the same rate at which blood arrives at the right atrium. When blood pressure rises at the right atrium, blood is arriving at the heart faster than it is being pumped out. The atrial baroreceptors solve the problem by stimulating the CV centers and increasing cardiac output until the backlog of venous blood is removed. Atrial pressure then returns to normal.

21

21

•**FIGURE 21-15** Baroreceptor Reflexes of the Carotid
and Aortic Sinuses

Chemoreceptor Reflexes   The **chemoreceptor reflexes**
respond to changes in the carbon dioxide, oxygen, or
pH levels in blood and cerebrospinal fluid (CSF)
(Figure 21-16•). The chemoreceptors involved are sensory neurons located in the **carotid bodies,** situated in
the neck near the carotid sinus, and the **aortic bodies,**
near the arch of the aorta. ∞ *p. 534* These receptors
monitor the composition of the arterial blood. Additional chemoreceptors are located on the ventrolateral surfaces of the medulla oblongata. These receptors
monitor the composition of CSF.

When chemoreceptors in the carotid bodies or aortic bodies detect a rise in the carbon dioxide content
or a fall in the pH of the arterial blood, the cardioacceleratory and vasomotor centers are stimulated and
the cardioinhibitory centers are inhibited (Figure
21-16•). This dual effect causes an increase in cardiac
output, peripheral vasoconstriction, and an elevation
in arterial blood pressure. A drop in the oxygen level
at the aortic bodies will have the same effects. Strong

stimulation of the carotid or aortic chemoreceptors
causes widespread sympathetic activation, with more-
dramatic increases in heart rate and cardiac output.

The chemoreceptors of the medulla oblongata are
primarily involved with the control of respiratory function and secondarily with regulating blood flow to the
brain. For example, a steep rise in CSF carbon dioxide
levels will trigger the vasodilation of cerebral vessels
but will produce vasoconstriction in most other organs. The result is increased blood flow—and hence
increased oxygen delivery—to the brain.

Arterial CO_2 levels can be reduced and O_2 levels
increased most effectively by coordinating cardiovascular and respiratory activities. Chemoreceptor stimulation also stimulates the respiratory centers, and the
rise in cardiac output and blood pressure is associated with an increased respiratory rate. Coordination of
cardiovascular and respiratory activities is vital, because accelerating tissue blood flow is useful only if
the circulating blood contains adequate oxygen. In ad-

•FIGURE 21-16 The Chemoreceptor Reflexes

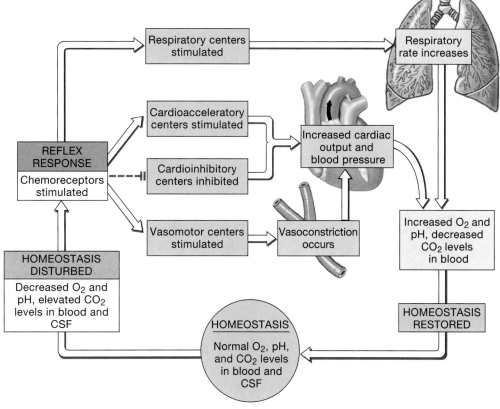

dition, a rise in the respiratory rate accelerates venous return through the action of the respiratory pump. (We shall consider other aspects of chemoreceptor activity and respiratory control in Chapter 23.)

CNS Activities and the Cardiovascular Centers

The output of the cardiovascular centers can also be influenced by activities in other areas of the brain. For example, the activation of either division of the autonomic nervous system will affect output from the cardiovascular centers:

- The cardioacceleratory and vasomotor centers are stimulated when a general sympathetic activation occurs. The result is an increase in cardiac output and blood pressure.
- When the parasympathetic division is activated, the cardioinhibitory centers are stimulated, producing a reduction in cardiac output. Parasympathetic activity does not directly affect the vasomotor centers, but vasodilation occurs as sympathetic activity declines.

The activities of higher brain centers can also affect blood pressure. Our thought processes or emotional states can produce significant changes in blood pressure by influencing cardiac output and vasomotor tone. For example, strong emotions of anxiety, fear, or rage are accompanied by an elevation in blood pressure, caused by cardiac stimulation and vasoconstriction.

HORMONES AND CARDIOVASCULAR REGULATION

The endocrine system provides both short-term and long-term regulation of cardiovascular performance. As we have seen, E and NE from the adrenal medullae stimulate cardiac output and peripheral vasoconstriction. Other hormones important in regulating cardiovascular function include (1) antidiuretic hormone (ADH), (2) angiotensin II, (3) erythropoietin (EPO), and (4) atrial natriuretic peptide (ANP). We introduced those hormones and their functions in Chapter 18. ∞ *pp. 591, 604, 605* Although ADH and angiotensin II affect blood pressure, all four hormones are concerned primarily with the long-term regulation of blood volume (Figure 21-17•).

Antidiuretic Hormone

Antidiuretic hormone (ADH) is released at the posterior pituitary gland in response to a decrease in blood volume or an increase in the osmotic concentration of the plasma or, secondarily, in response to circulating angiotensin II. The immediate result is a peripheral vasoconstriction that elevates blood pressure. This hormone also stimulates the conservation of water at the kidneys, thus preventing a reduction in blood volume that would further reduce blood pressure (Figure 21-17a•).

21

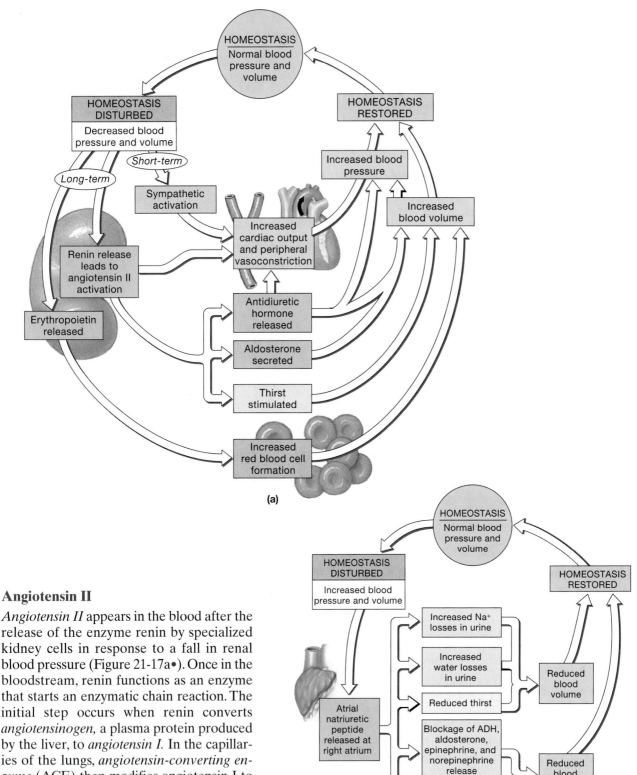

(a)

(b)

•FIGURE 21-17 The Regulation of Blood Pressure and Blood Volume. Factors that compensate for **(a)** decreased blood pressure and volume and for **(b)** increased blood pressure and volume.

Angiotensin II

Angiotensin II appears in the blood after the release of the enzyme renin by specialized kidney cells in response to a fall in renal blood pressure (Figure 21-17a•). Once in the bloodstream, renin functions as an enzyme that starts an enzymatic chain reaction. The initial step occurs when renin converts *angiotensinogen,* a plasma protein produced by the liver, to *angiotensin I.* In the capillaries of the lungs, *angiotensin-converting enzyme* (ACE) then modifies angiotensin I to angiotensin II, an active hormone with diverse effects.

Angiotensin II (1) stimulates the adrenal production of aldosterone, causing Na^+ retention and K^+ loss at the kidneys; (2) stimulates the secretion of ADH, in turn stimulating water reabsorption at the kidneys and complementing the effects of aldosterone; (3) stimulates thirst, resulting in increased fluid consumption (the presence of ADH and aldosterone ensures that the additional water consumed will be retained, elevating blood volume); and (4) stimulates cardiac output and triggers the constriction of arterioles, in turn elevating the systemic blood pressure. The effect of angiotensin II on blood pressure is 4–8 times greater than that produced by norepinephrine.

Erythropoietin

Erythropoietin (EPO) is released at the kidneys if the blood pressure declines or if the oxygen content of the blood becomes abnormally low (Figure 21-17a•). This hormone stimulates red blood cell production and maturation, elevating the blood volume and improving the oxygen-carrying capacity of blood.

Atrial Natriuretic Peptide

Atrial natriuretic peptide (nā-trē-ū-RET-ik; *natrium,* sodium + *ouresis,* making water), or *ANP,* is produced by cardiac muscle cells in the wall of the right atrium in response to excessive stretching during diastole. This peptide hormone reduces blood volume and blood pressure by (1) increasing sodium ion excretion at the kidneys, (2) promoting water losses by increasing the volume of urine produced; (3) reducing thirst; (4) blocking the release of ADH, aldosterone, epinephrine, and norepinephrine; and (5) stimulating peripheral vasodilation (Figure 21-17b•). As blood volume and blood pressure decline, the stress on the atrial walls is removed and ANP production ceases.

PATTERNS OF CARDIOVASCULAR RESPONSE

In this chapter and the preceding two chapters, we have considered the blood, the heart, and the cardiovascular system as individual entities. Yet, in our day-to-day lives, the cardiovascular system operates as an integrated complex. The interactions are fascinating and of considerable importance when physical or physiological conditions are changing rapidly.

   Two common stresses, exercise and blood loss, provide examples of the adaptability of this system and its ability to maintain homeostasis. The homeostatic responses involve an interplay among the cardiovascular system, the endocrine system, and other systems, and the central mechanisms are aided by autoregulation at the tissue level. We shall also consider the physiological mechanisms involved in *shock,* an important homeostatic disorder.

EXERCISE AND THE CARDIOVASCULAR SYSTEM

At rest, cardiac output averages about 5.6 liters per minute. That value changes dramatically during exercise. In addition, the pattern of blood distribution changes markedly, as detailed in Table 21-2.

Light Exercise

Before you begin to exercise, your heart rate increases slightly because a general rise in sympathetic activity occurs as you think about the workout ahead. As you begin light exercise, three interrelated changes occur:

1. *Extensive vasodilation occurs* as the rate of oxygen consumption in skeletal muscles increases. Peripheral resistance drops, blood flow through the capillaries increases, and blood enters the venous system at an accelerated rate.

2. *The venous return increases,* as skeletal muscle contractions squeeze blood along the peripheral veins and an increased breathing rate pulls blood into the venae cavae via the respiratory pump.

3. *Cardiac output rises,* primarily in response to (1) the rise in venous return (the Frank–Starling principle ∞ *p. 682*) and (2) atrial stretching (the atrial reflex). Some sympathetic stimulation occurs, leading to increases in heart rate and contractility, but massive sympathetic activation does not occur. The increased cardiac output keeps pace with the elevated demand, and arterial pressures are maintained despite the drop in peripheral resistance.

   This regulation by venous feedback produces a gradual increase in cardiac output to about double resting levels. Over this range, the blood flow to

21

TABLE 21-2 Distribution of Blood During Exercise

Organ	Tissue Blood Flow (ml/min)		
	Rest	**Light Exercise**	**Strenuous Exercise**
Skeletal muscles	1200	4500	12, 500
Heart	250	350	750
Brain	750	750	750
Skin	500	1500	1900
Kidney	1100	900	600
Abdominal viscera	1400	1100	600
Miscellaneous	600	400	400
Total cardiac output	5800	9500	17, 500

skeletal muscles, cardiac muscles, and skin increases. The increased flow to the muscles reflects the dilation of arterioles and precapillary sphincters in response to local factors; the increased blood flow to the skin occurs in response to the rise in body temperature.

Heavy Exercise

At higher levels of exertion, other physiological adjustments occur as the cardiac and vasomotor centers call for the general activation of the sympathetic nervous system. Cardiac output increases toward maximal levels, and major changes in the peripheral distribution of blood occur, facilitating the blood flow to active skeletal muscles.

Under massive sympathetic stimulation, your cardioacceleratory centers can increase cardiac output to levels as high as 20–25 liters per minute. Even at these rates, the increased circulatory demands of the skeletal muscles can be met only if the vasomotor centers severely restrict the blood flow to "nonessential" organs, such as those of your digestive system. During exercise at maximal levels, your blood essentially races between the skeletal muscles and the lungs and heart. Although most tissues experience reductions in blood flow, skin perfusion increases further, because body temperature continues to climb. Only the blood supply to the brain remains unaffected.

Exercise, Cardiovascular Fitness, and Health

Cardiovascular performance improves significantly with training. Table 21-3 compares the cardiac performance of athletes with that of nonathletes. Trained athletes have bigger hearts and larger stroke volumes than do nonathletes, and these are important functional differences.

Recall that cardiac output is equal to the stroke volume times the heart rate; for the same cardiac output, the person with a larger stroke volume has a slower heart rate. An athlete at rest can maintain normal blood flow to peripheral tissues at a heart rate as low as 50 bpm (beats per minute), and when necessary the cardiac output can increase to levels 50 percent higher than those of nonathletes. Thus, a trained athlete can tolerate sustained levels of activity that are well outside the capabilities of nonathletes.

Exercise and Cardiovascular Disease

Regular exercise has several beneficial effects. Even a moderate exercise routine (jogging 5 miles a week, for example) can lower total blood cholesterol levels. A high cholesterol level is one of the major risk factors for atherosclerosis, leading to cardiovascular disease and strokes. In addition, a healthy lifestyle—regular exercise, a balanced diet, weight control, and not smoking—reduces stress, lowers blood pressure, and slows plaque formation.

Regular moderate exercise may cut the incidence of heart attacks almost in half. However, at present only an estimated 8 percent of adults in the United States exercise at recommended levels. Exercise is also beneficial in accelerating recovery after a heart attack. Regular light-to-moderate exercise, such as walking, jogging, or bicycling, coupled with a low-fat diet and a low-stress lifestyle, not only reduces symptoms of CAD, such as angina, but also improves both mood and the overall quality of life. However, exercise does not remove the underlying medical problem, and atherosclerotic plaques do not disappear and seldom grow smaller with exercise.

There is no evidence that *intense* athletic training lowers the incidence of cardiovascular disease. On the contrary, the strains placed on all physiological systems, including the cardiovascular system, during an ultramarathon, iron-man triathlon, or other athletic extreme can be severe. Individuals with congenital aneurysms, cardiomyopathy, or cardiovascular disease risk fatal circulatory problems, such as an arrhythmia or heart attack, during severe exercise. Even healthy individuals can develop acute physiological disorders, such as kidney failure, after extreme exercise. We will discuss the effects of exercise on other systems in later chapters.

TABLE 21-3 **Effects of Training on Cardiovascular Performance**

Subject	Heart Weight (g)	Stroke Volume (ml)	Heart Rate (bpm)	Cardiac Output (l/min)	Blood Pressure (systolic/diastolic)
Nonathlete (rest)	300	60	83	5.0	120/80
Nonathlete (maximum)		104	192	19.9	187/75
Trained athlete (rest)	500	100	53	5.3	120/80
Trained athlete (maximum)		167	182	30.4	200/90*

* Diastolic pressures in athletes during maximal activity have not been accurately measured.

CARDIOVASCULAR RESPONSE TO HEMORRHAGING

In Chapter 19, we considered the local circulatory reaction to a break in the wall of a blood vessel. *pp. 645–646* When hemostasis fails to prevent a significant blood loss, the entire cardiovascular system makes adjustments to maintain blood pressure and restore blood volume (Figure 21-18•). The immediate problem is the maintenance of adequate blood pressure and peripheral blood flow. The long-term problem is the restoration of normal blood volume.

Short-Term Elevation of Blood Pressure

Almost as soon as the pressures start to decline, short-term adjustments begin. The steps include the following:

- The initial neural response occurs as carotid and aortic reflexes increase cardiac output and cause peripheral vasoconstriction (p. 713). With the blood volume reduced, cardiac output is maintained by increasing the heart rate, typically to rates of 180–200 bpm.
- The combination of stress and anxiety stimulates the sympathetic nervous system headquarters in the hypothalamus. This stimulation triggers a further increase in vasomotor tone, constricting the arterioles, and elevating blood pressure. At the same time, venoconstriction mobilizes the venous reserve and quickly improves venous return (p. 710).
- Long-term hormonal responses occur: The combination of ADH, angiotensin II, and EPO restrict fluid losses and take steps to increase blood volume (Figure 21-17a•, p. 716).
- Short-term hormonal effects also occur: For instance, (1) sympathetic activation causes the secretion of E and NE by the adrenal medulla, increasing cardiac output and extending peripheral vasoconstriction, and (2) the release of ADH by the posterior pituitary gland and the production of angiotensin II enhance vasoconstriction while participating in the long-term response.

This combination of adjustments elevates blood pressure and improves peripheral blood flow. After blood losses of up to 20 percent of the total blood volume, these short-term adjustments can restore normal arterial pressures and peripheral blood flow. For example, these adjustments are more than sufficient to compensate for the blood loss experienced when you donate blood. (Most blood banks collect 500 ml of whole blood, roughly 10 percent of the total blood volume.)

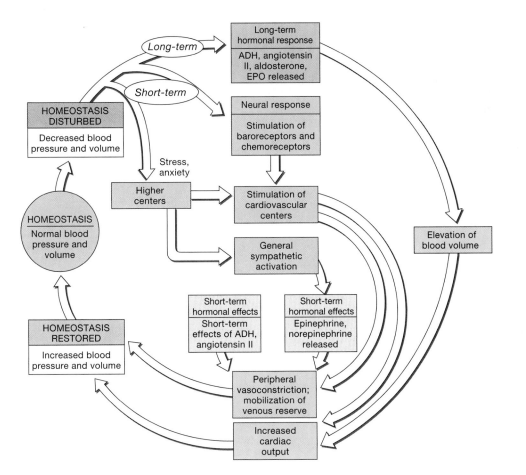

•**FIGURE 21-18** Cardiovascular Responses to Hemorrhaging and Blood Loss

Circulatory Shock

Shock is an acute circulatory crisis marked by low blood pressure (hypotension) and inadequate peripheral blood flow. Severe and potentially fatal symptoms develop as vital tissues become starved for oxygen and nutrients. Common causes of shock are (1) a drop in cardiac output after hemorrhaging or other fluid losses, (2) damage to the heart, (3) external pressure on the heart, and (4) extensive peripheral vasodilation. We shall focus on the cause, symptoms, and treatment of circulatory shock; the *Applications Manual* discusses other types. [AM] *Other Types of Shock*

Symptoms of **circulatory shock** appear after fluid losses of about 30 percent of the total blood volume. The cause can be hemorrhaging or fluid losses to the environment, as in dehydration or after severe burns. All cases of circulatory shock share the same six basic symptoms:

1. Hypotension, with systolic pressures below 90 mm Hg.
2. Pale, cool, and moist ("clammy") skin. The skin is pale and cool due to peripheral vasoconstriction; the moisture reflects the sympathetic activation of sweat glands.
3. Frequent confusion and disorientation, due to a drop in blood pressure at the brain.
4. A rise in heart rate and a rapid, weak pulse.
5. A cessation of urination, because the reduced blood flow to the kidneys slows or stops urine production.
6. A drop in blood pH *(acidosis)*, due to lactic acid generation in oxygen-deprived tissues.

In mild forms of circulatory shock, homeostatic adjustments can cope with the situation. The short-term and long-term responses detailed in Figure 21-18● are part of the adjustment process. During this period, peripheral blood flow is reduced but remains adequate to meet tissue demands.

When blood volume declines by more than 35 percent, homeostatic mechanisms are no longer able to cope with the situation. Despite sustained vasoconstriction and the mobilization of the venous reserve, blood pressure remains abnormally low, venous return is reduced, and cardiac output is inadequate (Figure 21-19a●). A vicious cycle begins when the low cardiac output damages the myocardium. This damage leads to a further reduction in cardiac output and subsequent reductions in blood pressure and venous return. As blood flow to the brain decreases, the individual becomes disoriented and confused.

When the mean arterial blood pressure (MAP) falls to about 50 mm Hg, carotid sinus baroreceptors trigger a massive activation of the vasomotor centers. In essence, the goal now is to preserve the circulation to the brain at any cost. The sympathetic output now causes a sustained and maximal vasoconstriction. This reflex, called the *central ischemic response,* reduces peripheral circulation to an absolute minimum, but it elevates blood pressure to about 70 mm Hg and improves blood flow to cerebral vessels.

The central ischemic response is a last-ditch effort that maintains adequate blood flow to the brain at the expense of other tissues. Unless prompt treatment is provided, the condition will soon prove fatal. Treatment must concentrate on (1) preventing further fluid losses and (2) giving a transfusion of whole blood, plasma expanders, or blood substitutes.

In the absence of treatment, the cardiovascular damage will soon become irreversible (Figure 21-19b●). At this point, conditions in the heart, liver, kidneys, and CNS are rapidly deteriorating to the point at which death will occur, even *with* medical treatment.

Irreversible shock begins when conditions in the tissues become so abnormal that the arteriolar smooth muscles and precapillary sphincters become unable to contract, despite the commands of the vasomotor centers. The result is a widespread peripheral vasodilation and an immediate and fatal decline in blood pressure. This event is called *circulatory collapse*. The blood pressure in many tissues then falls so low that the capillaries collapse like deflating balloons. Blood flow through these capillary beds then stops completely, and the cells in the affected tissues die. The dying cells release more abnormal chemicals, and the effect quickly spreads throughout the body.

•**FIGURE 21-19 Circulatory Shock. (a)** When blood
volume has decreased by more than 35 percent, systemic
blood pressure, tissue blood flow, and cardiac output
gradually drop. **(b)** Next, a series of integrated chain
reactions leads to a rapid decline in cardiac output, a
dramatic and irreversible fall in blood pressure, circulatory
collapse, and eventual death.

The vasoconstriction that occurs in response to a fall in blood pressure or a rise in CO_2 levels affects multiple tissues and organs simultaneously. The term *special circulation* refers to the circulation through organs where blood flow is controlled by separate mechanisms. We will note three important examples: the circulation to the brain, the heart, and the lungs.

THE BRAIN

In Chapter 14, we noted the existence of the blood–brain barrier, which isolates most CNS tissue from the general circulation. ∞ *p. 444* The brain has a very high demand for oxygen and receives a substantial supply of blood. Under a variety of conditions, the blood flow to the brain remains steady at about 750 ml/min. This supply is roughly 12 percent of the cardiac output, delivered to an organ that represents less than 2 percent of the body weight. Neurons do not maintain significant energy reserves, and in functional terms most of the adjustments made by the cardiovascular system treat blood flow to the brain as the number one priority. Even during a circulatory crisis, blood flow through the brain remains as near normal as possible. While the cardiovascular centers are calling for widespread peripheral vasoconstriction, the cerebral vessels are told to dilate.

Although total blood flow to the brain remains relatively constant, blood flow to specific regions of the brain changes from moment to moment. These changes occur in response to local changes in the composition of interstitial fluid that accompany neural activity. When you read, write, speak, or walk, specific regions of your brain become active. Blood flow to those regions increases almost instantaneously, ensuring that the active neurons will continue to receive the oxygen and nutrients they require.

The brain receives arterial blood from four arteries. Because these arteries form anastomoses inside the cranium, an interruption of any one vessel will not compromise the circulatory supply to the brain. If a plaque or a blood clot blocks an artery or if an artery ruptures, dependent tissues will be injured or killed. Symptoms of a *stroke,* or *cerebrovascular accident (CVA),* then appear. [AM] *The Causes and Treatment of Cerebrovascular Disease*

THE HEART

We described the anatomy of the coronary circulation in Chapter 20. ∞ *p. 664* The coronary arteries arise at the base of the ascending aorta, where systemic pressures are highest. Each time the heart contracts, it squeezes the coronary vessels, so blood flow is reduced. In the left ventricle, systolic pressures are high enough that blood can flow into the myocardium only during diastole. Normal cardiac muscle cells can tolerate these brief circulatory interruptions because they have substantial oxygen reserves.

When you are at rest, your coronary blood flow is about 250 ml/min. When the workload on your heart increases, local factors, such as reduced O_2 levels and lactic acid production, dilate the coronary vessels and increase blood flow. Epinephrine released during sympathetic stimulation promotes the vasodilation of coronary vessels while increasing your heart rate and the strength of cardiac contractions. As a result, coronary blood flow increases while vasoconstriction occurs in other tissues.

For uncertain reasons, some individuals experience *coronary spasms,* which can temporarily restrict coronary circulation and produce symptoms of angina. A permanent restriction or blockage of coronary vessels, as in coronary artery disease (CAD), and tissue damage, as caused by a myocardial infarction (MI), can limit the heart's ability to increase cardiac output, even under maximal stimulation. Individuals with these conditions will experience symptoms of heart failure when the cardiac workload increases much above resting levels. [AM] *Heart Failure*

THE LUNGS

The lungs contain roughly 300 million *alveoli* (al-VĒ-ō-lī; *alveolus,* sac), delicate epithelial pockets where gas exchange occurs. Each alveolus is surrounded by an extensive capillary network. Blood flow through the lungs is regulated primarily by local responses to the levels of oxygen within individual alveoli. When an alveolus contains oxygen in abundance, the associated vessels dilate, so blood flow increases. This increase promotes the absorption of oxygen from the alveolar air. When the oxygen content of the air is very low, the vessels constrict, so blood is shunted to alveoli that still contain significant levels of oxygen. This mechanism maximizes the efficiency of the respiratory system, because the circulation of blood through the capillaries of an alveolus has no benefit unless that alveolus contains oxygen.

This mechanism is precisely the opposite of the situation in other tissues, where a decline in oxygen levels causes local vasodilation rather than vasoconstriction. The difference makes functional sense, but its physiological basis remains a mystery.

Blood pressure in the pulmonary capillaries, averaging 10 mm Hg, is lower than the pressure in systemic capillaries. The BCOP (25 mm Hg) is the same as elsewhere in the bloodstream. As a result, reabsorption exceeds filtration in pulmonary capillaries. Fluid moves continuously into the pulmonary capillaries across the alveolar surfaces. This movement prevents fluid buildup in the alveoli that could interfere with the diffusion of respiratory gases. If the blood pressure in pulmonary capillaries rises above 25 mm Hg, fluid will enter the alveoli, causing *pulmonary edema.*

21

Long-Term Restoration of Blood Volume

Short-term compensation involves adjusting to the loss in blood volume. Long-term compensation involves restoring normal blood volume. After a serious hemorrhage, this process can take several days. The steps include the following:

- The decline in capillary blood pressure triggers the recall of fluids from the interstitial spaces (p. 710).

- Aldosterone and ADH promote fluid retention and reabsorption at the kidneys, preventing further reductions in blood volume.

- Thirst increases, and additional water is obtained by absorption across the digestive tract. This intake of fluid elevates plasma volume and ultimately replaces the interstitial fluids "borrowed" at the capillaries.

- Erythropoietin targets the bone marrow, stimulating the maturation of red blood cells, which increase blood volume and improve oxygen delivery to peripheral tissues.

☑ Why does blood pressure increase during exercise?

☑ How would applying a small pressure to the common carotid artery affect your heart rate?

☑ What effect would the vasoconstriction of the renal artery have on blood pressure and blood volume?

☑ Why does a person in circulatory shock have a rapid and weak pulse?

THE BLOOD VESSELS

You already know that the cardiovascular system is divided into the *pulmonary circuit* and the *systemic circuit*. The pulmonary circuit is composed of arteries and veins that transport blood between the heart and the lungs. This circuit begins at the right ventricle and ends at the left atrium. From the left ventricle, the arteries of the systemic circuit transport oxygenated blood and nutrients to all organs and tissues, ultimately returning deoxygenated blood to the right atrium. Figure 21-20• summarizes the primary circulatory routes within the pulmonary and systemic circuits.

In the pages that follow, we shall examine the vessels of the pulmonary and systemic circuits further. Two general functional patterns are worth noting at the outset:

1. The peripheral distributions of arteries and veins on the left and right sides are generally identical except near the heart, where the largest vessels connect to the atria or ventricles. For example, the distributions of the *left* and *right subclavian, axillary, brachial,* and *radial*

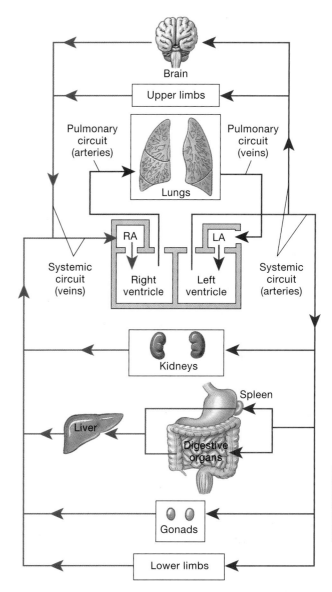

•**FIGURE 21-20** An Overview of the Pattern of Circulation. Here RA stands for right atrium; LA, for left atrium.

arteries parallel those of the *left* and *right subclavian, axillary, brachial,* and *radial veins,* respectively.

2. A single vessel may have several names as it crosses specific anatomical boundaries, making accurate anatomical descriptions possible when the vessel extends far into the periphery. For example, the *external iliac artery* becomes the *femoral artery* as it leaves the trunk and enters the lower limb.

THE PULMONARY CIRCUIT

Blood entering the right atrium has just returned from the peripheral capillary beds, where oxygen was released and carbon dioxide was absorbed. After traveling through the right atrium and ventricle, blood enters the pulmonary trunk, the start of the pulmonary circuit

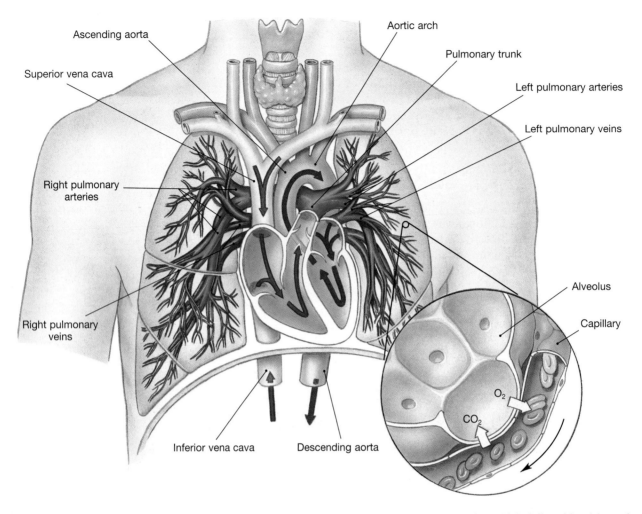

Ascending aorta

Superior vena cava

Right pulmonary arteries

Right pulmonary veins

Aortic arch

Pulmonary trunk

Left pulmonary arteries

Left pulmonary veins

Alveolus

Capillary

O_2

CO_2

Inferior vena cava

Descending aorta

•FIGURE 21-21 The Pulmonary Circuit. The pulmonary circuit consists of pulmonary arteries, which deliver blood from the right ventricle to the lungs; pulmonary capillaries, where gas exchange occurs; and pulmonary veins, which deliver blood to the left atrium. The pulmonary capillaries receive blood containing low levels of oxygen and higher levels of carbon dioxide. As the enlarged view shows, diffusion across the capillary walls at alveoli removes carbon dioxide and provides oxygen to the blood.
AM *Plates 7.4a, 7.6b; Scan 10*

(Figure 21-21•). At the lungs, oxygen is replenished, carbon dioxide is released, and the oxygenated blood is returned to the heart for distribution via the systemic circuit. Compared with the systemic circuit, the pulmonary circuit is short: The base of the pulmonary trunk and the lungs are only about 15 cm (6 in.) apart.

The arteries of the pulmonary circuit differ from those of the systemic circuit in that they carry deoxygenated blood. (For this reason, most color-coded diagrams show the pulmonary arteries in blue, the same color as systemic veins.) As it curves over the superior border of the heart, the pulmonary trunk gives rise to the **left** and **right pulmonary arteries.** These large arteries enter the lungs before branching repeatedly, giving rise to smaller and smaller arteries. The smallest branches, the *pulmonary arterioles,* provide blood to capillary networks that surround *alveoli.* The walls of these small air pockets are thin enough for gas exchange to occur between the capillary blood and in-

spired air. As it leaves the alveolar capillaries, oxygenated blood enters venules that in turn unite to form larger vessels carrying blood toward the **pulmonary veins.** These four veins, two from each lung, empty into the left atrium, completing the pulmonary circuit.

THE SYSTEMIC CIRCUIT

The systemic circuit supplies the capillary beds in all parts of the body not serviced by the pulmonary circuit. The systemic circuit, which at any moment contains about 84 percent of the total blood volume, begins at the left ventricle and ends at the right atrium.

Systemic Arteries

Figure 21-22• is an overview of the systemic arterial system. This figure indicates the relative locations of major systemic arteries. Figures 21-23 to 21-28• show the detailed distribution of these vessels and their branches.

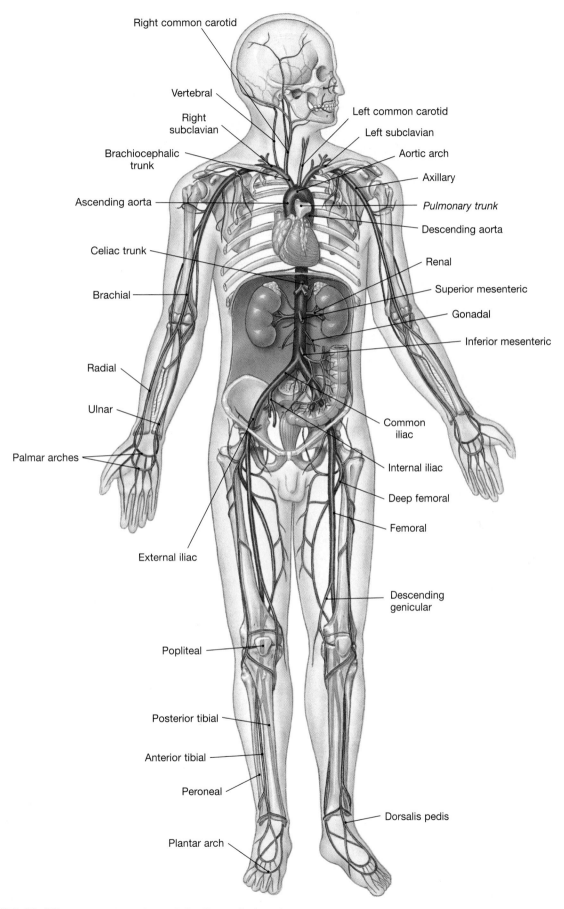

Right common carotid

Vertebral

Right subclavian

Brachiocephalic trunk

Ascending aorta

Celiac trunk

Brachial

Radial

Ulnar

Palmar arches

External iliac

Popliteal

Posterior tibial

Anterior tibial

Peroneal

Plantar arch

Left common carotid

Left subclavian

Aortic arch

Axillary

Pulmonary trunk

Descending aorta

Renal

Superior mesenteric

Gonadal

Inferior mesenteric

Common iliac

Internal iliac

Deep femoral

Femoral

Descending genicular

Dorsalis pedis

21

•**FIGURE 21-22** An Overview of the Major Systemic Arteries

•**FIGURE 21-23** Arteries of the Chest and Upper Limb. **(a)** A diagrammatic view. [AM] *Plates 6.1, 6.2*

The Ascending Aorta The **ascending aorta** begins at the aortic semilunar valve of the left ventricle (Figure 21-23•). The left and right coronary arteries originate at the base of the ascending aorta, just superior to the aortic semilunar valve. We explained the distribution of coronary vessels in Chapter 20 and illustrated them in Figure 20-9•, p. 666.

The Aortic Arch The **aortic arch** curves like a cane handle across the superior surface of the heart, connecting the ascending aorta with the *descending aorta.* Three elastic arteries originate along the aortic arch (Figures 21-22, 21-23, and 21-24•). These arteries, (1) the **brachiocephalic** (brā-kē-ō-se-FAL-ik) **trunk,** (2) the **left common carotid artery,** and (3) the **left subclavian artery,** deliver blood to the head, neck, shoulders, and upper limbs. The brachiocephalic trunk, also called the *innominate artery* (i-NOM-i-nāt; unnamed), ascends for a short distance before branching to form the **right subclavian artery** and the **right common carotid artery.**

We have only one brachiocephalic trunk, and the left common carotid and left subclavian arteries arise separately from the aortic arch. However, in terms of

By convention, several large arteries are called *trunks;* examples are the *pulmonary, brachiocephalic, thyrocervical,* and *celiac trunks.* Because most of the major arteries are paired, with one artery of each pair on either side of the body, the terms *right* and *left* will not be used unless the arteries to both sides are labeled.

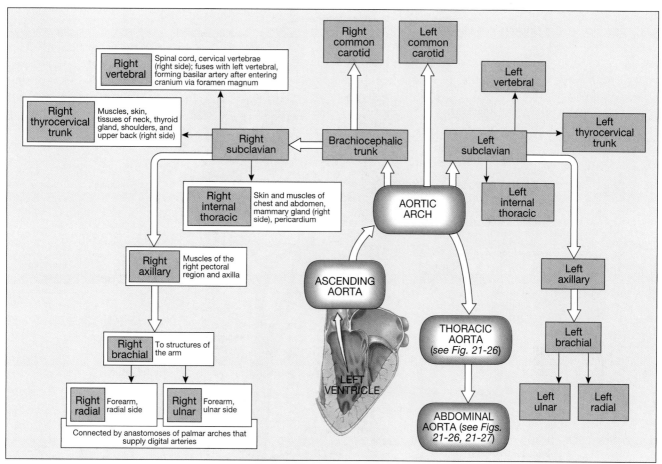

•**FIGURE 21-23** **Arteries of the Chest and Upper Limb** *(continued).* **(b)** A flow chart.

21

their peripheral distribution, the vessels on the left side are mirror images of those on the right side. Figures 21-23 and 21-24• illustrate the major branches of these arteries.

The subclavian arteries   The subclavian arteries supply blood to the arms, chest wall, shoulders, back, and CNS (Figures 21-22 and 21-23•). Three major branches arise before a subclavian artery leaves the thoracic cavity: (1) the **thyrocervical trunk,** which provides blood to muscles and other tissues of the neck, shoulder, and upper back, (2) the **internal thoracic artery,** supplying the pericardium and anterior wall of the chest, and (3) the **vertebral artery,** which provides blood to the brain and spinal cord.

After leaving the thoracic cavity and passing across the superior border of the first rib, the subclavian is called the **axillary artery.** This artery crosses the axilla to enter the arm, where it gives rise to *humeral circumflex arteries,* which supply structures next to the head of the humerus. Distally, it becomes the **brachial artery,** which supplies blood to the rest of the upper limb. As it approaches the coronoid fossa of the

humerus, the brachial artery divides into the **radial artery,** which follows the radius, and the **ulnar artery,** which follows the ulna to the wrist. These arteries supply blood to the forearm. At the wrist, they fuse to form the **superficial** and **deep palmar arches,** which supply blood to the hand and to the **digital arteries** of the thumb and fingers.

The carotid artery and the blood supply to the brain   The common carotid arteries ascend deep in the tissues of the neck. You can usually locate the carotid artery by pressing gently along either side of the windpipe (trachea) until you feel a strong pulse.

Each common carotid artery divides into an **external carotid artery** and an **internal carotid artery** (Figure 21-24•). The carotid sinus, located at the base of the internal carotid artery, may extend along a portion of the common carotid. The external carotid arteries supply blood to the structures of the neck, esophagus, pharynx, larynx, lower jaw, and face. The internal carotid arteries enter the skull through the carotid canals of the temporal bones, delivering blood to the brain (see Figures 7-3 and 7-4•, pp. 197–199).

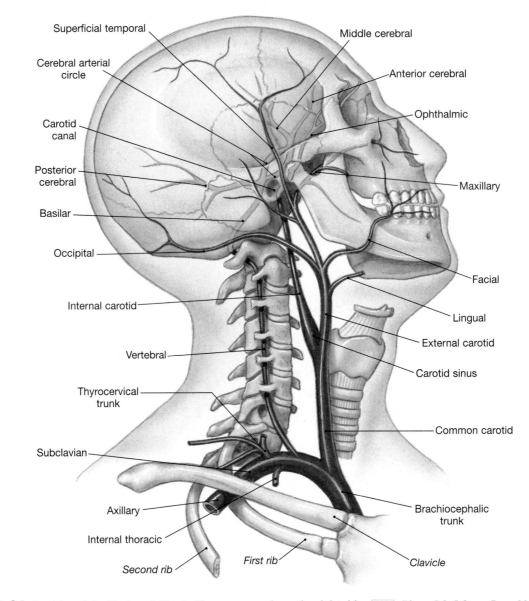

•**FIGURE 21-24** **Arteries of the Neck and Head.** Shown as seen from the right side. AM *Plates 5.2, 5.3a–c; Scan 11*

The internal carotid arteries ascend to the level of the optic nerves, where each artery divides into three branches: (1) an **ophthalmic artery,** which supplies the eyes, (2) an **anterior cerebral artery,** which supplies the frontal and parietal lobes of the brain, and (3) a **middle cerebral artery,** which supplies the mesencephalon and lateral surfaces of the cerebral hemispheres (Figures 21-24 and 21-25•).

The brain is extremely sensitive to changes in its circulatory supply. An interruption of blood flow for several seconds will produce unconsciousness, and after 4 minutes some permanent neural damage can occur. Such circulatory crises are rare, because blood reaches the brain through the vertebral arteries as well as by way of the internal carotids. The left and right vertebral arteries arise from the subclavian arteries

and ascend within the transverse foramina of the cervical vertebrae (see Figure 7-18b,c•, p. 215). The vertebral arteries enter the cranium at the foramen magnum, where they fuse along the ventral surface of the medulla oblongata to form the **basilar artery.** The vertebral arteries and basilar artery supply blood to the spinal cord, medulla oblongata, pons, and cerebellum before dividing into the **posterior cerebral arteries.** The **posterior communicating arteries** branch off the posterior cerebral arteries (Figure 21-25•).

The internal carotids normally supply the arteries of the anterior half of the cerebrum, and the rest of the brain receives blood from the vertebral arteries. But this circulatory pattern can easily change, because the internal carotid arteries and the basilar artery are interconnected in a ring-shaped anastomosis called the

Anterior cerebral

Internal carotid (cut)

Middle cerebral

Pituitary gland

Basilar

Vertebral

Anterior communicating

Anterior cerebral

Posterior communicating

Posterior cerebral

Cerebral arterial circle

Superior cerebellar

Anterior inferior cerebellar

Posterior inferior cerebellar

•**FIGURE 21-25** **Arteries of the Brain.** The major arteries on the inferior surface of the brain. [AM] *Plates 3.4, 3.5; Scan 11*

21

cerebral arterial circle, or *circle of Willis,* which encircles the infundibulum of the pituitary gland (Figure 21-25•). With this arrangement, the brain can receive blood from either the carotid or the vertebral arteries, and the chances for a serious interruption of circulation are reduced.

Strokes

Strokes, or *cerebrovascular accidents (CVAs),* are interruptions of the vascular supply to a portion of the brain. The *middle cerebral artery,* a major branch of the cerebral arterial circle, is the most common site of a stroke. Superficial branches deliver blood to the temporal lobe and large portions of the frontal and parietal lobes; deep branches supply the cerebral nuclei and portions of the thalamus. If a stroke blocks the middle cerebral artery on the left side of the brain, aphasia and a sensory and motor paralysis of the right side result. In a stroke affecting the middle cerebral artery on the right side, the individual experiences a loss of sensation and motor control over the left side and has difficulty drawing or interpreting spatial relationships. Strokes affecting vessels that supply the brain stem also produce distinctive symptoms; those affecting the lower brain stem are commonly fatal.

The Descending Aorta The **descending aorta** is continuous with the aortic arch. The diaphragm divides the descending aorta into a superior **thoracic aorta** and an inferior **abdominal aorta** (Figures 21-26 and 21-27, p. 731.)

The thoracic aorta The thoracic aorta begins at the level of vertebra T_5 and penetrates the diaphragm at the level of vertebra T_{12}. The thoracic aorta travels within the mediastinum, on the dorsal thoracic wall, slightly to the left of the vertebral column. It supplies blood to branches that service the tissues and organs of the mediastinum, the muscles of the chest and the diaphragm, and the thoracic spinal cord.

The branches of the thoracic aorta are anatomically grouped as either visceral or parietal. *Visceral branches* supply the organs of the chest: The **bronchial arteries** supply the nonrespiratory tissues of the lungs; the **pericardial arteries** supply the pericardium; the **esophageal arteries** supply the esophagus; and the **mediastinal arteries** supply the tissues of the mediastinum. *Parietal branches* supply the chest wall: The **intercostal arteries** supply the chest muscles and the vertebral column area, and the **superior phrenic** (FREN-ik) **arteries**

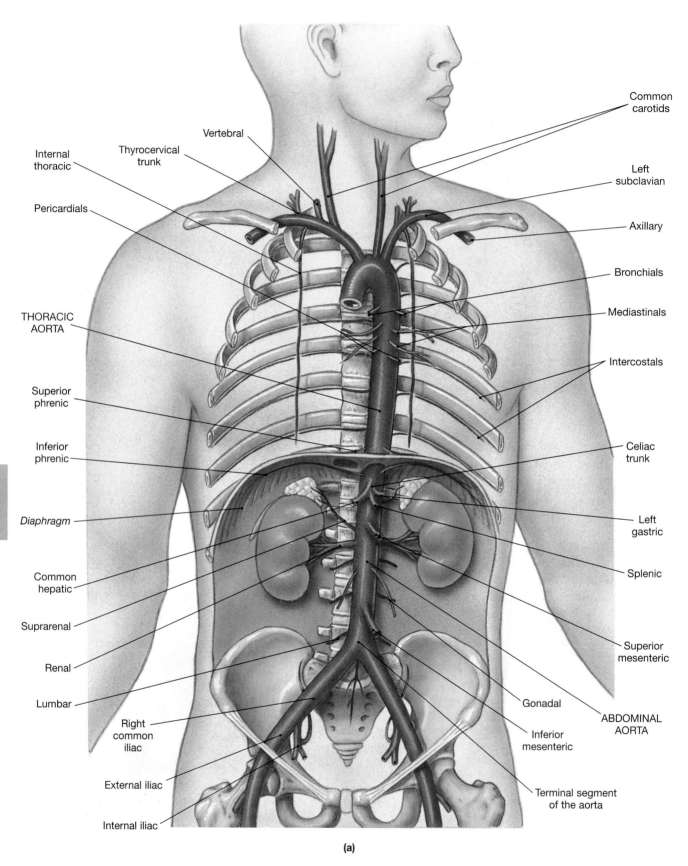

Common carotids

Vertebral

Internal thoracic

Thyrocervical trunk

Left subclavian

Pericardials

Axillary

Bronchials

Mediastinals

THORACIC AORTA

Intercostals

Superior phrenic

Inferior phrenic

Celiac trunk

Diaphragm

Left gastric

Common hepatic

Splenic

Suprarenal

Renal

Superior mesenteric

Lumbar

Gonadal

Right common iliac

ABDOMINAL AORTA

Inferior mesenteric

External iliac

Terminal segment of the aorta

Internal iliac

21

(a)

●**FIGURE 21-26 Major Arteries of the Trunk. (a)** A diagrammatic view, with most of the thoracic and abdominal organs removed. [AM] *Plates 7.6d, 7.10d,e; Scans 12, 13* deliver blood to the superior surface of the diaphragm, which separates the thoracic and abdominopelvic cavities. The branches of the thoracic aorta are diagrammed in Figure 21-26•.

The abdominal aorta   The abdominal aorta, which begins immediately inferior to the diaphragm, is a continuation of the thoracic aorta (Figure 21-26a•). The abdominal aorta descends slightly to the left of the vertebral column but posterior to the peritoneal cavity. It is commonly surrounded by a cushion of adipose tissue. At the level of vertebra L_4, the abdominal aorta splits into two major arteries—the *left* and *right common iliac arteries*—that supply deep pelvic structures and the lower limbs. The region where the aorta splits is called the *terminal segment of the aorta.*

The abdominal aorta delivers blood to all the abdominopelvic organs and structures. The major branches to visceral organs are unpaired, and they arise on the anterior surface of the abdominal aorta and extend into the mesenteries. Branches to the body wall, the kidneys, the urinary bladder, and other structures outside the abdominopelvic cavity are paired; they originate along the lateral surfaces of the abdominal aorta. Figure 21-26• shows the major arteries of the trunk after the removal of most thoracic and abdominal organs. Figure 21-27• gives the distribution of those arteries to abdominopelvic organs.

The abdominal aorta gives rise to three unpaired arteries (Figures 21-26 and 21-27•):

1. The **celiac** (SĒ-lē-ak) **trunk** delivers blood to the liver, stomach, and spleen. The celiac trunk divides into three branches: (a) the **left gastric artery,** which supplies the stomach and the inferior portion of the esophagus; (b) the **splenic artery,** which supplies the spleen and arteries to the stomach *(left gastroepiploic artery)* and pancreas *(pancreatic arteries);* and (c) the

21

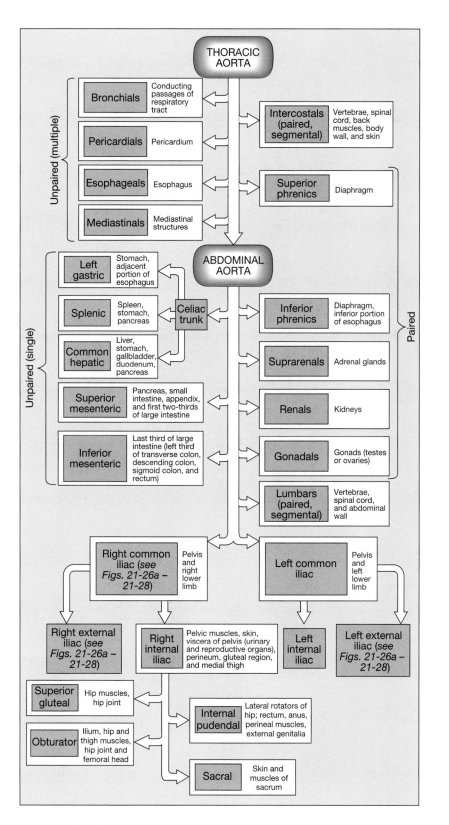

•**FIGURE 21-26** **Major Arteries of the Trunk** *(continued).* **(b)** A flow chart.

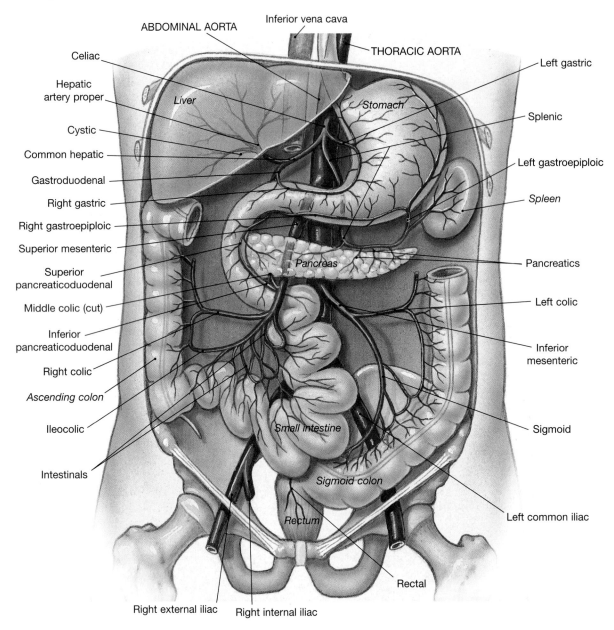

•**FIGURE 21-27** Arteries Supplying the Abdominopelvic Organs. *(See also Figure 24-23, p. 883.)* AM *Plate 7.10; Scans 12, 13*

common hepatic artery, which supplies arteries to the liver *(hepatic artery proper),* stomach *(right gastric artery),* gallbladder *(cystic artery),* and duodenal area *(gastroduodenal, right gastroepiploic,* and *superior pancreaticoduodenal arteries).*

2. The **superior mesenteric** (mez-en-TER-ik) **artery** arises about 2.5 cm inferior to the celiac trunk to supply arteries to the pancreas and duodenum *(inferior pancreaticoduodenal artery),* small intestine *(intestinal arteries),* and most of the large intestine *(right* and *middle colic* and the *ileocolic arteries).*

3. The **inferior mesenteric artery** arises about 5 cm superior to the terminal aorta and delivers blood to the

terminal portions of the colon *(left colic* and *sigmoid arteries)* and the rectum *(rectal arteries).*

The abdominal aorta also gives rise to five paired arteries (Figure 21-26•):

1. The **inferior phrenic arteries,** which supply the inferior surface of the diaphragm and the inferior portion of the esophagus.

2. The **suprarenal arteries,** which originate on either side of the aorta near the base of the superior mesenteric artery. Each suprarenal artery supplies one adrenal gland, which caps the superior part of a kidney.

3. The short (about 7.5 cm) **renal arteries,** which arise along the posterolateral surface of the abdominal

aorta, about 2.5 cm (1 in.) inferior to the superior mesenteric artery, and travel posterior to the peritoneal lining to reach the adrenal glands and kidneys. We shall consider the branches of the renal arteries in Chapter 26.

4. The **gonadal** (gō-NAD-al) **arteries,** which originate between the superior and inferior mesenteric arteries. In males, they are called *testicular arteries* and are long, thin arteries that supply blood to the testes and scrotum. In females, they are termed *ovarian arteries* and supply blood to the ovaries, uterine tubes, and uterus. The distribution of gonadal vessels (both arteries and veins) differs by gender; we shall describe the differences in Chapter 28.

5. Small **lumbar arteries,** which arise on the posterior surface of the aorta and supply the vertebrae, spinal cord, and abdominal wall.

Arteries of the Pelvis and Lower Limbs   Near the level of vertebra L_4, the terminal segment of the abdominal aorta divides to form a pair of elastic arteries, the **right** and **left common iliac** (IL-ē-ak) **arteries** (Figure 21-26•). These arteries carry blood to the pelvis and lower limbs (Figures 21-28•). Along the inner surface of the ilium, these arteries descend posterior to the cecum and sigmoid colon. At the level of the lumbosacral joint, each common iliac divides to form an **internal iliac artery** and an **external iliac artery** (Figure 21-27•). The internal iliac arteries enter the pelvic cavity to supply the urinary bladder, the internal and external walls of the pelvis, the external genitalia, the medial side of the thigh, and, in females, the uterus and vagina. The major tributaries of the internal iliac artery

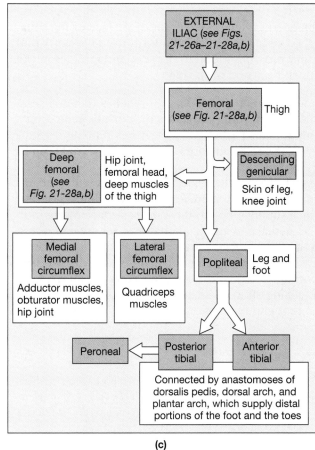

•**FIGURE 21-28** **Arteries of the Lower Limb.** **(a)** An anterior view. **(b)** A posterior view. **(c)** A flow chart of blood flow to a lower limb.   AM   *Plates 8.2, 8.7; Scans 5, 6*

are the *superior gluteal, internal pudendal, obturator,* and *sacral arteries.* The external iliac arteries supply blood to the lower limbs and are much larger in diameter than the internal iliac arteries.

Arteries of the thigh and leg Each external iliac artery crosses the surface of an iliopsoas muscle and penetrates the abdominal wall midway between the anterior superior iliac spine and the pubic symphysis on that side. It emerges on the anterior, medial surface of the thigh as the **femoral artery.** Roughly 5 cm distal to the emergence of the femoral artery, the **deep femoral artery** branches off its lateral surface (Figure 21-28a,b•). The deep femoral artery, which gives rise to the *femoral circumflex arteries,* supplies blood to the ventral and lateral regions of the skin and deep muscles of the thigh.

The femoral artery continues inferiorly and posterior to the femur. As it approaches the knee, it gives rise to the *descending genicular artery,* which supplies the area around the knee. At the popliteal fossa, posterior to the knee joint, the femoral artery becomes the **popliteal** (pop-LIT-ē-al) **artery.** The popliteal artery then branches to form the **posterior** and **anterior tibial arteries.** The posterior tibial artery gives rise to the **peroneal artery** before continuing inferiorly along the posterior surface of the tibia. The anterior tibial artery passes between the tibia and fibula, emerging on the anterior surface of the tibia. As it descends toward the foot, the anterior tibial artery provides blood to the skin and muscles of the anterior portion of the leg.

Arteries of the foot When it reaches the ankle, the anterior tibial artery becomes the **dorsalis pedis artery.** The dorsalis pedis artery branches repeatedly, supplying the ankle and dorsal portion of the foot (Figure 21-28a,b•).

As it reaches the ankle, the posterior tibial artery divides to form the **medial** and **lateral plantar arteries,** which supply blood to the plantar surface of the foot. The medial and lateral plantar arteries are connected to the dorsalis pedis artery through a pair of anastomoses. This arrangement produces a **dorsal arch** *(arcuate arch)* and a **plantar arch.** Small arteries branching off these arches supply the distal portions of the foot and the toes.

☑ A blockage of which branch from the aortic arch would interfere with blood flow to the left arm?

☑ Why would a compression of the common carotid arteries cause a person to lose consciousness?

☑ Grace is in an automobile accident, and her celiac trunk is ruptured. Which organs will be affected most directly by this injury?

Systemic Veins

Veins collect blood from each of the tissues and organs of the body by means of an elaborate venous network that drains into the right atrium of the heart via the superior and inferior venae cavae (Figure 21-29•). The branching pattern of peripheral veins is much more variable than is the branching pattern of arteries. The discussion that follows is based on the most common arrangement of veins. Complementary arteries and veins commonly run side by side, and in many cases they have comparable names.

One significant difference between the arterial and venous systems concerns the distribution of major veins in the neck and limbs. Arteries in these areas are located deep beneath the skin, protected by bones and surrounding soft tissues. In contrast, the neck and limbs generally have two sets of peripheral veins, one superficial and the other deep. This dual venous drainage is important for controlling body temperature. In hot weather, venous blood flows in superficial veins, where heat loss can occur; in cold weather, blood is routed to the deep veins to minimize heat loss.

The Superior Vena Cava All the body's systemic veins (except the cardiac veins) drain into either the superior vena cava or the inferior vena cava. The **superior vena cava (SVC)** receives blood from the tissues and organs of the head, neck, chest, shoulders, and upper limbs.

Venous return from the cranium Numerous veins drain the cerebral hemispheres. The *superficial cerebral veins* and small veins of the brain stem empty into a network of dural sinuses (Figure 21-30a•, p. 736), including the *superior* and *inferior sagittal sinuses,* the *petrosal sinuses,* the *occipital sinus,* the *left* and *right transverse sinuses,* and the *straight sinus* (Figure 21-30b•). The largest, the **superior sagittal sinus,** is in the falx cerebri (see Figure 14-4•, p. 441). Most of the *inferior cerebral veins* converge within the brain to form the **great cerebral vein,** which delivers blood from the interior of the cerebral hemispheres and the choroid plexus to the **straight sinus.** Other cerebral veins drain into the **cavernous sinus** with numerous small veins from the orbit.

The venous sinuses converge within the dura mater in the region of the lambdoidal suture. The left and right transverse sinuses converge at the base of the petrous portion of the temporal bone. There the transverse sinuses form the **sigmoid sinus,** which penetrates the jugular foramen and leaves the skull as the **internal jugular vein.** The internal jugular vein descends parallel to the common carotid artery in the neck (p. 728).

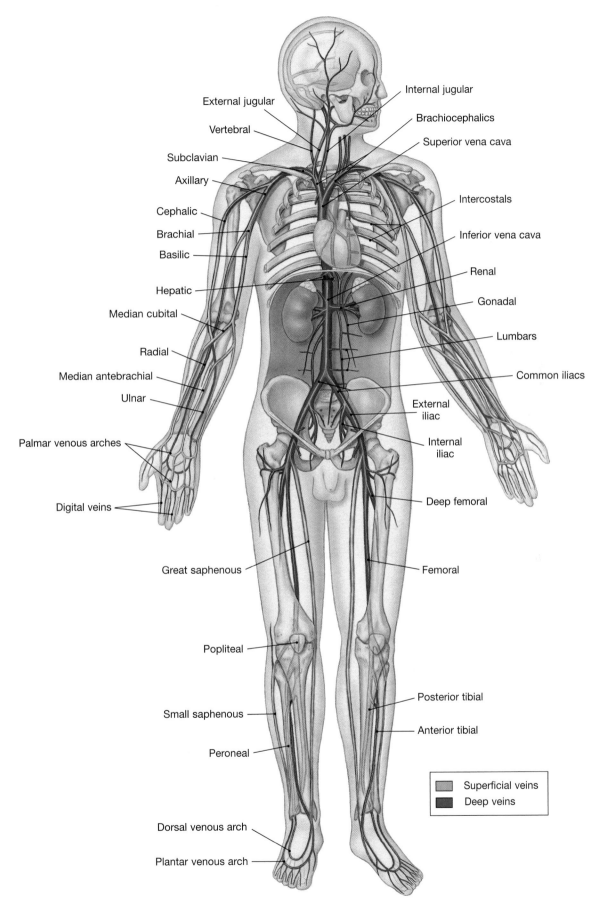

External jugular
Vertebral
Subclavian
Axillary
Cephalic
Brachial
Basilic
Hepatic
Median cubital
Radial
Median antebrachial
Ulnar
Palmar venous arches
Digital veins
Great saphenous
Popliteal
Small saphenous
Peroneal
Dorsal venous arch
Plantar venous arch

Internal jugular
Brachiocephalics
Superior vena cava
Intercostals
Inferior vena cava
Renal
Gonadal
Lumbars
Common iliacs
External iliac
Internal iliac
Deep femoral
Femoral
Posterior tibial
Anterior tibial

Superficial veins
Deep veins

21

•**FIGURE 21-29** **An Overview of the Major Systemic Veins**

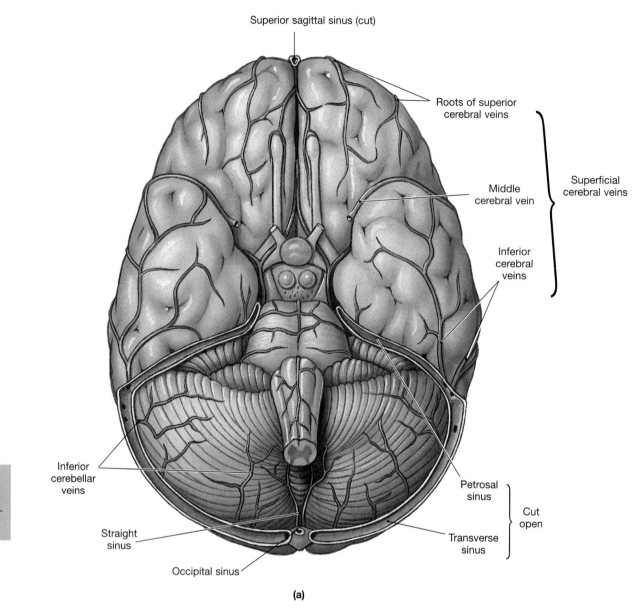

Superior sagittal sinus (cut)

Roots of superior
cerebral veins

Middle
cerebral vein

Superficial
cerebral veins

Inferior
cerebral
veins

Inferior
cerebellar
veins

Petrosal
sinus

Cut
open

Straight
sinus

Transverse
sinus

Occipital sinus

(a)

•**FIGURE 21-30 Major Veins of the Head, Neck, and Brain. (a)** An inferior view of the brain, showing the venous distribution. For the relationship of these veins to meningeal layers, *see Figure 14-4,* p. 441.

Vertebral veins drain the cervical spinal cord and the posterior surface of the skull. These vessels descend within the transverse foramina of the cervical vertebrae, in company with the vertebral arteries. The vertebral veins empty into the *brachiocephalic veins* of the chest (discussed later in the chapter).

Superficial veins of the head and neck The superficial veins of the head collect to form the **temporal, facial,** and **maxillary veins** (Figure 21-30b•). The temporal vein and the maxillary vein drain into the **external jugular**

vein. The facial vein drains into the internal jugular vein. A broad anastomosis between the external and internal jugular veins at the angle of the mandible provides dual venous drainage of the face, scalp, and cranium. The external jugular vein descends toward the chest just deep to the skin on the anterior surface of the sternocleidomastoid muscle. Posterior to the clavicle, the external jugular vein empties into the *subclavian vein.* In healthy individuals, the external jugular vein is easily palpable, and a *jugular venous pulse (JVP)* is sometimes detectable at the base of the neck.

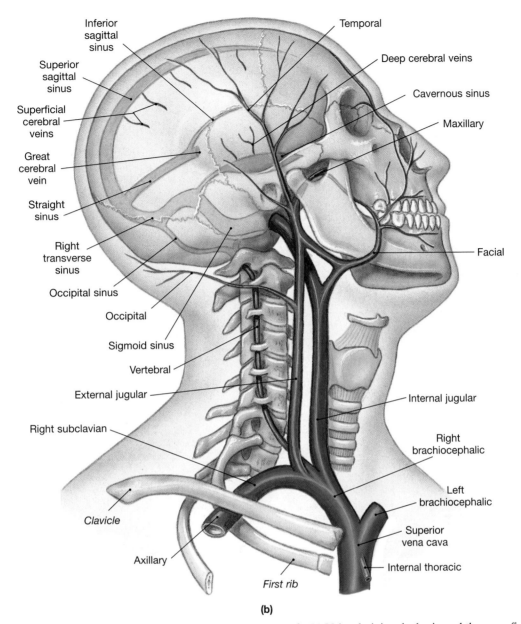

•FIGURE 21-30 Major Veins of the Head, Neck, and Brain *(continued).* **(b)** Veins draining the brain and the superficial and deep portions of the head and neck. AM *Plates 5.2, 5.3*

Venous return from the upper limbs The **digital veins** empty into the **superficial** and **deep palmar veins** of the hand, which are interconnected to form the **palmar venous arches.** The superficial arch empties into the **cephalic vein,** which ascends along the radial side of the forearm; the **median antebrachial vein;** and the **basilic vein,** which ascends on the ulnar side (Figure 21-31•). Anterior to the elbow is the superficial **median cubital vein.** This vein passes from the cephalic vein, medially and at an oblique angle, to connect to the basilic vein. (The median cubital is the vein from which venous blood samples are typically collected.) From the elbow, the basilic vein passes superiorly along the median surface of the biceps brachii muscle.

The deep palmar veins drain into the **radial vein** and the **ulnar vein.** After crossing the elbow, these veins fuse to form the **brachial vein.** The brachial vein is parallel to the brachial artery. As the brachial vein continues toward the trunk, it receives blood from the basilic vein before entering the axilla as the **axillary vein.**

Formation of the superior vena cava The cephalic vein joins the axillary vein on the lateral surface of the first rib, forming the **subclavian vein,** which continues into the chest. The subclavian vein passes superior to the first rib and along the superior margin of the clavicle. This vein then merges with the external and internal jugular veins of that side. This fusion creates the

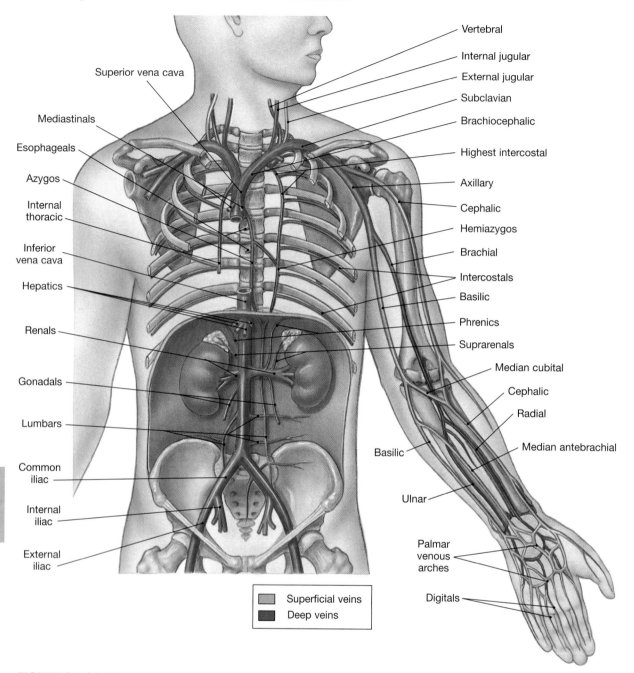

Vertebral

Internal jugular

External jugular

Subclavian

Brachiocephalic

Highest intercostal

Axillary

Cephalic

Hemiazygos

Brachial

Intercostals

Basilic

Phrenics

Suprarenals

Median cubital

Cephalic

Radial

Median antebrachial

Basilic

Ulnar

Palmar venous arches

Digitals

Superior vena cava

Mediastinals

Esophageals

Azygos

Internal thoracic

Inferior vena cava

Hepatics

Renals

Gonadals

Lumbars

Common iliac

Internal iliac

External iliac

Superficial veins

Deep veins

•FIGURE 21-31 The Venous Drainage of the Abdomen and Chest.   AM *Plates 6.1; 7.6b,d; 7.10a,d,e; 8.2, 8.7*

brachiocephalic vein, or *innominate vein,* which penetrates the body wall and enters the thoracic cavity.

Each brachiocephalic vein receives blood from the **vertebral vein** of the same side, which drains the back of the skull and spinal cord. Near the heart, at the level of the first and second ribs, the left and right brachiocephalic veins combine, creating the superior vena cava. Close to the point of fusion, the **internal thoracic vein** empties into the brachiocephalic vein.

The **azygos** (AZ-i-gos) **vein** is the major tributary of the superior vena cava. This vein ascends from the lumbar region over the right side of the vertebral column to invade the thoracic cavity through the di-

aphragm. The azygos vein joins the superior vena cava at the level of vertebra T_2. On the left side, the azygos receives blood from the smaller **hemiazygos vein,** which in many people also drains into the left brachiocephalic through the *highest intercostal vein.*

The azygos and hemiazygos veins are the chief collecting vessels of the thorax. They receive blood from (1) **intercostal veins,** which receive blood from the chest muscles; (2) **esophageal veins,** which drain blood from the inferior portion of the esophagus; and (3) smaller veins draining other mediastinal structures.

Figure 21-32a• diagrams the venous tributaries of the superior vena cava.

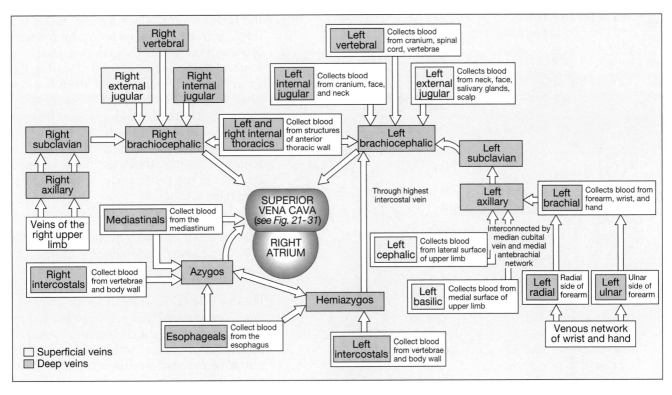

(a) Tributaries of the SVC

The Inferior Vena Cava   The **inferior vena cava (IVC)** collects most of the venous blood from organs inferior to the diaphragm. (A small amount reaches the superior vena cava via the azygos and hemiazygos veins.)

Veins draining the lower limbs   Blood leaving capillaries in the sole of each foot collects into a network of **plantar veins,** which supply the **plantar venous arch.** The plantar network provides blood to the deep veins of the leg: the **anterior tibial vein,** the **posterior tibial vein,** and the **peroneal vein** (Figure 21-33a•). The **dorsal venous arch** collects blood from capillaries on the superior surface of the foot and the **digital veins** of the toes. The plantar arch and the dorsal arch are extensively interconnected, and the path of blood flow can easily shift from superficial to deep veins.

The dorsal venous arch is drained by two superficial veins, the **great saphenous** (sa-FĒ-nus) **vein** (*saphenes,* prominent) and the **small saphenous vein.** The great saphenous vein ascends along the medial aspect of the leg and thigh, draining into the *femoral vein* near the hip joint. The small saphenous vein arises from the dorsal venous arch and ascends along the posterior and lateral aspect of the calf. This vein then enters the popliteal fossa, where it meets the **popliteal vein,** formed by the union of the peroneal and both tibial veins. The popliteal vein is easily palpated in the popliteal fossa adjacent to the adductor magnus muscle (Figure 21-33b•). At the femur, the popliteal vein becomes

21

(b) Tributaries of the IVC

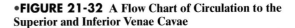
•**FIGURE 21-32 A Flow Chart of Circulation to the Superior and Inferior Venae Cavae**

(c)

the **femoral vein,** which ascends along the thigh, next to the femoral artery. Immediately before penetrating the abdominal wall, the femoral vein receives blood from (1) the great saphenous vein, (2) the **deep femoral vein,** which collects blood from deeper structures in the thigh, and (3) the *femoral circumflex vein,* which drains the region around the neck and head of the femur. The femoral vein penetrates the body wall and emerges in the pelvic cavity as the **external iliac vein.**

Veins draining the pelvis   The external iliac veins receive blood from the lower limbs, pelvis, and lower abdomen. As the left and right external iliac veins cross the inner surface of the ilium, they are joined by the **internal iliac veins,** which drain the pelvic organs. The internal iliac veins are formed by the fusion of the *superior gluteal, internal pudendal, obturator,* and *sacral veins.* The union of external and internal iliac veins forms the **common iliac vein.** The right and left common iliacs ascend at an oblique angle. Anterior to vertebra L_5, they unite to form the inferior vena cava (Figure 21-31•).

Veins draining the abdomen   The inferior vena cava ascends posterior to the peritoneal cavity, parallel to the aorta. The abdominal portion of the inferior vena cava collects blood from six major veins (Figures 21-31 and 21-32b•):

1. **Lumbar veins** drain the lumbar portion of the abdomen, including the spinal cord and body wall muscles. Superior branches of these veins are connected to the azygos vein (right side) and hemiazygos vein (left side), which empty into the superior vena cava.

2. **Gonadal** *(ovarian or testicular)* **veins** drain the ovaries or testes. The right gonadal empties into the inferior vena cava; the left gonadal generally drains into the left renal vein.

3. **Hepatic veins** from the liver empty into the inferior vena cava at the level of vertebra T_{10}.

4. **Renal veins,** the largest inferior vena cava tributaries, collect blood from the kidneys.

•**FIGURE 21-33 Venous Drainage from the Lower Limb. (a)** An anterior view. **(b)** A posterior view. **(c)** A flow chart of venous circulation to a lower limb. AM *Plate 8.7; Scans 5, 6*

5. **Suprarenal veins** drain the adrenal glands. In general, only the right suprarenal vein drains into the inferior vena cava, and the left into the left renal vein.

6. **Phrenic veins** drain the diaphragm. Only the right phrenic vein drains into the inferior vena cava; the left drains into the left renal vein.

Figure 21-32b• diagrams the tributaries of the IVC.

The Hepatic Portal System   The liver is the only digestive organ that drains directly into the inferior vena cava. Blood leaving the capillaries supplied by the celiac, superior, and inferior mesenteric arteries flows into the **hepatic portal system** (Figure 21-34•). A blood vessel connecting two capillary beds is called a *portal vessel;* the network is a *portal system.* ∞ *p. 587*

Blood flowing in the hepatic portal system is quite different from that in other systemic veins, because hepatic portal vessels contain substances absorbed by the stomach and intestines. For example, levels of blood glucose and amino acids in the hepatic portal vein often exceed those anywhere else in the circulatory system. The hepatic portal system delivers these and other absorbed compounds directly to the liver

for storage, metabolic conversion, or excretion. After passing through liver sinusoids, blood collects in the hepatic veins, which empty into the inferior vena cava. Blood from the digestive organs goes to the liver first, so the composition of the blood in the systemic circuit is fairly stable, regardless of any digestive activities.

The hepatic portal system begins in the capillaries of the digestive organs and ends as the hepatic portal vein discharges blood into liver sinusoids. Tributaries of the hepatic portal vein include (Figure 21-34•):

▪ The **inferior mesenteric vein,** which collects blood from capillaries along the inferior portion of the large intestine. Its tributaries include the *left colic vein* and the *superior rectal veins,* which drain the descending colon, sigmoid colon, and rectum.

▪ The **splenic vein,** formed by the union of the inferior mesenteric vein and veins from the spleen, the lateral border of the stomach *(left gastroepiploic vein),* and the pancreas *(pancreatic veins).*

▪ The **superior mesenteric vein,** which collects blood from veins draining the stomach *(right gastroepiploic*

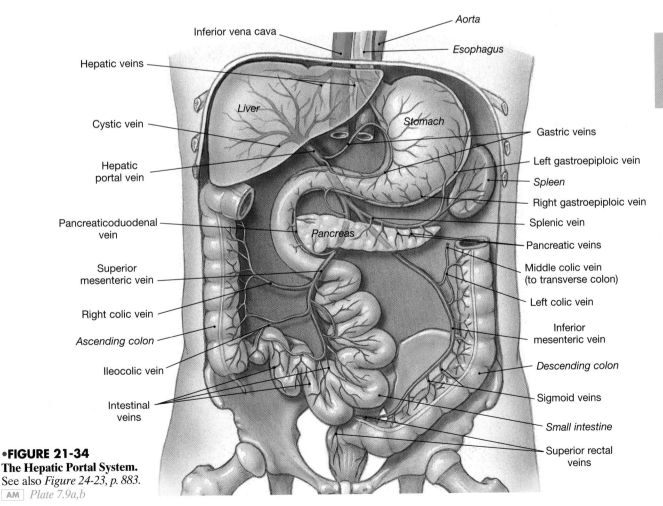

•**FIGURE 21-34**
The Hepatic Portal System.
See also *Figure 24-23,* p. 883.
AM *Plate 7.9a,b*

vein), the small intestine *(intestinal* and *pancreatico-duodenal veins),* and two-thirds of the large intestine *(ileocolic, right colic,* and *middle colic).*

The hepatic portal vein forms via the fusion of the superior mesenteric and splenic veins. The superior mesenteric normally contributes the greater volume of blood and most of the nutrients. As it proceeds, the hepatic portal vein receives blood from the **gastric veins,** which drain the medial border of the stomach, and from the **cystic vein** from the gallbladder.

FETAL CIRCULATION

Fetal and adult circulatory systems exhibit significant differences, reflecting different sources of respiratory and nutritional support. Embryonic lungs are collapsed and nonfunctional, and the digestive tract has nothing to digest. The nutritional and respiratory needs are provided by diffusion across the placenta.

Placental Blood Supply

Fetal circulation is diagrammed in Figure 21-35•. Blood flow to the placenta is provided by a pair of **umbilical arteries,** which arise from the internal iliac arteries and enter the umbilical cord. Blood returns from the placenta in the single **umbilical vein,** bringing oxygen and nutrients to the developing fetus. The umbilical vein drains into the **ductus venosus,** a vascular connection to an intricate network of veins within the developing liver. The ductus venosus collects blood from the veins of the liver and from the umbilical vein and empties into the inferior vena cava. When the placental connection is broken at birth, blood flow ceases along the umbilical vessels, and they soon degenerate.

Circulation in the Heart and Great Vessels

One of the most interesting aspects of circulatory development reflects the differences between the life of an embryo or fetus and that of an infant. Throughout embryonic and fetal life, the lungs are collapsed; yet, after delivery, the newborn infant must be able to extract oxygen from inspired air rather than across the placenta. AM *EMBRYOLOGY SUMMARY 16: The Development of the Cardiovascular System*

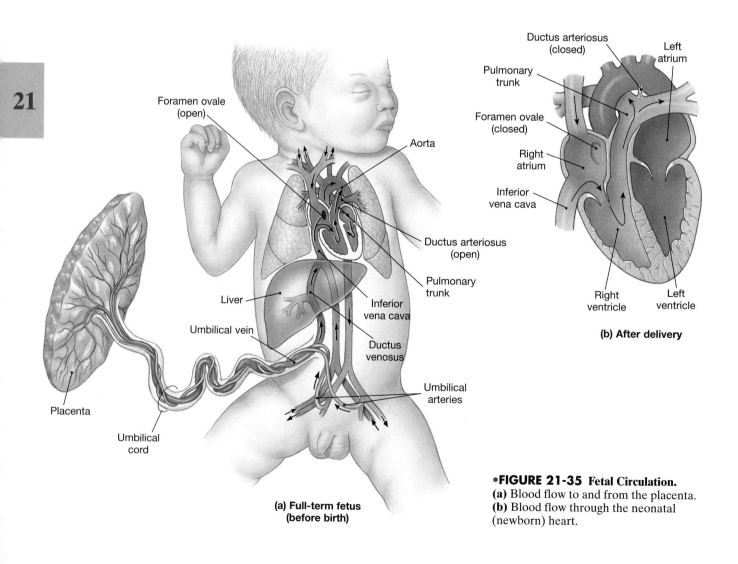

(a) **Full-term fetus (before birth)**

(b) **After delivery**

•**FIGURE 21-35 Fetal Circulation.**
(a) Blood flow to and from the placenta.
(b) Blood flow through the neonatal (newborn) heart.

Although the interatrial and interventricular septa develop early in fetal life, the interatrial partition remains functionally incomplete until birth. The **foramen ovale,** or *interatrial opening,* is associated with a long flap that acts as a valve. Blood can flow freely from the right atrium to the left atrium, but any backflow will close the valve and isolate the two chambers. Thus blood can enter the heart at the right atrium and bypass the pulmonary circuit. A second short-circuit exists between the pulmonary and aortic trunks. This connection, the **ductus arteriosus,** consists of a short, muscular vessel.

With the lungs collapsed, the capillaries are compressed and little blood flows through the lungs. During diastole, blood enters the right atrium and flows into the right ventricle, but it also passes into the left atrium through the foramen ovale. About 25 percent of the blood arriving at the right atrium bypasses the pulmonary circuit in this way. In addition, over 90 percent of the blood leaving the right ventricle passes through the ductus arteriosus and enters the systemic circuit rather than continuing to the lungs.

Circulatory Changes at Birth

At birth, dramatic changes occur. When the infant takes the first breath, the lungs expand, and so do the pulmonary vessels. The resistance in the pulmonary circuit declines suddenly, and blood rushes into the pulmonary vessels. Within a few seconds, rising O_2 levels stimulate the constriction of the ductus arteriosus, isolating the pulmonary and aortic trunks. As pressures rise in the left atrium, the valvular flap closes the foramen ovale. In adults, the interatrial septum bears the *fossa ovalis,* a shallow depression that marks the site of the foramen ovale (see Figure 20-6a,c•, p. 661). The remnants of the ductus arteriosus persist as the *ligamentum arteriosum,* a fibrous cord.

If the proper circulatory changes do not occur at birth or shortly thereafter, problems will eventually develop. The severity of the problem depends on which connection remains open and on the size of the opening. Treatment may involve the surgical closure of the foramen ovale, the ductus arteriosus, or both. Other forms of congenital heart defects result from abnormal cardiac development or inappropriate connections between the heart and major arteries and veins.

Congenital Circulatory Problems

Minor individual variations in the circulatory network are quite common. For example, very few individuals have identical patterns of venous distribution. Congenital circulatory problems serious enough to threaten homeostasis are relatively rare. They generally reflect abnormal formation of the heart or problems with the interconnections between the heart and the great vessels. Several examples of congenital circulatory defects are illustrated in Figure 21-36•. All these conditions can be surgically corrected, although multiple surgeries may be required.

The incomplete closure of the foramen ovale or ductus arteriosus results in similar types of problems. If the foramen ovale remains open, or *patent,* blood recirculates into the pulmonary circuit without entering the left ventricle (Figure 21-36a•). The movement, driven by the relatively high systemic pressure, is called a "left-to-right shunt." Arterial oxen content is normal, but the left ventricle must work much harder than usual to provide adequate blood flow through the systemic circuit. Hence pressures rise in the pulmonary circuit. The abnormality may not be immediately apparent, but pulmonary hypertension, pulmonary edema, and cardiac enlargement are eventual results. If the ductus arteriosus remains open, the same basic problems develop as blood ejected by the left ventricle reenters the pulmonary circuit. If valve defects, constricted pulmonary

vessels, or other abnormalities occur as well, pulmonary pressures can rise enough to force blood into the systemic circuit through the ductus arteriosus. This movement is called a "right-to-left shunt." Because normal blood oxygenation does not occur, the circulating blood develops a deep red color. The skin then develops the blue tones typical of *cyanosis,* (∞ *p. 149*), and the infant is known as a "blue baby."

Ventricular septal defects are openings in the interventricular septum (Figure 21-36b•). These are the most common congenital heart problems, affecting 0.12 percent of newborn infants. The opening between the left and right ventricles has the reverse effect of a connection between the atria: When the more powerful left ventricle beats, it ejects blood into the right ventricle and pulmonary circuit. The end results are the same as for a patent foramen ovale: a left-to-right shunt, with eventual pulmonary hypertension, pulmonary edema, and cardiac enlargement.

The *tetralogy of Fallot* (fa-LŌ) is a complex group of heart and circulatory defects that affect 0.10 percent of newborn infants. In this condition, (1) the pulmonary trunk is abnormally narrow, (2) the interventricular septum is incomplete, (3) the aorta originates where the interventricular septum normally ends, and (4) the right ventricle is enlarged and both ventricles are thickened owing to increased workloads (Figure 21-36c•).

(a) Patent foramen ovale and ductus arteriosus

(b) Ventricular septal defect

(c) Tetralogy of Fallot

(d) Transposition of great vessels

(e) Atrioventricular septal defect

•FIGURE 21-36 Congenital Circulatory Problems

In the *transposition of great vessels,* the aorta is connected to the right ventricle and the pulmonary artery is connected to the left ventricle (Figure 21-36d•). This malformation affects 0.05 percent of newborn infants.

In an *atrioventricular septal defect,* the atria and ventricles are incompletely separated (Figure 21-36e•).

The results are quite variable, depending on the extent of the defect and the effects on the atrioventricular valves. This type of defect most commonly affects infants with *Down syndrome,* a disorder caused by the presence of an extra copy of chromosome 21. ∞ *p. 499*

AGING AND THE CARDIOVASCULAR SYSTEM

The capabilities of your cardiovascular system gradually decline as you age. The major changes are as follows:

▪ *Age-related changes in blood.* Age-related changes in blood may include (1) a decreased hematocrit; (2) the constriction or blockage of peripheral veins by a *thrombus* (stationary blood clot), which can become detached, pass through the heart, and become wedged in a small artery, commonly in the lungs, causing a *pulmonary embolism;* and (3) a pooling of blood in the veins of the legs because valves are not working effectively.

▪ *The aging heart.* Age-related changes in the heart include (1) a reduction in the maximum cardiac output; (2) changes in the activities of the nodal and conducting cells; (3) a reduction in the elasticity of the fibrous skeleton; (4) a progressive atherosclerosis that can re-

strict coronary circulation; and (5) the replacement of damaged cardiac muscle cells by scar tissue.

▪ *Aging and blood vessels.* Age-related changes in blood vessels may be linked to arteriosclerosis: (1) The inelastic walls of arteries become less tolerant of sudden pressure increases, which can lead to an *aneurysm,* whose rupture may cause a stroke, myocardial infarction, or massive blood loss (depending on the vessel); (2) calcium salts can be deposited on weakened vascular walls, increasing the risk of a stroke or myocardial infarction; and (3) thrombi can form at atherosclerotic plaques.

☑ Whenever Tim gets angry, a large vein bulges in the lateral region of his neck. Which vein is this?

☑ A thrombus that blocks the popliteal vein would interfere with blood flow in which other veins?

☑ A blood sample taken from the umbilical cord contains a high concentration of oxygen and nutrients and a low concentration of carbon dioxide and waste products. Is this sample from an umbilical artery or from the umbilical vein? Explain.

INTEGRATION WITH OTHER SYSTEMS

The cardiovascular system is both anatomically and functionally linked to all other systems. In the section on vessel distribution, we demonstrated the extent of the anatomical connections. Figure 21-37• summarizes the physiological relationships between the cardiovascular system and other organ systems.

The most extensive communication occurs between the cardiovascular and lymphatic systems. Not only are the two systems physically interconnected, but cell populations of the lymphatic system use the cardiovascular system as a highway to move from one part of the body to another. In Chapter 22, we examine the lymphatic system in detail and consider its role in the immune response.

The human body contains miles of blood vessels, ranging from about an inch in diameter to a size barely wide enough for a single erythrocyte to pass. To gain a better understanding of the vessels that carry blood, use the resources available on both the CD-ROM and the Companion Website. The CD-ROM includes animations and tutorials; the website builds on those resources with text content and interactive problems as well as communication avenues for sharing ideas with your class. To complete the following Web Explorations, visit the MediaLab in Chapter 21 of your Companion Website.

Web Exploration 1

Estimated time for completion = 7 minutes

You have learned the names and locations of the body's major arteries and veins. Can the body create new blood vessels and therefore new circulatory patterns? This ability would benefit persons with blocked blood vessels. To read about the latest research into protein therapy for forming new vessels, select the keyword **NEW VESSELS** on the

Companion Website. Summarize the information in your own words. What disease states might be helped by this protein therapy? If you wish, share your thoughts with your classmates by using the communications tools on the Companion Website.

Web Exploration 2

Estimated time for completion = 10 minutes

You have been studying artists' renderings of arterial and venous pathways. Are these pathways the same in every person? To investigate normal variations in circulatory patterns, select the keyword **VARIANTS** on the Companion Website. From the list of arteries, view the femoral artery. Compare the descriptions and images on that site with Figure 21-28 of your text. Can this major artery follow different pathways? If so, how many? Back on the original Companion Website page, now select the femoral vein to view. Compare the venous variation with the arterial variation. From the information presented, what can you conclude about blood flow?

21

CHAPTER REVIEW

SELECTED CLINICAL TERMINOLOGY
Terms Discussed in This Chapter

aneurysm (AN-ū-rizm):   A bulge in the weakened wall of a blood vessel, generally an artery. *(p. 695* and *AM)*

arteriosclerosis (ar-tē-rē-ō-skle-RŌ-sis):   A thickening and toughening of arterial walls. *(p. 696)*

atherosclerosis (ath-er-ō-skle-RŌ-sis):   A type of arteriosclerosis characterized by changes in the endothelial lining and the formation of a plaque. *(p. 696)*

edema (e-DĒ-muh):   An abnormal accumulation of fluid in peripheral tissues. *(p. 710)*

hemorrhoids (HEM-o-roydz):   Varicose veins in the walls of the rectum, the anus, or both; commonly associated with frequent straining to force bowel movements. *(p. 700* and *AM)*

hypertension: Abnormally high blood pressure; usually defined in adults as blood pressure higher than 150/90. *(p. 706* and *AM)*

hypotension: Blood pressure so low that circulation to vital organs may be impaired. *(p. 706* and *AM)*

pressure points: Locations where muscular arteries can be compressed against skeletal elements to restrict or prevent the flow of blood in an emergency. *(p. 707)*

pulmonary embolism: A circulatory blockage caused by the trapping of a detached thrombus in a pulmonary artery. *(p. 744)*

shock: An acute circulatory crisis marked by hypotension and inadequate peripheral blood flow. *(p. 720* and *AM)*

sounds of Korotkoff: Distinctive sounds, caused by turbulent arterial blood flow, that are heard while measuring a person's blood pressure. *(p. 707)*

sphygmomanometer: A device that measures blood pressure by using an inflatable cuff placed around a limb. *(p. 707)*

thrombus: A stationary blood clot within a blood vessel. *(p. 744)*

varicose (VAR-i-kōs) **veins:**   Sagging, swollen veins distorted by gravity and by the failure of the venous valves. *(p. 700)*

INTEGUMENTARY SYSTEM

Stimulation of mast cells produces localized changes in blood flow and capillary permeability

Delivers immune system cells to injury sites; clotting response seals breaks in skin surface; carries away toxins from sites of infection; provides heat

THE CARDIOVASCULAR SYSTEM

FOR ALL SYSTEMS

Delivers oxygen, hormones, nutrients, and WBCs; removes carbon dioxide and metabolic wastes; transfers heat

Maintains healthy vessels and slows development of atherosclerosis with age by estrogens

Distributes reproductive hormones; provides nutrients, oxygen, and waste removal for developing fetus; local blood pressure changes responsible for physical changes during sexual arousal

REPRODUCTIVE SYSTEM

SKELETAL SYSTEM

Provides calcium needed for normal cardiac muscle contraction; protects blood cells developing in bone marrow

Provides Ca^{2+} and PO_4^{3-} for bone deposition; delivers EPO to bone marrow, parathyroid hormone and calcitonin to osteoblasts and osteoclasts

MUSCULAR SYSTEM

Skeletal muscle contractions assist in moving blood through veins; protects superficial blood vessels, especially in neck and limbs

Delivers oxygen and nutrients; removes carbon dioxide, lactic acid, and heat during skeletal muscle activity

NERVOUS SYSTEM

Controls patterns of circulation in peripheral tissues; modifies heart rate and regulates blood pressure; releases ADH

Endothelial cells maintain blood–brain barrier, help generate CSF

ENDOCRINE SYSTEM

Erythropoietin regulates production of RBCs; several hormones elevate blood pressure; epinephrine stimulates cardiac muscle, elevating heart rate and contractile force

Distributes hormones throughout the body; heart secretes ANP

LYMPHATIC SYSTEM

Defends against pathogens or toxins in blood; fights infections of cardiovascular organs; returns tissue fluid to circulation

Distributes WBCs; carries antibodies, which attack pathogens; clotting response assists in restricting spread of pathogens; granulocytes and lymphocytes produced in bone marrow

RESPIRATORY SYSTEM

Provides oxygen to cardiovascular organs and removes carbon dioxide; enzyme in lung capillaries converts angiotensin I to angiotensin II

RBCs transport oxygen and carbon dioxide between lungs and peripheral tissues

DIGESTIVE SYSTEM

Provides nutrients to cardiovascular organs; absorbs water and ions essential to maintenance of normal blood volume

Distributes digestive tract hormones; carries nutrients, water, and ions away from sites of absorption; delivers nutrients and toxins to liver

URINARY SYSTEM

Releases renin to elevate blood pressure and EPO to accelerate red blood cell production

Delivers blood to capillaries, where filtration occurs; accepts fluids and solutes reabsorbed during urine production

•FIGURE 21-37 Functional Relationships between the Cardiovascular System and Other Systems

Additional Terms Discussed in the *Applications Manual*

angiotensin-converting enzyme (ACE) inhibitors: Drugs that block the conversion of angiotensin I to angiotensin II; sometimes used in the treatment of chronic hypertension and congestive heart failure.

cardiogenic (kar-dē-ō-JEN-ik) **shock:** Shock that develops when the heart becomes unable to maintain normal cardiac output.

cerebral embolism: The occlusion of one of the cerebral arteries by a blood clot, fatty mass, or air bubble that originated at another site; a cause of strokes.

cerebral hemorrhage: The rupture of a cerebral blood vessel, commonly following the formation of an aneurysm; a cause of strokes.

cerebral thrombosis: The occlusion of one of the cerebral arteries by a clot formed at a plaque; a cause of strokes.

congestive heart failure (CHF): A condition that develops when the left ventricle can no longer keep pace with the right ventricle and blood backs up into the pulmonary circuit. Elevated pulmonary pressures lead to *pulmonary edema.*

distributive shock: Shock that results from a widespread, uncontrolled vasodilation, as in *neurogenic shock, septic shock,* or *anaphylactic shock.*

obstructive shock: Shock that develops when ventricular output is reduced by tissues or fluids pressing against the heart.

orthostatic hypotension: Hypotension that develops when a person is standing, due to an inability to make adjustments that increase blood pressure enough to maintain the blood supply to the brain.

phlebitis: The inflammation of a vein.

primary hypertension (essential hypertension): Hypertension without an obvious cause. Known risk factors include a hereditary history of hypertension, male gender, high plasma cholesterol, obesity, chronic stresses, and cigarette smoking.

secondary hypertension: Hypertension that develops as the result of kidney problems or abnormal production of ADH, aldosterone, renin, epinephrine, or other hormones.

transient ischemic attack (TIA): A temporary loss of cerebral function due to a momentary blockage in a cerebral artery; an indication of cerebrovascular disease.

STUDY OUTLINE

INTRODUCTION p. 692

1. Blood flows through a network of arteries, veins, and capillaries. All chemical and gaseous exchange between blood and interstitial fluid takes place across capillary walls.
2. **Arteries** and **veins** form an internal distribution system through which the heart propels blood. Arteries branch repeatedly, decreasing in size until they become **arterioles.** From the arterioles, blood enters the **capillary** networks. Blood flowing from the capillaries enters small **venules** before entering larger veins.

THE ANATOMY OF BLOOD VESSELS p. 692

The Structure of Vessel Walls p. 692

1. The walls of arteries and veins contain three layers: the innermost **tunica interna,** the **tunica media,** and the outermost **tunica externa.** *(Figure 21-1)*

Differences between Arteries and Veins p. 692

2. In general, the walls of arteries are thicker than those of veins. The endothelial lining of an artery cannot contract, so it is thrown into folds. Arteries constrict when blood pressure does not distend them; veins constrict very little. *(Figure 21-1)*

Arteries p. 693

3. The arterial system includes the large **elastic arteries,** medium-sized **muscular arteries,** and smaller arterioles. As we proceed toward the capillaries, the number of vessels increases, but the diameter of the individual vessels decreases and the walls become thinner. *(Figure 21-2)*
4. **Atherosclerosis,** a type of **arteriosclerosis,** is associated with changes in the endothelial lining of arteries. Fatty masses of tissue called **plaques** typically develop during atherosclerosis. *(Figure 21-4)*

Capillaries p. 695

5. Capillaries are the only blood vessels whose walls permit exchange between blood and interstitial fluid. Capillaries are **continuous** or **fenestrated. Sinusoids** have fenestrated walls and form elaborate networks that allow very slow blood flow. Sinusoids are located in the liver and in various endocrine organs. *(Figure 21-3)*

6. Capillaries form interconnected networks called **capillary beds (capillary plexuses). A precapillary sphincter** (a band of smooth muscle) adjusts the blood flow into each capillary. Blood flow in a capillary changes as **vasomotion** occurs. The entire capillary bed may be bypassed by blood flow through **arteriovenous anastomoses.** *(Figure 21-5)*

Veins p. 699

7. Venules collect blood from the capillaries and merge into **medium-sized veins** and then **large veins.** The arterial system is a high-pressure system; blood pressure in veins is much lower. **Valves** in veins prevent the backflow of blood. *(Figures 21-1, 21-2, 21-6)*

The Distribution of Blood p. 700

8. Peripheral **venoconstriction** helps maintain adequate blood volume in the arterial system after a hemorrhage. The **venous reserve** normally accounts for about 20 percent of the total blood volume. *(Figure 21-7)*

CARDIOVASCULAR PHYSIOLOGY p. 701

1. Cardiovascular regulation involves the manipulation of blood pressure and resistance to control the rates of blood flow and capillary exchange. *(Figure 21-8)*

Pressure p. 702

2. Flow is proportional to pressure difference; blood will flow from an area of higher pressure to one of lower pressure. The **circulatory pressure** is the pressure gradient across the systemic circuit.

Resistance p. 702

3. The **resistance (R)** determines the rate of blood flow through the systemic circuit. The major determinant of blood flow rate is the **peripheral resistance (PR),** the resistance of the arterial system. Neural and hormonal control mechanisms regulate blood pressure and peripheral resistance.

21

4. The most important determinant of peripheral resistance is the diameter of arterioles.

An Overview of Circulatory Pressures p. 704

5. The high arterial pressures overcome peripheral resistance and maintain blood flow through peripheral tissues. Capillary pressures are normally low; small changes in capillary pressure determine the rate of fluid movement into or out of the bloodstream. Venous pressure, normally low, determines *venous return* and affects cardiac output and peripheral blood flow. *(Figure 21-9; Table 21-1)*

Arterial Blood Pressure p. 705

6. Arterial blood pressure rises during ventricular systole and falls during ventricular diastole. The difference between these two blood pressures is the **pulse pressure.** *(Figures 21-10, 21-11)*

Capillary Exchange p. 706

7. At the capillaries, solute molecules diffuse across the capillary lining, and water-soluble materials diffuse through small spaces between endothelial cells. Water will move when driven by either **hydrostatic pressure** or **osmotic pressure.** The direction of water movement is determined by the balance between these two opposing pressures. Any change in hydrostatic or osmotic pressure in the blood or tissues will shift the point where the two forces are equal and determine whether there is a net loss or a net gain of fluid along the length of the capillary. *(Figures 21-12, 21-13)*

Venous Pressure and Venous Return p. 710

8. Valves, muscular compression, and the **respiratory pump** *(thoracoabdominal pump)* help the relatively low venous pressures propel blood toward the heart. *(Figure 21-6)*

CARDIOVASCULAR REGULATION p. 711

1. Homeostatic mechanisms ensure that **tissue perfusion** (blood flow) delivers adequate oxygen and nutrients.

2. Local, autonomic, and endocrine factors influence the coordinated regulation of cardiovascular function. Local factors change the pattern of blood flow within capillary beds in response to chemical changes in interstitial fluids. Central mechanisms respond to changes in arterial pressure or blood gas levels. Hormones can assist in short-term adjustments (changes in cardiac output and peripheral resistance) and long-term adjustments (changes in blood volume that affect cardiac output and gas transport). *(Figure 21-14)*

Autoregulation of Blood Flow within Tissues p. 712

3. Peripheral resistance is adjusted at the tissues by local factors that result in the dilation or constriction of precapillary sphincters.

Neural Mechanisms p. 712

4. **Baroreceptor reflexes** are autonomic reflexes that adjust cardiac output and peripheral resistance to maintain normal arterial pressures. *Baroreceptors* are in the **aortic sinuses,** the **carotid sinuses,** and the right atrium. *(Figure 21-15)*

5. **Chemoreceptor reflexes** respond to changes in the oxygen or carbon dioxide levels in the blood and cerebrospinal fluid. The stimulation of the *cardioacceleratory* and *vasomotor centers* leads to the stimulation of the heart rate and blood pressure by sympathetic activation; the *cardioinhibitory centers* exert their effects over the parasympathetic division. Epinephrine and norepinephrine stimulate cardiac output and peripheral vasoconstriction. *(Figure 21-16)*

Hormones and Cardiovascular Regulation p. 715

6. The endocrine system provides short-term regulation of cardiac output and peripheral resistance with epinephrine and norepinephrine from the adrenal medullae. Hormones involved in the long-term regulation of blood pressure and volume are *antidiuretic hormone (ADH), angiotensin II, erythropoietin (EPO),* and *atrial natriuretic peptide (ANP). (Figure 21-17)*

7. ADH and angiotensin II also promote peripheral vasoconstriction in addition to their other functions. ADH and aldosterone promote water and electrolyte retention and stimulate thirst. EPO stimulates red blood cell production. ANP encourages fluid loss, reduces blood pressure, inhibits thirst, and lowers peripheral resistance.

PATTERNS OF CARDIOVASCULAR RESPONSE p. 717

Exercise and the Cardiovascular System p. 717

1. During exercise, blood flow to skeletal muscles increases at the expense of blood flow to nonessential organs, so cardiac output rises. Cardiovascular performance improves with training. Athletes have larger stroke volumes, slower resting heart rates, and larger cardiac reserves than do nonathletes. *(Tables 21-2, 21-3)*

Cardiovascular Response to Hemorrhaging p. 719

2. Blood loss lowers blood volume and venous return and decreases cardiac output. Compensatory mechanisms include an increase in cardiac output, a mobilization of venous reserves, peripheral vasoconstriction, and the liberation of hormones that promote fluid retention and the manufacture of erythrocytes. *(Figure 21-18)*

3. **Shock** is an acute circulatory crisis marked by hypotension and inadequate peripheral blood flow. A severe drop in blood volume produces symptoms of **circulatory shock.** Causes of fluid loss may include hemorrhaging, dehydration, and severe burns. *(Figure 21-19)*

FOCUS: Special Circulation p. 722

4. The blood–brain barrier, the coronary circulation, and the circulation to alveolar capillaries in the lungs are examples of special circulation, in which circulatory dynamics and regulatory mechanisms differ from those in other tissues.

THE BLOOD VESSELS p. 723

1. The peripheral distributions of arteries and veins are generally identical on both sides of the body, except near the heart. *(Figure 21-20)*

The Pulmonary Circuit p. 723

2. The pulmonary circuit includes the pulmonary trunk, the **left** and **right pulmonary arteries,** and the **pulmonary veins,** which empty into the left atrium. *(Figure 21-21)*

The Systemic Circuit p. 724

3. The **ascending aorta** gives rise to the coronary circulation. The **aortic arch** communicates with the **descending aorta.** *(Figures 21-22–21-28)*

4. Arteries in the neck and limbs are deep beneath the skin; in contrast, there are generally two sets of peripheral veins, one superficial and one deep. This dual venous drainage is important for controlling body temperature.

5. The **superior vena cava** receives blood from the head, neck, chest, shoulders, and arms. *(Figures 21-29–21-32)*

6. The **inferior vena cava** collects most of the venous blood from organs inferior to the diaphragm. *(Figures 21-32, 21-33)*

7. The **hepatic portal system** directs blood from the other digestive organs to the liver before the blood returns to the heart. *(Figure 21-34)*

Fetal Circulation   p. 742

8. Major circulatory changes occur at birth as the lungs expand and the pulmonary circuit is activated. *(Figure 21-35)*
9. Congenital circulatory problems generally reflect abnormalities of the heart or of interconnections between the heart and great vessels. *(Figure 21-36)*

AGING AND THE CARDIOVASCULAR SYSTEM   p. 744

1. Age-related changes in the blood can include (1) a decreased hematocrit, (2) the constriction or blockage of peripheral veins by a *thrombus* (stationary blood clot), and (3) the pooling of blood in the veins of the legs because valves are not working effectively.

2. Age-related changes in the heart include (1) a reduction in the maximum cardiac output, (2) changes in the activities of the nodal and conducting cells, (3) a reduction in the elasticity of the fibrous skeleton, (4) a progressive atherosclerosis that can restrict coronary circulation, (5) the replacement of damaged cardiac muscle cells by scar tissue.

3. Age-related changes in blood vessels, commonly related to arteriosclerosis, include (1) a weakening in the walls of arteries, potentially leading to *aneurysm* formation, (2) calcium salt deposition on weakened vascular walls, increasing the risk of a stroke or myocardial infarction, and (3) thrombus formation at atherosclerotic plaques.

INTEGRATION WITH OTHER SYSTEMS   p. 745

1. The most extensive communication occurs between the cardiovascular and lymphatic systems. *(Figure 21-37)*

REVIEW QUESTIONS

LEVEL 1 REVIEWING FACTS AND TERMS

Match each numbered item with the most closely related lettered item. Use letters for answers in the spaces provided.

___ 1. veins	**a.** specialized fenestrated capillaries
___ 2. vasa vasorum	**b.** minimum blood pressure
___ 3. conducting arteries	**c.** vasomotor centers
___ 4. muscular arteries	**d.** drains the kidney
___ 5. sinusoids	**e.** vasoconstrictor fibers
___ 6. precapillary sphincter	**f.** ventricular stretching
___ 7. medulla oblongata	**g.** elastic arteries
___ 8. vascular resistance	**h.** blood supply to pelvis
___ 9. systolic pressure	**i.** vasodilator fibers
___ 10. diastolic pressure	**j.** largest superficial vein
___ 11. blood pressure	**k.** capacitance vessels
___ 12. arterioles	**l.** drains the liver
___ 13. cholinergic	**m.** largest artery in body
___ 14. adrenergic	**n.** stationary blood clot
___ 15. baroreceptors	**o.** distribution arteries
___ 16. Frank–Starling principle	**p.** foramen ovale
___ 17. aorta	**q.** vasomotion
___ 18. internal iliac artery	**r.** resistance vessels
___ 19. renal vein	**s.** migrating blood clot
___ 20. hepatic vein	**t.** carotid sinus
___ 21. great saphenous	**u.** friction
___ 22. interatrial opening	**v.** sounds of Korotkoff
___ 23. thrombus	**w.** "vessels of vessels"
___ 24. embolus	**x.** peak blood pressure

25. Blood vessels that carry blood away from the heart are
 (a) veins **(b)** arterioles
 (c) venules **(d)** arteries
26. The layer of the arteriole wall that provides the properties of contractility and elasticity is the
 (a) tunica adventitia **(b)** tunica media
 (c) tunica intima **(d)** tunica externa
27. Blood vessels that supply the walls of arteries and veins with blood are
 (a) coronary vessels **(b)** capillaries
 (c) vasa vasorum **(d)** metarterioles

28. Of the following arteries, the one that is an elastic artery is the
 (a) subclavian artery
 (b) external carotid artery
 (c) brachial artery
 (d) femoral artery
29. The two-way exchange of substances between blood and body cells occurs only through
 (a) arterioles **(b)** capillaries
 (c) venules **(d)** a, b, and c are correct
30. Large molecules such as peptides and proteins move into and out of the bloodstream by way of
 (a) continuous capillaries
 (b) fenestrated capillaries
 (c) thoroughfare channels
 (d) metarterioles
31. The alteration of blood flow due to the action of precapillary sphincters is
 (a) vasomotion **(b)** autoregulation
 (c) selective resistance **(d)** turbulence
32. The blood vessels that collect blood from all tissues and organs and return it to the heart are the
 (a) veins **(b)** arteries
 (c) capillaries **(d)** arterioles
33. Blood is compartmentalized within the veins because of the presence of
 (a) venous reservoirs **(b)** muscular walls
 (c) clots **(d)** valves
34. The most important factor in vascular resistance is
 (a) the viscosity of the blood
 (b) the diameter of blood vessel walls
 (c) turbulence due to irregular surfaces of blood vessels
 (d) the length of the blood vessels
35. The specialized exchange surfaces of the liver, bone marrow, and pituitary gland are the
 (a) continuous capillaries
 (b) fenestrated capillaries
 (c) sinusoids
 (d) arteriovenous anastomoses
36. Net hydrostatic pressure forces water _____ a capillary; net osmotic pressure forces water _____ a capillary.
 (a) into, out of
 (b) out of, into
 (c) out of, out of
 (d) a, b, and c are incorrect

21

37. The two arteries formed by the division of the brachiocephalic trunk are the
(a) aorta and internal carotid
(b) axillary and brachial
(c) external and internal carotid
(d) common carotid and subclavian

38. The unpaired arteries supplying blood to the visceral organs include the
(a) suprarenal, renal, lumbar
(b) iliac, gonadal, femoral
(c) celiac, superior and inferior mesenteric
(d) a, b, and c are correct

39. The paired arteries supplying blood to the body wall and other structures outside the abdominopelvic cavity include the
(a) left gastric, hepatic, splenic, phrenic
(b) suprarenals, renals, lumbars, gonadals
(c) iliacs, femorals, gonadals, ileocecals
(d) celiac, left gastric, superior and inferior mesenteric

40. The vein that drains the dural sinuses of the brain is the
(a) cephalic   (b) great saphenous
(c) internal jugular   (d) superior vena cava

41. The vein that drains the thorax and empties into the superior vena cava is the
(a) azygos   (b) basilic
(c) cardiac   (d) cephalic

42. The vein that collects most of the venous blood from below the diaphragm is the
(a) superior vena cava   (b) great saphenous
(c) inferior vena cava   (d) azygos

43. The tributaries of the hepatic portal vein include the
(a) lumbar, gonadal, renal, and suprarenal veins
(b) left colic, splenic, inferior and superior mesenteric veins
(c) phrenic, hepatic, renal, and suprarenal veins
(d) peroneal, iliac, saphenous, and femoral veins

44. Differentiate among the following types of pressure:
(a) systolic   (b) diastolic
(c) pulse   (d) mean arterial

45. What are the primary forces that cause fluid to move **(a)** out of a capillary and into the interstitial fluid at its arterial end? **(b)** into a capillary from the interstitial fluid at its venous end?

46. What two factors assist relatively low venous pressures in propelling blood toward the heart?

47. Give at least five examples of local vasodilators produced at the tissue level that accelerate blood flow through the tissue.

48. What two effects occur when the baroreceptor response to elevated blood pressure is triggered?

49. What factors affect the activity of chemoreceptors in the carotid and aortic bodies?

50. List six hormones important in the regulation of cardiovascular function.

51. What interrelated cardiovascular changes occur before and during light exercise?

52. What criteria are used to identify the basic symptoms of all forms of shock?

53. What circulatory changes occur at birth?

54. What age-related changes take place in the blood, heart, and blood vessels?

LEVEL 2 REVIEWING CONCEPTS

55. When dehydration occurs, there is
(a) accelerated reabsorption of water at the kidneys
(b) a recall of fluids
(c) an increase in the blood colloidal osmotic pressure
(d) a, b, and c are correct

56. Increased CO_2 levels in tissues would promote
(a) the constriction of precapillary sphincters
(b) an increase in the pH of the blood
(c) the dilation of precapillary sphincters
(d) a decrease in blood flow to tissues

57. Elevated levels of the hormones ADH and angiotensin II will produce
(a) increased peripheral vasodilation
(b) increased peripheral vasoconstriction
(c) increased peripheral blood flow
(d) increased venous return

58. The secretion of ADH and aldosterone are typical of the body's long-term compensation following
(a) a heart attack
(b) hypertension
(c) a serious hemorrhage
(d) heavy exercise

59. Relate the anatomical differences between arteries and veins to their functions.

60. What causes variations in the blood flow in individual capillaries?

61. Why do capillaries permit the diffusion of materials whereas arteries and veins do not?

62. How is blood pressure maintained in veins to cope with the force of gravity?

63. How do pressure and resistance affect cardiac output and peripheral blood flow?

64. Why is blood flow to the brain relatively continuous and constant?

65. Compare the effects of the cardioacceleratory and cardioinhibitory centers on cardiac output and blood pressure.

66. A nurse practitioner tells Mrs. B. that her blood pressure is 150/90. Explain what these numbers represent, and calculate Mrs. B.'s mean arterial pressure.

67. An accident victim displays the following symptoms: hypotension; pale, cool, moist skin; confusion and disorientation. Identify her condition, and explain why these symptoms occur. If you took her pulse, what would you probably find?

68. Emilio is in congestive heart failure. Because of this condition, his ankles and feet appear to be swollen. What is the relationship between congestive heart failure and the accumulation of fluid in the feet and ankles?

LEVEL 3 CRITICAL THINKING AND CLINICAL APPLICATIONS

69. Bob is sitting outside on a warm day and is sweating profusely. Mary wants to practice taking blood pressures, and he agrees to play patient. Mary finds that Bob's blood pressure is elevated, even though he is resting and has lost fluid from sweating. (She reasons that fluid loss should lower blood volume and thus blood pressure.) Why is Bob's blood pressure high instead of low?

70. The most common site of varicose veins is the greater saphenous vein of the leg. Why?

71. Who would have a higher pulse pressure: a resting athlete or a resting person who never exercises and has a sedentary job? Why?

72. People with allergies commonly take antihistamines with decongestants to relieve their symptoms. The container warns that the medication should not be taken by individuals who are being treated for high blood pressure. Why not?

73. Jolene awakens suddenly to the sound of her alarm clock. Realizing she is late for class, she jumps to her feet, feels lightheaded, and falls back on her bed. What probably caused this reaction? Why doesn't this happen all the time?

CHAPTER 22

The Lymphatic System and Immunity

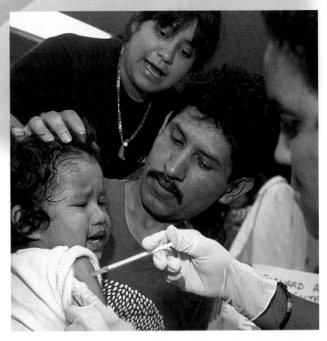

Every child's nightmare! As children, all we notice is the sting of a vaccination. As adults, we realize that the shot hurts only for an instant, whereas the disease it prevents can cripple or kill. We are constantly besieged by threats to our health from various sources—viruses, bacteria, fungi, toxins, and renegade body cells that have turned malignant. In this chapter, we shall examine the defense mechanisms that the body deploys to fight off these threats. We shall also see how modern medicine can manipulate the immune system to enhance the body's defenses.

CHAPTER OUTLINE AND OBJECTIVES

INTRODUCTION p. 752

AN OVERVIEW OF THE LYMPHATIC SYSTEM p. 752

ORGANIZATION OF THE LYMPHATIC SYSTEM p. 752
  Functions of the Lymphatic System p. 752
  Lymphatic Vessels p. 752
  Lymphocytes p. 756
  Lymphoid Tissues p. 758
  Lymphoid Organs p. 759
  The Lymphatic System and Body Defenses p. 763
① Identify the major components of the lymphatic system, and explain their functions.
② Discuss the importance of lymphocytes, and describe their distribution in the body.
③ Describe the structure of lymphoid tissues and organs, and explain their functions.

NONSPECIFIC DEFENSES p. 763
  Physical Barriers p. 765
  Phagocytes p. 765
  Immunological Surveillance p. 766
  Interferons p. 767
  Complement p. 767
  Inflammation p. 769
  Fever p. 770
④ List the body's nonspecific defenses, and describe the components and mechanisms of each.

SPECIFIC DEFENSES: THE IMMUNE RESPONSE p. 771
  Forms of Immunity p. 771
  Properties of Immunity p. 772
  Overview: The Immune Response p. 773
  T Cells and Cell-Mediated Immunity p. 773
  B Cells and Antibody-Mediated Immunity p. 777
  Primary and Secondary Responses to Antigen Exposure p. 781
  Hormones of the Immune System p. 782
⑤ Define specific resistance, and identify the forms and properties of immunity.
⑥ Distinguish between cell-mediated (cellular) immunity and antibody-mediated (humoral) immunity, and identify the cells responsible for each.
⑦ Discuss the types of T cells and the role played by each in the immune response.
⑧ Describe the structure of an antibody, and discuss the types of antibodies in body fluids and secretions.

⑨ Explain the effects of antibodies and how these effects are produced.
⑩ Discuss the primary and secondary responses to antigen exposure.
⑪ Discuss important hormones of the immune response and explain their significance.

SUMMARY: THE IMMUNE RESPONSE p. 784

THE DEVELOPMENT OF RESISTANCE p. 785
⑫ Describe the origin, development, activation, and regulation of resistance.

IMMUNE DISORDERS p. 787
  Autoimmune Disorders p. 787
  Immunodeficiency Diseases p. 788
  Allergies p. 788
⑬ Explain the origin of autoimmune disorders, immunodeficiency diseases, and allergies, and list important examples of each type of disorder.

STRESS AND THE IMMUNE RESPONSE p. 789

AGING AND THE IMMUNE RESPONSE p. 789

INTEGRATION WITH OTHER SYSTEMS p. 791
⑭ Give examples of interactions between the lymphatic system and each of the other organ systems.

MEDIALAB and CHAPTER REVIEW pp. 791, 793

The world is not always kind to the human body. Accidental bumps, cuts, and scrapes, chemical and thermal burns, extreme cold, and ultraviolet radiation are just a few of the hazards in our physical environment. Making matters worse, the world around us contains an assortment of *viruses, bacteria, fungi,* and *parasites* capable of not only surviving but thriving inside our bodies—and potentially causing us great harm. These organisms, called **pathogens,** are responsible for many human diseases. Each pathogen has a different lifestyle and attacks the body in a specific way. For example, viruses spend most of their time hiding within cells, whereas many bacteria multiply in interstitial fluids, and the largest parasites burrow through internal organs. ⟨AM⟩ *A Closer Look: The Nature of Pathogens*

Many organs and systems work together in an effort to keep us alive and healthy. In this ongoing struggle, the **lymphatic system** plays a central role. The lymphatic system consists of (1) *lymph,* a fluid; (2) a network of *lymphatic vessels;* (3) specialized cells called *lymphocytes;* and (4) an array of *lymphoid tissues* and *lymphoid organs* scattered throughout the body.

AN OVERVIEW OF THE LYMPHATIC SYSTEM

We introduced **lymphocytes,** the primary cells of the lymphatic system, in Chapters 4 and 19. ∞ *pp. 124, 637* These cells are vital to our ability to resist or overcome infection and disease. Lymphocytes respond to the presence of (1) invading pathogens, such as bacteria or viruses, (2) abnormal body cells, such as virus-infected cells or cancer cells, and (3) foreign proteins, such as the toxins released by some bacteria. They attempt to eliminate these threats or render them harmless by a combination of physical and chemical attacks.

The body has several physical and chemical barriers that either prevent or slow the entry of infectious organisms. These barriers are known as *nonspecific defenses,* because they do not distinguish one potential threat from another. In contrast, lymphocytes respond to specific threats. If bacteria invade peripheral tissues, lymphocytes organize a defense against that particular type of bacterium. For this reason, lymphocytes are said to provide a *specific defense,* known as the **immune response. Immunity** is the ability to resist infection and disease through the activation of specific defenses.

All the cells and tissues involved with the production of immunity are sometimes considered to be part of an *immune system,* a physiological system that includes not only the lymphatic system but also components of the integumentary, cardiovascular, respiratory, digestive, and other systems. For example, interactions between lymphocytes and Langerhans cells of the skin are important in mobilizing specific defenses against skin infections. We begin this chapter by examining the organization of the lymphatic system. We shall then consider how the lymphatic system interacts with cells and tissues of other systems to defend the body against infection and disease.

ORGANIZATION OF THE LYMPHATIC SYSTEM

The lymphatic system consists of the following:

1. A network of **lymphatic vessels,** often called **lymphatics,** which begin in peripheral tissues and end at connections to veins.
2. **Lymph,** a fluid that resembles plasma but contains a much lower concentration of suspended proteins.
3. **Lymphoid organs,** which are connected to lymphatic vessels and contain large numbers of lymphocytes.

Figure 22-1● provides a general overview of the components of this system. ⟨AM⟩ *EMBRYOLOGY SUMMARY 17: The Development of the Lymphatic System*

FUNCTIONS OF THE LYMPHATIC SYSTEM

The three primary functions of the lymphatic system are:

1. *The production, maintenance, and distribution of lymphocytes.* Lymphocytes are produced and stored within (1) lymphoid tissues and organs, such as the spleen and thymus, and (2) areas of red bone marrow.
2. *The return of fluid and solutes from peripheral tissues to the blood.* Capillaries normally deliver more fluid to the tissues than they carry away. ∞ *p. 709* The return of tissue fluids through lymphatic vessels maintains normal blood volume and eliminates local variations in the composition of the interstitial fluid.
3. *The distribution of hormones, nutrients, and waste products from their tissues of origin to the general circulation.* Substances that originate in the tissues but are for some reason unable to enter the bloodstream directly may do so by way of the lymphatic vessels. For example, lipids absorbed by the digestive tract commonly fail to enter the bloodstream through capillaries. They reach the bloodstream only after they have traveled along lymphatic vessels (a process we shall explore further in Chapter 24).

LYMPHATIC VESSELS

Lymphatic vessels carry lymph from peripheral tissues to the venous system. The smallest lymphatic vessels are called *lymphatic capillaries.*

Lymphatic Capillaries

The lymphatic network begins with the **lymphatic capillaries,** or *terminal lymphatics,* which branch through peripheral tissues. They differ from blood capillaries in

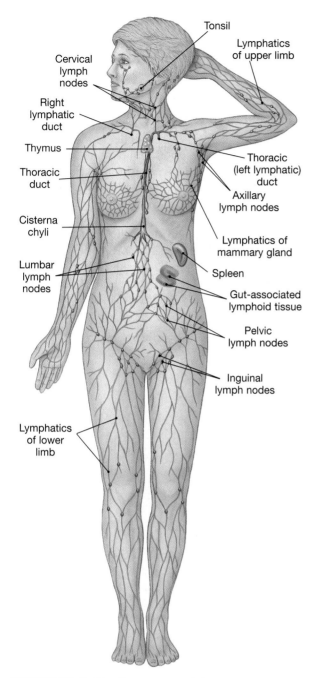

•**FIGURE 22-1** The Components of the Lymphatic System

Lymphatic capillaries are present in almost every tissue and organ in the body. Prominent lymphatic capillaries in the small intestine are called *lacteals;* these are important in the transport of lipids absorbed by the digestive tract. Lymphatic capillaries are absent in areas that lack a blood supply, such as the cornea of the eye. The bone marrow and the central nervous system also lack lymphatic vessels.

(a) Association of blood capillaries, tissue, and lymphatic capillaries

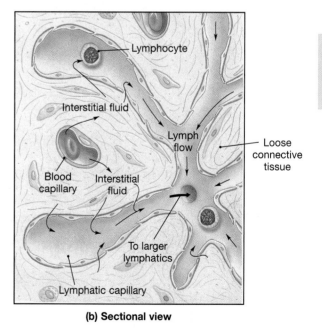

(b) Sectional view

•**FIGURE 22-2** Lymphatic Capillaries. **(a)** The interwoven network formed by blood capillaries and lymphatic capillaries. Arrows show the movement of fluid out of blood vessels and the net flow of interstitial fluid and lymph. **(b)** A sectional view showing fluid movement from the plasma through the interstitial fluid and into the lymphatic system.

that lymphatic capillaries (1) originate as blind pockets, (2) are larger in diameter, (3) have thinner walls, and (4) in sectional view typically have a flattened or irregular outline (Figure 22-2•). Although lymphatic capillaries are lined by endothelial cells, the basement membrane is incomplete or absent. The endothelial cells of a lymphatic capillary are not tightly bound together, but they do overlap. The region of overlap acts as a one-way valve. It permits the entry of fluids and solutes, even those as large as proteins, as well as viruses, bacteria, and cell debris, but it prevents their return to the intercellular spaces.

Small Lymphatic Vessels

From the lymphatic capillaries, lymph flows into larger lymphatic vessels that lead toward the trunk. The walls of these lymphatic vessels contain layers comparable to those of veins, and, like veins, the larger lymphatic vessels contain valves (Figure 22-3●). The valves are quite close together, and at each valve the lymphatic vessel bulges noticeably. As a result, large lymphatic vessels have a beaded appearance (Figure 22-3a●). The valves prevent the backflow of lymph within lymph vessels, especially those of the limbs. Pressures within the lymphatic system are minimal, and the valves are essential to maintaining normal lymph flow toward the thoracic cavity.

Lymphatic vessels commonly occur in association with blood vessels (Figure 22-3a●). Notice the differences in relative size, general appearance, and branching pattern that distinguish the lymphatic vessels from arteries and veins. Characteristic color differences are also apparent on examining living tissues. Most arteries are bright red; veins are dark red; and lymphatic vessels are a pale golden color. In general, a tissue will contain many more lymphatic vessels than veins, but the lymphatic vessels are much smaller.

Major Lymph-Collecting Vessels

Two sets of lymphatic vessels collect lymph from the lymphatic capillaries: superficial lymphatics and deep lymphatics. **Superficial lymphatics** are located in the subcutaneous layer deep to the skin; in the areolar tissues of the mucous membranes lining the digestive, respiratory, urinary, and reproductive tracts; and in the areolar tissues of the serous membranes lining the pleural, pericardial, and peritoneal cavities. **Deep lymphatics** are larger lymphatic vessels. They accompany deep arteries and veins supplying skeletal muscles and other organs of the neck, limbs, and trunk and the walls of visceral organs.

●**FIGURE 22-3 Lymphatic Vessels and Valves. (a)** A diagrammatic view of loose connective tissue, showing small blood vessels and a lymphatic vessel. The cross-sectional view emphasizes the structural differences among them. **(b)** A lymphatic valve. Lymphatic valves resemble those in veins. Each valve consists of a pair of flaps that permit fluid movement in only one direction. (LM × 43)

Superficial and deep lymphatics converge to form even larger vessels called *lymphatic trunks,* which in turn empty into two large collecting vessels: the thoracic duct and the right lymphatic duct. The *thoracic duct* collects lymph from the body inferior to the diaphragm and from the left side of the body superior to the diaphragm. The smaller *right lymphatic duct* collects lymph from the right side of the body superior to the diaphragm.

The **thoracic duct** begins inferior to the diaphragm at the level of vertebra L_2. The base of the thoracic duct is an expanded, saclike chamber called the **cisterna chyli** (KĪ-lē) (Figure 22-4•). The cisterna chyli receives lymph from the inferior part of the abdomen,

the pelvis, and the lower limbs by way of the *right* and *left lumbar trunks* and the *intestinal trunk.*

The inferior segment of the thoracic duct lies anterior to the vertebral column. From the second lumbar vertebra, it penetrates the diaphragm alongside the aorta and ascends along the left side of the vertebral column to the level of the left clavicle. After collecting lymph from the *left bronchomediastinal trunk,* the *left subclavian trunk,* and the *left jugular trunk,* it empties into the left subclavian vein near the left internal jugular vein (Figure 22-4b•). Lymph collected from the left side of the head, neck, and thorax, as well as lymph from the entire body inferior to the diaphragm, re-enters the venous circulation in this way.

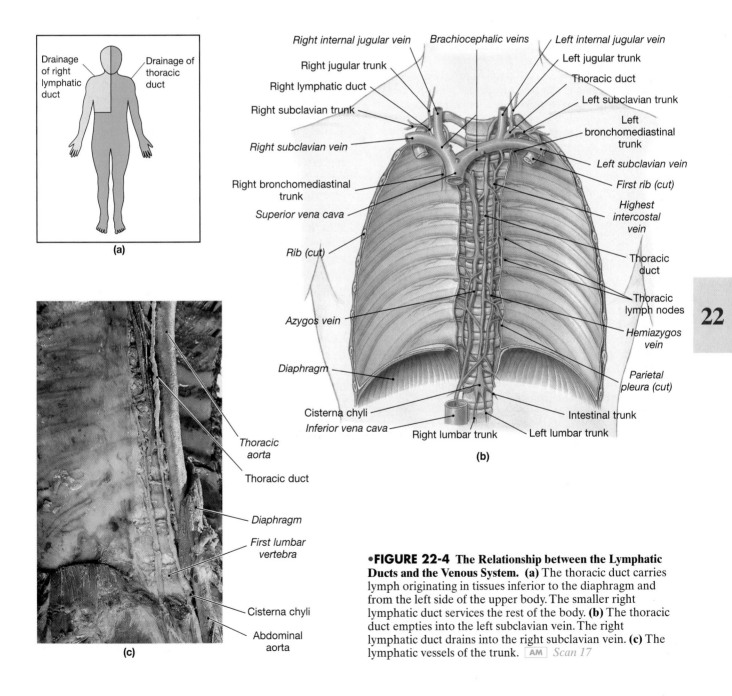

•**FIGURE 22-4** **The Relationship between the Lymphatic Ducts and the Venous System.** **(a)** The thoracic duct carries lymph originating in tissues inferior to the diaphragm and from the left side of the upper body. The smaller right lymphatic duct services the rest of the body. **(b)** The thoracic duct empties into the left subclavian vein. The right lymphatic duct drains into the right subclavian vein. **(c)** The lymphatic vessels of the trunk. [AM] *Scan 17*

22

The **right lymphatic duct** is formed by the merging of the *right jugular, right subclavian,* and *right bronchomediastinal trunks* in the area near the right clavicle. This duct empties into the right subclavian vein, delivering lymph from the right side of the body superior to the diaphragm (Figure 22-4b•).

Lymphedema

Blockage of the lymphatic drainage from a limb produces **lymphedema** (lim-fe-DĒ-muh). In this condition, interstitial fluids accumulate and the limb gradually becomes swollen and grossly distended. If the condition persists, the connective tissues lose their elasticity, and the swelling becomes permanent. Lymphedema by itself does not pose a major threat to life. The danger comes from the constant risk that an uncontrolled infection will develop in the affected area. Because the interstitial fluids are essentially stagnant, toxins and pathogens can accumulate and overwhelm the local defenses without fully activating the immune system.

Temporary lymphedema can result from tight clothing. Chronic lymphedema can result from scar tissue formation owing to repeated bacterial infections or from surgery that cuts or removes lymphatic vessels. Breast cancer surgery or radiation treatment can lead to lymphedema of the upper limb on the affected side. Lymphedema can also result from parasitic infections; in **filariasis** (fil-a-RĪ-a-sis), larvae of a parasitic roundworm, generally *Wuchereria bancrofti,* are transmitted by mosquitoes or black flies. The adult worms form massive colonies within lymphatic vessels and lymph nodes. Repeated scarring of the passageways eventually blocks lymphatic drainage and produces extreme lymphedema with permanent distension of tissues. The limbs or external genitalia typically become grossly distended, a condition known as **elephantiasis** (el-e-fan-TĪ-a-sis).

Therapy for chronic lymphedema consists of treating infections by the administration of antibiotics and (when possible) reducing the swelling. One possible treatment involves the application of elastic wrappings that squeeze the tissue. This external compression elevates the hydrostatic pressure of the interstitial fluids and opposes the entry of additional fluid from the capillaries.

LYMPHOCYTES

Lymphocytes account for 20–30 percent of the circulating white blood cell population. However, circulating lymphocytes are only a small fraction of the total lymphocyte population. The body contains some 10^{12} lymphocytes, with a combined weight of over a kilogram.

Types of Lymphocytes

Three classes of lymphocytes are in blood: (1) **T** (**t**hymus-dependent) **cells,** (2) **B** (**b**one marrow–derived) **cells,** and (3) **NK** (**n**atural **k**iller) **cells.** Each type has distinctive biochemical and functional characteristics.

T Cells   Approximately 80 percent of circulating lymphocytes are classified as T cells. Many types of T cells exist, including the following:

- **Cytotoxic T cells,** which attack foreign cells or body cells infected by viruses. Their attack commonly involves direct contact. These lymphocytes are the primary cells involved in the production of *cell-mediated immunity,* or *cellular immunity.*
- **Helper T cells,** which stimulate the activation and function of both T cells and B cells.
- **Suppressor T cells,** which inhibit the activation and function of both T cells and B cells.

The interplay between suppressor and helper T cells helps establish and control the sensitivity of the immune response. For this reason, these cells are also known as *regulatory T cells.*

These are the T cells that we will examine in the course of this chapter. It is not a complete list, however; other types of T cells participate in the immune response. For example, *inflammatory T cells* stimulate regional inflammation and local defenses in an injured tissue, and *suppressor/inducer T cells* suppress B cell activity but stimulate other T cells.

B Cells   B cells account for 10–15 percent of circulating lymphocytes. When stimulated, B cells can differentiate into **plasma cells.** Plasma cells are responsible for the production and secretion of *antibodies,* soluble proteins that are also known as *immunoglobulins.* ∞ *p. 119* These proteins bind to specific chemical targets called **antigens.** Most antigens are pathogens, parts or products of pathogens, or other foreign compounds. Most antigens are proteins, but some lipids, polysaccharides, and nucleic acids can also stimulate antibody production. When an antibody binds to its target antigen, a chain of events begins that leads to the destruction of the target compound or organism. B cells are responsible for *antibody-mediated immunity,* which is also known as *humoral* ("liquid") *immunity* because antibodies occur in body fluids.

NK Cells   The remaining 5–10 percent of circulating lymphocytes are NK cells, also known as **large granular lymphocytes.** These lymphocytes attack foreign cells, normal cells infected with viruses, and cancer cells that appear in normal tissues. Their continuous policing of peripheral tissues has been called *immunological surveillance.*

Life Span and Circulation of Lymphocytes

Lymphocytes are not evenly distributed in the blood, bone marrow, spleen, thymus, and peripheral lymphoid tissues. The ratio of B cells to T cells varies with the tissue or organ considered. For example, B cells are seldom found in the thymus, but in blood, T cells outnumber B cells by a ratio of 8:1. This ratio changes to 1:1 in the spleen and 1:3 in bone marrow.

The lymphocytes in these organs are visitors, not residents. All types of lymphocytes move throughout the body. They wander through a tissue and then enter a blood vessel or a lymphatic vessel for transport.

T cells move relatively quickly. For example, a wandering T cell may spend about 30 minutes in the blood, 5–6 hours in the spleen, and 15–20 hours in a lymph node. B cells, which are responsible for antibody production, move more slowly. A typical B cell spends about 30 hours in a lymph node before moving on.

Lymphocytes have relatively long life spans. Roughly 80 percent survive for 4 years, and some last 20 years or more. Throughout your life, you maintain normal lymphocyte populations by producing new lymphocytes in your bone marrow and lymphoid tissues.

Lymphocyte Production

In Chapter 19, we discussed *hemopoiesis,* the formation of the cellular elements of blood. ∞ *pp. 627–628 Erythropoiesis* (red blood cell formation) in adults is

normally confined to bone marrow, but lymphocyte production, or **lymphopoiesis** (lim-fō-poy-Ē-sis), involves the bone marrow, thymus, and peripheral lymphoid tissues (Figure 22-5•).

Bone marrow plays the primary role in the maintenance of normal lymphocyte populations. Hemocytoblast divisions in the bone marrow of adults generate the lymphoid stem cells that produce all types of lymphocytes. Two distinct populations of lymphoid stem cells are produced in the bone marrow.

One group of lymphoid stem cells remains in the bone marrow (Figure 22-5a•). Divisions of these cells produce immature B cells and NK cells. B cell development involves intimate contact with large **stromal cells** (*stroma,* a bed) in the bone marrow. The cytoplasmic extensions of stromal cells contact or even wrap around the developing B cells. Stromal cells produce an immune system hormone, or *cytokine,* called *interleukin-7,* which promotes the differentiation of B cells. (We will consider cytokines and their varied effects in a later section.)

As they mature, B cells and NK cells enter the bloodstream and migrate to peripheral tissues (Figure 22-5c•).

•**FIGURE 22-5 The Derivation and Distribution of Lymphocytes.** Hemocytoblast divisions produce stem cells with two fates. **(a)** One group remains in the bone marrow, producing daughter cells that mature into B cells or NK cells. **(b)** The other group migrates to the thymus, where subsequent divisions produce daughter cells that mature into T cells. The mature B cells, NK cells, and T cells circulate throughout the body in the bloodstream and then **(c)** temporarily reside in peripheral tissues.

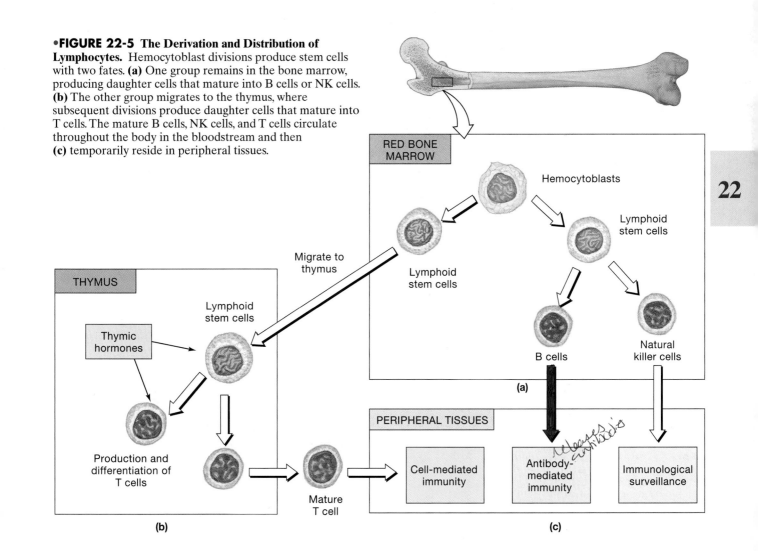

Most of the B cells move into lymph nodes, the spleen, or other lymphoid tissues. The NK cells migrate throughout the body, moving through peripheral tissues in search of abnormal cells.

The second group of lymphoid stem cells migrates to the thymus (Figure 22-5b•). While in the thymus, these cells and their descendants develop further in an environment that is isolated from the general circulation by the **blood–thymus barrier.** Under the influence of thymic hormones, the lymphoid stem cells divide repeatedly, producing the various kinds of T cells. At least seven thymic hormones have been identified. Their precise functions and interactions have yet to be determined.

When their development is nearing completion, T cells reenter the bloodstream and return to the bone marrow. They also travel to lymphoid tissues and organs, such as the spleen (Figure 22-5c•).

The T cells and B cells that migrate from their sites of origin retain the ability to divide. Their divisions produce daughter cells of the same type; for example, a dividing B cell produces other B cells, not T cells or NK cells. As we shall see, the ability to increase the number of lymphocytes of a specific type is important to the success of the immune response.

LYMPHOID TISSUES

Lymphoid tissues are connective tissues dominated by lymphocytes. In a **lymphoid nodule,** or *lymphatic nodule,* the lymphocytes are densely packed in an area of areolar tissue (Figure 22-6•). Lymphoid nodules occur in the connective tissue deep to the epithelia lining the respiratory, digestive, and urinary tracts. Typical nodules average about a millimeter in diameter, but the boundaries are not distinct, because no fibrous cap-

Intestinal lumen

Mucous membrane

Aggregate lymphoid nodule

Underlying connective tissue

(a) Lymphoid nodule

Pharyngeal epithelium

Pharyngeal tonsil

Palate

Palatine tonsil

Lingual tonsil

Germinal centers

•**FIGURE 22-6**
Lymphoid Nodules. (a) A typical nodule in section. (LM × 17) **(b)** The positions of the tonsils and a tonsil in section. Notice the relatively pale germinal centers, where lymphocyte cell divisions occur.

(b) Pharyngeal tonsil

22

sule surrounds them. They commonly have a central zone called a **germinal center,** which contains dividing lymphocytes (Figure 22-6b•).

MALT

The collection of lymphoid tissues linked with the digestive system is called the **mucosa-associated lymphoid tissue (MALT).** Clusters of lymphoid nodules deep to the epithelial lining of the intestine are known as **aggregate lymphoid nodules,** or *Peyer's patches* (Figure 22-6a•). In addition, the walls of the *vermiform appendix,* a blind pouch that originates near the junction between the small and large intestines, contain a mass of fused lymphoid nodules.

Tonsils

Large nodules in the walls of the pharynx are called **tonsils** (Figure 22-6b•). Most people have five tonsils:

1. A single **pharyngeal tonsil,** often called the *adenoids,* located in the posterior superior wall of the nasopharynx.

2, 3. A pair of **palatine tonsils,** located at the posterior margin of the oral cavity, along the boundary with the pharynx.

4, 5. A pair of **lingual tonsils,** which are usually not visible because they are located under the attached base of the tongue.

Infected Lymphoid Nodules

The lymphocytes in a lymphoid nodule are not always able to destroy bacterial or viral invaders that have crossed the adjacent epithelium. If pathogens become established in a lymphoid nodule, an infection develops. Two examples are probably familiar to you: *tonsillitis,* an infection of one of the tonsils (generally the pharyngeal or palatine), and *appendicitis,* an infection of the vermiform appendix that begins in the lymphoid nodules. Treatment commonly consists of antibiotic therapy, sometimes combined with the surgical removal of the infected tissues. AM *Infected Lymphoid Nodules*

LYMPHOID ORGANS

A fibrous connective tissue capsule separates lymphoid organs from surrounding tissues. These organs include the *lymph nodes,* the *thymus,* and the *spleen.*

Lymph Nodes

Lymph nodes are small, oval lymphoid organs ranging in diameter from 1 to 25 mm (to about 1 in.). Figure 22-1•, p. 753, shows the general pattern of lymph node distribution in the body. Each lymph node is covered by a capsule of dense connective tissue (Figure 22-7•). Bundles of collagen fibers extend from the capsule into the interior of the node. These fibrous partitions are called **trabeculae** (*trabecula,* a wall).

•FIGURE 22-7 The Structure of a Lymph Node

The shape of a typical lymph node resembles that of a kidney bean (Figure 22-7•). Blood vessels and nerves attach to the lymph node at the **hilus,** a shallow indentation. Two sets of lymphatic vessels are connected to each lymph node: afferent lymphatics and efferent lymphatics.

1. **Afferent lymphatics** carry lymph to the lymph node from peripheral tissues. The afferent lymphatics penetrate the capsule of the lymph node on the side opposite the hilus.
2. **Efferent lymphatics** are attached to the lymph node at the hilus. These vessels carry lymph away from the lymph node and toward the venous circulation.

Lymph Flow Lymph delivered by the afferent lymphatics flows through the lymph node within a network of sinuses, open passageways with incomplete walls (Figure 22-7•). Lymph first enters a *subcapsular sinus,* which contains a meshwork of branching reticular fibers, macrophages, and dendritic cells. **Dendritic cells** are involved in the initiation of the immune response; we shall consider their role in a later section. After passing through the subcapsular sinus, lymph flows through the **outer cortex** of the node. The outer cortex contains B cells within germinal centers that resemble those of lymphoid nodules.

Lymph then continues through lymph sinuses in the **deep cortex** *(paracortical area).* Lymphocytes leave the bloodstream and enter the lymph node by crossing the walls of blood vessels within the deep cortex. The deep cortical area is dominated by T cells.

After flowing through the sinuses of the deep cortex, lymph continues into the core, or **medulla,** of the lymph node. The medulla contains B cells and plasma cells organized into elongate masses known as **medullary cords.** After passing through a network of sinuses in the medulla, lymph enters the efferent lymphatics at the hilus.

Lymph Node Function A lymph node functions like a kitchen water filter: It filters and purifies lymph before that fluid reaches the venous circulation. As lymph flows through a lymph node, at least 99 percent of the antigens in the lymph are removed. Fixed macrophages in the walls of the lymphatic sinuses engulf debris or pathogens in lymph as it flows past. Antigens removed in this way are then processed by the macrophages and "presented" to nearby lymphocytes. Other antigens bind to receptors on the surfaces of dendritic cells, where they can stimulate lymphocyte activity. This process, *antigen presentation,* is generally the first step in the activation of the immune response.

In addition to filtering, lymph nodes provide an early-warning system. Any infection or other abnormality in a peripheral tissue will introduce abnormal antigens into the interstitial fluid and thus into the lymph leaving the area. These antigens will then stimulate macrophages and lymphocytes in nearby lymph nodes.

To protect a house against intrusion, you might guard all entrances and exits or place traps by the windows and doors. The distribution of lymphoid tissues and lymph nodes follows such a pattern. The largest lymph nodes are located where peripheral lymphatics connect with the trunk, such as in the groin, the axillae, and the base of the neck. These nodes are often called *lymph glands.* Because lymph is monitored in the cervical, inguinal, or axillary lymph nodes, potential problems can be detected and dealt with before they affect the vital organs of the trunk. Aggregations of lymph nodes also exist in the mesenteries of the gut, near the trachea and passageways leading to the lungs, and in association with the thoracic duct (Figure 22-4•, p. 755). These lymph nodes protect against pathogens and other antigens within the digestive and respiratory systems.

A minor injury commonly produces a slight enlargement of the nodes along the lymphatic vessels draining the region. This symptom, often called "swollen glands," typically indicates inflammation or infection of peripheral structures. The enlargement generally results from an increase in the number of lymphocytes and phagocytes in the node in response to a minor, localized infection. Chronic or excessive enlargement of lymph nodes constitutes **lymphadenopathy** (lim-fad-e-NOP-a-thē). This condition may occur in response to bacterial or viral infections, endocrine disorders, or cancer.

Cancer and the Lymphatic System

Lymphatic vessels are located in almost all portions of the body except the central nervous system, and lymphatic capillaries offer little resistance to the passage of cancer cells. As a result, metastasizing cancer cells commonly spread along lymphatic vessels. Under these circumstances, the lymph nodes serve as way stations for migrating cancer cells. Thus, an analysis of lymph nodes can provide information on the spread of the cancer cells, and such information helps determine the appropriate therapies. In Chapter 29, we shall discuss one example: identifying the stages of breast cancer by the degree of nodal involvement. *Lymphomas,* one group of cancers originating in the lymphatic system, are discussed in the *Applications Manual.* AM *Lymphomas.*

The Thymus

The **thymus** is located in the mediastinum, generally just posterior to the sternum. It is pink and has a grainy consistency. In newborn infants and young children, the thymus is relatively large, commonly extending from the base of the neck to the superior border of the heart. The thymus reaches its greatest size (relative to body size) in the first year or two after birth. Although the thymus continues to increase in size throughout childhood, the body as a whole grows even faster, so the size of the thymus relative to that of the other organs in the mediastinum gradually decreases. The thymus reaches its maximum absolute size, at a weight of about 40 g (1.4 oz), just before puberty. After puberty, it gradually diminishes in size and becomes increasingly fibrous; this process is called *involution*. By the time an individual reaches age 50, the thymus may weigh less than 12 g (0.3 oz). It has been suggested that the gradual decrease in the size and secretory abilities of the thymus may make the elderly more susceptible to disease.

The capsule that covers the thymus divides it into two **thymic lobes** (Figure 22-8a,b•). **Septae,** fibrous partitions that originate at the capsule, divide the lobes into **lobules,** which average 2 mm in width (Figure 22-8b,c•).

Each lobule consists of a densely packed outer **cortex** and a paler, central **medulla.** Lymphocytes in the cortex are dividing; as the T cells mature, they migrate into the medulla. After roughly 3 weeks, these T cells leave the thymus by entering one of the medullary blood vessels.

Lymphocytes in the cortex are arranged in clusters that are completely surrounded by **reticular epithelial cells.** These cells, which developed from epithelial cells of the embryo, also encircle the blood vessels of the cortex. The reticular epithelial cells (1) maintain the blood–thymus barrier and (2) secrete the thymic hormones that stimulate stem cell divisions and T cell differentiation.

As they mature, T cells leave the cortex and enter the medulla of the thymus. The medulla has no blood–thymus barrier. The reticular epithelial cells in

•**FIGURE 22-8 The Thymus. (a)** The appearance and position of the thymus in relation to other organs in the chest. **(b)** Anatomical landmarks on the thymus. **(c)** Fibrous septae divide the tissue of the thymus into lobules resembling interconnected lymphatic nodules. (LM × 40) **(d)** At higher magnification, we can see the unusual structure of Hassall's corpuscles. The small cells are lymphocytes in various stages of development. (LM × 532)

the medulla cluster together in concentric layers, forming distinctive structures known as **Hassall's corpuscles** (Figure 22-8d•). Despite their imposing appearance, the function of Hassall's corpuscles remains unknown. T cells in the medulla can enter or leave the bloodstream across the walls of blood vessels in this region or within one of the efferent lymphatics that collect lymph from the thymus.

Hormones of the Thymus

Hormones of the Thymus   The thymus produces several hormones that are important to the development and maintenance of normal immunological defenses. *Thymosin* (THĪ-mō-sin) is the name originally given to an extract from the thymus that promotes the development and maturation of lymphocytes. The thymic extract actually contains a blend of several complementary hormones: *thymosin-α, thymosin-β, thymosin V, thymopoietin, thymulin,* and others. The term *thymosins* now is sometimes used to refer to all thymic hormones.

The Spleen

The adult **spleen** contains the largest collection of lymphoid tissue in the body. It is about 12 cm (5 in.) long and weighs on average nearly 160 g (5.6 oz). The spleen lies along the curving lateral border of the stomach, extending between the ninth and eleventh ribs on the left side. It is attached to the lateral border of the stomach by the **gastrosplenic ligament,** a broad band of mesentery (Figure 22-9a•).

Functions of the Spleen   In gross dissection, the spleen is deep red owing to the blood it contains. In essence, the spleen performs the same functions for blood that lymph nodes perform for lymph. Functions of the spleen can be summarized as (1) the removal of abnormal blood cells and other blood components by phagocytosis, (2) the storage of iron from recycled red blood cells, and (3) the initiation of immune responses by B cells and T cells in response to antigens in circulating blood.

Surfaces of the Spleen   The spleen has a soft consistency, so its shape primarily reflects its association with the structures around it. It is in contact with the stomach, the left kidney, and the muscular diaphragm. The *diaphragmatic surface* is smooth and convex, conforming to the shape of the diaphragm and body wall. The *visceral surface* contains indentations of the shape of the stomach (the *gastric area*) and of the kidney (the *renal area*) (Figure 22-9b•). Splenic blood vessels and lymphatic vessels communicate with the spleen on the visceral surface at the **hilus,** a groove marking the border between the gastric and renal areas. The **splenic artery,** the **splenic vein,** and the

lymphatic vessels draining the spleen are attached at the hilus.

Histology of the Spleen   The spleen is surrounded by a capsule containing collagen and elastic fibers.[1] The cellular components within constitute the **pulp** of the spleen (Figure 22-9c•). **Red pulp** contains large quantities of red blood cells, whereas **white pulp** resembles lymphoid nodules.

The splenic artery enters at the hilus and branches to produce a number of arteries that radiate outward toward the capsule. These **trabecular arteries** branch extensively, and their finer branches are surrounded by areas of white pulp. Capillaries then discharge the blood into the red pulp.

The cell population of the red pulp includes all the normal components of circulating blood, plus fixed and free macrophages. The structural framework of the red pulp consists of a network of reticular fibers. The blood passes through this meshwork and enters large sinusoids, also lined by fixed macrophages. The sinusoids empty into small veins, and these ultimately collect into **trabecular veins** that continue toward the hilus.

This circulatory arrangement gives the phagocytes of the spleen an opportunity to identify and engulf any damaged or infected cells in circulating blood. Lymphocytes are scattered throughout the red pulp, and the area surrounding the white pulp has a high concentration of macrophages and dendritic cells. Thus any microorganism or other antigen in the blood will quickly come to the attention of the splenic lymphocytes.

Injury to the Spleen

An impact to the left side of the abdomen can distort or damage the spleen. Such injuries are known risks of contact sports, such as football and hockey, and of more-solitary athletic activities, such as skiing and sledding. However, the spleen tears so easily that a seemingly minor blow to the side can rupture the capsule. The result is serious internal bleeding and eventual circulatory shock. The spleen can also be damaged by infection, inflammation, or invasion by cancer cells.

Because the spleen is relatively fragile, it is very difficult to repair surgically. (Sutures typically tear out before they have been tensed enough to stop the bleeding.) A severely ruptured spleen is removed, a process called a **splenectomy** (sple-NEK-to-mē). A person without a spleen survives without difficulty but has a greater risk for bacterial infections than do individuals with a functional spleen. [AM] *Disorders of the Spleen*

---

[1] The spleens of dogs and cats have extensive layers of smooth muscle that can contract to eject blood into the bloodstream. The human spleen lacks those muscle layers and cannot contract.

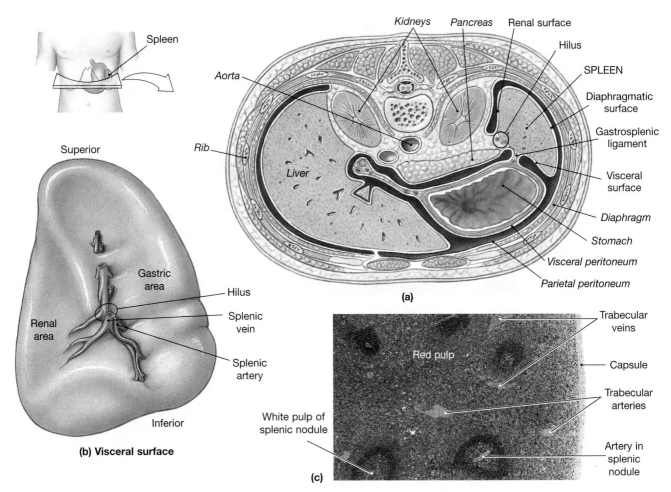

Kidneys Pancreas Renal surface
Hilus
Aorta
SPLEEN
Diaphragmatic surface
Gastrosplenic ligament
Rib
Liver
Visceral surface
Diaphragm
Stomach
Visceral peritoneum
Parietal peritoneum
(a)

Superior
Gastric area
Hilus
Splenic vein
Renal area
Splenic artery
Inferior
(b) Visceral surface

Red pulp
Trabecular veins
Capsule
Trabecular arteries
White pulp of splenic nodule
Artery in splenic nodule
(c)

•FIGURE 22-9 The Spleen. (a) A transverse section through the trunk, showing the typical position of the spleen within the abdominopelvic cavity. The shape of the spleen roughly conforms to the shapes of adjacent organs. **(b)** The external appearance of the intact spleen, showing major anatomical landmarks. Compare this view with that of part (a). **(c)** The histological appearance of the spleen. White pulp is dominated by lymphocytes; it appears blue because the nuclei of lymphocytes stain very darkly. Red pulp contains a preponderance of red blood cells. (LM × 38) AM *Plates 7.7d, 7.9*

22

THE LYMPHATIC SYSTEM AND BODY DEFENSES

The human body has multiple defense mechanisms, but these can be sorted into two general categories:

1. **Nonspecific defenses** do not distinguish one type of threat from another. Their response is the same, regardless of the type of invading agent. These defenses, which are present at birth, include *physical barriers, phagocytic cells, immunological surveillance, interferons, complement, inflammation,* and *fever.* They provide a defensive capability known as **nonspecific resistance.**

2. **Specific defenses** protect against particular threats. For example, a specific defense may protect against infection by one type of bacterium but ignore other bacteria and viruses. Many specific defenses develop after birth as a result of accidental or deliberate exposure to environmental hazards. *Specific defenses depend on the activities of lymphocytes.* The body's

specific defenses produce a state of protection known as immunity, or **specific resistance.**

Nonspecific and specific resistances are complementary. Both must function normally to provide adequate resistance to infection and disease.

☑ How would the blockage of the thoracic duct affect the circulation of lymph?

☑ If the thymus failed to produce thymic hormones, which population of lymphocytes would be affected?

☑ Why do lymph nodes enlarge during some infections?

NONSPECIFIC DEFENSES

Nonspecific defenses prevent the approach, deny the entrance, or limit the spread of microorganisms or other environmental hazards. Seven major categories of nonspecific defenses are summarized in Figure 22-10•:

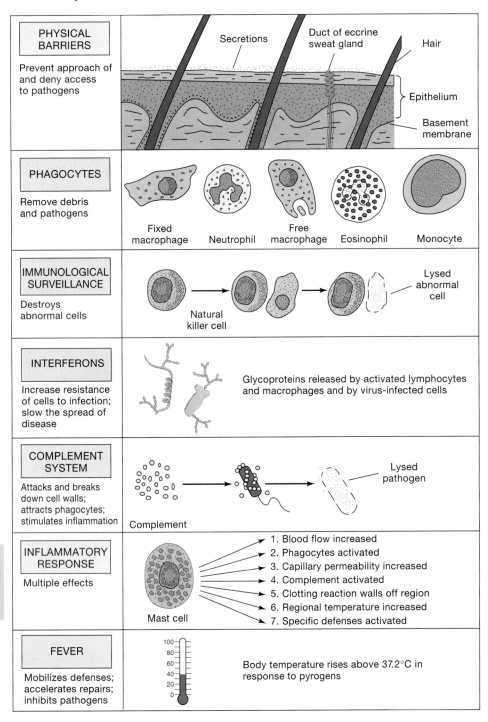

PHYSICAL BARRIERS	
Prevent approach of and deny access to pathogens	Secretions · Duct of eccrine sweat gland · Hair · Epithelium · Basement membrane
PHAGOCYTES	
Remove debris and pathogens	Fixed macrophage · Neutrophil · Free macrophage · Eosinophil · Monocyte
IMMUNOLOGICAL SURVEILLANCE	
Destroys abnormal cells	Natural killer cell · Lysed abnormal cell
INTERFERONS	
Increase resistance of cells to infection; slow the spread of disease	Glycoproteins released by activated lymphocytes and macrophages and by virus-infected cells
COMPLEMENT SYSTEM	
Attacks and breaks down cell walls; attracts phagocytes; stimulates inflammation	Complement · Lysed pathogen
INFLAMMATORY RESPONSE	
Multiple effects	Mast cell · 1. Blood flow increased · 2. Phagocytes activated · 3. Capillary permeability increased · 4. Complement activated · 5. Clotting reaction walls off region · 6. Regional temperature increased · 7. Specific defenses activated
FEVER	
Mobilizes defenses; accelerates repairs; inhibits pathogens	Body temperature rises above 37.2°C in response to pyrogens

•**FIGURE 22-10 Nonspecific Defenses.** Nonspecific defenses deny access to the body or destroy pathogens without distinguishing among specific types of pathogens.

1. *Physical barriers* keep hazardous organisms and materials outside the body. For example, a mosquito that lands on your head may be unable to reach the surface of the scalp if you have a full head of hair.

2. *Phagocytes* are cells that engulf pathogens and cell debris. Examples of phagocytes are the macrophages of peripheral tissues and the microphages of blood.

3. *Immunological surveillance* is the destruction of abnormal cells by NK cells in peripheral tissues.

4. *Interferons* are chemical messengers that coordinate the defenses against viral infection.

5. *Complement* is a system of circulating proteins that assist antibodies in the destruction of pathogens.

6. The *inflammatory response* is a local response to injury or infection that is directed at the tissue level. Inflammation tends to restrict the spread of an injury as well as combat an infection.

7. *Fever* is an elevation of body temperature that accelerates tissue metabolism and defenses.

PHYSICAL BARRIERS

To cause trouble, an antigenic compound or pathogen must enter the body tissues, and so must cross an epithelium. The epithelial covering of the skin has multiple layers, a keratin coating, and a network of desmosomes that lock adjacent cells together. *∞ pp. 146, 110* These barriers provide very effective protection for underlying tissues. Even along the more delicate internal passageways of the respiratory, digestive, and urinary tracts, the epithelial cells are tied together by tight junctions and generally are supported by a dense and fibrous basement membrane.

In addition to the barriers posed by the epithelial cells, most epithelia are protected by specialized accessory structures and secretions. The hairs on most areas of your body surface provide some protection against mechanical abrasion (especially on the scalp), and they often prevent hazardous materials or insects from contacting your skin surface. The epidermal surface also receives the secretions of sebaceous and sweat glands. These secretions flush the surface, washing away microorganisms and chemical agents. The secretions may also contain bactericidal chemicals, destructive enzymes *(lysozymes),* and antibodies.

The epithelia lining the digestive, respiratory, urinary, and reproductive tracts are more delicate, but they are equally well defended. Mucus bathes most surfaces of your digestive tract, and your stomach contains a powerful acid that can destroy many potential pathogens. Mucus moves across the respiratory tract lining; urine flushes the urinary passageways; and glandular secretions do the same for the reproductive tract. Special enzymes, antibodies, and an acidic pH add to the effectiveness of these secretions.

PHAGOCYTES

Phagocytes perform janitorial and police services in peripheral tissues, removing cellular debris and responding to invasion by foreign compounds or pathogens. These cells represent the "first line" of cellular defense against pathogenic invasion. Many phagocytes attack and remove the microorganisms before lymphocytes become aware of the incident. The human body has two general classes of phagocytic cells: *microphages* and *macrophages.*

Microphages

Microphages are the neutrophils and eosinophils that normally circulate in the blood. These phagocytic cells leave the bloodstream and enter peripheral tissues subjected to injury or infection. As we noted in Chapter 19, neutrophils are abundant, mobile, and quick to phagocytize cellular debris or invading bacteria. *∞ pp. 639–640* Eosinophils, which are less abundant, target foreign compounds or pathogens that have been coated with antibodies.

Macrophages

Macrophages are large, actively phagocytic cells. Your body contains several types of macrophages, and most are derived from the monocytes of the circulating blood. Macrophages either are fixed permanently or move freely. Although no organs or tissues are purely phagocytic, almost every tissue in the body shelters resident or visiting macrophages. This relatively diffuse collection of phagocytic cells has been called the **monocyte–macrophage system,** or the *reticuloendothelial system.*

An activated macrophage may respond to a pathogen in several ways. For example:

- It may engulf a pathogen or other foreign object and destroy it with lysosomal enzymes.
- It may bind to or remove a pathogen from the interstitial fluid but be unable to destroy the invader until assisted by other cells.
- It may destroy its target by releasing toxic chemicals, such as *tumor necrosis factor,* nitric oxide, or hydrogen peroxide, into the interstitial fluid.

We shall consider those responses further in a later section.

Fixed Macrophages **Fixed macrophages,** or *histiocytes,* are permanent residents of specific tissues and organs. These cells are normally incapable of movement, so the objects of their phagocytic attention must diffuse or otherwise move through the surrounding tissue until they are within range. Fixed macrophages are scattered among connective tissues, usually in close association with collagen or reticular fibers. Their presence has been noted in the papillary and reticular layers of the dermis, in the subarachnoid space of the meninges, and in bone marrow. In some organs, the fixed macrophages have special names: **Microglia** are macrophages in the central nervous system, and **Kupffer cells** are macrophages located in and around the liver sinusoids.

Free Macrophages **Free macrophages,** or *mobile macrophages,* travel throughout the body, arriving at an injury site by migration through adjacent tissues or by recruitment from the circulating blood. Some tissues contain free macrophages with distinctive characteristics. For example, the exchange surfaces of the lungs are patrolled by **alveolar macrophages,** also known as *phagocytic dust cells.*

Both classes of macrophages are derived from the monocytes of the blood. The primary difference is that fixed macrophages are always located within a given tissue, but the free macrophages come and go. During an infection, this distinction commonly breaks down, because the fixed macrophages may lose their attachments and begin roaming around the damaged tissue.

Movement and Phagocytosis

Free macrophages and microphages share a number of functional characteristics:

- Both can move through capillary walls by squeezing between adjacent endothelial cells, a process known as *emigration,* or *diapedesis.* ∞ *p. 638* The endothelial cells in an injured area develop membrane "markers" that let passing blood cells know that something is wrong. The cells then attach to the endothelial lining and migrate into the surrounding tissues.

- Both may be attracted to or repelled by chemicals in the surrounding fluids, a phenomenon called **chemotaxis.** They are particularly sensitive to cytokines released by other body cells and to the chemicals released by pathogens.

- For both, phagocytosis begins with **adhesion,** the attachment of the phagocyte to its target. Adhesion depends on binding of the surface of the target to receptors on the cell membrane of the phagocyte. Adhesion is followed by the formation of a vesicle containing the bound target, as detailed in Figure 3-12, p. 78. The contents of the vesicle are then digested when the vesicle fuses with lysosomes or peroxisomes.

All phagocytic cells function in much the same way, although the target of phagocytosis may differ from one cell type to another. The life span of an actively phagocytic cell can be rather brief. For example, most neutrophils expire before they have engulfed more than 25 bacteria, and in an infection a neutrophil may attack that many in an hour.

IMMUNOLOGICAL SURVEILLANCE

The immune system generally ignores the body's own cells unless they become abnormal in some way. Natural killer (NK) cells are responsible for recognizing and destroying abnormal cells when they appear in peripheral tissues. The constant monitoring of normal tissues by NK cells is called **immunological surveillance.**

The cell membrane of an abnormal cell generally contains antigens not found on the membranes of normal cells. NK cells recognize an abnormal cell by detecting the presence of those antigens. The recognition mechanism differs from that used by T cells or B cells. A T cell or a B cell can be activated only by exposure to a *specific* antigen at a *specific* site on a cell membrane. An NK cell responds to a variety of abnormal antigens that may appear anywhere on a cell membrane. NK cells are therefore much less selective about their targets than are other lymphocytes; if a membrane contains abnormal antigens, it will be attacked. As a result, NK cells are very versatile: A single NK cell can attack bacteria in the interstitial fluid, body cells infected with virus, or cancer cells.

NK cells also respond much more rapidly than T cells or B cells. The activation of T cells and B cells involves a relatively complex and time-consuming sequence of events. NK cells respond immediately on contact with an abnormal cell.

NK Cell Activation

Activated NK cells react in a predictable way:

1. If a cell has unusual surface proteins or other components in its cell membrane, an NK cell recognizes that other cell as abnormal. That recognition activates the NK cell, which then adheres to its target cell.
2. The Golgi apparatus moves around the nucleus until the maturing face points directly toward the abnormal cell. This process might be compared to the rotation of a tank turret to point the cannon toward the enemy. A flood of secretory vesicles is then produced at the Golgi apparatus. These vesicles, which contain proteins called **perforins,** travel through the cytoplasm toward the cell surface.
3. The perforins are then released at the cell surface by exocytosis. They diffuse across the gap between the NK cell and its target.
4. On reaching the opposing cell membrane, perforin molecules interact with one another and with the membrane. The result is the creation of a network of pores in the cell membrane (Figure 22-11●). These pores are large enough to permit the free passage of ions, proteins, and other intracellular materials. The target cell can no longer maintain its internal environment, and it quickly disintegrates.

It is not clear why perforin does not affect the membrane of the NK cell itself. It has been proposed that the NK cell membranes contain a second protein, called *protectin,* that binds and inactivates perforin.

NK Cells, Cancer, and Viral Infections

NK cells attack cancer cells and cells infected with viruses. Cancer cells probably appear throughout life, but their cell membranes generally contain unusual proteins called **tumor-specific antigens,** which NK cells recognize as abnormal. The affected cells are then destroyed, preserving tissue integrity. Unfortunately, some cancer cells avoid detection, perhaps because they lack tumor-specific antigens or because these antigens are covered in some way. Other cancer cells are able to destroy the NK cells that detect them. This process of avoiding detection or neutralizing body defenses is called **immunological escape.** Once immunological escape has occurred, cancer cells can multiply and spread without interference by NK cells.

NK cells are also important in fighting viral infections. Viruses reproduce inside cells, beyond the reach of circulating antibodies. However, infected cells incorporate viral antigens into their cell membranes, and NK cells recognize these infected cells as abnormal. By destroying them, NK cells can slow or prevent the spread of a viral infection.

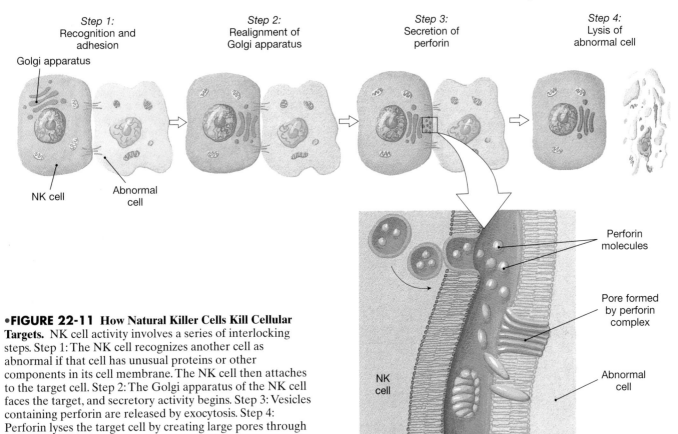

Step 1:
Recognition and
adhesion

Step 2:
Realignment of
Golgi apparatus

Step 3:
Secretion of
perforin

Step 4:
Lysis of
abnormal cell

Golgi apparatus

NK cell

Abnormal
cell

Perforin
molecules

Pore formed
by perforin
complex

NK
cell

Abnormal
cell

•**FIGURE 22-11 How Natural Killer Cells Kill Cellular Targets.** NK cell activity involves a series of interlocking steps. Step 1: The NK cell recognizes another cell as abnormal if that cell has unusual proteins or other components in its cell membrane. The NK cell then attaches to the target cell. Step 2: The Golgi apparatus of the NK cell faces the target, and secretory activity begins. Step 3: Vesicles containing perforin are released by exocytosis. Step 4: Perforin lyses the target cell by creating large pores through the cell membrane.

INTERFERONS

Interferons (in-ter-FĒR-onz) are small proteins released by activated lymphocytes and macrophages and by tissue cells infected with viruses. On reaching the membrane of a normal cell, an interferon binds to surface receptors on the cell and, via second messengers, triggers the production of **antiviral proteins** in the cytoplasm. Antiviral proteins do not interfere with the entry of viruses, but they do interfere with viral replication inside the cell. In addition to their role in slowing the spread of viral infections, interferons stimulate the activities of macrophages and NK cells.

At least three interferons exist, each of which has additional specialized functions: (1) **Alpha- (α) interferon,** produced by several types of leukocytes, attracts and stimulates NK cells; (2) **beta- (β) interferon,** secreted by fibroblasts, slows inflammation in a damaged area; and (3) **gamma- (γ) interferon,** secreted by T cells and NK cells, stimulates macrophage activity. Most cell types other than lymphocytes and macrophages respond to viral infection by secreting beta-interferon.

Interferons are examples of **cytokines** (SĪ-tō-kīnz), chemical messengers released by tissue cells to coordinate local activities. ∞ *p. 577* Cytokines produced by most cells are used only for paracrine communication, cell-to-cell communication within one tissue.

But the cytokines released by cellular defenders also act as hormones; they affect cells and tissues throughout the body. We shall discuss their role in the regulation of specific defenses in a later section.

COMPLEMENT

Your plasma contains 11 special **complement proteins (C),** which form the **complement system.** The term *complement* refers to the fact that this system complements, or supplements, the action of antibodies.

The complement proteins interact with one another in chain reactions, or *cascades,* comparable to those of the clotting system. Figure 22-12• provides an overview of the complement system. AM *A Closer Look: The Complement System*

Known effects of complement activation include:

▪ *The destruction of target cell membranes.* Five of the interacting complement proteins bind to the cell membrane. They form a functional unit called the **membrane attack complex (MAC),** which creates a pore in the membrane. The pores formed in this way are comparable to those produced by perforin and have the same effect. The target cell is soon destroyed.

▪ *The stimulation of inflammation.* Activated complement proteins enhance the release of histamine by

22

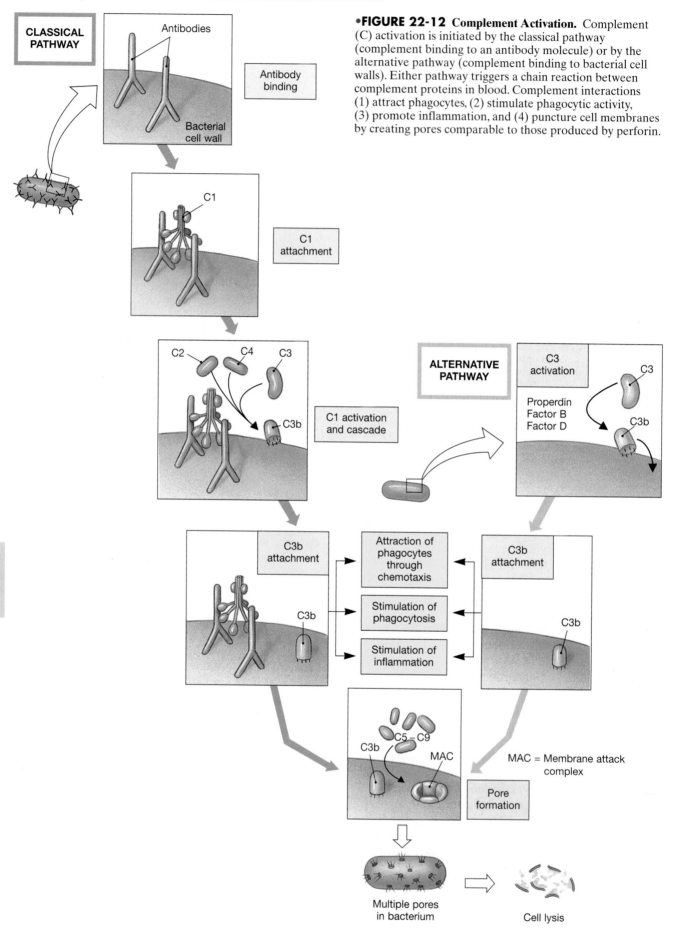

•FIGURE 22-12 Complement Activation. Complement (C) activation is initiated by the classical pathway (complement binding to an antibody molecule) or by the alternative pathway (complement binding to bacterial cell walls). Either pathway triggers a chain reaction between complement proteins in blood. Complement interactions (1) attract phagocytes, (2) stimulate phagocytic activity, (3) promote inflammation, and (4) puncture cell membranes by creating pores comparable to those produced by perforin.

mast cells and basophils. These substances increase the degree of local inflammation and accelerate blood flow to the region.

- *The attraction of phagocytes.* Activated complement proteins attract neutrophils and macrophages to the area, improving the chances that phagocytic cells will be able to cope with the injury or infection.

- *The enhancement of phagocytosis.* A coating of complement proteins and antibodies both attracts phagocytes and makes the target easier to engulf. Macrophage membranes contain receptors that can detect and bind to complement proteins and bound antibodies. After binding, the pathogens are easily engulfed. The antibodies involved are called **opsonins,** and this effect is called **opsonization.**

Complement Activation

Complement activation takes two routes: (1) the *classical pathway* and (2) the *alternative pathway.*

The Classical Pathway   The most rapid and effective activation of the complement system occurs through the **classical pathway** (Figure 22-12•). The process begins when one of the complement proteins *(C1)* binds to an antibody molecule already attached to its specific antigen—in this case, a bacterial cell wall. The bound complement protein then acts as an enzyme, catalyzing a series of reactions involving other complement proteins. The classical pathway ends with the conversion of an inactive complement protein, *C3,* to an active form, *C3b.* C3b binds the surface of the antigen and subsequently stimulates phagocytosis and promotes inflammation. It also triggers additional reactions that end with the creation of the membrane attack complex from complement factors *C5–C9.*

The Alternative Pathway   A less effective, slower activation of the complement system occurs in the absence of antibody molecules. This **alternative pathway,** or *properdin pathway,* is important in the defense against bacteria, some parasites, and virus-infected cells. The pathway begins when several complement proteins—including **properdin,** or *factor P, factor B,* and *factor D*—interact in the plasma (Figure 22-12•). This interaction can be triggered by exposure to foreign materials, such as the capsule of a bacterium. As does the classical pathway, the alternative pathway ends with the conversion of C3 to C3b, the stimulation of phagocytosis, and the subsequent formation of the membrane attack complex.

Some bacteria are unaffected by complement, because the complement proteins cannot interact with their cell membranes. These bacteria coat themselves in a protective capsule of carbohydrates. However, the classical pathway still provides a defense against these bacteria, once antibodies have bound to the capsule.

INFLAMMATION

Inflammation, or the *inflammatory response,* is a localized tissue response to injury. ∞ *p. 135* Inflammation produces local sensations of swelling, redness, heat, and pain. Figure 22-13• summarizes the events that occur during inflammation of the skin. Comparable events occur in almost any tissue subjected to physical damage or infection. Many stimuli can produce inflammation, including impact, abrasion,

22

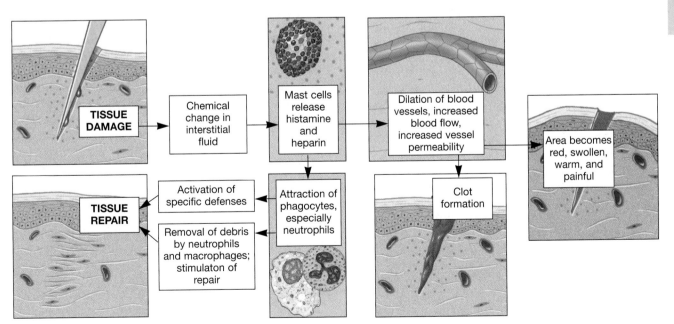

•**FIGURE 22-13** Inflammation

distortion, chemical irritation, infection by pathogens, or extreme temperatures (hot or cold). Each of these stimuli kills cells, damages connective tissue fibers, or injures the tissue in some other way. These changes alter the chemical composition of the interstitial fluid. Damaged cells release prostaglandins, proteins, and potassium ions, and the injury itself may have introduced foreign proteins or pathogens. The changes in the interstitial environment trigger the complex process of inflammation.

The three goals of inflammation are:

1. To perform a temporary repair at the injury site and prevent the access of additional pathogens.

2. To slow the spread of pathogens away from the injury site.

3. To mobilize local, regional, and systemic defenses that can overcome the pathogens and facilitate permanent repairs. The repair process is called *regeneration*.

The Response to Injury

Mast cells play a pivotal role in the inflammatory response. When stimulated by mechanical stress or chemical changes in the local environment, these cells release histamine, heparin, prostaglandins, and other chemicals into interstitial fluid. Events then proceed in a series of integrated steps:

1. The released histamine increases capillary permeability and accelerates blood flow through the area. The increased blood flow brings more cellular defenders to the site and carries away toxins and debris, diluting them and reducing their local impact.

2. Clotting factors and complement proteins leave the bloodstream and enter the injured or infected area. Clotting does not occur at the actual site of injury, due to the presence of heparin. However, a clot soon forms around the damaged area, and this clot both isolates the region and slows the spread of the chemical or pathogen into healthy tissues. Meanwhile, complement activation through the alternative pathway breaks down bacterial cell walls and attracts phagocytes.

3. The increased blood flow elevates the local temperature, increasing the rate of enzymatic reactions and accelerating the activity of phagocytes. The rise in temperature may also denature foreign proteins or vital enzymes of invading microorganisms.

4. Debris and bacteria are attacked by neutrophils drawn to the area by chemotaxis. As they circulate through a blood vessel in an injured area, neutrophils undergo *activation*. In this process, (1) they stick to the side of the vessel and move into the tissue by diapedesis; (2) their metabolic rate goes up dramatically, and, while this *respiratory burst* continues, they generate reactive compounds, such as nitric oxide and hydrogen peroxide, that can destroy engulfed pathogens; and (3) they secrete cytokines that attract other neutrophils and macrophages to the area.

5. Fixed and free macrophages engulf pathogens and cell debris. At first, these cells are outnumbered by neutrophils, but as the macrophages and neutrophils continue to secrete cytokines, the number of macrophages increases rapidly. Eosinophils may get involved if the foreign materials become coated with antibodies.

6. Other cytokines released by active phagocytes stimulate fibroblasts to begin barricading the area with scar tissue, reinforcing the clot and further slowing the invasion of adjacent tissues.

7. The combination of abnormal tissue conditions and chemicals released by mast cells stimulates local sensory neurons, producing sensations of pain. The individual is consciously aware of these sensations and may take steps to limit the damage, such as removing a splinter or cleaning a wound.

8. As inflammation is under way, the foreign proteins, toxins, microorganisms, and active phagocytes in the area activate the body's specific defenses.

Necrosis

After an injury, tissue conditions generally become even more abnormal before they begin to improve. **Necrosis** (ne-KRŌ-sis) is the tissue degeneration that occurs after cells have been injured or destroyed. The process begins several hours after the initial event, and the damage is caused by lysosomal enzymes. Lysosomes break down by autolysis, releasing digestive enzymes that first destroy the injured cells and then attack surrounding tissues. ∞ *p. 87*

As local inflammation continues, debris, fluid, dead and dying cells, and necrotic tissue components accumulate at the injury site. This viscous fluid mixture is known as **pus.** An accumulation of pus in an enclosed tissue space is called an **abscess.** [AM] *Complications of Inflammation*

FEVER

Fever is the maintenance of a body temperature greater than 37.2°C (99°F). In Chapter 14, we noted the presence of a temperature-regulating center in the preoptic area of the hypothalamus. ∞ *p. 459* Circulating proteins called **pyrogens** (PĪ-rō-jenz) can reset the body's thermostat and raise the temperature. A variety of stimuli, including pathogens, bacterial toxins, and antigen–antibody complexes, either act as pyrogens themselves or stimulate the release of pyrogens by macrophages. The pyrogen released by active macrophages is a cytokine called **endogenous pyrogen,** or **interleukin-1** (in-ter-LOO-kin), abbreviated **IL-1.**

Within limits, a fever can be beneficial. High body temperatures may inhibit some viruses and bacteria, but the most likely beneficial effect is on body metabolism. For each 1°C rise in temperature, your metabolic rate jumps by 10 percent. Your cells can move faster, so enzymatic reactions occur faster. The net results may be the quicker mobilization of tissue defenses and an accelerated repair process.

☑ What types of cells would be affected by a decrease in the number of monocyte-forming cells in bone marrow?

☑ A rise in the level of interferon in the body would suggest what kind of infection?

☑ What effects do pyrogens have in the body?

SPECIFIC DEFENSES: THE IMMUNE RESPONSE

Specific resistance, or immunity, is provided by the coordinated activities of T cells and B cells, which respond to the presence of specific antigens. The basic functional relationship can be summarized as follows:

1. *T cells* are responsible for **cell-mediated immunity** (or **cellular immunity**), our defense against abnormal cells and pathogens inside cells.
2. *B cells* provide **antibody-mediated immunity** (or **humoral immunity**), our defense against antigens and pathogens in body fluids.

Both mechanisms are important, because they come into play under different circumstances. Activated T cells do not respond to antigenic materials in solution, and the antibodies produced by activated B cells cannot cross cell membranes. Moreover, helper T cells play a crucial role in antibody-mediated immunity by stimulating the activity of B cells.

Our understanding of immunity has greatly improved in the past two decades, and a comprehensive discussion would involve hundreds of pages and thousands of details. The discussion that follows emphasizes important patterns and introduces general principles that will provide a foundation for future courses in microbiology and immunology.

FORMS OF IMMUNITY

Immunity is either *innate* or *acquired* (Figure 22-14•).

Innate Immunity

Innate immunity is genetically determined; it is present at birth and has no relationship to previous exposure to the antigen involved. For example, people do not get the same diseases goldfish do. Innate immunity breaks down only in the case of *AIDS* or other conditions that depress all aspects of specific resistance.

Acquired Immunity

Acquired immunity is not present at birth; you acquire immunity to a specific antigen only once you have been exposed to that antigen. Acquired immunity can be active or passive. **Active immunity** appears after exposure to an antigen, as a consequence of the immune response. The immune system is *capable* of

22

•**FIGURE 22-14** Types of Immunity

defending against a large number of antigens. However, the appropriate defenses are mobilized only after you encounter a particular antigen. Active acquired immunity can develop as a result of (1) natural exposure to an antigen in the environment (naturally acquired immunity) or (2) from deliberate exposure to an antigen (induced active immunity). **Naturally acquired immunity** normally begins to develop after birth, and it is continually enhanced as you encounter "new" pathogens or other antigens. You might compare this process to vocabulary development; a child begins with a few basic common words and learns new ones as needed. The purpose of **induced active immunity** is to stimulate antibody production under controlled conditions so that you will be able to overcome natural exposure to the pathogen some time in the future. This is the basic principle behind *immunization,* or *vaccination,* to prevent disease. A **vaccine** is a preparation that contains either a dead or an inactive pathogen or antigens derived from that pathogen. Vaccines are designed to induce an immune response.

Passive immunity is produced by the transfer of antibodies from another individual. **Natural passive immunity** results when a mother's antibodies protect her baby against infections during development (across the placenta) or in early infancy (through breast milk). In **induced passive immunity,** antibodies are administered to fight infection or prevent disease. For example, antibodies that attack the rabies virus are injected into a person bitten by a rabid animal.

PROPERTIES OF IMMUNITY

Regardless of the form, immunity exhibits four general properties: (1) *specificity,* (2) *versatility,* (3) *memory,* and (4) *tolerance.*

Specificity

A specific defense is activated by a specific antigen, and the immune response targets that particular antigen and no others. **Specificity** results from the activation of appropriate lymphocytes and the production of antibodies with targeted effects. Specificity occurs because T cells and B cells respond to the molecular structure of an antigen. The antigen shape and size determine which lymphocytes will respond to its presence. Each T cell or B cell has receptors that will bind to one specific antigen, ignoring all others. The response of an activated T cell or B cell is equally specific. Either lymphocyte will destroy or inactivate that antigen without affecting other antigens or normal tissues.

Versatility

Millions of antigens in the environment can pose a threat to health. Over a normal lifetime, an individual encounters only a fraction of that number—perhaps tens of thousands of antigens. Your immune system, however, has no way of anticipating which antigens it will encounter. It must be ready to confront *any* antigen at *any* time. **Versatility** results in part from the large diversity of lymphocytes present in the body and in part from variability in the structure of synthesized antibodies.

During development, differentiation of cells in the lymphatic system produces an enormous number of lymphocytes with varied antigen sensitivities. The trillion or more T cells and B cells in the human body include millions of different lymphocyte populations, distributed throughout the body. Each population consists of several thousand cells with receptors in their membranes that differ from those of other lymphocyte populations. As a result, each group of lymphocytes will respond to the presence of a different antigen.

Several thousand lymphocytes are not enough to overcome a pathogenic invasion. When activated in the presence of an appropriate antigen, however, a lymphocyte begins to divide, producing more lymphocytes with the same specificity. Thus, whenever an antigen is encountered, more lymphocytes are produced. The term **clone** refers to all the cells produced by the division of an activated lymphocyte. *All the members of a clone are sensitive to the same specific antigen.*

Think about running a snack shop with only samples on display. When customers come in and make a selection, you prepare the food on the spot. You can offer a wide selection, because the samples don't take up much space, and you do not have to expend energy preparing food that will not be eaten.

Memory

The immune system "remembers" antigens that it encounters. As a result of this **memory,** the immune response that occurs after a second exposure to an antigen is stronger and lasts longer than the response to the first exposure. During the initial response to an antigen, lymphocytes sensitive to its presence undergo repeated cycles of cell division. Two kinds of cells are produced: some that attack the invader, and others that remain inactive unless they are exposed to the same antigen again. These inactive *memory cells* enable your immune system to "remember" antigens it has previously encountered and to launch a faster, stronger counterattack if such an antigen reappears.

Tolerance

The immune system does not respond to all antigens. For example, all cells and tissues in the body contain some antigens that normally fail to stimulate an immune response. The immune system is said to exhibit **tolerance** toward such antigens.

The immune response targets foreign cells and compounds, but it generally ignores normal tissues. During their differentiation in the bone marrow (B cells) and thymus (T cells), cells that react to antigens normally present in the body are destroyed. As a result, mature B and T cells will ignore normal, or *self,* antigens and attack foreign, or *nonself,* antigens. Tolerance can also develop over time in response to chronic exposure to an antigen in the environment. Such tolerance lasts only as long as the exposure continues.

OVERVIEW: THE IMMUNE RESPONSE

Figure 22-15• presents an overview of the immune response. When an antigen triggers an immune response, it usually activates both T cells and B cells. The activation of T cells, which usually occurs first, involves active phagocytes exposed to the antigen. Once activated, T cells attack the antigen and stimulate the activation of B cells. Activated B cells mature into cells that produce antibodies; antibodies distributed in the bloodstream bind to and attack the antigen. We shall examine each of these processes more closely in the sections that follow.

T CELLS AND CELL-MEDIATED IMMUNITY

T cells play a key role in the initiation, maintenance, and control of the immune response. We have already introduced three major types of T cells:

1. *Cytotoxic T cells* (T_C), which are responsible for cell-mediated immunity. They enter peripheral tissues and directly attack antigens physically and chemically.
2. *Helper T cells* (T_H), which stimulate the responses of both T cells and B cells. They are absolutely vital to the immune response, because B cells must be activated by helper T cells before B cells can produce antibodies. The reduction in the helper T cell population that occurs in AIDS is largely responsible for the loss of immunity. (We shall discuss AIDS on p. 790.)
3. *Suppressor T cells* (T_S), which inhibit T cell and B cell activities and moderate the immune response.

T Cell Activation

Before an immune response can begin, T cells must be activated by exposure to an antigen. This activation seldom happens by direct lymphocyte–antigen interaction, and foreign compounds or pathogens entering a tissue commonly fail to stimulate an immediate immune response.

Antigen Presentation T cells recognize antigens when the antigens are bound to glycoproteins in cell membranes. Glycoproteins are integral membrane components. ∞ *p. 56* **Antigen presentation** occurs when an antigen–glycoprotein combination capable of activating T cells appears in a cell membrane. The structure of these glycoproteins is genetically determined. The genes controlling their synthesis are located along one portion of chromosome 6, in a region called the **major histocompatibility complex (MHC).** These membrane glycoproteins are called **MHC proteins,** or *human leukocyte antigens (HLAs).*

The amino acid sequences and shapes of MHC proteins differ among individuals. Each MHC molecule has a distinct three-dimensional shape with a relatively narrow central groove. An antigen that fits into this groove can be held in position by hydrogen bonding.

22

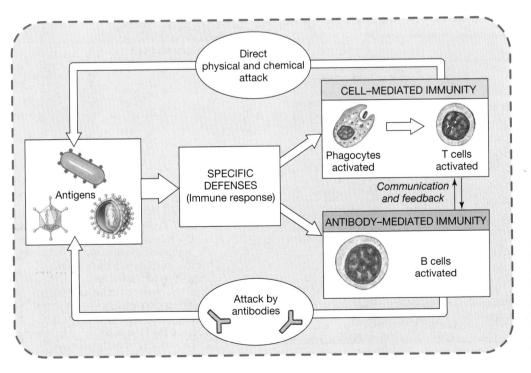

•**FIGURE 22-15 An Overview of the Immune Response.** This figure will be repeated, in reduced and simplified form as Navigator icons, in key figures throughout this chapter as we change topics.

Two major classes of MHC proteins are known: *Class I* and *Class II.*

MHC Class I **Class I** MHC proteins are in the membranes of all nucleated cells. These proteins are continuously being synthesized and exported to the cell membrane in vesicles created at the Golgi apparatus. As they form, Class I proteins pick up small peptides from the surrounding cytoplasm and carry them to the cell surface. If the cell is healthy and the peptides are normal, T cells will ignore them. If the cytoplasm contains abnormal (nonself) peptides or viral proteins (Figure 22-16a•), they will soon appear in the cell membrane, and T cells will be activated. Ultimately, their activation will lead to the destruction of the ab-

normal cell. This is the primary reason why donated organs are commonly rejected by the recipient; despite preliminary cross-match testing, the recipient's T cells recognize the transplanted tissue as foreign.

MHC Class II **Class II** MHC proteins are present only in the membranes of antigen-presenting cells and lymphocytes. **Antigen-presenting cells (APCs)** are specialized cells responsible for activating T cell defenses against foreign cells (including bacteria) and foreign proteins. Antigen-presenting cells include all the phagocytic cells of the monocyte–macrophage group discussed in other chapters, including (1) the free and fixed macrophages in connective tissues; (2) the Kupffer cells of the liver, and (3) the microglia in the cen-

(a) Infected cell

(b) Phagocytic antigen-presenting cell

•**FIGURE 22-16** **Antigens and Class II MHC Proteins.** The Navigator icon in the dashed box repeats Figure 22-15 but highlights the part that is relevant to *this* figure, so you can see how phagocyte activation relates to the rest of the immune response. **(a)** Viral or other foreign antigens appear in cell membranes bound to Class I MHC proteins. **(b)** Processed antigens appear on the surfaces of antigen-presenting cells bound to Class II MHC proteins.

22

tral nervous system (Chapter 12). ∞ *pp. 119, 370* The Langerhans cells of the skin and the dendritic cells of the lymph nodes and spleen are APCs that are not phagocytic. ∞ *p. 146*

Phagocytic APCs engulf and break down pathogens or foreign antigens. This **antigen processing** creates antigenic fragments, which are then bound to Class II MHC proteins and inserted into the cell membrane (Figure 22-16b•). *Class II MHC proteins appear in the cell membrane only when the cell is processing antigens.* Exposure to an APC membrane containing processed antigen can stimulate appropriate T cells.

The Langerhans cells and dendritic cells remove antigenic materials from their surroundings by pinocytosis rather than by phagocytosis. The cell membranes of these cells also present antigens that are bound to Class II MHC proteins.

Antigen Recognition Inactive T cells have receptors that recognize Class I or Class II MHC proteins. The receptors also have binding sites that detect the presence of specific bound antigens. If an MHC protein contains any antigen other than its specific target, the T cell remains inactive. If it contains the antigen that the T cell is programmed to detect, binding will occur. This process is called **antigen recognition,** because the T cell recognizes that it has found an appropriate target.

Some T cells can recognize antigens bound to Class I MHC proteins, whereas others can recognize antigens bound to Class II MHC proteins. Whether a T cell responds to antigens held by Class I or Class II proteins depends on the structure of the T cell membrane. The membrane proteins involved are members of a larger class of proteins called **CD** (*cluster of differentiation*) **markers.**

CD markers Lymphocytes, macrophages, and other related cells have CD markers. Each of the more than 50 types of CD markers is designated by an identifying number. All T cells have **CD3** markers in their membranes. Two other CD markers are associated with the MHC receptor complex in specific groups of T cells:

1. **CD8** markers are found on cytotoxic T cells and suppressor T cells. These cells are often called *CD8 T cells* or *CD8+ T cells.* CD8 T cells can be activated by exposure to antigens presented by Class I MHC proteins.
2. **CD4** markers are found on helper T cells. These cells are often called *CD4 T cells* or *CD4+ T cells.* CD4 T cells respond to antigens presented by Class II MHC proteins.

Costimulation ✹A T cell does not become active immediately after recognizing an antigen. Antigen recognition simply prepares the cell for activation. Before activation can occur, a T cell must bind to the stimulating cell at a second site. This vital secondary binding process is called *costimulation.* Costimulation

essentially confirms the "OK to activate" signal. Appropriate costimulation proteins appear in the presenting cell only if that cell has engulfed antigens or is infected by viruses. Many costimulation proteins are structurally related to cytokines released by activated lymphocytes. The effects these proteins have on the exposed T cell vary but typically include the stimulation of transcription at the nucleus and the promotion of cell division and differentiation.

Costimulation is like the safety on a gun: It helps prevent T cells from mistakenly attacking normal (self) tissues. If a cell displays an unusual antigen but does not display the "I am an active phagocyte" or "I am infected" signal, T cell activation will not occur. Costimulation is important only in determining whether a T cell will become activated. Once activation has occurred, the safety is off and the T cell will attack any cells that carry the target antigens.

We shall now consider the functions of CD8 and CD4 T cells in the immune response.

Cytotoxic T Cells

When activated by exposure to antigens bound to Class I MHC proteins, CD8 T cells undergo a series of divisions that generate cytotoxic T cells (T_C cells) and **memory T_C cells.** The cytotoxic T cells, also called *killer T cells,* roam throughout the injured tissue. When a cytotoxic T cell encounters its target antigens bound to Class I MHC proteins of another cell, it immediately destroys that cell. A cytotoxic T cell accomplishes this destruction in three ways (Figure 22-17•):

1. By rupturing the antigenic cell membrane through the release of perforin.
2. By killing the target cell by secreting a poisonous **lymphotoxin** (lim-fō-TOK-sin).
3. By activating genes in the cell's nucleus that tell the cell to die. We introduced genetically programmed cell death, called *apoptosis,* in Chapter 3. ∞ *p. 96*

The entire sequence of events, from the appearance of the antigen in a tissue to cell destruction by cytotoxic T cells, takes a significant amount of time. After the first exposure to an antigen, 2 days or more may pass before the concentration of cytotoxic T cells reaches effective levels at the site of injury or infection. Over this period, the damage or infection may spread, making it more difficult to control.

Memory T_C cells ensure that there will not be a delay in the response if the antigen reappears. These cells do not differentiate further the first time the antigen triggers an immune response, although thousands of them are produced. Instead, they remain in reserve. If the same antigen appears a second time, these cells will *immediately* differentiate into cytotoxic T cells, producing a swift, effective cellular response that can

22

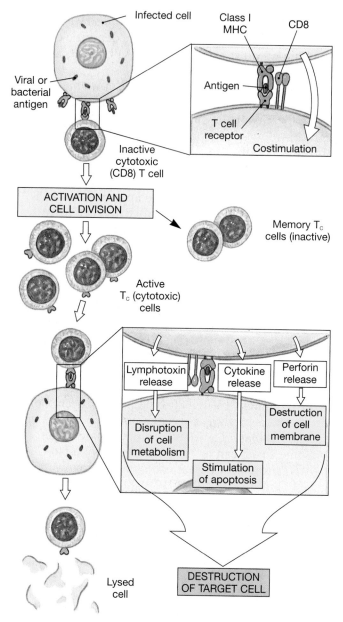

●**FIGURE 22-17** **Antigen Recognition and the Activation of Cytotoxic T Cells.** An inactive cytotoxic T cell must encounter an appropriate antigen bound to Class I MHC proteins. It must also receive costimulation from the membrane it contacts. It is then activated and undergoes divisions that produce memory T_C cells and active T_C cells. When one of these active cells encounters a membrane displaying the target antigen, the T_C cell will use one of several methods to destroy the cell. Notice that the dashed box now highlights the activation of T cells.

overwhelm an invading organism before it becomes well established in the tissues.

Suppressor T Cells

Suppressor T cells (T_S cells) depress the responses of other T cells and of B cells by secreting *suppression factors,* inhibitory cytokines of unknown structure. This suppression does not occur immediately: Suppressor T cell activation takes much longer than the activation of other types of T cells. In addition, most of the CD8 T cells in the bloodstream will on activation produce cytotoxic T cells rather than suppressor T cells. As a result, suppressor T cells act *after* the initial immune response. In effect, these cells put on the brakes and limit the degree of immune system activation from a single stimulus.

Helper T Cells

Upon activation, CD4 T cells undergo a series of cell divisions that produce helper T cells (T_H cells) and **memory T_H cells** (Figure 22-18●). The memory T_H cells remain in reserve, whereas the helper T cells secrete a variety of cytokines that coordinate specific and non-specific defenses and stimulate cell-mediated and antibody-mediated immunities. For example, activated helper T cells secrete cytokines that:

1. Stimulate the T cell divisions that produce memory T cells and accelerate the maturation of cytotoxic T cells.

2. Enhance nonspecific defenses by attracting macrophages to the area, preventing their departure, and stimulating their phagocytic activity and effectiveness.

3. Attract and stimulate the activity of NK cells, providing another mechanism for the destruction of abnormal cells and pathogens.

4. Promote B cell division, plasma cell maturation, and antibody production.

In addition, activated helper T cells participate in the direct activation of B cells sensitive to the antigen involved. We shall consider this aspect of helper T cell function in the next section.

Foreign antigen

Antigen-presenting cell (APC)

Class II MHC

CD4 protein

T cell receptor

Inactive T$_H$ (helper T) cell (CD4)

ACTIVATION AND CELL DIVISION

Memory T$_H$ cells (inactive)

Cytokines released

Active T$_H$ cells

To Figure 22-19

•**FIGURE 22-18** **Antigen Recognition and Activation of Helper T Cells.** Inactive helper T cells must be exposed to appropriate antigens bound to Class II MHC proteins. The T$_H$ cells then undergo activation, dividing to produce memory T$_H$ cells and active T$_H$ cells. Active T$_H$ cells secrete cytokines that stimulate cell-mediated and antibody-mediated immunities. They also interact with sensitized B cells, as Figure 22-19 shows.

B CELLS AND ANTIBODY-MEDIATED IMMUNITY

B cells are responsible for launching a chemical attack on antigens by producing appropriate *antibodies.* B cell activation proceeds in a series of steps diagrammed in Figure 22-19•.

B Cell Sensitization

As we noted earlier, the body has millions of populations of B cells. Each B cell carries its antibody molecules in its cell membrane. If antigens that will bind to those antibodies appear in the interstitial fluid, the B cell prepares to undergo activation. This process is called **sensitization.** Because B cells migrate throughout the body, pausing briefly in one lymphoid tissue or another, sensitization typically occurs at the lymph node closest to the site of infection or injury.

During sensitization, antigens enter the cell by endocytosis. They then appear on the surface of the B cell, bound to Class II MHC proteins. A sensitized B cell is now ready to go, but it generally will not undergo activation unless it receives the "OK" from a helper T cell.[2] The need for activation by a helper T cell helps prevent inappropriate activation, the same way that costimulation acts as a "safety" for cell-mediated immunity.

B Cell Activation

A sensitized B cell next encounters a helper T cell already activated by antigen presentation. The helper T cell binds to the MHC complex, recognizes the presence of an antigen, and begins secreting cytokines that (1) promote B cell activation, (2) stimulate B cell division, (3) accelerate plasma cell formation, and (4) enhance antibody production.

Graft Rejection and Immunosuppression

Organ transplantation is a treatment option for patients with severe disorders of the kidneys, liver, heart, lungs, or pancreas. Finding a suitable donor is the first major problem. In the United States, each day 6 people die while awaiting an organ transplant, and another 48 people are added to the transplant waiting list. After surgery has been performed, the major problem is **graft rejection.** In graft rejection, T cells are activated by contact with MHC proteins on cell membranes in the donated tissues. The cytotoxic T cells that develop then attack and destroy the foreign cells.

Significant improvements in transplant success can be made by **immunosuppression,** a reduction of the sensitivity of the immune system. Until recently, the drugs used to produce immunosuppression did not selectively target the immune response. For example, **prednisone** (PRED-ni-sōn), a corticosteroid, was used because it has anti-inflammatory effects that reduce the number of circulating white blood cells and depress the immune response. However, corticosteroid use also causes undesirable side effects. ∞ *p. 602*

An understanding of the communication among T cells, macrophages, and B cells has now led to the development of drugs with more selective effects. **Cyclosporin A (CsA),** a compound derived from a fungus, was the most important *immunosuppressive drug* developed in the 1980s. This compound depresses all aspects of the immune response primarily by suppressing helper T cell activity, while leaving suppressor T cells relatively unaffected. AM *Transplants and Immunosuppressive Drugs*

---

[2] Under certain circumstances, complex antigens can directly stimulate B cells. The B cell must bind multiple antigenic molecules simultaneously in a single, localized portion of its cell membrane. This process, called *capping,* is relatively rare.

22

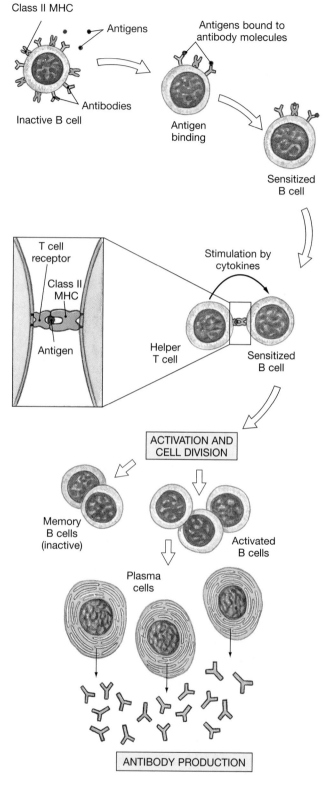

•**FIGURE 22-19** **The Sensitization and Activation of B Cells.** A B cell is sensitized by exposure to antigens that bind to antibodies in the B cell membrane. The B cell then displays those antigens in its cell membrane. Activated helper T cells encountering those antigens release cytokines that trigger the activation of the B cell. The activated B cell then divides, producing memory B cells and plasma cells that secrete antibodies.

Figure 22-19• diagrams the activation of a B cell by a helper T cell. The activated B cell typically divides several times, producing daughter cells that differentiate into plasma cells and *memory B cells.* The plasma cells begin synthesizing and secreting large numbers of antibodies into the interstitial fluid. These antibodies have the same target as the antibodies on the surface of the sensitized B cell. When stimulated by cytokines from helper T cells, a plasma cell can secrete up to 100 million antibody molecules each hour.

Memory B cells perform the same role for antibody-mediated immunity that memory T cells perform for cell-mediated immunity. Memory B cells do not respond to a threat on first exposure. Instead, they remain in reserve to deal with subsequent injuries or infections that involve the same antigens. On subsequent exposure, the memory B cells respond by dividing and differentiating into plasma cells that secrete antibodies in massive quantities.

Antibody Structure

Figure 22-20• presents different views of a single antibody molecule. An antibody consists of two parallel pairs of polypeptide chains: one pair of **heavy chains** and one pair of **light chains.** Each chain contains both *constant* and *variable segments.*

The Constant Segments The constant segments of the heavy chains form the base of the antibody molecule (Figure 22-20a,b•). B cells produce only five types of constant segments. These form the basis of a classification scheme that identifies antibodies as *IgG, IgE, IgD, IgM,* or *IgA,* as we shall discuss in the next section. The structure of the constant segments of the heavy chains determines the way the antibody is secreted and how it is distributed within the body. For ex-

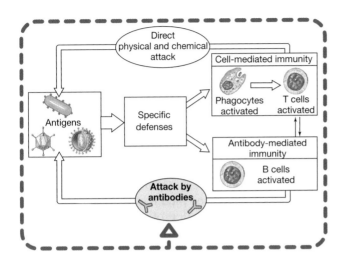

ample, antibodies in one class circulate in body fluids, whereas another type binds to the membranes of basophils and mast cells.

The heavy-chain constant segments, which are bound to constant segments of the light chains, also contain binding sites that can activate the complement system. These binding sites are covered when the antibody is secreted, but they become exposed when the antibody binds to an antigen.

The Variable Segments The specificity of the antibody molecule depends on the structure of the variable segments of the light and heavy chains. The free tips of the two variable segments contain the **antigen binding sites**

•**FIGURE 22-20 Antibody Structure. (a)** A diagrammatic view of antibody structure. **(b)** A computer-generated image of a typical antibody. **(c)** Antigen–antibody binding. **(d)** A hapten, or partial antigen, can become a complete antigen by binding to a carrier molecule.

of the antibody molecule (Figure 22-20a•). These sites can interact with an antigen in the same way that the active site of an enzyme interacts with a substrate molecule. ∞ *p. 54*

Small differences in the amino acid sequence of the variable segments affect the precise shape of the antigen binding site. These differences account for differences in specificity between the antibodies produced by different B cells. The distinctions are the result of minor genetic variations that occur during the production, division, and differentiation of B cells. The normal adult body contains roughly 10 trillion B cells. These cells can produce an estimated 100 million types of antibodies, each with a different specificity.

The antigen–antibody complex When an antibody molecule binds to its corresponding antigen molecule, an **antigen–antibody complex** is formed. Once the two molecules are in position, hydrogen bonding and other weak chemical forces lock them together.

Antibodies do not bind to the entire antigen. They bind to certain portions of its exposed surface, regions called **antigenic determinant sites** (Figure 22-20c•). The specificity of that binding depends initially on the three-dimensional "fit" between the variable segments of the antibody molecule and the corresponding antigenic determinant sites. A **complete antigen** is an antigen with at least two antigenic determinant sites, one for each of the antigen binding sites on an antibody molecule. Exposure to a complete antigen can lead to B cell sensitization and a subsequent immune response. Most environmental antigens have multiple antigenic determinant sites; entire microorganisms may have thousands.

Drug Reactions

Haptens, or *partial antigens,* do not ordinarily cause B cell activation and antibody production. Haptens include short peptide chains, steroids and other lipids, and several drugs, including antibiotics such as *penicillin.* Haptens may, however, become attached to carrier molecules, forming combinations that can function as complete antigens (Figure 22-20d•). In some cases, the carrier contributes an antigenic determinant site. The antibodies produced will attack both the hapten and the carrier molecule. If the carrier molecule is normally present in the tissues, the antibodies may begin attacking and destroying normal cells. This process is the basis for several drug reactions, including penicillin allergies.

Classes and Actions of Antibodies

Body fluids have five classes of antibodies, or **immunoglobulins (Igs):** *IgG, IgE, IgD, IgM,* and *IgA* (Table 22-1). The classes are determined by variations in the structure of the heavy-chain constant segments

and so have no effect on the antibody's specificity, which is determined by the antigen binding sites. The formation of an antigen–antibody complex may cause the elimination of the antigen in seven ways:

1. *Neutralization.* Both viruses and bacterial toxins have specific sites that must bind to target regions on body cells in order to enter or injure those cells. Antibodies may bind to those sites, making the virus or toxin incapable of attaching itself to a cell. This mechanism is known as **neutralization.**

2. *Agglutination and precipitation.* Each antibody molecule has two antigen binding sites, and most antigens have many antigenic determinant sites. If separate antigens (such as macromolecules or bacterial cells) are far apart, an antibody molecule will necessarily bind to two antigenic sites on the same antigen. If antigens are close together, however, an antibody can bind to antigenic determinant sites on two different antigens. In this way, antibodies can form "bridges" that tie large numbers of antigens together. The three-dimensional structure created in this way is known as an **immune complex.** When the antigen is a soluble molecule, such as a toxin, this process may create complexes too large to remain in solution. The formation of insoluble immune complexes is called **precipitation.** When the target antigen is on the surface of a cell or virus, the formation of large complexes is called **agglutination.** The clumping of red blood cells that occurs when incompatible blood types are mixed is an agglutination reaction. ∞ *p. 636*

3. *The activation of complement.* On binding to an antigen, portions of the antibody molecule change shape, exposing areas that bind complement proteins. The bound complement molecules then activate the complement system, destroying the antigen (as discussed previously).

4. *The attraction of phagocytes.* Antigens covered with antibodies attract eosinophils, neutrophils, and macrophages—cells that phagocytize pathogens and destroy foreign or abnormal cell membranes.

5. *Opsonization.* A coating of antibodies and complement proteins increases the effectiveness of phagocytosis. Recall from page 769 that this effect is called opsonization. Microorganisms such as bacteria have slick cell membranes or capsules, and a phagocyte must be able to hang onto its prey before it can engulf the prey. Phagocytes can bind more easily to antibodies and complement proteins on the surface of a pathogen than they can to the bare surface.

6. *The stimulation of inflammation.* Antibodies may promote inflammation through the stimulation of basophils and mast cells.

7. *The prevention of bacterial and viral adhesion.* Antibodies dissolved in saliva, mucus, and perspiration coat epithelia and provide an additional layer of defense. A covering of antibodies makes it difficult for pathogens to attach to body surfaces and penetrate their defenses.

TABLE 22-1 Classes of Antibodies

Structure	Description
IgG	**IgG** is the largest and most diverse class of antibodies. There are several types of IgG, but each type occurs as an individual molecule. Together they account for 80 percent of all antibodies. The IgG antibodies are responsible for resistance against many viruses, bacteria, and bacterial toxins. These antibodies can cross the placenta, and during embryological development, maternal IgG provides passive immunity to the fetus. However, the anti-Rh (anti-D) antibodies produced by Rh-negative mothers sensitized to Rh surface antigens are also IgG antibodies. These antibodies can cross the placenta and attack fetal Rh-positive red blood cells, producing the *hemolytic disease of the newborn.* ∞ *pp. 638–639*
IgE	**IgE** attaches as an individual molecule to the exposed surfaces of basophils and mast cells. When a suitable antigen appears and binds to the IgE molecules, the cell is stimulated to release histamine and other chemicals that accelerate inflammation in the immediate area. IgE is also important in the allergic response.
IgD	**IgD** is an individual molecule on the surfaces of B cells, where it can bind antigens in the extracellular fluid. This binding can play a role in the activation of the B cell involved.
IgM	**IgM** is the first antibody type secreted after an antigen arrives. The concentration of IgM then declines as IgG production accelerates. Although plasma cells secrete individual IgM molecules, IgM circulates as a five-antibody starburst. This configuration makes these antibodies particularly effective in forming immune complexes. The anti-A and anti-B antibodies responsible for the agglutination of cross-matched blood are IgM antibodies. ∞ *p. 636* IgM antibodies may also attack bacteria that are insensitive to IgG.
IgA (Secretory piece)	**IgA** is found primarily in glandular secretions such as mucus, tears, and saliva. These antibodies attack pathogens before they gain access to internal tissues. IgA antibodies circulate in blood as individual molecules or in pairs. Epithelial cells absorb them from the blood and attach a *secretory piece,* which confers solubility, before secreting the IgA molecules onto the epithelial surface.

22

☑ How can the presence of an abnormal peptide in the cytoplasm of a cell initiate an immune response?

☑ A decrease in the number of cytotoxic T cells would affect which type of immunity?

☑ How would a lack of helper T cells affect the antibody-mediated immune response?

☑ A sample of lymph contains an elevated number of plasma cells. Would you expect the amount of antibodies in the blood to be increasing or decreasing? Why?

PRIMARY AND SECONDARY RESPONSES TO ANTIGEN EXPOSURE

The initial response to antigen exposure is called the **primary response.** When an antigen appears again, it triggers a more extensive and prolonged **secondary response.** The secondary response reflects the presence of large numbers of memory cells that are primed for the arrival of the antigen. Primary and secondary responses are characteristic of both cell-mediated and antibody-mediated immunities. The differences between these responses are most easily demonstrated by following the production of antibodies over time.

The Primary Response

Because the antigen must activate the appropriate B cells, which must then differentiate into plasma cells, the primary response does not appear immediately (Figure 22-21a•). As plasma cells differentiate and begin secreting, the concentration of circulating antibodies undergoes a gradual, sustained rise.

During the primary response, the **antibody titer** ("standard"), or level of antibody activity, in the plasma does not peak until 1–2 weeks after the initial

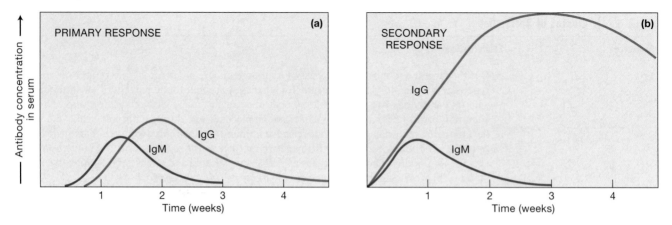

•**FIGURE 22-21** **The Primary and Secondary Immune Responses.** **(a)** The primary response takes about 2 weeks to develop peak antibody titers, and IgM and IgG antibody concentrations do not remain elevated. **(b)** The secondary response is characterized by a very rapid increase in IgG antibody titer, to levels much higher than those of the primary response. Antibody activity remains elevated for an extended period after the second exposure to the antigen.

exposure. If the individual is no longer exposed to the antigen, the antibody concentration thereafter declines. This reduction in antibody production occurs because (1) plasma cells have very high metabolic rates and survive for only a short time, and (2) further production of plasma cells is inhibited by suppression factors released by suppressor T cells. However, suppressor T cell activity does not begin immediately after antigen exposure, and under normal conditions helper cells outnumber suppressors by more than 3 to 1. As a result, many B cells are activated before suppressor T cell activity has a noticeable effect.

Two types of antibodies are involved in the primary response. Molecules of *immunoglobulin M, or IgM,* are the first to appear in the bloodstream. IgM is secreted by the plasma cells that form immediately after B cell activation. These lymphocytes do not pause to produce memory cells. Levels of *immunoglobulin G, or IgG,* rise more slowly, because the stimulated lymphocytes undergo repeated cell divisions and generate large numbers of memory cells as well as plasma cells. In effect, IgM provides an immediate but limited defense that fights the infection until massive quantities of IgG can be produced.

The Secondary Response

Unless they are exposed to the same antigen a second time, memory B cells do not differentiate into plasma cells. If and when that exposure occurs, the memory B cells respond immediately—faster than the B cells stimulated at initial exposure. This response is immediate in part because memory B cells are activated at relatively low antigen concentrations and in part because they synthesize more effective and destructive antibodies. Activated memory B cells divide and differentiate into plasma cells that secrete these anti-

bodies in massive quantities. This antibody secretion is the secondary response to antigen exposure.

During the secondary response, antibody titers increase more rapidly and reach levels many times higher than they did in the primary response (Figure 22-21b•). The secondary response appears even if the second exposure occurs years after the first, because memory cells may survive for 20 years or more.

The primary response to an antigen is not as effective a defense as the secondary response. The primary response develops slowly, and the antibodies are not produced in massive quantities. As a result, the primary response may not prevent an infection, whereas the secondary response is more than adequate. Therefore a person who survives the first infection can easily overcome a subsequent invasion by the same pathogen. The invading organisms will typically be destroyed before they have produced any disease symptoms. The effectiveness of the secondary response is one of the basic principles behind the use of immunization to prevent disease. AM *Immunization*

HORMONES OF THE IMMUNE SYSTEM

The specific and nonspecific defenses of the body are coordinated by physical interaction and by the release of chemical messengers. One example of physical interaction is antigen presentation by activated macrophages and helper T cells. An example of chemical messenger release is the release of cytokines by many of the cells involved in the immune response. Cytokines are classified according to their origins: *Lymphokines* are secreted by lymphocytes, and *monokines* by active macrophages and other antigen-presenting cells. These terms are misleading, however, because lymphocytes and macrophages may secrete the same

TABLE 22-2 Chemical Mediators (Cytokines) of the Immune Response

Compound	Functions
Interleukins	
Interleukin-1 (IL-1)	Stimulates T cells to produce IL-2, promotes inflammation; causes fever
IL-2	Stimulates growth and activation of other T cells and NK cells
IL-3	Stimulates production of mast cells and other blood cells
IL-4 (B cell differentiating factor); IL-5 (B cell growth factor); IL-6, IL-7 (B cell stimulating factors); IL-10; IL-11	Promote differentiation and growth of B cells and stimulate plasma cell formation and antibody production; each has somewhat different effects on macrophage and microphage activities
IL-8	Stimulates blood vessel formation (angiogenesis)
IL-9	Stimulates myeloid cell production (RBCs, platelets, granulocytes, monocytes)
IL-12	Stimulates T cell activity and cell-mediated immunity
IL-13	Suppresses production of several other cytokines (IL-1, IL-8, TNF); stimulates Class II MHC antigen presentation
Interferons (alpha, beta, gamma)	Activate other cells to prevent viral entry; inhibit viral replication; stimulate NK cells and macrophages
Tumor necrosis factors (TNFs)	Kill tumor cells; slow tumor growth; stimulate activities of T cells and eosinophils; inhibit parasites and viruses
Monocyte-chemotactic factor (MCF)	Attracts monocytes; activates them to macrophages
Migration-inhibitory factor (MIF)	Prevents macrophage migration from the area
Macrophage-activating factor (MAF)	Makes macrophages more active and aggressive
Microphage-chemotactic factor	Attracts microphages from the blood
Colony-stimulating factors (CSFs)	Stimulate RBC and WBC production
Defensins	Kill cells by piercing cell membrane
Growth-inhibitory factor (GIF)	Reduces or inhibits replication of target cells
Hemopoiesis-stimulating factor	Promotes blood cell production in bone marrow
Leukotrienes	Stimulate regional inflammation
Lymphotoxins	Kill cells; damage tissue; promote inflammation
Perforin	Destroys cell membranes by creating large pores
Transforming growth factor-β (TGF-β)	Stimulates production of IgA and of matrix proteins in connective tissues; inhibits macrophage activation and T_C maturation
Suppression factors	Inhibit T_C cell and B cell activity; depress immune response
Transfer factor	Sensitizes other T cells to same antigen

22

cytokines, and cytokines can also be secreted by cells involved with nonspecific defenses and tissue repair.

Table 22-2 summarizes the cytokines that have been identified to date. Five subgroups merit special attention: (1) *interleukins,* (2) interferons, (3) *tumor necrosis factors,* (4) chemicals that regulate phagocytic activities, and (5) *colony-stimulating factors.*

Interleukins

Interleukins may be the most diverse and important chemical messengers in the immune system. Nearly 20 types of interleukins have been identified; several are given in Table 22-2. Lymphocytes and macrophages are the primary sources of interleukins, but specific interleukins, such as interleukin-1 (IL-1), are also produced by endothelial cells, fibroblasts, and astrocytes. Interleukins have the following general functions:

1. *Increasing T cell sensitivity to antigens exposed on macrophage membranes.* Heightened sensitivity accelerates the production of cytotoxic and regulatory T cells.

2. *Stimulating B cell activity, plasma cell formation, and antibody production.* These events promote the production of antibodies and the development of antibody-mediated immunity.

3. *Enhancing nonspecific defenses.* Known effects of interleukin production include (1) the stimulation of inflammation, (2) the formation of scar tissue by fibroblasts, (3) the elevation of body temperature via the preoptic nucleus of the central nervous system, (4) the stimulation of mast cell formation, and (5) the promotion of adrenocorticotropic hormone (ACTH) secretion by the anterior pituitary gland.

4. *Moderating the immune response.* Some interleukins help suppress immune function and shorten the duration of an immune response.

Two interleukins, IL-1 and IL-2, are important in the stimulation and maintenance of the immune response. When released by activated macrophages and lymphocytes, these cytokines both stimulate the activities of other immune cells and further stimulate the secreting cell. The result is a positive feedback loop that promotes the recruitment of additional immune cells. Although mechanisms exist to control the degree of stimulation, the regulatory process sometimes breaks down, and massive production of interleukins can cause problems at least as severe as those of the primary infection. For example, in *Lyme disease* the release of IL-1 by activated macrophages in response to a localized bacterial infection produces symptoms of fever, pain, skin rash, and arthritis that affect the entire body. [AM] *Lyme Disease*

Some interleukins enhance the immune response, whereas others suppress it. The relative quantities secreted at any moment therefore have a significant effect on the nature and intensity of the response to an antigen. In the course of a typical infection, the pattern of interleukin secretion is constantly changing. Whether the individual succeeds in overcoming the infection is determined in part by whether stimulatory or suppressive interleukins predominate. As a result, interleukins and their interactions are now the focus of an intensive research effort.

Interferons

Interferons make the cell that synthesizes them, and that cell's neighbors, resistant to viral infection, thereby slowing the spread of the virus. These compounds may have other beneficial effects in addition to their antiviral activity. For example, alpha-interferons and gamma-interferons attract and stimulate NK cells, and beta-interferons slow the progress of inflammation associated with viral infection. Gamma-interferons also stimulate macrophages, making them more effective at killing bacterial or fungal pathogens.

Because they stimulate NK cell activity, interferons can be used to fight some cancers. For example, alpha-interferons have been used in the treatment of malignant melanoma, bladder cancer, ovarian cancer, and two forms of leukemia. Alpha- or gamma-interferons may be used to treat Kaposi's sarcoma, a cancer that typically develops in individuals with AIDS. [AM] *AIDS*

Tumor Necrosis Factors

Tumor necrosis factors (TNFs) slow tumor growth and kill sensitive tumor cells. Activated macrophages secrete one type of TNF and carry the molecules in their cell membranes. Cytotoxic T cells produce a different type of TNF. In addition to their effects on tumor cells, tumor necrosis factors stimulate granular leukocyte production, promote eosinophil activity, cause fever, and increase T cell sensitivity to interleukins.

Manipulating the Immune Response

As our understanding of the immune system grows, complex therapies involving combinations of cytokines, including interleukins, and *monoclonal antibodies*[3] are appearing. In one procedure, cytotoxic T cells were removed from patients with *malignant melanoma,* a particularly dangerous type of skin cancer. These lymphocytes were able to recognize tumor cells and had migrated to the tumor, but for some reason they appeared to be unable to kill the tumor cells.

The extracted lymphocytes were cultured, and viruses were used to insert multiple copies of the genes responsible for the production of tumor necrosis factor. The patients were then given periodic infusions of these "supercharged" T cells. To enhance T cell activity further, the researchers also administered doses of interleukin-2. Initial results are promising. It is clear that the ability to manipulate the immune response will revolutionize the treatment of many serious diseases.

Chemicals Regulating Phagocytic Activities

Several cytokines coordinate immune defenses by adjusting the activities of phagocytic cells. These cytokines include factors that attract free macrophages and microphages and prevent their premature departure.

Colony-Stimulating Factors

We introduced **colony-stimulating factors (CSFs)** in Chapter 19. ∞ *p. 644* These factors are produced by active T cells, cells of the monocyte–macrophage group, endothelial cells, and fibroblasts. CSFs stimulate the production of blood cells in bone marrow and lymphocytes in lymphoid tissues and organs.

SUMMARY: THE IMMUNE RESPONSE

We have now described the basic chemical and cellular interactions that follow the appearance of a foreign antigen. Figure 22-22• presents an integrated view of the immune response and its relationship to nonspecific defenses. Figure 22-23• provides an overview of the time course of the events responsible for overcoming bacterial infection.

In the early stages of bacterial infection, before antigen processing has occurred, neutrophils and NK cells migrate into the threatened area and destroy bacteria. Over time, cytokines draw increasing numbers of phagocytes to the region. Cytotoxic T cells appear as arriving T cells are activated by antigen presentation (Figure 22-23•). Last of all, the population of plasma cells rises as activated B cells differentiate. This rise is followed by a gradual, sustained increase in the level of circulating antibodies.

---

[3]Monoclonal antibodies are produced by a single clone of B cells under laboratory conditions. [AM] *Technology, Immunity, and Disease*

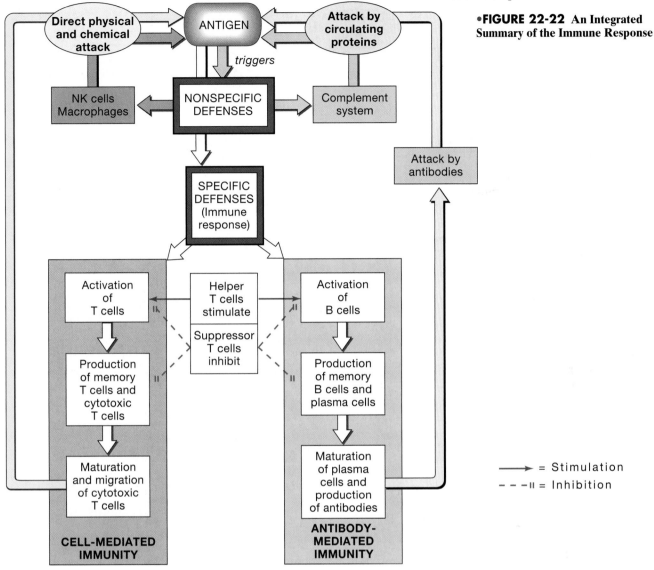

•**FIGURE 22-22** An Integrated Summary of the Immune Response

→ = Stimulation

-- ‖ = Inhibition

22

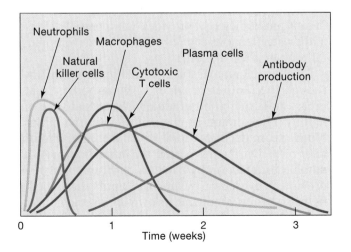

•**FIGURE 22-23 The Time Course of the Body's Response to Bacterial Infection.** An outline of the basic sequence of events that begins with the appearance of bacteria in peripheral tissues.

The basic sequence of events is very similar when a viral infection occurs. The initial steps are different, however, because cytotoxic T cells and NK cells can be activated by contact with infected cells. Figure 22-24• contrasts the steps involved in defense against bacterial infection with that against viral infection. Table 22-3 (p. 787) reviews the cells involved.

THE DEVELOPMENT OF RESISTANCE

The ability to produce an immune response after exposure to an antigen is **immunological competence.** Cell-mediated immunity can be demonstrated as early as the third month of fetal development, but active antibody-mediated immunity appears later.

The first cells that leave the fetal thymus migrate to the skin and into the epithelia lining the mouth, digestive tract, and, in females, the uterus and vagina. These

•**FIGURE 22-24** **Defenses against Bacterial and Viral Pathogens.** **(a)** Defenses against bacteria are usually initiated by active macrophages. **(b)** Defenses against viruses are usually activated after the infection of normal cells.

22

cells take up residence in these tissues as antigen-presenting cells, such as the Langerhans cells of the skin, whose primary function will be the activation of T cells. T cells that leave the thymus later in development populate lymphoid organs throughout the body.

The membranes of the first B cells to be produced in the liver and bone marrow carry IgM antibodies. Sometime after the fourth developmental month, the fetus may, if exposed to specific pathogens, produce IgM antibodies. Fetal antibody production is uncommon, however, because the developing fetus has natural passive immunity due to the transfer of IgG antibodies from the maternal bloodstream. These are the only antibodies that can cross the placenta, and they include the antibodies responsible for the clinical problems that accompany fetal–maternal Rh incompatibility, discussed in Chapter 19. ∞ *pp. 638–639* Because the anti-A and anti-B antibodies are IgM antibodies, and IgM cannot cross the placenta, problems with maternal–fetal incompatibilities involving the ABO blood groups rarely occur.

The natural immunity provided by maternal IgG may not be enough to protect the fetus if the maternal

defenses are overrun by a bacterial or viral infection. For example, the microorganisms responsible for syphilis and rubella ("German measles") can cross from the maternal to the fetal bloodstream, producing a congenital infection that leads to the production of fetal antibodies. IgM provides only a partial defense, and these infections can result in severe developmental problems for the fetus. AM *Fetal Infections*

Delivery eliminates the maternal supply of IgG. Although the mother provides IgA antibodies in her breast milk, the infant gradually loses its passive immunity. The amount of maternal IgG in the infant's bloodstream declines rapidly over the first 2 months after birth. During this period, the infant becomes vulnerable to infection by bacteria or viruses that were previously overcome by maternal antibodies. The infant also begins producing its own IgG, as the immune system begins to respond to infections, environmental antigens, and vaccinations. It has been estimated that, from birth to age 12, children encounter a "new" antigen every 6 weeks. (This fact explains why most parents, exposed to the same antigens when they were children, remain healthy while their children develop

TABLE 22-3 Cells That Participate in Tissue Defenses

Cell	Functions
Neutrophils	Phagocytosis; stimulation of inflammation
Eosinophils	Phagocytosis of antigen–antibody complexes; suppression of inflammation; involved in allergic response
Mast cells and basophils	Stimulation and coordination of inflammation by release of histamine, heparin, leukotrienes, prostaglandins
Antigen-Presenting Cells	
Macrophages (free and fixed macrophages, Kupffer cells, microglia, etc.)	Phagocytosis; antigen processing; antigen presentation with Class II MHC proteins; secretion of cytokines, especially interleukins and interferons
Dendritic cells, Langerhans cells	Antigen presentation bound to Class II MHC proteins
Lymphocytes	
NK cells	Destruction of cell membranes containing abnormal antigens
Cytotoxic T cells (T_C, CD8 marker)	Lysis of cell membranes containing antigens bound to Class I MHC proteins; secretion of perforins, defensins, lymphotoxins, and other cytokines
Helper T cells (T_H, CD4 marker)	Secretion of cytokines that stimulate cell-mediated and antibody-mediated immunity; activation of sensitized B cells
B cells	Differentiation into plasma cells that secrete antibodies and provide antibody-mediated immunity
Suppressor T cells (T_S, CD8 marker)	Secretion of suppression factors that inhibit the immune response
Memory cells (T_C, T_H, B)	Produced during the activation of T cells and B cells; remain in tissues awaiting rearrival of antigen

runny noses and colds.) Over this period, the concentration of circulating antibodies gradually rises toward normal adult levels and the populations of memory B cells and T cells continue to increase.

Skin tests can be used to determine whether an individual has developed resistance to a particular antigen. In this procedure, small quantities of antigen are injected into the skin, generally on the anterior surface of the forearm. If resistance exists, the region will become inflamed over the next 2–4 days. Many states require a tuberculosis test, called a *tuberculin skin test,* when someone applies for a food-service or health-service job. (When preparing or serving food or com-

ing in contact with patients, anyone with tuberculosis can accidentally transmit the bacteria.) If the test is positive, the individual has been exposed to the disease and has developed antibodies. Further tests must be performed to determine whether an infection is under way. Skin tests are also used to check for allergies to environmental antigens. [AM] *Delayed Hypersensitivity and Skin Tests*

IMMUNE DISORDERS

Because the immune response is so complex, many opportunities exist for things to go wrong. A variety of clinical conditions result from disorders of immune function. **Autoimmune disorders** develop when the immune response mistakenly targets normal body cells and tissues. In an **immunodeficiency disease,** either the immune system fails to develop normally or the immune response is blocked in some way. Autoimmune disorders and immunodeficiency diseases are relatively rare—clear evidence of the effectiveness of the immune system's control mechanisms. A far more common, and generally far less dangerous, class of immune disorders is the **allergies.** We shall consider examples of each type of disorder. [AM] *Immune Complex Disorders; Systemic Lupus Erythematosus*

AUTOIMMUNE DISORDERS

Autoimmune disorders affect an estimated 5 percent of adults in North America and Europe. We noted many examples of autoimmune disorders in previous chapters in discussions of their effects on the function of major systems. [AM] *Autoimmune Disorders*

The immune system usually recognizes and ignores the antigens normally found in the body—self antigens. The recognition system can malfunction, however. When it does, the activated B cells make antibodies against other body cells and tissues. These misguided antibodies are called **autoantibodies.** The trigger may be a reduction in suppressor T cell activity, the excessive stimulation of helper T cells, tissue damage that releases large quantities of antigenic fragments, haptens bound to compounds normally ignored, viral or bacterial toxins, or a combination of factors.

The symptoms produced depend on the specific antigen attacked by these autoantibodies. For example:

- The inflammation of *thyroiditis* results from the release of autoantibodies against thyroglobulin.
- *Rheumatoid arthritis* occurs when autoantibodies form immune complexes within connective tissues, especially around the joints.
- *Insulin-dependent diabetes mellitus* (IDDM) is generally caused by autoantibodies that attack cells in the pancreatic islets.

Many autoimmune disorders appear to be cases of mistaken identity. For example, proteins associated with the measles, Epstein–Barr, influenza, and other viruses contain amino acid sequences that are similar to those of myelin proteins. As a result, antibodies that target these viruses may also attack myelin sheaths. This mechanism accounts for the neurological complications that sometimes follow a vaccination or a viral infection. It is also the probable mechanism responsible for *multiple sclerosis.*

For unknown reasons, the risk of autoimmune problems increases if an individual has an unusual type of MHC proteins. At least 50 clinical conditions have been linked to specific variations in MHC structure.

IMMUNODEFICIENCY DISEASES

Immunodeficiency diseases result from (1) problems with the embryological development of lymphoid organs and tissues; (2) an infection with a virus, such as HIV, that depresses immune function; or (3) treatment with or exposure to immunosuppressive agents, such as radiation or drugs.

Individuals with **severe combined immunodeficiency disease (SCID)** fail to develop either cellular or antibody-mediated immunity. Lymphocyte populations are low, and normal B and T cells are absent. Such individuals cannot produce an immune response, and even a mild infection can prove fatal. Total isolation offers protection at great cost, with severe restrictions on lifestyle. Bone marrow transplants from compatible donors, normally a close relative, have been used to colonize lymphoid tissues with functional lymphocytes. Gene-splicing techniques have led to therapies that can treat at least one form of SCID. [AM] *Genetic Engineering and Gene Therapy*

AIDS, an immunodeficiency disease that we consider on page 790, is the result of a viral infection that targets primarily helper T cells. As the number of T cells declines, the normal immune control mechanism breaks down. When an infection occurs, suppressor factors released by suppressor T cells inhibit an immune response before the few surviving helper T cells can stimulate the formation of cytotoxic T cells or plasma cells in adequate numbers.

Immunosuppressive drugs have been used for many years to prevent graft rejection after transplant surgery. But immunosuppressive agents can destroy stem cells and lymphocytes, leading to a complete immunological failure. This outcome is one of the potentially fatal consequences of radiation exposure.

ALLERGIES

Allergies are inappropriate or excessive immune responses to antigens. The sudden increase in cellular activity or antibody titers can have a number of unpleasant side effects. For example, neutrophils or cytotoxic T cells may destroy normal cells while attacking the antigen, or the antigen–antibody complex may trigger a massive inflammatory response. Antigens that trigger allergic reactions are often called **allergens.**

There are several types of allergies. A complete classification recognizes four categories: *immediate hypersensitivity (Type I), cytotoxic reactions (Type II), immune complex disorders (Type III),* and *delayed hypersensitivity (Type IV).* We shall consider immediate hypersensitivity at this time, as it is probably the most common form of allergy. One form, *allergic rhinitis,* includes "hay fever" and environmental allergies that may affect 15 percent of the U.S. population. We discussed one example of a cytotoxic (Type II) reaction, the cross-reactions that follow the transfusion of an incompatible blood type, in Chapter 19. ∞ *p. 636* Other types of allergies are discussed in the *Applications Manual.* [AM] *Immune Complex Disorders; Delayed Hypersensitivity and Skin Tests*

Immediate Hypersensitivity

Immediate hypersensitivity begins with the process of sensitization. Sensitization is the initial exposure to an allergen that leads to the production of antibodies, specifically large quantities of IgE. The tendency to produce IgE antibodies in response to specific allergens may be genetically determined. In drug reactions, such as penicillin allergies, IgE antibodies are produced in response to a hapten (partial antigen) bound to a larger molecule inside the body.

Due to the lag time needed to activate B cells, produce plasma cells, and synthesize antibodies, the first exposure to an allergen does not produce symptoms. It merely sets the stage for the next encounter. After sensitization, the IgE antibodies become attached to the cell membranes of basophils and mast cells throughout the body. Later, when exposed to the same allergen, the bound antibodies stimulate these cells to release histamine, heparin, several cytokines, prostaglandins, and other chemicals into the surrounding tissues. A sudden, massive inflammation of the affected tissues results.

The cytokines and other mast cell secretions draw basophils, eosinophils, T cells, and macrophages to the area. These cells release their own chemicals, extending and exaggerating the responses initiated by mast cells.

The severity of the allergic reaction depends on the individual's sensitivity and on the location involved. If allergen exposure occurs at the body surface, the response may be restricted to that area. If the allergen enters the bloodstream, the response could be lethal.

Anaphylaxis In **anaphylaxis** (a-na-fi-LAK-sis; *ana-,* again + *phylaxis,* protection), a circulating allergen affects mast cells throughout the body (Figure 22-25•). The entire range of symptoms can develop within minutes. Changes in capillary permeabilities produce

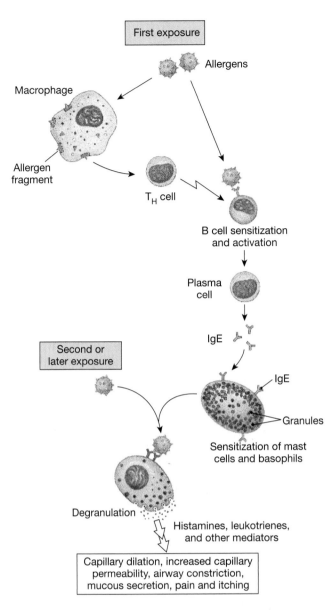

First exposure

Allergens

Macrophage

Allergen fragment

T$_H$ cell

B cell sensitization and activation

Plasma cell

IgE

Second or later exposure

IgE

Granules

Sensitization of mast cells and basophils

Degranulation

Histamines, leukotrienes, and other mediators

Capillary dilation, increased capillary permeability, airway constriction, mucous secretion, pain and itching

•**FIGURE 22-25** The Mechanism of Anaphylaxis

swelling and edema in the dermis, and raised welts, or *hives,* appear on the skin surface. Smooth muscles along the respiratory passageways contract; the narrowed passages make breathing extremely difficult. In severe cases, an extensive peripheral vasodilation occurs, producing a fall in blood pressure that can lead to a circulatory collapse. This response is **anaphylactic shock.**

Treatment of Anaphylaxis

Many of the symptoms of immediate hypersensitivity can be prevented by the prompt administration of **antihistamines** (an-tē-HIS-ta-mēnz), drugs that block the action of histamine. *Benadryl® (diphenhydramine hydrochloride)* is a popular antihistamine available over the counter. The treatment of severe anaphylaxis involves antihistamine, corticosteroid, and epinephrine injections.

STRESS AND THE IMMUNE RESPONSE

One of the normal effects of interleukin-1 secretion is the stimulation of adrenocorticotropic hormone (ACTH) production by the anterior pituitary gland. The production of ACTH in turn leads to the secretion of glucocorticoids by the adrenal cortex. ∞ *p. 602* The anti-inflammatory effects of the glucocorticoids may help control the size of the immune response. However, the long-term secretion of glucocorticoids, as in the resistance phase of the *general adaptation syndrome,* can inhibit the immune response and lower resistance to disease. ∞ *p. 612* Several of the tissue responses to glucocorticoids alter the effectiveness of specific and nonspecific defenses. Examples include the following:

▪ *The depression of the inflammatory response.* Glucocorticoids inhibit mast cells and reduce the permeability of capillaries. Inflammation is therefore less likely. When inflammation does occur, the reduced permeability of the capillaries slows the entry of fibrinogen, complement proteins, and cellular defenders.

▪ *A reduction in the activities and numbers of phagocytes in peripheral tissues.* This reduction further impairs nonspecific defense mechanisms and interferes with the processing and presentation of antigens to lymphocytes.

▪ *The inhibition of interleukin secretion.* A reduction in interleukin production depresses the response of lymphocytes, even to antigens bound to MHC proteins.

The mechanisms responsible for these changes are still under investigation. It is well known, however, that immune system depression due to chronic stress can be a serious threat to health.

AGING AND THE IMMUNE RESPONSE

With advancing age, your immune system becomes less effective at combating disease. T cells become less responsive to antigens; as a result, fewer cytotoxic T cells respond to an infection. This effect may, at least in part, be associated with the gradual involution of the thymus and a reduction in circulating levels of thymic hormones. Because the number of helper T cells is also reduced, B cells are less responsive, so antibody levels do not rise as quickly after antigen exposure. The net result is an increased susceptibility to viral and bacterial infection. For this reason, vaccinations for acute viral diseases, such as the flu (influenza), are strongly recommended for elderly individuals. The increased incidence of cancer in the elderly reflects the fact that immune surveillance declines, so tumor cells are not eliminated as effectively.

22

AIDS

Acquired immune deficiency syndrome (AIDS), or *late-stage HIV disease,* is caused by the **human immunodeficiency virus (HIV).** This virus is a *retrovirus,* which carries its genetic information in RNA rather than in DNA. The virus enters the cell by receptor-mediated endocytosis. ∞ *p. 77* In the body, the virus binds to CD4, the membrane protein characteristic of helper T cells. Several types of antigen-presenting cells, including those of the monocyte–macrophage line, are also infected by HIV, but it is the infection of helper T cells that leads to clinical problems.

Once the virus is inside a cell, the viral enzyme *reverse transcriptase* synthesizes a complementary strand of DNA, which is then incorporated into the cell's genetic material. When these inserted viral genes are activated, the infected cell begins synthesizing viral proteins. In effect, the introduced viral genes take over the cell's synthetic machinery and force the cell to produce additional viruses. These new viruses are then shed at the cell surface. AM *A Closer Look: The Nature of Pathogens*

Cells infected with HIV are ultimately killed. Several mechanisms may be responsible for cell destruction, including (1) the formation of pores in the cell membrane as the viruses are shed, (2) the cessation of cell maintenance due to the continuing synthesis of viral components, (3) autolysis, and (4) the stimulation of apoptosis.

The gradual destruction of helper T cells impairs the immune response, because these cells play a central role in coordinating cell-mediated and antibody-mediated responses to antigens. To make matters worse, suppressor T cells are relatively unaffected by the virus, and over time the excess of suppressing factors "turns off" the normal immune response. Circulating antibody levels decline, cell-mediated immunity is reduced, and the body is left without defenses against a wide variety of microbial invaders. With the affected person's immune function so reduced, ordinarily harmless pathogens can initiate lethal infections known as *opportunistic infections.* Because immune surveillance is also depressed, the risk of cancer increases.

Infection with HIV occurs by intimate contact with the body fluids of infected individuals. Although all body fluids carry the virus, the major routes of transmission involve contact with blood, semen, or vaginal secretions. Most individuals with AIDS become infected through sexual contact with an HIV-infected person (who may *not* necessarily be exhibiting the clinical symptoms of AIDS). The next largest group of infected persons consists of intravenous drug users who shared contaminated needles. Relatively few individuals have become infected with the virus after receiving a transfusion of contaminated blood or blood products. Finally, an increasing number of infants are born with AIDS, having acquired it from infected mothers.

AIDS is a public health problem of massive proportions. In 2000, an estimated 48,000 people in the United States will be diagnosed with AIDS. More than 423,000 people have died of AIDS in the United States alone. Estimates of the total number of individuals who are infected with HIV in the United States range from 650,000 to 900,000. The numbers worldwide are even more frightening: The World Health Organization estimates that 40 to 110 million individuals were infected when the year 2000 began. Every 15 seconds, another person becomes infected with the HIV virus.

The best defense against AIDS is to avoid sexual contact or sharing of needles with infected individuals. All forms of sexual intercourse carry the risk of viral transmission. The use of synthetic condoms greatly reduces (but does not eliminate) the chance of infection. Condoms that are not made of synthetic materials prevent pregnancy but do not block the passage of viruses.

Clinical symptoms of AIDS may not appear until 5–10 years or more after infection. When they do appear, they are commonly mild, consisting of lymphadenopathy and chronic but nonfatal infections. So far as is known, however, AIDS is almost always fatal, and most people who carry the virus will eventually die of the disease. (A handful of infected individuals have been able to tolerate the virus without apparent illness. For details, see the *Applications Manual.*)

Despite intensive efforts, a vaccine has yet to be developed that will provide immunity from HIV infection. While efforts to prevent the spread of HIV continue, the survival rate for AIDS patients has been steadily increasing, because new drugs are available that slow the progression of the disease and improved antibiotic therapies help combat secondary infections. This combination is extending the life span of patients while the search for more effective treatment continues. For more information on the distribution of HIV infection, current and future drug therapies, and additional details on HIV disease, consult the *Applications Manual.* AM *AIDS*

INTEGRATION WITH OTHER SYSTEMS

Figure 22-26• summarizes the interactions between the lymphatic system and other physiological systems. The relationships among the cells of the immune response and the nervous and endocrine systems are now the focus of intense research. It is apparent that there is substantial interaction among these systems.

For example, the cells of the immune system can influence central nervous system (CNS) and endocrine activity in the following ways:

- The thymus secretes oxytocin, antidiuretic hormone (ADH), and endorphins as well as the thymic hormones. The effects of these hormones on the CNS are not known, but the removal or destruction of the thymus leads to a decrease in brain endorphin levels.

- Thymic hormones as well as cytokines are involved in establishing the normal level of corticotropin-releasing hormone (CRH) produced by the hypothalamus. The removal or destruction of the thymus leads to a decrease in ACTH production and a decline in circulating adrenocortical hormones. Thymic hormones also stimulate the production of thyroid hormone–releasing hormone (TRH), leading to the secretion of thyroid-stimulating hormone (TSH) at the anterior pituitary gland. As a result, circulating thyroid hormone levels increase when an immune response is under way. The thyroid hormones further stimulate cell and tissue metabolism.

- Other thymic hormones affect the anterior pituitary gland directly, stimulating the secretion of prolactin and growth hormone.

Conversely, the nervous system can apparently adjust the sensitivity of the immune response. For example:

- The CNS innervates dendritic cells in the lymph nodes and spleen, Langerhans cells in the skin, and other antigen-presenting cells. The nerve endings release neurotransmitters that exaggerate and enhance local immune responses. For this reason, some skin conditions, such as *psoriasis,* worsen when a person is under stress.

- Neuroglia in the CNS produce cytokines that promote an immune response.

- A sudden decline in immune function can occur after even a brief period of emotional distress. Recent studies indicate that a heated argument can produce a steep reduction in circulating numbers of white blood cells over the following 24 hours.

The mechanisms involved and the functional significance of these interactions have yet to be determined. An explosive growth in our understanding of the immune response occurred during the last decade; that growth will continue in decades to come.

☑ Would the primary response or the secondary response be more affected by a lack of memory B cells for a particular antigen?

☑ Which kind of immunity protects a developing fetus, and how is that immunity produced?

☑ How does increased stress decrease the effectiveness of the immune response?

[EXPLORE] MediaLab

The lymphatic system is one of the most fascinating yet complicated systems in the body. To help you gain a better understanding of the physiology of immunity, use the resources available on both the CD-ROM and the Companion Website. The CD-ROM includes animations and tutorials; the website builds on those resources with text content and interactive problems as well as communication avenues for sharing ideas with your class. To complete the following Web Explorations, visit the MediaLab in Chapter 22 of your Companion Website.

Web Exploration 1

Estimated time for completion = 15 minutes

Lymphocytes travel from tissue to tissue and contact one another to initiate immune responses. Lymphocyte "traffic and homing"—traveling and knowing where to go—are current areas of research. To view the lymphatic system through the eyes of an artist, select the keyword **HOMING** on the Companion Website. Read this page and the next, and view the

linked images. Making rough drawings of lymphocytes, put together your own visual interpretation of lymphocyte actions. Incorporate information from the text, such as that in the section "T Cell Activation" (pp. 773–775), as well as that presented on the Web page.

Web Exploration 2

Estimated time for completion = 5 minutes

One of the most common ailments of the immune system is an allergic reaction. What are allergies? To test your knowledge, select the keyword **ALLERGIES** on the Companion Website. Answer the eight questions there, and keep track of your score. If you get any incorrect answers, jot down the correct answer and the provided explanation in your notes for further study. What did you learn from this quiz? If you wish, share your quiz score or questions about the material with your peers, using the communications tools on the Companion Website.

22

INTEGUMENTARY SYSTEM

Provides physical barriers to pathogen entry; Langerhans cells in epidermis and macrophages in dermis resist infection and present antigens to trigger immune response; mast cells trigger inflammation, mobilize cells of lymphatic system

Provides IgA for secretion onto integumentary surfaces

SKELETAL SYSTEM

Lymphocytes and other cells involved in the immune response are produced and stored in bone marrow

Assists in repair of bone after injuries; macrophages fuse to become osteoclasts

MUSCULAR SYSTEM

Protects superficial lymph nodes and the lymphatic vessels in the abdominopelvic cavity; muscle contractions help propel lymph along lymphatic vessels

Assists in repair after injuries

NERVOUS SYSTEM

Microglia present antigens that stimulate specific defenses; neuroglia secrete cytokines; innervation stimulates antigen-presenting cells

Cytokines affect hypothalamic production of CRH and TRH

ENDOCRINE SYSTEM

Glucocorticoids have anti-inflammatory effects; thymic hormones stimulate development and maturation of lymphocytes; many hormones affect immune function

Thymus secretes thymic hormones; cytokines affect cells throughout the body

CARDIOVASCULAR SYSTEM

Distributes WBCs; carries antibodies that attack pathogens; clotting response helps restrict spread of pathogens; granulocytes and lymphocytes produced in bone marrow

Fights infections of cardiovascular organs; returns tissue fluid to circulation

RESPIRATORY SYSTEM

Alveolar phagocytes present antigens and trigger specific defenses; provides O_2 required by lymphocytes and eliminates CO_2 generated during their metabolic activities

Tonsils protect against infection at entrance to respiratory tract

DIGESTIVE SYSTEM

Provides nutrients required by lymphatic tissues, digestive acids, and enzymes; provides nonspecific defense against pathogens

Tonsils and MALT defend against infection and toxins absorbed from tract; lymphatic vessels carry absorbed lipids to venous system

URINARY SYSTEM

Eliminates metabolic wastes generated by cellular activity; acid pH of urine provides nonspecific defense against urinary tract infection

THE LYMPHATIC SYSTEM

FOR ALL SYSTEMS

Provides specific defenses against infection; immune surveillance eliminates cancer cells; returns tissue fluid to circulation

Lysozymes and bactericidal chemicals in secretions provide nonspecific defense against reproductive tract infections

Provides IgA for secretion by epithelial glands

REPRODUCTIVE SYSTEM

22

•**FIGURE 22-26** Functional Relationships between the Lymphatic System and Other Systems

CHAPTER REVIEW

SELECTED CLINICAL TERMINOLOGY
Terms Discussed in This Chapter

acquired immune deficiency syndrome (AIDS): A disorder that develops following HIV infection, characterized by reduced circulating antibody levels and depressed cell-mediated immunity. *(p. 790 and AM)*

allergen: An antigen capable of triggering an allergic reaction. *(p. 788)*

allergy: An inappropriate or excessive immune response to antigens, triggered by the stimulation of mast cells bound to IgE. *(p. 787 and AM)*

anaphylactic shock: A drop in blood pressure that may lead to circulatory collapse, resulting from a severe case of anaphylaxis. *(p. 789)*

anaphylaxis (a-na-fi-LAK-sis): A type of allergy in which a circulating allergen affects mast cells throughout the body, producing numerous symptoms very quickly. *(p. 788)*

appendicitis: An infection and inflammation of the aggregate lymphoid nodules in the vermiform appendix. *(p. 759 and AM)*

autoimmune disorder: A disorder that develops when the immune response mistakenly targets normal body cells and tissues. *(p. 787)*

bacteria: Prokaryotic cells (cells lacking nuclei and other membranous organelles) that may be extracellular or intracellular pathogens. *(p. 752 and AM)*

filariasis: An infection by larval stages of a parasitic roundworm; the adult worms may scar and block lymphatic vessels, causing acute lymphedema, commonly affecting the external genitalia and lower limbs **(elephantiasis).** *(p. 756)*

fungi (singular, *fungus*): Eukaryotic organisms that absorb organic materials from the remains of dead cells; some fungi are pathogenic. *(p. 752 and AM)*

human immunodeficiency virus (HIV): The virus responsible for AIDS and related immunodeficiency disorders. *(p. 790 and AM)*

immune complex disorder: A disorder caused by the precipitation of immune complexes at sites such as the kidneys, where their presence disrupts normal tissue function. *(p. 788 and AM)*

immunodeficiency disease: A disease in which either the immune system fails to develop normally or the immune response is blocked. *(p. 787)*

immunosuppression: A reduction in the sensitivity of the immune system. *(p. 777)*

immunosuppressive drugs: Drugs administered to inhibit the immune response; examples include prednisone, cyclophosphamide, azathioprine, cyclosporin, and FK506. *(p. 777 and AM)*

lymphadenopathy (lim-fad-e-NOP-a-thē): A chronic or excessive enlargement of lymph nodes. *(p. 760)*

lymphedema: A painless accumulation of lymph in a region whose lymphatic drainage has been blocked. *(p. 756)*

lymphomas: Malignant cancers consisting of abnormal lymphocytes or lymphoid stem cells; examples include *Hodgkin's disease* and *non-Hodgkin's lymphoma.* *(p. 760 and AM)*

severe combined immunodeficiency disease (SCID): A congenital disorder resulting from the failure to develop cell-mediated or antibody-mediated immunity. *(p. 788 and AM)*

tonsillitis: An infection of one or more tonsils; symptoms include a sore throat, high fever, and leukocytosis (an abnormally high white blood cell count). *(p. 759 and AM)*

vaccine (vak-SĒN): A preparation of antigens derived from a specific pathogen; administered during *immunization,* or *vaccination.* *(p. 772)*

viruses: Noncellular pathogens that replicate by directing the synthesis of virus-specific proteins and nucleic acids inside tissue cells. *(p. 752 and AM)*

22

AM Additional Terms Discussed in the *Applications Manual*

active immunization: An immunization that intentionally stimulates a primary response to a pathogen before the pathogen is encountered in the environment.

bone marrow transplantation: The infusion of bone marrow from a compatible donor after the destruction of the host's marrow through radiation or chemotherapy; a treatment option for acute, late-stage lymphoma.

graft-versus-host disease (GVH): A condition that results when T cells in donor tissues, such as bone marrow, attack the tissues of the recipient.

hybridoma (hī-bri-DŌ-muh): A cell formed by the fusion of a cancer cell with a plasma cell; used to produce large amounts of a single antibody.

hypersplenism (hī-per-SPLĒN-izm): A condition caused by an overactive spleen; symptoms include anemia, leukopenia, and thrombocytopenia.

hyposplenism (hī-pō-SPLĒN-izm): A condition resulting from the absence or nonfunctional state of the spleen; there are few symptoms.

monoclonal (mo-nō-KLŌ-nal) **antibody:** An antibody produced in large quantities by a population of genetically identical cells.

mononucleosis: A condition resulting from chronic infection by the *Epstein–Barr virus (EBV);* symptoms include enlargement of the spleen, fever, sore throat, widespread swelling of lymph nodes, increased numbers of lymphocytes in the blood, and the presence of circulating antibodies to the virus.

passive immunization: An immunization in which the patient receives a dose of antibodies that will attack a pathogen immediately without involving the host's immune system.

protozoa: Unicellular eukaryotic organisms that are abundant in soil and water; some are pathogenic.

splenomegaly (splen-ō-MEG-a-lē): An enlargement of the spleen.

systemic lupus erythematosus (LOO-pus e-rith-ē-ma-TŌ-sis) **(SLE):** An autoimmune disorder resulting from a breakdown in the antigen recognition mechanism, leading to the production of antibodies that destroy healthy cells and tissues.

tonsillectomy: The removal of an inflamed tonsil.

STUDY OUTLINE

INTRODUCTION p. 752

1. The cells, tissues, and organs of the **lymphatic system** play a central role in the body's defenses against a variety of **pathogens,** or disease-causing organisms.

AN OVERVIEW OF THE LYMPHATIC SYSTEM p. 752

1. Lymphocytes, the primary cells of the lymphatic system, provide an **immune response** to specific threats to the body. **Immunity** is the ability to resist infection and disease through the activation of specific defenses.

ORGANIZATION OF THE LYMPHATIC SYSTEM p. 752

1. The lymphatic system includes a network of **lymphatic vessels,** or **lymphatics,** which carry **lymph** (a fluid similar to plasma but with a lower concentration of proteins). A series of **lymphoid organs** is connected to the lymphatic vessels. *(Figure 22-1)*

Functions of the Lymphatic System p. 752

2. The lymphatic system produces, maintains, and distributes lymphocytes (which attack invading organisms, abnormal cells, and foreign proteins). This system also helps maintain blood volume and eliminate local variations in the composition of interstitial fluid.

Lymphatic Vessels p. 752

3. Lymph flows along a network of lymphatic vessels, the smallest of which are the **lymphatic capillaries** (*terminal lymphatics*). The lymphatic vessels empty into the **thoracic duct** and the **right lymphatic duct.** *(Figures 22-1–22-4)*

Lymphocytes p. 756

4. The three classes of lymphocytes are **T cells** (**t**hymus-dependent), **B cells** (**b**one marrow–derived), and **NK cells** (**n**atural **k**iller).

5. Cytotoxic T cells attack foreign cells or body cells infected by viruses and provide **cell-mediated (cellular) immunity.** *Regulatory T cells* (**helper T cells** and **suppressor T cells**) regulate and coordinate the immune response.

6. B cells can differentiate into **plasma cells,** which produce and secrete *antibodies* that react with specific chemical targets called **antigens.** Antibodies in body fluids are called *immunoglobulins.* B cells are responsible for **antibody-mediated (humoral) immunity.**

7. NK cells (also called **large granular lymphocytes**) attack foreign cells, normal cells infected with viruses, and cancer cells. NK cells provide **immunological surveillance.**

8. Lymphocytes continuously migrate into and out of the blood through the lymphoid tissues and organs. **Lymphopoiesis** (lymphocyte production) involves the bone marrow, thymus, and peripheral lymphoid tissues. *(Figure 22-5)*

Lymphoid Tissues p. 758

9. Lymphoid tissues are connective tissues dominated by lymphocytes. In a **lymphoid nodule,** the lymphocytes are densely packed in an area of loose connective tissue. *(Figure 22-6)*

Lymphoid Organs p. 759

10. Important lymphoid organs include the **lymph nodes,** the **thymus,** and the **spleen.** Lymphoid tissues and organs are distributed in areas especially vulnerable to injury or invasion.

11. Lymph nodes are encapsulated masses of lymphoid tissue. The **deep cortex** is dominated by T cells; the **outer cortex** and **medulla** contain B cells. *(Figure 22-7)*

12. The thymus lies behind the sternum, in the anterior mediastinum. **Reticular epithelial cells** scattered among the lymphocytes produce thymic hormones. *(Figure 22-8)*

13. The adult spleen contains the largest mass of lymphoid tissue in the body. The cellular components form the **pulp** of the spleen. **Red pulp** contains large numbers of red blood cells, and **white pulp** resembles lymphoid nodules. *(Figure 22-9)*

The Lymphatic System and Body Defenses p. 763

14. The lymphatic system is a major component of the body's defenses. These defenses are either (1) **nonspecific defenses,** which protect without distinguishing one threat from another, or (2) **specific defenses,** which protect against particular threats only.

NONSPECIFIC DEFENSES p. 763

1. Nonspecific defenses prevent the approach, deny the entrance, or limit the spread of living or nonliving hazards. *(Figure 22-10)*

Physical Barriers p. 765

2. Physical barriers include hair, epithelia, and various secretions of the integumentary and digestive systems.

Phagocytes p. 765

3. The two types of phagocytic cells are **microphages** and **macrophages** (cells of the **monocyte–macrophage system**). Microphages are the neutrophils and eosinophils in circulating blood.

4. Phagocytes move among cells by *emigration,* or *diapedesis* (migration between adjacent endothelial cells), and exhibit **chemotaxis** (sensitivity and orientation to chemical stimuli).

Immunological Surveillance p. 766

5. Immunological surveillance involves constant monitoring of normal tissues by NK cells sensitive to abnormal antigens on the surfaces of otherwise normal cells. Cancer cells with **tumor-specific antigens** on their surfaces are killed. *(Figure 22-11)*

Interferons p. 767

6. Interferons, small proteins released by cells infected with viruses, trigger the production of **antiviral proteins,** which interfere with viral replication inside the cells. Interferons are **cytokines,** chemical messengers released by tissue cells to coordinate local activities.

Complement p. 767

7. At least 11 **complement proteins** make up the **complement system.** They interact with each other in chain reactions to destroy target cell membranes, stimulate inflammation, attract phagocytes, and/or enhance phagocytosis. *(Figure 22-12)*

Inflammation p. 769

8. Inflammation is a coordinated nonspecific response to tissue injury. *(Figure 22-13)*

Fever p. 770

9. A **fever** (body temperature greater than 37.2°C (99°F) can inhibit pathogens and accelerate metabolic processes.

SPECIFIC DEFENSES: THE IMMUNE RESPONSE p. 771

Forms of Immunity p. 771

1. Specific resistance, or immunity, involves **innate immunity** (genetically determined and present at birth) or **acquired immunity.** The two types of acquired immunity are **active immunity** (which appears after exposure to an antigen) and **passive immunity** (produced by the transfer of antibodies from another source). *(Figure 22-14)*

Properties of Immunity p. 772

2. Lymphocytes provide immunity, which has four general characteristics: **specificity, versatility, memory,** and **tolerance.** *Memory cells* enable the immune system to "remember" previous target antigens. Tolerance is the ability of the immune system to ignore some antigens, such as those of body cells.

Overview: The Immune Response p. 773

3. The immune response is triggered by the presence of an antigen; the response includes cell-mediated and antibody-mediated defenses. (*Figure 22-15*)

T Cells and Cell-Mediated Immunity p. 773

4. Foreign antigens must generally be processed by other cells (typically macrophages) and incorporated into cell membranes **(antigen presentation)** before the antigens can activate lymphocytes.

5. All body cells have membrane glycoproteins. The genes controlling their synthesis make up a chromosomal region called the **major histocompatibility complex (MHC).** The membrane glycoproteins are called **MHC proteins.** Lymphocytes are not activated by lone antigens but will respond to an antigen bound to either a **Class I** or a **Class II** MHC protein.

6. Class I MHC proteins are in all nucleated body cells. Class II MHC proteins are only in **antigen-presenting cells (APCs)** and lymphocytes. (*Figure 22-16*)

7. Whether a T cell responds to antigens held in Class I or II MHC proteins depends on the structure of the T cell membrane. T cell membranes contain proteins called **CD** (*cluster of differentiation*) **markers. CD3 markers** are present on all T cells. **CD8 markers** are on cytotoxic and suppressor T cells. **CD4 markers** are on all helper T cells.

8. Cell-mediated immunity (cellular immunity) results from the activation of CD8 T cells by antigens bound to Class I MHCs. When activated, most of these T cells divide to generate cytotoxic T cells and **memory T_C cells,** which remain on reserve to guard against future such attacks. Suppressor T cells depress the responses of other T cells and of B cells. (*Figure 22-17*)

9. Helper, or CD4, T cells respond to antigens presented by Class II MHC proteins. When activated, they secrete lymphokines that help coordinate specific and nonspecific defenses and regulate cell-mediated and antibody-mediated immunity. (*Figure 22-18*)

B Cells and Antibody-Mediated Immunity p. 777

10. B cells become **sensitized** when antigens bind to their membrane antibody molecules. The antigens are then displayed on the Class II MHC proteins of the B cells, which become activated by helper T cells activated by the same antigen.

11. An active B cell may differentiate into a plasma cell or produce daughter cells that differentiate into plasma cells and **memory B cells.** Antibodies are produced by the plasma cells. (*Figure 22-19*)

12. An antibody molecule consists of two parallel pairs of polypeptide chains containing *constant* and *variable segments.* (*Figure 22-20*)

13. When antibody molecules bind to an antigen, they form an **antigen–antibody complex.** Effects that appear after binding include **neutralization** (antibody binding such that viruses or bacterial toxins cannot bind); **precipitation** (formation of an insoluble **immune complex**) and **agglutination** (formation of large complexes); **opsonization** (coating of pathogens with antibodies and complement proteins to enhance phagocytosis); stimulation of inflammation; and prevention of bacterial or viral adhesion.

14. The five classes of antibodies in body fluids are **immunoglobulin** (1) **G (IgG),** responsible for resistance against many viruses, bacteria, and bacterial toxins; (2) **E (IgE),** which releases chemicals that accelerate local inflammation; (3) **D (IgD),** located on the surfaces of B cells;

(4) **M (IgM),** the first antibody type secreted after an antigen arrives; and (5) **A (IgA),** found in glandular secretions. (*Table 22-1*)

Primary and Secondary Responses to Antigen Exposure p. 781

15. The antibodies first produced by plasma cells are the agents of the **primary response.** The maximum **antibody titer** appears during the **secondary response** to antigen exposure. (*Figure 22-21*)

Hormones of the Immune System p. 782

16. **Interleukins** increase T cell sensitivity to antigens exposed on macrophage membranes; stimulate B cell activity, plasma cell formation, and antibody production; and enhance nonspecific defenses.

17. Interferons slow the spread of a virus by making the synthesizing cell and its neighbors resistant to viral infections.

18. **Tumor necrosis factors (TNFs)** slow tumor growth and kill tumor cells.

19. Several lymphokines adjust the activities of phagocytic cells to coordinate specific and nonspecific defenses. (*Table 22-2*)

SUMMARY: THE IMMUNE RESPONSE p. 784

1. The initial steps in the immune response to viral and bacterial infections differ. (*Figures 22-22–22-24; Table 22-3*)

THE DEVELOPMENT OF RESISTANCE p. 785

1. **Immunological competence** is the ability to demonstrate an immune response after exposure to an antigen. A developing fetus receives passive immunity from the maternal bloodstream. After delivery, the infant begins developing acquired immunity following exposure to environmental antigens.

IMMUNE DISORDERS p. 787

Autoimmune Disorders p. 787

1. **Autoimmune disorders** develop when the immune response mistakenly targets normal body cells and tissues.

Immunodeficiency Diseases p. 788

2. In an **immunodeficiency disease,** either the immune system does not develop normally or the immune response is blocked.

Allergies p. 788

3. **Allergies** are inappropriate or excessive immune responses to **allergens** (antigens that trigger allergic reactions). The four types of allergies are **immediate hypersensitivity** (*Type I*), *cytotoxic reactions (Type II), immune complex disorders (Type III),* and *delayed hypersensitivity (Type IV).*

4. In **anaphylaxis,** a circulating allergen affects mast cells throughout the body. (*Figure 22-25*)

STRESS AND THE IMMUNE RESPONSE p. 789

1. **Interleukin-1** released by active macrophages triggers the release of ACTH by the anterior pituitary gland. Glucocorticoids produced by the adrenal cortex moderate the immune response, but their long-term secretion can lower resistance to disease.

AGING AND THE IMMUNE RESPONSE p. 789

1. With aging, the immune system becomes less effective at combating disease.

INTEGRATION WITH OTHER SYSTEMS p. 791

1. There is substantial interaction between the cells and tissues involved with the immune response and the nervous and endocrine systems. (*Figure 22-26*)

REVIEW QUESTIONS

LEVEL 1 REVIEWING FACTS AND TERMS

1. Lymph from the lower abdomen, pelvis, and lower limbs is received by the
    (a) right lymphatic duct   (b) cisterna chyli
    (c) right thoracic duct   (d) aorta

2. Lymphoid stem cells that can form all types of lymphocytes occur in the
    (a) bloodstream   (b) thymus
    (c) bone marrow   (d) spleen

3. Lymphatic vessels are located everywhere except the
    (a) lower limbs
    (b) central nervous system
    (c) integument
    (d) digestive tract

4. The body's largest collection of lymphoid tissue is in the
    (a) adult spleen   (b) adult thymus
    (c) bone marrow   (d) tonsils

5. Red blood cells that are damaged or defective are removed from the bloodstream by the
    (a) thymus   (b) lymph nodes
    (c) spleen   (d) tonsils

6. Phagocytes move through capillary walls by squeezing between adjacent endothelial cells, a process known as
    (a) diapedesis   (b) chemotaxis
    (c) adhesion   (d) perforation

7. Perforins and protectin are proteins associated with the activity of
    (a) T cells   (b) B cells
    (c) NK cells   (d) plasma cells

8. Complement activation
    (a) stimulates inflammation
    (b) attracts phagocytes
    (c) enhances phagocytosis
    (d) a, b, and c are correct

9. Inflammation
    (a) aids in temporary repair at an injury site
    (b) slows the spread of pathogens
    (c) facilitates permanent repair
    (d) a, b, and c are correct

10. CD4 markers are associated with
    (a) cytotoxic T cells   (b) suppressor T cells
    (c) helper T cells   (d) a, b, and c are correct

11. Which two large collecting vessels are responsible for returning lymph to the veins of the circulatory system? What areas of the body does each serve?

12. Give a function for each of the following:
    (a) cytotoxic T cells
    (b) helper T cells
    (c) suppressor T cells
    (d) plasma cells
    (e) NK cells
    (f) stromal cells
    (g) reticular epithelial cells
    (h) interferons
    (i) pyrogens
    (j) T cells
    (k) B cells
    (l) interleukins
    (m) tumor necrosis factor
    (n) colony-stimulating factors

13. What are the three classes of lymphocytes, and where does each class originate?

14. What seven defenses, present at birth, provide the body with the defensive capability known as nonspecific resistance?

LEVEL 2 REVIEWING CONCEPTS

15. Compared with nonspecific defenses, specific defenses
    (a) do not distinguish between one threat and another
    (b) are always present at birth
    (c) protect against threats on an individual basis
    (d) deny entrance of pathogens to the body

16. T cells and B cells can be activated only by
    (a) pathogens
    (b) interleukins, interferons, and colony-stimulating factors
    (c) cells infected with viruses, bacteria, or cancer cells
    (d) exposure to a specific antigen at a specific site on a cell membrane

17. Class II MHC proteins appear in the cell membrane only when
    (a) the plasma cells are releasing antibodies
    (b) the cell is processing antigens
    (c) cytotoxic T cells are inhibited
    (d) NK cells are activated

18. List the four general properties of immunity; give an explanation of each.

19. How does a cytotoxic T cell destroy another cell displaying antigens bound to Class I MHC proteins?

20. How does the formation of an antigen–antibody complex cause the elimination of an antigen?

21. What effects follow the activation of the complement system?

22. An anesthesia technician is advised that she should be vaccinated against hepatitis B, which is caused by a virus. She is given one injection and is told to come back for a second injection in a month and a third injection after 6 months. Why is this series of injections necessary?

LEVEL 3 CRITICAL THINKING AND CLINICAL APPLICATIONS

23. An investigator at a crime scene discovers some body fluid on the victim's clothing. The investigator carefully takes a sample and sends it to the crime lab for analysis. On the basis of analysis of antibodies, could the crime lab determine whether the sample is blood plasma or semen? Explain.

24. Ted finds out that he has been exposed to the measles. He is concerned that he might have contracted the disease, so he goes to see his physician. The physician takes a blood sample and sends it to a lab for antibody titers. The results show an elevated level of IgM antibodies to rubella (measles) virus but very few IgG antibodies to the virus. Has Ted contracted the disease?

25. While walking along the street, you and your friend see an elderly woman whose left arm appears to be swollen to several times its normal size. Your friend remarks that the woman must have been in the tropics and contracted a form of filariasis that produces elephantiasis. You disagree, saying that it is more likely that the woman had a radical mastectomy (the removal of a breast because of cancer). Explain the rationale behind your answer.

26. Tilly has T cells that can respond to antigen A. Tilly's friend Harry has been exposed to antigen A, and Tilly offers to have some of her T cells transfused so that Harry will not come down with an infection. You overhear their conversation and observe that such a transfusion would not work. Why wouldn't it work?

27. You are a researcher interested in studying cells that can respond to the hormone FSH. How could you use antibodies to help you locate FSH-responsive cells?

CHAPTER 23

The Respiratory System

CHAPTER OUTLINE AND OBJECTIVES

INTRODUCTION p. 798

FUNCTIONS OF THE RESPIRATORY SYSTEM p. 798
① Describe the primary functions of the respiratory system.

ORGANIZATION OF THE RESPIRATORY SYSTEM p. 798
   The Respiratory Mucosa p. 798
② Explain how the delicate respiratory exchange surfaces are protected from pathogens, debris, and other hazards.

THE UPPER RESPIRATORY SYSTEM p. 801
   The Nose and Nasal Cavity p. 801
   The Pharynx p. 802
③ Identify the organs of the upper respiratory system, and describe their functions.

THE LARYNX p. 803
   Cartilages and Ligaments of the Larynx p. 803
   The Laryngeal Musculature p. 805
④ Describe the structure of the larynx, and discuss its role in normal breathing and sound production.

THE TRACHEA p. 805

THE PRIMARY BRONCHI p. 806
⑤ Discuss the structure of the extrapulmonary airways.

THE LUNGS p. 806
   Lobes and Surfaces of the Lungs p. 806
   The Bronchi p. 808
   The Bronchioles p. 809
   Alveolar Ducts and Alveoli p. 811
   The Blood Supply to the Lungs p. 813
⑥ Describe the superficial anatomy of the lungs, the structure of a pulmonary lobule, and the functional anatomy of the alveoli.

THE PLEURAL CAVITIES AND PLEURAL MEMBRANES p. 813

RESPIRATORY PHYSIOLOGY p. 814
   Pulmonary Ventilation p. 814
   Gas Exchange at the Respiratory Membrane p. 822
   Gas Pickup and Delivery p. 826
   Summary: Gas Transport p. 830
⑦ Describe the physical principles governing the movement of air into the lungs and the diffusion of gases into and out of the blood.

⑧ Describe the origins and actions of the respiratory muscles responsible for respiratory movements.
⑨ Differentiate between pulmonary ventilation and alveolar ventilation.
⑩ Describe how oxygen and carbon dioxide are transported in blood.

CONTROL OF RESPIRATION p. 830
   Local Regulation of Gas Transport and Alveolar Function p. 830
   The Respiratory Centers of the Brain p. 831
   Voluntary Control of Respiration p. 836
⑪ Describe the factors that influence the respiration rate.
⑫ Identify and discuss reflex respiratory activity and the brain centers involved in the control of respiration.

CHANGES IN THE RESPIRATORY SYSTEM AT BIRTH p. 836

AGING AND THE RESPIRATORY SYSTEM p. 837

INTEGRATION WITH OTHER SYSTEMS p. 837
⑬ Give examples of interactions between the respiratory system and each of the other organ systems.

MEDIALAB and CHAPTER REVIEW pp. 839, 840

Take a deep breath, and blow very hard to put out the candles. It sounds simple, but the process is very complex. Not only must a wide variety of skeletal muscles work together, but also the usually subconscious regulation of breathing must be consciously and forcefully overridden. In this chapter, we shall examine the homeostatic mechanisms responsible for the conscious and subconscious regulation of respiration. We shall also discuss the functional integration of respiratory and cardiovascular activities.

Cells need energy for maintenance, growth, defense, and replication. Our cells obtain that energy through aerobic mechanisms that require oxygen and produce carbon dioxide. Many aquatic organisms can obtain oxygen and excrete carbon dioxide by diffusion across the surface of the skin or in specialized structures, such as the gills of a fish. Such arrangements are poorly suited for life on land, because the exchange surfaces must be very thin and relatively delicate to permit rapid diffusion. In air, the exposed membranes collapse, evaporation and dehydration reduce blood volume, and the delicate surfaces become vulnerable to attack by pathogens. Our respiratory exchange surfaces are just as delicate as those of an aquatic organism, but they are confined to the inside of the lungs—in a warm, moist, protected environment. Under these conditions, diffusion can occur between the air and the blood.

The cardiovascular system provides the link between your interstitial fluids and the exchange surfaces of your lungs. Your circulating blood carries oxygen from the lungs to peripheral tissues; it also accepts and transports the carbon dioxide generated by those tissues, delivering it to the lungs.

Diffusion between blood and air occurs at **alveoli** (al-VĒ-ō-lī), air-filled pockets in the lungs. The distance between the blood in an alveolar capillary and the air inside an alveolus is generally less than 1 μm and in some cases as small as 0.1 μm. To meet the metabolic requirements of peripheral tissues, the exchange surfaces of the lungs must be very large. In fact, the total surface area for gas exchange in the lungs of adults is at least 35 times the surface area of the body. It is difficult to measure the exchange surfaces with precision; estimates of the surface area involved range from 70 m^2 to 140 m^2 (753 ft^2 to 1506 ft^2).

Our discussion of the respiratory system begins by following the air as it travels toward the alveoli. We shall then consider the mechanics of breathing and the physiology of respiration.

FUNCTIONS OF THE RESPIRATORY SYSTEM

The **respiratory system** has five basic functions:

1. Providing an extensive area for gas exchange between air and circulating blood.

2. Moving air to and from the exchange surfaces of the lungs.

3. Protecting respiratory surfaces from dehydration, temperature changes, or other environmental variations and defending the respiratory system and other tissues from invasion by pathogens.

4. Producing sounds involved in speaking, singing, and nonverbal communication.

5. Providing olfactory sensations to the central nervous system from the olfactory epithelium in the superior portions of the nasal cavity.

In addition, the capillaries of the lungs indirectly assist in the regulation of blood volume and blood pressure, through the conversion of angiotensin I to angiotensin II. ∞ *pp. 716–717*

ORGANIZATION OF THE RESPIRATORY SYSTEM

We can divide the respiratory system into an upper respiratory system and a lower respiratory system (Figure 23-1•). The **upper respiratory system** consists of the nose, nasal cavity, paranasal sinuses, and pharynx. These passageways filter, warm, and humidify the incoming air—protecting the more delicate surfaces of the lower respiratory system—and cool and dehumidify outgoing air. The **lower respiratory system** includes the larynx (voice box), trachea (windpipe), bronchi, bronchioles, and alveoli of the lungs.

Your **respiratory tract** consists of the airways that carry air to and from the exchange surfaces of your lungs. The respiratory tract consists of a *conducting portion* and a *respiratory portion*. The conducting portion begins at the entrance to the nasal cavity and extends through many passageways (the pharynx and larynx and along the trachea, bronchi, and bronchioles to the *terminal bronchioles*). The respiratory portion of the tract includes the delicate *respiratory bronchioles* and the sites of gas exchange, the alveoli.

Filtering, warming, and humidification of the inhaled air begin at the entrance to the upper respiratory system and continue throughout the rest of the conducting system. By the time air reaches the alveoli, most foreign particles and pathogens have been removed, and the humidity and temperature are within acceptable limits. The success of this "conditioning process" is due primarily to the properties of the *respiratory mucosa*.

THE RESPIRATORY MUCOSA

The **respiratory mucosa** (mū-KŌ-suh) lines the conducting portion of the respiratory system. A *mucosa* is a *mucous membrane,* one of the four types of membranes introduced in Chapter 4. It consists of an epithelium and an underlying layer of areolar tissue. ∞ *p. 129*

A pseudostratified ciliated columnar epithelium with numerous goblet cells lines the nasal cavity and the superior portion of the pharynx. ∞ *p. 115* The structure of the respiratory epithelium changes along the respiratory tract. The epithelium lining inferior portions of the pharynx is a stratified squamous epithelium similar to that of the oral cavity. These portions of the pharynx, which conduct air to the larynx,

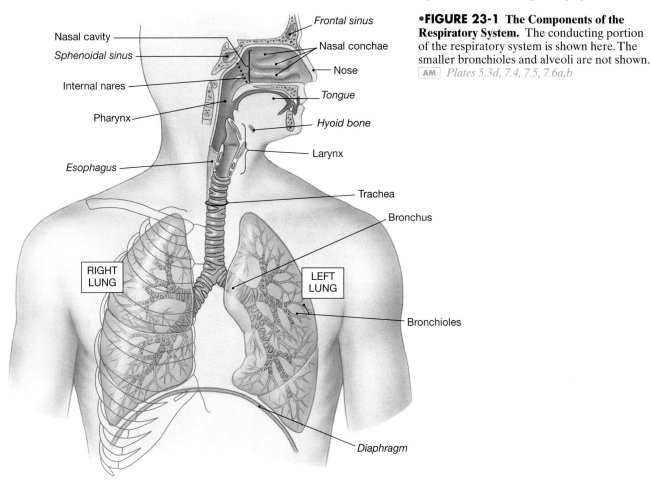

Nasal cavity

Sphenoidal sinus

Internal nares

Pharynx

Esophagus

Frontal sinus

Nasal conchae

Nose

Tongue

Hyoid bone

Larynx

Trachea

Bronchus

RIGHT LUNG

LEFT LUNG

Bronchioles

Diaphragm

•FIGURE 23-1 **The Components of the Respiratory System.** The conducting portion of the respiratory system is shown here. The smaller bronchioles and alveoli are not shown. AM *Plates 5.3d, 7.4, 7.5, 7.6a,b*

also convey food to the esophagus. The pharyngeal epithelium must therefore protect against abrasion and chemical attack.

At the beginning of the lower respiratory system is a pseudostratified ciliated columnar epithelium (Figure 23-2a,b•) comparable to that of the nasal cavity. In the smaller bronchioles, this pseudostratified epithelium is replaced by a cuboidal epithelium with scattered cilia. The exchange surfaces of the alveoli are lined by a very delicate simple squamous epithelium. Other, more specialized cells are scattered within that epithelium.

The **lamina propria** (LA-mi-nuh PRŌ-prē-uh) is the underlying layer of areolar tissue that supports the respiratory epithelium. In the upper respiratory system, trachea, and bronchi, the lamina propria contains mucous glands that discharge their secretions onto the epithelial surface. The lamina propria in the conducting portions of the lower respiratory system contains bundles of smooth muscle cells. At the bronchioles, the smooth muscles form relatively thick bands that encircle or spiral around the lumen.

The Respiratory Defense System

The delicate exchange surfaces of the respiratory system can be severely damaged if the inhaled air becomes contaminated with debris or pathogens. Such contamination is prevented by a series of filtration mechanisms that together make up the **respiratory defense system.**

Along much of the length of the respiratory tract, goblet cells in the epithelium and mucous glands in the lamina propria produce a sticky mucus that bathes exposed surfaces. In the nasal cavity, cilia sweep that mucus and any trapped debris or microorganisms toward the pharynx, where it will be swallowed and exposed to the acids and enzymes of the stomach. In the lower respiratory system, the cilia also beat toward the pharynx, moving a carpet of mucus toward the pharynx and cleaning the respiratory surfaces. This process is often described as a *mucus escalator* (Figure 23-2c•).

Filtration in the nasal cavity removes virtually all particles larger than about 10 μm from the inhaled air. Smaller particles may be trapped by the mucus of the nasopharynx or secretions of the pharynx before proceeding along the conducting system. Exposure to unpleasant stimuli, such as noxious vapors, large quantities of dust and debris, allergens, or pathogens, generally causes a rapid increase in the rate of mucus production in the nasal cavity and paranasal sinuses. (The familiar symptoms of the "common cold" result from the invasion of this respiratory epithelium by any of more than 200 types of viruses.)

23

(a) Superficial view

•**FIGURE 23-2** **The Respiratory Epithelium of the Nasal Cavity and Conducting System.** **(a)** A surface view of the epithelium. The cilia of the epithelial cells form a dense layer that resembles a shag carpet. The movement of these cilia propels mucus across the epithelial surface. (SEM × 1614) **(b)** The sectional appearance of the respiratory epithelium, a pseudostratified ciliated columnar epithelium. (LM × 1062) **(c)** A diagrammatic view of the respiratory epithelium of the trachea, showing the mechanism of mucus transport.

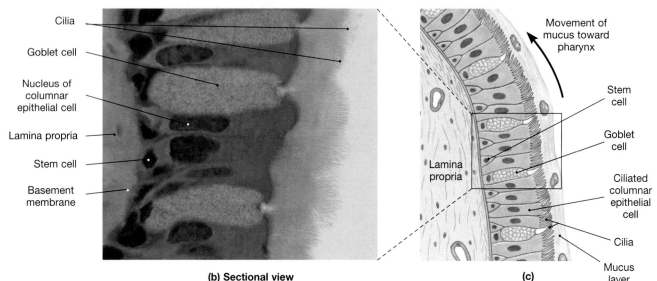

(b) Sectional view

(c)

Cilia
Goblet cell
Nucleus of columnar epithelial cell
Lamina propria
Stem cell
Basement membrane

Movement of mucus toward pharynx
Stem cell
Goblet cell
Lamina propria
Ciliated columnar epithelial cell
Cilia
Mucus layer

Most particles 1–5 μm in diameter are trapped in the mucus coating the respiratory bronchioles or in the liquid covering the alveolar surfaces. These areas are outside the boundaries of the mucus escalator, but the foreign particles can be engulfed by alveolar macrophages. Most particles smaller than about 0.5 μm remain suspended in the air.

Large quantities of airborne particles may overload the respiratory defenses and produce a variety of illnesses. For example, the presence of irritants in the lining of the conducting passageways can provoke the formation of abscesses that block airflow and reduce pulmonary function, and damage to the epithelium in the affected area may allow irritants to enter the surrounding tissues of the lung. The irritants then produce local inflammation. Airborne irritants are strongly linked to the development of lung cancer (p. 839). (We noted smoking-induced changes induced in the respiratory epithelium in Chapter 4, Figure 4-22•, p. 137.)

AM *Overloading the Respiratory Defenses*

Cystic Fibrosis

Cystic fibrosis (CF) is the most common lethal inherited disease affecting Caucasians of Northern European descent, occurring at a frequency of 1 birth in 2500. It occurs with less frequency in people of Southern European ancestry, in the Ashkenazic Jewish population, and in African Americans. The condition results from a defective gene located on chromosome 7. Individuals with CF seldom survive past age 30; death is generally the result of a massive bacterial infection of the lungs and associated heart failure.

The most serious symptoms appear because the respiratory mucosa in these individuals produces dense, viscous mucus that cannot be transported by the respiratory defense system. The mucus escalator stops working, and mucus blocks the smaller respiratory passageways. This blockage reduces the diameter of the airways, making breathing difficult, and the inactivation of the normal respiratory defenses leads to frequent bacterial infections.

AM *Cystic Fibrosis*

23

THE UPPER RESPIRATORY SYSTEM

As we have seen, the upper respiratory system consists of the nose, nasal cavity, paranasal sinuses, and pharynx (Figures 23-1 and 23-3•).

THE NOSE AND NASAL CAVITY

The nose is the primary passageway for air entering the respiratory system. Air normally enters through the paired **external nares** (NĀ-rēz), or *nostrils* (Figure 23-3a•), which open into the **nasal cavity.** The **vestibule** is the space contained within the flexible tissues of the nose (Figure 23-3c•). The epithelium of the vestibule contains coarse hairs that extend across the external nares. Large airborne particles, such as sand, sawdust, or even insects, are trapped in these hairs and are thereby prevented from entering the nasal cavity.

The *nasal septum* divides the nasal cavity into left and right portions (Figure 23-3b•). The bony portion of the nasal septum is formed by the fusion of the perpendicular plate of the ethmoid bone and the plate of the vomer (Figure 7-3d•, p. 198). The anterior portion of the nasal septum is formed of hyaline cartilage. This cartilaginous plate supports the *dorsum nasi* (DOR-sum NĀ-zī), or *bridge,* and *apex* (tip) of the nose.

The maxillary, nasal, frontal, ethmoid, and sphenoid bones form the lateral and superior walls of the nasal cavity. The mucous secretions produced in the associated *paranasal sinuses* (Figure 7-14•, p. 210), aided by the tears draining through the nasolacrimal ducts, help keep the surfaces of the nasal cavity moist and clean.

The *olfactory region,* or superior portion of the nasal cavity, includes the areas lined by olfactory epithelium: (1) the inferior surface of the cribriform plate, (2) the superior portion of the nasal septum, and (3) the superior nasal conchae. Receptors in the olfactory epithelium provide your sense of smell. ∞ *p. 535*

The *superior, middle,* and *inferior nasal conchae* project toward the nasal septum from the lateral walls of the nasal cavity. ∞ *pp. 204, 206* To pass from the vestibule to the internal nares, air tends to flow between adjacent conchae, through the **superior, middle,** and **inferior meatuses** (mē-Ā-tus-ez; *meatus,* a passage) (Figure 23-3b•). These are narrow grooves rather than open passageways; the incoming air bounces off the conchal surfaces and churns like a stream flowing over rapids. This turbulence serves a purpose: As the air eddies and swirls, small airborne particles are likely to come into contact with the mucus that coats the lining of the nasal cavity. In addition to promoting filtration, the turbulence allows extra time for warming and humidifying incoming air. It also creates eddy currents that bring olfactory stimuli to the olfactory receptors.

A bony **hard palate,** formed by portions of the maxillary and palatine bones, forms the floor of the nasal cavity and separates the oral and nasal cavities. A fleshy **soft palate** extends posterior to the hard palate, marking the boundary between the superior **nasopharynx** (nā-zō-FAR-inks) and the rest of the pharynx. The nasal cavity opens into the nasopharynx through a connection known as the **internal nares.**

•**FIGURE 23-3** The Nose, Nasal Cavity, and Pharynx. **(a)** The nasal cartilages and external landmarks on the nose. **(b)** The meatuses and the maxillary and ethmoidal sinuses.

(a)

(b)

23

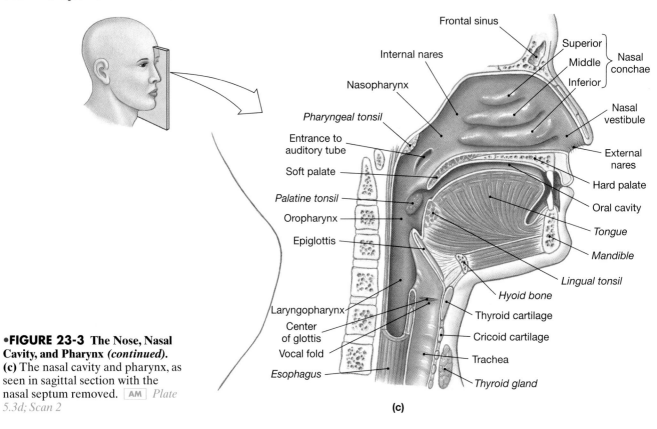

•FIGURE 23-3 The Nose, Nasal Cavity, and Pharynx *(continued)*. (c) The nasal cavity and pharynx, as seen in sagittal section with the nasal septum removed. [AM] *Plate 5.3d; Scan 2*

(c)

The Nasal Mucosa

The mucosa of the nasal cavity prepares the air you breathe for arrival at your lower respiratory system. Throughout much of the nasal cavity, the lamina propria contains an abundance of arteries, veins, and capillaries that bring nutrients and water to the secretory cells. The lamina propria of the nasal conchae also contains an extensive network of large and highly expandable veins. This vascularization provides a mechanism for warming and humidifying the incoming air (as well as for cooling and dehumidifying the outgoing air). As cool, dry air passes inward over the exposed surfaces of the nasal cavity, the warm epithelium radiates heat and the water in the mucus evaporates. Air moving from your nasal cavity to your lungs has been heated almost to body temperature, and it is nearly saturated with water vapor. This mechanism protects more-delicate respiratory surfaces from chilling or drying out—two potentially disastrous events. Breathing through your mouth eliminates much of the preliminary filtration, heating, and humidifying of the inhaled air. To avoid alveolar damage, patients breathing on a respirator, which utilizes a tube to provide air directly into the trachea, must receive air that has been externally filtered and humidified.

As air moves out of the respiratory tract, it again passes across the epithelium of the nasal cavity. This air is warmer and more humid than the air that enters; it warms the nasal mucosa, and moisture condenses on the epithelial surfaces. Thus breathing through your nose also helps prevent heat loss and water loss.

Nosebleeds

The extensive vascularization of the nasal cavity and the relatively vulnerable position of the nose make a nosebleed, or **epistaxis** (ep-i-STAK-sis), a fairly common event. This bleeding generally involves vessels of the mucosa covering the cartilaginous portion of the septum. A variety of factors may be responsible, including trauma, such as a punch in the nose, drying, infections, allergies, or clotting disorders. Hypertension can also provoke a nosebleed by rupturing small vessels of the lamina propria.

THE PHARYNX

The **pharynx** (FAR-inks) is a chamber shared by the digestive and respiratory systems. It extends between the internal nares and the entrances to the larynx and esophagus. The curving superior and posterior walls of the pharynx are closely bound to the axial skeleton, but the lateral walls are flexible and muscular.

The pharynx is divided into the nasopharynx, the oropharynx, and the laryngopharynx (Figure 23-3c•):

1. The **nasopharynx** is the superior portion of the pharynx. It is connected to the posterior portion of the nasal cavity through the internal nares and is separated from the oral cavity by the soft palate. The nasopharynx is lined by the same pseudostratified ciliated columnar epithelium as that in the nasal cavity. The *pharyngeal tonsil* is located on the posterior wall of the nasopharynx; on each side, one of the *auditory tubes* opens into the nasopharynx. ∞ *pp. 759, 558*

2. The **oropharynx** (*oris,* mouth) extends between the soft palate and the base of the tongue at the level of the hyoid bone. The posterior portion of the oral cavity communicates directly with the oropharynx, as does the posterior inferior portion of the nasopharynx. At the boundary between the nasopharynx and the oropharynx, the epithelium changes from pseudostratified columnar to stratified squamous.

3. The narrow **laryngopharynx** (la-rin-gō-FAR-inks), the inferior part of the pharynx, includes that portion of the pharynx between the hyoid bone and the entrance to the larynx and esophagus. Like the oropharynx, it is lined by a stratified squamous epithelium that resists abrasion, chemical attack, and pathogenic invasion.

THE LARYNX

Inhaled air leaves the pharynx and enters the larynx through the **glottis** (GLOT-is), a narrow opening. The **larynx** (LAR-inks) is a cartilaginous structure that surrounds and protects the glottis. The larynx begins at the level of vertebra C_4 or C_5 and ends at the level of vertebra C_6. Essentially a cylinder, the larynx has incomplete cartilaginous walls that are stabilized by ligaments and skeletal muscles (Figure 23-4•).

CARTILAGES AND LIGAMENTS OF THE LARYNX

Three large, unpaired cartilages form the larynx: the (1) thyroid cartilage, (2) cricoid cartilage, and (3) epiglottis (Figure 23-4•). The **thyroid cartilage** (*thyroid,* shield-shaped) is the largest laryngeal cartilage. Consisting of hyaline cartilage, it forms most of the anterior and lateral walls of the larynx. In section, this cartilage is U-shaped; it is incomplete posteriorly. The prominent anterior surface of this cartilage, which you can easily see and feel, is called the *laryngeal prominence* or *Adam's apple.* The inferior surface articulates with the cricoid cartilage. The superior surface has ligamentous attachments to the hyoid bone and to the epiglottis and smaller laryngeal cartilages.

The thyroid cartilage sits superior to the **cricoid** (KRĪ-koyd; ring-shaped) **cartilage,** another hyaline cartilage. The posterior portion of the cricoid is greatly expanded, providing support in the absence of the thyroid cartilage. The cricoid and thyroid cartilages protect the glottis and the entrance to the trachea, and their broad surfaces provide sites for the attachment

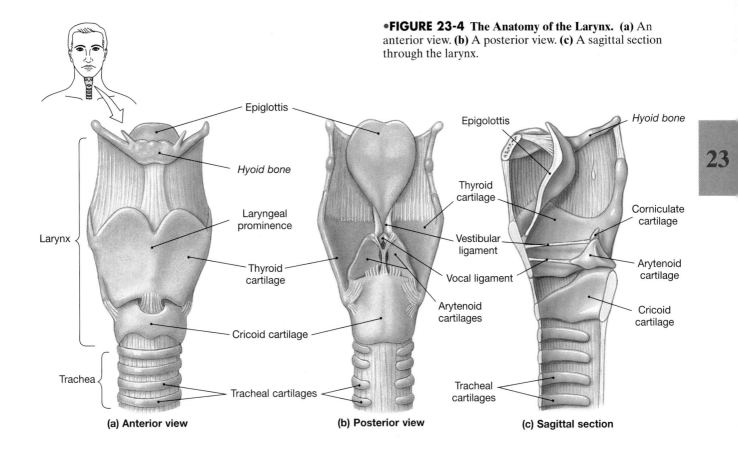

•**FIGURE 23-4 The Anatomy of the Larynx. (a)** An anterior view. **(b)** A posterior view. **(c)** A sagittal section through the larynx.

(a) Anterior view   **(b) Posterior view**   **(c) Sagittal section**

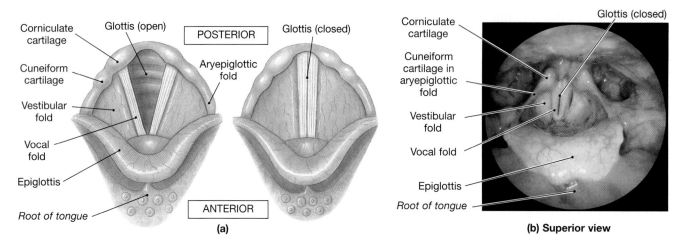

•**FIGURE 23-5** **The Glottis.** **(a)** A diagrammatic superior view of the larynx with the glottis open and closed. **(b)** A fiber-optic view of the entrance to the glottis, which is almost completely closed by the vocal folds.

of important laryngeal muscles and ligaments. Ligaments attach the inferior surface of the cricoid cartilage to the first cartilage of the trachea. The superior surface of the cricoid cartilage articulates with the small, paired *arytenoid cartilages.*

The shoehorn-shaped **epiglottis** (ep-i-GLOT-is) projects superior to the glottis and forms a lid over it. Composed of elastic cartilage, the epiglottis has ligamentous attachments to the anterior and superior borders of the thyroid cartilage and the hyoid bone. During swallowing, the larynx is elevated and the epiglottis folds back over the glottis, preventing the entry of liquids or solid food into the respiratory tract.

The larynx also contains three pairs of smaller hyaline cartilages: (1) The **arytenoid** (ar-i-TĒ-noyd; ladle-shaped) **cartilages** articulate with the superior border of the enlarged portion of the cricoid cartilage. (2) The **corniculate** (kor-NIK-ū-lāt; horn-shaped) **cartilages** articulate with the arytenoid cartilages. The corniculate and arytenoid cartilages are involved with the opening and closing of the glottis and the production of sound. (3) Elongate, curving **cuneiform** (kū-NĒ-i-form; wedge-shaped) **cartilages** lie within folds of tissue (the *aryepiglottic folds*) that extend between the lateral surface of each arytenoid cartilage and the epiglottis.

The various laryngeal cartilages are bound together by ligaments; additional ligaments attach the thyroid cartilage to the hyoid bone and the cricoid cartilage to the trachea (Figure 23-4a,b•). The **vestibular ligaments** and the **vocal ligaments** extend between the thyroid cartilage and the arytenoid cartilages.

The vestibular and vocal ligaments are covered by folds of laryngeal epithelium that project into the glottis. The vestibular ligaments lie within the superior pair of folds, known as the **vestibular folds.** These folds, which are relatively inelastic, help prevent foreign

objects from entering the glottis and protect the more delicate **vocal folds** (Figures 23-4b,c and 23-5•).

The vocal folds guard the entrance to the glottis. They are inferior to the vestibular folds. The vocal folds are highly elastic, because the vocal ligaments consist of elastic tissue. The vocal folds are involved with the production of sounds, and for this reason they are known as the **vocal cords.**

Sound Production

Air passing through the glottis vibrates the vocal folds and produces sound waves. The pitch (∞ *p. 566*) of the sound produced depends on the diameter, length, and tension in the vocal folds. The diameter and length are directly related to larynx size. The tension is controlled by the contraction of voluntary muscles that reposition the arytenoid cartilages relative to the thyroid cartilage. When the distance increases, the vocal folds tense and the pitch rises; when the distance decreases, the vocal folds relax and the pitch falls.

Children have slender, short vocal folds; their voices tend to be high-pitched. At puberty, the larynx of males enlarges much more than does that of females. The vocal cords of an adult male are thicker and longer, and produce lower tones, than those of an adult female.

Sound production at the larynx is called *phonation* (fō-NĀ-shun; *phone,* voice). Phonation is one component of speech production, but clear speech also requires *articulation,* the modification of those sounds by other structures. In a stringed instrument, such as a guitar, the quality of the sound produced does not depend solely on the nature of the vibrating string. The entire instrument becomes involved as the walls vibrate and the composite sound echoes within the hollow body. Similar amplification and resonance occur within your pharynx, oral cavity, nasal cavity, and paranasal sinuses. The combination determines the particular and distinctive sound of your voice. The final production of distinct words further depends on voluntary movements of the tongue, lips, and cheeks.

THE LARYNGEAL MUSCULATURE

The larynx is associated with (1) muscles of the neck and pharynx, which position and stabilize the larynx (p. 326), and (2) smaller intrinsic muscles that control tension in the vocal folds or open and close the glottis. These latter muscles insert on the thyroid, arytenoid, and corniculate cartilages. The opening or closing of the glottis involves rotational movements of the arytenoid cartilages that move the vocal folds.

When you swallow, both sets of muscles cooperate to prevent food or drink from entering the glottis. Before the material is swallowed, it is crushed and chewed into a pasty mass known as a *bolus*. Muscles of the neck and pharynx then elevate the larynx, bending the epiglottis over the glottis, so that the bolus can glide across the epiglottis rather than falling into the larynx. While this movement is under way, the glottis is closed.

Food or liquids that touch the vestibular or vocal folds trigger the *coughing reflex*. In a cough, the glottis is kept closed while the chest and abdominal muscles contract, compressing the lungs. When the glottis is opened suddenly, the resulting blast of air from the trachea ejects material that blocks the entrance to the glottis.

Laryngeal Problems

Infection or inflammation of the larynx is known as **laryngitis** (lar-in-JĪ-tis). It commonly affects the vibrational qualities of the vocal folds; hoarseness is the most familiar symptom. Mild cases are temporary and seldom serious. However, bacterial or viral infection of the epiglottis can be very dangerous; the resulting swelling may close the glottis and cause suffocation. This condition, acute **epiglottitis** (ep-i-glot-TĪ-tis), can develop rapidly after a bacterial infection of the throat. Young children are most likely to be affected.

☑ Why is the vascularization of the nasal cavity important?
☑ Why is the lining of the nasopharynx different from that of the oropharynx and the laryngopharynx?
☑ When the tension in your vocal folds increases, what happens to the pitch of your voice?

THE TRACHEA

The epithelium of the larynx is continuous with that of the **trachea** (TRĀ-kē-uh), or windpipe. The trachea is a tough, flexible tube with a diameter of about 2.5 cm (1 in.) and a length of about 11 cm (4.25 in.) (Figure 23-6•). The trachea begins anterior to vertebra C_6 in a ligamentous attachment to the cricoid cartilage. It ends in the mediastinum, at the level of vertebra T_5, where it branches to form the *right* and *left primary bronchi*.

The mucosa of the trachea resembles that of the nasal cavity and nasopharynx (Figure 23-2a•, p. 800). The **submucosa** (sub-mū-KŌ-suh), a thick layer of connective tissue, surrounds the mucosa. The submucosa contains mucous glands that communicate with the epithelial surface through a number of secretory ducts. The trachea contains 15–20 **tracheal cartilages** (Figure 23-6a•). Each tracheal cartilage is bound to neighboring cartilages by elastic **annular ligaments.** The tracheal cartilages stiffen the tracheal walls and protect the airway. They also prevent its collapse or overexpansion as pressures change in the respiratory system.

Each tracheal cartilage is C-shaped. The closed portion of the C protects the anterior and lateral surfaces of the trachea. The open portion of the C faces posteriorly, toward the esophagus (Figure 23-6b•). Because these cartilages are not continuous, the posterior tracheal wall can easily distort when you swallow, permitting large masses of food to pass through the esophagus.

An elastic ligament and the **trachealis muscle,** a band of smooth muscle, connect the ends of each tracheal cartilage (Figure 23-6b,c•). Contraction of the trachealis muscle alters the diameter of the trachea, changing the trachea's resistance to airflow. The normal diameter of the trachea changes from moment to moment, primarily under the control of the sympathetic division of the ANS. Sympathetic stimulation increases the diameter of the trachea and makes it easier to move large volumes of air along the respiratory passageways.

Tracheal Blockage

We sometimes inadvertently breathe in foreign objects; this process is called *aspiration*. Foreign objects that become lodged in the larynx or trachea are generally expelled by coughing. If the individual can speak or make a sound, the airway is still open, so no emergency measures should be taken. If the person can neither breathe nor speak, an immediate threat to life exists. Unfortunately, many victims become acutely embarrassed by this situation; instead of seeking assistance, they run to the nearest rest room and die there.

In the **Heimlich** (HĪM-lik) **maneuver,** or *abdominal thrust,* a rescuer applies compression to the victim's abdomen just inferior to the diaphragm. This action elevates the diaphragm forcefully and may generate enough pressure to remove the blockage. The maneuver must be performed properly to avoid damage to internal organs. Organizations such as the American Red Cross, local fire departments, and other charitable groups periodically hold brief training sessions in the proper performance of the Heimlich maneuver.

If blockage results from a swelling of the epiglottis or tissues surrounding the glottis, a professionally qualified rescuer may insert a curved tube through the pharynx and glottis to permit airflow. This procedure is called an *intubation*. If the blockage is immovable or the larynx has been crushed, a **tracheostomy** (trā-kē-OS-to-mē) may be performed. In this procedure, an incision is made through the anterior tracheal wall, and a tube is inserted. The tube bypasses the larynx and permits air to flow directly into the trachea.

23

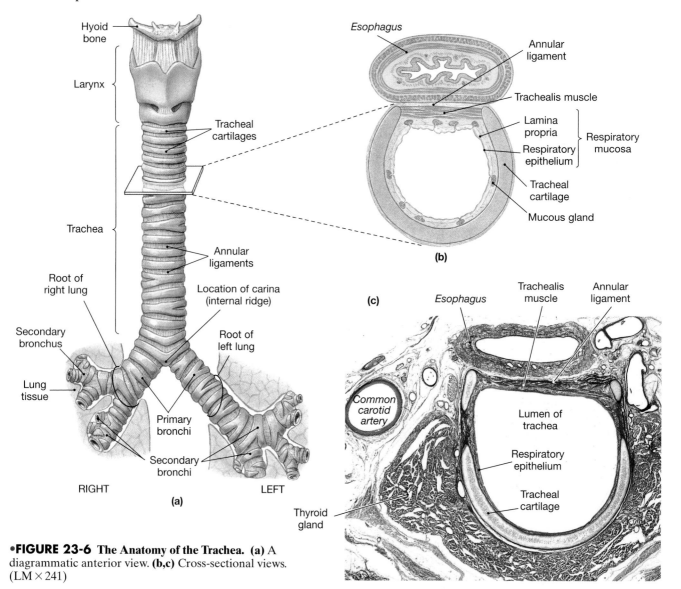

•**FIGURE 23-6** The Anatomy of the Trachea. **(a)** A diagrammatic anterior view. **(b,c)** Cross-sectional views. (LM × 241)

THE PRIMARY BRONCHI

The trachea branches within the mediastinum, giving rise to the **right** and **left primary bronchi** (BRONG-kī; singular, *bronchus*). A ridge called the **carina** (ka-RĪ-nuh) separates the two bronchi (Figure 23-6a•). Like the trachea, the primary bronchi have cartilaginous C-shaped supporting rings. The right primary bronchus supplies the right lung, and the left supplies the left lung. The right primary bronchus is larger in diameter, and descends toward the lung at a steeper angle, than the left. Thus most foreign objects that enter the trachea find their way into the right bronchus rather than the left.

Before branching further, each primary bronchus travels to a groove along the medial surface of its lung. This groove, the **hilus** of the lung, also provides access for entry to pulmonary vessels and nerves (Figure 23-7•). The entire array is firmly anchored in a mesh-

work of dense connective tissue. This complex, the **root** of the lung (Figure 23-6a•), attaches to the mediastinum and fixes the positions of the major nerves, vessels, and lymphatic vessels. The roots of the lungs are anterior to vertebrae T_5 (right) and T_6 (left).

THE LUNGS

The left and right **lungs** (Figure 23-7•) are in the left and right pleural cavities. Each lung is a blunt cone; the tip, or **apex,** points superiorly. The apex on each side extends into the base of the neck superior to the first rib. The broad concave inferior portion, or **base,** of each lung rests on the superior surface of the diaphragm.

LOBES AND SURFACES OF THE LUNGS

The lungs have distinct **lobes** that are separated by deep fissures (Figure 23-7•). The right lung has three

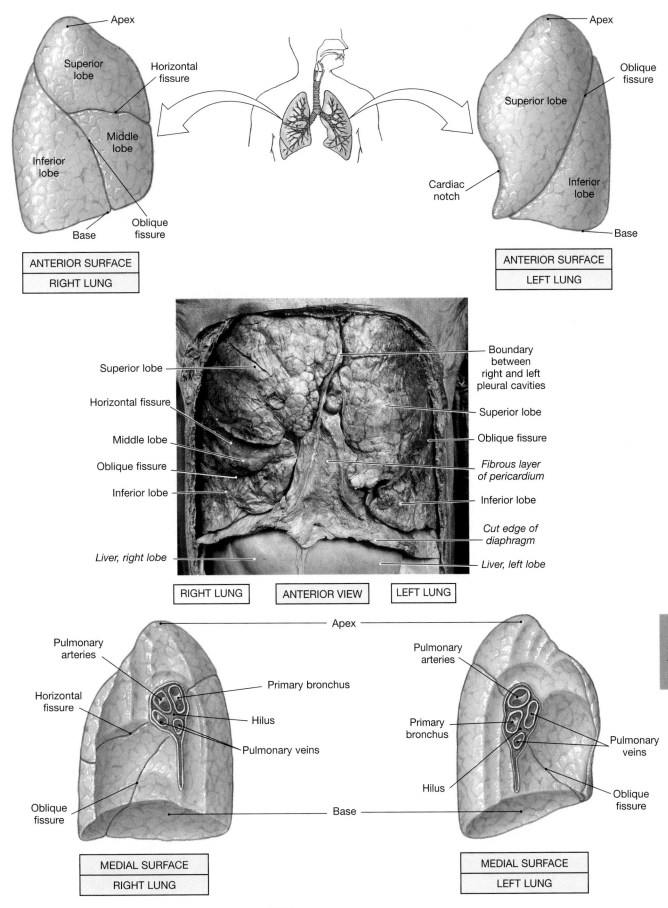

•**FIGURE 23-7** **The Gross Anatomy of the Lungs.**   AM   *Plates 7.4–7.6*

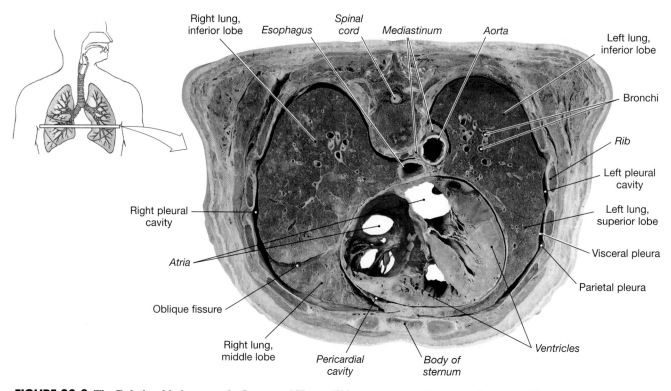

•**FIGURE 23-8** **The Relationship between the Lungs and Heart.** This transverse section was taken at the level of the cardiac notch.

lobes: *superior, middle,* and *inferior,* separated by the *horizontal* and *oblique fissures.* The left lung has only two lobes: *superior* and *inferior,* separated by the *oblique fissure.* The right lung is broader than the left, because most of the heart and great vessels project into the left thoracic cavity. However, the left lung is longer than the right lung, because the diaphragm rises on the right side to accommodate the mass of the liver.

The curving anterior and lateral surfaces of each lung follow the inner contours of the rib cage. The medial surface, which contains the hilus, has a more irregular shape. The medial surfaces of both lungs bear grooves that mark the positions of the great vessels and the heart (Figures 23-7 and 23-8•). The heart is located to the left of the midline, and the corresponding impression is larger in the left lung than in the right. In anterior view, the medial edge of the right lung forms a vertical line, whereas the medial margin of the left lung is indented at the **cardiac notch.**

Tuberculosis

Tuberculosis (too-ber-kū-LŌ-sis), or TB, results from a bacterial infection of the lungs; other organs may be invaded as well. The bacterium *Mycobacterium tuberculosis* may colonize the respiratory passageways, the interstitial spaces, the alveoli, or a combination of the three. Symptoms are variable but generally include coughing and chest pain, with fever, night sweats, fatigue, and weight loss. **AM** *Tuberculosis*

THE BRONCHI

The primary bronchi and their branches form the **bronchial tree** (Figure 23-9•). Because the left and right primary bronchi are outside the lungs, they are also called **extrapulmonary bronchi.** As the primary bronchi enter the lungs, they divide to form smaller passage-

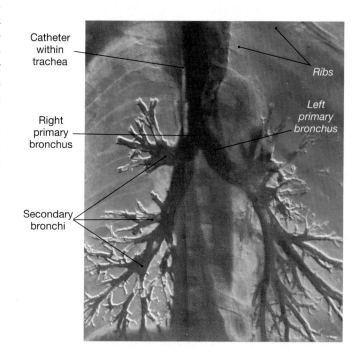

•**FIGURE 23-9** **The Bronchial Tree**

ways (Figures 23-6a, and 23-10a•). Those branches are collectively called the **intrapulmonary bronchi.**

Each primary bronchus divides to form **secondary bronchi,** also known as *lobar bronchi.* In each lung, one secondary bronchus goes to each lobe, so the right lung has three secondary bronchi and the left lung has two.

Figure 23-10• follows the branching pattern of the left primary bronchus as it enters the lung. (The number of branches has been reduced for clarity.) In each lung, the secondary bronchi branch to form **tertiary bronchi,** or *segmental bronchi.* The branching pattern differs between the two lungs, but each tertiary bronchus ultimately supplies air to a single **bronchopulmonary segment,** a specific region of one lung. The right lung has 10 bronchopulmonary segments (Figure 23-10b•). During development, the left lung also has 10 segments, but subsequent fusion of adjacent tertiary bronchi generally reduces that number to eight or nine.

The walls of the primary, secondary, and tertiary bronchi contain progressively lesser amounts of cartilage. In the secondary and tertiary bronchi, the cartilages form plates arranged around the lumen. These cartilages serve the same purpose as the rings of cartilage in the trachea and primary bronchi. As the amount of cartilage decreases, the relative amount of smooth muscle increases. With less cartilaginous support, the amount of tension in those smooth muscles has a greater effect on bronchial diameter and the resistance to airflow.

Bronchoscopy

During a respiratory infection, the bronchi and bronchioles can become inflamed. In this condition, called **bronchitis,** the smaller passageways may become greatly constricted, leading to difficulties in breathing. One method of investigating the status of the respiratory passageways is the use of a bronchoscope. A **bronchoscope** is a fiber-optic bundle small enough to be inserted into the trachea and steered along the conducting passageways to the level of the smaller bronchioles. This procedure is called **bronchoscopy** (brong-KOS-ko-pē). In addition to permitting the direct visualization of bronchial structures, the bronchoscope can collect tissue or mucus samples from the respiratory tract. In **bronchography** (brong-KOG-ra-fē), a bronchoscope or catheter introduces a radiopaque material into the bronchi. This technique can permit detailed X-ray analysis of bronchial masses, such as tumors, or other obstructions.

THE BRONCHIOLES

Each tertiary bronchus branches several times within the bronchopulmonary segment, giving rise to multiple **bronchioles.** These passageways branch further into the finest conducting branches, called **terminal bronchioles.** Roughly 6500 terminal bronchioles are supplied by each tertiary bronchus. The lumen of each terminal bronchiole has a diameter of 0.3–0.5 mm.

The walls of bronchioles, which lack cartilaginous supports, are dominated by smooth muscle tissue (Figure 23-10b•). In functional terms, the bronchioles are to the respiratory system what the arterioles are to the cardiovascular system. Varying the diameter of the bronchioles controls the amount of resistance to airflow and the distribution of air in the lungs.

The autonomic nervous system regulates the activity in this smooth muscle layer and thereby controls the diameter of the bronchioles. Sympathetic activation leads to **bronchodilation,** the enlargement of the airway diameter. Parasympathetic stimulation leads to **bronchoconstriction,** a reduction in the diameter of the airways. Bronchoconstriction also occurs during allergic reactions such as anaphylaxis, in response to histamine released by activated mast cells and basophils. *pp. 788–789*

Bronchodilation and bronchoconstriction adjust the resistance to airflow toward or away from the respiratory exchange surfaces. Tension in the smooth muscles commonly throws the bronchiolar mucosa into a series of folds, limiting airflow; excessive stimulation, as in *asthma* (AZ-muh), can almost completely prevent airflow along the terminal bronchioles. AM *Asthma*

Pulmonary Lobules

The connective tissues of the root of each lung extend into the lung's parenchyma. The fibrous partitions, or *trabeculae,* contain elastic fibers, smooth muscles, and lymphatic vessels. The trabeculae branch repeatedly, dividing the lobes into ever-smaller compartments. The branches of the conducting passageways, pulmonary vessels, and nerves of the lungs follow these trabeculae. The finest partitions, or **interlobular septa** (*septum,* a wall), divide the lung into **pulmonary lobules** (LOB-ūlz). Each pulmonary lobule is supplied by branches of the pulmonary arteries, pulmonary veins, and respiratory passageways (Figure 23-10•). The connective tissues of the septa are in turn continuous with those of the *visceral pleura,* the serous membrane covering the lungs.

Each terminal bronchiole delivers air to a single pulmonary lobule. Within the lobule, the terminal bronchiole branches to form several **respiratory bronchioles.** These are the thinnest and most delicate branches of the bronchial tree. They deliver air to the gas exchange surfaces of the lungs.

The preliminary filtration and humidification of incoming air are completed before incoming air leaves the terminal bronchioles. The epithelial cells of the terminal bronchioles and respiratory bronchioles are cuboidal, with only scattered cilia, and no goblet cells or underlying mucous glands are present.

23

•**FIGURE 23-10 The Bronchi and Lobules of the Lung.**
(a) The branching pattern of bronchi in the left lung, simplified. **(b)** The structure of a single pulmonary lobule, part of a bronchopulmonary segment.

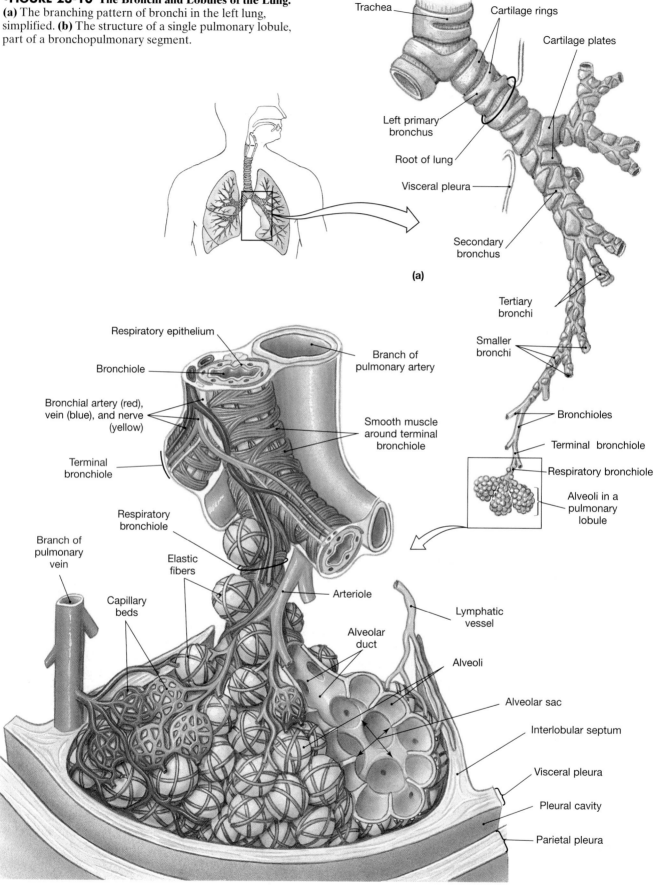

Trachea

Cartilage rings

Cartilage plates

Left primary bronchus

Root of lung

Visceral pleura

Secondary bronchus

(a)

Tertiary bronchi

Smaller bronchi

Bronchioles

Terminal bronchiole

Respiratory bronchiole

Alveoli in a pulmonary lobule

Respiratory epithelium

Bronchiole

Branch of pulmonary artery

Bronchial artery (red), vein (blue), and nerve (yellow)

Smooth muscle around terminal bronchiole

Terminal bronchiole

Respiratory bronchiole

Elastic fibers

Branch of pulmonary vein

Capillary beds

Arteriole

Lymphatic vessel

Alveolar duct

Alveoli

Alveolar sac

Interlobular septum

Visceral pleura

Pleural cavity

Parietal pleura

23

(b)

ALVEOLAR DUCTS AND ALVEOLI

Respiratory bronchioles are connected to individual alveoli and to multiple alveoli along regions called **alveolar ducts** (Figures 23-10b and 23-11•). Alveolar ducts end at **alveolar sacs,** common chambers connected to multiple individual alveoli. Each lung contains about 150 million alveoli, and their abundance gives the lung an open, spongy appearance (Figure 23-11a•). An extensive network of capillaries is associated with each alveolus; the capillaries are surrounded by a network of elastic fibers (Figure 23-12a•). This elastic tissue helps maintain the relative positions of the alveoli and respiratory bronchioles. Recoil of these fibers during exhalation reduces the size of the alveoli and helps push air out of the lungs.

The alveolar epithelium consists primarily of simple squamous epithelium (Figure 23-12b•). The squamous epithelial cells, called *Type I cells,* are unusually thin and delicate. Roaming **alveolar macrophages** (*dust cells*) patrol the epitheleal surface, phagocytizing any particulate matter that has eluded other respiratory defenses and reached the alveolar surfaces. **Septal cells,** also called *Type II cells,* are scattered among the squamous cells. These large cells produce **surfactant** (sur-FAK-tant), an oily secretion containing a mixture of phospholipids and proteins. Surfactant is secreted onto the alveolar surfaces, where it forms a superficial coating over a thin layer of water.

Surfactant reduces surface tension in the liquid coating the alveolar surface. Recall from Chapter 2 that *surface tension* results from the attraction between water molecules at an air–water boundary. ∞ *p. 36* Surface tension creates a barrier that keeps small objects from entering the water, but it also tends to collapse small bubbles. Alveolar walls are very delicate—rather like air bubbles; without surfactant, the surface tension would be so high that the alveoli would collapse. Surfactant forms a thin surface layer

(a)

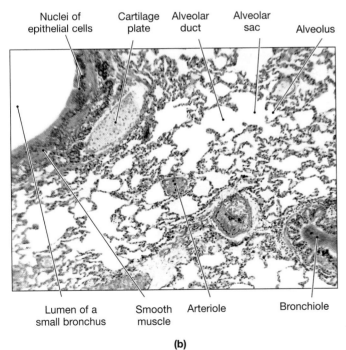

(b)

•**FIGURE 23-11 The Bronchioles. (a)** The distribution of a respiratory bronchiole supplying a portion of a lobule. **(b)** Alveolar sacs and alveoli. (LM × 42) **(c)** An SEM of the lung. Notice the open, spongy appearance of the lung tissue; compare with Figure 23-10b.

(c)

23

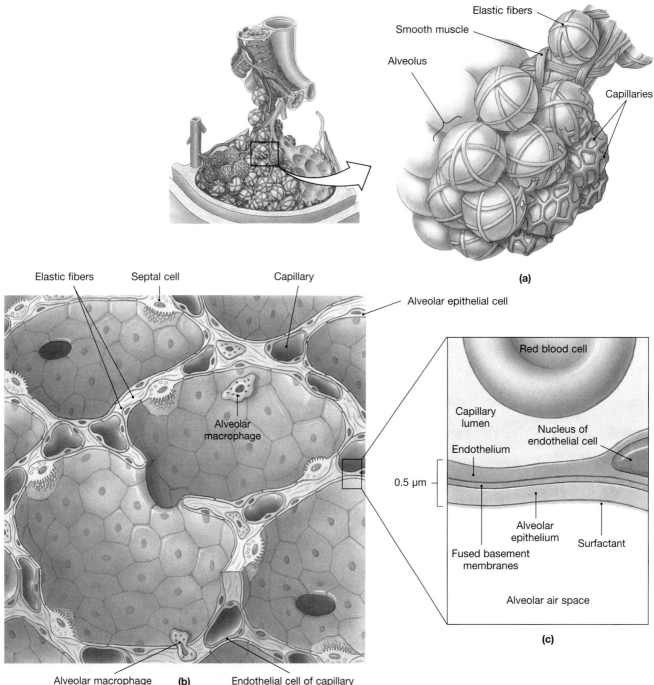

•**FIGURE 23-12 Alveolar Organization. (a)** The basic structure of a portion of a single lobule. A network of capillaries, supported by elastic fibers, surrounds each alveolus. Respiratory bronchioles also contain wrappings of smooth muscle that can change the diameter of these airways. **(b)** A diagrammatic view of alveolar structure. A single capillary may be involved with gas exchange across several alveoli simultaneously. **(c)** The respiratory membrane, which consists of an alveolar epithelial cell, a capillary endothelial cell, and their fused basement membranes.

that interacts with the water molecules, reducing the surface tension and keeping the alveoli open.

If septal cells produce inadequate amounts of surfactant due to injury or genetic abnormalities, the alveoli will collapse after each exhalation and respiration will become difficult. On each breath, the inhalation must then be forceful enough to pop open the alveoli.

A person who does not produce enough surfactant is soon exhausted by the effort required to keep inflating and deflating the lungs. This condition is called *respiratory distress syndrome.* ⟨AM⟩ *Respiratory Distress Syndrome (RDS)*

Gas exchange occurs across the **respiratory membrane** of the alveoli. The respiratory membrane is a composite structure consisting of three parts (Figure 23-12c•):

1. The squamous epithelial cell lining the alveolus.
2. The endothelial cell lining an adjacent capillary.
3. The fused basement membranes that lie between the alveolar and endothelial cells.

At the respiratory membrane, the total distance separating alveolar air from blood can be as little as 0.1 μm; it averages about 0.5 μm. Diffusion across the respiratory membrane proceeds very rapidly, because (1) the distance is small and (2) both oxygen and carbon dioxide are lipid-soluble. The membranes of the epithelial and endothelial cells thus do not pose a barrier to the movement of oxygen and carbon dioxide between blood and alveolar air spaces.

Pneumonia

Pneumonia (noo-MŌ-nē-uh) develops from a pathogenic infection or any other stimulus that causes inflammation of the lobules of the lung. As inflammation occurs, fluids leak into the alveoli and the respiratory bronchioles swell and constrict. Respiratory function deteriorates as a result. When bacteria are involved, they are generally types that normally inhabit the mouth and pharynx but have somehow managed to evade the respiratory defenses. Pneumonia becomes more likely when the respiratory defenses have already been compromised by other factors, such as epithelial damage from smoking or the breakdown of the immune system in AIDS. The most common pneumonia that develops in individuals with AIDS results from infection by the fungus *Pneumocystis carinii*. This fungus normally inhabits the alveoli, but the respiratory defenses of healthy individuals are able to prevent infection and tissue damage.

☑ Why are the cartilages that reinforce the trachea C-shaped?

☑ What would happen to the alveoli if surfactant were not produced?

☑ What path does air take in flowing from the glottis to the respiratory membrane?

THE BLOOD SUPPLY TO THE LUNGS

Your respiratory exchange surfaces receive blood from arteries of your pulmonary circuit. The pulmonary arteries enter the lungs at the hilus and branch with the bronchi as they approach the lobules. Each lobule receives an arteriole and a venule, and a network of capillaries surrounds each alveolus as part of the respiratory membrane. In addition to providing a mechanism for gas exchange, the endothelial cells of the alveolar capillaries are the primary source of *angiotensin-converting enzyme (ACE)*. This enzyme, which converts circulating angiotensin I to angiotensin

II, plays an important role in the regulation of blood volume and blood pressure. ∞ *p. 716*

Blood from the alveolar capillaries passes through the pulmonary venules and then enters the pulmonary veins, which deliver it to the left atrium. The conducting portions of your respiratory tract receive blood from the *external carotid arteries* (nasal passages and larynx), the *thyrocervical trunks* (the inferior larynx and trachea), and the *bronchial arteries* (the bronchi and bronchioles). (See Figures 21-23, 21-24, and 21-26•, pp. 726, 728, 730.) The capillaries supplied by the bronchial arteries provide oxygen and nutrients to the tissues of conducting passageways of your lungs. The venous blood from the bronchial capillaries ultimately flows into the pulmonary veins, bypassing the rest of the systemic circuit and diluting the oxygenated blood leaving the alveoli.

Pulmonary Embolisms

Blood pressure in the pulmonary circuit is usually relatively low, with systemic pressures of 30 mm Hg or less. With such pressures, pulmonary vessels can easily become blocked by small blood clots, fat masses, or air bubbles in the pulmonary arteries. Because the lungs receive the entire cardiac output, any such objects drifting in blood are likely to cause problems almost at once. The blockage of a branch of a pulmonary artery will stop blood flow to a group of lobules or alveoli. This condition is a **pulmonary embolism.** *Atherosclerosis* (damage to arterial walls and plaque formation; ∞, *p. 696*) and *venous thrombosis* (the blockage of a vein by a blood clot) can promote the development of a pulmonary embolism. In both cases, small blood clots tend to form, break loose, and drift in the bloodstream. If a pulmonary embolism is in place for several hours, the alveoli will permanently collapse. If the blockage occurs in a major pulmonary vessel rather than a minor tributary, pulmonary resistance increases. This resistance places extra strain on the right ventricle, which may be unable to maintain cardiac output. Congestive heart failure can result.

23

THE PLEURAL CAVITIES AND PLEURAL MEMBRANES

The thoracic cavity has the shape of a broad cone. Its walls are the rib cage, and the muscular diaphragm forms the floor. The two **pleural cavities** are separated by the mediastinum (Figure 23-8•, p. 808). Each lung occupies a single pleural cavity, which is lined by a serous membrane called the **pleura** (PLOO-ra; plural, *pleurae*). ∞ *p. 129* The pleura consists of two layers: the parietal pleura and the visceral pleura. The **parietal pleura** covers the inner surface of the thoracic wall and extends over the diaphragm and mediastinum. The **visceral pleura** covers the outer surfaces of the lungs, extending into the fissures between the lobes. Each pleural cavity actually represents a potential space rather than an open chamber, because the parietal and visceral pleurae are usually in close contact.

Both pleurae secrete a small amount of **pleural fluid.** Pleural fluid gives a moist, slippery coating that provides lubrication, thereby reducing friction between the parietal and visceral surfaces as you breathe. Samples of pleural fluid, obtained by means of a long needle inserted between the ribs, are sometimes obtained for diagnostic purposes. This sampling procedure is called *thoracentesis.* The fluid extracted is then examined for the presence of bacteria, blood cells, or other abnormal components. ∞ *p. 222*

In some disease states, the normal coating of pleural fluid is unable to prevent friction between the opposing pleural surfaces. The result is pain and pleural inflammation, a condition called **pleurisy.** When pleurisy develops, the secretion of pleural fluid may be excessive, or the inflamed pleurae may adhere to one another, limiting relative movement. In either case, breathing becomes difficult and prompt medical attention is required.

☑ Which arteries supply blood to the conducting portions and respiratory exchange surfaces of the lungs?

☑ List the functions of the pleura. What does it secrete?

RESPIRATORY PHYSIOLOGY

The general term *respiration* refers to two integrated processes: *external respiration* and *internal respiration.* The precise definitions of these terms vary from reference to reference. In this discussion, **external respiration** includes all the processes involved in the exchange of oxygen and carbon dioxide between the body's interstitial fluids and the external environment. The goal of external respiration, and the primary function of the respiratory system, is to meet the respiratory demands of cells. **Internal respiration** is the absorption of oxygen and the release of carbon dioxide by those cells. We shall consider the biochemical pathways responsible for oxygen consumption and carbon dioxide generation by mitochondria, often called *cellular respiration,* in Chapter 25.

Our discussion focuses on three integrated steps involved in external respiration (Figure 23-13●):

1. *Pulmonary ventilation,* or breathing, which involves the physical movement of air into and out of the lungs.
2. *Gas diffusion,* across the respiratory membrane between alveolar air spaces and alveolar capillaries and across capillary walls between blood and other tissues.
3. *The transport of oxygen and carbon dioxide* between alveolar capillaries and capillary beds in other tissues.

Abnormalities affecting any one of these steps will ultimately affect the gas concentrations of interstitial fluids and thereby cellular activities as well. If the oxygen content declines, the affected tissues will become oxygen-starved. **Hypoxia,** or low tissue oxygen levels, places severe limits on the metabolic activities of the affected area. For example, the effects of *coronary ischemia* result from chronic hypoxia affecting cardiac muscle cells. ∞ *p. 667* If the supply of oxygen is cut off completely, the condition of **anoxia** (a-NOKS-ē-uh; *a-,* without + *ox-,* oxygen) results. Anoxia kills cells very quickly. Much of the damage caused by strokes and heart attacks is the result of localized anoxia.

PULMONARY VENTILATION

Pulmonary ventilation is the physical movement of air into and out of the respiratory tract. The primary function of pulmonary ventilation is to maintain adequate *alveolar ventilation,* air movement into and out of the alveoli. Alveolar ventilation prevents the buildup of carbon dioxide in the alveoli and ensures a continuous supply of oxygen that keeps pace with absorption

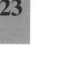

●**FIGURE 23-13 An Overview of Key Steps in Respiration.** This figure will be repeated, in reduced and simplified form as Navigator icons, in key figures throughout this chapter as we change topics.

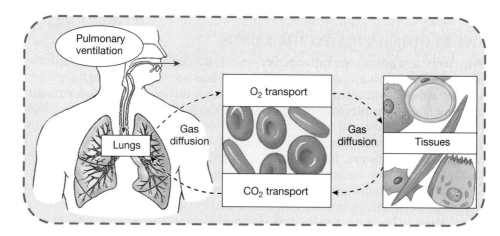

by the bloodstream. To understand this mechanical process, we need to look at basic physical principles governing the movement of air. One of the most basic principles is that our bodies and everything around us are compressed by the weight of Earth's atmosphere. Although we are seldom reminded of the fact, this **atmospheric pressure** has several important physiological effects.

Gas Pressure and Volume (Boyle's Law)

The primary differences between liquids and gases reflect the interactions between individual molecules. Although the molecules in a liquid are in constant motion, they are held closely together by weak interactions. In a polar liquid such as water, these interactions include hydrogen bonding between adjacent atoms. But, because the electrons of adjacent atoms tend to repel one another, liquids tend to resist compression. If you squeeze a balloon filled with water, it will distort into a different shape, but the volume of the new shape will be the same as that of the original.

In a gas, such as air, the molecules bounce around as independent entities. At normal atmospheric pressures, gas molecules are much farther apart than the molecules in a liquid, so the density of air is relatively low. The forces acting between gas molecules are minimal (the molecules are too far apart for weak interactions to occur), so an applied pressure can push them closer together. Consider a sealed container of air at atmospheric pressure. The pressure exerted by the enclosed gas results from the collision of gas molecules with the walls of the container. The more collisions, the higher the pressure.

You can change the gas pressure within a sealed container by changing the volume of the container, thereby giving the gas molecules more or less room in which to bounce around. If you decrease the volume of the container, the collisions will occur more frequently over a given time period, elevating the pressure of the gas (Figure 23-14a•). If you increase the volume, fewer collisions will occur per unit time, because it will take longer for a gas molecule to travel from one wall to another. As a result, the gas pressure inside the container will decline (Figure 23-14b•).

An inverse relationship thus exists between the pressure (P) and volume (V) of a gas in a closed container: Gas pressure is inversely proportional to volume. *If you decrease the volume of a gas, its pressure will rise; if you increase the volume of a gas, its pressure will fall.* In particular, the relationship between pressure and volume is reciprocal: If you double the external pressure on a flexible container, its volume will drop by half. If you reduce the external pressure by half, the volume of the container will double. This relationship, $P = 1/V$, first recognized by Robert Boyle in the 1600s, is called **Boyle's law.** [AM] *Boyle's Law and Air Overexpansion Syndrome*

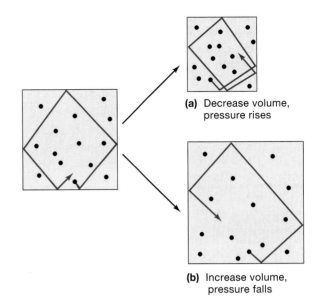

(a) Decrease volume, pressure rises

(b) Increase volume, pressure falls

•**FIGURE 23-14 Respiratory Pressure and Volume Relationships.** Gas molecules in a sealed container bounce off the walls and off one another, traveling the distance indicated in a given period of time. **(a)** If the volume decreases, each molecule will travel the same distance in that same time period but will strike the walls more frequently. The pressure exerted by the gas increases. **(b)** If the volume of the container increases, each molecule will strike the walls less often, lowering the pressure in the container.

Pressure and Airflow to the Lungs

Air will flow from an area of higher pressure to an area of lower pressure. This tendency for directed airflow, plus the pressure/volume relationship of Boyle's law, provide the basis for pulmonary ventilation. A single respiratory cycle consists of an *inspiration,* or inhalation, and an *expiration,* or exhalation. Inhalation and exhalation involve changes in the volume of the lungs. These changes create pressure gradients that move air into or out of the respiratory tract.

Each lung lies within a pleural cavity. The parietal and visceral pleurae are separated by only a thin film of pleural fluid. The two membranes can slide across one another, but they are held together by that fluid film. You encounter the same principle when you set a wet glass on a smooth tabletop. You can slide the glass easily, but when you try to lift it, you experience considerable resistance. As you pull the glass away from the tabletop, you create a powerful suction. The only way to defeat it is to tilt the glass so that air is pulled between the glass and the table, breaking the fluid bond.

A comparable fluid bond exists between the parietal pleura and the visceral pleura covering the lungs. As a result, the surface of each lung sticks to the inner wall of the chest and the superior surface of the diaphragm. Movements of the chest wall or diaphragm

23

thus directly affect the volume of the lungs. The basic principle is shown in Figure 23-15a•. The volume of the thoracic cavity changes when the diaphragm changes position or the rib cage moves:

- The diaphragm forms the floor of the thoracic cavity. The relaxed diaphragm has the shape of a dome that projects superiorly into the thoracic cavity. When the diaphragm contracts, it tenses and moves inferiorly. This movement increases the volume of the thoracic cavity and exerts pressure on the contents of the abdominopelvic cavity. When the diaphragm relaxes, it returns to its original position; the volume of the thoracic cavity decreases.

- Owing to the nature of the articulations between the ribs and the vertebrae, superior movement of the rib cage increases the depth and width of the thoracic cavity. Inferior movement of the rib cage reverses the process and reduces the volume of the thoracic cavity.

At the start of a breath, pressures inside and outside the thoracic cavity are identical, and no air moves into or out of the lungs (Figure 23-15b•). When the thoracic cavity enlarges, the pleural cavities and lungs expand to fill the additional space (Figure 23-15c•). This expansion lowers the pressure inside the lungs. Air now enters the respiratory passageways, because the pressure inside the lungs (P_{inside}) is lower than atmospheric pressure (pressure outside, or $P_{outside}$). Air continues to enter until the volume stops increasing and the internal pressure is the same as that outside. When the thoracic cavity decreases in volume, pressures rise inside the lungs, forcing air out of the respiratory tract (Figure 23-15d•).

Compliance   The **compliance** of the lungs is an indication of their expandability. The lower the compliance, the greater the force required to fill and empty the lungs. Factors affecting compliance include the following:

- *The connective tissue structure of the lungs.* The loss of supporting tissues resulting from alveolar damage, as in *emphysema*, increases compliance (p. 839).

- *The level of surfactant production.* On exhalation, the collapse of alveoli as a result of inadequate surfactant, as in respiratory distress syndrome, reduces compliance (p. 812).

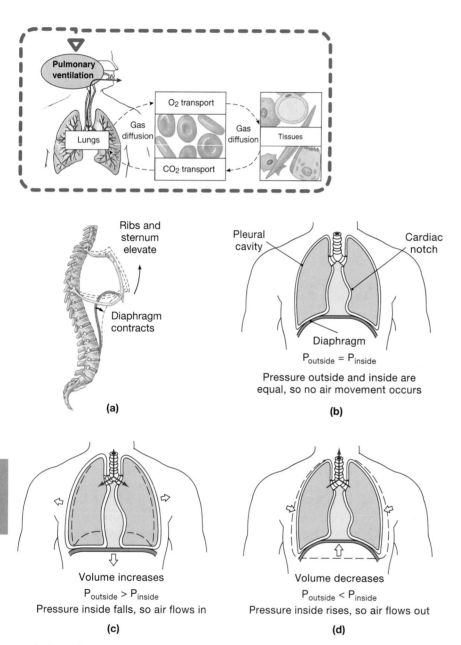

(a)

(b)
$P_{outside} = P_{inside}$
Pressure outside and inside are equal, so no air movement occurs

(c)
Volume increases
$P_{outside} > P_{inside}$
Pressure inside falls, so air flows in

(d)
Volume decreases
$P_{outside} < P_{inside}$
Pressure inside rises, so air flows out

•**FIGURE 23-15 Mechanisms of Pulmonary Ventilation.** The Navigator icon in the dashed box repeats Figure 23-13 but highlights the part that is relevant to *this* figure, so you can see how pulmonary ventilation relates to the entire process of respiration. **(a)** As the ribs are elevated or the diaphragm is depressed, the volume of the thoracic cavity increases. **(b)** An anterior view at rest, with no air movement. **(c)** *Inhalation:* Elevation of the rib cage and contraction of the diaphragm increase the size of the thoracic cavity. Pressure decreases, and air flows into the lungs. **(d)** *Exhalation:* When the rib cage returns to its original position, the volume of the thoracic cavity decreases. Pressure rises, and air moves out of the lungs.

■ *The mobility of the thoracic cage.* Arthritis or other skeletal disorders that affect the articulations of the ribs or spinal column will also reduce compliance.

At rest, the muscular activity involved in pulmonary ventilation accounts for 3–5 percent of the resting energy demand. If compliance is reduced, that figure climbs dramatically, and an individual may become exhausted simply trying to continue breathing.

Pressure Changes during Inhalation and Exhalation

To understand the mechanics of respiration and the principles of gas exchange, we must know the pressures recorded inside and outside the respiratory tract. We can report pressure readings in several ways (Table 23-1); we will use millimeters of mercury (mm Hg), as we did for blood pressure. Atmospheric pressure is also measured in *atmospheres;* one atmosphere of pressure *(1 atm)* is equivalent to 760 mm Hg.

The Intrapulmonary Pressure   The direction of airflow is determined by the relationship between atmospheric pressure and intrapulmonary pressure. **Intrapulmonary** (in-tra-PUL-mo-ner-ē) **pressure,** or **intra-alveolar** (in-tra-al-VĒ-ō-lar) **pressure,** is the pressure measured inside the respiratory tract, at the alveoli.

When you are relaxed and breathing quietly, the pressure difference between atmospheric and intrapulmonary pressures is relatively small. On inhalation, your lungs expand, and the intrapulmonary pressure drops to about 759 mm Hg. Because the intrapulmonary pressure is 1 mm Hg below atmospheric pressure, it is generally reported as −1 mm Hg. On exhalation, your lungs recoil and intrapulmonary pressure rises to 761 mm Hg, or +1 mm Hg (Figure 23-16a●).

The size of the pressure gradient increases when you breathe heavily. When a trained athlete breathes at maximum capacity, the pressure differentials can reach −30 mm Hg during inhalation and +100 mm Hg if the individual is straining with the glottis kept closed. This is one reason you are told to exhale while lifting weights; exhaling keeps your intrapulmonary

pressures and peritoneal pressure from climbing so high that an alveolar rupture or hernia could occur. Exhaling less forcefully against a closed glottis, a procedure known as the *Valsalva maneuver,* causes reflexive changes in blood pressure and cardiac output due to the compression of the aorta and venae cavae. When internal pressures rise, the venae cavae collapse and the venous return decreases. The resulting fall in cardiac output and blood pressure stimulates the aortic and carotid baroreceptors. ∞ *p. 713* Their stimulation causes a reflexive increase in heart rate and peripheral vasoconstriction. When the glottis opens and pressures return to normal, venous return and cardiac output rise. Because vasoconstriction has occurred, blood pressure then rises sharply. This rise inhibits the baroreceptors, and cardiac output, heart rate, and blood pressure quickly return to normal levels. The Valsalva maneuver is thus a simple way to check the cardiovascular responses to changes in arterial pressure and venous return.

The Intrapleural Pressure   **Intrapleural pressure** is the pressure measured in the space between the parietal and visceral pleurae. Intrapleural pressure averages about −4 mm Hg (Figure 23-16b●). It can reach −18 mm Hg during a powerful inhalation. This pressure is below atmospheric pressure due to the relationship between the lungs and the body wall. We noted earlier that the lungs are highly elastic. In fact, they would collapse to about 5 percent of their normal resting volume if the elastic fibers could recoil completely. The elastic fibers cannot recoil significantly, however, because they are not strong enough to overcome the fluid bond between the parietal and visceral pleurae.

The elastic fibers continuously fight that fluid bond and pull the lungs away from the chest wall and diaphragm. This pull lowers the intrapleural pressure slightly. Because the elastic fibers remain stretched even after a full exhalation, intrapleural pressure remains below atmospheric pressure throughout normal cycles of inhalation and exhalation. The cyclical changes in the intrapleural pressure are responsible

23

TABLE 23-1   The Four Most Common Methods of Reporting Gas Pressures

millimeters of mercury (mm Hg): This is the most common method of reporting blood pressure and gas pressures. Normal atmospheric pressure is approximately 760 mm Hg.

torr: This unit of measurement is preferred by many respiratory therapists; it is also commonly used in Europe and in some technical journals. One torr is equivalent to 1 mm Hg; in other words, normal atmospheric pressure is equal to 760 torr.

centimeters of water (cm H_2O): In a hospital setting, anesthetic gas pressures and oxygen pressures are commonly reported in centimeters of water. One cm H_2O is equivalent to 0.735 mm Hg; normal atmospheric pressure is 1033.6 cm H_2O.

pounds per square inch (psi): Pressures in compressed gas cylinders and other industrial applications are generally reported in psi. Normal atmospheric pressure at sea level is approximately 15 psi.

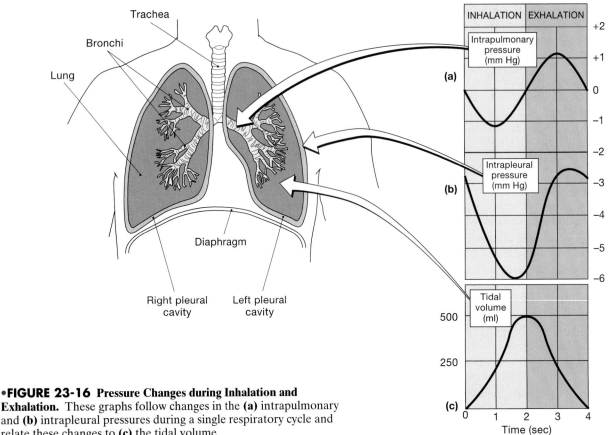

•**FIGURE 23-16** **Pressure Changes during Inhalation and Exhalation.** These graphs follow changes in the **(a)** intrapulmonary and **(b)** intrapleural pressures during a single respiratory cycle and relate these changes to **(c)** the tidal volume.

for the *respiratory pump,* the mechanism that assists the venous return to the heart. ∞ *pp. 710–711*

The Respiratory Cycle

A **respiratory cycle** is a single cycle of inhalation and exhalation. The **tidal volume** is the amount of air you move into or out of your lungs during a single respiratory cycle. The graphs in Figure 23-16a,b• follow the intrapulmonary and intrapleural pressures during a single respiratory cycle of an individual at rest. These graphs are also compared with the tidal volume (Figure 23-16c•).

At the start of the cycle, the intrapulmonary and atmospheric pressures are equal, and no air movement is occurring. Inhalation begins with the fall of intrapleural pressure that accompanies the expansion of the thoracic cavity. This pressure gradually falls to approximately −6 mm Hg. Over this period, intrapulmonary pressure drops to just under −1 mm Hg; it then begins to rise as air flows into the lungs. When exhalation begins, intrapleural and intrapulmonary pressures rise rapidly, forcing air out of the lungs. At the end of exhalation, air movement again ceases when the pressure difference between intrapulmonary and atmospheric pressures has been eliminated. The amount of air moved into the lungs during inhalation

is equal to the amount moved out of the lungs during exhalation. That amount is the tidal volume.

Pneumothorax

An injury to the chest wall that penetrates the parietal pleura or damages the alveoli and the visceral pleura can allow air into the pleural cavity. This **pneumothorax** (noo-mō-THO-raks) breaks the fluid bond between the pleurae and allows the elastic fibers to recoil. The result is called a "collapsed lung," or **atelectasis** (at-e-LEK-ta-sis; *ateles,* imperfect + *ektasis,* expansion). The treatment for a collapsed lung involves the removal of as much of the air as possible from the affected pleural cavity before the opening is sealed. This treatment lowers the intrapleural pressure and reinflates the lung.

Respiratory Muscles

In Chapter 11, we introduced the skeletal muscles involved in respiratory movements. Of those, the most important are the *diaphragm* and the *external intercostal muscles* (Figures 11-12, 11-14b, and 11-15b•, pp. 331, 337, and 338). They act in normal breathing at rest. The **accessory respiratory muscles** become active when the depth and frequency of respiration must be increased markedly. These muscles include the

AT REST   INHALATION   EXHALATION

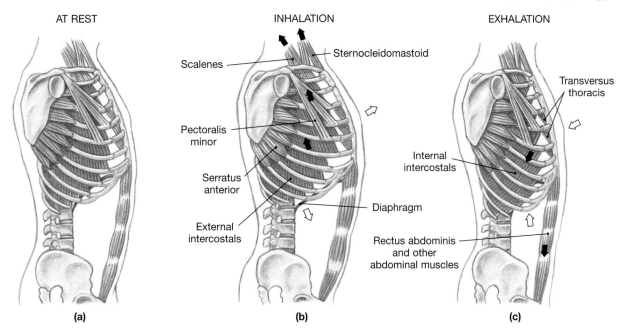

(a)   (b)   (c)

•**FIGURE 23-17** **The Respiratory Muscles.** The abdominal muscles that assist in exhalation are represented by a single muscle (the rectus abdominis). Diaphragmatic movements were also illustrated in Figure 23-15. **(a)** A lateral view at rest, with no air movement. **(b)** Inhalation, showing the muscles that elevate the ribs. **(c)** Exhalation, showing the muscles that depress the ribs.

internal intercostal, sternocleidomastoid, serratus anterior, pectoralis minor, scalene, transversus thoracis, transversus abdominis, external and *internal abdominal oblique,* and *rectus abdominis muscles* (Figures 11-11, pp. 328–329; 11-12; 11-14; and 23-17•).

Muscles Used in Inhalation Inhalation is an active process involving the contraction of one or more of these muscles:

▪ The contraction of the diaphragm increases the volume of the thoracic cavity by tensing and flattening its floor, and this increase draws air into the lungs. Diaphragmatic contraction is responsible for roughly 75 percent of the air movement in normal breathing at rest.

▪ The external intercostal muscles assist in inhalation by elevating the ribs. This action contributes roughly 25 percent to the volume of air in the lungs.

▪ Accessory muscles, including the sternocleidomastoid, serratus anterior, pectoralis minor, and scalene muscles, can assist the external intercostal muscles in elevating the ribs. These muscles increase the speed and amount of rib movement.

Muscles Used in Exhalation Exhalation is either passive or active, depending on the level of respiratory activity. When exhalation is active, it may involve one or more of the following muscles:

▪ The internal intercostal and transversus thoracis muscles depress the ribs and reduce the width and depth of the thoracic cavity.

▪ The abdominal muscles, including the external and internal abdominal oblique, transversus abdominis, and

rectus abdominis muscles, can assist the internal intercostal muscles in exhalation by compressing the abdomen and forcing the diaphragm upward.

Artificial Respiration

Artificial respiration is a technique to provide air to an individual whose respiratory muscles are no longer functioning. In **mouth-to-mouth resuscitation,** a rescuer provides ventilation by exhaling into the mouth or mouth and nose of the victim. After each breath, contact is broken to permit passive exhalation by the victim. Air provided in this way contains adequate oxygen to meet the needs of the victim. If the victim's cardiovascular system is nonfunctional as well, a technique called *cardiopulmonary resuscitation (CPR)* is required to maintain adequate blood flow and tissue oxygenation. ⃞ÁM *CPR*

23

Modes of Breathing

The respiratory muscles are used in various combinations, depending on the volume of air that must be moved into or out of the system. Respiratory movements are usually classified as *quiet breathing* or *forced breathing* by the pattern of muscle activity in the course of a single respiratory cycle.

Quiet Breathing In **quiet breathing,** or **eupnea** (ŪP-nē-uh), inhalation involves muscular contractions but exhalation is a passive process. Inhalation usually involves the contraction of both the diaphragm and the external intercostal muscles. The relative contributions of these muscles can vary:

■ During **diaphragmatic breathing,** or **deep breathing,** contraction of the diaphragm provides the necessary change in thoracic volume. Air is drawn into the lungs as the diaphragm contracts, and exhalation occurs passively when the diaphragm relaxes.

■ In **costal breathing,** or **shallow breathing,** the thoracic volume changes because the rib cage changes shape. Inhalation occurs when contractions of the external intercostal muscles elevate the ribs and enlarge the thoracic cavity. Exhalation occurs passively when these muscles relax.

During quiet breathing, expansion of the lungs stretches their elastic fibers. In addition, elevation of the rib cage stretches opposing skeletal muscles and elastic fibers in the connective tissues of the body wall. When the muscles of inhalation relax, these elastic components recoil, returning the diaphragm, the rib cage, or both to their original positions. This phenomenon is called **elastic rebound.**

Diaphragmatic breathing typically occurs at minimal levels of activity. As increased volumes of air are required, inspiratory movements become larger and the contribution of rib movement increases. Even when you are at rest, costal breathing can predominate when abdominal pressures, fluids, or masses restrict diaphragmatic movements. For example, pregnant women increasingly rely on costal breathing as the uterus enlarges and pushes the abdominal viscera against the diaphragm.

Forced Breathing **Forced breathing,** or **hyperpnea** (hī-PERP-nē-uh), involves active inspiratory and expiratory movements. Forced breathing calls on the accessory muscles to assist with inhalation, and exhalation involves contraction of the internal intercostal muscles. At absolute maximum levels of forced breathing, the abdominal muscles are used in exhalation. Their contraction compresses the abdominal contents, pushing them up against the diaphragm and further reducing the volume of the thoracic cavity.

Respiratory Rates

As you read this, you are probably breathing quietly, but when necessary your rate of respiration can change dramatically. Your **respiratory rate** is the number of breaths you take each minute. The normal respiratory rate of a resting adult ranges from 12 to 18 breaths each minute. Children breathe more rapidly, at rates of about 18–20 breaths per minute.

The Respiratory Minute Volume

We can calculate the amount of air moved each minute by multiplying the respiratory rate (f) by the tidal volume (V_T). This value is called the **respiratory minute volume,** $\dot{V}_E$. The tidal volume at rest varies from individual to individual, but it averages around

500 ml per breath. Therefore the respiratory minute volume at rest, at 12 breaths per minute, is approximately 6 liters per minute:

$$\underset{\left(\begin{array}{c}\text{volume of air moved}\\\text{each minute}\end{array}\right)}{\dot{V}_E} = \underset{\left(\begin{array}{c}\text{breaths per}\\\text{minute}\end{array}\right)}{f} \times \underset{\left(\begin{array}{c}\text{tidal}\\\text{volume}\end{array}\right)}{V_T}$$

$$= 12 \times 500 \text{ ml per minute}$$

$$= 6000 \text{ ml per minute}$$

$$= 6.0 \text{ liters per minute}$$

Alveolar Ventilation

The respiratory minute volume measures pulmonary ventilation and provides an indication of how much air is moving into and out of the respiratory tract. Only some of the inhaled air reaches the alveolar exchange surfaces. A typical inhalation pulls about 500 ml of air into your respiratory system. The first 350 ml inhaled travels along the conducting passageways and enters the alveolar spaces. The last 150 ml of inhaled air never gets farther than the conducting passageways and does not participate in gas exchange with blood. The volume of air in the conducting passages is known as the **anatomic dead space** (V_D). **Alveolar ventilation** ($\dot{V}_A$) is the amount of air reaching the alveoli each minute. The alveolar ventilation is less than the respiratory minute volume, because some of the air never reaches the alveoli but remains in the dead space of the lungs. We can calculate alveolar ventilation by subtracting the dead space from the tidal volume, using the following formula:

$$\underset{\left(\begin{array}{c}\text{alveolar}\\\text{ventilation}\end{array}\right)}{\dot{V}_A} = \underset{\left(\begin{array}{c}\text{breaths}\\\text{per minute}\end{array}\right)}{f} \times \underset{\left(\begin{array}{c}\text{tidal}\\\text{volume}\end{array} - \begin{array}{c}\text{anatomic}\\\text{dead space}\end{array}\right)}{(V_T - V_D)}$$

At rest, alveolar ventilation rates are approximately 4.2 liters per minute (12×350 ml). However, the gas arriving in the alveoli is significantly different from that of the surrounding atmosphere, because inhaled air always mixes with "used" air in the conducting passageways (the anatomic dead space) on its way to the exchange surfaces. The air in alveoli thus contains less oxygen and more carbon dioxide than does atmospheric air.

Relationships among V_T, $\dot{V}_E$, and $\dot{V}_A$

The respiratory minute volume can be increased by (1) increasing the tidal volume or (2) increasing the respiratory rate. Under maximum stimulation, the tidal volume can increase to roughly 4.8 liters. At peak respiratory rates of 40–50 breaths per minute and maximum cycles of inhalation and exhalation, the respiratory minute volume can approach 200 liters (about 55 gal) per minute.

23

In functional terms, the alveolar ventilation rate is more important than the respiratory minute volume, because it determines the rate of oxygen delivery to the alveoli:

- For a given respiratory rate, increasing the tidal volume will increase the alveolar ventilation rate.
- For a given tidal volume, increasing the respiratory rate will increase the alveolar ventilation rate.

The alveolar ventilation rate can change independently of the respiratory minute volume. In our previous example, the respiratory minute volume at rest was 6 liters and the alveolar ventilation rate was 4.2 l/min. If the respiratory rate rises to 20 breaths per minute but the tidal volume drops to 300 ml, the respiratory minute volume remains the same ($20 \times 300 = 6000$). However, the alveolar ventilation rate drops to only 3 l/min ($20 \times [300 - 150] = 3000$). Thus, whenever the demand for oxygen increases, both tidal volume *and* respiratory rate must be regulated closely. The mechanisms involved will be the focus of a later section.

Respiratory Performance and Volume Relationships

Only a small proportion of the air in the lungs is exchanged during a single quiet respiratory cycle; the tidal volume can be increased by inhaling more vigorously and exhaling more completely. We can divide the total volume of the lungs into *volumes* and *capacities* (Figure 23-18•). These values can be experimentally determined, and the values obtained are useful in diagnosing problems with pulmonary ventilation. Many respiratory volumes and capacities differ by gender, because adult females on average have smaller bodies, and thus smaller lung volumes, than do males. Representative values for both genders are indicated in Figure 23-18•.

Pulmonary volumes include the following:

- The **resting tidal volume** is the amount of air you move into or out of your lungs during a single respiratory cycle under resting conditions. The resting tidal volume averages about 500 ml in both males and females.
- The **expiratory reserve volume** (ERV) is the amount of air that you can voluntarily expel after you have completed a normal quiet respiratory cycle. If, with maximum use of the accessory muscles, you can expel an additional 1000 ml of air, your expiratory reserve volume is 1000 ml.
- The **residual volume** is the amount of air that remains in your lungs even after a maximal exhalation—typically about 1200 ml in males and 1100 ml in females. The **minimal volume,** a component of the residual volume, is the

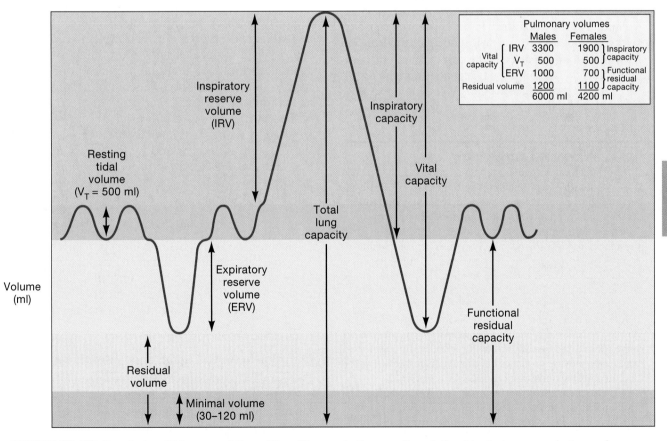

•**FIGURE 23-18** **Respiratory Volumes and Capacities.** The graph diagrams the relationships between respiratory volumes and capacities.

amount of air that would remain in your lungs if they were allowed to collapse. The minimal volume ranges from 30 to 120 ml, but, unlike other volumes, it cannot be measured in a healthy person. The minimal volume and the residual volume are very different, because the fluid bond between the lungs and the chest wall prevents the recoil of the elastic fibers. Some air remains in the lungs, even at minimal volume, because the surfactant coating the alveolar surfaces prevents their collapse.

■ The **inspiratory reserve volume** (IRV) is the amount of air that you can take in over and above the tidal volume. The inspiratory reserve volumes differ significantly by gender, because on average the lungs of males are larger than those of females. The inspiratory reserve volume of males averages 3300 ml versus 1900 ml in females.

We can determine respiratory capacities by adding the values of various volumes. Examples include the following:

■ The **inspiratory capacity** is the amount of air that you can draw into your lungs after you have completed a quiet respiratory cycle. It is the sum of the tidal volume and the inspiratory reserve volume.

■ The **functional residual capacity** (FRC) is the amount of air remaining in your lungs after you have completed a quiet respiratory cycle. It is the sum of the expiratory reserve volume and the residual volume.

■ The **vital capacity** is the maximum amount of air that you can move into or out of your lungs in a single respiratory cycle. It is the sum of the expiratory reserve, the tidal volume, and the inspiratory reserve. The vital capacity averages around 4800 ml in males and 3100 ml in females.

■ The **total lung capacity** is the total volume of your lungs. It is the sum of the vital capacity and the residual volume; it averages around 6000 ml in males and 4200 ml in females.

Pulmonary Function Tests

Pulmonary function tests monitor several aspects of respiratory function. A **spirometer** (spī-ROM-e-ter) measures parameters such as vital capacity, expiratory reserve, and inspiratory reserve. A **pneumotachometer** provides additional information by determining the rate of air movement throughout a respiratory cycle. A **peak flow meter** records the maximum rate of air movement during forced exhalation. These tests are relatively simple to perform, and they have considerable diagnostic significance. For example, in people with asthma, the constricted airways tend to close before an exhalation is completed. As a result, pulmonary function tests show a reduction in vital capacity, expiratory reserve volume, and peak flow rate. Air whistling through the constricted airways produces the characteristic "wheezing" that accompanies an asthma attack. [AM] *Asthma*

Conditions that restrict the maximum distensibility of the lungs have the same effect on vital capacity, because they lower the inspiratory reserve. However, because the airways are not affected, the expiratory reserve and expiratory flow rates are relatively normal.

☑ Mark breaks a rib that punctures the chest wall on his left side. What do you expect will happen to his left lung as a result?

☑ In pneumonia, fluid accumulates in the alveoli of the lungs. How would this accumulation affect vital capacity?

GAS EXCHANGE AT THE RESPIRATORY MEMBRANE

Pulmonary ventilation ensures that your alveoli are supplied with oxygen, and it removes the carbon dioxide arriving from your bloodstream. The actual process of gas exchange occurs between blood and alveolar air across the respiratory membrane. To understand these events, we consider (1) the *partial pressures* of the gases involved and (2) the diffusion of molecules between a gas and a liquid.

Dalton's Law and Partial Pressures

The air we breathe is not a single gas but a mixture of gases. Nitrogen molecules (N_2) are the most abundant, accounting for about 78.6 percent of the atmospheric gas molecules. Oxygen molecules (O_2), the second most abundant, account for roughly 20.9 percent of the atmospheric content. Most of the remaining 0.5 percent consists of water molecules, with carbon dioxide (CO_2) contributing a mere 0.04 percent.

Atmospheric pressure, 760 mm Hg, represents the combined effects of collisions involving each type of molecule in air. At any moment, 78.6 percent of those collisions will involve nitrogen molecules, 20.9 percent oxygen molecules, and so on. Thus each of the gases contributes to the total pressure in proportion to its relative abundance. This relationship is known as **Dalton's law.**

The **partial pressure** of a gas is the pressure contributed by a single gas in a mixture of gases. The partial pressure is abbreviated by the prefix *P* or *p*. All *the partial pressures added together equal the total pressure exerted by the gas mixture.* For the atmosphere, this relationship can be summarized as follows:

$$P_{N_2} + P_{O_2} + P_{H_2O} + P_{CO_2} = 760 \text{ mm Hg}$$

Because we know the individual percentages in air, we can easily calculate the partial pressure of each gas. For example, the partial pressure of oxygen, P_{O_2}, is 20.9 percent of 760 mm Hg, or roughly 159 mm Hg. The partial pressures for other atmospheric gases are included in Table 23-2.

Diffusion between Liquids and Gases (Henry's Law)

Differences in pressure move gas molecules from one place to another. Pressure differences also affect the movement of gas molecules into and out of solution. At a given temperature, *the amount of a particular gas in solution is directly proportional to the partial pressure of that gas.* This principle is known as **Henry's law.**

TABLE 23-2 Partial Pressures (mm Hg) and Normal Gas Concentrations (%) in Air

Source of Sample	Nitrogen (N_2)	Oxygen (O_2)	Carbon Dioxide (CO_2)	Water Vapor (H_2O)
Inhaled air (dry)	597 (78.6%)	159 (20.9%)	0.3 (0.04%)	3.7 (0.5%)
Alveolar air (saturated)	573 (75.4%)	100 (13.2%)	40 (5.2%)	47 (6.2%)
Exhaled air (saturated)	569 (74.8%)	116 (15.3%)	28 (3.7%)	47 (6.2%)

When a gas under pressure contacts a liquid, the pressure tends to force gas molecules into solution. At a given pressure, the number of dissolved gas molecules will rise until an equilibrium is established. At equilibrium, gas molecules diffuse out of the liquid as quickly as they enter it, so the total number of gas molecules in solution remains constant. If the partial pressure goes up, more gas molecules will go into solution; if the partial pressure goes down, gas molecules will come out of solution (Figure 23-19•).

You see Henry's law in action whenever you open a can of soda. The soda was put into the can under pressure, and the gas (carbon dioxide) is in solution.

When you open the can, the pressure falls and the gas molecules begin coming out of solution. The process will theoretically continue until an equilibrium develops between the surrounding air and the gas in solution. In fact, the volume of the can is so small, and the volume of the atmosphere so great, that within a half-hour or so, virtually all the carbon dioxide comes out of solution. You are then left with "flat" soda.

The actual *amount* of a gas in solution at a given partial pressure and temperature depends on the solubility of the gas in that particular liquid. Carbon dioxide is very soluble, oxygen is somewhat less soluble, and nitrogen has very limited solubility in body fluids.

•**FIGURE 23-19 Henry's Law and the Relationship between Solubility and Pressure.** Notice that the dashed box now highlights gas diffusion. **(a)** A solution containing dissolved gas molecules at equilibrium with air under a given pressure. **(b)** Increasing the pressure drives additional gas molecules into solution until a new equilibrium is established. **(c)** When the pressure decreases, some of the dissolved gas molecules leave the solution until a new equilibrium is reached.

23

The dissolved gas content is usually reported in milliliters of gas per 100 ml (1 dl) of solution. To see the differences in relative solubility, we can compare the partial pressure of each gas in the alveoli (Table 23-2) with the amount absorbed by plasma. The plasma in a pulmonary vein generally contains 2.62 ml/dl of dissolved CO_2 (P_{CO_2} = 40 mm Hg), 0.29 ml/dl of dissolved O_2 (P_{O_2} = 100 mm Hg), and 1.25 ml/dl of dissolved N_2 (P_{N_2} = 573 mm Hg).

Decompression Sickness

Decompression sickness is a painful condition that develops when a person is exposed to a sudden drop in atmospheric pressure. Nitrogen is the gas responsible for the problems experienced, owing to its high partial pressure in air. When the pressure drops, nitrogen comes out of solution, forming bubbles like those in a shaken can of soda. The bubbles may form in joint cavities, in the bloodstream, and in the cerebrospinal fluid. Individuals with decompression sickness typically curl up from the pain in affected joints; this reaction accounts for the common name for the condition, *the bends*. Decompression sickness most commonly affects scuba divers, who, while submerged, are breathing air under greater-than-normal pressures. It also can affect people in airplanes subject to sudden losses of cabin pressure. AM *Decompression Sickness*

The Composition of Alveolar Air

As soon as air enters the respiratory tract, its characteristics begin to change. In passing through the nasal cavity, inhaled air becomes warmer and the amount of water vapor increases. Humidification and filtration continue as the air travels through the pharynx, trachea, and bronchial passageways. On reaching the alveoli, the incoming air mixes with air that had remained in the alveoli from the previous respiratory cycle. The alveolar air thus contains more carbon dioxide and less oxygen than does atmospheric air.

The last 150 ml of inhaled air never gets farther than the conducting passageways and remains in the anatomic dead space of the lungs. During the subsequent exhalation, the departing alveolar air mixes with air in the dead space, producing yet another mixture that differs from both atmospheric and alveolar samples. The differences in composition between atmospheric (inhaled) and alveolar air are given in Table 23-2.

Diffusion at the Respiratory Membrane

Gas exchange at the respiratory membrane is efficient for the following five reasons:

1. *The differences in partial pressure across the respiratory membrane are substantial.* This fact is important, because the greater the difference in partial pressure, the faster the rate of gas diffusion. Conversely, if the P_{O_2} in alveoli decreases, the rate of oxygen diffusion into blood will drop. This is why many people feel light-headed at altitudes of 3000 m or more; the partial pressure of oxygen in their alveoli has dropped low enough that the rate of oxygen absorption is significantly reduced.

2. *The distances involved in gas exchange are small.* The fusion of capillary and alveolar basement membranes reduces the distance for gas exchange to an average of 0.5 μm. Inflammation of the lung tissue or fluid buildup in alveoli increases the diffusion distance and impairs alveolar gas exchange.

3. *The gases are lipid-soluble.* Both oxygen and carbon dioxide diffuse readily through the surfactant layer and the alveolar and endothelial cell membranes.

4. *The total surface area is large.* The combined alveolar surface area at peak inhalation may approach 140 m^2 (1506 ft^2). Damage to alveolar surfaces, which occurs in emphysema, reduces the available surface area and the efficiency of gas transfer.

5. *Blood flow and airflow are coordinated.* This arrangement improves the efficiency of both pulmonary ventilation and pulmonary circulation. For example, blood flow is greatest around alveoli with the highest P_{O_2} values, where oxygen uptake can proceed with maximum efficiency. If the normal blood flow is impaired, as it is in a *pulmonary embolism*, or if the normal airflow is interrupted, as it is in various forms of *pulmonary obstruction*, this coordination is lost and respiratory efficiency decreases. AM *Bronchitis, Emphysema, and COPD*

Partial Pressures in Alveolar Air and Alveolar Capillaries

Figure 23-20• illustrates the partial pressures of oxygen and carbon dioxide in the pulmonary and systemic circuits. Blood arriving in the pulmonary arteries has a lower P_{O_2} and a higher P_{CO_2} than does alveolar air. Diffusion between the alveolar mixture and the pulmonary capillaries thus elevates the P_{O_2} of blood while lowering its P_{CO_2}. By the time the blood enters the pulmonary venules, it has reached equilibrium with the alveolar air. Hence blood departs the alveoli with a P_{O_2} of about 100 mm Hg and a P_{CO_2} of roughly 40 mm Hg (Figure 23-20a•).

Diffusion between alveolar air and blood in the pulmonary capillaries occurs very rapidly. When you are at rest, a red blood cell moves through one of your pulmonary capillaries in about 0.75 second; when you exercise, that passage takes less than 0.3 second. This amount of time is usually sufficient to reach an equilibrium between the alveolar air and the blood.

Partial Pressures in the Systemic Circuit

The oxygenated blood now leaves the alveolar capillaries and returns to the heart, to be discharged into the systemic circuit. As this blood enters the pulmonary veins, it mixes with blood that flowed through capillaries around conducting passageways. Because gas ex-

•FIGURE 23-20 An Overview of Respiratory Processes and Partial Pressures in Respiration. **(a)** Partial pressures and diffusion at the respiratory membrane. **(b)** Partial pressures and diffusion in other tissues.

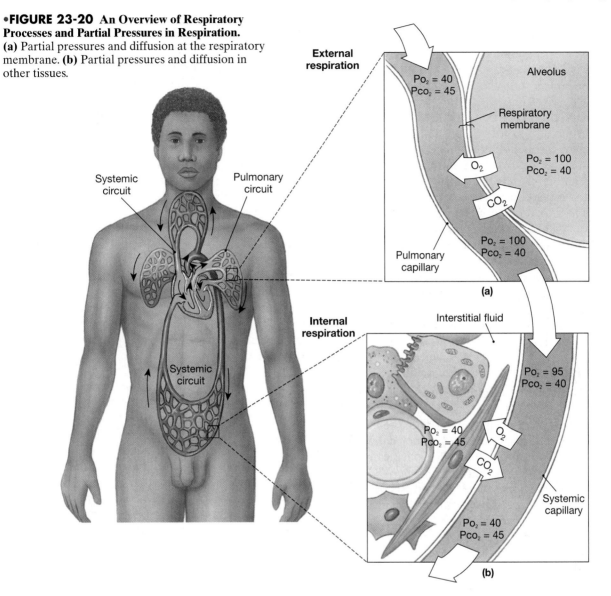

External respiration

Alveolus

$P_{O_2} = 40$
$P_{CO_2} = 45$

Respiratory membrane

O_2

$P_{O_2} = 100$
$P_{CO_2} = 40$

CO_2

$P_{O_2} = 100$
$P_{CO_2} = 40$

Pulmonary capillary

(a)

Systemic circuit

Pulmonary circuit

Systemic circuit

Internal respiration

Interstitial fluid

$P_{O_2} = 95$
$P_{CO_2} = 40$

$P_{O_2} = 40$
$P_{CO_2} = 45$

O_2

CO_2

Systemic capillary

$P_{O_2} = 40$
$P_{CO_2} = 45$

(b)

change occurs only at alveoli, the blood leaving the conducting passageways carries relatively little oxygen. The partial pressure of oxygen in the pulmonary veins therefore drops to about 95 mm Hg. This is the P_{O_2} in the blood that enters the systemic circuit, and no further changes in partial pressure occur until the blood reaches the peripheral capillaries (Figure 23-20b•).

Normal interstitial fluid has a P_{O_2} of 40 mm Hg. As a result, oxygen diffuses out of the capillaries and carbon dioxide diffuses in, until the capillary partial pressures are the same as those in the adjacent tissues. At a normal tissue P_{O_2}, blood entering the venous system still contains about 75 percent of its total oxygen content.

Blood entering the systemic circuit normally has a P_{CO_2} of 40 mm Hg. Inactive peripheral tissues normally have a P_{CO_2} of about 45 mm Hg. As a result, carbon dioxide diffuses into the blood as oxygen diffuses out (Figure 23-20b•).

Blood Gas Analysis

Blood samples can be analyzed to determine the concentration of dissolved gases. The usual tests include the determination of pH, P_{CO_2}, and P_{O_2} in an arterial sample. This analysis can be very helpful in monitoring patients after a heart attack or in chronic respiratory conditions such as obstructive disorders or asthma. A blood sample provides information about the degree of oxygenation in peripheral tissues. For example, if the P_{CO_2} is very high and the P_{O_2} is very low, tissues are not receiving adequate oxygen. This condition can be corrected by providing a gas mixture that has a high P_{O_2} (or even pure oxygen, with a P_{O_2} of 760 mm Hg). In addition, blood gas measurements determine the efficiency of gas exchange at the lungs. If the arterial P_{O_2} remains low despite the administration of oxygen or if the P_{CO_2} is very high, pulmonary exchange problems, such as pulmonary edema, emphysema, or pneumonia, must exist.

23

GAS PICKUP AND DELIVERY

Oxygen and carbon dioxide have limited solubilities in blood plasma. For example, at the normal P_{O_2} of alveoli, 100 ml of plasma will absorb only about 0.3 ml of oxygen. The limited solubilities of these gases are a problem, because peripheral tissues need more oxygen and generate more carbon dioxide than the plasma can absorb and transport.

The problem is solved by red blood cells (RBCs), which remove dissolved oxygen and CO_2 molecules from plasma and bind them (in the case of oxygen) or use them to manufacture soluble compounds (in the case of carbon dioxide). Because these reactions remove dissolved gases from blood plasma, gases continue to diffuse into the blood but never reach equilibrium.

The important thing about these reactions is that they are (1) *temporary* and (2) *completely reversible*. When plasma oxygen or carbon dioxide concentrations are high, the excess molecules are removed by RBCs. When plasma concentrations are falling, the RBCs release their stored reserves.

Oxygen Transport

Each 100 ml of blood leaving the alveolar capillaries carries away roughly 20 ml of oxygen. Of this amount, only about 0.3 ml (1.5 percent) consists of oxygen molecules in solution. The rest of the oxygen molecules are bound to *hemoglobin (Hb) molecules,* specifically to the iron ions in the center of heme units. ∞ *pp. 629–630* Recall that the hemoglobin molecule consists of four globular protein subunits, each containing a heme unit. Thus each hemoglobin molecule can bind four molecules of oxygen, forming oxyhemoglobin. This is a reversible reaction that can be summarized as

$$Hb + O_2 \rightleftharpoons HbO_2$$

Each red blood cell has approximately 280 million molecules of hemoglobin. Because a hemoglobin molecule contains four heme units, each RBC potentially can carry more than a billion molecules of oxygen.

Hemoglobin Saturation **Hemoglobin saturation** is the percentage of heme units containing bound oxygen. At most, each Hb molecule can carry four O_2 molecules. If all the Hb molecules in the blood are fully loaded with oxygen, saturation is 100 percent. If, on average, each Hb molecule carries two O_2 molecules, saturation is 50 percent.

In Chapter 2, we noted that the shape and functional properties of a protein change in response to changes in its environment. ∞ *p. 53* Hemoglobin is no exception; any shape changes that occur can affect oxygen binding. Under normal conditions, the most important environmental factors affecting hemoglobin are (1) the P_{O_2} of blood, (2) blood pH, (3) temperature, and (4) ongoing metabolic activity within RBCs.

Hemoglobin and P_{O_2} An **oxygen–hemoglobin saturation curve,** or *oxygen–hemoglobin dissociation curve,* is a graph that relates the saturation of hemoglobin to the partial pressure of oxygen (Figure 23-21•). The binding and dissociation of oxygen to hemoglobin is a typical reversible reaction. At equilibrium, oxygen molecules bind to heme at the same rate as other oxygen molecules are being released. If you increase the P_{O_2}, you shift the reaction to the right and more oxygen gets bound to hemoglobin. If you decrease the P_{O_2}, the reaction shifts to the left and more oxygen is released by hemoglobin. The graph of this relationship is a curve rather than a straight line, because the shape of the Hb molecule changes

P_{O_2} (mm Hg)	% saturation of Hb
10	13.5
20	35
30	57
40	75
50	83.5
60	89
70	92.7
80	94.5
90	96.5
100	97.5

•**FIGURE 23-21** **The Oxygen–Hemoglobin Saturation Curve.** The normal saturation characteristics of hemoglobin at various partial pressures of oxygen.

slightly each time it binds an oxygen molecule, and this change affects its ability to bind *another* oxygen molecule. In other words, the attachment of the first oxygen molecule makes it easier to bind the second; binding the second promotes binding of the third; and binding of the third enhances binding of the fourth.

Because each arriving oxygen molecule increases the affinity of hemoglobin for the *next* oxygen molecule, the saturation curve takes the form shown in Figure 23-21•. The slope is gradual until the first oxygen molecule binds to the hemoglobin; then the slope rises rapidly, with a plateau near 100 percent saturation. Over the steep initial slope, a very small decrease in plasma P_{O_2} will result in a large change in the amount of oxygen bound to or released from HbO_2. Because the curve rises quickly, hemoglobin will be more than 90 percent saturated if exposed to an alveolar P_{O_2} above 60 mm Hg. Thus near-normal oxygen transport can continue despite a decrease in the oxygen content of alveolar air. Without this ability, you could not survive at high altitudes, and conditions significantly reducing pulmonary ventilation would be immediately fatal.

At normal alveolar pressures (P_{O_2} = 100 mm Hg), the hemoglobin saturation is very high (97.5 percent), although complete saturation does not occur until the P_{O_2} reaches excessively high levels (about 250 mm Hg). In functional terms, the maximum saturation is not as important as the ability of hemoglobin to provide oxygen over the normal P_{O_2} range in body tissues. Over that range, from 100 mm Hg at the alveoli to perhaps 15 mm Hg in active tissues, the saturation drops from 97.5 percent to less than 20 percent, and a small change in P_{O_2} makes a big difference in terms of the amount of oxygen bound to hemoglobin.

Note that the relationship between P_{O_2} and hemoglobin saturation remains valid whether the P_{O_2} is rising or falling. *If the P_{O_2} increases, the saturation goes up and hemoglobin stores oxygen. If the P_{O_2} decreases, hemoglobin releases oxygen into its surroundings.* When oxygenated blood arrives in the peripheral capillaries, the blood P_{O_2} declines rapidly as a result of gas exchange with the interstitial fluid. As the P_{O_2} falls, hemoglobin gives up its oxygen.

The relationship between P_{O_2} and hemoglobin saturation provides a mechanism for automatic regulation of oxygen delivery. Inactive tissues have little demand for oxygen, and the local P_{O_2} is usually about 40 mm Hg. Under these conditions, hemoglobin will not release much oxygen. As it passes through the capillaries, it will go from 97 percent saturation (P_{O_2} = 95 mm Hg) to 75 percent saturation (P_{O_2} = 40 mm Hg). Because it still retains three-quarters of its oxygen, venous blood has a relatively large oxygen reserve. This reserve is important, because it can be mobilized if tissue oxygen demands increase.

Active tissues consume oxygen at an accelerated rate, so the P_{O_2} may drop to 15–20 mm Hg. Hemoglo-bin passing through these capillaries will then go from 97 percent saturation to about 20 percent saturation. In practical terms, this means that as blood circulates through peripheral capillaries, active tissues will receive 3.5 times as much oxygen as will inactive tissues.

Carbon Monoxide Poisoning

Murder or suicide victims who died in their cars inside a closed garage are popular characters for mystery writers. In real life, entire families are killed each winter by leaky furnaces or space heaters. The cause of death is **carbon monoxide poisoning.** The exhaust of automobiles and other petroleum-burning engines, of oil lamps, and of fuel-fired space heaters contains *carbon monoxide* (CO). Carbon monoxide competes with oxygen molecules for the binding sites on heme units. Unfortunately, the carbon monoxide usually wins, because at very low partial pressures it has a much stronger affinity for hemoglobin than does oxygen. The bond formed is extremely durable, and the attachment of a CO molecule essentially makes that heme unit inactive for respiratory purposes. Carbon monoxide will bind to hemoglobin at very low partial pressures. If CO molecules make up just 0.1 percent of the components of inhaled air, enough hemoglobin will be affected that survival will become impossible without medical assistance. Treatment may include (1) the administration of pure oxygen, because at sufficiently high partial pressures the oxygen molecules will gradually replace CO at the hemoglobin molecules, and, if necessary, (2) the transfusion of compatible red blood cells.

Hemoglobin and pH The oxygen–hemoglobin saturation curve in Figure 23-21• was determined in normal blood, with a pH of 7.4 and a temperature of 37°C. In addition to consuming oxygen, active tissues generate acids that lower the pH of the interstitial fluid. When the pH drops, the shape of hemoglobin molecules changes. As a result of this change, hemoglobin molecules release their oxygen reserves more readily, so the slope of the hemoglobin saturation curve changes (Figure 23-22a•). In other words, the saturation declines. Thus, at a tissue P_{O_2} of 40 mm Hg, hemoglobin molecules release 15 percent more oxygen at a pH of 7.2 than they do at a pH of 7.4. This effect of pH on the hemoglobin saturation curve is called the **Bohr effect.**

Carbon dioxide is the primary compound responsible for the Bohr effect. When CO_2 diffuses into the blood, the CO_2 rapidly diffuses into red blood cells. There, an enzyme called **carbonic anhydrase** catalyzes its reaction with water molecules:

$$CO_2 + H_2O \underset{}{\overset{carbonic\ anhydrase}{\rightleftharpoons}} H_2CO_3 \rightleftharpoons H^+ + HCO_3^-$$

The product of this enzymatic reaction, H_2CO_3, is called *carbonic acid,* because it dissociates into a hydrogen ion (H^+) and a bicarbonate ion (HCO_3^-). The

(a) Effect of pH

(b) Effect of temperature

•**FIGURE 23-22** **The Effects of pH and Temperature on Hemoglobin Saturation.** **(a)** When the pH drops below normal levels, more oxygen is released; the hemoglobin saturation curve shifts to the right. If the pH increases, less oxygen is released; the curve shifts to the left. **(b)** When the temperature rises, the saturation curve shifts to the right.

rate of carbonic acid formation depends on the amount of carbon dioxide in solution, which as we noted earlier depends on the P_{CO_2}. When the P_{CO_2} rises, the reaction proceeds from left to right and the rate of carbonic acid formation accelerates. The hydrogen ions generated diffuse out of the RBCs, and the pH of the plasma drops. When the P_{CO_2} declines, the reaction proceeds from right to left; hydrogen ions then diffuse into the RBCs, so the pH of the plasma rises.

Hemoglobin and temperature Temperature changes also affect the slope of the hemoglobin saturation curve (Figure 23-22b•). As the temperature rises, hemoglobin releases more oxygen; as the temperature declines, hemoglobin holds oxygen more tightly. Temperature effects are significant only in active tissues where large amounts of heat are being generated. For example, active skeletal muscles generate heat, and the heat warms blood that flows through these organs. As the blood warms, the Hb molecules release more oxygen than can be used by the active muscle fibers.

Hemoglobin and BPG Red blood cells produce ATP only by glycolysis, in which, as we saw in Chapter 10, lactic acid is formed. ∞ *p. 295* The metabolic pathways involved in glycolysis in an RBC also generate the compound **2,3-bisphosphoglycerate** (biz-fos-fō-GLIS-e-rāt), or **BPG.** Normal RBCs always contain BPG, which has a direct effect on oxygen binding and release. For any partial pressure of oxygen, the higher the concentration of BPG, the more oxygen will be released by the Hb molecules.

The concentration of BPG can be increased by thyroid hormones, growth hormone, epinephrine, androgens, and a high blood pH. These factors improve oxygen delivery to the tissues, because when BPG levels are elevated, hemoglobin will release about 10 percent more oxygen at a given P_{O_2} than it would otherwise. Both BPG synthesis and the Bohr effect improve oxygen delivery when the pH changes: BPG levels rise when the pH increases, and the Bohr effect appears when the pH decreases.

The production of BPG decreases as RBCs age. Thus, the level of BPG can determine how long a blood bank can hold fresh whole blood. When BPG levels get too low, hemoglobin becomes firmly bound to the available oxygen. The blood is then useless for transfusions, because the RBCs will no longer release oxygen to peripheral tissues, even at a disastrously low P_{O_2}.

Fetal Hemoglobin The RBCs of a developing fetus contain **fetal hemoglobin.** The structure of fetal hemoglobin, which differs from that of adult hemoglobin, gives it a much higher affinity for oxygen. At the same P_{O_2}, fetal hemoglobin binds more oxygen than does adult hemoglobin (Figure 23-23•). This trait is important in transferring oxygen across the placenta.

A fetus obtains oxygen from the maternal bloodstream. The maternal blood at the placenta has a relatively low P_{O_2}, ranging from 35 to 50 mm Hg. If maternal blood arrives at the placenta with a P_{O_2} of 40 mm Hg, hemoglobin saturation will be roughly 75 percent. The fetal blood arriving at the placenta has a P_{O_2} close to 20 mm Hg. However, because fetal

•**FIGURE 23-23** A Functional Comparison of Fetal and Adult Hemoglobins

hemoglobin has a higher affinity for oxygen, it is still 58 percent saturated.

As diffusion occurs between fetal blood and maternal blood, oxygen enters the fetal bloodstream until the P_{O_2} equilibrates at 30 mm Hg. At this P_{O_2} the maternal hemoglobin is less than 60 percent saturated, but the fetal hemoglobin is over 80 percent saturated. The steep slope of the saturation curve for fetal hemoglobin means that when fetal RBCs reach peripheral tissues, the Hb molecules will release a large amount of oxygen in response to a very small change in P_{O_2}.

 Adaptations to High Altitude
Atmospheric pressure decreases with increasing altitude, and so do the partial pressures of the component gases, including oxygen. People living in Denver or Mexico City function normally with alveolar oxygen pressures in the 80–90 mm Hg range. At higher elevations, the alveolar P_{O_2} continues to decline. At 3300 meters (10,826 ft), an altitude familiar to many hikers and skiers, the alveolar P_{O_2} falls to around 60 mm Hg. Despite the low alveolar P_{O_2}, millions of people live and work at altitudes this high or higher. Important physiological adjustments include an increased respiratory rate, an increased heart rate, and an elevated hematocrit. Thus, even though the hemoglobin is not fully saturated, the bloodstream holds more of it, and the round-trip between the lungs and the peripheral tissues takes less time. These responses represent an excellent example of the functional interplay between the respiratory and cardiovascular systems. However, most of these adaptations take days to weeks to appear. As a result, athletes planning to compete in events held at high altitude must begin training under such conditions well in advance.

Not everyone can tolerate high-altitude conditions. Roughly 20 percent of people who ascend to 2600 meters (8530 ft) or higher experience *mountain sickness*, or *altitude sickness*. Symptoms may include headache, disorientation, and fatal pulmonary or cerebral edema. AM *Mountain Sickness*

Carbon Dioxide Transport

Carbon dioxide is generated by aerobic metabolism in peripheral tissues. After entering the bloodstream, a CO_2 molecule is (1) converted to a molecule of carbonic acid, (2) bound to the protein portion of hemoglobin molecules within red blood cells, or (3) dissolved in plasma (Figure 23-24•). All three are completely reversible reactions. We shall consider the events under way as blood enters peripheral tissues in which the P_{CO_2} is 45 mm Hg.

Carbonic Acid Formation Most of the carbon dioxide absorbed by blood (roughly 70 percent of the total) is transported as molecules of carbonic acid. Carbon dioxide is converted to carbonic acid through the activity of the enzyme carbonic anhydrase in RBCs. The carbonic acid molecules immediately dissociate into a hydrogen ion and a bicarbonate ion, as described earlier (p. 827). Hence we can ignore the intermediate steps in this sequence and summarize the reaction as

$$CO_2 + H_2O \overset{\text{carbonic anhydrase}}{\rightleftharpoons} H^+ + HCO_3^-$$

This reaction is completely reversible. In peripheral capillaries this reaction proceeds vigorously, tying up large numbers of CO_2 molecules. The reaction continues as carbon dioxide diffuses out of the interstitial fluids.

The hydrogen ions and bicarbonate ions have different fates. Most of the hydrogen ions bind to hemoglobin molecules, forming $Hb \cdot H^+$. The Hb molecules thus function as buffers, tying up the released hydrogen ions before the ions leave the RBCs and affect the plasma pH. The bicarbonate ions move into the surrounding plasma with the aid of a countertransport mechanism that exchanges intracellular bicarbonate ions (HCO_3^-) for extracellular chloride ions (Cl^-). This exchange, which trades one anion for another, does not require ATP. The result is a mass movement of chloride ions into the RBCs, an event known as the **chloride shift.**

Hemoglobin Binding Roughly 23 percent of the carbon dioxide carried by your blood will be bound to the globular protein portions of the Hb molecules inside RBCs. These CO_2 molecules are attached to exposed amino groups ($-NH_2$) of the Hb molecules. The result is called **carbaminohemoglobin** (kar-ba-mē-nō-hē-mō-GLŌ-bin), $Hb \cdot CO_2$. This reversible reaction is summarized as follows:

$$CO_2 + HbNH_2 \rightleftharpoons HbNHCOOH$$

This reaction can be abbreviated without the amino groups as

$$CO_2 + Hb \rightleftharpoons Hb \cdot CO_2$$

23

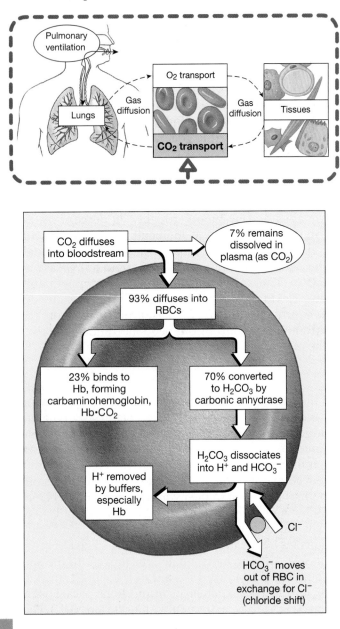

sult from the basic chemistry of the transport mechanisms. Other responses require coordinated adjustments in the activities of the cardiovascular and respiratory systems. We shall now consider those levels of control and regulation.

☑ As you exercise, hemoglobin releases more oxygen to your active skeletal muscles than it does when the muscles are at rest. Why?

☑ How would blockage of the trachea affect the blood pH?

•FIGURE 23-24 Carbon Dioxide Transport in Blood

CONTROL OF RESPIRATION

Peripheral cells continuously absorb oxygen from interstitial fluids and generate carbon dioxide. Under normal conditions, the cellular rates of absorption and generation are matched by the capillary rates of delivery and removal. These rates are identical to those of oxygen absorption and carbon dioxide excretion at the lungs. If these rates become unbalanced, homeostatic mechanisms intervene to restore equilibrium. Those mechanisms involve (1) changes in blood flow and oxygen delivery that are regulated at the local level and (2) changes in the depth and rate of respiration under the control of the brain's respiratory centers. The activities of the respiratory centers are coordinated with changes in cardiovascular function, such as fluctuations in blood pressure and cardiac output.

Plasma Transport Plasma becomes saturated with carbon dioxide quite rapidly, and only about 7 percent of the carbon dioxide absorbed by peripheral capillaries is transported in the form of dissolved gas molecules. The rest is absorbed by the RBCs for conversion by carbonic anhydrase or storage as carbaminohemoglobin.

SUMMARY: GAS TRANSPORT

Figure 23-25• summarizes the transportation of oxygen and carbon dioxide in the respiratory and cardiovascular systems. This transport is dynamic, capable of varying its responses to meet changing circumstances. Some of the responses are automatic and re-

LOCAL REGULATION OF GAS TRANSPORT AND ALVEOLAR FUNCTION

The rate of oxygen delivery in each tissue and the efficiency of oxygen pickup at the lungs are largely regulated at the local level. For example, if a peripheral tissue becomes more active, the interstitial P_{O_2} falls and the P_{CO_2} rises. This change increases the difference between the partial pressures in the tissues and in the arriving blood, so more oxygen is delivered and more carbon dioxide is carried away. In addition, the rising P_{CO_2} levels cause the relaxation of smooth muscles in the walls of arterioles and capillaries in the area, increasing local blood flow.

Local factors coordinate (1) *lung perfusion,* or blood flow to the alveoli, with (2) *alveolar ventilation,* or airflow, over a wide range of conditions and activity levels. As blood flows toward the alveolar capillaries, it is directed toward lobules in which the P_{O_2} is relatively high. This movement occurs because alveolar capillaries constrict when the local P_{O_2} is low. (We noted this response, the opposite of that seen in peripheral tissues, in Chapter 21. ∞ *p. 722*) Such a shift in blood flow tends to eliminate temporary differences

•**FIGURE 23-25 A Summary of the Primary Gas Transport Mechanisms.** (a) Oxygen transport. (b) Carbon dioxide transport.

in the oxygen and carbon dioxide contents of alveoli, lobules, or groups of lobules that could otherwise result from minor variations in local blood flow.

Smooth muscles in the walls of bronchioles are sensitive to the P_{CO_2} of the air they contain. When the P_{CO_2} goes up, the bronchioles increase in diameter (bronchodilation). When the P_{CO_2} declines, the bronchioles constrict (bronchoconstriction). Airflow is therefore directed to lobules in which the P_{CO_2} is high. Because the carbon dioxide is obtained from blood, these lobules are actively engaged in gas exchange. This response is especially important, because the improvement of airflow to functional alveoli can at least partially compensate for damage to pulmonary lobules.

By directing blood flow to alveoli with low CO_2 levels and improving airflow to alveoli with high CO_2 levels, local adjustments improve the efficiency of gas transport. When activity levels increase and the demand for oxygen rises, the cardiac output and respiratory rates increase under neural control, but the adjustments in alveolar blood flow and bronchiole diameter occur automatically.

THE RESPIRATORY CENTERS OF THE BRAIN

Respiratory control has both involuntary and voluntary components. Your brain's involuntary centers regulate the activities of the respiratory muscles and control the respiratory minute volume by adjusting the frequency and depth of pulmonary ventilation. They do so in response to sensory information arriving from your lungs and other portions of the respiratory tract as well as from a variety of other sites.

The voluntary control of respiration reflects activity in the cerebral cortex that affects either the output of the respiratory centers in the medulla oblongata and pons or motor neurons in the spinal cord that control respiratory muscles. The **respiratory centers** are three pairs of nuclei in the reticular formation of the medulla oblongata and pons. The motor neurons in the spinal cord are generally controlled by *respiratory reflexes,* but they can also be controlled voluntarily through commands delivered by the corticospinal pathway. ∞ *p. 488*

23

Respiratory Centers in the Medulla Oblongata

We introduced the *respiratory rhythmicity centers* of the medulla oblongata in Chapter 14. ∞ *p. 464* These paired centers set the pace for respiration. Each center can be subdivided into a **dorsal respiratory group (DRG)** and a **ventral respiratory group (VRG).** The DRG's *inspiratory center* contains neurons that control lower motor neurons innervating the external intercostal muscles and the diaphragm. This group functions in every respiratory cycle, whether quiet or forced. The VRG functions only during forced breathing. It includes neurons that innervate lower motor neurons controlling accessory respiratory muscles involved in active exhalation (an *expiratory center*) and maximal inhalation (an *inspiratory center*).

There is reciprocal inhibition between the neurons involved with inhalation and exhalation. ∞ *p. 428* When the inspiratory neurons are active, the expiratory neurons are inhibited, and vice versa. The pattern of interaction between these groups differs between quiet breathing and forced breathing. During quiet breathing (Figure 23-26a•):

- Activity in the DRG increases over a period of about 2 seconds, providing stimulation to the inspiratory muscles. Over this period, inhalation occurs.

- After 2 seconds, the DRG neurons become inactive. They remain quiet for the next 3 seconds and allow the inspiratory muscles to relax. Over this period, passive exhalation occurs.

During forced breathing (Figure 23-26b•):

- As the level of activity in the DRG increases, it stimulates neurons of the VRG that activate the accessory muscles involved in inhalation.

- At the end of each inhalation, active exhalation occurs as the neurons of the expiratory center stimulate the appropriate accessory muscles.

The basic pattern of respiration thus reflects a cyclic interaction between the DRG and VRG. The pace of this interaction is thought to be established by pacemaker cells that spontaneously undergo rhythmic patterns of activity. Attempts to locate the pacemaker, however, have been unsuccessful.

Central nervous system stimulants, such as amphetamines or even caffeine, increase your respiratory rate by facilitating the respiratory centers. These actions can be opposed by CNS depressants, such as barbiturates or opiates.

The Apneustic and Pneumotaxic Centers

The **apneustic** (ap-NOO-stik) **centers** and the **pneumotaxic** (noo-mō-TAKS-ik) **centers** of the pons are paired nuceli that adjust the output of the respiratory rhythmicity centers. Their activities adjust the

(a)

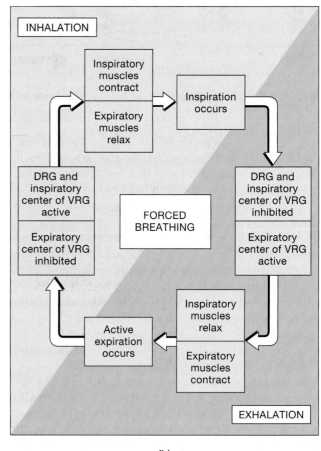

(b)

•**FIGURE 23-26** **Basic Regulatory Patterns of Respiration.**
(a) Quiet breathing. **(b)** Forced breathing.

respiratory rate and the depth of respiration in response to sensory stimuli or input from other centers in the brain. Each apneustic center provides continuous stimulation to the DRG on that side of the brain stem. During quiet breathing, stimulation from the apneustic center helps increase the intensity of inhalation over the following 2 seconds. Under normal conditions, after 2 seconds the apneustic center is inhibited by signals from the pneumotaxic center on that side. During forced breathing, the apneustic centers also respond to sensory input from the vagus nerves regarding the amount of lung inflation.

The pneumotaxic centers inhibit the apneustic centers and promote passive or active exhalation. Centers in the hypothalamus and cerebrum can alter the activity of the pneumotaxic centers as well as the respiratory rate and depth. However, essentially normal respiratory cycles continue even if the brain stem superior to the pons has been severely damaged. If the inhibitory output of the pneumotaxic centers is cut off by a stroke or other damage to the brain stem and if sensory innervation from the lungs is eliminated by cutting the vagus nerves, the person inhales to maximum capacity and maintains that state for 10–20 seconds at a time. Intervening exhalations are brief, and little pulmonary ventilation occurs.

The CNS regions involved with respiratory control are diagrammed in Figure 23-27•. Interactions between the DRG and VRG establish the basic pace and depth of respiration. The pneumotaxic centers modify that pace; an increase in pneumotaxic output quickens the pace of respiration by shortening the duration of each inhalation. A decrease in pneumotaxic output slows the respiratory pace but increases the depth of respiration, because the apneustic centers are more active.

Respiratory Reflexes

The activities of the respiratory centers are modified by sensory information from:

- Chemoreceptors sensitive to the P_{CO_2}, pH, and/or P_{O_2} of the blood or cerebrospinal fluid.
- Changes in blood pressure in the aortic or carotid sinuses.
- Stretch receptors that respond to changes in the volume of the lungs.

- Irritating physical or chemical stimuli in the nasal cavity, larynx, or bronchial tree.
- Other sensations, including pain, changes in body temperature, and abnormal visceral sensations.

Information from these receptors alters the pattern of respiration. The induced changes have been called *respiratory reflexes.*

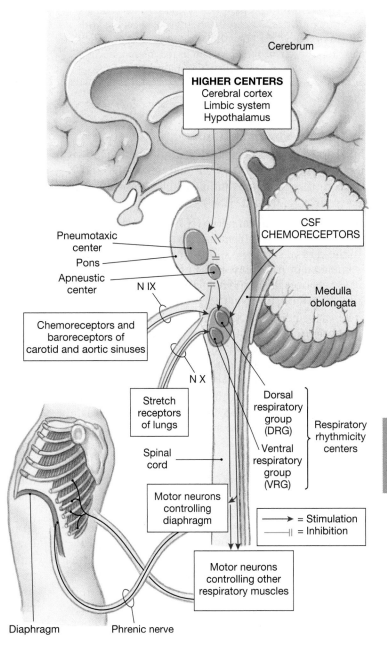

•**FIGURE 23-27 Respiratory Centers and Reflex Controls.** The positions of the major respiratory centers and other factors important to the reflex control of respiration. Pathways for conscious control over respiratory muscles are not shown.

SIDS

Sudden infant death syndrome (SIDS), also known as *crib death*, kills an estimated 10,000 infants each year in the United States alone. Most crib deaths occur between midnight and 9:00 A.M., in the late fall or winter, and involve infants 2–4 months old. Eyewitness accounts indicate that the sleeping infant suddenly stops breathing, turns blue, and relaxes. Genetic factors appear to be involved, but controversy remains as to the relative importance of other factors, such as laryngeal spasms, cardiac arrhythmias, upper respiratory system infections, viral infections, and CNS malfunctions. The age at the time of death corresponds with a period when the pacemaker complex and respiratory centers are establishing connections with other portions of the brain. It has recently been proposed that SIDS results from a problem in the interconnection process that disrupts the reflexive respiratory pattern.

The Chemoreceptor Reflexes

The respiratory centers are strongly influenced by chemoreceptor inputs from cranial nerves IX and X and from receptors that monitor the composition of the cerebrospinal fluid (CSF):

■ The glossopharyngeal nerve (N IX) carries chemoceptive information from the carotid bodies, adjacent to the carotid sinus. ∞ *p. 470* The carotid bodies are stimulated by a decrease in the pH or P_{O_2} of blood. Because changes in P_{CO_2} affect pH, these receptors are indirectly stimulated by a rise in the P_{CO_2}.

■ The vagus nerve (N X) monitors chemoreceptors in the aortic bodies, near the aortic arch. ∞ *p. 471* These receptors are sensitive to the same stimuli as the carotid bodies. Carotid and aortic body receptors are often called *peripheral chemoreceptors.*

■ Chemoreceptors are located on the ventrolateral surface of the medulla oblongata in a region known as the *chemosensitive area.* The neurons in that area respond only to the P_{CO_2} and pH of the CSF and are often called *central chemoreceptors.*

We discussed chemoreceptors and their effects on cardiovascular function in Chapters 17 and 21. ∞ *pp. 533–534, 714* Stimulation of these chemoreceptors leads to an increase in the depth and rate of respiration. Under normal conditions, a drop in arterial P_{O_2} has little effect on the respiratory centers until the arterial P_{O_2} drops by about 40 percent, to below 60 mm Hg. If the P_{O_2} of arterial blood drops to the level in peripheral tissues, 40 mm Hg, the respiratory rate will increase by only 50–70 percent. In contrast, a rise of just 10 percent in the arterial P_{CO_2} will cause the respiratory rate to double, even if the P_{O_2} remains completely normal. Carbon dioxide levels are therefore responsible for regulating respiratory activity under normal conditions.

Although the receptors monitoring CO_2 levels are more sensitive, oxygen and carbon dioxide receptors

work together in a crisis. Carbon dioxide is generated during oxygen consumption, so when oxygen concentrations are falling rapidly, CO_2 levels are usually increasing. This cooperation breaks down only under unusual circumstances. For example, an individual can hold his or her breath longer than normal by taking deep, full breaths, but the practice is very dangerous. The danger lies in the fact that the increased ability is due not to extra oxygen but to the loss of carbon dioxide. If the P_{CO_2} is driven down far enough, breath-holding ability can increase to the point at which the individual becomes unconscious from oxygen starvation in the brain without ever feeling the urge to breathe. AM *Shallow Water Blackout*

The chemoreceptors are subject to adaptation—a decrease in receptor sensitivity after chronic stimulation—if the P_{O_2} or P_{CO_2} remains abnormal for an extended period. This adaption can complicate the treatment of chronic respiratory disorders. For example, if the P_{O_2} remains low for an extended period and the P_{CO_2} remains chronically elevated, the chemoreceptors will reset to those values. They will accept those values as normal and will oppose any attempts to return the partial pressures to the proper range. AM *Chemoreceptor Accommodation and Opposition*

Because the chemoreceptors monitoring CO_2 levels are also sensitive to pH, any condition altering the pH of blood or CSF will affect respiratory performance. For example, the rise in lactic acid levels after exercise causes a drop in pH that helps stimulate respiratory activity.

Hypercapnia and Hypocapnia **Hypercapnia** is an increase in the P_{CO_2} of arterial blood. Figure 23-28a• diagrams the central response to hypercapnia, which is triggered by the stimulation of chemoreceptors in the carotid and aortic bodies and is reinforced by the stimulation of CNS chemoreceptors. Carbon dioxide crosses the blood–brain barrier quite rapidly, so a rise in arterial P_{CO_2} almost immediately elevates CSF CO_2 levels, lowering the pH of the CSF and stimulating the chemoreceptive neurons of the medulla oblongata.

These receptors stimulate your respiratory centers to increase the rate and depth of respiration. Your breathing becomes more rapid, and more air moves into and out of your lungs with each breath. Because more air moves into and out of the alveoli each minute, alveolar concentrations of carbon dioxide decline, accelerating the diffusion of carbon dioxide out of alveolar capillaries. Thus homeostasis is restored.

The most common cause of hypercapnia is hypoventilation. In **hypoventilation,** the respiratory rate remains abnormally low and is insufficient to meet the demands for normal oxygen delivery and carbon dioxide removal. Carbon dioxide then accumulates in the blood.

•FIGURE 23-28 **The Chemoreceptor Response to Changes in P_{CO_2}.** **(a)** A rise in arterial P_{CO_2} stimulates chemoreceptors that accelerate breathing cycles at the inspiratory center. This change increases the respiratory rate, encourages CO_2 loss at the lungs, and lowers arterial P_{CO_2}. **(b)** A drop in arterial P_{CO_2} inhibits these chemoreceptors. In the absence of stimulation, the rate of respiration decreases, slowing the rate of CO_2 loss and elevating arterial P_{CO_2}.

If the rate and depth of respiration exceed the demands for oxygen delivery and carbon dioxide removal, the condition of **hyperventilation** exists. Hyperventilation will gradually lead to **hypocapnia,** an abnormally low P_{CO_2}. If the arterial P_{CO_2} drops below normal levels, chemoreceptor activity decreases and the respiratory rate falls (Figure 23-28b•). This situation continues until the P_{CO_2} returns to normal and homeostasis is restored.

The Baroreceptor Reflexes

We described the effects of carotid and aortic baroreceptor stimulation on systemic blood pressure in Chapter 21. ∞ *pp. 713–714* The output from these baroreceptors also affects the respiratory centers. When blood pressure falls, the respiratory rate increases; when blood pressure rises, the respiratory rate declines. This adjustment results from the stimulation or inhibition of the respiratory centers by sensory fibers in the glossopharyngeal (N IX) and vagus (N X) nerves.

The Hering–Breuer Reflexes

The **Hering–Breuer reflexes** are named after the physiologists who described them in 1865. The sensory information from these reflexes is distributed to the apneustic centers and the ventral respiratory group. The Hering–Breuer reflexes are not involved in normal quiet breathing or in tidal volumes under 1000 ml. There are two such reflexes—the inflation reflex and the deflation reflex:

1. The **inflation reflex** prevents overexpansion of the lungs during forced breathing. The stretch receptors are located in the smooth muscle tissue around bronchioles and are stimulated by lung expansion. Sensory fibers leaving the stretch receptors of each lung reach the respiratory rhythmicity center on the same side over the vagus nerve. As lung volume increases, the dorsal respiratory group is gradually inhibited, and the expiratory center of the VRG is stimulated. Inhalation then stops as the lungs near maximum volume, and active exhalation then begins.

2. The **deflation reflex** inhibits the expiratory centers and stimulates the inspiratory centers when the lungs are deflating. The receptors, which are distinct from those of the inflation reflex, are located in the alveolar wall near the alveolar capillary network. The smaller the volume of the lungs, the greater the degree of inhibition, until exhalation stops and inhalation begins. This reflex normally functions only during forced exhalation, when both the inspiratory and expiratory centers are active.

23

Protective Reflexes

Protective reflexes operate when you are exposed to toxic vapors, chemical irritants, or mechanical stimulation of the respiratory tract. The receptors involved are located in the epithelium of the respiratory tract. Examples of protective reflexes include sneezing, coughing, and laryngeal spasms.

Sneezing is triggered by an irritation of the wall of your nasal cavity. Coughing is triggered by an irritation of your larynx, trachea, or bronchi. Both reflexes involve **apnea** (AP-nē-uh), a period in which respiration is suspended. They are usually followed by a forceful expulsion of air intended to remove the offending stimulus. The glottis is forcibly closed while the lungs are still relatively full. The abdominal and internal intercostal muscles then contract suddenly, creating pressures that will blast air out of your respiratory passageways when the glottis reopens. Air leaving the larynx can travel at 160 kph (99 mph), carrying mucus, foreign particles, and irritating gases out of the respiratory tract via the nose or mouth.

Laryngeal spasms result from the entry of chemical irritants, foreign objects, or fluids into the area around the glottis. This reflex generally closes your airway temporarily. A very strong stimulus, such as a toxic gas, could close the glottis so powerfully that you could lose consciousness and die without taking another breath. Fine chicken bones or fish bones that pierce the laryngeal walls can also stimulate laryngeal spasms, swelling, or both, restricting the airway.

Other Sensations That Affect Respiratory Function

Several other sensory stimuli can affect the activities of the respiratory centers. In some cases, the mechanism involved is not known. Examples include the following:

- Sudden pain or immersion in cold water can produce a temporary apnea. Chronic pain stimulates the sympathetic division of the autonomic nervous system, leading to an increase in the respiratory rate.

- Both vomiting and swallowing involve automatic adjustments in respiratory activity to prevent the entrance of foreign objects into the trachea.

- Both fever and an increase in body temperature due to exertion or overheating cause an increase in the respiratory rate. A reduction in body temperature leads to a decrease in the respiratory rate.

- Curiously, stretching the anal sphincter stimulates the respiratory centers and increases the rate of respiration. Although this reflex is occasionally used to stimulate respiration in an emergency situation, it is not clear which pathways are involved.

VOLUNTARY CONTROL OF RESPIRATION

Activity of your cerebral cortex has an indirect effect on your respiratory centers. For example:

- Conscious thought processes tied to strong emotions, such as rage or fear, affect the respiratory rate by stimulating centers in the hypothalamus.

- Emotional states can affect respiration through activation of the sympathetic or parasympathetic division of the autonomic nervous system. Sympathetic activation causes bronchodilation and increases the respiratory rate; parasympathetic stimulation has the opposite effects.

- An anticipation of strenuous exercise can trigger an automatic increase in the respiratory rate, along with increased cardiac output, by sympathetic stimulation.

We also have conscious control over our respiratory activities. This control may bypass the respiratory centers completely, using pyramidal fibers that innervate the same lower motor neurons that are controlled by the DRG and VRG. This control mechanism is an essential part of speaking, singing, and swimming, when respiratory activities must be precisely timed. Higher centers can also have an inhibitory effect on the apneustic centers and on the DRG and VRG; this effect is important when you hold your breath.

Our abilities to override the respiratory centers have limits, however. The chemoreceptor reflexes are extremely powerful respiratory stimulators, and they cannot be consciously suppressed. For example, you cannot kill yourself by holding your breath "till you turn blue." Once the P_{CO_2} rises to critical levels, you will be forced to take a breath.

CHANGES IN THE RESPIRATORY SYSTEM AT BIRTH

The respiratory system of a fetus and that of a newborn infant differ in several important ways. Before delivery, pulmonary arterial resistance is high, because the pulmonary vessels are collapsed. The rib cage is compressed, and the lungs and conducting passageways contain only small amounts of fluid and no air. During delivery, the lungs are compressed further, and as the placental connection is lost, blood oxygen levels fall and carbon dioxide levels climb rapidly. At birth, the newborn infant takes a truly heroic first breath through powerful contractions of the diaphragmatic and external intercostal muscles. The inhaled air must enter the respiratory passageways with enough force to overcome surface tension and inflate the bronchial tree and most of the alveoli. The same drop in pressure that pulls air into the lungs pulls blood into the pulmonary circulation. The changes in blood flow that occur lead to the closure of the *foramen ovale,* an interatrial connection, and the *ductus arteriosus,* the fetal connection between the pulmonary trunk and the aorta. ⚲ *p. 743* [AM] *EMBRYOLOGY SUMMARY 18: The Development of the Respiratory System*

The exhalation that follows fails to empty the lungs completely, because the rib cage does not return to its former, fully compressed state. Cartilages and connective tissues keep the conducting passageways open, and surfactant covering the alveolar surfaces prevents their collapse. Subsequent breaths complete the inflation of the alveoli. Pathologists sometimes use these physical changes to determine whether a newborn infant died before delivery or shortly thereafter. Before the first breath, the lungs are completely filled with fluid, and they will sink if placed in water. After the infant's first breath, even the collapsed lungs contain enough air to keep them afloat.

AGING AND THE RESPIRATORY SYSTEM

Many factors interact to reduce the efficiency of the respiratory system in elderly individuals. Three examples are particularly noteworthy:

1. As age increases, elastic tissue deteriorates throughout the body. This deterioration reduces the compliance of the lungs, lowering the vital capacity.

2. Chest movements are restricted by arthritic changes in the rib articulations and by decreased flexibility at the costal cartilages. Along with the changes in item 1, the stiffening and reduction in chest movement effectively limit the respiratory minute volume. This restriction contributes to the reduction in exercise performance and capabilities with increasing age.

3. Some degree of emphysema is normal in individuals over age 50. However, the extent varies widely with the lifetime exposure to cigarette smoke and other respiratory irritants. Figure 23-29• compares the res-

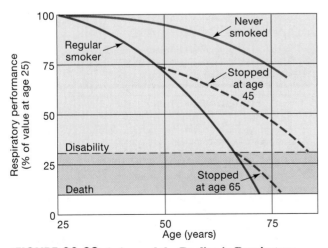

•**FIGURE 23-29 Aging and the Decline in Respiratory Performance.** The relative respiratory performances of individuals who have never smoked, individuals who quit smoking at age 45, individuals who quit smoking at age 65, and lifelong smokers.

piratory performance of individuals who have never smoked with individuals who have smoked for varying periods of time. The message is quite clear: Although some decrease in respiratory performance is inevitable, you can prevent serious respiratory deterioration by stopping smoking or never starting.

☑ What effect would exciting the pneumotaxic centers have on respiration?

☑ Are peripheral chemoreceptors as sensitive to the levels of carbon dioxide as they are to the levels of oxygen?

☑ Johnny is angry with his mother, so he tells her that he will hold his breath until he turns blue and dies. Should Johnny's mother worry?

INTEGRATION WITH OTHER SYSTEMS

The respiratory system has extensive anatomical connections to the cardiovascular system, and we have noted examples throughout this chapter. The functional ties are just as extensive.

The goal of respiratory activity is to maintain homeostatic oxygen and carbon dioxide levels in peripheral tissues. Changes in respiratory activity alone are seldom sufficient to accomplish this. For example, during exercise, when tissue oxygen demands are high, alveolar ventilation increases through the stimulation of the respiratory muscles. This response, which actually increases the demand for oxygen, serves no purpose unless cardiac output accelerates simultaneously, delivering more oxygen to active tissues.

We consider here four examples of the integration between the respiratory and cardiovascular systems:

1. At the local level, changes in lung perfusion in response to changes in alveolar P_{O_2} improve the efficiency of gas exchange within or among lobules.

2. Chemoreceptor stimulation not only increases respiratory drive; it also causes an elevation in blood pressure, through the stimulation of the vasomotor centers, and increased cardiac output.

3. The stimulation of baroreceptors in the lungs has secondary effects on cardiovascular function. For example, the stimulation of airway stretch receptors not only triggers the inflation reflex but also increases the heart rate. Thus, as the lungs fill, cardiac output rises and more blood flows through the alveolar capillaries.

4. The stimulation of baroreceptors in systemic blood vessels has secondary effects on respiratory function. When systemic blood pressure falls, stimulating carotid and aortic baroreceptors, the respiratory rate increases; when systemic blood pressure rises, the respiratory rate declines.

The respiratory system is functionally linked to all other systems as well. Figure 23-30• illustrates these interrelationships.

23

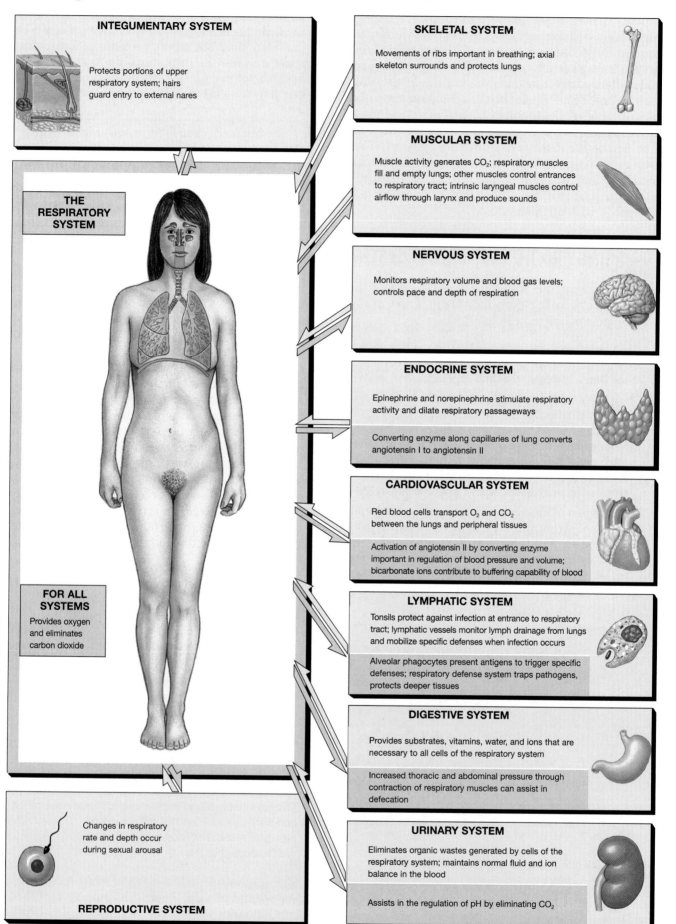

•FIGURE 23-30 Functional Relationships between the Respiratory System and Other Systems

Emphysema and Lung Cancer

Emphysema and lung cancer are two relatively common disorders that are both often associated with cigarette smoking. **Emphysema** (em-fi-SĒ-muh) is a chronic, progressive condition characterized by shortness of breath and an inability to tolerate physical exertion. The underlying problem is the destruction of alveolar surfaces and inadequate surface area for oxygen and carbon dioxide exchange. In essence, respiratory bronchioles and alveoli are functionally eliminated. The alveoli gradually expand, and adjacent alveoli merge to form larger air spaces supported by fibrous tissue without alveolar capillary networks. As connective tissues are eliminated, compliance increases; air moves into and out of the lungs more easily than before. However, the loss of respiratory surface area restricts oxygen absorption, so the individual becomes short of breath.

Emphysema has been linked to the inhalation of air that contains fine particulate matter or toxic vapors, such as those in cigarette smoke. Genetic factors also predispose individuals to this condition. Some degree of emphysema is a normal consequence of aging, however. An estimated 66 percent of adult males and 25 percent of adult females have detectable areas of emphysema in their lungs. AM *Bronchitis, Emphysema, and COPD*

Lung cancer, or *pleuropulmonary neoplasm,* is an aggressive class of malignancies originating in the bronchial passageways or alveoli. These cancers affect the epithelial cells that line conducting passageways, mucous glands, or alveoli. Symptoms generally do not appear until the condition has progressed to the point at which the tumor masses are restricting airflow or compressing adjacent structures. Chest pain, shortness of breath, a cough or a wheeze, and weight loss are common symptoms. Treatment programs vary with the cellular organization of the tumor and whether metastasis (cancer cell migration) has occurred, but surgery, radiation exposure, or chemotherapy may be involved.

Deaths from lung cancer were rare at the turn of the twentieth century, but 29,000 such deaths occurred in 1956, 105,000 in 1978, and 156,900 in 2000 in the United States. This rise coincides with the increased rate of smoking in the population. The number of diagnosed cases is also rising, doubling every 15 years. Each year, 22 percent of new cancers detected are lung cancers; in the United States in 2000, an estimated 89,500 men and 74,600 women will be diagnosed with this condition. Lung cancer is increasing markedly among women but declining among men; in 1989, 101,000 men and 54,000 women were diagnosed with lung cancer in the United States. AM *Lung Cancer*

[EXPLORE] MediaLab

The respiratory system is an integral part of our daily lives. We are told to breathe deeply when we are upset; women do breathing exercises when preparing for childbirth; the tobacco industry is struggling to remain competitive in light of mounting opposition to smoking. To gain a better understanding of the respiratory system, use the resources available on both the CD-ROM and the Companion Website. The CD-ROM includes animations and tutorials; the website builds on those resources with text content and interactive problems as well as communication avenues for sharing ideas with your class. To complete the following Web Explorations, visit the MediaLab in Chapter 23 of your Companion Website.

Web Exploration 1

Estimated time for completion = 15 minutes

Although asthma is one of the most common diseases associated with the respiratory system, not much is understood about its origins. The Centers for Disease Control and Prevention has noted an alarming increase in the number of people with asthma between the years 1980 and 1994. Select the keyword **ASTHMA** on the Companion Website to read about medically based theories for this increase in asthma incidence. What are your thoughts about this situation? Prepare a short essay describing the effects of asthma on the respiratory system and what you feel might be causing the increase in reported cases.

Web Exploration 2

Estimated time for completion = 5 minutes

Premature infants experience a host of problems as a result of their early birth. One such problem is the lack of production of surfactant in their lungs. Review the section in the text, "Alveolar Ducts and Alveoli" (pp. 811–813), that describes surfactant. Then select the keyword **SURFACTANT** on the Companion Website to learn about a breakthrough in surfactant production. Prepare an outline of the article's major points. How does surfactant work? Why does this chemical help premature infants?

CHAPTER REVIEW

SELECTED CLINICAL TERMINOLOGY
Terms Discussed in This Chapter

anoxia (a-NOKS-ē-uh):   A condition of tissue oxygen starvation caused by (1) circulatory blockage, (2) respiratory problems, or (3) cardiovascular problems. *(p. 814)*

asthma (AZ-muh):   An acute respiratory disorder characterized by unusually sensitive, irritated conducting passageways. *(p. 809 and AM)*

atelectasis (at-e-LEK-ta-sis):   A collapsed lung. *(p. 818)*

bronchitis (brong-KĪ-tis):   An inflammation of the bronchial lining. *(p. 809 and AM)*

bronchodilation (brong-kō-dī-LĀ-shun):   An enlargement of the respiratory passageways. *(p. 809)*

bronchography (brong-KOG-ra-fē):   A procedure in which radiopaque materials are introduced into the airways to improve X-ray imaging of the bronchial tree. *(p. 809)*

bronchoscope: A fiber-optic bundle small enough to be inserted into the trachea and finer airways; the procedure is called *bronchoscopy. (p. 809)*

cardiopulmonary resuscitation (CPR):   The application of cycles of compression to the rib cage and mouth-to-mouth breathing to maintain cardiovascular and respiratory function. *(p. 819 and AM)*

cystic fibrosis (CF):   A relatively common lethal inherited disease caused by an abnormal membrane channel protein; a major symptom is that mucous secretions become too thick to be transported easily, leading to respiratory problems. *(p. 800 and AM)*

decompression sickness, or *the bends:*   A condition caused by a drop in atmospheric pressure and the resulting formation of nitrogen gas bubbles in body fluids, tissues, and organs. *(p. 824 and AM)*

emphysema (em-fi-SĒ-muh):   A chronic, progressive condition characterized by shortness of breath and an inability to tolerate physical exertion. *(p. 839 and AM)*

epistaxis (ep-i-STAK-sis):   A nosebleed. *(p. 802)*

Heimlich (HĪM-lik) **maneuver,** or *abdominal thrust:*   Compression applied to the abdomen just inferior to the diaphragm, to force air out of the lungs to clear a blocked trachea or larynx. *(p. 805)*

hypercapnia (hī-per-KAP-nē-uh):   An increase in the P_{CO_2} of arterial blood. *(p. 834)*

hypocapnia: An abnormally low arterial P_{CO_2}. *(p. 835)*

hypoxia (hī-POKS-ē-uh):   A condition of reduced tissue P_{O_2}. *(p. 814)*

lung cancer *(pleuropulmonary neoplasm):*   A class of aggressive malignancies originating in the bronchial passageways or alveoli. *(p. 839 and AM)*

mountain sickness: An acute disorder resulting from CNS effects of the low gas partial pressures that occur at high altitudes. *(p. 829 and AM)*

pleurisy: An inflammation of the pleurae and secretion of excess amounts of pleural fluid. *(p. 814)*

pneumonia (noo-MŌ-nē-uh):   A respiratory disorder characterized by fluid leakage into the alveoli and/or swelling and constriction of the respiratory bronchioles. *(p. 813)*

pneumothorax (noo-mō-THO-raks):   The entry of air into the pleural cavity. *(p. 818)*

pulmonary embolism: A blockage of a branch of a pulmonary artery, with interruption of blood flow to a group of lobules and/or alveoli. *(p. 813)*

respiratory distress syndrome: A condition resulting from inadequate surfactant production and associated alveolar collapse. *(p. 812 and AM)*

sudden infant death syndrome (SIDS), or *crib death:*   The death of an infant due to respiratory arrest; the cause remains uncertain. *(p. 834)*

tracheostomy (trā-kē-OS-to-mē):   The insertion of a tube directly into the trachea to bypass a blocked or damaged larynx. *(p. 805)*

tuberculosis: A respiratory disorder caused by infection of the lungs by the bacterium *Mycobacterium tuberculosis.* *(p. 808 and AM)*

AM Additional Terms Discussed in the *Applications Manual*

23

chronic obstructive pulmonary disease (COPD):   A condition characterized by *chronic bronchitis* and *chronic airways obstruction.*

shallow water blackout: A potential risk for individuals who hyperventilate before swimming underwater; the CNS shuts down suddenly owing to hypoxia.

STUDY OUTLINE

INTRODUCTION   p. 798

1. Body cells must obtain oxygen and eliminate carbon dioxide. Gas exchange occurs at the **alveoli** of the lungs.

FUNCTIONS OF THE RESPIRATORY SYSTEM   p. 798

1. The functions of the **respiratory system** include (1) providing an area for gas exchange between air and circulating blood; (2) moving air to and from exchange surfaces; (3) protecting respiratory surfaces from environmental variations and defending the respiratory system and other tissues from invasion by pathogens; (4) producing sounds; and (5) providing olfactory sensations to the CNS.

ORGANIZATION OF THE RESPIRATORY SYSTEM   p. 798

1. The respiratory system includes the nose, nasal cavity, paranasal sinuses, pharynx, larynx, trachea, bronchi, bronchioles, and alveoli of the lungs. *(Figure 23-1)*

2. The **respiratory tract** consists of the conducting passageways that carry air to and from the alveoli. The passageways of the **upper respiratory system** filter and humidify incom-

ing air. The **lower respiratory system** includes delicate conduction passages and the alveolar exchange surfaces.

The Respiratory Mucosa   p. 798

3. The **respiratory mucosa** (respiratory epithelium and underlying connective tissue) lines the respiratory tract.

4. The respiratory epithelium changes in structure along the
respiratory tract. It is supported by the **lamina propria,** a
layer of areolar tissue. *(Figure 23-2)*

THE UPPER RESPIRATORY SYSTEM   p. 801

1. The components of the upper respiratory system consist of
the nose, nasal cavity, paranasal sinuses, and pharynx. *(Figures 23-1, 23-3)*

The Nose and Nasal Cavity   p. 801

2. Air normally enters the respiratory system through the **external nares,** which open into the **nasal cavity.** The **vestibule**
(entryway) is guarded by hairs that screen out large particles.

3. Incoming air flows through the **superior, middle,** and **inferior meatuses** (narrow grooves) and bounces off the conchal surfaces.

4. The **hard palate** separates the oral and nasal cavities. The
soft palate separates the superior nasopharynx from the
rest of the pharynx. The connections between the nasal cavity and nasopharynx are the **internal nares.**

5. The nasal mucosa traps particles, warms and humidifies incoming air, and cools and dehumidifies outgoing air.

The Pharynx   p. 802

6. The **pharynx** is a chamber shared by the digestive and respiratory systems. The **nasopharynx** is the superior part of the
pharynx. The **oropharynx** is continuous with the oral cavity.
The **laryngopharynx** includes the narrow zone between the
hyoid bone and the entrance to the esophagus. *(Figure 23-3)*

THE LARYNX   p. 803

1. Inhaled air passes through the **glottis** en route to the lungs;
the **larynx** surrounds and protects the glottis. *(Figure 23-4)*

Cartilages and Ligaments of the Larynx   p. 803

2. The cylindrical larynx is composed of three large cartilages
(the **thyroid cartilage, cricoid cartilage,** and **epiglottis**) and
three smaller pairs of cartilages (the **arytenoid, corniculate,** and **cuneiform cartilages**). The epiglottis projects into
the pharynx. *(Figure 23-4)*

3. Two pairs of folds span the glottis: the inelastic **vestibular
folds** and the more delicate **vocal folds.** Air passing
through the glottis vibrates the vocal folds, producing
sound. *(Figure 23-5)*

The Laryngeal Musculature   p. 805

4. The muscles of the neck and pharynx position and stabilize the larynx. The smaller intrinsic muscles regulate tension in the vocal folds or open and close the glottis. During
swallowing, both sets of muscles help prevent particles from
entering the glottis.

THE TRACHEA   p. 805

1. The **trachea** extends from the sixth cervical vertebra to the fifth
thoracic vertebra. The **submucosa** contains C-shaped **tracheal
cartilages,** which stiffen the tracheal walls and protect the airway. The posterior tracheal wall can distort to permit large
masses of food to pass through the esophagus. *(Figure 23-6)*

THE PRIMARY BRONCHI   p. 806

1. The trachea branches within the mediastinum to form the
right and **left primary bronchi.** Each bronchus enters a lung
at the **hilus** (a groove). The **root** is a connective tissue mass
including the bronchus, pulmonary vessels, and nerves.

THE LUNGS   p. 806

Lobes and Surfaces of the Lungs   p. 806

1. The **lobes** of the **lungs** are separated by fissures. The right
lung has three lobes, and the left lung has two. *(Figure 23-7)*

2. The anterior and lateral surfaces of the lungs follow the
inner contours of the rib cage. The concavity of the medial
surface of the left lung is the **cardiac notch,** which conforms
to the shape of the pericardium. *(Figure 23-8)*

The Bronchi   p. 808

3. The primary bronchi and their branches form the **bronchial
tree.** The **secondary** and **tertiary bronchi** are branches within the lungs. As they branch, the amount of cartilage in their
walls decreases and the amount of smooth muscle increases. *(Figures 23-9, 23-10)*

4. Each tertiary bronchus supplies air to a single **bronchopulmonary segment.** *(Figure 23-10)*

The Bronchioles   p. 809

5. **Bronchioles** within the bronchopulmonary segments ultimately branch into **terminal bronchioles.** Each terminal
bronchiole delivers air to a single **pulmonary lobule.** In the
lobule, the terminal bronchiole branches into **respiratory
bronchioles.** The connective tissues of the root of the lung
extend into the parenchyma of the lung as a series of *trabeculae* (partitions). These branch to form **interlobular
septa** that divide the lung into lobules. *(Figure 23-10)*

Alveolar Ducts and Alveoli   p. 811

6. The respiratory bronchioles open into **alveolar ducts;**
many alveoli are interconnected at each duct. The respiratory exchange surfaces are extensively connected to the
circulatory system via the vessels of the pulmonary circuit.
(Figure 23-11)

7. The **respiratory membrane** consists of a simple squamous
epithelium; **septal cells** scattered in it produce an oily secretion that keeps the alveoli from collapsing. **Alveolar
macrophages** patrol the epithelium and engulf foreign particles. *(Figure 23-12)*

The Blood Supply to the Lungs   p. 813

8. The conducting portions of the respiratory tract receive
blood from the external carotid arteries, the thyrocervical trunks, and the bronchial arteries. Venous blood flows
into the pulmonary veins, bypassing the rest of the systemic circuit and diluting the oxygenated blood leaving
the alveoli.

THE PLEURAL CAVITIES AND PLEURAL MEMBRANES   p. 813

1. Each lung occupies a single **pleural cavity** lined by a **pleura** (serous membrane). The two types of pleurae are the
parietal pleura, covering the inner surface of the thoracic
wall, and the **visceral pleura,** covering the lungs.

RESPIRATORY PHYSIOLOGY   p. 814

1. Respiratory physiology focuses on a series of integrated
processes. **External respiration** (the exchange of oxygen
and carbon dioxide between interstitial fluid and the external environment) includes **pulmonary ventilation**
(breathing). **Internal respiration** is the exchange of oxygen
and carbon dioxide between interstitial fluid and cells. If
oxygen content declines, the affected tissues will suffer from
hypoxia; if the oxygen supply is completely shut off, **anoxia** and tissue death result. *(Figure 23-13)*

Pulmonary Ventilation   p. 814

2. As pressure on a gas decreases, its volume expands; as pressure increases, gas volume contracts. This inverse relationship is **Boyle's law.** *(Figure 23-14; Table 23-1)*

23

3. The relationship between **intrapulmonary pressure** (the pressure inside the respiratory tract) and **atmospheric pressure (atm)** determines the direction of airflow. **Intrapleural pressure** is the pressure in the space between the parietal and visceral pleurae. *(Figures 23-14–23-16)*

4. The diaphragm and the external and internal intercostal muscles are involved in normal **quiet breathing,** or **eupnea.** Accessory muscles become active during the active inspiratory and expiratory movements of **forced breathing,** or **hyperpnea.** *(Figure 23-17)*

5. **Alveolar ventilation** is the amount of air reaching the alveoli each minute. The **vital capacity** includes the **tidal volume** plus the **expiratory** and **inspiratory reserve volumes.** The air left in the lungs at the end of maximum exhalation is the **residual volume.** *(Figure 23-18)*

Gas Exchange at the Respiratory Membrane p. 822

6. In a mixed gas, the individual gases exert a pressure proportional to their abundance in the mixture **(Dalton's law).** The pressure contributed by a single gas is its **partial pressure.**

7. The amount of a gas in solution is directly proportional to the partial pressure of that gas **(Henry's law).** *(Figure 23-19)*

8. Alveolar air and atmospheric air differ in composition. *(Table 23-2)*

Gas Pickup and Delivery p. 826

9. Blood entering peripheral capillaries delivers oxygen and absorbs carbon dioxide. The transport of oxygen and carbon dioxide in blood involves reactions that are completely reversible. *(Figure 23-20)*

10. Over the normal range of oxygen pressures in the body, a small change in plasma P_{O_2} will lead to a large change in the amount of oxygen bound or released. At alveolar P_{O_2}, hemoglobin is almost fully saturated; at the P_{O_2} of peripheral tissues, it retains a substantial oxygen reserve. When low plasma P_{O_2} continues for extended periods, red blood cells generate more **2,3-bisphosphoglycerate (BPG),** which reduces hemoglobin's affinity for oxygen. *(Figures 23-21, 23-22)*

11. **Fetal hemoglobin** has a stronger affinity for oxygen than does adult hemoglobin, aiding the removal of oxygen from maternal blood. *(Figure 23-23)*

12. Aerobic metabolism in peripheral tissues generates CO_2. About 7 percent of the CO_2 transported in blood is dissolved in the plasma; 23 percent is bound as **carbaminohemoglobin;** the rest is converted to carbonic acid, which dissociates into H^+ and HCO_3^-. *(Figure 23-24)*

Summary: Gas Transport p. 830

13. Driven by differences in partial pressure, oxygen enters the blood at the lungs and leaves in peripheral tissues; similar forces drive carbon dioxide into the blood at the tissues and into the alveoli at the lungs. *(Figure 23-25)*

CONTROL OF RESPIRATION p. 830

1. Local regulation compensates for small oscillations affecting individual tissues and organs; large-scale or extended changes require the integration of cardiovascular and respiratory responses.

Local Regulation of Gas Transport and Alveolar Function p. 830

2. Local factors regulate alveolar blood flow *(lung perfusion)* and airflow *(alveolar ventilation).* Alveolar capillaries constrict under conditions of low oxygen, and bronchioles dilate under conditions of high carbon dioxide.

The Respiratory Centers of the Brain p. 831

3. The **respiratory centers** include three pairs of nuclei in the reticular formation of the pons and medulla oblongata. The *respiratory rhythmicity centers* set the pace for respiration. The **apneustic centers** cause strong, sustained inspiratory movements, and the **pneumotaxic centers** inhibit the apneustic centers and promote exhalation. *(Figures 23-26, 23-27)*

4. The **inflation reflex** prevents overexpansion of the lungs during forced breathing. The **deflation reflex** stimulates inhalation when the lungs are collapsing. Chemoreceptor reflexes respond to changes in the P_{O_2} and P_{CO_2} of blood and cerebrospinal fluid. *(Figures 23-27, 23-28)*

Voluntary Control of Respiration p. 836

5. Conscious and unconscious thought processes can affect respiration by affecting the respiratory centers.

CHANGES IN THE RESPIRATORY SYSTEM AT BIRTH p. 836

1. Before delivery, the fetal lungs are fluid-filled and collapsed. At the first breath, the lungs inflate and never collapse completely thereafter.

AGING AND THE RESPIRATORY SYSTEM p. 837

1. The respiratory system is generally less efficient in the elderly because (1) elastic tissue deteriorates, lowering the vital capacity of the lungs; (2) movements of the chest are restricted by arthritic changes and decreased flexibility of costal cartilages; and (3) some degree of emphysema is generally present. *(Figure 23-29)*

INTEGRATION WITH OTHER SYSTEMS p. 837

1. The respiratory system has extensive anatomical and functional connections to the cardiovascular system. *(Figure 23-30)*

REVIEW QUESTIONS

LEVEL 1 REVIEWING FACTS AND TERMS

1. The C shape of the tracheal cartilages is important because
   (a) large masses of food can pass through the esophagus during swallowing
   (b) large masses of air can pass through the trachea
   (c) it allows greater tracheal elasticity and flexibility
   (d) a, b, and c are correct

2. Control over the amount of resistance to airflow and the distribution of air in the lungs is provided by the
   (a) diaphragm   (b) trachea
   (c) bronchioles   (d) alveoli

3. Pulmonary ventilation involves
   (a) internal respiration   (b) external respiration
   (c) cellular respiration   (d) air movement

4. The presence of an abnormally low oxygen content in tissue fluids is
    (a) anoxia   **(b)** edema
    (c) hypoxia   **(d)** emphysema

5. During inhalation, the lungs expand and the intrapulmonary pressure
    (a) rises to about 761 mm Hg
    (b) remains at 760 mm Hg
    (c) drops to about 759 mm Hg
    (d) does not change

6. During a normal inhalation, the intrapleural pressure is approximately
    (a) −1 mm Hg   **(b)** +6 mm Hg
    (c) +1 mm Hg   **(d)** −6 mm Hg

7. According to Henry's law, if the partial pressure of a gas increases,
    (a) gas molecules will come out of solution
    (b) more gas molecules will enter solution
    (c) the solubility of the gas will decrease
    (d) the volume of the gas will decrease

8. In tissues at a normal partial pressure of oxygen, blood entering the venous system contains about _____ of its total oxygen content.
    (a) 25 percent   **(b)** 50 percent
    (c) 75 percent   **(d)** 90 percent

9. Approximately 70 percent of the carbon dioxide absorbed by blood is transported
    (a) as bicarbonate ions
    (b) bound to hemoglobin
    (c) in the form of dissolved gas molecules
    (d) bound to oxygen molecules

10. The apneustic centers of the pons
    (a) inhibit the pneumotaxic and inspiratory centers
    (b) provide continuous stimulation to the inspiratory center
    (c) monitor blood gas levels
    (d) alter chemoreceptor sensitivity

11. All the following provide chemoreceptor input to the respiratory centers of the medulla oblongata except the
    (a) olfactory epithelium
    (b) medullary chemoreceptors
    (c) aortic body
    (d) carotid body

12. Sneezing and coughing are classic examples of
    (a) inflation reflexes   **(b)** deflation reflexes
    (c) protective reflexes   **(d)** Hering–Breuer reflexes

13. What are the five primary functions of the respiratory system?

14. Distinguish between the structures of the upper respiratory system and those of the lower respiratory system.

15. What defense mechanisms protect the respiratory system from becoming contaminated with debris or pathogens?

16. Name the three regions of the pharynx. Where is each region located?

17. What prevents you from swallowing food or liquids while you breathe?

18. How does the parietal pleura differ from the visceral pleura?

19. What four integrated steps are involved in external respiration?

20. What important physiological adjustments are necessary to adapt to and tolerate high-altitude conditions?

21. By what three mechanisms is carbon dioxide transported in the bloodstream?

22. What effect does P_{CO_2} have on smooth muscles in the walls of bronchioles?

LEVEL 2 REVIEWING CONCEPTS

23. Parasympathetic stimulation to the smooth muscle tissue layer in the bronchioles causes
    (a) bronchoconstriction
    (b) bronchodilation
    (c) a relaxation of muscle tone
    (d) an increase in tidal volume

24. If you have a respiration rate of 15 breaths per minute and a tidal volume of 500 ml of air, your respiratory minute volume is
    (a) 7.5 l/min
    (b) 75 l/min
    (c) 750 l/min
    (d) 7500 l/min

25. If an inhalation in one respiratory cycle pulls in 1000 ml of air, the amount of air reaching the alveolar spaces is about
    (a) 300 ml
    (b) 850 ml
    (c) 150 ml
    (d) 700 ml

26. Gas exchange at the respiratory membrane is efficient because
    (a) the differences in partial pressure are substantial
    (b) the gases are lipid-soluble
    (c) the total surface area is large
    (d) a, b, and c are correct

27. For any partial pressure of oxygen, if the concentration of 2,3-bisphosphoglycerate (BPG) increases,
    (a) the amount of oxygen released by hemoglobin will decrease
    (b) the oxygen levels in hemoglobin will be unaffected
    (c) the amount of oxygen released by hemoglobin will increase
    (d) the amount of carbon dioxide carried by hemoglobin will increase

28. The primary physiological adjustment necessary for an athlete to compete at high altitudes is
    (a) increased respiratory rate
    (b) increased heart rate
    (c) elevated hematocrit
    (d) a, b, and c are correct

29. An increase in the partial pressure of carbon dioxide in arterial blood causes chemoreceptors to stimulate the respiratory centers, causing
    (a) decreased respiratory rate
    (b) increased respiratory rate
    (c) hypocapnia
    (d) hypercapnia

30. What is the functional significance of the decrease in the amount of cartilage and increase in the amount of smooth muscle in the lower respiratory passageways?

31. Why can't you swallow solid food or liquid properly while you are talking?

32. Why is breathing through the nasal cavity more desirable than breathing through the mouth?

33. How would you justify the statement "the bronchioles are to the respiratory system what the arterioles are to the cardiovascular system"?

34. How are septal cells involved with keeping the alveoli from collapsing?

35. Why is diffusion across the respiratory membrane rapid and efficient?

36. How does pulmonary ventilation differ from alveolar ventilation, and what is the function of each type of ventilation?

23

37. What is the significance of **(a)** Boyle's law, **(b)** Dalton's law, and **(c)** Henry's law to the process of respiration?

38. If the pleural cavity is penetrated due to an injury, why does the lung collapse?

39. What happens to the process of respiration when a person is sneezing or coughing?

40. What are the differences between pulmonary volumes and respiratory capacities? How are pulmonary volumes and respiratory capacities determined?

41. What are the primary effects of pH, temperature, and 2,3-bisphosphoglycerate (BPG) on hemoglobin saturation?

42. What is the functional difference between the dorsal respiratory group (DRG) and the ventral respiratory group (VRG) of the medulla oblongata?

43. What types of sensory information modify the activities of the respiratory centers?

LEVEL 3 CRITICAL THINKING AND CLINICAL APPLICATIONS

44. Billy's normal alveolar ventilation rate (AVR) during mild exercise is 6.0 l/min. While at the beach on a warm summer day, he goes snorkeling. The snorkel has a volume of 50 ml. Assuming that the water is not too cold and that snorkeling for Billy is mild exercise, what would his respiratory rate have to be for him to maintain an AVR of 6.0 l/min while snorkeling?

(Assume a constant tidal volume of 500 ml and an anatomic dead space of 150 ml.)

45. A decrease in blood pressure triggers a baroreceptor reflex that leads to increased ventilation. What is the possible advantage of this reflex?

46. Mr. B. has had chronic advanced emphysema for 15 years. While hospitalized with a respiratory infection, he goes into respiratory distress. Without thinking, his nurse immediately administers pure oxygen, which causes Mr. B. to stop breathing. Why?

47. You spend the night at a friend's house during the winter. Your friend's home is quite old, and the hot-air furnace lacks a humidifier. When you wake up in the morning, you have a fair amount of nasal congestion and decide you might be coming down with a cold. After a steamy shower and some juice for breakfast, the nasal congestion disappears. Explain.

48. Why would a person with kyphosis (see Chapter 7, p. 212) exhibit a lower-than-normal vital capacity?

49. Why do individuals who are anemic generally not exhibit an increase in respiratory rate or tidal volume even though their blood is not carrying enough oxygen?

50. Doris has an obstruction of her right primary bronchus. As a result, how would you expect the oxygen–hemoglobin saturation curve for her right lung to compare with that for her left?

CHAPTER 24

The Digestive System

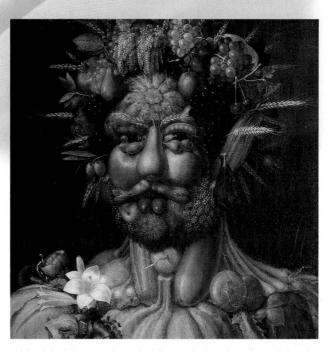

Although Giuseppe Arcimboldo may have taken the notion a bit too literally, modern physiology has confirmed that, in a sense, "you are what you eat." We all need a regular supply of nutrients in our diet. Unfortunately, the nutrients in food are not ready for use by our cells. Without processing by the digestive system, food would be of no more use to us than a lump of coal in the gas tank of a car. This chapter introduces the organs of the digestive system and their varied functions in digestion.

CHAPTER OUTLINE AND OBJECTIVES

INTRODUCTION   p. 846

AN OVERVIEW OF THE STRUCTURE AND FUNCTION
OF THE DIGESTIVE TRACT   p. 846
  Histological Organization of the Digestive Tract   p. 846
  The Movement of Digestive Materials   p. 849
  The Control of Digestive Function   p. 849
  The Peritoneum   p. 851
① Identify the organs of the digestive system.
② List the functions of the digestive system.
③ Describe the functional histology of the digestive tract.
④ Describe the processes by which materials move
through the digestive tract.
⑤ Describe the mechanisms that regulate digestion.

THE ORAL CAVITY   p. 853
  The Tongue   p. 854
  Salivary Glands   p. 854
  The Teeth   p. 856

THE PHARYNX   p. 858

THE ESOPHAGUS   p. 858
  Histology of the Esophagus   p. 860
  Swallowing   p. 860
⑥ Describe the anatomy and functions of the oral cavity,
pharynx, and esophagus.

THE STOMACH   p. 861
  Anatomy of the Stomach   p. 861
  Regulation of Gastric Activity   p. 865
  Digestion and Absorption in the Stomach   p. 868
⑦ Describe the anatomy of the stomach, its histological
features, and its roles in digestion and absorption.

THE SMALL INTESTINE AND ASSOCIATED GLANDULAR
ORGANS   p. 868
  The Small Intestine   p. 869
  Intestinal Movements   p. 872
  The Pancreas   p. 872
  The Liver   p. 874
  The Gallbladder   p. 879
  The Coordination of Secretion and Absorption   p. 880
⑧ Describe the anatomical and histological
characteristics of the small intestine.
⑨ Explain the functions of the intestinal secretions, and
discuss the regulation of secretory activities.

⑩ Describe the structure, functions, and regulation of the
accessory digestive organs.

THE LARGE INTESTINE   p. 882
  The Cecum   p. 882
  The Colon   p. 882
  The Rectum   p. 884
  Histology of the Large Intestine   p. 884
  Physiology of the Large Intestine   p. 885
⑪ Describe the structure of the large intestine, its
movements, and its absorptive processes.

DIGESTION AND ABSORPTION   p. 887
  The Processing and Absorption of Nutrients   p. 887
  Carbohydrate Digestion and Absorption   p. 888
  Lipid Digestion and Absorption   p. 890
  Protein Digestion and Absorption   p. 890
  Water Absorption   p. 891
  Ion Absorption   p. 891
  Vitamin Absorption   p. 892
⑫ Describe the digestion and absorption of organic and
inorganic nutrients.

AGING AND THE DIGESTIVE SYSTEM   p. 893

INTEGRATION WITH OTHER SYSTEMS   p. 893
⑬ Give examples of interactions between the digestive
system and each of the other organ systems.

MEDIALAB and CHAPTER REVIEW   p. 895

F ew people give any serious thought to the digestive system unless it malfunctions. Yet we spend hours of conscious effort filling and emptying it. References to this system are part of our everyday language. Think of how often we use expressions relating to the digestive system: We may "have a gut feeling," "want to chew on" something, or find someone's opinions "hard to swallow." When something does go wrong with the digestive system, even something minor, most of us seek treatment immediately. For this reason, television advertisements promote toothpaste and mouthwash, diet supplements, antacids, and laxatives.

Metabolic reactions within cells provide the energy cells need to continue functioning. These reactions require two essential ingredients: (1) oxygen and (2) organic molecules, such as carbohydrates, fats, or proteins, that can be broken down by intracellular enzymes. The respiratory system, working in concert with the cardiovascular system, supplies the necessary oxygen. The **digestive system,** working with the cardiovascular and lymphatic systems, provides the organic molecules. In effect, the digestive system provides both the fuel that keeps all the body's cells running and the building blocks needed for cell growth and repair.

The digestive system consists of a muscular tube, the **digestive tract,** also called the *gastrointestinal (GI) tract* or *alimentary canal,* and various **accessory organs.** The *oral cavity* (mouth), *pharynx, esophagus, stomach, small intestine,* and *large intestine* make up the digestive tract. Accessory digestive organs include the teeth, tongue, and various *glandular organs,* such as the salivary glands, liver, and pancreas, which secrete into ducts emptying into the digestive tract. Food enters the digestive tract and passes along its length. On the way, the secretions of the glandular organs, which contain water, enzymes, buffers, and other components, assist in preparing organic and inorganic nutrients for absorption across the epithelium of the digestive tract.

We can consider digestive functions to be a series of integrated steps:

1. **Ingestion** occurs when materials enter the digestive tract via the mouth. Ingestion is an active process involving conscious choice and decision making.
2. **Mechanical processing** is physical manipulation and distortion that makes materials easier to propel along the digestive tract and increases the surface area for enzymatic attack. Mechanical processing may or may not be required before ingestion; you can swallow liquids immediately but must process most solids first. Tearing and crushing with the teeth, followed by squashing and compaction by the tongue, are examples of preliminary mechanical processing. Swirling, mixing, and churning motions of the stomach and intestines provide mechanical processing after ingestion.

3. **Digestion** refers to the chemical breakdown of food into small organic fragments suitable for absorption by the digestive epithelium. Simple molecules in food, such as glucose, can be absorbed intact, but epithelial cells have no way to deal with molecules the size and complexity of proteins, polysaccharides, or triglycerides. These molecules must be disassembled by digestive enzymes prior to absorption. For example, the starches in a potato are of no value until enzymes have broken them down to simple sugars that your digestive epithelium can absorb and distribute to your cells.
4. **Secretion** is the release of water, acids, enzymes, buffers, and salts by the epithelium of the digestive tract and by glandular organs.
5. **Absorption** is the movement of organic substrates, electrolytes (inorganic ions), vitamins, and water across the digestive epithelium and into the interstitial fluid of the digestive tract.
6. **Excretion** is the removal of waste products from body fluids. The digestive tract and glandular organs discharge waste products in secretions that enter the lumen of the tract. Most of these waste products, after mixing with the indigestible residue of the digestive process, will leave the body. The ejection of materials from the digestive tract, a process called **defecation** (def-e-KĀ-shun), or *egestion,* eliminates materials as **feces.**

The lining of the digestive tract also plays a protective role by safeguarding surrounding tissues against (1) the corrosive effects of digestive acids and enzymes; (2) mechanical stresses, such as abrasion; and (3) bacteria that either are swallowed with food or reside in the digestive tract. The digestive epithelium and its secretions provide a nonspecific defense against these bacteria. Bacteria reaching the lamina propria are attacked by macrophages and other cells of the immune system.

AN OVERVIEW OF THE STRUCTURE AND FUNCTION OF THE DIGESTIVE TRACT

Figure 24-1• shows the major components of the digestive system. The digestive tract begins at the oral cavity, continues through the pharynx, esophagus, stomach, small intestine, and large intestine, and ends at the anus. These structures have overlapping functions, but each has certain areas of specialization and shows distinctive histological characteristics.

HISTOLOGICAL ORGANIZATION OF THE DIGESTIVE TRACT

The major layers of the digestive tract include (1) the *mucosa,* (2) the *submucosa,* (3) the *muscularis externa,* and (4) the *serosa.* Sectional, diagrammatic views of

24

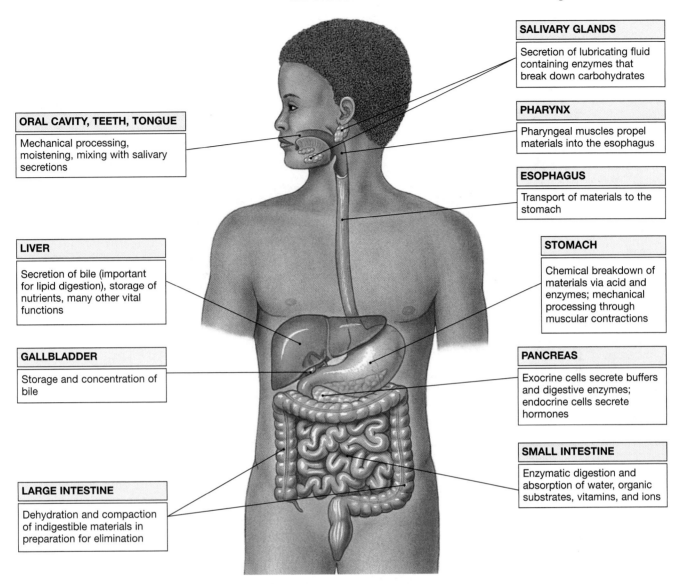

•FIGURE 24-1 **The Components of the Digestive System.** The major regions and accessory organs of the digestive tract, together with their primary functions.

these layers are presented in Figure 24-2•. The structure of the layers varies by region; Figure 24-2 is a composite view that most closely resembles the appearance of the small intestine, the longest segment of the digestive tract.

The Mucosa

The inner lining, or **mucosa,** of the digestive tract is a *mucous membrane.* Mucous membranes, introduced in Chapter 4, consist of a layer of an epithelium moistened by glandular secretions and an underlying layer of areolar tissue, the *lamina propria.* ∞ *p. 129*

The Digestive Epithelium The mucosal epithelium is either simple or stratified, depending on the location and the stresses to which it is most often subjected. The oral cavity, pharynx, and esophagus (where

mechanical stresses are most severe) are lined by a stratified squamous epithelium, whereas the stomach, the small intestine, and almost the entire length of the large intestine (where absorption occurs) have a simple columnar epithelium that contains goblet cells. Scattered among the columnar cells are **enteroendocrine cells,** which secrete hormones that coordinate the activities of the digestive tract and the accessory glands.

The lining of the digestive tract is often thrown into longitudinal folds, which disappear as the tract fills, and permanent transverse folds, or *plicae* (PLĪ-sē; singular, *plica* [PLĪ-ka]) (Figure 24-2•). The folding dramatically increases the surface area available for absorption. The secretions of gland cells located in the mucosa and submucosa—or in accessory glandular organs—are carried to the epithelial surfaces by ducts.

24

•FIGURE 24-2 The Structure of the Digestive Tract. A diagrammatic view of a representative portion of the digestive tract. The features illustrated are those of the small intestine.

Epithelial Renewal and Repair

The life span of a typical epithelial cell varies from 2 to 3 days in the esophagus to 6 days in the large intestine. The lining of the entire digestive tract is continuously renewed through the divisions of epithelial stem cells, keeping pace with the rate of cell destruction and loss at epithelial surfaces. This high rate of cell division explains why radiation and anticancer drugs that inhibit mitosis have drastic effects on the digestive tract. Lost epithelial cells are no longer replaced, and the cumulative damage to the epithelial lining quickly leads to problems in nutrient absorption. In addition, the exposure of the lamina propria to digestive enzymes can cause internal bleeding and other serious problems.

The Lamina Propria   The lamina propria consists of a layer of areolar tissue that also contains blood vessels, sensory nerve endings, lymphatic vessels, smooth muscle cells, and scattered areas of lymphoid tissue. In the oral cavity, pharynx, esophagus, stomach, and *duodenum* (the proximal portion of the small intestine), the lamina propria also contains the secretory cells of mucous glands.

In most areas of the digestive tract, the lamina propria contains a narrow band of smooth muscle and elastic fibers. This band is called the **muscularis** (mus-kū-LAR-is) **mucosae.** The smooth muscle cells in the muscularis mucosae are arranged in two concentric layers (Figure 24-2•). The inner layer encircles the

lumen (the *circular muscle*), and the outer layer contains muscle cells oriented parallel to the long axis of the tract (the *longitudinal layer*). Contractions in these layers alter the shape of the lumen and move the epithelial pleats and folds.

The Submucosa

The **submucosa** is a layer of dense irregular connective tissue that surrounds the muscularis mucosae. This layer has large blood vessels and lymphatic vessels, and in some regions it also contains exocrine glands that secrete buffers and enzymes into the lumen of the digestive tract. Along its outer margin, the submucosa contains a network of nerve fibers and scattered neurons. This **submucosal plexus,** or *plexus of Meissner,* contains sensory neurons, parasympathetic ganglionic neurons, and sympathetic postganglionic fibers that innervate the mucosa and submucosa (Figure 24-2•).

The Muscularis Externa

The submucosal plexus lies along the inner border of the **muscularis externa,** a region dominated by smooth muscle cells. Like the smooth muscle cells in the muscularis mucosae, those in the muscularis externa are arranged in an inner, circular layer and an outer, longitudinal layer. These layers play an essential role in mechanical processing and in the movement of materials along the digestive tract. These movements are coordinated primarily by neurons of the **myenteric** (mī-en-TER-ik) **plexus** (*mys,* muscle + *enteron,* intestine), or *plexus of Auerbach.* This network of parasympathetic ganglia and sympathetic postganglionic fibers lies sandwiched between the circular and longitudinal muscle layers. Parasympathetic stimulation increases muscle tone and activity; sympathetic stimulation promotes muscular inhibition and relaxation.

The Serosa

Along most portions of the digestive tract inside the peritoneal cavity, the muscularis externa is covered by a serous membrane known as the **serosa** (Figure 24-2•). ∞ *p. 129* There is no serosa covering the muscularis externa of the oral cavity, pharynx, esophagus, and rectum, where a dense network of collagen fibers firmly attaches the digestive tract to adjacent structures. This fibrous sheath is called an *adventitia.*

THE MOVEMENT OF DIGESTIVE MATERIALS

Your digestive tract contains layers of *visceral smooth muscle tissue,* a type of smooth muscle introduced in Chapter 10. ∞ *p. 305* The smooth muscle along the digestive tract shows rhythmic cycles of activity due to the presence of *pacesetter cells.* These smooth muscle cells undergo spontaneous depolarization, and their contraction triggers a wave of contraction that spreads through the entire muscular sheet. Pacesetter cells are located in the muscularis mucosae and muscularis externa, whose layers surround the lumen of the digestive tract. The coordinated contractions of the muscularis externa play a vital role in the movement of materials along the tract, through *peristalsis,* and in mechanical processing, through *segmentation.*

Peristalsis

The muscularis externa propels materials from one portion of the digestive tract to another by contractions known as **peristalsis** (per-i-STAL-sis). Peristalsis consists of waves of muscular contractions that move a **bolus,** or small oval mass of digestive contents, along the length of the digestive tract. During a peristaltic movement, the circular muscles contract behind the bolus. Longitudinal muscles ahead of the bolus contract next, shortening adjacent segments. A wave of contraction in the circular muscles then forces the bolus in the desired direction (Figure 24-3•).

Segmentation

Most areas of the small intestine and some portions of the large intestine undergo cycles of contraction

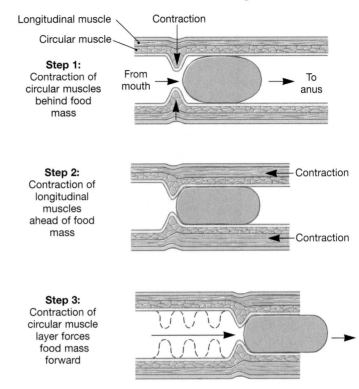

•**FIGURE 24-3 Peristalsis.** Peristalsis propels materials along the length of the digestive tract.

that produce **segmentation.** These movements churn and fragment the bolus, mixing the contents with intestinal secretions. Because they do not follow a set pattern, segmentation movements do not push materials along the tract in any one direction.

THE CONTROL OF DIGESTIVE FUNCTION

The activities of the digestive system are regulated by neural, hormonal, and local mechanisms (Figure 24-4•).

Neural Mechanisms

The movement of materials along your digestive tract and many secretory functions are controlled primarily by neural mechanisms. For example, peristaltic movements limited to a few centimeters of the digestive tract are triggered by sensory receptors in the walls of the digestive tract. The motor neurons that control smooth muscle contraction and glandular secretion are located in the myenteric plexus. These neurons are usually considered to be parasympathetic, because they are innervated by parasympathetic preganglionic fibers. However, the sensory neurons and interneurons that are responsible for local reflexes are also located in the myenteric plexus rather than in the central nervous system. As we noted in Chapter 16, the reflexes controlled by these

24

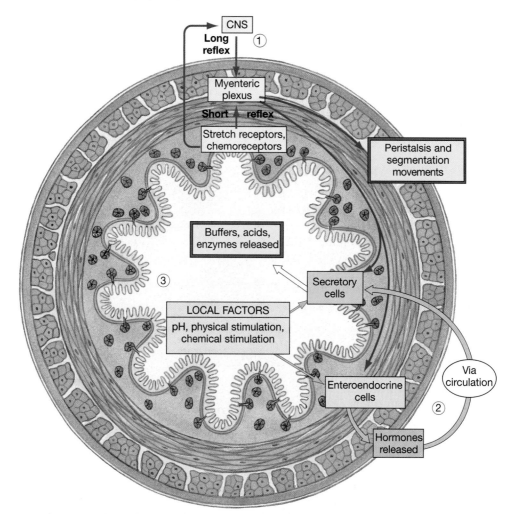

•**FIGURE 24-4 The Regulation of Digestive Activities.** The major factors responsible for regulating digestive activities: ① neural mechanisms, ② hormonal mechanisms, and ③ local mechanisms.

neurons are called *short reflexes.* ∞ *p. 520* These reflexes are also called *myenteric reflexes,* and the term *enteric nervous system* is often used to refer to the neural network that coordinates these reflexes along the digestive tract.

In general, short reflexes control activities in one region of the digestive tract. This control may involve coordinating local peristalsis and triggering the secretion of digestive glands. Many neurons are involved—the enteric nervous system has roughly as many neurons and neurotransmitters as the spinal cord has. The specific functions and interactions of these neurotransmitters in the enteric nervous system remain largely unknown.

Sensory information from receptors in the digestive tract is also distributed to the CNS, where it can trigger *long reflexes.* ∞ *p. 520* Long reflexes, which involve interneurons and motor neurons in the CNS, provide a higher level of control over digestive and glandular activities. These reflexes generally control large-scale peristaltic waves that move materials from one region of the digestive tract to another. Long reflexes may involve motor fibers in the glossopharyngeal, vagus, or pelvic nerves that synapse in the myenteric plexus. The sensitivity of the smooth muscle cells to neural commands can be enhanced or inhibited by digestive hormones.

Hormonal Mechanisms

Your digestive tract is a hormone factory that produces at least 18 hormones. These hormones affect almost every aspect of digestive function, and some of them also affect the activities of other systems. The hormones (*gastrin, secretin,* and others) are peptides produced by enteroendocrine cells in the digestive tract. These hormones reach their target organs by distribution in the bloodstream. We shall consider each of these hormones further as we proceed down the length of the digestive tract.

Local Mechanisms

Prostaglandins, histamine, and other chemicals released into interstitial fluid may affect adjacent cells within a small segment of the digestive tract. These local messengers are important in coordinating a response to conditions (such as a change in the local pH or other chemical or physical stimulation) that affect only a portion of the tract. For example, histamine release in the lamina propria of the stomach stimulates the secretion of acid by cells in the adjacent epithelium.

THE PERITONEUM

The abdominopelvic cavity contains the *peritoneal cavity.* The peritoneal cavity is lined by a serous membrane that consists of a superficial mesothelium covering a layer of areolar tissue. We can divide the serous membrane into the serosa, or *visceral peritoneum,* which covers organs within the peritoneal cavity, and the *parietal peritoneum,* which lines the inner surfaces of the body wall.

The serous membrane lining the peritoneal cavity continuously produces peritoneal fluid that lubricates the peritoneal surfaces. About 7 liters of fluid is secreted and reabsorbed each day, although the volume within the peritoneal cavity at any one time is very small. Liver disease, kidney disease, and heart failure can cause an increase in the rate of fluid movement into the peritoneal cavity. The accumulation of fluid creates a characteristic abdominal swelling called **ascites** (a-SĪ-tēz). The distortion of internal organs by the contained fluid can result in symptoms such as heartburn, indigestion, and low back pain.

Mesenteries

Portions of your digestive tract are suspended within the peritoneal cavity by sheets of serous membrane that connect the parietal peritoneum with the visceral peritoneum. These **mesenteries** (MEZ-en-ter-ēz) are double sheets of peritoneal membrane (Figure 24-2•). The areolar tissue between the mesothelial surfaces provides an access route for the passage of blood vessels, nerves, and lymphatic vessels to and from the digestive tract. Mesenteries also stabilize the positions of the attached organs and prevent your intestines from becoming entangled during digestive movements or sudden changes in body position.

In embryonic development, the digestive tract and accessory organs are suspended within the peritoneal cavity by *dorsal* and *ventral mesenteries* (Figure 24-5a•). The ventral mesentery later disappears along most of the digestive tract, persisting in adults only (1) on the ventral surface of the stomach, between the stomach and liver (the lesser omentum); and (2) between the liver and the anterior abdominal wall (the falciform ligament) (Figure 24-5b,c,d•). The **lesser omentum** (ō-MEN-tum; *omentum,* fat skin) stabilizes the position

of the stomach and provides an access route for blood vessels and other structures entering or leaving the liver. The **falciform** (FAL-si-form) **ligament** helps stabilize the position of the liver relative to the diaphragm and abdominal wall. [AM] *EMBRYOLOGY SUMMARY 19: The Development of the Digestive System*

As the digestive tract elongates, it twists and turns within the crowded peritoneal cavity. The dorsal mesentery of the stomach becomes greatly enlarged and forms an enormous pouch that extends inferiorly between the body wall and the anterior surface of the small intestine. This pouch, the **greater omentum** (Figure 24-5b,d•), hangs like an apron from the lateral and inferior borders of the stomach. Adipose tissue in the greater omentum conforms to the shapes of the surrounding organs, providing padding and protection across the anterior and lateral surfaces of the abdomen. The lipids in the adipose tissue are an important energy reserve, and the greater omentum provides insulation that reduces heat loss across the anterior abdominal wall.

All but the first 25 cm of the small intestine is suspended by the **mesentery proper,** a thick mesenterial sheet that provides stability but permits some independent movement. The mesentery associated with the initial portion of the small intestine (the *duodenum*) and pancreas fuses with the posterior abdominal wall, locking those structures in position. Their anterior surfaces are covered by peritoneum, but the rest of the organs lie outside the peritoneal cavity. Organs that lie posterior to, rather than within, the peritoneal cavity are called **retroperitoneal** (*retro,* behind).

A **mesocolon** is a mesentery associated with a portion of the large intestine. During normal development, the mesocolon of the *ascending colon,* the *descending colon,* and the *rectum* of the large intestine fuse to the dorsal body wall. These regions are thus locked in place. Thereafter these organs are retroperitoneal, with the visceral peritoneum covering only their anterior surfaces and portions of their lateral surfaces (Figure 24-5b,c,d•). The **transverse mesocolon,** which supports the transverse colon, and the **sigmoid mesocolon,** which supports the sigmoid colon, are all that remains of the original mesocolon.

24

Peritonitis

An inflammation of the peritoneal membrane produces symptoms of **peritonitis** (per-i-tō-NĪ-tis), a painful condition that interferes with the normal functioning of the affected organs. Physical damage, chemical irritation, and bacterial invasion of the peritoneum can lead to severe and even fatal cases of peritonitis. Peritonitis may be caused in untreated appendicitis by the rupturing of the appendix and the subsequent release of bacteria into the peritoneal cavity. Peritonitis is a potential complication of any surgery in which the peritoneal cavity is opened.

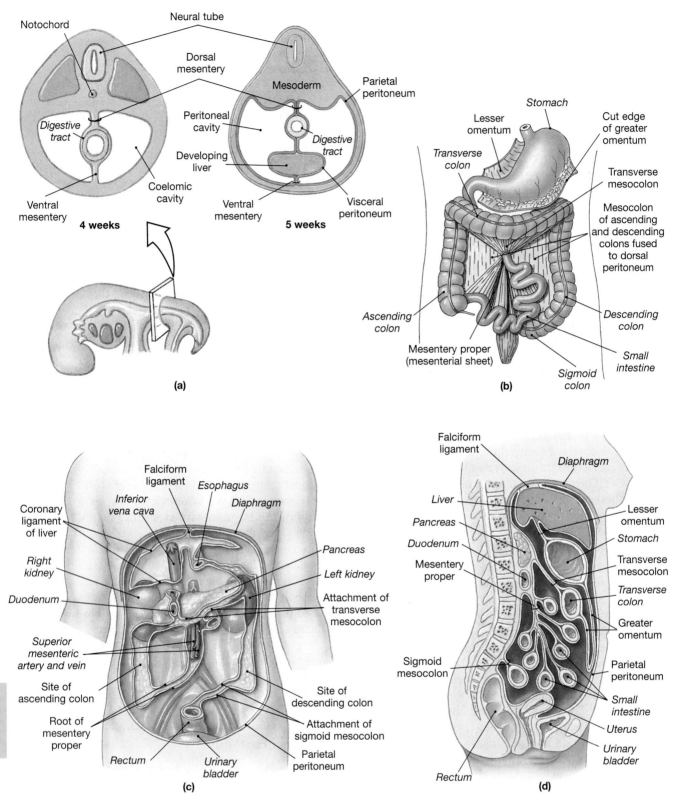

•**FIGURE 24-5** **Mesenteries.** **(a)** During embryonic development, the digestive tube is initially suspended by dorsal and ventral mesenteries. The ventral mesentery in adults is lost except where it connects the stomach to the liver *(lesser omentum)* and the liver to the anterior body wall and diaphragm *(falciform ligament)*. **(b)** A diagrammatic view of the organization of mesenteries in an adult. As the digestive tract enlarges, mesenteries associated with the proximal portion of the small intestine, the pancreas, and the ascending and descending portions of the colon fuse to the body wall. **(c)** An anterior view of the empty peritoneal cavity, showing attachment sites where fusion occurs. **(d)** A sagittal section of the mesenteries of an adult. Notice that the pancreas, duodenum, and rectum are retroperitoneal.

✓ What is the importance of the mesenteries?

✓ Which would be more efficient in propelling intestinal contents from one place to another—peristalsis or segmentation?

✓ What effect would a drug that blocks the parasympathetic stimulation of the digestive tract have on peristalsis?

THE ORAL CAVITY

Our exploration of the digestive tract will follow the path of ingested materials from the mouth to the anus. The mouth opens into the **oral cavity,** or **buccal** (BUK-al) **cavity** (Figure 24-6•). We can summarize the functions of the oral cavity as follows: (1) *analysis* of material before swallowing; (2) *mechanical processing* through the actions of the teeth, tongue, and palatal surfaces; (3) *lubrication* by mixing with mucus and salivary gland secretions; and (4) limited *digestion* of carbohydrates and lipids.

The oral cavity is lined by the **oral mucosa,** which has a stratified squamous epithelium. Only the regions exposed to severe abrasion (such as the superior surface of the tongue and the opposing surfaces of the hard palate, part of the roof of the mouth) are covered by a layer of keratinized cells. The epithelial lining of the cheeks, lips, and inferior surface of the tongue is relatively thin, nonkeratinized, and delicate. Although nutrient absorption does not occur in the oral cavity, the mucosa inferior to the tongue is thin enough and vascular enough to permit the rapid absorption of lipid-soluble drugs. *Nitroglycerin* may be administered via this route to treat acute angina attacks.

The mucosae of the **cheeks,** or lateral walls of the oral cavity, are supported by pads of fat and the buccinator muscles. ∞ *p. 321* Anteriorly, the mucosa of each cheek is continuous with that of the lips, or **labia** (LĀ-bē-uh; singular, *labium*). The **vestibule** is the space between the cheeks (or lips) and the teeth. The **gingivae** (JIN-ji-vē), or *gums,* are ridges of oral mucosa that surround the base of each tooth on the alveolar processes of the maxillary bones and mandible. ∞ *pp. 205, 207* The gingivae in most regions are firmly bound to the periostea of the underlying bones.

The roof of the oral cavity is formed by the hard and soft palates; the tongue dominates its floor (Figure 24-6b•). The floor of the mouth inferior to the tongue receives extra support from the mylohyoid muscle. ∞ *p. 328* The *hard palate* is formed by the palatine processes of the maxillary bones and the horizontal plates of the palatine bones. A prominent central ridge, or *raphe,* extends along the midline of the hard palate. The mucosa lateral and anterior to the raphe is thick, with complex ridges. When your tongue compresses food against your hard palate, these ridges provide traction. The *soft palate* lies posterior to the hard palate. A thinner and more-delicate mucosa covers the posterior margin of the hard palate and extends onto the soft palate.

The posterior margin of the soft palate supports the **uvula** (Ū-vū-luh), a dangling process that helps prevent food from entering the pharynx prematurely. On either side of the uvula are two pairs of muscular *pharyngeal arches* (Figure 24-6•). The more anterior **palatoglossal** (pal-a-tō-GLOS-al) **arch** extends between the soft palate and the base of the tongue. A

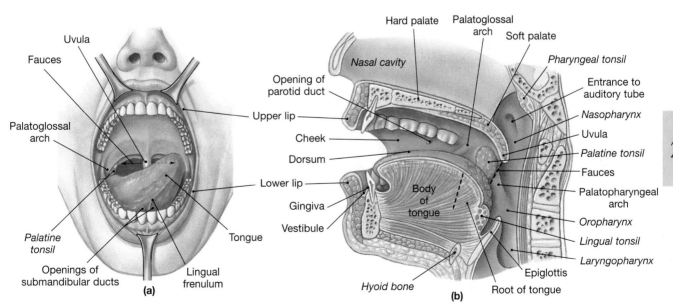

•**FIGURE 24-6 The Oral Cavity. (a)** An anterior view of the oral cavity, as seen through the open mouth. **(b)** A sagittal section.
AM *Plate 5.3d*

curving line that connects the palatoglossal arches and uvula forms the boundaries of the **fauces** (FAW-sēz), the passageway between the oral cavity and the oropharynx. The more posterior **palatopharyngeal** (pal-a-tō-fa-RIN-jē-al) **arch** extends from the soft palate to the pharyngeal wall. A palatine tonsil lies between the palatoglossal and palatopharyngeal arches on either side.

THE TONGUE

The **tongue** (Figure 24-6●) manipulates materials inside the mouth and is occasionally used to bring foods (such as ice cream) into the oral cavity. The primary functions of the tongue are (1) mechanical processing by compression, abrasion, and distortion; (2) manipulation to assist in chewing and to prepare the material for swallowing; (3) sensory analysis by touch, temperature, and taste receptors, and (4) secretion of mucus and the enzyme lingual lipase.

We can divide the tongue into an anterior **body,** or *oral portion,* and a posterior **root,** or *pharyngeal portion.* The superior surface, or *dorsum,* of the body contains a forest of fine projections, the *lingual papillae.* ∞ *p. 536* The thickened epithelium covering each papilla assists in the movement of materials by the tongue. A V-shaped line of circumvallate papillae roughly indicates the boundary between the body and the root of the tongue, which is situated in the oropharynx (Figure 24-6b●).

The epithelium covering the inferior surface of the tongue is thinner and more delicate than that of the dorsum. Along the inferior midline is the **lingual frenulum** (FREN-ū-lum; *frenulum,* a small bridle), a thin fold of mucous membrane that connects the body of the tongue to the mucosa covering the floor of the oral cavity. Ducts from two pairs of salivary glands open on either side of the lingual frenulum. The lingual frenulum prevents extreme movements of the tongue. However, if your lingual frenulum is *too* restrictive, you cannot eat or speak normally. When properly diagnosed, this condition, called **ankyloglossia** (ang-ki-lō-GLOS-ē-uh), can be corrected surgically.

The tongue's epithelium is flushed by the secretions of small glands that extend into the lamina propria of the tongue. These secretions contain water, mucus, and the enzyme **lingual lipase.** Lingual lipase begins the enzymatic breakdown of lipids, specifically triglycerides, before you have swallowed the food.

Your tongue contains two groups of skeletal muscles: (1) **intrinsic tongue muscles** and (2) **extrinsic tongue muscles.** All gross movements of the tongue are performed by the relatively large extrinsic muscles. ∞ *p. 325* The smaller intrinsic muscles change the shape of your tongue and assist the extrinsic muscles during precise movements, as in speech. Both intrinsic and extrinsic tongue muscles are under the control of the hypoglossal nerve (N XII).

SALIVARY GLANDS

Three pairs of salivary glands secrete into the oral cavity (Figure 24-7a●). Each pair of salivary glands has a distinctive cellular organization and produces *saliva,* a

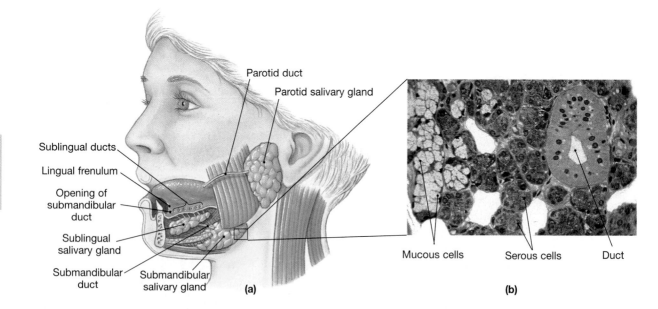

●**FIGURE 24-7 The Salivary Glands. (a)** A lateral view, showing the relative positions of the salivary glands and ducts on the left side of the head. For clarity, the left ramus and body of the mandible have been removed. For the positions of the ducts in the oral cavity, see Figure 24-6. **(b)** The submandibular gland secretes a mixture of mucins, produced by mucous cells, and enzymes, produced by serous cells. (LM × 303) AM *Plates 5.2, 5.3a,b*

mixture of glandular secretions, with slightly different properties:

1. The large **parotid** (pa-ROT-id) **salivary glands** lie inferior to the zygomatic arch deep to the skin that covers the lateral and posterior surface of the mandible. Each gland has an irregular shape, extending from the mastoid process of the temporal bone across the outer surface of the masseter muscle. ∞ *p. 324* The parotid salivary glands produce a thick, serous secretion containing large amounts of *salivary amylase,* an enzyme that breaks down starches (complex carbohydrates). The secretions of each parotid gland are drained by a **parotid duct** *(Stensen's duct),* which empties into the vestibule at the level of the second upper molar.

2. The **sublingual** (sub-LING-gwal) **salivary glands** are covered by the mucous membrane of the floor of the mouth. These glands produce a watery, mucous secretion that acts as a buffer and lubricant. Numerous **sublingual ducts** *(Rivinus' ducts)* open along either side of the lingual frenulum.

3. The **submandibular salivary glands** are situated in the floor of the mouth along the inner surfaces of the mandible within a depression called the *mandibular groove.* The submandibular glands (Figure 24-7b•) secrete a mixture of buffers, glycoproteins called *mucins,* and salivary amylase. The **submandibular ducts** *(Wharton's ducts)* open into the mouth on either side of the lingual frenulum immediately posterior to the teeth (Figure 24-6a•).

Saliva

Your salivary glands produce 1.0–1.5 liters of saliva each day. Saliva is 99.4 percent water, and the remaining 0.6 percent includes an assortment of electrolytes (principally Na^+, Cl^-, and HCO_3^-), buffers, glycoproteins, antibodies, enzymes, and waste products. The glycoproteins, called **mucins,** are primarily responsible for the lubricating action of saliva. ∞ *p. 56* About 70 percent of saliva originates in the submandibular salivary glands, 25 percent in the parotids, and the remaining 5 percent in the sublingual salivary glands.

A continuous background level of secretion flushes the oral surfaces, helping keep them clean. Buffers in the saliva keep the pH of your mouth near 7.0 and prevent the buildup of acids produced by bacterial action. In addition, saliva contains antibodies (IgA) and lysozymes that help control populations of oral bacteria. A reduction or elimination of salivary secretions, caused by radiation exposure, emotional distress, or other factors, triggers a bacterial population explosion in the oral cavity. This proliferation rapidly leads to recurring infections and the progressive erosion of the teeth and gums.

The saliva produced when you eat has a variety of functions, including:

- Lubricating the mouth.
- Moistening and lubricating materials in the mouth.
- Dissolving chemicals that can stimulate the taste buds and provide sensory information about the material.
- Initiating the digestion of complex carbohydrates before the material is swallowed. The enzyme involved is **salivary amylase,** which is also known as *ptyalin* or *alpha-amylase.* Although the digestive process begins in the oral cavity, it is not completed there, and no absorption of nutrients occurs across the lining of the oral cavity.

Mumps

The **mumps virus** most often targets the salivary glands, especially the parotid salivary glands, although other organs can also become infected. Infection typically occurs at 5–9 years of age. The first exposure stimulates antibody production and in most cases confers permanent immunity; active immunity can be conferred by immunization. In postadolescent males, the mumps virus can also infect the testes and cause sterility. Infection of the pancreas by the mumps virus can produce temporary or permanent diabetes; other organ systems, including the central nervous system, are affected in severe cases. An effective mumps vaccine is available. Widespread distribution of that vaccine has reduced the incidence of the disease in the United States.

Control of Salivary Secretion

Salivary secretions are normally controlled by the autonomic nervous system. Each salivary gland receives parasympathetic and sympathetic innervation. The parasympathetic outflow originates in the **salivatory nuclei** of the medulla oblongata and synapses in the submandibular and otic ganglia. ∞ *pp. 469, 471, 514* Any object in your mouth can trigger a salivary reflex by stimulating receptors monitored by the trigeminal nerve (N V) or by stimulating taste buds innervated by cranial nerves N VII, IX, or X. Parasympathetic stimulation accelerates secretion by all the salivary glands, resulting in the production of large amounts of saliva. The role of the sympathetic innervation remains uncertain; evidence suggests that it provokes the secretion of small amounts of very thick saliva.

The salivatory nuclei are also influenced by other brain stem nuclei as well as by the activities of higher centers. For example, chewing with an empty mouth, the smell of food, or even thinking about food will initiate an increase in salivary secretion rates; that is why chewing gum is so effective at keeping your mouth moist. The presence of irritating stimuli in the esophagus, stomach, or intestines will also accelerate saliva production, as will the sensation of nausea. Increased saliva production in response to unpleasant stimuli helps reduce the magnitude of the stimulus by

24

dilution, by a rinsing action, or by buffering strong acids or bases.

THE TEETH

Movements of the tongue are important in passing food across the opposing surfaces of the **teeth.** The **occlusal,** or opposing, **surfaces** of your teeth perform chewing, or **mastication** (mas-ti-KĀ-shun), of food. Mastication breaks down tough connective tissues in meat and the plant fibers in vegetable matter, and it helps saturate the materials with salivary secretions and enzymes.

Figure 24-8a• is a sectional view through an adult tooth. The bulk of each tooth consists of a mineralized matrix similar to that of bone. This material, called **dentin,** differs from bone in that it does not contain cells. Instead, cytoplasmic processes extend into the dentin from cells in the central **pulp cavity,** an interior chamber. The pulp cavity receives blood vessels and nerves from the **root canal,** a narrow tunnel located at the **root,** or base, of the tooth. Blood vessels and nerves enter the root canal through an opening called the **apical foramen** to supply the pulp cavity.

The root of each tooth sits in a bony socket called an *alveolus.* Collagen fibers of the **periodontal ligament** extend from the dentin of the root to the bone of the alveolus, creating a strong articulation known as a gomphosis. ∞ *p. 247* A layer of **cementum** (se-MEN-tum) covers the dentin of the root, providing protection and firmly anchoring the periodontal ligament. Cementum is very similar in histological structure to bone, and it is less resistant to erosion than is dentin.

The **neck** of the tooth marks the boundary between the root and the **crown,** the exposed portion of the tooth. A shallow groove called the **gingival sulcus** surrounds the neck of each tooth. The mucosa of the gingival sulcus is very thin, and it is not tightly bound to the periosteum. The epithelium is bound to the tooth over an extensive area. This epithelial attachment prevents bacterial access to the lamina propria of the gingiva and the relatively soft cementum of the root. When you brush and massage your gums, you stimulate the epithelial cells and strengthen the attachment. If this attachment breaks down, a condition called **gingivitis,** a bacterial infection of the gingivae, can occur.

The dentin of the crown is covered by a layer of **enamel.** Enamel contains calcium phosphate in a crystalline form; it is the hardest biologically manufactured substance. Adequate amounts of calcium, phosphates, and vitamin D during childhood are essential if the enamel coating is to be complete and resistant to decay.

Tooth decay generally results from the action of bacteria that inhabit your mouth. Bacteria adhering to the surfaces of the teeth produce a sticky matrix that traps food particles and creates deposits of **plaque.** Over time, this organic material can become calcified, forming a hard layer of *tartar,* or *dental calculus,* which can be difficult to remove. Tartar deposits most commonly develop at or near the gingival sulcus, where brushing cannot remove the relatively soft plaque deposits.

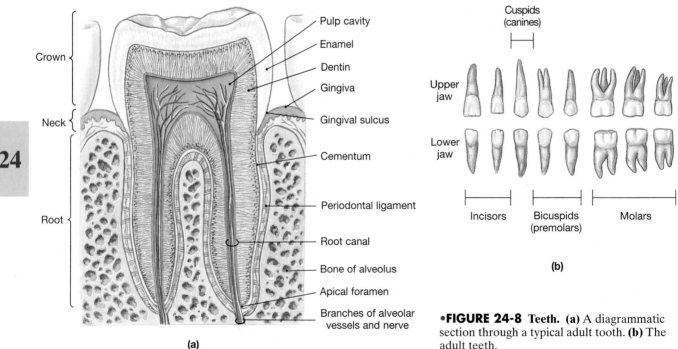

•**FIGURE 24-8 Teeth. (a)** A diagrammatic section through a typical adult tooth. **(b)** The adult teeth.

Types of Teeth

The alveolar processes of the maxillary bones and the mandible form the *upper* and *lower dental arches,* respectively. These arches contain four types of teeth, each with specific functions (Figure 24-8b•):

1. **Incisors** (in-SĪ-zerz) are blade-shaped teeth located at the front of the mouth. Incisors are useful for clipping or cutting, as when you nip off the tip of a carrot stick. These teeth have a single root.
2. The **cuspids** (KUS-pidz), or *canines,* are conical, with a sharp ridgeline and a pointed tip. They are used for tearing or slashing. You might weaken a tough piece of celery by the clipping action of the incisors but then take advantage of the shearing action provided by the cuspids. Cuspids have a single root.
3. **Bicuspids** (bī-KUS-pidz), or *premolars,* have flattened crowns with prominent ridges. They crush, mash, and grind. Bicuspids have one or two roots.
4. **Molars** have very large, flattened crowns with prominent ridges. They excel at crushing and grinding. You can usually shift a tough nut or sparerib to your bicuspids and molars for successful crunching. Molars typically have three or more roots.

Dental Succession

During embryonic development, two sets of teeth begin to form. The first to appear are the **deciduous teeth** (de-SID-ū-us; *deciduus,* falling off), the temporary teeth of the **primary dentition.** Deciduous teeth are also called *primary teeth, milk teeth,* or *baby teeth.* Most children have 20 deciduous teeth—five on each side of the upper and lower jaws (Figure 24-9a•). These teeth will later be replaced by the adult **secondary dentition,** or *permanent dentition* (Figure 24-9b•). Adult jaws are larger and can accommodate more than 20 permanent teeth. Three additional molars appear on each side of the upper and lower jaws as the individual ages. These teeth extend the length of the tooth rows posteriorly and bring the permanent tooth count to 32.

On each side of the upper or lower jaw, the primary dentition consists of two incisors, one cuspid, and a pair of deciduous molars. These deciduous teeth are gradually replaced by the secondary dentition.

As replacement proceeds, the periodontal ligaments and roots of the primary teeth erode until the deciduous teeth either fall out or are pushed aside by the **eruption,** or emergence, of the secondary teeth. The adult premolars take the place of the deciduous molars, and the adult molars extend the tooth row as the jaw enlarges. The third molars, or *wisdom teeth,* may not erupt before age 21. When wisdom teeth fail to erupt, it is because they develop in inappropriate positions or because space on the dental arch is inadequate. Any teeth that develop in locations that do not permit their eruption are called *impacted teeth.* Impacted wisdom teeth can be surgically removed to prevent the formation of abscesses.

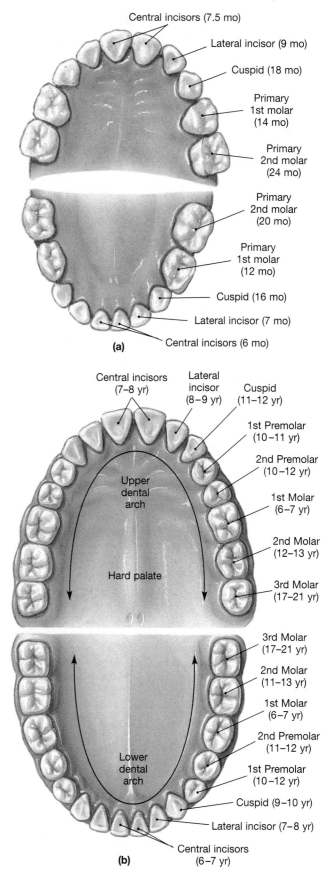

•**FIGURE 24-9 Primary and Secondary Teeth. (a)** The primary teeth, with the age at eruption given in months. **(b)** The adult teeth, with the age at eruption given in years.

24

Dental Problems and Solutions

The mass of a plaque deposit protects the bacteria that normally reside in the mouth from salivary secretions. As the bacteria digest nutrients, these pathogens generate acids that erode the structure of the teeth. The result is **dental caries,** or "cavities." *Streptococcus mutans* is the most abundant bacterium at these sites, and vaccines are now being developed to promote resistance to it and thereby prevent dental caries. If *S. mutans* (or another bacterium) reaches the pulp and infects it, *pulpitis* (pul-PĪ-tis) results. Treatment generally involves the complete removal of the pulp tissue, especially the sensory innervation and all areas of decay; the pulp cavity is then packed with appropriate materials. This procedure is called a *root canal.*

Brushing the exposed surfaces of your teeth after you eat helps prevent the settling of bacteria and the entrapment of food particles, but bacteria between your teeth, in the region known as the *interproximal space,* and within the gingival sulcus may elude the brush. Dentists therefore recommend the daily use of dental floss to clean these spaces and stimulate the gingival epithelium. If bacteria and plaque remain within the gingival sulcus for extended periods, the acids generated will begin eroding the connections between the neck of the tooth and the gingiva. The gums appear to recede from the teeth, and **periodontal disease** develops. As this disease progresses, the bacteria attack the cementum, eventually destroying the periodontal ligament and eroding the bone of the alveolus. This deterioration loosens the tooth, and it falls out or must be pulled. Periodontal disease is the most common cause of the loss of teeth.

Lost or broken teeth have commonly been replaced by "false teeth" attached to a plate or frame inserted into the mouth. In the 1980s, an alternative was developed that uses *dental implants.* A ridged titanium cylinder is inserted into the alveolus, and osteoblasts lock the ridges into the surrounding bone. After 4–6 months, an artificial tooth is screwed into the cylinder.

Dental implants are not suitable for everyone. For example, enough alveolar bone must be present to provide a firm attachment, and complications may occur during or after surgery. Nevertheless, as the technique evolves, dental implants will become increasingly important. Roughly 42 percent of individuals over age 65 have lost all their teeth; the rest have lost an average of 10 teeth.

Mastication

The *muscles of mastication* close your jaws and slide or rock your lower jaw from side to side. ∞ *p. 324* Chewing is not a simple process; it can involve any combination of mandibular elevation, depression, protraction, retraction, and medial–lateral movement. (Try classifying the movements involved the next time you eat.)

During mastication, you force food from the oral cavity to the vestibule and back, crossing and recrossing the occlusal surfaces. This movement results in part from the action of the muscles of mastication, but control would be impossible without the aid of the muscles of the cheeks, lips, and tongue. Once you have shredded or torn the material to a satisfactory consistency and have moistened it with salivary secretions, your tongue begins compacting the debris into a bolus. You can swallow a compact, moist, cohesive bolus relatively easily.

☑ Which type of epithelium lines the oral cavity?

☑ The digestion of which nutrient would be affected by damage to the parotid salivary glands?

☑ Which type of tooth is most useful for chopping off bits of relatively rigid foods?

THE PHARYNX

The **pharynx** serves as a common passageway for solid food, liquids, and air. We described the epithelial lining and divisions of the pharynx—the nasopharynx, the oropharynx, and the laryngopharynx—in Chapter 23. ∞ *pp. 802–803* Food normally passes through the oropharynx and laryngopharynx on its way to the esophagus. Both of these divisions have a stratified squamous epithelium similar to that of the oral cavity. The lamina propria contains scattered mucous glands and the lymphoid tissue of the pharyngeal, palatal, and lingual tonsils. Deep to the lamina propria lies a dense layer of elastic fibers, bound to the underlying skeletal muscles.

We described the specific pharyngeal muscles involved in swallowing in Chapter 11. ∞ *p. 326* In summary:

- The *pharyngeal constrictor muscles* push the bolus toward the esophagus.
- The *palatopharyngeus* and *stylopharyngeus muscles* elevate the larynx.
- The *palatal muscles* elevate the soft palate and adjacent portions of the pharyngeal wall.

These muscles cooperate with muscles of the oral cavity and esophagus to initiate swallowing, which pushes the bolus along the esophagus and into the stomach.

THE ESOPHAGUS

The **esophagus** (Figure 24-10•) is a hollow muscular tube with a length of approximately 25 cm (1 ft) and a diameter of about 2 cm (0.75 in.) at its widest point. The primary function of the esophagus is to carry solid food and liquids to the stomach.

The esophagus begins posterior to the cricoid cartilage, at the level of vertebra C_6. From this point, the narrowest portion of the esophagus, it descends toward the thoracic cavity posterior to the trachea,

•**FIGURE 24-10 The Esophagus. (a)** A transverse section through the esophagus. **(b)** The esophageal mucosa. (LM × 77) **(c)** A color-enhanced SEM of the transition between the esophageal and gastric mucosae at the lower esophageal sphincter.

passes inferiorly along the dorsal wall of the mediastinum, and enters the abdominopelvic cavity through the **esophageal hiatus** (hī-Ā-tus), an opening in the diaphragm. The esophagus then empties into the stomach anterior to vertebra T_7.

In the neck, the esophagus receives blood from the external carotid arteries and the thyrocervical trunks. In the mediastinum, it is supplied by the esophageal arteries and branches of the bronchial arteries. As it passes through the esophageal hiatus, the esophagus receives blood from the inferior phrenic arteries, and the portion of the esophagus that is adjacent to the stomach is supplied by the left gastric artery. Blood from esophageal capillaries collects into the esophageal, inferior thyroid, azygos, and gastric veins.

The esophagus is innervated by parasympathetic and sympathetic fibers from the esophageal plexus. ∞ *p. 517* Resting muscle tone in the circular muscle layer in the superior 3 cm (1 in.) of the esophagus normally prevents air from entering your esophagus. A comparable zone at the inferior end of the esophagus normally remains in a state of active contraction. This condition prevents the backflow of materials from the stomach into the esophagus. Neither region has a well-defined sphincter muscle comparable to those located elsewhere along the digestive tract. Nevertheless, the terms *upper esophageal sphincter* and *lower esophageal sphincter (cardiac sphincter)* are often used to describe these regions at either end of the esophagus, because they are similar in function to other sphincters.

24

HISTOLOGY OF THE ESOPHAGUS

The wall of the esophagus contains mucosal, submucosal, and muscularis layers comparable to those described in Figure 24-2• (p. 848). Distinctive features of the esophageal wall (Figure 24-10•) include the following:

■ The mucosa of the esophagus contains a nonkeratinized, stratified squamous epithelium similar to that of the pharynx and oral cavity.

■ The mucosa and submucosa are thrown into large folds that extend the length of the esophagus. These folds allow for expansion during the passage of a large bolus; except when you swallow, muscle tone in the walls keeps the lumen closed.

■ The muscularis mucosae consists of an irregular layer of smooth muscle.

■ The submucosa contains scattered *esophageal glands,* which produce a mucous secretion that reduces friction between the bolus and the esophageal lining.

■ The muscularis externa has inner circular and outer longitudinal layers. In the superior third of the esophagus, these layers contain skeletal muscle fibers; in the middle third, there is a mixture of skeletal and smooth muscle tissue; along the inferior third, only smooth muscle occurs.

■ There is no serosa, but an adventitia of connective tissue outside the muscularis externa anchors the esophagus in position against the dorsal body wall. Over the 1–2 cm between the diaphragm and stomach, the esophagus is retroperitoneal, with peritoneum covering the anterior and left lateral surfaces.

Esophageal Varices

The veins draining the inferior portion of the esophagus empty into tributaries of the hepatic portal vein. ∞ *p. 741* If the venous pressure in the hepatic portal vein becomes abnormally high, due to liver damage or the constriction of the vessels, blood will pool in the submucosal veins of the esophagus. The veins become grossly distended and create bulges in the esophageal wall. These distorted **esophageal varices** (VAR-i-sēz; *varices,* dilated veins) may constrict the esophageal passageway. They may also rupture, bleeding into the submucosal tissues or onto the epithelial surface. Esophageal varices commonly develop in individuals with advanced *cirrhosis,* a chronic liver disorder that restricts hepatic blood flow.

SWALLOWING

Swallowing, or **deglutition,** is a complex process whose initiation can be voluntarily controlled, but it proceeds automatically once it begins. Although you are consciously aware of, and voluntarily control, swallowing when you eat or drink, swallowing can also occur unconsciously, as saliva collects at the back of the mouth.

Each day you swallow approximately 2400 times. We can divide swallowing into buccal, pharyngeal, and esophageal phases:

1. The **buccal phase** begins with the compression of the bolus against the hard palate. Subsequent retraction of the tongue then forces the bolus into the oropharynx and assists in the elevation of the soft palate, thereby isolating the nasopharynx (Figure 24-11a,b•). The buccal phase is strictly voluntary. Once the bolus enters the oropharynx, reflex responses are initiated and the bolus is moved toward the stomach.

(a)   **BUCCAL PHASE**   **(b)**

(c)   **PHARYNGEAL PHASE**   **(d)**

(e)   **(f)**

(g)   **ESOPHAGEAL PHASE**   **(h)**

•**FIGURE 24-11 The Swallowing Process.** This sequence, based on a series of X-rays, shows the stages of swallowing and the movement of materials from the mouth to the stomach.

2. The **pharyngeal phase** begins as the bolus comes into contact with the palatoglossal and palatopharyngeal arches and the posterior pharyngeal wall (Figure 24-11c,d•). The **swallowing reflex** begins as tactile receptors on the palatal arches and uvula are stimulated by the passage of the bolus. The information is relayed to the **swallowing center** of the medulla oblongata over the trigeminal and glossopharyngeal nerves. Motor commands originating at this center then target the pharyngeal musculature, producing a coordinated and stereotyped pattern of muscle contraction. Elevation of the larynx and folding of the epiglottis direct the bolus past the closed glottis, while the uvula and soft palate block passage back to the nasopharynx. It takes less than a second for the pharyngeal muscles to propel the bolus into the esophagus. During this period, the respiratory centers are inhibited and breathing stops.

3. The **esophageal phase** of swallowing begins as the contraction of pharyngeal muscles forces the bolus through the entrance to the esophagus (Figure 24-11e–h•). Once in the esophagus, the bolus is pushed toward the stomach by a peristaltic wave. The approach of the bolus triggers the opening of the lower esophageal sphincter, and the bolus then continues into the stomach.

Primary peristaltic waves are peristaltic movements that are coordinated by afferent and efferent fibers in the glossopharyngeal and vagus nerves. For a typical bolus, the entire trip takes about 9 seconds to complete. Liquids may make the journey in a few seconds, arriving ahead of the peristaltic contractions with the assistance of gravity.

A dry or poorly lubricated bolus travels much more slowly, and a series of *secondary peristaltic waves* may be required to push it all the way to the stomach. Secondary peristaltic waves are local reflexes triggered by the stimulation of sensory receptors in the esophageal walls. These receptors relay information by way of the submucosal and myenteric plexuses, producing peristaltic contractions in the absence of CNS instructions. You cannot swallow a completely dry bolus, because friction with the walls of the esophagus will make peristalsis ineffective. (For this reason, you cannot swallow an entire slice of processed white bread without taking a drink.) Other problems with bolus movement may involve conditions such as *achalasia,* a constriction of the lower esophageal sphincter, and *esophagitis,* an inflammation of the esophagus.
AM *Achalasia and Esophagitis*

☑ What is unusual about the muscularis externa of the esophagus?

☑ Where in the human body would you find the fauces?

☑ What is occurring when the soft palate and larynx elevate and the glottis closes?

THE STOMACH

The **stomach** performs four major functions: (1) the bulk storage of ingested food, (2) the mechanical breakdown of ingested food, (3) the disruption of chemical bonds in food material through the action of acids and enzymes, and (4) the production of *intrinsic factor,* a glycoprotein whose presence in the digestive tract is required for the absorption of vitamin B_{12}. Ingested substances mix with the secretions of the glands of the stomach. This mixing produces a viscous, highly acidic, soupy mixture of partially digested food called **chyme** (kīm).

ANATOMY OF THE STOMACH

The stomach has the shape of an expanded J (Figure 24-12•). A short **lesser curvature** forms the medial surface of the organ, and a long **greater curvature** forms the lateral surface. The anterior and posterior surfaces are smoothly rounded. The shape and size of the stomach are extremely variable from individual to individual and even from one meal to the next. In an "average" stomach, the lesser curvature has a length of approximately 10 cm (4 in.), and the greater curvature measures about 40 cm (16 in.). The stomach typically extends between the levels of vertebrae T_7 and L_3.

We can divide the stomach into four regions (Figure 24-12•):

1. *The cardia.* The **cardia** (KAR-dē-uh) is the smallest part of the stomach. It consists of the superior, medial portion of the stomach within 3 cm (1.2 in.) of the junction between the stomach and the esophagus. The cardia contains abundant mucous glands whose secretions coat the connection with the esophagus and help protect that tube from the acids and enzymes of the stomach.

2. *The fundus.* The portion of the stomach that is superior to the junction between the stomach and esophagus is the **fundus** (FUN-dus). The fundus contacts the inferior, posterior surface of the diaphragm (Figure 24-12a•).

3. *The body.* The area between the fundus and the curve of the J is the **body** of the stomach. The body is the largest region of the stomach, and it functions as a mixing tank for ingested food and secretions produced in the stomach. *Gastric glands* (*gaster,* stomach) in the fundus and body secrete most of the acids and enzymes involved in gastric digestion.

4. *The pylorus.* The **pylorus** (pī-LOR-us) is the curve of the J. The pylorus is divided into a **pyloric antrum** (*antron,* cavity), which is connected to the body, and a **pyloric canal,** which empties into the *duodenum,* the proximal segment of the small intestine. As mixing movements occur during digestion, the pylorus frequently changes shape. A muscular **pyloric sphincter**

24

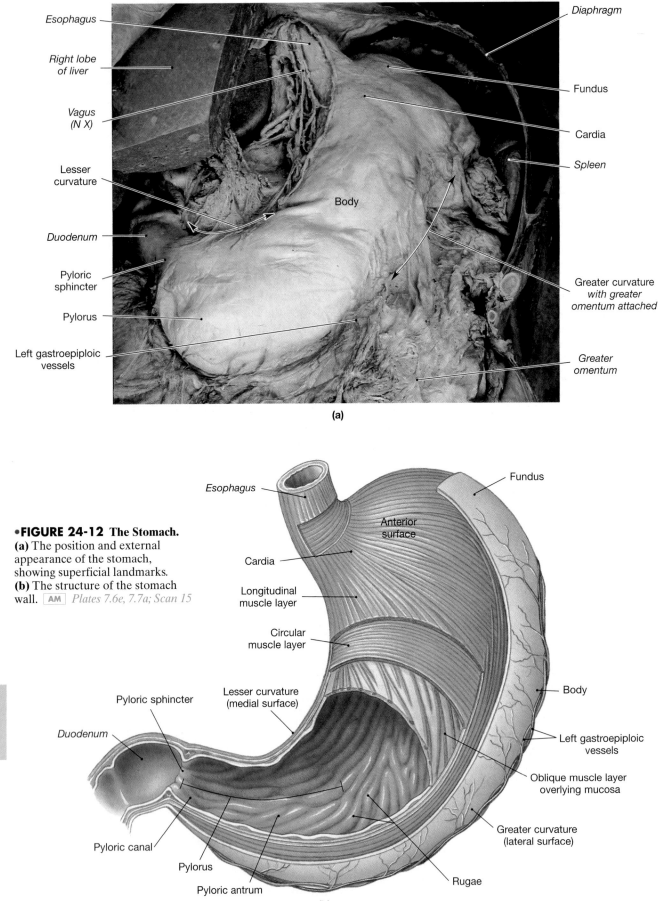

•FIGURE 24-12 The Stomach.
(a) The position and external appearance of the stomach, showing superficial landmarks.
(b) The structure of the stomach wall. AM Plates 7.6e, 7.7a; Scan 15

regulates the release of chyme into the duodenum. Glands in the pylorus secrete mucus and important digestive hormones, including *gastrin,* a hormone that stimulates the activity of gastric glands.

The stomach's volume increases while you eat, then decreases as chyme enters the small intestine. When your stomach is relaxed (empty), the mucosa is thrown into prominent folds called **rugae** (ROO-gē; wrinkles) (Figure 24-12b•). Rugae are temporary features that let the gastric lumen expand. As your stomach fills, the rugae flatten out. When your stomach is full, the rugae almost disappear. When empty, your stomach resembles a muscular tube with a narrow, constricted lumen. When full, it can expand to contain 1–1.5 liters of material.

Musculature of the Stomach

The muscularis mucosae and muscularis externa of the stomach contain extra layers of smooth muscle cells in addition to the usual circular and longitudinal layers. The muscularis mucosae generally contains an outer, circular layer of muscle cells. The muscularis externa has an inner, **oblique layer** of smooth muscle (Figure 24-12b•). The extra layers of smooth muscle strengthen the stomach wall and assist in the mixing and churning activities essential to the formation of chyme.

Histology of the Stomach

A simple columnar epithelium lines all portions of the stomach (Figure 24-13a•). The epithelium is a *secretory sheet,* which produces a carpet of mucus that covers the interior surfaces of the stomach. The alkaline mucous layer protects epithelial cells against the acids and enzymes in the gastric lumen.

The stomach receives blood from (1) the *left gastric artery;* (2) the *splenic artery,* which supplies the *left gastroepiploic artery;* and (3) the *common hepatic artery,* which supplies the *right gastric, gastroduodenal,* and *right gastroepiploic arteries.* It is drained by the *gastric* and *gastroepiploic veins.* ∞ *pp. 732, 741*

Shallow depressions called **gastric pits** open onto the gastric surface (Figure 24-13b,c,d•). The mucous cells at the base, or *neck,* of each gastric pit actively divide, replacing superficial cells that are shed into the chyme. A typical gastric epithelial cell has a life span of 3–7 days, but exposure to strong alcohol or other chemicals that damage or kill epithelial cells will increase the rate of cell turnover.

Gastric Glands In the fundus and body of the stomach, each gastric pit communicates with several **gastric glands,** which extend deep into the underlying lamina propria (Figure 24-13b,c•). Gastric glands are dominated by two types of secretory

•**FIGURE 24-13 The Stomach Lining. (a)** A surface view of the gastric mucosa of the full stomach, showing the entrances to the gastric pits. (SEM × 35) **(b)** A section through gastric pits and gastric glands. (LM × 300)

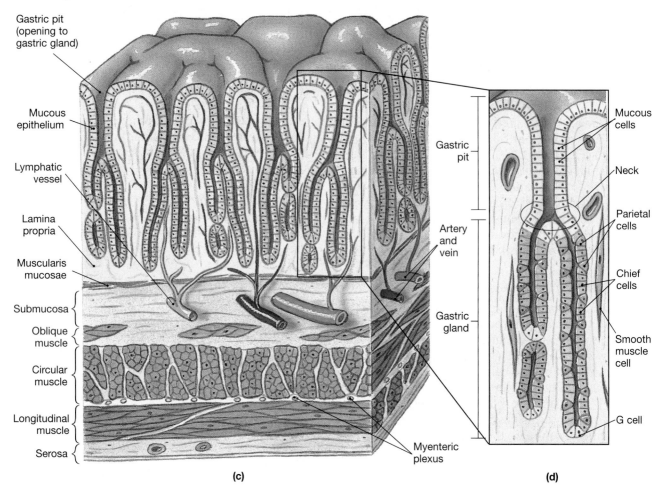

•**FIGURE 24-13** **The Stomach Lining** (*continued*). **(c)** The organization of the stomach wall. **(d)** A gastric gland.

cells: (1) *parietal cells* and (2) *chief cells.* Together they secrete about 1500 ml of **gastric juice** each day.

Parietal cells **Parietal cells** are especially common along the proximal portions of each gastric gland (Figure 24-13d•). These cells secrete **intrinsic factor,** a glycoprotein that facilitates the absorption of **vitamin B₁₂** across the intestinal lining. Recall from Chapter 19 that this vitamin is essential for normal erythropoiesis. ∞ *p. 634*

Parietal cells also secrete *hydrochloric acid* (HCl). They do not produce HCl in the cytoplasm, however, because it is such a strong acid that it would erode a secretory vesicle and destroy the cell. Instead, H$^+$ and Cl$^-$, the two ions that form HCl, are transported independently by different mechanisms (Figure 24-14•). Hydrogen ions are generated inside a parietal cell as the enzyme carbonic anhydrase converts carbon dioxide and water to carbonic acid (H$_2$CO$_3$). The carbonic acid promptly dissociates into hydrogen ions and bicarbonate ions (HCO$_3^-$). The hydrogen ions are actively transported into the lumen of the gastric gland. The bicarbonate ions are ejected

•**FIGURE 24-14** **The Secretion of Hydrochloric Acid.** An active parietal cell generates H$^+$ by the dissociation of carbonic acid within the cell. The bicarbonate is exchanged for Cl$^-$ in the interstitial fluid; the chloride ions diffuse into the lumen of the gastric gland as the hydrogen ions are transported out of the cell.

into the interstitial fluid by a countertransport mechanism that exchanges intracellular bicarbonate ions for extracellular chloride ions. The chloride ions then diffuse across the cell and through open chloride channels in the cell membrane into the lumen of the gastric gland.

The bicarbonate ions released by the parietal cell diffuse through the interstitial fluid into the bloodstream. When gastric glands are actively secreting, enough bicarbonate ions enter the bloodstream to increase the pH of the blood significantly. This sudden influx of bicarbonate ions has been called the *alkaline tide*.

The secretory activities of the parietal cells can keep the stomach contents at a pH of 1.5–2.0. This highly acidic environment does not by itself digest chyme but has four important functions:

1. The acidity of gastric juice kills most of the microorganisms ingested with food.
2. The acidity denatures proteins and inactivates most of the enzymes in food.
3. The acidity helps break down plant cell walls and the connective tissues in meat.
4. An acidic environment is essential for the activation and function of *pepsin*, a protein-digesting enzyme secreted by chief cells.

Chief cells **Chief cells** are most abundant near the base of a gastric gland (Figure 24-13d•). These cells secrete **pepsinogen** (pep-SIN-ō-jen), an inactive proenzyme. Pepsinogen is converted by the acid in the gastric lumen to **pepsin**, an active *proteolytic* (protein-digesting) enzyme. Pepsin functions most effectively at a strongly acidic pH of 1.5–2.0. The stomachs of newborn infants (but not of adults) also produce **rennin**, also known as *chymosin*, and **gastric lipase**, enzymes important for the digestion of milk. Rennin coagulates milk proteins; gastric lipase initiates the digestion of milk fats.

Pyloric Glands Glands in the pylorus produce primarily a mucous secretion rather than enzymes or acid. In addition, several types of enteroendocrine cells are scattered among the mucus-secreting cells. These cells produce at least seven hormones, most notably the hormone **gastrin** (GAS-trin). Gastrin is produced by *G cells*. G cells are most abundant in the gastric pits of the pyloric antrum. Gastrin stimulates (1) the secretion of both parietal and chief cells and (2) contractions of the gastric wall that mix and stir the gastric contents. The pyloric glands also contain *D cells*, which release **somatostatin**, a hormone that inhibits gastrin release. D cells continuously release their secretions into the interstitial fluid adjacent to the G cells. This inhibition of gastrin production can be overpowered by neural and hormonal stimuli when the stomach is preparing for digestion or is already engaged in digestion.

Gastritis and Peptic Ulcers

An inflammation of the gastric mucosa is called **gastritis** (gas-TRĪ-tis). This condition can develop after a person has swallowed drugs, including alcohol and aspirin. Gastritis can also appear after severe emotional or physical stress, bacterial infection of the gastric wall, or the ingestion of strongly acidic or alkaline chemicals.

A **peptic ulcer** develops when the digestive acids and enzymes manage to erode their way through the defenses of the stomach lining or proximal portions of the small intestine. The specific locations are indicated by the terms **gastric ulcer** (stomach) or **duodenal ulcer** (duodenum). Peptic ulcers result from the excessive production of acid or the inadequate production of the alkaline mucus that defends the epithelium against that acid. Since the late 1970s, drugs such as *cimetidine (Tagamet®)* have been used to inhibit acid production by parietal cells. Infections involving the bacterium *Helicobacter pylori* are responsible for at least 80 percent of peptic ulcers. Hence treatment for gastric ulcers commonly involves the administration of antibiotic drugs. [AM] *Peptic Ulcers*

REGULATION OF GASTRIC ACTIVITY

The production of acid and enzymes by the gastric mucosa can be (1) controlled by the central nervous system, (2) regulated by short reflexes coordinated in the wall of the stomach, and (3) regulated by digestive tract hormones. Three phases of gastric control are identified, although considerable overlap exists among them: the *cephalic phase*, the *gastric phase*, and the *intestinal phase*. These phases are summarized in Figure 24-15•.

The Cephalic Phase

The **cephalic phase** of gastric secretion begins when you see, smell, taste, or think of food (Figure 24-15a•). This stage, which is directed by the CNS, prepares your stomach to receive food. The neural output proceeds by way of the parasympathetic division of the autonomic nervous system, and the vagus nerves innervate the submucosal plexus of the stomach. Next, postganglionic parasympathetic fibers innervate mucous cells, chief cells, parietal cells, and G cells of the stomach. In response to stimulation, the production of gastric juice accelerates, reaching rates of about 500 ml/h. This phase generally lasts only minutes before the gastric phase commences. Emotional states can exaggerate or inhibit the cephalic phase. For example, anger or hostility leads to excessive gastric secretion, whereas anxiety, stress, or fear decreases gastric secretion and gastric contractions, or *motility*.

24

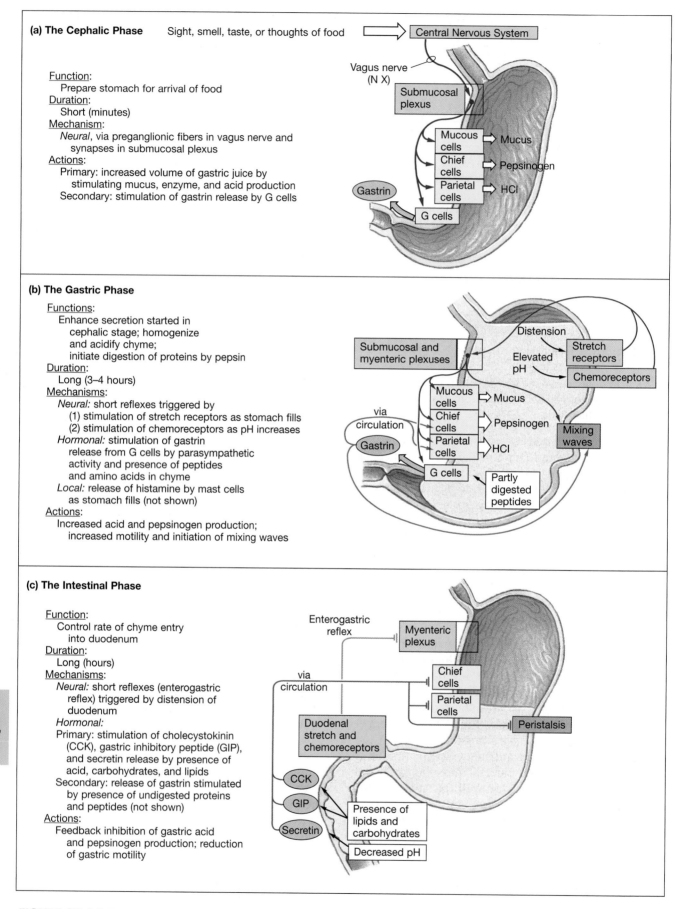

(a) The Cephalic Phase   Sight, smell, taste, or thoughts of food

Function:
    Prepare stomach for arrival of food
Duration:
    Short (minutes)
Mechanism:
    Neural, via preganglionic fibers in vagus nerve and
        synapses in submucosal plexus
Actions:
    Primary: increased volume of gastric juice by
        stimulating mucus, enzyme, and acid production
    Secondary: stimulation of gastrin release by G cells

(b) The Gastric Phase

Functions:
    Enhance secretion started in
        cephalic stage; homogenize
        and acidify chyme;
        initiate digestion of proteins by pepsin
Duration:
    Long (3–4 hours)
Mechanisms:
    Neural: short reflexes triggered by
        (1) stimulation of stretch receptors as stomach fills
        (2) stimulation of chemoreceptors as pH increases
    Hormonal: stimulation of gastrin
        release from G cells by parasympathetic
        activity and presence of peptides
        and amino acids in chyme
    Local: release of histamine by mast cells
        as stomach fills (not shown)
Actions:
    Increased acid and pepsinogen production;
        increased motility and initiation of mixing waves

(c) The Intestinal Phase

Function:
    Control rate of chyme entry
        into duodenum
Duration:
    Long (hours)
Mechanisms:
    Neural: short reflexes (enterogastric
        reflex) triggered by distension of
        duodenum
    Hormonal:
    Primary: stimulation of cholecystokinin
        (CCK), gastric inhibitory peptide (GIP),
        and secretin release by presence of
        acid, carbohydrates, and lipids
    Secondary: release of gastrin stimulated
        by presence of undigested proteins
        and peptides (not shown)
Actions:
    Feedback inhibition of gastric acid
        and pepsinogen production; reduction
        of gastric motility

•**FIGURE 24-15** The Phases of Gastric Secretion

The Gastric Phase

The **gastric phase** begins with the arrival of food in the stomach and builds on the stimulation provided during the cephalic phase (Figure 24-15b•). The stimuli that initiate the gastric phase are (1) distension of the stomach, (2) an increase in the pH of the gastric contents, and (3) the presence of undigested materials in the stomach, especially proteins and peptides. The gastric phase consists of the following responses:

1. *Neural response.* The stimulation of stretch receptors in the stomach wall and chemoreceptors in the mucosa triggers short reflexes coordinated in the submucosal and myenteric plexuses. The postganglionic fibers leaving the submucosal plexus innervate parietal cells and chief cells, and the release of ACh stimulates their secretion. Proteins, alcohol in small doses, and caffeine enhance gastric secretion markedly by stimulating chemoreceptors in the gastric lining. The stimulation of the myenteric plexus produces mixing waves in the muscularis externa.

2. *Hormonal response.* Neural stimulation and the presence of peptides and amino acids in chyme stimulate the secretion of gastrin, primarily by G cells of the pyloric antrum. Gastrin entering the interstitial fluid of the stomach must enter capillaries and complete a round-trip of the bloodstream before the hormone stimulates parietal and chief cells of the fundus and body. Both parietal and chief cells respond to the presence of gastrin by accelerating their secretion rates. The effect on the parietal cells is the most pronounced, and the pH of the gastric juice declines as a result. Gastrin also stimulates gastric motility.

3. *Local response.* A distortion of the gastric wall also stimulates the release of histamine in the lamina propria. The source of the histamine is thought to be mast cells in the connective tissue of this layer. Histamine binds to receptors on the parietal cells and stimulates acid secretion.

The gastric phase may continue for 3–4 hours while the ingested materials are processed by the acids and enzymes. During this period, gastrin stimulates contractions in the muscularis externa of the stomach and intestinal tract. The effects are strongest in the stomach, where stretch receptors are stimulated as well. The initial contractions are weak pulsations in the gastric walls. These *mixing waves* occur several times per minute, and they gradually increase in intensity. After an hour, the material in the stomach is churning like clothing in a washing machine.

When the contractions begin, the pH of the gastric contents is high; only the material in contact with the gastric epithelium is exposed to undiluted digestive acids and enzymes. As mixing occurs, the acid is diluted but the pH remains elevated until a large volume of gastric juice has been secreted and the contents thoroughly mixed. This process generally takes several hours. As the pH throughout the chyme reaches 1.5–2.0 and the amount of undigested protein decreases, gastrin production declines, and so do the rates of acid and enzyme secretion by parietal and chief cells.

The Intestinal Phase

The **intestinal phase** of gastric secretion begins when chyme starts to enter the small intestine (Figure 24-15c•). The intestinal phase generally starts after several hours of mixing contractions, when waves of contraction begin sweeping down the length of the stomach. Each time the pylorus contracts, a small quantity of chyme squirts through the pyloric sphincter. The purpose of the intestinal phase is to control the rate of gastric emptying and to ensure that the secretory, digestive, and absorptive functions of the small intestine can proceed with reasonable efficiency. Although we shall consider the intestinal phase now as it affects stomach activity, the arrival of chyme in the small intestine also triggers other neural and hormonal events that coordinate the activities of the intestinal tract and the pancreas, liver, and gallbladder.

The intestinal phase involves a combination of neural and hormonal responses:

1. *Neural response.* Chyme leaving the stomach relieves some of the distension in the stomach wall, thus reducing the stimulation of stretch receptors. At the same time, the distension of the duodenum by chyme stimulates stretch receptors and chemoreceptors that trigger the **enterogastric reflex.** This reflex temporarily inhibits both central and local stimulation of gastrin production and gastric contractions. It also stimulates the contraction of the pyloric sphincter. The net result is that immediately after chyme enters the small intestine, gastric contractions decrease in strength and frequency and further discharge of chyme is prevented. This result gives the duodenum time to deal with the arriving acids before the next wave of gastric contraction occurs. At the same time, local reflexes at the duodenum stimulate mucus production, which helps protect the intestinal lining from the arriving acids and enzymes.

2. *Hormonal response.* Several hormonal responses are triggered by the arrival of chyme in the duodenum:
   a. The arrival of lipids (especially triglycerides and fatty acids) and carbohydrates in the duodenum stimulates the secretion of the hormones *cholecystokinin* (kō-lē-sis-tō-KĪ-nin), or CCK, and *gastric inhibitory peptide* (GIP). The CCK inhibits gastric secretion of acids and enzymes. The GIP, which also targets the pancreas, inhibits gastric secretion and reduces the rate and force of gastric contractions. As a result, a meal high in fats stays in your stomach longer, and enters the duodenum at a more leisurely pace, than a low-fat meal. This delay allows more time for lipids to be digested and absorbed in the small intestine.

24

b. A drop in pH below 4.5 stimulates the secretion of the hormone *secretin* (se-KRĒ-tin) by enteroendocrine cells of the duodenum. Secretin inhibits parietal cell and chief cell activity in the stomach. It also targets (1) the pancreas, where it stimulates the production of buffers that will protect the duodenum by neutralizing the acid in chyme, and (2) the liver, where it stimulates bile secretion.

c. The arrival of partially digested proteins in the duodenum stimulates G cells in the duodenal wall. These cells secrete gastrin, which circulates to the stomach and accelerates acid and enzyme production. In effect, this is a feedback mechanism that regulates the amount of gastric processing to meet the requirements of a specific meal.

In general, the rate of chyme movement into the small intestine is highest when your stomach is greatly distended and the meal contains relatively little protein. A large meal containing small amounts of protein, large amounts of carbohydrates (such as rice or pasta), wine (alcohol), and after-dinner coffee (caffeine) will leave your stomach extremely quickly because both alcohol and caffeine stimulate gastric secretion and motility.

DIGESTION AND ABSORPTION IN THE STOMACH

The stomach performs preliminary digestion of proteins by pepsin and, for a variable period, permits the digestion of carbohydrates and lipids by salivary amylase and lingual lipase. Until the pH throughout the contents of the stomach falls below 4.5, salivary amylase and lingual lipase continue to digest carbohydrates and lipids in the meal. These enzymes generally remain active 1–2 hours after a meal.

As the stomach contents become more fluid and the pH approaches 2.0, pepsin activity increases and protein disassembly begins. Protein digestion is not completed in the stomach, because time is limited and pepsin attacks only specific types of peptide bonds, not all peptide bonds. However, pepsin generally has enough time to break down complex proteins into smaller peptide and polypeptide chains before the chyme enters the duodenum.

Although digestion does occur in the stomach, nutrient absorption does not occur there. There are several reasons: (1) the epithelial cells are covered by a blanket of alkaline mucus and are not directly exposed to chyme, (2) the epithelial cells lack the specialized transport mechanisms of cells that line the small intestine, (3) the gastric lining is relatively impermeable to water, and (4) digestion has not proceeded to completion by the time chyme leaves the stomach. At this stage, most carbohydrates, lipids, and proteins are only partially broken down.

Some drugs can be absorbed in the stomach. For example, ethyl alcohol can diffuse through the mucous barrier and penetrate the lipid membranes of the epithelial cells. As a result, you absorb alcohol in your stomach before any nutrients in a meal reach the bloodstream. Meals containing large amounts of fat will slow the rate of alcohol absorption, because alcohol is lipid-soluble, and some of it will be dissolved in fat droplets in the chyme. Aspirin is another lipid-soluble drug that can enter the bloodstream across the gastric mucosa. Such drugs alter the properties of the mucous layer and can promote epithelial damage by stomach acids and enzymes. Prolonged aspirin use can cause gastric bleeding, and individuals with stomach ulcers usually avoid aspirin.

Stomach Cancer

Stomach, or *gastric*, **cancer** is one of the most common lethal cancers, responsible for roughly 14,000 deaths in the United States each year. The incidence is higher in countries such as Japan or Korea, where the typical diet includes large quantities of pickled foods. Because the symptoms can resemble those of gastric ulcers, the condition may not be reported in its early stages. Diagnosis generally involves X-rays of the stomach at various degrees of distension. The mucosa can also be visually inspected by using a flexible instrument called a *gastroscope*. Attachments permit the collection of tissue samples for analysis. Treatment of stomach cancer involves the surgical removal of part or all of the stomach, *gastrectomy* (gas-TREK-to-mē). People can survive even a total gastrectomy, because the only absolutely vital function of the stomach is the secretion of intrinsic factor. Protein breakdown can still be performed by the small intestine, although at reduced efficiency, and the loss of such functions as food storage and acid production is not life-threatening.

How would a large meal affect the pH of blood that leaves the stomach?

When a person suffers from chronic ulcers in the stomach, the branches of the vagus nerve that serve the stomach are sometimes severed. Why?

THE SMALL INTESTINE AND ASSOCIATED GLANDULAR ORGANS

Your stomach is a holding tank where food is saturated with gastric juices and exposed to stomach acids and the digestive effects of pepsin. These are preliminary steps; most important digestive and absorptive functions occur in your small intestine, where chemical digestion is completed and the products of digestion are absorbed. The mucosa of the small intestine produces only a few of the enzymes involved. The pancreas provides digestive enzymes as well as buffers that help neutralize chyme. The liver secretes *bile,* a solution stored in the gallbladder for subsequent discharge into the small intestine. Bile contains buffers and *bile salts,* compounds that facilitate lipid digestion and absorption.

THE SMALL INTESTINE

The **small intestine** plays the key role in the digestion and absorption of nutrients. Ninety percent of nutrient absorption occurs in the small intestine, and most of the rest occurs in the large intestine. The small intestine averages 6 m (20 ft) in length (range: 4.5–7.5 m; 14.8–24.6 ft) and has a diameter ranging from 4 cm (1.6 in.) at the stomach to about 2.5 cm (1 in.) at the junction with the large intestine. It occupies all abdominal regions except the right and left hypochondriac and epigastric regions (Figure 1-7•, p. 17). The small intestine has three subdivisions: (1) the duodenum, (2) the jejunum, and (3) the ileum.

The **duodenum** (doo-AH-de-num), 25 cm (10 in.) in length, is the section closest to the stomach. This portion of the small intestine is a "mixing bowl" that receives chyme from the stomach and digestive secretions from the pancreas and liver. From its connection with the stomach, the duodenum curves in a C that encloses the pancreas. Except for the proximal 2.5 cm (1 in.), the duodenum is in a retroperitoneal position between vertebrae L_1 and L_4 (Figure 24-5•, p. 852).

A rather abrupt bend marks the boundary between the duodenum and the **jejunum** (je-JOO-num). At this junction, the small intestine reenters the peritoneal cavity, supported by a sheet of mesentery. The jejunum is about 2.5 meters (8 ft) long. The bulk of chemical digestion and nutrient absorption occurs in the jejunum.

The **ileum** (IL-ē-um) is the final segment of the small intestine. It is also the longest, averaging 3.5 meters (12 ft) in length. The ileum ends at the **ileocecal valve,** a sphincter that controls the flow of materials from the ileum into the *cecum* of the large intestine.

The small intestine fills much of the peritoneal cavity, and its position is stabilized by a broad mesentery attached to the dorsal body wall (Figure 24-5•, p. 852). Movement of the small intestine during digestion is restricted by the stomach, the large intestine, the abdominal wall, and the pelvic girdle. Blood vessels, lymphatic vessels, and nerves reach these segments of the small intestine within the connective tissue of the mesentery. The primary blood vessels involved are branches of the superior mesenteric artery and the superior mesenteric vein. ∞ *pp. 732, 741–742*

The subdivisions of the small intestine—the duodenum, jejunum, and ileum—are distinguished by both histological specialization and primary functions. The locations of these segments in the peritoneal cavity are indicated in Figure 24-16a•.

Histology of the Small Intestine

The intestinal lining bears a series of transverse folds called **plicae,** or *plicae circulares* (Figures 24-16b and 24-17a,b•). Unlike the rugae in the stomach, each plica is a permanent feature that does not disappear when the small intestine fills. The small intestine contains roughly 800 plicae; their presence greatly increases the surface area available for absorption.

Figure 24-17c,d,e• presents a more detailed view of the intestinal wall.

Intestinal Villi The mucosa of the small intestine is thrown into a series of fingerlike projections, the **intestinal villi.** The intestinal villi are covered by a simple columnar epithelium that is carpeted with microvilli. Because the microvilli project from the epithelium like the bristles on a brush, these cells are said to have a *brush border* (Figure 24-17c,d,e•).

If the small intestine were a simple tube with smooth walls, it would have a total absorptive area of roughly

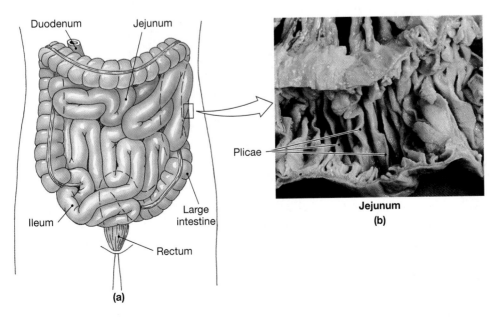

Duodenum   Jejunum

Ileum

Large intestine

Rectum

(a)

Plicae

Jejunum
(b)

•**FIGURE 24-16 Regions of the Small Intestine. (a)** The positions of the duodenum, jejunum, and ileum in the abdominopelvic cavity. **(b)** A representative view of the jejunum. *Plates 7.6e, 7.7c*

24

•**FIGURE 24-17 The Intestinal Wall. (a)** A section of the intestine. (LM × 2.5) **(b)** A single plica and multiple villi. **(c)** The organization of the intestinal wall. **(d)** Internal structures in a single villus, showing the capillary and lymphatic supplies. **(e)** A villus in sectional view. (LM × 252)

3300 cm² (3.6 ft²). Instead, the mucosa contains plicae, each plica supports a forest of villi, and each villus is covered by epithelial cells whose exposed surfaces contain microvilli. This arrangement increases the total area for absorption by a factor of more than 600, to approximately 2 million cm² (more than 2200 ft²).

The lamina propria of each villus contains an extensive network of capillaries. These capillaries originate in a vascular network within the submucosa. They carry absorbed nutrients to the hepatic portal circulation for delivery to the liver. ∞ *p. 741* The liver adjusts the nutrient concentrations of blood before the blood reaches the general systemic circulation.

In addition to capillaries and nerve endings, each villus contains a lymphatic capillary called a **lacteal** (LAK-tē-al; *lacteus,* milky) (Figure 24-17d,e•). Lacteals transport materials that are unable to enter blood capillaries. For example, absorbed fatty acids are assembled into protein–lipid packages too large to diffuse into the bloodstream. These packets, called *chylomicrons,* reach the venous circulation by way of the thoracic duct, which delivers lymph into the left subclavian vein. ∞ *p. 755* The name *lacteal* refers to the pale, cloudy appearance of lymph that contains large quantities of lipids.

Contractions of the muscularis mucosae and smooth muscle cells within the intestinal villi move the villi back and forth, exposing the epithelial surfaces to the liquefied intestinal contents. This movement improves the efficiency of absorption, because local differences in nutrient concentration will be quickly eliminated. Movements of the villi also squeeze the lacteals, thus assisting in the movement of lymph out of the villi.

Intestinal Glands Between the columnar epithelial cells, goblet cells eject mucins onto the intestinal surfaces. At the bases of the villi are the entrances to the **intestinal glands,** or *crypts of Lieberkuhn.* These glandular pockets extend deep into the underlying lamina propria (Figure 24-17c•). Near the base of each intestinal gland, stem cell divisions produce new generations of epithelial cells. These new cells are continuously displaced toward the intestinal surface. In a few days, they will have reached the tip of a villus, where they are shed into the intestinal lumen. This ongoing process renews the epithelial surface, and the subsequent disintegration of the shed cells adds enzymes to the lumen.

Several important brush border enzymes enter the intestinal lumen in this way. *Brush border enzymes* are integral membrane proteins located on the surfaces of intestinal microvilli. The enzymes have important digestive functions: These enzymes attack materials in contact with the brush border, and the breakdown products are absorbed by the epithelial cells. Once the epithelial cells are shed, they disintegrate within the

lumen, and the intracellular and brush border enzymes are released. *Enterokinase* is a brush border enzyme that enters the lumen in this way. Enterokinase does not directly participate in digestion, but it activates proenzymes secreted by the pancreas. (We shall consider the functions of enterokinase and other brush border enzymes in a later section.) Intestinal glands also contain enteroendocrine cells responsible for the production of several intestinal hormones, including gastrin, cholecystokinin, and secretin.

The duodenum has numerous mucous glands, both in the epithelium and deep to it. In addition to intestinal glands, the submucosa contains **submucosal glands,** or *Brunner's glands,* which produce copious quantities of mucus when chyme arrives from the stomach. Mucus produced by these glands protects the epithelium from the acidity of chyme. It also contains buffers that help elevate the pH of the chyme. Along the length of the duodenum, the pH of chyme goes from 1–2 to 7–8. The submucosal glands also secrete the hormone *urogastrone,* which inhibits gastric acid production and stimulates the division of epithelial stem cells along the digestive tract.

Regional Specializations The duodenum has few plicae and their villi are small. The primary function of the duodenum is to receive chyme from the stomach and neutralize the acids before they can damage the absorptive surfaces of the small intestine. Over the proximal half of the jejunum, however, plicae and villi are very prominent. The plicae and villi thereafter gradually decrease in size. This reduction parallels the reduction in absorptive activity; most nutrient absorption has occurred before ingested materials reach the ileum. One rather drastic surgical method of promoting weight loss is the removal of a significant portion of the jejunum. The reduction in absorptive area causes a marked weight loss and may not prevent adequate nutrition, but the side effects can be very troublesome.
AM *Drastic Weight-Loss Techniques*

The distal portions of the ileum lack plicae, and the lamina propria there contains 20–30 masses of lymphoid tissue called *aggregate lymphoid nodules,* or *Peyer's patches.* ∞ *p. 759* The lymphocytes in these nodules protect the small intestine from bacteria that are normal inhabitants of the large intestine. Lymphoid nodules are most abundant in the terminal portion of the ileum, near the entrance to the large intestine.

24

Intestinal Secretions

Roughly 1.8 liters of watery **intestinal juice** enters your intestinal lumen each day. Intestinal juice moistens chyme, assists in buffering acids, and liquefies both the digestive enzymes provided by the pancreas and the products of digestion. Much of this fluid volume arrives by osmosis, as water flows out of the mucosa and into the relatively concentrated chyme. The rest is secreted by intestinal glands, stimulated by the activation of touch receptors and stretch receptors in the intestinal walls.

The submucosal glands help protect the duodenal epithelium from gastric acids and enzymes. These glands increase their secretory activities in response to (1) local reflexes, (2) the release of the hormone *enterocrinin* by enteroendocrine cells of the duodenum, and (3) parasympathetic stimulation via the vagus nerves. Mechanisms 1 and 2 operate only after chyme arrives in the duodenum. However, because vagus nerve activity triggers their secretion, the submucosal glands begin secreting during the cephalic phase of gastric secretion, long before the chyme reaches the pyloric sphincter. Thus the duodenal lining has protection in advance.

Sympathetic stimulation inhibits the activation of the submucosal glands, leaving the duodenal lining relatively unprepared for the arrival of chyme. This fact probably accounts for the common observation that duodenal ulcers can be caused by chronic stress or other factors that promote sympathetic activation.

INTESTINAL MOVEMENTS

After chyme has arrived in the duodenum, weak peristaltic contractions move it slowly toward the jejunum. These contractions are myenteric reflexes not under central nervous system control. Their effects are limited to within a few centimeters of the site of the original stimulus. These short reflexes are controlled by motor neurons in the submucosal and myenteric plexuses. In addition, some of the smooth muscle cells contract periodically, even without stimulation, establishing a basic contractile rhythm that then spreads from cell to cell.

The stimulation of the parasympathetic system increases the sensitivity of these myenteric reflexes and accelerates both local peristalsis and segmentation. More-elaborate reflexes coordinate activities along the entire length of the small intestine. Two examples are triggered by the stimulation of stretch receptors in the stomach as it fills. The **gastroenteric reflex** stimulates motility and secretion along the entire small intestine; the **gastroileal** (gas-trō-IL-ē-al) **reflex** triggers the relaxation of the ileocecal valve. The net result is that materials pass from the small intestine into the large intestine. Thus the gastroenteric and gastroileal reflexes accelerate movement along the small intestine—the opposite effect of the enterogastric reflex.

Hormones released by the digestive tract can enhance or suppress reflex responses. For example, the gastroileal reflex is triggered by stretch receptor stimulation, but the degree of ileocecal valve relaxation is enhanced by gastrin, which is secreted in large quantities when food enters the stomach.

Vomiting

The responses of the digestive tract to chemical or mechanical irritation are rather predictable. Gastric secretion accelerates all along the digestive tract, and the intestinal contents are eliminated as quickly as possible. The *vomiting reflex* occurs in response to irritation of the fauces, pharynx, esophagus, stomach, or proximal portions of the small intestine. These sensations are relayed to the *vomiting center* of the medulla oblongata, which coordinates the motor responses. In preparation, the pylorus relaxes and the contents of the duodenum and proximal jejunum are discharged into the stomach by strong peristaltic waves that travel toward the stomach rather than toward the ileum. **Vomiting,** or *emesis* (EM-e-sis), then occurs as the stomach regurgitates its contents through the esophagus and pharynx. As regurgitation occurs, the uvula and soft palate block the entrance to the nasopharynx. Increased salivary secretion assists in buffering the stomach acids, thereby preventing erosion of the teeth. In conditions marked by repeated vomiting, severe tooth damage can occur; we shall discuss one example, the eating disorder *bulimia,* in Chapter 25. Most of the force of vomiting comes from expiratory movements that elevate intra-abdominal pressures and force the stomach against the tensed diaphragm.

THE PANCREAS

Your **pancreas** lies posterior to your stomach, extending laterally from the duodenum toward the spleen (Figure 24-18a•). The pancreas is an elongate, pinkish gray organ with a length of about 15 cm (6 in.) and a weight of about 80 g (3 oz). The broad **head** of the pancreas lies within the loop formed by the duodenum as it leaves the pylorus. The slender **body** of the pancreas extends toward the spleen, and the **tail** is short and bluntly rounded. The pancreas is retroperitoneal and is firmly bound to the posterior wall of the abdominal cavity.

The surface of the pancreas has a lumpy, lobular texture. A thin, transparent connective tissue capsule wraps the entire organ. You can see the pancreatic lobules, associated blood vessels, and excretory ducts through the anterior capsule and the overlying layer of peritoneum. Arterial blood reaches the pancreas by way of branches of the splenic, superior mesenteric, and common hepatic arteries. The pancreatic arteries and pancreaticoduodenal arteries are the major branches from these vessels. The splenic vein and its branches drain the pancreas.

The pancreas is primarily an exocrine organ, producing digestive enzymes and buffers. The large **pancreatic duct** (*duct of Wirsung*) delivers these secretions to the duodenum. A small **accessory pancreatic duct,** or *duct of Santorini,* may branch from the pancreatic duct. The pancreatic duct extends within the attached mesentery to reach the duodenum, where it meets the *common bile duct* from the liver

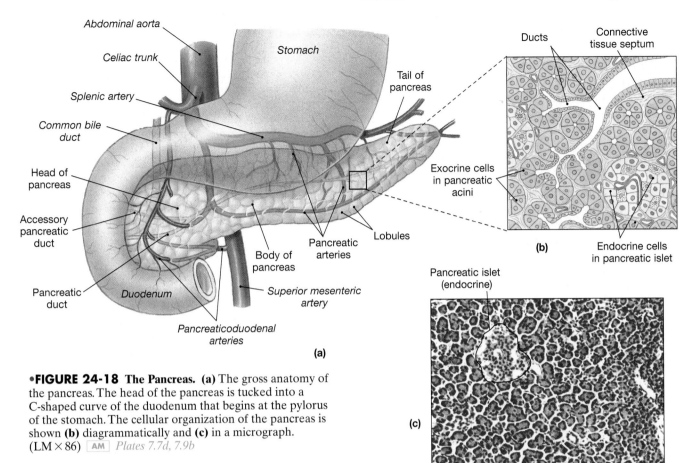

•**FIGURE 24-18 The Pancreas. (a)** The gross anatomy of the pancreas. The head of the pancreas is tucked into a C-shaped curve of the duodenum that begins at the pylorus of the stomach. The cellular organization of the pancreas is shown **(b)** diagrammatically and **(c)** in a micrograph. (LM × 86) AM *Plates 7.7d, 7.9b*

and gallbladder. The two ducts then empty into the *duodenal ampulla,* a chamber located roughly halfway along the length of the duodenum (Figure 24-21b•, p. 878). When present, the accessory pancreatic duct generally empties into the duodenum independently, outside the duodenal ampulla.

Histological Organization

Partitions of connective tissue divide the interior of the pancreas into distinct lobules. The blood vessels and tributaries of the pancreatic ducts are situated within these connective tissue septa (Figure 24-18b•). The pancreas is an example of a *compound tubuloalveolar gland,* a gland structure that we described in Chapter 4. ∞ *p. 117* In each lobule, the ducts branch repeatedly before ending in blind pockets called the **pancreatic acini** (AS-i-nī). Each pancreatic acinus is lined by a simple cuboidal epithelium. *Pancreatic islets,* the endocrine tissues of the pancreas, are scattered among the acini (Figure 24-18b,c•). The islets account for only about 1 percent of the cellular population of the pancreas.

The pancreas has two distinct functions, one endocrine and the other exocrine. The endocrine cells of the pancreatic islets secrete insulin and glucagon into the bloodstream. ∞ *p. 605* The exocrine cells include the acinar cells and the epithelial cells that line the duct system. Together they secrete **pancreatic juice,** an alkaline mixture of digestive enzymes, water, and ions, into the small intestine. Pancreatic enzymes are secreted by the acinar cells. These enzymes do most of the digestive work in the small intestine, breaking down ingested materials into small molecules suitable for absorption. The water and ions, secreted primarily by the cells lining the pancreatic ducts, assist in diluting and buffering the acids in the chyme.

Physiology of the Pancreas

Each day your pancreas secretes about 1000 ml (1 qt) of pancreatic juice. The secretory activities are controlled primarily by hormones from the duodenum. When chyme arrives in the duodenum, secretin is

24

released. This hormone triggers the pancreatic secretion of a watery buffer solution with a pH of 7.5–8.8. Among its other components, this secretion contains bicarbonate and phosphate buffers that help elevate the pH of the chyme. A different duodenal hormone, cholecystokinin, stimulates the production and secretion of pancreatic enzymes. Pancreatic enzyme secretion also increases under stimulation by the vagus nerves. As we noted earlier, this stimulation occurs during the cephalic phase of gastric regulation, so the pancreas starts to synthesize enzymes before food even reaches the stomach. Such a head start is important, because enzyme synthesis takes much longer than buffer production. By starting early, the pancreatic cells will be ready to meet the demand when chyme arrives in the duodenum.

The specific pancreatic enzymes involved include the following:

■ **Pancreatic alpha-amylase,** a **carbohydrase** (kar-bō-HĪ-drās)—an enzyme that breaks down certain starches. Pancreatic alpha-amylase is almost identical to salivary amylase.

■ **Pancreatic lipase,** which breaks down certain complex lipids, releasing products, such as fatty acids, that can be easily absorbed.

■ **Nucleases,** which break down nucleic acids.

■ **Proteolytic enzymes,** which break certain proteins apart. The proteolytic enzymes of the pancreas include proteases and peptidases. **Proteases** break apart large protein complexes, whereas **peptidases** break small peptide chains into individual amino acids.

Proteolytic enzymes account for about 70 percent of the total pancreatic enzyme production. The enzymes are secreted as inactive proenzymes that are activated only after they reach the small intestine. Proenzymes discussed earlier in the text include pepsinogen, angiotensinogen, plasminogen, fibrinogen, and many of the clotting factors and enzymes of the complement system. ∞ *pp. 767–769* As in the stomach, the release of a proenzyme rather than an active enzyme in the pancreas protects the secretory cells from the destructive effects of their own products. Among the proenzymes secreted by the pancreas are **trypsinogen** (trip-SIN-ō-jen), **chymotrypsinogen** (kī-mō-trip-SIN-ō-jen), **procarboxypeptidase** (prō-kar-bok-sē-PEP-ti-dās), and **proelastase** (pro-ē-LAS-tās).

Once inside the duodenum, enterokinase located on the brush border and in the lumen triggers the conversion of trypsinogen to **trypsin,** an active protease. Trypsin then activates the other proenzymes, producing **chymotrypsin, carboxypeptidase,** and **elastase.** Each enzyme attacks peptide bonds linking specific amino acids and ignores others. Together, they break down proteins into a mixture of dipeptides, tripeptides, and amino acids.

Pancreatitis

Pancreatitis (pan-krē-a-TĪ-tis) is an inflammation of the pancreas. A blockage of the excretory ducts, bacterial or viral infections, ischemia, and drug reactions, especially those involving alcohol, are among the factors that may produce this condition. These stimuli provoke a crisis by injuring exocrine cells in at least a portion of the organ. Lysosomes in the damaged cells then activate the proenzymes, and autolysis begins. The proteolytic enzymes digest the surrounding, undamaged cells, activating their enzymes and starting a chain reaction. In most cases, only a portion of the pancreas will be affected, and the condition subsides in a few days. In 10–15 percent of pancreatitis cases, the process does not subside; the enzymes can ultimately destroy the pancreas.

THE LIVER

The **liver,** the largest visceral organ, is one of the most versatile organs in the body. Most of its mass lies in the right hypochondriac and epigastric regions, but it may extend into the left hypochondriac and umbilical regions as well. The liver weighs about 1.5 kg (3.3 lb). This large, firm, reddish brown organ provides essential metabolic and synthetic services.

Anatomy of the Liver

The liver is wrapped in a tough fibrous capsule and covered by a layer of visceral peritoneum. On the anterior surface, the **falciform ligament** marks the division between the **left lobe** and the **right lobe** of the liver (Figure 24-19a,b•). A thickening in the posterior margin of the falciform ligament is the **round ligament,** or *ligamentum teres,* a fibrous band that marks the path of the fetal umbilical vein.

On the posterior surface of the liver, the impression left by the inferior vena cava marks the division between the right lobe and the small **caudate** (KAW-dāt) **lobe** (Figure 24-19a,c•). Inferior to the caudate lobe lies the **quadrate lobe,** sandwiched between the left lobe and the gallbladder. Afferent blood vessels and other structures reach the liver by traveling within the connective tissue of the lesser omentum. They converge at a region called the **hilus,** or *porta hepatis* ("doorway to the liver").

We discussed the circulation to the liver in Chapter 21 and summarized that information in Figures 21-27 and 21-34•, pp. 732, 741. Roughly one-third of the blood supply to the liver is arterial blood from the hepatic artery proper. The rest is venous blood from the hepatic portal vein, which begins in the capillaries of the esophagus, stomach, small intestine, and most of the large intestine. Liver cells, called **hepatocytes** (he-PAT-ō-sīts), adjust circulating levels of nutrients by selective absorption and secretion. Blood leaving the liver returns to the systemic circuit through the hepatic veins, which open into the inferior vena cava.

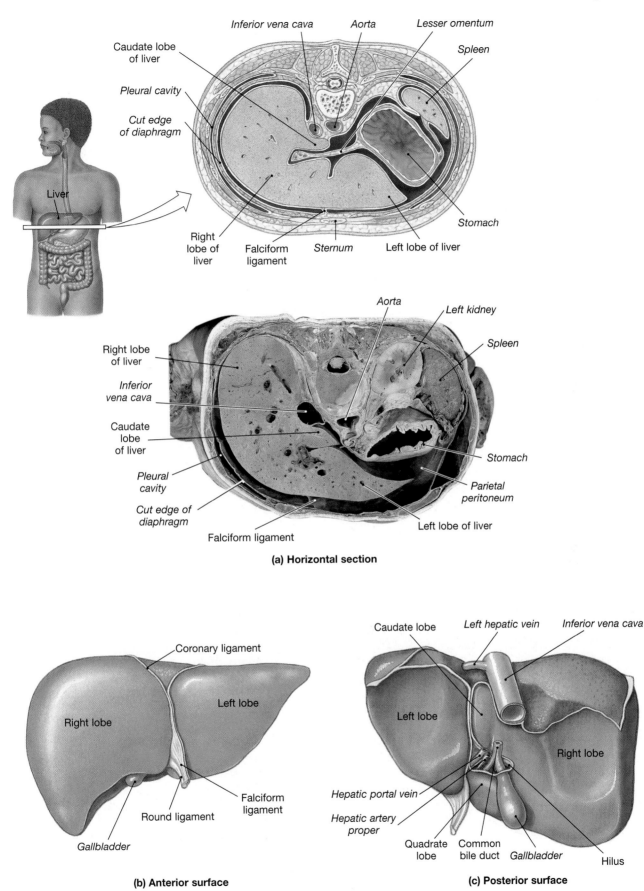

Caudate lobe of liver

Pleural cavity

Cut edge of diaphragm

Liver

Inferior vena cava   Aorta   Lesser omentum

Spleen

Right lobe of liver   Falciform ligament   Sternum   Left lobe of liver

Stomach

Aorta   Left kidney

Spleen

Right lobe of liver

Inferior vena cava

Caudate lobe of liver

Pleural cavity

Cut edge of diaphragm

Falciform ligament

Stomach

Parietal peritoneum

Left lobe of liver

(a) Horizontal section

Coronary ligament

Left lobe

Right lobe

Round ligament

Falciform ligament

Gallbladder

(b) Anterior surface

Caudate lobe   Left hepatic vein   Inferior vena cava

Left lobe

Right lobe

Hepatic portal vein

Hepatic artery proper

Quadrate lobe   Common bile duct   Gallbladder

Hilus

(c) Posterior surface

24

•**FIGURE 24-19 The Anatomy of the Liver. (a)** Horizontal sectional views (diagrammatic and actual) through the superior abdomen. **(b)** The anterior surface of the liver. **(c)** The posterior surface of the liver. *Plates 7.6e, 7.7, 7.8*

Histological Organization of the Liver

Each lobe of the liver is divided by connective tissue into approximately 100,000 **liver lobules,** the basic functional units of the liver. The histological organization and structure of a typical liver lobule are shown in Figure 24-20•.

Adjacent lobules are separated from each other by an *interlobular septum.* The hepatocytes in a liver lobule form a series of irregular plates arranged like the spokes of a wheel (Figure 24-20a,b•). The plates are only one cell thick, and exposed hepatocyte surfaces are covered with short microvilli. Within a lobule, sinusoids between adjacent plates empty into the

•**FIGURE 24-20 Liver Histology.** **(a)** A single liver lobule and its cellular components. **(b)** A diagrammatic view of liver structure, showing relationships among lobules. **(c)** A section through liver lobules from a pig liver. (LM × 38) (The interlobular septa in a human liver are very difficult to see at comparable magnification.) **(d)** A portal area. (LM × 31)

central vein. (We introduced sinusoids in Chapter 21. ∞ *p. 698*) The liver sinusoids lack a basement membrane, so large openings between the endothelial cells allow solutes—even those as large as plasma proteins—to pass out of the bloodstream and into the spaces surrounding the hepatocytes.

In addition to typical endothelial cells, the sinusoidal lining includes a large number of **Kupffer** (KOOP-fer) **cells,** also known as *stellate reticuloendothelial cells.* ∞ *p. 765* These phagocytic cells, part of the monocyte–macrophage system, engulf pathogens, cell debris, and damaged blood cells. Kupffer cells are also responsible for storing (1) iron, (2) some lipids, and (3) heavy metals, such as tin or mercury, that are absorbed by the digestive tract.

Blood enters the liver sinusoids from small branches of the hepatic portal vein and hepatic artery proper. A typical liver lobule has a hexagonal shape in cross section (Figure 24-20b•). There are six **portal areas,** or *hepatic triads,* one at each corner of the lobule. A portal area contains three structures: (1) a branch of the hepatic portal vein, (2) a branch of the hepatic artery proper, and (3) a small branch of the bile duct (Figure 24-20a,b,d•).

Branches from the arteries and veins deliver blood to the sinusoids of adjacent liver lobules (Figure 24-20a,b•). As blood flows through the sinusoids, hepatocytes absorb solutes from the plasma and secrete materials such as plasma proteins. Blood then leaves the sinusoids and enters the central vein of the lobule. The central veins ultimately merge to form the hepatic veins, which then empty into the inferior vena cava. Liver diseases, such as the various forms of *hepatitis,* and conditions such as alcoholism can lead to degenerative changes in the liver tissue and constriction of the circulatory supply. [AM] *Liver Disease*

Portal Hypertension

Pressures in the hepatic portal system are usually low, averaging 10 mm Hg or less. This pressure can increase markedly if blood flow through the liver becomes restricted as a result of liver damage or a blood clot. A rise in portal pressure is called **portal hypertension.** As pressures rise, small peripheral veins and capillaries in the portal system become distended and are likely to rupture; intestinal bleeding can occur. Under these conditions, esophageal varices (p. 860) may develop, and fluid may leak into the peritoneal cavity across the serosal surfaces of the liver and viscera, producing ascites (p. 887).

The Bile Duct System

The liver secretes a fluid called **bile** into a network of narrow channels between the opposing membranes of adjacent liver cells. These passageways, called **bile canaliculi,** extend outward, away from the central vein.

Eventually they connect with fine **bile ductules** (DUK-tūlz), which carry bile to bile ducts in the nearest portal area. The **right** and **left hepatic ducts** collect bile from all the bile ducts of the liver lobes. These ducts unite to form the **common hepatic duct,** which leaves the liver (Figure 24-21•). The bile in the common hepatic duct either (1) flows into the *common bile duct,* which empties into the duodenal ampulla, or (2) enters the *cystic duct,* which leads to the gallbladder.

The **common bile duct** is formed by the union of the **cystic duct** and the common hepatic duct. The common bile duct passes within the lesser omentum toward the stomach, turns, and penetrates the wall of the duodenum to meet the pancreatic duct at the duodenal ampulla.

The Physiology of the Liver

The liver is responsible for three general categories of functions: (1) *metabolic regulation,* (2) *hematological regulation,* and (3) *bile production.* The liver has more than 200 functions; in this discussion, we shall provide only a general overview.

Metabolic Regulation The liver is the primary organ involved in regulating the composition of circulating blood. All blood leaving the absorptive surfaces of the digestive tract enters the hepatic portal system and flows into the liver. Liver cells extract absorbed nutrients or toxins from blood before it reaches the systemic circulation through the hepatic veins. Excess nutrients are removed and stored, and deficiencies are corrected by mobilizing stored reserves or performing synthetic activities. For example:

- *Carbohydrate metabolism.* The liver stabilizes blood glucose levels at about 90 mg/dl. If blood glucose levels drop, hepatocytes break down glycogen reserves and release glucose into the bloodstream. They also synthesize glucose from other carbohydrates or from available amino acids. The synthesis of glucose from other compounds is a process called *gluconeogenesis.* If blood glucose levels climb, hepatocytes remove glucose from the bloodstream and either store it as glycogen or use it to synthesize lipids that can be stored in the liver or other tissues. These metabolic activities are regulated by circulating hormones, such as insulin and glucagon. ∞ *pp. 606–608*

- *Lipid metabolism.* The liver regulates circulating levels of triglycerides, fatty acids, and cholesterol. When those levels decline, the liver breaks down its lipid reserves and releases them into the bloodstream. When the levels are high, the lipids are removed for storage. However, because most lipids absorbed by the digestive tract bypass the hepatic portal circulation, this regulation occurs only after lipid levels have risen within the general circulation.

- *Amino acid metabolism.* The liver removes excess amino acids from the bloodstream. These amino acids

24

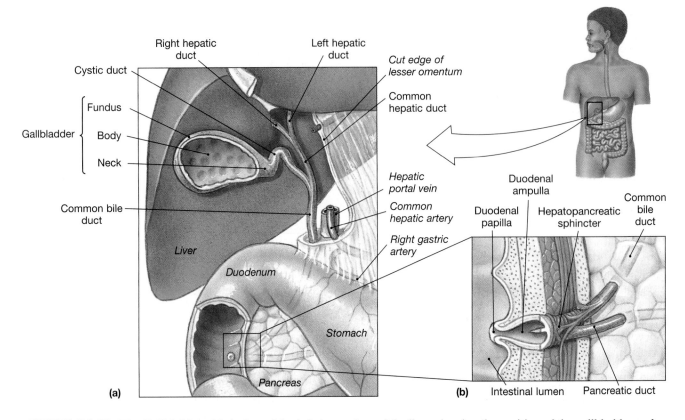

•**FIGURE 24-21** **The Gallbladder.** (a) A view of the inferior surface of the liver, showing the position of the gallbladder and ducts that transport bile from the liver to the gallbladder and duodenum. A portion of the lesser omentum has been cut away. (b) An interior view of the duodenum, showing the duodenal ampulla and related structures. [AM] *Plates 7.7b,d, 7.8b; Scan 14*

can be used to synthesize proteins or can be converted to lipids or glucose for storage.

■ *The removal of waste products.* When converting amino acids to lipids or carbohydrates, or when breaking down amino acids to get energy, the liver strips off the amino groups, a process called *deamination.* Ammonia, a toxic waste product, is formed. The liver neutralizes ammonia by conversion to *urea,* a fairly harmless compound excreted at the kidneys. Other waste products, circulating toxins, and drugs are also removed from the blood for inactivation, storage, or excretion.

■ *Vitamin storage.* Fat-soluble vitamins (A, D, E, and K) and vitamin B_{12} are absorbed from the blood and stored in the liver. These reserves are called on when your diet contains inadequate amounts of those vitamins.

■ *Mineral storage.* The liver converts iron reserves to ferritin and stores this protein–iron complex. ∞ *p. 633*

■ *Drug inactivation.* The liver removes and breaks down circulating drugs, thereby limiting the duration of their effects. When they prescribe drugs, physicians must take into account the rate at which the liver removes a particular drug. For example, a drug that is absorbed relatively quickly must be administered every few hours to keep the plasma concentrations at therapeutic levels.

Liver Disease

Any condition that severely damages the liver represents a serious threat to life. The liver has a limited ability to regenerate itself after injury, but liver function will not fully recover unless the normal vascular pattern returns. Examples of important types of liver disease include **cirrhosis,** which is characterized by the replacement of lobules by fibrous tissue, and various forms of **hepatitis** as a result of viral infections. Liver transplants are in some cases used to combat liver disease, but the supply of suitable donor tissue is limited and the success rate is highest in young, otherwise healthy individuals. Clinical trials are now under way to test an artificial liver known as *ELAD* (extracorporeal *l*iver *a*ssist *d*evice) that may prove suitable for the long-term support of persons with chronic liver disease. [AM] *Liver Disease*

Hematological Regulation The liver, the largest blood reservoir in your body, receives about 25 percent of the cardiac output. As blood passes through, the liver performs the following functions:

■ *Phagocytosis and antigen presentation.* Kupffer cells in the liver sinusoids engulf old or damaged red blood cells, cellular debris, and pathogens from the bloodstream. Kupffer cells are antigen-presenting cells that can stimulate an immune response. ∞ *p. 774*

- *The synthesis of plasma proteins.* Hepatocytes synthesize and release most of the plasma proteins. These include the albumins, which contribute to the osmotic concentration of the blood; the various types of transport proteins; clotting proteins; and complement proteins.

- *The removal of circulating hormones.* The liver is the primary site for the absorption and recycling of epinephrine, norepinephrine, insulin, thyroid hormones, and steroid hormones, such as the sex hormones (estrogens and androgens) and corticosteroids. The liver also absorbs cholecalciferol (vitamin D_3) from the blood. Liver cells then convert cholecalciferol, which may be synthesized in the skin or absorbed in the diet, into an intermediary product, 25-hydroxy-D_3, that is released back into the bloodstream. This intermediary is absorbed by the kidneys and used to generate calcitriol, a hormone important to Ca^{2+} metabolism. ∞ *p. 604*

- *The removal of antibodies.* The liver absorbs and breaks down antibodies, releasing amino acids to be recycled.

- *The removal or storage of toxins.* Lipid-soluble toxins in the diet, such as DDT, are absorbed by the liver and stored in lipid deposits, where they do not disrupt cellular functions. Other toxins are removed from the bloodstream and are either broken down or excreted in the bile.

- *The synthesis and secretion of bile.* Bile is synthesized in the liver and excreted into the lumen of the duodenum. Bile consists mostly of water, with minor amounts of ions, *bilirubin* (a pigment derived from hemoglobin), cholesterol, and an assortment of lipids collectively known as the **bile salts.** The water and ions assist in the dilution and buffering of acids in chyme as it enters the small intestine.

Bile salts are synthesized from cholesterol in the liver. Several related compounds are involved; the most abundant are derivatives of the steroids *cholate* and *chenodeoxycholate.*

The Functions of Bile Most dietary lipids are not water-soluble. Mechanical processing in the stomach creates large drops containing a variety of lipids. Pancreatic lipase is not lipid-soluble, so the enzymes can interact with lipids only at the surface of a lipid drop. The larger the droplet, the more lipids are inside, isolated and protected from these enzymes. Bile salts break the droplets apart, a process called **emulsification** (ē-mul-si-fi-KĀ-shun).

Emulsification creates tiny *emulsion droplets* with a superficial coating of bile salts. The formation of tiny droplets increases the surface area available for enzymatic attack. In addition, the layer of bile salts facilitates interaction between the lipids and lipid-digesting enzymes supplied by the pancreas. After lipid digestion has been completed, bile salts promote the absorption of lipids by the intestinal epithelium. More than 90 percent of the bile salts are themselves reabsorbed, primarily in the ileum, as lipid digestion is completed. The reabsorbed bile salts enter the hepatic portal circulation and are collected and recycled by the liver. The cycling of bile salts from the liver to the small intestine and back is called the **enterohepatic circulation of bile.**

THE GALLBLADDER

The **gallbladder** is a hollow, pear-shaped organ that stores and concentrates bile prior to its excretion into the small intestine. This muscular sac is located in a fossa, or recess, in the posterior surface of the liver's right lobe. The gallbladder is divided into three regions: (1) the **fundus,** (2) the **body,** and (3) the **neck** (Figure 24-21a•). The cystic duct leads from the gallbladder toward its union with the common hepatic duct to form the common bile duct. At the duodenum, the common bile duct meets the pancreatic duct before emptying into a chamber called the **duodenal ampulla** (am-PUL-a) (Figure 24-21b•). The duodenal ampulla receives buffers and enzymes from the pancreas and bile from the liver and gallbladder. It opens into the duodenum at the **duodenal papilla,** a small mound.

The **hepatopancreatic sphincter** *(sphincter of Oddi),* a muscular sphincter, encircles the lumen of the common bile duct and generally the pancreatic duct and duodenal ampulla as well. This sphincter remains contracted unless stimulated by the intestinal hormone cholecystokinin (CCK).

Physiology of the Gallbladder

The gallbladder has two major functions: (1) *bile storage* and (2) *bile modification.* Liver cells produce roughly 1 liter of bile each day, but the hepatopancreatic sphincter remains closed until chyme enters the duodenum. In the meantime, when bile cannot flow along the common bile duct, it enters the cystic duct for storage within the expandable gallbladder. When full, the gallbladder contains 40–70 ml of bile. The composition of bile gradually changes as it remains in the gallbladder. Much of the water is absorbed, and the bile salts and other components of bile become increasingly concentrated.

Bile secretion occurs continuously, but bile release into the duodenum occurs only under the stimulation of cholecystokinin. In the absence of CCK, the hepatopancreatic sphincter remains closed and bile exiting the liver in the common hepatic duct reaches the gallbladder by way of the cystic duct. Cholecystokinin (1) relaxes the hepatopancreatic sphincter

24

and (2) stimulates contractions in the walls of the gallbladder. These contractions push bile into the small intestine. Whenever chyme enters the duodenum, CCK is released, but the amount secreted increases markedly when the chyme contains large amounts of lipids.

⚕ Problems with Bile Storage and Excretion

If bile becomes too concentrated, crystals of insoluble minerals and salts begin to appear. These deposits are called **gallstones.** Merely having them, a condition termed **cholelithiasis** (kō-lē-li-THĪ-a-sis; *chole*, bile), is not a problem as long as the stones remain small. Small gallstones are normally flushed down the bile duct and excreted. In **cholecystitis,** the gallstones are so large that they can damage the wall of the gallbladder or block the cystic or common bile duct. A recent therapy for cholecystitis involves immersing the individual in water and shattering the stones with focused sound waves. The apparatus used is called a *lithotripter.* The particles produced are then small enough to pass through the bile duct without difficulty. In severe cases of cholecystitis, the gallbladder can become infected, inflamed, or perforated. Under these conditions, it may be surgically removed in a procedure known as a *cholecystectomy.* This loss does not seriously impair digestion, because bile production continues at normal levels. However, the bile is more dilute, and its entry into the small intestine is not as closely tied to the arrival of food in the duodenum. AM *Cholecystitis*

THE COORDINATION OF SECRETION AND ABSORPTION

A combination of neural and hormonal mechanisms coordinates the activities of the digestive glands. These regulatory mechanisms are centered around the duodenum, where the acids must be neutralized and the appropriate enzymes added.

Neural mechanisms involving the central nervous system (CNS) deal with (1) preparing the digestive tract for activity (parasympathetic innervation) or inhibiting gastrointestinal activity (sympathetic innervation) and (2) coordinating the movement of materials along the length of the digestive tract (the enterogastric, gastroenteric, and gastroileal reflexes).

In addition, motor neurons synapsing in the digestive tract release a variety of neurotransmitters. Many of these chemicals are also released in the CNS, but in general their functions are poorly understood. Examples of potentially important neurotransmitters include Substance P, enkephalins, and endorphins.

We introduced hormones important to the regulation of intestinal and glandular function in the course of our discussion. We will now summarize the introductory information and will consider some additional details about the regulatory mechanisms involved. Table 24-1 also summarizes this information.

Intestinal Hormones

The intestinal tract secretes a variety of hormones, but it has proved very difficult to determine the primary effects of these hormones, largely because all are peptide hormones with similar chemical structures. Careful analyses have led to a marked increase in the number of identified intestinal hormones, but their specific functions have yet to be sorted out to everyone's satisfaction. Many of these hormones have multiple effects that target several regions of the digestive tract as well as affecting the accessory glandular organs.

Duodenal enteroendocrine cells produce the following hormones known to coordinate digestive functions:

- *Secretin.* **Secretin** is released when chyme arrives in the duodenum. The primary effect of secretin is to cause an increase in the secretion of bile and buffers by the liver and pancreas. Among its secondary effects, secretin reduces gastric motility and secretory rates.

- *Cholecystokinin.* **Cholecystokinin (CCK)** is secreted when chyme arrives in the duodenum, especially when the chyme contains lipids and partially digested proteins. In the pancreas, CCK accelerates the production and secretion of all types of digestive enzymes. It also causes a relaxation of the hepatopancreatic sphincter and contraction of the gallbladder, resulting in the ejection of bile and pancreatic juice into the duodenum. The net effects of CCK are thus to increase the secretion of pancreatic enzymes and to push pancreatic secretions and bile into the duodenum. The presence of CCK in high concentrations has two additional effects: (1) It inhibits gastric activity, and (2) it appears to have CNS effects that reduce the sensation of hunger.

- *Gastric inhibitory peptide.* **Gastric inhibitory peptide (GIP)** is secreted when fats and carbohydrates, especially glucose, enter the small intestine. The inhibition of gastric activity is accompanied by the stimulation of insulin release at the pancreatic islets, so GIP is also known as *glucose-dependent insulinotropic peptide.* This hormone has a number of secondary effects. For instance, it stimulates the activity of the submucosal glands of the duodenum, stimulates lipid synthesis in adipose tissue, and increases glucose use by skeletal muscles.

- *Vasoactive intestinal peptide.* **Vasoactive intestinal peptide (VIP)** stimulates the secretion of intestinal glands, dilates regional capillaries, and inhibits acid production in the stomach. By dilating capillaries in active areas of the intestinal tract, VIP provides an efficient mechanism for removing absorbed nutrients.

- *Gastrin.* Gastrin is secreted by G cells in the duodenum when they are exposed to large quantities of incompletely digested proteins.

- *Enterocrinin.* **Enterocrinin,** a hormone released when chyme enters the small intestine, stimulates mucin production by the submucosal glands of the duodenum.

TABLE 24-1   Important Digestive Hormones and Their Primary Effects

Hormone	Stimulus	Origin	Target	Effects
Cholecystokinin (CCK)	Arrival of chyme containing lipids and partially digested proteins	Duodenum	Pancreas	Stimulates production of pancreatic enzymes
			Gallbladder	Stimulates contraction of gallbladder
			Duodenum	Causes relaxation of hepato-pancreatic sphincter
			Stomach	Inhibits gastric secretion and motion
			CNS	May reduce hunger
Enterocrinin	Arrival of chyme in the duodenum	Duodenum	Submucosal glands	Stimulates production of alkaline mucus
Gastric inhibitory peptide (GIP)	Arrival of chyme containing large quantities of fats and glucose	Duodenum	Pancreas	Stimulates release of insulin by pancreatic islets
			Stomach	Inhibits gastric secretion and motility
			Adipose tissue	Stimulates lipid synthesis
			Skeletal muscle	Stimulates glucose use
Gastrin	Vagus nerve stimulation or arrival of food in the stomach	Stomach	Stomach	Stimulates production of acids and enzymes; increases motility
	Arrival of chyme containing large quantities of undigested proteins	Duodenum	Stomach	As above
Secretin	Arrival of chyme in the duodenum	Duodenum	Pancreas	Stimulates production of alkaline buffers
			Stomach	Inhibits gastric secretion and motility
			Liver	Increases rate of bile secretion
Vasoactive intestinal peptide (VIP)	Arrival of chyme in the duodenum	Duodenum	Duodenal glands, stomach	Stimulates buffer secretion; inhibits acid production; dilates intestinal capillaries

■ *Other intestinal hormones.* Several other hormones are produced in relatively small quantities. Examples include *motilin,* which stimulates intestinal contractions; *villikinin,* which promotes the movement of villi and associated lymph flow; and *somatostatin,* which inhibits gastric secretion.

Functional interactions among gastrin, secretin, CCK, GIP, and VIP are diagrammed in Figure 24-22•.

Intestinal Absorption

On average it takes about 5 hours for materials to pass from your duodenum to the end of your ileum, so the first of the materials to enter the duodenum after you eat breakfast may leave the small intestine at lunchtime. Along the way, absorptive effectiveness is enhanced by the fact that so much of the mucosa is movable. The microvilli can be moved by their supporting microfilaments, the individual villi by smooth muscle cells, groups of villi by the muscularis mucosae, and the plicae by the muscularis mucosae and the muscularis externa. These movements stir and mix the intestinal contents. As a result, the environment around each epithelial cell changes from moment to moment.

24

☑ How is the small intestine adapted for the absorption of nutrients?

☑ How would a meal that is high in fat affect the level of cholecystokinin in the blood?

☑ How would the pH of intestinal contents be affected if the small intestine did not secrete the hormone secretin?

☑ The digestion of which nutrient would be most impaired by damage to the exocrine pancreas?

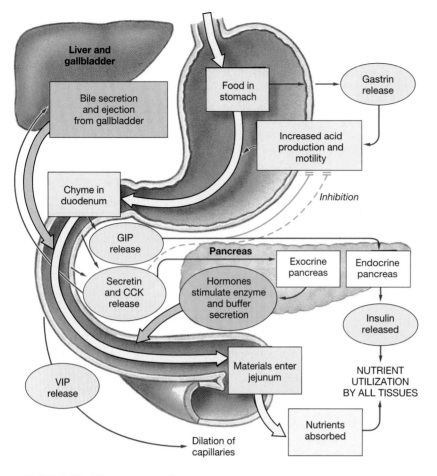

●**FIGURE 24-22** **The Activities of Major Digestive Tract Hormones.** The primary actions of gastrin, secretin, CCK, GIP, and VIP.

THE LARGE INTESTINE

The horseshoe-shaped **large intestine** begins at the end of the ileum and ends at the anus. The large intestine lies inferior to the stomach and liver and almost completely frames the small intestine (Figure 24-16a●, p. 869). The major functions of the large intestine include (1) the reabsorption of water and compaction of intestinal contents into feces, (2) the absorption of important vitamins liberated by bacterial action, and (3) the storage of fecal material prior to defecation.

The large intestine, also known as the **large bowel,** has an average length of about 1.5 meters (5 ft) and a width of 7.5 cm (3 in.). We can divide it into three parts: (1) the pouchlike *cecum,* the first portion of the large intestine; (2) the *colon,* the largest portion; and (3) the *rectum,* the last 15 cm (6 in.) of the large intestine and the end of the digestive tract (Figure 24-23●).

THE CECUM

Material arriving from the ileum first enters an expanded pouch called the **cecum** (SĒ-kum). The ileum attaches to the medial surface of the cecum and opens into the cecum at the **ileocecal** (il-ē-ō-SĒ-kal) **valve** (Figure 24-23a●). The cecum collects and stores materials from the ileum and begins the process of compaction.

The slender, hollow **vermiform appendix** (*vermis,* worm), more commonly known simply as the *appendix,* is attached to the posteromedial surface of the cecum (Figure 24-23●). The appendix is generally about 9 cm (3.5 in.) long, but its size and shape are quite variable. A small mesentery called the **mesoappendix** connects the appendix to the ileum and cecum. The mucosa and submucosa of the appendix are dominated by lymphoid nodules, and the primary function of the appendix is as an organ of the lymphatic system. Inflammation of the appendix is known as *appendicitis.* ⟶ *p. 759*

THE COLON

The **colon** has a larger diameter and a thinner wall than the small intestine. Distinctive features of the colon include the following (Figure 24-23a,c●):

- The wall of the colon forms a series of pouches, or **haustra** (HAWS-truh; singular, *haustrum*). Cutting into the intestinal lumen reveals that the creases between the haustra affect the mucosal lining as well, producing a series of internal folds. Haustra permit the expansion and elongation of the colon, rather like the bellows that allow an accordion to lengthen.

- Three separate longitudinal bands of smooth muscle—called the **taenia coli** (TĒ-nē-a KŌ-lī)—are on the outer surfaces of the colon just deep to the serosa. These bands correspond to the outer layer of the muscularis externa in other portions of the digestive tract. Muscle tone within the taenia coli is what creates the haustra.

- The serosa of the colon contains numerous teardrop-shaped sacs of fat called **epiploic** (e-pip-LŌ-ik) **appendages.**

We can subdivide the colon into four regions: the ascending colon, transverse colon, descending colon, and sigmoid colon (Figure 24-23a●).

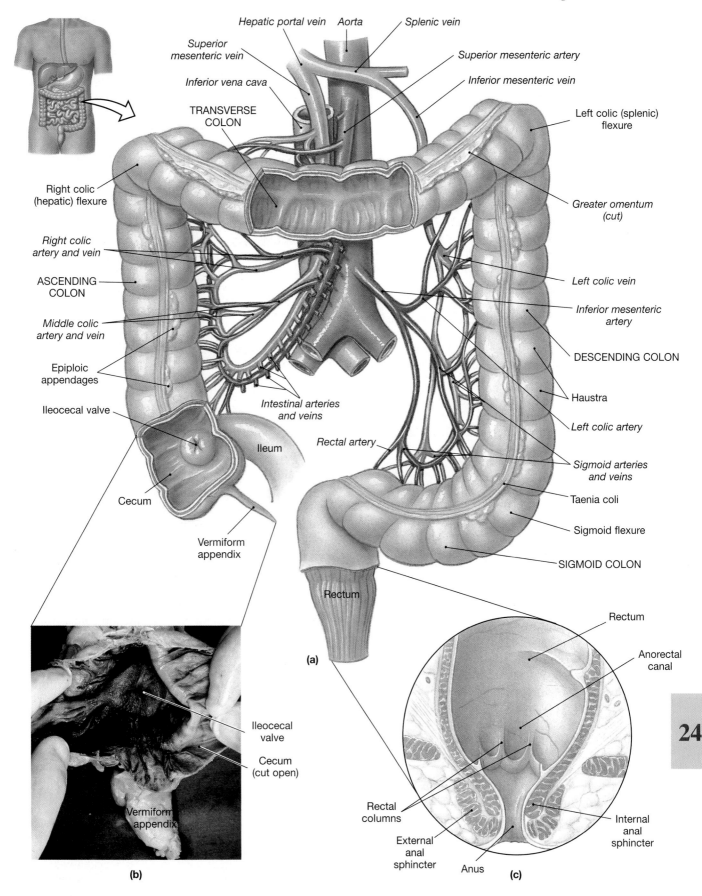

Hepatic portal vein   Aorta   Splenic vein

Superior mesenteric vein

Inferior vena cava

Superior mesenteric artery

Inferior mesenteric vein

Left colic (splenic) flexure

TRANSVERSE COLON

Right colic (hepatic) flexure

Greater omentum (cut)

Right colic artery and vein

Left colic vein

ASCENDING COLON

Inferior mesenteric artery

Middle colic artery and vein

DESCENDING COLON

Epiploic appendages

Haustra

Intestinal arteries and veins

Left colic artery

Ileocecal valve

Ileum

Rectal artery

Sigmoid arteries and veins

Cecum

Taenia coli

Sigmoid flexure

Vermiform appendix

SIGMOID COLON

Rectum

(a)

Ileocecal valve

Cecum (cut open)

Vermiform appendix

(b)

Rectum

Anorectal canal

Rectal columns

External anal sphincter

Anus

Internal anal sphincter

(c)

24

•**FIGURE 24-23 The Large Intestine. (a)** The gross anatomy and regions of the large intestine. **(b)** The vermiform appendix. **(c)** The rectum and anus. *Plates 7.6e, 7.7a,b; Scan 16*

1. The **ascending colon** begins at the superior border of the cecum and ascends along the right lateral and posterior wall of the peritoneal cavity to the inferior surface of the liver. There the colon bends sharply to the left at the **right colic flexure,** or *hepatic flexure.* This flexure marks the end of the ascending colon and the beginning of the transverse colon (Figure 24-5b,c•, p. 852).

2. The **transverse colon** curves anteriorly from the right colic flexure and crosses the abdomen from right to left (Figure 24-5b,d•). It is supported by the transverse mesocolon and is separated from the anterior abdominal wall by the layers of the greater omentum. As the transverse colon reaches the left side of the body, it passes inferior to the greater curvature of the stomach. Near the spleen, the colon makes a 90° turn at the **left colic flexure,** or *splenic flexure,* and becomes the descending colon.

3. The **descending colon** proceeds inferiorly along the left side (Figure 24-5b,c•) until reaching the iliac fossa. The descending colon is retroperitoneal and firmly attached to the abdominal wall. At the iliac fossa, the descending colon curves at the **sigmoid flexure** and becomes the sigmoid colon.

4. The sigmoid flexure is the start of the **sigmoid** (SIG-moyd) **colon** (*sigmeidos,* the Greek letter *S*), an S-shaped segment that is only about 15 cm (6 in.) long. It lies posterior to the urinary bladder, suspended from the sigmoid mesocolon (Figure 24-5b–d•). The sigmoid colon empties into the *rectum.*

The large intestine receives blood from tributaries of the superior mesenteric and inferior mesenteric arteries. Venous blood is collected from the large intestine by the superior mesenteric and inferior mesenteric veins. ∞ *pp. 732, 741–742*

THE RECTUM

The **rectum** (REK-tum) forms the last 15 cm (6 in.) of the digestive tract (Figure 24-23a,c•). The rectum is an expandable organ for the temporary storage of fecal material. The movement of fecal materials into the rectum triggers the urge to defecate.

The last portion of the rectum, the **anorectal** (ā-nō-REK-tal) **canal,** contains small longitudinal folds called **rectal columns.** The distal margins of these columns are joined by transverse folds that mark the boundary between the columnar epithelium of the proximal rectum and a stratified squamous epithelium like that in the oral cavity. The **anus,** or *anal orifice,* is the exit of the anorectal canal. There the epidermis becomes keratinized and identical to the surface of the skin. The circular muscle layer of the muscularis externa in this region forms the **internal anal sphincter** (Figure 24-23c•). The smooth muscle cells of the internal anal sphincter are not under voluntary control. The **external anal sphincter** guards the anus. This sphincter,

which consists of a ring of skeletal muscle fibers, is under voluntary control.

The lamina propria and submucosa of the anorectal canal bear a network of veins. If venous pressures there rise too high due to straining during defecation, the veins can become distended, producing *hemorrhoids.*

Colon Cancer

Colon cancers are relatively common. Approximately 94,700 cases are diagnosed in the United States each year (in addition to 34,700 cases of rectal cancers). In 1999 an estimated 47,900 deaths occurred from colon cancers (and another 8700 deaths from rectal cancers). The mortality rate for these cancers remains high, and the best defense appears to be early detection and prompt treatment. The standard screening test involves checking the feces for blood. This is a simple procedure that can easily be performed on a stool (fecal) sample as part of a routine physical. [AM] *Colon Inspection and Cancer*

HISTOLOGY OF THE LARGE INTESTINE

Although the diameter of the colon is roughly three times that of the small intestine, its wall is much thinner. The major characteristics of the colon are the lack of villi, the abundance of goblet cells, and the presence of distinctive intestinal glands (Figure 24-24•). The glands in the large intestine are deeper than those of the small intestine, and they are dominated by goblet cells. The mucosa of the large intestine does not produce enzymes; any digestion that occurs results from enzymes introduced in the small intestine or from bacterial action. The mucus provides lubrication as the fecal material becomes less moist and more compact. Mucous secretion occurs as local stimuli, such as friction or exposure to harsh chemicals, trigger short reflexes involving local nerve plexuses. Large lymphoid nodules are scattered throughout the lamina propria and submucosa.

The muscularis externa of the large intestine is unusual, because the longitudinal layer has been reduced to the muscular bands of the taenia coli. However, the mixing and propulsive contractions of the colon resemble those of the small intestine.

Diverticulosis

In **diverticulosis** (dī-ver-tik-ū-LŌ-sis), pockets *(diverticula)* form in the mucosa, generally in the sigmoid colon. These pockets get forced outward, probably by the pressures generated during defecation. If the pockets push through weak points in the muscularis externa, they form semi-isolated chambers that are subject to recurrent infection and inflammation. The infections cause pain and occasional bleeding, a condition known as *diverticulitis* (dī-ver-tik-ū-LĪ-tis). Inflammation of other portions of the colon is called *colitis* (kō-LĪ-tis). [AM] *Inflammatory Bowel Disease*

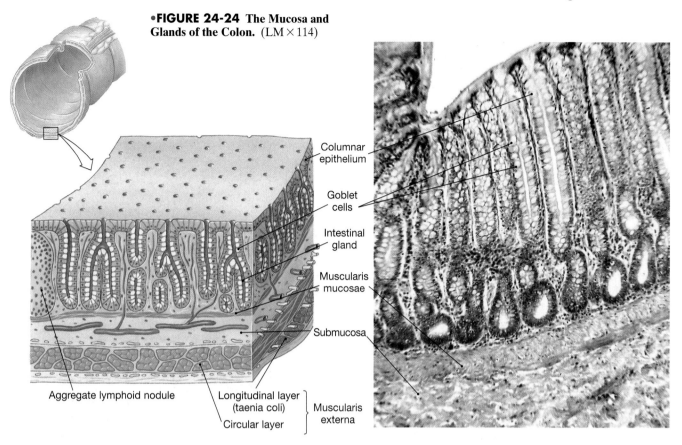

•**FIGURE 24-24** The Mucosa and Glands of the Colon. (LM × 114)

Columnar epithelium
Goblet cells
Intestinal gland
Muscularis mucosae
Submucosa
Aggregate lymphoid nodule
Longitudinal layer (taenia coli)
Circular layer
Muscularis externa

PHYSIOLOGY OF THE LARGE INTESTINE

Less than 10 percent of the nutrient absorption under way in the digestive tract occurs in the large intestine. Nevertheless, the absorptive operations in this segment of the digestive tract are important. The large intestine also prepares fecal material for ejection from the body.

Absorption in the Large Intestine

The reabsorption of water is an important function of the large intestine. Although roughly 1500 ml of material enters your colon each day, only about 200 ml of feces is ejected. The remarkable efficiency of digestion can best be appreciated by considering the average composition of feces: 75 percent water, 5 percent bacteria, and the rest a mixture of indigestible materials, small quantities of inorganic matter, and the remains of epithelial cells.

In addition to reabsorbing water, the large intestine absorbs a number of other substances that remain in the feces or that were secreted into the digestive tract along its length. Examples include useful compounds, such as bile salts and vitamins, organic waste products, such as urobilinogen, and various toxins generated by bacterial action. Most of the bile salts entering the large intestine are promptly reabsorbed in the cecum and are transported in blood to the liver for secretion into bile.

Vitamins   Vitamins are organic molecules that are important as cofactors or coenzymes in many metabolic pathways. Bacteria that reside in the colon generate three vitamins that supplement our dietary supply:

1. *Vitamin K,* a fat-soluble vitamin that the liver needs in order to synthesize four clotting factors, including prothrombin. Intestinal bacteria produce roughly half of our daily vitamin K requirements.

2. *Biotin,* a water-soluble vitamin important to various reactions, notably those of glucose metabolism.

3. *Vitamin B$_5$* (pantothenic acid), a water-soluble vitamin required in the manufacture of steroid hormones and some neurotransmitters.

Vitamin K deficiencies, which lead to impaired blood clotting, result from (1) a deficiency of lipids in the diet, which impairs the absorption of all fat-soluble vitamins, or (2) problems affecting lipid processing and absorption, such as inadequate bile production or chronic diarrhea. Disorders resulting from deficiencies of biotin or vitamin B$_5$ are extremely rare after infancy, because the intestinal bacteria produce sufficient amounts to supplement any dietary shortage.

Organic Wastes   In Chapter 19, we discussed the fate of bilirubin, a breakdown product of heme. ∞ *p. 632* In the large intestine, bacteria convert bilirubin to

24

urobilinogens and *stercobilinogens.* Some urobilinogens are absorbed into the bloodstream and then excreted in urine. On exposure to oxygen, some urobilinogens and stercobilinogens are converted to **urobilins** and **stercobilins.** These pigments in various proportions give feces a yellow-brown or brown coloration. Bacterial action breaks down peptides that remain in the feces and generates (1) ammonia, in the form of soluble *ammonium ions* (NH_4^+); (2) *indole* and *skatole,* two nitrogen-containing compounds that are primarily responsible for the odor of feces; and (3) hydrogen sulfide (H_2S), a gas that produces a "rotten egg" odor. Significant amounts of ammonia and smaller amounts of other toxins cross the colonic epithelium and enter the hepatic portal circulation. These toxins are removed by the liver and converted to relatively nontoxic compounds that can be released into the blood and excreted at the kidneys.

Indigestible carbohydrates are not altered by intestinal enzymes, so they arrive in the colon virtually intact. These complex polysaccharides provide a reliable nutrient source for colonic bacteria, whose metabolic activities are responsible for the small quantities of **flatus,** or intestinal gas, in the large intestine. Meals containing large numbers of indigestible carbohydrates (such as franks and beans) stimulate bacterial gas production, leading to distension of the colon, cramps, and the frequent discharge of intestinal gases.

Movements of the Large Intestine

The gastroileal and gastroenteric reflexes move materials into the cecum while you eat. Movement from the cecum to the transverse colon is very slow, allowing hours for water absorption to convert the already thick material into a sludgy paste. Peristaltic waves move material along the length of the colon, and segmentation movements, called *haustral churning,* mix the contents of adjacent haustra. Movement from the transverse colon through the rest of the large intestine results from powerful peristaltic contractions called **mass movements,** which occur a few times each day. The stimulus is distension of the stomach and duodenum; the commands are relayed over the intestinal nerve plexuses. The contractions force feces into the rectum and produce the conscious urge to defecate.

The rectal chamber is usually empty except when one of those powerful peristaltic contractions forces feces out of the sigmoid colon. Distension of the rectal wall then triggers the **defecation reflex,** which involves two positive feedback loops (Figure 24-25●). Both loops are triggered by stretch receptor stimulation in the walls of the rectum. The first loop is a short reflex that triggers a series of peristaltic contractions in the rectum that move feces toward the anus (Steps 1, 2, 3a, and 4a). The second loop is a long reflex coordinated by the sacral parasympathetic system. This reflex stimulates mass movements that push feces

toward the rectum from the descending colon and sigmoid colon (Steps 3b and 4b).

Rectal stretch receptors also trigger two reflexes important to the *voluntary* control of defecation. One is a long reflex mediated by parasympathetic innervation within the pelvic nerves. This reflex causes the

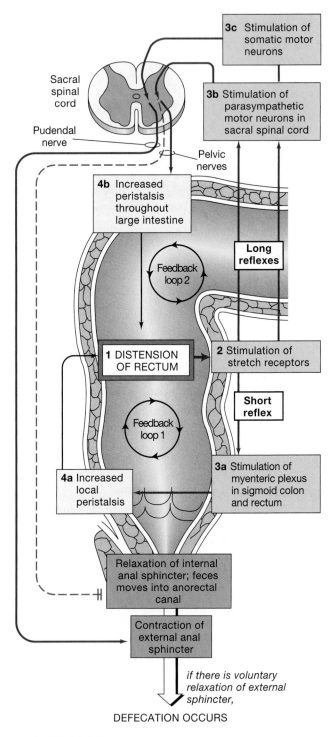

●**FIGURE 24-25 The Defecation Reflex.** Short and long reflexes (Steps 1, 2, 3a, and 4a and Steps 3b and 4b, respectively) promote movement of fecal material toward the anus. Another long reflex triggered by rectal stretch receptor stimulation (Step 3c) prevents involuntary defecation.

relaxation of the internal anal sphincter, the smooth muscle sphincter that controls the movement of feces into the anorectal canal. The second (Step 3c in Figure 24-25•) is a somatic reflex that stimulates the immediate contraction of the external anal sphincter, a skeletal muscle. ∞ *p. 334* The motor commands are carried by the pudendal nerves.

The elimination of feces requires that both the internal and external anal sphincters be relaxed, but these reflexes open the internal sphincter and close the external sphincter. The actual release of feces requires conscious effort to open the external sphincter. In addition to opening the external sphincter, consciously directed activities such as tensing the abdominal muscles or making expiratory movements while closing the glottis elevate intra-abdominal pressures and help force fecal materials out of the rectum.

If the external anal sphincter remains constricted, the peristaltic contractions cease until additional rectal expansion triggers the defecation reflex a second time. The urge to defecate usually develops when rectal pressure reaches about 15 mm Hg. If pressure inside the rectum exceeds 55 mm Hg, the external anal sphincter will relax and defecation will occur. This mechanism regulates defecation in infants and in adults with severe spinal cord injuries.

Diarrhea and Constipation

Diarrhea (dī-a-RĒ-uh) exists when an individual has frequent, watery bowel movements. This condition results when the mucosa of the colon becomes unable to maintain normal levels of absorption or the rate of fluid entry into the colon exceeds its maximum reabsorptive capacity. Bacterial, viral, or protozoan infection of the colon or small intestine can cause acute bouts of diarrhea lasting several days. Severe diarrhea is life-threatening due to cumulative losses of fluids and ions. In *cholera* (KOL-e-ruh), bacteria bound to the intestinal lining release toxins that stimulate a massive fluid secretion across the intestinal epithelium. Without treatment, a person with cholera can die of acute dehydration in a matter of hours. [AM] *Diarrhea*

Constipation is infrequent defecation, generally involving dry, hard feces. Constipation occurs when fecal materials move through the colon so slowly that excessive water reabsorption occurs. The feces become extremely compact, difficult to move, and highly abrasive. Inadequate dietary fiber and fluids, coupled with a lack of exercise, are common causes. Constipation can generally be treated by the oral administration of stool softeners, such as *Colace®*, laxatives, or **cathartics** (ka-THAR- tiks), which promote defecation. These compounds promote water movement into the feces, increase fecal mass, or irritate the lining of the colon to stimulate peristalsis. For example, indigestible fiber adds bulk to the feces, retaining moisture and stimulating stretch receptors that promote peristalsis. The promotion of peristalsis is a benefit of high-fiber cereals. Active movement during exercise also assists in the movement of fecal materials through the colon.

DIGESTION AND ABSORPTION

A typical meal contains carbohydrates, proteins, lipids, water, electrolytes, and vitamins. Your digestive system handles each component differently. Large organic molecules must be broken down by digestion before absorption can occur. Water, electrolytes, and vitamins can be absorbed without preliminary processing, but special transport mechanisms may be involved.

THE PROCESSING AND ABSORPTION OF NUTRIENTS

Food contains large organic molecules, many of them insoluble. The digestive system first breaks down the physical structure of the ingested material and then proceeds to disassemble the component molecules into smaller fragments. This disassembly eliminates any antigenic properties, so after absorption the fragments don't trigger an immune response. The molecules released into the bloodstream will be absorbed by cells and either (1) broken down to provide energy for the synthesis of ATP or (2) used to synthesize carbohydrates, proteins, and lipids. This section will focus on the mechanics of digestion and absorption; the fate of the compounds inside cells will be the focus of Chapter 25.

Most ingested organic materials are complex chains of simpler molecules. In a typical dietary carbohydrate, the basic molecules are simple sugars. In a protein, the building blocks are amino acids; in lipids, they are generally fatty acids; and in nucleic acids, they are nucleotides. Digestive enzymes break the bonds between the component molecules of carbohydrates, proteins, lipids, and nucleic acids in a process called *hydrolysis*. ∞ *p. 38*

The classes of digestive enzymes differ with respect to their targets. *Carbohydrases* break the bonds between simple sugars, *proteases* split the linkages between amino acids, and *lipases* separate the fatty acids from glycerides. Specific enzymes in each class may be even more selective, breaking bonds between specific molecules. For example, a particular carbohydrase might break bonds between glucose molecules but not those between glucose and another simple sugar.

Digestive enzymes secreted by the salivary glands, tongue, stomach, and pancreas are mixed into the ingested material as it passes along the digestive tract. These enzymes break down large carbohydrates, proteins, lipids, and nucleic acids into smaller fragments, which in turn must typically be broken down further before absorption can occur. The final enzymatic steps involve brush border enzymes, which are attached to the exposed surfaces of microvilli.

Nucleic acids are broken down into their component nucleotides. Brush border enzymes digest these nucleotides into sugars, phosphates, and nitrogenous bases that are absorbed by active transport. However, nucleic acids represent only a small fraction of the total nutrients absorbed each day. The digestive fates of

24

carbohydrates, lipids, and proteins, the major dietary components, are summarized in Figure 24-26•. Table 24-2 reviews the major digestive enzymes and their functions. We shall now take a closer look at the digestion and absorption of carbohydrates, lipids, and proteins.

CARBOHYDRATE DIGESTION AND ABSORPTION

The digestion of complex carbohydrates (simple polysaccharides and starches) proceeds in two steps. One

step involves carbohydrases produced by the salivary glands and pancreas; the other, brush border enzymes.

Salivary and Pancreatic Enzymes

The digestion of complex carbohydrates involves two enzymes—salivary amylase and pancreatic alpha-amylase (Figure 24-26a•)—that function effectively at a pH of 6.7–7.5. Carbohydrate digestion begins in the mouth during mastication, by the action of salivary amylase from the parotid and submandibular salivary

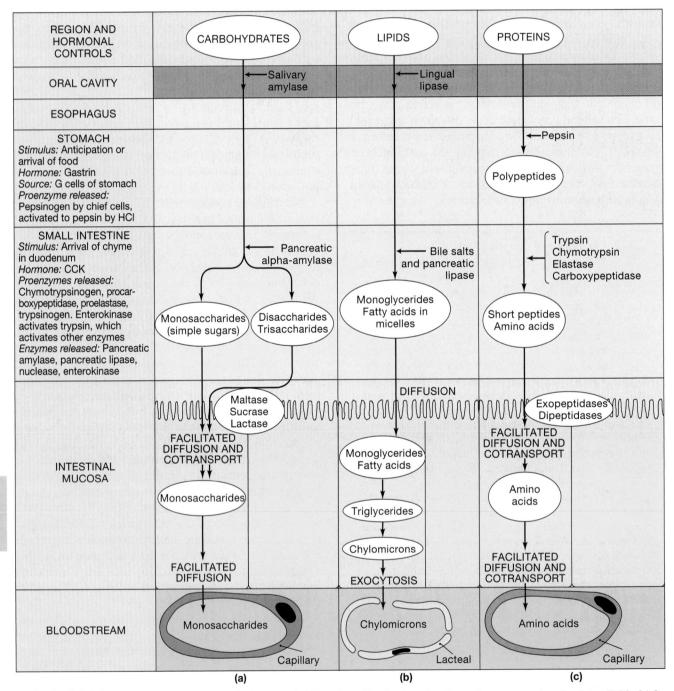

•**FIGURE 24-26 A Summary of the Chemical Events in Digestion.** For further details on the enzymes involved, see Table 24-2.

TABLE 24-2 Digestive Enzymes and Their Functions

Enzyme (proenzyme)	Source	Optimal pH	Target	Products	Remarks
Carbohydrases					
Maltase, sucrase, lactase	Brush border of small intestine	7–8	Maltose, sucrose, lactose	Monosaccharides	Found in membrane surface of microvilli
Pancreatic alpha-amylase	Pancreas	6.7–7.5	Complex carbohydrates	Disaccharides and trisaccharides	Breaks bonds between monomers
Salivary amylase	Salivary glands	6.7–7.5	Complex carbohydrates	Disaccharides and trisaccharides	Breaks bonds between monomers
Proteases					
Carboxypeptidase (procarboxypeptidase)	Pancreas	7–8	Proteins, polypeptides, amino acids	Short-chain peptides	Activated by trypsin
Chymotrypsin (chymotrypsinogen)	Pancreas	7–8	Proteins, polypeptides	Short-chain peptides	Activated by trypsin
Dipeptidases, peptidases	Brush border of small intestine	7–8	Dipeptides, tripeptides	Amino acids	Found in membrane surface of brush border
Elastase (proelastase)	Pancreas	7–8	Elastin	Short-chain peptides	Activated by trypsin
Enterokinase	Brush border and lumen of small intestine	7–8	Trypsinogen	Trypsin	Reaches lumen through disintegration of shed epithelial cells
Pepsin (pepsinogen)	Chief cells of stomach	1.5–2.0	Proteins, polypeptides	Short-chain polypeptides	Secreted as proenzyme, pepsinogen; activated by H^+ in stomach acid
Rennin	Stomach	3.5–4.0	Milk proteins		Secreted only in infants; causes protein coagulation
Trypsin (trypsinogen)	Pancreas	7–8	Proteins, polypeptides	Short-chain peptides	Proenzyme activated by enterokinase; activates other pancreatic proteases
Lipases					
Lingual lipase	Glands of tongue	6.7–7.5	Triglycerides	Fatty acids and monoglycerides	
Pancreatic lipase	Pancreas	7–8	Triglycerides	Fatty acids and monoglycerides	Bile salts must be present for efficient action
Nucleases	Pancreas	7–8	Nucleic acids	Nitrogenous bases and simple sugars	Includes ribonuclease for RNA and deoxyribonuclease for DNA

24

glands. Salivary amylase breaks down starches (complex carbohydrates), producing a mixture composed primarily of *disaccharides* (two simple sugars) and *trisaccharides* (three simple sugars). Salivary amylase continues to digest the starches and glycogen in the food for 1–2 hours before stomach acids render it inactive. Because the enzymatic content of saliva is not high, only a small amount of digestion occurs over this period.

In the duodenum, the remaining complex carbohydrates are broken down by the action of pancreatic alpha-amylase. Any disaccharides or trisaccharides produced, as well as any already present in the food, are ignored by both salivary and pancreatic amylases. Additional hydrolysis does not occur until these molecules contact the intestinal mucosa.

Brush Border Enzymes

Prior to absorption, disaccharides and trisaccharides are fragmented into *monosaccharides* (simple sugars) by brush border enzymes of the intestinal microvilli.

The enzyme **maltase** splits bonds between the two glucose molecules of the disaccharide **maltose. Sucrase** breaks the disaccharide **sucrose** into glucose and *fructose,* another six-carbon sugar. **Lactase** hydrolyzes the disaccharide **lactose** into a molecule of glucose and one of *galactose.* Lactose is the primary carbohydrate in milk, so by breaking down lactose, lactase provides essential services throughout infancy and early childhood. If the intestinal mucosa stops producing lactase by the time of adolescence, the individual becomes *lactose-intolerant.* After ingesting milk and other dairy products, individuals who are lactose-intolerant can have a variety of unpleasant digestive problems.

AM *Lactose Intolerance*

Absorption of Monosaccharides

The intestinal epithelium then absorbs the monosaccharides by facilitated diffusion and cotransport mechanisms (see Figure 3-9•, p. 75). Both methods involve a carrier protein. Facilitated diffusion and cotransport differ in three major ways:

1. *Facilitated diffusion moves only one molecule or ion through the cell membrane, whereas cotransport moves more than one molecule or ion through the membrane at the same time.* In cotransport, the transported materials move in the same direction: down the concentration gradient for at least one of the transported substances.

2. *Facilitated diffusion does not require ATP.* Although cotransport by itself does not consume ATP, the cell must often expend ATP to preserve homeostasis. For example, the process may introduce sodium ions that must later be pumped out of the cell.

3. *Facilitated diffusion will not occur if there is an opposing concentration gradient for the particular molecule or ion.* Cotransport can occur despite an opposing concentration gradient for one of the transported substances. For example, cells lining the small intestine will continue to absorb glucose when glucose concentrations inside the cells are much higher than they are in the intestinal contents.

The cotransport system responsible for the uptake of glucose also brings sodium ions into the cell. This passive process resembles facilitated diffusion except that both a sodium ion and a glucose molecule must bind to the carrier protein before they can move into the cell. Glucose cotransport is an example of sodium-linked cotransport. Comparable cotransport mechanisms exist for other simple sugars and for some amino acids. Although these mechanisms deliver valuable nutrients to the cytoplasm, they also bring in sodium ions that must be ejected by the sodium–potassium exchange pump.

The simple sugars that are transported into the cell at the apical surface diffuse through the cytoplasm and reach the interstitial fluid by facilitated diffusion across the basolateral surfaces. These monosaccharides then diffuse into the capillaries of the villus for eventual transport to the liver in the hepatic portal vein.

LIPID DIGESTION AND ABSORPTION

Lipid digestion involves lingual lipase from glands of the tongue and pancreatic lipase from the pancreas (Figure 24-26b•). The most important and abundant dietary lipids are triglycerides, which consist of three fatty acids attached to a single molecule of glycerol (see Figure 2-15•, p. 49). The lingual and pancreatic lipases break off two of the fatty acids, leaving monoglycerides.

Lipases are water-soluble enzymes, and lipids tend to form large drops that exclude water molecules. As a result, lipases can attack only the exposed surfaces of the lipid drops. Lingual lipase begins breaking down triglycerides in the mouth and stomach (until inactivated), but the lipid drops are so large, and the available time so short, that very little lipid digestion has occurred by the time the chyme enters the duodenum.

Bile salts improve chemical digestion by emulsifying the lipid drops into tiny emulsion droplets, thus providing better access for pancreatic lipase. The emulsification occurs only after the chyme has been mixed with bile in the duodenum. Pancreatic lipase then breaks apart the triglycerides to form a mixture of fatty acids and monoglycerides. As these molecules are released, they interact with bile salts in the surrounding chyme to form small lipid–bile salt complexes called **micelles** (mī-SELZ). A micelle is only about 2.5 nm (0.0025 μm) in diameter.

When a micelle contacts the intestinal epithelium, the lipids diffuse across the cell membrane and enter the cytoplasm. The intestinal cells synthesize new triglycerides from the monoglycerides and fatty acids. These triglycerides, in company with absorbed steroids and phospholipids, are then coated with proteins, creating complexes known as **chylomicrons** (kī-lō-MĪ-kronz; *chylos,* milky lymph + *mikros,* small).

The intestinal cells then secrete the chylomicrons into interstitial fluid by exocytosis. The superficial protein coating of chylomicrons keeps them suspended in the interstitial fluid, but their size generally prevents them from diffusing into capillaries. Most of the chylomicrons released diffuse into the intestinal lacteals, which lack basement membranes and have large gaps between adjacent endothelial cells. From the lacteals they proceed along the lymphatic vessels and through the thoracic duct. Chylomicrons finally enter the bloodstream at the left subclavian vein.

Most of the bile salts within micelles are reabsorbed by sodium-linked cotransport. Only about 5% of the bile salts secreted by the liver enters the colon, and only about 1% is lost in feces.

PROTEIN DIGESTION AND ABSORPTION

Proteins have very complex structures, so protein digestion is both complex and time-consuming. The first problem to overcome is the disruption of the three-dimensional organization of the food so that proteolytic enzymes can attack individual proteins. This step involves mechanical processing in the oral cavity, through mastication, and chemical processing in the stomach, through the action of hydrochloric acid. Exposure of the bolus to a strongly acidic environment breaks down plant cell walls and the connective tissues in animal products and also most pathogens.

The acidic contents of the stomach also provide the proper environment for the activity of pepsin, the proteolytic enzyme secreted by chief cells of the stomach (Figure 24-26c•). Pepsin, which works effectively at a pH of 1.5–2.0, breaks the peptide bonds within a polypeptide chain. When chyme enters the duodenum, enterokinase produced in the small intestine triggers the conversion of trypsinogen to **trypsin,** and the pH is adjusted to 7–8. Pancreatic proteases can now begin working. Trypsin, chymotrypsin, and elastase are like pepsin in that they break specific peptide bonds within a polypeptide. Each enzyme has a different specialty. For example, trypsin breaks peptide bonds involving the amino acids *arginine* or *lysine;* chymotrypsin targets peptide bonds involving *tyrosine* or *phenylalanine.*

Carboxypeptidase also acts in the small intestine. This enzyme chops off the last amino acid of a polypeptide chain, ignoring the identities of the amino acids involved. Thus, while the other peptidases are generating amino acid chains of varying length, carboxypeptidase is producing free amino acids.

Absorption of Amino Acids

The epithelial surfaces of the small intestine contain several peptidases, including **dipeptidases,** enzymes that break short peptide chains into individual amino acids. (Dipeptidases break apart *dipeptides.*) These amino acids, as well as those produced by the pancreatic enzymes, are absorbed through both facilitated diffusion and cotransport mechanisms.

After diffusing to the basal surface of the cell, the amino acids are released into interstitial fluid by facilitated diffusion and cotransport. Once in the interstitial fluid, the amino acids diffuse into intestinal capillaries for transport to the liver by means of the hepatic portal vein.

WATER ABSORPTION

Our cells cannot actively absorb or secrete water. All water movements across the digestive tract lining and the production of glandular secretions, involve passive water flow along osmotic gradients. When two fluids are separated by a selectively permeable membrane, water tends to flow into the solution that has the higher concentration of solutes. ∞ *p. 73* Osmotic movements are rapid, so interstitial fluid and the fluids in the intestinal lumen always have the same osmolarity (osmotic concentration of solutes).

Our epithelial cells continuously absorb nutrients and ions from the intestinal contents; these activities gradually lower the solute concentration in the lumen. As the solute concentration drops, water moves into the surrounding tissues, maintaining osmotic equilibrium.

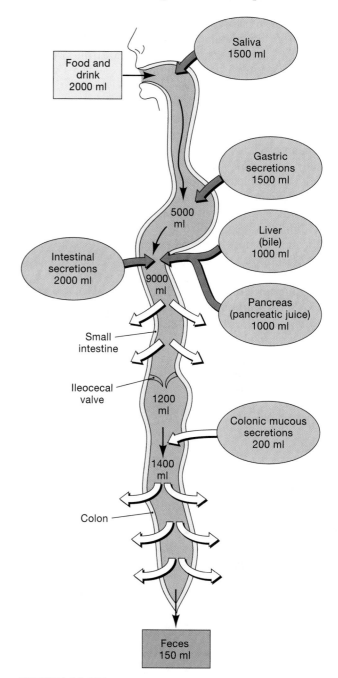

•**FIGURE 24-27** **Digestive Secretion and Absorption of Water.** The white arrows indicate water reabsorption.

24

Each day roughly 2000 ml of water enters your digestive tract in the form of food or drink. The salivary, gastric, intestinal, pancreatic, and bile secretions provide about 7000 ml. Of that total, only about 150 ml is lost in feces. The sites of secretion and absorption of water are indicated in Figure 24-27•.

ION ABSORPTION

Osmosis does not distinguish among solutes; all that matters is the total concentration of solutes. To maintain homeostasis, however, the concentrations of specific ions must be closely regulated. Thus each ion must be handled individually, and the rate of intestinal absorption of each ion must be tightly controlled (Figure 24-28•). Many of the regulatory mechanisms controlling the rates of absorption are poorly understood.

Sodium ions (Na^+), usually the most abundant cations in food, may enter intestinal cells by diffusion, by cotransport with another nutrient, or by active transport. These ions are then pumped into interstitial fluid across the base of the cell. The rate of Na^+ uptake from the lumen is generally proportional to the concentration of Na^+ in the intestinal contents. As a result, eating heavily salted foods leads to increased sodium ion absorption and an associated gain of water through osmosis. The rate of sodium ion absorption by the digestive tract is increased by aldosterone, a steroid hormone from the adrenal cortex. ∞ *p. 601*

Calcium ion (Ca^{2+}) absorption involves active transport at the epithelial surface. The rate of transport is accelerated by parathyroid hormone (PTH) and calcitriol. ∞ *pp. 182, 598, 604*

As other solutes move out of the lumen, the concentration of potassium ions (K^+) increases. These ions can diffuse into the epithelial cells along the concentration gradient. The absorption of magnesium (Mg^{2+}), iron (Fe^{2+}), and other cations involves specific carrier proteins; the cell must use ATP to obtain and transport these ions to interstitial fluid. Regulatory factors controlling the absorption of these cations are poorly understood.

The anions chloride (Cl^-), iodide (I^-), bicarbonate (HCO_3^-), and nitrate (NO_3^-) are absorbed by diffusion or carrier-mediated transport. Phosphate (PO_4^{3-}) and sulfate (SO_4^{2-}) ions enter epithelial cells only by active transport.

VITAMIN ABSORPTION

Vitamins are organic compounds required in very small quantities. There are two major groups of vitamins: fat-soluble vitamins and water-soluble vitamins. Vitamins A, D, E, and K are **fat-soluble vitamins;** their structure allows them to dissolve in lipids. The nine

•**FIGURE 24-28 Ion and Vitamin Absorption by the Digestive Tract**

water-soluble vitamins include the B vitamins, common in milk and meats, and vitamin C, found in citrus fruits. In Chapter 25, we shall consider the functions of vitamins and associated nutritional problems.

All but one of the water-soluble vitamins are easily absorbed by diffusion across the digestive epithelium. Vitamin B_{12} cannot be absorbed by the intestinal mucosa in normal amounts unless this vitamin has been bound to intrinsic factor, a glycoprotein secreted by the parietal cells of the stomach (p. 864). The combination is then absorbed via active transport.

Fat-soluble vitamins in the diet enter the duodenum in fat droplets, mixed with triglycerides. The vitamins remain in association with these lipids as they form emulsion droplets and, after further digestion, micelles. The fat-soluble vitamins are then absorbed from the micelles along with the fatty acids and monoglycerides. Vitamin K produced in the colon is absorbed with other lipids released through bacterial action. Taking supplements of fat-soluble vitamins while you have an empty stomach, are fasting, or are on a low-fat diet will be relatively ineffective, because proper absorption requires the presence of other lipids.

Figure 24-28• summarizes the digestive fates of the major electrolytes and vitamins. Review this figure carefully to become familiar with the outlined mechanisms and events. In Chapter 25, we shall examine the interplay between the respiratory and digestive systems.

Malabsorption Syndromes

Malabsorption is a disorder characterized by abnormal nutrient absorption. Difficulties in the absorption of all classes of compounds will result from damage to the accessory glands or the intestinal mucosa. If the accessory organs are functioning normally but their secretions cannot reach the duodenum, the condition is called *biliary obstruction* (bile duct blockage) or *pancreatic obstruction* (pancreatic duct blockage). Alternatively, the ducts may remain open but the glandular cells may be damaged and unable to continue normal secretory activities. We noted two examples, pancreatitis and cirrhosis, earlier in this chapter.

Even with the normal enzymes present in the lumen, absorption will not occur if the mucosa cannot function properly. A genetic inability to manufacture specific enzymes will result in discrete patterns of malabsorption; lactose intolerance (p. 890) is a good example. Mucosal damage due to ischemia, radiation exposure, toxic compounds, or infection will adversely affect absorption and as a result will deplete nutrient and fluid reserves.

AGING AND THE DIGESTIVE SYSTEM

Essentially normal digestion and absorption occur in elderly individuals. However, many changes occur in the digestive system that parallel age-related changes we have already described for other systems:

■ *The division rate of epithelial stem cells declines.* The digestive epithelium becomes more susceptible to damage by abrasion, acids, or enzymes. Peptic ulcers therefore become more likely. Stem cells in the epithelium divide less frequently with age, so tissue repair is less efficient. In the mouth, esophagus, and anus, the stratified epithelium becomes thinner and more fragile.

■ *Smooth muscle tone decreases.* General motility decreases, and peristaltic contractions are weaker as a result of a decrease in smooth muscle tone. These changes slow the rate of fecal movement and promote constipation. Sagging and inflammation of the haustra in the colon can produce symptoms of diverticulosis (p. 884). Straining to eliminate compacted fecal materials can stress the less-resilient walls of blood vessels, producing hemorrhoids. Problems are not restricted to the lower digestive tract. For example, weakening of muscular sphincters can lead to esophageal reflux and frequent bouts of "heartburn."

■ *The effects of cumulative damage become apparent.* A familiar example is the gradual loss of teeth due to dental caries or gingivitis. Cumulative damage can involve internal organs as well. Toxins such as alcohol and other injurious chemicals that are absorbed by the digestive tract are transported to the liver for processing. The cells of the liver are not immune to these toxic compounds, and chronic exposure can lead to cirrhosis or other types of liver disease.

■ *Cancer rates increase.* Cancers are most common in organs in which stem cells divide to maintain epithelial cell populations. Rates of colon cancer and stomach cancer rise with age; oral and pharyngeal cancers are particularly common among elderly smokers.

■ *Changes in other systems have direct or indirect effects on the digestive system.* For example, the reduction in bone mass and calcium content in the skeleton is associated with erosion of the tooth sockets and eventual tooth loss. The decline in olfactory and gustatory sensitivities with age can lead to dietary changes that affect the entire body.

☑ What component of food would increase the number of chylomicrons in the lacteals?

☑ The absorption of which vitamin would be impaired by the removal of the stomach?

☑ Why is diarrhea potentially life-threatening but constipation is not?

INTEGRATION WITH OTHER SYSTEMS

The digestive system is functionally linked to all other systems. It has extensive anatomical connections to the nervous, cardiovascular, endocrine, and lymphatic systems. Figure 24-29• summarizes the physiological relationships between the digestive system and other organ systems.

24

INTEGUMENTARY SYSTEM

Provides vitamin D₃ needed for the absorption of calcium and phosphorus

Provides lipids for storage by adipocytes in subcutaneous layer

SKELETAL SYSTEM

Skull, ribs, vertebrae, and pelvic girdle support and protect parts of digestive tract; teeth important in mechanical processing of food

Absorbs calcium and phosphate ions for incorporation into bone matrix; provides lipids for storage in yellow marrow

MUSCULAR SYSTEM

Protects and supports digestive organs in abdominal cavity; controls entrances and exits to digestive tract; liver metabolizes lactic acid from active muscles

Liver regulates blood glucose and fatty acid levels, removes lactic acid from bloodstream

NERVOUS SYSTEM

ANS regulates movement and secretion; reflexes coordinate passage of materials along tract; control over skeletal muscles regulates ingestion and defecation; hypothalamic centers control hunger, satiation, and feeding behaviors

Provides substrates essential for neurotransmitter synthesis

ENDOCRINE SYSTEM

Epinephrine and norepinephrine stimulate constriction of sphincters and depress digestive activity; hormones coordinate activity along tract

Provides nutrients and substrates to endocrine cells; endocrine cells of pancreas secrete insulin and glucagon; liver produces angiotensinogen

THE DIGESTIVE SYSTEM

FOR ALL SYSTEMS

Absorbs organic substrates, vitamins, ions, and water required by all cells

CARDIOVASCULAR SYSTEM

Distributes hormones of the digestive tract; carries nutrients, water, and ions from sites of absorption; delivers nutrients and toxins to liver

Absorbs fluid to maintain normal blood volume; absorbs vitamin K; liver excretes heme (as bilirubin), synthesizes coagulation proteins

LYMPHATIC SYSTEM

Tonsils and MALT defend against infection; MALT defends against toxins absorbed from the tract; lymphatic vessels carry absorbed lipids to venous system

Secretions of digestive tract (acids and enzymes) provide nonspecific defense against pathogens

RESPIRATORY SYSTEM

Increased thoracic and abdominal pressures through contraction of respiratory muscles can assist in defecation

Pressure of digestive organs against the diaphragm can assist exhalation and limit inhalation

REPRODUCTIVE SYSTEM

Provides additional nutrients required to support gamete production and (in pregnant women) embryonic and fetal development

URINARY SYSTEM

Excretes toxins absorbed by the digestive epithelium; excretes some bilirubin produced by liver

Absorbs water needed to excrete waste products at the kidneys; absorbs ions needed to maintain normal body fluid concentrations

24

•**FIGURE 24-29** Functional Relationships between the Digestive System and Other Systems

[EXPLORE]
MediaLab

Food drives our culture; we are constantly subjected to advertisements promoting the taste, aroma, and nutrient content of food. Our enjoyment of food depends on the proper functioning of our sensory mechanisms and the digestive system. To gain a better understanding of the anatomy and physiology of the digestive system, use the resources available on both the CD-ROM and the Companion Website. The CD-ROM includes animations and tutorials; the website builds on those resources with text content and interactive problems as well as communication avenues for sharing ideas with your class. To complete the following Web Explorations, visit the Media-Lab in Chapter 24 of your Companion Website.

Web Exploration 1

Estimated time for completion = 5 minutes

Although the gastrointestinal tract is very long, it includes relatively few organs. Viewing these organs on paper is not quite the same as seeing them intact in an organism. To test

your ability to identify the digestive organs in situ, select the keyword **GI QUIZ** on the Companion Website. Match the organs that are indicated with the proper term, and check your grade. Then take the multiple-choice quiz, using the "test yourself" link. How good are you at transferring the information from your text to other images?

Web Exploration 2

Estimated time for completion = 10 minutes

Hepatitis can be a serious threat to public health. The name *hepatitis* literally means inflammation of the liver. This disease has several variations: A, B, C, D, E, and G. To learn more about the differences among these viruses, select the keyword **HEPATITIS** on the Companion Website. Read the brief introduction, and then move through the first four pages in this site. Create a table depicting the characteristics of each type of hepatitis. Include modes of transmission, symptoms, and any other information you can find.

CHAPTER REVIEW

SELECTED CLINICAL TERMINOLOGY
Terms Discussed in This Chapter

ascites (a-SĪ-tēz): Fluid leakage into the peritoneal cavity across the serous membranes of the liver and viscera. *(p. 851)*

cathartics (ka-THAR-tiks): Drugs that promote defecation. *(p. 887)*

cholecystitis (kō-lē-sis-TĪ-tis): An inflammation of the gallbladder due to a blockage of the cystic or common bile duct by gallstones. *(p. 880 and AM)*

cholelithiasis (kō-lē-li-THĪ-a-sis): The presence of gallstones in the gallbladder. *(p. 880 and AM)*

cholera (KOL-e-ruh): A bacterial infection of the digestive tract that causes massive fluid losses through diarrhea. *(p. 887 and AM)*

cirrhosis (sir-Ō-sis): A disease characterized by the widespread destruction of hepatocytes by exposure to drugs (especially alcohol), viral infection, ischemia, or blockage of the hepatic ducts. *(pp. 860, 878 and AM)*

colitis (kō-LĪ-tis): A general term for a condition characterized by an inflammation of the colon. *(p. 884 and AM)*

constipation: Infrequent defecation, generally involving dry, hard feces. *(p. 887 and AM)*

diarrhea (dī-a-RĒ-uh): Frequent, watery bowel movements. *(p. 887 and AM)*

diverticulitis (dī-ver-tik-ū-LĪ-tis): An infection and inflammation of mucosal pockets of the large intestine (diverticula). *(p. 884 and AM)*

diverticulosis (dī-ver-tik-ū-LŌ-sis): The formation of diverticula, generally along the sigmoid colon. *(p. 884)*

esophageal varices (VAR-i-sēz): Swollen and fragile esophageal veins that result from portal hypertension. *(p. 860)*

gallstones: Deposits of minerals, bile salts, and cholesterol that form if bile becomes too concentrated. *(p. 880 and AM)*

gastrectomy (gas-TREK-to-mē): The surgical removal of the stomach, generally to treat advanced stomach cancer. *(p. 868)*

gastritis (gas-TRĪ-tis): An inflammation of the gastric mucosa. *(p. 865)*

gastroscope: A fiber-optic instrument inserted into the mouth and directed along the esophagus and into the stomach; used to examine the interior of the stomach and to perform minor surgical procedures. *(p. 868)*

hepatitis (hep-a-TĪ-tis): A virus-induced disease of the liver; forms include *hepatitis A, B,* and *C. (p. 878 and AM)*

lactose intolerance: A malabsorption syndrome that results from the lack of the enzyme *lactase* at the brush border of the intestinal epithelium. *(p. 890 and AM)*

mumps: A viral infection that in children tends to focus in the salivary glands, primarily the parotid. *(p. 855)*

pancreatitis (pan-krē-a-TĪ-tis): An inflammation of the pancreas. *(p. 874)*

peptic ulcer: Erosion of the gastric lining or duodenal lining by stomach acids and enzymes. *(p. 865 and AM)*

periodontal disease: A loosening of the teeth within the alveolar sockets caused by erosion of the periodontal ligaments by acids produced through bacterial action. *(p. 858)*

peritonitis (per-i-tō-NĪ-tis): An inflammation of the peritoneal membrane. *(p. 851)*

portal hypertension: High venous pressures in the hepatic portal system. *(p. 877)*

pulpitis (pul-PĪ-tis): An infection of the pulp of a tooth; may be treated by a *root canal* procedure. *(p. 858)*

24

Additional Terms Discussed in the *Applications Manual*

achalasia (ak-a-LĀ-zē-uh): A condition that results when a bolus cannot reach the stomach due to the constriction of the lower esophageal sphincter.

colectomy (kō-LEK-to-mē): The removal of all or a portion of the colon.

colonoscope (kō-LON-ō-skōp): A fiber-optic instrument inserted into the anus and moved into the colon to look for problems in the rectum, sigmoid colon, and ascending colon.

colostomy (kō-LOS-to-mē): The attachment of the cut end of the colon to an opening in the body wall after a colectomy.

esophagitis (ē-sof-a-JĪ-tis): An inflammation of the esophagus.

gastric stapling: The surgical reduction in the size of the stomach by stapling closed a portion of the gastric lumen; an attempt to correct an overeating problem.

gastroenteritis (gas-trō-en-ter-Ī-tis): A condition characterized by vomiting and diarrhea and resulting from bacterial toxins, viral infections, or various poisons.

giardiasis (jē-ar-DĪ-a-sis): An infection of the digestive tract by the protozoan *Giardia intestinalis*. Some infected individuals do not develop symptoms, but others develop violent diarrhea, abdominal pains, cramps, nausea, and vomiting.

ileostomy (il-ē-OS-to-mē): The attachment of the cut end of the ileum to an opening in the body wall after a colectomy.

inflammatory bowel disease *(ulcerative colitis):* A chronic inflammation of the digestive tract, most commonly affecting the colon.

irritable bowel syndrome: A disorder characterized by diarrhea, constipation, or both alternately. When constipation is the primary problem, this condition may be called a *spastic colon* or *spastic colitis.*

laparoscopy (lap-a-ROS-ko-pē): The use of a flexible fiber-optic instrument introduced through the abdominal wall to permit direct visualization of the viscera, tissue sampling, and limited surgical procedures.

liver biopsy: A sample of liver tissue, generally taken through the anterior abdominal wall by means of a long needle.

liver scans: Radiological procedures that involve the injection of radioisotope labeled compounds that will be selectively absorbed by Kupffer cells, liver cells, or abnormal liver tissues.

perforated ulcer: A particularly dangerous ulcer in which the gastric acids erode through the wall of the digestive tract and enter the peritoneal cavity.

polyps (PAH-lips): Small mucosal tumors that grow from the intestinal wall.

STUDY OUTLINE

INTRODUCTION   p. 846

1. The **digestive system** consists of the muscular **digestive tract** and various **accessory organs.**
2. Digestive functions are **ingestion, mechanical processing, digestion, secretion, absorption,** and **excretion.**

AN OVERVIEW OF THE STRUCTURE AND FUNCTION OF THE DIGESTIVE TRACT   p. 846

1. The digestive tract consists of the *oral cavity, pharynx, esophagus, stomach, small intestine, large intestine,* and *anus. (Figure 24-1)*

Histological Organization of the Digestive Tract   p. 846

2. The digestive tract is lined by a mucous epithelium moistened by glandular secretions of the epithelial and accessory organs.
3. The *lamina propria* and epithelium form the **mucosa** (mucous membrane) of the digestive tract. Proceeding outward, we encounter the **submucosa,** the **muscularis externa,** and a layer of areolar tissue called the *adventitia.* Within the peritoneal cavity, the muscularis externa is covered by a serous membrane called the **serosa.** *(Figure 24-2)*

The Movement of Digestive Materials   p. 849

4. Many smooth muscle cells are not innervated by motor neurons, and the neurons that do innervate smooth muscles are not under voluntary control.
5. The muscularis externa propels materials through the digestive tract by the contractions of **peristalsis. Segmentation** movements in the small intestine churn digestive materials. *(Figure 24-3)*

The Control of Digestive Function   p. 849

6. Digestive tract activities are controlled by neural, hormonal, and local mechanisms. *(Figure 24-4)*

The Peritoneum   p. 851

7. Double sheets of peritoneal membrane called **mesenteries** suspend the digestive tract. The **greater omentum** lies anterior to the abdominal viscera. Its adipose tissue provides padding, protection, insulation, and an energy reserve. *(Figure 24-5)*

THE ORAL CAVITY   p. 853

1. The functions of the **oral cavity,** or **buccal cavity,** are (1) *analysis* of foods; (2) *mechanical processing* by the teeth, tongue, and palatal surfaces; (3) *lubrication* by mixing with mucus and salivary secretions; and (4) initiating the *digestion* of carbohydrates and lipids.
2. The oral cavity is lined by **oral mucosa.** The **hard** and **soft palates** form its roof, and the **tongue** forms its floor. *(Figure 24-6)*

The Tongue   p. 854

3. **Intrinsic** and **extrinsic tongue muscles** are controlled by the hypoglossal nerve. *(Figure 24-6)*

Salivary Glands   p. 854

4. The **parotid, sublingual,** and **submandibular salivary glands** discharge their secretions into the oral cavity. *(Figure 24-7)*

The Teeth   p. 856

5. **Mastication** (chewing) of the **bolus** occurs through the contact of the **occlusal** (opposing) **surfaces** of the **teeth.** The **periodontal ligament** anchors each tooth in an *alveolus,* or bony socket. **Dentin** forms the basic structure of a tooth. The **crown** is coated with **enamel,** and the **root** with **cementum.** *(Figure 24-8)*
6. The 20 primary teeth, or **deciduous teeth,** are replaced by the 32 teeth of the **secondary dentition** during development. *(Figure 24-9)*

THE PHARYNX   p. 858

1. Propulsion of the bolus through the **pharynx** results from the contractions of the *pharyngeal constrictor muscles* and the *palatal muscles* and from elevation of the larynx.

THE ESOPHAGUS   p. 858

1. The **esophagus** carries solids and liquids from the pharynx to the stomach through the **esophageal hiatus,** an opening in the diaphragm.

Histology of the Esophagus   p. 860

2. The esophageal mucosa consists of a stratified epithelium. Mucous secretion by esophageal glands of the submucosa reduces friction during the passage of foods. The proportions of skeletal and smooth muscle of the muscularis externa change from the pharynx to the stomach. *(Figure 24-10)*

Swallowing   p. 860

3. Swallowing **(deglutition)** can be divided into **buccal, pharyngeal,** and **esophageal phases.** Swallowing begins with the compaction of a bolus and its movement into the pharynx, followed by the elevation of the larynx, reflection of the *epiglottis,* and closure of the *glottis.* After the *upper esophageal sphincter* is opened, peristalsis moves the bolus down the esophagus to the *lower esophageal sphincter.* *(Figure 24-11)*

THE STOMACH   p. 861

1. The **stomach** has four major functions: the (1) bulk storage of ingested food, (2) mechanical breakdown of food, (3) disruption of chemical bonds via acids and enzymes, and (4) production of *intrinsic factor.*

Anatomy of the Stomach   p. 861

2. The four regions of the stomach are the **cardia, fundus, body,** and **pylorus.** The **pyloric sphincter** guards the exit from the stomach. In a relaxed state, the stomach lining contains numerous **rugae** (ridges and folds). *(Figure 24-12)*

3. Within the **gastric glands, parietal cells** secrete **intrinsic factor** and *hydrochloric acid.* **Chief cells** secrete **pepsinogen,** which is converted by acids in the gastric lumen to the enzyme **pepsin. Enteroendocrine** cells of the stomach secrete several compounds, notably the hormone **gastrin.** *(Figures 24-13, 24-14)*

Regulation of Gastric Activity   p. 865

4. Gastric secretion consists of (1) the **cephalic phase,** which prepares the stomach to receive ingested materials; (2) the **gastric phase,** which begins with the arrival of food in the stomach; and (3) the **intestinal phase,** which controls the rate of gastric emptying. *(Figure 24-15)*

Digestion and Absorption in the Stomach   p. 868

5. The preliminary digestion of proteins by pepsin begins in the stomach. Very little nutrient absorption occurs in the stomach.

THE SMALL INTESTINE AND ASSOCIATED GLANDULAR ORGANS   p. 868

1. Most of the important digestive and absorptive functions occur in the **small intestine.** Digestive secretions and buffers are provided by the pancreas, liver, and gallbladder.

The Small Intestine   p. 869

2. The small intestine consists of the **duodenum,** the **jejunum,** and the **ileum.** A sphincter, the **ileocecal valve,** marks the transition between the small and large intestines. *(Figure 24-16)*

3. The intestinal mucosa bears transverse folds called **plicae** and small projections called **intestinal villi.** These folds and projections increase the surface area for absorption. Each villus contains a terminal lymphatic called a **lacteal.** Pockets called **intestinal glands** are lined by enteroendocrine, goblet, and stem cells. *(Figure 24-17)*

4. The **submucosal (Brunner's) glands** of the duodenum produce mucus, buffers, and the hormone **urogastrone.** The ileum contains masses of lymphoid tissue called *aggregate lymphoid nodules,* or *Peyer's patches,* near the entrance to the large intestine.

5. **Intestinal juice** moistens chyme, helps buffer acids, and holds digestive enzymes and digestive products in solution.

Intestinal Movements   p. 872

6. Some of the smooth muscle cells contract periodically, without stimulation, to produce a pattern of peristaltic contractions that move the intestinal contents.

The Pancreas   p. 872

7. The **pancreatic duct** penetrates the wall of the duodenum. Within each lobule of the **pancreas,** ducts branch repeatedly before ending in the **pancreatic acini** (blind pockets). *(Figure 24-18)*

8. The pancreas has two functions: endocrine (secreting insulin and glucagon into the blood) and exocrine (secreting water, ions, and digestive enzymes into the small intestine). Pancreatic enzymes include **carbohydrases, lipases, nucleases,** and **proteolytic enzymes.**

The Liver   p. 874

9. The **liver** performs metabolic and hematological regulation and produces **bile.** The bile ducts from all the **liver lobules** unite to form the **common hepatic duct.** That duct meets the **cystic duct** to form the **common bile duct,** which empties into the duodenum. *(Figure 24-19)*

10. The liver lobule is the organ's basic functional unit. **Hepatocytes** form irregular plates arranged as wheel spokes. **Bile canaliculi** carry bile to the **bile ductules,** which lead to **portal areas.** *(Figure 24-20)*

11. In **emulsification, bile salts** break apart large drops of lipids, making them accessible to lipases secreted by the pancreas.

The Gallbladder   p. 879

12. The **gallbladder** stores and concentrates bile. *(Figure 24-21)*

The Coordination of Secretion and Absorption   p. 880

13. Neural and hormonal mechanisms coordinate the activities of the digestive glands. Gastrointestinal activity is stimulated by parasympathetic innervation and inhibited by sympathetic innervation. The **enterogastric, gastroenteric,** and **gastroileal reflexes** coordinate movement from the stomach to the large intestine.

14. Intestinal hormones include **enterocrinin, secretin, cholecystokinin (CCK), gastric inhibitory peptide (GIP), vasoactive intestinal peptide (VIP),** and **gastrin.** *(Figure 24-22; Table 24-1)*

THE LARGE INTESTINE   p. 882

1. The main functions of the **large intestine** are to (1) reabsorb water and compact materials into feces, (2) absorb vitamins produced by bacteria, and (3) store fecal material prior to defecation. The large intestine consists of the *cecum, colon,* and *rectum.* *(Figure 24-23)*

The Cecum   p. 882

2. The **cecum** collects and stores material from the ileum and begins the process of compaction. The **vermiform appendix** is attached to the cecum. *(Figure 24-23)*

The Colon   p. 882

3. The **colon** has a larger diameter and a thinner wall than the small intestine. It bears **haustra** (pouches) and the **taenia coli** (longitudinal bands of muscle). *(Figure 24-23)*

24

The Rectum p. 884

4. The **rectum** terminates in the **anorectal canal** leading to the **anus.** *(Figure 24-23)*

Histology of the Large Intestine p. 884

5. Characteristics of the colon include an absence of villi and the presence of goblet cells and deep intestinal glands. *(Figure 24-24)*

Physiology of the Large Intestine p. 885

6. The large intestine reabsorbs water and other substances such as vitamins, urobilinogen, bile salts, and toxins. Bacteria are responsible for intestinal gas, or **flatus.**

7. The gastroileal reflex moves materials from the ileum into the cecum while you eat. Distension of the stomach and duodenum stimulates **mass movements** of materials from the transverse colon through the rest of the large intestine and into the rectum. Muscular sphincters control the passage of fecal materal to the anus. Distension of the rectal wall triggers the **defecation reflex.** *(Figure 24-25)*

DIGESTION AND ABSORPTION p. 887

The Processing and Absorption of Nutrients p. 887

1. The digestive system breaks down the physical structure of the ingested material and then disassembles the component molecules into smaller fragments by *hydrolysis.* *(Figure 24-26; Table 24-2)*

Carbohydrate Digestion and Absorption p. 888

2. Salivary and pancreatic amylases break down complex carbohydrates into *disaccharides* and *trisaccharides.* These are broken down into *monosaccharides* by enzymes at the epithelial surface. The monosaccharides are then absorbed by the intestinal epithelium by facilitated diffusion or cotransport. *(Figure 24-26)*

Lipid Digestion and Absorption p. 890

3. *Triglycerides* are emulsified into lipid droplets. The resulting fatty acids and *monoglycerides* interact with bile salts to form **micelles,** from which they diffuse across the intestinal epithelium. *(Figure 24-26)*

Protein Digestion and Absorption p. 890

4. Protein digestion involves the gastric enzyme pepsin and the various pancreatic proteases. Peptidases liberate amino acids that are absorbed and exported to interstitial fluid. *(Figure 24-26)*

Water Absorption p. 891

5. About 2000 ml of water is ingested each day, and digestive secretions provide about 7000 ml. Nearly all is reabsorbed by osmosis. *(Figure 24-27)*

Ion Absorption p. 891

6. Various processes such as diffusion, cotransport, and carrier-mediated and active transports are responsible for the movements of cations (sodium, calcium, potassium, and so on) and anions (chloride, iodide, bicarbonate, and so on) into epithelial cells. *(Figure 24-28)*

Vitamin Absorption p. 892

7. The **water-soluble vitamins** (except for B_{12}) easily diffuse across the digestive epithelium. **Fat-soluble vitamins** are enclosed within fat droplets and are absorbed with the products of lipid digestion. *(Figure 24-28)*

AGING AND THE DIGESTIVE SYSTEM p. 893

1. Age-related changes include a thinner and more fragile epithelium due to a reduction in epithelial stem cell divisions, weaker peristaltic contractions as smooth muscle tone decreases, effects of cumulative damage, increased cancer rates, and related changes in other systems.

INTEGRATION WITH OTHER SYSTEMS p. 893

1. The digestive system has extensive anatomical connections to the nervous, cardiovascular, endocrine, and lymphatic systems. *(Figure 24-29)*

REVIEW QUESTIONS

LEVEL 1 REVIEWING FACTS AND TERMS

1. The enzymatic breakdown of large molecules into their basic building blocks is
   (a) absorption
   (b) secretion
   (c) mechanical digestion
   (d) chemical digestion

2. The inner layer of the digestive tract is the
   (a) mucosa   (b) submucosa
   (c) serosa   (d) muscularis

3. The muscularis externa propels materials from one portion of the digestive tract to another by the contractions of
   (a) segmentation   (b) propulsion
   (c) mass movements (d) peristalsis

4. The activities of the digestive system are regulated by
   (a) hormonal mechanisms
   (b) local mechanisms
   (c) neural mechanisms
   (d) a, b, and c are correct

5. Double sheets of peritoneum that provide support and stability for the organs of the peritoneal cavity are
   (a) mediastinums   (b) mucous membranes
   (c) omenta   (d) mesenteries

6. The layer of the peritoneum that lines the inner surfaces of the body wall is the
   (a) visceral peritoneum   (b) parietal peritoneum
   (c) greater omentum   (d) lesser omentum

7. The peritoneal fold that stabilizes and supports the small intestine is the
   (a) serosa   (b) lesser omentum
   (c) mesentery   (d) parietal peritoneum

8. Intrinsic factor and hydrochloric acid are secreted by cells in the stomach wall called
   (a) parietal cells (b) chief cells
   (c) acinar cells   (d) G cells

9. Protein digestion in the stomach results primarily from secretions released by
   (a) G cells   (b) parietal cells
   (c) chief cells   (d) D cells

10. The part of the digestive tract that plays the primary role in the digestion and absorption of nutrients is, or are, the
    (a) large intestine (b) small intestine
    (c) stomach   (d) cecum and colon

11. The duodenal hormone that stimulates the production and secretion of pancreatic enzymes is
    (a) enterokinase   (b) gastrin
    (c) secretin   (d) cholecystokinin

12. The essential metabolic or synthetic service provided by the liver is
   (a) metabolic regulation
   (b) hematological regulation
   (c) bile production
   (d) a, b, and c are correct

13. Bile release from the gallbladder into the duodenum occurs only under the stimulation of
   (a) cholecystokinin   **(b)** secretin
   (c) gastrin   **(d)** enterokinase

14. The major function(s) of the large intestine is (are) the
   (a) reabsorption of water and compaction of feces
   (b) absorption of vitamins liberated by bacterial action
   (c) storage of fecal material prior to defecation
   (d) a, b, and c are correct

15. The part of the colon that empties into the rectum is the
   (a) ascending colon   **(b)** descending colon
   (c) transverse colon   **(d)** sigmoid colon

16. Three vitamins generated by bacteria in the colon are
   (a) vitamins A, D, and E
   (b) B complex vitamins and vitamin C
   (c) vitamin K, biotin, and pantothenic acid
   (d) niacin, thiamine, and riboflavin

17. The final enzymatic steps in the digestive process are accomplished by
   (a) brush border enzymes of the microvilli
   (b) enzymes secreted by the stomach
   (c) enzymes secreted by the pancreas
   (d) the action of bile from the gallbladder

18. What are the six steps involved in digestion?

19. What is the purpose of the transverse or longitudinal folds of the epithelium in the digestive tract?

20. Name and describe the layers of the digestive tract, proceeding from the innermost to the outermost layer.

21. What three basic mechanisms regulate the activities of the digestive tract?

22. What is the relationship among the parietal peritoneum, visceral peritoneum, and mesenteries?

23. What are the four primary functions of the oral cavity?

24. List the three pairs of salivary glands, and identify the ducts used by their glandular secretions.

25. What are the functions of saliva, and how are salivary secretions controlled?

26. What specific function does each of the four types of teeth perform in the oral cavity?

27. What role do the pharyngeal muscles play in swallowing?

28. What are the three phases of swallowing, and how are they controlled?

29. What contributions do the gastric and pyloric glands of the stomach make to the digestive process?

30. What three subdivisions of the small intestine are involved in the digestion and absorption of food?

31. What are the primary functions of the pancreas, liver, and gallbladder in digestion?

32. Which hormones produced by duodenal enteroendocrine cells effectively coordinate digestive functions?

33. What are the three primary functions of the large intestine?

34. What two positive feedback loops are involved in the defecation reflex?

35. Describe five age-related changes that occur in the digestive system.

LEVEL 2 REVIEWING CONCEPTS

36. If the lingual frenulum is too restrictive, an individual
   (a) has difficulty tasting food
   (b) cannot swallow properly
   (c) cannot control movements of the tongue
   (d) cannot eat or speak normally

37. Increased secretion by all the salivary glands results from
   (a) sympathetic stimulation
   (b) hormonal stimulation
   (c) parasympathetic stimulation
   (d) myenteric reflexes

38. The production of acid and enzymes by the gastric mucosa is controlled and regulated by
   (a) the central nervous system
   (b) short reflexes coordinated in the stomach wall
   (c) digestive tract hormones
   (d) a, b, and c are correct

39. The gastric phase of secretion is initiated by
   (a) distension of the stomach
   (b) an increase in the pH of the gastric contents
   (c) the presence of undigested materials in the stomach
   (d) a, b, and c are correct

40. Chyme reaches the small intestine most quickly when
   (a) the stomach is greatly distended
   (b) a large amount of gastrin is released
   (c) the meal contains a relatively small quantity of proteins
   (d) a and c are correct

41. A drop in pH below 4.5 in the duodenum stimulates the secretion of
   (a) secretin   **(b)** cholecystokinin
   (c) gastrin   **(d)** a, b, and c are correct

42. If you wanted to nip the tip off a carrot stick, which teeth would you most likely use?

43. Differentiate peristalsis and segmentation by their actions and outcomes.

44. How does the stomach facilitate and assist in digestion?

45. How do the three phases of gastric secretion promote and facilitate gastric control?

46. Why should a person who has stomach ulcers avoid aspirin?

47. Nutritionists have found that after a heavy meal, the pH of blood increases slightly, especially in the veins that carry blood away from the stomach. What causes this "postenteric alkaline tide"?

LEVEL 3 CRITICAL THINKING AND CLINICAL APPLICATIONS

48. Some persons with gallstones develop pancreatitis. How could this occur?

49. Barb suffers from Crohn's disease, a regional inflammation of the intestine. This disease is thought to have some genetic basis, but the actual cause is as yet unknown. When the disease flares up, she experiences abdominal pain, weight loss, and anemia. Which part(s) of the intestine is (are) probably involved, and what is the cause of her symptoms?

50. What symptoms would you expect to observe in a person whose small intestine is blocked at the level of the jejunum?

24

CHAPTER 25

Metabolism and Energetics

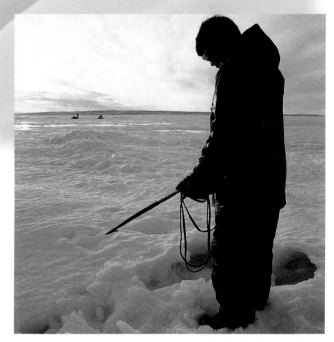

Humans are remarkably adaptable animals; we can maintain homeostasis near the poles, where the temperature remains below 0°C (32°F) most of the time. The key to survival under varied environmental conditions is the preservation of normal body temperature. In part, we can accomplish this by shedding or adding layers of clothing, but hormonal and metabolic adjustments are also crucial. Thus, our metabolic rate rises in cold climates, generating additional heat that warms the body. This chapter will examine several aspects of metabolism, including nutrition, energy use, heat production, and the regulation of body temperature under normal and abnormal conditions.

CHAPTER OUTLINE AND OBJECTIVES

INTRODUCTION p. 901

AN OVERVIEW OF METABOLISM p. 901
① Define *metabolism*, and explain why cells need to synthesize new organic components.

CARBOHYDRATE METABOLISM p. 903
  Glycolysis p. 904
  Mitochondrial ATP Production p. 904
  Energy Yield of Glycolysis and Cellular Respiration p. 910
  Other Catabolic Pathways p. 911
  Gluconeogenesis p. 912
② Describe the basic steps in glycolysis, the TCA cycle, and the electron transport chain.
③ Summarize the energy yield of glycolysis and cellular respiration.

LIPID METABOLISM p. 913
  Lipid Catabolism p. 913
  Lipid Synthesis p. 914
  Lipid Transport and Distribution p. 914
④ Describe the pathways involved in lipid metabolism and the mechanisms necessary for lipid transport and distribution.

PROTEIN METABOLISM p. 916
  Amino Acid Catabolism p. 916
  Protein Synthesis p. 919
⑤ Discuss protein metabolism and the use of proteins as an energy source.

NUCLEIC ACID METABOLISM p. 919
  RNA Catabolism p. 919
  Nucleic Acid Synthesis p. 920
⑥ Discuss nucleic acid metabolism.

METABOLIC INTERACTIONS p. 920
  The Absorptive State p. 921
  The Postabsorptive State p. 923
⑦ Differentiate between the absorptive and postabsorptive metabolic states, and summarize the characteristics of each.

DIET AND NUTRITION p. 926
  Food Groups and Food Pyramids p. 926
  Nitrogen Balance p. 926
  Minerals p. 928
  Vitamins p. 929
  Diet and Disease p. 931
⑧ Explain what constitutes a balanced diet and why such a diet is important.

BIOENERGETICS p. 932
  Food and Energy p. 933
  Metabolic Rate p. 933
  Thermoregulation p. 933
⑨ Define *metabolic rate*, and discuss the factors involved in determining an individual's BMR.
⑩ Discuss the homeostatic mechanisms that maintain a constant body temperature.

MEDIALAB and CHAPTER REVIEW pp. 937, 938

C ells are chemical factories that break down organic molecules to obtain energy. This energy can be used to generate ATP. Reactions within mitochondria provide most of the energy needed by a typical cell. ∞ *pp. 87–89, 293* To carry out these metabolic reactions, our cells must have a reliable supply of oxygen and nutrients, including water, vitamins, mineral ions, and organic substrates. (Recall that *substrates* are the reactants in enzymatic reactions. ∞ *p. 54*) Oxygen is absorbed at the lungs; the other substances are obtained by absorption at the digestive tract. The cardiovascular system ensures prompt distribution of all those substances to interstitial fluids throughout the body, where they become accessible to our cells.

The energy released inside the cell supports growth, cell division, contraction, secretion, and a variety of other special functions that vary from cell to cell and tissue to tissue. Because different tissue types contain different populations of cells, the energy and nutrient requirements of any two tissues, such as loose connective tissue and cardiac muscle, are quite different. When cells, tissues, and organs modify their patterns of activity, the metabolic needs of the body change. There are short-term and long-term changes. For example, nutrient requirements vary from moment to moment (resting versus active), hour to hour (asleep versus awake), and year to year (growing child versus adult).

Differentiation creates tissues and organs that are specialized to store excess nutrients; the storage of lipids in adipose tissue is one familiar example. When nutrients are abundant, energy reserves are built up. These reserves, generally glycogen deposits or lipid inclusions, can then be called on whenever the digestive tract cannot provide the right quantity or quality of nutrients. The endocrine system, with the assistance of the nervous system, adjusts and coordinates the metabolic activities of the body's tissues and controls the storage and mobilization of nutrient reserves.

AN OVERVIEW OF METABOLISM

The term **metabolism** (me-TAB-ō-lizm) refers to all chemical reactions that occur in an organism. At the cellular level, chemical reactions provide the energy needed to maintain homeostasis and to perform essential functions. The intracellular reactions constitute *cellular metabolism*. The essential functions performed by this process include (1) *metabolic turnover*, the periodic breakdown and replacement of the organic components of a cell; (2) growth and cell division; and (3) special processes, such as secretion, contraction, and action potential propagation.

Figure 25-1• provides a broad overview of the processes involved in cellular metabolism. Amino

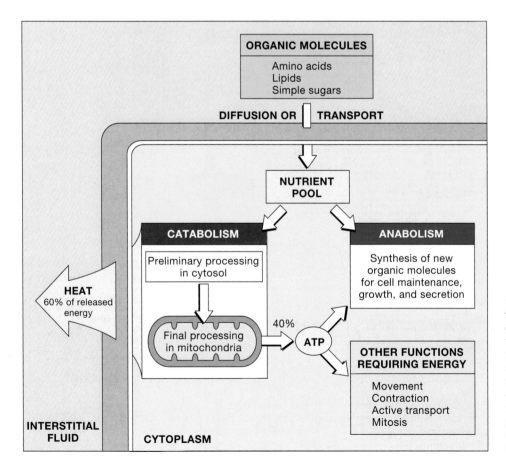

25

•**FIGURE 25-1 An Introduction to Cellular Metabolism.** The cell obtains organic molecules from the interstitial fluid and breaks them down to obtain ATP. Only about 40 percent of the energy released by catabolism is captured in ATP; the rest is radiated as heat. The ATP generated by catabolism provides energy for all vital cellular activities, including anabolism.

acids, lipids, and simple sugars cross the cell membrane to join nutrients already in the cytoplasm. All the cell's organic building blocks collectively form a *nutrient pool.*

Catabolism breaks down organic substrates, releasing energy that can be used to synthesize ATP or other high-energy compounds. ∞ *p. 38* Catabolism proceeds in a series of steps. In general, the initial steps occur in the cytosol, where enzymes break down large organic molecules into smaller fragments. For example, carbohydrates are broken down to short carbon chains, triglycerides are split into fatty acids and glycerol, and proteins are broken down to individual amino acids.

Relatively little ATP is produced during these preparatory steps. However, the simple molecules formed can be absorbed and processed by mitochondria, and the mitochondrial steps release significant amounts of energy. As mitochondrial enzymes break the covalent bonds that hold these molecules together, they capture roughly 40 percent of the energy released. The captured energy is used to convert ADP to ATP. The other 60 percent escapes as heat that warms the interior of the cell and the surrounding tissues.

The ATP produced by mitochondria provides energy to support **anabolism,** the synthesis of new organic molecules (∞ *p. 38*), as well as other cell functions. Those additional functions, such as ciliary or cell movement, contraction, active transport, and cell division, vary from one cell to another. For example, muscle fibers need ATP to provide energy for contraction, and gland cells need ATP to synthesize and transport their secretions. We have considered such specialized functions in other chapters, so we shall restrict our focus here to anabolic processes.

Anabolism is an "uphill" process that involves the formation of new chemical bonds. Cells synthesize new organic components for four basic reasons (Figure 25-2●):

1. *To perform structural maintenance or repairs.* All cells must expend energy to perform ongoing maintenance and repairs, because most structures in the cell are temporary rather than permanent. Their removal and replacement are part of the process of *metabolic turnover.* ∞ *p. 60*

2. *To support growth.* Cells preparing to divide increase in size and synthesize extra proteins and organelles.

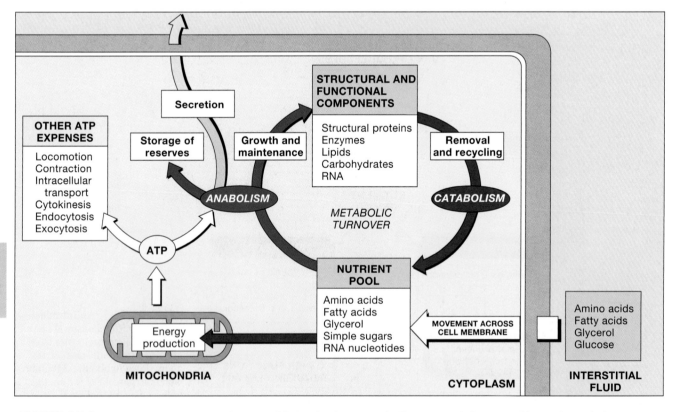

●**FIGURE 25-2 Metabolic Turnover and Cellular ATP Production.** Metabolic turnover is integrated into other cellular functions. Organic molecules released during metabolic turnover can be recycled by means of anabolic pathways and used to build other large molecules. They can also be broken down for energy production. Through catabolism, the cell must provide ATP for anabolism and for other cellular functions.

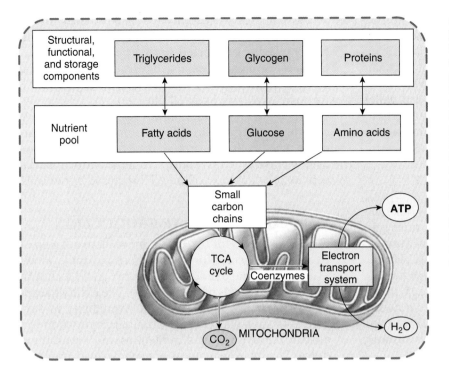

•**FIGURE 25-3** **Nutrient Use in Cellular Metabolism.** Cells use the nutrient pool to build up reserves and to synthesize cellular structures. Catabolism within mitochondria provides the ATP necessary to sustain cell functions. Mitochondria are "fed" small carbon chains that are produced by the breakdown of carbohydrates (especially glucose, stored as glycogen), lipids (especially fatty acids from triglycerides) and proteins (amino acids). The mitochondria absorb these breakdown products for further catabolism by means of the tricarboxylic acid (TCA) cycle and the electron transport system (ETS). This figure will be repeated, in reduced and simplified form as Navigator icons, in key figures throughout this chapter as we change topics.

3. *To produce secretions.* Secretory cells must synthesize their products and deliver them to the interstitial fluid.

4. *To build nutrient reserves.* Most cells prepare for a rainy day—a period of emergency, an interval of extreme activity, or a time when the supply of nutrients in the bloodstream is inadequate. Cells prepare for such times by building up reserves, nutrients stored in a form that can be mobilized as needed. The most abundant storage form of carbohydrate is glycogen, a branched chain of glucose molecules; the most abundant lipids are triglycerides, consisting primarily of fatty acids; and proteins, the most abundant organic components in the body, consist of complex chains of amino acids. For example, muscle cells and liver cells store glucose in the form of glycogen, and adipocytes and liver cells store triglycerides.

The nutrient pool is the source of the substrates for both catabolism and anabolism. As you might expect, the cell tends to conserve the materials needed to build new compounds and breaks down the rest. The cell continuously replaces membranes, organelles, enzymes, and structural proteins. These anabolic activities require more amino acids than lipids and few carbohydrates. In general, when a cell with excess carbohydrates, lipids, and amino acids needs energy, it will break down carbohydrates first. Lipids are the second choice, and amino acids are seldom broken down if other energy sources are available.

Mitochondria are important, because they provide the energy that supports cellular operations. The cell feeds its mitochondria from its nutrient pool, and in return the cell gets the ATP it needs. However, mitochondria are picky eaters; they will accept only specific organic molecules for processing and energy production. Thus, chemical reactions in the cytoplasm take whichever organic nutrients are available and break them down into smaller fragments acceptable to the mitochondria. The mitochondria then break the fragments down further, generating carbon dioxide, water, and ATP (Figure 25-3•). This mitochondrial activity involves two pathways: the *TCA cycle* and the *electron transport system.* In the next section, we shall describe these important catabolic and anabolic reactions that occur in our cells.

CARBOHYDRATE METABOLISM

Most cells generate ATP and other high-energy compounds by breaking down carbohydrates, especially glucose. The complete reaction sequence can be summarized as follows:

$$\underset{\text{glucose}}{C_6H_{12}O_6} + \underset{\text{oxygen}}{6\,O_2} \longrightarrow \underset{\text{carbon dioxide}}{6\,CO_2} + \underset{\text{water}}{6\,H_2O}$$

The breakdown occurs in a series of small steps, and several of the steps release sufficient energy to support the conversion of ADP to ATP. The complete catabolism of one molecule of glucose will provide a typical cell in the body with a net gain of 36 molecules of ATP.

25

Although most of the actual ATP production occurs inside mitochondria, the first steps take place in the cytosol. We outlined the steps in Chapter 10; because those steps do not require oxygen, they are said to be *anaerobic*. The subsequent reactions, which occur in mitochondria, consume oxygen and are called *aerobic*. ∞ *pp. 293–294* The mitochondrial activity responsible for ATP production is called **aerobic metabolism,** or **cellular respiration.**

GLYCOLYSIS

Glycolysis (glī-KOL-i-sis; *glykus,* sweet + *lysis,* dissolution) is the breakdown of glucose to **pyruvic acid.** ∞ *p. 294* In this process, a series of enzymatic steps breaks the six-carbon glucose molecule ($C_6H_{12}O_6$) into two three-carbon molecules of pyruvic acid (CH_3—CO—COOH). At the normal pH inside the cell, each pyruvic acid molecule loses a hydrogen ion and exists as the negatively charged ion CH_3—CO—COO⁻. This ionized form is usually called *pyruvate* rather than pyruvic acid.

Glycolysis requires (1) glucose molecules, (2) appropriate cytoplasmic enzymes, (3) ATP and ADP, (4) inorganic phosphates, and (5) **NAD (n**icotinamide **a**denine **d**inucleotide), a coenzyme that removes hydrogen atoms during one of the enzymatic reactions. (Recall from Chapter 2 that coenzymes are organic molecules essential to enzyme function. ∞ *p. 55*) If any of these participants is missing, glycolysis cannot take place.

The steps in glycolysis are diagrammed in Figure 25-4•. This diagram is an overview rather than a detailed description of the pathway involved. Glycolysis begins when an enzyme *phosphorylates*—that is, attaches a phosphate group—to the last (sixth) carbon atom on a glucose molecule, creating **glucose-6-phosphate.** Although this step "costs" the cell one ATP molecule, it has two important results: (1) It traps the glucose molecule within the cell, because phosphorylated glucose cannot cross the cell membrane; and (2) it prepares the glucose molecule for further biochemical reactions.

A second phosphorylation occurs in the cytosol before the six-carbon chain is broken into two three-carbon fragments. Energy benefits begin to appear as these fragments are converted to pyruvic acid. Two of the steps release enough energy to generate ATP from ADP and inorganic phosphate (PO_4^{3-}, or P_i). In addition, two molecules of NAD are converted to NADH. The net reaction looks like this:

Glucose + 2 NAD + 2 ADP + 2 P_i ⟶
(6-carbon)

2 Pyruvic acid + 2 NADH + 2 ATP
(3-carbon)

This anaerobic reaction sequence provides for the cell a net gain of two molecules of ATP for each glucose molecule converted to two pyruvic acid molecules. A few highly specialized cells, such as red blood cells, lack mitochondria and derive all their ATP through glycolysis. Skeletal muscle fibers rely on glycolysis for ATP production during periods of active contraction; using the ATP provided by glycolysis alone, most cells can survive for brief periods. However, when oxygen is readily available, mitochondrial activity provides most of the ATP required by our cells.

MITOCHONDRIAL ATP PRODUCTION

Glycolysis yields for the cell an immediate net gain of two ATP molecules for each glucose molecule. However, a great deal of additional energy is still locked in the chemical bonds of pyruvic acid. The ability to capture that energy depends on the availability of oxygen. If oxygen supplies are adequate, mitochondria absorb the pyruvic acid molecules and break them down. The hydrogen atoms of each pyruvic acid molecule (CH_3—CO—COOH) are removed by coenzymes and will ultimately be the source of most of the energy gain for the cell. The carbon and oxygen atoms are removed and released as carbon dioxide, a process called **decarboxylation** (dē-kar-boks-i-LĀ-shun).

Each mitochondrion has two membranes surrounding it. ∞ *p. 87* The *outer membrane* contains large-diameter pores that are permeable to ions and small organic molecules such as pyruvic acid. The ions and molecules easily enter the *intermembrane space,* which separates the outer membrane from the *inner membrane.* The inner membrane contains a carrier protein that moves pyruvic acid into the mitochondrial matrix.

In the mitochondrion, a pyruvic acid molecule participates in a complex reaction involving NAD and another coenzyme, **coenzyme A (CoA).** This reaction yields one molecule of carbon dioxide, one of NADH, and one of **acetyl-CoA** (as-Ē-til-KŌ-ā). Acetyl-CoA consists of a two-carbon **acetyl group** (CH_3CO) bound to coenzyme A. Next, the acetyl group is transferred from CoA to a four-carbon molecule of *oxaloacetic acid,* producing *citric acid.*

The TCA Cycle

The formation of citric acid from acetyl-CoA is the first step in a sequence of enzymatic reactions called the **tricarboxylic** (trī-kar-bok-SIL-ik) **acid (TCA) cycle,** or **citric acid cycle.** This reaction sequence is sometimes called the *Krebs cycle* in honor of Hans Krebs, the biochemist who described these reactions in 1937. The function of the cycle is to remove hydrogen atoms from organic molecules and transfer them to coenzymes. The overall pattern of the TCA cycle is

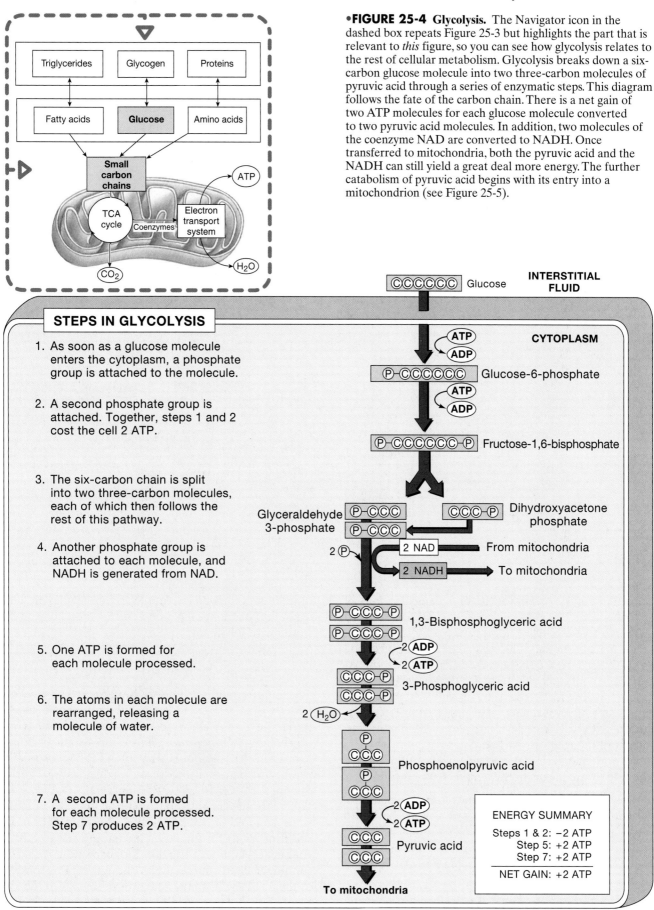

•**FIGURE 25-4 Glycolysis.** The Navigator icon in the dashed box repeats Figure 25-3 but highlights the part that is relevant to *this* figure, so you can see how glycolysis relates to the rest of cellular metabolism. Glycolysis breaks down a six-carbon glucose molecule into two three-carbon molecules of pyruvic acid through a series of enzymatic steps. This diagram follows the fate of the carbon chain. There is a net gain of two ATP molecules for each glucose molecule converted to two pyruvic acid molecules. In addition, two molecules of the coenzyme NAD are converted to NADH. Once transferred to mitochondria, both the pyruvic acid and the NADH can still yield a great deal more energy. The further catabolism of pyruvic acid begins with its entry into a mitochondrion (see Figure 25-5).

STEPS IN GLYCOLYSIS

1. As soon as a glucose molecule enters the cytoplasm, a phosphate group is attached to the molecule.

2. A second phosphate group is attached. Together, steps 1 and 2 cost the cell 2 ATP.

3. The six-carbon chain is split into two three-carbon molecules, each of which then follows the rest of this pathway.

4. Another phosphate group is attached to each molecule, and NADH is generated from NAD.

5. One ATP is formed for each molecule processed.

6. The atoms in each molecule are rearranged, releasing a molecule of water.

7. A second ATP is formed for each molecule processed. Step 7 produces 2 ATP.

INTERSTITIAL FLUID

Glucose

CYTOPLASM

Glucose-6-phosphate

Fructose-1,6-bisphosphate

Glyceraldehyde 3-phosphate

Dihydroxyacetone phosphate

2 NAD — From mitochondria

2 NADH → To mitochondria

1,3-Bisphosphoglyceric acid

3-Phosphoglyceric acid

Phosphoenolpyruvic acid

Pyruvic acid

To mitochondria

ENERGY SUMMARY

Steps 1 & 2: −2 ATP
Step 5: +2 ATP
Step 7: +2 ATP

NET GAIN: +2 ATP

25

•**FIGURE 25-5 The TCA Cycle.** Notice that the dashed box now highlights the use of small carbon chains in the TCA cycle. This cycle completes the breakdown of organic molecules begun by glycolysis and other catabolic pathways. **(a)** An overview of the TCA cycle and the distribution of carbon, hydrogen, and oxygen atoms.

(a)

shown in Figure 25-5•. For the biochemical details of glycolysis and the TCA cycle, consult the *Applications Manual.* [AM] *A Closer Look: Aerobic Metabolism*

At the start of the TCA cycle, the two-carbon acetyl group carried by CoA is attached to a four-carbon ox-aloacetic acid molecule to make the six-carbon compound citric acid. Coenzyme A is released intact and can thus bind another acetyl group. A complete revolution of the TCA cycle removes two carbon atoms, regenerating the four-carbon chain. (This is why the reaction sequence is called a *cycle.*) We can summarize the fate of the atoms in the acetyl group as follows:

■ The two carbon atoms are removed in enzymatic reactions that incorporate four oxygen atoms and form two molecules of carbon dioxide, a waste product.

■ The hydrogen atoms are removed by the coenzyme NAD or a related coenzyme called **FAD** (*flavin ade-nine dinucleotide*).

Several of the steps involved in a revolution of the TCA cycle involve more than one reaction and re-

quire more than one enzyme. Water molecules are tied up in two of those steps. The entire sequence can be summarized as

$$CH_3CO-CoA + 3\,NAD + FAD + GDP + P_i + 2\,H_2O \longrightarrow$$
$$CoA + 2\,CO_2 + 3\,NADH + FADH_2 + 2\,H^+ + GTP$$

The only immediate energy benefit of one revolution of the TCA cycle is the formation of a single molecule of GTP *(guanosine triphosphate)* from GDP *(guanosine diphosphate)* and P_i. In practical terms, GTP is the equivalent of ATP, because GTP readily transfers a phosphate group to ADP, producing ATP: GTP + ADP $\longrightarrow$ GDP + ATP.

The formation of GTP from GDP in the TCA cycle is an example of **substrate-level phosphorylation.** In this process, an enzyme uses the energy released by a chemical reaction to transfer a phosphate group to a suitable acceptor molecule. Although GTP is formed in the TCA cycle, many reaction pathways in the cytosol phosphorylate ADP and form ATP directly. For example, the ATP produced during glycolysis is gen-

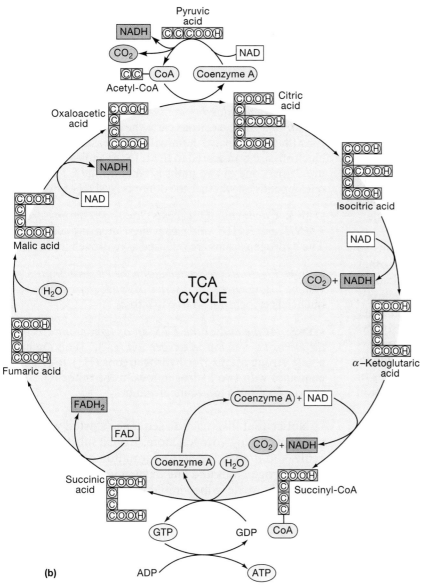

25

•**FIGURE 25-5 The TCA Cycle**
(continued). **(b)** A more complete view, showing the fate of the carbon chains. The TCA cycle begins with the transfer of an acetyl group from coenzyme A to oxaloacetic acid, a four-carbon molecule in the mitochondrial matrix. In a series of enzymatic reactions, the two added carbon atoms, together with oxygen atoms, are eliminated as carbon dioxide. One molecule of ATP is produced indirectly (via GTP) for each turn of the cycle, but much more ATP is ultimately obtained from the hydrogen atoms that are removed by the coenzymes NAD and FAD.

ecules, and oxygen is an atmospheric gas. The only problem is that the reaction releases a tremendous amount of energy all at once. In fact, this reaction releases so much energy that it is used to launch the space shuttle into orbit. Cells cannot handle energy explosions; energy release must be gradual, as it is in oxidative phosphorylation. During oxidative phosphorylation, this powerful reaction proceeds in a series of small, enzymatically controlled steps. Under these conditions, energy can be captured and ATP generated.

erated through substrate-level phosphorylation. Normally, however, substrate-level phosphorylation provides a relatively small amount of energy compared with *oxidative phosphorylation.*

Oxidative Phosphorylation and the ETS

Oxidative phosphorylation is the generation of ATP within mitochondria in a reaction sequence that requires coenzymes and consumes oxygen. This process produces over 90 percent of the ATP used by our cells. The key reactions take place in the *electron transport system (ETS),* a series of integral and peripheral proteins in the inner membrane of the mitochondrion. The basis of oxidative phosphorylation is very simple:

$$2\,H_2 + O_2 \longrightarrow 2\,H_2O$$

Cells can easily obtain the ingredients for this reaction. Hydrogen is a component of all organic mol-

Oxidation, Reduction, and Energy Transfer   The enzymatic steps of oxidative phosphorylation involve oxidation and reduction. (There are different types of oxidation and reduction reactions, but the most important for our purposes are those involving the transfer of electrons.) The loss of electrons is a form of **oxidation;** the acceptance of electrons is a form of **reduction.** The two reactions are always paired. When electrons pass from one molecule to another, the electron donor is oxidized and the electron recipient reduced. Oxidation and reduction are important because electrons carry chemical energy. In a typical oxidation–reduction reaction, *the reduced molecule gains energy at the expense of the oxidized molecule.*

In such an exchange, the reduced molecule does not acquire all the energy released by the oxidized molecule. Some energy is always released as heat, and additional energy may be used to perform physical or chemical work, such as the formation of ATP. By leading the electrons through a series of oxidation–reduction reactions before they ultimately combine with oxygen atoms, cells can capture and use much of the energy released in the formation of water.

Coenzymes play a key role in this process. A coenzyme acts as an intermediary that accepts electrons from one molecule and transfers them to another molecule. In the TCA cycle, NAD and FAD remove hydrogen atoms from organic substrates. Each hydrogen atom consists of an electron (e^-) and a proton (a hydrogen ion, H^+). Thus when a coenzyme accepts hydrogen atoms, the coenzyme is reduced and gains energy. The donor molecule loses electrons and energy as it gives up its hydrogen atoms.

Next NADH and $FADH_2$, the reduced forms of NAD and FAD, transfer their hydrogen atoms to other coenzymes. The protons are subsequently released, and the electrons, which carry the chemical energy, enter a sequence of oxidation–reduction reactions. This sequence ends with the electrons' transfer to oxygen and the formation of a water molecule. At several steps along the oxidation–reduction sequence, enough energy is released to support the synthesis of ATP from ADP. We shall now consider that reaction sequence in greater detail.

The coenzyme FAD accepts two hydrogen atoms from the TCA cycle and in doing so gains two electrons. The coenzyme NAD also gains two electrons as two hydrogen atoms are removed from the donor molecule. However, NAD binds one electron as part of a hydrogen atom, forming NADH; it removes the electron from the other hydrogen atom and releases the proton (H^+). For this reason, the reduced form of NAD is often described as "NADH + H^+."

Other coenzymes involved in the initial steps of oxidative phosphorylation are **FMN** (*flavin mononucleotide*) and **coenzyme Q** (*ubiquinone*). Coenzymes are either free in the mitochondrial matrix (NAD) or attached to the inner mitochondrial membrane (FAD, FMN, coenzyme Q).

The Electron Transport System

The **electron transport system (ETS)**, or *respiratory chain*, is a sequence of proteins called **cytochromes** (SĪ-tō-krōmz; *cyto-*, cell + *chroma*, color). Each cytochrome has two components: a protein and a pigment. The protein, embedded in the inner membrane of a mitochondrion, surrounds the pigment complex, which contains a metal ion (either iron, Fe^{3+}, or copper, Cu^{2+}). We shall consider four cytochromes: *b, c, a,* and *a_3.*

Figure 25-6● summarizes the basic steps in oxidative phosphorylation. We shall first consider the path taken by the electrons that are captured and delivered by coenzymes (Figure 25-6a●):

STEP 1: *A coenzyme strips a pair of hydrogen atoms from a substrate molecule.* As we have seen, different coenzymes are used for different substrate molecules. During glycolysis, which occurs in the cytoplasm, NAD is reduced to NADH. Within mitochondria, both NAD and FAD are reduced through reactions of the TCA cycle, producing NADH and $FADH_2$, respectively.

STEP 2: *NADH and $FADH_2$ deliver hydrogen atoms to coenzymes embedded in the inner membrane of a mitochondrion.* The electrons carry the energy, and the protons that accompany them will be released before the electrons are transferred to the ETS. As indicated in Figure 25-6a●, one of two paths is taken to the ETS; which one depends on whether the donor is NADH or $FADH_2$.

STEP 3: *Coenzyme Q accepts hydrogen atoms from $FMNH_2$ and $FADH_2$ and passes electrons to cytochrome b. The hydrogen atoms are released as hydrogen ions (H^+).*

STEP 4: *Electrons are passed along the electron transport system, losing energy in a series of small steps.* The sequence is cytochrome *b* to *c* to *a* to *a_3.*

STEP 5: *At the end of the ETS, an oxygen atom accepts the electrons, creating an oxygen ion (O^-).* This ion has a very strong affinity for hydrogen ions (H^+); it quickly combines with two hydrogen ions in the mitochondrial matrix, forming a molecule of water.

Notice that this reaction sequence started with the removal of two hydrogen atoms from a substrate molecule, and it ended with the formation of water from two hydrogen ions and one oxygen ion. This is the reaction that we described initially as releasing too much energy if performed in a single step. Because the reaction has occurred in a series of small steps, however, the combination of hydrogen and oxygen can take place quietly rather than explosively.

The electron transport system does not produce ATP directly. Instead, it creates the conditions necessary for ATP production. The electrons release energy as they pass from coenzyme to cytochrome and from cytochrome to cytochrome. The energy released at several steps drives hydrogen ion pumps that move hydrogen ions from the mitochondrial matrix into the intermembrane space, between the inner and outer membranes of a mitochondrion. These hydrogen ion pumps create a large concentration gradient for hydrogen ions across the inner membrane. This concentration gradient is a form of potential energy, and it is used to provide the energy to convert ADP to ATP. Despite the concentration gradient, the hydrogen ions cannot diffuse into the matrix because they are not lipid-soluble. However, hydrogen ion channels in the inner membrane permit the diffusion of H^+ into the matrix. These ion channels and their attached *coupling factors* use the kinetic energy of passing hydrogen ions to generate ATP. This process is *chemiosmosis* (kem-ē-oz-MŌ-sis), or *chemiosmotic phosphorylation.*

•FIGURE 25-6 Oxidative Phosphorylation. (a) The sequence of oxidation–reduction reactions involved. **(b)** The location of the coenzymes and the electron transport system. Notice the sites where hydrogen ions are pumped into the intermembrane space, providing the concentration gradient essential to the generation of ATP.

25

ATP Generation and the ETS Figure 25-6b• diagrams the mechanism of ATP generation. Hydrogen ion pumping occurs (1) as FMN reduces coenzyme Q, (2) as cytochrome *b* reduces cytochrome *c*, and (3) as electrons are passed from cytochrome *a* to cytochrome a_3.

For each pair of electrons removed by NAD from a substrate in the TCA cycle, six hydrogen ions are pumped across the inner membrane of the mitochondrion and into the intermembrane space (Figure 25-6b•). Their reentry into the matrix provides the energy to generate three molecules of ATP. For each pair of electrons removed by FAD from a substrate in the TCA cycle, four hydrogen ions are pumped across the inner membrane and into the intermembrane space. Their reentry into the matrix provides the energy to generate two molecules of ATP.

Oxidative phosphorylation is the most important mechanism for the generation of ATP. In most cases, a chronic suspension, or even a significant reduction, of the rate of oxidative phosphorylation will kill the cell. If many cells are affected, the individual may die. Oxidative phosphorylation requires both oxygen and electrons; the rate of ATP generation is thus limited by the availability of either oxygen or electrons.

Cells obtain oxygen by diffusion from the extracellular fluid. If the supply of oxygen is cut off, mitochondrial ATP production will cease, because reduced cytochrome a_3 will have no acceptor for its electrons. With the last reaction stopped, the entire ETS comes to a halt, like cars at a washed-out bridge. Oxidative phosphorylation can no longer take place, and cells quickly succumb to energy starvation. Because neurons have a high demand for energy, the brain is one of the first organs to be affected.

Hydrogen cyanide gas is sometimes used as a pesticide to kill rats or mice; in some states where capital punishment is legal, that gas is used to execute criminals. The cyanide ion (CN^-) binds to cytochrome a_3 and prevents the transfer of electrons to oxygen. As a result, cells die from energy starvation.

ENERGY YIELD OF GLYCOLYSIS AND CELLULAR RESPIRATION

For most cells, the complete reaction pathway that begins with glucose and ends with carbon dioxide and water is the main method of generating ATP. Figure 25-7• reviews the process in terms of energy production:

■ *Glycolysis.* During glycolysis, the cell gains 2 molecules of ATP directly for each glucose molecule broken down anaerobically to pyruvic acid. Two molecules of NADH are also produced. In most cells, electrons are passed from NADH to FAD by means of an intermediate in the intermembrane space and then to CoQ and the electron transport system. This sequence of events ultimately provides an additional 4 ATP molecules.

■ *The electron transport system.* The TCA cycle breaks down the 2 pyruvic acid molecules, transferring hydrogen atoms to NADH and $FADH_2$. These coenzymes provide electrons to the ETS; each of the 8 molecules of NADH yields 3 ATP and 1 water molecule; each of the 2 $FADH_2$ molecules yields 2 ATP and 1 water molecule. Thus the shuffling from the TCA cycle to the ETS yields 28 molecules of ATP.

■ *The TCA cycle.* Each of the two revolutions of the TCA cycle required to break down both pyruvic acid molecules completely yields 1 molecule of ATP by way of GTP. This cycling provides an additional gain of 2 molecules of ATP.

Summing up, for each glucose molecule processed, the cell gains 36 molecules of ATP: two from glycolysis, 4 from the NADH generated in glycolysis, 2 from the TCA cycle (by means of GTP) and a total of 28 from the ETS.

Your cardiac muscle cells and liver cells are able to gain an additional 2 ATP molecules for each glucose molecule broken down. This gain is accomplished by increasing the energy yield from the NADH generated during glycolysis from 4 ATP to 6 ATP molecules. In these cells, each NADH molecule passes electrons to an intermediate that generates NADH, not $FADH_2$, in the mitochondrial matrix. The subsequent transfer of electrons to FMN, CoQ, and the ETS results in the production of 6 ATP molecules, just as if the 2 NADH molecules had been generated in the TCA cycle.

Carbohydrate Loading

Although other nutrients can be broken down to provide substrates for the TCA cycle, carbohydrates require the least processing and preparation. It is not surprising, therefore, that athletes have tried to devise ways of exploiting these compounds as ready sources of energy.

Eating carbohydrates *just before* you exercise does not improve your performance and can decrease your endurance by slowing the mobilization of existing energy reserves. Runners or swimmers preparing for lengthy endurance contests, such as a marathon or 5-km swim, do not eat immediately before competing, and for 2 hours before the race they limit their intake to drinking water. However, these athletes often eat carbohydrate-rich meals for 3 days before the event. This **carbohydrate loading** increases the carbohydrate reserves of muscle tissue that will be called on during the competition.

You can obtain maximum effects by exercising to exhaustion for 3 days before you start a high-carbohydrate diet; this practice is called *carbohydrate depletion/loading*. Carbohydrate depletion/loading has a number of potentially unpleasant side effects, including muscle and kidney damage. Sports physiologists recommend that athletes use this routine fewer than three times a year.

•FIGURE 25-7 A Summary of the Energy Yield of Aerobic Metabolism. For each glucose molecule broken down by glycolysis, only two molecules of ATP (net) are produced. However, glycolysis, the formation of acetyl-CoA, and the TCA cycle all yield molecules of reduced coenzymes (NADH and/or $FADH_2$). Many additional ATP molecules are produced when electrons from these coenzymes pass through the electron transport system. In most cells, each of the two NADH molecules produced in glycolysis provides another two ATP molecules. Each of the eight NADH molecules produced in the mitochondria yields three ATP molecules, for a total of 24. Another two ATP molecules are gained from each of the two $FADH_2$ molecules generated in the mitochondria. The TCA cycle generates an additional two ATP molecules in the form of GTP.

☑ What is the primary role of the TCA cycle in the production of ATP?

☑ The NADH produced by glycolysis in skeletal muscle fibers leads to the production of 2 ATP molecules in the mitochondria, but the NADH produced by glycolysis in cardiac muscle cells leads to the production of 3 ATP molecules. Why?

☑ How would a decrease in the level of cytoplasmic NAD affect ATP production in mitochondria?

OTHER CATABOLIC PATHWAYS

Aerobic metabolism is relatively efficient and capable of generating large amounts of ATP. It is the cor-

nerstone of normal metabolism, but it has one obvious limitation: The cell must have adequate supplies of both oxygen and glucose. Cells can survive only for brief periods without adequate oxygen. Low glucose concentrations have a much lesser effect on most cells, because they can break down other nutrients to provide substrates for the TCA cycle (Figure 25-3, p. 903). Many cells can switch from one nutrient source to another as needed. For example, many cells shift from glucose-based to lipid-based ATP production when necessary. When actively contracting, skeletal muscles catabolize glucose, but at rest they utilize fatty acids.

Cells break down proteins for energy only when lipids or carbohydrates are unavailable, primarily

25

because proteins make up the enzymes and organelles that the cell needs. Nucleic acids are present only in small amounts; they are seldom catabolized for energy, even when the cell is dying of acute starvation. This restraint makes sense, because the DNA in the nucleus determines all the structural and functional characteristics of the cell. We shall consider the catabolism of other compounds in later sections as we discuss lipid, protein, and nucleic acid metabolism.

GLUCONEOGENESIS

Because some of the steps in glycolysis—glucose breakdown—are essentially irreversible, cells cannot generate glucose by performing glycolysis in reverse, using the same enzymes (Figure 25-8•). Therefore, glycolysis and the production of glucose involve a different set of regulatory enzymes, and the two processes are independently regulated. Pyruvic acid or other three-carbon molecules can be used to synthesize glucose; as a result, a cell can create glucose molecules from other carbohydrates, lactic acid, glycerol, or some amino acids. However, acetyl-CoA cannot be used to make glucose, because the decarboxylation step between pyruvic acid and acetyl-CoA cannot be reversed. **Gluconeogenesis** (gloo-kō-nē-ō-JEN-e-sis) is the synthesis of glucose from noncarbohydrate precursors, such as lactic acid, glycerol, or amino acids. Fatty acids and many amino acids cannot be used for gluconeogenesis, because their catabolic pathways produce acetyl-CoA.

Glucose molecules created by gluconeogenesis can be used to manufacture other simple sugars, complex carbohydrates, proteoglycans, or nucleic acids. In the liver and in skeletal muscle, glucose molecules are stored as glycogen. The process of glycogen formation from glucose is known as **glycogenesis.** It is a complex process that involves several steps and requires the high-energy compound *uridine triphosphate (UTP).* Glycogen is an important energy reserve that can be broken down when the cell cannot obtain enough glucose from interstitial fluid. The breakdown of glycogen is called **glycogenolysis.** Glycogenolysis occurs quickly and involves a single enzymatic step. Although glycogen molecules are large, glycogen reserves take up very little space because they form compact, insoluble granules.

25

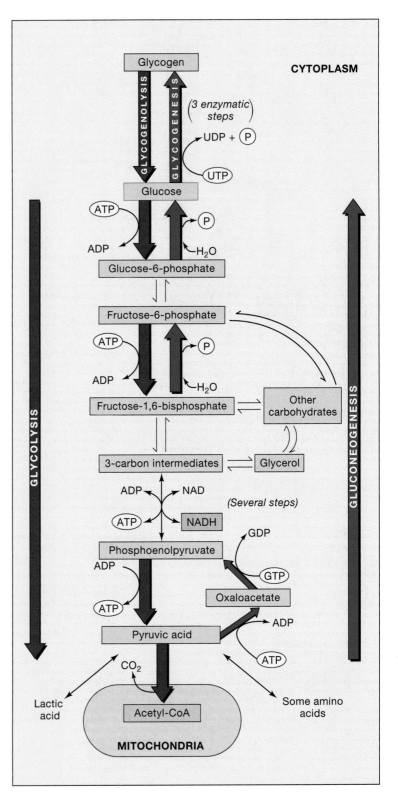

•**FIGURE 25-8 Carbohydrate Breakdown and Synthesis.** A flow chart of the pathways for glycolysis and gluconeogenesis. Many of the reactions are freely reversible, but separate regulatory enzymes control the key steps, which are indicated by colored arrows. Some amino acids, carbohydrates other than glucose, and glycerol can be converted to glucose. The enzymatic reaction that converts pyruvic acid to acetyl-CoA cannot be reversed.

LIPID METABOLISM

Like carbohydrates, lipid molecules contain carbon, hydrogen, and oxygen, but the atoms are present in different proportions. Triglycerides are the most abundant lipid in the body, so our discussion shall focus on pathways for triglyceride breakdown and synthesis.

LIPID CATABOLISM

During lipid catabolism, or **lipolysis,** lipids are broken down into pieces that can be either converted to pyruvic acid or channeled directly into the TCA cycle. A triglyceride is first split into its component parts by hydrolysis. This step yields one molecule of glycerol and three fatty acid molecules. Glycerol enters the TCA cycle after enzymes in the cytosol convert it to pyruvic acid. The catabolism of fatty acids involves a completely different set of enzymes.

Beta-Oxidation

Fatty acid molecules are broken down into two-carbon fragments in a sequence of reactions known as **beta-oxidation.** This process occurs inside mitochondria, so the carbon chains can enter the TCA cycle immediately. Figure 25-9• diagrams one step in the process of beta-oxidation. Each step generates molecules of acetyl-CoA, NADH, and $FADH_2$, leaving a shorter carbon chain bound to coenzyme A.

Beta-oxidation provides substantial energy benefits. For each two-carbon fragment removed from the fatty acid, the cell gains 12 ATP molecules from the processing of acetyl-CoA in the TCA cycle, plus 5 ATP molecules from the NADH and $FADH_2$. The cell can therefore gain 144 ATP molecules from the breakdown of one 18-carbon fatty acid molecule. This result is almost 1.5 times the energy obtained by the complete breakdown of three six-carbon glucose molecules. The catabolism of other lipids follows similar patterns, generally ending with the formation of acetyl-CoA.

Lipids and Energy Production

Lipids are important as an energy reserve, because they can provide large amounts of ATP. They are insoluble, so

25

•**FIGURE 25-9 Beta-Oxidation.** During beta-oxidation, the carbon chains of fatty acids are broken down to yield molecules of acetyl-CoA, which can be used in the TCA cycle. The reaction also donates hydrogen atoms to coenzymes that deliver them to the electron transport system.

lipids can be stored in compact droplets in the cytosol. This storage method saves space, but when the lipid droplets are large, it is difficult for water-soluble enzymes to get at them. Lipid reserves are therefore more difficult to access than carbohydrate reserves. Also, most lipids are processed inside mitochondria, and mitochondrial activity is limited by the availability of oxygen. The net result is that lipids cannot provide large amounts of ATP in a short time. However, cells with modest energy demands can shift to lipid-based energy production when glucose supplies are limited. Skeletal muscle fibers normally cycle between lipid metabolism and carbohydrate metabolism. When at rest (when energy demands are low), these cells break down fatty acids. When active (when energy demands are large and immediate), skeletal muscle fibers shift to glucose metabolism.

LIPID SYNTHESIS

The synthesis of lipids is known as **lipogenesis** (li-pō-JEN-e-sis). Figure 25-10• shows the major pathways of lipogenesis. Glycerol is synthesized from *dihydroxyacetone phosphate,* one of the intermediate products of glycolysis. The synthesis of most other types of lipids, including nonessential fatty acids and steroids, begins with acetyl-CoA. Lipogenesis can use almost any organic substrate, because lipids, amino acids, and carbohydrates can be converted to acetyl-CoA.

Fatty acid synthesis involves a reaction sequence quite distinct from that of beta-oxidation. Your cells cannot *build* every fatty acid they can break down. **Linoleic acid** and **linolenic acid,** both 18-carbon unsaturated fatty acids, cannot be synthesized. These are called **essential fatty acids,** because they must be included in your diet. They are synthesized by plants. A deficiency of essential fatty acids generally occurs only among hospitalized individuals receiving nutrients in an intravenous solution. A diet poor in linoleic acid slows growth and alters the appearance of the skin. These fatty acids are also needed to synthesize prostaglandins and some of the phospholipids incorporated into cell membranes throughout the body.

LIPID TRANSPORT AND DISTRIBUTION

Like glucose, lipids are needed throughout the body. All cells need lipids to maintain their cell membranes, and important steroid hormones must reach target cells in many different tissues. Because most lipids are not soluble in water, special transport mechanisms are required to carry them from one region of the body to another. Most lipids circulate through the bloodstream as lipoproteins. **Lipoproteins** are lipid–protein complexes that contain large insoluble glycerides and cholesterol with a superficial coating of phospholipids and proteins. The proteins and phospholipids make the entire complex soluble. Exposed proteins, which can bind to specific membrane receptors, determine which cells absorb the associated lipids.

Lipoproteins are usually classified according to size and the relative proportions of lipid versus protein.

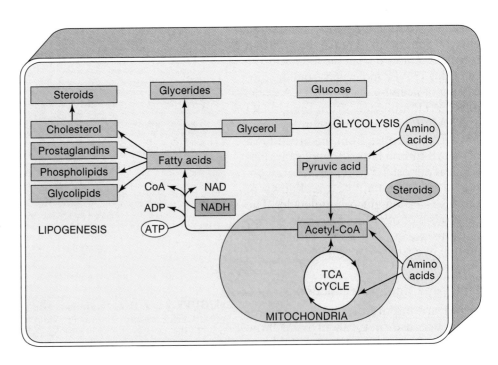

•**FIGURE 25-10 Lipid Synthesis.** Pathways of lipid synthesis begin with acetyl-CoA. Molecules of acetyl-CoA can be strung together in the cytosol, yielding fatty acids. Those fatty acids can be used to synthesize glycerides or other lipid molecules. Lipids can be synthesized from amino acids or carbohydrates by using acetyl-CoA as an intermediate.

25

Dietary Fats and Cholesterol

Elevated cholesterol levels are associated with the development of atherosclerosis (Chapter 21) and coronary artery disease (CAD; Chapter 20). ∞ *pp. 696, 667* Current nutritional advice suggests that you reduce cholesterol intake to under 300 mg per day. This amount represents a 40 percent reduction for the average American adult. As a result of rising concerns about cholesterol, such phrases as "low-cholesterol," "contains no cholesterol," and "cholesterol-free" are now widely used in the advertising and packaging of foods. Cholesterol content alone does not tell the entire story; we must consider some basic information about cholesterol and about lipid metabolism in general:

- *Cholesterol has many vital functions in the human body.* It serves as a waterproofing for the epidermis, a lipid component of all cell membranes, a key constituent of bile, and the precursor of several steroid hormones and one vitamin (vitamin D_3). Because cholesterol is so important, dietary restrictions should have the goal of keeping cholesterol levels within acceptable limits. The goal is not to eliminate cholesterol from the diet or from the circulating blood.

- *The cholesterol content of the diet is not the only source of circulating cholesterol.* The human body can manufacture cholesterol from the acetyl-CoA obtained through glycolysis or the beta-oxidation of other lipids. Dietary cholesterol probably accounts for only about 20 percent of the cholesterol in the bloodstream; the rest is the result of metabolism of saturated fats in the diet. If the diet contains an abundance of saturated fats, cholesterol levels in the blood will rise because excess lipids are broken down to acetyl-CoA and used to synthesize cholesterol. Consequently, a person trying to lower serum cholesterol by dietary control must restrict other lipids—especially saturated fats—as well.

- *Genetic factors affect each individual's cholesterol level.* If you reduce your dietary supply of cholesterol, your body will synthesize more to maintain "acceptable" concentrations in the blood. The acceptable level depends on your genetic programming. Because individuals differ in genetic makeup, their cholesterol levels can vary even on similar diets. In virtually all instances, however, dietary restrictions can lower blood cholesterol significantly.

- *Cholesterol levels vary with age and physical condition.* At age 19, three out of four males have fasting cholesterol levels (levels measured 8–12 hours after a meal) below 170 mg/dl. Cholesterol levels in females of this age are slightly higher, typically at or below 175 mg/dl. As age increases, the cholesterol values gradually climb; over age 70 the values are 230 mg/dl (males) and 250 mg/dl (females). Cholesterol levels are considered unhealthy if they are higher than those of 90 percent of the population in that age group. For males, this value ranges from 185 mg/dl at age 19 to 250 mg/dl at age 70. For females, the comparable values are 190 mg/dl and 275 mg/dl, respectively.

To determine whether you may need to do anything about your cholesterol level, remember three simple rules:

1. Individuals of any age with total cholesterol values below 200 mg/dl probably do not need to change their lifestyle unless they have a family history of CAD and atherosclerosis.
2. Those with cholesterol levels between 200 and 239 mg/dl should modify their diet, lose weight (if overweight), and have annual checkups.
3. Cholesterol levels over 240 mg/dl warrant drastic changes in dietary lipid consumption, perhaps coupled with drug treatment. Drug therapies are always recommended in cases in which the serum cholesterol level exceeds 350 mg/dl. Examples of drugs used to lower cholesterol levels include *cholestyramine, colestipol,* and *lovastatin.*

Most physicians, when ordering a blood test for cholesterol, also request information about circulating triglycerides. In fasting individuals, triglycerides are usually present at levels of 40–150 mg/dl. (After a person has consumed a fatty meal, triglyceride levels may be temporarily elevated.)

When cholesterol levels are high, or when an individual has a family history of atherosclerosis or CAD, further tests and calculations may be performed. The HDL level is measured, and the LDL level is calculated as

$$LDL = Cholesterol - HDL - \left[\frac{Triglycerides}{5}\right]$$

A high total cholesterol value linked to a high LDL level spells trouble. In effect, an unusually large amount of cholesterol is being exported to peripheral tissues. Problems can also exist if the individual has high total cholesterol—or even normal total cholesterol—but low HDL levels (below 35 mg/dl). In this case, excess cholesterol delivered to the tissues cannot easily be returned to the liver for excretion. In either event, the amount of cholesterol in peripheral tissues, and especially in arterial walls, is likely to increase. For years, LDL:HDL ratios were taken to be valid predictors of the risk of developing atherosclerosis. Risk-factor analysis and LDL levels are now thought to be more-accurate indicators. For males with more than one risk factor, many clinicians recommend dietary and drug therapy if LDL levels exceed 130 mg/dl, regardless of the total cholesterol or HDL levels.

25

The following five major groups of lipoproteins are recognized:

1. **Chylomicrons.** Roughly 95 percent of the weight of a chylomicron consists of triglycerides. Chylomicrons are the largest lipoproteins, ranging in diameter from 0.03 to 0.5 μm. They are produced by intestinal epithelial cells. ∞ *p. 890* Chylomicrons carry absorbed lipids from the intestinal tract to the bloodstream. The other lipoproteins shuttle lipids among various tissues. The liver is the primary source of all other types of lipoproteins.

2. **Very low-density lipoproteins (VLDLs).** Very low-density lipoproteins contain triglycerides manufactured by the liver plus small amounts of phospholipids and cholesterol. The primary function of VLDLs is to transport these triglycerides to peripheral tissues. The VLDLs range in diameter from 25 to 75 nm (0.025–0.075 μm).

3. **Intermediate-density lipoproteins (IDLs).** Intermediate-density lipoproteins are intermediate in size and lipid composition between VLDLs and low-density lipoproteins (LDLs). They contain smaller amounts of triglycerides than do VLDLs and relatively more phospholipids and cholesterol than do LDLs.

4. **Low-density lipoproteins (LDLs).** Low-density lipoproteins contain cholesterol, lesser amounts of phospholipids, and very few triglycerides. These lipoproteins, which are about 25 nm in diameter, deliver cholesterol to peripheral tissues. Because this cholesterol may wind up in arterial plaques, LDL cholesterol is often called "bad cholesterol."

5. **High-density lipoproteins (HDLs).** High-density lipoproteins, about 10 nm in diameter, have roughly equal amounts of lipid and protein. The lipids are largely cholesterol and phospholipids. The primary function of HDLs is transporting excess cholesterol from peripheral tissues back to the liver for storage or excretion in the bile. Because HDL cholesterol is returning from peripheral tissues and will not cause circulatory problems, it is called "good cholesterol." Actually, applying the terms *good* and *bad* to cholesterol can be misleading, for cholesterol metabolism is complex and variable. (For more details, see the discussion "Dietary Fats and Cholesterol" on p. 915.)

Figure 25-11• diagrams the probable relationships among these lipoproteins. Chylomicrons produced in the intestinal tract reach the venous circulation by entering lymphatic capillaries and traveling through the thoracic duct (Figure 25-11a•). Although chylomicrons are too large to diffuse across capillary walls, the endothelial lining of capillaries in skeletal muscle, cardiac muscle, adipose tissue, and the liver contains the enzyme **lipoprotein lipase,** which breaks down complex lipids. When chylomicrons contact these endothelial walls, enzymatic activity releases fatty acids and monoglycerides that can diffuse across the endothelium and into interstitial fluid.

The liver controls the distribution of other lipoproteins (Figure 25-11b•):

STEP 1: Liver cells synthesize VLDLs for discharge into the bloodstream.

STEP 2: In peripheral capillaries, lipoprotein lipase removes many of the triglycerides from VLDLs, leaving IDLs; the triglycerides are broken down into fatty acids and monoglycerides.

STEP 3: When IDLs reach the liver, additional triglycerides are removed and the protein content is altered. This process creates LDLs, which then returns to peripheral tissues to deliver cholesterol.

STEP 4: LDLs leave the bloodstream through capillary pores or cross the endothelium by vesicular transport.

STEP 5: Once in peripheral tissues, the LDLs are absorbed by means of receptor-mediated endocytosis. ∞ *p. 77* The amino acids and cholesterol then enter the cytoplasm.

STEP 6: The cholesterol not used by the cell in the synthesis of lipid membranes or other products diffuses out of the cell.

STEP 7: The cholesterol then reenters the bloodstream, where it is absorbed by HDLs and returned to the liver.

STEP 8: In the liver, the HDLs are absorbed and their cholesterol is extracted. Some of the cholesterol recovered will be used in the synthesis of LDLs, and the rest will be excreted in bile salts.

STEP 9: The HDLs stripped of their cholesterol are released into the bloodstream to travel into peripheral tissues and absorb additional cholesterol.

PROTEIN METABOLISM

The body can synthesize 100,000 to 140,000 different proteins with various forms, functions, and structures. Yet, each protein contains some combination of the same 20 amino acids. (Appendix IV shows the structures of these amino acids.) Under normal conditions, cellular proteins are continuously recycled. Peptide bonds are broken, and the free amino acids are used in new proteins. This recycling occurs in the cytosol.

If other energy sources are inadequate, mitochondria can break down amino acids in the TCA cycle to generate ATP. Not all amino acids enter the TCA cycle at the same point, so the ATP benefits vary. However, the average ATP yield per gram is comparable to that of carbohydrate catabolism.

AMINO ACID CATABOLISM

The first step in amino acid catabolism is the removal of the amino group ($-NH_2$). This process requires a coenzyme derivative of **vitamin B_6** *(pyridoxine).* The

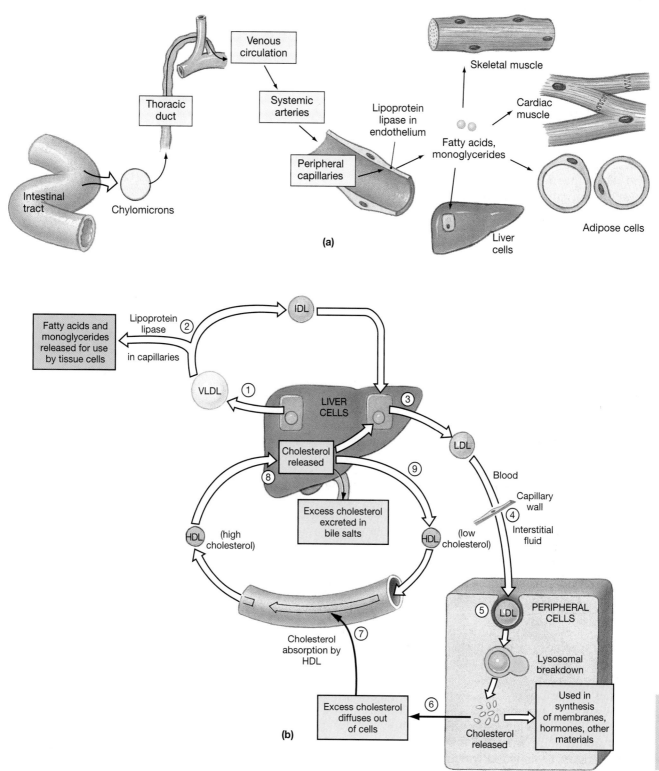

•**FIGURE 25-11 Lipid Transport and Utilization. (a)** Chylomicrons synthesized at the intestinal epithelium reach the bloodstream by way of the thoracic duct. They are broken down by lipoprotein lipase in capillaries that supply blood to skeletal muscle, cardiac muscle, adipose tissue, and the liver. **(b)** ① Liver cells synthesize VLDLs that deliver triglycerides to peripheral tissues. ② Lipoprotein lipase in endothelial cells breaks down these triglycerides and releases fatty acids and monoglycerides that diffuse into the surrounding tissues. ③ The IDLs that remain returns to the liver, where they are absorbed and converted to LDLs that contain cholesterol. ④ The LDLs circulate to peripheral tissues, cross the endothelium, and are absorbed by cells by endocytosis. ⑤ Some of the cholesterol is used in cellular processes. ⑥ The excess cholesterol diffuses back into the bloodstream. ⑦ In plasma, the cholesterol is absorbed by HDLs produced by the liver. ⑧ On returning to the liver, the HDLs are absorbed and the cholesterol is extracted. Some of the cholesterol will be exported once again, in LDLs; excess cholesterol is excreted in bile salts. ⑨ The HDLs stripped of their cholesterol are released into the bloodstream.

25

amino group is removed by *transamination* (tranz-am-i-NĀ-shun) or *deamination* (dē-am-i-NĀ-shun). We shall consider other aspects of amino acid catabolism in a later section.

Transamination

Transamination attaches the amino group of an amino acid to a **keto acid** (Figure 25-12a•). A keto acid resembles an amino acid except that the second carbon binds an oxygen atom rather than an amino group. Transamination converts the keto acid into an amino acid that can enter the cytosol and be used for protein synthesis. In the process, the original amino acid becomes a keto acid that can be broken down in the TCA cycle. Many different tissues perform transaminations. These reactions enable a cell to synthesize many of the amino acids needed for protein synthesis. Cells of the liver, skeletal muscles, heart, lung, kidney, and brain, which are particularly active in protein synthesis, perform many transaminations.

Deamination

Deamination is performed in preparing an amino acid for breakdown in the TCA cycle (Figure 25-12b•). Deamination is the removal of an amino group and a hydrogen atom in a reaction that generates an ammonia (NH_3) molecule or an ammonium ion (NH_4^+). Ammonia molecules are highly toxic, even in low concentrations. Your liver, the primary site of deamination, has the enzymes needed to deal with the problem of ammonia generation. Liver cells convert the ammonia to **urea**, a relatively harmless, water-soluble compound excreted in urine. The **urea cycle** is the reaction sequence involved in urea production (Figure 25-12c•).

When glucose supplies are low and lipid reserves are inadequate, liver cells break down internal proteins and absorb additional amino acids from the blood. The amino acids are deaminated, and the carbon chains are broken down to provide ATP.

Proteins and ATP Production

Three factors make protein catabolism an impractical source of quick energy:

1. Proteins are more difficult to break apart than are complex carbohydrates or lipids.

2. One of the byproducts, ammonia, is a toxin that can damage cells.

(a) Transamination

(b) Deamination

(c) Urea cycle

•**FIGURE 25-12 Amino Acid Catabolism. (a)** In transamination, an enzyme removes the amino group ($—NH_2$) from one molecule and attaches it to a keto acid. **(b)** In deamination, an enzyme strips the amino group and a hydrogen atom from an amino acid and produces a keto acid and an ammonium ion. **(c)** The urea cycle takes two metabolic waste products, ammonia and carbon dioxide, and produces a molecule of urea. Urea is a relatively harmless, soluble compound that is excreted in the urine.

3. Proteins form the most important structural and functional components of any cell. Extensive protein catabolism therefore threatens homeostasis at the cellular and systems levels.

PROTEIN SYNTHESIS

We discussed the basic mechanism for protein synthesis in Chapter 3 (Figures 3-24 and 3-25•, pp. 93, 94–95). Your body can synthesize roughly half of the various amino acids needed to build proteins. There are 10 **essential amino acids.** You cannot synthesize eight of them (*isoleucine, leucine, lysine, threonine, tryptophan, phenylalanine, valine,* and *methionine*); the other two (*arginine* and *histidine*) can be synthesized but in amounts that are insufficient for growing children. Because the body can make other amino acids on demand, they are called the **nonessential amino acids.** Your body can readily synthesize the carbon frameworks of the nonessential amino acids, and a nitrogen group can be attached by transamination or amination. **Amination** is the attachment of an amino group (Figure 25-13•).

Protein deficiency diseases develop when an individual does not consume adequate amounts of all essential amino acids. All amino acids must be available if protein synthesis is to occur. Every transfer RNA molecule must appear at the active ribosome at the proper time bearing its individual amino acid. If that does not happen, the entire process comes to a halt. Regardless of its energy content, if the diet is deficient in essential amino acids, the individual will be malnourished to some degree. Examples of protein deficiency diseases include *marasmus* and *kwashiorkor*. More than 100 million children worldwide have symptoms of these disorders, although neither condition is common in the United States today. AM *Protein Deficiency Diseases*

Several inherited metabolic disorders result from an inability to produce specific enzymes involved with amino acid metabolism. *Phenylketonuria* (fen-il-kē-tō-NOO-rē-uh), or PKU, is an example. Individuals with PKU cannot convert phenylalanine to tyrosine, due to a defect in the enzyme *phenylalanine hydroxylase.* This reaction is an essential step in the synthesis of norepinephrine, epinephrine, dopamine, and melanin. If PKU is not detected in infancy, central nervous system development is inhibited and severe brain damage results. AM *Phenylketonuria*

NUCLEIC ACID METABOLISM

Cells contain both DNA and RNA. In Chapter 3, we considered the replication of DNA, the mechanics of cell division, and the importance of DNA in regulating the structural and functional characteristics of the cell. ∞ *pp. 91–99* The genetic information contained in the DNA of the nucleus is essential to the cell's long-term survival. As a result, the DNA in the nucleus is never catabolized for energy, even if the cell is dying of starvation. But the RNA in the cell is involved in protein synthesis, and RNA molecules are broken down and replaced regularly.

RNA CATABOLISM

In the breakdown of RNA, the bonds between nucleotides are broken and the molecule is disassembled into individual nucleotides. The nucleotides are usually recycled into new nucleic acids. However, they can be catabolized to simple sugars and nitrogenous bases.

RNA catabolism makes a relatively insignificant contribution to the total energy budget of the cell. Proteins account for 30 percent of the cell's weight, and much more energy can be provided through the catabolism of nonessential proteins. Even when RNA is broken down, only the sugars and pyrimidines provide energy. The sugars can enter the glycolytic pathways. The pyrimidines (cytosine and uracil, in RNA) are converted to acetyl-CoA and metabolized through the TCA cycle. The purines (adenine and guanine) cannot be catabolized. Instead, they are deaminated and excreted as **uric acid.** Like urea, uric acid is a relatively nontoxic waste product, but it is far less soluble than urea. Urea and uric acid are called **nitrogenous wastes,** because they are waste products that contain nitrogen atoms.

Normal uric acid concentrations in plasma average 2.7–7.4 mg/dl, depending on gender and age. When plasma concentrations exceed 7.4 mg/dl, *hyperuricemia* (hī-per-ū-ri-SĒ-mē-uh) exists. This condition may affect 18 percent of the U.S. population. At concentrations over 7.4 mg/dl, body fluids are saturated with uric acid. Although symptoms may not appear at once, uric acid crystals may begin to form in body fluids. The condition that then develops is called *gout*. Most cases of hyperuricemia and gout are linked to problems with the excretion of uric acid by the kidneys. AM *Gout*

25

α –Ketoglutarate   Glutamic acid

•**FIGURE 25-13** **Amination.** Amination attaches an amino group to a keto acid. This is an important step in the synthesis of nonessential amino acids. Amino groups can also be attached through transamination (see Figure 25-12a).

•**FIGURE 25-14** A Summary of the Pathways of Catabolism and Anabolism. An overview of major catabolic and anabolic pathways for lipids, carbohydrates, and proteins.

NUCLEIC ACID SYNTHESIS

Most cells synthesize RNA, but DNA synthesis occurs only in cells preparing for mitosis and cell division or for meiosis (nuclear events involved in gamete production). ∞ *p. 96* Messenger RNA (mRNA), transfer RNA (tRNA), and ribosomal RNA (rRNA) are transcribed by different forms of RNA polymerase:

■ Messenger RNA is manufactured only when specific genes are activated. A strand of mRNA has a life span of minutes to hours.

■ Ribosomal RNA and tRNA are more durable than mRNA strands. For example, the average life span of a strand of rRNA is just over 5 days. However, because a typical cell contains roughly 100,000 ribosomes and many times that number of tRNA molecules, their replacement involves a considerable amount of synthetic activity.

Figure 25-14• summarizes the metabolic pathways for lipids, carbohydrates, and proteins. The diagram follows the reactions in a "typical" cell. Yet no one cell can perform all the anabolic and catabolic operations and interconversions required by the body as a whole. As differentiation proceeds, each cell type develops its own complement of enzymes, and this enzyme complement determines the cell's metabolic capabilities. In the presence of such cellular diversity, homeostasis

can be preserved only when the metabolic activities of tissues, organs, and organ systems are coordinated.

☑ How would a diet that is deficient in pyridoxine (vitamin B_6) affect protein metabolism?

☑ Elevated levels of uric acid in the blood can be an indicator of increased metabolism of which organic compound?

☑ Why are high-density lipoproteins (HDLs) considered to be beneficial?

METABOLIC INTERACTIONS

The nutrient requirements of each tissue vary with the types and quantities of enzymes present in the cytoplasm. From a metabolic standpoint, we can consider the body in terms of five distinctive components: the liver, adipose tissue, skeletal muscle, neural tissue, and other peripheral tissues:

1. *The liver.* The liver is the focal point of metabolic regulation and control. Liver cells contain a great diversity of enzymes, and they can break down or synthesize most of the carbohydrates, lipids, and amino acids needed by other body cells. Liver cells have an extensive circulatory supply, so they are in an excellent po-

sition to monitor and adjust the nutrient composition of circulating blood. The liver also contains significant energy reserves in the form of glycogen deposits.

2. *Adipose tissue.* Adipose tissue stores lipids, primarily as triglycerides. Adipocytes are located in many areas; in previous chapters, we noted the presence of fat cells in areolar tissue, in mesenteries, within red and yellow marrows, in the epicardium, and around the eyes.

3. *Skeletal muscle.* Skeletal muscle accounts for almost half of an individual's body weight, and these cells maintain substantial glycogen reserves. In addition, their contractile proteins can be broken down and the amino acids used as an energy source if other nutrients are unavailable.

4. *Neural tissue.* Neural tissue has a high demand for energy, but the cells do not maintain reserves of carbohydrates, lipids, or proteins. Neurons must be provided with a reliable supply of glucose, because they are generally unable to metabolize other molecules. If blood glucose becomes too low, neural tissue in the central nervous system cannot continue to function, and the individual becomes unconscious.

5. *Other peripheral tissues.* Other peripheral tissues do not maintain large metabolic reserves, but they are able to metabolize glucose, fatty acids, or other substrates. Their preferred source of energy varies with the instructions provided by the endocrine system.

The interrelationships among these five components can best be understood by considering the events that occur over a 24-hour period. During this time, the body experiences two broad patterns of metabolic activity: (1) the *absorptive state* and (2) the *postabsorptive state.*

THE ABSORPTIVE STATE

The **absorptive state** is the period following a meal, when nutrient absorption is under way. After a typical meal, the absorptive state continues for about 4 hours. If you are fortunate enough to eat three meals a day, you spend 12 out of every 24 hours in the absorptive state.

A typical meal contains proteins, lipids, and carbohydrates in varying proportions. While you are in the absorptive state, your intestinal mucosa busily absorbs these nutrients. Glucose and amino acids enter the bloodstream, and the hepatic portal vein carries them to the liver. Most of the absorbed fatty acids are packaged in chylomicrons, which enter the lacteals.

Some of the carbohydrates, lipids, and amino acids are broken down at once to provide energy for cellular operations. The remainder is stored, lessening the impact of future shortages. Insulin is the primary hormone of the absorptive state, although various other hormones stimulate amino acid uptake (growth hormone, or GH) and protein synthesis (GH, androgens, and estrogens). We shall now consider the activities under way in specific sites, with particular reference to Table 25-1 and Figure 25-15•.

TABLE 25-1 Regulatory Hormones and Their Effects on Peripheral Metabolism

Hormone	Effect on General Peripheral Tissues	Selective Effects on Target Tissues
Absorptive State		
Insulin	Increased glucose uptake and utilization	*Liver:* Glycogenesis *Adipose tissue:* Lipogenesis *Skeletal muscle:* Glycogenesis
Insulin and growth hormone	Increased amino acid uptake and protein synthesis	*Skeletal muscle:* Fatty acid catabolism
Androgens, estrogens	Increased amino acid use in protein synthesis	*Skeletal muscle:* Muscle hypertrophy (especially androgens)
Postabsorptive State		
Glucagon		*Liver:* Glycogenolysis
Epinephrine		*Liver:* Glycogenolysis *Adipose tissue:* Lipolysis
Glucocorticoids	Decreased use of glucose; increased reliance on ketone bodies and fatty acids	*Liver:* Glycogenolysis *Adipose tissue:* Lipolysis, gluconeogenesis *Skeletal muscle:* Glycogenolysis, protein breakdown, amino acid release
Growth hormone	Complements effects of glucocorticoids	Acts with glucocorticoids

25

Nutrient Source   Distributed in Bloodstream   Destinations

•**FIGURE 25-15 The Absorptive State.** During the absorptive state, the primary metabolic goal is anabolic activity, especially growth and the storage of energy reserves.

The Liver

The liver regulates the levels of glucose and amino acids in the blood arriving in the hepatic portal vein before that blood reaches the inferior vena cava. Despite the continuous absorption of glucose at the intestinal mucosa, blood glucose levels do not sky-rocket, because the liver cells, under insulin stimulation, remove glucose from blood delivered by the hepatic portal circulation. Blood glucose levels do rise, but only from about 90 mg/dl to perhaps 150 mg/dl, even after a meal rich in carbohydrates. The liver uses some of the absorbed glucose to generate the ATP required to perform synthetic operations. Glycogenesis (glycogen formation) continues until glycogen accounts for about 5 percent of the total liver weight. If excess glucose still remains in the bloodstream, hepatocytes (liver cells) use glucose to synthesize triglycerides. Although small quantities of lipids are normally stored in the liver, most of the synthesized triglycerides are bound to transport proteins and released into the bloodstream as VLDLs. Peripheral tissues, primarily adipose tissues, then absorb these lipids for storage.

The liver does not control circulating levels of amino acids as precisely as it does glucose concentrations. Plasma amino acid levels normally range between 35 and 65 mg/dl, but they may become elevated after a protein-rich meal. The absorbed amino acids are used to support the synthesis of proteins, including plasma proteins and the proenzymes of the clotting system. Liver cells can also synthesize many amino acids, and an amino acid present in abundance may be converted to another, less common type and released into the bloodstream.

Most of the lipids absorbed by the digestive tract do not reach the liver. Triglycerides, cholesterol, and large fatty acids reach the general venous circulation in chylomicrons that are transported in the thoracic duct. Most of these lipids are absorbed by other tissues.

Adipose Tissue

During the absorptive state, adipocytes remove fatty acids and glycerol from the bloodstream. Lipids continue to be removed from the blood for 4–6 hours after you have eaten a fatty meal. Over this period, the presence of chylomicrons may give the plasma a milky appearance, a characteristic called **lipemia** (lip-Ē-mē-uh).

Adipocytes are particularly active in absorbing these lipids and in synthesizing new triglycerides for storage. At normal blood glucose concentrations, any glucose entering these cells will be catabolized to provide the energy needed for lipogenesis (lipid synthesis). Adipocytes also absorb amino acids as needed for protein synthesis. Although these cells can use glucose or amino acids to manufacture triglycerides, they do so only if circulating concentrations are unusually high.

If on a daily basis you take in more nutrients during the absorptive state than you catabolize during the postabsorptive state, the fat deposits in your adipose tissue will enlarge. Most of the increase represents an increase in the size of individual adipocytes. An increase in the total number of adipocytes does not ordinarily occur, except in children before puberty and in extremely obese adults.

> ### Obesity
> A useful definition of **obesity** is 20 percent above ideal weight, because that is the point at which serious health risks appear. On the basis of that criterion, 20–30 percent of men and 30–40 percent of women in the United States can be considered obese. Simply stated, obese individuals are taking in more food energy than they use. Unfortunately, there is very little agreement as to the underlying cause of this situation. The two major categories of obesity, *regulatory obesity* and *metabolic obesity*, are described in the *Applications Manual.* AM *Obesity*

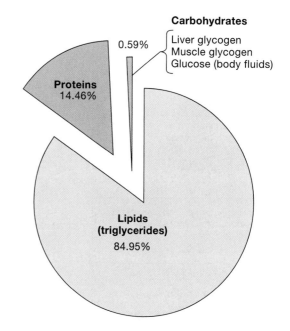

•**FIGURE 25-16** **Metabolic Reserves.** The distribution of the estimated metabolic reserves of a 70-kg individual.

Skeletal Muscle, Neural Tissue, and Other Peripheral Tissues

When blood glucose and amino acid concentrations are elevated, insulin is released from the pancreatic islets, and all tissues increase their rates of absorption and utilization. Glucose is catabolized for energy, and the amino acids are used to build proteins.

Glucose is normally retained in the body, because the kidneys prevent the loss of glucose molecules in urine. The kidneys' ability to conserve glucose breaks down only when blood glucose concentrations are extraordinarily high, somewhere in excess of 180 mg/dl. The amino acid content of urine is not as carefully regulated; amino acids commonly appear in the urine after a protein-rich meal.

When blood glucose levels are elevated, most cells ignore the circulating lipids, so the adipocytes have little competition. In resting skeletal muscles, a significant portion of the metabolic demand is met through the catabolism of fatty acids. Glucose molecules are used to build glycogen reserves, which may account for 0.5–1 percent of the weight of each muscle fiber.

THE POSTABSORPTIVE STATE

The **postabsorptive state** is the period when nutrient absorption is not under way and your body must rely on internal energy reserves to continue meeting its energy demands. You spend roughly 12 hours each day in the postabsorptive state, although a person who is skipping meals can extend it considerably. Metabolic activity in the postabsorptive state is focused on the mobilization of energy reserves and the maintenance of normal blood glucose levels. These activities are coordinated by several hormones, including glucagon, epinephrine, glucocorticoids, and growth hormone (Table 25-1).

The metabolic reserves of a typical 70-kg (154-lb) individual include carbohydrates, lipids, and proteins (Figure 25-16•). Due to its high energy content, the adipose tissue represents a disproportionate percentage of the total reserve in the form of triglycerides. Most of the available protein reserve is located in the contractile proteins of skeletal muscle. Carbohydrate reserves are relatively small and sufficient for only a few hours or, at most, overnight.

We shall now examine the events under way in specific tissues and organs during this period, as diagrammed in Figure 25-17•.

The Liver

As the absorptive state ends, your intestinal cells stop providing glucose to the portal circulation. At first, the peripheral tissues continue to remove glucose from the blood, and blood glucose levels begin to decline. The liver responds by reducing its synthetic activities. When plasma concentrations fall below 80 mg/dl, liver cells begin breaking down glycogen reserves and releasing glucose into the bloodstream. This glycogenolysis occurs in response to a rise in circulating levels of glucagon and epinephrine. The liver contains 75–100 g of glycogen that is readily available,

25

•**FIGURE 25-17 The Postabsorptive State.** In the postabsorptive state, energy reserves are mobilized, and peripheral tissues (except neural tissues) shift from glucose catabolism to fatty acid or ketone body catabolism to obtain energy.

and this reserve is adequate to maintain blood glucose levels for about 4 hours.

As glycogen reserves decline and plasma glucose levels fall to about 70 mg/dl, liver cells begin to make glucose in an attempt to stabilize blood glucose levels. The shift from glycogenolysis to gluconeogenesis occurs under stimulation by *glucocorticoids,* steroid hormones from the adrenal cortex. ∞ *p. 602*

Gluconeogenesis By gluconeogenesis, liver cells synthesize glucose molecules from smaller carbon fragments. In effect, any carbon fragment that can be converted to pyruvic acid or one of the three-carbon compounds involved in glycolysis in the cytoplasm can be used to synthesize glucose. (We discussed the conversion of lactic acid to glucose in the liver in Chapter 10. (∞ *p. 295*) With glucose already in short supply, lipids and amino acids must be catabolized to provide the ATP molecules needed for these syntheses.

Utilization of Lipids In the postabsorptive state, your liver absorbs fatty acids and glycerol from the blood. The glycerol molecules are converted to glucose. Fatty acids are broken down through beta-oxidation to produce large quantities of acetyl-CoA. However, because the enzymatic reaction that converts pyruvic acid to acetyl-CoA cannot be reversed, acetyl-CoA cannot be used to synthesize glucose. Instead,

■ Some of the acetyl-CoA molecules deliver their two-carbon acetyl fragments to the TCA cycle, where they are broken down. The ATP generated can then be used to support gluconeogenesis.

■ In addition, some of the molecules of acetyl-CoA are converted to special compounds that can be utilized by peripheral tissues. These compounds, called **ketone bodies,** are organic acids that are also produced during the catabolism of amino acids.

Utilization of Amino Acids Before an amino acid can be used for either gluconeogenesis or energy production by means of breakdown in the TCA cycle, the amino group ($-NH_2$) must be removed. The structure of the remaining carbon chain determines its subsequent fate. After deamination, some amino acids can be converted to molecules of pyruvic acid or to one of the intermediary molecules of the TCA cycle. Other amino acids—including most of the essential amino acids—can be converted only to acetyl-CoA and must be either broken down further or converted to ketone bodies.

The liver is the most active site of amino acid breakdown. There the ammonia generated by deamination is converted to urea. This relatively harmless, water-soluble compound circulates in the bloodstream until it is excreted in urine. The urea concentration of blood rises during the postabsorptive period, because the rate of amino acid catabolism increases.

Ketone Bodies During the postabsorptive state, liver cells conserve glucose and break down lipids and amino acids. Both lipid catabolism and amino acid catabolism generate acetyl-CoA. As the concentration of acetyl-CoA rises, ketone bodies begin to form. There are three ketone bodies: (1) **acetoacetate** (as-ē-tō-AS-e-tāt), (2) **acetone** (AS-e-tōn), and (3) **betahydroxybutyrate** (bā-ta-hī-droks-ē-BŪ-te-rāt). Liver cells do not metabolize any of the ketone bodies, and these compounds diffuse through the cytoplasm and into the general circulation. Cells in peripheral tissues then absorb the ketone bodies and reconvert them to acetyl-CoA for introduction into the TCA cycle.

The normal concentration of ketone bodies in the blood is about 30 mg/dl, and very few of these compounds appear in urine. During even a brief period of fasting, the increased production of ketone bodies results in **ketosis** (kē-TŌ-sis), a high concentration of ketone bodies in body fluids. In ketosis, the concentration of ketone bodies is elevated in blood, a condition called **ketonemia** (kē-tō-NĒ-mē-uh), and in urine, a condition called **ketonuria** (kē-tō-NOO-rē-uh). Ketonemia and ketonuria are clear indications that the catabolism of proteins and lipids is under way. Acetone, which diffuses out of the pulmonary capillaries and into the alveoli very readily, can be smelled on the individual's breath.

Summary In summary, during the postabsorptive state your liver attempts to stabilize blood glucose concentrations, first by the breakdown of glycogen reserves and later by gluconeogenesis. Over the remainder of the postabsorptive state, the combination of lipid and amino acid catabolism provides the necessary ATP and generates large quantities of ketone bodies that diffuse into the bloodstream.

Ketoacidosis

A ketone body is an acid that dissociates in solution, releasing a hydrogen ion. As a result, the appearance of ketone bodies in the bloodstream—ketonemia—a threat to the plasma pH that must be controlled by buffers. During prolonged starvation, ketone levels continue to rise. Eventually, the buffering capacities are exceeded and a dangerous drop in pH occurs. This acidification of the blood is called **ketoacidosis** (kē-tō-as-i-DŌ-sis). In severe ketoacidosis, the circulating concentration of ketone bodies can reach 200 mg/dl and the pH may fall below 7.05. A pH that low can disrupt tissue activities and cause coma, cardiac arrhythmias, and death.

In *diabetes mellitus,* most peripheral tissues cannot use glucose, because they lack insulin. ∞ *pp. 606–607* Under these circumstances, cells survive by catabolizing lipids and proteins. The result is the production of large numbers of ketone bodies. This condition leads to *diabetic ketoacidosis,* the most common and life-threatening form of ketoacidosis.

Adipose Tissue

Adipose tissue contains a tremendous storehouse of energy in the form of triglycerides. Fat accounts for approximately 15 percent of the body weight of the average individual—enough to provide a 1–2 month reserve of ATP. Although some areas, including the eyelids, nose, and the backs of the hands and feet, rarely contain adipose tissue, other regions are preferential sites of deposition. Typically, an individual's adipose tissue is distributed among the hypodermis (50 percent), the greater omentum (10–15 percent), between muscles (5–8 percent), and packed around the kidneys (12 percent) and reproductive organs (15–20 percent).

As blood glucose levels decline, the rate of triglyceride synthesis falls. Under the stimulation of epinephrine, glucocorticoids, and growth hormone, adipocytes soon begin breaking down their lipid reserves, releasing fatty acids and glycerol into the bloodstream. This process, called *fat mobilization,* continues for the duration of the postabsorptive state. A normal individual retains about 2 months' supply of energy in the triglycerides of adipose tissue. The evolutionary advantages are obvious: The retention of an energy reserve provides a buffer against daily, monthly, and even seasonal changes in the available food supply.

Skeletal Muscle

At the start of the postabsorptive state, skeletal muscles obtain energy by breaking down their glycogen reserves and catabolizing the glucose released. As the concentrations of fatty acids and ketone bodies in the bloodstream increase, these substrates become increasingly important as an energy source.

Skeletal muscle as a whole contains twice as much glycogen as the liver, but it is distributed throughout your muscular system. These glucose reserves are not directly available to other tissues, because the lack of a key enzyme makes skeletal muscle cells unable to release glucose into the bloodstream. Once skeletal muscle fibers metabolize fatty acids as an energy source, they continue to break down their glycogen reserves and convert the pyruvic acid molecules to lactic acid. Lactic acid then diffuses out of the muscle fibers and into the bloodstream. However, even if all of the available glycogen reserves in your body were mobilized as glucose or as lactic acid, the energy provided would be only enough to get you through a good night's sleep. If the postabsorptive state continues for an unusually long period, long enough that lipid reserves are being depleted, muscle proteins will be broken down. The amino acids that are released diffuse into the blood for use by the liver in gluconeogenesis.

Other Peripheral Tissues

With rising plasma concentrations of lipids and ketone bodies and falling blood glucose levels, peripheral

25

tissues gradually decrease their reliance on glucose. Circulating ketone bodies and fatty acids are absorbed and converted to acetyl-CoA for entry into the TCA cycle.

Neural Tissue

Neurons are unusual in that they continue "business as usual" during the postabsorptive state. Neurons are dependent on a reliable supply of glucose, and changes in the activity of the liver, adipose tissue, skeletal muscle, and other peripheral tissues are intended to ensure that the supply of glucose to the nervous system continues unaffected, despite daily or even weekly changes in the availability of nutrients. Only after a prolonged period of starvation will neural tissue begin to metabolize ketone bodies and lactic acid molecules as well as glucose. AM *Adaptations to Starvation*

☑ What process in the liver would you expect to increase after you have eaten a high-carbohydrate meal?

☑ Why does the amount of urea in blood increase during the postabsorptive state?

☑ If a cell accumulates more acetyl-CoA than it can metabolize by way of the TCA cycle, what products are likely to form?

DIET AND NUTRITION

The postabsorptive state can be maintained for a considerable period, but homeostasis can be maintained indefinitely only if your digestive tract regularly absorbs fluids, organic substrates, minerals, and vitamins, keeping pace with cellular demands. The absorption of nutrients from food is called **nutrition.**

The individual requirement for each nutrient varies from day to day and from person to person. *Nutritionists* attempt to analyze a diet in terms of its ability to meet the needs of a specific individual. A **balanced diet** contains all the ingredients necessary to maintain homeostasis, including adequate substrates for energy generation, essential amino acids and fatty acids, minerals, and vitamins. In addition, your diet must include enough water to replace losses in urine, feces, and evaporation. A balanced diet prevents **malnutrition,** an unhealthy state resulting from inadequate or excessive intake of one or more nutrients.

FOOD GROUPS AND FOOD PYRAMIDS

One method of avoiding malnutrition is to include members of each of the six **basic food groups** in the diet. These are (1) the *milk, yogurt, and cheese group;* (2) the *meat, poultry, fish, dry beans, eggs, and nuts group;* (3) the *vegetable group;* and (4) the *fruit group;* (5) the *bread, cereal, rice, and pasta group;* and (6) the *fats, oils, and sweets group.* Each group differs from the others in the typical balance of proteins, carbohydrates, and

lipids contained as well as in the amount and identity of vitamins and minerals. The six groups are arranged in a *food pyramid,* which has the bread, cereal, rice, and pasta group at the bottom (Figure 25-18●). The aim of this display is to emphasize the need to restrict dietary fats, oils, and sugar and to increase your consumption of breads and such, which are rich in complex carbohydrates (polysaccharides such as starch).

These are rather artificial groupings at best, and they are downright misleading at worst. For instance, many processed foods belong in a combination of these groups. What is important is that you obtain nutrients in sufficient *quantity* (adequate to meet your energy needs) and *quality* (including essential amino acids, fatty acids, vitamins, and minerals). How these nutrients are packaged is a secondary concern. There is nothing magical about the number six; since 1940, the U.S. government has advocated 11, 7, 4, or 6 food groups at various times. The key is making intelligent choices about what you eat. The wrong choices can lead to malnutrition even if all six groups are represented.

For example, consider the case of the essential amino acids. Your liver cannot synthesize any of these amino acids, and you must obtain them from your diet. Some members of the meat and dairy groups, such as beef, fish, poultry, eggs, and milk, contain all the essential amino acids in sufficient quantities. They are said to contain **complete proteins.** Many plants also contain adequate *amounts* of protein, but these are **incomplete proteins,** which are deficient in one or more of the essential amino acids. Vegetarians, who restrict themselves to the fruit and vegetable groups (with or without the bread group), must become adept at juggling the constituents of their meals to include a combination of ingredients that will meet all their amino acid requirements. Even with a proper balance of amino acids, vegetarians face a significant problem, because vitamin B_{12} is obtained only from animal products or from fortified cereals or tofu. (Although some health-food products, such as *Spirulina*®, are marketed as sources of this vitamin, the B_{12} present is in a form that humans cannot utilize.)

NITROGEN BALANCE

A variety of important compounds in the body contain nitrogen atoms. These **N compounds** include:

- Amino acids, which are part of the framework of all proteins and protein derivatives, such as glycoproteins and lipoproteins.

- Purines and pyrimidines, the nitrogenous bases of RNA and DNA.

- *Creatine,* important in energy storage in muscle tissue (as creatine phosphate).

- *Porphyrins,* complex ring-shaped molecules that bind metal ions and are essential to the function of hemoglobin, myoglobin, and the cytochromes.

•**FIGURE 25-18**
The Food Pyramid

A Guide to Daily Food Choices

Nutrient Group	Provides	Deficiencies
Fats, oil, sweets	Calories	The majority are deficient in most minerals and vitamins
Milk, yogurt, cheese	Complete proteins; fats; carbohydrates; calcium; potassium; magnesium; sodium; phosphorus; vitamins A, B_{12}, pantothenic acid, thiamine, riboflavin	Dietary fiber, vitamin C
Meat, poultry, fish, dry beans, eggs, nuts	Complete proteins; fats; potassium; phosphorus; iron; zinc; vitamins E, thiamine, B_6	Carbohydrates, dietary fiber, several vitamins
Fruits	Carbohydrates; vitamins A, C, E, folacin; dietary fiber; potassium	Many are low in fats, calories, and protein
Vegetables	Carbohydrates; vitamins A, C, E, folacin; dietary fiber; potassium	Many are low in fats, calories, and protein
Bread, cereal, rice, pasta	Carbohydrates; vitamins E, thiamine, niacin, folacin; calcium; phosphorus; iron; sodium; dietary fiber	Fats

Despite the importance of nitrogen to these compounds, your body neither stores nitrogen nor maintains reserves of N compounds as it does carbohydrates (glycogen) and lipids (triglycerides). Your body can synthesize the carbon chains of the N compounds, but you must obtain nitrogen atoms either by recycling N compounds already in the body or by absorbing nitrogen from your diet. You are in **nitrogen balance** if the amount of nitrogen you absorb from the diet balances the amount you lose in urine and feces. This is the normal condition, and it means that the rates of N compound synthesis and breakdown are equivalent.

Growing children, athletes, persons recovering from an illness or injury, and pregnant or lactating women actively synthesize N compounds, so these individuals must absorb more nitrogen than they excrete. Such individuals are in a state of **positive nitrogen balance.** When excretion exceeds ingestion, a **negative nitrogen balance** exists. This is an extremely unsatisfactory situation: The body contains only about a kilogram of nitrogen tied up in N compounds, and a decrease of one-third can be fatal. Even when energy reserves are mobilized, as during starvation, carbohydrates and lipid reserves are broken down first and N compounds are conserved.

Like N compounds, minerals and vitamins are essential components of the diet. Your body cannot synthesize minerals, and your cells can generate only a small quantity of very few vitamins. We shall consider them next.

25

MINERALS

Minerals are inorganic ions released through the dissociation of electrolytes. Minerals are important for three reasons:

1. *Ions such as sodium and chloride determine the osmotic concentrations of body fluids.* Potassium is important in maintaining the osmotic concentration of the cytoplasm inside body cells.

2. *Ions in various combinations play major roles in important physiological processes,* as we have discussed. These processes include the maintenance of transmembrane potentials, the construction and maintenance of the skeleton, muscle contraction, action potential generation, neurotransmitter release, hormone production, blood clotting, the transport of respiratory gases, buffer systems fluid absorption, and waste removal, to name but a few examples encountered in earlier chapters.

3. *Ions are essential cofactors in a variety of enzymatic reactions.* For example, the calcium-dependent ATPase

in skeletal muscle also requires the presence of magnesium ions, and another ATPase required for the conversion of glucose to pyruvic acid needs both potassium and magnesium ions. Carbonic anhydrase, important in CO_2 transport, buffering systems, and gastric acid secretion, requires the presence of zinc ions. Finally, each component of the electron transport system requires an iron atom, and the final cytochrome (a_3) of the ETS must bind a copper ion as well.

The major minerals and a summary of their functional roles are given in Table 25-2. Your body contains significant reserves of several important minerals; these reserves help reduce the effects of variations in dietary supply. The reserves are often small, however, and chronic dietary reductions can lead to a variety of clinical problems. Alternatively, because storage capabilities are limited, a dietary excess of mineral ions can be equally dangerous.

Problems involving iron are particularly common. The body of a healthy male contains about 3.5 g of

TABLE 25-2 Minerals and Mineral Reserves*

Mineral	Significance	Total Body Content	Primary Route of Excretion	Recommended Daily Intake
Bulk Minerals				
Sodium	Major cation in body fluids; essential for normal membrane function	110 g, primarily in body fluids	Urine, sweat, feces	0.5–1.0 g
Potassium	Major cation in cytoplasm; essential for normal membrane function	140 g, primarily in cytoplasm	Urine	1.9–5.6 g
Chloride	Major anion in body fluids	89 g, primarily in body fluids	Urine, sweat	0.7–1.4 g
Calcium	Essential for normal muscle and neuron function, and normal bone structure	1.36 kg, primarily in skeleton	Urine, feces	0.8–1.2 g
Phosphorus	As phosphate in high-energy compounds, nucleic acids, and bone matrix	744 g, primarily in skeleton	Urine, feces	0.8–1.2 g
Magnesium	Cofactor of enzymes, required for normal membrane functions	29 g (skeleton, 17 g; cytoplasm and body fluids, 12 g)	Urine	0.3–0.4 g
Trace Minerals				
Iron	Component of hemoglobin, myoglobin, cytochromes	3.9 g (1.6 g stored as ferritin or hemosiderin)	Urine (traces)	10–18 mg
Zinc	Cofactor of enzyme systems, notably carbonic anhydrase	2 g	Urine, hair (traces)	15 mg
Copper	Required as cofactor for hemoglobin synthesis	127 mg	Urine, feces (traces)	2–3 mg
Manganese	Cofactor for some enzymes	11 mg	Feces, urine (traces)	2.5–5 mg
Cobalt	Cofactor for transaminations	1.1 g	Feces, urine	0.0001 mg
Selenium	Antioxidant	Variable	Feces, urine	0.05–0.2 mg
Chromium	Cofactor for glucose metabolism	0.0006	Feces, urine	0.005–0.2 mg

*For information on the effects of deficiencies and excesses, see Table 27-2, p. 996.

iron in the ionic form Fe^{2+}. Of that amount, 2.5 g is bound to the hemoglobin of circulating red blood cells, and the rest is stored in the liver and bone marrow. In females, the total body iron content averages 2.4 g, with roughly 1.9 g incorporated into red blood cells. Thus a woman's iron reserves consist of only 0.5 g, half that of a typical man. If the diet contains inadequate amounts of iron, women are therefore more likely to develop signs of iron deficiency than are men.

VITAMINS

Vitamins are assigned to either of two groups, according to their chemical structure and characteristics: (1) *fat-soluble vitamins* or (2) *water-soluble vitamins.*

Fat-Soluble Vitamins

Vitamins A, D, E, and K are the **fat-soluble vitamins.** These vitamins are absorbed primarily from your digestive tract along with the lipid contents of micelles. However, when exposed to sunlight, your skin can synthesize small amounts of vitamin D, and intestinal bacteria produce some vitamin K. The mode of action of these vitamins is unknown. Vitamin A has long been recognized as a structural component of the visual pigment retinal, but its more general metabolic effects are not well understood. Vitamin D is ultimately converted to calcitriol, which binds to cytoplasmic receptors within the intestinal epithelium and promotes an increase in the rate of intestinal calcium and phosphorus absorption. Vitamin E is thought to stabilize intracellular membranes. Vitamin K is a necessary participant in a reaction essential to the synthesis of several proteins, including at least three of the clotting factors. Current information about the fat-soluble vitamins is summarized in Table 25-3.

Because they dissolve in lipids, fat-soluble vitamins normally diffuse into cell membranes and other lipids in the body, including the lipid inclusions in the liver and adipose tissue. Your body therefore contains a significant reserve of these vitamins, and normal metabolic operations can continue for several months after dietary sources have been cut off. For this reason, symptoms of **avitaminosis** (ā-vī-ta-min-Ō-sis), or **vitamin deficiency disease,** rarely result from dietary insufficiency of fat-soluble vitamins. However, avitaminosis involving either fat-soluble or water-soluble vitamins can be caused by a variety of factors other than dietary deficiencies. An inability to absorb a vitamin from the digestive tract, inadequate storage, or excessive demand may each produce the same result.

As Table 25-3 points out, *too much* of a vitamin can produce effects just as unpleasant as *too little.* **Hypervitaminosis** (hi-per-vī-ta-min-Ō-sis) occurs when the dietary intake exceeds the body's abilities to store, utilize, or excrete a particular vitamin. This condition most commonly involves one of the fat-soluble vitamins because the excess is retained and stored in body lipids.

TABLE 25-3 The Fat-Soluble Vitamins

Vitamin	Significance	Sources	Daily Requirement	Effects of Deficiency	Effects of Excess
A	Maintains epithelia; required for synthesis of visual pigments	Leafy green and yellow vegetables	1 mg	Retarded growth, night blindness, deterioration of epithelial membranes	Liver damage, skin peeling, CNS effects (nausea, anorexia)
D (steroids including cholecalciferol, or D_3)	Required for normal bone growth, calcium and phosphorus absorption at gut and retention at kidneys	Synthesized in skin exposed to sunlight	None*	Rickets, skeletal deterioration	Calcium deposits in many tissues, disrupting functions
E (tocopherols)	Prevents breakdown of vitamin A and fatty acids	Meat, milk, vegetables	12 mg	Anemia, other problems suspected	None reported
K	Essential for liver synthesis of prothrombin and other clotting factors	Vegetables; production by intestinal bacteria	0.7–0.14 mg	Bleeding disorders	Liver dysfunction, jaundice

*Unless sunlight exposure is inadequate for extended periods and alternative sources (fortified milk products) are unavailable.

25

Hypervitaminosis

"If a little is good, a lot must be better" is a common but dangerously incorrect attitude about vitamins. When the dietary supply of fat-soluble vitamins is excessive, the tissue lipids absorb the additional vitamins. Because these vitamins will later diffuse back into the bloodstream, once the symptoms of hypervitaminosis appear, they are likely to persist.

When absorbed in massive amounts (from ten to thousands of times the recommended daily allowance), fat-soluble vitamins can produce acute symptoms of *vitamin toxicity*. Vitamin A toxicity is the most common condition; it afflicts some children whose parents are overanxious about proper nutrition and vitamins. A single enormous overdose can produce nausea, vomiting, headache, dizziness, lethargy, and even death. Chronic overdose can lead to hair loss, joint pain, hypertension, weight loss, and liver enlargement.

At least 19 cases of vitamin D toxicity were reported in the Boston area during 1992. Symptoms included fatigue, weight loss, and potentially severe damage to the kidneys and cardiovascular system. The problems resulted from drinking milk fortified with vitamin D. Due to problems at one dairy, some of the milk sold had over 230,000 units of vitamin D per quart instead of the usual 400 units per quart. The incident highlighted the need for quality control in the production, and care in the consumption, of vitamin supplements.

Water-Soluble Vitamins

Most of the **water-soluble vitamins** are components of coenzymes (Table 25-4). For example, NAD is derived from niacin, FAD from vitamin B_2 (riboflavin), and coenzyme A from vitamin B_5 (pantothenic acid).

Water-soluble vitamins are rapidly exchanged between the fluid compartments of the digestive tract and the circulating blood, and excessive amounts are readily excreted in urine. For this reason, hypervitaminosis involving water-soluble vitamins is relatively uncommon except among individuals taking large doses of vitamin supplements. However, only vitamins B_{12} and C are stored in significant quantities, so insufficient intake of other water-soluble vitamins can lead to initial symptoms of vitamin deficiency within a period of days to weeks.

The bacterial inhabitants of our intestines help prevent deficiency diseases by producing small amounts of five of the nine water-soluble vitamins, in addition to fat-soluble vitamin K. Your intestinal epithelium can easily absorb all the water-soluble vitamins except B_{12}. The B_{12} molecule is large, and, as we discussed in Chapter 24, it must be bound to the *intrinsic factor* from the gastric mucosa before absorption can occur. ∞ *p. 864*

TABLE 25-4 The Water-Soluble Vitamins

Vitamin	Significance	Sources	Daily Requirement	Effects of Deficiency	Effects of Excess
B_1 (thiamine)	Coenzyme in decarboxylations	Milk, meat, bread	1.9 mg	Muscle weakness, CNS and cardiovascular problems, including heart disease; called *beriberi*	Hypotension
B_2 (riboflavin)	Part of FMN and FAD	Milk, meat	1.5 mg	Epithelial and mucosal deterioration	Itching, tingling
Niacin (nicotinic acid)	Part of NAD	Meat, bread, potatoes	14.6 mg	CNS, GI, epithelial, and mucosal deterioration; called *pellagra*	Itching, burning; vasodilation; death after large dose
B_5 (pantothenic acid)	Part of acetyl-CoA	Milk, meat	4.7 mg	Retarded growth, CNS disturbances	None reported
B_6 (pyridoxine)	Coenzyme in amino acid and lipid metabolisms	Meat	1.42 mg	Retarded growth, anemia, convulsions, epithelial changes	CNS alterations, perhaps fatal
Folacin (folic acid)	Coenzyme in amino acid and nucleic acid metabolisms	Vegetables, cereal, bread	0.1 mg	Retarded growth, anemia, gastrointestinal disorders, developmental abormalities	Few noted except at massive doses
B_{12} (cobalamin)	Coenzyme in nucleic acid metabolism	Milk, meat	4.5 μg	Impaired RBC production, causing *pernicious anemia*	Polycythemia
Biotin	Coenzyme in decarboxylations	Eggs, meat, vegetables	0.1–0.2 mg	Fatigue, muscular pain, nausea, dermatitis	None reported
C (ascorbic acid)	Coenzyme; delivers hydrogen ions, antioxidant	Citrus fruits	60 mg	Epithelial and mucosal deterioration; called *scurvy*	Kidney stones

DIET AND DISEASE

Diet has a profound influence on general health. We have already considered the effects of too many or too few nutrients, above-normal or below-normal concentrations of minerals, and hypervitaminosis or avitaminosis. More-subtle, long-term problems can occur when the diet includes the wrong proportions or combinations of nutrients. The average diet in the United States contains too many calories, and lipids provide too great a proportion of those calories. This diet increases the incidence of obesity, heart disease, atherosclerosis, hypertension, and diabetes in the U.S. population.

☑ Would an athlete in intensive training try to maintain a positive or a negative nitrogen balance?

☑ How would a decrease in the amount of bile salts in the bile affect the amount of vitamin A in the body?

Alcohol: A Risky Diversion

Alcohol production and sales are big business throughout the Western world. Beer commercials on television, billboards advertising various brands of liquor, TV or movie characters enjoying a drink—all demonstrate the significance of alcohol in our society. Many people are unaware of the medical consequences of this cultural fondness for alcohol. Problems with alcohol are usually divided into those stemming from alcohol abuse and those involving alcoholism. The boundary between these conditions is hazy. *Alcohol abuse* is the general term for overuse and the resulting behavioral and physical effects of overindulgence. *Alcoholism* is chronic alcohol abuse with the physiological changes associated with addiction to other CNS-active drugs. Alcoholism has received the most attention in recent years, although alcohol abuse—especially when combined with driving an automobile—is also in the limelight.

Consider the following statistics:

- Alcoholism affects more than 10 million people in the United States alone. The lifetime risk of developing alcoholism is estimated at 10 percent.

- Alcoholism is probably the most expensive health problem today, with an annual estimated direct cost of more than $136 billion. Indirect costs, in terms of damage to automobiles, property, and innocent accident victims, are unknown.

- An estimated 25–40 percent of U.S. hospital patients are undergoing treatment related to alcohol consumption. Approximately 200,000 deaths occur annually due to alcohol-related medical conditions. Some major clinical conditions are caused almost entirely by alcohol consumption. For example, alcohol is responsible for 60–90 percent of all liver disease in the United States.

- Alcohol affects all physiological systems. Major clinical symptoms of alcoholism include (1) disorientation and confusion (nervous system); (2) ulcers, diarrhea, and cirrhosis (digestive system); (3) cardiac arrhythmias, cardiomyopathy, and anemia (cardiovascular system); (4) depressed sexual drive and testosterone levels (reproductive system); and (5) itching and angiomas (integumentary system).

- The toll on newborn infants has risen steadily since the 1960s as the number of women drinkers has increased. Women consuming 1 ounce of alcohol per day during pregnancy have a higher rate of spontaneous abortion and produce children with lower birthweights than do women who consume no alcohol. Women who drink heavily may bear children with *fetal alcohol syndrome (FAS)*. This condition is marked by characteristic facial abnormalities, a small head, slow growth, and mental retardation.

- Alcohol abuse is considerably more widespread than alcoholism. Although the medical effects are less well-documented, they are certainly significant.

Several factors interact to produce alcoholism. The primary risk factors are gender (males are more likely to become alcoholics than are females) and a family history of alcoholism. There does appear to be a genetic component; a gene on chromosome 11 has been implicated in some inherited forms. The relative importance of genes versus social environment has been difficult to assess. It is likely that alcohol abuse and alcoholism result from a variety of factors. Treatment may consist of counseling and behavior modification. To be successful, treatment must include the total avoidance of alcohol. Supporting groups, such as Alcoholics Anonymous, can be very helpful in providing a social framework for abstinence. The use of the drug *disulfiram (Antabuse®)* has not proved to be as successful as originally anticipated. Antabuse sensitizes the individual to alcohol so that a drink produces intense nausea; it was anticipated that this would be an effective deterrent. Clinical tests indicated that it could increase the time between drinks but could not prevent drinking altogether.

25

Eating Disorders

Eating disorders are psychological problems that result in either inadequate or excessive food consumption. The most common conditions are **anorexia nervosa,** characterized by self-induced starvation, and **bulimia,** characterized by feeding binges followed by vomiting, laxative use, or both. Adolescent females account for most cases of anorexia nervosa and bulimia. Males account for only 5–10 percent of anorexia or bulimia cases. A common thread in the two conditions is an obsessive concern about food and body weight.

According to current estimates, the incidence of anorexia nervosa in the United States ranges from 0.4 to 1.5 per 100,000 population. The incidence among Caucasian women age 12–18 is estimated to be 1 percent. A typical person with this condition is an adolescent Caucasian woman whose weight is roughly 30 percent below normal levels. Although underweight, she is convinced that she is too fat, so she refuses to eat normal amounts of food.

The psychological factors responsible for anorexia are complex. Young women with this condition tend to be high achievers who are attempting to reach an "ideal" weight that will be envied and admired and thereby provide a sense of security and accomplishment. The factors thus tend to be a combination of their view of society ("thin is desirable or demanded"), their view of themselves ("I am not yet thin enough"), and a desire to be able to control their fate ("I can decide when to eat"). Female models, figure skaters, gymnasts, and theater and arts majors of any age may feel forced to drop weight to remain competitive. The few male anorexics diagnosed typically face comparable stresses. They tend to be athletes, such as jockeys or wrestlers, who need to maintain a minimal weight to succeed in their careers.

Young anorexic women may continue to starve themselves down to a weight of 30–35 kg (66–77 lb). Dry skin, peripheral edema, an abnormally low heart rate and blood pressure, a reduction in bone and muscle mass, and a cessation of menstrual cycles are relatively common symptoms. Some of the changes, especially in bone mass, can be permanent. Treatment is difficult, and only 50–60 percent of anorexics who regain normal weight stay there for 5 years or more. Death rates from severe anorexia nervosa range from 10 to 15 percent.

Bulimia is more common than anorexia nervosa. In this condition, the individual goes on an "eating binge" that may involve a meal that lasts 1–2 hours and may include 20,000 or more calories. The meal is followed by induced vomiting, commonly accompanied by the use of laxatives (drugs that promote the movement of the material through the digestive tract) and diuretics (drugs that promote fluid loss in urine). These often-expensive binges may occur several times each week, separated by periods of either normal eating or fasting.

Bulimia generally involves women of the same age group as anorexia nervosa. The actual incidence is difficult to determine; published estimates for young college-age women range from 5 to 18 percent. However, many bulimics are not diagnosed until they are age 30–40. Because they ingest plenty of food, bulimics may have normal body weight, so the condition is harder to diagnose than anorexia nervosa. The health risks of bulimia result from (1) cumulative damage to the stomach, esophagus, oral cavity, and teeth by repeated exposure to stomach acids; (2) electrolyte imbalances resulting from the loss of sodium and potassium ions in the gastric juices, diarrhea, and urine; (3) edema; and (4) cardiac arrhythmias.

The underlying cause of bulimia remains uncertain. Societal factors are certainly involved, but bulimia has also been strongly correlated with depression and with elevated levels of antidiuretic hormone (ADH) in cerebrospinal fluid.

25

BIOENERGETICS

The study of **bioenergetics** examines the acquisition and use of energy by organisms. When chemical bonds are broken, energy is released. Inside cells, a significant amount of energy may be used to synthesize ATP, but much of it is lost to the environment as heat. The process of **calorimetry** (kal-o-RIM-e-trē) determines the total amount of energy released when the bonds of organic molecules are broken. The unit of measurement is the **calorie** (KAL-o-rē), defined as the amount of energy required to raise the temperature of 1 g of water one degree centigrade. One gram of water is not a very practical measure when you are interested in the metabolic operations that keep a 70-kg human alive, so the **kilocalorie** (KIL-ō-kal-o-rē) (kc), or **Calorie** (with a capital *C*), also known as "large calorie," is used instead. One Calorie is the amount of energy needed to raise the temperature of 1 *kilo*gram of water one degree centigrade. Calorie-counting guides that give the caloric value of various foods indicate Calories, not calories.

FOOD AND ENERGY

In cells, organic molecules are oxidized to carbon dioxide and water. Oxidation also occurs when something burns, and this process can be experimentally controlled. A known amount of food is placed in a chamber called a **calorimeter** (kal-o-RIM-e-ter), which is filled with oxygen and surrounded by a known volume of water. Once the food is inside, the chamber is sealed and the contents are electrically ignited. When the material has completely oxidized and only ash remains in the chamber, the number of Calories released can be determined by comparing the water temperatures before and after the test.

The energy potential of food is usually expressed in Calories per gram (C/g). The catabolism of lipids entails the release of a considerable amount of energy, roughly 9.46 C/g. The catabolism of carbohydrates or proteins is not as rewarding, because many of the carbon and hydrogen atoms are already bound to oxygen. Their average yields are comparable: 4.18 C/g for carbohydrate and 4.32 C/g for protein. Most foods are mixtures of fats, proteins, and carbohydrates, and the values in a "Calorie counter" vary as a result.

METABOLIC RATE

Clinicians can examine your metabolic state and determine how many Calories you are utilizing. The result can be expressed as Calories per hour, Calories per day, or Calories per unit of body weight per day. What is actually measured is the sum of all the varied anabolic and catabolic processes occurring in your body—your **metabolic rate** at that time. The metabolic rate will change according to the activity under way; for instance, sprinting and sleeping measurements are quite different. In an attempt to reduce the variations, clinicians standardize the testing conditions so as to determine the **basal metabolic rate (BMR).** Ideally, the BMR is the minimum, resting energy expenditures of an awake, alert person.

A direct method of determining the BMR simply monitors respiratory activity, because in resting persons energy utilization is proportional to oxygen consumption. If we assume that average amounts of carbohydrates, lipids, and proteins are being catabolized, the ratio gives 4.825 Calories per liter of oxygen consumed.

An average individual has a BMR of 70 C per hour, or about 1680 C per day. Although the test conditions are standardized, many uncontrollable factors can influence the BMR. These include age, gender, physical condition, body weight, and genetic differences such as ethnic variations.

Because the BMR is technically difficult to measure, and because circulating thyroid hormone levels have a profound effect on the BMR, clinicians usually monitor the concentration of thyroid hormones rather than the actual metabolic rate. The results are then compared with normal values, to obtain an index of metabolic activity. One such test, the **T_4 assay,** measures the amount of thyroxine in the blood.

Daily energy expenditures for a given individual vary widely with the activities undertaken. For example, a person leading a sedentary life may have near-basal energy demands, but a single hour of swimming can increase the daily caloric requirements by 500 C or more. If your daily energy intake exceeds your total energy demands, you will store the excess, primarily as triglycerides in adipose tissue. If your daily caloric expenditures exceed your dietary supply, there will be a net reduction in your body's energy reserves and a corresponding loss in weight. This relationship accounts for the significance of Calorie counting and exercise in a weight-control program.

The control of appetite is poorly understood. Stretch receptors along the digestive tract, especially in the stomach, do play a role, but other factors are probably more important. Social factors, psychological pressures, and dietary habits are important. Evidence also indicates that complex hormonal stimuli interact to affect appetite. For example, the hormones *cholecystokinin* and *adrenocorticotropic hormone (ACTH)* suppress appetite. The hormone *leptin,* released by adipose tissues, also plays a role. During the absorptive state, adipose tissues release leptin into the bloodstream as they synthesize triglycerides. Leptin binds to neurons in the central nervous system that deal with emotion and the control of appetite. The result is a sense of satiation and the suppression of appetite.

THERMOREGULATION

The BMR estimates the rate of energy use by the body. The energy not captured and harnessed by cells is released as heat. This heat loss serves an important homeostatic purpose. Humans are subject to vast changes in environmental temperatures, but our complex biochemical systems have a major limitation: Enzymes operate over only a relatively narrow range of temperatures. Our bodies have anatomical and physiological mechanisms that keep body temperatures within acceptable limits, regardless of the environmental conditions. This homeostatic process is called **thermoregulation.** ∞ *p. 12* Failure to control body temperature can result in a series of

25

physiological changes. For example, a body temperature below 36°C (97°F) or above 40°C (104°F) can cause disorientation, and a temperature above 42°C (108°F) can cause convulsions and permanent cell damage.

We continuously produce heat as a byproduct of metabolism. When energy use increases, due to physical activity, or when our cells are more active metabolically, as they are during the absorptive state, additional heat is generated. The heat produced by biochemical reactions is retained in water, which accounts for nearly 66 percent of your body weight. Water is an excellent conductor of heat, so the heat produced in one region of the body is rapidly distributed by diffusion as well as through the bloodstream. If body temperature is to remain constant, that heat must be lost to the environment at the same rate that it is generated. When the environmental conditions vary from "ideal," becoming too warm or too cold, the body must control the gains or losses to maintain homeostasis.

Mechanisms of Heat Transfer

Heat exchange with the environment involves four basic processes: (1) *radiation,* (2) *conduction,* (3) *convection,* and (4) *evaporation.* These mechanisms are illustrated in Figure 25-19•.

Warm objects lose heat energy as infrared **radiation.** When you feel the heat from the sun, you are expe-

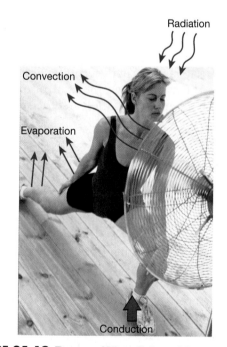

•**FIGURE 25-19** **Routes of Heat Gain and Loss**

riencing radiant heat. Your body loses heat the same way but in proportionately smaller amounts. More than 50 percent of the heat you lose is attributable to radiation. The exact amount varies with both body temperature and skin temperature.

Conduction is the direct transfer of energy through physical contact. When you arrive in an air-conditioned classroom and sit on a cold plastic chair, you are immediately aware of this process. Conduction is generally not an effective mechanism for gaining or losing heat. We cannot estimate the value of its contribution, because it varies with the temperature of the object and with the amount of surface area involved. When lying on cool sand in the shade, conductive losses can be considerable; when standing, conductive losses are negligable.

Convection is the result of conductive heat loss to the air that overlies the surface of the body. Warm air rises, because it is lighter than cool air. As your body conducts heat to the air next to your skin, that air warms and rises, moving away from the skin surface. Cooler air replaces it; as it in turn becomes warmed, the pattern repeats. Convection accounts for roughly 15 percent of your heat loss; convection is insignificant as a mechanism of heat gain.

When water evaporates, it changes from a liquid to a vapor. **Evaporation** absorbs energy, roughly 0.58 C per gram of water evaporated. The rate of evaporation occurring at your skin is highly variable. Each hour, 20–25 ml of water crosses epithelia and evaporates from the alveolar surfaces and the surface of the skin. This insensible perspiration remains relatively constant; at rest it accounts for roughly 20 percent of your average heat loss. The sweat glands responsible for sensible perspiration have a tremendous scope of activity, ranging from virtual inactivity to secretory rates of 2–4 liters per hour.

The Regulation of Heat Gain and Heat Loss

Heat loss and heat gain involve the coordinated activity of many systems. That activity is coordinated by the **heat-loss center** and **heat-gain center** in the preoptic area of the anterior hypothalamus. ∞ *p. 459* These centers modify the activities of other hypothalamic nuclei. The overall effect is to control temperature by influencing two events: the rate of heat production and the rate of heat loss to the environment. These events may be further supported by behavioral modifications.

Mechanisms for Increasing Heat Loss When the temperature at the preoptic nucleus exceeds its thermostat setting, the heat-loss center is stimulated. The stimulation of this center has three major effects:

1. *The inhibition of the vasomotor center causes peripheral vasodilation, and warm blood flows to the surface of the body.* The skin takes on a reddish color, skin temperatures rise, and radiational and convective losses increase.

2. *As blood flow to the skin increases, sweat glands are stimulated to increase their secretory output.* The perspiration flows across the body surface, and evaporative losses accelerate. A maximal secretion rate would, if it were completely evaporated, remove 2320 C per hour.

3. *The respiratory centers are stimulated, and the depth of respiration increases.* Often the individual begins respiring through an open mouth rather than through the nasal passageways, increasing evaporative losses through the lungs.

Heat Exhaustion and Heat Stroke

Heat exhaustion and heat stroke are malfunctions of the thermoregulatory system. In **heat exhaustion,** also known as *heat prostration*, the individual has difficulty in maintaining blood volume. The heat-loss center is stimulating sweat glands, whose secretions moisten the surface of the skin to provide evaporative cooling. As fluid losses mount, blood volume decreases. The resulting decline in blood pressure is not countered by peripheral vasoconstriction, because the heat-loss center is actively stimulating peripheral vasodilation. As blood flow to the brain declines, headache, nausea, and eventual collapse follow. Treatment is straightforward: Provide fluids, salts, and a cooler environment.

Heat stroke is more serious and can follow an untreated case of heat exhaustion. Predisposing factors include any preexisting condition, such as heart disease or diabetes, that affects peripheral circulation. The thermoregulatory center ceases to function, the sweat glands are inactive, and the skin becomes hot and dry. Unless the situation is recognized in time, body temperature may climb to 41°–45°C (106°–113°F). Temperatures in this range quickly disrupt a variety of vital physiological systems and destroy brain, liver, skeletal muscle, and kidney cells. Effective treatment involves lowering body temperature as rapidly as possible.

Mechanisms for Promoting Heat Gain The function of the heat-gain center of the brain is to prevent **hypothermia** (hī-pō-THER-me-uh), or below-normal body temperature. When the temperature at the preoptic nucleus drops below acceptable levels, the heat-loss center is inhibited and the heat-gain center is activated.

Heat conservation The sympathetic vasomotor center decreases blood flow to the dermis of the skin, thus reducing losses by radiation, convection, and conduction. The skin cools, and with blood flow restricted, it may take on a bluish or pale coloration. The epithelial cells are not damaged, because they can tolerate extended periods at temperatures as low as 25°C (77°F) or as high as 49°C (120°F).

In addition, blood returning from the limbs is shunted into a network of deep veins. ∞ *p. 734* In warm weather, blood flows in a superficial venous network. In cold weather, blood is diverted to a network of deep veins that lie deep to an insulating layer of subcutaneous fat. This venous network wraps around the deep arteries, and heat diffuses from the warm blood flowing outward to the limbs into the cooler blood returning from the periphery. This arrangement traps the heat close to the body core and restricts heat loss. Diffusion between fluids that are moving in opposite directions is called *countercurrent exchange.* (We shall return to this topic in Chapter 26.)

Heat generation The mechanisms available to generate heat can be divided into two broad categories: (1) shivering thermogenesis and (2) nonshivering thermogenesis. In **shivering thermogenesis** (ther-mō-JEN-e-sis), a gradual increase in muscle tone increases the energy consumption of skeletal muscle tissue throughout your body. Both agonists and antagonists are involved, and the degree of stimulation varies with the demand.

If the heat-gain center is extremely active, muscle tone increases to the point at which stretch receptor stimulation will produce brief, oscillatory contractions of antagonistic muscles. In other words, you begin to **shiver.** Shivering increases the workload of the muscles and further elevates oxygen and energy consumption. The heat that is produced warms the deep vessels, to which blood has been shunted by the sympathetic vasomotor center. Shivering can elevate body temperature quite effectively; it increases the rate of heat generation by as much as 400 percent.

In **nonshivering thermogenesis,** hormones are released that increase the metabolic activity of all tissues:

- The heat-gain center stimulates the adrenal medullae, via the sympathetic division of the autonomic nervous system, and epinephrine is released. Epinephrine increases the rates of glycogenolysis in liver and skeletal muscle and the metabolic rate of most tissues. These effects are immediate.

- The preoptic nucleus regulates the production of thyrotropin-releasing hormone (TRH) by the hypothalamus. In children, when body temperatures are below normal, additional TRH is released, stimulating the release of thyroid-stimulating hormone (TSH) by the anterior pituitary gland. The thyroid gland responds to this release of TSH by increasing the rate of thyroxine release into the blood. Thyroxine increases not only the rate of carbohydrate catabolism but also the rate of catabolism of all other nutrients. These effects develop gradually, over a period of days to weeks.

Induced Hypothermia

Hypothermia may be intentionally produced during surgery to reduce the metabolic rate of a particular organ or of the patient's entire body. In controlled hypothermia, the individual is first anesthetized to prevent the shivering that would otherwise fight the induction of hypothermia.

During open-heart surgery, the body is typically cooled to 25°–32°C (79°–89°F). This cooling reduces the metabolic demands of the body, which will be receiving blood from an external pump or oxygenator. The heart must be stopped completely during the operation, and it cannot be well supplied with blood over this period. So the heart is exposed to an *arresting solution* at 0°–4°C (32°–39°F) and maintained at a temperature below 15°C (60°F) for the duration of the operation. At these temperatures, the cardiac muscle can tolerate several hours of ischemia without damage.

When cardiac surgery is performed on infants, a deep hypothermia may be produced by cooling the entire body to temperatures as low as 11°C (52°F) for an hour or more. In effect, this procedure duplicates the conditions experienced by accidental drowning victims. AM *Accidental Hypothermia*

Sources of Individual Variation

The timing of thermoregulatory responses differs from individual to individual. A person may undergo **acclimatization** (a-klī-ma-ti-ZĀ-shun), the physiological adjustment to a particular environment over time. For example, natives of Tierra del Fuego once lived naked in the snow, but Hawaii residents unpack their sweaters when the temperature drops below 22°C (72°F).

Another interesting source of variation is body size. Although heat *production* occurs within the mass of the body, heat *loss* must occur across a body surface. As an object (or person) gets larger, its surface area increases at a much slower rate than does its total volume. This relationship affects thermoregulation, because heat is generated by the "volume" (that is, by internal tissues) and lost at the body surface. Hence small individuals lose heat more readily than do large individuals.

Thermoregulatory Problems of Infants During embryonic development, the maternal surroundings are at normal body temperature. At birth, the infant's temperature-regulating mechanisms are not fully functional; also, infants lose heat quickly as a result of their small size. Newborns must be dried promptly and kept bundled up; for those born prematurely, a thermally regulated incubator is required. Infants' body temperatures are also less stable than those of adults. Their metabolic rates decline while they sleep, then rise after they awake.

Infants cannot shiver, but they have a different mechanism for raising body temperature rapidly. The adipose tissue between the shoulder blades, around the neck, and possibly elsewhere in the upper body is histologically and functionally different from most of the adipose tissue in adults. The tissue is highly vascularized, and the individual adipocytes contain numerous mitochondria. Together these characteristics give the tissue a deep, rich color responsible for the name **brown fat.** *∞ p. 122* The individual adipocytes are innervated by sympathetic autonomic fibers. When these nerves are stimulated, lipolysis accelerates in the adipocytes. The cells do not capture the energy released through fatty acid catabolism, and it radiates into the surrounding tissues as heat. This heat quickly warms the blood that passes through the surrounding network of vessels, and it is then distributed throughout the body. In this way, an infant can accelerate metabolic heat generation by 100 percent very quickly, whereas nonshivering thermogenesis in an adult raises heat production by only 10–15 percent after a period of weeks.

With increasing age and size, body temperature becomes more stable, so the importance of this thermoregulatory mechanism declines. Adults have little if any brown fat; with increased body size, skeletal muscle mass, and insulation, shivering thermogenesis is significantly more effective in elevating body temperature.

Thermoregulatory Variations Among Adults Adults of the same body weight may differ in their thermal responses if their weight is distributed differently. Which tissues account for their weight is also a factor. Adipose tissue is an excellent insulator, conducting heat at only about one-third the rate of other tissues. As a result, individuals with a more substantial layer of subcutaneous fat may not begin to shiver until long after their more slender companions.

Two otherwise similar individuals may also differ in their response to temperature changes because their hypothalamic thermostats are at different settings. We experience daily oscillations in body temperature, with temperatures falling 1°–2°C (1.8°–3.6°F) at night and peaking during the day or early evening. Individuals vary in terms of their time of maximum temperature setting; some have a series of peaks, with an afternoon low. The origin of these patterns is uncertain. It is not the result of daily activity regimens—people who work at night still show their temperature peaks over the same range of times as the rest of the population.

Fevers

Any elevated body temperature is called **pyrexia** (pī-REK-sē-uh). Pyrexia is usually temporary. A **fever** is a body temperature maintained at greater than 37.2°C (99°F). We discussed fevers when we examined non-specific defenses. ∞ *p. 770* Fevers occur for a variety of reasons, not all of them pathological. In young children, transient fevers with no ill effects can result from exercise in warm weather. Similar exercise-related elevations were rarely encountered in adults until running marathons became popular. Temperatures from 39° to 41°C (103° to 106°F) may result. For this reason, competitions are usually held when the air temperature is below 28°C (82°F).

Fevers can also result from such factors as the following:

- Abnormalities affecting the entire thermoregulatory mechanism, such as heat exhaustion or heat stroke.
- Clinical problems that restrict blood flow, such as congestive heart failure.
- Conditions that impair sweat gland activity, such as drug reactions and some skin conditions.
- The resetting of the hypothalamic thermostat by circulating *pyrogens,* most notably interleukin-1.

We can classify fevers as *chronic* or *acute*. The classification and treatment of fevers are discussed in the *Applications Manual.* AM *Fevers*

☑ How would the BMR (basal metabolic rate) of a pregnant woman compare with her own BMR while she is in the non-pregnant state?

☑ What effect would the vasoconstriction of peripheral blood vessels have on an individual's body temperature on a hot day?

☑ Why do infants have greater problems with thermoregulation than do adults?

[EXPLORE]
[MediaLab]

Jenny Craig®, Richard Simmons, Weight Watchers®—these names conjure up thoughts of dieting and, therefore, metabolic control. The term *metabolism* refers to many biochemical reactions, all centered around the production and burning of energy. To help you understand these reactions, use the resources available on both the CD-ROM and the Companion Website. The CD-ROM includes animations and tutorials; the website builds on those resources with text content and interactive problems as well as communication avenues for sharing ideas with your class. To complete the following Web Exercises, visit the MediaLab in Chapter 25 of your Companion Website.

Web Exploration 1

Estimated time for completion = 12 minutes

The TCA cycle, or Krebs cycle, is at the heart of carbohydrate metabolism. Until you become familiar with this cycle, its many reactions can be daunting. To gain practice with the reactions of the TCA cycle, select the keyword **TCA** on the Companion Website. Compare the cycle shown there with that in Figure 25-5 of your text, and watch the animated cycle in the lower right corner. Link to "Another animated Krebs Cycle" to view this cycle with highlights that explain each reaction. Reproduce the cycle in your notes; include the reaction types and product names as seen on these Web pages.

Web Exploration 2

Estimated time for completion = 5 minutes

What is your personal basal metabolic rate (BMR)? In other words, how much energy does your body require while you are at rest and while you are exercising? To calculate your BMR, select the keyword **BMR** on the Companion Website. Read the explanation there, and fill in the information requested on the table. Record your BMR. Now go back to the table and change some values. What happens to the BMR when you change your age? your weight? your percentage of heavy or moderate exercise? Prepare a short essay describing how this information applies to your daily routine.

25

CHAPTER REVIEW

SELECTED CLINICAL TERMINOLOGY
Terms Discussed in This Chapter

avitaminosis (ā-vī-ta-min-Ō-sis): A vitamin deficiency disease. *(p. 929)*

carbohydrate loading: Eating large quantities of carbohydrates in the days preceding an athletic competition to increase endurance. *(p. 910)*

eating disorders: Psychological problems that result in inadequate or excessive food consumption. Examples include anorexia nervosa and bulimia. *(p. 932)*

heat exhaustion: A malfunction of the thermoregulatory system caused by excessive fluid loss in perspiration. *(p. 933)*

heat stroke: A condition in which the thermoregulatory center stops functioning and body temperature rises uncontrollably. *(p. 933)*

hyperuricemia (hī-per-ū-ri-SĒ-mē-uh): Levels of plasma uric acid above 7.4 mg/dl; can result in the condition called *gout*. *(p. 919)*

hypervitaminosis (hī-per-vī-ta-min-Ō-sis): A disorder caused by the ingestion of excessive quantities of one or more vitamins. *(p. 930)*

hypothermia (hī-pō-THER-mē-uh): Below-normal body temperature. *(p. 936 and AM)*

ketoacidosis (kē-tō-as-i-DŌ-sis): The acidification of blood due to the presence of ketone bodies. *(p. 925 and AM)*

ketonemia (kē-tō-NE-mē-uh): Elevated levels of ketone bodies in blood. *(p. 925)*

ketonuria (kē-tō-NOO-rē-uh): The presence of ketone bodies in urine. *(p. 925)*

ketosis (kē-TŌ-sis): Abnormally high concentration of ketone bodies in body fluids. *(p. 925)*

obesity: A body weight more than 20 percent above the ideal weight for a given individual. *(p. 923 and AM)*

phenylketonuria (fen-il-kē-tō-NOO-rē-uh): An inherited metabolic disorder resulting from an inability to convert phenylalanine to tyrosine. *(p. 919 and AM)*

protein deficiency diseases: Nutritional disorders resulting from a lack of essential amino acids. *(p. 919 and AM)*

AM Additional Terms Discussed in the *Applications Manual*

antipyretic drugs: Drugs administered to control or reduce fever.

gout: A metabolic disorder characterized by the precipitation of uric acid crystals within joint cavities.

liposuction: The removal of adipose tissue by suction through an inserted tube.

STUDY OUTLINE

INTRODUCTION   p. 901
1. Cells in the human body are chemical factories that break down organic substrates to obtain energy.

AN OVERVIEW OF METABOLISM   p. 901
1. In general, during *cellular metabolism,* cells break down excess carbohydrates first, then lipids, while conserving amino acids. Only about 40 percent of the energy released through catabolism is captured in ATP; the rest is released as heat. *(Figure 25-1)*
2. Cells synthesize new compounds: (1) to perform structural maintenance or repairs, (2) to support growth, and (3) to produce secretions. *(Figure 25-2)*
3. Cells feed small organic molecules to their mitochondria; in return, they obtain the ATP they need to perform cellular functions. *(Figure 25-3)*

CARBOHYDRATE METABOLISM   p. 903
1. Most cells generate ATP and other high-energy compounds through the breakdown of carbohydrates.

Glycolysis   p. 904
2. **Glycolysis** and **aerobic metabolism,** or **cellular respiration,** provide most of the ATP used by typical cells. Glycogen can be broken down to glucose molecules. In glycolysis,

each molecule of glucose yields 2 molecules of pyruvic acid (as pyruvate ions), a net of 2 molecules of ATP, and 2 of NADH. *(Figure 25-4)*

Mitochondrial ATP Production   p. 904
3. In the presence of oxygen, the pyruvic acid molecules enter mitochondria, where they are broken down completely in the **tricarboxylic acid (TCA) cycle.** Carbon and oxygen atoms are lost as carbon dioxide; hydrogen atoms are passed to coenzymes, which initiate the oxygen-consuming and ATP-generating reaction **oxidative phosphorylation.** *(Figure 25-5)*
4. **Cytochromes** of the **electron transport system (ETS)** pass electrons to oxygen, resulting in the formation of water. As this transfer occurs, the ETS generates ATP. *(Figure 25-6)*

Energy Yield of Glycolysis and Cellular Respiration   p. 910
5. For each glucose molecule processed through glycolysis, the TCA cycle, and the ETS, most cells gain 36 molecules of ATP. *(Figure 25-7)*

Other Catabolic Pathways   p. 911
6. Cells can break down other nutrients to provide substrates for the TCA cycle if supplies of glucose are limited.

Gluconeogenesis   p. 912
7. **Gluconeogenesis,** the synthesis of glucose from noncarbohydrate precursors, such as lactic acid, glycerol, or amino

acids, enables a liver cell to synthesize glucose molecules when carbohydrate reserves are depleted. **Glycogenesis** is the process of glycogen formation. Glycogen is an important energy reserve when the cell cannot obtain enough glucose from interstitial fluid. *(Figure 25-8)*

LIPID METABOLISM p. 913

Lipid Catabolism p. 913

1. During **lipolysis** (lipid catabolism), lipids are broken down into pieces that can be converted into pyruvic acid or channeled into the TCA cycle.
2. Triglycerides, the most abundant lipids in the body, are split into glycerol and fatty acids. The glycerol enters the glycolytic pathways, and the fatty acids enter the mitochondria.
3. **Beta-oxidation** is the breakdown of a fatty acid molecule into two-carbon fragments that can be used in the TCA cycle. The steps of beta-oxidation cannot be reversed, and the body cannot manufacture all the fatty acids needed for normal metabolic operations. *(Figure 25-9)*
4. Lipids cannot provide large amounts of ATP in a short amount of time. However, cells can shift to lipid-based energy production when glucose reserves are limited.

Lipid Synthesis p. 914

5. In **lipogenesis,** the synthesis of lipids, almost any organic substrate can be used to form glycerol. **Essential fatty acids** cannot be synthesized and must be included in the diet. *(Figure 25-10)*

Lipid Transport and Distribution p. 914

6. Most lipids circulate as **lipoproteins** (lipid–protein complexes that contain large glycerides and cholesterol). The largest lipoproteins, chylomicrons, carry absorbed lipids from the intestinal tract to the bloodstream. All other lipoproteins are derived from the liver and carry lipids to and from various tissues of the body. *(Figure 25-11)*
7. Capillary walls of adipose tissue, skeletal muscle, cardiac muscle, and the liver contain **lipoprotein lipase,** an enzyme that breaks down complex lipids, releasing a mixture of fatty acids and monoglycerides. *(Figure 25-11)*

PROTEIN METABOLISM p. 916

Amino Acid Catabolism p. 916

1. If other energy sources are inadequate, mitochondria can break down amino acids in the TCA cycle to generate ATP. In the mitochondria, the amino group can be removed by **transamination** or by **deamination.** *(Figure 25-12)*
2. Protein catabolism is impractical as a source of quick energy.

Protein Synthesis p. 919

3. Roughly half the amino acids needed to build proteins can be synthesized. There are 10 **essential amino acids,** which must be acquired through the diet. **Amination,** the attachment of an amino acid group to a carbon framework, is an important step in the synthesis of **nonessential amino acids.** *(Figure 25-13)*

NUCLEIC ACID METABOLISM p. 919

1. DNA in the nucleus is never catabolized for energy.

RNA Catabolism p. 919

2. RNA molecules are broken down and replaced regularly. They are generally recycled as new nucleic acids, but the nucleotides can be catabolized to simple sugars and nitrogenous bases. In general, nucleic acids do not contribute significantly to the cell's energy reserves.

Nucleic Acid Synthesis p. 920

3. Most cells synthesize RNA, but DNA synthesis occurs only in cells preparing for mitosis or meiosis.

METABOLIC INTERACTIONS p. 920

1. No one cell of a human can perform all the anabolic and catabolic operations necessary to support life. Homeostasis can be preserved only when metabolic activities of different tissues are coordinated. *(Figure 25-14)*
2. The body has five metabolic components: the liver, adipose tissue, skeletal muscle, neural tissue, and other peripheral tissues. The liver is the focal point for metabolic regulation and control. Adipose tissue stores lipids, primarily in the form of triglycerides. Skeletal muscle contains substantial glycogen reserves, and the contractile proteins can be mobilized and the amino acids used as an energy source. Neural tissue does not contain energy reserves; glucose must be supplied to it for energy. Other peripheral tissues are able to metabolize glucose, fatty acids, or other substrates under the direction of the endocrine system.

The Absorptive State p. 921

3. For about 4 hours after a meal, nutrients enter the blood as intestinal absorption proceeds. *(Figure 25-15; Table 25-1)*
4. The liver closely regulates the glucose content of blood and the circulating levels of amino acids.
5. **Lipemia** (milky appearance of the plasma due to the presence of lipids) commonly marks the **absorptive state.** Adipocytes remove fatty acids and glycerol from the bloodstream and synthesize new triglycerides to be stored for later use.
6. During the absorptive state, glucose molecules are catabolized and amino acids are used to build proteins. Skeletal muscles may also catabolize circulating fatty acids, and the energy is used to increase glycogen reserves.

The Postabsorptive State p. 923

7. The **postabsorptive state** extends from the end of the absorptive state to the next meal. *(Figures 25-16, 25-17)*
8. When blood glucose falls, the liver begins breaking down glycogen reserves and releasing the glucose into the bloodstream. As the time between meals increases, liver cells synthesize glucose molecules from smaller carbon fragments and from glycerol molecules. Fatty acids undergo beta-oxidation; the fragments enter the TCA cycle or combine to form **ketone bodies.** *(Table 25-1)*
9. Some amino acids can be converted to pyruvic acid and used for gluconeogenesis; others, including most of the essential amino acids. are converted to acetyl-CoA and are either catabolized or converted to ketone bodies.
10. The average individual carries a 1–2 month energy reserve in adipose tissue. During the postabsorptive state, lipolysis increases and the fatty acids are released into the bloodstream for catabolism.
11. Skeletal muscles metabolize ketone bodies and fatty acids. Their glycogen reserves are broken down to yield lactic acid, which diffuses into the bloodstream. After a prolonged fast, muscle proteins are broken down.
12. Neural tissue continues to be supplied with glucose as an energy source until blood glucose levels become very low.

DIET AND NUTRITION p. 926

1. **Nutrition** is the absorption of nutrients from food. A **balanced diet** contains all the ingredients necessary to maintain homeostasis; it prevents **malnutrition.**

Food Groups and Food Pyramids p. 926

2. The six **basic food groups** are milk, yogurt, and cheese; meat, poultry, fish, dry beans, eggs, and nuts; vegetable; fruit; bread, cereal, rice, and pasta; and fats, oils, and sweets. These are arranged in a *food pyramid* with the bread and cereal group forming the base. *(Figure 25-18)*

25

Nitrogen Balance p. 926

3. Amino acids, purines, pyrimidines, creatine, and porphyrins are **N compounds,** which contain nitrogen atoms. An adequate dietary supply of nitrogen is essential, because the body does not maintain large nitrogen reserves.

Minerals p. 928

4. **Minerals** act as cofactors in a variety of enzymatic reactions. They also contribute to the osmotic concentration of body fluids and play a role in transmembrane potentials, action potentials, neurotransmitter release, muscle contraction, skeletal construction and maintenance, gas transport, buffer systems, fluid absorption, and waste removal. *(Table 25-2)*

Vitamins p. 929

5. Vitamins are needed in very small amounts. Vitamins A, D, E, and K are the **fat-soluble vitamins;** taken in excess, they can lead to **hypervitaminosis. Water-soluble vitamins** are not stored in the body; a lack of adequate dietary supplies may lead to **deficiency disease (avitaminosis).** *(Tables 25-3, 25-4)*

Diet and Disease p. 931

6. A balanced diet can improve general health.

BIOENERGETICS p. 932

1. The energy content of food is usually expressed as **Calories** per gram (C/g). Our cells can capture less than half the energy content of glucose or any other nutrient.

Food and Energy p. 933

2. The catabolism of lipids releases 9.46 C/g, about twice the amount released by equivalent weights of carbohydrates or proteins.

Metabolic Rate p. 933

3. The total of all the anabolic and catabolic processes under way is an individual's **metabolic rate.** The **basal metabolic rate (BMR)** is the rate of energy utilization at rest.

Thermoregulation p. 933

4. The homeostatic regulation of body temperature is **thermoregulation.** Heat exchange with the environment involves four processes: **radiation, conduction, convection,** and **evaporation.** *(Figure 25-19)*

5. The *preoptic area* of the hypothalamus acts as the body's thermostat, affecting the **heat-loss center** and the **heat-gain center.**

6. Mechanisms for increasing heat loss include both physiological mechanisms (peripheral vasodilation, increased perspiration, and increased respiration) and behavioral modifications.

7. Responses that conserve heat include decreased blood flow to the dermis and *countercurrent exchange.*

8. Heat is generated by **shivering thermogenesis** and **nonshivering thermogenesis.**

9. Thermoregulatory responses differ among individuals. One important source of variation is **acclimatization** (adjusting physiologically to an environment over time).

10. **Pyrexia** is an elevated body temperature. **Fever,** a body temperature above 37.2°C (99°F), can result from problems with the thermoregulatory mechanism, circulation, or sweat gland activity or from the resetting of the hypothalamic thermostat by circulating pyrogens.

REVIEW QUESTIONS

LEVEL 1 REVIEWING FACTS AND TERMS

1. Cells synthesize new organic components to
   (a) perform structural maintenance and repairs
   (b) support growth
   (c) produce secretions
   (d) a, b, and c are correct

2. During the complete catabolism of one molecule of glucose, a typical cell gains
   (a) 4 ATP   (b) 18 ATP
   (c) 36 ATP   (d) 144 ATP

3. The breakdown of glucose to pyruvic acid is
   (a) glycolysis
   (b) gluconeogenesis
   (c) cellular respiration
   (d) oxidative phosphorylation

4. Glycolysis yields an *immediate* net gain of _____ for the cell.
   (a) 1 ATP   (b) 2 ATP
   (c) 4 ATP   (d) 36 ATP

5. The only *immediate* energy benefit of one turn of the TCA cycle is the formation of a single molecule of
   (a) GTP   (b) NAD
   (c) FAD   (d) CoA

6. The process that produces over 90 percent of the ATP used by our cells is
   (a) glycolysis
   (b) the TCA cycle
   (c) substrate-level phosphorylation
   (d) oxidative phosphorylation

7. The electron transport system and phosphorylation associated with it generally yield a total of _____ molecules of ATP in the complete catabolism of one glucose molecule.
   (a) 2   (b) 4
   (c) 32   (d) 36

8. The synthesis of glucose from nonglucose precursors is
   (a) glycolysis   (b) glycogenesis
   (c) gluconeogenesis (d) beta-oxidation

9. The sequence of reactions responsible for the breakdown of fatty acid molecules is
   (a) beta-oxidation   (b) the TCA cycle
   (c) lipogenesis   (d) a, b, and c are correct

10. The essential fatty acids that cannot be synthesized by the body but must be included in the diet are
    (a) linoleic and linolenic
    (b) leucine and lysine
    (c) cholesterol and glycerol
    (d) HDLs and LDLs

11. The lipoproteins that transport excess cholesterol from peripheral tissues back to the liver for storage or excretion in the bile are the
    (a) chylomicrons   (b) VLDLs
    (c) LDLs   (d) HDLs

12. The removal of an amino group in a reaction that generates an ammonia molecule is
    (a) ketoacidosis   (b) transamination
    (c) deamination   (d) denaturation

13. The part of the RNA molecule that *cannot* be catabolized to provide energy is the
(a) phosphate   (b) sugar
(c) pyrimidine   (d) purine

14. The focal point of metabolic regulation and control is the
(a) brain   (b) liver
(c) heart   (d) kidneys

15. When the body is relying on internal energy reserves to continue meeting its energy demands, it is in the
(a) postabsorptive state
(b) absorptive state
(c) starvation state
(d) deprivation state

16. A complete protein contains
(a) the proper balance of amino acids
(b) all the essential amino acids in sufficient quantities
(c) a combination of nutrients selected from the food pyramid
(d) N compounds produced by the body

17. All minerals and most vitamins
(a) are fat-soluble
(b) cannot be stored by the body
(c) cannot be synthesized by the body
(d) must be synthesized by the body because they are not present in adequate amounts in the diet

18. The vitamins generally associated with vitamin toxicity are
(a) fat-soluble vitamins
(b) water-soluble vitamins
(c) the B complex vitamins
(d) vitamins C and B_{12}

19. The basal metabolic rate (BMR) represents the
(a) maximum energy expenditure when exercising
(b) minimum, resting energy expenditure of an awake, alert person
(c) minimum energy expenditure during light exercise
(d) muscular energy expenditure added to the resting energy expenditure

20. The abuse of laxatives and diuretics is commonly associated with
(a) avitaminosis   (b) kwashiorkor
(c) bulimia   (d) marasmus

21. Over half the heat loss from our bodies is attributable to
(a) radiation   (b) conduction
(c) convection   (d) evaporation

22. The most effective mechanism for elevating body temperature in adults is
(a) lipolysis in brown fat
(b) carbohydrate loading
(c) fatty acid catabolism
(d) shivering thermogenesis

23. The resetting of the hypothalamic thermostat by circulating pyrogens such as interleukin-1 produces
(a) hypothermia
(b) fever
(c) maintenance of normal body temperature
(d) shivering

24. Define the terms *metabolism*, *anabolism*, and *catabolism*.

25. Write the complete reaction sequence for carbohydrate metabolism.

26. What is a lipoprotein? What are the major groups of lipoproteins, and how do they differ?

27. Why are vitamins and minerals essential components of the diet?

28. What energy yields, in Calories per gram, are associated with the catabolism of carbohydrates, lipids, and proteins?

29. What is the basal metabolic rate (BMR)?

30. What four basic mechanisms are involved in heat exchange between the body and the environment?

31. What causative factors or conditions are associated with the production of fever?

LEVEL 2 REVIEWING CONCEPTS

32. The function of the TCA cycle is to
(a) produce energy during periods of active muscle contraction
(b) break six-carbon chains into three-carbon fragments
(c) prepare the glucose molecule for further reactions
(d) remove hydrogen atoms from organic molecules and transfer them to coenzymes

33. During periods of fasting or starvation, the presence of ketone bodies in the bloodstream causes
(a) an increase in blood pH
(b) a decrease in blood pH
(c) lipemia
(d) diabetes insipidus

34. What happens during glycolysis? What conditions are necessary for this process to take place?

35. Why is the TCA cycle called a cycle? What substance(s) enter(s) the cycle, and what substances leave it?

36. What is oxidative phosphorylation? Explain how the electron transport system is involved in this process.

37. How are lipids catabolized in the body? How is beta-oxidation involved with lipid catabolism?

38. How is RNA catabolized to produce energy?

39. How do the absorptive and postabsorptive states maintain normal blood glucose levels?

40. Why is the liver the focal point for metabolic regulation and control?

41. How can the food pyramid be used as a tool to obtain nutrients in sufficient quantity and quality? Why are the dietary fats, oils, and sugars at the top of the pyramid and bread, cereal, rice, and pasta at the bottom?

42. How is the brain involved in the regulation of body temperature?

43. Some articles in popular magazines refer to "good cholesterol" and "bad cholesterol." To which types and functions of cholesterol might these terms refer? Explain your answer.

LEVEL 3 CRITICAL THINKING AND CLINICAL APPLICATIONS

44. While resting and alert, Mary has a respiratory rate of 10 breaths per minute and a tidal volume of 300 ml. The air she is breathing contains 18 percent oxygen. Estimate Mary's basal metabolic rate for a 24-hour period.

45. Why is a starving person more susceptible to infectious disease than one who is well nourished?

46. Individuals with anorexia nervosa typically exhibit bradycardia, hypotension, and decreased heart size. These problems can eventually lead to death from heart failure. How does anorexia cause these symptoms?

47. The drug *colestipol* binds bile salts in the intestine, forming complexes that cannot be absorbed. How would this drug affect cholesterol levels in blood?

25

CHAPTER 26

The Urinary System

CHAPTER OUTLINE AND OBJECTIVES

INTRODUCTION   p. 943

ORGANIZATION OF THE URINARY SYSTEM   p. 943
① Identify the components of the urinary system, and describe the vital functions performed by this system.

THE KIDNEYS   p. 943
Superficial Anatomy of the Kidneys   p. 944
Sectional Anatomy of the Kidneys   p. 944
The Blood Supply to the Kidneys   p. 946
Innervation of the Kidneys   p. 946
The Nephron   p. 946
② Describe the structural features of the kidneys.
③ Describe the structure of the nephron and the processes involved in the formation of urine.
④ Identify the major blood vessels associated with each kidney, and trace the path of blood flow through a kidney.

RENAL PHYSIOLOGY   p. 953
Basic Principles of Urine Formation   p. 953
Filtration   p. 954
Reabsorption and Secretion   p. 958
The Function of the Vasa Recta   p. 967
Summary: Renal Function   p. 968
The Composition of Normal Urine   p. 969
⑤ List and describe the factors that influence filtration pressure and the rate of filtrate formation.
⑥ Identify the types of transport mechanisms found along the nephron, and discuss the reabsorptive or secretory functions of each segment of the nephron and collecting system.
⑦ Explain the role of countercurrent multiplication in the formation of a concentration gradient in the renal medulla.
⑧ Describe how antidiuretic hormone and aldosterone levels influence the volume and concentration of urine.
⑨ Describe the normal characteristics, composition, and solute concentrations of a representative urine sample.

Urinating in public is considered bad form. But this whimsical fountain in Brussels, Belgium, depicts a very natural and necessary process. After our digestive system assimilates a meal and our body harvests and utilizes the nutrients it needs, we must discharge the resulting waste products. The urinary system performs this service while efficiently conserving water and other valuable substances. As we shall see, it also carries out a variety of other vital but less obvious functions (none of which has been immortalized in sculpture, so far as we know).

URINE TRANSPORT, STORAGE, AND ELIMINATION   p. 972
The Ureters   p. 972
The Urinary Bladder   p. 974
The Urethra   p. 976
The Micturition Reflex and Urination   p. 976
⑩ Describe the structures and functions of the ureters, urinary bladder, and urethra.
⑪ Discuss the voluntary and involuntary regulation of urination, and describe the micturition reflex.

AGING AND THE URINARY SYSTEM   p. 977
⑫ Describe the effects of aging on the urinary system.

INTEGRATION WITH OTHER SYSTEMS   p. 978
⑬ Give examples of interactions between the urinary system and each of the other organ systems.

MEDIALAB and CHAPTER REVIEW   pp. 978, 980

The human body contains trillions of cells bathed in extracellular fluid. In previous chapters, we compared these cells to factories that burn nutrients to obtain energy. Imagine what would happen if *real* factories were built as close together as cells in the body. What a mess they would make! Each would generate piles of garbage, and the smoke they produced (while depleting the oxygen supply) would drastically reduce air quality. In short, a serious pollution problem would result.

The coordinated activities of the digestive, cardiovascular, respiratory, and urinary systems prevent the development of similar pollution problems in the body. The digestive tract absorbs nutrients from food, and the liver adjusts the nutrient concentration of the circulating blood. The cardiovascular system delivers these nutrients and oxygen from the respiratory system to peripheral tissues. As blood leaves these tissues, it carries the carbon dioxide and waste products to sites of excretion. The carbon dioxide is removed at the lungs. Most of the organic waste products are removed by the urinary system, the focus of this chapter.

The **urinary system** performs the vital function of removing the organic waste products generated by cells throughout the body. It also has other essential functions that are often overlooked. A more complete list of urinary system functions includes the following:

- *Regulating blood volume and blood pressure* by (1) adjusting the volume of water lost in urine, (2) releasing erythropoietin, and (3) releasing renin.
- *Regulating plasma concentrations of sodium, potassium, chloride, and other ions* by controlling the quantities lost in urine and controlling calcium ion levels by the synthesis of calcitriol.
- *Helping to stabilize blood pH* by controlling the loss of hydrogen ions and bicarbonate ions in urine.
- *Conserving valuable nutrients* by preventing their excretion in urine while excreting organic waste products, especially nitrogenous wastes, such as *urea* and *uric acid*.
- *Assisting the liver* in detoxifying poisons and, during starvation, deaminating amino acids so that other tissues can break them down.

These activities are carefully regulated to keep the composition of blood within acceptable limits. A disruption of any one of these functions will have immediate and potentially fatal consequences.

In this chapter, we will consider the functional organization of the urinary system and will describe the major regulatory mechanisms that control urine production and concentration.

ORGANIZATION OF THE URINARY SYSTEM

The urinary system (Figure 26-1•) has two major functions: (1) *excretion,* the removal of waste products from body fluids; and (2) *elimination,* the discharge of these waste products into the environment. The excretory functions of the urinary system are performed by the two **kidneys.** These organs produce **urine,** a fluid containing water, ions, and small soluble compounds. Urine leaving the kidneys travels along the **urinary tract,** which consists of paired tubes called **ureters** (ū-RĒ-terz), to the **urinary bladder,** a muscular sac for temporary urine storage. On leaving the urinary bladder, urine passes through the **urethra** (ū-RĒ-thra), which conducts the urine to the exterior. The urinary bladder and urethra are responsible for urine elimination, a process called **urination,** or **micturition** (mik-tū-RI-shun). This process occurs when the contraction of the muscular urinary bladder forces urine through the urethra and out of the body. ⓐⓜ *EMBRYOLOGY SUMMARY 20: The Development of the Urinary System*

THE KIDNEYS

The kidneys are located on either side of the vertebral column, between vertebrae T_{12} and L_3. The left kidney lies slightly superior to the right kidney (Figures 26-1, anterior view, and 26-2a•, posterior view).

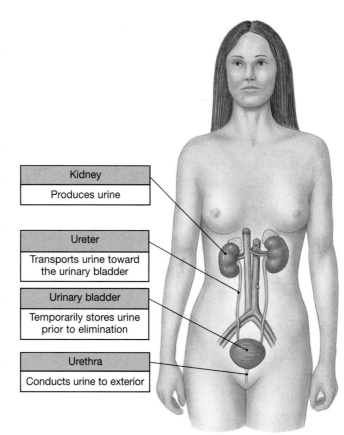

Kidney
Produces urine

Ureter
Transports urine toward the urinary bladder

Urinary bladder
Temporarily stores urine prior to elimination

Urethra
Conducts urine to exterior

Anterior view

•**FIGURE 26-1 An Introduction to the Urinary System.** An anterior view of the urinary system, showing the positions of the kidneys and other components.

26

(a) Posterior view

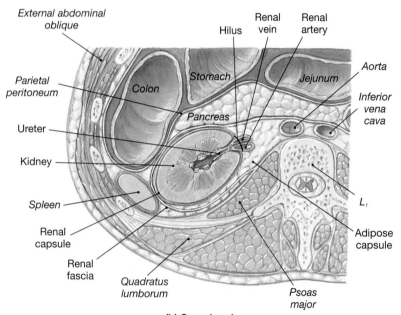

(b) Superior view

•FIGURE 26-2 The Position of the Kidneys. **(a)** A posterior view of the trunk. **(b)** A superior view of a section at the level indicated in (a). *Plates 7.9b, 7.10*

The superior surface of each kidney is capped by an adrenal gland (Figures 26-2a and 26-3•). The kidneys and adrenal glands lie between the muscles of the dorsal body wall and the parietal peritoneum in a retroperitoneal position (Figure 26-2b•).

The position of the kidneys in the abdominal cavity is maintained by (1) the overlying peritoneum, (2) contact with adjacent visceral organs, and (3) sup-

porting connective tissues. Each kidney is protected and stabilized by three concentric layers of connective tissue (Figure 26-2b•):

1. The **renal capsule,** a layer of collagen fibers that covers the outer surface of the entire organ.

2. The **adipose capsule,** a thick layer of adipose tissue that surrounds the renal capsule.

3. The **renal fascia,** a dense, fibrous outer layer. Collagen fibers extend outward from the renal capsule through the adipose capsule to this layer. The renal fascia anchors the kidney to surrounding structures. Posteriorly, the renal fascia fuses with the deep fascia surrounding the muscles of the body wall. Anteriorly, the renal fascia forms a thick layer that fuses with the peritoneum.

In effect, each kidney hangs suspended by collagen fibers from the renal fascia and is packed in a soft cushion of adipose tissue. This arrangement prevents the jolts and shocks of day-to-day existence from disturbing normal kidney function. If the suspensory fibers break or become detached, a slight bump or jar can displace the kidney and stress the attached vessels and ureter. This condition, called a *floating kidney,* is especially dangerous, because the ureters or renal blood vessels can become twisted or kinked during movement.

SUPERFICIAL ANATOMY OF THE KIDNEYS

Each reddish brown kidney is shaped like a kidney bean. A typical adult kidney (Figures 26-3 and 26-4•) is about 10 cm (4 in.) in length, 5.5 cm (2.2 in.) in width, and 3 cm (1.2 in.) in thickness. Each kidney weighs about 150 g (5.25 oz). The **hilus,** a prominent medial indentation, is the point of entry for the *renal artery* and *renal nerves* and the point of exit for the *renal vein* and ureter.

SECTIONAL ANATOMY OF THE KIDNEYS

The fibrous renal capsule covers the outer surface of the kidney and lines the **renal sinus,** an internal cavity within the kidney (Figure 26-4a•). Renal blood vessels and the ureter draining the kidney pass through the hilus and branch within the renal sinus. The renal

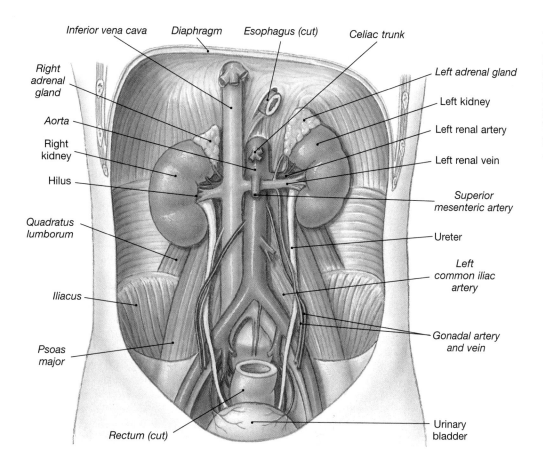

•**FIGURE 26-3** The **Urinary System in Gross Dissection.** The abdominopelvic cavity (with the digestive organs removed), showing the kidneys, ureters, urinary bladder, and blood supply to the urinary structures. *Plate 7.10d,e*

Inferior vena cava

Diaphragm

Esophagus (cut)

Celiac trunk

Right adrenal gland

Aorta

Right kidney

Hilus

Quadratus lumborum

Iliacus

Psoas major

Rectum (cut)

Left adrenal gland

Left kidney

Left renal artery

Left renal vein

Superior mesenteric artery

Ureter

Left common iliac artery

Gonadal artery and vein

Urinary bladder

Medulla

Cortex

Renal capsule

Renal pyramid

Medulla

Renal pyramids

Connection to minor calyx

Renal sinus

Renal sinus

Minor calyx

Renal pelvis

Adipose tissue in renal sinus

Major calyx

Major calyx

Renal pelvis

Hilus

Minor calyx

Hilus

Renal lobe

Renal papilla

Ureter

Ureter

Renal papilla

Renal columns

Renal lobe

Renal capilla

Renal capsule

(a)

(b)

•**FIGURE 26-4** **The Structure of the Kidney.** **(a)** A diagrammatic view of a frontal section through the left kidney. **(b)** A frontal section of the left kidney. *Plate 7.10a,b*

26

capsule is bound to the surfaces of these structures, stabilizing their positions.

The kidney itself has two layers: An outer cortex and an inner medulla. The renal **cortex** is the superficial portion of the kidney; it is in contact with the renal capsule. The cortex is reddish brown and granular. The renal **medulla** consists of 6 to 18 distinct conical or triangular structures called **renal pyramids.** The base of each pyramid faces the cortex, and the tip of each pyramid, a region known as the **renal papilla,** projects into the renal sinus. Each pyramid has a series of fine grooves that converge at the papilla. Adjacent renal pyramids are separated by bands of cortical tissue called **renal columns,** which extend into the medulla. The columns have a distinctly granular texture, similar to that of the cortex. A **renal lobe** consists of a renal pyramid, the overlying area of renal cortex, and adjacent tissues of the renal columns.

Urine production occurs in the renal lobes. Ducts within each renal papilla discharge urine into a cup-shaped drain called a **minor calyx** (KĀ-liks). Four or five minor calyces (KAL-i-sēz) merge to form a **major calyx,** and two or three major calyces combine to form the **renal pelvis,** a large, funnel-shaped chamber. The renal pelvis, which fills most of the renal sinus, is connected to the ureter, which drains the kidney.

Urine production begins in microscopic, tubular structures called **nephrons** (NEF-ronz) in the cortex of each renal lobe. Each kidney has roughly 1.25 million nephrons, with a combined length of about 145 km (85 miles).

THE BLOOD SUPPLY TO THE KIDNEYS

Your kidneys receive 20–25 percent of your total cardiac output. In normal, healthy individuals, about 1200 ml of blood flows through the kidneys each minute. That is a phenomenal amount of blood for organs with a combined weight of less than 300 g (10.5 oz)!

Each kidney receives blood from a **renal artery,** which originates along the lateral surface of the abdominal aorta near the level of the superior mesenteric artery (Figure 21-26a•, p. 730). As it enters the renal sinus, the renal artery provides blood to the **segmental arteries** (Figure 26-5a•). Segmental arteries further divide into a series of **interlobar arteries,** which radiate outward through the renal columns between the renal pyramids. The interlobar arteries supply blood to the **arcuate** (AR-kū-āt) **arteries,** which arch along the boundary between the cortex and medulla of the kidney. Each arcuate artery gives rise to a number of **interlobular arteries,** which supply the cortical portions of the adjacent renal lobes. Branching from each interlobular artery are numerous **afferent arterioles,** which deliver blood to the capillaries supplying individual nephrons (Figure 26-5b•).

From the capillaries of the nephrons, blood enters a network of venules and small veins that converge on the **interlobular veins.** The interlobular veins deliver blood to **arcuate veins,** which empty into **interlobar veins.** The interlobar veins drain into the **renal vein** (Figure 26-5a,c•).

✚ Analysis of Renal Blood Flow

The rate of blood flow through the kidneys can be determined by administering the compound *para-aminohippuric acid (PAH),* which is removed at the nephrons and eliminated in urine. Virtually all the PAH contained in the blood that arrives at the kidneys is removed before the blood departs in the renal veins. Renal blood flow can thus be calculated by comparing plasma concentrations of PAH with the amount secreted in urine. [AM] *PAH and the Calculation of Renal Blood Flow*

INNERVATION OF THE KIDNEYS

The kidneys and ureters are innervated by **renal nerves.** Most of the nerve fibers involved are sympathetic postganglionic fibers from the superior mesenteric ganglion. ∞ *p. 510* A renal nerve enters each kidney at the hilus and follows the tributaries of the renal arteries to reach individual nephrons. The sympathetic innervation (1) adjusts rates of urine formation by changing blood flow and blood pressure at the nephron, and (2) stimulates the release of renin, which ultimately restricts losses of water and salt in the urine by stimulating reabsorption at the nephron.

THE NEPHRON

Each nephron consists of a renal tubule roughly 50 mm long and a renal corpuscle. The **renal tubule,** a long tubular passageway, begins at the **renal corpuscle** (KOR-pus-ul), a cup-shaped chamber. The renal corpuscle is approximately 200 μm (0.2 mm) in diameter. It contains a capillary network called the **glomerulus** (glo-MER-ū-lus; plural, *glomeruli*), which consists of about 50 intertwining capillaries. Blood arrives at the glomerulus by way of an **afferent arteriole** and departs in an **efferent arteriole.** These structures of the nephron are shown in Figure 26-5•.

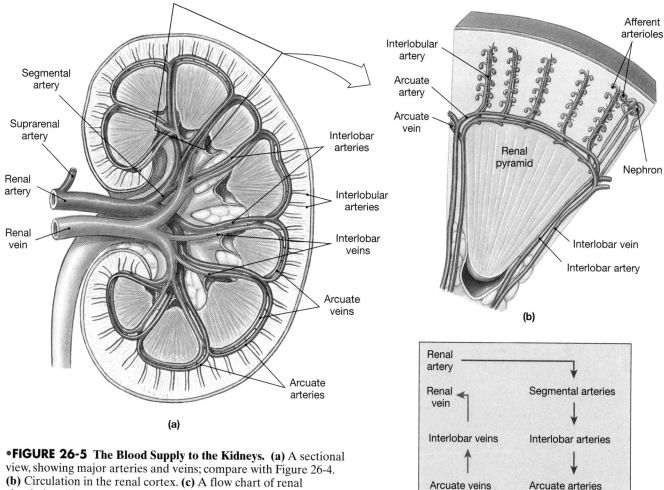

•FIGURE 26-5 The Blood Supply to the Kidneys. (a) A sectional view, showing major arteries and veins; compare with Figure 26-4. **(b)** Circulation in the renal cortex. **(c)** A flow chart of renal circulation. *Plate 7.10a,b,c; Scans 12, 13*

In the renal corpuscle, blood pressure forces fluid and dissolved solutes out of the glomerular capillaries and into the *capsular space.* This process is called *filtration.* ∞ *p. 75* Filtration produces an essentially protein-free solution, known as a **filtrate,** that is otherwise very similar to blood plasma.

From the renal corpuscle, filtrate enters the renal tubule. The renal tubule is responsible for:

▪ Reabsorbing all the useful organic substrates that enter the filtrate.

▪ Reabsorbing over 90 percent of the water present in filtrate.

▪ Secreting into the tubular fluid any waste products that were missed by the filtration process.

The renal tubule has two convoluted (coiled or twisted) segments—the *proximal convoluted tubule* (PCT) and the *distal convoluted tubule* (DCT)—separated by a simple U-shaped tube, the *loop of Henle* (HEN-lē). The convoluted segments are in the

cortex, and the loop extends partially or completely into the medulla. For clarity, the nephron shown in Figure 26-6• has been shortened and straightened. The regions of the nephron vary by structure and function. As it travels along the tubule, the filtrate, now called **tubular fluid,** gradually changes in composition. The changes that occur and the characteristics of the urine that results vary with the activities under way in each segment of the nephron. Figure 26-6• and Table 26-1 (p. 949) survey the regional specializations.

Each nephron empties into the **collecting system,** a series of tubes that carry tubular fluid away from

26

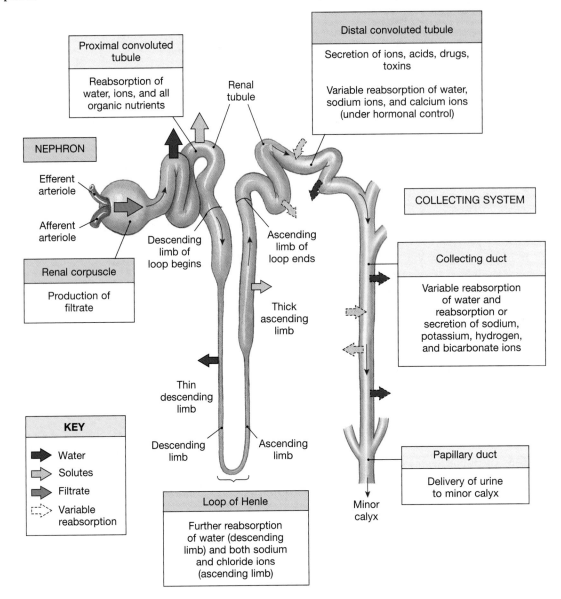

Proximal convoluted tubule

Reabsorption of water, ions, and all organic nutrients

NEPHRON

Efferent arteriole

Afferent arteriole

Renal corpuscle

Production of filtrate

Renal tubule

Descending limb of loop begins

Ascending limb of loop ends

Distal convoluted tubule

Secretion of ions, acids, drugs, toxins

Variable reabsorption of water, sodium ions, and calcium ions (under hormonal control)

COLLECTING SYSTEM

Collecting duct

Variable reabsorption of water and reabsorption or secretion of sodium, potassium, hydrogen, and bicarbonate ions

Thick ascending limb

Thin descending limb

Descending limb

Ascending limb

KEY

- ➡ Water
- ➡ Solutes
- ➡ Filtrate
- ⇢ Variable reabsorption

Loop of Henle

Further reabsorption of water (descending limb) and both sodium and chloride ions (ascending limb)

Papillary duct

Delivery of urine to minor calyx

Minor calyx

•**FIGURE 26-6** **A Representative Nephron.** The major functions of each segment of the nephron and collecting system. For clarity, the nephron has been shortened and straightened considerably.

the nephron. *Collecting ducts* receive this fluid from many nephrons. Each collecting duct begins in the cortex and descends into the medulla, carrying fluid to a *papillary duct.* This duct drains into a minor calyx.

Nephrons from different locations differ slightly in structure. Roughly 85 percent of all nephrons are **cortical nephrons,** located almost entirely within the superficial cortex of the kidney (Figure 26-7a,b•, p. 950). In a cortical nephron, the loop of Henle is relatively short, and the efferent arteriole delivers blood to a network of **peritubular capillaries,** which surround the entire renal tubule. These capillaries drain into small venules that carry blood to the interlobular veins (Figure 26-5c•).

The remaining 15 percent of nephrons, termed **juxtamedullary** (juks-ta-MED-ū-lar-ē) **nephrons** (*juxta,* near), have long loops of Henle that extend deep into the medulla (Figure 26-7a,c•). In these nephrons, the peritubular capillaries are connected to the **vasa recta** (*vasa,* vessel + *recta,* straight), long, straight capillaries that parallel the loop of Henle.

Because they are more numerous than juxtamedullary nephrons, cortical nephrons perform most of the reabsorptive and secretory functions of the kidneys. However, as you will see later in the chapter, it is the juxtamedullary nephrons that give the kidneys the ability to produce concentrated urine.

TABLE 26-1 The Organization of the Nephron and Collecting System in the Kidney

Region	Length	Diameter	Primary Function	Histological Characteristics
NEPHRON				
Renal corpuscle Glomerulus Parietal epithelium Visceral epithelium	150–250 μm (spherical)	150–250 μm	Filtration of plasma	Glomerulus (capillary knot), supporting cells, and lamina densa, enclosed by Bowman's capsule; visceral epithelium (podocytes) and parietal epithelium separated by capsular space
Renal tubule				
Proximal convoluted tubule (PCT) Microvilli	14 mm	60 μm	Reabsorption of ions, organic molecules, vitamins, water; secretion of drugs, toxins, acids	Cuboidal cells with microvilli
Loop of Henle Thin segment	30 mm	15 μm	Descending limb: reabsorption of water from tubular fluid	Squamous or low cuboidal cells
Thick segment		30 μm	Ascending limb: reabsorption of ions; assists in creation of a concentration gradient in the medulla	
Distal convoluted tubule (DCT)	5 mm	30–50 μm	Reabsorption of sodium ions and calcium ions; secretion of acids, ammonia, drugs, toxins	Cuboidal cells with few if any microvilli
COLLECTING SYSTEM				
Collecting duct	15 mm	50–100 μm	Reabsorption of water, sodium ions; secretion or reabsorption of bicarbonate ions or hydrogen ions	Cuboidal to columnar cells
Papillary duct	5 mm	100–200 μm	Conduction of tubular fluid to minor calyx; contributes to concentration gradient of the medulla	Columnar cells

We shall now examine the structure of each segment of a representative nephron.

The Renal Corpuscle

The renal corpuscle has a diameter of 150–250 μm on average. It includes (1) a region known as **Bowman's capsule** and (2) the capillary network of the glomerulus (Figure 26-8•, p. 951). Connected to the initial segment of the renal tubule, Bowman's capsule forms the outer wall of the renal corpuscle and encapsulates the glomerular capillaries.

The glomerulus is surrounded by Bowman's capsule, much as the heart is surrounded by the pericardial cavity. The outer wall of the capsule is lined by a simple squamous **parietal epithelium.** This layer is continuous with the **visceral epithelium,** which covers the glomerular capillaries. The **capsular space** separates the parietal and visceral epithelia. The two epithelial layers are continuous where the glomerular

26

•**FIGURE 26-7** Cortical and Juxtamedullary Nephrons. **(a)** The general appearance and location of nephrons in the kidneys. **(b)** The circulation to a cortical nephron. **(c)** The circulation to a juxtamedullary nephron. The length of the loop of Henle is not drawn to scale.

(b) Cortical nephron

(c) Juxtamedullary nephron

capillaries are connected to the afferent arteriole and efferent arteriole.

The visceral epithelium consists of large cells with complex processes, or "feet," that wrap around the **lamina densa,** the specialized basement membrane of the glomerular capillaries (Figure 26-8b,c,d●). These unusual cells are called **podocytes** (PŌ-do-sīts; *podos,* foot + -*cyte,* cell). The podocyte feet are known as **pedicels.** Materials passing out of the blood at the glomerulus must be small enough to pass between the narrow gaps, or **filtration slits,** between adjacent pedicels.

The glomerular capillaries are fenestrated capillaries—that is, their endothelium contains large-diameter pores (Figure 26-8d●). The lamina densa differs from the basement membrane of other capillary networks in that it may encircle more than one capillary. Special

supporting cells that lie between adjacent capillaries play a role in controlling their diameter and thus in the rate of capillary blood flow. Together, the fenestrated endothelium, the lamina densa, and the filtration slits form the *filtration membrane.* During filtration, blood pressure forces water and small solutes across this membrane and into the capsular space. The larger solutes, especially plasma proteins, are excluded. Filtration at the renal corpuscle is both effective and passive, but it has one major limitation: In addition to metabolic wastes and excess ions, compounds such as glucose, free fatty acids, amino acids, vitamins, and other solutes enter the capsular space. These potentially useful materials are recaptured before filtrate leaves the kidneys, with much of the reabsorption occurring in the proximal convoluted tubule.

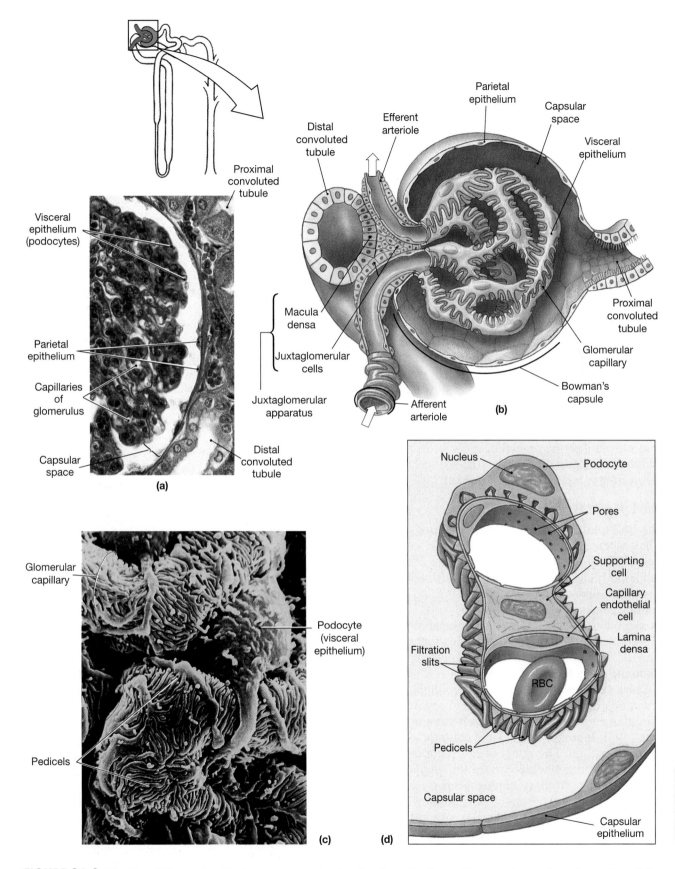

•FIGURE 26-8 The Renal Corpuscle. **(a)** A renal corpuscle, showing the epithelium of Bowman's capsule and a portion of the glomerular capillary network. (LM × 1120) **(b)** Important structural features of a renal corpuscle. **(c)** The glomerular surface, showing individual podocytes and their processes. (SEM × 27,248) **(d)** A podocyte with pedicels covering the adjacent surfaces of the lamina densa.

26

Glomerulonephritis

Glomerulonephritis (glo-mer-ū-lō-nef-RĪ-tis) is an inflammation of the renal cortex that affects the filtration mechanism of the kidneys. This condition, which can develop after an infection involving *Streptococcus* bacteria, is an *immune complex disorder.* ∞ *p. 788* The primary infection may not occur in or near the kidneys. However, as the immune system responds to the infection, the number of circulating antigen–antibody complexes skyrockets. These complexes are small enough to pass through the lamina densa but too large to fit between the filtration slits of the filtration membrane. As a result, the filtration mechanism clogs up and filtrate production drops. Any condition that leads to a massive immune response can cause glomerulonephritis, including viral infections and autoimmune disorders.

The Proximal Convoluted Tubule

The **proximal convoluted tubule (PCT)** is the first segment of the renal tubule (Figure 26-6●). The entrance to the PCT lies almost directly opposite the point where the afferent and efferent arterioles connect to the glomerulus. The lining of the PCT is a simple cuboidal epithelium whose apical surfaces bear microvilli. The tubular cells absorb organic nutrients, ions, water, and plasma proteins (if any) from the tubular fluid and release them into the **peritubular fluid,** the interstitial fluid surrounding the renal tubule. Reabsorption is the primary function of the PCT, but the epithelial cells can also secrete substances into the lumen.

The Loop of Henle

The PCT makes an acute bend that turns the renal tubule toward the renal medulla. This turn leads to the **loop of Henle,** or *nephron loop* (Figures 26-6, p. 948, and 26-7●). The loop of Henle can be divided into a **descending limb** and an **ascending limb.** Fluid in the descending limb travels toward the renal pelvis, and that in the ascending limb travels toward the renal cortex. Each limb contains a **thick segment** and a **thin segment.** (The terms *thick* and *thin* refer to the height of the epithelium, not to the diameter of the lumen.)

Thick segments have a cuboidal epithelium; a thin squamous epithelium lines the thin segments. The thick descending limb has functions similar to those of the PCT: It pumps sodium and chloride ions out of the tubular fluid. The effect of this pumping is most noticeable in the medulla, where the long ascending limbs of juxtamedullary nephrons create unusually high solute concentrations in peritubular fluid. The thin segments are freely permeable to water but not to solutes; water movement out of these segments helps concentrate the tubular fluid.

The Distal Convoluted Tubule

The thick ascending limb of the loop of Henle ends where it forms a sharp angle near the renal corpuscle.

The **distal convoluted tubule (DCT),** the third segment of the renal tubule, begins there. The initial portion of the DCT passes between the afferent and efferent arterioles (Figure 26-8b●).

In sectional view, the DCT differs from the PCT in that the DCT has a smaller diameter and its epithelial cells lack microvilli (Table 26-1). The DCT is an important site for:

- The active secretion of ions, acids, and other materials.
- The selective reabsorption of sodium ions and calcium ions from tubular fluid.
- The selective reabsorption of water, which assists in concentrating the tubular fluid.

The Juxtaglomerular Apparatus The epithelial cells of the DCT near the renal corpuscle are taller than those elsewhere along the DCT, and their nuclei are clustered together. This region is called the **macula densa** (MAK-ū-la DEN-sa) (Figure 26-8b●). The cells of the macula densa are closely associated with unusual smooth muscle fibers in the wall of the afferent arteriole. These fibers are known as **juxtaglomerular cells.** Together, the macula densa and juxtaglomerular cells form the **juxtaglomerular apparatus (JGA).** The juxtaglomerular apparatus is an endocrine structure that secretes the hormone *erythropoietin* and the enzyme *renin.* ∞ *p. 604*

The Collecting System

The distal convoluted tubule, the last segment of the nephron, opens into the collecting system. The collecting system consists of collecting ducts and papillary ducts (Figure 26-6●, p. 948). Individual nephrons drain into a nearby **collecting duct.** Several collecting ducts converge to empty into a larger **papillary duct,** which in turn empties into a minor calyx. The epithelium lining the collecting system is typically a columnar epithelium.

In addition to transporting tubular fluid from the nephron to the renal pelvis, the collecting system adjusts its composition and determines the final osmotic concentration and volume of urine.

Polycystic Kidney Disease

Polycystic (po-lē-SIS-tik) **kidney disease** is an inherited condition affecting the structure of kidney tubules. Swellings (cysts) develop along the length of the tubules, some growing large enough to compress adjacent nephrons and vessels. Kidney function deteriorates, and the nephrons can eventually become nonfunctional. The process is so gradual that serious problems seldom appear before the individual is 30–40 years of age. Common symptoms include sharp pain in the sides, recurrent urinary infections, and the presence of blood in the urine. Treatment is symptomatic, focusing on the prevention of infection and reduction of pain with analgesics. In severe cases, hemodialysis or kidney transplantation may be required.

☑ Which portions of the nephron are in the renal cortex?

☑ Why don't plasma proteins pass into the capsular space under normal circumstances?

☑ Damage to which part of the nephron would interfere with the control of blood pressure?

RENAL PHYSIOLOGY

BASIC PRINCIPLES OF URINE FORMATION

The goal of urine production is to maintain homeostasis by regulating the volume and composition of blood. This process involves the excretion of solutes, specifically metabolic waste products. Three organic waste products are noteworthy:

1. *Urea.* Urea is the most abundant organic waste. You generate roughly 21 g of urea each day. Most of it is produced during the breakdown of amino acids.

2. *Creatinine.* Creatinine is generated in skeletal muscle tissue by the breakdown of *creatine phosphate,* a high-energy compound that plays an important role in muscle contraction. ∞ *p. 293* Your body generates roughly 1.8 g of creatinine each day, and virtually all of it is excreted in urine.

3. *Uric acid.* Uric acid is formed by the recycling of nitrogenous bases from RNA molecules. ∞ *p. 919* You produce approximately 480 mg of uric acid each day.

These waste products are dissolved in the bloodstream and can be eliminated only while dissolved in urine. As a result, their removal is accompanied by an unavoidable water loss. The kidneys are usually capable of producing concentrated urine with an osmotic concentration of 1200–1400 mOsm/l, more than four times that of plasma. (See the box "Expressing Solute Concentrations" on p. 954.) If the kidneys were not able to concentrate the filtrate produced by glomerular filtration, fluid losses would lead to fatal dehydration in a matter of hours. At the same time, your kidneys ensure that the fluid that *is* lost does not contain potentially useful organic substrates, such as sugars or amino acids, that are contained in blood plasma. These valuable materials must be reabsorbed and retained for use by other tissues.

To accomplish these goals, your kidneys rely on three distinct processes:

1. *Filtration.* In **filtration,** blood pressure forces water across a filtration membrane at the renal corpuscle. Solute molecules small enough to pass through the filtration membrane are carried by the surrounding water molecules.

2. *Reabsorption.* **Reabsorption** is the removal of water and solutes from the filtrate, across the tubular epithelium, and into the peritubular fluid. Reabsorption occurs after filtrate has left the renal corpuscle. Most of the reabsorbed materials are nutrients that your body can use. Whereas filtration occurs solely on the basis of size, reabsorption is a selective process. Solute reabsorption may involve simple diffusion or the activity of carrier proteins in the tubular epithelium. The reabsorbed substances pass into the peritubular fluid and eventually reenter the blood. Water reabsorption occurs passively, through osmosis.

3. *Secretion.* **Secretion** is the transport of solutes from the peritubular fluid, across the tubular epithelium, and into the tubular fluid. Secretion is necessary because filtration does not force all the dissolved materials out of the plasma. Tubular secretion provides a backup for the filtration process that can further lower the plasma concentration of undesirable materials. Secretion can be the primary method of excretion for some compounds, including many drugs.

Together, these processes create a fluid that is very different from other body fluids. Table 26-2 provides an indication of the efficiency of the renal system by comparing the concentrations of some of the substances present in urine and plasma.

All segments of the nephron and collecting system participate in the formation of urine. Most regions of the nephron perform a combination of reabsorption and secretion, but the balance between the two shifts from one region to another:

- Filtration occurs exclusively in the renal corpuscle, across the filtration membrane.

- Nutrient reabsorption occurs primarily along the proximal convoluted tubules, but it also occurs elsewhere along the renal tubule and within the collecting system.

TABLE 26-2 Significant Differences between Solute Concentrations of Urine and Plasma

Component	Urine	Plasma
Ions (mEq/l)*		
Sodium (Na^+)	147.5	138.4
Potassium (K^+)	47.5	4.4
Chloride (Cl^-)	153.3	106
Bicarbonate (HCO_3^-)	1.9	27
Metabolites and Nutrients (mg/dl)		
Glucose	0.009	90
Lipids	0.002	600
Amino acids	0.188	4.2
Proteins	0.000	7.5 g/dl
Nitrogenous Wastes (mg/dl)		
Urea	1800	10–20
Creatinine	150	1–1.5
Ammonia	60	<0.1
Uric acid	40	3

*See the box "Expressing Solute Concentrations" on page 954.

26

▪ Active secretion occurs primarily at the proximal and distal convoluted tubules.

▪ The loops of Henle, especially the long loops of the juxtamedullary nephrons, and the collecting system interact to regulate the final volume and solute concentration of the urine.

Normal kidney function can continue only as long as filtration, reabsorption, and secretion function within relatively narrow limits. A disruption in kidney function has immediate effects on the composition of the circulating blood. If both kidneys are affected, death will occur within a few days unless medical assistance is provided.

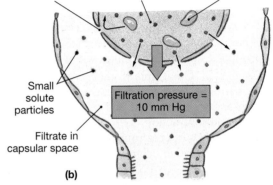

•**FIGURE 26-9 Glomerular Filtration. (a)** The filtration membrane. **(b)** Filtration pressure.

Expressing Solute Concentrations

The osmotic concentration, or *osmolarity*, of a solution is the total number of solute particles in each liter. ∞ *p. 74* Osmolarity is usually expressed in **osmoles** per liter (Osm/l) or **milliosmoles** per liter (mOsm/l). If each liter of a fluid contains one mole of dissolved particles, the solute concentration is 1 Osm/l, or 1000 mOsm/l. Body fluids have an osmotic concentration of about 300 mOsm/l. In comparison, that of seawater is about 1000 mOsm/l, and fresh water about 5 mOsm/l. Ion concentrations are often reported in *millequivalents* per liter (mEq/l), whereas the concentrations of large organic molecules are usually reported in grams or milligrams per unit volume of solution (typically mg or g per dl). AM *Solutions and Concentrations*

FILTRATION

Filtration occurs in the renal corpuscle as fluids move across the wall of the glomerulus and into the capsular space. The process of **glomerular filtration** involves passage across the filtration membrane. This membrane has three components: (1) the capillary endothelium, (2) the lamina densa, and (3) the filtration slits (Figures 26-8d and 26-9a•).

Glomerular capillaries are fenestrated capillaries with pores ranging from 60 to 100 nm (0.06 to 0.1 μm) in diameter. These openings are small enough to prevent the passage of blood cells, but they are too large to restrict the diffusion of solutes, even those the size of plasma proteins. The lamina densa is more selective; only small plasma proteins, nutrients, and ions can cross it. The filtration slits are the finest filters of all. The gaps are only 6–9 nm wide, which is small enough to block the passage of most small plasma proteins. As a result, under normal circumstances no plasma proteins other than a few albumin molecules (average diameter 7 nm) can cross the filtration membrane and enter the capsular space. However, plasma proteins are all that stay behind, so the filtrate contains dissolved ions and small organic molecules in roughly the same concentrations as in plasma.

Filtration Pressures

We discussed the major forces that act across capillary walls in Chapters 21 and 22. (You may find it helpful to review Figures 21-12 and 21-13• before you proceed.) ∞ *pp. 708, 709* The primary factor involved in glomerular filtration is basically the same as that governing fluid and solute movement across capillaries throughout the body: the balance between hydrostatic pressure (fluid pressure) and colloid osmotic pressure (pressure due to materials in solution) on either side of the capillary walls.

Hydrostatic Pressure The **glomerular hydrostatic pressure (GHP)** is the blood pressure in the glomerular capillaries. It tends to push water and solute molecules out of the plasma and into the filtrate. The GHP is significantly higher than capillary pressures elsewhere in the systemic circuit due to the arrangement of vessels at the glomerulus.

Blood pressure is low in typical systemic capillaries, because capillary blood flows into the venous system, where resistance is relatively low. However, at the glomerulus, blood leaving the glomerular capillaries flows into an efferent arteriole, whose diameter is *smaller* than that of the afferent arteriole. The efferent arteriole offers considerable resistance, so relatively high pressures are needed to force blood into it. As a result, glomerular pressures are similar to those of small arteries, averaging 45–55 mm Hg instead of the 35 mm Hg typical of peripheral capillaries.

Glomerular hydrostatic pressure is opposed by the **capsular hydrostatic pressure (CsHP).** The CsHP tends to push water and solutes out of the filtrate and into the plasma. This pressure results from the resistance to flow along the nephron and conducting system. (Before additional filtrate can enter the capsule, some of the filtrate already present must be forced into the PCT.) The CsHP averages about 15 mm Hg.

The *net hydrostatic pressure (NHP)* is the difference between glomerular hydrostatic pressure, which tries to push water and solutes out of the bloodstream, and capsular hydrostatic pressure, which tries to push water and solutes into the bloodstream.

Because the glomerular hydrostatic pressure averages 50 mm Hg and capsular hydrostatic pressure averages 15 mm Hg, the net hydrostatic pressure is 35 mm Hg. This can be written in an equation as:

$$NHP = GHP - CsHP = 50 \, mm \, Hg - 15 \, mm \, Hg = 35 \, mm \, Hg$$

Colloid Osmotic Pressure   Recall from Chapter 21 that the colloid osmotic pressure of a solution is the osmotic pressure resulting from the presence of suspended proteins. ∞ *pp. 707–708* The **blood colloid osmotic pressure (BCOP)** tends to draw water out of the filtrate and into the plasma. It thus opposes filtration. Over the entire length of the glomerular capillary bed, the BCOP averages about 25 mm Hg. Under normal conditions, very few plasma proteins enter the capsular space, so there is no opposing colloid osmotic pressure within the capsule. However, if the glomeruli are damaged by disease or injury and plasma proteins begin passing into the capsular space, a *capsular colloid osmotic pressure* is created that promotes filtration and increases fluid losses in urine.

Filtration Pressure   The **filtration pressure (FP)** at the glomerulus is the difference between the hydrostatic pressure and the colloid osmotic pressure acting across the glomerular capillaries. Under normal circumstances, this relationship can be summarized as

$$FP = NHP - BCOP$$
filtration pressure = 35 mm Hg − 25 mm Hg = 10 mm Hg

This value represents the average pressure forcing water and dissolved materials out of the glomerular capillaries and into the capsular spaces (Figure 26-9b•). Problems that affect filtration pressure can seriously disrupt kidney function and cause a variety of clinical symptoms. [AM] *Conditions Affecting Filtration*

Glomerular Filtration Rate

The **glomerular filtration rate (GFR)** is the amount of filtrate your kidneys produce each minute. Each kidney contains about 6 m² of filtration surface, and the GFR averages an astounding *125 ml per minute.* This means that roughly 10 percent of the fluid delivered to your kidneys by the renal arteries leaves the bloodstream and enters the capsular spaces.

A *creatinine clearance test* is often used to estimate the GFR. Creatinine, which results from the breakdown of creatine phosphate in muscle tissue, is normally eliminated in urine. Creatinine enters the filtrate at the glomerulus, and it is not reabsorbed in significant amounts. By monitoring the creatinine concentrations in blood and the amount excreted in urine, a clinician can easily estimate the GFR. Consider a person who eliminates 84 mg of creatinine each hour and has a plasma creatinine concentration of 1.4 mg/dl. The GFR is equal to the amount secreted, divided by the plasma concentration:

$$\frac{84 \, mg/h}{1.4 \, mg/dl} = 60 \, dl/h = 100 \, ml/min$$

The GFR is usually reported in milliliters per minute.

This value is only an approximation of the GFR, because up to 15 percent of creatinine enters the urine by means of active tubular secretion. When necessary, a more accurate GFR determination can be performed by using the complex carbohydrate *inulin,* which is not metabolized in the body and is neither reabsorbed nor secreted by the kidney tubules.

In the course of a single day, your glomeruli generate about 180 liters (50 gal) of filtrate, roughly 70 times the total plasma volume. But as filtrate passes through the renal tubules, about 99 percent of it is reabsorbed. You should now appreciate the significance of tubular reabsorption!

Blood Pressure and Glomerular Filtration

The glomerular filtration rate depends on the filtration pressure across glomerular capillaries. Any factor that alters the filtration pressure will therefore alter the GFR and thereby affect kidney function. One of the most significant factors is a drop in renal blood pressure.

Despite the large volume of filtrate generated, the filtration pressure is relatively low. Whenever the mean arterial pressure falls by 10 percent (from a normal value of about 100 mm Hg to 90 mm Hg), the GFR is severely restricted. If blood pressure at the glomeruli drops by 20 percent, from 50 mm Hg to 40 mm Hg, kidney filtration will cease, because the filtration pressure will be 0 mm Hg. The kidneys are therefore sensitive to changes in blood pressure that have little or no effect on other organs. Hemorrhaging, shock, and dehydration are relatively common clinical conditions that can cause a dangerous decline in the GFR.

26

Controlling the GFR

Glomerular filtration is the vital first step essential to all other kidney functions. If filtration does not occur, waste products are not excreted, pH control is jeopardized, and an important mechanism of blood volume regulation is lost. It should be no surprise to find that a variety of regulatory mechanisms exist to ensure that your GFR remains within normal limits.

Filtration depends on adequate blood flow to the glomerulus and the maintenance of normal filtration pressures. Three interacting levels of control stabilize your GFR: (1) *autoregulation* at the local level, (2) *hormonal regulation* initiated by the kidneys, and (3) *autonomic regulation,* primarily by the sympathetic division of the autonomic nervous system.

Autoregulation of the GFR The goal of autoregulation is to maintain an adequate GFR despite changes in local blood pressure and blood flow. This goal is accomplished by changing the diameters of the afferent arterioles, the efferent arterioles, and the glomerular capillaries. The most important regulatory mechanisms stabilize the GFR when systemic blood pressure declines. A reduction in blood flow and a decline in glomerular blood pressure trigger (1) the dilation of the afferent arteriole, (2) the relaxation of supporting cells and the dilation of the glomerular capillaries, and (3) the constriction of the efferent arteriole. This combination increases blood flow and elevates glomerular blood pressure to normal levels. As a result, filtration rates remain relatively constant. The GFR also remains relatively constant when systemic blood pressure rises. A rise in renal blood pressure stretches the walls of the afferent arterioles, and the smooth muscle cells respond by contracting. The reduction in the diameter of the afferent arterioles decreases glomerular blood flow and keeps the GFR within normal limits.

Hormonal Regulation of the GFR The GFR is regulated by the hormones of the renin–angiotensin system and atrial natriuretic peptide (ANP). We introduced those hormones and their actions in Chapters 18 and 21. ∞ *pp. 604–605, 716–717* Renin is an enzyme released by the juxtaglomerular apparatus (JGA). This release occurs when (1) renal blood flow declines as a result of a decrease in blood volume, a fall in systemic pressures, or a blockage in the renal artery or its tributaries; or (2) the osmotic concentration of the tubular fluid that reaches the distal convoluted tubule decreases.[1]

In the renin–angiotensin system, renin converts the inactive protein angiotensinogen to angiotensin I, and the angiotensin-converting enzyme in the capillaries of the lungs then converts angiotensin I to the active form, angiotensin II. Figure 26-10▪ diagrams the primary effects of angiotensin II. These effects include the following:

- *In peripheral capillary beds,* angiotensin II causes a brief but powerful vasoconstriction of arterioles and precapillary sphincters, elevating pressures in the renal arteries and their tributaries.

- *At the nephron,* angiotensin II causes the constriction of the efferent arteriole, further elevating glomerular pressures and filtration rates. It also directly stimulates the reabsorption of sodium ions and water at the PCT.

- *At the adrenal glands,* angiotensin II stimulates the secretion of aldosterone by the adrenal cortex. The aldosterone accelerates sodium reabsorption in the DCT and cortical portion of the collecting system.

- *In the CNS,* angiotensin II (1) causes the sensation of thirst, (2) triggers the release of antidiuretic hormone (ADH), stimulating the reabsorption of water in the distal portion of the DCT and the collecting system, and (3) increases sympathetic motor tone, further stimulating peripheral vasoconstriction.

The final results are an increase in systemic blood volume and blood pressure and the restoration of normal GFR.

Atrial natriuretic peptide is released in response to the stretching of the atrial walls of the heart by increased blood volume or blood pressure. Among its other effects, ANP triggers the dilation of the afferent arteriole and constriction of the efferent arteriole. This effect elevates glomerular pressures and increases the GFR, leading to increased urine production and decreased blood volume and pressure.

Autonomic Regulation of the GFR Most of the autonomic innervation of the kidneys consists of sympathetic postganglionic fibers. (The role of the few parasympathetic fibers in regulating kidney function is not known.) Sympathetic activation has one direct effect on the GFR: It produces a powerful vasoconstriction of the afferent arterioles, decreasing the GFR and slowing the production of filtrate. The sympathetic activation triggered by an acute fall in blood pressure or a heart attack will override the local regulatory mechanisms that act to stabilize the GFR. As the crisis passes and sympathetic tone decreases, the filtration rate gradually returns to normal levels.

When the sympathetic division alters regional patterns of blood circulation, blood flow to the kidneys is often affected. For example, the dilation of superficial vessels in warm weather shunts blood away from your kidneys, so glomerular filtration declines temporarily. The effect becomes especially pronounced during periods of strenuous exercise. As the blood flow

---

[1] When the rate of glomerular filtration decreases, so does the rate of fluid movement along the nephron. Because the tubular fluid then spends more time in the ascending limb of the loop of Henle, the concentration of sodium and chloride ions becomes abnormally low.

The T_m and Renal Threshold Normally, any plasma proteins and nutrients, such as amino acids and glucose, are removed from the tubular fluid by cotransport or facilitated diffusion. If the concentrations of these nutrients rise in the tubular fluid, the rates of reabsorption increase until the carrier proteins are saturated. A concentration higher than the tubular maximum will exceed the reabsorptive abilities of the nephron, so some of the material will remain in the tubular fluid and appear in the urine. The transport maximum thus determines the **renal threshold,** the plasma concentration at which a specific compound or ion will begin appearing in urine.

The renal threshold varies with the substance involved. The renal threshold for glucose is approximately 180 mg/dl. When plasma glucose concentrations remain higher than 180 mg/dl, glucose concentrations in tubular fluid will exceed the T_m of the tubular cells, so glucose will appear in urine. The presence of glucose in urine is a condition called *glycosuria.* After you have eaten a meal rich in carbohydrates, your plasma glucose levels may exceed the T_m for a brief period. Your liver will quickly lower circulating glucose levels, and very little glucose will be lost in your urine. However, chronically elevated plasma and urinary glucose concentrations are very unusual. (Glycosuria is one of the key symptoms of diabetes mellitus. ∞ *p. 607*)

The renal threshold for amino acids is lower than that for glucose; amino acids appear in urine when plasma concentrations exceed 65 mg/dl. Your plasma amino acid levels commonly exceed the renal threshold after you have eaten a protein-rich meal, causing some amino acids to appear in your urine. This condition is termed *aminoaciduria.*

The T_m values for water-soluble vitamins are relatively low; as a result, excess quantities of these vitamins are excreted in urine. (This is typically the fate of daily water-soluble vitamin supplements.) Cells of the renal tubule ignore a number of other compounds in the tubular fluid. As water and other compounds are removed, the concentration of the ignored materials gradually rises in the tubular fluid. Table 26-3 contains a partial listing of substances actively reabsorbed, secreted, or ignored by the renal tubules.

We shall now proceed along the nephron to consider the changes in the composition and concentration of the filtrate that is produced at the renal corpuscle. Most of what follows applies equally to cortical and juxtamedullary nephrons. The major differences between the two types of nephron are that the loop of Henle of a cortical nephron is shorter and does not extend as far into the medulla as does the loop of Henle of a juxtamedullary nephron (Figure 26-7a•, p. 950). The long loop of Henle in a juxtamedullary nephron extends deep into the renal pyramids, where it plays a vital role in water conservation and the for-

TABLE 26-3 Tubular Reabsorption and Secretion

Reabsorbed	Secreted	No Transport Mechanism
Ions	**Ions**	Urea
Na^+, Cl^-, K^+,	K^+, H^+, Ca^{2+},	Water
Ca^{2+}, Mg^{2+},	PO_4^{3-}	Urobilinogen
SO_4^{2-}, HCO_3^-		Bilirubin
	Wastes	
	Creatinine	
Metabolites	Ammonia	
Glucose	Organic acids	
Amino acids	and bases	
Proteins		
Vitamins	**Miscellaneous**	
	Neurotransmitters	
	(ACh, NE, E, dopamine)	
	Histamine	
	Drugs (penicillin, atropine, morphine, many others)	

mation of concentrated urine. Because this process is so important, affecting the tubular fluid produced by every nephron in the kidney, and because the functions of the renal corpuscle and of the proximal and distal convoluted tubules are the same in all nephrons, we shall use a juxtamedullary nephron as our example. Table 26-4 surveys the functions of the various parts of the nephron.

The Proximal Convoluted Tubule

The cells of the proximal convoluted tubule normally reabsorb 60–70 percent of the volume of the filtrate produced in the renal corpuscle. The reabsorbed materials enter the peritubular fluid and diffuse into peritubular capillaries.

The PCT has five major functions:

1. *The reabsorption of organic nutrients.* Under normal circumstances, before the tubular fluid enters the loop of Henle, the PCT reabsorbs more than 99 percent of the glucose, amino acids, and other organic nutrients. This reabsorption involves a combination of facilitated transport and cotransport.

2. *The active reabsorption of ions.* The PCT actively transports ions, including sodium, potassium, magnesium, bicarbonate, phosphate, and sulfate ions. The ion pumps involved are individually regulated and may be influenced by circulating ion or hormone levels. For example, angiotensin II stimulates Na^+ reabsorption along the PCT. By absorbing carbon dioxide, the PCT

26

TABLE 26-4 Renal Segments and Their Functions

Segment	General Functions	Specific Functions	Mechanisms
Renal corpuscle	*Filtration* of plasma; generates approximately 180 liters/day of filtrate similar in composition to blood plasma without plasma proteins	*Filtration* of water, inorganic and organic solutes from plasma; retention of plasma proteins and blood cells	Glomerular hydrostatic (blood) pressure working across capillary endothelium, lamina densa, and filtration slits
Proximal convoluted tubule (PCT)	*Reabsorption* of 60–70% of the water (108–116 liters/day), 99–100% of the organic substrates, and 60–70% of the sodium and chloride ions in the original filtrate	*Reabsorption:* Active: glucose, other simple sugars, amino acids, vitamins, ions (including sodium, potassium, calcium, magnesium, phosphate, and bicarbonate) Passive: urea, chloride ions, lipid-soluble materials, water *Secretion:* Hydrogen ions, ammonium ions, creatinine, drugs, and toxins (as at DCT)	Carrier-mediated transport, including facilitated transport (glucose, amino acids), cotransport (glucose, ions), or countertransport (with secretion of H^+) Diffusion (solutes) or osmosis (water) Countertransport with sodium ions
Loop of Henle	*Reabsorption* of 25% of the water (45 liters/day) and 20–25% of the sodium and chloride ions present in the original filtrate; creation of the concentration gradient in the medulla	*Reabsorption:* Sodium and chloride ions Water	Active transport via Na^+–K^+/2 Cl^- transporter Osmosis
Distal convoluted tubule (DCT)	*Reabsorption* of a variable amount of water (usually 5%, or 9 liters/day), under ADH stimulation, and a variable amount of sodium ions, under aldosterone stimulation	*Reabsorption:* Sodium and chloride ions Sodium ions (variable) Calcium ions (variable) Water (variable) *Secretion:* Hydrogen ions, ammonium ions Creatinine, drugs, toxins	Cotransport Countertransport with potassium ions; aldosterone-regulated Carrier-mediated transport stimulated by parathyroid hormone and calcitriol Osmosis; ADH-regulated Countertransport with sodium ions Carrier-mediated transport
Collecting system	*Reabsorption* of a variable amount of water (usually 9.3%, or 16.8 liters/day), under ADH stimulation, and a variable amount of sodium ions, under aldosterone stimulation	*Reabsorption:* Sodium ions (variable) Bicarbonate ions (variable) Water (variable) Urea (distal portions only) *Secretion:* Potassium and hydrogen ions (variable)	Countertransport with potassium or hydrogen ions; aldosterone-regulated Diffusion, generated within tubular cells Osmosis; ADH-regulated Diffusion Carrier-mediated transport
Peritubular capillaries	*Redistribution* of water and solutes reabsorbed in the cortex	Return of water and solutes to the general circulation	Osmosis and diffusion
Vasa recta	*Redistribution* of water and solutes reabsorbed in the medulla and stabilization of the concentration gradient of the medulla	Return of water and solutes to the general circulation	Osmosis and diffusion

26

•**FIGURE 26-11** **Transport Activities at the PCT.** Sodium ions may enter a tubular cell from the filtrate by diffusion, cotransport, or countertransport. The sodium ions are then pumped into peritubular fluid by the sodium–potassium exchange pump. Other reabsorbed solutes may be ejected into the peritubular fluid by separate active transport mechanisms.

indirectly recaptures roughly 90 percent of the bicarbonate ions from tubular fluid (Figure 26-11•). Bicarbonate is important in stabilizing blood pH, a process we shall examine further in Chapter 27.

3. *The reabsorption of water.* The reabsorptive processes have a direct effect on the solute concentrations inside and outside the tubules. The filtrate entering the PCT has the same osmotic concentration as that of surrounding peritubular fluid. As transport activities proceed, the solute concentration of tubular fluid decreases and that of peritubular fluid and adjacent capillaries increases. Osmosis then pulls water out of the tubular fluid and into the peritubular fluid. Along the PCT, this mechanism results in the reabsorption of roughly 108 liters of water each day.

4. *The passive reabsorption of ions.* As active reabsorption of ions occurs and water leaves tubular fluid by osmosis, the concentration of other solutes in tubular fluid increases above that in peritubular fluid. If the tubular cells are permeable to them, those solutes will move through passive diffusion across the tubular cells and into the peritubular fluid. Urea, chloride ions, and lipid-soluble materials may diffuse out of the PCT in this way. As this diffusion occurs, it further reduces the solute concentration of the tubular fluid and promotes additional water reabsorption by osmosis.

5. *Secretion.* Active secretion also occurs along the PCT. Because the DCT performs comparatively little reabsorption, we will consider secretory mechanisms when we discuss the DCT.

Sodium ion reabsorption plays an important role in these processes. Sodium ions may enter the tubular cells by diffusion through Na^+ leak channels; by the sodium-linked cotransport of glucose, amino acids, or

26

other organic solutes; or by countertransport for hydrogen ions (Figure 26-11●). In the tubular cell, the sodium ions diffuse toward the basement membrane. The cell membrane in this area contains sodium–potassium exchange pumps that eject sodium ions in exchange for extracellular potassium ions. Reabsorbed sodium ions then diffuse into the adjacent peritubular capillaries.

The reabsorption of ions and compounds along the PCT involves many different carrier proteins. Some people have an inherited inability to manufacture one or more of these carrier proteins and are therefore unable to recover specific solutes from tubular fluid. For example, in *renal glycosuria* (glī-kō-SOO-rē-uh), a defective carrier protein makes it impossible for the PCT to reabsorb glucose from tubular fluid. [AM] *Inherited Problems with Tubular Function*

☑ How does the reabsorption of calcium ions by the kidney affect the sodium ion concentration of filtrate?

☑ What occurs when the plasma concentration of a substance exceeds its tubular maximum?

☑ How would a decrease in blood pressure affect the GFR?

The Loop of Henle and Countercurrent Exchange

Roughly 60–70 percent of the volume of filtrate produced at the glomerulus has been reabsorbed before the tubular fluid reaches the loop of Henle. In the process, the useful organic substrates have been reclaimed, along with many mineral ions. The loop of Henle will reabsorb roughly half of the water as well as two-thirds of the sodium and chloride ions remaining in the tubular fluid.

Countercurrent Multiplication   The loop of Henle accomplishes this reabsorption in a very efficient way. It uses the principle of countercurrent exchange, which we introduced in Chapter 25 in our discussion of heat conservation techniques. ∞ *p. 935*

The thin descending limb and the thick ascending limb of the loop of Henle are very close together, separated by peritubular fluid. The exchange that occurs between these segments is called **countercurrent multiplication.** *Countercurrent* refers to the fact that the exchange occurs between fluids moving in opposite directions; tubular fluid in the descending limb flows toward the renal pelvis, whereas tubular fluid in the ascending limb flows toward the cortex. *Multiplication* refers to the fact that the effect of the exchange increases as fluid movement continues.

The two parallel segments of the loop of Henle have very different permeability characteristics. The thin descending limb is permeable to water but relatively impermeable to solutes. The thick ascending limb, which is relatively impermeable to both water

and solutes, contains active transport mechanisms that pump sodium and chloride ions from the tubular fluid into the peritubular fluid of the medulla. Countercurrent multiplication operates as follows:

- *Sodium and chloride are pumped out of the thick ascending limb and into the peritubular fluid.*

- *This pumping elevates the osmotic concentration in the peritubular fluid around the thin descending limb.*

- *The result is an osmotic flow of water out of the thin descending limb and into the peritubular fluid, increasing the solute concentration in the thin descending limb.*

- *The arrival of the highly concentrated solution in the thick ascending limb accelerates the transport of sodium and chloride ions into the peritubular fluid of the medulla.*

Notice that this arrangement is a simple positive feedback loop: Solute pumping at the ascending limb leads to higher solute concentrations in the descending limb, which then result in accelerated solute pumping in the ascending limb.

Figure 26-12a● diagrams ion transport across the epithelium of the thick ascending limb. Active transport at the apical surface moves sodium, potassium, and chloride ions out of the tubular fluid. The carrier is called a **Na^+–K^+/2 Cl^- transporter,** because each cycle of the pump carries a sodium ion, a potassium ion, and two chloride ions into the tubular cell. Potassium and chloride ions are pumped into the peritubular fluid by cotransport carriers. However, potassium ions are removed from the peritubular fluid as the sodium–potassium exchange pump pumps sodium ions out of the tubular cell. These potassium ions diffuse back into the lumen of the tubule through potassium leak channels. The net result is that Na^+ and Cl^- enter the peritubular fluid of the renal medulla.

The removal of sodium and chloride ions from the tubular fluid in the ascending limb elevates the osmotic concentration of the peritubular fluid around the thin descending limb (Figure 26-12b●). The thin descending limb is permeable to water but impermeable to solutes. Thus, as tubular fluid travels into the medulla along the thin descending limb, osmosis moves water into the peritubular fluid. Solutes remain behind, so the tubular fluid reaching the turn of the loop of Henle has a higher osmotic concentration than it did at the start.

The pumping mechanism of the thick ascending limb is very effective: Almost two-thirds of the sodium and chloride ions that enter it are pumped out of the tubular fluid before that fluid reaches the DCT. In other tissues, differences in solute concentration are quickly resolved by osmosis. Osmosis cannot occur across the wall of the thick ascending limb, however, because the tubular epithelium there is impermeable to water. Thus, as Na^+ and Cl^- are removed, the solute concentration in the tubular fluid declines. Tubular

(a) The mechanism of sodium and chloride ion transport involves (1) a Na⁺–K⁺/ 2 Cl⁻ carrier at the apical surface and (2) two carriers at the basal surface of the tubular cell: a potassium–chloride cotransport pump and a sodium–potassium exchange pump. Because the potassium ions removed from the lumen of the tubule ultimately diffuse back in through leak channels, the net result is the transport of sodium and chloride ions into the peritubular fluid.

(b) Active transport of NaCl along the ascending thick limb results in the movement of water from the descending limb.

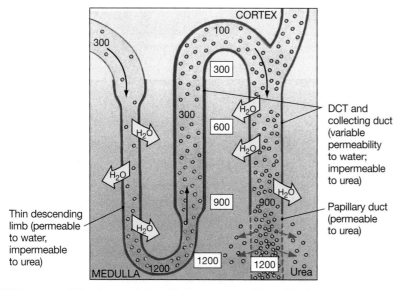

KEY	
●	= Cotransport
○	= Exchange pump
⊣ ⊢	= Leak channel
------	= Diffusion
---▸	= Reabsorption
---▸	= Secretion

(c) The permeability characteristics of both the loop and the collecting duct tend to concentrate urea in the tubular fluid and in the medulla. The loop of Henle, DCT, and collecting duct are impermeable to urea. As water reabsorption occurs, the urea concentration rises. The papillary ducts' permeability to urea accounts for roughly one-third of the solutes in the deepest portions of the medulla.

•**FIGURE 26-12** Countercurrent Multiplication and Concentration of Urine

fluid arrives at the DCT with an osmotic concentration of only about 100 mOsm/l, one-third of the concentration of the peritubular fluid of the renal cortex.

The rate of ion transport by the thick ascending limb is proportional to the ion's concentration in tubular fluid. As a result, more sodium and chloride ions are pumped into the medulla at the start of the thick ascending limb, where NaCl concentrations are highest, than near the cortex. This regional difference in the rate of ion transport is the basis of the concentration gradient within the medulla.

The Concentration Gradient of the Medulla Normally, the maximum solute concentration of the peritubular fluid near the turn of the loop of Henle is about 1200 mOsm/l. Sodium and chloride ions pumped out of the loop's ascending limb account for roughly two-thirds of that gradient (750 mOsm/l). The rest of the concentration gradient results from the presence of urea. To understand how the urea arrived in the medulla, we must look ahead to events in the last segments of the collecting system (Figure 26-12c•). The thick ascending limb of the loop of Henle, the DCT,

26

and the collecting ducts are impermeable to urea. As water reabsorption occurs, the concentration of urea gradually rises. The tubular fluid reaching the papillary duct typically contains urea at a concentration of about 450 mOsm/l. Because the papillary ducts are permeable to urea, the urea concentration in the deepest parts of the medulla also averages 450 mOsm/l.

Benefits of Countercurrent Multiplication
The countercurrent mechanism performs two services:

1. It is an efficient way to reabsorb solutes and water before the tubular fluid reaches the DCT and collecting system.
2. It establishes a concentration gradient that permits the passive reabsorption of water from the tubular fluid in the collecting system. This reabsorption is regulated by circulating levels of antidiuretic hormone (ADH).

The tubular fluid arriving at the descending limb of the loop of Henle has an osmotic concentration of roughly 300 mOsm/l, primarily due to the presence of ions such as Na^+ and Cl^-. The concentration of organic wastes, such as urea, is low. Roughly half of the tubular fluid entering the loop of Henle is reabsorbed along the thin descending limb, and two-thirds of the Na^+ and Cl^- is reabsorbed along the thick ascending limb. As a result, the DCT receives a reduced volume of

tubular fluid with an osmotic concentration of about 100 mOsm/l. Urea and other organic wastes, which were not pumped out of the thick ascending limb, now represent a significant percentage of the dissolved solutes.

The Distal Convoluted Tubule

The composition and volume of tubular fluid change dramatically as it travels from the capsular space to the distal convoluted tubule. Only 15–20 percent of the initial filtrate volume reaches the DCT, and the concentrations of electrolytes and organic wastes in the arriving tubular fluid no longer resemble the concentrations in blood plasma. Selective reabsorption or secretion, primarily along the DCT (Figure 26-13•)

KEY:

⚪	= Exchange pump
⚫	= Cotransport
🔘	= Countertransport
-----	= Diffusion
----▶	= Reabsorption
----▶	= Secretion
⊣⊢	= Leak channel

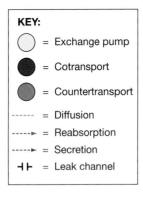

•**FIGURE 26-13 Tubular Secretion and Solute Reabsorption at the DCT.** (a) The basic pattern of the absorption of sodium and chloride ions and the secretion of potassium ions. (b) Aldosterone-regulated absorption of sodium ions, linked to the passive loss of potassium ions.

(a) Sodium and chloride reabsorption   (b) Sodium–potassium exchange

26

makes the final adjustments in the solute composition and volume of the tubular fluid.

Reabsorption Throughout most of the DCT, the tubular cells actively transport Na^+ and Cl^- out of the tubular fluid (Figure 26-13a•). Tubular cells along the distal portions of the DCT also contain ion pumps that reabsorb tubular Na^+ in exchange for another cation (usually K^+) (Figure 26-13b•). The ion pump and the Na^+ channels involved are controlled by the hormone *aldosterone,* produced by the adrenal cortex. Aldosterone stimulates the synthesis and incorporation of sodium ion pumps and sodium channels in cell membranes along the DCT and collecting duct. The net result is a reduction in the number of sodium ions lost in urine. However, sodium ion conservation is associated with potassium ion loss (Figure 26-13b•). Prolonged aldosterone stimulation can therefore produce *hypokalemia,* a dangerous reduction in the plasma K^+ concentration.

The DCT is also the primary site of Ca^{2+} reabsorption. This process is regulated by circulating levels of parathyroid hormone and calcitriol. ∞ *p. 598*

Secretion Filtration does not force all the dissolved materials out of plasma, and blood entering peritubular capillaries still contains a number of potentially undesirable substances. In most cases, the concentrations of these materials are too low to cause physiological problems. However, any ions or compounds present in peritubular capillaries will diffuse into the peritubular fluid. If those concentrations become too high, the tubular cells may absorb these materials from the peritubular fluid and secrete them into the tubular fluid. Table 26-3, p. 959, gives a partial listing of substances secreted into tubular fluid by the distal and proximal convoluted tubules.

The rate of K^+ and H^+ secretion rises or falls in response to changes in their concentrations in peritubular fluid. The higher the concentration in the peritubular fluid, the higher the rate of secretion. Potassium and hydrogen ions merit special attention, because their concentrations in body fluids must be maintained within relatively narrow limits.

Potassium ion secretion Figure 26-13a,b• diagrams the mechanism of K^+ secretion. Potassium ions are removed from peritubular fluid in exchange for sodium ions obtained from the tubular fluid. These potassium ions diffuse into the lumen through potassium channels at the apical surfaces of the tubule cells. In effect, the tubular cells trade sodium ions in the tubular fluid for excess potassium ions in body fluids.

Hydrogen ion secretion Hydrogen ion secretion is also associated with the reabsorption of sodium. Figure 26-13c• shows two secretory routes. Both involve the generation of carbonic acid by the enzyme *carbonic anhydrase.* ∞ *pp. 829, 864* Hydrogen ions generated by the dissociation of the carbonic acid are secreted by means of sodium-linked countertransport in exchange for Na^+ in the tubular fluid. The bicarbonate ions diffuse into the peritubular fluid and then into the bloodstream, where they help prevent changes in plasma pH.

(c) H^+ secretion and HCO_3^- reabsorption

•**FIGURE 26-13** **Tubular Secretion and Solute Reabsorption at the DCT** *(continued).* **(c)** Hydrogen ion secretion and the acidification of urine occur by two routes. The central theme is the exchange of hydrogen ions in the cytoplasm for sodium ions in the tubular fluid and the reabsorption of the bicarbonate ions generated in the process.

26

Hydrogen ion secretion acidifies the tubular fluid while elevating the pH of the blood. Hydrogen ion secretion accelerates when the pH of the blood falls, as in *lactic acidosis,* which can develop after exhaustive muscle activity or *ketoacidosis,* which can develop in starvation or diabetes mellitus. ∞ *p. 925* The combination of H^+ removal and HCO_3^- production at the kidneys plays an important role in the control of blood pH. Because one of the secretory pathways is aldosterone-sensitive, aldosterone stimulates H^+ secretion. Prolonged aldosterone stimulation can cause *alkalosis,* abnormally high blood pH.

In Chapter 25, we noted that the production of lactic acid and ketone bodies during the postabsorptive state can cause acidosis. Under these conditions, the PCT and DCT will deaminate amino acids in reactions that strip off the amino groups ($-NH_2$). The reaction sequence ties up H^+ and yields both **ammonium ions** (NH_4^+) and HCO_3^-. As indicated in Figure 26-13c•, the ammonium ions are then pumped into tubular fluid by sodium-linked countertransport and the bicarbonate ions enter the bloodstream by way of the peritubular fluid.

Tubular deamination thus has two major benefits: (1) It provides carbon chains suitable for catabolism, and (2) it generates bicarbonate ions that add to the buffering capabilities of plasma.

The Collecting System

The collecting ducts receive tubular fluid from many nephrons and carry it toward the renal sinus along the concentration gradient in the medulla. The normal amount of water and solute loss in the collecting system is regulated in two ways: (1) by aldosterone, which controls sodium ion pumps along most of the DCT and the proximal portion of the collecting system; and (2) by ADH, which controls the permeability of the DCT and collecting system to water. The collecting system also has other reabsorptive and secretory functions, many of which are important to the control of body fluid pH.

Reabsorption Important examples of solute reabsorption in the collecting system include the following:

■ *Sodium ion reabsorption.* The collecting system contains aldosterone-sensitive ion pumps that exchange Na^+ in tubular fluid for K^+ in peritubular fluid (Figure 26-13b•).

■ *Bicarbonate reabsorption.* Bicarbonate ions are reabsorbed in exchange for chloride ions in the peritubular fluid (Figure 26-13c•).

■ *Urea reabsorption.* The concentration of urea in the tubular fluid entering the collecting duct is relatively high. The fluid entering the papillary duct generally has the same osmotic concentration as that of interstitial fluid of the medulla—about 1200 mOsm/l—but contains a much higher concentration of urea. As a result, urea tends to diffuse out of the tubular fluid and into the peritubular fluid of the deepest portion of the medulla.

Secretion The collecting system is an important site for the control of body fluid pH by means of the secretion of hydrogen or bicarbonate ions. If the pH of the peritubular fluid declines, carrier proteins pump hydrogen ions into the tubular fluid and reabsorb bicarbonate ions that help restore normal pH. If the pH of the peritubular fluid rises (a much less common event), the collecting system secretes bicarbonate ions and pumps hydrogen ions into the peritubular fluid. The net result is that the body eliminates a buffer and gains hydrogen ions that will lower the pH. We shall examine those responses in more detail in Chapter 27 when we consider acid–base balance.

The Control of Urine Volume and Osmotic Concentration

Urine volume and osmotic concentration are regulated by controlling water reabsorption. Water reabsorption by osmosis occurs in the proximal convoluted tubule and the descending limb of the loop of Henle. The water permeabilities of these regions cannot be adjusted, and water reabsorption occurs whenever the osmotic concentration of the peritubular fluid exceeds that of the tubular fluid. The ascending limb of the loop of Henle is impermeable to water, but 1–2 percent of the volume of water in the original filtrate is recovered during sodium ion reabsorption in the distal convoluted tubule and collecting system. Because these water movements cannot be prevented, they represent the *obligatory water reabsorption.* Obligatory water reabsorption usually recovers 85 percent of the volume of filtrate produced.

The volume of water lost in urine depends on how much of the water in the remaining tubular fluid (15 percent of the filtrate volume, or roughly 27 liters per day) is reabsorbed along the DCT and the collecting system. The amount can be precisely controlled by a process called *facultative water reabsorption.* Precise control is possible because these segments are relatively impermeable to water except in the presence of ADH. The hormone ADH causes the appearance of special *water channels* in the apical cell membranes. Water channels dramatically enhance the rate of osmotic water movement. The higher the circulating levels of ADH, the greater the number of water channels and the greater the water permeability of these segments.

As we noted earlier in this chapter, the tubular fluid arriving at the DCT has an osmotic concentration of only about 100 mOsm/l. In the presence of ADH, osmosis occurs and water moves out of the DCT until the osmotic concentration of the tubular fluid equals that of the surrounding cortex (roughly 300 mOsm/l). The tubular fluid then travels through the collecting duct, which passes along the concentration gradient of the medulla. Additional water is then reabsorbed, and the urine reaching the minor calyx has an osmotic concentration closer to 1200 mOsm/l. Just how closely the osmotic concentration approaches 1200 mOsml/l depends on how much ADH is present.

Figure 26-14• diagrams the effects of ADH on the DCT and collecting system. In the absence of ADH, water reabsorption does not occur in these segments, so all the fluid reaching the DCT is lost in the urine. The individual then produces large amounts of very dilute urine (Figure 26-14a•). That is what happens in cases of *diabetes insipidus* (∞ *p. 591*), when urinary water losses may reach 24 liters per day, and urine osmotic concentration is 30–400 mOsm/l. As ADH levels rise, the DCT and collecting system become more permeable to water, the amount of water reabsorbed increases, and urine osmotic concentration climbs. Under maximum ADH stimulation, the DCT and collecting system become so permeable to water that the osmotic concentration of the urine is equal to that of the deepest portion of the medulla (Figure 26-14b•). Notice that the concentration of urine can never *exceed* that of the medulla, because the concentrating mechanism relies on osmosis.

You continuously secrete ADH at low levels, and your DCT and collecting system always have a significant degree of water permeability. The DCT normally reabsorbs roughly 9 liters of water per day, or about 5 percent of the original volume of filtrate produced by your glomeruli. At normal levels of ADH, the collecting system reabsorbs roughly 16.8 liters per day, or about 9.3 percent of the original volume of filtrate. A healthy adult typically produces 1200 ml of urine per day (about 0.6 percent of the filtrate volume), with an osmotic concentration of 800–1000 mOsm/l.

The effects of ADH are opposed by those of atrial natriuretic peptide (ANP), which is released by cardiac muscle cells in the atrial walls when the walls are excessively stretched. This hormone stimulates the production of a large volume of relatively dilute urine, soon restoring normal plasma volume.

THE FUNCTION OF THE VASA RECTA

The solutes and water reabsorbed in the medulla must be returned to the general circulation without

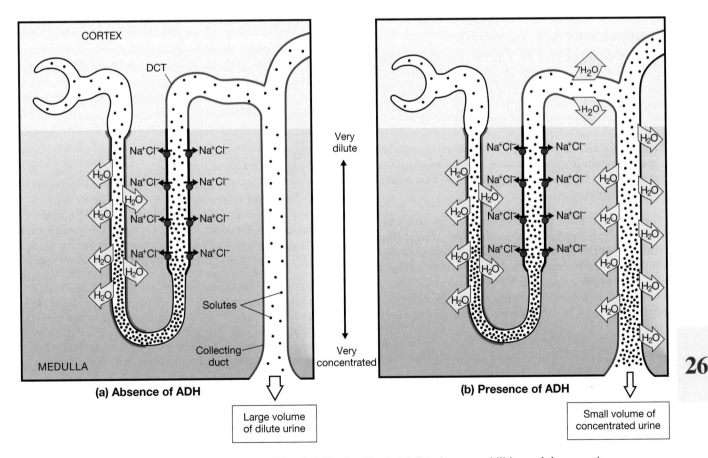

•**FIGURE 26-14** **The Effects of ADH on the DCT and Collecting Duct.** **(a)** Tubule permeabilities and the osmotic concentration of urine without ADH. **(b)** Tubule permeabilities and the osmotic concentration of urine at maximal ADH levels.

disrupting the concentration gradient. This return is the function of the vasa recta. Blood enters the vasa recta with an osmotic concentration of approximately 300 mOsm/l. The blood descending into the medulla gradually increases in osmotic concentration as the solute concentration in the peritubular fluid rises. This increase in blood osmotic concentration involves both solute absorption and water loss, but solute absorption predominates, because the plasma proteins limit osmotic flow out of the blood. ∞ *p. 709* Blood flowing toward the cortex gradually decreases in osmotic concentration as the solute concentration of the peritubular fluid declines. Again, this decrease involves a combination of solute diffusion and osmosis, but in this case osmosis predominates, because the presence of plasma proteins does not oppose the osmotic flow of water into the blood.

The net results are that (1) some of the solutes absorbed in the descending portion of the vasa recta do not diffuse out in the ascending portion, and (2) more water moves into the ascending portion of the vasa recta than is moved out in the descending portion. Thus the vasa recta carries both water and solutes out of the medulla. Under normal conditions, the removal of solutes and water by the vasa recta precisely balances the rates of solute reabsorption and osmosis in the medulla.

SUMMARY: RENAL FUNCTION

Table 26-4, p. 960, lists the functions of each segment of the nephron and collecting system. Figure 26-15• provides a functional overview that summarizes the major steps involved in the reabsorption of water and the production of concentrated urine:

STEP 1: The filtrate produced at the renal corpuscle has the same osmotic concentration as does plasma, about 300 mOsm/l. It has the composition of blood plasma without the plasma proteins.

STEP 2: In the PCT, the active removal of ions and organic substrates produces a continuous osmotic flow of water out of the tubular fluid. This process reduces the volume of filtrate but keeps the solutions inside and outside the tubule isotonic. Between 60 and 70 percent of the filtrate volume has been reabsorbed before the tubular fluid reaches the descending limb of the loop of Henle.

STEP 3: In the PCT and descending limb of the loop of Henle, water moves into the surrounding peritubular fluids, leaving a small volume (15–20 percent of the original filtrate) of highly concentrated tubular fluid. The volume reduction has occurred by obligatory water reabsorption.

STEP 4: The thick ascending limb is impermeable to water and solutes. The tubular cells actively transport Na+ and Cl− out of the tubular fluid. This transport lowers the osmotic concentration of tubular fluid without affecting the volume of the tubular fluid. The tubular fluid reaching the distal convoluted tubule is hypotonic relative to the peritubular fluid, with an osmotic concentration of only about 100 mOsm/l. Because only Na+ and Cl− are removed, urea now accounts for a significantly higher proportion of the total osmotic concentration at the end of the loop than it did at the start.

STEP 5: The final adjustments in the composition of the tubular fluid are made in the DCT and the collecting system. Although the DCT and collecting duct are generally impermeable to solutes, the osmotic concentration of tubular fluid can be adjusted through active transport (reabsorption or secretion). Some of these transport activities are hormonally regulated.

STEP 6: The final adjustments in the volume and osmotic concentration of the tubular fluid are made by controlling the water permeabilities of the distal portions of the DCT and the collecting system. These segments are relatively impermeable to water unless exposed to ADH. Under maximum ADH stimulation, urine volume is at a minimum and urine osmotic concentration is equal to that of the peritubular fluid in the deepest portion of the medulla (roughly 1200 mOsm/l).

STEP 7: The vasa recta absorbs solutes and water reabsorbed by the loop of Henle and the collecting ducts and thereby maintains the concentration gradient of the medulla.

Diuretics

Diuresis (dī-ū-RĒ-sis; *dia*, through + *ouresis*, urination) is the elimination of urine. Whereas *urination* is an equivalent term in a general sense, *diuresis* typically indicates the production of a large volume of urine. **Diuretics** (dī-ū-RET-iks) are drugs that promote the loss of water in urine. The usual goal in diuretic administration is a reduction of blood volume, blood pressure, or both. The ability to control renal water losses with relatively safe and effective diuretics has saved the lives of many individuals, especially those with high blood pressure or congestive heart failure.

Diuretics have many mechanisms of action, but all such drugs affect transport activities or water reabsorption along the nephron and collecting system. For example, a class of diuretics called *thiazides* (THĪ-a-zīdz) promotes water loss by reducing sodium and chloride ion transport in the proximal and distal convoluted tubules.
[AM] *Diuretics*

Diuretic use for nonclinical reasons is on the rise. For example, large doses of diuretics are taken by some bodybuilders to improve muscle definition and by some fashion models or jockeys to reduce body weight. This practice is extremely dangerous and has caused several deaths due to severe dehydration or cardiac arrest.

26

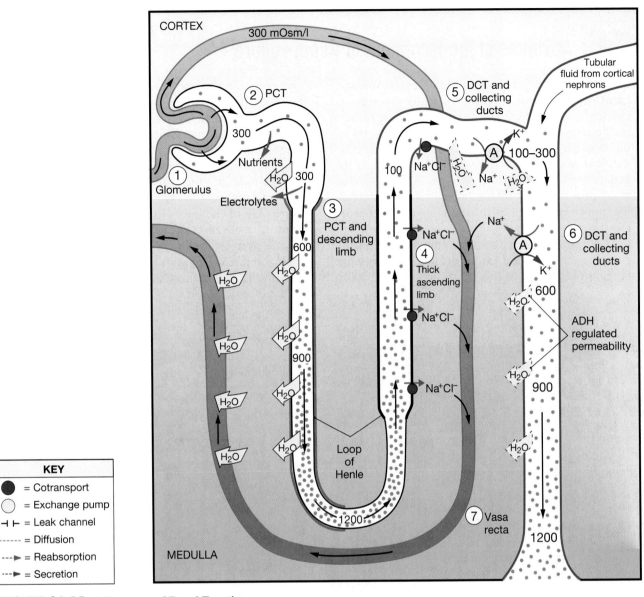

•**FIGURE 26-15** A Summary of Renal Function

☑ What effect would increased amounts of aldosterone have on the K$^+$ concentration of urine?

☑ What effect would a decrease in the Na$^+$ concentration of filtrate have on the pH of tubular fluid?

☑ How would the lack of juxtamedullary nephrons affect the volume and osmotic concentration of urine?

☑ Why does a decrease in the amount of Na$^+$ in the distal convoluted tubule lead to an increase in blood pressure?

THE COMPOSITION OF NORMAL URINE

More than 99 percent of the 180 liters of filtrate produced each day by the glomeruli is reabsorbed and never reaches the renal pelvis. General characteristics of normal urine are listed in Table 26-5. However, the

TABLE 26-5 **General Characteristics of Normal Urine**

Characteristic	Normal Range
pH	6.0 (range: 4.5–8)
Specific gravity	1.003–1.030
Osmotic concentration (osmolarity)	855–1335 mOsm/l
Water content	93–97 percent
Volume	1200 ml/day
Color	Clear yellow
Odor	Varies with composition
Bacterial content	Sterile

26

Advances in the Treatment of Renal Failure

Many conditions can result in renal failure. Management of chronic renal failure typically involves restricting water and salt intake and minimizing protein intake, with few dietary proteins. This combination reduces strain on the urinary system by (1) minimizing the volume of urine produced and (2) preventing the generation of large quantities of nitrogenous wastes. Acidosis, a common problem in persons with renal failure, can be countered by the ingestion of bicarbonate ions.

If drugs and dietary controls cannot stabilize the composition of blood, more-drastic measures are taken. In **hemodialysis** (hē-mō-dī-AL-i-sis), an artificial membrane is used to regulate the composition of blood by means of a *dialysis machine* (Figure 26-16a●). The basic principle involved in this process, called **dialysis**, involves passive diffusion across a selectively permeable membrane. The patient's blood flows past an artificial *dialysis membrane*, which contains pores large enough to permit the diffusion of small ions but small enough to prevent the loss of plasma proteins. On the other side of the membrane flows a special **dialysis fluid.**

The composition of typical dialysis fluid is indicated in Table 26-6. As diffusion takes place across the membrane, the composition of the blood changes.

●**FIGURE 26-16 Hemodialysis. (a)** A patient is hooked up to a dialysis machine. **(b)** The dialysis procedure. Preparation for hemodialysis typically involves the implantation of a pair of shunts connected by a loop that permits normal blood flow when the patient is not hooked up to the machine.

TABLE 26-6 The Composition of Dialysis Fluid

Component	Plasma	Dialysis Fluid
Electrolytes (mEq/l)		
Potassium	4	3
Bicarbonate	27	36
Phosphate	3	0
Sulfate	0.5	0
Nutrients (mg/dl)		
Glucose	80–100	125
Nitrogenous wastes (mg/dl)		
Urea	20	0
Creatinine	1	0
Uric acid	3	0

Note: Only the significant variations are given; values for other electrolytes are usually similar. Although these values are representative, the precise composition can be tailored to meet specific clinical needs. For example, if plasma potassium levels are too low, the dialysis fluid concentration can be elevated to remedy the situation. Changes in the osmotic concentration of the dialysis fluid can also be used to adjust an individual's blood volume, generally by adjusting the glucose content of the dialysis fluid.

Potassium ions, phosphate ions, sulfate ions, urea, creatinine, and uric acid diffuse across the membrane into the dialysis fluid. Bicarbonate ions and glucose diffuse into the bloodstream. In effect, diffusion across the dialysis membrane takes the place of normal glomerular filtration, and the characteristics of the dialysis fluid ensure that important metabolites remain in the bloodstream rather than diffusing across the membrane.

In practice, silicone rubber tubes called *shunts* are inserted into a medium-sized artery and vein. (The typical location is the forearm, although the lower leg is sometimes used.) The two shunts are then connected as shown in Figure 26-16b•. The connection acts like a short-circuit that does not impede blood flow, and the shunts can be used like taps in a wine barrel, to draw a blood sample or to connect the individual to a dialysis machine. For long-term dialysis, a surgically created arteriovenous anastomosis provides access.

When connected to the dialysis machine, the individual sits quietly while blood circulates from the arterial shunt, through the machine, and back through the venous shunt. In the machine, the blood flows within a tube composed of dialysis membrane, and diffusion occurs between the blood and the surrounding dialysis fluid.

The use of a dialysis machine is suggested when a patient's *BUN (blood urea nitrogen)* value exceeds 100 mg/dl (the normal value is up to 30 mg/dl). Dialysis techniques can maintain patients who are awaiting a transplant or those whose kidney function has been temporarily disrupted. Hemodialysis does have drawbacks, however: (1) The patient must sit by the machine about 15 hours a week; (2) between treatments, the symptoms of uremia gradually develop; (3) hypotension can develop as a result of fluid loss during dialysis; (4) air bubbles in the tubing can cause an embolism to form in the bloodstream; (5) anemia commonly develops; and (6) the shunts can serve as sites of recurring infections.

One alternative to the use of a dialysis machine is peritoneal dialysis. In **peritoneal dialysis,** the peritoneal lining is used as a dialysis membrane. Dialysis fluid is introduced into the peritoneum through a catheter in the abdominal wall, and at intervals the fluid is removed and replaced. For example, one procedure involves cycling 2 liters of fluid in an hour—15 minutes for infusion, 30 minutes for exchange, and 15 minutes for fluid reclamation. This process is usually performed in a hospital. An interesting variation on this procedure is called **continuous ambulatory peritoneal dialysis (CAPD).** In this procedure, the patient administers 2 liters of dialysis fluid through the catheter and then continues with life as usual until 4–6 hours later, when the fluid is removed and replaced.

Probably the most satisfactory solution, in terms of overall quality of life, is *kidney transplantation.* This procedure involves the implantation of a new kidney obtained from a living donor or from a cadaver. One-third of the estimated 14,000 kidneys transplanted in 1999 were obtained from living, related donors. In most cases, the damaged kidney is removed and its blood supply is connected to the transplant. When the original kidney is left in place, an arterial graft is inserted to carry blood from the iliac artery or the aorta to the transplant, which is placed in the pelvis or lower abdomen.

The success rate for kidney transplantation varies, depending on how aggressively the recipient's T cells attack the donated organ and whether an infection develops. The 1-year success rate for transplantation is now 85–95 percent. The use of kidneys taken from close relatives significantly improves the chances that the transplant will succeed. Immunosuppressive drugs are administered to reduce tissue rejection, but unfortunately this treatment also lowers the individual's resistance to infection.

26

composition of the 1.2 liters of urine produced varies with the metabolic and hormonal events under way.

The composition and concentration of urine are two integrated but distinct properties. The *composition* of urine reflects the filtration, absorption, and secretion activities of the nephrons. Some compounds, such as urea, are neither actively excreted nor reabsorbed along the nephron. In contrast, organic nutrients are completely reabsorbed, and other compounds, such as creatinine, that are missed by the filtration process are actively secreted into tubular fluid.

These processes determine the identities and amounts of materials excreted in urine. The *concentration* of these components in a given urine sample depends on the osmotic movement of water across the walls of the tubules and collecting ducts. Because the composition and concentration of urine vary independently, you can produce a small quantity of concentrated urine or a large quantity of dilute urine and still excrete the same amount of dissolved materials. As a result, physicians interested in a detailed assessment of renal function commonly rely on a 24-hour urine collection rather than on a single urine sample.

Urinalysis

Normal urine is a noncloudy, sterile solution. Its yellow color results from the presence of urobilin, generated in the kidneys from the urobilinogens produced by intestinal bacteria and absorbed in the colon (see Figure 19-5●, p. 632). The evaporation of small molecules, such as ammonia, accounts for the characteristic odor of urine. Other substances not normally present, such as acetone or other ketone bodies, can also impart a distinctive smell. **Urinalysis,** the analysis of a urine sample, is a diagnostic tool of considerable importance, even in high-technology medicine. A standard urinalysis involves an assessment of urine color and appearance, two characteristics that can be determined without specialized equipment. In the seventeenth century, physicians classified the taste of the urine as sweet, salty, and so on, but quantitative chemical tests have long since replaced the taste bud assay. Table 26-7 gives values for urinalysis. AM *Urinalysis*

URINE TRANSPORT, STORAGE, AND ELIMINATION

Filtrate modification and urine production end when the fluid enters the renal pelvis. The urinary tract (the ureters, urinary bladder, and urethra) is responsible for the transport, storage, and elimination of urine. A *pyelogram* (PĪ-el-ō-gram) is an image of the urinary system, obtained by taking an X-ray of the kidneys after a radiopaque compound has been administered. Such an image provides an orientation to the relative sizes and positions of these organs (Figure 26-17●). The sizes of the minor and major calyces, the renal pelvis, the ureters,

•**FIGURE 26-17** A Radiographic View of the Urinary System. A color-enhanced pyelogram (posterior view).

the urinary bladder, and the proximal portion of the urethra are somewhat variable, because these regions are lined by a *transitional epithelium* that can tolerate cycles of distension and contraction without damage. ∞ *p. 112*

We shall now examine these components of the urinary system.

THE URETERS

The ureters are a pair of muscular tubes that extend from the kidneys to the urinary bladder—a distance of about 30 cm (12 in.). Each ureter begins at the funnel-shaped renal pelvis (Figure 26-4●, p. 945). The ureters extend inferiorly and medially, passing over the anterior surfaces of the *psoas major muscles* (Figures 26-2b, p. 944, and 26-3●, p. 945). The ureters are retroperitoneal and are firmly attached to the posterior abdominal wall. The paths taken by the ureters in men and women are different due to variations in the nature, size, and position of the reproductive organs.

26

TABLE 26-7 Typical Values from Standard Urine Testing

Compound	Primary Source	Daily Elimination*	Concentration	Remarks
Nitrogenous wastes				
Urea	Deamination of amino acids at liver and kidneys	21 g	1.8 g/dl	Rises if negative nitrogen balance exists
Creatinine	Breakdown of creatinine phosphate in skeletal muscle	1.8 g	150 mg/dl	Proportional to muscle mass; decreases during atrophy or muscle disease
Ammonia	Deamination by liver and kidney, absorption from intestinal tract	0.68 g	60 mg/dl	
Uric acid	Breakdown of purines	0.53 g	40 mg/dl	Increases in gout, liver diseases
Hippuric acid	Breakdown of dietary toxins	4.2 mg	350 μg/dl	
Urobilin	Urobilinogens absorbed at colon	1.5 mg	125 μg/dl	Gives urine its yellow color
Bilirubin	Hemoglobin breakdown product	0.3 mg	20 μg/dl	Increase may indicate problem with liver elimination or excess production; causes yellow skin color in jaundice
Nutrients and metabolites				
Carbohydrates		0.11 g	9 μg/dl	Primarily glucose; *glycosuria* develops if T_m is exceeded
Ketone bodies		0.21 g	17 μg/dl	Ketonuria may occur during postabsorptive state
Lipids		0.02 g	1.6 μg/dl	May increase in some kidney diseases
Amino acids		2.25 g	287.5 μg/dl	Note relatively high loss compared with other metabolites due to low T_m; excess (*aminoaciduria*) indicates T_m problem
Ions				
Sodium		4.0 g	333 mg/dl	Varies with diet, urine pH, hormones, etc.
Chloride		6.4 g	533 mg/dl	
Potassium		2.0 g	166 mg/dl	Varies with diet, urine pH, hormones, etc.
Calcium		0.2 g	17 mg/dl	Hormonally regulated (PTH/CT)
Magnesium		0.15 g	13 mg/dl	
Blood cells[†]				
RBCs		130,000/day	100/ml	Excess (*hematuria*) indicates vascular damage
WBCs		650,000/day	500/ml	Excess (*pyuria*) indicates renal infection or inflammation

26

*Representative values for a 70-kg male.
[†] Usually estimated by counting the cells in a sample of sediment after urine centrifugation.

As Figure 26-18a• shows, in males, the base of the urinary bladder lies between the rectum and the pubic symphysis; in females, the base of the urinary bladder sits inferior to the uterus and anterior to the vagina (Figure 26-18b•).

The ureters penetrate the posterior wall of the urinary bladder without entering the peritoneal cavity. They pass through the bladder wall at an oblique angle, and the **ureteral openings** are slit-like rather than rounded (Figure 26-18c•). This shape helps prevent the backflow of urine toward the ureter and kidneys when the urinary bladder contracts.

Histology of the Ureters

The wall of each ureter consists of three layers: (1) an inner mucosa, consisting of a transitional epithelium and the surrounding lamina propria; (2) a middle muscular layer made up of longitudinal and circular bands of smooth muscle; and (3) an outer connective tissue layer that is continuous with the fibrous renal capsule and peritoneum (Figure 26-19a•). About every 30 seconds, a peristaltic contraction begins at the renal pelvis and sweeps along the ureter, forcing urine toward the urinary bladder.

Problems with the Conducting System

Local blockages of the collecting ducts or ureter can result from the formation of *casts*—small blood clots, epithelial cells, lipids, or other materials. Casts are commonly eliminated in urine and are visible in microscopic analysis of urine samples. **Calculi** (KAL-kū-lī), or *kidney stones,* form within the urinary tract from calcium deposits, magnesium salts, or crystals of uric acid. This condition is called *nephrolithiasis* (nef-rō-li-THĪ-a-sis). The blockage of the urinary passage by a stone or by other factors, such as external compression, results in **urinary obstruction.** Urinary obstruction is a serious problem because, in addition to causing pain, it reduces or prevents filtration in the affected kidney by elevating the capsular hydrostatic pressure.

Calculi are generally visible on an X-ray. If peristalsis and fluid pressures are insufficient to dislodge them, they must be surgically removed or destroyed. One interesting nonsurgical procedure involves breaking these stones apart with a *lithotripter,* the same apparatus used to destroy gallstones. ∞ *p. 880* Another nonsurgical approach is the insertion of a catheter that is armed with a laser that can shatter calculi with intense light beams.

THE URINARY BLADDER

The urinary bladder is a hollow, muscular organ that functions as a temporary storage reservoir for urine (Figure 26-18c•). The dimensions of the urinary bladder vary with the state of distension, but the full urinary bladder can contain about a liter of urine.

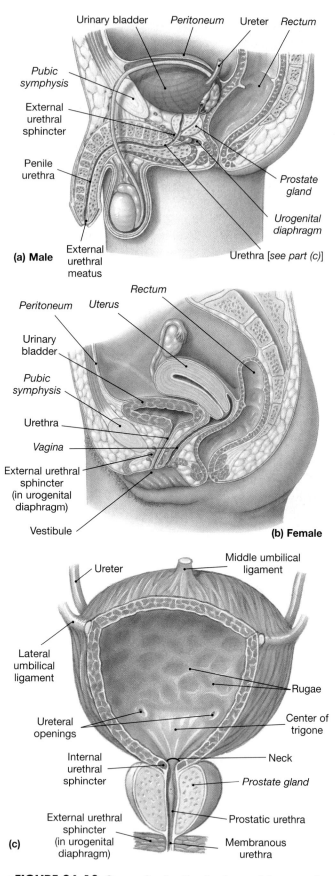

(a) Male

(b) Female

(c)

•**FIGURE 26-18 Organs for the Conduction and Storage of Urine. (a)** The ureter, urinary bladder, and urethra in the male and **(b)** in the female. **(c)** The urinary bladder of a male.
AM *Plates 7.10e, 7.11a,b*

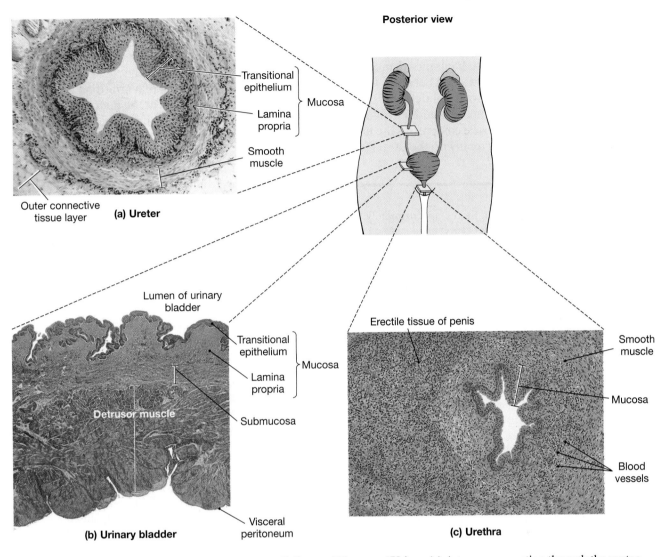

Posterior view

(a) Ureter
- Transitional epithelium ⎤
- Lamina propria ⎦ Mucosa
- Smooth muscle
- Outer connective tissue layer

(b) Urinary bladder
- Lumen of urinary bladder
- Transitional epithelium ⎤
- Lamina propria ⎦ Mucosa
- Submucosa
- Detrusor muscle
- Visceral peritoneum

(c) Urethra
- Erectile tissue of penis
- Smooth muscle
- Mucosa
- Blood vessels

•**FIGURE 26-19** **The Histology of the Organs that Collect and Transport Urine. (a)** A transverse section through the ureter. A thick layer of smooth muscle surrounds the lumen. (For a close-up of transitional epithelium, review Figure 4-4c•, pp. 112–113.) (LM × 53) **(b)** The wall of the urinary bladder. (LM × 29) **(c)** A transverse section through the male urethra. (LM × 49)

The superior surfaces of the urinary bladder are covered by a layer of peritoneum; several peritoneal folds assist in stabilizing its position. The **middle umbilical ligament** extends from the anterior, superior border toward the umbilicus (navel). The **lateral umbilical ligaments** pass along the sides of the bladder to the umbilicus. These fibrous cords contain the vestiges of the two *umbilical arteries,* which supplied blood to the placenta during embryonic and fetal development. ∞ *p. 742* The urinary bladder's posterior, inferior, and anterior surfaces lie outside the peritoneal cavity. In these areas, tough ligamentous bands anchor the urinary bladder to the pelvic and pubic bones.

In sectional view, the mucosa lining the urinary bladder is usually thrown into folds, or **rugae,** that disappear as the bladder fills. The triangular area bounded by the ureteral openings and the entrance to the urethra constitutes a region called the **trigone** (TRĪ-gōn) of the urinary bladder. The mucosa there is smooth and very thick. The trigone acts as a funnel that channels urine into the urethra when the urinary bladder contracts.

The urethral entrance lies at the apex of the trigone, at the most inferior point in the urinary bladder. The region surrounding the urethral opening, known as the **neck** of the urinary bladder, contains a muscular **internal urethral sphincter,** or *sphincter vesicae.* The smooth muscle fibers of the internal urethral sphincter provide involuntary control over the discharge of urine from the urinary bladder. The urinary bladder is innervated by postganglionic fibers from ganglia in the hypogastric plexus and by parasympathetic fibers from intramural ganglia that are controlled by branches of the pelvic nerves.

26

Histology of the Urinary Bladder

The wall of the urinary bladder contains mucosa, submucosa, and muscularis layers. The muscularis layer consists of inner and outer longitudinal smooth muscle layers, with a circular layer sandwiched between. Collectively, these layers form the powerful **detrusor** (de-TROO-sor) **muscle** of the urinary bladder (Figure 26-19b•). Contraction of this muscle compresses the urinary bladder and expels its contents into the urethra.

Bladder Cancer

Each year in the United States, approximately 52,000 new cases of **bladder cancer** are diagnosed, and there are 9500 deaths from this condition. The incidence among males is three times that among females, and most patients are age 60–70. Environmental factors, especially exposure to *2-naphthylamine* or related compounds, are responsible for most bladder cancers. For this reason, the bladder cancer rate is highest among cigarette smokers and employees of chemical and rubber companies. The mechanism appears to involve damage to tumor suppressor genes, such as *p53*, that regulate cell division. The prognosis is reasonably good for localized superficial cancers (88 percent 5-year survival), but it is poor for persons with severe metastatic bladder cancer (9 percent 5-year survival). Treatment of metastasized bladder cancer is very difficult, because the cancer spreads rapidly through adjacent lymphatic vessels and through the bone marrow of the pelvis.

THE URETHRA

The urethra extends from the neck of the urinary bladder to the exterior. The urethrae of males and females differ in length and in function. In males, the urethra extends from the neck of the urinary bladder to the tip of the penis, a distance that may be 18–20 cm (7–8 in.). The male urethra can be subdivided into three portions (Figure 26-18a,c•): (1) the prostatic urethra, (2) the membranous urethra, and (3) the penile urethra. The **prostatic urethra** passes through the center of the prostate gland. The **membranous urethra** includes the short segment that penetrates the *urogenital diaphragm,* the muscular floor of the pelvic cavity. The **penile** (PĒ-nīl) **urethra** extends from the distal border of the urogenital diaphragm to the external opening, or **external urethral meatus,** at the tip of the penis. In females, the urethra is very short, extending 3–5 cm (1–2 in.) from the bladder to the vestibule (Figure 26-18b•). The external urethral meatus is near the anterior wall of the vagina.

In both genders, as the urethra passes through the urogenital diaphragm, a circular band of skeletal muscle forms the **external urethral sphincter.** This muscular band acts as a valve. The external urethral sphincter is under voluntary control, via the perineal branch of the pudendal nerve. ∞ *p. 420* This sphincter has a resting muscle tone and must be voluntarily relaxed to permit micturition.

Histology of the Urethra

The urethral lining consists of a stratified epithelium that varies from transitional at the neck of the urinary bladder, through stratified columnar at the midpoint, to stratified squamous near the external urethral meatus. The lamina propria is thick and elastic, and the mucous membrane is thrown into longitudinal folds (Figure 26-19c•). Mucin-secreting cells are located in the epithelial pockets. In males, the epithelial mucous glands may form tubules that extend into the lamina propria. Connective tissues of the lamina propria anchor the urethra to surrounding structures. In females, the lamina propria contains an extensive network of veins, and the entire complex is surrounded by concentric layers of smooth muscle.

Urinary Tract Infections

Urinary tract infections (UTIs) result from the colonization of the urinary tract by bacterial or fungal invaders. The intestinal bacterium *Escherichia coli* is most commonly involved. Women are particularly susceptible to UTIs, because the external urethral orifice is so close to the anus. Sexual intercourse can push bacteria into the urethra; because the female urethra is relatively short, the urinary bladder can then become infected.

The condition may have no symptoms, but it can be detected by the presence of bacteria and blood cells in urine. If inflammation of the urethral wall occurs, the condition is termed *urethritis;* inflammation of the lining of the bladder is called *cystitis.* Many infections, including sexually transmitted diseases (STDs) such as gonorrhea, cause a combination of urethritis and cystitis. These conditions cause painful urination, a symptom known as *dysuria* (dis-ū-rē-uh), and the urinary bladder becomes tender and sensitive to pressure. Despite the discomfort produced, the individual feels the urge to urinate frequently. Urinary tract infections generally respond to antibiotic therapies, although reinfections can occur.

In untreated cases, the bacteria may proceed along the ureters to the renal pelvis. The resulting inflammation of the walls of the renal pelvis produces *pyelitis* (pī-e-LĪ-tis). If the bacteria invade the renal cortex and medulla as well, *pyelonephritis* (pī-e-lō-nef-RĪ-tis) results. Signs and symptoms of pyelonephritis include a high fever, intense pain on the affected side, vomiting, diarrhea, and the presence of blood cells and pus in the urine.

THE MICTURITION REFLEX AND URINATION

Urine reaches the urinary bladder by the peristaltic contractions of the ureters. The process of urination is coordinated by the **micturition reflex.** Components of this reflex are diagrammed in Figure 26-20•.

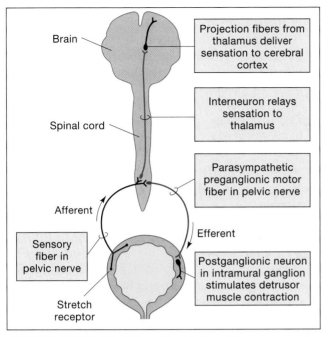

•**FIGURE 26-20 The Micturition Reflex.** The components of the reflex arc that stimulates smooth muscle contractions in the urinary bladder. Micturition occurs after voluntary relaxation of the external urethral sphincter.

Stretch receptors in the wall of the urinary bladder are stimulated as the bladder fills with urine. Afferent fibers in the pelvic nerves carry the impulses generated to the sacral spinal cord. Their increased level of activity (1) facilitates parasympathetic motor neurons in the sacral spinal cord and (2) stimulates interneurons that relay sensations to the thalamus and then, through projection fibers, to the cerebral cortex. As a result, you become consciously aware of the fluid pressure in your urinary bladder.

The urge to urinate generally appears when the bladder contains about 200 ml of urine. The micturition reflex begins to function when the stretch receptors have provided adequate stimulation to parasympathetic preganglionic motor neurons. Action potentials carried by efferent fibers within the pelvic nerves then stimulate ganglionic neurons in the wall of the urinary bladder. These neurons in turn stimulate sustained contraction of the detrusor muscle.

This contraction elevates fluid pressures in the urinary bladder, but urine ejection does not occur unless both the internal and external urethral sphincters are relaxed. The relaxation of the external urethral sphincter occurs under voluntary control. When the external urethral sphincter relaxes, so does the internal urethral sphincter. If the external urethral sphincter does not relax, the internal urethral sphincter remains closed and the urinary bladder gradually relaxes.

A further increase in bladder volume begins the cycle again, usually within an hour. Each increase in urinary volume leads to an increase in stretch receptor

stimulation that makes the sensation more acute. Once the volume exceeds 500 ml, the micturition reflex may generate enough pressure to force open the internal urethral sphincter. This opening leads to a reflexive relaxation of the external urethral sphincter, and urination occurs despite voluntary opposition or potential inconvenience. At the end of a normal micturition, less than 10 ml of urine remains in the bladder.

Infants lack voluntary control over urination, because the necessary corticospinal connections have yet to be established. Toilet training before age 2 years typically involves training the parent to anticipate the timing of the reflex rather than training the child to exert conscious control.

Incontinence

Incontinence (in-KON-ti-nens) is the inability to control urination voluntarily. Trauma to the internal or external urethral sphincter can contribute to incontinence in otherwise healthy adults. For example, childbirth can stretch and damage the sphincter muscles, so some mothers then develop *stress incontinence*. In this condition, elevated intra-abdominal pressures, caused for example by a cough or sneeze, can overwhelm the sphincter muscles, causing urine to leak out. Incontinence can also develop in older individuals due to a general loss of muscle tone.

Damage to the central nervous system, the spinal cord, or the nerve supply to the urinary bladder or external urethral sphincter can also produce incontinence. For example, incontinence commonly accompanies Alzheimer's disease or a stroke. In most cases, the affected individual develops an *automatic bladder*. The micturition reflex remains intact, but voluntary control of the external urethral sphincter is lost; the individual cannot prevent the reflexive emptying of the urinary bladder. Damage to the pelvic nerves can abolish the micturition reflex entirely, because those nerves carry both afferent and efferent fibers of this reflex arc. The urinary bladder becomes greatly distended with urine. It remains filled to capacity, and the excess trickles into the urethra in an uncontrolled stream. The insertion of a catheter is commonly required to facilitate the discharge of urine.

AGING AND THE URINARY SYSTEM

In general, aging is associated with an increased incidence of kidney problems. We described examples, such as *nephrolithiasis*—the formation of calculi, or kidney stones—earlier in the chapter (page 974). Other age-related changes in the urinary system include:

26

- *A decline in the number of functional nephrons.* The total number of kidney nephrons drops by 30–40 percent between ages 25 and 85.

- *A reduction of the GFR.* This reduction results from decreased numbers of glomeruli, cumulative damage to the filtration apparatus in the remaining glomeruli, and reductions in renal blood flow.

▪ *A reduced sensitivity to ADH.* With age, the distal portions of the nephron and collecting system become less responsive to ADH. Less reabsorption of water and sodium ions occurs, and more sodium ions are lost in urine.

▪ *Problems with the micturition reflex.* The following three factors are involved in such problems:

1. The sphincter muscles lose muscle tone and become less effective at voluntarily retaining urine. *Incontinence* results, commonly involving a slow leakage of urine.

2. The ability to control micturition can be lost after a stroke, Alzheimer's disease, or other CNS problems affecting the cerebral cortex or hypothalamus.

3. In males, **urinary retention** may develop secondary to enlargement of the prostate gland *(prostatic hypertrophy)*. In this condition, swelling and distortion of surrounding prostatic tissues compress the urethra, restricting or preventing the flow of urine.

INTEGRATION WITH OTHER SYSTEMS

The urinary system is not the only organ system concerned with excretion. The urinary, integumentary, respiratory, and digestive systems are considered to form an anatomically diverse **excretory system.** Its components perform all the excretory functions of the body that affect the composition of body fluids. Examples of excretory activities discussed in earlier chapters include these:

▪ *Integumentary system.* Water and electrolyte losses in sensible perspiration can affect plasma volume and composition. The effects are most apparent when loss-es are extreme, as when you are at peak sweat production. You also eliminate small amounts of metabolic wastes, including urea, in perspiration.

▪ *Respiratory system.* The lungs remove the carbon dioxide generated by cells. Small amounts of other compounds, such as acetone and water, evaporate into the alveoli and are eliminated when you exhale.

▪ *Digestive system.* Your liver excretes small amounts of metabolic waste products in bile, and you lose a variable amount of water in feces.

These excretory activities have an impact on the composition of body fluids. The respiratory system, for example, is the primary site of carbon dioxide excretion. But the excretory functions of these systems are not regulated as closely as are those of the kidneys. Under normal circumstances, the effects of integumentary and digestive excretory activities are minor compared with those of the urinary system.

Figure 26-21• summarizes the functional relationships between the urinary system and other systems. We shall explore many of these relationships further in Chapter 27 when we consider major aspects of fluid, pH, and electrolyte balance.

☑ What effect would a high-protein diet have on the composition of urine?

☑ An obstruction of a ureter by a kidney stone would interfere with the flow of urine between which two points?

☑ The ability to control the micturition reflex depends on your ability to control which muscle?

Filtering toxins out of our blood is so vital that the body maintains more filtering units than are needed to ensure nonstop cleansing. Those filtering units are the microscopic nephrons within each kidney. To help you understand the function of nephrons and the filtering process, use the resources available on both the CD-ROM and the Companion Website. The CD-ROM includes animations and tutorials; the website builds on those resources with text content and interactive problems as well as communication avenues for sharing ideas with your class. To complete the following Web Explorations, visit the MediaLab in Chapter 26 of your Companion Website.

Web Exploration 1

Estimated time for completion = 10 minutes

The anatomy of the kidney, and of the nephron in particular, determines the functioning of the urinary system. To test your understanding of these structures, select the keyword **URINARY QUIZZES** on the Companion Website. You will get a menu of 12 quizzes. After studying the text's diagrams of the kidney and the nephron, choose the "Cross Section of a Kidney #3" quiz and try to identify the structures presented. Then take the "Nephron 3" quiz. If you wish, share your answers with your classmates, using the communication tools on the Companion Website.

Web Exploration 2

Estimated time for completion = 10 minutes

Urinalysis is a common medical procedure. What information can it provide? Review the clinical discussion "Urinalysis" (p. 972) and Table 26-7 (p. 973) in your text and the "Urinalysis" section of the *Applications Manual* about normal values for urinalysis tests. Then select the keyword **URINALYSIS** on the Companion Website. Read the objectives of the exercise there, and try to solve the first two case studies. Which values are out of normal range? What do those abnormal values signify? Write brief summaries of your diagnoses, and then link to the answers to check yourself. If you wish, discuss these cases in a group.

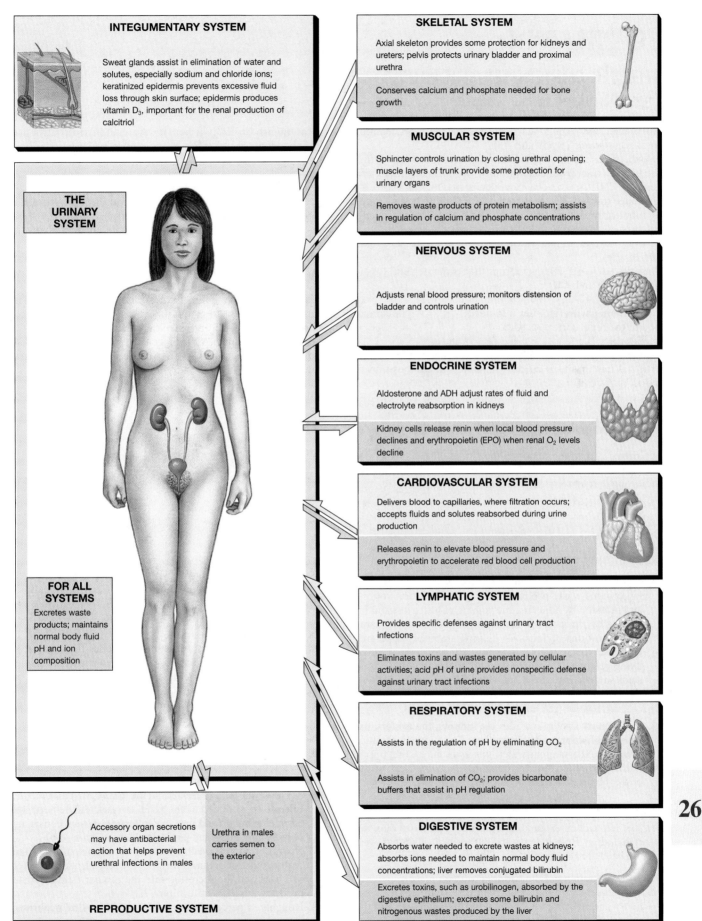

INTEGUMENTARY SYSTEM

Sweat glands assist in elimination of water and solutes, especially sodium and chloride ions; keratinized epidermis prevents excessive fluid loss through skin surface; epidermis produces vitamin D_3, important for the renal production of calcitriol

THE URINARY SYSTEM

FOR ALL SYSTEMS

Excretes waste products; maintains normal body fluid pH and ion composition

Accessory organ secretions may have antibacterial action that helps prevent urethral infections in males

Urethra in males carries semen to the exterior

REPRODUCTIVE SYSTEM

SKELETAL SYSTEM

Axial skeleton provides some protection for kidneys and ureters; pelvis protects urinary bladder and proximal urethra

Conserves calcium and phosphate needed for bone growth

MUSCULAR SYSTEM

Sphincter controls urination by closing urethral opening; muscle layers of trunk provide some protection for urinary organs

Removes waste products of protein metabolism; assists in regulation of calcium and phosphate concentrations

NERVOUS SYSTEM

Adjusts renal blood pressure; monitors distension of bladder and controls urination

ENDOCRINE SYSTEM

Aldosterone and ADH adjust rates of fluid and electrolyte reabsorption in kidneys

Kidney cells release renin when local blood pressure declines and erythropoietin (EPO) when renal O_2 levels decline

CARDIOVASCULAR SYSTEM

Delivers blood to capillaries, where filtration occurs; accepts fluids and solutes reabsorbed during urine production

Releases renin to elevate blood pressure and erythropoietin to accelerate red blood cell production

LYMPHATIC SYSTEM

Provides specific defenses against urinary tract infections

Eliminates toxins and wastes generated by cellular activities; acid pH of urine provides nonspecific defense against urinary tract infections

RESPIRATORY SYSTEM

Assists in the regulation of pH by eliminating CO_2

Assists in elimination of CO_2; provides bicarbonate buffers that assist in pH regulation

DIGESTIVE SYSTEM

Absorbs water needed to excrete wastes at kidneys; absorbs ions needed to maintain normal body fluid concentrations; liver removes conjugated bilirubin

Excretes toxins, such as urobilinogen, absorbed by the digestive epithelium; excretes some bilirubin and nitrogenous wastes produced by the liver

26

•**FIGURE 26-21** Functional Relationships between the Urinary System and Other Systems

CHAPTER REVIEW

SELECTED CLINICAL TERMINOLOGY
Terms Discussed in This Chapter

aminoaciduria: Amino acid loss in urine; the most common form is *cystinuria*. *(p. 959 and AM)*

calculi (KAL-kū-lī): Insoluble deposits that form within the urinary tract from calcium salts, magnesium salts, or uric acid. *(p. 974)*

clearance test: A test that permits the GFR to be calculated by monitoring plasma and renal concentrations of a specific solute, such as creatinine. *(p. 955 and AM)*

cystitis: An inflammation of the lining of the urinary bladder. *(p. 976)*

diuretics (dī-ū-RET-iks): Drugs that promote fluid loss in urine. *(p. 968 and AM)*

dysuria (dis-ū-rē-uh): Painful urination. *(p. 976)*

glomerulonephritis (glo-mer-ū-lō-nef-RĪ-tis): An inflammation of the renal cortex. *(p. 952)*

hematuria: Blood loss in urine. *(p. 957 and AM)*

hemodialysis (hē-mō-dī-AL-i-sis): A technique in which an artificial membrane is used to regulate the composition of blood. *(p. 970)*

incontinence (in-KON-ti-nens): An inability to control urination voluntarily. *(p. 977)*

polycystic kidney disease: An inherited abnormality that affects the development and structure of kidney tubules. *(p. 952)*

proteinuria: Protein loss in urine. *(p. 957 and AM)*

pyelitis (pī-e-LĪ-tis): An inflammation of the walls of the renal pelvis. *(p. 976)*

pyelogram (PĪ-el-ō-gram): An image obtained by taking an X-ray of the kidneys after a radiopaque compound has been administered. *(p. 972 and AM)*

pyelonephritis (pī-e-lō-nef-RĪ-tis): An inflammation of the kidney tissues. *(p. 976)*

renal failure: An inability of the kidneys to excrete wastes in sufficient quantities to maintain homeostasis. *(p. 958)*

urinalysis: A physical and chemical assessment of urine. *(p. 972)*

urinary obstruction: A blockage of the urinary tract. *(p. 974)*

urinary tract infection (UTI): An infection that results from the colonization of the urinary tract by bacterial or fungal invaders. *(p. 976)*

AM Additional Terms Discussed in the *Applications Manual*

nephritis: An inflammation of the kidneys.

uremia (ū-RĒ-mē-uh): A change in the composition of blood, indicating that all renal functions are abnormal.

urethritis: An inflammation of the urethral wall.

STUDY OUTLINE

INTRODUCTION p. 943

1. The functions of the **urinary system** include (1) eliminating organic waste products; (2) regulating blood volume and pressure by adjusting the volume of water lost and releasing erythropoietin and renin; (3) regulating plasma concentrations of ions; (4) helping stabilize blood pH; (5) conserving nutrients; and (6) assisting the liver in detoxifying poisons and, during starvation, deaminating amino acids so that they can be catabolized by other tissues.

ORGANIZATION OF THE URINARY SYSTEM p. 943

1. The urinary system includes the **kidneys,** the **ureters,** the **urinary bladder,** and the **urethra.** The kidneys produce **urine,** a fluid containing water, ions, and soluble compounds. During **urination (micturition)** urine is forced out of the body. *(Figures 26-1, 26-2)*

THE KIDNEYS p. 943

1. The left kidney extends superiorly slightly more than the right kidney. Both kidneys and the adrenal gland that overlies each are retroperitoneal. *(Figure 26-3)*

Superficial Anatomy of the Kidneys p. 944

2. The **hilus,** a medial indentation, provides entry for the *renal artery* and *renal nerves* and exit for the *renal vein* and the ureter. *(Figures 26-3, 26-4)*

Sectional Anatomy of the Kidneys p. 944

3. The ureter communicates with the **renal pelvis,** a chamber that branches into two **major calyces.** Each major calyx is connected to four or five **minor calyces,** which enclose the **renal papillae.** *(Figure 26-4)*

The Blood Supply to the Kidneys p. 946

4. The blood supply to the kidneys includes the **renal, segmental, interlobar, arcuate,** and **interlobular arteries** and the **interlobular, arcuate, interlobar,** and **renal veins.** *(Figure 26-5)*

Innervation of the Kidneys p. 946

5. The **renal nerves,** which innervate the kidneys and ureters, are dominated by sympathetic postganglionic fibers.

The Nephron p. 946

6. The **nephron** (the basic functional unit in the kidney) consists of the **renal corpuscle** and **renal tubule.** The renal tubule is long and narrow and divided into the *proximal convoluted tubule,* the *loop of Henle,* and the *distal convoluted tubule.* **Filtrate** is produced at the renal corpuscle. The nephron empties **tubular fluid** into the **collecting system,** consisting of **collecting ducts** and **papillary ducts.** *(Figure 26-6)*

7. Nephrons are responsible for the production of filtrate, the reabsorption of organic nutrients, the reabsorption of water and ions, and the secretion into the tubular fluid of waste products missed by filtration. *(Table 26-1)*

8. Roughly 85 percent of the nephrons are **cortical nephrons,** located in the renal **cortex.** The **juxtamedullary nephrons**

26

are closer to the renal **medulla,** with their *loops of Henle* extending deep into the **renal pyramids.** *(Figure 26-7)*

9. Blood travels from the efferent arteriole to the **peritubular capillaries** and the **vasa recta.** *(Figure 26-8)*

10. The renal tubule begins at the renal corpuscle. It includes a knot of intertwined capillaries called the **glomerulus,** surrounded by **Bowman's capsule.** Blood arrives at the glomerulus via the **afferent arteriole** and departs in the **efferent arteriole.** *(Figure 26-8)*

11. At the glomerulus, **podocytes** cover the **lamina densa** of the capillaries that project into the **capsular space.** The **pedicels** of the podocytes are separated by narrow **filtration slits.** *(Figure 26-8)*

12. The **proximal convoluted tubule (PCT)** actively reabsorbs nutrients, plasma proteins, and ions from filtrate. These substances are released into the **peritubular fluid,** which surrounds the nephron.

13. The **loop of Henle** includes a **descending limb** and an **ascending limb;** each limb contains a **thick segment** and a **thin segment.** The ascending limb delivers fluid to the **distal convoluted tubule (DCT),** which actively secretes ions, toxins, and drugs and reabsorbs sodium ions from the tubular fluid. *(Figures 26-6, 26-8)*

RENAL PHYSIOLOGY   p. 953

Basic Principles of Urine Formation   p. 953

1. Urine formation involves **filtration, reabsorption,** and **secretion.** *(Table 26-2)*

Filtration   p. 954

2. **Glomerular filtration** occurs as fluids move across the wall of the glomerulus into the capsular space, in response to the **glomerular hydrostatic pressure (GHP)**—the hydrostatic (blood) pressure in the glomerular capillaries. This movement is opposed by the **capsular hydrostatic pressure (CsHP)** and by the **blood colloid osmotic pressure (BCOP).** The **filtration pressure (FP)** at the glomerulus is the difference between the blood pressure and the opposing capsular and osmotic pressures. *(Figure 26-9)*

3. The **glomerular filtration rate (GFR)** is the amount of filtrate produced in the kidneys each minute. Any factor that alters the filtration pressure acting across the glomerular capillaries will change the GFR and affect kidney function. *(Figure 26-10)*

4. A drop in filtration pressures stimulates the **juxtaglomerular apparatus (JGA)** to release renin and erythropoietin. *(Figure 26-10)*

5. Sympathetic activation (1) produces a powerful vasoconstriction of the afferent arterioles, decreasing the GFR and slowing the production of filtrate, (2) alters the GFR by changing the regional pattern of blood circulation, and (3) stimulates the release of renin by the juxtaglomerular apparatus. *(Figure 26-10)*

Reabsorption and Secretion   p. 958

6. Four types of *carrier-mediated transport (facilitated diffusion, active transport, cotransport,* and *countertransport)* are involved in modifying filtrate. The saturation limit of a carrier protein is its **transport maximum.** The transport maximum determines the **renal threshold,** the plasma concentration at which various compounds will appear in urine. *(Table 26-3, Table 26-4)*

7. Glomerular filtration produces a filtrate with a composition similar to blood plasma but with few, if any, plasma proteins.

8. The cells of the PCT normally reabsorb sodium and other ions, water, and almost all the organic nutrients that enter

filtrate. These cells also secrete various substances into the tubular fluid. *(Figure 26-11)*

9. Water and ions are reclaimed from tubular fluid by the loop of Henle. A concentration gradient in the renal medulla encourages the osmotic flow of water out of the tubular fluid. The **countercurrent multiplication** between the ascending and descending limbs of the loop of Henle helps create the osmotic gradient in the medulla. As water is lost by osmosis and the volume of tubular fluid decreases, the urea concentration rises. *(Figure 26-12)*

10. The DCT performs final adjustments by actively secreting or absorbing materials. Sodium ions are actively absorbed, in exchange for potassium or hydrogen ions discharged into tubular fluid. Aldosterone secretion increases the rate of sodium reabsorption and potassium loss. *(Figure 26-13)*

11. The amount of water and solutes in the tubular fluid of the collecting ducts is further regulated by aldosterone and ADH secretions. *(Figure 26-14)*

The Function of the Vasa Recta   p. 967

12. Normally, the removal of solutes and water by the vasa recta precisely balances the rates of reabsorption and osmosis in the renal medulla.

Summary: Renal Function   p. 968

13. Each segment of the nephron and collecting system contributes to the production of concentrated urine. *(Figure 26-15; Table 26-4)*

The Composition of Normal Urine   p. 969

14. **Hemodialysis** is a technique used to regulate the composition of blood artificially. *(Figure 26-16; Table 26-6)*

15. More than 99 percent of the filtrate produced each day is reabsorbed before reaching the renal pelvis. *(Tables 26-5, 26-7)*

URINE TRANSPORT, STORAGE, AND ELIMINATION   p. 972

1. Urine production ends when tubular fluid enters the renal pelvis. The rest of the urinary system is responsible for transporting, storing, and eliminating urine. *(Figures 26-17, 26-18)*

The Ureters   p. 972

2. The ureters extend from the renal pelvis to the urinary bladder. Peristaltic contractions by smooth muscles move the urine. *(Figure 26-19)*

The Urinary Bladder   p. 974

3. The urinary bladder is stabilized by the **middle umbilical ligament** and the **lateral umbilical ligaments.** Internal features include the **trigone,** the **neck,** and the **internal urethral sphincter.** The mucosal lining contains prominent **rugae** (folds). Contraction of the **detrusor muscle** compresses the bladder and expels urine into the urethra. *(Figures 26-18, 26-19)*

The Urethra   p. 976

4. In both genders, as the urethra passes through the *urogenital diaphragm,* a circular band of skeletal muscles forms the **external urethral sphincter,** which is under voluntary control. *(Figures 26-18, 26-19)*

The Micturition Reflex and Urination   p. 976

5. Urination is coordinated by the **micturition reflex,** which is initiated by stretch receptors in the wall of the urinary bladder. Voluntary urination involves coupling this reflex with the voluntary relaxation of the external urethral sphincter, which allows the opening of the **internal urethral sphincter.** *(Figure 26-20)*

26

AGING AND THE URINARY SYSTEM p. 977

1. Aging is generally associated with increased kidney problems. Age-related changes in the urinary system include (1) declining numbers of functional nephrons, (2) reduced GFR, (3) reduced sensitivity to ADH, (4) problems with the micturition reflex. (**Urinary retention** may develop in men whose prostate gland is enlarged.)

INTEGRATION WITH OTHER SYSTEMS p. 978

1. The urinary, integumentary, respiratory, and digestive systems are sometimes considered an anatomically diverse **excretory system,** whose components work together to perform all the excretory functions that affect the composition of body fluids. *(Figure 26–21)*

REVIEW QUESTIONS

LEVEL 1 REVIEWING FACTS AND TERMS

1. The point of entry for the renal artery and exit for the renal vein and ureter is a medial indentation called the
- (a) renal column
- (b) medulla
- (c) hilus
- (d) renal cortex

2. The glomerulus is located within the
- (a) renal corpuscle
- (b) renal tubule
- (c) renal pelvis
- (d) renal column

3. Specialized cells that engulf organic materials that might otherwise clog the filter at the lamina densa are
- (a) supporting cells
- (b) juxtamedullary cells
- (c) podocytes
- (d) macula densa

4. Large cells with complex processes, or "feet," that wrap around the glomerular capillaries are
- (a) vasa recta
- (b) podocytes
- (c) astrocytes
- (d) supporting cells

5. After it leaves Bowman's capsule, filtrate empties into the
- (a) distal convoluted tubule
- (b) loop of Henle
- (c) proximal convoluted tubule
- (d) collecting duct

6. The distal convoluted tubule is an important site for
- (a) active secretion of ions
- (b) active secretion of acids and other materials
- (c) selective reabsorption of sodium ions from the tubular fluid
- (d) a, b, and c are correct

7. The endocrine structure that secretes renin and erythropoietin is the
- (a) juxtaglomerular apparatus
- (b) vasa recta
- (c) Bowman's capsule
- (d) adrenal gland

8. The primary purpose of the collecting system is to
- (a) transport urine from the urinary bladder to the urethra
- (b) selectively reabsorb sodium ions from tubular fluid
- (c) transport urine from the renal pelvis to the ureters
- (d) make final adjustments to the osmotic concentration and volume of urine

9. Pickup or delivery of substances that are reabsorbed or secreted by the PCT and the DCT is provided by the
- (a) afferent arteriole
- (b) peritubular capillaries
- (c) renal artery
- (d) efferent arteriole

10. The most abundant organic waste in urine is
- (a) uric acid
- (b) creatinine
- (c) urea
- (d) creatine phosphate

11. The removal of water and solute molecules from filtrate after it enters the renal tubules is
- (a) filtration
- (b) secretion
- (c) reabsorption
- (d) elimination

12. The transport of solutes from peritubular fluid into tubular fluid is
- (a) reabsorption
- (b) filtration
- (c) elimination
- (d) secretion

13. The force that tends to drive water and solutes into interstitial fluid is the
- (a) glomerular hydrostatic pressure
- (b) net hydrostatic pressure
- (c) capsular hydrostatic pressure
- (d) net colloid osmotic pressure

14. An inflammation of the lining of the urinary bladder is
- (a) urethritis
- (b) cystitis
- (c) dysuria
- (d) pyelitis

15. What is the primary function of the urinary system?

16. What structures are components of the urinary system?

17. Trace the pathway of the protein-free filtrate from the time it is produced in the renal corpuscle until it drains into the renal pelvis in the form of urine. (Use arrows to indicate the direction of flow.)

18. Name the segments of the nephron distal to the renal corpuscle, and state the function(s) of each.

19. What role does the lamina densa play in the renal corpuscle?

20. What is the function of the juxtaglomerular apparatus?

21. Using arrows, trace a drop of blood from its entry into the renal artery until its exit via a renal vein.

22. Name and define the three distinct processes involved in the production of urine.

23. What are the primary effects of angiotensin II on kidney function and regulation?

24. Which parts of the urinary system are responsible for the transport, storage, and elimination of urine?

LEVEL 2 REVIEWING CONCEPTS

25. The urinary system regulates blood volume and pressure by
- (a) adjusting the volume of water lost in urine
- (b) releasing erythropoietin
- (c) releasing renin
- (d) a, b, and c are correct

26. The balance of solute and water reabsorption in the medulla is maintained by the
- (a) segmental arterioles and veins
- (b) interlobar arteries and veins
- (c) vasa recta
- (d) arcuate arteries

27. Sympathetic activation of the nerve fibers in the nephron causes
- (a) the regulation of glomerular blood flow and pressure
- (b) the stimulation of renin release from the juxtaglomerular apparatus
- (c) the direct stimulation of water and Na^+ reabsorption
- (d) a, b, and c are correct

26

28. An increase in the colloid osmotic pressure of the filtrate, caused by damage to the glomerulus, would
  (a) decrease the amount of plasma delivered to the kidney
  (b) enhance the movement of water and solutes into the capsular space
  (c) cause a decrease in the renal blood pressure
  (d) decrease movement of water and solutes into the capsular space

29. Sodium reabsorption in the DCT and cortical portion of the collecting system is accelerated by the secretion of
  (a) ADH   **(b)** renin
  (c) aldosterone   **(d)** erythropoietin

30. When ADH levels rise,
  (a) the amount of water reabsorbed increases
  (b) the DCT becomes impermeable to water
  (c) the amount of water reabsorbed decreases
  (d) sodium ions are exchanged for potassium ions

31. The control of blood pH by the kidneys involves
  (a) the addition of hydrogen ions and removal of bicarbonate ions from filtrate
  (b) a decrease in the amount of water reabsorbed
  (c) hydrogen ion removal and bicarbonate ion production
  (d) potassium ion secretion

32. How does a cortical nephron differ from a juxtamedullary nephron?

33. How can you determine the filtration pressure (FP) at the glomerulus?

34. What interacting controls stabilize the glomerular filtration rate (GFR)?

35. What is the renal threshold?

36. What primary changes occur in the composition and concentration of filtrate as a result of activity in the proximal convoluted tubule?

37. What two functions does the countercurrent mechanism perform for the kidney?

38. What events in the distal convoluted tubule and collecting system determine the final composition and concentration of filtrate?

39. Describe the micturition reflex.

LEVEL 3 CRITICAL THINKING AND CLINICAL APPLICATIONS

40. Why do long-haul trailer truck drivers commonly experience kidney problems?

41. Physicians often ask for urine samples collected over a 24-hour period rather than a single sample. Why?

42. Randy enjoys his four or five cups of coffee a day and his six or seven beers a few nights a week. What are the effects of his caffeine and alcohol consumption on his urinary system?

43. For the past week, Susan has felt a burning sensation in the urethral area when she urinates. She checks her temperature and finds that she has a low-grade fever. What unusual substances are likely to be present in her urine?

44. *Lasix*^{TM} is a diuretic that acts by decreasing the amounts of sodium and chloride ions actively transported by the ascending limb of the loop of Henle. Why would this medication be given to someone with high blood pressure (hypertension)?

45. Carlos has advanced arteriosclerosis. An analysis of his blood indicates elevated levels of aldosterone and decreased levels of ADH. Explain.

46. *Mannitol* is a sugar that is filtered but not reabsorbed by the kidneys. What effect would drinking a solution of mannitol have on the volume of urine produced?

47. The drug *Diamox*^{TM} is sometimes used in the treatment of epilepsy. It inhibits the action of carbonic anhydrase in the proximal convoluted tubule. Polyuria (the elimination of an unusually large volume of urine) is a side effect associated with this medication. Why does this symptom occur?

26

CHAPTER 27

Fluid, Electrolyte, and Acid–Base Balance

CHAPTER OUTLINE AND OBJECTIVES

INTRODUCTION   p. 985

FLUID BALANCE AND ELECTROLYTE BALANCE   p. 985
   Basic Concepts in the Regulation of Fluids and Electrolytes   p. 987
   An Overview of the Primary Regulatory Hormones   p. 988
   The Interplay between Fluid Balance and Electrolyte Balance   p. 989
   Fluid Balance   p. 989
   Electrolyte Balance   p. 991
① Compare the composition of intracellular and extracellular fluids.
② Explain the basic concepts involved in the regulation of fluids and electrolytes.
③ Identify the hormones that play important roles in regulating fluid balance and electrolyte balance, and describe their effects.
④ Discuss the mechanisms by which sodium, potassium, calcium, and chloride ion concentrations are regulated to maintain electrolyte balance.

ACID–BASE BALANCE   p. 997
   The Importance of pH Control   p. 997
   Types of Acids in the Body   p. 998
   Mechanisms of pH Control   p. 998
   Maintenance of Acid–Base Balance   p. 1001
⑤ Explain the buffering systems that balance the pH of the intracellular and extracellular fluids.
⑥ Describe the compensatory mechanisms involved in the maintenance of acid–base balance.

DISTURBANCES OF ACID–BASE BALANCE   p. 1004
   Respiratory Acidosis   p. 1005
   Respiratory Alkalosis   p. 1006
   Metabolic Acidosis   p. 1007
   Metabolic Alkalosis   p. 1008
   The Detection of Acidosis and Alkalosis   p. 1008
⑦ Identify the most frequent threats to acid–base balance, and explain how the body responds when the pH of body fluids varies outside normal limits.

AGING AND FLUID, ELECTROLYTE, AND ACID–BASE BALANCE   p. 1009

MEDIALAB and CHAPTER REVIEW   p. 1011

After a severe accident, the treatment given in the first hour can mean the difference between life and death. Emergency measures generally must start before the patient arrives at a hospital, even when the trip is made by Med-Evac helicopter. After serious trauma, the body's regulatory mechanisms may be unable to maintain homeostasis without assistance. The patient's internal conditions must therefore be monitored and stabilized while in transit. In virtually all medical emergencies, fluid and electrolyte levels must be adjusted, generally by the administration of intravenous solutions containing water, electrolytes, and buffers. This simple therapy can replace lost blood, provide nutrients, and control the pH of body fluids. In this chapter, we shall examine the mechanisms responsible for the maintenance of water, electrolyte, and acid–base balance under normal and abnormal conditions.

The next time you see a small pond, think about the fish it contains. They live out their lives totally dependent on the quality of that isolated environment. Severe water pollution will kill the fish, but even subtle changes can have equally grave effects. Changes in the volume of the pond, for example, can be quite important. If evaporation removes too much of the pond water, the fish become overcrowded; oxygen and food supplies run out, and the fish suffocate or starve. The ionic concentration of the pond water is also crucial. Most of the fish in a freshwater pond will die if the water becomes too salty; those in a saltwater pond will die if their environment becomes too dilute. The pH of the pond water, too, is a vital factor; that is one reason acid rain is such a problem.

Your cells live in a pond whose shores are the exposed surfaces of your skin. Most of your body weight is water. Water accounts for up to 99 percent of the volume of the fluid outside cells, and it is an essential ingredient of cytoplasm. All of a cell's operations rely on water as a diffusion medium for the distribution of gases, nutrients, and waste products. If the water content of the body changes, cellular activities are jeopardized. For example, when the water content reaches very low levels, proteins denature, enzymes cease functioning, and cells die. To survive, we must maintain a normal volume and composition in the **extracellular fluid** (**ECF;** interstitial fluid, plasma, and other body fluids) and **intracellular fluid** (**ICF;** cytosol).

The ionic concentrations and pH of the body water are as important as its absolute quantity. If concentrations of calcium or potassium ions in the ECF become too high, cardiac arrhythmias develop and death can result. A pH outside the normal range can lead to a variety of dangerous problems. Low pH is especially dangerous, because hydrogen ions break chemical bonds, change the shapes of complex molecules, disrupt cell membranes, and impair tissue functions.

In this chapter, we shall consider the dynamics of exchange among the various body fluids and between the body and the external environment. Stabilizing the volumes, solute concentrations, and pH of the ECF and ICF involves three interrelated processes:

1. *Fluid balance.* You are in **fluid balance** when the amount of water you gain each day is equal to the amount you lose to the environment. The maintenance of normal fluid balance involves regulating body water content and distribution in the ECF and ICF. The digestive system is the primary source for water gains. A small amount of water is generated by metabolic activity. The urinary system is the primary route for water loss under normal conditions, but as we noted in Chapter 25, sweat gland activity can become important when body temperature is elevated. ∞ *p. 934* Because your cells and tissues cannot transport water, fluid balance reflects primarily the control of *electrolyte balance.*

2. *Electrolyte balance.* **Electrolytes** are ions released through the dissociation of inorganic compounds; they are so named because they will conduct an electrical current in a solution. ∞ *pp. 40–41* Each day, your body fluids gain electrolytes from the food and drink you consume and lose electrolytes in urine, sweat, and feces. For each ion, daily gains must balance daily losses. For example, if you lose 500 mg of Na^+ in urine and insensible perspiration, you will need to gain 500 mg of Na^+ from food and drink to remain in sodium balance. If the gains and losses for every electrolyte are in balance, you are said to be in **electrolyte balance.** Electrolyte balance primarily involves balancing the rates of absorption across the digestive tract with rates of loss at the kidneys, although losses at sweat glands and other sites can play a secondary role.

3. *Acid–base balance.* You are in **acid–base balance** when the production of hydrogen ions is precisely offset by their loss. When acid–base balance exists, the pH of body fluids remains within normal limits. Preventing a reduction in pH is the primary problem, because your body generates a variety of acids during normal metabolic operations. The kidneys play a major role by secreting hydrogen ions into the urine and generating buffers that enter the bloodstream. This secretion occurs primarily in the distal segments of the distal convoluted tubule (DCT) and along the collecting system. ∞ *pp. 965–966* The lungs also play a key role through the elimination of carbon dioxide.

Much of the material in this chapter was introduced in earlier chapters that focused on aspects of fluid, electrolyte, or acid–base balance that affected specific systems. This chapter provides an overview that integrates those discussions to highlight important functional patterns. Few other chapters have such wide-ranging clinical importance: The treatment of any serious illness affecting the nervous, cardiovascular, respiratory, urinary, or digestive system must include steps to restore normal fluid, electrolyte, and acid–base balances. Because this chapter builds on information provided in earlier chapters, you will encounter many references to relevant discussions and figures that you should use when you need a quick review.

FLUID BALANCE AND ELECTROLYTE BALANCE

Figure 27-1a• provides an overview of the composition of the body of a 70-kg individual with a minimum of body fat. This distribution was obtained by averaging values for males and females, age 18–40 years. Water accounts for roughly 60 percent of the total body weight of an adult male and 50 percent of that of an adult female. The gender difference primarily reflects the proportionately larger mass of adipose tissue in adult females and the greater average muscle mass in adult males. (Adipose tissue is 10 percent water, whereas

27

(a)

Body Composition of Adult Males and Females

Material/Compartment	Percentage of Total Body Weight	
	Adult Males	Adult Females
Solids (protein, lipids, minerals, carbo-hydrates, organic and inorganic materials)	40	50
Water	60	50
ICF	33	27
ECF	27	23
Interstitial fluid	21.5	18
Plasma	4.5	4.0
Minor components	1.0	1.0

(b)

•**FIGURE 27-1** **The Composition of the Human Body.**
(a) The body composition (averaged for both genders) and major body fluid compartments of a 70-kg individual. For technical reasons, it is extremely difficult to determine the precise volume of any of these fluid compartments; estimates of their relative sizes vary widely. **(b)** A comparison of the body compositions of adult males and females, age 18–40 years.

skeletal muscle is 75 percent water.) In both genders, intracellular fluid contains more of the total body water than does extracellular fluid. Exchange between the ICF and the ECF occurs across cell membranes by osmosis, diffusion, and carrier-mediated transport. (To review the mechanisms involved, see Table 3-2, p. 79.)

The largest subdivisions of the ECF are the *interstitial fluid* of peripheral tissues and the *plasma* of circulating blood. Minor components of the ECF include lymph, cerebrospinal fluid (CSF), synovial fluid, serous fluids (pleural, pericardial, and peritoneal fluids), aqueous humor, perilymph, and endolymph. Precise measurements of total body water provide additional information on gender differences in the distribution of body water (Figure 27-1b•). The greatest variation is in the ICF as a result of differences in the intracellular water content of fat versus muscle. Less-striking differences occur in the ECF values due to variations in the interstitial fluid volume of various tissues and the larger blood volume in males versus females.

In clinical situations, it is customary to estimate that two-thirds of the total body water is in the ICF and one-third in the ECF. This ratio underestimates the real volume of the ECF, but that underestimation is appropriate when we consider rapid fluid and solute movements. Portions of the ECF, including the water in bone, in many dense connective tissues, and in many of the minor ECF components, are relatively isolated. Exchange between these fluid volumes and the rest of the ECF occurs more slowly than does exchange between plasma and other interstitial fluids.

Exchange among the subdivisions of the ECF occurs primarily across the endothelial lining of capillaries. Fluid may also travel from the interstitial spaces to plasma through lymphatic vessels that drain into the venous system. ∞ *pp. 752–756* The identity and quantity of dissolved electrolytes, proteins, nutrients, and waste products in the ECF vary regionally. (For a chemical analysis of the composition of ECF compartments, see Appendix VI.) Yet the variations among the segments of the ECF seem minor when compared with the major differences between the ECF and the ICF.

The ECF and ICF are called **fluid compartments,** because they commonly behave as distinct entities.

27

The presence of a cell membrane and active transport at the membrane surface enable cells to maintain internal environments with a composition different from that of their surroundings. The principal ions in the ECF are sodium, chloride, and bicarbonate. The ICF contains an abundance of potassium, magnesium, and phosphate ions, plus large numbers of negatively charged proteins. Figure 27-2• compares the ICF with the two major subdivisions of the ECF.

If the cell membrane were freely permeable, diffusion would continue until these ions were evenly distributed across the membrane. But it does not, because cell membranes are selectively permeable: Ions can enter or leave the cell only via specific membrane channels. In addition, carrier mechanisms move specific ions into or out of the cell.

Despite the differences in the concentration of specific substances, the ICF and ECF osmotic concentrations are identical. Osmosis eliminates minor concentration differences almost at once, because most cell membranes are freely permeable to water. (The only noteworthy exceptions are the apical surfaces of epithelial cells along the ascending limb of the loop of Henle, distal convoluted tubule, and collecting system.) ∞ *p. 966* Because changes in

solute concentrations lead to immediate changes in water distribution, the regulation of fluid balance and that of electrolyte balance are tightly intertwined.

Physiologists and clinicians pay particular attention to ionic distributions across membranes and to the electrolyte composition of body fluids. Appendix VI reports normal values in the units most often encountered in clinical reports.

BASIC CONCEPTS IN THE REGULATION OF FLUIDS AND ELECTROLYTES

You must understand four basic principles before we can proceed to a discussion of fluid balance and electrolyte balance:

1. *All the homeostatic mechanisms that monitor and adjust the composition of body fluids respond to changes in the ECF, not in the ICF.* Receptors monitoring the composition of two key components of the ECF—plasma and cerebrospinal fluid—detect significant changes in composition or volume and trigger appropriate neural and endocrine responses. This arrangement makes functional sense, because a change in one ECF component will spread rapidly throughout the extracellular compartment and will affect all the

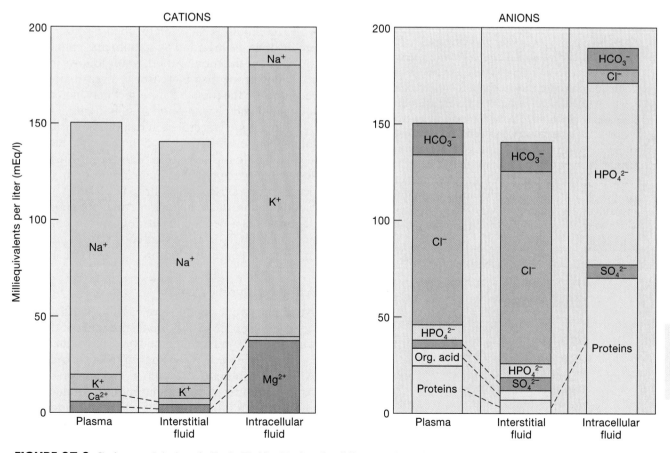

•**FIGURE 27-2 Cations and Anions in Body Fluids.** Notice the differences in cation and anion concentrations in the various body fluid compartments. For information about the composition of other body fluids, see Appendix VI.

body's cells. In contrast, the ICF is contained within trillions of individual cells physically and chemically isolated from one another by their cell membranes. Thus changes in the ICF in one cell have no direct effect on the composition of the ICF in distant cells and tissues unless those changes affect the ECF.

2. *No receptors directly monitor fluid or electrolyte balance.* In other words, our receptors cannot detect how many liters of water or grams of sodium, chloride, or potassium the body contains or count how many liters or grams we gain or lose in the course of a day. But our receptors *can* monitor *plasma volume* and *osmotic concentration.* Because fluid continuously circulates between interstitial fluid and plasma and exchange occurs between the ECF and ICF, plasma volume and osmotic concentration are good indicators of the state of fluid balance and electrolyte balance for the body as a whole.

3. *Our cells cannot move water molecules by active transport.* All water movement across cell membranes and epithelia occurs passively, in response to osmotic gradients. These gradients can be established by the active transport of specific ions, such as sodium and chloride. You may find it useful to remember the simple phrase *water follows salt.* As we saw in earlier chapters, when sodium and chloride ions (or other solutes) are actively transported across a membrane or epithelium, water follows by osmosis. This basic principle accounts for water absorption across the digestive epithelium and water conservation in the kidneys.

4. *The body's content of water or electrolytes will rise if dietary gains exceed losses to the environment and will fall if losses exceed gains.* This basic rule is important when you consider the mechanics of fluid balance and electrolyte balance. Homeostatic adjustments generally affect the balance between urinary excretion and dietary absorption. As we saw in Chapter 26, the physiological adjustments in renal function are regulated primarily by circulating hormones. These hormones can also produce complementary changes in behavior. For example, the combination of angiotensin II and aldosterone can give you a sensation of thirst—which stimulates you to drink fluids—and a taste for heavily salted foods.

AN OVERVIEW OF THE PRIMARY REGULATORY HORMONES

Major physiological adjustments affecting fluid balance and electrolyte balance are mediated by three hormones: (1) *antidiuretic hormone (ADH),* (2) *aldosterone,* and (3) *atrial natriuretic peptide (ANP).* Interactions among these hormones were shown in Figures 21-17, 21-18, and 26-10●. ∞ *pp. 716, 719, 957*

Antidiuretic Hormone

The hypothalamus contains special cells known as **osmoreceptors,** which monitor the osmotic concentration of the ECF. These cells are sensitive to subtle changes: A 2 percent change in osmotic concentration (approximately 6 mOsm/l) is sufficient to alter osmoreceptor activity.

The population of osmoreceptors includes neurons that secrete ADH. These neurons are located in the anterior hypothalamus, and their axons release ADH near fenestrated capillaries in the posterior pituitary gland. The rate of ADH release varies directly with the osmotic concentration: The higher the osmotic concentration, the more ADH is released.

Increased release of ADH has two important effects: (1) It stimulates water conservation at the kidneys, reducing urinary water losses and concentrating the urine; and (2) it stimulates the thirst center, promoting the intake of fluids. As we saw in Chapter 21, the combination of decreased water loss and increased water gain gradually restores normal plasma osmotic concentration. ∞ *pp. 715–716*

Aldosterone

The secretion of aldosterone by the adrenal cortex plays a major role in determining the rate of Na^+ absorption and K^+ loss along the distal convoluted tubule (DCT) and collecting system of the kidneys. The higher the plasma concentration of aldosterone, the more efficiently the kidneys conserve Na^+. Because "water follows salt," the conservation of Na^+ also stimulates water retention; as Na^+ is reabsorbed, Cl^- follows (Figure 26-13a●, p. 964), and as sodium and chloride ions move out of the tubular fluid, water follows by osmosis. Aldosterone also increases the sensitivity of salt receptors on the tongue. This effect may increase your interest in and consumption of salty foods.

Aldosterone is secreted in response to rising K^+ or falling Na^+ levels in the blood that reaches the adrenal cortex or to the activation of the renin–angiotensin system. As we noted in earlier chapters, renin release occurs in response to (1) a drop in plasma volume or blood pressure at the juxtaglomerular apparatus of the nephron, (2) a decline in filtrate osmotic concentration at the DCT, or, as we shall soon see, (3) falling Na^+ or rising K^+ concentrations in the renal circulation.

Atrial Natriuretic Peptide

Atrial natriuretic peptide is released by cardiac muscle fibers in response to abnormal stretching of the atrial walls, caused by elevated blood pressure or an increase in blood volume. We considered the many effects of ANP in Chapters 18, 21, and 26. ∞ *pp. 605, 717, 967* Among its other effects, this hormone (1) reduces thirst and (2) blocks the release of ADH and aldosterone that might otherwise lead to the conservation of water and salt. The resulting diuresis (fluid loss at the kidneys) lowers both blood pressure and plasma volume, eliminating the source of the stimulation.

THE INTERPLAY BETWEEN FLUID BALANCE AND ELECTROLYTE BALANCE

At first glance, it can be very difficult to distinguish between water balance and electrolyte balance. For example, when you lose body water, your plasma volume decreases and electrolyte concentrations rise. Conversely, when you gain or lose excess electrolytes, there is an associated water gain or loss due to osmosis. However, because the regulatory mechanisms involved are quite different, it is often useful to consider fluid balance and electrolyte balance as distinct entities. This distinction is absolutely vital in a clinical setting, where problems with fluid balance and electrolyte balance must be identified and corrected promptly.

FLUID BALANCE

Water circulates freely within the ECF compartment. At capillary beds throughout the body, hydrostatic pressure forces water out of plasma and into interstitial spaces. Some of that water is reabsorbed along the distal portion of the capillary bed, and the rest enters lymphatic vessels for transport to the venous circulation. This circulation is diagrammed in Figure 27-3•, which indicates two additional important relationships between components of the ECF:

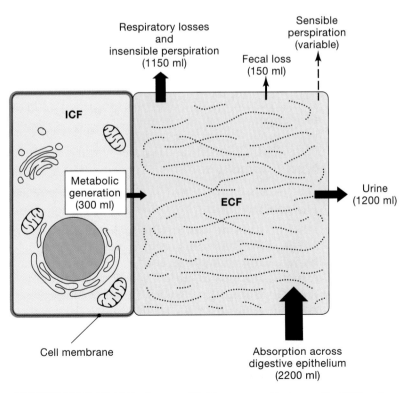

1. Water moves back and forth across the mesothelial surfaces that line the peritoneal, pleural, and pericardial cavities and through the synovial membranes that line joint capsules. The flow rate is significant; for example, roughly 7 liters of peritoneal fluid is produced and reabsorbed each day. The actual volume present at any time in the peritoneal cavity, however, is very small—less than 35 ml.

2. Water also moves between blood and cerebrospinal fluid (CSF), the aqueous humor and vitreous humor of the eye, and the perilymph and endolymph of the inner ear. The volumes involved in these water movements are very small, and the volume and composition of these fluids are closely regulated. For those reasons, we shall largely ignore them in the discussion that follows.

Fluid Movement within the ECF

In the discussion of capillary dynamics in Chapter 21, we considered the basic principles that determine fluid movement among the divisions of the ECF. ∞ *pp. 708–710* The exchange between plasma and interstitial fluid, by far the largest components of the ECF, is determined by the relationship between the net hydrostatic pressure, which tends to push water out of the plasma and into the interstitial fluid, and the net colloid osmotic pressure, which tends to draw water out of the interstitial fluid and into the plasma. The interaction between these opposing forces, diagrammed in Figure 21-13• (p. 709), results in the continuous filtration of fluid from the capillaries into the interstitial fluid. This volume of fluid is then redistributed: After passing through the channels of the lymphatic system, the fluid returns to the venous system. At any moment, interstitial fluid and minor fluid compartments contain roughly 80 percent of the ECF volume, and plasma contains the other 20 percent.

Any factor that affects the net hydrostatic pressure or the net colloid osmotic pressure will alter the distribution of fluid within the ECF. The movement of abnormal amounts of water from plasma into interstitial fluid is called **edema.** ∞ *p. 710* For example, pulmonary edema can result from an increase in the blood pressure in pulmonary capillaries, and a generalized edema can result from a decrease in blood colloid osmotic pressure, as in advanced starvation, when plasma protein concentrations decline. Localized edema can result from damage to capillary walls, as in bruising, the constriction of regional venous circulation, or a blockage of the lymphatic drainage, as in *lymphedema,* discussed in Chapter 22. ∞ *p. 756*

27

•**FIGURE 27-3 Fluid Exchanges.** Fluid movement between the ICF and ECF and between the ECF and the environment. The volumes are not drawn to scale; functionally, the ICF is twice as large as the ECF.

Fluid Exchange with the Environment

Figure 27-3• and Table 27-1 indicate the major routes of exchange with the environment:

- *Water losses.* You lose roughly 2500 ml of water each day through urine, feces, and *insensible perspiration,* the gradual movement of water across the epithelia of the skin and respiratory tract. The losses due to *sensible perspiration,* the secretory activities of the sweat glands, vary with the activities you undertake. Sensible perspiration can cause significant water deficits, with maximum perspiration rates reaching 4 liters per hour. ∞ *p. 934*

- *The temperature rise accompanying a fever.* Fever can also increase water losses. For each degree your temperature rises above normal, your daily insensible water loss increases by 200 ml. The advice "drink plenty of fluids" for anyone who is sick has a definite physiological basis.

- *Water gains.* A water gain of roughly 2500 ml/day is required to balance your average water losses. This value amounts to roughly 40 ml/kg body weight per day. You obtain water through eating (1000 ml), drinking (1200 ml), and *metabolic generation* (300 ml).

Metabolic generation of water is the production of water within cells, primarily as a result of oxidative phosphorylation in mitochondria. (We described the synthesis of water at the end of the electron transport system in Chapter 25.) ∞ *p. 908* When a cell breaks down 1 g of lipid, 1.7 ml of water is generated. Breaking down proteins or carbohydrates results in much lower values (0.41 ml/g protein; 0.55 ml/g carbohydrate). A typical mixed diet in the United States contains 46 percent carbohydrates, 40 percent lipids, and 14 percent protein. This diet produces roughly 300 ml of water per day, about 12 percent of your average daily requirement.

TABLE 27-1 Water Balance

Source	Daily Input (ml)
Water content of food	1000
Water consumed as liquid	1200
Metabolic water during catabolism	300
Total	2500

Method of Elimination	Daily Output (ml)
Urination	1200
Evaporation at skin	750
Evaporation at lungs	400
Loss in feces	150
Total	2500

Water Excess and Water Depletion   The body's water content cannot easily be determined. However, the concentration of Na^+, the most abundant ion in the ECF, provides useful clues to the state of water balance. When the body water content rises enough to reduce the Na^+ concentration of the ECF below 130 mEq/l, the state of **hyponatremia** (*natrium,* sodium) exists. When the body water content declines, the Na^+ concentration rises; when that concentration exceeds 150 mEq/l, **hypernatremia** exists.

Hyponatremia is a sign of **overhydration,** or *water excess.* Hyponatremia results from (1) the ingestion of a large volume of fresh water or the *infusion* (injection into the bloodstream) of a hypotonic solution; (2) an inability to eliminate excess water in urine, due to chronic renal failure (p. 958), heart failure (p. 677), cirrhosis (p. 878), or other disorders; or (3) endocrine disorders, such as excessive ADH production. The reduction in Na^+ concentrations leads to a shift of fluid into the ICF. The first signs are the effects on central nervous system function. In the early stages of hyponatremia, the individual behaves as if drunk on alcohol. This condition, *water intoxication,* may sound odd but is extremely dangerous. Untreated cases can rapidly progress from confusion to hallucinations, convulsions, coma, and then death. Treatment of severe water intoxication generally involves diuretics and infusing a concentrated salt solution that elevates Na^+ levels to near-normal levels.

Dehydration, also known as *volume depletion* or *water depletion,* develops when water losses outpace water gains. The osmotic concentration of plasma gradually increases, so hypernatremia results. The loss of body water is associated with severe thirst, dryness and wrinkling of the skin, and a fall in plasma volume and blood pressure. Eventually, circulatory shock develops, generally with fatal consequences. Treatment for dehydration entails administering hypotonic fluids by mouth or intravenous infusion. This treatment increases ECF volume and restores normal electrolyte concentrations. [AM] *Water and Weight Loss*

Fluid Shifts

Water movement between the ECF and ICF is called a **fluid shift.** Fluid shifts occur rapidly, reaching equilibrium within minutes to hours. These shifts occur in response to changes in the osmotic concentration of the ECF. We explored the basic relationships in Chapter 3; see Figure 3-8•, p. 74, which shows the effects of hypertonic and hypotonic solutions on cells:

- *If the osmotic concentration of your ECF increases, that fluid will become hypertonic with respect to your ICF.* Water will then move from the cells into the ECF until osmotic equilibrium is restored. The osmotic concentration of the ECF will increase if you lose water but retain electrolytes.

■ *If the osmotic concentration of your ECF decreases, that fluid will become hypotonic with respect to your ICF.* Water will then move from the ECF into the cells, and the ICF volume will increase. The osmotic concentration of the ECF will decrease if you gain water but do not gain electrolytes.

In summary, if the osmotic concentration of the ECF changes, a fluid shift between the ICF and ECF will tend to oppose the change. Because the volume of the ICF is much greater than that of the ECF, the ICF acts as a water reserve. In effect, instead of a large change in the osmotic concentration of the ECF, smaller changes occur in both the ECF and ICF. Two examples will demonstrate the dynamic exchange of water between the ECF and ICF.

Allocation of Water Losses When you lose water but retain electrolytes, the osmotic concentration of the ECF rises. Osmosis then moves water out of the ICF and into the ECF until the two solutions are again isotonic. At this point, both the ECF and ICF are somewhat more concentrated than normal, and both volumes are lower than they were before the fluid loss. Because the ICF has roughly twice the functional volume of the ECF, the net change in the ECF is relatively small.

Conditions that cause severe water losses include excessive perspiration (exercising in hot weather), inadequate water consumption, repeated vomiting, and diarrhea. These conditions promote water losses far in excess of electrolyte losses, so body fluids become increasingly concentrated. Responses that attempt to restore homeostasis include ADH and renin secretion and (as soon as possible) an increase in fluid intake.

Distribution of Water Gains When you drink a glass of pure water, or when you are given hypotonic solutions intravenously, the water content of your body increases without a corresponding increase in the concentration of electrolytes. As a result, the ECF increases in volume but becomes hypotonic with respect to the ICF. A fluid shift then occurs, and the volume of the ICF increases at the expense of the ECF. Once again, the larger volume of the ICF limits the amount of osmotic change. After the fluid shift, the ECF and ICF have slightly larger volumes and slightly lower osmotic concentrations than they did originally.

Normally, this situation will be promptly corrected. The reduced plasma osmotic concentration depresses the secretion of ADH, discouraging fluid intake and increasing water losses in urine. If the situation is *not* corrected, a variety of clinical problems will develop as water shifts into the intracellular fluid, distorting cells, changing the solute concentrations around enzymes, and disrupting normal cell functions.

ELECTROLYTE BALANCE

You are in electrolyte balance when the rates of gain and loss are equal for each electrolyte in your body. Electrolyte balance is important because:

■ *Total electrolyte concentrations directly affect water balance,* as we described previously.

■ *The concentrations of individual electrolytes can affect cell functions.* We saw many examples in earlier chapters, such as the effect of abnormal Na^+ concentrations on neuron activity and the effects of high or low Ca^{2+} and K^+ concentrations on cardiac muscle tissue.

Two cations, Na^+ and K^+, merit attention, because (1) they are major contributors to the osmotic concentrations of the ECF and ICF, respectively, and (2) they directly affect the normal functioning of all cells. Sodium is the dominant cation in the ECF. More than 90 percent of the osmotic concentration of the ECF results from the presence of sodium salts, mainly sodium chloride (NaCl) and sodium bicarbonate ($NaHCO_3$), so changes in the osmotic concentration of body fluids generally reflect changes in Na^+ concentration. Normal Na^+ concentrations in the ECF average 136–142 mEq/l, versus 10 mEq/l or less in the ICF. Potassium is the dominant cation in the ICF, where concentrations reach 160 mEq/l. Extracellular K^+ concentrations are generally very low, from 3.8 to 5.0 mEq/l.

Two general rules about sodium balance and potassium balance are worth noting:

1. *The most common problems with electrolyte balance are caused by an imbalance between gains and losses of sodium ions.*

2. *Problems with potassium balance are less common but significantly more dangerous than are those related to sodium balance.*

Sodium Balance

The total amount of sodium in the ECF represents a balance between two factors:

1. *Sodium ion uptake across the digestive epithelium.* Sodium ions enter the ECF by crossing the digestive epithelium through diffusion and carrier-mediated transport. The rate of absorption varies directly with the amount of sodium in the diet.

2. *Sodium ion excretion at the kidneys and other sites.* Sodium losses occur primarily by excretion in urine and through perspiration. The kidneys are the most important sites of Na^+ regulation. We discussed the mechanisms for sodium reabsorption at the kidneys in Chapter 26. ∞ *pp. 959–966*

27

A person in sodium balance typically gains and loses 48–144 mEq (1.1–3.3 g) of Na^+ each day. When sodium gains exceed sodium losses, the total Na^+ content of the ECF goes up; when losses exceed gains, the Na^+

content declines. However, a change in the Na$^+$ *content* of the ECF does not produce a change in the Na$^+$ *concentration*. When sodium intake or output changes, a corresponding gain or loss of water tends to keep the Na$^+$ concentration constant. For example, if you eat a very salty meal, the osmotic concentration of your ECF will not increase. When sodium ions are pumped across your digestive epithelium, the solute concentration in that portion of the ECF increases, whereas that of the intestinal contents decreases. Osmosis then occurs. Additional water enters the ECF from the digestive tract, elevating blood volume and blood pressure. Thus, persons with high blood pressure are advised to restrict the amount of salt in their diets.

Sodium Balance and ECF Volume The regulatory mechanism, diagrammed in Figure 27-4●, changes the ECF volume but keeps the Na$^+$ concentration relatively stable. If large amounts of salt are consumed *without* adequate fluid, as when you eat salty potato chips without taking a drink, the plasma Na$^+$ concentration will rise temporarily. A change in ECF volume will soon follow, however. Fluid will exit the ICF, increasing ECF volume and lowering Na$^+$ concentrations somewhat. The secretion of ADH restricts water loss and stimulates thirst, promoting additional water consumption. Due to the inhibition of water receptors in the pharynx, ADH secretion begins even before the Na$^+$ absorption occurs; the secretion rate rises further after Na$^+$ absorption due to osmoreceptor stimulation.

When sodium losses exceed gains, the volume of the ECF decreases. This reduction occurs without a significant change in the osmotic concentration of the ECF. Thus, if you perspire heavily but consume only pure water, you will lose sodium and the osmotic concentration of your ECF will drop briefly. However, as soon as the osmotic concentration drops by 2 percent or more, ADH secretion will decrease, so water losses at your kidneys will increase. As water leaves the ECF, the osmotic concentration will return to normal.

Minor changes in ECF volume do not matter, because they do not cause adverse physiological effects. If the regulation of Na$^+$ concentrations results in a large change in ECF volume, the situation will be corrected by the same homeostatic mechanisms responsible for regulating blood volume and blood pressure. This is the case because when the ECF volume changes, so does plasma volume and, in turn, blood volume. If the ECF volume rises, blood volume goes up; if the ECF volume drops, blood volume goes down. As we saw in Chapter 21, blood volume has a direct effect on blood pressure. A rise in blood volume elevates blood pressure; a drop in blood volume lowers blood pressure. The net result is that homeostatic mechanisms can monitor ECF volume indirectly by monitoring blood pressure. The receptors involved are baroreceptors at the carotid sinus, the aortic sinus, and the right atrium. ∞ *p. 713* The regulatory steps are reviewed in Figure 27-5●.

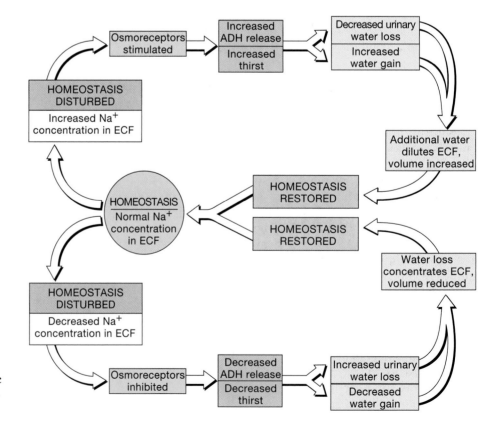

●**FIGURE 27-4** The Homeostatic Regulation of Normal Sodium Ion Concentrations in Body Fluids

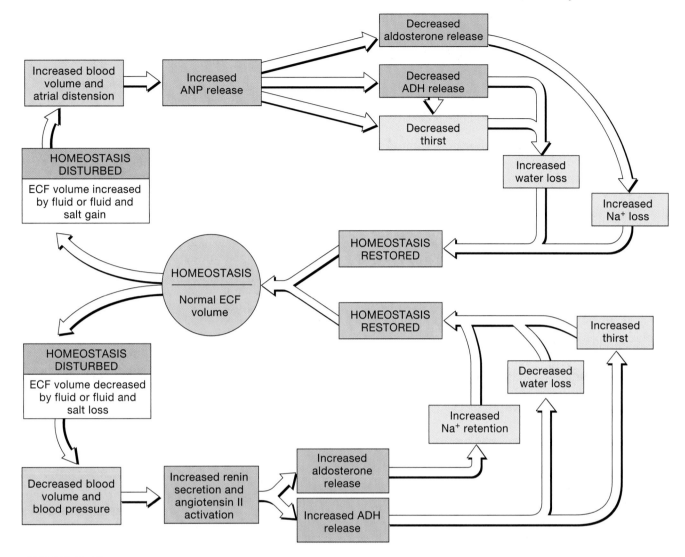

•**FIGURE 27-5** The Integration of Fluid Volume Regulation and Sodium Ion Concentrations in Body Fluids

If the ECF volume is inadequate, both blood volume and blood pressure decline and the renin–angiotensin system is activated. In response, losses of water and Na^+ are reduced, and gains of water and Na^+ are increased. The net result is that the ECF volume increases. Although the total amount of Na^+ in the ECF is increasing (gains exceed losses), the Na^+ concentration in the ECF remains unchanged, because Na^+ absorption is accompanied by osmotic water movement.

If the plasma volume becomes abnormally large, venous return increases, stretching the atrial wall and stimulating the release of ANP. Atrial natriuretic peptide reduces thirst and blocks the secretion of ADH and aldosterone, which together promote water or salt conservation. As a result, salt and water loss at the kidneys increases and the volume of the ECF declines.

Potassium Balance

Roughly 98 percent of the potassium content of the human body is in the ICF. Cells expend energy to recover potassium ions as they diffuse out of the cytoplasm and into the ECF. The K^+ concentration outside the cell is relatively low, and the concentration in the ECF at any moment represents a balance between (1) the rate of entry across the digestive epithelium and (2) the rate of loss into urine. Potassium loss in urine is regulated by controlling the activities of ion pumps along the distal portions of the nephron and collecting system. Whenever a sodium ion is reabsorbed from the tubular fluid, it generally is exchanged for a cation, typically K^+, in the peritubular fluid.

Urinary K^+ losses are usually limited to the amount gained by absorption across the digestive epithelium, typically 50–150 mEq (1.9–5.8 g) per day.

27

Athletes and Salt Loss

Unfounded notions and rumors about water and salt requirements during exercise abound. Sweat is a hypotonic solution that contains Na^+ in lower concentration than the ECF. As a result, a person who is sweating profusely loses more water than salt, and this loss leads to a rise in the Na^+ concentration of the ECF. The water content of the ECF decreases as the water loss occurs, so blood volume drops. As we noted on page 990, this condition is called *volume depletion*. Because this volume depletion occurs at the same time that blood is being shunted away from the kidneys, waste products accumulate in the blood and kidney function is impaired.

To prevent volume depletion, athletes should pause at brief intervals to drink liquids. The primary problem in volume depletion is water loss, and research has shown no basis for the rumor that cramps will result if you drink while exercising. Salt pills and the various sports beverages promoted for "faster absorption" and "better electrolyte balance" have no apparent benefits despite their relatively high cost. Body reserves of electrolytes are sufficient to tolerate extended periods of strenuous activity, and problems with Na^+ balance are extremely unlikely except during ultramarathons or other activities that involve maximal exertion for more than 6 hours.

Some sports beverages contain sugars and vitamins as well as electrolytes. During endurance events (marathons, ultramarathons, distance cycling), solutions containing less than 10 g/dl of glucose may improve performance when consumed late in the competition, as metabolic reserves are exhausted. However, high sugar concentrations (above 10 g/dl) can cause cramps, diarrhea, and other problems. The benefit of "glucose polymers" in sports drinks has yet to be proved. Drinking beverages "fortified" with vitamins is actively discouraged. Vitamins are not lost during exercise, and the consumption of these beverages in large volumes could, over time, cause hypervitaminosis.

(The potassium losses in feces and perspiration are negligible.) The K^+ concentration in the ECF is controlled by adjustments in the rate of active secretion along the distal convoluted tubule and collecting system of the nephron.

The rate of tubular secretion of K^+ varies in response to the following three factors:

1. *Changes in the K^+ concentration of the ECF.* In general, the higher the extracellular concentration of potassium, the higher the rate of secretion.

2. *Changes in pH.* When the pH of the ECF falls, so does the pH of peritubular fluid. The rate of potassium secretion then declines, because hydrogen ions, rather than potassium ions, are secreted in exchange for sodium ions in tubular fluid. The mechanisms for H^+ secretion were summarized in Figure 26-13c• (p. 965).

3. *Aldosterone levels.* The rate at which K^+ is lost in urine is strongly affected by aldosterone, because the ion pumps sensitive to this hormone reabsorb Na^+ from filtrate in exchange for K^+ from peritubular fluid. Aldosterone secretion is stimulated by angiotensin II as part of the regulation of blood volume. High plasma K^+ concentrations also stimulate aldosterone secretion directly. Either way, under the influence of aldosterone, the amount of sodium conserved and the amount of potassium excreted in urine are directly related.

Other Electrolytes

The ECF concentrations of other electrolytes are regulated as well. We shall note the most important ions involved. Additional information about sodium, potassium, and these other ions is in Table 27-2 (p. 996).

Calcium Balance Calcium is the most abundant mineral in the body. A typical body has 1–2 kg (2.2–4.4 lb) of it; 99 percent is deposited in the skeleton. In addition to forming the crystalline component of bone, calcium ions play key roles in the control of muscular and neural activities, in blood clotting, as cofactors for enzymatic reactions, and as second messengers.

As we noted in Chapters 6 and 18, calcium homeostasis reflects primarily an interplay between the reserves in bone, the rate of absorption across the digestive tract, and the rate of loss at the kidneys. The hormones parathyroid hormone (PTH), calcitriol, and, to a lesser degree, calcitonin maintain calcium homeostasis in the ECF. Parathyroid hormone and calcitriol raise Ca^{2+} concentrations; their actions are opposed by calcitonin. ∞ *pp. 597, 598*

A small amount of Ca^{2+} is lost in the bile, and under normal circumstances, very little Ca^{2+} escapes in urine or feces. To keep pace with biliary, urinary, and fecal Ca^{2+} losses, an adult must absorb only 0.8–1.2 g/day of Ca^{2+}. That represents only about 0.03 percent of the calcium reserve in the skeleton. Calcium absorption at the digestive tract and reabsorption along the distal convo-

Hypokalemia and Hyperkalemia

When the plasma concentration of potassium falls below 2 mEq/l, extensive muscular weakness develops, followed by eventual paralysis. We discussed this condition, called **hypokalemia** (*kalium*, potassium), in connection with ion effects on cardiac function. ∞ *p. 686* Causes of hypokalemia include:

- *Inadequate dietary K^+ intake.* If K^+ gains from the diet do not keep pace with the rate of K^+ loss in urine, K^+ concentrations in the ECF will drop.

- *The administration of diuretic drugs.* Several diuretics, including *Lasix*, can produce hypokalemia by increasing the volume of urine produced. Although the concentration of K^+ in urine is low, the greater the total volume, the more potassium is lost.

- *Excessive aldosterone secretion.* The condition of *aldosteronism*, characterized by excessive aldosterone secretion, results in hypokalemia, because the reabsorption of Na^+ is tied to the secretion of K^+. ∞ *p. 965*

- *An increase in the pH of the ECF.* When the H^+ concentration of the ECF declines, cells exchange intracellular H^+ for extracellular K^+. This ion swap helps stabilize the extracellular pH but gradually lowers the K^+ concentration of the ECF.

Treatment for hypokalemia generally includes increasing dietary intake of potassium by salting food with potassium salts (KCl) or by taking potassium tablets, such as *Slow-K®*. Severe cases are treated by the infusion of a solution containing K^+ at a concentration of 40–60 mEq/l.

High K^+ concentrations in the ECF produce an equally dangerous condition known as **hyperkalemia.** Severe cardiac arrhythmias appear when the K^+ concentration exceeds 8 mEq/l. ∞ *p. 686* Hyperkalemia results from factors such as:

- *Renal failure.* Kidney failure due to damage or disease will prevent normal K^+ secretion and thereby produce hyperkalemia.

- *The administration of diuretic drugs that block Na^+ reabsorption.* When sodium reabsorption slows down, so does potassium secretion. Hyperkalemia can result.

- *A decline in the pH of the ECF.* When the pH of the ECF declines, hydrogen ions move into the ICF in exchange for intracellular potassium ions. In addition, potassium secretion at the kidney tubules slows down, because hydrogen ions are secreted instead of potassium ions. The combination of increased K^+ entry into the ECF and decreased K^+ secretion can produce a dangerous hyperkalemia very rapidly.

Treatment for hyperkalemia includes the (1) elevation of ECF volume with a solution low in K^+; (2) stimulation of K^+ loss in urine by using appropriate diuretics, such as Lasix; (3) administration of buffers (generally sodium bicarbonate) that can control the pH of the ECF; (4) restriction of dietary K^+ intake; and (5) administration of enemas or laxatives that contain compounds, such as *kayexolate*, that promote K^+ loss across the digestive lining. In cases resulting from renal failure, kidney dialysis may also be required.

---

luted tubule are stimulated by PTH from the parathyroid glands and calcitriol from the kidneys.

Hypercalcemia and Hypocalcemia

Hypercalcemia exists when the Ca^{2+} concentration of the ECF is above 11 mEq/l. The primary cause of hypercalcemia in adults is *hyperparathyroidism.* ∞ *p. 598* Less-common causes include malignant cancers of the breast, lung, kidney, or bone marrow. Severe hypercalcemia (12–13 mEq/l) causes such symptoms as fatigue, confusion, cardiac arrhythmias, and calcification of the kidneys and soft tissues throughout the body.

The condition **hypocalcemia** (a Ca^{2+} concentration under 4 mEq/l) is much less common than hypercalcemia. Hypoparathyroidism, vitamin D deficiency, or chronic renal failure is typically responsible for hypocalcemia. Symptoms include muscle spasms, sometimes including generalized convulsions, weak heartbeats, cardiac arrhythmias, and osteoporosis.

Magnesium Balance The adult body contains about 29 g of magnesium, almost 60 percent of it deposited in the skeleton. The magnesium in body fluids is contained primarily in the ICF, where the concentration averages about 26 mEq/l. Magnesium is required as a cofactor for several important enzymatic reactions, including the phosphorylation of glucose within cells and the use of ATP by contracting muscle fibers. Magnesium is also important as a structural component of bone.

The Mg^{2+} concentration of the ECF averages 1.5–2.5 mEq/l, considerably lower than levels in the ICF. The proximal convoluted tubule reabsorbs magnesium very effectively. Keeping pace with the daily urinary loss requires a minimum dietary intake of only 24–32 mEq (0.3–0.4 g) per day.

Phosphate Balance Phosphate ions are required for bone mineralization, and roughly 744 g of PO_4^{3-} is bound up in the mineral salts of the skeleton. In body

27

TABLE 27-2 Electrolyte Balance

Ion and Normal ECF Range (mEq/l)	Disorder (mEq/l)	Symptoms	Causes	Treatment
Sodium (136–142)	Hypernatremia (>150)	Thirst, dryness and wrinkling of skin, reduced blood volume and pressure, eventual circulatory collapse	Dehydration; loss of hypotonic fluid	Ingestion of water or intravenous infusion of hypotonic solution
	Hyponatremia (<130)	Disturbed CNS function (water intoxication): confusion, hallucinations, convulsions, coma; death in severe cases	Infusion or ingestion of large volumes of hypotonic solution	Diuretic use and infusion of hypertonic salt solution
Potassium (3.8–5.0)	Hyperkalemia (>8)	Severe cardiac arrhythmias	Renal failure; use of diuretics; chronic acidosis	Infusion of hypotonic solution; selection of different diuretics; infusion of buffers; dietary restrictions
	Hypokalemia (<2)	Muscular weakness and paralysis	Low-potassium diet; diuretics; hypersecretion of aldosterone; chronic alkalosis	Increase in dietary K^+ content; ingestion of K^+ tablets or solutions; infusion of potassium solution
Calcium (4.5–5.3)	Hypercalcemia (>11)	Confusion, muscle pain, cardiac arrhythmias, kidney stones, calcification of soft tissues	Hyperparathyroidism; cancer; vitamin D toxicity; calcium supplement overdose	Infusion of hypotonic fluid to lower Ca^{2+} levels; surgery to remove parathyroid gland; administration of calcitonin
	Hypocalcemia (<4)	Muscle spasms, convulsions, intestinal cramps, weak heartbeats, cardiac arrhythmias, osteoporosis	Poor diet; lack of vitamin D; renal failure; hypoparathyroidism; hypomagnesemia	Calcium supplements; administration of vitamin D
Magnesium (1.5–2.5)	Hypermagnesemia (>4)	Confusion, lethargy, respiratory depression, hypotension	Overdose of magnesium supplements or antacids (rare)	Infusion of hypotonic solution to lower plasma concentration
	Hypomagnesemia (<0.8)	Hypocalcemia, muscle weakness, cramps, cardiac arrhythmias, hyptertension	Poor diet; alcoholism; severe diarrhea; kidney disease; malabsorption syndrome; ketoacidosis	Intravenous infusion of solution high in Mg^{2+}
Phosphate (1.8–2.6)	Hyperphosphatemia (>6)	No immediate symptoms; chronic elevation leads to calcification of soft tissues	High dietary phosphate intake; hypoparathyroidism	Dietary reduction
	Hypophosphatemia (<1)	Anorexia, dizziness, muscle weakness, cardiomyopathy, osteoporosis	Poor diet; kidney disease; malabsorption syndrome; hyperparathyroidism; vitamin D deficiency	Dietary improvement; hormone or vitamin supplements
Chloride (100–108)	Hyperchloremia (>112)	Acidosis, hyperkalemia	Dietary excess; increased chloride retention	Infusion of hypotonic solution to lower plasma concentration
	Hypochloremia (<95)	Alkalosis, anorexia, muscle cramps, apathy	Vomiting; hypokalemia	Diuretic use and infusion of hypertonic salt solution

27

fluids, the most important functions of PO_4^{3-} involve the ICF, where the ions are required for the formation of high-energy compounds, the activation of enzymes, and the synthesis of nucleic acids.

The PO_4^{3-} concentration of the plasma is usually 1.8–2.6 mEq/l. Phosphate ions are reabsorbed from tubular fluid along the proximal convoluted tubule; urinary and fecal losses of PO_4^{3-} amount to 30–45 mEq (0.8–1.2 g) per day. Phosphate ion reabsorption along the PCT is stimulated by calcitriol.

Chloride Balance Chloride ions are the most abundant anions in the ECF. The plasma concentration ranges from 100–108 mEq/l. In the ICF, Cl^- concentrations are usually low (3 mEq/l). Chloride ions are absorbed across the digestive tract with sodium ions; several carrier proteins along the renal tubules reabsorb Cl^- with Na^+. ∞ *pp. 962–964* The rate of Cl^- loss is small; a gain of 48–146 mEq (1.7–5.1 g) per day will keep pace with losses in urine and perspiration.

☑ What effect would drinking a pitcher of distilled water have on your level of ADH?

☑ How would eating a meal high in salt affect the amount of fluid in the ICF?

☑ What effect would being in the desert without water for a day have on your plasma osmotic concentration?

ACID–BASE BALANCE

In Chapter 2, we introduced the topic of pH and the chemical nature of acids, bases, and buffers. Table 27-3 reviews key terms important to the discussion that follows. If you need a more detailed review, refer to the appropriate sections of Chapter 2 before you proceed. ∞ *pp. 42–43*

The pH of body fluids can be altered by the introduction of either acids or bases. A general classification of acids and bases sorts them into *strong* versus *weak:*

■ *Strong acids and bases. Strong acids* and *strong bases* dissociate completely in solution. For example, hydrochloric acid (HCl), a strong acid, dissociates in solution by the reaction

$$HCl \longrightarrow H^+ + Cl^-$$
hydrochloric
acid

■ *Weak acids and bases.* When *weak acids* or *weak bases* enter a solution, a significant number of molecules remain intact; dissociation is not complete. This means that if you place molecules of a weak acid in one solution and the same number of molecules of a strong acid in another solution, the weak acid will liberate fewer hydrogen ions and have less of an impact on the pH of the solution than the strong acid will. Carbonic acid is a weak acid. At the normal pH of the ECF, an equilib-

TABLE 27-3 A Review of Important Terms Relating to Acid–Base Balance

pH	The negative exponent (negative logarithm) of the hydrogen ion concentration
Neutral	A solution with a pH of 7, which contains equal numbers of hydrogen ions and hydroxide ions
Acidic	A solution with a pH below 7, in which hydrogen ions predominate
Basic, or alkaline	A solution with a pH above 7, in which hydroxide ions predominate
Acid	A substance that dissociates to release hydrogen ions, decreasing pH
Base	A substance that dissociates to release hydroxide ions or to tie up hydrogen ions, increasing pH
Salt	An ionic compound consisting of a cation other than hydrogen and an anion other than a hydroxide ion
Buffer	A substance that tends to oppose changes in the pH of a solution by removing or replacing hydrogen ions; in body fluids, buffers maintain pH within normal limits (7.35–7.45)

rium state exists, and the reaction can be diagrammed as follows:

$$H_2CO_3 \rightleftharpoons H^+ + HCO_3^-$$
carbonic acid   bicarbonate ion

THE IMPORTANCE OF pH CONTROL

The pH of your body fluids reflects interactions among all the acids, bases, and salts in solution. The pH of the ECF normally remains within relatively narrow limits, usually from 7.35–7.45. Any deviation outside the normal range is extremely dangerous, because changes in H^+ concentrations disrupt the stability of cell membranes, alter protein structure, and change the activities of important enzymes. You could not survive for long with an ECF pH below 6.8 or above 7.7.

When the pH of plasma falls below 7.35, **acidemia** exists. The physiological state that results is called **acidosis.** When the pH of plasma rises above 7.45, **alkalemia** exists. The physiological state that results is called **alkalosis.** Acidosis and alkalosis affect virtually all body systems, but the nervous system and cardiovascular system are particularly sensitive to pH fluctuations. For example, severe acidosis (pH below 7.0) can be deadly, because (1) central nervous system function deteriorates, and the individual may become comatose; (2) cardiac contractions grow weak and irregular, and symptoms of heart failure may develop; and (3) peripheral vasodilation produces a dramatic

27

drop in blood pressure, and circulatory collapse can occur.

The control of pH is therefore an extremely important homeostatic process of great physiological and clinical significance. Although both acidosis and alkalosis are dangerous, in practice, problems with acidosis are much more common than are problems with alkalosis. This is so because several acids, including carbonic acid, are generated by normal cellular activities.

TYPES OF ACIDS IN THE BODY

The body contains three general categories of acids: (1) *volatile acids,* (2) *fixed acids,* and (3) *organic acids.*

A **volatile acid** is an acid that can leave solution and enter the atmosphere. Carbonic acid (H_2CO_3) is an important volatile acid in body fluids. At the lungs, carbonic acid breaks down into carbon dioxide and water; the carbon dioxide diffuses into the alveoli. In peripheral tissues, carbon dioxide in solution interacts with water to form carbonic acid, which dissociates to release hydrogen ions and bicarbonate ions. The complete reaction sequence is

$$\underset{\substack{\text{carbon}\\\text{dioxide}}}{CO_2} + \underset{\text{water}}{H_2O} \rightleftharpoons \underset{\substack{\text{carbonic}\\\text{acid}}}{H_2CO_3} \rightleftharpoons \underset{}{H^+} + \underset{\substack{\text{bicarbonate}\\\text{ion}}}{HCO_3^-}$$

This reaction occurs spontaneously in body fluids, but it occurs most rapidly in the presence of *carbonic anhydrase (CA),* an enzyme found in the cytoplasm of red blood cells, liver and kidney cells, parietal cells of the stomach, and in many other cell types.

Because most of the carbon dioxide in solution is converted to carbonic acid, and most of the carbonic acid dissociates, the partial pressure of carbon dioxide (P_{CO_2}) and the pH are inversely related (Figure 27-6•). When carbon dioxide levels rise, additional hydrogen ions and bicarbonate ions are released, so the pH goes down. (Recall that the pH is a *negative exponent;* when the concentration of hydrogen ions goes up, the pH value goes down.) The P_{CO_2} is the most important factor affecting the pH in body tissues.

At the alveoli, carbon dioxide diffuses into the atmosphere, the number of hydrogen ions and bicarbonate ions drops, and the pH rises. We shall consider this process, which effectively removes hydrogen ions from solution, in more detail later in the chapter.

Fixed acids are acids that do not leave solution; once produced, they remain in body fluids until they are eliminated at the kidneys. Sulfuric acid and phosphoric acid are the most important fixed acids. They are generated in small amounts during the catabolism of amino acids and compounds that contain phosphate groups, including phospholipids and nucleic acids.

Organic acids are acid participants in or byproducts of aerobic metabolism. Lactic acid, produced by the anaerobic metabolism of pyruvate, and ketone bodies, synthesized from acetyl-CoA, are important organic acids. Under normal conditions, most organic acids are metabolized rapidly, so significant accumulations do not occur. But relatively large amounts of organic acids are produced (1) during periods of anaerobic metabolism, because lactic acid builds up rapidly; and (2) during starvation or excessive lipid catabolism, because ketone bodies accumulate.

MECHANISMS OF pH CONTROL

To maintain acid–base balance over long periods of time, your body must balance gains and losses of hydrogen ions. Hydrogen ions are gained at the digestive tract and through metabolic activities within your cells. You must eliminate these hydrogen ions at the kidneys, by secret-

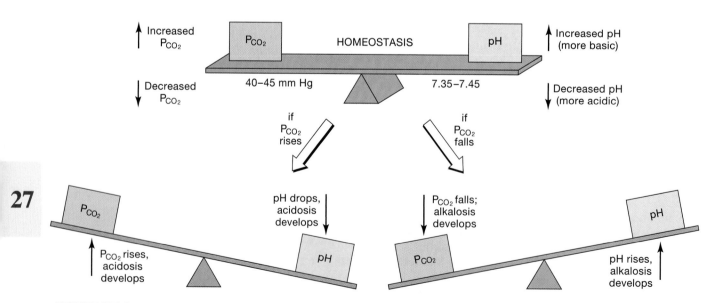

•**FIGURE 27-6** **The Basic Relationship between P_{CO_2} and Plasma pH.** The P_{CO_2} is inversely related to the pH.

ing H^+ into urine, and at the lungs, by forming water and carbon dioxide from H^+ and HCO_3^-. The sites of elimination are far removed from the sites of acid production. As these hydrogen ions travel through the body, they must be neutralized to avoid tissue damage.

The acids produced in the course of normal metabolic operations are temporarily neutralized by buffers and buffer systems in body fluids. *Buffers* are dissolved compounds that can provide or remove H^+ and thereby stabilize the pH of a solution. ∞ *p. 43* Buffers include weak acids that can donate H^+ and weak bases that can absorb H^+. A **buffer system** in body fluids generally consists of a combination of a weak acid and its dissociation products. The anion released functions as a weak base. Adding H^+ to the solution will upset the equilibrium, and the resulting formation of additional molecules of the weak acid will remove some of the H^+ from the solution.

The body has three major buffer systems, each with slightly different characteristics and distribution:

1. *Protein buffer systems.* Protein buffer systems contribute to the regulation of pH in the ECF and ICF. These buffer systems interact extensively with the other buffer systems.

2. *The carbonic acid–bicarbonate buffer system.* The carbonic acid–bicarbonate buffer system is most important in the ECF.

3. *The phosphate buffer system.* The phosphate buffer system has an important role in buffering the pH of the ICF and of urine.

Figure 27-7• shows the locations of these buffer systems.

Protein Buffer Systems

Protein buffer systems depend on the ability of amino acids to respond to pH changes by accepting or releasing H^+. The underlying mechanism is shown in Figure 27-8•:

▪ If the pH climbs, the carboxyl group ($-COOH$) of the amino acid can dissociate, acting as a weak acid and releasing a hydrogen ion. The carboxyl group then becomes a carboxylate ion ($-COO^-$). At the normal pH of body fluids (7.35–7.45), the carboxyl groups of most amino acids have already given up their hydrogen ions. (Proteins carry negative charges primarily for that reason.) However, some amino acids, notably *histidine*

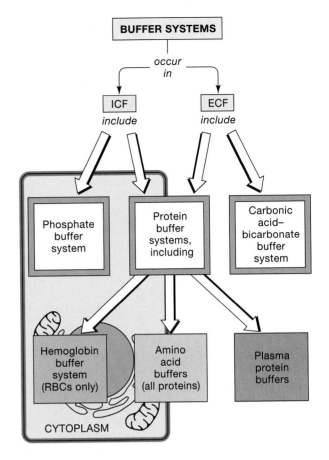

•**FIGURE 27-7** **Buffer Systems in Body Fluids.** Phosphate buffers occur primarily in the ICF, whereas the carbonic acid–bicarbonate buffer system occurs primarily in the ECF. Protein buffer systems are in both the ICF and the ECF. Extensive interactions take place among these systems.

and *cysteine,* have R groups (side chains) that will donate hydrogen ions if the pH climbs outside the normal range. Their buffering effects are very important in both the ECF and ICF.

▪ If the pH drops, the amino group ($-NH_2$) can act as a weak base and accept an additional hydrogen ion, forming an amino ion ($-NH_3^+$). This effect is primarily limited to free amino acids and the last amino acid in a polypeptide chain, because the amino groups in peptide bonds cannot function as buffers.

Plasma proteins contribute to the buffering capabilities of blood. Interstitial fluid contains extracellular

In alkaline medium, amino acid acts as an acid

In acidic medium, amino acid acts as a base

•**FIGURE 27-8** **Amino Acid Buffers.** Amino acids can either donate a hydrogen ion (as at left) or accept one (as at right), depending on the pH of their surroundings.

protein fibers and dissolved amino acids that also assist in the regulation of pH. In the ICF of active cells, structural and other proteins provide an extensive buffering capability that prevents destructive pH changes when organic acids, such as lactic acid or pyruvic acid, are produced by cellular metabolism.

Because there is an exchange between the ECF and ICF, the protein buffer system can help stabilize the pH of the ECF. For example, when the pH of the ECF decreases, cells pump H^+ out of the ECF and into the ICF, where they can be buffered by intracellular proteins. When the pH of the ECF rises, exchange pumps in cell membranes exchange H^+ in the ICF for K^+ in the ECF.

These mechanisms can assist in buffering the pH of the ECF. The process is slow, however, because hydrogen ions must be individually transported across the cell membrane. As a result, the protein buffer system in most cells cannot make rapid, large-scale adjustments in the pH of the ECF.

The Hemoglobin Buffer System   The situation is somewhat different for red blood cells. Red blood cells, which contain roughly 5.5 percent of the ICF, are normally suspended in the plasma. These cells are densely packed with hemoglobin, and their cytoplasm contains large amounts of carbonic anhydrase. These cells have a significant effect on the pH of the ECF, because they absorb carbon dioxide from the plasma and convert it to carbonic acid. Carbon dioxide can diffuse across the RBC membrane very quickly, so no transport mechanism is needed. As the carbonic acid dissociates, the bicarbonate ions move into the plasma via the *chloride shift.* ∞ *p. 829* The hydrogen ions

are buffered by hemoglobin molecules. At the lungs, the entire reaction sequence, diagrammed in Figure 23-25•, p. 831, proceeds in reverse. This is known as the **hemoglobin buffer system.**

The hemoglobin buffer system is the only intracellular buffer system that can have an immediate effect on the pH of the ECF. *The hemoglobin buffer system helps prevent drastic changes in pH when the plasma P_{CO_2} is rising or falling.*

The Carbonic Acid–Bicarbonate Buffer System

With the exception of red blood cells, some cancer cells, and tissues temporarily deprived of oxygen, your cells generate carbon dioxide virtually 24 hours a day. As we described above, most of the carbon dioxide is converted to carbonic acid, which then dissociates into a hydrogen ion and a bicarbonate ion. The carbonic acid and its dissociation products form the **carbonic acid–bicarbonate buffer system.**

This buffer system consists of the following reaction, introduced in our discussion of volatile acids (Figure 27-9a•):

$$CO_2 + H_2O \rightleftharpoons H_2CO_3 \rightleftharpoons H^+ + HCO_3^-$$

carbon   water   carbonic   bicarbonate
dioxide   acid   ion

Because the reaction is freely reversible, changing the concentrations of any participant will affect the concentrations of all other participants. For example, if hydrogen ions are added, most of them will be removed by interactions with HCO_3^-, forming H_2CO_3 (carbonic acid). In the process, the HCO_3^- acts as a weak base that buffers the excess H^+. The H_2CO_3

•**FIGURE 27-9 The Carbonic Acid–Bicarbonate Buffer System. (a)** Basic components of the carbonic acid–bicarbonate buffer system, showing their relationships to carbon dioxide and the bicarbonate reserve. **(b)** The response of the carbonic acid–bicarbonate buffer system to hydrogen ions generated by fixed or organic acids in body fluids.

formed in this way will in turn dissociate into CO_2 and water (Figure 27-9b•). The extra CO_2 can then be excreted at the lungs. In effect, this reaction takes the H^+ released by a strong organic or fixed acid and generates a volatile acid that can easily be eliminated.

This buffer system can also protect against increases in pH, although such changes are relatively rare. If hydrogen ions are removed from plasma, the reaction is driven to the right: The P_{CO_2} declines, and the dissociation of H_2CO_3 replaces the missing H^+.

The primary role of the carbonic acid–bicarbonate buffer system is to prevent pH changes caused by organic acids and fixed acids in the ECF. This buffer system has three important limitations:

1. *It cannot protect the ECF from pH changes that result from elevated or depressed levels of CO_2.* A buffer system cannot protect against changes in the concentration of its own weak acid. In the case above, the addition of excess H^+ drove the reaction to the left. But if we had added excess CO_2 instead of excess H^+, the elevated CO_2 would have driven the reaction to the right. Additional H_2CO_3 would have formed and dissociated into H^+ and HCO_3^-. This reaction would have reduced the pH of the plasma.

2. *It can function only when the respiratory system and the respiratory control centers are working normally.* Normally, the elevation in P_{CO_2} that occurs when fixed or organic acids are buffered will stimulate an increase in the respiratory rate. This increase accelerates the removal of CO_2 at the lungs. If the respiratory passageways are blocked, blood flow to the lungs will be impaired, or if the respiratory centers do not respond normally, the efficiency of the buffer system will be reduced.

3. *The ability to buffer acids is limited by the availability of bicarbonate ions.* Every time a hydrogen ion is removed from plasma, a bicarbonate ion goes with it. When all the bicarbonate ions have been tied up, buffering capabilities are lost.

Problems due to a lack of bicarbonate ions are rare for several reasons. Body fluids contain a large reserve of HCO_3^-, primarily in the form of dissolved molecules of the weak base *sodium bicarbonate* ($NaHCO_3$). This readily available supply of HCO_3^- is known as the **bicarbonate reserve.** The reaction sequence involved (Figure 27-9a•) is

$$Na^+ \ + \ HCO_3^- \ \rightleftharpoons \ NaHCO_3$$

bicarbonate   sodium
ion   bicarbonate

When hydrogen ions enter the ECF, the bicarbonate ions tied up in H_2CO_3 molecules are replaced by HCO_3^- from the bicarbonate reserve. This relationship is diagrammed in Figure 27-9b•.

Additional HCO_3^- can be generated at the kidneys, through mechanisms we introduced in Chapter 26

(Figure 26-13c•, p. 965). In the distal convoluted tubule and collecting system, carbonic anhydrase converts CO_2 within tubular cells into H_2CO_3, which then dissociates. The hydrogen ion is pumped into tubular fluid in exchange for a sodium ion, and the bicarbonate ion is transported into peritubular fluid in exchange for a chloride ion. In effect, tubular cells remove HCl from peritubular fluid in exchange for $NaHCO_3$.

The Phosphate Buffer System

The **phosphate buffer system** consists of the anion $H_2PO_4^-$, which is a weak acid. The operation of the phosphate buffer system resembles that of the carbonic acid–bicarbonate buffer system. The reversible reaction involved is

$$H_2PO_4^- \ \rightleftharpoons \ H^+ \ + \ HPO_4^{2-}$$

dihydrogen   monohydrogen
phosphate   phosphate

The weak acid is *dihydrogen phosphate* ($H_2PO_4^-$), and the anion released is *monohydrogen phosphate* (HPO_4^{2-}). In the ECF, the phosphate buffer system plays only a supporting role in the regulation of pH, primarily because the concentration of HCO_3^- far exceeds that of HPO_4^{2-}. However, the phosphate buffer system is quite important in buffering the pH of the ICF. In addition, cells contain a *phosphate reserve* in the form of the weak base *sodium monohydrogen phosphate* (Na_2HPO_4). The phosphate buffer system is also important in stabilizing the pH of urine. The dissociation of Na_2HPO_4 provides additional HPO_4^{2-} for use by this buffer system:

$$2\,Na^+ \ + \ HPO_4^{2-} \ \rightleftharpoons \ Na_2HPO_4$$

monohydrogen   sodium
phosphate   monohydrogen
               phosphate

MAINTENANCE OF ACID–BASE BALANCE

Although buffer systems can tie up excess H^+, they provide only a temporary solution. The hydrogen ions have not been eliminated but merely rendered harmless. For homeostasis to be preserved, the captured H^+ must ultimately be removed from body fluids. The problem is that the supply of buffer molecules is limited. Suppose that a buffer molecule prevents a pH change by binding a hydrogen ion that enters the ECF. The buffer is then tied up, reducing the capacity of the ECF to cope with any more H^+. Eventually, all the buffer molecules will be bound to H^+, and pH control will be impossible.

The situation can be resolved only by removing the H^+ from the ECF, thereby freeing the buffer molecules, or by replacing the buffer molecules. Similarly, if a buffer provides a hydrogen ion to maintain normal

27

pH, either another hydrogen ion must be obtained or the buffer must be replaced.

The maintenance of acid–base balance thus includes balancing H$^+$ gains and losses. This balancing act involves coordinating the actions of buffer systems with respiratory mechanisms and renal mechanisms. The respiratory and renal mechanisms support the buffer systems by (1) secreting or absorbing H$^+$, (2) controlling the excretion of acids and bases, and, when necessary, (3) generating additional buffers. It is the *combination* of buffer systems and these respiratory and renal mechanisms that maintains your pH within narrow limits.

Respiratory Compensation

Respiratory compensation is a change in the respiratory rate that helps stabilize the pH of the ECF. Respiratory compensation occurs whenever your pH strays outside normal limits. This compensation is effective, because respiratory activity has a direct effect on the carbonic acid–bicarbonate buffer system. Increasing or decreasing the rate of respiration alters pH by lowering or raising the P_{CO_2}. When the P_{CO_2} rises, the pH falls, because the addition of CO_2 drives the carbonic acid–bicarbonate buffer system to the right. When the P_{CO_2} falls, the pH rises because the removal of CO_2 drives that buffer system to the left.

We described the mechanisms responsible for the control of respiratory rate in Chapter 23 and shall present only a brief summary here. (If necessary, review Figures 23-27, p. 833, and 23-28•, p. 835.) Chemoreceptors of the carotid and aortic bodies are sensitive to the P_{CO_2} of the circulating blood. Other receptors, located on the ventrolateral surfaces of the medulla oblongata, monitor the P_{CO_2} of the CSF. A rise in P_{CO_2} stimulates the receptors; a fall in the P_{CO_2} inhibits them. The stimulation of the chemoreceptors leads to an increase in the respiratory rate. As the rate of respiration increases, more CO_2 is lost at the lungs, so the P_{CO_2} returns to normal levels. When the P_{CO_2} of the blood or CSF declines, respiratory activity becomes depressed and the breathing rate decreases. This decrease causes an elevation of the P_{CO_2} in the ECF.

Renal Compensation

Renal compensation is a change in the renal rates of H$^+$ and HCO$_3^-$ secretion or reabsorption in response to changes in plasma pH. Under normal conditions, the body generates enough organic and fixed acids to add about 100 mEq of H$^+$ to the ECF each day. An equivalent number of hydrogen ions must therefore be excreted in urine to maintain acid–base balance. In addition, the kidneys assist the lungs by eliminating CO_2 that enters the renal tubules during filtration or that diffuses into the tubular fluid as it travels toward the renal pelvis.

Hydrogen ions are secreted into the tubular fluid along the proximal convoluted tubule (PCT), the distal convoluted tubule (DCT), and the collecting system. The basic mechanisms involved were detailed in Figures 26-11 and 26-13c• (pp. 961, 965). The ability to eliminate a large number of hydrogen ions in a normal volume of urine depends on the presence of buffers in the urine. The secretion of H$^+$ can continue only until the pH of the tubular fluid reaches 4.0–4.5. (At that point, the H$^+$ concentration gradient is so great that hydrogen ions leak out of the tubule as fast as they are pumped in.) If there were no buffers in the tubular fluid, the kidneys could secrete less than 1 percent of the acid produced each day before the pH reached this limit. To maintain acid balance under these conditions, the kidneys would have to produce about 1000 liters of urine each day just to keep pace with the generation of H$^+$ in the body. Buffers in tubular fluid are therefore extremely important, because they keep the pH high enough for H$^+$ secretion to continue. Metabolic acids are continuously being generated; without these buffering mechanisms, the kidneys would be unable to maintain homeostasis.

Figure 27-10• diagrams the primary routes of H$^+$ secretion and shows the buffering mechanisms that stabilize the pH of tubular fluid. The three major buffers involved (Figure 27-10a•) are ① the carbonic acid–bicarbonate buffer system, ② the phosphate buffer system, and ③ ammonia. Glomerular filtration puts components of the carbonic acid–bicarbonate and phosphate buffer systems into the filtrate. The ammonia is generated by tubule cells (primarily those of the PCT).

Figure 27-10a• shows the secretion of H$^+$, which relies on carbonic anhydrase (CA) activity within the tubular cells. The hydrogen ions generated may be pumped into the lumen in exchange for sodium ions, individually, or with chloride ions. The net result is the secretion of H$^+$, accompanied by the removal of CO_2 (from the tubular fluid, the tubule cells, and the ECF), and the release of sodium bicarbonate into the ECF.

Figure 27-10b• shows the generation of ammonia within the tubules. As the tubule cells use the enzyme *glutaminase* to break down the amino acid *glutamine*, amino groups are released as ammonium ions (NH$_4^+$) or as ammonia (NH$_3$). The ammonium ions are transported into the lumen in exchange for Na$^+$ in the tubular fluid. The NH$_3$, which is highly volatile and also toxic to cells, rapidly diffuses into the tubular fluid. There it reacts with a hydrogen ion, forming NH$_4^+$. This reaction buffers the tubular fluid and removes a potentially dangerous compound from body fluids. The carbon chains of the glutamine molecules are ultimately converted to HCO$_3^-$, which is cotransported with Na$^+$ into the ECF. The generation of ammonia by tubule cells thus ties up H$^+$ in the tubular fluid and releases sodium bicarbonate into the ECF, where it

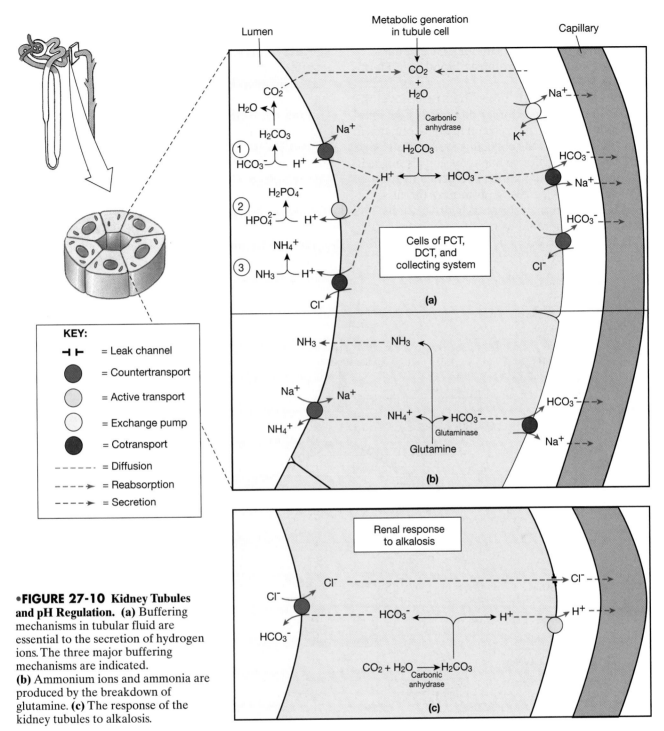

•**FIGURE 27-10 Kidney Tubules and pH Regulation. (a)** Buffering mechanisms in tubular fluid are essential to the secretion of hydrogen ions. The three major buffering mechanisms are indicated. **(b)** Ammonium ions and ammonia are produced by the breakdown of glutamine. **(c)** The response of the kidney tubules to alkalosis.

contributes to the bicarbonate reserve. These mechanisms (low body fluid pH) of H⁺ secretion and buffering are always functioning, but their levels of activity vary widely with the pH of the ECF.

The Renal Responses to Acidosis and Alkalosis   Acidosis (low body fluid pH) develops when the normal plasma buffer mechanisms are stressed by excessive numbers of hydrogen ions. The kidney tubules do not distinguish among the various acids that may cause acidosis. Whether the fall in pH results from the production of volatile, fixed, or organic acids, the renal contribution remains limited to the (1) secretion of H⁺, (2) activity of buffers in the tubular fluid, (3) removal of CO_2, and (4) reabsorption of $NaHCO_3$.

The tubular cells thus bolster the capabilities of the carbonic acid–bicarbonate buffer system. They do so by increasing the concentration of bicarbonate ions in the ECF, replacing those already used to remove hydrogen ions from solution. In a starving individual,

27

the tubular cells break down amino acids, yielding ammonium ions that are pumped into the tubular fluid, carbon chains for catabolism, and bicarbonates to help buffer ketone bodies in the blood (see Figure 26-13c•, p. 965).

When alkalosis (high body fluid pH) develops, (1) the rate of H^+ secretion at the kidneys declines, (2) the tubule cells do not reclaim the bicarbonates in tubular fluid, and (3) the collecting system transports HCO_3^- into tubular fluid while releasing a stong acid (HCl) into peritubular fluid. The concentration of HCO_3^- in plasma decreases, promoting the dissociation of H_2CO_3 and the release of hydrogen ions. The additional H^+ generated at the kidneys helps return the pH to normal levels. The renal response to alkalosis is diagrammed in Figure 27-10c•.

DISTURBANCES OF ACID–BASE BALANCE

Figure 27-11• summarizes the interactions among buffer systems, respiration, and renal function in normal acid–base balance. In combination, these mechanisms are generally able to control pH very precisely, so the pH of the ECF seldom varies more than 0.1 pH units, from 7.35 to 7.45. When buffering mechanisms are severely stressed, however, the pH wanders outside these limits, producing symptoms of alkalosis or acidosis.

If you are considering a career in a health-related field, an understanding of acid–base dynamics will be essential to you for clinical diagnosis and patient management under a variety of conditions. Temporary shifts in the pH of body fluids occur frequently. Rapid

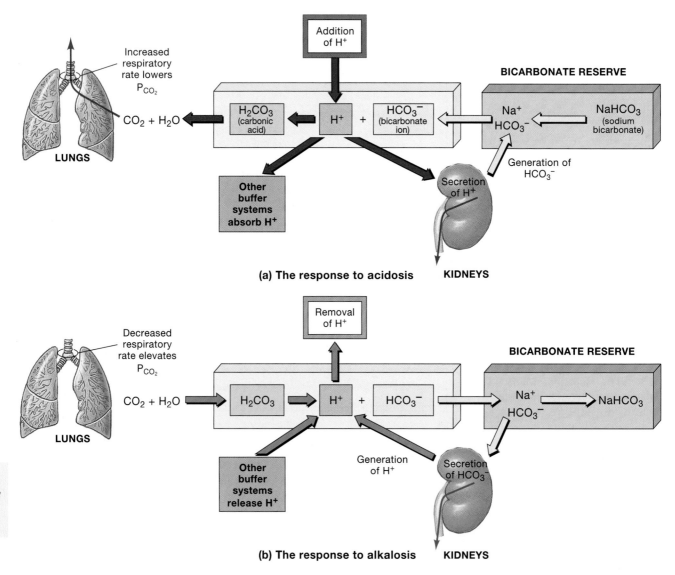

(a) The response to acidosis

(b) The response to alkalosis

•**FIGURE 27-11** **The Central Role of the Carbonic Acid–Bicarbonate Buffer System in the Regulation of Plasma pH.**
Interactions between the carbonic acid–bicarbonate buffer system and other buffer systems and compensation mechanisms. **(a)** The response to a fall in pH caused by the addition of H^+. **(b)** The response to a rise in pH caused by the removal of H^+.

and complete recovery occurs through a combination of buffer system activity and the respiratory and renal responses. More-serious and prolonged disturbances of acid–base balance can result from any factor that affects one of the principal regulatory mechanisms:

- *Any disorder affecting circulating buffers, respiratory performance, or renal function.* Several conditions described in earlier chapters, including *emphysema* (Chapter 23) and *renal failure* (Chapter 26), are associated with dangerous changes in pH. ∞ *pp. 839, 958*
- *Cardiovascular conditions.* Conditions such as *heart failure* or *hypotension* can affect the pH of internal fluids by causing fluid shifts and by changing glomerular filtration rates and respiratory efficiency.
- *Conditions affecting the central nervous system.* For example, neural damage or disease that affects the CNS will affect the respiratory and cardiovascular reflexes that are essential to normal pH regulation.

Serious abnormalities in acid–base balance generally show an initial *acute phase,* in which the pH moves rapidly away from the normal range. If the condition persists, physiological adjustments occur; the individual then enters the *compensated phase.* Unless the underlying problem is corrected, compensation cannot be completed, and blood chemistry will remain abnormal. The pH typically remains outside normal limits, even after compensation has occurred. Even if the pH is within the normal range, the P_{CO_2} or HCO_3^- concentrations can be abnormal.

The primary source of the problem is usually indicated by the name given to the resulting condition. For example:

- *Respiratory acid–base disorders* result from a mismatch between carbon dioxide generation in peripheral tissues and carbon dioxide excretion at the lungs. When a respiratory acid–base disorder is present, the carbon dioxide level of the ECF is abnormal.
- *Metabolic acid–base disorders* are caused by the generation of organic or fixed acids or by conditions affecting the concentration of HCO_3^- in the ECF.

Respiratory compensation alone may restore normal acid–base balance in individuals with respiratory acid–base disorders. In contrast, compensation mechanisms for metabolic acid–base disorders may be able to stabilize pH, but other aspects of acid–base balance (buffer system function, bicarbonate levels, and P_{CO_2}) remain abnormal until the underlying metabolic cause is corrected.

We can subdivide the respiratory and metabolic categories to create four major classes of acid–base disturbances: (1) *respiratory acidosis,* (2) *respiratory alkalosis,* (3) *metabolic acidosis,* and (4) *metabolic alkalosis.*

RESPIRATORY ACIDOSIS

Respiratory acidosis develops when the respiratory system cannot eliminate all the carbon dioxide generated by peripheral tissues. The primary symptom is low plasma pH due to **hypercapnia,** an elevated plasma P_{CO_2}. The usual cause is hypoventilation, an abnormally low respiratory rate. ∞ *p. 834* When the P_{CO_2} in the ECF rises, H^+ and HCO_3^- concentrations also begin rising as H_2CO_3 forms and dissociates. Other buffer systems can tie up some of the H^+, but once the combined buffering capacity has been exceeded, the pH begins to fall rapidly. The effects are diagrammed in Figure 27-12a•.

Respiratory acidosis is the most common challenge to acid–base equilibrium. Your tissues generate carbon dioxide rapidly. Even a few minutes of hypoventilation can cause acidosis, reducing the pH of the ECF to as low as 7.0. Under normal circumstances, the chemoreceptors that monitor the P_{CO_2} of plasma and of cerebrospinal fluid (CSF) will eliminate the problem by calling for an increase in breathing (pulmonary ventilation) rates.

If the chemoreceptors fail to respond, if the breathing rate cannot be increased, or if the circulatory supply to the lungs is inadequate, the pH will continue to decline. If the decline is severe, **acute respiratory acidosis** will develop. Acute respiratory acidosis is an immediate, life-threatening condition. It is especially dangerous in persons whose tissues are generating large amounts of carbon dioxide or who are incapable of normal respiratory activity. For this reason, the reversal of acute respiratory acidosis is probably the major goal in the resuscitation of cardiac arrest or drowning victims. Thus first-aid, CPR, and life-saving courses always stress the "ABCs" of emergency care: Airway, Breathing, and Circulation.

Chronic respiratory acidosis develops when normal respiratory function has been compromised but the compensatory mechanisms have not failed completely. Individuals with CNS injuries, or those whose respiratory centers have been desensitized by drugs such as alcohol or barbiturates, do not respond to the warning signals from the chemoreceptors. As a result, these people are prone to developing acidosis due to chronic hypoventilation.

Even when respiratory centers are intact and functional, damage to some components of the respiratory system can prevent increased pulmonary exchange. Examples of conditions fostering chronic respiratory acidosis include emphysema, congestive heart failure, and pneumonia (in which alveolar damage or blockage typically occurs). Pneumothorax and respiratory muscle paralysis have a similar effect, because they, too, limit the ability to maintain adequate breathing rates.

When a normal pulmonary response does not occur, the kidneys respond by increasing the rate of H^+ secretion into tubular fluid. This response slows the rate of pH change. However, renal mechanisms alone cannot return the pH to normal until the underlying respiratory or circulatory problems are corrected.

27

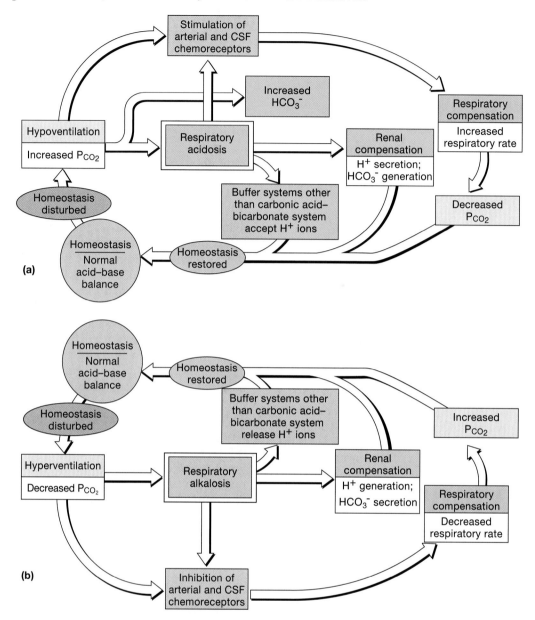

•**FIGURE 27-12 Respiratory Acid–Base Regulation.** (a) Respiratory acidosis and (b) respiratory alkalosis result from inadequate or excessive breathing. In healthy individuals, respiratory responses, combined with the responses of the kidneys, are generally able to restore normal acid–base balance.

The primary problem in respiratory acidosis is that the rate of pulmonary exchange is inadequate to keep the arterial P_{CO_2} within normal limits. The efficiency of breathing can typically be improved temporarily by inducing bronchodilation or by using mechanical aids that provide air under positive pressure. If breathing has ceased, artificial respiration or a mechanical ventilator will be required. These measures may restore normal pH if the respiratory acidosis was neither severe nor prolonged. Treatment of acute respiratory acidosis is complicated by the fact that it causes a complementary *metabolic acidosis,* as we shall soon see, due to the generation of lactic acid in oxygen-starved tissues.

RESPIRATORY ALKALOSIS

Problems with **respiratory alkalosis** (Figure 27-12b•) are relatively uncommon. Respiratory alkalosis develops when respiratory activity lowers plasma P_{CO_2} to below-normal levels, a condition called **hypocapnia.** A temporary hypocapnia can be produced by *hyperventilation,* when increased respiratory activity leads to a reduction in the arterial P_{CO_2}. ∞ *p. 835* Continued hyperventilation can elevate the pH to levels as high as 7.8–8. This condition generally corrects itself, because the reduction in P_{CO_2} removes the stimulation for the chemoreceptors, so the urge to breathe fades until carbon dioxide levels have returned to normal. *Respiratory alkalosis caused by hyperventilation seldom persists long enough to cause a clinical emergency.*

Common causes of hyperventilation include physical stresses, such as pain, or psychological stresses, such as extreme anxiety. Hyperventilation gradually elevates the pH of the cerebrospinal fluid, and central nervous system function is affected. The initial symptoms involve tingling sensations in the hands, feet, and lips. A light-headed feeling may also be noted. If the hyperventilation continues, the individual may lose consciousness. When unconsciousness occurs, any contributing psychological stimuli are removed, and the breathing rate declines. The P_{CO_2} then rises until the pH returns to normal.

A simple treatment for respiratory alkalosis caused by hyperventilation consists of having the individual breathe and rebreathe the air contained in a small paper bag. As the P_{CO_2} in the bag rises, so do the person's alveolar and arterial CO_2 concentrations. This change eliminates the problem and restores the pH to normal levels. Other problems with respiratory alkalosis are rare and involve primarily (1) individuals adapting to high altitudes, where the low P_{O_2} promotes hyperventilation, (2) patients on mechanical respirators, or (3) individuals with brain stem injuries who are incapable of responding to shifts in plasma CO_2 concentrations.

METABOLIC ACIDOSIS

Metabolic acidosis is the second most common type of acid–base imbalance. It has three major causes:

1. The most widespread cause of metabolic acidosis is the production of a large number of fixed or organic acids. The hydrogen ions liberated by these acids overload the carbonic acid–bicarbonate buffer system, so the pH begins to decline (Figure 27-13a•). We introduced two examples of metabolic acidosis earlier:

   a. **Lactic acidosis** can develop after strenuous exercise or prolonged tissue hypoxia (oxygen starvation) as active cells rely on anaerobic respiration. (See Figure 10-18•, p. 294.)

   b. **Ketoacidosis** results from the generation of large quantities of ketone bodies during the postabsorptive state of metabolism. Ketoacidosis is a problem in starvation and a potential complication of poorly controlled diabetes mellitus. In either case, peripheral tissues are unable to obtain adequate glucose from the bloodstream and begin metabolizing lipids and ketone bodies. ∞ *p. 925*

2. A less common cause of metabolic acidosis is an impaired ability to excrete H^+ at the kidneys. For example, conditions marked by severe kidney damage, such as *glomerulonephritis* (Chapter 26), typically result in a severe metabolic acidosis. ∞ *p. 952* Metabolic acidosis is also caused by diuretics that turn off the sodium–hydrogen transport system in the kidney

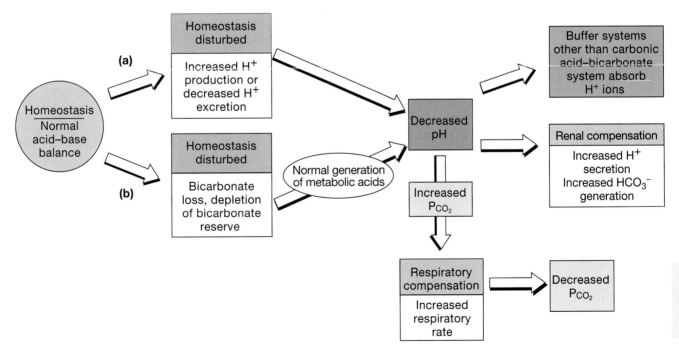

•**FIGURE 27-13 The Response to Metabolic Acidosis.** Metabolic acidosis can result from either **(a)** increased acid production or decreased acid secretion, leading to a buildup of H^+ in body fluids, or **(b)** a loss of bicarbonate ions that makes the carbonic acid–bicarbonate buffer system incapable of preventing a fall in pH. Respiratory and renal compensation mechanisms can stabilize pH, but the blood chemistry remains abnormal until the levels of acid production, acid secretion, and bicarbonate ion concentration return to normal.

27

tubules. The secretion of H^+ is directly or indirectly linked to the reabsorption of Na^+. When Na^+ reabsorption stops, so does H^+ secretion.

3. Metabolic acidosis occurs after a severe bicarbonate loss (Figure 27-13b•). The carbonic acid–bicarbonate buffer system relies on bicarbonate ions to balance hydrogen ions that threaten pH balance. A drop in the HCO_3^- concentration in the ECF thus reduces the effectiveness of this buffer system, and acidosis soon develops. The most common cause of HCO_3^- depletion is chronic diarrhea. Under normal conditions, most of the bicarbonate ions secreted into the digestive tract in pancreatic, hepatic, and mucous secretions are reabsorbed before the feces are eliminated. In diarrhea, these bicarbonates are lost, and the HCO_3^- concentration of the ECF drops accordingly.

The nature of the problem must be understood before treatment can begin. Potential causes are so varied that a clinician must piece together relevant clues to make a diagnosis. In some cases, the diagnosis is very straightforward; for example, a patient in metabolic acidosis after a bicycle race probably has lactic acidosis. In other cases, the clinician must be a detective.

Compensation for metabolic acidosis generally involves a combination of respiratory and renal mechanisms. Hydrogen ions interacting with bicarbonate ions form carbon dioxide molecules that are eliminated at the lungs, whereas the kidneys excrete additional hydrogen ions into the urine and generate bicarbonate ions that are released into the ECF.

Combined Respiratory and Metabolic Acidosis

Respiratory acidosis and metabolic acidosis are typically linked, because oxygen-starved tissues generate large quantities of lactic acid and because sustained hypoventilation leads to decreased arterial P_{O_2}. The problem can be particularly serious in cases of near-drowning, in which the body fluids have high P_{CO_2}, low P_{O_2}, and large amounts of lactic acid generated by the muscles of the struggling person. Prompt emergency treatment is essential. The usual procedure involves some form of artificial or mechanical respiratory assistance coupled with the intravenous infusion of an isotonic solution that contains sodium lactate, sodium gluconate, or sodium bicarbonate.

METABOLIC ALKALOSIS

Metabolic alkalosis occurs when HCO_3^- concentrations become elevated (Figure 27-14•). The bicarbonate ions then interact with hydrogen ions in solution, forming H_2CO_3.

The reduction in H^+ concentrations then causes symptoms of alkalosis.

Metabolic alkalosis is relatively rare, but we noted one interesting cause in Chapter 24. ∞ p. 865 The secretion of hydrochloric acid (HCl) by the gastric mucosa is associated with the influx of large numbers of bicarbonate ions into the ECF. This phenomenon, known as the *alkaline tide,* temporarily elevates the HCO_3^- concentration in the ECF during meals. A person who begins vomiting repeatedly will continue to generate stomach acids to replace those that are lost, so HCO_3^- concentration of the ECF will continue to rise.

Compensation for metabolic alkalosis involves a reduction in breathing rate, coupled with the increased loss of HCO_3^- in urine. Treatment of mild cases addresses the primary cause, generally by the control of vomiting, and may involve the administration of solutions that contain NaCl or KCl.

Treatment of acute cases of metabolic alkalosis may involve the administration of ammonium chloride (NH_4Cl). Metabolism of the ammonium ion in the liver results in the liberation of a hydrogen ion, so in effect the introduction of NH_4Cl leads to the internal generation of HCl, a strong acid. As the HCl diffuses into the bloodstream, the pH falls toward normal levels.

THE DETECTION OF ACIDOSIS AND ALKALOSIS

Virtually anyone who has a problem that affects the cardiovascular, respiratory, urinary, digestive, or nervous system may develop potentially dangerous acid–base imbalances. For this reason, most diagnostic blood tests include tests designed to provide informa-

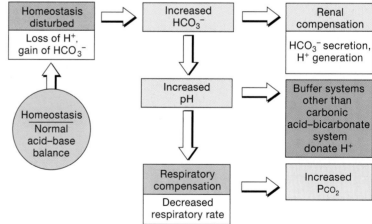

•**FIGURE 27-14 Metabolic Alkalosis.** Metabolic alkalosis most commonly results from the loss of acids, especially stomach acid lost through vomiting. As replacement gastric acids are produced, the alkaline tide introduces a great many bicarbonate ions into the bloodstream, so pH increases.

TABLE 27-4 Changes in Blood Chemistry Associated with the Major Classes of Acid–Base Disorders

Disorder	pH (normal = 7.35–7.45)	HCO_3^- (normal = 24–28 mEq/l)	P_{CO_2} (mm Hg) (normal = 35–45)	Remarks	Treatment
Respiratory acidosis	Decreased (below 7.35)	*Acute:* normal *Compensated:* increased (above 28)	Increased (above 50)	Generally caused by hypoventilation and CO_2 buildup in tissues and blood	Improve ventilation; in some cases, with bronchodilation and mechanical assistance
Metabolic acidosis	Decreased (below 7.35)	Decreased (below 24)	*Acute:* normal *Compensated:* decreased (below 35)	Caused by buildup of organic or fixed acid, impaired H^+ elimination at kidneys, or HCO_3^- loss in urine or feces	Administration of bicarbonate (gradual) with other steps as needed to correct primary cause
Respiratory alkalosis	Increased (above 7.45)	*Acute:* normal *Compensated:* decreased (below 24)	Decreased (below 35)	Generally caused by hyperventilation and reduction in plasma CO_2 levels	Reduce respiratory rate, allow rise in P_{CO_2}
Metabolic alkalosis	Increased (above 7.45)	Increased (above 28)	Increased (above 45)	Generally caused by prolonged vomiting and associated acid loss	pH below 7.55: no treatment; pH above 7.55: may require administration of NH_4Cl

tion about pH and buffer function. Standard tests include blood pH, P_{CO_2}, and HCO_3^- levels. These data make the recognition of acidosis or alkalosis and the classification of a particular condition as respiratory or metabolic in nature relatively straightforward. Table 27-4 indicates the patterns that characterize the four major categories of acid–base disorders. Additional steps, such as determining the anion gap, plotting blood test results on a graph, or *nomogram,* or using a diagnostic chart can help in identifying possible causes of the problem and in distinguishing compensated from uncompensated conditions. Details and diagnostic exercises are included in the *Applications Manual.* AM
Diagnostic Classification of Acid–Base Disorders

☑ What effect would a decrease in the pH of the body fluids have on the respiratory rate?

☑ Why does the tubular fluid in nephrons need to be buffered?

☑ How would a prolonged fast affect the body's pH?

☑ Why can prolonged vomiting produce metabolic alkalosis?

⧗ AGING AND FLUID, ELECTROLYTE, AND ACID–BASE BALANCE

Aging affects many aspects of fluid, electrolyte, and acid–base balance, including the following:

▪ Total body water content gradually decreases with age. Between ages 40 and 60, total body water content averages 55 percent for males and 47 percent for fe-

males. After age 60, the values decline to roughly 50 percent for males and 45 percent for females. Among other effects, this decrease reduces the dilution of waste products, toxins, and administered drugs.

▪ A reduction in the GFR and in the number of functional nephrons reduces the ability to regulate pH through renal compensation.

▪ The ability to concentrate urine declines, so more water is lost in urine. In addition, the rate of insensible perspiration increases as the skin becomes thinner and more delicate. Maintaining fluid balance therefore requires a higher daily water intake. A reduction in ADH and aldosterone sensitivity makes older people less able than younger people to conserve body water when losses exceed gains.

▪ Many people over age 60 experience a net loss in body mineral content as muscle mass and skeletal mass decrease. This loss can, at least in part, be prevented by a combination of exercise and increased dietary mineral supply.

▪ The reduction in vital capacity that accompanies aging reduces the ability to perform respiratory compensation, increasing the risk of respiratory acidosis. This problem can be compounded by arthritis, which can reduce vital capacity; and by emphysema, another condition that, to some degree, develops with aging.

▪ Disorders affecting major systems become more common with increasing age. Most if not all of these disorders have some impact on either fluid, electrolyte, or acid–base balance.

27

Fluid, Electrolyte, and Acid–Base Balance in Infants

Fetuses and infants have very different requirements for the maintenance of fluid, electrolyte, and acid–base balance than do adults. A fetus obtains the water, organic nutrients, and electrolytes it needs from the maternal bloodstream. Buffers in the fetal bloodstream provide short-term pH control, and the maternal kidneys eliminate the generated H^+. The body water content is high; at birth, water accounts for roughly 75 percent of the newborn infant's body weight, as compared with 50–60 percent in adults. Several factors contribute to this difference, including (1) the larger infant blood volume (10–12.5 percent of body weight versus 6–7 percent in adults), (2) a proportionately larger volume of cerebrospinal fluid, (3) a larger interstitial fluid volume (in part due to existing for 9 months under weightless conditions; body water content also changes in orbiting astronauts), and (4) differences in the water content and in the proportions, in terms of body mass, of organs and tissues; for example, compared with adults, newborn infants have a proportionately larger heart and brain, which are 75–79 percent water, but less than half as much adipose tissue, which is 10 percent water.

During the first 2–3 days after delivery, roughly 6 percent of the infant's excess water is lost. The loss thereafter is more gradual, with the typical adult body water content appearing after age 2 years.

At birth, the distribution of body water is also different from that of an adult. In a newborn, the ECF accounts for roughly 35 percent of total body weight, versus 40 percent for the ICF. Because the ECF and ICF in infants are similar in volume, the ICF is less effective than in adults at buffering changes in the ECF volume. As a result, less water must be lost from the ECF before the ICF volume is reduced enough to damage cells. As growth occurs, the ICF volume (as a percentage of total body weight) remains relatively stable, whereas the ECF volume gradually decreases. The relative decline in the ECF volume becomes evident after a few months; the adult relationship between the ECF and ICF volumes is reached after approximately 2 years.

Basic aspects of electrolyte balance are the same in newborn infants as in adults, but the effects of fluctuations in the diet are much more immediate in newborns because reserves of minerals and energy sources are much smaller. (At any age, the bone mass and adipose tissue mass are relatively small.) The problem is compounded by the fact that the metabolic rate (per unit body mass) of infants is twice that of adults. Thus, infants have an elevated demand for nutrients, and that demand must be met promptly. This is one reason that infants require frequent feedings; another is that they have a much higher demand for water than do adults.

The elevated metabolic rate also means an accelerated production of waste products that must be eliminated at the kidneys. But the kidneys of a newborn are unable to produce urine with an osmotic concentration above 450 mOsm/l. So a newborn must produce more urine to eliminate the metabolic waste products. Water losses at other sites are higher as well. Because the surface-to-volume ratio and the respiratory rate are relatively high, the rate of insensible perspiration is much higher in infants than in adults. To keep pace with rates of water loss in urine and insensible perspiration, newborns must consume, on a proportional basis, seven times as much water as adults do. If water intake is inadequate, waste products accumulate, and metabolic acidosis may develop. Although the kidneys become effective at concentrating urine after about 1 month, the elevated metabolic rate and water loss remain. As a result, infants continue to consume (and lose) roughly twice as much water as do adults by body weight. Infants are therefore at greater risk of dehydration; they can survive for only 1–3 days without water, whereas adults can survive for a week or more.

The increased metabolic rate of infants also means an accelerated demand for oxygen and more-rapid generation of CO_2. The respiratory rate in newborns is relatively high—roughly 40 breaths per minute. And the breaths are deep; in proportion to body weight, newborns must move twice as much air as adults do. As a result, the functional residual capacity of the lungs (the amount of air in the lungs after one quiet respiratory cycle) is about half that of adults. ∞ *p. 822* In adults, the functional residual capacity is large relative to the tidal volume (the amount of air moving into or out of the lungs in one respiratory cycle), and this proportion helps prevent rapid changes in the P_{O_2} and P_{CO_2} of alveolar air. In infants, as soon as the respiratory rate changes, the alveolar air composition changes. Because changes in the respiratory rate can cause sudden changes in the P_{O_2} and P_{CO_2} of arterial blood, newborns are at greater risk of developing respiratory acidosis or respiratory alkalosis.

Many newborn infants have some degree of respiratory acidosis caused by the interruption of placental circulation during labor. This condition is known as *fetal stress*. It is therefore important that the newborn begin breathing as soon as possible after delivery so that the excess CO_2 can be eliminated. Most full-term newborns are given oxygen briefly to improve blood oxygenation and to help them eliminate excess CO_2. (Oxygen administration to premature infants can have unwanted side effects.) The pH of infants in fetal stress typically does not fall below 7.25; in most cases, no treatment other than oxygen administration is required.

If the fetus struggles during labor or the delivery is prolonged, *fetal distress* can develop. This condition results from the combination of respiratory acidosis, due to decreased placental circulation, and metabolic acidosis, due to lactic acid production. The pH can fall as low as 7.0, with potentially fatal effects on the heart and central nervous system. Because fetal distress is one of the risks of vaginal delivery, the fetus is commonly monitored during labor and delivery. Early in labor, the heart rate is checked by ultrasound; once delivery is under way, blood samples can be taken from the scalp. If the fetal heart rate becomes too rapid and irregular early in labor, or if the arterial pH drops below 7.2, prompt surgical delivery (by cesarean section) may be recommended. Treatment of a newborn who is in fetal distress typically includes the administration of oxygen and the intravenous infusion of a sodium bicarbonate solution.

Normal body functioning requires fluid, electrolyte, and pH values to remain within narrow ranges. Stresses, including exercise, alter these values in interstitial fluid. Advertisements for sports drinks tout the ability to replace electrolytes after exercise; the body has its own methods for handling small shifts in fluid homeostasis. To understand these methods better, use the resources available on both the CD-ROM and the Companion Website. The CD-ROM includes animations and tutorials; the website builds on those resources with text content and interactive problems as well as communication avenues for sharing ideas with your class. To complete the following Web Explorations, visit the MediaLab in Chapter 27 of your Companion Website.

Web Exploration 1

Estimated time for completion = 10 minutes

Dehydration is a serious concern for people who exercise in hot climates. It is important to be able to recognize the signs and symptoms of dehydration and to know what actions to take should this condition arise. To learn more about the significance of dehydration and of fluid balance, select the keyword **DEHYDRATION** on the Companion Website. After reading the article, take the fluid balance test. Determine your personal fluid needs. Prepare a short essay explaining dehydration. Include your fluid balance calculations.

Web Exploration 2

Estimated time for completion = 5 minutes

Replacement fluids are designed to provide electrolytes and fluid volume to individuals who have lost fluids due to illness. There are five popular brands of replacement fluids; each has a slightly different ion concentration. Select the keyword **REHYDRATION** on the Companion Website. Analyze the chart of ion concentrations by brand. What are the main ions in each fluid? How do the brands differ? Why do sodium and chloride levels vary across the chart while the potassium level is constant? What is the significance of the calcium and magnesium levels in Resol® and of the low levels of citrate in WHO®? You may wish to work with your classmates as a group.

CHAPTER REVIEW

SELECTED CLINICAL TERMINOLOGY
Terms Discussed in This Chapter

hyperkalemia: Plasma K^+ levels above 8 mEq/l. *(p. 995)*
hypernatremia: Plasma Na^+ levels above 150 mEq/l. *(p. 990)*
hypokalemia: Plasma K^+ levels below 2 mEq/l. *(p. 995)*
hyponatremia: Plasma Na^+ levels below 130 mEq/l. *(p. 990)*
metabolic acidosis: A type of acidosis caused by the inability to excrete hydrogen ions, the production of numerous fixed and/or organic acids, or a severe bicarbonate loss. *(p. 1007 and AM)*

metabolic alkalosis: A rare form of alkalosis due to high concentrations of bicarbonate ions in body fluids. *(p. 1008 and AM)*
respiratory acidosis: Acidosis resulting from inadequate respiratory activity, characterized by elevated levels of carbon dioxide (hypercapnia) in body fluids. *(p. 1005 and AM)*
respiratory alkalosis: Alkalosis due to excessive respiratory activity, which depresses carbon dioxide levels and elevates the pH of body fluids. *(p. 1006 and AM)*

27

Additional Terms Discussed in the *Applications Manual*

anion gap: The difference between the concentration of Na^+ and the combined concentrations of HCO_3^- and Cl^-.

nomogram: A graph that can be used to assess acid–base balance in a clinical setting.

STUDY OUTLINE

INTRODUCTION p. 985

1. The maintenance of normal volume and normal composition of extracellular and intracellular fluids is vital to life. Three types of homeostasis are involved: **fluid balance, electrolyte balance,** and **acid–base balance.**

FLUID BALANCE AND ELECTROLYTE BALANCE p. 985

1. The **intracellular fluid (ICF)** contains nearly two-thirds of the total body water; the **extracellular fluid (ECF)** contains the rest. Exchange occurs between the ICF and ECF, but the two **fluid compartments** retain their distinctive characteristics. *(Figures 27-1, 27-2)*

Basic Concepts in the Regulation of Fluids and Electrolytes p. 987

2. Homeostatic mechanisms that monitor and adjust the composition of body fluids respond to changes in the ECF.

3. No receptors directly monitor fluid or electrolyte balance; receptors involved in fluid balance and in electrolyte balance respond to changes in plasma volume or osmotic concentration.

4. Our cells cannot move water molecules by active transport; all water movements across cell membranes and epithelia occur passively in response to osmotic gradients.

5. The body content of water or electrolytes will rise if intake exceeds outflow and will fall if losses exceed gains.

An Overview of the Primary Regulatory Hormones p. 988

6. ADH encourages water reabsorption at the kidneys and stimulates thirst. Aldosterone increases the rates of sodium reabsorption at the kidneys. ANP opposes those actions and promotes fluid and electrolyte losses in urine.

The Interplay between Fluid Balance and Electrolyte Balance p. 989

7. The regulatory mechanisms are quite different between fluid balance and electrolyte balance, and the distinction is important clinically.

Fluid Balance p. 989

8. Water circulates freely within the ECF compartment.

9. Water losses are normally balanced by gains through eating, drinking, and **metabolic generation.** *(Figure 27-3; Table 27-1)*

10. Water movement between the ECF and ICF is called a **fluid shift.** If the ECF becomes hypertonic relative to the ICF, water will move from the ICF into the ECF until osmotic equilibrium has been restored. If the ECF becomes hypotonic relative to the ICF, water will move from the ECF into the cells, and the volume of the ICF will increase.

Electrolyte Balance p. 991

11. Problems with electrolyte balance generally result from an imbalance between sodium gains and losses. Problems with potassium balance are less common but more dangerous.

12. The rate of sodium uptake across the digestive epithelium is directly proportional to the amount of sodium in the diet. Sodium losses occur mainly in urine and through perspiration. *(Figure 27-4)*

13. Shifts in sodium balance result in the expansion or contraction of the ECF. Large variations in the ECF volume are corrected by the homeostatic mechanisms triggered by changes in blood volume. If the volume becomes too low, ADH and aldosterone are secreted; if the volume becomes too high, ANP is secreted. *(Figure 27-5)*

14. Potassium ion concentrations in the ECF are very low and not as closely regulated as are sodium ion concentrations. Potassium excretion increases as ECF concentrations rise, under aldosterone stimulation, and when the pH rises. Potassium retention occurs when the pH falls.

15. The ECF concentrations of other electrolytes such as calcium, magnesium, phosphate, and chloride are also regulated. *(Table 27-2)*

ACID–BASE BALANCE p. 997

1. Acids and bases are either *strong* or *weak. (Table 27-3)*

The Importance of pH Control p. 997

2. The pH of normal body fluids ranges from 7.35 to 7.45; variations outside this relatively narrow range produce symptoms of **acidosis** or **alkalosis.**

Types of Acids in the Body p. 998

3. **Volatile acids** can leave solution and enter the atmosphere; **fixed acids** remain in body fluids until excreted at the kidneys; **organic acids** are acid participants in or byproducts of aerobic metabolism.

4. Carbonic acid is the most important factor affecting the pH of the ECF. In solution, CO_2 reacts with water to form carbonic acid. An inverse relationship exists between pH and the concentration of CO_2. *(Figure 27-6)*

5. Sulfuric acid and phosphoric acid, the most important fixed acids, are generated during the catabolism of amino acids and compounds containing phosphate groups.

6. Organic acids include metabolic products such as lactic acid and ketone bodies.

Mechanisms of pH Control p. 998

7. A **buffer system** typically consists of a weak acid and the anion released by its dissociation, which functions as a weak base. The three major buffer systems are (1) **protein buffer systems** in the ECF and ICF; (2) the **carbonic acid–bicarbonate buffer system,** most important in the ECF; and (3) the **phosphate buffer system** in the ICF and urine. *(Figure 27-7)*

8. In protein buffer systems, the component amino acids respond to changes in pH by accepting or releasing hydrogen ions. The **hemoglobin buffer system** is a protein buffer system that helps prevent drastic changes in pH when the P_{CO_2} is rising or falling. *(Figure 27-8)*

9. The carbonic acid–bicarbonate buffer system prevents pH changes caused by organic acids and fixed acids in the ECF. The readily available supply of bicarbonate ions is the **bicarbonate reserve.** *(Figure 27-9)*

10. The phosphate buffer system plays a supporting role in regulating the pH of the ECF but is important in buffering the pH of the ICF and of urine.

Maintenance of Acid–Base Balance p. 1001

11. The lungs help regulate pH by affecting the carbonic acid–bicarbonate buffer system. Changing the respiratory rate can raise or lower the P_{CO_2} of body fluids, affecting the buffering capacity. This process is **respiratory compensation.**

12. In **renal compensation,** the kidneys vary their rates of hydrogen ion secretion and bicarbonate ion reabsorption, depending on the pH of the ECF. *(Figure 27-10)*

DISTURBANCES OF ACID–BALANCE BALANCE p. 1004

1. Interactions among buffer systems, respiration, and renal function normally maintain the pH of the ECF with very tight control: generally within a pH range of 7.35–7.45. *(Figure 27-11)*

2. **Respiratory acid–base disorders** result when abnormal respiratory function causes an extreme rise or fall in CO_2 levels in the ECF. **Metabolic acid–base disorders** are caused by the generation of organic or fixed acids or by conditions affecting the concentration of bicarbonate ions in the ECF.

Respiratory Acidosis p. 1005

3. **Respiratory acidosis** results from excessive levels of CO_2 in body fluids. *(Figure 27-12)*

Respiratory Alkalosis p. 1006

4. **Respiratory alkalosis** is a relatively rare condition associated with hyperventilation. *(Figure 27-12)*

Metabolic Acidosis p. 1007

5. **Metabolic acidosis** results from the depletion of the bicarbonate reserve. It is caused by an inability to excrete hydrogen ions at the kidneys, the production of large numbers of fixed and organic acids, or the bicarbonate loss that accompanies chronic diarrhea. *(Figure 27-13)*

Metabolic Alkalosis p. 1008

6. **Metabolic alkalosis** occurs when bicarbonate ion concentrations become elevated, as from extended periods of vomiting. *(Figure 27-14)*

The Detection of Acidosis and Alkalosis p. 1008

7. Standard diagnostic blood tests such as blood pH, P_{CO_2}, and bicarbonate levels are used to recognize and classify acidosis and alkalosis conditions as respiratory or metabolic in nature. *(Figure 27-15; Table 27-4)*

AGING AND FLUID, ELECTROLYTE, AND ACID–BASE BALANCE p. 1009

1. Changes affecting fluid, electrolyte, and acid–base balance in the elderly include (1) reduced total body water content, (2) impaired ability to perform renal compensation, (3) increased water demands due to reduced ability to concentrate urine and reduced sensitivity to ADH and aldosterone, (4) a net loss of minerals, (5) reductions in respiratory efficiency that affect the ability to perform respiratory compensation, and (6) increased incidence of conditions that secondarily affect fluid, electrolyte, or acid–base balance.

REVIEW QUESTIONS

LEVEL 1 REVIEWING FACTS AND TERMS

Match each numbered item with the most closely related lettered item. Use letters for answers in the spaces provided.

___ 1. ECF
___ 2. ADH
___ 3. aldosterone
___ 4. ANP
___ 5. oxidative phosphorylation
___ 6. overhydration
___ 7. dehydration
___ 8. sodium
___ 9. potassium
___ 10. carotid sinus
___ 11. calcitonin
___ 12. hydrochloric acid
___ 13. carbonic acid
___ 14. carbon dioxide
___ 15. fixed acid
___ 16. organic acid
___ 17. hypoventilation
___ 18. hyperventilation
___ 19. elevated bicarbonate ion concentration
___ 20. severe bicarbonate loss

a. reduces thirst
b. dominant cation in ECF
c. hyponatremia
d. weak acid
e. baroreceptors
f. promotes calcium loss
g. respiratory alkalosis
h. stimulates water conservation
i. sulfuric acid
j. volatile acid
k. ketone bodies
l. interstitial fluid
m. respiratory acidosis
n. metabolic acidosis
o. hypernatremia
p. metabolic alkalosis
q. dominant cation in ICF
r. conservation of sodium ions
s. strong acid
t. metabolic generation of water

21. A person is in fluid balance when
  (a) the ECF and ICF are isotonic
  (b) no fluid moves between compartments
  (c) the amount of water gained each day is equal to the amount lost to the environment
  (d) a, b, and c are correct

22. The primary components of the extracellular fluid are
  (a) lymph and cerebrospinal fluid
  (b) plasma and serous fluids
  (c) interstitial fluid and plasma
  (d) a, b, and c are correct

23. The principal ions in the ECF are
  (a) sodium and chloride
  (b) sodium and potassium
  (c) phosphate and chloride
  (d) phosphate and bicarbonate

24. All the homeostatic mechanisms that monitor and adjust the composition of body fluids respond to changes
  (a) in the ICF   (b) in the ECF
  (c) inside the cell   (d) a, b, and c are correct

25. Homeostatic adjustments to monitor fluid and electrolyte balance occur in response to changes in
  (a) the pH of body fluids
  (b) autonomic nervous system activity
  (c) the concentration of dissolved gases
  (d) plasma volume or osmotic concentration

26. A rise in body temperature accompanying a fever
  (a) decreases water losses
  (b) increases water losses
  (c) does not affect water gain or loss
  (d) causes the body to retain water

27

27. The most common problems with electrolyte balance are caused by an imbalance between gains and losses of
(a) calcium ions   (b) chloride ions
(c) potassium ions   (d) sodium ions

28. Angiotensin II produces a coordinated elevation in the ECF volume by
(a) stimulating thirst
(b) causing the release of ADH
(c) triggering production and secretion of aldosterone
(d) a, b, and c are correct

29. The rate of tubular secretion of potassium ions changes in response to
(a) changes in pH
(b) changes in aldosterone levels
(c) changes in the potassium ion concentration in the ECF
(d) a, b, and c are correct

30. Respiratory acidosis develops when plasma pH is
(a) elevated due to a decreased plasma P_{CO_2}
(b) decreased due to an elevated plasma P_{CO_2}
(c) elevated due to an elevated plasma P_{CO_2}
(d) decreased due to a decreased plasma P_{CO_2}

31. Metabolic alkalosis occurs when
(a) bicarbonate ion concentrations become elevated
(b) there is a severe bicarbonate loss
(c) the kidneys fail to excrete hydrogen ions
(d) ketone bodies are generated in abnormally large quantities

32. What are fluid shifts? What is their function, and what factors can cause them?

33. Which three major hormones mediate major physiological adjustments that affect fluid and electrolyte balance? What are the primary effects of each hormone?

34. What influence does aldosterone have on the relationship between sodium and potassium?

35. Define and give an example of (a) a volatile acid, (b) a fixed acid, and (c) an organic acid. Which represent(s) the greatest threat to acid–base balance? Why?

LEVEL 2 REVIEWING CONCEPTS

36. The higher the plasma concentration of aldosterone, the more efficiently the kidney will
(a) conserve sodium ions
(b) retain potassium ions
(c) stimulate urinary water loss
(d) secrete greater amounts of ADH

37. If the ECF is hypertonic with respect to the ICF, _____ until osmotic equilibrium is reached.
(a) water will move from the ECF into the cells
(b) ions will move from the ECF into the cells
(c) ions will move from the cells into the ICF
(d) water will move from the cells into the ECF

38. The osmotic concentration of the ECF decreases if the individual gains water without a corresponding
(a) gain of electrolytes
(b) loss of water
(c) fluid shift from the ECF to the ICF
(d) a, b, and c are correct

39. Chronic renal failure and congestive heart failure ultimately result in
(a) hypernatremia   (b) hyponatremia
(c) hyperkalemia   (d) hypokalemia

40. When pure water is consumed,
(a) the ECF becomes hypertonic with respect to the ICF
(b) the ECF becomes hypotonic with respect to the ICF
(c) the ICF becomes hypotonic with respect to the plasma
(d) water moves from the ICF into the ECF

41. When the plasma concentration of potassium decreases to a dangerous level, the result is
(a) kidney failure
(b) extensive muscular weakness, followed by paralysis
(c) decreased blood-clotting capability
(d) the cessation of enzyme activity in cells

42. In a protein buffer system, if the pH rises,
(a) the protein acquires a hydrogen ion from carbonic acid
(b) hydrogen ions are buffered by hemoglobin molecules
(c) a hydrogen ion is released, and a carboxyl ion is formed
(d) a chloride shift occurs

43. Increasing or decreasing the rate of respiration alters the pH by
(a) lowering or raising the partial pressure of CO_2
(b) lowering or raising the partial pressure of O_2
(c) lowering or raising the partial pressure of N_2
(d) a, b, and c are correct

44. Differentiate among fluid balance, electrolyte balance, and acid–base balance, and explain why each is important to homeostasis.

45. A sample of plasma contains chloride ions at a concentration of 250 mg/dl. How many mmol/l of chloride ions are in the sample of plasma?

46. Why should a person with a fever drink plenty of fluids?

47. Describe the effects of losses and gains of sodium on the ECF.

48. What are the three major buffer systems in body fluids? How does each system work?

49. How do respiratory and renal mechanisms support the buffer systems?

50. Differentiate between respiratory compensation and renal compensation.

51. Distinguish between respiratory and metabolic disorders that disturb acid–base balance.

52. What is the difference between metabolic alkalosis and respiratory alkalosis? What can cause these conditions?

53. The most recent advice from medical and nutritional experts is to decrease the intake of salt so that it does not exceed the amount needed to maintain a constant ECF composition. What effect does excessive salt and water ingestion have on (a) urine volume, (b) urine concentration, and (c) blood pressure?

54. Exercise physiologists recommend that adequate amounts of fluid be ingested before, during, and after exercise. Why is fluid replacement during extensive sweating important?

LEVEL 3 CRITICAL THINKING AND CLINICAL APPLICATIONS

55. After falling into an abandoned stone quarry filled with water and nearly drowning, a young boy is rescued. His rescuers assess his condition. They find that his body fluids have high P_{CO_2} and low P_{O_2}, and that large amounts of lactic acid had been generated by the boy's muscles as he struggled in the water. As a clinician, diagnose the boy's condition and recommend the necessary treatment to restore his body to homeostasis.

56. A person who is hyperventilating and disoriented is brought into the emergency room. Blood analysis indicates that the person has elevated levels of potassium (hyperkalemia). Is the most likely cause acidosis or alkalosis? Explain.

57. Willy has chronic emphysema. A blood analysis indicates that his plasma pH is 7.4, although his plasma bicarbonate level is significantly elevated. How can this be?

58. While visiting a foreign country, Milly inadvertently drinks the water, even though she had been advised not to. She contracts an intestinal disease that causes severe diarrhea. How would you expect her condition to affect her blood pH, urine pH, and pattern of ventilation?

59. Yuka is dehydrated, so her physician prescribes intravenous fluids. The attending nurse becomes distracted and erroneously gives Yuka a hypertonic glucose solution instead of normal saline. What effect will this mistake have on Yuka's plasma levels of ADH and urine volume?

60. Refer to the diagnostic flow chart in Figure 27-15•. Use information from the blood test results in the accompanying table

to categorize the acid–base disorders that affect the patients represented there.

Results	Patient 1	Patient 2	Patient 3	Patient 4
pH	7.5	7.2	7.0	7.7
P_{CO_2}	32	45	60	50
Na^+	138	140	140	136
HCO_3^-	22	20	28	34
Cl^-	106	102	101	91
Anion gap*	10	18	12	11

* Anion gap = Na^+ concentration − (HCO_3^- concentration + Cl^- concentration). Refer to the *Applications Manual* for additional information.   AM   *Diagnostic Classification of Acid–Base Disorders*

•**FIGURE 27-15** **A Diagnostic Chart for Acid–Base Disorders**

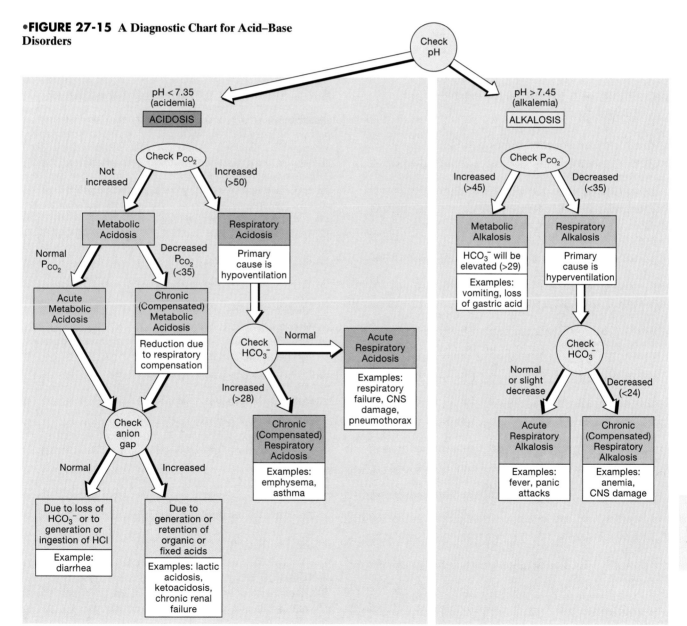

27

CHAPTER 28

The Reproductive System

CHAPTER OUTLINE AND OBJECTIVES

INTRODUCTION   p. 1017

AN OVERVIEW OF THE REPRODUCTIVE SYSTEM   p. 1017
① Summarize the functions of the human reproductive system and its principal components.

THE REPRODUCTIVE SYSTEM OF THE MALE   p. 1017
The Testes   p. 1017
The Anatomy of a Spermatozoon   p. 1025
The Male Reproductive Tract   p. 1025
The Accessory Glands   p. 1028
Semen   p. 1029
The Penis   p. 1030
Hormones and Male Reproductive Function   p. 1032
② Describe the components of the male reproductive system.
③ Discuss the processes of meiosis and spermatogenesis in the testes.
④ Describe the roles the male reproductive tract and accessory glands play in the functional maturation, nourishment, storage, and transport of spermatozoa.
⑤ Discuss the normal composition of semen.
⑥ Describe the male external genitalia.
⑦ Describe the hormonal mechanisms that regulate male reproductive functions.

THE REPRODUCTIVE SYSTEM OF THE FEMALE   p. 1034
The Ovaries   p. 1035
The Uterine Tubes   p. 1039
The Uterus   p. 1040
The Vagina   p. 1045
The External Genitalia   p. 1046
The Mammary Glands   p. 1047
Hormones and the Female Reproductive Cycle   p. 1049
⑧ Describe the components of the female reproductive system.
⑨ Discuss the processes of meiosis and oogenesis in the ovaries.
⑩ Identify the phases and events of the ovarian and uterine cycles.
⑪ Describe the structure, histology, and functions of the vagina.
⑫ Name and describe the parts of the female external genitalia and mammary glands.
⑬ Describe the anatomical, physiological, and hormonal aspects of the female reproductive cycle.

THE PHYSIOLOGY OF SEXUAL INTERCOURSE   p. 1053
Male Sexual Function   p. 1053
Female Sexual Function   p. 1056
⑭ Discuss the physiology of sexual intercourse as it affects the reproductive systems of males and females.

AGING AND THE REPRODUCTIVE SYSTEM   p. 1056
Menopause   p. 1056
The Male Climacteric   p. 1057
⑮ Describe the changes in the reproductive system that occur with aging.

INTEGRATION WITH OTHER SYSTEMS   p. 1057
⑯ Give examples of interactions between the reproductive system and each of the other organ systems.

MEDIALAB and CHAPTER REVIEW   p. 1059

In a sense, our reproductive systems do not exist for us as individuals; they exist for the human race. But how can we measure the value to us—to the part of us that is not just cells and tissues and organs—of the ability to have children? Of contributing something to the next generation and, in the process, making a little part of ourselves live on? In this chapter, we shall study the system that has enabled humans to exist on Earth for a very long time.

The reproductive system is the only system that is not essential to the life of the individual, although its activities do impact other systems. Instead, this system ensures the continued existence of the human species. Many primitive societies failed to discover the basic link between sexual activity and childbirth and assumed that cosmic forces are what produces new individuals. Although our society has a much clearer view of the reproductive process, a sense of wonder remains. Sexually mature males and females produce individual reproductive cells that are brought together by the sex act. The fusion of these cells starts a chain of events that leads to the appearance of an infant who will mature as part of the next generation.

In this and the next chapter, we shall consider the mechanics of this remarkable process. We begin by examining the anatomy and physiology of the reproductive system in this chapter. The human reproductive system produces, stores, nourishes, and transports functional male and female reproductive cells, or **gametes** (GAM-ēts). Chapter 29 begins with **fertilization,** also known as *conception,* in which a male gamete and female gamete unite. All the cells in a human body are the mitotic descendants of a **zygote** (ZĪ-gōt), the single cell created by the fusion of a gamete from the father and a gamete from the mother. The gradual transformation of that single cell into a functional adult, over a span of 15–20 years, is the process of *development.* During this period, hormones produced by the reproductive system direct gender-specific patterns of development.

AN OVERVIEW OF THE REPRODUCTIVE SYSTEM

The **reproductive system** includes the following:

- **Gonads** (GŌ-nadz), or reproductive organs that produce gametes and hormones.
- Ducts that receive and transport the gametes.
- Accessory glands and organs that secrete fluids into the ducts of the reproductive system or into other excretory ducts.
- Perineal structures, associated with the reproductive system, that are collectively known as the **external genitalia** (jen-i-TĀ-lē-uh).

In both males and females, the ducts are connected to chambers and passageways that open to the exterior of the body. The structures involved constitute the *reproductive tract.* The male and female reproductive systems are functionally quite different, however. In adult males, the **testes** (TES-tēz; singular, *testis*), or male gonads, secrete sex hormones called *androgens,* (principally *testosterone*). (We introduced testosterone and other sex hormones in Chapter 18.) *pp. 608–609* The testes also produce the male

gametes, called **spermatozoa** (sper-ma-tō-ZŌ-uh; singular, *spermatozoon*), or *sperm*—one-half billion each day. During *emission,* mature spermatozoa travel along a lengthy duct system, where they are mixed with the secretions of accessory glands. The mixture created is known as **semen** (SĒ-men). During *ejaculation,* semen is expelled from the body.

In adult females, the **ovaries,** or female gonads, typically release only one immature gamete, an **oocyte,** per month. This immature gamete travels along short *uterine tubes,* which end in the muscular organ called the *uterus* (Ū-ter-us). If a sperm reaches the occyte and initiates the process of fertilization, the oocyte matures into an **ovum** (plural, *ova*). A short passageway, the *vagina* (va-JĪ-nuh), connects the uterus with the exterior. During *sexual intercourse,* ejaculation introduces semen into the vagina, and the spermatozoa then ascend the female reproductive tract. If fertilization occurs, the uterus will enclose and support a developing *embryo* as the embryo grows into a *fetus* and prepares for birth.

We will now examine the anatomy of the male and female reproductive systems further and will consider the physiological and hormonal mechanisms responsible for the regulation of reproductive function. In earlier chapters, we introduced the anatomical reference points used in the following discussions; you may find it helpful to review the figures on the pelvic girdle (Figures 8-8 and 8-10•, pp. 235, 236), perineal musculature (Figure 11-13•, p. 333), pelvic innervation (Figure 13-12•, p. 421), and regional blood supply (Figures 21-28 and 21-33•, pp. 733, 740).

THE REPRODUCTIVE SYSTEM OF THE MALE

The principal structures of the male reproductive system are shown in Figure 28-1•. Proceeding from a testis, the spermatozoa travel within the *epididymis* (ep-i-DID-i-mus); the *ductus deferens* (DUK-tus DEF-e-renz), or *vas deferens;* the *ejaculatory duct;* and the *urethra* before leaving the body. Accessory organs—the *seminal* (SEM-i-nal) *vesicles,* the *prostate* (PROS-tāt) *gland,* and the *bulbourethral* (bul-bō-ū-RĒ-thral) *glands*—secrete into the ejaculatory ducts and urethra. The external genitalia consist of the *scrotum* (SKRŌ-tum), which encloses the testes, and the *penis* (PĒ-nis), an erectile organ through which the distal portion of the urethra passes.

THE TESTES

Each testis has the shape of a flattened egg that is roughly 5 cm (2 in.) long, 3 cm (1.2 in.) wide, and 2.5 cm (1 in.) thick. Each has a weight of 10–15 g (0.35–0.53 oz). The testes hang within the **scrotum,** a fleshy pouch suspended inferior to the perineum,

28

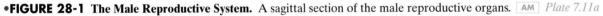

•**FIGURE 28-1** **The Male Reproductive System.** A sagittal section of the male reproductive organs. AM *Plate 7.11a*

anterior to the anus and posterior to the base of the penis (Figure 28-1). AM *EMBRYOLOGY SUMMARY 21: The Development of the Reproductive System*

Descent of the Testes

During development, the testes form inside the body cavity adjacent to the kidneys. A bundle of connective tissue fibers extends from each testis to the posterior wall of a small, anterior and inferior pocket of the peritoneum. These fibers constitute the **gubernaculum testis.** As growth proceeds, the gubernacula do not get any longer, so they lock the testes in position. As a result, the relative position of each testis changes as the rest of the body enlarges: The testis gradually moves inferiorly and anteriorly toward the anterior abdominal wall (Figure 28-2a•). During the seventh developmental month, fetal growth continues at a rapid pace, and circulating hormones stimulate a contraction of the gubernaculum testis. Over this period, each testis moves through the abdominal musculature, accompanied by small pockets of the peritoneal cavity. This process is called the **descent of the testes.**

As it moves through the body wall, each testis is accompanied by the ductus deferens and the testicular blood vessels, nerves, and lymphatic vessels. Together, these structures form the body of the spermatic cord, which we shall discuss next.

Cryptorchidism

In **cryptorchidism** (kript-OR-ki-dizm; *crypto*, hidden), one or both of the testes have not descended into the scrotum by the time of birth. Typically, the cryptorchid testes are lodged in the abdominal cavity or within the inguinal canal. Cryptorchidism occurs in about 3 percent of full-term deliveries and in roughly 30 percent of premature births. In most instances, normal descent occurs a few weeks later, but the condition can be surgically corrected if it persists. Corrective measures should be taken before *puberty* (sexual maturation), because cryptorchid (abdominal) testes will not produce spermatozoa. Thus the individual will be *sterile (infertile),* or unable to bear children. In most cases, if the testes cannot be moved into the scrotum, they will be removed, because about 10 percent of males with uncorrected cryptorchid testes eventually develop testicular cancer. This surgical procedure is called a *bilateral orchiectomy* (or-kē-EK-to-mē; *orchis,* testis).

The Spermatic Cords

The **spermatic cords** are paired structures extending between the abdominopelvic cavity and the testes. Each spermatic cord consists of layers of fascia and muscle enclosing the ductus deferens and the blood vessels, nerves, and lymphatic vessels that supply the testes. The blood vessels include the *deferential artery,*

•**FIGURE 28-2** **The Descent of the Testes.** **(a)** Sagittal sectional views of the positional changes involved in the descent of the testes. Because the size of the gubernaculum testis remains constant (see the scale bar at the left) while the rest of the fetus grows, the relative position of the testis shifts. **(b)** Frontal views during the descent of the testes and the formation of the spermatic cords.

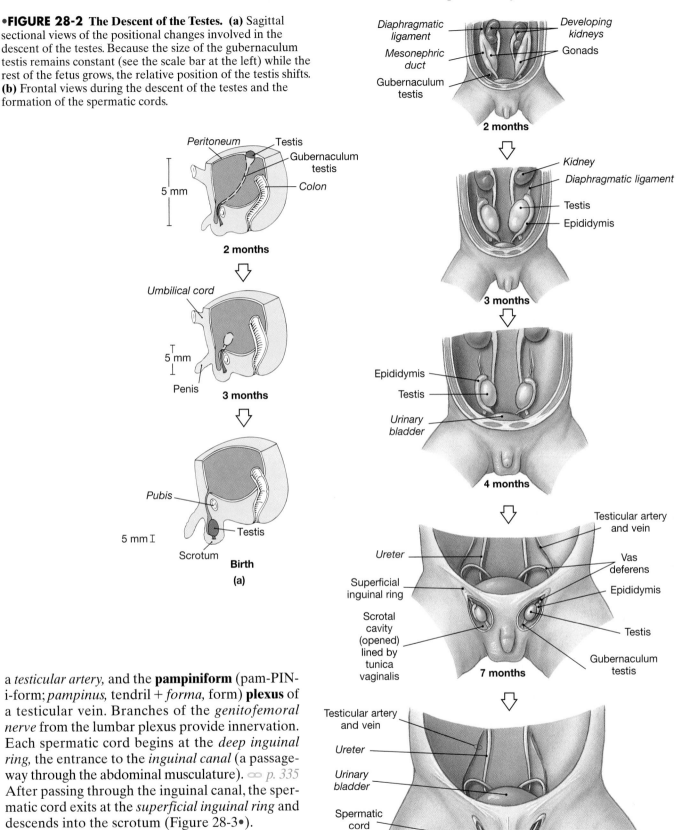

a *testicular artery,* and the **pampiniform** (pam-PIN-i-form; *pampinus,* tendril + *forma,* form) **plexus** of a testicular vein. Branches of the *genitofemoral nerve* from the lumbar plexus provide innervation. Each spermatic cord begins at the *deep inguinal ring,* the entrance to the *inguinal canal* (a passageway through the abdominal musculature). ∞ *p. 335* After passing through the inguinal canal, the spermatic cord exits at the *superficial inguinal ring* and descends into the scrotum (Figure 28-3•).

The inguinal canals form during development as the testes descend into the scrotum; at that time, these canals link the scrotal cavities with the peritoneal cavity. In normal adult males, the inguinal canals are closed, but the presence of the spermatic cords creates weak points in the abdominal wall that remain

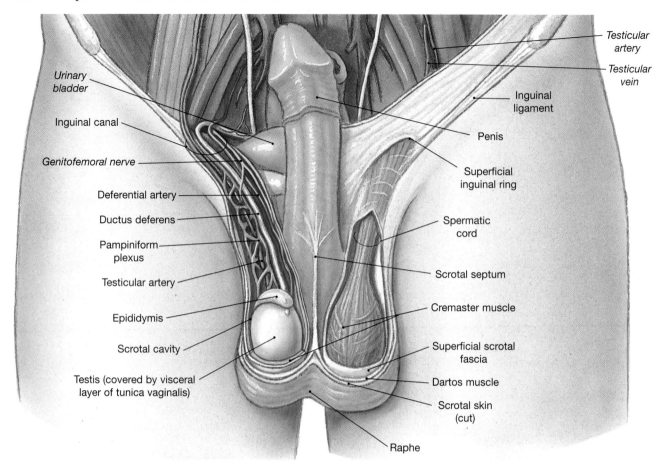

Urinary bladder

Inguinal canal

Genitofemoral nerve

Deferential artery

Ductus deferens

Pampiniform plexus

Testicular artery

Epididymis

Scrotal cavity

Testis (covered by visceral layer of tunica vaginalis)

Testicular artery

Testicular vein

Inguinal ligament

Penis

Superficial inguinal ring

Spermatic cord

Scrotal septum

Cremaster muscle

Superficial scrotal fascia

Dartos muscle

Scrotal skin (cut)

Raphe

•**FIGURE 28-3** The Male Reproductive System in Frontal View

throughout life. As a result, *inguinal hernias,* protrusions of a portion of the intestine into the inguinal canal, are relatively common in males. ∞ *p. 335* The inguinal canals in females are very small, containing only the *ilioinguinal nerves* and the *round ligaments* of the uterus. The abdominal wall is nearly intact, so inguinal hernias in women are very rare.

The Scrotum and the Position of the Testes

The scrotum is divided internally into two chambers. The partition between the two is marked by a raised thickening in the scrotal surface known as the **raphe** (RĀ-fē) (Figure 28-3•). Each testis lies in a separate chamber, or **scrotal cavity.** Because the scrotal cavities are separated by a partition, infection or inflammation of one testis does not ordinarily spread to the other. A narrow space separates the inner surface of the scrotum from the outer surface of the testis. The **tunica vaginalis** (TOO-ni-ka vaj-i-NAL-is), a serous membrane, lines the scrotal cavity and reduces friction between the opposing parietal (scrotal) and visceral (testicular) surfaces. The tunica vaginalis is an isolated portion of the peritoneum that lost its connection with the peritoneal cavity after the testes descended, when the inguinal canal closed.

The scrotum consists of a thin layer of skin and the underlying superficial fascia. The dermis contains a layer of smooth muscle, the **dartos** (DAR-tōs) **muscle.** Resting muscle tone in the dartos muscle causes the characteristic wrinkling of the scrotal surface. A layer of skeletal muscle, the **cremaster** (krē-MAS-ter) **muscle,** lies deep to the dermis. Contraction of the cremaster during sexual arousal, or in response to changes in temperature, tenses the scrotum and pulls the testes closer to the body. Normal spermatozoa development in the testes requires temperatures about 1.1°C (2°F) lower than those elsewhere in the body. The cremaster muscle relaxes or contracts to move the testes away from or toward the body as needed to maintain acceptable testicular temperatures. When air or body temperature rises, the cremaster muscle relaxes and the testes move away from the body. Sudden cooling of the scrotum, as occurs when a man enters a cold swimming pool, results in cremasteric contractions that pull the testes closer to the body and keep testicular temperatures from falling.

28

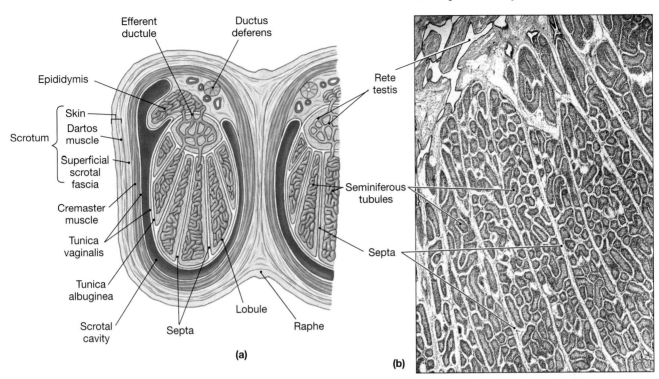

•**FIGURE 28-4** The Structure of the Testes. **(a)** A frontal section. **(b)** A section through a testis. (LM × 26)

The Structure of the Testes

Deep to the tunica vaginalis covering the testis is the **tunica albuginea** (al-bū-JIN-ē-uh), a dense layer of connective tissue rich in collagen fibers. These fibers are continuous with those surrounding the adjacent epididymis. The collagen fibers of the tunica albuginea also extend into the substance of the testis, forming fibrous partitions, or *septa* (Figure 28-4•). These septa converge toward the region nearest the entrance to the epididymis. The connective tissues in this region support the blood vessels and lymphatic vessels that supply the testis and the *efferent ductules,* which transport spermatozoa to the epididymis.

Histology of the Testes

The septa subdivide the testis into a series of **lobules.** Roughly 800 slender, tightly coiled **seminiferous** (se-mi-NIF-e-rus) **tubules** are distributed among the lobules (Figures 28-4 and 28-5•). Each tubule averages about 80 cm (31 in.) in length, and a typical testis contains nearly one-half mile of seminiferous tubules. Sperm production occurs within these tubules.

Each seminiferous tubule forms a loop that is connected to a maze of passageways known as the **rete** (RĒ-tē; *rete,* a net) **testis** (Figure 28-4•). Fifteen to twenty large **efferent ductules** connect the rete testis to the epididymis.

Because the seminiferous tubules are tightly coiled, most histological preparations show them in transverse section (Figure 28-5a•). Each tubule is surrounded by a delicate capsule, and areolar tissue fills the spaces between the tubules. Within those spaces are numerous blood vessels and large **interstitial cells** *(cells of Leydig)* (Figure 28-5b•). Interstitial cells are responsible for the production of *androgens,* the dominant sex hormones in males. *Testosterone* is the most important androgen.

Spermatozoa are produced by the process of **spermatogenesis** (sper-ma-tō-JEN-e-sis). Spermatogenesis begins at the outermost layer of cells in the seminiferous tubules and proceeds toward the lumen (Figure 28-5b,c•). At each step in this process, the daughter cells move closer to the lumen. First, stem cells called **spermatogonia** (sper-ma-tō-GŌ-nē-uh) divide by mitosis to produce daughter cells, some of which differentiate into primary spermatocytes. **Primary spermatocytes** (sper-MA-tō-sīts) are the cells that begin *meiosis,* a specialized form of cell division involved only in the production of gametes (spermatozoa in males, ova in females). Spermatocytes give rise to spermatids. **Spermatids** (SPER-ma-tidz) are immature gametes that subsequently differentiate into spermatozoa. The spermatozoa lose contact with the wall of the seminiferous tubule and enter the fluid in the lumen.

Each seminiferous tubule contains spermatogonia, spermatocytes at various stages of meiosis, spermatids, spermatozoa, and large **sustentacular** (sus-ten-TAK-ū-lar) **cells** (or *Sertoli cells*). Sustentacular cells are attached to the tubular capsule and extend toward the lumen between the other cell types (Figure 28-5b,c•).

28

•**FIGURE 28-5** **The Seminiferous Tubules.** **(a)** A transverse section through a coiled seminiferous tubule. **(b)** A cross section through a single tubule. (LM × 786) **(c)** The lining of a tubule. Sustentacular cells surround the stem cells of the tubule and support the developing spermatocytes and spermatids.

Seminiferous tubule containing spermatids completing spermiogenesis

Seminiferous tubule containing spermatids beginning spermiogenesis

Seminiferous tubules containing spermatozoa about to be released into the lumen

(a)

LUMEN

Spermatids completing spermiogenesis

Initial spermiogenesis

Secondary spermatocyte in meiosis II

Level of blood–testis barrier

Fibroblast

Connective tissue capsule

Interstitial cells

Spermatids beginning spermiogenesis

Sustentacular cells

Secondary spermatocyte

Primary spermatocyte preparing for meiosis I

Capillary

Spermatogonium

(c)

Interstitial cells

Spermatogonia

Dividing spermatocytes

Spermatids

Sustentacular cells

Lumen

Connective tissue capsule

Heads of maturing spermatozoa

(b)

Dividing spermatocytes

Sustentacular cell

Capillary

Lumen

Spermatozoa

Interstitial cells

Spermatogonium

Spermatogenesis

Spermatogenesis involves three integrated processes:

1. *Mitosis.* Spermatogonia undergo cell divisions throughout adult life. (You may wish to review the description of mitosis and cell division in Chapter 3. ∞ *pp. 98–99*) One daughter cell from each division is pushed toward the lumen of the seminiferous tubule.

These cells differentiate into primary spermatocytes, which prepare to begin meiosis.

2. *Meiosis.* **Meiosis** (mī-Ō-sis) is a special form of cell division involved in gamete production. Gametes contain half the normal set of chromosomes. As a result, the fusion of the nuclei of a male gamete and a female gamete produces a cell that has the normal number of chromosomes (46) rather than twice that number.

In the seminiferous tubules, meiotic divisions that begin with primary spermatocytes produce spermatids, undifferentiated male gametes.

3. *Spermiogenesis.* Spermatids are small, relatively unspecialized cells. In **spermiogenesis,** spermatids differentiate into physically mature spermatozoa. Spermatozoa are among the most highly specialized cells in the body. Spermiogenesis involves major changes in a spermatid's internal and external structures.

Mitosis and Meiosis In both males and females, mitosis and meiosis differ significantly in terms of the events that take place in the nucleus. Mitosis is part of the process of somatic cell division, which produces two daughter cells, each containing 23 *pairs* of chromosomes (Figure 28-6a•). Each pair consists of one chromosome provided by the father and another by the mother at the time of fertilization. Because the daughter cells contain both members of each chromosome pair (for a total of 46 chromosomes), they are called **diploid** (DIP-loyd; *diplo,* double) cells. Meiosis involves two cycles of cell division (*meiosis I* and *meiosis II*) and produces four cells, each of which contains 23 individual chromosomes (Figure 28-6b•). Because these cells contain only one member of each pair of chromosomes, they are called **haploid** (HAP-loyd; *haplo,* single) cells. The events in the nucleus shown in Figure 28-6b• are the same whether you consider the formation of spermatozoa or ova.

As a cell prepares to begin meiosis, DNA replication occurs within the nucleus as if the cell were about to undergo mitosis. As prophase of the first meiotic division, **meiosis I,** arrives, the chromosomes condense and become visible with a light microscope. As in mitosis, each chromosome consists of two duplicate *chromatids.*

The corresponding maternal and paternal chromosomes now come together, an event known as **synapsis** (sin-AP-sis). Synapsis involves 23 pairs of chromosomes; each member of each pair consists of two chromatids. A matched set of four chromatids is called a **tetrad** (TET-rad; *tetras,* four). Some exchange of genetic material can occur between the chromatids of a chromosome pair at this stage of meiosis. This exchange, called *crossing-over,* increases genetic variation among offspring; we shall discuss it in Chapter 29.

The nuclear envelope disappears at the end of prophase I. As metaphase I begins, the tetrads line up along the metaphase plate. As anaphase I begins, the tetrads break up—the maternal and paternal chromosomes separate. This is a major difference between mitosis and meiosis: In mitosis, each daughter cell receives one of the two copies of every chromosome, maternal and paternal, but in meiosis I each daughter cell receives both copies of *either* the maternal chromosome *or* the paternal chromosome from each tetrad. (Compare the two parts of Figure 28-6•.)

As anaphase proceeds, the maternal and paternal components are randomly and independently distributed. That is, as each tetrad splits, you cannot predict which daughter cell will receive copies of the maternal chromosome rather than copies of the paternal

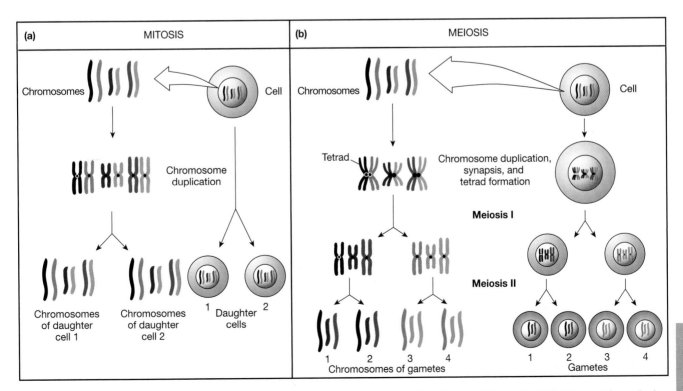

•**FIGURE 28-6 Chromosomes in Mitosis and Meiosis. (a)** Steps in mitosis. (See Figure 3-28, pp. 98–99.) **(b)** Steps in meiosis.

28

chromosome. As a result, telophase I ends with the formation of two daughter cells containing unique combinations of maternal and paternal chromosomes. Both cells contain 23 chromosomes. Because the first meiotic division reduces the number of chromosomes from 46 to 23, it is called a **reductional division.** Each of these chromosomes still consists of two duplicate chromatids. These duplicates will separate during **meiosis II.**

The interphase separating meiosis I and meiosis II is very brief, and no DNA is replicated during this period. The cell then proceeds through prophase II, metaphase II, and anaphase II. During anaphase II, the duplicate chromatids separate. Telophase II thus yields *four cells,* each containing 23 chromosomes. Because the number of chromosomes has not changed, meiosis II is an **equational division.**

In males, the mitotic division of a spermatogonium produces two daughter cells. One is a spermatogonium that remains in contact with the basement membrane, and the other is a primary spermatocyte that is displaced toward the lumen (Figure 28-7•). As meiosis begins, each primary spermatocyte contains 46 individual chromosomes. At the end of meiosis I, the daughter cells are called **secondary spermatocytes.** Every secondary spermatocyte contains 23 chromosomes, each of which consists of a pair of duplicate chromatids. The secondary spermatocytes soon enter prophase II. The completion of metaphase II, anaphase II, and telophase II yields four spermatids, each containing 23 chromosomes.

For each primary spermatocyte that enters meiosis, four spermatids are produced. Because cytokinesis (cytoplasmic division) is not completed in meiosis I or meiosis II, the four spermatids initially remain interconnected by bridges of cytoplasm. These connections assist in the transfer of nutrients and hormones between the cells, thus helping ensure that the cells develop in synchrony. The bridges are not broken until the last stages of physical maturation.

Spermiogenesis   In **spermiogenesis,** the last step of spermatogenesis, each spermatid matures into a single spermatozoon, or sperm (Figure 28-7•). Developing spermatocytes undergoing meiosis and spermatids undergoing spermiogenesis are not free in the seminiferous tubules. Instead, they are surrounded by the cytoplasm of the sustentacular cells. As spermiogenesis proceeds, the spermatids gradually develop the appearance of mature spermatozoa. At *spermiation,* a spermatozoon loses its attachment to the sustentacular cell and enters the lumen of the seminiferous tubule. The entire process, from spermatogonial division to spermiation, takes approximately 9 weeks.

Sustentacular Cells   Sustentacular cells play a key role in spermatogenesis. These cells have six important functions that directly or indirectly affect mitosis, meiosis, and spermiogenesis within the seminiferous tubules:

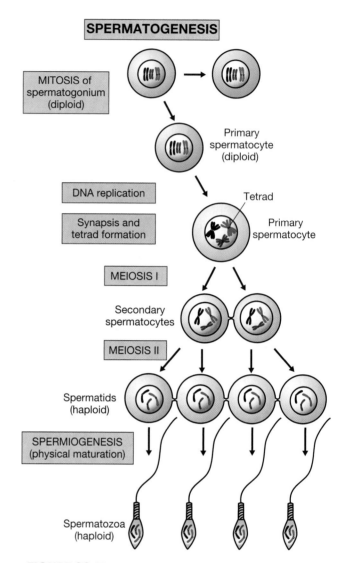

SPERMATOGENESIS

MITOSIS of spermatogonium (diploid)

Primary spermatocyte (diploid)

DNA replication

Synapsis and tetrad formation

Tetrad

Primary spermatocyte

MEIOSIS I

Secondary spermatocytes

MEIOSIS II

Spermatids (haploid)

SPERMIOGENESIS (physical maturation)

Spermatozoa (haploid)

•**FIGURE 28-7** **Spermatogenesis.** Meiosis in the seminiferous tubules, showing the distribution of only a few chromosomes.

1. *The maintenance of the blood–testis barrier.* The seminiferous tubules are isolated from the general circulation by a **blood–testis barrier,** comparable in function to the blood–brain barrier. ∞ *p. 444* Sustentacular cells are joined by tight junctions, forming a layer that divides the seminiferous tubule into an outer *basal compartment,* which contains the spermatogonia, and an inner *lumenal compartment* (or *adlumenal compartment*), where meiosis and spermiogenesis occur. Transport across the sustentacular cells is tightly regulated so that conditions in the lumenal compartment remain very stable. The fluid in the lumen of a seminiferous tubule is produced by sustentacular cells, which also regulate the fluid's composition. This fluid is very different from the surrounding interstitial fluid. For example, the fluid in seminiferous tubules is high in androgens, estrogens, potassium, and amino acids. The blood–testis barrier is essential to preserving the differences between this tubular fluid and interstitial

fluid. In addition, developing spermatozoa contain sperm-specific antigens in their cell membranes. These antigens, not found in somatic cell membranes, would be attacked by the immune system if the blood–testis barrier did not prevent their detection.

2. *The support of mitosis and meiosis.* Spermatogenesis depends on the stimulation of sustentacular cells by circulating follicle-stimulating hormone (FSH) and testosterone. Stimulated sustentacular cells then promote the division of spermatogonia and the meiotic divisions of spermatocytes.

3. *The support of spermiogenesis.* Spermiogenesis requires the presence of sustentacular cells. These cells surround and enfold the spermatids, providing nutrients and chemical stimuli that promote their development.

4. *The secretion of inhibin.* Sustentacular cells secrete *inhibin* (in-HIB-in), a peptide hormone, in response to factors released by developing spermatozoa. Inhibin depresses the pituitary production of FSH and perhaps the hypothalamic secretion of gonadotropin-releasing hormone (GnRH). The faster the rate of sperm production, the more inhibin is secreted. By regulating FSH and GnRH secretion, sustentacular cells provide feedback control of spermatogenesis.

5. *The secretion of androgen-binding protein. Androgen-binding protein (ABP)* binds androgens (primarily testosterone) in the fluid contents of the seminiferous tubules. This protein is thought to be important in elevating the concentration of androgens within the seminiferous tubules and in stimulating spermiogenesis. The production of ABP is stimulated by FSH.

6. *The secretion of Müllerian-inhibiting factor. Müllerian-inhibiting factor (MIF)* is secreted by sustentacular cells in the developing testes. This hormone causes regression of the fetal *Müllerian ducts,* passageways that in females participate in the formation of the uterine tubes and the uterus. In males, inadequate MIF production during development leads to the retention of these ducts and the failure of the testes to descend into the scrotum.

Testicular Cancer

Testicular cancer occurs at a relatively low rate of about 3 cases per 100,000 males per year. Although only about 7200 new cases occur each year in the United States, testicular cancer is the most common cancer among males age 15–35. The incidence among Caucasian-American males has more than doubled since the 1930s, but the incidence among African-American males has remained unchanged. The reason for this difference is not known.

More than 95 percent of testicular cancers are the result of abnormal spermatogonia or spermatocytes rather than abnormal sustentacular cells, interstitial cells, or other cells of the testes. Treatment generally consists of a combination of orchiectomy and chemotherapy. The survival rate has increased from about 10 percent in 1970 to about 95 percent in 1999, primarily as a result of earlier diagnosis and improved treatment protocols.

THE ANATOMY OF A SPERMATOZOON

Each spermatozoon has three distinct regions: (1) the head, (2) the middle piece, and (3) the tail (Figure 28-8●). The **head** is a flattened ellipse containing a nucleus with densely packed chromosomes. At the tip of the head is the **acrosomal** (ak-rō-SŌ-mal) **cap,** a membranous compartment containing enzymes essential to fertilization. During spermiogenesis, saccules of the spermatid's Golgi apparatus fuse and flatten into an *acrosomal vesicle,* which ultimately forms the acrosomal cap of the spermatozoon.

A short **neck** attaches the head to the **middle piece.** The neck contains both centrioles of the original spermatid. The microtubules of the distal centriole are continuous with those of the middle piece and tail. Mitochondria in the middle piece are arranged in a spiral around the microtubules. Mitochondrial activity provides the ATP that is needed to move the tail.

The **tail** is the only flagellum in the human body. A *flagellum,* a whiplike organelle, moves a cell from one place to another. ∞ *p. 83* Whereas cilia beat in a predictable, waving fashion, the flagellum of a spermatozoon has a complex, corkscrew motion.

Unlike other, less specialized cells, a mature spermatozoon lacks an endoplasmic reticulum, Golgi apparatus, lysosomes, peroxisomes, inclusions, and many other intracellular structures. The loss of these organelles reduces the cell's size and mass; it is essentially a mobile carrier for the enclosed chromosomes, and extra weight would slow it down. Because the cell does not contain glycogen or other energy reserves, however, it must absorb nutrients (primarily fructose) from the surrounding fluid.

THE MALE REPRODUCTIVE TRACT

The testes produce physically mature spermatozoa that are, as yet, incapable of successful fertilization of an oocyte. The other portions of the male reproductive system are responsible for the functional maturation, nourishment, storage, and transport of spermatozoa.

The Epididymis

Late in their development, spermatozoa detach from the sustentacular cells and lie within the lumen of the seminiferous tubule. They have most of the physical characteristics of mature spermatozoa but are functionally immature and incapable of coordinated locomotion or fertilization. Fluid currents, created by cilia lining the efferent ductules, transport the immobile gametes into the epididymis (Figure 28-4a●). The **epididymis,** the start of the male reproductive

•**FIGURE 28-8** **Spermiogenesis and Spermatozoon Structure.** **(a)** The differentiation of a spermatid into a spermatozoon. This differentiation process is completed in approximately 5 weeks. **(b)** Human spermatozoa. (SEM × 1688)

tract, is a coiled tube bound to the posterior border of the testis (Figure 28-9a●).

The epididymis can be felt through the skin of the scrotum. Consisting of a tubule almost 7 meters (23 ft) long, the epididymis is coiled and twisted so as to take up very little space. It has (1) a head, (2) a body, and (3) a tail. The superior **head** is the portion of the epididymis proximal to the testis. The head receives spermatozoa from the efferent ductules.

The **body** begins distal to the last efferent ductule and extends inferiorly along the posterior margin of the testis. Near the inferior border of the testis, the number of coils decreases, marking the start of the **tail.** The tail recurves and ascends to its connection with the ductus deferens. Spermatozoa are stored primarily within the tail of the epididymis.

The epididymis has the following three functions:

1. *It monitors and adjusts the composition of the fluid produced by the seminiferous tubules.* The pseudostratified columnar epithelial lining of the epididymis bears distinctive stereocilia, which increase the surface area available for absorption from and secretion into the fluid in the tubule (Figure 28-9b●).

2. *It acts as a recycling center for damaged spermatozoa.* Cellular debris and damaged spermatozoa are absorbed in the epididymis, and the products of enzymatic breakdown are released into the surrounding interstitial fluids for pickup by the epididymal blood vessels.

3. *It stores and protects spermatozoa and facilitates their functional maturation.* A spermatozoon passes through the epididymis in about 2 weeks and completes its functional maturation at this time. Over this period, the spermatozoa exist in a sheltered environment that is precisely regulated by the surrounding epithelial cells. Although spermatozoa leaving the epididymis are

(b)

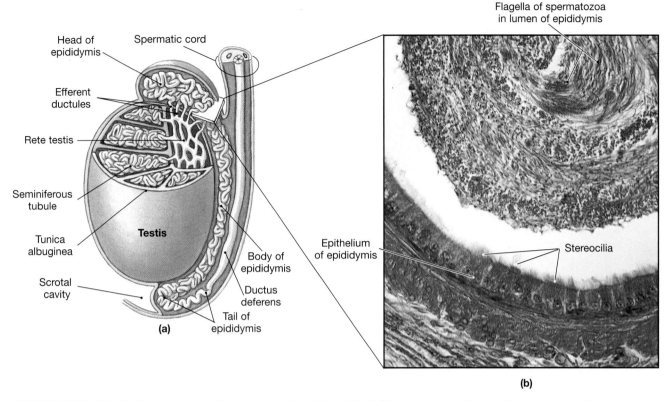

•**FIGURE 28-9 The Epididymis. (a)** A diagrammatic view. **(b)** Epithelial features, especially the elongate stereocilia characteristic of the epididymis. (LM × 1304)

mature, they remain immobile. To become *motile* (actively swimming) and fully functional, spermatozoa must undergo **capacitation.** Capacitation normally occurs in two steps: (1) Spermatozoa become motile when mixed with secretions of the seminal vesicles, and (2) they become capable of successful fertilization when exposed to conditions in the female reproductive tract. The epididymis secretes a substance (as yet unidentified) that prevents premature capacitation.

Transport along the epididymis involves a combination of fluid movement and peristaltic contractions of smooth muscle in the walls of the epididymis. After passing along the tail of the epididymis, the spermatozoa enter the ductus deferens.

The Ductus Deferens

Each **ductus deferens,** or *vas deferens,* is 40–45 cm (16–18 in.) long. It begins at the tail of the epididymis (Figures 28-1, p. 1018, and 28-9a•) and, as part of the spermatic cord, ascends through the inguinal canal (Figure 28-3•, p. 1020). Inside the abdominal cavity, the ductus deferens passes posteriorly, curving inferiorly along the lateral surface of the urinary bladder toward the superior and posterior margin of the prostate gland. Just before the ductus deferens reaches the prostate gland and seminal vesicles, its lumen

enlarges. This expanded portion is known as the **ampulla** (am-PŪL-uh) (Figure 28-10a•).

The wall of the ductus deferens contains a thick layer of smooth muscle (Figure 28-10b•). Peristaltic contractions in this layer propel spermatozoa and fluid along the duct, which is lined by a pseudostratified ciliated columnar epithelium. In addition to transporting spermatozoa, the ductus deferens can store spermatozoa for several months. During this time, the spermatozoa remain in a state of suspended animation and have low metabolic rates.

The junction of the ampulla with the duct of the seminal vesicle marks the start of the **ejaculatory duct.** This short passageway (2 cm, or less than 1 in.) penetrates the muscular wall of the prostate gland and empties into the urethra near the opening of the ejaculatory duct from the opposite side (Figures 28-1 and 28-10a•).

The Urethra

The **urethra** males extends 18–20 cm (7–8 in.) from the urinary bladder to the tip of the penis (Figure 28-1•). It is divided into *prostatic, membranous,* and *penile regions.* ∞ *p. 976* In males, the urethra is a passageway used by both the urinary and reproductive systems.

28

•**FIGURE 28-10 The Ductus Deferens and Accessory Glands.** (a) A posterior view of the prostate gland, showing subdivisions of the ductus deferens in relation to surrounding structures. (b) The ductus deferens, showing the smooth muscle around the lumen. (LM [top] × 34, SEM × 42) (Reproduced from R. G. Kessel and R. H. Kardon, *Tissues and Organs: A Text-Atlas of Scanning Electron Microscopy,* W. H. Freeman & Co., 1979.) Sections of (c) the seminal vesicle (LM × 44), (d) the prostate gland (LM × 49), and (e) a bulbourethral gland. (LM × 177)

THE ACCESSORY GLANDS

The fluids contributed by the seminiferous tubules and the epididymis account for only about 5 percent of the volume of semen. The fluid component of semen is a mixture of the secretions of many glands, each with distinctive biochemical characteristics. Important glands include the *seminal vesicles,* the *prostate gland,* and the *bulbourethral glands.* Major functions of these glands, which occur only in males, include (1) activating the spermatozoa; (2) providing the nutrients spermatozoa need for motility; (3) propelling spermatozoa and fluids along the reproductive tract, mainly by peristaltic contractions; and (4) producing buffers that counteract the acidity of urethral and vaginal environments.

The Seminal Vesicles

The ductus deferens on each side ends at the junction between the ampulla and the duct that drains the seminal vesicle (Figure 28-10a•). The **seminal vesicles** are glands embedded in connective tissue on either side of the midline, sandwiched between the posterior wall of the urinary bladder and the rectum. Each seminal vesicle is a tubular gland with a total length of about 15 cm (6 in.). The body of the gland has many short side branches. The entire assemblage is coiled and folded into a compact, tapered mass roughly 5 cm × 2.5 cm (2 in. × 1 in.). The location of the seminal vesicles is shown in Figures 28-1, 28-10a, and 28-11a•, p. 1031).

Seminal vesicles are extremely active secretory glands with an epithelial lining that contains extensive folds (Figure 28-10c•). The seminal vesicles contribute about 60 percent of the volume of semen. Although the vesicular fluid generally has the same osmotic concentration as that of blood plasma, the composition of the two fluids is quite different. In particular, the secretion of the seminal vesicles contains (1) higher concentrations of fructose, which is easily metabolized by spermatozoa; (2) prostaglandins, which can stimulate smooth muscle contractions along the male and female reproductive tracts; and (3) fibrinogen, which after ejaculation forms a temporary clot within the vagina. The secretions of the seminal vesicles are slightly alkaline. This alkalinity helps neutralize acids in the secretions of the prostate gland and within the vagina. When mixed with the secretions of the seminal vesicles, previously inactive but functional spermatozoa begin beating their flagella, becoming highly motile.

The secretions of the seminal vesicles are discharged into the ejaculatory duct at *emission,* when peristaltic contractions are under way in the ductus deferens, seminal vesicles, and prostate gland. These contractions are under control of the sympathetic nervous system.

The Prostate Gland

The **prostate gland** is a small, muscular, rounded organ about 4 cm (1.6 in.) in diameter. The prostate gland encircles the proximal portion of the urethra as it leaves the urinary bladder (Figures 28-10a and 28-11a•). The glandular tissue of the prostate consists of a cluster of 30–50 compound tubuloalveolar glands (Figure 28-10d•). ∞ *p. 117* These glands are surrounded by and wrapped in a thick blanket of smooth muscle fibers.

The prostate gland produces **prostatic fluid,** a slightly acidic solution that contributes 20–30 percent

of the volume of semen. In addition to several other compounds of uncertain significance, prostatic secretions contain **seminalplasmin** (sem-i-nal-PLAZ-min), an antibiotic that may help prevent urinary tract infections in males. These secretions are ejected into the prostatic urethra by peristaltic contractions of the muscular wall.

Prostatitis

Prostatic inflammation, or **prostatitis** (pros-ta-TĪ-tis), can occur in males at any age, but it most commonly afflicts older men. Prostatitis can result from bacterial infections but also occurs in the apparent absence of pathogens. Symptoms can resemble those of prostate cancer. Individuals with prostatitis complain of pain in the lower back, perineum, or rectum, in some cases accompanied by painful urination and the discharge of mucous secretions from the external urethral meatus. Antibiotic therapy is effective in treating most cases that result from bacterial infection.

The Bulbourethral Glands

The paired **bulbourethral glands,** or *Cowper's glands,* are situated at the base of the penis, covered by the fascia of the urogenital diaphragm (Figures 28-10a and 28-11a•). The bulbourethral glands are round, with diameters approaching 10 mm (less than 0.5 in.). The duct of each gland travels alongside the penile urethra for 3–4 cm (1.2–1.6 in.) before emptying into the urethral lumen. These are compound, tubuloalveolar mucous glands (Figure 28-10e•) that secrete a thick, alkaline mucus. The secretion helps neutralize any urinary acids that may remain in the urethra and lubricates the *glans,* or tip of the penis.

SEMEN

A typical ejaculation releases 2–5 ml of semen. This volume of fluid, called an **ejaculate,** contains:

- *Spermatozoa.* A normal **sperm count** ranges from 20 million to 100 million spermatozoa per cubic milliliter of sperm.
- *Seminal fluid.* **Seminal fluid,** the fluid component of semen, is a mixture of glandular secretions with a distinct ionic and nutrient composition. A typical sample of seminal fluid contains the combined secretions of the seminal vesicles (60 percent), the prostate gland (30 percent), the sustentacular cells and epididymis (5 percent), and the bulbourethral glands (less than 5 percent).
- *Enzymes.* Several important enzymes are present in seminal fluid, including (1) a protease that may help dissolve mucous secretions in the vagina; (2) seminalplasmin, an antibiotic enzyme from the prostate gland that kills a variety of bacteria, including

28

Escherichia coli; (3) a prostatic enzyme that converts fibrinogen to fibrin after ejaculation; and (4) *fibrinolysin,* which subsequently liquefies the clotted semen.

Semen Analysis

Semen analysis is commonly performed to assess male fertility. A semen sample is donated after a 36-hour period of sexual abstinence. Standard tests include the following:

- *Volume.* Normal ejaculate volume is 2–5 ml. An abnormally low volume may indicate problems with the prostate gland or seminal vesicles.

- *Sperm count.* Each ejaculate should contain more than 60 million spermatozoa. The concentration of spermatozoa should be over 20 million per milliliter of semen. Most individuals with lower sperm counts are infertile, because too few spermatozoa survive the ascent of the female reproductive tract to perform fertilization. A low sperm count may reflect inflammation of the epididymis, ductus deferens, or prostate gland.

- *Motility.* At least 60 percent of the spermatozoa in the sample should be beating their flagella and swimming actively.

- *Shape.* At least 60 percent of the spermatozoa should have normal shapes. Common abnormalities are malformed heads and "twin" spermatozoa that did not separate at the time of spermiation.

- *Liquefaction.* Within a few minutes after ejaculation, semen coagulates, liquefying again some time later. The function of this clotting is unknown. Normal semen liquefies after 15–30 minutes. An extended liquefaction time may indicate problems with the secretions of accessory glands.

The determination of male fertility problems in the absence of abnormal semen analysis results may require additional tests. In what is often called the "hamster test," a semen sample is placed on a slide with the oocyte of a hamster. Normal human spermatozoa will fertilize the oocyte, although further development is impossible. If fertilization does not occur, there may be problems with the enzymes in the acrosomal cap.

A complete chemical analysis of semen appears in Appendix VI.

THE PENIS

The **penis** is a tubular organ through which the distal portion of the urethra passes (Figure 28-11a●). It conducts urine to the exterior and introduces semen into the female's vagina during sexual intercourse. The penis is divided into three regions: (1) the root, (2) the body, and (3) the glans (Figure 28-11c●). The **root** of the penis is the fixed portion that attaches the penis to the body wall. This connection occurs within the urogenital triangle immediately inferior to the pubic symphysis. ∞ *p. 234* The **body (shaft)** of the penis is the tubular, movable portion. The **glans** of the penis is the expanded distal end that surrounds the external urethral meatus. The *neck* is the narrow portion of the penis between the shaft and the glans.

The skin overlying the penis resembles that of the scrotum. The dermis contains a layer of smooth muscle, and the underlying areolar tissue allows the thin skin to move without distorting underlying structures. The subcutaneous layer also contains superficial arteries, veins, and lymphatic vessels.

A fold of skin called the **prepuce** (PRĒ-pūs), or *foreskin,* surrounds the tip of the penis. The prepuce attaches to the relatively narrow neck of the penis and continues over the glans. **Preputial** (pre-PŪ-shal) **glands** in the skin of the neck and the inner surface of the prepuce secrete a waxy material known as **smegma** (SMEG-ma). Unfortunately, smegma can be an excellent nutrient source for bacteria. Mild inflammation and infections in this region are common, especially if the area is not washed thoroughly and frequently. One way to avoid trouble is to perform a **circumcision** (ser-kum-SIZH-un), the surgical removal of the prepuce. In Western societies (especially the United States), this procedure is generally performed shortly after birth. Although the practice of circumcision remains controversial, strong religious and cultural biases and epidemiological evidence suggest that it will continue.

Deep to the areolar tissue, a dense network of elastic fibers encircles the internal structures of the penis. Most of the body of the penis consists of three cylindrical columns of **erectile tissue** (Figure 28-11a,b●). Erectile tissue consists of a three-dimensional maze of vascular channels incompletely separated by partitions of elastic connective tissue and smooth muscle fibers. In the resting state, the arterial branches are constricted and the muscular partitions are tense. This combination restricts blood flow into the erectile tissue. The parasympathetic innervation of the penile arteries involves neurons that release nitric oxide (NO) at their synaptic knobs. The smooth muscles in the arterial walls relax when NO is released, at which time (1) the vessels dilate, (2) blood flow increases, (3) the vascular channels become engorged with blood, and (4) **erection** of the penis occurs. The flaccid (nonerect) penis hangs inferior to the pubic symphysis and anterior to the scrotum, but during erection the penis stiffens and assumes a more upright position.

The anterior surface of the flaccid penis covers two cylindrical masses of erectile tissue, the **corpora cavernosa** (KOR-po-ruh ka-ver-NŌ-suh; singular, *cavernosum*). The two are separated by a thin septum

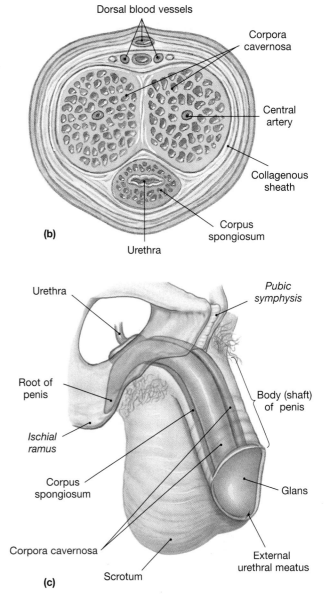

•**FIGURE 28-11** **The Penis.** **(a)** A frontal section through the penis and associated organs. **(b)** A sectional view through the penis. **(c)** An anterior and lateral view of the penis, showing positions of the erectile tissues.

and encircled by a dense collagenous sheath. The corpora cavernosa diverge at their bases, forming the **crura** (*crura,* legs; singular, *crus*) of the penis. Each crus is bound to the ramus of the ischium and pubis by tough connective tissue ligaments. The corpora cavernosa extend along the length of the penis as far as the neck. The erectile tissue within each corpus cavernosum surrounds a central artery (Figure 28-11b•).

The relatively slender **corpus spongiosum** (spon-jē-Ō-sum) surrounds the penile urethra (Figure 28-11a,b•). This erectile body extends from the superficial fascia of the urogenital diaphragm to the tip of the penis, where it expands to form the glans. The sheath surrounding the corpus spongiosum contains more elastic fibers than does that of the corpora cavernosa, and the erectile tissue contains a pair of small arteries.

The Circumcision Debate

Worldwide, roughly 25 percent of males are circumcised. However, the practice of neonatal male circumcision continues to be hotly debated. Strong epidemiological evidence favors circumcision: Circumcised males have a lower incidence of neonatal and adult urinary tract infections, genital ulcerative disease, penile carcinoma, and HIV infection (in some groups) than are uncircumcised males. Circumcision also carries serious negative aspects and risks, however, such as pain, bleeding, penile injury, and postoperative infection. Many health professionals, as well as parents, feel that these neonatal problems outweigh the statistical advantages in adulthood. Neither the American Academy of Pediatrics nor the Canadian Paediatric Society advocates routine neonatal circumcision; both support parental choice on the basis of an individual assessment of the potential risks and benefits.

28

Prostatic Hypertrophy and Prostate Cancer

In most cases, enlargement of the prostate gland, or **benign prostatic hypertrophy,** occurs spontaneously in men over age 50. The increase in size occurs while testosterone production by the interstitial cells decreases. At the same time, the interstitial cells begin releasing small quantities of estrogens into the bloodstream. The combination of lower testosterone levels and the presence of estrogen probably stimulates prostatic growth. In severe cases, prostatic swelling constricts and blocks the urethra and constricts the rectum. If not corrected, the urinary obstruction can cause permanent kidney damage. Partial surgical removal is the most effective treatment. In the procedure known as a **TURP** (transurethral prostatectomy), an instrument pushed along the urethra restores normal function by cutting away the swollen prostatic tissue. Most of the prostate gland remains in place, and no external scars result.

Prostate cancer, a malignant, metastasizing cancer of the prostate gland, is for males the second most common cancer and the second most common cause of cancer deaths. In 1999, approximately 334,500 new cases of prostate cancer were diagnosed in the United States, with about 41,800 deaths. Most patients are elderly (average age 72 at diagnosis). There are racial differences in susceptibility that are poorly understood. The prostate cancer rates in the 50–54 age group are twice as high for African-American males as for Caucasian-American males. (The rates at all ages are about one-third higher for African Americans.) The prostate cancer rates for Asian-American males are relatively low compared with those of either Caucasian Americans or African Americans. For all age groups and all races, the rates of prostate cancer are rising sharply. The reason for the increase is not known.

Prostate cancer normally originates in one of the secretory glands. As the cancer progresses, it produces a nodular lump or swelling on the surface of the prostate gland. Palpation of this gland through the rectal wall, a procedure known as a *digital rectal exam* (DRE), is the easiest diagnostic screening procedure. *Transrectal prostatic ultrasound* (TRUS) can be used to obtain more-detailed information about the status of the prostate but at significantly higher cost to the patient. Blood tests are also used for screening purposes. The most sensitive is a blood test for *prostate-specific antigen (PSA).* Elevated levels of this antigen, normally present in low concentrations, may indicate the presence of prostate cancer. The *serum enzyme assay,* which checks the level of the isozyme *prostatic acid phosphatase,* detects prostate cancer in later stages of development. Screening with periodic PSA tests is now being recommended for men over age 50.

If the condition is detected before the cancer cells have spread to other organs, the usual treatment is localized radiation or the surgical removal of the prostate gland. This operation, **prostatectomy** (prosta-TEK-to-mē), can be effective in controlling the condition, but it can have undesirable side effects, including urinary incontinence and the loss of sexual function. Modified surgical procedures can reduce the risks and maintain normal sexual function in almost three out of four patients.

The prognosis is much worse for prostate cancer diagnosed after metastasis has occurred, because metastasis rapidly involves the lymphatic system, lungs, bone marrow, liver, or adrenal glands. The survival rates at this stage become relatively low. Potential treatments for metastasized prostate cancer include widespread irradiation, hormonal manipulation, lymph node removal, and aggressive chemotherapy. Because the cancer cells are stimulated by testosterone, treatment may involve castration or hormones that depress GnRH or LH production. Three treatment options exist: (1) *an estrogen:* the usual choice is *diethylstilbestrol (DES);* (2) *drugs that mimic GnRH:* These drugs are given in high doses, producing a surge in LH production followed by a sharp decline to very low levels, presumably as the endocrine cells adapt to the excessive stimulation; and (3) *drugs that block the action of androgens:* Several new drugs, including *flutamide* and *finasateride,* prevent the stimulation of cancer cells by testosterone. Despite these interesting advances in treatment, however, the average survival time for patients diagnosed with advanced prostatic cancer is only 2.5 years.

HORMONES AND MALE REPRODUCTIVE FUNCTION

The hormonal interactions in males are diagrammed in Figure 28-12●. We introduced major reproductive hormones in Chapter 18. *pp. 608–609* The anterior pituitary gland releases *follicle-stimulating hormone* (FSH) and *luteinizing hormone* (LH). The pituitary release of these hormones occurs in the presence of *gonadotropin-releasing hormone* (GnRH), a peptide synthesized in the hypothalamus and carried to the anterior pituitary by the hypophyseal portal system.

The hormone GnRH is secreted in pulses rather than continuously. In adult males, small pulses occur at

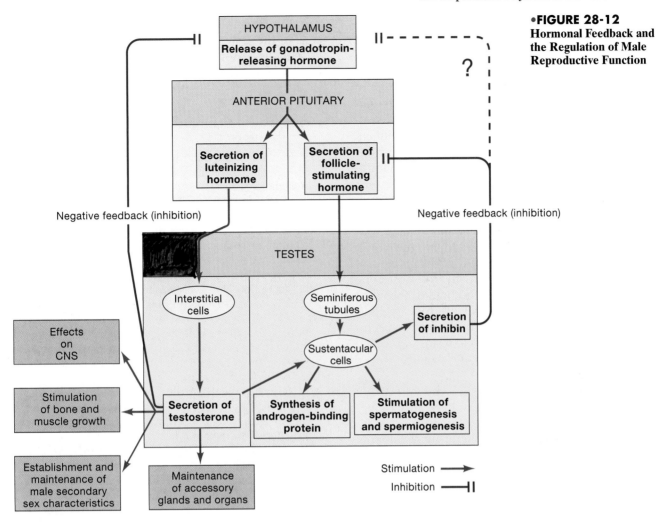

60–90 minute intervals. As levels of GnRH change, so do the rates of secretion of FSH and LH (and testosterone, which is released in response to LH). Unlike the situation in women, which we will consider later in the chapter, the GnRH pulse frequency in adult males remains relatively steady from hour to hour, day to day, and year to year. As a result, plasma levels of FSH, LH, and testosterone remain within a relatively narrow range throughout adult life.

FSH and Spermatogenesis

In males, FSH targets primarily the sustentacular cells of the seminiferous tubules. Under FSH stimulation, and in the presence of testosterone from the interstitial cells, sustentacular cells (1) promote spermatogenesis and spermiogenesis and (2) secrete androgen-binding protein (ABP).

The rate of spermatogenesis is regulated by a negative-feedback mechanism involving GnRH, FSH, and inhibin. Under GnRH stimulation, FSH promotes spermatogenesis along the seminiferous tubules. As spermatogenesis accelerates, however, so does the rate of inhibin secretion by the sustentacular cells of the

testes. Inhibin inhibits FSH production in the anterior pituitary gland and may also suppress the secretion of GnRH at the hypothalamus.

The net effect is that when FSH levels become elevated, inhibin production increases until the FSH levels return to normal. If FSH levels decline, inhibin production falls, so the rate of FSH production accelerates.

LH and Androgen Production

In males, LH causes the secretion of testosterone and other androgens by the interstitial cells of the testes. Testosterone, the most important androgen, has numerous functions, such as (1) stimulating spermatogenesis and promoting the functional maturation of spermatozoa, through its effects on sustentacular cells; (2) affecting central nervous system (CNS) function, including the influence of libido (sexual drive) and related behaviors; (3) stimulating metabolism throughout the body, especially pathways concerned with protein synthesis and muscle growth; (4) establishing and maintaining the secondary sex characteristics, such as the distribution of facial hair, increased muscle mass and body size, and the quantity and location

28

of characteristic adipose tissue deposits; and (5) maintaining the accessory glands and organs of the male reproductive tract.

Testosterone functions like other steroid hormones. ∞ *pp. 583–584* It circulates while bound to one of two types of transport proteins: (1) *gonadal steroid-binding globulin* (GBG), which carries roughly two-thirds of circulating testosterone, and (2) the albumins, which bind the remaining one-third. Testosterone diffuses across the cell membrane and binds to an intracellular receptor. The hormone–receptor complex then binds to the DNA in the nucleus. In many target tissues, some of the arriving testosterone is converted to **dihydrotestosterone (DHT).** A small amount of DHT diffuses back out of the cell and into the bloodstream, and DHT levels are usually about 10 percent of circulating testosterone levels. Dihydrotestosterone can also enter peripheral cells and bind to the same hormone receptors targeted by testosterone. In addition, some tissues, notably those of the external genitalia, respond to DHT rather than to testosterone, and other tissues, including the prostate gland, are more sensitive to DHT than to testosterone.

Testosterone production begins around the seventh week of development and reaches a peak after roughly 6 months of development. Over this period, the secretion of Müllerian-inhibiting factor by developing sustentacular cells leads to the regression of the Müllerian ducts. The early surge in testosterone levels stimulates the differentiation of the male duct system and accessory organs and affects CNS development. The best-known CNS effects occur in the developing hypothalamus, where testosterone apparently programs the hypothalamic centers that are involved with (1) GnRH production and with the regulation of pituitary FSH and LH secretion, (2) sexual behaviors, and (3) sexual drive. As a result of this prenatal exposure to testosterone, the hypothalamic centers will respond appropriately when the individual becomes sexually mature. The factors responsible for regulating the fetal production of testosterone are not known.

Testosterone levels are low at birth. Up to puberty, background testosterone levels, although still relatively low, are higher in males than in females. Testosterone secretion accelerates markedly at puberty, initiating sexual maturation and the appearance of secondary sex characteristics. In adult males, negative feedback controls the level of testosterone production. Above-normal testosterone levels inhibit the release of GnRH by the hypothalamus. This inhibition causes a reduction in LH secretion and lowers testosterone levels.

The plasma of adult males also contains relatively small amounts of estradiol (2 ng/dl, versus 525 ng/dl of testosterone). Seventy percent of the estradiol is formed from circulating testosterone. The rest is secreted, primarily by the interstitial and sustentacular cells of the testes. The conversion of testosterone to estradiol is performed by an enzyme called *aromatase*. For unknown reasons, estradiol production increases in older men.

DHEA

Dehydroepiandrosterone, or **DHEA,** is the primary androgen secreted by the zona reticularis of the adrenal cortex. ∞ *p. 602* As we noted in Chapter 18, these androgens, which are secreted in small amounts, are converted to testosterone (or estrogens) by other tissues. The significance of this small adrenal androgen contribution in both genders remains uncertain, but some people have suggested that DHEA could be used to treat a variety of conditions, including diabetes, heart disease, and depression. The effects of long-term, high-dose DHEA administration remain largely unknown. However, you will recall from Chapter 18 that the long-term effects of androgen abuse can be quite serious. ∞ *p. 611* High levels of DHEA in women have been linked to masculinization, due to the conversion of DHEA to testosterone, and to an increased risk of ovarian cancer.

Nevertheless, DHEA is being promoted as a wonder drug to increase vitality, strength, and muscle mass. Direct sales of the hormone were banned in 1984, because the side-effects were unknown, but food supplements prepared from wild Mexican yams are now being advertised as containing "DHEA precursors." These claims are false; the compounds contained in these supplements have no effect on circulating levels of DHEA. The current recommendations are that (1) DHEA use be restricted to controlled, supervised clinical trials; and (2) that no one under age 30 use DHEA.

☑ On a warm day, would the cremaster muscle be contracted or relaxed? Why?

☑ What will occur if the arteries within the penis dilate?

☑ What effect would low levels of FSH have on sperm production?

THE REPRODUCTIVE SYSTEM OF THE FEMALE

A woman's reproductive system produces sex hormones and functional gametes and also must be able to protect and support a developing embryo and nourish the newborn infant. The principal organs of the female reproductive system are the *ovaries,* the *uterine tubes,* the *uterus,* the *vagina,* and the components of the external genitalia (Figure 28-13•). As in males, a variety of accessory glands secrete into the female reproductive tract.

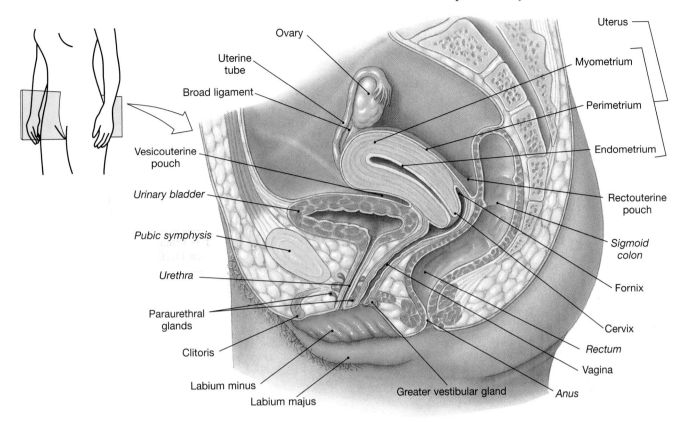

Ovary

Uterine tube

Broad ligament

Vesicouterine pouch

Urinary bladder

Pubic symphysis

Urethra

Paraurethral glands

Clitoris

Labium minus

Labium majus

Greater vestibular gland

Uterus

Myometrium

Perimetrium

Endometrium

Rectouterine pouch

Sigmoid colon

Fornix

Cervix

Rectum

Vagina

Anus

•**FIGURE 28-13** **The Female Reproductive System.** A sagittal section of the female reproductive organs. [AM] *Plate 7.11b*

The ovaries, uterine tubes, and uterus are enclosed within an extensive mesentery known as the **broad ligament**. The uterine tubes run along the superior border of the broad ligament and open into the pelvic cavity lateral to the ovaries. The **mesovarium** (mes-ō-VAR-ē-um), a thickened fold of mesentery, supports and stabilizes the position of each ovary. The broad ligament attaches to the sides and floor of the pelvic cavity, where it becomes continuous with the parietal peritoneum. The broad ligament thus subdivides this part of the peritoneal cavity. The pocket formed between the posterior wall of the uterus and the anterior surface of the colon is the **rectouterine** (rek-tō-Ū-ter-in) **pouch;** the pocket formed between the uterus and the posterior wall of the bladder is the **vesicouterine** (ves-i-kō-Ū-ter-in) **pouch.** These subdivisions are most apparent in sagittal section (Figure 28-13•).

Several other ligaments assist the broad ligament in supporting and stabilizing the position of the uterus and associated reproductive organs. These ligaments lie within the mesentery sheet of the broad ligament and are connected to the ovaries or uterus. The broad ligament limits side-to-side movement and rotation, and the other ligaments (described with the ovaries and uterus) prevent superior–inferior movement.

THE OVARIES

The paired ovaries are small, almond-shaped organs near the lateral walls of the pelvic cavity (Figures 28-13 and 28-14•). The ovaries (1) produce immature female gametes, or oocytes; (2) secrete female sex hormones, including estrogens and progestins; and (3) secrete inhibin, involved in the feedback control of pituitary FSH production.

The position of each ovary is stabilized by the mesovarium and by a pair of supporting ligaments: the ovarian ligament and the suspensory ligament (Figure 28-14•). The **ovarian ligament** extends from the uterus, near the attachment of the uterine tube, to the medial surface of the ovary. The **suspensory ligament** extends from the lateral surface of the ovary past the open end of the uterine tube to the pelvic wall. The suspensory ligament contains the major blood vessels of the ovary, the **ovarian artery** and **ovarian vein.** These vessels are connected to the ovary at the **ovarian hilum,** where the ovary attaches to the mesovarium.

28

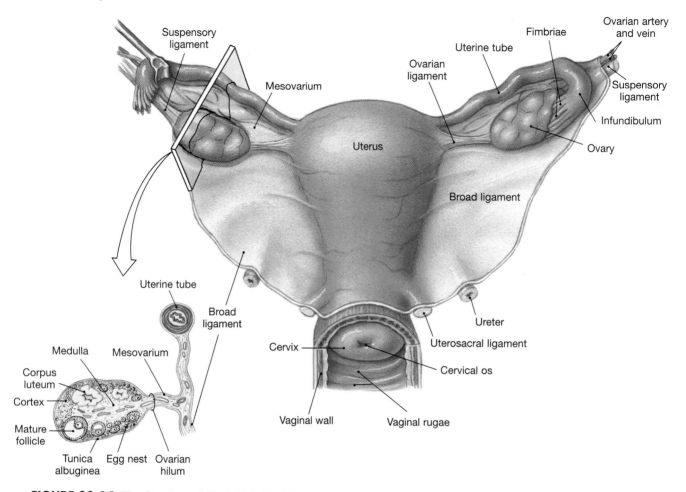

•**FIGURE 28-14 The Ovaries and Their Relationships to the Uterine Tube and Uterus.** A posterior view of the uterus, uterine tubes, and ovaries and a sectional view of the ovary, uterine tube, and associated mesenteries.

A typical ovary is a flattened oval about 5 cm in length, 2.5 cm in width, and 8 mm in thickness (2 in. × 1 in. × 0.33 in.) and weighs 6–8 g (roughly 0.25 oz). An ovary is pink or yellowish and has a nodular consistency. The visceral peritoneum, or *germinal epithelium,* covering the surface of each ovary consists of a layer of columnar epithelial cells that overlies a dense connective tissue layer called the **tunica albuginea.** We can divide the interior tissues, or **stroma,** of the ovary into a superficial *cortex* and a deeper *medulla.* Gametes are produced in the cortex.

Oogenesis

Ovum production, or **oogenesis** (ō-ō-JEN-e-sis), begins before a woman's birth, accelerates at puberty, and ends at *menopause.* Between puberty and menopause, oogenesis occurs on a monthly basis as part of the *ovarian cycle.*

Unlike spermatogonia, the **oogonia** (ō-ō-GŌ-nē-uh), or stem cells of females, complete their mitotic di-

visions before birth. Between the third and seventh months of fetal development, the daughter cells, or **primary oocytes** (Ō-ō-sīts), prepare to undergo meiosis. They proceed as far as prophase of meiosis I, but at that time the process comes to a halt. The primary oocytes then remain in a state of suspended development until the individual reaches puberty. At that time, rising levels of FSH trigger the start of the ovarian cycle. Each month thereafter, some of the primary oocytes will be stimulated to undergo further development. Not all primary oocytes produced during development survive until puberty. The ovaries have roughly 2 million *primordial follicles* at birth, each containing a primary oocyte. By the time of puberty, that number has dropped to about 400,000. The rest of the primordial follicles degenerate in a process called **atresia** (a-TRĒ-zē-uh).

Although the nuclear events under way during meiosis in the ovaries are the same as those in the testes, the process differs in two important details:

1. The cytoplasm of the primary oocyte is unevenly distributed during the two meiotic divisions. Oogenesis produces one functional ovum, which contains most of the original cytoplasm, and two or three **polar bodies,** nonfunctional cells that later disintegrate (Figure 28-15•).

2. The ovary releases a **secondary oocyte** rather than a mature ovum. The secondary oocyte is suspended in metaphase of meiosis II; meiosis will not be completed unless and until fertilization occurs.

The Ovarian Cycle

Ovarian follicles are specialized structures in which both oocyte growth and meiosis I occur. The ovarian follicles are located in the cortex of the ovaries. Primary oocytes are located in the outer portion of the ovarian cortex near the tunica albuginea in clusters called *egg nests.* Each primary oocyte within an egg nest is surrounded by a single squamous layer of *follicular cells.* The primary oocyte and its follicular cells form a **primordial follicle.** After sexual maturation, a different group of primordial follicles is activated each month. This monthly process is known as the **ovarian cycle.**

The ovarian cycle can be divided into a **follicular phase,** or *preovulatory phase,* and a **luteal phase,** or *postovulatory phase.* Important steps in the ovarian cycle can be summarized as follows (Figure 28-16•):

STEP 1: *The formation of primary follicles.* Follicle formation is stimulated by FSH from the anterior pituitary gland. The ovarian cycle begins as activated primordial follicles develop into **primary follicles.** In a primary follicle, the follicular cells enlarge and undergo repeated cell divisions. The divisions create several layers of follicular cells around the oocyte. These follicle cells are now called **granulosa cells.**

As layers of granulosa cells develop around the primary oocyte, microvilli from the surrounding granulosa cells intermingle with those of the primary oocyte. The microvilli are surrounded by a layer of glycoproteins; the entire region is called the **zona pellucida** (ZŌ-na pe-LOO-si-duh). The microvilli increase the surface area available for the transfer of materials from the granulosa cells to the rapidly enlarging oocyte.

The conversion from primordial to primary follicles and subsequent follicular development occurs under FSH stimulation. As the granulosa cells enlarge and multiply, adjacent cells in the ovarian stroma form a layer of **thecal cells** around the follicle. Thecal cells and granulosa cells work together to produce sex hormones called *estrogens.*

STEP 2: *The formation of secondary follicles.* Although many primordial follicles develop into primary follicles, only a few will proceed to the next step. The transformation begins as the wall of the follicle thickens and the granulosa cells begin secreting small amounts of fluid. This **follicular fluid,** or *liquor folliculi,* accumulates in small pockets that gradually expand and separate the inner and outer layers of the follicle. At this stage, the complex is known as a **secondary follicle.** Although the primary oocyte continues to grow at a steady pace, the follicle as a whole now enlarges rapidly because follicular fluid accumulates.

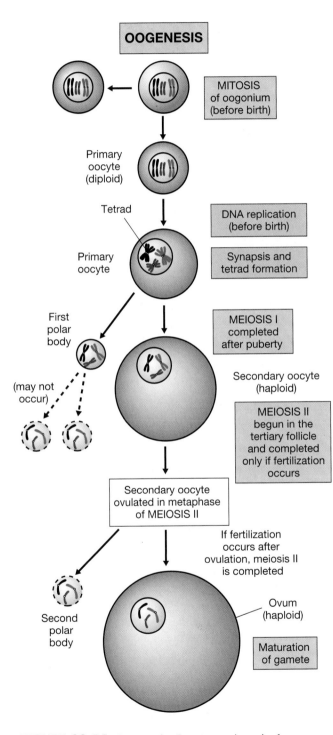

•FIGURE 28-15 Oogenesis. In oogenesis, a single primary oocyte produces an ovum and two or three nonfunctional polar bodies. Compare this schematic diagram with Figure 28-7, p. 1024.

28

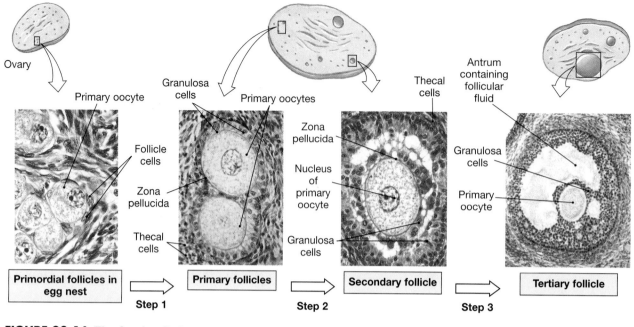

•**FIGURE 28-16** The Ovarian Cycle

STEP 3: *The formation of a tertiary follicle.* Eight to ten days after the start of the ovarian cycle, the ovaries generally contain only a single secondary follicle destined for further development. By the tenth to fourteenth day of the cycle, that follicle has formed a **tertiary follicle,** or *mature Graafian* (GRAF-ē-an) *follicle,* roughly 15 mm in diameter. This complex spans the entire width of the ovarian cortex and distorts the ovarian capsule, creating a prominent bulge in the surface of the ovary. The oocyte projects into the **antrum** (AN-trum), or expanded central chamber of the follicle. The antrum is surrounded by a mass of granulosa cells.

Until this time, the primary oocyte has been suspended in prophase of meiosis I. As the development of the tertiary follicle ends, LH levels begin rising, prompting the primary oocyte to complete meiosis I. Instead of producing two secondary oocytes, the first meiotic division yields a secondary oocyte and a small, nonfunctional polar body. The secondary oocyte then enters meiosis II but stops once again on reaching metaphase. Meiosis II will not be completed unless fertilization occurs.

Generally on day 14 of a 28-day cycle, the secondary oocyte and the surrounding granulosa cells lose their connections with the follicular wall. The granulosa cells immediately surrounding the secondary oocyte now drift free within the antrum and are known as the **corona radiata** (ko-RŌ-nuh rā-dē-A-tuh).

STEP 4: *Ovulation.* At **ovulation,** the tertiary follicle releases the secondary oocyte. The distended follicular wall ruptures, releasing the follicular contents, including the secondary oocyte and corona radiata, into the pelvic cavity. The sticky follicular fluid keeps the corona radiata attached to the surface of the ovary, where direct contact with projections of the uterine tube or with fluid currents established by their ciliated epithelium can transfer the secondary oocyte to the uterine tube.

STEP 5: *The formation and degeneration of the corpus luteum.* The empty tertiary follicle initially collapses, and ruptured vessels bleed into the antrum. The remaining granulosa cells then invade the area, proliferating to create an endocrine structure known as the **corpus luteum** (LOO-tē-um), named for its yellow color (*lutea,* yellow). This process occurs under LH stimulation.

The lipids contained in the corpus luteum are used to manufacture steroid hormones known as **progestins** (prō-JES-tinz), principally the steroid **progesterone** (prō-JES-ter-ōn). Although moderate amounts of estrogens are also secreted by the corpus luteum, levels are not as high as they were at ovulation, and progesterone is the principal hormone in the interval after ovulation. Its primary function is to prepare the uterus for pregnancy by stimulating the maturation of the uterine lining and the secretions of uterine glands.

STEP 6: *Unless fertilization occurs, the corpus luteum begins to degenerate roughly 12 days after ovulation.* Progesterone and estrogen levels then fall markedly. Fibroblasts invade the nonfunctional corpus luteum, producing a knot of pale scar tissue called a **corpus albicans** (AL-bi-kanz). The disintegration, or *involution,* of the corpus luteum marks the end of the ovarian cycle. A new ovarian cycle then begins with the activation of another group of primordial follicles.

Released secondary oocyte

Corona radiata

Ruptured tertiary follicle

Secondary oocyte within corona radiata

Follicular fluid

Outer surface of ovary

Ruptured follicle wall

OVULATION   Step 4   Step 5   **Corpus luteum**   Step 6   **Corpus albicans**

Age and Oogenesis

Although many primordial follicles may have developed into primary follicles, and several primary follicles may have been converted to secondary follicles, generally only a single oocyte will be released into the pelvic cavity at ovulation. The rest undergo atresia. At puberty, each ovary contains about 200,000 primordial follicles. Forty years later, few if any follicles remain, although only about 500 will have been ovulated during the interim.

Ovarian Cancer

A woman in the United States has a 1 in 70 chance of developing **ovarian cancer** in her lifetime. In 2000, an estimated 23,100 ovarian cancers were diagnosed, and 14,000 women died from this condition. Although ovarian cancer is the third most common reproductive cancer among women, it is the most dangerous because ovarian cancer is seldom diagnosed in its early stages. The prognosis is relatively good for cancers that originate in the general ovarian tissues or from abnormal oocytes. These cancers respond well to some combination of chemotherapy, radiation, and surgery. However, 85 percent of ovarian cancers develop from epithelial cells, and sustained remission can be obtained in only about one-third of the cases of this type. AM *The Diagnosis and Treatment of Ovarian Cancer*

THE UTERINE TUBES

Each **uterine tube** (*Fallopian tube* or *oviduct*) is a hollow, muscular tube measuring roughly 13 cm (5 in.) in length (Figures 28-13, sagittal view; 28-14 and 28-17a•,

posterior views). Each uterine tube is divided into three segments (Figure 28-17a,b•):

1. *The infundibulum.* The end closest to the ovary forms an expanded funnel, or **infundibulum,** with numerous fingerlike projections that extend into the pelvic cavity. The projections are called **fimbriae** (FIM-brē-ē). The inner surfaces of the infundibulum are lined with cilia that beat toward the middle segment of the uterine tube, the *ampulla.*

2. *The ampulla.* The thickness of the smooth muscle layers in the wall of the middle segment, or **ampulla,** of the uterine tube gradually increases as it approaches the uterus.

3. *The isthmus.* The ampulla leads to the **isthmus** (IS-mus), a short segment connected to the uterine wall.

Histology of the Uterine Tube

The epithelium lining the uterine tube is composed of ciliated columnar epithelial cells with scattered mucin-secreting cells (Figure 28-17c•). The mucosa is surrounded by concentric layers of smooth muscle (Figure 28-17b•). Oocyte transport involves a combination of ciliary movement and peristaltic contractions in the walls of the uterine tube. A few hours before ovulation, sympathetic and parasympathetic nerves from the hypogastric plexus "turn on" this beating pattern and initiate peristalsis. The uterine tubes transport a secondary oocyte for final maturation and fertilization. It normally takes 3–4 days for an oocyte to travel from the infundibulum to the *uterine cavity.* If fertilization is to occur, the secondary oocyte must encounter spermatozoa during the first 12–24 hours of its passage. Fertilization typically occurs near the

28

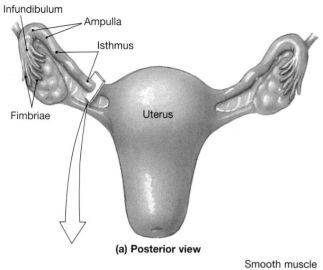

Infundibulum
Ampulla
Isthmus
Fimbriae
Uterus

(a) Posterior view

Smooth muscle

Columnar epithelium   **(b)**   Lamina propria

Cilia   Microvilli of mucin-secreting cells

(c)

•**FIGURE 28-17 The Uterine Tubes. (a)** Regions of the uterine tubes. **(b)** A sectional view of the isthmus. (LM × 122) **(c)** A colorized SEM of the ciliated lining of the uterine tube.

boundary between the ampulla and isthmus of the uterine tube.

Along with its transport function, the uterine tube also provides a nutrient-rich environment that contains lipids and glycogen. This mixture supplies nutrients to both spermatozoa and a developing *pre-embryo* (the cell cluster produced by the initial divisions that follow fertilization). Unfertilized oocytes degenerate in the terminal portions of the uterine tubes or within the uterus.

Pelvic Inflammatory Disease (PID)

Pelvic inflammatory disease (PID) in women is a major cause of sterility (infertility). An infection of the uterine tubes, PID affects an estimated 850,000 women each year in the United States. In many cases, sexually transmitted pathogens are involved. As much as 50–80 percent of all first cases may be due to infection by *Neisseria gonorrhoeae,* the organism responsible for symptoms of **gonorrhea** (gon-ō-RĒ-uh), a sexually transmitted disease. Invasion of the region by bacteria that normally reside in the vagina can also cause PID. Symptoms of PID include fever, lower abdominal pain, and elevated white blood cell counts. In severe cases, the infection can spread to other visceral organs or produce a generalized peritonitis.

Sexually active women in the 15–24 age group have the highest incidence of PID. Whereas the use of an oral contraceptive decreases the risk of infection, the presence of an intrauterine device (IUD) may increase the risk by 1.4–7.3 times. Treatment with antibiotics may control the condition, but chronic abdominal pain may persist. In addition, damage and scarring of the uterine tubes can cause infertility by preventing the passage of a zygote to the uterus. Recently, another sexually transmitted bacterium, belonging to the genus *Chlamydia,* has been identified as the probable cause of up to 50 percent of all cases of PID. Despite the fact that women with this infection may develop few if any symptoms, scarring of the uterine tubes can still produce infertility.

THE UTERUS

The **uterus** provides mechanical protection, nutritional support, and waste removal for the developing *embryo* (weeks 1–8) and *fetus* (from week 9 to delivery). In addition, contractions in the muscular wall of the uterus are important in ejecting the fetus at the time of birth.

The uterus is a small, pear-shaped organ about 7.5 cm (3 in.) long with a maximum diameter of 5 cm (2 in.). It weighs 30–40 g (1–1.4 oz). In its normal position, the uterus bends anteriorly near its base, a condition known as *anteflexion* (an-tē-FLEK-shun). In this position, the uterus tilts anteriorly, covering across the superior and posterior surfaces of the urinary bladder (Figure 28-13•, p. 1035). If the uterus bends backward toward the sacrum, the condition is termed *retroflexion* (re-trō-FLEK-shun). Retroflexion, which occurs in about 20 percent of adult women, has no clinical significance. (A retroflexed uterus generally becomes anteflexed spontaneously during the third month of pregnancy.)

Suspensory Ligaments of the Uterus

In addition to the broad ligament, three pairs of suspensory ligaments stabilize the position of the uterus and limit its range of movement (Figure 28-18a,b•). The **uterosacral** (ū-te-rō-SĀ-kral) **ligaments** extend from the lateral surfaces of the uterus to the anterior face of the sacrum, keeping the body of the uterus from moving inferiorly and anteriorly. The **round ligaments** arise on the lateral margins of the uterus just posterior and inferior to the attachments of the uterine tubes. These ligaments extend through the inguinal canal and end in the connective tissues of the external genitalia. The round ligaments restrict primarily posterior movement of the uterus. The **lateral** *(cardinal)* **ligaments** extend from the base of the uterus and vagina to the lateral walls of the pelvis. These ligaments tend to prevent inferior movement of the uterus. Additional mechanical support is provided by the muscles and fascia of the pelvic floor. ∞ *pp. 332–334*

Internal Anatomy of the Uterus

We can divide the uterus into two anatomical regions (Figure 28-18c•, p. 1042): the body and the cervix. The uterine **body,** or *corpus,* is the largest region of the uterus. The **fundus** is the rounded portion of the body superior to the attachment of the uterine tubes. The body ends at a constriction known as the uterine **isthmus.** The **cervix** (SER-viks) is the inferior portion of the uterus that extends from the isthmus to the vagina.

The tubular cervix projects about 1.25 cm (0.5 in.) into the vagina. Within the vagina, the distal end of the cervix forms a curving surface that surrounds the **cervical os** (*os,* an opening or mouth), or *external orifice* of the uterus. The cervical os leads into the **cervical canal,** a constricted passageway that opens into the **uterine cavity** of the body at the **internal os,** or *internal orifice.*

The uterus receives blood from branches of the **uterine arteries,** which arise from branches of the *internal iliac arteries,* and the *ovarian arteries,* which arise from the abdominal aorta inferior to the renal arteries. The arteries to the uterus are extensively interconnected. This arrangement helps ensure a reliable flow of blood to the organ despite changes in

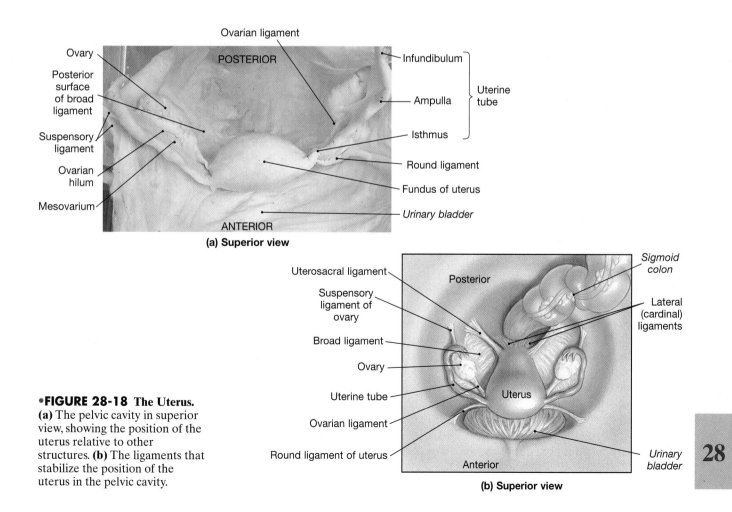

(a) Superior view

•FIGURE 28-18 The Uterus.
(a) The pelvic cavity in superior view, showing the position of the uterus relative to other structures. **(b)** The ligaments that stabilize the position of the uterus in the pelvic cavity.

(b) Superior view

28

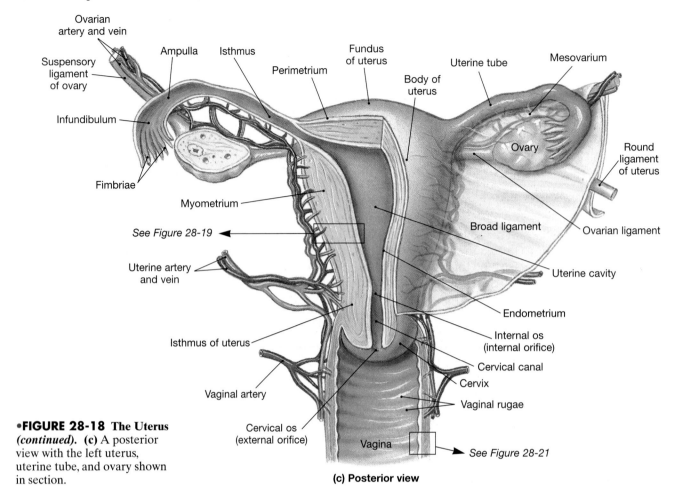

•FIGURE 28-18 The Uterus
(continued). (c) A posterior
view with the left uterus,
uterine tube, and ovary shown
in section.

(c) Posterior view

position and the changes in uterine shape that accompany pregnancy. Numerous veins and lymphatic vessels also supply each portion of the uterus. The uterus is innervated by autonomic fibers from the hypogastric plexus (sympathetic) and from sacral segments S_3 and S_4 (parasympathetic). Sensory information reaches the central nervous system within the dorsal roots of spinal nerves T_{11} and T_{12}. The most-delicate anesthetic procedures used during labor and delivery, known as *segmental blocks,* target only spinal nerves T_{10}–L_1.

The Uterine Wall

The dimensions of the uterus are highly variable. In adult women of reproductive age who have not given birth, the uterine wall is about 1.5 cm (0.5 in.) thick. The uterine wall has a thick, outer, muscular **myometrium** (mī-ō-MĒ-trē-um; *myo-,* muscle + *metra,* uterus) and a thin, inner, glandular **endometrium** (en-dō-MĒ-trē-um), or *mucosa.* The fundus and the posterior surface of the uterine body and isthmus are covered by a serous membrane continuous with the peritoneal lining. This incomplete serosa is called the **perimetrium** (Figure 28-18c•).

The endometrium contributes about 10 percent to the mass of the uterus. The glandular and vascular tissues of the endometrium support the physiological demands of the growing fetus. Vast numbers of uterine glands open onto the endometrial surface. These glands extend deep into the lamina propria almost to the myometrium. Under the influence of estrogen, the uterine glands, blood vessels, and epithelium change with the phases of the monthly *uterine cycle.*

The myometrium, the thickest portion of the uterine wall, forms almost 90 percent of the mass of the uterus. Smooth muscle in the myometrium is arranged into longitudinal, circular, and oblique layers. The smooth muscle tissue of the myometrium provides much of the force needed to move a large fetus out of the uterus and into the vagina.

Cervical Cancer

Cervical cancer is the most common reproductive system cancer in women age 15–34. Roughly 12,800 new cases of invasive cervical cancer are diagnosed each year in the United States, and approximately 33 percent of the individuals will eventually die of this condition. Another 34,900 patients are diagnosed with a less aggressive form of cervical cancer. Cervical and other uterine tumors and cancers are discussed in the *Applications Manual.*
AM *Uterine Tumors and Cancers*

Histology of the Uterus We can divide the endometrium into (1) a **functional zone,** the layer closest to the uterine cavity; and (2) an outer **basilar zone,** adjacent to the myometrium (Figure 28-19a•). The functional zone contains most of the uterine glands and contributes most of the endometrial thickness. The basilar zone attaches the endometrium to the myometrium and contains the terminal branches of the tubular endometrial glands.

Endometrium

Simple columnar epithelium | Endometrial glands | Functional zone | Basilar zone

Uterine cavity

Myometrium

(a)

Spiral artery   Endometrium   Radial artery

Lumen

Straight artery

Myometrium

Uterine artery

(b)

Uterine glands   Arcuate arteries

•**FIGURE 28-19** **The Uterine Wall. (a)** The basic structure of the endometrium. (LM × 32) **(b)** A sectional view of the uterine wall, showing the endometrial regions and the circulatory supply to the endometrium.

Within the myometrium, branches of the uterine arteries form **arcuate arteries,** which encircle the endometrium. **Radial arteries** supply **straight arteries,** which deliver blood to the basilar zone of the endometrium and **spiral arteries,** which supply the functional zone (Figure 28-19b•).

The structure of the basilar zone remains relatively constant over time, but that of the functional zone undergoes cyclical changes in response to sex hormone levels. These cyclical changes produce the characteristic histological features of the uterine cycle.

The Uterine Cycle

The **uterine cycle,** or *menstrual* (MEN-stroo-al) *cycle,* is a repeating series of changes in the structure of the endometrium. The uterine cycle averages 28 days in length, but it can range from 21 to 35 days in healthy women of reproductive age. We can divide the cycle into three phases: (1) *menses,* (2) the *proliferative phase,* and (3) the *secretory phase.* The histological appearance of the endometrium during each phase is shown in Figure 28-20•. The phases occur in response to hormones associated with the regulation of the ovarian cycle. Menses and the proliferative phase occur during the follicular phase of the ovarian cycle. The secretory phase corresponds to the luteal phase of the ovarian cycle. We shall consider the regulatory mechanism in a later section.

Menses The uterine cycle begins with the onset of **menses** (MEN-sēz), an interval marked by the degeneration of the functional zone of the endometrium. This degeneration occurs in patches. It is caused by the constriction of the spiral arteries, which reduces blood flow to areas of endometrium. Deprived of oxygen and nutrients, the secretory glands and other tissues in the functional zone begin to deteriorate. Eventually, the weakened arterial walls rupture, and blood pours into the connective tissues of the functional zone. Blood cells and degenerating tissues then break away and enter the uterine lumen, to be lost by passage through the cervical os and into the vagina. Only the functional zone is affected, because the deeper, basilar zone is provided with blood from the straight arteries, which remain unconstricted.

The sloughing of tissue is gradual, and at each site, repairs begin almost at once. Nevertheless, before menses has ended, the entire functional zone has been lost (Figure 28-20a•). This process of endometrial sloughing is called **menstruation** (men-stroo-Ā-shun). Menstruation generally lasts from 1 to 7 days, and over this period roughly 35 to 50 ml of blood is lost. The process can be relatively painless. Painful menstruation, or **dysmenorrhea,** can result from uterine inflammation and contraction or from conditions involving adjacent pelvic structures.

28

•**FIGURE 28-20 The Uterine Cycle.** The appearance of the endometrium **(a)** at menses (LM × 63) and during the **(b)** proliferative (LM × 66) and **(c)** secretory (LM × 52; blowup, LM × 150) phases of the uterine cycle.

The Proliferative Phase   The basilar zone, including the basal parts of the uterine glands, survives menses intact. In the days after menses, the epithelial cells of these glands multiply and spread across the endometrial surface, restoring the integrity of the uterine epithelium (Figure 28-20b•). Further growth and vascularization result in the complete restoration of the functional zone. As this reorganization proceeds, the endometrium is in the **proliferative phase.** This restoration occurs at the same time as the enlargement of primary and secondary follicles in the ovary. The proliferative phase is stimulated and sustained by estrogens secreted by the developing ovarian follicles.

By the time ovulation occurs, the functional zone is several millimeters thick, and prominent mucous glands extend to the border with the basilar zone. At this time, the endometrial glands are manufacturing a mucus rich in glycogen. The entire functional zone is highly vascularized, with small arteries spiraling toward the inner surface from larger arteries in the myometrium.

The Secretory Phase   During the **secretory phase** of the uterine cycle, the endometrial glands enlarge, accelerating their rates of secretion, and the arteries that supply the uterine wall elongate and spiral through the tissues of the functional zone (Figure 28-20c•). This activity occurs under the combined stimulatory effects of progestins and estrogens from the corpus luteum. This phase begins at the time of ovulation and persists as long as the corpus luteum remains intact.

Secretory activities peak about 12 days after ovulation. Over the next day or two, the glandular activity declines, and the uterine cycle comes to a close as the corpus luteum stops producing stimulatory hormones. A new cycle then begins with the onset of menses and the disintegration of the functional zone. The secretory phase generally lasts 14 days. As a result, you can determine the date of ovulation by counting backward 14 days from the first day of menses.

Endometriosis

In **endometriosis** (en-dō-mē-trē-Ō-sis), an area of endometrial tissue begins to grow outside the uterus. The cause is unknown; because this condition is most common in the inferior portion of the peritoneum, one possibility is that pieces of endometrium sloughed during menstruation are in some way forced through the uterine tubes into the peritoneal cavity, where they have reattached. The severity of the condition depends on the size of the abnormal mass and its location. Abdominal pain, bleeding, pressure on adjacent structures, and infertility are common symptoms. As the island of endometrial tissue enlarges, the symptoms become more severe.

Diagnosis can generally be made by inserting a laparoscope through a small opening in the abdominal wall. Using this device, a physician can inspect the outer surfaces of the uterus and uterine tubes, the ovaries, and the lining of the pelvic cavity. Treatment of endometriosis may involve hormonal therapy or surgical removal of the endometrial mass. If the condition is widespread, a *hysterectomy* (removal of the uterus) or *oophorectomy* (removal of the ovaries) may be required.

Menarche and Menopause The uterine cycle begins at puberty. The first cycle is known as **menarche** (me-NAR-kē; *mēn*, month + *archē*, beginning). It typically occurs at age 11–12. The cycles continue until age 45–55, when **menopause** (MEN-ō-pawz), the last uterine cycle, occurs. In the interim, the regular appearance of uterine cycles is interrupted only by circumstances such as illness, stress, starvation, or pregnancy.

If menarche does not appear by age 16, or if the normal uterine cycle of an adult woman becomes interrupted for 6 months or more, the condition of **amenorrhea** (ā-men-ō-RĒ-uh) exists. *Primary amenorrhea* is the failure to initiate menses. This condition may indicate developmental abnormalities, such as nonfunctional ovaries, the absence of a uterus, or an endocrine or genetic disorder. It can also result from malnutrition; puberty is delayed if leptin levels are too low. ∞ *p. 609* Transient *secondary amenorrhea* can be caused by severe physical or emotional stresses. In effect, the reproductive system gets "switched off." Factors that cause either type of amenorrhea include drastic weight loss, anorexia nervosa, and severe depression or grief. Amenorrhea has also been observed in marathon runners and other women engaged in training programs that require sustained high levels of exertion and severely reduce body lipid reserves.

THE VAGINA

The **vagina** is an elastic, muscular tube. It extends between the cervix and the *vestibule,* a space bounded by the female external genitalia. The vagina has an average length of 7.5–9 cm (3–3.5 in.) but is highly distensible, so its size varies.

At the proximal end of the vagina, the cervix projects into the **vaginal canal.** The shallow recess surrounding the cervical protrusion is known as the **fornix** (FOR-niks). The vagina lies parallel to the rectum, and the two are in close contact posteriorly. Anteriorly, the urethra extends along the superior wall of the vagina from the urinary bladder to the external urethral meatus, which opens into the vestibule. The primary blood supply of the vagina is via the **vaginal branches** of the internal iliac (or uterine) arteries and veins. Innervation is from the hypogastric plexus, sacral nerves S$_2$–S$_4$, and branches of the pudendal nerve. ∞ *p. 420*

The vagina has three major functions:

1. It serves as a passageway for the elimination of menstrual fluids.

2. It receives the penis during sexual intercourse and holds spermatozoa prior to their passage into the uterus.

3. It forms the inferior portion of the *birth canal,* through which the fetus passes during delivery.

Histology of the Vagina

In sectional view, the lumen of the vagina appears constricted, forming a rough H. The vaginal walls contain a network of blood vessels and layers of smooth muscle (Figure 28-21•). The lining is moistened by the secretions of the cervical glands and by the movement of

•**FIGURE 28-21** The Histology of the Vagina. (LM × 27)

28

water across the permeable epithelium. The vagina and vestibule are separated by the **hymen** (HĪ-men), an elastic epithelial fold that partially or completely blocks the entrance to the vagina before the onset of sexual intercourse. The two *bulbospongiosus muscles* extend along either side of the vaginal entrance; their contractions constrict the entrance. ∞ *p. 333* These muscles cover the *vestibular bulbs,* masses of erectile tissue on either side of the vaginal entrance. The vestibular bulbs have the same embryological origins as the corpus spongiosum of the penis in males.

The vaginal lumen is lined by a nonkeratinized stratified squamous epithelium (Figure 28-21•). In the relaxed state, this epithelium is thrown into folds called *rugae.* The underlying lamina propria is thick and elastic, and it contains small blood vessels, nerves, and lymph nodes. The vaginal mucosa is surrounded by an elastic **muscularis** layer, with layers of smooth muscle fibers arranged in circular and longitudinal bundles continuous with the uterine myometrium. The portion of the vagina adjacent to the uterus has a serosal covering continuous with the pelvic peritoneum. Along the rest of the vagina, the muscularis layer is surrounded by an *adventitia* of fibrous connective tissue.

The vagina contains a population of resident bacteria, usually harmless, supported by nutrients in the cervical mucus. The metabolic activity of these bacteria creates an acidic environment, which restricts the growth of many pathogens. *Vaginitis* (va-jin-Ī-tis), an inflammation of the vaginal canal, is caused by fungi, bacteria, or parasites. In addition to any discomfort that may result, the condition may affect the survival of spermatozoa and thereby reduce fertility. ☐ᴬᴹ *Vaginitis* An acidic environment also inhibits sperm motility; for this reason, the buffers in semen are important to successful fertilization.

The hormonal changes associated with the ovarian cycle also affect the vaginal epithelium. A *vaginal smear* is a sample of epithelial cells shed at the surface of the vagina. By examining these cells, a clinician can estimate the corresponding stage in the ovarian and uterine cycles. This technique is an example of *exfoliative cytology,* a diagnostic procedure. ∞ *p. 115*

THE EXTERNAL GENITALIA

The area containing the female external genitalia is the **vulva** (VUL-vuh), or *pudendum* (Figure 28-22•). The vagina opens into the **vestibule,** a central space bounded by small folds known as the **labia minora** (LĀ-bē-uh mi-NOR-uh; singular, *labium minus*). The labia minora are covered with a smooth, hairless skin. The urethra opens into the vestibule just anterior to the vaginal entrance. The **paraurethral glands,** or *Skene's glands,* discharge into the urethra near the external urethral opening. Anterior to this opening, the **clitoris** (KLI-to-ris) projects into the vestibule.

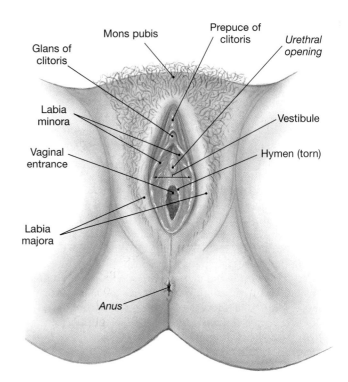

Mons pubis — Prepuce of clitoris — *Urethral opening* — Glans of clitoris — Labia minora — Vestibule — Vaginal entrance — Hymen (torn) — Labia majora — *Anus*

•**FIGURE 28-22** The Female External Genitalia

The clitoris, a small, rounded tissue projection, is the female equivalent of the penis; it is derived from the same embryonic structures. Internally, it contains erectile tissue comparable to the corpora cavernosa of the penis. The clitoris engorges with blood during sexual arousal. A small erectile *glans* sits atop it; extensions of the labia minora encircle the body of the clitoris, forming the **prepuce,** or *hood,* of the clitoris. ☐ᴬᴹ *EMBRYOLOGY SUMMARY 21: The Development of the Reproductive System*

A variable number of small **lesser vestibular glands** discharge their secretions onto the exposed surface of the vestibule, keeping it moist. During sexual arousal, a pair of ducts discharges the secretions of the **greater vestibular glands** *(Bartholin's glands)* into the vestibule near the posterolateral margins of the vaginal entrance. These mucous glands have the same embryological origins as the bulbourethral glands of males.

The outer limits of the vulva are established by the mons pubis and the labia majora. The bulge of the **mons pubis** is created by adipose tissue deep to the skin anterior to the pubic symphysis. Adipose tissue also accumulates within the **labia majora** (singular, *labium majus*), prominent folds of skin that encircle and partially conceal the labia minora and adjacent structures. The outer margins of the labia majora and the mons pubis are covered with coarse hair, but the inner surfaces of the labia majora are relatively hairless. Sebaceous glands and scattered apocrine sweat glands secrete onto the inner surface of the labia majora, moistening and lubricating them.

THE MAMMARY GLANDS

A newborn infant cannot fend for itself, and several of its key systems have yet to complete their development. Over the initial period of adjustment to an independent existence, the infant gains nourishment from the milk secreted by the maternal **mammary glands.** Milk production, or **lactation** (lak-TĀ-shun), occurs in these glands. In females, mammary glands are specialized organs of the integumentary system that are controlled mainly by hormones of the reproductive system and by the *placenta,* a temporary structure that provides the embryo or fetus with nutrients.

On each side, a mammary gland lies in the subcutaneous tissue of the **pectoral fat pad** deep to the skin of the chest (Figure 28-23a•). Each breast bears a **nipple,** a small conical projection where the ducts of the underlying mammary gland open onto the body surface. The reddish-brown skin around each nipple is the **areola** (a-RĒ-ō-luh). Large sebaceous glands deep to the areolar surface give it a grainy texture.

The glandular tissue of the mammary gland consists of separate lobes, each containing several secretory lobules. Ducts leaving the lobules converge, giving rise to a single **lactiferous** (lak-TIF-e-rus) **duct** in each lobe (Figure 28-23•). Near the nipple, that lactiferous duct enlarges, forming an expanded chamber called a **lactiferous sinus.** Typically, 15–20 lactiferous sinuses open onto the surface of each nipple. Dense connective tissue surrounds the duct system and forms partitions that extend between the lobes and the lobules. These bands of connective tissue, the *suspensory ligaments of the breast,* originate in the dermis of the overlying skin. A layer of areolar tissue separates the mammary gland complex from the underlying pectoralis muscles. Branches of the *internal thoracic artery* supply blood to each mammary gland (Figure 21-23•, p. 726).

Figure 28-23b,c• compares the histological organizations of inactive and active mammary glands. The inactive, or *resting,* mammary gland is dominated by a duct system rather than by active glandular cells. The size of the mammary glands in a nonpregnant woman reflects primarily the amount of adipose tissue present, not the amount of glandular tissue. The secretory apparatus does not complete its development unless pregnancy occurs. The active mammary gland is a tubuloalveolar gland, consisting of multiple glandular tubes that end in secretory alveoli. We shall discuss the hormonal mechanisms involved in Chapter 29.

☑ As the result of infections such as gonorrhea, scar tissue can block the lumen of each uterine tube. How would this blockage affect a woman's ability to conceive?

☑ What is the advantage of the acidic pH of the vagina?

☑ Which layer of the uterus sloughs off during menstruation?

☑ Would the blockage of a single lactiferous sinus interfere with the delivery of milk to the nipple? Explain.

(a)

(b) Inactive mammary gland

(c) Active mammary gland

•**FIGURE 28-23 The Mammary Glands. (a)** The mammary gland of the left breast. **(b)** An inactive mammary gland of a nonpregnant woman. (LM × 53) **(c)** An active mammary gland of a nursing woman. (LM × 119)

28

Breast Cancer

The mammary glands are cyclically stimulated by the changing levels of circulating reproductive hormones that accompany the uterine cycle. The effects usually go unnoticed, but occasional discomfort or even inflammation of mammary gland tissues can occur late in the cycle. If inflamed lobules become walled off with scar tissue, **cysts** are created. Clusters of cysts can be felt in the breast as discrete masses, a condition known as **fibrocystic disease.** Because the symptoms are similar to those of breast cancer, biopsies may be needed to distinguish between this benign condition and breast cancer.

Breast cancer is a malignant, metastasizing cancer of the mammary gland. It is the leading cause of death in women between the ages of 35 and 45, but it is most common in women over age 50. Approximately 41,200 deaths occurred in the United States from breast cancer in 2000, and approximately 184,200 new cases were reported. An estimated 12 percent of women in the United States will develop breast cancer at some point in their lifetime, and the rate is steadily rising. The incidence is highest among Caucasian Americans, somewhat lower in African Americans, and lowest in Asian Americans and American Indians. Notable risk factors include (1) a family history of breast cancer, (2) a pregnancy after age 30, and (3) early menarche (first menstrual period) or late menopause (last menstrual period). Breast cancers in males are very rare, but about 400 men die from breast cancer each year in the United States.

Despite repeated studies, no links have been proven between oral contraceptive use, estrogen therapy, fat consumption, or alcohol use and breast cancer. It appears likely that multiple factors are involved. Most women never develop breast cancer—even women in families with a history of the disease. Adequate amounts of nutrients and vitamins and a diet rich in fruits and vegetables appear to offer some protection against the development of breast cancer. Mothers who breast-fed (nursed) their babies have a 20-percent lower incidence of breast cancer after menopause than do mothers who did not breast-feed. The reason is not known. Adding to the mystery, nursing does not appear to affect the incidence of premenopausal breast cancer.

Early detection of breast cancer is the key to reducing mortalities. *Most cases of breast cancer are found through self-examination,* but the use of clinical screening techniques has increased in recent years.

Mammography involves the use of X-rays to examine breast tissues. The radiation dosage can be low, because only soft tissues must be penetrated. This procedure gives the clearest picture of conditions in the breast tissues. Ultrasound can provide some information, but the resulting images lack the detail of standard mammograms.

For treatment to be successful, breast cancer must be identified while it is still relatively small and localized. Once it has grown larger than 2 cm (0.78 in.), the chances of long-term survival worsen. A poor prognosis also follows if the cancer cells have spread through the lymphatic system to the axillary lymph nodes. If the nodes are not yet involved, the chances of 5-year survival are about 82 percent, but if four or more nodes are involved, the survival rate drops to 21 percent.

Treatment of breast cancer begins with the removal of the tumor. Because in most cases the cancer cells begin to spread before the condition is diagnosed, part or all of the affected mammary gland is surgically removed:

- In a **segmental mastectomy,** or *lumpectomy,* only a portion of the mammary gland is removed.
- In a **total mastectomy,** the entire mammary gland is removed, but other tissues are left intact.
- In a **radical mastectomy,** the pectoralis muscles, the mammary gland, and the axillary lymph nodes are removed. In a *modified radical mastectomy,* the most common operation, the mammary gland and nodes are removed but the muscular tissue remains intact.

A combination of chemotherapy, radiation treatments, and hormone treatments may be used to supplement the surgical procedures. *Tamoxifen*™ is an anti-estrogen drug that is used to treat some cases of breast cancer. It is more effective than conventional chemotherapy for treating breast cancer in women over 50, and it has fewer unpleasant side effects. It can also be used in addition to regular chemotherapy in the treatment of advanced-stage disease. As an added bonus, tamoxifen prevents and even reverses the osteoporosis of aging. This drug has risks, however. When given to premenopausal women, tamoxifen can cause amenorrhea and hot flashes similar to those of menopause. Tamoxifen has also been linked to an increased risk of endometrial cancer and potentially to liver cancer as well. It has been proposed that this drug and others like it be used to prevent breast cancer as well as to treat it.

HORMONES AND THE FEMALE REPRODUCTIVE CYCLE

The activity of the female reproductive tract is under hormonal control that involves an interplay between secretions of both the pituitary gland and gonads. But the regulatory pattern in females is much more complicated than in males, because it must coordinate the ovarian and uterine cycles. Circulating hormones control the **female reproductive cycle,** coordinating the ovarian and uterine cycles to ensure proper reproductive function. If the two cycles cannot be coordinated normally, infertility results. A woman who fails to ovulate cannot conceive, even if her uterus is perfectly normal. A woman who ovulates normally but whose uterus is not ready to support an embryo will be just as infertile. Because the processes are complex and difficult to study, many of the biochemical details of the female reproductive cycle still elude us, but the general patterns are reasonably clear.

As in males, GnRH from the hypothalamus regulates reproductive function in females. However, in females the GnRH pulse frequency and amplitude (amount secreted per pulse) change throughout the course of the ovarian cycle. If the hypothalamus were a radio, the pulse frequency would correspond to the radio frequency, and the amplitude would be the volume control. We will consider changes in pulse frequency, as their effects are both dramatic and reasonably well-understood. Changes in GnRH pulse frequency are essential to normal FSH and LH production and thus to normal ovulation. If GnRH is absent or supplied at a constant rate (without pulses), FSH and LH secretion will stop in a matter of hours.

When you shift from one radio frequency to another, you change stations. You may then hear a very different message—from talk radio to jazz or from hard rock to classical. When the GnRH pulse frequency shifts, the pituitary gland hears a different mes-sage. The cells responsible for FSH and LH production are called *gonadotropes*. When the GnRH pulse frequency shifts, these cells change their pattern of FSH and LH production. For example, at one pulse frequency, the gonadotropes respond by preferentially secreting FSH, whereas at another frequency, LH is the primary hormone released. FSH and LH production also occurs in pulses that follow the rhythm of GnRH pulses. The precise effects of these pulses on target cells are not known. Changes in GnRH pulse frequency are controlled primarily by circulating levels of estrogens and progestins. Estrogens increase the GnRH pulse frequency, and progestins decrease it.

Hormones and the Follicular Phase

Follicular development begins under FSH stimulation; each month some of the primordial follicles begin to develop into primary follicles. As the follicles enlarge, thecal cells start producing *androstenedione,* a steroid hormone that is a key intermediate in the synthesis of most sex hormones (Figure 28-24•). Androstenedione is absorbed by the granulosa cells and converted to estrogens. In addition, small quantities of estrogens are secreted by interstitial cells scattered throughout the ovarian stroma. Circulating estrogens are bound primarily to albumins, with lesser amounts carried by GBG (gonadal steroid-binding globulin).

Three estrogens are in circulation: (1) estradiol, (2) estrone, and (3) estriol. All have similar effects on their target tissues. **Estradiol** (es-tra-DĪ-ol) is the most abundant estrogen, and its effects on target tissues are most pronounced. It is the dominant hormone prior to ovulation. In estrogen synthesis, androstenedione is first converted to testosterone, which the enzyme aromatase converts to estradiol. The synthesis of both *estrone* and *estriol* proceeds directly from androstenedione (Figure 28-24•).

•**FIGURE 28-24** **Pathways of Steroid Hormone Synthesis.** All gonadal steroids are derived from cholesterol. In men, the pathway ends with the synthesis of testosterone, which may subsequently be converted to dihydrotestosterone. In women, an additional step past testosterone leads to estradiol synthesis. The synthesis of progesterone and the estrogens other than estradiol involve alternative pathways.

28

•**FIGURE 28-25** The Hormonal
Regulation of Ovarian Activity

Estrogens have multiple functions, affecting the activities of many tissues and organs throughout the body. Important general functions include (1) stimulating bone and muscle growth; (2) maintaining female secondary sex characteristics, such as body hair distribution and the location of adipose tissue deposits; (3) affecting central nervous system (CNS) activity (especially in the hypothalamus, where estrogens increase sexual drive); (4) maintaining functional accessory reproductive glands and organs; and (5) initiating the repair and growth of the endometrium. Figure 28-25•, which diagrams the hormonal regulation of ovarian activity, includes an overview of the effects of estrogen on various aspects of reproductive function.

Figure 28-26• shows the changes in circulating hormone levels that accompany the ovarian cycle. Early in the follicular phase, estrogen levels are low and the GnRH pulse frequency is 16–24 per day (one pulse every 60–90 minutes). At this pulse frequency, FSH is the dominant hormone released by the anterior pituitary gland; the estrogen released by developing follicles inhibits LH secretion. As secondary follicles develop, FSH levels decline due to the negative feedback effects of inhibin. Follicular development and maturation continues, however, supported by the combination of estrogens, FSH, and LH.

As one or more tertiary follicles begin forming, the concentration of circulating estrogens rises steeply. As

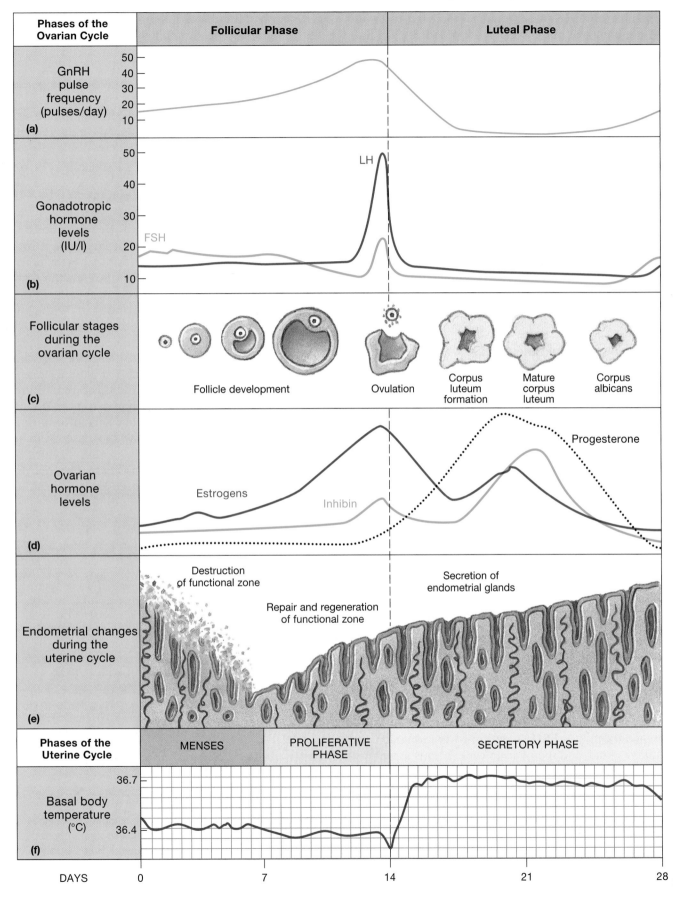

•FIGURE 28-26 The Hormonal Regulation of the Female Reproductive Cycle a result, the GnRH pulse frequency increases to about 36 per day (one pulse every 30–60 minutes). The increased pulse frequency stimulates LH secretion. In addition, at roughly day 10 of the cycle, the effect of estrogen on LH secretion changes from inhibition to stimulation. The switchover occurs only after rising estrogen levels have exceeded a threshold value for about 36 hours. (The threshold value and the time required vary among individuals.) High estrogen levels also increase gonadotrope sensitivity to GnRH. At about day 14, the estrogen level has peaked, the gonadotropes are at maximum sensitivity, and the GnRH pulses are arriving about every 30 minutes. The result is a massive release of LH from the anterior pituitary gland. This sudden surge in LH concentration triggers (1) the completion of meiosis I by the primary oocyte, (2) the rupture of the follicular wall, and (3) ovulation. Ovulation typically occurs 34–38 hours after the LH surge begins, roughly 9 hours after the LH peak.

Hormones and the Luteal Phase

The high LH levels that trigger ovulation also promote progesterone secretion and the formation of the corpus luteum. As progesterone levels rise and estrogen levels fall, the GnRH pulse frequency declines sharply, soon reaching 1–4 pulses per day. This frequency of GnRH pulses stimulates LH secretion more than it does FSH secretion, and the LH maintains the structure and secretory function of the corpus luteum.

Although moderate amounts of estrogens are secreted by the corpus luteum, progesterone is the main hormone of the luteal phase. Its primary function is to continue the preparation of the uterus for pregnancy by enhancing the blood supply to the functional zone and stimulating the secretion of the endometrial glands. Progesterone levels remain high for the next week, but unless pregnancy occurs, the corpus luteum then begins to degenerate. Roughly 12 days after ovulation, the corpus luteum becomes nonfunctional, and progesterone and estrogen levels fall markedly. The blood supply to the functional zone is restricted, and the endometrial tissues begin to deteriorate. As progesterone and estrogen levels drop, the GnRH pulse frequency increases. This increase stimulates FSH secretion by the anterior pituitary gland, and the ovarian cycle begins again.

The hormonal changes involved with the ovarian cycle in turn affect the activities of other reproductive tissues and organs. At the uterus, the hormonal changes maintain the uterine cycle.

Hormones and the Uterine Cycle

Figure 28-26e• follows changes in the endometrium during a single uterine cycle. The declines in progesterone and estrogen levels that accompany the degeneration of the corpus luteum (Figure 28-26c,d•) result in menses. The sloughing of endometrial tissue continues for several days, until rising estrogen levels stimulate the repair and regeneration of the functional zone of the endometrium. The proliferative phase continues until rising progesterone levels mark the arrival of the secretory phase. The combination of estrogen and progesterone then causes the enlargement of the endometrial glands and an increase in their secretory activities.

Hormones and Body Temperature

The monthly hormonal fluctuations cause physiological changes that affect the core body temperature. During the follicular phase, when estrogen is the dominant hormone, *basal body temperature,* or the resting body temperature measured on awakening in the morning, is about 0.3°C lower than it is during the luteal phase, when progesterone dominates. At the time of ovulation, basal body temperature declines noticeably, making the rise in temperature over the following day even more noticeable (Figure 28-26f•). As a result, by keeping records of body temperature over a few uterine cycles, a woman can often determine the precise day of ovulation. This information can be important for individuals who wish to avoid or promote a pregnancy, because fertilization typically occurs within a day of ovulation. Thereafter, oocyte viability and the likelihood of successful fertilization decrease markedly.

Premenstrual Syndrome

Several physical and physiological changes occur in women 7–10 days before the start of menses. Fluid retention, breast enlargement, headaches, pelvic pain, and bloating are common symptoms. These sensations may be associated with psychological changes producing irritability, anxiety, and depression. This combination of symptoms has been called **premenstrual syndrome (PMS).**

The mechanism responsible for PMS has yet to be determined. Changes in sex hormone levels may be involved directly, by action on peripheral organ systems, or indirectly, by the modification of neurotransmitter release in the CNS. There are no laboratory tests or procedures to diagnose PMS, but tracking the appearance of symptoms over a 2–3 month period can reveal characteristic patterns. Treatment is symptomatic and may involve exercise, dietary change, or medication, depending on the nature of the primary symptom. For example, if headache is the major problem, analgesics are prescribed; diuretics may be used to combat bloating and fluid retention. For severe PMS, drugs can be administered that block GnRH secretion and stop uterine cycles for 6 months or more. During the interim, estrogens can be administered to prevent symptoms of premature menopause.

☑ What changes would you expect to observe in the ovarian cycle if the LH surge did not occur?

☑ What effect would a blockage of progesterone receptors in the uterus have on the endometrium?

☑ What event occurs in the uterine cycle when the levels of estrogens and progesterone decline?

THE PHYSIOLOGY OF SEXUAL INTERCOURSE

Sexual intercourse, also known as *coitus* (KŌ-i-tus) or *copulation,* introduces semen into the female reproductive tract. We shall now consider the process as it affects the reproductive systems of males and females.

MALE SEXUAL FUNCTION

Sexual function in males is coordinated by complex neural reflexes that we do not yet understand completely. The reflex pathways utilize the sympathetic and parasympathetic divisions of the autonomic nervous system. During sexual **arousal,** erotic thoughts, the stimulation of sensory nerves in the genital region, or both lead to an increase in the parasympathetic outflow over the pelvic nerves. This outflow leads to erection of the penis (as discussed on p. 1030). The integument covering the glans of the penis contains numerous sensory receptors, and erection tenses the skin and increases their sensitivity. Subsequent stimulation can initiate the secretion of the bulbourethral glands, lubricating the penile urethra and the surface of the glans.

During intercourse, the sensory receptors of the penis are rhythmically stimulated. This stimulation eventually results in the coordinated processes of emission and ejaculation. **Emission** occurs under sympathetic stimulation. The process begins when the peristaltic contractions of the ampulla push fluid and spermatozoa into the prostatic urethra. The seminal vesicles then begin contracting, and the contractions increase in force and duration over the next few seconds. Peristaltic contractions also appear in the walls of the prostate gland. The combination moves the seminal mixture into the membranous and penile portions of the urethra. While the contractions are proceeding, sympathetic commands also cause the contraction of the urinary bladder and the internal urethral sphincter. The combination of elevated pressure inside the bladder and the contraction of this sphincter effectively prevents the passage of semen into the bladder.

Ejaculation occurs as powerful, rhythmic contractions appear in the *ischiocavernosus* and *bulbospongiosus muscles,* two superficial skeletal muscles of the pelvic floor. The ischiocavernosus muscles insert along the sides of the penis; their contractions serve primarily to stiffen that organ. The bulbospongiosus muscle wraps around the base of the penis; the contraction of this muscle pushes semen toward the external urethral opening. These contractions are controlled by somatic motor neurons in the inferior lumbar and superior sacral segments of the spinal cord. (The positions of these muscles are shown in Figure 11-13•, p. 333.)

Ejaculation is associated with intensely pleasurable sensations, an experience known as male **orgasm** (OR-gazm). Several other noteworthy physiological changes occur at this time, including pronounced but temporary increases in heart rate and blood pressure. After ejaculation has been completed, blood begins to leave the erectile tissue, and the erection begins to subside. This subsidence, called **detumescence** (de-tū-MES-ens), is mediated by the sympathetic nervous system.

In summary, arousal, erection, emission, and ejaculation are controlled by a complex interplay between the sympathetic and parasympathetic divisions of the autonomic nervous system. Higher centers, including the cerebral cortex, can facilitate or inhibit many of the important reflexes, thereby modifying the patterns of sexual function. Any physical or psychological factor that affects a single component of the system can result in male sexual dysfunction, also called **impotence.**

Impotence

Impotence is an inability to achieve or maintain an erection. Various physical causes may be responsible for impotence, because erection involves vascular changes as well as neural commands. For example, low blood pressure in the arteries servicing the penis, due to a circulatory blockage such as a plaque, will affect the ability to obtain an erection. Drugs, alcohol, trauma, or illnesses that affect the autonomic nervous system or the central nervous system can have the same effect. But male sexual performance can also be strongly affected by the psychological state of the individual. Temporary periods of impotence are relatively common in healthy individuals who are experiencing severe stresses or emotional problems. Depression, anxiety, and fear of impotence are examples of emotional factors that can result in sexual dysfunction. The prescription drug Viagra® has proven useful in treating many cases of impotence; it enhances and prolongs the effects of nitric oxide on the erectile tissue of the penis.

28

Birth Control Strategies

For physiological, logistical, financial, or emotional reasons, most adults practice some form of conception control during their reproductive years. Well over 50 percent of U.S. women age 15–44 carry out some method of contraception. In 1996, an estimated 50 million American women were taking oral birth control pills. When the simplest and most obvious method, sexual abstinence, is unsatisfactory for some reason, another method of contraception must be used to avoid unwanted pregnancies. Many methods are available, so the selection process can be quite involved. Because each has specific strengths and weaknesses, the potential risks and benefits must be carefully analyzed on an individual basis. We shall consider only a few of the available contraception methods here.

Sterilization is a surgical procedure that makes an individual unable to provide functional gametes for fertilization. Either sexual partner may be sterilized. In a **vasectomy** (vaz-EK-to-mē), a segment of the ductus deferens is removed, making it impossible for spermatozoa to pass from the epididymis to the distal portions of the reproductive tract. The surgery can be performed in a physician's office in a matter of minutes. The spermatic cords are located as they ascend from the scrotum on either side; after each cord is opened, the ductus deferens is severed. A 1-cm section is removed, and the cut ends are tied shut. The cut ends cannot reconnect, and in time, scar tissue forms a permanent seal. Alternatively, the cut ends of the ductus deferens are blocked with silicone plugs that can later be removed. This more recent vasectomy procedure makes it possible to restore fertility at a later date. After a vasectomy, the man experiences normal sexual function, because the secretions of the epididymis and testes normally account for only about 5 percent of the volume of semen. Spermatozoa continue to develop, but they remain within the epididymis until they degenerate. The failure rate for this procedure is 0.08 percent. (*Failure* for a birth control method is defined as a resulting pregnancy.)

The uterine tubes can be blocked by a surgical procedure known as a **tubal ligation.** The failure rate for this procedure is estimated at 0.45 percent. Because the surgery requires that the abdominopelvic cavity be opened, complications are more likely than with vasectomy. As in a vasectomy, attempts may be made to restore fertility after a tubal ligation.

Oral contraceptives manipulate the female hormonal cycle so that ovulation does not occur. The contraceptive pills produced in the 1950s contained relatively large amounts of progestins. These concentrations were adequate to suppress pituitary production of GnRH, so FSH was not released and ovulation did not occur. In most of the oral contraceptive products developed subsequently, small amounts of estrogens have been added. Current *combination pills* differ significantly from the earlier products in that the hormonal doses are much lower, with only one-tenth the progestins and less than half the estrogens. The hormones are administered in a cyclic fashion, beginning 5 days after the start of menses and continuing for 3 weeks. Then, over the fourth week, the woman takes placebo pills or no pills at all. Low-dosage combination pills are sometimes prescribed for women who experience irregular uterine cycles, because these pills create a regular 28-day cycle.

At least 20 brands of combination oral contraceptives are now available, and more than 200 million women are using them worldwide. In the United States, 25 percent of women under age 45 use a combination pill to prevent conception. The failure rate for the combination oral contraceptives, when used as prescribed, is 0.24 percent over a 2-year period. Birth control pills are not risk-free. Combination pills can worsen problems associated with severe hypertension, diabetes mellitus, epilepsy, gallbladder disease, heart trouble, and acne. Women taking oral contraceptives are also at increased risk of venous thrombosis, strokes, pulmonary embolism, and (for women over 35) heart disease. Pregnancy has similar risks, however.

Two progesterone-only forms of birth control are now available: Depo-provera® and the Norplant® system. *Depo-provera* is injected every 3 months. Uterine cycles are initially irregular; in roughly 50 percent of women using this product, uterine cycles then cease. The major problems with this contraceptive method are (1) a tendency to gain weight after 3 or more years of use, and (2) a slow return to fertility (up to 18 months) after injections are discontinued. The Silastic® (silicone rubber) tubes of the *Norplant system* are saturated with progesterone and inserted under the skin. This method provides birth control for approximately 5 years, but to date the relatively high cost has limited its use. Because it does not supply estrogen, it produces fewer hormonal side-effects than combination pills do. Fertility returns immediately after the removal of a Norplant device. Both Depo-provera and the Norplant system are easy to use and are extremely convenient.

The **condom,** also called a *prophylactic* or "rubber," covers the body of the penis during intercourse and keeps spermatozoa from reaching the female reproductive tract. Condoms are also used to prevent the transmission of sexually transmitted diseases, such as syphilis, gonorrhea, and AIDS. The reported condom failure rate varies from 6 to 17 percent, depending on the study criteria.

Vaginal barriers, such as the *diaphragm* and *cervical cap,* rely on similar principles. A diaphragm, the most popular form of vaginal barrier in use today, consists of a dome of latex rubber with a small metal hoop supporting the rim. Because vaginas vary in size, women choosing this method must be individually fitted. Before intercourse, the diaphragm is inserted so that it covers the cervical os. The diaphragm is usually coated with a small amount of spermicidal (sperm-killing) jelly or cream, adding to the effectiveness of the barrier. The failure rate of a properly fitted diaphragm is estimated at 5–6 percent. The cervical cap is smaller and lacks the metal rim. It, too, must be fitted carefully, but unlike the diaphragm, it can be left in place for several days. The failure rate (8 percent) is higher than that for diaphragm use.

An **intrauterine device (IUD)** consists of a small plastic loop or a T that can be inserted into the uterine cavity. The mechanism of action remains uncertain, but it is known that IUDs stimulate prostaglandin production in the uterus. The resulting change in the chemical composition of uterine secretions lowers the chances of fertilization and subsequent *implantation* of the zygote into the uterine lining. (We shall discuss implantation in Chapter 29.) In the United States, IUDs are in limited use, but they remain popular in many other countries. The failure rate is estimated at 5–6 percent.

The **rhythm method** involves abstaining from sexual activity on the days ovulation might be occurring. The timing is estimated on the basis of previous patterns of menstruation and in some cases by monitoring changes in basal body temperature and cervical mucus. The failure rate of the rhythm method is very high—almost 25 percent.

Sterilization, oral contraceptives, condoms, and vaginal barriers are the primary contraception methods for all age groups. But the proportion of the population using a particular method varies by age group.

Sterilization is most popular among older women, who may already have had children. Relative availability also plays a role. For example, a sexually active female under age 18 can buy a condom more easily than she can obtain a prescription for an oral contraceptive. But many of the differences are attributable to the relationship between risks and benefits for each age group.

When considering the use and selection of contraceptives, many people simply examine the list of potential complications and make the "safest" choice. For example, media coverage of the risks associated with oral contraceptives made many women reconsider their use. But complex decisions should not be made on such a simplistic basis, and the risks associated with use of contraceptives must be considered in light of their relative efficiencies. Although pregnancy is a natural phenomenon, it has risks, and the mortality rate for pregnant women in the United States averages about 8 deaths per 100,000 pregnancies. That average incorporates a broad range; the rate is 5.4 per 100,000 women under 20 but 27 per 100,000 among women over 40. Although these risks are small, for pregnant women over age 35 the chances of dying from complications related to pregnancy are almost twice as great as the chances of being killed in an automobile accident and are many times greater than the risks associated with the use of oral contraceptives. For women in developing nations, the comparison is even more striking. The mortality rate for pregnant women in parts of Africa is approximately 1 per 150 pregnancies. In addition to preventing pregnancy, combination birth control pills have also been shown to reduce the risks of ovarian and endometrial cancers and fibrocystic breast disease.

Before age 35, *the risks associated with contraceptive use are lower than the risks associated with pregnancy.* The notable exception involves individuals who take the pill but also smoke cigarettes. Younger women are more fertile, so despite a lower mortality rate for each pregnancy, they are likely to have more pregnancies. As a result, birth control failures imply a higher risk in the younger age groups.

After age 35, the risks of complications associated with oral contraceptive use increase, but the risks of using other methods remain relatively stable. Women over age 35 (smokers) or 40 (nonsmokers) are therefore typically advised to seek other forms of contraception. Because each contraceptive method has its own advantages and disadvantages, research on contraception control continues. [AM] *Experimental Contraceptive Methods*

FEMALE SEXUAL FUNCTION

The phases of female sexual function are comparable to those of male sexual function. During sexual arousal, parasympathetic activation leads to an engorgement of the erectile tissues of the clitoris and increased secretion of cervical mucous glands and the greater vestibular glands. Clitoral erection increases the receptors' sensitivity to stimulation, and the cervical and vestibular glands lubricate the vaginal walls. A network of blood vessels in the vaginal walls becomes filled with blood at this time, and the vaginal surfaces are also moistened by the fluid that moves across the epithelium from underlying connective tissues. (This process accelerates during intercourse as the result of mechanical stimulation.) Paraympathetic stimulation also causes engorgement of blood vessels at the nipples, making them more sensitive to touch and pressure.

During sexual intercourse, rhythmic contact of the penis with the clitoris and vaginal walls, reinforced by touch sensations from the breasts and other stimuli (visual, olfactory, and auditory), provides stimulation that leads to orgasm. Female orgasm is accompanied by peristaltic contractions of the uterine and vaginal walls and, by means of impulses traveling over the pudendal nerves, rhythmic contractions of the bulbospongiosus and ischiocavernosus muscles. The latter contractions give rise to the sensations of orgasm.

Sexually Transmitted Diseases

Sexually transmitted diseases (STDs) are transferred from individual to individual, primarily or exclusively by sexual intercourse. A variety of bacterial, viral, and fungal infections are included in this category. At least two dozen STDs are currently recognized. The bacterium *Chlamydia* can cause PID and infertility. Other types of STDs are also quite dangerous, and a few, including AIDS, are deadly. The incidence of STDs has been increasing in the United States since 1984; an estimated 15 million cases were diagnosed during 1999. Poverty, coupled with drug use, prostitution, and the appearance of drug-resistant pathogens all contribute to the problem. The *Applications Manual* contains a detailed discussion of the most common forms of STD, including *gonorrhea, syphilis, herpes, genital warts,* and *chancroid.*

AM *Sexually Transmitted Diseases*

 # AGING AND THE REPRODUCTIVE SYSTEM

Just as the aging process affects our other systems, it also affects the reproductive systems of men and women. These systems become fully functional at puberty, as noted earlier in the chapter. Thereafter, the most striking age-related changes in the female reproductive system occur at menopause. Comparable age-related changes in the male reproductive system occur more gradually and over a longer period of time.

MENOPAUSE

Menopause is usually defined as the time that ovulation and menstruation cease. Menopause typically occurs at age 45–55, but in the years preceding it, the ovarian and uterine cycles become irregular. This interval is called *perimenopause*. A shortage of primordial follicles is the underlying cause of the cycle irregularities. It has been estimated that almost 7 million potential oocytes are in fetal ovaries after 5 months of development, but the number drops to about 2 million at birth and to a few hundred thousand at puberty. As perimenopause arrives, the number of follicles responding each month begins to drop markedly. As the number of available follicles decreases, levels of estrogen decline and may fail to rise enough to trigger ovulation. By age 50, there are often no primordial follicles left to respond to FSH. In **premature menopause,** this depletion occurs before age 40.

Menopause is accompanied by a decline in circulating concentrations of estrogen and progesterone and a sharp and sustained rise in the production of GnRH, FSH, and LH. The decline in estrogen levels leads to reductions in the size of the uterus and breasts, accompanied by a thinning of the urethral and vaginal epithelia. The reduced estrogen concentrations have also been linked to the development of osteoporosis, presumably because bone deposition proceeds at a slower rate. A variety of neural effects are also reported, including "hot flashes," anxiety, and depression. Hot flashes typically begin while estrogen levels are declining and cease when estrogen levels reach minimal values. These intervals of elevated body temperature are associated with surges in LH production. The hormonal mechanisms involved in other CNS effects of menopause are poorly understood. In addition, the risk of atherosclerosis and other forms of cardiovascular disease increases after menopause.

The majority of women experience only mild symptoms, but some individuals experience acutely unpleasant symptoms in perimenopause or during or after menopause. For most of those individuals, hormone replacement therapies involving a combination of estrogens and progestins can prevent osteoporosis and the neural and vascular changes associated with menopause. The hormones are administered as pills, by injection, or by transdermal "estrogen patches." The synthetic hormone *etidronate* inhibits osteoporosis by suppressing osteoclast activity. When given at intervals over a 2-year period, it increases bone mass and reduces the rate of fracture incidence in postmenopausal women.

THE MALE CLIMACTERIC

Changes in the male reproductive system occur more gradually than do those in the female reproductive system. The period of change is known as the **male climacteric.** Levels of circulating testosterone begin to decline between ages 50 and 60, and levels of circulating FSH and LH increase. Although sperm production continues (men well into their eighties can father children), older men experience a gradual reduction in sexual activity. This decrease may be linked to declining testosterone levels. Some clinicians suggest the use of testosterone replacement therapy to enhance libido (sexual drive) in elderly men.

INTEGRATION WITH OTHER SYSTEMS

Normal human reproduction is a complex process that requires the participation of multiple systems. The hormones discussed in this chapter play a major role in coordinating reproductive events (Table 28-1). The reproductive process depends on a variety of physical, physiological, and psychological factors, many of which require intersystem cooperation. Figure 28-27• summarizes the relationships between the reproductive system and other physiological systems. The man's sperm count must be adequate, the semen must have the correct pH and nutrients, and erection

TABLE 28-1 Hormones of the Reproductive System

Hormone	Source	Regulation of Secretion	Primary Effects
Gonadotropin-releasing hormone (GnRH)	Hypothalamus	*Males:* inhibited by testosterone and possibly by inhibin *Females:* GnRH pulse frequency increased by estrogens, decreased by progestins	Stimulates FSH secretion, LH synthesis As above
Follicle-stimulating hormone (FSH)	Anterior pituitary	*Males:* stimulated by GnRH, inhibited by inhibin *Females:* stimulated by GnRH, inhibited by inhibin	*Males:* stimulates spermatogenesis and spermiogenesis through effects on sustentacular cells *Females:* stimulates follicle development, estrogen production, and oocyte maturation
Luteinizing hormone (LH)	Anterior pituitary	*Males:* stimulated by GnRH *Females:* production stimulated by GnRH, secretion by the combination of high GnRH pulse frequencies and high estrogen levels	*Males:* stimulates interstitial cells to secrete testosterone *Females:* stimulates ovulation, formation of corpus luteum, and progestin secretion
Androgens (primarily testosterone and dihydrotestosterone)	Interstitial cells of testes	Stimulated by LH	Establish and maintain secondary sex characteristics and sexual behavior; promote maturation of spermatozoa; inhibit GnRH secretion
Estrogens (primarily estradiol)	Granulosa and thecal cells of developing follicles; corpus luteum	Stimulated by FSH	Stimulate LH secretion (at high levels); establish and maintain secondary sex characteristics and sexual behavior; stimulate repair and growth of endometrium; increase frequency of GnRH pulses
Progestins (primarily progesterone)	Granulosa cells from mid-cycle through functional life of corpus luteum	Stimulated by LH	Stimulate endometrial growth and glandular secretion; reduce frequency of GnRH pulses
Inhibin	Sustentacular cells of testes and granulosa cells of ovaries	Stimulated by factors released by developing spermatozoa (male) and developing follicles (female)	Inhibits secretion of FSH (and possibly of GnRH)

28

INTEGUMENTARY SYSTEM

Covers external genitalia; provides sensations that stimulate sexual behaviors; mammary gland secretions provide nourishment for newborn

Reproductive hormones affect distribution of body hair and subcutaneous fat deposits

THE REPRODUCTIVE SYSTEM

FOR ALL SYSTEMS

Secretion of hormones with effects on growth and metabolism

Urethra in males carries semen to exterior; kidneys remove wastes generated by reproductive tissues

Accessory organ secretions may have antibacterial action that helps prevent urethral infections in males

URINARY SYSTEM

SKELETAL SYSTEM

Pelvis protects reproductive organs of females, portion of ductus deferens and accessory glands in males

Sex hormones stimulate growth and maintenance of bones; sex hormones at puberty accelerate growth and closure of epiphyseal cartilages

MUSCULAR SYSTEM

Contractions of skeletal muscles eject semen from male reproductive tract; muscle contractions during sex act produce pleasurable sensations in both sexes

Reproductive hormones, especially testosterone, accelerate skeletal muscle growth

NERVOUS SYSTEM

Controls sexual behaviors and sexual function

Sex hormones affect CNS development and sexual behaviors

ENDOCRINE SYSTEM

Hypothalamic regulatory factors and pituitary hormones regulate sexual development and function; oxytocin stimulates smooth muscle contractions in uterus and mammary glands

Steroid sex hormones and inhibin inhibit secretory activities of hypothalamus and pituitary gland

CARDIOVASCULAR SYSTEM

Distributes reproductive hormones; provides nutrients, oxygen, and waste removal for fetus; local blood pressure changes responsible for physical changes during sexual arousal

Estrogens may maintain healthy vessels and slow development of atherosclerosis

LYMPHATIC SYSTEM

Provides IgA for secretion by epithelial glands; assists in repairs and defense against infection

Lysozymes and bactericidal chemicals in secretions provide nonspecific defense against reproductive tract infections

RESPIRATORY SYSTEM

Provides oxygen and removes carbon dioxide generated by tissues of reproductive system

Changes in respiratory rate and depth occur during sexual arousal, under control of the nervous system

DIGESTIVE SYSTEM

Provides additional nutrients required to support gamete production and (in pregnant women) embryonic and fetal development

•**FIGURE 28-27** Functional Relationships between the Reproductive System and Other Systems

and ejaculation must occur in the proper sequence; the woman's ovarian and uterine cycles must be properly coordinated, ovulation and oocyte transport must occur normally, and her reproductive tract must provide a hospitable environment for sperm survival, movement, and subsequent fertilization. For these steps to occur, the reproductive, digestive, endocrine, nervous, cardiovascular, and urinary systems must all be functioning normally.

Even when all else is normal and fertilization occurs at the proper time and place, a healthy infant will not be produced unless the zygote, a single cell the size of a pinhead, manages to develop into a full-term fetus that

typically weighs about 3 kg. In Chapter 29, we shall consider the process of development, focusing on the mechanisms that determine both the structure of the body and the distinctive characteristics of each individual.

☑ An inability to contract the ischiocavernosus and bulbospongiosus muscles would interfere with which part(s) of the male sex act?

☑ What changes occur in females during sexual arousal as the result of increased parasympathetic stimulation?

☑ Why does the level of FSH rise and remain high during menopause?

Men Are from Mars, Women Are from Venus is the title of a popular book. Although men and women are not from two separate planets, our reproductive systems are anatomically and physiologically different. To gain a better understanding of these differences, use the resources available on both the CD-ROM and the Companion Website. The CD-ROM includes animations and tutorials; the website builds on those resources with text content and interactive problems as well as communication avenues for sharing ideas with your class. To complete the following Web Explorations, visit the Media-Lab in Chapter 28 of your Companion Website.

Web Exploration 1

Estimated time for completion = 10 minutes

To understand the reproductive system, you must learn two anatomies, two physiological cycles, and two similar—yet not identical—hormonal mechanisms. Select the keyword

REPRODUCTION QUIZ on the Companion Website to find a quiz that covers all aspects of this system. Each question has a tutorial button that provides help in answering the posed question. Test your knowledge of the reproductive system by working through all 12 questions, taking note of those that you miss. Using these questions as an outline, create a reference sheet for your studies.

Web Exploration 2

Estimated time for completion = 5 minutes

Impotence is a serious concern for males as they age. Honest information about this topic is difficult to find other than from physicians. To learn more about impotence, including the various causes and remedies, select the keyword **IMPOTENCE** on the Companion Website. Read the article, and outline the important information presented in it. Using your notes, prepare an informational brochure that would be suitable for the general public.

CHAPTER REVIEW

SELECTED CLINICAL TERMINOLOGY
Terms Discussed in This Chapter

amenorrhea: The failure of menarche to appear before age 16, or a cessation of menstruation for 6 months or more in an adult female of reproductive age. *(p. 1045)*

breast cancer: A malignant, metastasizing cancer of the mammary gland that is the primary cause of death for women age 35–45. *(p. 1048)*

cervical cancer: A malignant, metastasizing cancer of the cervix that is the most common reproductive cancer in women. *(p. 1042 and AM)*

cryptorchidism (kript-OR-ki-dizm): The failure of one or both testes to descend into the scrotum by the time of birth. *(p. 1018)*

dysmenorrhea: Painful menstruation. *(p. 1043)*

endometriosis (en-dō-mē-trē-Ō-sis): The growth of endometrial tissue outside the uterus. *(p. 1045)*

fibrocystic disease: Clusters of lobular cysts within the tissues of the mammary gland. *(p. 1048)*

gonorrhea (gon-ō-RĒ-uh): A sexually transmitted disease. *(p. 1040 and AM)*

28

impotence: An inability to achieve or maintain an erection. *(p. 1053)*

mammography: The use of X-rays to examine breast tissue. *(p. 1048)*

mastectomy: The surgical removal of part or all of a cancerous mammary gland. *(p. 1048)*

orchiectomy (or-kē-EK-to-mē): The surgical removal of a testis. *(p. 1018)*

ovarian cancer: A malignant, metastasizing cancer of the ovaries that is the most dangerous reproductive cancer in women. *(p. 1039 and AM)*

pelvic inflammatory disease (PID): An infection of the uterine tubes. *(p. 1040)*

prostate cancer: A malignant, metastasizing cancer of the prostate gland; the second most common cause of cancer deaths in males. *(p. 1032)*

prostatectomy (pros-ta-TEK-to-mē): The surgical removal of the prostate gland. *(p. 1032)*

prostate-specific antigen (PSA): An antigen whose level in blood increases in persons with prostate cancer. *(p. 1032)*

sexually transmitted diseases (STDs): Diseases transferred from one individual to another primarily or exclusively through sexual contact. Examples include gonorrhea, syphilis, herpes, and AIDS. *(p. 1056 and AM)*

vaginitis (va-jin-Ī-tis): An infection of the vaginal canal by fungal or bacterial pathogens. *(p. 1046 and AM)*

vasectomy (vaz-EK-to-mē): The surgical removal of a segment of the ductus deferens, making it impossible for spermatozoa to reach the distal portions of the male reproductive tract. *(p. 1054)*

AM **Additional Terms Discussed in the** *Applications Manual*

endometrial polyps: Benign epithelial tumors of the uterine lining; they can be asymptomatic and so typically go unnoticed.

leiomyomas (lī-ō-mī-Ō-maz), or **fibroids:** Benign smooth muscle tumors; myometrial leiomyomas are the most common tumors in women.

STUDY OUTLINE

INTRODUCTION   p. 1017

1. The human reproductive system produces, stores, nourishes, and transports functional **gametes** (reproductive cells). **Fertilization** is the fusion of male and female gametes to create a **zygote.**

AN OVERVIEW OF THE REPRODUCTIVE SYSTEM   p. 1017

1. The **reproductive system** includes **gonads** (**testes** or **ovaries**), ducts, accessory glands and organs, and the **external genitalia.**
2. In males, the testes produce **spermatozoa,** which are expelled from the body in **semen** during *ejaculation.* The ovaries of a sexually mature female produce an **oocyte** (immature **ovum**) that travels along *uterine tubes* toward the *uterus.* The *vagina* connects the uterus with the exterior.

THE REPRODUCTIVE SYSTEM OF THE MALE   p. 1017

1. The spermatozoa travel along the *epididymis,* the *ductus deferens,* the *ejaculatory duct,* and the *urethra* before leaving the body. Accessory organs (notably the *seminal vesicles, prostate gland,* and *bulbourethral glands*) secrete into the ejaculatory ducts and the urethra. The *scrotum* encloses the testes, and the *penis* is an erectile organ. *(Figure 28-1)*

The Testes   p. 1017

2. The **descent of the testes** through the *inguinal canals* occurs during development. The testes remain connected to internal structures via the **spermatic cords.** The **raphe** marks the boundary between the two chambers in the **scrotum.** *(Figures 28-2, 28-3)*
3. The **dartos** muscle gives the scrotum a wrinkled appearance; the **cremaster muscle** pulls the testes close to the body. The **tunica albuginea** surrounds each testis. Septa extend from the tunica albuginea to the region of the testis closest to the entrance to the epididymis, creating a series of **lobules.** *(Figure 28-4)*

4. **Seminiferous tubules** within each lobule are the sites of sperm production. From there, spermatozoa pass through the **rete testis. Efferent ductules** connect the rete testis to the epididymis. Between the seminiferous tubules are **interstitial cells,** which secrete sex hormones. *(Figures 28-4, 28-5)*
5. Seminiferous tubules contain **spermatogonia,** stem cells involved in **spermatogenesis** (the production of spermatozoa), and **sustentacular cells,** which sustain and promote the development of spermatozoa. *(Figures 28-5–28-7)*

The Anatomy of a Spermatozoon   p. 1025

6. Each **spermatozoon** has a **head, middle piece,** and **tail.** *(Figure 28-8)*

The Male Reproductive Tract   p. 1025

7. From the testis, the spermatozoa enter the **epididymis,** an elongate tubule with **head, body,** and **tail** regions. The epididymis monitors and adjusts the composition of the fluid in the seminiferous tubules and serves as a recycling center for damaged spermatozoa. *(Figure 28-9)*
8. The **ductus deferens,** or *vas deferens,* begins at the epididymis and passes through the inguinal canal as part of the spermatic cord. Near the prostate gland, the ductus deferens enlarges to form the **ampulla.** The junction of the base of the seminal vesicle and the ampulla creates the **ejaculatory duct,** which empties into the urethra. *(Figure 28-10)*
9. The **urethra** extends from the urinary bladder to the tip of the penis. It can be divided into *prostatic, membranous,* and *penile regions.*

The Accessory Glands   p. 1028

10. Each **seminal vesicle** is an active secretory gland that contributes about 60 percent of the volume of semen; its secretions contain fructose, which is easily metabolized by spermatozoa. The **prostate gland** secretes slightly acidic **prostatic fluid.** Alkaline mucus secreted by the **bulbourethral glands** has lubricating properties. *(Figure 28-10)*

28

Semen p. 1029

11. A typical ejaculation releases 2–5 ml of semen (an **ejaculate**), which contains 20–100 million spermatozoa per milliliter. The fluid component of semen is **seminal fluid.**

The Penis p. 1030

12. The skin overlying the **penis** resembles that of the scrotum. Most of the **body** of the penis consists of three masses of **erectile tissue.** Beneath the superficial fascia are two **corpora cavernosa** and a single **corpus spongiosum,** which surrounds the urethra. Dilation of the erectile tissue with blood produces an **erection.** *(Figure 28-11)*

Hormones and Male Reproductive Function p. 1032

13. Important regulatory hormones include **FSH** *(follicle-stimulating hormone),* **LH** *(luteinizing hormone),* and **GnRH** *(gonadotropin-releasing hormone).* *Testosterone* is the most important androgen. *(Figure 28-12)*

THE REPRODUCTIVE SYSTEM OF THE FEMALE p. 1034

1. Principal organs of the female reproductive system include the ovaries, *uterine tubes, uterus, vagina,* and external genitalia. *(Figure 28-13)*

2. The ovaries, uterine tubes, and uterus are enclosed within the **broad ligament.** The **mesovarium** supports and stabilizes each ovary.

The Ovaries p. 1035

3. The ovaries are held in position by the **ovarian ligament** and the **suspensory ligament.** Major blood vessels enter the ovary at the **ovarian hilum.** Each ovary is covered by a **tunica albuginea.** *(Figure 28-14)*

4. **Oogenesis** (ovum production) occurs monthly in **ovarian follicles** as part of the **ovarian cycle.** The ovarian cycle is divided into a **follicular** *(preovulatory)* **phase** and a **luteal** *(postovulatory)* **phase.**

5. As development proceeds, **primordial, primary, secondary,** and **tertiary follicles** are produced in turn. At **ovulation,** a **secondary oocyte** and the surrounding follicular cells of the **corona radiata** are released through the ruptured ovarian wall. The follicular cells remaining within the ovary form the **corpus luteum,** which later degenerates into a **corpus albicans** of scar tissue. *(Figures 28-15, 28-16)*

The Uterine Tubes p. 1039

6. Each **uterine tube** has an **infundibulum** with **fimbriae** (fingerlike projections), an **ampulla,** and an **isthmus.** Each uterine tube opens into the *uterine cavity.* For fertilization to occur, the secondary oocyte must encounter spermatozoa during the first 12–24 hours of its passage from the infundibulum to the uterus. *(Figure 28-17)*

The Uterus p. 1040

7. The **uterus** provides mechanical protection and nutritional support to the developing embryo. Normally, the uterus bends anteriorly near its base *(anteflexion).* It is stabilized by the broad ligament, **uterosacral ligaments, round ligaments,** and **lateral ligaments.** *(Figure 28-18)*

8. Major anatomical landmarks of the uterus include the **body, isthmus, cervix, cervical os** *(external orifice),* **uterine cavity, cervical canal,** and **internal os** *(internal orifice).* The uterine wall consists of an inner **endometrium,** a muscular **myometrium,** and a superficial **perimetrium** (an incomplete serous layer). *(Figures 28-18, 28-19)*

9. A typical 28-day **uterine,** or *menstrual,* **cycle** begins with the onset of **menses** and the destruction of the **functional zone** of the endometrium. This process of **menstruation** continues from 1 to 7 days. *(Figure 28-20)*

10. After menses, the **proliferative phase** begins, and the functional zone undergoes repair and thickens. Menstrual activity begins at **menarche** and continues until **menopause.** *(Figure 28-20)*

The Vagina p. 1045

11. The **vagina** is a muscular tube extending between the uterus and the external genitalia. Before the first sexual intercourse, a thin epithelial fold, the **hymen,** partially blocks the entrance to the vagina. *(Figures 28-21, 28-22)*

The External Genitalia p. 1046

12. The components of the **vulva** are the **vestibule, labia minora, paraurethral glands, clitoris, labia majora,** and **lesser** and **greater vestibular glands.** *(Figures 28-13, 28-22)*

The Mammary Glands p. 1047

13. A newborn infant gains nourishment from milk secreted by maternal **mammary glands.** *(Figure 28-23)*

Hormones and the Female Reproductive Cycle p. 1049

14. Hormonal regulation of the **female reproductive cycle** involves the coordination of the ovarian and uterine cycles.

15. **Estradiol,** the most important *estrogen,* is the dominant hormone of the follicular phase. Ovulation occurs in response to a mid-cycle surge in LH. *(Figures 28-24, 28-25)*

16. The hypothalamic secretion of GnRH occurs in pulses that trigger the pituitary secretion of FSH and LH. FSH initiates follicular development, and activated follicles and ovarian interstitial cells produce estrogens. High levels of estrogen stimulate LH secretion, increase pituitary sensitivity to GnRH, and increase the GnRH pulse frequency. **Progesterone,** one of the **progestins,** is the principal hormone of the luteal phase. Changes in estrogen and progesterone levels are responsible for the maintenance of the uterine cycle. *(Figures 28-25, 28-26)*

THE PHYSIOLOGY OF SEXUAL INTERCOURSE p. 1053

Male Sexual Function p. 1053

1. During sexual **arousal** in males, erotic thoughts, sensory stimulation, or both lead to parasympathetic activity that produces erection. Stimuli accompanying **sexual intercourse** lead to **emission** and **ejaculation.** Contractions of the bulbospongiosus muscles are associated with **orgasm.**

Female Sexual Function p. 1056

2. The phases of female sexual function resemble those of male sexual function, with parasympathetic arousal and skeletal muscle contractions associated with orgasm.

AGING AND THE REPRODUCTIVE SYSTEM p. 1056

Menopause p. 1056

1. Menopause (the time that ovulation and menstruation cease) typically occurs at age 45–55. The production of GnRH, FSH, and LH rise, whereas circulating concentrations of estrogen and progesterone decline.

The Male Climacteric p. 1057

2. During the **male climacteric,** at age 50–60, circulating testosterone levels fall, whereas levels of FSH and LH rise.

INTEGRATION WITH OTHER SYSTEMS p. 1057

1. Hormones play a major role in coordinating reproduction. *(Table 28-1)*

2. Reproduction requires the normal functioning of the digestive, nervous, cardiovascular, and urinary systems. *(Figure 28-27)*

28

REVIEW QUESTIONS

LEVEL 1 REVIEWING FACTS AND TERMS

1. Perineal structures associated with the reproductive system are collectively known as
(a) gonads   (b) sex gametes
(c) external genitalia   (d) accessory glands

2. Interstitial cells are responsible for the
(a) production of spermatozoa
(b) maintenance of the blood–testis barrier
(c) support of spermiogenesis
(d) production of androgens

3. Sustentacular cells are responsible for the secretion of
(a) inhibin
(b) androgen-binding protein
(c) Müllerian-inhibiting factor
(d) a, b, and c are correct

4. During meiosis, when synapsis occurs, corresponding maternal and paternal chromosomes come together to produce
(a) 46 pairs of chromosomes
(b) 23 chromosomes
(c) 23 pairs of chromosomes
(d) the haploid number of chromosomes

5. The completion of meiosis in males produces four spermatids, each containing
(a) 23 chromosomes
(b) 23 pairs of chromosomes
(c) the diploid number of chromosomes
(d) 46 pairs of chromosomes

6. Erection of the penis occurs when
(a) sympathetic activation of penile arteries occurs
(b) arterial branches are constricted, and muscular partitions are tense
(c) the vascular channels become engorged with blood
(d) a, b, and c are correct

7. In males, the primary target of FSH is the
(a) sustentacular cells of the seminiferous tubules
(b) interstitial cells of the seminiferous tubules
(c) cells of Leydig
(d) epididymis

8. Testosterone and other androgens are secreted by the
(a) hypothalamus
(b) anterior pituitary gland
(c) sustentacular cells
(d) interstitial cells

9. The ovaries are responsible for
(a) the production of female gametes
(b) the secretion of female sex hormones
(c) the secretion of inhibin
(d) a, b, and c are correct

10. Gametes are produced in the _____ of the ovary.
(a) germinal epithelium   (b) medulla
(c) cortex   (d) tunica albuginea

11. Ovum production, or oogenesis, begins
(a) before birth   (b) after birth
(c) at puberty   (d) after puberty

12. In females, meiosis is not completed until
(a) birth
(b) puberty
(c) fertilization occurs
(d) uterine implantation occurs

13. If fertilization is to occur, the ovum must encounter spermatozoa during the first _____ of its passage.
(a) 1 to 5 hours   (b) 6 to 11 hours
(c) 12 to 24 hours   (d) 25 to 36 hours

14. The part of the endometrium that undergoes cyclical changes in response to levels of sex hormones is the
(a) serosa
(b) basilar zone
(c) muscular myometrium
(d) functional zone

15. The secretory phase of the uterine cycle is influenced primarily by the stimulatory effects of progestins and estrogens from the
(a) anterior pituitary gland   (b) hypothalamus
(c) endometrium   (d) corpus luteum

16. A sudden surge in LH secretion causes the
(a) onset of menses
(b) rupture of the follicular wall and ovulation
(c) beginning of the proliferative phase
(d) end of the uterine cycle

17. The principal hormone of the postovulatory phase is
(a) progesterone   (b) estradiol
(c) estrogen   (d) luteinizing hormone

18. At ovulation, the basal body temperature
(a) is not affected
(b) increases noticeably
(c) declines sharply
(d) may increase or decrease a few degrees

19. Impotence is the inability to
(a) produce sufficient amounts of spermatozoa for fertilization
(b) achieve or maintain an erection
(c) achieve orgasm
(d) ejaculate

20. Erection of the clitoris is caused by the influence of
(a) parasympathetic activation
(b) sympathetic activation
(c) increased output of estrogen
(d) increased output of progesterone

21. Menopause is accompanied by
(a) sustained rises in GnRH, FSH, and LH
(b) drops in circulating levels of estrogen and progesterone
(c) a thinning of the urethral and vaginal walls
(d) a, b, and c are correct

22. Which reproductive structures are common to both males and females?

23. Trace the duct system that the sperm traverses from the site of production to the exterior.

24. Which accessory organs and glands contribute to the composition of semen? What are the functions of each?

25. What are the two primary cell populations in the testes that are responsible for functions related to reproductive activity? What are the functions of these cells?

26. What are the three primary functions of the epididymis?

27. Identify the three regions of the male urethra.

28. Describe the composition of a typical sample of ejaculate.

29. List the functions of testosterone in males.

30. What are the primary functions of the ovaries?

28

31. List and summarize the important steps in the ovarian cycle.

32. Describe the histological composition of the uterine wall.

33. What are the three major functions of the vagina?

34. What is the functional significance of the normally acidic pH of the vagina?

35. What is the role of the clitoris in the female reproductive system?

36. What is the function of the lesser and greater vestibular glands in females?

37. Trace the route that milk takes from its site of production to the outside of the female.

LEVEL 2 REVIEWING CONCEPTS

38. How does the human reproductive system differ functionally from all other systems in the body?

39. How are the male and female reproductive systems functionally different?

40. How is the process of meiosis involved in the development of the spermatozoon and the ovum?

41. Describe the erectile tissues of the penis. How does erection occur?

42. Using an average cycle of 28 days, describe each of the three phases of the uterine cycle.

43. Describe the hormonal events associated with the ovarian cycle.

44. Describe the hormonal events associated with the uterine cycle.

45. Summarize the steps that occur in sexual arousal and orgasm. Do these processes differ in males and females?

46. How does the aging process affect the reproductive systems of men and women?

47. How do birth control pills prevent conception?

LEVEL 3 CRITICAL THINKING AND CLINICAL APPLICATIONS

48. Diane has peritonitis (an inflammation of the peritoneum), which her physician says resulted from a urinary tract infection. Why could this situation occur more readily in females than in males?

49. Rod suffers an injury to the sacral region of his spinal cord. Will he still be able to achieve an erection? Explain.

50. A 7-year-old girl develops an ovarian tumor that involves granulosa cells. What symptoms would you expect to observe?

51. How would a blockage of the enzyme required for the production of testosterone affect a woman's ovarian cycle?

52. Women bodybuilders and women with eating disorders such as anorexia nervosa commonly experience amenorrhea. What does this fact suggest about the relation between body fat and menstruation? What might be the benefit of amenorrhea under such circumstances?

28

CHAPTER 29

Development and Inheritance

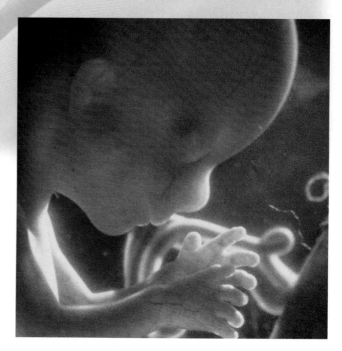

The physiological processes we have studied thus far have been relatively brief in duration. Many last only a fraction of a second; others may take hours at most. But some important processes are measured in months, years, or decades. A human develops in the womb for 9 months, grows to maturity in 15 or 20 years, and may live the better part of a century. During that time, he or she is always changing. Birth, growth, maturation, aging, and death are all parts of a single, continuous process. That process does not end with the individual, because humans can pass at least some of their characteristics on to their offspring. Thus, each generation gives rise to a new generation that will repeat the cycle. In this chapter, we shall explore the continuity of life, from conception to death.

CHAPTER OUTLINE AND OBJECTIVES

INTRODUCTION   p. 1065

AN OVERVIEW OF TOPICS IN DEVELOPMENT   p. 1065

FERTILIZATION   p. 1065
  The Oocyte at Ovulation   p. 1066
  Oocyte Activation   p. 1067
① Describe the process of fertilization.
② Explain how developmental processes are regulated.

AN OVERVIEW OF PRENATAL DEVELOPMENT   p. 1067
③ List the three prenatal periods, and describe the major events associated with each.

THE FIRST TRIMESTER   p. 1068
  Cleavage and Blastocyst Formation   p. 1068
  Implantation   p. 1068
  Placentation   p. 1072
  Embryogenesis   p. 1075
④ Explain how the germ layers participate in the formation of extraembryonic membranes.
⑤ Discuss the importance of the placenta as an endocrine organ.

THE SECOND AND THIRD TRIMESTERS   p. 1081
  Pregnancy and Maternal Systems   p. 1081
  Structural and Functional Changes in the Uterus   p. 1083
⑥ Describe the interplay between the maternal organ systems and the developing fetus.
⑦ Discuss the structural and functional changes in the uterus during gestation.

LABOR AND DELIVERY   p. 1084
  Stages of Labor   p. 1084
  Premature Labor   p. 1084
  Multiple Births   p. 1086
⑧ List and discuss the events that occur during labor and delivery.

POSTNATAL DEVELOPMENT   p. 1086
  The Neonatal Period, Infancy, and Childhood   p. 1086
  Adolescence and Maturity   p. 1088
  Senescence   p. 1089

⑨ Identify the features and functions associated with the various life stages.

GENETICS, DEVELOPMENT, AND INHERITANCE   p. 1090
  Genes and Chromosomes   p. 1090
  The Human Genome Project   p. 1095
⑩ Relate basic principles of genetics to the inheritance of human traits.

MEDIALAB and CHAPTER REVIEW   p. 1098

Time refuses to stand still; today's infant will be tomorrow's adult. The gradual modification of anatomical structures during the period from fertilization to maturity is called **development.** The changes that occur during development are truly remarkable. What begins as a single cell slightly larger than the period at the end of this sentence becomes an individual whose body contains trillions of cells organized into highly specialized tissues, organs, and organ systems. The creation of different cell types required by this process is called **differentiation.** Differentiation occurs through selective changes in genetic activity. As development proceeds, some genes are turned off and others are turned on. The identities of these genes vary from one cell type to another.

A basic understanding of human development provides important insights into anatomical structures. In addition, many of the mechanisms of development and growth are similar to those responsible for the repair of injuries. In this chapter, we shall focus on major aspects of development. We shall consider highlights of the developmental process rather than describe the events in great detail. We shall also consider the regulatory mechanisms and how developmental patterns can be modified—for good or ill. Few topics in the biological sciences hold such fascination, and fewer still confront the investigator with so dazzling an array of technological, moral, and logistical challenges.

AN OVERVIEW OF TOPICS IN DEVELOPMENT

Development involves (1) the division and differentiation of cells and (2) the changes that produce and modify anatomical structures. Development begins at **fertilization,** or **conception.** We can divide development into periods characterized by specific anatomical changes. **Embryological development** comprises the events that occur during the first 2 months after fertilization. The study of these events is called **embryology** (em-brē-OL-o-jē). **Fetal development** begins at the start of the ninth week and continues until birth. Embryological and fetal development are sometimes referred to collectively as **prenatal development,** the primary focus of this chapter. **Postnatal development** commences at birth and continues to **maturity,** when the aging process begins.

Although all humans go through the same developmental stages, differences in their genetic makeup produce distinctive individual characteristics. **Inheritance** refers to the transfer of genetically determined characteristics from generation to generation. **Genetics** is the study of the mechanisms responsible for inheritance. In this chapter, we shall consider basic genetics as it applies to inherited characteristics, such as gender, hair color, and various diseases.

FERTILIZATION

Fertilization involves the fusion of two haploid gametes, producing a zygote that contains 46 chromosomes, the normal number for a somatic cell. The functional roles and contributions of the male and female gametes are very different. The spermatozoon simply delivers the paternal chromosomes to the site of fertilization. It must travel a relatively large distance and is small, efficient, and highly streamlined. In contrast, the female gamete must provide all the nourishment and genetic programming to support development of the embryo for nearly a week after conception. The volume of this gamete is therefore much greater than that of the spermatozoon (Figure 29-1a•). Recall from Chapter 28 that ovulation releases a secondary oocyte suspended in metaphase of meiosis II. At fertilization, the diameter of the secondary oocyte is more than twice the entire length of the spermatozoon. The ratio of their volumes is even more striking—roughly 2000:1.

The spermatozoa arriving in the vagina are already motile, as a result of mixing with secretions of the seminal vesicles—the first step of *capacitation.* ∞ *p. 1027* (A substance secreted by the epididymis appears to prevent premature capacitation.) But they cannot perform fertilization until they have been exposed to conditions in the female reproductive tract. The mechanism responsible for this second step of capacitation remains unknown.

Fertilization typically occurs near the junction between the ampulla and isthmus of the uterine tube, generally within a day after ovulation. By this time, a secondary oocyte has traveled a few centimeters, but spermatozoa must cover the distance between the vagina and the ampulla of the uterine tube. A spermatozoon can propel itself at speeds of only about 34 μm per second, roughly equivalent to 12.5 cm (5 in.) per hour, so in theory it should take spermatozoa several hours to reach the upper portions of the uterine tubes. The actual passage time, however, ranges from 2 hours to as little as 30 minutes. Contractions of the uterine musculature and ciliary currents in the uterine tubes have been suggested as likely mechanisms for accelerating the movement of spermatozoa from the vagina to the fertilization site.

Even with transport assistance and available nutrients, this is not an easy passage. Of the roughly 200 million spermatozoa introduced into the vagina in a typical ejaculation, only about 10,000 enter the uterine tube, and fewer than 100 reach the isthmus. In general, a male with a sperm count below 20 million per milliliter is functionally sterile because too few spermatozoa survive to reach and fertilize an oocyte. Dozens of spermatozoa are required for successful fertilization, because a single sperm cannot penetrate the *corona radiata,* a layer of follicle cells that surrounds the oocyte. ∞ *p. 1039*

29

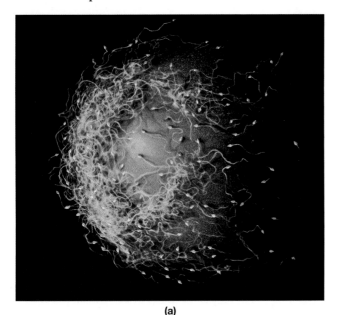

(a)

THE OOCYTE AT OVULATION

Ovulation occurs before the oocyte is completely mature. The secondary oocyte leaving the follicle is in metaphase of meiosis II. The cell's metabolic operations have been discontinued, and the oocyte drifts in a sort of suspended animation, awaiting the stimulus for further development. If fertilization does not occur, the oocyte disintegrates without completing meiosis.

Fertilization is complicated by the fact that when the secondary oocyte leaves the ovary, it is surrounded by the corona radiata. Fertilization and the events that follow are diagrammed in Figure 29-1b•. The cells of the corona radiata protect the secondary oocyte as it passes through the ruptured follicular wall, across the surface of the ovary, and into the infundibulum of the uterine tube. Although the physical process of fertilization requires only a single spermatozoon in contact

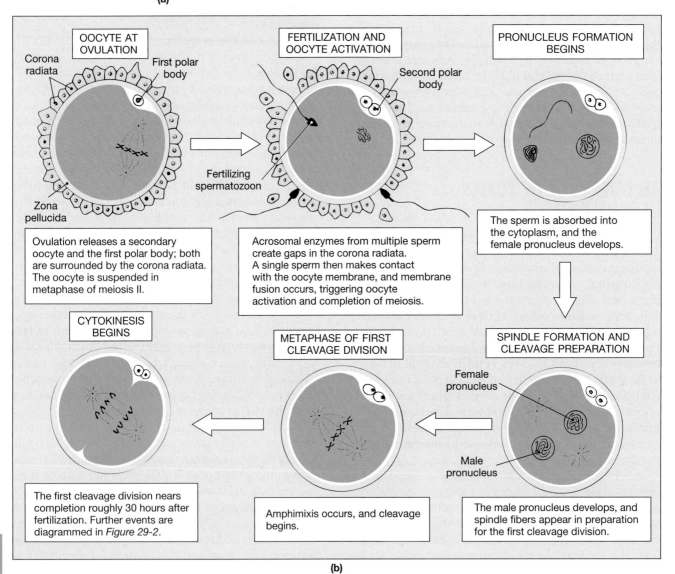

(b)

•**FIGURE 29-1 Fertilization. (a)** An oocyte and numerous sperm at the time of fertilization. Notice the difference in size between the gametes. **(b)** Fertilization and the preparations for cleavage.

with the oocyte membrane, that spermatozoon must first penetrate the corona radiata. The acrosomal cap of the sperm contains several enzymes, including **hyaluronidase** (hī-al-ū-RON-a-dās), which breaks down the bonds between adjacent follicle cells. Dozens of spermatozoa must release hyaluronidase before the connections between the follicular cells break down enough to permit fertilization.

No matter how many spermatozoa slip through the gap in the corona radiata, normally only a single spermatozoon will accomplish fertilization and activate the oocyte. The first step is the binding of the spermatozoon to *sperm receptors* in the zona pellucida. This step triggers the rupture of the acrosomal cap. The hyaluronidase and **acrosin,** another proteolytic enzyme, then digest a path through the zona pellucida toward the surface of the oocyte. When the sperm contacts that surface, the sperm and oocyte membranes begin to fuse. This step is the trigger for *oocyte activation,* a complex process we shall discuss in the next section. As the membranes fuse, the entire sperm enters cytoplasm of the oocyte.

OOCYTE ACTIVATION

Oocyte activation involves a series of changes in the metabolic activity of the oocyte. The trigger for activation is contact and fusion of the cell membranes of the sperm and oocyte. This process is accompanied by the depolarization of the oocyte membrane due to an increased permeability to sodium ions. The entry of sodium ions in turn causes the release of calcium ions from the smooth endoplasmic reticulum. The sudden rise in Ca^{2+} levels has important effects, including:

- *The exocytosis of vesicles located just interior to the oocyte membrane.* This process, called the *cortical reaction,* releases enzymes that inactivate the sperm receptors and harden the zona pellucida. This combination prevents **polyspermy** (fertilization by more than one sperm). Polyspermy creates a zygote that is incapable of normal development. (Prior to the completion of the cortical reaction, the depolarization of the oocyte membrane probably prevents fertilization by any sperm that penetrate the zona pellucida.)
- *The completion of meiosis II and formation of the second polar body.*
- *The activation of enzymes that cause a rapid increase in the cell's metabolic rate.* The cytoplasm contains large numbers of mRNA strands, but until now they have been inactivated by special proteins. The mRNA strands are now activated, so protein synthesis accelerates rapidly. Most of the proteins synthesized will be needed if development is to proceed.

After oocyte activation has occurred and meiosis has been completed, the nuclear material remaining within the ovum reorganizes as the **female pronucleus** (Figure 29-1b•). While these changes are under way, the nucleus of the spermatozoon swells, becoming the **male pronucleus.** The male pronucleus then migrates toward the center of the cell, and spindle fibers form. The two pronuclei fuse, a process called *amphimixis* (am-fi-MIK-sis). Fertilization is now complete, with the formation of a zygote that contains the normal complement of 46 chromosomes.

AN OVERVIEW OF PRENATAL DEVELOPMENT

During prenatal development, a single cell ultimately forms a 3–4 kg infant who in postnatal development will grow through adolescence and maturity toward old age and eventual death. One of the most fascinating aspects of development is its apparent order and simplicity. Continuity exists at all levels and at all times. Nothing leaps into existence, unheralded and without apparent precursors; differentiation and increasing structural complexity occur hand in hand.

Differentiation involves changes in the genetic activity of some cells but not others. There is a continuous exchange of information between the nucleus and the cytoplasm in a cell. ∞ *p. 89* Activity in the nucleus varies in response to chemical messages that arrive from the surrounding cytoplasm. In turn, ongoing nuclear activity alters conditions within the cytoplasm by directing the synthesis of specific proteins. In this way, the nucleus can affect enzyme activity, cell structure, and membrane properties.

In development, differences in the cytoplasmic composition of individual cells trigger changes in genetic activity. These changes in turn lead to further changes in the cytoplasm, and the process continues in a sequential fashion. But if all the cells of the embryo are derived from cell divisions of a zygote, how do the cytoplasmic differences originate? What sets this process in motion? The important first step occurs before fertilization, while the oocyte is in the ovary.

Before ovulation, the growing oocyte accepts amino acids, nucleotides, and glucose, as well as more-complex materials such as phospholipids, mRNA molecules, and proteins, from the surrounding granulosa cells. Because not all follicle cells manufacture and deliver the same nutrients and instructions to the oocyte, the contents of the cytoplasm are not evenly distributed. After fertilization, subsequent divisions divide the cytoplasm of the zygote into ever smaller cells that differ from one another in their cytoplasmic composition. These differences alter genetic activity, creating cell lines with increasingly diverse fates.

As development proceeds, some of these cells will release chemical substances, such as RNA molecules, polypeptides, and small proteins, that affect the differentiation of other embryonic cells. This type of chemical interplay among developing cells is called *induction* (in-DUK-shun). Induction can work over very short distances, as when two types of cells are in direct contact. It may also operate over longer distances, with the inducing chemicals functioning as hormones.

This type of regulation, which involves an integrated series of interacting steps, can control very complex processes. The mechanism is not risk-free, however. The appearance of an abnormal or inappropriate inducer can throw the entire development plan off course. [AM] *Teratogens and Abnormal Development*

The time spent in prenatal development is known as the period of **gestation** (jes-TĀ-shun). For convenience, we usually think of the gestation period as three integrated **trimesters,** each 3 months in duration:

1. The **first trimester** is the period of embryological and early fetal development. During this period, the rudiments of all the major organ systems appear.

2. The **second trimester** is dominated by the development of organs and organ systems, and this process nears completion. The body proportions change; by the end of this trimester, the fetus looks distinctively human.

3. The **third trimester** is characterized by rapid fetal growth. Early in the third trimester, most of the major organ systems become fully functional. An infant born 1 month or even 2 months prematurely has a reasonable chance of survival.

The *Applications Manual* contains EMBRYOLOGY SUMMARIES, in which we introduce key steps in embryological and fetal development and trace the development of specific organ systems. We shall refer to those Summaries in the discussions that follow. As we proceed, you should review the material as indicated. Doing so will help you understand the "big picture" as well as the specific details.

THE FIRST TRIMESTER

At the moment of conception, the fertilized ovum is a single cell that has a diameter of about 0.135 mm (0.005 in.) and a weight of approximately 150 mg. By the end of the first trimester (twelfth developmental week), the fetus is almost 75 mm (3 in.) long and weighs perhaps 14 g (0.5 oz).

Many important and complex developmental events occur during the first trimester. We shall focus on four general processes: cleavage, implantation, placentation, and embryogenesis:

1. **Cleavage** (KLĒV-ij) is a sequence of cell divisions that begins immediately after fertilization and ends at the first contact with the uterine wall. During cleavage, the zygote becomes a **pre-embryo,** which develops into a multicellular complex known as a *blastocyst.* (We introduce cleavage and blastocyst formation in [AM] *EMBRYOLOGY SUMMARY 1: The Development of Tissues*.)

2. **Implantation** begins with the attachment of the blastocyst to the endometrium of the uterus and continues as the blastocyst invades maternal tissues. During implan-

tation, other important events take place that set the stage for the formation of vital embryonic structures.

3. **Placentation** (plas-en-TĀ-shun) occurs as blood vessels form around the periphery of the blastocyst and the **placenta** develops. The placenta is a complex organ that permits exchange between the maternal and embryonic circulatory systems. It supports the fetus in the second and third trimesters, but it is lost at birth as the newborn becomes physically independent of the mother.

4. **Embryogenesis** (em-brē-ō-JEN-e-sis) is the formation of a viable embryo. This process establishes the foundations for all major organ systems.

These processes are both complex and vital to the survival of the embryo. Perhaps because the events in the first trimester are so complex, this is the most dangerous period in prenatal life. Only about 40 percent of conceptions produce embryos that survive the first trimester. For that reason, pregnant women are warned to take great care to avoid drugs and other disruptive stresses during the first trimester, in the hopes of preventing an error in the delicate processes under way.

CLEAVAGE AND BLASTOCYST FORMATION

Cleavage is a series of cell divisions that subdivides the cytoplasm of the zygote (Figure 29-2•). The first cleavage division produces a pre-embryo consisting of two identical cells. Identical cells produced by cleavage divisions are called **blastomeres** (BLAS-tō-mērz). The first division is completed roughly 30 hours after fertilization, and subsequent cleavage divisions occur at intervals of 10–12 hours. During the initial cleavage divisions, all the blastomeres undergo mitosis simultaneously. As the number of blastomeres increases, the timing becomes less predictable.

After 3 days of cleavage, the pre-embryo is a solid ball of cells, resembling a mulberry. This stage is called the **morula** (MOR-ū-la; *morula,* mulberry). The morula typically reaches the uterus on day 4. Over the next 2 days, the blastomeres form a **blastocyst,** a hollow ball with an inner cavity known as the **blastocoele** (BLAS-tō-sēl). The blastomeres are no longer identical in size and shape. The outer layer of cells, which separates the outside world from the blastocoele, is called the **trophoblast** (TRŌ-fō-blast). The function is implied by the name: *trophos,* food + *blast,* precursor. These cells are responsible for providing nutrients to the developing embryo. A second group of cells, the **inner cell mass,** lies clustered at one end of the blastocyst. These cells are exposed to the blastocoele but are insulated from contact with the outside environment by the trophoblast. In time, the inner cell mass will form the embryo.

IMPLANTATION

During blastocyst formation, enzymes released by the trophoblast erode a hole through the zona pellucida, which is then shed in a process known as *hatching.* The

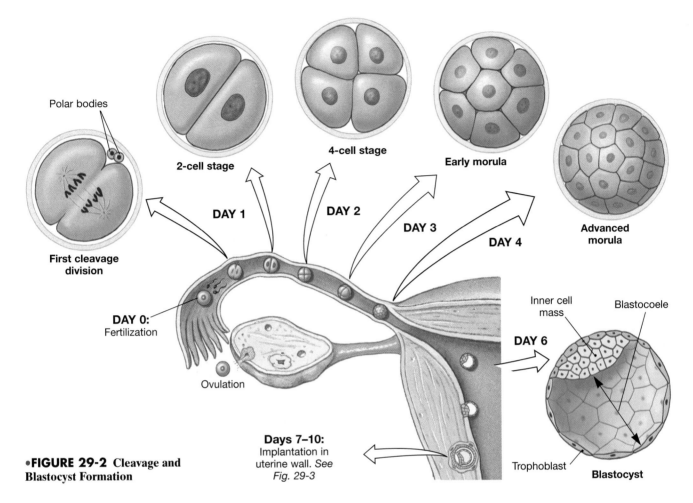

Polar bodies

First cleavage
division

DAY 0:
Fertilization

Ovulation

DAY 1

2-cell stage

DAY 2

4-cell stage

DAY 3

Early morula

DAY 4

Advanced
morula

Inner cell
mass

Blastocoele

DAY 6

Trophoblast

Blastocyst

Days 7–10:
Implantation in
uterine wall. *See
Fig. 29-3*

●**FIGURE 29-2** Cleavage and
Blastocyst Formation

blastocyst is now freely exposed to the fluid contents of the uterine cavity. This fluid, rich in glycogen, is secreted by the endometrial glands of the uterus. Over the past few days, the pre-embryo and then early blastocyst has been absorbing fluid and nutrients from its surroundings; that process now accelerates, and the blastocyst enlarges. When fully formed, the blastocyst contacts the endometrium, and implantation occurs. Stages in the implantation process are illustrated in Figure 29-3●.

Implantation begins as the surface of the blastocyst closest to the inner cell mass touches and adheres to the uterine lining (see day 7, Figure 29-3●). At the point of contact, the trophoblast cells divide rapidly, making the trophoblast several layers thick. The cells closest to the interior of the blastocyst remain intact, forming a layer of **cellular trophoblast,** or *cytotrophoblast.* Near the endometrial wall, the cell membranes separating the trophoblast cells disappear, creating a layer of cytoplasm containing multiple nuclei (day 8). This outer layer is called the **syncytial** (sin-SISH-al) **trophoblast,** or *syncytiotrophoblast.*

The syncytial trophoblast erodes a path through the uterine epithelium by secreting hyaluronidase. This enzyme dissolves the intercellular cement between adjacent epithelial cells, just as the hyaluronidase released by spermatozoa dissolves the connections between cells of the corona radiata. At first, this erosion creates a gap in the uterine lining, but the migration and divisions of epithelial cells soon repair the surface. When the repairs are completed, the blastocyst loses contact with the uterine cavity. Development occurs entirely within the functional zone of the endometrium. ∞ *p. 1043*

In most cases, implantation occurs in the fundus or elsewhere in the body of the uterus. In an **ectopic pregnancy,** implantation occurs somewhere other than within the uterus, such as in one of the uterine tubes. Approximately 0.6 percent of pregnancies are ectopic pregnancies, which do not produce a viable embryo and can be life-threatening. AM *Ectopic Pregnancies*

As implantation proceeds, the syncytial trophoblast continues to enlarge and spread into the surrounding endometrium (day 9). The erosion of uterine glands releases nutrients that are absorbed by the syncytial trophoblast and distributed by diffusion through the underlying cellular trophoblast to the inner cell mass. These nutrients provide the energy needed to support the early stages of embryo formation. Trophoblastic extensions grow around endometrial capillaries. As the capillary walls are destroyed, maternal blood begins to percolate through trophoblastic channels known as

29

lacunae. Fingerlike **villi** extend away from the trophoblast into the surrounding endometrium and gradually increase in size and complexity until about day 21. By about day 9, the syncytial trophoblast begins breaking down larger endometrial veins and arteries, and blood flow through the lacunae accelerates.

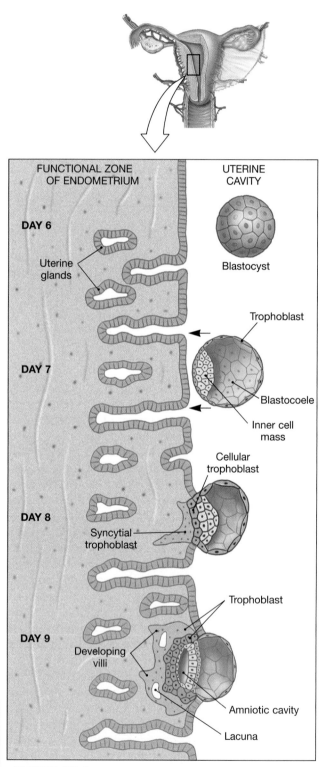

•**FIGURE 29-3** **Stages in Implantation**

Gestational Trophoblastic Neoplasias

The trophoblast undergoes repeated nuclear divisions, shows extensive and rapid growth, has a very high demand for energy, invades and spreads through adjacent tissues, and fails to activate the maternal immune system. In short, the trophoblast has many of the characteristics of cancer cells. In about 0.1 percent of pregnancies, something goes wrong with the regulatory mechanisms, and a normal placenta does not develop. Instead, the syncytial trophoblast behaves like a tumor. This condition is called **gestational trophoblastic neoplasia.** The least dangerous form, a *hydatidiform* (hī-da-TID-i-form) *mole,* is not malignant. However, about 20 percent of gestational trophoblastic neoplasias will metastasize, invading other tissues, with potentially fatal results. As a result, prompt surgical removal of the mass is essential, sometimes followed by chemotherapy.

Formation of the Amniotic Cavity

The inner cell mass has little apparent organization early in the blastocyst stage. Yet, by the time of implantation, the inner cell mass has separated from the trophoblast. The separation gradually increases, creating a fluid-filled chamber called the **amniotic** (am-nē-OT-ik) **cavity.** The trophoblast will later be separated from the amniotic cavity by layers of cells that originate at the inner cell mass and line the amniotic cavity. These layers form the *amnion,* a membrane we shall examine later in the chapter. The amniotic cavity can be seen in day 9 of Figure 29-3•; additional details from day 10 to day 12 are shown in Figure 29-4•. When the amniotic cavity first appears, the cells of the inner cell mass are organized into an oval sheet that is two cell layers thick: a superficial layer that faces the amniotic cavity, and a deeper layer that is exposed to the fluid contents of the blastocoele.

Gastrulation and Germ Layer Formation

By day 12, a third layer begins to form through **gastrulation** (gas-troo-LĀ-shun) (Figure 29-4•). During gastrulation, cells in specific areas of the surface move toward a central line known as the **primitive streak.** At the primitive streak, these migrating cells leave the surface and move between the two existing layers. This movement creates three distinct embryonic layers: (1) the **ectoderm,** consisting of superficial cells that did not migrate into the interior of the inner cell mass; (2) the **endoderm,** consisting of the cells that face the blastocoele; and (3) the **mesoderm,** consisting of the poorly organized layer of migrating cells between the ectoderm and the endoderm. Collectively, these three embryonic layers are called **germ layers.** The formation of mesoderm and the fates of each germ layer are summarized in AM *EMBRYOLOGY SUMMARY 4: The Development of Organ Systems.* Table 29-1 contains a more comprehensive listing of the contributions each germ layer makes to the body systems described in earlier chapters.

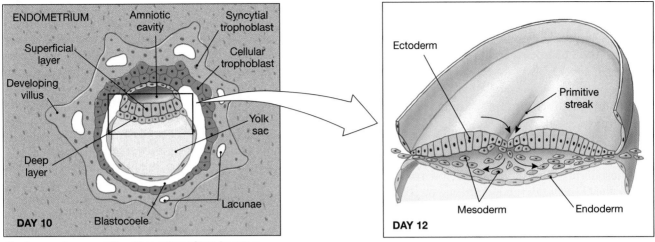

The inner cell mass at Day 10 consists of two layers: a superficial layer, facing the amniotic cavity, and a deep layer, exposed to the blastocoele. Migration of cells around the amniotic cavity is the first step in the formation of the amnion. Migration of cells around the edges of the blastocoele is the first step in yolk sac formation.

Migration of superficial cells into the interior creates a third layer. From the time this process (gastrulation) begins, the superficial layer is called *ectoderm*, the deep layer *endoderm*, and the migrating cells *mesoderm*.

•**FIGURE 29-4** The Inner Cell Mass and Gastrulation

TABLE 29-1 The Fates of the Germ Layers

Ectodermal Contributions

Integumentary system: epidermis, hair follicles and hairs, nails, and glands communicating with the skin (apocrine and merocrine sweat glands, mammary glands, and sebaceous glands)

Skeletal system: pharyngeal cartilages and their derivatives in adults (portion of sphenoid bone, the auditory ossicles, the styloid processes of the temporal bones, the cornu and superior rim of the hyoid bone)*

Nervous system: all neural tissue, including brain and spinal cord

Endocrine system: pituitary gland and adrenal medullae

Respiratory system: mucous epithelium of nasal passageways

Digestive system: mucous epithelium of mouth and anus, salivary glands

Mesodermal Contributions

Skeletal system: all components except some pharyngeal derivatives

Muscular system: all components

Endocrine system: adrenal cortex, endocrine tissues of heart, kidneys, and gonads

Cardiovascular system: all components

Lymphatic system: all components

Urinary system: the kidneys, including the nephrons and the initial portions of the collecting system

Reproductive system: the gonads and the adjacent portions of the duct systems

Miscellaneous: the lining of the body cavities (thoracic, pericardial, and peritoneal) and the connective tissues that support all organ systems

Endodermal Contributions

Endocrine system: thymus, thyroid gland, and pancreas

Respiratory system: respiratory epithelium (except nasal passageways) and associated mucous glands

Digestive system: mucous epithelium (except mouth and anus), exocrine glands (except salivary glands), liver, and pancreas

Urinary system: urinary bladder and distal portions of the duct system

Reproductive system: distal portions of the duct system, stem cells that produce gametes

*The neural crest is derived from ectoderm and contributes to the formation of the skull and the skeletal derivatives of the embryonic pharyngeal arches.

29

Gastrulation produces an oval, three-layered sheet known as the **embryonic disc.** The embryonic disc will form the body of the embryo, whereas the rest of the blastocyst will be involved in forming the *extraembryonic membranes.*

The Formation of Extraembryonic Membranes

Germ layers also participate in the formation of four **extraembryonic membranes:** (1) the *yolk sac* (endoderm and mesoderm), the (2) *amnion* (ectoderm and mesoderm), the (3) *allantois* (endoderm and mesoderm), and (4) the *chorion* (mesoderm and trophoblast). Although these membranes support embryological and fetal development, they leave few traces of their existence in adult systems. Figure 29-5● shows representative stages in the development of the extraembryonic membranes.

The Yolk Sac   The **yolk sac** begins as a layer of cells spread out around the outer edges of the blastocoele to form a complete pouch. This pouch is already visible 10 days after fertilization (Figure 29-4●). As gastrulation proceeds, mesodermal cells migrate around this pouch and complete the formation of the yolk sac (Figure 29-5a●). Blood vessels soon appear within the mesoderm, and the yolk sac becomes an important site of blood cell formation.

The Amnion   The ectodermal layer enlarges, and ectodermal cells spread over the inner surface of the amniotic cavity. Mesodermal cells soon follow, creating a second, outer layer (Figure 29-5a●). This combination of mesoderm and ectoderm is the **amnion** (AM-nē-on). As development proceeds, the amnion and the amniotic cavity continue to enlarge. The amniotic cavity contains **amniotic fluid,** which surrounds and cushions the developing embryo or fetus (Figure 29-5b–e●).

The Allantois   The third extraembryonic membrane begins as an outpocketing of the endoderm near the base of the yolk sac (Figure 29-5b●). The free endodermal tip then grows toward the wall of the blastocyst, surrounded by a mass of mesodermal cells. This sac of endoderm and mesoderm is the **allantois** (a-LAN-tō-is). The base of the allantois later gives rise to the urinary bladder. (We illustrate the formation of the allantois and its relationship to the urinary bladder in AM *EMBRYOLOGY SUMMARY 20: The Development of the Urinary System.*)

The Chorion   The mesoderm associated with the allantois spreads around the blastocyst, separating the cellular trophoblast from the blastocoele. This combination of mesoderm and trophoblast is the **chorion** (KOR-ē-on) (Figure 29-5a,b●).

When implantation first occurs, the nutrients absorbed by the trophoblast can easily reach the inner cell mass by simple diffusion. But as the embryo and the trophoblast enlarge, the distance between them increases, so diffusion alone can no longer keep pace with the demands of the developing embryo. Blood vessels now begin to develop within the mesoderm of the chorion, creating a rapid-transit system that links the embryo with the trophoblast.

The appearance of blood vessels in the chorion is the first step in the creation of a functional placenta. By the third week of development, the mesoderm extends along the core of each trophoblastic villus, forming **chorionic villi** in contact with maternal tissues (Figures 29-5b and 29-6●, p. 1074). These villi continue to enlarge and branch, creating an intricate network within the endometrium. Embryonic blood vessels develop within each villus. Blood flow through those chorionic vessels begins early in the third week of development, when the embryonic heart starts beating. The blood supply to the chorionic villi arises from the allantoic arteries and veins.

As the chorionic villi enlarge, more maternal blood vessels are eroded. Maternal blood now moves slowly through complex lacunae lined by the syncytial trophoblast. Chorionic blood vessels pass close by, and gases and nutrients diffuse between the embryonic and maternal circulations across the trophoblast layers. Maternal blood then reenters the venous system of the mother through the broken walls of small uterine veins. No mixing of maternal and fetal blood occurs, because the two are always separated by layers of trophoblast.

PLACENTATION

At first the entire blastocyst is surrounded by chorionic villi. The chorion continues to enlarge, expanding like a balloon within the endometrium. By the fourth week, the embryo, amnion, and yolk sac are suspended within an expansive, fluid-filled chamber (Figure 29-5c●). The **body stalk,** the connection between embryo and chorion, contains the distal portions of the allantois and blood vessels that carry blood to and from the placenta. The narrow connection between the endoderm of the embryo and the yolk sac is called the **yolk stalk.** (We illustrate the formation of the yolk stalk and body stalk in AM *EMBRYOLOGY SUMMARY 19: The Development of the Digestive System.*)

The placenta does not continue to enlarge indefinitely. Regional differences in placental organization begin to develop as placental expansion creates a prominent bulge in the endometrial surface. This relatively thin portion of the endometrium, called the **decidua capsularis** (dē-SID-ū-a kap-sū-LA-ris; *deciduus,* a falling off), no longer participates in nutrient exchange, and the

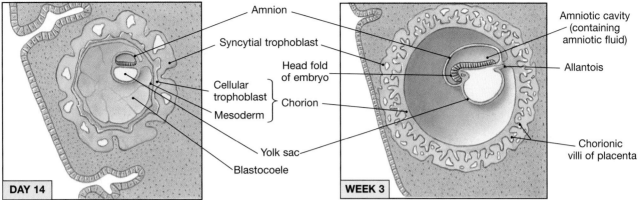

DAY 14

Amnion

Syncytial trophoblast

Cellular trophoblast

Mesoderm

Yolk sac

Blastocoele

(a) Migration of mesoderm around the inner surface of the trophoblast creates the chorion. Mesodermal migration around the outside of the amniotic cavity, between the ectodermal cells and the trophoblast, creates the amnion. Mesodermal migration around the endodermal pouch creates the yolk sac.

WEEK 3

Amniotic cavity (containing amniotic fluid)

Head fold of embryo

Chorion

Allantois

Chorionic villi of placenta

(b) The embryonic disc bulges into the amniotic cavity at the head fold. The allantois, an endodermal extension surrounded by mesoderm, extends toward the trophoblast.

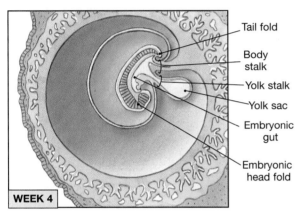

WEEK 4

Tail fold

Body stalk

Yolk stalk

Yolk sac

Embryonic gut

Embryonic head fold

(c) The embryo now has a head fold and a tail fold. Constriction of the connection between the embryo and the surrounding trophoblast constricts the yolk stalk and body stalk.

Uterus

Myometrium

Decidua parietalis

Decidua capsularis

Decidua basalis

Umbilical stalk

Placenta

Chorionic villi of placenta

Yolk sac

Uterine lumen

WEEK 5

(d) The developing embryo and extraembryonic membranes bulge into the uterine cavity. The trophoblast pushing out into the uterine lumen remains covered by endometrium but no longer participates in nutrient absorption and embryo support. The embryo moves away from the placenta, and the body stalk and yolk stalk fuse to form an umbilical stalk.

Decidua parietalis

Decidua basalis

Decidua capsularis

Chorion

Placenta

Amnion

Umbilical cord

Amniotic cavity

WEEK 10

(e) The amnion has expanded greatly, filling the uterine cavity. The fetus is connected to the placenta by an elongated umbilical cord that contains a portion of the allantois, blood vessels, and the remnants of the yolk stalk.

•**FIGURE 29-5** Extraembryonic Membranes and Placenta Formation

29

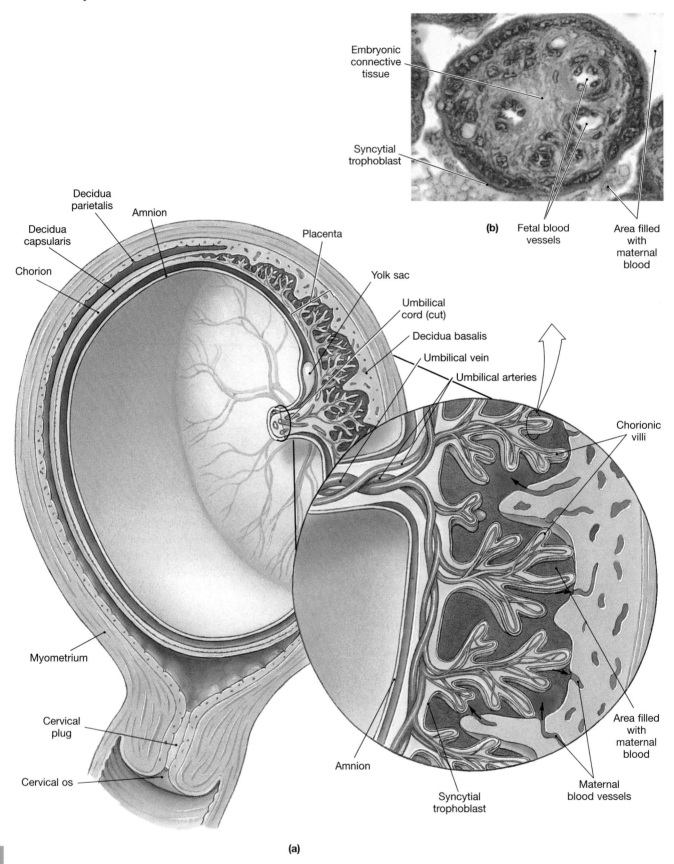

Embryonic connective tissue

Syncytial trophoblast

(b) Fetal blood vessels

Area filled with maternal blood

Decidua parietalis

Amnion

Decidua capsularis

Chorion

Placenta

Yolk sac

Umbilical cord (cut)

Decidua basalis

Umbilical vein

Umbilical arteries

Chorionic villi

Myometrium

Area filled with maternal blood

Cervical plug

Amnion

Maternal blood vessels

Cervical os

Syncytial trophoblast

(a)

•**FIGURE 29-6 A Three-Dimensional View of Placental Structure.** **(a)** For clarity, the uterus is shown after the embryo has been removed and the umbilical cord cut. Arrows indicate the direction of blood flow. Blood flows into the placenta through ruptured maternal arteries. It then flows around chorionic villi, which contain fetal blood vessels. **(b)** A cross section through a chorionic villus, showing the syncytial trophoblast exposed to the maternal blood space. (LM × 2045)

29

chorionic villi in this region disappear (Figures 29-5d, 29-6a●). Placental functions are now concentrated in a disc-shaped area in the deepest portion of the endometrium, a region called the **decidua basalis** (ba-SA-lis). The rest of the uterine endometrium, which has no contact with the chorion, is called the **decidua parietalis.**

As the end of the first trimester approaches, the fetus moves farther from the placenta (Figure 29-5d,e●). The fetus and placenta remain connected by the **umbilical cord,** or *umbilical stalk,* which contains the allantois, the placental blood vessels, and the yolk stalk.

Placental Circulation

Figure 29-6a● diagrams circulation at the placenta near the end of the first trimester. Blood flows to the placenta through the paired **umbilical arteries** and returns in a single **umbilical vein.** ∞ *p. 742* The chorionic villi provide the surface area for active and passive exchanges of gases, nutrients, and waste products between the fetal and maternal bloodstreams (Figure 29-6b●).

The placenta places a considerable demand on the maternal cardiovascular system, and blood flow to the uterus and placenta is extensive. If the placenta is torn or otherwise damaged, the consequences may prove fatal for both fetus and mother. [AM] *Problems with Placentation*

The Endocrine Placenta

In addition to its role in the nutrition of the fetus, the placenta acts as an endocrine organ. Hormones are synthesized by the syncytial trophoblast and released into the maternal bloodstream. The hormones produced include *human chorionic gonadotropin, human placental lactogen, placental prolactin, relaxin, progesterone,* and *estrogens.*

Human Chorionic Gonadotropin The hormone **human chorionic gonadotropin (hCG)** appears in the maternal bloodstream soon after implantation has occurred. The presence of hCG in blood or urine samples provides a reliable indication of pregnancy. Kits sold for the early detection of pregnancy are sensitive to the presence of this hormone.

In function, hCG resembles luteinizing hormone (LH), because it maintains the integrity of the corpus luteum and promotes the continued secretion of progesterone. ∞ *p. 1052* As a result, the endometrial lining remains perfectly functional, and menses does not occur. In the absence of hCG, the pregnancy would end because another uterine cycle would begin and the functional zone of the endometrium would disintegrate.

In the presence of hCG, the corpus luteum persists for 3–4 months before gradually decreasing in size and

secretory function. The decline in luteal function does not trigger the return of uterine cycles, because by the end of the first trimester, the placenta actively secretes both estrogens and progesterone.

Human Placental Lactogen and Placental Prolactin **Human placental lactogen (hPL),** or *human chorionic somatomammotropin* (hCS), helps prepare the mammary glands for milk production. It also has a stimulatory effect on other tissues comparable to that of growth hormone (GH). At the mammary glands, the conversion from resting to active status requires the presence of placental hormones (hPL, **placental prolactin,** estrogen, and progesterone) as well as several maternal hormones (GH, prolactin [PRL], and thyroid hormones). We shall consider the hormonal control of mammary gland function in a later section.

Relaxin **Relaxin** is a peptide hormone that is secreted by the placenta as well as by the corpus luteum during pregnancy. Relaxin (1) increases the flexibility of the pubic symphysis, permitting the pelvis to expand during delivery; (2) causes the dilation of the cervix, making it easier for the fetus to enter the vaginal canal; and (3) suppresses the release of oxytocin by the hypothalamus and delays the onset of labor contractions.

Progesterone and Estrogens After the first trimester, the placenta produces sufficient amounts of progesterone to maintain the endometrial lining and continue the pregnancy. As the end of the third trimester approaches, estrogen production by the placenta accelerates. As we shall see in a later section, the rising estrogen levels play a role in stimulating labor and delivery.

EMBRYOGENESIS

Shortly after gastrulation begins, the body of the embryo begins to separate itself from the rest of the embryonic disc. The body of the embryo and its internal organs now start to form. This process, called **embryogenesis,** begins as folding and differential growth of the embryonic disc produce a bulge that projects into the amniotic cavity (Figure 29-5b●). This projection is known as the **head fold.** Similar movements lead to the formation of a **tail fold** (Figure 29-5c●).

The embryo is now physically as well as developmentally distinct from the embryonic disc and the extraembryonic membranes. The definitive orientation of the embryo can now be seen, complete with dorsal and ventral surfaces and left and right sides. Table 29-2 provides an overview of the subsequent development of the major organs and body systems. The changes in proportions and appearance that occur

29

TABLE 29-2 An Overview of Prenatal and Early Postnatal Development

Background Material in *Applications Manual*
AM *EMBRYOLOGY SUMMARY 1: The Development of Tissues*
AM *EMBRYOLOGY SUMMARY 2: The Development of Epithelia*
AM *EMBRYOLOGY SUMMARY 3: The Development of Connective Tissues*
AM *EMBRYOLOGY SUMMARY 4: The Development of Organ Systems*

Gestational Age (Months)	Size and Weight	Integumentary System	Skeletal System	Muscular System	Nervous System	Special Sense Organs
1	5 mm 0.02 g		(b) Formation of somites	(b) Formation of somites	(b) Formation of neural tube	(b) Formation of eyes and ears
2	28 mm 2.7 g	(b) Formation of nail beds, hair follicles, sweat glands	(b) Formation of axial and appendicular cartilages	(c) Rudiments of axial musculature	(b) CNS, PNS organization, growth of cerebrum	(b) Formation of taste buds, olfactory epithelium
3	78 mm 26 g	(b) Epidermal layers appear	(b) Spreading of ossification centers	(c) Rudiments of appendicular musculature	(c) Basic spinal cord and brain structure	
4	133 mm 0.15 kg	(b) Formation of hair, sebaceous glands (c) Sweat glands	(b) Articulations (c) Facial and palatal organization	Fetus starts moving	(b) Rapid expansion of cerebrum	(c) Basic eye and ear structure (b) Formation of peripheral receptors
5	185 mm 0.46 kg	(b) Keratin production, nail production			(b) Myelination of spinal cord	
6	230 mm 0.64 kg			(c) Perineal muscles	(b) Formation of CNS tract (c) Layering of cortex	
7	270 mm 1.492 kg	(b) Keratinization, formation of nails, hair				(c) Eyelids open, retinae sensitive to light
8	310 mm 2.274 kg		(b) Formation of epiphyseal cartilages			(c) Taste receptors functional
9	346 mm 3.2 kg					
Postnatal development		Hair changes in consistency and distribution	Formation and growth of epiphyseal cartilages continue	Muscle mass and control increase	Myelination, layering, CNS tract formation continue	
Location of relevant text and illustrations		AM Embryology Summary 5	Ch 6: pp. 175–179 AM Embryology Summaries 6, 7, 8	AM Embryology Summary 9	Ch. 13: pp. 407–408 Ch. 14: pp. 438–439 AM Embryology Summaries 10, 11, 12	AM Embryology Summary 13

Note: (b) = beginning; (c) = completion

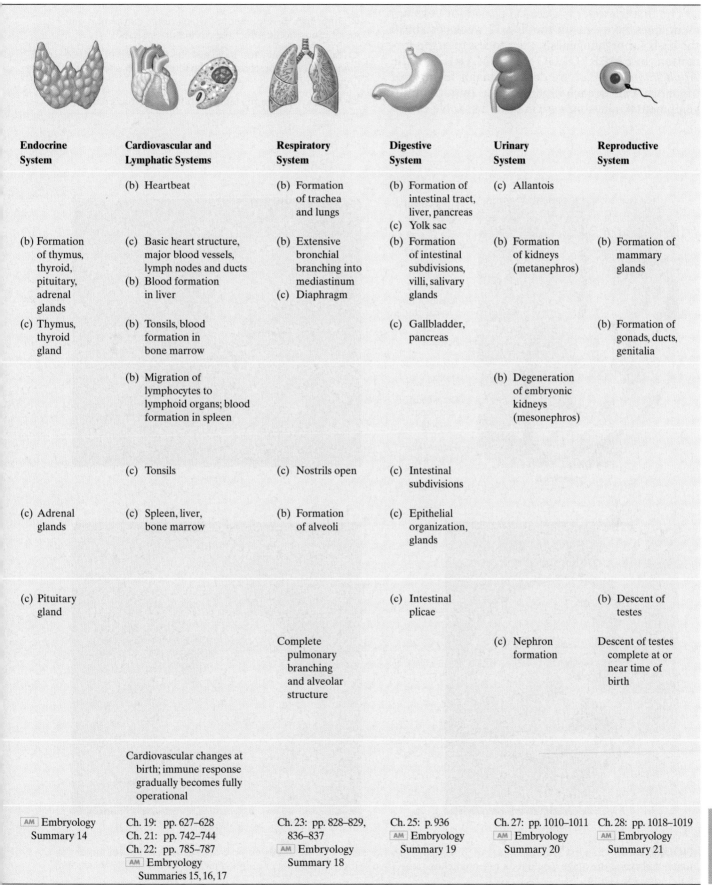

Endocrine System	Cardiovascular and Lymphatic Systems	Respiratory System	Digestive System	Urinary System	Reproductive System
	(b) Heartbeat	(b) Formation of trachea and lungs	(b) Formation of intestinal tract, liver, pancreas	(c) Allantois	
			(c) Yolk sac		
(b) Formation of thymus, thyroid, pituitary, adrenal glands	(c) Basic heart structure, major blood vessels, lymph nodes and ducts	(b) Extensive bronchial branching into mediastinum	(b) Formation of intestinal subdivisions, villi, salivary glands	(b) Formation of kidneys (metanephros)	(b) Formation of mammary glands
	(b) Blood formation in liver	(c) Diaphragm			
(c) Thymus, thyroid gland	(b) Tonsils, blood formation in bone marrow		(c) Gallbladder, pancreas		(b) Formation of gonads, ducts, genitalia
	(b) Migration of lymphocytes to lymphoid organs; blood formation in spleen			(b) Degeneration of embryonic kidneys (mesonephros)	
	(c) Tonsils	(c) Nostrils open	(c) Intestinal subdivisions		
(c) Adrenal glands	(c) Spleen, liver, bone marrow	(b) Formation of alveoli	(c) Epithelial organization, glands		
(c) Pituitary gland			(c) Intestinal plicae		(b) Descent of testes
		Complete pulmonary branching and alveolar structure		(c) Nephron formation	Descent of testes complete at or near time of birth
	Cardiovascular changes at birth; immune response gradually becomes fully operational				
AM Embryology Summary 14	Ch. 19: pp. 627–628 Ch. 21: pp. 742–744 Ch. 22: pp. 785–787 AM Embryology Summaries 15, 16, 17	Ch. 23: pp. 828–829, 836–837 AM Embryology Summary 18	Ch. 25: p. 936 AM Embryology Summary 19	Ch. 27: pp. 1010–1011 AM Embryology Summary 20	Ch. 28: pp. 1018–1019 AM Embryology Summary 21

between the second developmental week and the end of the first trimester are summarized in Figure 29-7•.

The first trimester is a critical period for development, because events in the first 12 weeks establish the basis for **organogenesis,** the process of organ formation. In EMBRYOLOGY SUMMARIES in the *Applications Manual,* we describe major features of organogenesis in each organ system. Important developmental milestones are indicated in Table 29-2.

☑ What is the developmental fate of the inner cell mass of the blastocyst?

☑ Improper development of which of the extraembryonic membranes would affect the circulatory system?

☑ Sue's pregnancy test indicates elevated levels of the hormone hCG (human chorionic gonadotropin). Is she pregnant?

☑ What are two important functions of the placenta?

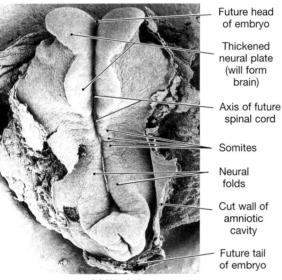

Future head of embryo

Thickened neural plate (will form brain)

Axis of future spinal cord

Somites

Neural folds

Cut wall of amniotic cavity

Future tail of embryo

(a) Week 2

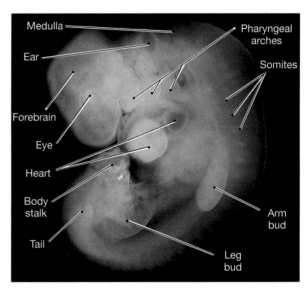

Medulla

Ear

Forebrain

Eye

Heart

Body stalk

Tail

Pharyngeal arches

Somites

Arm bud

Leg bud

(b) Week 4

Chorionic villi

Amnion

Umbilical cord

Placenta

(c) Week 8

(d) Week 12

•**FIGURE 29-7 The First Trimester.** **(a)** An SEM of the superior surface of a monkey embryo after 2 weeks of development. A human embryo at this stage would look essentially the same. **(b–d)** Fiber-optic views of human embryos at 4, 8, and 12 weeks. For actual sizes, see Figure 29-14, p. 1088.

Technology and the Treatment of Infertility

Infertility *(sterility)* is usually defined as an inability to achieve pregnancy after 1 year of appropriately timed intercourse. Problems with fertility are relatively common. An estimated 10–15 percent of U.S. married couples are infertile, and another 10 percent are unable to have as many children as they desire. It is thus not surprising that reproductive physiology has become a popular field and that the treatment of infertility has become a major medical industry.

An infertile woman is unable to produce functional oocytes or to support a developing embryo. An infertile man is incapable of providing a sufficient number of motile sperm capable of successful fertilization. Because the infertility of either sexual partner has the same result, the diagnosis and treatment of infertility must involve the evaluation of both partners. Approximately 40 percent of infertility cases are attributed to the female partner, 40 percent to the male partner, and 20 percent to both partners.

Recent advances in our understanding of reproductive physiology are providing new solutions to fertility problems. These approaches, called **assisted reproductive technologies (ART),** are diagrammed in Figure 29-8•:

- *Low sperm count.* In cases of male infertility due to low sperm counts, semen from several ejaculates can be pooled, concentrated, and introduced into the female reproductive tract. This technique, *artificial insemination,* may lead to normal fertilization and pregnancy. In cases in which a male's spermatozoa are unable to penetrate the oocyte, single-sperm fertilization has been accomplished with the microscopic manipulation of the oocyte and corona radiata.

- *Abnormal spermatozoa.* If the male cannot produce functional spermatozoa, they can be obtained from a *sperm bank,* a storage facility for donor sperm.

- *Hormonal problems.* If the problem involves the woman's inability to ovulate because her gonadotropin or estrogen levels are low or she is unable to maintain adequate progesterone levels after ovulation, these hormones can be provided.

- *Problems with oocyte production.* Fertility drugs, such as clomiphene *(Clomid®),* stimulate ovarian oocyte production. Clomiphene blocks the feedback inhibition of the hypothalamus and pituitary gland by estrogens. As a result, circulating FSH levels rise, so more follicles are stimulated to complete their development. Injected purified gonadotropins, such as *Pergonal®* (FSH and LH) and *Metrodine®* (FSH), are also used to accelerate ovum development. The chance that a single oocyte will be fertilized through sexual intercourse is about 1 in 3. Increasing the number of oocytes released increases the odds of fertilization and therefore the odds of a pregnancy. It is not easy to determine just how much ovarian stimulation will be needed, however, so treatment with fertility drugs commonly results in multiple births. Careful

•**FIGURE 29-8** **The Treatment of Infertility**

29

monitoring of follicle development somewhat reduces the chances of a multiple birth.

■ *Problems with oocyte transport from the ovary to the uterine tube.* When problems occur with the transport of the oocyte from the ovary to the uterine tube (for instance, due to scarring of the fimbriae), a procedure called **GIFT,** or *gamete intrafallopian tube transfer,* can be used. (*Fallopian tube* is another name for the uterine tube.) In this procedure, the ovaries are stimulated with injected hormones, and a large "crop" of mature oocytes is removed from tertiary follicles. The individual oocytes are examined for defects, inserted into the uterine tubes, and exposed to high concentrations of spermatozoa from the husband or donor. The success rate of this procedure is less than that of natural fertilization (4 percent versus 33 percent), and not every pregnancy produces an infant. The cost of a single procedure (successful or not) averages $5000.

■ *Blocked uterine tubes.* The blockage of the uterine tubes or damage to the uterine lining can interfere with oocyte, sperm, and zygote transport. In the GIFT procedure, fertilization occurs in its normal location, within the uterine tube. This site is not essential, and fertilization can also take place in a test tube or petri dish. This process is called **in vitro fertilization,** or **IVF** (*vitro,* glass). If a carefully controlled fluid environment is provided, early development will proceed normally. One variation on the GIFT procedure, called **ZIFT** (*zygote intrafallopian tube transfer),* exposes selected oocytes to spermatozoa outside the body and inserts zygotes or pre-embryos, rather than oocytes, into the uterine tubes. If multiple zygotes are available, some can be frozen and stored for later insertion if the initial attempt fails. The cost for a single ZIFT procedure, which has a 4 percent success rate, ranges between $8000 and $10,000.

Alternatively, the zygote can be maintained in an artificial environment through the first 2–3 days of development. This procedure is commonly selected if the uterine tubes are damaged or blocked. The cleavage-stage embryo is then placed directly into the uterus rather than into one of the uterine tubes. The cost of this procedure is comparable to that for ZIFT.

■ *Abnormal oocytes.* If the oocytes released by the ovaries are abnormal in some way or if menopause has already occurred, viable oocytes can be obtained from a suitable donor. Through treatment with fertility drugs, the donor's ovaries are stimulated to produce a large crop of oocytes. These are collected and fertilized in vitro, generally by the spermatozoa of the recipient's husband. After cleavage has begun, the pre-embryo is placed in the recipient's uterus, which has been "primed" by progesterone therapy. The pregnancy rate of this procedure is roughly 33 percent for women over age 40, using oocytes donated by women in their early twenties. Oocyte donation has a much higher success rate for these women than does ZIFT or GIFT. This difference suggests that age-related changes in the characteristics and quality of the oocytes, rather than changes in hormone levels or uterine responsiveness, are typically the primary cause of infertility in older women.

■ *An abnormal uterine environment.* If fertilization and transport occur normally but the uterus cannot maintain a pregnancy, the problem may involve low levels of progesterone secretion by the corpus luteum. Hormone therapy may solve this problem. If the maternal uterus simply cannot support development, the zygote or cleavage-stage embryo can be introduced into the uterus of a *surrogate mother.* If the embryo survives and makes contact with the endometrium, development will proceed normally despite the fact that the mother has no genetic relationship with the embryo.

Surrogate motherhood, which sounds relatively simple and straightforward, has proven to be one of the most explosive solutions in terms of ethics and legality. Since 1990, several court cases have resulted from disputes over surrogate motherhood and who merits legal custody of the infant. Legal battles have also broken out over a variety of other complex questions, and some of them will take years to sort out. To understand the problem, consider the following questions:

■ Do parents share property rights over frozen and stored zygotes? Can a husband have any of the stored zygotes implanted into the uterus of his second wife without the consent of his first wife, who provided the oocytes? Can a wife use her husband's stored spermatozoa to become pregnant after his death?

■ If both donor oocyte and donor sperm are used, do adoption laws apply?

■ If the husband provided the spermatozoa that fertilized the oocyte of a donor who is not his wife, for implantation into a surrogate mother, can the wife, the surrogate mother, or the oocyte donor sue for custody of the child after a divorce?

■ If a hospital stores frozen pre-embryos or spermatozoa but the freezer breaks down, what is the hospital's liability? What is the monetary value of a frozen pre-embryo?

If you use your imagination, you can probably think of even more complex problems, many of which will probably be debated in courtrooms over the next decade.

THE SECOND AND THIRD TRIMESTERS

By the end of the first trimester (Figure 29-7d•), the rudiments of all the major organ systems have formed. Over the next 3 months, the fetus will grow to a weight of about 0.64 kg (1.4 lb). During this second trimester, the fetus, encircled by the amnion, grows faster than the surrounding placenta. When the mesoderm on the outer surface of the amnion contacts the mesoderm on the inner surface of the chorion, these layers fuse, creating a compound *amniochorionic membrane*. Figure 29-9a• shows a 4-month fetus; Figure 29-9b• shows a 6-month fetus.

During the third trimester, most of the organ systems become ready to fulfill their normal functions without medical assistance. The rate of growth starts to decrease, but in absolute terms this trimester sees the largest weight gain. In the last 3 months of gestation, the fetus gains about 2.6 kg (5.7 lb), reaching a full-term weight of approximately 3.2 kg (7 lb). EMBRYOLOGY SUMMARIES in the *Applications Manual* illustrate organ system development in the second and third trimesters, and highlights are noted in Table 29-2.

At the end of gestation, a typical uterus will undergo a tremendous size increase. Figure 29-10a–c• shows the positions of the uterus, fetus, and placenta from 16 weeks to *full term* (9 months). When the pregnancy is at full term, the uterus and fetus push many of the maternal abdominal organs out of their normal positions (Figure 29-10c,d•).

PREGNANCY AND MATERNAL SYSTEMS

The developing fetus is totally dependent on maternal organ systems for nourishment, respiration, and waste removal. These functions must be performed by maternal systems in addition to their normal operations. For example, the mother must absorb enough oxygen, nutrients, and vitamins for herself *and* for her fetus, and she must eliminate all the generated wastes. Although this is not a burden over the initial weeks of gestation, the demands placed on the mother become significant as the fetus grows. For the mother to survive under these conditions, the maternal systems must make major compensatory adjustments. In practical terms, the mother must breathe, eat, and excrete for two. Among the major changes that occur in maternal systems are the following:

- *The maternal respiratory rate goes up and the tidal volume increases.* As a result, the mother's lungs deliver the extra oxygen required, and remove the excess carbon dioxide generated, by the fetus.
- *The maternal blood volume increases.* This increase occurs because (1) blood flowing into the placenta reduces the volume in the rest of the systemic circuit, and (2) fetal activity lowers the blood P_{O_2} and elevates

(a)

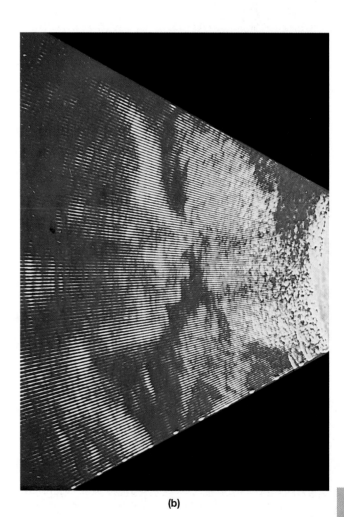

(b)

•**FIGURE 29-9** **The Second and Third Trimesters.**
(a) A 4-month fetus, as seen through a fiber-optic endoscope.
(b) Head of a 6-month fetus, as seen through ultrasound.

29

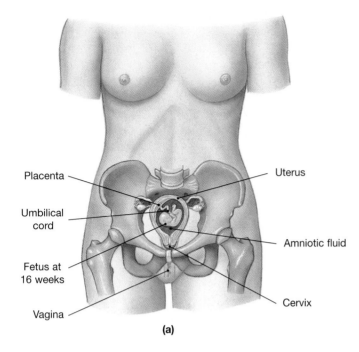

Placenta

Umbilical cord

Fetus at 16 weeks

Vagina

Uterus

Amniotic fluid

Cervix

(a)

9
8 months
7 months
6 months
5 months
4 months
3 months

After dropping, in preparation for delivery

(b)

•**FIGURE 29-10** **Growth of the Uterus and Fetus.** **(a)** Pregnancy at 16 weeks, showing the positions of the uterus, fetus, and placenta. **(b)** Pregnancy at 3 months to 9 months (full term), showing the position of the uterus within the abdomen. **(c)** Pregnancy at full term. Note the position of the uterus and full-term fetus within the abdomen and the displacement of abdominal organs, as compared with **(d)**, a sectional view through the abdominopelvic cavity of a woman who is not pregnant.

Stomach

Transverse colon

Fundus of uterus

Placenta

Umbilical cord

Liver

Small intestine

Pancreas

Aorta

Common iliac vein

Urinary bladder

Pubic symphysis

Urethra

Vagina

Mucus plug in cervical canal

Rectum

(c) Pregnant female (full-term infant)

(d) Nonpregnant female

29

the P_{CO_2}. The combination stimulates the production of renin and erythropoietin, leading to an increase in maternal blood volume through mechanisms detailed in Chapter 21 (see Figure 21-17a•, p. 716). By the end of gestation, the maternal blood volume has increased by almost 50 percent.

■ *The maternal requirements for nutrients and vitamins climb 10–30 percent.* Pregnant women must nourish both themselves and their fetus and tend to have increased sensations of hunger.

■ *The maternal glomerular filtration rate increases by roughly 50 percent.* This increase, which corresponds to the increase in blood volume, accelerates the excretion of metabolic wastes generated by the fetus. Because the volume of urine produced increases and the weight of the uterus presses down on the urinary bladder, pregnant women need to urinate frequently.

■ *The uterus undergoes a tremendous increase in size.* Structural and functional changes in the expanding uterus are so important that we shall discuss them in a separate section.

■ *The mammary glands increase in size, and secretory activity begins.* Mammary gland development requires a combination of hormones, including human placental lactogen and placental prolactin from the placenta, and PRL, estrogens, progesterone, GH, and thyroxine from maternal endocrine organs. By the end of the sixth month of pregnancy, the mammary glands are fully developed and begin to produce secretions that are stored in the duct system of those glands.

Abortion

Abortion is the termination of a pregnancy. Most references distinguish among spontaneous, therapeutic, and induced abortions. **Spontaneous abortions,** or *miscarriages,* occur as a result of developmental or physiological problems. For example, spontaneous abortions can result from chromosomal defects in the embryo or from hormonal problems, such as inadequate LH production by the maternal pituitary gland (or reduced LH sensitivity at the corpus luteum), inadequate progesterone sensitivity in the endometrium, or placental failure to produce adequate levels of hCG. Spontaneous abortions occur in roughly 15 percent of all pregnancies. **Therapeutic abortions** are performed when continuing the pregnancy represents a threat to the life of the mother. AM *Problems with the Maintenance of a Pregnancy*

Induced abortions, or *elective abortions,* are performed at the woman's request. Induced abortions remain the focus of considerable controversy. Most induced abortions involve unmarried and/or adolescent women. The ratio between abortions and deliveries for married women is 1:10, whereas it is nearly 2:1 for unmarried women and adolescents. In most states, induced abortions are legal during the first three months after conception; with restrictions, induced abortions may be permitted until the fifth or sixth month.

STRUCTURAL AND FUNCTIONAL CHANGES IN THE UTERUS

At the end of gestation, a typical uterus has grown from 7.5 cm (3 in.) in length and 30–40 g (1–1.4 oz) in weight to 30 cm (12 in.) in length and 1100 g (2.4 lb) in weight. It may then contain almost 5 liters of fluid, so the organ with contents has a total weight of roughly 10 kg (22 lb). This remarkable expansion occurs through the enlargement (hypertrophy) and elongation of existing cells, especially smooth muscle fibers, rather than by an increase in the total number of cells.

The tremendous stretching of the uterus is associated with a gradual increase in the rates of spontaneous smooth muscle contractions in the myometrium. In the early stages of pregnancy, the contractions are weak, painless, and brief. Evidence indicates that the progesterone released by the placenta has an inhibitory effect on the uterine smooth muscle, preventing more-extensive and more-powerful contractions.

Three major factors oppose the calming action of progesterone:

1. *Rising estrogen levels.* Estrogens produced by the placenta increase the sensitivity of the uterine smooth muscles and make contractions more likely. Throughout pregnancy, progesterone exerts the dominant effect, but as delivery approaches, the production of estrogen accelerates and the myometrium becomes more sensitive to stimulation. Estrogens also increase the sensitivity of smooth muscle fibers to oxytocin.

2. *Rising oxytocin levels.* Rising oxytocin levels stimulate an increase in the force and frequency of uterine contractions. Oxytocin release is stimulated by high estrogen levels and by the distortion of the cervix. Uterine distortion occurs as the weight of the fetus increases.

3. *Prostaglandin production.* Estrogens and oxytocin stimulate the production of prostaglandins in the endometrium. These prostaglandins further stimulate smooth muscle contractions.

Late in pregnancy, some women experience occasional spasms in the uterine musculature, but the contractions are neither regular nor persistent. These contractions are called **false labor. True labor** begins when the biochemical and mechanical factors reach the point of no return. After 9 months of gestation, multiple factors interact to initiate true labor. Once the **labor contractions** have begun in the myometrium, positive feedback ensures that the contractions will continue until delivery has been completed.

Figure 29-11• diagrams important factors that stimulate and sustain labor. The actual trigger for the onset of true labor may be events in the fetus rather than in the mother. When labor commences, the fetal pituitary gland secretes oxytocin that is released into the maternal bloodstream at the placenta. The resulting increase in myometrial contractions and prostaglandin production, on top of the priming effects of estrogens and maternal oxytocin, may be the "last straw."

29

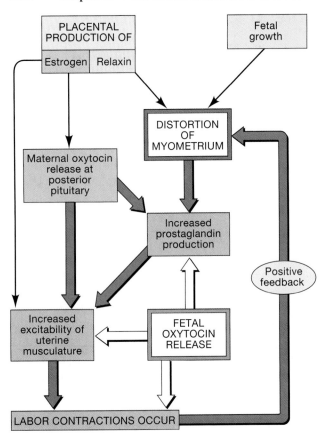

•**FIGURE 29-11** Factors Involved in the Initiation of Labor and Delivery

ly. Late in the process, the amniochorionic membrane ruptures, an event sometimes referred to as having the "water break." If this event occurs before other events of the dilation stage, the life of the fetus is at risk; medical steps must then be taken to accelerate labor.

The Expulsion Stage

The **expulsion stage** begins as the cervix, pushed open by the approaching fetus, completes its dilation (Figure 29-12b•). In this stage, contractions reach maximum intensity; they may occur at 2-minute or 3-minute intervals and last a full minute. Expulsion continues until the fetus has emerged from the vagina; in most cases, it lasts less than 2 hours. The arrival of the newborn infant into the outside world is **delivery,** or birth.

If the vaginal canal is too small to permit the passage of the fetus and there is acute danger of perineal tearing, a clinician may temporarily enlarge the passageway by making an incision through the perineal musculature. After delivery, this **episiotomy** (e-pēz-ē-OT-o-mē) can be repaired with sutures, a much simpler procedure than dealing with the bleeding and tissue damage associated with an extensive perineal tear. If complications arise during the dilation or expulsion stages, the infant can be removed by **cesarean section,** or "*C-section.*" In such cases, an incision is made through the abdominal wall and the uterus is opened just enough to allow passage of the infant's head. This procedure is performed during 15–25 percent of the deliveries in the United States—more often than necessary, according to some studies. Over the last decade, efforts have been made to reduce the frequency of both episiotomies and cesarean sections.

The Placental Stage

During the **placental stage** of labor, muscle tension builds in the walls of the partially empty uterus, and the organ gradually decreases in size (Figure 29-12c•). This uterine contraction tears the connections between the endometrium and the placenta. In general, within an hour of delivery, the placental stage ends with the ejection of the placenta, or *afterbirth.* The disruption of the placenta is accompanied by a loss of blood. Because the maternal blood volume has increased greatly during pregnancy, this loss can be tolerated without difficulty.

LABOR AND DELIVERY

The goal of labor is **parturition** (par-tū-RISH-un), the forcible expulsion of the fetus. During true labor, each labor contraction begins near the top of the uterus and sweeps in a wave toward the cervix. These contractions are strong and occur at regular intervals. As parturition approaches, the contractions increase in force and frequency, changing the position of the fetus and moving it toward the cervical canal.

STAGES OF LABOR

Labor has traditionally been divided into three stages: (1) the *dilation stage,* (2) the *expulsion stage,* and (3) the *placental stage* (Figure 29-12•).

The Dilation Stage

The **dilation stage** begins with the onset of true labor, as the cervix dilates and the fetus begins to move toward the cervical canal (Figure 29-12a•). This stage is highly variable in length but typically lasts 8 or more hours. At the start of this stage, the labor contractions last up to half a minute and occur at intervals of once every 10–30 minutes; their frequency increases steadi-

PREMATURE LABOR

Premature labor occurs when true labor begins before the fetus has completed normal development. The chances of newborn survival are directly related to the infant's body weight at delivery. Even with massive supportive efforts, infants weighing less than 400 g (14 oz) at birth will not survive, primarily because the respiratory, cardiovascular, and urinary

Umbilical cord

Pubic symphysis

Cervix

Vagina

Cervical canal

Placenta

Fully developed fetus

Sacral promontory

(a) Dilation stage

(b) Expulsion stage

Uterus

Ejection of the placenta

(c) Placental stage

•**FIGURE 29-12** The Stages of Labor

systems are unable to support life without the aid of maternal systems. As a result, the dividing line between spontaneous abortion and **immature delivery** is usually set at 500 g (17.6 oz), the normal weight near the end of the second trimester.

Most fetuses born at 25–27 weeks of gestation (birth weight under 600 g) die despite intensive neonatal care; survivors have a high risk of developmental abnormalities. **Premature delivery** usually refers to infants born at 28–36 weeks (birth weight over 1 kg). With care, they have a good chance of survival and normal development. AM *Complexity and Perfection*

MULTIPLE BIRTHS

Multiple births (twins, triplets, quadruplets, and so forth) can occur for several reasons. The ratio of twin births to single births in the U.S. population is roughly 1:89. "Fraternal," or **dizygotic** (dī-zī-GOT-ik), twins develop when two separate eggs were ovulated and subsequently fertilized. Because chromosomes shuffle during meiosis, the odds against any two zygotes from the same parents having identical genes exceeds 1:8.4 million. Seventy percent of twins are dizygotic.

"Identical," or **monozygotic,** twins result either from the separation of blastomeres early in cleavage or by the splitting of the inner cell mass before gastrulation. In either event, the genetic makeup of the pair is identical because both formed from the same pair of gametes. Triplets, quadruplets, and larger multiples can result from multiple ovulations, blastomere splitting, or some combination of the two. For unknown reasons, the statistics for naturally occurring multiple births fall into a pattern: Twins occur at a rate of 1:89, triplets at a rate of $1:89^2$, quadruplets at $1:89^3$, and so forth. The incidence of multiple births can be increased by exposure to fertility drugs that stimulate the maturation of abnormally large numbers of oocytes. (See the discussion "Technology and the Treatment of Infertility" on pp. 1079–1080.)

Multiple pregnancies pose special problems for maternal systems, because the strains on the mother are multiplied proportionately. The chances of premature labor are increased, and the risks to the mother are higher than for single births. The risks for the fetuses are also increased, both during gestation and after delivery, because even at full term the newborn infants have a lower average birthweight. They are also more likely to have problems during delivery. For example, in more than half of twin deliveries, one or both fetuses enter the vaginal canal in an abnormal position.

Complete splitting of the blastomeres or the embryonic disc can produce identical twins. If the separation is not complete, **conjoined** *(Siamese)* **twins** may develop. These infants typically share some skin, a portion of the liver, and perhaps other internal organs as well. When the fusion is minor, the infants can be surgically separated with some success. Most conjoined twins with more extensive fusions fail to survive delivery.

Forceps Deliveries and Breech Births

In most pregnancies, by the end of gestation, the fetus has rotated within the uterus to transit the birth canal head first, facing the mother's sacrum. In about 6 percent of deliveries, the fetus faces the mother's pubis instead. These infants can be delivered normally, given enough time, but risks to infant and mother are reduced by a *forceps delivery*. The forceps resemble large, curved salad tongs that can be separated for insertion into the vaginal canal one side at a time. Once in place, they are reunited and used to grasp the head of the fetus. An intermittent pull is applied, so that the forces on the head resemble those of normal delivery.

In 3–4 percent of deliveries, the legs or buttocks of the fetus enter the vaginal canal first. Such deliveries are **breech births.** Risks to the infant are higher in breech births than in normal deliveries, because the umbilical cord can become constricted and cut off placental blood flow. The head is normally the widest part of the fetus; the mother's cervix may dilate enough to pass the baby's legs and body but not the head. This entrapment compresses the umbilical cord, prolongs delivery, and subjects the fetus to severe distress and potential injury. If attempts to reposition the fetus or promote further dilation are unsuccessful over the short term, delivery by cesarean section may be required.

POSTNATAL DEVELOPMENT

Developmental processes do not cease at delivery, because newborn infants have few of the anatomical, functional, or physiological characteristics of mature adults. In the course of postnatal development, every individual passes through five **life stages:** (1) the *neonatal period,* (2) *infancy,* (3) *childhood,* (4) *adolescence,* and (5) *maturity.* Each stage is typified by a distinctive combination of characteristics and abilities.

These stages are a familiar part of human experience. Although each stage has distinctive features, the transitions between them are gradual and the boundaries are indistinct. At maturity, development ends and the process of aging, or *senescence,* begins.

THE NEONATAL PERIOD, INFANCY, AND CHILDHOOD

The **neonatal period** extends from birth to 1 month thereafter. **Infancy** then continues to 2 years of age, and **childhood** lasts until **adolescence,** the period of sexual and physical maturation. Two major events are under way during these developmental stages:

1. The organ systems other than those associated with reproduction become fully operational and gradually acquire the functional characteristics of adult structures.
2. The individual grows rapidly, and body proportions change significantly.

Pediatrics is a medical specialty focusing on postnatal development from infancy through adolescence.

Infants and young children cannot clearly describe the problems they are experiencing, so pediatricians and parents must be skilled observers. Standardized tests are used to assess developmental progress relative to average values. [AM] *Monitoring Postnatal Development*

The Neonatal Period

Physiological and anatomical changes occur as the fetus completes the transition to the status of newborn infant, or **neonate.** Before delivery, dissolved gases, nutrients, wastes, hormones, and antibodies were transferred across the placenta. At birth, the neonate must become relatively self-sufficient, with respiration, digestion, and excretion performed by its own specialized organs and organ systems. The transition from fetus to neonate can be summarized as follows:

- The lungs at birth are collapsed and filled with fluid. Filling them with air involves a massive and powerful inhalation. ∞ *p. 836*

- When the lungs expand, the pattern of cardiovascular circulation changes due to alterations in blood pressure and flow rates. The ductus arteriosus closes, isolating the pulmonary and systemic trunks. Closure of the foramen ovale separates the atria of the heart, completing the separation of the pulmonary and systemic circuits. ∞ *pp. 742–744*

- The typical heart rates (120–140 beats per minute) and respiratory rate (30 breaths per minute) of neonates are considerably higher than those of adults. In conjunction, a neonate's metabolic rate is roughly twice that of an adult for equivalent body weights.

- Before birth, the digestive system remains relatively inactive, although it does accumulate a mixture of bile secretions, mucus, and epithelial cells. This collection of debris is excreted during the first few days of life. Over that period, the newborn infant begins to nurse.

- As waste products build up in the arterial blood, they are excreted at the kidneys. Glomerular filtration is normal, but the neonate cannot concentrate urine to any significant degree. As a result, urinary water losses are high, and neonatal fluid requirements are much greater than those of adults. ∞ *pp. 1010–1011*

- The neonate has little ability to control body temperature, particularly in the first few days after delivery. As the infant grows larger and its insulating subcutaneous adipose "blanket" gets thicker, its metabolic rate also rises. Daily and even hourly shifts in body temperature continue throughout childhood. ∞ *p. 936*

Throughout the neonatal period, the newborn is dependent on nutrients contained in the milk secreted by the maternal mammary glands.

Lactation and the Mammary Glands By the end of the sixth month of pregnancy, the mammary glands are fully developed, and the gland cells begin to produce a secretion known as **colostrum** (ko-LOS-trum). Colostrum, which will be provided to the infant during the first 2 or 3 days of life, contains more proteins and far less fat than breast milk contains. Many of the proteins are antibodies that may help the infant ward off infections until its own immune system becomes fully functional. In addition, the mucins present in both colostrum and milk can inhibit the replication of a family of viruses *(rotaviruses)* that can cause dangerous forms of gastroenteritis and infant diarrhea.

As colostrum production drops, the mammary glands convert to milk production. Breast milk consists of water, proteins, amino acids, lipids, sugars, and salts. It also contains large quantities of *lysozymes,* enzymes with antibiotic properties. Human milk provides roughly 750 Calories per liter. The secretory rate varies with the demand, but a 5–6 kg (11–13 lb) infant usually requires about 850 ml of milk per day.

Mammary gland secretion is triggered when the infant sucks on the nipple. The stimulation of tactile receptors there leads to the stimulation of secretory neurons in the paraventricular nucleus of the mother's hypothalamus. These neurons release oxytocin at the posterior pituitary. When oxytocin reaches the mammary gland, this hormone causes the contraction of *myoepithelial cells,* contractile cells in the walls of the lactiferous ducts and sinuses. The result is milk ejection, or the **milk let-down reflex** (Figure 29-13•).

3. Stimulation of hypothalamic nuclei

Posterior pituitary gland

4. Oxytocin released

5. Milk ejected

1. Receptors in nipples stimulated

2. Impulses propagated to spinal cord

•**FIGURE 29-13** The Milk Let-Down Reflex

The milk let-down reflex continues to function until *weaning,* typically 1–2 years after birth. Milk production ceases soon after, and the mammary glands gradually return to a resting state.

Infancy and Childhood

The most rapid growth occurs during prenatal development, and the rate of growth declines after delivery. Postnatal growth during infancy and childhood occurs under the direction of circulating hormones, notably pituitary growth hormone, adrenal steroids, and thyroid hormones. These hormones affect each tissue and organ in specific ways, depending on the sensitivities of the individual cells. As a result, growth does not occur uniformly, so the body proportions gradually change. The head, for example, is relatively large at birth but decreases in proportion with the rest of the body to adulthood (Figure 29-14●).

ADOLESCENCE AND MATURITY

Adolescence begins at **puberty,** the period of sexual maturation, and ends when growth is completed. Three major hormonal events interact at the onset of puberty:

1. The hypothalamus increases its production of gonadotropin-releasing hormone (GnRH). Evidence indicates that this increase is dependent on adequate levels of *leptin,* a hormone released by adipose tissues. ∞ *p. 609*

2. The anterior pituitary gland becomes more sensitive to the presence of GnRH, and the circulating levels of FSH and LH rise rapidly.

3. Ovarian or testicular cells become more sensitive to FSH and LH. These changes initiate (1) gamete formation, (2) the production of sex hormones that stimulate the appearance of secondary sex characteristics and behaviors, and (3) a sudden acceleration in the growth rate, culminating in the closure of the epiphyseal cartilages.

The age at which puberty begins varies. In the United States today, puberty generally occurs at about age 14 in boys and 13 in girls, but the normal ranges are broad (9–14 in boys, 8–13 in girls). Many body systems alter their activities in response to circulating sex hormones and to the presence of growth hormone, thyroid hormones, PRL, and adrenocortical hormones, so gender-specific differences in structure and function develop. The following are some of the changes induced by the endocrine system at puberty:

4 WEEKS **8 WEEKS**

16 WEEKS

—5 ft
—4 ft
—3 ft
—2 ft
—1 ft

Newborn 6 years Adult

●**FIGURE 29-14 Growth and Changes in Body Form.** The views at 4, 8, and 16 weeks are presented at actual size. Notice the changes in body form and proportions as development proceeds. These changes do not stop at birth. For example, the head, which contains the brain and sense organs, is relatively large at birth.

29

■ *Integumentary system.* Testosterone stimulates the development of terminal hairs on the face and chest, whereas those follicles continue to produce fine hairs under estrogen stimulation. The hairline recedes under testosterone stimulation. Both testosterone and estrogen stimulate terminal hair growth in the axillae and in the genital area. These hormones also stimulate sebaceous gland secretion and may cause acne. Adipose tissues respond differently to testosterone than to estrogens, and this difference produces changes in the subcutaneous distribution of body fat. In women, the combination of estrogens, PRL, growth hormone, and thyroid hormones promotes the initial development of the mammary glands. Although the duct system becomes more elaborate, true secretory alveoli do not develop, and much of the growth of the breasts during this period reflects increased deposition of fat rather than glandular tissue.

■ *Skeletal system.* Both testosterone and estrogen accelerate bone deposition and skeletal growth. In the process, they promote the closure of the epiphyses and thus place a limit on growth in height. Estrogens cause more-rapid epiphyseal closure than does testosterone. In addition, the period of skeletal growth is shorter in girls than in boys, and girls generally do not grow as tall as boys. Girls grow most rapidly between ages 10 and 13, whereas boys grow most rapidly between ages 12 and 15.

■ *Muscular system.* Sex hormones stimulate the growth of skeletal muscle fibers, increasing strength and endurance. The effects of testosterone greatly exceed those of the estrogens, and the increased muscle mass accounts for significant gender differences in body mass, even for males and females of the same height. The stimulatory effects of testosterone on muscle mass have produced an interest in anabolic steroids among competitive athletes of both genders.

■ *Nervous system.* Sex hormones affect central nervous system centers concerned with sexual drive and sexual behaviors. These centers differentiated in gender-specific ways during the second and third trimesters, when the fetal gonads secreted either testosterone (in males) or estrogens (in females). The surge in sex hormone secretion at puberty activates these centers.

■ *Cardiovascular system.* Testosterone stimulates erythropoiesis and thereby increases the blood volume and the hematocrit. In females, once uterine cycles have begun, the iron loss associated with menses increases the risk of developing iron-deficiency anemia. Late in each uterine cycle, estrogens and progesterone promote the movement of water from plasma into interstitial fluid, leading to an increase in tissue water content. Estrogens decrease plasma cholesterol levels and slow plaque formation. As a result, premenopausal women have a lower risk of atherosclerosis than do adult men.

■ *Respiratory system.* Testosterone stimulates disproportionate growth of the larynx and a thickening and lengthening of the vocal cords. These changes cause a gradual deepening of the voice of males as compared with that of females.

■ *Reproductive system.* In males, testosterone stimulates the functional development of the accessory reproductive glands, such as the prostate gland and seminal vesicles, and helps promote spermatogenesis. In females, estrogens target the uterus, promoting a thickening of the myometrium, increasing blood flow to the endometrium, and stimulating cervical mucus production. Estrogens also promote the functional development of the female accessory reproductive organs. Once uterine cycles have begun, the first few may not be accompanied by ovulation. After that initial stage, the woman will be fertile, even though growth and physical maturation will continue for several years.

After puberty, the continued background secretion of estrogens or androgens maintains these gender-specific differences. In both genders, growth continues at a slower pace until age 18–21. By that time most of the epiphyseal cartilages have closed. The boundary between adolescence and maturity is hazy, because it has physical, emotional, behavioral, and legal aspects. Adolescence is often said to be over when growth stops in the late teens or early twenties. The individual is then considered mature.

SENESCENCE

Although growth may cease at maturity, physiological changes continue. The gender-specific differences produced at puberty are retained, but further changes occur when sex hormone levels decline at menopause or the male climacteric. *pp. 1056, 1057* All these changes are part of the process of **senescence,** or aging. Aging reduces the efficiency and capabilities of the individual. Even in the absence of other factors, such as disease or injury, senescence-related changes at the molecular level ultimately lead to death.

Table 29-3 summarizes the age-related changes in physiological systems that we have discussed in earlier chapters. Taken together, these changes reduce the functional abilities of the individual. They also affect homeostatic mechanisms. As a result, the elderly are less able to make homeostatic adjustments in response to internal or environmental stresses. The risks of contracting a variety of bacterial or viral diseases are proportionately increased as immune function deteriorates. This deterioration leads to drastic physiological changes that affect all internal systems. Death ultimately occurs when some combination of stresses cannot be countered by existing homeostatic mechanisms.

Physicians attempt to forestall death by adjusting homeostatic mechanisms or removing the sources of stress. **Geriatrics** is a medical specialty that deals with the mechanics of aging; physicians trained in geriatrics are known as **geriatricians.** Problems commonly encountered by geriatricians include infections, cancers, heart disease, strokes, arthritis, and anemia. These

29

TABLE 29-3 Effects of Aging on Organ Systems

The characteristic physical and functional changes that are part of the aging process affect all organ systems. Examples discussed in previous chapters include:

- A loss of elasticity in the skin that produces sagging and wrinkling. ∞ *p. 160*
- A decline in the rate of bone deposition, leading to weak bones, and degenerative changes in joints that make them less mobile. ∞ *pp. 185, 187*
- Reductions in muscular strength and ability. ∞ *pp. 299–300*
- Impairment of coordination, memory, and intellectual function. ∞ *p. 499*
- Reductions in the production of and sensitivity to circulating hormones. ∞ *p. 614*
- Appearance of cardiovascular problems and a reduction in peripheral blood flow that can affect a variety of vital organs. ∞ *p. 744*
- Reduced sensitivity and responsiveness of the immune system, leading to infection, cancer, or both. ∞ *p. 789*
- Reduced elasticity in the lungs, leading to decreased respiratory function. ∞ *p. 837*
- Decreased peristalsis and muscle tone along the digestive tract. ∞ *p. 893*
- Decreased peristalsis and muscle tone in the urinary system, coupled with a reduction of the glomerular filtration rate. ∞ *pp. 977–978*
- Functional impairment of the reproductive system, which eventually becomes inactive when menopause or the male climacteric occurs. ∞ *pp. 1056, 1057*

conditions are directly related to the age-induced changes in vital systems. [AM] *Death and Dying*

☑ Why does a mother's blood volume increase during pregnancy?

☑ What effect would a decrease in progesterone have on the uterus during late pregnancy?

☑ An increase in the levels of GnRH, FSH, LH, and sex hormones in blood mark the onset of which stage of development?

GENETICS, DEVELOPMENT, AND INHERITANCE

We introduced chromosome structure and the functions of genes in Chapter 3. ∞ *pp. 89–95* Chromosomes contain DNA, and genes are functional segments of DNA. Each gene carries the information needed to direct the synthesis of a specific polypeptide. Every nucleated somatic cell in your body carries copies of the original 46 chromosomes present when you were a zygote. Those chromosomes and their component genes constitute your **genotype** (JĒN-ō-tīp).

Through development and differentiation, the instructions contained in the genotype are expressed in many ways. No single cell or tissue uses all the information and instructions contained in the genotype. For example, in muscle fibers the genes important for the formation of excitable membranes and contractile proteins are active, whereas a different set of genes is operating in cells of the pancreatic islets. Collectively, however, the instructions contained in your genotype determine the anatomical and physiological characteristics that make you a unique individual. Those anatomical and physiological characteristics constitute your **phenotype** (FĒN-ō-tīp). In architectural terms, the genotype is a set of plans and the phenotype is the finished building. Specific elements in your phenotype, such as your hair color and eye color, skin tone, and foot size, are called phenotypic *characters,* or *traits.*

Your genotype is derived from the genotypes of your parents. Yet you are not an exact copy of either parent; nor are you an easily identifiable mixture of their characteristics. Our discussion of genetics will begin with the basic patterns of inheritance and their implications. We will then examine the mechanisms responsible for regulating the activities of the genotype during prenatal development.

GENES AND CHROMOSOMES

The 46 chromosomes carried by each somatic cell occur in pairs; every somatic cell contains 23 pairs of chromosomes. At amphimixis, one member of each pair was contributed by the spermatozoon, and the other by the ovum. The two members of each pair are known as **homologous** (hō-MOL-o-gus) **chromosomes.** Twenty-two of those pairs are called **autosomal** (aw-tō-SŌ-mal) **chromosomes.** Most of the genes of the autosomal chromosomes affect somatic characteristics, such as hair color and skin pigmentation. The chromosomes of the twenty-third pair are called the **sex chromosomes;** one of their functions is to determine whether the individual will be genetically male or female. Figure 29-15● shows the **karyotype,** or entire set of chromosomes, of a normal male.

Autosomal Chromosomes

The two chromosomes in a homologous autosomal pair have the same structure and carry genes that af-

•**FIGURE 29-15 Human Chromosomes.** The 23 pairs of somatic-cell chromosomes of a normal male.

fect the same traits. Suppose that one member of the pair contains three genes in a row, with number 1 determining hair color; number 2, eye color; and number 3, skin pigmentation. The other chromosome will carry genes that affect the same traits, and the genes will be in the same sequence. The genes will also be located at equivalent positions on their respective chromosomes. A **locus** (LŌ-kus; plural, *loci*) is a gene's position on a chromosome.

The two chromosomes in a pair may not carry the same *form* of each gene, however. The various forms of any gene are called **alleles** (a-LĒLZ). These *alternate forms* determine the precise effect of the gene on your phenotype. If the two chromosomes of a homologous pair carry the same allele of a particular gene, you are **homozygous** (hō-mō-ZĪ-gus) for the trait affected by that gene. That allele will indeed be expressed in your phenotype. For example, if you receive a gene for curly hair from your father and a gene for curly hair from your mother, you will be homozygous for curly hair— and you will have curly hair. About 80 percent of an individual's genome consists of homozygous alleles.

Interactions Between Alleles   Because the chromosomes of a homologous pair have different origins, one paternal and the other maternal, they do not necessarily carry the same alleles. When you have two different alleles of the same gene, you are **heterozygous** (het-er-ō-ZĪ-gus) for the trait determined by that gene. The phenotype that results from a heterozygous genotype depends on the nature of the interaction between the corresponding alleles. For example, if you received a gene for curly hair from your father but a gene for

straight hair from your mother, whether *you* will have curly hair, straight hair, or even wavy hair depends on the relationship between the alleles for those traits:

- An allele that is **dominant** will be expressed in the phenotype *regardless of any conflicting instructions carried by the other allele.* For instance, an individual with only one allele for freckles will have freckles, because that allele is dominant over the "nonfreckle" allele.

- An allele that is **recessive** will be expressed in the phenotype only if that same allele is present on *both* chromosomes of a homologous pair. For example, in Chapter 5 we described the albino condition, characterized by an inability to synthesize the yellow-brown pigment *melanin*. ∞ *p. 147* The presence of one allele that allows for melanin production will result in normal coloration. Two recessive alleles must be present to produce an albino individual. A single gene can have many different alleles, some dominant and others recessive.

- In **incomplete dominance,** heterozygous alleles produce a phenotype that is distinct from the phenotypes of individuals homozygous for one allele or the other. For instance, consider a gene that affects red blood cell shape. Individuals with homozygous alleles that carry instructions for normal adult hemoglobin A have red blood cells of normal shape. Individuals with homozygous alleles for hemoglobin S, an abnormal form, have red blood cells that will sickle in peripheral capillaries when the P_{O_2} decreases. These individuals develop *sickle cell anemia.* ∞ *p. 631* Individuals who are heterozygous for this trait do not develop the anemia; they have red blood cells that will sickle only when tissue oxygen levels are extremely low.

- In **codominance,** an individual with a heterozygous allele for a given trait exhibits both the dominant and recessive phenotypes for that trait. Blood type in humans is determined by codominance. ∞ *pp. 634–635* The alleles for types A and B blood are dominant over the allele for type O blood, but a person with one type A allele and one type B allele will have type AB blood, not A or B. Type AB blood has *both* type A and type B antigens. The distinction between incomplete dominance and codominance can be difficult. For example, a person heterozygous for hemoglobin A and hemoblogin S shows incomplete dominance for RBC shape but codominance for RBC hemoglobin; each RBC contains a mixture of hemoglobin A and hemoglobin S.

Penetrance and Expressivity   Differences in genotype lead to distinctive variations in phenotype, but the relationships are not always easily predictable. The presence of a particular pair of alleles does not affect the phenotype in the same way in every individual. **Penetrance** is the percentage of individuals with a particular genotype that show the "expected" phenotype. The effects of the genotype in the other individuals may be overridden by the activity of other genes or by environmental factors. For example, *emphysema,* a respiratory disorder discussed in Chapter 23, has been

29

linked to a specific abnormal genotype. ∞ *p. 839* Roughly 20 percent of the individuals with this genotype do not develop emphysema. The penetrance of this genotype is therefore approximately 80 percent. The effects of environmental factors are apparent; most people who develop emphysema are cigarette smokers.

If a given genotype *does* affect the phenotype, it can do so to varying degrees, again depending on the activity of other genes or environmental stimuli. For example, identical twins do not have exactly the same fingerprints, even though they have the same genotype. The extent to which a particular allele is expressed when it is present is termed its **expressivity.**

Environmental effects on genetic expression are particularly evident during embryological and fetal development. Drugs, including certain antibiotics, alcohol, and the nicotine in cigarette smoke, can disrupt fetal development. Stimuli that result in abnormal development are called **teratogens** (TER-a-tō-jenz).

[AM] *Teratogens and Abnormal Development*

Predicting Inheritance   Not every allele can be neatly characterized as dominant or recessive. Several that can be are included in Table 29-4. If you consider the traits listed there, you can predict the characteristics of individuals on the basis of the parents' alleles.

In such calculations, dominant alleles are traditionally indicated by capitalized abbreviations, and recessive alleles are abbreviated in lowercase letters. For a given trait, the possibilities are indicated by *AA* (homozygous dominant), *Aa* (heterozygous), or *aa* (homozygous recessive). Each gamete involved in fertilization contributes a single allele for a given trait. That allele must be one of the two contained by all cells in the parent's body. Consider, for example, the offspring of an albino mother and a father with normal skin pigmentation. Because albinism is a recessive trait, the maternal alleles are abbreviated *aa*. No matter which of her oocytes is fertilized, it will carry the recessive *a* allele. The father has normal pigmentation, a dominant trait. He is therefore either homozygous *or* heterozygous for this trait, because both *AA* and *Aa* will produce the same phenotype for normal skin pigmentation. Every sperm produced by a homozygous father will carry the *A* allele. In contrast, half the sperm produced by a heterozygous father will carry the dominant allele *A,* and the other half will carry the recessive allele *a.*

A simple box diagram known as a **Punnett square** lets us predict the probabilities that children will have particular characteristics by showing the possible combinations of parental alleles they can inherit. In the Punnett squares shown in Figure 29-16•, the maternal alleles for normal versus albino skin pigmentation are listed along the horizontal axis, and the paternal ones along the vertical axis. The possible combinations are indicated in the small boxes. Figure 29-16a• considers the possible offspring of an *aa* mother and an *AA* father. All

TABLE 29-4   The Inheritance of Selected Traits

DOMINANT TRAITS

One allele determines phenotype; the other is suppressed:
normal skin pigmentation
lack of freckles
brachydactyly (short fingers)
ability to taste phenylthiocarbamide (PTC)
free earlobes
curly hair
presence of Rh factor on red blood cell membranes
ability to roll the tongue into a U-shape
color vision
Both dominant alleles are expressed (codominance):
presence of A or B antigens on red blood cell membranes
structure of serum proteins (albumins, transferrins)
Two alleles produce intermediate traits (incomplete dominance):
hemoglobin A and hemoglobin S production

RECESSIVE TRAITS

albinism
freckles
normal digits
inability to taste phenylthiocarbamide (PTC)
attached earlobes
straight hair
blond hair
red hair (expressed only if individual is also homozygous for blond hair)
lack of A or B surface antigens (Type O blood)
absence of Rh surface antigen
inability to roll the tongue into a U-shape

SEX-LINKED TRAITS

color blindness
hemophilia

POLYGENIC TRAITS

eye color
hair colors other than pure blond or red

Note: For a listing of inherited clinical conditions, see Table 29-5.

the children must have the genotype *Aa*, so all will have normal skin pigmentation. Compare these results with those of Figure 29-16b•, for a heterozygous father *(Aa)*. The heterozygous male produces two types of gametes, *A* and *a*, and the secondary oocyte may be fertilized by either one. As a result, the probability is 50 percent that a child of such a father will inherit the genotype *Aa* and so have normal skin pigmentation. The probability of inheriting the genotype *aa*, and thus having the albino phenotype, is also 50 percent.

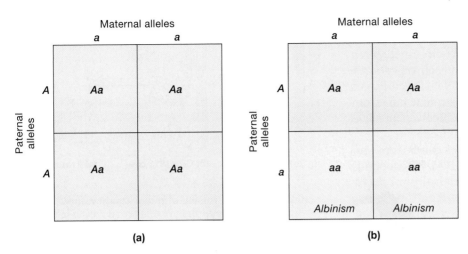

Maternal alleles

(a)

Maternal alleles

(b)

•**FIGURE 29-16** **Predicting Phenotypic Characteristics by Using Punnett Squares.** **(a)** All the offspring of a homozygous dominant father *(AA)* and a homozygous recessive mother *(aa)* will be heterozygous *(Aa)* for that trait. Their phenotype for that trait will be the same as that of the father. **(b)** The offspring of a heterozygous father *(Aa)* and a homozygous recessive mother *(aa)* will be either heterozygous or homozygous for the recessive trait. In this example, half the offspring will have normal skin coloration, and the other half will be albinos.

The Punnett square can also be used to draw conclusions about the identity and genotype of a parent. For example, a man with the genotype *AA* cannot be the father of an albino child *(aa)*.

Simple Inheritance In **simple inheritance,** phenotypes are determined by interactions between a single pair of alleles. (Our examples thus far involve simple inheritance.) We can predict the frequency of appearance of an inherited disorder that results from simple inheritance by using a Punnett square. Although they are rare in terms of overall numbers, more than 1200 inherited conditions have been identified that reflect the presence of one or two abnormal alleles for a single gene. A partial listing is given in Table 29-5, along with sites with more information.

Polygenic Inheritance Many phenotypic characters are determined by interactions among several genes. Such interactions are called **polygenic inheritance.** The resulting phenotype depends on how the genes interact. Because multiple alleles are involved, the frequency of occurrence cannot easily be predicted by using a simple Punnett square. In *suppression,* one gene suppresses the other and the second gene has no effect on the phenotype. In *complementary gene action,* dominant alleles on two genes interact to produce a phenotype different from that seen when one gene contains recessive alleles. The risks of developing several important adult disorders, including hypertension and coronary artery disease, are linked to polygenic inheritance.

Many of the developmental disorders responsible for fetal deaths and congenital malformations result from polygenic inheritance. In these cases, the genetic composition of the individual does not by itself determine the onset of the disease. Instead, the conditions regulated by these genes establish a susceptibility to particular environmental influences. Thus, not every individual with the genetic tendency for a certain condition will develop that condition. It is therefore difficult to track polygenic conditions through successive generations. However, because many inherited polygenic conditions are *likely* but not *guaranteed* to occur, steps can be taken to prevent a crisis. For example, you can reduce hypertension by controlling your diet and fluid volume, and you can prevent coronary artery disease by lowering your serum cholesterol levels.

Sources of Individual Variation Just as you are not a copy of either parent, you are not a 50:50 mixture of

TABLE 29-5 **Fairly Common Inherited Disorders**

Disorder	Text Page and/or *Applications Manual*
Autosomal dominants	
Polycystic kidney disease	p. 952 and *AM*
Marfan's syndrome	pp. 120, 181 and *AM*
Huntington's disease	p. 498 and *AM*
Autosomal recessives	
Deafness	p. 571 and *AM*
Albinism	pp. 148, 1091
Sickle cell anemia	p. 631 and *AM*
Cystic fibrosis	p. 800 and *AM*
Phenylketonuria	p. 919 and *AM*
Tay-Sachs disease	p. 398 and *AM*
X-linked	
Duchenne's muscular dystrophy	p. 277 and *AM*
Myotonic muscular dystrophy	*AM*
Hemophilia (A and B)	p. 649
Testicular feminization syndrome	*AM*
Color blindness	p. 555

Note: The genetic bases have been identified for at least some forms of the diseases in blue.

29

their characteristics. We noted one reason in Chapter 28: During meiosis, maternal and paternal chromosomes are randomly distributed, so each gamete has a unique blend of maternal and paternal chromosomes. ∞ *pp. 1023–1024* You may have an allele for curly hair from your father and an allele for straight hair from your mother even though your sister received an allele for straight hair from each of your parents.[1]

In addition, parts of chromosomes can become rearranged during synapsis (Figure 29-17•). When tetrads form, adjacent chromatids may overlap; this process is called **crossing-over.** Under these conditions, the chromatids may break, and the overlapping segments trade places. This process is called **translocation.** All the changes in chromosome structure that occur during meiosis produce gametes whose chromosomes differ from those of each parent. This reshuffling process is called **genetic recombination.** Genetic recombination, which occurs during meiosis in both males and females, greatly increases the range of possible variation among gametes, and thus among members of successive generations, whose genotypes are formed by the combination of gametes in fertilization. It can also complicate the tracing of inheritance of genetic disorders.

During recombination, portions of chromosomes may break away and be lost, or *deleted.* The effects of a deletion on a zygote depend on the nature of the lost genes. In some cases, the effects depend on whether the abnormal gamete is produced through oogenesis or spermatogenesis. This phenomenon is **genomic imprinting.** For example, the deletion of a specific segment of chromosome 15 causes two very different disorders, *Angelman syndrome* and *Prader–Willi syndrome.* Symptoms of Angleman syndrome include hyperactivity, severe mental retardation, and seizures; this condition results when the abnormal chromosome is provided by the oocyte. Symptoms of Prader–Willi syndrome include short stature, reduced muscle tone and skin pigmentation, underdeveloped gonads, and a mental retardation varying from slight to severe; this condition results when the abnormal chromosome is delivered by the sperm.

Recombination that produces abnormal chromosome shapes or numbers is lethal for the zygote in almost all cases. Roughly 10 percent of zygotes have **chromosomal abnormalities**—that is, damaged, broken, missing, or extra copies of chromosomes—but only about 0.5 percent of newborn infants have such abnormalities. Chromosomal abnormalities produce a variety of serious clinical conditions, in addition to contributing to prena-

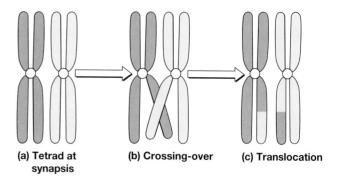

(a) Tetrad at synapsis   (b) Crossing-over   (c) Translocation

•**FIGURE 29-17 Crossing-over and Translocation. (a)** Synapsis, with the formation of a tetrad during meiosis. **(b)** Crossing-over of homologous portions of two chromosomes. **(c)** The breakage and exchange of corresponding sections on the chromosomes in (b).

tal mortality. *Few individuals with chromosomal abnormalities survive to full term.* The high mortality rate and severity of the problems reflect the fact that large numbers of genes have been added or deleted.

Variations at the level of the individual gene can result from mutations that affect the nucleotide sequence of one allele. **Spontaneous mutations** are the result of random errors in DNA replication. Such errors are relatively common, but in most cases the error is detected and repaired by enzymes in the nucleus. Those that go undetected and unrepaired have the potential to change the phenotype in some way.

Mutations occurring during meiosis can produce gametes that contain abnormal alleles. These alleles may be dominant or recessive, and they may occur on autosomal chromosomes or on sex chromosomes. The vast majority of mutations make the zygote incapable of completing normal development. Mutation, rather than chromosomal abnormalities, is probably the primary cause of the high mortality rate among pre-embryos and embryos. (Roughly 50 percent of all zygotes fail to complete cleavage, and another 10 percent fail to reach the fifth month of gestation.) AM *Complexity and Perfection*

If the abnormal allele is dominant but does not affect gestational survival, the individual's phenotype will show the effects of the mutation. If the abnormal allele is recessive and is on an autosomal chromosome, it will not affect the individual's phenotype as long as the zygote contains a normal allele contributed by the other parent at fertilization. Over generations, a recessive autosomal allele can spread through the population, remaining undetected until a fertilization occurs in which the two gametes contribute identical recessive alleles. This individual, who will be homozygous for the abnormal allele, will be the first to show the phenotypic effects of the original mutation. Individuals who are heterozygous for the abnormal allele but who do not show the effects of the mutation are called **carriers.** Genetic tests are available to determine if an individual is

---

[1] In very rare cases, an individual will receive both alleles from one parent. The few documented cases appear to have resulted when duplicate maternal chromatids failed to separate during meiosis II and the corresponding chromosome provided by the sperm did not participate in amphimixis. This condition is called *uniparental disomy.* The event generally remains undetected, because the individuals are phenotypically normal.

a carrier for one of several autosomal recessive disorders, including Tay-Sachs disease and Huntington's disease. The information can be useful in counseling prospective parents. For example, if both parents are carriers for the same disorder, they have a 25 percent probability of producing a child with the disease. This information may affect their decision to conceive.

Sex Chromosomes

Unlike the other 22 chromosomal pairs, the sex chromosomes may not by identical in appearance and gene content. There are two types of sex chromosomes: an **X chromosome** and a **Y chromosome.** X chromosomes are considerably larger, and have more genes, than do Y chromosomes. The Y chromosome includes dominant alleles that specify that an individual with that chromosome will be male. The normal pair of sex chromosomes in males is XY. Females do not have a Y chromosome; their sex chromosome pair is XX.

All oocytes carry an X chromosome, because the only sex chromosomes females have are X chromosomes. But each sperm carries either an X or a Y chromosome, because males have one of each and can pass along either one. Thus, with a Punnett square you can show that the ratio of males to females in offspring should be 1:1. The birth statistics differ slightly from that prediction, with 106 males born for every 100 females. It has been suggested that more males are born because the sperm that carries the Y chromosome can reach the oocyte first, since they do not have to carry the extra weight of the larger X chromosome.

The X chromosome also carries genes that affect somatic structures. These characteristics are called **X-linked** (or *sex-linked*), because in most cases there are no corresponding alleles on the Y chromosome. The inheritance of characteristics regulated by these genes does not follow the pattern of alleles on autosomal chromosomes. The best known of these single-allele characters are those associated with noticeable diseases or functional deficits.

The inheritance of color blindness exemplifies the differences between sex-linked inheritance and autosomal inheritance. ∞ *p. 555* Normal color vision is determined by the presence of a dominant allele, *C,* on the X chromosome, whereas red–green color blindness results from the absence of *C* and the presence of a recessive allele, *c,* on the X chromosome. A woman, with her two X chromosomes, can be either homozygous dominant *(CC)* or heterozygous *(Cc)* and still have normal color vision. She will be unable to distinguish reds from greens only if she carries two recessive alleles, *cc.* But a male has only one X chromosome, so whichever allele that chromosome carries will determine whether he has normal color vision or is red–green color-blind. A Punnett square for an X-linked trait, as in Figure 29-18•, reveals that the sons produced by a father with normal vision and a heterozygous mother will have a

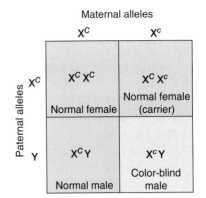

•**FIGURE 29-18 X-Linked Inheritance.** Recessive alleles on X but not on Y chromosomes produce genetic disorders in males at a higher frequency than in females. Females have two X chromosomes, so a female with a recessive allele will not develop such a disorder if she has a normal allele on the other X chromosome. A male with an X chromosome bearing the recessive allele will develop the disorder.

50 percent chance of being red–green color-blind, whereas any daughters will have normal color vision.

A number of other clinical disorders noted earlier in the text are X-linked traits, including certain forms of hemophilia, diabetes insipidus, and muscular dystrophy. In several instances, advances in techniques of molecular genetics have enabled geneticists to localize the specific genes on the X chromosome. These techniques provide a relatively direct method of screening for the presence of a particular condition before the symptoms appear and even before birth.

THE HUMAN GENOME PROJECT

Few of the genes responsible for inherited disorders have been identified or even associated with a specific chromosome. That situation is changing rapidly, however, due to the **Human Genome Project.** This project, funded by the National Institutes of Health and the Department of Energy, is attempting to transcribe the entire human **genome**—that is, the full complement of genetic material—chromosome by chromosome and gene by gene. The project began in October 1990 and was expected to take 15 years. Progress has been more rapid than expected, and it is now expected to be completed by the end of 2003.

The first step in understanding the human genome is to prepare a map of the individual chromosomes. **Karyotyping** (KAR-ē-ō-tīp-ing; *karyon,* nucleus + *typos,* mark) is the determination of an individual's chromosomal complement. Figure 29-15• shows a set of normal human chromosomes. Each chromosome has characteristic banding patterns, and segments can be stained with special dyes. The patterns are useful as reference points for the preparation of more-detailed genetic maps. The banding patterns themselves can be useful, as abnormal banding patterns are characteristic of some genetic disorders and several cancers.

29

As of 2000, a progress report includes the following:

■ Nine chromosomes—chromosomes 3, 5, 11, 12, 16, 19, 21, 22, and the Y chromosome—have been mapped completely, and preliminary maps have been made of all other chromosomes.

■ More than 38,000 genes have been tentatively identified, and over 11,000 have been mapped. That is a significant number, but it is only a small fraction of the 100,000–140,000 genes in the human genome.

■ The genes responsible for more than 60 inherited disorders have been identified, including several listed in Table 29-5 and those shown in Figure 29-19•. Genetic screenings are now done for many of these disorders.

The Human Genome Project is attempting to determine the normal genetic composition of a "typical" human. Yet we all are variations on a basic theme. As we improve our abilities to manipulate our own ge-

netic foundations, we will face troubling ethical and legal decisions. For example, few people object to the insertion of a "correct" gene into somatic cells to cure a specific disease. (See the section "Genetic Engineering and Gene Therapy" in the *Applications Manual*.) But what if we could insert that modified gene into a gamete and change not only that individual but his or her descendants as well? And what if the gene did not correct or prevent a disorder but "improved" the individual by increasing intelligence, height, or vision or altering some other phenotypic characteristic? Such difficult questions will not go away. In the years to come, we will have to find answers acceptable to us all.

☑ Curly hair is an autosomal dominant trait. What would be the phenotype of a person who is heterozygous for this trait?

☑ Why are children not identical copies of their parents?

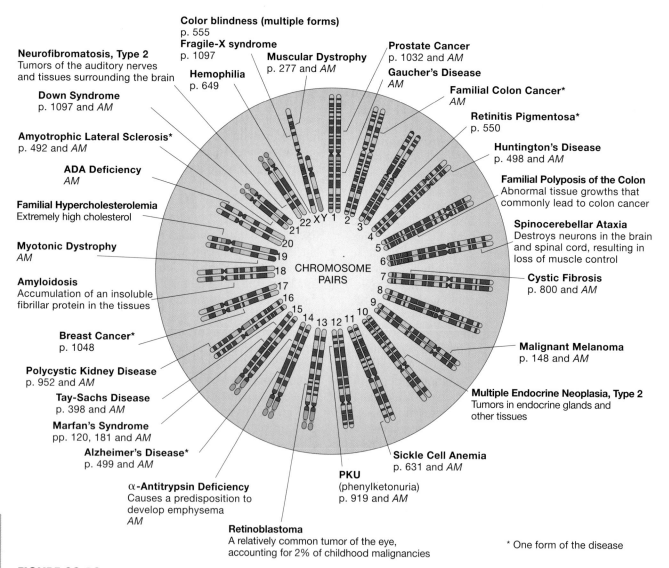

Color blindness (multiple forms)
p. 555
Fragile-X syndrome
p. 1097
Muscular Dystrophy
p. 277 and *AM*
Prostate Cancer
p. 1032 and *AM*
Gaucher's Disease
AM
Hemophilia
p. 649
Familial Colon Cancer*
AM
Neurofibromatosis, Type 2
Tumors of the auditory nerves
and tissues surrounding the brain
Retinitis Pigmentosa*
p. 550
Down Syndrome
p. 1097 and *AM*
Huntington's Disease
p. 498 and *AM*
Amyotrophic Lateral Sclerosis*
p. 492 and *AM*
Familial Polyposis of the Colon
Abnormal tissue growths that
commonly lead to colon cancer
ADA Deficiency
AM
Spinocerebellar Ataxia
Destroys neurons in the brain
and spinal cord, resulting in
loss of muscle control
Familial Hypercholesterolemia
Extremely high cholesterol
Myotonic Dystrophy
AM
Cystic Fibrosis
p. 800 and *AM*
Amyloidosis
Accumulation of an insoluble
fibrillar protein in the tissues
Breast Cancer*
p. 1048
Malignant Melanoma
p. 148 and *AM*
Polycystic Kidney Disease
p. 952 and *AM*
Multiple Endocrine Neoplasia, Type 2
Tumors in endocrine glands and
other tissues
Tay-Sachs Disease
p. 398 and *AM*
Marfan's Syndrome
pp. 120, 181 and *AM*
Sickle Cell Anemia
p. 631 and *AM*
Alzheimer's Disease*
p. 499 and *AM*
PKU
(phenylketonuria)
p. 919 and *AM*
α-Antitrypsin Deficiency
Causes a predisposition to
develop emphysema
AM
Retinoblastoma
A relatively common tumor of the eye,
accounting for 2% of childhood malignancies
* One form of the disease

CHROMOSOME PAIRS

•**FIGURE 29-19 A Map of Human Chromosomes.** The banding patterns of typical chromosomes in a male individual and the locations of the genes responsible for specific inherited disorders. The chromosomes are not drawn to scale.

29

Chromosomal Abnormalities and Genetic Analysis

Embryos that have abnormal autosomal chromosomes rarely survive. Two types of abnormalities in autosomal chromosomes, however, do not invariably kill the individual before birth. These are translocation defects and trisomy.

In a **translocation defect,** crossing-over occurs between different chromosome pairs. For example, a piece of chromosome 8 may become attached to chromosome 14 rather than to the other chromatid of chromosome 8. The genes moved to their new position may function abnormally, becoming inactive or overactive. A translocation between chromosomes 8 and 14 is responsible for *Burkitt's lymphoma,* a type of lymphatic system cancer.

In **trisomy,** something goes wrong in meiosis: One of the gametes at fertilization contributes an extra copy of one chromosome, so the zygote then has three copies of this chromosome rather than two. The nature of the trisomy is indicated by the number of the chromosome involved. Zygotes with extra copies of chromosomes seldom survive. Individuals with trisomy 13 and trisomy 18 may survive until delivery but rarely live longer than a year. The notable exception is trisomy 21.

Trisomy 21, or **Down syndrome,** is the most common viable chromosomal abnormality. Estimates of the frequency of appearance range from 1.5 to 1.9 per 1000 births for the U.S. population. Affected individuals exhibit mental retardation and characteristic physical malformations, including a facial appearance that gave rise to the term *mongolism,* once used to describe this condition. The degree of mental retardation ranges from moderate to severe. Few individuals with this condition lead independent lives. Anatomical problems affecting the cardiovascular system commonly prove fatal during childhood or early adulthood. Although some individuals survive to moderate old age, many develop Alzheimer's disease while still relatively young (before age 40).

For unknown reasons, maternal age is linked to the risk of having a child with trisomy 21. For a maternal age below 25, the incidence of Down syndrome approaches 1 in 2000 births, or 0.05 percent. For maternal ages 30–34, the odds increase to 1:900, and for ages 35–44 they go from 1:290 to 1:46, or more than 2 percent. These statistics are becoming increasingly significant, because many women have delayed childbearing until their mid-thirties or later.

Abnormal numbers of sex chromosomes do not produce effects as severe as those induced by extra or missing autosomal chromosomes. In **Klinefelter syndrome,** the individual carries the sex chromosome pattern XXY. The phenotype is male, but the extra X chromosome causes reduced androgen production. As a result, the breasts are slightly enlarged, the testes fail to mature, and affected individuals are sterile. The incidence of this condition among newborn males averages 1:750.

Individuals with **Turner syndrome** have only a single, female sex chromosome; their sex chromosome complement is abbreviated XO. This kind of chromosomal deletion is known as **monosomy.** The incidence of this condition at delivery has been estimated as 1:10,000. At birth, the condition may not be recognized, because the phenotype is normal female. But maturational changes do not appear at puberty. The ovaries are nonfunctional, and estrogen production occurs at negligible levels.

Fragile-X syndrome causes mental retardation, abnormal facial development, and increased testicular size in affected males. The cause is an abnormal X chromosome that contains a *genetic stutter,* an abnormal repetition of a single triplet. In this case, the problem is the repetition of the nucleotide sequence CCG at one site on the X chromosome. (A genetic stutter on another chromosome is responsible for Huntington's disease. ∞ *p. 498*) The presence of the stutter in some way disrupts the normal functioning of adjacent genes and so produces the symptoms of the disorder.

Many of these conditions can be detected prior to birth by the analysis of fetal cells. In **amniocentesis,** a sample of amniotic fluid is removed and the fetal cells that it contains are analyzed. This procedure permits the identification of over 20 congenital conditions, including Down syndrome. The needle inserted to obtain a fluid sample is guided into position by using ultrasound. Unfortunately, amniocentesis has two major drawbacks:

1. Because the sampling procedure represents a potential threat to the health of the fetus and mother, amniocentesis is performed only when known risk factors are present. Examples of risk factors would include a family history of specific conditions or, in the case of Down syndrome, a maternal age over 35.

2. Sampling cannot safely be performed until the volume of amniotic fluid is large enough that the fetus will not be injured during the sampling process. The usual time for amniocentesis is at a gestational age of 14–15 weeks. It can take several weeks to obtain results once samples have been collected; by the time the results are received, the option of therapeutic abortion may no longer be available.

In an alternative procedure known as **chorionic villus sampling,** cells collected from the chorionic villi during the first trimester are analyzed. Although it can be performed at an earlier gestational age than can amniocentesis, chorionic villus sampling has largely been abandoned because the risks of misdiagnosis are high and placental damage can cause spontaneous abortion.

29

[EXPLORE]
[MediaLab]

"You have your father's nose and your mother's build, but where did you get those curls?" Human inheritance and subsequent development are fascinating areas of study. To further your understanding of genetics and development, use the resources available on both the CD-ROM and the Companion Website. The CD-ROM includes animations and tutorials; the website builds on those resources with text content and interactive problems as well as communication avenues for sharing ideas with your class. To complete the following Web Explorations, visit the MediaLab in Chapter 29 of your Companion Website.

Web Exploration 1

Estimated time for completion = 10 minutes

Human development from embryo to neonate is a sequential process of change that includes many steps and structures. To help you keep this material in order, select the keyword **TIME LINE** on the Companion Website. There you will see a diagram of a fertilized egg and a time line. Identify the structures shown by clicking on the name in the box on the

upper right. Information about each structure is given in the box on the lower right. Use the forward arrow at the bottom of the page to view a series of diagrams. Create your own time line, adding information from both this website and your text.

Web Exploration 2

Estimated time for completion = 5 minutes

As the text indicates, the goals of the government-funded Human Genome Project are to identify all the genes in human DNA, to determine and store the sequence of bases in those genes, and to use this information to further medical research. Create a list of the benefits this information could produce. Then select the keyword **HUMAN GENOME PROJECT** on the Companion Website to read about the potential benefits as seen by the project's developers. Does your list match this one? Could this project have any negative implications? Discuss your thoughts about both positive and negative implications of the Human Genome Project, and prepare a short essay describing your position.

CHAPTER REVIEW

SELECTED CLINICAL TERMINOLOGY
Terms Discussed in This Chapter

amniocentesis: An analysis of fetal cells taken from a sample of amniotic fluid. *(p. 1097)*
breech birth: A delivery during which the legs or buttocks of the fetus enter the vaginal canal first. *(p. 1086)*
chorionic villus sampling: An analysis of cells collected from the chorionic villi during the first trimester. *(p. 1097)*
ectopic pregnancy: A pregnancy in which implantation occurs somewhere other than the uterus. *(p. 1069 and AM)*

gestational trophoblastic neoplasia: A tumor formed by undifferentiated, rapid growth of the syncytial trophoblast; if untreated, the neoplasm may become malignant. *(p. 1070)*
infertility: The inability to achieve pregnancy after 1 year of appropriately timed intercourse. *(p. 1079)*
in vitro fertilization: Fertilization outside the body, generally in a petri dish. *(p. 1080)*

AM Additional Terms Discussed in the *Applications Manual*

abruptio placentae (ab-RUP-shē-ō pla-SEN-tē): A tearing of the placenta after the fifth gestational month.
Apgar rating: A method of evaluating newborn infants for developmental problems and neurological damage.
congenital malformation: A severe structural abnormality, present at birth, that affects major systems.
DDST: The Denver Developmental Screening Test, given to infants and children up to age 5 to monitor their development.
fetal alcohol syndrome (FAS): A neonatal condition resulting from maternal alcohol consumption; characterized by developmental defects typically involving the skeletal, nervous, and/or cardiovascular system.
placenta previa: A condition, resulting from implantation in or near the cervix, in which the placenta covers the cervix and prevents normal birth.

pseudohermaphrodite (soo-dō-her-MAF-rō-dīt): An individual who has the genotype of one gender but has some or most of the phenotypic characteristics of the opposite gender.
teratogens (TER-a-tō-jenz): Stimuli that disrupt normal development by damaging cells, by altering chromosome structure, or by altering the chemical environment of the embryo.
testicular feminization syndrome: A form of male pseudohermaphroditism resulting from an inability to respond to testosterone.
toxemia (tok-SĒ-mē-uh) **of pregnancy:** Disorders affecting the maternal cardiovascular system; includes *eclampsia* and *preeclampsia*.

29

STUDY OUTLINE

INTRODUCTION   p. 1065

1. **Development** is the gradual modification of physical and physiological characteristics from **conception** to maturity. The creation of different cell types is **differentiation.**

AN OVERVIEW OF TOPICS IN DEVELOPMENT   p. 1065

1. **Inheritance** is the transfer of genetically determined characteristics from generation to generation. **Genetics** is the study of the mechanisms of inheritance. **Prenatal development** occurs before birth; **postnatal development** begins at birth and continues to **maturity,** when aging begins.

FERTILIZATION   p. 1065

1. **Fertilization,** or *conception,* normally occurs in the uterine tube within a day after ovulation. Spermatozoa cannot fertilize a secondary oocyte until they have undergone *capacitation.*

The Oocyte at Ovulation   p. 1066

2. The acrosomal caps of the spermatozoa release **hyaluronidase** and **acrosin,** enzymes required to penetrate the corona radiata and zona pellucida of the oocyte. When a single spermatozoon contacts the oocyte membrane, fertilization begins and **oocyte activation** follows.

Oocyte Activation   p. 1067

3. During activation, the oocyte completes meiosis II and thus becomes a functionally mature ovum. **Polyspermy** is prevented by membrane depolarization and the *cortical reaction.*
4. After activation, the **female pronucleus** and the **male pronucleus** fuse in a process called *amphimixis. (Figure 29-1)*

AN OVERVIEW OF PRENATAL DEVELOPMENT   p. 1067

1. During prenatal development, differences in the cytoplasmic composition of individual cells trigger changes in genetic activity. The chemical interplay among developing cells is **induction.**
2. The 9-month **gestation** period can be divided into three **trimesters.**

THE FIRST TRIMESTER   p. 1068

1. In the **first trimester, cleavage** subdivides the cytoplasm of the zygote in a series of mitotic divisions; the zygote becomes a **pre-embryo** and then a **blastocyst.** During **implantation,** the blastocyst burrows into the uterine endometrium. **Placentation** occurs as blood vessels form around the blastocyst and the **placenta** develops. **Embryogenesis** is the formation of a viable embryo.

Cleavage and Blastocyst Formation   p. 1068

2. The blastocyst consists of an outer **trophoblast** and an **inner cell mass.** *(Figure 29-2)*

Implantation   p. 1068

3. Implantation occurs about 7 days after fertilization as the blastocyst adheres to the uterine lining. *(Figure 29-3)*
4. As the trophoblast enlarges and spreads, maternal blood flows through open **lacunae.** After **gastrulation,** there is an **embryonic disc** composed of **endoderm, ectoderm,** and an intervening **mesoderm.** It is from these **germ layers** that the body systems differentiate. *(Figure 29-4; Table 29-1)*
5. Germ layers help form four **extraembryonic membranes:** the yolk sac, amnion, allantois, and chorion. *(Figure 29-5)*
6. The **yolk sac** is an important site of blood cell formation. The **amnion** encloses fluid that surrounds and cushions the developing embryo. The base of the **allantois** later gives rise to the urinary bladder. Circulation within the vessels of the

chorion provides a rapid-transit system that links the embryo with the trophoblast.

Placentation   p. 1072

7. **Chorionic villi** extend outward into the maternal tissues, forming an intricate, branching network through which maternal blood flows. As development proceeds, the **umbilical cord** connects the fetus to the placenta. The trophoblast synthesizes **human chorionic gonadotropin (hCG),** estrogens, progesterone, **human placental lactogen (hPL), placental prolactin,** and **relaxin.** *(Figure 29-6)*

Embryogenesis   p. 1075

8. The first trimester is critical, because events in the first 12 weeks establish the basis for **organogenesis** (organ formation). *(Figure 29-7; Table 29-2)*
9. **Infertility** is relatively common, but recent advances are providing new solutions. *(Figure 29-8)*

THE SECOND AND THIRD TRIMESTERS   p. 1081

1. In the **second trimester,** the organ systems increase in complexity. During the **third trimester,** many of the organ systems become fully functional. *(Figure 29-9; Table 29-2)*
2. The fetus undergoes its largest weight gain in the third trimester. At the end of gestation, the fetus and enlarged uterus push many of the mother's abdominal organs out of their positions. *(Figure 29-10)*

Pregnancy and Maternal Systems   p. 1081

3. The developing fetus is totally dependent on maternal organs for nourishment, respiration, and waste removal. Maternal adaptations include increased respiratory rate, tidal volume, blood volume, nutrient and vitamin intake, glomerular filtration rate, and changes in the size of the uterus and mammary glands.

Structural and Functional Changes in the Uterus   p. 1083

4. Progesterone produced by the placenta has an inhibitory effect on uterine muscles; its calming action is opposed by estrogens, oxytocin, and prostaglandins. At some point, multiple factors interact to produce **labor contractions** in the uterine wall. *(Figure 29-11)*

LABOR AND DELIVERY   p. 1084

1. The goal of **true labor** is **parturition,** forcible expulsion of the fetus.

Stages of Labor   p. 1084

2. Labor can be divided into three stages: the **dilation stage,** the **expulsion stage,** and the **placental stage.** *(Figure 29-12)*

Premature Labor   p. 1084

3. **Premature labor** may result in **immature delivery** or **premature delivery.**

Multiple Births   p. 1086

4. Twin births are either **dizygotic** ("fraternal") or **monozygotic** ("identical").

POSTNATAL DEVELOPMENT   p. 1086

1. Postnatal development involves a series of five **life stages:** the neonatal period, infancy, childhood, adolescence, and maturity. *Senescence* (aging) begins at maturity and ends in the death of the individual.

The Neonatal Period, Infancy, and Childhood   p. 1086

2. The **neonatal period** extends from birth to 1 month after. In the transition from fetus to **neonate,** the respiratory, circulatory, digestive, and urinary systems begin functioning independently. The newborn must also begin thermoregulation.

29

3. Mammary gland cells produce protein-rich **colostrum** during the neonate's first few days of life and then convert to milk production. These secretions are released as a result of the **milk let-down reflex.** *(Figure 29-13)*

4. Body proportions gradually change during **infancy** (from age 1 month to 2 years) and **childhood** (from 2 years to puberty). *(Figure 29-14)*

Adolescence and Maturity p. 1088

5. **Adolescence** begins at **puberty:** (1) The hypothalamus increases its production of GnRH; (2) circulating levels of FSH and LH rise rapidly; and (3) ovarian or testicular cells become more sensitive to FSH and LH. These changes initiate gamete formation, the production of sex hormones, and a sudden rise in growth rate. The hormonal changes at puberty, especially changes in sex hormone levels, produce gender-specific differences in the structure and function of many systems; these differences will be retained. Adolescence continues until growth is completed. Further changes occur when sex hormone levels decline at menopause or the male climacteric.

Senescence p. 1089

6. **Senescence** then begins, producing gradual changes in the functional capabilities of all systems. *(Table 29-3)*

GENETICS, DEVELOPMENT, AND INHERITANCE p. 1090

1. Every somatic cell carries copies of the original 46 chromosomes in the zygote; these chromosomes and their component genes constitute the individual's **genotype.** The physical expression of the genotype is the individual's **phenotype.**

Genes and Chromosomes p. 1090

2. Every somatic human cell contains 23 pairs of chromosomes; each pair consists of **homologous chromosomes.**

Twenty-two pairs are **autosomal chromosomes.** The chromosomes of the twenty-third pair are the **sex chromosomes;** they differ between the sexes. *(Figure 29-15)*

3. Chromosomes contain DNA, and genes are functional segments of DNA. The various forms of a gene are called **alleles.** If both homologous chromosomes carry the same allele of a particular gene, the individual is **homozygous;** if they carry different alleles, the individual is **heterozygous.**

4. Alleles are either **dominant** or **recessive,** depending on how their traits are expressed. *(Table 29-4)*

5. Combining maternal and paternal alleles in a **Punnett square** diagram helps us predict the characteristics of offspring. *(Figure 29-16)*

6. In **simple inheritance,** phenotypic characteristics are determined by interactions between a single pair of alleles. **Polygenic inheritance** involves interactions among alleles on several genes.

7. **Genetic recombination,** the gene reshuffling, or **crossing-over** and **translocation,** that occurs during meiosis, increases the genetic variation of male and female gametes. *(Figure 29-17)*

8. The two types of sex chromosomes are an **X chromosome** and a **Y chromosome.** The normal sex chromosome complement of males is XY; that of females is XX. The X chromosome carries **X-linked** *(sex-linked)* **genes,** which affect somatic structures but have no corresponding alleles on the Y chromosome. *(Figure 29-18)*

The Human Genome Project p. 1095

9. The **Human Genome Project** has mapped more than 11,000 of our 100,000–140,000 genes, including some of those responsible for inherited disorders. *(Figure 29-19; Table 29-5)*

REVIEW QUESTIONS

LEVEL 1 REVIEWING FACTS AND TERMS

1. The gradual modification of anatomical structures during the period from conception to maturity is
(a) development (b) differentiation
(c) embryogenesis (d) capacitation

2. Human fertilization involves the fusion of two haploid gametes, producing a zygote that contains
(a) 23 chromosomes
(b) 46 chromosomes
(c) the normal haploid number of chromosomes
(d) 46 pairs of chromosomes

3. The secondary oocyte leaving the follicle is in
(a) interphase
(b) metaphase of the first meiotic division
(c) telophase of the second meiotic division
(d) metaphase of the second meiotic division

4. Chemical interplay among developing cells is
(a) gestation (b) induction
(c) capacitation (d) placentation

5. The stage of development that follows cleavage is the
(a) blastocyst (b) morula
(c) trophoblast (d) blastocoele

6. The process that establishes the foundation of all major organ systems is
(a) cleavage (b) implantation
(c) placentation (d) embryogenesis

7. What begins as a zygote arrives in the uterine cavity as a
(a) morula (b) trophoblast
(c) lacuna (d) blastomere

8. The first extraembryonic membrane to appear is the
(a) allantois (b) yolk sac
(c) amnion (d) chorion

9. The primitive streak appears during
(a) cleavage (b) implantation
(c) gastrulation (d) placentation

10. The part of the uterine endometrium that has no contact with the chorion is the
(a) decidua capsularis (b) decidua basalis
(c) decidua parietalis (d) allantois

11. The surface that provides for active and passive exchange between the fetal and maternal bloodstreams is the
(a) yolk stalk (b) chorionic villi
(c) umbilical veins (d) umbilical arteries

12. The placental hormone that appears in the maternal bloodstream soon after implantation is
(a) human chorionic gonadotropin
(b) human placental lactogen
(c) placental prolactin
(d) relaxin

13. The placental stage of labor ends with
(a) the afterbirth
(b) rupture of the amniotic sac
(c) delivery of the newborn
(d) parturition

14. Milk let-down is associated with
(a) events occurring in the uterus
(b) placental hormonal influences
(c) circadian rhythms
(d) reflex action triggered by nursing

15. An individual who has two different alleles for the same gene is _____ for that trait.
   (a) heterozygous   (b) homozygous
   (c) dominant   (d) recessive

16. If an allele must be present on both the maternal and paternal chromosomes to affect the phenotype, the allele is said to be
   (a) dominant   (b) recessive
   (c) complementary   (d) heterozygous

17. The percentage of individuals with a particular genotype who show the expected phenotype reflects
   (a) dominance
   (b) polygenic inheritance
   (c) spontaneous mutation
   (d) penetrance

18. For a trait *A*, the genotype of a homozygous recessive individual is represented as
   (a) *AA*   (b) *Aa*
   (c) *aa*   (d) *Ab*

19. The normal sex chromosome composition of males is
   (a) XXXX   (b) XYXY
   (c) XXYXXY   (d) XY

20. Describe the changes that occur in the sperm and egg immediately after fertilization.

21. What is induction? Explain its significance.

22. Summarize the developmental changes that occur during the first, second, and third trimesters.

23. (a) What are the four extraembryonic membranes?
   (b) How do they form, and what are their functions?

24. Indicate when each of the following appears during development, and describe each: embryonic disc, blastocyst, morula, zygote.

25. What is the trophoblast, and what are its three major functions?

26. Identify the three stages of labor, and describe the events that characterize each stage.

27. What factors are involved in initiating labor contractions?

28. Identify the three life stages that occur between birth and approximately age 10. Describe the characteristics of each stage and when it occurs.

29. What hormonal events are responsible for puberty? Which life stage does it initiate?

30. What occurs during the life stage of maturity?

31. What is senescence? Give some examples of how it affects organ systems throughout the human body.

LEVEL 2 REVIEWING CONCEPTS

32. Relaxin is a peptide hormone that
   (a) increases the flexibility of the pubic symphysis
   (b) causes dilation of the cervix
   (c) suppresses the release of oxytocin by the hypothalamus
   (d) a, b, and c are correct

33. The production of prostaglandins in the endometrium
   (a) initiates the release of oxytocin for parturition
   (b) stimulates smooth muscle contractions
   (c) initiates secretory activity in the mammary glands
   (d) a, b, and c are correct

34. During adolescence, the events that interact to promote increased hormone production and sexual maturation result from activity of the
   (a) hypothalamus
   (b) anterior pituitary gland
   (c) ovaries and testicular cells
   (d) a, b, and c are correct

35. What activity during oocyte activation prevents penetration by additional sperm?

36. In addition to its role in the nutrition of the fetus, what are the primary endocrine functions of the placenta?

37. Discuss the changes that occur in maternal systems during pregnancy. Why are these changes functionally significant?

38. Discuss the changes that occur in the uterus during pregnancy. How do these changes affect uterine tissues, and which hormones are involved?

39. Why are uterine contractions in the early stages of pregnancy weak, painless, and brief?

40. During true labor, what physiological mechanisms ensure that uterine contractions continue until delivery has been completed?

41. To what does the phrase "having the water break" refer during the process of labor?

42. What physiological adjustments must an infant make during the neonatal period to survive?

43. Distinguish between the following paired terms:
   (a) monozygotic and dizygotic
   (b) genotype and phenotype
   (c) heterozygous and homozygous
   (d) simple inheritance and polygenic inheritance

44. What would you conclude about a trait in each of the following situations?
   (a) Children who exhibit this trait have at least one parent who exhibits the same trait.
   (b) Children exhibit this trait even though neither parent exhibits it.
   (c) The trait is expressed more commonly in sons than in daughters.
   (d) The trait is expressed equally in daughters and sons.

45. Explain why more men than women are red–green colorblind. Which type of inheritance is involved?

46. Explain the impact of genetic recombination on the production of gametes and on the traits of individuals in future generations.

47. Explain the goals and possible benefits of the Human Genome Project.

LEVEL 3 CRITICAL THINKING AND CLINICAL APPLICATIONS

48. Hemophilia A, a condition in which blood does not clot properly, is a recessive trait located on the X chromosome (X^h). A woman heterozygous for this trait (XX^h) mates with a normal male (XY). What is the probability that this couple will have hemophiliac daughters? What is the probability that this couple will have hemophiliac sons?

49. Joan is a 27-year-old nurse who is in labor with her first child. She remembers from her anatomy and physiology class that calcium ions can increase the force of smooth muscle contractions, and because the labor is prolonged, she asks her physician for a calcium injection. The surprised physician informs Joan that such an injection is definitely out of the question. Why?

50. A new mother tells you that when she nurses her baby, she feels as if she is having menstrual cramps. How would you explain this phenomenon?

51. Explain why the normal heart and respiratory rates of neonates are so much higher than those of adults, even though adults are much larger.

52. Pregnant women are advised against taking aspirin during their third trimester. Why?

53. Sally gives birth to a baby with a congenital deformity of the stomach. Sally swears that it is the result of a viral infection that she suffered during her third trimester. Is this a possibility? Explain.

29

Answers to Concept Check Questions

CHAPTER 1

Page 12

1. White blood cells, or any cells or organisms, that respond to changes in its environment are exhibiting responsiveness, or irritability.
2. A histologist investigates the structure and properties of the tissue level of organization.
3. The study of the physiology of specific organs is called special physiology. In this particular case, the field of study is cardiac physiology (the study of heart function). Because heart failure is typically caused by disease, this specialty would overlap or be closely related to pathological physiology.

Page 14

1. Physiological systems can function normally only under carefully controlled conditions. Homeostatic regulation prevents potentially disruptive changes in the body's internal environment.
2. When homeostasis fails, organ systems function less efficiently or malfunction. The result is the state that we call disease. If the situation is not corrected, death can result.
3. Positive feedback is useful in processes, such as blood clotting, that must move quickly to completion once they have begun. It is harmful in situations in which a stable condition must be maintained, because it tends to increase any departure from the desired condition. For example, positive feedback in the regulation of body temperature would cause a slight fever to spiral out of control, with fatal results. For this reason, physiological systems normally exhibit negative feedback, which tends to oppose any departure from the norm.

Page 22

1. The two eyes would be separated by a midsagittal section.
2. The body cavity inferior to the diaphragm is the abdominopelvic cavity.

CHAPTER 2

Page 36

1. Atoms combine with each other so as to gain a complete set of eight electrons in their outer energy level. Oxygen atoms do not have a full outer energy level, so they readily react with many other elements to attain this stable arrangement. Neon already has a full outer energy level and thus has little tendency to combine with other elements.
2. Hydrogen has three isotopes: hydrogen-1, with a mass of 1; deuterium, with a mass of 2; and tritium, with a mass of 3. The heavier sample must contain a higher proportion of one or both of the heavier isotopes.
3. The atoms in a water molecule are held together by polar covalent bonds. Water molecules are attracted to one another by hydrogen bonds.

Page 39

1. Because this reaction involves a large molecule being broken down into two smaller ones, it is a decomposition reaction. Because energy is released in the process, the reaction can also be classified as exergonic.
2. Enzymes in our cells promote chemical reactions by lowering the activation energy requirements. Enzymes make it possible for chemical reactions to proceed under the conditions compatible with life.

Page 44

1. When it dissolves in water, salt dissociates into ions, charged particles that are capable of conducting an electrical current. Sugar molecules are held together by covalent bonds and so do not dissociate in solution; thus there are no ions to carry a current.
2. Stomach discomfort is commonly the result of excess stomach acidity ("acid indigestion"). Antacids contain a weak base that neutralizes the excess acid.

Page 51

1. A compound with a C:H:O ratio of 1:2:1 is a carbohydrate. The body uses carbohydrates chiefly as an energy source.
2. When two monosaccharides undergo a dehydration synthesis reaction, they form a disaccharide.
3. The most abundant lipid in a sample taken from beneath the skin would be a triglyceride.
4. An analysis of the lipid content of human cell membranes would indicate the presence of mostly phospholipids and small amounts of cholesterol and glycolipids.

Page 56

1. Proteins are chains of small organic molecules called amino acids.
2. An agent that breaks hydrogen bonds would affect the secondary level of protein structure.
3. The heat of boiling will break bonds that maintain the protein's tertiary structure, quaternary structure, or both. The resulting change in shape will affect the ability of the protein molecule to perform its normal biological functions. These changes are known as denaturation.
4. If the active site of an enzyme changes so that the site better fits its substrate, the level of enzyme activity will increase. But if the change alters the active site to the extent that the enzyme's substrate can no longer bind or binds poorly, the enzyme's activity will decrease or be inhibited.

Page 60

1. The nucleic acid RNA (ribonucleic acid) contains the sugar ribose. The nucleic acid DNA (deoxyribonucleic acid) contains the sugar deoxyribose instead; both contain nitrogenous bases and phosphate groups.
2. Phosphorylation of an ADP molecule would yield a molecule of ATP.

CHAPTER 3

Page 70

1. The phospholipids of the bilayer membrane form a physical barrier between the internal environment of the cell and the external environment.
2. Channel proteins are integral proteins that allow water and small ions to pass through the cell membrane.

Page 75

1. Diffusion is driven by a concentration gradient. The larger the concentration gradient, the faster the rate of diffusion. The smaller the concentration gradient, the slower the rate of diffusion. If the concentration of oxygen in the lungs were to decrease, the concentration gradient between oxygen in the lungs and oxygen in the blood would decrease (as long as the oxygen level of the blood remained constant). Thus oxygen would diffuse more slowly into the blood.
2. The 10 percent salt solution would be hypertonic with respect to the cells lining the nasal cavity, because this solution contains a higher concentration of salt than do the cells. The hypertonic solution would draw water out of the cells, causing the cells to shrink and loosening mucus, thus relieving the congestion.

Page 80

1. To transport hydrogen ions against their concentration gradient—that is, from a region where they are less concentrated (the cells lining the stomach) to a region where they are more concentrated (the interior of the stomach)—energy must be expended. An active transport process must be involved.
2. If the cell membrane were freely permeable to sodium ions, more of these positively charged ions would move into the cell and the transmembrane potential would move closer to zero.
3. This process is an example of phagocytosis.

Page 84

1. The fingerlike projections on the surface of the intestinal cells are microvilli. They increase the cells' surface area, enhancing the cells' ability to absorb nutrients.
2. Because the flagellum is an organelle of locomotion, if sperm lacked a flagellum, these cells would be unable to move. (As a result, they would be unable to reach an oocyte to perform fertilization.)

Page 89

1. The SER functions in the synthesis of lipids such as steroids. Ovaries and testes produce large amounts of steroid hormones, which are lipids, and thus need large amounts of SER.
2. Mitochondria produce energy, in the form of ATP molecules, for the cell. A large number of mitochondria in a cell indicates a high demand for energy.

Page 95

1. The nucleus of a cell contains DNA, which codes for the production of all the cell's polypeptides and proteins. Some of these proteins are structural proteins, which are responsible for the shape and other physical characteristics of the cell. Other proteins are enzymes, which govern cellular metabolism, direct the production of cell proteins, and control all the cell's activities.

2. If a cell lacked the enzyme RNA polymerase, the cell would not be able to transcribe RNA from DNA.

Page 102

1. This cell would be in the G_1 phase of its life cycle.
2. The deletion of a single nucleotide from a coding sequence of DNA would alter the entire base sequence after the point of deletion. As a result, different codons would be lined up on the messenger RNA that was transcribed from the affected region. These codons, in turn, would result in the incorporation of a different series of amino acids into the polypeptide. Almost certainly, the polypeptide product would be not functional.
3. If spindle fibers failed to form during mitosis, the cell would not be able to separate the chromosomes into two sets. If cytokinesis occurred, the result would be one cell with two sets of chromosomes and one cell with none.

CHAPTER 4

Page 110

1. The presence of many microvilli on the free surface of epithelial cells greatly increases the cell's surface area, allowing for increased absorption.
2. Gap junctions allow small molecules and ions to pass from cell to cell. Among epithelial cells, they help coordinate functions such as the beating of cilia. In cardiac and smooth muscle tissues, they are essential to the coordination of muscle cell contractions.
3. All these regions are subject to mechanical trauma and abrasion—by food (pharynx and esophagus), feces (anus), and intercourse or childbirth (vagina).

Page 117

1. No. A simple squamous epithelium does not provide enough protection against infection, abrasion, or dehydration and so is not located on the skin surface.
2. This process is holocrine secretion.
3. The gland is an endocrine gland.

Page 124

1. Collagen fibers add strength to connective tissue. A vitamin C deficiency thus might result in the production of connective tissue that is weak and prone to damage.
2. Antihistamines act against the molecule histamine, which is produced by mast cells.
3. The tissue is adipose (fat) tissue.

Page 128

1. Cartilage lacks a direct blood supply, which is necessary for proper healing to occur. Materials that are needed to repair damaged cartilage must diffuse from the blood to the chondroblasts; this process takes a long time and retards the healing process.
2. Intervertebral discs are composed of fibrocartilage.
3. The two connective tissues that contain a fluid matrix are blood and lymph.

Page 132

1. The pleural, peritoneal, and pericardial cavities are all lined by serous membranes.
2. This is an example of a mucous membrane.
3. This tissue is probably fascia, a type of dense regular connective tissue that attaches muscles to skin and bones.

Page 135

1. This muscle tissue is smooth muscle tissue; both cardiac and skeletal muscle tissues are striated.
2. The cells are probably neurons.
3. New skeletal muscle is produced by the division and fusion of satellite cells, mesenchymal cells that persist in adult skeletal muscle tissue.

CHAPTER 5

Page 147

1. This outer layer of skin is the stratum corneum.
2. The splinter is lodged in the stratum granulosum.
3. Fresh water is hypotonic with respect to skin cells, so water will move into the cells by osmosis, causing them to swell.
4. Sanding the tips of the fingers will not permanently remove fingerprints. The ridges of the fingerprints are formed in layers of the skin that are constantly regenerated, so these ridges will eventually reappear. The pattern of the ridges is determined by the arrangement of tissue in the dermis, which is not affected by sanding.

Page 149

1. When exposed to the ultraviolet radiation in sunlight or sunlamps, melanocytes in the epidermis and dermis synthesize the pigment melanin, darkening the color of the skin.
2. When skin is warm, blood is diverted to the superficial dermis for the purpose of eliminating heat. Because blood is red, it imparts a red cast to the skin.
3. The hormone cholecalciferol (vitamin D_3) is needed to form strong bones and teeth. The first step in this hormone's production involves the exposure of the skin to specific wavelengths of UV light. When the body surface is covered, UV light cannot penetrate to the blood in the skin to begin vitamin D_3 production, resulting in fragile bones.

Page 152

1. The capillaries and sensory neurons that supply the epidermis are located in the papillary layer of the dermis.
2. The presence of elastic fibers and the resilience of skin turgor allow the dermis to undergo repeated cycles of stretching and recoil.

Page 155

1. The contraction of the arrector pili muscles pulls the hair follicles erect, depressing the area at the base of the hair and making the surrounding skin appear higher. The result is known as "goose bumps."
2. Hair is a derivative of the epidermis, but the follicles are in the dermis. Where the epidermis and deep dermis are destroyed, new hair will not grow.

Page 157

1. Sebaceous glands produce a secretion called sebum. Sebum lubricates and protects the keratin of the hair shaft, lubricates and conditions the surrounding skin, and inhibits the growth of bacteria.
2. Apocrine sweat glands produce a secretion that contains several kinds of organic compounds. Some of these compounds have an odor, and others produce an odor when metabolized by skin bacteria. Deodorants are used to mask the odor of such secretions.
3. Apocrine sweat glands enlarge and increase their secretion production in response to the increase in sex hormones that occurs at puberty.

Page 160

1. The combination of fibrin clot, fibroblasts, and the extensive network of capillaries in tissue that is healing is called granulation tissue.
2. Skin can regenerate effectively even after considerable damage has occurred because stem cells persist in both the epithelial and connective tissue components of skin. When injury occurs, cells of the stratum germinativum replace epithelial cells while mesenchymal cells replace cells lost from the dermis.
3. As a person ages, the blood supply to the dermis decreases and merocrine sweat glands become less active. These changes make it more difficult for the elderly to cool themselves in hot weather.

CHAPTER 6

Page 174

1. If the ratio of collagen to hydroxyapatite in a bone increased, the bone would become less strong (as well as more flexible).
2. The presence of concentric layers of bone around a central canal is indicative of an osteon. Osteons make up compact bone. Because the ends (epiphyses) of long bones are primarily cancellous (spongy) bone, this sample is most likely from the marrow cavity of the shaft (diaphysis) of a long bone.
3. Because osteoclasts break down or demineralize bone, the bone would have a reduced mineral content (less mass); as a result, it would also be weaker.

Page 178

1. During intramembranous ossification, fibrous connective tissue is replaced by bone.
2. In endochondral ossification, cells of the inner layer of the perichondrium differentiate into osteoblasts.
3. Long bones of the body, such as the femur, have an epiphyseal cartilage, a plate of cartilage that separates the epiphysis from the diaphysis as long as the bone is still growing lengthwise. An X-ray would indicate whether the epiphyseal cartilage is still present. If it is, growth is still occurring; if it is not, the bone has reached its adult length.

Page 182

1. The larger arm muscles of the weight lifter would apply more mechanical stress to the bones of the upper limbs. In response to that stress, the bones would grow thicker. For similar reasons, we would expect the jogger to have heavier thigh bones.

2. Growth continues throughout childhood. At puberty, a growth spurt occurs and is followed by the closure of the epiphyseal cartilages. The later puberty begins, the taller the child will be when the growth spurt begins, so the taller the individual will be when growth is completed.
3. Increased levels of growth hormone prior to puberty will result in excessive bone growth, making the individual taller.

Page 184

1. The bones of children who have rickets are poorly mineralized and as a result are quite flexible. Under the weight of the body, the leg bones bend. The instability makes walking difficult and can lead to other problems of the legs and feet.
2. Parathyroid hormone (PTH) stimulates osteoclasts to release calcium ions from bone. Increased PTH secretion would result in an increase in the level of calcium ions in the blood.
3. Calcitonin lowers blood calcium levels by (1) inhibiting osteoclast activity and (2) increasing the rate of calcium excretion at the kidneys.

Page 187

1. An external callus forms early in the healing process, when cells from the endosteum and periosteum migrate to the area of the fracture. These cells form an enlarged collar (external callus) that circles the bone in the area of the fracture.
2. The sex hormones known as estrogens play an important role in moving calcium into bones. After menopause, the level of these hormones decreases dramatically; as a result, older women have difficulty replacing the calcium in bones that is being lost due to normal aging. In males, the level of sex hormones (androgens) does not decrease until much later in life.
3. Bones that form in the tendons of joints, such as the knee and wrist, are sesamoid bones.

CHAPTER 7

Page 208

1. The foramen magnum is located in the base of the occipital bone.
2. Tomás has fractured his right parietal bone.
3. The internal jugular veins pass through openings between the occipital and the temporal bones.
4. The sphenoid bone contains the sella turcica, which in turn contains the pituitary gland.
5. Olfactory nerves, which deal with the sense of smell (olfaction), pass through the olfactory foramina of the cribriform plate from the nasal cavity to the olfactory bulbs. If the cribriform plate failed to form, these sensory nerves could not reach the olfactory bulbs, and the sense of smell would be lost.

Page 214

1. The adult vertebral column has fewer vertebrae because the five sacral vertebrae fuse to form a single sacrum and the four coccygeal vertebrae fuse to form a single coccyx.
2. The secondary curves of the spine allow us to balance our body weight on our legs with minimal muscular effort. Without the secondary curves, we would not be able to stand upright for extended periods.
3. When you run your finger along a person's spine, you can feel the spinous processes of the vertebrae.

Page 219

1. The dens is located on the axis, or second cervical vertebra, which is in the neck.
2. The presence of transverse foramina indicate that this vertebra is a cervical vertebra.
3. The lumbar vertebrae must support a great deal more weight than do vertebrae that are more superior in the spinal column. The large vertebral bodies allow the weight to be distributed over a larger area.

Page 222

1. True ribs are attached directly to the sternum by their own costal cartilage. False ribs either do not attach to the sternum (as in the floating ribs) or attach by means of a common costal cartilage (as in the vertebrochondral ribs).
2. Improper compression of the chest during CPR can—and commonly does—result in a fracture of the sternum or ribs.

CHAPTER 8

Page 230

1. The clavicle attaches the scapula to the sternum and thus restricts the scapula's range of movement. When the clavicle is broken, the scapula has a greater range of movement and is less stable.

2. The head of the humerus articulates with the scapula at the glenoid cavity.

Page 233

1. The two rounded projections on either side of the elbow are the lateral and medial epicondyles of the humerus.
2. The radius is lateral when the forearm is pronated, as in the anatomical position.
3. The first distal phalanx is located at the tip of the thumb; Bill's thumb is broken.

Page 237

1. The three bones that make up the ossa coxae are the ilium, ischium, and pubis.
2. The pelvis of females is generally smoother and lighter than that of males and has less-prominent markings. The pelvic outlet is enlarged, and there is less curvature on the sacrum and coccyx. The pelvic inlet is wider and more circular. The pelvis as a whole is relatively broad and low. The ilia project farther laterally. The inferior angle between the pubic bones is greater than 100°, as opposed to 90° or less for the pelvis of males. These differences adapt the pelvis of females for supporting the weight of the developing fetus and enable the newborn to pass through the pelvic outlet during delivery.
3. When you are seated, your body weight is borne by the ischial tuberosities of the pelvis.

Page 241

1. Although the fibula is not part of the knee joint and does not bear weight, it is an important point of attachment for many leg muscles. When the fibula is fractured, these muscles cannot function properly to move the leg, and walking is difficult and painful. The fibula also helps stabilize the ankle joint.
2. Joey has most likely fractured the calcaneus (heel bone).
3. The talus transmits the weight of the body from the tibia toward the toes.
4. Muscles that pull the femur toward the axis of the body are attached at the linea aspera. When these muscles are strengthened, they apply more stress to the femur at the linea aspera, so this ridge will get larger and heavier.

CHAPTER 9

Page 250

1. All these joints (other than synostoses) consist of bony regions separated by fibrous or cartilaginous connective tissue.
2. Originally, each of these joints is a syndesmosis. As the bones interlock, they form sutural joints.
3. Articular cartilages lack a blood supply; they rely on synovial fluid to supply nutrients and eliminate wastes. Impairing the circulation of synovial fluid would have the same effect as impairing a tissue's blood supply. Nutrients would not be delivered to meet the tissue's needs, and wastes would accumulate. Damage to and ultimately the death of the cells in the tissue would result.

Page 256

1. (a) abduction; (b) supination; (c) flexion
2. When you do jumping jacks and move your lower limbs away from the midline of the body, the movement is abduction. When you bring the lower limbs back together, the movement is adduction.
3. Flexion and extension are the movements associated with hinge joints.

Page 257

1. Intervertebral discs are not found between the first and second cervical vertebrae, between sacral vertebrae in the sacrum, or between coccygeal vertebrae in the coccyx. An intervertebral disc between the first and second cervical vertebrae would prohibit rotation. The vertebrae in the sacrum and coccyx are fused.
2. (a) flexion; (b) lateral flexion; (c) rotation

Page 260

1. Ligaments and muscles provide most of the stability for the shoulder joint.
2. Because the subscapular bursa is located in the shoulder joint, the tennis player would be more likely to develop inflammation of this structure (bursitis). The condition is associated with repetitive motion that occurs at the shoulder, such as swinging a tennis racket. The jogger would be more at risk for injuries to the knee joint.
3. A shoulder separation is an injury involving partial or complete dislocation of the acromioclavicular joint.
4. Terry has most likely damaged his annular ligament.

Page 263

1. The iliofemoral, pubofemoral, and ischiofemoral ligaments are at the hip joint.
2. Damage to the menisci of the knee joint would result in a decrease in the joint's stability. The individual would have a hard time locking the knee in place while standing and would have to use muscle contractions to stabilize the joint. If the person had to stand for a long period, the muscles would fatigue and the knee would "give out." We would also expect the individual to feel pain.
3. "Clergyman's knee" is a bursitis that commonly affects members of the clergy, who spend a great deal of time kneeling. The work of carpet layers and roofers necessitates kneeling and sliding along on their knees, causing a similar inflammation of the bursae in the knee joint.

CHAPTER 10

Page 278

1. Because tendons attach muscles to bones, severing the tendon would disconnect the muscle from the bone. If the muscle were to contract, nothing would happen.
2. Skeletal muscle appears striated when viewed through a microscope because the Z lines and thick filaments of the myofibrils within the muscle fibers are aligned.
3. You would expect the greatest concentration of calcium ions to be in the cisternae of the sarcoplasmic reticulum in resting skeletal muscle.

Page 286

1. A muscle's ability to contract depends on the formation of cross-bridges between the myosin and actin myofilaments in the muscle. A drug that interferes with cross-bridge formation would prevent muscle contraction.
2. Because the amount of cross-bridge formation is proportional to the amount of available calcium ions, increased permeability of the sarcolemma to Ca^{2+} would lead to an increased intracellular concentration of Ca^{2+} and a greater degree of contraction. In addition, because the amount of calcium ions in the sarcoplasm must be reduced for relaxation to occur, an increase in the permeability of the sarcolemma to Ca^{2+} could make the muscle unable to relax completely.
3. Without acetylcholinesterase, the motor end plate would be continuously stimulated by acetylcholine; the muscle would lock in a state of contraction.

Page 293

1. A muscle's ability to contract depends on the formation of cross-bridges between the myosin and actin myofilaments in the muscle,. In a muscle that is overstretched, the myofilaments would overlap very little. Very few cross-bridges would form, and the contraction would be weak. If the myofilaments did not overlap at all, then no cross-bridges would form and the muscle could not contract.
2. During treppe, there is not enough time between successive contractions to reabsorb all the calcium ions that were released during the prior contraction event. As a result, calcium ions accumulate in the sarcoplasm at higher than normal levels, allowing more cross-bridges to form and tension to increase.
3. Yes. Contraction occurs as thick and thin filaments interact. The entire muscle can shorten (isotonic, concentric), elongate (isotonic, eccentric), or remain the same length (isometric), depending on the relationship between the resistance and the tension produced by actin and myosin interactions.

Page 300

1. A sprinter requires large amounts of energy for a short burst of activity. To supply this demand for energy, the sprinter's muscles switch to anaerobic metabolism. Anaerobic metabolism is not as efficient in producing energy as aerobic metabolism is, and the process also produces acidic waste products; this combination contributes to fatigue. Conversely, marathon runners derive most of their energy from aerobic metabolism, which is more efficient and does not produce the amount of waste products that anaerobic metabolism does.
2. Activities that require short periods of strenuous activity produce a greater oxygen debt, because this type of activity relies heavily on energy production by anaerobic metabolism. Because lifting weights is more strenuous over the short term than swimming is, weight lifting would likely produce a greater oxygen debt than would swimming laps, which is an aerobic activity.
3. Individuals who excel at endurance activities have a higher than normal percentage of slow fibers. Those fibers are physiologically better adapted to this type of activity than are fast fibers, which are less vascular and fatigue faster.

Page 305

1. Cardiac muscle cells are joined by gap junctions, which allow ions and small molecules to flow directly from one cell to another. This type of junction allows for action potentials that are generated in one cell to spread rapidly to adjacent cells. Thus all the cells contract simultaneously as if they were a single unit (a syncytium).
2. Cardiac and smooth muscle contractions are more affected by changes in the concentration of Ca^{2+} in the extracellular fluid than are skeletal muscle contractions because in cardiac and smooth muscles, most of the calcium ions that trigger a contraction come from the extracellular fluid. In skeletal muscle, most of the calcium ions come from the sarcoplasmic reticulum.
3. The actin and myosin filaments of smooth muscle are not as rigidly organized as those of skeletal muscle. This loose organization allows smooth muscle to contract over a wider range of resting lengths.

CHAPTER 11

Page 314

1. A pennate muscle contains more muscle fibers than does a parallel muscle of the same size. A muscle that has more muscle fibers has more myofibrils and sarcomeres. As a result, the contraction of a pennate muscle generates more tension than would the contraction of a parallel muscle of the same size.
2. The opening between the stomach and the small intestine would be guarded by a circular muscle, or sphincter. The concentric circles of muscle fibers found in sphincters are ideally suited for opening and closing openings and for acting as valves in the body.
3. The joint between the occipital bone and the first cervical vertebra is part of a first-class lever system. The joint between the two bones, the fulcrum, lies between the skull, which provides the resistance, and the neck muscles, which provide the applied force.

Page 320

1. The origin of a muscle is the end that remains stationary during an action. Because the gracilis muscle moves the tibia, the origin of this muscle must be on the pelvis (pubis and ischium).
2. Muscles A and B are antagonists, because they perform opposite actions.
3. The name *flexor carpi radialis longus* tells you that this muscle is a long muscle that lies next to the radius and flexes the wrist.

Page 328

1. Contraction of the masseter muscle elevates the mandible; relaxation of this muscle depresses the mandible. You would probably be chewing something.
2. You would expect the buccinator muscle, which forms the mouth for blowing, to be well developed in a trumpet player.
3. Swallowing involves contractions of the palatal muscles, which elevate the soft palate as well as portions of the superior pharyngeal wall. Elevation of the superior portion of the pharynx enlarges the opening to the auditory tube, permitting airflow to the middle ear and the inside of the eardrum. Making this opening larger by swallowing facilitates airflow into or out of the middle ear cavity.

Page 334

1. Damage to the external intercostal muscles would interfere with breathing.
2. A blow to the rectus abdominis muscle would cause that muscle to contract forcefully, resulting in flexion of the vertebral column. In other words, you would "double over."
3. The sore muscles are most likely the erector spinae muscles, especially the longissimus and the iliocostalis muscles of the lumbar region. These muscles would have to contract harder to counterbalance the increased anterior weight you would bear when carrying heavy boxes.

Page 346

1. When you shrug your shoulders, you are contracting your levator scapulae muscles.
2. The rotator cuff muscles include the supraspinatus, infraspinatus, subscapularis, and teres minor muscles. The tendons of these muscles help enclose and stabilize the shoulder joint.
3. Injury to the flexor carpi ulnaris muscle would impair the ability to perform flexion and adduction at the wrist.

Page 353

1. Injury to the obturator muscle would impair your ability to perform lateral rotation at the hip.

2. "Hamstring" refers to a group of muscles that collectively flex the knee. These muscles are the biceps femoris, semimembranosus, and semi-tendinosus muscles.

3. The calcanean tendon attaches the soleus and gastrocnemius muscles to the calcaneus (heel bone). When these muscles contract, they produce plantar flexion at the ankle. A torn calcanean tendon would make extension of the foot difficult.

CHAPTER 12

Page 373

1. The afferent division of the PNS is composed of nerves that carry sensory information to the brain and spinal cord. Damage to this division would interfere with a person's ability to experience a variety of sensory stimuli.

2. Most sensory neurons of the PNS are unipolar. Thus these neurons are most likely sensory neurons.

3. Microglial cells are small phagocytic cells that occur in increased number in damaged and diseased areas of the CNS.

Page 386

1. The depolarization of the neuron cell membrane involves the opening of gated sodium channels and the rapid influx of sodium ions into the cell. If the sodium channels were blocked, a neuron would not be able to depolarize.

2. If the extracellular concentration of potassium ions were to decrease, more potassium ions would leave the cell; hence the electrical gradient across the membrane (transmembrane potential) would increase. This condition is called hyperpolarization.

3. Action potentials are propagated along myelinated axons by saltatory propagation at speeds much higher than those along unmyelinated axons. The axon with a propagation speed of 50 meters per second must be the myelinated axon.

Page 392

1. When an action potential reaches the presynaptic terminal of a cholinergic synapse, voltage-regulated calcium ion channels open and the influx of Ca^{2+} triggers the release of ACh into the synapse to stimulate the next neuron. If these calcium channels were blocked, the ACh would not be released and transmission across the synapse would cease.

2. Because of synaptic delay, the pathway with fewer neurons (three) will transmit impulses more rapidly.

3. The effect of a neurotransmitter on postsynaptic membranes depends on the receptor molecule that binds it. In this case, although the neurotransmitter and target tissues are the same, the smooth muscle cells in blood vessels of skeletal muscle have different NE receptors than do those in the blood vessels of intestines. Because different receptors are involved, the effects of neurotransmitter binding can be totally different; in this example, they have opposite effects.

Page 399

1. This arrangement represents a convergent circuit.

2. For a severed axon to heal, it must come into contact with and grow into the new column of Schwann cells that forms distal to the injury site. If the axon fails to make this connection, the column of Schwann cells degenerates, the axon stops growing, and the connection will not be reestablished. Closely aligning the two ends of the axons after an injury increases the chances that the connection will be made and that innervation reestablished.

3. Interneurons are responsible for analyzing sensory input and coordinating motor output. Without interneurons, the spinal cord would not be able to process sensory information or to make appropriate motor responses.

CHAPTER 13

Page 412

1. The ventral root of a spinal nerve is composed of both visceral and somatic motor fibers. Damage to this root would interfere with motor function.

2. The cerebrospinal fluid that surrounds the spinal cord is located in the subarachnoid space, which lies beneath the epithelium of the arachnoid layer and superficial to the pia mater.

Page 413

1. Because the virus that causes polio would be located in somatic motor neurons, it would be in the anterior gray horns of the spinal cord, where the cell bodies of these neurons are located.

2. A disease that damages myelin sheaths would affect the columns of the spinal cord, because that part of the cord is composed of bundles of myelinated axons.

Page 420

1. The dorsal rami of spinal nerves innervate the skin and muscles of the back. The skin and muscles of the back of the neck and of the shoulders will be affected by such an anesthetic.

2. The phrenic nerves that innervate the diaphragm originate in the cervical plexus. Damage to this plexus, or more specifically to the phrenic nerves, would greatly interfere with the ability to breathe and might result in death.

3. Compression of the sciatic nerve produces the characteristic sensation that you perceive when your leg has "fallen asleep."

Page 428

1. The minimum number of neurons required for a reflex arc is two. One must be a sensory neuron, to bring impulses to the CNS, and the other a motor neuron, to bring about a response to the sensory input.

2. The suckling reflex is an innate reflex.

3. When stretch receptors are stimulated by gamma motor neurons, the muscle spindles become more sensitive. As a result, little if any stretching stimulus would be needed to stimulate the contraction of the quadriceps muscles in the patellar reflex. Thus the reflex response would appear more quickly.

Page 431

1. This response is the tendon reflex.

2. During a withdrawal reflex, the limb on the opposite side is extended. This response is called a crossed extensor reflex.

3. A positive Babinski reflex is abnormal for an adult; it indicates possible damage of descending tracts in the spinal cord.

CHAPTER 14

Page 438

1. The three primary brain vesicles are the prosencephalon, the mesencephalon, and the rhombencephalon. The prosencephalon gives rise to the cerebrum and diencephalon; the mesencephalon does not subdivide further; and the rhombencephalon develops into the cerebellum, the pons, and the medulla oblongata.

2. The response is controlled by the mesencephalon.

Page 445

1. If an interventricular foramen became blocked, cerebrospinal fluid could not flow from the first or second ventricle into the third ventricle. Cerebrospinal fluid would continue to form within that ventricle, so the blocked ventricle would swell with fluid—a condition known as hydrocephalus.

2. Diffusion across the arachnoid granulations is the means by which cerebrospinal fluid reenters the bloodstream. If this process decreased, excess fluid would accumulate in the ventricles; the volume of fluid in the ventricles would increase.

3. The blood–brain barrier, consisting of capillary endothelial cells stimulated by adjacent astrocytes, restricts and regulates the movement of water-soluble molecules from the blood to the extracellular fluid of the brain.

Page 451

1. The primary motor cortex is located in the precentral gyrus of the frontal lobe of the cerebrum.

2. Damage to the temporal lobes of the cerebrum would interfere with the processing of olfactory (smell) and auditory (sound) impulses.

3. The stroke has damaged Jake's speech center, located in the frontal lobe.

Page 454

1. The cerebral nuclei control muscle tone and coordinate learned movement patterns. A person who has damage to these nuclei would exhibit difficulty in walking and in making fluid, precise movements of the upper limbs.

2. The temporal lobe of Paul's cerebrum is probably involved, specifically the hippocampus and the amygdaloid body. His problems may also involve other parts of the limbic system that act as a gate for loading and retrieving long-term memories.

3. Terri's symptoms imply injury to the mamillary bodies of the hypothalamus. She may also have injured the motor areas of the brain stem that control the muscles involved in these processes.

Page 459

1. The lateral geniculate nuclei are involved with processing visual information. Damage to these nuclei would interfere with the sense of sight.

2. Changes in body temperature would stimulate the preoptic area of the hypothalamus, a division of the diencephalon.

Page 464

1. The vermis and arbor vitae are part of the cerebellum.
2. Although the medulla oblongata is small, it contains many vital reflex centers, including those that control breathing and regulate the heart rate and blood pressure. Damage to the medulla oblongata can result in a cessation of breathing or in lethal changes in heart rate and blood pressure.
3. Cells of the substantia nigra release the neurotransmitter dopamine at synapses with neurons in the cerebral nuclei. Dopamine produces IPSPs and decreases the activity of motor neurons. A decrease in the release of dopamine would result in the motor neurons' being more active, producing symptoms including twitches, tremors, and other types of involuntary contractions.

CHAPTER 15

Page 487

1. The fasciculus gracilis in the posterior column of the spinal cord carries information about touch and pressure from the lower limbs to the brain. It is this tract that is being compressed.
2. Nociceptors are stimulated by pain. The action potentials generated by these receptors are carried by the lateral spinothalamic tracts.
3. Impulses carried along the right fasciculus gracilis are destined for the primary sensory cortex of the left cerebral hemisphere.

Page 492

1. The anatomical basis for opposite-side motor control is that crossing-over (decussation) of axons occurs, so the motor fibers of the corticospinal pathway innervate lower motor neurons on the opposite side of the body.
2. The superior portion of the motor cortex controls the upper limb and upper portion of the lower limb. An injury to this area would affect the ability to control the muscles in those regions of the body.
3. Motor neurons of the red nucleus help control the muscle tone of skeletal muscles. Increased stimulation of these neurons would increase stimulation of the skeletal muscles, producing increased muscle tone.

Page 496

1. We would expect Tina's brain waves to be beta waves, which are characteristic of adults who are experiencing stress and/or tension.
2. An inability to comprehend the spoken or written word indicates a problem with the general interpretive area of the brain, which in most individuals is located in the left temporal lobe of the cerebrum.
3. Recalling information for an A & P test involves fact memory, specifically secondary memory (a type of long-term memory).

Page 499

1. The reticular activating system is responsible for rousing the cerebrum to a state of consciousness. If your RAS were suddenly stimulated, you would wake up.
2. A drug that increases the amount of serotonin released in the brain would produce a heightened perception of certain sensory stimuli, such as auditory or visual stimuli, and hallucinations.
3. Some possible reasons for slower recall and for loss of memory in the elderly include a loss of neurons (possibly those involved in specific memories), changes in synaptic organization of the brain, changes in the neurons themselves, and decreased blood flow, which would affect the metabolic rate of neurons and perhaps slow the retrieval of information from memory.

CHAPTER 16

Page 506

1. Two neurons are required to conduct an action potential from the spinal cord to smooth muscles in the wall of the intestine. One neuron is required to carry the action potential from the spinal cord to the autonomic ganglion, and a second to carry the action potential from the autonomic ganglion to the smooth muscle.
2. The sympathetic division of the autonomic nervous system is responsible for the physiological changes that occur in response to stress and increased activity.
3. The sympathetic division of the autonomic nervous system includes preganglionic fibers from the lumbar and thoracic portions of the spinal cord, whereas the parasympathetic division includes preganglionic fibers from the cranial and sacral portions.

Page 512

1. The nerves that synapse in collateral ganglia originate in the inferior thoracic and superior lumbar portion of the spinal cord; they pass through the chain ganglia to the collateral ganglia.
2. Acetylcholine is the neurotransmitter released by the preganglionic fibers of the sympathetic nervous system. A drug that stimulates ACh receptors would stimulate the postganglionic fibers of the sympathetic nerves, resulting in increased sympathetic activity.
3. Blocking the beta receptors on cells would decrease or prevent sympathetic stimulation of the tissues that contain those cells. The heart rate and the force of contraction and relaxation of the smooth muscle in the walls of blood vessels would decrease. These changes would contribute to a lowering of the blood pressure.

Page 516

1. The vagus nerve (N X) carries the parasympathetic fibers that innervate the lungs, heart, stomach, liver, pancreas, and parts of the small and large intestines (as well as several other visceral organs).
2. Muscarinic receptors are a type of acetylcholine receptor located in postganglionic synapses of the parasympathetic nervous system. The stimulation of these receptors at the heart would cause potassium ion channels to open, resulting in hyperpolarization of the cell membrane and a decreased heart rate.
3. The parasympathetic division is sometimes referred to as the anabolic system because parasympathetic stimulation leads to a general increase in the nutrient content of the blood. Cells throughout the body respond to the increase by absorbing the nutrients and using them to support growth and other anabolic activities.

Page 522

1. Most blood vessels receive sympathetic stimulation, so a decrease in sympathetic tone would lead to a relaxation of the muscles in the walls of these vessels and hence to vasodilation. Blood flow to the tissue would, in turn, increase.
2. A patient who is anxious about an impending root canal would probably exhibit some or all of the following changes: a dry mouth, an increased heart rate, an increase in blood pressure, an increased rate of breathing, cold sweats, an urge to urinate or defecate, a change in the motility of the digestive tract (that is, "butterflies in the stomach"), and dilated pupils. These changes would be the result of anxiety or stress causing an increase in sympathetic stimulation.
3. A brain tumor that interferes with hypothalamic function would interfere with autonomic function as well. Centers in the posterior and lateral hypothalamus coordinate and regulate sympathetic function, whereas centers in the anterior and medial hypothalamus control parasympathetic function.

CHAPTER 17

Page 534

1. The receptor with the smaller receptor field will provide more-precise sensory information—thus, receptor A.
2. Nociceptors are pain receptors. When they are stimulated, you perceive a painful sensation in the affected hand.
3. Proprioceptors relay information about limb position and movement to the CNS, especially to the cerebellum. If this information were blocked, your movements would be uncoordinated and you would probably not be able to walk.

Page 538

1. By the end of the lab period, adaptation has occurred. In response to the constant level of stimulation, your receptor neurons have become less active, partially as the result of synaptic fatigue.
2. Your taste receptors (taste buds) are sensitive only to molecules and ions that are in solution. If you dry the surface of your tongue, the salt ions or sugar molecules have no moisture in which to dissolve, so they will not stimulate the taste receptors.
3. The difference in the taste of your grandfather's food is the result of several age-related factors. The number of taste buds declines dramatically after age 50, and those that remain are not as sensitive as they once were. In addition, the loss of olfactory receptors contributes to the perception of less flavor in foods.

Page 550

1. The first layer of the eye that would be affected by inadequate tear production would be the conjunctiva. Drying of this layer would produce an irritated, scratchy feeling.
2. When the lens is round, you are looking at an object that is close to you.

3. Renee will probably not be able to see at all. The fovea of the eye contains cones but no rods. Rods respond to light of low intensity, but cones need high light intensity to be stimulated. In a dimly lit room, the light would not be strong enough to stimulate these photoreceptors.
4. If the canal of Schlemm were blocked, the aqueous humor would not be able to drain; glaucoma would develop. As the quantity of fluid increased, the pressure within the eye would increase, distorting soft tissues and interfering with vision. If untreated, the condition would ultimately cause blindness.

Page 557

1. Even with a congenital lack of cones in your eyes, you would still be able to see—as long as you had functioning rods. But because cone cells function in color vision, you would see only black and white.
2. A deficiency of vitamin A in the diet would affect the quantity of retinal the body could produce and thus would interfere with night vision.
3. A decrease in phosphodiesterase activity would lead to a higher level of intracellular cGMP. This rise would, in turn, keep the gated sodium channels open and decrease the ability of receptor neurons to respond to photons.

Page 571

1. Without the movement of the round window, the perilymph would not be moved by the vibration of the stapes at the oval window. There would be little or no perception of sound.
2. The loss of stereocilia (as a result of constant exposure to loud noises, for instance) would reduce hearing sensitivity and could lead to deafness.
3. The auditory tube allows pressure to equalize on both sides of the tympanic membrane. If this tube were blocked and external pressures then declined, the pressure on the inside of the tympanic membrane would be greater than that on the outside, forcing the membrane outward and producing pain.

CHAPTER 18

Page 586

1. Neural responses occur within fractions of a second and do not last long (short duration). Conversely, endocrine responses may be slow to appear but will last for minutes to days (long duration).
2. Adenylate cyclase is the enzyme that converts ATP to cAMP. A molecule that blocks this enzyme would block the action of any hormone that required cAMP as a second messenger.
3. A cell's hormonal sensitivities are determined by the presence or absence of the necessary receptor complex for a given hormone.

Page 592

1. Dehydration increases the osmotic pressure of the blood. The increase in blood osmotic pressure would stimulate the posterior pituitary to release more ADH.
2. Somatomedins mediate the action of growth hormone. When the levels of somatomedins are elevated, the level of growth hormone would be elevated as well.
3. Elevated circulating levels of cortisol would inhibit the cells that control the release of ACTH from the pituitary gland; therefore, the level of ACTH would decrease. This is an example of a negative feedback mechanism.

Page 603

1. An individual whose diet lacks iodine would not be able to form the hormone thyroxine. As a result, we would expect to see the symptoms associated with thyroxine deficiency, such as a decreased rate of metabolism, a decreased body temperature, a poor response to physiological stress, and an increase in the size of the thyroid gland (goiter).
2. Most of the thyroid hormone in the blood is bound to proteins called thyroid-binding globulins. These compounds represent a large reservoir of the thyroid hormone thyroxine that guards against rapid fluctuations in the level of this important hormone. Because such a large amount is stored in this way, it takes several days to deplete the supply of thyroxine, even after the thyroid gland has been removed.
3. The removal of the parathyroid glands would result in a decrease in the blood concentration of calcium ions. This decrease could be counteracted by increasing the amounts of vitamin D and calcium in the diet.
4. One of the functions of cortisol is to decrease the cellular use of glucose while increasing the available glucose by promoting the breakdown of glycogen and the conversion of amino acids to carbohydrates. The net result of elevated cortisol levels would be an elevation in the level of glucose in the blood.

Page 610

1. An individual with Type 1 or Type 2 diabetes has such elevated levels of glucose in the blood that the kidneys cannot reabsorb all the glucose; some glucose is lost in urine. Because the urine contains high concentrations of glucose, less water can be reclaimed by osmosis; the volume of urine production increases. The water losses reduce blood volume and elevate blood osmotic pressure, promoting thirst and triggering the secretion of ADH.
2. Glucagon stimulates the conversion of glycogen to glucose in the liver. Increased levels of glucagon would lead to decreased amounts of glycogen in the liver.
3. Increased amounts of light would inhibit the production (and release) of melatonin from the pineal gland, which receives neural input from the optic tracts. The secretion of melatonin by the pineal gland is influenced by light–dark cycles.

Page 614

1. The type of hormonal interaction exemplified by the effects of insulin and glucagon is antagonism. In this type of hormonal interaction, two hormones have opposite effects on their target tissues.
2. The hormones GH, thyroid hormone, PTH, and the gonadal hormones all play a role in formation and development of the skeletal system.
3. During the resistance phase of the general adaptation syndrome, there is a high demand for glucose, especially by the nervous system. The hormones GH-RH and CRH increase the levels of GH and ACTH, respectively. Growth hormone mobilizes fat reserves and promotes the catabolism of protein; ACTH increases cortisol, which stimulates the conversion of glycogen to glucose as well as the catabolism of fat and protein.

CHAPTER 19

Page 627

1. Venipuncture is a common sampling technique because (1) superficial veins are easy to locate; (2) the walls of veins are thinner than those of arteries; and (3) blood pressure in veins is relatively low, so the puncture wound seals quickly.
2. A decrease in the amount of plasma proteins in the blood would cause a decrease in (1) plasma osmotic pressure; (2) the ability to fight infection; and (3) the transport and binding of some ions, hormones, and other molecules.
3. During a viral infection, you would expect the level of immunoglobulins (antibodies) in the blood to be elevated.

Page 634

1. The hematocrit measures the amount of formed elements (mostly red blood cells) as a percentage of the total blood. In hemorrhage, the loss of blood, especially of red blood cells, would lower the hematocrit.
2. A decreased blood flow to the kidneys will trigger the release of erythropoietin. The elevated erythropoietin will lead to an increase in erythropoiesis (red blood cell formation). Thus, Dave's hematocrit will increase.
3. The liver processes bilirubin prior to excretion in the bile. Diseases that cause damage to the liver, such as hepatitis or cirrhosis, would impair the liver's ability to perform this function. As a result, bilirubin would accumulate in the blood, producing jaundice.

Page 637

1. Surface antigens on RBCs are glycolipids in the cell membrane.
2. A person with Type O blood can accept only Type O blood.
3. If a person with Type A blood receives a transfusion of Type B blood, the red blood cells will clump, or agglutinate, potentially blocking blood flow to various organs and tissues.

Page 644

1. In an infected cut, we would find a large number of neutrophils, phagocytic white blood cells that are generally the first to arrive at the site of an injury. They specialize in dealing with infectious bacteria.
2. The cells that produce circulating antibodies are B lymphocytes; these white blood cells would be found in elevated numbers.
3. During inflammation, basophils release a variety of chemicals, such as histamine and heparin, that exaggerate the inflammation and attract other types of white blood cells.

Page 648

1. Megakaryocytes are the precursors of platelets, which play an important role in hemostasis and the clotting process. A decreased number of megakaryocytes would result in fewer platelets, which in turn would interfere with the ability to clot properly.
2. The use of antibiotics would lower the number of intestinal bacteria and thus the amount of vitamin K produced. This decrease in vitamin K would lead to a decrease in the production of several clotting factors—most notably, prothrombin. As a result, clotting time would increase.
3. The activation of Factor XII initiates the intrinsic pathway.

App.
I

CHAPTER 20

Page 668

1. The semilunar valves on the right side of the heart guard the opening to the pulmonary artery. Damage to these valves would interfere with the blood flow to this vessel.
2. When the ventricles begin to contract, they force the AV valves to close, tensing the chordae tendineae. The chordae tendineae are attached to the papillary muscles, which begin contracting just before the rest of the ventricular myocardium does. Their contraction prevents the AV valves from opening back into the atria.
3. The left ventricle is more muscular than the right ventricle because the left ventricle must generate enough force to propel blood throughout all body systems except the lungs. The right ventricle must generate only enough force to propel blood a few centimeters to the lungs.

Page 673

1. If these cells failed to function, the heart would still continue to beat but at a slower rate.
2. If the impulses from the atria were not delayed at the AV node, they would be conducted through the ventricles so quickly by the bundle branches and Purkinje cells that the ventricles would begin contracting immediately, before the atria had finished their contraction. As a result, the ventricles would not be as full of blood as they could be, and the pumping of the heart would not be as efficient, especially during activity.
3. The cardioinhibitory center of the medulla oblongata is part of the parasympathetic branch of the autonomic nervous system. Damage to this center would result in fewer parasympathetic action potentials to the heart and an increase in heart rate due to sympathetic dominance.

Page 678

1. When pressure in the left ventricle is rising, the heart is contracting but no blood is leaving the heart. During this initial phase of contraction, both the AV valves and the semilunar valves are closed. The increase in pressure is the result of increased tension as the cardiac muscle contracts. When the pressure in the ventricle exceeds the pressure in the aorta, the aortic semilunar valves are forced open and blood is rapidly ejected from the ventricle.
2. An increase in the size of the QRS complex indicates a larger-than-normal amount of electrical activity during ventricular depolarization. One possible cause is an increase in the size of the heart. Because more cardiac muscle is depolarizing, the magnitude of the electrical event would be greater.

Page 683

1. If the heart beats too rapidly, it does not have sufficient time to fill completely between beats. The heart pumps in proportion to the amount of blood that enters: The less blood that enters, the less the heart can pump. If it beats too fast, very little blood will enter the bloodstream; tissues will suffer damage from the lack of blood supply.
2. Stimulating the acetylcholine receptors of the heart would cause the heart to slow down. The cardiac output is the product of stroke volume and the heart rate; if the heart rate decreases, so will the cardiac output (assuming that the stroke volume doesn't change).
3. The venous return fills the heart with blood, stretching the heart muscle. According to the Frank–Starling principle, the more the heart muscle is stretched, the more forcefully it will contract (to a point). The more forceful the contraction, the more blood the heart will eject with each beat (stroke volume). Therefore, increased venous return would increase the stroke volume (if all other factors are constant).
4. An increase in sympathetic stimulation of the heart would result in an increased heart rate and increased force of contraction. The end-systolic volume (ESV) is the amount of blood that remains in a ventricle after a contraction (systole). The more forcefully the heart contracts, the more blood it ejects and the lower the ESV is. Therefore, increased sympathetic stimulation should result in a lower ESV.
5. SV = EDV − ESV, so SV = 125 ml − 40 ml = 85 ml

Page 687

1. Caffeine acts directly on the conducting system and contractile cells of the heart, increasing the rate at which they depolarize. Drinking large amounts of caffeinated coffee would increase the heart rate.
2. The cardioinhibitory center of the medulla oblongata is part of the parasympathetic division of the autonomic nervous system. Damage to this center would result in fewer parasympathetic action potentials to the heart and an increase in heart rate due to sympathetic dominance.
3. An increase in the extracellular concentration of calcium ions would result in an increased force of cardiac contraction. More calcium ions

would enter heart cells when they were stimulated, due to the increased concentration gradient. The increased amount of intracellular Ca^{2+} would lead to increased binding of troponin, more myosin heads binding to actin, and a stronger and longer-lasting contraction.
4. A drug that increases the length of time required for the repolarization of pacemaker cells would decrease the heart rate, because the pacemaker cells would generate fewer action potentials per minute.

CHAPTER 21

Page 701

1. These blood vessels are veins. Arteries and arterioles have a large amount of smooth muscle tissue in a thick, well-developed tunica media.
2. Blood pressure in the arterial system pushes blood into the capillaries. Blood pressure in the venous system is very low, and other forces help keep the blood moving. Valves in veins prevent blood from flowing backward whenever the venous pressure drops; this is not a problem in arteries.
3. You would find fenestrated capillaries in organs and tissues where small peptides move freely into and out of the blood: for instance, endocrine glands, the choroid plexus of the brain, absorptive areas of the intestine, and filtration areas of the kidneys.

Page 711

1. In a healthy individual, the blood pressure should be greater at the aorta than at the inferior vena cava. Blood, like other fluids, moves along a pressure gradient from areas of high pressure to areas of low pressure. If the pressure were higher in the inferior vena cava, the blood would flow backward.
2. While a person stands for periods of time, blood pools in the lower limbs. This pooling decreases the venous return to the heart. In turn, the cardiac output decreases, sending less blood to the brain, causing lightheadedness and fainting. A hot day adds to this effect, because body water is lost and blood volume is reduced, through sweating.
3. The nurse first heard the sounds of Korotkoff when the pressure in the cuff reached 125 mm Hg. At this point, the pressure in the vessel during systole is just enough to overcome the pressure in the cuff. Turbulent flow produced in the constricted vessel then produces the audible sounds.

Page 723

1. During exercise, blood pressure increases because (1) blood flow to muscles increases, (2) cardiac output increases, and (3) resistance in visceral tissues increases.
2. Pressure at this site would decrease blood pressure at the carotid sinus, where the carotid baroreceptors are located. This decrease would cause a decreased frequency of action potentials along the glossopharyngeal nerve (IX) to the medulla oblongata, and more sympathetic impulses would be sent to the heart. The net result would be an increase in the heart rate.
3. The vasoconstriction of the renal artery would decrease both blood flow and blood pressure at the kidney. In response, the kidney would increase the amount of renin that it releases, which in turn would lead to an increase in the level of angiotensin II. The angiotensin II would bring about increased blood pressure and increased blood volume.
4. A person in circulatory shock has a decreased venous return to the heart and, in turn, a decreased cardiac output. The decreased cardiac output accounts for the weak pulse. Because the cardiac output is decreased, the baroreceptors are stimulated. As a result, sympathetic stimulation to the heart increases, causing the rapid heart rate. Although the heart is beating faster, there is less blood to pump and pulse pressure remains low.

Page 734

1. The left subclavian artery is the branch of the aorta that sends blood to the left arm.
2. The common carotid arteries carry blood to the head. A compression of the common carotid arteries would result in decreased blood pressure at the carotid sinus and a rapid fall in blood flow to the brain, resulting in unconsciousness. An immediate reflexive increase in heart rate and blood pressure would follow.
3. Organs served by the celiac trunk include the stomach, spleen, liver, and pancreas.

Page 744

1. The vein that is bulging is the external jugular vein.
2. A blockage of the popliteal vein would interfere with blood flow in the tibial and peroneal veins (which form the popliteal vein) and the small saphenous vein (which joins the popliteal vein).
3. This blood sample must have come from the umbilical vein, which carries oxygenated, nutrient-rich blood from the placenta to the fetus.

CHAPTER 22

Page 763

1. The thoracic duct drains lymph from the area beneath the diaphragm and from the left side of the head and thorax. Most of the lymph enters the venous blood by way of this duct. A blockage of this duct would impair the circulation of lymph through most of the body and would promote the accumulation of fluid in the limbs (lymphedema).

2. The thymic hormones play a role in the differentiation of lymphoid stem cells into T lymphocytes. A lack of these hormones would result in an absence of T lymphocytes.

3. During an infection, the lymphocytes and phagocytes in the lymph nodes in the affected region undergo cell division to deal with the infectious agent more effectively. This increase in the number of cells in the node causes the node to become enlarged or swollen.

Page 771

1. A decrease in the number of monocyte-forming cells in the bone marrow would result in a decreased number of macrophages in the body, because all types of macrophages are derived from monocytes. These include the Kupffer's cells of the liver, Langerhans cells in the skin and digestive tract, and alveolar macrophages.

2. A rise in the level of interferon would suggest a viral infection. Interferon is released from cells that are infected with viruses. It does not help an infected cell but "interferes" with the virus's ability to infect other cells.

3. Pyrogens stimulate the temperature control area of the preoptic nucleus of the hypothalamus. The result is an increase in body temperature, or fever.

Page 781

1. Abnormal peptides in the cytoplasm of a cell can become attached to MHC (major histocompatibility complex) proteins and are then displayed on the surface of the cell's membrane. Peptides presented in this manner are then recognized by T cells, which can initiate an immune response.

2. Cytotoxic T cells function in cell-mediated immunity. A decrease in the number of cytotoxic T cells would interfere with the ability to kill foreign cells and tissues as well as cells infected by viruses.

3. Helper T cells promote B cell division, the maturation of plasma cells, and the production of antibodies by the plasma cells. Without helper T cells, the antibody-mediated immune response would probably not occur.

4. Plasma cells produce and secrete antibodies. In a sample containing an elevated number of plasma cells, we would expect the amount of antibodies in the blood to increase.

Page 791

1. The secondary response would be more affected by the lack of memory B cells for a particular antigen. The ability to produce a secondary response depends on the presence of memory B cells and T cells that are formed during the primary response to an antigen. These cells are not involved in the primary response but are held in reserve against future contact with the same antigen.

2. The developing fetus is protected primarily by natural passive immunity, the product of IgG antibodies that cross the placenta from the mother's bloodstream.

3. Stress can interfere with the immune response by depressing the inflammatory response, reducing the number and activity of phagocytes, and inhibiting interleukin secretion.

CHAPTER 23

Page 805

1. The rich blood supply to the nose delivers body heat to the nasal cavity, so the inhaled air is warmed before it leaves the nasal cavity. The heat also evaporates moisture from the epithelium to humidify the incoming air. The moisture is derived from the blood supply as well. The blood supply also brings nutrients and water to the secretory cells of the nasal mucosa.

2. The nasopharynx receives only air from the nasal cavity. The oropharynx and laryngopharynx receive both air from the nasal cavity and food from the oral cavity. Ingested solids and liquids can damage delicate cells; thus, the areas that are in contact with food have a highly protective stratified squamous epithelium, like that of the exterior skin. The lining of the nasopharynx is the same as that of the nasal cavity, a pseudostratified ciliated columnar epithelium.

3. Increased tension in the vocal folds will raise the pitch of your voice.

Page 813

1. The tracheal cartilages are C-shaped to allow room for esophageal expansion when large pieces of food or volumes of liquid are swallowed.

2. Without surfactant, the alveoli would collapse as a result of surface tension in the thin layer of water that moistens the alveolar surfaces.

3. Air that passes through the glottis flows into the larynx and through the trachea. From there, the air flows into a primary bronchus, which supplies the lungs. In the lungs, the air passes to bronchi, bronchioles, a terminal bronchiole, a respiratory bronchiole, an alveolar duct, an alveolar sac, an alveolus, and ultimately to the respiratory membrane.

Page 814

1. The pulmonary arteries supply the exchange surfaces; the external carotid arteries, the thyrocervical trunks, and the bronchial arteries supply the conducting portions of the respiratory system.

2. The pleura is a serous membrane. Pleural surfaces secrete pleural fluid, which lubricates the opposing parietal and visceral surfaces to prevent friction during breathing.

Page 822

1. Because the rib penetrates Mark's chest wall, atmospheric air will enter his thoracic cavity. This condition is a pneumothorax. Pressure within the pleural cavity is normally lower than atmospheric pressure, but when air enters the pleural cavity, the natural elasticity of the lung may cause the lung to collapse. The resulting condition is atelectasis, or a collapsed lung.

2. Because the fluid produced in pneumonia takes up space that would normally be occupied by air, the vital capacity would decrease.

Page 830

1. As skeletal muscles become more active, they generate more heat and more acidic waste products, which lower the pH of the surrounding fluid. The combination of increased temperature and reduced pH causes the hemoglobin to release more oxygen than it would otherwise.

2. Blockage of the trachea would interfere with the body's ability to gain oxygen and to eliminate carbon dioxide. Because most carbon dioxide is carried in blood as bicarbonate ion that is formed from the dissociation of carbonic acid, an inability to eliminate carbon dioxide would result in an excess of hydrogen ions, thus lowering the body's pH.

Page 837

1. The pneumotaxic centers inhibit the inspiratory center and the apneustic center. Exciting the pneumotaxic centers would result in shorter breaths and a more rapid rate of breathing.

2. Peripheral chemoreceptors are more sensitive to carbon dioxide levels than to oxygen levels. When carbon dioxide dissolves, it produces hydrogen ions, thereby lowering pH and altering cell or tissue activity.

3. Johnny's mother shouldn't worry. When Johnny holds his breath, the level of carbon dioxide in his blood will increase. This increase will in turn lead to increased stimulation of the inspiratory center, forcing Johnny to breathe again.

CHAPTER 24

Page 853

1. The mesenteries are double layers of serous membrane that support and stabilize the positions of organs in the abdominopelvic cavity and provide a route for the associated blood vessels, nerves, and lymphatic vessels.

2. Peristalsis would be more efficient in propelling intestinal contents. Segmentation is essentially a churning action that mixes intestinal contents with digestive fluids. Peristalsis consists of waves of contractions that not only mix the contents but also propel them along the digestive tract.

3. Parasympathetic stimulation increases muscle tone and activity in the digestive tract. A drug that blocks this activity would decrease the rate of peristalsis.

Page 858

1. The oral cavity is lined by a stratified squamous epithelium. This type of lining is very protective and so is located in areas of the body that receive a great deal of friction or abrasion.

2. Because the parotid salivary glands secrete salivary amylase, an enzyme that digests complex carbohydrates, damage to these glands would interfere with the digestion of carbohydrates.

3. The incisors are the type of tooth best suited for chopping (or cutting or shearing) pieces of relatively rigid food, such as raw vegetables.

Page 861

1. The muscularis externa of the esophagus is unusual because (1) it contains skeletal muscle cells along most of the length of the esophagus and (2) it is surrounded by an adventitia rather than a serosa.
2. The fauces is the opening between the oral cavity and the pharynx.
3. Swallowing (deglutition) is occurring.

Page 868

1. The larger the meal (especially in terms of protein), the more stomach acid is secreted. The hydrogen ions for the acid come from the blood that enters the stomach; therefore, the blood leaving the stomach will have fewer than normal hydrogen ions and will be decidedly alkaline—that is, have a higher pH. This phenomenon is referred to as the alkaline tide.
2. The vagus nerve contains parasympathetic motor fibers that can stimulate gastric secretions. This stimulation can occur even if food is not present in the stomach (the cephalic phase of gastric digestion). Cutting the branches of the vagus that supply the stomach would prevent this type of secretion from occurring and thereby decrease the chance of ulcer formation.

Page 881

1. The small intestine has several adaptations that increase surface area to increase its absorptive capacity. The walls of the small intestine are thrown into folds, the plicae circulares. The tissue that covers the plicae forms fingerlike projections, the villi. The cells that cover the villi have an exposed surface covered by small fingerlike projections, the microvilli. In addition, the small intestine has a very rich supply of blood vessels and lymphatic vessels, which transport the nutrients that are absorbed.
2. The cholecystokinin level in the blood would increase.
3. The hormone secretin, among other things, stimulates the pancreas to release fluid high in buffers to neutralize the chyme that enters the duodenum from the stomach. If the small intestine did not secrete secretin, the pH of the intestinal contents would be lower than normal.
4. Damage to the exocrine pancreas would most affect the digestion of fats (lipids). Enzymes for carbohydrate digestion are produced by salivary glands and the small intestine as well as by the pancreas. Enzymes for protein digestion are produced by the stomach and the small intestine as well as by the pancreas. Even though the digestion of carbohydrates and proteins would not be as complete as it is when the pancreas is functioning, some digestion would still take place. Because the pancreas is the primary source of lipases, lipid digestion would be most impaired.

Page 893

1. Chylomicrons are formed from the fats digested in a meal. A meal that is high in fat would increase the number of chylomicrons in the lacteals.
2. The removal of the stomach would interfere with the absorption of vitamin B_{12}. This vitamin requires intrinsic factor, which is produced by the parietal cells of the stomach.
3. A person can lose fluid and electrolytes faster than these substances can be replaced. This loss would result in dehydration and possibly death. Although it can be quite uncomfortable, constipation is not potentially life-threatening, because it does not interfere with any major body process that supports life. The few toxic waste products that are normally eliminated by way of the digestive system can move into the blood and can ultimately be eliminated by the kidneys.

CHAPTER 25

Page 911

1. The primary role of the TCA cycle in ATP production is to transfer electrons from substrates to coenzymes. These electrons carry energy that can then be used as an energy source for the production of ATP by the electron transport system.
2. The NADH produced by glycolysis cannot enter the mitochondria, where the enzymes of the electron transport chain are located. An intermediary in the mitochondrial membrane can, however, transfer the electrons from the NADH to a coenzyme within the mitochondria. In skeletal muscle cells, the intermediary transfers the electrons to FAD, whereas in cardiac muscle cells, a different intermediary is used that transfers the electrons to another NAD. In mitochondria, each NADH yields 3 molecules of ATP, whereas each $FADH_2$ yields just 2 molecules of ATP. The different intermediaries account for the difference in ATP yield.
3. A decrease in the level of cytoplasmic NAD would lead to a decrease in the amount of ATP production in mitochondria. The mitochondria depend on a supply of pyruvic acid from glycolysis. Glycolysis, in turn, requires NAD. A decrease in NAD would decrease the available pyruvic acid for the TCA cycle and thus would decrease overall ATP production.

Page 920

1. Pyridoxine (vitamin B_6) is an important coenzyme in deamination and transamination, the first steps in processing amino acids in the cell. A diet

deficient in this vitamin would interfere with the ability to metabolize proteins.
2. Uric acid is the product of purine degradation in the body. The macromolecules that contain purines are the nucleic acids. Elevated levels of uric acid can indicate increased breakdown of nucleic acids.
3. High-density lipoproteins are considered to be beneficial because they reduce the amount of fat (including cholesterol) in the bloodstream by transporting fat back to the liver for storage or excretion in the bile.

Page 926

1. After you have eaten a high-carbohydrate meal, you would expect increased glycogenesis (the formation of glycogen) to occur in the liver.
2. Urea is formed from byproducts of protein metabolism. During the postabsorptive state, many amino acids are being metabolized and the ammonia produced by deamination is converted to urea in the liver. Thus, the amount of urea in the blood increases.
3. Excess acetyl-CoA is likely converted into ketone bodies.

Page 931

1. An athlete in extensive training—adding muscle mass—would try to maintain a positive nitrogen balance.
2. Bile salts are necessary for the digestion and absorption of fats and fat-soluble vitamins. Vitamin A is a fat-soluble vitamin. A decrease in the amount of bile salts in the bile would result in a decreased ability to absorb vitamin A from food and hence in a vitamin A deficiency.

Page 937

1. The BMR of a pregnant woman would be higher than her own BMR when she is not pregnant, due to the increased metabolism associated with support of the fetus as well as the added effect of fetal metabolism.
2. The vasoconstriction of peripheral vessels would decrease both blood flow to the skin and the amount of heat that the body can lose. As a result, the body temperature would increase.
3. Infants have higher surface-to-volume ratios than do adults, and the temperature-regulating mechanisms of the body are not fully functional at birth. As a result, infants must expend more energy to maintain body temperature, and they get cold more easily than do healthy adults.

CHAPTER 26

Page 953

1. The renal corpuscle, proximal convoluted tubule, distal convoluted tubule, and the proximal portions of the loop of Henle and collecting duct are all in the renal cortex.
2. The pores of the glomerular capillaries will not allow substances the size of plasma proteins to pass into the capsular space, and the filtration slits of the podocyte will allow only the smallest plasma proteins to pass.
3. Damage to the juxtaglomerular apparatus of the nephron would interfere with the hormonal control of blood pressure.

Page 962

1. The reabsorption of calcium ions by the kidney is a countertransport mechanism in which sodium ions are traded for calcium ions. As more calcium is reabsorbed, more sodium enters the filtrate, increasing the sodium ion concentration of the filtrate.
2. When the plasma concentration of a substance exceeds its tubular maximum, the excess is not reabsorbed but is excreted in urine.
3. A decrease in blood pressure would reduce the blood hydrostatic pressure within the glomerulus and hence decrease the GFR.

Page 969

1. Aldosterone promotes Na^+ retention and K^+ secretion at the kidneys. In response to increased amounts of aldosterone, the K^+ concentration of urine would increase.
2. The secretion of H^+ by the nephron involves a countertransport mechanism with Na^+. If the concentration of Na^+ in the filtrate decreased, fewer hydrogen ions could be secreted. The result would be a tubular fluid with a higher pH.
3. With no juxtamedullary nephrons, there would not be a large osmotic gradient in the medulla. The kidneys would not be able to form concentrated urine.
4. When the amount of Na^+ in the tubular fluid passing through the distal convoluted tubule decreases, the cells of the macula densa are stimulated to release renin. Renin activates angiotensin, and this activation brings about an increase in blood pressure.

Page 978

1. Urea, a nitrogenous waste, is formed during the metabolism of amino acids, which are obtained by the breakdown of proteins. Thus, a high-protein diet would lead to increased urea production and an increased amount of urea in the urine. Fluid volume might also increase as a result of the need to flush the excess urea.

2. An obstruction of a ureter would interfere with the passage of urine between the renal pelvis and the urinary bladder.

3. To control the micturition reflex, you must be able to control the external urinary sphincter, a ring of skeletal muscle formed by the urogenital diaphragm, that acts as a valve.

CHAPTER 27

Page 997

1. Drinking a pitcher of distilled water would temporarily lower your blood osmolarity (osmotic concentration). Because ADH release is triggered by increases in osmolarity, a decrease in osmolarity would lead to a decrease in the level of ADH in your blood.

2. Eating a meal high in salt would temporarily increase the osmolarity of your ECF. As a result, some of the water in the ICF would shift to the ECF.

3. Being in the desert without water, you would lose fluid through perspiration, urine formation, and respiration. As a result, the osmotic concentration of your plasma (and other body fluids) would increase.

Page 1009

1. A decrease in the pH of body fluids would stimulate the respiratory centers of the medulla oblongata. The result would be an increase in the respiratory rate.

2. The kidney tubules modify the pH of the filtrate by secreting H^+ or reabsorbing HCO_3^-. The pH of the tubular fluid must be kept above about 4.5, because H^+ secretion cannot continue against a large concentration gradient. The buffers allow the filtrate to take more H^+ without decreasing the pH below the critical level.

3. In a prolonged fast, fatty acids are mobilized and large numbers of ketone bodies are formed. These molecules are acids that lower the body's pH. (The lowered pH would eventually lead to ketoacidosis.)

4. In vomiting, large amounts of stomach acid are lost from the body. This acid is formed by the parietal cells of the stomach by taking H^+ from blood. Prolonged vomiting would lead to the excessive removal of H^+ from the blood to produce the acid, thus raising the body's pH and leading to metabolic alkalosis.

CHAPTER 28

Page 1034

1. On a warm day, the cremaster muscle (as well as the dartos muscle) would be relaxed so that the scrotal sac could descend away from the warmth of the body and cool the testes.

2. The dilation of the arteries within the penis allows blood flow to increase and the vascular chambers to become engorged with blood, resulting in erection.

3. FSH is required for the production of ABP, a protein that binds testosterone and keeps a high level of that hormone available to support spermatogenesis. Low levels of FSH would lead to low levels of testosterone in the seminiferous tubules and thus a lower rate of sperm production and a low sperm count.

Page 1047

1. A blockage of the uterine tube would cause sterility.

2. The acidic pH of the vagina helps prevent bacterial, fungal, and parasitic infections in this area.

3. The functional layer of the endometrium sloughs off during menstruation.

4. The blockage of a single lactiferous sinus would not interfere with the delivery of milk to the nipple, because each breast generally has 15–20 lactiferous sinuses.

Page 1053

1. If the LH surge did not occur during an ovarian cycle, ovulation and corpus luteum formation would not occur.

2. Progesterone is responsible for the functional maturation and secretion of the endometrium. A blockage of progesterone receptors would inhibit the development of the endometrium (and would make the uterus unprepared for pregnancy).

3. A decline in the levels of estrogens and progesterone signals the beginning of the menses, the end of the uterine cycle.

Page 1059

1. An inability to contract the ischiocavernosus and bulbospongiosus muscles would interfere with a male's ability to ejaculate and to experience orgasm.

2. As the result of parasympathetic stimulation in females during sexual arousal, the erectile tissues of the clitoris engorge with blood, the secretion of cervical and vaginal glands increases, blood flow to the walls of the vagina increases, and the blood vessels in the nipples engorge.

3. At menopause, circulating estrogen levels begin to drop. Estrogen has an inhibitory effect on FSH (and on GnRH). As the level of estrogen declines, the levels of FSH rise and remain high.

CHAPTER 29

Page 1078

1. The inner cell mass of the blastocyst eventually develops into the embryo.

2. Cells of the mesoderm migrate to form the yolk sac, in which blood vessels appear. The yolk sac becomes an important site of blood cell formation. Improper development of this extraembryonic membrane would thus affect the development and function of the circulatory system.

3. After fertilization, the developing trophoblasts and, later, the placenta produce and release the hormone hCG. Sue is pregnant.

4. Placental functions include (1) supplying the developing fetus with a route for gas exchange, nutrient transfer, and waste product elimination; and (2) producing hormones that affect maternal systems.

Page 1090

1. During pregnancy, blood flow through the placenta reduces the volume of blood in the mother's systemic circuit. This reduction stimulates an increase in maternal blood volume to compensate.

2. Progesterone reduces uterine contractions. A decrease in progesterone at any time during the pregnancy could lead to uterine contractions and, in late pregnancy, labor.

3. An increase in the levels of GnRH, FSH, LH, and sex hormones in blood mark the onset of puberty.

Page 1096

1. A person who is heterozygous for curly hair would have one dominant allele and one recessive allele for that trait. The person's phenotype would be "curly hair."

2. One reason that children are not identical copies of their parents is that during meiosis, parental chromosomes are randomly distributed such that each gamete has a unique set of chromosomes. Also, mutations and the crossing-over that occurs during meiosis introduce new variations.

Answers to End-of-Chapter Questions

CHAPTER 1

Level 1 Reviewing Facts and Terms

1. g **2.** d **3.** a **4.** j **5.** b **6.** l **7.** n **8.** f **9.** h **10.** e **11.** c **12.** o **13.** k **14.** i **15.** m **16.** c **17.** b **18.** b **19.** a **20.** d **21.** a **22.** b **23.** c **24.** a **25.** d **26.** c **27.** d **28.** b **29.** c **30.** c **31.** b

Level 2 Reviewing Concepts

32. responsiveness, adaptability, growth, differentiation, reproduction, movement, metabolism, absorption, respiration, and excretion

33. (a) Anatomy is the study of internal and external structures and the physical relationships among body parts. **(b)** Physiology is the study of how organisms perform their vital functions.

34. The organization at each level determines the characteristics and functions of higher levels.
35. chemical (atoms and molecules) → cellular → tissue → organ → organ system → organism
36. Homeostatic regulation refers to adjustments in physiological systems that are responsible for the preservation of homeostasis.
37. Autoregulation occurs when the activities of a cell, tissue, organ, or organ system change automatically (that is, without neural or endocrine input) when faced with some environmental change. Extrinsic regulation results from the activities of the nervous or endocrine system. It causes more-extensive and potentially more-effective adjustments in activities.
38. Receptor: receives the stimulus; control or integration center: determines whether the stimulus has upset homeostasis, and also initiates an appropriate response; effector: responds to the stimulus according to instructions from the control center.
39. In negative feedback, a variation outside the normal range triggers an automatic response that corrects the situation. In positive feedback, the initial stimulus produces a response that exaggerates the stimulus.
40. The body is erect, and the hands are at the sides with the palms facing forward.
41. Serous membranes surround the organs. They secrete a watery fluid that reduces friction when the organs move within the internal cavities.
42. stomach (You'd cut the pericardium to access the heart.)
43. (a) dorsal cavity (b) ventral cavity (c) thoracic cavity (d) abdominopelvic cavity
44. (a) pericardium (b) pleura (c) peritoneum

Level 3 Critical Thinking and Clinical Applications

45. Calcitonin is released when calcium levels are elevated. This hormone should bring about a decrease in blood calcium levels, thus decreasing the stimulus for its own release.
46. The initial increase in blood flow to active muscles is an example of autoregulation. The increased demands for oxygen and for waste removal cause a local increase in blood flow even before responses from the nervous or endocrine system take place. Autoregulation is an immediate and automatic response to environmental change (in this case, low oxygen levels and increased wastes) that does not require the nervous or endocrine system.
47. b

CHAPTER 2

Level 1 Reviewing Facts and Terms

1. a **2.** b **3.** d **4.** a **5.** h **6.** n **7.** e **8.** k **9.** a **10.** i **11.** c
12. o **13.** f **14.** b **15.** m **16.** d **17.** g **18.** j **19.** l **20.** b **21.** c
22. d **23.** a **24.** b **25.** d **26.** a **27.** b
28. protons, neutrons, and electrons
29. Potential energy is stored energy. Kinetic energy is the energy of motion.
30. Enzymes are specialized protein catalysts that lower the activation energy for chemical reactions. Enzymes speed up chemical reactions but are not used up or changed in the process.
31. Carbon dioxide, oxygen, water, inorganic acids, inorganic bases, inorganic salts.
32. carbohydrates, lipids, proteins, and nucleic acids
33. (1) They provide a significant energy reserve. (2) They serve as insulation and thus act in heat preservation. (3) They protect organs by cushioning them.
34. (1) support (structural proteins); (2) movement (contractile proteins); (3) transport (transport proteins); (4) buffering; (5) metabolic regulation; (6) coordination and control (hormones and neurotransmitters); and (7) defense (antibodies)
35. (a) DNA: deoxyribose, phosphate, and nitrogenous bases (A, T, C, G); (b) RNA: ribose, phosphate, and nitrogenous bases (A, U, C, G)
36. (1) adenosine, (2) phosphate groups, and (3) appropriate enzymes

Level 2 Reviewing Concepts

37. c **38.** c **39.** d
40. (a) 20 (b) 20 (c) 20 (d) 20 (e) 40 (f) first level: 2; second level: 8; third level: 8; fourth level: 2
41. Nonpolar covalent bonds have an equal sharing of electrons. Polar covalent bonds have an unequal sharing of electrons. Ionic bonds lose and/or gain electrons.
42. As a result of thermal inertia, a great deal of energy must be provided to raise or lower the temperature of a volume of liquid water by 1°C. Once a large volume of water has reached a particular temperature, its temperature will change further only slowly. Because water accounts for roughly 66 percent of our body weight, thermal inertia helps stabilize body temperature.

43. Pure water, with a pH of 7, is neutral because it contains equal numbers of hydrogen ions and hydroxyl ions.
44. Buffer systems maintain pH within normal limits by removing or replacing hydrogen ions as needed.
45. The molecule is a nucleic acid. Carbohydrates and lipids do not contain nitrogen. Although both proteins and nucleic acids contain nitrogen, only nucleic acids normally contain phosphorus.

Level 3 Critical Thinking and Clinical Applications

46. The number of neutrons in an atom is equal to the atomic weight minus the atomic number. Thus sulfur has $32 - 16 = 16$ neutrons. The atomic number indicates the number of protons, so a neutral sulfur atom contains 16 protons—plus 16 electrons to balance the protons electrically. The electrons would be distributed as follows: 2 in the first level, 8 in the second level, and the remaining 6 in the third level. To achieve a full 8 electrons in the third (outermost) level, the sulfur atom can accept 2 electrons in an ionic bond or can share 2 electrons in a covalent bond. Because hydrogen atoms can share 1 electron in a covalent bond, the sulfur atom can form two covalent bonds with hydrogen, one with each of two hydrogen atoms.
47. If a person exhales large amounts of CO_2, the equilibrium will shift to the left and the level of H^+ in the blood will decrease. A decrease in the amount of H^+ will cause the pH to rise.
48. Because NaCl is an ionic compound, it will dissociate in water, producing equal numbers of sodium ions and chloride ions. Thus each mole of NaCl will produce 2 moles of ions. To prepare a solution containing 1.2 moles of ions, the student should dissolve 0.6 moles of NaCl. The molecular weight of NaCl is the sum of its atomic weights: 23 g/mol + 35.5 g/mol = 58.5 g/mol. Therefore, 0.6 mol × 58.5 g/mol = 35 g of NaCl should be added to the solution.

CHAPTER 3

Level 1 Reviewing Facts and Terms

1. c **2.** c **3.** d **4.** a **5.** a **6.** d **7.** a **8.** c **9.** d **10.** b **11.** b
12. (1) Cells are the building blocks of all plants and animals. (2) Cells are produced by the division of pre-existing cells. (3) Cells are the smallest units that perform all vital physiological functions. (4) Each cell maintains homeostasis at the cellular level. (5) Homeostasis at the tissue, organ, organ system, and organism levels reflect the combined and coordinated actions of many cells.
13. (1) physical isolation, (2) regulation of exchange with the environment, (3) sensitivity, and (4) structural support
14. Membrane proteins function as receptors, channels, carriers, enzymes, anchors, and identifiers.
15. (1) diffusion, (2) filtration, (3) carrier-mediated transport, and (4) vesicular transport
16. (1) distance (2) size of the gradient, (3) molecule size, and (4) temperature
17. (1) Osmosis is the diffusion of water molecules across a membrane. (2) Osmosis occurs across a selectively permeable membrane that is freely permeable to water but not freely permeable to solutes. (3) In osmosis, water will flow across a membrane toward the solution that has the higher concentration of solutes.
18. (a) nonmembranous: cytoskeleton, microvilli, centrioles, cilia, flagella, and ribosomes;
(b) membranous: nucleus, mitochondria, endoplasmic reticulum, Golgi apparatus, liposomes, and peroxisomes
19. (1) synthesis of proteins, carbohydrates, and lipids; (2) storage of absorbed or synthesized molecules; (3) transport of materials; and (4) detoxification of drugs or toxins
20. (1) prophase, (2) metaphase, (3) anaphase, and (4) telophase

Level 2 Reviewing Concepts

21. b **22.** a **23.** b **24.** c **25.** c
26. specificity, saturation, regulation, type of transport (active or passive), and number of ions, molecules, or solutes transported simultaneously
27. (a) Both processes use carrier proteins and exhibit saturation.
(b) Facilitated diffusion is passive, does not expend ATP, and has concentration gradient, whereas active transport is active, expends ATP, and has no concentration gradient.
28. By ejecting sodium ions from the cytosol and reclaiming potassium ions from the extracellular fluid in a ratio of 3:2, the sodium–potassium pump maintains the negative resting membrane potential.
29. Cytosol is a colloid that has a high concentration of K^+, and suspended proteins; it also contains small quantities of carbohydrates and large reserves of amino acids and lipids and may contain insoluble materials known as inclusions. Interstitial fluid has a high concentration of Na^+ and fewer suspended proteins and organic nutrients.

30. In transcription, RNA polymerase uses genetic information to assemble a strand of mRNA. In translation, ribosomes use information carried by the mRNA strand to assemble functional proteins.

31. G_0: normal cell functions; G_1: cell growth, duplication of organelles, and protein synthesis; S: DNA replication and synthesis of histones; G_2: protein synthesis

32. Prophase: Chromatin condenses and chromosomes become visible; centrioles migrate to opposite poles of the cell and spindle fibers develop; and the nuclear membrane disintegrates. Metaphase: Chromatids attach to spindle fibers and line up along the metaphase plate. Anaphase: Chromatids separate and migrate toward opposite poles of the cell. Telophase: The nuclear membrane forms; chromosomes disappear as the chromatin relaxes; and nucleoli appear.

33. (a) Cytokinesis is the cytoplasmic movement that separates two daughter cells. **(b)** It completes mitosis.

Level 3 Critical Thinking and Clinical Applications

34. This process is facilitated diffusion, which requires a carrier molecule but not cellular energy. The energy for this process is provided by the concentration gradient of the substance being transported. When all the carriers are actively involved in transport, the rate of transport reaches a saturation point.

35. Solution A must have initially had more solutes than solution B. As a result, water moved by osmosis across the selectively permeable membrane from side B to side A, increasing the fluid level on side A.

36. For the dialysis fluid to remove urea without removing water, it should not contain urea. Because urea is a small molecule, it will diffuse through the dialysis membrane from an area of high concentration (the blood) to an area of low concentration (the dialysis fluid). To prevent an associated osmotic water movement, the dialysis fluid should have an osmotic concentration similar to that of blood plasma but with higher concentrations of solutes such as bicarbonate ions or glucose. As urea diffuses into the dialysis fluid, glucose and bicarbonate diffuse into the blood; as a result, the solute concentrations remain in balance and no osmotic water movement occurs.

CHAPTER 4

Level 1 Reviewing Facts and Terms

1. j **2.** g **3.** o **4.** m **5.** a **6.** t **7.** c **8.** p **9.** b **10.** f **11.** i
12. e **13.** k **14.** h **15.** s **16.** n **17.** d **18.** l **19.** r **20.** q **21.** a
22. c **23.** d **24.** c **25.** c **26.** b **27.** d **28.** d **29.** b **30.** c **31.** c
32. a **33.** d **34.** s **35.** a **36.** b

37. (1) It consists mainly of cells rather than extracellular materials; (2) it has a free surface exposed to the external environment or to some internal chamber or passageway; (3) it is avascular—that is, it has no blood vessels; and (4) damaged or lost epithelial cells are continuously being replaced.

38. (1) It provides physical protection; (2) it controls permeability; (3) it provides sensations; and (4) it produces specialized secretions.

39. (1) tight junctions; (2) desmosomes; and (3) gap junctions

40. (1) simple; (2) stratified; and (3) transitional

41. (1) squamous; (2) cuboidal; and (3) columnar

42. (1) merocrine secretion; (2) apocrine secretion; (3) holocrine secretion

43. (1) specialized cells; (2) extracellular protein fibers; and (3) a fluid ground substance

44. (1) They establish a structural framework for the body; (2) they transport fluids in the body; (3) they provide protection; (4) they support, surround, and interconnect other tissue types; (5) they store energy; and (6) they defend the body against invasion by microorganisms.

45. Fluid connective tissues are blood and lymph. They have a liquid matrix, and the proteins are held in solution. Supporting connective tissues are cartilage and bone. They have a firm, dense matrix that contains large, insoluble fibers.

46. (1) collagen fibers; (2) reticular fibers; and (3) elastic fibers

47. (1) tendons; (2) aponeuroses; (3) elastic tissue; and (4) ligaments

48. (1) hyaline cartilage; (2) elastic cartilage; and (3) fibrocartilage

49. (1) mucous membranes; (2) serous membranes; (3) cutaneous membranes; and (4) synovial membranes

50. Transudate is a fluid formed on the surface of a serous membrane; special types are pleural fluid, pericardial fluid, peritoneal fluid. It prevents friction between the walls of cavities and surfaces of organs.

51. (1) smooth muscle tissue; (2) skeletal muscle tissue; and (3) cardiac muscle tissue

52. (1) neurons, which transmit electrical impulses in the form of changes in the transmembrane potential; and (2) neuroglia, which comprise several kinds of supporting cells and play a role in providing nutrients to neurons

Level 2 Reviewing Concepts

53. a

54. Substances are moved over the epithelial surface by the synchronized beating of cilia, like a continuously moving escalator. The ciliated epithelium that lines the respiratory tract moves mucus, which has trapped dust particles from inhaled air, up from the lungs and toward the throat, where it can then be swallowed and eventually eliminated from the body.

55. Holocrine secretion destroys the gland cell. The entire cell becomes packed with secretory products and then bursts, releasing the secretion but killing the cell. Gland cells must be replaced by the division of stem cells.

56. Exocrine secretions are secreted onto a surface or outward through a duct. Endocrine secretions are secreted by ductless glands into surrounding tissues. The secretions are called hormones, which usually diffuse into the bloodstream for distribution to other parts of the body.

57. Tight junctions block the passage of water or solutes between the cells. In the digestive system, these junctions keep enzymes, acids, and waste products from damaging delicate underlying tissues.

58. Intercellular connections called desmosomes are most abundant in superficial layers of skin. Desmosomes create links that lock cells together, causing dead skin cells to be shed in thick sheets.

59. The extensive connections between cells formed by cell junctions, intercellular cement, and physical interlocking hold skin cells together and can deny access to chemicals or pathogens that cover their free surfaces. If the skin is damaged and the connections are broken, infection can easily occur.

60. In infants and young children, the adipose tissue in specific areas of the upper body is different from most of the adipose tissue in adults. This tissue, called brown fat, is highly vascularized, and the fat cells contain many mitochondria. When stimulated by the nervous system, lipid breakdown accelerates and the cells do not capture the released energy; the energy radiates into the surrounding tissues as heat. That heat warms the circulating blood, which distributes it throughout the body. In this way, an infant can accelerate metabolic heat generation by 100 percent very quickly.

61. Adipocytes (fat cells) are metabolically active cells. Their lipids are continuously being broken down and replaced. During a weight-loss program, nutrients are scarce, causing the fat cells to decrease in size. The cells are not killed but merely reduced in size. Once the individual is off the weight-loss program, these cells increase in size and the lost weight will be regained in the same locations.

62. Cartilage is avascular, because chondrocytes produce a chemical that discourages the formation of blood vessels. This property makes nutrient and oxygen delivery difficult. In adults, interstitial and appositional growth do not occur, so there is no division of chondrocytes or differentiation of the inner cells of the perichondrium into chondrocytes. Both the avascularity and the lack of division and differentiation contribute to the poor healing of cartilage.

63. Cutaneous membranes are thick, relatively waterproof, and usually dry.

64. Cardiac muscle fibers are incapable of dividing, because they do not contain satellite cells. Therefore, the regeneration of damaged or diseased cardiac muscle tissue is not possible.

65. Elastic ligaments are found along the spinal column, where they are important in stabilizing the position of the vertebrae. It is the elasticity of the elastic ligaments that allows the spine to bend and straighten easily.

Level 3 Critical Thinking and Clinical Applications

66. Because apocrine secretions are released by pinching off a portion of the secreting cell, you could test for the presence of cell membranes, specifically for the phospholipids in cell membranes. Merocrine secretions do not contain a portion of the secreting cell, so they would lack membrane constituents.

67. The respiratory passages are lined by ciliated pseudostratified columnar epithelium. This type of epithelium contains many goblet cells that produce mucus. The mucus traps debris and foreign material that is moved by the beating cilia to the pharynx to be swallowed. Chronic smoking initially paralyzes the cilia, resulting in a buildup of mucus in the airways. The hot air and particulates sear and burn off the cilia. The epithelium is also damaged by the heat and responds by producing even more mucus; the mucus cannot be moved due to the lack of cilia. The only way that the respiratory system can clear the debris and mucus is with forceful bursts of air: coughing.

68. Because animal intestine is modified for absorption, you would look for a slide that shows a single layer of epithelium lining the cavity. The cells would be cuboidal or columnar and would likely have microvilli on the surface to increase surface area. Because the esophagus receives undigested food, it would have a stratified epithelium consisting of squamous cells to protect it.

69. Step 1: Check for striations. (If striations are present, the choices are skeletal muscle or cardiac muscle. If striations are absent, the tissue is smooth muscle.) Step 2: Check for the presence of intercalated discs. (If these discs are present, the tissue is cardiac muscle. If they are absent, it is skeletal muscle.)

CHAPTER 5

Level 1 Reviewing Facts and Terms

1. a **2.** c **3.** d **4.** b **5.** c **6.** a **7.** d **8.** b **9.** b **10.** a **11.** d **12.** b **13.** a **14.** a

15. Epidermal cell division occurs in the stratum germinativum.
16. Keratin production occurs in keratinocytes of the stratum granulosum.
17. carotene and melanin
18. Epidermal growth factor promotes the divisions of germinal cells in the stratum germinativum and stratum spinosum. It also accelerates the production of keratin in differentiating epidermal cells and stimulates both epidermal development and epidermal repair after injury and synthetic activity and secretion by epithelial cells.
19. (1) papillary layer, which consists of loose connective tissue and contains capillaries and sensory neurons; (2) reticular layer, which consists of dense irregular connective tissue and bundles of collagen fibers. Both layers contain networks of blood vessels, lymphatic vessels, and nerve fibers.
20. (1) internal root sheath, which surrounds the root; (2) external root sheath, which extends from the skin surface to hair matrix; (3) glassy membrane, the thickened basement membrane
21. (1) apocrine sweat glands; (2) merocrine sweat glands

Level 2 Reviewing Concepts

22. Insensible perspiration is water loss via evaporation through the stratum corneum of the skin. Sensible perspiration is produced by active sweat glands.
23. Substances that are fat soluble pass through the permeability barrier easily, because the barrier is composed primarily of lipids surrounding the epidermal cells. Water-soluble drugs are hydrophobic and thus don't penetrate the permeability barrier easily.
24. A tan is a result of the synthesis of melanin in the skin. Melanin helps prevent skin damage by absorbing UV radiation before it reaches the deep layers of the epidermis and dermis. Within the epidermal cells, melanin concentrates around the outer wall of the nucleus, so it absorbs the UV light before it can damage nuclear DNA.
25. Lines of cleavage represent the orientation of the collagen and elastin fibers of the dermis, an orientation that resists normal stresses on the skin. Incisions along the lines of cleavage are more likely to remain closed, and thus will heal more quickly, than would incisions not along lines of cleavage.
26. Inflammation of the skin is painful because sensory receptors are abundant in the skin. Changes associated with swelling stimulate the sensory receptors and bare nerve endings, resulting in a painful sensation.
27. The subcutaneous layer is not highly vascular, and it does not contain major organs; thus the potential for tissue damage is reduced.
28. The cells that produce the nails can be affected by conditions that alter body metabolism; thus the nails hint at possible disease conditions.
29. Scabs temporarily restore epidermal integrity, restricting further entry of microorganisms.
30. The dermis becomes thinner and the elastic fiber network decreases in size, weakening the integument and causing a loss of resilience.

Level 3 Critical Thinking and Clinical Applications

31. The child probably has a fondness for vegetables that are high in carotene, such as sweet potatoes, squash, and carrots. It is not uncommon for parents to feed babies what they will eat best. If the child consumes large amounts of carotene, the yellow-orange pigment will be stored in the skin, producing a yellow-orange skin color.
32. Most elderly people have poor blood flow to the skin. Thus, temperature receptors in the skin do not sense as much warmth as when there is a rich blood supply. The sensory information is relayed to the brain, and the brain interprets the temperature as cool or cold.
33. (a) Ultraviolet radiation in sunlight converts a cholesterol-related steroid into vitamin D_3, or cholecalciferol. This compound is then converted to calcitriol, which is essential for normal calcium and phosphorus absorption by the small intestine. Calcium and phosphorus are necessary for normal bone maintenance and growth. (b) Milk is routinely fortified with cholecalciferol, normally identified as "vitamin D," which is easily absorbed by the intestines.
34. Sweating from merocrine glands is precisely regulated, and one influencing factor is emotional state. Presumably, a person who is lying is nervous and sweats noticeably; this sweating can be measured by the lie detector machine.

CHAPTER 6

Level 1 Reviewing Facts and Terms

1. b **2.** a **3.** a **4.** c **5.** d **6.** b **7.** b **8.** c **9.** a **10.** d
11. (1) support; (2) storage of minerals and lipids; (3) blood cell production; (4) protection; and (5) leverage
12. (1) osteocytes; (2) osteoblasts; (3) osteoclasts; and (4) osteoprogenitor cells
13. Osteoblasts are responsible for the production of new bone (osteogenesis). Osteoclasts contain lysosomes that secrete enzymes and acids that dissolve the bony matrix and release the stored minerals of calcium and phosphate (osteolysis).
14. (1) It isolates the bone from surrounding tissues; (2) provides a route for the circulatory and nervous supply; and (3) actively participates in bone growth and repair.
15. In intramembranous ossification, bone develops from mesenchyme or fibrous connective tissue. In endochondral ossification, bone develops from a cartilage model.
16. (1) the nutrient artery and vein; (2) metaphyseal vessels; and (3) periosteal vessels
17. (a) calcium salts and phosphate salts and vitamins A, C, and D_3; (b) calcitriol, growth hormone, thyroxine, estrogens (in females) or androgens (in males,) calcitonin, and parathyroid hormone
18. (1) the bones; (2) the intestinal tract; and (3) the kidneys
19. It stimulates osteoclast activity, increases the rate of intestinal absorption, and decreases the rate of excretion of calcium ions.
20. It inhibits osteoclast activity, decreases the rate of calcium absorption, and increases the rate of excretion of calcium ions.

Level 2 Reviewing Concepts

21. Nutrients reach the osteocytes by diffusion along canaliculi that open onto the surface of the trabeculae.
22. The osteons are parallel to the long axis of the shaft, which does not bend when forces are applied to either end. Stresses or impacts to the side of the shaft can lead to a fracture.
23. Inactivity in unstressed bones leads to the removal of calcium salts. Up to one-third of the bone mass can be lost in this manner, causing the bones to become thin and brittle.
24. The digestive and urinary systems (kidneys) play important roles in providing the calcium and phosphate minerals needed for bone growth. In return, the skeleton provides protection and acts as a reserve of calcium, phosphate, and other minerals that can compensate for changes in the dietary supplies of these ions.
25. The chondrocytes of the epiphyseal cartilage enlarge and divide, increasing the thickness of the cartilage. On the shaft side, the chondrocytes become ossified, "chasing" the expanding epiphyseal cartilage away from the shaft.
26. At puberty, the stimulation of ossification by the sex hormones causes the epiphyseal cartilages to gets narrower and narrower until they virtually disappear.
27. The thyroid gland secretes calcitonin, which stimulates calcium excretion and inhibits calcium absorption and osteoclast activity. Without calcitonin, blood calcium levels might become abnormally high.
28. When a bone fracture is repaired, the bone tends to be stronger and thicker than normal at the fracture site.
29. People get shorter with age due in part to a loss of bone density in the vertebrae.
30. Bone markings give clues as to the size, weight, gender, and general appearance of an individual.
31. Elevations and projections on bones are sites of attachment for tendons and ligaments.

Level 3 Critical Thinking and Clinical Applications

32. The fracture might have damaged the epiphyseal cartilage in Sally's right leg. Even though the bone healed properly, the damaged leg did not produce as much cartilage as did the undamaged leg. The result would be a shorter bone on the side of the injury.
33. Ossification can begin in the fetus as early as 6 weeks. By 8–10 weeks, the process is well under way and requires large amounts of calcium and phosphate to make the new bone. If the mother's diet does not supply sufficient amounts of calcium for the fetus, then hormonal regulation will cause minerals to be taken from the mother's bones for use by her fetus. Because Sherry has a diet deficient in calcium, the demineralization of her bone is supplying the calcium needs of her fetus. As a result, Sherry has weak bones that fractured easily.
34. A person deficient in vitamin D_3 would not be able to absorb calcium effectively from the digestive tract, leading to a shortage of calcium in the blood. To maintain homeostasis, the decrease in blood calcium would trigger the release of PTH. The PTH, in turn, would stimulate osteoclasts to release enough calcium from the bone to maintain proper calcium levels

in the blood. Levels of calcitonin would probably decrease, because this hormone lowers blood calcium levels and would aggravate the situation caused by the vitamin D_3 deficiency.

35. In kidney failure, the kidneys are not able to retrieve important ions that have been filtered from the blood, including calcium. As a result, calcium would be lost from the body and blood levels of calcium would drop. The reduced levels of calcium in the blood would trigger higher levels of PTH, which, in turn, would trigger an increase in the release of minerals from bone by the osteoclasts. Over time, this mineral release would result in demineralized bones similar to those of an individual with osteoporosis.

CHAPTER 7

Level 1 Reviewing Facts and Terms

1. i **2.** e **3.** l **4.** a **5.** g **6.** j **7.** b **8.** k **9.** d **10.** h **11.** c **12.** f **13.** b **14.** d **15.** a **16.** a **17.** d **18.** c **19.** a **20.** d **21.** a **22.** b **23.** c **24.** a **25.** d **26.** c
27. (1) lacrimal bone; (2) nasal bone; (3) maxillary bone; (4) zygomatic bone; and (5) mandible
28. (1) lambdoidal suture; (2) coronal suture; (3) sagittal suture; and (4) squamous suture
29. The vomer forms a bony nasal septum that separates the right and left nasal cavities and separates the external nares anteriorly in the fleshy portion of the nose.
30. the frontal, sphenoid, ethmoid, and maxillary bones
31. It is the only bone in the body that does not articulate with another bone.
32. Because the fontanels are made of fibrous connective tissue, they are quite flexible; they allow the skull to be distorted without damage. Such distortion normally occurs during delivery and eases the passage of the infant along the birth canal.
33. cervical (7), thoracic (12), lumbar (5), sacral (1), and coccygeal (1)
34. (1) cervical curve; (2) thoracic curve; (3) lumbar curve; and (4) sacral curve
35. (1) It protects the heart, lungs, thymus, and other structures in the thoracic cavity. (2) It serves as an attachment point for muscles involved with respiration, the position of the vertebral column, and movements of the pectoral girdle and upper limb.
36. The upper seven pairs are connected directly to the sternum by the costal cartilages. Ribs 8–12 are called false ribs because they do not attach directly to the sternum.

Level 2 Reviewing Concepts

37. d
38. The petrous portion of the temporal bone encloses the structures of the inner ear. The middle ear is located in the tympanic cavity within the petrous portion. The external auditory canal ends at the tympanic membrane, which leads to the inner ear. Mastoid air cells within the mastoid process are connected to the tympanic cavity.
39. The pituitary gland rests in the sella turcica of the sphenoid bone.
40. Human skulls require less effort by muscles and ligaments to balance the head on the vertebral column. The neck muscles and ligamentum nuchae in humans are weaker than those of comparable structures in other animals. Due to the weakness of these muscles and ligaments, the human bony processes and lines of the posterior skull are not as well developed as are those in other animals. The location of the human foramen magnum facilitates the balancing of the skull above the vertebral column and allows an upright posture. The presence of a cervical curvature allows for support of the weight of the head.
41. Because it is accessible, the body of the sternum is often used as a site for taking red bone marrow samples. The xiphoid process is used as a landmark in CPR; this part of the sternum is easily broken, because its only attachment is superior, to the sternum. A broken xiphoid process can cause injury to nearby organs.
42. The lumbar vertebrae have massive bodies and carry a lot of weight; these factors contribute to the rupturing of a disc. The cervical vertebrae are more delicate and have small bodies; these factors increase the possibility of dislocations and fractures in the cervical vertebrae.
43. A "runny" nose is caused by an increase in the mucus draining from the membranes of the nasal cavity. The increased production of mucus is a response to many kinds of stimuli, including irritants and bacterial or viral infections. Its purpose is to flush out the offending substances or microorganisms from the nasal cavity.
44. Keeping your back straight keeps the weight aligned along the axis of your vertebral column, where it can be transferred to your lower limbs. Bending your back would strain the muscles and ligaments of the back, increasing the risk of injury.

Level 3 Critical Thinking and Clinical Applications

45. The virus could have been inhaled through the nose and passed into the cranium by way of the cribriform plate of the ethmoid bone.
46. Joe has lordosis. The extra weight that he is carrying in his abdomen shifts his center of gravity anteriorly. To compensate for this shift, the muscles of the spine exaggerate the lumbar curvature to prevent Joe from falling forward. The situation is similar to that occurring in women in late pregnancy.
47. On the basis of this evidence, you would believe the defense. If the child died as the result of violent shaking, you would expect the same type of damage as in whiplash, a dislocation that typically involves the superior cervical vertebrae. The damage described by the medical examiner is consistent with a child having his head stuck and pulling or twisting to one side to try to free himself. If the child pulled or twisted hard enough, he could fracture the articular processes, leading to a damaged spinal cord.
48. The large bones of a child's cranium are not yet fused; they are connected by fontanels, areas of connective tissue. By examining the bones, the archaeologist could readily see if sutures had formed. By knowing approximately how long it takes for the various fontanels to close and by determining their sizes, she could estimate the ages of the bones.

CHAPTER 8

Level 1 Reviewing Facts and Terms

1. b **2.** b **3.** a **4.** a **5.** d **6.** b **7.** d **8.** b **9.** d **10.** d **11.** b **12.** a **13.** c **14.** d
15. the clavicle (collar bone) and the scapula (shoulder blade)
16. the acromion and the coracoid process
17. the humerus (arm) and the ulna and radius (forearm)
18. extension and flexion
19. (1) scaphoid bone; (2) lunate bone; (3) triquetrum; (4) pisiform bone; (5) trapezium; (6) trapezoid bone; (7) capitate bone; and (8) hamate bone
20. femur (thigh), tibia and fibula (leg), tarsal bones (ankle), metatarsal bones (foot), and phalanges (toes)
21. ischium, ilium, and pubis
22. The false pelvis consists of the expanded, blade-like portions of each ilium superior to the iliopectineal line. Structures inferior to that line—including the inferior portions of each ilium, both pubic bones, the ischia, the sacrum, and the coccyx—form the true pelvis.
23. (1) talus; (2) calcaneus; (3) cuboid bone; (4) navicular bone; and (5–7) three cuneiform bones
24. The pollex is the thumb, whereas the hallux is the great toe.

Level 2 Reviewing Concepts

25. The clavicles are small and fragile, so fractures are quite common. The position of the clavicles also makes them vulnerable to injury and damage.
26. The pelvic girdle consists of the ossa coxae. The pelvis is a composite structure; it consists of the ossa coxae of the appendicular skeleton and the sacrum and coccyx of the axial skeleton.
27. The slender fibula parallels the tibia of the lower leg and provides an important site for muscle attachment. It does not help in transferring weight to the ankle and foot, however, because it is excluded from the knee joint.
28. tendons
29. The clavicles are small and fragile, so they are easy to break. Once this part of the pectoral girdle is broken, the assailant would no longer have efficient use of the arms.
30. The pelvic girdle transmits the weight of the body to the legs. The pectoral girdle doesn't support the body's weight and so its bones needn't be as massive as those of the pelvic girdle.

Level 3 Critical Thinking and Clinical Applications

31. You could feel the inferior iliac notch, which is at the same level as the head of the femur. You could feel the medial malleolus of the tibia at the ankle.
32. Fred probably dislocated his shoulder, which is quite a common injury due to the weak nature of the glenohumeral joint.
33. Cindy's pelvis might not be broad enough, or the angle inferior to the pubic symphysis might not be large enough, for Cindy to deliver the baby by natural childbirth.

CHAPTER 9

Level 1 Reviewing Facts and Terms

1. d **2.** a **3.** c **4.** c **5.** d **6.** a **7.** b **8.** c **9.** b **10.** b **11.** d **12.** c **13.** c **14.** a **15.** b **16.** a **17.** d **18.** a

Level 2 Reviewing Concepts

19. d **20.** c

21. Menisci may subdivide a synovial cavity, channel the flow of synovial fluid, and allow variations in the shape of the articular surfaces.

22. A joint cannot be both highly mobile and very strong. The shoulder (glenohumeral) joint demonstrates the principle that strength and stability must be sacrificed to obtain mobility; it permits the greatest range of motion of any joint in the body. The relatively loose, oversized, poorly reinforced articular capsule lessens the stability of the joint but allows a great range of motion.

23. Articular cartilages do not have perichondrium, and their matrix contains more water than do other cartilages.

24. The slight mobility of the pubic symphysis facilitates childbirth, by allowing the pelvis to spread easily while the baby moves through the birth canal.

25. Height decreases during adulthood in part as a result of osteoporosis in the vertebrae and in part as a result of the decline in water content of the nucleus pulposus region of the intervertebral discs.

26. Contraction of the biceps brachii muscle causes flexion of the forearm at the elbow joint.

27. When the knee is "locked," it is more stable than when it is not locked. Locking occurs when the knee is fully extended and the fibular collateral ligaments are tight.

Level 3 Critical Thinking and Clinical Applications

28. Dave sprained his ankle. This condition occurs when ligaments are stretched to the point at which some of the collagen fibers are torn. Stretched ligaments in joints can cause the release of synovial fluid, which results in swelling and pain.

29. Bob has probably damaged his menisci, which act as cushions to absorb the force of movements such as jumping. When Bob landed off-balance on his right knee, the excessive force could have pushed the menisci out of place. In turn, the synovial membranes and/or the ligaments were damaged, resulting in swelling.

30. Her pain is probably caused by bursitis, which can result from repetitive motion, infection, trauma, chemical irritation, or pressure over the joint. Given the location of the pain, her pain probably results from the repetitive motion of practicing pitches.

CHAPTER 10

Level 1 Reviewing Facts and Terms

1. d **2.** c **3.** b **4.** c **5.** d **6.** d **7.** c **8.** d

9. (1) skeletal muscle; (2) cardiac muscle; and (3) smooth muscle

10. (1) producing skeletal movement; (2) maintaining posture and body position; (3) supporting soft tissue; (4) guarding entrances and exits; and (5) maintaining body temperature

11. (1) epimysium: surrounds entire muscle; (2) perimysium: surrounds muscle bundles (fascicles); and (3) endomysium: surrounds skeletal muscle fibers

12. all the muscle fibers controlled by a single motor neuron

13. transverse (T) tubules

14. tropomyosin and troponin

15. (1) exposure of active sites; (2) attachment of cross-bridges; (3) pivoting of myosin head (power stroke); (4) detachment of cross-bridges; and (5) activation of myosin heads (cocking)

16. (1) the frequency of motor unit stimulation and (2) the number of motor units involved

17. intrafusal fibers of the muscle spindle

18. ATP, creatine phosphate, and glycogen

19. (1) aerobic metabolism and (2) glycolysis

20. (1) substantial intracellular energy reserves; (2) a normal circulatory supply; and (3) a normal oxygen concentration in the blood

21. calmodulin

Level 2 Reviewing Concepts

22. d **23.** b **24.** d **25.** a

26. Acetylcholine released by the motor neuron at the neuromuscular junction changes the permeability of the cell membrane at the motor end plate. The permeability change allows the influx of positive charge, which triggers an action potential. The action potential spreads across the entire surface of the muscle fiber and into the interior via T tubules. The cytoplasmic concentration of calcium ions (released from the sarcoplasmic reticulum) increases, triggering the start of a contraction. The contraction ends when AChE removes the ACh from the synaptic cleft and motor end plate.

27. Skeletal muscle cells are large and have a high metabolic turnover. The genes contained in their nuclei direct the production of the enzymes and structural proteins required for normal contraction. The presence of multiple gene copies speeds up this process.

28. In an initial latent period (after the stimulus arrives and before the tension begins to increase), the action potential in the muscle is generated and triggers the release of calcium ions from the SR. In the contraction phase, the calcium binds to troponin (cross-bridges form) and the tension begins to increase. In the relaxation phase, tension drops because cross-bridges have detached and because calcium levels have fallen; the active sites are once again covered by the troponin–tropomyosin complex.

29. (1) O_2 for aerobic respiration is consumed by liver cells, which need to make a great deal of ATP to convert lactic acid to glucose. (2) O_2 for aerobic respiration is consumed by skeletal muscle fibers as they restore ATP, creatine phosphate, and glycogen concentrations to their former levels. (3) The normal O_2 concentration in blood and peripheral tissues is replenished.

30. Skeletal muscle fibers become smaller in diameter; skeletal muscles become smaller and less elastic; tolerance for exercise decreases; and the ability to recover from muscle injuries decreases.

31. Cardiac muscle fibers are mechanically, chemically, and electrically connected to one another, causing the entire tissue to resemble a single, enormous muscle fiber that performs as a functional syncytium.

32. Cardiac muscle tissue has the property of automaticity. The contractions are timed by pacemaker cells, specialized cardiac muscle fibers.

33. At the motor end plates of neuromuscular junctions, the blocking of the binding process by atracurium would inhibit the ability of the muscle to contract. Thus the muscle would remain relaxed. Atracurium could be useful if administered before surgery.

34. They will atrophy. Even a temporary reduction in muscle use, such as when you wear a cast, can cause some degree of atrophy. Physical therapy can offset this effect in cases of temporary inactivity.

35. In rigor mortis, the membranes of the dead cells are no longer selectively permeable; the SR is no longer able to retain calcium ions. As calcium ions enter the sarcoplasm, a sustained contraction develops. The muscles lock in the contracted position, making the body extremely stiff. Contraction persists because the dead muscle cells can no longer make ATP, which is necessary for cross-bridge detachment from the active sites. Rigor mortis begins a few hours after death and lasts 15–25 hours, until the lysosomal enzymes released by autolysis break down the myofilaments.

36. Visceral smooth muscle fibers are interconnected by gap junctions. When one fiber fires an action potential, it is conducted quickly to neighboring cells. The action potential in uninnervated smooth muscle fibers may be the result of exposure to chemicals, hormones, oxygen, carbon dioxide, stretching, or irritation.

37. Powerful, gross movements; each time the motor unit is activated, all the fibers will contract maximally. More-powerful contractions result from the activity of large motor units than from the activity of small motor units. Fine movements require numerous small motor units that can be individually activated.

Level 3 Critical Thinking and Clinical Applications

38. Because organophosphates block the action of acetylcholinesterase, the acetylcholine released into the synaptic cleft would not be removed. It would continue to stimulate the motor end plate, causing a state of persistent contraction (spastic paralysis). If the muscles of respiration were affected (which is likely), Ivan would die of suffocation. Prior to death, the most obvious symptoms would be uncontrolled tetanic contractions of the skeletal muscles.

39. This is a case of competitive inhibition; you would want to increase the level of acetylcholine at the synapse. You would want to use the drug that inhibits acetylcholinesterase, allowing the acetylcholine concentration at the synapse to increase. In turn, the likelihood that the acetylcholine would bind its receptor and lead to muscle stimulation would increase, thus alleviating the symptoms. Blocking acetylcholine would aggravate the symptoms and make the condition worse.

40. If muscle damage had occurred, the doctor would find enzymes and other molecules that normally are present only inside the skeletal muscle cells. These molecules would be in Mary's blood if the cell membranes were broken, allowing the cell contents to escape. Particularly significant would be enzymes associated with energy metabolism, such as CPK.

CHAPTER 11

Level 1 Reviewing Facts and Terms

1. b **2.** d **3.** d **4.** d **5.** c **6.** d **7.** a **8.** b **9.** a **10.** b **11.** b **12.** b **13.** c **14.** a **15.** b **16.** d **17.** a

18. (1) the anatomical arrangement of the muscle fibers; and (2) the type of muscle attachment to the skeleton

19. (1) parallel; (2) convergent; (3) pennate; and (4) circular

20. unipennate: All muscle cells are on the same side of the tendon; bipennate: Muscle fibers are on both sides of the tendon; multipennate: The tendon branches within the muscles.

21. In first-class levers, the fulcrum lies between the applied force and the resistance. In second-class levers, the resistance lies between the applied force and the fulcrum. In third-class levers, the force is applied between the resistance and the fulcrum.

22. (1) agonist (prime mover): a muscle whose contraction is chiefly responsible for producing a particular movement; (2) antagonist: a muscle whose actions oppose that of an agonist; and (3) synergist: a muscle that assists the agonist in performing an action and in preventing movement at a joint, thereby stabilizing the origin of the agonist

23. The axial musculature positions the head and vertebral column and moves the rib cage, assisting in the movements that make breathing possible. The appendicular musculature stabilizes or moves the components of the appendicular skeleton.

24. (1) muscles of the head and neck; (2) muscles of the vertebral column; (3) oblique and rectus muscles; and (4) muscles of the pelvic floor

25. the masseter, temporalis, and pterygoid muscles

26. (1) supporting the organs of the pelvic cavity; (2) flexion at joints of the sacrum and coccyx; and (3) controlling movement of materials through the urethra and anus

27. (1) muscles that position the pectoral girdle; (2) muscles that move the arm; (3) muscles that move the forearm and hand; and (4) muscles that move the hand and fingers

28. (1) muscles that move the thigh; (2) muscles that move the leg; and (3) muscles that move the foot and toes

29. (1) the deltoid muscle (shoulder); (2) the gluteus medius and maximus; and (3) the vastus lateralis of the thigh. Injections into these thick muscle masses are preferred sites because the likelihood of piercing a blood vessel or nerve is minimized there.

Level 2 Reviewing Concepts

30. b 31. b 32. b

33. The vertebral column does not need a massive series of flexors, because many of the large trunk muscles flex the vertebral column when they contract. In addition, most of the body weight lies anterior to the vertebral column and gravity tends to flex the intervertebral joints.

34. The urethral and anal sphincter muscles are usually constricted, preventing the passage of urine and feces. The muscles are innervated by nerves that are under conscious control, so the sphincters normally relax only to allow the passage of wastes when the individual chooses to.

35. In a convergent muscle, the direction of pull can be changed by stimulating only one group of muscle cells at any one time. When all the fibers contract at once, they do not pull as hard on the tendon as would a parallel muscle of the same size. The reason is that the muscle fibers on opposite sides of the tendon are pulling in different directions rather than working together.

36. A pennate muscle contains more muscle fibers than does a parallel muscle of the same size. A muscle that has more muscle fibers also has more myofibrils and sarcomeres, resulting in a contraction that generates more tension.

37. Lifting heavy objects becomes easier as the elbow approaches a 90° angle. As you decrease the angle at or near full flexion, tension production declines, so movement becomes more difficult.

38. The muscles of mastication will be affected; therefore, chewing will be inhibited.

39. flexion at the knee and extension at the hip

Level 3 Critical Thinking and Clinical Applications

40. Contraction of the frontalis muscle would wrinkle Mary's brow; contraction of the procerus muscle would flare her nostrils. Contraction of the levator labii muscle on the right side would raise the right side of her lip, as in sneering. Mary is not happy to see Jill.

41. Jeff should do squatting exercises. If he places a weight on his shoulders as he does these, he will notice better results, because the quadriceps muscles would be working against a greater resistance.

42. As the child grasps the ice cream cone, he uses flexor muscles, principally the flexor digitorum superficialis, flexor digitorum profundus, and flexor pollicis longus muscles. Opening his mouth requires the contraction of the platysma and the relaxation of the masseter, temporalis, and pterygoid muscles. The licking action would involve the genioglossus muscle (to depress and protract the tongue) and the palatoglossus muscle (to elevate the tongue).

CHAPTER 12

Level 1 Reviewing Facts and Terms

1. c 2. b 3. a 4. d 5. c 6. c 7. d 8. c 9. c 10. b 11. a
12. b 13. a 14. b 15. a 16. a 17. c 18. b 19. d 20. a

21. (a) brain and spinal cord (b) all other nerve fibers, divided between the efferent division (which consists of the somatic nervous system and the autonomic nervous system) and the afferent division (which consists of receptors and sensory neurons)

22. (1) neurons: transmit nerve impulses; and (2) neuroglia: act as supporting cells

23. (a) astrocytes (b) microglia

24. (1) satellite cells and (2) Schwann cells

25. (1) anaxonic; (2) unipolar; (3) bipolar; and (4) multipolar

26. (1) sensory neurons: transmit impulses from the PNS to the CNS; (2) motor neurons: transmit impulses from the CNS to peripheral effectors; and (3) interneurons: analyze sensory inputs and coordination of motor outputs

27. (1) the conduction of one or more action potentials along an axon and (2) the chemical transmission of signals across one or more synapses

28. Voltage-regulated channels open or close in response to changes in the transmembrane potential. Chemically regulated channels open or close when they bind specific extracellular chemicals. Mechanically regulated channels open or close in response to physical distortion of the membrane surface.

29. (1) The change in the transmembrane potential decreases with distance. (2) The graded potential spreads passively due to local currents. (3) The graded potential may involve either depolarization or hyperpolarization. (4) The stronger the stimulus, the greater the change in the transmembrane potential and the larger the area affected.

30. The properties of the action potential are independent of the relative strength of the depolarizing stimulus.

31. The membrane depolarizes to threshold. Next, voltage-regulated sodium channels are activated, and the membrane rapidly depolarizes. These sodium channels are then inactivated, and potassium channels are activated. Finally, normal permeability returns. The voltage-regulated sodium channels become activated once the repolarization is complete; the voltage-regulated potassium channels begin closing as the transmembrane potential reaches threshold levels.

32. An action potential arrives, and the synaptic knob depolarizes. Next, extracellular calcium ions enter the synaptic knob and trigger the exocytosis of ACh. The ACh binds, and the postsynaptic membrane depolarizes. Finally, ACh is removed by AChE.

33. pH changes, ionic changes, and temperature changes

Level 2 Reviewing Concepts

34. b 35. d 36. c 37. b 38. a 39. a 40. c 41. a

42. Neurons lack a centrosome. Neural stem cells are not present in the CNS, and CNS neurons usually lose their centrioles during differentiation, making cell division impossible.

43. Collaterals enable a single neuron to innervate several other cells—thus they allow divergence to occur.

44. Axoplasmic transport is the movement of products that are synthesized in the cell body out to the synaptic knobs. Retrograde flow is the movement of materials toward the cell body.

45. In continuous conduction, which occurs in unmyelinated axons, an action potential appears to move across the membrane surface in a series of tiny steps. In saltatory conduction, which occurs in myelinated axons, only the nodes along the axon can respond to a depolarizing stimulus.

46. Action potentials last longer in muscle fibers than in nerve fibers. Muscle fibers conduct action potentials at a slower speed.

47. Neurotransmitters are chemicals that are often classified as excitatory or inhibitory on the basis of their effects on postsynaptic membranes. Neuromodulators are compounds that influence neurotransmitter release or the postsynaptic cell's response to the neurotransmitter.

48. Temporal summation is the addition of stimuli that arrive in rapid succession. It occurs at a single synapse and is active repeatedly. Spatial summation occurs when simultaneous stimuli have a cumulative effect on the transmembrane potential. It involves multiple synapses that are active simultaneously.

49. Divergence permits the broad distribution of a specific input; in convergence, several neurons synapse on the postsynaptic neuron, permitting the variable control of motor neurons by providing a mechanism for their voluntary and involuntary control.

50. the synthesis, release, and recycling of neurotransmitter molecules; the movement of materials to and from the cell body via axoplasmic transport; and the recovery from action potentials

51. Electrical synapses occur between neighboring cells that are connected by gap junctions. Gap junctions allow for the direct passage of electrical current through a continuous cytoplasm, so activity in electrically

coupled cells is nearly simultaneous. Chemical synapses involve the release, diffusion, and binding of a neurotransmitter in response to an action potential in the presynaptic cell, before electrical activity can be generated in the postsynaptic cell. Chemical synaptic transmission thus takes longer than electrical synaptic transmission.

52. During long-term strenuous physical activity, endorphins are released in the brain. Endorphins are structurally similar to morphine and relieve pain by suppressing the release of the neurotransmitter Substance P from pain neurons.

53. In MS, the myelin sheath is destroyed. Action potentials travel slower in unmyelinated fibers than in myelinated fibers. The destruction of myelin slows the time it takes for motor neurons to communicate with their effector muscles. This delay results in varying degrees of uncoordinated muscle activity and paralysis. Cumulative sensory and motor losses may eventually lead to generalized sensory deficiencies and muscular paralysis.

Level 3 Critical Thinking and Clinical Applications

54. Brain tumors result from uncontrolled division of neuroglia. Unlike neurons, neuroglia are still capable of cell division. In addition, cells of the meninges can give rise to tumors.

55. The kidney condition is causing Harry to retain potassium ions. As a result, the K^+ concentration of the extracellular fluid is higher than normal. Under these conditions, less potassium diffuses from heart muscle cells than normal, resulting in a resting potential that is less negative (more positive). This change in resting potential moves the transmembrane potential closer to threshold, so it is easier to stimulate the muscle. The ease of stimulation accounts for the increased number of contractions that produces the rapid heart rate.

56. To reach threshold, the postsynaptic membrane must receive enough neurotransmitter to produce an EPSP of 20 mV (10 mV to reach threshold and 10 mV to cancel the IPSPs produced by the inhibitory neurons). Each neuron releases enough neurotransmitter to produce a change of 2 mV, so 10 of the 15 excitatory neurons must be stimulated to produce this effect by spatial summation.

57. Without myelination, information about limb movement and body position moves slowly to the brain, and motor responses move slowly to the muscles. By the time the brain is aware of limb movement or position and issues a motor command, the limb has already moved. When the motor response reaches the skeletal muscle, the response is no longer appropriate. As the peripheral neurons gradually become myelinated, information flow and processing speed up, so we see improved balance, coordination, and capabilities.

58. Initially, the recording would show normal action potentials in response to the constant stimulation. After 50,000–100,000 action potentials have passed, the recording would start to indicate less responsiveness from the neuron; ultimately, the neuron would fail to respond. Because very few ions actually move across the membrane during an action potential, neurons can carry thousands of action potentials before the concentrations of sodium and potassium ions must be reestablished. The activity of the sodium–potassium exchange pump is required. Blocking ATPase activity would shut down the pump, and the cell could not maintain the necessary gradients of sodium and potassium ions. As a result, the membrane could not respond to stimulation until the ion concentration gradients were reestablished.

CHAPTER 13

Level 1 Reviewing Facts and Terms

1. d **2.** a **3.** a **4.** c **5.** c **6.** d **7.** b **8.** c **9.** c **10.** a **11.** b **12.** a **13.** b **14.** c **15.** c **16.** d

17. (1) the dura mater, the outermost covering; (2) the arachnoid membrane, the second meningeal layer; and (3) the pia mater, the innermost meningeal layer

18. the epineurium, which surrounds the spinal nerve; the perineurium, which surrounds fascicles; and the endoneurium, which surrounds individual axons

Level 2 Reviewing Concepts

19. d

20. The first cervical nerve exits anterior to vertebra C_1 (between the skull and vertebra); the last cervical nerve exits posterior to vertebra C_7 (between the last cervical vertebra and the first thoracic vertebra). There are thus 8 cervical nerves but only 7 cervical vertebrae.

21. The cell bodies of spinal motor neurons are located in the anterior gray horns, so damage to these horns would result in a loss of motor control.

22. 1: walls of vertebral canal; 2: epidural space; 3: dura mater; 4: subdural space; 5: arachnoid membrane; 6: subarachnoid space; 7: pia mater; 8: spinal cord

23. Cerebrospinal fluid fills the subarachnoid space. The CSF acts as a shock absorber as well as a diffusion medium for dissolved gases, nutrients, chemical messengers, and waste products.

24. (1) involvement of pools of interneurons; (2) intersegmental distribution; (3) involvement of reciprocal innervation; (4) motor response prolonged by reverberating circuits; and (5) cooperation of reflexes to produce a coordinated, controlled response

25. **(a)** ventral ramus **(b)** ventral ramus **(c)** dorsal ramus **(d)** ventral ramus

26. **(a)** cervical plexus **(b)** lumbosacral plexus **(c)** brachial plexus **(d)** lumbosacral plexus

27. Transmission across a chemical synapse always involves a synaptic delay, but with only one synapse (a monosynaptic reflex), the delay between stimulus and response is minimized. In a polysynaptic reflex, the length of delay is proportional to the number of synapses involved.

28. (1) arrival of stimulus and activation of receptor; (2) activation of sensory neuron; (3) information processing; (4) activation of a motor neuron; and (5) response by an effector (muscle or gland)

29. The ventral roots contain axons of motor neurons, so muscles controlled by the neurons of the damaged root would be paralyzed.

30. increases muscle tone

31. The spinal cord ends at the level of L_2, so there is little chance of damaging it below that level. A needle can be inserted through the meninges inferior to the conus medullaris into the subarachnoid space with minimal risk to the cauda equina.

Level 3 Critical Thinking and Clinical Applications

32. the median nerve

33. the radial nerve

34. the genitofemoral nerve

35. Ramon has a damaged femoral nerve. This nerve also supplies the sensory innervation of the skin on the anteromedial surface of the thigh and medial surfaces of the leg and foot, so he may also experience numbness in those regions.

CHAPTER 14

Level 1 Reviewing Facts and Terms

1. i **2.** n **3.** f **4.** a **5.** k **6.** q **7.** o **8.** d **9.** r **10.** b **11.** h **12.** c **13.** j **14.** e **15.** l **16.** m **17.** p **18.** g **19.** h **20.** e **21.** a **22.** k **23.** c **24.** l **25.** b **26.** d **27.** j **28.** f **29.** g **30.** i **31.** a **32.** k **33.** d **34.** b **35.** a **36.** d **37.** b **38.** c **39.** a **40.** b **41.** a **42.** b **43.** b **44.** a

45. The cerebellum adjusts voluntary and involuntary motor activities on the basis of sensory information and stored memories of previous experiences.

46. The hypothalamus contains centers involved with emotions, autonomic function, and the regulation of body temperature and hormone production. It is the primary link between the nervous and endocrine systems.

47. (1) cushioning delicate neural structures; (2) supporting the brain; and (3) transporting nutrients, chemical messengers, and waste products

48. (1) portions of the hypothalamus where the capillary endothelium is extremely permeable; (2) capillaries in the pineal gland; and (3) capillaries at the choroid plexus

49. The limbic system consists of nuclei and tracts along the border between the cerebrum and diencephalon. Its functions include establishing emotional states and related behavioral drives; linking the conscious, intellectual functions of the cerebral cortex with the unconscious and autonomic functions of the brain stem; and facilitating memory storage and retrieval.

50. I: olfactory nerve; II: optic nerve; III: oculomotor nerve; IV: trochlear nerve; V: trigeminal nerve; VI: abducens nerve; VII: facial nerve; VIII: vestibulocochlear nerve; IX: glossopharyngeal nerve; X: vagus nerve; XI: accessory nerve; and XII: hypoglossal nerve

Level 2 Reviewing Concepts

51. d **52.** a

53. the hippocampus, which is part of the limbic system

54. the feeding and thirst centers of the hypothalamus

55. c

56. the substantia nigra, which releases the neurotransmitter dopamine at the cerebral nuclei

57. The left hemisphere contains the general interpretive and speech centers and is responsible for language-based skills. Reading, writing, speaking, performing analytical tasks, and logical decision making are dependent on processing done in the left hemisphere. The right hemisphere analyzes sensory information and relates the body to the sensory environment. Interpretive centers in this hemisphere permit the identification of familiar objects by touch, smell, sight, taste, or feel. It is important in understanding three-dimensional relationships and in analyzing the emotional context of a conversation.

58. centers in the medulla oblongata

59. The brain has more neurons and neuronal pools than the spinal cord does, and the brain has more-complex interconnections among the neurons and neuronal pools.

60. The general intrepretive area (Wernicke's area, sensory) and the speech center (Broca's area, motor) are involved in speech. Aphasia results in the absence of or defects in speech and the inability to comprehend language. Lesions in the general interpretive area produce defective visual and auditory comprehension of language, repetition of spoken sentences, and defective naming of objects. Lesions in the speech center result in hesitant and distorted speech.

Level 3 Critical Thinking and Clinical Applications

61. The sensory innervation of the nasal mucosa is by way of the maxillary branch of the trigeminal nerve (N V). Irritation of the nasal lining increases the frequency of action potentials along the maxillary branch of the trigeminal nerve to the semilunar ganglion. Second-order neurons from the semilunar ganglion carry increased frequency of action potentials to centers in the mesencephalon, which in turn excite the neurons of the reticular activating system (RAS). Increased activity by the RAS can raise the cerebrum back to consciousness.

62. The officer is testing the function of Bill's cerebellum. Many drugs, including alcohol, have pronounced effects on the function of the cerebellum. A person who is under the influence of alcohol cannot properly anticipate the range and speed of limb movement, because processing and correction by the cerebellum are slow. As a result, Bill might have a difficult time walking a straight line or touching his finger to his nose.

63. The local anesthetic temporarily numbs portions of the hypoglossal nerve (N XII), which controls movements of the tongue. If Tyler tries to eat before the anesthetic wears off, he could bite or otherwise injure his tongue without being aware of it. Movements of the tongue manipulate food during chewing and help initiate the swallowing reflex, so the inability to move food within the mouth could result in choking.

CHAPTER 15

Level 1 Reviewing Facts and Terms

1. a **2.** d **3.** c **4.** d

5. (1) nociceptors: a variety of stimuli associated with tissue damage; (2) thermoreceptors: respond to changes in temperature; (3) mechanoreceptors: stimulated or inhibited by physical distortion, contact, or pressure; and (4) chemoreceptors: monitor the chemical composition of body fluids and respond to the presence of specific molecules

6. (1) posterior column pathway: provides conscious sensations of highly localized ("fine") touch, pressure, vibration, and proprioception; (2) spinothalamic pathway: provides conscious sensations of poorly localized ("crude") touch, pressure, pain, and temperature; and (3) spinocerebellar pathway: carries proprioceptive information about the position of skeletal muscles, tendons, and joints to the cerebellum

7. (1) corticobulbar tracts; (2) lateral corticospinal tracts; and (3) anterior corticospinal tracts

8. (1) vestibulospinal pathway; (2) tectospinal pathway; and (3) reticulospinal pathway

9. (1) It integrates proprioceptive sensations with visual information from the eyes and equilibrium-related sensations from the inner ear; and (2) it adjusts the activities of the voluntary and involuntary motor centers on the basis of sensory information and the stored memories of previous experiences.

10. (1) The number of neurons in the cerebral cortex continues to increase until at least age 1; (2) the brain as a whole grows in size and complexity until at least age 4; and (2) myelination of CNS neurons continues at least until puberty, reducing the delay between stimulus and response and improving motor control.

11. (1) They are performed by neurons of the cerebral cortex and involve complex interactions between areas of the cortex and between the cerebral cortex and other parts of the brain; (2) they involve both conscious and unconscious information processing; and (3) their functions are subject to modification and adjustment over time.

12. increased neurotransmitter release, facilitation of synapses, and the formation of additional synaptic connections

13. During non-REM sleep, the entire body relaxes and activity at the cerebral cortex is at a minimum. The heart rate, blood pressure, respiratory rate, and energy utilization decline. During REM sleep, active dreaming occurs, accompanied by alterations in blood pressure and respiratory rates. Muscle tone decreases markedly, and there is less response to outside stimuli.

14. a reduction in brain volume and weight, a reduction in the number of neurons, a decrease in blood flow to the brain, changes in the synaptic organization, and intracellular and extracellular changes in CNS neurons

Level 2 Reviewing Concepts

15. d

16. The sensory neuron that delivers the sensations to the CNS is a first-order neuron. Within the CNS, the axon of the first-order neuron synapses on a second-order neuron, an interneuron located in the spinal cord or brain stem. The second-order neuron synapses on a third-order neuron in the thalamus. The axons of third-order neurons synapse on neurons of the primary sensory cortex of the cerebral hemispheres.

17. The autonomic nervous system controls visceral effectors, such as smooth and cardiac muscles, glands, and fat cells. The somatic nervous system, which is under voluntary control, controls the contractions of skeletal muscles.

18. A motor homunculus, a mapped-out area of the primary motor cortex, provides an indication of the degree of fine motor control available. A sensory homunculus indicates the degree of sensitivity of peripheral sensory receptors.

19. Injury to the primary motor cortex affects the ability to exert fine control over motor units. Gross movements are still possible, however, because they are controlled by the cerebral nuclei that use the reticulospinal or rubrospinal tracts. Thus, walking and other voluntary and involuntary movements can still be performed, but the movements are imprecise and awkward.

20. Muscle tone is controlled by the red nuclei of the mesencephalon, whose axons descend in the rubrospinal tract. The motor neurons of the red nucleus are controlled by neurons in the cerebral nuclei, cerebellum, and reticular formation. Stimulation of the motor neurons increases muscle tone, whereas inhibition decreases muscle tone.

21. A CVA is more likely to be devastating if it occurs in the left hemisphere, which in most people is the site of the general interpretive and speech centers. That is the hemisphere responsible for language-based skills, analytical tasks, and logical decision making. If a CVA occurred in the right hemisphere, it could affect the patient's ability to analyze sensory information and to relate the body to the sensory environment.

Level 3 Critical Thinking and Clinical Applications

22. The tumor is most likely adjacent to the corticobulbar tracts. The axons of those tracts carry action potentials to motor nuclei of the cranial nerves, which control eye muscles and facial expression.

23. You would expect to observe some degree of paralysis on the left side of Doris' body, relative to the amount of motor cortex deprived of blood. You might also observe some degree of memory loss, mood changes, and impairment of planning abilities.

24. The drug is mimicking the effects of the neurotransmitter serotonin. The areas of the brain that are being affected include the occipital lobe (visual hallucinations, color enhancement), the temporal lobe (auditory hallucinations), and the limbic system (increased sexual appetite).

CHAPTER 16

Level 1 Reviewing Facts and Terms

1. a **2.** d **3.** c **4.** b **5.** a **6.** c **7.** c **8.** a **9.** d

10. Sympathetic preganglionic fibers emerge from the thoracolumbar area (T_1 through L_2) of the spinal cord. Parasympathetic fibers emerge from the brain stem and the sacral region of the spinal cord (craniosacral).

11. Preganglionic neuron (T_5–L_2) → collateral ganglia → postganglionic fibers → visceral effector in abdominopelvic cavity

12. (1) celiac ganglion; (2) superior mesenteric ganglion; and (3) inferior mesenteric ganglion

13. (1) the release of norepinephrine at specific locations and (2) the secretion of epinephrine (and modest amounts of norepinephrine) into the bloodstream

14. When stimulated, the presynaptic membrane at a cholinergic synapse releases acetylcholine, whereas an adrenergic synapse releases norepinephrine.

15. N III, N VII, N IX, and N X

16. (1) ciliary ganglion; (2) sphenopalatine ganglion; (3) submandibular ganglion; and (4) otic ganglion

17. (1) cardiac plexus: heart rate–increase (sympathetic)/decrease (parasympathetic); heart strength–increase (sympathetic)/decrease (parasympathetic); blood pressure–increase (sympathetic)/decrease (parasympathetic); (2) pulmonary plexus: respiratory passageways dilate/constrict; (3) esophageal plexus: respiratory rate increase/decrease; (4) celiac plexus: digestion inhibited/stimulated; (5) inferior mesenteric plexus: digestion inhibited/stimulated; and (6) hypogastric plexus:

defecation inhibited/stimulated; urination inhibited/stimulated; sexual organs: stimulate secretion/erection
18. receptor, sensory neuron, interneuron or visceral motor neuron, and two visceral motor neurons
19. Processing centers in the medulla oblongata coordinate complex sympathetic and parasympathetic reflexes. These medullary centers are regulated by the hypothalamus. Centers in the posterior and lateral hypothalamus coordinate and regulate sympathetic function, and portions of the anterior and medial hypothalamus control the parasympathetic division.

Level 2 Reviewing Concepts
20. d **21.** a
22. mental alertness: increased (sympathetic)/decreased (parasympathetic); metabolic rate: increased/decreased; digestive and urinary functions: inhibited/stimulated; use of energy reserves: stimulated/inhibited; respiratory rate: increased/decreased; and heart rate and blood pressure: increased/decreased
23. The preganglionic fibers innervating the cervical ganglia originate in the white rami of thoracic segments, which are undamaged.
24. Preganglionic fibers entering the adrenal gland proceed to the adrenal medulla, where they synapse on neuroendocrine cells that release the neurotransmitters norepinephrine and epinephrine into the bloodstream.
25. The stimulation of alpha receptors activates enzymes on the inside of the cell membrane; the stimulation of beta receptors in organs and tissues triggers changes in the metabolic activity of the target cell. Beta-receptor stimulation results in the formation of a second messenger that activates or deactivates key enzymes.
26. Most of the acetylcholine released is inactivated by acetylcholinesterase within the synapse. Any acetylcholine diffusing into the surrounding tissues will be inactivated by the enzyme tissue cholinesterase.
27. If autonomic motor neurons maintain a background level of activity at all times, they can either increase or decrease their activity, providing a range of control options.
28. Due to the stimulation of sympathetic nervous activity, you would experience an increased respiratory rate, which would facilitate gas exchange at the lungs; an increase in vasoconstriction of peripheral blood vessels, causing an increase in blood pressure; an increase in heart rate and force of contraction, causing an increased rate of blood delivery; and an increase in the release of glucose into the blood, increasing the nutrient supply for energy metabolism.

Level 3 Critical Thinking and Clinical Applications
29. Stress-induced stomach ulcers are due to excessive sympathetic stimulation. The sympathetic division causes the vasoconstriction of blood vessels that supply the digestive organs, causing an almost total shutdown of blood supply to the stomach. Lack of blood leads to tissue death and necrosis, which causes the ulcers.
30. You would want to use a beta-blocker, because the stimulation of beta-1 receptors increases heart rate. By blocking these receptors, parasympathetic influence will predominate and the heart rate will slow down.
31. epinephrine, because it would reduce inflammation and relax the smooth muscle of the airways, making it easier for Billy to breathe
32. norepinephrine, binding to alpha-1 receptors

CHAPTER 17

Level 1 Reviewing Facts and Terms
1. b **2.** d **3.** d **4.** a **5.** b **6.** a **7.** d **8.** b **9.** b **10.** a **11.** d
12. d **13.** a **14.** c **15.** c **16.** b **17.** c **18.** d **19.** d **20.** b **21.** d
22. d **23.** c
24. (1) An arriving stimulus alters the transmembrane potential of the receptor membrane. (2) The receptor potential directly or indirectly affects a sensory neuron. (3) Action potentials travel to the CNS along an afferent fiber.
25. (1) tactile receptors; (2) baroreceptors; and (3) proprioceptors
26. (1) free nerve endings: sensitive to touch and pressure; (2) root hair plexus: monitors distortions and movements across the body surface; (3) tactile discs: detect fine touch and pressure; (4) tactile corpuscles: detect fine touch and pressure; (5) lamellated corpuscles: sensitive to pulsing or vibrating stimuli (deep pressure); and (6) Ruffini corpuscles: sensitive to pressure and distortion of the skin
27. Axons leaving the olfactory epithelium collect into twenty or more bundles that penetrate the cribriform plate of the ethmoid bone to reach the olfactory bulbs of the cerebrum. Axons leaving the olfactory bulb travel along the olfactory tract to reach the olfactory cortex, the hypothalamus, and portions of the limbic system.
28. (1) filiform papillae; (2) fungiform papillae; and (3) circumvallate papillae

29. (a) the sclera and the cornea **(b)** It (1) provides mechanical support and some physical protection, (2) serves as an attachment site for the extrinsic eye muscles, and (3) contains structures that assist in the focusing process.
30. iris, ciliary body, and choroid
31. A nerve impulse passes along the optic nerve, portions of which cross those of the other eye at the optic chiasm. There the impulse may cross to the opposite side of the brain (along the optic tract). The impulse either ends at nuclei that control eye reflexes or continues to the thalamus. On reaching the thalamus, an impulse is transmitted to another neuron, which conducts it to the occipital lobe of the cerebral cortex.
32. (1) malleus, (2) incus, and (3) stapes, which transmit a mechanical vibration (amplified along the way) from the tympanic membrane to the oval window
33. (1) Sound waves arrive at the tympanic membrane. (2) Movement of the tympanic membrane causes displacement of the auditory ossicles. (3) Movement of the stapes at the oval window establishes pressure waves in the perilymph of the vestibular duct. (4) The pressure waves distort the basilar membrane on their way to the round window of the tympanic duct. (5) Vibration of the basilar membrane causes vibration of hair cells against the tectorial membrane. (6) Information about the region and intensity of stimulation is relayed to the CNS over the cochlear branch of N VIII.

Level 2 Reviewing Concepts
34. c **35.** c
36. The general senses include somatic and visceral sensation. The special senses are senses whose receptors are confined to the head.
37. as action potentials, regardless of the type of stimulus
38. information about the strength, duration, variation, and movement of the stimulus
39. The olfactory system has extensive connections to the limbic system, which is responsible for memories and emotions.
40. a sty, a painful swelling
41. A visual acuity of 20/15 means that Jane can discriminate images at a distance of 20 feet, whereas someone with "normal" vision must be 5 feet closer (15 feet away) to see the same details. Jane's visual acuity is better than that of someone with 20/20 vision.

Level 3 Critical Thinking and Clinical Applications
42. Your medial rectus muscles would contract, directing your gaze more medially. In addition, your pupils would constrict and the lenses would become more spherical.
43. Sally is apparently nearsighted (myopic) and needs convex lenses to correct the condition.
44. The noise from the fireworks has transferred so much energy to the endolymph in the cochlea of her ears that this fluid continues to move for a long time. The vibrations associated with normal conversation are not strong enough to overcome the currents already moving through the endolymph, making these vibrations difficult to discern against the background "noise." And as long as the endolymph is moving, it will vibrate the tectorial membrane and will stimulate the hair cells, producing a "ringing" sensation.
45. The rapid descent in the elevator causes the maculae in the saccule of your vestibule to slide upward, producing the sensation of downward vertical motion. When the elevator abruptly stops, the maculae do not. It takes a few seconds for them to come to rest in the normal position. As long as the maculae are displaced, you will perceive movement.
46. When Juan closes his eyes, the visual cues are gone and his brain must rely solely on proprioceptive information and information from the static equilibrium centers (saccule and utricle). As a result of his problem with the saccules or utricles, his brain does not receive sufficient information to maintain balance. The movement of the arms toward the side of the impaired receptors is due to the deficit of information arriving from that side of the body.

CHAPTER 18

Level 1 Reviewing Facts and Terms
1. a **2.** b **3.** d **4.** a **5.** c **6.** d **7.** b **8.** d **9.** c **10.** b **11.** a
12. (1) the activation of adenyl cyclase; (2) the release or entry of calcium ions; (3) the activation of phosphodiesterase; and (4) the activation of other intracellular enzymes
13. (1) The hypothalamus secretes regulatory hormones that control endocrine cells in the pituitary gland; (2) the hypothalamus contains autonomic centers that exert direct neural control over the endocrine cells of the adrenal medulla; and (3) the hypothalamus releases hormones into the bloodstream at the posterior pituitary.
14. (1) thyroid-stimulating hormone (TSH); (2) adrenocorticotropic hormone (ACTH); (3) follicle-stimulating hormone (FSH); (4) luteinizing

hormone (LH); (5) prolactin (PRL); (6) growth hormone (GH); and (7) melanocyte-stimulating hormone (MSH)

15. (1) antidiuretic hormone (ADH) and (2) oxytocin

16. (1) an increase in the rate of energy consumption and utilization in cells; (2) an acceleration in the production of sodium–potassium ATPase; (3) the activation of genes coding for the synthesis of enzymes involved in glycolysis and energy production; (4) the acceleration of ATP production by mitochondria; and (5) in growing children, the normal development of the skeletal, muscular and nervous systems

17. calcitonin: decreases the concentration of calcium ions in body fluids; parathyroid hormone: causes an increase in the concentration of calcium ions in body fluids

18. (1) zona glomerulosa: mineralocorticoids; (2) zona fasciculata: glucocorticoids; and (3) zona reticularis: androgens

19. The activation of the adrenal medulla results in the secretion of epinephrine and norepinephrine, which trigger a mobilization of glycogen reserves in skeletal muscles and accelerate the breakdown of glucose to provide ATP. Stored fats are broken down to fatty acids in adipose tissue. Glycogen molecules are broken down in the liver, and the resulting glucose molecules are released into the bloodstream. In the heart, the stimulation of beta-1 receptors triggers an increase in the rate and force of cardiac muscle contraction.

20. (1) calcitriol, which stimulates calcium and phosphate ion absorption along the digestive tract; and (2) erythropoietin, which stimulates the production of red blood cells by the bone marrow and is involved in the regulation of blood pressure and blood volume

21. (1) It promotes the loss of sodium ions and water at the kidneys; (2) inhibits the secretion of water-conserving hormones, such as ADH and aldosterone; (3) suppresses thirst; and (4) blocks the effects of angiotensin II and norepinephrine on arterioles

22. (1) alpha cells: glucagon; (2) beta cells: insulin; (3) delta cells: somatostatin; and (4) F cells: pancreatic polypeptide

23. (a) (1) alarm phase, in which energy reserves are mobilized and the body prepares itself with "fight or flight" responses; (2) resistance phase, in which glucocorticoids are released to mobilize lipid and protein reserves, conserve glucose for neural tissues, elevate and stabilize blood glucose concentrations, and conserve salts and water and lose K^+ and H^+; and (3) exhaustion phase, in which homeostatic regulation breaks down **(b)** epinephrine in the alarm phase and glucocorticoids in the resistance phase

Level 2 Reviewing Concepts

24. In the nervous system, the source and destination of communication are quite specific and the effects are short lived. In endocrine communication, the effects are slow to appear and commonly persist for days. A single hormone can alter the metabolic activities of multiple tissues and organs simultaneously.

25. Hormones direct the synthesis of an enzyme (or other protein) not already present in the cytoplasm. They turn an existing enzyme "on" or "off" and increase the rate of synthesis of a particular enzyme or other protein.

26. The two hormones may have opposing, or antagonistic, effects; they may have additive or synergistic effects; one hormone may have a permissive effect on another (that is, the first hormone is needed for the second to produce its effect); or they may produce different but complementary results in specific tissues and organs.

27. In endocrine reflexes, the functional counterpart of neural reflexes, a stimulus triggers the production of a hormone. Both neural and endocrine reflexes are controlled by negative feedback mechanisms in most cases.

28. Because phosphodiesterase converts cAMP to AMP, the inactivation of this enzyme would prolong the effect of the hormone.

Level 3 Critical Thinking and Clinical Applications

29. Extreme thirst and frequent urination are characteristics of both diabetes insipidus and diabetes mellitus. To distinguish between the two, glucose levels in the blood and urine could be measured. A high glucose concentration would indicate diabetes mellitus.

30. Julie's poor diet would not supply enough Ca^{2+} for her developing fetus, which would remove large amounts of Ca^{2+} from the maternal blood. A lowering of the mother's blood Ca^{2+} would lead to an increase in parathyroid hormone levels and increased mobilization of stored Ca^{2+} from maternal skeletal reserves.

31. Sherry's symptoms suggest hyperthyroidism. Blood tests could be performed to assay the levels of TSH, T_3, and T_4. From these tests, the physician could make a positive diagnosis (hormone levels would be elevated in hyperthyroidism) and also determine whether the condition is primary (a problem with the thyroid gland) or secondary (a problem with hypothalamo-pituitary control of the thyroid gland).

32. All steroid hormones are quite similar in structure. When there is an excess of one steroid, some is converted to another form, commonly a

form similar to aldosterone. Increased levels of aldosterone lead to the retention of sodium and water and can produce bloating, as in patients receiving steroid therapy.

CHAPTER 19

Level 1 Reviewing Facts and Terms

1. c **2.** c **3.** a **4.** c **5.** d **6.** a **7.** b **8.** d **9.** d **10.** b

11. (1) the transportation of dissolved gases, nutrients, hormones, and metabolic wastes; (2) the regulation of pH and electrolyte composition of interstitial fluids throughout the body; (3) the restriction of fluid losses through damaged vessels or at other injury sites; (4) defense against toxins and pathogens; and (5) the stabilization of body temperature

12. albumins, which maintain the osmotic pressure of plasma and are important in the transport of fatty acids; globulins, which bind small ions, hormones, or compounds that might otherwise be filtered out of the blood at the kidneys or have very low solubility in water (transport globulins) or attack foreign proteins and pathogens (immunoglobulins); and fibrinogen, which functions in blood clotting

13. (a) anti-B antibodies **(b)** anti-A antibodies **(c)** neither anti-A nor anti-B antibodies **(d)** anti-A and anti-B antibodies

14. (1) ameboid movement, the extension of a cellular process; (2) emigration, squeezing between adjacent endothelial cells in the capillary wall; and (3) positive chemotaxis, the attraction to specific chemical stimuli

15. neutrophils, eosinophils, basophils, and monocytes

16. (1) T cells, which are responsible for cell-mediated immunity; (2) B cells, which are responsible for humoral immunity; and (3) NK cells, which are responsible for immune surveillance

17. (1) M-CSF: monocytes and macrophages; (2) G-CSF: granulocytes; (3) GM-CSF: granulocytes and monocytes; and (4) multi-CSF: granulocytes, monocytes, and megakaryocytes

18. (1) the transport of chemicals important to clotting; (2) the formation of a temporary patch in the walls of damaged blood vessels; and (3) active contraction after the clot has formed

19. (1) thrombopoietin (TPO); (2) interleukin-6 (IL-6); and (3) multi-CSF

20. (1) the vascular phase; (2) the platelet phase; (3) the coagulation phase; (4) clot retraction; and (5) fibrinolysis

21. The common pathway begins when thromboplastin from either the extrinsic or intrinsic pathway appears in the plasma.

22. An embolus is a drifting blood clot. A thrombus is a blood clot that sticks to the wall of an intact blood vessel.

Level 2 Reviewing Concepts

23. a **24.** d **25.** b **26.** d

27. Red blood cells are biconcave discs that lack mitochondria, ribosomes, and nuclei, and they contain a large amount of hemoglobin.

28. White blood cells defend against toxins and pathogens. Neutrophils, eosinophils, and monocytes engulf and digest bacteria, protozoa, fungi, viruses, and cellular debris. Lymphocytes specialize to attack and destroy specific foreign cells, proteins, and cancerous cells, directly or through the production of antibodies.

29. Blood stabilizes and maintains body temperature by absorbing and redistributing the heat of the body. Heat is absorbed from active skeletal muscles. Dermal capillaries dilate when body temperature rises, thereby increasing blood flow to the skin and dissipating the excess heat to the air. Dermal capillaries constrict when body temperature falls, thereby decreasing blood flow to the skin and thus conserving heat for organs that are more temperature sensitive.

30. Arterial puncture is required for checking the efficiency of gas exchange at the lungs. Samples are usually drawn from the radial artery at the wrist or from the brachial artery at the elbow.

31. RhoGam is given to prevent an Rh-negative mother from producing anti-Rh–positive antibodies, which could attack the blood of an Rh-positive fetus. It can be administered soon after the delivery of the first Rh-positive baby. RhoGam contains anti-Rh–positive antibodies that remove fetal Rh-positive antigens from the mother's bloodstream before the mother's immune system recognizes their presence. Thus, the mother's immune system is not induced to produce anti-Rh–positive antibodies.

32. Aspirin inactivates platelet enzymes involved in the production of thromboxanes and prostaglandins, and it inhibits endothelial cell production of prostacyclin. Thus, aspirin inhibits clotting.

Level 3 Critical Thinking and Clinical Applications

33. A prolonged prothrombin time and a normal partial thromboplastin time indicate a deficiency in the extrinsic system but not in the intrinsic system or common pathway. Factor VII would be deficient.

34. reticulocytes, to maintain a constant number of formed elements since there is not enough time for large numbers of erythrocytes to mature
35. multiple bruises throughout the skin, prolonged bleeding time, and prolonged clotting time (as a result of antibiotic-induced thrombocytopenia)
36. In many cases of kidney disease, the cells responsible for producing erythropoietin are damaged or destroyed. The reduction in erythropoietin levels leads to reduced erythropoiesis and fewer red blood cells, resulting in anemia.
37. Intrinsic factor, an essential part of the transport system for the absorption of vitamin B_{12} by the intestinal cells, is produced by specialized stomach cells. When most of the stomach was removed, intrinsic factor was no longer available to facilitate the absorption of vitamin B_{12}, so injection directly into the bloodstream was necessary. (If Randy did not take the B_{12} shots, he would develop pernicious anemia.)
38. Antibiotics can kill the bacteria that normally reside in the digestive tract. These bacteria are the body's primary source of vitamin K, which is necessary for the synthesis of prothrombin and some other clotting factors. Thus, a reduction in clotting ability could result.

CHAPTER 20

Level 1 Reviewing Facts and Terms

1. c 2. b 3. b 4. b 5. a 6. a 7. b 8. a 9. c
10. During ventricular contraction, tension in the papillary muscles pulls against the chordae tendineae, which keep the cusps of the AV valve from swinging into the atrium. This action prevents the backflow, or regurgitation, of blood into the atrium as the ventricle contracts.
11. (1) The epicardium is the visceral pericardium, which covers the outer surface of the heart. (2) The myocardium is the muscular wall of the heart, which forms both atria and ventricles. It contains cardiac muscle tissue and associated connective tissues, blood vessels, and nerves. (3) The endocardium is a squamous epithelium that covers the inner surfaces of the heart, including the valves.
12. The atrioventricular (AV) valves prevent the backflow of blood from the ventricles into the atria. The right AV valve is the tricuspid valve; the left AV valve is the bicuspid valve. The pulmonary and aortic semilunar valves prevent the backflow of blood from the pulmonary trunk and aorta into the right and left ventricles.
13. SA node → AV node → AV bundle → right and left bundle branches → Purkinje fibers (into the mass of ventricular muscle tissue)
14. The cardiac cycle is the period between the end of one heartbeat and the beginning of the next heartbeat. The cycle begins with atrial systole as the atria contract and push blood into the relaxed ventricles. As the atria relax and atrial diastole begins, the ventricles contract (ventricular systole), forcing blood through the semilunar valves into the pulmonary trunk and aorta. The ventricles then relax (ventricular diastole). For the rest of the cardiac cycle, both the atria and ventricles are in diastole; passive filling occurs.
15. (1) preload, the stretch on the heart before it contracts; (2) contractility, the forcefulness of contraction of individual ventricular muscle fibers; and (3) afterload, the pressure that must be exceeded before blood can be ejected from the ventricles

Level 2 Reviewing Concepts

16. a 17. c 18. d 19. a
20. The right atrium receives blood from the systemic circuit and passes it to the right ventricle. The right ventricle discharges blood into the pulmonary circuit. The left atrium collects blood from the pulmonary circuit. Contraction of the left ventricle ejects blood into the systemic circuit.
21. The first sound ("lubb"), which marks the start of ventricular contraction, is produced as the AV valves close and the semilunar valves open. The second sound ("dupp") occurs when the semilunar valves close, marking the start of ventricular diastole. The third heart sound is associated with blood flow into the atria, and the fourth with atrial contraction. Listening to the heart sounds (auscultation) is a simple, effective method of cardiac diagnosis.
22. Stroke volume (SV) is the volume of blood ejected by a ventricle in a single contraction. Cardiac output (CO) is the amount of blood pumped by a ventricle in 1 minute. CO (in ml/min) = SV (in ml) × HR (in beats/min)
23. stroke volume and heart rate
24. Preload is the degree of stretching of the ventricular muscle cells during ventricular diastole. There are no antagonistic muscle groups to extend the cardiac muscle tissue after each contraction. The stretching force is provided by the pouring of blood into the heart and by the elasticity of the fibrous skeleton.
25. Sympathetic activation causes the release of NE by postganglionic fibers and the secretion of NE and E by the adrenal medullae. These hormones stimulate the metabolism of cardiac muscle cells and increase the

force and degree of contraction. They also depolarize nodal cells toward threshold, which in turn increases the heart rate. Parasympathetic stimulation causes the release of ACh at membrane surfaces, where it produces hyperpolarization and inhibition. The result is a decrease in the heart rate and the force of cardiac contractions.
26. All these hormones have positive inotropic effects, which means that they increase the strength of contraction of the heart.

Level 3 Critical Thinking and Clinical Applications

27. In skeletal muscles, increased accumulation of lactic acid during exercise decreases the efficiency of contraction and can ultimately prevent contraction due to muscle fatigue. If this situation occurred in the heart, the results could be fatal. Fortunately, the heart can metabolize lactic acid, which is usually broken down before it can interfere with cardiac function. Thus, the heart can continue to beat normally despite a temporary reduction in the local oxygen supply.
28. This patient has a second-degree heart block, a condition in which some signals from the SA node do not reach the ventricular muscle. In this case, for every two action potentials generated by the SA node, only one is reaching the ventricles. This accounts for the two P waves for each QRS complex.
29. Person 1 has a cardiac output of 4500 ml (CO = HR × SV). Person 2 has a cardiac output of 8550 ml. According to Starling's law, in a normal heart the cardiac output is directly proportional to the venous return. Thus, person 2 has the greater venous return. Ventricular filling decreases with increased heart rate; person 1 has the lower heart rate and therefore the longer ventricular filling time.
30. It will decrease the force of cardiac contraction, which is directly proportional to stroke volume. Verapamil should lower Karen's stroke volume.
31. As a result of a myocardial infarction, cardiac muscle cells die. As they die, their cell membranes break down and their cytoplasmic contents leak into the surrounding interstitial fluid. Because the cells contain relatively high concentrations of K^+, the concentration of K^+ in the surrounding fluid would increase as K^+ leaks from the dead and dying cells. The interstitial fluid would contain more K^+, so less K^+ would diffuse through the membranes of living cells, decreasing the potential difference across the membrane and making them more excitable.
32. The extra contraction that is generated by the ectopic pacemaker is accompanied by an extra refractory period. The next normal action potential from the AV node usually arrives while the ventricles are still in the refractory period that accompanies the action potential generated by the ectopic pacemaker. As a result, the action potential does not trigger a contraction and the ventricles are not stimulated again until the next normal signal along the conducting pathways. This longer-than-normal period between subsequent contractions produces the feeling of the heart "skipping" a beat.

CHAPTER 21

Level 1 Reviewing Facts and Terms

1. k 2. w 3. g 4. o 5. a 6. q 7. c 8. u 9. x 10. b 11. v
12. r 13. e 14. i 15. t 16. f 17. m 18. h 19. d 20. l 21. j
22. p 23. n 24. s 25. d 26. b 27. c 28. a 29. b 30. b 31. a
32. a 33. d 34. b 35. c 36. b 37. d 38. c 39. b 40. c 41. a
42. c 43. b
44. **(a)** the peak blood pressure measured during ventricular systole **(b)** the minimum blood pressure at the end of ventricular diastole **(c)** the difference between the systolic and diastolic pressures **(d)** a single value for blood pressure, calculated by adding one-third of the pulse pressure to the diastolic pressure
45. **(a)** hydrostatic pressure **(b)** osmotic pressure
46. (1) muscular compression of peripheral veins, and (2) the respiratory pump
47. (1) decreased tissue O_2 levels/increased CO_2 levels; (2) the generation of lactic acid or other acids by tissue cells; (3) increased concentrations of K^+ or H^+ in interstitial fluid; (4) chemicals released during local inflammation; and (5) elevated local temperature
48. (1) a decrease in the cardiac output due to parasympathetic stimulation and inhibition of sympathetic activity, and (2) widespread peripheral vasodilation due to the inhibition of excitatory neurons in the vasomotor center
49. changes in the carbon dioxide, oxygen, or pH levels in the blood and the carbon dioxide level or pH of the cerebrospinal fluid
50. (1) epinephrine/norepinephrine; (2) aldosterone; (3) antidiuretic hormone; (4) angiotensin II; (5) erythropoietin; and (6) atrial natriuretic peptide
51. Before exercise begins, the heart rate increases slightly due to a general rise in sympathetic activity as you think about the workout ahead.

Extensive vasodilation occurs; venous return increases; and cardiac output rises.

52. hypotension: systolic pressures below 90 mm Hg; pale, cool, and moist ("clammy") skin; confusion and disorientation; a rapid, weak pulse; the cessation of urination; and a drop in blood pH (acidosis)

53. When an infant takes its first breath, the lungs expand and pulmonary vessels dilate. The smooth muscles in the ductus arteriosus contract, isolating the pulmonary and aortic trunks, and blood begins flowing through the pulmonary circuit. As pressure rises in the left atrium, the valvular flap closes the foramen ovale, completing the circulatory remodeling.

54. blood: decreased hematocrit, the formation of thrombin, and valvular malfunction; heart: a reduction in maximum cardiac output, changes in the activities of the nodal and conducting fibers, a reduction in the elasticity of the fibrous skeleton, progressive atherosclerosis, and the replacement of damaged cardiac muscle fibers by scar tissue; blood vessels: a progressive inelasticity in arterial walls, the deposition of calcium salts on weakened vascular walls, and the formation of thrombi at atherosclerotic plaques

Level 2 Reviewing Concepts

55. d **56.** c **57.** b **58.** c

59. Artery walls are generally thicker, and they contain more smooth muscle and elastic fibers, enabling them to resist and adjust to the pressure generated by the heart. Venous walls are thinner; the pressure in veins is less than that in arteries. Arteries constrict more than veins do when not expanded by blood pressure, due to a greater degree of elastic tissue. Finally, the endothelial lining of an artery has a pleated appearance because it cannot contract and so forms folds. The lining of a vein looks like a typical endothelial layer.

60. Precapillary sphincters at entrances to capillaries cause alternating contractions and relaxations that result in a series of pulses, producing variations in blood flow (vasomotion).

61. Capillary walls are thin, so distances for diffusion are small. Continuous capillaries have small gaps between adjacent endothelial cells that permit the diffusion of water and small solutes into the surrounding interstitial fluid but prevent the loss of blood cells and plasma proteins. Fenestrated capillaries contain pores that permit very rapid exchange of fluids and solutes between interstitial fluid and plasma. The walls of arteries and veins are several cell layers thick and are not specialized for diffusion.

62. Movement in the surrounding skeletal muscles squeezes venous blood toward the heart. This mechanism, the muscular pump, is assisted by the presence of valves in the veins, which prevent backflow of the blood. The respiratory pump results from the increase in internal pressure of the thoracic cavity on exhalation, which pushes venous blood into the right atrium.

63. Cardiac output and peripheral blood flow are directly proportional to blood pressure. Blood pressure is closely regulated by a combination of neural and hormonal mechanisms. The resistance of the circulatory system opposes the movement of blood, so blood flow is inversely proportional to the resistance. Sources of peripheral resistance include vascular resistance, viscosity, and turbulence.

64. The brain receives arterial blood via four arteries that form anastomoses within the cranium. An interruption of any one vessel will not compromise the circulatory supply to the brain.

65. The cardioacceleratory and vasomotor centers are stimulated when a general sympathetic activation occurs. The result is an increase in cardiac output and blood pressure. When the parasympathetic division is activated, the cardioinhibitory center is stimulated, reducing cardiac output.

66. The top number (150) is the systolic pressure, the peak blood pressure measured during ventricular systole. The bottom number (90) is the diastolic pressure, the minimum blood pressure at the end of ventricular diastole. The mean arterial pressure (MAP) is calculated by adding one-third of the pulse pressure to the diastolic pressure: $(150 - 90)/3 + 90 = 110$. (Note: pulse pressure = systolic pressure − diastolic pressure.)

67. The accident victim is in circulatory shock. The hypotension results from a loss of blood volume and decreased cardiac output. Her skin is pale and cool due to peripheral vasoconstriction; the moisture results from the sympathetic activation of sweat glands. Falling blood pressure to the brain causes confusion and disorientation. If you took her pulse, you would find it to be rapid and weak, reflecting the heart's response to reduced blood flow and volume.

68. In congestive heart failure, the heart cannot produce enough force to circulate the blood properly. As fluid accumulates in the capillaries, the blood hydrostatic pressure increases and the blood osmotic pressure decreases. The result is a fluid shift from the blood to the interstitial space. The fluid accumulation exceeds the ability of the lymphatic system to drain it, resulting in peripheral edema (swelling in the limbs).

Level 3 Critical Thinking and Clinical Applications

69. There are three contributing factors to Bob's elevated blood pressure: (1) The loss of water by sweating increases blood viscosity. The number of red blood cells remains about the same, but because there is less plasma volume, the concentration of red cells is increased, thus increasing the blood viscosity. Increased viscosity increases peripheral resistance and contributes to increased blood pressure. (2) To cool Bob's body, there is increased blood flow to the skin. This in turn increases venous return, which increases stroke volume and cardiac output (Starling's law of the heart). The increased cardiac output can also contribute to increased blood pressure. (3) The heat stress that Bob is experiencing leads to increased sympathetic stimulation (the reason for the sweating). Increased sympathetic stimulation of the heart will increase heart rate and stroke volume, thus increasing cardiac output and blood pressure.

70. The greater saphenous vein of the leg is a large vein that lies just beneath the skin, with no muscle to support it as with the deeper leg veins. Also, due to its location, the vessel is more susceptible to the effects of gravity than are vessels in the arms. When there is any impediment to blood flow from the legs, blood tends to collect in the greater saphenous vein and to stretch it; the skin does very little to resist the stretching. Even if the valves in the vein are closed, their edges won't meet due to the stretching, allowing more blood to collect inferior to the valves and causing even more stretching. This positive feedback mechanism leads to the black-and-blue nodules of varicose veins.

71. Pulse pressure is the difference between the systolic and diastolic pressures. The heart rate of a resting athlete is slower than that of a nonathlete, due to the size and condition of the heart muscle. The cardiac output of an athlete would be the same or even slightly higher than that of a nonathlete, however, because the athlete's stroke volume is greater. The larger stroke volume, the greater the difference between the systolic and diastolic pressures. (The systolic pressure is proportional to the volume of blood ejected per beat.) Thus, since the athlete has the larger stroke volume, he or she would have the larger pulse pressure.

72. Antihistamines and decongestants are sympathomimetic drugs; they have the same effects on the body as does stimulation of the sympathetic nervous system. In addition to the desired effects of counteracting the symptoms of the allergy, these medications can produce an increased heart rate, increased stroke volume, and increased peripheral resistance, all of which will contribute to elevating blood pressure. In a person with hypertension (high blood pressure), these drugs would aggravate this condition, with potentially hazardous consequences.

73. When Jolene rapidly moved from a lying position to a standing position, gravity caused her blood volume to move to the lower parts of her body away from the heart, decreasing venous return. The decreased venous return resulted in a decreased EDV, leading to a decreased stroke volume and cardiac output. In turn, blood flow to the brain decreased, so the diminished oxygen supply caused her to be light-headed and feel faint. This reaction doesn't happen all the time because as soon as the pressure drops due to inferior movement of blood, the baroreceptor reflex should be triggered. Normally, a rapid change in blood pressure is sensed by baroreceptors in the aortic arch and carotid sinus. Action potentials from these areas are carried to the medulla oblongata, where appropriate responses are integrated. In this case, we would expect an increase in peripheral resistance to compensate for the decreased blood pressure. If this doesn't compensate enough for the drop, then an increase in heart rate and force of contraction would occur. In normal individuals, these responses occur so quickly that changes in pressure following changes in body position go unnoticed.

CHAPTER 22

Level 1 Reviewing Facts and Terms

1. b **2.** c **3.** b **4.** a **5.** c **6.** a **7.** c **8.** d **9.** d **10.** c

11. (1) thoracic duct: collects lymph from the body below the diaphragm and from the left side of the body above the diaphragm; (2) right thoracic duct: collects lymph from the right side of the body above the diaphragm

12. **(a)** lymphocytes responsible for cell-mediated immunity **(b)** stimulate the activation and function of T cells and B cells **(c)** inhibit the activation and function of both T cells and B cells **(d)** produce and secrete antibodies **(e)** recognize and destroy abnormal cells **(f)** produce interleukin-7, which promotes the differentiation of B cells **(g)** maintain the blood–thymus barrier and secrete the thymic hormones that stimulate stem cell division and T cell differentiation **(h)** interfere with viral replication inside the cell and stimulate the activities of macrophages and NK cells **(i)** reset the body's thermostat, causing a rise in body temperature (fever) **(j)** provide cell-mediated immunity, which defends against abnormal cells and pathogens inside cells **(k)** provide humoral immunity, which defends against antigens and pathogens in the body (but not inside cells)

(l) enhance nonspecific defenses and increase T cell sensitivity and stimulate B cell activity (m) slow tumor growth and kill sensitive tumor cells (n) stimulate the production of blood cells in the bone marrow and lymphocytes in lymphoid tissues and organs
13. (1) T cells, derived from the thymus; (2) B cells, derived from bone marrow; and (3) NK cells, derived from bone marrow
14. (1) physical barriers; (2) phagocytic cells; (3) immunological surveillance; (4) interferons; (5) complement; (6) inflammation; and (7) fever

Level 2 Reviewing Concepts

15. c **16.** d **17.** b
18. (1) specificity: The immune response is triggered by a specific antigen and defends against only that antigen; (2) versatility: The immune system can differentiate among tens of thousands of antigens it may encounter during a normal lifetime; (3) memory: The immune response following the second exposure to a particular antigen is stronger and lasts longer than before; and (4) tolerance: Some antigens, such as those on your own cells, don't elicit an immune response.
19. It may rupture the cell membrane through the release of perforin, kill the target cell by secreting a poisonous lymphotoxin, or activate genes within the nucleus of the cell that tell the cell to die.
20. by neutralization, agglutination and precipitation, the activation of complement, the attraction of phagocytes, opsonization, the stimulation of inflammation, or the prevention of bacterial and viral adhesion
21. the destruction of target cell membranes, the stimulation of inflammation, the attraction of phagocytes, and the enhancement of phagocytosis
22. The injections are timed to trigger the primary and secondary responses of the immune system. When the technician is first exposed to the hepatitis antigens, B cells produce daughter cells that differentiate into plasma cells and memory B cells. The plasma cells begin producing antibodies, which represent the primary response to exposure. However, the primary response does not maintain elevated antibody levels for long periods, so the second and third injections are necessary to trigger the secondary (anamnestic) response, when memory B cells differentiate into plasma cells and produce antibody concentrations that remain high much longer.

Level 3 Critical Thinking and Clinical Applications

23. IgA antibodies are found in body secretions such as tears, saliva, semen, vaginal secretions but not in blood plasma. Plasma contains IgM, IgG, IgD, and IgE antibodies. By testing for the presence or absence of IgA and IgG, the lab could determine to whether the sample is blood plasma or semen.
24. Yes. On initial contact with a virus, the immune system first produces IgM antibodies. The response is fairly rapid but short-lived. About the time that IgM peaks, IgG levels are beginning to rise. IgG plays the more important role in eventually controlling the disease. The fact that Ted's blood sample has an elevated level of IgM antibodies indicates that he is in the early stages of a primary response to the measles virus.
25. In a radical mastectomy, the neighboring lymph glands in the axilla and surrounding region are removed along with the cancerous breast. (The lymph nodes are removed as a precautionary measure to try to prevent the spread of cancer cells spread by way of the lymphatic system.) The lymphatic vessels from the limb on the affected side are tied off, because there is no place for the lymph to drain to. Over time, lymphedema occurs, causing swelling of the limb.
26. For Harry's body to produce an immune response to antigen A, the antigen would have to be presented on his MHC proteins on the surface of his cells. For T cells to respond to an antigen, those antigens must be bound to MHC proteins of antigen-presenting cells. Tilly and Harry almost certainly have different MHC proteins, so introducing Tilly's cells would not help Harry. Her cells would not recognize the MHC proteins on the antigen-presenting cells and so would not respond to the antigen. In addition, her cells would be recognized as foreign and would be attacked and destroyed by Harry's macrophages.
27. Because FSH is a protein hormone, it binds to receptors on the surface of its target cells. If you could isolate these receptors, you could inject them into an animal, such as a rabbit, that would give an immune response to the human protein. By isolating from the rabbit the plasma cells that produced antibodies to the FSH receptor, you could clone them to produce large amounts of antibodies. Each plasma cell produces only one type of antibody, so this process would produce monoclonal antibodies. You could then tag the antibodies with a marker, such as the fluorescent molecule fluorescein, to make them visible when viewed under a fluorescent microscope. You could then introduce the antibodies into cell populations that you wished to test. If the receptors for FSH were present, the antibodies would bind to them and you could locate them by locating the fluorescent marker.

CHAPTER 23

Level 1 Reviewing Facts and Terms

1. a **2.** c **3.** d **4.** c **5.** c **6.** d **7.** b **8.** c **9.** a **10.** b **11.** a
12. c
13. (1) providing an extensive area for gas exchange between inhaled air and circulating blood; (2) moving air to and from the exchange surfaces of the lungs; (3) protecting the respiratory surfaces from dehydration, temperature changes, and other environmental variations and defending the respiratory system and other tissues from invasion by pathogens; (4) producing sounds involved in speaking, singing, or nonverbal communication; and (5) providing olfactory sensations to the CNS from the olfactory epithelium in the superior portions of the nasal cavity (The respiratory system also indirectly assists in the regulation of blood volume, blood pressure, and the control of body fluid pH.)
14. The upper respiratory system consists of the nose, nasal cavity, paranasal sinuses, and pharynx. The lower respiratory system consists of the larynx, trachea, bronchi, bronchioles, and alveoli of the lungs.
15. Goblet cells in the epithelium and mucous glands in the lamina propria of the respiratory tract produce a sticky mucus that bathes exposed surfaces. In the nasal cavity, cilia sweep mucus and any trapped debris or microorganisms toward the pharynx, where it will be swallowed and exposed to the acids and enzymes of the stomach. In the lower respiratory system, the cilia beat toward the pharynx, creating a mucus escalator that cleans the respiratory passageways. Filtration occurs in the nasal cavity. Particles may become trapped in the liquid covering the alveolar surfaces. Finally, foreign particles can be engulfed by alveolar macrophages.
16. (1) nasopharynx, the superior portion, where the nasal cavity opens into the pharynx; (2) oropharynx, the middle portion, posterior to the oral cavity; and (3) laryngopharynx, the inferior portion that is posterior to the hyoid bone and glottis
17. During swallowing, the larynx is elevated and the epiglottis folds back over the glottis, preventing the entry of liquids or solid food into the respiratory passageways.
18. The parietal pleura covers the inner surface of the thoracic wall and extends over the diaphragm and mediastinum. The visceral pleura covers the outer surfaces of the lungs, extending into the fissures between the lobes.
19. (1) pulmonary ventilation (breathing); (2) gas diffusion across the respiratory membrane; (3) the storage and transport of oxygen and carbon dioxide; and (4) the exchange of dissolved gases between blood and interstitial fluids
20. increased respiratory rate, increased heart rate, and elevated hematocrit
21. (1) conversion to carbonic acid; (2) binding to the hemoglobin of red blood cells; and (3) dissolution in plasma
22. When the P_{CO_2} goes up, the bronchioles increase in diameter (bronchodilation). When the P_{CO_2} declines, the bronchioles constrict (bronchoconstriction). Airflow is therefore directed to lobules where the P_{CO_2} is high, which helps improve gas exchange, because these will also be the lobules where the P_{O_2} is very low.

Level 2 Reviewing Concepts

23. a **24.** a **25.** d **26.** d **27.** c **28.** d **29.** b
30. With less cartilaginous support, the amount of tension in the smooth muscles has a greater effect on bronchial diameter and the resistance to air flow.
31. During talking, the airways are open and the food passages are closed. Trying to swallow while talking is therefore likely to lead to inadvertent entry of food particles into the airways, which will then trigger coughing to expel the food.
32. The nasal cavity is designed to cleanse, moisten, and warm inhaled air, whereas the mouth is not. Air that has entered through the mouth is drier than air that has entered through the nose and as a result can irritate the trachea, causing soreness of the throat.
33. The walls of bronchioles, like the walls of arterioles, are dominated by smooth muscle tissue. Varying the diameter of the bronchioles (bronchodilation or bronchoconstriction) provides control over the amount of resistance to airflow and the distribution of air within the lungs, just as vasodilation and vasoconstriction of the arterioles regulate blood flow and blood distribution.
34. Septal cells produce surfactant, which reduces surface tension in the fluid coating the alveolar surface. The alveolar walls are so delicate that without surfactant, the surface tension would be so high that the alveoli would collapse.
35. The differences in partial pressure are substantial; the diffusion distances involved are small; the gases are lipid soluble; the total surface area is large; and blood flow and airflow are coordinated.
36. Pulmonary ventilation is the physical movement of air into and out of the respiratory tract. Its primary function is to maintain adequate alveolar

ventilation. Alveolar ventilation is air movement into and out of the alveoli. This process prevents the buildup of carbon dioxide in the alveoli and ensures a continuous supply of oxygen that keeps pace with absorption by the bloodstream.

37. **(a)** Boyle's law describes the inverse relationship between pressure and volume: If volume decreases, pressure rises; if volume increases, pressure falls. **(b)** Dalton's law states that each of the gases that make up a mixture of gases contributes to the total pressure in proportion to its relative abundance; that is, all the partial pressures added together equal the total pressure exerted by the gas mixture. **(c)** Henry's law states that, at a given temperature, the amount of a particular gas that dissolves in a liquid is directly proportional to the partial pressure of that gas.

38. Such an injury can allow air into the pleural cavity. This pneumothorax breaks the fluid bond between the parietal pleura and the visceral pleura and allows the elastic fibers to recoil, causing the lung to collapse (atelectasis).

39. Both sneezing and coughing involve a temporary cessation of respiration (apnea).

40. Pulmonary volumes include resting tidal volume (averaging 500 ml), expiratory reserve volume (approximately 1200 ml), residual volume (averaging 1200 ml), minimal volume (30–120 ml), and inspiratory reserve volume (approximately 3600 ml). These values are determined experimentally. Respiratory capacities include inspiratory capacity, functional residual capacity, vital capacity, and total lung capacity. Respiratory capacities are determined by adding the values of various volumes.

41. The effect of pH on the hemoglobin saturation curve is called the Bohr effect. When the partial pressure of CO_2 rises, the pH drops; when the partial pressure of CO_2 declines, the pH rises. Hemoglobin molecules release more oxygen when the pH drops than they would when the pH rises. Hydrogen ions tend to "bump" oxygen molecules from the heme units. As the temperature rises, hemoglobin releases more oxygen; as the temperature declines, hemoglobin holds oxygen more tightly. For any partial pressure of oxygen, the higher the concentration of BPG, the more oxygen will be released by the hemoglobin molecules.

42. The DRG is the inspiratory center that contains neurons that control lower motor neurons innervating the external intercostal muscles and the diaphragm. The DRG functions in every respiratory cycle, whether quiet or forced. The VRG functions only during forced respiration—active exhalation and maximal inhalation. The neurons involved with active exhalation are sometimes said to form an expiratory center.

43. chemoreceptors sensitive to the P_{CO_2}, pH, and/or P_{O_2} of the blood or cerebrospinal fluid; changes in blood pressure in the aorta or carotid sinuses; stretch receptors that respond to changes in the volume of the lungs; irritating physical or chemical stimuli in the nasal cavity, larynx, or bronchial tree; and pain, changes in body temperature, and abnormal visceral sensations.

Level 3 Critical Thinking and Clinical Applications

44. AVR = respiratory rate × (tidal volume − dead air space). In this case, the dead air space is 200 ml (the anatomical dead air space plus the volume of the snorkel), therefore AVR = respiratory rate × (500 ml − 200 ml). We want to maintain an AVR of 6.0 l/min, we have 6.0 l/min = RR × 300 ml and so respiratory rate = (6.0 l/min) ÷ 300 ml = 20 breaths/min.

45. An increase in ventilation will increase the movement of venous blood back to the heart (recall the respiratory pump). This movement would increase venous return and thus help to increase blood pressure (Starling's law of the heart).

46. A person with chronic emphysema has constantly elevated levels of CO_2 in the blood, due to an inability to eliminate CO_2 efficiently as a result of the physical damage to the lungs. Over time, the brain ignores the stimulatory signals produced by the increased CO_2 and begins to rely on information from the peripheral chemoreceptors to set the pace of breathing (in other words, accommodation has occurred). The peripheral chemoreceptors also accommodate to the elevated CO_2 and respond primarily to the level of O_2 in the blood, increasing breathing when O_2 levels are low and decreasing breathing when O_2 levels are high. When pure O_2 was administered, chemoreceptors responded with fewer action potentials to the medulla oblongata, so Mr. B. stopped breathing.

47. While you were sleeping, the air that you were breathing was so dry that it absorbed more than the normal amount of moisture as it passed through the nasal cavity. The loss of moisture made the mucous secretions quite viscous and harder for the cilia to move. Your nasal epithelia continued to secrete mucus, but very little of it moved; ultimately, the nasal congestion resulted. After the shower and juice, more moisture was transferred to the mucus, loosening it and making it easier to move, thus clearing up the problem.

48. A person with kyphosis has an abnormal thoracic curvature of the spine. The abnormal curvature interferes with rib movement during ventilation as well as lung expansion. As a result, the lungs cannot expand to

their full potential. This restriction would decrease the inspiratory reserve and possibly the tidal volume, thus leading to a decreased vital capacity.

49. In anemia, the decreased ability of blood to carry oxygen is due to the lack of functional hemoglobin, red blood cells, or both. The disease does not interfere with the exchange of carbon dioxide within the alveoli nor with the amount of oxygen that will dissolve in the plasma. Because the chemoreceptors respond to dissolved gases and pH, as long as the concentrations of dissolved carbon dioxide, oxygen, and pH are normal, ventilation patterns should not change significantly.

50. The obstruction in Doris' right lung would not allow for gas exchange. Thus, the blood moving through the right lung would not oxygenate and would retain carbon dioxide. The retention of carbon dioxide in the blood would lead to a lower pH than the blood leaving the left lung. The lower pH would shift the oxygen–hemoglobin saturation curve to the left (the Bohr effect) for the right lung, compared with that for the left lung.

CHAPTER 24

Level 1 Reviewing Facts and Terms

1. d **2.** a **3.** d **4.** d **5.** d **6.** b **7.** b **8.** a **9.** c **10.** b **11.** d **12.** d **13.** a **14.** d **15.** d **16.** c **17.** a

18. (1) ingestion; (2) mechanical processing; (3) secretion; (4) digestion; (5) absorption; and (6) excretion

19. The folds increase the surface area available for absorption and may permit the expansion of the lumen after a large meal.

20. mucosa: the epithelial layer that performs chemical digestion and absorption of nutrients; submucosa: the connective tissue layer containing lymphatic and blood vessels and the submucosal nerve plexus; muscularis externa: the smooth muscle layer containing the myenteric nerve plexus; and serosa: the outermost layer, connective tissue that forms the visceral peritoneum

21. (1) neural mechanisms; (2) hormonal mechanisms; and (3) local mechanisms

22. The visceral peritoneum is continuous with the parietal peritoneum, which lines the inner surfaces of the body wall. The mesenteries are double sheets of peritoneal serous membranes that connect the parietal peritoneum with the visceral peritoneum. The mesenteries provide an access route for the passage of the blood vessels, nerves, and lymphatic vessels to and from the digestive tract. They also stabilize the position of the attached organs and prevent the intestines from becoming entangled during digestive movements or sudden changes in body positions.

23. (1) sensory analysis of material before swallowing; (2) mechanical processing through the actions of the teeth, tongue, and palatal surfaces; (3) lubrication by mixing with mucus and salivary secretions; and (4) limited digestion of carbohydrates and lipids

24. (1) parotid salivary glands: Stensen's duct; (2) sublingual salivary glands: Rivunus' ducts; and (3) submandibular salivary glands: Wharton's ducts

25. lubricating the mouth; moistening and lubricating materials in the mouth; dissolving chemicals that can stimulate the taste buds and provide sensory information about the material in the mouth; and initiating the digestion of complex carbohydrates before the material is swallowed. Salivary secretions are usually controlled by the ANS. Parasympathetic stimulation accelerates salivary secretion, whereas sympathetic innervation is believed to promote the secretion of small amounts of very thick saliva.

26. (1) incisors: clipping or cutting; (2) cuspids: tearing or slashing; (3) bicuspids: crushing, mashing, grinding; and (4) molars: crushing and grinding

27. The pharyngeal constrictors push the bolus toward the esophagus. The palatopharyngeus and stylopharyngeus muscles elevate the larynx. The palatal muscles raise the soft palate and adjacent portions of the pharyngeal wall.

28. (1) buccal phase; (2) pharyngeal phase; and (3) esophageal phase. Swallowing is controlled by the swallowing center of the medulla oblongata via the trigeminal and glossopharyngeal nerves. The motor commands originating at the swallowing center are distributed by cranial nerves V, IX, X, and XII. Along the esophagus, primary peristaltic contractions are coordinated by afferent and efferent fibers within the glossopharyngeal and vagus nerves, but secondary peristaltic contractions occur in the absence of CNS instructions.

29. The gastric glands are dominated by two types of secretory cells: parietal and chief cells. Parietal cells secrete intrinsic factor and hydrochloric acid. Intrinsic factor facilitates the absorption of vitamin B_{12} across the intestinal lining. The low pH of gastric juice kills most of the microorganisms ingested with food. The acid helps break down plant cell walls and the connective tissues in meat. The acidic environment is essential for the activation and function of pepsin, a protein-digesting enzyme. Chief cells secrete pepsinogen, an inactive proenzyme that is converted by the acid in the lumen to its active form, pepsin. The pyloric glands consist of G cells, which produce gastrin, and D cells, which release somatostatin, a hormone that inhibits gastrin release. Gastrin stimulates the secretion of both parietal and chief cells as well as contractions of the gastric wall that mix and stir the gastric contents.

30. (1) duodenum; (2) jejunum; and (3) ileum

31. The pancreas provides digestive enzymes as well as buffers that assist in the neutralization of chyme. The liver and gallbladder provide bile, which contains additional buffers and bile salts that facilitate the digestion and absorption of lipids. The liver is responsible for metabolic regulation, hematological regulation, and bile production. It is the primary organ involved with regulating the composition of circulating blood.

32. enterocrinin, which stimulates the submucosal glands of the duodenum; secretin, which stimulates the pancreas and liver to increase the secretion of water and buffers; cholecystokinin (CCK), which causes an increase in the release of pancreatic secretions and bile into the duodenum, inhibits gastric activity, and appears to have CNS effects that reduce the sensation of hunger; gastric inhibitory peptide (GIP), which stimulates insulin release at pancreatic islets and the activity of the duodenal submucosal glands; vasoactive intestinal peptide (VIP), which stimulates the secretion of intestinal glands, dilates regional capillaries, and inhibits acid production in the stomach; gastrin, which is secreted by G cells in the duodenum when they are exposed to large quantities of incompletely digested proteins; and, in small quantities, motilin, which stimulates intestinal contractions, villikinin, which promotes the movement of villi and associated lymph flow, and somatostatin, which inhibits gastric secretion

33. (1) resorption of water and compaction of the intestinal contents into feces; (2) absorption of important vitamins liberated by bacterial action; and (3) storage of fecal material prior to defecation

34. (1) Stretch receptors in the rectal walls promote a series of peristaltic contractions in the colon and rectum, moving feces toward the anus; and (2) the sacral parasympathetic system, also activated by the stretch receptors, stimulates peristalsis via motor commands distributed by the pelvic nerves.

35. (1) The rate of epithelial stem cell division declines; (2) smooth muscle tone decreases; (3) the effects of cumulative damage become apparent; (4) cancer rates increase; and (5) changes in other systems have direct or indirect effects on the digestive system.

Level 2 Reviewing Concepts

36. d   **37.** c   **38.** d   **39.** d   **40.** d   **41.** a

42. incisors

43. Peristalsis consists of waves of muscular contractions that move along the length of the digestive tract. During a peristaltic movement, the circular muscles contract behind the digestive contents. Longitudinal muscles contract next, shortening adjacent segments. A wave of contraction in the circular muscles then forces the materials in the desired direction. Segmentation movements churn and fragment the digestive materials, mixing the contents with intestinal secretions. Because they do not follow a set pattern, segmentation movements do not produce directional movement of materials along the tract.

44. The stomach stores ingested food, mechanically breaks down ingested food, disrupts chemical bonds through the actions of acids and enzymes, and produces intrinsic factor.

45. (1) The cephalic phase begins with the sight or thought of food. Directed by the CNS, this phase prepares the stomach to receive food. (2) The gastric phase begins with the arrival of food in the stomach; this phase is initiated by distension of the stomach, an increase in the pH of the gastric contents, and the presence of undigested materials in the stomach. (3) The intestinal phase begins when chyme starts to enter the small intestine. This phase controls the rate of gastric emptying and ensures that the secretory, digestive, and absorptive functions of the small intestine can proceed at reasonable efficiency.

46. Aspirin is acidic, and acids aggravate stomach ulcers. Also, aspirin promotes bleeding, interfering with the healing of ulcers.

47. After a heavy meal, bicarbonate ions pass from the parietal cells of the stomach into the extracellular fluid, causing the pH of the extracellular fluid to rise. As the extracellular fluid exchanges ions with the blood, the blood pH also increases.

Level 3 Critical Thinking and Clinical Applications

48. If a gallstone is small enough, it can pass through the common bile duct and block the pancreatic duct. Enzymes from the pancreas then cannot reach the small intestine. As they accumulate, the enzymes irritate the duct and ultimately the exocrine pancreas, producing pancreatitis.

49. The small intestine, especially the jejunum and ileum, are probably involved. Regional inflammation is the cause of Barb's pain. The inflamed tissue will not absorb nutrients; hence she is losing weight. Among the nutrients that are not absorbed are iron and vitamin B_{12}, which are necessary for formation of hemoglobin and red blood cells; hence, she has developed anemia.

50. You would expect to observe pain as the intestinal contents continued to accumulate and stretch the intestinal wall. The abdomen would be distended in the region proximal to the obstruction. Because material cannot leave the small intestine, increased distension and irritation would lead to vomiting, further increasing pressure on the intestine. The vomited material would contain bile, because little of the bile would be absorbed by the blocked intestine. No feces or intestinal gas would be produced because the blockage would prevent movement into the large intestine. Because the blockage is in the region of the jejunum, some absorption of nutrients would take place, but not as much as normal due to the distention and the irritation of the intestinal lining. Various symptoms of malnutrition would result.

CHAPTER 25

Level 1 Reviewing Facts and Terms

1. d   **2.** c   **3.** a   **4.** c   **5.** a   **6.** d   **7.** c   **8.** c   **9.** a   **10.** a   **11.** d
12. c   **13.** d   **14.** b   **15.** a   **16.** b   **17.** c   **18.** a   **19.** b   **20.** c   **21.** a
22. d   **23.** b

24. Metabolism is all of the chemical reactions occurring in the cells of the body. Anabolism is those chemical reactions that result in the synthesis of complex molecules from simpler reactants; products of anabolism are used for maintenance/repair, growth, and secretion. Catabolism is the breakdown of complex molecules into their building block molecules, resulting in the release of energy for the synthesis of ATP and related molecules.

25. $C_6H_{12}O_6 + 6\ O_2 \longrightarrow 6\ CO_2 + 6\ H_2O$

26. Lipoproteins are lipid–protein complexes that contain large insoluble glycerides and cholesterol, with a superficial coating of phospholipids and proteins. The major groups are chylomicrons, which consist of 95 percent triglyceride, are the largest lipoproteins, and carry absorbed lipids from the intestinal tract to the bloodstream; very low-density lipoproteins (VLDLs), which consist of triglyceride, phospholipid, and cholesterol and transport triglycerides to peripheral tissues; intermediate-density lipoproteins (IDLs), which are intermediate in size and composition between VLDLs and LDLs; low-density lipoproteins (LDLs, or "bad cholesterols"), which are mostly cholesterol and deliver cholesterol to peripheral tissues; and high-density lipoproteins (HDLs, or "good cholesterols"), which are equal parts protein and lipid (cholesterol and phospholipids) and transport excess cholesterol back to the liver for storage or excretion in bile.

27. The body cannot synthesize these nutrients.

28. carbohydrates: 4.18 C/g; lipids: 9.46 C/g; and proteins: 4.32 C/g

29. The BMR is the minimum, resting energy expenditure of an awake, alert person.

30. (1) radiation (heat loss as infrared waves); (2) conduction (heat loss to surfaces in physical contact); (3) convection (heat loss to the air); and (4) evaporation (heat loss with water being vaporized on the skin)

31. abnormalities of thermoregulation, such as heat exhaustion or heat stroke; clinical problems that restrict blood flow; conditions that impair sweat gland activity; and the resetting of the hypothalamic thermostat by circulating pyrogens

Level 2 Reviewing Concepts

32. d   **33.** b

34. Glycolysis results in the breakdown of glucose to pyruvic acid through a series of enzymatic steps, and 4 ATP and 2 NADH are also produced. Glycolysis requires glucose, specific cytoplasmic enzymes, ATP and ADP, inorganic phosphates, and NAD.

35. The TCA reaction sequence is a cycle because the four-carbon starting compound (oxaloacetic acid) is regenerated at the end. Acetyl-CoA and oxaloacetic acid enter the cycle, and CO_2, NADH, ATP, $FADH_2$, and oxaloacetic acid leave the cycle.

36. Oxidative phosphorylation is the generation of ATP within mitochondria, through a reaction sequence that requires coenzymes and consumes oxygen. The electron transport system consists of a sequence of metalloproteins called cytochromes, which pass electrons (from H atoms) along in small steps, gradually releasing energy for the formation of ATP and producing water as a byproduct.

37. A triglyceride is hydrolyzed, yielding glycerol and fatty acids. Glycerol is converted to pyruvic acid and enters the TCA cycle. Fatty acids are broken into two-carbon fragments by beta-oxidation inside mitochondria. The two-carbon compounds then enter the TCA cycle.

38. RNA is broken into its nucleotide building blocks, which are either recycled into new nucleic acids or catabolized to sugars (which enter glycolysis) and nitrogenous bases. The pyrimidines are converted into acetyl-CoA for the TCA cycle, and the purines are excreted.

39. The primary hormone of the absorptive state is insulin, which prevents a large surge in blood glucose after a meal. Insulin causes the liver to remove glucose from the hepatic portal circulation. During the postabsorptive state, blood glucose begins to decline, triggering the liver to release glucose via glycogenolysis and gluconeogenesis.

40. Liver cells can break down or synthesize most carbohydrates, lipids, and amino acids. The liver has an extensive blood supply and thus can easily monitor blood composition of these nutrients and regulate accordingly. The liver also stores energy in the form of glycogen.

41. The food pyramid indicates how much of each food group an individual should consume per day to ensure adequate intake of nutrients and calories. The placement of fats, oils, and sugars at the top of the food pyramid indicates that such foods are to be consumed very sparingly, whereas carbohydrates, placed at the bottom of the pyramid, are to be consumed in largest relative quantities.

42. The brain contains the hypothalamus, the thermostat of the body. The hypothalamus regulates the ANS control of such mechanisms as sweating and shivering thermogenesis via negative feedback homeostatic mechanisms.

43. These terms refer to the high-density lipoproteins (HDL) and low-density lipoproteins (LDL), lipoproteins in the blood that transport cholesterol. HDL ("good cholesterol") transports excess cholesterol to the liver for storage or breakdown, whereas LDL ("bad cholesterol") transports cholesterol to peripheral tissues, which unfortunately may include the arteries. The buildup of cholesterol in the arteries is linked to cardiovascular disease.

Level 3 Critical Thinking and Clinical Applications

44. Mary's AVR = 10 breaths/min × (300 ml/breath − 150 ml/breath).
10 breaths/min × 150 ml/breath = 1500 ml/min
1500 ml/min × 1l/1000 ml = 1.5 l/min
daily gas consumption = 1.5 l/min × 60 min/h × 24 h/day = 2160 l/day
daily oxygen consumption = 2160 l air/day × 0.18 O_2 = 388.8 l O_2/day
BMR = 388.8 l O_2/day × 4.825 kcal/l O_2 consumed = 1875 kcal (Calories)/day

45. During starvation, the body must use fat and protein reserves to supply the energy necessary to sustain life. Some of the protein that is metabolized for energy is the gamma globulin fraction of the blood, which is mostly composed of antibodies. This loss of antibodies, coupled with a lack of amino acids to synthesize new ones (as well as protective molecules such as interferon and complement proteins), renders an individual more susceptible than usual to contracting a disease and less likely to recover from it.

46. The response to anorexia is the same as in starvation. To supply the needed energy, the body utilizes reserves of fat and protein. Once the heart has exhausted its available fat reserves, it begins to break down contractile fibers to use the amino acids from the protein as an energy source. Eventually, the size of the heart and the force of cardiac contraction decrease. The weakened heart cannot contract as quickly as before, and bradycardia results. The combination of bradycardia and a decreased force of contraction contribute to hypotension. Ultimately, the heart will pump so inefficiently that it will not be able to serve the needs of the body's tissues; death will occur as the result of heart failure.

47. Colestipol would lead to a decrease in the plasma levels of cholesterol. Bile salts are necessary for the absorption of fats. If bile salts could not be absorbed, the absorption of fats—namely, cholesterol and triglycerides—would decrease. The result would be a decrease in cholesterol from a dietary source and a decrease in fatty acids that could be used to synthesize new cholesterol. In addition, the body would have to replace the bile salts that are being lost with the feces. The drop in cholesterol levels would be further enhanced because bile salts are formed from cholesterol.

CHAPTER 26

Level 1 Reviewing Facts and Terms

1. c **2.** a **3.** a **4.** b **5.** c **6.** d **7.** a **8.** d **9.** b **10.** c **11.** c **12.** d **13.** b **14.** b

15. The urinary system performs vital excretory functions and eliminates the organic waste products generated by cells throughout the body.

16. the kidneys, ureters, urinary bladder, and urethra

17. renal corpuscle (glomerulus/Bowman's capsule) → proximal convoluted tubule → loop of Henle → distal convoluted tubule → collecting duct → papillary duct → renal pelvis

18. proximal convoluted tubule: reabsorbs all the useful organic substrates from the filtrate; loop of Henle: reabsorbs over 90 percent of the water in the filtrate; and distal convoluted tubule: secretes into the filtrate waste products that were missed by filtration

19. The lamina densa is an elaborate basement membrane surrounding the capillaries of the glomerulus. During filtration, it restricts the passage of large plasma proteins but permits the movement of smaller molecules, including albumin, many organic nutrients, and ions.

20. The juxtaglomerular apparatus secretes the enzyme renin and the hormone erythropoietin.

21. renal artery → segmental arteries → interlobar arteries → arcuate arteries → interlobular arteries → afferent arterioles → nephrons → interlobular veins → arcuate veins → interlobar veins → renal vein

22. (1) filtration: the selective removal of large solutes and suspended materials from a solution on the basis of size; requires a filtration membrane and hydrostatic pressure, as provided by gravity or by blood pressure; (2) reabsorption: the removal of water and solute molecules from the filtrate after it enters the renal tubules; and (3) secretion: the transport of solutes from the peritubular fluid, across the tubular epithelium, and into the tubular fluid

23. In peripheral capillary beds, angiotensin II causes powerful vasoconstriction, elevating pressures in the renal arteries and their tributaries. At the nephron, angiotensin II causes the efferent arteriole to constrict, elevating glomerular pressures and filtration rates. At the PCT, it stimulates the reabsorption of sodium ions and water. In the CNS, angiotensin II triggers the release of ADH, stimulating the reabsorption of water in the distal portion of the DCT and the collecting system, and it causes the sensation of thirst. At the adrenal gland, angiotensin II stimulates the secretion of aldosterone by the cortex and epinephrine by the medulla oblongata. The aldosterone accelerates sodium reabsorption in the DCT and the cortical portion of the collection system. Epinephrine causes the heart rate and force of contraction to increase, elevating renal blood pressure.

24. ureters, urinary bladder, and urethra

Level 2 Reviewing Concepts

25. d **26.** c **27.** d **28.** b **29.** c **30.** a **31.** a

32. Cortical nephrons are located in the superficial cortex of the kidney. Juxtamedullary nephrons are located closer to the medulla, and their loops of Henle extend deep into the renal pyramids.

33. Net filtration pressure = glomerular hydrostatic pressure − capsular hydrostatic pressure − blood colloid osmotic pressure: NFP = (GHP − CsHP) − BCOP

34. autoregulation at the local level; hormonal regulation initiated by the kidneys; and autonomic regulation (sympathetic division of the ANS)

35. The renal threshold is the plasma concentration at which a specific compound or ion will begin appearing in urine.

36. As a result of facilitated diffusion and cotransport mechanisms, 99 percent of the glucose, amino acids, and other nutrients are reabsorbed before the filtrate leaves the PCT. A reduction of the solute concentration of the tubular fluid occurs due to active ion reabsorption of sodium, potassium, calcium, magnesium bicarbonate, phosphate, and sulfate ions. The passive diffusion of urea, chloride ions, and lipid-soluble materials further reduces the solute concentration of the tubular fluid and promotes additional water reabsorption.

37. (1) It is an efficient way to reabsorb solutes and water before the tubular fluid reaches the DCT and collecting system; and (2) it establishes a concentration gradient that will permit the passive reabsorption of water from urine in the collecting system.

38. In the distal convoluted tubule (DCT) and collecting duct, sodium–chloride cotransport and aldosterone-stimulated reabsorption of sodium ions adjust the osmolarity of the filtrate. The osmotic concentration of urine is controlled by variations in the water permeabilities of the distal portion of the DCT and the collecting ducts. The segments are impermeable to water unless they are exposed to antidiuretic hormone (ADH) from the posterior pituitary. At normal concentrations of ADH, the distal portions of the DCT and the collecting ducts are somewhat permeable to water, and there is an osmotic flow of water out of the urine as it passes along the collecting ducts. Under these conditions, urine entering the minor calyx has an osmolarity approaching 1200 milliosmoles/liter.

39. The urge to urinate usually appears when the bladder contains about 200 ml of urine. The micturition reflex begins to function when the stretch receptors have provided adequate stimulation to the parasympathetic motor neurons. The activity in the motor neurons generates action potentials that reach the smooth muscle in the wall of the urinary bladder. These efferent impulses travel over the pelvic nerves, producing a sustained contraction of the urinary bladder.

Level 3 Critical Thinking and Clinical Applications

40. Truck drivers may not urinate as frequently as they should. Resisting the urge to urinate can result in urine backing up into the kidney. This backup puts pressure on the kidney tissues, which can lead to tissue death and ultimately to kidney failure.

41. A single sample is useful in detecting the presence or absence of specific solutes (such as glucose), but it is less useful in detecting the amount of fluid lost via urination or the amount of some substance (urea, creatinine, etc.) lost per day. The amount of water reabsorbed from the filtrate can change from moment to moment, making the concentration in any one sample variable. The concentration of a solute in a single sample is much more variable than the amount excreted in a 24-hour period.

42. Caffeine increases the glomerular filtration rate by causing vasodilation of the glomerular capillaries. The increase in rate of flow through the renal tubule reduces the time available for the reabsorption of solutes. The alcohol acts as a diuretic, inhibiting ADH secretion and resulting in excess loss of water by the formation of dilute urine. The resulting dehydration leads to decreased secretion by body glands, including the salivary glands, which makes the mouth dry.

43. Susan may have a urinary tract infection, so her urine is likely to contain blood cells and bacteria.

44. By decreasing the active transport of sodium and chloride ions from the loop of Henle, the drug produces a more concentrated filtrate and a less concentrated interstitial fluid surrounding the loop of Henle. This combination of factors would lead to the production of a large volume of urine for two reasons: First, the more concentrated filtrate would retain water due to the increased osmolarity; second, less water would leave the descending limb because the osmotic gradient would not be as great. The increase in urine volume would lower blood volume. Because blood volume and blood pressure are directly related, if the blood volume decreases, the blood pressure would decrease (assuming no other critical factors change).

45. Arteriosclerosis contributes to hypertension (high blood pressure), which would trigger a baroreceptor reflex that would lead to a decrease in the level of ADH in the blood. Assuming that the arteriosclerosis is affecting all of Carlos' large arteries, including the renal arteries, the stiffening of the vessels would decrease the blood flow to the kidneys, thus triggering the release of renin from the juxtaglomerular apparatus. The renin would catalyze the conversion of angiotensinogen into angiotensin I, and the angiotensin I would be converted into angiotensin II at the lungs. The angiotensin II would stimulate the secretion of aldosterone, which would increase sodium and water reabsorption, thus increasing blood volume. The increase in blood volume would add to the hypertension, further decreasing the levels of ADH.

46. Because mannitol is filtered but not reabsorbed, drinking a mannitol solution would lead to an increase in the osmolarity of the filtrate. Less water would be reabsorbed, and an increased volume of urine would be produced.

47. Carbonic anhydrase catalyzes the reaction that forms carbonic acid, a source of hydrogen ions that are excreted by the kidneys. Hydrogen ion excretion is accomplished by an antiport system in which sodium ions are exchanged for hydrogen ions. Fewer hydrogen ions would be available, so less sodium would be reabsorbed, contributing to an increased osmolarity of the filtrate. In turn, an increased volume of urine and more-frequent urination would result.

CHAPTER 27

Level 1 Reviewing Facts and Terms

1. l **2.** h **3.** r **4.** a **5.** t **6.** c **7.** o **8.** b **9.** q **10.** e **11.** f
12. s **13.** d **14.** j **15.** i **16.** k **17.** m **18.** g **19.** p **20.** n **21.** c
22. c **23.** a **24.** b **25.** d **26.** b **27.** d **28.** d **29.** d **30.** b **31.** a
32. A fluid shift is a water movement between the ECF and ICF. This movement helps prevent drastic variations in the volume of the ECF. Changes in the osmolarity of the ECF can cause fluid shifts. If the ECF becomes more concentrated (hypertonic) with respect to the ICF, water will move from the cells into the ECF until equilibrium is restored. If the ECF becomes more dilute (hypotonic), water will move from the ECF into the cells. Changes in osmolarity can result from water loss (such as excessive perspiration, dehydration, vomiting, or diarrhea), water gain (drinking pure water, administering hypotonic solutions intravenously), and changes in electrolyte concentrations (such as sodium).

33. (1) antidiuretic hormone (ADH): stimulates water conservation at the kidneys and stimulates the thirst center; (2) aldosterone: determines the rate of sodium reabsorption and potassium secretion along the DCT and collecting system of the kidney; and (3) atrial natriuretic peptide: reduces thirst and blocks the release of ADH and aldosterone

34. Aldosterone promotes sodium conservation in exchange for potassium excretion in the urine.

35. (a) A volatile acid is an acid that can leave solution and enter the atmosphere. Carbon dioxide is an important volatile acid. (b) A fixed acid is an acid that does not leave solution; once produced, it remains in body fluids until excreted at the kidneys. Sulfuric acid and phosphoric acid are the most important fixed acids. (c) An organic acid contains carbon and

is usually a participant in or byproduct of cellular metabolism. Important examples are lactic acid and ketone bodies. Volatile acids, such as carbon dioxide, are the greatest threat to acid–base balance, because the CO_2 concentration is the most important factor affecting the pH of body tissues.

Level 2 Reviewing Concepts

36. a **37.** d **38.** d **39.** a **40.** b **41.** b **42.** c **43.** a
44. Fluid balance is a state in which the amount of water gained each day is equal to the amount lost to the environment. It is vital that the water content of the body remain stable, because water is an essential ingredient of cytoplasm and accounts for about 99 percent of the volume of extracellular fluid. Electrolyte balance exists when there is neither a net gain nor a net loss of any ion in body fluids. It is important that the ionic concentrations in body water remain within normal limits; if levels of calcium or potassium become too high, for instance, cardiac arrhythmias can develop. Acid–base balance exists when the production of hydrogen ions precisely offsets their loss. The pH of body fluids must remain within a relatively narrow range; variations outside this range can be life-threatening.

45. To go from mg/dl to mmol/l, multiply by 10 and divide by the atomic weight (35.5 g/mol) of chlorine: $(250 \text{ mg/dl} \times 10 \text{ dl/l})/(35.5 \text{ g/mol}) = 70.42$ mmol/l.

46. The temperature rise accompanying a fever can increase water losses. For each degree the temperature rises above normal, the daily water loss increases by 200 ml.

47. Whenever sodium gains exceed sodium losses, the volume of the ECF increases but its osmolarity remains the same. When sodium losses exceed gains, the ECF volume decreases but there is no significant change in the osmolarity of the ECF.

48. (1) protein buffer systems: These depend on the ability of amino acids to respond to changes in pH by accepting or releasing hydrogen ions. If the pH rises, the carboxyl group of the amino acid dissociates to release a hydrogen ion. If the pH drops, the amino group accepts an additional hydrogen ion to form an NH_3^+ group. The plasma proteins contribute to the buffering capabilities of the blood; inside cells, protein buffer systems stabilize the pH of the ECF by absorbing extracellular hydrogen ions or exchanging intracellular hydrogen ions for extracellular potassium. (2) carbonic acid–bicarbonate system: Most carbon dioxide generated in tissues is converted to carbonic acid, which dissociates into a hydrogen ion and a bicarbonate ion. Hydrogen ions released by dissociation of organic or fixed acids combine with bicarbonate ions, elevating the P_{CO_2}; additional CO_2 is lost at the lungs. (3) phosphate buffer systems: This buffer system consists of $H_2PO_4^-$, a weak acid that, in solution, reversibly dissociates into a hydrogen ion and HPO_4^{2-}. The phosphate buffer system plays a relatively small role in regulating the pH of the ECF, because the ECF contains far higher concentrations of bicarbonate ions than phosphate ions; however, it is important in buffering the pH of the ICF.

49. by secreting or absorbing hydrogen ions, controlling the excretion of acids and bases, and generating additional buffers

50. Respiratory compensation is a change in the respiratory rate that helps stabilize the pH of the ECF. Increasing or decreasing the rate of respiration alters pH by lowering or raising the P_{CO_2}. When the P_{CO_2} goes down, the pH rises; when the P_{CO_2} increases, the pH decreases. Renal compensation is a change in the rates of hydrogen and bicarbonate ion secretion or reabsorption in response to changes in plasma pH. Tubular hydrogen ion secretion results in the diffusion of bicarbonate ions into the ECF.

51. Respiratory disorders result from abnormal carbon dioxide levels in the ECF. An imbalance exists between the rate of CO_2 removal at the lungs and its generation in other tissues. Metabolic disorders are caused by the generation of organic or fixed acids or by conditions affecting the concentration of bicarbonate ions in the ECF.

52. Alkalosis is a rise in the pH of body fluids. Respiratory alkalosis (hypocapnia) results from an abnormally low level of carbon dioxide, usually caused by hyperventilation. Metabolic alkalosis occurs when bicarbonate ion levels rise and the bicarbonate ions interact with hydrogen ions to form carbonic acid. It can result from episodes of prolonged vomiting when stomach acids are being generated, causing the ECF bicarbonate concentration to rise.

53. (a) Excessive salt and water intake causes an increase in total blood volume and blood pressure. Increased blood pressure results in decreased ADH secretion and reduced renin secretion from the kidneys. Decreased renin results in a decreased production of angiotensin II, which reduces the rate at which aldosterone is secreted. Increased sodium ions and increased blood pressure cause the secretion of atrial natriuretic peptide, inhibiting ADH secretion and sodium ion reabsorption in the DCT. These changes cause increased loss of sodium in the urine and an increase in the volume of urine produced. (b) If the amount of salt ingested is excessive, the urine volume will rise and the concentration of salt in the urine will be high. If the amount of water is excessive, the

urine volume will increase and the concentration of sodium ions in the urine will decrease. (c) If the amount of salt and water ingested in food exceeds the amount needed to maintain a constant ECF composition, the total blood volume and the blood pressure will increase.
54. Since sweat is usually hypotonic, the loss of a large volume of sweat causes hypertonicity in body fluids. The loss of fluid volume is primarily from the interstitial space, which leads to a reduction in plasma volume and an increase in the hematocrit. Severe dehydration can cause the blood viscosity to increase substantially, resulting in an increased work load on the heart, ultimately increasing the probability of heart failure.

Level 3 Critical Thinking and Clinical Applications

55. The young boy has metabolic and respiratory acidosis. Sustained hypoventilation during drowning leads to decreased arterial P_{O_2}, and oxygen-starved tissues generate large quantities of lactic acid. Prompt emergency treatment is essential; the usual procedure involves some form of artificial or mechanical respiratory assistance coupled with the intravenous infusion of an isotonic solution containing sodium lactate, sodium gluconate, or sodium bicarbonate.
56. The most likely cause is acidosis. The disorientation could be the result of decreased pH in the ICF of the brain. The hyperventilation would be a response to acidosis, because hyperventilation removes carbon dioxide, the source of a volatile acid, carbonic acid. Finally, in acidosis, hydrogen ions move into the ICF and potassium ions move out in exchange. This exchange, coupled with increased potassium ion retention at the kidney due to increased hydrogen ion excretion, could account for the hyperkalemia.
57. As long as the ratio of bicarbonate ion to carbonic acid in the blood is 20:1, the pH of the blood will be 7.4. If the bicarbonate is significantly elevated without a change in pH, then the concentration of carbon dioxide and carbonic acid in the blood must also be increased. This condition is consistent with emphysema. Persons with emphysema have lost surface area and ventilation capacity and cannot efficiently clear the carbon dioxide from the blood. The result is a chronic elevation in the level of carbon dioxide and carbonic acid. To maintain homeostasis, the body makes adjustments, primarily at the kidneys, to raise the bicarbonate level appropriately to keep the pH in the normal range.
58. Digestive secretions contain high levels of bicarbonate, so persons with diarrhea can lose significant amounts of this important ion, leading to acidosis. We would expect Milly's blood pH to be lower than 7.4 and the pH of her urine to be low (due to increased renal excretion of hydrogen ion). We would also expect an increase in the rate and depth of breathing as the respiratory system tries to compensate by eliminating carbon dioxide.
59. The hypertonic solution will cause fluid to move from the ICF to the ECF, further aggravating Yuka's dehydration. The slight increase in pressure and osmolarity of the blood should lead to an increase in ADH, although ADH levels are probably quite high already. Urine volume would probably increase, because much of the glucose would not be reabsorbed. The osmolarity of the tubular filtrate would increase, decreasing water reabsorption and increasing urine volume.
60. Patient 1 has compensated respiratory alkalosis. Patient 2 has acute metabolic acidosis due to generation or retention of organic or fixed acids. Patient 3 has acute respiratory acidosis. Patient 4 has metabolic alkalosis.

CHAPTER 28

Level 1 Reviewing Facts and Terms

1. c **2.** d **3.** d **4.** c **5.** a **6.** c **7.** a **8.** d **9.** d **10.** c **11.** a **12.** c **13.** c **14.** d **15.** d **16.** b **17.** a **18.** c **19.** b **20.** a **21.** d
22. reproductive organs (gonads) that produce gametes and hormones; ducts that receive and transport gametes; accessory glands and organs that secrete fluids into these or other excretory ducts; and external genitalia
23. Seminiferous tubules → rete testis → efferent ducts → epididymis → ductus deferens → ejaculatory duct → urethra
24. The accessory organs/glands include the seminal vesicles, prostate gland, and the bulbourethral glands. The major functions of these glands are activating the spermatozoa, providing the nutrients sperm need for motility, propelling sperm and fluids along the reproductive tract, and producing buffers that counteract the acidity of the urethral and vaginal contents.
25. interstitial cells (cells of Leydig): produce male sex hormones (androgens), the most important one being testosterone; sustentacular cells: maintain the blood–testis barrier, support spermatogenesis and spermiogenesis, and secrete inhibin, androgen-binding protein, and Müllerian-inhibiting substance

26. (1) monitoring and adjusting the composition of tubular fluid; (2) acting as a recycling center for damaged spermatozoa; and (3) storing spermatozoa and facilitating their functional maturation
27. (1) prostatic urethra; (2) membranous urethra; and (3) penile urethra
28. spermatozoa: normal count is 20–100 million/ml; seminal fluid: a mixture of glandular secretions with distinctive ionic and nutrient composition; enzymes: a protease that may help dissolve mucous secretions in the vagina and seminalplasmin, an antibiotic enzyme that kills a variety of bacteria
29. stimulating spermatogenesis and promoting the functional maturation of spermatozoa; maintaining the male reproductive accessory organs; determining male secondary sex characteristics; stimulating metabolic operations, especially those concerned with protein synthesis and muscle growth; and influencing brain development by stimulating sexual behaviors and sexual drive
30. producing and releasing immature female gametes (oocytes); secreting female sex hormones, including estrogens and progestins; and secreting inhibin, involved in the feedback control of pituitary FSH production
31. (1) the formation of primary follicles; (2) the formation of secondary follicles; (3) the formation of a tertiary follicle; (4) ovulation; and (5) the formation and degeneration of the corpus luteum
32. The myometrium is the outer muscular layer; the endometrium is the inner glandular layer; and the perimetrium is an incomplete serosal layer.
33. (1) It serves as a passageway for the elimination of menstrual fluids; (2) it receives the penis during sexual intercourse and holds spermatozoa prior to their passage into the uterus; and (c) in childbirth, it forms the lower portion of the birth canal through which the fetus passes during delivery.
34. The acids, which are secreted by bacteria, inhibit the growth of many pathogens.
35. The clitoris is a structural component of the external genitalia of the female. It is the female equivalent of the penis; they are derived from the same embryonic structures. Internally, the clitoris contains erectile tissue comparable to the corpus spongiosum of the penis. The clitoris becomes engorged with blood during sexual arousal.
36. The lesser vestibular glands discharge their secretions onto the exposed surface of the vestibule, keeping it moistened. During arousal, a pair of ducts discharges the secretions of the greater vestibular glands into the vestibule near the posterolateral margins of the vaginal entrance.
37. secretory lobules of glandular tissue (lobes) → ducts → lactiferous duct → lactiferous sinus → open onto the surface of each nipple

Level 2 Reviewing Concepts

38. The reproductive system is the only physiological system that isn't required for survival of the individual.
39. In males, the gonads (testes) secrete androgens and produce one-half billion gametes (sperm) each day. The sperm are mixed with secretions, forming semen, and are expelled from the body during ejaculation. In females, the gonads (ovaries) secrete estrogen and progesterone and release one ovum (gamete) every month. The ovum may be fertilized within the oviduct.
40. Meiosis is the two-step nuclear division resulting in four haploid cells from one diploid cell. In males, four sperm are produced from each diploid cell. In females, only 1 oocyte (plus up to three polar bodies) is produced from each diploid precursor.
41. The corpora cavernosa extend along the length of the penis as far as the neck of the penis, and the erectile tissue within each corpus cavernosum surrounds a central artery. The slender corpus spongiosum surrounds the urethra. This erectile body extends from the superficial fascia of the urogenital diaphragm to the tip of the penis, where it expands to form the glans. The sheath surrounding the corpus spongiosum contains more elastic fibers than do the corpora cavernosa, and the erectile tissue contains a pair of arteries. Erection occurs when the parasympathetic neurons of the penile arteries release nitric oxide, causing the smooth muscles in the arterial walls to relax. The vessels dilate, blood flow increases, the vascular channels become engorged with blood, and erection of the penis occurs.
42. (1) menses: the interval marked by the degeneration and loss of the functional zone of the endometrium; lasts from 1–7 days, and 35–50 ml of blood is lost; (2) proliferative phase: growth and vascularization result in the complete restoration of the functional zone; lasts from the end of menses until the beginning of ovulation, around day 14; (3) secretory phase: the endometrial glands enlarge, accelerating their rates of secretion and the arteries elongate and spiral through the tissues of the functional zone; occurs under the combined stimulatory effects of progestins and estrogens from the corpus luteum; begins at ovulation and persists as long as the corpus luteum remains intact
43. As follicular development proceeds, the concentration of circulating estrogen rises. Secondary follicles contain increased numbers of granulosa cells, and the level of circulating inhibin rises. The rising estrogen and

inhibin levels inhibit hypothalamic secretion of GnRH and pituitary production and release of FSH. As the follicles develop and estrogen levels rise, the pituitary output of LH gradually increases. Estrogens, FSH, and LH continue to support follicular development and maturation despite a gradual decline in FSH levels. In the second week of the ovarian cycle, estrogen levels sharply increase—and the tertiary follicle enlarges in preparation for ovulation. By day 14, estrogen levels peak, triggering a massive outpouring of LH from the anterior pituitary. The rupture of the follicular wall results in ovulation. Next, LH stimulates the formation of the corpus luteum, which secretes moderate amounts of estrogens but large amounts of progesterone, the principal hormone of the postovulatory period. About 12 days after ovulation, declining progesterone and estrogen levels stimulate hypothalamic receptors and GnRH production increases, leading to increased FSH and LH production in the anterior pituitary; the cycle begins again.

44. The corpus luteum degenerates and progesterone and estrogen levels drop, resulting in the endometrial breakdown of menses. Next, rising levels of FSH, LH, and estrogen stimulate the repair and regeneration of the functional zone of the endometrium. During the postovulatory phase, the combination of estrogen and progesterone cause the enlargement of the endometrial glands and an increase in their secretory activity.

45. During sexual arousal, erotic thoughts or physical stimulation of sensory nerves in the genital region increases the parasympathetic outflow over the pelvic nerve, leading to erection of the clitoris or penis. Orgasm is the intensely pleasurable sensation associated with perineal muscle contraction and ejaculation in males, and with uterine and vaginal contractions and perineal muscle contraction in females. These processes are comparable in males and females.

46. Women age 45–55 undergo menopause, the time that ovulation and menstruation cease, accompanied by a sharp and sustained rise in the production of GnRH, FSH, and LH and a drop in the concentrations of circulating estrogen and progesterone. The decline in estrogen levels leads to reductions in the size of the uterus and breasts, accompanied by a thinning of the urethral and vaginal walls. In addition to neural and cardiovascular effects, reduced estrogen concentrations have been linked to the development of osteoporosis, presumably because bone deposition proceeds at a slower rate. Men age 50–60+ undergo the male climacteric, a time when circulating testosterone levels begin to decline and circulating levels of FSH and LH rise. Although sperm production continues, there is a gradual reduction in sexual activity in older men.

47. Birth control pills contain progestins and estrogens, which are administered in a cyclic fashion, beginning 5 days after the start of menses and continuing for the next 3 weeks. Over the fourth week, placebo pills or no pills are taken. The slightly elevated levels of progestin and estrogen inhibit GnRH release at the hypothalamus and the release of FSH and LH from the anterior pituitary. Without FSH, primordial follicles do not initiate development and an LH surge is necessary for ovulation to occur.

Level 3 Critical Thinking and Clinical Applications

48. There is no direct entry into the abdominopelvic cavity in males as there is in females. In females, the urethral opening is in close proximity to the vaginal orifice, so infectious organisms can exit from the urethral meatus and enter the vagina. They can then proceed through the vagina to the uterus, into the uterine tubes, and finally into the peritoneal cavity.

49. The sacral region of the spinal cord contains the parasympathetic centers that control the external genitalia, so damage to this area of the spinal cord would interfere with the ability to achieve an erection by way of parasympathetic stimuli. However, erection can also occur by way of sympathetic centers in the lower thoracic region of the spinal cord. Visual, auditory, or cerebral stimuli can result in decreased tone in the arteries serving the penis; increased blood flow and erection result. Tactile stimulation of the penis, however, would not generate an erection.

50. Granulosa cells produce estrogens, so a tumor involving these cells would lead to elevated levels of estrogens similar to or higher than those of an adult female. We would expect to observe the development of secondary sex characteristics.

51. Ovarian thecal cells and granulosa cells depend on androgens as a substrate for the production of female hormones. A blockage of the testosterone-producing enzyme would decrease the available substrate for estrogens and progesterone; the levels of these hormones would decline. Without sufficient estrogen levels, follicles would not develop properly and ovulation would not occur.

52. It suggests that a certain amount of body fat is necessary for menstrual cycles to occur. The nervous system appears to respond to circulating levels of the hormone leptin; when leptin levels fall below a certain set point, menstruation ceases. Without proper fat reserves, a woman might not be able to have a successful pregnancy. To avoid harm to the mother and the death of a fetus, the body prevents pregnancy by shutting down the ovarian cycle and thus the menstrual cycle. When appropriate energy reserves are available, the cycles begin again.

CHAPTER 29

Level 1 Reviewing Facts and Terms

1. a **2.** b **3.** d **4.** b **5.** b **6.** d **7.** a **8.** b **9.** c **10.** c **11.** b **12.** a **13.** a **14.** d **15.** a **16.** b **17.** d **18.** c **19.** d

20. When a sperm contacts the oocyte, their plasma membranes fuse. The oocyte is then activated: Its metabolic rate rises; it completes meiosis II; and a cortical reaction occurs that prevents additional sperm from entering. (Vesicles beneath the oocyte surface fuse with the cell membrane and discharge their contents.) The male and female pronuclei fuse (amphimixis), and the zygote begins preparing for the first cleavage division.

21. Induction is the chemical interplay among developing cells. Some cells release chemical substances that affect the differentiation of other cells. Induction controls the differentiation of embryonic cells into specialized cells—a crucial part of human development.

22. During the first trimester, the rudiments of all major organ systems appear. During the second trimester, the organs and organ systems complete most of their development. During the third trimester, the fetus grows rapidly and most of the major organ systems become fully functional.

23. **(a)** (1) yolk sac; (2) amnion; (3) allantois; and (4) chorion **(b)** The yolk sac forms from endoderm and mesoderm; it is an important site of blood cell formation. The amnion forms from ectoderm and mesoderm; it encloses the fluid that surrounds and cushions the developing embryo and fetus. The allantois forms from endoderm and mesoderm; its base gives rise to the urinary bladder. The chorion forms from mesoderm and trophoblast; circulation through chorionic vessels provides a "rapid transit system" for blood and nutrients

24. embryonic disc: an oval sheet of cells, develops at days 9–10 from the inner cell mass of the blastocyst; blastocyst: a hollow ball of cells formed after 5 days of cleavage; morula: a solid ball of cells formed during the first 5 days after fertilization; zygote: the fertilized ovum, prior to the start of cleavage, which begins immediately after fertilization and ends with the formation of the blastocyst

25. The trophoblast is the outer layer of cells that surrounds the blastocoele. Its functions include eroding a path through the uterine epithelium to merge the blastocyst with the functional zone of the endometrium; absorbing nutrients from uterine cells to support the early stages of embryo formation; creating lacunae, through which maternal blood percolates; participating in the formation of the placenta; and secreting placental hormones.

26. (1) dilation stage: It begins with the onset of true labor, as the cervix dilates and the fetus begins to move toward the cervical canal; late in this stage, the amniochorionic membrane ruptures. (2) expulsion stage: It begins as the cervix dilates completely and continues till the fetus has completely emerged from the vagina (delivery). (3) placental stage: The uterus gradually contracts, tearing the connections between the endometrium and the placenta and ejecting the placenta.

27. Relaxin produced by the placenta softens the pubic symphysis, and the weight of the fetus then deforms the cervical orifice. Deformation of the cervix and the rising estrogen levels promote the release of oxytocin, and the already stretched muscles become even more excitable.

28. (1) neonatal period (birth to 1 month): The newborn becomes relatively self-sufficient and begins performing respiration, digestion, and excretion on its own. Heart rates and fluid requirements are higher than those of adults. Neonates have little ability to thermoregulate. (2) infancy (1 month to 2 years): Major organ systems (other than those related to reproduction) become fully operational. (3) Childhood (2 years to puberty): Growth continues; body proportions change significantly.

29. Three events interact to promote increased hormone production and sexual maturation at puberty: (1) The hypothalamus increases its production of GnRH; (2) the anterior pituitary becomes more sensitive to the presence of GnRH and the circulating levels of FSH and LH rise rapidly; and (3) ovarian or testicular cells become more sensitive to FSH and LH. Puberty initiates adolescence, which includes gametogenesis and the production of male or female sex hormones that stimulate the appearance of secondary sexual characteristics and behaviors.

30. Maturity is the cessation of growth, normally occurring in the late teens or early twenties. An individual is reproductively mature immediately after puberty, but skeletal maturation continues until the last epiphyseal cartilages close, sometime after age 20.

31. Senescence is the process of aging, which affects all organ systems. Examples include the loss of elasticity in skin; the declining rate of bone deposition and degenerative changes in joints; reduced muscular strength; impaired coordination, memory, and intellectual function; reduced production of and sensitivity to hormones; cardiovascular problems; reduced immune system function; reduced elasticity in the lungs; decreased muscle tone along the digestive tract; reduced muscle tone in the urinary system; and the functional impairment of the reproductive system.

Level 2 Reviewing Concepts

32. d   **33.** b   **34.** d

35. Polyspermy is prevented by the cortical reaction, in which oocyte vesicles release their sperm-inhibiting contents via exocytosis.

36. The placenta produces human chorionic gonadotropin, which maintains the integrity of the corpus luteum and promotes the continued secretion of progesterone (keeping the endometrial lining functional; human placental lactogen and placental prolactin, which help prepare the mammary glands for milk production; and relaxin, which increases the flexibility of the pubic symphysis, causes dilation of the cervix, and suppresses the release of oxytocin by the hypothalamus, delaying the onset of labor contractions.

37. The respiratory rate and tidal volume increase, allowing the lungs to obtain the extra oxygen and remove the excess carbon dioxide generated by the fetus. Maternal blood volume increases, compensating for blood that will be lost during delivery. Requirements for nutrients and vitamins climb 10–30 percent, reflecting the fact that some of the mother's nutrients go to nourish the fetus. The glomerular filtration rate increases by about 50 percent, which corresponds to the increased blood volume and accelerates the excretion of metabolic wastes generated by the fetus.

38. By delivery, the uterus has quadrupled in length and grown to almost 20 times its original weight. The myometrium is stretched tremendously, in association with an increase in smooth muscle contractions. Progesterone released by the placenta inhibits the uterine smooth muscle to prevent more-violent contractions. The calming action of progesterone is opposed by estrogens, oxytocin, and prostaglandins; later in pregnancy, the production of estrogens and prostaglandins increases.

39. Progesterone released by the placenta has an inhibitory effect on the uterine smooth muscle.

40. Positive feedback mechanisms ensure that labor contractions continue until delivery has been completed.

41. the rupturing of the amnion, generally late in the dilation stage of labor

42. The neonate must fill its lungs (which are collapsed and filled with fluid at birth) with air, changing the pattern of cardiovascular circulation due to changes in blood pressure and flow rates. It must excrete the mixture of debris that has collected in the fetal digestive system. The neonate must obtain nourishment from a new source—the mother's mammary glands. Neonatal fluid requirements are high, because the infant cannot concentrate its urine significantly. The infant also has little ability to thermoregulate, although as it grows its insulating adipose tissue increases and its metabolic rate also rises.

43. **(a)** Monozygotic ("identical") twins result from either the separation of the blastomeres early in cleavage or from the splitting of the inner cell mass prior to gastrulation. The genetic makeup of the pair is almost identical, because both formed from the same gamete. Dizygotic twins result from two separate oocytes that were subsequently fertilized. **(b)** Genotype refers to all the chromosomes and their genes of the individual. Phenotype refers to the physical and physiological characteristics of the individual. The genotype is a primary determinant of the phenotype, although other factors (injury, disease, environment) can also play a role. **(c)** If both chromosomes of a homologous pair carry the same allele of a particular gene, the individual is homozygous for that trait. If the two chromosomes carry different alleles for that gene, the individual is het-

erozygous for that trait. **(d)** In simple inheritance, phenotypes are determined by interactions between a single pair of alleles. In polygenic inheritance, interactions occur among multiple genes.

44. **(a)** dominant **(b)** recessive **(c)** X-linked **(d)** autosomal

45. The trait of red–green color blindness is carried on the X chromosome. Men are thus more likely to inherit it because they have only one X chromosome, so whatever that chromosome carries will determine whether they are color-blind. Women have two X chromosomes, so they are color-blind only if they are homozygous-recessive. Red–green color blindness is an example of X-linked inheritance.

46. Genetic recombination, which occurs at synapsis during meiosis, results in increased variability in traits carried by different gametes and therefore in offspring.

47. The Human Genome Project is attempting to transcribe the entire human genome. The project will help identify the genes that are responsible for inherited disorders and will localize the specific chromosomes involved.

Level 3 Critical Thinking and Clinical Applications

48. The probability that this couple's daughters will have hemophilia is 0, because each daughter will receive a normal allele from her father. There is a 50 percent chance that a son will have hemophilia, because each son has a 50 percent chance of receiving the mother's normal allele and a 50 percent chance of receiving the father's recessive allele.

49. Excess calcium could cause tetanic contractions of the smooth muscle of the uterus. These contractions would cut off the blood supply to the placenta and deprive the fetus of oxygen.

50. During nursing, the mechanical stimulus of suckling triggers a neural reflex that leads to the release of oxytocin from the posterior pituitary. Oxytocin enters the bloodstream and stimulates smooth muscle in the ducts of the mammary glands to contract and move milk into the lactiferous sinuses—the milk let-down reflex. Oxytocin receptors are also located on smooth muscle cells of the uterus, so the rise in oxytocin sometimes leads to uterine contractions as well, producing a feeling similar to menstrual cramping.

51. In adults, heat is lost across the skin. In infants, the ratio of surface area to volume is very high, so infants lose heat very quickly. To maintain a constant body temperature despite the heat loss, cellular metabolism must be high. Cellular metabolism requires oxygen, so increased metabolism demands an increased respiratory rate. Cardiac output must increase as well to move the blood from the lungs to the tissues. The range of contraction in the neonate heart is limited, so the greatest increase in cardiac output is achieved by increasing the heart rate.

52. The active ingredient of aspirin blocks enzymes necessary for the production of prostaglandins. Prostaglandins promote uterine contractions during parturition; blocking their synthesis would interfere with normal labor. Prostaglandins are also involved in blood clotting. Taking aspirin could increase the time it would take for blood to clot following delivery, leading to excess blood loss and the possible need for a transfusion.

53. It is very unlikely that the baby's condition is the result of a viral infection contracted during the third trimester. The development of organ systems occurs during the first trimester. By the end of the second trimester, most organ systems are fully formed. During the third trimester, the fetus undergoes tremendous growth but very little new organ formation.

Weights and Measures

Accurate descriptions of physical objects would be impossible without a precise method of reporting the pertinent data. Dimensions such as length and width are reported in standardized units of measurement, such as inches or centimeters. These values can be used to calculate the **volume** of an object, a measurement of the amount of space the object fills. Mass is another important physical property. The **mass** of an object is determined by the amount of matter the object contains; on Earth the mass of an object determines the object's weight.

In the United States, length and width are typically described in inches, feet, or yards; volumes in pints, quarts, or gallons; and weights in ounces, pounds, or tons. These are units of the **U.S. system** of measurement. Table 1 summarizes the familiar and unfamiliar terms used in the U.S. system. For reference purposes, this table also includes a definition of the "household units," popular in recipes. The U.S. system can be very difficult to work with, because there is no logical relationship among the various units. For example, there are 12 inches in a foot, 3 feet in a yard, and 1760 yards in a mile. Without a clear pattern of organization, the conversion of feet to inches or miles to feet can be confusing and time-consuming. The relationships among ounces, pints, quarts, and gallons are no more logical than those among ounces, pounds, and tons.

In contrast, the **metric system** has a logical organization based on powers of 10, as indicated in Table 2. For example, a **meter (m)** is the basic unit for the measurement of size. For measurements of larger objects, data can be reported in **dekameters** (*deka,* ten),

TABLE 1 The U.S. System of Measurement

Physical Property	Unit	Relationship to Other U.S. Units	Relationship to Household Units
Length	inch (in.)	1 in. = 0.083 ft	
	foot (ft)	1 ft = 12 in.	
		= 0.33 yd	
	yard (yd)	1 yd = 36 in.	
		= 3 ft	
	mile (mi)	1 mi = 5280 ft	
		= 1760 yd	
Volume	fluidram (fl dr)	1 fl dr = 0.125 fl oz	
	fluid ounce (fl oz)	1 fl oz = 8 fl dr	= 6 teaspoons (tsp)
		= 0.0625 pt	= 2 tablespoons (tbsp)
	pint (pt)	1 pt = 128 fl dr	= 32 tbsp
		= 16 fl oz	= 2 cups (c)
		= 0.5 qt	
	quart (qt)	1 qt = 256 fl dr	= 4 c
		= 32 fl oz	
		= 2 pt	
		= 0.25 gal	
	gallon (gal)	1 gal = 128 fl oz	
		= 8 pt	
		= 4 qt	
Mass	grain (gr)	1 gr = 0.002 oz	
	dram (dr)	1 dr = 27.3 gr	
		= 0.063 oz	
	ounce (oz)	1 oz = 437.5 gr	
		= 16 dr	
	pound (lb)	1 lb = 7000 gr	
		= 256 dr	
		= 16 oz	
	ton (t)	1 t = 2000 lb	

TABLE 2 The Metric System of Measurement

Physical Property	Unit	Relationship to Standard Metric Units	Conversion to U.S. Units	
Length	nanometer (nm)	1 nm = 0.000000001 m (10^{-9})	= 3.94 × 10^{-8} in.	25,400,000 nm = 1 in.
	micrometer (µm)	1 µm = 0.000001 m (10^{-6})	= 3.94 × 10^{-5} in.	25,400 µm = 1 in.
	millimeter (mm)	1 mm = 0.001 m (10^{-3})	= 0.0394 in.	25.4 mm = 1 in.
	centimeter (cm)	1 cm = 0.01 m (10^{-2})	= 0.394 in.	2.54 cm = 1 in.
	decimeter (dm)	1 dm = 0.1 m (10^{-1})	= 3.94 in.	0.254 dm = 1 in.
	meter (m)	standard unit of length	= 39.4 in.	0.0254 m = 1 in.
			= 3.28 ft	0.3048 m = 1 ft
			= 1.093 yd	0.914 m = 1 yd
	kilometer (km)	1 km = 1000 m	= 3280 ft	
			= 1093 yd	
			= 0.62 mi	1.609 km = 1 mi
Volume	microliter (µl)	1 µl = 0.000001 l (10^{-6}) = 1 cubic millimeter (mm^3)		
	milliliter (ml)	1 ml = 0.001 l (10^{-3}) = 1 cubic centimeter $(cm^3$ or cc)	= 0.0338 fl oz	5 ml = 1 tsp
				15 ml = 1 tbsp
				30 ml = 1 fl oz
	centiliter (cl)	1 c = 0.01 l (10^{-2})	= 0.338 fl oz	2.95 c = 1 fl oz
	deciliter (dl)	1 dl = 0.1 l (10^{-1})	= 3.38 fl oz	0.295 dl = 1 fl oz
	liter (l)	standard unit of volume	= 33.8 fl oz	0.0295 l = 1 fl oz
			= 2.11 pt	0.473 l = 1 pt
			= 1.06 qt	0.946 l = 1 qt
Mass	picogram (pg)	1 pg = 0.000000000001 g (10^{-12})		
	nanogram (ng)	1 ng = 0.000000001 g (10^{-9})	= 0.000000015 gr	66,666,666 mg = 1 gr
	microgram (µg)	1 µg = 0.000001 g (10^{-6})	= 0.000015 gr	66,666 mg = 1 gr
	milligram (mg)	1 mg = 0.001 g (10^{-3})	= 0.015 gr	66.7 mg = 1 gr
	centigram (cg)	1 cg = 0.01 g (10^{-2})	= 0.15 gr	6.67 cg = 1 gr
	decigram (dg)	1 dg = 0.1 g (10^{-1})	= 1.5 gr	0.667 dg = 1 gr
	gram (g)	standard unit of mass	= 0.035 oz	28.4 g = 1 oz
			= 0.0022 lb	454 g = 1 lb
	dekagram (dag)	1 dag = 10 g		
	hectogram (hg)	1 hg = 100 g		
	kilogram (kg)	1 kg = 1000 g	= 2.2 lb	0.454 kg = 1 lb
	metric ton (kt)	1 mt = 1000 kg	= 1.1 t	
			= 2205 lb	0.907 kt = 1 t

Temperature	Centigrade	Fahrenheit
Freezing point of pure water	0°	32°
Normal body temperature	36.8°	98.6°
Boiling point of pure water	100°	212°
Conversion	°C ⟶ °F: °F = (1.8 × °C) + 32	°F ⟶ °C: °C = (°F − 32) × 0.56

hectometers (*hekaton,* hundred), or **kilometers** (**km;** *chilioi,* thousand); for smaller objects, data can be reported in **decimeters** (0.1 m; *decem,* ten), **centimeters** (**cm** = 0.01 m; *centum,* hundred), **millimeters** (**mm** = 0.001 m; *mille,* thousand), and so forth. In the metric system, the same prefixes are used to report weights, based on the **gram (g),** and volumes, based on the **liter (l).** This text reports data in metric units, in most cases with U.S. system equivalents. Use this opportunity to become familiar with the metric system, because most technical sources report data only in metric units; most of the world outside the United

States uses the metric system exclusively. Conversion factors are included in Table 2.

The U.S. and metric systems also differ in their methods of reporting temperatures. In the United States, temperatures are usually reported in degrees Fahrenheit (°F), whereas scientific literature and individuals in most other countries report temperatures in degrees centigrade or Celsius (°C). The relationship between temperatures in degrees Fahrenheit and those in degrees centigrade is indicated in Table 2.

The following illustration spans the entire range of measurements that we will consider in this book. Gross anatomy traditionally deals with structural organization as seen with the naked eye or with a simple hand lens. A microscope can provide higher levels of magnification and can reveal finer details. Before the 1950s, most information was provided by *light microscopy*. A photograph taken through a *light microscope* is called a **light micrograph (LM).** Light microscopy can magnify cellular structures up to about 1000 times and can show details as fine as 0.25 μm. The symbol **μm** stands for **micrometer;** 1 μm = 0.001 mm, or 0.00004 inches. With a light microscope, we can iden-

tify cell types, such as muscle fibers or neurons, and can see large structures within a cell. Because individual cells are relatively transparent, thin sections cut through a cell are treated with dyes that stain specific structures to make them easier to see.

Although special staining techniques can show the general distribution of proteins, lipids, carbohydrates, and nucleic acids in the cell, many fine details of intracellular structure remained a mystery until investigators began using *electron microscopy*. This technique uses a focused beam of electrons, rather than a beam of light, to examine cell structure. In *transmission electron microscopy*, electrons pass through an ultrathin section to strike a photographic plate. The result is a **transmission electron micrograph (TEM).** Transmission electron microscopy shows the fine structure of cell membranes and intracellular structures. In *scanning electron microscopy*, electrons bouncing off exposed surfaces create a **scanning electron micrograph (SEM).** Although it cannot achieve as much magnification as transmission microscopy, scanning microscopy provides a three-dimensional perspective of cell structure.

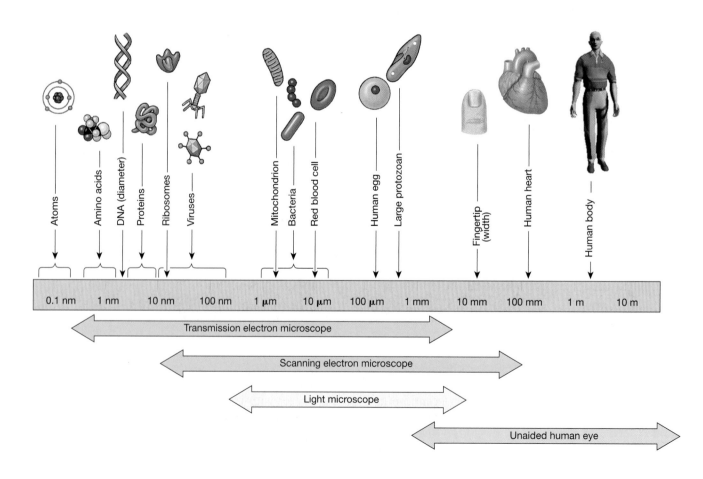

The **periodic table** presents the known elements in order of their atomic weights. Each horizontal row represents a single electron shell. The number of elements in that row is determined by the maximum number of electrons that can be stored at that energy level. The element at the left end of each row contains a single electron in its outermost electron shell; the element at the right end of the row has a filled outer electron shell. Organizing the elements in this fashion highlights similarities that reflect the composition of the outer electron shell. These similarities are evident when you examine the vertical columns. All the gases of the right-most column—helium, neon, argon, krypton, xenon, and radon—have full electron shells; each is a gas at normal atmospheric temperature and pressure, and none reacts readily with other elements. These elements, highlighted in blue, are known as the *noble*, or *inert*, *gases*. In contrast, the elements of the left-most column below hydrogen—lithium, sodium, potassium, rubidium, cesium, and francium—are silvery, soft metals that are so highly reactive that pure forms cannot be found in nature. The fourth and fifth electron levels can store up to 18 electrons. Table inserts are used for the so-called *lanthanide* and *actinide series* to save space, as higher levels can store up to 32 electrons. Elements of particular importance to our discussion of human anatomy and physiology are highlighted in pink.

Legend:
Atomic number — 1
Chemical symbol — H
Element name — Hydrogen
Atomic weight — 1.01

Main table:

1	2	3	4	5	6	7	8	9	10	11	12	13	14	15	16	17	18
1 H Hydrogen 1.01																	2 He Helium 4.00
3 Li Lithium 6.94	4 Be Beryllium 9.01											5 B Boron 10.81	6 C Carbon 12.01	7 N Nitrogen 14.01	8 O Oxygen 16.00	9 F Fluorine 19.00	10 Ne Neon 20.18
11 Na Sodium 22.99	12 Mg Magnesium 24.31											13 Al Aluminum 26.98	14 Si Silicon 28.09	15 P Phosphorus 30.97	16 S Sulfur 32.07	17 Cl Chlorine 35.45	18 Ar Argon 39.95
19 K Potassium 39.10	20 Ca Calcium 40.08	21 Sc Scandium 44.96	22 Ti Titanium 47.88	23 V Vanadium 50.94	24 Cr Chromium 52.00	25 Mn Manganese 54.94	26 Fe Iron 55.85	27 Co Cobalt 58.93	28 Ni Nickel 58.69	29 Cu Copper 63.55	30 Zn Zinc 65.39	31 Ga Gallium 69.72	32 Ge Germanium 72.61	33 As Arsenic 74.92	34 Se Selenium 78.96	35 Br Bromine 79.90	36 Kr Krypton 83.80
37 Rb Rubidium 85.47	38 Sr Strontium 87.62	39 Y Yttrium 88.91	40 Zr Zirconium 91.22	41 Nb Niobium 92.91	42 Mo Molybdenum 95.94	43 Tc Technetium (98)	44 Ru Ruthenium 101.07	45 Rh Rhodium 102.91	46 Pd Palladium 106.42	47 Ag Silver 107.87	48 Cd Cadmium 112.41	49 In Indium 114.82	50 Sn Tin 118.71	51 Sb Antimony 121.76	52 Te Tellurium 127.60	53 I Iodine 126.90	54 Xe Xenon 131.29
55 Cs Cesium 132.91	56 Ba Barium 137.33	57 La* Lanthanum 138.91	72 Hf Hafnium 178.49	73 Ta Tantalum 180.95	74 W Tungsten 183.85	75 Re Rhenium 186.21	76 Os Osmium 190.2	77 Ir Iridium 192.22	78 Pt Platinum 195.08	79 Au Gold 196.97	80 Hg Mercury 200.59	81 Tl Thallium 204.38	82 Pb Lead 207.2	83 Bi Bismuth 208.98	84 Po Polonium (209)	85 At Astatine (210)	86 Rn Radon (222)
87 Fr Francium (223)	88 Ra Radium 226.03	89 Ac† Actinium 227.03	104 Rf Rutherfordium (261)	105 Db Dubnium (262)	106 Jl Joliotium (263)	107 Bh Bohrium (262)	108 Hn Hahnium (265)	109 Mt Meitnerium (266)	110 Unnamed (269)	111 Unnamed (272)	112 Unnamed (272)						

*Lanthanide series

58 Ce Cerium 140.12	59 Pr Praseodymium 140.91	60 Nd Neodymium 144.24	61 Pm Promethium (145)	62 Sm Samarium 150.36	63 Eu Europium 151.96	64 Gd Gadolinium 157.25	65 Tb Terbium 158.93	66 Dy Dysprosium 162.50	67 Ho Holmium 164.93	68 Er Erbium 167.26	69 Tm Thulium 168.93	70 Yb Ytterbium 173.04	71 Lu Lutetium 174.97

†Actinide series

90 Th Thorium 232.04	91 Pa Protactinium 231.04	92 U Uranium 238.03	93 Np Neptunium 237.05	94 Pu Plutonium (244)	95 Am Americium (243)	96 Cm Curium (247)	97 Bk Berkelium (247)	98 Cf Californium (251)	99 Es Einsteinium (252)	100 Fm Fermium (257)	101 Md Mendelevium (258)	102 No Nobelium (259)	103 Lr Lawrencium (260)

Condensed Structural Formulas of the Amino Acids

Amino acids with acidic or basic side chains (R groups)

Aspartic acid (asp)   Glutamic acid (glu)   Tyrosine (tyr)

Lysine (lys)   Arginine (arg)   Histidine (his)

Amino acids with uncharged but polar side chains

Serine (ser)   Threonine (thr)   Methionine (met)   Cysteine (cys)

Amino acids with hydrocarbon side chains

Glycine (gly)   Alanine (ala)   Valine (val)   Tryptophan (trp)   Asparagine (asn)

Glutamine (gln)   Leucine (leu)   Isoleucine (ile)   Phenylalanine (phe)   Proline (pro)

Eponyms in Common Use

Eponym	Equivalent Term(s)	Individual Referenced
The Cellular Level of Organization (Chapter 3)		
Golgi apparatus		Camillo Golgi (1844–1926), Italian histologist; shared Nobel Prize in 1906
Krebs cycle	Tricarboxylic, or citric acid, cycle	Hans Adolph Krebs (1900–1981), British biochemist; shared Nobel Prize in 1953
The Skeletal System (Chapters 6–9)		
Colles' fracture		Abraham Colles (1773–1843), Irish surgeon
Haversian canals	Central canals	Clopton Havers (1650–1702), English anatomist and microscopist
Haversian systems	Osteons	Clopton Havers
Pott's fracture		Percivall Pott (1713–1788), English surgeon
Sharpey's fibers	Perforating fibers	William Sharpey (1802–1880), Scottish histologist and physiologist
Volkmann's canals	Perforating canals	Alfred Wilhelm Volkmann (1800–1877), German surgeon
Wormian bones	Sutural bones	Olas Worm (1588–1654), Danish anatomist
The Muscular System (Chapters 10, 11)		
Achilles' tendon	Calcanean tendon	Achilles, hero of Greek mythology
Cori cycle		Carl Ferdinand Cori (1896–1984) and Gerty Theresa Cori (1896–1957), American biochemists; shared Nobel Prize in 1947
The Nervous System (Chapters 12–16)		
Broca's center	Speech center	Pierre Paul Broca (1824–1880), French surgeon
Foramen of Munro	Interventricular foramen	John Cummings Munro (1858–1910), American surgeon
Nissl bodies		Franz Nissl (1860–1919), German neurologist
Purkinje cells		Johannes E. Purkinje (1781–1869), Czechoslovakian physiologist
Nodes of Ranvier		Louis Antoine Ranvier (1835–1922), French physiologist
Island of Reil	Insula	Johann Christian Reil (1759–1813), German anatomist
Fissure of Rolando	Central sulcus	Luigi Rolando (1773–1831), Italian anatomist
Schwann cells	Neurolemmocytes	Theodor Schwann (1810–1882), German anatomist
Aqueduct of Sylvius	Mesencephalic aqueduct	Jacobus Sylvius (Jacques Dubois, 1478–1555), French anatomist
Sylvian fissure	Lateral sulcus	Franciscus Sylvius (Franz de le Boë, 1614–1672), Dutch anatomist
Pons varolii	Pons	Costanzo Varolio (1543–1575), Italian anatomist
Sensory Function (Chapter 17)		
Organ of Corti	Spiral organ	Alfonso Corti (1822–1888), Italian anatomist
Eustachian tube	Auditory tube	Bartolomeo Eustachio (1520–1574), Italian anatomist
Golgi tendon organs	Tendon organs	Camillo Golgi (1844–1926), Italian histologist; shared Nobel Prize in 1906
Hertz (Hz)		Heinrich Hertz (1857–1894), German physicist
Meibomian glands	Tarsal glands	Heinrich Meibom (1638–1700), German anatomist
Meissner's corpuscles	Tactile corpuscles	Georg Meissner (1829–1905), German physiologist
Merkel's discs	Tactile discs	Friedrich Siegismund Merkel (1845–1919), German anatomist
Pacinian corpuscles	Lamellated corpuscles	Fillippo Pacini (1812–1883), Italian anatomist
Ruffini's corpuscles		Angelo Ruffini (1864–1929), Italian anatomist
Canal of Schlemm	Scleral venous sinus	Friedrich S. Schlemm (1795–1858), German anatomist
The Endocrine System (Chapter 18)		
Islets of Langerhans	Pancreatic islets	Paul Langerhans (1847–1888), German pathologist
Interstitial cells of Leydig	Interstitial cells	Franz von Leydig (1821–1908), German anatomist

Eponym	Equivalent Term(s)	Individual Referenced
The Cardiovascular System (Chapters 19–21)		
Bundle of His		Wilhelm His (1863–1934), German physician
Purkinje fibers		Johannes E. Purkinje (1781–1869), Czechoslovakian physiologist
Starling's law		Ernest Henry Starling (1866–1927), English physiologist
Circle of Willis	Cerebral arterial circle	Thomas Willis (1621–1675), English physician
The Lymphatic System (Chapter 22)		
Hassall's corpuscles		Arthur Hill Hassall (1817–1894), English physician
Kupffer cells	Stellate reticuloendothelial cells	Karl Wilhelm Kupffer (1829–1902), German anatomist
Langerhans cells		Paul Langerhans (1847–1888), German pathologist
Peyer's patches	Aggregate lymphoid nodules	Johann Conrad Peyer (1653–1712), Swiss anatomist
The Respiratory System (Chapter 23)		
Bohr effect		Niels Bohr (1885–1962), Danish physicist; won Nobel Prize in 1922
Boyle's law		Robert Boyle (1621–1691), English physicist
Charles' law		Jacques Alexandre César Charles (1746–1823), French physicist
Dalton's law		John Dalton (1766–1844), English physicist
Henry's law		William Henry (1775–1837), English chemist
The Digestive System (Chapter 24)		
Plexus of Auerbach	Myenteric plexus	Leopold Auerbach (1827–1897), German anatomist
Brunner's glands	Submucosal glands	Johann Conrad Brunner (1653–1727), Swiss anatomist
Kupffer cells	Stellate reticuloendothelial cells	Karl Wilhelm Kupffer (1829–1902), German anatomist
Crypts of Lieberkuhn	Intestinal glands	Johann Nathaniel Lieberkuhn (1711–1756), German anatomist
Plexus of Meissner	Submucosal plexus	Georg Meissner (1829–1905), German physiologist
Sphincter of Oddi	Hepaticopancreatic sphincter	Ruggero Oddi (1864–1913), Italian physician
Peyer's patches		Johann Conrad Peyer (1653–1712), Swiss anatomist
Duct of Santorini	Accessory pancreatic duct	Giovanni Domenico Santorini (1681–1737), Italian anatomist
Stensen's duct	Parotid duct	Niels Stensen (1638–1686), Danish physician/priest
Ampulla of Vater	Duodenal ampulla	Abraham Vater (1684–1751), German anatomist
Wharton's duct	Submandibular duct	Thomas Wharton (1614–1673), English physician
Duct of Wirsung	Pancreatic duct	Johann Georg Wirsung (1600–1643), German physician
The Urinary System (Chapter 26)		
Bowman's capsule	Glomerular capsule	Sir William Bowman (1816–1892), English physician
Loop of Henle	Nephron loop	Friedrich Gustav Jakob Henle (1809–1885), German histologist
The Reproductive System (Chapters 28, 29)		
Bartholin's glands	Greater vestibular glands	Casper Bartholin, Jr. (1655–1738), Danish anatomist
Cowper's glands	Bulbourethral glands	William Cowper (1666–1709), English surgeon
Fallopian tube	Uterine tube/oviduct	Gabriele Fallopio (1523–1562), Italian anatomist
Graafian follicle	Tertiary follicle	Reijnier de Graaf (1641–1673), Dutch physician
Interstitial cells of Leydig	Interstitial cells	Franz von Leydig (1821–1908), German anatomist
Glands of Littre	Lesser vestibular glands	Alexis Littre (1658–1726), French surgeon
Sertoli cells	Sustentacular cells	Enrico Sertoli (1842–1910), Italian histologist

Normal Physiological Values

Tables 3 and 4 present normal averages or ranges for the chemical composition of body fluids. These values are approximations rather than absolute values, because test results vary from laboratory to laboratory owing to differences in procedures, equipment, normal solutions, and so forth. Blanks in the tabular data appear where data are not available; sources used in the preparation of these tables follow. The following locations in the text contain additional information about body fluid analysis:

Table 19-3 (p. 642) presents data on the cellular composition of whole blood.

Table 26-2 (p. 953) compares the average compositions of urine and plasma.

Tables 26-5 (p. 969) and 26-7 (p. 973) give the general characteristics of normal urine.

Sources

Braunwauld, Eugene, Kurt J. Isselbacher, Dennis L. Kasper, Jean D. Wilson, Joseph B. Martin, and Anthony S. Fauci, eds. 1998. *Harrison's Principles of Internal Medicine,* 14th ed. New York: McGraw-Hill.

Lentner, Cornelius, ed. 1981. *Geigy Scientific Tables,* 8th ed. Basel, Switzerland: Ciba–Geigy Limited.

Halsted, James A. 1976. *The Laboratory in Clinical Medicine: Interpretation and Application.* Philadelphia: W.B. Saunders Company.

Wintrobe, Maxwell, G. Richard Lee, Dane R. Boggs, Thomas C. Bitnell, John Foerster, John W. Athens, and John N. Lukens. 1981. *Clinical Hematology,* Philadelphia: Lea and Febiger.

App. VI

TABLE 3 The Composition of Minor Body Fluids

Test	Perilymph	Endolymph	Synovial Fluid	Sweat	Saliva	Semen
			Normal Averages or Ranges			
pH			7.4	4–6.8	6.4*	7.19
Specific gravity			1.008–1.015	1.001–1.008	1.007	1.028
Electrolytes (mEq/l)						
Potassium	5.5–6.3	140–160	4.0	4.3–14.2	21	31.3
Sodium	143–150	12–16	136.1	0–104	14*	117
Calcium	1.3–1.6	0.05	2.3–4.7	0.2–6	3	12.4
Magnesium	1.7	0.02		0.03–4	0.6	11.5
Bicarbonate	17.8–18.6	20.4–21.4	19.3–30.6		6*	24
Chloride	121.5	107.1	107.1	34.3	17	42.8
Proteins (total) (mg/dl)	200	150	1.72 g/dl	7.7	386†	4.5 g/dl
Metabolites (mg/dl)						
Amino acids				47.6	40	1.26 g/dl
Glucose	104		70–110	3.0	11	224 (fructose)
Urea				26–122	20	72
Lipids (total)	12		20.9	‡	25–500§	188

* Increases under salivary stimulation.
† Primarily alpha-amylase, with some lysozomes.
‡ Not present in eccrine secretions.
§ Cholesterol.

App.
VI

TABLE 4 The Chemistry of Blood, Cerebrospinal Fluid, and Urine

| Test | Normal Range | | |
	Blood*	CSF	Urine
pH	S: 7.38–7.44	7.31–7.34	4.6–8.0
Osmolarity (mOsm/l)	S: 280–295	292–297	500–800
Electrolytes		(mEq/l unless noted)	(urinary loss per 24-hour period†)
Bicarbonate	P: 21–28	20–24	
Calcium	S: 4.5–5.3	2.1–3.0	6.5–16.5 mEq
Chloride	S: 100–108	116–122	120–240 mEq
Iron	S: 50–150 μg/l	23–52 μg/l	40–150 μg
Magnesium	S: 1.5–2.5	2–2.5	4.9–16.5 mEq
Phosphorus	S: 1.8–2.6	1.2–2.0	0.8–2 g
Potassium	P: 3.8–5.0	2.7–3.9	35–80 mEq
Sodium	P: 136–142	137–145	120–220 mEq
Sulfate	S: 0.2–1.3		1.07–1.3 g
Metabolites		(mg/dl unless noted)	(urinary loss per 24-hour period‡)
Amino acids	P/S: 2.3–5.0	10.0–14.7	41–133 mg
Ammonia	P: 20–150 μg/dl	25–80 μg/dl	340–1200 mg
Bilirubin	S: 0.5–1.0	<0.2	0.02–1.9 mg
Creatinine	P/S: 0.6–1.2	0.5–1.9	1.01–2.5
Glucose	P/S: 70–110	40–70	16–132 mg
Ketone bodies	S: 0.3–2.0	1.3–1.6	10–100 mg
Lactic acid	WB: 5–20§	10–20	100–600 mg
Lipids (total)	S: 400–1000	0.8–1.7	0–31.8 mg
Cholesterol (total)	S: 150–300	0.2–0.8	1.2–3.8 mg
Triglycerides	S: 40–150	0–0.9	
Urea	P/S: 23–43	13.8–36.4	12.6–28.6
Uric acid	S: 2.0–7.0	0.2–0.3	80–976 mg
Proteins	(g/dl)	(mg/dl)	(urinary loss per 24-hour period‡)
Total	S: 6.0–7.8	2.0–4.5	47–76.2 mg
Albumin	S: 3.2–4.5	10.6–32.4	10–100 mg
Globulins (total)	S: 2.3–3.5	2.8–15.5	7.3 mg (average)
Immunoglobulins	S: 1.0–2.2	1.1–1.7	3.1 mg (average)
Fibrinogen	P: 0.2–0.4	0.65 (average)	

*S = serum, P = plasma, WB = whole blood.
†Because urinary output averages just over 1 liter per day, these electrolyte values are comparable to mEq/l.
‡Because urinary metabolite and protein data approximate mg/l or g/l, these data must be divided by 10 for comparison with CSF or blood concentrations.
§Venous blood sample.

GLOSSARY OF KEY TERMS

Pronunciation Guide:

The pronunciations given in this text correspond to those given in medical and college dictionaries published in the United States. International pronunciations may vary; some of these pronunciations are included in the Glossary as alternatives. Pronunciation symbols are as follows:

"ā" as in "tray" (trā)

"ē" as in "ear" (ēr)

"ī" as in "fly" (flī)

"ō" as in "bone" (bōn)

"ū" as in "you" (ū)

Emphasized syllables are shown in capital letters, as in united (ū-NĪ-ted) or factor (FAK-tor).

A

abdomen: The region of the trunk bounded by the diaphragm and pelvis.

abdominopelvic cavity: The portion of the ventral body cavity that contains abdominal and pelvic subdivisions; also contains the peritoneal cavity.

abducens (ab-DOO-senz): Cranial nerve VI, which innervates the lateral rectus muscle of the eye.

abduction: Movement away from the midline of the body, as viewed in the anatomical position.

abortion: The premature loss or expulsion of an embryo or fetus.

abruptio placentae (ab-RUP-shē-ō pla-SEN-tē): The premature loss of a placental connection to the uterus, leading to maternal hemorrhaging and shock.

abscess: A localized collection of pus within a damaged tissue.

absorption: The active or passive uptake of gases, fluids, or solutes.

acclimatization: The physical adaption to a long-term environmental change, such as adaptions that accompany a change in season or in latitude.

accommodation: An alteration in the curvature of the lens of the eye to focus an image on the retina.

acetabulum (a-se-TAB-ū-lum): The fossa on the lateral aspect of the pelvis that accommodates the head of the femur.

acetylcholine (ACh) (as-ē-til-KŌ-lēn): A chemical neurotransmitter in the brain and peripheral nervous system; the dominant neurotransmitter in the peripheral nervous system, released at neuromuscular junctions and synapses of the parasympathetic division.

acetylcholinesterase (AChE): An enzyme found in the synaptic cleft, bound to the postsynaptic membrane, and in tissue fluids; breaks down and inactivates acetylcholine molecules.

acetyl-CoA: An acetyl group bound to coenzyme A, a participant in the anabolic and catabolic pathways for carbohydrates, lipids, and many amino acids.

acetyl group: $-CH_3C=O$.

achalasia (āk-a-LĀ-zē-uh): A condition that develops when the lower esophageal sphincter fails to dilate and ingested materials cannot enter the stomach.

Achilles tendon: See **calcanean tendon.**

acid: A compound whose dissociation in solution releases a hydrogen ion and an anion; an acidic solution has a pH below 7.0 and contains an excess of hydrogen ions.

acidosis (a-sid-Ō-sis): An abnormal physiological state characterized by a plasma pH below 7.35.

acinus/acini (AS-i-nī): A histological term referring to a blind pocket, pouch, or sac.

acne: A condition characterized by the inflammation of sebaceous glands and follicles; commonly affects adolescents and in most cases involves the face.

acoustic: Pertaining to sound or the sense of hearing.

acquired immune deficiency syndrome (AIDS): A disease caused by the human immunodeficiency virus (HIV); characterized by the destruction of helper T cells and a resulting severe impairment of the immune response.

acromegaly: A condition caused by the overproduction of growth hormone in adults, characterized by a thickening of bones and an enlargement of cartilages and other soft tissues.

acromion (a-KRŌ-mē-on): A continuation of the scapular spine that projects above the capsule of the scapulohumeral joint.

acrosomal cap (ak-rō-SŌ-mal): A membranous sac at the tip of a spermatozoon that contains hyaluronidase.

actin: The protein component of microfilaments that forms thin filaments in skeletal muscles and produces contractions of all muscles through interaction with thick (myosin) filaments; see also **sliding filament theory.**

action potential: A conducted change in the transmembrane potential of excitable cells, initiated by a change in the membrane permeability to sodium ions; see also **nerve impulse.**

activation energy: The energy required to initiate a specific chemical reaction.

active transport: The ATP-dependent absorption or excretion of solutes across a cell membrane.

acute: Sudden in onset, severe in intensity, and brief in duration.

adaptation: A change in pupillary size in response to changes in light intensity; a decrease in receptor sensitivity or perception after chronic stimulation; physiological responses that produce acclimatization.

Addison's disease: A condition resulting from the hyposecretion of glucocorticoids; characterized by lethargy, weakness, hypotension, and increased skin pigmentation.

adduction: Movement toward the axis or midline of the body, as viewed in the anatomical position.

adenine: A purine; one of the nitrogenous bases in the nucleic acids RNA and DNA.

adenohypophysis (ad-e-nō-hī-POF-i-sis): The anterior portion of the pituitary gland; also called **anterior pituitary.**

adenoids: The pharyngeal tonsil.

adenosine: A combination of adenine and ribose.

adenosine diphosphate (ADP): A compound consisting of adenosine with two phosphate groups attached.

adenosine monophosphate (AMP): A nucleotide consisting of adenine plus a phosphate group (PO_4^{3-}); also called *adenosine phosphate.*

adenosine triphosphate (ATP): A high-energy compound consisting of adenosine with three phosphate groups attached; the third is attached by a high-energy bond.

adenylate cyclase: An enzyme bound to the inner surfaces of cell membranes that can convert ATP to cyclic-AMP; also called *adenylyl cyclase.*

adhesion: The fusion of two mesenterial layers after damage or irritation of their opposing surfaces; this process restricts relative movement of the organs involved; the binding of a phagocyte to its target.

adipocyte (AD-i-pō-sīt): A fat cell.

adipose tissue: Loose connective tissue dominated by adipocytes.

adrenal cortex: The superficial portion of the adrenal gland that produces steroid hormones.

adrenal gland: A small endocrine gland that secretes steroids and catecholamines and is located superior to each kidney; also called *suprarenal gland.*

adrenal medulla: The core of the adrenal gland; a modified sympathetic ganglion that secretes catecholamines into the blood after sympathetic activation.

adrenergic (ad-ren-ER-jik): A synaptic terminal that, when stimulated, releases norepinephrine.

adrenocortical hormone: Any steroid produced by the adrenal cortex.

adrenocorticotropic hormone (ACTH): The hormone that stimulates the production and secretion of glucocorticoids by the zona fasciculata of the adrenal cortex; released by the anterior pituitary in response to corticotropin-releasing hormone.

adventitia (ad-ven-TISH-a): The superficial layer of connective tissue surrounding an internal organ; fibers are continuous with those of surrounding tissues, providing support and stabilization.

aerobic: Requiring the presence of oxygen.

aerobic metabolism: The complete breakdown of organic substrates into carbon dioxide and water, via pyruvic acid; a process that yields large amounts of ATP but requires mitochondria and oxygen.

afferent: Toward.

afferent arteriole: An arteriole that carries blood to a glomerulus of the kidney.

afferent fiber: An axon that carries sensory information to the central nervous system.

afterbirth: The distal portions of the umbilical cord and placenta that are ejected from the uterus during the placental stage of labor.

agglutination (a-gloo-ti-NĀ-shun): The aggregation of red blood cells due to interactions between surface antigens and plasma antibodies.

agglutinins (a-GLOO-ti-ninz): Immunoglobulins in plasma that react with antigens on the surfaces of foreign red blood cells when donor and recipient differ in blood type.

agglutinogens (a-gloo-TIN-ō-jenz): Antigens on the surfaces of red blood cells whose presence and structure are genetically determined.

aggregate lymphoid nodules: Lymphoid nodules beneath the epithelium of the small intestine; also called *Peyer's patches.*

agonist: A muscle responsible for a specific movement.

agranular: Without granules; *agranular leukocytes* are monocytes and lymphocytes.

AIDS: *See* **acquired immune deficiency syndrome.**

alba: White.

albicans (AL-bi-kanz): White.

albinism: The absence of pigment in hair and skin, caused by the inability of melanocytes to produce melanin.

albuginea (al-bū-JIN-ē-uh): White.

aldosterone: A mineralocorticoid produced by the zona glomerulosa of the adrenal cortex; stimulates sodium and water conservation at the kidneys; secreted in response to the presence of angiotensin II.

aldosteronism: The condition caused by the oversecretion of aldosterone, characterized by fluid retention, edema, and hypertension.

alkalosis (al-ka-LŌ-sis): The condition characterized by a plasma pH greater than 7.45; associated with a relative deficiency of hydrogen ions or an excess of bicarbonate ions.

allantois (a-LAN-tō-is): One of the four extraembryonic membranes; provides vascularity to the chorion and is therefore essential to placenta formation; the proximal portion becomes the urinary bladder.

alleles (a-LĒLZ): Alternate forms of a particular gene.

allergen: An antigenic compound that produces a hypersensitivity response.

alpha-blockers: Drugs that prevent the stimulation of alpha receptors.

alpha cells: Cells in the pancreatic islets that secrete glucagon.

alpha receptors: Membrane receptors sensitive to norepinephrine or epinephrine; stimulation normally results in the excitation of the target cell.

alveolar sac: An air-filled chamber that supplies air to several alveoli.

alveolus/alveoli (al-VĒ-o-lī): Blind pockets at the end of the respiratory tree, lined by a simple squamous epithelium and surrounded by a capillary network; sites of gas exchange with the blood; a bony socket that holds the root of a tooth.

Alzheimer's disease: A disorder resulting from degenerative changes in populations of neurons in the cerebrum, causing dementia characterized by problems with attention, short-term memory, and emotions.

amacrine cells (AM-a-krīn): Modified neurons in the retina that facilitate or inhibit communication between bipolar cells and ganglion cells.

amenorrhea (ā-men-ō-RĒ-uh): A failure to commence menstruation at adolescence or the cessation of menstruation prior to menopause.

amination: The attachment of an amino group to a carbon chain; performed by a variety of cells and important in the synthesis of amino acids.

amino acids: Organic compounds whose chemical structure can be summarized as R—$CHNH_2$—$COOH$.

amino group: —NH_2.

amnesia: A temporary or permanent memory loss.

amniocentesis: The sampling of amniotic fluid for analytical purposes; used to detect certain genetic abnormalities.

amnion (AM-nē-on): One of the four extraembryonic membranes; surrounds the developing embryo or fetus.

amniotic fluid (am-nē-OT-ik): Fluid that fills the amniotic cavity; cushions and supports the embryo or fetus.

amphiarthrosis (am-fē-ar-THRŌ-sis): An articulation that permits a small degree of independent movement; *see* **interosseous membrane** and **pubic symphysis.**

amphicytes (AM-fi-sīts): Supporting cells that surround neurons in the peripheral nervous system; also called *satellite cells.*

amphimixis (am-fi-MIK-sis): The fusion of male and female pronuclei after fertilization.

ampulla/ampullae (am-PŪL-la): A localized dilation in the lumen of a canal or passageway.

amygdaloid body (ah-MIG-da-loyd): A cerebral nucleus that is a component of the limbic system and acts as an interface between that system, the cerebrum, and sensory systems.

amylase: An enzyme that breaks down polysaccharides; produced by the salivary glands and pancreas.

anabolism (a-NAB-ō-lizm): The synthesis of complex organic compounds from simpler precursors.

anaerobic: Without oxygen.

analgesic: A substance that relieves pain.

anal triangle: The posterior subdivision of the perineum.

anamnestic response (an-am-NES-tik): The sudden and exaggerated production of antibodies after a second exposure to a specific antigen; due to the activation of memory B cells.

anaphase (AN-a-fāz): The mitotic stage in which the paired chromatids separate and move toward opposite ends of the spindle apparatus.

anaphylaxis (a-na-fi-LAK-sis): A hypersensitivity reaction due to the binding of antigens to immunoglobulins (IgE) on the surfaces of mast cells; the release of histamine, serotonin, and prostaglandins by mast cells then causes widespread inflammation; a sudden decline in blood pressure may occur, producing anaphylactic shock.

anastomosis (a-nas-to-MŌ-sis): The joining of two tubes, usually referring to a connection between two peripheral vessels without an intervening capillary bed.

anatomical position: An anatomical reference position; the body viewed from the anterior surface with the palms facing forward; supine.

anatomy (a-NAT-o-mē): The study of the structure of the body.

anaxonic neuron (an-ak-SON-ik): A central nervous system neuron that has many processes but no apparent axon.

androgen (AN-drō-jen): A steroid sex hormone primarily produced by the interstitial cells of the testis and manufactured in small quantities by the adrenal cortex in either gender.

anemia (a-NĒ-mē-uh): The condition marked by a reduction in the hematocrit, the hemoglobin content of the blood, or both.

anencephaly (an-en-SEF-a-lē): A developmental defect characterized by incomplete development of the cerebral hemispheres and cranium.

anesthesia: The total or partial loss of sensation from a region of the body.

aneurysm (AN-ū-rizm): A weakening and localized dilation in the wall of a blood vessel.

angiogram (AN-jē-ō-gram): An X-ray image of circulatory pathways.

angiography: The X-ray examination of vessel distribution after the introduction of radiopaque substances into the bloodstream.

angiotensin I: The hormone produced by the activation of angiotensinogen by renin; angiotensin-converting enzyme converts angiotensin I into angiotensin II in lung capillaries.

angiotensin II: A hormone that causes an elevation in systemic blood pressure, stimulates the secretion of aldosterone, promotes thirst, and causes the release of antidiuretic hormone; angiotensin-converting enzyme in lung capillaries converts angiotensin I into angiotensin II.

angiotensinogen: The blood protein produced by the liver that is converted to angiotensin I by the enzyme renin.

anion (AN-ī-on): An ion bearing a negative charge.

ankyloglossia (ang-ki-lō-GLOS-ē-uh): A condition characterized by an overly robust and restrictive lingual frenulum.

anorectal canal (ā-nō-REK-tal): The distal portion of the rectum that contains the rectal columns and ends at the anus.

anorexia nervosa: An eating disorder marked by a loss of appetite and pronounced weight loss.

anoxia (a-NOKS-ē-uh): Tissue oxygen deprivation.

antagonist: A muscle that opposes the movement of an agonist.

antebrachium: The forearm.

anteflexion (an-tē-FLEK-shun): The normal position of the uterus, with the superior surface bent forward.

anterior: On or near the front, or ventral surface, of the body.

anterior pituitary: *See* **pituitary gland** or **adenohypophysis.**

anterograde amnesia: The inability to store memories of events that occur after a specific incident or time.

anthracosis (an-thra-KŌ-sis): "Black lung disease," a deterioration of respiratory exchange efficiency due to the chronic inhalation of coal dust.

antibiotic: A chemical agent that selectively kills pathogens.

antibody (AN-ti-bod-ē): A globular protein produced by plasma cells that will bind to specific antigens and promote their destruction or removal from the body.

antibody-mediated immunity: The form of immunity resulting from the presence of circulating antibodies produced by plasma cells; also called *humoral immunity.*

anticholinesterase: A chemical compound that blocks the action of acetylcholine and causes prolonged and intensive stimulation of postsynaptic membranes.

anticoagulant: A compound that slows or prevents clot formation by interfering with the clotting system.

anticodon: Three nitrogenous bases on a tRNA molecule that interact with an appropriate codon on a strand of mRNA.

antidiuretic hormone (ADH) (an-tī-dī-ū-RET-ik): A hormone synthesized in the hypothalamus and secreted at the posterior pituitary; causes water retention at the kidneys and an elevation of blood pressure.

antigen: A substance capable of inducing the production of antibodies.

antigen–antibody complex: The combination of an antigen and a specific antibody.

antigenic determinant site: A portion of an antigen that can interact with an antibody molecule.

antigen-presenting cell (APC): A cell that processes antigens and displays them, bound to MHC proteins; essential to the initiation of a normal immune response.

antihistamine (an-tī-HIS-ta-mēn): A chemical agent that blocks the action of histamine on peripheral tissues.

antipyretic agents: Chemicals that reduce fever.

antrum (AN-trum): A chamber or pocket.

anulus (AN-ū-lus): A cartilage or bone shaped like a ring; also spelled *annulus.*

anuria (a-NŪ-rē-uh): A cessation of urine production.

anus: The external opening of the anorectal canal.

aorta: The large, elastic artery that carries blood away from the left ventricle and into the systemic circuit.

aortic reflex: A baroreceptor reflex triggered by increased aortic pressures; leads to a reduction in cardiac output and a fall in systemic pressure.

Apgar test: A test used to assess the neurological status of a newborn infant.

aphasia: The inability to speak.

apnea (AP-nē-uh): The cessation of breathing.

apneustic center (ap-NOO-stik): A respiratory center whose chronic activation would lead to apnea at full inhalation.

apocrine secretion: A mode of secretion in which the glandular cell sheds portions of its cytoplasm.

aponeurosis/aponeuroses (ap-ō-nū-RŌ-sis): A broad tendinous sheet that may serve as the origin or insertion of a skeletal muscle.

apoplexy: A cerebrovascular accident (stroke).

appendicitis: An inflammation of the appendix.

appendicular: Pertaining to the upper or lower limbs.

appendix: *See* **vermiform appendix.**

appositional growth: The enlargement of a bone by the addition of cartilage or bony matrix at its surface.

aqueous humor: A fluid similar to perilymph or cerebrospinal fluid that fills the anterior chamber of the eye.

arachidonic acid: One of the essential fatty acids.

arachnoid (a-RAK-noyd): The middle meninx that encloses cerebrospinal fluid and protects the central nervous system.

arachnoid granulations: Processes of the arachnoid that project into the superior sagittal sinus; sites where cerebrospinal fluid enters the venous circulation.

arbor vitae: The central, branching mass of white matter inside the cerebellum.

arcuate (AR-kū-āt): Curving.

areflexia (ā-re-flek-sē-uh): The absence of normal reflex responses to stimulation.

areola (a-RĒ-ō-la): The pigmented area that surrounds the nipple of a breast.

areolar: Containing minute spaces, as in areolar tissue.

areolar tissue: Loose connective tissue with an open framework.

arrector pili (ar-REK-tor PI-lē): Smooth muscles whose contractions force hairs to stand erect.

arrhythmias (ā-RITH-mē-az): Abnormal patterns of cardiac contractions.

arteriole (ar-TĒ-rē-ōl): A small arterial branch that delivers blood to a capillary network.

artery: A blood vessel that carries blood away from the heart and toward a peripheral capillary.

arthritis (ar-THRĪ-tis): An inflammation of a joint.

arthroscope: A fiber-optic device intended for visualizing the interior of joints; also used for certain forms of joint surgery.

articular: Pertaining to a joint.

articular capsule: The dense collagen fiber sleeve that surrounds a joint and provides protection and stabilization.

articular cartilage: The cartilage pad that covers the surface of a bone inside a joint cavity.

articulation (ar-tik-ū-LĀ-shun): A joint; the formation of words.

arytenoid cartilages (ar-i-TĒ-noyd): A pair of small cartilages in the larynx.

ascending tract: A tract carrying information from the spinal cord to the brain.

ascites (a-SĪ-tēz): The overproduction and accumulation of peritoneal fluid.

asphyxia: Unconsciousness due to oxygen deprivation at the central nervous system.

aspirate: To remove or obtain by suction; to inhale.

association areas: Cortical areas of the cerebrum that are responsible for the integration of sensory inputs and/or motor commands.

association neuron: *See* **interneuron.**

asthma (AZ-ma): A reversible constriction of smooth muscles around respiratory passageways, commonly caused by an allergic response.

astigmatism: A visual disturbance due to an irregularity in the shape of the cornea.

astrocyte (AS-trō-sīt): One of the four types of neuroglia in the central nervous system; responsible for maintaining the blood–brain barrier by the stimulation of endothelial cells.

ataxia: A failure to coordinate muscular activities normally.

atelectasis (at-e-LEK-ta-sis): The collapse of a lung or a portion of a lung.

atherosclerosis (ath-er-ō-skle-RŌ-sis): The formation of fatty plaques in the walls of arteries, leading to circulatory impairment.

atom: The smallest stable unit of matter.

atomic number: The number of protons in the nucleus of an atom.

atomic weight: Roughly, the average total number of protons and neutrons in the atoms of a particular element.

atresia (a-TRĒ-zē-uh): The closing of a cavity, or its incomplete development; refers to the degeneration of developing ovarian follicles.

atria: Thin-walled chambers of the heart that receive venous blood from the pulmonary or systemic circuit.

atrial natriuretic peptide (ANP) (nā-trē-ū-RET-ik): The hormone released by specialized atrial cardiocytes when they are stretched by an abnormally large venous return; promotes fluid loss and reductions in blood pressure and in venous return.

atrial reflex: The reflexive increase in heart rate after an increase in venous return; due to mechanical and neural factors; also called *Bainbridge reflex.*

atrioventricular (AV) node (ā-trē-ō-ven-TRIK-ū-lar): Specialized cardiocytes that relay the contractile stimulus to the bundle of His, the bundle branches, the Purkinje fibers, and the ventricular myocardium; located at the boundary between the atria and ventricles.

atrioventricular (AV) valve: One of the valves that prevent backflow into the atria during ventricular systole.

atrophy (AT-rō-fē): The wasting away of tissues from a lack of use, ischemia, or nutritional abnormalities.

auditory: Pertaining to the sense of hearing.

auditory ossicles: The bones of the middle ear: malleus, incus, and stapes.

auditory tube: A passageway that connects the nasopharynx with the middle ear cavity; also called *Eustachian tube* or *pharyngotympanic tube.*

auricle: A broad, flattened process that resembles the external ear; in the ear, the expanded, projecting portion that surrounds the external auditory canal, also called *pinna;* in the heart, the externally visible flap formed by the collapse of the outer wall of a relaxed atrium.

autoantibodies: Antibodies that react with antigens on the surfaces of a person's own cells and tissues.

autodigestion: The digestion of tissues by digestive acids or enzymes from the stomach or pancreas.

autoimmunity: The immune system's sensitivity to normal cells and tissues, resulting in the production of autoantibodies.

autolysis: The destruction of a cell due to the rupture of lysosomal membranes in its cytoplasm.

automaticity: The spontaneous depolarization to threshold, characteristic of cardiac pacemaker cells.

autonomic ganglion: A collection of visceral motor neurons outside the central nervous system.

autonomic nerve: A peripheral nerve consisting of preganglionic or postganglionic autonomic fibers.

autonomic nervous system (ANS): Centers, nuclei, tracts, ganglia, and nerves involved in the unconscious regulation of visceral functions; includes components of the central nervous system and the peripheral nervous system.

autopsy: The detailed examination of a body after death, normally performed by a pathologist.

autoregulation: Changes in activity that maintain homeostasis in direct response to changes in the local environment; does not require neural or endocrine control.

autosomal (aw-to-SŌ-mal): Chromosomes other than the X or Y chromosome.

avascular (ā-VAS-kū-lar): Without blood vessels.

avitaminosis (ā-vī-ta-min-Ō-sis): A condition caused by the inadequate intake of one or more essential vitamins.

avulsion: An injury involving the violent tearing away of body tissues.

axilla: The armpit.

axolemma: The cell membrane of an axon, continuous with the cell membrane of the cell body and dendrites and distinct from any neuroglial coverings.

axon: The elongate extension of a neuron that conducts an action potential.

axon hillock: In a multipolar neuron, the portion of the cell body adjacent to the initial segment.

axoplasm (AK-sō-plazm): The cytoplasm within an axon.

B

Babinski sign: The reflexive extension and abduction of the toes after the medial, plantar surface of the foot is stroked; positive reflex (Babinski sign) is normal up to age 1.5 years; thereafter, a positive reflex indicates damage to descending tracts.

bacteria: Single-celled microorganisms, some pathogenic, that are common in the environment and in and on the body.

Bainbridge reflex: *See* **atrial reflex.**

baroreception: The ability to detect changes in pressure.

baroreceptor reflex: A reflexive change in cardiac activity in response to changes in blood pressure.

baroreceptors (bar-ō-rē-SEP-torz): The receptors responsible for baroreception.

basal metabolic rate (BMR): The resting metabolic rate of a normal fasting individual under homeostatic conditions.

base: A compound whose dissociation releases a hydroxide ion (OH^-) or removes a hydrogen ion (H^+) from the solution.

basement membrane: A layer of filaments and fibers that attach an epithelium to the underlying connective tissue.

basilar membrane: The membrane that supports the organ of Corti and separates the cochlear duct from the scala tympani in the inner ear.

basophils (BĀ-sō-filz): Circulating granulocytes (white blood cells) similar in size and function to tissue mast cells.

B cells: Lymphocytes capable of differentiating into plasma cells, which produce antibodies.

benign: Not malignant.

beta cells: Cells of the pancreatic islets that secrete insulin in response to elevated blood sugar concentrations.

beta oxidation: Fatty acid catabolism that produces molecules of acetyl-CoA.

beta receptors: Membrane receptors sensitive to epinephrine; stimulation may result in the excitation or inhibition of the target cell.

bicarbonate ions: HCO_3^-; anion components of the carbonic acid–bicarbonate buffer system.

bicuspid (bī-KUS-pid): Having two cusps or leafs; refers to a premolar tooth, which has two roots, or to the left AV valve, which has two cusps.

bicuspid valve: The left atrioventricular (AV) valve, also called *mitral valve.*

bifurcate: To branch into two parts.

bile: The exocrine secretion of the liver; stored in the gallbladder and ejected into the duodenum.

bile salts: Steroid derivatives in bile; responsible for the emulsification of ingested lipids.

bilirubin (bil-ē-ROO-bin): A pigment that is the byproduct of hemoglobin catabolism.

bioenergetics: The analysis of energy production and utilization by cells.

biofeedback: The use of artificial signals to provide feedback about unconscious, visceral motor activities.

biopsy: The removal of a small sample of tissue for pathological analysis.

bipennate: A muscle whose fibers are arranged on either side of a common tendon.

bladder: A muscular sac that distends as fluid is stored and whose contraction ejects the fluid at an appropriate time; used alone, the term usually refers to the urinary bladder.

blastocoele (BLAS-tō-sēl): A fluid-filled cavity within a blastocyst.

blastocyst (BLAS-tō-sist): An early stage in the developing embryo, consisting of an outer trophoblast and an inner cell mass.

blastomere (BLAS-tō-mēr): One of the cells in the morula; a collection of cells produced by the division of the zygote.

blockers/blocking agents: Drugs that block membrane pores or prevent binding to membrane receptors.

blood–brain barrier: The isolation of the central nervous system from the general circulation; primarily the result of astrocyte regulation of capillary permeabilities.

blood clot: A network of fibrin fibers and trapped blood cells.

blood–CSF barrier: The isolation of the cerebrospinal fluid from the capillaries of the choroid plexus; primarily the result of specialized ependymal cells.

blood pressure: A force exerted against vessel walls by the blood in the vessels, due to the push exerted by cardiac contraction and the elasticity of the vessel walls; usually measured along one of the muscular arteries, with systolic pressure measured during ventricular systole and diastolic pressure during ventricular diastole.

blood–testis barrier: The isolation of the interior of the seminiferous tubules from the general circulation, due to the activities of the sustentacular (Sertoli) cells.

Bohr effect: The increased oxygen release by hemoglobin in the presence of elevated carbon dioxide levels.

boil: An abscess of the skin, normally involving a sebaceous gland.

bolus: A compact mass; usually refers to compacted ingested material on its way to the stomach.

bone: *See* **osseous tissue.**

botulinus toxin (bot-ū-LĪ-nus): A toxin, produced by the anaerobic bacterium *Clostridium botulinum,* that can cause severe food poisoning.

bowel: The intestinal tract.

Bowman's capsule: The cup-shaped initial portion of the renal tubule; surrounds the glomerulus and receives the glomerular filtrate.

Boyle's law: The principle that, in a gas, pressure and volume are inversely related.

brachial: Pertaining to the arm.

brachial plexus: A network formed by branches of spinal nerves C_5–T_1 en route to innervating the upper limb.

brachium: The arm.

bradycardia (brad-ē-KAR-dē-uh): A slow heart rate, below 50 bpm.

brain stem: The brain minus the cerebrum, diencephalon, and cerebellum.

brevis: Short.

Broca's area: The speech center of the brain, normally located on the neural cortex of the left cerebral hemisphere.

bronchial tree: The trachea, bronchi, and bronchioles.

bronchitis (brong-KĪ-tis): An inflammation of the bronchial passageways.

bronchodilation: The dilation of the bronchial passages; can be caused by sympathetic stimulation.

bronchodilators (brong-kō-dī-LĀ-torz): Drugs that produce bronchodilation; some are used clinically in treating asthma.

bronchoscope: A fiber-optic instrument used to examine the bronchial passageways.

bronchus/bronchi: A branch of the bronchial tree between the trachea and bronchioles.

buccal (BUK-al): Pertaining to the cheeks.

buffer: A compound that stabilizes the pH of a solution by removing or releasing hydrogen ions.

buffer system: Interacting compounds that prevent increases or decreases in the pH of body fluids; includes the carbonic acid–bicarbonate buffer system, the phosphate buffer system, and the protein buffer system.

bulbar: Pertaining to the brain stem.

bulbourethral glands (bul-bō-ū-RĒ-thral): Mucous glands at the base of the penis that secrete into the penile urethra; the equivalent of the greater vestibular glands of females; also called *Cowper's glands.*

bundle branches: Specialized conducting cells in the ventricles that carry the contractile stimulus from the bundle of His to the Purkinje fibers.

bundle of His (hiss): Specialized conducting cells in the interventricular septum that carry the contracting stimulus from the AV node to bundle branches and then to Purkinje fibers.

bursa: A small sac filled with synovial fluid that cushions adjacent structures and reduces friction.

bursitis: A painful inflammation of one or more bursae.

C

calcanean tendon: The large tendon that inserts on the calcaneus; tension on this tendon produces plantar flexion of the foot; also called *Achilles tendon.*

calcaneus (kal-KĀ-nē-us or kal-KAN-ē-us): The heelbone, the largest of the tarsal bones.

calcification: The deposition of calcium salts within a tissue.

calcitonin (kal-si-TŌ-nin): The hormone secreted by C cells of the thyroid when calcium ion concentrations are abnormally high; restores homeostasis by increasing the rate of bone deposition and the rate of calcium loss at the kidneys.

calculus/calculi (KAL-kū-lī): Concretions of insoluble materials that form within body fluids, especially the gallbladder, kidneys, or urinary bladder.

callus: A localized thickening of the epidermis due to chronic mechanical stresses; a thickened area that forms at the site of a bone break as part of the repair process.

calorie (c) (KAL-o-rē): The amount of heat that is required to raise the temperature of 1 gram of water 1°C.

Calorie (C): The amount of heat that is required to raise the temperature of 1 kilogram of water 1°C; also called *kilocalorie.*

calorigenic effect: The stimulation of energy production and heat loss by thyroid hormones.

calvaria (kal-VAR-ē-uh or kal-VĀ-rē-ah): The skullcap, consisting of the superior portions of the frontal, parietal, and occipital bones.

calyx/calyces (KĀL-i-sēz): Cup-shaped divisions of the renal pelvis.

canaliculi (kan-a-LIK-ū-lī): Microscopic passageways between cells; bile canaliculi carry bile to bile ducts in the liver; in bone, canaliculi permit the diffusion of nutrients and wastes to and from osteocytes.

cancellous bone (KAN-sel-us): Spongy bone, composed of a network of bony struts.

cancer: A malignant tumor that can undergo metastasis.

cannula: A tube that can be inserted into the body; commonly placed in blood vessels prior to transfusion or dialysis.

canthus, medial and lateral (KAN-thus): The angles formed at either corner of the eye between the upper and lower eyelids.

capacitation: The activation process that must occur before a spermatozoon can successfully fertilize an oocyte; occurs in the vagina after ejaculation.

capillary (KAP-i-lār-ē or ka-PIL-a-rē): A small blood vessel, located between an arteriole and a venule, whose thin wall permits the diffusion of gases, nutrients, and wastes between plasma and interstitial fluids.

capitulum (ka-PIT-ū-lum): A general term for a small, elevated articular process; refers to the rounded distal surface of the humerus that articulates with the head of the radius.

caput: The head.

carbaminohemoglobin (kar-bam-ē-nō-HĒ-mō-GLŌ-bin): Hemoglobin bound to carbon dioxide molecules.

carbohydrase (kar-bō-HĪ-drās): An enzyme that breaks down carbohydrate molecules.

carbohydrate (kar-bō-HĪ-drāt): An organic compound containing carbon, hydrogen, and oxygen in a ratio that approximates 1:2:1.

carbon dioxide: CO_2; a compound produced by the decarboxylation reactions of aerobic metabolism.

carbonic anhydrase: An enzyme that catalyzes the reaction $H_2O + CO_2 \rightleftharpoons H_2CO_3$; important in carbon dioxide transport, gastric acid secretion, and renal pH regulation.

carboxypeptidase (kar-bok-sē-PEP-ti-dās): A protease that breaks down proteins and releases amino acids.

carcinogenic (kar-sin-ō-JEN-ik): Stimulating cancer formation in affected tissues.

cardia (KAR-dē-uh): The area of the stomach surrounding its connection with the esophagus.

cardiac: Pertaining to the heart.

cardiac cycle: One complete heartbeat, including atrial and ventricular systole and diastole.

cardiac glands: Mucous glands characteristic of the cardia of the stomach.

cardiac output: The amount of blood ejected by the left ventricle each minute; normally about 5 liters.

cardiac reserve: The potential percentage increase in cardiac output above resting levels.

cardiac tamponade: A compression of the heart due to fluid accumulation in the pericardial cavity.

cardiocyte (KAR-dē-ō-sīt): A cardiac muscle cell.

cardiomyopathy (kar-dē-ō-mī-OP-a-thē): A progressive disease characterized by damage to the cardiac muscle tissue.

cardiopulmonary resuscitation (CPR): A method of artificially maintaining respiratory and circulatory functions.

cardiovascular: Pertaining to the heart, blood, and blood vessels.

cardiovascular centers: Poorly localized centers in the reticular formation of the medulla oblongata of the brain; includes cardioacceleratory, cardioinhibitory, and vasomotor centers.

cardium: The heart.

carina (ka-RĪ-na): A ridge on the inner surface of the base of the trachea that runs anteroposteriorly, between two primary bronchi.

carotene (KAR-ō-tēn): A yellow-orange pigment, found in carrots and in green and orange leafy vegetables, that the body can convert to vitamin A.

carotid artery: The principal artery of the neck, servicing cervical and cranial structures; one branch, the internal carotid, provides a major blood supply to the brain.

carotid body: A group of receptors, adjacent to the carotid sinus, that are sensitive to changes in the carbon dioxide levels, pH, and oxygen concentrations of arterial blood.

carotid sinus: A dilated segment at the base of the internal carotid artery whose walls contain baroreceptors sensitive to changes in blood pressure.

carotid sinus reflex: Reflexive changes in blood pressure that maintain homeostatic pressures at the carotid sinus, stabilizing blood flow to the brain.

carpus/carpal: The wrist.

cartilage: A connective tissue with a gelatinous matrix that contains an abundance of fibers.

castration: The removal of the testes; also called *bilateral orchiectomy.*

catabolism (ka-TAB-ō-lizm): The breakdown of complex organic molecules into simpler components, accompanied by the release of energy.

catalyst (KAT-uh-list): A substance that accelerates a specific chemical reaction but that is not altered by the reaction.

cataract: A reduction in lens transparency that causes visual impairment.

catecholamine (kat-e-KŌL-am-ēn): Epinephrine, norepinephrine, dopamine, and related compounds.

cathepsins (ka-THEP-sinz): Enzymes, present in the sarcoplasm of skeletal muscle cells, that can break down contractile proteins, thereby providing amino acids that can act as a supplemental energy source.

catheter (KATH-e-ter): A tube surgically inserted into a body cavity or along a blood vessel or excretory passageway for the collection of body fluids, monitoring of blood pressure, or introduction of medications or radiographic dyes.

cation (KAT-ī-on): An ion that bears a positive charge.

cauda equina (KAW-da ek-WĪ-na): Spinal nerve roots distal to the tip of the adult spinal cord; they extend caudally inside the vertebral canal en route to lumbar and sacral segments.

caudal/caudally: Closest to or toward the tail (coccyx).

caudate nucleus (KAW-dāt): One of the cerebral nuclei of the extrapyramidal system; involved with the unconscious control of muscular activity.

cavernous tissue: Erectile tissue that can be engorged with blood; located in the penis (males) and clitoris (females).

cecum (SĒ-kum): An expanded pouch at the start of the large intestine.

cell: The smallest living unit in the human body.

cell body: Body; the body of a neuron; also called *soma.*

cell-mediated immunity: Resistance to disease through the activities of sensitized T cells that destroy antigen-bearing cells by direct contact or through the release of lymphotoxins; also called *cellular immunity.*

cellulitis (sel-ū-LĪ-tis): Diffuse inflammation, normally involving areas of loose connective tissue, such as the subcutaneous layer.

cementum (se-MEN-tum): Bony material that covers the root of a tooth and is not shielded by a layer of enamel.

center of ossification: The site in a connective tissue where bone formation begins.

central canal: Longitudinal canal in the center of an osteon that contains blood vessels and nerves; also called *Haversian canal;* a passageway along the longitudinal axis of the spinal cord that contains cerebrospinal fluid.

central nervous system (CNS): The brain and spinal cord.

central sulcus: A groove in the surface of a cerebral hemisphere, between the primary sensory and primary motor areas of the cortex.

centriole: A cylindrical intracellular organelle composed of nine groups of microtubules, three in each group; functions in mitosis or meiosis by organizing the microtubules of the spindle apparatus.

centromere (SEN-trō-mēr): The localized region where two chromatids remain connected after the chromosomes have replicated; site of spindle fiber attachment.

centrosome: A region of cytoplasm that contains a pair of centrioles oriented at right angles to one another.

centrum: The body of a vertebra.

cephalic: Pertaining to the head.

cerebellum (ser-e-BEL-um): The posterior portion of the metencephalon, containing the cerebellar hemispheres; includes the arbor vitae, cerebellar nuclei, and cerebellar cortex.

cerebral cortex: An extensive area of neural cortex covering the surfaces of the cerebral hemispheres.

cerebral hemispheres: A pair of expanded portions of the cerebrum covered in neural cortex.

cerebral nuclei: Nuclei of the cerebrum that are important in the subconscious control of skeletal muscle activity.

cerebral palsy: A chronic condition resulting from damage to motor areas of the brain during development or at delivery.

cerebral peduncle: A mass of nerve fibers on the ventrolateral surface of the mesencephalon; contains ascending tracts that terminate in the thalamus and descending tracts that originate in the cerebral hemispheres.

cerebrospinal fluid (CSF): Fluid bathing the internal and external surfaces of the central nervous system; secreted by the choroid plexus.

cerebrovascular accident (CVA): The occlusion of a blood vessel that supplies a portion of the brain, resulting in damage to the dependent neurons; also called *stroke.*

cerebrum (SER-ē-brum): The largest portion of the brain, composed of the cerebral hemispheres; includes the cerebral cortex, the cerebral nuclei, and the internal capsule.

cerumen: The waxy secretion of the ceruminous glands along the external auditory canal.

ceruminous glands (se-ROO-mi-nus): Integumentary glands that secrete cerumen.

cervical enlargement: A thickening of the cervical portion of the spinal cord caused by the abundance of central nervous system neurons involved with motor control of the upper limbs.

cervix: The inferior portion of the uterus.

cesarean section: The surgical delivery of an infant via an incision through the lower abdominal wall and uterus.

chalazion (kah-LĀ-zē-on): An inflammation and distension of a Meibomian gland on the eyelid; also called *sty.*

chancre (SHANG-ker): A skin lesion that develops at the primary site of a syphilis infection.

chemoreception: The detection of changes in the concentrations of dissolved compounds or gases.

chemotaxis (kē-mō-TAK-sis): The attraction of phagocytic cells to the source of abnormal chemicals in tissue fluids.

chemotherapy: The treatment of illness through the administration of specific chemicals.

chloride shift: The movement of plasma chloride ions into red blood cells in exchange for bicarbonate ions generated by the intracellular dissociation of carbonic acid.

cholecystitis (kō-lē-sis-TĪ-tis): An inflammation of the gallbladder.

cholecystokinin (CCK) (kō-lē-sis-tō-KĪ-nin): A duodenal hormone that stimulates the contraction of the gallbladder and the secretion of enzymes by the exocrine pancreas; also called *pancreozymin.*

cholelithiasis (kō-lē-li-THĪ-a-sis): The formation or presence of gallstones.

cholesterol: A steroid component of cell membranes and a substrate for the synthesis of steroid hormones and bile salts.

choline: A breakdown product or precursor of acetylcholine.

cholinergic synapse (kō-lin-ER-jik): A synapse where the presynaptic membrane releases acetylcholine on stimulation.

cholinesterase (kō-li-NES-ter-ās): The enzyme that breaks down and inactivates acetylcholine.

chondrocyte (KON-drō-sīt): A cartilage cell.

chondroitin sulfate (kon-DROY-tin): The predominant proteoglycan in cartilage, responsible for the gelatinous consistency of the matrix.

chordae tendineae (KOR-dē TEN-di-nē-ē): Fibrous cords that stabilize the position of the AV valves in the heart, preventing backflow during ventricular systole.

chorion/chorionic (KOR-ē-on/ko-rē-ON-ik): An extraembryonic membrane, consisting of the trophoblast and underlying mesoderm, that forms the placenta.

choroid: The middle, vascular layer in the wall of the eye.

choroid plexus: The vascular complex in the roof of the third and fourth ventricles of the brain, responsible for the production of cerebrospinal fluid.

chromatid (KRŌ-ma-tid): One complete copy of a DNA strand and its associated nucleoproteins.

chromatin (KRŌ-ma-tin): A histological term referring to the grainy material visible in cell nuclei during interphase; the appearance of the DNA content of the nucleus when the chromosomes are uncoiled.

chromosomes: Dense structures, composed of tightly coiled DNA strands and associated histones, that become visible in the nucleus when a cell prepares to undergo mitosis or meiosis; normal human somatic cells contain 46 chromosomes apiece.

chronic: Habitual or long term.

chylomicrons (kī-lō-MĪ-kronz): Relatively large droplets that may contain triglycerides, phospholipids, and cholesterol in association with proteins; synthesized and released by intestinal cells and transported to the venous blood by the lymphatic system.

chyme (kīm): A semifluid, acidic mixture of ingested food and digestive secretions that forms in the stomach during the early phases of digestion.

chymotrypsin (kī-mō-TRIP-sin): A protease in the small intestine.

chymotrypsinogen: The inactive proenzyme, secreted by the pancreas, that is subsequently converted to chymotrypsin.

ciliary body: A thickened region of the choroid that encircles the lens of the eye; includes the ciliary muscle and the ciliary processes that support the suspensory ligaments of the lens.

cilium/cilia: A slender organelle that extends above the free surface of an epithelial cell and generally undergoes cycles of movement; composed of a basal body and microtubules in a 9×2 array.

circulatory system: The network of blood vessels and lymphatic vessels that facilitate the distribution and circulation of extracellular fluid.

circumduction (sir-kum-DUK-shun): A movement at a synovial joint in which the distal end of the bone describes a circle but the shaft does not rotate.

circumvallate papilla (sir-kum-VAL-āt pa-PIL-la): One of the large, dome-shaped papillae on the superior surface of the tongue that form a V, separating the body of the tongue from the root.

cirrhosis (sir-RŌ-sis): A liver disorder characterized by the degeneration of hepatocytes and their replacement by fibrous connective tissue.

cisterna (sis-TUR-na): An expanded chamber.

citric acid cycle: *See* **TCA cycle.**

cleavage (KLĒ-vij): Mitotic divisions that follow the fertilization of an ovum and lead to the formation of a blastocyst.

cleavage lines: Stress lines in the skin that follow the orientation of major bundles of collagen fibers in the dermis.

clitoris (KLI-to-ris): A small erectile organ of females that is the developmental equivalent of the penis in males.

clone: As a verb, the production of genetically identical cells; as a noun, a genetically identical copy.

clonus (KLŌ-nus): Rapid cycles of muscle contraction and relaxation.

clot: A network of fibrin fibers and trapped blood cells; also called a *thrombus* if it occurs within the circulatory system.

clotting factors: Plasma proteins, synthesized by the liver, that are essential to the clotting response.

clotting response: The series of events that result in the formation of a clot.

coccygeal ligament: The fibrous extension of the dura mater and filum terminale; provides longitudinal stabilization to the spinal cord.

coccyx (KOK-siks): The terminal portion of the spinal column, consisting of relatively tiny, fused vertebrae.

cochlea (KOK-lē-uh): The spiral portion of the bony labyrinth of the inner ear that surrounds the organ of hearing.

cochlear duct (KOK-lē-ar): The central membranous tube within the cochlea that is filled with endolymph and contains the organ of Corti; also called *scala media.*

codon (KŌ-don): A sequence of three nitrogenous bases along an mRNA strand that will specify the location of a single amino acid in a peptide chain.

coelom (SĒ-lōm): The ventral body cavity, lined by a serous membrane and subdivided during fetal development into the pleural, pericardial, and abdominopelvic (peritoneal) cavities.

coenzymes (kō-EN-zīmz): Complex organic cofactors; most are structurally related to vitamins.

cofactor: Ions or molecules that must be attached to the active site before an enzyme can function; examples include mineral ions and several vitamins.

colectomy (kō-LEK-to-mē): The surgical removal of part or all of the colon.

colitis: An inflammation of the colon.

collagen: A strong, insoluble protein fiber common in connective tissues.

collateral ganglion (kō-LAT-er-al): A sympathetic ganglion situated anterior to the spinal column and separate from the sympathetic chain.

Colles' fracture (KOL-lēz): A fracture of the distal end of the radius and possibly the ulna, with posterior and dorsal displacement of the distal bone fragments.

colliculus/colliculi (kol-IK-ū-lus): A little mound; in the brain, refers to one of the thickenings in the roof of the mesencephalon; the superior colliculus is associated with the visual system, and the inferior colliculi with the auditory system.

colloid/colloidal suspension: A solution containing large organic molecules in suspension.

colon: The large intestine.

colonoscope (kō-LON-ō-skōp): A fiber-optic device for examining the interior of the colon.

colostomy (kō-LOS-to-mē): The surgical connection of a portion of the colon to the body wall, sometimes performed after a colectomy to permit the discharge of fecal materials.

colostrum (ko-LOS-trum): The secretion of the mammary glands at the time of childbirth and for a few days thereafter; contains more protein and less fat than does the milk secreted later.

coma (kō-ma): An unconscious state from which an individual cannot be aroused, even by strong stimuli.

comedo (kō-MĒ-dō): An inflamed sebaceous gland.

comminuted: Broken or crushed into small pieces.

commissure: A crossing over from one side to another.

common bile duct: The duct formed by the union of the cystic duct from the gallbladder and the bile ducts from the liver; terminates at the duodenal ampulla, where it meets the pancreatic duct.

common pathway: In the clotting response, the events that begin with the appearance of thromboplastin and end with the formation of a clot.

compact bone: Dense bone that contains parallel osteons.

compensation curves: The cervical and lumbar curves that develop to center the body weight over the legs.

complement: A system of 11 plasma proteins that interact in a chain-reaction after exposure to activated antibodies or the surfaces of certain pathogens; complement proteins promote cell lysis, phagocytosis, and other defense mechanisms.

compliance: Distensibility; the ability of certain organs to tolerate changes in volume; indicates the presence of elastic fibers and smooth muscles.

compound: A molecule containing two or more elements in combination.

concentration: The amount (in grams) or number of atoms, ions, or molecules (in moles) per unit volume.

concentration gradient: Regional differences in the concentration of a particular substance.

conception: Fertilization.

concha/conchae (KONG-kē): Three pairs of thin, scroll-like bones that project into the nasal cavities; the superior and medial conchae are part of the ethmoid bone, and the inferior conchae are separate bones.

concussion: A violent blow or shock; loss of consciousness due to a violent blow to the head.

condyle: A rounded articular projection on the surface of a bone.

cone: A photoreceptor of the retina, responsible for color vision.

congenital (kon-JEN-i-tal): Present at birth.

congestive heart failure (CHF): The failure to maintain adequate cardiac output due to circulatory problems or myocardial damage.

conjunctiva (kon-junk-TĪ-va): A layer of stratified squamous epithelium that covers the inner surfaces of the eyelids and the anterior surface of the eye to the edges of the cornea.

conjunctivitis: An inflammation of the conjunctiva.

connective tissue: One of the four primary tissue types; provides a structural framework that stabilizes the relative positions of the other tissue types; includes connective tissue proper, cartilage, bone, and blood; contains cell products, cells, and ground substance.

continuous propagation: The propagation of an action potential along an unmyelinated axon or a muscle cell membrane, wherein the action potential affects every portion of the membrane surface.

contractility: The ability to contract; possessed by skeletal, smooth, and cardiac muscle cells.

contralateral reflex: A reflex that affects the opposite side of the body from the stimulus.

conus medullaris: The conical tip of the spinal cord that gives rise to the filum terminale.

convergence: In the nervous system, the innervation of a single neuron by axons from several neurons; most common along motor pathways.

coracoid process (KOR-uh-koyd): A hook-shaped process of the scapula that projects above the anterior surface of the capsule of the shoulder joint.

Cori cycle: The metabolic exchange of lactic acid from skeletal muscle for glucose from the liver; performed during the recovery period after muscular exertion.

cornea (KOR-nē-uh): The transparent portion of the fibrous tunic of the anterior surface of the eye.

corniculate cartilages (kor-NIK-ū-lāt): A pair of small laryngeal cartilages.

cornu: Horn-shaped.

corona radiata (ko-RŌ-na rā-dē-A-ta): A layer of follicle cells surrounding a secondary oocyte at ovulation.

coronoid (ko-RŌ-noyd): Hooked or curved.

corpora cavernosa (ka-ver-NŌ-suh): Two parallel masses of erectile tissue within the body of the penis (males) or clitoris (females).

corpora quadrigemina (KOR-po-ra quad-ri-JEM-i-na): The superior and inferior colliculi of the mesencephalic tectum (roof) in the brain.

corpus/corpora: Body.

corpus callosum: The bundle of axons that links centers in the left and right cerebral hemispheres.

corpus luteum (LOO-tē-um): The progestin-secreting mass of follicle cells that develops in the ovary after ovulation.

corpus spongiosum (spon-jē-Ō-sum): The mass of erectile tissue that surrounds the urethra in the penis and expands distally to form the glans.

cortex: The outer layer or portion of an organ.

Corti, organ of: The receptor complex in the cochlear duct that includes the inner and outer hair cells, supporting cells and structures, and the tectorial membrane; provides the sensation of hearing.

corticobulbar tracts (kor-ti-kō-BUL-bar): Descending tracts that carry information or commands from the cerebral cortex to nuclei and centers in the brain stem.

corticospinal tracts: Descending tracts that carry motor commands from the cerebral cortex to the anterior gray horns of the spinal cord.

corticosteroid: A steroid hormone produced by the adrenal cortex.

corticosterone (kor-ti-KOS-te-rōn): A corticosteroid secreted by the zona fasciculata of the adrenal cortex; a glucocorticoid.

corticotropin-releasing hormone (CRH): The releasing hormone, secreted by the hypothalamus, that stimulates secretion of adrenocorticotropic hormone by the anterior pituitary.

cortisol (KOR-ti-sol): A corticosteroid secreted by the zona fasciculata of the adrenal cortex; a glucocorticoid.

costa/costae: A rib.

cotransport: The membrane transport of a nutrient, such as glucose, in company with the movement of an ion, normally sodium; transport requires a carrier protein but does not involve direct ATP expenditure and can occur regardless of the concentration gradient for the nutrient.

countercurrent exchange: The diffusion between two solutions that travel in opposite directions.

countercurrent multiplication: Active transport between two limbs of a loop that contains a fluid moving in one direction;

responsible for the concentration of urine in the kidney tubules.

covalent bond (kō-VĀ-lent): A chemical bond between atoms that involves the sharing of electrons.

cranial: Pertaining to the head.

cranial nerves: Peripheral nerves originating at the brain.

craniosacral division (krā-nē-ō-SAK-ral): *See* **parasympathetic division.**

craniostenosis (krā-nē-ō-sten-Ō-sis): A skull deformity caused by the premature closure of the cranial sutures.

cranium: The braincase; the skull bones that surround and protect the brain.

creatine: A nitrogenous compound, synthesized in the body, that can form a high-energy bond by connecting to a phosphate group and that serves as an energy reserve.

creatine phosphate: A high-energy compound in muscle cells; during muscle activity, the phosphate group is donated to ADP, regenerating ATP; also called *phosphorylcreatine.*

creatinine: A breakdown product of creatine metabolism.

crenation: Cellular shrinkage due to an osmotic movement of water out of the cytoplasm.

cribriform plate: A portion of the ethmoid bone that contains the foramina used by the axons of olfactory receptors en route to the olfactory bulbs of the cerebrum.

cricoid cartilage (KRĪ-koyd): A ring-shaped cartilage that forms the inferior margin of the larynx.

crista/cristae: A ridge-shaped collection of hair cells in the ampulla of a semicircular canal; the crista and cupula form a receptor complex sensitive to movement along the plane of the canal.

cross-bridge: A myosin head that projects from the surface of a thick filament and that can bind to an active site of a thin filament in the presence of calcium ions.

cruciate ligaments: A pair of intracapsular ligaments (anterior and posterior) in the knee.

cryosurgery: A surgical technique that involves freezing and killing cells in a localized area.

cryptorchid testis: An undescended testis that is in the abdominopelvic cavity rather than in the scrotum.

cuneiform cartilages (kū-NĒ-i-form): A pair of small cartilages in the larynx.

cupula (KŪ-pū-la): A gelatinous mass that is located in the ampulla of a semicircular canal in the inner ear and whose movement stimulates the hair cells of the crista.

curare: A toxin that prevents neural stimulation of neuromuscular junctions.

Cushing's disease: A condition caused by the oversecretion of adrenal steroids.

cutaneous membrane: The epidermis and papillary layer of the dermis.

cuticle: The layer of dead, keratinized cells that surrounds the shaft of a hair; for nails, *see* **eponychium.**

cyanosis: A bluish coloration of the skin due to the presence of deoxygenated blood in vessels near the body surface.

cyst: A fibrous capsule containing fluid or other material.

cystic duct: A duct that carries bile between the gallbladder and the common bile duct.

cystitis: An inflammation of the urinary bladder.

cytochrome (SĪ-tō-krōm): A pigment component of the electron transport system; a structural relative of heme.

cytokinesis (sī-tō-ki-NĒ-sis): The cytoplasmic movement that separates two daughter cells at the completion of mitosis.

cytology (sī-TOL-o-jē): The study of cells.

cytoplasm: The material between the cell membrane and the nuclear membrane; cell contents.

cytosine: A pyrimidine; one of the nitrogenous bases in the nucleic acids RNA and DNA.

cytoskeleton: A network of microtubules and microfilaments in the cytoplasm.

cytosol: The fluid portion of the cytoplasm.

cytotoxic: Poisonous to cells.

cytotoxic T cells: Lymphocytes involved in cell-mediated immunity that kill target cells by direct contact or by the secretion of lymphotoxins; also called *killer T cells* and T_C *cells*.

D

daughter cells: Genetically identical cells produced by somatic cell division.

deamination (dē-am-i-NĀ-shun): The removal of an amino group from an amino acid.

decarboxylation (dē-kar-boks-i-LĀ-shun): The removal of a molecule of carbon dioxide.

decerebrate: Lacking a cerebrum.

decomposition reaction: A chemical reaction that breaks a molecule into smaller fragments.

decubitis ulcers: Ulcers that form where chronic pressure interrupts blood flow to a portion of the skin.

decussate: To cross over to the opposite side, usually referring to the crossover of the descending tracts of the corticospinal pathway on the ventral surface of the medulla oblongata.

defecation (def-e-KĀ-shun): The elimination of fecal wastes.

deglutition: Swallowing.

degradation: Breakdown, catabolism.

dehydration: A reduction in the water content of the body that threatens homeostasis.

dehydration synthesis: The joining of two molecules associated with the removal of a water molecule.

delta cell: A pancreatic islet cell that secretes growth hormone-inhibiting hormone.

dementia: A loss of mental abilities.

demyelination: The loss of the myelin sheath of an axon, normally due to chemical or physical damage to Schwann cells or oligodendrocytes.

denaturation: A temporary or permanent change in the three-dimensional structure of a protein.

dendrite (DEN-drīt): A sensory process of a neuron.

denticulate ligaments: Supporting fibers that extend laterally from the surface of the spinal cord, tying the pia mater to the dura mater and providing lateral support for the spinal cord.

dentin (DEN-tin): The bonelike material that forms the body of a tooth; differs from bone in that it lacks osteocytes and osteons.

deoxyribonucleic acid (DNA) (dē-ok-sē-rī-bō-noo-KLĀ-ik): A nucleic acid consisting of a chain of nucleotides that contain the sugar deoxyribose and the nitrogenous bases adenine, guanine, cytosine, and thymine.

deoxyribose: A five-carbon sugar resembling ribose but lacking an oxygen atom.

depolarization: A change in the transmembrane potential from a negative value toward 0 mV.

depression: Inferior (downward) movement of a body part.

dermatitis: An inflammation of the skin.

dermatome: A sensory region monitored by the dorsal rami of a single spinal segment.

dermis: The connective tissue layer beneath the epidermis of the skin.

detrusor muscle (de-TROO-sor): A smooth muscle in the wall of the urinary bladder.

detumescence (dē-tū-MES-ens): The loss of a penile erection.

development: Growth and the acquisition of increasing structural and functional complexity; includes the period from conception to maturity.

diabetes insipidus: Polyuria due to inadequate production of antidiuretic hormone.

diabetes mellitus (mel-Ī-tus): Polyuria and glycosuria, most commonly due to the inadequate production of insulin with a resulting elevation of blood glucose levels.

diabetogenic effect: An elevation in blood sugar concentrations after the secretion of growth hormone or glucagon.

dialysis: Diffusion between two solutions of differing solute concentrations across a selectively permeable membrane, which permits the passage of some solutes but not others; regulates the composition of blood.

diapedesis (dī-a-pe-DĒ-sis): The movement of white blood cells through the walls of blood vessels by migration between adjacent endothelial cells.

diaphragm (DĪ-a-fram): Any muscular partition; the respiratory muscle that separates the thoracic cavity from the abdominopelvic cavity.

diaphysis (dī-A-fi-sis): The shaft of a long bone.

diarrhea (dī-a-RĒ-uh): Abnormally frequent defecation, associated with the production of unusually fluid feces.

diarthrosis (dī-ar-THRŌ-sis): A synovial joint.

diastolic pressure: Pressure measured in the walls of a muscular artery when the left ventricle is in diastole.

diencephalon (dī-en-SEF-a-lon): A division of the brain that includes the epithalamus, thalamus, and hypothalamus.

differential count: The determination of the relative abundance of each type of white blood cell on the basis of a random sampling of 100 white blood cells.

differentiation: The gradual appearance of characteristic cellular specializations during development as the result of gene activation or repression.

diffusion: Passive molecular movement from an area of higher concentration to an area of lower concentration.

digestion: The chemical breakdown of ingested materials into simple molecules that can be absorbed by the cells of the digestive tract.

digestive system: The digestive tract and associated glands.

digestive tract: An internal passageway that begins at the mouth, ends at the anus, and is lined by a mucous membrane; also called *gastrointestinal tract*.

dilate: To increase in diameter; to enlarge or expand.

diploid (DIP-loyd): Having a complete somatic complement of chromosomes (23 pairs in human cells).

disaccharide (di-SAK-a-rīd): A compound formed by the joining of two simple sugars by dehydration synthesis.

dislocation: The forceful displacement of an articulating bone to an abnormal position, generally accompanied by damage to tendons, ligaments, the articular capsule, or other structures.

dissociation (di-sō-sē-Ā-shun): *See* **ionization.**

distal: Movement away from the point of attachment or origin; for a limb, away from its attachment to the trunk.

distal convoluted tubule (DCT): The portion of the nephron closest to the connecting tubules and collecting duct; an important site of active secretion.

diuresis (dī-ūr-Ē-sis): Fluid loss at the kidneys; the production of urine.

divergence: In neural tissue, the spread of information from one neuron to many neurons; an organizational pattern common along sensory pathways of the central nervous system.

diverticulitis (dī-ver-tik-ū-LĪ-tis): An inflammation of a diverticulum.

diverticulosis (dī-ver-tik-ū-LŌ-sis): The formation of diverticula in the wall of an organ.

diverticulum: A sac or pouch in the wall of the colon or other organ.

dizygotic twins (dī-zī-GOT-ik): Twins that result from the fertilization of two oocytes.

DNA molecule: Two DNA strands wound in a double helix and held together by weak bonds between complementary nitrogenous base pairs.

dominant trait: An allele whose presence will determine the phenotype, regardless of the nature of its companion allele.

dopamine (DŌ-pa-mēn): An important neurotransmitter in the central nervous system.

dorsal: Toward the back, posterior.

dorsal root ganglion: A peripheral nervous system ganglion containing the cell bodies of sensory neurons.

dorsiflexion: The elevation of the superior surface of the foot through movement at the ankle.

Down syndrome: A genetic abnormality resulting from the presence of three copies of chromosome 21; individuals with this condition have characteristic physical and intellectual deficits.

duct: A passageway that delivers exocrine secretions to an epithelial surface.

ductus arteriosus (ar-tē-rē-Ō-sus): A vascular connection between the pulmonary trunk and the aorta that functions throughout fetal life; normally closes at birth or shortly thereafter and persists as the ligamentum arteriosum.

ductus deferens (DUK-tus DEF-e-renz): A passageway that carries spermatozoa from the epididymis to the ejaculatory duct.

duodenal ampulla: A chamber that receives bile from the common bile duct and pancreatic secretions from the pancreatic duct.

duodenal papilla: A conical projection from the inner surface of the duodenum that contains the opening of the duodenal ampulla.

duodenum (doo-AH-de-num or doo-ō-DĒ-num): The proximal 25 cm of the small intestine that contains short villi and submucosal glands.

dura mater (DOO-ra MĀ-ter or DOO-ra MAH-ter): The outermost component of the cranial and spinal meninges.

dynorphin (dī-NOR-fin): A powerful neuromodulator, produced in the central nervous system, that blocks pain perception by inhibiting pain pathways.

dyslexia: The impaired ability to comprehend written words.

dysmenorrhea: Painful menstruation.

dysmetria (dis-MĒT-rē-uh): Difficulty in performing movements due to problems with the interpretation and anticipation of the distance to be covered.

dysuria (dis-ū-rē-uh): Painful urination.

E

eccrine glands (EK-rin): Sweat glands of the skin that produce a watery secretion.

echocardiography (ek-ō-kar-dē-OG-ra-fē): The examination of the heart by using modified ultrasound techniques.

ectoderm: One of the three primary germ layers; covers the surface of the embryo and gives rise to the nervous system, the epidermis and associated glands, and a variety of other structures.

ectopic (ek-TOP-ik): Outside the normal location.

effector: A peripheral gland or muscle cell innervated by a motor neuron.

efferent: Away from.

efferent arteriole: An arteriole carrying blood away from a glomerulus of the kidney.

efferent fiber: An axon that carries impulses away from the central nervous system.

egestion: *See* defecation.

ejaculation (ē-jak-ū-LĀ-shun): The ejection of semen from the penis as the result of muscular contractions of the bulbospongiosus and ischiocavernosus muscles.

ejaculatory ducts (ē-JAK-ū-la-to-rē): Short ducts that pass within the walls of the prostate gland and connect the ductus deferens with the prostatic urethra.

elastase (ē-LAS-tās): A pancreatic enzyme that breaks down elastin fibers.

elastin: Connective tissue fibers that stretch and recoil, providing elasticity to connective tissues.

electrical coupling: A connection between adjacent cells that permits the movement of ions and the transfer of graded or conducted changes in the transmembrane potential from cell to cell.

electrocardiogram (ECG, EKG) (e-lek-trō-KAR-dē-ō-gram): A graphic record of the electrical activities of the heart, as monitored at specific locations on the body surface.

electroencephalogram (EEG): A graphic record of the electrical activities of the brain.

electrolytes (ē-LEK-trō-līts): Soluble inorganic compounds whose ions will conduct an electrical current in solution.

electron: One of the three fundamental subatomic particles; bears a negative charge and normally orbits the protons of the nucleus.

electron transport system (ETS): The cytochrome system responsible for most of the energy production in cells; a complex bound to the inner mitochondrial membrane.

element: All the atoms with the same atomic number.

elephantiasis (el-e-fan-TĪ-a-sis): A lymphedema caused by the infection and blockage of lymphatic vessels by mosquito-borne parasites.

elevation: Movement in a superior, or upward, direction.

elimination: The ejection of wastes from the body through urination or defecation.

embolism (EM-bō-lizm): The obstruction or closure of a vessel by an embolus.

embolus (EM-bō-lus): An air bubble, fat globule, or blood clot drifting in the bloodstream.

embryo (EM-brē-ō): The developmental stage beginning at fertilization and ending at the start of the third developmental month.

embryology (em-brē-OL-o-jē): The study of embryonic development, focusing on the first 2 months after fertilization.

emesis (EM-e-sis): Vomiting.

emmetropia: Normal vision.

emulsification (ē-mul-si-fi-KĀ-shun): The physical breakup of fats in the digestive tract, forming smaller droplets accessible to digestive enzymes; normally the result of mixing with bile salts.

enamel: Crystalline material similar in mineral composition to bone, but harder and without osteocytes, that covers the exposed surfaces of the teeth.

encephalitis: An inflammation of the brain.

endocarditis: An inflammation of the endocardium of the heart.

endocardium (en-dō-KAR-dē-um): The simple squamous epithelium that lines the heart and is continuous with the endothelium of the great vessels.

endochondral ossification (en-dō-KON-dral): The conversion of a cartilaginous model to bone; the characteristic mode of formation for skeletal elements other than the bones of the cranium, the clavicles, and sesamoid bones.

endocrine gland: A gland that secretes hormones into the blood.

endocrine system: The endocrine glands of the body.

endocytosis: The movement of relatively large volumes of extracellular material into the cytoplasm via the formation of a membranous vesicle at the cell surface; includes pinocytosis and phagocytosis.

endoderm: One of the three primary germ layers; the layer on the undersurface of the embryonic disc; gives rise to the epithelia and glands of the digestive system, the respiratory system, and portions of the urinary system.

endogenous: Produced within the body.

endolymph (EN-dō-limf): The fluid contents of the membranous labyrinth (the saccule, utricle, semicircular canals, and cochlear duct) of the inner ear.

endometrial glands: The secretory glands of the endometrium.

endometrium (en-dō-MĒ-trē-um): The mucous membrane lining the uterus.

endomysium (en-dō-MĪZ-ē-um): A delicate network of connective tissue fibers that surrounds individual muscle cells.

endoneurium: A delicate network of connective tissue fibers that surrounds individual nerve fibers.

endoplasmic reticulum (en-dō-PLAZ-mik re-TIK-ū-lum): A network of membranous channels in the cytoplasm of a cell that function in intracellular transport, synthesis, storage, packaging, and secretion.

endorphins (en-DOR-finz): Neuromodulators, produced in the central nervous system, that inhibit activity along pain pathways.

endosteum: An incomplete cellular lining on the inner (medullary) surfaces of bones.

endothelium (en-dō-THĒ-lē-um): The simple squamous epithelium that lines blood and lymphatic vessels.

enkephalins (en-KEF-a-linz): Neuromodulators, produced in the central nervous system, that inhibit activity along pain pathways.

enteritis (en-ter-Ī-tis): An inflammation of the intestinal tract.

enterocrinin: A hormone secreted by the lining of the duodenum after exposure to chyme; stimulates the secretion of the submucosal glands.

enteroendocrine cells (en-ter-ō-EN-dō-krin): Endocrine cells scattered among the epithelial cells that line the digestive tract.

enterogastric reflex: The reflexive inhibition of gastric secretion; initiated by the arrival of chyme in the small intestine.

enterohepatic circulation: The excretion of bile salts by the liver, followed by the absorption of bile salts by intestinal cells for return to the liver by the hepatic portal vein.

enterokinase: An enzyme in the lumen of the small intestine that activates the proenzymes secreted by the pancreas.

enzyme: A protein that catalyzes a specific biochemical reaction.

eosinophil (ē-ō-SIN-ō-fil): A microphage (white blood cell) with a lobed nucleus and red-staining granules; participates in the immune response and is especially important during allergic reactions.

ependyma (ep-EN-di-mah): The layer of cells lining the ventricles and central canal of the central nervous system.

epicardium: A serous membrane covering the outer surface of the heart; also called *visceral pericardium.*

epidermis: The epithelium covering the surface of the skin.

epididymis (ep-i-DID-i-mus): A coiled duct that connects the rete testis to the ductus deferens; site of functional maturation of spermatozoa.

epidural block: Anesthesia caused by the elimination of sensory inputs from dorsal nerve roots after drugs are introduced into appropriate regions of the epidural space.

epidural space: The space between the spinal dura mater and the walls of the vertebral foramen; contains blood vessels and adipose tissue; a common site of injection for regional anesthesia.

epiglottis (ep-i-GLOT-is): A blade-shaped flap of tissue, reinforced by cartilage, that is attached to the dorsal and superior surface of the thyroid cartilage; folds over the entrance to the larynx during swallowing.

epimysium (ep-i-MĪZ-ē-um): A dense layer of collagen fibers that surrounds a skeletal muscle and is continuous with the tendons/aponeuroses of the muscle and with the perimysium.

epineurium: A dense layer of collagen fibers that surrounds a peripheral nerve.

epiphyseal cartilage (e-pi-FI-zē-al): The cartilaginous region between the epiphysis and diaphysis of a growing bone.

epiphysis (e-PIF-i-sis): The head of a long bone.

epistaxis (ep-i-STAK-sis): A nosebleed.

epithelium (e-pi-THĒ-lē-um): One of the four primary tissue types; a layer of cells that forms a superficial covering or an internal lining of a body cavity or vessel.

eponychium (ep-ō-NIK-ē-um): A narrow zone of stratum corneum that extends across the surface of a nail at its exposed base; also called *cuticle.*

equational division: The second division of meiosis.

equilibrium (ē-kwi-LIB-rē-um): A dynamic state in which two opposing forces or processes are in balance.

erection: The stiffening of the penis due to the engorgement of the erectile tissues of the corpora cavernosa and corpus spongiosum.

erythema (er-i-THĒ-ma): Redness and inflammation at the surface of the skin.

erythrocyte (e-RITH-rō-sīt): A red blood cell; has no nucleus and contains large quantities of hemoglobin.

erythrocytosis (e-rith-rō-sī-TŌ-sis): An abnormally large number of erythrocytes in the circulating blood.

erythropoietin (e-rith-rō-POY-ē-tin): A hormone released by tissues, especially the kidneys, exposed to low oxygen concentrations; stimulates erythropoiesis (red blood cell formation) in bone marrow.

Escherichia coli: A normal bacterial resident of the large intestine.

esophagus: A muscular tube that connects the pharynx to the stomach.

essential amino acids: Amino acids that cannot be synthesized in the body in adequate amounts and must be obtained from the diet.

essential fatty acids: Fatty acids that cannot be synthesized in the body and must be obtained from the diet.

estrogens (ES-trō-jenz): A class of steroid sex hormones that includes estradiol.

eupnea (ŪP-nē-uh): Normal quiet breathing.

evaporation: A movement of molecules from the liquid state to the gaseous state.

eversion (ē-VER-zhun): A turning outward.

excitable membranes: Membranes that conduct action potentials, a characteristic of muscle cells and nerve cells.

excitatory postsynaptic potential (EPSP): The depolarization of a postsynaptic membrane by a chemical neurotransmitter released by the presynaptic cell.

excretion: A removal from body fluids.

exocrine gland: A gland that secretes onto the body surface or into a passageway connected to the exterior.

exocytosis (ek-sō-sī-TŌ-sis): The ejection of cytoplasmic materials by the fusion of a membranous vesicle with the cell membrane.

expiration: Exhalation; breathing out.

expiratory reserve: The amount of additional air that can be voluntarily moved out of the respiratory tract after one normal exhalation.

extension: An increase in the angle between two articulating bones; the opposite of flexion.

external auditory canal: A passageway in the temporal bone that leads to the tympanic membrane of the inner ear.

external ear: The auricle, external auditory meatus, external auditory canal, and tympanic membrane.

external nares (NA-rēz): The nostrils; the external openings into the nasal cavity.

external respiration: The diffusion of gases between the alveolar air and the alveolar capillaries and between the systemic capillaries and peripheral tissues.

exteroceptors: General sensory receptors in the skin, mucous membranes, and special sense organs that provide information about the external environment and about our position within it.

extracellular fluid: All body fluids other than that contained within cells; includes plasma and interstitial fluid.

extraembryonic membranes: The yolk sac, amnion, chorion, and allantois.

extrafusal fibers: Contractile muscle fibers (as opposed to the sensory intrafusal fibers, or muscle spindles).

extrinsic pathway: A clotting pathway that begins with damage to blood vessels or surrounding tissues and ends with the formation of tissue thromboplastin.

F

fabella: A sesamoid bone commonly located in the gastrocnemius muscle, just behind the knee.

facilitated: Brought closer to threshold, as in the depolarization of a nerve cell membrane toward threshold; making the cell more sensitive to depolarizing stimuli.

facilitated diffusion: The passive movement of a substance across a cell membrane by means of a protein carrier.

falciform ligament (FAL-si-form): A sheet of mesentery that contains the ligamentum teres, the fibrous remains of the umbilical vein of the fetus.

falx (falks): Sickle-shaped.

falx cerebri (falks ser-Ē-brē): The curving sheet of dura mater that extends between the two cerebral hemispheres; encloses the superior sagittal sinus.

fasciae (FASH-ē-ē): Connective tissue fibers, primarily collagenous, that form sheets or bands beneath the skin to attach, stabilize, enclose, and separate muscles and other internal organs.

fasciculus (fa-SIK-ū-lus): A small bundle; usually refers to a collection of nerve axons or muscle fibers.

fatty acids: Hydrocarbon chains that end in a carboxylic acid group.

fauces (FAW-sēz): The passage from the mouth to the pharynx, bounded by the palatal arches, the soft palate, and the uvula.

febrile: Characterized by or pertaining to a fever.

feces: Waste products eliminated by the digestive tract at the anus; contains indigestible residue, bacteria, mucus, and epithelial cells.

fenestra: An opening.

fertilization: The fusion of an oocyte and a spermatozoon to form a zygote.

fetus: The developmental stage lasting from the start of the third developmental month to delivery.

fibrillation (fi-bri-LĀ-shun): Uncoordinated contractions of individual muscle cells that impair or prevent normal function.

fibrin (FĪ-brin): Insoluble protein fibers that form the basic framework of a blood clot.

fibrinogen (fī-BRIN-ō-jen): A plasma protein that is the soluble precursor of the fibrous protein fibrin.

fibrinolysis (fī-brin-OL-i-sis): The breakdown of the fibrin strands of a blood clot by a proteolytic enzyme.

fibroblasts (FĪ-brō-blasts): Cells of connective tissue proper that are responsible for the production of extracellular fibers and the secretion of the organic compounds of the extracellular matrix.

fibrocartilage: Cartilage containing an abundance of collagen fibers; located around the edges of joints, in the intervertebral discs, the menisci of the knee, and so on.

fibrous tunic: The outermost layer of the eye, composed of the sclera and cornea.

fibula (FIB-ū-luh): The lateral, slender bone of the leg.

filariasis (fil-a-RĪ-a-sis): A condition resulting from infection by mosquito-borne parasites; can cause elephantiasis.

filiform papillae: Slender conical projections from the dorsal surface of the anterior two-thirds of the tongue.

filtrate: The fluid produced by filtration at a glomerulus in the kidney.

filtration: The movement of a fluid across a membrane whose pores restrict the passage of solutes on the basis of size.

filtration pressure: The hydrostatic pressure responsible for filtration.

filum terminale: A fibrous extension of the spinal cord, from the conus medullaris to the coccygeal ligament.

fimbriae (FIM-brē-ē): Fringes; the fingerlike processes that surround the entrance to the uterine tube.

fissure: An elongate groove or opening.

fistula: An abnormal passageway between two organs or from an internal organ or space to the body surface.

flaccid: Limp, soft, flabby; a muscle without muscle tone.

flagellum/flagella (fla-JEL-uh): An organelle that is structurally similar to a cilium but is used to propel a cell through a fluid.

flatus: Intestinal gas.

flavin adenine dinucleotide (FAD): A coenzyme important in oxidative phosphorylation; cycles between the oxidized ($FADH_2$) and reduced (FAD) states.

flavin adenine mononucleotide (FMN): A coenzyme important in oxidative phosphorylation; cycles between the oxidized ($FMNH_2$) and reduced (FMN) states.

flexion (FLEK-shun): A movement that reduces the angle between two articulating bones; the opposite of extension.

flexor: A muscle that produces flexion.

flexor reflex: A reflex contraction of the flexor muscles of a limb in response to an unpleasant stimulus.

flexure: A bending.

fluoroscope: An instrument that permits the body to be examined in real time with X-rays rather than with fixed images on photographic plates.

folia (FŌ-lē-uh): Leaflike folds; the slender folds in the surface of the cerebellar cortex.

follicle (FOL-i-kul): A small secretory sac or gland.

follicle-stimulating hormone (FSH): A hormone secreted by the anterior pituitary; stimulates oogenesis (female) and spermatogenesis (male).

folliculitis (fo-lik-ū-LĪ-tis): An inflammation of a follicle, such as a hair follicle of the skin.

fontanel (fon-tuh-NEL): A relatively soft, flexible, fibrous region between two flat bones in the developing skull; also spelled *fontanelle*.

foramen/foramina (fo-RĀ-men/fo-RAM-in-uh): An opening or passage through a bone.

forearm: The distal portion of the upper limb between the elbow and wrist.

forebrain: The cerebrum.

fornix (FOR-niks): An arch or the space bounded by an arch; in the brain, an arching tract that connects the hippocampus with the mamillary bodies; in the eye, a slender pocket situated where the epithelium of the ocular conjunctiva folds back on itself as the palpebral conjunctiva.

fossa: A shallow depression or furrow in the surface of a bone.

fourth ventricle: An elongate ventricle of the metencephalon (pons and cerebellum) and the myelencephalon (medulla oblongata) of the brain; the roof contains a region of choroid plexus.

fovea (FŌ-vē-uh): The portion of the retina that provides the sharpest vision because it has the highest concentration of cones; also called *macula lutea*.

fracture: A break or crack in a bone.

frenulum (FREN-ū-lum): A bridle; *see* **lingual frenulum.**

frontal plane: A sectional plane that divides the body into an anterior portion and a posterior portion; also called *coronal plane.*

fructose: A hexose (six-carbon simple sugar) in foods and in semen.

fundus (FUN-dus): The base of an organ.

fungiform papillae: The mushroom-shaped papillae on the dorsal and dorsolateral surfaces of the tongue.

furuncle (FUR-ung-kl): A boil, resulting from the invasion and inflammation of a hair follicle or sebaceous gland.

G

gallbladder: The pear-shaped reservoir for bile after it is secreted by the liver.

gametes (GAM-ēts): Reproductive cells (spermatozoa or oocytes) that contain half the normal chromosome complement.

gametogenesis (ga-mē-tō-JEN-e-sis): The formation of gametes.

gamma aminobutyric acid (GABA) (GAM-ma a-MĒ-nō-bū-TIR-ik): A neurotransmitter of the central nervous system whose effects are generally inhibitory.

gamma motor neurons: Motor neurons that adjust the sensitivities of muscle spindles (intrafusal fibers).

ganglion/ganglia: A collection of neuron cell bodies outside the central nervous system.

gangliosides: Glycolipids that are important components of cell membranes in the central nervous system.

gap junctions: Connections between cells that permit electrical coupling.

gaster (GAS-ter): The stomach; the body, or belly, of a skeletal muscle.

gastrectomy (gas-TREK-to-mē): The partial or total surgical removal of the stomach.

gastric: Pertaining to the stomach.

gastric glands: The tubular glands of the stomach whose cells produce acid, enzymes, intrinsic factor, and hormones.

gastric inhibitory peptide (GIP): A duodenal hormone released when the arriving chyme contains large quantities of carbohydrates; triggers the secretion of insulin and a slowdown in gastric activity.

gastrin (GAS-trin): A hormone produced by enteroendocrine cells of the stomach, after exposure to mechanical stimuli or stimulation of the vagus nerve, and of the duodenum, after exposure to chyme that contains undigested proteins.

gastritis (gas-TRĪ-tis): An inflammation of the stomach.

gastroenteric reflex (gas-trō-en-TER-ik): An increase in peristalsis along the small intestine; triggered by the arrival of food in the stomach.

gastroileal reflex (gas-trō-IL-ē-al): Peristaltic movements that shift materials from the ileum to the colon; triggered by the arrival of food in the stomach.

gastrointestinal (GI) tract: *See* **digestive tract.**

gastroscope: A fiber-optic instrument that permits visual inspection of the stomach lining.

gastrulation (gas-troo-LĀ-shun): The movement of cells of the inner cell mass that creates the three primary germ layers of the embryo.

gene: A portion of a DNA strand that functions as a hereditary unit, is located at a particular site on a specific chromosome, and codes for a specific protein.

genetic engineering: Research and experiments involving the manipulation of the genetic makeup of an organism.

genetics: The study of mechanisms of heredity.

geniculate (je-NIK-ū-lāt): Like a little knee; the medial geniculates and the lateral geniculates are nuclei in the walls of the thalamus of the brain.

genitalia (jen-i-TĀ-lē-uh): The reproductive organs.

genotype (JĒN-ō-tīp): An individual's genetic complement, which determines the individual's phenotype.

germinal centers: Pale regions in the interior of lymphoid tissues or nodules, where cell divisions are under way.

gestation (jes-TĀ-shun): The period of intrauterine development.

gingivae (JIN-ji-vē): The gums.

gingivitis: An inflammation of the gums.

gland: Cells that produce exocrine or endocrine secretions.

glans: The expanded tip of the penis that surrounds the external urethral meatus; continuous with the corpus spongiosum.

glaucoma: An eye disorder characterized by rising intraocular pressures due to inadequate drainage of aqueous humor at the canal of Schlemm.

glenoid cavity: A rounded depression that forms the articular surface of the scapula at the shoulder joint.

glial cells (GLĒ-al): *See* **neuroglia.**

globular proteins: Proteins whose tertiary structure makes them rounded and compact.

glomerular capsule: The expanded initial portion of the nephron that surrounds the glomerulus.

glomerular filtration rate: The rate of filtrate formation at the glomerulus.

glomerulonephritis (glo-mer-ū-lō-nef-RĪ-tis): An inflammation of the glomeruli of the kidneys.

glomerulus (glo-MER-ū-lus): A ball or knot; in the kidneys, a knot of capillaries that projects into the enlarged, proximal end of a nephron; the site of filtration, the first step in the production of urine.

glossopharyngeal nerve (glos-ō-fa-RIN-jē-al): Cranial nerve IX.

glottis (GLOT-is): The passageway from the pharynx to the larynx.

glucagon (GLOO-ka-gon): A hormone secreted by the alpha cells of the pancreatic islets; elevates blood glucose concentrations.

glucocorticoids: Hormones secreted by the zona fasciculata of the adrenal cortex to modify glucose metabolism; cortisol, cortisone, and corticosterone are important examples.

glucogenic amino acids: Amino acids that can be broken down, converted to pyruvic acid, and used in gluconeogenesis.

gluconeogenesis (gloo-kō-nē-ō-JEN-e-sis): The synthesis of glucose from protein or lipid precursors.

glucose (GLOO-kōs): A six-carbon sugar, $C_6H_{12}O_6$; the preferred energy source for most cells and normally the only energy source for neurons.

glycerides: Lipids composed of glycerol bound to fatty acids.

glycogen (GLĪ-kō-jen): A polysaccharide that is an important energy reserve; a polymer consisting of a long chain of glucose molecules.

glycogenesis: The synthesis of glycogen from glucose molecules.

glycogenolysis: Glycogen breakdown and the liberation of glucose molecules.

glycolipids (glī-cō-LIP-idz): Compounds created by the combination of carbohydrate and lipid components.

glycolysis (glī-KOL-i-sis): The anaerobic cytoplasmic breakdown of glucose into lactic acid by way of pyruvic acid, with a net gain of two ATP molecules.

glycoprotein (GLĪ-kō-prō-tēn): A compound containing a relatively small carbohydrate group attached to a large protein.

glycosuria (glī-kō-SOO-rē-uh): The presence of glucose in urine.

goblet cell: A goblet-shaped, mucus-producing, unicellular gland in certain epithelia of the digestive and respiratory tracts.

goiter: An enlargement of the thyroid gland.

Golgi apparatus (gol-jē): A cellular organelle consisting of a series of membranous plates that give rise to lysosomes and secretory vesicles.

Golgi tendon organ: A receptor sensitive to tension in a tendon.

gomphosis (gom-FŌ-sis): A fibrous synarthrosis that binds a tooth to the bone of the jaw; *see* **periodontal ligament.**

gonadotropin-releasing hormone (GnRH): A hypothalamic releasing hormone that causes the secretion of both follicle-stimulating hormone and luteinizing hormone by the anterior pituitary gland.

gonadotropins (gō-nad-ō-TRŌ-pinz): Follicle-stimulating hormone and luteinizing hormone, hormones that stimulate gamete development and sex hormone secretion.

gonads (GŌ-nadz): Reproductive organs that produce gametes and hormones.

gout: A condition resulting from elevated uric acid concentrations in blood and in peripheral tissues.

granulocytes (GRAN-ū-lō-sīts): White blood cells containing granules that are visible with the light microscope; includes eosinophils, basophils, and neutrophils; also called *granular leukocytes.*

gray matter: Areas in the central nervous system that are dominated by neuron cell bodies, neuroglia, and unmyelinated axons.

gray ramus: A bundle of postganglionic sympathetic nerve fibers that are distributed to effectors in the body wall, skin, and limbs by way of a spinal nerve.

greater omentum: A large fold of the dorsal mesentery of the stomach; hangs anterior to the intestines.

greater vestibular glands: Mucous glands in the vaginal walls that secrete into the vestibule; the equivalent of the bulbourethral glands of males.

greenstick fracture: A fracture in which a bone cracks and bends, most commonly involving the long bones of young children.

groin: The inguinal region.

gross anatomy: The study of the structural features of the body without the aid of a microscope.

growth hormone (GH): An anterior pituitary hormone that stimulates tissue growth and anabolism when nutrients are abundant and restricts tissue glucose dependence when nutrients are in short supply.

growth hormone–inhibiting hormone (GH-IH): A hypothalamic regulatory hormone that inhibits growth hormone secretion by the anterior pituitary; also called *somatostatin.*

guanine: A purine; one of the nitrogenous bases in the nucleic acids RNA and DNA.

gustation (gus-TĀ-shun): Taste.

gynecologist (gī-ne-KOL-o-jist): A physician specializing in the female reproductive system.

gyrus (JĪ-rus): A prominent fold or ridge of neural cortex on the surfaces of the cerebral hemispheres.

H

hair: A keratinous strand produced by epithelial cells of the hair follicle.

hair cells: Sensory cells of the inner ear.

hair follicle: An accessory structure of the integument; a tube lined by a stratified squamous epithelium that begins at the surface of the skin and ends at the hair papilla.

hair root: A thickened, conical structure consisting of a connective tissue papilla and the overlying matrix, a layer of epithelial cells that produces the hair shaft.

hallux: The big toe.

haploid (HAP-loyd): Possessing half the normal number of chromosomes; a characteristic of gametes.

hapten: A partial antigen that can bind to an antibody but cannot stimulate antibody production; a foreign compound that has only one antigenic determinant site.

hard palate: The bony roof of the oral cavity, formed by the maxillary and palatine bones.

Hassall's corpuscles: Aggregations of epithelial cells in the thymus whose functions are unknown.

haustra (HAWS-truh): Saclike pouches along the length of the large intestine that result from tension in the taenia coli.

heat exhaustion: A condition that is characterized by excessive perspiration and results in dangerous losses of fluids and salts.

heat stroke: A condition resulting from the failure of the body's normal temperature control mechanisms; characterized by a cessation of sweating and a potentially fatal elevation of body temperature.

Heimlich maneuver (HĪM-lik): A technique for removing an airway blockage by external compression of the abdomen and forceful elevation of the diaphragm.

helper T cells: Lymphocytes whose secretions and other activities coordinate cell-mediated and antibody-mediated immunities; also called T_H *cells.*

hematocrit (he-MAT-ō-krit): The percentage of the volume of whole blood contributed by cells; also called *volume of packed red cells (VPRC)* or *packed cell volume (PCV).*

hematoma: A tumor or swelling filled with blood.

hematuria (hē-ma-TOO-rē-uh): The presence of abnormal numbers of red blood cells in urine.

heme (hēm): A porphyrin ring containing a central iron atom that can reversibly bind oxygen molecules; a component of the hemoglobin molecule.

hemocytoblasts: Stem cells whose divisions produce each of the various populations of blood cells.

hemodialysis (hē-mō-dī-AL-i-sis): Dialysis of the blood.

hemoglobin (HĒ-mō-glō-bin): A protein composed of four globular subunits, each bound to a heme molecule; gives red blood cells the ability to transport oxygen in the blood.

hemolysis: The breakdown (lysis) of red blood cells.

hemophilia (hē-mō-FĒL-ē-uh): A congenital condition due to the inadequate synthesis of one of the clotting factors.

hemopoiesis (hē-mō-poy-Ē-sis): Blood cell formation and differentiation.

hemorrhage: Blood loss.

hemorrhoids (HEM-o-roydz): Swollen, varicose veins that protrude from the walls of the rectum and/or the anorectal canal.

hemostasis: The cessation of bleeding.

hemothorax: The entry of blood into one of the pleural cavities.

heparin (HEP-a-rin): An anticoagulant released by activated basophils and mast cells.

hepatic duct: The duct that carries bile away from the liver lobes and toward the union with the cystic duct.

hepatic portal vein: The vessel that carries blood between the intestinal capillaries and the sinusoids of the liver.

hepatitis (hep-a-TĪ-tis): An inflammation of the liver, resulting from exposure to toxic chemicals, drugs, or viruses.

hepatocyte (he-PAT-ō-sīt): A liver cell.

hernia: The protrusion of a loop or portion of a visceral organ through the abdominopelvic wall or into the thoracic cavity.

herniated disc: The rupture of the connective tissue sheath of the nucleus pulposus of an intervertebral disc.

heterotopic: Ectopic; outside the normal location.

heterozygous (het-er-ō-ZĪ-gus): Possessing two different alleles at corresponding sites on a chromosome pair; the individual's phenotype is determined by one or both of the alleles.

hexose: A six-carbon simple sugar.

hiatus (hī-Ā-tus): A gap, cleft, or opening.

high-density lipoprotein (HDL): A lipoprotein with a relatively small lipid content; thought to be responsible for the movement of cholesterol from peripheral tissues to the liver.

hilum/hilus (HĪ-lum): A localized region where blood vessels, lymphatic vessels, nerves, and/or other anatomical structures are attached to an organ.

hippocampus: A region, beneath the floor of a lateral ventricle, involved with emotional states and the conversion of short-term to long-term memories.

histamine (HIS-tuh-mēn): The chemical released by stimulated mast cells or basophils to initiate or enhance an inflammatory response.

histology (his-TOL-o-jē): The study of tissues.

histones: Proteins associated with the DNA of the nucleus; the DNA strands are wound around them.

holocrine (HO-lō-krin): A form of exocrine secretion in which the secretory cell becomes swollen with vesicles and then ruptures.

homeostasis (hō-mē-ō-STĀ-sis): The maintenance of a relatively constant internal environment.

homologous chromosomes (hō-MOL-o-gus): The members of a chromosome pair.

homozygous (hō-mō-ZĪ-gus): Having the same allele for a given phenotypic character on two homologous chromosomes.

hormone: A compound that is secreted by one cell and travels through the circulatory system to affect the activities of cells in another portion of the body.

human chorionic gonadotropin (hCG): The placental hormone that maintains the corpus luteum for the first 3 months of pregnancy.

human immunodeficiency virus (HIV): The infectious agent that causes acquired immune deficiency syndrome (AIDS).

human leukocyte antigen (HLA): *See* **MHC protein.**

human placental lactogen (hPL): The placental hormone that stimulates the functional development of the mammary glands.

humoral immunity: *See* **antibody-mediated immunity.**

hyaluronan: A carbohydrate component of proteoglycans in the matrix of many connective tissues.

hyaluronidase: An enzyme that breaks down the bonds between adjacent follicle cells; produced by some bacteria and found in the acrosomal cap of a spermatozoon.

hydrocephalus: A condition resulting from excessive production or inadequate drainage of cerebrospinal fluid.

hydrogen bond: A weak interaction between the hydrogen atom on one molecule and a negatively charged portion of another molecule.

hydrolysis (hī-DROL-i-sis): The breakage of a chemical bond through the addition of a water molecule; the reverse of dehydration synthesis.

hydrophilic (hī-drō-PHIL-ik): Freely associating with water; readily entering into solution.

hydrophobic: Incapable of freely associating with water molecules; insoluble.

hydrostatic pressure: Fluid pressure.

hydroxide ion (hī-DROKS-īd): OH^-.

hypercapnia (hī-per-KAP-nē-uh): High plasma carbon dioxide concentrations, commonly as a result of hypoventilation or inadequate tissue perfusion.

hyperglycemia: Elevated plasma glucose concentrations.

hyperkalemia (hī-per-kā-LĒ-mē-uh): Abnormally high potassium concentrations in extracellular fluid.

hypernatremia: Abnormally high sodium concentrations in extracellular fluid.

hyperopia: Farsightedness, characterized by an inability to focus on nearby objects.

hyperplasia: An abnormal enlargement of an organ due to an increase in the number of cells.

hyperpnea (hī-PERP-nē-uh): Abnormal increases in the rate and depth of respiration.

hyperpolarization: The movement of the transmembrane potential away from the normal resting potential and farther from 0 mV.

hyperreflexia: Abnormally exaggerated reflex responses to stimulation.

hypersecretion: The overactivity of glands that produce exocrine or endocrine secretions.

hypersensitivity: An overreaction to an allergen that results in tissue damage and inflammation.

hypertension: Abnormally high blood pressure.

hyperthermia: Excessively high body temperature.

hyperthyroidism: An excessive production of thyroid hormones.

hypertonic: In comparing two solutions, the solution with the higher osmolarity.

hypertrophy (hī-PER-trō-fē): An increase in the size of tissue without cell division.

hyperventilation (hī-per-ven-ti-LĀ-shun): A rate of respiration sufficient to reduce plasma P_{CO_2} to levels below normal.

hypervitaminosis (hī-per-vī-ta-min-Ō-sis): A clinical condition caused by the excessive ingestion and uptake of vitamins.

hypesthesia: An abnormally decreased sensitivity to stimuli.

hypocapnia: An abnormally low plasma P_{CO_2}, commonly as a result of hyperventilation.

hypodermic needle: A needle inserted through the skin to introduce drugs into the subcutaneous layer.

hypodermis: *See* **subcutaneous layer.**

hypokalemia (hī-pō-ka-LĒ-mē-uh): Abnormally low plasma potassium concentrations.

hyponatremia: Abnormally low plasma sodium concentrations.

hyponychium (hī-pō-NIK-ē-um): A thickening in the epidermis beneath the free edge of a nail.

hypophyseal portal system (hī-pō-FI-sē-al): The network of vessels that carry blood from capillaries in the hypothalamus to capillaries in the anterior pituitary gland (hypophysis).

hypophysis (hī-POF-i-sis): The anterior pituitary gland, which is subdivided into the pars distalis and the pars intermedia.

hyporeflexia: Abnormally depressed reflex responses to stimuli.

hyposecretion: Abnormally low rates of exocrine or endocrine secretion.

hypothalamus: The floor of the diencephalon; the region of the brain containing centers involved with the unconscious regulation of visceral functions, emotions, drives, and the coordination of neural and endocrine functions.

hypothermia (hī-pō-THER-mē-uh): An abnormally low body temperature.

hypothesis: A prediction that can be subjected to scientific analysis and review.

hypotonic: In comparing two solutions, the solution with the lower osmolarity.

hypoventilation: A respiratory rate that is insufficient to keep plasma P_{CO_2} within normal levels.

hypovitaminosis: A clinical condition resulting from inadequate vitamin ingestion and uptake; vitamin deficiency.

hypovolemic (hī-pō-vō-LĒ-mik): An abnormally low blood volume.

hypoxia (hī-POKS-ē-uh): A low tissue oxygen concentration.

I

ileocecal valve (il-ē-ō-SĒ-kal): A fold of mucous membrane that guards the connection between the ileum and the cecum.

ileostomy (il-ē-OS-to-mē): The surgical creation of an opening into the ileum; the opening created when the ileum is surgically attached to the abdominal wall.

ileum (IL-ē-um): The last 2.5 m of the small intestine.

ilium (IL-ē-um): The largest of the three bones whose fusion creates an os coxae.

immunity: Resistance to injuries and diseases caused by foreign compounds, toxins, or pathogens.

immunization: The production of immunity by the deliberate exposure to antigens under conditions that prevent the development of illness but stimulate the production of memory B cells.

immunodeficiency: An inability to produce normal numbers and types of antibodies and sensitized lymphocytes.

immunoglobulin (i-mū-nō-GLOB-ū-lin): A circulating antibody.

immunosuppression (i-mū-nō-su-PRE-shun): The suppression of immune responses by the administration of drugs or exposure to toxic chemicals, radiation, or infection.

implantation (im-plan-TĀ-shun): The erosion of a blastocyst into the uterine wall.

impotence: The inability to obtain or maintain an erection.

inclusions: Aggregations of insoluble pigments, nutrients, or other materials in cytoplasm.

incontinence (in-KON-ti-nens): The inability to control micturition (or defecation) voluntarily.

incus (IN-kus): The central auditory ossicle, situated between the malleus and the stapes in the middle ear cavity.

inducer: A stimulus that promotes the activity of a specific gene.

inexcitable: Incapable of conducting an action potential.

infarct: An area of dead cells that results from an interruption of blood flow.

infection: The invasion and colonization of body tissues by pathogens.

inferior: Below, in reference to a particular structure, with the body in the anatomical position.

inferior vena cava: The vein that carries blood from the parts of the body inferior to the heart to the right atrium.

infertility: The inability to conceive; also called *sterility.*

inflammation: A nonspecific defense mechanism that operates at the tissue level; characterized by swelling, redness, warmth, pain, and some loss of function.

inflation reflex: A reflex mediated by the vagus nerve (N X) that prevents overexpansion of the lungs.

infundibulum (in-fun-DIB-ū-lum): A tapering, funnel-shaped structure; in the brain, the connection between the pituitary gland and the hypothalamus; in the uterine tube, the entrance bounded by fimbriae that receives the oocytes at ovulation.

ingestion: The introduction of materials into the digestive tract by way of the mouth.

inguinal canal: A passage through the abdominal wall that marks the path of testicular descent and that contains the testicular arteries, veins, and ductus deferens.

inguinal region: The area near the junction of the trunk and the thighs that contains the external genitalia.

inhibin (in-HIB-in): A hormone, produced by the sustentacular cells, that inhibits the pituitary secretion of follicle-stimulating hormone.

inhibitory postsynaptic potential (IPSP): A hyperpolarization of the postsynaptic membrane after the arrival of a neurotransmitter.

initial segment: The proximal portion of the axon where an action potential first appears.

injection: The forcing of fluid into a body part or organ.

inner cell mass: Cells of the blastocyst that will form the body of the embryo.

inner ear: *See* **internal ear.**

innervation: The distribution of sensory and motor nerves to a specific region or organ.

insensible perspiration: Evaporative water loss by diffusion across the epithelium of the skin or evaporation across the alveolar surfaces of the lungs.

insertion: A point of attachment of a muscle; the end that is easily movable.

insoluble: Incapable of dissolving in solution.

insomnia: A sleep disorder characterized by the inability to fall asleep.

inspiration: Inhalation; the movement of air into the respiratory system.

inspiratory reserve: The maximum amount of air that can be drawn into the lungs over and above the normal tidal volume.

insulin (IN-su-lin): A hormone secreted by the beta cells of the pancreatic islets; causes a reduction in plasma glucose concentrations.

integument (in-TEG-ū-ment): The skin.

intercalated discs (in-TER-ka-lā-ted): Regions where adjacent cardiocytes interlock and where gap junctions permit electrical coupling between the cells.

intercellular cement: Proteoglycans situated between adjacent epithelial cells.

intercellular fluid: *See* **interstitial fluid.**

interdigitate: To interlock.

interferons (in-ter-FĒR-onz): Peptides released by virus-infected cells, especially lymphocytes, that slow viral replication and make other cells more resistant to viral infection.

interleukins (in-ter-LOO-kinz): Peptides, released by activated monocytes and lymphocytes, that assist in the coordination of cell-mediated and antibody-mediated immunities.

internal capsule: The collection of afferent and efferent fibers of the white matter of the cerebral hemispheres, visible on gross dissection of the brain.

internal ear: The membranous labyrinth that contains the organs of hearing and equilibrium.

internal nares: The entrance to the nasopharynx from the nasal cavity.

internal respiration: The diffusion of gases between interstitial fluid and cytoplasm.

interneuron: An association neuron; central nervous system neurons that are between sensory and motor neurons.

interoceptors: Sensory receptors monitoring the functions and status of internal organs and systems.

interosseous membrane: The fibrous connective tissue membrane between the shafts of the tibia and fibula and between the radius and ulna; an example of a fibrous amphiarthrosis.

interphase: The stage in the life cycle of a cell during which the chromosomes are uncoiled and all normal cellular functions except mitosis are under way.

intersegmental reflex: A reflex that involves several segments of the spinal cord.

interstitial fluid (in-ter-STISH-al): The fluid in the tissues that fills the spaces between cells.

interstitial growth: A form of cartilage growth through the growth, mitosis, and secretion of chondrocytes in the matrix.

interventricular foramen: The opening that permits fluid movement between the lateral and third ventricles of the brain.

intervertebral disc: A fibrocartilage pad between the bodies of successive vertebrae that absorbs shocks.

intestinal crypt: A tubular epithelial pocket that is lined by secretory cells and opens into the lumen of the digestive tract; also called *intestinal gland.*

intestine: The tubular organ of the digestive tract.

intracellular fluid: The cytosol.

intrafusal fibers: Muscle spindle fibers.

intramembranous ossification (in-tra-MEM-bra-nus): The formation of bone within a connective tissue without the prior development of a cartilaginous model.

intramuscular injection: The injection of medication into the bulk of a skeletal muscle.

intraocular pressure: The hydrostatic pressure exerted by the aqueous humor of the eye.

intrapleural pressure: The pressure measured in a pleural cavity; also called *intrathoracic pressure.*

intrapulmonary pressure (in-tra-PUL-mo-ner-ē): The pressure measured in an alveolus of the lungs; also called *intraalveolar pressure.*

intrauterine: Within the uterus; during prenatal development.

intrinsic factor: A glycoprotein, secreted by the parietal cells of the stomach, that facilitates the intestinal absorption of vitamin B$_{12}$.

intrinsic pathway: A pathway of the clotting system that begins with the activation of platelets and ends with the formation of platelet thromboplastin.

inversion: A turning inward.

in vitro: Outside the body, in an artificial environment.

in vivo: In the living body.

involuntary: Not under conscious control.

ion: An atom or molecule bearing a positive or negative charge due to the donation or acceptance, respectively, of an electron.

ionic bond (ī-ON-ik): A molecular bond created by the attraction between ions with opposite charges.

ionization (ī-on-i-ZĀ-shun): Dissociation; the breakdown of a molecule in solution to form ions.

ipsilateral: A reflex response that affects the same side as the stimulus.

iris: A contractile structure, made up of smooth muscle, that forms the colored portion of the eye.

ischemia (is-KĒ-mē-uh): An inadequate blood supply to a region of the body.

ischium (IS-kē-um or is-SHĒ-um): One of the three bones whose fusion creates the os coxae.

islets of Langerhans: *See* **pancreatic islets.**

isometric contraction: A muscle contraction characterized by rising tension production but no change in length.

isotonic: A solution with an osmolarity that does not result in water movement across cell membranes.

isotonic contraction: A muscle contraction during which tension climbs and then remains stable as the muscle shortens.

isotopes: Forms of an element whose atoms contain the same number of protons but different numbers of neutrons (and thus differ in atomic weight).

isthmus (IS-mus): A narrow band of tissue connecting two larger masses.

J

jaundice (JAWN-dis): A condition characterized by yellowing of connective tissues due to elevated tissue bilirubin levels; normally associated with damage to the liver or bile ducts.

jejunum (je-JOO-num): The middle part of the small intestine.

joint: An area where adjacent bones interact; also called *articulation.*

juxtaglomerular apparatus: The macula densa and the juxtaglomerular cells; a complex responsible for the release of renin and erythropoietin.

juxtaglomerular cells: Modified smooth muscle cells in the walls of the afferent and efferent arterioles adjacent to the glomerulus and the macula densa.

K

karyotyping (KAR-ē-ō-tī-ping): The determination of the chromosomal characteristics of an individual or cell.

keratin (KER-a-tin): The tough, fibrous protein component of nails, hair, calluses, and the general integumentary surface.

keratinization (KER-a-tin-i-zā-shun): The production of keratin by a stratified squamous epithelium; also called *cornification.*

keto acid: A molecule that ends in —COCOOH; the carbon chain that remains after the deamination or transamination of an amino acid.

ketoacidosis (kē-tō-as-i-DŌ-sis): A reduction in the pH of body fluids due to the presence of large numbers of ketone bodies.

ketogenic amino acids: Amino acids whose catabolism yields ketone bodies rather than pyruvic acid.

ketone bodies: Keto acids produced during the catabolism of lipids and ketogenic amino acids; specifically, acetone, acetoacetate, and beta-hydroxybutyrate.

ketonemia (kē-tō-NĒ-mē-uh): Abnormal concentrations of ketone bodies in blood.

ketonuria (kē-tō-NŪ-rē-uh): Abnormal concentrations of ketone bodies in urine.

ketosis (kē-TŌ-sis): A condition characterized by the abnormal production of ketone bodies.

kidney: A component of the urinary system; an organ functioning in the regulation of plasma composition, including the excretion of wastes and the maintenance of normal fluid and electrolyte balances.

killer T cells: *See* **cytotoxic T cells.**

kilocalorie (KIL-ō-kal-o-rē): *See* **Calorie.**

Krebs cycle: *See* **TCA cycle.**

Kupffer cells (KOOP-fer): Stellate reticular cells of the liver; phagocytic cells of the liver sinusoids.

kyphosis (kī-FŌ-sis): An exaggerated thoracic curvature.

L

labium/labia (LĀ-bē-uh): Lips; the labia majora and labia minora are components of the female external genitalia.

labrum: A lip or rim.

labyrinth: A maze of passageways; the structures of the inner ear.

lacrimal gland (LAK-ri-mal): A tear gland on the dorsolateral surface of the eye.

lactase: An enzyme that breaks down milk proteins.

lactation (lak-TĀ-shun): The production of milk by the mammary glands.

lacteal (LAK-tē-al): A terminal lymphatic within an intestinal villus.

lactic acid: A compound produced from pyruvic acid under anaerobic conditions.

lactiferous duct (lak-TIF-e-rus): A duct draining one lobe of the mammary gland.

lactiferous sinus: An expanded portion of a lactiferous duct adjacent to the nipple of a breast.

lacuna (la-KOO-nuh): A small pit or cavity.

lambdoidal suture (lam-DOYD-al): The synarthrosis between the parietal and occipital bones of the cranium.

lamellae (la-MEL-ē): Concentric layers; the concentric layers of bone within an osteon.

lamina (LA-min-uh): A thin sheet or layer.

lamina propria (LA-min-uh PRŌ-prē-uh): The loose connective tissue that underlies a mucous epithelium and forms part of a mucous membrane.

laminectomy: The removal of the spinous processes of a vertebra to gain access and treat a herniated disc.

Langerhans cells (LAN-ger-hanz): Cells in the epithelium of the skin and digestive tract that participate in the immune response by presenting antigens to T cells.

laparoscope (LAP-a-ro-skōp): A fiber-optic instrument used to visualize the contents of the abdominopelvic cavity.

large intestine: The terminal portions of the intestinal tract, consisting of the colon, the rectum, and the anorectal canal.

laryngopharynx (la-rin-gō-FAR-inks): The division of the pharynx that is inferior to the epiglottis and superior to the esophagus.

larynx (LAR-inks): A complex cartilaginous structure that surrounds and protects the glottis and vocal cords; the superior margin is bound to the hyoid bone, and the inferior margin is bound to the trachea.

latent period: The time between the stimulation of a muscle and the start of the contraction phase.

lateral: Pertaining to the side.

lateral apertures: Openings in the roof of the fourth ventricle that permit the circulation of cerebrospinal fluid into the subarachnoid space.

lateral ventricle: A fluid-filled chamber within a cerebral hemisphere.

laxatives: Compounds that promote defecation through increased peristalsis or an increase in the water content and volume of feces.

lens: The transparent body that is inferior to the iris and pupil and superior to the vitreous humor.

lesion: A localized abnormality in tissue organization.

lesser omentum: A small pocket in the mesentery that connects the lesser curvature of the stomach to the liver.

leukemia (loo-KĒ-mē-uh): A malignant disease of the blood-forming tissues.

leukocyte (LOO-kō-sīt): A white blood cell.

leukocytosis (loo-kō-sī-TŌ-sis): Abnormally high numbers of circulating white blood cells.

leukopenia (loo-kō-PĒ-nē-uh): Abnormally low numbers of circulating white blood cells.

ligament (LI-ga-ment): A dense band of connective tissue fibers that attaches one bone to another.

ligamentum arteriosum: The fibrous strand in adults that is the remnant of the ductus arteriosus of the fetal stage.

ligamentum nuchae (li-guh-MEN-tum NOO-kē or NOO-kā): An elastic ligament between the vertebra prominens and the occipital bone.

ligamentum teres: The fibrous strand in the falciform ligament of adults that is the remnant of the umbilical vein of the fetal stage.

ligate: To tie off.

limbic system (LIM-bik): The group of nuclei and centers in the cerebrum and diencephalon that are involved with emotional states, memories, and behavioral drives.

limbus (LIM-bus): The edge of the cornea, marked by the transition from the corneal epithelium to the ocular conjunctiva.

linea alba: The tendinous band along the midline of the rectus abdominis muscle.

lingual: Pertaining to the tongue.

lingual frenulum: An epithelial fold that attaches the inferior surface of the tongue to the floor of the mouth.

lipase (LĪ-pās): A pancreatic enzyme that breaks down triglycerides.

lipemia (lip-Ē-mē-uh): An elevated concentration of lipids in the blood.

lipid: An organic compound containing carbons, hydrogens, and oxygens in a ratio that does not approximate 1:2:1; includes fats, oils, and waxes.

lipofuscin (li-pō-FŪ-shun): A pigment inclusion in aging nerve cells that is of unknown significance.

lipogenesis (li-pō-JEN-e-sis): The synthesis of lipids from non-lipid precursors.

lipoids: Prostaglandins, steroids, phospholipids, glycolipids, and so on.

lipolysis (lī-POL-i-sis): The catabolism of lipids as a source of energy.

lipoprotein (lī-pō-PRŌ-tēn): A compound containing a relatively small lipid bound to a protein.

liver: An organ of the digestive system that has varied and vital functions, including the production of plasma proteins, the excretion of bile, the storage of energy reserves, the detoxification of poisons, and the interconversion of nutrients.

lobule (LOB-ūl): Histologically, the basic organizational unit of the liver.

local hormone: *See* **prostaglandin.**

long-term memories: Memories that persist for an extended period.

loop of Henle: The portion of the nephron that creates the concentration gradient in the renal medulla.

loose connective tissue: A loosely organized, easily distorted connective tissue that contains several fiber types, a varied population of cells, and a viscous ground substance.

lordosis (lor-DŌ-sis): An exaggeration of the lumbar curvature.

lumbar: Pertaining to the lower back.

lumen: The central space within a duct or other internal passageway.

lungs: The paired organs of respiration, situated in the pleural cavities.

luteinizing hormone (LH) (LOO-tē-in-ī-zing): An anterior pituitary hormone that, in females, assists follicle-stimulating hormone in follicle stimulation, triggers ovulation, and promotes the maintenance and secretion of the endometrial glands; in males, stimulates spermatogenesis; also called *lutropin;* formerly called *interstitial cell–stimulating hormone* in males.

luxation (luks-Ā-shun): The dislocation of a joint.

lymph: The fluid contents of lymphatic vessels, similar in composition to interstitial fluid.

lymphadenopathy (lim-fad-e-NOP-a-thē): The pathological enlargement of the lymph nodes.

lymphatic vessels: The vessels of the lymphatic system; also called *lymphatics.*

lymphedema (lim-fe-DĒ-ma): The swelling of peripheral tissues as a result of excessive lymph production or inadequate drainage.

lymph nodes: Lymphatic organs that monitor the composition of lymph.

lymphocyte (LIM-fō-sīt): A cell of the lymphatic system that participates in the immune response.

lymphokines: Chemicals secreted by activated lymphocytes.

lymphopoiesis (lim-fō-poy-Ē-sis): The production of lymphocytes from lymphoid stem cells.

lymphotoxin (lim-fō-TOK-sin): A secretion of lymphocytes that kills the target cells.

lysis (LĪ-sis): The destruction of a cell through the rupture of its cell membrane.

lysosome (LĪ-sō-sōm): An intracellular vesicle containing digestive enzymes.

lysozyme: An enzyme, present in some exocrine secretions, that has antibiotic properties.

M

macrophage (MAK-rō-fāj): A phagocytic cell of the monocyte–macrophage system.

macula (MAK-ū-la): A receptor complex, located in the saccule or utricle of the inner ear, that responds to linear acceleration or gravity.

macula densa (MAK-ū-la DEN-sa): A group of specialized secretory cells that is located in a portion of the distal convoluted tubule, adjacent to the glomerulus and the juxtaglomerular cells; a component of the juxtaglomerular apparatus.

macula lutea (LOO-tē-uh): *See* **fovea.**

major histocompatibility complex: *See* **MHC protein.**

male climacteric: The age-related cessation of gametogenesis in males as a result of reduced sex hormone production.

malignant tumor: A form of cancer characterized by rapid cell growth and the spread of cancer cells throughout the body.

malleus (MAL-ē-us): The first auditory ossicle, bound to the tympanic membrane and the incus.

malnutrition: An unhealthy state produced by inadequate dietary intake of nutrients, calories, and/or vitamins.

mamillary bodies (MAM-i-lar-ē): Nuclei in the hypothalamus that affect eating reflexes and behaviors; a component of the limbic system.

mammary glands: Milk-producing glands of the female breast.

manus: The hand.

marrow: A tissue that fills the internal cavities in bone; dominated by hemopoietic cells (red bone marrow) or by adipose tissue (yellow bone marrow).

mass peristalsis: A powerful peristaltic contraction that moves fecal materials along the colon and into the rectum.

mass reflex: A hyperreflexia in an area innervated by spinal cord segments that are distal to an area of injury.

mast cell: A connective tissue cell that, when stimulated, releases histamine, serotonin, and heparin, initiating the inflammatory response.

mastectomy: The surgical removal of part or all of a mammary gland.

mastication (mas-ti-KĀ-shun): Chewing.

mastoid sinus: Air-filled spaces in the mastoid process of the temporal bone.

matrix: The extracellular fibers and ground substance of a connective tissue.

maxillary sinus (MAK-si-ler-ē): One of the paranasal sinuses; an air-filled chamber lined by a respiratory epithelium that is located in a maxillary bone and opens into the nasal cavity.

meatus (mē-Ā-tus): An opening or entrance into a passageway.

mechanoreception: The detection of mechanical stimuli, such as touch, pressure, or vibration.

medial: Toward the midline of the body.

mediastinum (mē-dē-as-TĪ-um or mē-dē-AS-ti-num): The central tissue mass that divides the thoracic cavity into two pleural cavities; includes the aorta and other great vessels, the esophagus, trachea, thymus, the pericardial cavity and heart, and a host of nerves, small vessels, and lymphatic vessels; in males, the area of connective tissue attaching a testis to the epididymis, proximal portion of ductus deferens, and associated vessels.

medulla: The inner layer or core of an organ.

medulla oblongata: The most caudal of the five brain regions, also called *myelencephalon*.

medullary cavity: The space within a bone that contains the marrow.

medullary rhythmicity center: The center in the medulla oblongata that sets the background pace of respiration; includes inspiratory and expiratory centers.

megakaryocytes (meg-a-KAR-ē-ō-sīts): Bone marrow cells responsible for the formation of platelets.

meiosis (mī-Ō-sis): Cell division that produces gametes with half the normal somatic chromosome complement.

melanin (ME-la-nin): The yellow-brown pigment produced by the melanocytes of the skin.

melanocyte (me-LAN-ō-sīt): A specialized cell in the deeper layers of the stratified squamous epithelium of the skin; responsible for the production of melanin.

melanocyte-stimulating hormone (MSH): A hormone, produced by the pars intermedia of the anterior pituitary, that stimulates melanin production.

melanomas (mel-a-NŌ-maz): Dangerous malignant skin cancers that involve melanocytes.

melatonin (mel-a-TŌ-nin): A hormone secreted by the pineal gland; inhibits secretion of MSH and GnRH.

membrane: Any sheet or partition; a layer consisting of an epithelium and the underlying connective tissue.

membrane flow: The movement of sections of membrane surface to and from the cell surface and components of the endoplasmic reticulum, the Golgi apparatus, and vesicles.

membrane potential: *See* **transmembrane potential.**

membranous labyrinth: Endolymph-filled tubes that enclose the receptors of the inner ear.

memory: The ability to recall information or sensations; can be divided into short-term and long-term memories.

menarche (me-NAR-kē): The beginning of menstrual function; the first menstrual period, which normally occurs at puberty.

meninges (men-IN-jēz): Three membranes that surround the surfaces of the central nervous system; the dura mater, the pia mater, and the arachnoid.

meningitis: An inflammation of the spinal or cranial meninges.

meniscectomy: The removal of a meniscus.

meniscus (men-IS-kus): A fibrocartilage pad between opposing surfaces in a joint.

menses (MEN-sēz): The first portion of the uterine cycle, the portion in which the endometrial lining sloughs away.

merocrine (MER-ō-krin): A method of secretion in which the cell ejects materials through exocytosis.

mesencephalic aqueduct: The passageway that connects the third ventricle (diencephalon) with the fourth ventricle (metencephalon).

mesencephalon (mez-en-SEF-a-lon): The midbrain; the region between the diencephalon and pons.

mesenchyme: Embryonic or fetal connective tissue.

mesentery (MEZ-en-ter-ē): A double layer of serous membrane that supports and stabilizes the position of an organ in the abdominopelvic cavity and provides a route for the associated blood vessels, nerves, and lymphatic vessels.

mesoderm: The middle germ layer, between the ectoderm and endoderm of the embryo.

mesothelium (mez-ō-THĒ-lē-um): A simple squamous epithelium that lines one of the divisions of the ventral body cavity.

messenger RNA (mRNA): RNA formed at transcription to direct protein synthesis in the cytoplasm.

metabolic turnover: The continuous breakdown and replacement of organic materials within cells.

metabolism (me-TAB-ō-lizm): The sum of all biochemical processes under way within the human body at any moment; includes anabolism and catabolism.

metabolites (me-TAB-ō-līts): Compounds produced in the body as a result of metabolic reactions.

metacarpal bones: The five bones of the palm of the hand.

metalloproteins (met-al-ō-PRŌ-tēnz): Plasma proteins that transport metal ions.

metaphase (MET-a-fāz): The stage of mitosis in which the chromosomes line up along the equatorial plane of the cell.

metaphysis (me-TA-fi-sis): The region of a long bone between the epiphysis and diaphysis, corresponding to the location of the epiphyseal cartilage of the developing bone.

metarteriole (met-ar-TĒ-rē-ōl): A vessel that connects an arteriole to a venule and that provides blood to a capillary plexus.

metastasis (me-TAS-ta-sis): The spread of cancer cells from one organ to another, leading to the establishment of secondary tumors.

metatarsal bone: One of the five bones of the foot that articulate with the tarsal bones (proximally) and the phalanges (distally).

metencephalon (met-en-SEF-a-lon): The pons and cerebellum of the brain.

MHC protein: A surface antigen that is important to the recognition of foreign antigens and that plays a role in the coordination and activation of the immune response; also called *human leukocyte antigen (HLA)*.

micelle (mī-SEL): A droplet with hydrophilic portions on the outside; a spherical aggregation of bile salts, monoglycerides, and fatty acids in the lumen of the intestinal tract.

microcephaly (mī-krō-SEF-a-lē): An abnormally small cranium as a result of the premature closure of one or more fontanels.

microfilaments: Fine protein filaments visible with the electron microscope; components of the cytoskeleton.

microglia (mī-KROG-lē-uh): Phagocytic neuroglia in the central nervous system.

microphages: Neutrophils and eosinophils.

microtubules: Microscopic tubules that are part of the cytoskeleton and are a component in cilia, flagella, the centrioles, and spindle fibers.

microvilli: Small, fingerlike extensions of the exposed cell membrane of an epithelial cell.

micturition (mik-tū-RI-shun): Urination.

midbrain: The mesencephalon.

middle ear: The space between the external and internal ears that contains auditory ossicles.

midsagittal plane: A plane passing through the midline of the body that divides it into left and right halves.

mineralocorticoid: Corticosteroids produced by the zona glomerulosa of the adrenal cortex; steroids such as aldosterone that affect mineral metabolism.

miscarriage: A spontaneous abortion.

mitochondrion (mī-tō-KON-drē-on): An intracellular organelle responsible for generating most of the ATP required for cellular operations.

mitosis (mī-TŌ-sis): The division of a single cell nucleus that produces two identical daughter cell nuclei; an essential step in cell division.

mitral valve (MĪ-tral): *See* **bicuspid valve.**

mixed gland: A gland that contains exocrine and endocrine cells, or an exocrine gland that produces serous and mucous secretions.

mixed nerve: A peripheral nerve that contains sensory and motor fibers.

mole: A quantity of an element or compound having a mass in grams equal to the element's atomic weight or to the compound's molecular weight.

molecular weight: The sum of the atomic weights of all the atoms in a molecule.

molecule: A chemical structure containing two or more atoms that are held together by chemical bonds.

monoclonal antibodies (mo-nō-KLŌ-nal): Antibodies produced by genetically identical cells under laboratory conditions.

monocytes (MON-ō-sīts): Phagocytic agranulocytes (white blood cells) in the circulating blood.

monoglyceride (mo-nō-GLI-se-rīd): A lipid consisting of a single fatty acid bound to a molecule of glycerol.

monokines: Secretions released by activated cells of the monocyte–macrophage system to coordinate various aspects of the immune response.

monosaccharide (mon-ō-SAK-uh-rīd): A simple sugar, such as glucose or ribose.

monosynaptic reflex: A reflex in which the sensory afferent neuron synapses directly on the motor efferent neuron.

monozygotic twins: Twins produced by the splitting of a single fertilized egg (zygote).

morula (MOR-ū-la): A mulberry-shaped collection of cells produced by the mitotic divisions of a zygote.

motor unit: All of the muscle cells controlled by a single motor neuron.

mucins (MŪ-sinz): Proteoglycans responsible for the lubricating properties of mucus.

mucosa (mū-KŌ-sa): A mucous membrane; the epithelium plus the lamina propria.

mucosa-associated lymphoid tissue (MALT): The extensive collection of lymphoid tissues linked with the digestive system.

mucous: Indicating the presence or production of mucus.

mucous membrane: *See* **mucosa.**

mucus: A lubricating fluid that is composed of water and mucins and is produced by unicellular and multicellular glands along the digestive, respiratory, urinary, and reproductive tracts.

multipennate: A muscle whose internal fibers are organized around several tendons.

multipolar neuron: A neuron with many dendrites and a single axon; the typical form of a motor neuron.

multiunit smooth muscle: A smooth muscle tissue whose muscle cells are innervated in motor units.

muscarinic receptors (mus-kar-IN-ik): Membrane receptors sensitive to acetylcholine and to muscarine, a toxin produced by certain mushrooms; located at all parasympathetic neuromuscular and neuroglandular junctions and at a few sympathetic neuromuscular and neuroglandular junctions.

muscle: A contractile organ composed of muscle tissue, blood vessels, nerves, connective tissues, and lymphatic vessels.

muscle tissue: A tissue characterized by the presence of cells capable of contraction; includes skeletal, cardiac, and smooth muscle tissues.

muscularis externa (mus-kū-LAR-is): Concentric layers of smooth muscle responsible for peristalsis.

muscularis mucosae: The layer of smooth muscle beneath the lamina propria; responsible for moving the mucosal surface.

mutagens (MŪ-ta-jenz): Chemical agents that induce mutations and may be carcinogenic.

mutation: A change in the nucleotide sequence of the DNA in a cell.

myalgia (mī-AL-jē-uh): Muscle pain.

myasthenia gravis (mī-as-THĒ-nē-a GRA-vis): A muscular weakness due to a reduction in the number of acetylcholine receptor sites on the sarcolemmal surface; suspected to be an autoimmune disorder.

myelencephalon (mī-el-en-SEF-a-lon): *See* **medulla oblongata.**

myelin (MĪ-e-lin): An insulating sheath around an axon; consists of multiple layers of neuroglial membrane; significantly increases conduction rate along the axon.

myelination: The formation of myelin.

myenteric plexus (mī-en-TER-ik): Parasympathetic motor neurons and sympathetic postganglionic fibers located between the circular and longitudinal layers of the muscularis externa.

myocardial infarction (mī-ō-KAR-dē-al): A heart attack; damage to the heart muscle due to an interruption of regional coronary circulation.

myocarditis: An inflammation of the myocardium.

myocardium: The cardiac muscle tissue of the heart.

myofibril: Organized collections of myofilaments in skeletal and cardiac muscle cells.

myofilaments: Fine protein filaments composed primarily of the proteins actin (thin filaments) and myosin (thick filaments).

myoglobin (MĪ-ō-glō-bin): An oxygen-binding pigment that is especially common in slow skeletal muscle fibers and cardiac muscle cells.

myogram: A recording of the tension produced by muscle fibers on stimulation.

myometrium (mī-ō-MĒ-trē-um): The thick layer of smooth muscle in the wall of the uterus.

myopia: Nearsightedness, an inability to accommodate for distant vision.

myosepta: Connective tissue partitions that separate adjacent skeletal muscles.

myosin: The protein component of thick filaments.

myositis (mī-ō-SĪ-tis): An inflammation of muscle tissue.

N

nail: A keratinous structure produced by epithelial cells of the nail root.

narcolepsy: A sleep disorder characterized by falling asleep at inappropriate moments.

nares, external (NA-rēz): The entrance from the exterior to the nasal cavity.

nares, internal: The entrance from the nasal cavity to the nasopharynx.

nasal cavity: A chamber in the skull that is bounded by the internal and external nares.

nasolacrimal duct: The passageway that transports tears from the nasolacrimal sac to the nasal cavity.

nasolacrimal sac: A chamber that receives tears from the lacrimal ducts.

nasopharynx (nā-zō-FAR-inks): A region that is posterior to the internal nares and superior to the soft palate and ends at the oropharynx.

N compound: An organic compound containing nitrogen atoms.

necrosis (nek-RŌ-sis): The death of cells or tissues from disease or injury.

negative feedback: A corrective mechanism that opposes or negates a variation from normal limits.

neonate: A newborn infant, or baby.

neoplasm: A tumor, or mass of abnormal tissue.

nephritis (nef-RĪ-tis): An inflammation of the kidney.

nephrolithiasis (nef-rō-li-THĪ-a-sis): A condition resulting from the formation of kidney stones.

nephron (NEF-ron): The basic functional unit of the kidney.

nerve impulse: An action potential in a neuron cell membrane.

neural cortex: An area of gray matter at the surface of the central nervous system.

neurilemma (noo-ri-LEM-muh): The outer surface of a neuroglia that encircles an axon.

neurofibrils: Microfibrils in the cytoplasm of a neuron.

neurofilaments: Microfilaments in the cytoplasm of a neuron.

neuroglandular junction: A cell junction at which a neuron controls or regulates the activity of a secretory (gland) cell.

neuroglia (noo-RŌG-lē-ah, noo-RŌ-glī-ah, or noo-rō-GLĪ-ah): Cells of the central nervous system and peripheral nervous system that support and protect neurons; also called *glial cells.*

neurohypophysis (NOO-rō-hī-pof-i-sis): The posterior pituitary, or pars nervosa.

neuromodulator (noo-rō-MOD-ū-la-tor): A compound, released by a neuron, that adjusts the sensitivities of another neuron to specific neurotransmitters.

neuromuscular junction: A synapse between a neuron and a muscle cell.

neuron (NOO-ron) or **neurone** (NOO-rōn): A cell in neural tissue that is specialized for intercellular communication through (1) changes in membrane potential and (2) synaptic connections.

neurotransmitter: A chemical compound released by one neuron to affect the transmembrane potential of another.

neurotubules: Microtubules in the cytoplasm of a neuron.

neurulation: The embryological process responsible for the formation of the central nervous system.

neutron: A fundamental particle that does not carry a positive or a negative charge.

neutropenia: An abnormally low number of neutrophils in the circulating blood.

neutrophil (NOO-trō-fil): A microphage that is very numerous and normally the first of the mobile phagocytic cells to arrive at an area of injury or infection.

nicotinic receptors (nik-ō-TIN-ik): Acetylcholine receptors on the surfaces of sympathetic and parasympathetic ganglion cells; respond to the compound nicotine.

nipple: An elevated epithelial projection on the surface of the breast; contains the openings of the lactiferous sinuses.

Nissl bodies: The ribosomes, Golgi apparatus, rough endoplasmic reticulum, and mitochondria of the perikaryon of a typical nerve cell.

nitrogenous wastes: Organic waste products of metabolism that contain nitrogen, such as urea, uric acid, and creatinine.

nociception (nō-sē-SEP-shun): Pain perception.

node of Ranvier: The area between adjacent neuroglia where the myelin covering of an axon is incomplete.

nodose ganglion (NŌ-dōs): A sensory ganglion of cranial nerve X.

noradrenaline: *See* **norepinephrine.**

norepinephrine (NE) (nor-ep-i-NEF-rin): A catecholamine neurotransmitter in the peripheral nervous system and central nervous system, released at most sympathetic neuromuscular and neuroglandular junctions, and a hormone secreted by the adrenal medulla; also called *noradrenaline.*

nucleic acid (noo-KLĀ-ik): A polymer of nucleotides that contains a pentose sugar, a phosphate group, and one of four nitrogenous bases that regulate the synthesis of proteins and make up the genetic material in cells.

nucleolus (noo-KLĒ-ō-lus): The dense region in the nucleus that is the site of RNA synthesis.

nucleoplasm: The fluid content of the nucleus.

nucleoproteins: Proteins of the nucleus that are generally associated with DNA.

nucleotide: A compound consisting of a nitrogenous base, a simple sugar, and a phosphate group.

nucleus: A cellular organelle that contains DNA, RNA, and proteins; in the central nervous system, a mass of gray matter.

nucleus pulposus (pul-PŌ-sus): The gelatinous central region of an intervertebral disc.

nutrient: An inorganic or organic compound that can be broken down in the body to produce energy.

nystagmus: An unconscious, continuous movement of the eyes as if to adjust to constant motion.

O

obesity: Body weight 10–20 percent above standard values as a result of body fat accumulation.

occlusal surface (o-KLOO-sal): The opposing surfaces of the teeth that come into contact when processing food.

ocular: Pertaining to the eye.

oculomotor nerve (ok-ū-lō-MŌ-ter): Cranial nerve III, which controls the extrinsic oculomotor muscles other than the superior oblique and the lateral rectus muscles.

olecranon (ō-LEK-ruh-non or ō-lah-KRŌ-non): The proximal end of the ulna that forms the prominent point of the elbow.

olfaction: The sense of smell.

olfactory bulb (ōl-FAK-tor-ē): The expanded ends of the olfactory tracts; the sites where the axons of the first cranial nerves (N I) synapse on central nervous system interneurons that lie beneath the frontal lobes of the cerebrum.

oligodendrocytes (o-li-gō-DEN-drō-sīts): Central nervous system neuroglia that maintain cellular organization within gray matter and provide a myelin sheath in areas of white matter.

oligopeptide (ol-i-gō-PEP-tīd): A short chain of amino acids.

oncogene (ON-kō-jēn): A gene that can turn a normal cell into a cancer cell.

oncologists (on-KOL-o-jists): Physicians specializing in the study and treatment of tumors.

oocyte (Ō-ō-sīt): A cell whose meiotic divisions will produce a single ovum and three polar bodies.

oogenesis (ō-ō-JEN-e-sis): Ovum production.

oogonia (ō-ō-GŌ-nē-uh): Stem cells in the ovaries whose divisions give rise to oocytes.

oophorectomy (ō-of-ō-REK-to-mē): The surgical removal of the ovaries.

ooplasm: The cytoplasm of the ovum.

opsin: A protein; one structural component of the visual pigment rhodopsin.

opsonization: An effect of coating an object with antibodies; the attraction and enhancement of phagocytosis.

optic chiasm (KĪ-azm): The crossing point of the optic nerves.

optic nerve: The second cranial nerve (N II), which carries signals from the eye to the optic chiasm.

optic tract: The tract over which nerve impulses from the retina are transmitted between the optic chiasm and the thalamus.

orbit: The bony cavity of the skull that contains the eyeball.

orchiectomy (or-kē-EK-to-mē): The surgical removal of one or both testes.

orchitis: An inflammation of the testes.

organelle (or-gan-EL): An intracellular structure that performs a specific function or group of functions.

organic compound: A compound containing carbon, hydrogen, and in most cases oxygen.

organogenesis: The formation of organs during embryological and fetal development.

organs: Combinations of tissues that perform complex functions.

origin: A point of attachment of a muscle; the end that is not easily movable.

oropharynx: The middle portion of the pharynx, bounded superiorly by the nasopharynx, anteriorly by the oral cavity, and inferiorly by the laryngopharynx.

os coxae/ossa coxae: The hip bone(s).

osmolarity (oz-mō-LAR-i-tē): The total concentration of dissolved materials in a solution, regardless of their specific identities, expressed in moles; also called *osmotic concentration.*

osmoreceptor: A receptor sensitive to changes in the osmolarity of plasma.

osmosis (oz-MŌ-sis): The movement of water across a selectively permeable membrane from one solution to another solution that contains a higher solute concentration.

osmotic pressure: The force of osmotic water movement; the pressure that must be applied to prevent osmosis across a membrane.

osseous tissue: A strong connective tissue containing specialized cells and a mineralized matrix of crystalline calcium phosphate and calcium carbonate; also called **bone.**

ossicles: Small bones.

ossification: The formation of bone.

osteoblast: (OS-tē-ō-blast): A cell that produces the fibers and matrix of bone.

osteoclast (OS-tē-ō-klast): A cell that dissolves the fibers and matrix of bone.

osteocyte (OS-tē-ō-sīt): A bone cell responsible for the maintenance and turnover of the mineral content of the surrounding bone.

osteogenic layer (os-tē-ō-JEN-ik): The inner, cellular layer of the periosteum that participates in bone growth and repair.

osteolysis (os-tē-OL-i-sis): The breakdown of the mineral matrix of bone.

osteon (OS-tē-on): The basic histological unit of compact bone, consisting of osteocytes organized around a central canal and separated by concentric lamellae.

otic: Pertaining to the ear.

otitis media: An inflammation of the middle ear cavity.

otolith: A complex formed by the combination of a gelatinous matrix and statoconia, aggregations of calcium carbonate crystals; located above one of the maculae of the vestibule.

oval window: An opening in the bony labyrinth where the stapes attaches to the membranous wall of the vestibular duct.

ovarian cycle (ō-VAR-ē-an): The monthly chain of events that leads to ovulation.

ovary: The female reproductive organ that produces gametes.

ovulation (ov-ū-LĀ-shun): The release of a secondary oocyte, surrounded by cells of the corona radiata, after the rupture of the wall of a tertiary follicle.

ovum/ova (Ō-vum): The functional product of meiosis II, produced after the fertilization of a secondary oocyte.

oxidation: The loss of electrons or hydrogen atoms or the acceptance of an oxygen atom.

oxidative phosphorylation: The capture of energy as ATP during a series of oxidation–reduction reactions; a reaction sequence that occurs in the mitochondria and involves coenzymes and the electron transport system.

oxytocin (oks-i-TŌ-sin): A hormone produced by hypothalamic cells and secreted into capillaries at the posterior pituitary; stimulates smooth muscle contractions of the uterus or mammary glands in females and the prostate gland in males.

P

pacemaker cells: Cells of the sinoatrial node that set the pace of cardiac contraction.

Pacinian corpuscle (pa-SIN-ē-an): A receptor sensitive to vibration.

palate: The horizontal partition separating the oral cavity from the nasal cavity and nasopharynx; divided into an anterior bony (hard) palate and a posterior fleshy (soft) palate.

palatine: Pertaining to the palate.

palpate: To examine by touch.

palpebrae (pal-PĒ-brē): Eyelids.

pancreas: A digestive organ containing exocrine and endocrine tissues; the exocrine portion secretes pancreatic juice, and the endocrine portion secretes hormones, including insulin and glucagon.

pancreatic duct: A tubular duct that carries pancreatic juice from the pancreas to the duodenum.

pancreatic islets: Aggregations of endocrine cells in the pancreas; also called *islets of Langerhans.*

pancreatic juice: A mixture of buffers and digestive enzymes that is discharged into the duodenum under the stimulation of the enzymes secretin and cholecystokinin.

pancreatitis (pan-krē-a-TĪ-tis): An inflammation of the pancreas.

Papanicolaou (Pap) test: A test for the detection of malignancies based on the cytological appearance of epithelial cells, especially those of the cervix and uterus.

papilla (pa-PIL-la): A small, conical projection.

paralysis: The loss of voluntary motor control over a portion of the body.

paranasal sinuses: Bony chambers, lined by respiratory epithelium, that open into the nasal cavity; the frontal, ethmoidal, sphenoidal, and maxillary sinuses.

parasagittal: A section or plane that parallels the midsagittal plane but that does not pass along the midline.

parasympathetic division: One of the two divisions of the autonomic nervous system; also called *craniosacral division;* generally responsible for activities that conserve energy and lower the metabolic rate.

parasympathomimetic drugs: Drugs that mimic the actions of parasympathetic stimulation.

parathyroid glands: Four small glands embedded in the posterior surface of the thyroid gland; secrete parathyroid hormone.

parathyroid hormone (PTH): A hormone secreted by the parathyroid glands when plasma calcium levels fall below the normal range; causes increased osteoclast activity, increased intestinal calcium uptake, and decreased calcium ion loss at the kidneys.

parenchyma (pa-RENG-ki-ma): The cells of a tissue or organ that are responsible for fulfilling its functional role; distinguished from the stroma of that tissue or organ.

paresthesia: A sensory abnormality that produces a tingling sensation.

parietal: Referring to the body wall or outer layer.

parietal cells: Cells of the gastric glands that secrete hydrochloric acid and intrinsic factor.

Parkinson's disease: A progressive motor disorder caused by the degeneration of cerebral nuclei.

parotid salivary glands (pa-ROT-id): Large salivary glands that secrete a saliva containing high concentrations of salivary (alpha) amylase.

pars distalis (dis-TAL-is): The large, anterior portion of the anterior pituitary gland.

pars intermedia (in-ter-MĒ-dē-uh): The portion of the anterior pituitary that is immediately adjacent to the posterior pituitary and the infundibulum.

pars nervosa: The posterior pituitary gland.

pars tuberalis: The portion of the anterior pituitary that wraps around the infundibulum superior to the posterior pituitary.

parturition (par-tū-RISH-un): Childbirth, delivery.

patella (pa-TEL-uh): The sesamoid bone of the kneecap.

pathogen: A disease-causing organism.

pathogenic: Disease-causing.

pathologist (pa-THO-lo-jist): An M.D. specializing in the identification of diseases on the basis of characteristic structural and functional changes in tissues and organs.

pedicel (PED-i-sel): A slender process of a podocyte that forms part of the filtration apparatus of the kidney glomerulus.

pedicles (PE-di-kulz): Thick, bony struts that connect the vertebral body with the articular and spinous processes.

pelvic cavity: The inferior subdivision of the abdominopelvic (peritoneal) cavity; encloses the urinary bladder, the sigmoid colon and rectum, and male or female reproductive organs.

pelvis: A bony complex created by the articulations among the coxae, the sacrum, and the coccyx.

penis (PĒ-nis): A component of the male external genitalia; a copulatory organ that surrounds the urethra and serves to introduce semen into the female vagina; the developmental equivalent of the female clitoris.

pepsin: A proteolytic enzyme secreted by the chief cells of the gastric glands in the stomach.

peptidases: Enzymes that split peptide bonds and release amino acids.

peptide: A chain of amino acids linked by peptide bonds.

peptide bond: A covalent bond between the amino group of one amino acid and the carboxyl group of another.

perforating canal: A passageway in compact bone that is at right angles to the axes of the osteons, between the periosteum and endosteum.

perfusion: The blood flow through a tissue.

pericardial cavity (per-i-KAR-dē-al): The space between the parietal pericardium and the epicardium (visceral pericardium) that covers the outer surface of the heart.

pericarditis: An inflammation of the pericardium.

pericardium (per-i-KAR-dē-um): The fibrous sac that surrounds the heart; its inner, serous lining is continuous with the epicardium.

perichondrium (per-i-KON-drē-um): The layer that surrounds a cartilage, consisting of an outer fibrous region and an inner cellular region.

perikaryon (per-i-KAR-ē-on): The cytoplasm that surrounds the nucleus in the cell body of a neuron.

perilymph (PER-ē-limf): A fluid similar in composition to cerebrospinal fluid; located in the spaces between the bony labyrinth and the membranous labyrinth of the inner ear.

perimysium (pe-ri-MĪZ-ē-um): A connective tissue partition that separates adjacent fasciculi in a skeletal muscle.

perineum (pe-ri-NĒ-um): The pelvic floor and its associated structures.

perineurium: A connective tissue partition that separates adjacent bundles of nerve fibers in a peripheral nerve.

periodontal ligament (per-ē-ō-DON-tal): Collagen fibers that bind the cementum of a tooth to the periosteum of the surrounding alveolus.

periosteum (per-ē-OS-tē-um): The layer that surrounds a bone, consisting of an outer fibrous region and inner cellular region.

peripheral nervous system (PNS): All neural tissue outside the central nervous system.

peripheral resistance: The resistance to blood flow; primarily caused by friction with the vascular walls.

peristalsis (per-i-STAL-sis): A wave of smooth muscle contractions that propels materials along the axis of a tube such as the digestive tract, the ureters, or the ductus deferens.

peritoneal cavity: *See* **abdominopelvic cavity.**

peritoneum (per-i-tō-NĒ-um): The serous membrane that lines the peritoneal (abdominopelvic) cavity.

peritonitis (per-i-tō-NĪ-tis): An inflammation of the peritoneum.

peritubular capillaries: A network of capillaries that surrounds the proximal and distal convoluted tubules of the kidneys.

permeability: The ease with which dissolved materials can cross a membrane; if the membrane is freely permeable, any molecule can cross it; if impermeable, nothing can cross; most biological membranes are selectively permeable.

peroxisome: A membranous vesicle containing enzymes that break down hydrogen peroxide (H_2O_2).

pes: The foot.

petrosal ganglion: A sensory ganglion of the glossopharyngeal nerve (N IX).

petrous: Stony; usually refers to the thickened portion of the temporal bone that encloses the inner ear.

pH: The negative exponent of the hydrogen ion concentration, expressed in moles per liter.

phagocyte: A cell that performs phagocytosis.

phagocytosis (fa-gō-sī-TŌ-sis): The engulfing of extracellular materials or pathogens; the movement of extracellular materials into the cytoplasm by enclosure in a membranous vesicle.

phalanx/phalanges (fa-LAN-gēz): Bones of the fingers or toes.

pharmacology: The study of drugs, their physiological effects, and their clinical uses.

pharynx (FAR-inks): The throat; a muscular passageway shared by the digestive and respiratory tracts.

phasic response: A pattern of response to stimulation by sensory neurons that are normally inactive; stimulation causes a burst of neural activity that ends when the stimulus either stops or stops changing in intensity.

phenotype (FĒN-ō-tīp): Physical characteristics that are genetically determined.

phonation (fō-NĀ-shun): Sound production at the larynx.

phosphate group: PO_4^{3-}; a functional group that can be attached to an organic molecule; required for the formation of high-energy bonds.

phospholipid (fos-fō-LIP-id): An important membrane lipid whose structure includes both hydrophilic and hydrophobic regions.

phosphorylation (fos-for-i-LĀ-shun): The addition of a high-energy phosphate group to a molecule.

photoreception: Sensitivity to light.

physiology (fiz-ē-OL-o-jē): The study of function; deals with the ways organisms perform vital activities.

pia mater: The tough, outer meningeal layer that surrounds the central nervous system.

pigment: A compound with a characteristic color.

pineal gland (PIN-ē-ul): Neural tissue in the posterior portion of the roof of the diencephalon; secretes melatonin.

pinna: *See* **auricle.**

pinocytosis (pi-nō-sī-TŌ-sis): The introduction of fluids into the cytoplasm by enclosing them in membranous vesicles at the cell surface.

pituitary gland: An endocrine organ that is situated in the sella turcica of the sphenoid bone and is connected to the hypothalamus by the infundibulum; includes the posterior pituitary (pars nervosa) and the anterior pituitary (pars intermedia and pars distalis).

placenta (pla-SENT-uh): A temporary structure in the uterine wall that permits diffusion between the fetal and maternal circulatory systems; *see* **afterbirth.**

plantar: Referring to the sole of the foot.

plasma (PLAZ-muh): The fluid ground substance of whole blood; what remains after the cells have been removed from a sample of whole blood.

plasma cell: An activated B cell that secretes antibodies.

plasmalemma (plaz-ma-LEM-a): A cell membrane.

platelets (PLĀT-lets): Small packets of cytoplasm that contain enzymes important in the clotting response; manufactured in bone marrow by megakaryocytes.

pleura (PLOO-ra): The serous membrane that lines the pleural cavities.

pleural cavities: Subdivisions of the thoracic cavity that contain the lungs.

pleuritis (ploor-Ī-tis): An inflammation of the pleura; also called *pleurisy.*

plexus (PLEK-sus): A network or braid.

plica (PLĪ-ka): A permanent transverse fold in the wall of the small intestine.

pneumotaxic center (noo-mō-TAKS-ik): A center in the reticular formation of the pons that regulates the activities of the apneustic and respiratory rhythmicity centers to adjust the pace of respiration.

pneumothorax (noo-mō-THO-raks): The introduction of air into the pleural cavity.

podocyte (PŌ-do-sīt): A cell whose processes surround the kidney glomerular capillaries and assist in filtration.

polar body: A nonfunctional packet of cytoplasm that contains chromosomes eliminated from an oocyte during meiosis.

polar bond: A covalent bond in which electrons are shared unequally.

polarized: Referring to cells that have regional differences in organelle distribution or cytoplasmic composition along a specific axis, such as between the basement membrane and free surface of an epithelial cell.

pollex: The thumb.

polymer: A large molecule consisting of a long chain of subunits.

polymorph: A polymorphonuclear leukocyte; a neutrophil.

polypeptide: A chain of amino acids strung together by peptide bonds; those containing more than 100 peptides are called *proteins.*

polyribosome: Several ribosomes linked by their translation of a single mRNA strand.

polysaccharide (pol-ē-SAK-uh-rīd): A complex sugar, such as glycogen or a starch.

polysynaptic reflex: A reflex in which interneurons are interposed between the sensory fiber and the motor neuron(s).

polyunsaturated fats: Fatty acids containing carbon atoms that are linked by double bonds.

polyuria (pol-ē-Ū-rē-uh): Excessive urine production.

pons: The portion of the metencephalon that is anterior to the cerebellum.

popliteal (pop-LIT-ē-al): Pertaining to the back of the knee.

porphyrins (POR-fi-rinz): Ring-shaped molecules that form the basis of important respiratory and metabolic pigments, including heme and the cytochromes.

porta hepatis: A region of mesentery between the duodenum and liver that contains the hepatic artery proper, the hepatic portal vein, and the common bile duct; also called *hilus.*

positive feedback: A mechanism that increases a deviation from normal limits after an initial stimulus.

postabsorptive state: A period that begins 4 hours after a meal; characterized by falling blood glucose concentrations and the mobilization of metabolic reserves.

postcentral gyrus: The primary sensory cortex, where touch, vibration, pain, temperature, and taste sensations arrive and are consciously perceived.

posterior: Toward the back; dorsal.

postganglionic neuron: An autonomic neuron in a peripheral ganglion, whose activities control peripheral effectors.

postovulatory phase: The secretory phase of the menstrual cycle.

postsynaptic membrane: The portion of the cell membrane of a postsynaptic cell that is part of a synapse.

potential difference: The separation of opposite charges; requires a barrier that prevents ion migration.

precentral gyrus: The primary motor cortex of a cerebral hemisphere, located anterior to the central sulcus.

prefrontal cortex: The anterior portion of each cerebral hemisphere; thought to be involved with higher intellectual functions, predictions, calculations, and so forth.

preganglionic neuron: A visceral motor neuron in the central nervous system whose output controls one or more ganglionic motor neurons in the peripheral nervous system.

premolars: Bicuspids; teeth with flattened occlusal surfaces; located anterior to the molar teeth.

premotor cortex: The motor association area between the precentral gyrus and the prefrontal area.

preoptic nucleus: The hypothalamic nucleus that coordinates thermoregulatory activities.

prepuce (PRĒ-pūs): The loose fold of skin that surrounds the glans penis (in males) or the clitoris (in females).

preputial glands (prē-PŪ-shal): Glands on the inner surface of the prepuce that produce a viscous, odorous secretion called *smegma.*

presbyopia: Farsightedness; an inability to accommodate for near vision.

presynaptic membrane: The synaptic surface where neurotransmitter release occurs.

prevertebral ganglion: *See* **collateral ganglion.**

prime mover: A muscle that performs a specific action.

proenzyme: An inactive enzyme secreted by an epithelial cell.

progesterone (prō-JES-ter-ōn): The most important progestin secreted by the corpus luteum after ovulation.

progestins (prō-JES-tinz): Steroid hormones structurally related to cholesterol; progesterone is an example.

prognosis: A prediction about the possibility or time course of recovery from a specific disease.

projection fibers: Axons carrying information from the thalamus to the cerebral cortex.

prolactin (prō-LAK-tin): The hormone that stimulates functional development of the mammary gland in females; a secretion of the anterior pituitary gland.

prolapse: The abnormal descent or protrusion of a portion of an organ, such as the vagina or anorectal canal.

proliferative phase: A portion of the uterine cycle; the interval of estrogen-induced repair of the functional zone of the endometrium through the growth and proliferation of epithelial cells in the uterine glands.

pronation (prō-NĀ-shun): The rotation of the forearm that makes the palm face posteriorly.

prone: Lying face down with the palms facing the floor.

pronucleus: An enlarged ovum or spermatozoon nucleus that forms after fertilization but before amphimixis.

properdin: The complement factor that prolongs and enhances non-antibody-dependent complement binding to bacterial cell walls.

prophase (PRŌ-fāz): The initial phase of mitosis; characterized by the appearance of chromosomes, the breakdown of the nuclear membrane, and the formation of the spindle apparatus.

proprioception (prō-prē-ō-SEP-shun): The awareness of the positions of bones, joints, and muscles.

prostaglandin (pros-ta-GLAN-din): A fatty acid secreted by one cell that alters the metabolic activities or sensitivities of adjacent cells; also called *local hormone*.

prostatectomy (pros-ta-TEK-to-mē): The surgical removal of the prostate gland.

prostate gland (PROS-tāt): An accessory gland of the male reproductive tract, contributing roughly one-third of the volume of semen.

prostatitis (pros-ta-TĪ-tis): An inflammation of the prostate gland.

prosthesis: An artificial substitute for a body part.

protease: *See* **proteinase.**

protein: A large polypeptide with a complex structure

proteinase: An enzyme that breaks down proteins into peptides and amino acids.

proteinuria (prō-tēn-ŪR-ē-uh): Abnormal amounts of protein in urine.

proteoglycan (prō-tē-ō-GLĪ-kan): A compound containing a large polysaccharide complex attached to a relatively small protein; examples include hyaluronan and chondroitin sulfate.

prothrombin: A circulating proenzyme of the common pathway of the clotting system; converted to thrombin by the enzyme thromboplastin.

proton: A fundamental particle bearing a positive charge.

protraction: Movement anteriorly in the horizontal plane.

proximal: Toward the attached base of an organ or structure.

proximal convoluted tubule (PCT): The portion of the nephron that is situated between Bowman's capsule and the loop of Henle; the major site of active reabsorption from filtrate.

pruritis (proo-RĪ-tus): Itching.

pseudopodia (soo-dō-PŌ-dē-ah): Temporary cytoplasmic extensions typical of mobile or phagocytic cells.

pseudostratified epithelium: An epithelium that contains several layers of nuclei but whose cells are all in contact with the underlying basement membrane.

psoriasis (sō-RĪ-uh-sis): A skin condition characterized by excessive keratin production and the formation of dry, scaly patches on the body surface.

psychosomatic condition: An abnormal physiological state with a psychological origin.

puberty: A period of rapid growth, sexual maturation, and the appearance of secondary sexual characteristics; normally occurs at ages 10–15 years.

pubic symphysis: The fibrocartilaginous amphiarthrosis between the pubic bones of the ossa coxae.

pubis (PŪ-bis): The anterior, inferior component of the os coxae.

pudendum (pū-DEN-dum): The external genitalia.

pulmonary circuit: Blood vessels between the pulmonary semilunar valve of the right ventricle and the entrance to the left atrium; the blood flow through the lungs.

pulmonary ventilation: The movement of air into and out of the lungs.

pulp cavity: The internal chamber in a tooth, containing blood vessels, lymphatic vessels, nerves, and the cells that maintain the dentin.

pulpitis (pul-PĪ-tis): An inflammation of the tissues of the pulp cavity.

pulvinar: The thalamic nucleus involved in the integration of sensory information prior to projection to the cerebral hemispheres.

pupil: The opening in the center of the iris through which light enters the eye.

purine: A nitrogen compound with a double ring-shaped structure; examples include adenine and guanine, two nitrogenous bases that are common in nucleic acids.

Purkinje cell (pur-KIN-jē): A large, branching neuron of the cerebellar cortex.

Purkinje fibers: Specialized conducting cardiocytes in the ventricles of the heart.

pus: An accumulation of debris, fluid, dead and dying cells, and necrotic tissue.

P wave: A deflection of the ECG corresponding to atrial depolarization.

pyelogram (PĪ-el-ō-gram): A radiographic image of the kidneys and ureters.

pyelonephritis (pī-e-lō-nef-RĪ-tis): An inflammation of the kidneys.

pyloric sphincter (pī-LOR-ik): A sphincter of smooth muscle that regulates the passage of chyme from the stomach to the duodenum.

pylorus (pī-LOR-us): The gastric region between the body of the stomach and the duodenum; includes the pyloric sphincter.

pyrexia (pī-REK-sē-uh): A fever.

pyrimidine: A nitrogen compound with a single ring-shaped structure; examples include cytosine, thymine, and uracil, nitrogenous bases that are common in nucleic acids.

pyrogen (PĪ-rō-jen): A compound that promotes a fever.

pyruvic acid (pī-ROO-vik): A three-carbon compound produced by glycolysis.

Q

quadriplegia: Paralysis of the upper and lower limbs.

quaternary structure: The three-dimensional protein structure produced by interactions between protein subunits.

R

radiodensity: The relative resistance to the passage of X-rays.

radiographic techniques: Methods of visualizing internal structures by using various forms of radiational energy.

radiopaque: Having a high radiodensity.

rami communicantes: Axon bundles that link the spinal nerves with the ganglia of the sympathetic chain.

ramus/rami: A branch.

raphe (RĀ-fē): A seam.

receptive field: The area monitored by a single sensory receptor.

recessive trait: An allele that affects the phenotype only when the individual is homozygous for that trait.

recombinant DNA: DNA created by splicing together a specific gene from one organism into the DNA strand of another organism.

rectal columns: Longitudinal folds in the walls of the anorectal canal.

rectouterine pouch (rek-tō-Ū-te-rin): The peritoneal pocket between the anterior surface of the colon and the posterior surface of the uterus.

rectum (REK-tum): The inferior 15 cm (6 in.) of the digestive tract.

rectus: Straight.

red blood cell (RBC): *See* **erythrocyte.**

reduction: The gain of hydrogen atoms or electrons or the loss of an oxygen molecule.

reductional division: The first meiotic division, which reduces the chromosome number from 46 to 23.

reflex: A rapid, automatic response to a stimulus.

reflex arc: The receptor, sensory neuron, motor neuron, and effector involved in a particular reflex; interneurons may be present, depending on the reflex considered.

refraction: The bending of light rays as they pass from one medium to another.

refractory period: The period between the initiation of an action potential and the restoration of the normal resting potential; during this period, the membrane will not respond normally to stimulation.

relaxation phase: The period after a contraction when the tension in the muscle fiber returns to resting levels.

relaxin: A hormone that loosens the pubic symphysis; secreted by the placenta.

renal: Pertaining to the kidneys.

renal corpuscle: The initial portion of the nephron, consisting of an expanded chamber that encloses the glomerulus.

renin: The enzyme released by cells of the juxtaglomerular apparatus when renal blood flow declines; converts angiotensinogen to angiotensin I.

rennin: A gastric enzyme that breaks down milk proteins.

replication: Duplication.

repolarization: The movement of the transmembrane potential away from a positive value and toward the resting potential.

residual volume: The amount of air remaining in the lungs after maximum forced exhalation.

respiration: The exchange of gases between cells and the environment; includes pulmonary ventilation, external respiration, internal respiration, and cellular respiration.

respiratory minute volume (V_E): The amount of air moved into and out of the respiratory system each minute.

respiratory pump: A mechanism by which changes in the intrapleural pressures during the respiratory cycle assist the venous return to the heart; also called *thoracoabdominal pump*.

resting potential: The transmembrane potential of a normal cell under homeostatic conditions.

rete (RĒ-tē): An interwoven network of blood vessels or passageways.

reticular activating system (RAS): The mesencephalic portion of the reticular formation; responsible for arousal and the maintenance of consciousness.

reticular formation: A diffuse network of gray matter that extends the entire length of the brain stem.

reticulospinal tracts: Descending tracts of the medial pathway that carry involuntary motor commands issued by neurons of the reticular formation.

retina: The innermost layer of the eye, lining the vitreous chamber; also called *neural tunic*.

retinal (RET-i-nal): A visual pigment derived from vitamin A.

retraction: Movement posteriorly in the horizontal plane.

retroflexion (re-trō-FLEK-shun): A posterior tilting of the uterus that has no clinical significance.

retrograde flow (RET-rō-grād): The transport of materials from the telodendria to the cell body of a neuron.

retroperitoneal (re-trō-per-i-tō-NĒ-al): Behind or outside the peritoneal cavity.

reverberation: A positive feedback along a chain of neurons such that they remain active once stimulated.

rheumatism (ROO-muh-tizm): A condition characterized by pain in muscles, tendons, bones, or joints.

Rh factor: An agglutinogen that may be present (Rh-positive) or absent (Rh-negative) from the surfaces of red blood cells.

rhizotomy: The surgical transection of a dorsal root, normally performed to relieve pain.

rhodopsin (rō-DOP-sin): The visual pigment in the membrane disks of the distal segments of rods.

rhythmicity center: A medullary center responsible for the pace of respiration; includes inspiratory and expiratory centers.

ribonucleic acid (rī-bō-noo-KLĀ-ik): A nucleic acid consisting of a chain of nucleotides that contain the sugar ribose and the nitrogenous bases adenine, guanine, cytosine, and uracil.

ribose: A five-carbon sugar that is a structural component of RNA.

ribosome: An organelle that contains rRNA and proteins and is essential to mRNA translation and protein synthesis.

rigor mortis: The extended muscular contraction and rigidity that occurs after death; the result of calcium ion release from the sarcoplasmic reticulum and the exhaustion of cytoplasmic ATP reserves.

rod: A photoreceptor responsible for vision in dim lighting.

rough endoplasmic reticulum (RER): A membranous organelle that is a site of protein synthesis and storage.

round window: An opening in the bony labyrinth of the inner ear that exposes the membranous wall of the tympanic duct to the air of the middle ear cavity.

rubrospinal tracts: Descending tracts of the lateral pathway that carry involuntary motor commands issued by the red nucleus of the mesencephalon.

Ruffini corpuscles (roo-FĒ-nē): Receptors sensitive to tension and stretch in the dermis of the skin.

rugae (ROO-gē): Mucosal folds in the lining of the empty stomach that disappear as gastric distension occurs.

S

saccule (SAK-ūl): A portion of the vestibular apparatus of the inner ear; contains a macula important for static equilibrium.

sagittal plane: A sectional plane that divides the body into left and right portions.

salivatory nucleus (SAL-i-va-to-rē): The medullary nucleus that controls the secretory activities of the salivary glands.

salt: An inorganic compound consisting of a cation other than H^+ and an anion other than OH^-.

saltatory propagation: The relatively rapid propagation of an action potential between successive nodes of a myelinated axon.

sarcolemma: The cell membrane of a muscle cell.

sarcoma (sar-KŌ-ma): A tumor of connective tissues.

sarcomere: The smallest contractile unit of a striated muscle cell.

sarcoplasm: The cytoplasm of a muscle cell.

satellite cells: *See* **amphicytes.**

scala media: *See* **cochlear duct.**

scala tympani: *See* **tympanic duct.**

scala vestibuli: *See* **vestibular duct.**

scar tissue: The thick, collagenous tissue that forms at an injury site.

Schlemm, canal of: The passageway that delivers aqueous humor from the anterior chamber of the eye to the venous circulation.

Schwann cells: Neuroglia responsible for the neurilemma that surrounds axons in the peripheral nervous system.

sciatica (sī-AT-i-ka): Pain felt along the peripheral distribution of the sciatic nerve.

sciatic nerve (sī-A-tik): A nerve innervating the posteromedial portions of the thigh and leg.

sclera (SKLER-uh): The fibrous, outer layer of the eye that forms the white area of the anterior surface; a portion of the fibrous tunic of the eye.

sclerosis: A hardening and thickening that commonly occurs secondary to tissue inflammation.

scoliosis (skō-lē-Ō-sis): An abnormal, exaggerated lateral curvature of the spine.

scrotum (SKRŌ-tum): The loose-fitting, fleshy pouch that encloses the testes of the male.

sebaceous glands (sē-BĀ-shus): Glands that secrete sebum; normally associated with hair follicles.

sebum (SĒ-bum): A waxy secretion that coats the surfaces of hairs.

secondary sex characteristics: Physical characteristics that appear at puberty in response to sex hormones but are not involved in the production of gametes.

secretin (sē-KRĒ-tin): A hormone, secreted by the duodenum, that stimulates the production of buffers by the pancreas and inhibits gastric activity.

semen (SĒ-men): The fluid ejaculate that contains spermatozoa and the secretions of accessory glands of the male reproductive tract.

semicircular ducts: The tubular components of the membranous labyrinth of the inner ear; responsible for dynamic equilibrium.

semilunar valve: A three-cusped valve guarding the exit from one of the cardiac ventricles; the pulmonary and aortic valves.

seminal vesicles (SEM-i-nal): Glands of the male reproductive tract that produce roughly 60 percent of the volume of semen.

seminiferous tubules (se-mi-NIF-e-rus): Coiled tubules where spermatozoon production occurs in the testis.

senescence: Aging.

sensible perspiration: Water loss due to secretion by sweat glands.

septae (SEP-tē): Partitions that subdivide an organ.

serosa: *See* **serous membrane.**

serotonin (ser-ō-TŌ-nin): A neurotransmitter in the central nervous system; a compound that enhances inflammation and is released by activated mast cells and basophils.

serous cell: A cell that produces a serous secretion.

serous membrane: A squamous epithelium and the underlying loose connective tissue; the lining of the pericardial, pleural, and peritoneal cavities.

serous secretion: A watery secretion that contains high concentrations of enzymes; produced by serous cells.

serum: The ground substance of blood plasma from which clotting agents have been removed.

sesamoid bone: A bone that forms within a tendon.

sigmoid colon (SIG-moyd): The S-shaped 18-cm-long portion of the colon between the descending colon and the rectum.

sign: The visible evidence of the presence of a disease.

simple epithelium: An epithelium containing a single layer of cells above the basement membrane.

sinoatrial (SA) node: The natural pacemaker of the heart; situated in the wall of the right atrium.

sinus: A chamber or hollow in a tissue; a large, dilated vein.

sinusitis: An inflammation of a nasal sinus.

sinusoid (SĪ-nus-oyd): An extensive network of vessels in the liver, adrenal cortex, spleen, and pancreas; similar in histological structure to capillaries.

skeletal muscle: A contractile organ of the muscular system.

skeletal muscle tissue: A contractile tissue dominated by skeletal muscle fibers; characterized as striated, voluntary muscle.

sliding filament theory: The concept that a sarcomere shortens as the thick and thin filaments slide past one another.

small intestine: The duodenum, jejunum, and ileum; the digestive tract between the stomach and the large intestine.

smegma (SMEG-ma): The secretion of the preputial glands of the penis or clitoris.

smooth endoplasmic reticulum (SER): A membranous organelle in which lipid and carbohydrate synthesis and storage occur.

smooth muscle tissue: Muscle tissue in the walls of many visceral organs; characterized as nonstriated, involuntary muscle.

soft palate: The fleshy posterior extension of the hard palate, separating the nasopharynx from the oral cavity.

sole: The inferior surface of the foot.

solute: Any materials dissolved in a solution.

solution: A fluid containing dissolved materials.

solvent: The fluid component of a solution.

soma (SŌ-ma): Body; the body of a neuron; also called *cell body.*

somatic (sō-MAT-ik): Pertaining to the body.

somatic nervous system (SNS): The efferent division of the nervous system that innervates skeletal muscles.

somatomedins: Compounds stimulating tissue growth; released by the liver after the secretion of growth hormone; also called *insulin-like growth factors.*

somatotropin: Growth hormone; produced by the anterior pituitary in response to growth hormone–releasing hormone (GH-RH).

sperm: *See* **spermatozoon.**

spermatic cord: Collectively, the spermatic vessels, nerves, lymphatic vessels, and the ductus deferens, extending between the testes and the proximal end of the inguinal canal.

spermatids (SPER-ma-tidz): The product of meiosis in males; cells that differentiate into spermatozoa.

spermatocyte (sper-MA-tō-sīt): A cell of the seminiferous tubules that is engaged in meiosis.

spermatogenesis (sper-ma-tō-JEN-e-sis): Spermatozoon production.

spermatogonia (sper-ma-tō-GŌ-nē-uh): Stem cells whose mitotic divisions give rise to other stem cells and to primary spermatocytes.

spermatozoon/spermatozoa (sper-ma-to-ZŌ-a): A male gamete; also called *sperm.*

spermicide: A compound toxic to spermatozoa; used as a contraceptive method.

spermiogenesis: The process of spermatid differentiation that leads to the formation of physically mature spermatozoa.

sphincter (SFINK-ter): A muscular ring that contracts to close the entrance or exit of an internal passageway.

spinal nerve: One of 31 pairs of nerves that originate on the spinal cord from anterior and posterior roots.

spindle apparatus: A muscle spindle (intrafusal fibers) and its sensory and motor innervation.

spinocerebellar tracts: Ascending tracts that carry sensory information to the cerebellum.

spinothalamic tracts: Ascending tracts that carry poorly localized touch, pressure, pain, vibration, and temperature sensations to the thalamus.

spinous process: The prominent posterior projection of a vertebra; formed by the fusion of two laminae.

spleen: A lymphoid organ important for the phagocytosis of red blood cells, the immune response, and lymphocyte production.

splenectomy (splē-NEK-to-mē): The surgical removal of the spleen.

sprain: A forceful distortion of an articulation that produces damage to the capsule, ligaments, or tendons but not dislocation.

sputum (SPŪ-tum): A viscous mucus that is transported to the pharynx by the mucus escalator of the respiratory tract and is ejected from the mouth.

squama: A broad, flat surface.

squamous (SKWĀ-mus): Flattened.

squamous epithelium: An epithelium whose superficial cells are flattened and platelike.

stapedius (sta-PĒ-dē-us): A muscle of the middle ear whose contraction tenses the auditory ossicles and reduces the forces transmitted to the oval window.

stapes (STĀ-pēz): The auditory ossicle attached to the tympanic membrane.

statoconia: Densely packed calcium carbonate crystals forming masses that are attached, by a gelatinous matrix, to hair cell receptors of maculae in the vestibule of the inner ear.

stenosis (ste-NŌ-sis): A constriction or narrowing of a passageway.

stereocilia: Elongate microvilli characteristic of the epithelium of the epididymis, portions of the ductus deferens, and the inner ear.

steroid: A ring-shaped lipid structurally related to cholesterol.

stimulus: An environmental change that produces a change in cellular activities; often used to refer to events that alter the transmembrane potentials of excitable cells.

stratified: Containing several layers.

stratum (STRA-tum): A layer.

stretch receptors: Sensory receptors that respond to stretching of the surrounding tissues.

stroma: The connective tissue framework of an organ; distinguished from the functional cells (parenchyma) of that organ.

subarachnoid space: A meningeal space containing cerebrospinal fluid; the area between the arachnoid membrane and the pia mater.

subclavian (sub-KLĀ-vē-an): Pertaining to the region immediately posterior and inferior to the clavicle.

subcutaneous layer: The layer of loose connective tissue below the dermis; also called *hypodermis* or *superficial fascia.*

sublingual salivary glands (sub-LING-gwal): Mucus-secreting salivary glands inferior to the tongue.

submandibular salivary glands: Salivary glands nestled in depressions on the medial surfaces of the mandible; salivary glands that produce a mixture of mucins and enzymes (salivary amylase).

submucosa (sub-mū-KŌ-sa): The region between the muscularis mucosae and the muscularis externa.

submucosal glands: Mucous glands in the submucosa of the duodenum; also called *Brunner's glands.*

subserous fascia: The loose connective tissue layer beneath the serous membrane that lines the ventral body cavity.

substrate: A participant (product or reactant) in an enzyme-catalyzed reaction.

sulcus (SUL-kus): A groove or furrow.

summation: The temporal or spatial addition of stimuli.

superficial fascia: *See* **subcutaneous layer.**

superior: Above, in reference to a portion of the body in the anatomical position.

superior vena cava (SVC): The vein that carries blood to the right atrium from parts of the body that are superior to the heart.

supination (soo-pi-NĀ-shun): The rotation of the forearm such that the palm faces anteriorly.

supine (SOO-pīn): Lying face up, with palms facing anteriorly.

suppressor T cells: Lymphocytes that inhibit B cell activation and the secretion of antibodies by plasma cells.

suprarenal gland (soo-pra-RĒ-nal): *See* **adrenal gland.**

surfactant (sur-FAK-tant): A lipid secretion that coats the alveolar surfaces of the lungs and prevents their collapse.

sustentacular cells (sus-ten-TAK-ū-lar): Supporting cells of the seminiferous tubules of the testis; responsible for the differentiation of spermatids, the maintenance of the blood–testis barrier, and the secretion of inhibin, androgen-binding protein, and Müllerian-inhibiting factor.

sutural bones: Irregular bones that form in fibrous tissue between the flat bones of the developing cranium; also called *Wormian bones.*

suture: A fibrous joint between flat bones of the skull.

sympathectomy (sim-path-EK-to-mē): The transection of the sympathetic innervation to a region.

sympathetic division: The division of the autonomic nervous system that is responsible for "fight or flight" reactions; primarily concerned with the elevation of metabolic rate and increased alertness.

sympathomimetic drugs: Drugs that mimic the actions of sympathetic stimulation.

symphysis (SIM-fi-sis): A fibrous amphiarthrosis, such as that between adjacent vertebrae or between the pubic bones of the ossa coxae.

symptom: An abnormality of function as a result of disease.

synapse (SIN-aps): The site of communication between a nerve cell and some other cell; if the other cell is not a neuron, the term *neuromuscular* or *neuroglandular junction* is often used.

synaptic delay (sin-AP-tik): The period between the arrival of an impulse at the presynaptic membrane and the initiation of an action potential in the postsynaptic membrane.

synarthrosis (sin-ar-THRŌ-sis): A joint that does not permit relative movement between the articulating elements; *see* **lambdoidal suture.**

synchondrosis (sin-kon-DRŌ-sis): A cartilaginous synarthrosis, such as the articulation between the epiphysis and diaphysis of a growing bone.

syncope (SIN-kō-pē): A sudden, transient loss of consciousness; a faint.

syncytial trophoblast (sin-SISH-al): The multinucleate cytoplasmic layer that covers the blastocyst; responsible for uterine erosion and implantation.

syncytium (sin-SI-shē-um): A multinucleate mass of cytoplasm, produced by the fusion of cells or repeated mitoses without cytokinesis.

syndesmosis (sin-dez-MŌ-sis): A fibrous amphiarthrosis.

syndrome: A discrete set of symptoms that occur together.

syneresis (sin-ER-ē-sis): Clot retraction.

synergist (SIN-er-jist): A muscle that assists a prime mover in performing its primary action.

synostosis (sin-os-TŌ-sis): A synarthrosis formed by the fusion of the articulating elements.

synovial cavity (si-NŌ-vē-ul): A fluid-filled chamber in a synovial joint.

synovial fluid: The substance secreted by synovial membranes that lubricates joints.

synovial joint: A freely movable joint where the opposing bone surfaces are separated by synovial fluid; a diarthrosis.

synovial membrane: An incomplete layer of fibroblasts confronting the synovial cavity, plus the underlying loose connective tissue.

synthesis (SIN-the-sis): Manufacture; anabolism.

system: An interacting group of organs that performs one or more specific functions.

systemic circuit: The vessels between the aortic semilunar valve and the entrance to the right atrium; the circulatory system other than the vessels of the pulmonary circuit.

systole (SIS-tō-lē): The period of cardiac contraction.

systolic pressure: The peak arterial pressure measured during ventricular systole.

T

tachycardia (tak-ē-KAR-dē-uh): An abnormally rapid heart rate.

tactile: Pertaining to the sense of touch.

taenia coli (TĒ-nē-a KŌ-lī): Three longitudinal bands of smooth muscle in the muscularis externa of the colon.

tarsal bones: The bones of the ankle (the talus, calcaneus, navicular, and cuneiform bones).

tarsus: The ankle.

TCA (tricarboxylic acid) cycle: The aerobic reaction sequence that occurs in the matrix of mitochondria; in the process, organic molecules are broken down, carbon dioxide molecules are released, and hydrogen molecules are transferred to coenzymes that deliver them to the electron transport system; also called *citric acid cycle* or *Krebs cycle.*

T cells: Lymphocytes responsible for cell-mediated immunity and for the coordination and regulation of the immune response; includes regulatory T cells (helpers and suppressors) and cytotoxic (killer) T cells.

tears: The fluid secretion of the lacrimal glands that bathes the anterior surfaces of the eyes.

tectorial membrane (tek-TOR-ē-al): The gelatinous membrane suspended over the hair cells of the organ of Corti.

tectospinal tracts: Descending tracts of the medial pathway that carry involuntary motor commands issued by the colliculi.

tectum: The roof of the mesencephalon of the brain.

telencephalon (tel-en-SEF-a-lon): The forebrain or cerebrum, including the cerebral hemispheres, the internal capsule, and the cerebral nuclei.

telodendria (te-lō-DEN-drē-uh): Terminal axonal branches that end in synaptic knobs.

telophase (TĒL-ō-fāz): The final stage of mitosis, characterized by the disappearance of the spindle apparatus, the reappearance of the nuclear membrane, the disappearance of the chromosomes, and the completion of cytokinesis.

temporal: Pertaining to time (temporal summation) or to the temples (temporal bone).

tendinitis: A painful inflammation of a tendon.

tendon: A collagenous band that connects a skeletal muscle to an element of the skeleton.

tentorium cerebelli (ten-TOR-ē-um ser-e-BEL-ē): A dural partition that separates the cerebral hemispheres from the cerebellum.

teratogen (TER-a-tō-jen): A stimulus that causes developmental defects.

teres: Long and round.

terminal: Toward the end.

tertiary follicle: A mature ovarian follicle, containing a large, fluid-filled chamber.

tertiary structure: The protein structure that results from interactions among distant portions of the same molecule; complex coiling and folding.

testes (TES-tēz): The male gonads, sites of gamete production and hormone secretion.

testosterone (tes-TOS-te-rōn): The principal androgen produced by the interstitial cells of the testes.

tetanic contraction: A sustained skeletal muscle contraction due to repeated stimulation at a frequency that prevents muscle relaxation.

tetanus: A tetanic contraction; also refers to a disease state that results from the stimulation of muscle cells by certain bacterial toxins.

tetany: A tetanic contraction; also refers to abnormally prolonged contractions that result from disturbances in electrolyte balance.

tetrad (TET-rad): Paired, duplicated chromosomes visible at the start of meiosis I.

tetraiodothyronine (tet-ra-ī-ō-dō-THĪ-rō-nēn): T_4, or thyroxine, a thyroid hormone.

thalamus: The walls of the diencephalon.

thalassemia (thal-ah-SĒ-mē-uh): A hereditary disorder that affects hemoglobin synthesis and produces anemia.

theory: A hypothesis that makes valid predictions, as demonstrated by evidence that is testable, unbiased, and repeatable.

therapy: The treatment of disease.

thermogenesis (ther-mō-JEN-e-sis): Heat production.

thermography: A diagnostic procedure involving the production of an infrared image.

thermoreception: Sensitivity to temperature changes.

thermoregulation: Homeostatic maintenance of body temperature.

thick filament: A cytoskeletal filament in a skeletal or cardiac muscle cell; composed of myosin, with a core of titin.

thin filament: A cytoskeletal filament in a skeletal or cardiac muscle cell; consists of actin, troponin, and tropomyosin.

thoracolumbar division (tho-ra-kō-LUM-bar): The sympathetic division of the autonomic nervous system.

thorax: The chest.

threshold: The transmembrane potential at which an action potential begins.

thrombin (THROM-bin): The enzyme that converts fibronogen to fibrin.

thrombocyte (THROM-bō-sīt): *See* **platelets;** in nonmammalian vertebrates, nucleated cells that are the equivalent of platelets in humans.

thrombocytopenia (throm-bō-sī-tō-PĒ-nē-uh): An abnormally low platelet count in the circulating blood.

thromboembolism (throm-bō-EM-bō-lizm): The occlusion of a blood vessel by a drifting blood clot.

thromboplastin: An enzyme that converts prothrombin to thrombin.

thrombus: A blood clot that develops at the lumenal wall of a blood vessel.

thymine: A pyrimidine; one of the nitrogenous bases in the nucleic acid DNA.

thymosins (THĪ-mō-sinz): Thymic hormones essential to the development and differentiation of T cells.

thymus: A lymphoid organ, the site of T cell formation.

thyroglobulin (thī-rō-GLOB-ū-lin): A circulating transport globulin that binds thyroid hormones.

thyroid gland: An endocrine gland whose lobes are lateral to the thyroid cartilage of the larynx.

thyroid hormones: Thyroxine (T_4) and triiodothyronine (T_3), hormones of the thyroid gland; stimulate tissue metabolism, energy utilization, and growth.

thyroid-stimulating hormone (TSH): The hormone, produced by the anterior pituitary, that triggers the secretion of thyroid hormones by the thyroid gland.

thyroxine (thī-ROKS-ēn): A thyroid hormone; also called T_4 or *tetraiodothyronine*.

tidal volume: The volume of air moved into and out of the lungs during a normal quiet respiratory cycle.

tissue: A collection of specialized cells and cell products that performs a specific function.

titer: The plasma antibody concentration.

tolerance: The failure to produce antibodies against antigens normally present in the body.

tonic response: An increase or decrease in the frequency of action potentials by sensory receptors that are chronically active.

tonsil: A lymphoid nodule in the wall of the pharynx; the palatine, pharyngeal, and lingual tonsils.

topical: Applied to the body surface.

toxic: Poisonous.

trabecula (tra-BEK-ū-la): A connective tissue partition that subdivides an organ.

trabeculae carneae (tra-BEK-ū-lē KAR-nē-ē): Muscular ridges projecting from the walls of the ventricles of the heart.

trachea (TRĀ-kē-a): The windpipe, an airway extending from the larynx to the primary bronchi.

tracheal ring: A C-shaped supporting cartilage of the trachea.

tracheostomy (trā-kē-OS-to-mē): The surgical opening of the anterior tracheal wall to permit airflow.

trachoma: An infectious disease of the conjunctiva and cornea.

tract: A bundle of axons in the central nervous system.

transamination (tranz-am-i-NĀ-shun): The enzymatic transfer of an amino group from an amino acid to another carbon chain.

transcription: The encoding of genetic instructions on a strand of mRNA.

transdermal medication: The administration of medication by absorption through the skin.

transection: The severing or cutting of an object in the transverse plane.

transfusion: The transfer of blood from a donor directly into the bloodstream of another person.

transient ischemic attack (TIA): A temporary loss of consciousness due to the occlusion of a small blood vessel in the brain.

translation: The process of peptide formation from the instructions carried by an mRNA strand.

transmembrane potential: The potential difference, measured across a cell membrane and expressed in millivolts, that results from the uneven distribution of positive and negative ions across the cell membrane.

transudate (TRAN-sū-dāt): A fluid that diffuses across a serous membrane and lubricates opposing surfaces.

transverse tubules: The transverse, tubular extensions of the sarcolemma that extend deep into the sarcoplasm, contacting cisternae of the sarcoplasmic reticulum; also called *T tubules.*

treppe (TREP-ē): A steplike increase in tension production after the repeated stimulation of a muscle, even though the muscle is allowed to complete each relaxation phase.

triad: In a skeletal muscle fiber, the combination of a transverse tubule and two cisternae of the sarcoplasmic reticulum; in the liver, a combination of branches of the hepatic portal vein, hepatic artery proper, and bile duct at a liver lobule.

tricarboxylic acid cycle (trī-kar-bok-SIL-ik): *See* **TCA cycle.**

tricuspid valve (trī-KUS-pid): The right atrioventricular valve, which prevents the backflow of blood into the right atrium during ventricular systole.

trigeminal nerve (trī-JEM-i-nal): Cranial nerve V, which provides sensory information from the lower portions of the face (including the upper and lower jaws) and delivers motor commands to the muscles of mastication.

triglyceride (trī-GLIS-e-rīd): A lipid that is composed of a molecule of glycerol attached to three fatty acids.

trigone (TRĪ-gōn): The triangular region of the urinary bladder that is bounded by the exits of the ureters and the entrance to the urethra.

triiodothyronine: T_3, a thyroid hormone.

trisomy (trī-SŌ-mē): The abnormal possession of three copies of a chromosome; trisomy 21 is responsible for Down syndrome.

trochanter (trō-KAN-ter): Large processes near the head of the femur.

trochlea (TRŌK-lē-uh): A pulley; the spool-shaped medial portion of the condyle of the humerus.

trochlear nerve (TRŌK-lē-ar): Cranial nerve IV, controlling the superior oblique muscle of the eye.

trophoblast (TRŌ-fō-blast): The superficial layer of the blastocyst that will be involved in implantation, hormone production, and placenta formation.

tropomyosin (trō-pō-MĪ-ō-sin): A protein on thin filaments that covers the active sites in the absence of free calcium ions.

troponin (TRŌ-pō-nin): A protein on thin filaments that binds to tropomyosin and to G actin.

trunk: The thoracic and abdominopelvic regions; a major arterial branch.

trypsin (TRIP-sin): One of the pancreatic proteases.

trypsinogen (trip-SIN-ō-jen): The inactive proenzyme that is secreted by the pancreas and is converted to trypsin in the duodenum.

T tubules: *See* **transverse tubules.**

tuberculum (too-BER-kū-lum): A small, localized elevation on a bony surface.

tuberosity: A large, roughened elevation on a bony surface.

tubulin: A protein subunit of microtubules.

tumor: A tissue mass formed by the abnormal growth and replication of cells.

tunica (TOO-ni-ka): A layer or covering.

tunica externa: The outermost layer of connective tissue fibers that stabilizes the position of a blood vessel.

tunica intima: The innermost layer of connective tissue fibers in a blood vessel; consists of the endothelium plus an underlying elastic membrane.

tunica media: The middle layer of connective tissue fibers in a blood vessel; contains collagen, elastin, and smooth muscle fibers in varying proportions.

T wave: A deflection of the ECG corresponding to ventricular repolarization.

twitch: A single stimulus–contraction–relaxation cycle in a skeletal muscle.

tympanic duct: The perilymph-filled chamber of the inner ear, adjacent to the basilar membrane; pressure changes there distort the round window; also called *scala tympani.*

tympanic membrane (tim-PAN-ik): The membrane that separates the external auditory canal from the middle ear; the membrane whose vibrations are transferred to the auditory ossicles and ultimately to the oval window; also called *eardrum* or *tympanum.*

type A axons: Large myelinated axons.

type B axons: Small myelinated axons.

type C axons: Small unmyelinated axons.

U

ulcer: An area of epithelial sloughing associated with damage to the underlying connective tissues and blood vessels.

ultrasound: A diagnostic visualization procedure that uses high-frequency sound waves.

umbilical cord (um-BIL-i-kal): The connecting stalk between the fetus and the placenta; contains the allantois, the umbilical arteries, and the umbilical vein.

umbilicus: The navel.

unicellular gland: Goblet cells.

unipennate muscle: A muscle whose fibers are on one side of the tendon.

unipolar neuron: A sensory neuron whose cell body is in a dorsal root ganglion or a sensory ganglion of a cranial nerve.

unmyelinated axon: An axon whose neurilemma does not contain myelin and across which continuous conduction occurs.

uracil: A pyrimidine; one of the nitrogenous bases in the nucleic acid RNA.

uremia (ū-RĒ-mē-uh): An abnormal condition caused by impaired kidney function; characterized by the retention of wastes and the disruption of many other organ systems.

ureters (ū-RĒ-terz): Muscular tubes, lined by transitional epithelium, that carry urine from the renal pelvis to the urinary bladder.

urethra (ū-RĒ-thra): A muscular tube that carries urine from the urinary bladder to the exterior.

urethritis: An inflammation of the urethra.

urinalysis: An analysis of the physical and chemical characteristics of urine.

urinary bladder: The muscular, distensible sac that stores urine prior to micturition.

urination: The voiding of urine; micturition.

urobilin: A compound derived from urobilinogen and ultimately from the bilirubin excreted in bile.

uterus (Ū-ter-us): The muscular organ of the female reproductive tract in which implantation, placenta formation, and fetal development occur.

utricle (Ū-tri-kul): The largest chamber of the vestibular apparatus of the inner ear; contains a macula important for static equilibrium.

uvea: The vascular tunic of the eye.

uvula (Ū-vū-luh): A dangling, fleshy extension of the soft palate.

V

vagina (va-JĪ-na): A muscular tube extending between the uterus and the vestibule.

varicose veins (VAR-i-kōs): Distended superficial veins.

vasa recta: Long, straight capillaries that parallel the loop of Henle in juxtamedullary nephrons.

vasa vasorum: Blood vessels that supply the walls of large arteries and veins.

vascular: Pertaining to blood vessels.

vascularity: The blood vessels in a tissue.

vascular spasm: A contraction of the wall of a blood vessel at an injury site; may slow the rate of blood loss.

vasoconstriction: A reduction in the diameter of arterioles due to the contraction of smooth muscles in the tunica media; elevates peripheral resistance; may occur in response to local factors, through the action of hormones, or from the stimulation of the vasomotor center.

vasodilation: An increase in the diameter of arterioles due to the relaxation of smooth muscles in the tunica media; reduces peripheral resistance; may occur in response to local factors, through the action of hormones, or after decreased stimulation of the vasomotor center.

vasomotion: Changes in the pattern of blood flow through a capillary bed in response to changes in the local environment.

vasomotor center: The center in the medulla oblongata whose stimulation produces vasoconstriction and an elevation of peripheral resistance.

vein: A blood vessel carrying blood from a capillary bed toward the heart.

vena cava (VĒ-na KĀ-va): One of the major veins delivering systemic blood to the right atrium; superior and inferior venae cavae.

ventilation: Air movement into and out of the lungs.

ventral: Pertaining to the anterior surface.

ventricle (VEN-tri-kul): A fluid-filled chamber; in the heart, one of the large chambers discharging blood into the pulmonary or systemic circuits; in the brain, one of four fluid-filled interior chambers.

ventricular escape: The initiation of ventricular contractions after a pause caused by impaired conduction of the contractile stimulus from the AV node.

ventricular folds: Mucosal folds in the laryngeal walls that do not play a role in sound production; the false vocal cords.

venule (VEN-ūl): Thin-walled veins that receive blood from capillaries.

vermiform appendix: A blind tube connected to the cecum of the large intestine; also called *appendix*.

vermis (VER-mis): A midsagittal band of neural cortex on the surface of the cerebellum.

vertebral canal: The passageway that encloses the spinal cord; a tunnel bounded by the neural arches of adjacent vertebrae.

vertebral column: The cervical, thoracic, and lumbar vertebrae, the sacrum, and the coccyx.

vertebrochondral ribs: Ribs 8–10; false ribs, connected to the sternum by shared cartilaginous bars.

vertebrosternal ribs: Ribs 1–7; true ribs, connected to the sternum by individual cartilaginous bars.

vertigo: Dizziness.

vesicle: A membranous sac in the cytoplasm of a cell.

vestibular complex: The utricle, saccule, and semicircular canals of the inner ear.

vestibular duct: The perilymph-filled chamber of the inner ear, superior to the vestibular membrane; pressure changes there result from distortions of the oval window; also called *scala vestibuli*.

vestibular membrane: The membrane that separates the cochlear duct from the vestibular duct of the inner ear.

vestibular nucleus: The processing center for sensations that arrive from the vestibular apparatus of the inner ear; located near the border between the pons and the medulla oblongata.

vestibule (VES-ti-būl): A chamber; in the inner ear, the utricle and saccule.

vestibulospinal tracts: Descending tracts of the medial pathway that carry involuntary motor commands issued by the vestibular nucleus to stabilize the position of the head.

villus/villi: A slender projection of the mucous membrane of the small intestine.

virus: A noncellular pathogen.

viscera (VIS-e-rah): Organs in the ventral body cavity.

visceral: Pertaining to viscera or their outer coverings.

visceral smooth muscle: A smooth muscle tissue that forms sheets or layers in the walls of visceral organs; the cells may not be innervated, and the layers often show automaticity (rhythmic contractions).

viscosity: The resistance to flow that a fluid exhibits as a result of molecular interactions within the fluid.

viscous: Thick, syrupy.

vital capacity: The maximum amount of air that can be moved into or out of the respiratory system; the sum of the inspiratory reserve, the expiratory reserve, and the tidal volume.

vitamin: An essential organic nutrient that functions as a coenzyme in vital enzymatic reactions.

vitreous humor: The gelatinous mass in the vitreous chamber of the eye.

vocal folds: Folds in the laryngeal wall that contain elastic ligaments whose tension can be voluntarily adjusted; the true vocal cords, responsible for phonation.

voluntary: Controlled by conscious thought processes.

vulva (VUL-vuh): The female pudendum (external genitalia).

W

Wallerian degeneration: The disintegration of an axon and its myelin sheath distal to an injury site.

white blood cells (WBCs): Leukocytes; the granulocytes and agranulocytes of blood.

white matter: Regions in the central nervous system that are dominated by myelinated axons.

white ramus: A nerve bundle containing the myelinated preganglionic axons of sympathetic motor neurons en route to the sympathetic chain or to a collateral ganglion.

Wormian bones: *See* **sutural bones.**

X

X chromosome: The sex chromosome that is present both in genetic males and in genetic females.

xiphoid process (ZĪ-foyd): The slender, inferior extension of the sternum.

Y

Y chromosome: The sex chromosome whose presence indicates that the individual is a genetic male.

yolk sac: One of the four extraembryonic membranes, composed of an inner layer of endoderm and an outer layer of mesoderm.

Z

zona fasciculata (ZŌ-na fa-sik-ū-LA-ta): The region of the adrenal cortex that secretes glucocorticoids.

zona glomerulosa (glo-mer-ū-LŌ-sa): The region of the adrenal cortex that secretes mineralocorticoids.

zona pellucida (pe-LOO-si-duh): The region between a developing oocyte and the surrounding follicular cells of the ovary.

zona reticularis (re-ti-kū-LAR-is): The region of the adrenal cortex that secretes androgens.

zygote (ZĪ-gōt): The fertilized ovum, prior to the start of cleavage.

ILLUSTRATION CREDITS

Chapter 1 **CO-1** Kai-Uwe Waerner/AP/Wide World Photos **1-7a,b,c** Custom Medical Stock Photo, Inc. **1-12aL,aR,bL** Science Source/Photo Researchers, Inc. **1-12bR** Custom Medical Stock Photo, Inc. **1-13b** CNRI/Photo Researchers, Inc. **1-13c** Photo Researchers, Inc. **1-13d** Diagnostic Radiological Health Sciences Learning Lab **1-14b** Alexander Tsiaras/Science Source/Photo Researchers, Inc. **1-15bL** Science Source/Photo Researchers, Inc.

Chapter 2 **CO-2** Martin Dohrn/Photo Researchers, Inc.

Chapter 3 **CO-3** Prof. P. Motta, Department of Anatomy, University *La Sapienza,* Rome/Science Photo Library/Photo Researchers, Inc. **3-8a,b,c** David M. Philips/Visuals Unlimited **3-13a** M. Sahliwa/Visuals Unlimited **3-14b** Fawcett/Hirokawa/Heuser/Science Source/Photo Researchers, Inc. **3-15a** Fawcett/de Harven/Kalnins/Photo Researchers, Inc. **3-17** R. Bollender, Don Fawcett/Visuals Unlimited **3-18b** Biophoto Associates/Photo Researchers, Inc. **3-19b** Dr. Birgit H. Satir **3-20aR** CNRI/Science Source/Photo Researchers, Inc. **3-21b** Biophoto Associates/Photo Researchers, Inc. **3-28a,b,d,e,f** Ed Reschke/Peter Arnold, Inc. **3-28c** James Solliday/Biological Photo Service

Chapter 4 **CO-4** David M. Phillips/Photo Researchers, Inc. **4-3a** Ward's Natural Science Establishment, Inc. **4-3b** Frederic H. Martini **4-4b** Courtesy of Gregory N. Fuller, M.D., Ph.D., Section of Neuropathology, M. D. Anderson Cancer Center, Houston, Texas **4-4cL,cR, 4-5a,b** Frederic H. Martini **4-5c** Courtesy of Gregory N. Fuller, M.D., Ph.D., Section of Neuropathology, M. D. Anderson Cancer Center, Houston, Texas **4-6a** Phototake/Carolina Biological Supply Company **4-9a** Ward's Natural Science Establishment, Inc. **4-10a,b** Project Masters, Inc./The Bergman Collection **4-11aL** Frederic H. Martini **4-11aR** Prof. P. Motta, Dept. of Anatomy, University *La Sapienza,* Rome/Science Photo Library/Photo Researchers, Inc. **4-11b** Ward's Natural Science Establishment, Inc. **4-12a** John D. Cunningham/Visuals Unlimited **4-12b** Frederic H. Martini **4-12c** Bruce Iverson/Visuals Unlimited **4-15a** Frederic H. Martini **4-15b** Robert Brons/Biological Photo Service **4-15c** Science Source/Photo Researchers, Inc. **4-15d** Ed Reschke/Peter Arnold, Inc. **4-16** Frederic H. Martini **4-19a,b** G. W. Willis, M.D. Biological Photo Service

Chapter 5 **CO-5** Steve Mason/PhotoDisc, Inc. **5-2b,c** Frederic H. Martini **5-3** John D. Cunningham/Visuals Unlimited **5-4b** Biophoto Associates/Photo Researchers, Inc. **5-7b** John D. Cunningham/Visuals Unlimited **5-8, 5-9a,b** Frederic H. Martini

Chapter 6 **CO-6** SuperStock, Inc. **6-2a** Ralph T. Hutchings **6-3a,b** © R. G. Kessel and R. H. Kardon, *Tissues and Organs: A Text-Atlas of Scanning Electron Microscopy,* W. H. Freeman & Co., 1979. All Rights Reserved. **6-4b** Ralph T. Hutchings **6-7L,R** Frederic H. Martini **6-7R** Frederic H. Martini **6-9a,b** Project Masters, Inc./The Bergman Collection **6-15a,b** Southern Illinois University/Visuals Unlimited **6-15d** Southern Illinois University/Peter Arnold, Inc. **6-15e** Custom Medical Stock Photo, Inc. **6-15f** Scott Camazine/Photo Researchers, Inc. **6-15g** Frederic H. Martini **6-15h** Southern Illinois University/Visuals Unlimited **6-15i** Project Masters, Inc./The Bergman Collection **6-16a,b** Prof. P. Motta, Dept. of Anatomy, University *La Sapienza,* Rome/Science Photo Library/Photo Researchers, Inc.

Chapter 7 **CO-7** Frans Lanting/Stone **7-1aL,aR, 7-5a,b, 7-6a,b, 7-7a,b** Ralph T. Hutchings **7-7c** Michael J. Timmons **7-8a,b, 7-9a,b, 7-10a, 7-11, 7-12a,b,c, 7-13R** Ralph T. Hutchings **7-14b** Michael Siegfried, M.D. **7-18a,b,c, 7-19a,b,c, 7-20a,b,c, 7-21a,b,c, 7-22a,b, 7-23a,b T7-2C,L,R** Ralph T. Hutchings

Chapter 8 **CO-8** David Madison/Stone **8-2b** Bates/Custom Medical Stock Photo, Inc. **8-2c, 8-3a,b,c 8-4a,b, 8-5a,b, 8-6a,b, 8-7a,b, 8-8a,b, 8-11a,b, 8-12a,b, 8-13a,b, 8-14a,b** Ralph T. Hutchings

Chapter 9 **CO-9** Kevin Horan/Stone **9-3aT,cL,cR,d, 9-4a,c, 9-5B1,2,3,4,T1,2,3, 9-10a,b** Ralph T. Hutchings

Chapter 10 **CO-10** NASA/Johnson Space Center **10-02** Ward's Natural Science Establishment, Inc. **10-04b** Dr. Don Fawcett/Photo Researchers, Inc. **10-17a** Comack, D. (ed): *Ham's Histology,* 9th ed. Philadelphia: J. B. Lippincott, 1987. By Permission. **10-17bB,bT** Frederic H. Martini **10-21a** Phototake NYC **10-22a** Frederic H. Martini

Chapter 11 **CO-11** CMCD/PhotoDisc, Inc

Chapter 12 **CO-12** Sy Napier/Science Photo Library/Photo Researchers, Inc. **12-03** David Scott David Scott/Phototake **12-07a** Frederic H. Martini **12-08a** Biophoto Associates/Photo Researchers

Chapter 13 **CO-13** Richard Ellis/Agence France Presse/Corbis **13-3, 13-4a** Ralph T. Hutchings **13-4b** Hinerfeld/Custom Medical Stock Photo, Inc **13-5** Michael J. Timmons **13-6a** © R. G. Kessel and R. H. Kardon, *Tissues and Organs: A Text-Atlas of Scanning Electron Microscopy,* W. H. Freeman & Co., 1979. All Rights Reserved.

Chapter 14 **CO-14** Will Hart **14-1a,b,c,d** Ralph T. Hutchings **14-6** Visuals Unlimited **14-10b** Michael J. Timmons **14-10c** Pat Lynch/Photo Researchers, Inc. **14-12a,b, 14-15b** Ralph T. Hutchings **14-16b** Daniel P. Perl, M.D., Mount Sinai School of Medicine **14-17a,b, 14-19a** Ralph T. Hutchings

Chapter 15 **CO-15** Max Hilaire/The Image Bank **15-9e** Larry Mulvehill/Photo Researchers, Inc.

Chapter 16 **CO-16** Gary Gladstone/The Image Bank

Chapter 17 **CO-17** Escher, Maurits Cornelius, *Convex and Concave,* 1955/Cordon Art B.V./ Art Resource, Inc. **17-2d** Frederic H. Martini **17-4c** G. W. Willis/Terraphotographics/Biological Photo Service **17-5a** Ralph T. Hutchings **17-8a** Ed Reschke/Peter Arnold, Inc. **17-8c** Custom Medical Stock Photo, Inc. **17-19** Richmond International, Inc. **17-23b** Lennart Nilsson/Bonnier Alba AB, *Behold Man,* pp. 206–207. **17-27b** Michael J. Timmons **17-28b** Ward's Natural Science Establishment, Inc.

Chapter 18 **CO-18** Focus on Sports, Inc. **18-6b** Manfred Kage/Peter Arnold, Inc. **18-10c, 18-12b,c** Frederic H. Martini **18-14c, 18-15b** Ward's Natural Science Establishment, Inc. **18-19a,b** Project Masters, Inc./The Bergman Collection **18-19c** John Paul Kay/Peter Arnold, Inc. **18-19d** Custom Medical Stock Photo, Inc. **18-19e** Biophoto Associates/Science Source/Photo Researchers, Inc.

Chapter 19 **CO-19** Lester Lefkowitz/Stone **19-1a** Martin M. Rotker **19-2a** David Scharf/Peter Arnold, Inc. **19-2b** Ed Reschke/Peter Arnold, Inc. **19-2c** David Scharf/Peter Arnold, Inc. **19-4a,b** Stanley Flegler/Visuals Unlimited **19-9a,b,c,d,e** Ed Reschke/Peter Arnold, Inc. **19-12b** Custom Medical Stock Photo, Inc.

Chapter 20 **CO-20** Lunagrafix, Inc./Photo Researchers, Inc. **20-3a** Ralph T. Hutchings **20-5b** Peter Arnold, Inc. **20-6b** Copyright Lennart Nilsson, *Behold Man,* Little, Brown and Company. **20-6c, 20-7b** Ralph T. Hutchings **20-8cL** Science Photo Library/Photo Researchers, Inc. **20-8cR** Biophoto Associates/Science Source/Photo Researchers, Inc. **20-9c** Ralph T. Hutchings **20-10a,b** Howard Sochurek/Medichrome/The Stock Shop, Inc. **20-10c** Larry Mulvehill/Photo Researchers, Inc. **20-14a** Michael J. Timmons

Chapter 21   CO-21 Professor P. M. Motta, A. Caggiati, and G. Macchiarelli/Science Photo Library/Photo Researchers, Inc. **21-1** Biophoto Associates/Photo Researchers, Inc. **21-4a** B & B Photos/Custom Medical Stock Photo, Inc. **21-4b** William Ober/Visuals Unlimited **21-5b** Biophoto Associates/Photo Researchers, Inc.

Chapter 22   CO-22 Steven Rubin/The Image Works **22-3b** Frederic H. Martini **22-4c** Ralph T. Hutchings **22-6a** D. M. Phillips/Visuals Unlimited **22-6b** Biophoto Associates/Photo Researchers, Inc. **22-8c,d, 22-9c** Frederic H. Martini **22-20b** Leonard Lessin/Peter Arnold, Inc.

Chapter 23   CO-23 Richard Hutchings/PhotoEdit **23-2a** Photo Researchers, Inc. **23-2b** Frederic H. Martini **23-3b** Ralph T. Hutchings **23-5b** Phototake NYC **23-6c** John D. Cunningham/Visuals Unlimited **23-7, 23-8** Ralph T. Hutchings **23-9** CNRI/Science Photo Library/Photo Researchers, Inc. **23-11b** Ward's Natural Science Establishment, Inc. **23-11c** Don W. Fawcett, M.D. Micrograph by P. Gehr, from Bloom & Fawcett, *Textbook of Histology,* W. B. Saunders Co.

Chapter 24   CO-24 Erich Lessing/Art Resource, N.Y. **24-7b** Frederic H. Martini **24-10a** Alfred Pasieka/Peter Arnold, Inc. **24-10b** Astrid and Hanns-Frieder Michler/Science Photo Library/Photo Researchers, Inc. **24-10c** Prof. P. Motta, Department of Anatomy, University *La Sapienza,* Rome/Science Photo Library/Photo Researchers, Inc. **24-12a** Ralph T. Hutchings **24-13a** Prof. P. Motta/Department of Anatomy, University *La Sapienza,* Rome/Science Photo Library/ Photo Researchers, Inc. **24-13b** John D. Cunningham/Visuals Unlimited **24-16b** Ralph T. Hutchings **24-17a** Phototake NYC **24-17e** M. I. Walker/Photo Researchers, Inc. **24-18c** Frederic H. Martini **24-19** Ralph T. Hutchings **24-20c,d** Michael J. Timmons **24-23b** Ralph T. Hutchings **24-24** Ward's Natural Science Establishment, Inc.

Chapter 25   CO-25 Robert Semeniuk/The Stock Market **25-18** Christopher Bissell/Stone

Chapter 26   CO-26 Kavaler/Art Resource, N.Y. **26-4b** Ralph T. Hutchings **26-8c** David M. Phillips/Visuals Unlimited **26-16a** Peter Arnold, Inc. **26-17** Photo Researchers, Inc. **26-19a** Ward's Natural Science Establishment, Inc. **26-19b** Frederic H. Martini **26-19c** G. W. Willis, M.D./ Biological Photo Service

Chapter 27   CO-27 Textron Inc.

Chapter 28   CO-28 Barros & Barros/The Image Bank **28-4b** Frederic H. Martini **28-5a** Don W. Fawcett, M.D., Harvard Medical School **28-5b** Ward's Natural Science Establishment, Inc. **28-8b** David M. Phillips/Visuals Unlimited **28-9b** Frederic H. Martini **28-10a** Ward's Natural Science Establishment, Inc. **28-10b** © R. G. Kessel and R. H. Kardon, *Tissues and Organs: A Text-Atlas of Scanning Electron Microscopy,* W. H. Freeman & Co., 1979. All Rights Reserved. **28-10c,d,e, 28-16a,b,c,d** Frederic H. Martini **28-16e** C. Edelmann/La Villete/Photo Researchers, Inc. **28-16f** G.W. Willis, M.D./ Biological Photo Service **28-16g** G. W. Willis/Terraphotographics/Biological Photo Service **28-17b** Frederic H. Martini **28-17c** Custom Medical Stock Photo, Inc. **28-18a** Ralph T. Hutchings **28-19a** Ward's Natural Science Establishment, Inc. **28-20a,b** Frederic H. Martini **28-20cB** Michael J. Timmons **28-20cT** Frederic H. Martini **28-21** Michael J. Timmons **28-23b** Fred E. Hossler/ Visuals Unlimited **28-23c** Frederic H. Martini

Chapter 29   CO-29 Petit Format/Nestle/Photo Researchers, Inc. **29-1a** Francis Leroy, Biocosmos/Science Photo Library/Custom Medical Stock Photo, Inc. **29-6b** Frederic H. Martini **29-7a** Dr. Arnold Tamarin **29-7b,c,d, 29-8a** Photo Lennart Nilsson/Albert Bonniers Forlag **29-8b** Photo Researchers, Inc. **29-15** Science Photo Library/Photo Researchers, Inc.

INDEX

Selected page references to the
Applications Manual appear in blue type.

A

A band, 275–276
Abdomen, 16
  veins of, 740–741
Abdominal aorta, 729, 731–733
Abdominal cavity, 20, 21, 22
Abdominal reflex, 432
Abdominal region, 16
Abdominal thrust, 805
Abdominis, 317
Abdominopelvic cavity, 20, 21, 22
Abdominopelvic quadrant, 15–17, 26
Abdominopelvic region, 17, 26
Abducens nerve, 467, 474
Abduction, joints, 251, 252
Abductor, 317
Abductor digiti minimi, 354
Abductor hallucis, 354
Abductor pollicis brevis, 344, 346
Abortion, 1083
Abrasion, 158; 48 AM
Abruptio placentae, 1098; 194 AM
Abscess, 135, 138, 770; 134 AM
Absolute refractory period, 381, 669–670
Absorption, 846
  of amino acids, 891, 922
  of carbohydrates, 888–890, 922
  ion, 891–892
  and large intestine, 885–886
  of lipids, 922
  malabsorption syndromes, 893
  and metabolism, 2
  and small intestine, 881
  and stomach, 868
Absorptive state, 921–923
  and adipose tissue, 922–923
  and liver, 922
Accessory duct, 872
Accessory ligaments, 249
Accessory nerve, 473, 474
Accessory respiratory muscles, 818–819
Accessory structures
  digestive system, 846
  eye, 538–541
  integumentary system, 144
Acclimatization, 936
Accommodation, visual, 547–549
Accommodation curves, of spine, 212
ACE (angiotensin-converting enzyme), 813; 125, 127, 179 AM
Acetabular notch, 234, 260
Acetabulum, 234
Acetate, 388
Acetoacetate, 925
Acetone, 925
Acetylcholine (ACh), 280, 281, 391, 490, 506
  release of, 387–389
  sympathetic stimulation of, 512
Acetylcholinesterase (AChE), 388
  and skeletal muscle contraction, 281
Acetyl-CoA, 904
Acetyl group, 904
ACh. *See* Acetylcholine (ACh)
Achalasia, 861, 896; 162 AM
Achilles tendon, 239, 351
Achondroplasia, 190; 55 AM
Acid–base balance, 997–1011
  aging and, 1009
  buffer systems, 999–1001
  disturbances of, 1004–1009

maintenance of, 1001–1004
meaning of, 985
in newborn infants, 1010–1011
and renal compensation, 1002–1004
and respiratory compensation, 1002
Acid–base disorders, 181 AM
  blood chemistry changes related to, 1009
  causes of, 1005
  detection of, 1008–1009
  metabolic acidosis, 1007–1008
  metabolic alkalosis, 1008
  respiratory acidosis, 1005–1006
  respiratory alkalosis, 1006–1007
Acidemia, 997
Acidic solution, 42, 997
Acidophils, 640
Acidosis, 42, 720, 997
  detection of, 1008–1009
  metabolic, 1007–1008, 1009
  renal response to, 1003–1004
  respiratory, 1005–1006, 1009
Acids, 40, 42–43, 997
  fixed acids, 998
  organic acids, 998
  strong acid, 42–43, 997
  volatile acids, 998
  weak acid, 43, 997
Acinar glands, 117
Acinus, 117
Acne, 156, 163
Acoustic nerve, 470
Acquired immune deficiency syndrome (AIDS), 788, 790, 793; 138–142 AM
  nature of, 788
Acquired immunity, 771–772
Acquired reflexes, 424
Acromegaly, 181, 190, 590, 616; 56 AM
Acromial end, clavicle, 228
Acromioclavicular joint, 230
Acromioclavicular ligament, 259
Acromion process, 230
Acrosin, 1067
Acrosomal cap, 1025
Actin, 81, 275
Actin filaments, 81
Actinic keratosis, 45 AM
Action potentials, 275, 281, 363, 380–386, 395–396
  all-or-none principle, 380
  and axons, 384–386
  compared to graded potentials, 384
  continuous propagation, 383–384
  generation of, 380–382
  and heartbeat, 668–670
  in muscle tissue, 386
  propagation of, 383–384
  propagation speed of, 384–386
  rate of generation, 395–396
  refractory period, 381–382
  saltatory propagation, 384, 385
  and sodium–potassium ATPase, 382
Actions, skeletal muscle, 315–316
Activation energy, 39
Activators, 22 AM
Active (gated) channels, 69, 70, 376–377
Active immunity, 771
Active immunization, 793; 136, 137 AM
Active processes, 71

Active site
  and enzymes, 54
  and G actin, 277
Active transport, 76, 79
  secondary, 76–77, 79
Actose, 890
Acute mountain sickness, 154 AM
Acute otitis media, 94–95 AM
Acute renal failure, 958
Adam's apple, 803
Adaptability, of organisms, 2
Adaptation
  central adaptation, 529–530
  peripheral adaptation, 529
Addison's disease, 149, 602, 616, 618; 102 AM
Adduction, joints, 251, 252
Adductor brevis, 347, 349
Adductor hallucis, 354
Adductor longus, 347, 349
Adductor magnus, 347, 349
Adductor pollicis, 346
Adductors, 317, 347, 349
Adenine, 56, 91
Adenohypophysis, 587–591
Adenoids, 759
Adenosine, 57
Adenosine diphosphate (ADP), 57, 645
Adenosine monophosphate (AMP), 57
Adenosine triphosphatase (ATPase), 57
Adenosine triphosphate (ATP), 57, 87–89, 294–295
  and active transport, 76
  and aerobic metabolism, 293–294
  and carbohydrate metabolism, 903–911
  and cell division, 99
  and creatine phosphate (CP), 293
  and electron transport system, 910
  energy storage, 57–59
  and glycolysis, 294
  mitochondrial energy production, 87–89, 904–910
  and oxygen debt, 296
  and proteins, 916, 918–919
  and skeletal muscle contractions, 293–294
  structure of, 58
Adenylate cyclase, 392, 582
ADH. *See* Antidiuretic hormone (ADH)
Adenylyl cyclase, 582
Adhesions, 130, 138
  in phagocytosis, 766
  prevention of bacterial and viral, 780
Adipocytes, 118, 119
Adipose capsule, 944
Adipose cells, 119
Adipose tissue, 118, 121–122
  and absorptive state, 922–923
  autonomic control of, 519
  brown fat, 122
  and metabolism, 921, 922–923, 925
  and postabsorptive state, 925
  and weight loss, 122
  white fat, 121
Adolescence, 1086, 1088–1089
ADP. *See* Adenosine diphosphate (ADP)
Adrenal cortex, 589, 600–603
Adrenal gland, 8, 600–603

adrenal cortex, 600–603
adrenal medulla, 600
autonomic control of, 519
disorders of, 601
zona fasciculata, 600, 602
zona glomerulosa, 600, 601
Adrenal hormones
  aldosterone, 601
  corticosteroids, 600
  epinephrine, 603
  glucocorticoids, 602
  mineralocorticoids, 601
  norepinephrine, 603
Adrenaline. *See* Epinephrine (E)
Adrenal medulla, 507, 508, 510, 600
Adrenergic receptors, 515
Adrenergic synapses, 389, 511, 712
Adrenocortical steroids, 600
Adrenocorticotropic hormone (ACTH), 149, 588, 589, 933
Adrenogenital syndrome, 602
Adult respiratory distress syndrome (ARDS), 152 AM
Adventitia, 849
Adventitious bursae, 249
Aerobic endurance, 299
  activities related to, 299
Aerobic metabolism, 88, 904
  alternate pathways for, 911–912
  and ATP production, 293–294, 904–911
  energy yield of, 910–911
Afferent arteriole, 946–947, 950
Afferent division, peripheral nervous system, 361, 366
Afferent fibers, 366
Afferent lymphatics, 760
Afferent vessels. *See* Veins
Afterbirth, 1084
Afterload, cardiodynamics, 683
Agglutination
  immune response, 780
  of red blood cells, 636
Agglutinins, 636
Agglutinogens A, B, and Rh (D), 635
Aggregate lymphoid nodules (Peyer's patches), 759, 871
Aging
  acid–base balance and, 1009
  and blood, 744
  and blood vessels, 744
  and bone, 185–187
  and brain, 499
  and cancer, 136
  and digestive system, 893
  and exercise, 299
  and female reproductive system, 1056
  and fluid and electrolyte balance, 1009
  and gustation, 538
  and heart, 744
  and hormones, 614
  and immunity, 789
  and integumentary system, 160
  and joints, 265
  and male reproductive system, 1057
  and olfaction, 536
  and oogenesis, 1039
  process of, 1089, 1090
  and respiratory system, 837
  and skeletal muscles, 299–300
  and skeletal system, 242
  and tissues, 136
  and urinary system, 977–978

Agonists, skeletal muscle, 315
Agranular leukocytes (agranulo-
    cytes), 637
AIDS, 784, 788, 790, 793;
    138–142 AM
Ala, sacral (wing), 219
Alanine, 52, 93
Alarm phase, general adaptation
    syndrome (GAS), 612, 613
Alba, 317
Albinism, 61, 148; 22 AM
Albumins, 445, 596, 626–627
Alcoholism
    facts about, 931
    risk factors, 931
Aldosterone, 601, 965, 994
    and fluid/electrolyte balance, 988
Aldosteronism, 601, 995; 102 AM
Alkalemia, 997
Alkaline phosphatase, 175
Alkaline solution, 42, 997
Alkaline tide, 865, 1008
Alkalosis, 42, 966, 997
    detection of, 1008–1009
    metabolic, 1008, 1009, 1011
    renal response to, 1003–1004
    respiratory, 1006–1007, 1009
Allantois, 1072
Alleles, 1091
Allergens, 641, 788, 793
Allergies, 787, 788–789, 793
    anaphylaxis, 788
    categories of, 788
    immediate hypersensitivity, 788
All-or-none principle
    action potentials, 380
    muscles, 286
Allosteric
    activation, 22 AM
    effects, 61, 21 AM
    inhibition, 22 AM
Alopecia areata; 47 AM
Alpha-2-macroglobulin, 647
Alpha-amylase, 855
Alpha-blockers, 522, 524; 91 AM
Alpha cells, 606, 608
Alpha chains, hemoglobin, 629
Alpha-helix, 52, 53
Alpha-interferon, 767, 784
Alpha particles, 15 AM
Alpha receptors, 511
Alpha-thalessemia, 109–111 AM
Alpha waves, 493, 494
Alternative pathway, complement
    system, 768, 769
Altitude, adaptation to, 829
Altitude sickness, 829
Alveolar air
    composition of, 824
    partial pressures in, 824
Alveolar ducts, 811–813
Alveolar glands, 117
Alveolar macrophages, 765, 811
Alveolar processes, 205, 207
Alveolar sacs, 811
Alveolar ventilation, 814–815, 820,
    821
    changes in, 830–831
Alveoli, 9, 111, 722, 724, 798,
    811–813
    and respiratory membrane,
    812–813
    surfactant, 811–812
Alveolus, and tooth, 856
Alzheimer's disease, 499, 500;
    89 AM
Amacrine cells, 544
Amenorrhea, 1045, 1059
Amination, 919
Amino acid derivatives, 579
    types of, 579
Amino acids, 51–52
    absorption of, 891, 922
    catabolism, 916–919
    essential amino acids, 919

linking of, 52
and liver, 877–878, 922
nonessential amino acids, 919
in protein metabolism, 919
Aminoaciduria, 959, 980; 179 AM
Amino group, 44, 51
Ammonium ions, 886, 966
Amnesia, 496, 500; 85 AM
Amniocentesis, 115, 1097, 1098
Amniochorionic membrane, 1081
Amnion, 1070, 1072
Amniotic cavity, 1070
Amniotic fluid, 115, 1072
Amoeboid movement, 637
Amphetamines, 498; 88 AM
Amphiarthroses (slightly movable
    joints), 246, 247
    symphysis, 247
    syndesmosis, 247
Amphicytes (satellite cells), 132,
    272, 367, 370
Amphimixis, 1067
Amplification, 582
Ampulla, 561, 1027, 1039, 1040
Amygdaloid body, 451, 452–454
Amylin, 606
Amyloid, 499
Amyotrophic lateral sclerosis
    (ALS), 492, 500; 83–84 AM
Anabolic steroids, abuse by ath-
    letes, 611
Anabolic system. See
    Parasympathetic division of
    ANS
Anabolism, 38, 902
Anaerobic endurance, 298–299
    activities related to, 299
Anaerobic process, glycolysis as,
    294
Analgesics, 87 AM
Anal orifice, 884
Anal sphincter, 884
Anal triangle, 332, 333, 334
Anaphase, 98
Anaphylactic shock, 789, 793;
    128 AM
    treatment of, 789
Anaphylaxis, 788, 789, 793
Anaplasia, 137, 138
Anastomoses, 666
Anatomical directions, 17–19
Anatomical landmarks, 15, 16
Anatomical neck, 230
Anatomical position, 15
Anatomical regions, 15–17
Anatomic dead space, 820
Anatomy
    anatomical position, 15
    gross anatomy, 3
    microscopic anatomy, 3
    sectional anatomy, 19–25
    study of, 2–3
    superficial anatomy, 15–19
Anaxonic neurons, 365
Anchoring proteins, 70
Anconeus, 317, 340, 342
Androgen-binding protein, 1025
Androgens, 85, 580, 589, 602, 608,
    610, 611, 617, 1017, 1021
    abuse by athletes, 611
    and bone, 181, 182
    luteinizing hormone and produc-
    tion of, 1033–1034
    and male reproductive system,
    1057
Androstenedione, 1049
Anemia, 630, 650; 5, 11,
    113–115 AM
    iron-deficiency, 633
    pernicious, 634
    and vitamin B_{12}, 634
Anencephaly, 493, 500
Anesthesia, 572
Aneurysms, 695, 744, 745; 125 AM
Angelman syndrome, 1094

Angina pectoris, 667, 687
Angiography, 667
Angiotensin-converting enzyme,
    (ACE), 813;
Angiotensin-converting enzyme
    inhibitors, 747; 125, 127,
    179 AM
Angiotensin I, 604
Angiotensin II, 601, 604
    and cardiovascular regulation, 716
Angiotensinogen, 604
Angle
    mandibular, 207
    ribs, 221
Angular motion, joints, 251, 252
Anion gap, 1012
Anions, 33
Ankle, 239
    bones of, 239
Ankyloglossia, 854
Ankylosis, 267; 60, 61 AM
Annular ligament, 260, 805
Anorectal canal, 884
Anorexia nervosa, 932
Anoxia, 814, 840
ANP. See Atrial natriuretic peptide
    (ANP)
ANS. See Autonomic nervous
    system (ANS)
Antabuse, 931
Antacids, 43
Antagonistic effects, hormones, 610
Antagonists, skeletal muscle, 315
Antebrachial region, 16
Antebrachial vein, 737
Antebrachium, 16
Anteflexion, uterus, 1040
Anterior, 18, 19, 317
Anterior cardiac vein, 668
Anterior cavity, eye, 541
Anterior cerebral artery, 728
Anterior chamber, eye, 545
Anterior clinoid process, 203
Anterior commissure, 446
Anterior corticospinal tracts, 488,
    491
Anterior cruciate ligament, 263
Anterior fontanel, 211
Anterior gray horns, 413
Anterior group, 457
Anterior inferior iliac spine, 234
Anterior interventricular branch,
    666
Anterior interventricular sulcus,
    657
Anterior lobes, 461
Anterior longitudinal ligament,
    256, 257
Anterior margin, tibia, 238, 239
Anterior median fissure, 407
Anterior nucleus, 454, 457
Anterior pituitary, 587–591
Anterior section, 20
Anterior spinocerebellar tracts, 486
Anterior spinothalamic tracts, 484,
    486, 532
Anterior superior iliac spine, 234
Anterior vertebral arches, 215
Anterior white column, 413
Anterior white commissure, 413
Anterograde amnesia, 500; 85 AM
Anterograde flow, 365
Anthracosis, 149 AM
Antiangiogenesis factor, 101, 125,
    138
Antibodies, 51, 54, 119, 624, 626,
    756, 777–781
    antigen–antibody complex, 780
    fetal antibody production, 786
    immunoglobulins, classification
    of, 780–781
    and liver, 879
    structure of, 778–780
Antibody-mediated immunity
    (humoral immunity), 756, 771

Antibody titer, 781–782
Anticholinesterase drugs, 390, 401
Anticoagulants, 647
Anticodon, 93
Anticonvulsants, 88 AM
Antidiuretic hormone (ADH), 459,
    579, 591, 617
    and cardiovascular regulation,
    715
    and fluid/electrolyte balance,
    988
Antigen–antibody complex, 780
Antigen binding sites, 779–780
Antigenic determinant sites, 780
Antigen presentation, 760, 773–775
    antigen-presenting cells, 774–775,
    787
Antigen processing, 775
Antigen recognition, 775
Antigens, 634
    and liver, 878–879
    nature of, 756
    partial, 780
    response to exposure, 781–782,
    788–789
    and T cell activation, 774–775
Antihistamines, 789
Anti-inflammatory drugs, 49 AM
Anti-inflammatory effects, gluco-
    corticoids, 602
Antioxidants, 603
Antiport, 75
Antipsychotic drugs, 87–88 AM
Antipyretic drugs, 938; 173 AM
Antithrombin-III, 647
Antiviral proteins, 767
Antrum, 188, 1038
Anulus fibrosus, 256
Anuria, 174, 180 AM
Anus, 884
Aortic arch, 662, 726–729
    carotid artery, 727–729
    subclavian arteries, 726, 727
Aortic bodies, 534, 714
Aortic reflex, 713
Aortic semilunar valve, 662
Aortic sinuses, 533, 662, 713
Aortic stenosis, 119–120 AM
Apex
    heart, 658
    lung, 806
    nose, 801
    patella, 238, 239
    sacrum, 219
Apgar rating, 1098; 195, 197 AM
Aphasia, 449, 500; 81 AM
Apical foramen, 856
Aplastic anemia, 115 AM
Apnea, 836
Apneustic center, 464, 832–833
Apocrine secretion, 115–116
Apocrine sweat glands, 157
Apoenzyme, 61; 21 AM
Apolipoproteins, 627
    apolipoprotein-E, 696
Aponeuroses, 7, 122, 273, 312
Apoptosis, 96, 775
Appendectomy, 132 AM
Appendicitis, 759, 793, 882; 132 AM
Appendicular musculature, 317,
    335–355. See also Skeletal
    muscle
Appendicular skeleton, 6, 194,
    226–244
    limb, lower, 237–241
    limb, upper, 230–233
    pectoral girdle, 228–230
    pelvic girdle, 233, 234–236
Appendix, 882
Appetite, control of, 933
Appositional growth, 125, 126
    bone growth, 176, 178, 179
Aqueduct of Sylvius, 439
Aqueous humor, 541, 545
Aqueous solution, 40–42

Arachidonic acid, 48, 580
Arachnoid, 409, 410, 440
Arachnoid granulations, 443
Arachnoid membrane, 410
Arachnoid trabeculae, 410
Arachnoid villi, 443
Arbor vitae, 461
Arches, of foot, 240
Arcuate arch (dorsal arch), 734
Arcuate arteries, 946, 1043
Arcuate fibers, 446
Arcuate line, 235
Arcuate veins, 946
Areflexia, 432; 79 AM
Areola, 1047
Areolar tissue, 121
Arginine, 891, 919
Arginine vasopressin (AVP), 591, 617
Arm. See Limbs, upper
Aromatase, 1034, 1049
Arousal, 497–498
    and reticular activating system, 497–498
    sexual intercourse, 1053, 1056
Arrector pili, 153
Arrhythmias, cardiac, 674, 687
Arterial anastomosis, 698
Arterial puncture, 626
Arteries, 8, 124, 655, 692, 693–695.
    See also specific arteries
    arterioles, 692, 695
    axillary, 727
    basilar artery, 728
    brachial arteries, 626, 695, 727
    brachiocephalic trunk, 726
    bronchial, 729
    carotid artery, 726, 727–729
    cerebral artery, 728, 729
    compared to veins, 692–693
    digital arteries, 727
    disorders of, 695, 696–697
    dorsalis pedis artery, 734
    elastic arteries, 694–695
    esophageal, 729
    femoral artery, 695, 723, 734
    gastric artery, 731, 732
    gonadal arteries, 733
    hepatic artery, 732, 877
    iliac arteries, common, 731, 733
    intercostal arteries, 729
    internal thoracic artery, 727
    lumbar arteries, 733
    mediastinal, 729
    mesenteric arteries, 732
    muscular arteries, 695
    ophthalmic artery, 728
    pericardial, 729
    peroneal artery, 734
    phrenic arteries, 729–731, 732
    plantar arteries, 734
    popliteal artery, 734
    posterior communicating
        arteries, 728
    pulmonary arteries, 724
    radial artery, 727
    renal arteries, 732–733
    splenic artery, 731, 762
    subclavian arteries, 726, 727
    suprarenal arteries, 732
    systemic arteries, 724–734
    thyrocervical trunk, 727
    tibial artery, 734
    ulnar artery, 727
    umbilical arteries, 742
    vertebral artery, 727
Arterioles, 692, 695
    blood pressure in, 706
    pulmonary arterioles, 724
Arteriosclerosis, 499, 695, 696–697, 745
    formation of, 696–697
    forms of, 696
    treatment of, 697
Arteriovenous anastomoses, 698

Arthritis, 265, 266; 60–61 AM
    types of, 265
Arthropods, 30 AM
Arthroscope, 267; 59 AM
Arthroscopic surgery, 267; 59 AM
Articular capsule, 248, 259
    hip joint, 260–261
    knee joint, 262–263
Articular cartilage, 126, 169, 177, 178, 248
Articular discs, 231, 249
Articular facet, 214
Articular processes, vertebrae, 214
Articulation, in speech, 804
Articulations. See Joints
Artificial insemination, 1079
Artificial respiration, 819
Artificial sweeteners, 46
Aryepiglottic folds, 804
Arytenoid cartilages, 804
Asbestosis, 149 AM
Ascending aorta, 662, 726
Ascending colon, 851, 884
Ascending limb, 952
Ascending tracts, 413
Ascites, 130, 138, 851, 895
Aspartate, 391, 392
Aspartate aminotransferase, 679
Aspiration, 805
Aspirin, and blood clotting, 649
Assisted reproductive technologies
    See Infertility treatment
Assisted reproductive technologies
    (ART), 1079
Association areas, cerebral cortex, 447–448
Association fibers, 446
Association neurons, 367
Asthma, 809, 822, 840; 8, 10, 151–152 AM
Astigmatism, 549
Astral rays, 98
Astrocytes, 367, 368
    functions of, 368
Asynchronous motor unit summation, 289
Ataxia, 462, 475
Atelectasis, 818, 840
Atenolol, 683
Atherosclerosis, 696, 745, 813
Athletes
    and anabolic steroid abuse, 611
    and fluid and electrolyte balance, 994
Atlas, 215
Atmospheric pressure, 815
Atomic mass unit (amu), 31
Atoms, 4, 5, 30–36
    atomic number, 30
    atomic weight, 31–32
    and elements, 30–31
    energy levels, 32
    isotopes, 30–31
    structure of, 30–32
    subatomic particles of, 30
ATP. See Adenosine triphosphate
    (ATP)
ATPase (adenosine triphosphatase), 57
Atresia, 1036
Atrial appendage, 657
Atrial baroreceptors, 713
Atrial diastole, 675
Atrial fibrillation, 688; 122 AM
Atrial flutter, 688; 122 AM
Atrial natriuretic peptide (ANP), 605, 956
    and cardiovascular regulation, 717
    and fluid/electrolyte balance, 988
Atrial reflex, 685, 713
Atrial systole, 675, 676–677
Atrioventricular (AV) valves, 660, 664, 665
Atrioventricular node, 664, 671, 672

Atrioventricular septal defect, 744
Atrophy, skeletal muscles, 298
Atropine, 390, 401
Attenuated bacteria and viruses, 136 AM
Audiogram, 95, 96 AM
Auditory association area, 447
Auditory cortex, 447
Auditory nerve, 470
Auditory ossicles, 202, 558–560
Auditory tube, 126, 326, 558, 803
Auricle (pinna), 126, 127, 557
Auricular surface, 218, 235
Auricularis, 317
Auscultation, 26, 677; 8 AM
Autoantibodies, 787
Autoimmune disorders, 265, 787–788, 793; 6, 143, 144 AM
Autoimmune response, 602
Autologous marrow transplant, 651; 116 AM
Autologous transfusion, 112 AM
Autolysis, 87
Automatic bladder, 977
Automaticity, 303, 670
Autonomic function, hypothalamus
    and control of, 457
Autonomic ganglia, 367, 414, 504
Autonomic nervous system (ANS), 362, 366, 487, 503–526. See
    also Parasympathetic division
    of ANS; Sympathetic division
    of ANS
    autonomic tone, 518–519
    comparison of sympathetic/
        parasympathetic divisions, 517–519
    drug effects, 522
    dual innervation, 516–517
    and heart, 682, 683–685
    higher levels of control, 521–523
    innervation patterns, 506
    organization of, 504–506
    parasympathetic division, 362, 504, 505–506, 512–516
    sympathetic division, 362, 504–505, 506–512
    visceral reflexes, 520–521
Autonomic reflexes. See Visceral
    reflexes
Autonomic tone, 518
    in dual innervation, 518
Autoregulation
    capillary, 698
    homeostasis, 12
    at tissue level, 711–712
Autorhythmicity, 670
Autosomal chromosomes, 1090–1095
Autosomal dominants, 1093
Autosomal recessives, 1093
Avascular tissue, 107
Avitaminosis, 929, 938
Avogadro's number, 32
Avulsions, 48 AM
Axial muscles, 7
Axial musculature, 317, 320–335.
    See also Skeletal muscle
Axial skeleton, 6, 193–225
    components of, 194, 195
    functions of, 194
    skull, 195–211
    thoracic cage, 219–222
    vertebral column, 212–219
Axillary artery, 727
Axillary border (lateral border),
    scapula, 229
Axillary nerve, 419
Axillary vein, 737
Axis, 215–216
Axoaxonal synapse, 395
Axolemma, 363, 370
Axon hillock, 363
Axons, 134, 361, 363–365
    and action potentials, 384–386

classification of, 384
Axoplasm, 363
Axoplasmic transport, 364–365; 72, 74 AM
Azathioprine, 136 AM
Azotemia, 180 AM
Azygos vein, 738

B
Babinski sign, 430, 431
Baby teeth, 857
Bacillus, 24 AM
Bacteria, 752, 793; 23–24 AM
Bacterial adhesion, prevention of, 780
Bacterial infection
    defenses against bacterial
        pathogens, 785–787
    time course of body's response
        to, 785
Bainbridge reflex, 685
Balanced diet, 926
Baldness, 155, 163
Ball-and-socket joints, 256
Balloon angioplasty, 668, 687
Band cells, 643
Barbiturates, 683
Barium-contrast X-ray, 23
Baroreceptor reflexes, 521
    and cardiovascular regulation, 713–714
    and respiration, 835
Baroreceptors, 532, 533, 534
    locations of, 713
Bartholin's glands, 1046
Basal body, 82–83
Basal body temperature, 1052
Basal cell carcinoma, 164; 45 AM
Basal ganglia, 451
Basal (germinative) cells, 110, 146, 534
Basal lamina, 110
Basal metabolic rate, 933
Basal nuclei, 451
Basal surface, 107
Base
    heart, 657–658
    lung, 806
    patella, 238
    sacral, 219
Basement membrane, 107, 110, 950
Bases, 40, 42, 43, 997
    strong base, 43
    weak base, 43
Basic solution, 42, 997
Basilar artery, 728
Basilar membrane, 565
Basilar zone, uterus, 1043
Basilic vein, 737
Basophils, 119, 124, 637, 640, 641, 644, 787
B cells, 641, 756, 771, 777–781, 786, 787
    activation of, 777–778
    memory B cells, 778
    sensitization, 777
bcl-2 (gene), 96
Bedsores (decubitus ulcers), 151, 163
Behavior
    and brain chemistry, 498
    and hormones, 614
Bell's palsy, 469, 475
Belt desmosome, 110
Benadryl, 789
Bends, the, 824
Benign prostatic hypertrophy, 1032
Benign tumors, 100–101, 102
Beta-amyloid, 89 AM
Beta-blockers, 522, 524; 91 AM
Beta cells, 606, 608
Beta chains, hemoglobin, 629
Betahydroxybutyrate, 925

Beta-interferon, 767, 784
Beta-oxidation, lipid metabolism, 913
Beta particles, 15 [AM]
Beta receptors, 512
Beta-thalassemia, 109, 111 [AM]
Beta waves, 493, 494
Bias, avoiding, in research, 3–4 [AM]
Biaxial articulation, 251
Bicarbonate reabsorption, 966
Bicarbonate reserve, 1001
Biceps, 317
Biceps brachii, 231, 260, 285, 293, 312, 316, 339–340, 342
Biceps femoris, 349, 351
Bicuspids, 857
Bicuspid valve, 662
Bifid, 214
Bilateral orchiectomy, 1018
Bile, 597, 868
   circulation of, 879
   functions of, 879
   secretion of, 877, 879, 880
   synthesis of, 879
Bile canaliculi, 877
Bile duct, 877
Bile ductules, 877
Bile modification, 879
Bile salts, 49, 869, 879, 885, 890
Bile storage, 879, 880
Biliary obstruction, 893
Bilirubin, 632, 879
Biliverdin, 632
Bioenergetics
   energy and food, 933
   metabolic rate, 933
   thermoregulation, 933–937
Biofeedback, 523
Biogenic amines, 391, 579
Biohybrid artificial pancreas, 104 [AM]
Biology, cellular, 23–33 [AM]
Biology, molecular, 34–40 [AM]
Biology, study of, 2
Biomechanics, 312–314
Biopsies, 137, 411
Biotin
   facts about, 930
   production of, 885
Bipennate muscle, 313
Bipolar cells, 544
Bipolar neurons, 365
Birth control, methods of, 1054–1055
   experimental, 189 [AM]
Birthmark, 151
2,3-Bisphosphoglycerate (BPG), 828
Blast cells, 643
Blastocoele, 1068
Blastocyst, 1068, 1069
Blastomeres, 1068
Bleaching, vision, 553, 554
Bleeding time, 646
Blindness, 549
   color, 555, 572, 1095
Blind spot, 544
Blisters, 147
Blocking agents, 522
Blood, 8, 118, 124, 623–653
   and aging, 744
   collection/analysis of, 626
   disorders of, 631, 632, 638, 649
   distribution of, 700–701
   formed elements, 624
   functions of, 624
   hemopoiesis, 627–628
   hemostasis, 645–649
   and liver, 878–879
   plasma, 124, 624
   platelets, 644–645
   red blood cells, 124, 624
   white blood cells, 124, 637–644
   whole blood, 625, 626
Blood–brain barrier, 368, 440, 444–445, 698

Blood cells, production, 168
Blood clotting, 624, 626, 646
   abnormalities of, 649
   clot retraction, 648
   clotting factors, 645
   clotting system, 116–118 [AM]
   drugs effecting coagulation, 649
   extrinsic pathway, 646
   feedback control of, 14, 15, 666
   fibrinolysis, 648
   intrinsic pathway, 646
   procoagulants, 646
   tests for, 651
   and vitamin K, 647–648
Blood colloid osmotic pressure, 708, 955
Blood–CSF barrier, 445
Blood doping, 634; 114 [AM]
Blood flow, 704
Blood gas analysis, 825
Blood pressure (BP), 704, 705–706
   diastolic pressure, 705
   disorders of, 706
   elastic rebound, 705
   and glomerular filtration rate, 955
   and hemorrhaging, 719
   measurement of, 707
   in small arteries/arterioles, 706
   systolic pressure, 705
Blood smear, 626
Blood supply
   to bone, 178–180
   to brain, 443–445, 701, 710, 713, 714, 718, 720, 722, 727–729
   to esophagus, 859
   to heart, 664–668, 722
   to kidneys, 946
   to liver, 874
   to lungs, 722, 813
   to placenta, 742, 1075
   to skeletal muscle, 273
   to skin, 151
   to stomach, 863
   to uterus, 1041–1042, 1043
Blood tests, types of, 634, 650
Blood–thymus barrier, 758
Blood types, 634–637
   agglutinins, 636
   compatibility testing, 636–637
   cross-reactions, 636
   Rh factor, 635
Blood vessels, 8
   and aging, 744
   arteries, 124, 655, 692, 693–695
   capillaries, 124, 655, 692, 695–699
   cardiovascular regulation, 701
   distribution of blood, 700–701
   exchange vessels, 655
   veins, 124, 655, 692, 699–700
   walls, structure of, 692, 695
Blood volume, and hemorrhaging, 723
Blue cones, 554
Body
   epididymis, 1026
   gallbladder, 879
   hyoid bone, 207
   mandible, 207
   muscle, 312
   pancreas, 872
   penis, 1030
   ribs, 221
   scapula, 229
   sphenoid bone, 203, 209
   sternum, 222
   stomach, 861
   tongue, 854
   uterus, 1041
   vertebra, 213
Body cavities, 20–22
   abdominal cavity, 21
   abdominopelvic cavity, 21
   dorsal body cavities, 20, 21
   pelvic cavity, 20, 21
   pericardial cavity, 20, 21, 22

peritoneal cavity, 22
   pleural cavity, 21
   relationships of, 20
   thoracic cavity, 21
   ventral body cavities, 20–22
Body stalk, 1072
Body temperature. See also Thermoregulation
   and blood, 624, 626
   and heart, 686
   and heart rate, 686
   and hormones, 1052
   and perspiration, 157
   regulation of, 459
   and skeletal muscles, 271
Bohr effect, 827
Boil, 156
Bolus, 805, 849, 858
Bone bruise, 356
Bone conduction tests, 95 [AM]
Bone grafts, 184–185
Bone marrow, 6, 169
   lymphocyte production, 757
   red, 168, 169, 633
   transplantation of, 793; 116, 134 [AM]
   yellow, 168, 169, 633
Bone(s), 6, 128–129, 167–192
   abnormal growth, 181
   and aging, 185–187
   blood supply, 178–180
   bone markings (surface features), 187
   as calcuim reservoir, 182–184
   cells in, 171–172
   classification of, 168–169
   compact bone, 168, 169, 172–173
   compared to cartilage, 129
   composition of, 128
   diameter increases in, 178, 179
   disorders of, 181, 184, 185–187, 190
   endochondral ossification, 175, 176–178
   endosteum, 174
   exercise effects, 180–181
   flat bones, 168, 169
   fractures, 184–185
   growth of, 175–180
   hormone effects, 181–182
   innervation, 180
   integration with other systems, 187, 189
   intramembranous ossification, 175–176
   irregular bones, 168, 169
   length increases in, 171
   long bones, 168, 169
   matrix of, 170–171, 173
   nutritional effects, 181–182
   organization of bone tissue, 128
   periosteum, 174
   remodeling, 180
   sesamoid bones, 168, 169
   short bones, 168, 169
   spongy bone, 168, 169, 173–174
   sutural bones, 168, 169
Bony crepitus, 267
Bony fusion, 247
Bony labyrinth, 560–561
Boron neutron capture therapy (BNCT), 40 [AM]
Botulinus toxin, 67 [AM]
Botulism, 281, 306, 390; 67 [AM]
Bowman's capsule, 949
Bowman's glands, 534
Boyle's law, 815
BPG (2,3-bisphosphoglycerate), 828
Brachial arteries, 626, 695, 727
Brachialis, 317, 340, 342
Brachial palsies, 78 [AM]
Brachial plexus, 419–420
Brachial region, 16
Brachial vein, 737

Brachiocephalic trunk, 726
Brachiocephalic vein, 736, 738
Brachioradialis, 340, 342
Brachium, 16
Bradycardia, 672, 683, 687
Brain, 7
   and aging, 499
   behavior and brain chemistry, 498
   blood–brain barrier, 444–445
   blood supply to, 443–445, 701, 710, 713, 714, 718, 720, 722, 727–729
   cerebellum, 436, 438, 460–462
   cerebral cortex, 445–451
   cerebral nuclei, 451–452, 453
   cerebrospinal fluid, 441–443
   cerebrum, 436, 438, 445–452
   and consciousness, 496–498
   cranial meninges, 440–441
   cranial nerves, 464–474
   development of, 438–439
   diencephalon, 436, 438, 439, 454–459
   disorders of, 443, 444, 449, 475, 493, 498, 499
   electroencephalogram monitoring, 493
   hypothalamus, 456, 457–459
   limbic system, 452–454
   medulla oblongata, 438, 462–464
   and memory, 494–496
   mesencephalon, 436–438, 439, 459–460
   motor pathways, 487–493
   organization of, 436–440
   pons, 438, 462
   protection and support of, 440–445
   respiratory centers of, 831–836
   sensory pathways, 481–487
   thalamus, 456–457
   ventricles of, 439–440
   views of, 437, 455
Brain-derived neutrophic factor (BDNF), 77 [AM]
Brain stem, 436–438, 456
   and vision, 556–557
Brain waves
   and sleep, 496–497
   types of, 493, 494
Branched glands, 117
Breast cancer, 1048, 1059
Breasts. See Mammary glands
Breathing, modes of, 832
Breech birth, 1086, 1098
Brevis, 316, 317
Broad ligament, 1035
Broca's area, 449
Brodmann's areas, 449–450
Bronchi, 9
   components of, 806
   inflammation of, 809
Bronchial arteries, 729, 813
Bronchial tree, 808
Bronchioles
   components of, 809–810, 811
   dilation and constriction of, 809
   respiratory bronchioles, 809
   terminal bronchioles, 809
Bronchitis, 809, 840; 153 [AM]
Bronchoconstriction, 809
Bronchodilation, 809, 840
Bronchodilators, 152 [AM]
Bronchography, 809, 840
Bronchomediastinal trunk, 755
Bronchopulmonary segment, 809
Bronchoscope, 809, 840
Bronchoscopy, 809
Brown fat, 122, 936
Bruit, 703
Brunner's glands, 871
Brush border, 869
Brush border enzymes, 871, 889
Buccal cavity. See Oral cavity
Buccal phase, swallowing, 860

Buccinator, 317, 321, 322, 853
Buccinator muscle, 316
Buffers, 43, 51, 54, 997
   and neutralization, 43
Buffer systems, 999–1001
   carbonic acid–bicarbonate buffer
      system, 999, 1000–1001
   functions of, 999
   hemoglobin buffer system, 1000
   phosphate buffer system, 999,
      1001
   protein buffer systems, 999–1000
Buffy coat, 628
Bulbar conjunctiva, 538
Bulbospongiosus muscles, 334,
   1046, 1053
Bulbourethral glands, 1017, 1029
Bulimia, 871, 932
Bulk flow, 74
BUN (blood urea nitrogen), 971
Bundle branches, 671, 672
Bundle of His, 672
Bunions, 249, 266
Burkitt's lymphoma, 1097; 133 AM
Burning pain, 530
Burns, 162
   classification of, 162
   effects of, 162
   treatment of, 162
Bursae
   disorders of, 249
   and joints, 249
   of shoulder joint, 259
   subacromial bursa, 259
   subcoracoid bursa, 259
   subdeltoid bursa, 259
   subscapular bursa, 259
Bursitis, 249, 259, 266, 356
Button desmosomes, 110
Butyrate, 630; 113 AM

C

C_1 inactivator, 647
C3, 769
C3b, 769
Caffeine, 496
Calcaneal tendon, 239, 351
Calcaneus, 239
Calcification, bone, 118, 175
Calcitonin, 597
   and bone, 181, 182, 184
Calcitriol, 49, 149, 181, 182, 580,
   598, 604
   and growth, 610
Calcium, 31
   bone storage of, 168, 182–184
   in fluid and electrolyte balance,
      994–995, 996
   nutritional factors, 928
Calcium channel blockers, 667, 687
Calcium ions, 582, 647–648
   and heartbeat, 670
Calculi, 974, 980
Calmodulin, 305, 582
Calories, 596, 932–933
Caloriogenic effect, 596
Calorimeter, 933
Calorimetry, 932
Calsequestrin, 275
Calvaria, 196
Canal, 188
Canaliculi, 128, 170, 171
Canal of Schlemm, 545
Canals of Volkmann, 173
Cancellous bone. See Spongy bone
Cancer, 100–101, 102; 37–40 AM
   and aging, 136
   bladder, 976
   breast, 1048, 1059
   and cell division, 100–101
   cervical, 1042, 1059
   colon, 884
   lungs, 839, 840

lymphomas, 760
   metastasis, 101, 102
   and natural killer (NK) cells, 766
   ovarian, 1039, 1060
   prostate gland, 1032, 1060
   skin, 148, 164, 784
   stomach, 868
   testicular, 1025
   tissue changes in, 137
Candidiasis, 29, 46, 189 AM
Canines, 857
Capacitance vessels, 700
Capacitation, 1027, 1065
Capillaries, 8, 79, 124, 655, 692,
   695–699
   capillary beds, 698
   capillary exchange, 706–710
   continuous capillaries, 695–698
   fenestrated capillaries, 698
   filtration across walls of, 75
   lymphatic capillaries, 752–753
   vasomotion, 698–699
Capillary autoregulation, 698
Capillary beds, 698, 699
Capillary endothelium, 950
Capillary exchange, 701, 706–710
   diffusion, 706
   filtration, 707, 708–710
   net colloid osmotic pressure, 708
   net filtration pressure, 708
   reabsorption, 707–710
Capillary hemangioma, 151, 163
Capillary hydrostatic pressure
   (CHP), 708
Capillary plexus, 698
Capitate bone, 232
Capitis, 317
Capitulum, 221, 230
Capsular colloid osmotic pressure,
   955
Capsular hydrostatic pressure
   (CsHP), 955
Capsular space, 947, 949
Capsule, 124
Carbaminohemoglobin, 630, 829
Carbohydrase, 874, 887
Carbohydrate depletion/loading,
   910, 938
Carbohydrate metabolism, 903–912
   and ATP generation, 903–911
   gluconeogenesis, 912
   glycolysis, 904, 905
   oxidative phosphorylation,
      907–910
   tricarboxylic acid (TCA) cycle,
      903, 904–907
Carbohydrates, 44–46
   absorption of, 888–890, 922
   of cell membrane, 70
   digestion of, 888–890
   disaccharides, 45–46
   and liver, 877
   monosaccharides, 44–45
   polysaccharides, 46
Carbon, 31
Carbon dioxide, 34, 40
   and respiration, 825, 834
Carbon dioxide transport, 829–830,
   831
   carbonic acid formation, 829
   hemoglobin binding, 829
   plasma transport, 830
Carbonic acid, 43, 827–828, 998
   formation of, 829
Carbonic acid–bicarbonate buffer
   system, 43, 999, 1000–1001
Carbonic anhydrase, 827, 829, 965,
   998
Carbon monoxide, 390, 391
Carbon monoxide poisoning, 827
Carboxyl group, 44
Carboxylic acid group, 47, 51
Carboxypeptidase, 874, 891
Carcinogens, 103; 38 AM
Cardia, 861

Cardiac arrest, 122 AM
Cardiac arrhythmias, 674, 687
Cardiac centers, 463–464, 683, 712
Cardiac cycle, 674–678
   diastole, 674
   energy for, 678
   heart sounds, 677–678
   phases of, 675
   pressure/volume changes in,
      676–677
   systole, 674
Cardiac impressions, 808
Cardiac muscle, 271
Cardiac muscle cells, 660
Cardiac muscle tissue, 4, 132, 133,
   302–303, 660
   cells of, 302
   compared to skeletal muscle tis-
      sue, 302–303, 660, 669
   contractions of, 303
   intercalated discs, 302–303
   structure of, 302–303
Cardiac myocytes, 132, 302
Cardiac notch, 808
Cardiac output, 681
   abnormal conditions affecting,
      686
   exercise and, 685–686
   and heart rate, 686
Cardiac pacemaker, 671
Cardiac physiology, 4
Cardiac plexus, 517, 683
Cardiac reserve, 685
Cardiac sphincter, 859
Cardiac tamponade, 657, 687
Cardiac veins, 668
Cardioacceleratory center, 683, 712
Cardioacceleratory reflex, 521
Cardiocytes, 132, 302
Cardiodynamics, 680–686
   contractility, 681, 682–683
   drug effects, 683
   end-diastolic volume, 680
   end-systolic volume, 680
   Frank–Starling principle, 682
   and hormones, 682
   ion concentration, effects of, 683
   preload, 681–682
Cardiogenic shock, 747; 127 AM
Cardioinhibitory center, 683, 712
Cardiomyopathies, 688; 118 AM
Cardiopulmonary resuscitation
   (CPR), 222, 819, 840; 153 AM
Cardiovascular centers, 463–464,
   712
   and CNS activities, 715
Cardiovascular physiology, 4,
   701–711
   arterial blood pressure, 705–706
   capillary exchange, 706–710
   circulatory pressure, 702, 704–705
   resistance, 702–703
   venous pressure and return,
      710–711
Cardiovascular regulation, 711–717
   and baroceptor reflexes, 713–714
   and chemoreceptor reflexes, 713,
      714–715
   hormonal control, 715–717
   local factors, 711–712
   neural control, 712–715
   vasomotor tone, 712–713
Cardiovascular response, 717–723
Cardiovascular system, 2, 3, 8, 624,
   654–750; 105–128 AM
   and aging, 744
   autonomic control of, 518
   cardiodynamics, 680–686
   cardiovascular physiology, 4,
      701–711
   cardiovascular regulation,
      711–717
   cardiovascular response, 717–723
   disorders of, 108–128 AM
   examination of, 106–108 AM

and exercise, 299, 300, 717–718
   functions of, 4, 8
   heart, anatomy of, 655–668
   heartbeat, 668–679
   hemorrhaging, effects of, 719–723
   integration with other systems,
      745, 746
   organs of, 8
   overview of, 655
   and puberty, 1089
   shock, effects of, 720–721
   smooth muscles of, 304
Carditis, 687; 118 AM
Carina, 806
Carotene, 147
Carotid artery, 443, 533, 726,
   727–729
   common, 726
   external, 695, 727
   internal, 443, 727–728
Carotid bodies, 534, 714
Carotid canals, 202, 727
Carotid sinus, 533, 713
Carpal bones, 168, 232–233
   distal carpals, 232
   proximal carpals, 232
Carpal tunnel syndrome, 233, 243,
   344, 356
Carpus, 232, 317
Carrier-mediated transport, 71,
   75–77, 79, 958–959
   active transport, 76, 79, 958
   characteristics of, 75, 958
   cotransport, 958
   countertransport, 958
   facilitated diffusion, 75–76, 79,
      958
   secondary active transport, 76–77,
      79
   sodium–potassium exchange
      pump, 76
Carrier proteins, 69, 70
Carriers, 1094–1095
Cartilage, 6, 125–128, 129
   compared to bone, 129
   elastic cartilage, 126, 127
   fibrocartilage, 126–128
   growth of, 125–126
   hyaline cartilage, 126, 127, 128
   and joints, 249
   and knee injuries, 128
   of larynx, 803–805
Cartilaginous joint, 247
Cascade, 646
Casts, in urine, 974; 179–180 AM
Catabolism, 38, 902
   amino acids, 916–919
   lipids, 913–914
   RNA, 919
Catalysts, 39
Cataracts, 546, 572
Catecholamines, 579
Catechol-O-methyltransferase
   (COMT), 511
Categorical hemisphere, 451
Cathartics, 887, 895
Catheter, 667–668
Cations, 33
CAT (computerized axial tomogra-
   phy) scan, 23, 24
Cauda equina, 409
Caudal, 18
Caudal anesthesia, 432; 76 AM
Caudate lobe, 874
Caudate nucleus, 451
Cavernous hemangioma, 151, 163
Cavernous sinus, 734
Cavities, 858
C cells (parafollicular cells), 184,
   597, 599
Cdc2 protein, 100
CD markers, 775
Cecum, 869, 882
Celiac artery, 510
Celiac ganglion, 510

Celiac plexus, 517
Celiac trunk, branches of, 731–732
Cell adhesion molecules (CAMs), 108
Cell biology, 65
Cell body (soma), 134, 362–363
Cell division, 96–102
　and cancer, 100–101
　cytokinesis, 99
　DNA replication, 96, 97
　and energy, 99
　interphase, 96–97, 98
　mitosis, 96, 98–99
　mitotic rate, 99
Cell junctions, 109–110
　desmosomes, 109, 110
　gap junctions, 109–110
Cell-mediated immunity (cellular immunity), 756, 771
Cell membrane, 66, 67, 68–80
　carbohydrates of, 70
　carrier-mediated transport, 75–77
　diffusion, 71–73
　filtration, 75
　and hormones, 582–583
　lipids of, 68–69
　membrane channels, 376–377
　osmosis, 73–74
　permeability, 71–80
　proteins of, 69–70
　transmembrane potential, 79–80, 373–377
　vesicular transport, 77–79
Cell physiology, 4
Cells, 3, 59
　anatomy of, 66–67
　cell membrane, 66, 67, 68–80
　cytoplasm, 66, 80–89
　differentiation, 101–102
　first description of, 59–60, 65
　fluids related to, 66
　nature of, 59–60, 65
　nucleus, 66, 89–95
　organelles, 4, 66, 67, 80–89
　sex cells, 66
　somatic cells, 66
　study of, 65–67
Cells of Leydig, 1021
Cell structure and function, 31–33 AM
Cell theory, 65
Cellular biology, 23–33 AM
Cellular immunity (cell-mediated immunity), 641
Cellular level, of organization, 4, 5
Cellular metabolism, 901–903
Cellular respiration, 88, 814, 904
Cellular trophoblast, 1069
Cellulitis, 134–135 AM
Cellulose, 46
Cementum, 856
Center, neurons, 399
Centimeters of water, 815, 817
Central adaptation, 530
Central canal (Haversian canal), 171, 172–173, 368, 412
Central chemoreceptors, 834
Central ischemic response, 720
Central nervous system (CNS), 7, 361. See also Brain; Nervous system; Spinal cord
　and cardiovascular centers, 715
　demyelination of, 370
　neuroglia of, 367–370
　tumors of, 370
Central retinal artery, 544
Central retinal vein, 544
Central sulcus, 446
Central vein, 877
Centrioles, 82, 83, 363
Centromere, 91, 98
Centrosome, 66, 67, 82
Centrum, vertebral, 213
Cephalic, 18

Cephalic phase, of gastric secretion, 865, 866
Cephalic region, 16
Cephalic vein, 737
Cephalon, 16
Cerebellar hemispheres, 461
Cerebellar nuclei, 462, 490–491
Cerebellum, 436, 438, 439, 460–462
　components of, 461–462
　functions of, 460–461
　and motor pathways, 491–492
Cerebral aqueduct, 439
Cerebral arterial circle, 729
Cerebral artery, 728, 729
Cerebral cortex, 436, 445–451, 490, 492
　association areas, 447–448
　Brodmann's areas, 449–450
　cerebral lobes, 445, 447
　general interpretive area, 449
　hemispheric lateralization, 450–451
　integrative centers of, 448–450
　motor areas, 447
　prefrontal cortex, 448, 449
　sensory areas, 447
　speech center, 449
　white matter, 446–447
Cerebral embolism, 747; 128 AM
Cerebral hemispheres, 436
　hemispheric lateralization, 450–451
　and vision, 556
Cerebral hemorrhage, 747; 128 AM
Cerebral meningitis, 409
Cerebral nuclei, 451–452, 453, 490–491, 492
　functions of, 451–452
　structure of, 451, 453
Cerebral palsy, 489, 500
Cerebral peduncles, 460, 462
Cerebral thrombosis, 747; 128 AM
Cerebral veins, 734–736
Cerebrospinal fluid (CSF), 368, 410, 411, 439, 441–443
　circulation of, 443
　formation of, 442–443
　functions of, 441–442
　and hydrocephalus, 444
Cerebrovascular accidents (CVAs), 398, 443, 475, 729
Cerebrovascular diseases, 443
Cerebrum, 436, 438, 439, 445–452
　cerebral cortex, 445–451
　white matter, 446–447
Cerumen, 157, 558
Ceruminous glands, 157, 558
Cervical canal, 1041
Cervical cancer, 1042, 1059; 189 AM
Cervical cap, 1055
Cervical curve, spine, 212
Cervical enlargement, of spinal cord, 407
Cervical os, 1041
Cervical plexus, 418
Cervical region, 16
Cervical vertebrae, 212, 214–216
　atlas, 215
　axis, 215–216
　vertebra prominens, 216
Cervicis, 16, 317
Cervix, 115, 1041
　cancer of, 1042, 1059
Cesarean section, 1084
Chalazion, 538
Chalones, 100
Chancroid, 1056; 191 AM
Channels, 70
　gated channels, 69, 70
　leak channels, 69, 70
Cheeks, 853
Chemical bonds, 32–36
　covalent bonds, 34–35
　high-energy bond, 57

hydrogen bonds, 35–36
ionic bonds, 33–34
polar covalent bonds, 35
Chemical compounds. See Compounds
Chemical gradient, 72
Chemical level of organization, 4, 5, 29–63
　atoms, 30–36
　chemical bonds, 32–36
　chemical reactions, 36–39
　inorganic compounds, 39–43
　and living cells, 59–60
　organic compounds, 39, 44–59
Chemically regulated channels, 376, 377
Chemical notation, rules for, 37
Chemical reactions, 36–39
　anabolism, 38
　catabolism, 38
　decomposition, 38, 39
　endergonic reactions, 38
　energy concepts in, 36–37
　and enzymes, 39
　exchange reactions, 38
　exergonic reactions, 38
　reversible reactions, 38
　synthesis, 38
Chemical sense, 533–534
Chemical synapses, 387–390
Chemiosmosis, 908
Chemistry, study of, 30
　applying principles of, 15–22 AM
Chemoreceptor reflexes
　and cardiovascular regulation, 713, 714–715
　and respiration, 834–835
Chemoreceptors, 482, 533–534
　locations of, 714
Chemosensitive area, 834
Chemotaxis, 766
Chemotherapy, 138; 40 AM
Chenodeoxycholate, 879
Chest tube, 222, 223
Chief cells, 598, 599, 864, 865
Childhood, 1086, 1088
Chiropractors (D.C.), 59 AM
Chlamydia, 1040, 1056; 190 AM
Chlamydia pneumoniae, 697
Chloride
　in fluid and electrolyte balance, 996, 997
　nutritional factors, 928
Chloride channels, EPSP/IPSP summation and, 395
Chloride ion, 34
Chloride shift, 829, 1000
Chlorine, 31
Cholate, 879
Cholecalciferol, 149, 181, 604
Cholecystectomy, 880
Cholecystitis, 880, 895; 167 AM
Cholecystokinin, 867, 871, 874, 879, 880, 933
Cholelithiasis, 880, 895
Cholera, 887, 895
Cholesterol, 47, 49, 60, 718
　facts about, 915
　good and bad, 916
　high levels, 915
Choline, 388
Cholinergic receptors, 515
Cholinergic synapses, 387–389, 511, 712
Cholinesterase, 281, 388
Cholinesterase inhibitor, 307, 390; 68 AM
Chondrocytes, 125, 128
Chondroitin sulfates, 125
Chordae tendineae, 662
Chorion, 1072
Chorionic villi, 1072
Chorionic villus sampling, 1097, 1098
Choroid, 542

Choroid plexus, 442, 698
Chromatids, 91, 98, 1023
Chromatin, 90
Chromium, 928
Chromosomal abnormalities, 1094, 1097
　disorders from, 1097
　monosomy, 1097
　nature of, 1097
　testing for, 1097
　translocaton defect, 1097
　trisomy, 1097
Chromosomal microtubules, 98
Chromosomes, 90
　alleles, interactions between, 1091
　autosomal chromosomes, 1090–1095
　daughter chromosomes, 98
　deoxyribonucleic acid (DNA), 90–93
　homologous chromosomes, 1090
　sex chromosomes, 1090, 1095
　structure of, 9, 91
　X chromosome, 1095
　Y chromosome, 1095
Chronic airways obstruction, 153 AM
Chronic fatigue syndrome (CFS) 307; 66 AM
Chronic obstructive pulmonary disease (COPD); 840; 154 AM
Chronic renal failure, 958
Chylomicrons, 871, 890, 916
Chyme, 861, 863, 865, 867, 868
Chymosin, 865
Chymotrypsin, 874, 891
Chymotrypsinogen, 874, 889
Ciguatoxin (CTX), 390; 75 AM
Cilia, 66, 67, 82–83
Ciliary body, 542
Ciliary ganglia, 386, 467, 468
Ciliary muscle, 547, 548
Ciliary neurotrophic factor, 77 AM
Ciliary processes, 542
Ciliated epithelium, 108
Cimetidine, 865
Cingulate gyrus, 454
Circadian rhythms, 557, 603
　control of, 458, 459
Circle of Willis, 729
Circular muscle (sphincter), 313, 848
Circulatory collapse, 720
Circulatory pressure, 702, 704–705
　blood pressure, 702, 704
　capillary hydrostatic pressure, 702
　venous pressure, 702, 704
Circulatory shock, 720–721
Circulatory system, 624. See also Blood vessels; Heart
　blood pressure, 705–706
　capillary exchange, 706–710
　components of, 655
　development of, 743
　disorders of, 743–744, 747
　fetal circulation, 742–743
　homeostatic mechanisms, 711–717
　neural control of, 712–715
　pulmonary circulation, 723–724
　special circulation, 722
　systemic circulation, 724–742
　terms related to, 704
　vasoconstrictors, 712
　vasodilators, 712
　venous pressure, 710–711
Circumcision, 1030, 1031
Circumduction, joints, 251, 252
Circumferential lamellae, 173, 178
Circumflex branch, 666
Circumvallate papillae, 536
Cirrhosis, 860, 878, 893, 895; 166–167 AM
11-Cis, 551
Cisterna chyli, 755

Cisternae, 84, 85
Citratephosphate dextrose, 649
Citric acid cycle (tricarboxylic acid, or Krebs, cycle), 293, 904–907
Classical pathway, complement system, 768, 769
Clavicle, 175, 222, 228–229
-clavius, 317
Clearance test, 980
Cleavage, 1068, 1069
Cleft palate, 205
Cleido-, 317
Clenbuterol, 611
Clinical diagnosis, 1 AM
Clitoris, 11, 1046
Clomiphene (Clomid), 1079
Clone, 772; 142 AM
Clonus, 432; 80 AM
Closed fractures, 186
Closed wounds, 48 AM
Clostridium, 289
Closure, epiphyseal growth, 178, 181
Clotting factors, 645, 646
Clotting proteins, 51
Clotting system, 644
Clubfoot, 240
Club hair, 155
CNS. See Central Nervous system (CNS)
Coagulation phase, hemostasis, 645, 646–648. See also Blood clotting
Coagulation time, 646
Coated vesicles, 77
Cobalt, 928
Coccus, 24 AM
Coccygeal cornua, 219
Coccygeal ligament, 407, 410
Coccygeus, 317, 334
Coccyx, 212, 219
Cochlea, 561, 565
Cochlear branch, 470, 568
Cochlear duct, 561, 564–565
Cochlear implants, 95–96 AM
Cochlear nucleus, 470, 568
Coding strand, 92
Codominance, 1091
Codon, 92, 93–94
Coelom, 20–22
Coenzyme A (CoA), 904, 906
Coenzyme Q, 908
Coenzymes, 55. See also Vitamins
Cofactors, 55; 21 AM
Coitus, 1053–1056
Colace, 887
Colectomy, 896; 167 AM
Colic vein, 741
Colitis, 884, 895; 167 AM
Collagen, 52, 53, 54, 170, 174
Collagen fibers, 119–120
Collagenous tissues. See Dense connective tissue
Collateral ganglia, 507, 508, 510
  anatomy of, 510
  functions of, 510
Collaterals, 363, 698
Collecting ducts, 948, 949, 952
Collecting system, 960, 966
  kidney, 947–948, 966
  nephron, 948, 949, 952–953
Collecting tubule, 947–948
Colles' fracture, 186
Colliculi, 490
Colloid, 42, 594
Colloid osmotic pressure (NCOP), 955
Colon, 882–884
  regions of, 882–884
  structure of, 882–883
Colon cancer, 884
Colonoscope, 896; 167 AM
Colony-stimulating factors (CSFs), 644, 783, 784
Color vision, 554

color blindness, 555, 572, 1095
Colostomy, 896; 167 AM
Colostrum, 1087
Columnar epithelium, 112–115
  pseudostratified columnar epithelium, 114, 115
  simple columnar epithelium, 112–115
  stratified columnar epithelium, 114, 115
Columns, spinal cord, 399, 413
Comedos, 48 AM
Comminuted fractures, 186
Commissural fibers, 446
Commissures
  anterior white, 413
  gray, 413
Common bile duct, 872, 877
Common hepatic artery, 732
Common hepatic ducts, 877
Common pathway, 646
Compact bone, 168, 169, 172–173
  function of, 173
  histology of, 171
  structure of, 172–173
Compartments, 355
Compartment syndromes, 355, 356
Compazine, 563
Compensation curves, spine, 212
Competitive inhibition, 21 AM
Complement, activation of, 780
Complementary base pairs, 57
Complementary gene action, 1093
Complement proteins, 640, 767
Complement system, 764, 767–769
  alternative pathway, 768, 769
  classic pathway, 768, 769
  effects of activation, 769
Complete antigen, 780
Complete heart block, 121 AM
Complete proteins, 926
Complete tetanus, 287
Compliance, of lungs, 816–817
Compound glands, 117
Compounds, 33
  high-energy compounds, 57–59, 87
  inorganic, 39–43
  organic, 39, 44–59
Compound tubuloacinar gland, 873
Compression, 271
Compression fractures, 186
Concentration gradient, 71, 706
  size of, 72
Concentrations, 41
  solute, 41–42
  solutions, 17–18 AM
Concentric contraction, 290–291
Concentric lamellae, 171
Conception, 1017, 1065
Condensation (dehydration synthesis), 38, 45, 52
Condensation reaction, 45
Condom, 1055
Conducting arteries, 694–695
Conducting cells, 671
Conducting system of heart, 662, 668, 670–672
  atrioventricular node, 671, 672
  bundle branches, 671, 672
  disorders of, 679
  sinoatrial node, 671
Conduction, of heat, 934
Conduction deficits, 672, 687; 121 AM
Conductive deafness, 571, 572
Condylar processes, 207
Condyle, 188, 230
Condyloid joint (ellipsoidal joint), 254
Cones, eye, 544, 550, 551, 554
Congenital malformation, 1098; 195 AM
Congenital talipes equinovarus, 240, 243

Congestive heart failure (CHF), 747; 123–125 AM
Conjoined twins, 1086
Conjunctiva, 538
Conjunctivitis, 538–539
Connecting tubules, 947–948
Connectins, 276
Connective tissue, 4, 107, 117–129
  components of, 117, 131–132
  connective tissue proper, 118–124
  fluid connective tissue, 118, 124
  functions of, 118
  of heart, 664
  supporting connective tissue, 118, 125–129
Connective tissue proper, 118–124
  cell types of, 118–119
  dense connective tissue, 122–124
  embryonic connective tissue, 120
  fibers of, 119–120
  ground substance, 120
  loose connective tissue, 120–122
Connexons, 110, 387
Consciousness, 496–498
  arousal, 497–498
  sleep, 496–497
Consensual light reflex, 520, 521
Constant segments, antibodies, 778
Constipation, 887, 895
Contact dermatitis, 152, 163
Continuous ambulatory peritoneal dialysis, 971
Continuous capillaries, 695–698
Continuous passive motion, 265, 266
Continuous propagation, action potentials, 383–384
Contraception, 1054–1055
Contraceptive methods; experimental; 189 AM
Contractile cells, 668–670
Contractile proteins, 51, 54
Contractility, cardiodynamics, 681, 682–683
Contraction phase, muscle twitch, 286
Contractions, pregnancy, 1083
Contracture, 355
Contralateral reflex arc, 428
Control center, 12
Contusion, 151; 48, 80 AM
Conus arteriosus, 662
Conus medullaris, 407
Convection, of heat, 934
Convergence, neurons, 396–397
Convergent muscle, 312, 313
Converting enzymes, 813
Convulsive seizures, 84 AM
Copper, nutritional factors, 928
Copulation, 1053–1056
Coracoacromial ligament, 259
Coracobrachialis, 339, 340
Coracoclavicular ligament, 259
Coracohumeral ligament, 259
Coracoid process, 230
Cords, brachial plexus, 419
Cori cycle, 296
Cornea, 538, 541
Corneal epithelium, 538
Corneal transplant, 541
Corniculate cartilages, 804
Cornification, 147
Coronal plane, 19
Coronal suture, 196, 197
Corona radiata, 1038, 1065
Coronary angiogram, 123 AM
Coronary arteries, 664–668
  diseases of, 667–668
  left, 666
  right, 664
Coronary arteriography, 688; 123 AM
Coronary artery bypass graft, 668, 688

Coronary artery disease, 667–668, 688, 696, 722
Coronary circulation
  cardiac veins, 668
  coronary arteries, 664–668
Coronary ischemia, 667, 688, 814
Coronary sinus, 660, 668
Coronary spasms, 722
Coronary sulcus, 657
Coronary thrombosis, 679, 688
Coronary veins, 660
Coronoid fossa, 230
Coronoid processes, 207, 231
Corpora quadrigemina, 459
Corposa cavernosa, 1030–1031
Corpus albicans, 1038
Corpus callosum, 446
  surgical cutting of, 451
Corpus luteum, 609, 1038
Corpus spongiosum, 1031
Corpus striatum, 451
Corrugator supercilii, 322
Cortex
  adrenal, 600–603
  bone, 169
  hair, 153
  kidney, 946
  thymus, 761
Cortical nephrons, 948, 950
Cortical reaction, 1067
Corticobulbar tracts, 488, 489, 491
Corticospinal tracts, 481, 488, 491
Corticosteroids, 49, 580, 600
Corticosterone, 602
Corticotropin, 589
Corticotropin-releasing hormone (CRH), 588, 589
Cortisol, 602
Cortisone, 602
Costae. See Ribs
Costal breathing, 820
Costal cartilages, 221
Costal groove, 221
Costalis, 317
Costal processes, 214
Costimulation, 775
Cotransport, 75, 958
  features of, 890
Coughing, 836
Coughing reflex, 521, 805
Coumadin, 649
Countercurrent heat exchange, 935
Countercurrent multiplication, 962–963
Countertransport, 75, 958
Coupling factors, 908
Covalent bonds, 34–35
  nonpolar, 35, 41
  polar, 35
  single and double, 34
Cowper's glands. See Bulbourethral glands
Coxal joint (hip joint), 260–261
Cradle cap, 156
Cramps, muscle, 356
Cranial, 18
Cranial bones, 195, 200–204
  ethmoid bone, 204, 209
  frontal bone, 201, 209
  occipital bone, 200, 209
  parietal bone, 200
  sphenoid bone, 203, 209
  temporal bones, 202, 209
Cranial cavity, 20, 21, 195
Cranial meninges, 409, 440–441
  arachnoid, 440
  dura mater, 440
  functions of, 440–441
  pia mater, 440
Cranial nerves, 361, 464–474
  abducens nerve, 467
  accessory nerve, 473
  facial nerve, 468, 469
  glossopharyngeal nerve, 470–471
  hypoglossal nerve, 473

oculomotor nerve, 467
olfactory nerve, 466
optic nerve, 466
origins of, 465
summary of, 474
trigeminal nerve, 468
trochlear nerve, 467
vagus nerve, 471–472
vestibulocochlear nerve, 470
Cranial reflexes, 424; 82–83 AM
Cranial trauma, 441, 475; 80 AM
Craniosacral division. *See*
Parasympathetic division of
ANS
Craniostenosis, 211, 223
Cranium, 20, 195
veins of, 734–736
Creatine, 293, 926
Creatine phosphate (CP), 293, 294,
953
Creatine phosphokinase (CPK),
293, 679
Creatinine, 953, 955
Creatinine clearance test, 955
Cremaster muscle, 1020
Crenation, 74
Crest, 188
Cretinism, 597, 616, 618
Crib death. *See* Sudden infant
death syndrome (SIDS)
Cribriform plate, 204
Cricoid cartilage, 803–804
Crisis phase; of fevers, 173 AM
Cristae, 87, 561
Crista galli, 204
Cross-bridges, 278, 282–283, 285
Crossed extensor reflexes, 428–429
Crossing-over, 1023, 1094
Cross-match testing, 636, 637
Cross-reactions, 636
Cross-training, 299
Crown, of tooth, 856
Crude touch and pressure recep-
tors, 532
Crura, 1031
Crural palsies, 78 AM
Crural region, 16
Crus, 16
Crusts, 43 AM
Cryosurgery, 49 AM
Cryptorchidism, 1018, 1059
Crypts of Lieberkuhn, 871
Crystallins, 546
CSF. *See* Cerebrospinal fluid (CSF)
CSFs (colony stimulating factors),
644, 783, 784
CT (computerized tomography)
scan, 23, 24
Cubic millimeter, 628
Cubital vein, 737
Cuboidal epithelium, 112·
simple cuboidal epithelium,
112–113
stratified cuboidal epithelium,
112–113
transitional epithelium, 112–113
Cuboid bone, 239
Cumulative trauma disorder, 344
Cuneiform cartilages, 126, 804
Cuneiform bones, 239
Cupula, 561
Current, 375
Cushing's disease, 602, 616, 618;
102 AM
Cuspids, 857
Cusps, 662
Cutaneous, 317
Cutaneous membrane, 6, 129, 131,
144
Cutaneous plexus, 151
Cuticle
hair, 153
nail, 158
Cyanosis, 149, 163, 743
Cycles, sound waves, 566

Cyclic-AMP (cAMP), 392, 511,
535, 582
Cyclic-GMP (cGMP), 392, 551,
553, 582
Cyclin, 100
Cyclophosphamide, 136 AM
Cyclosporin A, 777
Cysteine, 52
Cystic duct, 877
Cystic fibrosis (CF), 800, 840; 6,
150–151 AM
Cystic vein, 742
Cystinuria, 179 AM
Cystitis, 976, 980
Cysts, 134 AM
Cysts, breasts, 1048
Cytochromes, 908
Cytokines, 757, 767
Cytokinesis, 98, 99
Cytokines (local hormones), 577
Cytology, 3, 65
Cytoplasm, 66, 80–89
cytosol, 80
organelles, 80–89
Cytosine, 56, 91
Cytoskeleton, 66, 67, 70, 81–82
intermediate filaments, 81–82
microfilaments, 81
microtubules, 82
thick filaments, 81
Cytosol, 66, 67, 80
Cytotoxic T cells, 756, 773, 775–776,
785, 787
Cytotrophoblast, 1069

D

Dalton, 31
Dalton's law, 822
Dancer's fracture, 243; 58 AM
Dark-adapted state, 554
Dark current, 551
Dartos, 1020
Daughter cells, 96, 99
Daughter chromosomes, 98
D cells, 865
Deafness, 571, 572
Deamination, 878, 918
Death and dying, 198–199 AM
Death, rigor mortis, 283
Decarboxylation, 904
Decay, 31
Decerebrate rigidity, 475; 81 AM
Decibels, sound, 567
Decidua basalis, 1075
Decidua capsularis, 1072–1075
Decidua parietalis, 1075
Deciduous teeth, 857
Decomposition, 38, 39
Decompression sickness, 824, 840;
153 AM
Decubitis ulcers, 151, 163
Decussation, 83, 464
Deep, directional term, 18
Deep breathing, 820
Deep cortex, 760
Deep fascia, 131–132, 272
Deep femoral artery, 734
Deep femoral vein, 740
Deep lymphatics, 754
Deep palmar arch, 727
Deep sleep, 496
Defecation, 846, 886–887
reflex in, 886–887
Defensins, 640, 783
Defibrillator, 688; 122 AM
Deficiency disease, vitamin, 929
Deflation reflex, 835
Degenerative arthritis, 265, 266;
60 AM
Degenerative joint disease (DJD),
265; 60 AM
Deglutition. *See* Swallowing
Degranulation, 640

Dehydration, 990
Dehydration synthesis (condensa-
tion), 38, 45, 52
Dehydroepiandrosterone
(DHEA), 1034
Delayed hypersensitivity, 143 AM
Delayed-onset muscle soreness
(DOMS), 300
Delirium, 500; 85 AM
Delivery (birth), 1084
Delta cells, 606, 608
Delta waves, 493, 494
Deltoid, 230, 313, 317, 339, 340
Deltoid tuberosity, 230
Dementia, 499, 500; 85 AM
Demifacets, 217
Demyelination, 370, 401
disorders, 74 AM
Denaturation, 55
Dendrites, 134, 363
Dendritic cells, 760, 787
Dendritic spines, 363
Dens, 215
Dense area, 110
Dense bodies, 304
Dense bone. *See* Compact bone
Dense connective tissue, 118,
122–124
dense irregular connective tis-
sue, 122–124
dense regular connective tissue,
122
elastic tissue, 122–123, 124
ligaments, 122, 124. *See also*
Ligaments
tendons, 122. *See also* Tendons
Dense irregular connective tissue,
122–124
Dense regular connective tissue,
124
Densitometer, 179 AM
Dental calculus, 856
Dental caries, 858
Dental implants, 858
Dentate gyrus, 454
Denticulate ligaments, 410
Dentin, 856
Denver Developmental Screening
Test (DDST), 1098; 198 AM
Deoxyhemoglobin, 630
Deoxyribonucleic acid (DNA), 56,
57, 58, 90–93
characteristics of, 57, 58
and control of cell structure/func-
tion, 95
DNA fingerprinting, 91
recombinant DNA, 103
replication of, 96, 97
and transcription, 91–93
and translation, 93–95
Deoxyribose, 56
Depolarization, 379, 380
in heartbeat, 669
Depo-provera, 1054
Depression, 87 AM
Depression, joints, 253, 254
Depressor muscles, 317
depressor anguli oris, 322
depressor labii superioris, 322
Dermal bones, 175
Dermal ossification, 175–176
Dermal papillae, 145
Dermatitis, 152, 163; 44 AM
Dermatomes, 414
Dermis, 6, 144, 150–152, 160
blood supply to, 160
lines of cleavage, 151
papillary layer, 150
reticular layer, 150
wrinkles/stretch marks, 150–151
Descending aorta, 662, 729–733
abdominal aorta, 29, 731–733
thoracic aorta, 729–731
Descending colon, 851, 884
Descending limb, 952

Descending tracts, 413
Desmin, 277, 304
Desmopressin acetate (DDAVP),
591
Desmosomes, 110
Detached retina, 543–544
Detrusor, 975, 976
Detumescence, 1053
Deuterium, 31
Development, 1065
embryological development, 1065
fetal development, 1065
and induction, 1067–1068
postnatal development, 1065
prenatal development, 1065
Developmental anatomy, 3
Developmental disorders,
191–199 AM
Developmental dyslexia, 449
Deviated septum, 210, 223
Dextran, 74, 102
DHEA (dehydroepiandrosterone),
1034
Diabetes, 590, 591; 103–104 AM
characteristics of, 591, 607
diabetes insipidus, 591, 618, 967,
1095
diabetes mellitus, 76, 590, 607,
618, 959
insulin-dependent (Type 1) diabetes, 607, 618, 787
medical problems related to, 607
non-insulin-dependent (Type 2)
diabetes, 607, 618
Diabetic ketoacidosis, 925
Diabetic nephropathy, 607, 618
Diabetic neuropathy, 607, 618
Diabetic retinopathy, 545, 607,
618
Diabetogenic effect, 590
Diacylglycerol, 582
Diagnosis, 1, 5–13 AM
Diagnostic tests, 8–10, 11–13 AM
Dialysis, 970–971
continuous ambulatory peritoneal dialysis, 971
dialysis fluid, 970–971
hemodialysis, 970
peritoneal dialysis, 971
Diapedesis, 638, 766
Diaper rash, 152, 163
Diaphragm, 20, 21, 222, 331, 332,
418
birth control, 1055
Diaphragma sellae, 587
Diaphragmatic breathing, 820
Diaphragmatic hernia, 335, 356;
68–69 AM
Diaphragmatic surface of spleen,
762
Diaphysis, 169, 170, 173
Diarrhea, 887, 895
Diarthroses (freely movable
joints), 246, 247–250
accessory structures, 249
articular cartilages, 248
ball-and-socket joints, 256, 257
and bursae, 249
and cartilages, 249
dynamic motion of, 250–254
ellipsoidal joints, 254, 257
gliding joints, 254, 257
hinge joints, 254
and ligaments, 249
pivot joints, 254, 257
saddle joints, 254–256, 257
synovial fluid, 248–249
synovial membrane, 248
and tendons, 249
Diastole, 674
atrial diastole, 675
ventricular diastole, 675, 677
Diastolic pressure, 705
Dicrotic notch, 677
Dicumerol, 649

Diencephalon, 436, 438, 439, 454–459
 hypothalamus, 456, 457–459
 thalamus, 456–457
Dietary fats, 915
Diethylstilbestrol (DES), 1032
Differential count, 641
Differentiation, 1065
 cells, 101–102
 organisms, 2
Diffusion, 71–74, 79
 capillary exchange, 706
 across cell membranes, 72–74
 facilitated diffusion, 75–76
 influencing factors, 72
 net diffusion, 71
 osmosis, 73–74
Digastric, 326–328
Digestion, 2, 846
 of carbohydrates, 887
 digestive enzymes, 887, 890–891
 gastric secretion, phases of, 863–865
 hormones in, 850, 864–865, 880–881
 intestinal absorption, 881
 intestinal movements, 872
 ion absorption, 891–892
 of lipids, 890
 of protein, 891
 and stomach, 868
 vitamin absorption, 892–893
 water absorption, 891
Digestive enzymes, 890
 brush border enzymes, 889
 classification of, 889
 hydrolysis, 887
 of pancreas, 874, 888–889, 890
 of salivary glands, 855, 888–889
 of small intestine, 889, 890
 of stomach, 864–865, 868, 890
 of tongue, 854, 890
Digestive epithelium, 847
 renewal and repair of, 848
Digestive system, 5, 10, 845–899; 157–173 AM
 accessory organs, 846
 and aging, 893
 autonomic control of, 519
 components of, 847
 disorders of, 860, 865, 868, 880, 884, 893, 896; 162–173 AM
 esophagus, 858–861
 examination of, 157–162 AM
 functions of, 10, 846
 gallbladder, 879–880
 integration with other systems, 893–894
 large intestine, 882–887
 liver, 874–879
 oral cavity, 853–858
 organs of, 10
 pancreas, 872–874
 peritoneum, 851–852
 pharynx, 858
 regulation of, 849–851
 salivary glands, 854–856
 small intestine, 869–872
 smooth muscles of, 304
 stomach, 861–865
 symptoms of, 157 AM
 teeth, 856–858
 tongue, 854
Digestive tract, 846–853
 histological organization of, 846–849
 mesenteries, 851, 852
 mucosa, 847–848
 muscularis externa, 848–849
 peristalsis, 849
 segmentation, 849
 serosa, 849
 submucosa, 848
Digital arteries, 727
Digital extensors, 353

Digital flexors, 353
Digitalis, 683
Digital rectal exam (DRE), 1032
Digital subtraction angiography (DSA), 23, 25
Digital veins, 737
Diglyceride, 47, 48
Dihydrogen phosphate, 1001
Dihydrotestosterone (DHT), 1034
Dihydroxyacetone phosphate, 914
Dilation, 136
Dilation stage, labor, 1084, 1085
Dipeptidases, 891
Dipeptides, 52, 891
Diphtheria, 370, 401; 74 AM
Diploë, 169
Diploid cells, 1023
Dipole, 40
Direct communication, 577
Direct light reflex, 521
Disaccharides, 45–46, 889
Disconnection syndrome, 451
Discs, tactile (Merkel's), 532
Discs, photoreceptors, 550
Disease, 5 AM
 and homeostasis, 14, 26
 and tissue changes, 137
Dislocation, 250, 259
Displaced fractures, 186
Disseminated intravascular coagulation, 649
Dissociation, 40, 41
Distal, 18
Distal carpal bones, 232
Distal convoluted tubule (DCT), 947, 952, 960, 965–966, 967
 juxtaglomerular apparatus, 952
 reabsorption, 965
 secretion, 965
Distal radioulnar joint, 231
Distress, fetal, 1011
Distribution arteries, 695
Distributive shock, 747; 127 AM
Disulfide bonds, 52
Disulfiram, 931
Diuresis, 968
Diuretics, 968, 980, 995; 125, 179 AM
Divalent ions, 17 AM
Divergence, neurons, 396, 397
Diverticulitis, 884, 895
Diverticulosis, 884, 893, 895
Dizygotic twins, 1086
DMSO (dimethyl sulfoxide), 150
DNA. See Deoxyribonucleic acid (DNA)
DNA fingerprinting, 91, 102
DNA polymerase, 96–97
Dobutamine, 683
Dominant hemisphere, 451
Dominant traits, 1091, 1092
Dopamine, 389, 391, 460, 579, 589, 591, 683
Dorsal, 18
Dorsal arch (arcuate arch), 734
Dorsal body cavity, 20, 21
Dorsal interossei, 346
Dorsalis pedis artery, 734
Dorsal ramus, 414
Dorsal respiratory group, 832
Dorsal root ganglia, 407
Dorsal roots, 407
Dorsiflexion, 240
 joints, 253, 254
Dorsiflexors, 353
Dorsum, tongue, 854
Dorsum nasi, 801
Double covalent bond, 34
Double helix, 57
Down-regulation, 582
Down syndrome, 115, 498, 744, 1097
Dramamine, 563
Drives, and hypothalamus, 459
Drugs

absorption in stomach, 868
abuse, 88 AM
addiction, 88 AM
and autonomic nervous system, 522
 blocking agents, 522
 intramuscular injection, 349
 and liver, 878
 mimetic drugs, 522
 reactions to, 780
 and renal elimination, 958
 and synaptic function, 390
 transdermal administration, 150
DSA (digital subtraction angiography) images, 23, 25
DSR (dynamic spatial reconstruction) images, 23, 26
D-tubocurarine, 390, 401
Dual innervation, 516–517
 anatomy of, 516–517
 autonomic tone in, 518
Duchenne's muscular dystrophy (DMD), 277, 306; 66–67 AM
Ductless glands, 115
Duct of Santorini, 872
Duct of Wirsung, 872
Ducts, 115
Ductus arteriosus, 743, 744, 814, 836
Ductus deferens, 11, 1017, 1027, 1028
Ductus venosus, 742
Duodenal ampulla, 873, 879
Duodenal papilla, 879
Duodenal ulcer, 865
Duodenum, 848, 851, 861, 869, 871, 872. See also Small intestine
Dural folds, 440, 441
Dural sinuses, 441
Dura mater, 409–410, 440
Dust cells, 811
Dwarfism, 181, 190, 590
Dynamic spatial reconstruction (DSR) images, 23, 26
Dynein, 82
Dynorphin, 391, 392
Dyslexia, 449, 475
Dysmenorrhea, 1043, 1059
Dysmetria, 475; 82 AM
Dysplasia, 137, 138
Dysuria, 976, 980

E

E. See Epinephrine (E)
Ear. See also Hearing
 auditory ossicles, 558–560
 and equilibrium, 557, 561–564
 external ear, 557–558
 inner ear, 202, 557, 560–561
 middle ear, 557, 558–560
 sensory receptors of, 560, 561
Ear bones, 202
Eardrum (tympanic membrane), 202
Ear wax, 157
Eating disorders, 938
 anorexia nervosa, 932
 bulimia, 932
Eccentric contraction, 291–292
Eccrine sweat glands, 157
ECG. See Electrocardiogram
Echocardiography, 688; 9, 123 AM
Echogram, 24, 26
Eclampsia, 194 AM
Ectoderm, 1070
Ectopic bones, 175; 54 AM
Ectopic pacemaker, 672; 193–194 AM
Ectopic pregnancy, 1069, 1098; 193–194 AM
Eczema, 152, 163
Edema, 710, 745, 989
 causes of, 710

EDTA (ethylenediaminetetroacetic acid), 649
EEG (electroencephalogram), 493, 500
Effectors, 12, 361
Efferent arteriole, 947, 948, 950, 952
Efferent division, nervous system, 361–362
Efferent ductules, 1021
Efferent fibers, 366
Efferent lymphatics, 760
Efferent vessels. See Arteries
Egg nests, 1037
Eicosanoids, 47, 48, 580
Ejaculate, semen, 1029
Ejaculation, 521, 1017, 1053
Ejaculatory duct, 1017
Ejection fraction, 677, 680
EKG. See Electrocardiogram
ELAD (extracorporeal liver assist device), 878
Elastase, 874, 891
Elastic arteries, 694–695
Elastic cartilage, 126, 127
Elastic fibers, 120
Elastic ligaments, 120, 124
Elastic membrane, 692
Elastic rebound, 820
 blood pressure, 705
Elastic tissue, 122–123, 124
Elastin, 120
Elbow joint, 230, 260
 ligaments of, 260
 stability of, 260
 structure of, 260
Elective abortions, 1083
Electrical force, 30
Electrical synapses, 386–387
Electrocardiogram (ECG/EKG), 673–674, 688
 intervals of, 674
 waves of, 673
Electrochemical gradient, 72, 375
Electroencephalogram (EEG), 493, 500
Electrolytes, 40–41. See also Fluid and electrolyte balance
 balance, 985, 991–997
Electron cloud, 30
Electron microscopy, 65; 31 AM
Electrons, 30
 and energy levels, 32
Electron shell, 30, 34
Electron transport system (ETS), 903, 907, 908–910
 and ATP generation, 910, 911
Electrosurgery, 49 AM
Elements, 30–31
 atomic weight, 31–32
 in human body, 31
 trace, 31
Elephantiasis, 710, 756, 793
Elevation, joints, 253, 254
Elimination, 943
ELISA (enzyme-linked immunosorbent assay) test, 141 AM
Ellipsoidal joints, 254, 257
Embolism, 649, 650, 745
Embolus, 649
Embryo, 1017
Embryogenesis, 1068, 1075–1078
Embryological development, 1065
Embryology, 3, 26
 brain, 438–439
 circulatory system, 742–743
Embryonic connective tissue, 120
Embryonic disc, 1072
Emesis, 871
Emigration, 638, 766
Emission, 592, 1053
Emmetropia, 548
Emotions
 and hypothalamus, 459
 and limbic system, 454

Emphysema, 816, 837, 839, 840, 1005, 1091–1092; 5, 42, 153–154 [AM]
Emulsification, of lipids, 879
Emulsion droplets, 879
Enamel, of tooth, 856
End-diastolic volume (EDV), 677, 680
Endergonic reactions, 38
Endocarditis, 688; 118 [AM]
Endocardium, 658
Endochondral ossification, 175, 176–178
Endocrine cells, 578
Endocrine communication, 577, 578
Endocrine factors, 712
Endocrine glands, 115
　disorders of, 602
Endocrine pancreas, 605
Endocrine placenta, 1075
Endocrine reflexes, 584–586
Endocrine system, 5, 8, 361, 576–622; 97–104 [AM] See also Hormones; specific organs
　adrenal gland, 8, 600–603
　disorders of, 590, 591, 597, 598, 607, 616–617; 99–104 [AM]
　　classification of, 97–98 [AM]
　　symptoms of, 98–99 [AM]
　endocrine functions of heart, 605
　endocrine functions of kidney, 604
　endocrine functions of pancreas, 605–608
　endocrine reflexes, 584–586
　functions of, 8
　gonads, 608–610
　and hypothalamus, 585
　organs of, 8, 579
　parathyroid glands, 597–600
　pineal gland, 603
　pituitary gland, 586–593
　reproductive system hormones, 589, 602, 608–610
　thymus, 604, 605
　thyroid gland, 593–597
Endocytosis, 77, 79
　phagocytosis, 77–79
　pinocytosis, 77, 78
　receptor-mediated endocytosis, 77, 78
Endoderm, 1070
Endogenous pyrogen, 770
Endolymph, 560
Endolymphatic duct, 563
Endolymphatic sac, 563
Endometrial cancer, 188–189 [AM]
Endometrial polyps, 1060
Endometriosis, 1045, 1059
Endometrium, 1042
Endomysium, 272
Endoneurium, 414
Endoplasmic reticulum (ER), 66, 67, 84–85
　functions of, 84
　rough endoplasmic reticulum, 85
　smooth endoplasmic reticulum, 85
　in specialized cells, 85
Endorphins, 391, 392, 401
Endosomes, 77
Endosteum, 172, 174
Endothelins, 645
Endothelium, 111
Endothelium-derived relaxation factor, 712
End-systolic volume (ESV), 677, 680
Endurance, 296
Energy, 36
　activation energy, 39
　kinetic energy, 36, 37
　potential energy, 36–37
Energy levels, atoms, 32

Energy production
　and adenosine triphosphate (ATP), 57–59, 87–89
　Calories, 932–933
　and carbohydrate metabolism, 903–912
　and lipid metabolism, 913–914
　and mitochondria, 87–89
　and stress response, 612–614
Enkephalins, 391, 392
Enlargements, of spinal cord, 407
Enteric nervous system, 850
Enteritis, 168 [AM]
Enterocrinin, 872, 880, 881
Enteroendocrine cells, 847
Enterogastric reflex, 867
Enterohepatic circulation, of bile, 879
Enterokinase, 871, 891
Environmental factors, and nervous system, 398
Enzyme activity control, 19–22 [AM]
Enzymes, 51, 54–56
　in cell membrane, 69, 70
　and chemical reactions, 39
　coenzymes, 55
　functions of, 54–56
　isozymes, 54
　respiratory, 88
　in seminal fluid, 1029–1030
　synthesis reaction, 54, 55
　and temperature and pH, 55–56
Eosin, 640
Eosinophils, 119, 124, 637, 638, 640–641, 644, 787
Ependyma, 368, 439
Ependymal cells, 367, 368
Epicardium, 657, 658
Epicondyles, 230
Epicranius, 321
Epidermal derivatives, 152–158
Epidermal growth factor, 100
Epidermal ridges, 145
Epidermis, 6, 144, 145–150
　aging and, 160
　circulation in, 148–149
　epidermal growth factor, 150
　pigmentation of, 149
　stratum corneum, 146–147
　stratum germinativum, 145–146
　stratum granulosum, 146
　stratum lucidum, 146
　stratum spinosum, 146
　and sun, 147–148
　and vitamin D, 149
Epididymis, 11, 1017, 1025–1027
　components of, 1026
　functions of, 1026–1027
Epididymitis, 183 [AM]
Epidural block, 410, 431
Epidural hemorrhage, 443, 475
Epidural space, 409
Epiglottis, 804
Epiglottitis, 805
Epilepsy, 500; 84 [AM]
Epimysium, 272
Epinephrine (E), 391, 510, 511–512, 579, 603, 617, 935
　and skeletal muscles, 296
　and stress reaction, 612
Epineurium, 414
Epiphyseal cartilage (epiphyseal plate), 178
Epiphyseal fractures, 186
Epiphyseal line, 178, 180
Epiphyseal plate (epiphyseal cartilage), 178, 181
Epiphysis, 169, 170, 173, 177, 178
Epiploic appendages, 882
Episiotomy, 1084
Epistaxis, 802, 840
Epithalamus, 454
Epithelial tissue, 107–117
　characteristics of, 107
　columnar epithelium, 112–115

cuboidal epithelium, 112–113
　exfoliative cytology, 115
　functions of, 107–108
　intercellular connections in, 108–110
　maintaining integrity of, 108–110
　maintenance and repair of, 110
　simple epithelium, 110
　specialization of, 108
　squamous epithelium, 111–112
　stratified epithelium, 110–111
Epithelium, 107
EPO. See Erythropoietin
Eponychium, 158
Equational division, 1024
Equilibrium, 528, 557, 561–564
　in chemical reactions, 38
　and inner ear, 560
　pathways for, 563–564
　and saccule, 563
　semicircular ducts, 561–563
　and utricle, 563
Equivalent (Eq), 61; 17 [AM]
ER (endoplasmic reticulum), 66, 67, 84–85
Erectile tissue, 1030
Erection, penis, 1030
Erector spinae, 328, 329
Erosions, 43 [AM]
Eruption, of secondary teeth, 857
Erysipelas, 164; 134–135 [AM]
Erythema, 162
Erythroblastosis fetalis, 638
Erythroblasts, 633
Erythrocytes, 124. See also Red blood cells (RBCs)
Erythrocytosis, 651; 114 [AM]
Erythromycin, 445
Erythropoiesis, 633–634
　regulation of, 633–634
　stages of, 633
Erythropoietin (EPO; erythropoiesis-stimulating hormone), 100, 604, 611, 628, 634, 952
　and cardiovascular regulation, 717
Esophageal arteries, 729
Esophageal glands, 860
Esophageal hiatus, 859
Esophageal phase, swallowing, 860, 861
Esophageal plexus, 517, 859
Esophageal varices, 860, 895
Esophageal veins, 738
Esophagitis, 861, 896; 162 [AM]
Esophagus, 10, 858–861
　blood supply to, 859
　disorders of, 860
　histology of, 860
　structure of, 858–859
　swallowing, 860–861
Essential amino acids, 919, 924
Essential fatty acids, 914
Estradiol, 589, 609, 1049
Estriol, 1049
Estrogens, 49, 85, 580, 589, 602, 609, 610, 617, 1037
　and bone, 181, 182
　estradiol, 589, 609, 1049
　functions of, 1050
　and ovarian cycle, 1049–1052, 1057
　placental, 1075
　and pregnancy, 1083
Estrone, 1049
Ethmoidal sinuses, 204, 210
Ethmoid bone, 204, 209
Ethylenediaminetetraacetic acid (EDTA), 649
Etidronate, 1056
Eupnea, 819–820
Eustachian tube (pharyngotympanic tube), 558
Evaporation, 934
Eversion, 253, 254

Excess postexercise oxygen consumption (EPOC), 296
Exchange pump, 76
Exchange reactions, 38
Exchange transfusions, 111–112 [AM]
Exchange vessels, 655
Excitable membrane, 376
Excitation–contraction coupling, 281, 305
Excitatory neurotransmitters, 387
Excitatory postsynaptic potentials (EPSPs), 393
　summation of, 394–395
Excretion, 2, 846, 943
Excretory system, 978
Exercise
　aerobic endurance, 299
　and aging, 299
　anaerobic endurance, 298–299
　and bone, 180–181
　and carbohydrate loading, 910
　cardiac output and, 685–686
　and cardiovascular system, 299, 300, 717–718
　and heat loss, 296
　and integumentary system, 300
　interval training, 299
　and nervous system, 300
　and respiratory system, 300
　and skeletal muscles, 298–299
Exergonic reactions, 38
Exfoliative cytology, 115, 138, 1046
Exhaustion phase, general adaptation syndrome (GAS), 612, 614
Exocrine cells, 578
Exocrine glands, 115–117
　ceruminous glands, 157
　mixed exocrine glands, 116
　mucous glands, 116
　multicellular exocrine glands, 116, 117
　sebaceous glands, 155–156
　secretory modes, 115–116
　serous glands, 116
　sweat glands, 156–157
　types of secretions, 116–117
　unicellular exocrine glands, 116–117
Exocrine pancreas, 605
Exocytosis, 79
　and Golgi apparatus, 86
Exons, 92
Exophthalmos, 619; 99 [AM]
Expiratory center, 832
Expiratory reserve volume (ERV), 821
Expressivity, 1092
Expulsion stage, labor, 1084, 1085
Extension, joints, 251, 252
Extensor, 317
Extensor carpi radialis, 316, 342, 343
Extensor carpi ulnaris, 342, 343
Extensor digiti minimi, 344
Extensor digitorum, 312, 344
Extensor digitorum brevis, 354
Extensor digitorum longus, 349, 353
Extensor hallucis longus, 349, 353
Extensor indicis, 344
Extensor pollicis, 344
Extensor retinacula, 344, 353
External abdominal obliques, 330, 331, 332
External anal sphincter, 334, 884
External auditory canal, 202, 557
External callus, 184, 190
External carotid arteries, 695, 727, 813
External ear, 557–558
External elastic membrane, 692
External genitalia, 1017
　female, 1046

External iliac artery, 723, 733
External intercostals, 330, 331, 332
External nares, 206, 801
External occipital protuberance, 200
External orifice, uterus, 1041
External respiration, 814
External root sheath, 154, 155
External tension, 288–289
External urethral meatus, 976
External urethral sphincter, 334, 976
Externus, 316, 317
Exteroceptors, 366, 530
Extracapsular ligaments, 249
Extracellular fluid, 66, 985, 986
Extracorporeal liver assist device (ELAD), 878
Extraembryonic membranes, prenatal development, 1072, 1073
Extrafusal muscle fibers, 426
Extrapulmonary bronchi, 808
Extrapyramidal system, 489
Extrinsic, 316, 317
Extrinsic muscles, 316
    of eye, 323
    of foot and toes, 351–353
    of hand, 344
    of tongue, 854
Extrinsic pathway, blood clotting, 646
Extrinsic regulation, homeostasis, 12
Eye, 539–550. See also Vision
    accessory structures, 538–541
    aqueous humor, 541, 545
    autonomic control of, 518
    chambers of, 545–546
    choroid, 542
    ciliary body, 542
    cornea, 541
    disorders of, 545, 546, 563
    extrinsic eye muscles, 323
    fibrous tunic, 541
    insulation of, 541
    iris, 542
    lens, 546–549
    neural tunic (retina), 541–544
    optic disc, 544
    pupil, 542
    sclera, 541
    vascular tunic (uvea), 541–542
    vitreous body, 545, 546
Eyelashes, 538
Eyelids, 538–539
Eye sockets (orbits), 201

F

Fabella, 351
Facets, 188, 238
Facial bones, 196, 205–210
    hyoid bone, 207
    inferior nasal conchae, 206
    lacrimal bones, 206, 209
    mandible, 207, 209
    maxillary bones, 205, 209
    nasal bones, 206
    nasal complex, 208, 210
    orbital complex, 208
    palatine bones, 205
    vomer, 206
    zygomatic bones, 206, 209
Facial complex, 196
Facial expression, muscles of, 320–322
Facial nerve, 202, 320, 468, 469, 474
Facial veins, 736
Facilitated diffusion, 75–76, 79, 890
Facilitation, of neurons, 394
F actin, 277, 279
Fact memories, 494
Factor B, 769
Factor D, 769

Factor III, 646
Factor P, 769
Factors, 585
Facultative water reabsorption, 966
FAD (flavin adenine dinucleotide), 906, 908
Fainting, 710
Falciform ligament, 851, 874
Fallopian tubes. See Uterine tubes
False labor, 1083
False pelvis, 235, 236
False ribs, 221
Falx cerebelli, 441
Falx cerebri, 204, 441
Familial hypercholesterolemia, 61; 22 AM
Fasciae, 131–132
Fascicles, 272, 312, 316, 414
Fasciculus, 413
Fasciculus cuneatus, 482, 486
Fasciculus gracilis, 482, 486
Fast-adapting receptors, 529
Fast channels, 669
Fast fibers, 296, 297, 298
Fast pain, 530
Fat cells (adipocytes), 118, 119
    turnover rate, 60
Fatigued muscle, 295
Fat mobilization, 925
Fat pads, and joints, 249
Fats, dietary, 915
Fat-soluble vitamins, 878, 893, 929
Fat substitutes, 47
Fatty acids, 47–48
    essential fatty acids, 914
    and health, 48
    monounsaturated fatty acids, 48
    omega-3 fatty acids, 48
    polyunsaturated fatty acids, 48
    prostaglandins, 48
    saturated fatty acids, 47, 48
    unsaturated fatty acids, 47, 48
Fauces, 854
F cells, 606, 608
Feces, 846
Feedback control mechanism, negative, 423
Feeding center, 459
Female pronucleus, 1067
Female reproductive system, 1034–1052
    and aging, 1056
    autonomic control of, 519
    disorders of, 1039, 1040, 1045, 1059
    external genitalia, 1046
    hormones of, 1049–1052
    mammary glands, 1047
    menopause, 1056
    ovaries, 589, 604, 609, 1035–1039
    uterine tubes, 1039–1040
    uterus, 1040–1045
    vagina, 1045–1046
Femoral artery, 695, 723, 734
Femoral circumflex vein, 740
Femoral head, 237
    ligament of, 261
Femoral nerve, 420, 422
Femoral region, 16
Femoral vein, 739, 740
Femoris, 317
Femur, 16, 178, 237–238
Fenestrated capillaries, 587, 698
Ferritin, 633
Fertility drugs, 1079–1080
Fertilization, 609, 1017, 1065–1067
    conditions for, 1039–1040
    events of, 1065–1067
    oocyte activation, 1067
    pronucleus formation, 1067
Fetal alcohol syndrome, 931, 1098; 192–193 AM
Fetal antibody production, 787
Fetal circulation, 742–743
    changes at birth, 743

in heart and great vessels, 742–743
    placental blood supply, 742
Fetal development, 1065
Fetal distress, 1011
Fetal hemoglobin, 630, 828–829
Fetal stress, 1010–1011
Fetus, 1017
Fever, 764, 770, 937; 173 AM
    water losses and, 990
Fibrillin, 120
Fibrin, 158, 627
Fibrinogen, 626–627
Fibrinolysin, 1030
Fibrinolysis, 648
Fibroblast growth factor, 100
Fibroblasts, 118
Fibrocartilage, 126–128
Fibrocystic disease, 1048, 1059
Fibroids, 1060
Fibromyalgia syndrome, 307; 65–66 AM
Fibrosis, 136, 299, 306
Fibrous joint, 247
Fibrous pericardium, 657
Fibrous protein, 52, 53
Fibrous skeleton, 664
Fibrous tunic, 541
Fibula, 238–239
Fibular collateral ligament, 263
Fight or flight response, 504–505, 612, 613
Filariasis, 710, 756, 793
Filgrastim, 644
Filiform papillae, 536
Filling time, 681
Filtrate, 947
Filtration, 71, 75, 79, 947
    capillary exchange, 707, 708–710
    glomerular filtration rate, 955–957
    by kidneys, 953, 954–958
Filtration membrane, 950
Filtration pressures, 954–955
    colloid osmotic pressure, 955
    hydrostatic pressure, 954–955
Filtration slits, 950, 954
Filum terminale, 407
Fimbriae, 1039
Final diagnosis, 8 AM
Finasateride, 1032
Fine touch and pressure receptors, 532
Fingers
    bones of, 233
    muscles of, 343–346
First-class lever, 313, 314
First-degree burn, 162
First-degree heart block, 121 AM
First heart sound, 677
First messenger, 582
First-order neurons, 482
Fissures, 188, 445; 43 AM
Fixators, 316
Fixed acids, 998
Fixed macrophages, 119, 760, 762, 765
Fixed ribosomes, 84, 85
Flagella, 83
Flagellum, spermatozoon, 1025
Flat bones, 168, 169
Flat feet, 240, 243
Flatus, 886
Flavin adenine dinucleotide (FAD), 906, 908
Flavin mononucleotide (FMN), 908
Flatworms, 29 AM
Flexion, joints, 251, 252
Flexor, 317
Flexor carpi radialis, 342, 343
Flexor carpi ulnaris, 342, 343
Flexor digiti minimi brevis, 346, 354
Flexor digitorum brevis, 354
Flexor digitorum longus, 353

Flexor digitorum profundus, 344
Flexor digitorum superficialis, 344
Flexor hallucis brevis, 354
Flexor muscles, 428
Flexor pollicis brevis, 346
Flexor pollicis longus, 344
Flexor reflex, 428, 429
Flexor retinaculum, 233, 344
Floaters, 550
Floating kidney, 944
Floating ribs, 221
Flocculonodular lobe, 461
Fluid and electrolyte balance, 985–997
    and aging, 1009
    and athletes, 994
    basic concepts in, 987–988
    and calcium, 994–995, 996
    and chloride, 996, 997
    disorders of, 990, 995
    electrolyte balance, 985, 991–997
    fluid balance, 985, 989–991
    fluid exchange with environment, 990
    fluid movement, intracellular and extracellular, 989
    fluid shifts, 990–991
    hormones in, 988
    interplay in, 988
    and magnesium, 995, 996
    and phosphates, 995–997
    potassium balance, 993–994, 996
    sodium balance, 991–993, 996
Fluid compartments, 986–987
Fluid connective tissue, 118, 124
    blood, 124
    lymph, 124
Fluid shifts, 990–991
Fluoxetine (Prozac), 498
Flutamide, 1032
FMN (flavin mononucleotide), 908
Foam cells, 697
Focal calcification, 696
Focal distance, 547
Focal point, 547
Focal seizures; 84 AM
Folacin, 930
Folia, 461
Folic acid, 930
Follicle cavity, 594
Follicles, 608–609
Follicle-stimulating hormone (FSH), 580, 588, 589, 1025
    and female reproductive system, 1037, 1049–1052, 1057
    and male reproductive system, 1032, 1033, 1057
Follicular cells, 609, 1037
Follicular epithelium, thyroid, 597, 599
Follicular fluid, 1037
Follicular phase, 1037
Folliculitis, 156, 163
Follitropin. See Follicle-stimulating hormone (FSH)
Fontanels, 211
Food groups, basic, 926
Food pyramid, 926, 927. See also Nutrition
Foot, 239–240
    arches of, 240
    arteries of, 734
    bones of, 239–240
    clubfoot, 240
    muscles of, 351–355
Foramen lacerum, 202
Foramen magnum, 200
Foramen of Munro, 439
Foramen ovale, 203, 662, 743, 744, 836
Foramen rotundum, 203
Foramen spinosum, 203
Foramina (foramen), 188, 196
Forced breathing, 820, 832
Forceps delivery, 1086

Foreskin, 1030
Formed elements of blood, 124
Forming face, 86
Fornix
  of brain, 454
  of eye, 539
  of vagina, 1045
Fossa, 188
Fossa ovalis, 662, 743
Fourth heart sound, 678
Fourth ventricle, 439
Fovea, 544
Fovea capitis, 237, 261
Fovea centralis, 544
Fractionated blood, 626
Fracture hematoma, 184, 190
Fractures, 172, 190
  classification of, 186
  hip, 261
  repair of, 184–185
Fragile-X syndrome, 1097
Frank–Starling principle, 682
Freckles, 149
Free edge, 158
Freely permeable membrane, 71
Free macrophages, 119, 765
Free nerve endings, 528, 532
Free radicals, 34–35, 603
Free ribosomes, 84
Freeze-fracture technique, 89
Frequency
  light, 550
  sound, 564, 566
Frontal bone, 197, 201, 209
Frontal eye field, 448
Frontalis, 321, 322, 323
Frontal lobe, 446, 447
Frontal plane, 19
Frontal section, 20
Frontal sinuses, 201, 210
Frontal squama, 201
Fructose, 45, 890
FSH. See Follicle-stimulating hormone (FSH)
Fulcrum, 313
Full-thickness burns, 162
Full-thickness graft, 162
Functional groups, 44
Functional residual capacity, 822
Functional syncytium, 303
Functional zone, uterus, 1043
Fundus
  gallbladder, 879
  stomach, 861
  uterus, 1041
Fungi, 752, 793; 29 AM
Fungiform papillae, 536
Funny bone, 230
Furuncle, 156, 163

G

G₀ phase, 96
G₁ phase, 96
G₂ phase, 97
GABA (gamma aminobutyric acid), 389, 391, 395, 490
G actin, 277, 279
Galactose, 890
Galactosemia, 61; 22 AM
Galea aponeurotica, 321
Gallbladder, 10, 877, 879–880
  and bile, 880
  disorders of, 880
  functions of, 879
  structure of, 879
Gallstones, 880, 895
Gamete intrafallopian transfer (GIFT), 1080
Gametes, 1017
Gamma aminobutyric acid (GABA), 389, 391, 395, 490
Gamma efferents, 426, 430
Gamma-hydroxybutyrate, 611

Gamma-interferon, 767, 784
Gamma motor neurons, 426
Gamma rays, 15 AM
Ganglia, 370
Ganglion cells, 544, 555, 556
Ganglionic neurons, 504, 512
Gangliosides, 398
Gap junctions, 109–110, 386
Gases, 36
Gaster, 312
Gastrectomy, 868, 895
Gastric area, 762
Gastric artery, 731, 732
Gastric glands, 861, 863–865
  chief cells, 864, 865
  parietal cells, 864–865
Gastric inhibitory peptide (GIP), 867, 880, 881
Gastric juice, 864
Gastric lipase, 865
Gastric phase, of gastric secretion, 866, 867
Gastric pits, 863
Gastric stapling, 896; 165 AM
Gastric ulcer, 865
Gastric veins, 742
Gastrin, 850, 863, 880, 881, 896
Gastritis, 865, 895
Gastrocnemius, 286, 315, 351, 353
Gastroenteric reflex, 872
Gastroenteritis, 896; 168 AM
Gastroileal reflex, 872
Gastroscope, 868, 895
Gastrosplenic ligament, 762
Gastrulation, 1070–1072
Gated channels, 69, 70, 376–377
G cells, 865
G-CSF, 644
Gemelli, 347
Gender differences, skeletal system, 241
General adaptation syndrome (GAS), 612–614, 618, 789
  alarm phase, 612, 613
  exhaustion phase, 613–614
  resistance phase, 612–614
General interpretive area, 449
Generalized seizures, 84 AM
General senses, 482, 528, 530–534
  chemical sense, 533–534
  pain, 530–531
  pressure, 533
  temperature, 531
  touch, 530, 532–533
Generator potential, 529
Genes, 91
  activation of, 91–95
  genetic recombination, 1094
  hormone effects on, 583, 584
  mutations, 99
  mutations of, 1094
  transcription, 91–93
  translation, 93–95
  and triplet code, 91
Gene splicing, 35, 36 AM
Gene therapy, 34–37 AM
Genetic code, 91
Genetic disorders. See Chromosomal abnormalities
Genetic engineering, 103; 34–37 AM
Genetic recombination, 1094
Genetics. See also Inheritance
  chromosomal abnormalities, 1094, 1097
  chromosomes, 1090–1095
  dominant traits, 1091, 1092
  expressivity, 1092
  genotype, 1090, 1091–1092
  Human Genome Project, 1095–1096
  penetrance, 1091–1092
  phenotype, 1090, 1091–1092, 1093
  recessive traits, 1091, 1092
  study of, 1065

Genetic stutter, 1097
Geniculate ganglion, 469
Genio-, 317
Genioglossus, 325
Geniohyoid, 327, 328
Genital warts, 1056; 46, 191 AM
Genitofemoral nerve, 420, 422, 1019
Genome, 1095
Genomic imprinting, 1094
Genotype, 1090, 1091–1092
Gentamicin, 571
Geriatricians, 1089–1090
Geriatrics, 1089
Germ cells, 66
Germinal center, 759
Germinal epithelium, 1036
Germinative (basal) cells, 110, 146, 534
Germ layers, prenatal development, 1070–1072
Gestation, 1068
Gestational neoplasms, 1070, 1098
GH. See Growth hormone (GH)
GHP (glomerular hydrostatic pressure), 954
Giardiasis, 896; 28, 168 AM
Giemsa stains, 637
GIF (growth inhibitory factor), 873
GIFT (gamete intrafallopian transfer), 1080
Gigantism, 181, 190
Gingivae, 853
Gingival sulcus, 856
Gingivitis, 856
Gland cells, 107
Glands, 107
Glandular areas, 117
Glandular epithelium, 107–108
  endocrine glands, 115
  exocrine glands, 115–117
Glans
  clitoris, 1047
  penis, 1029, 1030
Glassy membrane, 154, 155
Glaucoma, 546, 572
Glenohumeral joint. See Shoulder joint
Glenohumeral ligaments, 259
Glenoid cavity, 230
Glenoid labrum, 258
Glial cells. See Neuroglia
Glial growth factor, 77 AM
Glial maturation factor, 77 AM
Gliding joints, 254, 257
Gliding movement, joints, 250
Global aphasia, 449
Globular protein, 52
Globulins, 626–627
Globus pallidus, 451
Glomerular capillaries, 950
Glomerular filtration, 954–958
Glomerular filtration rate, 955–957
  and blood pressure, 955
  maternal, 1083
  regulation of, 956–957
Glomerular hydrostatic pressure (GHP), 954
Glomerulonephritis, 952, 980, 1007
Glomerulus, 601, 946
Glosso-/-glossal, 317
Glossopharyngeal nerve, 470–471, 474
Glottis, 803, 804
Glucagon, 606, 608
  effects of, 606, 608
Glucocorticoids, 344, 589, 602, 617, 924
  abnormalities of, 602
  effects of, 602
  and stress response, 612
  types of, 602
Gluconeogenesis, 608, 877, 912
  and liver, 924

Glucose, 41, 44, 45
  absorption of, 922
  regulation by liver, 922
Glucose-dependent insulinotropic peptide, 880
Glucose-6-phosphate, 904
Glucose-sparing effect, 590, 602
Glutamate, 391, 392, 530
Gluteal group, 346–349
Gluteal lines, 234
Gluteal nerves, 422
Gluteal region, 16
Gluteus, 16
Gluteus maximus, 347, 348
Gluteus medius, 347, 349
Gluteus minimus, 347, 349
Glycerides, 47, 48–49
Glycerol, 48
Glycine, 52, 391
Glycocalyx, 70
Glycogen, 46, 294
Glycogenesis, 912
Glycogenolysis, 912
Glycolipids, 47, 50–51, 70
Glycolysis, 88, 275, 904, 905
  and ATP production, 294
  energy yield of, 910–911
  process of, 904, 905
  skeletal muscle contractions, 294
Glyconeogenesis, and liver, 923–924
Glycoproteins, 56, 70
Glycosaminoglycans, 109
Glycosuria, 607, 618, 959
GM-CSF, 644
Gnostic area, 449
Goblet cells, 116–117
Goiter, 616, 618; 99 AM
Golgi apparatus, 66, 67, 85–86
  and exocytosis, 86
  vesicles of, 86
Golgi tendon organs, 428, 533
Gomphosis, 247
Gonadal arteries, 733
Gonadal steroid-binding globulin (GBG), 1034
Gonadal veins, 740
Gonadotropes, 1049
Gonadotropin-releasing hormone, 588, 589, 1025
  and female reproductive cycle, 1049, 1050–1052, 1057
  and male reproductive system, 1032–1033, 1057
Gonadotropins, 589
  follicle-stimulating hormone (FSH), 580, 589, 590
  luteinizing hormone (LH), 580, 589
Gonads, 8, 608–610, 1017
Gonorrhea, 1040, 1056, 1059; 23, 190 AM
"Goose bumps," 153
Gout, 919, 938; 12, 171 AM
G-proteins, 553, 582
Gracilis, 317, 347
Graded potentials, 377–380
  characteristics of, 379
  compared to action potentials, 384
  depolarization, 379
  hyperpolarization, 379
  postsynaptic potentials, 392–395
  repolarization, 379
Graft rejection, 777
Graft-versus-host (GVH) disease, 793; 134 AM
Grams per deciliter (g/dl), 17 AM
Grand mal seizures, 84
Granular leukocytes (granulocytes), 637
Granulation tissue, 160, 163
Granulosa cells, 1037
Graves' disease, 597; 99 AM

Gravity, and muscle contractions, 293
Gray commissures, 413
Gray matter, 363, 370, 399
  spinal cord, 413
Gray ramus, 414, 508
Great cardiac veins, 668
Great cerebral vein, 734
Greater curvature, 861
Greater horns (greater cornua), 207
Greater omentum, 851
Greater sciatic notch, 234
Greater trochanter, 237–238
Greater tubercle, 230
Greater vestibular glands, 1046
Greater wings, 203
Great saphenous vein, 668
Great vessels, 656
Green cones, 554
Greenstick fracture, 186
Gross anatomy, 3
  specialties of, 3
Ground substance, 117, 120
Growth
  hormones of, 610
  of organisms, 2
Growth-associated protein 43 (GAP-43), 77 AM
Growth factors, 100, 577
Growth hormone (GH), 100, 126, 580, 588, 589–591, 610, 617
  abnormalities of, 590
  actions of, 589–591
  and bone, 181, 182
  control of production, 590
  and skeletal muscles, 296
  stimulation of, 590
Growth hormone–inhibiting hormone, 590
Growth hormone–releasing hormone, 590
Growth-inhibitory factor (GIF), 783
GTP (guanosine triphosphate), 59, 551, 906
Guanine, 56, 91
Guanosine diphosphate, 906
Guanosine triphosphate (GTP), 59, 551, 906
Gubernaculum testis, 1018
Guillain–Barré syndrome, 370, 401; 74 AM
Gums, 853
Gustation, 528, 536–538
  and aging, 538
  gustatory discrimination, 537–538
  gustatory pathways, 536
  gustatory receptors, 536
  taste sensations, types of, 537–538
Gustatory cells, 536
Gustatory cortex, 447
Gynecomastia, 602
Gyri, 445–446

H

Hair, 152–155
  color of, 153
  follicle structure, 154–155
  functions of, 153
  growth cycle, 155
  hair follicles, 152, 153, 154–155, 160
  hair loss, 155
  hair production, 154–155
  types of, 153
Hair bulb, 153
Hair cells, 557, 561, 564, 568, 571
Hair follicles, 6
Hair matrix, 154
Hair papilla, 153
Half-life, 31; 15 AM
Hallucis, 317

Hallux, 16, 240
Hamate bone, 232
Hamstrings, 349
Hand
  bones of, 233
  muscles of, 339–346
Hansen's disease (leprosy), 432; 42, 78 AM
Haploid cells, 1023
Haptens, 780
Hard keratin, 153
Hard palate, 205, 801, 853
Hassall's corpuscles, 762
Hatching, 1068
Haustra, 882
Haustral churning, 886
Haversian canal (central canal), 172–173
Haversian system (osteon), 172
Hct (hematocrit), 628, 634, 650
HDLs (high-density lipoproteins), 916
Head, 188
  bones of. See Skull
  epididymis, 1026
  femur, 237
  fibula, 238
  humerus, 230
  muscles of, 320–325
  pancreas, 872
  radius, 231
  ribs, 221
  spermatozoon, 1025
  thick filaments, 278
  ulna, 231
  veins of, 736
Headaches, 72 AM
Head fold, 1075
Hearing, 528, 557, 564–571
  auditory pathways, 568–570
  auditory sensitivity, 570–571
  cochlear duct and, 564–565
  disorders of, 571
  and organ of Corti, 565
  process of, 568
  sensory receptors, 564
  and sound, 568, 569
Heart, 8, 654–690
  and aging, 744
  atrioventricular node, 671, 672
  autonomic innervation, 682, 683–685, 686
  blood supply to, 664–668, 722
  and body temperature changes, 686
  bundle branches, 671, 672
  cardiac cycle, 674–678
  cardiac muscle tissue, 302–303, 660
  cardiodynamics, 680–686
  conducting system, 662, 668, 670–672
  connective tissues of, 664
  contractility. See Cardiodynamics
  disorders of, 657, 667–668, 673, 677, 679, 686, 688
  electrocardiogram (ECG/EKG), 673–674
  endocrine cells of, 605
  exercise and cardiac output, 685–686
  fibrous skeleton, 664
  heartbeat, 668–678
  heart rate, 683–685
  heart wall, 658–660
  hormonal effects on, 685
  left atrium, 662
  pericardium, 656–657
  right atrium, 660–662
  right ventricle, 662
  sinoatrial node, 671–672
  sounds of, 677–678
  superficial anatomy of, 657–658
  valves of, 663–664, 665
Heart attacks, 649, 679

coronary thrombosis, 679
  enzymes tested for, 679
  myocardial infarction, 679
  treatment for, 679
Heartbeat, 668–678
  and action potentials, 668–670
  and calcium ions, 670
  contractile cells, 668–670
  refractory period, 668
Heart block, 688; 121 AM
Heartburn, 851
Heart failure, 677, 683, 688, 722
Heart murmur, 678
Heart rate, 683–685
  abnormal rate, 686
  atrial reflex, 686
  autonomic innervation, 686
  and cardiac output, 686
  and hormones, 686
  and ion concentration, 686
  and sinoatrial node, 686
Heart wall, 658–660
  endocardium, 658
  epicardium, 657, 658
  myocardium, 658
Heat, meaning of, 37
Heat capacity, 40
Heat exhaustion, 935, 938
Heat-gain center, 934
Heat-loss center, 934
Heat production. See Thermoregulation
Heat stroke, 935, 938
Heavy chains, 778
Heavy metal poisoning, 365, 370
Heimlich maneuver, 805, 840
Helicobacter pylori, 865
Helium, 32
Helminths, 29 AM
Helper T cells, 756, 773, 776–777, 787
Hemangioma
  capillary, 151, 163
  cavernous, 151, 163
Hematocrit (Hct), 628, 634, 650
Hematologists, 633
Hematuria, 631, 650, 957, 980; 178 AM
Heme units, 629
Hemiazygos vein, 738
Hemidesmosome, 109, 110
Hemispheric lateralization, 450–451
  categorical hemisphere, 451
Hemocytoblasts, 628
Hemodialysis, 970, 980
Hemoglobin, 52, 53, 148, 149, 628, 629–630
  abnormalities of, 631
  and BPG, 828
  and carbon dioxide transport, 829
  fetal, 828–829
  function of, 630
  heme units, 629
  and oxygen transport, 826–829
  and pH, 827–828
  recycling of, 631–633
  structure of, 629–630
  and temperature, 828
Hemoglobin buffer system, 1000
Hemoglobin concentration (Hb), 634
Hemoglobin saturation, 826–828
Hemoglobinuria, 631, 650
Hemolysis, 74
Hemolytic anemia, 115 AM
Hemolytic disease of the newborn, 638, 650
Hemolytic jaundice, 113 AM
Hemolyze, 631, 636
Hemophilia, 649, 650, 1095
Hemopoiesis, 627–628, 757
  process of, 627–628
Hemopoiesis-stimulating factor, 783
Hemorrhage

epidural, 443, 475
  subdural, 443, 475
Hemorrhagic anemia, 115 AM
Hemorrhaging, cardiovascular response to, 719–723
Hemorrhoids, 700, 745, 884; 126 AM
Hemosiderin, 633
Hemostasis, 645–649
  abnormal, 649
  coagulation phase, 645, 646–648
  drugs effecting coagulation, 649
  platelet phase, 645–646
  vascular phase, 645
Henry's law, 822–824
Heparin, 119, 641, 647
Hepatic artery, 732, 877
Hepatic ducts, 877
Hepatic flexure, 884
Hepatic portal system, 741–742
Hepatic portal vein, 877
Hepatic triads, 877
Hepatic veins, 740, 877
Hepatitis, 877, 878, 895; 165–166 AM
Hepatocytes, 874
Hepatopancreatic sphincter, 879
Hering–Breuer reflexes, and respiration, 835
Hernia, 356, 1020; 68–69 AM
  types of, 335
Herniated disc, 257, 266
Herpes, 1056; 191 AM
Herpes varicella-zoster virus, 431
Hertz (Hz), 566
Heterologous marrow transplant, 651; 116 AM
Heterotopic bones, 175; 54 AM
Heterozygous, 1091
High-ceiling diuretics, 179 AM
High-density lipoproteins (HDLs), 916
High-energy bond, 57, 87
High-energy compounds, 44, 57–59, 87, 391
  adenosine triphosphate (ATP), 87
Higher-order functions
  consciousness, 496–498
  memory, 494–496
High-frequency sound, 566
Hilus
  bronchi, 806
  kidney, 944
  liver, 874
  lymph nodes, 760
  spleen, 762
Hinge joints, 254, 257
Hip fractures, 265
Hip joint, 260–261
  articular capsule, 260–261
  fractures of, 261
  ligaments of, 261
  stabilization of hip, 261
  structure of, 261
Hippocampus, 363, 454
Hirsutism, 48 AM
Histamine, 119, 391, 641
Histidine, 919
Histology, 3, 26, 107
Histones, 90
HIV (human immunodeficiency virus), 790, 793
Hives, 152, 789
HLAs (human leukocyte antigens), 773
Hodgkin's disease (HD); 133 AM
Holocrine secretion, 116, 156
Holoenzyme, 61; 21 AM
Homeostasis, 4, 12–15
  autoregulation, 12
  and disease, 14; 5–6 AM
  extrinsic regulation, 12
  negative feedback, 12–14
  positive feedback, 14–15
  regulation of, 12–15

Homologous chromosomes, 1090
Homozygous, 1091
Horizontal cells, 544
Horizontal plate, palatine bones, 205
Hormonal stimuli, 584
Hormones, 12, 54, 115, 577, 578–584, 585. *See also specific hormones*
 abuse in athletics, 611
 adrenocorticotropic hormone (ACTH), 149, 588, 589, 933
 and aging, 614
 aldosterone, 601, 965, 988, 994
 amino acid derivatives, 579
 androgens, 589, 602, 608, 610, 611
 antidiuretic hormone (ADH), 591
 atrial natriuretic peptide (ANP), 605
 and bone growth/maintenance, 181–182
 calcitriol, 604
 and cardiovascular regulation, 715–717
 and cell membrane, 582–583
 classification of, 578–580
 corticosteroids, 600
 and digestion, 850, 864–865, 880–881
 epinephrine (E), 603
 erythropoiesis-stimulating hormone, 634
 erythropoietin (EPO), 604
 estrogens, 609, 611
 of female reproductive system, 1049–1052
 in fluid and electrolyte balance, 988
 follicle-stimulating hormone (FSH), 580, 588, 589
 functions of, 580–584
 gastrointestinal, 880–881
 and gene activity, 583, 584
 glomerular filtration rate and, 956
 glucocorticoids, 589, 602
 gonadotropin-releasing hormone, 589
 gonadotropins, 589
 and G-proteins, 582
 and growth, 610
 growth hormone, 588, 589–591
 growth hormone–inhibiting hormone, 590
 growth hormone–releasing hormone, 590
 and heart, 682, 685
 heart, effects on, 685
 of immune system, 782–784
 inhibiting hormones, 588
 insulin, 606
 insulin-like growth factors (IGFs), 590
 integration with other systems, 614–615
 and intercellular receptors, 583–584
 leptin, 609, 933
 lipid derivatives, 580
 and liver, 879
 local, 48
 luteinizing hormone (LH), 589
 of male reproductive system, 1032–1034
 mechanisms of action, 580–584
 melanocyte-stimulating hormone, 591
 melatonin, 603
 mineralocorticoids, 601
 neurotransmitters, 391
 norepinephrine (NE), 603
 oxytocin, 591–592
 patterns of hormonal interaction, 610

peptide hormones, 579–581
 of placenta, 1075
 prolactin, 589
 prolactin-inhibiting hormone, 589
 regulatory hormones, 459, 585
 releasing hormones, 588
 renin, 604–605
 and skeletal muscles, 296
 and stress, 612–614
 testosterone, 608, 611
 of thymus, 762
 thyroid-stimulating hormone (TSH), 580, 588, 589
Horner's syndrome, 510; 89 AM
Horns, spinal cord, 412
Howship's lacunae, 174
Human body, levels of organization, 4–11
Human chorionic gonadotropin (hCG), 1075
Human Genome Project (HGP), 1095–1096
Human immunodeficiency virus (HIV), 790, 793
Human leukocyte antigens (HLAs), 773
Human physiology, 3
Human placental lactogen (hPL), 1075
Human tetanus immune globulin, 289
Humeral circumflex arteries, 727
Humeroradial joint, 260
Humeroulnar joint, 230, 260
Humerus, 178, 180, 230, 231
Humoral immunity (antibody-mediated immunity), 641, 756, 771
Humoral stimuli, 584
Huntington's disease, 498, 500; 87 AM
Hyaline cartilage, 126, 127, 128
Hyaluronan (hyaluronic acid), 109
Hyaluronidase, 1067, 1068
Hybridoma, 793; 142 AM
Hydatidiform mole, 1070
Hydration sphere, 40
Hydro cephalus, 443, 444, 475
Hydrochloric acid, 42–43, 864
Hydrocortisone, 602
Hydrogen, 30, 31
Hydrogen bonds, 35–36
Hydrogen ions, 42
 in body fluids, 42
Hydrogen ion secretion, 965–966
Hydrogen peroxide, 640
Hydrolysis, 38, 40, 45, 46, 48, 887
Hydrophilic molecules, 41
Hydrophobic molecules, 41
Hydrostatic pressure, 74, 704, 708
 of interstitial fluid, 708
 in kidney, 954–955
Hydroxide ion, 42
Hydroxyapatite, 170
Hydroxyl group, 44
Hydroxyurea, 630
Hymen, 1046
Hyoglossus, 325
Hyoid bone, 207
Hypercalcemia, 686, 995, 996
Hypercapnia, 834, 840, 1005
Hyperchloremia, 996
Hypercholesterolemia, 697
Hyperchromic red blood cells, 115 AM
Hyperextension, joints, 251, 252
Hyperglycemia, 607, 618
Hyperkalemia, 398, 401, 686, 995, 996, 1011
Hyperkeratosis, 44 AM
Hypermagnesemia, 996
Hypernatremia, 990, 996, 1011
Hyperopia, 548, 572
Hyperostosis, 190; 55–56 AM

Hyperparathyroidism, 598, 995; 102 AM
Hyperphosphatemia, 996
Hyperpnea, 820
Hyperpolarization, 379, 387
Hyperreflexia, 432; 79–80 AM
Hypersomnia, 500; 86 AM
Hypersplenism, 793; 134 AM
Hypertension, 706, 745; 126–127 AM
 portal, 877
Hyperthyroidism, 597; 99 AM
Hypertonic solution, 74
Hypertrophic cardiomyopathy (HCM),118 AM
Hypertrophy, skeletal muscles, 298
Hyperuricemia, 919, 938
Hyperventilation, 835, 1006–1007
Hypervitaminosis, 929, 930, 938
Hypervolemic, blood volume, 637, 650
Hypesthesia, 572
Hypnotic drugs, 87 AM
Hypoaldosteronism, 601; 102 AM
Hypocalcemia, 686, 995, 996
Hypocalcemic tetany, 598, 618
Hypocapnia, 835, 840, 1006
Hypochloremia, 996
Hypochromic red blood cells, 115 AM
Hypodermic needle, 152, 163
Hypodermis. *See* Subcutaneous layer
Hypogastric plexus, 517, 1042
Hypoglossal canals, 200
Hypoglossal nerves, 200, 325, 473, 474
Hypoglycemia, 619
Hypogonadism, 589, 617
Hypokalemia, 686, 965, 995, 996, 1011
Hypomagnesemia, 996
Hyponatremia, 990, 996, 1011
Hyponychium, 158
Hypoparathyroidism, 598; 99 AM
Hypophosphatemia, 996
Hypophyseal fossa, 203
Hypophyseal portal system, 587–588
Hypophysis. *See* Pituitary gland
Hyporeflexia, 432; 79 AM
Hyposplenism, 793; 134 AM
Hypotension, 706, 745; 127 AM
Hypothalamus, 12, 436, 456, 457–459
 and endocrine activity, 585
 functions of, 457–459
 and thermoregulation, 934
Hypothermia, 935, 938
 accidental, 173 AM
 induced, 936
Hypothesis, 2–3 AM
Hypothyroidism, 597; 99 AM
 congenital, 616, 618
Hypotonic solution, 74
Hypoventilation, 834, 1005
Hypovolemic, blood volume, 637, 650
Hypoxia, 634, 650, 814, 840
Hysterectomy, 1045
H zone, 275, 276

I

I band, 275, 276–277
IGFs (insulin-like growth factors), 590
Ileocecal valve, 869, 882
Ileostomy, 896; 167 AM
Ileum, 869, 872
Iliac arteries, 731, 733
Iliac crest, 235
Iliac fossa, 235
Iliac spines, 234

Iliac tuberosity, 235
Iliacus, 347, 349
Iliac vein, 740
Ilio-, 317
Iliococcygeus, 334
Iliocostalis group, spine, 328, 329
Iliofemoral ligament, 261
Iliohypogastric nerve, 422
Ilioinguinal nerve, 422, 1020
Iliopsoas, 347, 349
Iliotibial tract, 348
Ilium, 234
Image reversal, visual, 549
Immature delivery, 1086
Immediate hypersensitivity, allergies, 788
Immovable joints. *See* Synarthroses (immovable joints)
Immune complex, 780
Immune complex disorders, 793, 952; 143 AM
Immune disorders, 793
 allergies, 787, 788–789
 autoimmune disorders, 787–788
 immune complex disorders, 793, 952; 143 AM
 immunodeficiency diseases, 787
Immune response, 634, 752
Immune surveillance, 641
Immune system, manipulation of, 784
Immunity, 752
 acquired immunity, 771–772
 and aging, 789
 antibodies, 777–781
 B cells, 777–781
 cell-mediated, 756, 771
 and colony-stimulating factors, 784
 complement system, 767–769
 fever, 770
 immunological surveillance, 766
 inflammation, 767–770
 innate immunity, 771
 integration with other systems, 791, 792
 interferons, 767
 and interferons, 783
 and interleukins, 783–784
 macrophages, 765
 and memory cells, 772
 microphages, 765
 natural killer (NK) cell activation, 766
 nonspecific defenses, 638, 752, 763–771
 passive immunity, 772
 patterns of immune response, 784–785
 phagocytes, 764, 765, 769
 phagocytosis, 766
 physical barriers, 764, 765
 proteins of immune system, 51
 resistance, development of, 785–787
 response to antigens, 781–782, 788–789
 specific defenses, 639, 752, 763
 specificity of, 772
 and stress, 789
 T cells, 773–777
 tolerance, 772–773
 and tumor necrosis factors, 784
 versatility of, 772
Immunization, 136–138 AM
 active and passive, 793
 tetanus, 289
Immunodeficiency diseases, 787, 793
 AIDS, 788, 790, 793
 severe combined immunodeficiency disease (SCID), 788
Immunoglobulins, 626, 756. *See also* Antibodies
 IgA, 778, 780, 781, 786

IgD, 778, 780, 781
IgE, 778, 780, 781
IgG, 778, 780, 781, 782, 786–787
IgM, 778, 780, 781, 782
Immunological competence, 785
Immunological escape, 766
Immunological surveillance, 756, 764
Immunosuppression, 777, 793
Immunosuppressive drugs, 777, 793
Immunotherapy, 138; 40 AM
Impacted teeth, 857
Impermeable membrane, 71
Implantation, prenatal develop-
    ment, 1068–1072
Impotence, 1053, 1060
Inactivated vaccines, 136, 137 AM
Incision, 158; 48 AM
Incisors, 857
Inclusions, 80
Incomplete dominance, 1091
Incomplete proteins, 926
Incomplete tetanus, 287
Incontinence, 977, 980
Incus, 559
Indole, 886
Induced abortions, 1083
Induced active immunity, 772
Induced passive immunity, 772
Inducer T cells, 756
Induction, and development,
    1067–1068
Inert gases, 32
Infancy, 1086, 1088
Infants, and thermoregulation, 936
Infarct, 649
Infection, 135; 6 AM
    sepsis, 162
Infectious agents; 24 AM
Inferior, 18
Inferior abdominal oblique, 323
Inferior angle, scapula, 230
Inferior articular facets, 221
Inferior articular processes, 214
Inferior cerebellar peduncles, 462
Inferior colliculus, 460, 490
Inferior demifacets, 217
Inferior ganglion, 471
Inferioris, 317
Inferior meatuses, 801
Inferior mesenteric artery, 510, 732
Inferior mesenteric ganglion, 510
Inferior mesenteric plexus, 517
Inferior nasal conchae, 206
Inferior nuchal line, 200
Inferior orbital fissure, 205
Inferior ramus, 234
Inferior rectus, 323
Inferior sagittal sinus, 441
Inferior section, 19
Inferior temporal lines, 200
Inferior vena cava, 660, 739–741
Infertility, 1040, 1079, 1098
    and sperm count, 1065
Infertility treatment, 1079–1080
    artificial insemination, 1079
    fertility drugs, 1079–1080
    gamete intrafallopian transfer
        (GIFT), 1080
    surrogate mother, 1080
    in vitro fertilization, 1080
    zygote intrafallopian tube trans-
        fer (ZIFT), 1080
Infestations, 6, 29 AM
Inflammation, 135–136, 764,
    767–770
    effects of, 769–770
    glucocorticoids as anti-inflamma-
        tory, 602
    process of, 769–770
    stimulation of, 767–769, 780
Inflammatory bowel disease, 896;
    167 AM
Inflammatory diseases (arthritis),
    60 AM
Inflammatory T cells, 756

Inflation reflex, 835
Information processing, nervous
    system, 392–396
    reflex arc, 423
Infraorbital foramen, 205
Infraspinatus muscles, 259, 339, 340
Infraspinous fossa, 230
Infundibulum, 436, 457, 587, 1039,
    1040
Ingestion, 846
Inguen, 16
Inguinal, 317
Inguinal canal, 335, 1019–1020;
    68 AM
Inguinal hernia, 335, 356, 1020
Inguinal region, 16
Inheritance, 1092–1095
    individual variations in,
        1093–1095
    meaning of, 1065
    polygenic inheritance, 1092, 1093
    prediction of, 1092–1093
    simple inheritance, 1093
Inhibin, 589, 608, 1025
    and reproductive system, 1057
Inhibiting hormones, 588
    growth hormone–inhibiting hor-
        mone, 590
    prolactin-inhibiting hormone, 589
Inhibition, of spinal reflexes, 430
Inhibitors, 22 AM
Inhibitory neurotransmitters, 387
Inhibitory postsynaptic potentials
    (IPSPs), 393
    summation of, 394–395
Initial segment, of axon, 363
Innate immunity, 771
Innate reflexes, 424
Inner cell mass, 1068, 1071
Inner ear, 202, 557, 560–561
    and equilibrium, 560
Inner segment, photoreceptors, 550
Innervation, 320
    bone, 180
    heart, 683–685
    kidneys, 946
    skeletal muscles, 273, 280, 320
    skin, 152
    uterus, 1042
Innominate artery, 726
Innominate bones (ossa coxae),
    232, 234–235
Innominate vein, 736, 738
Inorganic compounds, 39–43
    acids, 42–43
    bases, 42, 43
    buffers, 43
    salts, 43
    types in human body, 40
    water, 40–42
Inositol triphosphate, 392, 582
Insensible perspiration, 147, 934,
    990
Insertion, skeletal muscle, 316
Insomnia, 500; 86 AM
Inspection, 26; 7 AM
Inspiratory capacity, 822
Inspiratory center, 832
Inspiratory reserve volume (IRV),
    822
Insula, 446
Insulin, 70, 76, 586, 606–608, 617
    effects of, 606
    and growth, 610
Insulin-dependent cells, 607
Insulin-dependent (Type 1) dia-
    betes, 607, 618, 787
Insulin-dependent diabetes melli-
    tus (IDDM), 103–104 AM
Insulin-independent cells, 607
Insulin-like growth factors (IGFs),
    590
Insulin pump, 103 AM
Integral proteins, 69
Integration center, 12

Integrative centers, cerebral cortex,
    448–450
Integrative effects, hormones, 610
Integument, 136, 144
Integumentary system, 5, 6,
    143–166; 42–50 AM
    accessory structures, 144, 152–158
    and aging, 160
    autonomic control of, 518
    burns of, 162
    components of, 144
    cutaneous membrane, 144
    dermatitis, 152
    dermis, 150–152
    disorders of, 47–50 AM
        treatment of, 49–50 AM
    drug administration through, 150
    epidermis, 144, 145–150
    examination of, 42–47 AM
    and exercise, 300
    functions of, 6, 145
    glands of, 155–158
    hair, 152–155
    injury and repair of, 158–160
    innervation of, 152
    nails, 158
    organs of, 6
    and puberty, 1089
    relationships with other systems,
        160–161
    skin cancer, 148
    smooth muscles of, 304
    subcutaneous layer, 144, 152
Intensity, sound, 564, 567
Intention tremors, 82 AM
Intention tumor, 475
Interatrial septum, 660
Intercalated discs, 132, 302–303,
    660
Intercellular cement, 108–109
Intercondylar eminence, 238, 239
Intercondylar fossa, 238
Intercostal arteries, 729
Intercostal muscles, 221, 330, 331,
    332
Intercostal veins, 738
Interferons, 764, 767, 783, 784
    types of, 767, 783
Interleukin-1, 770
Interleukin-6, 645
Interleukin-7, 757
Interleukins, 783–784
    functions of, 784
Interlobar arteries, 946
Interlobar veins, 946
Interlobular arteries, 946
Interlobular septa, 809, 876
Interlobular veins, 946
Intermediate cuneiform bone, 239
Intermediate-density lipoproteins
    (IDLs), 916
Intermediate fibers, 297, 298
Intermediate filaments, 81–82
Intermediate mass, 457
Intermuscular septa, 315
Internal abdominal obliques, 330,
    331, 332
Internal acoustic canal, 202
Internal anal sphincter, 884
Internal callus, 184, 190
Internal capsule, 447
Internal carotid arteries, 443,
    727–728
Internal elastic membrane, 692
Internal iliac arteries, 733, 1041
Internal intercostals, 330, 331
Internal jugular veins, 443, 734
Internal nares, 801
Internal occipital crest, 441
Internal os (internal orifice), 1041
Internal respiration, 814
Internal root sheath, 154, 155
Internal tension, 288–289
Internal thoracic artery, 668, 727,
    1047

Internal urethral sphincter, 975
Interneurons, 367, 428, 429, 430
Internodal pathways, 671
Internodes, 370
Internus, 316, 317
Interoceptors, 366, 530
Interossei dorsal, 354
Interossei plantar, 354
Interosseous membrane, 231, 238
Interphase, 96–97, 98
    G_0 phase, 96
    G_1 phase, 96
    G_2 phase, 97
    S phase, 96–97
Interproximal space, 858
Intersegmental reflex arcs, 425
Interspinales, 328, 329, 330
Interspinous ligament, 256, 257
Interstitial cells, 589, 608, 1021
Interstitial cell–stimulating hor-
    mone, 589
Interstitial fluid, 66, 124, 986
    compared to plasma, 624
Interstitial fluid colloid osmotic
    pressure, 708
Interstitial growth, 125–126
    bone growth, 176
Interstitial lamellae, 173
Intertransversarii, 328, 329, 330
Intertrochanteric crest, 237, 238
Intertrochanteric line, 237, 238
Intertubercular groove, 230
Intertubercular sulcus, 230
Interval training, 299
Interventricular foramen, 439
Interventricular septum, 660
Intervertebral articulations,
    256–258
    movements, 257
Intervertebral discs, 213, 247, 256
    problems of, 257
Intervertebral foramina, 214
Intestinal crypts, 871
Intestinal glands, 871
Intestinal juice, 871
Intestinal phase, of gastric secre-
    tion, 865, 866, 867–868
Intestinal trunks, 755
Intestinal villi, 869–871
Intracapsular ligaments, 249
Intracellular fluid, 80, 985, 986
Intrafusal muscle fibers, 426
Intramembranous ossification,
    175–176
Intramural ganglia, 505
Intramuscular injection, 349, 356
Intraocular pressure, 545
Intrapleural pressure, 817–818
Intrapulmonary bronchi, 809
Intrapulmonary pressure, 817
Intrauterine device (IUD), 1040,
    1055
Intravenous pyelogram (IVP), 175
    AM
Intrinsic, 317
Intrinsic factor, 634, 861, 864, 893,
    930
Intrinsic muscles, 316
    of foot, 353–355
    of hand, 344, 345–346
    of tongue, 854
Intrinsic pathway, 646
Introns, 92
Intubation, 805
Inulin, 955
Invasion, by malignant cells, 101,
    102
Inversion, joints, 253, 254
In vitro fertilization (IVF), 1080,
    1098
Involuntary contractions, 361
Involution, 761, 1038
Iodide ions, 595
Iodine, 31
    and thyroid hormones, 596–597

Ion absorption, 891–892
Ion concentration, and heart rate, 686
Ionic bonds, 33–34
Ionization, 40
Ion pumps, 76
Ions, 33
    ammonium, 886, 966
    iodide, 595
    reabsorption of, 961
    sodium, 34, 375, 966
Ipsilateral reflex arcs, 428
Iris, eye, 542
Iron, 31
    nutritional factors, 928–929
    and red blood cells, 633
Iron-deficiency anemia, 633; 11, 113–114, 115 [AM]
Irregular bones, 168, 169
Irreversible shock, 720, 721
Irritable bowel syndrome (IBS), 896; 167 [AM]
Ischemia, 355
Ischial ramus, 234
Ischial spine, 234
Ischial tuberosity, 234
Ischiocavernosus muscles, 333, 334, 1053
Ischiofemoral ligament, 261
Ischium, 234
Islets of Langerhans, 605
Isoleucine, 919
Isomers, 45
    pharmaceutical use of, 18 [AM]
Isometric contractions, 291, 292
Isoproterenol, 683
Isosmotic solution, 74
Isotonic contractions, 290–292
Isotonic solution, 74
Isotopes, 30–31
Isovolumetric contraction, 677
Isovolumetric relaxation, 677
Isozymes, 54
Isthmus
    thyroid gland, 593
    uterine tube, 1039, 1040
    uterus, 1041
Itching sensation, 533

J

Jaundice, 149, 632, 650; 42, 113 [AM]
Jejunum, 869, 871, 872
Joints, 6, 196, 245–269
    and aging, 265
    amphiarthroses (slightly movable joints), 246, 247
    articular form and function, 250–256
    classification of, 246–250
    diarthroses (freely movable joints), 246, 247–250
    dislocation of, 250
    disorders of, 265, 266
    elbow joint, 260
    factors stabilizing, 249–250
    hip joint, 260–261
    knee joint, 261–263
    mobility of, 250–254
    repair of, 265
    replacement of, 265
    shoulder joint, 258–259
    synarthroses (immovable joints), 246–247
Jugular foramen, 200, 202
Jugular ganglion, 471
Jugular notch, 222
Jugular trunk, 755
Jugular vein, 200, 202, 440, 734, 736
Jugular venous pulse, 736
Junctional folds, 281
Juvenile-onset diabetes, 618; 104 [AM]
Juxtaglomerular apparatus, 952, 956

K

Juxtaglomerular cells, 952
Juxtamedullary nephrons, 948, 950

Karyotyping, 103, 1090, 1095; 34 [AM]
Kayexolate, 995
Keloid, 160, 163
Keratin, 52, 53, 54, 112, 146
    soft and hard, 153
Keratinization, 147, 154
Keratinocytes, 145
Keratohyalin, 146
Keto acid, 918
Ketoacidosis, 619, 925, 938, 1007; 103 [AM]
Ketone bodies, 619, 924, 925, 1007; 103 [AM]
Ketonemia, 925, 938
Ketonuria, 925, 938
Ketosis, 925, 938
Kidneys, 8, 10, 604–605, 943–967
    blood flow analysis, 946
    blood supply to, 946
    collecting system, 966
    connective tissue of, 944
    disorders of, 952, 957–958, 974
    filtration, 953, 954–958
    functions of, 967–968
    hormones of, 604–605
    innervation of, 946
    nephron, 946–953
    reabsorption, 953, 959, 965, 966
    renal failure, treatment of, 970–971
    response to acidosis and alkalosis, 1003–1004
    secretion, 953–954, 965, 966
    sectional anatomy of, 944–946, 947
    superficial anatomy of, 944
    urine formation, 953–954
    vasa recta, 967–968
Kidney stones, 974
Kidney transplantation, 971
Killer T cells, 775
Kilocalorie, 932
Kinase, 582
Kinesins, 82, 365
Kinetic energy, 36, 37
Kinetochore, 91, 98
Kinocilium, 561
Klinefelter syndrome, 1097
Kneecaps, 168
Knee extensors, 349
Knee flexors, 349
Knee-jerk reflex, 426, 431
Knee joint, 262
    injuries to, 263
    knee injuries and cartilage, 128
    ligaments of, 262, 263
    locking of, 263
    structure of, 262
Krebs cycle (tricarboxylic acid, or citric acid, cycle), 293, 904
Kupffer cells, 765, 877
Kwashiorkor, 919; 6, 42, 170 [AM]
Kyphosis, 212, 223; 57 [AM]

L

Labeled line, 529
Labetalol, 683
Labia, 11
    oral cavity, 853
Labia majora, 1046
Labia minora, 1046
Labor
    contractions, 1083
    false labor, 1083
    premature, 1084–1086
    stages of, 1084, 1085

Lacerations, 48, 80 [AM]
Lacrimal apparatus, 539–541
Lacrimal bones, 206, 209
Lacrimal canals, 541
Lacrimal caruncle, 538
Lacrimal fossa, 201
Lacrimal gland, 539
Lacrimal lake, 541
Lacrimal puncta, 541
Lacrimal sac, 205, 539, 541
Lacrimal sulcus, 206
Lacrimal (tear) gland, 201
Lactase, 889, 890
Lactate dehydrogenase (LDH), 679
Lactate ion, 295
Lactation, 1047
Lacteals, 753, 871
Lactic acid, 44, 295–296
Lactic acidosis, 1007
Lactiferous duct, 1047
Lactiferous sinus, 1047
Lactose, 890
Lactose intolerance, 890, 895; 168–169 [AM]
Lacunae, 125, 170, 171, 1070
Lacus lacrimalis, 541
Lambdoidal suture, 196, 197
Lamellae, 171, 173
Lamellated (pacinian) corpuscles, 152, 532
Lamina densa, 950
Laminae, 213–214
Lamina propria, 129, 534, 799, 848
Laminectomy, 267
Langerhans cells, 146, 160, 787
Lanugo, 153
Laparoscopy, 896
Large bowel. See Large intestine
Large granular lymphocytes, 641, 756
Large intestine, 10, 882–887
    absorption in, 885–886
    cecum, 882
    colon, 882–884
    histology of, 884
    movements of, 886–887
    rectum, 882, 884
Large ribosomal subunit, 84, 94
Large veins, 699
Laryngeal elevators, 326
Laryngeal prominence, 803
Laryngeal spasms, 836
Laryngitis, 805
Laryngopharynx, 803
Larynx, 9, 803–805
    cartilages of, 803–805
    infection of, 805
    ligaments of, 803–805
    muscles of, 805
    sound production, 804–805
Lata, 317
Latent period, muscle twitch, 286
Lateral, 18
Lateral angle, scapula, 230
Lateral apertures, 443
Lateral border, scapula, 229
Lateral canthus, 538
Lateral condyles, 238
Lateral corticospinal tracts, 488, 491
Lateral cuneiform bone, 239
Lateral epicondyles, 230, 237, 238
Lateral excursion, 324
Lateral femoral cutaneous nerve, 420, 422
Lateral flexion, 253, 254
Lateral geniculate nucleus, 457
Lateral gray horns, 413
Lateral group, 457
Lateralis, 317
Lateral ligaments, 1041
Lateral malleolus, 238
Lateral masses, 204
Lateral menisci, 263

Lateral rectus, 323
Lateral rotation, 253
Lateral rotators, 347, 349
Lateral sacral crest, 218
Lateral spinothalamic tracts, 484, 485, 486, 530
Lateral sulcus, 446
Lateral supracondylar ridge, 237
Lateral tibial condyles, 238, 239
Lateral umbilical ligaments, 975
Lateral ventricles, 439
Lateral white column, 413
Late-stage HIV disease, 138 [AM]
Latissimus, 317
Latissimus dorsi, 316, 339, 340
Lauric acid, 47
Laxatives, 887
LDH (lactate dehydrogenase), 679
LDLs (low-density lipoproteins), 679, 696, 916
L-dopa, 445
Leads, 673
Leak channels, 69, 70, 373, 376
Lecithin, 47, 50
Left anterior descending artery, 666
Left atrioventricular (AV) valve, 662
Left atrium, 655
    structure of, 662
Left bundle branches, 672
Left colic flexure, 884
Left common iliac arteries, 733
Left coronary artery, 666
Left gastroepiploic artery, 731
Left primary bronchus, 806
Left pulmonary arteries, 662
Left pulmonary veins, 662
Left rotation, 253
Left thalamus, 436
Left ventricle, 655
    compared to right ventricle, 662–663
    structure of, 662
Legs, 237–239
Leiomyomas, 1060
Lens, eye, 546–549
Lens fibers, 546
Lentiform nucleus, 451
Lentigos, 149
Leprosy (Hansen's disease), 432; 42, 78 [AM]
Leptin, 609, 933
Lesser curvature, 861
Lesser horns (lesser cornua), 207
Lesser omentum, 851
Lesser sciatic notch, 234
Lesser trochanter, 237–238
Lesser tubercle, 230
Lesser vestibular glands, 1046
Lesser wings, 203
Lesion, 6, 42, 43–44 [AM]
Leucine, 919
Leukemia, 641, 650; 115–116 [AM]
Leukocytes. See White blood cells (WBCs)
Leukocytosis, 641, 650
Leukopenia, 641, 650
Leukotrienes, 47, 48, 580, 640, 783
Levator, 317
Levator ani, 334
Levator palpebrae superioris, 322, 538
Levator scapulae, 335, 336
Levator veli palatini, 326
Levels of organization, 4–11
    cellular level, 4, 5
    molecular level, 4, 5
    organism level, 4, 5
    organ system level, 4, 5
    tissue level, 4, 5
Levers, 313–314
    classes of, 313–314
LH. See Luteinizing hormone (LH)
Life stages, 1086–1090

Ligaments, 6, 119, 122
  acromioclavicular ligament, 259
  annular ligament, 260
  anterior cruciate ligament, 263
  of breast, 1047
  coracoacromial ligament, 259
  coracoclavicular ligament, 259
  coracohumeral ligament, 259
  elastic, 124
  of elbow joint, 260
  of femoral head, 261
  fibular collateral ligament, 263
  glenohumeral ligaments, 259
  of hip joint, 261
  iliofemoral ligament, 261
  injury to, 249
  intracapsular, 249
  ischiofemoral ligament, 261
  and joints, 249
  of knee joint, 262, 263
  of larynx, 803–805
  ligamentum teres, 261
  of ovaries, 1035
  patellar ligament, 263
  popliteal ligament, 263
  posterior cruciate ligament, 263
  pubofemoral ligament, 261
  radial collateral ligament, 260
  of shoulder joint, 259
  tibial collateral ligament, 263
  transverse acetabular ligament, 261
  types of, 249
  of uterus, 1041
Ligamentum arteriosum, 662, 743
Ligamentum flavum, 256, 257
Ligamentum nuchae, 216, 256, 257
Ligamentum teres, 261, 874
Ligands, 70, 77, 78
Ligases, 97
Light, and vision, 550
Light-adapted state, 554
Light chains, 778
Light micrograph (LM), 30 AM
Light microscopy, 65; 30 AM
Limbic lobe, 454
Limbic system, 452–454
  components of, 452–454
  functions of, 452
Limbs, lower, 237–241
  ankle, 239
  arteries of, 733–734
  femur, 237–238
  fibula, 238–239
  foot, 239–240
  leg, 237–239
  muscles of, 346–355
  patella, 238
  thigh, 237–238
  tibia, 238
  veins of, 739–740
Limbs, upper, 230–233
  arm, 230–232
  arteries of, 727
  carpals, 232–233
  forearm, 231–232
  hand, 233
  humerus, 230, 231
  muscles of, 339–346
  radius, 231–232
  ulna, 231, 232
  veins of, 737
Limbus, 541
Line, 188
Linea alba, 331
Linea aspera, 237, 238
Linear motion, joints, 250
Linea terminalis (pelvic brim), 235
Lines of cleavage, 151
Lingual frenulum, 854
Lingual lipase, 854, 890
Lingual papillae, 536, 854
Lingual tonsils, 759
Linoleic acid, 914
Linolenic acid, 914

Lipases, 887, 889, 890
Lipemia, 922
Lipid, 50
Lipid derivatives, 58, 580
  eicosanoids, 580
  steroid hormones, 580
Lipid metabolism, 913–916
  beta-oxidation, 913
  and energy production, 913–914
  lipid catabolism, 913–914
  lipid transport/distribution, 914–916, 917
  lipoproteins, 914–916
  synthesis of lipids, 914
Lipids, 46–51
  absorption of, 922
  of cell membrane, 68–69
  digestion of, 890–891
  eicosanoids, 47, 580
  fatty acids, 47–48
  glycerides, 47, 48–49
  glycolipids, 47, 50–51
  and liver, 877, 922, 924
  phospholipids, 47, 50–51
  prostaglandins, 47, 48
  steroids, 47, 49
  storage of, 168
  structural lipids, 51
Lipid solubility, 73
Lipofuscin, 499
Lipogenesis, 914
Lipolysis, 913
Lipoprotein lipase, 916
Lipoproteins, 627, 914–916
  chylomicrons, 916
  high-density lipoproteins (HDLs), 916
  intermediate-density lipoproteins (IDLs), 916
  low-density lipoproteins (LDLs), 916
  very low-density lipoproteins (VLDLs), 916
Liposomes, 150
Liposuction, 122, 138, 938; 171–172 AM
Lips, 853
Liquids, 36
Liquor folliculi, 1037
Lithium, 32
Lithotripter, 880, 974
Liver, 10, 874–879
  and absorptive state, 922
  and amino acids, 922, 924
  bile secretion, 877, 879, 880
  blood supply to, 874
  diseases of, 877, 878
  and gluconeogenesis, 924
  and glucose regulation, 922
  hematological activities of, 878–879
  histology of, 876–877
  and hormones, 879
  and ketone bodies, 924, 925
  and lipids, 922, 924
  liver lobules, 876–877
  metabolic activities of, 877–878
  and metabolism, 920–921, 922, 923–925
  and postabsorptive state, 923–925
  structure of, 874–875
Liver biopsy, 896; 165 AM
Liver cells, turnover rate, 60
Liver function tests, 165 AM
Liver scans, 896; 165 AM
Liver spots, 149
Lobar bronchi, 809
Lobes
  cerebral, 446, 447
  kidney, 946
  liver, 874
  lungs, 806–808
  mammary glands, 1047
  thyroid gland, 593
Lobotomy, 449

Lobules, 761
  of liver, 876–877
  of lung, 809–810
  of mammary glands, 1047
  of testes, 1021
  of thymus, 761
Local current, 379
Local hormones, 48, 577
Local potentials, 377–380
Local vasoconstrictors, 712
Local vasodilators, 712
Locking of knee, 263
Lockjaw, 289
Locus of gene, 1091
Long bones, 168, 169
Longissimus, 316
Longissimus group, 328, 329
Longitudinal arch, 240
Longitudinal fasciculi, 446
Longitudinal fissure, 446
Longitudinal layer, 848
Long reflexes, 520, 850
Long-term memories, 494, 495
Longus, 316
Longus capitis, 329, 330
Longus colli, 329, 330
Loop of Henle, 947, 949, 952, 960, 962–964, 968
Loose connective tissue, 118, 120–122
  adipose tissue, 121–122
  reticular tissue, 122
  structure of, 120–122
Lordosis, 212, 223; 57 AM
Low-density lipoproteins (LDLs), 679, 696, 916
Lower motor neuron, 487
Lower respiratory system, 798
Low-frequency sound, 566
LSD (lysergic acid diethylamide), 498
Lumbar arteries, 733
Lumbar curve, spine, 212
Lumbar enlargement, of spinal cord, 407
Lumbar plexus, 420–422
Lumbar puncture, 411, 431
Lumbar region, 16
Lumbar trunks, 755
Lumbar veins, 740
Lumbar vertebrae, 212, 214, 217–218
Lumborum, 317
Lumbricals, 346, 354
Lumbus, 16
Lumen, 109, 156
Lunate bone, 232
Lunate surface, 234
Lung cancer, 839, 840; 155–156 AM
Lungs, 9, 806–813
  alveolar ducts, 811–813
  blood supply to, 722, 813
  bronchi, 806
  bronchioles, 809–810, 811
  cancer of, 839, 840; 155–156 AM
  changes in lung perfusion, 831–832
  compliance of, 816–817
  lobes of, 806–808
  lobules of, 809–810
  surfaces of, 806–808
Lunula, 158
Luteal phase, 1037, 1038–1039
Luteinizing hormone (LH), 580, 588, 589
  and female reproductive cycle, 1049, 1052, 1057
  and male reproductive system, 1032, 1033–1034, 1057
Lutropin. See Luteinizing hormone (LH)
Luxation, 250, 266
Lyme disease, 784; 138 AM
Lymph, 118, 124, 752
Lymphadenopathy, 760, 793

Lymphatic capillaries, 752–753
Lymphatic system, 5, 9, 624, 706, 751–796; 129–144 AM
  cancer of, 760
  disorders of, 756, 762; 129, 130–144 AM
  examination of, 129–130 AM
  functions of, 9, 752
  lymphatic vessels, 752–756
  lymph nodes, 759–760
  lymphocytes, 756–758
  lymphoid tissues, 758–759
  organs of, 9, 752
  spleen, 762–763
  thymus, 760–762
Lymphatic vessels, 9, 124, 752–756
  deep lymphatics, 754
  lymphatic capillaries, 752–753
  right lymphatic duct, 755
  superficial lymphatics, 754
  thoracic duct, 755
  valves of, 754
Lymphedema, 756, 793, 989
Lymph glands, 760
Lymph nodes, 9, 759–760
  flow of lymph, 760
  functions of, 760
  lymphatic vessels connected to, 760
Lymphocytes, 118, 119, 124, 604, 605, 627, 637, 640, 641, 644, 752, 756–758, 787
  B cells, 641, 756
  circulation of, 756–757
  natural killer (NK) cells, 641, 756, 758
  production of, 757–758
  T cells, 641, 756
Lymphocytosis, 641
Lymphogranuloma venereum (LGV), 23, 190 AM
Lymphoid leukemias, 115–116 AM
Lymphoid nodule, 758–759
Lymphoid organs, 752
Lymphoid stem cells, 628, 633
Lymphoid tissues, 644, 752, 758–759
  infection of, 759
  mucosa-associated lymphoid tissue (MALT), 759
  tonsils, 759
Lymphokines, 782
Lymphomas, 760, 793; 133–134 AM
Lymphopenia, 641
Lymphopoiesis, 757
Lymphotoxin, 775, 783
Lysine, 891, 919
Lysosomal storage diseases, 87
Lysosomes, 66, 67, 77, 78, 86–87
Lysozymes, 539, 765, 1087

M
MAC (membrane attack complex), 767
Macrocytic cells, 115 AM
Macrophage-activating factor (MAF), 783
Macrophages, 79, 118–119, 638, 641, 765, 787
  fixed, 119, 760, 762, 765
  free, 119, 765
Macroscopic anatomy, 3
Macula densa, 952
Maculae, 563
Macula lutea, 544
Macules, 43 AM
Magnesium, 31
  in fluid and electrolyte balance, 996
  nutritional factors, 928
Magnocells (M cells), 555
Magnus, 316, 317
Major, 316, 317
Major calyx, 946

Major histocompatibility complex (MHC), 773–775
  proteins of, 70, 773
Major motor aphasia, 81 [AM]
Malabsorption syndromes, 893
Malathion, 390
Male climacteric, 1057
Male pattern baldness, 155, 163; 47 [AM]
Male pronucleus, 1067
Male reproductive system
  and aging, 1057
  autonomic control of, 519
  bulbourethral glands, 1029
  climacteric, 1057
  disorders of, 1018, 1029, 1060
  epididymis, 1025–1027
  hormones of, 1032–1034
  penis, 1030–1031
  prostate gland, 1029
  semen, 1029–1030
  seminal vesicles, 1029
  testes, 608, 609, 1017–1025
  urethra, 1027
Male sexual function, 1053
Malignant melanoma, 148, 163, 784; 46 [AM]
Malignant tumors, 101, 102. See also Cancer
Malleus, 558
Malnutrition, 919, 926
MALT (mucosa-associated lymphoid tissue), 759
Maltase, 890
Maltose, 890
Mamillary bodies, 454, 458
Mammary glands, 11, 157, 1047
  cancer of, 1048, 1059
  development in pregnancy, 1047
  hormonal stimulation of, 589
  milk production, 1047, 1087–1088
  pregnancy and development of, 1083
  structure of, 1047
Mammography, 1048, 1060
Mammotropin. See Prolactin (mammotropin)
Mandible, 175, 207, 209
Mandibular angle, 207
Mandibular body, 207
Mandibular canal, 207
Mandibular foramen, 207
Mandibular fossa, 202
Mandibular groove, 855
Mandibular notch, 207
Manganese, nutritional factors, 928
Mania, 87 [AM]
Mannitol, 179 [AM]
Manual region, 16
Manubrium, 222
Manus, 16
MAP (mean arterial pressure), 705
Marasmus, 919; 170 [AM]
Marfan's syndrome, 120, 181, 190, 695; 55 [AM]
Marginal branches, 664
Margination, 638
Marrow cavity, 169, 172, 173, 178
Mass, 30
Masseter, 324
Mass movement, large intestine, 886
Mass number, 31
Mass reflex, 432; 80 [AM]
Mast cells, 118, 119, 787
Mastectomy, 1048, 1060
Mastication, 856
  muscles of, 324
  process of, 858
Mastoid air cells, 202
Mastoid fontanel, 211
Mastoiditis, 202; 95 [AM]
Mastoid process, 202
Matrix
  of bone, 170–171, 173

of connective tissue, 117
of mitochondria, 87
Matter, 30
  states of, 36
Maturation-promoting factor (m-phase promoting factor), 100
Mature Graafian follicle, 1038
Maturing face, 86
Maturity, 1065
Maturity-onset diabetes, 104 [AM]
Maxillae, 205
Maxillary bones, 205, 209
Maxillary sinuses, 205, 210
Maxillary veins, 736
Maximus, 316, 317
M cells (magnocells), 555
M-CSF, 644
Mean arterial pressure (MAP), 705
Mean corpuscular hemoglobin concentration (MCHC), 634
Mean corpuscular volume (MCV), 634
Mechanically regulated channels, 376, 377
Mechanical processing, 846, 853
Mechanoreceptors, 482, 531–533
  baroreceptors, 532, 533, 534
  proprioceptors, 532, 533, 534
  tactile receptors, 532–533
Medastinum, 656
Medial, 18
Medial border, scapula, 229
Medial canthus, 538
Medial condyles, 238
Medial cuneiform bone, 239
Medial epicondyles, 230, 238
Medial geniculate nucleus, 457
Medial group, 457
Medialis, 317
Medial lemniscus, 483
Medial malleolus, 238
Medial menisci, 263
Medial rectus, 323
Medial rotation, 253
Medial supracondylar ridge, 237
Medial surface of lung, 808
Medial tibial condyles, 238, 239
Median aperture, 443
Median cubital vein, 626
Median eminence, 457, 458
Median nerve, 344, 419
Median sacral crest, 218
Mediastinal arteries, 729
Mediastinum, 20, 21, 22
Medical anatomy, 3
Medium-sized veins, 699
Medius, 317
Medulla
  adrenal, 600, 601, 602, 603
  concentration gradient of, 963–964
  hair, 153
  kidney, 946
  lymph node, 760
  thymus, 760
Medulla oblongata, 438, 439, 462–464
  components of, 463–464
  functions of, 463
  and respiration, 832
  vasomotor center of, 700
Medullary branch, 473
Medullary cavity, 169, 172, 173, 178
Medullary cords, 760
Megakaryocytes, 644
Meibomian glands, 538
Meiosis, 96
  meiosis I, 1023
  meiosis II, 1024
  nature of, 1023
  spermatogenesis, 1021, 1022–1023, 1025
Meissner's corpuscles, 152, 532
Melanin, 119, 147, 1091

Melanocytes, 119, 146, 147–148, 160
Melanocyte-stimulating hormone, 149, 589, 591
Melanoma, malignant, 148, 163, 784
Melanosomes, 147
Melanotropin, 591
Melatonin, 454–456, 579, 603
Membrane attack complex (MAC), 767
Membrane channels, 376–377
  chemically regulated channels, 376, 377
  gated channels, 376–377
  mechanically regulated channels, 376, 377
  passive channels, 376
  voltage-regulated channels, 376, 377
Membrane flow, 87
Membrane renewal vesicles, 86
Membranes, 51
  cutaneous membrane, 129, 130
  mucous membranes, 129, 130
  serous membranes, 129–130
  synovial membranes, 129, 130, 131
Membranous labyrinth, 560
Membranous organelles, 80–81, 84–89
Membranous urethra, 976
Memory, 494–496
  brain regions related to, 495
  formation/storage, 495–496
  immune system, 772
  loss of, 495, 496
  types of, 494–495
Memory B cells, 778
Memory cells, 787
  and immunity, 772
Memory consolidation, 494, 495
Memory engram, 496
Memory T cells, 775, 776
Menarche, 1045
Ménière's disease, 95 [AM]
Meninges, 214
  cranial meninges, 409, 440–441
  spinal meninges, 409–412
Meningitis, 409, 431; 76 [AM]
Meniscectomy, 267; 59–60 [AM]
Meniscus, 249
Menopause, 1036, 1045, 1056
Menses, 1043
Menstrual cycle. See Uterine cycle
Menstruation, 1043
Mental foramina, 207
Mentalis, 322
Mental protuberance, 207
Merkel cells, 145, 532
Merkel's (tactile) discs, 152, 532
Merocrine secretion, 115
Merocrine sweat glands, 157, 160
Mesencephalic aqueduct, 439
Mesencephalon, 436–438, 439, 459–460
Mesenchymal cells, 118, 119
Mesenchyme, 120
Mesenteric arteries, 732
Mesenteric vein, 741–742
Mesenteries, 851, 852
Mesentery proper, 851
Mesoappendix, 882
Mesocolon, 851
Mesoderm, 1070
Mesothelium, 111, 129
Mesovarium, 1035
Messenger RNA (mRNA), 57, 84, 91–93
Metabolic acidosis, 1007–1008, 1009, 1011
  causes of, 1007–1008
  combined with respiratory acidosis, 1008
  compensation for, 1008
  ketoacidosis, 1007
  lactic acidosis, 1007
  treatment of, 1008

Metabolic alkalosis, 1008, 1009, 1011
Metabolic anomalies, 22 [AM]
Metabolic obesity, 923; 171 [AM]
Metabolic rate, 933
Metabolic regulation, 51
Metabolic turnover, 60, 901, 902
Metabolism, 2, 36; 18–19 [AM]
  absorptive state, 921–923
  and adipose tissue, 921, 922–923, 925
  aerobic. See Aerobic metabolism
  anabolism, 38, 902
  carbohydrate metabolism, 903–911
  catabolism, 38, 902
  cellular metabolism, 901–903
  lipid metabolism, 913–916
  and liver, 877–878, 920–921, 922, 923–925
  and nervous system, 398
  and neurons, 921, 926
  nucleic acid metabolism, 919–920
  postabsorptive state, 923–925
  protein metabolism, 916–919
  and skeletal muscle, 921, 923, 925
  water generated from, 990
Metabolites, 39, 953, 959, 971
Metacarpals, 233
Metalloproteins, 627
Metaphase, 98
Metaphase plate, 98
Metaphyseal vessels, 180
Metaphysis, 169, 170, 177
Metaplasia, 137, 138
Metarteriole, 698
Metastasis, 101, 102
Metatarsal bones, 239–240
Metencephalon, 439
Methionine, 94, 919
Methotrexate, 136 [AM]
Metopic suture, 201, 247
Metoprolol, 667, 683
Metrodine, 1079
MHC (major histocompatibility complex) proteins, 70
Micelles, 51, 68, 890
Microanatomy, methods of, 30–31 [AM]
Microcephaly, 211, 223
Microcytic cells, 115 [AM]
Microfilaments, 67, 81
Microglia, 367, 370, 765
Microliter (μl), 628
Micromoles (μmol), 32
Microns (μ), 30 [AM]
Microphage-chemotactic factor, 783
Microphages, 118, 119, 638, 765
Microscopic anatomy, 3
  specialties of, 3
Microscopy
  electron microscopy, 65
  light microscopy, 65
Microtubules, 67, 82
Microvilli, 66, 67, 82
Micturition. See Urination
Micturition reflex, 976
Middle cardiac veins, 668
Middle cerebellar peduncles, 462
Middle cerebral artery, 728, 729
Middle clinoid process, 203
Middle ear, 202, 557, 558–560
Middle meatuses, 801
Middle nasal conchae, 204
Middle piece, spermatozoon, 1025
Middle umbilical ligaments, 975
Midsagittal section, 19, 20
Migraine headaches, 72 [AM]
Migration-inhibitory factor (MIF), 783
Milk let-down reflex, 1087–1088
Milk production, mammary glands, 1047, 1087–1088
Milliequivalents per liter (mEq/l), 17 [AM]

Milligrams per deciliter (mg/dl), 17 AM
Millimeters of mercury (mm Hg), 817
Millimole (mmol), 32, 60
Milliosmole (mOsm), 954
Millimoles per liter (mmol/l or mM), 17 AM
Milliosmoles per liter (mOsm/l), 17 AM
Mimetic drugs, 522
Mineralocorticoids, 601, 617
Minerals, 928–929
  bulk minerals, listing of, 928
  functions of, 928
  and liver, 878
  storage of, 168
  trace minerals, listing of, 928
Minimal volume, 821–822
Minimus, 316, 317
Minor, 316, 317
Minor calyx, 946
Minoxidil, 155
Miscarriages, 1083
Mitochondria, 67, 87–89
  ATP production, 87–89, 904–910
  membranes of, 904
  structure of, 87
Mitosis, 96, 98–99
  anaphase, 98
  metaphase, 98
  nature of, 1023
  prophase, 98
  spermatogenesis, 1022
  telophase, 98–99
Mitotic rate, 99
Mitral stenosis, 119–120 AM
Mitral valve, 662
Mitral valve prolapse, 678
Mixed exocrine glands, 116
Mixed nerves, 407
M line, 275, 276
Mobile macrophages, 765
Moderator band, 662
Modulator, 22 AM
Modified radical mastectomy, 1048
Molarity, 41
Molars, 857
Mole (mol), 32, 60
Molecular biology, 34–40 AM
Molecular level of organization, 4, 5. See also Chemical level of organization
Molecular motors, 82
Molecular weight, calculation of, 36
Molecules, 4, 30–36
Moles per liter (mol/l or M), 17 AM
Monaxial articulation, 251
Monoclonal antibodies (MoAbs), 17, 142 AM
Mongolism, 1097
Monoamine oxidase (MAO), 511
Monoarthritic, 267
Monoclonal antibodies, 784, 793
Monocyte-chemotactic factor (MCF), 783
Monocyte–macrophage system, 765
Monocytes, 124, 172, 637, 638, 641, 644
Monoglycerides, 47, 48, 890
Monokines, 782
Mononucleosis, 793; 134 AM
Monosaccharides, 44–45, 46
  absorption of, 890
Monosomy, 1097
Monosynaptic reflexes, 424, 425
Monounsaturated fatty acid, 48
Monovalent ions, 17 AM
Monozygotic twins, 1086
Mons pubis, 1046
MOPP (drug combination), 134 AM
Morula, 1068
Motilin, 881

Motion, range of, 246
Motion sickness, 563
Motivational system. See Limbic system
Motor areas, cerebral cortex, 447
Motor end plate, 281
Motor homunculus, 489
Motor neurons, 366–367, 487–488
  activation of, 423
  types of, 367
Motor nuclei, 413
Motor pathways, 399, 487–493
  and cerebellum, 491–492
  levels of processing, 492–493
  medial and lateral, 489–490, 491
  motor homunculus, 489
  pyramidal system, 488–489
Motor speech area, 449
Motor units, 289–290
Mountain sickness, 829, 840; 154 AM
Mouth-to-mouth resuscitation, 819
Movement
  of organisms, 2
  of synovial joints, 250–254
M phase, 97
m-phase promoting factor (MPF, maturation-promoting factor), 100
MPTP (drug), 81–82 AM
MRI scan, 23, 24, 26
MS (multiple sclerosis), 370, 401, 413, 788
Mucins, 56, 116, 855
Mucosa, digestive tract, 847–848
Mucosa-associated lymphoid tissue (MALT), 759
Mucous connective tissue, 120
Mucous glands, 116
Mucous membranes, 129, 130, 847
Mucus, 56, 115
Mucus escalator, 799, 800
Müllerian ducts, 1025
Müllerian-inhibiting factor, 1025
Multicellular glands, 116
  exocrine glands, 116, 117
  organization of, 117
Multi-CSF, 644
Multifidus, 328, 330
Multinucleated cells, 132, 273
Multipennate muscle, 313
Multiple births, 1086
Multiple motor unit summation, 289
Multiple sclerosis (MS), 370, 401, 413, 788; 74, 76 AM
Multipolar neurons, 362, 363, 365, 366
Multiunit smooth muscle cells, 305
Mumps, 855, 895
Muscarine, 515
Muscarinic receptors, 515, 516
Muscle cells, turnover rate, 60
Muscle cramps, 356
Muscle fibers, 132
Muscles. See Muscle tissue; specific muscles
Muscle spindles, 533
  and stretch reflex, 426
Muscle tissue, 4, 107, 132–134, 271–310
  action potentials in, 386
  cardiac muscle tissue, 132, 133
  skeletal muscle tissue, 132, 133
  smooth muscle tissue, 132–134
Muscle tone, 290
Muscular arteries, 695
Muscular dystrophies, 277, 306, 1095; 66–67 AM
Muscularis externa, digestive tract, 848–849
Muscularis layer, 1046
Muscularis mucosae, 848
Muscular system, 5, 7, 311–359; 62–69 AM

disorders, 63–69 AM
  symptoms of, 62–63 AM
  functions of, 7
  organs of, 7
  and puberty, 1089
Musculi pectinati, 660
Musculocutaneous nerve, 419
Musculoskeletal system, 265. See also Skeletal muscle; Skeletal system
Mutagens, 103
Mutations, 97
Myasthenia gravis, 281, 306; 67–68 AM
Mycobacterium tuberculosis, 808
Myelencephalon, 439
Myelin, 370
Myelination, 370, 386
Myelin sheath, 370
Myelocytes, 643
Myelography, 411, 431
Myeloid leukemias, 115–116 AM
Myeloid stem cells, 628, 633, 642
Myeloid tissue, 633
Myenteric plexus, 849
Myenteric reflexes, 850
Mylohyoid, 207, 327, 853
Mylohyoid line, 207
Myoblasts, 273–274
Myocardial infarction, 649, 679, 688, 722
Myocarditis, 688; 118 AM
Myocardium, 658
Myoepithelial cells, 157, 1087
Myofacial pain syndrome, 208
Myofibrils, 275
Myofilaments, 275
Myoglobin, 52, 297, 629
Myogram, 286, 288
Myometrium, 1042
Myoneural junction, 280–281
Myopia, 548, 572
Myosin, 81, 275, 279
  reactivation of, 283
Myosin light chain kinase, 305
Myotonic dystrophy, 67 AM
Myxedema, 597, 618

N
NAD (nicotinamide adenine dinucleotide), 904
Nail bed, 158
Nail body, 158
Nail folds, 158
Nail grooves, 158
Nail root, 158
Nails, 6, 158
  structure of, 158
Na^+-K^+/2 Cl^- transporter, 962, 963
Nanometer, 30
2-Naphthylamine, 976
Narcolepsy, 86 AM
Nasal bones, 206
Nasal cavity, 9, 801
Nasal complex, 208, 210
Nasal conchae, 204
Nasalis, 317, 322
Nasal mucosa, 802
Nasal septum, 196, 801
Nasolacrimal canal, 205
Nasolacrimal duct, 205, 539, 541
Nasopharynx, 801, 803
Natural killer (NK) cells, 641, 756, 758, 766, 781
  activation of, 766
  and cancer, 766
  and viral infections, 766
Naturally acquired immunity, 772
Natural pacemaker, 671
Natural passive immunity, 772
Navicular bone, 239
N compounds, 926–927
NE. See Norepinephrine (NE)

Near point of vision, 549
Neck, 188
  bladder, 975
  femur, 237
  gallbladder, 879
  muscles of, 320, 326–328
  penis, 1030
  radius, 231
  ribs, 221
  spermatozoon, 1025
  tooth, 856
  veins of, 736
Necrosis, 135, 138, 151, 770
Necrotizing fasciitis, 64 AM
Negative feedback, 12–14
Negative feedback control mechanism, 423
Negative inotropic action, 682
Negative nitrogen balance, 927
Neisseria gonorrhoeae, 1040
Neohematocytes, 112 AM
Neomycin, 571
Neon, 32
Neonatal period, 1086, 1087–1088
Neonatal respiratory distress syndrome (NRDS), 152 AM
Neonate, acid–base balance in, 1010–1011
Neoplasm. See Tumors
Nephritis, 980; 178 AM
Nephrogenic diabetes insipidus, 591
Nephrolithiasis, 974
Nephrons, 946–953
  collecting system, 952–953
  distal convoluted tubule, 952
  proximal convoluted tubule, 952
  renal corpuscle, 949–952
  types of, 948, 950
Nerve deafness, 571, 572
Nerve fibers, 134, 273, 361
Nerve grafts, 79 AM
Nerve growth factors (NGF), 100, 416, 431; 77 AM
Nerve impulses. See Synapse
Nerve palsies, 416
Nerve plexuses, 416–422
  brachial plexus, 419–420
  cervical plexus, 418
  lumbar plexus, 420–422
  sacral plexus, 420–422
Nerves. See also specific nerves
  cranial nerves, 361, 464–474
  peripheral nerves, 414–416
  spinal nerves, 361, 414–422
Nervous system, 5, 7; 70–96 AM
  See also Brain; Spinal cord;
  action potentials, 395–396
  afferent division, 361
  and aging, 499
  anatomical organization, 399
  autonomic nervous system, 362
  central nervous system, 361
  disorders of, 370, 401; 72–96 AM
    symptoms of, 70–71 AM
  efferent division, 361–362
  and environmental factors, 398
  and exercise, 300
  functional organization, 396–397
  functional overview, 362
  functions of, 7
  information processing, 392–396
  injury/regeneration of, 372
  integration with other systems, 399–400
  and metabolic processes, 398
  neurophysiological concepts, 373–386
  organs of, 7
  peripheral nervous system, 361
  postsynaptic potentials, 392–395
  presynaptic facilitation, 395
  presynaptic inhibition, 395
  and puberty, 1089
  somatic nervous system, 361
  synapses, 386–392

Net colloid osmotic pressure (NCOP), 708
Net diffusion, 71
Net filtration pressure (NFP), 708
Net hydrostatic pressure (NHP), 955
Neupogen, 644
Neural cortex, 399, 436
Neural mechanisms, 712
Neural retina, 542, 543, 545
Neural stimuli, 584
Neural tissue, 107, 134, 360–405
  neuroglia, 361, 367–371
  neurons, 362–367
Neural tube, 438
Neural tunic, 541–544
Neurilemma, 370
Neurocoel, 438
Neuroendocrine cells, 510
Neuroendocrine reflexes, 591
Neuroepithelium, 107
Neurofibrillary tangles, 499
Neurofibrils, 363
Neurofilaments, 362–363
Neurogenic shock, 127 [AM]
Neuroglandular junction, 364
Neuroglia, 134, 361, 367–371
  astrocytes, 367, 368
  of central nervous system, 367–370
  ependymal cells, 367, 368
  ganglia, 370
  microglia, 367, 370
  oligodendrocytes, 367, 368–370
  of peripheral nervous system, 370–371
  satellite cells, 367, 370
  Schwann cells, 367, 370, 371
Neurohypophysis, 591–592
Neurolemmocytes, 370
Neurological examination, 71–72 [AM]
Neuromodulators, 392, 395
  endorphins, 392
  functions of, 392
Neuromuscular junction, 280–281, 364
Neuronal pools, 396
Neurons, 134, 361, 362–367, 399
  anaxonic neurons, 365
  axon, 363–365
  axoplasmic transport, 364–365
  bipolar neurons, 365
  cell body (soma), 134, 362–363
  convergence, 396–397
  dendrites, 363
  divergence, 396, 397
  injury of, 372
  and metabolism, 921, 926
  motor neurons, 366–367
  multipolar neurons, 362, 363, 365, 366
  neuronal pools, 396
  parallel processing, 397
  and postabsorptive state, 926
  reverberation, 397
  sensory neurons, 366, 482
  serial processing, 397
  synapse, 364–365
  turnover rates, 60
  unipolar neurons, 365–366
Neuropeptides, 391, 392
Neurophysiology
  action potentials, 380–386
  graded potentials, 377–380
  membrane channels, 376–377
  transmembrane potential, 373–377
Neurosyphilis, 191 [AM]
Neurotoxins, 382, 401
Neurotransmitters, 280, 364, 578
  acetylcholine (ACh), 387–389
  categories of, 391
  dopamine, 389
  drug effects, 498

excitatory, 387
gamma aminobutyric acid (GABA), 389
imbalance, effects of, 498
inhibitory, 387
norepinephrine, 389
and parasympathetic stimulation, 515
serotonin, 389
and sympathetic function, 511–512
Neurotrophin-3 (NT-3), 77 [AM]
Neurotrophin-4 (NT-4), 77 [AM]
Neurotubules, 362
Neutral fats, 48–49
Neutralization
  of antigens, 780
  and buffers, 43
Neutral solution, 42, 997
Neutrons, 30
Neutrophilic band cell, 643
Neutrophilic myelocyte, 643
Neutrophils, 119, 124, 637, 638, 639–640, 644, 787
  activation of, 770
Newborn infants, acid–base balance in, 1010–1011. See also Neonatal period
Niacin, 930
Nicotinamide adenine dinucleotide (NAD), 904
Nicotine, 390, 401, 496
Nicotinic receptors, 515, 516
Night blindness, 554
9 + 0 array, 82
9 + 2 array, 82
Nipples, 1047
Nissl bodies, 363
Nitric oxide, 34–35, 390, 391
  sympathetic stimulation of, 512
Nitrogen, 31
Nitrogen balance, and nutrition, 927
Nitrogenous base, 56
Nitrogenous wastes, 919, 953
Nitroglycerin, 667, 853
Nitroxidergic synapses, 512, 712
NK cells. See Natural killer (NK) cells
Nociceptors, 482, 530–531
Nodes of Ranvier, 370
Nodose ganglion, 471
Nodules, 43 [AM]
Nomogram, 1009, 1012
Nondisplaced fractures, 186
Nonessential amino acids, 919
Non-Hodgkin's lymphoma (NHL), 133 [AM]
Non-insulin-dependent (Type 2) diabetes, 607, 618; 104 [AM]
Nonmembranous organelles, 80–84
Nonpolar covalent bonds, 35, 41
Nonshivering thermogenesis, 935–936
Nonspecific defenses, 638, 752, 763–771
Nonspecific resistance, 763
Nonstriated involuntary muscle, 134
Nonstriated muscle. See Smooth muscle tissue
Noradrenaline, 389
Norepinephrine (NE), 389, 391, 506, 510, 579, 603, 617
  stimulation of, 511–512
Normal saline, 74, 102
Normoblast, 633
Normochromic, 650
Normochromic red blood cells, 115 [AM]
Normocytic, 650
Normocytic cells, 115 [AM]
Normovolemic blood volume, 637, 650
Norplant, 1054–1055

Nose, 801–802
  components of, 801
  nasal mucosa, 802
Nosebleeds, 802
Nuchal, 317
Nuchal lines, inferior and superior, 200
Nuclear envelope, 89
Nuclear imaging, 61; 15 [AM]
Nuclear matrix, 90
Nuclear pores, 89–90
Nucleases, 874, 889
Nucleic acid metabolism, 919–920
  nucleic acid synthesis, 920
  RNA catabolism, 919
Nucleic acids, 56–57, 58, 59
  deoxyribonucleic acid (DNA), 56, 57, 58
  ribonucleic acid (RNA), 56, 57, 58
  structure of, 56–77
Nucleoli, 66, 67, 90
Nucleoplasm, 90
Nucleoside, 56
Nucleosomes, 90
Nucleotides, 56–57
Nucleus, 66, 67, 89–95
  of atom, 30
  center of neurons, 399
  chromosomes, 90–921
  genes, 91
Nucleus basalis, 495
Nucleus cuneatus, 464, 483
Nucleus gracilis, 464, 483
Nucleus pulposus, 256, 257
Nucleus solitarius, 536
Nursemaid's elbow, 260
Nutrient artery, 180
Nutrient foramina, 180
Nutrient pool, 902, 903
Nutrients, 2, 39
  maternal requirements for, 1083
Nutrient vein, 180
Nutrition
  and balanced diet, 926
  and bone growth/maintenance, 181–182
  cholesterol levels, 915
  diet and disease, 931
  food pyramid, 926, 927
  meaning of, 926
  minerals, 928–929
  nitrogen balance, 927
  vitamins, 929–930
Nutritionists, 926
Nystagmus, 564, 572

O

Obesity, 923, 938; 171–172 [AM]
Obligatory water reabsorption, 966
Oblique group, 320, 323, 330–332, 863
Obliques, 316, 317
Obstructive jaundice, 113 [AM]
Obstructive shock, 747; 127 [AM]
Obturator, 347, 349
Obturator foramen, 235
Obturator nerve, 422
Occipital bone, 200, 209
Occipital condyles, 200
Occipital crest, 200
Occipital fontanel, 211
Occipitalis, 321, 322
Occipital lobe, 446, 447
Occlusal surfaces of teeth, 856
Ocular conjunctiva, 538
Oculo-, 317
Oculomotor muscles, 323
Oculomotor nerve, 467, 474
Odontoid process, 216
Odorant binding proteins, 535
Off-center neurons, 555
Oil glands. See Sebaceous glands
Olecranal joint. See Elbow joint

Olecranon, 231
Olecranon fossa, 230
Olfaction, 528, 534–536
  and aging, 536
  olfactory discrimination, 536
  olfactory organs, 534, 535
  olfactory pathways, 535–536
  olfactory receptors, 534, 535
Olfactory bulbs, 466, 535
Olfactory cortex, 447, 535
Olfactory epithelium, 534
Olfactory foramina, 204
Olfactory glands, 534
Olfactory nerves, 204, 466, 474
Olfactory organs, 534, 535
Olfactory receptors, 534, 535
Olfactory region, 801
Olfactory (smell) receptors, 204
Olfactory tracts, 466
Oligodendrocytes, 367, 370
Oliguria, 174 [AM]
Olivary nuclei, 464
Olives, 464
Omega-3 fatty acids, 48, 60
Omohyoid, 327, 328
On-center neurons, 555
Oncogenes, 101, 102; 37 [AM]
Oncologists, 138; 38 [AM]
Oncotic pressure, 708n
Oocytes, 608, 1017, 1036
  abnormal, 1080
  and fertilization, 1065, 1067
  at ovulation, 1066–1067
Oogenesis, 1036–1037
  and aging, 1039
Oogonia, 1036
Oophorectomy, 1045
Open fractures, 186
Open wounds, 48 [AM]
Ophthalmic artery, 728
Opioids, 391, 392
Opponens digiti minimi, 346
Opponens pollicis, 346
Opportunistic infections, 790
Opposition, joints, 253, 254
Opsin, 550, 552, 553
Opsonins, 769
Opsonization, 769
  immune response, 780
Optic canals, 203
Optic chiasm, 457, 466
Optic disc, 544
Optic nerve, 466, 474
Optic radiation, 556
Optic tracts, 457, 466
Oral cavity, 803, 853–858
  functions of, 853–854
  structures of, 853–854
Oral contraceptives, 1054
Oral mucosa, 853
Ora serrata, 542
Orbicularis, 317
Orbicularis oculi, 322, 538
Orbicularis oris, 313, 320–321, 322
Orbital complex, 208
Orbital fat, 541
Orbital process, 205
Orbital rim, 205
Orbits, 201, 208
Orchiectomy, 1060
Orchitis, 183 [AM]
Organelles, 4, 66, 67, 80–89
  centrioles, 82
  cilia, 82–83
  cytoskeleton, 81–82
  endoplasmic reticulum, 84–85
  flagella, 83
  Golgi apparatus, 85–86
  lysosomes, 86–87
  membrane flow, 87
  membranous, 80–81
  microvilli, 82
  mitochondria, 87–89
  nonmembranous, 80–84

peroxisomes, 87
ribosomes, 83–84
Organic acids, 998
Organic compounds, 39, 44–59
    carbohydrates, 44–46
    lipids, 46–51
    nucleic acids, 56–57, 58, 59
    proteins, 51–56
Organism level, of organization, 4, 5
Organization
    levels of, 4–11
    of organism, 2
Organ level, of organization, 4, 5
Organ of Corti, 565
Organogenesis, 1078
Organ systems, 3, 4, 5, 6–11
    cardiovascular system, 8
    digestive system, 10
    endocrine system, 8
    integumentary system, 6
    lymphatic system, 9
    muscular system, 7
    nervous system, 7
    reproductive system, 11
    respiratory system, 9
    skeletal system, 6
    urinary system, 10
Organ transplantation, graft rejection, 777
Orgasm, 1053
Origin, skeletal muscles, 316
Oris, 317
Oropharynx, 803
Orthopedists/orthopedic surgeons, 58 AM
Orthostasis, 127 AM
Orthostatic hypotension, 747; 127 AM
Os coxae, 232, 234–235
Osmolarity, 74
Osmoles (Osm), 954
Osmoreceptors, 591, 988
Osmosis, 73–74, 79
    across cell membranes, 73–74
    characteristics of, 73
    hydrostatic pressure, 74
    osmotic concentration, 74
    osmotic pressure, 74
    and tonicity, 74
Osmotic pressure, 707–708
Osseous tissue. See Bone(s)
Ossification, 175–180
    endochondral, 176–178
    intramembranous, 175–176
Ossification center, 176
    primary, 177
    secondary, 178
Osteitis deformans (Paget's disease), 190; 57 AM
Osteoarthritis, 265, 266; 60 AM
Osteoblasts, 171–172, 176
Osteoclast-activating factor, 187, 190
Osteoclasts, 172
Osteocytes, 128, 170, 171, 176
Osteogenesis, 171–172, 175
Osteogenesis imperfecta, 190; 55 AM
Osteoid, 171
Osteolysis (resorption), 172
Osteomalacia, 184, 190; 55 AM
Osteomyelitis, 190; 57 AM
Osteon (Haversian system), 172
Osteopaths (O.D.)/osteopathic surgeons, 59 AM
Osteopenia, 185, 190
Osteopetrosis, 190
Osteoporosis, 136, 185–187, 190; 57 AM
Osteoprogenitor cells, 172
Otic ganglia, 468, 471
Otitis media, 558
Otoliths, 563
Outer cortex, 760
Outer segment, photoreceptors, 550

Oval window, 559, 561
Ovarian artery, 733, 1035, 1041
Ovarian cancer, 1039, 1060
Ovarian cycle, 1036, 1037–1039
    corpus luteum, formation of, 1038
    hormones of, 1049–1052
    and menarche, 1045
    ovulation, 1038
    primary follicle formation, 1037
    secondary follicle formation, 1037
    tertiary follicle formation, 1038
Ovarian follicles, 1037
Ovarian hilum, 1035
Ovarian ligament, 1035
Ovarian vein, 1035
Ovaries, 8, 11, 1017, 1035–1039
    hormones of, 609
    oogenesis, 1036–1037
    ovarian cycle, 1036, 1037–1039
    structure of, 1035–1036
Overhydration, 990
Overuse syndrome, 344
Ovulation, 1038
    and oocyte, 1066–1067
    tracking with body temperature, 1052
Ovum, 1017
Oxidation, in oxidative phosphorylation, 907–908
Oxidative phosphorylation, 907–910
    electron transport system, 907, 908–910
    oxidation in, 907–908
    reduction in, 907–908
    steps in, 909
Oxygen, 31, 40
    and respiration, 824–825
Oxygen debt, 296
Oxygen–hemoglobin saturation curve, 826–827
Oxygen transport, 826–829, 830, 831
    fetal hemoglobin, 828–829
    hemoglobin saturation, 826–828
Oxyhemoglobin, 629
Oxyphils, 598
Oxytocin, 445, 459, 579–580, 591–592
    and pregnancy, 1083

P

p53 gene, 100, 976
Pacemaker cells, 132, 303, 671
Pacesetter cells, 305, 849
Pacinian (lamellated) corpuscles, 152, 532
Packed cell volume, 628
Packed red blood cells (PRBCs), 651; 112 AM
Paget's disease (osteitis deformans), 190; 57 AM
Pain, 6, 7, 93–94 AM
    fast and slow pain, 530
    referred pain, 485
    sensation of, 530–531
Pain receptors, 423
Palatal muscles, 326, 858
Palate, 801, 853
Palatine bones, 205
Palatine processes, 205
Palatine tonsils, 759
Palatoglossal arch, 853
Palatoglossus, 325
Palatopharyngeal arch, 853
Palatopharyngeus, 858
Palmar interossei, 346
Palmaris brevis, 346
Palmaris longus, 342, 343
Palmar veins, 737
Palmar venous arches, 737
Palpation, 26; 7 AM
Palpebrae, 317, 538–539

Palpebral conjunctiva, 538
Palpebral fissure, 538
Palsies, 416, 431
Pampiniform plexus, 1019
Pancreas, 8, 10, 604, 605–608, 872–874
    endocrine system, 605–608
    endocrine tissue of, 605–608
    enzymes of, 874, 888–889, 890
    exocrine pancreas, 605
    glucagon, 606, 608
    histology of, 873
    inflammation of, 874
    insulin, 606–608
    structure of, 872–873
Pancreatic acini, 605, 873
Pancreatic alpha-amylase, 874, 889
Pancreatic arteries, 731, 872
Pancreatic duct, 872
Pancreatic islets, 605, 608, 873
Pancreatic juice, 873
Pancreatic lipase, 874
Pancreatic obstruction, 893
Pancreatic polypeptide, 606
Pancreatitis, 874, 893, 895
Pantothenic acid, 885
Papillary duct, 948, 949, 952
Papillary layer, dermis, 150
Papillary muscles, 662
Papillary plexus, 151
Pap test, 115
Papules, 43 AM
Para-aminohippuric acid, 946
Paracortical area, 760
Paracrine communication, 577
Paracrine factors, 577
Parafollicular cells (C cells), 184, 597, 599
Parahippocampal gyrus, 454
Parallel muscle, 312
Parallel processing, neurons, 397
Paralysis
    muscular, 281, 298
    and spinal cord, 413
Paralytic shellfish poisoning (PSP), 75 AM
Paranasal sinuses, 9, 210, 801
Paraplegia, 413, 431
Parasagittal section, 19, 20
Parasites, 752
Parasomnias, 500; 86 AM
Parasympathetic blocking agents, 522, 524; 91 AM
Parasympathetic division of ANS, 362, 504, 505–506, 512–516
    compared to sympathetic division of ANS, 517–519
    functions of, 513–514
    muscarinic receptors, 515, 516
    and neurotransmitter release, 514–516
    nicotinic receptors, 515, 516
    organization of, 513
Parasympathomimetic drugs, 91 AM
Parasympathetic reflexes, 521
Parasympathomimetic drugs, 522, 524
Parathyroid glands, 8, 181, 182, 597–600
    disorders of, 598
    structure of, 598
Parathyroid hormones, 597, 598, 599, 604, 617
    and bone, 181, 182
    and growth, 610
Paraurethral glands, 1046
Paraventricular nucleus, 458, 459, 591
Paravertebral ganglia, 506–507
Parenchyma, 120
Paresthesia, 432, 572; 78 AM
Parietal bones, 169, 200
Parietal cells, 864
Parietal epithelium, 949

Parietal lobe, 446, 447
Parietal pericardium, 21, 22, 657
Parietal peritoneum, 22, 851
Parietal pleura, 22, 813
Parieto-occipital sulcus, 446
Parkinson's disease, 389, 445, 460, 475, 490; 81–82 AM
Parotid duct, 855
Parotid salivary glands, 855
Paroxysmal atrial tachycardia (PAT), 688; 122 AM
Pars distalis, 587, 589
Pars intermedia, 587, 591
Pars nervosa, 591–592
Pars tuberalis, 587
Partial antigens, 780
Partial pressure, of gas, 822, 823, 824–825
Partial-thickness burns, 162
Parturition, 1084
Parvo cells, 555
Passive channels, 376
Passive immunity, 772
Passive immunization, 793; 136–138 AM
Passive processes, 71
Patella, 168, 238
    injury to, 263
Patellar ligament, 238, 263
Patellar reflex, 426, 431
Patellar retinaculae, 263
Patellar surface, 237, 238
Pathogens, 752; 23 AM
Pathological physiology, 4
Pathologists, 137, 138
Pathology, 5 AM
Pathos, 4
Pathways, 399, 481, 482
    in chemical reactions, 39
    motor pathways, 399, 487–493
    sensory pathways, 399, 481–487
P cells, 555
Peak flow meter, 822
Pectinate, 317
Pectinate muscles, 660
Pectineal line, 235
Pectineus, 347, 349
Pectoral fat pad, 1047
Pectoral girdle, 228–230
    clavicle, 228–229
    scapula, 228, 229–230
Pectoralis group, 312
Pectoralis major, 339, 340
Pectoralis minor, 335, 336, 339
Pedal region, 16
Pediatrics, 1086–1087; 198 AM
Pedicels, 950
Pedicles, 213–214
Pelvic brim, 235
Pelvic cavity, 20, 21, 22, 235
Pelvic diaphragm, 332, 333, 334
Pelvic floor, muscles of, 320, 332–334
Pelvic girdle, 233, 234–236
    coxa, 234–235
    pelvis, 235–236
Pelvic inflammatory disease (PID), 1040, 1060
Pelvic inlet, 235, 236
Pelvic nerves, 513
Pelvic outlet, 235, 236
Pelvic region, 16
Pelvis, 16, 235–236
    arteries of, 733–734
    and childbirth, 236
    male vs. female, 236
    veins of, 740
Penetrance, 1091–1092
Penicillin, 445, 780
Penile urethra, 976
Penis, 11, 1017, 1030–1031
    components of, 1030
    erection of, 1030
Pennate muscle, 312–313
Pentose, 56

Pepsin, 55, 865, 889
Pepsinogen, 865, 889
Peptic ulcer, 865, 895; 162, 165 AM
Peptidases, 874
Peptide bond, 52
Peptide hormones, 579–581
Peptides, 52
Perception, 482
Percussion, 26; 8 AM
Perfluorochemical (PFC) emulsions, 112 AM
Perforated ulcer, 896; 165 AM
Perforating canals, 173
Perforating fibers, 174
Perforins, 766, 783
Pergonal, 1079
Pericardial arteries, 729
Pericardial cavity, 20, 21, 22, 655–656
Pericardial effusions, 130
Pericardial fluid, 130, 657
Pericardial sac, 657
Pericarditis, 130, 138, 657, 688
Pericardium, 21, 22, 129, 656–657
segments of, 656–657
Perichondrium, 124, 125, 177
Perikaryon, 362, 363
Perilymph, 561
Perimenopause, 1056
Perimetrium, 1042
Perimysium, 272
Perineum, 235, 320, 332
Perineurium, 414
Perinuclear space, 89
Periodic table, 31
Periodontal disease, 858, 895
Periodontal ligament, 247, 856
Periosteal vessels, 180
Periosteum, 124, 128, 170, 174
Peripheral adaptation, 530
Peripheral chemoreceptors, 834
Peripheral effector, 423
Peripheral nerves, 414–416
axillary nerve, 419
femoral nerve, 420, 422
genitofemoral nerve, 420, 422
lateral femoral cutaneous nerve, 420, 422
median nerve, 419
musculocutaneous nerve, 419
peroneal nerve, 420, 422
phrenic nerve, 418
pudendal nerve, 420, 422
radial nerve, 419
sciatic nerve, 420, 422
tibial nerve, 420, 422
ulnar nerve, 419
Peripheral nervous system (PNS), 7, 361
afferent division of, 366
neuroglia of, 370–371
Peripheral neuropathies, 416, 431, 607
Peripheral proteins, 69
Peripheral resistance, 703, 704
Peristalsis, 861
digestive tract, 849
Peritoneal cavity, 22, 851
Peritoneal dialysis, 971
Peritoneal fluid, 130
Peritoneum, 22, 129, 851–852
Peritonitis, 130, 138, 851, 895
Peritubular capillaries, 960
Peritubular fluid, 952
Permeability
cell membrane, 71–80
epithelial tissue and control of, 107
Permissive effects, of hormones, 610
Pernicious anemia, 634; 115 AM
Peroneal artery, 734
Peroneal palsy, 79 AM
Peroneal vein, 739
Peroneus, 351, 353

Peroxidases, 595
Peroxisomes, 66, 67, 87
Perpendicular plate, 204
palatine bones, 205
Personality, 498
Perspiration, sensible and insensible, 147, 157, 934, 990
Pertibular capillaries, 948
Pes, 16
PET (positron emission tomography) scans, 23, 26; 16 AM
Petit mal seizures, 84 AM
Petrous portion, temporal bone, 202
Peyer's patches (aggregate lymphoid nodules), 759, 871
pH, 42, 43, 997. See also Acid–base balance
of blood, 626
and buffer systems, 999–1001
control mechanisms, 998–1001
enzymes and, 55–56
and hemoglobin, 827–828
importance in body fluids, 997–998
potassium balance and, 994, 995
Phagocytes, 764, 765, 769, 780
macrophages, 765
microphages, 765
phagocytosis, 766
Phagocytic activities, chemicals regulating, 784
Phagocytic dust cells, 765
Phagocytosis, 77–79, 638, 766, 769
and liver, 878
Phagosome, 77
Phalanges, 233, 240
Phantom limb pain, 485
Pharyngeal arches, 854
Pharyngeal constrictors, 326, 858
Pharyngeal phase, swallowing, 861
Pharyngeal tonsil, 759, 803
Pharyngotympanic tube (Eustachian tube), 558
Pharynx, 9, 10, 802–803, 858
laryngopharynx, 803
muscles of, 326, 858
nasopharynx, 803
oropharynx, 803
Phasic receptors, 529
Phenotype, 1090, 1091–1092, 1093
Phenylalanine, 891, 919
Phenylalanine hydroxylase, 919
Phenylketonuria (PKU), 61, 919, 938; 12, 22, 170–171 AM
Phenylthiocarbamide (PTC), 538
Phenylthiourea, 538
Pheochromocytoma, 103 AM
Phimosis, 183 AM
Phlebitis, 747
Phonation, 804
Phosphate buffer system, 999, 1001
Phosphate group, 5, 44
Phosphates, in fluid and electrolyte balance, 995–997
Phosphodiesterase (PDE), 553, 582–583
Phospholipase C, 582
Phospholipid bilayers, 68
Phospholipids, 47, 50–51
of cell membrane, 68
Phosphoric acid, 998
Phosphorus, 31
nutritional factors, 928
Phosphorylation, 57, 582, 904; 22 AM
oxidative, 907–910
substrate-level, 906–907
Phosphorylcreatine, 293
Photons, 550
Photoreceptors, 541, 543, 545, 546, 550–553
Photorefractive keratectomy, 548
Phrenic arteries, 729–731, 732
Phrenic nerve, 418

Phrenic veins, 741
Physical barriers, and immunity, 764, 765
Physical conditioning. See Exercise Physiology
specialties of, 4
study of, 2, 3–4
Pia mater, 409, 410, 440
Pigmented layer, 542, 543
Pineal gland, 8, 445, 454, 603
hormone of, 603
Pinealocytes, 603
Pinna (auricle), 126, 127, 557
Pinocytosis, 77, 78
Piriformis, 317, 347, 349
Pisiform bone, 232
Pitch, sound, 566
Pituitary gland, 8, 203, 436, 586–593
anterior pituitary, 587–591, 593
hypophyseal portal system, 587–588
hypothalamic control, 588
posterior pituitary, 591–592
Pituitary growth failure (pituitary dwarfism), 181, 190, 590
Pituitary hormones
adrenocorticotropic hormone (ACTH), 588, 589
antidiuretic hormone, 591
gonadotropins, 589
growth hormone, 588, 589–591
melanocyte-stimulating hormone, 591
oxytocin, 591–592
prolactin, 589
thyroid-stimulating hormone, 580, 588, 589, 590
Pivot joints, 254, 257
Placenta, 1068, 1072–1075
blood supply to, 742, 1075
delivery of, 1084
hormones of, 1075
placentation, 1068, 1072–1075
structure of, 1073, 1074
Placental prolactin, 1075
Placental stage, labor, 1084, 1085
Placenta previa, 1098; 194 AM
Planar joints, 254
Planes, 19–20
types of, 19–20
Planta, 16
Plantar arch, 734
Plantar arteries, 734
Plantar flexion, 240
joints, 253
Plantar flexors, 353
Plantaris, 353
Plantar reflex, 430, 431
Plantar region, 16
Plantar veins, 739
Plantar venous arch, 739
Plaque
artery, 649, 650, 667–668, 697
in the CNS, 499
in teeth, 856
Plasma, 124, 624, 627
albumins, 626–627
and carbon dioxide transport, 830
compared to interstitial fluid, 624
compared to urine, 953
fibrinogen, 626–627
globulins, 626–627
proteins, 624, 627
Plasma cells, 118, 119, 627, 641, 756
Plasma expanders, 627
Plasmalemma, 68
Plasma membrane. See Cell membrane
Plasma proteins, and liver, 879
Plasmin, 647, 648
Plasminogen, 648
Plasticity, 305
Plateau, in heartbeat, 669
Platelet adhesion, 645
Platelet aggregation, 645

Platelet-derived growth factor, 645
Platelet phase, hemostasis, 645–646
Platelet plug, 645
Platelets, 124, 626, 627, 644–645
functions of, 644
platelet factors, 646
production of, 644–645
Platys-, 317
Platysma, 321, 322
Pleated sheet, 52, 53
Pleura, 21, 22, 129, 330, 813–814
pleural cavity, 20, 21, 22, 813
pleural fluid, 130, 814
Pleural effusions, 130, 138
Pleural rub, 130
Pleuripotent stem cells, 628
Pleurisy, 814, 840
Pleuritis, 130, 138
Plexus of Auerbach, 849
Plexus of Meissner, 848
Plicae, 847
Plicae circulares, 869
PMNs (polymorphonuclear leukocytes), 639
PMS (premenstrual syndrome), 1052
Pneumocystis carinii, 813
Pneumonia, 813, 840
Pneumotachometer, 822
Pneumotaxic center, 464, 832–833
Pneumothorax, 818, 840
PNS. See Peripheral nervous system (PNS)
Podiatrists (D.P.M.), 59 AM
Podocytes, 950
Point mutations, 97
Polar bodies, 1037
Polar covalent bonds, 35
Polarity, epithelial cells, 107, 108
Polar molecule, 40
Polio, 298, 306, 413; 68 AM
Pollex, 233
Pollicis, 317
Polyarthritic, 267
Polycystic kidney disease, 952, 980
Polycythemia, 651; 108 AM
Polygenic inheritance, 1092, 1093
Polygenic traits, 1092, 1093
Polymorphonuclear leukocytes (PMNs), 639
Polypeptide, 52
Polyps, 167 AM
endometrial, 1060
intestinal, 896
Polyribosomes, 95
Polysaccharides, 46
Polysome, 95
Polyspermy, 1067
Polysynaptic reflexes, 425, 428–429
Polyunsaturated fatty acids, 48
Polyuria, 591, 607, 618; 174, 180 AM
Pons, 438, 462
and respiration, 832–833
Popliteal artery, 734
Popliteal ligament, 263
Popliteal surface, 237
Popliteal vein, 739
Popliteus, 317, 349, 350
Porphyrin, 826
Portal areas, 877
Portal hypertension, 877, 895
Portal system, 587, 741–742
Portal vessels, 587
Port-wine stains, 151
Positive chemotaxis, 638
Positive end-expiratory pressure (PEEP), 152 AM
Positive feedback, 14–15
Positive feedback loop, 14
Positive inotropic action, 682
Positive nitrogen balance, 927
Positron emission tomography (PET) scans, 23, 26

Postabsorptive state, 921, 923–926
  and adipose tissue, 925
  and liver, 923–925
  and neurons, 926
  and skeletal muscle, 925
Postcentral gyrus, 447
Posterior, 18, 317
Posterior cardiac vein, 668
Posterior cavity, eye, 541
Posterior cerebral artery, 728
Posterior chamber, eye, 545
Posterior clinoid process, 203
Posterior column pathway,
  482–484, 485, 486, 532, 533
Posterior communicating arteries,
  728
Posterior cruciate ligament, 263
Posterior descending artery, 664
Posterior gray horns, 413
Posterior group, 457
Posterior inferior iliac spine, 234
Posterior interventricular branch,
  664
Posterior interventricular sulcus,
  657
Posterior lobes, 461
Posterior longitudinal ligament,
  256, 257
Posterior median sulcus, 407
Posterior pituitary, 589, 591–592
Posterior section, 20
Posterior spinocerebellar tracts,
  486
Posterior superior iliac spine, 234
Posterior vertebral arches, 215
Posterior white column, 413
Postganglionic fibers, 367, 504
Postnatal development, 1065,
  1086–1090
  adolescence, 1086, 1088–1089
  aging, 1089, 1090
  childhood, 1086, 1088
  infancy, 1086, 1088
  neonatal period, 1086, 1087–1088
Postovulatory phase, 1037,
  1038–1039
Postsynaptic cell, 364
Postsynaptic neurons, 386
Postsynaptic potentials, 392–395
  inhibitory postsynaptic potentials,
  393
  summation, 393–395
Post-traumatic amnesia (PTA),
  500; 85 [AM]
Postural reflex, 427
Potassium, 31
  in fluid and electrolyte balance,
  993–994, 996
  nutritional factors, 928
Potassium ions, and resting poten-
  tial, 375–376
Potassium ion secretion, 965
Potential difference, 80
Potential energy, 36–37, 80, 375
Pott's fracture, 186, 239
Pounds per square inch, 817
Power, 296
Power stroke, 282
Prader–Willi syndrome, 1094
Precapillary sphincter, 698
Precentral gyrus, 447
Precipitation, immune response,
  780
Precocious puberty, 614
Prednisone, 777
Preeclampsia, 194 [AM]
Pre-embryo, 1040, 1068
Prefrontal cortex, 448, 449
Prefrontal lobotomy, 449
Preganglionic fibers, 367
Preganglionic neurons, 504, 512
Pregnancy
  contractions, 1083
  ectopic, 193–194 [AM]
  hormones in, 1083

infertility, 1079
  labor, stages of, 1084, 1085
  mammary gland development,
  1047
  maternal physiological changes,
  1081–1083
  multiple births, 1086
  and Rh factors, 639
  toxemia of, 194 [AM]
  uterine changes, 1083
Preliminary diagnosis, 8 [AM]
Preload, cardiodynamics, 681–682
Premature atrial contractions
  (PACs), 688; 121–122 [AM]
Premature delivery, 1086
Premature labor, 1084–1086
Premature menopause, 1056
Premature ventricular contractions
  (PVCs), 688; 122 [AM]
Premenstrual syndrome (PMS),
  1052
Premolars, 857
Premotor cortex, 448
Pre-mRNA, 92
Prenatal development, 1065
  allantois, 1072
  amnion, 1072
  blastocyst, 1068, 1069
  chorion, 1072
  cleavage, 1068, 1069
  extraembryonic membranes,
  1072, 1073
  first trimester, 1068–1078
  gastrulation, 1070–1072
  germ layers, 1070–1072
  implantation, 1068–1072
  organogenesis, 1078
  overview of, 1076–1077
  placenta, 1072–1075
  second trimester, 1076–1077,
  1081–1083
  third trimester, 1076–1077,
  1081–1083
  yolk sac, 1072
Prenatal testing, 1097
Preoptic areas, 458, 459
Preovulatory phase, 1037
Prepotential, 671
Prepuce, 1030, 1046
Preputial glands, 1030
Presbyopia, 548, 572
Pressure, sensation of, 532–533
Pressure points, 707, 745
Presynaptic cell, 364
Presynaptic facilitation, 395
Presynaptic inhibition, 395
Presynaptic neurons, 386
Prevertebral ganglia, 507
Pricking pain, 530
Primary amenorrhea, 1045
Primary brain vesicles, 438
Primary bronchi, 806
Primary center of ossification, 177
Primary curves, spine, 212
Primary fissure, 461
Primary follicles, 1037
Primary hypertension, 747;
  126–127 [AM]
Primary lysosomes, 77, 78, 86
Primary motor cortex, 447
Primary oocytes, 1036
Primary response, to antigens,
  781–782
Primary sensory cortex, 447
Primary spermatocytes, 1021
Primary structure, of protein, 52, 53
Primary taste sensations, 537
Primary teeth, 857
Primary tumor (primary neo-
  plasm), 101, 102
Prime mover, skeletal muscles, 315
Primitive streak, 1070
Primordial follicle, 1036, 1037
Prions, 27–28 [AM]
P–R interval, 674

Procarboxypeptidase, 874, 889
Procerus, 322
Process, 188
Procoagulants, 646
Products, of chemical reactions, 36
Proelastase, 874, 889
Proenzymes, 646
Proerythroblasts, 633
Profundus, 316
Progenitor cells, 642
Progesterone, 589
  and ovarian cycle, 1038, 1057
  placental, 1075
Progestins, 580, 589, 609, 1038
Prognosis, 10 [AM]
Prohormones, 579
Projection fibers, 447
Prokaryotic cells, 23 [AM]
Prolactin-inhibiting hormone
  (PIH), 589
Prolactin (mammotropin), 100, 580,
  589, 1075
Prolactin-releasing factor, 589
Proliferative phase of uterine cycle,
  1044
Proline, 93
Prometaphase, 98
Promoter (control segment), 91
Pronation, 343
  joints, 253
Pronator, 317
Pronator quadratus, 342, 343
Pronator teres, 342, 343
Prone position, 15
Pronucleus, formation, 1067
Properdin, 769
Prophase, 98
Propranolol, 667, 683
Proprioceptors, 366, 482, 530, 532,
  533, 534
Prosencephalon, 438
Prostacyclins, 580, 645, 646
Prostaglandins, 47, 48, 391, 580
  and pregnancy, 1083
Prostatectomy, 1032, 1060
Prostate gland, 11, 1017, 1029
  cancer of, 1032, 1060
  disorders of, 1029, 1032
  secretions of, 1029
Prostate-specific antigen (PSA),
  1032, 1060
Prostatic acid phosphatase, 1032
Prostatic fluid, 1029
Prostatic hypertrophy, 978
Prostatic urethra, 976
Prostatitis, 1029
Proteases, 874, 887, 889
  inhibitors, 141 [AM]
Protectin, 766
Protein buffer systems, 999–1000
Protein C, 647
Protein deficiency disease, 919, 938;
  6, 170 [AM]
Protein kinase C, 582
Protein metabolism, 916–919
  amino acid catabolism, 916–919
  protein synthesis, 919
Proteins, 51–56
  and amino acids, 51–52
  antibodies, 51
  and ATP, 916, 918–919
  buffers, 51
  of cell membrane, 69–70
  complete and incomplete proteins, 926
  contractile proteins, 51
  digestion of, 891
  enzymes, 54–56
  functions of, 51
  glycoproteins, 56
  proteoglycans, 56
  shape and function, 52–54
  structural proteins, 51
  structure of, 51–52
  transport proteins, 51

Protein synthesis, 91
  overview of process, 93
  translation, 93–95
Proteinuria, 957, 980; 178 [AM]
Proteoglycans, 56, 70
Proteolytic enzymes, 874
Prothrombin time (PT), 118 [AM]
Prothrombinase, 646
Proton acceptor, 43
Proton donors, 42
Protons, 30
Proto-oncogenes, 37 [AM]
Protozoa, 793; 28 [AM]
Protraction, joints, 253, 254
Proximal, 18
Proximal carpal bones, 232
Proximal convoluted tubule (PCT),
  947, 952, 959–962, 968
  functions of, 959–961
Proximal radioulnar joint, 231
Prozac (fluoxetine), 498
Pruritis, 164; 42 [AM]
PSA (prostate-specific antigen),
  1032, 1060
Pseudohermaphrodite, 1098;
  193 [AM]
Pseudopodia, 79
Pseudostratified columnar epitheli-
  um, 115
Pseudounipolar neurons, 365–366
Psoas, 317
Psoas major, 347, 349
Psoriasis, 158, 163, 791; 44 [AM]
Psychoses, 87 [AM]
Psychotropic drugs, 87 [AM]
Pterygoid, 324
Pterygoid plates, 203
Pterygoid processes, 203
Ptyalin, 855
Puberty, 1088–1089
  behavioral changes in, 614
Pubic angle, 236
Pubic region, 16
Pubic symphysis, 234, 247
Pubic tubercle, 234–235
Pubis, 16, 234
Pubococcygeus, 334
Pubofemoral ligament, 261
Pudendal nerve, 420, 422
Pulled groin, 349
Pulmonary arteries, 724
Pulmonary arterioles, 724
Pulmonary circuit, 655, 723–724
Pulmonary circulation
  pulmonary arteries, 724
  pulmonary veins, 724
Pulmonary edema, 124–125 [AM]
Pulmonary embolism, 649, 744,
  745, 813, 824, 840
Pulmonary function tests, 822
Pulmonary lobules, 809–810
Pulmonary obstruction, 824
Pulmonary plexus, 517
Pulmonary semilunar valve, 662
Pulmonary trunk, 662
Pulmonary ventilation, 814–822
  alveolar ventilation, 820, 821
  intrapleural pressure, 817–818
  intrapulmonary pressure, 817
  modes of breathing, 819–820
  pressure and airflow to lungs,
  815–817
  pulmonary volumes, 821–822
  respiratory cycle, 818
  respiratory minute volume, 820,
  821
  and respiratory muscles, 818–819
  respiratory rates, 820
Pulmonary volumes, 821–822
  expiratory reserve volume, 821
  inspiratory reserve volume, 822
  minimal volume, 821–822
  residual volume, 821–822
  and respiratory capacities, 822
  resting tidal volume, 821

Pulp, of spleen, 762
Pulp cavity, 856
Pulpitis, 858, 895
Pulse pressure, 705
Pulses, 586
Pulvinar, 457
Punctures, 48 [AM]
Punnett square, 1092–1093
Pupil, eye, 542
Pupillary constrictor muscles, 542
Pupillary dilator muscles, 542
Pupillary reflex, 520, 521, 542
Purines, 56
Purkinje cells, 461, 486
Purkinje fibers, 671, 672
Pus, 135, 770; 134 [AM]
Pustules, 43 [AM]
Putamen, 451
P wave, 673
Pyelitis, 976, 980
Pyelogram, 980
Pyelonephritis, 976, 980
Pyloric antrum, 861
Pyloric canal, 861
Pyloric glands, 865
Pyloric sphincter, 861–863
Pylorus, 861–863, 865
Pyramidal, 317
Pyramidal cells, 447, 488
Pyramidal system, 488–489
  motor homunculus, 489
  tracts of, 488, 490
Pyramids, 488
Pyrexia, 937
Pyridoxine, 916
Pyrimidines, 56
Pyrogens, 770, 937
Pyruvate, 904
Pyruvic acid, 88, 294, 295, 296, 904

Q
QRS complex, 673
Q–T interval, 674
Quadrate lobe, 874
Quadratus femoris, 347
Quadratus lumborum, 328, 330
Quadratus plantae, 354
Quadriceps, 317
Quadriceps femoris, 238, 349
Quadriceps group, 316
Quadriplegia, 413, 431
Quaternary structure, of protein, 52, 53
Quiet breathing, 819–820, 832

R
Rabies, 365, 401; 72, 74 [AM]
Radial artery, 626, 707, 727, 1043
Radial collateral ligament, 260
Radial fossa, 230
Radial groove, 230, 260
Radial head, 231
Radialis, 317
Radial keratotomy, 548
Radial nerve, 230, 260, 419
Radial nerve palsy, 78 [AM]
Radial notch, 231
Radial tuberosity, 231, 260
Radiant energy, 550
Radiation, of heat, 934
Radical mastectomy, 1048
Radiodensity, 23
Radiographic anatomy, 3
Radioisotopes, 31, 60; 15–16 [AM]
Radiologists, 23, 26
Radiology
  CAT scan, 24
  CT scan, 23, 24
  DSA images, 23, 25
  echogram, 24
  MRI scan, 23, 24

spiral–CT scan, 23, 25
  ultrasound, 23, 24
  X–rays, 23
Radiopharmaceuticals, 61; 16–17 [AM]
Radioulnar joint, 231
Radius, 231–232
RAIU (radioactive iodine uptake measurement), 16 [AM]
Rami communicantes, 414
Ramus, 188
  of ischium, 234
  mandibular, 207
Range of motion, 246
Raphe, 312, 853, 1020
Rapid eye movement (REM) sleep, 497
Ras proteins, 100
Raynaud's disease, 89 [AM]
RBC (red blood cell) count, 634
RDS (respiratory distress syndrome), 812, 840
Reabsorption
  capillary exchange, 707–710
  kidneys, 953, 959
Reactants, 36
Reactive atoms, 33
Recall of fluids, 710
Receptive field, 528
Receptive speech area, 449
Receptor-mediated endocytosis, 77, 78
Receptor potential, 529; 91 [AM]
Receptor proteins, 69, 70
Receptors, 12, 361, 423. See also Sensory receptors
  exteroceptors, 366
  interoceptors, 366
  parasympathetic, 515
  proprioceptors, 366
  specificity of, 528
  sympathetic, 511–512
Receptor site, 75
Recessive traits, 1091, 1092
Reciprocal inhibition, 428, 429
Recognition proteins, 69, 70
Recombinant DNA, 103
Recovery heat, 296
Recovery period, skeletal muscle contractions, 295–296
Recruitment, 289
Rectal columns, 884
Rectal veins, 741
Rectouterine pouch, 1035
Rectum, 851, 882, 884
  structure of, 884
Rectus, 316, 317
Rectus abdominis, 222, 330–331, 332, 819
Rectus femoris, 313, 349, 350
Rectus group, 320, 322, 330–332
Red blood cells (RBCs), 124, 624, 628–637
  abundance of, 628
  agglutinins, 636
  and blood tests, 634
  blood types, 634–637
  hemoglobin, 629–630
  and iron, 633
  life span of, 630–633
  production of, 633–634
  structure of, 628–629
Red cones, 554
Red marrow, 168, 169, 633
Red muscles, 298
Red nucleus, 460, 490
Red pulp, of spleen, 762
Reduction, in oxidative phosphorylation, 907–908
Reductional division, 1024
Referred pain, 485
Reflex arc, 422–423
  intersegmental, 425
  visceral, 520
Reflex centers, 463

Reflexes, 362, 422–431. See also Cranial reflexes; Spinal reflexes; Visceral reflexes
  acquired reflexes, 424
  cranial reflexes, 424
  development of, 424
  endocrine reflexes, 584–586
  gastrointestinal, 872
  and homeostasis, 422, 423
  innate reflexes, 424
  monosynaptic reflex, 424, 425
  and neural circuitry, 424–425
  polysynaptic reflex, 425
  processing sites, 424
  reflex arc, 422–423
  respiratory, 831, 833
  somatic reflexes, 424
  spinal reflexes, 424, 425–431
  steps in activation of, 423
  visceral reflexes, 424, 520–521
Refraction, vision, 546–547
Refractory period
  action potentials, 381–382
  heartbeat, 669–670
Regeneration, 770
  of epithelial cells, 107
  of tissue, 135, 136, 138
Regional anatomy, 3
Regulation
  of carrier-mediated transport, 75
  of cell division, 100–101
  of enzymes, 55
Regulatory hormones, 459, 585
Regulatory obesity, 923; 171 [AM]
Regulatory T cells, 756
Regurgitation, in heart, 664
Reinforcement, of spinal reflexes, 430
Relative refractory period, 381, 670
Relaxation phase, muscle twitch, 286
Relaxin, 236, 1075
Releasing hormones, 588
  angiotensin II, 601
  corticotropin-releasing hormone, 588, 589
  gonadotropin-releasing hormone, 588, 589
  growth hormone–releasing hormone, 590
  prolactin-releasing factor, 589
  thyrotropin-releasing hormone, 588, 589
Remission, 138; 40 [AM]
Remodeling, bone, 180
REM (rapid eye movement) sleep, 497
Renal area, 762
Renal arteries, 732–733, 944, 946
Renal capsule, 944
Renal columns, 946
Renal compensation, and acid–base balance, 1002–1004
Renal corpuscle, 946, 949–952, 960, 968
  Bowman's capsule, 949
  glomerular capillaries, 950
Renal failure, 957, 958, 980, 1005
  treatment of, 970–971
Renal fascia, 944
Renal glycosuria, 962; 178–179 [AM]
Renal lobe, 946
Renal nerves, 944, 946
Renal papilla, 946
Renal pelvis, 946
Renal pyramids, 946
Renal sinus, 944
Renal threshold, 959
Renal tubule, 946, 949, 952
Renal veins, 740, 944, 946
Renin, 601, 604–605, 865, 889, 952
  functions of, 956
Repeatability, need for, in research, 4 [AM]
Repolarization, 379

Repressor genes, 100
Reproduction, 2. See also Reproductive system
Reproductive cells, 66
Reproductive hormones, 610
Reproductive system, 5, 11, 1016–1063; 183–199 [AM]. See also Female reproductive system; Male reproductive system
  components of, 1017
  disorders, 187–199 [AM]
  examination of, female, 185–187 [AM]
  examination of, male, 183–184 [AM]
  functions of, 11
  integration with other systems, 1057–1059
  organs of, 11
  and puberty, 1089
  smooth muscles of, 304
Reproductive tract, 1017
Residual volume, 821
Resistance, 271, 695, 702–703, 704
  cell membrane as, 375
  in circulation, 703, 704
  development of, 785–787
  and muscle contraction, 292
  nonspecific, 763
  specific, 763
Resistance phase, general adaptation syndrome (GAS), 612–614
Resistance vessels, 695
Resonance, sound, 567
Resorption (osteolysis), 172
Respiration, 2
  and baroreceptor reflexes, 835
  cellular, 88, 814, 904
  and chemoreceptor reflexes, 834–835
  external respiration, 814
  forced breathing, 820
  gas exchange. See Respiratory gases
  and Hering–Breuer reflexes, 835
  internal respiration, 814
  local regulation of blood flow, 831–832
  muscles used in, 818–819
  protective reflexes, 836
  pulmonary ventilation, 814–822
  quiet breathing, 819–820
  respiratory centers, 831–836
  respiratory cycle, 818
  voluntary control of, 836
Respiratory acidosis, 1005–1006, 1009, 1011
  acute, 1005
  causes of, 1005
  chronic, 1005
  treatment of, 1006
Respiratory alkalosis, 1006–1007, 1009, 1011
Respiratory bronchioles, 798, 809
Respiratory burst, 640, 770
Respiratory centers, 831–836
Respiratory chain. See Electron transport system
Respiratory compensation, and acid–base balance, 1002
Respiratory cycle, 818
Respiratory defense system, 799–800
Respiratory distress syndrome (RDS), 812, 840
Respiratory epithelium, 798–799, 800
Respiratory gases
  abnormalities, 834–835
  alveolar air, composition of, 824
  blood gas analysis, 825
  carbon dioxide pressure, 825
  carbon dioxide transport, 829–830, 831

Dalton's law, 822
  gas exchange at respiratory membrane, 822–825
  Henry's law, 822–824
  oxygen pressure, 824–825
  oxygen transport, 826–829, 830, 831
  partial pressure, 822, 823, 824–825
Respiratory membrane, 812–813
  gas exchange at, 822–825
Respiratory minute volume, 820, 821
Respiratory mucosa, 798–800
  and cystic fibrosis, 800
  lamina propria, 799
  respiratory defense system, 799–800
  respiratory epithelium, 798–799, 800
Respiratory pump, 710–711, 818
Respiratory rates, 820
Respiratory reflexes, 831, 833
Respiratory rhythmicity centers, 464
Respiratory system, 5, 9, 797–844; 145–156 AM
  and aging, 837
  autonomic control of, 519
  changes at birth, 836–837
  disorders of, 800, 805, 812, 840; 146–156 AM
    symptoms of, 145 AM
  examination of, 145–146 AM
  and exercise, 300
  functions of, 9, 798
  interaction with other systems, 837–838
  larynx, 803–805
  lower, 798
  lungs, 806–813
  nose, 801–802
  organization of, 798–800
  organs of, 9
  pharynx, 802–803
  pleura, 813–814
  primary bronchi, 806
  and puberty, 1089
  respiratory mucosa, 798–800
  smooth muscles of, 304
  trachea, 805–806
  upper, 798, 801–803
Respiratory tract, 798
Responsiveness, of organisms, 2
Resting potential, 80, 375–376
  and potassium ions, 375–376
  and sodium ions, 375
Resting tidal volume, 821
Rete testis, 1021
Reticular activating system (RAS), 460
  and arousal, 497–498
Reticular epithelial cells, 761
Reticular fibers, 120, 122
Reticular formation, 454, 490, 491
Reticular lamina, 110
Reticular layer, dermis, 150
Reticular tissue, 122
Reticulocyte, 633
Reticulocyte count, 634
Reticuloendothelial system, 765
Reticulospinal tracts, 490, 491
Retina, 541–544
  detached, 543–544
  organization of, 543–544
  retinal processing, 555–556
Retin-A, and wrinkles, 151
Retinal (pigment), 550
Retinitis pigmentosa, 550, 572
Retinopathy, 545
Retraction
  blood clots, 648
  joints, 253, 254
Retroflexion, uterus, 1040–1041
Retrograde amnesia, 500; 85 AM
Retrograde flow, 365

Retroperitoneal organs, 22, 851
Reverberation, neurons, 397
Reverse transcriptase, 790
Reversible reactions, 38–39
R group, 51
Rheumatic fever, 664; 119 AM
Rheumatic heart disease (RHD), 664, 688; 119–120 AM
Rheumatism, 265, 266; 60 AM
Rheumatoid arthritis, 265, 266, 787; 60 AM
Rh factor, 635
  Rh-negative, 635
  Rh-positive, 635
Rhinitis, allergic, 788
Rhizotomy, 82 AM
Rhodopsin, 550, 551, 552
RhoGam, 638
Rhombencephalon, 438
Rhomboideus, 317, 335, 336
Rhythm method, 1055
Ribonucleic acid (RNA), 56, 57
  catabolism of, 919
  characteristics of, 57, 58
  messenger RNA (mRNA), 91–93
  ribosomal RNA (rRNA), 57, 84, 90
  RNA processing, 92–93
  transcription of, 920
  transfer RNA (tRNA), 93–95
Ribose, 56
Ribosomal RNA (rRNA), 57, 84, 90
Ribosomes, 66, 67, 83–84
  free and fixed, 84, 85
Ribs, 219–222
  mobility of, 221–222
  structure of, 221
Rickets, 184, 190, 610
Right atrioventricular (AV) valve, 662
Right atrium, 655, 660–662
  structure of, 660–662
Right bundle branches, 672
Right colic flexure, 884
Right common iliac arteries, 733
Right coronary artery, 664
Right lymphatic duct, 755, 756
Right primary bronchus, 806
Right pulmonary arteries, 662
Right pulmonary veins, 662
Right rotation, 253
Right thalamus, 436
Right ventricle, 655, 662
  compared to left ventricle, 662–663
  structure of, 662
Rigor mortis, 283, 306
Ringer's solution, 627
Risorius, 316, 317, 322
Rivinus' ducts, 855
RNA. See Ribonucleic acid (RNA)
RNA polymerase, 91
Rods, eye, 544, 550, 551
Rogaine, 155
Root
  hair, 153
  lung, 806
  penis, 1030
  tongue, 854
  tooth, 856
Root canal, 856, 858
Root hair plexus, 153, 532
Rotation, joints, 251, 253
Rotator, 317
Rotator cuff, 259, 339, 356
Rotatores, 328, 329, 330
Rotaviruses, 1087
Rough endoplasmic reticulum (RER), 6, 67, 85
Round ligaments, 874, 1020, 1041
Round window, 561
Roundworms, 29 AM
Rubrospinal tracts, 490, 491
Ruffini corpuscles, 532

Rugae
  bladder, 975
  stomach, 863
  vagina, 1046
Rule of nines, 162
R wave, 673

S

Saccules, 561, 563
Sacral body, 219
Sacral canal, 218
Sacral cornua, 218
Sacral curve, spine, 212
Sacral foramina, 218
Sacral hiatus, 218
Sacral plexus, 420–422
Sacral promontory, 219
Sacral tuberosity, 218–219
Sacroiliac joint, 218, 235
Sacrum, 212, 218–219
Saddle joints, 254–256, 257
Sagittal plane, 19
Sagittal section, 20
Sagittal sinus, superior, 734
Sagittal suture, 196, 197
Saline, normal, 74, 102
Saliva, 855
Salivary amylase, 855, 889
Salivary glands, 10, 854–856
  automatic control of, 518
  control of secretion, 855–856
  enzymes of, 855, 888–889
  pairs of, 854
  saliva, 855
Salivatory nuclei, 855
Saltatory propagation, action potentials, 384, 385
Salt loss, 994
Salts, 40, 43, 997
Saphenous nerve, 422
Saphenous veins, 739
Sarcolemma, 274, 280, 281
Sarcomeres, 275–279, 281–284, 313
Sarcoplasm, 274
Sarcoplasmic reticulum, 275
Sartorius, 316, 317, 349, 350
Satellite cells, 132, 272, 367, 370
Saturated fatty acids, 47, 48
Saturated solution, 21 AM
Saturation curve, 21 AM
Saturation limits, 55, 75
Saxitoxin (STX), 390; 75 AM
Scab, 158–160, 163
Scalenes, 330, 332, 819
Scales, 43 AM
Scanning electron micrograph (SEM), 31 AM
Scanning electron microscopy, 31 AM
Scaphoid bone, 232
Scapula, 228, 229–230
Scapularis, 317
Scapular spine, 230
Scapulohumeral joint. See Shoulder joint
Scar tissue, 136, 160
Schizophrenia, 498, 500
Schwann cells, 367, 370, 371, 372
Sciatica, 432; 79 AM
Sciatic compression, 78–79 AM
Sciatic nerve, 234, 420, 422
Scientific method, 3; 1, 2–4 AM
Sclera, 541
Scoliosis, 212, 223; 57 AM
Scotomas, 550
Scrotal cavity, 1020
Scrotum, 11, 1017–1018, 1020
Scurvy, 181, 190
Seafood, toxins found in, 390
Seasonal affective disorder (SAD), 603, 618; 103 AM
Sebaceous follicles, 156

Sebaceous glands, 6, 155–156
  control of, 157
Seborrheic dermatitis, 156, 163
Sebum, 156
Second-degree heart block, 121 AM
Secondary active transport, 76–77, 79
Secondary amenorrhea, 1045
Secondary brain vesicles, 439
Secondary bronchi, 809
Secondary curves, spine, 212
Secondary dentition, 857
Secondary follicles, 1037
Secondary hypertension, 747; 126–127 AM
Secondary lysosomes, 77, 78, 86
Secondary memories, 494
Secondary oocytes, 1037
Secondary ossification centers, 178
Secondary peristaltic waves, 861
Secondary response, to antigens, 782
Secondary spermatocytes, 1024
Secondary structure, of protein, 52, 53
Secondary tumors, 101, 102
Second-class lever, 314
Second-degree burn, 162
Second heart sound, 677–678
Second messengers, 392, 511, 582
Second-order neurons, 482
Secretin, 850, 868, 873–874, 880, 881
Secretion, 846
  kidneys, 953–954, 959, 961
Secretory phase of uterine cycle, 1044
Secretory sheet, 117
Secretory vesicles, 86
Sectional anatomy, 19–25
  body cavities, 20–22
  planes, 19–20
  sections, 19–20
Sections, 19–20
Sedatives, 87 AM
Segmental arteries, 946
Segmental blocks, 1041
Segmental bronchi, 809
Segmental mastectomy, 1048
Segmental veins, 946
Segmentation, digestive tract, 849
Seizure, 493, 500; 84 AM
Selectively permeable membrane, 71
Selenium, 928
Sellaris (saddle) joints, 254–256
Sella turcica, 203, 586
Semen, 1017, 1029–1030
  analysis of, 1030
  components of, 1029–1030
Semicircular canals, 561
Semicircular ducts, 561–563
Semilunar ganglion, 468
Semilunar valves, 664, 665
Semimembranosus, 349, 350
Seminal fluid, 1029
  enzymes in, 1029–1030
Seminalplasmin, 1029
Seminal vesicles, 11, 1017, 1029
Seminiferous tubules, 1021, 1022
Semispinalis muscles, 328, 330
Semitendinosus, 349, 350
Senescence, 1089–1090
Senile cataracts, 546
Senile dementia, 499, 500
Senile lentigos, 149
Sensation, 482
  adaptation, 529–530
  and epithelial tissue, 107
  general senses, 482, 528, 530–534
  special senses, 528, 534–572
Sense organs, 528
Sensible perspiration, 147, 157, 934, 990

Sensitization, 636
  and allergies, 788
Sensory areas, cerebral cortex, 447
Sensory association areas, 447
Sensory coding, 529
Sensory disorders, 92–96 AM
Sensory ganglia, 366
Sensory homunculus, 483
Sensory loss, 533
Sensory neurons, 366, 482
  activation of, 423
  types of, 482
  types of receptors, 366
Sensory nuclei, 413
Sensory pathways, 399, 481–487
  organization of, 482
  posterior column pathway,
    482–484, 485
  sensory homunculus, 483
  sensory neurons, 482
  spinocerebellar pathway, 486–487
  spinothalamic pathway, 484–485,
    486
Sensory receptors, 6, 482
  chemoreceptors, 482, 533–534
  exteroceptors, 530
  gustatory receptors, 536
  interoceptors, 530
  interpretation of sensory infor-
    mation, 529–530
  mechanoreceptors, 482, 531–533
  nociceptors, 482, 530–531
  olfactory receptors, 534, 535
  proprioceptors, 530
  specificity of, 528
  thermoreceptors, 482, 530, 531
  transduction, 528
Sepsis, 162, 163
Septae, 761
Septal cells, 811–812
Septal defects, 210
Septic shock, 127 AM
Septicemia, 651
Septum pellucidum, 439
Serial processing, neurons, 397
Series elastic elements, 288
Serosa, 129, 849
Serotonin, 389, 391, 395, 603, 645
Serous glands, 116
Serous membranes, 22, 129–130
Serratus, 317
Serratus anterior, 335, 337, 819
Sertoli cells, 1021
Serum, 627
Serum and plasma protein assays,
  165 AM
Serum bilirubin assay, 165 AM
Serum enzyme assay, 1032
Serum enzyme tests, 165 AM
Serum glutamic oxaloacetic
  transaminase, 679
Sesamoid bones, 168, 169
Set point, 12
Severe combined immunodeficien-
  cy disease (SCID), 788, 793
Sex cells, 66
Sex chromosomes, 1090, 1095
Sex hormones. See also Androgens;
  Estrogens
Sex-linked traits, 1092
Sexual arousal, 521
Sexual intercourse, 1017,
  1053–1056
  female sexual function, 1056
  male sexual function, 1053
  phases of, 1053
Sexually transmitted diseases
  (STDs), 1040, 1056, 1060;
  190–191 AM
Shaft
  bone, 173
  femur, 237
  hair, 153
  humerus, 230
  radius, 231

ribs, 221
  ulna, 231
Shallow breathing, 820
Shallow water blackout, 834, 840
Sharpey's fibers, 174
Shingles, 416, 431
Shivering, 935
Shivering thermogenesis, 935
Shock, 745
  cardiovascular response to,
    720–721
  circulatory shock, 720–721
  irreversible, 720, 721
Short bones, 168, 169
Short reflexes, 850
Short-term memories, 494
Shoulder girdle. See Pectoral girdle
Shoulder joint, 230, 258–259
  bursae, 259
  ligaments, 259
  muscles of, 259
  structure of, 259
  tendons of, 259
Shoulders, muscles of, 335–339
Shoulder separation, 259, 266
Shunts, 444, 971
SI (Système Internationale), 18 AM
Siamese twins, 1086
Sickle cell anemia, 52, 97, 630, 631,
  650, 1091; 6, 11, 34, 112–113 AM
Sigmoid colon, 884
Sigmoid flexure, 884
Sigmoid mesocolon, 851
Sigmoid sinus, 734
Sign, 6 AM
Sign, of disease, 14
Silent heart attack, 679
Silicosis, 149 AM
Simple columnar epithelium,
  112–115
Simple cuboidal epithelium,
  112–113
Simple epithelium, 110
Simple glands, 117
Simple inheritance, 1093
Simple squamous epithelium, 111
Simple sugar, 44–45, 46
Single covalent bond, 34
Sinoatrial node, 664, 671–672,
  684–685, 686
Sinuses, 188, 196
  disorders of, 210
  frontal, 201
  paranasal sinuses, 210
Sinusitis, 210, 223
Sinusoids, 698
Skatole, 886
Skeletal muscle, 7, 132, 133,
  271–310. See also specific muscles
  actions, 315–316
  and aerobic endurance, 299
  and aging, 299–300
  agonists, 315
  all-or-none principle, 286
  and anaerobic endurance,
    298–299
  antagonists, 315
  atrophy, 298
  autonomic control of, 519
  blood supply, 273
  circular muscle, 313
  complete tetanus, 287
  connective tissue organization,
    271–273
  convergent muscle, 312, 313
  disorders of, 277, 306, 356
  extrinsic eye muscles, 323
  of facial expression, 320–322
  functions of, 271
  of hand, 339–346
  of head, 320–325
  and hormones, 296
  hypertrophy, 298
  innervation, 273, 280, 320

insertion, 316
  integration with other systems,
    300–301
  levers, 313–314
  of limbs, lower, 346–355
  of limbs, upper, 339–346
  of mastication, 324
  and metabolism, 921, 923, 925
  motor units, 289–290
  muscle tone, 290
  of neck, 320, 326–328
  origins, 316
  parallel muscle, 312, 313
  of pelvic floor, 320, 332–334
  pennate muscle, 312–313
  of pharynx, 326, 858
  and postabsorptive state, 925
  red muscles, 298
  and respiration, 818–819
  of shoulder joint, 259
  of shoulders, 335–339
  skeletal muscle contraction, 28–86
  skeletal muscle fibers, 273–279
  sphincter muscle, 313
  of spine, 320
  of stomach, 863
  synergist, 315–316
  terms related to, 316–317
  of tongue, 325, 854
  treppe, 287, 288
  twitch, 286–287
  of vertebral column, 328–330
  views of, 318–319, 336–337
  as voluntary muscles, 273
  wave summation, 287
  white muscles, 297
Skeletal muscle contractions,
  280–284
  and acetylcholinesterase (AChE),
    281
  and adenosine triphosphate
    (ATP), 293–294
  and aerobic metabolism, 293–294
  concentric contraction, 290–291
  contraction cycle, 281, 282–283,
    284
  eccentric contraction, 291–292
  elastic forces, 292
  fatigued muscle, 295
  glycolysis, 294
  and gravity, 293
  and heat loss, 296
  isometric contractions, 291, 292
  isotonic contractions, 290–292
  and lactic acid, 295–296
  length–tension relationships, 285
  and moderate activity, 294–295
  and muscle length, 292–293
  opposing muscle contractions, 292
  and oxygen debt, 295
  and peak activity levels, 295
  recovery period, 295–296
  and relaxation, 281
  speed of contraction, 292
  subconscious control of, 457
Skeletal muscle fibers, 273–279
  compared to cardiac muscle,
    302–303, 660, 665
  distribution of, 297–298
  energy stored in, 293
  excitation–contraction coupling,
    281
  fast fibers, 296, 297, 298
  intermediate fibers, 297, 298
  myofibrils, 275
  neuromuscular junction, 280–281
  paralysis, 281
  sarcolemma, 274
  sarcomeres, 275–279
  sarcoplasmic reticulum, 275
  sliding filament theory, 279
  slow fibers, 296–297, 298
  thick filaments, 275, 278, 279
  thin filaments, 275, 277–278, 279
Skeletal muscle system

appendicular musculature, 317,
    335–355
  axial musculature, 317, 320–335
Skeletal system, 5, 6, 193–269. See
    also Appendicular skeleton;
    Axial skeleton; 51–61 AM
  and age, 242
  appendicular skeleton, 226–244
  axial skeleton, 193–225
  bone, 167–192
  disorders; 57–61 AM
    symptoms of; 52 AM
  examination of; 52–57 AM
  functions of, 6, 168
  gender differences, 241
  individual variation in, 241
  integration with other systems,
    187, 189
  organs of, 6
  and puberty, 1089
Skene's glands, 1046
Skill memories, 494
Skin, 42–50 AM. See also
    Integumentary system
  examination of, 42–47 AM
  secondary disorders of, 47–
    49 AM
Skin cancer, 148, 164, 784
Skin color, 147–149
  abnormal, 149
  melanocytes, 147–148, 149
  pigments, 147–149
Skin conditions, treatment of,
    49–50 AM
Skin graft, 162, 163; 49 AM
Skin signs, 164; 42, 43 AM
Skin tests, 787
Skin turgor, 150
Skull, 195–211
  cranial bones, 200–204
  disorders of, 211
  facial bones, 205–210
  fontanels, 211
  foramina and fissures of, summa-
    ry of, 208, 209
  of infants/children, 211
  sectional anatomy of, 199
  sutures of, 196, 197
  views of, 197–199
SLE (systemic lupus erythemato-
    sus), 793
Sleep, 496–497
  and brain waves, 496–497
Sleep apnea, 500; 86–87 AM
Sleep disorders, 497; 86–87 AM
Sliding filament theory, 279
Slightly movable joints. See
    Amphiarthroses (slightly
    movable joints)
Slipped disc, 257, 266
Slow-adapting receptors, 529–530
Slow calcium channels, 669
Slow fibers, 296–297, 298
Slow-K, 995
Slow pain, 530
Slow potassium channels, 669
Small cardiac vein, 668
Small intestine, 10, 851, 861,
    869–872
  absorption in, 881
  duodenum, 869, 871, 872
  enzymes of, 889, 890
  ileum, 869, 872
  intestinal crypts, 871
  intestinal villi, 869–871
  jejunum, 869, 871, 872
  regional specializations in, 871
  secretions of, 871–872
Small ribosomal subunit, 84
Smegma, 1030
Smell. See Olfaction
Smooth endoplasmic reticulum
    (SER), 6, 67, 85
Smooth muscle, 271
Smooth muscle cells, 303

Smooth muscle tissue, 132–134, 303–305
  compared to other muscle tissue, 304–305
  control of contractions, 305
  excitation–contraction coupling, 305
  functions of, 305
  length–tension relationships, 305
  smooth muscle cells, 303
  visceral smooth muscle cells, 305
Smooth muscle tone, 305
Sneezing, 836
SNS (somatic nervous system), 361, 366
Sodium, 31
  in fluid and electrolyte balance, 991–993, 996
  nutritional factors, 928
Sodium bicarbonate, 43, 1001
Sodium channel inactivation, 381
Sodium chloride, 33, 34
Sodium hydroxide, 43
Sodium ions, 34
  reabsorption of, 966
  and resting potential, 375
Sodium-linked cotransport, 890
Sodium monohydrogen phosphate, 1001
Sodium–potassium ATPase, 76, 375
  and action potentials, 382
Sodium–potassium exchange pump, 76
Soft keratin, 153
Soft palate, 801, 853
Solar plexus, 517
Soleus, 351
Solids, 36
Solitary nucleus, 536
Solute, 40
  reabsorption of, 966
Solute concentrations, 41–42
Solutions, 40–42
  colloids, 42
  concentrations of, 17–18 AM
  hyperosmotic solution, 74
  hypertonic solution, 74
  hypotonic solution, 74
  isosmotic solution, 74
  isotonic solution, 74
  pH, 42
  solute concentrations, 41–42
  suspensions, 42
Solvent, 40
Soma (cell body), 134, 362–363
Somatic association area, 447
Somatic cells, 66
Somatic motor association area, 448
Somatic motor neurons, 366
Somatic motor system, 487–488
Somatic nervous system (SNS), 361, 366
Somatic reflexes, 424
Somatic sensory neurons, 366
Somatocrinin, 590
Somatomedins, 590
Somatostatin, 590, 865, 881
Somatotropin. See Growth hormone (GH)
Sound
  and hearing, 568, 569
  nature of, 565–567
  production of, 804–805
Sounds of Korotkoff, 707, 745
Spasticity, 475; 81 AM
Spatial summation, 394
Special circulation, 722
Special physiology, 4
Special senses, 528, 534–572
  equilibrium, 557, 561–564
  gustation, 536–538
  hearing, 557, 564–571
  olfaction, 534–536
  vision, 538–557

Specific defenses, 639, 752, 763
Specificity, 54
  in carrier-mediated transport, 75
  of immunity, 772
  of receptors, 528
Specific resistance, 763
Speech center, cerebral cortex, 449
Speech production, 804–805
Sperm, 1017
Spermatic cord, 1018–1020
Spermatids, 1021, 1024
Spermatocytes, 1021, 1024
Spermatogenesis, 1021, 1022–1025
  and follicle-stimulating hormone, 1033
  meiosis, 1021, 1022–1023, 1024, 1025
  mitosis, 1022
  spermiogenesis, 1024, 1026
  and sustentacular cells, 1024–1025
Spermatogonia, 1021
Spermatozoa, 1017, 1024
  abnormal, 1079
  and fertilization, 1065, 1067
  spermiogenesis and, 1026
  structure of, 1025
Sperm bank, 1079
Sperm count, 1029, 1030
  low, 1079
  and sterility, 1065
Spermiation, 1024
Spermicide, 1055
Spermiogenesis, 1023, 1024, 1025, 1026
Sperm receptors, 1067
S phase, 96–97
Sphenoidal fontanel, 211
Sphenoidal sinus, 203, 210
Sphenoidal spine, 203
Sphenoid bone, 203, 209
Sphenomandibularis muscle, 325
Sphenopalatine ganglion, 468, 469
Sphincter (circular muscle), 303, 313
Sphincter of Oddi, 879
Sphincter vesicae, 975
Sphygmomanometer, 707, 745
Spicules, 176
Spina bifida, 214, 223
Spinal accessory nerve. See Accessory nerve
Spinal branch, 473
Spinal cavity, 20, 21
Spinal compression, 77 AM
Spinal concussion, 76 AM
Spinal contusion, 77 AM
Spinal cord, 7
  disorders of, 409, 413, 431, 432
  gray matter, 413
  gross anatomy of, 407–413
  reflexes, 422–431
  sectional anatomy of, 412–413
  spinal meninges, 409–412
  spinal nerves, 414–422
  white matter, 413
Spinal curvature, 212
  abnormalities of, 212
  types of curves, 212
Spinal flexors, 330
Spinalis group, 328, 329
Spinal laceration, 77 AM
Spinal meninges, 409–412
  arachnoid, 409, 410
  dura mater, 409–410
  pia mater, 409, 410
Spinal meningitis, 409
Spinal nerves, 361, 407, 414–422
  arrangement of, 407
  brachial plexus, 419–420
  cervical plexus, 418
  distribution of, 414–416
  lumbar plexus, 420–422
  as mixed nerves, 407
  nerve plexuses, 416–422

  peroneal nerve, 420, 422
  sacral plexus, 420–422
  tissue of, 414
Spinal reflexes, 424, 425–431
  Babinski sign, 430, 431
  contralateral reflex arc, 428
  crossed extensor reflexes, 428–429
  flexor reflex, 428, 429
  knee-jerk reflex, 426
  monosynaptic reflexes, 425–427
  plantar reflex, 430
  polysynaptic reflexes, 428–429
  postural reflex, 427
  reinforcement and inhibition of, 430
  stretch reflex, 425–427
  tendon reflex, 428
  and voluntary movements, 430–431
  withdrawal reflex, 428
Spinal shock, 432; 76 AM
Spinal tap, 410, 411, 431, 442
Spinal transection, 77 AM
Spindle apparatus, 82
Spindle fibers, 98
Spine. See Vertebral column
Spine (process), 188
Spinocerebellar pathway, 486–487, 533
Spinocerebellar tracts, 486
Spinothalamic pathway, 484–485, 486
Spinothalamic tracts, 481, 484
Spinous process, 214, 216
Spiral arteries, 1043
Spiral-CT scans, 23, 25
Spiral fractures, 186
Spiral ganglion, 470, 568
Spirillum, 24 AM
Spirochete, 24 AM
Spirometer, 822
Splanchnic nerves, 510
Spleen, 9, 762–763
  functions of, 762
  histology of, 762
  injury to, 762
  surface of, 762
Splenectomy, 762
Splenic artery, 731, 762
Splenic flexure, 884
Splenic vein, 741, 762, 872
Splenius, 317, 329
Splenomegaly, 793; 134 AM
Split-thickness graft, 162
Spongy bone, 168, 169, 173–174
  function of, 173–174
  structure of, 173
Spontaneous abortion, 1083, 1086
Spontaneous mutations, 1094; 194–195
Sports injuries, 339
Sports medicine, 339
Sprained ankle, 57 AM
Sprains, 249, 266, 356
Squama, 202
Squamosal suture, 196, 197
Squamous cell carcinoma, 164; 46 AM
Squamous epithelium, 111–112
  simple squamous epithelium, 111
  stratified squamous epithelium, 111–112
Squamous portion, 202
Stapedius muscle, 560
Stapes, 559
Starches, 46
Starling's law of the heart, 682
Start codon, 94
Starvation, 172–173 AM
Statoacoustic nerve. See Vestibulocochlear nerve
Stellate reticuloendothelial cells, 877
Stem cells, 96, 99, 110, 628, 633

Stenosis, of heart valves, 119–120 AM
Stensen's duct, 855
Stercobilinogens, 632
Stercobilins, 632, 886
Stereocilia, 108, 561
Sterile. See Infertility
Sterilization, male and female, 1054, 1055
Sternal end, clavicle, 228
Sternocleidomastoid, 327, 328, 819
Sternohyoid, 327, 328
Sternothyroid, 327, 328
Sternum, 222
  components of, 222
Steroid-binding proteins, 627
Steroid hormones, 580
  estrogens, 589, 602, 609, 610
  glucocorticoids, 602
  synthesis pathways, 1049
  testosterone, 608
Steroids, 47, 49
  cholesterol, 47, 49
Stethoscope, 677
Stimulants, 88 AM
Stomach, 10, 861–865
  absorption in, 868
  blood supply to, 863
  cancer of, 868
  digestion in, 868
  digestive enzymes of, 864–865, 868, 889, 891
  gastric glands, 863–865
  gastric secretion, 863–865
  muscles of, 863
  pyloric glands, 865
  regions of, 861–863
Stop codon, 94, 95
Straight arteries, 1043
Straight sinus, 734
Strains, 349, 356
Stratified columnar epithelium, 114, 115
Stratified cuboidal epithelium, 112–113
Stratified epithelium, 110–111
Stratified squamous epithelium, 111–112
Stratum basale, 145–146
Stratum corneum, 146–147
Stratum germinativum, 145–146
Stratum granulosum, 146
Stratum lucidum, 146
Stratum spinosum, 146
Streptococcus, 858, 952
Streptokinase, 649
Stress
  fetal, 1010–1011
  fight or flight response, 504–505, 612, 613
  general adaptation syndrome (GAS), 612–614
  and hormones, 612–614
  and immunity, 789
  stress response, 612–614
Stress fractures, 356
Stress incontinence, 977
Stretch marks, 151
Stretch reflex, 425–427
  and muscle spindles, 426–427
  and passive muscle movement, 427
Striated involuntary muscle, 132
Striated muscle, 277
Striated skeletal muscle, 132
Striated voluntary muscle, 132
Striations, 277
Stroke, 398, 443, 475, 649, 729
Stroke volume (SV), 677, 680
Stroma, 120, 122, 1036
Stromal cells, 757
Strong acid, 42–43, 997
Strong base, 43
Structural formula, 34
Structural lipids, 51

Structural proteins, 51, 54
Sty, 538
Styloglossus, 325
Stylohyoid, 327, 328
Stylohyoid ligaments, 207
Styloid process, 202, 231, 232
Stylomastoid foramen, 202
Stylopharyngeus, 858
Subacromial bursa, 259
Subarachnoid space, 410, 439
Subatomic particles, 30
Subcapsular sinus, 760
Subclavian artery, 727
Subclavian trunk, 755
Subclavian vein, 736, 737
Subclavius, 335, 337
Subcoracoid bursa, 259
Subcutaneous injection, 152
Subcutaneous layer, 6, 131, 144, 152
Subdeltoid bursa, 259
Subdural hemorrhage, 443, 475
Subdural space, 410
Sublingual ducts, 855
Sublingual salivary glands, 855
Subluxation, 250, 266
Submandibular ducts, 855
Submandibular ganglion, 468, 469
Submandibular salivary glands,
    207, 855
Submucosa, 805
    digestive tract, 848
Submucosal glands, 871
Submucosal plexus, 848
Subscapular bursa, 259
Subscapular fossa, 230
Subscapularis, 259, 339, 340
Subserous fascia, 132
Substance P, 391, 392, 530
Substantia nigra, 460
Substrate-level phosphorylation,
    906–907
Substrates, 54
Sucrase, 889, 890
Sucrose, 45, 890
Sudden infant death syndrome
    (SIDS), 834, 840
Sudoriferous glands, 156–157
Sugars. See also Carbohydrates
    and carbohydrate digestion, 890
    simple, 44–45, 46
    types of, 890
Sulci (sulcus), 188, 445
Sulfadiazine, 445
Sulfisoxazole, 445
Sulfur, 31
Sulfuric acid, 998
Summation, 393–395
    of excitatory postsynaptic poten-
        tials, 394–395
    facilitation, 394
    of inhibitory postsynaptic poten-
        tials, 394–395
    spatial summation, 394
    temporal summation, 393–394
    of twitches, 287
Sun
    and epidermis, 147–148
    ultraviolet radiation, 147–148
Superficial, 18
Superficial anatomy, 15–19
    anatomical directions, 17–19
    anatomical landmarks, 15
    anatomical regions, 15–17
Superficial fascia, 131
Superficialis, 316
Superficial lymphatics, 754
Superficial palmar arch, 727
Superficial transverse perineus, 334
Superior, 18
Superior angle, scapula, 230
Superior articular facets, 221
Superior articular processes, 216
Superior border, scapula, 229
Superior cerebellar peduncles, 462
Superior colliculus, 459–460, 490, 564

Superior demifacets, 217
Superior ganglion, 471
Superior hypophyseal artery, 587
Superioris, 317
Superior meatuses, 801
Superior mesenteric artery, 510, 732
Superior mesenteric ganglion, 510
Superior nasal conchae, 204
Superior nuchal line, 200
Superior oblique, 323
Superior orbital fissure, 203
Superior phrenic arteries, 729–731
Superior ramus, 235
Superior rectus, 323
Superior sagittal sinus, 441
Superior section, 19
Superior temporal lines, 200, 201
Superior vena cava, 660, 734–739
Superoxide anions, 640
Supination, 343
    joints, 253
Supinator, 317, 342, 343
Supine position, 15
Supporting cells, 561
Supporting connective tissue, 118,
    125–129
    bone, 128–129
    cartilage, 125–129
Suppression, gene, 1093
Suppression factors, 783
Suppressor T cells, 756, 773, 776, 787
Suprachiasmatic nucleus, 458, 557
Supraoptic nucleus, 458, 459, 591
Supraorbital foramen, 201
Supraorbital margins, 201
Supraorbital notch, 201
Suprarenal arteries, 732
Suprarenal gland. See Adrenal
    gland
Suprarenal veins, 741
Supraspinatus, 339, 340
Supraspinatus muscles, 259
Supraspinous fossa, 230
Supraspinous ligament, 256, 257
Sura, 16
Sural nerve, 420
Sural region, 16
Surface anatomy, 3
Surface antigens, 635
Surface tension, 36
Surfactant, 811–812, 816
Surgical anatomy, 3
Surgical neck, 230
Surrogate mother, 1080
Suspensions, 42
Suspensory ligaments, 542, 1035,
    1041
    breast, 1047
Sustentacular cells, 589, 608, 1021
    and spermatogenesis, 1024–1025
Sutural bones, 168, 169, 196
Sutures, 196, 197
    coronal suture, 196, 197
    as immovable joint, 247
    lambdoidal suture, 196, 197
    sagittal suture, 196, 197
    squamosal suture, 196, 197
    sutural bones, 196
SV (stroke volume), 677, 680
Swallowing
    esophagus, 860–861
    movements of larynx during, 805
    pharynx, 805, 858
    phases of, 860–861
Swallowing center, 861
Swallowing reflex, 521, 861
Sweat glands, 6, 156–157
    apocrine sweat glands, 157
    control of, 157
    functions of, 157
    merocrine sweat glands, 157
Sympathectomy, 89 AM
Sympathetic blocking agents, 522,
    524
Sympathetic chain ganglia, 506–510

anatomy of, 508
functions of, 507–509
Sympathetic division of ANS, 362,
    504–505, 506–512
    activation of, 510
    adrenal medullae, 507, 508, 510
    alpha receptors, 511
    beta receptors, 512
    collateral ganglia, 507, 508, 510
    compared to parasympathetic
        division of ANS, 517–519
    and neurotransmitters, 511–512
    sympathetic chain ganglia,
        506–510
Sympathetic nerves, 414, 508
Sympathetic reflexes, 521
Sympathomimetic drugs, 522, 524
Symphysis, 247
Symport, 75
Symptoms of disease, 14; 6 AM
Synapse, 364–365, 386–392
    chemical synapses, 387–390
    cholinergic synapses, 387–389
    drug effects, 390
    electrical synapses, 386–387
    postsynaptic potentials, 392–395
    presynaptic inhibition and facili-
        tation, 395
    synaptic delay, 388
    synaptic fatigue, 389
Synapsis, 1023
Synaptic cleft, 280–281
Synaptic communication, 577, 578
Synaptic knobs, 364
Synaptic terminals, 280, 363
Synaptic vesicles, 364
Synarthroses (immovable joints),
    246–247
    gomphosis, 247
    suture, 247
    synchondrosis, 247
Synchondrosis, 247
Syncytial trophoblast, 1069
Syndesmosis, 247
Syneresis, blood clots, 648
Synergist, skeletal muscle, 315–316
Synergistic effects, hormones, 610
Synostosis, 247
Synovial fluid, 131, 248–249
    functions of, 248–249
Synovial joints, 246, 247–250
Synovial membranes, 130, 131, 248
Synovial tendon sheaths, 249, 343
Synovitis, 60 AM
Synthesis, 38
Syphilis, 1056; 23, 190–191 AM
Système Internationale (SI),
    18 AM
Systemic anatomy, 3
Systemic arteries, 724–734
Systemic circuit, 655
Systemic circulation, 724–742
    abdominal aorta, 731–733
    aortic arch, 726–729
    to brain, 727–729
    carotid arteries, 727–729
    descending aorta, 729–733
    of foot, 734
    of limbs, upper, 727
    of pelvis, 733–734
    systemic arteries, 724–734
Systemic lupus erythematosus
    (SLE), 793; 143 AM
Systemic physiology, 4
Systole, 674
    atrial systole, 675, 676–677
    ventricular systole, 675, 677
Systolic pressure, 705

T

T_3, 595, 596
T_4, 595, 596
T_4 assay, 951

Tabes dorsalis, 191 AM
Tachycardia, 672, 683, 688
Tactile discs (Merkel's discs), 152
Tactile (Meissner's) corpuscles, 152
Tactile receptors, 482, 531, 532–533
Taenia coli, 882
Tagamet, 865
Tail
    epididymis, 1026
    pancreas, 872
    spermatozoon, 1025
    thick filaments, 278
Tail fold, 1075
Talus, 239, 240
Tamoxifen, 1048
Target cells, 578
Tarsal bones, 168, 239
Tartar, 856
Taste. See Gustation
Taste buds, 537, 538
Taste hairs, 536
Taste pore, 536
Tay–Sachs disease, 398, 401; 33,
    75–76 AM
TCA (tricarboxylic acid) cycle, 293,
    904–907
T cells, 641, 756, 771, 773–777
    activation of, 773–775
    costimulation, 775
    cytotoxic T cells, 756, 773,
        775–776, 787
    helper T cells, 756, 773, 776–777,
        787
    memory T cells, 775, 776
    suppressor T cells, 756, 773, 776,
        787
Tears, lacrimal apparatus, 539–541
Tectorial membrane, 565
Tectospinal tracts, 490, 491
Tectum, 459
Teeth, 856–858
    disorders of, 858
    and mastication, 858
    primary and secondary, 857
    structure of, 856
    types of, 857
Telencephalon, 439
Telodendria, 363
Telophase, 98–99
Temperature
    diffusion rate and, 72
    enzymes and, 55–56
    and hemoglobin, 828
    sensation of, 530, 531
Template strand, 92
Temporal bones, 195, 202, 209
Temporalis, 200, 317, 324
Temporalis muscle, 207
Temporal lobe, 446, 447
    seizures, 84 AM
Temporal process, 202, 206
Temporal summation, 393–394
Temporal veins, 736
Temporomandibular joint, 208
    TMJ syndrome, 208, 223
Temporoparietalis, 322
Tendinitis, 356
-tendinosus, 317
Tendinous inscriptions, 331
Tendon reflex, 428
Tendons, 7, 118, 119, 122, 273
    and joints, 249
    of shoulder joint, 259
Tenosynovitis, 344
Tension, 271
    external, 288–289
    internal, 288–289
Tension headaches, 72 AM
Tensor, 317
Tensor fasciae latae, 347, 348
Tensor tympani muscle, 559
Tensor veli palatini, 326
Tentorium cerebelli, 441
Teratogens, 1092, 1098;
    191–193 AM

Teratology, 191 AM
Teres, 316, 317
Teres major, 316, 339, 340
Teres minor, 259, 339, 340
Terminal bronchioles, 798, 809
Terminal cisternae, 275
Terminal ganglia, 505
Terminal hairs, 153
Terminal lymphatics, 752
Terminal segment of aorta, 731
Terminal web, 81
Tertiary bronchi, 809
Tertiary follicles, 1038
Tertiary memories, 494–495
Tertiary structure, of protein, 52, 53
Testes, 8, 11, 1017–1025
    blood–testis barrier, 1024–1025
    cancer of, 1025
    descent of, 1018, 1019
    histology of, 1021–1022
    hormones of, 608, 609
    and scrotum, 1020
    spermatic cord, 1018–1020
    spermatogenesis, 1022–1025
    structure of, 1021
    undescended, 1018
Testicular arteries, 733
Testicular feminization syndrome,
    1098; 193 AM
Testosterone, 49, 296, 583, 589, 608,
    611, 1017, 1021
    functions of, 1034, 1057
Testosterone-binding globulin
    (TeBG), 627
Tetanus, 306
    effects of, 289
Tetracycline, 445
Tetrad, 1023
Tetraiodothyronine (T$_4$), 595
Tetralogy of Fallot, 743, 744
Tetrodotoxin (TTX), 382, 390; 32,
    74–75 AM
Thalamic nuclei, 456, 457
Thalamus, 456–457
Thalassemia, 97, 630, 631, 650; 109,
    111 AM
Thalidomide, 18, 192 AM
Thecal cells, 1037
Theory, 4 AM
Therapeutic abortions, 1083
Thermal inertia, 40
Thermogenesis, shivering and non-
    shivering, 935–936
Thermoreceptors, 482, 530, 531
Thermoregulation, 12–14, 162,
    933–937
    fever, 937
    heat conservation, 935
    heat gain mechanisms, 935–936
    heat generation, 935–936
    heat-loss/heat-gain centers, 934
    heat-loss mechanisms, 934–935
    heat-transfer mechanisms, 934
    individual differences, 936–937
    and infants, 936
Theta waves, 493, 494
Thiazides, 968; 179 AM
Thick filaments, 81, 275, 278, 279
Thick segment, 952
Thick skin, 145
Thigh, muscles of, 346–349
Thin filaments, 275, 277–278, 279
Thin segment, 952
Thin skin, 145
Third-class lever, 314
Third-degree burns, 162
Third-degree heart block, 121 AM
Third heart sound, 678
Third-order neuron, 482
Third ventricle, 439
Thirst center, 459
Thoracentesis, 222, 223, 814
Thoracic aorta, 729–731
Thoracic cage, 219–222
    functions of, 219

Thoracic cavity, 20, 21, 22
Thoracic curve spine, 212
Thoracic duct, 755
Thoracic region, 16
Thoracic vein, 738
Thoracic vertebrae, 212, 214,
    216–217
Thoracis, 16, 317
Thoracoabdominal pump, 711
Thoracolumbar division. See
    Sympathetic division of ANS
Thoroughfare channel, 698
Threonine, 919
Threshold, 380
Thrombin, 646
Thrombocytes. See Platelets
Thrombocyte-stimulating factor
    (thrombopoietin), 645
Thrombocytopenia, 644
Thrombocytopenic purpura, 651
Thrombocytopoiesis, 644
Thrombocytosis, 644
Thrombomodulin, 647
Thrombopoietin, 645
Thromboxane A$_2$, 645
Thromboxanes, 645
Thrombus, 649, 650, 744, 745
Thymic lobes, 761
Thymine, 56, 91
Thymopoietin, 762
Thymosin-α, 762
Thymosin-β, 762
Thymosins, 100, 604, 605, 762
Thymosin V, 762
Thymulin, 762
Thymus, 8, 9, 604, 760–762
    cortex, 761
    hormones of, 762
    lymphocyte production, 758
    medulla, 761
    thymosins, 604, 605
Thymus hormones, 605
Thyrocervical trunks, 727, 813
Thyroglobulin, 594–595
Thyrohyoid, 327
Thyroid-binding globulins (TBGs),
    595, 627
Thyroid-binding prealbumin,
    595–596
Thyroid cartilage, 593, 803
Thyroid follicles, 112, 594
Thyroid gland, 8, 593–597
    calcitonin, 597
    C cells of, 597
    disorders of, 597
    structure of, 593–594
    thyroid follicles, 594
Thyroid hormones, 579, 583–584,
    599
    function of, 595, 596–597
    and growth, 610
    and iodine, 596–597
    and skeletal muscles, 296
    tetraiodothyronine (T$_4$), 595
    thyroxine, 595, 617
    triiodothyronine (T$_3$), 595, 617
Thyroiditis, 787
Thyroid peroxidase, 595
Thyroid scan, 16 AM
Thyroid–stimulating hormone
    (TSH), 580, 588, 589
Thyrotoxic crisis, 619
Thyrotoxicosis, 597, 618
Thyrotropin, 589
Thyrotropin-releasing hormone
    (TRH), 588, 589
Thyroxine, 181, 182, 595, 936
Tibia, 178, 238
Tibial artery, 734
Tibial collateral ligament, 263
Tibialis, 317, 351
Tibialis anterior, 351, 353

Tibialis posterior, 353
Tibial nerve, 420, 422
Tibial tuberosity, 238, 239
Tibial veins, 739
Tic douloureux, 468, 475
Tickle sensation, 533
Tidal volume, 818
Tight junction, 109
Timolol, 683
Tissue cholinestrase, 515
Tissue culture, 149
Tissue factor, 646
Tissue level, of organization, 4, 5
Tissue perfusion, 711
Tissue plasminogen activator
    (t-PA), 648, 649
Tissues, 102
    and aging, 136
    connective tissue, 107, 117–129
    epithelial tissue, 107–117
    histological changes in disease,
        137
    inflammation of, 135–136
    muscle tissue, 107, 132–134
    nature of, 3, 107
    neural tissue, 107, 134
    regeneration of, 136
    study of, 107
Titin, 276
TMJ syndrome, 208, 223
TNFs (tumor necrosis factors), 783,
    784
Tolerance, to drugs, 88 AM
Tolerance, immunity, 772–773
Tongue, 854
    digestive enzymes of, 855, 890
    muscles of, 325, 854
Tonicity, and osmosis, 74
Tonic receptors, 529
Tonsillectomy, 793; 130, 132 AM
Tonsillitis, 759, 793; 130, 132 AM
Tonsils, 759
Torr, 817
Total force, 289
Total lung capacity, 822
Total mastectomy, 1048
Total peripheral resistance, 703,
    704
Total placenta previa, 194 AM
Touch, sensation of, 532–533
Toxemia of pregnancy, 1098;
    194 AM
Toxic shock syndrome (TSS),
    127 AM
Toxins
    and large intestine, 886
    and liver, 879
Trabeculae, 173, 174, 759, 809
Trabeculae carneae, 662
Trabecular arteries, 762
Trabecular veins, 762
Trace elements, 31
Trace minerals, 928
Tracer, 61; 15 AM
Trachea, 9, 805–806
    blockage of, 805
Tracheal cartilages, 805
Trachealis, 805
Tracheostomy, 805, 840
Tracts, in nervous system, 399, 413
11-Trans, 551
Transamination, 918
Transcalciferin, 627
Transcortins, 600, 627
Transcription, 91–93
Transdermal administration, drugs,
    150
Transderm-Scop, 563
Transducin, 553
Transduction, sensory receptors,
    528
Trans fatty acids, 48
Transfer factor, 783
Transferrin, 627, 633
Transfer RNA (tRNA), 57, 93–95

Transforming growth factor-β, 783
Transfusion, blood, 631, 637, 650;
    111–112 AM
Transfusion reactions, 636
Transient ischemic attack (TIA),
    747; 128 AM
Transitional epithelium, 112–113
Translation, 93–95
    and polyribosomes, 95
Translocation, 1094
Translocaton defect, 1097
Transmembrane potential, 79–80,
    373–377
    active forces, 375
    chemical gradients, 374
    electrical gradients, 374–375
    electrochemical gradient, 375
    passive forces, 374–375
    resting potential, 80
    sodium–potassium ATPase, 375
Transmembrane proteins, 69
Transmission electron micrograph
    (TEM), 31 AM
Transmission electron microscopy,
    31 AM
Transport gobulins, 626–627
Transport maximum, 958
Transport proteins, 51, 54
Transport vesicles, 85, 86
Transposition of great vessels, 744
Transrectal prostatic ultrasound
    (TRUS), 1032
Transthyretin, 595–596, 627
Transudate, 129
Transurethral prostatectomy
    (TURP), 1032
Transverse acetabular ligament, 261
Transverse arch, 240
Transverse colon, 884
Transverse costal facets, 217
Transverse fibers, 462
Transverse foramina, 214
Transverse fractures, 186
Transverse lines, 219
Transverse mesentery, 851
Transverse perineus muscles, 334
Transverse plane, 19
Transverse processes, 214, 216
Transverse section, 19
Transverse sinus, 441
Transverse tubules, 274–275
Transversospinalis, 330
Transversus, 316, 317
Transversus abdominis, 330, 331, 332
Transversus muscles, 330, 331
Transversus thoracis, 330, 331, 332,
    819
Trapezium, 232
Trapezius, 317, 335, 337
Trapezoid bone, 232
Trauma, 48 AM
Tremor, 475; 81, 82 AM
Treppe, 287, 288
Tretinoin, 151
Triad, 275
Triaxial articulation, 251
Tricarboxylic acid (TCA) cycle
    (citric acid, or Krebs, cycle),
    88, 293, 903, 904–907, 910
Triceps, 317
Triceps brachii, 315, 339, 340, 342
Trichinosis, 307; 64 AM
Trichomoniasis; 28; 189 AM
Trick knee, 263
Tricuspid valve, 662
Trigeminal nerve, 324, 468, 474
Trigeminal neuralgia, 82 AM
Triglycerides, 47, 48–49, 889
Trigone, 975
Triiodothyronine (T$_3$), 595
Trimesters in pregnancy
    first, 1068–1078
    second, 1068, 1076–1077,
        1081–1083
    third, 1076–1077, 1081–1083

Tripeptide, 52
Triple covalent bond, 34
Triplet code, 91, 93
Triquetrum, 232
Trisaccharides, 889
Trisomy, 1097
Tritium, 31
Trochanter, 188
Trochlea, 188, 230, 239
Trochlear nerve, 467, 474
Trochlear notch, 231
Trophic hormones, 589
Trophoblast, 1068, 1069
Tropic hormones, 589
Tropomyosin, 277
Troponin, 277
Troponin–tropomyosin complex, 277–278, 279
True labor, 1083
True pelvis, 235, 236
True ribs, 219–221
Trunks, brachial plexus, 419
Trypsin, 55–56, 874, 891
Trypsinogen, 874
Tryptophan, 579, 919
TSH (thyroid-stimulating hormone), 580, 588, 589
T tubules, 274–275
Tubal ligation, 1054
Tuberal area, 457, 458
Tubercle, 188, 221
Tuberculin skin test, 787
Tuberculosis, 602, 808, 840; 151 AM
Tuberculum, 221
Tuberosity, 188
Tubular fluid, 947
Tubular glands, 117
Tubular maximum, 958
Tubulin, 82
Tubuloacinar, 117
Tubuloalveolar glands, 117, 1047
Tumor necrosis factors (TNFs), 783, 784
Tumors, 100–101, 102
  benign, 100–101, 102
  of central nervous system, 370
  malignant, 101, 102
Tumor-specific antigens, 766
Tunica albuginea, 1021, 1036
Tunica externa, 692, 693
Tunica interna, 692, 693
Tunica media, 692, 693
Tunica vaginalis, 1020
Turbulence, resistance in circulation, 703, 704
Turner syndrome, 1097
Turnover rate, 60
T wave, 673
Twins, 1086
Twitch, skeletal muscle, 286–287
Two-point discrimination test, 533; 94 AM
Two-to-one (2:1) heart block, 121 AM
Tympanic cavity, 202, 558–560
Tympanic duct, 565
Tympanic membrane, 202, 558
Tympanum, 558
Type AB blood, 635
Type A blood, 635
Type A fibers, axons, 384, 386
Type B blood, 635
Type B fibers, axons, 384, 386
Type C fibers, axons, 384, 386
Type O blood, 635
Type 1 diabetes, 103–104 AM
Type 2 diabetes, 104 AM
Tyrosine, 147, 579, 595, 891

U

Ubiquinone, 908
Ulcers
  decubitis, 151

digestive system, 865
  skin, 151, 163
Ulna, 231, 232
Ulnar artery, 727
Ulnar collateral ligament, 260
Ulnar head, 231
Ulnaris, 317
Ulnar nerve, 230, 419
Ulnar notch, 231
Ulnar palsy, 78 AM
Ulnar vein, 737
Ultrasound, 23, 24, 26
Ultraviolet (UV) radiation, and skin, 147–148
Umami, 537
Umbilical arteries, 742, 975, 1075
Umbilical cord, 1075
Umbilical ligaments, 975
Umbilical stalk, 1075
Umbilical vein, 742, 1075
Unicellular exocrine glands, 116–117
Unicellular glands, 116
Uniparental disomy, 1094
Unipennate muscle, 312
Unipolar neurons, 365–366
Universal donors, 636
Universal recipients, 636
Unmyelinated axons, 370
Unsaturated fatty acids, 47, 48
Upper motor neuron, 487
Upper respiratory system, 798, 801–803
Up-regulation, 582
Uracil, 56
Urea, 878, 918, 943, 953
Urea cycle, 918
Urea reabsorption, 966
Uremia, 980; 180 AM
Ureteral opening, 974
Ureters, 10, 943, 972–974
  histology of, 974
  structure of, 972–974
Urethra, 10, 11, 943, 976
  histology of, 976
  infections of, 976
  male, 1017, 1027
  structure of, 976
Urethral sphincter, 975
Urethritis, 976, 980
Uric acid, 919, 943, 953
Uridine triphosphate (UTP), 59, 912
Urinalysis, 972, 980
Urinary bladder, 10, 943, 974–976
  cancer of, 976
  histology of, 976
  structure of, 974–975
Urinary obstruction, 974, 980
Urinary retention, 978
Urinary system, 5, 10, 942–983; 174–182 AM
  and aging, 977–978
  autonomic control of, 519
  disorders of, 980; 177–182 AM
  examination of, 174–176 AM
  functions of, 10, 943–944
  integration with other systems, 978, 979
  kidneys, 943–967
  micturation reflex, 976
  organs of, 10, 943, 945
  smooth muscles of, 304
  ureters, 972–974
  urethra, 976
  urinary bladder, 974–976
Urinary tract, 943
  infections of, 976, 980
Urination, 521, 943, 968
  incontinence, 977
  micturation reflex, 976
Urine, 943
  buffers in, 1002
  compared to plasma, 953
  composition of, 969–972

storage of, 972–977
  transport of, 972
  urination, 976–977
  values in urine testing, 972
  waste products in, 953
Urine formation, 953–954
  carrier-mediated transport, 958–959
  countercurrent multiplication, 962–963, 966
  filtration, 953, 954–958
  reabsorption, 953, 966–967
  secretion, 953–954, 966
Urinometer, 179 AM
Uro-, 317
Urobilinogens, 632, 886
Urobilins, 632, 886
Urogastrone, 871
Urogenital diaphragm, 332, 333, 334, 976
Urogenital triangle, 332, 333, 334
Urokinase, 649
Urticaria, 152, 163
Uterine arteries, 1041
Uterine cavity, 1041
Uterine cycle, 1042, 1043–1045
  hormones of, 1052
  and menopause, 1045
  menses, 1043
  proliferative phase, 1044
  secretory phase, 1044
Uterine environment, abnormal, 1080
Uterine tubes, 11, 1017, 1039–1040
  blocked, 1080
  histology of, 1039–1040
  structure of, 1039, 1040
Uterosacral ligaments, 1041
Uterus, 11, 1017, 1040–1045
  blood supply to, 1041–1042, 1043
  changes in pregnancy, 1083
  histology of, 1043
  innervation of, 1042
  ligaments of, 1041
  structure of, 1042
  uterine cycle, 1043–1045
  uterine wall, 1042–1043
Utricle, 561, 563
Uvea, 541–542
Uvula, 853

V

Vaccine, 793; 136, 137 AM
Vagina, 11, 1017, 1045–1046
  functions of, 1045
  histology of, 1045–1046
  structure of, 1045
Vaginal barriers, 1055
Vaginal branches, 1045
Vaginal canal, 1045
Vaginal smear, 1046
Vaginitis, 1046, 1060
Vagus nerve, 471–472, 474
Valine, 919
Valsalva maneuver, 817
Valves
  lymphatic vessels, 754
  veins, 699–700
Valves of heart, 663–664, 665
  atrioventricular (AV) valves, 660, 664, 665
  disorders of, 664, 678, 679
  semilunar valves, 664, 665
Valvular heart disease, 664, 688
Valvular stenosis, 688; 119–120 AM
Variable segments, antibodies, 778
Varicose veins, 700, 745; 125–126 AM
Varicosities, 511
Vasa recta, 948, 960, 967–968
Vasa vasorum, 692
Vascular phase, hemostasis, 645
Vascular resistance, 703, 704

Vascular spasm, 645
Vascular tunic, 541–542
Vas deferens, 1027, 1028
Vasectomy, 1054, 1060
Vasoactive intestinal peptide (VIP), 880, 881
Vasoconstriction, 591, 694
Vasoconstrictors, 712
Vasodilation, 694
Vasodilators, 712
Vasomotion, 698–699
Vasomotor centers, 459, 464, 712
  of medulla oblongata, 700
Vasomotor reflexes, 521
Vasomotor tone, 712–713
Vasopressin, 591
Vastus, 317, 349, 350
Vegetarians, 926
Veins, 8, 124, 655, 692, 699–700. See also specific nerves
  of abdomen, 740–741
  antebrachial vein, 737
  axillary vein, 737
  azygos vein, 738
  basilic vein, 737
  brachial vein, 737
  brachiocephalic vein, 736, 738
  cephalic vein, 737
  cerebral veins, 734–736
  colic vein, 741
  compared to arteries, 692–693
  of cranium, 734–736
  cubital vein, 737
  cystic vein, 742
  esophageal veins, 738
  facial veins, 736
  femoral vein, 739, 740
  gastric veins, 742
  gonadal veins, 740
  of head and neck, 736
  hemiazygos vein, 738
  hepatic portal system, 741–742
  hepatic veins, 740, 877
  iliac vein, 740
  inferior vena cava, 739–741
  intercostal veins, 738
  jugular vein, 734, 736
  large veins, 699
  of limbs, lower, 739–740
  of limbs, upper, 737
  lumbar veins, 740
  maxillary veins, 736
  medium-sized veins, 699
  mesenteric vein, 741
  palmar veins, 737
  of pelvis, 740
  peroneal, 739
  phrenic veins, 741
  plantar veins, 739
  popliteal vein, 739
  pulmonary veins, 724
  rectal veins, 741
  renal veins, 740
  saphenous veins, 739
  splenic vein, 741
  subclavian vein, 736, 737
  superior vena cava, 734–739
  suprarenal veins, 741
  temporal veins, 736
  thoracic vein, 738
  tibial veins, 739
  ulnar vein, 737
  umbilical vein, 742
  valves of, 699–700
  venous pressure and return, 710–711
  venules, 698, 699
  vertebral veins, 736, 738
Vellus hairs, 153
Venipuncture, 626, 650
Venoconstriction, 700
Venous pressure, 704, 710–711
  and muscular compression, 710
  and respiratory pump, 710–711
Venous reserve, 701

Venous return, 681, 700
Venous thrombosis, 813
Ventral, 18, 19
Ventral body cavity, 20–22
Ventral group, 457
Ventral nuclei, 457
Ventral ramus, 414
Ventral respiratory group (VRG), 832
Ventral roots, 407
Ventricles of brain, 368, 439–440
Ventricular diastole, 675, 677
Ventricular ejection, 677
Ventricular escape, 688; 121 AM
Ventricular fibrillation (VF), 688; 122 AM
Ventricular septal defects, 743, 744
Ventricular systole, 675, 677
Ventricular tachycardia (VT), 688; 122 AM
Venules, 173, 692, 698, 699
Vermiform appendix, 759, 882
Vermis, 461
Vernix caseosa, 156
Vertebrae, 20, 212
  articular processes, 214
  vertebral arch, 213–214
  vertebral body, 213, 216
Vertebral arch, 213–214
Vertebral arteries, 214, 443, 727
Vertebral body, 213, 216
Vertebral border (medial border), scapula, 229
Vertebral canal, 213
Vertebral column, 212–219
  cervical vertebrae, 214–216
  coccyx, 219
  lumbar vertebrae, 212, 214, 217–218
  muscles of, 320, 328–330
  sacrum, 218–219
  spinal curvature, 212–213
  thoracic vertebrae, 212, 214, 216–217
  vertebra, structure of, 213–214
  vertebral articulation, 214
Vertebral foramen, 213, 216
Vertebral ribs, 221
Vertebral veins, 214, 736, 738
Vertebra prominens, 216
Vertebrochondral ribs, 219–221
Vertebrosternal ribs, 221
Vertigo, 95 AM
Very low-density lipoproteins (VLDLs), 916
Vesicles, 77; 43 AM
Vesicouterine pouch, 1035
Vesicular transport, 71, 77–79
  endocytosis, 77, 79
  exocytosis, 79
Vestible, oral cavity, 853
Vestibular branch, 470, 563
Vestibular bulbs, 1046
Vestibular complex, 561, 562. See also Equilibrium
Vestibular duct, 564–565
Vestibular folds, 804
Vestibular ganglia, 563
Vestibular ligaments, 804
Vestibular nuclei, 386, 470, 490, 563
Vestibule, 561
  nasal cavity, 801
  vagina, 1045, 1046
Vestibulocochlear nerve, 470, 474
Vestibulospinal tracts, 490, 491, 564
Viagra, 1053
Vibrio, 24 AM
Villi, 1070
Villikinin, 881
Viral adhesion, prevention of, 780
Viral infections
  defenses against viral pathogens, 784, 785, 786
  and natural killer (NK) cells, 766
Virtual Human database, 325

Viruses, 752, 793; 24–27 AM
Viscera, 22
Visceral epithelium, 949
Visceral motor neurons, 367
Visceral motor system. See Autonomic nervous system (ANS)
Visceral pericardium, 22, 657
Visceral peritoneum, 22, 851
Visceral pleura, 22, 809, 813
Visceral reflex arc, 520
Visceral reflexes, 424, 520–521
  cardioacceleratory reflex, 521
  consensual light reflex, 521
  coughing reflex, 521
  pupillary reflex, 521
  swallowing reflex, 521, 861
  urination, 521
  vasomotor reflexes, 521
Visceral sensory neurons, 366
Visceral smooth muscle cells, 305
Visceral smooth muscle tissue, 849
Visceral surface of spleen, 762
Viscosity
  of blood, 626
  resistance in circulation, 703, 704
Vision, 528, 538–557
  accommodation, 547–549
  bleaching, 553, 554
  and brain stem, 556–557
  central processing, 556–557
  and cerebral hemispheres, 556
  color vision, 554
  disorders of, 548, 549, 550, 554, 555
  image reversal, 549
  light/dark adaptation, 554
  night blindness, 554
  photoreception, steps in, 552–553
  photoreceptors, 541, 543, 545, 546, 550–553
  refraction, 546–547
  retinal processing, 555–556
  visual acuity, 549
  visual field, 556
  visual pathways, 554–557
  visual pigments, 550–552
Visual acuity, 549
Visual association area, 447
Visual axis, of eye, 544
Visual cortex, 447
Visual field, 556
Visual pigments, 550–552, 554
Vital capacity, 822
Vital signs, 8 AM
Vitamin A, 929, 930
  and bone, 181
  and night blindness, 554
  and visual pigments, 550, 554
Vitamin B$_1$, 930
Vitamin B$_2$, 930
Vitamin B$_5$, 930
  production of, 885
Vitamin B$_6$, 930
  and amino acid catabolism, 916–918
Vitamin B$_{12}$, 181, 930
  absorption of, 864, 893
  and anemia, 634
Vitamin C, 930
  and bone, 181
Vitamin D, 929, 930
  and bone, 181, 184
  and calcitriol, 604
  and epidermis, 149
Vitamin E, 929
Vitamin K, 181, 929
  absorption of, 893
  and blood clotting, 647–648
  production of, 885
Vitamins, 55, 149, 929–930. See also specific vitamins
  absorption of, 892–893
  deficiency disease, 929
  fat-soluble, 878, 893, 929

hypervitaminosis, 928, 929
  listing of, 929, 930
  and liver, 878
  maternal requirements for, 1083
  produced by colonic bacteria, 885
  water-soluble, 893, 930
Vitamin toxicity, 930
Vitiligo, 149
Vitreous body (vitreous humor), 545, 546
Vocal cords, 804
Vocal folds, 804
Vocal ligaments, 804
Volatile acids, 998
Voltage-regulated channels, 376, 377
Volt (V), 80
Volume depletion, 994
Volume of packed red cells, 628
Voluntary contractions, 361
Vomer, 206
Vomiting, 872
Vomiting reflex, 872
Von Willebrand disease, 649
Von Willebrand factor (vWF), 649
Vulva, 1046

W
Wallerian degeneration, 372
Warfarin, 649
Waste products
  and liver, 878
  organic, 885–886
Water, 36, 40–42
  absorption of, 891
  hydrogen ions in, 42
  metabolic generation of, 990
  overhydration and depletion, 990
  properties of, 40
  reabsorption of, 961, 966–967
  solutions, 40–42
  and weight loss, 990
Water channels, 966
Water intoxication, 990
Water receptors, 537
Water-soluble vitamins, 893, 930
Wavelengths, light, 550, 554
Wave summation, 287
  skeletal muscle, 287
Weak acid, 43, 997
Weak base, 43
Weight loss
  and adipose tissue, 122
  and water, 990
Wernicke's area, 449
Western blot procedure, 141 AM
Wharton's ducts, 855
Wharton's jelly, 120
Wheal, 43 AM
Whiplash, 214, 223, 257
White blood cells (WBCs), 124, 580, 624, 637–644
  basophils, 637, 640, 641
  circulation of, 637–638
  differential count, 641
  eosinophils, 637, 638, 640–641
  functions of, 638–639
  lymphocytes, 637, 640, 641
  monocytes, 637, 638, 640, 641
  neutrophils, 637, 639–640
  production of, 642–644
White fat, 121
White matter, 370
  cerebral, 446–447
  spinal cord, 413
White muscles, 297
White pulp, of spleen, 762
White ramus, 414, 507
Whole blood, 625, 626
Whole blood substitutes, 112 AM
Whole facet, 217
Wing, sacral, 219
Wisdom teeth, 857

Withdrawal, from drugs, 88 AM
Withdrawal reflex, 428
Work, 36
Wormian bones, 168, 196
Wounds, healing of, 158–160
Wright's stain, 637
Wrinkles, skin, 150–151
Wrist, bones of, 232–233
Wuchereria bancrofti, 756

X
X chromosome, 1095
Xerosis, 147, 164; 44 AM
Xiphoid process, 222
X-linked traits, 1093, 1095
X-rays, 23, 24, 26

Y
Y chromosome, 1095
Yellow marrow, 168, 169, 633
Yolk sac, 1072
Yolk stalk, 1072

Z
Z disc, 278
ZIFT (zygote intrafallopian tube transfer), 1080
Zinc, nutritional factors, 928
Z lines, 276–277, 278, 279, 285
Zona fasciculata, 600, 602
Zona glomerulosa, 600, 601
Zona pellucida, 1037
Zona reticularis, 600, 602
Zone of overlap, 275, 276
Zygomatic arch, 202
Zygomatic bones, 206, 209
Zygomaticofacial foramen, 206
Zygomatic process, 202
Zygomaticus major and minor, 322
Zygote, 1017
  and fertilization, 1067
Zygote intrafallopian tube transfer (ZIFT), 1080

SOME ABBREVIATIONS USED IN THIS TEXT

ACh — acetylcholine
AChE — acetylcholinesterase
ACTH — adrenocorticotropic hormone
ADH — antidiuretic hormone
ADP — adenosine diphosphate
AIDS — acquired immune deficiency syndrome
ALS — amyotrophic lateral sclerosis
AMP — adenosine monophosphate
ANP — atrial natriuretic peptide
ANS — autonomic nervous system
AP — arterial pressure
ARDS — adult respiratory distress syndrome
atm — atmospheric pressure
ATP — adenosine triphosphate
ATPase — adenosine triphosphatase
AV — atrioventricular
AVP — arginine vasopressin
BMR — basal metabolic rate
BCOP — blood colloid osmotic pressure
BPG — bisphosphoglycerate
bpm — beats per minute
BUN — blood urea nitrogen
C — kilocalorie; centigrade
CABG — coronary artery bypass graft
CAD — coronary artery disease
cAMP — cyclic-AMP
CAPD — continuous ambulatory peritoneal dialysis
CCK — cholecystokinin
CD — cluster of differentiation
CF — cystic fibrosis
CHF — congestive heart failure
CHP — capillary hydrostatic pressure
CsHP — capsular hydrostatic pressure
CNS — central nervous system
CO — cardiac output; carbon monoxide
CoA — coenzyme A
COMT — catecol-O-methyltransferase
COPD — chronic obstructive pulmonary disease
CP — creatine phosphate
CPK, CK — creatine phosphokinase
CPM — continous passive motion
CPR — cardiopulmonary resuscitation
CRF — chronic renal failure
CRH — corticotropin-releasing hormone
CSF — cerebrospinal fluid; colony-stimulating factors
CT — computerized tomography; calcitonin
CVA — cerebrovascular accident
CVS — cardiovascular system
DAG — diacylglycerol
D.C. — Doctor of Chiropractic
DCT — distal convoluted tubule
DDST — Denver Developmental Screening Test
DIC — disseminated intravascular coagulation
DJD — degenerative joint disease
DMD — Duchenne's muscular dystrophy
DNA — deoxyribonucleic acid
D.O. — Doctor of Osteopathy
D.P.M. — Doctor of Podiatric Medicine
DSA — digital subtraction angiography
E — epinephrine
ECF — extracellular fluid
ECG — electrocardiogram
EDV — end-diastolic volume
EEG — electroencephalogram
EKG — electrocardiogram
ELISA — enzyme-linked immunosorbent assay
EPSP — excitatory postsynaptic potential
ERV — expiratory reserve volume
ESV — end-systolic volume
ETS — electron transport system
FAD — flavin adenine dinucleotide

FAS — fetal alcohol syndrome
FES — functional electrical stimulation
FMN — flavin mononucleotide
FRC — functional residual capacity
FSH — follicle-stimulating hormone
GABA — gamma aminobutyric acid
GAS — general adaptation syndrome
GC — glucocorticoid
GFR — glomerular filtration rate
GH — growth hormone
GH-IH — growth hormone–inhibiting hormone
GHP — glomerular hydrostatic pressure
GH-RH — growth hormone–releasing hormone
GIP — gastric inhibitory peptide
GnRH — gonadotropin-releasing hormone
GTP — guanosine triphosphate
Hb — hemoglobin
hCG — human chorionic gonadotropin
HCl — hydrochloric acid
HDL — high-density lipoprotein
HDN — hemolytic disease of the newborn
hGH — human growth hormone
HIV — human immunodeficiency virus
HLA — human leukocyte antigen
HMD — hyaline membrane disease
HP — hydrostatic pressure
hPL — human placental lactogen
HR — heart rate
Hz — Hertz
ICF — intracellular fluid
ICOP — interstitial fluid colloid osmotic pressure
IGF — insulin-like growth factor
IH — inhibiting hormone
IM — intramuscular
IP_3 — inositol triphosphate
IPSP — inhibitory postsynaptic potential
IRV — inspiratory reserve volume
ISF — interstitial fluid
IUD — intrauterine device
IVC — inferior vena cava
IVF — in vitro fertilization
kc — kilocalorie
LDH — lactate dehydrogenase
LDL — low-density lipoprotein
L-DOPA — levodopa
LH — luteinizing hormone
LLQ — left lower quadrant
LM — light micrograph
LSD — lysergic acid diethylamide
LUQ — left upper quadrant
MAO — monoamine-oxidase
MAP — mean arterial pressure
MC — mineralocorticoid
M.D. — Doctor of Medicine
mEq — milliequivalent
MHC — major histocompatibility complex
MI — myocardial infarction
mm Hg — millimeters of mercury
mmol — millimole
mOsm — milliosmole
MRI — magnetic resonance imaging
mRNA — messenger RNA
MS — multiple sclerosis
MSH — melanocyte-stimulating hormone
MSH-IH — melanocyte-stimulating hormone–inhibiting hormone
NAD — nicotinamide adenine dinucleotide
NE — norepinephrine
NFP — net filtration pressure
NHP — net hydrostatic pressure
NO — nitric oxide
NRDS — neonatal respiratory distress syndrome
OP — osmotic pressure
Osm — osmoles
OT — oxytocin
PAC — premature atrial contraction
PAT — paroxysmal atrial tachycardia

PCT — proximal convoluted tubule
PCV — packed cell volume
PEEP — positive end-expiratory pressure
PET — positron emission tomography
PFC — perfluorochemical emulsion
PG — prostaglandin
PID — pelvic inflammatory disease
PIH — prolactin-inhibiting hormone
PIP — phosphatidylinositol
PKC — protein kinase C
PKU — phenylketonuria
PLC — phospholipase C
PMN — polymorphonuclear leukocyte
PNS — peripheral nervous system
PR — peripheral resistance
PRF — prolactin-releasing factor
PRL — prolactin
psi — pounds per square inch
PT — prothrombin time
PTA — post-traumatic amnesia; plasma thromboplastin antecedent
PTC — phenylthiocarbamide
PTH — parathyroid hormone
PVC — premature ventricular contraction
RAS — reticular activating system
RBC — red blood cell
RDA — recommended daily allowance
RDS — respiratory distress syndrome
REM — rapid eye movement
RER — rough endoplasmic reticulum
RH — releasing hormone
RHD — rheumatic heart disease
RLQ — right lower quadrant
RNA — ribonucleic acid
rRNA — ribosomal RNA
RUQ — right upper quadrant
SA — sinoatrial
SCA — sickle cell anemia
SCID — severe combined immunodeficiency disease
SEM — scanning electron micrograph
SER — smooth endoplasmic reticulum
SGOT — serum glutamic oxaloacetic transaminase
SIADH — syndrome of inappropriate ADH secretion
SIDS — sudden infant death syndrome
SLE — systemic lupus erythematosus
SNS — somatic nervous system
STD — sexually transmitted disease
SV — stroke volume
SVC — superior vena cava
T_3 — triiodothyronine
T_4 — tetraiodothyronine, or thyroxine
TB — tuberculosis
TBG — thyroid-binding globulin
TEM — transmission electron micrograph
TIA — transient ischemic attack
T_m — transport (tubular) maximum
TMJ — temporomandibular joint
t-PA — tissue plasminogen activator
TRH — thyrotropin-releasing hormone
tRNA — transfer RNA
TSH — thyroid-stimulating hormone
TSS — toxic shock syndrome
TX — thyroxine
U.S. — United States
UTI — urinary tract infection
UTP — uridine triphosphate
UV — ultraviolet
$\dot{V}_A$ — alveolar ventilation
$\dot{V}_D$ — anatomic dead space
$\dot{V}_E$ — respiratory minute volume
V_T — tidal volume
VF — ventricular fibrillation
VLDL — very low-density lipoprotein
VPRC — volume of packed red cells
VT — ventricular tachycardia
WBC — white blood cell

Minimum System Requirements:

Windows
Pentium 133 Mhz
32 MB RAM
8x CD-ROM drive
Windows 95 or newer

Macintosh
Power PC 133 Mhz
32 MBRAM
8x CD-ROM drive
OS 8.0

Includes Netscape Navigator®
Netscape®, Netscape Navigator®, and the Netscape N and Ship's Wheel logos are registered trademarks of Netscape Communications Corporation in the United States and other countries.

Includes QuickTime™
QuickTime™ and the QuickTime logo are trademarks of Apple Computer, Inc., registered in the United States and other countries.

Includes Macromedia Shockwave™ Player and Macromedia Flash™ Player
Macromedia Shockwave™ Player and Macromedia Flash™ Player software by Macromedia, Inc., Copyright © 1995-2000 Macromedia, Inc. All rights reserved. Macromedia, Shockwave, and Flash are trademarks of Macromedia, Inc.

VISIBLE PRODUCTIONS

Martini Interactive
Fundamentals of Anatomy and Physiology, 5/E
Martini
Prentice Hall, Inc.
LICENSE AGREEMENT
YOU SHOULD CAREFULLY READ THE FOLLOWING TERMS AND CONDITIONS BEFORE BREAKING THE SEAL ON THE PACKAGE. AMONG OTHER THINGS, THIS AGREEMENT LICENSES THE ENCLOSED SOFTWARE TO YOU AND CONTAINS WARRANTY AND LIABILITY DISCLAIMERS. BY BREAKING THE SEAL ON THE PACKAGE, YOU ARE ACCEPTING AND AGREEING TO THE TERMS AND CONDITIONS OF THIS AGREEMENT. IF YOU DO NOT AGREE TO THE TERMS OF THIS AGREEMENT, DO NOT BREAK THE SEAL. YOU SHOULD PROMPTLY RETURN THE PACKAGE UNOPENED.
LICENSE.
Subject to the provisions contained herein, Prentice-Hall, Inc. ("PH") hereby grants to you a non-exclusive, non-transferable license to use the object code version of the computer software product ("Software") contained in the package on a single computer of the type identified on the package.
SOFTWARE AND DOCUMENTATION.
PH shall furnish the Software to you on media in machine-readable object code form and may also provide the standard documentation ("Documentation") containing instructions for operation and use of the Software.
LICENSE TERM AND CHARGES.
The term of this license commences upon delivery of the Software to you and is perpetual unless earlier terminated upon default or as otherwise set forth herein.
TITLE.
Title, and ownership right, and intellectual property rights in and to the Software and Documentation shall remain in PH and/or in suppliers to PH of programs contained in the Software. The Software is provided for your own internal use under this license. This license does not include the right to sublicense and is personal to you and therefore may not be assigned (by operation of law or otherwise) or transferred without the prior written consent of PH. You acknowledge that the Software in source code form remains a confidential trade secret of PH and/or its suppliers and therefore you agree not to attempt to decipher or decompile, modify, disassemble, reverse engineer or prepare derivative works of the Software or develop source code for the Software or knowingly allow others to do so. Further, you may not copy the Documentation or other written materials accompanying the Software.
UPDATES.
This license does not grant you any right, license, or interest in and to any improvements, modifications, enhancements, or updates to the Software and Documentation. Updates, if available, may be obtained by you at PH's then current standard pricing, terms, and conditions.
LIMITED WARRANTY AND DISCLAIMER.
PH warrants that the media containing the Software, if provided by PH, is free from defects in material and workmanship under normal use for a period of sixty (60) days from the date you purchased a license to it.
THIS IS A LIMITED WARRANTY AND IT IS THE ONLY WARRANTY MADE BY PH. THE SOFTWARE IS PROVIDED 'AS IS' AND PH SPECIFICALLY DISCLAIMS ALL WARRANTIES OF ANY KIND, EITHER EXPRESS OR IMPLIED, INCLUDING, BUT NOT LIMITED TO, THE IMPLIED WARRANTY OF MERCHANTABILITY AND FITNESS FOR A PARTICULAR PURPOSE. FURTHER, COMPANY DOES NOT WARRANT, GUARANTY OR MAKE ANY REPRESENTATIONS REGARDING THE USE, OR THE RESULTS OF THE USE, OF THE SOFTWARE IN TERMS OF CORRECTNESS, ACCURACY, RELIABILITY, CURRENTNESS, OR OTHERWISE AND DOES NOT WARRANT THAT THE OPERATION OF ANY SOFTWARE WILL BE UNINTERRUPTED OR ERROR FREE. COMPANY EXPRESSLY DISCLAIMS ANY WARRANTIES NOT STATED HEREIN. NO ORAL OR WRITTEN INFORMATION OR ADVICE GIVEN BY PH, OR ANY PH DEALER, AGENT, EMPLOYEE OR OTHERS SHALL CREATE, MODIFY OR EXTEND A WARRANTY OR IN ANY WAY INCREASE THE SCOPE OF THE FOREGOING WARRANTY, AND NEITHER SUBLICENSEE OR PURCHASER MAY RELY ON ANY SUCH INFORMATION OR ADVICE. If the media is subjected to accident, abuse, or improper use; or if you violate the terms of this Agreement, then this warranty shall immediately be terminated. This warranty shall not apply if the Software is used on or in conjunction with hardware or programs other than the unmodified version of hardware and programs with which the Software was designed to be used as described in the Documentation.
LIMITATION OF LIABILITY.
Your sole and exclusive remedies for any damage or loss in any way connected with the Software are set forth below. UNDER NO CIRCUMSTANCES AND UNDER NO LEGAL THEORY, TORT, CONTRACT, OR OTHERWISE, SHALL PH BE LIABLE TO YOU OR ANY OTHER PERSON FOR ANY INDIRECT, SPECIAL, INCIDENTAL, OR CONSEQUENTIAL DAMAGES OF ANY CHARACTER INCLUDING, WITHOUT LIMITATION, DAMAGES FOR LOSS OF GOODWILL, LOSS OF PROFIT, WORK STOPPAGE, COMPUTER FAILURE OR MALFUNCTION, OR ANY AND ALL OTHER COMMERCIAL DAMAGES OR LOSSES, OR FOR ANY OTHER DAMAGES EVEN IF PH SHALL HAVE BEEN INFORMED OF THE POSSIBILITY OF SUCH DAMAGES, OR FOR ANY CLAIM BY ANY OTHER PARTY. PH'S THIRD PARTY PROGRAM SUPPLIERS MAKE NO WARRANTY, AND HAVE NO LIABILITY WHATSOEVER, TO YOU. PH's sole and exclusive obligation and liability and your exclusive remedy shall be: upon PH's election, (i) the replacement of your defective media; or (ii) the repair or correction of your defective media if PH is able, so that it will conform to the above warranty; or (iii) if PH is unable to replace or repair, you may terminate this license by returning the Software. Only if you inform PH of your problem during the applicable warranty period will PH be obligated to honor this warranty. You may contact PH to inform PH of the problem as follows:
SOME STATES OR JURISDICTIONS DO NOT ALLOW THE EXCLUSION OF IMPLIED WARRANTIES OR LIMITATION OR EXCLUSION OF CONSEQUENTIAL DAMAGES, SO THE ABOVE LIMITATIONS OR EXCLUSIONS MAY NOT APPLY TO YOU. THIS WARRANTY GIVES YOU SPECIFIC LEGAL RIGHTS AND YOU MAY ALSO HAVE OTHER RIGHTS WHICH VARY BY STATE OR JURISDICTION.
MISCELLANEOUS.
If any provision of this Agreement is held to be ineffective, unenforceable, or illegal under certain circumstances for any reason, such decision shall not affect the validity or enforceability (i) of such provision under other circumstances or (ii) of the remaining provisions hereof under all circumstances and such provision shall be reformed to and only to the extent necessary to make it effective, enforceable, and legal under such circumstances. All headings are solely for convenience and shall not be considered in interpreting this Agreement. This Agreement shall be governed by and construed under New York law as such law applies to agreements between New York residents entered into and to be performed entirely within New York, except as required by U.S. Government rules and regulations to be governed by Federal law.
YOU ACKNOWLEDGE THAT YOU HAVE READ THIS AGREEMENT, UNDERSTAND IT, AND AGREE TO BE BOUND BY ITS TERMS AND CONDITIONS. YOU FURTHER AGREE THAT IT IS THE COMPLETE AND EXCLUSIVE STATEMENT OF THE AGREEMENT BETWEEN US THAT SUPERSEDES ANY PROPOSAL OR PRIOR AGREEMENT, ORAL OR WRITTEN, AND ANY OTHER COMMUNICATIONS BETWEEN US RELATING TO THE SUBJECT MATTER OF THIS AGREEMENT.
U.S. GOVERNMENT RESTRICTED RIGHTS.
Use, duplication or disclosure by the Government is subject to restrictions set forth in subparagraphs (a) through (d) of the Commercial Computer-Restricted Rights clause at FAR 52.227-19 when applicable, or in subparagraph (c) (1) (ii) of the Rights in Technical Data and Computer Software clause at DFARS 252.227-7013, and in similar clauses in the NASA FAR Supplement.